CHARACTERISTIC INFRARED ABSORPTION FREQUENCIES[a]

Bond	Compound type	Frequency range, cm^{-1}	Reference
C—H	Alkanes	2850–2960	Sec. 13.15
		1350–1470	
C—H	Alkenes	3020–3080 (*m*)	Sec. 13.15
		675–1000	
C—H	Aromatic rings	3000–3100 (*m*)	Sec. 13.15
		675–870	
C—H	Alkynes	3300	Sec. 13.15
C═C	Alkenes	1640–1680 (*v*)	Sec. 13.15
C≡C	Alkynes	2100–2260 (*v*)	Sec. 13.15
C⩦C	Aromatic rings	1500, 1600 (*v*)	Sec. 13.15
C—O	Alcohols, ethers, carboxylic acids, esters	1080–1300	Sec. 16.13
			Sec. 17.17
			Sec. 18.22
			Sec. 20.25
C═O	Aldehydes, ketones, carboxylic acids, esters	1690–1760	Sec. 19.17
			Sec. 18.22
			Sec. 20.25
O—H	Monomeric alcohols, phenols	3610–3640 (*v*)	Sec. 16.13
			Sec. 24.13
	Hydrogen-bonded alcohols, phenols	3200–3600(*broad*)	Sec. 16.13
			Sec. 24.13
	Carboxylic acids	2500–3000 (*broad*)	Sec. 18.22
N—H	Amines	3300–3500 (*m*)	Sec. 23.20
C—N	Amines	1180–1360	Sec. 23.20
C≡N	Nitriles	2210–2260 (*v*)	
$—NO_2$	Nitro compounds	1515–1560	
		1345–1385	

[a] All bands strong unless marked: *m*, moderate; *w*, weak; *v*, variable.

Third Edition

ORGANIC CHEMISTRY

ROBERT THORNTON MORRISON

ROBERT NEILSON BOYD

New York University

ALLYN AND BACON, INC.

BOSTON
LONDON
SYDNEY
TORONTO

1 20

470 ATLANTIC AVENUE, BOSTON

LIBRARY OF CONGRESS CATALOG CARD NUMBER: 72-91904

ISBN 0-205-03239-7

Fourteenth printing . . . . . . June, 1979

PRINTED IN THE UNITED STATES OF AMERICA

Contents

PART II Special Topics

PART III Biomolecules

Preface

In preparing this third edition, we have done just about all the things that one can do to revise a textbook: added new material, deleted old, corrected mistakes, rewritten, and reorganized. To increase the book's value as a teaching aid, we have made two major, related changes: a change in organization, and a change in content. Our aim was to bring the book up-to-date—not just in the chemistry of each topic, but in the *selection* of topics, so that it would reflect, to the extent that a beginning textbook can, the directions that organic chemistry is taking today.

We have divided the book into three parts, and thus have explicitly recognized what has always been our practice and that of most other teachers: to assign for study the first twenty-odd chapters of our book and then, of the last ten or twelve chapters, to pick three or four or five.

In the twenty-four chapters of Part I, the student is introduced to the fundamentals of organic chemistry. As before, these chapters are tightly woven together; although certain sections—or even chapters—can be omitted, or their sequence altered, the organization is necessarily a fairly rigid one.

In Parts II and III the student reinforces his understanding of the fundamentals by applying them to more complicated systems. It is not really so important just which of these later chapters are chosen—although each of us may consider one or another of these topics absolutely essential—as that *some* of them are studied. What is learned at the beginning of the course can evaporate pretty rapidly if the lid is not screwed down at the end.

The new edition is about as long as the previous one, and contains the same number of chapters. But, through deletion and transfer of material, about 100 pages have been cut from the early part of the book. Using these pages, and rearranging some old material, we have written seven quite new chapters: Carbanions I, Carbanions II, Macromolecules, Rearrangements and Neighboring Group Effects, Molecular Orbitals and Orbital Symmetry, Fats, and Biochemical Processes and Molecular Biology.

The teacher will most probably assign all or most of Part I. From Part II he may then select topics to take the student more deeply into "straight" organic chemistry: use of carbanions in organic synthesis; conjugate addition; polynuclear and heterocyclic compounds; rearrangements and neighboring group effects; the application of the orbital symmetry concept to concerted reactions of various kinds. In Part III, he may have the student learn something of the organic chemistry of biomolecules: fats, carbohydrates, proteins, and nucleic acids. And, as before, the brighter or more ambitious student may dip into chemistry not studied by the rest of the class.

There are other changes in organization. Glycols are introduced with alcohols, epoxides with ethers, and dicarboxylic acids with monocarboxylic acids. The aldol and Claisen condensations and the Wittig and Reformatsky reactions appear in Chapter 21, just after aldehydes and ketones and esters. These changes do not, we have found, place an undue burden on the student, but they do stimulate him and keep him busy in the middle part of the course: after the onslaught of new ideas in the beginning and before the complexities of the later topics. The student is *ready* for polyfunctional compounds at this time. He certainly finds epoxides more exciting than ordinary ethers, and can do more with phthalic and succinic anhydrides than with acetic anhydride. With carbanion chemistry in Chapter 21, he has opened to him the really important routes to carbon–carbon bond formation.

As before, we use problems as the best way to help the student to *learn* what he has been exposed to, and to let him broaden his acquaintance with organic chemistry beyond the bounds of the text. We have extended the practice—introduced in our first edition—of inserting problems *within the chapter* as checkpoints on the student's progress; nearly half of the more than 1300 problems are thus used to provide a kind of programmed instruction.

Spectroscopic analysis, chiefly nmr and infrared, is again introduced in Chapter 13, and in subsequent chapters the spectral characteristics of each class of compound is outlined. Emphasis is on the study of spectra themselves—there are 97 nmr and infrared spectra to be analyzed in problems—or of spectral data.

There is much that is entirely new: new reactions, like thallation, solvomercuration, Corey-House hydrocarbon synthesis, organoborane synthesis of acids and ketones; the use of enamines, 2-oxazolines, tetrahydropyranyl esters; and, of course, electrocyclic reactions, cycloaddition, and sigmatropic shifts. In one of the new chapters, rearrangements and neighboring group effects are discussed as related, often indistinguishable, kinds of intramolecular nucleophilic attack. In another, the concept of orbital symmetry is applied to concerted reactions; the treatment is based on the roles played by the highest occupied and lowest unoccupied molecular orbitals, and the student learns the enormous power of the simple Woodward-Hoffmann rules by applying them to dozens of examples: in the text, and in problems of graded difficulty.

In teaching today, one must recognize that many organic chemists will end up working in biological fields, and even those calling themselves biologists must know much organic chemistry. It seems clear, then, that we must do an even better job of teaching the fundamentals of organic chemistry to all students, whatever their ultimate goal. At the same time, the student should be made aware of the role of organic chemistry in biology, and it is with this in mind that we approach the study of biomolecules. Emphasis is on their structure and their chemistry in the test tube—the foundation for any further study. In addition, we have tried to give the student some idea of the ways in which their properties as organic molecules underlie their functions in biological systems, and how all this ultimately goes back to our fundamental ideas of structure. Biology, on the molecular level, *is* organic chemistry, and we try to let the student see this.

No changes have been made just for the sake of change. In our rewriting and in the selection of new topics, we have stuck to the principle we have always held: *these are beginning students, and they need all the help they can get*. Discussion of neighboring group effects or the Woodward-Hoffmann rules is pitched at the same level as the chlorination of methane in Chapter 2. New material is introduced at the rate at which we have found students can absorb it. Once presented, a principle is used and re-used. In a beginning course, we cannot hope to cover more than a tiny fraction of this enormous field; but what we *can* hope for is to make a good job of what we do teach.

ROBERT THORNTON MORRISON

ROBERT NEILSON BOYD

Acknowledgments

Our thanks to Sadtler Research Laboratories for the infrared spectra labeled "Sadtler" and to the Infrared Data Committee of Japan for those labeled "IRDC," and to the following people for permission to reproduce material: Professor George A. Olah, Figure 5.7; Irving Geis and Harper and Row, Publishers, Figure 37.1; the editors of The Journal of the American Chemical Society, Figures 5.7, 13.17, and 13.18; and Walt Disney Productions, Figure 9.13.

Acknowledgments

Our thanks to Sadtler Research Laboratories for the infrared spectra labeled "Sadtler" and to the Infrared Data Committee of Japan for those labeled "IRDC," and to the following people for permission to reproduce material: Professor George A. Olah, Figure 5.7; Irving Geis and Harper and Row, Publishers, Figure 37.1; the editors of The Journal of the American Chemical Society, Figures 5.7, 13.17, and 13.18; and Walt Disney Productions, Figure 9.13.

PART I

The Fundamentals

Chapter 1 Structure and Properties

1.1 Organic chemistry

Organic chemistry is the chemistry of the **compounds of carbon.**

The misleading name "organic" is a relic of the days when chemical compounds were divided into two classes, inorganic and organic, depending upon where they had come from. Inorganic compounds were those obtained from minerals; organic compounds were those obtained from vegetable or animal sources, that is, from material produced by living organisms. Indeed, until about 1850 many chemists believed that organic compounds *must* have their origin in living organisms, and consequently could never be synthesized from inorganic material.

These compounds from organic sources had this in common: they all contained the element carbon. Even after it had become clear that these compounds did not have to come from living sources but could be made in the laboratory, it was convenient to keep the name *organic* to describe them and compounds like them. The division between inorganic and organic compounds has been retained to this day.

Today, although many compounds of carbon are still most conveniently isolated from plant and animal sources, most of them are synthesized. They are sometimes synthesized from inorganic substances like carbonates or cyanides, but more often from other organic compounds. There are two large reservoirs of organic material from which simple organic compounds can be obtained: *petroleum* and *coal.* (Both of these are "organic" in the old sense, being products of the decay of plants and animals.) These simple compounds are used as building blocks from which larger and more complicated compounds can be made.

We recognize petroleum and coal as the *fossil fuels*, laid down over millenia and non-replaceable, that are being consumed at an alarming rate to meet our constantly increasing demands for power. There *is*, fortunately, an alternative source of power—nuclear energy—but where are we to find an alternative reservoir of organic raw material?

What is so special about the compounds of carbon that they should be separated from compounds of all the other hundred-odd elements of the Periodic Table? In part, at least, the answer seems to be this: there are so very many compounds of carbon, and their molecules can be so large and complex.

The number of compounds that contain carbon is many times greater than the number of compounds that do not contain carbon. These organic compounds have been divided into families, which generally have no counterparts among the inorganic compounds.

Organic molecules containing thousands of atoms are known, and the arrangement of atoms in even relatively small molecules can be very complicated. One of the major problems in organic chemistry is to find out how the atoms are arranged in molecules, that is, to determine the structures of compounds.

There are many ways in which these complicated molecules can break apart, or rearrange themselves, to form new molecules; there are many ways in which atoms can be added to these molecules, or new atoms substituted for old ones. Much of organic chemistry is devoted to finding out what these reactions are, how they take place, and how they can be used to synthesize compounds we want.

What is so special about carbon that it should form so many compounds? The answer to this question came to August Kekulé in 1854 during a London bus ride.

> "One fine summer evening, I was returning by the last omnibus, 'outside' as usual, through the deserted streets of the metropolis, which are at other times so full of life. I fell into a reverie and lo! the atoms were gambolling before my eyes. . . . I saw how, frequently, two smaller atoms united to form a pair, how a larger one embraced two smaller ones; how still larger ones kept hold of three or even four of the smaller; whilst the whole kept whirling in a giddy dance. I saw how the larger ones formed a chain. . . . I spent part of the night putting on paper at least sketches of these dream forms."—August Kekulé, 1890.

Carbon atoms can attach themselves to one another to an extent not possible for atoms of any other element. Carbon atoms can form chains thousands of atoms long, or rings of all sizes; the chains and rings can have branches and cross-links. To the carbon atoms of these chains and rings there are attached other atoms, chiefly hydrogen, but also fluorine, chlorine, bromine, iodine, oxygen, nitrogen, sulfur, phosphorus, and many others. (Look, for example, at cellulose on page 1126, chlorophyll on page 1004, and oxytocin on page 1143.)

Each different arrangement of atoms corresponds to a different compound, and each compound has its own characteristic set of chemical and physical properties. It is not surprising that close to a million compounds of carbon are known today and that thousands of new ones are being made each year. It is not surprising that the study of their chemistry is a special field.

Organic chemistry is a field of immense importance to technology: it is the chemistry of dyes and drugs, paper and ink, paints and plastics, gasoline and rubber tires; it is the chemistry of the food we eat and the clothing we wear.

Organic chemistry is fundamental to biology and medicine. Aside from water, living organisms are made up chiefly of organic compounds; the molecules of "molecular biology" are organic molecules. Ultimately, biological processes are a matter of organic chemistry.

1.2 The structural theory

"Organic chemistry nowadays almost drives me mad. To me it appears like a primeval tropical forest full of the most remarkable things, a dreadful endless jungle into which one does not dare enter for there seems to be no way out."—Friedrich Wöhler, 1835.

How can we even begin to study a subject of such enormous complexity? Is organic chemistry today as Wöhler saw it a century and a half ago? The jungle is still there—largely unexplored—and in it are more remarkable things than Wöhler ever dreamed of. But, so long as we do not wander too far too fast, we can enter without fear of losing our way, for we have a chart: the **structural theory**.

The structural theory is the basis upon which millions of facts about hundreds of thousands of individual compounds have been brought together and arranged in a systematic way. It is the basis upon which these facts can best be accounted for and understood.

The structural theory is the framework of ideas about how atoms are put together to make molecules. The structural theory has to do with the order in which atoms are attached to each other, and with the electrons that hold them together. It has to do with the shapes and sizes of the molecules that these atoms form, and with the way that electrons are distributed over them.

A molecule is often represented by a picture or a model—sometimes by several pictures or several models. The atomic nuclei are represented by letters or wooden balls, and the electrons that join them by lines or dots or wooden pegs. These crude pictures and models are useful to us only if we understand what they are intended to mean. Interpreted in terms of the structural theory, they tell us a good deal about the compound whose molecules they represent: how to go about making it; what physical properties to expect of it—melting point, boiling point, specific gravity, the kind of solvents the compound will dissolve in, even whether it will be colored or not; what kind of chemical behavior to expect—the kind of reagents the compound will react with and the kind of products that will be formed, whether it will react rapidly or slowly. We would know all this about a compound that we had never encountered before, simply on the basis of its structural formula and what we understand its structural formula to mean.

1.3 The chemical bond before 1926

Any consideration of the structure of molecules must begin with a discussion of *chemical bonds*, the forces that hold atoms together in a molecule.

We shall discuss chemical bonds first in terms of the theory as it had developed prior to 1926, and then in terms of the theory of today. The introduction of quantum mechanics in 1926 caused a tremendous change in ideas about how molecules are formed. For convenience, the older, simpler language and pictorial representations are often still used, although the words and pictures are given a modern interpretation.

In 1916 two kinds of chemical bond were described: the *ionic bond* by Walther Kossel (in Germany) and the *covalent bond* by G. N. Lewis (of the University of

California). Both Kossel and Lewis based their ideas on the following concept of the atom.

A positively charged nucleus is surrounded by electrons arranged in concentric shells or energy levels. There is a maximum number of electrons that can be accommodated in each shell: two in the first shell, eight in the second shell, eight or eighteen in the third shell, and so on. The greatest stability is reached when the outer shell is full, as in the noble gases. Both ionic and covalent bonds arise from the tendency of atoms to attain this stable configuration of electrons.

The **ionic bond** results from **transfer of electrons**, as, for example, in the formation of lithium fluoride. A lithium atom has two electrons in its inner shell

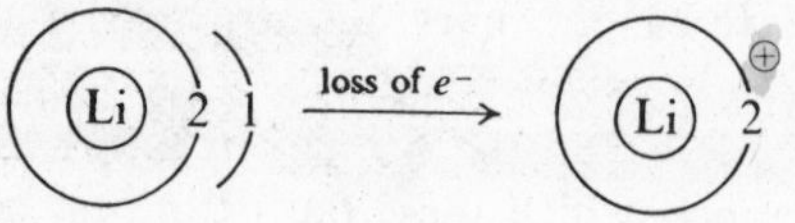

$$Li \longrightarrow Li^{+} + e^{-}$$

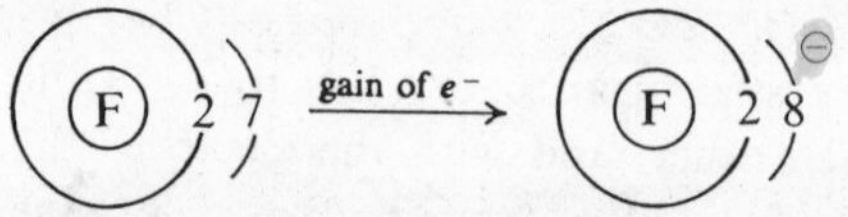

$$F + e^{-} \longrightarrow F^{-}$$

and one electron in its outer or valence shell; the loss of one electron would leave lithium with a full outer shell of two electrons. A fluorine atom has two electrons in its inner shell and seven electrons in its valence shell; the gain of one electron would give fluorine a full outer shell of eight. Lithium fluoride is formed by the transfer of one electron from lithium to fluorine; lithium now bears a positive charge and fluorine bears a negative charge. The electrostatic attraction between the oppositely charged ions is called an ionic bond. Such ionic bonds are typical of the salts formed by combination of the metallic elements (electropositive elements) on the far left side of the Periodic Table with the non-metallic elements (electronegative elements) on the far right side.

The **covalent bond** results from **sharing of electrons**, as, for example, in the formation of the hydrogen molecule. Each hydrogen atom has a single electron; by sharing a pair of electrons, both hydrogens can complete their shells of two. Two fluorine atoms, each with seven electrons in the valence shell, can complete their octets by sharing a pair of electrons. In a similar way we can visualize the formation of HF, H_2O, NH_3, CH_4, and CF_4. Here, too, the bonding force is electrostatic attraction: this time between each electron and *both* nuclei.

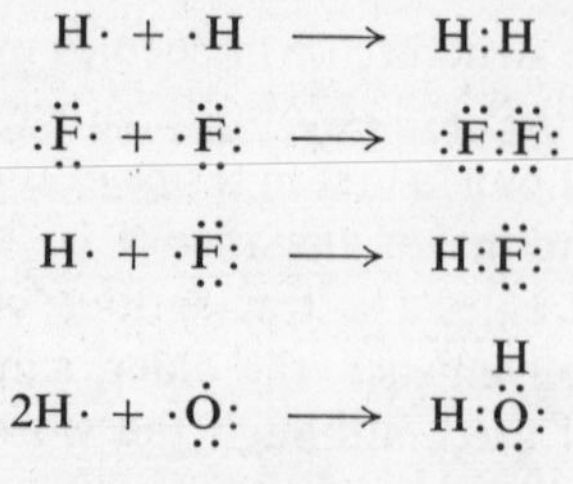

$$H\cdot + \cdot H \longrightarrow H:H$$

$$:F\cdot + \cdot F: \longrightarrow :F:F:$$

$$H\cdot + \cdot F: \longrightarrow H:F:$$

$$2H\cdot + \cdot O: \longrightarrow H:O:$$

$$3H\cdot + \cdot N: \longrightarrow H:N:$$

$$4\text{H}\cdot + \cdot\dot{\underset{\cdot}{\text{C}}}\cdot \longrightarrow \begin{array}{c}\text{H}\\ \text{H}:\ddot{\underset{\cdot\cdot}{\text{C}}}:\text{H}\\ \text{H}\end{array}$$

$$4:\ddot{\underset{\cdot\cdot}{\text{F}}}\cdot + \cdot\dot{\underset{\cdot}{\text{C}}}\cdot \longrightarrow \begin{array}{c}:\ddot{\text{F}}:\\ :\ddot{\underset{\cdot\cdot}{\text{F}}}:\ddot{\underset{\cdot\cdot}{\text{C}}}:\ddot{\underset{\cdot\cdot}{\text{F}}}:\\ :\underset{\cdot\cdot}{\text{F}}:\end{array}$$

The covalent bond is typical of the compounds of carbon; it is the bond of chief importance in the study of organic chemistry.

Problem 1.1 Which of the following would you expect to be ionic, and which non-ionic? Give a simple electronic structure for each, showing only valence shell electrons.

(a) KBr	(c) NF_3	(e) $CaSO_4$	(g) PH_3
(b) H_2S	(d) $CHCl_3$	(f) NH_4Cl	(h) CH_3OH

Problem 1.2 Give a likely simple electronic structure for each of the following, assuming them to be completely covalent. Assume that every atom (except hydrogen, of course) has a complete octet, and that two atoms may share more than one pair of electrons.

(a) H_2O_2	(c) $HONO_2$	(e) HCN	(g) H_2CO_3
(b) N_2	(d) NO_3^-	(f) CO_2	(h) C_2H_6

1.4 Quantum mechanics

In 1926 there emerged the theory known as *quantum mechanics*, developed, in the form most useful to chemists, by Erwin Schrödinger (of the University of Zurich). He worked out mathematical expressions to describe the motion of an electron in terms of its energy. These mathematical expressions are called *wave equations*, since they are based upon the concept that electrons show properties not only of particles but also of waves.

A wave equation has a series of solutions, called *wave functions*, each corresponding to a different energy level for the electron. For all but the simplest of systems, doing the mathematics is so time-consuming that at present—and super-high-speed computers will some day change this—only approximate solutions can be obtained. Even so, quantum mechanics gives answers agreeing so well with the facts that it is accepted today as the most fruitful approach to an understanding of atomic and molecular structure.

> "Wave mechanics has shown us what is going on, and at the deepest possible level . . . it has taken the concepts of the experimental chemist—the imaginative perception that came to those who had lived in their laboratories and allowed their minds to dwell creatively upon the facts that they had found—and it has shown how they all fit together; how, if you wish, they all have one single rationale; and how this hidden relationship to each other can be brought out."—C. A. Coulson, London, 1951.

1.5 Atomic orbitals

A wave equation cannot tell us exactly where an electron is at any particular moment, or how fast it is moving; it does not permit us to plot a precise orbit about the nucleus. Instead, it tells us the *probability* of finding the electron at any particular place.

The region in space where an electron is likely to be found is called an **orbital.** There are different kinds of orbitals, which have different sizes and different shapes, and which are disposed about the nucleus in specific ways. The particular kind of orbital that an electron occupies depends upon the energy of the electron. It is the shapes of these orbitals and their disposition with respect to each other that we are particularly interested in, since these determine—or, more precisely, can conveniently be *thought of* as determining—the arrangement in space of the atoms of a molecule, and even help determine its chemical behavior.

It is convenient to picture an electron as being smeared out to form a cloud. We might think of this cloud as a sort of blurred photograph of the rapidly moving electron. The shape of the cloud is the shape of the orbital. The cloud is not uniform, but is densest in those regions where the probability of finding the electron is highest, that is, in those regions where the average negative charge, or *electron density*, is greatest.

Let us see what the shapes of some of the atomic orbitals are. The orbital at the lowest energy level is called the 1*s* orbital. It is a sphere with its center at the nucleus of the atom, as represented in Fig. 1.1. An orbital has no definite

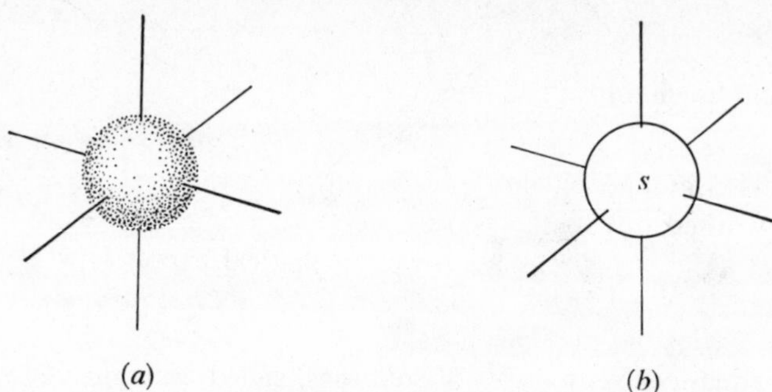

Figure 1.1. Atomic orbitals: *s* orbital. Nucleus at center.

boundary since there is a probability, although a very small one, of finding the electron essentially separated from the atom—or even on some other atom! However, the probability decreases very rapidly beyond a certain distance from the nucleus, so that the distribution of charge is fairly well represented by the electron cloud in Fig. 1.1*a*. For simplicity, we may even represent an orbital as in Fig. 1.1*b*, where the solid line encloses the region where the electron spends most (say 95%) of its time.

At the next higher energy level there is the 2*s* orbital. This, too, is a sphere with its center at the atomic nucleus. It is—*naturally*—larger than the 1*s* orbital: the higher energy (lower stability) is due to the greater average distance between electron and nucleus, with the resulting decrease in electrostatic attraction. (Consider the work that must be done—the energy put into the system—to move an electron away from the oppositely charged nucleus.)

Next there are three orbitals of equal energy called 2*p* orbitals, shown in

Fig. 1.2. Each $2p$ orbital is dumbbell-shaped. It consists of two lobes with the atomic nucleus lying between them. The axis of each $2p$ orbital is perpendicular to the axes of the other two. They are differentiated by the names $2p_x$, $2p_y$, and $2p_z$, where the x, y, and z refer to the corresponding axes.

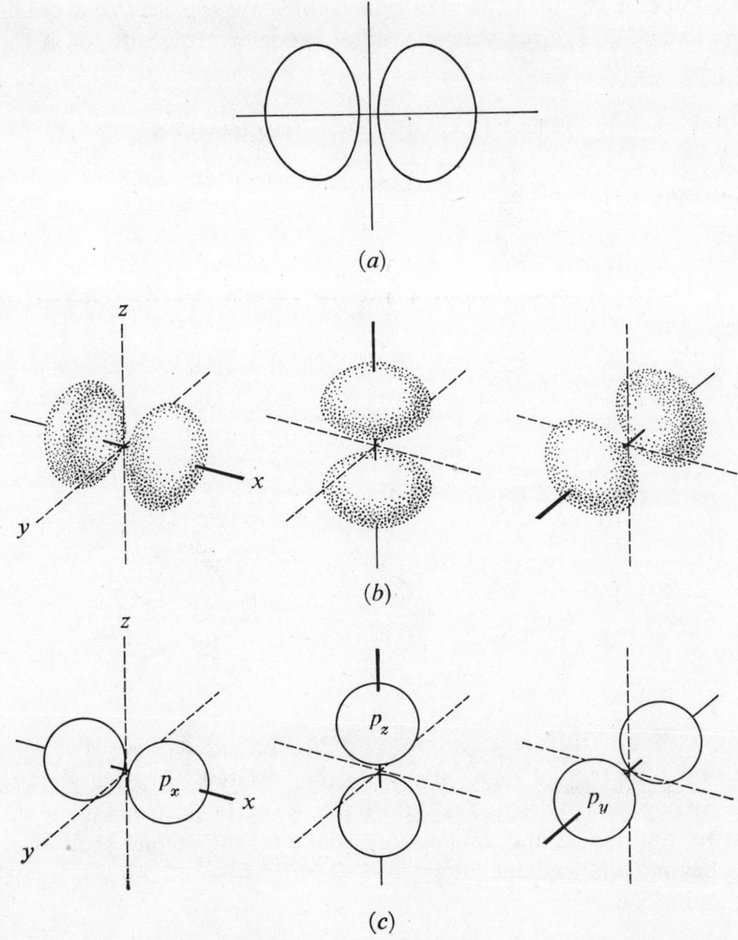

Figure 1.2. Atomic orbitals: p orbitals. Axes mutually perpendicular. (*a*) Cross-section showing the two lobes of a single orbital. (*b*) Approximate shape as pairs of distorted ellipsoids. (*c*) Representation as pairs of not-quite-touching spheres.

1.6 Electronic configuration. Pauli exclusion principle

There are a number of "rules" that determine the way in which the electrons of an atom may be distributed, that is, that determine the *electronic configuration* of an atom.

The most fundamental of these rules is the **Pauli exclusion principle**: *only two electrons can occupy any atomic orbital, and to do so these two must have* **opposite** *spins.* These electrons of opposite spins are said to be *paired. Electrons of* **like** *spin tend to get as far from each other as possible.* This tendency is the most important of all the factors that determine the shapes and properties of molecules.

The exclusion principle, advanced in 1925 by Wolfgang Pauli, Jr. (of the Institute for Theoretical Physics, Hamburg, Germany), has been called the cornerstone of chemistry.

The first ten elements of the Periodic Table have the electronic configurations shown in Table 1.1. We see that an orbital becomes occupied only if the orbitals

Table 1.1 ELECTRONIC CONFIGURATIONS

	1s	2s	2p		
H	(·)				
He	(··)				
Li	(··)	(·)	○	○	○
Be	(··)	(··)	○	○	○
B	(··)	(··)	(·)	○	○
C	(··)	(··)	(·)	(·)	○
N	(··)	(··)	(·)	(·)	(·)
O	(··)	(··)	(··)	(·)	(·)
F	(··)	(··)	(··)	(··)	(·)
Ne	(··)	(··)	(··)	(··)	(··)

of lower energy are filled (e.g., 2*s* after 1*s*, 2*p* after 2*s*). We see that an orbital is not occupied by a pair of electrons until other orbitals of equal energy are each occupied by one electron (e.g., the 2*p* orbitals). The 1*s* electrons make up the first shell of two, and the 2*s* and 2*p* electrons make up the second shell of eight. For elements beyond the first ten, there is a third shell containing a 3*s* orbital, 3*p* orbitals, and so on.

Problem 1.3 (a) Show the electronic configurations for the next eight elements in the Periodic Table (from sodium through argon). (b) What relationship is there between electronic configuration and periodic family? (c) Between electronic configuration and chemical properties of the elements?

1.7 Molecular orbitals

In molecules, as in isolated atoms, electrons occupy orbitals, and in accordance with much the same "rules." These *molecular orbitals* are considered to be centered about many nuclei, perhaps covering the entire molecule; the distribution of nuclei and electrons is simply the one that results in the most stable molecule.

To make the enormously complicated mathematics more workable, two simplifying assumptions are commonly made: (a) that each pair of electrons is essentially localized near just two nuclei, and (b) that the shapes of these localized molecular orbitals and their disposition with respect to each other are related in a simple way to the shapes and disposition of atomic orbitals in the component atoms.

The idea of localized molecular orbitals—or what we might call *bond orbitals*—is evidently not a bad one, since mathematically this method of approximation is successful with most (although *not all*) molecules. Furthermore, this idea closely parallels the chemist's classical concept of a bond as a force acting between two atoms and pretty much independent of the rest of the molecule; it can hardly be accidental that this concept has worked amazingly well for a hundred years. Significantly, the exceptional molecules for which classical formulas do not work are just those for which the localized molecular orbital approach does not work, either. (Even these cases, we shall find, can be handled by a rather simple adaptation of classical formulas, an adaptation which again parallels a method of mathematical approximation.)

The second assumption, of a relationship between atomic and molecular orbitals, is a highly reasonable one, as discussed in the following section. It has proven so useful that, when necessary, atomic orbitals of certain kinds have been *invented* just so that the assumption can be retained.

1.8 The covalent bond

Now let us consider the formation of a molecule. For convenience we shall picture this as happening by the coming together of the individual atoms, although most molecules are not actually made this way. We make physical models of molecules out of wooden or plastic balls that represent the various atoms; the location of holes or snap fasteners tells us how to put them together. In the same way, we shall make *mental* models of molecules out of mental atoms; the location of atomic orbitals—some of them imaginary—will tell us how to put these together.

For a covalent bond to form, two atoms must be located so that an orbital of one *overlaps* an orbital of the other; each orbital must contain a single electron. When this happens, the two atomic orbitals merge to form a single *bond orbital* which is occupied by both electrons. The two electrons that occupy a bond orbital must have opposite spins, that is, must be paired. Each electron has available to it the entire bond orbital, and thus may be considered to "belong to" both atomic nuclei.

This arrangement of electrons and nuclei contains less energy—that is, is more stable—than the arrangement in the isolated atoms; as a result, formation of a bond is accompanied by evolution of energy. The amount of energy (per mole) that is given off when a bond is formed (or the amount that must be put in to break the bond) is called the *bond dissociation energy*. For a given pair of atoms, the greater the overlap of atomic orbitals, the stronger the bond.

What gives the covalent bond its strength? It is the increase in electrostatic attraction. In the isolated atoms, each electron is attracted by—and attracts—one positive nucleus; in the molecule, each electron is attracted by *two* positive nuclei.

It is the concept of "overlap" that provides the mental bridge between atomic orbitals and bond orbitals. Overlap of atomic orbitals means that the bond orbital occupies much of the same region in space that was occupied by *both* atomic orbitals. Consequently, an electron from one atom can, to a considerable extent, remain in its original, favorable location with respect to "its" nucleus, and at the same time occupy a similarly favorable location with respect to the second nucleus; the same holds, of course, for the other electron.

The principle of *maximum overlap*, first stated in 1931 by Linus Pauling (at the California Institute of Technology), has been ranked only slightly below the exclusion principle in importance to the understanding of molecular structure.

As our first example, let us consider the formation of the hydrogen molecule, H_2, from two hydrogen atoms. Each hydrogen atom has one electron, which occupies the 1*s* orbital. As we have seen, this 1*s* orbital is a sphere with its center at the atomic nucleus. For a bond to form, the two nuclei must be brought closely enough together for overlap of the atomic orbitals to occur (Fig. 1.3). For hydrogen, the system is most stable when the distance between the nuclei is 0.74 A;

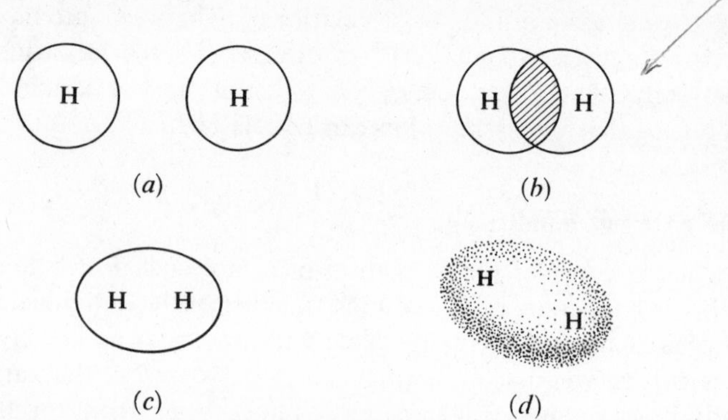

Figure 1.3. Bond formation: H_2 molecule. (*a*) Separate *s* orbitals. (*b*) Overlap of *s* orbitals. (*c*) and (*d*) The σ bond orbital.

this distance is called the **bond length**. At this distance the stabilizing effect of overlap is exactly balanced by repulsion between the similarly charged nuclei. The resulting hydrogen molecule contains 104 kcal/mole less energy than the hydrogen atoms from which it was made. We say that the hydrogen–hydrogen bond has a length of 0.74 A and a strength of 104 kcal.

This bond orbital has roughly the shape we would expect from the merging of two *s* orbitals. As shown in Fig. 1.3, it is sausage-shaped, with its long axis lying along the line joining the nuclei. It is cylindrically symmetrical about this long axis; that is, a slice of the sausage is circular. Bond orbitals having this shape are called *σ orbitals* (*sigma orbitals*) and the bonds are called *σ bonds*. We may visualize the hydrogen molecule as two nuclei embedded in a single sausage-shaped electron cloud. This cloud is densest in the region between the two nuclei, where the negative charge is attracted most strongly by the two positive charges.

The size of the hydrogen molecule—as measured, say, by the volume inside the 95% probability surface—is considerably *smaller* than that of a single hydrogen atom. Although surprising at first, this shrinking of the electron cloud is actually what would be expected. It is the powerful attraction of the electrons by *two* nuclei that gives the molecule greater stability than the isolated hydrogen atoms; this must mean that the electrons are held tighter, *closer*, than in the atoms.

Next, let us consider the formation of the fluorine molecule, F_2, from two fluorine atoms. As we can see from our table of electronic configurations (Table 1.1), a fluorine atom has two electrons in the 1*s* orbital, two electrons in the 2*s* orbital, and two electrons in each of two 2*p* orbitals. In the third 2*p* orbital there is a single electron which is unpaired and available for bond formation. Overlap of this *p* orbital with a similar *p* orbital of another fluorine atom permits electrons to pair and the bond to form (Fig. 1.4). The electronic charge is concentrated between the two nuclei, so that the back lobe of each of the overlapping orbitals

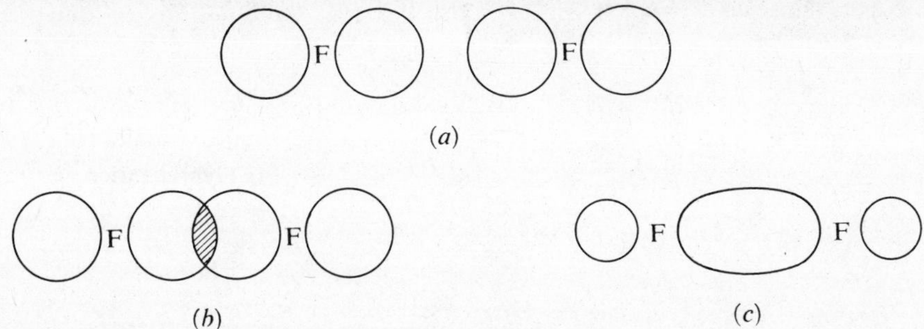

Figure 1.4. Bond formation: F_2 molecule. (*a*) Separate *p* orbitals. (*b*) Overlap of *p* orbitals. (*c*) The σ bond orbital.

shrinks to a comparatively small size. Although formed by overlap of atomic orbitals of a different kind, the fluorine–fluorine bond has the same general shape as the hydrogen–hydrogen bond, being cylindrically symmetrical about a line joining the nuclei; it, too, is given the designation of σ bond. The fluorine–fluorine bond has a length of 1.42 A and a strength of about 38 kcal.

As the examples show, a covalent bond results from the overlap of two atomic orbitals to form a bond orbital occupied by a pair of electrons. *Each kind of covalent bond has a characteristic length and strength.*

1.9 Hybrid orbitals: *sp*

Let us next consider beryllium chloride, $BeCl_2$.

Beryllium (Table 1.1) has no unpaired electrons.

	1*s*	2*s*	2*p*		
Be	⊙	⊙	○	○	○

How are we to account for its combining with two chlorine atoms? Bond formation is an energy-releasing (stabilizing) process, and the tendency is to form bonds—and as many as possible—even if this results in bond orbitals that bear little resemblance to the atomic orbitals we have talked about. If our method of mental molecule-building is to be applied here, it must be modified. We must invent an imaginary kind of beryllium atom, one that is about to become bonded to two chlorine atoms.

To arrive at this divalent beryllium atom, let us do a little electronic bookkeeping. First, we "promote" one of the 2*s* electrons to an empty *p* orbital:

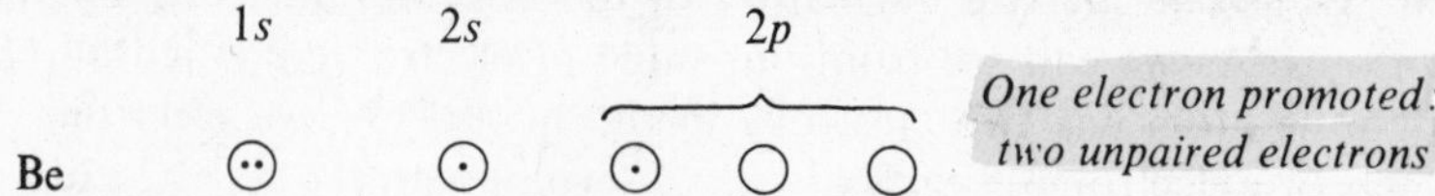

This provides two unpaired electrons, which are needed for bonding to two chlorine atoms. We might now expect beryllium to form one bond of one kind, using the *p* orbital, and one bond of another kind, using the *s* orbital. Again, this is contrary to fact: the two bonds in beryllium chloride are known to be equivalent.

Next, then, we *hybridize* the orbitals. Various combinations of one *s* and

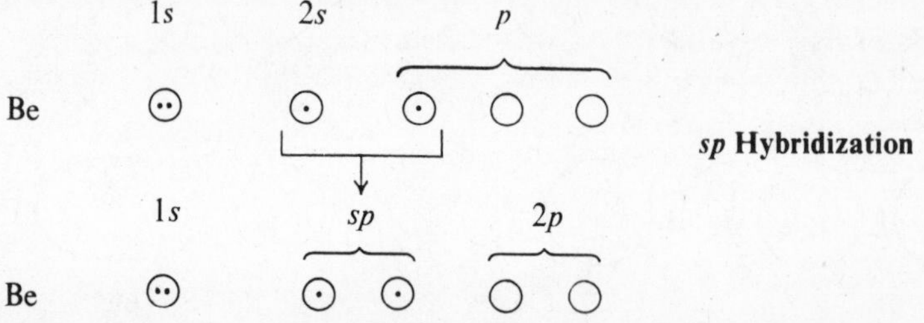

one *p* orbitals are taken mathematically, and the mixed (*hybrid*) orbitals with the greatest degree of *directional character* are found (Fig. 1.5). The more an atomic orbital is concentrated in the direction of the bond, the greater the overlap and the stronger the bond it can form. Three highly significant results emerge from the

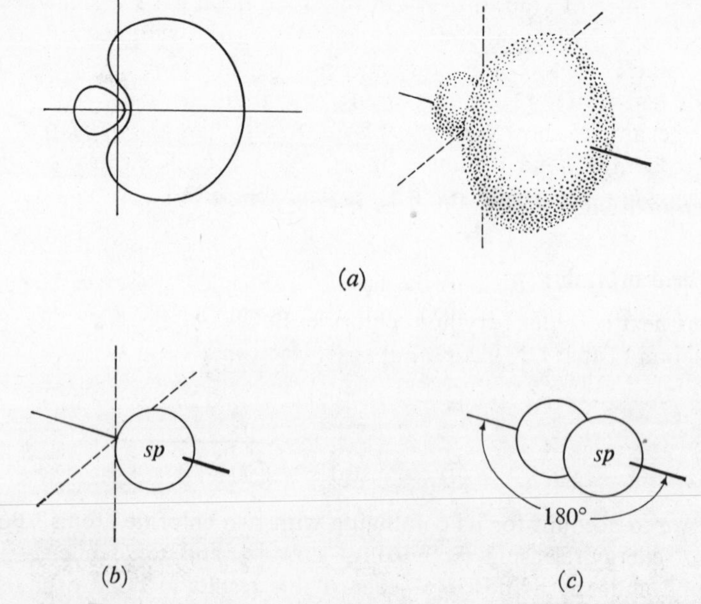

Figure 1.5. Atomic orbitals: hybrid *sp* orbitals. (*a*) Cross-section and approximate shape of a single orbital. Strongly directed along one axis. (*b*) Representation as a sphere, with small back lobe omitted. (*c*) Two orbitals, with axes lying along a straight line.

calculations: (a) the "best" hybrid orbital is much more strongly directed than either the *s* or *p* orbital; (b) the two best orbitals are exactly equivalent to each other; and (c) these orbitals point in exactly opposite directions—*the arrangement that permits them to get as far away from each other as possible* (remember the Pauli exclusion principle). The angle between the orbitals is thus 180°.

These particular hybrid orbitals are called *sp* orbitals, since they are considered to arise from the mixing of *one s* orbital and *one p* orbital. They have the shape shown in Fig. 1.5*a*; for convenience we shall neglect the small back lobe and represent the front lobe as a sphere.

Using this *sp-hybridized* beryllium, let us construct beryllium chloride. An extremely important concept emerges here: **bond angle**. For maximum overlap between the *sp* orbitals of beryllium and the *p* orbitals of the chlorines, the two chlorine nuclei must lie along the axes of the *sp* orbitals; that is, they must be located on exactly opposite sides of the beryllium atom (Fig. 1.6). The angle between the beryllium–chlorine bonds must therefore be 180°.

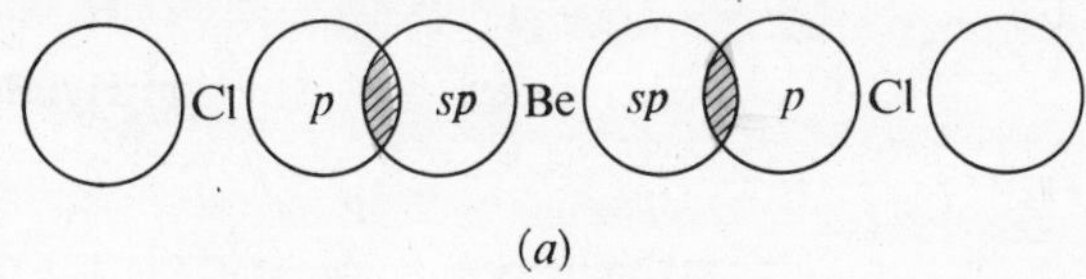

(*a*)

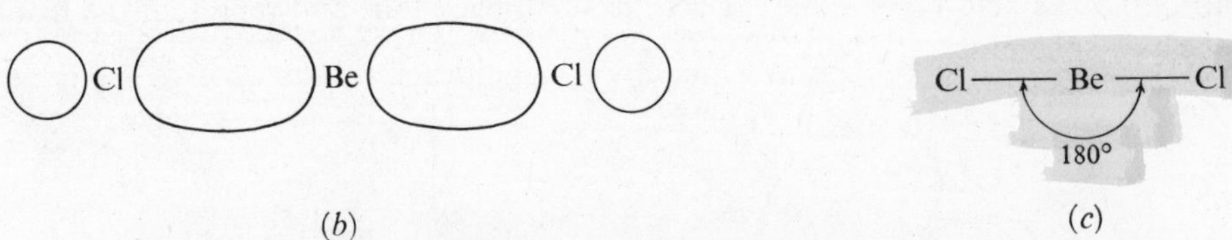

(*b*) (*c*)

Figure 1.6. Bond formation: $BeCl_2$ molecule. (*a*) Overlap of *sp* and *p* orbitals. (*b*) The σ bond orbitals. (*c*) Shape of molecule.

Experiment has shown that, as calculated, beryllium chloride is a *linear molecule*, all three atoms lying along a single straight line.

There is nothing magical about the increase in directional character that accompanies hybridization. The two lobes of the *p* orbital are of opposite *phase* (Sec. 29.2); combination with an *s* orbital amounts to *addition* on one side of the nucleus, but *subtraction* on the other.

1.10 Hybrid orbitals: sp^2

Next, let us look at boron trifluoride, BF_3. Boron (Table 1.1) has only one unpaired electron, which occupies a 2*p* orbital. For three bonds we need three

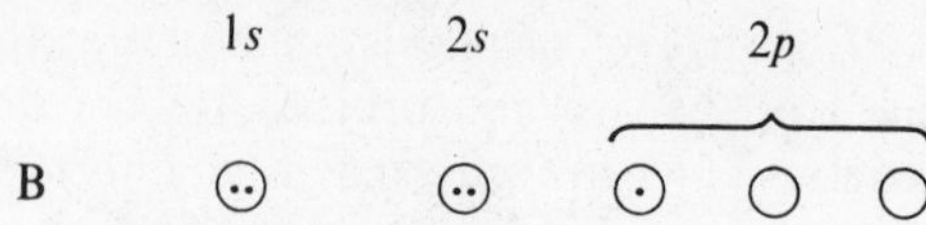

unpaired electrons, and so we promote one of the 2*s* electrons to a 2*p* orbital:

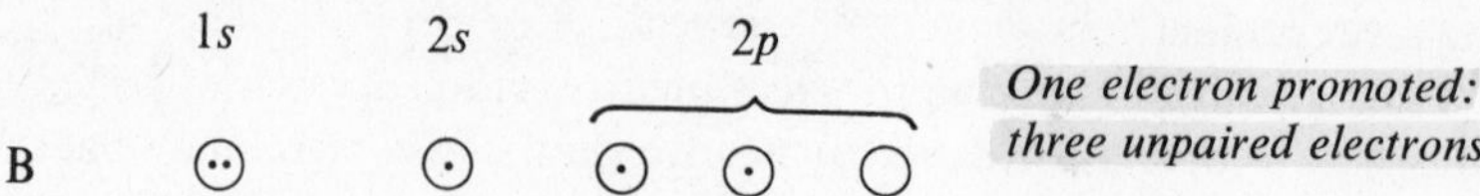

If, now, we are to "make" the most stable molecule possible, we must "make" the strongest bonds possible; for these we must provide the most strongly directed atomic orbitals that we can. Again, hybridization provides such orbitals: three hybrid orbitals, exactly equivalent to each other. Each one has the shape shown

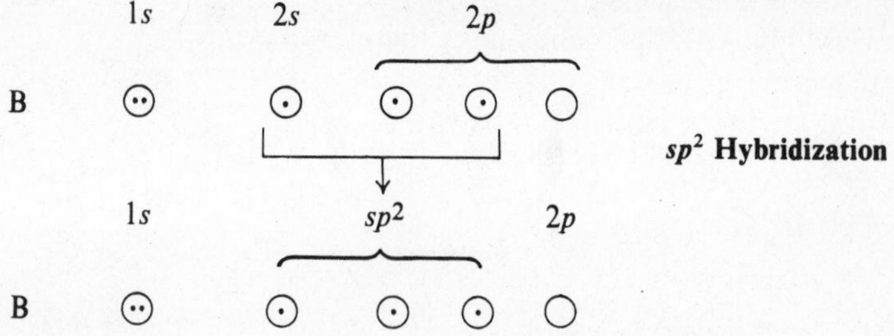

in Fig. 1.7; as before, we shall neglect the small back lobe and represent the front lobe as a sphere.

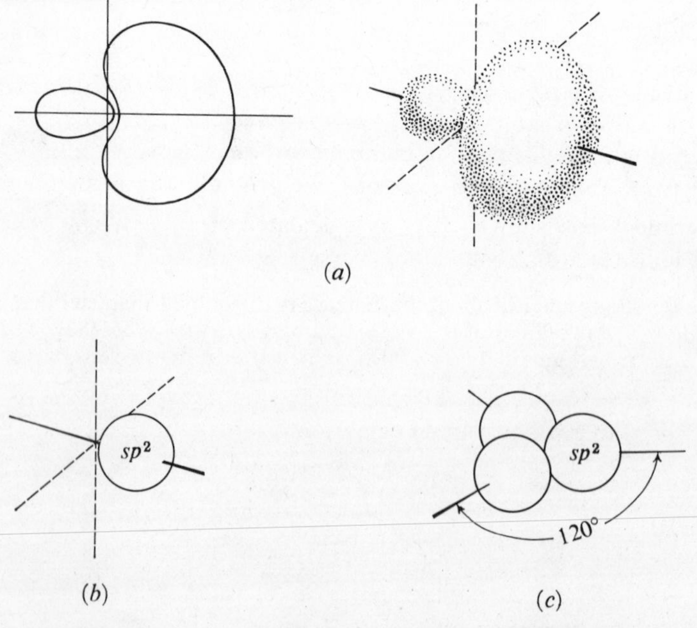

Figure 1.7. Atomic orbitals: hybrid sp^2 orbitals. (*a*) Cross-section and approximate shape of a single orbital. Strongly directed along one axis. (*b*) Representation as a sphere, with small back lobe omitted. (*c*) Three orbitals, with axes directed toward corners of equilateral triangle.

These hybrid orbitals are called sp^2 orbitals, since they are considered to arise from the mixing of *one s* orbital and *two p* orbitals. They lie in a plane, which includes the atomic nucleus, and are directed to the corners of an equilateral triangle; the angle between any two orbitals is thus 120°. Again we see the geometry that permits the orbitals to be as far apart as possible: here, a *trigonal* (three-cornered) arrangement.

When we arrange the atoms for maximum overlap of each of the sp^2 orbitals of boron with a p orbital of fluorine, we obtain the structure shown in Fig. 1.8: a *flat* molecule, with the boron atom at the center of a triangle and the three fluorine atoms at the corners. Every bond angle is 120°.

Figure 1.8. BF_3 molecule.

F sp^2 sp^2 B sp^2 F F 120°

Experiment has shown that boron fluoride has exactly this flat, symmetrical structure calculated by quantum mechanics.

1.11 Hybrid orbitals: sp^3

Now, let us turn to one of the simplest of organic molecules, *methane*, CH_4.

Carbon (Table 1.1) has an unpaired electron in each of the two p orbitals, and on this basis might be expected to form a compound CH_2. (It *does*, but

	1s	2s	2p		
C	⊙(··)	⊙(··)	⊙	⊙	○

CH_2 is a highly reactive molecule whose properties center about the need to provide carbon with two more bonds.) Again, we see the tendency to form as many bonds as possible: in this case, to combine with *four* hydrogen atoms.

To provide four unpaired electrons, we promote one of the 2s electrons to the empty p orbital:

	1s	2s	2p		
C	⊙(··)	⊙	⊙	⊙	⊙

One electron promoted: four unpaired electrons

Once more the most strongly directed orbitals are hybrid orbitals: this time, sp^3 orbitals, from the mixing of *one s* orbital and *three p* orbitals. Each one has the

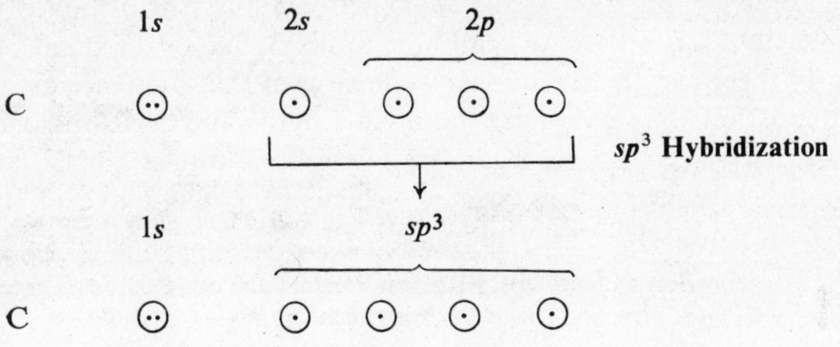

shape shown in Fig. 1.9; as with *sp* and sp^2 orbitals, we shall neglect the small back lobe and represent the front lobe as a sphere.

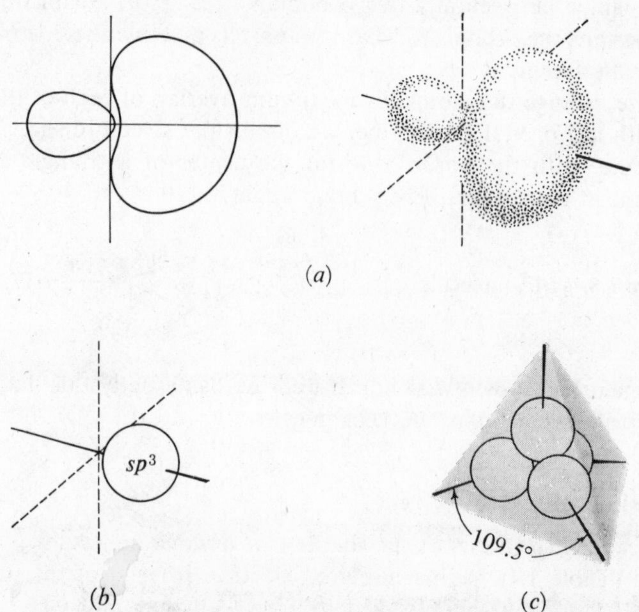

Figure 1.9. Atomic orbitals: hybrid sp^3 orbitals. (*a*) Cross-section and approximate shape of a single orbital. Strongly directed along one axis. (*b*) Representation as a sphere, with small back lobe omitted. (*c*) Four orbitals, with axes directed toward corners of tetrahedron.

Now, how are sp^3 orbitals arranged in space? The answer is no surprise to us: in the way that lets them get as far away from each other as possible. They are directed to the corners of a regular tetrahedron. The angle between any two orbitals is the tetrahedral angle 109.5° (Fig. 1.9). Just as mutual repulsion among orbitals gives two linear bonds or three trigonal bonds, so it gives four tetrahedral bonds.

Overlap of each of the sp^3 orbitals of carbon with an *s* orbital of hydrogen results in methane: carbon at the center of a regular tetrahedron, and the four hydrogens at the corners (Fig. 1.10).

Experimentally, methane has been found to have the highly symmetrical tetrahedral structure we have assembled. Each carbon–hydrogen bond has exactly the same length, 1.10 A; the angle between any pair of bonds is the tetrahedral angle 109.5°. It takes 104 kcal/mole to break one of the bonds of methane.

Thus, in these last three sections, we have seen that there are associated with covalent bonds not only characteristic bond lengths and bond dissociation energies but also characteristic bond *angles*. These bond angles can be conveniently related to the arrangement of atomic orbitals—including hybrid orbitals—involved in bond formation; they ultimately go back to the Pauli exclusion principle and the tendency for unpaired electrons to get as far from each other as possible.

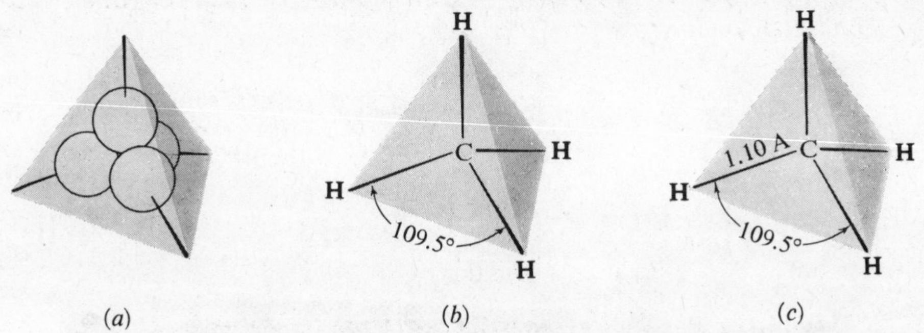

Figure 1.10. Bond formation: CH_4 molecule. (*a*) Tetrahedral sp^3 orbitals. (*b*) Predicted shape: H nuclei located for maximum overlap. (*c*) Shape and size.

Unlike the ionic bond, which is equally strong in all directions, *the covalent bond is a directed bond.* We can begin to see why the chemistry of the covalent bond is so much concerned with molecular size and shape.

1.12 Unshared pairs of electrons

Two familiar compounds, ammonia (NH_3) and water (H_2O), show how *unshared pairs of electrons* can affect molecular structure.

In ammonia, nitrogen resembles the carbon of methane. Nitrogen is sp^3-hybridized, but (Table 1.1) has only three unpaired electrons; they occupy three

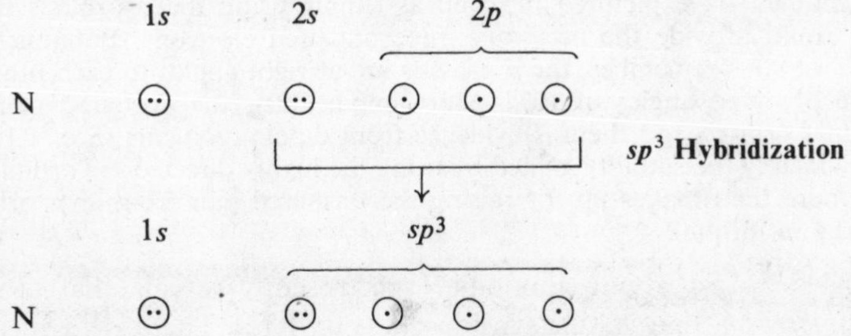

of the sp^3 orbitals. Overlap of each of these orbitals with the *s* orbital of a hydrogen atom results in ammonia (Fig. 1.11). The fourth sp^3 orbital of nitrogen contains a pair of electrons.

If there is to be maximum overlap and hence maximum bond strength, the hydrogen nuclei must be located at three corners of a tetrahedron; the fourth corner is occupied by an unshared pair of electrons. Considering only atomic nuclei, we would expect ammonia to be shaped like a pyramid with nitrogen at the apex and hydrogen at the corners of a triangular base. Each bond angle should be the tetrahedral angle 109.5°.

Experimentally, ammonia is found to have the pyramidal shape calculated by quantum mechanics. The bond angles are 107°, slightly smaller than the predicted value; it has been suggested that the unshared pair of electrons occupies more space than any of the hydrogen atoms, and hence tends to compress the bond

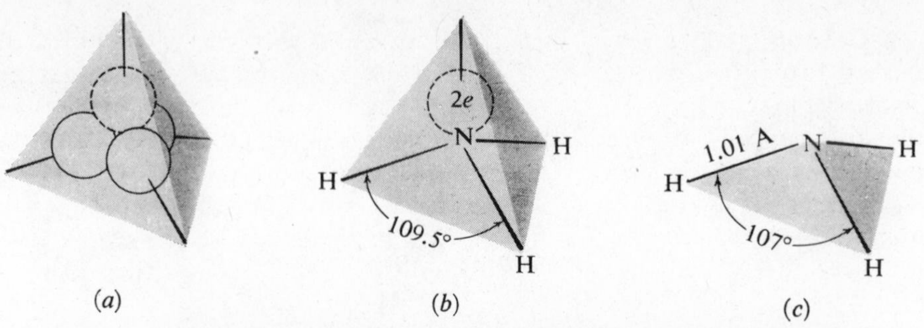

Figure 1.11. Bond formation: NH_3 molecule. (*a*) Tetrahedral sp^3 orbitals. (*b*) Predicted shape, showing unshared pair: H nuclei located for maximum overlap. (*c*) Shape and size.

angles slightly. The nitrogen–hydrogen bond length is 1.01 A; it takes 103 kcal/mole to break one of the bonds of ammonia.

The sp^3 orbital occupied by the unshared pair of electrons is a region of high electron density. This region is a source of electrons for electron-seeking atoms and molecules, and thus gives ammonia its basic properties (Sec. 1.22).

There are two other conceivable electronic configurations for ammonia, but neither fits the facts.

(a) Since nitrogen is bonded to three other atoms, we might have pictured it as using sp^2 orbitals, as boron does in boron trifluoride. But ammonia is *not* a flat molecule, and so we must reject this possibility. It is the unshared pair of electrons on nitrogen that makes the difference between NH_3 and BF_3—these electrons need to stay away from those in the carbon–hydrogen bonds, and the tetrahedral shape makes this possible.

(b) We might have pictured nitrogen as simply using the *p* orbitals for overlap, since they would provide the necessary three unpaired electrons. But this would give bond angles of 90°—remember, the *p* orbitals are at right angles to each other—in contrast to the observed angles of 107°. More importantly, the unshared pair would be buried in an *s* orbital, and there is evidence from dipole moments (Sec. 1.16) that this is not so. Evidently the stability gained by using the highly directed sp^3 orbitals for bond formation more than makes up for raising the unshared pair from an *s* orbital to the higher-energy sp^3 orbital.

One further fact about ammonia: spectroscopy reveals that the molecule undergoes *inversion*, that is, turns inside-out (Fig. 1.12). There is an energy barrier

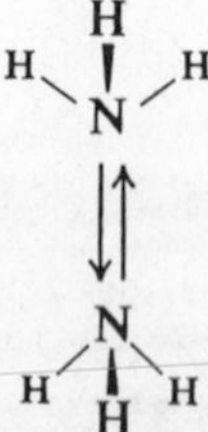

Figure 1.12. Inversion of ammonia.

of only 6 kcal/mole between one pyramidal arrangement and the other, equivalent one. This energy is provided by molecular collisions, and even at room temperature the fraction of collisions hard enough to do the job is so large that a rapid transformation between pyramidal arrangements occurs.

Compare ammonia with methane, which does *not* undergo inversion. The unshared pair plays the role of a carbon–hydrogen bond in determining the most stable shape of the molecule, tetrahedral. But, unlike a carbon–hydrogen bond, the unshared pair cannot maintain a *particular* tetrahedral arrangement; the pair points now in one direction, and the next instant in the opposite direction.

Finally, let us consider water, H_2O. The situation is similar to that for ammonia, except that oxygen has only two unpaired electrons, and hence it bonds

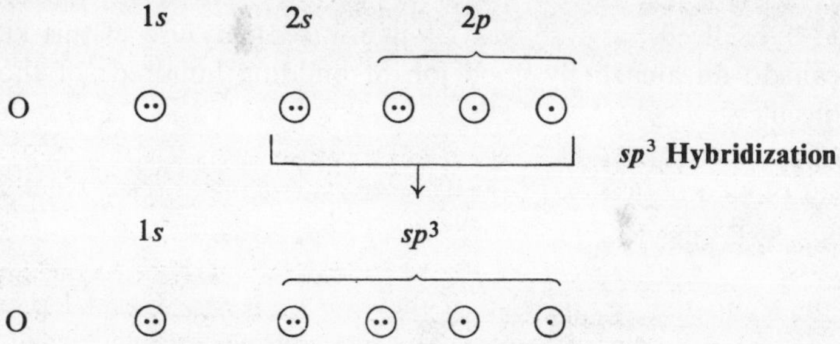

with only two hydrogen atoms, which occupy two corners of a tetrahedron. The other two corners of the tetrahedron are occupied by unshared pairs of electrons (Fig. 1.13).

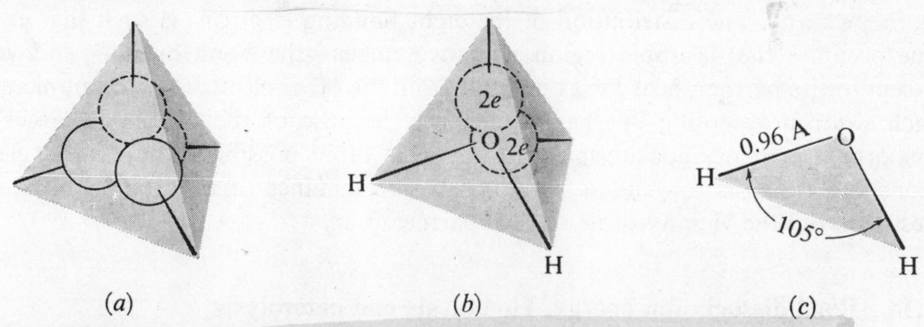

Figure 1.13. Bond formation: H_2O molecule. (*a*) Tetrahedral sp^3 orbitals. (*b*) Predicted shape, showing unshared pairs: H nuclei located for maximum overlap. (*c*) Shape and size.

As actually measured, the H—O—H angle is 105°, smaller than the calculated tetrahedral angle, and even smaller than the angle in ammonia. Here there are two bulky unshared pairs of electrons compressing the bond angles. The oxygen–hydrogen bond length is 0.96 A; it takes 118 kcal/mole to break one of the bonds of water.

Because of the unshared pairs of electrons on oxygen, water is basic, although less strongly so than ammonia (Sec. 1.22).

Problem 1.4 Predict the shape of each of the following molecules, and tell how you arrived at your prediction: (a) the ammonium ion, NH_4^+; (b) the hydronium ion, H_3O^+; (c) methyl alcohol, CH_3OH; (d) methylamine, CH_3NH_2.

1.13 Intramolecular forces

We must remember that the particular method of mentally building molecules that we are learning to use is artificial: it is a purely intellectual process involving imaginary overlap of imaginary orbitals. There are other, equally artificial ways that use different mental or physical models. Our method is the one that so far has seemed to work out best for the organic chemist. Our kit of mental atomic models will contain just three "kinds" of carbon: *tetrahedral* (sp^3-hybridized), *trigonal* (sp^2-hybridized), and *digonal* (sp-hybridized). By use of this kit, we shall find, one can do an amazingly good job of building hundreds of thousands of organic molecules.

But, however we arrive at it, we see the actual structure of a molecule to be the net result of a combination of *repulsive* and *attractive* forces, which are related to *charge* and *electron spin*.

(a) *Repulsive forces.* Electrons tend to stay as far apart as possible because they have the same charge and also, if they are unpaired, because they have the same spin (Pauli exclusion principle). The like-charged atomic nuclei, too, repel each other.

(b) *Attractive forces.* Electrons are attracted by atomic nuclei—as are the nuclei by the electrons—because of their opposite charge, and hence tend to occupy the region between two nuclei. Opposite spin *permits* (although, in itself, probably does not actually *encourage*) two electrons to occupy the same region.

In methane, for example, the four hydrogen nuclei are as widely separated as they can be. The distribution of the eight bonding electrons is such that each one occupies the desirable region near two nuclei—the bond orbital—and yet, except for its partner, is as far as possible from the other electrons. We can picture each electron accepting—perhaps reluctantly because of their similar charges—one orbital-mate of opposite spin, but staying as far as possible from all other electrons and even, as it wanders within the loose confines of its orbital, doing its best to avoid the vicinity of its restless partner.

1.14 Bond dissociation energy. Homolysis and heterolysis

We have seen that energy is liberated when atoms combine to form a molecule. For a molecule to break into atoms, an equivalent amount of energy must be consumed. *The amount of energy consumed or liberated when a bond is broken or formed is known as the* **bond dissociation energy**, **D**. It is characteristic of the particular bond. Table 1.2 lists bond dissociation energies that have been measured for a number of bonds. As can be seen, they vary widely, from weak bonds like I—I (36 kcal/mole) to very strong bonds like H—F (136 kcal/mole). Although the accepted values may change as experimental methods improve, certain trends are clear.

We must not confuse *bond dissociation energy* (D) with another measure of bond strength called *bond energy* (E). If one begins with methane, for example, and breaks, successively, four carbon–hydrogen bonds, one finds four different bond dissociation energies:

$$CH_4 \longrightarrow CH_3 + H\cdot \qquad D(CH_3\text{—}H) = 104 \text{ kcal/mole}$$

$$CH_3 \longrightarrow CH_2 + H\cdot \qquad D(CH_2\text{—}H) = 106$$

$$CH_2 \longrightarrow CH + H\cdot \qquad D(CH—H) = 106$$

$$CH \longrightarrow C + H\cdot \qquad D(C—H) = 81$$

The carbon–hydrogen bond energy in methane, E(C—H), on the other hand, is a single average value:

$$CH_4 \longrightarrow C + 4H\cdot \qquad \Delta H = 397 \text{ kcal/mole}, \quad E(C—H) = 397/4 = 99 \text{ kcal/mole}$$

We shall generally find bond dissociation energies more useful for our purposes.

Table 1.2 Bond Dissociation Energies, Kcal/Mole

$A{:}B \longrightarrow A\cdot + \cdot B \qquad \Delta H$ = Bond Dissociation Energy or D(A–B)

H—H 104			CH_3—H 104
H—F 136	F—F 38		CH_3—F 108
H—Cl 103	Cl—Cl 58		CH_3—Cl 84
H—Br 88	Br—Br 46		CH_3—Br 70
H—I 71	I—I 36		CH_3—I 56
CH_3—H 104	CH_3—CH_3 88	CH_3—Cl 84	CH_3—Br 70
C_2H_5—H 98	C_2H_5—CH_3 85	C_2H_5—Cl 81	C_2H_5—Br 69
n-C_3H_7—H 98	*n*-C_3H_7—CH_3 85	*n*-C_3H_7—Cl 82	*n*-C_3H_7—Br 69
i-C_3H_7—H 95	*i*-C_3H_7—CH_3 84	*i*-C_3H_7—Cl 81	*i*-C_3H_7—Br 68
t-C_4H_9—H 91	*t*-C_4H_9—CH_3 80	*t*-C_4H_9—Cl 79	*t*-C_4H_9—Br 63
H_2C=CH—H 104	H_2C=CH—CH_3 92	H_2C=CH—Cl 84	
H_2C=$CHCH_2$—H 88	H_2C=$CHCH_2$—CH_3 72	H_2C=$CHCH_2$—Cl 60	H_2C=$CHCH_2$—Br 47
C_6H_5—H 112	C_6H_5—CH_3 93	C_6H_5—Cl 86	C_6H_5—Br 72
$C_6H_5CH_2$—H 85	$C_6H_5CH_2$—CH_3 70	$C_6H_5CH_2$—Cl 68	$C_6H_5CH_2$—Br 51

So far, we have spoken of breaking a molecule into two atoms or into an atom and a group of atoms. Thus, of the two electrons making up the covalent bond, one goes to each fragment; such bond-breaking is called *homolysis*. We shall also encounter reactions involving bond-breaking of a different kind: *heterolysis*, in which both bonding electrons go to the same fragment.

$$A{:}B \longrightarrow A\cdot + B\cdot$$ **Homolysis:** *one electron to each fragment*

$$A{:}B \longrightarrow A + {:}B$$ **Heterolysis:** *both electrons to one fragment*

(These words are taken from the Greek: *homo* and *hetero*, the same and different; and *lysis*, a loosing. To a chemist *lysis* means "cleavage" as in, for example, *hydro-lysis*, "cleavage by water.")

Simple heterolysis of a neutral molecule yields, of course, a positive ion and a negative ion. Separation of these oppositely charged particles takes a great deal of energy: 100 kcal/mole or so *more* than separation of neutral particles. In the gas phase, therefore, bond dissociation generally takes place by the easier route, homolysis. In an ionizing solvent (Sec. 1.21), on the other hand, heterolysis is the preferred kind of cleavage.

1.15 Polarity of bonds

Besides the properties already described, certain covalent bonds have another property: **polarity**. Two atoms joined by a covalent bond share electrons; their nuclei are held by the same electron cloud. But in most cases the two nuclei do not share the electrons equally; the electron cloud is denser about one atom than the other. One end of the bond is thus relatively negative and the other end is relatively positive; that is, there is a *negative pole* and a *positive pole*. Such a bond is said to be a **polar bond**, or to *possess polarity*.

We can indicate polarity by using the symbols δ_+ and δ_-, which indicate *partial* + and − charges. For example:

$$\overset{\delta_+}{H}-\overset{\delta_-}{F} \qquad \overset{\delta_+}{H}\overset{\delta_-}{O}\overset{\delta_+}{H} \qquad \overset{\delta_+}{H}\,\overset{\delta_-}{N}(\underset{\delta_+}{H})\,\overset{\delta_+}{H}$$

Polar bonds

We can expect a covalent bond to be polar if it joins atoms that differ in their tendency to attract electrons, that is, atoms that differ in *electronegativity*. Furthermore, the greater the difference in electronegativity, the more polar the bond will be.

The most electronegative elements are those located in the upper right-hand corner of the Periodic Table. Of the elements we are likely to encounter in organic chemistry, fluorine has the highest electronegativity, then oxygen, then nitrogen and chlorine, then bromine, and finally carbon. Hydrogen does not differ very much from carbon in electronegativity; it is not certain whether it is more or less electronegative.

Electronegativity $\qquad F > O > Cl, N > Br > C, H$

Bond polarities are intimately concerned with both physical and chemical properties. The polarity of bonds can lead to polarity of molecules, and thus profoundly affect melting point, boiling point, and solubility. The polarity of a bond determines the kind of reaction that can take place at that bond, and even affects reactivity at nearby bonds.

1.16 Polarity of molecules

A molecule is polar if the center of negative charge does not coincide with the center of positive charge. Such a molecule constitutes a *dipole*: two equal and opposite charges separated in space. A dipole is often symbolized by $+\!\!\!\longrightarrow$, where the arrow points from positive to negative. The molecule possesses a dipole moment, μ, which is equal to the magnitude of the charge, e, multiplied by the distance, d, between the centers of charge:

$$\underset{\substack{\text{in}\\\text{Debye}\\\text{units, D}}}{\mu} = \underset{\substack{\text{in}\\\text{e.s.u.}}}{e} \times \underset{\substack{\text{in}\\\text{cm}}}{d}$$

In a way that cannot be gone into here, it is possible to measure the dipole moments of molecules; some of the values obtained are listed in Table 1.3. We shall be interested in the values of dipole moments as indications of the relative polarities of different molecules.

Table 1.3 DIPOLE MOMENTS, D

H_2	0	HF	1.75	CH_4	0
O_2	0	H_2O	1.84	CH_3Cl	1.86
N_2	0	NH_3	1.46	CCl_4	0
Cl_2	0	NF_3	0.24	CO_2	0
Br_2	0	BF_3	0		

It is the *fact* that some molecules are polar which has given rise to the *speculation* that some bonds are polar. We have taken up bond polarity first simply because it is convenient to consider that the polarity of a molecule is a composite of the polarities of the individual bonds.

Molecules like H_2, O_2, N_2, Cl_2, and Br_2 have zero dipole moments, that is, are non-polar. The two identical atoms of each of these molecules have, of course, the same electronegativity and share electrons equally; e is zero and hence μ is zero, too.

A molecule like hydrogen fluoride has the large dipole moment of 1.75 D. Although hydrogen fluoride is a small molecule, the very high electronegative fluorine pulls the electrons strongly; although d is small, e is large, and hence μ is large, too.

Methane and carbon tetrachloride, CCl_4, have zero dipole moments. We certainly would expect the individual bonds—of carbon tetrachloride at least—to be polar; because of the very symmetrical tetrahedral arrangement, however, they exactly cancel each other out (Fig. 1.14). In methyl chloride, CH_3Cl, the polarity

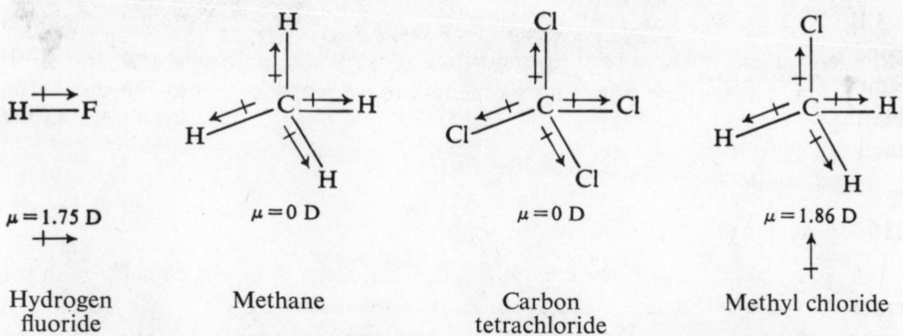

Figure 1.14. Dipole moments of some molecules. Polarity of bonds and of molecules.

of the carbon–chlorine bond is not canceled, however, and methyl chloride has a dipole moment of 1.86 D. Thus the polarity of a molecule depends not only upon the polarity of its individual bonds but also upon the way the bonds are directed, that is, upon the shape of the molecule.

Ammonia has a dipole moment of 1.46 D. This could be accounted for as a net dipole moment (*a vector sum*) resulting from the three individual bond moments, and would be in the direction shown in the diagram. In a similar way, we could account for water's dipole moment of 1.84 D.

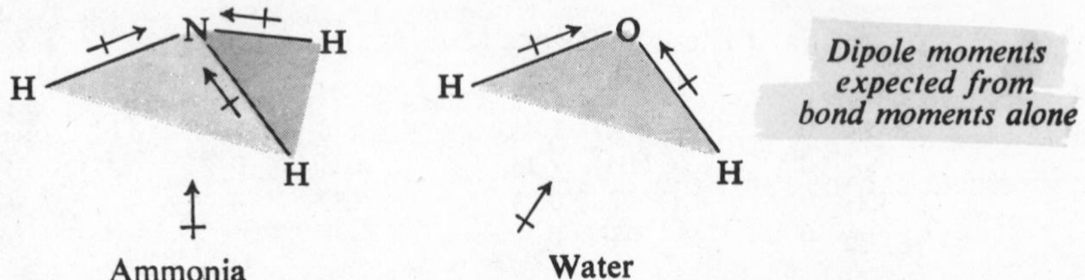

Now, what kind of dipole moment would we expect for nitrogen trifluoride, NF_3, which, like ammonia, is pyramidal? Fluorine is the most electronegative element of all and should certainly pull electrons strongly from nitrogen; the N—F bonds should be highly polar, and their vector sum should be large—far larger than for ammonia with its modestly polar N—H bonds.

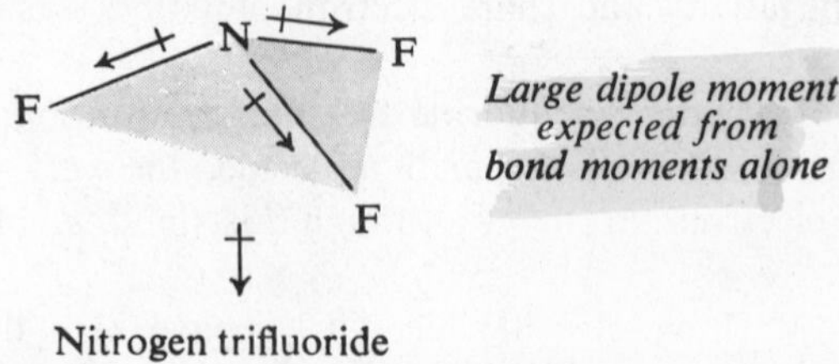

What are the facts? Nitrogen trifluoride has a dipole moment of only 0.24 D. It is not larger than the moment for ammonia, but rather is *much smaller*.

How are we to account for this? We have forgotten the *unshared pair of electrons*. In NF_3 (as in NH_3) this pair occupies an sp^3 orbital and must contribute a dipole moment in the direction opposite to that of the net moment of the N—F bonds (Fig. 1.15); these opposing moments are evidently of about the same size,

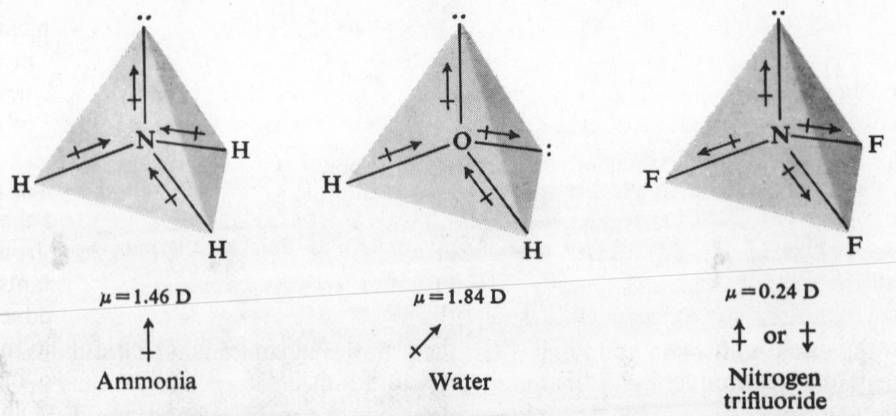

Figure 1.15. Dipole moments of some molecules. Contribution from unshared pairs. In NF_3, moment due to unshared pair opposes vector sum of bond moments.

and the result is a small moment, in which direction we cannot say. In ammonia the observed moment is probably due chiefly to the unshared pair, augmented by the sum of the bond moments. In a similar way, unshared pairs of electrons must contribute to the dipole moments of water and, indeed, of any molecules in which they appear.

Dipole moments can give valuable information about the structure of molecules. For example, any structure for carbon tetrachloride that would result in a polar molecule can be ruled out on the basis of dipole moment alone. The evidence of dipole moment thus supports the tetrahedral structure for carbon tetrachloride. (However, it does not prove this structure, since there are other conceivable structures that would also result in a non-polar molecule.)

Problem 1.5 Which of the following conceivable structures of CCl_4 would also have a zero dipole moment? (a) Carbon at the center of a square with a chlorine at each corner. (b) Carbon at the apex of a pyramid with a chlorine at each corner of a square base.

Problem 1.6 Suggest a shape for the CO_2 molecule that would account for its zero dipole moment.

Problem 1.7 In Sec. 1.12 we rejected two conceivable electronic configurations for ammonia. (a) If nitrogen were sp^2-hybridized, what dipole moment would you expect for ammonia? What *is* the dipole moment of ammonia? (b) If nitrogen used p orbitals for bonding, how would you expect the dipole moments of ammonia and nitrogen trifluoride to compare? How *do* they compare?

The dipole moments of most compounds have never been measured. For these substances we must predict polarity from structure. From our knowledge of electronegativity, we can estimate the polarity of bonds; from our knowledge of bond angles, we can then estimate the polarity of molecules, taking into account any unshared pairs of electrons.

1.17 Structure and physical properties

We have just discussed one physical property of compounds: dipole moment. Other physical properties—like melting point, boiling point, or solubility in a particular solvent—are also of concern to us. The physical properties of a new compound give valuable clues about its structure. Conversely, the structure of a compound often tells us what physical properties to expect of it.

In attempting to synthesize a new compound, for example, we must plan a series of reactions to convert a compound that we have into the compound that we want. In addition, we must work out a method of separating our product from all the other compounds making up the reaction mixture: unconsumed reactants, solvent, catalyst, by-products. Usually the *isolation* and *purification* of a product take much more time and effort than the actual making of it. The feasibility of isolating the product by distillation depends upon its boiling point and the boiling points of the contaminants; isolation by recrystallization depends upon its solubility in various solvents and the solubility of the contaminants. Success in the laboratory often depends upon making a good prediction of physical properties from structure.

We have seen that there are two extreme kinds of chemical bonds: ionic bonds, formed by the transfer of electrons, and covalent bonds, formed by the sharing of electrons. The physical properties of a compound depend largely upon which kind of bonds hold its atoms together in the molecule.

1.18 Melting point

In a crystalline solid the particles acting as structural units—ions or molecules—are arranged in some very regular, symmetrical way; there is a geometric pattern repeated over and over within a crystal.

Melting is the change from the highly ordered arrangement of particles in the crystalline lattice to the more random arrangement that characterizes a liquid (see Figs. 1.16 and 1.17). Melting occurs when a temperature is reached at which the thermal energy of the particles is great enough to overcome the intracrystalline forces that hold them in position.

An **ionic compound** forms crystals in which the structural units are *ions*. Solid sodium chloride, for example, is made up of positive sodium ions and negative chloride ions alternating in a very regular way. Surrounding each positive

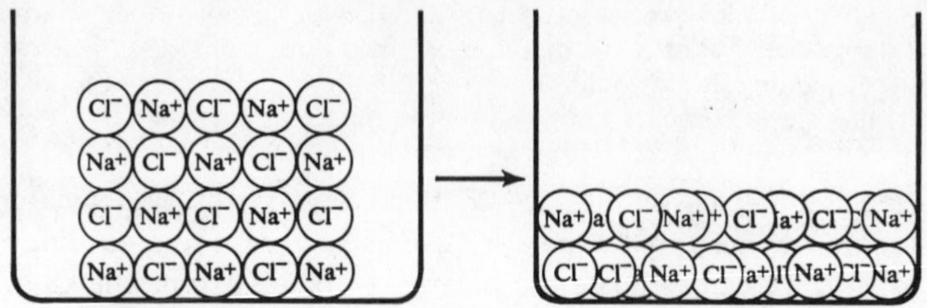

Figure 1.16. Melting of an ionic crystal. Units are ions.

ion and equidistant from it are six negative ions: one on each side of it, one above and one below, one in front and one in back. Each negative ion is surrounded in a similar way by six positive ions. There is nothing that we can properly call a *molecule* of sodium chloride. A particular sodium ion does not "belong" to any one chloride ion; it is equally attracted to six chloride ions. The crystal is an extremely strong, rigid structure, since the electrostatic forces holding each ion in position are powerful. These powerful *interionic* forces are overcome only at a very high temperature; sodium chloride has a melting point of 801°.

Crystals of other ionic compounds resemble crystals of sodium chloride in having an ionic lattice, although the exact geometric arrangement may be different. As a result, these other ionic compounds, too, have high melting points. Many molecules contain both ionic and covalent bonds. Potassium nitrate, KNO_3, for example, is made up of K^+ ions and NO_3^- ions; the oxygen and nitrogen atoms of the NO_3^- ion are held to each other by covalent bonds. The physical properties of compounds like these are largely determined by the ionic bonds; potassium nitrate has very much the same sort of physical properties as sodium chloride.

A **non-ionic compound**, one whose atoms are held to each other entirely by covalent bonds, forms crystals in which the structural units are *molecules*. It is the forces holding these molecules to each other that must be overcome for melting to occur. In general, these *intermolecular* forces are very weak compared with the

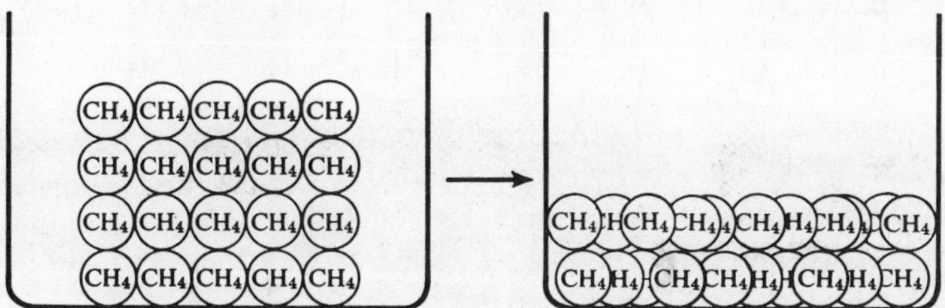

Figure 1.17. Melting of a non-ionic crystal. Units are molecules.

forces holding ions to each other. To melt sodium chloride we must supply enough energy to break ionic bonds between Na^+ and Cl^-. To melt methane, CH_4, we do not need to supply enough energy to break covalent bonds between carbon and hydrogen; we need only supply enough energy to break CH_4 molecules away from each other. In contrast to sodium chloride, methane melts at $-183°$.

1.19 Intermolecular forces

What kind of forces hold neutral molecules to each other? Like interionic forces, these forces seem to be electrostatic in nature, involving attraction of positive charge for negative charge. There are two kinds of intermolecular forces: *dipole–dipole interactions* and *van der Waals forces.*

Dipole–dipole interaction is the attraction of the positive end of one polar molecule for the negative end of another polar molecule. In hydrogen chloride, for example, the relatively positive hydrogen of one molecule is attracted to the relatively negative chlorine of another:

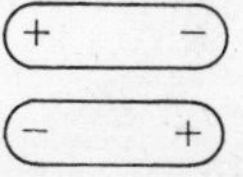

As a result of dipole–dipole interaction, polar molecules are generally held to each other more strongly than are non-polar molecules of comparable molecular weight; this difference in strength of intermolecular forces is reflected in the physical properties of the compounds concerned.

An especially strong kind of dipole–dipole attraction is **hydrogen bonding**, in which *a hydrogen atom serves as a bridge between two electronegative atoms, holding one by a covalent bond and the other by purely electrostatic forces.* When hydrogen is attached to a highly electronegative atom, the electron cloud is greatly distorted toward the electronegative atom, exposing the hydrogen nucleus. The strong positive charge of the thinly shielded hydrogen nucleus is strongly attracted by the negative charge of the electronegative atom of a second molecule. This attraction has a strength of about 5 kcal/mole, and is thus much weaker than the covalent bond—about 50–100 kcal/mole—that holds it to the first electronegative atom. It is, however, much stronger than other dipole–dipole attractions.

Hydrogen bonding is generally indicated in formulas by a broken line:

```
                                        H       H         H     H
                                        |       |         |     |
H—F---H—F     H—O---H—O      H—N---H—N     H—N---H—O
                 |       |          |       |         |
                 H       H          H       H         H
```

For hydrogen bonding to be important, both electronegative atoms must come from the group: **F**, **O**, **N**. Only hydrogen bonded to one of these three elements is positive enough, and only these three elements are negative enough, for the necessary attraction to exist. These three elements owe their special effectiveness to the concentrated negative charge on their small atoms.

Hydrogen bonding, we shall find, not only exerts profound effects on the boiling point and solubility properties of compounds, but also plays a key role in determining the *shapes* of large molecules like proteins and nucleic acids, shapes that in a very direct way determine, in turn, their biological properties: the size of the "pockets" in the hemoglobin molecule, just big enough to hold heme groups with their oxygen-carrying iron atoms (p. 1152); the helical shape of α-keratin and collagen molecules that makes wool and hair strong, and tendons and skin tough (p. 1158). It is hydrogen bonding that makes the double helix of DNA *double*—and thus permits the self-duplication of molecules that is the basis of heredity (p. 1179).

There must be forces between the molecules of a non-polar compound, since even such compounds can solidify. Such attractions are called **van der Waals forces**. The existence of these forces is accounted for by quantum mechanics. We can roughly visualize them arising in the following way. The average distribution of charge about, say, a methane molecule is symmetrical, so that there is no net dipole moment. However, the electrons move about, so that at any instant of time the distribution will probably be distorted, and a small dipole will exist. This momentary dipole will affect the electron distribution in a second methane molecule nearby. The negative end of the dipole tends to repel electrons, and the positive end tends to attract electrons; the dipole thus *induces* an oppositely oriented dipole in the neighboring molecule:

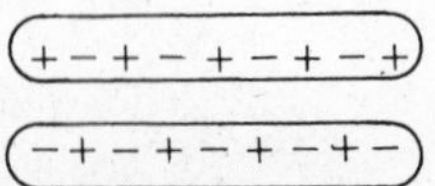

Although the momentary dipoles and induced dipoles are constantly changing, the net result is attraction between the two molecules. These van der Waals forces have a very short range; they act only between the portions of different molecules that are in close contact, that is, between the surfaces of molecules. As we shall see, the relationship between the strength of van der Waals forces and the surface areas of molecules (Sec. 3.12) will help us to understand the effect of molecular size and shape on physical properties. We must not underestimate the power of these weakest intermolecular forces; acting between non-polar chains of phospholipids, for example, they are the mortar in the walls of living cells.

With respect to other atoms to which it is not bonded—whether in another molecule or in another part of the same molecule—every atom has an effective "size," called its *van der Waals radius.* As two non-bonded atoms are brought together the attraction between them steadily increases, and reaches a maximum when they are just "touching"—that is to say, when the distance between the nuclei is equal to the sum of the van der Waals radii. Now, if the atoms are forced still closer together, van der Waals attraction is very rapidly replaced by van der Waals *repulsion.* Thus, non-bonded atoms welcome each other's touch, but strongly resist crowding.

We shall find both attractive and repulsive van der Waals forces important to our understanding of molecular structure.

1.20 Boiling point

Although the particles in a liquid are arranged less regularly and are freer to move about than in a crystal, each particle is attracted by a number of other particles. Boiling involves the breaking away from the liquid of individual molecules or pairs of oppositely charged ions (see Figs. 1.18 and 1.19). This occurs

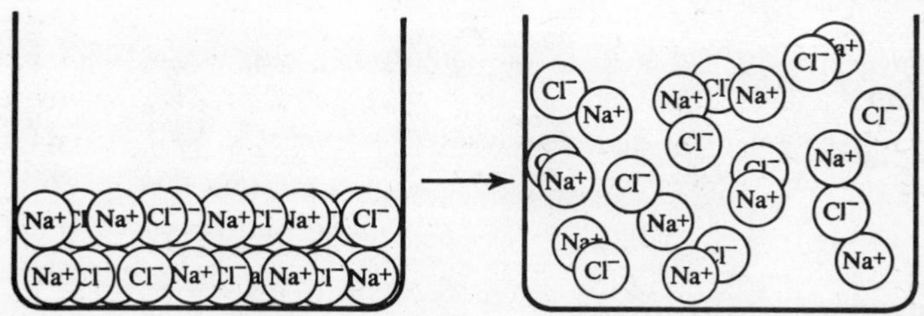

Figure 1.18. Boiling of an ionic liquid. Units are ions and ion pairs.

when a temperature is reached at which the thermal energy of the particles is great enough to overcome the cohesive forces that hold them in the liquid.

In the liquid state the unit of an ionic compound is again the ion. Each ion is still held strongly by a number of oppositely charged ions. Again there is nothing we could properly call a molecule. A great deal of energy is required for a pair of oppositely charged ions to break away from the liquid; boiling occurs only at a very high temperature. The boiling point of sodium chloride, for example, is 1413°. In the gaseous state we have an *ion pair*, which can be considered a sodium chloride molecule.

In the liquid state the unit of a non-ionic compound is again the molecule. The weak intermolecular forces here—dipole–dipole interactions and van der Waals forces—are more readily overcome than the strong interionic forces of ionic compounds, and boiling occurs at a very much lower temperature. Non-polar methane boils at $-161.5°$, and even polar hydrogen chloride boils at only $-85°$.

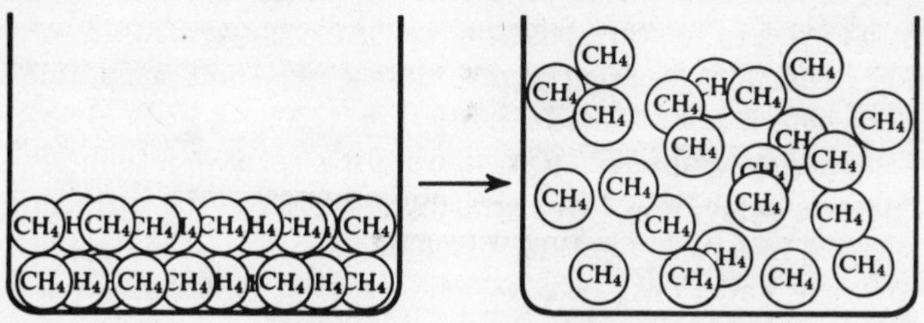

Figure 1.19. Boiling of a non-ionic liquid. Units are molecules.

Liquids whose molecules are held together by hydrogen bonds are called *associated liquids.* Breaking these hydrogen bonds takes considerable energy, and so an associated liquid has a boiling point that is abnormally high for a compound of its molecular weight and dipole moment. Hydrogen fluoride, for example, boils 100 degrees higher than the heavier, non-associated hydrogen chloride; water boils 160 degrees higher than hydrogen sulfide.

The bigger the molecules, the stronger the van der Waals forces. Other things being equal—polarity, hydrogen bonding—boiling point rises with increasing molecular size. Boiling points of organic compounds range upward from that of tiny, non-polar methane, but we seldom encounter boiling points much above 350°; at higher temperatures, covalent bonds *within* the molecules start to break, and decomposition competes with boiling. It is to lower the boiling point and thus minimize decomposition that distillation of organic compounds is often carried out under reduced pressure.

Problem 1.8 Which of the following organic compounds would you predict to be *associated* liquids? Draw structures to show the hydrogen bonding you would expect. (a) CH_3OH; (b) CH_3OCH_3; (c) CH_3F; (d) CH_3Cl; (e) CH_3NH_2; (f) $(CH_3)_2NH$; (g) $(CH_3)_3N$.

1.21 Solubility

When a solid or liquid dissolves, the structural units—ions or molecules—become separated from each other, and the spaces in between become occupied by solvent molecules. In dissolution, as in melting and boiling, energy must be supplied to overcome the interionic or intermolecular forces. Where does the necessary energy come from? The energy required to break the bonds between solute particles is supplied by the formation of bonds between the solute particles and the solvent molecules: the old attractive forces are replaced by new ones.

A great deal of energy is necessary to overcome the powerful electrostatic forces holding together an ionic lattice. Only water or other highly polar solvents are able to dissolve ionic compounds appreciably. What kind of bonds are formed between ions and a polar solvent? By definition, a polar molecule has a positive end and a negative end. Consequently, there is electrostatic attraction between a positive ion and the negative end of the solvent molecule, and between a negative ion and the positive end of the solvent molecule. These attractions are called **ion-dipole** bonds. Each ion–dipole bond is relatively weak, but in the aggregate they supply enough energy to overcome the interionic forces in the crystal. In solution each ion is surrounded by a cluster of solvent molecules, and is said to be *solvated*; if the solvent happens to be water, the ion is said to be *hydrated.* In solution, as in the solid and liquid states, the unit of a substance like sodium chloride is the ion, although in this case it is a solvated ion (see Fig. 1.20).

To dissolve ionic compounds a solvent must also have a high *dielectric constant*, that is, have high insulating properties to lower the attraction between oppositely charged ions once they are solvated.

Water owes its superiority as a solvent for ionic substances not only to its polarity and its high dielectric constant, but to another factor as well: it contains the —OH group and thus can form hydrogen bonds. Water solvates both cations

Figure 1.20. Ion–dipole interactions: solvated cation and anion.

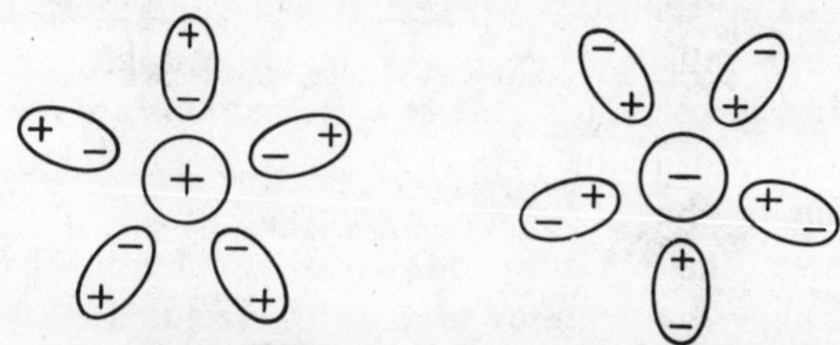

and anions: cations, at its negative pole (its unshared electrons, essentially); anions, through hydrogen bonding.

The solubility characteristics of non-ionic compounds are determined chiefly by their polarity. Non-polar or weakly polar compounds dissolve in non-polar or weakly polar solvents; highly polar compounds dissolve in highly polar solvents. "Like dissolves like" is an extremely useful rule of thumb. Methane dissolves in carbon tetrachloride because the forces holding methane molecules to each other and carbon tetrachloride molecules to each other are replaced by very similar forces holding methane molecules to carbon tetrachloride molecules.

Neither methane nor carbon tetrachloride is readily soluble in water. The highly polar water molecules are held to each other by very strong dipole–dipole interactions—hydrogen bonds; there could be only very weak attractive forces between water molecules on the one hand and the non-polar methane or carbon tetrachloride molecules on the other.

In contrast, the highly polar organic compound methanol, CH_3OH, is quite soluble in water. Hydrogen bonds between water and methanol molecules can readily replace the very similar hydrogen bonds between different methanol molecules and different water molecules.

As we shall see, much of organic chemistry is concerned with reactions between non-ionic compounds (generally organic) and ionic compounds (inorganic and organic), and it is necessary to select a solvent in which both the reagents will dissolve. Water is a poor solvent for most organic compounds, but this difficulty can be overcome by addition of a second solvent like methanol.

Solvents like water or methanol are called *protic solvents*: solvents containing hydrogen that is attached to oxygen or nitrogen, and hence is appreciably acidic (Sec. 1.22). Through hydrogen bonding such solvents tend to solvate anions particularly strongly; and anions, as it turns out, are usually the important half of an ionic reagent. Thus, although protic solvents dissolve the reagent and bring it into contact with the organic molecule, at the same time they stabilize the anions and lower their reactivity drastically; their basicity is weakened and, with it, the related property, *nucleophilic* power (Sec. 14.5).

Recent years have seen the development and widespread use of *aprotic solvents*: polar solvents of moderately high dielectric constants, which do not contain acidic hydrogen. For example:

$$CH_3-\underset{\underset{O}{\|}}{S}-CH_3 \qquad H-\underset{\underset{O}{\|}}{C}-N\begin{matrix}CH_3\\CH_3\end{matrix}$$

Dimethyl sulfoxide
DMSO

N,N-Dimethylformamide
DMF

Sulfolane

These solvents dissolve both organic and inorganic reagents but, in dissolving ionic compounds, solvate *cations* most strongly, and leave the anions relatively unencumbered and highly reactive; anions are more basic and more nucleophilic.

Since about 1958, reports of dramatic solvent effects on a wide variety of reactions have appeared, first about dimethylformamide (DMF) and more recently about dimethyl sulfoxide (DMSO): reactions that, in most solvents, proceed slowly at high temperatures to give low yields may be found, in an aprotic solvent, to proceed rapidly—often at room temperature—to give high yields. A change of solvent may cause a million-fold change in reaction rate. A solvent is not simply a place—a kind of gymnasium—where solute molecules may gambol about and occasionally collide; the solvent is intimately *involved* in any reaction that takes place in it, and we are just beginning to find out how much it is involved, and in what way.

Individual molecules may have both polar and non-polar parts and, if the molecules are big enough, these parts display their individual solubility properties. The polar parts dissolve in water; the non-polar parts dissolve in a non-polar solvent or, if there is none about, cluster together—in effect, dissolve in each other. Such dual solubility behavior gives soaps and detergents their cleansing power, and controls the alignment of molecules in cell membranes; a globular protein molecule—an enzyme, say—coils up to expose its polar parts to the surrounding water and to hide its non-polar parts, and in doing this takes on the particular shape needed for its characteristic biological properties.

1.22 Acids and bases

Turning from physical to chemical properties, let us review briefly one familiar topic that is fundamental to the understanding of organic chemistry: acidity and basicity.

The terms *acid* and *base* have been defined in a number of ways, each definition corresponding to a particular way of looking at the properties of acidity and basicity. We shall find it useful to look at acids and bases from two of these viewpoints; the one we select will depend upon the problem at hand.

According to the **Lowry-Brønsted** definition, *an acid is a substance that gives up a proton*, and *a base is a substance that accepts a proton.* When sulfuric acid dissolves in water, the acid H_2SO_4 gives up a proton (hydrogen nucleus) to the base H_2O to form a new acid H_3O^+ and a new base HSO_4^-. When hydrogen chloride reacts with ammonia, the acid HCl gives up a proton to the base NH_3 to form the new acid NH_4^+ and the new base Cl^-.

$$\underset{\text{Stronger acid}}{H_2SO_4} + \underset{\text{Stronger base}}{H_2O} \rightleftarrows \underset{\text{Weaker acid}}{H_3O^+} + \underset{\text{Weaker base}}{HSO_4^-}$$

$$\underset{\text{Stronger acid}}{HCl} + \underset{\text{Stronger base}}{NH_3} \rightleftarrows \underset{\text{Weaker acid}}{NH_4^+} + \underset{\text{Weaker base}}{Cl^-}$$

According to the Lowry-Brønsted definition, the strength of an acid depends upon its tendency to give up a proton, and the strength of a base depends upon its tendency to accept a proton. Sulfuric acid and hydrogen chloride are strong acids since they tend to give up a proton very readily; conversely, bisulfate ion, HSO_4^-, and chloride ion must necessarily be weak bases since they have little tendency to hold on to protons. In each of the reactions just described, the equilibrium favors the formation of the weaker acid and the weaker base.

If aqueous H_2SO_4 is mixed with aqueous NaOH, the acid H_3O^+ (hydronium ion) gives up a proton to the base OH^- to form the new acid H_2O and the new base H_2O. When aqueous NH_4Cl is mixed with aqueous NaOH, the acid NH_4^+

$$\underset{\text{Stronger acid}}{H_3O^+} + \underset{\text{Stronger base}}{OH^-} \rightleftarrows \underset{\text{Weaker acid}}{H_2O} + \underset{\text{Weaker base}}{H_2O}$$

$$\underset{\text{Stronger acid}}{NH_4^+} + \underset{\text{Stronger base}}{OH^-} \rightleftarrows \underset{\text{Weaker acid}}{H_2O} + \underset{\text{Weaker base}}{NH_3}$$

(ammonium ion) gives up a proton to the base OH^- to form the new acid H_2O and the new base NH_3. In each case the strong base, hydroxide ion, has accepted a proton to form the weak acid H_2O. If we arrange these acids in the order shown, we must necessarily arrange the corresponding (conjugate) bases in the opposite order.

Acid strength $\begin{matrix} H_2SO_4 \\ HCl \end{matrix} > H_3O^+ > NH_4^+ > H_2O$

Base strength $\begin{matrix} HSO_4^- \\ Cl^- \end{matrix} < H_2O < NH_3 < OH^-$

Like water, many organic compounds that contain oxygen can act as bases and accept protons; ethyl alcohol and ethyl ether, for example, form the *oxonium ions* I and II. For convenience, we shall often refer to a structure like I as a *protonated alcohol* and a structure like II as a *protonated ether*.

$$\underset{\text{Ethyl alcohol}}{C_2H_5\ddot{\underset{..}{O}}H} + H_2SO_4 \rightleftarrows \underset{\substack{\text{I} \\ \text{An oxonium ion} \\ \text{Protonated ethyl alcohol}}}{C_2H_5\overset{\oplus}{\ddot{O}}H} + HSO_4^-$$
$$\qquad\qquad\qquad\qquad\qquad\qquad\ddot{H}$$

$$\underset{\text{Ethyl ether}}{(C_2H_5)_2\ddot{O}:} + HCl \rightleftarrows \underset{\substack{\text{II} \\ \text{An oxonium ion} \\ \text{Protonated ethyl ether}}}{(C_2H_5)_2\overset{\oplus}{\ddot{O}}:H} + Cl^-$$

According to the **Lewis** definition, *a base is a substance that can furnish an electron pair to form a covalent bond*, and *an acid is a substance that can take up an electron pair to form a covalent bond.* Thus **an acid is an electron-pair acceptor** and **a base is an electron-pair donor**. This is the most fundamental of the acid-base concepts, and the most general; it includes all the other concepts.

A proton is an acid because it is deficient in electrons, and needs an electron pair to complete its valence shell. Hydroxide ion, ammonia, and water are bases because they contain electron pairs available for sharing. In boron trifluoride, BF_3, boron has only six electrons in its outer shell and hence tends to accept another pair to complete its octet. Boron trifluoride is an acid and combines with

$$\mathrm{F{-}\underset{\underset{F}{|}}{\overset{\overset{F}{|}}{B}}} + \mathrm{:NH_3} \rightleftharpoons \mathrm{F{-}\underset{\underset{F}{|}}{\overset{\overset{F}{\ominus|}}{B}}{:}\overset{\oplus}{N}H_3}$$

Acid Base

$$\mathrm{F{-}\underset{\underset{F}{|}}{\overset{\overset{F}{|}}{B}}} + \mathrm{:\ddot{O}(C_2H_5)_2} \rightleftharpoons \mathrm{F{-}\underset{\underset{F}{|}}{\overset{\overset{F}{\ominus|}}{B}}{:}\overset{\oplus}{\ddot{O}}(C_2H_5)_2}$$

Acid Base

such bases as ammonia or ethyl ether. Aluminum chloride, $AlCl_3$, is an acid, and for the same reason. In stannic chloride, $SnCl_4$, tin has a complete octet, but can accept additional pairs of electrons (e.g., in $SnCl_6^{--}$) and hence it is an acid, too.

We write a formal negative charge on boron in these formulas because it has one more electron—half-interest in the pair shared with nitrogen or oxygen—than is balanced by the nuclear charge; correspondingly, nitrogen or oxygen is shown with a formal positive charge.

We shall find the Lewis concept of acidity and basicity fundamental to our understanding of organic chemistry. To make it clear that we are talking about this kind of acid or base, we shall often use the expression *Lewis acid* (or *base*), or sometimes *acid* (or *base*) *in the Lewis sense.*

Chemical properties, like physical properties, depend upon molecular structure. Just what features in a molecule's structure tell us what to expect about its acidity or basicity? We can try to answer this question in a general way now, although we shall return to it many times later.

To be acidic in the Lowry-Brønsted sense, a molecule must, of course, contain hydrogen. The degree of acidity is determined largely by the kind of atom that holds the hydrogen and, in particular, by that atom's ability to accommodate the electron pair left behind by the departing hydrogen ion. This ability to accommodate the electron pair seems to depend upon several factors, including (a) the atom's *electronegativity*, and (b) its *size*. Thus, within a given row of the Periodic Table, acidity increases as electronegativity increases:

Acidity

$$\mathrm{H{-}CH_3 < H{-}NH_2 < H{-}OH < H{-}F}$$
$$\mathrm{H{-}SH < H{-}Cl}$$

And within a given family, acidity increases as the size increases:

Acidity

$$\mathrm{H{-}F < H{-}Cl < H{-}Br < H{-}I}$$
$$\mathrm{H{-}OH < H{-}SH < H{-}SeH}$$

Among organic compounds, we can expect appreciable Lowry-Brønsted acidity from those containing O—H, N—H, and S—H groups.

To be acidic in the Lewis sense, a molecule must be electron-deficient; in particular, we would look for an atom bearing only a sextet of electrons.

Problem 1.9 Predict the relative acidity of: (a) methyl alcohol (CH_3OH) and methylamine (CH_3NH_2); (b) methyl alcohol (CH_3OH) and methanethiol (CH_3SH); (c) H_3O^+ and NH_4^+.

Problem 1.10 Which is the stronger acid of each pair: (a) H_3O^+ or H_2O; (b) NH_4^+ or NH_3; (c) H_2S or HS^-; (d) H_2O or OH^-? (e) What relationship is there between *charge* and acidity?

To be basic in either the Lowry-Brønsted or the Lewis sense, a molecule must have an electron pair available for sharing. The availability of these unshared electrons is determined largely by the atom that holds them: its electronegativity, its size, its charge. The operation of these factors here is necessarily opposite to what we observed for acidity; the better an atom accommodates the electron pair, the less available the pair is for sharing.

Problem 1.11 Arrange the members of each group in order of basicity: (a) F^-, OH^-, NH_2^-, CH_3^-; (b) HF, H_2O, NH_3; (c) Cl^-, SH^-; (d) F^-, Cl^-, Br^-, I^-; (e) OH^-, SH^-, SeH^-.

Problem 1.12 Predict the relative basicity of methyl fluoride (CH_3F), methyl alcohol (CH_3OH), and methylamine (CH_3NH_2).

Problem 1.13 Arrange the members of each group in order of basicity: (a) H_3O^+, H_2O, OH^-; (b) NH_3, NH_2^-; (c) H_2S, HS^-, S^{--}. (d) What relationship is there between charge and basicity?

1.23 Electronic and steric effects

Like acidity and basicity, other chemical properties, too, depend upon molecular structure. Indeed, most of this book will be concerned with finding out what this relationship is.

A particular compound is found to undergo a particular reaction. Not surprisingly, other compounds of similar structure are also found to undergo the same reaction—but faster or slower, or with the equilibrium lying farther to the right or left. We shall, first of all, try to see how it is that a particular kind of structure predisposes a compound to a particular reaction. Then—and much of our time will be spent with this—we shall try to see how it is that variations in molecular structure give rise to variations in reactivity: to differences in rate of reaction or in position of equilibrium.

To do all this is a complicated business, and to help us we shall mentally analyze the molecule: we shall consider that a molecule consists of a reaction center to which are attached various substituents. The nature of the reaction center determines what reaction occurs. The nature of the substituents determines the reactivity.

A substituent affects reactivity in two general ways: (a) by its *electronic effect*, that is, by its effect on the availability of electrons at the reaction center; and (b) by its *steric effect*, that is, by its effect on crowding in the molecule. Since hydrogen is the element most commonly attached to carbon, it is used as the standard of reference. We consider a substituent G, which may be an atom or a group of atoms, to be attached to a carbon in place of a hydrogen, and we ask the question: How does G—C〰 compare with H—C〰?

Let us look first at **electronic effects**. At some stage of most reactions, a positive or negative charge develops in the reacting molecule. Reactivity usually depends upon how easily the molecule can accommodate that charge. Accommodation of charge depends, in turn, on the electronic effects of the substituents. Compared with hydrogen, a substituent may tend either to *withdraw electrons* (G←C~) or to *release electrons* (G→C~). An electron-withdrawing substituent will help to accept the surplus of electrons that constitutes a negative charge; an electron-releasing substituent will help offset the deficit of electrons that constitutes a positive charge.

Just how does a substituent exert its electronic effect? Despite the vast amount of work that has been done—and is still being done—on this problem, there is no general agreement, except that at least two factors must be at work. We shall consider electron withdrawal and release to result from the operation of two factors: the *inductive effect* and the *resonance effect*.

The **inductive effect** depends upon the "intrinsic" tendency of a substituent to withdraw electrons—by definition, its electronegativity—acting either through the molecular chain or through space. The effect weakens steadily with increasing distance from the substituent. Most elements likely to be substituted for hydrogen in an organic molecule are more electronegative than hydrogen, so that most substituents exert electron-withdrawing inductive effects: for example, —F, —Cl, —Br, —I, —OH, $—NH_2$, $—NO_2$.

The **resonance effect** involves *delocalization* of electrons—typically, those called π (pi) electrons. It depends upon the overlap of certain orbitals, and therefore can operate only when the substituent is located in certain special ways relative to the charge center. By its very nature, as we shall see (Sec. 6.25), the resonance effect is a stabilizing effect, and so it amounts to electron withdrawal from a negatively charged center, and electron release to a positively charged center.

A substituent can influence reactivity not only by its electronic effect (inductive and/or resonance), but also, in some cases, by its **steric effect**: an effect due to crowding at some stage of the reaction, and dependent therefore on the *size* of the substituent.

Problem 1.14 Predict the relative basicity of NH_3 and NF_3.

1.24 Isomerism

Before we start our systematic study of the different kinds of organic compounds, let us look at one further concept which illustrates especially well the fundamental importance of molecular structure: the concept of **isomerism**.

The compound *ethyl alcohol* is a liquid boiling at 78°. Analysis (by the methods described later, Sec. 2.26) shows that it contains carbon, hydrogen, and oxygen in the proportions 2C:6H:1O. Measurement of its mass spectrum shows that it has a molecular weight of 46. The molecular formula of ethyl alcohol must therefore be C_2H_6O. Ethyl alcohol is a quite reactive compound. For example, if a piece of sodium metal is dropped into a test tube containing ethyl alcohol, there is a vigorous bubbling and the sodium metal is consumed; hydrogen gas is evolved and there is left behind a compound of formula C_2H_5ONa. Ethyl alcohol reacts with hydriodic acid to form water and a compound of formula C_2H_5I.

The compound *methyl ether* is a gas with a boiling point of −24°. It is clearly a different substance from ethyl alcohol, differing not only in its physical properties but also in its chemical properties. It does not react at all with sodium metal. Like ethyl alcohol, it reacts with hydriodic acid, but it yields a compound of formula CH_3I. Analysis of methyl ether shows that it contains carbon, hydrogen, and oxygen in the same proportions as ethyl alcohol, 2C:6H:1O. It has the same molecular weight as ethyl alcohol, 46. We conclude that it has the same molecular formula, C_2H_6O.

Here we have two substances, ethyl alcohol and methyl ether, which have the same molecular formula, C_2H_6O, and yet quite clearly are different compounds. How can we account for the existence of these two compounds? The answer is: *they differ in molecular structure*. Ethyl alcohol has the structure represented by I, and methyl ether the structure represented by II. As we shall see, the differences in physical and chemical properties of these two compounds can readily be accounted for on the basis of the difference in structure.

```
    H H               H     H
    | |               |     |
  H-C-C-O-H         H-C-O-C-H
    | |               |     |
    H H               H     H
     I                  II
 Ethyl alcohol      Methyl ether
```

Different compounds that have the same molecular formula are called **isomers** (Gr.: *isos*, equal; *meros*, part). They contain the same numbers of the same kinds of atoms, but the atoms are attached to one another in different ways. Isomers are different compounds because they have different molecular structures.

This difference in molecular structure gives rise to a difference in properties; it is the difference in properties which tells us that we are dealing with different compounds. In some cases, the difference in structure—and hence the difference in properties—is so marked that the isomers are assigned to different chemical families, as, for example, ethyl *alcohol* and methyl *ether*. In other cases the difference in structure is so subtle that it can be described only in terms of three-dimensional models. Other kinds of isomerism fall between these two extremes.

PROBLEMS

1. Which of the following would you expect to be ionic, and which non-ionic? Give a simple electronic structure (Sec. 1.3) for each, showing only valence shell electrons.

(a) $MgCl_2$ (b) CH_2Cl_2 (c) ICl (d) NaOCl (e) $KClO_4$ (f) $SiCl_4$ (g) $BaSO_4$ (h) CH_3NH_2

2. Give a likely simple electronic structure (Sec. 1.3) for each of the following, assuming them to be completely covalent. Assume that every atom (except hydrogen, of course) has a complete octet, and that two atoms may share more than one pair of electrons.

(a) N_2H_4 (b) H_2SO_4 (c) HSO_4^- (d) $COCl_2$ (e) HONO (f) NO_2^- (g) CO_3^{--} (h) C_2H_4 (i) C_2H_2 (j) CH_2O (k) CH_2O_2 (l) C_3H_8

3. What shape would you expect each of the following to have?

(a) $(CH_3)_3B$
(b) the methyl anion, $CH_3{:}^-$
(c) the methyl cation, CH_3^+
(d) H_2S
(e) the amide ion, NH_2^-
(f) methyl ether
(g) the fluoborate ion, BF_4^-
(h) $(CH_3)_3N$

4. In many complex ions, e.g., $Co(NH_3)_6^{+++}$, the bonds to the central atom can be pictured as utilizing six equivalent sp^3d^2 (or d^2sp^3) hybrid orbitals. On the basis of maximum separation of orbitals, what geometry would you expect these complexes to have?

5. Indicate the direction of the dipole moment, *if any*, that you would expect for each of the following:

(a) HBr	(d) CH_2Cl_2	(g) methyl ether
(b) ICl	(e) $CHCl_3$	(h) $(CH_3)_3N$
(c) I_2	(f) CH_3OH	(i) CF_2Cl_2

6. (a) Although HCl (1.27 A) is a longer molecule than HF (0.92 A), it has a *smaller* dipole moment (1.03 D compared to 1.75 D). How do you account for this fact? (b) The dipole moment of CH_3F is 1.847 D, and of CD_3F, 1.858 D. (D is ^2H, deuterium.) Compared with the C—H bond, what is the direction of the C—D dipole?

7. What do the differences in properties between lithium acetylacetonate (m.p. very high, insoluble in chloroform) and beryllium acetylacetonate (m.p. 108°, b.p. 270°, soluble in chloroform) suggest about their structures?

8. *n*-Butyl alcohol (b.p. 118°) has a much higher boiling point than its isomer ethyl ether (b.p. 35°), yet both compounds show the same solubility (8 g per 100 g) in water.

```
   H  H  H  H                    H  H     H  H
   |  |  |  |                    |  |     |  |
H—C—C—C—C—O—H               H—C—C—O—C—C—H
   |  |  |  |                    |  |     |  |
   H  H  H  H                    H  H     H  H
```

n-Butyl alcohol   Ethyl ether

How do you account for these facts?

9. Rewrite the following equations to show the Lowry-Brønsted acids and bases actually involved. Label each as stronger or weaker, as in Sec. 1.22.

(a) $HCl(aq) + NaHCO_3(aq) \rightleftarrows H_2CO_3 + NaCl$
(b) $NaOH(aq) + NaHCO_3(aq) \rightleftarrows Na_2CO_3 + H_2O$
(c) $NH_3(aq) + HNO_3(aq) \rightleftarrows NH_4NO_3(aq)$
(d) $NaCN(aq) \rightleftarrows HCN(aq) + NaOH(aq)$
(e) $NaH + H_2O \longrightarrow H_2 + NaOH$
(f) $CaC_2 + H_2O \longrightarrow Ca(OH)_2 + C_2H_2$
Calcium carbide   Acetylene

10. What is the Lowry-Brønsted acid in (a) HCl dissolved in water; (b) HCl (unionized) dissolved in benzene? (c) Which solution is the more strongly acidic?

11. Account for the fact that nearly every organic compound containing oxygen dissolves in cold concentrated sulfuric acid to yield a solution from which the compound can be recovered by dilution with water.

12. For each of the following molecular formulas, draw structures like those in Sec. 1.24 (a line for each shared pair of electrons) for all the isomers you can think of. Assume that every atom (except hydrogen) has a complete octet, and that two atoms may share more than one pair of electrons.

(a) C_2H_7N	(c) C_4H_{10}	(e) C_3H_8O
(b) C_3H_8	(d) C_3H_7Cl	(f) C_2H_4O

13. In ordinary distillation, a liquid is placed in a flask and heated, at ordinary or reduced pressure, until distillation is complete. In the modification called *flash distillation*, the liquid is dripped into a heated flask at the same rate that it distills out, so that there is little liquid in the flask at any time. What advantage might flash distillation have, and under what conditions might you use it?

About Working Problems

Working problems is a necessary part of your work for two reasons: it will guide your study in the right direction, and, after you have studied a particular chapter, it will show whether or not you have reached your destination.

You should work all the problems that you can; you should get help with the ones you cannot work yourself. The first problems in each set are easy, but provide the drill in drawing formulas, naming compounds, and using reactions that even the best student needs. The later problems in each set are the kind encountered by practicing chemists, and test your ability to *use* what you have learned.

You can check your answers to many of the problems in the answer section in the back of the book, and by use of the index.

C_2H_7N

Chapter 2 Methane
Energy of Activation. Transition State

2.1 Hydrocarbons

Certain organic compounds contain only two elements, hydrogen and carbon, and hence are known as **hydrocarbons**. On the basis of structure, hydrocarbons are divided into two main classes, **aliphatic** and **aromatic**. Aliphatic hydrocarbons are further divided into families: alkanes, alkenes, alkynes, and their cyclic analogs (cycloalkanes, etc.). We shall take up these families in the order given.

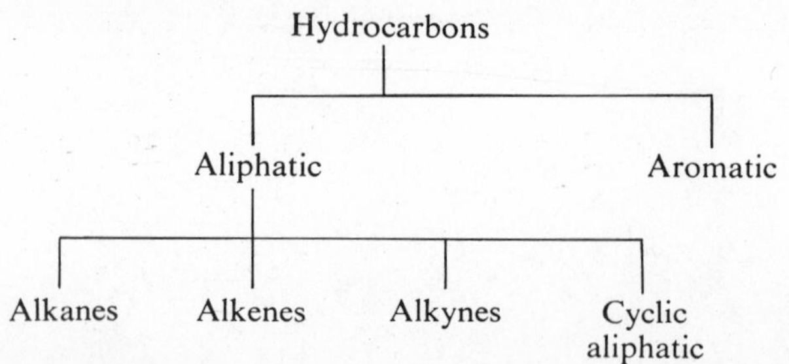

The simplest member of the alkane family and, indeed, one of the simplest of all organic compounds is **methane**, CH_4. We shall study this single compound at some length, since most of what we learn about it can be carried over with minor modifications to any alkane.

2.2 Structure of methane

As we discussed in the previous chapter (Sec. 1.11), each of the four hydrogen atoms is bonded to the carbon atom by a covalent bond, that is, by the sharing of a pair of electrons. When carbon is bonded to four other atoms, its bonding orbitals (sp^3 orbitals, formed by the mixing of one s and three p orbitals) are directed to the corners of a tetrahedron (Fig. 2.1a). This tetrahedral arrangement is the one that permits the orbitals to be as far apart as possible. For each of these orbitals to overlap most effectively the spherical *s* orbital of a hydrogen atom, and thus to form the strongest bond, each hydrogen nucleus must be located at a corner of this tetrahedron (Fig. 2.1*b*).

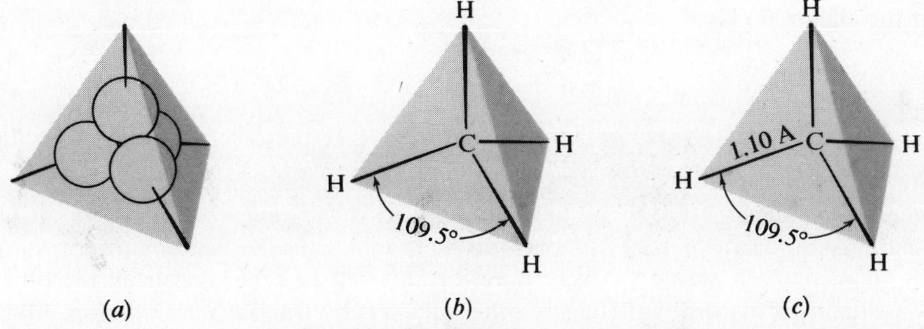

Figure 2.1. Methane molecule. (*a*) Tetrahedral sp^3 orbitals. (*b*) Predicted shape: H nuclei located for maximum overlap. (*c*) Shape and size.

The tetrahedral structure of methane has been verified by electron diffraction (Fig. 2.1*c*), which shows beyond question the arrangement of atoms in such simple molecules. Later on, we shall examine some of the evidence that led chemists to accept this tetrahedral structure long before quantum mechanics or electron diffraction was known.

We shall ordinarily write methane with a dash to represent each pair of electrons shared by carbon and hydrogen (I). To focus our attention on individual electrons, we may sometimes indicate a pair of electrons by a pair of dots (II). Finally, when we wish to consider the actual shape of the molecule, we shall use a simple three-dimensional picture (III).

H | H—C—H | H	H ·· H:C:H ·· H	H H—○—H H
I	II	III

2.3 Physical properties

As we discussed in the previous chapter (Sec. 1.18), the unit of such a non-ionic compound, whether solid, liquid, or gas, is the molecule. Because the methane molecule is highly symmetrical, the polarities of the individual carbon–hydrogen bonds cancel out; as a result, the molecule itself is non-polar.

Attraction between such non-polar molecules is limited to van der Waals forces; for such small molecules, these attractive forces must be tiny compared with the enormous forces between, say, sodium and chloride ions. It is not surprising, then, that these attractive forces are easily overcome by thermal energy, so that melting and boiling occur at very low temperatures: m.p. −183°, b.p. −161.5°. (Compare these values with the corresponding ones for sodium chloride: m.p. 801°, b.p. 1413°.) As a consequence, methane is a gas at ordinary temperatures.

Methane is colorless and, when liquefied, is less dense than water (sp.gr. 0.4). In agreement with the rule of thumb that "like dissolves like," it is only slightly soluble in water, but very soluble in organic liquids such as gasoline, ether, and alcohol. In its physical properties methane sets the pattern for the other members of the alkane family.

2.4 Source

Methane is an end product of the anaerobic ("without air") decay of plants, that is, of the breakdown of certain very complicated molecules. As such, it is the major constituent (up to 97%) of **natural gas**. It is the dangerous *firedamp* of the coal mine, and can be seen as *marsh gas* bubbling to the surface of swamps.

If methane is wanted in very pure form, it can be separated from the other constituents of natural gas (mostly other alkanes) by fractional distillation. Most of it, of course, is consumed as fuel without purification.

According to one theory, the origins of life go back to a primitive earth surrounded by an atmosphere of methane, water, ammonia, and hydrogen. Energy—radiation from the sun, lightning discharges—broke these simple molecules into reactive fragments (free radicals, Sec. 2.12); these combined to form larger molecules which eventually yielded the enormously complicated organic compounds that make up living organisms. (Recent detection of organic molecules in space has even led to the speculation that "organic seeds for life could have existed in interstellar clouds.")

Evidence that this *could* have happened was found in 1953 by the Nobel Prize winner Harold C. Urey and his student Stanley Miller at the University of Chicago. They showed that an electric discharge converts a mixture of methane, water, ammonia, and hydrogen into a large number of organic compounds, including amino acids, the building blocks from which proteins, the "stuff of life" (Chap. 36), are made. (It is perhaps appropriate that we begin this study of organic chemistry with methane and its conversion into free radicals.)

The methane generated in the final decay of a once-living organism may well be the very substance from which—in the final analysis—the organism was derived. "*. . . earth to earth, ashes to ashes, dust to dust. . . .*"

2.5 Reactions

In its chemical properties as in its physical properties, methane sets the pattern for the alkane family (Sec. 3.18). Typically, it reacts only with highly reactive substances—or under very vigorous conditions, which, as we shall see, amounts to the same thing. At this point we shall take up only its oxidation: by oxygen, by halogens, and even by water.

2.6 Oxidation. Heat of combustion

Combustion to carbon dioxide and water is characteristic of organic compounds; under special conditions it is used to determine their content of carbon and hydrogen (Sec. 2.26).

Combustion of methane is the principal reaction taking place during the

REACTIONS OF METHANE

1. Oxidation

$$CH_4 + 2O_2 \xrightarrow{\text{flame}} CO_2 + 2H_2O + \text{heat (213 kcal/mole)}$$ *Combustion*

$$6CH_4 + O_2 \xrightarrow{1500^\circ} \underset{\text{Acetylene}}{2HC{\equiv}CH} + 2CO + 10H_2$$ Discussed in Sec. 8.5.

$$CH_4 + H_2O \xrightarrow[\text{Ni}]{850^\circ} CO + 3H_2$$

2. Halogenation

$$CH_4 \xrightarrow{X_2} \underset{}{CH_3X} \xrightarrow{X_2} CH_2X_2 \xrightarrow{X_2} CHX_3 \xrightarrow{X_2} CX_4$$

(each step also gives HX) *Heat or light required*

Reactivity of X_2 $F_2 > Cl_2 > Br_2$ ($> I_2$)
Unreactive

burning of natural gas. It is hardly necessary to emphasize its importance in the areas where natural gas is available; the important product is not carbon dioxide or water but *heat*.

Burning of hydrocarbons takes place only at high temperatures, as provided, for example, by a flame or a spark. Once started, however, the reaction gives off heat which is often sufficient to maintain the high temperature and to permit burning to continue. *The quantity of heat evolved when one mole of a hydrocarbon is burned to carbon dioxide and water is called the* **heat of combustion**; for methane its value is 213 kcal.

Through controlled *partial* oxidation of methane and the high-temperature catalytic reaction with water, methane is an increasingly important source of products other than heat: of hydrogen, used in the manufacture of ammonia; of mixtures of carbon monoxide and hydrogen, used in the manufacture of *methanol* and other alcohols; and of *acetylene* (Sec. 8.5), itself the starting point of large-scale production of many organic compounds.

Oxidation by halogens is of particular interest to us—partly because we know more about it than the other reactions of methane—and, in one way or another, is the topic of discussion throughout the remainder of this chapter.

2.7 Chlorination: a substitution reaction

Under the influence of ultraviolet light or at a temperature of 250–400° a mixture of the two gases, methane and chlorine, reacts vigorously to yield hydrogen chloride and a compound of formula CH_3Cl. We say that methane has undergone **chlorination**, and we call the product, CH_3Cl, *chloromethane* or *methyl chloride* (CH_3 = **methyl**).

Chlorination is a typical example of a broad class of organic reactions known as **substitution**. A chlorine atom has been substituted for a hydrogen atom of methane, and the hydrogen atom thus replaced is found combined with a second atom of chlorine.

$$\underset{\text{Methane}}{\text{H}-\overset{\text{H}}{\underset{\text{H}}{\text{C}}}-\text{H}} + \underset{\text{Chlorine}}{\text{Cl}-\text{Cl}} \xrightarrow{\text{light or heat}} \underset{\substack{\text{Methyl chloride}\\ \text{(Chloromethane)}}}{\text{H}-\overset{\text{H}}{\underset{\text{H}}{\text{C}}}-\text{Cl}} + \underset{\substack{\text{Hydrogen}\\ \text{chloride}}}{\text{H}-\text{Cl}}$$

The methyl chloride can itself undergo further substitution to form more hydrogen chloride and CH_2Cl_2, *dichloromethane* or *methylene chloride* (CH_2 = **methylene**).

$$\text{H}-\overset{\text{H}}{\underset{\text{H}}{\text{C}}}-\text{Cl} + \text{Cl}-\text{Cl} \xrightarrow{\text{light or heat}} \underset{\substack{\text{Methylene chloride}\\ \text{(Dichloromethane)}}}{\text{H}-\overset{\text{H}}{\underset{\text{Cl}}{\text{C}}}-\text{Cl}} + \text{H}-\text{Cl}$$

In a similar way, chlorination may continue to yield $CHCl_3$, *trichloromethane* or *chloroform*, and CCl_4, *tetrachloromethane* or *carbon tetrachloride*. These last two compounds are already familiar to us, chloroform as an anesthetic, and carbon tetrachloride as a non-flammable cleaning agent and the fluid in certain fire extinguishers.

$$\underset{\text{Methane}}{CH_4} \xrightarrow{Cl_2} \underset{\substack{\text{Methyl}\\ \text{chloride}}}{\overset{\substack{HCl\\ +}}{CH_3Cl}} \xrightarrow{Cl_2} \underset{\substack{\text{Methylene}\\ \text{chloride}}}{\overset{\substack{HCl\\ +}}{CH_2Cl_2}} \xrightarrow{Cl_2} \underset{\text{Chloroform}}{\overset{\substack{HCl\\ +}}{CHCl_3}} \xrightarrow{Cl_2} \underset{\substack{\text{Carbon}\\ \text{tetrachloride}}}{\overset{\substack{HCl\\ +}}{CCl_4}}$$

Heat or light required

2.8 Control of chlorination

Chlorination of methane may yield any one of four organic products, depending upon the stage to which the reaction is carried. Can we control this reaction so that methyl chloride is the principal organic product? That is, can we limit the reaction to the first stage, *mono*chlorination?

We might at first expect—naïvely, as it turns out—to accomplish this by providing only one mole of chlorine for each mole of methane. But let us see what happens if we do so. At the beginning of the reaction there is only methane for the chlorine to react with, and consequently only the first stage of chlorination takes place. This reaction, however, yields methyl chloride, so that as the reaction proceeds methane disappears and methyl chloride takes its place.

As the proportion of methyl chloride grows, it competes with the methane for the available chlorine. By the time the concentration of methyl chloride exceeds that of methane, chlorine is more likely to attack methyl chloride than methane, and the second stage of chlorination becomes more important than the first. A large amount of methylene chloride is formed, which in a similar way is chlorinated to chloroform and this, in turn, is chlorinated to carbon tetrachloride. When we finally work up the reaction product, we find that it is a mixture of all four chlorinated methanes together with some unreacted methane.

The reaction may, however, be limited almost entirely to monochlorination if we use a large excess of methane. In this case, even at the very end of the reaction unreacted methane greatly exceeds methyl chloride. Chlorine is more likely to attack methane than methyl chloride, and thus the first stage of chlorination is the principal reaction.

Because of the great difference in their boiling points, it is easy to separate the excess methane (b.p. $-161.5°$) from the methyl chloride (b.p. $-24°$) so that the methane can be mixed with more chlorine and put through the process again. While there is a low **conversion** of methane into methyl chloride in each cycle, the **yield** of methyl chloride based on the chlorine consumed is quite high.

The use of a large excess of one reactant is a common device of the organic chemist when he wishes to limit reaction to only one of a number of reactive sites in the molecule of that reactant.

2.9 Reaction with other halogens: halogenation

Methane reacts with bromine, again at high temperatures or under the influence of ultraviolet light, to yield the corresponding bromomethanes: methyl bromide, methylene bromide, bromoform, and carbon tetrabromide.

$$\underset{\text{Methane}}{CH_4} \xrightarrow{Br_2} \underset{\text{Methyl bromide}}{\overset{\displaystyle HBr \atop +}{CH_3Br}} \xrightarrow{Br_2} \underset{\text{Methylene bromide}}{\overset{\displaystyle HBr \atop +}{CH_2Br_2}} \xrightarrow{Br_2} \underset{\text{Bromoform}}{\overset{\displaystyle HBr \atop +}{CHBr_3}} \xrightarrow{Br_2} \underset{\text{Carbon tetrabromide}}{\overset{\displaystyle HBr \atop +}{CBr_4}}$$

Heat or light required

Bromination takes place somewhat less readily than chlorination.

Methane does not react with iodine at all. With fluorine it reacts so vigorously that, even in the dark and at room temperature, the reaction must be carefully controlled: the reactants, diluted with an inert gas, are mixed at low pressure.

We can, therefore, arrange the halogens in order of reactivity.

Reactivity of halogens $F_2 > Cl_2 > Br_2\ (> I_2)$

This same order of reactivity holds for the reaction of the halogens with other alkanes and, indeed, with most other organic compounds. The spread of reactivities is so great that only chlorination and bromination proceed at such rates as to be generally useful.

2.10 Relative reactivity

Throughout our study of organic chemistry, we shall constantly be interested in *relative reactivities*. We shall compare the reactivities of various reagents toward the same organic compound, the reactivities of different organic compounds toward the same reagent, and even the reactivities of different sites in an organic molecule toward the same reagent.

It should be understood that when we compare reactivities we compare rates of reaction. When we say that chlorine is *more reactive* than bromine toward methane, we mean that under the same conditions (same concentration, same temperature, etc.) chlorine reacts with methane *faster* than does bromine. From another point of view, we mean that the bromine reaction must be carried out under more vigorous conditions (higher concentration or higher temperature) if it is to take place as fast as the chlorine reaction. When we say that methane and iodine do not react at all, we mean that the reaction is too slow to be significant.

We shall want to know not only what these relative reactivities are, but also, whenever possible, how to account for them. To see what factors cause one reaction to be faster than another, we shall take up in more detail this matter of the different reactivities of the halogens toward methane. Before we can do this, however, we must understand a little more about the reaction itself.

2.11 Reaction mechanisms

It is important for us to know not only *what* happens in a chemical reaction but also *how* it happens, that is, to know not only the *facts* but also the *theory*.

For example, we know that methane and chlorine under the influence of heat or light form methyl chloride and hydrogen chloride. Just how is a molecule of methane converted into a molecule of methyl chloride? Does this transformation involve more than one step, and, if so, what are these steps? Just what is the function of heat or light?

The answer to questions like these, that is, *the detailed, step-by-step description of a chemical reaction, is called a* **mechanism**. It is only a hypothesis; it is advanced to account for the facts. As more facts are discovered, the mechanism must also account for them, or else be modified so that it does account for them; it may even be necessary to discard a mechanism and to propose a new one.

It would be difficult to say that a mechanism had ever been *proved.* If, however, a mechanism accounts satisfactorily for a wide variety of facts; if we make predictions based upon this mechanism and find these predictions borne out; if the mechanism is consistent with mechanisms for other, related reactions; then the mechanism is said to be *well established*, and it becomes part of the theory of organic chemistry.

Why are we interested in the mechanisms of reactions? As an important part of the theory of organic chemistry, they help make up the framework on which we hang the facts we learn. An understanding of mechanisms will help us to see a pattern in the complicated and confusing multitude of organic reactions. We shall find that many apparently unrelated reactions proceed by the same or similar mechanisms, so that most of what we have already learned about one reaction may be applied directly to many new ones.

By knowing how a reaction takes place, we can make changes in the experimental conditions—not by trial and error, but logically—that will improve the yield of the product we want, or that will even alter the course of the reaction completely and give us an entirely different product. As our understanding of reactions grows, so does our power to control them.

2.12 Mechanism of chlorination. Free radicals

It will be worthwhile to examine the mechanism of chlorination of methane in some detail. The same mechanism holds for bromination as well as chlorination, and for other alkanes as well as methane; it even holds for many compounds which, while not alkanes, contain alkane-like portions in their molecules. Closely related mechanisms are involved in oxidation (combustion) and other reactions of alkanes. More important, this mechanism illustrates certain general principles that can be carried over to a wide range of chemical reactions. Finally, by studying the evidence that supports the mechanism, we can learn something of how a chemist finds out what goes on during a chemical reaction.

Among the facts that must be accounted for are these: (a) Methane and chlorine do not react in the dark at room temperature. (b) Reaction takes place readily, however, in the dark at temperatures over 250°, or (c) under the influence of ultraviolet light at room temperature. (d) When the reaction is induced by light, many (several thousand) molecules of methyl chloride are obtained for each photon of light that is absorbed by the system. (e) The presence of a small amount of oxygen slows down the reaction for a period of time, after which the reaction proceeds normally; the length of this period depends upon how much oxygen is present.

The mechanism that accounts for these facts most satisfactorily, and hence is generally accepted, is shown in the following equation:

$$(1) \qquad Cl_2 \xrightarrow{\text{heat or light}} 2Cl\cdot$$

$$(2) \qquad Cl\cdot + CH_4 \longrightarrow HCl + CH_3\cdot$$

$$(3) \qquad CH_3\cdot + Cl_2 \longrightarrow CH_3Cl + Cl\cdot$$

then (2), (3), (2), (3), etc.

The first step is the breaking of a chlorine molecule into two chlorine atoms. Like the breaking of any bond, this requires energy, the *bond dissociation energy*, and in Table 1.2 (p. 21) we find that in this case the value is 58 kcal/mole. The energy is supplied as either heat or light.

$$\text{energy} + :\ddot{\underset{..}{Cl}}:\ddot{\underset{..}{Cl}}: \longrightarrow :\ddot{\underset{..}{Cl}}\cdot + \cdot\ddot{\underset{..}{Cl}}:$$

The chlorine molecule undergoes *homolysis* (Sec. 1.14): that is, cleavage of the chlorine–chlorine bond takes place in a symmetrical way, so that each atom retains one electron of the pair that formed the covalent bond. This **odd electron** is not *paired* as are all the other electrons of the chlorine atom; that is, it does not have a partner of opposite spin (Sec. 1.6). *An atom or group of atoms possessing an odd (unpaired) electron is called a* **free radical**. In writing the symbol for a free radical, we generally include a dot to represent the odd electron just as we include a plus or minus sign in the symbol of an ion.

Once formed, what is a chlorine atom most likely to do? Like most free radicals, it is extremely reactive because of its tendency to gain an additional electron and thus have a complete octet; from another point of view, energy was supplied to each chlorine atom during the cleavage of the chlorine molecule, and this energy-rich particle tends strongly to lose energy by the formation of a new chemical bond.

To form a new chemical bond, that is, to react, the chlorine atom must collide with some other molecule or atom. What is it most likely to collide with? Obviously, it is most likely to collide with the particles that are present in the highest concentration: chlorine molecules and methane molecules. Collision with another chlorine atom is quite unlikely simply because there are very few of these reactive, short-lived particles around at any time. Of the likely collisions, that with a chlorine molecule causes no net change; reaction may occur, but it can result only in the exchange of one chlorine atom for another:

$$:\underset{\cdot\cdot}{\overset{\cdot\cdot}{\mathrm{Cl}}}\cdot + :\underset{\cdot\cdot}{\overset{\cdot\cdot}{\mathrm{Cl}}}:\underset{\cdot\cdot}{\overset{\cdot\cdot}{\mathrm{Cl}}}: \longrightarrow :\underset{\cdot\cdot}{\overset{\cdot\cdot}{\mathrm{Cl}}}:\underset{\cdot\cdot}{\overset{\cdot\cdot}{\mathrm{Cl}}}: + :\underset{\cdot\cdot}{\overset{\cdot\cdot}{\mathrm{Cl}}}\cdot \qquad \textit{Collision probable but not productive}$$

Collision of a chlorine atom with a methane molecule is both *probable* and *productive*. The chlorine atom abstracts a hydrogen atom, with one electron, to form a molecule of hydrogen chloride:

$$\underset{\text{Methane}}{\mathrm{H}:\underset{\underset{\mathrm{H}}{\cdot\cdot}}{\overset{\overset{\mathrm{H}}{\cdot\cdot}}{\mathrm{C}}}:\mathrm{H}} + \cdot\underset{\cdot\cdot}{\overset{\cdot\cdot}{\mathrm{Cl}}}: \longrightarrow \mathrm{H}:\underset{\cdot\cdot}{\overset{\cdot\cdot}{\mathrm{Cl}}}: + \underset{\text{Methyl radical}}{\mathrm{H}:\underset{\underset{\mathrm{H}}{\cdot\cdot}}{\overset{\overset{\mathrm{H}}{\cdot\cdot}}{\mathrm{C}}}\cdot} \qquad \textit{Collision probable and productive}$$

Now the methyl group is left with an odd, unpaired electron; the carbon atom has only seven electrons in its valence shell. One free radical, the chlorine atom, has been consumed, and a new one, the methyl radical, $CH_3\cdot$, has been formed in its place. This is step (2) in the mechanism.

Now, what is this methyl radical most likely to do? Like the chlorine atom, it is extremely reactive, and for the same reason: the tendency to complete its octet, to lose energy by forming a new bond. Again, collisions with chlorine molecules or methane molecules are the probable ones, not collisions with the relatively scarce chlorine atoms or methyl radicals. But collision with a methane molecule could at most result only in the exchange of one methyl radical for another:

$$\mathrm{H}:\underset{\underset{\mathrm{H}}{\cdot\cdot}}{\overset{\overset{\mathrm{H}}{\cdot\cdot}}{\mathrm{C}}}:\mathrm{H} + \cdot\underset{\underset{\mathrm{H}}{\cdot\cdot}}{\overset{\overset{\mathrm{H}}{\cdot\cdot}}{\mathrm{C}}}:\mathrm{H} \longrightarrow \mathrm{H}:\underset{\underset{\mathrm{H}}{\cdot\cdot}}{\overset{\overset{\mathrm{H}}{\cdot\cdot}}{\mathrm{C}}}\cdot + \mathrm{H}:\underset{\underset{\mathrm{H}}{\cdot\cdot}}{\overset{\overset{\mathrm{H}}{\cdot\cdot}}{\mathrm{C}}}:\mathrm{H} \qquad \textit{Collision probable but not productive}$$

The collision of a methyl radical with a chlorine molecule is, then, the important one. The methyl radical abstracts a chlorine atom, with one of the bonding electrons, to form a molecule of methyl chloride:

$$\underset{\text{Methyl radical}}{\mathrm{H}:\underset{\underset{\mathrm{H}}{\cdot\cdot}}{\overset{\overset{\mathrm{H}}{\cdot\cdot}}{\mathrm{C}}}\cdot} + :\underset{\cdot\cdot}{\overset{\cdot\cdot}{\mathrm{Cl}}}:\underset{\cdot\cdot}{\overset{\cdot\cdot}{\mathrm{Cl}}}: \longrightarrow \underset{\text{Methyl chloride}}{\mathrm{H}:\underset{\underset{\mathrm{H}}{\cdot\cdot}}{\overset{\overset{\mathrm{H}}{\cdot\cdot}}{\mathrm{C}}}:\underset{\cdot\cdot}{\overset{\cdot\cdot}{\mathrm{Cl}}}:} + :\underset{\cdot\cdot}{\overset{\cdot\cdot}{\mathrm{Cl}}}\cdot \qquad \textit{Collision probable and productive}$$

The other product is a chlorine atom. This is step (3) in the mechanism.

Here again the consumption of one reactive particle has been accompanied by the formation of another. The new chlorine atom attacks methane to form a methyl radical, which attacks a chlorine molecule to form a chlorine atom, and so the sequence is repeated over and over. Each step produces not only a new reactive particle but also a molecule of product: methyl chloride or hydrogen chloride.

This process cannot, however, go on forever. As we saw earlier, union of two short-lived, relatively scarce particles is not likely; but every so often it does happen, and when it does, this particular sequence of reactions stops. Reactive particles are consumed but not generated.

$$:\ddot{\underset{..}{Cl}}\cdot + \cdot\ddot{\underset{..}{Cl}}: \longrightarrow :\ddot{\underset{..}{Cl}}:\ddot{\underset{..}{Cl}}:$$

$$CH_3\cdot + \cdot CH_3 \longrightarrow CH_3:CH_3$$

$$CH_3\cdot + \cdot\ddot{\underset{..}{Cl}}: \longrightarrow CH_3:\ddot{\underset{..}{Cl}}:$$

It is clear, then, how the mechanism accounts for facts (a), (b), (c), and (d) on page 47: either light or heat is required to cleave the chlorine molecule and form the initial chlorine atoms; once formed, each atom may eventually bring about the formation of many molecules of methyl chloride.

2.13 Chain reactions

The chlorination of methane is an example of a **chain reaction**, *a reaction that involves a series of steps, each of which generates a reactive substance that brings about the next step.* While chain reactions may vary widely in their details, they all have certain fundamental characteristics in common.

(1) $Cl_2 \xrightarrow{\text{heat or light}} 2Cl\cdot$ **Chain-initiating step**

(2) $Cl\cdot + CH_4 \longrightarrow HCl + CH_3\cdot$

(3) $CH_3\cdot + Cl_2 \longrightarrow CH_3Cl + Cl\cdot$

(2)–(3): **Chain-propagating steps**

then (2), (3), (2), (3), etc., *until finally*:

(4) $Cl\cdot + \cdot Cl \longrightarrow Cl_2$

or

(5) $CH_3\cdot + \cdot CH_3 \longrightarrow CH_3CH_3$

or

(6) $CH_3\cdot + \cdot Cl \longrightarrow CH_3Cl$

(4)–(6): **Chain-terminating steps**

First in the chain of reactions is a **chain-initiating step**, in which energy is absorbed and a reactive particle generated; in the present reaction it is the cleavage of chlorine into atoms (step 1).

There are one or more **chain-propagating steps**, each of which consumes a reactive particle and generates another; here they are the reaction of chlorine atoms with methane (step 2), and of methyl radicals with chlorine (step 3).

Finally, there are **chain-terminating steps**, in which reactive particles are consumed but not generated; in the chlorination of methane these would involve the union of two of the reactive particles, or the capture of one of them by the walls of the reaction vessel.

Under one set of conditions, about 10,000 molecules of methyl chloride are formed for every quantum (photon) of light absorbed. Each photon cleaves one chlorine molecule to form two chlorine atoms, each of which starts a chain. On the average, each chain consists of 5000 repetitions of the chain-propagating cycle before it is finally stopped.

2.14 Inhibitors

Finally, how does the mechanism of chlorination account for fact (e), that a small amount of oxygen slows down the reaction for a period of time, which depends upon the amount of oxygen, after which the reaction proceeds normally?

Oxygen is believed to react with a methyl radical to form a new free radical:

$$CH_3\cdot + O_2 \longrightarrow CH_3{-}O{-}O\cdot$$

The $CH_3OO\cdot$ radical is much less reactive than the $CH_3\cdot$ radical, and can do little to continue the chain. By combining with a methyl radical, one oxygen molecule breaks a chain, and thus prevents the formation of thousands of molecules of methyl chloride; this, of course, slows down the reaction tremendously. After all the oxygen molecules present have combined with methyl radicals, the reaction is free to proceed at its normal rate.

A substance that slows down or stops a reaction even though present in small amount is called an **inhibitor**. *The period of time during which inhibition lasts, and after which the reaction proceeds normally, is called the inhibition period.* Inhibition by a relatively small amount of an added material is quite characteristic of chain reactions of any type, and is often one of the clues that first leads us to suspect that we are dealing with a chain reaction. It is hard to see how else a few molecules could prevent the reaction of so many. (We shall frequently encounter the use of oxygen to inhibit free-radical reactions.)

2.15 Heat of reaction

In our consideration of the chlorination of methane, we have so far been concerned chiefly with the particles involved—molecules and atoms—and the changes that they undergo. As with any reaction, however, it is important to consider also the energy changes involved, since these changes determine to a large extent how fast the reaction will go, and, in fact, whether it will take place at all.

By using the values of bond dissociation energies given in Table 1.2 (p. 21), we can calculate the energy changes that take place in a great number of reactions. In the conversion of methane into methyl chloride, two bonds are broken, $CH_3{-}H$ and Cl—Cl, consuming 104 + 58, or a total of 162 kcal/mole. At the same time two new bonds are formed, $CH_3{-}Cl$ and H—Cl, liberating 84 + 103, or a total of 187 kcal/mole. The result is the liberation of 25 kcal of heat for every mole of

$$\underbrace{\underset{104}{CH_3{-}H} + \underset{58}{Cl{-}Cl}}_{162} \longrightarrow \underbrace{\underset{84}{CH_3{-}Cl} + \underset{103}{H{-}Cl}}_{187} \qquad \Delta H = -25 \text{ kcal}$$

methane that is converted into methyl chloride; this is, then, an **exothermic reaction.** (This calculation, we note, does not depend on our knowing the mechanism of the reaction.)

When heat is liberated, the heat content (enthalpy), H, of the molecules themselves must decrease; the change in heat content, ΔH, is therefore given a negative sign. (In the case of an endothermic reaction, where heat is absorbed, the increase in heat content of the molecules is indicated by a positive ΔH.)

Problem 2.1 Calculate ΔH for the corresponding reaction of methane with: (a) bromine, (b) iodine, (c) fluorine.

The value of -25 kcal that we have just calculated is the *net* ΔH for the overall reaction. A more useful picture of the reaction is given by the ΔH's of the individual steps. These are calculated below:

(1) $\underset{(58)}{\text{Cl—Cl}} \longrightarrow 2\text{Cl}\cdot \qquad \Delta H = +58 \text{ kcal}$

(2) $\text{Cl}\cdot + \underset{(104)}{\text{CH}_3\text{—H}} \longrightarrow \text{CH}_3\cdot + \underset{(103)}{\text{H—Cl}} \qquad \Delta H = +1$

(3) $\text{CH}_3\cdot + \underset{(58)}{\text{Cl—Cl}} \longrightarrow \underset{(84)}{\text{CH}_3\text{—Cl}} + \text{Cl}\cdot \qquad \Delta H = -26$

It is clear why this reaction, even though exothermic, occurs only at a high temperature (in the absence of light). The chain-initiating step, without which reaction cannot occur, is highly *endothermic*, and takes place (at a significant rate) only at a high temperature. Once the chlorine atoms are formed, the two chain-propagating steps—one only slightly endothermic, and the other exothermic—occur readily many times before the chain is broken. The difficult cleavage of chlorine is the barrier that must be surmounted before the subsequent easy steps can be taken.

Problem 2.2 Calculate ΔH for the corresponding steps in the reaction of methane with: (a) bromine, (b) iodine, (c) fluorine.

We have assumed so far that exothermic reactions proceed readily, that is, are reasonably fast at ordinary temperatures, whereas endothermic reactions proceed with difficulty, that is, are slow except at very high temperatures. This assumed relationship between ΔH and rate of reaction is a useful rule of thumb when other information is not available; it is *not*, however, a *necessary* relationship, and there are many exceptions to the rule. We shall go on, then, to a discussion of another energy quantity, the *energy of activation*, which is related in a more exact way to rate of reaction.

2.16 Energy of activation

To see what actually happens during a chemical reaction, let us look more closely at a specific example, the attack of chlorine atoms on methane:

$$\text{Cl}\cdot + \underset{(104)}{\text{CH}_3\text{—H}} \longrightarrow \underset{(103)}{\text{H—Cl}} + \text{CH}_3\cdot \qquad \Delta H = +1 \text{ kcal} \quad E_{\text{act}} = 4 \text{ kcal}$$

This reaction is comparatively simple: it occurs in the gas phase, and is thus not complicated by the presence of a solvent; it involves the interaction of a single atom and the simplest of organic molecules. Yet from it we can learn certain principles that apply to any reaction.

Just what must happen if this reaction is to occur? First of all, a chlorine atom and a methane molecule must **collide**. Since chemical forces are of extremely short range, a hydrogen–chlorine bond can form only when the atoms are in close contact.

Next, to be *effective*, the collision must provide a certain *minimum amount of energy*. Formation of the H—Cl bond liberates 103 kcal/mole; breaking the

CH_3—H bond requires 104 kcal/mole. We might have expected that only 1 kcal/mole additional energy would be needed for reaction to occur; however, this is not so. Bond-breaking and bond-making evidently are not perfectly synchronized, and the energy liberated by the one process is not completely available for the other. Experiment has shown that if reaction is to occur, an additional 4 kcal/mole of energy must be supplied.

The minimum amount of energy that must be provided by a collision for reaction to occur is called the **energy of activation**, E_{act}. Its source is the kinetic energy of the moving particles. Most collisions provide less than this minimum quantity and are fruitless, the original particles simply bouncing apart. Only solid collisions between particles one or both of which are moving unusually fast are energetic enough to bring about reaction. In the present example, at 275°, only about one collision in 40 is sufficiently energetic.

Finally, in addition to being sufficiently energetic, the collisions must occur when the particles are properly **oriented**. At the instant of collision, the methane molecule must be turned in such a way as to present a hydrogen atom to the full force of the impact. In the present example, only about one collision in eight is properly oriented.

In general, then, *a chemical reaction requires collisions of sufficient energy* (E_{act}) *and of proper orientation*. There is an energy of activation for nearly every reaction where bonds are broken, even for exothermic reactions, in which bond-making liberates more energy than is consumed by bond-breaking.

The attack of bromine atoms on methane is more highly endothermic, with a ΔH of +16 kcal.

$$\underset{}{Br\cdot} + \underset{(104)}{CH_3\text{—}H} \longrightarrow \underset{(88)}{H\text{—}Br} + CH_3\cdot \qquad \Delta H = +16\text{ kcal} \quad E_{act} = 18\text{ kcal}$$

Breaking the CH_3—H bond, as before, requires 104 kcal/mole, of which only 88 kcal is provided by formation of the H—Br bond. It is evident that, even if this 88 kcal were completely available for bond-breaking, at least an additional 16 kcal/mole would have to be supplied by the collision. In other words, the E_{act} of an endothermic reaction must be at least as large as the ΔH. As is generally true, the E_{act} of the present reaction (18 kcal) is actually somewhat larger than the ΔH.

2.17 Progress of reaction: energy changes

These energy relationships can be seen more clearly in diagrams like Figs. 2.2 and 2.3. Progress of reaction is represented by horizontal movement from reactants on the left to products on the right. Potential energy (that is, all energy except kinetic) at any stage of reaction is indicated by the height of the curve.

Let us follow the course of reaction in Fig. 2.2. We start in a potential energy valley with a methane molecule and a chlorine atom. These particles are moving, and hence possess kinetic energy in addition to the potential energy shown. The exact amount of kinetic energy varies with the particular pair of particles, since some move faster than others. They collide, and kinetic energy is converted into potential energy. With this increase in potential energy, reaction begins, and we move up the energy hill. If enough kinetic energy is converted, we reach the top of the hill and start down the far side.

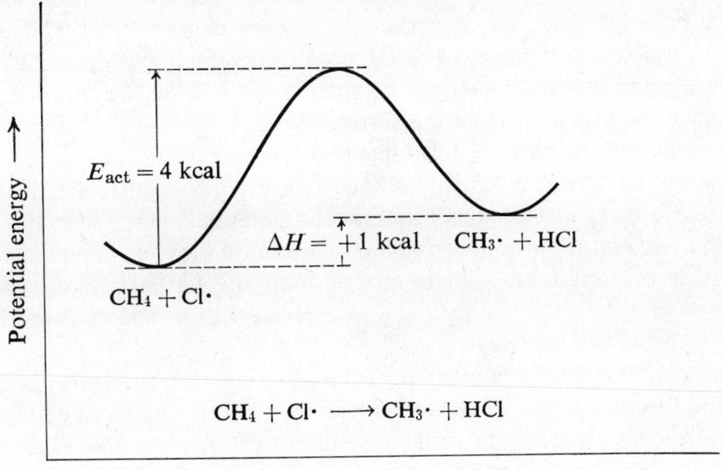

Figure 2.2. Potential energy changes during progress of reaction: the methane–chlorine atom reaction.

During the descent, potential energy is converted back into kinetic energy, until we reach the level of the products. The products contain a little more potential energy than did the reactants, and we find ourselves in a slightly higher valley than the one we left. With this net increase in potential energy there must be a corresponding decrease in kinetic energy. The new particles break apart, and since they are moving more slowly than the particles from which they were formed,

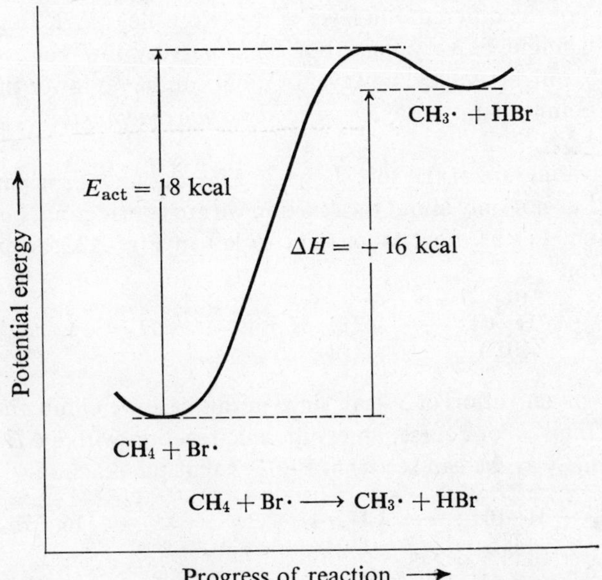

Figure 2.3. Potential energy changes during progress of reaction: the methane–bromine atom reaction.

we observe a drop in temperature. Heat will be *taken up* from the surroundings.

In the bromine reaction, shown in Fig. 2.3, we climb a much higher hill and end up in a much higher valley. The increase in potential energy—and the corresponding decrease in kinetic energy—is much larger than in the chlorine reaction; more heat will be taken up from the surroundings.

An exothermic reaction follows much the same course. (Take, for example, the reverse of the bromine reaction; that is, read from right to left in Fig. 2.3.) In this case, however, the products contain less potential energy than did the reactants so that we end up in a lower valley than the one we left. Since this time the new particles contain more kinetic energy than the particles from which they were formed, and hence move faster, we observe a rise in temperature. Heat will be *given off* to the surroundings.

In any reaction there are many collisions that provide too little energy for us to reach the top of the hill. These collisions are fruitless, and we slide back to our original valley. Many collisions provide sufficient energy, but take place when the molecules are improperly oriented. We then climb an energy hill, but we are off the road; we may climb very high without finding the pass that leads over into the next valley.

The difference in level between the two valleys is, of course, the ΔH; the difference in level between the reactant valley and the top of the hill is the E_{act}. We are concerned only with these differences, and not with the absolute height at any stage of the reaction. We are not even concerned with the relative levels of the reactant valleys in the chlorine and bromine reactions. We need only to know that in the chlorine reaction we climb a hill 4 kcal high and end up in a valley 1 kcal higher than our starting point; and that in the bromine reaction we climb a hill 18 kcal high and end up in a valley 16 kcal higher than our starting point.

As we shall see, it is the height of the hill, the E_{act}, that determines the rate of reaction, and not the difference in level of the two valleys, ΔH. In going to a lower valley, the hill might be very high, but *could* be very low—or even non-existent. In climbing to a higher valley, however, the hill can be no lower than the valley to which we are going; that is to say, *in an endothermic reaction the E_{act} must be at least as large as the ΔH.*

An energy diagram of the sort shown in Figs. 2.2 and 2.3 is particularly useful because it tells us not only about the reaction we are considering, but also about the reverse reaction. Let us move from right to left in Fig. 2.2, for example. We see that the reaction

$$\underset{}{CH_3\cdot} + \underset{(103)}{H\text{—}Cl} \longrightarrow \underset{(104)}{CH_3\text{—}H} + Cl\cdot \qquad \Delta H = -1, \quad E_{act} = 3$$

has an energy of activation of 3 kcal, since in this case we climb the hill from the higher valley. This is, of course, an exothermic reaction with a ΔH of -1 kcal.

In the same way we can see from Fig. 2.3 that the reaction

$$CH_3\cdot + \underset{(88)}{H\text{—}Br} \longrightarrow \underset{(104)}{CH_3\text{—}H} + Br\cdot \qquad \Delta H = -16, \quad E_{act} = 2$$

has an energy of activation of 2 kcal, and is exothermic with a ΔH of -16 kcal. (We notice that, even though exothermic, these last two reactions have energies of activation.)

Reactions like the cleavage of chlorine into atoms fall into a special category:

$$\underset{(58)}{\text{Cl—Cl}} \longrightarrow \text{Cl}\cdot + \cdot\text{Cl} \qquad \Delta H = +58, \quad E_{\text{act}} = 58$$

a bond is broken but no bonds are formed. The reverse of this reaction, the union of chlorine atoms, involves no bond-breaking and hence would be expected to

$$\text{Cl}\cdot + \cdot\text{Cl} \longrightarrow \underset{(58)}{\text{Cl—Cl}} \qquad \Delta H = -58, \quad E_{\text{act}} = 0$$

take place very easily, in fact, with no energy of activation at all. This is considered to be generally true for reactions involving the union of two free radicals.

If there is no hill to climb in going from chlorine atoms to a chlorine molecule, but simply a slope to descend, the cleavage of a chlorine molecule must involve simply the ascent of a slope as shown in Fig. 2.4. The E_{act} for the cleavage of a chlorine molecule, then, must equal the ΔH, that is, 58 kcal. This equality of E_{act} and ΔH is believed to hold generally for reactions in which molecules dissociate into radicals.

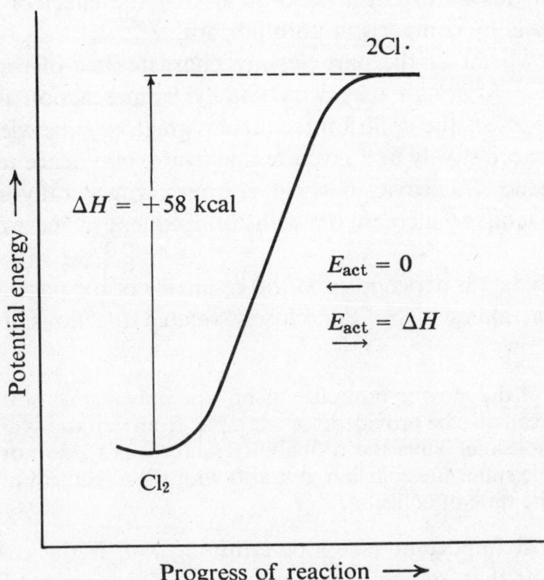

Figure 2.4. Potential energy changes during progress of reaction: simple dissociation.

2.18 Rate of reaction

A chemical reaction is the result of collisions of sufficient energy and proper orientation. The rate of reaction, therefore, must be the rate at which these effective collisions occur, the number of effective collisions, let us say, that occur during each second within each cc of reaction space. We can then express the rate as the product of three factors. (The number expressing the probability that a collision will have the proper orientation is commonly called the **probability factor**.) Anything that affects any one of these factors affects the rate of reaction.

number of effective collisions per cc per sec	=	total number of collisions per cc per sec	×	fraction of collisions that have sufficient energy	×	fraction of collisions that have proper orientation
rate	=	collision frequency	×	energy factor	×	probability factor (orientation factor)

The **collision frequency** depends upon (a) how closely the particles are crowded together, that is, concentration or pressure; (b) how large they are; and (c) how fast they are moving, which in turn depends upon their weight and the temperature.

We can change the concentration and temperature, and thus change the rate. We are familiar with the fact that an increase in concentration causes an increase in rate; it does so, of course, by increasing the collision frequency. A rise in temperature increases the collision frequency; as we shall see, it also increases the energy factor, and this latter effect is so great that the effect of temperature on collision frequency is by comparison unimportant.

The size and weight of the particles are characteristic of each reaction and cannot be changed. Although they vary widely from reaction to reaction, this variation does not affect the collision frequency greatly. A heavier weight makes the particle move more slowly at a given temperature, and hence tends to decrease the collision frequency. A heavier particle is, however, generally a larger particle, and the larger size tends to increase the collision frequency. These two factors thus tend to cancel out.

The **probability factor** depends upon the geometry of the particles and the kind of reaction that is taking place. For closely related reactions it does not vary widely.

Kinetic energy of the moving molecules is not the only source of the energy needed for reaction; energy can also be provided, for example, from vibrations among the various atoms within the molecule. Thus the probability factor has to do not only with what atoms in the molecule suffer the collision, but also with the alignment of the other atoms in the molecule at the time of collision.

By far the most important factor determining rate is the **energy factor**: the fraction of collisions that are sufficiently energetic. This factor depends upon the temperature, which we can control, and upon the energy of activation, which is characteristic of each reaction.

At a given temperature the molecules of a particular compound have an average velocity and hence an average kinetic energy that is characteristic of this system; in fact, the temperature is a measure of this average kinetic energy. But the individual molecules do not all travel with the same velocity, some moving faster than the average and some slower. The distribution of kinetic energy is shown in Fig. 2.5 by the familiar bell-shaped curve that describes the distribution among individuals of so many qualities, for example, height, intelligence, income, or even life expectancy. The number of molecules with a particular kinetic energy is greatest

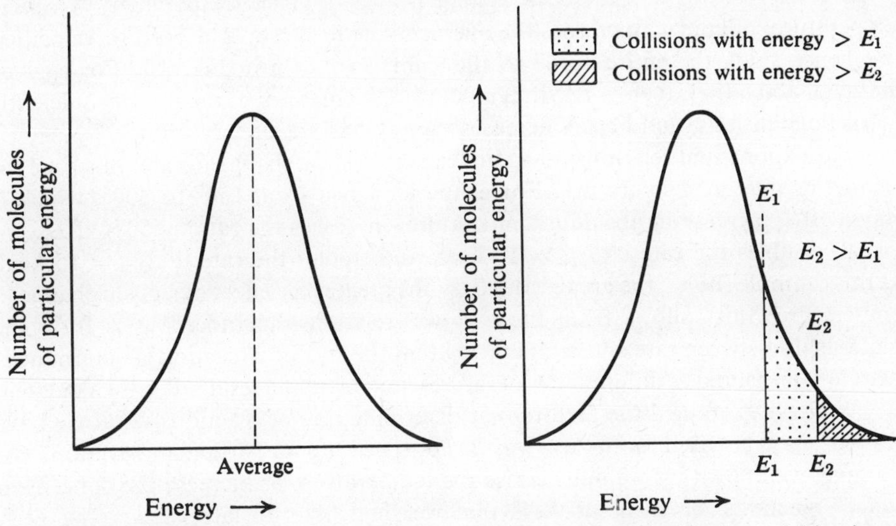

Figure 2.5. Distribution of kinetic energy among molecules.

Figure 2.6. Distribution of kinetic energy among collisions.

for an energy near the average and decreases as the energy becomes larger or smaller than the average.

The distribution of collision energies, as we might expect, is described by a similar curve, Fig. 2.6. Let us indicate collisions of a particular energy, E_{act}, by a vertical line. The number of collisions with energy equal to or greater than E_{act} is indicated by the shaded area under the curve to the right of the vertical line. The fraction of the total number of collisions that have this minimum energy, E_{act}, is then the fraction of the total area that is shaded. It is evident that *the greater the value of E_{act}, the smaller the fraction of collisions that possess that energy.*

The exact relationship between energy of activation and fraction of collisions with that energy is:

$$e^{-E_{act}/RT} = \text{fraction of collisions with energy greater than } E_{act}$$

where

$$e = 2.718 \text{ (base of natural logarithms)}$$
$$R = 1.986 \text{ (gas constant)}$$
$$T = \text{absolute temperature.}$$

Using P for the probability factor and Z for the collision frequency, we arrive at the rate equation:

$$\text{rate} = PZe^{-E_{act}/RT}$$

This exponential relationship is important to us in that it indicates that a small difference in E_{act} has a large effect on the fraction of sufficiently energetic collisions, and hence on the rate of reaction. For example, at 275°, out of every million collisions, 10,000 provide sufficient energy if E_{act} = 5 kcal, 100 provide sufficient energy if E_{act} = 10 kcal, and only one provides sufficient energy if E_{act} = 15 kcal. This means that (all other things being equal) a reaction with E_{act} = 5 kcal will go 100 times as fast as one with E_{act} = 10 kcal, and 10,000 times as fast as one with E_{act} = 15 kcal.

We have so far considered a system held at a given temperature. A rise in temperature, of course, increases the average kinetic energy and average velocities, and hence shifts the entire curve to the right, as shown in Fig. 2.7. For a given energy of activation, then, a rise in temperature increases the fraction of sufficiently energetic collisions, and hence increases the rate, as we already know.

The exponential relationship again leads to a large change in rate, this time for a small change in temperature. For example, a rise from 250° to 300°, which is only a 10% increase in absolute temperature, increases the rate by 50% if E_{act} = 5 kcal, doubles the rate if E_{act} = 10 kcal, and trebles the rate if E_{act} = 15 kcal. As this example shows, the greater the E_{act}, the greater the effect of a given change in temperature; this follows from the $e^{-E_{act}/RT}$ relationship. Indeed, it is from the relationship between rate and temperature that the E_{act} of a reaction is determined: the rate is measured at different temperatures, and from the results E_{act} is calculated.

We have examined the factors that determine rate of reaction. What we have learned may be used in many ways. To speed up a particular reaction, for example, we know that we might raise the temperature, or increase the concentration of reactants, or even (in ways that we shall take up later) lower the E_{act}.

Of immediate interest, however, is the matter of relative reactivities. Let us see, therefore, how our knowledge of reaction rates can help us to account for the fact that one reaction proceeds faster than another, even though conditions for the two reactions are identical.

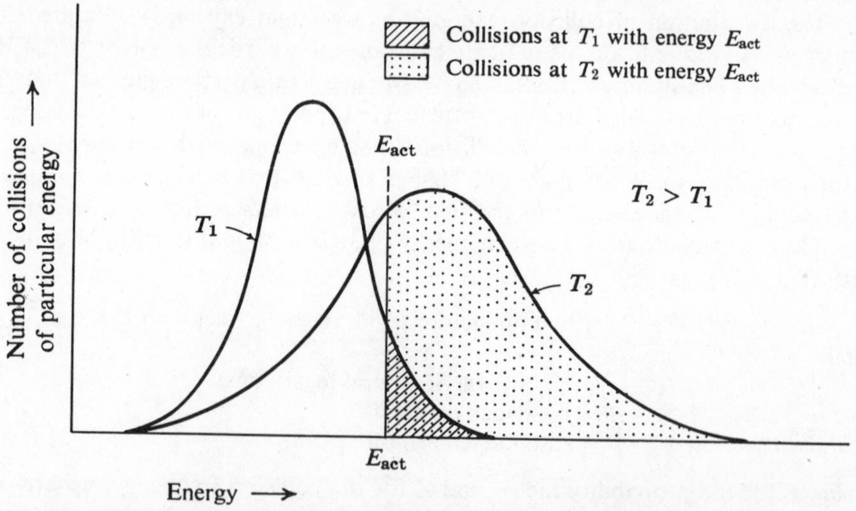

Figure 2.7. Change in collision energies with change in temperature.

2.19 Relative rates of reaction

We have seen that the rate of a reaction can be expressed as a product of three factors:

$$\text{rate} = \text{collision frequency} \times \text{energy factor} \times \text{probability factor}$$

Two reactions could proceed at different rates because of differences in any or all these factors. To account for a difference in rate, we must first see in which of these factors the difference lies.

As an example, let us compare the reactivities of chlorine and bromine atoms toward methane; that is, let us compare the rates, under the same conditions, of the two reactions:

$$Cl\cdot + CH_3{-}H \longrightarrow H{-}Cl + CH_3\cdot \qquad \Delta H = +1, \quad E_{act} = 4$$

$$Br\cdot + CH_3{-}H \longrightarrow H{-}Br + CH_3\cdot \qquad \Delta H = +16, \quad E_{act} = 18$$

Since temperature and concentration must be the same for the two reactions if we are to compare them under the same conditions, any difference in **collision frequency** would have to arise from differences in particle weight or size. A bromine atom is heavier than a chlorine atom, and it is also larger; as we have seen, the effects of these two properties tend to cancel out. In actuality, the collision frequencies differ by only a few per cent. It is generally true that for the same temperature and concentration, two closely related reactions differ but little in collision frequency. A difference in collision frequency therefore cannot be the cause of a large difference in reactivity.

The nature of the **probability factor** is very poorly understood. Since our two reactions are quite similar, however, we might expect them to have similar probability factors. Experiment has shown this to be true: whether chlorine or bromine atoms are involved, about one in every eight collisions with methane has the proper orientation for reaction. In general, where closely related reactions are concerned, we may assume that a difference in probability factor is *not likely* to be the cause of a large difference in reactivity.

We are left with a consideration of the **energy factor**. At a given temperature, the fraction of collisions that possess the amount of energy required for reaction depends upon how large that amount is, that is, depends upon the E_{act}. In our example E_{act} is 4 kcal for the chlorine reaction, 18 kcal for the bromine reaction. As we have seen, a difference of this size in the E_{act} causes an enormous difference in the energy factor, and hence in the rate. At 275°, of every 10 million collisions, 250,000 are sufficiently energetic when chlorine atoms are involved, and only *one* when bromine atoms are involved. Because of the difference in E_{act} alone, then, chlorine atoms are 250,000 times as reactive as bromine atoms toward methane.

As we encounter, again and again, differences in reactivity, we shall in general attribute them to differences in E_{act}; in many cases we shall be able to account for these differences in E_{act} on the basis of differences in molecular structure. *It must be understood that we are justified in doing this only when the reactions being compared are so closely related that differences in collision frequency and in probability factor are comparatively insignificant.*

2.20 Relative reactivities of halogens toward methane

With this background, let us return to the reaction between methane and the various halogens, and see if we can account for the order of reactivity given before, $F_2 > Cl_2 > Br_2 > I_2$, and in particular for the fact that iodine does not react at all.

From the table of bond dissociation energies (Table 1.2, p. 21) we can calculate for each of the four halogens the ΔH for each of the three steps of halogenation. Since E_{act} has been measured for only a few of these reactions, let us see what tentative conclusions we can reach using only ΔH.

			X =	F	Cl	Br	I
(1)	$X_2 \longrightarrow 2X\cdot$		$\Delta H =$	+38	+58	+46	+36
(2)	$X\cdot + CH_4 \longrightarrow HX + CH_3\cdot$			−32	+1	+16	+33
(3)	$CH_3\cdot + X_2 \longrightarrow CH_3X + X\cdot$			−70	−26	−24	−20

Since step (1) involves simply dissociation of molecules into atoms, we may quite confidently assume (Sec. 2.17 and Fig. 2.4) that ΔH in this case is equal to E_{act}. Chlorine has the largest E_{act}, and should dissociate most slowly; iodine has the smallest E_{act}, and should dissociate most rapidly. Yet this does not agree with the observed order of reactivity. Thus, except possibly for fluorine, dissociation of the halogen into atoms cannot be the step that determines the observed reactivities.

Step (3), attack of methyl radicals on halogen, is exothermic for all four halogens, and for chlorine, bromine, and iodine it has very nearly the same ΔH. For these reactions, E_{act} *could* be very small, and does indeed seem to be so; probably only a fraction of a kcal. Even iodine has been found to react readily with methyl radicals generated in another way, e.g., by the heating of tetramethyllead. In fact, iodine is sometimes employed as a free-radical "trap" or "scavenger" in the study of reaction mechanisms. The third step, then, cannot be the cause of the observed relative reactivities.

This leaves step (2), abstraction of hydrogen from methane by a halogen atom. Here we see a wide spread of ΔH's, from the highly exothermic reaction with the fluorine atom to the highly endothermic reaction with the iodine atom. The endothermic bromine atom reaction must have an E_{act} of at least 16 kcal; as we have seen, it is actually 18 kcal. The slightly endothermic chlorine atom reaction could have a very small E_{act}; it is actually 4 kcal. At a given temperature, then, the fraction of collisions of sufficient energy is much larger for methane and chlorine atoms than for methane and bromine atoms. To be specific, at 275° the fraction is about 1 in 40 for chlorine and 1 in 10 million for bromine.

A bromine atom, on the average, collides with many methane molecules before it succeeds in abstracting hydrogen; a chlorine atom collides with relatively few. During its longer search for the proper methane molecule, a bromine atom is more likely to encounter another scarce particle—a second halogen atom or a methyl radical—or be captured by the vessel wall; the chains should therefore be much shorter than in chlorination. Experiment has shown this to be so: where the average chain length is several thousand for chlorination, it is less than 100 for bromination. Even though bromine atoms are formed more rapidly than chlorine atoms at a given temperature because of the lower E_{act} of step (1), overall bromination is slower than chlorination because of the shorter chain length.

For the endothermic reaction of an iodine atom with methane, E_{act} can be no less than 33 kcal, and is probably somewhat larger. Even for this minimum value of 33 kcal, an iodine atom must collide with an enormous number of methane molecules (10^{12} or a million million at 275°) before reaction is likely to occur. Virtually no iodine atoms last this long, but instead recombine to form iodine molecules; the reaction therefore proceeds at a negligible rate. Iodine atoms are easy to form; it is their inability to abstract hydrogen from methane that prevents iodination from occurring.

We cannot predict the E_{act} for the highly exothermic attack of fluorine atoms on methane, but we would certainly not expect it to be any larger than for the attack of chlorine atoms on methane. It appears actually to be smaller (about 1 kcal), thus permitting even longer chains. Because of the surprising weakness of the fluorine–fluorine bond, fluorine atoms should be formed faster than chlorine atoms; thus there should be not only longer chains in fluorination but also *more* chains. The overall reaction is extremely exothermic, with a ΔH of -102 kcal, and the difficulty of removing this heat is one cause of the difficulty of control of fluorination.

Of the two chain-propagating steps, then, step (2) is more difficult than step (3) (see Fig. 2.8). Once formed, methyl radicals react easily with any of the halogens; it is how fast methyl radicals are formed that limits the rate of overall reaction. Fluorination is fast because fluorine atoms rapidly abstract hydrogen atoms from methane; E_{act} is only 1 kcal. Iodination does not take place because iodine atoms find it virtually impossible to abstract hydrogen from methane; E_{act} is more than 33 kcal.

Values of E_{act} for step (2), we notice, parallel the values of ΔH. Since the same bond, CH_3—H, is being broken in every case, the differences in ΔH reflect differences in bond dissociation energy among the various hydrogen–halogen bonds. Ultimately, it appears, the reactivity of a halogen toward methane depends upon the strength of the bond which that halogen forms with hydrogen.

One further point requires clarification. We have said that an E_{act} of 33 kcal is too great for the reaction between iodine atoms and methane to proceed at a significant rate; yet the initial step in each of these halogenations requires an even

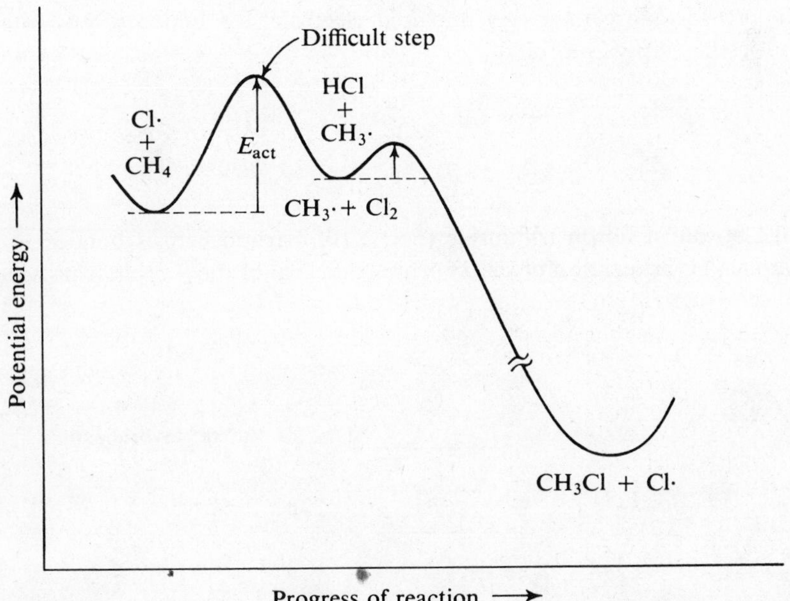

Figure 2.8. Potential energy changes during progress of reaction: chlorination of methane. Formation of radical is difficult step.

greater E_{act}. The difference is this: since halogenation is a chain reaction, dissociation of each molecule of halogen gives rise ultimately to many molecules of methyl halide; hence, even though dissociation is very slow, the overall reaction can be fast. The attack of iodine atoms on methane, however, is a chain-carrying step and if it is slow the entire reaction must be slow; under these circumstances chain-terminating steps (e.g., union of two iodine atoms) become so important that effectively there is *no* chain.

2.21 Structure of the methyl radical. sp^2 Hybridization

We have spent a good part of this chapter discussing the formation and reactions of the methyl free radical $CH_3\cdot$. Just what is this molecule like? What is its shape? How are the electrons distributed and, in particular, where is the odd electron?

These are important questions, for the answers apply not only to this simple radical but to any free radical, however complicated, that we shall encounter. The *shape*, naturally, underlies the three-dimensional chemistry—the stereochemistry—of free radicals. The *location of the odd electron* is intimately involved with the stabilization of free radicals by substituent groups.

As we did when we "made" methane (Sec. 1.11), let us start with the electronic configuration of carbon,

	1s	2s	2p		
C	⊙(··)	⊙(··)	⊙	⊙	○

and, to provide more than two unpaired electrons for bonding, promote a 2s electron to the empty 2p orbital:

	1s	2s	2p			
C	⊙(··)	⊙	⊙	⊙	⊙	*One electron promoted: four unpaired electrons*

Like boron in boron trifluoride (Sec. 1.10), carbon here is bonded to three other atoms. Hybridization of the 2s orbital and two of the *p* orbitals provides the

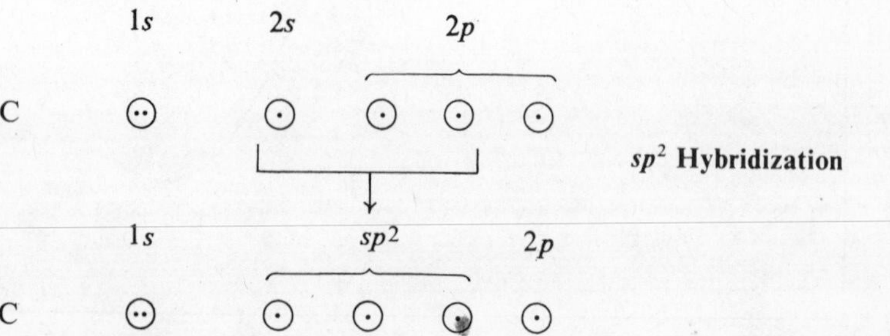

necessary orbitals: three strongly directed sp^2 orbitals which, as we saw before, lie in a plane that includes the carbon nucleus, and are directed to the corners of an equilateral triangle.

If we arrange the carbon and three hydrogens of a methyl radical to permit maximum overlap of orbitals, we obtain the structure shown in Fig. 2.9*a*. It is

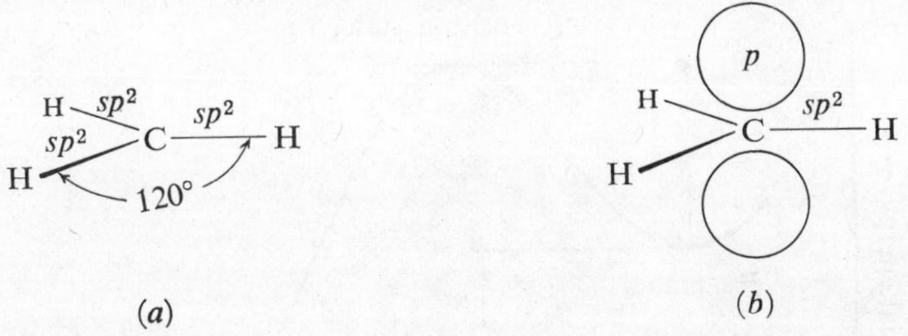

Figure 2.9. Methyl radical. (*a*) Only σ bonds shown. (*b*) Odd electron in *p* orbital above and below plane of σ bonds.

flat, with the carbon atom at the center of a triangle and the three hydrogen atoms at the corners. Every bond angle is 120°.

Now where is the odd electron? In forming the sp^2 orbitals, the carbon atom has used only two of its three *p* orbitals. The remaining *p* orbital consists of two equal lobes, one lying above and the other lying below the plane of the three sp^2 orbitals (Fig. 2.9*b*); it is occupied by the odd electron.

This is not the only conceivable electronic configuration for the methyl radical: an alternative treatment would lead to a pyramidal molecule like that of ammonia, except that the fourth sp^3 orbital contains the odd electron instead of an electron pair (Sec. 1.12). Quantum mechanical calculations do not offer a clear-cut decision between the two configurations. Spectroscopic studies indicate that the methyl radical is actually flat, or nearly so. Carbon is trigonal, or not far from it; the odd electron occupies a *p* orbital, or at least an orbital with much *p* character.

Compare the shapes of three molecules in which the central atom is bonded to three other atoms: (a) boron trifluoride, with no unshared electrons, trigonal; (b) ammonia, with an unshared *pair*, tetrahedral; and (c) the methyl radical, with a *single* unshared electron, trigonal or intermediate between trigonal and tetrahedral.

There is stereochemical evidence (for example, Sec. 7.10) that most other free radicals are either flat or, if pyramidal, undergo rapid *inversion* like that of the ammonia molecule (Sec. 1.12).

Problem 2.3 Besides free radicals, we shall encounter two other kinds of reactive particles, carbonium ions (positive charge on carbon) and carbanions (negative charge on carbon). Suggest an electronic configuration, and from this predict the shape, of the methyl cation, CH_3^+; of the methyl anion, $CH_3{:}^-$.

2.22 Transition state

Clearly, the concept of E_{act} is to be our key to the understanding of chemical reactivity. To make it *useful*, we need a further concept: *transition state*.

A chemical reaction is presumably a continuous process involving a gradual transition from reactants to products. It has been found extremely helpful, however, to consider the arrangement of atoms at an intermediate stage of reaction as though it were an actual molecule. This intermediate structure is called the **transition state**; its energy content corresponds to the top of the energy hill (Fig. 2.10).

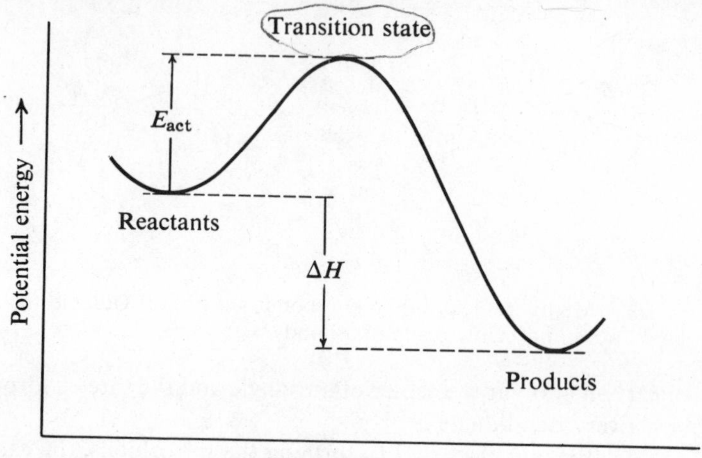

Figure 2.10. Potential energy changes during progress of reaction: transition state at top of energy hump.

The reaction sequence is now:

$$\text{reactants} \longrightarrow \text{transition state} \longrightarrow \text{products}$$

Just as ΔH is the difference in energy content between reactants and products, so E_{act} *is the difference in energy content between reactants and transition state.*

The transition state concept is useful for this reason: we can analyze the structure of the transition state very much as though it were a molecule, and attempt to estimate its stability. Any factor that stabilizes the transition state relative to the reactants tends to lower the energy of activation; that is to say, any factor that lowers the top of the energy hill more than it lowers the reactant valley reduces the net height we must climb during reaction. Transition state stability will be the basis—whether explicit or implicit—of almost every discussion of reactivity in this book.

But the transition state is only a fleeting arrangement of atoms which, by its very nature—lying at the top of an energy hill—cannot be isolated and examined. How can we possibly know anything about its structure? Well, let us take as an example the transition state for the abstraction of hydrogen from methane by a halogen atom, and see where a little thinking will lead us.

To start with, we can certainly say this: the carbon–hydrogen bond is stretched but not entirely broken, and the hydrogen–halogen bond has started to form but is not yet complete. This condition could be represented as

$$\underset{\text{Reactants}}{\mathrm{CH_4} + \cdot\mathrm{X}} \longrightarrow \underset{\text{Transition state}}{\left[\mathrm{H_3C}\text{---}\overset{\delta\cdot}{\mathrm{H}}\text{---}\overset{\delta\cdot}{\mathrm{X}}\right]} \longrightarrow \underset{\text{Products}}{\mathrm{H_3C}\cdot + \mathrm{H{-}X}}$$

where the dashed lines indicate partly broken or partly formed bonds.

Now, what can we say about the shape of the methyl group in this transition state? In the reactant, where methyl holds the hydrogen, carbon is tetrahedral (sp^3-hybridized); in the product, where methyl has lost the hydrogen, carbon is trigonal (sp^2-hybridized). In the transition state, where the carbon–hydrogen bond is partly broken, hybridization of carbon is somewhere between sp^3 and sp^2. The methyl group is partly but not completely flattened; bond angles are greater than 109.5° but less than 120°.

—H + X· ⟶ δ· --H--- δ· X ⟶ · + H—X

Reactant — *Tetrahedral* | Transition state — *Becoming trigonal* | Product — *Trigonal*

Finally, where is the odd electron? It is on chlorine in the reactants, on the methyl group in the products, and divided between the two in the transition state. (Each atom's share is represented by the symbol δ·.) The methyl group *partly* carries the odd electron it will have in the product, and to this extent has taken on some of the character of the free radical it will become.

Thus, in a straightforward way, we have drawn a picture of the transition state that shows the bond-making and bond-breaking, the spatial arrangement of the atoms, and the distribution of the electrons.

(This particular transition state is intermediate between reactants and products not only in the time sequence but also in structure. Not *all* transition states are intermediate in structure: as shown on page 462, reactant and product in S_N2 reactions are tetrahedral, whereas the transition state contains pentavalent carbon.)

In Sec. 2.18, we looked at the matter of reaction rates from the standpoint of the *collision theory*. An alternative, more generally useful approach is the *transition state* (or *thermodynamic*) *theory* of reaction rates. An equilibrium is considered to exist between the reactants and the transition state, and this is handled in the same way as true equilibria of reversible reactions (Sec. 18.11). Energy of activation (E_{act}) and probability factor are replaced by, respectively, *heat* (*enthalpy*) *of activation* ($\Delta H\ddagger$) and *entropy of activation* ($\Delta S\ddagger$), which together make up *free energy of activation* ($\Delta G\ddagger$).

$$\Delta G\ddagger = \Delta H\ddagger - T\Delta S\ddagger$$

The smaller (the less positive) the $\Delta H\ddagger$ and the larger (the more positive) the $\Delta S\ddagger$, the smaller $\Delta G\ddagger$ will be, and the faster the reaction.

Entropy corresponds, roughly, to the randomness of a system; equilibrium tends to favor the side in which fewer restrictions are placed on the atoms and molecules. Entropy of activation, then, is a measure of the relative randomness of reactants and transition state; the fewer the restrictions that are placed on the arrangement of atoms in the transition state—relative to the reactants—the faster the reaction will go. We can see, in a general way, how probability factor and entropy of activation measure much the same thing. A low probability factor means that a rather special orientation of atoms is required on collision. In the other language, an unfavorable (low) entropy of activation means that rather severe restrictions are placed on the positions of atoms in the transition state.

2.23 Reactivity and development of the transition state

For the abstraction of hydrogen from methane by a halogen atom, we have just seen that the transition state differs from the reactants—and this difference is, of course, what we are looking for—chiefly in being like the products. This is generally true for reactions in which free radicals (or, for that matter, carbonium ions or carbanions) are formed.

But just *how much* does this particular transition state resemble the products? How far have bond-breaking and bond-making gone? How flat has the methyl group become, and to what extent does it carry the odd electron?

Surprisingly, we can answer even questions like these, at least in a relative way. **In a set of similar reactions, the higher the E_{act}, the later the transition state is reached in the reaction process.** Of the theoretical considerations underlying this postulate, we shall mention only this: the difference in electronic distribution that we call a difference in structure corresponds to a difference in energy; the greater the difference in structure, the greater the difference in energy. If E_{act} is high, the transition state differs greatly from the reactants in energy and, presumably, also in electronic structure; if E_{act} is low, the transition state differs little from the reactants in energy and, presumably, also in electronic structure (see Fig. 2.11).

Practically, this postulate has been found extremely useful in the interpretation of experimental results; among other things, as we shall see, it enables us to account for the relationship between reactivity and selectivity (Sec. 3.28).

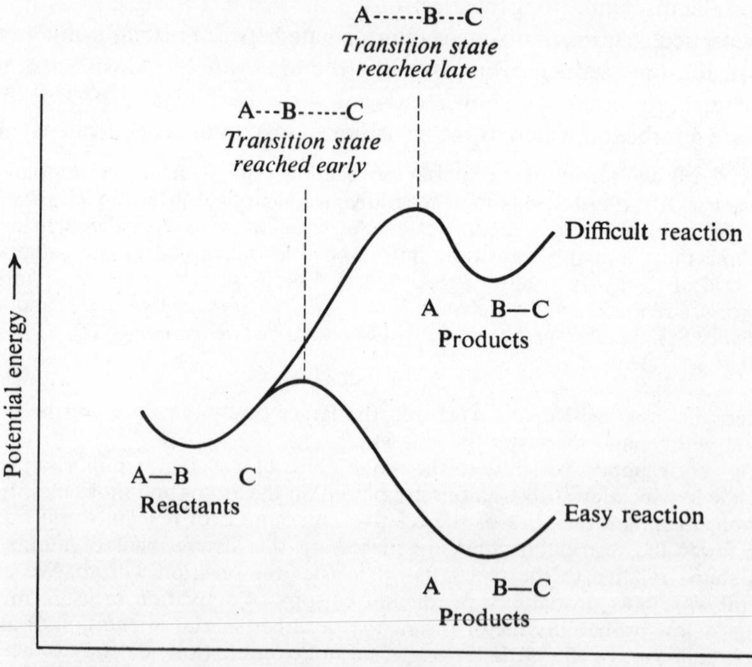

Figure 2.11. Potential energy changes during progress of reaction: reactivity and development of the transition state. Difficult reaction: transition state reached late, resembles products. Easy reaction: transition state reached early, resembles reactants.

Abstraction of hydrogen by the highly reactive chlorine atom has a low E_{act}. According to the postulate, then, the transition state is reached before the reaction has proceeded very far, and when the carbon–hydrogen bond is only slightly stretched. Atoms and electrons are still distributed much as they were in the reactants; carbon is still nearly tetrahedral. The methyl group has developed little free-radical character.

Abstraction of hydrogen by the less reactive bromine atom, in contrast, has a very high E_{act}. The transition state is reached only after reaction is well along toward completion and when the carbon–hydrogen bond is more nearly broken. The geometry and electron distribution has begun to approach that of the products, and carbon may well be almost trigonal. The methyl group has developed much free-radical character.

Thus, *in the attack by a reagent of high reactivity, the transition state tends to resemble the reactant; in the attack by a reagent of low reactivity, the transition state tends to resemble the products.*

2.24 Molecular formula: its fundamental importance

In this chapter we have been concerned with the structure of methane: the way in which atoms are put together to form a molecule of methane. But first we had to know what kinds of atoms these are and how many of them make up the molecule; we had to know that methane is CH_4. Before we can assign a structural formula to a compound, we must first know its molecular formula.

Much of the chapter has been spent in discussing the substitution of chlorine for the hydrogen of methane. But first we had to know that there *is* substitution, that each step of the reaction yields a product that contains one less hydrogen atom and one more chlorine atom than the reactant; we had to know that CH_4 is converted successively into CH_3Cl, CH_2Cl_2, $CHCl_3$, and CCl_4. Before we can discuss the reactions of an organic compound, we must first know the molecular formulas of the products.

Let us review a little of what we know about the assigning of a molecular formula to a compound. We must carry out:

(a) a *qualitative elemental analysis*, to find out what kinds of atoms are present in the molecule;

(b) a *quantitative elemental analysis*, to find out the relative numbers of the different kinds of atoms, that is, to determine the *empirical formula*;

(c) a *molecular weight determination*, which (combined with the empirical formula) shows the actual numbers of the different kinds of atoms, that is, gives us the *molecular formula.*

Most of this should be familiar to the student from previous courses in chemistry. What we shall concentrate on here will be the application of these principles to organic analysis.

2.25 Qualitative elemental analysis

The presence of carbon or hydrogen in a compound is detected by **combustion**: heating with copper oxide, which converts carbon into carbon dioxide and hydrogen into water. (*Problem:* How could each of these products be identified?)

$$(C,H) + CuO \xrightarrow{\text{heat}} Cu + CO_2 + H_2O$$

Covalently bonded halogen, nitrogen, and sulfur must be converted into inorganic ions, which can then be detected in already familiar ways. This conversion is accomplished in either of two ways: (a) through **sodium fusion**, treatment with hot molten sodium metal;

$$(C,H,X,N,S) + Na \xrightarrow{\text{heat}} Na^+X^- + Na^+CN^- + Na^+S^{--}Na^+$$

or (b) through **Schöniger oxidation** by oxygen gas.

$$(C,H,X,N,S) + O_2 \xrightarrow{\text{NaOH}} Na^+X^- + Na^+NO_2^- + Na^+SO_3^{--}Na^+$$

(A simpler method of detecting halogen in *some* organic compounds is discussed in Sec. 14.24.)

By these methods, we could show, for example, that methane contains carbon and hydrogen, or that methyl chloride contains carbon, hydrogen, and chlorine.

Further tests would show the absence of any other element in these compounds, except possibly oxygen, for which there is no simple chemical test; presence or absence of oxygen would be shown by a quantitative analysis.

Problem 2.4 (a) How would you detect halide ion as a product of sodium fusion or oxidation? (b) If sulfur and/or nitrogen is also present in an organic molecule, this test cannot be carried out on a sodium fusion mixture until it has been acidified and boiled. Why is this so?

Problem 2.5 Only carbon and hydrogen were detected by a qualitative elemental analysis of the compound ethyl alcohol; quantitative analysis gave 52.1% carbon and 13.1% hydrogen. (a) Why would it be assumed that ethyl alcohol contains oxygen? (b) What percentage of oxygen would be assumed?

2.26 Quantitative elemental analysis: carbon, hydrogen, and halogen

Knowing what elements make up a compound, we must next determine the proportions in which they are present. To do this, we carry out very much the same analysis as before, only this time on a quantitative basis. To find out the relative amounts of carbon and hydrogen in methane, for example, we would completely oxidize a measured amount of methane and weigh the carbon dioxide and water formed.

In a quantitative combustion, a weighed sample of the organic compound is passed through a *combustion train*: a tube packed with copper oxide heated to 600–800°, followed by a tube containing a drying agent (usually Dehydrite, magnesium perchlorate) and a tube containing a strong base (usually Ascarite, sodium hydroxide on asbestos). The water formed is absorbed by the drying agent, and the carbon dioxide is absorbed by the base; the increase in weight of each tube gives the weight of product formed.

For example, we might find that a sample of methane weighing 9.67 mg produced 26.53 mg of CO_2 and 21.56 mg of H_2O. Now, only the fraction $C/CO_2 = 12.01/44.01$ of the carbon dioxide is carbon, and only the fraction $2H/H_2O = 2.016/18.02$ of the water is hydrogen. Therefore

$$\text{wt. C} = 26.53 \times 12.01/44.01 \qquad \text{wt. H} = 21.56 \times 2.016/18.02$$

wt. C (in sample) = 7.24 mg wt. H (in sample) = 2.41 mg

and the percentage composition is

$$\% \text{ C} = 7.24/9.67 \times 100 \qquad \% \text{ H} = 2.41/9.67 \times 100$$

$$\% \text{ C (in sample)} = 74.9 \qquad \% \text{ H (in sample)} = 24.9$$

Since the total of carbon and hydrogen is 100%, within the limits of error of the analysis, oxygen (or any other element) must be absent.

In quantitative, as in qualitative, analysis, covalently bonded halogen must be converted into halide ion. The organic compound is heated either (a) in a bomb with sodium peroxide or (b) in a sealed tube with nitric acid (*Carius method*). The halide ion thus formed is converted into silver halide, which can be weighed.

Problem 2.6 When 7.36 mg of methyl chloride was heated in a bomb with sodium peroxide, the chloride ion liberated yielded 20.68 mg of silver chloride. (a) What percentage of chlorine is indicated by this analysis? (b) What percentage of chlorine would be expected from a compound of formula CH_3Cl? (c) What weight of silver chloride would you expect from 7.36 mg of methylene chloride? (d) Of chloroform? (e) Of carbon tetrachloride?

(We shall take up other quantitative analytical methods when we need them: nitrogen and sulfur analysis, Sec. 10.12; methoxyl determination, Sec. 17.16; neutralization equivalent, Sec. 18.21; saponification equivalent, Sec. 20.24.)

2.27 Empirical formula

Knowing the percentage composition of a compound, we can now calculate the **empirical formula**: *the simplest formula that shows the relative numbers of the different kinds of atoms in a molecule*. For example, in 100 g (taken for convenience) of methane there are 74.9 g of carbon and 24.9 g of hydrogen, according to our quantitative analysis. Dividing each quantity by the proper atomic weight gives the number of gram-atoms of each element.

$$\text{C:} \ \frac{74.9}{12.01} = 6.24 \text{ gram-atoms}$$

$$\text{H:} \ \frac{24.9}{1.008} = 24.7 \text{ gram-atoms}$$

Since a gram-atom of one element contains the same number of atoms as a gram-atom of any other element, we now know the relative number of carbon and hydrogen atoms in methane: $C_{6.24}H_{24.7}$. Conversion to smallest whole numbers gives the empirical formula CH_4 for methane.

$$\text{C: } 6.24/6.24 = 1$$

$$\text{H: } 24.7/6.24 = 3.96\text{, approximately } 4$$

Problem 2.7 Calculate the percentage composition and then the empirical formula for each of the following compounds: (a) Combustion of a 3.02-mg sample of a compound gave 8.86 mg of carbon dioxide and 5.43 mg of water. (b) Combustion of an 8.23-mg sample of a compound gave 9.62 mg of carbon dioxide and 3.94 mg of water. Analysis of a 5.32-mg sample of the same compound by the Carius method gave 13.49 mg of silver chloride.

2.28 Molecular weight. Molecular formula

At this stage we know what kinds of atoms make up the molecule we are studying, and in what ratio they are present. This knowledge is summarized in the empirical formula.

But this is not enough. On the basis of just the empirical formula, a molecule of methane, for example, might contain one carbon and four hydrogens, or two carbons and eight hydrogens, or *any* multiple of CH_4. We still have to find the **molecular formula**: *the formula that shows the actual number of each kind of atom in a molecule.*

To find the molecular formula, we must determine the molecular weight: today, almost certainly by mass spectrometry, which gives an exact value (Sec. 13.2). *Ethane*, for example, has an empirical formula of CH_3. A molecular weight of 30 is found, indicating that, of the possible molecular formulas, C_2H_6 must be the correct one.

Problem 2.8 Quantitative elemental analysis shows that the empirical formula of a compound is CH. The molecular weight is found to be 78. What is the molecular formula?

Problem 2.9 Combustion of a 5.17-mg sample of a compound gives 10.32 mg of carbon dioxide and 4.23 mg of water. The molecular weight is 88. What is the molecular formula of the compound?

PROBLEMS

1. Calculate the percentage composition of A, B, and C from the following analytical data:

	wt. sample	wt. CO_2	wt. H_2O	wt. AgCl
A	4.37 mg	15.02 mg	2.48 mg	—
B	5.95 mg	13.97 mg	2.39 mg	7.55 mg
C	4.02 mg	9.14 mg	3.71 mg	—

2. What is the percentage composition of:

(a) C_3H_7Cl (c) $C_4H_8O_2$ (e) CH_4ON_2
(b) C_2H_6O (d) $C_6H_8O_2N_2S$ (f) C_6H_8NCl

3. What is the empirical formula of an organic compound whose percentage composition is:

(a) 85.6% C, 14.4% H (d) 29.8% C, 6.3% H, 44.0% Cl
(b) 92.2% C, 7.8% H (e) 48.7% C, 13.6% H, 37.8% N
(c) 40.0% C, 6.7% H (f) 25.2% C, 2.8% H, 49.6% Cl

(*Note:* remember that oxygen often is not determined directly.)

4. A qualitative analysis of *papaverine*, one of the alkaloids in opium, showed carbon, hydrogen, and nitrogen. A quantitative analysis gave 70.8% carbon, 6.2% hydrogen, and 4.1% nitrogen. Calculate the empirical formula of papaverine.

5. *Methyl orange*, an acid-base indicator, is the sodium salt of an acid that contains carbon, hydrogen, nitrogen, sulfur, and oxygen. Quantitative analysis gave 51.4% carbon, 4.3% hydrogen, 12.8% nitrogen, 9.8% sulfur, and 7.0% sodium. What is the empirical formula of methyl orange?

6. Combustion of 6.51 mg of a compound gave 20.47 mg of carbon dioxide and

8.36 mg of water. The molecular weight was found to be 84. Calculate: (a) percentage composition; (b) empirical formula; and (c) molecular formula of the compound.

7. A liquid of molecular weight 60 was found to contain 40.0% carbon and 6.7% hydrogen. What is the molecular formula of the compound?

8. A gas of the same empirical formula as the compound in Problem 7 has a molecular weight of 30. What is its molecular formula?

9. *Indigo*, an important dyestuff, gave an analysis of 73.3% carbon, 3.8% hydrogen, and 10.7% nitrogen. Molecular weight determination gave a value of 262. What is the molecular formula of indigo?

10. The hormone *insulin* contains 3.4% sulfur. (a) What is the minimum molecular weight of insulin? (b) The actual molecular weight is 5734; how many sulfur atoms are probably present per molecule?

11. Calculate ΔH for:

(a)–(d) $H_2 + X_2 \longrightarrow 2HX$, where X = F, Cl, Br, I
(e) $C_2H_6 + Br_2 \longrightarrow C_2H_5Br + HBr$
(f) $C_6H_5CH_3 + Br_2 \longrightarrow C_6H_5CH_2Br + HBr$
(g) $H_2C{=}CHCH_3 + Br_2 \longrightarrow H_2C{=}CHCH_2Br + HBr$
(h) Reactions (e), (f), and (g) proceed by the same free radical mechanism as halogenation of methane. Calculate ΔH for each step in these three reactions.

12. A conceivable mechanism for the chlorination of methane involves the following steps:

(1) $$Cl_2 \longrightarrow 2Cl\cdot$$

(2) $$Cl\cdot + CH_4 \longrightarrow CH_3Cl + H\cdot$$

(3) $$H\cdot + Cl_2 \longrightarrow HCl + Cl\cdot$$

then (2), (3), (2), (3), etc.

(a) Calculate ΔH for each of these steps. (b) Why does this mechanism seem less likely than the accepted one given in Sec. 2.12? (Additional, conclusive evidence against this alternative mechanism will be presented in Sec. 7.10.)

13. (a) Free methyl radicals react with methane as follows:

$$(i) \quad CH_3\cdot + CH_4 \longrightarrow CH_4 + CH_3\cdot$$

On the basis of the bond strengths involved, show why the above reaction takes place rather than the following:

$$(ii) \quad CH_3\cdot + CH_4 \longrightarrow CH_3{-}CH_3 + H\cdot$$

(b) Reaction (*i*) has an E_{act} of 13 kcal. In Sec. 2.12 it was listed as probable (but unproductive) on grounds of collision probability. In actuality, how probable is reaction (*i*) in, say, a 50:50 mixture of CH_4 and Cl_2? (*Hint:* See Secs. 2.20 and 2.18.)

14. Bromination of methane is slowed down by addition of fairly large amounts of HBr. (a) Suggest a possible explanation for this. (*Hint:* See Sec. 2.17.) (b) Account for the fact that HCl does not have a similar effect upon chlorination. (c) Any reaction tends to slow down as reactants are used up and their concentrations decrease. How do you account for the fact that bromination of methane slows down to an unusually great extent, more than, say, chlorination of methane?

15. A mixture of H_2 and Cl_2 does not react in the dark at room temperature. At high temperatures or under the influence of light (of a wavelength absorbed by chlorine) a violent reaction occurs and HCl is formed. The photochemical reaction yields as many as a million molecules of HCl for each photon absorbed. The presence of a small amount of oxygen slows down the reaction markedly. (a) Outline a possible mechanism to account for these facts. (b) Account for the fact that a mixture of H_2 and I_2 does not behave in the same way. (Hydrogen iodide is actually formed, but by an entirely different mechanism.)

16. A stream of tetramethyllead vapor, $(CH_3)_4Pb$, was passed through a quartz tube which was heated at one spot; a mirror of metallic lead was deposited at the hot point, and the gas escaping from the tube was found to be chiefly ethane. The tube was next heated upstream of the lead mirror while more tetramethyllead was passed through; a new mirror appeared at the hot point, the old mirror disappeared, and the gas escaping from the tube was now found to be chiefly tetramethyllead. Experiments like this, done by Fritz Paneth at the University of Berlin, were considered the first good evidence for the existence of short-lived free radicals like methyl. (a) Show how these experimental results can be accounted for in terms of intermediate free radicals. (b) The farther upstream the tube was heated, the more slowly the old mirror disappeared. Account for this.

17. When a small amount (0.02%) of tetraethyllead, $(C_2H_5)_4Pb$, is added to a mixture of methane and chlorine, chlorination takes place at only 140° instead of the usual minimum of 250°. In light of Problem 16, show how this fact strengthens the mechanism of Sec. 2.12.

Chapter 3 Alkanes

Free-Radical Substitution

3.1 Classification by structure: the family

The basis of organic chemistry, we have said, is the structural theory. We separate all organic compounds into a number of families on the basis of structure. Having done this, we find that we have at the same time classified the compounds as to their physical and chemical properties. A particular set of properties is thus characteristic of a particular kind of structure.

Within a family there are variations in properties. All members of the family may, for example, react with a particular reagent, but some may react more readily than others. Within a single compound there may be variations in properties, one part of a molecule being more reactive than another part. These variations in properties correspond to variations in structure.

As we take up each family of organic compounds, we shall first see what structure and properties are characteristic of the family. Next we shall see how structure and properties vary within the family. We shall not simply memorize these facts, but, whenever possible, shall try to understand properties in terms of structure, and to understand variations in properties in terms of variations in structure.

Having studied methane in some detail, let us now look at the more complicated members of the alkane family. These hydrocarbons have been assigned to the same family as methane on the basis of their structure, and on the whole their properties follow the pattern laid down by methane. However, certain new points will arise simply because of the greater size and complexity of these compounds.

3.2 Structure of ethane

Next in size after methane is **ethane**, C_2H_6. If we connect the atoms of this molecule by covalent bonds, following the rule of one bond (one pair of electrons) for each hydrogen and four bonds (four pairs of electrons) for each carbon, we arrive at the structure

```
  H H              H  H
  ·· ··            |  |
H:C:C:H          H—C—C—H
  ·· ··            |  |
  H H              H  H
```

Ethane

Each carbon is bonded to three hydrogens and to the other carbon.

Since each carbon atom is bonded to four other atoms, its bonding orbitals (sp^3 orbitals) are directed toward the corners of a tetrahedron. As in the case of methane, the carbon–hydrogen bonds result from overlap of these sp^3 orbitals with the *s* orbitals of the hydrogens. The carbon–carbon bond arises from overlap of two sp^3 orbitals.

The carbon–hydrogen and carbon–carbon bonds have the same general electron distribution, being cylindrically symmetrical about a line joining the atomic nuclei (see Fig. 3.1); because of this similarity in shape, the bonds are given the same name, *σ bonds* (*sigma bonds*).

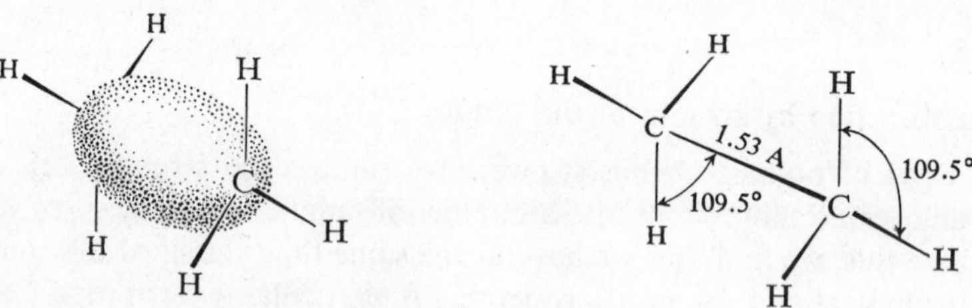

Figure 3.1. Ethane molecule. Carbon–carbon single bond: σ bond.

Figure 3.2. Ethane molecule: shape and size.

In ethane, then, the bond angles and carbon–hydrogen bond lengths should be very much the same as in methane, that is, about 109.5° and about 1.10 A, respectively. Electron diffraction and spectroscopic studies have verified this structure in all respects, giving (Fig. 3.2) the following measurements for the molecule: bond angles, 109.5°; C—H length, 1.10 A; C—C length, 1.53 A. Similar studies have shown that, with only slight variations, these values are quite characteristic of carbon–hydrogen and carbon–carbon bonds and of carbon bond angles in alkanes.

3.3 Free rotation about the carbon–carbon single bond. Conformations. Torsional strain

This particular set of bond angles and bond lengths still does not limit us to a single arrangement of atoms for the ethane molecule, since the relationship between the hydrogens of one carbon and the hydrogens of the other carbon is not specified. We could have an arrangement like I in which the hydrogens exactly oppose each other, an arrangement like II in which the hydrogens are perfectly staggered, or an infinity of intermediate arrangements. Which of these is the actual structure of ethane? The answer is: *all of them.*

We have seen that the σ bond joining the carbon atoms is cylindrically symmetrical about a line joining the two carbon nuclei; overlap and hence bond

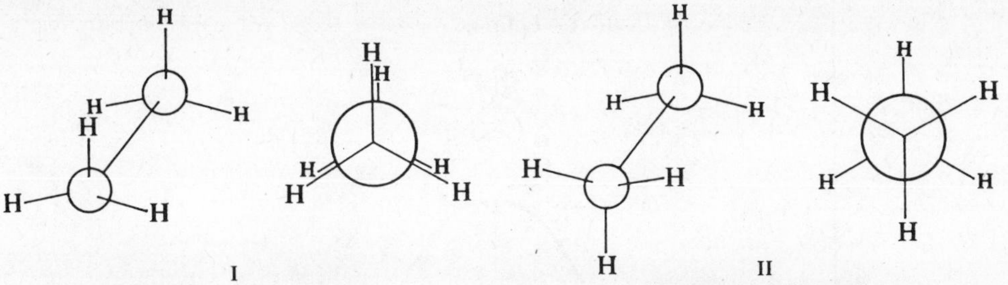

I
Eclipsed conformation

II
Staggered conformation

Ethane

strength should be the same for all these possible arrangements. If the various arrangements do not differ in energy, then the molecule is not restricted to any one of them, but can change freely from one to another. Since the change from one to another involves rotation about the carbon–carbon bond, we describe this freedom to change by saying that *there is free rotation about the carbon–carbon single bond.*

Different arrangements of atoms that can be converted into one another by rotation about single bonds are called **conformations**. I is called the *eclipsed conformation;* II is called the *staggered conformation.* (The infinity of intermediate conformations are called *skew conformations.*)

The highly useful representations of the kind

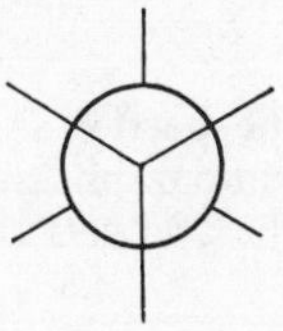

are called *Newman projections,* after M. S. Newman, of The Ohio State University, who first proposed their use.

The picture is not yet complete. Certain physical properties show that rotation is *not quite free*: there is an energy barrier of about 3 kcal/mole. The potential energy of the molecule is at a minimum for the staggered conformation, increases with rotation, and reaches a maximum at the eclipsed conformation (Fig. 3.3). Most ethane molecules, naturally, exist in the most stable, staggered conformation; or, put differently, any molecule spends most of its time in the most stable conformation.

How free are ethane molecules to rotate from one staggered arrangement to another? The 3-kcal barrier is not a very high one; even at room temperature the fraction of collisions with sufficient energy is large enough that a rapid interconversion between staggered arrangements occurs. For most practical purposes, we may still consider that the carbon–carbon single bond permits free rotation.

The nature of the rotational barrier in ethane is not understood or—what is not exactly the same thing—is not readily explained. It is too high to be due merely to van der Waals forces (Sec. 1.19): although thrown closer together in the eclipsed conformation than in the staggered conformation, the hydrogens on opposite carbons are not big enough for this to cause appreciable crowding. The barrier is considered to arise in some way from interaction among the electron clouds of the carbon–hydrogen bonds. Quantum

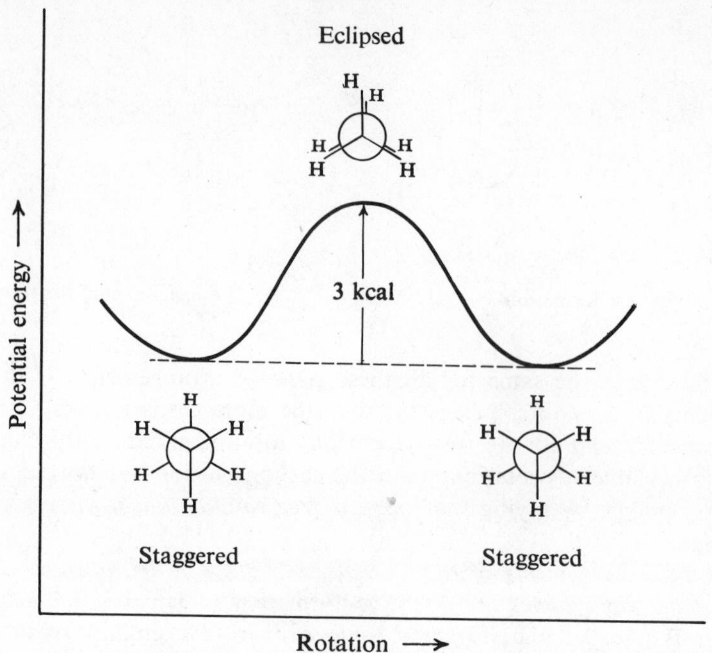

Figure 3.3. Potential energy changes during rotation about carbon–carbon single bond of ethane.

mechanical calculations show that the barrier should exist, and so perhaps "lack of understanding" amounts to difficulty in paraphrasing the mathematics in physical terms. Like the bond orbitals in methane, the two sets of orbitals in ethane tend to be as far apart as possible—to be *staggered.*

The energy required to rotate the ethane molecule about the carbon–carbon bond is called *torsional energy*. We speak of the relative instability of the eclipsed conformation—or any of the intermediate skew conformations—as being due to *torsional strain.*

As the hydrogens of ethane are replaced by other atoms or groups of atoms, other factors affecting the relative stability of conformations appear: van der Waals forces, dipole–dipole interactions, hydrogen bonding. But the tendency for the bond orbitals on adjacent carbons to be staggered remains, and any rotation away from the staggered conformation is accompanied by torsional strain.

3.4 Propane and the butanes

The next member of the alkane family is **propane,** C_3H_8. Again following the rule of one bond per hydrogen and four bonds per carbon, we arrive at structure I.

Here, rotation can occur about two carbon–carbon bonds, and again is essentially free. Although the methyl group is considerably larger than hydrogen, the rotational barrier (3.3 kcal/mole) is only a little higher than for ethane. Evidently there is still not significant crowding in the eclipsed conformation, and the rotational barrier is due chiefly to the same factor as the barrier in ethane: *torsional strain.*

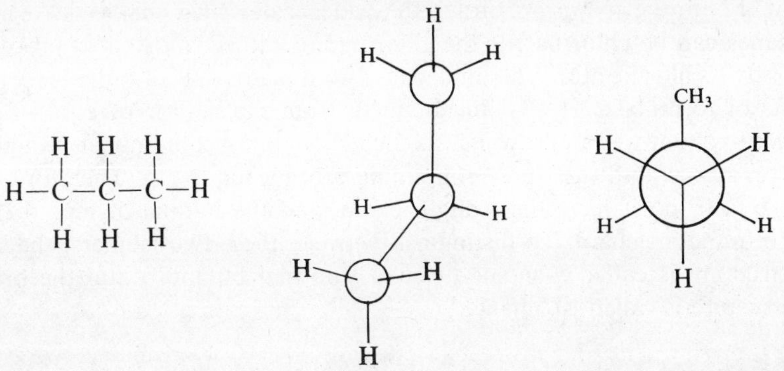

```
    H  H  H
    |  |  |
H—C—C—C—H
    |  |  |
    H  H  H
```

I
Propane

When we consider **butane**, C_4H_{10}, we find that there are two possible structures, II and III. II has a four-carbon chain and III has a three-carbon chain with a

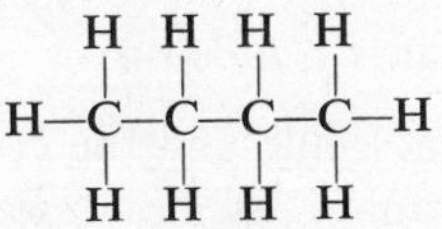

```
    H  H  H  H
    |  |  |  |
H—C—C—C—C—H
    |  |  |  |
    H  H  H  H
```

II
n-Butane

```
    H     H     H
    |     |     |
H—C———C———C—H
    |     |     |
    H     |     H
        H—C—H
          |
          H
```

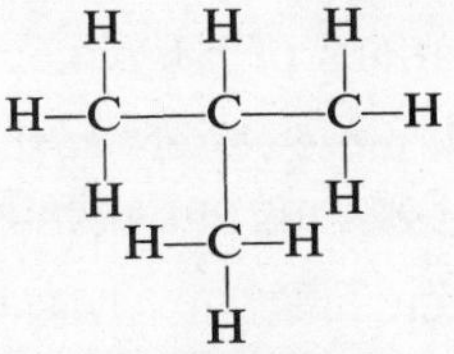

III
Isobutane

one-carbon branch. There can be no doubt that these represent different structures, since no amount of moving, twisting, or rotating about carbon–carbon bonds will cause these structures to coincide. We can see that in the *straight-chain* structure (II) each carbon possesses at least two hydrogens, whereas in the *branched-chain* structure (III) one carbon possesses only a single hydrogen; or we may notice that in the branched-chain structure (III) one carbon is bonded to three other carbons, whereas in the straight-chain structure (II) no carbon is bonded to more than two other carbons.

In agreement with this prediction, we find that two compounds of the same formula, C_4H_{10}, have been isolated. There can be no doubt that these two substances are different compounds, since they show definite differences in their physical and chemical properties (see Table 3.1); for example one boils at 0° and the other at −12°. By definition, they are *isomers* (Sec. 1.24).

Table 3.1 PHYSICAL CONSTANTS OF THE ISOMERIC BUTANES

	n-Butane	Isobutane
b.p.	0°	−12°
m.p.	−138°	−159°
sp.gr. at −20°	0.622	0.604
solub. in 100 ml alcohol	1813 ml	1320 ml

Two compounds of formula C_4H_{10} are known and we have drawn two structures to represent them. The next question is: which structure represents which compound? For the answer we turn to the evidence of **isomer number**. Like methane, the butanes can be chlorinated; the chlorination can be allowed to proceed until there are two chlorine atoms per molecule. From the butane of b.p. 0°, *six* isomeric products of formula $C_4H_8Cl_2$ are obtained, from the butane of b.p. −12°, only *three*. We find that we can draw just six dichlorobutanes containing a straight chain of carbon atoms, and just three containing a branched chain. Therefore, the butane of b.p. 0° must have the straight chain, and the butane of b.p. −12° must have the branched chain. To distinguish between these two isomers, the straight-chain structure is called ***n*-butane** (spoken "normal butane") and the branched-chain structure is called **isobutane**.

Problem 3.1 Draw the structures of all possible dichloro derivatives of: (a) *n*-butane; (b) isobutane.

Problem 3.2 Could we assign structures to the isomeric butanes on the basis of the number of isomeric *mono*chloro derivatives?

3.5 Conformations of *n*-butane. Van der Waals repulsion

Let us look more closely at the *n*-butane molecule and the conformations in which it exists. Focusing our attention on the middle C—C bond, we see a molecule

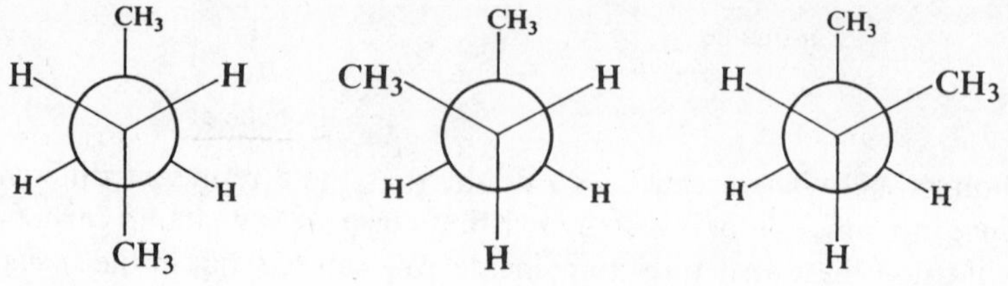

I
Anti conformation

II III
Gauche conformations

n-Butane

similar to ethane, but with a methyl group replacing one hydrogen on each carbon. As with ethane, staggered conformations have lower torsional energies and hence are more stable than eclipsed conformations. But, due to the presence of the methyl groups, two new points are encountered here: first, there are several *different* staggered conformations; and second, a factor besides torsional strain comes into play to affect conformational stabilities.

There is the *anti* conformation, I, in which the methyl groups are as far apart as they can be (dihedral angle 180°). There are two *gauche* conformations, II and III, in which the methyl groups are only 60° apart. (Conformations II and III are mirror images of each other, and are of the same stability; nevertheless, they *are* different. Make models and convince yourself that this is so.)

The *anti* conformation, it has been found, is more stable (by 0.8 kcal/mole) than the *gauche* (Fig. 3.4). Both are free of torsional strain. But in a *gauche* conformation, the methyl groups are crowded together, that is, are thrown together closer than the sum of their van der Waals radii; under these conditions, van der Waals forces are *repulsive* (Sec. 1.19) and raise the energy of the conformation. We say that there is *van der Waals repulsion* (or *steric repulsion*) between the

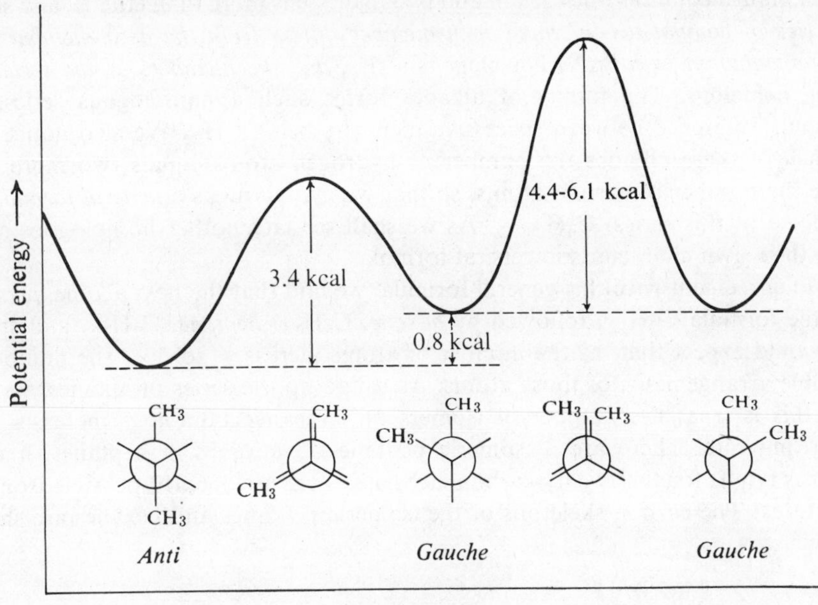

Figure 3.4. Potential energy changes during rotation about C_2—C_3 bond of *n*-butane.

methyl groups, and that the molecule is less stable because of *van der Waals strain* (or *steric strain*).

Van der Waals strain can affect not only the relative stabilities of various staggered conformations, but also the heights of the barriers between them. The energy maximum reached when two methyl groups swing past each other—rather than past hydrogens—is the highest rotational barrier of all, and has been estimated at 4.4–6.1 kcal/mole. Even so, it is low enough that—at ordinary temperatures, at least—the energy of molecular collisions causes rapid rotation; a given molecule exists now in a *gauche* conformation, and the next instant in the *anti* conformation.

We shall return to the relationships among conformations like these of *n*-butane in Sec. 4.20.

Problem 3.3 Both calculations and experimental evidence indicate that the dihedral angle between the methyl groups in the *gauche* conformation of *n*-butane is actually somewhat *larger* than 60°. How would you account for this?

Problem 3.4 Considering only rotation about the bond shown, draw a potential energy *vs.* rotation curve like Fig. 3.4 for: (a) $(CH_3)_2CH$—$CH(CH_3)_2$; (b) $(CH_3)_2CH$—CH_2CH_3; (c) $(CH_3)_3C$—$C(CH_3)_3$. (d) Compare the heights of the various energy barriers with each other and with those in Fig. 3.4.

3.6 Higher alkanes. The homologous series

If we examine the molecular formulas of the alkanes we have so far considered, we see that butane contains one carbon and two hydrogens more than propane, which in turn contains one carbon and two hydrogens more than ethane, and so on. *A series of compounds in which each member differs from the next member by a constant amount is called a* **homologous series**, *and the members of the series are called* **homologs**. The family of alkanes forms such a homologous series, the constant difference between successive members being CH_2. We also notice that in each of these alkanes the number of hydrogen atoms equals two more than twice the number of carbon atoms, so that we may write as a *general formula* for members of this series, C_nH_{2n+2}. As we shall see later, other homologous series have their own characteristic general formulas.

In agreement with this general formula, we find that the next alkane, *pentane*, has the formula C_5H_{12}, followed by *hexane*, C_6H_{14}, *heptane*, C_7H_{16}, and so on. We would expect that, as the number of atoms increases, so does the number of possible arrangements of those atoms. As we go up the series of alkanes, we find that this is true: the number of isomers of successive homologs increases at a surprising rate. There are 3 isomeric pentanes, 5 hexanes, 9 heptanes, and 75 decanes (C_{10}); for the twenty-carbon eicosane, there are 366,319 possible isomeric structures! The carbon skeletons of the isomeric pentanes and hexanes are shown below.

```
                                          C
                                          |
C—C—C—C—C         C—C—C—C               C—C—C          Pentanes
                      |                   |
                      C                   C
n-Pentane         Isopentane           Neopentane
b.p. 36°          b.p. 28°             b.p. 9.5°

C—C—C—C—C—C      C—C—C—C—C        C—C—C—C—C
                        |              |
                        C              C
b.p. 69°          b.p. 60°         b.p. 63°            Hexanes

          C
          |
      C—C—C—C           C—C—C—C
          |               | |
          C               C C
      b.p. 50°          b.p. 58°
```

It is important to practice drawing the possible isomeric structures that correspond to a single molecular formula. In doing this, a set of molecular models is especially helpful since it will show that many structures which appear to be different when drawn on paper are actually identical.

Problem 3.5 Draw the structures of: (a) the nine isomeric heptanes (C_7H_{16}); (b) the eight chloropentanes ($C_5H_{11}Cl$); (c) the nine dibromobutanes ($C_4H_8Br_2$).

3.7 Nomenclature

We have seen that the names *methane*, *ethane*, *propane*, *butane*, and *pentane* are used for alkanes containing respectively one, two, three, four, and five carbon atoms. Table 3.2 gives the names of many larger alkanes. Except for the first

Table 3.2 NAMES OF ALKANES

CH_4	methane	C_9H_{20}	nonane
C_2H_6	ethane	$C_{10}H_{22}$	decane
C_3H_8	propane	$C_{11}H_{24}$	undecane
C_4H_{10}	butane	$C_{12}H_{26}$	dodecane
C_5H_{12}	pentane	$C_{14}H_{30}$	tetradecane
C_6H_{14}	hexane	$C_{16}H_{34}$	hexadecane
C_7H_{16}	heptane	$C_{18}H_{38}$	octadecane
C_8H_{18}	octane	$C_{20}H_{42}$	eicosane

four members of the family, the name is simply derived from the Greek (or Latin) prefix for the particular number of carbons in the alkane; thus **pent**ane for five, **hex**ane for six, **hept**ane for seven, **oct**ane for eight, and so on.

The student should certainly memorize the names of at least the first ten alkanes. Having done this, he has at the same time essentially learned the names of the first ten alkenes, alkynes, alcohols, etc., since the names of many families of compounds are closely related. Compare, for example, the names *propane*, *propene*, and *propyne* for the three-carbon alkane, alkene, and alkyne.

But nearly every alkane can have a number of isomeric structures, and there must be an unambiguous name for each of these isomers. The butanes and pentanes are distinguished by the use of prefixes: ***n***-butane and **iso**butane; ***n***-pentane, **iso**pentane, and **neo**pentane. But there are 5 hexanes, 9 heptanes, and 75 decanes; it would be difficult to devise, and even more difficult to remember, a different prefix for each of these isomers. It is obvious that some systematic method of naming is needed.

As organic chemistry has developed, several different methods have been devised to name the members of nearly every class of organic compounds; each method was devised when the previously used system had been found inadequate for the growing number of increasingly complex organic compounds. Unfortunately for the student, perhaps, several systems have survived and are in current use. Even if we are content ourselves to use only one system, we still have to understand the names used by other chemists; hence it is necessary for us to learn more than one system of nomenclature. But before we can do this, we must first learn the names of certain organic groups.

3.8 Alkyl groups

In our study of inorganic chemistry, we found it useful to have names for certain groups of atoms that compose only part of a molecule and yet appear many times as a unit. For example, NH_4^+ is called *ammonium*; NO_3^-, *nitrate*; SO_3^{--}, *sulfite*; and so on.

In a similar way names are given to certain groups that constantly appear as structural units of organic molecules. We have seen that chloromethane, CH_3Cl, is also known as *methyl chloride*. The CH_3 group is called **methyl** wherever it appears, CH_3Br being *methyl* bromide, CH_3I, *methyl* iodide, and CH_3OH, *methyl* alcohol. In an analogous way, the C_2H_5 group is **ethyl**; C_3H_7, **propyl**; C_4H_9, **butyl**; and so on.

These groups are named simply by dropping *-ane* from the name of the corresponding alkane and replacing it by *-yl*. They are known collectively as

alkyl groups. The general formula for an alkyl group is C_nH_{2n+1}, since it contains one less hydrogen than the parent alkane, C_nH_{2n+2}.

Among the alkyl groups we again encounter the problem of isomerism. There is only one methyl chloride or ethyl chloride, and correspondingly only one methyl group or ethyl group. We can see, however, that there are two propyl chlorides, I and II, and hence that there must be two propyl groups. These groups

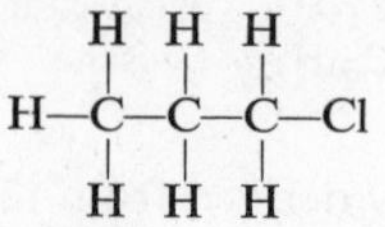

```
     H  H  H              H  H  H
     |  |  |              |  |  |
  H—C—C—C—Cl          H—C—C—C—H
     |  |  |              |  |  |
     H  H  H              H  Cl H
        I                    II
  n-Propyl chloride    Isopropyl chloride
```

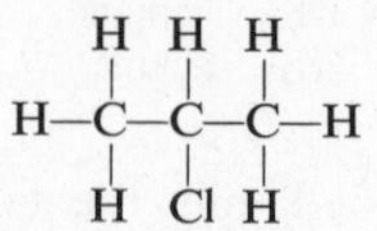

both contain the propane chain, but differ in the point of attachment of the chlorine; they are called ***n*-propyl** and **isopropyl**. We can distinguish the two

$CH_3CH_2CH_2—$ *n*-Propyl

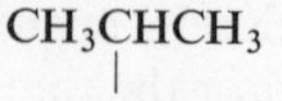

$CH_3\underset{|}{C}HCH_3$ Isopropyl

chlorides by the names *n-propyl chloride* and *isopropyl chloride*; we distinguish the two propyl bromides, iodides, alcohols, and so on in the same way.

We find that there are four butyl groups, two derived from the straight-chain *n*-butane, and two derived from the branched-chain isobutane. These are given the designations ***n***- (*normal*), ***sec***- (*secondary*), ***iso***-, and ***tert***- (*tertiary*), as shown below. Again the difference between *n*-butyl and *sec*-butyl and between isobutyl and *tert*-butyl lies in the point of attachment of the alkyl group to the rest of the molecule.

$CH_3CH_2CH_2CH_2—$ *n*-Butyl

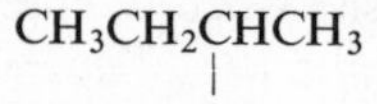

$CH_3CH_2\underset{|}{C}HCH_3$ *sec*-Butyl

```
CH3                      CH3
   \                      |
    CHCH2—          CH3—C—
   /                      |
CH3                      CH3
   Isobutyl            tert-Butyl
```

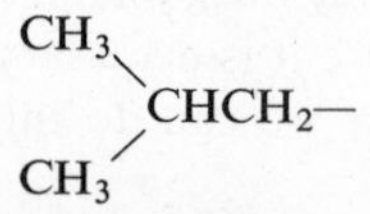

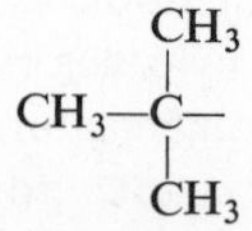

Beyond butyl the number of isomeric groups derived from each alkane becomes so great that it is impracticable to designate them all by various prefixes. Even though limited, this system is so useful for the small groups just described that it is widely used; a student must therefore memorize these names and learn to recognize these groups at a glance in whatever way they happen to be represented.

However large the group concerned, one of its many possible arrangements can still be designated by this simple system. The prefix *n*- is used to designate any alkyl group in which all carbons form a single continuous chain and in which the point of attachment is the very end carbon. For example:

$CH_3CH_2CH_2CH_2CH_2Cl$ *n*-Pentyl chloride

$CH_3(CH_2)_4CH_2Cl$ *n*-Hexyl chloride

The prefix *iso*- is used to designate any alkyl group (of six carbons or less) that has a single one-carbon branch on the next-to-last carbon of a chain and has the point of attachment at the opposite end of the chain. For example:

$$\begin{matrix} CH_3 \\ & \diagdown \\ & & CHCH_2CH_2Cl \\ & \diagup \\ CH_3 \end{matrix} \qquad\qquad \begin{matrix} CH_3 \\ & \diagdown \\ & & CH(CH_2)_2CH_2Cl \\ & \diagup \\ CH_3 \end{matrix}$$

Isopentyl chloride Isohexyl chloride

If the branching occurs at any other position, or if the point of attachment is at any other position, this name does not apply.

Now that we have learned the names of certain alkyl groups, let us return to the original problem: the naming of alkanes.

3.9 Common names of alkanes

As we have seen, the prefixes *n*-, *iso*-, and *neo*- are adequate to differentiate the various butanes and pentanes, but beyond this point an impracticable number of prefixes would be required. However, the prefix *n*- has been retained for any alkane, no matter how large, in which all carbons form a continuous chain with no branching:

$$CH_3CH_2CH_2CH_2CH_3 \qquad\qquad CH_3(CH_2)_4CH_3$$

n-Pentane *n*-Hexane

An *isoalkane* is a compound of six carbons or less in which all carbons except one form a continuous chain and that one carbon is attached to the next-to-end carbon:

$$\begin{matrix} CH_3 \\ & \diagdown \\ & & CHCH_2CH_3 \\ & \diagup \\ CH_3 \end{matrix} \qquad\qquad \begin{matrix} CH_3 \\ & \diagdown \\ & & CH(CH_2)_2CH_3 \\ & \diagup \\ CH_3 \end{matrix}$$

Isopentane Isohexane

In naming any other of the higher alkanes, we make use of the IUPAC system, outlined in the following section.

(It is sometimes convenient to name alkanes as derivatives of methane; see, for example, I on p. 129.)

3.10 IUPAC names of alkanes

To devise a system of nomenclature that could be used for even the most complicated compounds, various committees and commissions representing the chemists of the world have met periodically since 1892. In its present modification, the system so devised is known as the **IUPAC system** (International Union of Pure and Applied Chemistry). Since this system follows much the same pattern for all families of organic compounds, we shall consider it in some detail as applied to the alkanes.

Essentially the rules of the IUPAC system are:

1. Select as the parent structure the longest continuous chain, and then consider the compound to have been derived from this structure by the replacement of hydrogen by various alkyl groups. Isobutane (I) can be considered to arise

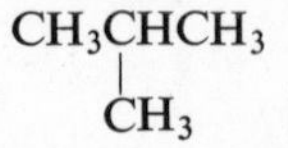

CH_3CHCH_3
$\quad | $
$\quad CH_3$
I
Methylpropane
(Isobutane)

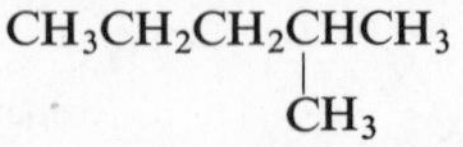

$CH_3CH_2CH_2CHCH_3$
$\qquad\qquad | $
$\qquad\qquad CH_3$
II
2-Methylpentane

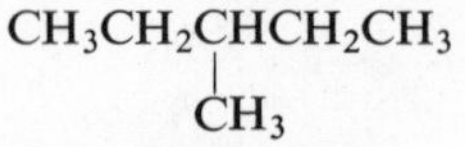

$CH_3CH_2CHCH_2CH_3$
$\qquad\quad | $
$\qquad\quad CH_3$
III
3-Methylpentane

from propane by the replacement of a hydrogen atom by a methyl group, and thus may be named *methylpropane*.

2. Where necessary, as in the isomeric methylpentanes (II and III), indicate by a number the carbon to which the alkyl group is attached.

3. In numbering the parent carbon chain, start at whichever end results in the use of the lowest numbers; thus II is called *2-methylpentane* rather than 4-methylpentane.

4. If the same alkyl group occurs more than once as a side chain, indicate this by the prefix *di-*, *tri-*, *tetra-*, etc., to show how many of these alkyl groups there are, and indicate by various numbers the positions of *each* group, as in *2,2,4-trimethylpentane* (IV).

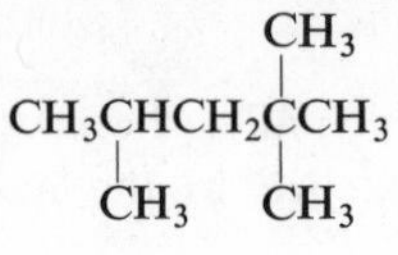

$\qquad\qquad\quad CH_3$
$\qquad\qquad\quad | $
$CH_3CHCH_2CCH_3$
$\quad | \qquad\quad | $
$\quad CH_3 \quad CH_3$
IV
2,2,4-Trimethylpentane

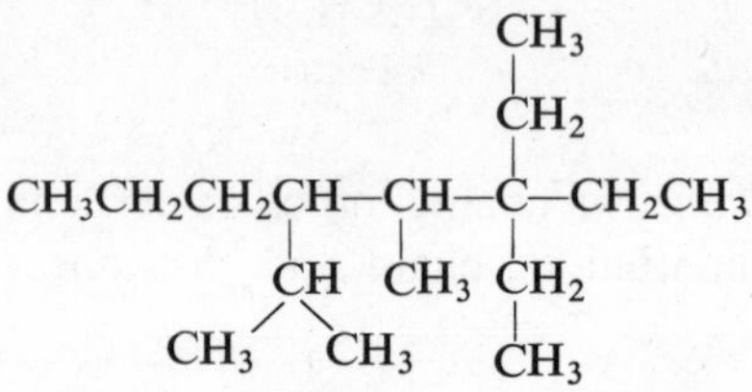

$\qquad\qquad\qquad\qquad\qquad\quad CH_3$
$\qquad\qquad\qquad\qquad\qquad\quad | $
$\qquad\qquad\qquad\qquad\qquad\quad CH_2$
$\qquad\qquad\qquad\qquad\qquad\quad | $
$CH_3CH_2CH_2CH{-}CH{-}C{-}CH_2CH_3$
$\qquad\qquad\qquad | \qquad | \qquad | $
$\qquad\qquad\quad CH \quad CH_3 \; CH_2$
$\qquad\qquad\quad / \; \backslash \qquad\quad | $
$\qquad\quad CH_3 \quad CH_3 \quad CH_3$
V
4-Methyl-3,3-diethyl-5-isopropyloctane

5. If there are several different alkyl groups attached to the parent chain, name them in order of increasing size or in alphabetical order; as in *4-methyl-3,3-diethyl-5-isopropyloctane* (V).

There are additional rules and conventions used in naming very complicated alkanes, but the five fundamental rules mentioned here will suffice for the compounds we are likely to encounter.

Problem 3.6 Give the IUPAC names for: (a) the isomeric hexanes shown on page 80; (b) the nine isomeric heptanes (see Problem 3.5, p. 80).

Problem 3.7 The IUPAC names for *n*-propyl and isopropyl chlorides are *1-chloropropane* and *2-chloropropane*. On this basis name: (a) the eight isomeric chloropentanes; (b) the nine isomeric dibromobutanes (see Problem 3.5, p. 80).

3.11 Classes of carbon atoms and hydrogen atoms

It has been found extremely useful to classify each carbon atom of an alkane with respect to the number of other carbon atoms to which it is attached.

A **primary** *(1°) carbon atom is attached to only one other carbon atom; a* **secondary** *(2°) is attached to two others; and a* **tertiary** *(3°) to three others.* For example:

```
  1° 2° 2° 1°           1°   3°   1°            1°   3°   2°   1°
  H  H  H  H            H    H    H             H    H    H    H
  |  |  |  |            |    |    |             |    |    |    |
H-C--C--C--C-H        H-C----C----C-H         H-C----C----C----C-H
  |  |  |  |            |    |    |             |    |    |    |
  H  H  H  H            H    |    H             H    |    H    H
                           H-C-H                    H-C-H
                             |                        |
                             H                        H
                             1°                       1°
```

Each hydrogen atom is similarly classified, being given the same designation of *primary*, *secondary*, or *tertiary* as the carbon atom to which it is attached.

We shall make constant use of these designations in our consideration of the relative reactivities of various parts of an alkane molecule.

3.12 Physical properties

The physical properties of the alkanes follow the pattern laid down by methane, and are consistent with the alkane structure. An alkane molecule is held together entirely by covalent bonds. These bonds either join two atoms of the same kind and hence are non-polar, or join two atoms that differ very little in electronegativity and hence are only slightly polar. Furthermore, these bonds are directed in a very symmetrical way, so that the slight bond polarities tend to cancel out. As a result an alkane molecule is either non-polar or very weakly polar.

As we have seen (Sec. 1.19), the forces holding non-polar molecules together (van der Waals forces) are weak and of very short range; they act only between the portions of different molecules that are in close contact, that is, between the surfaces of molecules. Within a family, therefore, we would expect that the larger the molecule—and hence the larger its surface area—the stronger the intermolecular forces.

Table 3.3 lists certain physical constants for a number of the *n*-alkanes. As we can see, the boiling points and melting points rise as the number of carbons increases. The processes of boiling and melting require overcoming the intermolecular forces of a liquid and a solid; the boiling points and melting points rise because these intermolecular forces increase as the molecules get larger.

Except for the very small alkanes, *the boiling point rises 20 to 30 degrees for each carbon that is added to the chain*; we shall find that this increment of 20–30° per carbon holds not only for the alkanes but also for each of the homologous series that we shall study.

The increase in melting point is not quite so regular, since the intermolecular forces in a crystal depend not only upon the size of the molecules but also upon how well they fit into a crystal lattice.

The first four *n*-alkanes are gases, but, as a result of the rise in boiling point and melting point with increasing chain length, the next 13 (C_5—C_{17}) are liquids, and those containing 18 carbons or more are solids.

Table 3.3 ALKANES

Name	Formula	M.p., °C	B.p., °C	Density (at 20°)
Methane	CH_4	−183	−162	
Ethane	CH_3CH_3	−172	− 88.5	
Propane	$CH_3CH_2CH_3$	−187	− 42	
n-Butane	$CH_3(CH_2)_2CH_3$	−138	0	
n-Pentane	$CH_3(CH_2)_3CH_3$	−130	36	0.626
n-Hexane	$CH_3(CH_2)_4CH_3$	− 95	69	.659
n-Heptane	$CH_3(CH_2)_5CH_3$	− 90.5	98	.684
n-Octane	$CH_3(CH_2)_6CH_3$	− 57	126	.703
n-Nonane	$CH_3(CH_2)_7CH_3$	− 54	151	.718
n-Decane	$CH_3(CH_2)_8CH_3$	− 30	174	.730
n-Undecane	$CH_3(CH_2)_9CH_3$	− 26	196	.740
n-Dodecane	$CH_3(CH_2)_{10}CH_3$	− 10	216	.749
n-Tridecane	$CH_3(CH_2)_{11}CH_3$	− 6	234	.757
n-Tetradecane	$CH_3(CH_2)_{12}CH_3$	5.5	252	.764
n-Pentadecane	$CH_3(CH_2)_{13}CH_3$	10	266	.769
n-Hexadecane	$CH_3(CH_2)_{14}CH_3$	18	280	.775
n-Heptadecane	$CH_3(CH_2)_{15}CH_3$	22	292	
n-Octadecane	$CH_3(CH_2)_{16}CH_3$	28	308	
n-Nonadecane	$CH_3(CH_2)_{17}CH_3$	32	320	
n-Eicosane	$CH_3(CH_2)_{18}CH_3$	36		
Isobutane	$(CH_3)_2CHCH_3$	−159	− 12	
Isopentane	$(CH_3)_2CHCH_2CH_3$	−160	28	.620
Neopentane	$(CH_3)_4C$	− 17	9.5	
Isohexane	$(CH_3)_2CH(CH_2)_2CH_3$	−154	60	.654
3-Methylpentane	$CH_3CH_2CH(CH_3)CH_2CH_3$	−118	63	.676
2,2-Dimethylbutane	$(CH_3)_3CCH_2CH_3$	− 98	50	.649
2,3-Dimethylbutane	$(CH_3)_2CHCH(CH_3)_2$	−129	58	.668

Problem 3.8 Using the data of Table 3.3, make a graph of: (a) b.p. *vs.* carbon number for the *n*-alkanes; (b) m.p. *vs.* carbon number; (c) density *vs.* carbon number.

There are somewhat smaller differences among the boiling points of alkanes that have the same carbon number but different structures. On pages 77 and 80 the boiling points of the isomeric butanes, pentanes, and hexanes are given. We see that in every case *a branched-chain isomer has a lower boiling point than a straight-chain isomer*, and further, that the more numerous the branches, the lower the boiling point. Thus *n*-butane has a boiling point of 0° and isobutane −12°. *n*-Pentane has a boiling point of 36°, isopentane with a single branch 28°, and neopentane with two branches 9.5°. This effect of branching on boiling point is observed within all families of organic compounds. That branching should lower the boiling point is reasonable: with branching the shape of the molecule tends to approach that of a sphere; and as this happens the surface area decreases, with the result that the intermolecular forces become weaker and are overcome at a lower temperature.

In agreement with the rule of thumb, "like dissolves like," the alkanes are soluble in non-polar solvents such as benzene, ether, and chloroform, and are insoluble in water and other highly polar solvents. Considered themselves as solvents, the liquid alkanes dissolve compounds of low polarity and do not dissolve compounds of high polarity.

The density increases with size of the alkanes, but tends to level off at about 0.8; thus all alkanes are less dense than water. It is not surprising that nearly all organic compounds are less dense than water since, like the alkanes, they consist chiefly of carbon and hydrogen. In general, to be denser than water a compound must contain a heavy atom like bromine or iodine, or several atoms like chlorine.

3.13 Industrial source

The principal source of alkanes is **petroleum**, together with the accompanying **natural gas**. Decay and millions of years of geological stresses have transformed the complicated organic compounds that once made up living plants or animals into a mixture of alkanes ranging in size from one carbon to 30 or 40 carbons. Formed along with the alkanes, and particularly abundant in California petroleum, are *cycloalkanes* (Chap. 9), known to the petroleum industry as *naphthenes*.

The other fossil fuel, coal, is a potential second source of alkanes: processes are being developed to convert coal, through hydrogenation, into gasoline and fuel oil, and into synthetic gas to offset anticipated shortages of natural gas.

Natural gas contains, of course, only the more volatile alkanes, that is, those of low molecular weight; it consists chiefly of methane and progressively smaller amounts of ethane, propane, and higher alkanes. For example, a sample taken from a pipeline supplied by a large number of Pennsylvania wells contained methane, ethane, and propane in the ratio of 12:2:1, with higher alkanes making up only 3% of the total. The propane–butane fraction is separated from the more volatile components by liquefaction, compressed into cylinders, and sold as *bottled gas* in areas not served by a gas utility.

Petroleum is separated by distillation into the various fractions listed in Table 3.4; because of the relationship between boiling point and molecular weight, this amounts to a rough separation according to carbon number. Each fraction is still a very complicated mixture, however, since it contains alkanes of a range of carbon numbers, and since each carbon number is represented by numerous isomers. The use that each fraction is put to depends chiefly upon its volatility or viscosity, and it matters very little whether it is a complicated mixture or a

Table 3.4 PETROLEUM CONSTITUENTS

Fraction	Distillation Temperature, °C	Carbon Number
Gas	Below 20°	C_1–C_4
Petroleum ether	20–60°	C_5–C_6
Ligroin (light naphtha)	60–100°	C_6–C_7
Natural gasoline	40–205°	C_5–C_{10}, and cycloalkanes
Kerosine	175–325°	C_{12}–C_{18}, and aromatics
Gas oil	Above 275°	C_{12} and higher
Lubricating oil	Non-volatile liquids	Probably long chains attached to cyclic structures
Asphalt or petroleum coke	Non-volatile solids	Polycyclic structures

single pure compound. (In gasoline, as we shall see in Sec. 3.30, the structures of the components are of key importance.)

The chief use of all but the non-volatile fractions is as fuel. The gas fraction, like natural gas, is used chiefly for heating. Gasoline is used in those internal combustion engines that require a fairly volatile fuel, kerosine is used in tractor and jet engines, and gas oil is used in Diesel engines. Kerosine and gas oil are also used for heating purposes, the latter being the familiar "furnace oil."

The lubricating oil fraction, especially that from Pennsylvania crude oil (*paraffin-base petroleum*), often contains large amounts of long-chain alkanes (C_{20}–C_{34}) that have fairly high melting points. If these remained in the oil, they might crystallize to waxy solids in an oil line in cold weather. To prevent this, the oil is chilled and the wax is removed by filtration. After purification this is sold as solid *paraffin wax* (m.p. 50–55°) or used in *petrolatum jelly* (Vaseline). Asphalt is used in roofing and road building. The coke that is obtained from paraffin-base crude oil consists of complex hydrocarbons having a high carbon-to-hydrogen ratio; it is used as a fuel or in the manufacture of carbon electrodes for the electrochemical industries. Petroleum ether and ligroin are useful solvents for many organic materials of low polarity.

In addition to being used directly as just described, certain petroleum fractions are converted into other kinds of chemical compounds. Catalytic **isomerization** changes straight-chain alkanes into branched-chain ones. The **cracking** process (Sec. 3.31) converts higher alkanes into smaller alkanes and alkenes, and thus increases the gasoline yield; it can even be used for the production of "natural" gas. In addition, the alkenes thus formed are perhaps the most important raw materials for the large-scale synthesis of aliphatic compounds. The process of **catalytic reforming** (Sec. 12.4) converts alkanes and cycloalkanes into aromatic hydrocarbons and thus helps provide the raw material for the large-scale synthesis of another broad class of compounds.

3.14 Industrial source *vs.* laboratory preparation

We shall generally divide the methods of obtaining a particular kind of organic compound into two categories: *industrial source* and *laboratory preparation.* We may contrast the two in the following way, although it must be realized that there are many exceptions to these generalizations.

An industrial source must provide large amounts of the desired material at the lowest possible cost. A laboratory preparation may be required to produce only a few hundred grams or even a few grams; cost is usually of less importance than the time of the investigator.

For many industrial purposes a mixture may be just as suitable as a pure compound; even when a single compound is required, it may be economically feasible to separate it from a mixture, particularly when the other components may also be marketed. In the laboratory a chemist nearly always wants a single pure compound. Separation of a single compound from a mixture of related substances is very time-consuming and frequently does not yield material of the required purity. Furthermore, the raw material for a particular preparation may well be the hard-won product of a previous preparation or even series of preparations, and hence he wishes to convert it as completely as possible into his desired compound. On an industrial scale, if a compound cannot be isolated from naturally occurring material, it may be synthesized along with a number of related compounds by some inexpensive reaction. In the laboratory, whenever possible, a reaction is selected that forms a single compound in high yield.

In industry it is frequently worth while to work out a procedure and design apparatus that may be used in the synthesis of only one member of a chemical family. In the laboratory a chemist is seldom interested in preparing the same compound over and over again, and hence he makes use of methods that are applicable to many or all members of a particular family.

In our study of organic chemistry, we shall concentrate our attention on versatile laboratory preparations rather than on limited industrial methods. In learning these we may, for the sake of simplicity, use as examples the preparation of compounds that may actually never be made by the method shown. We may discuss the synthesis of ethane by the hydrogenation of ethylene, even though we can buy all the ethane we need from the petroleum industry. However, if we know how to convert ethylene into ethane, then, when the need arises, we also know how to convert 2-methyl-1-hexene into 2-methylhexane, or cholesterol into cholestanol, or, for that matter, cottonseed oil into oleomargarine.

3.15 Preparation

Each of the smaller alkanes, from methane through *n*-pentane and isopentane, can be obtained in pure form by fractional distillation of petroleum and natural gas; neopentane does not occur naturally. Above the pentanes the number of isomers of each homolog becomes so large and the boiling point differences become so small that it is no longer feasible to isolate individual, pure compounds; these alkanes must be synthesized by one of the methods outlined below.

In some of these equations, the symbol **R** is used to represent **any alkyl group**. This convenient device helps to summarize reactions that are typical of an entire family, and emphasizes the essential similarity of the various members.

In writing these generalized equations, however, we must not lose sight of one important point. An equation involving RCl, to take a specific example, has meaning only in terms of a reaction that we can carry out in the laboratory using a real compound, like methyl chloride or *tert*-butyl chloride. Although *typical* of alkyl halides, a reaction may differ widely in rate or yield depending upon the particular alkyl group actually concerned. We may use quite different experimental conditions for methyl chloride than for *tert*-butyl chloride; in an extreme case, a reaction that goes well for methyl chloride might go so slowly or give so many side products as to be completely useless for *tert*-butyl chloride.

PREPARATION OF ALKANES

1. Hydrogenation of alkenes. Discussed in Sec. 6.3.

$$\underset{\text{Alkene}}{C_nH_{2n}} \xrightarrow{H_2 + \text{Pt, Pd, or Ni}} \underset{\text{Alkane}}{C_nH_{2n+2}}$$

2. Reduction of alkyl halides

(a) Hydrolysis of Grignard reagent. Discussed in Sec. 3.16.

$$RX + Mg \longrightarrow \underset{\text{Grignard reagent}}{RMgX} \xrightarrow{H_2O} RH$$

Example:

$$\underset{\textit{sec}\text{-Butyl bromide}}{CH_3CH_2\underset{|}{\overset{}{C}}HCH_3 \atop \quad\; Br} \xrightarrow{Mg} \underset{\textit{sec}\text{-Butylmagnesium bromide}}{CH_3CH_2\underset{|}{C}HCH_3 \atop \quad\; MgBr} \xrightarrow{H_2O} \underset{n\text{-Butane}}{CH_3CH_2\underset{|}{C}HCH_3 \atop \quad\; H}$$

(b) Reduction by metal and acid. Discussed in Sec. 3.15.

$$RX + Zn + H^+ \longrightarrow RH + Zn^{++} + X^-$$

Example:

$$\underset{\textit{sec}\text{-Butyl bromide}}{CH_3CH_2\underset{|}{C}HCH_3 \atop \quad\; Br} \xrightarrow{Zn,\ H^+} \underset{n\text{-Butane}}{CH_3CH_2\underset{|}{C}HCH_3 \atop \quad\; H}$$

3. Coupling of alkyl halides with organometallic compounds. Discussed in Sec. 3.17.

$$\underset{\substack{\textit{May be} \\ 1°,\ 2°,\ 3°}}{RX} \xrightarrow{Li} \underset{\text{Alkyllithium}}{RLi} \xrightarrow{CuX} \underset{\substack{\text{Lithium} \\ \text{dialkylcopper}}}{R_2CuLi} \;+\; \underset{\textit{Should be } 1°}{R'X} \longrightarrow R{-}R'$$

Examples:

$$\underset{\substack{\text{Ethyl} \\ \text{chloride}}}{CH_3CH_2Cl} \xrightarrow{Li} \underset{\text{Ethyllithium}}{CH_3CH_2Li} \xrightarrow{CuI} \underset{\substack{\text{Lithium} \\ \text{diethylcopper}}}{(CH_3CH_2)_2CuLi} \;+\; \underset{n\text{-Heptyl bromide}}{CH_3(CH_2)_5CH_2Br} \longrightarrow \underset{n\text{-Nonane}}{CH_3(CH_2)_7CH_3}$$

$$\underset{\textit{tert}\text{-Butyl chloride}}{CH_3\overset{CH_3}{\underset{Cl}{C}}CH_3} \xrightarrow{Li} \xrightarrow{CuI} (t\text{-}C_4H_9)_2CuLi \;+\; \underset{n\text{-Pentyl bromide}}{CH_3CH_2CH_2CH_2CH_2Br} \longrightarrow \underset{\text{2,2-Dimethylheptane}}{CH_3\overset{CH_3}{\underset{CH_3}{C}}CH_2CH_2CH_2CH_2CH_3}$$

By far the most important of these methods is the hydrogenation of alkenes. When shaken under a slight pressure of hydrogen gas in the presence of a small amount of catalyst, alkenes are converted smoothly and quantitatively into alkanes of the same carbon skeleton. The method is limited only by the availability of the proper alkene. This is not a very serious limitation; as we shall see (Sec. 5.11), alkenes are readily prepared, chiefly from alcohols, which in turn can be readily synthesized (Sec. 15.7) in a wide variety of sizes and shapes.

Reduction of an alkyl halide, either via the Grignard reagent or directly with metal and acid, involves simply the replacement of a halogen atom by a hydrogen atom; the carbon skeleton remains intact. This method has about the same applicability as the previous method, since, like alkenes, alkyl halides are generally prepared from alcohols. Where either method could be used, the hydrogenation of alkenes would probably be preferred because of its simplicity and higher yield.

The coupling of alkyl halides with organometallic compounds is the only one of these methods in which carbon–carbon bonds are formed and a new, bigger carbon skeleton is generated.

3.16 The Grignard reagent: an organometallic compound

When a solution of an alkyl halide in dry ethyl ether, $(C_2H_5)_2O$, is allowed to stand over turnings of metallic magnesium, a vigorous reaction takes place: the solution turns cloudy, begins to boil, and the magnesium metal gradually disappears. The resulting solution is known as a **Grignard reagent**, after Victor Grignard (of the University of Lyons) who received the Nobel prize in 1912 for its discovery. It is one of the most useful and versatile reagents known to the organic chemist.

$$\underset{\text{Methyl iodide}}{CH_3I} + Mg \xrightarrow{\text{ether}} \underset{\text{Methylmagnesium iodide}}{CH_3MgI}$$

$$\underset{\text{Ethyl bromide}}{CH_3CH_2Br} + Mg \xrightarrow{\text{ether}} \underset{\text{Ethylmagnesium bromide}}{CH_3CH_2MgBr}$$

The Grignard reagent has the general formula RMgX, and the general name **alkylmagnesium halide**. The carbon–magnesium bond is covalent but highly polar, with carbon pulling electrons from electropositive magnesium; the magnesium–halogen bond is essentially ionic.

$$R{:}Mg^{+}{:}\ddot{\underset{\cdot\cdot}{X}}{:}^{-}$$

Since magnesium becomes bonded to the same carbon that previously held halogen, the alkyl group remains intact during the preparation of the reagent. Thus *n*-propyl chloride yields *n*-propylmagnesium chloride, and isopropyl chloride yields isopropylmagnesium chloride.

$$\underset{n\text{-Propyl chloride}}{CH_3CH_2CH_2Cl} + Mg \xrightarrow{\text{ether}} \underset{n\text{-Propylmagnesium chloride}}{CH_3CH_2CH_2MgCl}$$

$$\underset{\text{Isopropyl chloride}}{CH_3CHClCH_3} + Mg \xrightarrow{\text{ether}} \underset{\text{Isopropylmagnesium chloride}}{CH_3CHMgClCH_3}$$

The Grignard reagent is the best-known member of a broad class of substances, called **organometallic** compounds, in which carbon is bonded to a metal: lithium, potassium, sodium, zinc, mercury, lead, thallium—almost any metal known. Each kind of organometallic compound has, of course, its own set of properties, and its particular uses depend on these. But, whatever the metal, it is less electronegative than carbon, and the carbon–metal bond—like the one in the

$$\overset{\delta-}{R}—\overset{\delta+}{M}$$

Grignard reagent—is highly polar. Although the organic group is not a full-fledged *carbanion*—an anion in which carbon carries negative charge—it nevertheless has considerable carbanion character. As we shall see, organometallic compounds owe their enormous usefulness chiefly to one common quality: they can serve as a source from which carbon is readily transferred *with its electrons.*

The Grignard reagent is highly reactive. It reacts with numerous inorganic compounds including water, carbon dioxide, and oxygen, and with most kinds of organic compounds; in many of these cases the reaction provides the best way to make a particular class of organic compound.

The reaction with water to form an alkane is typical of the behavior of the Grignard reagent—and many of the more reactive organometallic compounds—toward acids. In view of the marked carbanion character of the alkyl group, we may consider the Grignard reagent to be the magnesium salt, RMgX, of the extremely weak acid, R—H. The reaction

$$\mathrm{RMgX} + \underset{\text{Stronger acid}}{\mathrm{HOH}} \longrightarrow \underset{\text{Weaker acid}}{\mathrm{R{-}H}} + \mathrm{Mg(OH)X}$$

is simply the displacement of the weaker acid, R—H, from its salt by the stronger acid, HOH.

An alkane is such a weak acid that it is displaced from the Grignard reagent by compounds that we might ordinarily consider to be very weak acids themselves, or possibly not acids at all. Any compound containing hydrogen attached to oxygen or nitrogen is tremendously more acidic than an alkane, and therefore can decompose the Grignard reagent: for example, ammonia or methyl alcohol.

$$\mathrm{RMgX} + \underset{\text{Stronger acid}}{\mathrm{NH_3}} \longrightarrow \underset{\text{Weaker acid}}{\mathrm{R{-}H}} + \mathrm{Mg(NH_2)X}$$

$$\mathrm{RMgX} + \underset{\text{Stronger acid}}{\mathrm{CH_3OH}} \longrightarrow \underset{\text{Weaker acid}}{\mathrm{R{-}H}} + \mathrm{Mg(OCH_3)X}$$

For the preparation of an alkane, one acid is as good as another, so we naturally choose water as the most available and convenient.

Problem 3.9 (a) Which alkane would you expect to get by the action of water on *n*-propylmagnesium chloride? (b) On isopropylmagnesium chloride? (c) Answer (a) and (b) for the action of deuterium oxide ("heavy water," D_2O).

Problem 3.10 On conversion into the Grignard reagent followed by treatment with water, how many alkyl bromides would yield: (a) *n*-pentane; (b) 2-methylbutane; (c) 2,3-dimethylbutane; (d) neopentane? Draw the structures in each case.

3.17 Coupling of alkyl halides with organometallic compounds

To make an alkane of higher carbon number than the starting material requires formation of carbon–carbon bonds, most directly by the coupling together of two alkyl groups. The most versatile method of doing this is through a synthesis developed during the late 1960s by E. J. Corey and Herbert House, working independently at Harvard University and Massachusetts Institute of Technology.

Coupling takes place in the reaction between a *lithium dialkylcopper*, R_2CuLi, and an alkyl halide, R′X. (R′ stands for an alkyl group that may be the same as, or different from, R.)

$$\underset{\text{Lithium dialkylcopper}}{R_2CuLi} + \underset{\text{Alkyl halide}}{R'X} \longrightarrow \underset{\text{Alkane}}{R\text{—}R'} + RCu + LiX$$

An alkyllithium, RLi, is prepared from an alkyl halide, RX, in much the same way as a Grignard reagent. To it is added cuprous halide, CuX, and then, finally, the second alkyl halide, R′X. Ultimately, the alkane is synthesized from the two alkyl halides, RX and R′X.

$$RX \xrightarrow{Li} \underset{\text{Alkyl lithium}}{RLi} \xrightarrow{CuX} \underset{\text{Lithium dialkylcopper}}{R_2CuLi} \xrightarrow{R'X} R\text{—}R'$$

For good yields, R′X should be a *primary* halide; the alkyl group R in the organometallic may be primary, secondary, or tertiary. For example:

$$\underset{\text{Methyl bromide}}{CH_3Br} \xrightarrow{Li} \underset{\text{Methyllithium}}{CH_3Li} \xrightarrow{CuI} \underset{\text{Lithium dimethylcopper}}{(CH_3)_2CuLi} \xrightarrow[n\text{-Octyl iodide}]{CH_3(CH_2)_6CH_2I} \underset{n\text{-Nonane}}{CH_3(CH_2)_7CH_3}$$

$$\underset{\text{sec-Butyl chloride}}{CH_3CH_2CH(Cl)CH_3} \xrightarrow{Li} \xrightarrow{CuI} (CH_3CH_2CH(CH_3)\text{—})_2CuLi \xrightarrow[n\text{-Pentyl bromide}]{CH_3CH_2CH_2CH_2CH_2Br} \underset{\text{3-Methyloctane}}{CH_3CH_2CH(CH_3)(CH_2)_4CH_3}$$

The choice of organometallic reagent is crucial. Grignard reagents or organolithium compounds, for example, couple with only a few unusually reactive organic halides. Organosodium compounds couple, but are so reactive that they couple, as they are being formed, with their parent alkyl halide; the reaction of sodium with alkyl halides (*Wurtz reaction*) is thus limited to the synthesis of symmetrical alkanes, R—R.

Organocopper compounds were long known to be particularly good at the formation of carbon–carbon bonds, but are unstable. Here, they are generated *in situ* from the organolithium, and then combine with more of it to form these relatively stable organometallics. They exist as complex aggregates but are believed to correspond roughly to $R_2Cu^-Li^+$. The anion here is an example of an *ate* complex, the negative counterpart of an *onium* complex (amm*onium*, ox*onium*).

Although the mechanism is not understood, evidence strongly suggests this much: the alkyl group R is transferred from copper, taking a pair of electrons with it, and attaches itself to the alkyl group R′ by pushing out halide ion (*nucleophilic aliphatic substitution*, Sec. 14.9).

Problem 3.11 (a) Outline two conceivable syntheses of 2-methylpentane from three-carbon compounds. (b) Which of the two would you actually use? Why?

3.18 Reactions

The alkanes are sometimes referred to by the old-fashioned name of *paraffins*. This name (Latin: *parum affinis*, not enough affinity) was given to describe what appeared to be the low reactivity of these hydrocarbons.

But reactivity depends upon the choice of reagent. If alkanes are inert toward hydrochloric and sulfuric acids, they react readily with acids like $HF–SbF_5$ and $FSO_3H–SbF_5$ ("magic acid") to yield a variety of products. If alkanes are inert toward oxidizing agents like potassium permanganate or sodium dichromate, most of this chapter is devoted to their oxidation by halogens. Certain yeasts feed happily on alkanes to produce proteins—certainly a chemical reaction. As Professor M. S. Kharasch (p. 189) used to put it, consider the "inertness" of a room containing natural gas, air, and a lighted match.

Still, on a comparative basis, reactivity *is* limited. "Magic acid" is, after all, one of the strongest acids known; halogenation requires heat or light; combustion needs a flame or spark to get it started.

Much of the chemistry of alkanes involves free-radical chain reactions, which take place under vigorous conditions and usually yield mixtures of products. A reactive particle—typically an atom or free radical—is needed to begin the attack on an alkane molecule. It is the generation of this reactive particle that requires the vigorous conditions: the dissociation of a halogen molecule into atoms, for example, or even (as in pyrolysis) dissociation of the alkane molecule itself.

In its attack, the reactive particle abstracts hydrogen from the alkane; the alkane itself is thus converted into a reactive particle which continues the reaction sequence, that is, carries on the chain. But an alkane molecule contains many hydrogen atoms and the particular product eventually obtained depends upon *which* of these hydrogen atoms is abstracted. Although an attacking particle may show a certain selectivity, it can abstract a hydrogen from any part of the molecule, and thus bring about the formation of many isomeric products.

REACTIONS OF ALKANES

1. Halogenation. Discussed in Secs. 3.19–3.22.

$$-\overset{|}{\underset{|}{C}}-H + X_2 \xrightarrow{250\text{–}400^\circ\text{, or light}} -\overset{|}{\underset{|}{C}}-X + HX$$

Usually a mixture

Reactivity X_2: $Cl_2 > Br_2$

H: $3^\circ > 2^\circ > 1^\circ > CH_3—H$

Example:

$$\underset{\text{Isobutane}}{CH_3-\overset{CH_3}{\overset{|}{C}H}-CH_3} \xrightarrow[250\text{–}400^\circ]{Cl_2} \underset{\text{Isobutyl chloride}}{CH_3-\overset{CH_3}{\overset{|}{C}H}-CH_2Cl} \text{ and } \underset{\textit{tert}\text{-Butyl chloride}}{CH_3-\overset{CH_3}{\overset{|}{\underset{\underset{Cl}{|}}{C}}}-CH_3}$$

2. Combustion. Discussed in Sec. 3.30.

$$C_nH_{2n+2} + \text{excess } O_2 \xrightarrow{\text{flame}} nCO_2 + (n+1)H_2O \qquad \Delta H = \text{heat of combustion}$$

Example:

$$n\text{-}C_5H_{12} + 8\,O_2 \xrightarrow{\text{flame}} 5CO_2 + 6H_2O \qquad \Delta H = -845 \text{ kcal}$$

3. Pyrolysis (cracking). Discussed in Sec. 3.31.

$$\text{alkane} \xrightarrow{400\text{–}600^\circ\text{; with or without catalysts}} H_2 + \text{smaller alkanes} + \text{alkenes}$$

3.19 Halogenation

As we might expect, halogenation of the higher alkanes is essentially the same as the halogenation of methane. It can be complicated, however, by the formation of mixtures of isomers.

Under the influence of ultraviolet light, or at 250–400°, chlorine or bromine converts alkanes into chloroalkanes (alkyl chlorides) or bromoalkanes (alkyl bromides); an equivalent amount of hydrogen chloride or hydrogen bromide is formed at the same time. When diluted with an inert gas, and in an apparatus designed to carry away the heat produced, fluorine has recently been found to give analogous results. As with methane, iodination does not take place at all.

Depending upon which hydrogen atom is replaced, any of a number of isomeric products can be formed from a single alkane. Ethane can yield only one haloethane; propane, *n*-butane, and isobutane can yield two isomers each; *n*-pentane can yield three isomers, and isopentane, four isomers. Experiment has shown that on halogenation an alkane yields a mixture of all possible isomeric products, indicating that all hydrogen atoms are susceptible to replacement. For example, for chlorination:

$$CH_3CH_3 \xrightarrow[\text{light, }25^\circ]{Cl_2} CH_3CH_2\text{—}Cl$$

Ethane → b.p. 13°, Chloroethane, Ethyl chloride

$$CH_3CH_2CH_3 \xrightarrow[\text{light, }25^\circ]{Cl_2} CH_3CH_2CH_2\text{—}Cl \quad \text{and} \quad CH_3\underset{\displaystyle Cl}{\underset{|}{C}}HCH_3$$

Propane → b.p. 47°, 1-Chloropropane, *n*-Propyl chloride, *45%*; and b.p. 36°, 2-Chloropropane, Isopropyl chloride, *55%*

$$CH_3CH_2CH_2CH_3 \xrightarrow[\text{light, }25^\circ]{Cl_2} CH_3CH_2CH_2CH_2\text{—}Cl \quad \text{and} \quad CH_3CH_2\underset{\displaystyle Cl}{\underset{|}{C}}HCH_3$$

n-Butane → b.p. 78.5°, 1-Chlorobutane, *n*-Butyl chloride, *28%*; and b.p. 68°, 2-Chlorobutane, *sec*-Butyl chloride, *72%*

$$\underset{\text{Isobutane}}{CH_3\overset{\overset{CH_3}{|}}{C}HCH_3} \xrightarrow[\text{light, 25°}]{Cl_2} \underset{\substack{\text{b.p. 69°} \\ \text{1-Chloro-2-} \\ \text{methylpropane} \\ \text{Isobutyl chloride} \\ 64\%}}{CH_3\overset{\overset{CH_3}{|}}{C}HCH_2{-}Cl} \quad \text{and} \quad \underset{\substack{\text{b.p. 51°} \\ \text{2-Chloro-2-} \\ \text{methylpropane} \\ \textit{tert}\text{-Butyl chloride} \\ 36\%}}{CH_3\overset{\overset{CH_3}{|}}{\underset{\underset{Cl}{|}}{C}}CH_3}$$

Bromination gives the corresponding bromides but in different proportions:

$$\underset{\text{Ethane}}{CH_3CH_3} \xrightarrow[\text{light, 146°}]{Br_2} CH_3CH_2Br$$

$$\underset{\text{Propane}}{CH_3CH_2CH_3} \xrightarrow[\text{light, 146°}]{Br_2} \underset{3\%}{CH_3CH_2CH_2Br} \quad \text{and} \quad \underset{97\%}{CH_3\underset{\underset{Br}{|}}{C}HCH_3}$$

$$\underset{n\text{-Butane}}{CH_3CH_2CH_2CH_3} \xrightarrow[\text{light, 146°}]{Br_2} \underset{2\%}{CH_3CH_2CH_2CH_2Br} \quad \text{and} \quad \underset{98\%}{CH_3CH_2\underset{\underset{Br}{|}}{C}HCH_3}$$

$$\underset{\text{Isobutane}}{CH_3\overset{\overset{CH_3}{|}}{C}HCH_3} \xrightarrow[\text{light, 146°}]{Br_2} \underset{\textit{trace}}{CH_3\overset{\overset{CH_3}{|}}{C}HCH_2Br} \quad \text{and} \quad \underset{\textit{over } 99\%}{CH_3\overset{\overset{CH_3}{|}}{\underset{\underset{Br}{|}}{C}}CH_3}$$

Problem 3.12 Draw the structures of: (a) the three monochloro derivatives of *n*-pentane; (b) the four monochloro derivatives of isopentane.

Although both chlorination and bromination yield mixtures of isomers, the results given above show that the *relative amounts* of the various isomers differ markedly depending upon the halogen used. Chlorination gives mixtures in which no isomer greatly predominates; in bromination, by contrast, one isomer may predominate to such an extent as to be almost the only product, making up 97–99% of the total mixture. In bromination, there is a high degree of *selectivity* as to which hydrogen atoms are to be replaced. (As we shall see in Sec. 3.28, this characteristic of bromination is due to the relatively low reactivity of bromine atoms, and is an example of a general relationship between *reactivity* and *selectivity*.)

Chlorination of an alkane is not usually suitable for the laboratory preparation of an alkyl chloride; any one product is necessarily formed in low yield, and is difficult to separate from its isomers, whose boiling points are seldom far from its own. Bromination, on the other hand, often gives a nearly pure alkyl bromide in high yield. As we shall see, it is possible to predict just which isomer will predominate; if this product is the one desired, direct bromination could be a feasible synthetic route.

On an industrial scale, chlorination of alkanes is important. For many purposes, for example, use as a solvent, a mixture of isomers is just as suitable as, and much cheaper than, a pure compound. It may be even worthwhile, when necessary, to separate a mixture of isomers if each isomer can then be marketed.

Problem 3.13 How do you account for the fact that not only bromination but also chlorination is a feasible laboratory route to a neopentyl halide, $(CH_3)_3CCH_2X$?

3.20 Mechanism of halogenation

Halogenation of alkanes proceeds by the same mechanism as halogenation of methane:

(1) $X_2 \xrightarrow[\text{or ultraviolet light}]{250\text{–}400°} 2X\cdot$ **Chain-initiating step**

(2) $X\cdot + RH \longrightarrow HX + R\cdot$

(3) $R\cdot + X_2 \longrightarrow RX + X\cdot$

(2) and (3): **Chain-propagating steps**

then (2), (3), (2), (3), *etc., until finally a chain is terminated* (Sec. 2.13)

A halogen atom abstracts hydrogen from the alkane (RH) to form an alkyl radical (R·). The radical in turn abstracts a halogen atom from a halogen molecule to yield the alkyl halide (RX).

Which alkyl halide is obtained depends upon which alkyl radical is formed.

$CH_4 \xrightarrow{X\cdot} CH_3\cdot \xrightarrow{X_2} CH_3X$

Methane, Methyl radical, Methyl halide

$CH_3CH_3 \xrightarrow{X\cdot} CH_3CH_2\cdot \xrightarrow{X_2} CH_3CH_2X$

Ethane, Ethyl radical, Ethyl halide

$CH_3CH_2CH_3$ (Propane) $\xrightarrow{X\cdot}$

- $\xrightarrow{\text{abstraction of 1° H}} CH_3CH_2CH_2\cdot$ (*n*-Propyl radical) $\xrightarrow{X_2} CH_3CH_2CH_2X$ (*n*-Propyl halide)
- $\xrightarrow{\text{abstraction of 2° H}} CH_3\dot{C}HCH_3$ (Isopropyl radical) $\xrightarrow{X_2} CH_3CH(X)CH_3$ (Isopropyl halide)

This in turn depends upon the alkane and which hydrogen atom is abstracted from it. For example, *n*-propyl halide is obtained from a *n*-propyl radical, formed from propane by abstraction of a primary hydrogen; isopropyl halide is obtained from an isopropyl radical, formed by abstraction of a secondary hydrogen.

How fast an alkyl halide is formed depends upon how fast the alkyl radical is formed. Here also, as was the case with methane (Sec. 2.20), of the two chain-propagating steps, step (2) is more difficult than step (3), and hence controls the rate of overall reaction. Formation of the alkyl radical is difficult, but once formed the radical is readily converted into the alkyl halide (see Fig. 3.5).

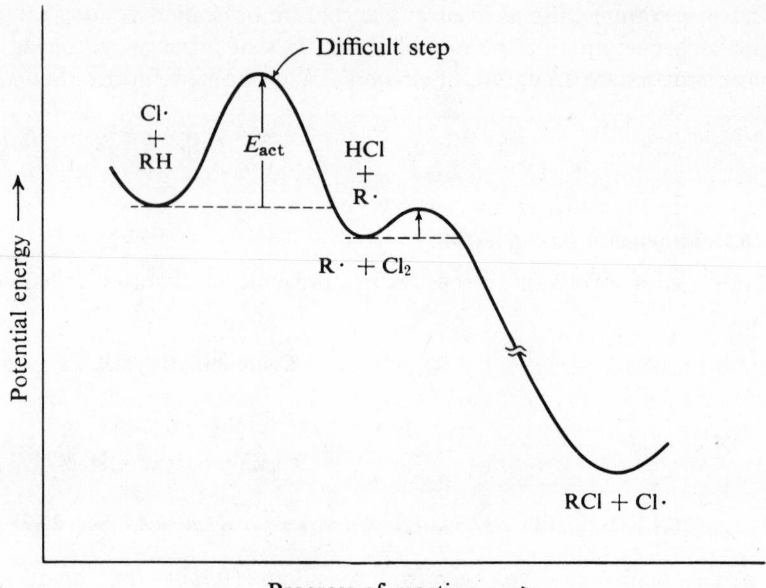

Figure 3.5. Potential energy changes during progress of reaction: chlorination of an alkane. Formation of radical is rate-controlling step.

3.21 Orientation of halogenation

With this background let us turn to the problem of **orientation**; that is, let us examine the factors that determine *where* in a molecule reaction is most likely to occur. It is a problem that we shall encounter again and again, whenever we study a compound that offers more than one reactive site to attack by a reagent. It is an important problem, because orientation determines what product we obtain.

As an example let us take chlorination of propane. The relative amounts of *n*-propyl chloride and isopropyl chloride obtained depend upon the relative rates at which *n*-propyl radicals and isopropyl radicals are formed. If, say, isopropyl radicals are formed faster, then isopropyl chloride will be formed faster, and will make up a larger fraction of the product. As we can see, *n*-propyl radicals are formed by abstraction of primary hydrogens, and isopropyl radicals by abstraction of secondary hydrogens.

```
                              H  H  H                  H  H  H
                abstraction   |  |  |                  |  |  |
                 of 1° H   → H—C—C—C·    —Cl₂→    H—C—C—C—Cl
                ┌             |  |  |                  |  |  |
                │             H  H  H                  H  H  H
   H  H  H      │            n-Propyl                 n-Propyl
   |  |  |      │             radical                 chloride
H—C—C—C—H  —Cl·—┤
   |  |  |      │
   H  H  H      │             H  H  H                  H  H  H
   Propane      │ abstraction |  |  |                  |  |  |
                └ of 2° H  → H—C—Ċ—C—H   —Cl₂→    H—C—C—C—H
                              |     |                  |  |  |
                              H     H                  H  Cl H
                             Isopropyl                Isopropyl
                              radical                 chloride
```

Thus *orientation is determined by the relative rates of competing reactions.* In this case we are comparing the rate of abstraction of primary hydrogens with the rate of abstraction of secondary hydrogens. What are the factors that determine the rates of these two reactions, and in which of these factors may the two reactions differ?

First of all, there is the collision frequency. This must be the same for the two reactions, since both involve collisions of the same particles: a propane molecule and a chlorine atom.

Next, there is the probability factor. If a primary hydrogen is to be abstracted, the propane molecule must be so oriented at the time of collision that the chlorine atom strikes a primary hydrogen; if a secondary hydrogen is to be abstracted, the propane must be so oriented that the chlorine collides with a secondary hydrogen. Since there are six primary hydrogens and only two secondary hydrogens in each molecule, we might estimate that the probability factor favors abstraction of primary hydrogens by the ratio of 6:2, or 3:1.

Considering only collision frequency and our guess about probability factors, we predict that chlorination of propane would yield *n*-propyl chloride and isopropyl chloride in the ratio of 3:1. As shown on page 95, however, the two chlorides are formed in roughly equal amounts, that is, in the ratio of about 1:1, or 3:3. The proportion of isopropyl chloride is about three times as great as predicted. Evidently, about three times as many collisions with secondary hydrogens are successful as collisions with primary hydrogens. If our assumption about the probability factor is correct, this means that E_{act} is less for abstraction of a secondary hydrogen than for abstraction of a primary hydrogen.

Chlorination of isobutane presents a similar problem. In this case, abstraction of one of the nine primary hydrogen leads to the formation of isobutyl chloride, whereas abstraction of the single tertiary hydrogen leads to the formation of *tert*-butyl chloride. We would estimate, then, that the probability factor favors

```
                                      H                    H
                                      |                    |
                                    H-C-H                H-C-H
                                    H | H                H | H
                     abstraction    | | |                | | |
                      of 1° H  -->  H-C-C-C·   --Cl2-->  H-C-C-C-Cl
                    |               | | |                | | |
        H           |               H H H                H H H
        |           |              Isobutyl             Isobutyl
      H-C-H         |              radical              chloride
      H | H         |
      | | |   Cl·   |
    H-C-C-C-H ------|
      | | |         |
      H H H         |                 H                    H
    Isobutane       |                 |                    |
                    |               H-C-H                H-C-H
                    |               H | H                H | H
                    | abstraction   | | |                | | |
                    |  of 3° H -->  H-C-C-C-H  --Cl2-->  H-C-C-C-H
                                    |  ·  |              | | |
                                    H     H              H Cl H
                                  tert-Butyl           tert-Butyl
                                   radical              chloride
```

formation of isobutyl chloride by the ratio of 9:1. The experimental results given on page 96 show that the ratio is roughly 2:1, or 9:4.5. Evidently, about 4.5 times as many collisions with the tertiary hydrogen are successful as collisions with the primary hydrogens. This, in turn, probably means that E_{act} is less for abstraction of a tertiary hydrogen than for abstraction of a primary hydrogen, and, in fact, even less than for abstraction of a secondary hydrogen.

Study of the chlorination of a great many alkanes has shown that these are typical results. After allowance is made for differences in the probability factor, the rate of abstraction of hydrogen atoms is always found to follow the sequence 3° > 2° > 1°. At room temperature, for example, the relative rates *per hydrogen atom* are 5.0:3.8:1.0. Using these values we can predict quite well the ratio of isomeric chlorination products from a given alkane. For example:

$$\underset{n\text{-Butane}}{CH_3CH_2CH_2CH_3} \xrightarrow[\text{light, 25°}]{Cl_2} \underset{n\text{-Butyl chloride}}{CH_3CH_2CH_2CH_2Cl} \quad \text{and} \quad \underset{sec\text{-Butyl chloride}}{CH_3CH_2CHClCH_3}$$

$$\frac{n\text{-butyl chloride}}{sec\text{-butyl chloride}} = \frac{\text{no. of 1° H}}{\text{no. of 2° H}} \times \frac{\text{reactivity of 1° H}}{\text{reactivity of 2° H}}$$

$$= \frac{6}{4} \times \frac{1.0}{3.8}$$

$$= \frac{6}{15.2} \quad \textit{equivalent to} \quad \frac{28\%}{72\%}$$

In spite of these differences in reactivity, chlorination rarely yields a great preponderance of any single isomer. In nearly every alkane, as in the examples we have studied, the less reactive hydrogens are the more numerous; their lower reactivity is compensated for by a higher probability factor, with the result that appreciable amounts of every isomer are obtained.

Problem 3.14 Predict the proportions of isomeric products from chlorination at room temperature of: (a) propane; (b) isobutane; (c) 2,3-dimethylbutane; (d) *n*-pentane (*Note:* There are *three* isomeric products); (e) isopentane; (f) 2,2,3-trimethylbutane; (g) 2,2,4-trimethylpentane. For (a) and (b) check your calculations against the experimental values given on pages 95 and 96.

The same sequence of reactivity, 3° > 2° > 1°, is found in bromination, but with enormously larger reactivity ratios. At 146°, for example, the relative rates per hydrogen atom are 1600:82:1. Here, differences in reactivity are so marked as vastly to outweigh probability factors.

Problem 3.15 Answer Problem 3.14 for bromination at 146°.

3.22 Relative reactivities of alkanes toward halogenation

The best way to measure the relative reactivities of different compounds toward the same reagent is by the **method of competition**, since this permits an exact quantitative comparison under identical reaction conditions. Equimolar amounts of two compounds to be compared are mixed together and allowed to react with a limited amount of a particular reagent. Since there is not enough reagent for both compounds, the two compete with each other. Analysis of the reaction products shows which compound has consumed more of the reagent and hence is more reactive.

For example, if equimolar amounts of methane and ethane are allowed to react with a small amount of chlorine, about 400 times as much ethyl chloride as methyl chloride is obtained, showing that ethane is 400 times as reactive as methane. When allowance is made for the relative numbers of hydrogens in the two kinds of molecules, we see that each hydrogen of ethane is about 270 times as reactive as each hydrogen of methane.

$$\underset{1}{CH_3Cl} \xleftarrow{CH_4} Cl_2 \xrightarrow[\text{light, } 25^\circ]{C_2H_6} \underset{400}{C_2H_5Cl}$$

Problem 3.16 Because of the rather large difference in reactivity between ethane and methane, competition experiments have actually used mixtures containing more methane than ethane. If the molar ratio of methane to ethane were 10:1, what ratio of ethyl chloride to methyl chloride would you expect to obtain? What practical advantage would this experiment have over one involving a 1:1 ratio?

Data obtained from similar studies of other compounds are consistent with this simple generalization: *the reactivity of a hydrogen depends chiefly upon its class, and not upon the alkane to which it is attached.* Each primary hydrogen of propane, for example, is about as easily abstracted as each primary hydrogen in *n*-butane or isobutane; each secondary hydrogen of propane, about as easily as each secondary hydrogen of *n*-butane or *n*-pentane; and so on.

The hydrogen atoms of methane, which fall into a special class, are even less reactive than primary hydrogens, as shown by the above competition with ethane.

Problem 3.17 On chlorination, an equimolar mixture of ethane and neopentane yields neopentyl chloride and ethyl chloride in the ratio of 2.3:1. How does the reactivity of a primary hydrogen in neopentane compare with that of a primary hydrogen in ethane?

3.23 Ease of abstraction of hydrogen atoms. Energy of activation

At this stage we can summarize the effect of structure on halogenation of alkanes in the following way. The controlling step in halogenation is abstraction of hydrogen by a halogen atom:

$$R{-}H + X\cdot \longrightarrow H{-}X + R\cdot$$

The relative ease with which the different classes of hydrogen atoms are abstracted is:

Ease of abstraction of hydrogen atoms $3^\circ > 2^\circ > 1^\circ > CH_4$

This sequence applies (a) to the various hydrogens within a single alkane and hence governs **orientation** of reaction, and (b) to the hydrogens of different alkanes and hence governs **relative reactivities**.

Earlier, we concluded that these differences in ease of abstraction—like most differences in rate between closely related reactions (Sec. 2.19)—are probably due to differences in E_{act}. By study of halogenation at a series of temperatures (Sec. 2.18), the values of E_{act} listed in Table 3.5 were measured. In agreement with our tentative conclusions, the increasing rate of reaction along the series, methyl, 1°,

Table 3.5 ENERGIES OF ACTIVATION, KCAL/MOLE

$$R{-}H + X\cdot \longrightarrow R\cdot + H{-}X$$

R	X = Cl	X = Br
CH_3	4	18
1°	1	13
2°	0.5	10
3°	0.1	7.5

2°, 3°, is paralleled by a decreasing E_{act}. In chlorination the differences in E_{act}, like the differences in rate, are small; in bromination both differences are large.

We have seen (Sec. 2.18) that the larger the E_{act} of a reaction, the larger the increase in rate brought about by a given rise in temperature. We have just found that the differences in rate of abstraction among primary, secondary, and tertiary hydrogens are due to differences in E_{act}. We predict, therefore, that a rise in temperature should speed up abstraction of primary hydrogens (with the largest E_{act}) most, and abstraction of tertiary hydrogens (with the smallest E_{act}) least; the three classes of hydrogen should then display more nearly the same reactivity.

This leveling-out effect has indeed been observed: as the temperature is raised, the relative rates per hydrogen atom change from 5.0:3.8:1.0 toward 1:1:1. At very high temperatures virtually every collision has enough energy for abstraction of even primary hydrogens. It is generally true that *as the temperature is raised a given reagent becomes less selective in the position of its attack*; conversely, as the temperature is lowered it becomes more selective.

How can we account for the effect of structure on ease of abstraction of hydrogen atoms? Since this is a matter of E_{act}, we must look for our answer, as always, in the transition state. To do this, however, we must first shift our focus from the hydrogen atom being abstracted to the radical being formed.

3.24 Stability of free radicals

In Table 1.2 (p. 21) we find the dissociation energies of the bonds that hold hydrogen atoms to a number of groups. These values are the ΔH's of the following reactions:

$$CH_3{-}H \longrightarrow CH_3\cdot + H\cdot \qquad \Delta H = 104 \text{ kcal}$$

$$CH_3CH_2{-}H \longrightarrow \underset{\text{A 1° radical}}{CH_3CH_2\cdot} + H\cdot \qquad \Delta H = 98$$

$$CH_3CH_2CH_2{-}H \longrightarrow \underset{\text{A 1° radical}}{CH_3CH_2CH_2\cdot} + H\cdot \qquad \Delta H = 98$$

$$CH_3\underset{\displaystyle H}{\underset{|}{C}}HCH_3 \longrightarrow \underset{\text{A 2° radical}}{CH_3\dot{C}HCH_3} + H\cdot \qquad \Delta H = 95$$

$$CH_3\overset{\displaystyle CH_3}{\overset{|}{\underset{\displaystyle H}{\underset{|}{C}}}}CH_3 \longrightarrow \underset{\text{A 3° radical}}{CH_3\overset{\displaystyle CH_3}{\overset{|}{\dot{C}}}CH_3} + H\cdot \qquad \Delta H = 91$$

By definition, bond dissociation energy is the amount of energy that must be supplied to convert a mole of alkane into radicals and hydrogen atoms. As we can see, the amount of energy needed to form the various classes of radicals decreases in the order: $CH_3\cdot > 1° > 2° > 3°$.

$$\text{R—H} \longrightarrow \text{R}\cdot + \text{H}\cdot \qquad \Delta H = \text{bond dissociation energy}$$

If less energy is needed to form one radical than another, it can only mean that, *relative to the alkane from which it is formed*, the one radical contains less energy than the other, that is to say, is *more stable* (see Fig. 3.6).

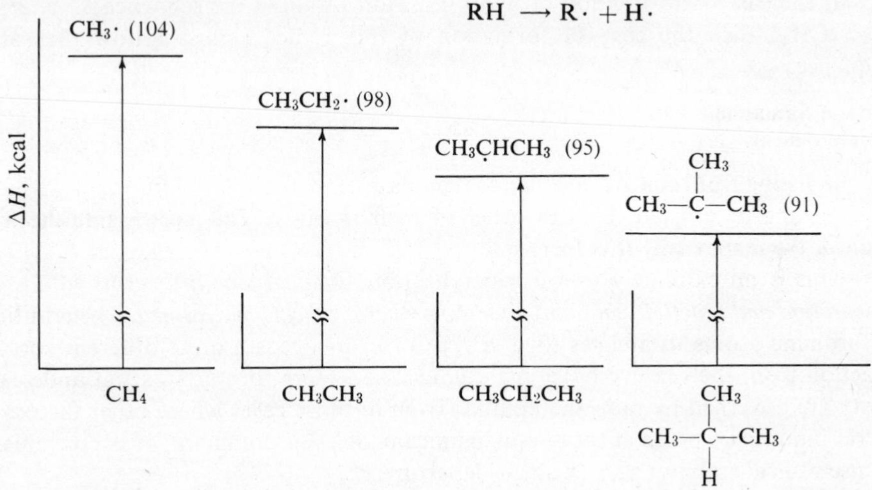

Figure 3.6. Relative stabilities of free radicals. (Plots aligned with each other for easy comparison.)

We are not attempting to compare the absolute energy contents of, say, methyl and ethyl radicals; we are simply saying that the difference in energy between methane and methyl radicals is greater than the difference between ethane and ethyl radicals. *When we compare stabilities of free radicals, it must be understood that our standard for each radical is the alkane from which it is formed.* As we shall see, this is precisely the kind of stability that we are interested in.

Relative to the alkane from which each is formed, then, the order of stability of free radicals is:

Stability of free radicals $$3° > 2° > 1° > CH_3\cdot$$

3.25 Ease of formation of free radicals

Let us return to the halogenation of alkanes. Orientation and reactivity, we have seen (Sec. 3.23), are governed by the relative ease with which the different classes of hydrogen atoms are abstracted. But by definition, the hydrogen being abstracted and the radical being formed belong to the same class. Abstraction of a primary hydrogen yields a primary radical, abstraction of a secondary hydrogen yields a secondary radical, and so on. For example:

$$\underset{\text{A 1° hydrogen}}{CH_3CH_2CH_2\text{—H}} + Br\cdot \longrightarrow \text{H—Br} + \underset{\text{A 1° radical}}{CH_3CH_2CH_2\cdot}$$

$$CH_3\underset{\underset{H}{|}}{C}HCH_3 + Br\cdot \longrightarrow H{-}Br + CH_3\dot{C}HCH_3$$

A 2° hydrogen — A 2° radical

$$CH_3\overset{\overset{CH_3}{|}}{\underset{\underset{H}{|}}{C}}CH_3 + Br\cdot \longrightarrow H{-}Br + CH_3\overset{\overset{CH_3}{|}}{\dot{C}}CH_3$$

A 3° hydrogen — A 3° radical

If the ease of abstraction of hydrogen atoms follows the sequence 3° > 2° > 1° > CH_4, then the ease of formation of free radicals must follow the same sequence:

Ease of formation of free radicals $3° > 2° > 1° > CH_3\cdot$

In listing free radicals in order of their ease of formation, we find that we have at the same time listed them in order of their stability. **The more stable the free radical, the more easily it is formed.**

This is an extremely useful generalization. *Radical stability seems to govern orientation and reactivity in many reactions where radicals are formed.* The addition of bromine atoms to alkenes (Sec. 6.17), for example, is a quite different sort of reaction from the one we have just studied; yet, there too, orientation and reactivity are governed by radical stability. (Even in those cases where other factors—steric hindrance, polar effects—are significant or even dominant, it is convenient to use radical stability as a point of departure.)

3.26 Transition state for halogenation

Is it reasonable that the more stable radical should be formed more easily?

We have already seen that the differences in reactivity toward halogen atoms are due chiefly to differences in E_{act}: the more stable the radical, then, the lower the E_{act} for its formation. This, in turn, means that the more stable the radical, the more stable the transition state leading to its formation—both stabilities being measured, as they must be, against the same standard, the reactants. (*Remember*: E_{act} is the difference in energy content between reactants and transition state.)

Examination of the transition state shows that this is exactly what we would expect. As we saw before (Sec. 2.22), the hydrogen–halogen bond is partly formed and the carbon–hydrogen bond is partly broken. To the extent that the bond is

$$-\overset{|}{\underset{|}{C}}-H + \cdot X \longrightarrow \left[-\overset{|\delta\cdot}{\underset{|}{C}}\cdots H\cdots \overset{\delta\cdot}{X}\right] \longrightarrow -\overset{|}{\underset{|}{C}}\cdot + H{-}X$$

Reactants	Transition state	Products
Halogen has odd electron	*Carbon acquiring free-radical character*	*Carbon has odd electron*

broken, the alkyl group possesses character of the free radical it will become. *Factors that tend to stabilize the free radical tend to stabilize the incipient free radical in the transition state.*

We have seen that the stabilities of free radicals follow the sequence 3° > 2° > 1° > $CH_3\cdot$. A certain factor (*delocalization of the odd electron*, Sec. 6.28) causes the energy difference between isobutane and the *tert*-butyl radical, for example, to be smaller than between propane and the isopropyl radical. It is not unreasonable that this same factor should cause the energy difference between isobutane and the *incipient tert*-butyl radical in the transition state to be smaller than between propane and the *incipient* isopropyl radical in its transition state (Fig. 3.7).

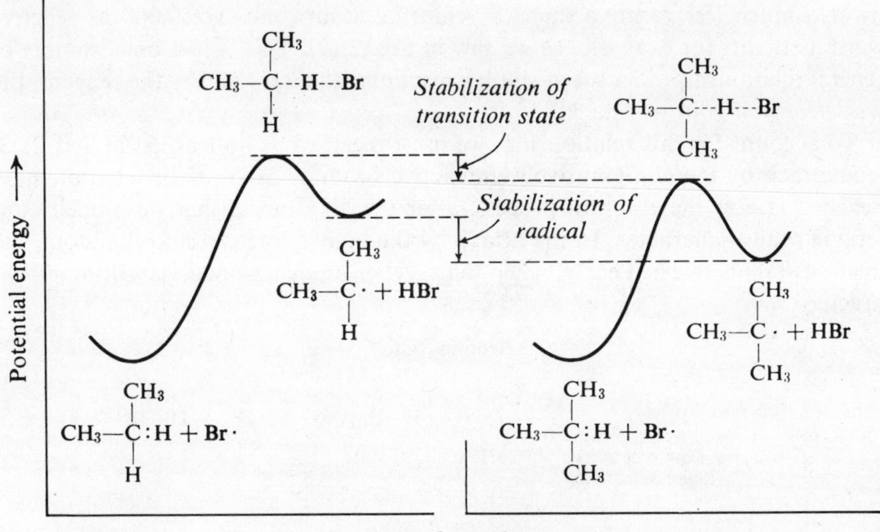

Figure 3.7. Molecular structure and rate of reaction. Stability of transition state parallels stability of radical: more stable radical formed faster. (Plots aligned with each other for easy comparison.)

3.27 Orientation and reactivity

Throughout our study of organic chemistry, we shall approach the problems of orientation and reactivity in the following way.

Both problems involve comparing the rates of closely related reactions: in the case of orientation, reactions at different sites in the same compound; in the case of reactivity, reactions with different compounds. For such closely related reactions, variations in rate are due mostly to differences in E_{act}; by definition, E_{act} is the difference in energy content between reactants and transition state.

We shall examine the most likely structure for the transition state, then, to see what structural features affect its stability without at the same time affecting by an equal amount the stability of the reactants; that is, we shall look for factors that tend to increase or decrease the energy difference between reactants and transition state. Having decided what structural features affect the E_{act}, we shall compare the transition states for the reactions whose rates we wish to compare: the more stable the transition state, the faster the reaction.

In many, if not most, reactions where a free radical is formed, as in the present case, the transition state differs from the reactants chiefly in being like the product. It is reasonable, then, that the factor most affecting the E_{act} should be the *radical character* of the transition state. Hence we find that the more stable the radical, the more stable the transition state leading to its formation, and the faster the radical is formed.

3.28 Reactivity and selectivity

In its attack on alkanes, the bromine atom is much more selective than the chlorine atom (with relative rate factors of 1600:82:1 as compared with 5.0:3.8:1). It is also much less reactive than the chlorine atom (only 1/250,000 as reactive toward methane, for example, as we saw in Sec. 2.19). This is just one example of a general relationship: in a set of similar reactions, the *less reactive* the reagent, the *more selective* it is in its attack.

To account for this relationship, we must recall what we learned in Sec. 2.23. In the attack by the comparatively unreactive bromine atom, the transition state is reached late in the reaction process, after the alkyl group has developed considerable radical character. In the attack by the highly reactive chlorine atom, the transition state is reached early, when the alkyl group has gained very little radical character.

Bromination

$$\text{R—H} + \text{Br}\cdot \longrightarrow \left[\overset{\delta\cdot}{\text{R}}\text{----H--}\overset{\delta\cdot}{\text{Br}}\right] \longrightarrow \text{R}\cdot + \text{H—Br}$$

Low reactivity; high selectivity

Transition state

Reached late: much radical character

Chlorination

$$\text{R—H} + \text{Cl}\cdot \longrightarrow \left[\overset{\delta\cdot}{\text{R}}\text{--H----}\overset{\delta\cdot}{\text{Cl}}\right] \longrightarrow \text{R}\cdot + \text{H—Cl}$$

High reactivity; low selectivity

Transition state

Reached early: little radical character

Now, by "selectivity" we mean here the differences in rate at which the various classes of free radicals are formed; a more stable free radical is formed faster, we said, because the factor that stabilizes it—delocalization of the odd electron (Sec. 6.28)—also stabilizes the incipient radical in the transition state. If this is so, then the more fully developed the radical character in the transition state, the more effective delocalization will be in stabilizing the transition state. The isopropyl radical, for example, is 3 kcal more stable than the *n*-propyl radical; if the radicals were *completely* formed in the transition state, the difference in E_{act} would be 3 kcal. Actually, in bromination the difference in E_{act} is 3 kcal: equal, within the limits of experimental error, to the maximum potential stabilization, indicating, as we expected, a great deal of radical character. In chlorination, by contrast, the difference in E_{act} is only 0.5 kcal, indicating only very slight radical character.

A similar situation exists for reactions of other kinds. Whatever the factor responsible for differences in stability among a set of transition states—whether it is delocalization of an odd electron, or accommodation of a positive or negative charge, or perhaps a change in crowding of the atoms—the factor will operate more effectively when the transition state is more fully developed, that is, when the reagent is less reactive.

3.29 Non-rearrangement of free radicals. Isotopic tracers

Our interpretation of orientation (Sec. 3.21) was based on an assumption that we have not yet justified: that the relative amounts of isomeric halides we find in the product reflect the relative rates at which various free radicals were formed from the alkane. From isobutane, for example, we obtain twice as much isobutyl chloride as *tert*-butyl chloride, and we assume from this that, by abstraction of hydrogen, isobutyl radicals are formed twice as fast as *tert*-butyl radicals.

Yet how do we know, in this case, that every isobutyl radical that is formed ultimately yields a molecule of isobutyl chloride? Suppose some isobutyl radicals were to change—by *rearrangement* of atoms—into *tert*-butyl radicals, which then react with chlorine to yield *tert*-butyl chloride. This supposition is not so far-

$$\underset{\text{Isobutane}}{CH_3-\overset{\overset{\large CH_3}{|}}{\underset{\underset{\large H}{|}}{C}}-CH_3} \xrightarrow{Cl\cdot} \underset{\text{Isobutyl radical}}{CH_3-\overset{\overset{\large CH_3}{|}}{\underset{\underset{\large H}{|}}{C}}-CH_2\cdot} \xrightarrow[\textit{Does not happen}]{\text{rearrangement}} \underset{\textit{tert}\text{-Butyl radical}}{CH_3-\overset{\overset{\large CH_3}{|}}{\dot{C}}-CH_3}$$

$$\downarrow Cl_2$$

$$\underset{\textit{tert}\text{-Butyl chloride}}{CH_3-\overset{\overset{\large CH_3}{|}}{\underset{\underset{\large Cl}{|}}{C}}-CH_3}$$

fetched as we, in our present innocence, might think; the doubt it raises is a very real one. We shall shortly see that another kind of reactive intermediate particle, the carbonium ion, is very prone to rearrange, with less stable ions readily changing into more stable ones (Sec. 5.22).

H. C. Brown (of Purdue University) and Glen Russell (now of Iowa State University) decided to test the possibility that free radicals, like carbonium ions, might rearrange, and chose the chlorination of isobutane as a good test case, because of the large difference in stability between *tert*-butyl and isobutyl radicals. If rearrangement of alkyl radicals can indeed take place, it should certainly happen here.

What the problem comes down to is this: does every abstraction of primary hydrogen lead to isobutyl chloride, and every abstraction of tertiary hydrogen lead to *tert*-butyl chloride? This, we might say, we could never know, because all hydrogen atoms are exactly alike. But are they? Actually, three isotopes of hydrogen exist: ^1H, *protium*, ordinary hydrogen; ^2H or D, *deuterium*, heavy hydrogen; and ^3H or T, *tritium*. Protium and deuterium are distributed in nature in the ratio of 5000:1. (Tritium, the unstable, radioactive isotope, is present in traces, but can be made by neutron bombardment of ^6Li.) Modern methods of separation of isotopes have made very pure deuterium available, at moderate prices, in the form of deuterium oxide, D_2O, heavy water.

Brown and Russell prepared the deuterium-labeled isobutane I,

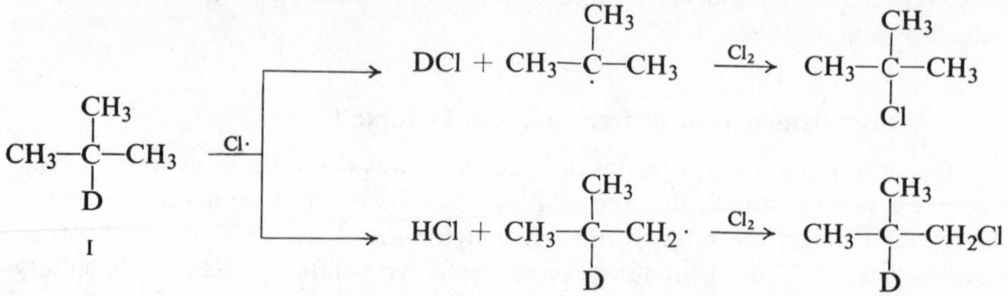

$$\underset{\text{I}}{CH_3-\underset{D}{\overset{CH_3}{\overset{|}{\underset{|}{C}}}}-CH_3} \xrightarrow{Cl\cdot} \begin{cases} DCl + CH_3-\overset{CH_3}{\overset{|}{\dot{C}}}-CH_3 \xrightarrow{Cl_2} CH_3-\underset{Cl}{\overset{CH_3}{\overset{|}{\underset{|}{C}}}}-CH_3 \\ HCl + CH_3-\underset{D}{\overset{CH_3}{\overset{|}{\underset{|}{C}}}}-CH_2\cdot \xrightarrow{Cl_2} CH_3-\underset{D}{\overset{CH_3}{\overset{|}{\underset{|}{C}}}}-CH_2Cl \end{cases}$$

photochemically chlorinated it, and analyzed the products. The DCl:HCl ratio (determined by the mass spectrometer) was found to be equal (within experimental error) to the *tert*-butyl chloride:isobutyl chloride ratio. Clearly, every abstraction of a tertiary hydrogen (*deuterium*) gave a molecule of *tert*-butyl chloride, and every abstraction of a primary hydrogen (*protium*) gave a molecule of isobutyl chloride. *Rearrangement of the intermediate free radicals did not occur.*

All the existing evidence indicates quite strongly that, although rearrangement of free radicals occasionally happens, it is not very common and does not involve simple alkyl radicals.

Problem 3.18 (a) What results would have been obtained if some isobutyl radicals *had* rearranged to *tert*-butyl radicals? (b) Suppose that, instead of rearranging, isobutyl radicals were, in effect, converted into *tert*-butyl radicals by the reaction

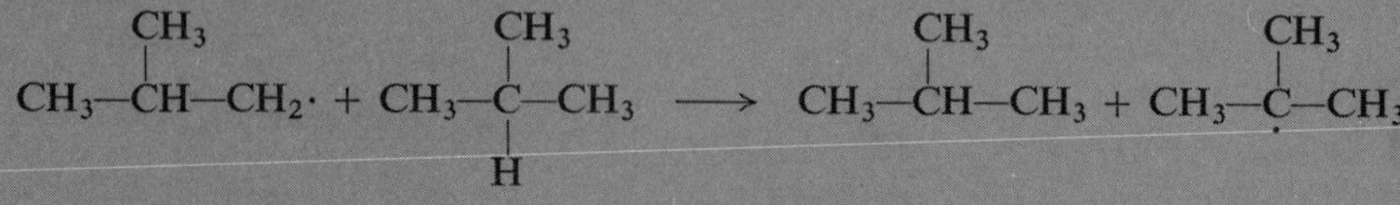

$$CH_3-\overset{CH_3}{\overset{|}{CH}}-CH_2\cdot + CH_3-\underset{H}{\overset{CH_3}{\overset{|}{\underset{|}{C}}}}-CH_3 \longrightarrow CH_3-\overset{CH_3}{\overset{|}{CH}}-CH_3 + CH_3-\overset{CH_3}{\overset{|}{\dot{C}}}-CH_3$$

What results would Brown and Russell have obtained?

Problem 3.19 Keeping in mind the availability of D_2O, suggest a way to make I from *tert*-butyl chloride. (*Hint:* See Sec. 3.16.)

The work of Brown and Russell is just one example of the way in which we can gain insight into a chemical reaction by using isotopically labeled compounds. We shall encounter many other examples in which isotopes, used either as *tracers*, as in this case, or for the detection of *isotope effects* (Sec. 11.15), give us information about reaction mechanisms that we could not get in any other way.

Besides deuterium and tritium, isotopes commonly used in organic chemistry include: ^{14}C, available as $^{14}CH_3OH$ and $Ba^{14}CO_3$; ^{18}O, as $H_2{}^{18}O$; ^{15}N, as $^{15}NH_3$, $^{15}NO_3{}^-$, and $^{15}NO_2{}^-$; ^{36}Cl, as chlorine or chloride; ^{131}I, as iodide.

Problem 3.20 Bromination of methane is slowed down by the addition of HBr (Problem 14, p. 71); this is attributed to the reaction

$$CH_3\cdot + HBr \longrightarrow CH_4 + Br\cdot$$

which, as the reverse of one of the chain-carrying steps, slows down bromination. How might you test whether or not this reaction actually occurs in the bromination mixture?

Problem 3.21 In Sec. 2.12 the reaction

$$Cl\cdot + Cl_2 \longrightarrow Cl_2 + Cl\cdot$$

was listed as probable but unproductive. Given ordinary chlorine (made up of ^{35}Cl and ^{37}Cl) and $^{36}Cl_2$, and a mass spectrometer, how would you go about finding out whether or not the reaction actually occurs?

3.30 Combustion

The reaction of alkanes with oxygen to form carbon dioxide, water, and—most important of all—*heat*, is the chief reaction occurring in the internal combustion engine; its tremendous practical importance is obvious.

The mechanism of this reaction is extremely complicated and is not yet fully understood. There seems to be no doubt, however, that it is a free-radical chain reaction. The reaction is extremely exothermic and yet requires a very high temperature, that of a flame, for its initiation. As in the case of chlorination, a great deal of energy is required for the bond-breaking that generates the initial reactive particles; once this energy barrier is surmounted, the subsequent chain-carrying steps proceed readily and with the evolution of energy.

A higher compression ratio has made the modern gasoline engine more efficient than earlier ones, but has at the same time created a new problem. Under certain conditions the smooth explosion of the fuel-air mixture in the cylinder is replaced by **knocking**, which greatly reduces the power of the engine.

The problem of knocking has been successfully met in two general ways: (a) proper selection of the hydrocarbons to be used as fuel, and (b) addition of tetraethyllead.

Experiments with pure compounds have shown that hydrocarbons of differing structures differ widely in knocking tendency. The relative antiknock tendency of a fuel is generally indicated by its **octane number**. An arbitrary scale has been set up, with *n*-heptane, which knocks very badly, being given an octane number of zero, and 2,2,4-trimethylpentane ("iso-octane") being given the octane number of 100. There are available today fuels with better antiknock qualities than "iso-octane."

The gasoline fraction obtained by direct distillation of petroleum (*straight-run gasoline*) is improved by addition of compounds of higher octane number; it is sometimes entirely replaced by these better fuels. Branched-chain alkanes and alkenes, and aromatic hydrocarbons generally have excellent antiknock qualities; these are produced from petroleum hydrocarbons by *catalytic cracking* (Sec. 3.31) and *catalytic reforming* (Sec. 12.4). Highly branched alkanes are synthesized from alkenes and alkanes by *alkylation* (Sec. 6.16).

In 1922 T. C. Midgley, Jr., and T. A. Boyd (of the General Motors Research Laboratory) found that the octane number of a fuel is greatly improved by addition of a small amount of tetraethyllead, $(C_2H_5)_4Pb$. Gasoline so treated is called *ethyl* gasoline or *leaded* gasoline. Nearly 50 years of research has finally shown that tetraethyllead probably works by producing tiny particles of lead oxides, on whose surface certain reaction chains are broken.

In addition to carbon dioxide and water, however, the gasoline engine discharges other substances into the atmosphere, substances that are either smog-producing or downright poisonous: unburned hydrocarbons, carbon monoxide, nitrogen oxides, and, from leaded gasoline, various compounds of lead—in the United States, hundreds of tons of lead a day. Growing public concern about these pollutants has caused a minor revolution in the petroleum and auto industries. *Converters* are being developed to clean up exhaust emissions: by catalytic oxidation of hydrocarbons and carbon monoxide, and by the breaking down of nitrogen oxides into nitrogen and oxygen. But most of these oxidation catalysts contain platinum, which is poisoned by lead; there has been a move to get the lead out of gasoline—not, initially, to cut down on lead pollution, but to permit converters to function. This has, in turn, brought back the problem of knocking, which is being met in two ways: (a) by lowering the compression ratio of the new automobiles being built; and (b) by increasing the octane number of gasoline through changes in hydrocarbon composition—through addition of aromatics and through increased use of isomerization (Sec. 3.13).

3.31 Pyrolysis: cracking

Decomposition of a compound by the action of heat alone is known as **pyrolysis**. This word is taken from the Greek *pyr*, fire, and *lysis*, a loosing, and hence to chemists means "cleavage by heat"; compare *hydro-lysis*, "cleavage by water."

The pyrolysis of alkanes, particularly when petroleum is concerned, is known as **cracking**. In *thermal cracking* alkanes are simply passed through a chamber heated to a high temperature. Large alkanes are converted into smaller alkanes, alkenes, and some hydrogen. This process yields predominantly ethylene (C_2H_4) together with other small molecules. In a modification called *steam cracking*, the hydrocarbon is diluted with steam, heated for a fraction of a second to 700–900°, and rapidly cooled. Steam cracking is of growing importance in the production of hydrocarbons as chemicals, including ethylene, propylene, butadiene, isoprene, and cyclopentadiene. Another source of smaller hydrocarbons is *hydrocracking*, carried out in the presence of hydrogen at high pressure and at much lower temperatures (250–450°).

The low-molecular-weight alkenes obtained from these cracking processes can be separated and purified, and are the most important raw materials for the large-scale synthesis of aliphatic compounds.

Most cracking, however, is directed toward the production of fuels, not chemicals, and for this *catalytic cracking* is the major process. Higher boiling petroleum fractions (typically, gas oil), are brought into contact with a finely divided silica–alumina catalyst at 450–550° and under slight pressure. Catalytic cracking not only increases the yield of gasoline by breaking large molecules into smaller ones, but also improves the quality of the gasoline: this process involves *carbonium ions* (Sec. 5.15), and yields alkanes and alkenes with the highly branched structures desirable in gasoline.

Through the process of *alkylation* (Sec. 6.16) some of the smaller alkanes and alkenes are converted into high-octane synthetic fuels.

Finally, by the process of *catalytic reforming* (Sec. 12.4) enormous quantities of the aliphatic hydrocarbons of petroleum are converted into *aromatic* hydrocarbons which are used not only as superior fuels but as the starting materials in the synthesis of most aromatic compounds (Chap. 10).

3.32 Determination of structure

One of the commonest and most important jobs in organic chemistry is to determine the structural formula of a compound just synthesized or isolated from a natural source.

The compound will fall into one of two groups, although at first we probably shall not know *which* group. It will be either (a) a previously reported compound, which we must identify, or (b) a new compound, whose structure we must prove.

If the compound has previously been encountered by some other chemist who determined its structure, then a description of its properties will be found somewhere in the chemical literature, together with the evidence on which its structure was assigned. In that case, we need only to show that our compound is identical with the one previously described.

If, on the other hand, our compound is a new one that has never before been reported, then we must carry out a much more elaborate proof of structure.

Let us see—in a general way now, and in more detail later—just how we would go about this job. We are confronted by a flask filled with gas, or a few milliliters of liquid, or a tiny heap of crystals. We must find the answer to the question: *what is it?*

First, we purify the compound and determine its physical properties: melting point, boiling point, density, refractive index, and solubility in various solvents. In the laboratory today, we would measure various spectra of the compound (Chap. 13), in particular the infrared spectrum and the nmr spectrum; indeed, because of the wealth of information to be gotten in this way, spectroscopic examination might well be the first order of business after purification. From the mass spectrum we would get a very accurate molecular weight.

We would carry out a qualitative elemental analysis to see what elements are present (Sec. 2.25). We might follow this with a quantitative analysis, and from this and the molecular weight we could calculate a molecular formula (Sec. 2.26); we would certainly do this if the compound is suspected of being a new one.

Next, we study systematically the behavior of the compound toward certain reagents. This behavior, taken with the elemental analysis, solubility properties, and spectra, generally permits us to *characterize* the compound, that is, to decide what family the unknown belongs to. We might find, for example, that the compound is an alkane, or that it is an alkene, or an aldehyde, or an ester.

Now the question is: *which* alkane is it? Or which alkene, or which aldehyde, or which ester? To find the answer, we first go to the chemical literature and look up compounds of the particular family to which our unknown belongs.

If we find one described whose physical properties are identical with those of our unknown, then the chances are good that the two compounds are identical. For confirmation, we generally convert the unknown by a chemical reaction into a new compound called a **derivative**, and show that this derivative is identical with the product derived in the same way from the previously reported compound.

If, on the other hand, we do not find a compound described whose physical properties are identical with those of our unknown, then we have a difficult job on our hands: we have a new compound, and must prove its structure. We may carry out a *degradation*: break the molecule apart, identify the fragments, and deduce what the structure must have been. To clinch any proof of structure, we attempt to *synthesize* the unknown by a method that leaves no doubt about its structure.

Problem 3.22 The final step in the proof of structure of an unknown alkane was its synthesis by the coupling of lithium di(*tert*-butyl)copper with *n*-butyl bromide. What was the alkane?

In Chap. 13, after we have become familiar with more features of organic structure, we shall see how spectroscopy fits into the general procedure outlined above.

3.33 Analysis of alkanes

An unknown compound is characterized as an alkane on the basis of negative evidence.

Upon qualitative elemental analysis, an alkane gives negative tests for all elements except carbon and hydrogen. A quantitative combustion, if one is carried out, shows the absence of oxygen; taken with a molecular weight determination, the combustion gives the molecular formula, C_nH_{2n+2}, which is that of an alkane.

An alkane is insoluble not only in water but also in dilute acid and base and in concentrated sulfuric acid. (As we shall see, most kinds of organic compounds dissolve in one or more of these solvents.)

An alkane is unreactive toward most chemical reagents. Its infrared spectrum lacks the absorption bands characteristic of groups of atoms present in other families of organic compounds (like OH, C=O, C=C, etc.).

Once the unknown has been characterized as an alkane, there remains the second half of the problem: finding out *which* alkane.

On the basis of its physical properties—boiling point, melting point, density, refractive index, and, most reliable of all, its infrared and mass spectra—it may be identified as a previously studied alkane of known structure.

If it turns out to be a new alkane, the proof of structure can be a difficult job. Combustion and molecular weight determination give its molecular formula. Clues about the arrangement of atoms are given by its infrared and nmr spectra. (For compounds like alkanes, it may be necessary to lean heavily on x-ray diffraction and mass spectrometry.)

Final proof lies in synthesis of the unknown by a method that can lead only to the particular structure assigned.

(The spectroscopic analysis of alkanes will be discussed in Secs. 13.15–13.16.)

PROBLEMS

1. Give the structural formula of:

(a) 2,2,3,3-tetramethylpentane
(b) 2,3-dimethylbutane
(c) 3,4,4,5-tetramethylheptane
(d) 3,4-dimethyl-4-ethylheptane
(e) 2,4-dimethyl-4-ethylheptane
(f) 2,5-dimethylhexane
(g) 2-methyl-3-ethylpentane
(h) 2,2,4-trimethylpentane

2. Draw out the structural formula and give the IUPAC name of:

(a) $(CH_3)_2CHCH_2CH_2CH_3$
(b) $CH_3CH_2C(CH_3)_2CH_2CH_3$
(c) $(C_2H_5)_2C(CH_3)CH_2CH_3$
(d) $CH_3CH_2CH(CH_3)CH(CH_3)CH(CH_3)_2$
(e)
$$\begin{array}{l} CH_3CH_2\underset{\displaystyle CH_3}{\underset{|}{C}}HCH_2\underset{\displaystyle CH_2CH_2CH_3}{\underset{|}{C}}HCH_2CH_3 \end{array}$$
(f) $(CH_3)_3CCH_2C(CH_3)_3$
(g) $(CH_3)_2CHCH_2CH_2CH(C_2H_5)_2$
(h) $(CH_3)_2CHCH(CH_3)CH_2C(C_2H_5)_2CH_3$
(i) $(CH_3)_2CHC(C_2H_5)_2CH_2CH_2CH_3$
(j)
$$\begin{array}{l} CH_3CH_2\overset{\displaystyle CH_3}{\overset{|}{C}}HCH_2\underset{\displaystyle CH_2CH_2CH_3}{\underset{|}{C}}H\overset{\displaystyle CH_3}{\overset{|}{C}}HCH_3 \end{array}$$

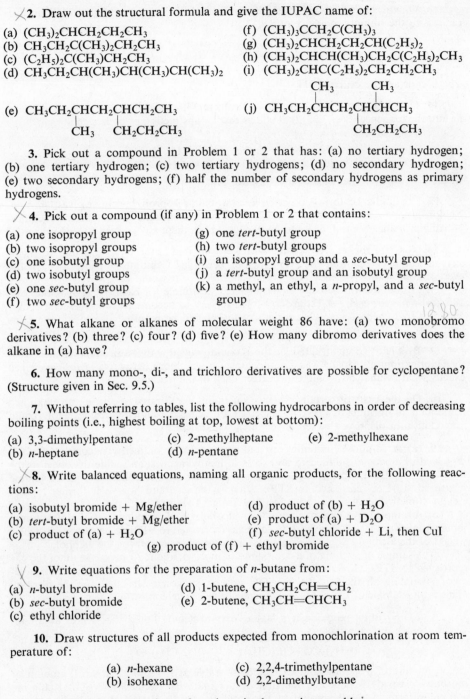

3. Pick out a compound in Problem 1 or 2 that has: (a) no tertiary hydrogen; (b) one tertiary hydrogen; (c) two tertiary hydrogens; (d) no secondary hydrogen; (e) two secondary hydrogens; (f) half the number of secondary hydrogens as primary hydrogens.

4. Pick out a compound (if any) in Problem 1 or 2 that contains:

(a) one isopropyl group
(b) two isopropyl groups
(c) one isobutyl group
(d) two isobutyl groups
(e) one *sec*-butyl group
(f) two *sec*-butyl groups
(g) one *tert*-butyl group
(h) two *tert*-butyl groups
(i) an isopropyl group and a *sec*-butyl group
(j) a *tert*-butyl group and an isobutyl group
(k) a methyl, an ethyl, a *n*-propyl, and a *sec*-butyl group

5. What alkane or alkanes of molecular weight 86 have: (a) two monobromo derivatives? (b) three? (c) four? (d) five? (e) How many dibromo derivatives does the alkane in (a) have?

6. How many mono-, di-, and trichloro derivatives are possible for cyclopentane? (Structure given in Sec. 9.5.)

7. Without referring to tables, list the following hydrocarbons in order of decreasing boiling points (i.e., highest boiling at top, lowest at bottom):

(a) 3,3-dimethylpentane
(b) *n*-heptane
(c) 2-methylheptane
(d) *n*-pentane
(e) 2-methylhexane

8. Write balanced equations, naming all organic products, for the following reactions:

(a) isobutyl bromide + Mg/ether
(b) *tert*-butyl bromide + Mg/ether
(c) product of (a) + H_2O
(d) product of (b) + H_2O
(e) product of (a) + D_2O
(f) *sec*-butyl chloride + Li, then CuI
(g) product of (f) + ethyl bromide

9. Write equations for the preparation of *n*-butane from:

(a) *n*-butyl bromide
(b) *sec*-butyl bromide
(c) ethyl chloride
(d) 1-butene, $CH_3CH_2CH{=}CH_2$
(e) 2-butene, $CH_3CH{=}CHCH_3$

10. Draw structures of all products expected from monochlorination at room temperature of:

(a) *n*-hexane
(b) isohexane
(c) 2,2,4-trimethylpentane
(d) 2,2-dimethylbutane

11. Predict the proportions of products in the previous problem.

12. (a) Reaction of an aldehyde with a Grignard reagent is an important way of making alcohols. Why must one scrupulously dry the aldehyde before adding it to the Grignard reagent? (b) Why would one not prepare a Grignard reagent from $BrCH_2CH_2OH$?

13. On the basis of bond strengths in Table 1.2, page 21, add the following free radicals to the stability sequence of Sec. 3.24:

(a) *vinyl*, $H_2C{=}CH\cdot$
(b) *allyl*, $H_2C{=}CHCH_2\cdot$
(c) *benzyl*, $C_6H_5CH_2\cdot$

Check your answer on page 211.

14. On the basis of your answer to Problem 13, predict how the following would fit into the sequence (Sec. 3.23) that shows ease of abstraction of hydrogen atoms:

(a) *vinylic* hydrogen, $H_2C{=}CH{-}H$
(b) *allylic* hydrogen, $H_2C{=}CHCH_2{-}H$
(c) *benzylic* hydrogen, $C_6H_5CH_2{-}H$

Check your answer against the facts on page 210.

15. Free-radical chlorination of *either n*-propyl or isopropyl bromide gives 1-bromo-2-chloropropane, and of *either* isobutyl or *tert*-butyl bromide gives 1-bromo-2-chloro-2-methylpropane. What appears to be happening? Is there any pattern to this behavior?

16. (a) If a rocket were fueled with kerosine and liquid oxygen, what weight of oxygen would be required for every liter of kerosine? (Assume kerosine to have the average composition of n-$C_{14}H_{30}$.) (b) How much heat would be evolved in the combustion of one liter of kerosine? (Assume 157 kcal/mole for each $-CH_2-$ group and 186 kcal/mole for each $-CH_3$ group.) (c) If it were to become feasible to fuel a rocket with free hydrogen atoms, what weight of fuel would be required to provide the same heat as a liter of kerosine and the necessary oxygen? (Assume H_2 as the sole product.)

17. By what two quantitative methods could you show that a product isolated from the chlorination of propane was a monochloro or a dichloro derivative of propane? Tell exactly what results you would expect from each of the methods.

18. On the basis of certain evidence, including its infrared spectrum, an unknown compound of formula $C_{10}H_{22}$ is suspected of being 2,7-dimethyloctane. How could you confirm or disprove this tentatively assigned structure?

19. (a) A solution containing an unknown amount of methyl alcohol (CH_3OH) dissolved in *n*-octane is added to an excess of methylmagnesium iodide dissolved in the high-boiling solvent, *n*-butyl ether. A gas is evolved, and is collected and its volume measured: 1.04 cc (corrected to STP). What is the gas, and how is it formed? What weight of methyl alcohol was added to the Grignard reagent?

(b) A sample of 4.12 mg of an unknown alcohol, ROH, is added to methylmagnesium iodide as above; there is evolved 1.56 cc of gas (corrected to STP). What is the molecular weight of the alcohol? Suggest a possible structure or structures for the alcohol.

(c) A sample of 1.79 mg of a compound of mol. wt. about 90 gave 1.34 ml of the gas (corrected to STP). How many "active (that is, acidic) hydrogens" are there per molecule? Assuming all these to be in —OH groups, suggest a structure for the alcohol. (This is an example of the *Zerewitinoff active hydrogen determination.*)

20. (a) *tert*-Butyl peroxide is a stable, easy-to-handle liquid that serves as a convenient source of free radicals:

$$(CH_3)_3CO{-}OC(CH_3)_3 \xrightarrow[\text{or light}]{130^\circ} 2(CH_3)_3CO\cdot$$

A mixture of isobutane and CCl_4 is quite stable at 130–140°. If a small amount of *tert*-butyl peroxide is added, a reaction occurs that yields (chiefly) *tert*-butyl chloride and chloroform. A small amount of *tert*-butyl alcohol ($(CH_3)_3COH$, equivalent to the peroxide used) is also isolated. Give all steps in a likely mechanism for this reaction.

(b) When irradiated with ultraviolet light, or in the presence of a small amount of peroxides, *tert*-butyl hypochlorite, $(CH_3)_3C{-}O{-}Cl$, reacts with alkanes to form, in equimolar amounts, alkyl chlorides and *tert*-butyl alcohol. Outline all steps in a likely mechanism for this reaction.

Chapter 4 Stereochemistry I. Stereoisomers

4.1 Stereochemistry and stereoisomerism

The science of organic chemistry, we said, is based on the relationship between molecular structure and properties. That part of the science which deals with structure *in three dimensions* is called **stereochemistry** (Gr.: *stereos*, solid).

One aspect of stereochemistry is *stereoisomerism*. Isomers, we recall, are different compounds that have the same molecular formula. The particular kind of isomers that are different from each other *only* in the way the atoms are oriented in space (but are like one another with respect to which atoms are joined to which other atoms) are called **stereoisomers.**

Pairs of stereoisomers exist that differ so little in structure—and hence in properties—that of all the physical measurements we can make, only one, involving a special instrument and an unusual kind of light, can distinguish between them. Yet, despite this close similarity, the existence of such stereoisomers provides us with one of our most sensitive probes into mechanisms of chemical reactions; very often, one of these isomers is selected for study, not because it is different from ordinary compounds in its three-dimensional chemistry, but because it can be made to reveal what ordinary compounds hide. And, again despite their close similarity, one isomer of such a pair may serve as a nourishing food, or as an antibiotic, or as a powerful heart stimulant, and the other isomer may be useless.

In this chapter, we shall learn how to predict the existence of the kinds of stereoisomers called *enantiomers* and *diastereomers*, how to represent and designate their structures, and, in a general way, how their properties will compare. Then, in following chapters, we shall begin to use what we learn in this one. In Secs. 5.5–5.6, we shall learn about the kind of stereoisomers called *geometric isomers*. In Chapter 7, the emphasis will shift from what these stereoisomers *are*, to how they are formed, what they do, and what they can tell us.

We have already (Secs. 3.3 and 3.5) begun our study of the branch of stereochemistry called *conformational analysis*; we shall return to it, especially in Chap. 9, and make use of it throughout the rest of the book.

4.2 Isomer number and tetrahedral carbon

Let us begin our study of stereochemistry with methane and some of its simple substitution products. Any compound, however complicated, that contains carbon bonded to four other atoms can be considered to be a derivative of methane; and whatever we learn about the shape of the methane molecule can be applied to the shapes of vastly more complicated molecules.

The evidence of electron diffraction, x-ray diffraction, and spectroscopy shows that when carbon is bonded to four other atoms its bonds are directed toward the corners of a tetrahedron. But as early as 1874, years before the direct determination of molecular structure was possible, the tetrahedral carbon atom was proposed by J. H. van't Hoff, while he was still a student at the University of Utrecht. His proposal was based upon the evidence of **isomer number.**

For any atom Y, *only one substance of formula* CH_3Y *has ever been found.* Chlorination of methane yields only one compound of formula CH_3Cl; brominations yields only one compound of formula CH_3Br. Similarly, only one CH_3F is known, and only one CH_3I. Indeed, the same holds true if Y represents, not just an atom, but a group of atoms (unless the group is so complicated that in itself it brings about isomerism); there is only one CH_3OH, only one CH_3COOH, only one CH_3SO_3H.

What does this suggest about the arrangement of atoms in methane? It suggests that every hydrogen atom in methane is equivalent to every other hydrogen atom, so that replacement of any one of them gives rise to the same product. If the hydrogen atoms of methane were not equivalent, then replacement of one would yield a different compound than replacement of another, and isomeric substitution products would be obtained.

In what ways can the atoms of methane be arranged so that the four hydrogen atoms are equivalent? There are three such arrangements: (a) a *planar* arrangement (I) in which carbon is at the center of a rectangle (or square) and a hydrogen

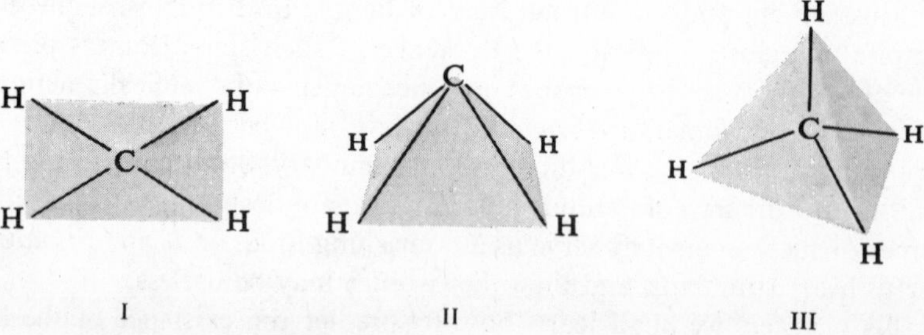

atom is at each corner; (b) a *pyramidal* arrangement (II) in which carbon is at the apex of a pyramid and a hydrogen atom is at each corner of a square base; (c) a *tetrahedral* arrangement (III) in which carbon is at the center of a tetrahedron and a hydrogen atom is at each corner.

How do we know that each of these arrangements could give rise to only one substance of formula CH_3Y? As always for problems like this, the answer lies in the use of molecular models. (Gumdrops and toothpicks can be used to make structures like I and II, for which the bond angles of ordinary molecular models are not suited.) For example, we make two identical models of I. In one model we replace, say, the upper right-hand H with a different atom Y, represented by a differently colored ball or gumdrop; in the other model we similarly replace, say, the lower right-hand H. We next see whether or not the two resulting models are *superimposable*; that is, we see whether or not, by any manipulations except bending or breaking bonds, we can make the models coincide in all their parts. If the two models are superimposable, they simply represent two molecules of the same compound; if the models are not superimposable, they represent molecules of different compounds which, since they have the same molecular formula, are by definition *isomers* (p. 37). Whichever hydrogen we replace in I (or in II or III), we get the same structure. From any arrangement other than these three, we would get more than one structure.

As far as compounds of the formula CH_3Y are concerned, the evidence of isomer number limits the structure of methane to one of these three possibilities.

Problem 4.1 How many isomers of formula CH_3Y would be possible if methane were a pyramid with a *rectangular* base? What are they? (*Hint:* If you have trouble with this question now, try it again after you have studied Sec. 4.7.)

For any atom Y *and for any atom* Z, *only one substance of formula* CH_2YZ *has ever been found.* Halogenation of methane, for example, yields only one compound of formula CH_2Cl_2, only one compound of formula CH_2Br_2, and only one compound of formula CH_2ClBr.

Of the three possible structures of methane, only the tetrahedral one is consistent with this evidence.

Problem 4.2 How many isomers of formula CH_2YZ would be expected from each of the following structures for methane? (a) Structure I with carbon at the center of a rectangle; (b) structure I with carbon at the center of a square; (c) structure II; (d) structure III.

Thus, only the tetrahedral structure for methane agrees with the evidence of isomer number. It is true that this is negative evidence; one might argue that isomers exist which have never been isolated or detected simply because the experimental techniques are not good enough. But, as we said before, any compound that contains carbon bonded to four other atoms can be considered to be a derivative of methane; in the preparation of hundreds of thousands of compounds of this sort, the number of isomers obtained has always been consistent with the concept of the tetrahedral carbon atom.

There is additional, positive evidence for the tetrahedral carbon atom: the finding of just the kind of isomers—*enantiomers*—that are predicted for compounds of formula CWXYZ. It was the existence of enantiomers that convinced van't Hoff that the carbon atom is tetrahedral. But to understand what enantiomers are, we must first learn about the property called *optical activity*.

4.3 Optical activity. Plane-polarized light

Light possesses certain properties that are best understood by considering it to be a wave phenomenon in which the vibrations occur at right angles to the direction in which the light travels. There are an infinite number of planes passing through the line of propagation, and ordinary light is vibrating in all these planes. If we consider that we are looking directly into the beam of a flashlight, Fig. 4.1

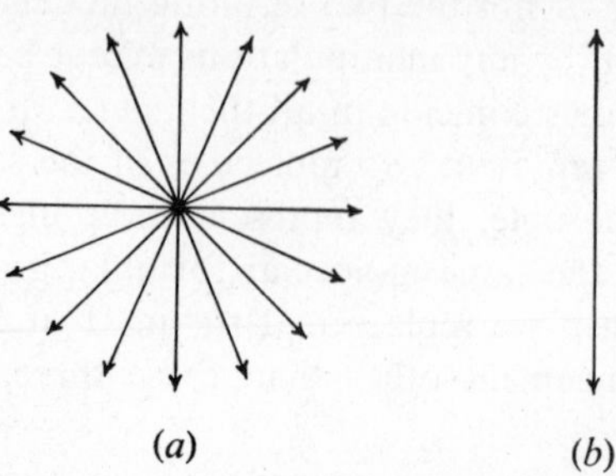

Figure 4.1. Schematic representation of (*a*) ordinary light and (*b*) plane-polarized light. Light traveling perpendicular to page; vibrations in plane of page.

shows schematically the sort of vibrations that are taking place, all perpendicular to a line between our eye and the paper (flashlight). **Plane-polarized light** *is light whose vibrations take place in only one of these possible planes.* Ordinary light is turned into plane-polarized light by passing it through a lens made of the material known as Polaroid or more traditionally through pieces of *calcite* (a particular crystalline form of $CaCO_3$) so arranged as to constitute what is called a *Nicol prism.*

An **optically active substance** *is one that rotates the plane of polarized light.* When polarized light, vibrating in a certain plane, is passed through an optically active substance, it emerges vibrating in a different plane.

4.4 The polarimeter

How can this rotation of the plane of polarized light—this optical activity—be detected? It is both detected and measured by an instrument called the **polarimeter**, which is represented schematically in Fig. 4.2. It consists of a light source, two lenses (Polaroid or Nicol), and between the lenses a tube to hold the substance that is being examined for optical activity. These are arranged so that the light

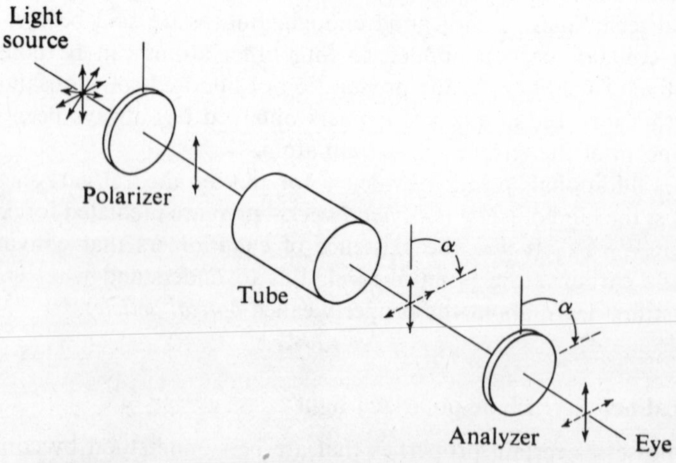

Light source
Polarizer
Tube
Analyzer
Eye

Figure 4.2. Schematic representation of a polarimeter. Solid lines: before rotation. Broken lines: after rotation. α is angle of rotation.

passes through one of the lenses (*polarizer*), then the tube, then the second lens (*analyzer*), and finally reaches our eye. When the tube is empty, we find that the maximum amount of light reaches our eye when the two lenses are so arranged that they pass light vibrating in the same plane. If we rotate the lens that is nearer our eye, say, we find that the light dims, and reaches a minimum when the lens is at right angles to its previous position.

Let us adjust the lenses so that a maximum amount of light is allowed to pass. (In practice, it is easier to detect a minimum than a maximum; the principle remains the same.) Now let us place the sample to be tested in the tube. If the substance does not affect the plane of polarization, light transmission is still at a maximum and the substance is said to be **optically inactive**. If, on the other hand, the substance rotates the plane of polarization, then the lens nearer our eye must be rotated to conform with this new plane if light transmission is again to be a maximum, and the substance is said to be **optically active**. If the rotation of the plane, and hence our rotation of the lens, is to the right (clockwise), the substance is **dextrorotatory** (Latin: *dexter*, right); if the rotation is to the left (counterclockwise), the substance is **levorotatory** (Latin: *laevus*, left).

We can determine not only that the substance has rotated the plane, and in which direction, but also *by how much*. The amount of rotation is simply the number of degrees that we must rotate the lens to conform with the light. The symbols **+** and **–** are used to indicate rotations to the right and to the left, respectively.

The lactic acid (p. 121) that is extracted from muscle tissue rotates light to the right, and hence is known as *dextrorotatory* lactic acid, or (+)-lactic acid. The 2-methyl-1-butanol that is obtained from fusel oil (a by-product of the fermentation of starch to ethyl alcohol) rotates light to the left, and is known as *levorotatory* 2-methyl-1-butanol, or (–)-2-methyl-1-butanol.

4.5 Specific rotation

Since optical rotation of the kind we are interested in is caused by individual molecules of the active compound, *the amount of rotation depends upon how many molecules the light encounters in passing through the tube.*

The light will encounter twice as many molecules in a tube 20 cm long as in a tube 10 cm long, and the rotation will be twice as large. If the active compound is in solution, the number of molecules encountered by the light will depend upon the concentration. For a given tube length, light will encounter twice as many molecules in a solution of 2 g per 100 cc of solvent as in a solution containing 1 g per 100 cc of solvent, and the rotation will be twice as large. When allowances are made for the length of tube and the concentration, it is found that the amount of rotation, as well as its direction, is a characteristic of each individual optically active compound.

Specific rotation is the number of degrees of rotation observed if a 1-decimeter tube is used, and the compound being examined is present to the extent of 1g/cc. This is usually calculated from observations with tubes of other lengths and at different concentrations by means of the equation

$$[\alpha] = \frac{\alpha}{l \times d}$$

$$\text{specific rotation} = \frac{\text{observed rotation (degrees)}}{\text{length (dm)} \times \text{g/cc}}$$

where d represents density for a pure liquid or concentration for a solution.

The specific rotation is as much a property of a compound as its melting point, boiling point, density, or refractive index. Thus the specific rotation of the 2-methyl-1-butanol obtained from fusel oil is

$$[\alpha]_{D}^{20} = -5.756°$$

Here 20 is the temperature and D is the wavelength of the light used in the measurement (*D* line of sodium, 5893 A).

Problem 4.3 The concentration of cholesterol dissolved in chloroform is 6.15 g per 100 ml of solution. (a) A portion of this solution in a 5-cm polarimeter tube causes an observed rotation of −1.2°. Calculate the specific rotation of cholesterol. (b) Predict the observed rotation if the same solution were placed in a 10-cm tube. (c) Predict the observed rotation if 10 ml of the solution were diluted to 20 ml and placed in a 5-cm tube.

Problem 4.4 A sample of a pure liquid in a 10-cm tube is placed in a polarimeter, and a reading of +45° is made. How could you establish that [α] is really +45° and not −315°? That it is +45° and not +405° or, for that matter, +765°?

4.6 Enantiomerism: the discovery

The optical activity we have just described was discovered in 1815 at the Collège de France by the physicist Jean-Baptiste Biot.

In 1848 at the École normale in Paris the chemist Louis Pasteur made a set of observations which led him a few years later to make a proposal that is the foundation of stereochemistry. Pasteur, then a young man, had come to the École normale from the Royal College of Besançon (where he had received his *baccalaurent ès sciences* with the rating of *médiocre* in chemistry), and had just won his *docteur ès sciences*. To gain some experience in crystallography, he was repeating another chemist's earlier work on salts of tartaric acid when he saw something that no one had noticed before: optically inactive sodium ammonium tartrate existed as a mixture of two different kinds of crystals, which were *mirror images* of each other. Using a hand lens and a pair of tweezers, he carefully and laboriously separated the mixture into two tiny piles—one of right-handed crystals and the other of left-handed crystals—much as one might separate right-handed and left-handed gloves lying jumbled together on a shop counter. Now, although the original mixture was optically inactive, each set of crystals dissolved in water was found to be *optically active*! Furthermore, the specific rotations of the two solutions were exactly *equal, but of opposite sign*; that is to say, one solution rotated plane-polarized light to the right, and the other solution an equal number of degrees to the left. In all other properties the two substances were identical.

Since the difference in optical rotation was observed *in solution*, Pasteur concluded that it was characteristic, not of the crystals, but of the *molecules*. He proposed that, like the two sets of crystals themselves, the molecules making up the crystals were *mirror images of each other*. He was proposing the existence of isomers whose structures differ only in being mirror images of each other, and whose properties differ only in the direction of rotation of polarized light.

There remained only for van't Hoff to point out that a *tetrahedral* carbon atom would account not only for the absence of isomers of formula CH_3Y and CH_2YZ, but also for the existence of mirror-image isomers—*enantiomers*—like Pasteur's tartaric acids.

4.7 Enantiomerism and tetrahedral carbon

Let us convince ourselves that such mirror-image isomers should indeed exist. Starting with the actual, tetrahedral arrangement for methane, let us make a model of a compound CWXYZ, using a ball of a different color for each different atom or group represented as W, X, Y, and Z. Let us then imagine that we are holding this model before a mirror, and construct a second model of what its mirror image would look like. We now have two models which look something like this:

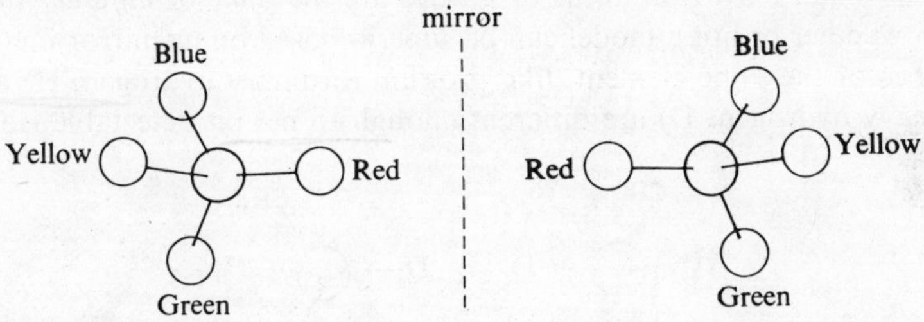

mirror

Blue, Yellow, Red, Green | Blue, Red, Yellow, Green

which are understood to stand for this:

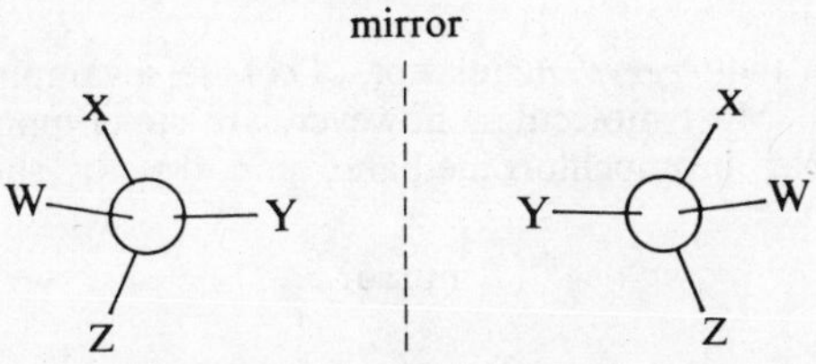

mirror

X, W, Y, Z | X, Y, W, Z

Not superimposable: isomers

Are these two models superimposable? *No*. We may twist and turn them as much as we please (so long as no bonds are broken), but although two groups of each may coincide, the other two do not. The models are not superimposable, and therefore must represent two isomers of formula CWXYZ.

As predicted, mirror-image isomers do indeed exist, and thousands of instances besides the tartaric acids are known. There are, for example, two isomeric *lactic*

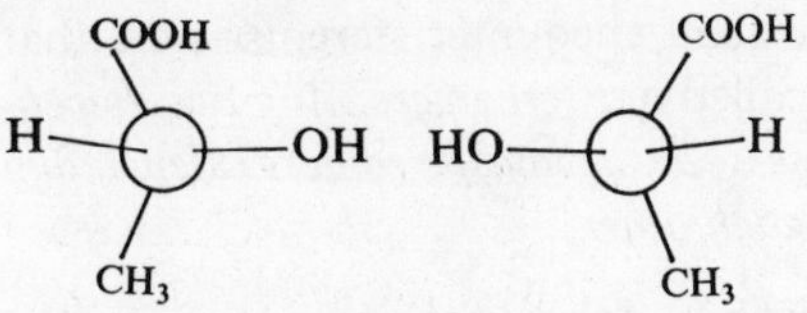

COOH, H, OH, CH_3 | COOH, HO, H, CH_3

Lactic acid

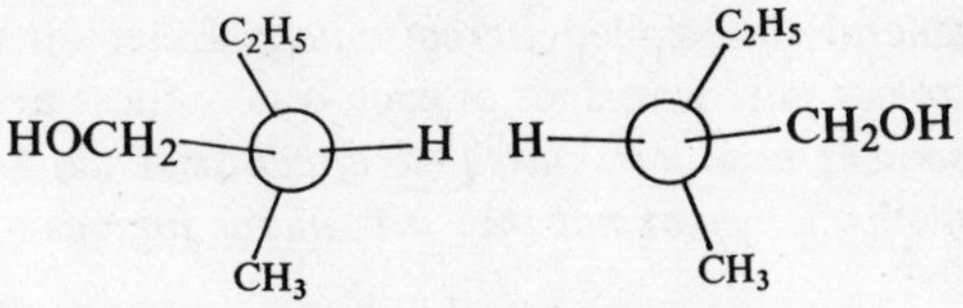

C_2H_5, $HOCH_2$, H, CH_3 | C_2H_5, H, CH_2OH, CH_3

2-Methyl-1-butanol

acids and two *2-methyl-1-butanols*, two *chloroiodomethanesulfonic acids* and two *sec-butyl chlorides*.

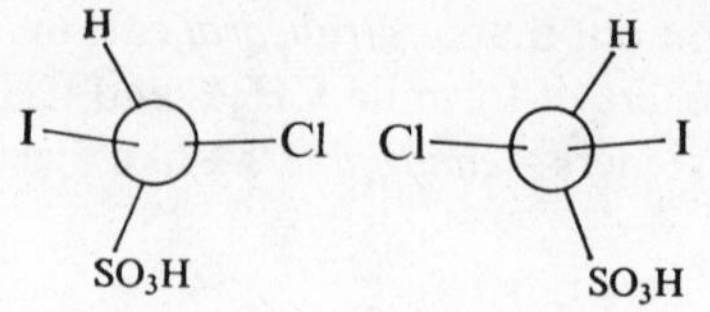

Chloroiodomethanesulfonic acid

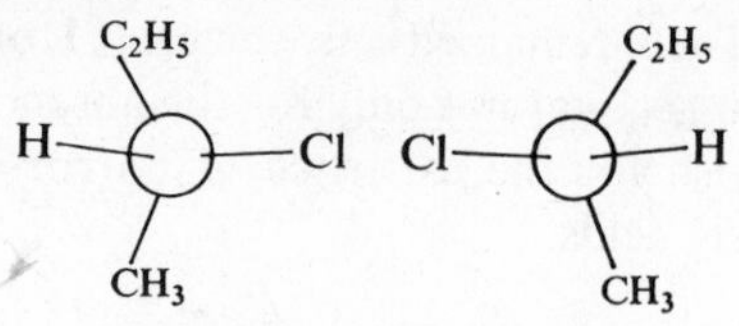

sec-Butyl chloride

As we can see, the structures of each pair are mirror images; as we can easily verify by use of models, the structures of each pair are not superimposable and therefore represent isomers. (In fact, we have *already* verified this, since the models we made for CWXYZ can, of course, stand for any of these.)

At this point we do not need to know the chemistry of these compounds, or even what structure a particular collection of letters (—COOH, say, or —CH_2OH) stands for; we can tell when atoms or groups are the *same* or *different* from each other, and whether or not a model can be superimposed on its mirror image. Even two isotopes of the same element, like protium (ordinary hydrogen, H) and deuterium (heavy hydrogen, D) are different enough to permit detectable isomerism:

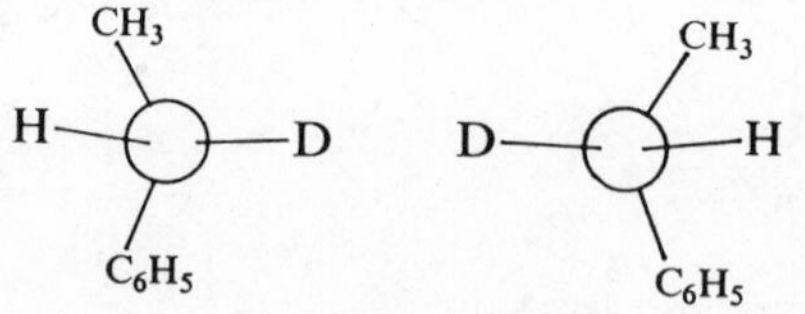

α-Deuterioethylbenzene

We must remember that *everything* (except, of course, a vampire) has a mirror image, including all molecules. Most molecules, however, are superimposable on their mirror images, as, for example, bromochloromethane, and do not show this mirror-image isomerism.

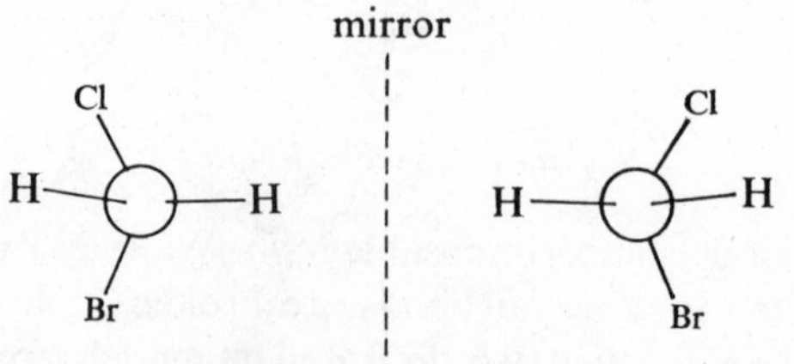

mirror

Bromochloromethane
Superimposable: no isomerism

Mirror-image isomers are called *enantiomers*. Since they differ from one another only in the way the atoms are oriented in space, enantiomers belong to the general class called *stereoisomers*. Later on we shall encounter stereoisomers that are *not* mirror images of each other; these are called *diastereomers*. *Any two stereoisomers are thus classified either as enantiomers or as diastereomers, depending upon whether or not they are mirror images of each other.*

The non-superimposability of mirror images that brings about the existence of enantiomers also, as we shall see, gives them their optical activity, and hence enantiomers are often referred to as (one kind of) *optical isomers*. We shall make no use of the term *optical isomer*, since it is hard to define—indeed, is often used undefined—and of doubtful usefulness.

4.8 Enantiomerism and optical activity

Most compounds do not rotate the plane of polarized light. How is it that *some* do? It is not the particular chemical family that they belong to, since optically active compounds are found in all families. To see what special structural feature gives rise to optical activity, let us look more closely at what happens when polarized light is passed through a sample of a single pure compound.

When a beam of polarized light passes through an individual molecule, in nearly every instance its plane is rotated a tiny amount by interaction with the charged particles of the molecule; the direction and extent of rotation varies with the orientation of the particular molecule in the beam. For most compounds, because of the random distribution of the large number of molecules that make up even the smallest sample of a single pure compound, for every molecule that the light encounters, there is another (identical) molecule oriented *as the mirror image of the first*, which exactly cancels its effect. The net result is no rotation, that is, optical inactivity. Thus optical inactivity is not a property of individual molecules, but rather of the *random distribution of molecules that can serve as mirror images of each other*.

Optical inactivity requires, then, that one molecule of a compound act as the mirror image of another. But in the special case of CWXYZ, we have found (Sec. 4.7) a molecule whose mirror image is not just another, identical molecule, but rather a molecule of a different, isomeric compound. In a pure sample of a single enantiomer, no molecule can serve as the mirror image of another; there is no exact canceling-out of rotations, and the net result is optical activity. Thus, the same non-superimposability of mirror images that gives rise to enantiomerism also is responsible for optical activity.

4.9 Prediction of enantiomerism. Chirality

Molecules that are not superimposable on their mirror images are **chiral**.

Chirality is the necessary and sufficient condition for the existence of enantiomers. That is to say: *a compound whose molecules are chiral can exist as enantiomers; a compound whose molecules are achiral* (without chirality) *cannot exist as enantiomers*.

When we say that a molecule and its mirror image are superimposable, we mean that if—in our mind's eye—we were to bring the image from behind the mirror where it seems to be, it could be made to coincide in all its parts with the molecule. To decide whether or not a molecule is chiral, therefore, we make a model of it and a model of its mirror image, and see if we can superimpose them. This is the safest way, since properly handled it must give us the right answer. It is the method that we should use until we have become quite familiar with the ideas involved; even then, it is the method we should use when we encounter a new type of compound.

After we have become familiar with the models themselves, we can draw pictures of the models, and *mentally* try to superimpose them. Some, we find, are not superimposable, like these:

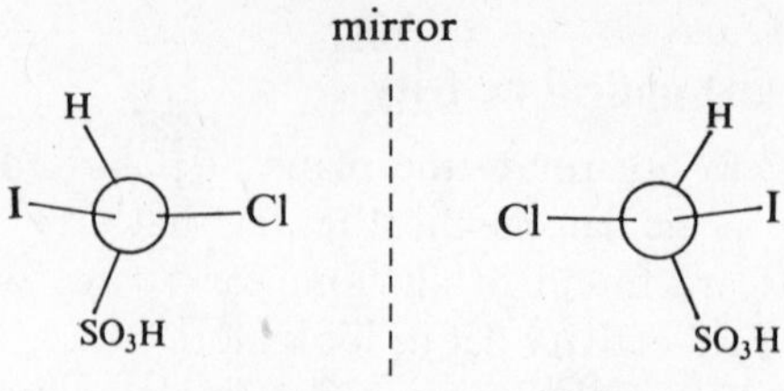

mirror

H — I — Cl — SO_3H | H — Cl — I — SO_3H

Chloroiodomethanesulfonic acid
Not superimposable: enantiomers

These molecules are chiral, and we know that chloroiodomethanesulfonic acid can exist as enantiomers, which have the structures we have just made or drawn.

Others, we find, are superimposable, like these:

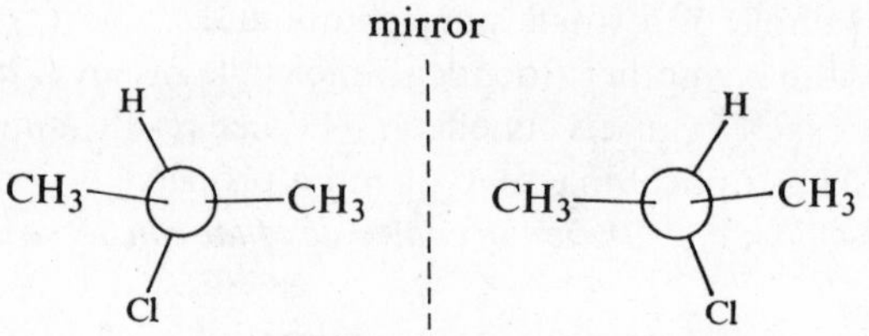

mirror

H — CH_3 — CH_3 — Cl | H — CH_3 — CH_3 — Cl

Isopropyl chloride
Superimposable: no enantiomers

These molecules are achiral, and so we know that isopropyl chloride cannot exist as enantiomers.

> "I call any geometrical figure, or any group of points, *chiral*, and say it has *chirality*, if its image in a plane mirror, ideally realized, cannot be brought to coincide with itself."—Lord Kelvin, 1893.

In 1964, Cahn, Ingold, and Prelog (see p. 130) proposed that chemists use the terms "chiral" and "chirality" as defined by Kelvin. Based on the Greek word for "hand" (*cheir*), chirality means "handedness," in reference to that pair of non-superimposable mirror images we constantly have before us: our two hands. There has been wide-spread acceptance of Kelvin's terms, and they have largely displaced the earlier "dissymmetric" and "dissymmetry" (and the still earlier—and less accurate—"asymmetric" and "asymmetry"), although one must expect to encounter the older terms in the older chemical literature.

Whatever one calls it, it is non-superimposability-on-mirror-image that is the necessary and sufficient condition for enantiomerism; it is also a necessary—but *not* sufficient—condition for optical activity (see Sec. 4.13).

4.10 The chiral center

So far, all the chiral molecules we have talked about happen to be of the kind CWXYZ; that is, in each molecule there is a carbon (C*) that holds four different groups.

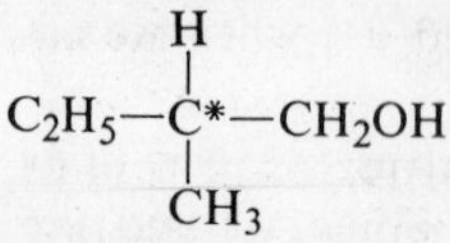

$C_2H_5—C^*(H)(CH_3)—CH_2OH$	$CH_3—C^*(H)(OH)—COOH$	$C_2H_5—C^*(H)(Cl)—CH_3$	$C_6H_5—C^*(H)(D)—CH_3$
2-Methyl-1-butanol	Lactic acid	*sec*-Butyl chloride	α-Deuterioethylbenzene

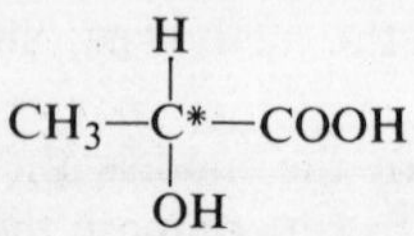

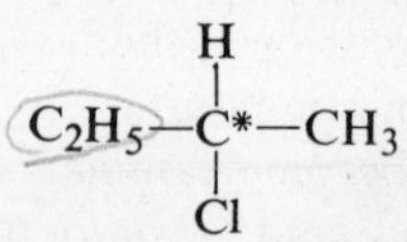

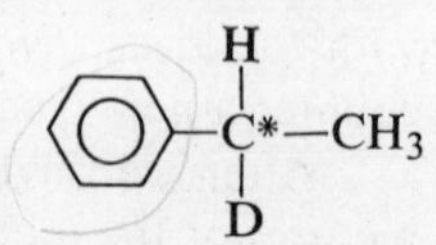

A carbon atom to which four different groups are attached is a **chiral center**. (Sometimes it is called *chiral carbon*, when it is necessary to distinguish it from *chiral nitrogen, chiral phosphorus*, etc.)

Many—*but not all*—molecules that contain a chiral center are chiral. Many—*but not all*—chiral molecules contain a chiral center. There are molecules that contain chiral centers—more than one—and yet are achiral (Sec. 4.18). There are chiral molecules that contain no chiral centers (see, for example, Problem 6, p. 315).

The presence or absence of a chiral center is thus no criterion of chirality. However, most of the chiral molecules that we shall take up do contain chiral centers, and it will be useful for us to look for such centers; if we find a chiral center, then we should consider the *possibility* that the molecule is chiral, and hence can exist in enantiomeric forms. We shall later (Sec. 4.18) learn to recognize the kind of molecule that may be achiral in spite of the presence of chiral centers; such molecules contain more than one chiral center.

After becoming familiar with the use of models and of pictures of models, the student can make use of even simpler representations of molecules containing chiral centers, which can be drawn much faster. This is a more dangerous method, however, and must be used properly to give the right answers. We simply draw a cross and attach to the four ends the four groups that are attached to the chiral center. The chiral center is understood to be located where the lines cross. Chemists have agreed that such a diagram stands for a particular structure: *the horizontal lines represent bonds coming toward us out of the plane of the paper, whereas the vertical lines represent bonds going away from us behind the plane of the paper.* That is to say:

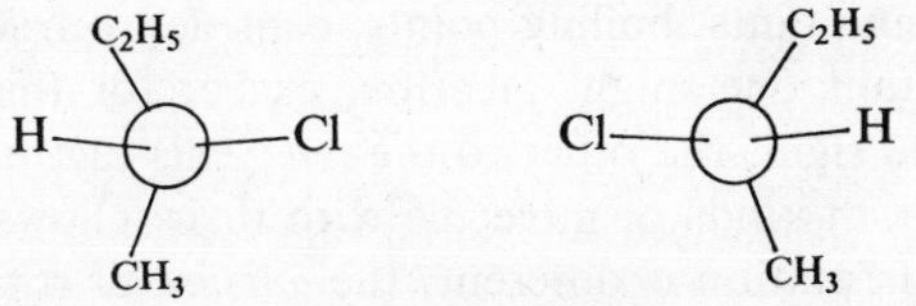

C_2H_5 C_2H_5
H— —Cl Cl— —H
CH_3 CH_3

can be represented by

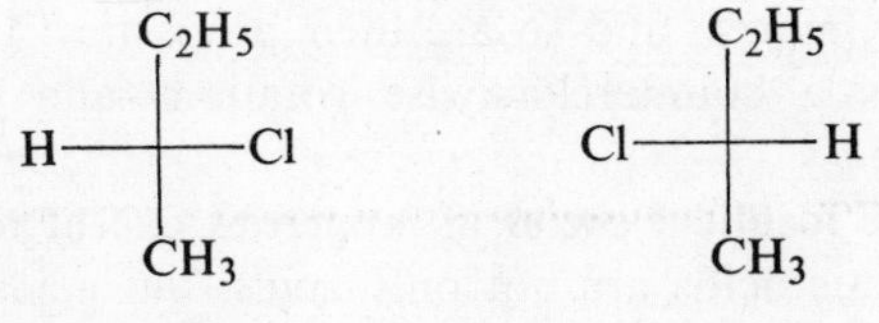

C_2H_5 C_2H_5
H— —Cl Cl— —H
CH_3 CH_3

In testing the superimposability of two of these flat, two-dimensional representations of three-dimensional objects, we must follow a certain procedure and obey certain rules. First, we use these representations only for molecules that contain a chiral center. Second, we draw one of them, and then draw the other as its mirror image. (Drawing these formulas *at random* can lead to some interesting but quite *wrong* conclusions about isomer numbers.) Third, in our mind's eye we may slide these formulas or rotate them end for end, *but we may not remove them from the plane of the paper*. Used with caution, this method of representation is convenient; it is not foolproof, however, and in doubtful cases models or pictures of models should be used.

Problem 4.5 Using cross formulas, decide which of the following compounds are chiral. Check your answers by use of stick-and-ball formulas, and finally by use of models.

(a) 1-chloropentane
(b) 2-chloropentane
(c) 3-chloropentane
(d) 1-chloro-2-methylpentane
(e) 2-chloro-2-methylpentane
(f) 3-chloro-2-methylpentane
(g) 4-chloro-2-methylpentane
(h) 1-chloro-2-bromobutane

Problem 4.6 (a) Neglecting stereoisomers for the moment, draw all isomers of formula C_3H_6DCl. (b) Decide, as in Problem 4.5, which of these are chiral.

4.11 Enantiomers

Isomers that are mirror images of each other are called **enantiomers.** The two different lactic acids whose models we made in Sec. 4.7 are enantiomers (Gr.: *enantio-*, opposite). So are the two 2-methyl-1-butanols, the two *sec*-butyl chlorides, etc. How do the properties of enantiomers compare?

Enantiomers have identical physical properties, except for the direction of rotation of the plane of polarized light. The two 2-methyl-1-butanols, for example,

	(+)-2-Methyl-1-butanol	(−)-2-Methyl-1-butanol (Fermentation Product)
Specific rotation	+5.756°	−5.756°
Boiling point	128.9°	128.9°
Density	0.8193	0.8193
Refractive index	1.4107	1.4107

have identical melting points, boiling points, densities, refractive indices, and any other physical constant one might measure, except for this: one rotates plane-polarized light to the right, the other to the left. This fact is not surprising, since the interactions of both kinds of molecule with their fellows should be the same. Only the *direction* of rotation is different; the *amount* of rotation is the same, the specific rotation of one being +5.756°, the other −5.756°. It is reasonable that these molecules, being so similar, can rotate light by the same amount. The molecules are mirror images, and so are their properties: the mirror image of a clockwise rotation is a counterclockwise rotation—and of exactly the same *magnitude.*

Enantiomers have identical chemical properties except toward optically active reagents. The two lactic acids are not only acids, but acids of exactly the same strength; that is, dissolved in water at the same concentration, both ionize to exactly the same degree. The two 2-methyl-1-butanols not only form the same products—*alkenes* on treatment with hot sulfuric acid, *alkyl bromides* on treatment with HBr, *esters* on treatment with acetic acid—but also form them at exactly the same rate. This is quite reasonable, since the atoms undergoing attack in each case are influenced in their reactivity by exactly the same combination of substituents. The reagent approaching either kind of molecule encounters the same environment, except, of course, that one environment is the mirror image of the other.

In the special case of a reagent that is itself optically active, on the other hand, the influences exerted on the reagent are *not* identical in the attack on the two enantiomers, and reaction rates will be different—so different, in some cases, that reaction with one isomer does not take place at all. In biological systems, for example, such stereochemical specificity is the rule rather than the exception, since the all-important catalysts, *enzymes*, and most of the compounds they work on, are optically active. The sugar (+)-glucose plays a unique role in animal metabolism (Sec. 34.3) and is the basis of a multimillion-dollar fermentation industry (Sec. 15.5); yet (−)-glucose is neither metabolized by animals nor fermented by yeasts. When the mold *Penicillium glaucum* feeds on a mixture of enantiomeric tartaric acids, it consumes only the (+)-enantiomer and leaves (−)-tartaric acid behind. The hormonal activity of (−)-adrenaline is many times that of its enantiomer; only one stereoisomer of chloromycetin is an antibiotic. (+)-Ephedrine not only has no activity as a drug, but actually interferes with the action of its enantiomer. Among amino acids, only one asparagine and one leucine are sweet, and only one glutamic acid enhances the flavor of food. It is (−)-carvone that gives oil of spearmint its characteristic odor; yet the enantiomeric (+)-carvone is the essence of caraway.

Consider, as a crude analogy, a right and left hand of equal strength (the enantiomers) hammering a nail (an optically inactive reagent) and inserting a right-handed screw (an optically active reagent). Hammering requires exactly corresponding sets of muscles in the two hands, and can be done at identical rates. Inserting the screw uses different sets of muscles: the right thumb pushes, for example, whereas the left thumb pulls.

Or, let us consider reactivity in the most precise way we know: by the transition-state approach (Sec. 2.22).

Take first the reactions of two enantiomers with an optically inactive reagent. The reactants in both cases are of exactly the same energy: one enantiomer plus the reagent, and the other enantiomer plus the same reagent. The two transition states for the reactions are mirror images (they are enantiomeric, Sec. 4.17), and hence are of exactly the same energy, too. Therefore, the energy differences between reactants and transition states—the E_{act}'s—are identical, and so are the rates of reaction.

Now take the reactions of two enantiomers with an optically *active* reagent. Again the reactants are of the same energy. The two transition states, however, are *not* mirror images of each other (they are diastereomeric), and hence are of *different* energies; the E_{act}'s are different, and so are the rates of reaction.

4.12 The racemic modification

A mixture of equal parts of enantiomers is called a **racemic modification**. *A racemic modification is optically inactive:* when enantiomers are mixed together, the rotation caused by a molecule of one isomer is exactly canceled by an equal and opposite rotation caused by a molecule of its enantiomer.

The prefix **±** is used to specify the racemic nature of the particular sample, as, for example, (±)-lactic acid or (±)-2-methyl-1-butanol.

It is useful to compare a racemic modification with a compound whose molecules are superimposable on their mirror images, that is, with an achiral compound. They are both optically inactive, and for exactly the same reason. Because of the random distribution of the large number of molecules, for every molecule that the light encounters there is a second molecule, a mirror image of the first, aligned just right to cancel the effect of the first one. In a racemic modification this second molecule happens to be an isomer of the first; for an achiral compound it is not an isomer, but another, identical molecule (Sec. 4.8).

(For an optically active substance uncontaminated by its enantiomer, we have seen, such cancellation of rotation cannot occur since no other molecule can serve as the mirror image of another, no matter how random the distribution.)

Problem 4.7 To confirm the statements of the three preceding paragraphs, make models of: (a) a pair of enantiomers, e.g., CHClBrI; (b) a pair of identical achiral molecules, e.g., CH_2ClBr; (c) a pair of identical chiral molecules, e.g., CHClBrI. (d) Which pairs are mirror images?

The identity of most physical properties of enantiomers has one consequence of great practical significance. They cannot be separated by ordinary methods: not by fractional distillation, because their boiling points are identical; not by fractional crystallization, because their solubilities in a given solvent are identical (unless the solvent is optically active); not by chromatography, because they are held equally strongly on a given adsorbent (unless it is optically active). The separation of a racemic modification into enantiomers—the *resolution* of a racemic modification—is therefore a special kind of job, and requires a special kind of approach (Sec. 7.9).

The first resolution was, of course, the one Pasteur carried out with his hand lens and tweezers (Sec. 4.6). But this method can almost never be used, since racemic modifications seldom form mixtures of crystals recognizable as mirror images. Indeed, even sodium ammonium tartrate does not, unless it crystallizes at a temperature below 28°. Thus partial credit for Pasteur's discovery has been given to the cool Parisian climate—and, of course, to the availability of tartaric acid from the winemakers of France.

The method of resolution nearly always used—one also discovered by Pasteur—involves the use of optically active reagents, and is described in Sec. 7.9.

Although popularly known chiefly for his great work in bacteriology and medicine, Pasteur was by training a chemist, and his work in chemistry alone would have earned him a position as an outstanding scientist.

4.13 Optical activity: a closer look

We have seen (Sec. 4.8) that, like enantiomerism, optical activity results from—and *only* from—chirality: the non-superimposability of certain molecules on their mirror images. Whenever we observe (molecular) optical activity, we know we are dealing with chiral molecules.

Is the reverse true? Whenever we deal with chiral molecules—with compounds that exist as enantiomers—must we always observe optical activity? *No.* We have just seen that a 50:50 mixture of enantiomers is optically inactive. Clearly, if we are to *observe* optical activity, the material we are dealing with must contain an *excess* of one enantiomer: enough of an excess that the net optical rotation can be detected by the particular polarimeter at hand.

Furthermore, this excess of one enantiomer must persist long enough for the optical activity to be measured. If the enantiomers are rapidly interconverted, then before we could measure the optical activity due to one enantiomer, it would be converted into an equilibrium mixture, which—since enantiomers are of exactly the same stability—must be a 50:50 mixture and optically inactive.

Even if all these conditions are met, the magnitude—and hence the detectability—of the optical rotation depends on the structure of the particular molecule concerned. In compound I, for example, the four groups attached to the chiral center differ only in chain length.

$$CH_3CH_2CH_2CH_2CH_2CH_2-\overset{\displaystyle CH_2CH_3}{\underset{\displaystyle CH_2CH_2CH_3}{\overset{|}{\underset{|}{C}}}}-CH_2CH_2CH_2CH_3$$

I

Ethyl-*n*-propyl-*n*-butyl-*n*-hexylmethane

It has been calculated that this compound should have the tiny specific rotation of 0.00001° —far below the limits of detection by any existing polarimeter. In 1965, enantiomerically pure samples of both enantiomers of I were prepared (see Problem 19, p. 1026), and each was found to be optically inactive.

At our present level of study, the matter of speed of interconversion will give us no particular trouble. Nearly all the chiral molecules we encounter in this book lie at either of two extremes, which we shall easily recognize: (a) molecules—like those described in this chapter—which owe their chirality to chiral centers; here interconversion of enantiomers (*configurational* enantiomers) is so slow—because bonds have to be broken—that we need not concern ourselves at all about interconversion; (b) molecules whose enantiomeric forms (*conformational* enantiomers) are interconvertible simply by rotations about single bonds; here—for the compounds we shall encounter—interconversion is so fast that ordinarily we need not concern ourselves at all about the existence of the enantiomers.

4.14 Configuration

The arrangement of atoms that characterizes a particular stereoisomer is called its **configuration.**

Using the test of superimposability, we conclude, for example, that there are two stereoisomeric *sec*-butyl chlorides; their *configurations* are I and II. Let us

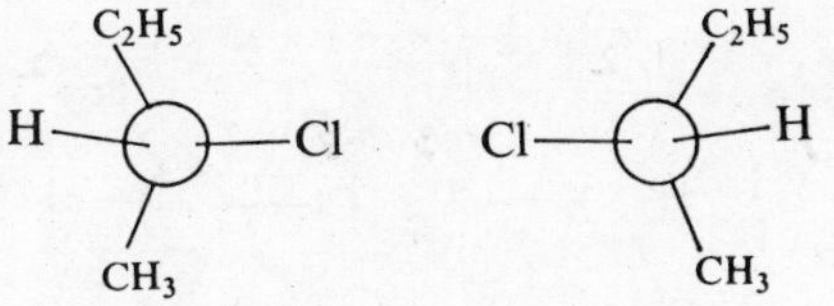

C_2H_5 / H—C—Cl / CH_3 (I) C_2H_5 / Cl—C—H / CH_3 (II)

I II

sec-Butyl chloride

say that, by methods we shall take up later (Sec. 7.9), we have obtained in the laboratory samples of two compounds of formula $C_2H_5CHClCH_3$. We find that one rotates the plane of polarized light to the right, and the other to the left; we put them into two bottles, one labeled "(+)-*sec*-butyl chloride" and the other "(−)-*sec*-butyl chloride."

We have made two models to represent the two configurations of this chloride. We have isolated two isomeric compounds of the proper formula. Now the question arises, which configuration does each isomer have? Does the (+)-isomer, say, have configuration I or configuration II? How do we know which structural formula, I or II, to draw on the label of each bottle? That is to say, how do we *assign configuration*?

Until 1951 the question of configuration could not be answered in an absolute sense for any optically active compound. But in that year J. M. Bijvoet—most fittingly Director of the van't Hoff Laboratory at the University of Utrecht (Sec. 4.2)—reported that, using a special kind of x-ray analysis (the method of anomalous scattering), he had determined the actual arrangement in space of the atoms of an optically active compound. The compound was a salt of (+)-tartaric acid, the same acid that—almost exactly 100 years before—had led Pasteur to his discovery of optical isomerism. Over the years prior to 1951, the relationships between the configuration of (+)-tartaric acid and the configurations of hundreds of optically active compounds had been worked out (by methods that we shall take up later, Sec. 7.5); when the configuration of (+)-tartaric acid became known, these other configurations, too, immediately became known. (In the case of the *sec*-butyl chlorides, for example, the (−)-isomer is known to have configuration I, and the (+)-isomer configuration II.)

4.15 Specification of configuration: R and S

Now, a further problem arises. How can we specify a particular configuration in some simpler, more convenient way than by always having to draw its picture? The most generally useful way yet suggested is the use of the prefixes R and S. According to a procedure proposed by R. S. Cahn (The Chemical Society, London), Sir Christopher Ingold (University College, London), and V. Prelog (Eidgenössiche Technische Hochschule, Zurich), two steps are involved.

Step 1. Following a set of *sequence rules* (Sec. 4.16), we assign a sequence of priority to the four atoms or groups of atoms attached to the chiral center.

In the case of CHClBrI, for example, the four atoms attached to the chiral center are all different and priority depends simply on atomic number, the atom of higher number having higher priority. Thus I, Br, Cl, H.

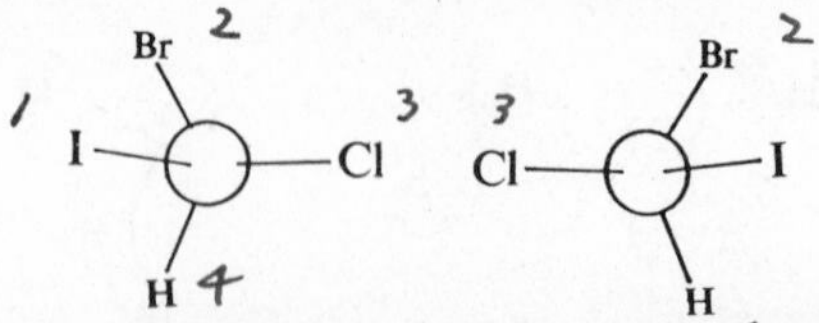

I II

Bromochloroiodomethane

Step 2. We visualize the molecule oriented so that the group of *lowest* priority is directed *away* from us, and observe the arrangement of the remaining groups. If, in proceeding from the group of highest priority to the group of second priority and thence to the third, our eye travels in a clockwise direction, the configuration is specified **R** (Latin: *rectus*, right); if counterclockwise, the configuration is specified S (Latin: *sinister*, left).

Thus, configurations I and II are viewed like this:

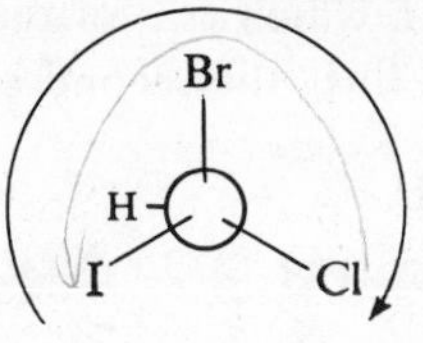

Br
H
I Cl
R

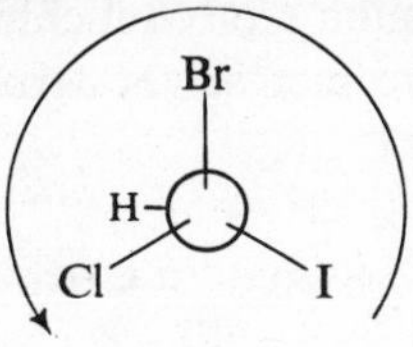

Br
H
Cl I
S

and are specified R and S, respectively.

A complete name for an optically active compound reveals—if they are known—both configuration and direction of rotation, as, for example, (S)-(+)-*sec*-butyl chloride. A racemic modification can be specified by the prefix RS, as, for example, (RS)-*sec*-butyl chloride.

(Specification of compounds containing more than one chiral center is discussed in Sec. 4.19.)

We must not, of course, confuse the direction of optical rotation of a compound—a physical property of a real substance, like melting point or boiling point—with the direction in which our eye happens to travel when we imagine a molecule held in an arbitrary manner. So far as we are concerned, unless we happen to know what has been established experimentally for a specific compound, we have no idea whether (+) or (−) rotation is associated with the (R)- or the (S)-configuration.

4.16 Sequence rules

For ease of reference and for convenience in reviewing, we shall set down here those sequence rules we shall have need of. The student should study Rules 1 and 2 now, and Rule 3 later when the need for it arises.

Sequence Rule 1. If the four atoms attached to the chiral center are all different, priority depends on atomic number, with the atom of higher atomic number getting higher priority. If two atoms are isotopes of the same element, the atom of higher mass number has the higher priority.

For example, in chloroiodomethanesulfonic acid the sequence is I, Cl, S, H; in α-deuterioethyl bromide it is Br, C, D, H.

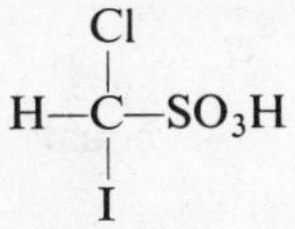

```
     Cl                    H
     |                     |
  H—C—SO3H            H3C—C—Br
     |                     |
     I                     D
```

Chloroiodomethanesulfonic acid 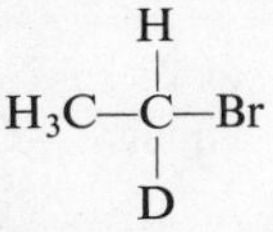 α-Deuterioethyl bromide

Problem 4.8 Make models and then draw both stick-and-ball pictures and cross formulas for the enantiomers of: (a) chloroiodomethanesulfonic acid and (b) α-deuterioethyl bromide. Label each as R or S.

Sequence Rule 2. If the relative priority of two groups cannot be decided by Rule 1, it shall be determined by a similar comparison of the next atoms in the groups (and so on, if necessary, working outward from the chiral center). That is to say, if two atoms attached to the chiral center are the same, we compare the atoms attached to each of these first atoms.

For example, take *sec*-butyl chloride, in which two of the atoms attached to the chiral center are themselves carbon. In CH_3 the second atoms are H, H, H;

$$\begin{array}{c} \mathrm{H} \\ | \\ \mathrm{CH_3{-}CH_2{-}C{-}CH_3} \\ | \\ \mathrm{Cl} \end{array}$$

sec-Butyl chloride

in C_2H_5 they are C, H, H. Since carbon has a higher atomic number than hydrogen, C_2H_5 has the higher priority. A complete sequence of priority for *sec*-butyl chloride is therefore Cl, C_2H_5, CH_3, H.

In 3-chloro-2-methylpentane the C, C, H of isopropyl takes priority over the C, H, H of ethyl, and the complete sequence of priority is Cl, isopropyl, ethyl, H.

$$\begin{array}{c} \mathrm{CH_3\ \ H} \\ |\quad\ | \\ \mathrm{CH_3{-}CH{-}C{-}CH_2{-}CH_3} \\ | \\ \mathrm{Cl} \end{array} \qquad\qquad \begin{array}{c} \mathrm{CH_3\ \ H} \\ |\quad\ | \\ \mathrm{CH_3{-}CH{-}C{-}CH_2Cl} \\ | \\ \mathrm{Cl} \end{array}$$

3-Chloro-2-methylpentane 1,2-Dichloro-3-methylbutane

In 1,2-dichloro-3-methylbutane the Cl, H, H of CH_2Cl takes priority over the C, C, H of isopropyl. Chlorine has a higher atomic number than carbon, and the fact that there are *two* C's and only *one* Cl does not matter. (One higher number is worth more than two—or three—of a lower number.)

Problem 4.9 Into what sequence of priority must these alkyl groups always fall: CH_3, 1°, 2°, 3°?

Problem 4.10 Specify as R or S each of the enantiomers you drew: (a) in Problem 4.5 (p. 126); (b) in Problem 4.6 (p. 126).

Sequence Rule 3. (*The student should defer study of this rule until he needs it.*) Where there is a double or triple bond, both atoms are considered to be duplicated or triplicated. Thus,

$$\mathrm{{-}\overset{|}{C}{=}A} \text{ equals } \begin{array}{c} | \\ \mathrm{{-}C{-}A} \\ |\quad | \\ \mathrm{A\quad C} \end{array} \text{ and } \mathrm{{-}C{\equiv}A} \text{ equals } \begin{array}{c} \mathrm{A\quad C} \\ |\quad | \\ \mathrm{{-}C{-}A} \\ |\quad | \\ \mathrm{A\quad C} \end{array}$$

For example, in glyceraldehyde the OH group has the highest priority of all,

$$\begin{array}{c} \mathrm{H} \\ | \\ \mathrm{C{=}O} \\ | \\ \mathrm{H{-}C{-}OH} \\ | \\ \mathrm{CH_2OH} \end{array} \qquad \begin{array}{c} \mathrm{H} \\ | \\ \mathrm{{-}C{=}O} \end{array} \text{ equals } \begin{array}{c} \mathrm{H} \\ | \\ \mathrm{{-}C{-}O} \\ |\quad | \\ \mathrm{O\quad C} \end{array}$$

Glyceraldehyde

and the O, O, H of —CHO takes priority over the O, H, H of $-CH_2OH$. The complete sequence is then —OH, —CHO, $-CH_2OH$, —H.

The phenyl group, C_6H_5—, is handled as though it had one of the Kekulé structures:

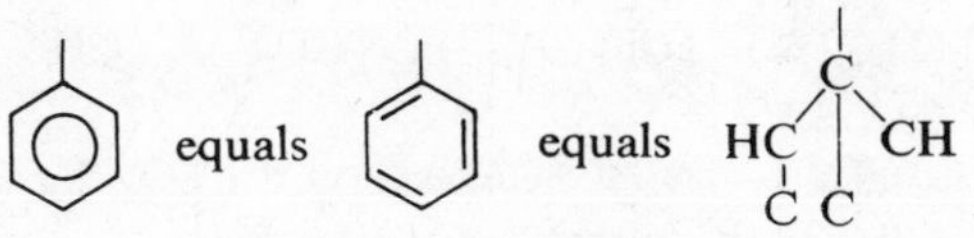

[benzene ring] equals [Kekulé structure] equals HC(—C)—C(—C)(—)—CH

In 1-amino-2-methyl-1-phenylpropane, for example, the C, C, C, of phenyl takes

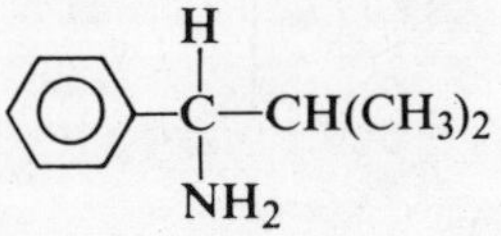

C_6H_5—C(H)(NH$_2$)—CH(CH$_3$)$_2$

priority over the C, C, H of isopropyl, but not over N, which has a higher atomic number. The entire sequence is then NH_2, C_6H_5, C_3H_7, H.

The vinyl group, CH_2=CH—, takes priority over isopropyl.

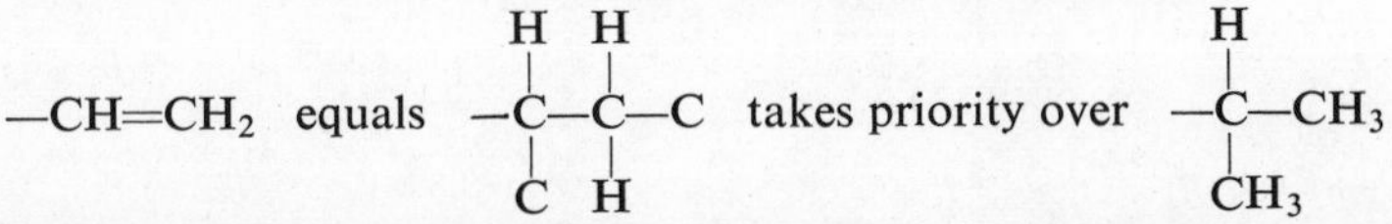

—CH=CH$_2$ equals —C(H)(C)—C(H)(H)—C takes priority over —C(H)(CH$_3$)—CH$_3$

Following the "senior" branch, —CH_2—C, we arrive at C in vinyl as compared with H in the —CH_2—H of isopropyl.

Problem 4.11 Draw and specify as R or S the enantiomers (if any) of:

(a) 3-chloro-1-pentene
(b) 3-chloro-4-methyl-1-pentene
(c) $HOOCCH_2CHOHCOOH$, malic acid
(d) $C_6H_5CH(CH_3)NH_2$
(e) methylethyl-*n*-propylisopropylmethane
(f) $C_6H_5CHOHCOOH$, mandelic acid
(g) $CH_3CH(NH_2)COOH$, alanine

4.17 Diastereomers

Next, we must learn what stereoisomers are possible for compounds whose molecules contain, not just one, but *more than one* chiral center. (In Chapter 34, we shall be dealing regularly with molecules that contain *five* chiral centers.)

Let us start with 2,3-dichloropentane. This compound contains two chiral

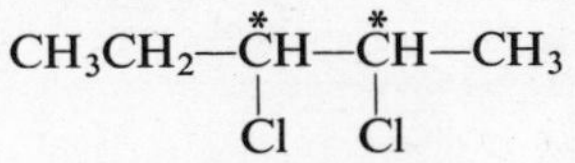

CH_3CH_2—C*H(Cl)—C*H(Cl)—CH_3

2,3-Dichloropentane

centers, C-2 and C-3. (What four groups are attached to each of these carbon atoms?) How many stereoisomers are possible?

Using models, let us first make structure I and its mirror image II, and see if these are superimposable. We find that I and II are not superimposable, and hence must be enantiomers. (As before, we may represent the structures by pictures, and mentally try to superimpose these. Or, we may use the simple "cross" representations, being careful, as before (Sec. 4.10), not to remove the drawings from the plane of the paper or blackboard.)

Next, we try to interconvert I and II by rotations about carbon–carbon bonds. We find that they are not interconvertible in this way, and hence each of them is capable of retaining its identity and, if separated from its mirror image, of showing optical activity.

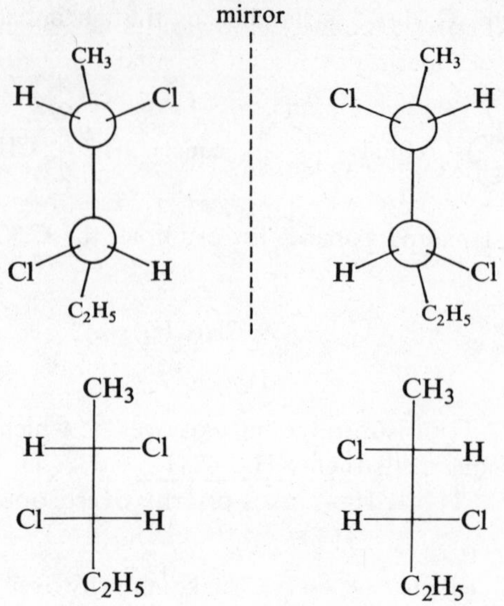

mirror

CH_3 H Cl | CH_3 Cl H

Cl H | H Cl

C_2H_5 | C_2H_5

CH_3 | CH_3

H—Cl | Cl—H

Cl—H | H—Cl

C_2H_5 | C_2H_5

I | II

Not superimposable
Enantiomers

Are there any other stereoisomers of 2,3-dichloropentane? We can make structure III, which we find to be non-superimposable on either I or II; it is not, of

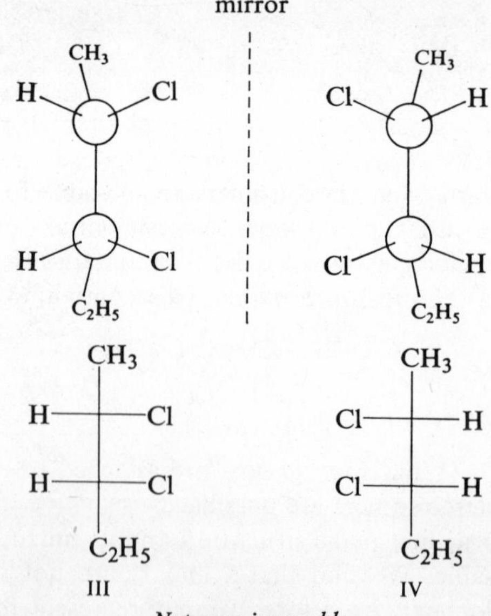

Not superimposable
Enantiomers

course, the mirror image of either. What is the relationship between III and I? Between III and II? They are stereoisomers but not enantiomers. *Stereoisomers that are not mirror images of each other are called* **diastereomers**. Compound III is a diastereomer of I, and similarly of II.

Now, is III chiral? Using models, we make its mirror image, structure IV, and find that this is not superimposable on (or interconvertible with) III. Structures III and IV represent a second pair of enantiomers. Like III, compound IV is a diastereomer of I and of II.

How do the properties of diastereomers compare?

Diastereomers have similar chemical properties, since they are members of the same family. Their chemical properties are *not identical*, however. In the reaction of two diastereomers with a given reagent, neither the two sets of reactants nor the two transition states are mirror images, and hence—except by sheer coincidence—will not be of equal energies. E_{act}'s will be different and so will the rates of reaction.

Diastereomers have different physical properties: different melting points, boiling points, solubilities in a given solvent, densities, refractive indexes, and so on. Diastereomers differ in specific rotation; they may have the same or opposite signs of rotation, or some may be inactive.

As a result of their differences in boiling point and in solubility, they can, in principle at least, be separated from each other by fractional distillation or fractional crystallization; as a result of differences in molecular shape and polarity, they differ in adsorption, and can be separated by chromatography.

Given a mixture of all four stereoisomeric 2,3-dichloropentanes, we could separate it, by distillation, for example, into two fractions but no further. One fraction would be the racemic modification of I plus II; the other fraction would be the racemic modification of III plus IV. Further separation would require *resolution* of the racemic modifications by use of optically active reagents (Sec. 7.9).

Thus the presence of two chiral centers can lead to the existence of as many as four stereoisomers. For compounds containing three chiral centers, there could be as many as eight stereoisomers; for compounds containing four chiral centers, there could be as many as sixteen stereoisomers, and so on. The maximum number of stereoisomers that can exist is equal to 2^n, where n is the number of chiral centers. (In any case where *meso* compounds exist, as discussed in the following section, there will be fewer than this maximum number.)

4.18 *Meso* structures

Now let us look at 2,3-dichlorobutane, which also has two chiral centers. Does this compound, too, exist in four stereoisomeric forms?

$$CH_3-\overset{*}{C}H(Cl)-\overset{*}{C}H(Cl)-CH_3$$

2,3-Dichlorobutane

Using models as before, we arrive first at the two structures V and VI. These are mirror images that are not superimposable or interconvertible; they are therefore enantiomers, and each should be capable of optical activity.

mirror

V VI

Not superimposable
Enantiomers

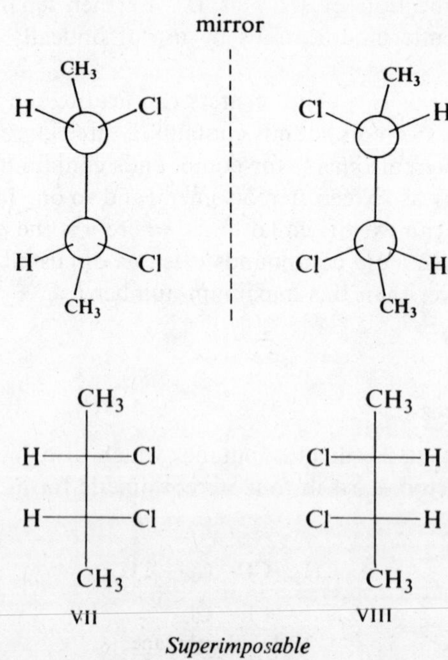

Superimposable
A *meso* compound

Next, we make VII, which we find to be a diastereomer of V and of VI. We now have three stereoisomers; is there a fourth? *No*. If we make VIII, the mirror image of VII, we find the two to be superimposable; turned end-for-end,

VII coincides in every respect with VIII. In spite of its chiral centers, VII is not chiral. It cannot exist in two enantiomeric forms, and it cannot be optically active. It is called a *meso* compound.

A **meso compound** *is one whose molecules are superimposable on their mirror images even though they contain chiral centers.* A *meso* compound is optically inactive for the same reason as any other compound whose molecules are achiral: the rotation caused by any one molecule is cancelled by an equal and opposite rotation caused by another molecule that is the mirror image of the first (Sec. 4.8).

We can often recognize a *meso* structure on sight by the fact that (in at least one of its conformations) one half of the molecule is the mirror image of the other half. This can be seen for *meso*-2,3-dichlorobutane by imagining the molecule to be cut by a plane lying where the dotted line is drawn. The molecule has *a plane of symmetry*, and cannot be chiral. (*Caution:* If we do not see a plane of symmetry, however, this does not necessarily mean that the molecule is chiral.)

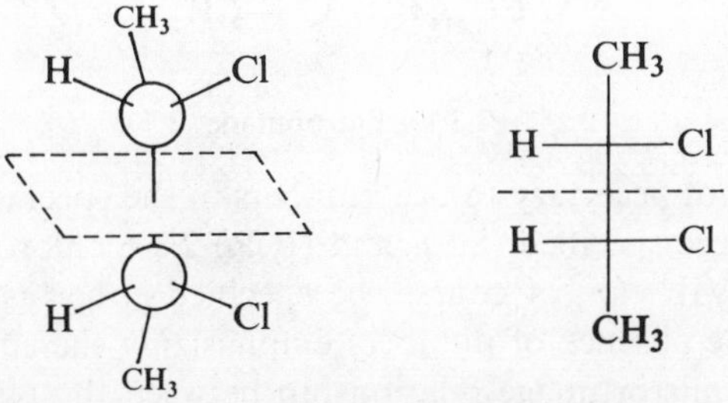

CH_3 H Cl — H Cl CH_3

$$\begin{array}{c} CH_3 \\ H\!-\!\!\!+\!\!\!-\!Cl \\ \text{- - - - - - -} \\ H\!-\!\!\!+\!\!\!-\!Cl \\ CH_3 \end{array}$$

Problem 4.12 Draw stereochemical formulas for all the possible stereoisomers of the following compounds. Label pairs of enantiomers, and *meso* compounds. Tell which isomers, if separated from all other stereoisomers, will be optically active. Pick out several examples of diastereomers.

(a) 1,2-dibromopropane
(b) 3,4-dibromo-3,4-dimethylhexane
(c) 2,4-dibromopentane
(d) 2,3,4-tribromohexane
(e) 1,2,3,4-tetrabromobutane
(f) 2-bromo-3-chlorobutane
(g) 1-chloro-2-methylbutane
(h) 1,3-dichloro-2-methylbutane

4.19 Specification of configuration: more than one chiral center

Now, how do we specify the configuration of compounds which, like these, contain more than one chiral center? They present no special problem; we simply specify the configuration about *each* of the chiral centers, and by use of numbers tell which specification refers to which carbon.

Consider, for example, the 2,3-dichloropentanes (Sec. 4.17). We take each of the chiral centers, C–2 and C–3, in turn—ignoring for the moment the existence

$$CH_3CH_2-\overset{3}{\overset{*}{C}}H(Cl)-\overset{2}{\overset{*}{C}}H(Cl)-CH_3$$

2,3-Dichloropentane

of the other—and follow the steps of Sec. 4.15 and use the Sequence Rules of

Sec. 4.16. In order of priority, the four groups attached to C–2 are Cl, CH_3CH_2CHCl—, CH_3, H. On C–3 they are Cl, CH_3CHCl—, CH_3CH_2—, H. (Why is CH_3CHCl— "senior" to CH_3CH_2—?)

Taking in our hands—or in our mind's eye—a model of the particular stereoisomer we are interested in, we focus our attention first on C–2 (ignoring C–3), and then on C–3 (ignoring C–2). Stereoisomer I (p. 134), for example, we specify (2S,3S)-2,3-dichloropentane. Similarly, II is (2R,3R), III is (2S,3R), and IV is (2R,3S). These specifications help us to analyze the relationships among the stereoisomers. As enantiomers, I and II have opposite—that is, mirror-image—configurations about both chiral centers: 2S,3S and 2R,3R. As diastereomers, I and III have opposite configurations about one chiral center, and the same configuration about the other: 2S,3S and 2S,3R.

We would handle 2,3-dichlorobutane (Sec. 4.18) in exactly the same way. Here it happens that the two chiral centers occupy equivalent positions along the

$$CH_3-\overset{*}{C}H(Cl)-\overset{*}{C}H(Cl)-CH_3$$

2,3-Dichlorobutane

chain, and so it is not necessary to use numbers in the specifications. Enantiomers V and VI (p. 136) are specified (S,S)- and (R,R)-2,3-dichlorobutane, respectively. The *meso* isomer, VII, can, of course, be specified either as (R,S)- or (S,R)-2,3-dichlorobutane—the absence of numbers emphasizing the equivalence of the two specifications. The mirror-image relationship between the two ends of this molecule is consistent with the *opposite* designations of R and S for the two chiral centers. (Not all (R,S)-isomers, of course, are *meso* structures—only those whose two halves are chemically equivalent.)

Problem 4.13 Give the R/S specification for each stereoisomer you drew in Problem 4.12 (p. 137).

4.20 Conformational isomers

In Sec. 3.5, we saw that there are several different staggered conformations of *n*-butane, each of which lies at the bottom of an energy valley—at an *energy minimum*—separated from the others by energy hills (see Fig. 3.4, p. 79). *Different conformations corresponding to energy minima are called* **conformational isomers,** *or* **conformers.** Since conformational isomers differ from each other only in the way their atoms are oriented in space, they, too, are stereoisomers. Like stereoisomers of any kind, a pair of conformers can either be mirror images of each other or not.

n-Butane exists as three conformational isomers, one *anti* and two *gauche* (Sec. 3.5). The *gauche* conformers, II and III, are mirror images of each other, and hence are (conformational) enantiomers. Conformers I and II (or I and III) are *not* mirror images of each other, and hence are (conformational) diastereomers.

Although the barrier to rotation in *n*-butane is a little higher than in ethane, it is still low enough that—at ordinary temperatures, at least—interconversion of conformers is easy and rapid. Equilibrium exists, and favors a higher population of the more stable *anti* conformer; the populations of the two *gauche* conformers—

mirror images, and hence of exactly equal stability—are, of course, equal. Put differently, any given molecule spends the greater part of its time as the *anti* conformer, and divides the smaller part equally between the two *gauche* conformers. As a result of the rapid interconversion, these isomers cannot be separated.

Problem 4.14 Return to Problem 3.4 (p. 79) and, for each compound: (a) tell how many conformers there are, and label pairs of (conformational) enantiomers; (b) give the order of relative abundance of the various conformers.

Easy interconversion is characteristic of nearly every set of conformational isomers, and is the quality in which such isomers differ most from the kind of stereoisomers we have encountered so far in this chapter. This difference in interconvertibility is due to a difference in height of the energy barrier separating stereoisomers, which is, in turn, due to a difference in origin of the barrier. By definition, interconversion of conformational isomers involves rotation about single bonds; the rotational barrier is—in most cases—a very low one and interconversion is easy and fast. The other kind of stereoisomers, *configurational isomers*, or *inversional isomers*, differ from one another in configuration about a chiral center. Interconversion here involves the breaking of a covalent bond, for which there is a very high barrier: 50 kcal/mole or more (Sec. 1.14). Interconversion is difficult, and—unless one deliberately provides conditions to bring it about—is negligibly slow.

Interconvertibility of stereoisomers is of great practical significance because it limits their *isolability*. Hard-to-interconvert stereoisomers can be separated (with special methods, of course, for resolution of enantiomers) and studied individually; among other things, their optical activity can be measured. Easy-to-interconvert isomers cannot be separated, and single isolated isomers cannot be studied; optical activity cannot be observed, since any chiral molecules are present only as non-resolvable racemic modifications.

Our general approach to stereoisomers involves, then, two stages; first, we test the *superimposability* of possible isomeric structures, and then we test their *interconvertibility*. Both tests are best carried out with models. We make models of the two molecules and, without allowing any rotations about single bonds, we try to superimpose them: if they cannot be superimposed, they represent isomers. Next, we allow the models all possible rotations about single bonds, and repeatedly try to superimpose them: if they still cannot be superimposed, they are non-interconvertible, and represent *configurational isomers*; but if they can be superimposed after rotation, they are interconvertible and represent *conformational isomers*.

In dealing with those aspects of stereochemistry that depend on isolation of stereoisomers—isomer number or optical activity, for example, or study of the reactions of a single stereoisomer—we can ignore the existence of easy-to-interconvert isomers, which means *most* conformational isomers. For convenience the following "ground rule" will hold for discussions and problems in this book: unless specifically indicated otherwise, *the terms "stereoisomers," "enantiomers," and "diastereomers" will refer only to configurational isomers, including geometric isomers* (Sec. 5.6), and will exclude conformational isomers. The latter will be referred to as "conformational isomers," "conformers," "conformational enantiomers," and "conformational diastereomers."

There is no sharp boundary between easy-to-interconvert and hard-to-interconvert stereoisomers. Although we can be sure that interconversion of configurational isomers will be hard, we cannot be sure that interconversion of conformational isomers will be easy. Depending upon the size and nature of substituents, the barrier to rotation about single bonds can be of any height, from the low one in ethane to one comparable to that for breaking a covalent bond. Some conformational isomers exist that are readily isolated, kept, and studied; indeed, study of such isomers (*atropisomers*) makes up a large and extremely important part of stereochemistry, one which, unfortunately, we shall not be able to take up in this beginning book. Other conformational isomers exist that can be isolated, not at ordinary temperatures, but at lower temperatures, where the average collision energy is lower. The conformational isomers that we shall encounter in this book, however, have low rotational barriers, and we may assume—until we learn otherwise—that when we classify stereoisomers as configurational or conformational, we at the same time classify them as hard-to-interconvert or easy-to-interconvert.

Problem 4.15 At low temperatures, where collision energies are small, two isomeric forms of the badly crowded $CHBr_2CHBr_2$ have been isolated by crystallization. (a) Give a formula or formulas (Newman projections) corresponding to each of the separable forms. (b) Which, if either, of the materials, as actually isolated at low temperatures, would be optically active? Explain.

PROBLEMS

1. What is meant by each of the following?

(a) optical activity
(b) dextrorotatory
(c) levorotatory
(d) specific rotation
(e) chirality
(f) chiral molecule
(g) chiral center
(h) superimposable
(i) enantiomers
(j) diastereomers
(k) *meso* compound
(l) racemic modification
(m) configuration
(n) conformations
(o) R
(p) S
(q) +
(r) −
(s) configurational isomers
(t) conformational isomers

2. (a) What is the necessary and sufficient condition for enantiomerism? (b) What is a necessary but not a sufficient condition for optical activity? (c) What conditions must be met for the observation of optical activity? (d) How can you tell from its formula whether or not a compound can exist as enantiomers? (e) What restrictions, if any, must be applied to the use of planar formulas in (d)? To the use of models in (d)? (f) Exactly how do you go about deciding whether a molecule should be specified as R or as S?

3. Compare the dextrorotatory and levorotatory forms of *sec*-butyl alcohol, $CH_3CH_2CHOHCH_3$, with respect to:

(a) boiling point
(b) melting point
(c) specific gravity
(d) specific rotation
(e) refractive index
(f) solubility in 100 g of water
(g) rate of reaction with HBr
(h) infrared spectrum
(i) nmr spectrum
(j) adsorption on alumina
(k) retention time in gas chromatography
(l) specification as R or S

4. Which of the following objects are chiral?

(a) nail, screw, pair of scissors, knife, spool of thread;
(b) glove, shoe, sock, pullover sweater, coat sweater, scarf tied around your neck;
(c) child's block, rubber ball, Pyramid of Cheops, helix (p. 1157), double helix (p. 1179);

(d) basketball, football, tennis racket, golf club, baseball bat, shotgun barrel, rifle barrel;
(e) your hand, your foot, your ear, your nose, yourself.

5. Assuming both your hands to be of equal strength and skill, which of the following operations could you perform with equal speed and efficiency?
(a) driving a screw, sawing a board, drilling a hole;
(b) opening a door, opening a milk bottle, opening a coffee jar, turning on the hot water;
(c) signing your name, sharpening a pencil, throwing a ball, shaking hands with another right hand, turning to page 142.

6. Draw and specify as R or S the enantiomers (if any) of:
(a) 3-bromohexane
(b) 3-chloro-3-methylpentane
(c) 1,2-dibromo-2-methylbutane
(d) 1,3-dichloropentane
(e) 3-chloro-2,2,5-trimethylhexane
(f) 1-deuterio-1-chlorobutane, $CH_3CH_2CH_2CHDCl$

7. (a) What is the lowest molecular weight alkane that is chiral? Draw stereochemical formulas of the enantiomers and specify each as R or S. (b) Is there another alkane of the same molecular weight that is also chiral? If there is, give its structure and name, and specify the enantiomers as R or S.

8. Draw stereochemical formulas for all the possible stereoisomers of the following compounds. Label pairs of enantiomers, and *meso* compounds. Tell which isomers, if separated from all other stereoisomers, will be optically active. Give one isomer of each set its R/S specification.
(a) $CH_3CHBrCHOHCH_3$
(b) $CH_3CHBrCHBrCH_2Br$
(c) $C_6H_5CH(CH_3)CH(CH_3)C_6H_5$
(d) $CH_3CH_2CH(CH_3)CH_2CH_2CH(CH_3)CH_2CH_3$
(e) $CH_3CH(C_6H_5)CHOHCH_3$
(f) $CH_3CHOHCHOHCHOHCH_2OH$
(g) $HOCH_2(CHOH)_3CH_2OH$
(h) CH_2—$CHCl$ / CH_2—$CHCl$ (*Make models.*)
(i) CH_2—$CHCl$ / $CHCl$—CH_2
(j) methylethyl-*n*-propyl-*n*-butylammonium chloride, $(RR'R''R'''N)^+Cl^-$. (See Sec. 1.12.)
(k) methylethyl-*n*-propyl-*sec*-butylammonium chloride

9. (a) In a study of chlorination of propane, four products (A, B, C, and D) of formula $C_3H_6Cl_2$ were isolated. What are their structures?

(b) Each was chlorinated further, and the number of trichloro products ($C_3H_5Cl_3$) obtained from each was determined by gas chromatography. A gave one trichloro product; B gave two; and C and D each gave three. What is the structure of A? Of B? Of C and D?

(c) By another synthetic method, compound C was obtained in optically active form. Now what is the structure of C? Of D?

(d) When optically active C was chlorinated, one of the trichloropropanes (E) obtained was optically active, and the other two were optically inactive. What is the structure of E? Of the other two?

10. Draw configurational isomers (if any) of: (a) CH_2BrCH_2Cl; (b) $CH_3CHBrCH_2Cl$. (c) For each substance of (a) and (b), draw all conformers. Label pairs of conformational enantiomers.

11. The more stable conformer of *n*-propyl chloride, CH_3CH_2—CH_2Cl, is the *gauche*. What does this indicate about the interaction between —Cl and —CH_3? How do you account for this interaction? (*Hint:* See Sec. 1.19.)

12. (a) What must be the dipole moment of the *anti* conformation of 1,2-dichloroethane, CH_2Cl—CH_2Cl? (b) At 32° in the gas phase, the measured dipole moment of

1,2-dichloroethane is 1.12 D. What does this single fact tell you about the conformational make-up of the compound? (c) The dipole moment of a mixture of X and Y is given by the expression

$$\mu^2 = N_X\mu_X{}^2 + N_Y\mu_Y{}^2$$

where N is the mole fraction of each kind of molecule. From bond moments, it has been estimated that the *gauche* conformation of 1,2-dichloroethane should have a dipole moment of about 3.2 D. Calculate the conformational composition of 1,2-dichloroethane at 32° in the gas phase.

Chapter 5 Alkenes I. Structure and Preparation

Elimination

5.1 Unsaturated hydrocarbons

In our discussion of the alkanes we mentioned briefly another family of hydrocarbons, the **alkenes**, which contain less hydrogen, carbon for carbon, than the alkanes, and which can be converted into alkanes by addition of hydrogen. The alkenes were further described as being obtained from alkanes by loss of hydrogen in the cracking process.

Since alkenes evidently contain less than the maximum quantity of hydrogen, they are referred to as **unsaturated hydrocarbons**. This unsaturation can be satisfied by reagents other than hydrogen and gives rise to the characteristic chemical properties of alkenes.

5.2 Structure of ethylene. The carbon–carbon double bond

The simplest member of the alkene family is **ethylene**, C_2H_4. In view of the ready conversion of ethylene into ethane, we can reasonably expect certain structural similarities between the two compounds.

To start, then, we connect the carbon atoms by a covalent bond, and then attach two hydrogen atoms to each carbon atom. At this stage we find that each carbon atom possesses only six electrons in its valence shell, instead of the required eight, and that the entire molecule needs an additional pair of electrons if it is to be neutral. We can solve both these problems by assuming that the carbon atoms can share two pairs of electrons. To describe this sharing of two pairs of electrons, we say that the carbon atoms are joined by a *double bond. The* **carbon–carbon double bond** *is the distinguishing feature of the alkene structure.*

```
 H H            H     H
H:C̈::C̈:H          \   /
                   C=C
                  /   \
                 H     H
```

Ethylene

Quantum mechanics gives a more detailed picture of ethylene and the carbon–carbon double bond. To form bonds with three other atoms, carbon makes use of three equivalent hybrid orbitals: sp^2 orbitals, formed by the mixing of *one s* and *two p* orbitals. As we have seen (Sec. 1.10), sp^2 orbitals lie in one plane, that of the carbon nucleus, and are directed toward the corners of an equilateral triangle; the angle between any pair of orbitals is thus 120°. This **trigonal** arrangement (Fig. 5.1) permits the hybrid orbitals to be as far apart as possible. Just as mutual

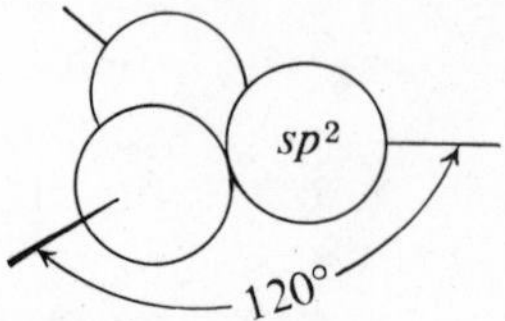

Figure 5.1. Atomic orbitals: hybrid sp^2 orbitals. Axes directed toward corners of equilateral triangle.

repulsion among orbitals gives four tetrahedral bonds, so it gives three trigonal bonds.

If we arrange the two carbons and four hydrogens of ethylene to permit maximum overlap of orbitals, we obtain the structure shown in Fig. 5.2. Each

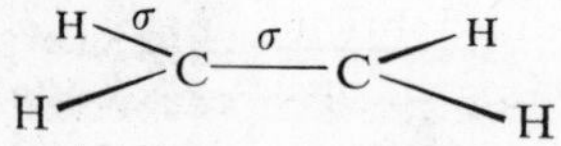

H σ σ H
C—C
H H

Figure 5.2. Ethylene molecule: only σ bonds shown.

carbon atom lies at the center of a triangle, at whose corners are located the two hydrogen atoms and the other carbon atom. Every bond angle is 120°. Although distributed differently about the carbon nucleus, these bonds individually are very similar to the bonds in ethane, being cylindrically symmetrical about a line joining the nuclei, and are given the same designation: *σ bond* (*sigma bond*).

The molecule is not yet complete, however. In forming the sp^2 orbitals, each carbon atom has used only two of its three *p* orbitals. The remaining *p* orbital consists of two equal lobes, one lying above and the other lying below the plane of the three sp^2 orbitals (Fig. 5.3); it is occupied by a single electron. If the *p* orbital

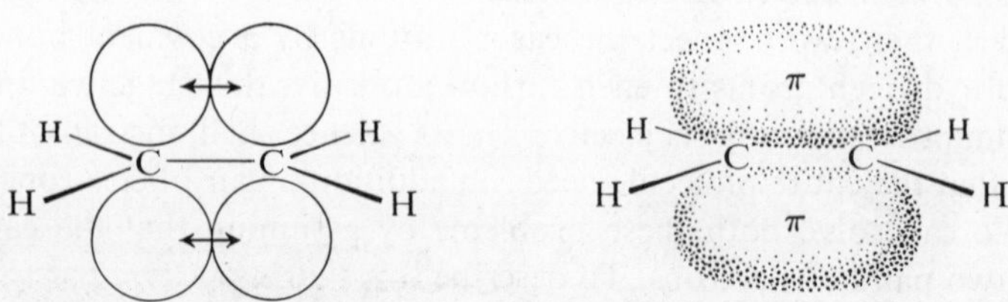

Figure 5.3. Ethylene molecule: carbon–carbon double bond. Overlap of *p* orbitals gives π bond; π cloud above and below plane.

of one carbon atom overlaps the *p* orbital of the other carbon atom, the electrons pair up and an additional bond is formed.

Because it is formed by the overlap of *p* orbitals, and to distinguish it from the differently shaped σ bonds, this bond is called a π *bond* (*pi bond*). It consists of two parts, one electron cloud that lies above the plane of the atoms, and another electron cloud that lies below. Because of less overlap, the π bond is weaker than the carbon–carbon σ bond. As we can see from Fig. 5.3, this overlap can occur only when all six atoms lie in the same plane. Ethylene, then, is a *flat molecule*.

The carbon–carbon "double bond" is thus made up of a strong σ bond and a weak π bond. The total bond energy of 163 kcal is greater than that of the carbon–carbon single bond of ethane (88 kcal). Since the carbon atoms are held more tightly together, the C—C distance in ethylene is less than the C—C distance in ethane; that is to say, the carbon–carbon double bond is shorter than the carbon–carbon single bond.

The σ bond in ethylene has been estimated to have a strength of about 95 kcal: stronger than the one in ethane because it is formed by overlap of sp^2 orbitals (Sec. 5.4). On this basis, we would estimate the strength of the π bond to be 68 kcal.

This quantum mechanical structure of ethylene is verified by direct evidence. Electron diffraction and spectroscopic studies show ethylene (Fig. 5.4) to be a flat molecule, with bond angles very close to 120°. The C—C distance is 1.34 A as compared with the C—C distance of 1.53 A in ethane.

Figure 5.4. Ethylene molecule: shape and size.

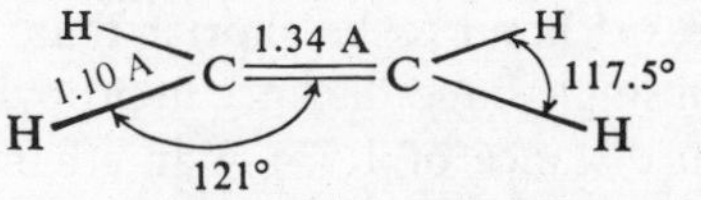

In addition to these direct measurements, we shall soon see that two important aspects of alkene chemistry are consistent with the quantum mechanical picture of the double bond, and are most readily understood in terms of that picture. These are (a) the concept of *hindered rotation* and the accompanying phenomenon of *geometric isomerism* (Sec. 5.6), and (b) the kind of reactivity characteristic of the carbon–carbon double bond (Sec. 6.2).

5.3 Propylene

The next member of the alkene family is **propylene**, C_3H_6. In view of its great similarity to ethylene, it seems reasonable to assume that this compound also contains a carbon–carbon double bond. Starting with two carbons joined by a double bond, and attaching the other atoms according to our rule of one bond per hydrogen and four bonds per carbon, we arrive at the structure

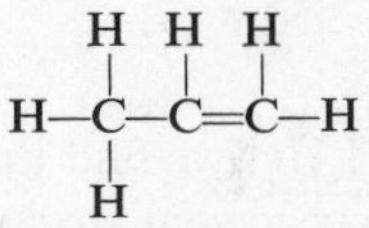

```
     H  H  H
     |  |  |
  H—C—C=C—H
     |
     H
```

Propylene

5.4 Hybridization and orbital size

The carbon–carbon double bond in alkenes is shorter than the carbon–carbon single bond in alkanes because four electrons bind more tightly than two. But, in addition, certain other bonds in alkenes are significantly shorter than their counterparts in alkanes: for example, the C—H distance is 1.103 A in ethylene compared with 1.112 A in ethane. To account for this and other differences in bond length, we must consider differences in hybridization of carbon.

The carbon–hydrogen bonds of ethylene are single bonds just as in, say, ethane, but they are formed by overlap of sp^2 orbitals of carbon, instead of sp^3 orbitals as in ethane. Now, compared with an sp^3 orbital, an sp^2 orbital has less *p* character and more *s* character. A *p* orbital extends some distance from the nucleus; an *s* orbital, on the other hand, lies close about the nucleus. As the *s* character of a hybrid orbital increases, the effective size of the orbital decreases and, with it, the length of the bond to a given second atom. Thus an sp^2–s carbon–hydrogen bond should be shorter than an sp^3–s carbon–hydrogen bond.

Benzene, in most ways a quite different kind of molecule from ethylene (Sec. 10.1), also contains sp^2–s carbon–hydrogen bonds; the C—H bond distance is 1.10 A, almost exactly the same as in ethylene. Acetylene (Sec. 8.2) contains *sp*-hybridized carbon which, in view of the even greater *s* character of the orbitals, should form even shorter bonds than in ethylene; this expectation is correct, the sp–s bond being only 1.079 A.

A consideration of hybridization and orbital size would lead one to expect an sp^2–sp^3 bond to be shorter than an sp^3–sp^3 bond. In agreement, the carbon–carbon single bond distance in propylene is 1.501 A, as compared with the carbon–carbon distance of 1.534 A in ethane. The sp–sp^3 carbon–carbon single bond in methylacetylene (Sec. 8.19) is even shorter, 1.459 A. These differences in carbon–carbon single bond lengths are greater than the corresponding differences in carbon–hydrogen bond lengths; however, another factor (Sec. 8.18) besides the particular hybridization of carbon may be at work here.

Consideration of hybridization and orbital size helps us to understand other properties of molecules besides bond length: the relative acidities of certain hydrocarbons (Sec. 8.10), for example, and the relative basicities of certain amines (Sec. 31.11). We might reasonably expect shorter bonds to be stronger bonds, and in agreement Table 1.2 (p. 21) shows that the C—H bond dissociation energy in ethylene (104 kcal) is larger than that in ethane (98 kcal), and the C—C (single) bond dissociation energy in propylene (92 kcal) is greater than that in ethane (88 kcal). Indeed, as will be discussed in Sec. 8.19, by affecting the stability of molecules, changes in hybridization may be of more fundamental importance than has been generally recognized.

5.5 The butylenes

Going on to the **butylenes**, C_4H_8, we find that there are a number of possible arrangements. First of all, we may have a straight-chain skeleton as in *n*-butane, or a branched-chain structure as in isobutane. Next, even when we restrict ourselves to the straight-chain skeleton, we find that there are two possible arrangements that differ in position of the double bond in the chain. So far, then, we have a total of three structures; as indicated, these are given the names *1-butene*, *2-butene*, and *isobutylene*.

```
  H H H H              H H H H              H     H
  | | | |              | | | |              |     |
H—C—C—C=C—H          H—C—C=C—C—H          H—C—C=C—H
  | |                  |     |              |  |
  H H                  H     H              H  |
                                             H—C—H
                                               |
                                               H
 1-Butene              2-Butene          Isobutylene
```

How do the facts agree with the prediction of three isomeric butylenes? Experiment has shown that not three but *four* alkenes of the formula C_4H_8 exist; they have the physical properties shown in Table 5.1.

Table 5.1 PHYSICAL PROPERTIES OF THE BUTYLENES

Name	b.p., °C	m.p., °C	Density (−20°)	Refractive Index (−12.7°)
Isobutylene	−7	−141	0.640	1.3727
1-Butene	−6	< −195	.641	1.3711
trans-2-Butene	+1	−106	.649	1.3778
cis-2-Butene	+4	−139	.667	1.3868

On hydrogenation, the isomer of b.p. −7° yields isobutane; this butylene evidently contains a branched chain, and has therefore the structure we have designated isobutylene.

On hydrogenation, the other three isomers all yield the same compound, *n*-butane; they evidently have a straight-chain skeleton. In ways that we shall study later (Sec. 6.29), it is possible to break an alkene molecule apart at the double bond, and from the fragments obtained deduce the position of the double bond in the molecule. When this procedure is carried out, the isomer of b.p. −6° yields products indicating clearly that the double bond is at the end of the chain; this butylene has therefore the structure we have designated 1-butene. When the same procedure is carried out on the two remaining isomers, both yield the same mixture of products; these products show that the double bond is in the middle of the chain.

Judging from the products of hydrogenation and the products of cleavage, we would conclude that the butylenes of b.p. +1° and +4° *both* have the structure we have designated 2-butene. Yet the differences in boiling point, melting point, and other physical properties show clearly that they are not the same compound, that is, that they are isomers. In what way can their structures differ?

To understand the kind of isomerism that gives rise to two 2-butenes, we must examine more closely the structure of alkenes and the nature of the carbon–carbon double bond. Ethylene is a flat molecule. We have seen that this flatness is a result of the geometric arrangement of the bonding orbitals, and in particular the overlap that gives rise to the π orbital. For the same reasons, a portion of any alkene must also be flat, the two doubly-bonded carbons and the four atoms attached to them lying in the same plane.

If we examine the structure of 2-butene more closely, and particularly if we use molecular models, we find that there are two quite different ways, I and II, in which the atoms can be arranged (aside from the infinite number of possibilities arising from rotation about the single bonds). In one of the structures the methyl groups lie on the same side of the molecule (I), and in the other structure they lie on opposite sides of the molecule (II).

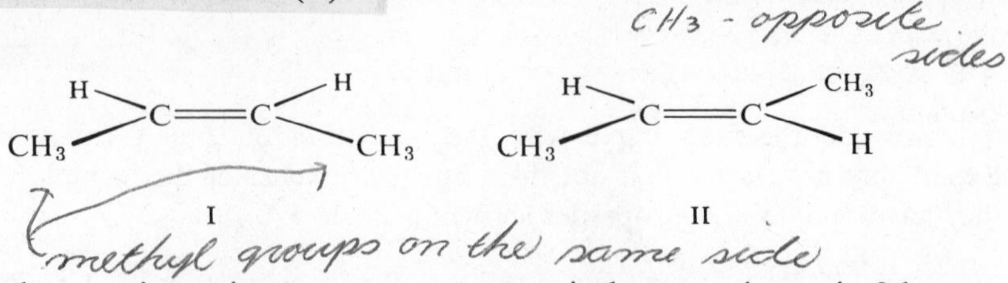

I II

Now the question arises: can we expect to isolate two isomeric 2-butenes corresponding to these two different structures, or are they too readily interconverted—like, say, the conformations of *n*-butane (Sec. 3.5)?

Conversion of I into II involves rotation about the carbon–carbon double bond. The possibility of isolating isomers depends upon the energy required for this rotation. We have seen that the formation of the π bond involves overlap of the *p* orbitals that lie above and below the plane of the σ orbitals. To pass from one of these 2-butenes to the other, the molecule must be twisted so that the *p* orbitals no longer overlap; that is, the π bond must be broken (see Fig. 5.5).

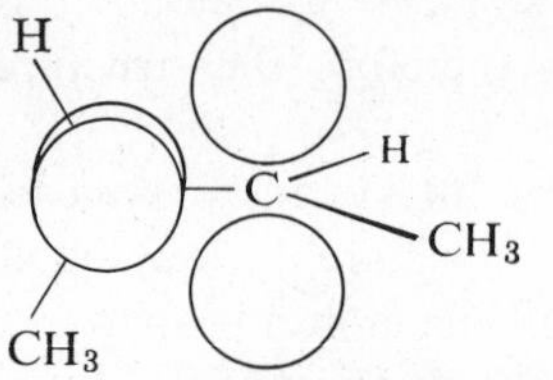

Figure 5.5. Hindered rotation about carbon–carbon double bond. Rotation would prevent overlap of *p* orbitals and would break π bond.

Breaking the π bond requires about 70 kcal of energy; at room temperature an insignificant proportion of collisions possess this necessary energy, and hence the rate of this interconversion is extremely small. Because of this 70-kcal energy barrier, then, *there is* **hindered rotation** *about the carbon–carbon double bond.* As a result of this hindered rotation, two isomeric 2-butenes can be isolated. These are, of course, the butylenes of b.p. $+1°$ and b.p. $+4°$.

5.6 Geometric isomerism

Since the isomeric 2-butenes differ from one another *only* in the way the atoms are oriented in space (but are like one another with respect to which atoms are attached to which other atoms), they belong to the general class we have called *stereoisomers* (Sec. 4.1). They are not, however, mirror images of each other, and hence are not enantiomers. As we have already said, *stereoisomers that are not mirror images of each other are called* **diastereomers**.

The particular kind of diastereomers that owe their existence to hindered rotation about double bonds are called **geometric isomers**. The isomeric 2-butenes, then, are diastereomers, and more specifically, geometric isomers.

We recall that the arrangement of atoms that characterizes a particular stereoisomer is called its *configuration.* The configurations of the isomeric 2-butenes are the structures I and II. These configurations are differentiated in their names by the prefixes **cis-** (Latin: on this side) and **trans-** (Latin: across), which indicate that the methyl groups are on the same side or on opposite sides of the molecule. In a way that we shall take up shortly (Sec. 5.9), the isomer of b.p. +4° has been assigned the *cis* configuration and the isomer of b.p. +1° the *trans* configuration.

I	II	
$CH_3(H)C{=}C(H)CH_3$	$CH_3(H)C{=}C(CH_3)H$	**Geometric isomers**
cis-2-Butene	*trans*-2-Butene	
b.p. +4°	b.p. +1°	

There is hindered rotation about *any* carbon–carbon double bond, but it gives rise to geometric isomerism only if there is a certain relationship among the groups attached to the doubly-bonded carbons. We can look for this isomerism by drawing the possible structures (or better yet, by constructing them from molecular models), and then seeing if these are indeed isomeric, or actually identical. On this basis we find that propylene, 1-butene, and isobutylene should not show

$H_2C{=}C(H)CH_3$	$H_2C{=}C(H)C_2H_5$	$H_2C{=}C(CH_3)CH_3$	*No geometric isomerism*
Propylene	1-Butene	Isobutylene	

isomerism; this conclusion agrees with the facts. Many higher alkenes may, of course, show geometric isomerism.

If we consider compounds other than hydrocarbons, we find that 1,1-dichloro- and 1,1-dibromoethene should not show isomerism, whereas the 1,2-dichloro- and 1,2-dibromoethenes should. In every case these predictions have been found correct. Isomers of the following physical properties have been isolated.

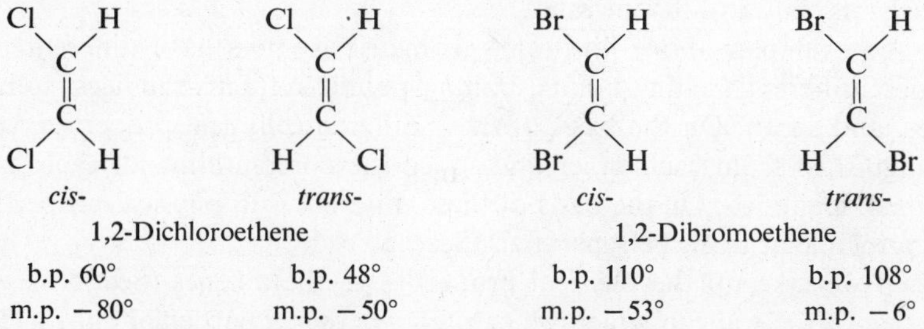

cis-	*trans-*	*cis-*	*trans-*
1,2-Dichloroethene		1,2-Dibromoethene	
b.p. 60°	b.p. 48°	b.p. 110°	b.p. 108°
m.p. −80°	m.p. −50°	m.p. −53°	m.p. −6°

As we soon conclude from our examination of these structures, geometric isomerism cannot exist if either carbon carries two identical groups. Some possible combinations are shown below.

Isomerism

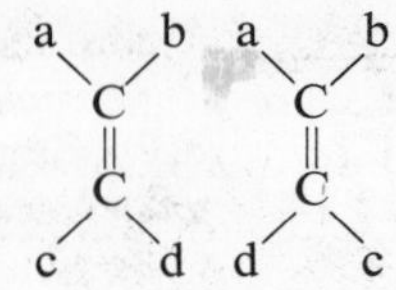

Isomerism

No isomerism

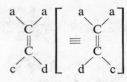

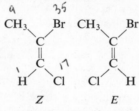

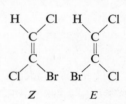

The phenomenon of geometric isomerism is a general one and can be encountered in any class of compounds that contain carbon–carbon double bonds (or even double bonds of other kinds).

The prefixes *cis* and *trans* work very well for disubstituted ethylenes and some trisubstituted ethylenes. But how are we to specify configurations like these?

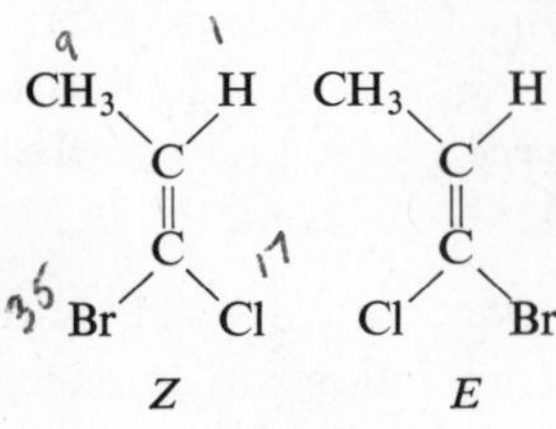

Z *E*	*Z* *E*	*Z* *E*
1-Bromo-1-chloropropene	2-Bromo-1-chloropropene	1-Bromo-1,2-dichloroethene
$CH_3 > H$	$Br > CH_3$	$Cl > H$
$Br > Cl$	$Cl > H$	$Br > Cl$

Which groups are our reference points? Looking at each doubly-bonded carbon in turn, we arrange its two atoms or groups in their Cahn-Ingold-Prelog sequence. We then take the group of higher priority on the one carbon and the group of higher priority on the other carbon, and tell whether they are on the same side of the molecule or on opposite sides. So that it will be clear that we are using this method of specification, we use the letter ***Z*** to mean *on the same side*, and the letter ***E*** to mean *on opposite sides.* (From the German: *zusammen*, together, and *entgegen*, opposite.)

In so far as chemical and physical properties are concerned, geometric isomers show the same relationship to each other as do the other diastereomers we have encountered (Sec. 4.17). They contain the same functional groups and hence show similar chemical properties. Their chemical properties are *not identical*, however, since their structures are neither identical nor mirror images; they react with the same reagents, but at different rates.

As the examples above illustrate, geometric isomers have different physical properties: different melting points, boiling points, refractive indices, solubilities, densities, and so on. On the basis of these different physical properties, they can be distinguished from each other and, once the configuration of each has been determined, identified. On the basis of these differences in physical properties they can, in principle at least, be separated. (See Sec. 4.17.)

When we take up the physical properties of the alkenes (Sec. 5.9), we shall discuss one of the ways in which we can tell whether a particular substance is the *cis*- or *trans*-isomer, that is, one of the ways in which we *assign configuration.*

A pair of geometric isomers are, then, diastereomers. Where do they fit into the other classification scheme, the one based on how the stereoisomers are interconverted

(Sec. 4.20)? We shall discuss this question in more detail later (Sec. 7.1), but for the moment we can say this. In the important quality of *isolability*, geometric isomers resemble configurational isomers, and for a very good reason: in both cases interconversion requires bond breaking—a π bond in the case of geometric isomers.

5.7 Higher alkenes

As we can see, the butylenes contain one carbon and two hydrogens more than propylene, which in turn contains one carbon and two hydrogens more than ethylene. The alkenes, therefore, form another homologous series, the increment being the same as for the alkanes: CH_2. The general formula for this family is C_nH_{2n}.

As we ascend the series of alkenes, the number of isomeric structures for each member increases even more rapidly than in the case of the alkane series; in addition to variations in the carbon skeletons, there are variations in the position of the double bond for a given skeleton, and the possibility of geometric isomerism.

Problem 5.1 Neglecting enantiomerism, draw structures of: (a) the six isomeric pentylenes (C_5H_{10}); (b) the four chloropropylenes (C_3H_5Cl); (c) the eleven chlorobutylenes (C_4H_7Cl). Specify as *Z* or *E* each geometric isomer.

5.8 Names of alkenes

Common names are seldom used except for three simple alkenes: *ethylene*, *propylene*, and *isobutylene*. The various alkenes of a given carbon number are, however, sometimes referred to collectively as the *pentylenes* (*amylenes*), *hexylenes*, *heptylenes*, and so on. (One sometimes encounters the naming of alkenes as derivatives of ethylene: as, for example, *tetramethylethylene* for $(CH_3)_2C{=}C(CH_3)_2$.) Most alkenes are named by the IUPAC system.

The rules of the IUPAC system are:

1. Select as the parent structure the longest continuous chain *that contains the carbon–carbon double bond*; then consider the compound to have been derived from this structure by replacement of hydrogen by various alkyl groups. The parent structure is known as *ethene*, *propene*, *butene*, *pentene*, and so on, depending upon the number of carbon atoms; each name is derived by changing the ending *-ane* of the corresponding alkane name to **-ene**:

$H_2C{=}CH_2$ — Ethene

$CH_3{-}CH{=}CH_2$ — Propene

$CH_3CH_2CH{=}CH_2$ — 1-Butene

$CH_3CH{=}CHCH_3$ — 2-Butene (*cis-* or *trans-*)

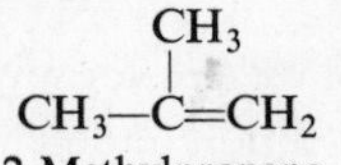

$CH_3{-}C(CH_3){=}CH_2$ — 2-Methylpropene

$CH_3{-}C(CH_3)_2{-}CH{=}CH_2$ — 3,3-Dimethyl-1-butene

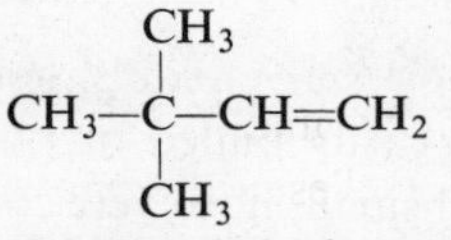

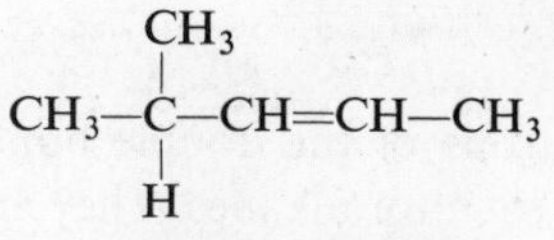

$CH_3{-}CH(CH_3){-}CH{=}CH{-}CH_3$ — 4-Methyl-2-pentene (*cis-* or *trans-*)

2. Indicate by a number the position of the double bond in the parent chain. Although the double bond involves two carbon atoms, designate its position by the number of the *first* doubly-bonded carbon encountered when numbering from the end of the chain nearest the double bond; thus *1-butene* and *2-butene*.

3. Indicate by numbers the positions of the alkyl groups attached to the parent chain.

Problem 5.2 Give the structural formula of: (a) 2,3-dimethyl-2-butene; (b) 3-bromo-2-methylpropene; (c) *cis*-2-methyl-3-heptene; (d) (*E*)-2-chloro-2-butene.

Problem 5.3 Referring to your answer to Problem 5.1 (p. 151), give IUPAC names for: (a) the isomeric pentylenes; (b) the isomeric chloropropenes.

5.9 Physical properties

As a class, the alkenes possess physical properties that are essentially the same as those of the alkanes. They are insoluble in water, but quite soluble in non-polar solvents like benzene, ether, chloroform, or ligroin. They are less dense than water. As we can see from Table 5.2, the boiling point rises with increasing carbon number; as with the alkanes, the boiling point rise is 20–30° for each added carbon, except for the very small homologs. As before, branching lowers the boiling point. A comparison of Table 5.2 with Table 3.3 (p. 86) shows that the boiling point of an alkene is very nearly the same as that of the alkane with the corresponding carbon skeleton.

Table 5.2 ALKENES

Name	Formula	M.p., °C	B.P., °C	Density (at 20°C)
Ethylene	$CH_2{=}CH_2$	−169	−102	
Propylene	$CH_2{=}CHCH_3$	−185	−48	
1-Butene	$CH_2{=}CHCH_2CH_3$		−6.5	
1-Pentene	$CH_2{=}CH(CH_2)_2CH_3$		30	0.643
1-Hexene	$CH_2{=}CH(CH_2)_3CH_3$	−138	63.5	.675
1-Heptene	$CH_2{=}CH(CH_2)_4CH_3$	−119	93	.698
1-Octene	$CH_2{=}CH(CH_2)_5CH_3$	−104	122.5	.716
1-Nonene	$CH_2{=}CH(CH_2)_6CH_3$		146	.731
1-Decene	$CH_2{=}CH(CH_2)_7CH_3$	−87	171	.743
cis-2-Butene	*cis*-$CH_3CH{=}CHCH_3$	−139	4	
trans-2-Butene	*trans*-$CH_3CH{=}CHCH_3$	−106	1	
Isobutylene	$CH_2{=}C(CH_3)_2$	−141	−7	
cis-2-Pentene	*cis*-$CH_3CH{=}CHCH_2CH_3$	−151	37	.655
trans-2-Pentene	*trans*-$CH_3CH{=}CHCH_2CH_3$		36	.647
3-Methyl-1-butene	$CH_2{=}CHCH(CH_3)_2$	−135	25	.648
2-Methyl-2-butene	$CH_3CH{=}C(CH_3)_2$	−123	39	.660
2,3-Dimethyl-2-butene	$(CH_3)_2C{=}C(CH_3)_2$	−74	73	.705

Like alkanes, alkenes are at most only weakly polar. Since the loosely held π electrons of the double bond are easily pulled or pushed, dipole moments are larger than for alkanes. They are still small, however: compare the dipole moments shown for propylene and 1-butene, for example, with the moment of 1.83 D for methyl chloride. The bond joining the alkyl group to the doubly-bonded carbon has a small polarity, which is believed to be in the direction shown, that is, with the alkyl group releasing electrons to the doubly-bonded carbon. Since this polarity is not canceled by a corresponding polarity in the opposite direction, it gives a net dipole moment to the molecule.

$CH_3(H)C{=}CH_2$	$C_2H_5(H)C{=}CH_2$
$\mu = 0.35$ D	$\mu = 0.37$ D

cis-2-Butene, with two methyl groups on one side of the molecule and two hydrogens on the other, should have a small dipole moment. In *trans*-2-butene, on the other hand, with one methyl and one hydrogen on each side of the molecule, the bond moments should cancel out. Although the dipole moments have not

$CH_3(H)C{=}C(H)CH_3$	$CH_3(H)C{=}C(CH_3)H$
cis-2-Butene	*trans*-2-Butene
expect small $\leftrightarrow$	expect $\mu = 0$
b.p. +4°	b.p. +1°
m.p. −139°	m.p. −106°

been measured directly, a small difference in polarity is reflected in the higher boiling point of the *cis*-isomer.

This same relationship exists for many pairs of geometric isomers. Because of its higher polarity the *cis*-isomer is generally the higher boiling of a pair; because of its lower symmetry it fits into a crystalline lattice more poorly, and thus generally has the lower melting point.

The differences in polarity, and hence the differences in melting point and boiling point, are greater for alkenes that contain elements whose electronegativities differ widely from that of carbon. For example:

$H(Cl)C{=}C(Cl)H$	$H(Cl)C{=}C(H)Cl$	$H(Br)C{=}C(Br)H$	$H(Br)C{=}C(H)Br$	$H(I)C{=}C(I)H$	$H(I)C{=}C(H)I$
cis	*trans*	*cis*	*trans*	*cis*	*trans*
$\mu = 1.85$ D	$\mu = 0$	$\mu = 1.35$ D	$\mu = 0$	$\mu = 0.75$ D	$\mu = 0$
b.p. 60°	b.p. 48°	b.p. 110°	b.p. 108°	b.p. 188°	b.p. 192°
m.p. −80°	m.p. −50°	m.p. −53°	m.p. −6°	m.p. −14°	m.p. +72°

The relationship between configuration and boiling point or melting point is only a rule of thumb, to which there are many exceptions (for example, the boiling points of the diiodoethenes). Measurement of dipole moment, on the other hand, frequently enables us positively to designate a particular isomer as *cis* or *trans*.

Problem 5.4 (a) Indicate the direction of the net dipole moment for each of the dihaloethenes. (b) Would *cis*-2,3-dichloro-2-butene have a larger or smaller dipole moment than *cis*-1,2-dichloroethene? (c) Indicate the direction of the net dipole moment of *cis*-1,2-dibromo-1,2-dichloroethene. Will it be larger or smaller than the dipole moment of *cis*-1,2-dichloroethene? Why?

5.10 Industrial source

Alkenes are obtained in industrial quantities chiefly by the cracking of petroleum (Sec. 3.31). The smaller alkenes can be obtained in pure form by fractional distillation and are thus available for conversion into a large number of important aliphatic compounds. Higher alkenes, which cannot be separated from the complicated cracking mixture, remain as valuable components of gasoline.

1-Alkenes of even carbon number, consumed in large quantities in the manufacture of detergents, are available through controlled ionic polymerization of ethylene by the Ziegler-Natta method (Sec. 32.6).

5.11 Preparation

Alkenes containing up to five carbon atoms can be obtained in pure form from the petroleum industry. Pure samples of more complicated alkenes must be prepared by methods like those outlined below.

The introduction of a carbon–carbon double bond into a molecule containing only single bonds must necessarily involve the **elimination** of atoms or groups from two adjacent carbons:

$$-\underset{\text{Y}}{\overset{|}{\underset{|}{\text{C}}}}-\underset{\text{Z}}{\overset{|}{\underset{|}{\text{C}}}}- \longrightarrow -\overset{|}{\text{C}}=\overset{|}{\text{C}}- \qquad \textbf{Elimination}$$

In the cracking process already discussed, for example, the atoms eliminated are both hydrogen atoms:

$$-\underset{\text{H}}{\overset{|}{\underset{|}{\text{C}}}}-\underset{\text{H}}{\overset{|}{\underset{|}{\text{C}}}}- \xrightarrow{\text{heat}} -\overset{|}{\text{C}}=\overset{|}{\text{C}}- + H_2$$

The elimination reactions described below not only can be used to make simple alkenes, but also—and this is much more important—provide the best general ways to introduce carbon–carbon double bonds into molecules of all kinds.

PREPARATION OF ALKENES

1. Dehydrohalogenation of alkyl halides. Discussed in Sec. 5.12–5.14.

$$-\underset{\text{H}}{\overset{|}{\underset{|}{\text{C}}}}-\underset{\text{X}}{\overset{|}{\underset{|}{\text{C}}}}- + \text{KOH} \xrightarrow{\text{alcohol}} -\overset{|}{\text{C}}=\overset{|}{\text{C}}- + \text{KX} + H_2O$$

Ease of dehydrohalogenation of alkyl halides
3° > 2° > 1°

Examples:

$$\underset{n\text{-Butyl chloride}}{CH_3CH_2CH_2CH_2Cl} \xrightarrow{\text{KOH (alc)}} \underset{\text{1-Butene}}{CH_3CH_2CH{=}CH_2}$$

$$\underset{sec\text{-Butyl chloride}}{CH_3CH_2CHClCH_3} \xrightarrow{\text{KOH (alc)}} \underset{\substack{\text{2-Butene}\\ 80\%}}{CH_3CH{=}CHCH_3} + \underset{\substack{\text{1-Butene}\\ 20\%}}{CH_3CH_2CH{=}CH_2}$$

2. Dehydration of alcohols. Discussed in Sec. 5.19–5.23.

$$\underset{\text{Alcohols}}{-\overset{|}{\underset{\underset{H}{|}}{C}}-\overset{|}{\underset{\underset{OH}{|}}{C}}-} \xrightarrow{\text{acid}} \underset{\text{Alkenes}}{-\overset{|}{C}{=}\overset{|}{C}-} + H_2O$$

Ease of dehydration of alcohols
3° > 2° > 1°

Examples:

$$\underset{\text{Ethyl alcohol}}{H-\overset{\overset{H}{|}}{\underset{\underset{H}{|}}{C}}-\overset{\overset{H}{|}}{\underset{\underset{OH}{|}}{C}}-H} \xrightarrow{\text{acid}} \underset{\text{Ethylene}}{H-\overset{\overset{H}{|}}{C}{=}\overset{\overset{H}{|}}{C}-H} + H_2O$$

$$\underset{n\text{-Butyl alcohol}}{CH_3CH_2CH_2CH_2OH} \xrightarrow{\text{acid}} \underset{\text{1-Butene}}{CH_3CH_2CH{=}CH_2} + \underset{\substack{\text{2-Butene}\\ \textit{Chief product}}}{CH_3CH{=}CHCH_3}$$

$$\underset{sec\text{-Butyl alcohol}}{CH_3CH_2-\underset{\underset{OH}{|}}{CH}-CH_3} \xrightarrow{\text{acid}} \underset{\substack{\text{2-Butene}\\ \textit{Chief product}}}{CH_3CH{=}CHCH_3} + \underset{\text{1-Butene}}{CH_3CH_2CH{=}CH_2}$$

3. Dehalogenation of vicinal dihalides. Discussed in Sec. 5.11.

$$-\overset{|}{\underset{\underset{X}{|}}{C}}-\overset{|}{\underset{\underset{X}{|}}{C}}- + Zn \longrightarrow -\overset{|}{C}{=}\overset{|}{C}- + ZnX_2$$

Example:

$$\underset{\text{2,3-Dibromobutane}}{CH_3CHBrCHBrCH_3} \xrightarrow{\text{Zn}} \underset{\text{2-Butene}}{CH_3CH{=}CHCH_3}$$

4. Reduction of alkynes. Discussed in Sec. 8.9.

$$\underset{\text{An alkyne}}{R-C{\equiv}C-R} \xrightarrow[\text{Pd or Ni-B (P-2)}]{H_2} \begin{matrix} R & & R \\ & C{=}C & \\ H & & H \end{matrix} \quad \textit{Cis}$$

$$\underset{\text{An alkyne}}{R-C{\equiv}C-R} \xrightarrow{\text{Na or Li, } NH_3} \begin{matrix} R & & H \\ & C{=}C & \\ H & & R \end{matrix} \quad \textit{Trans}$$

The most important of these methods of preparation—since they are the most generally applicable—are the **dehydrohalogenation of alkyl halides** and the **dehydration of alcohols**. Both methods suffer from the disadvantage that, where the structure permits, hydrogen can be eliminated from the carbon on either side of the carbon bearing the —X or —OH; this frequently produces isomers. Since the isomerism usually involves only the position of the double bond, it is not important in the cases where we plan to convert the alkene into an alkane.

As we shall see later, alkyl halides are generally prepared from the corresponding alcohols, and hence both these methods ultimately involve preparation from alcohols; however, dehydrohalogenation generally leads to fewer complications and is often the preferred method despite the extra step in the sequence.

Dehalogenation of vicinal (Latin: *vicinalis*, neighboring) dihalides is severely limited by the fact that these dihalides are themselves generally prepared from the alkenes. However, it is sometimes useful to convert an alkene to a dihalide while we perform some operation on another part of the molecule, and then to regenerate the alkene by treatment with zinc; this procedure is referred to as *protecting the double bond.*

When a pure *cis*- or *trans*-alkene is wanted, uncontaminated with its stereoisomer, it can often be prepared by reduction of an alkyne with the proper reagent (Sec. 8.9).

5.12 Dehydrohalogenation of alkyl halides

Alkyl halides are converted into alkenes by **dehydrohalogenation**: *elimination of the elements of hydrogen halide.* Dehydrohalogenation involves removal of the halogen atom together with a hydrogen atom from a carbon adjacent to the one

Dehydrohalogenation: elimination of HX

$$\underset{\text{Alkyl halide}}{-\overset{|}{\underset{\underset{H}{|}}{C}}-\overset{|}{\underset{\underset{X}{|}}{C}}-} + \text{KOH (alcoholic)} \longrightarrow \underset{\text{Alkene}}{-\overset{|}{C}=\overset{|}{C}-} + KX + H_2O$$

bearing the halogen. It is not surprising that the reagent required for the elimination of what amounts to a molecule of acid is a strong base.

The alkene is prepared by simply heating together the alkyl halide and a solution of potassium hydroxide in alcohol. For example:

$$\underset{n\text{-Propyl chloride}}{CH_3CH_2CH_2Cl} \xrightarrow{\text{KOH (alc)}} \underset{\text{Propylene}}{CH_3CH{=}CH_2} \xleftarrow{\text{KOH (alc)}} \underset{\text{Isopropyl chloride}}{CH_3\underset{\underset{Cl}{|}}{C}HCH_3}$$

$$\underset{n\text{-Butyl chloride}}{CH_3CH_2CH_2CH_2Cl} \xrightarrow{\text{KOH (alc)}} \underset{\text{1-Butene}}{CH_3CH_2CH{=}CH_2}$$

$$\underset{sec\text{-Butyl chloride}}{CH_3CH_2\underset{\underset{Cl}{|}}{C}HCH_3} \xrightarrow{\text{KOH (alc)}} \underset{\substack{\text{2-Butene}\\ 80\%}}{CH_3CH{=}CHCH_3} + \underset{\substack{\text{1-Butene}\\ 20\%}}{CH_3CH_2CH{=}CH_2}$$

As we can see, in some cases this reaction yields a single alkene, and in other cases yields a mixture. *n*-Butyl chloride, for example, can eliminate hydrogen only from C–2 and hence yields only 1-butene. *sec*-Butyl chloride, on the other hand, can eliminate hydrogen from either C–1 or C–3 and hence yields both 1-butene and 2-butene. Where the two alkenes can be formed, 2-butene is the chief product; this fact fits into a general pattern for dehydrohalogenation which is discussed in Sec. 5.14.

Problem 5.5 Give structures of all alkenes expected from dehydrohalogenation of: (a) 1-chloropentane, (b) 2-chloropentane, (c) 3-chloropentane, (d) 2-chloro-2-methylbutane, (e) 3-chloro-2-methylbutane, (f) 2-chloro-2,3-dimethylbutane, (g) 1-chloro-2,2-dimethylpropane.

Problem 5.6 What alkyl halide (*if any*) would yield each of the following pure alkenes upon dehydrohalogenation? (a) isobutylene, (b) 1-pentene, (c) 2-pentene, (d) 2-methyl-1-butene, (e) 2-methyl-2-butene, (f) 3-methyl-1-butene.

5.13 Mechanism of dehydrohalogenation

The function of hydroxide ion is to pull a hydrogen ion away from carbon; simultaneously a halide ion separates and the double bond forms. We should

$$-\overset{|}{\underset{|}{C}}(:\ddot{X}:)\,:\,\overset{|}{C}(H)- \;+\; :\ddot{O}H^- \longrightarrow -\overset{|}{C}::\overset{|}{C}- + :\ddot{X}:^- + H_2O$$

represented as

$$-\overset{X}{\overset{|}{\underset{|}{C}}}-\underset{\underset{H}{|}}{\overset{|}{C}}- \;\; (OH^-) \longrightarrow \;\; \rangle C{=}C\langle \; + X^- + H_2O$$

where arrows show the direction of electron shift

notice that, in contrast to free radical reactions, the breaking of the C—H and C—X bonds occurs in an unsymmetrical fashion: hydrogen relinquishes *both* electrons to carbon, and halogen retains *both* electrons. The electrons left behind by hydrogen are now available for formation of the second bond (the π bond) between the carbon atoms.

What supplies the energy for the breaking of the carbon–hydrogen and carbon–halogen bonds?

(a) First, there is formation of the bond between the hydrogen ion and the very strong base, hydroxide ion.

(b) Next, there is formation of the π bond which, although weak, does supply about 70 kcal/mole of energy.

(c) Finally—and this is extremely important—there is the energy of solvation of the halide ions. Alcohol, like water, is a polar solvent. A liberated halide ion is surrounded by a cluster of these polar molecules; each solvent molecule is oriented so that the positive end of its dipole is near the negative ion (Fig. 5.6).

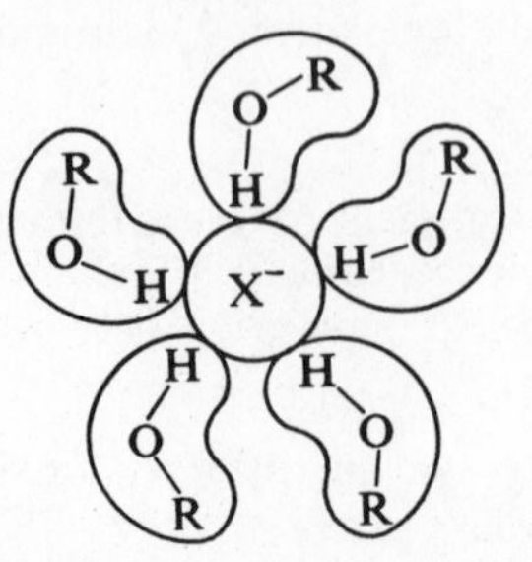

Figure 5.6. Ion-dipole interaction: solvated halide ion.

Although each of these *ion–dipole bonds* (Sec. 1.21) is weak, in the aggregate they supply a great deal of energy. (We should recall that the ion–dipole bonds in hydrated sodium and chloride ions provide the energy for the breaking down of the sodium chloride crystalline lattice, a process which in the absence of water requires a temperature of 801°.) *Just as a hydrogen ion is pulled out of the molecule by a hydroxide ion, so a halide ion is pulled out by solvent molecules.*

The free-radical reactions of the alkanes, which we studied in Chap. 3, are chiefly gas phase reactions. It is significant that ionic reactions (like the one just discussed) occur chiefly in solution.

(We shall return to dehydrohalogenation after we have learned a little more chemistry (Sec. 14.20), and have a look at the *evidence* for this mechanism.)

5.14 Orientation and reactivity in dehydrohalogenation

In cases where a mixture of isomeric alkenes can be formed, which isomer, if any, will predominate? Study of many dehydrohalogenation reactions has shown that one isomer generally does predominate, and that it is possible to predict which isomer this will be—that is, to predict the *orientation* of elimination—on the basis of molecular structure.

$$CH_3CH_2CHBrCH_3 \xrightarrow{\text{KOH (alc)}} \underset{81\%}{CH_3CH{=}CHCH_3} \text{ and } \underset{19\%}{CH_3CH_2CH{=}CH_2}$$

$$CH_3CH_2CH_2CHBrCH_3 \xrightarrow{\text{KOH (alc)}} \underset{71\%}{CH_3CH_2CH{=}CHCH_3} \text{ and } \underset{29\%}{CH_3CH_2CH_2CH{=}CH_2}$$

$$CH_3CH_2\overset{CH_3}{\overset{|}{C}}BrCH_3 \xrightarrow{\text{KOH (alc)}} \underset{71\%}{CH_3CH{=}\overset{CH_3}{\overset{|}{C}}CH_3} \text{ and } \underset{29\%}{CH_3CH_2\overset{CH_3}{\overset{|}{C}}{=}CH_2}$$

Once more, orientation is determined by the relative rates of competing reactions. For *sec*-butyl bromide, attack by base at any one of three hydrogens (those on C–1) can lead to the formation of 1-butene; attack at either of two hydrogens (on C–3) can lead to the formation of 2-butene. We see that 2-butene is the preferred product—that is, is formed faster—*despite* a probability factor of 3:2 working against its formation. The other examples fit the same pattern: the preferred product is the alkene that has the greater number of alkyl groups attached to the doubly-bonded carbon atoms.

Ease of formation of alkenes

$$R_2C{=}CR_2 > R_2C{=}CHR > R_2C{=}CH_2,\ RCH{=}CHR > RCH{=}CH_2$$

In Sec. 6.4 we shall find evidence that the stability of alkenes follows exactly the same sequence.

Stability of alkenes

$$R_2C{=}CR_2 > R_2C{=}CHR > R_2C{=}CH_2,\ RCH{=}CHR > RCH{=}CH_2 > CH_2{=}CH_2$$

In dehydrohalogenation, the more stable the alkene the more easily it is formed.

Examination of the transition state involved shows that it is reasonable that the more stable alkene should be formed faster:

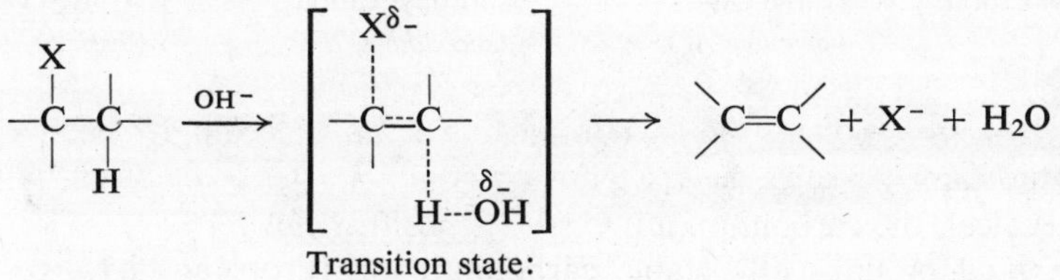

$$-\overset{X}{\underset{|}{\overset{|}{C}}}-\underset{H}{\underset{|}{\overset{|}{C}}}- \xrightarrow{OH^-} \left[-\overset{X^{\delta-}}{\overset{\vdots}{C}}\!\cdots\!\underset{H\cdots\overset{\delta-}{OH}}{\underset{\vdots}{C}}- \right] \longrightarrow \;\rangle C{=}C\langle\; + X^- + H_2O$$

Transition state:
partly formed double bond

The double bond is partly formed, and the transition state has thus acquired alkene character. Factors that stabilize an alkene also stabilize an *incipient* alkene in the transition state.

Alkene stability not only determines *orientation* of dehydrohalogenation, but also is an important factor in determining the *reactivity* of an alkyl halide toward elimination, as shown below.

Reactant	⟶	*Product*	*Relative rates*	*Relative rates per* H
CH_3CH_2Br	⟶	$CH_2{=}CH_2$	1.0	1.0
$CH_3CH_2CH_2Br$	⟶	$CH_3CH{=}CH_2$	3.3	5.0
$CH_3CHBrCH_3$	⟶	$CH_3CH{=}CH_2$	9.4	4.7
$(CH_3)_3CBr$	⟶	$(CH_3)_2C{=}CH_2$	120	40

As one proceeds along a series of alkyl halides from 1° to 2° to 3°, the structure by definition becomes more branched at the carbon carrying the halogen. This increased branching has two results: it provides a greater number of hydrogens for attack by base, and hence a more favorable probability factor toward elimination; and it leads to a more highly branched, more stable alkene, and hence a more stable transition state and lower E_{act}. As a result of this combination of factors, **in dehydrohalogenation the order of reactivity of RX is 3° > 2° > 1°.**

Problem 5.7 Predict the major product of each dehydrohalogenation in Problem 5.5, page 157.

5.15 Carbonium ions

To account for the observed facts, we saw earlier, a certain mechanism was advanced for the halogenation of alkanes; the heart of this mechanism is the fleeting existence of free radicals, highly reactive neutral particles bearing an odd electron.

Before we can discuss the preparation of alkenes by dehydration of alcohols, we must first learn something about another kind of reactive particle: the **carbonium ion**, *a group of atoms that contains a carbon atom bearing only six electrons.* Carbonium ions are classified as primary, secondary, or tertiary after the carbon bearing the positive charge. They are named by use of the word *cation.* For example:

H	H	H	$\mathrm{CH_3}$
$\mathrm{H{:}\ddot{C}\oplus}$	$\mathrm{CH_3{:}\ddot{C}\oplus}$	$\mathrm{CH_3{:}\ddot{C}{:}CH_3}$	$\mathrm{CH_3{:}\ddot{C}{:}CH_3}$
$\mathrm{\ddot{H}}$	$\mathrm{\ddot{H}}$	$\oplus$	$\oplus$
Methyl cation	Ethyl cation (*primary*, 1°)	Isopropyl cation (*secondary*, 2°)	*tert*-Butyl cation (*tertiary*, 3°)

Like the free radical, the carbonium ion is an exceedingly reactive particle, and for the same reason: the tendency to complete the octet of carbon. Unlike the free radical, the carbonium ion carries a positive charge.

One kind of unusually stable carbonium ion (Problem 12.12, p. 398) was recognized as early as 1902 by the salt-like character of certain organic compounds. But direct observation of simple alkyl cations should be exceedingly difficult, by virtue of the very reactivity—and hence short life—that we attribute to them. Nevertheless, during the 1920's and 1930's, alkyl cations were proposed as intermediates in many organic reactions, and their existence was generally accepted, due largely to the work of three chemists: Hans Meerwein of Germany, "the father of modern carbonium ion chemistry;" Sir Christopher Ingold of England; and Frank Whitmore of the United States. The evidence consisted of a wide variety of observations made in studying the chemistry of alkenes, alcohols, alkyl halides, and many other kinds of organic compounds: observations that revealed a basically similar pattern of behavior most logically attributed to intermediate carbonium ions. A sizable part of this book will be devoted to seeing what that pattern is.

In 1963, George Olah (now at Case Western Reserve University) reported the *direct observation* of simple alkyl cations. Dissolved in the extremely powerful Lewis acid SbF_5, alkyl fluorides (and, later, other halides) were found to undergo ionization to form the cation, which could be studied at leisure. There was a

$$\mathrm{RF + SbF_5 \longrightarrow R^+\ SbF_6^-}$$

dramatic change in the nmr spectrum (Chap. 13), from the spectrum of the alkyl fluoride to the spectrum of a molecule that contained no fluorine but instead sp^2-hybridized carbon with a very low electron density. Figure 5.7 shows what was observed for the *tert*-butyl fluoride system: a simple spectrum but, by its very simplicity, enormously significant. Although potentially very reactive, the *tert*-butyl cation can do little in this environment except try to regain the fluoride ion—and the SbF_5 is an even stronger Lewis acid than the cation.

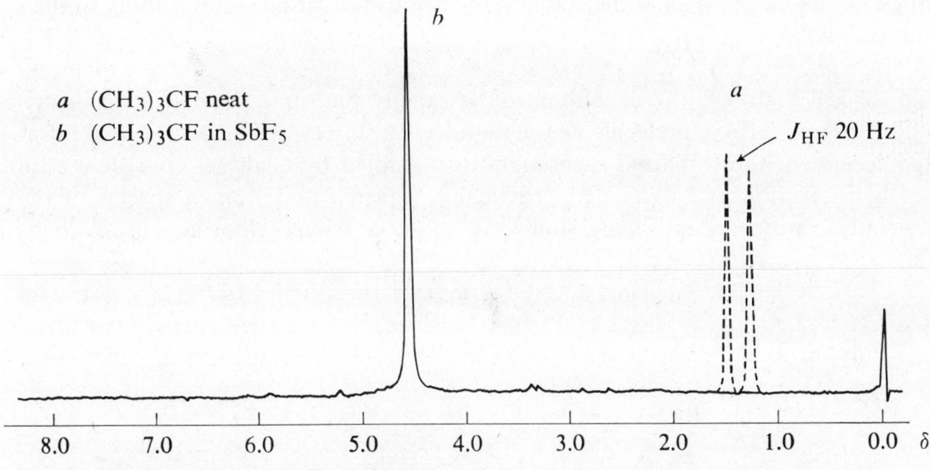

Courtesy of *The Journal of American Chemical Society*

Figure 5.7. Proton nmr spectrum of (*a*) *tert*-butyl fluoride and (*b*) *tert*-butyl cation. In (*a*), proton signal split into two peaks by coupling with nearby fluorine. In (*b*), single peak, shifted far downfield; strong deshielding due to low electron density on positive carbon.

By methods like this, Olah has opened the door to the study not just of the existence of organic cations of many kinds, but of intimate details of their structure.

5.16 Structure of carbonium ions

In a carbonium ion, the electron-deficient carbon is bonded to three other atoms, and for this bonding uses sp^2 orbitals; the bonds are trigonal, directed to the corners of an equilateral triangle. This part of a carbonium ion is therefore

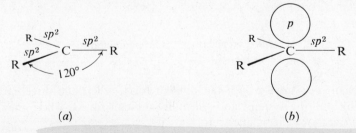

Figure 5.8. A carbonium ion. (*a*) Only σ bonds shown. (*b*) Empty *p* orbital above and below plane of σ bonds.

flat, the electron-deficient carbon and the three atoms attached to it lying in the same plane (Fig. 5.8*a*).

But our description of the molecule is not yet quite complete. Carbon has left a *p* orbital, with its two lobes lying above and below the plane of the σ bonds (Fig. 5.8*b*); in a carbonium ion, the *p* orbital is *empty*. Although formally empty, this *p* orbital, we shall find, is intimately involved in the chemistry of carbonium ions: in their stability, and the stability of various transition states leading to their formation.

There can be little doubt that carbonium ions actually are flat. The quantum mechanical picture of a carbonium ion is exactly the same as that of boron trifluoride (Sec. 1.10), a molecule whose flatness is firmly established. Nmr and infrared spectra of the stabilized carbonium ions studied by Olah are consistent with sp^2 hybridization and flatness: in particular, infrared and Raman spectra of the *tert*-butyl cation are strikingly similar to those of trimethylboron, known to be flat.

Evidence of another kind indicates that carbonium ions not only normally *are* flat, but have a strong *need to be* flat. Consider the three tertiary alkyl bromides: *tert*-butyl bromide; and I and II, which are *bicyclic* (two-ringed) compounds with

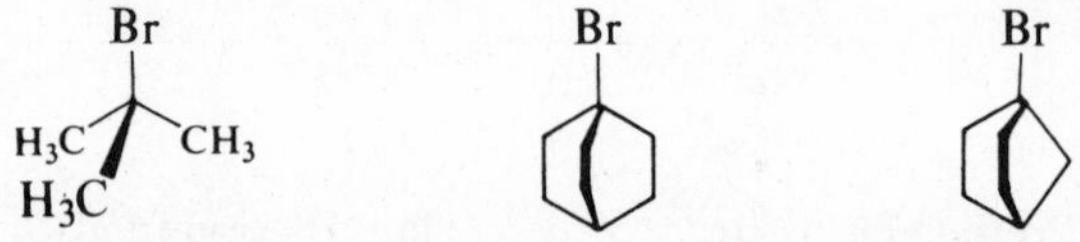

Br, H_3C, CH_3, H_3C — *tert*-Butyl bromide

Br — I — 1-Bromobicyclo-[2.2.2]octane

Br — II — 1-Bromobicyclo-[2.2.1]heptane

bromine at the bridgehead. The impact of a high-energy electron can remove bromine from an alkyl bromide and generate a carbonium ion; the energy of the electron required to do the job can be measured. On electron impact, I requires 5 kcal/mole more energy to form the carbonium ion than does *tert*-butyl bromide, and II requires 20 kcal/mole more energy.

How are we to interpret these facts? On conversion into a carbonium ion, three carbons must move into the plane of the electron-deficient carbon: easy for

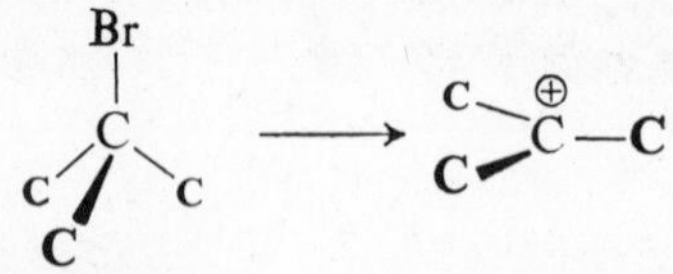

Br, C, C, C, C $\longrightarrow$ C, $\oplus$C—C, C

Alkyl bromide
Tetrahedral

Carbonium ion
Trigonal

the open-chain *tert*-butyl group; but difficult for I, where the three carbons are tied back by the ring system; and still more difficult for II, where they are tied back more tightly by the smaller ring.

Imagine—or, better, *make*—a model of I or II. You could squash the top of the molecule flat, but only by distorting the angles of the other bonds away from their normal tetrahedral angle, and thus introducing *angle strain* (Sec. 9.7).

Now, why is there this need to be flat? Partly, to permit formation of the strongest possible σ bonds through sp^2 hybridization. But there is a second advantage of flatness, one which is related to the major factor determining carbonium ion stability, *accommodation of charge*.

5.17 Stability of carbonium ions. Accommodation of charge

The characteristic feature of a carbonium ion is, by definition, the electron-deficient carbon and the attendant positive charge. The relative stability of a carbonium ion is determined chiefly by how well it *accommodates* that charge.

According to the laws of electrostatics, **the stability of a charged system is increased by dispersal of the charge**. Any factor, therefore, that tends to spread out the positive charge of the electron-deficient carbon and distribute it over the rest of the ion must stabilize a carbonium ion.

Consider a substituent, G, attached to an electron-deficient carbon in place of a hydrogen atom. Compared with hydrogen, G may either release electrons or withdraw electrons (Sec. 1.23).

Carbonium Ion Stability

$G\rightarrow C^{\oplus}$	$G\leftarrow C^{\oplus}$
G releases electrons: *disperses charge,* *stabilizes cation*	G withdraws electrons: *intensifies charge,* *destabilizes cation*

An electron-releasing substituent tends to reduce the positive charge at the electron-deficient carbon; in doing this, the substituent itself becomes somewhat positive. This dispersal of the charge stabilizes the carbonium ion.

An electron-withdrawing substituent tends to intensify the positive charge on the electron-deficient carbon, and hence makes the carbonium ion less stable.

We consider (Sec. 1.23) electronic effects to be of two kinds: *inductive effects*, related to the electronegativity of substituents; and *resonance effects*. In the case of carbonium ions, we shall see (Sec. 8.21), a resonance effect involves overlap of the "empty" *p* orbital of the electron-deficient carbon with orbitals on other, nearby atoms; the result is, of course, that the *p* orbital is no longer empty, and the electron-deficient carbon no longer so positive. Maximum overlap depends on coplanarity in this part of the molecule, and it is here that we find the second advantage of flatness in a carbonium ion.

So far, we have discussed only factors operating *within* a carbonium ion to make it more or less stable than another carbonium ion. But what is *outside* the carbonium ion proper—its environment—can be even more important in determining how fast a carbonium ion is formed, how long it lasts, and what happens to it. There are *anions*, one of which may stay close by to form an ion pair. There is the *solvent*: a cluster of solvent molecules, each with the negative end of its dipole turned toward the cation; possibly one solvent molecule—or two—playing a special role through overlap of one or both lobes of the *p* orbital. There may be a *neighboring group effect* (Chap. 28), in which a substituent on a neighboring carbon approaches closely enough to share its electrons and form a covalent bond: an internal factor, actually, but in its operation much like an external factor.

In all this we see the characteristic of carbonium ions that underlies their whole pattern of behavior: a *need for electrons to complete the octet of carbon.*

5.18 Relative stabilities of alkyl cations

The amount of energy required to remove an electron from a molecule or atom is called the *ionization potential.* (It is really the ionization *energy.*) The ionization potential of a free radical is, by definition, the ΔH for the conversion of the radical into a carbonium ion:

$$\text{R}\cdot \longrightarrow \text{R}^+ + e^- \qquad \Delta H = \text{ionization potential}$$

In ways that we cannot go into, the ionization potentials of many free radicals have been measured. For example:

$$CH_3\cdot \longrightarrow CH_3^+ + e^- \qquad \Delta H = 229 \text{ kcal/mole}$$

$$CH_3CH_2\cdot \longrightarrow CH_3CH_2^+ + e^- \qquad \Delta H = 202$$

$$CH_3\underset{\bullet}{C}HCH_3 \longrightarrow CH_3\underset{+}{C}HCH_3 + e^- \qquad \Delta H = 182$$

$$CH_3-\underset{\bullet}{\overset{CH_3}{\overset{|}{C}}}-CH_3 \longrightarrow CH_3-\underset{+}{\overset{CH_3}{\overset{|}{C}}}-CH_3 + e^- \qquad \Delta H = 171$$

As we can see, the values decrease in the order: $CH_3\cdot > 1° > 2° > 3°$.

Bond dissociation energies have already shown (Sec. 3.24) that the amount of energy required to form free radicals from alkanes decreases in the same order: $CH_3\cdot > 1° > 2° > 3°$. If we combine these two sets of data—ionization potentials and bond dissociation energies—we see (Fig. 5.9) that, relative to the various alkanes concerned, the order of stability of carbonium ions is:

Stability of carbonium ions $3° > 2° > 1° > CH_3^+$

Differences in stability between carbonium ions are much larger than between free radicals. The *tert*-butyl free radical, for example, is 13 kcal more stable than the methyl free radical; the *tert*-butyl cation is 71 kcal more stable than the methyl cation.

What we really want as standards for stability of carbonium ions are, of course, the kinds of compounds they are generated from: alcohols at this particular point or, later, alkyl halides (Chap. 14). However, the relative stabilities of most ordinary neutral molecules closely parallel the relative stabilities of the alkanes, so that the relative order of stabilities that we have arrived at is certainly valid whatever the source of the carbonium ions. To take an extreme example, the difference in stability between methyl and *tert*-butyl cations relative to the alkanes, as we have just calculated it, is 71 kcal. Relative to other standards, the difference in stability is: alcohols, 57 kcal; chlorides, 74 kcal; bromides, 78 kcal; and iodides, 76 kcal.

Now, by definition, the distinction among primary, secondary, and tertiary cations is the number of alkyl groups attached to the electron-deficient carbon. The facts are, then, *that the greater the number of alkyl groups, the more stable the carbonium ion.*

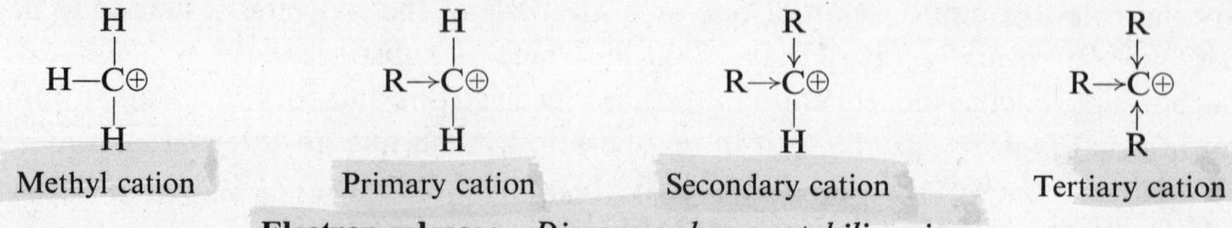

Electron release: *Disperses charge, stabilizes ion*

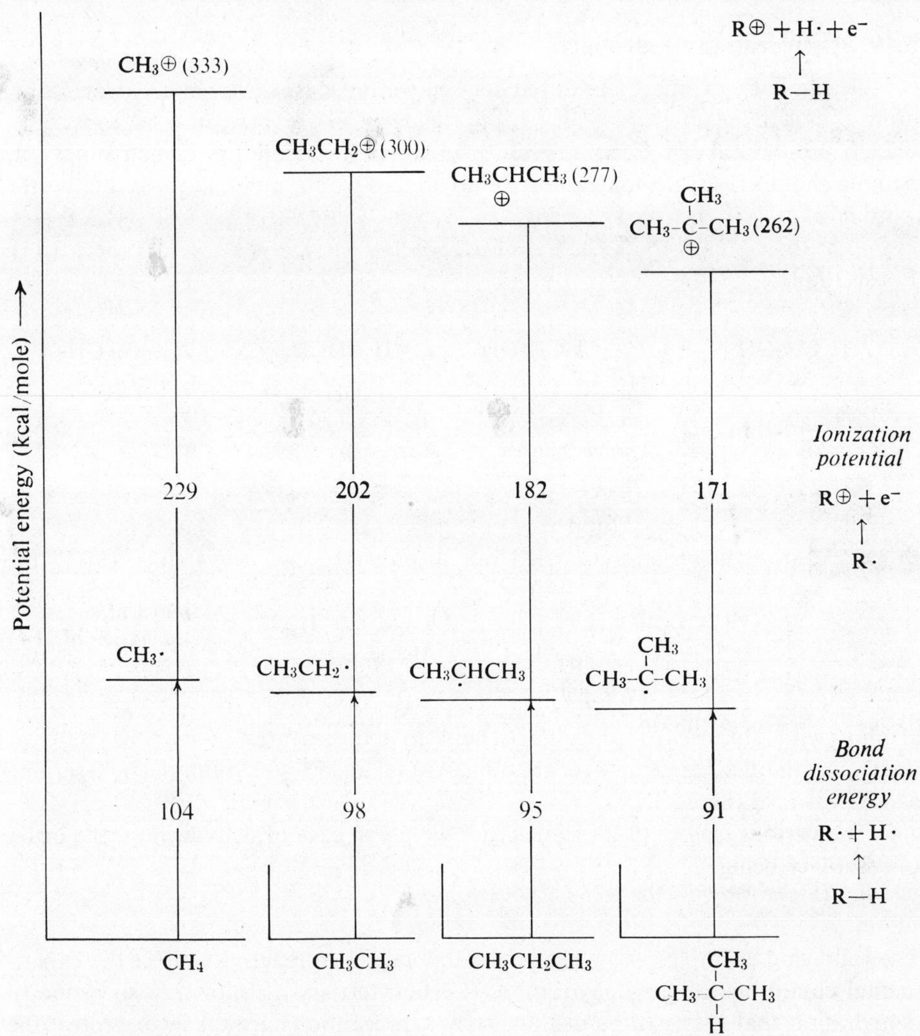

Figure 5.9. Relative stabilities of alkyl cations. (Plots aligned with each other for easy comparison.)

If our generalization about dispersal of charge applies in this case, alkyl groups must *release electrons* here: possibly through an inductive effect, possibly through resonance (*hyperconjugation*, Sec. 8.21).

The electronic effects of alkyl groups are the most poorly understood of all such effects. Dipole moments indicate that they are electron-releasing when attached to a π electron system, but weakly electron-withdrawing in saturated hydrocarbons. Ingold (p. 160) has suggested that alkyl groups can do pretty much what is demanded of them by other groups in the molecule. This much seems clear: (a) in general, carbonium ions and incipient carbonium ions are stabilized by electron release; (b) alkyl groups stabilize carbonium ions; (c) alkyl groups affect a wide variety of reactions in a manner consistent with their being electron-releasing.

5.19 Dehydration of alcohols

Alcohols are compounds of the general formula, ROH, where R is any alkyl group; the hydroxyl group, —OH, is characteristic of alcohols, just as the carbon–carbon double bond is characteristic of alkenes. An alcohol is named simply by naming the alkyl group that holds the hydroxyl group and following this by the word *alcohol*. It is classified as *primary* (1°), *secondary* (2°), or *tertiary* (3°), depending upon the nature of the carbon atom holding the hydroxyl group (Sec. 3.11). For example:

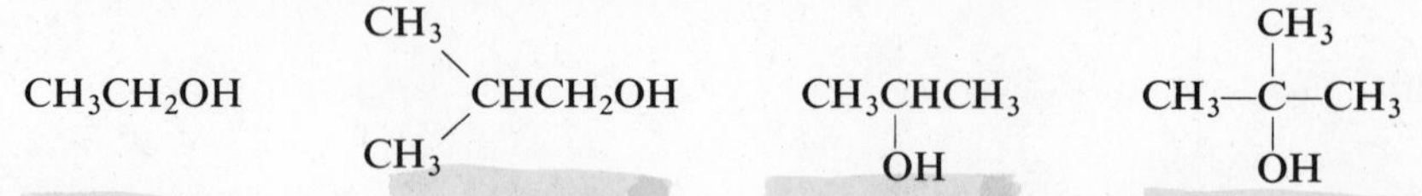

CH_3CH_2OH	$(CH_3)_2CHCH_2OH$	$CH_3CH(OH)CH_3$	$CH_3-C(CH_3)(OH)-CH_3$
Ethyl alcohol	Isobutyl alcohol	Isopropyl alcohol	*tert*-Butyl alcohol
A primary alcohol	*A primary alcohol*	*A secondary alcohol*	*A tertiary alcohol*

An alcohol is converted into an alkene by **dehydration**: *elimination of a molecule of water*. Dehydration requires the presence of an acid and the application of heat. It is generally carried out in either of two ways: (a) heating the

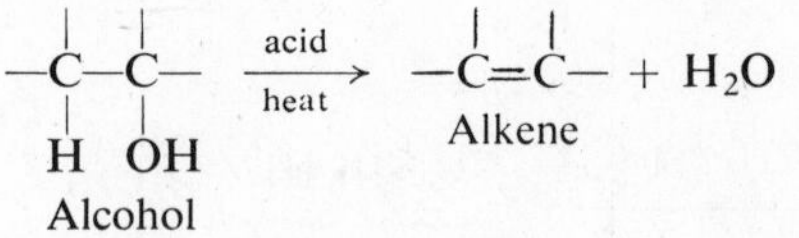

$$\underset{\text{Alcohol}}{\underset{\text{H}\quad\text{OH}}{-\overset{|}{\underset{|}{C}}-\overset{|}{\underset{|}{C}}-}} \xrightarrow[\text{heat}]{\text{acid}} \underset{\text{Alkene}}{-\overset{|}{C}=\overset{|}{C}-} + H_2O$$

Dehydration: elimination of H_2O

alcohol with sulfuric or phosphoric acid to temperatures as high as 200°, or (b) passing the alcohol vapor over alumina, Al_2O_3, at 350–400°, alumina here serving as a Lewis acid (Sec. 1.22).

The various classes of alcohols differ widely in ease of dehydration, the order of reactivity being

Ease of dehydration of alcohols $3° > 2° > 1°$

The following examples show how these differences in reactivity affect the experimental conditions of the dehydration. (Certain tertiary alcohols are so prone to dehydration that they can be distilled only if precautions are taken to protect the system from the acid fumes in the ordinary laboratory.)

$$\underset{\text{Ethyl alcohol}}{CH_3CH_2OH} \xrightarrow[170°]{95\%\ H_2SO_4} \underset{\text{Ethylene}}{CH_2{=}CH_2}$$

$$\underset{n\text{-Butyl alcohol}}{CH_3CH_2CH_2CH_2OH} \xrightarrow[140°]{75\%\ H_2SO_4} \underset{\substack{\text{2-Butene}\\ \textit{Chief product}}}{CH_3CH{=}CHCH_3}$$

$$\underset{sec\text{-Butyl alcohol}}{CH_3CH_2CHOHCH_3} \xrightarrow[100°]{60\%\ H_2SO_4} \underset{\substack{\text{2-Butene}\\ \textit{Chief product}}}{CH_3CH{=}CHCH_3}$$

$$\underset{tert\text{-Butyl alcohol}}{CH_3-\overset{\overset{CH_3}{|}}{\underset{\underset{OH}{|}}{C}}-CH_3} \xrightarrow[85\text{–}90°]{20\%\ H_2SO_4} \underset{\text{Isobutylene}}{CH_3-\overset{\overset{CH_3}{|}}{C}{=}CH_2}$$

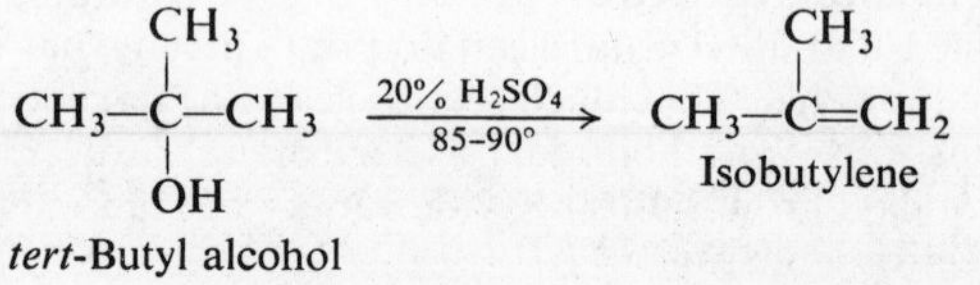

Where isomeric alkenes can be formed, we again find the tendency for one isomer to predominate. Thus, *sec*-butyl alcohol, which might yield both 2-butene and 1-butene, actually yields almost exclusively the 2-isomer (see Sec. 5.23).

The formation of 2-butene from *n*-butyl alcohol illustrates a characteristic of dehydration that is not shared by dehydrohalogenation: the double bond can be formed at a position remote from the carbon originally holding the —OH group. This characteristic is accounted for later (Sec. 5.22). It is chiefly because of the greater certainty as to where the double bond will appear that dehydrohalogenation is often preferred over dehydration as a method of making alkenes.

5.20 Mechanism of dehydration of alcohols

The generally accepted mechanism for the dehydration of alcohols is summarized in the following equations; for the sake of simplicity, ethyl alcohol is used as the example. The alcohol unites (step 1) with a hydrogen ion to form the protonated alcohol, which dissociates (step 2) into water and a carbonium ion; the carbonium ion then loses (step 3) a hydrogen ion to form the alkene.

(1)
$$\begin{array}{cccc} & \text{H} & & \text{H} \\ & | & & | \\ \text{H—} & \text{C} & : & \text{C—H} \\ & \ddot{} & & \ddot{} \\ & \text{H} & & :\ddot{\text{O}}:\text{H} \end{array} + \text{H}^+ \rightleftharpoons \begin{array}{cccc} & \text{H} & & \text{H} \\ & | & & | \\ \text{H—} & \text{C} & : & \text{C—H} \\ & \ddot{} & & \ddot{} \\ & \text{H} & & :\text{O}:\text{H} \\ & & & \ddot{}\ \oplus \\ & & & \text{H} \end{array}$$

Alcohol — Protonated alcohol

(2)
$$\begin{array}{cccc} & \text{H} & & \text{H} \\ & | & & | \\ \text{H—} & \text{C} & : & \text{C—H} \\ & \ddot{} & & \ddot{} \\ & \text{H} & & :\text{O}:\text{H} \\ & & & \ddot{\text{H}}\ \oplus \end{array} \rightleftharpoons \begin{array}{cccc} & \text{H} & & \text{H} \\ & | & & | \\ \text{H—} & \text{C} & : & \text{C—H} \\ & \ddot{} & & \oplus \\ & \text{H} & & \end{array} + \text{H}:\ddot{\underset{\cdot\cdot}{\text{O}}}: \overset{\text{H}}{}$$

Carbonium ion

(3)
$$\begin{array}{cccc} & \text{H} & & \text{H} \\ & | & & | \\ \text{H—} & \text{C} & : & \text{C—H} \\ & \cdot\cdot & & \oplus \\ & \text{H} & & \end{array} \rightleftharpoons \begin{array}{cccc} & \text{H} & & \text{H} \\ & | & & | \\ \text{H—} & \text{C} & :: & \text{C—H} \end{array} + \text{H}^+$$

Alkene

The double bond is thus formed in two stages, —OH being lost (as H_2O) in step (2) and —H being lost in step (3). This is in contrast to dehydrohalogenation (Sec. 5.13), where the halogen and hydrogen are lost simultaneously.

The first step is simply an acid–base equilibrium in the Lowry-Brønsted sense (Sec. 1.22). When sulfuric acid, for example, is dissolved in water, the following reaction occurs:

$$\text{H}:\underset{\cdot\cdot}{\overset{\text{H}}{\ddot{\text{O}}}}: + \text{HO—}\underset{\text{O}}{\overset{\text{O}}{\underset{|}{\overset{|}{\text{S}}}}}\text{—OH} \rightleftharpoons \text{H}:\underset{\oplus}{\overset{\text{H}}{\ddot{\text{O}}}}:\text{H} + \text{HSO}_4^-$$

Stronger base — Weaker base

The proton held by the very weak base, HSO_4^-, is transferred to the stronger base, H_2O, to form the oxonium ion, H_3O^+; the basic properties of each are due of course, to the unshared electrons that are available for sharing with the hydrogen ion. An alcohol also contains an oxygen atom with unshared electrons and hence displays basicity comparable to that of water. The first step of the mechanism is more properly represented as

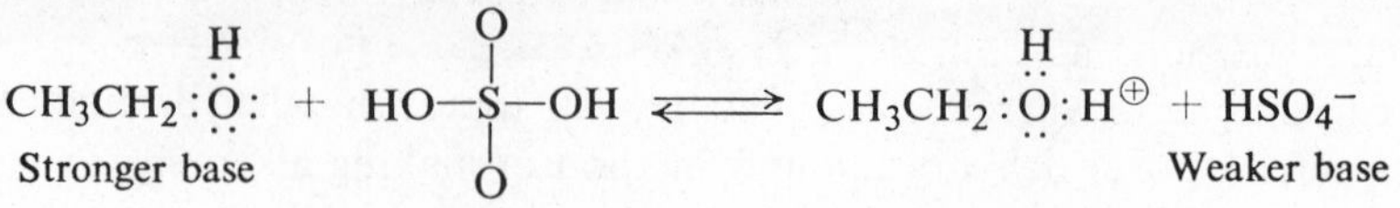

$$\underset{\text{Stronger base}}{CH_3CH_2:\overset{H}{\ddot{O}}:} + HO-\overset{O}{\underset{O}{S}}-OH \rightleftharpoons CH_3CH_2:\overset{H}{\ddot{O}}:H^{\oplus} + \underset{\text{Weaker base}}{HSO_4^-}$$

where the proton held by the bisulfate ion is transferred to the stronger base, ethyl alcohol, to yield the substituted oxonium ion, $C_2H_5OH_2^+$, the protonated alcohol.

In a similar way, step (3) does not actually involve the expulsion of a naked hydrogen ion, but rather a transfer of the hydrogen ion to a base, the strongest one around, C_2H_5OH.

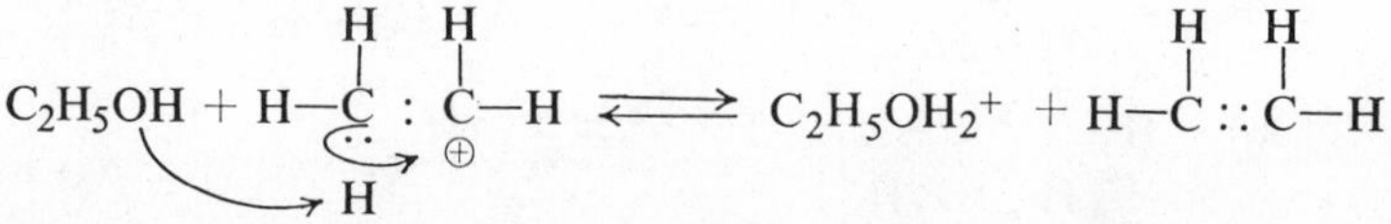

$$C_2H_5OH + H-\overset{H}{\underset{H}{C}}:\overset{H}{\underset{\oplus}{C}}-H \rightleftharpoons C_2H_5OH_2^+ + H-\overset{H}{C}::\overset{H}{C}-H$$

For convenience we shall frequently show the addition or expulsion of a hydrogen ion, H^+, but it should be understood that in all cases this actually involves the transfer of a proton from one base to another.

In summary, the mechanism for dehydration is the following:

Dehydration

(1) 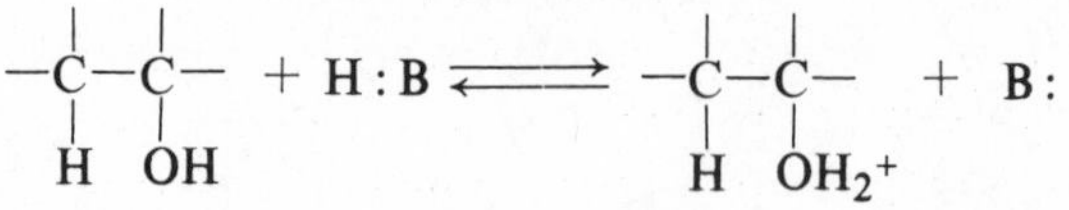 $$-\underset{H}{\overset{|}{C}}-\underset{OH}{\overset{|}{C}}- + H:B \rightleftharpoons -\underset{H}{\overset{|}{C}}-\underset{OH_2^+}{\overset{|}{C}}- + B:$$

(2) $$-\underset{H}{\overset{|}{C}}-\underset{OH_2^+}{\overset{|}{C}}- \rightleftharpoons -\underset{H}{\overset{|}{C}}-\underset{\oplus}{\overset{|}{C}}- + H_2O$$

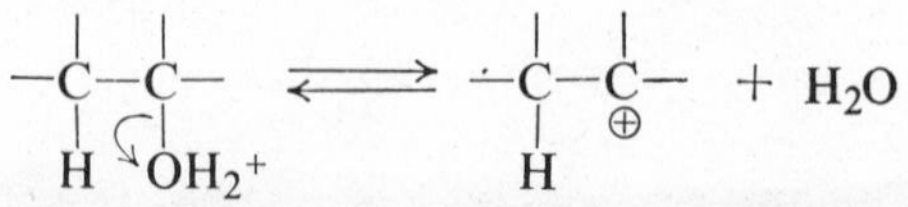

(3) 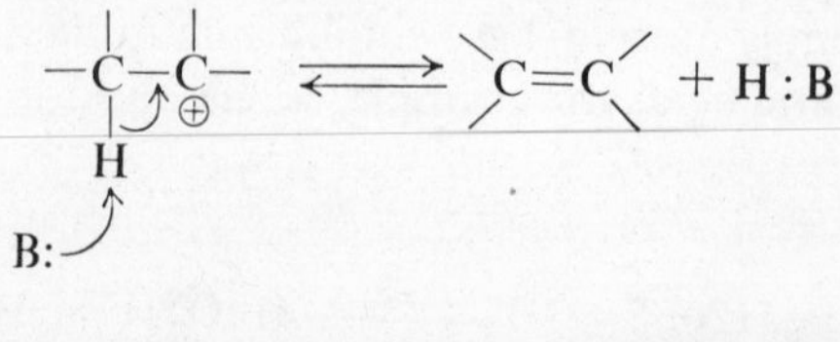 $$-\underset{\underset{B:}{H}}{\overset{|}{C}}-\underset{\oplus}{\overset{|}{C}}- \rightleftharpoons \;\rangle C=C\langle\; + H:B$$

All three reactions are shown as equilibria, since each step is readily reversible; as we shall soon see, the exact reverse of this reaction sequence is involved in the formation of alcohols from alkenes (Sec. 6.10). Equilibrium (1) lies very far to the right; sulfuric acid, for example, is known to be nearly completely ionized in alcohol solution. Since there is a very low concentration of carbonium ions present at any time, equilibrium (2) undoubtedly lies very far to the left. Occasionally one of these few carbonium ions undergoes reaction (3) to form the alkene. Under the conditions of dehydration the alkene, being quite volatile, is generally driven from the reaction mixture, and thus equilibrium (3) is shifted to the right. As a consequence the entire reaction system is forced toward completion.

The carbonium ion is formed by dissociation of the protonated alcohol; this involves separation of a charged particle, R^+, from a neutral particle, H_2O. It is obvious that this process requires much less energy than would formation of a carbonium ion from the alcohol itself, since the latter process involves separation of a positive particle from a negative particle. Viewed in another way, the carbonium ion (a Lewis acid) releases the weak base, water, much more readily than it

$$ROH_2^{\oplus} \longrightarrow R^{\oplus} + H_2O$$

Weak base: *good leaving group* — **Easy**

$$ROH \longrightarrow R^{\oplus} + OH^-$$

Strong base: *poor leaving group* — **Difficult**

releases the extremely strong base, hydroxide ion; that is to say, water is a much better *leaving group* than hydroxide ion. Indeed, the evidence indicates that separation of a hydroxide ion from an alcohol almost never occurs; reactions involving cleavage of the C—O bond of an alcohol seem in nearly every case to require an acidic catalyst, the function of which, as in the present case, is to form the protonated alcohol.

Finally, we must realize that even dissociation of the protonated alcohol is made possible only by solvation of the carbonium ion (compare Sec. 5.13). Energy for the breaking of the carbon–oxygen bond is supplied by the formation of many ion–dipole bonds between the carbonium ion and the polar solvent.

As we shall see, a carbonium ion can undergo a number of different reactions; just which one occurs depends upon experimental conditions. All reactions of a carbonium ion have a common end: *they provide a pair of electrons to complete the octet of the positively charged carbon.* In the present case a hydrogen ion is eliminated from the carbon adjacent to the positive, electron-deficient carbon; the pair of electrons formerly shared by this hydrogen are available for formation of a π bond.

$$-\overset{|}{\underset{\underset{H}{|}}{C}}-\overset{|}{\underset{\oplus}{C}}- \longrightarrow \;>C=C<\; + H{:}B$$

B: (base removes H)

We can see how the mechanism accounts for the fact that dehydration is catalyzed by acids. Now let us see how the mechanism also accounts for the fact that the ease with which alcohols undergo dehydration follows the sequence $3° > 2° > 1°$.

Problem 5.8 According to the **principle of microscopic reversibility**, a reaction and its reverse follow exactly the same path but in opposite directions. On this basis write a detailed mechanism for the *hydration of alkenes*, a reaction that is the exact reverse of the dehydration of alcohols. (Check your answer in Sec. 6.10.)

5.21 Ease of formation of carbonium ions

The ease with which alcohols undergo dehydration follows the sequence 3° > 2° > 1°. There is evidence that a controlling factor in dehydration is the formation of the carbonium ion, and that one alcohol is dehydrated more easily than another chiefly because it forms a carbonium ion more easily.

Carbonium ions can be formed from compounds other than alcohols, and in reactions other than elimination. In all these cases the evidence indicates that the ease of formation of carbonium ions follows the same sequence:

Ease of formation of carbonium ions 3° > 2° > 1° > CH_3^+

In listing carbonium ions in order of their ease of formation, we find that we have at the same time listed them in order of their stability. **The more stable the carbonium ion, the more easily it is formed.**

Is it reasonable that the more stable carbonium ion should be formed more easily? To answer this question, we must look at a reaction in which a carbonium ion is formed, and consider the nature of the transition state.

In the dehydration of an alcohol, the carbonium ion is formed by loss of water from the protonated alcohol, ROH_2^+, that is, by breaking of the carbon–oxygen bond. In the reactant the positive charge is mostly on oxygen, and in the product it is on carbon. In the transition state the C—O bond must be partly broken, oxygen having partly pulled the electron pair away from carbon. The positive charge originally on oxygen is now divided between carbon and oxygen. Carbon has partly gained the positive charge it is to carry in the final carbonium ion.

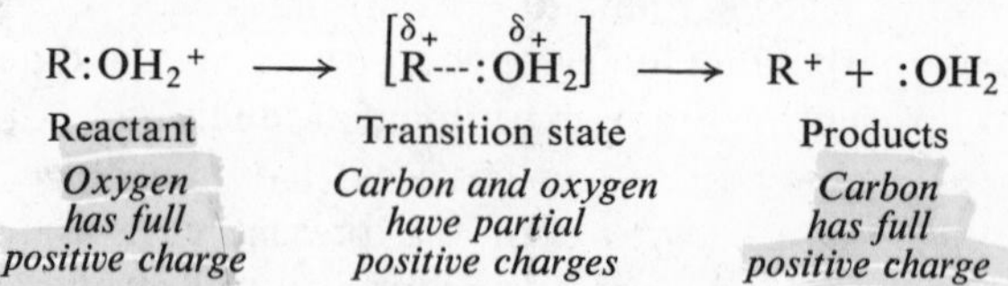

Electron-releasing groups tend to disperse the partial positive charge (δ_+) developing on carbon, and in this way stabilize the transition state. Stabilization of the transition state lowers E_{act} and permits a faster reaction (see Fig. 5.10).

Thus the same factor, electron release, that stabilizes the carbonium ion also stabilizes the *incipient* carbonium ion in the transition state. The more stable carbonium ion is formed faster.

We shall return again and again to the relationship between electronic effects and dispersal of charge, and between dispersal of charge and stability. We shall find that these relationships will help us to understand carbonium ion reactions of many kinds, and, in fact, all reactions in which a charge—positive or negative—develops or disappears. These will include reactions as seemingly different from dehydra-

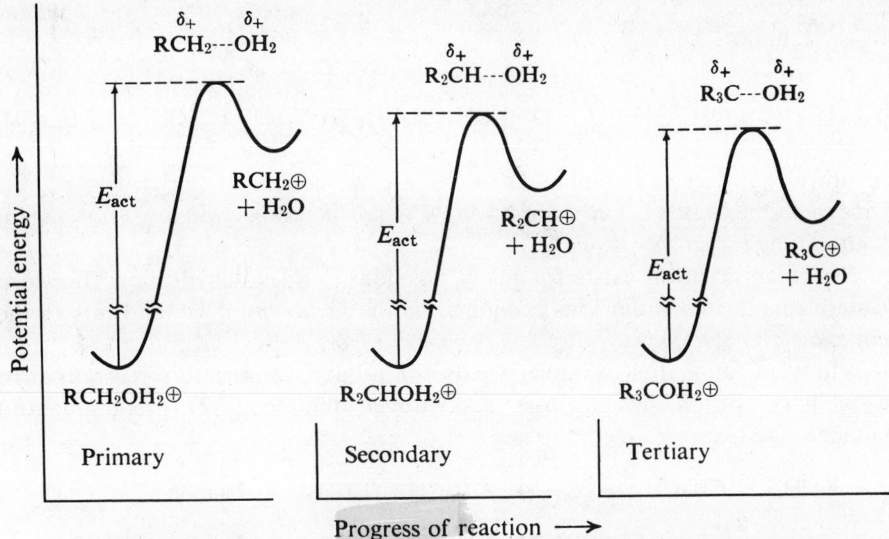

Figure 5.10. Molecular structure and rate of reaction. Stability of transition state parallels stability of carbonium ion: more stable carbonium ion formed faster. (Plots aligned with each other for easy comparison.)

tion of alcohols as: addition to alkenes, aromatic and aliphatic substitution, rearrangements, acidity and basicity.

5.22 Rearrangement of carbonium ions

Very often, dehydration gives alkenes that do not fit the mechanism as we have so far seen it. The double bond appears in unexpected places; sometimes the carbon skeleton is even changed. For example:

$$\underset{n\text{-Butyl alcohol}}{CH_3CH_2CH_2CH_2OH} \xrightarrow{H^+} \underset{\substack{\text{2-Butene} \\ \textit{Chief product}}}{CH_3CH{=}CHCH_3}$$

$$\underset{\text{2-Methyl-1-butanol}}{CH_3CH_2\overset{\overset{\large CH_3}{|}}{C}HCH_2OH} \xrightarrow{H^+} \underset{\substack{\text{2-Methyl-2-butene} \\ \textit{Chief product}}}{CH_3CH{=}\overset{\overset{\large CH_3}{|}}{C}CH_3}$$

$$\underset{\text{3,3-Dimethyl-2-butanol}}{CH_3\overset{\overset{\large CH_3}{|}}{\underset{\underset{\large CH_3}{|}}{C}}CHOHCH_3} \xrightarrow{H^+} \underset{\substack{\text{2,3-Dimethyl-2-butene} \\ \textit{Chief product}}}{CH_3\overset{\overset{\large CH_3}{|}}{C}{=}\underset{\underset{\large CH_3}{|}}{C}CH_3} \quad \text{and} \quad \underset{\text{2,3-Dimethyl-1-butene}}{CH_3\overset{\overset{\large CH_3}{|}}{C}H\underset{\underset{\large CH_3}{|}}{C}{=}CH_2}$$

Take the formation of 2-butene from *n*-butyl alcohol. Loss of water from the protonated alcohol gives the *n*-butyl carbonium ion. Loss of the proton from the carbon adjacent to the positive carbon could give 1-butene but *not* the 2-butene that is the major product.

$$CH_3CH_2CH_2CH_2\overset{\oplus}{O}H_2 \longrightarrow H_2O + CH_3CH_2CH_2CH_2\oplus \longrightarrow CH_3CH_2CH{=}CH_2$$

The other examples are similar. In each case we conclude that if, indeed, the alkene is formed from a carbonium ion, *it is not the same carbonium ion that is initially formed from the alcohol.*

A similar situation exists for many reactions besides dehydration. The idea of intermediate carbonium ions accounts for the facts *only* if we add this to the theory: *a carbonium ion can rearrange to form a more stable carbonium ion.*

n-Butyl alcohol, for example, yields the *n*-butyl cation; this rearranges to the *sec*-butyl cation, which loses a hydrogen ion to give (predominantly) 2-butene:

$$CH_3CH_2CH_2CH_2\overset{\oplus}{O}H_2 \longrightarrow CH_3CH_2CH_2CH_2\oplus + H_2O$$

$$\underset{(1°)}{CH_3CH_2CH_2CH_2\oplus} \longrightarrow \underset{(2°)}{CH_3CH_2\underset{\oplus}{C}HCH_3} \qquad \textit{Rearrangement}$$

$$CH_3CH_2\underset{\oplus}{C}HCH_3 \longrightarrow \underset{\substack{\text{2-Butene}\\ \textit{Chief product}}}{CH_3CH{=}CHCH_3} + H^+$$

In a similar way, the 2-methyl-1-butyl cation rearranges to the 2-methyl-2-butyl cation,

$$\underset{(1°)}{CH_3CH_2\overset{\overset{CH_3}{|}}{C}HCH_2\oplus} \longrightarrow \underset{(3°)}{CH_3CH_2\underset{\oplus}{\overset{\overset{CH_3}{|}}{C}}CH_3} \qquad \textit{Rearrangement}$$

and the 3,3-dimethyl-2-butyl cation rearranges to the 2,3-dimethyl-2-butyl cation.

$$\underset{(2°)}{CH_3\overset{\overset{CH_3}{|}}{\underset{\underset{CH_3}{|}}{C}}{-}\underset{\oplus}{C}HCH_3} \longrightarrow \underset{(3°)}{CH_3\underset{\oplus}{\overset{\overset{CH_3}{|}}{C}}{-}\overset{\overset{CH_3}{|}}{\underset{\underset{H}{|}}{C}}CH_3} \qquad \textit{Rearrangement}$$

We notice that in each case rearrangement occurs in the way that yields the more stable carbonium ion: primary to a secondary, primary to a tertiary, or secondary to a tertiary.

Just how does this rearrangement occur? Frank Whitmore (of The Pennsylvania State University) pictured rearrangement as taking place in this way: a hydrogen atom or alkyl group migrates *with a pair of electrons* from an adjacent carbon to the carbon bearing the positive charge. The carbon that loses the migrating group acquires the positive charge. A migration of hydrogen with a pair of electrons is known as a **hydride shift**; a similar migration of an alkyl group is known as an **alkyl shift**. These are just two examples of the most common kind

of rearrangement, the **1,2-shifts**: *rearrangements in which the migrating group moves from one atom to the very next atom.*

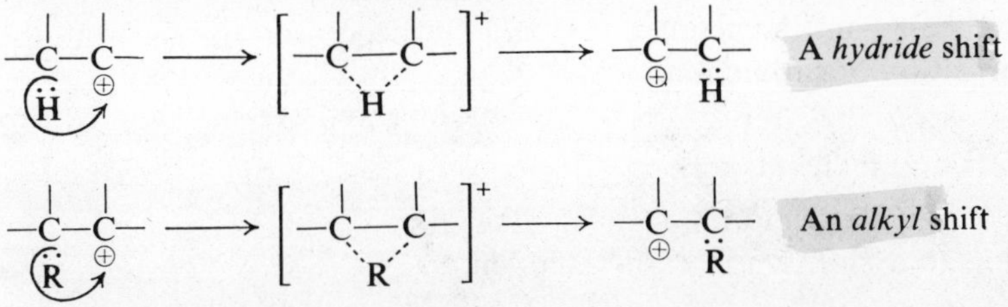

$$-\overset{|}{C}-\underset{\oplus}{\overset{|}{C}}- \text{ (with } \ddot{H} \text{ on first C)} \longrightarrow \left[-\overset{|}{C}\underset{H}{\cdots}\overset{|}{C}-\right]^+ \longrightarrow -\underset{\oplus}{\overset{|}{C}}-\underset{\ddot{H}}{\overset{|}{C}}- \quad \text{A } \textit{hydride} \text{ shift}$$

1,2 Shifts

$$-\overset{|}{C}-\underset{\oplus}{\overset{|}{C}}- \text{ (with } \ddot{R} \text{ on first C)} \longrightarrow \left[-\overset{|}{C}\underset{R}{\cdots}\overset{|}{C}-\right]^+ \longrightarrow -\underset{\oplus}{\overset{|}{C}}-\underset{\ddot{R}}{\overset{|}{C}}- \quad \text{An } \textit{alkyl} \text{ shift}$$

We can account for rearrangements in dehydration in the following way. A carbonium ion is formed by the loss of water from the protonated alcohol. **If a 1,2-shift of hydrogen or alkyl can form a more stable carbonium ion, then such a rearrangement takes place.** The new carbonium ion now loses a proton to yield an alkene.

In the case of the *n*-butyl cation, a shift of hydrogen yields the more stable *sec*-butyl cation; migration of an ethyl group would simply form a different *n*-butyl cation. In the case of the 2-methyl-1-butyl cation, a hydride shift yields a tertiary cation, and hence is preferred over a methyl shift, which would only yield a secondary cation. In the case of the 3,3-dimethyl-2-butyl cation, on the other hand, a methyl shift can yield a tertiary cation and is the rearrangement that takes place.

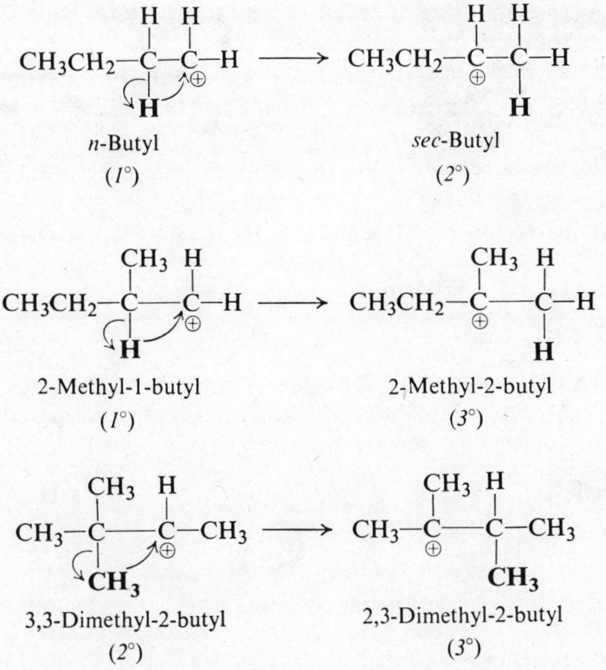

$$CH_3CH_2-\underset{H}{\overset{H}{C}}-\underset{\oplus}{\overset{H}{C}}-H \longrightarrow CH_3CH_2-\underset{\oplus}{\overset{H}{C}}-\underset{H}{\overset{H}{C}}-H$$

n-Butyl (*1°*) → *sec*-Butyl (*2°*)

$$CH_3CH_2-\underset{H}{\overset{CH_3}{C}}-\underset{\oplus}{\overset{H}{C}}-H \longrightarrow CH_3CH_2-\underset{\oplus}{\overset{CH_3}{C}}-\underset{H}{\overset{H}{C}}-H$$

2-Methyl-1-butyl (*1°*) → 2-Methyl-2-butyl (*3°*)

$$CH_3-\underset{CH_3}{\overset{CH_3}{C}}-\underset{\oplus}{\overset{H}{C}}-CH_3 \longrightarrow CH_3-\underset{\oplus}{\overset{CH_3}{C}}-\underset{CH_3}{\overset{H}{C}}-CH_3$$

3,3-Dimethyl-2-butyl (*2°*) → 2,3-Dimethyl-2-butyl (*3°*)

Historically, it was the occurrence of rearrangements that was chiefly responsible for the development of the carbonium ion theory. Reactions of seemingly quite different kinds involved rearrangements that followed the same general pattern; the search for a common basis led to the concept of the carbonium ion. Today, the occurrence (or non-occurrence) of rearrangements of the kind we have seen here is the best—and sometimes the *only*—evidence for (or against) the intermediate formation of carbonium ions.

In our short acquaintance with the carbonium ion, we have encountered two of its reactions. **A carbonium ion may:**

(a) eliminate a hydrogen ion to form an alkene;
(b) rearrange to a more stable carbonium ion.

This list will grow rapidly.

In rearrangement, as in every other reaction of a carbonium ion, the electron-deficient carbon atom gains a pair of electrons, this time at the expense of a neighboring carbon atom, one that can better accommodate the positive charge.

5.23 Orientation and reactivity in dehydration

At this point, we know this much about dehydration of alcohols.

(a) It involves the formation of a carbonium ion. How fast dehydration takes place depends chiefly upon how fast this carbonium ion is formed, which, in turn, depends upon how stable the carbonium ion is. The stability of the carbonium ion depends upon the dispersal of the positive charge, which is determined by electron-release or electron-withdrawal by the attached groups.

(b) If this initially formed carbonium ion can rearrange via a 1,2-shift to form a more stable carbonium ion, it will do so.

This brings us to the last step of dehydration. (c) The carbonium ion—either the original one or the one formed by rearrangement—loses a proton to form an alkene. Now, if isomeric alkenes can be formed in this step, which, if any, will predominate? The examples we have already encountered give us the answer:

$$CH_3CH_2\underset{\oplus}{C}HCH_3 \longrightarrow CH_3CH{=}CHCH_3 \quad \text{and} \quad CH_3CH_2CH{=}CH_2$$

2-Butene — *Preferred product*; 1-Butene

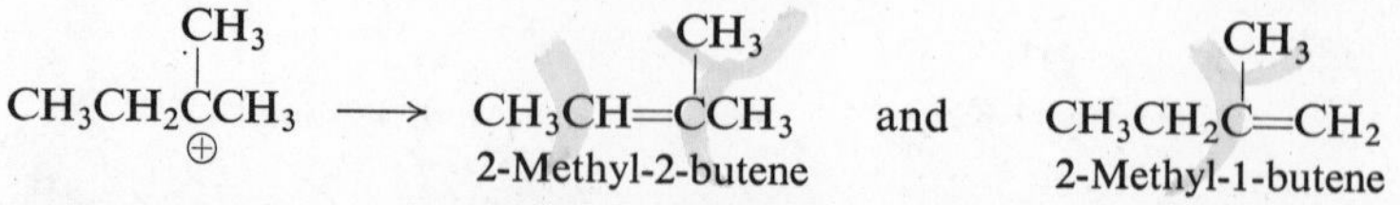

$$CH_3CH_2\overset{CH_3}{\underset{\oplus}{C}}CH_3 \longrightarrow CH_3CH{=}\overset{CH_3}{C}CH_3 \quad \text{and} \quad CH_3CH_2\overset{CH_3}{C}{=}CH_2$$

2-Methyl-2-butene — *Preferred product*; 2-Methyl-1-butene

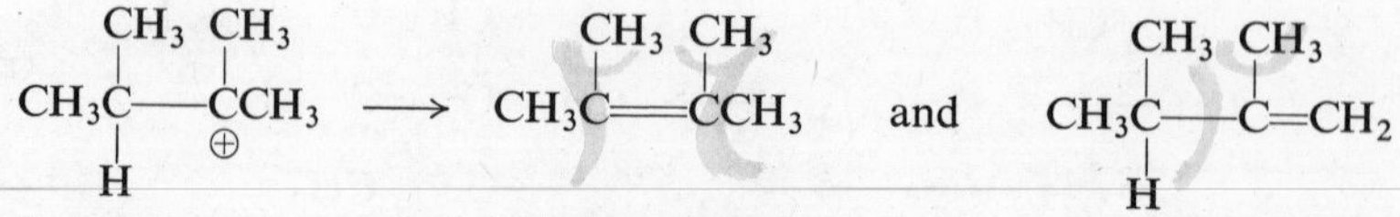

$$CH_3\overset{CH_3}{\underset{H}{C}}{-}\overset{CH_3}{\underset{\oplus}{C}}CH_3 \longrightarrow CH_3\overset{CH_3}{C}{=}\overset{CH_3}{C}CH_3 \quad \text{and} \quad CH_3\overset{CH_3}{\underset{H}{C}}{-}\overset{CH_3}{C}{=}CH_2$$

2,3-Dimethyl-2-butene — *Preferred product*; 2,3-Dimethyl-1-butene

Here, as in dehydrohalogenation, the preferred alkene is the more highly substituted one, that is, the *more stable* one (Sec. 6.4). **In dehydration, the more stable alkene is the preferred product.**

Once more, examination of the transition state involved shows that it is reasonable that the more stable alkene should be formed faster:

$$\underset{\oplus\ \ \ \ \text{H}}{-\overset{|}{\text{C}}-\overset{|}{\text{C}}-} \xrightarrow{\text{ROH}} \left[\begin{array}{c} -\overset{|}{\text{C}}\text{---}\overset{|}{\text{C}}- \\ \delta_+ \quad \vdots \\ \quad \delta_+ \\ \text{H---}\underset{\text{H}}{\text{OR}} \end{array}\right] \longrightarrow \;\rangle\text{C=C}\langle\; + \text{ROH}_2^+$$

Transition state:
partly formed double bond

As the proton is pulled away by the base (the solvent), the electrons it leaves behind become shared by the two carbons, and the carbon–carbon bond acquires double-bond character. Factors that stabilize an alkene also stabilize an *incipient* alkene in the transition state.

Problem 5.9 Predict the *major* product of dehydration of each of the following: (a) $(CH_3)_2C(OH)CH_2CH_3$, (b) $(CH_3)_2CHCHOHCH_3$, (c) $(CH_3)_2C(OH)CH(CH_3)_2$.

PROBLEMS

1. Give the structural formula of:

(a) 3,6-dimethyl-1-octene
(b) 3-chloropropene
(c) 2,4,4-trimethyl-2-pentene
(d) *trans*-3,4-dimethyl-3-hexene
(e) (*Z*)-3-chloro-4-methyl-3-hexene
(f) (*E*)-1-deuterio-2-chloropropene
(g) (R)-3-bromo-1-butene
(h) (S)-*trans*-4-methyl-2-hexene

2. Draw out the structural formula and give the IUPAC name of:

(a) isobutylene
(b) *cis*-$CH_3CH_2CH{=}CHCH_2CH_3$
(c) $(CH_3)_3CCH{=}CH_2$
(d) *trans*-$(CH_3)_2CHCH{=}CHCH(CH_3)_2$
(e) $(CH_3)_2CHCH_2CH{=}C(CH_3)_2$
(f) $(CH_3CH_2)_2C{=}CH_2$

3. Indicate which of the following compounds show geometric (*cis-trans*) isomerism, draw the isomeric structures, and specify each as *Z* or *E*.

(a) 1-butene
(b) 2-butene
(c) 1,1-dichloroethene
(d) 1,2-dichloroethene
(e) 2-methyl-2-butene
(f) 1-pentene
(g) 2-pentene
(h) 1-chloropropene
(i) 1-chloro-2-methyl-2-butene
(j) 3-methyl-4-ethyl-3-hexene
(k) 2,4-hexadiene ($CH_3CH{=}CHCH{=}CHCH_3$)

4. There are 13 isomeric hexylenes (C_6H_{12}) disregarding geometric isomerism. (a) Draw the structure and give the IUPAC name for each. (b) Indicate which ones show geometric isomerism, draw the isomeric structures, and specify each as *Z* or *E*. (c) One of the hexylenes is chiral. Which one is it? Draw structures of the enantiomers, and specify each as R or S.

5. In which of the following will *cis*-3-hexene differ from *trans*-3-hexene?

(a) b.p.
(b) m.p.
(c) adsorption on alumina
(d) infrared spectrum
(e) dipole moment
(f) refractive index
(g) rate of hydrogenation
(h) product of hydrogenation
(i) solubility in ethyl alcohol
(j) density
(k) retention time in gas chromatography

(1) Which *one* of the above would absolutely prove the configuration of each isomer?

6. Write balanced equations for preparation of propylene from:

(a) $CH_3CH_2CH_2OH$ (*n*-propyl alcohol)
(b) $CH_3CHOHCH_3$ (isopropyl alcohol)
(c) isopropyl chloride
(d) the alkyne, $CH_3C{\equiv}CH$
(e) propylene bromide (1,2-dibromopropane)

7. Give structures of the products expected from dehydrohalogenation of:

(a) 1-bromohexane
(b) 2-bromohexane
(c) 1-bromo-2-methylpentane
(d) 2-bromo-2-methylpentane
(e) 3-bromo-2-methylpentane
(f) 4-bromo-2-methylpentane
(g) 1-bromo-4-methylpentane
(h) 3-bromo-2,3-dimethylpentane

8. In those cases in Problem 7 where more than one product can be formed, predict the *major* product.

9. Which alcohol of each pair would you expect to be more easily dehydrated?

(a) $CH_3CH_2CH_2CH_2CH_2OH$ or $CH_3CH_2CH_2CHOHCH_3$
(b) $(CH_3)_2C(OH)CH_2CH_3$ or $(CH_3)_2CHCHOHCH_3$
(c) $(CH_3)_2CHC(OH)(CH_3)_2$ or $(CH_3)_2CHCH(CH_3)CH_2OH$

10. (a) Show all steps in the synthesis of propylene from propane by ordinary laboratory methods (*not* cracking). (b) If the steps in (a) were carried out starting with *n*-butane, would a single product or a mixture be expected?

11. When dissolved in SbF_5, *n*-propyl fluoride and isopropyl fluoride give solutions with identical nmr spectra, indicating that identical species are present. How do you account for this?

12. (a) When neopentyl alcohol, $(CH_3)_3CCH_2OH$, is heated with acid, it is slowly converted into an 85:15 mixture of two alkenes of formula C_5H_{10}. What are these alkenes, and how are they formed? Which one would you think is the major product, and why?

(b) Would you expect neopentyl bromide, $(CH_3)_3CCH_2Br$, to undergo the kind of dehydrohalogenation described in Sec. 5.13? Actually, when heated in aqueous alcohol, neopentyl bromide slowly reacts to yield, among other products, the same alkenes as those in (a). Suggest a mechanism for this particular kind of dehydrohalogenation. Why does this reaction, unlike that in (a), *not* require acid catalysis?

13. When 3,3-dimethyl-1-butene is treated with hydrogen chloride there is obtained a mixture of 3-chloro-2,2-dimethylbutane and 2-chloro-2,3-dimethylbutane. What does the formation of the second product suggest to you? Propose a likely mechanism for this reaction, which is an example of *electrophilic addition*. Check your answer in Secs. 6.10 and 6.12.

Chapter 6 Alkenes II. Reactions of the Carbon–Carbon Double Bond

Electrophilic and Free-Radical Addition

6.1 The functional group

The characteristic feature of the alkene structure is the carbon–carbon double bond. The characteristic reactions of an alkene are those that take place at the double bond. *The atom or group of atoms that defines the structure of a particular family of organic compounds and, at the same time, determines their properties is called the* **functional group.**

In alkyl halides the functional group is the halogen atom, and in alcohols the —OH group; in alkenes it is the carbon–carbon double bond. We must not forget that an alkyl halide, alcohol, or alkene has alkyl groups attached to these functional groups; under the proper conditions, the alkyl portions of these molecules undergo the reactions typical of alkanes. However, the reactions that are *characteristic* of each of these compounds are those that occur at the halogen atom or the hydroxyl group or the carbon–carbon double bond.

A large part of organic chemistry is therefore the chemistry of the various functional groups. We shall learn to associate a particular set of properties with a particular group wherever we may find it. When we encounter a complicated molecule, which contains a number of different functional groups, we may expect the properties of this molecule to be roughly a composite of the properties of the various functional groups. The properties of a particular group may be modified, of course, by the presence of another group and it is important for us to understand these modifications, but our point of departure is the chemistry of individual functional groups.

6.2 Reactions of the carbon–carbon double bond: addition

Alkene chemistry is the chemistry of the carbon–carbon double bond.

What kind of reaction may we expect of the double bond? The double bond consists of a strong σ bond and a weak π bond; we might expect, therefore, that reaction would involve the breaking of this weaker bond. This expectation is correct; the typical reactions of the double bond are of the sort,

$$-\overset{|}{C}=\overset{|}{C}- + YZ \longrightarrow -\underset{Y}{\overset{|}{\underset{|}{C}}}-\underset{Z}{\overset{|}{\underset{|}{C}}}- \qquad \textbf{Addition}$$

where the π bond is broken and two strong σ bonds are formed in its place.

A reaction in which two molecules combine to yield a single molecule of product is called an **addition reaction**. The reagent is simply *added to* the organic molecule, in contrast to a substitution reaction where part of the reagent is *substituted for* a portion of the organic molecule. Addition reactions are necessarily limited to compounds that contain atoms sharing more than one pair of electrons, that is, to compounds that contain multiply-bonded atoms.

What kind of reagent may we expect to add to the carbon–carbon double bond? In our structure of the bond there is a cloud of π electrons above and below the plane of the atoms (see Fig. 6.1). These π electrons are less involved than the

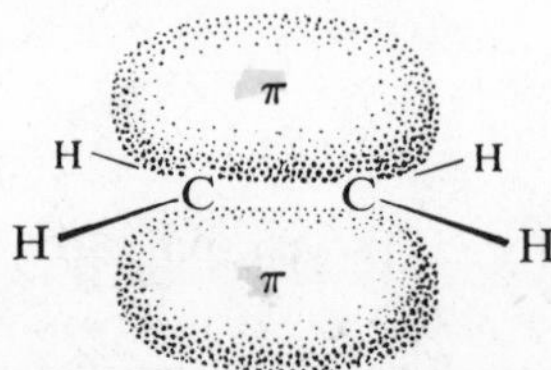

Figure 6.1. Carbon–carbon double bond: π bond is source of electrons.

σ electrons in holding together the carbon nuclei. As a result, they are themselves held less tightly. These loosely held π electrons are particularly available to a reagent that is seeking electrons. It is not surprising, then, that in many of its reactions the carbon–carbon double bond serves as a **source of electrons**: that is, it acts as a **base**. The compounds with which it reacts are those that are deficient in electrons, that is, are *acids*. *These acidic reagents that are seeking a pair of electrons are called* **electrophilic reagents** (Greek: electron-loving). *The typical reaction of an alkene is* **electrophilic addition**, or, in other words, addition of acidic reagents.

Reagents of another kind, *free radicals*, seek electrons—or, rather, seek *an* electron. And so we find that alkenes also undergo **free-radical addition.**

Most alkenes contain not only the carbon–carbon double bond but also alkyl groups, which have essentially the alkane structure. Besides the addition reactions characteristic of the carbon–carbon double bond, therefore, alkenes may undergo the free-radical substitution characteristic of alkanes. The most important of these addition and substitution reactions are summarized below, and will be discussed in detail in following sections.

There are reagents that can add either as acids or as free radicals, and with strikingly different results; there are reagents that are capable both of adding to the double bond and of bringing about substitution. We shall see how, by our choice of conditions, we can lead these reagents along the particular reaction path —electrophilic or free-radical, addition or substitution—we want them to follow.

The alkyl groups attached to the doubly-bonded carbons modify the reactions of the double bond; the double bond modifies the reactions of the alkyl groups.

We shall be concerned with seeing what these modifications are and, where possible, how they can be accounted for.

REACTIONS OF ALKENES

Addition Reactions

$$-\overset{|}{C}=\overset{|}{C}- + YZ \longrightarrow -\underset{Y}{\overset{|}{\underset{|}{C}}}-\underset{Z}{\overset{|}{\underset{|}{C}}}-$$

1. Addition of hydrogen. Catalytic hydrogenation. Discussed in Sec. 6.3.

$$-\overset{|}{C}=\overset{|}{C}- + H_2 \xrightarrow{\text{Pt, Pd, or Ni}} -\underset{H}{\overset{|}{\underset{|}{C}}}-\underset{H}{\overset{|}{\underset{|}{C}}}-$$

Example:

$$\underset{\substack{\text{Propene}\\ \text{(Propylene)}}}{CH_3CH=CH_2} \xrightarrow{H_2,\ Ni} \underset{\text{Propane}}{CH_3CH_2CH_3}$$

2. Addition of halogens. Discussed in Secs. 6.5, 6.13, and 7.11–7.12.

$$-\overset{|}{C}=\overset{|}{C}- + X_2 \longrightarrow -\underset{X}{\overset{|}{\underset{|}{C}}}-\underset{X}{\overset{|}{\underset{|}{C}}}- \qquad X_2 = Cl_2,\ Br_2$$

Example:

$$\underset{\substack{\text{Propene}\\ \text{(Propylene)}}}{CH_3CH=CH_2} \xrightarrow{Br_2 \text{ in } CCl_4} \underset{\substack{\text{1,2-Dibromopropane}\\ \text{(Propylene bromide)}}}{CH_3CHBrCH_2Br}$$

3. Addition of hydrogen halides. Discussed in Secs. 6.6–6.7 and 6.17.

$$-\overset{|}{C}=\overset{|}{C}- + HX \longrightarrow -\underset{H}{\overset{|}{\underset{|}{C}}}-\underset{X}{\overset{|}{\underset{|}{C}}}- \qquad HX = HCl,\ HBr,\ HI$$

Examples:

$$\underset{\text{Propene}}{CH_3CH=CH_2} \xrightarrow{HI} \underset{\substack{\text{2-Iodopropane}\\ \text{(Isopropyl iodide)}}}{CH_3CHICH_3}$$

$$CH_3CH=CH_2 \xrightarrow{HBr} \begin{cases} \xrightarrow{\text{no peroxides}} \underset{\substack{\text{2-Bromopropane}\\ \text{(Isopropyl bromide)}}}{CH_3CHBrCH_3} & \textbf{Markovnikov addition} \\ \xrightarrow{\text{peroxides}} \underset{\substack{\text{1-Bromopropane}\\ (n\text{-Propyl bromide})}}{CH_3CH_2CH_2Br} & \textbf{Anti-Markovnikov addition} \end{cases}$$

4. Addition of sulfuric acid. Discussed in Sec. 6.8.

$$-\overset{|}{C}=\overset{|}{C}- + H_2SO_4 \longrightarrow -\underset{H}{\overset{|}{\underset{|}{C}}}-\underset{OSO_3H}{\overset{|}{\underset{|}{C}}}-$$

Example:

$$\underset{\text{Propene}}{CH_3CH{=}CH_2} \xrightarrow{\text{conc. } H_2SO_4} \underset{\text{Isopropyl hydrogen sulfate}}{CH_3\underset{OSO_3H}{\underset{|}{C}}HCH_3}$$

5. Addition of water. Hydration. Discussed in Sec. 6.9.

$$-\overset{|}{C}=\overset{|}{C}- + HOH \xrightarrow{H^+} -\underset{H}{\overset{|}{\underset{|}{C}}}-\underset{OH}{\overset{|}{\underset{|}{C}}}-$$

Example:

$$\underset{\text{Propene}}{CH_3CH{=}CH_2} \xrightarrow{H_2O,\ H^+} \underset{\substack{\text{Isopropyl alcohol} \\ \text{(2-Propanol)}}}{CH_3\underset{OH}{\underset{|}{C}}HCH_3}$$

6. Halohydrin formation. Discussed in Sec. 6.14.

$$-\overset{|}{C}=\overset{|}{C}- + X_2 + H_2O \longrightarrow -\underset{X}{\overset{|}{\underset{|}{C}}}-\underset{OH}{\overset{|}{\underset{|}{C}}}- + HX \qquad X_2 = Cl_2, Br_2$$

Example:

$$\underset{\substack{\text{Propylene} \\ \text{(Propene)}}}{CH_3CH{=}CH_2} \xrightarrow{Cl_2,\ H_2O} \underset{\substack{\text{Propylene chlorohydrin} \\ \text{(1-Chloro-2-propanol)}}}{CH_3\underset{OH}{\underset{|}{CH}}-\underset{Cl}{\underset{|}{CH_2}}}$$

7. Dimerization. Discussed in Sec. 6.15.

Example:

$$\underset{\text{Isobutylene}}{CH_3-\overset{CH_3}{\overset{|}{C}}{=}CH_2 + CH_3-\overset{CH_3}{\overset{|}{C}}{=}CH_2} \xrightarrow{\text{acid}} \underset{\text{2,4,4-Trimethyl-2-pentene}}{CH_3-\underset{CH_3}{\underset{|}{\overset{CH_3}{\overset{|}{C}}}}-CH{=}\overset{CH_3}{\overset{|}{C}}-CH_3}$$

$$\text{and} \quad \underset{\text{2,4,4-Trimethyl-1-pentene}}{CH_3-\underset{CH_3}{\underset{|}{\overset{CH_3}{\overset{|}{C}}}}-CH_2-\overset{CH_3}{\overset{|}{C}}{=}CH_2}$$

8. Alkylation. Discussed in Sec. 6.16.

$$-\overset{|}{C}=\overset{|}{C}- + R-H \xrightarrow{\text{acid}} -\underset{H}{\overset{|}{\underset{|}{C}}}-\underset{R}{\overset{|}{\underset{|}{C}}}-$$

Example:

$$\underset{\text{Isobutylene}}{CH_3-\overset{CH_3}{\overset{|}{C}}=CH_2} + \underset{\text{Isobutane}}{CH_3-\underset{CH_3}{\overset{CH_3}{\overset{|}{\underset{|}{C}}}}-H} \xrightarrow{H_2SO_4} \underset{\text{2,2,4-Trimethylpentane}}{CH_3\underset{H}{\overset{CH_3}{\overset{|}{\underset{|}{C}}}}-CH_2-\underset{CH_3}{\overset{CH_3}{\overset{|}{\underset{|}{C}}}}-CH_3}$$

9. Oxymercuration-demercuration. Discussed in Sec. 15.8.

$$-\overset{|}{C}=\overset{|}{C}- + H_2O + Hg(OAc)_2 \longrightarrow -\underset{HO}{\overset{|}{\underset{|}{C}}}-\underset{HgOAc}{\overset{|}{\underset{|}{C}}}- \xrightarrow{NaBH_4} \underset{\substack{\textit{Markovnikov}\\ \textit{orientation}}}{-\underset{HO}{\overset{|}{\underset{|}{C}}}-\underset{H}{\overset{|}{\underset{|}{C}}}-}$$

10. Hydroboration-oxidation. Discussed in Secs. 15.9–15.11.

$$-\overset{|}{C}=\overset{|}{C}- + \underset{\text{Diborane}}{(BH_3)_2} \longrightarrow -\underset{H}{\overset{|}{\underset{|}{C}}}-\underset{B-}{\overset{|}{\underset{|}{C}}}- \xrightarrow[OH^-]{H_2O_2} \underset{\substack{\textit{Anti-Markovnikov}\\ \textit{orientation}}}{-\underset{H}{\overset{|}{\underset{|}{C}}}-\underset{OH}{\overset{|}{\underset{|}{C}}}-}$$

11. Addition of free radicals. Discussed in Secs. 6.17 and 6.18.

$$-\overset{|}{C}=\overset{|}{C}- + Y-Z \xrightarrow[\text{or light}]{\text{peroxides}} -\underset{Y}{\overset{|}{\underset{|}{C}}}-\underset{Z}{\overset{|}{\underset{|}{C}}}-$$

Example:

$$\underset{\text{1-Octene}}{n\text{-}C_6H_{13}CH=CH_2} + \underset{\text{Bromotrichloromethane}}{BrCCl_3} \xrightarrow{\text{peroxides}} \underset{\text{3-Bromo-1,1,1-trichlorononane}}{n\text{-}C_6H_{13}\underset{Br}{\underset{|}{C}H}-CH_2-CCl_3}$$

12. Polymerization. Discussed in Secs. 6.19 and 32.3–32.6.

13. Addition of carbenes. Discussed in Secs. 9.15–9.16.

14. Hydroxylation. Glycol formation. Discussed in Secs. 6.20 and 17.12.

$$-\overset{|}{C}=\overset{|}{C}- + KMnO_4 \text{ or } HCO_2OH \longrightarrow -\underset{OH}{\overset{|}{\underset{|}{C}}}-\underset{OH}{\overset{|}{\underset{|}{C}}}-$$

Example:

$$\underset{\substack{\text{Propylene}\\\text{(Propene)}}}{CH_3CH{=}CH_2} \xrightarrow{KMnO_4 \text{ or } HCO_2OH} \underset{\substack{\text{Propylene glycol}\\\text{(1,2-Propanediol)}}}{CH_3{-}\underset{OH}{\underset{|}{CH}}{-}\underset{OH}{\underset{|}{CH_2}}}$$

Substitution Reactions

15. Halogenation. Allylic substitution. Discussed in Sec. 6.21.

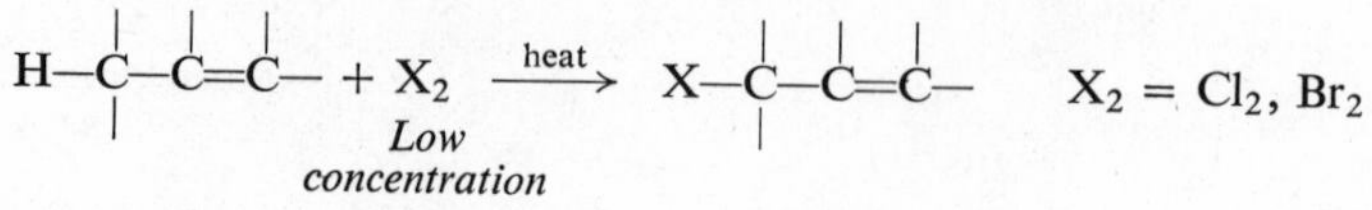

$$H{-}\overset{|}{\underset{|}{C}}{-}\overset{|}{C}{=}\overset{|}{C}{-} + \underset{\substack{\textit{Low}\\\textit{concentration}}}{X_2} \xrightarrow{\text{heat}} X{-}\overset{|}{\underset{|}{C}}{-}\overset{|}{C}{=}\overset{|}{C}{-} \qquad X_2 = Cl_2, Br_2$$

Examples:

$$\underset{\substack{\text{Propylene}\\\text{(Propene)}}}{CH_3CH{=}CH_2} \xrightarrow{Cl_2,\ 600^\circ} \underset{\substack{\text{Allyl chloride}\\\text{(3-Chloro-l-propene)}}}{Cl{-}CH_2CH{=}CH_2}$$

Cyclohexene + NBS ⟶ 3-Bromocyclohexene (Br on the allylic carbon)

Cyclohexene   N-Bromosuccinimide   3-Bromocyclohexene

Cleavage Reactions

16. Ozonolysis. Discussed in Sec. 6.29.

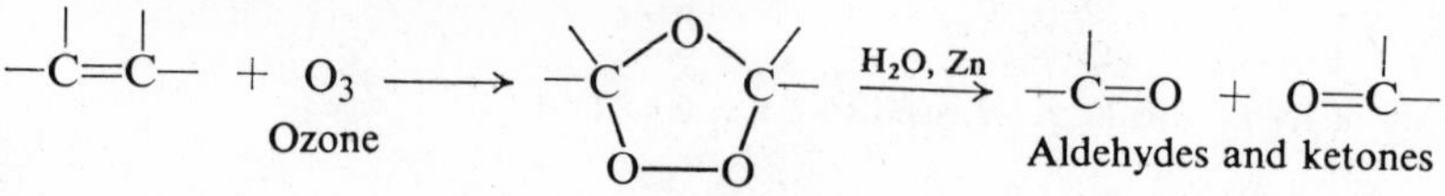

$$-\overset{|}{C}{=}\overset{|}{C}{-} + \underset{\text{Ozone}}{O_3} \longrightarrow \underset{\text{Ozonide}}{\text{(ozonide ring: }{-}C{<}^{O}_{O-O}{>}C{-})} \xrightarrow{H_2O,\ Zn} \underset{\text{Aldehydes and ketones}}{-\overset{|}{C}{=}O + O{=}\overset{|}{C}{-}}$$

Used to determine structure

Examples:

$$\underset{\text{1-Butene}}{CH_3CH_2CH{=}CH_2} \xrightarrow{O_3} \xrightarrow{H_2O,\ Zn} CH_3CH_2\overset{H}{\overset{|}{C}}{=}O + O{=}\overset{H}{\overset{|}{C}}H$$

$$\underset{\text{Isobutylene}}{CH_3{-}\overset{CH_3}{\overset{|}{C}}{=}CH_2} \xrightarrow{O_3} \xrightarrow{H_2O,\ Zn} CH_3\overset{CH_3}{\overset{|}{C}}{=}O + O{=}\overset{H}{\overset{|}{C}}H$$

6.3 Hydrogenation. Heat of hydrogenation

We have already encountered hydrogenation as the most useful method for preparing alkanes (Sec. 3.15). It is not limited to the synthesis of alkanes, but is a general method for the conversion of a carbon–carbon double bond into a carbon–carbon single bond: using the same apparatus, the same catalyst, and very nearly the same conditions, we can convert an alkene into an alkane, an unsaturated alcohol into a saturated alcohol, or an unsaturated ester into a saturated ester. Since the reaction is generally quantitative, and since the volume of hydrogen consumed can be easily measured, hydrogenation is frequently used as an analytical tool; it can, for example, tell us the number of double bonds in a compound.

$$-\overset{|}{C}=\overset{|}{C}- + H-H \longrightarrow -\underset{H}{\overset{|}{\underset{|}{C}}}-\underset{H}{\overset{|}{\underset{|}{C}}}- \qquad \Delta H = \textbf{heat of hydrogenation}$$

Hydrogenation is exothermic: the two σ bonds (C—H) being formed are, together, stronger than the σ bond (H—H) and π bond being broken. *The quantity of heat evolved when one mole of an unsaturated compound is hydrogenated is called the* **heat of hydrogenation**; it is simply ΔH of the reaction, but the minus sign is not included. The heat of hydrogenation of nearly every alkene is fairly close to an approximate value of 30 kcal for each double bond in the compound (see Table 6.1).

Table 6.1 HEATS OF HYDROGENATION OF ALKENES

Alkene	Heat of hydrogenation, kcal/mole
Ethylene	32.8
Propylene	30.1
1-Butene	30.3
1-Pentene	30.1
1-Heptene	30.1
3-Methyl-1-butene	30.3
3,3-Dimethyl-1-butene	30.3
4,4-Dimethyl-1-pentene	29.5
cis-2-Butene	28.6
trans-2-Butene	27.6
Isobutylene	28.4
cis-2-Pentene	28.6
trans-2-Pentene	27.6
2-Methyl-1-butene	28.5
2,3-Dimethyl-1-butene	28.0
2-Methyl-2-butene	26.9
2,3-Dimethyl-2-butene	26.6

Although hydrogenation is an exothermic reaction, it proceeds at a negligible rate in the absence of a catalyst, even at elevated temperatures. The uncatalyzed reaction must have, therefore, a very large energy of activation. The function of the catalyst is to lower the energy of activation (E_{act}) so that the reaction can proceed rapidly at room temperature. The catalyst does not, of course, affect the net energy change of the overall reaction; it simply lowers the energy hill between the reactants and products (see Fig. 6.2).

A catalyst lowers E_{act} by permitting reaction to take place in a different way, that is, by a different mechanism. In this case, the reactants are adsorbed on the enormous surface of the finely divided metal, where reaction actually occurs. Reaction between the adsorbed molecules is very different from the reaction that would have to take place otherwise; it is believed, for example, that the catalytic surface breaks the π bond of the alkene prior to reaction with hydrogen.

Lowering the energy hill, as we can see, decreases the energy of activation of the reverse reaction as well, and thus increases the rate of *de*hydrogenation. We might expect, therefore, that platinum, palladium, and nickel, under the proper conditions, should serve as dehydrogenation catalysts; this is indeed the case. We are familiar with the fact that, although a catalyst speeds up a reaction, it does

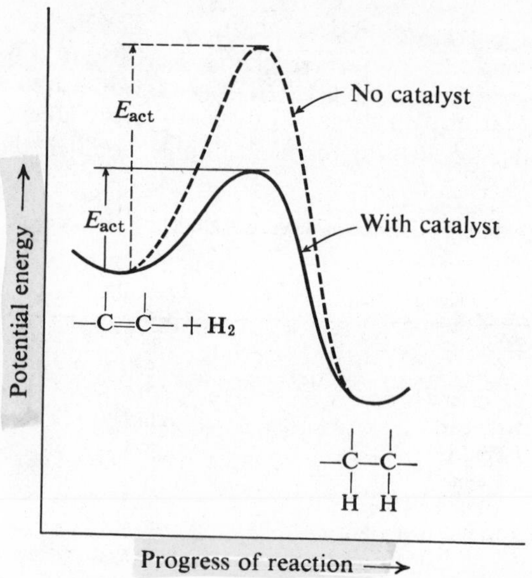

Figure 6.2. Potential energy changes during progress of reaction: effect of catalyst.

not shift the position of equilibrium; this is, of course, because it speeds up both the forward and reverse reactions (See Sec. 30.7).

Like hydrogenation, the addition of other reagents to the double bond is generally exothermic. The energy consumed by the breaking of the Y—Z and π bonds is almost always less than that liberated by formation of the C—Y and C—Z bonds.

$$-\overset{|}{C}=\overset{|}{C}- + \text{Y—Z} \longrightarrow -\underset{\text{Y}}{\overset{|}{\underset{|}{C}}}-\underset{\text{Z}}{\overset{|}{\underset{|}{C}}}- + \text{heat}$$

6.4 Heat of hydrogenation and stability of alkenes

Heats of hydrogenation can often give us valuable information about the relative stabilities of unsaturated compounds. For example, of the isomeric 2-butenes, the *cis*-isomer has a heat of hydrogenation of 28.6 kcal, the *trans*-isomer one of 27.6 kcal. Both reactions consume one mole of hydrogen and yield the same product, *n*-butane. Therefore, if the *trans*-isomer *evolves* 1 kcal less energy than the *cis*-isomer, it can only mean that it *contains* 1 kcal less energy; in other words, the *trans*-isomer is *more stable* by 1 kcal than the *cis*-isomer (see Fig. 6.3). In a similar way, *trans*-2-pentene (heat of hydrogenation = 27.6 kcal) must be more stable by 1.0 kcal than *cis*-2-pentene (heat of hydrogenation = 28.6 kcal).

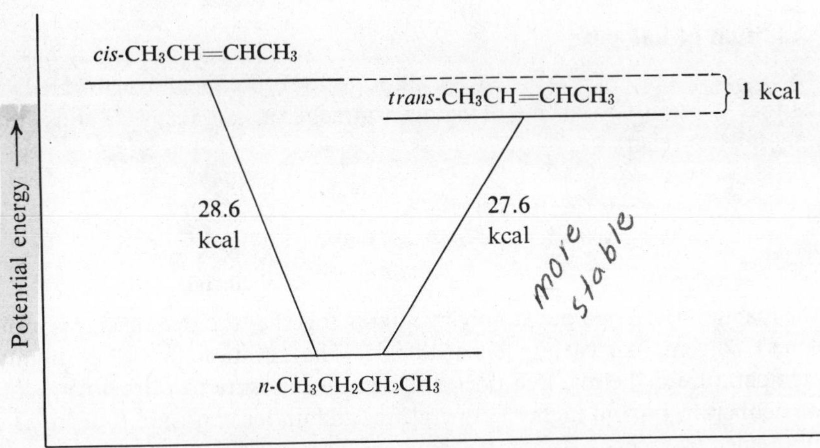

Figure 6.3. Heats of hydrogenation and stability: *cis*- and *trans*-2-butene.

Of simple disubstituted ethylenes, it is usually the *trans*-isomer that is the more stable. The two larger substituents are located farther apart than in the *cis*-isomer; there is less crowding, and less van der Waals strain (Sec. 3.5).

Heats of hydrogenation show that the stability of an alkene also depends upon the position of the double bond. The following examples are typical:

$CH_3CH_2CH{=}CH_2$
30.3 kcal

$CH_3CH{=}CHCH_3$
cis 28.6; *trans* 27.6

$CH_3CH_2CH_2CH{=}CH_2$
30.1 kcal

$CH_3CH_2CH{=}CHCH_3$
cis 28.6; *trans* 27.6

$CH_3CH(CH_3)CH{=}CH_2$
30.3 kcal

$CH_2{=}C(CH_3)CH_2CH_3$
28.5

$CH_3C(CH_3){=}CHCH_3$
26.9

Each set of isomeric alkenes yields the same alkane. The differences in heat of hydrogenation must therefore be due to differences in stability. In each case, **the greater the number of alkyl groups attached to the doubly-bonded carbon atoms, the more stable the alkene.**

Stability of alkenes

$$R_2C{=}CR_2 > R_2C{=}CHR > R_2C{=}CH_2,\ RCH{=}CHR > RCH{=}CH_2 > CH_2{=}CH_2$$

We have seen (Secs. 5.14, 5.23) that the stability of alkenes determines orientation in dehydrohalogenation and dehydration.

Problem 6.1 (a) Write a balanced equation for combustion of 1-butene. (b) How does this equation compare with the corresponding one for *cis*-2-butene? For *trans*-2-butene? (c) The following heats of combustion have been measured for these three butenes: 648.1, 647.1, 649.8 kcal. Which heat of combustion do you think applies to each butene? (d) Assign the following heats of combustion to 1-pentene, and *cis*- and *trans*-2-pentene: 804.3, 806.9, 805.3.

6.5 Addition of halogens

Alkenes are readily converted by chlorine or bromine into saturated compounds that contain two atoms of halogen attached to adjacent carbons; iodine generally fails to react.

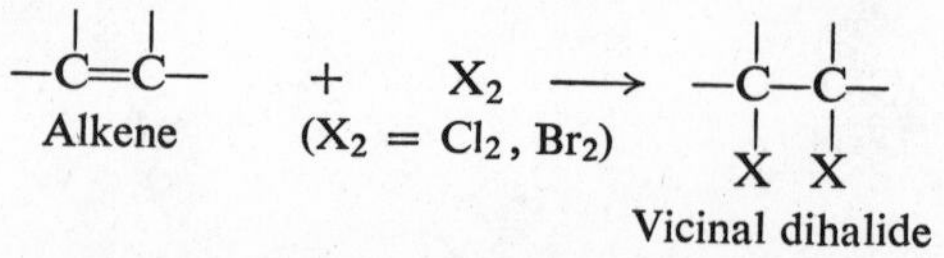

$$-\overset{|}{C}=\overset{|}{C}- \quad + \quad X_2 \longrightarrow -\overset{|}{\underset{X}{\underset{|}{C}}}-\overset{|}{\underset{X}{\underset{|}{C}}}-$$

Alkene $(X_2 = Cl_2, Br_2)$ Vicinal dihalide

The reaction is carried out simply by mixing together the two reactants, usually in an inert solvent like carbon tetrachloride. The addition proceeds rapidly at room temperature or below, and does not require exposure to ultraviolet light; in fact, we deliberately avoid higher temperatures and undue exposure to light, as well as the presence of excess halogen, since under those conditions substitution might become an important side reaction.

This reaction is by far the best method of preparing **vicinal dihalides.** For example:

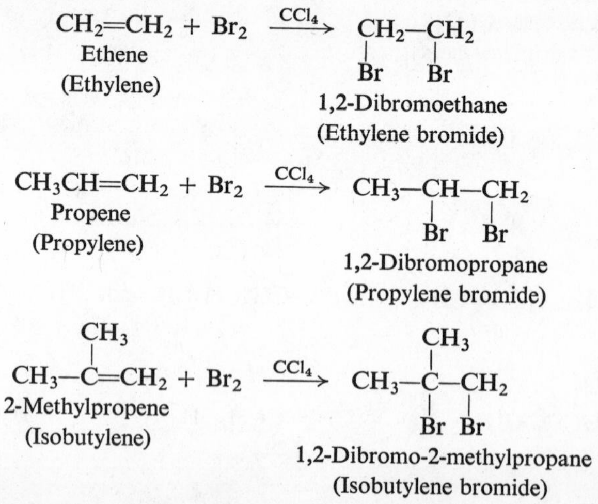

$$CH_2{=}CH_2 + Br_2 \xrightarrow{CCl_4} \underset{Br}{\underset{|}{CH_2}}-\underset{Br}{\underset{|}{CH_2}}$$

Ethene (Ethylene) → 1,2-Dibromoethane (Ethylene bromide)

$$CH_3CH{=}CH_2 + Br_2 \xrightarrow{CCl_4} CH_3-\underset{Br}{\underset{|}{CH}}-\underset{Br}{\underset{|}{CH_2}}$$

Propene (Propylene) → 1,2-Dibromopropane (Propylene bromide)

$$CH_3-\overset{CH_3}{\overset{|}{C}}{=}CH_2 + Br_2 \xrightarrow{CCl_4} CH_3-\underset{Br}{\underset{|}{\overset{CH_3}{\overset{|}{C}}}}-\underset{Br}{\underset{|}{CH_2}}$$

2-Methylpropene (Isobutylene) → 1,2-Dibromo-2-methylpropane (Isobutylene bromide)

Addition of bromine is extremely useful for detection of the carbon–carbon double bond. A solution of bromine in carbon tetrachloride is red; the dihalide, like the alkene, is colorless. Rapid decolorization of a bromine solution is characteristic of compounds containing the carbon–carbon double bond. (However, see Sec. 6.30.)

A common method of naming alkene derivatives is illustrated here. As we see, the product of the reaction between ethylene and bromine has the IUPAC name of 1,2-dibromoethane. It is also frequently called *ethylene bromide*, the word *ethylene* forming part of the name even though the compound is actually saturated. This is an old-fashioned name, and is meant to indicate the product of the reaction between ethylene and bromine, just as, for example, *sodium bromide* would indicate the product of the reaction between sodium and bromine. It should not be confused with the different compound, 1,2-dibromoethene, BrCH=CHBr. In a similar way, we have *propylene bromide, isobutylene bromide,* and so on.

We shall shortly encounter other saturated compounds that are named in a similar way, as, for example, *ethylene bromohydrin* and *ethylene glycol.* These names have in common the use of two words, the first of which is the name of the alkene; in this way they can be recognized as applying to compounds no longer containing the double bond.

6.6 Addition of hydrogen halides. Markovnikov's rule

An alkene is converted by hydrogen chloride, hydrogen bromide, or hydrogen iodide into the corresponding alkyl halide.

$$\underset{\text{Alkene}}{-\overset{|}{C}=\overset{|}{C}-} + \underset{(HX\,=\,HCl,\,HBr,\,HI)}{HX} \longrightarrow \underset{\text{Alkyl halide}}{-\underset{H}{\overset{|}{\underset{|}{C}}}-\underset{X}{\overset{|}{\underset{|}{C}}}-}$$

The reaction is frequently carried out by passing the dry gaseous hydrogen halide directly into the alkene. Sometimes the moderately polar solvent, acetic acid, which will dissolve both the polar hydrogen halide and the non-polar alkene, is used. The familiar aqueous solutions of the hydrogen halides are not generally used; in part, this is to avoid the addition of water to the alkene (Sec. 6.9).

Problem 6.2 (a) What is the acid in an aqueous solution of HBr? In dry HBr? (b) Which is the stronger acid? (c) Which can better transfer a hydrogen ion to an alkene?

In this way, ethylene is converted into an ethyl halide, the hydrogen becoming attached to one doubly-bonded carbon and the halogen to the other.

$$\underset{\text{Ethylene}}{CH_2{=}CH_2} + HI \longrightarrow \underset{\text{Ethyl iodide}}{CH_3CH_2I}$$

Propylene could yield either of two products, the *n*-propyl halide or the isopropyl halide, depending upon the orientation of addition, that is, depending upon which carbon atoms the hydrogen and halogen become attached to. Actually, it is found that the isopropyl halide greatly predominates.

$$\underset{H-I}{CH_3-CH{=}CH_2} \not\longrightarrow \underset{n\text{-Propyl iodide}}{CH_3-\underset{H}{\underset{|}{CH}}-\underset{I}{\underset{|}{CH_2}}}$$

$$\underset{I-H}{CH_3-CH{=}CH_2} \longrightarrow \underset{\text{Isopropyl iodide}}{CH_3-\underset{I}{\underset{|}{CH}}-\underset{H}{\underset{|}{CH_2}}} \qquad \textit{Actual product}$$

In the same way, isobutylene could yield either of two products, isobutyl halide or *tert*-butyl halide; here the orientation of addition is such that the *tert*-butyl halide greatly predominates.

$$CH_3-\underset{|}{\overset{CH_3}{\overset{|}{C}}}=CH_2 + H-I \;\not\longrightarrow\; CH_3-\underset{H}{\overset{CH_3}{\overset{|}{\underset{|}{C}}}}-\underset{I}{\underset{|}{CH_2}}$$

Isobutyl iodide

$$CH_3-\overset{CH_3}{\overset{|}{C}}=CH_2 + I-H \longrightarrow CH_3-\underset{I}{\overset{CH_3}{\overset{|}{\underset{|}{C}}}}-\underset{H}{\underset{|}{CH_2}}$$

tert-Butyl iodide

Actual product

Orientation in alkane substitutions (Sec. 3.21) depends upon which hydrogen is replaced; orientation in alkene additions depends upon which doubly-bonded carbon accepts Y and which accepts Z of a reagent YZ.

Examination of a large number of such additions showed the Russian chemist Vladimir Markovnikov (of the University of Kazan) that where two isomeric products are possible, one product usually predominates. He pointed out in 1869 that the orientation of addition follows a pattern which we can summarize as: *In the ionic addition of an acid to the carbon–carbon double bond of an alkene, the hydrogen of the acid attaches itself to the carbon atom that already holds the greater number of hydrogens.* This statement is generally known as **Markovnikov's rule.** Thus: "Unto everyone that hath shall be given," or "Them as has, gits."

Thus, in the addition to propylene we see that the hydrogen goes to the carbon bearing two hydrogen atoms rather than to the carbon bearing one. In the addition to isobutylene, the hydrogen goes to the carbon bearing two hydrogens rather than to the carbon bearing none.

Using Markovnikov's rule, we can correctly predict the principal product of many reactions. For example:

$$CH_3CH_2CH{=}CH_2 + HI \longrightarrow CH_3CH_2CHICH_3$$

1-Butene → *sec*-Butyl iodide (2-Iodobutane)

$$CH_3\overset{CH_3}{\overset{|}{C}}{=}CH-CH_3 + HI \longrightarrow CH_3-\underset{I}{\underset{|}{\overset{CH_3}{\overset{|}{C}}}}-CH_2-CH_3$$

2-Methyl-2-butene → *tert*-Pentyl iodide (2-Iodo-2-methylbutane)

$$CH_3CH{=}CHCH_3 + HI \longrightarrow CH_3CHICH_2CH_3$$

2-Butene → *sec*-Butyl iodide (2-Iodobutane)

$$CH_2{=}CHCl + HI \longrightarrow CH_3CHICl$$

Vinyl chloride (Chloroethene) → 1-Chloro-1-iodoethane

$$CH_3CH_2CH{=}CHCH_3 + HI \longrightarrow CH_3CH_2CHICH_2CH_3 + CH_3CH_2CH_2CHICH_3$$

2-Pentene → 3-Iodopentane + 2-Iodopentane

In 2-pentene each of the doubly-bonded carbons holds one hydrogen, so that according to the rule we should expect neither product to predominate. Here again the prediction is essentially correct, roughly equal quantities of the two isomers actually being obtained.

The examples have involved the addition of hydrogen iodide; exactly similar results are obtained in the addition of hydrogen chloride and, except for special conditions indicated in the following section, of hydrogen bromide.

Reactions that, from the standpoint of orientation, give exclusively or nearly exclusively *one* of several possible isomeric products are called **regiospecific.** (From the Latin *regio*, direction, and pronounced "reejio.")

Addition of hydrogen halides to alkenes can be used to make alkyl halides. The fact that addition occurs with a specific orientation, as summarized by Markovnikov's rule, rather than at random, is an advantage since a fairly pure product can generally be obtained. At the same time, the synthesis is, of course, limited to those products that are formed in agreement with Markovnikov's rule; for example, we can make isopropyl iodide in this way, but not *n*-propyl iodide. (As we shall see later, there are other, more important ways to prepare alkyl halides.)

6.7 Addition of hydrogen bromide. Peroxide effect

Addition of hydrogen chloride and hydrogen iodide to alkenes follows Markovnikov's rule. Until 1933 the situation with respect to hydrogen bromide was exceedingly confused. It had been reported by some workers that addition of hydrogen bromide to a particular alkene yields a product in agreement with Markovnikov's rule; by others, a product in contradiction to Markovnikov's rule; and by still others, a mixture of both products. It had been variously reported that the product obtained depended upon the presence or absence of water, or of light, or of certain metallic halides; it had been reported that the product obtained depended upon the solvent used, or upon the nature of the surface of the reaction vessel.

In 1933, M. S. Kharasch and F. R. Mayo at the University of Chicago brought order to this chemical chaos by discovering that the orientation of addition of hydrogen bromide to the carbon–carbon double bond is determined solely by the presence or absence of **peroxides.**

Organic peroxides are compounds containing the —O—O— linkage. They are encountered, generally in only very small amounts, as impurities in many organic compounds, where they have been slowly formed by the action of oxygen. Certain peroxides are deliberately synthesized, and used as reagents.

Kharasch and Mayo found that if one carefully excludes peroxides from the reaction system, or if one adds certain **inhibitors**—*hydroquinone* (p. 878), for example, or *diphenylamine* (p. 728)—the addition of HBr to alkenes follows Markovnikov's rule. On the other hand, if one does not exclude peroxides, or if one

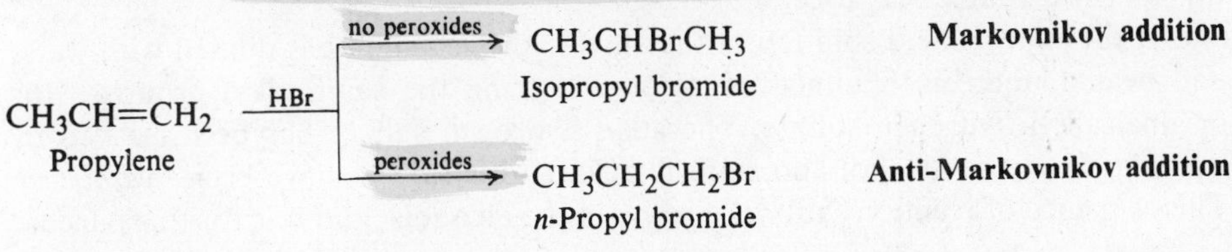

$$\underset{\text{Isobutylene}}{CH_3-\overset{\overset{CH_3}{|}}{C}=CH_2} \xrightarrow{HBr} \begin{cases} \xrightarrow{\text{no peroxides}} \underset{\textit{tert}\text{-Butyl bromide}}{CH_3-\overset{\overset{CH_3}{|}}{\underset{\underset{Br}{|}}{C}}-CH_3} & \textbf{Markovnikov addition} \\ \xrightarrow{\text{peroxides}} \underset{\text{Isobutyl bromide}}{CH_3-\overset{\overset{CH_3}{|}}{\underset{\underset{H}{|}}{C}}-CH_2Br} & \textbf{Anti-Markovnikov addition} \end{cases}$$

deliberately puts peroxides into the reaction system, HBr adds to alkenes in exactly the reverse direction.

This reversal of the orientation of addition caused by the presence of peroxides is known as the **peroxide effect.** Of the reactions we are studying, *only* the addition of hydrogen bromide shows the peroxide effect. The presence or absence of peroxides has no effect on the orientation of addition of hydrogen chloride, hydrogen iodide, sulfuric acid, water, etc. As we shall see (Secs. 6.11 and 6.17), both Markovnikov's rule and the peroxide effect can readily be accounted for in ways that are quite consistent with the chemistry we have learned so far.

6.8 Addition of sulfuric acid

Alkenes react with cold, concentrated sulfuric acid to form compounds of the general formula $ROSO_3H$, known as **alkyl hydrogen sulfates.** These products are formed by addition of hydrogen ion to one side of the double bond and bisulfate

$$\underset{\text{Alkene}}{-\overset{|}{C}=\overset{|}{C}-} + \underset{\text{Sulfuric acid}}{H-O-\overset{\overset{O}{|}}{\underset{\underset{O}{|}}{S}}-O-H} \longrightarrow \underset{\text{Alkyl hydrogen sulfate}}{-\overset{|}{\underset{\underset{H}{|}}{C}}-\overset{|}{\underset{|}{C}}-O-\overset{\overset{O}{|}}{\underset{\underset{O}{|}}{S}}-O-H}$$

ion to the other. It is important to notice that carbon is bonded to oxygen and not to sulfur.

Reaction is carried out simply by bringing the reactants into contact: a gaseous alkene is bubbled through the acid, and a liquid alkene is stirred or shaken with the acid. Since alkyl hydrogen sulfates are soluble in sulfuric acid, a clear solution results. The alkyl hydrogen sulfates are deliquescent solids, and are difficult to isolate. As the examples below show, the concentration of sulfuric acid required for reaction depends upon the particular alkene involved; we shall later account for this in a reasonable way (Sec. 6.11).

If the sulfuric acid solution of the alkyl hydrogen sulfate is diluted with water and heated, there is obtained an alcohol bearing the same alkyl group as the original alkyl hydrogen sulfate. The alkyl hydrogen sulfate has been cleaved by water to form the alcohol and sulfuric acid, and is said to have been *hydrolyzed.* This sequence of reactions affords a route to the alcohols, and it is for this purpose that addition of sulfuric acid to alkenes is generally carried out. This is an excellent method for the large-scale manufacture of alcohols, since alkenes are readily

$$\underset{\text{Ethylene}}{CH_2{=}CH_2} \xrightarrow{98\%\ H_2SO_4} \underset{\text{Ethyl hydrogen sulfate}}{CH_3CH_2OSO_3H} \xrightarrow{H_2O,\ \text{heat}} \underset{\text{Ethyl alcohol}}{CH_3CH_2OH} + H_2SO_4$$

$$\underset{\text{Propylene}}{CH_3CH{=}CH_2} \xrightarrow{80\%\ H_2SO_4} \underset{\text{Isopropyl hydrogen sulfate}}{CH_3CH(OSO_3H)CH_3} \xrightarrow{H_2O,\ \text{heat}} \underset{\text{Isopropyl alcohol}}{CH_3CH(OH)CH_3}$$

$$\underset{\text{Isobutylene}}{CH_3{-}C(CH_3){=}CH_2} \xrightarrow{63\%\ H_2SO_4} \underset{\textit{tert}\text{-Butyl hydrogen sulfate}}{CH_3{-}C(CH_3)(OSO_3H){-}CH_3} \xrightarrow{H_2O,\ \text{heat}} \underset{\textit{tert}\text{-Butyl alcohol}}{CH_3{-}C(CH_3)(OH){-}CH_3}$$

obtained by the cracking of petroleum. Because the addition of sulfuric acid follows Markovnikov's rule, certain alcohols cannot be obtained by this method. For example, isopropyl alcohol can be made but not *n*-propyl alcohol; *tert*-butyl alcohol, but not isobutyl alcohol.

The fact that alkenes dissolve in cold, concentrated sulfuric acid to form the alkyl hydrogen sulfates is made use of in the purification of certain other kinds of compounds. Alkanes or alkyl halides, for example, which are insoluble in sulfuric acid, can be freed from alkene impurities by washing with sulfuric acid. A gaseous alkane is bubbled through several bottles of sulfuric acid, and a liquid alkane is shaken with sulfuric acid in a separatory funnel.

6.9 Addition of water. Hydration

Water adds to the more reactive alkenes in the presence of acids to yield alcohols. Since this addition, too, follows Markovnikov's rule, the alcohols are

$$\underset{\text{Alkene}}{-\overset{|}{C}{=}\overset{|}{C}-} + H_2O \xrightarrow{H^+} \underset{\text{Alcohol}}{-\overset{|}{\underset{H}{C}}-\overset{|}{\underset{OH}{C}}-}$$

the same as those obtained by the two-step synthesis just described; this direct hydration is, of course, the simpler and cheaper of the two processes. Hydration of alkenes is the principal industrial source of those lower alcohols whose formation is consistent with Markovnikov's rule.

$$\underset{\text{Isobutylene}}{CH_3{-}C(CH_3){=}CH_2} \xrightarrow{H_2O,\ H^+} \underset{\textit{tert}\text{-Butyl alcohol}}{CH_3{-}C(CH_3)(OH){-}CH_3}$$

6.10 Electrophilic addition: mechanism

Before we consider other reactions of alkenes, it will be helpful to examine the mechanism of some of the reactions we have already discussed. After we have done this, we shall return to our systematic consideration of alkene reactions, prepared to understand them better in terms of these earlier reactions.

We shall take up first the addition of those reagents which contain ionizable hydrogen: the hydrogen halides, sulfuric acid, and water. The generally accepted mechanism will be outlined, and then we shall see how this mechanism accounts for certain facts. Like dehydration of alcohols, addition is pictured as involving carbonium ions. We shall notice certain resemblances between these two kinds of reaction; these resemblances are evidence that a common intermediate is involved.

Addition of the acidic reagent, HZ, is believed to proceed by two steps:

(1) $-\overset{|}{C}=\overset{|}{C}- + H{:}Z \longrightarrow -\underset{H}{\overset{|}{C}}-\overset{|}{\underset{\oplus}{C}}- + {:}Z$ $\quad$ $HZ = HCl, HBr, HI, H_2SO_4, H_3O^+$

(2) $-\underset{H}{\overset{|}{C}}-\overset{|}{\underset{\oplus}{C}}- + {:}Z \longrightarrow -\underset{H}{\overset{|}{C}}-\underset{Z}{\overset{|}{C}}-$ $\quad$ ${:}Z = Cl^-, Br^-, I^-, HSO_4^-, H_2O$

Step (1) involves transfer of hydrogen ion from :Z to the alkene to form a carbonium ion; this is a transfer of a proton from one base to another.

Electrophilic addition

$$-\overset{|}{C}{:}{:}\overset{|}{C}- \quad Z{:}H \longrightarrow Z{:} + -\underset{H}{\overset{|}{C}}{:}\underset{\oplus}{\overset{|}{C}}-$$

represented as

$$-\overset{|}{C}=\overset{|}{C}- \quad Z-H \longrightarrow Z{:} + -\underset{H}{\overset{|}{C}}-\underset{\oplus}{\overset{|}{C}}-$$

Step (2) is the union of the carbonium ion with the base :Z.

Step (1) is the difficult step, and its rate largely or entirely controls the overall rate of addition. This step involves attack by an acidic, electron-seeking reagent—that is, an *electrophilic* reagent—and hence the reaction is called **electrophilic addition.** The electrophile need not necessarily be a Lowry–Brønsted acid transferring a proton, as shown here, but, as we shall see, can be almost any kind of electron-deficient molecule (Lewis acid).

On the basis of step (2), we can add another reaction to our list of Sec. 5.22. **A carbonium ion may:**

(c) combine with a negative ion or other basic molecule to form a halide, a bisulfate, an alcohol, etc.

This reaction, like the earlier ones, provides the electron-deficient carbon with a pair of electrons.

The general mechanism is illustrated by specific examples: addition of hydrogen chloride,

(1) $CH_3-CH=CH_2 + H{:}\ddot{C}l{:} \longrightarrow CH_3-\underset{\oplus}{C}H-CH_3 + {:}\ddot{C}l{:}^-$

(2) $$CH_3{-}\underset{\oplus}{CH}{-}CH_3 + :\ddot{Cl}:^- \longrightarrow CH_3{-}\underset{\underset{Cl}{|}}{CH}{-}CH_3$$

of sulfuric acid,

(1) $$CH_3{-}CH{=}CH_2 + H{:}OSO_3H \longrightarrow CH_3{-}\underset{\oplus}{CH}{-}CH_3 + :OSO_3H^-$$

(2) $$CH_3{-}\underset{\oplus}{CH}{-}CH_3 + :OSO_3H^- \longrightarrow CH_3{-}\underset{\underset{OSO_3H}{|}}{CH}{-}CH_3$$

and of water.

(1) $$CH_3{-}CH{=}CH_2 + H{:}OH_2^+ \rightleftarrows CH_3{-}\underset{\oplus}{CH}{-}CH_3 + :OH_2$$

(2a) $$CH_3{-}\underset{\oplus}{CH}{-}CH_3 + :OH_2 \rightleftarrows CH_3{-}\underset{\underset{\oplus OH_2}{|}}{CH}{-}CH_3$$

(2b) $$CH_3{-}\underset{\underset{\oplus OH_2}{|}}{CH}{-}CH_3 + :OH_2 \rightleftarrows CH_3{-}\underset{\underset{OH}{|}}{CH}{-}CH_3 + H{:}OH_2^+$$

We notice that the carbonium ion combines with water to form not the alcohol but the protonated alcohol; in a subsequent reaction this protonated alcohol releases a hydrogen ion to another base to form the alcohol. This sequence of reactions, we can see, is just the reverse of that proposed for the dehydration of alcohols (Sec. 5.20). In dehydration, the equilibria are shifted in favor of the alkene chiefly by the removal of the alkene from the reaction mixture by distillation: in hydration, the equilibria are shifted in favor of the alcohol partly by the high concentration of water.

Let us see how this mechanism accounts for some of the facts.

First, the mechanism is consistent with (a) *the acidic nature of the reagents.* According to the mechanism, the first step in all these reactions is the transfer of a hydrogen ion to the alkene. This agrees with the fact that all these reagents except water are strong acids in the classical sense; that is, they can readily supply hydrogen ions. The exception, water, requires the presence of a strong acid for reaction to occur.

Next, the mechanism is consistent with (b) *the basic nature of alkenes.* The mechanism pictures the alkene as a base, supplying electrons to an attacking acid. This agrees with the structure of the carbon–carbon double bond: basicity is due to the loosely held, mobile π electrons.

In the following sections we shall see that the mechanism is also consistent with (c) *the orientation of addition*, (d) *the relative reactivities of alkenes*, and (e) *the occurrence of rearrangements.*

Problem 6.3 Addition of D_2O to 2-methyl-2-butene (in the presence of D^+) was found (as we might expect) to yield the alcohol $(CH_3)_2C(OD)CHDCH_3$. When the reaction was about half over, it was interrupted and the unconsumed alkene was isolated; mass spectrometric analysis showed that it contained almost no deuterium. This fact is considered to be evidence that formation of the carbonium ion is rate-determining: that as soon as a carbonium ion is formed, it rapidly reacts with water to yield the alcohol. Show how this conclusion is justified. (*Hint:* What results would you expect if the carbonium ions were formed rapidly and reversibly, and only every so often combined with water?)

6.11 Electrophilic addition: orientation and reactivity

The mechanism is consistent with the orientation of addition of acidic reagents, and with the effect of structure on relative reactivities.

Addition of hydrogen chloride to three typical alkenes is outlined below, with the two steps of the mechanism shown. In accord with Markovnikov's rule, propylene yields isopropyl chloride, isobutylene yields *tert*-butyl chloride, and 2-methyl-2-butene yields *tert*-pentyl chloride.

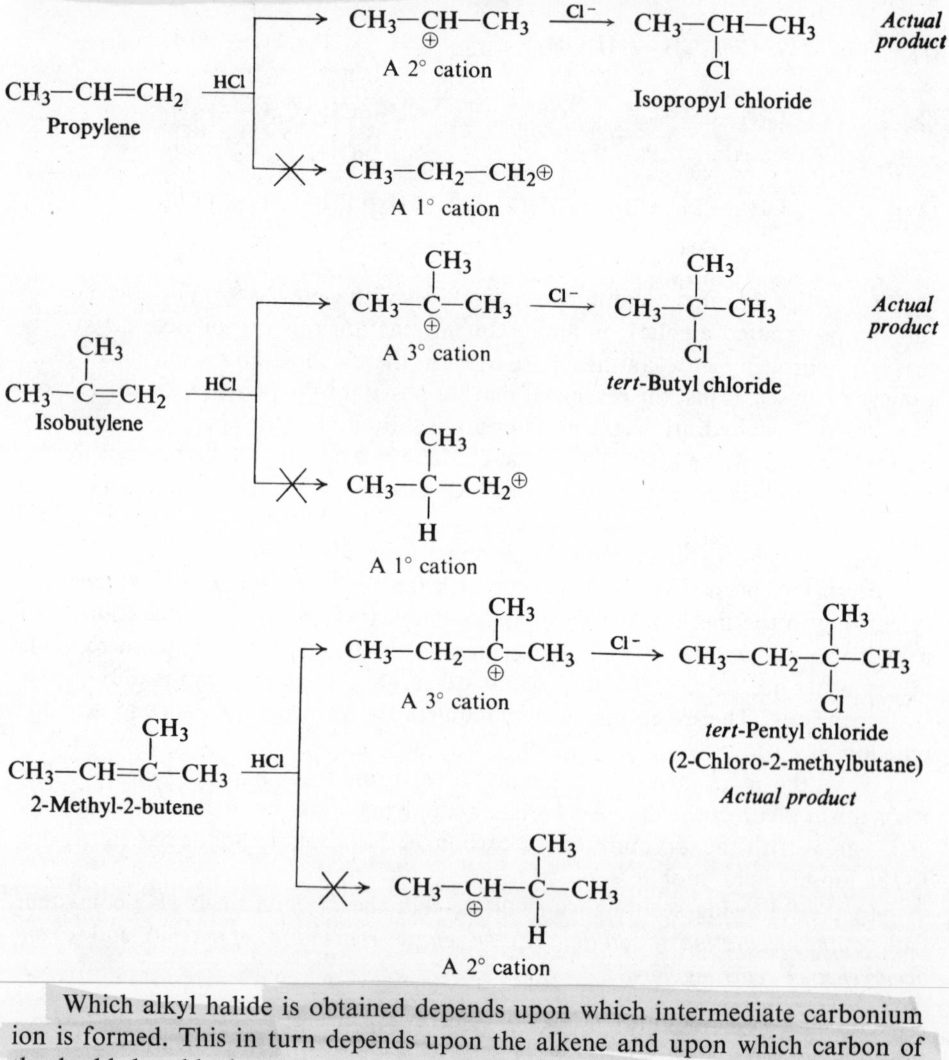

$$CH_3{-}CH{=}CH_2 \xrightarrow{HCl} CH_3{-}\overset{\oplus}{C}H{-}CH_3 \xrightarrow{Cl^-} CH_3{-}CH(Cl){-}CH_3$$

Propylene → A 2° cation → Isopropyl chloride (*Actual product*)

$$CH_3{-}CH{=}CH_2 \xrightarrow{HCl} \not\to CH_3{-}CH_2{-}CH_2^{\oplus}$$

A 1° cation

$$(CH_3)_2C{=}CH_2 \xrightarrow{HCl} CH_3{-}\overset{\oplus}{C}(CH_3){-}CH_3 \xrightarrow{Cl^-} CH_3{-}C(CH_3)(Cl){-}CH_3$$

Isobutylene → A 3° cation → *tert*-Butyl chloride (*Actual product*)

$$(CH_3)_2C{=}CH_2 \xrightarrow{HCl} \not\to CH_3{-}C(CH_3)(H){-}CH_2^{\oplus}$$

A 1° cation

$$CH_3{-}CH{=}C(CH_3){-}CH_3 \xrightarrow{HCl} CH_3{-}CH_2{-}\overset{\oplus}{C}(CH_3){-}CH_3 \xrightarrow{Cl^-} CH_3{-}CH_2{-}C(CH_3)(Cl){-}CH_3$$

2-Methyl-2-butene → A 3° cation → *tert*-Pentyl chloride (2-Chloro-2-methylbutane) *Actual product*

$$CH_3{-}CH{=}C(CH_3){-}CH_3 \xrightarrow{HCl} \not\to CH_3{-}\overset{\oplus}{C}H{-}C(CH_3)(H){-}CH_3$$

A 2° cation

Which alkyl halide is obtained depends upon which intermediate carbonium ion is formed. This in turn depends upon the alkene and upon which carbon of the double bond hydrogen goes to. Propylene, for example, could yield an *n*-propyl cation if hydrogen went to C–2 or an isopropyl cation if hydrogen went to C–1.

Orientation is thus determined by the relative rates of two competing reactions: formation of one carbonium ion or another. The fact that propylene is converted into the isopropyl cation instead of the *n*-propyl cation means that the isopropyl cation is formed *faster* than the *n*-propyl cation.

In each of the examples given above, the product obtained shows that in the initial step a secondary cation is formed faster than a primary, or a tertiary faster than a primary, or a tertiary faster than a secondary. Examination of many cases of addition of acids to alkenes shows that this is a general rule: orientation is governed by the ease of formation of carbonium ions, which follows the sequence 3° > 2° > 1°.

In listing carbonium ions in order of their ease of formation from alkenes, we find that once more (compare Sec. 5.21) we have listed them in order of their stability (Sec. 5.18).

Stability of carbonium ions $3° > 2° > 1° > CH_3^+$

We can now replace Markovnikov's rule by a more general rule: **electrophilic addition to a carbon–carbon double bond involves the intermediate formation of the more stable carbonium ion.**

Is it reasonable that the more stable carbonium ion should be formed more easily? We answered this question in Sec. 5.21 by considering the transition state leading to a carbonium ion; let us do the same here.

In addition reactions, the carbonium ion is formed by attachment of hydrogen ion to one of the doubly-bonded carbons. In the reactant the positive charge is entirely on the hydrogen ion; in the product it is on the carbon atom. In the transition state, the C—H bond must be partly formed, and the double bond partly broken. As a result the positive charge is divided between hydrogen and carbon.

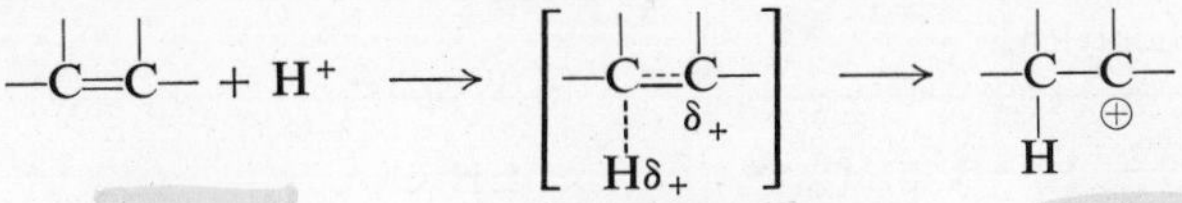

$$-\overset{|}{C}=\overset{|}{C}- + H^+ \longrightarrow \left[-\overset{|}{C}\overset{\delta_+}{\text{---}}\underset{\underset{H\delta_+}{\vdots}}{\overset{|}{C}}- \right] \longrightarrow -\overset{|}{\underset{\underset{H}{|}}{C}}-\overset{|}{\underset{\oplus}{C}}-$$

Electrophilic addition

Reactants	Transition state	Product
Hydrogen has full positive charge	*Carbon and hydrogen have partial positive charges*	*Carbon has full positive charge*

Electron-releasing groups tend to disperse the partial positive charge (δ_+) developing on carbon and in this way stabilize the transition state. Stabilization of the transition state lowers E_{act} and permits a faster reaction (see Fig. 6.4). As before, the electron release that stabilizes the carbonium ion also stabilizes the *incipient* carbonium ion in the transition state. The more stable carbonium ion is formed faster.

Thus, the rate of addition of a hydrogen ion to a double bond depends upon the stability of the carbonium ion being formed. As we might expect, this factor determines not only the **orientation** of addition to a simple alkene, but also the **relative reactivities** of different alkenes.

Alkenes generally show the following order of reactivity toward addition of acids:

Reactivity of alkenes toward acids

$$(CH_3)_2C=CH_2 > CH_3CH=CHCH_3,\ CH_3CH_2CH=CH_2,\ CH_3CH=CH_2 > CH_2=CH_2 > CH_2=CHCl$$

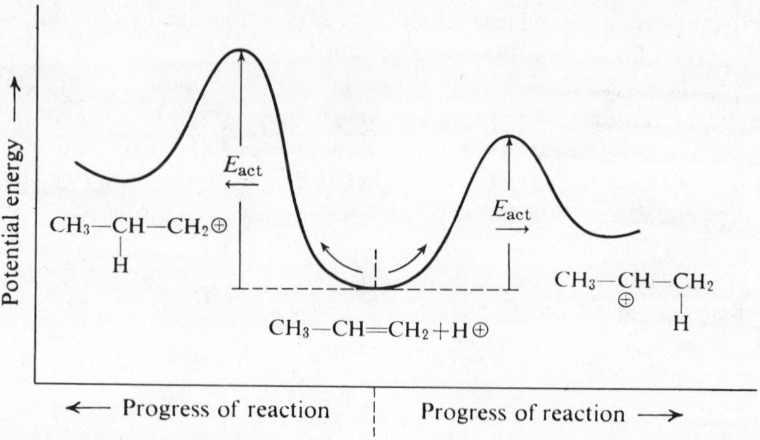

Figure 6.4. Molecular structure and orientation of reaction. Stability of transition state parallels stability of carbonium ion: more stable carbonium ion formed faster.

Isobutylene, which forms a tertiary cation, reacts faster than 2-butene, which forms a secondary cation. 1-Butene, 2-butene, and propylene, which form secondary cations, react faster than ethylene, which forms a primary cation.

$$\underset{\text{Isobutylene}}{CH_3-\overset{CH_3}{\overset{|}{C}}=CH_2} + H^+ \longrightarrow \underset{\text{A 3° cation}}{CH_3-\overset{CH_3}{\overset{|}{\underset{\oplus}{C}}}-CH_3}$$

$$\underset{\text{2-Butene}}{CH_3CH=CHCH_3} + H^+ \longrightarrow \underset{\text{A 2° cation}}{CH_3CH_2\underset{\oplus}{C}HCH_3}$$

$$\underset{\text{1-Butene}}{CH_3CH_2CH=CH_2} + H^+ \longrightarrow \underset{\text{A 2° cation}}{CH_3CH_2\underset{\oplus}{C}HCH_3}$$

$$\underset{\text{Propylene}}{CH_3CH=CH_2} + H^+ \longrightarrow \underset{\text{A 2° cation}}{CH_3\underset{\oplus}{C}HCH_3}$$

$$\underset{\text{Ethylene}}{CH_2=CH_2} + H^+ \longrightarrow \underset{\text{A 1° cation}}{CH_3CH_2\oplus}$$

Halogens, like other elements in the upper right-hand corner of the Periodic Table, tend to attract electrons. Just as electron release by alkyl groups disperses the positive charge and stabilizes a carbonium ion, so electron withdrawal by halogens intensifies the positive charge and destabilizes the carbonium ion. It is not surprising that vinyl chloride, $CH_2=CHCl$, is *less* reactive than ethylene.

We can begin to see what a powerful weapon we have for attacking the problems that arise in connection with a wide variety of reactions that involve carbonium ions. We know that the more stable the carbonium ion, the faster it is formed; that its stability depends upon dispersal of the charge; and that dispersal of charge is determined by the electronic effects of the attached groups. We have already found that this same approach enables us to deal with such seemingly different facts as (a) the relative ease of dehydration of alcohols; (b) the relative reactivities of alkenes toward addition of acids; and (c) the orientation of addition of acids to alkenes.

6.12 Electrophilic addition: rearrangement

The mechanism of electrophilic addition is consistent with the occurrence of rearrangements.

If carbonium ions are intermediates in electrophilic addition, then we should expect the reaction to be accompanied by the kind of rearrangement that we said earlier is highly characteristic of carbonium ions (Sec. 5.22). Rearrangements are not only observed, but they occur according to just the pattern that would be predicted.

For example, addition of hydrogen chloride to 3,3-dimethyl-1-butene yields not only 2-chloro-3,3-dimethylbutane, but also 2-chloro-2,3-dimethylbutane:

$$CH_3-\underset{CH_3}{\overset{CH_3}{C}}-CH{=}CH_2 \xrightarrow{HCl} CH_3-\underset{CH_3}{\overset{CH_3}{C}}-\underset{\oplus}{CH}-CH_3 \xrightarrow{Cl^-} CH_3-\underset{CH_3}{\overset{CH_3}{C}}-\underset{Cl}{CH}-CH_3$$

3,3 Dimethyl-1-butene; 2-Chloro-3,3-dimethylbutane

$\downarrow$ rearrangement

$$CH_3-\underset{\oplus}{\overset{CH_3}{C}}-\underset{CH_3}{CH}-CH_3 \xrightarrow{Cl^-} CH_3-\underset{Cl}{\overset{CH_3}{C}}-\underset{CH_3}{CH}-CH_3$$

2-Chloro-2,3-dimethylbutane

Since a 1,2-shift of a methyl group can convert the initially formed secondary cation into the more stable tertiary cation, such a rearrangement does occur, and much of the product is derived from this new ion. (If we compare this change in carbon skeleton with the one accompanying dehydration of 3,3-dimethyl-2-butanol (p. 171), we can begin to see how the idea arose that these apparently unrelated reactions proceed through the same intermediate.)

Problem 6.4 Addition of HCl to 3-methyl-1-butene yields a mixture of two alkyl chlorides. What are they likely to be, and how is each formed? Give detailed equations.

Problem 6.5 The reaction of aqueous HCl with 3,3-dimethyl-2-butanol yields 2,3-dimethyl-2-chlorobutane. *Using only reaction steps that you have already encountered*, propose a detailed mechanism for this reaction. (Check your answer in Sec. 16.5.)

6.13 Mechanism of addition of halogens

Electrophilic addition of acids to alkenes involves two steps, the first being attachment of hydrogen ion to form the carbonium ion. What is the mechanism of the addition of chlorine and bromine?

From the structure of the double bond we might expect that here again it is an electron source, a base, and hence that the halogen acts as an electrophilic reagent, an acid. This idea is supported by the fact that alkenes usually show the same order of reactivity toward halogens as toward the acids already studied: electron-releasing substituents activate an alkene, and electron-withdrawing substituents deactivate an alkene.

The commonly accepted mechanism for addition of halogens to alkenes has two steps, and is quite analogous to the mechanism for addition of hydrogen-containing acids (protic acids). In step (1) halogen adds as a positive halogen ion

(1) $$-\overset{|}{C}=\overset{|}{C}- + :\ddot{X}:\ddot{X}: \longrightarrow -\overset{|}{C}-\overset{|}{\underset{:\ddot{X}:}{C}}\oplus- + :\ddot{X}:^-$$

(2) $$-\overset{|}{\underset{:\ddot{X}:}{C}}-\overset{|}{C}\oplus- + :\ddot{X}:^- \longrightarrow -\overset{|}{\underset{:\ddot{X}:}{C}}-\overset{|}{\underset{:\ddot{X}:}{C}}-$$

to the double bond to form a carbonium ion. In step (2) the carbonium ion combines with a negative halide ion. (The mechanism is somewhat simplified for our present purpose, and will be modified in Sec. 7.12.)

It seems reasonable that an alkene should abstract hydrogen ion from the very polar hydrogen halide molecule. Is it reasonable that an alkene should abstract a positive halogen ion from the non-polar halogen molecule? Let us look at this problem more closely.

It is true that a halogen molecule is non-polar, since the two identical atoms share electrons equally. This is certainly not true, however, for a halogen molecule while it is under the influence of the powerful electric field of a nearby carbon–carbon double bond. The dense electron cloud of the double bond tends to repel the similarly charged electron cloud of the halogen molecule; this repulsion makes the halogen atom that is nearer the double bond relatively positive and its partner

$$>C{=}C< \qquad \overset{\delta_+}{Br}-\overset{\delta_-}{Br}$$

Polarization of Br_2 by a double bond

relatively negative. *The distortion of the electron distribution in one molecule caused by another molecule is called* **polarization**. Here, we would say that the alkene has *polarized* the halogen molecule.

The more positive halogen of this polarized molecule is then abstracted by the alkene to form a carbonium ion, leaving a negative halide ion. This halide ion, or more probably another just like it, finally collides with the carbonium ion to yield the product, a dihalide.

Let us look at some of the evidence for this mechanism. If a carbonium ion is the intermediate, we might expect it to react with almost any negative ion or basic molecule that we care to provide. For example, the carbonium ion formed in the reaction between ethylene and bromine should be able to react not only with bromide ion but also—if these are present—with chloride ion, iodide ion, nitrate ion, or water.

The facts are in complete agreement with this expectation. When ethylene is bubbled into an aqueous solution of bromine and sodium chloride, there is formed not only the dibromo compound but also the bromochloro compound and the bromoalcohol. Aqueous sodium chloride *alone* is completely inert toward ethylene; chloride ion or water can react only after the carbonium ion has been formed by the action of bromine. In a similar way bromine and aqueous sodium iodide or sodium nitrate convert ethylene into the bromoiodo compound or the bromonitrate, as well as into the dibromo compound and the bromoalcohol.

$$CH_2{=}CH_2 \xrightarrow{Br_2} CH_2Br{-}CH_2^{\oplus}$$

$\xrightarrow{Br^-}$ $CH_2Br{-}CH_2Br$ 1,2-Dibromoethane

$\xrightarrow{Cl^-}$ $CH_2Br{-}CH_2Cl$ 2-Bromo-1-chloroethane

$\xrightarrow{I^-}$ $CH_2Br{-}CH_2I$ 2-Bromo-1-iodoethane

$\xrightarrow{NO_3^-}$ $CH_2Br{-}CH_2ONO_2$ 2-Bromoethyl nitrate

$\xrightarrow{H_2O}$ $CH_2Br{-}CH_2\overset{\oplus}{O}H_2$ $\xrightarrow{-H^+}$ $CH_2Br{-}CH_2OH$ 2-Bromoethanol

Bromine in water with no added ions yields the dibromo compound and the bromoalcohol.

In addition to the elegant work just described, the stereochemistry of the reaction provides powerful support for a two-step addition of halogen. At the same time, as we shall see in Sec. 7.12, it requires a modification in the mechanism.

6.14 Halohydrin formation

As we have just seen, addition of chlorine or bromine in the presence of water can yield compounds containing halogen and hydroxyl groups on adjacent carbon atoms. These compounds are commonly referred to as **halohydrins**. Under proper conditions, they can be made the major products. For example:

$$\underset{\text{Ethylene}}{CH_2{=}CH_2} \xrightarrow{Br_2,\ H_2O} \underset{\text{OH}\quad\text{Br}}{CH_2{-}CH_2}$$

Ethylene bromohydrin (2-Bromoethanol)

$$\underset{\text{Propylene}}{CH_3{-}CH{=}CH_2} \xrightarrow{Cl_2,\ H_2O} CH_3{-}\underset{OH}{CH}{-}\underset{Cl}{CH_2}$$

Propylene chlorohydrin (1-Chloro-2-propanol)

There is evidence, of a kind we are not prepared to go into here, that these compounds are formed by reaction of halogen and water (as shown in Sec. 6.13) rather than by addition of preformed hypohalous acid, HOX. Whatever the mechanism, the result is addition of the elements of hypohalous acid (HO— and —X), and the reaction is often referred to in that way.

We notice that in propylene chlorohydrin chlorine is attached to the terminal carbon. This orientation is, we say, quite understandable in light of the mechanism and what we know about formation of carbonium ions: the initial addition of chlorine occurs in the way that yields the more stable secondary cation. However, we shall have to modify (Sec. 17.15) this interpretation of the orientation to fit the modified mechanism of Sec. 7.12.

6.15 Addition of alkenes. Dimerization

Under proper conditions, isobutylene is converted by sulfuric or phosphoric acid into a mixture of two alkenes of molecular formula C_8H_{16}. Hydrogenation of either of these alkenes produces the same alkane, 2,2,4-trimethylpentane (Sec. 3.30). The two alkenes are isomers, then, and differ only in position of the double bond. (*Problem:* Could they, instead, be *cis–trans* isomers?) When studied by the methods discussed at the end of this chapter (Sec. 6.29), these two alkenes are found to have the structures shown:

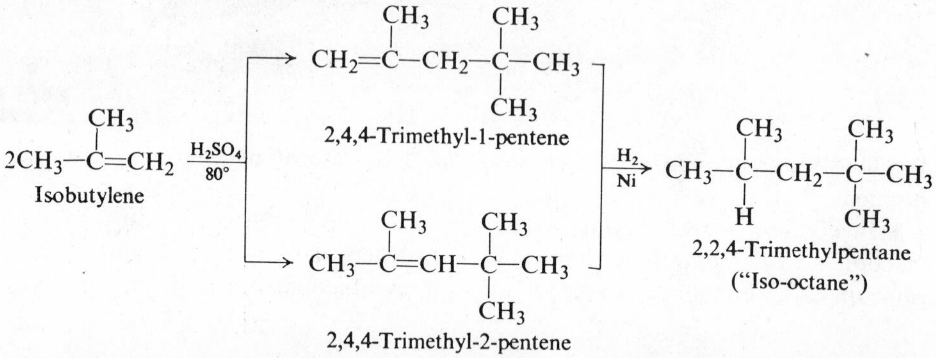

$$2CH_3-\overset{CH_3}{\overset{|}{C}}=CH_2 \xrightarrow[80°]{H_2SO_4} \begin{cases} CH_2=\overset{CH_3}{\overset{|}{C}}-CH_2-\underset{CH_3}{\underset{|}{\overset{CH_3}{\overset{|}{C}}}}-CH_3 \quad \text{2,4,4-Trimethyl-1-pentene} \\ CH_3-\overset{CH_3}{\overset{|}{C}}=CH-\underset{CH_3}{\underset{|}{\overset{CH_3}{\overset{|}{C}}}}-CH_3 \quad \text{2,4,4-Trimethyl-2-pentene} \end{cases} \xrightarrow[Ni]{H_2} CH_3-\underset{H}{\underset{|}{\overset{CH_3}{\overset{|}{C}}}}-CH_2-\underset{CH_3}{\underset{|}{\overset{CH_3}{\overset{|}{C}}}}-CH_3$$

Isobutylene

2,2,4-Trimethylpentane ("Iso-octane")

Since the alkenes produced contain exactly twice the number of carbon and hydrogen atoms as the original isobutylene, they are known as **dimers** (*di* = two, *mer* = part) of isobutylene, and the reaction is called **dimerization**. Other alkenes undergo analogous dimerizations.

Let us see if we can devise an acceptable mechanism for this dimerization. There are a great many isomeric octenes; if our mechanism should lead us to just the two that are actually formed, this in itself would provide considerable support for the mechanism.

Since the reaction is catalyzed by acid, let us write as step (1) addition of a hydrogen ion to isobutylene to form the carbonium ion; the tertiary cation would, of course, be the preferred ion.

$$(1) \qquad CH_3-\overset{CH_3}{\overset{|}{C}}=CH_2 + H^+ \longrightarrow CH_3-\underset{\oplus}{\overset{CH_3}{\overset{|}{C}}}-CH_3$$

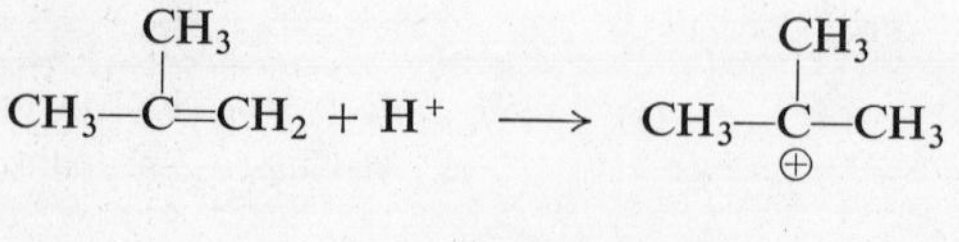

A carbonium ion undergoes reactions that provide electrons to complete the octet of the positively charged carbon atom. But a carbon–carbon double bond is an excellent electron source, and a carbonium ion might well go there in its quest for electrons. Let us write as step (2), then, addition of the *tert*-butyl cation to isobutylene; again, the orientation of addition is such as to yield the more stable

$$(2)\qquad CH_3-\overset{\overset{CH_3}{|}}{C}=CH_2 + \oplus\overset{\overset{CH_3}{|}}{\underset{\underset{CH_3}{|}}{C}}-CH_3 \longrightarrow CH_3-\overset{\overset{CH_3}{|}}{\underset{\oplus}{C}}-CH_2-\overset{\overset{CH_3}{|}}{\underset{\underset{CH_3}{|}}{C}}-CH_3$$

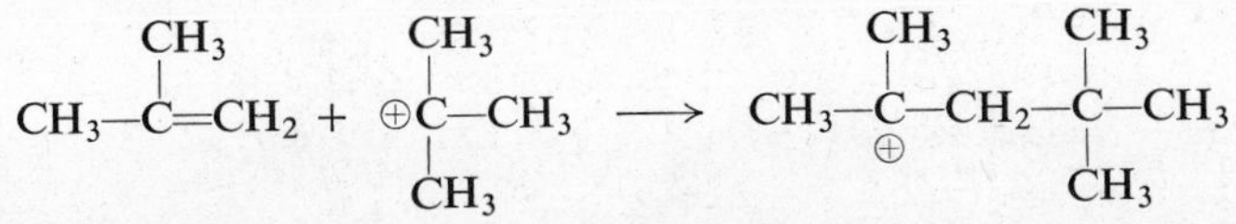

tertiary cation. Step (2) brings about the union of two isobutylene units, which is, of course, necessary to account for the products.

What is this new carbonium ion likely to do? We might expect that it could add to another molecule of alkene and thus make an even larger molecule; under certain conditions this does indeed happen. Under the present conditions, however, we know that this reaction stops at eight-carbon compounds, and that these compounds are alkenes. Evidently, the carbonium ion undergoes a reaction familiar to us: loss of a hydrogen ion (step 3). Since the hydrogen ion can be lost from a carbon on either side of the positively charged carbon, two products should be possible.

$$(3)\qquad CH_3-\overset{\overset{CH_3}{|}}{\underset{\oplus}{C}}-CH_2-\overset{\overset{CH_3}{|}}{\underset{\underset{CH_3}{|}}{C}}-CH_3 \longrightarrow \begin{cases} H^+ + CH_2=\overset{\overset{CH_3}{|}}{C}-CH_2-\overset{\overset{CH_3}{|}}{\underset{\underset{CH_3}{|}}{C}}-CH_3 \\[2ex] H^+ + CH_3-\overset{\overset{CH_3}{|}}{C}=CH-\overset{\overset{CH_3}{|}}{\underset{\underset{CH_3}{|}}{C}}-CH_3 \end{cases}$$

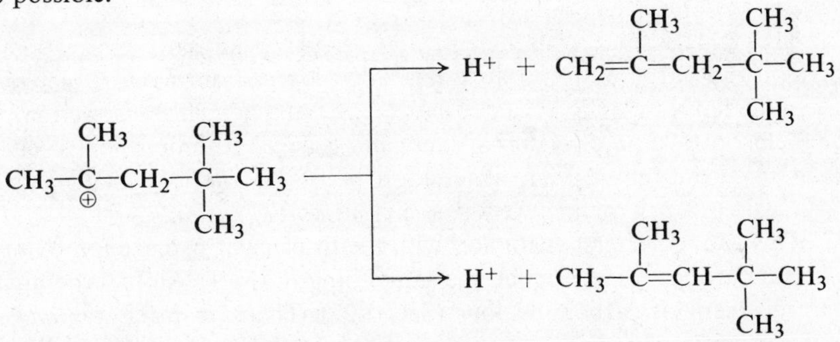

We find that the products expected on the basis of our mechanism are just the ones that are actually obtained. The fact that we can make this prediction simply on the basis of the fundamental properties of carbonium ions as we understand them is, of course, powerful support for the entire carbonium ion theory.

From what we have seen here, we can add one more reaction to those undergone by carbonium ions. **A carbonium ion may:**

(d) add to an alkene to form a larger carbonium ion.

6.16 Addition of alkanes. Alkylation

The large amounts of 2,2,4-trimethylpentane consumed as high-test gasoline are not made today by the dimerization reaction just described, but in another, cheaper way. Isobutylene and isobutane are allowed to react in the presence of an

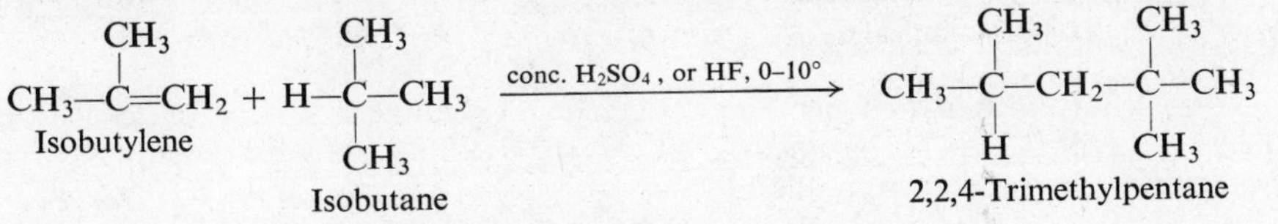

$$\underset{\text{Isobutylene}}{CH_3-\overset{\overset{CH_3}{|}}{C}=CH_2} + \underset{\text{Isobutane}}{H-\overset{\overset{CH_3}{|}}{\underset{\underset{CH_3}{|}}{C}}-CH_3} \xrightarrow{\text{conc. } H_2SO_4\text{, or HF, 0–10°}} \underset{\text{2,2,4-Trimethylpentane}}{CH_3-\overset{\overset{CH_3}{|}}{\underset{\underset{H}{|}}{C}}-CH_2-\overset{\overset{CH_3}{|}}{\underset{\underset{CH_3}{|}}{C}}-CH_3}$$

acidic catalyst, to form directly 2,2,4-trimethylpentane, or "iso-octane." This reaction is, in effect, addition of an alkane to an alkene.

The commonly accepted mechanism of this **alkylation** is based on the study of many related reactions and involves in step (3) a reaction of carbonium ions that we have not previously encountered.

(1) $$CH_3-\underset{}{\overset{CH_3}{\overset{|}{C}}}=CH_2 + H^+ \longrightarrow CH_3-\underset{\oplus}{\overset{CH_3}{\overset{|}{C}}}-CH_3$$

(2) $$CH_3-\overset{CH_3}{\overset{|}{C}}=CH_2 + \oplus\underset{\underset{CH_3}{|}}{\overset{CH_3}{\overset{|}{C}}}-CH_3 \longrightarrow CH_3-\underset{\oplus}{\overset{CH_3}{\overset{|}{C}}}-CH_2-\underset{\underset{CH_3}{|}}{\overset{CH_3}{\overset{|}{C}}}-CH_3$$

(3) $$CH_3-\underset{\oplus}{\overset{CH_3}{\overset{|}{C}}}-CH_2-\underset{\underset{CH_3}{|}}{\overset{CH_3}{\overset{|}{C}}}-CH_3 + H-\underset{\underset{CH_3}{|}}{\overset{CH_3}{\overset{|}{C}}}-CH_3 \longrightarrow CH_3-\underset{\underset{H}{|}}{\overset{CH_3}{\overset{|}{C}}}-CH_2-\underset{\underset{CH_3}{|}}{\overset{CH_3}{\overset{|}{C}}}-CH_3 + \oplus\underset{\underset{CH_3}{|}}{\overset{CH_3}{\overset{|}{C}}}-CH_3$$

then (2), (3), (2), (3), etc.

The first two steps are identical with those of the dimerization reaction. In step (3) a carbonium ion abstracts a hydrogen atom *with its pair of electrons* (a **hydride ion**, essentially) from a molecule of alkane. This abstraction of hydride ion yields an alkane of eight carbons, and a new carbonium ion to continue the chain. As we might expect, abstraction occurs in the way that yields the *tert*-butyl cation rather than the less stable (1°) isobutyl cation.

This is not our first encounter with the transfer of hydride ion to an electron-deficient carbon; we saw much the same thing in the 1,2-shifts accompanying the rearrangement of carbonium ions (Sec. 5.22). There, transfer was *intramolecular* (within a molecule); here, it is *intermolecular* (between molecules). We shall find hydride transfer playing an important part in the chemistry of carbonyl compounds (Chap. 19).

Let us now bring our list of carbonium ion reactions up to date. **A carbonium ion may:**

(a) eliminate a hydrogen ion to form an alkene;
(b) rearrange to a more stable carbonium ion;
(c) combine with a negative ion or other basic molecule;
(d) add to an alkene to form a larger carbonium ion;
(e) abstract a hydride ion from an alkane.

A carbonium ion formed by (b) or (d) can subsequently undergo any of the reactions.

As we see, all reactions of a carbonium ion have a common end: *they provide a pair of electrons to complete the octet of the positively charged carbon.*

Problem 6.6 When ethylene is alkylated by isobutane in the presence of acid, there is obtained, not neohexane, $(CH_3)_3CCH_2CH_3$, but chiefly 2,3-dimethylbutane. Account in detail for the formation of this product.

6.17 Free-radical addition. Mechanism of the peroxide-initiated addition of HBr

In the absence of peroxides, hydrogen bromide adds to alkenes in agreement with Markovnikov's rule; in the presence of peroxides, the direction of addition is exactly reversed (see Sec. 6.7).

To account for this *peroxide effect*, Kharasch and Mayo proposed that addition can take place by two entirely different mechanisms: Markovnikov addition by the ionic mechanism that we have just discussed, and anti-Markovnikov addition by a free-radical mechanism. Peroxides initiate the free-radical reaction; in their absence (or if an inhibitor, p. 189, is added), addition follows the usual ionic path.

The essence of the mechanism is that hydrogen and bromine add to the double bond as *atoms* rather than as ions; the intermediate is a *free radical* rather than a carbonium ion. Like halogenation of alkanes, this is a chain reaction, this time involving addition rather than substitution.

(1) peroxides ⟶ Rad·

(2) Rad· + H:Br ⟶ Rad:H + Br·

} **Chain-initiating steps**

(3) Br· + —C(|)=C(|)— ⟶ —C(|)(Br)—C(|)(·)—

(4) —C(|)(Br)—C(|)(·)— + H:Br ⟶ —C(|)(Br)—C(|)(H)— + Br·

} **Chain-propagating steps**

then (3), (4), (3), (4), etc.

Decomposition of the peroxide (step 1) to yield free radicals is a well-known reaction. The free radical thus formed abstracts hydrogen from hydrogen bromide (step 2) to form a bromine atom. The bromine atom adds to the double bond (step 3), and, in doing so, converts the alkene into a free radical.

—C(|)::C(|)— + Br· ⟶ —C(|)(:Br)—:—C(|)(·)— **Free-radical addition**

This free radical, like the free radical initially generated from the peroxide, abstracts hydrogen from hydrogen bromide (step 4). Addition is now complete, and a new bromine atom has been generated to continue the chain. As in halogenation of alkanes, every so often a reactive particle combines with another one, or is captured by the wall of the reaction vessel, and a chain is terminated.

The mechanism is well supported by the facts. The fact that a very few molecules of peroxide can change the orientation of addition of many molecules of hydrogen bromide strongly indicates a chain reaction. So, too, does the fact that a very few molecules of inhibitor can prevent this change in orientation. It is not surprising to find that these same compounds are efficient inhibitors of many other chain reactions. Although their exact mode of action is not understood, it seems clear that they break the chain, presumably by forming unreactive radicals.

We must not confuse the effects of peroxides, which may have been formed by the action of oxygen, with the effects of oxygen itself. Peroxides *initiate* free-radical reactions; oxygen *inhibits* free-radical reactions (see Sec. 2.14).

The mechanism involves addition of a bromine atom to the double bond. It is supported, therefore, by the fact that anti-Markovnikov addition is caused not only by the presence of peroxides but also by irradiation with light of a wavelength known to dissociate hydrogen bromide into hydrogen and bromine atoms.

Recently, the light-catalyzed addition of hydrogen bromide to several alkenes was studied by means of esr (electron spin resonance) spectroscopy, which not only can detect the presence of free radicals at extremely low concentrations, but also can tell something about their structure (see Sec. 13.14). Organic free radicals were shown to be present at appreciable concentration, in agreement with the mechanism.

Is it reasonable that free-radical addition of hydrogen bromide should occur with orientation opposite to that of ionic addition? Let us compare the two kinds of addition to propylene.

Ionic addition: *Markovnikov orientation*

$$\underset{\text{Propylene}}{CH_3{-}CH{=}CH_2} \xrightarrow{HBr} \underset{\text{A 2° cation}}{CH_3{-}\overset{}{\underset{\oplus}{C}H}{-}CH_3} \xrightarrow{Br^-} \underset{\text{Isopropyl bromide}}{CH_3{-}\underset{Br}{\underset{|}{C}H}{-}CH_3}$$

$$\underset{\text{Propylene}}{CH_3{-}CH{=}CH_2} \xrightarrow{HBr} \not\rightarrow \underset{\text{A 1° cation}}{CH_3{-}CH_2{-}CH_2^{\oplus}}$$

Free-radical addition: *Anti-Markovnikov orientation*

$$\underset{\text{Propylene}}{CH_3{-}CH{=}CH_2} \xrightarrow{Br\cdot} \underset{\text{A 2° free radical}}{CH_3{-}\dot{C}H{-}CH_2Br} \xrightarrow{HBr} \underset{n\text{-Propyl bromide}}{CH_3{-}CH_2{-}CH_2Br}$$

$$\underset{\text{Propylene}}{CH_3{-}CH{=}CH_2} \xrightarrow{Br\cdot} \not\rightarrow \underset{\text{A 1° free radical}}{CH_3{-}\underset{Br}{\underset{|}{C}H}{-}CH_2\cdot}$$

Ionic addition yields isopropyl bromide because a secondary cation is formed faster than a primary. Free-radical addition yields *n*-propyl bromide because a secondary free radical is formed faster than a primary. Examination of many cases of anti-Markovnikov addition shows that orientation is governed by the ease of formation of free radicals, which follows the sequence $3° > 2° > 1°$.

In listing free radicals in order of their ease of formation from alkenes, we find that once more (compare Sec. 3.25) we have listed them in order of their stability (Sec. 3.24):

Stability of free radicals $3° > 2° > 1° > CH_3\cdot$

Free-radical addition to a carbon–carbon double bond involves the intermediate formation of the more stable free radical.

Thus we find the chemistry of free radicals and the chemistry of carbonium ions following much the same pattern: the more stable particle is formed more easily, whether by abstraction or dissociation, or by addition to a double bond. Even the order of stability of the two kinds of particle is the same: 3° > 2° > 1° > CH_3. In this particular case orientation is reversed simply because the hydrogen adds first in the ionic reaction, and bromine adds first in the radical reaction.

6.18 Other free-radical additions

In the years since the discovery of the peroxide effect, dozens of reagents besides HBr have been found (mostly by Kharasch) to add to alkenes in the presence of peroxides or light. Exactly analogous free-radical mechanisms are generally accepted for these reactions, too.

For the addition of carbon tetrachloride to an alkene, for example,

$$RCH{=}CH_2 + CCl_4 \xrightarrow{\text{peroxides}} RCH(Cl){-}CH_2{-}CCl_3$$

the following mechanism has been proposed:

(1) $$\text{peroxide} \longrightarrow \text{Rad}\cdot$$

(2) $$\text{Rad}\cdot + Cl{:}CCl_3 \longrightarrow \text{Rad}{:}Cl + \cdot CCl_3$$

(3) $$\cdot CCl_3 + RCH{=}CH_2 \longrightarrow R\dot{C}H{-}CH_2{-}CCl_3$$

(4) $$R\dot{C}H{-}CH_2{-}CCl_3 + Cl{:}CCl_3 \longrightarrow RCH(Cl){-}CH_2{-}CCl_3 + \cdot CCl_3$$

then (3), (4), (3), (4), etc.

In the next section, we shall encounter another example of free-radical addition—*polymerization*—which has played a key part in the creation of this age of plastics.

Problem 6.7 In the presence of a trace of peroxide or under the influence of ultraviolet light, 1-octene reacts:

(a) with $CHCl_3$ to form 1,1,1-trichlorononane;
(b) with $CHBr_3$ to form 1,1,3-tribromononane;
(c) with $CBrCl_3$ to form 1,1,1-trichloro-3-bromononane;
(d) with $H{-}S{-}CH_2COOH$ (thioglycolic acid) to yield $n\text{-}C_8H_{17}{-}S{-}CH_2COOH$;
(e) with aldehydes, $R{-}\underset{\displaystyle H}{C}{=}O$, to yield ketones, $n\text{-}C_8H_{17}{-}\underset{\displaystyle \overset{\|}{O}}{C}{-}R$.

Show all steps of a likely mechanism for these reactions.

Problem 6.8 From the addition of CCl_4 to alkenes, $RCH{=}CH_2$, there is obtained not only $RCHClCH_2CCl_3$, but also $RCHClCH_2{-}\underset{\displaystyle R}{CH}CH_2CCl_3$. *Using only the kinds of reactions you have already encountered*, suggest a mechanism for the formation of this second product.

Problem 6.9 In the dark at room temperature, a solution of chlorine in tetrachloroethylene can be kept for long periods with no sign of reaction. When irradiated with ultraviolet light, however, the chlorine is rapidly consumed, with the formation of hexachloroethane; many molecules of product are formed for each photon of light absorbed; this reaction is slowed down markedly when oxygen is bubbled through the solution.

(a) How do you account for the absence of reaction in the dark? (b) Outline all steps in the most likely mechanism for the photochemical reaction. Show how it accounts for the facts, including the effect of oxygen.

Free-radical addition is probably even commoner than has been suspected. Recent work indicates that free-radical chains do not always require light or decomposition of highly unstable compounds like peroxides for their initiation. Sometimes a change from a polar solvent—which can stabilize a polar transition state—to a non-polar solvent causes a change from a heterolytic reaction to a homolytic one (Sec. 1.14). In some cases, it may even be that chains are started by *concerted homolysis*, in which cleavage of comparatively stable molecules (halogens, for example) is aided by the simultaneous breaking and making of other bonds. In the absence of the clue usually given by the method of initiation, the free-radical nature of such reactions is harder to detect; one depends upon inhibition by oxygen, detailed analysis of reaction kinetics, or a change in orientation or stereochemistry.

6.19 Free-radical polymerization of alkenes

When ethylene is heated under pressure with oxygen, there is obtained a compound of high molecular weight (about 20,000), which is essentially an alkane with a very long chain. This compound is made up of many ethylene units and

$$n\mathrm{CH_2{=}CH_2} \xrightarrow{\mathrm{O_2,\ heat,\ pressure}} \sim\mathrm{CH_2{-}CH_2{-}CH_2{-}CH_2{-}CH_2{-}CH_2}\sim \quad \text{or} \quad (\mathrm{{-}CH_2CH_2{-}})_n$$

Polyethylene

hence is called *polyethylene* (**poly** = many). It is familiar to most of us as the plastic material of packaging films.

The formation of polyethylene is a simple example of the process called **polymerization**: *the joining together of many small molecules to make very large molecules.* The compound composed of these very large molecules is called a **polymer** (Greek: *poly* + *meros*, many parts). The simple compounds from which polymers are made are called **monomers** (*mono* = one).

Polymerization of substituted ethylenes yields compounds whose structures contain the long chain of polyethylene, with substituents attached at more or less regular intervals. For example, vinyl chloride yields *poly(vinyl chloride)*, used to

$$n\mathrm{CH_2{=}CH(Cl)} \xrightarrow{\text{peroxides}} \sim\mathrm{CH_2{-}CH(Cl){-}CH_2{-}CH(Cl){-}CH_2{-}CH(Cl)}\sim \quad \text{or} \quad (\mathrm{{-}CH_2{-}CH(Cl){-}})_n$$

Vinyl chloride → Poly(vinyl chloride)

make phonograph records, plastic pipe, and—when plasticized with high-boiling esters—raincoats, shower curtains, and coatings for metals and upholstery fabrics.

Many other groups (e.g., $-COOCH_3$, $-CN$, $-C_6H_5$) may be attached to the doubly-bonded carbons. These substituted ethylenes polymerize more or less readily, and yield plastics of widely differing physical properties and uses, but the polymerization process and the structure of the polymer are basically the same as for ethylene or vinyl chloride.

Polymerization requires the presence of a small amount of an **initiator**. Among the commonest of these initiators are peroxides, which function by breaking down to form a free radical. This radical adds to a molecule of alkene, and in doing so generates another free radical. This radical adds to another molecule of alkene to generate a still larger radical, which in turn adds to another molecule of alkene, and so on. Eventually the chain is terminated by steps, such as union of two radicals, that consume but do not generate radicals.

$$\left.\begin{array}{l}\text{Peroxide} \longrightarrow \text{Rad}\cdot \\ \text{Rad}\cdot + CH_2{=}\underset{\displaystyle R}{\underset{|}{CH}} \longrightarrow \text{Rad}CH_2{-}\underset{\displaystyle R}{\underset{|}{CH}}\cdot\end{array}\right\} \textbf{Chain-initiating steps}$$

$$\text{Rad}CH_2{-}\underset{\displaystyle R}{\underset{|}{CH}}\cdot + CH_2{=}\underset{\displaystyle R}{\underset{|}{CH}} \longrightarrow \text{Rad}CH_2{-}\underset{\displaystyle R}{\underset{|}{CH}}{-}CH_2{-}\underset{\displaystyle R}{\underset{|}{CH}}\cdot \longrightarrow \textit{etc.} \quad \textbf{Chain-propagating step}$$

This kind of polymerization, each step of which consumes a reactive particle and produces another, similar particle, is an example of *chain-reaction polymerization*. In Chap. 32, we shall encounter chain-reaction polymerization that takes place, not by way of free radicals, but by way of organic ions. We shall also encounter *step-reaction polymerization*, which involves a series of reactions each of which is essentially independent of the others.

Problem 6.10 Give the structure of the monomer from which each of the following polymers would most likely be made:
(a) Orlon (fibers, fabrics), $\sim CH_2CH(CN)CH_2CH(CN)\sim$;
(b) Saran (packaging film, seat covers), $\sim CH_2CCl_2CH_2CCl_2\sim$;
(c) Teflon (chemically resistant articles), $\sim CF_2CF_2CF_2CF_2\sim$.

Problem 6.11 Can you suggest a reason why polymerization should take place in a way ("head-to-tail") that yields a polymer with regularly alternating groups?

6.20 Hydroxylation. Glycol formation

Certain oxidizing agents convert alkenes into compounds known as **glycols**. Glycols are simply dihydroxy alcohols; their formation amounts to the addition of two hydroxyl groups to the double bond.

$$-\overset{|}{C}{=}\overset{|}{C}- \xrightarrow[\text{or } HCO_2OH]{\text{cold alkaline } KMnO_4} -\underset{\displaystyle OH}{\underset{|}{\overset{|}{C}}}{-}\underset{\displaystyle OH}{\underset{|}{\overset{|}{C}}}-$$

A glycol

Of the numerous oxidizing agents that cause hydroxylation, two of the most commonly used are (a) cold alkaline $KMnO_4$, and (b) peroxyformic acid, HCO_2OH.

Hydroxylation with permanganate is carried out by stirring together at room temperature the alkene and the aqueous permanganate solution: either neutral—the reaction produces OH^-—or, better, slightly alkaline. Heat and the addition of acid are avoided, since these more vigorous conditions promote further oxidation of the glycol, with cleavage of the carbon–carbon double bond (Sec. 6.29).

Hydroxylation with peroxyformic acid is carried out by allowing the alkene to stand with a mixture of hydrogen peroxide and formic acid, HCOOH, for a few hours, and then heating the product with water to hydrolyze certain intermediate compounds.

A glycol is frequently named by adding the word *glycol* to the name of the alkene from which it is formed. For example:

$$\underset{\text{Ethylene}}{3CH_2{=}CH_2} + 2KMnO_4 + 4H_2O \longrightarrow \underset{\text{Ethylene glycol}}{3\underset{OH}{\underset{|}{C}H_2}{-}\underset{OH}{\underset{|}{C}H_2}} + 2MnO_2 + 2KOH$$

$$\underset{\text{Propylene}}{CH_3{-}CH{=}CH_2} \xrightarrow{HCO_2OH} \xrightarrow{H_2O} \underset{\text{Propylene glycol}}{CH_3{-}\underset{OH}{\underset{|}{C}H}{-}\underset{OH}{\underset{|}{C}H_2}}$$

Hydroxylation of alkenes is the most important method for the synthesis of glycols. Moreover, oxidation by permanganate is the basis of a very useful analytical test known as the **Baeyer test** (Sec. 6.30).

(We shall discuss the stereochemistry and mechanism of glycol formation in Sec. 17.12.)

6.21 Substitution by halogen. Allylic hydrogen

So far in our discussion of alkenes, we have concentrated on the carbon–carbon double bond, and on the addition reactions that take place there. Now let us turn to the alkyl groups that are present in most alkene molecules.

Since these alkyl groups have the alkane structure, they should undergo alkane reactions, for example, substitution by halogen. But an alkene molecule presents *two* sites where halogen can attack, the double bond and the alkyl groups. Can we direct the attack to just one of these sites? The answer is yes, *by our choice of reaction conditions.*

We know that alkanes undergo substitution by halogen at high temperatures or under the influence of ultraviolet light, and generally in the gas phase: conditions that favor formation of free radicals. We know that alkenes undergo addition of halogen at low temperatures and in the absence of light, and generally in the liquid phase: conditions that favor ionic reactions, or at least do not aid formation of radicals.

$$-\overset{|}{C}{=}\overset{|}{C}{-}\underset{H}{\underset{|}{\overset{|}{C}}}{-}$$

$\overset{\delta-}{X}{-}\overset{\delta+}{X}$ (arrow to C=C) — Ionic attack, *Addition*

$X\cdot$ (arrow to H) — Free-radical attack, *Substitution*

If we wish to direct the attack of halogen to the alkyl portion of an alkene molecule, then, we choose conditions that are favorable for the free-radical reaction and unfavorable for the ionic reaction. Chemists of the Shell Development Company found that, at a temperature of 500–600°, a mixture of gaseous propylene and chlorine yields chiefly the substitution product, 3-chloro-1-propene, known as *allyl chloride* ($CH_2{=}CH{-}CH_2{-}$ = **allyl**). Bromine behaves similarly.

$$CH_3{-}CH{=}CH_2 \xrightarrow{Cl_2} \begin{cases} \xrightarrow[\text{CCl}_4\text{ soln.}]{\text{low temp.}} CH_3{-}\underset{\large Cl}{\underset{|}{CH}}{-}\underset{\large Cl}{\underset{|}{CH_2}} & \text{1,2-Dichloropropane (Propylene chloride)} \\ \xrightarrow[\text{gas phase}]{500\text{–}600^\circ} Cl{-}CH_2{-}CH{=}CH_2 + HCl & \text{3-Chloro-1-propene (Allyl chloride)} \end{cases}$$

Propylene

Ionic: *addition*

Free-radical: *substitution*

In view of Secs. 6.17–6.18, we might wonder why a halogen atom does not add to a double bond, instead of abstracting a hydrogen atom. H. C. Brown (of Purdue University) has suggested that the halogen atom *does* add but, at high temperatures, is expelled before the second step of free-radical addition can occur.

Free-radical addition

$$X\cdot + CH_3{-}CH{=}CH_2 \rightleftarrows CH_3{-}\dot{C}H{-}CH_2X \;(\text{I}) \xrightarrow{X_2} CH_3{-}\underset{X}{\underset{|}{CH}}{-}CH_2X + X\cdot$$

Free-radical substitution

$$X\cdot + CH_3{-}CH{=}CH_2 \longrightarrow \dot{C}H_2{-}CH{=}CH_2 + HX \xrightarrow{X_2} X{-}CH_2{-}CH{=}CH_2 + X\cdot$$

Allyl radical + HX

Allyl halide
Actual product at high temperature or low halogen concentration
(X = Cl, Br)

Consistent with Brown's explanation is the finding that *low concentration* of halogen can be used instead of high temperature to favor substitution over (free-radical) addition. Addition of the halogen atom gives radical I, which falls apart (to regenerate the starting material) if the temperature is high or if it does not soon encounter a halogen molecule to complete the addition. The allyl radical, on the other hand, once formed, has little option but to wait for a halogen molecule, whatever the temperature or however low the halogen concentration.

Problem 6.12 (a) What would the allyl radical have to do to return to the starting material? (b) From bond dissociation energies, calculate the minimum E_{act} for this reaction.

The compound **N-bromosuccinimide (NBS)** is a reagent used *for the specific purpose of brominating alkenes at the allylic position*; NBS functions simply by providing a constant, low concentration of bromine. As each molecule of HBr is formed by the halogenation, NBS converts it into a molecule of Br_2.

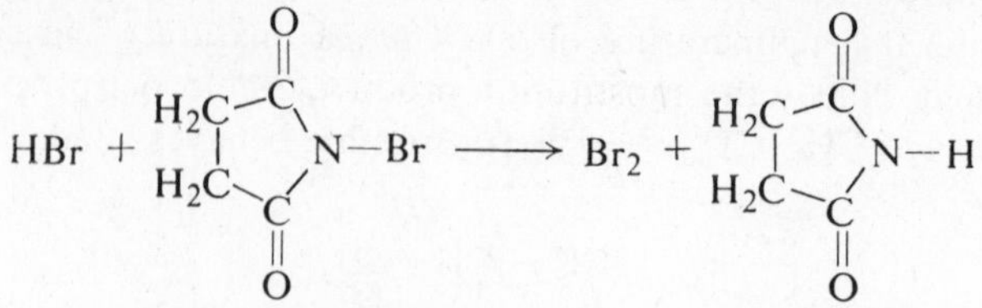

$$HBr + \begin{matrix} & & O \\ & & \| \\ H_2C & - & C \\ | & & & \diagdown \\ & & & & N{-}Br \\ | & & & \diagup \\ H_2C & - & C \\ & & \| \\ & & O \end{matrix} \longrightarrow Br_2 + \begin{matrix} & & O \\ & & \| \\ H_2C & - & C \\ | & & & \diagdown \\ & & & & N{-}H \\ | & & & \diagup \\ H_2C & - & C \\ & & \| \\ & & O \end{matrix}$$

N-Bromosuccinimide (NBS) — Succinimide

6.22 Orientation and reactivity in substitution

Thus alkenes undergo substitution by halogen in exactly the same way as do alkanes. Furthermore, just as the alkyl groups affect the reactivity of the double bond toward addition, so the double bond affects the reactivity of the alkyl groups toward substitution.

Halogenation of many alkenes has shown that: (a) hydrogens attached to doubly-bonded carbons undergo very little substitution; and (b) hydrogens attached to carbons adjacent to doubly-bonded carbons are particularly reactive toward substitution. Examination of reactions which involve attack not only by halogen atoms but by other free radicals as well has shown that this is a general rule: hydrogens attached to doubly-bonded carbons, known as **vinylic** hydrogens, are harder to abstract than ordinary primary hydrogens; hydrogens attached to a carbon atom adjacent to a double bond, known as **allylic** hydrogens, are even easier to abstract than tertiary hydrogens.

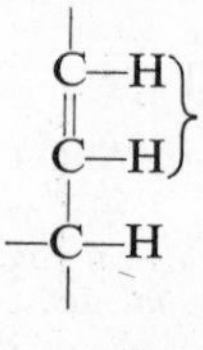

$$\begin{matrix} | \\ C{-}H \\ \| \\ C{-}H \\ | \\ -C{-}H \\ | \end{matrix} \quad \begin{matrix} \left.\begin{matrix} \\ \\ \end{matrix}\right\} \textbf{Vinylic hydrogen: } \textit{hard to abstract} \\ \\ \textbf{Allylic hydrogen: } \textit{easy to abstract} \end{matrix}$$

We can now expand the reactivity sequence of Sec. 3.23.

Ease of abstraction of hydrogen atoms	allylic > 3° > 2° > 1° > CH_4, vinylic

Substitution in alkenes seems to proceed by the same mechanism as substitution in alkanes. For example:

$$\underset{\text{Ethylene}}{CH_2{=}CH{-}H} \xrightarrow{Cl\cdot} \underset{\text{Vinyl radical}}{CH_2{=}CH\cdot} \xrightarrow{Cl_2} \underset{\text{Vinyl chloride}}{CH_2{=}CH{-}Cl}$$

$$\underset{\text{Propylene}}{CH_2{=}CH{-}CH_2{-}H} \xrightarrow{Cl\cdot} \underset{\text{Allyl radical}}{CH_2{=}CH{-}CH_2\cdot} \xrightarrow{Cl_2} \underset{\text{Allyl chloride}}{CH_2{=}CH{-}CH_2Cl}$$

Evidently the vinyl radical is formed very slowly and the allyl radical is formed very rapidly. We can now expand the sequence of Sec. 3.25.

Ease of formation of free radicals allyl > 3° > 2° > 1° > $CH_3\cdot$, vinyl

Are these findings in accord with our rule that *the more stable the radical, the more rapidly it is formed*? Is the slowly formed vinyl radical relatively unstable, and the rapidly formed allyl radical relatively stable?

The bond dissociation energies in Table 1.2 (p. 21) show that 104 kcal of energy is needed to form vinyl radicals from a mole of ethylene, as compared with 98 kcal for formation of ethyl radicals from ethane. Relative to the hydrocarbon from which each is formed, then, the vinyl radical contains more energy and is less stable than a primary radical, and about the same as a methyl radical.

On the other hand, bond dissociation energies show that only 88 kcal is needed for formation of allyl radicals from propylene, as compared with 91 kcal for formation of *tert*-butyl radicals. Relative to the hydrocarbon from which each is formed, the allyl radical contains less energy and is more stable than the *tert*-butyl radical.

We can now expand the sequence of Sec. 3.24; relative to the hydrocarbon from which each is formed, the order of stability of free radicals is:

Stability of free radicals allyl > 3° > 2° > 1° > $CH_3\cdot$, vinyl

In some way, then, the double bond affects the stability of certain free radicals; it exerts a similar effect on the incipient radicals of the transition state, and thus affects the rate of their formation. We have already seen (Sec. 5.4) a possible explanation for the unusually strong bond to vinylic hydrogen. The high stability of the allyl radical is readily accounted for by the structural theory: specifically, by the concept of *resonance*.

6.23 Resonance theory

It will be helpful first to list some of the general principles of the concept of resonance, and then to discuss these principles in terms of a specific example, the structure of the allyl radical.

(a) *Whenever a molecule can be represented by two or more structures that differ only in the arrangement of electrons—that is, by structures that have the same arrangement of atomic nuclei—there is* **resonance**. The molecule is a **hybrid** of all these structures, and cannot be represented satisfactorily by any one of them. Each of these structures is said to **contribute** to the hybrid.

(b) *When these contributing structures are of about the same stability* (*that is, have about the same energy content*), *then* **resonance is important**. The contribution of each structure to the hybrid depends upon the relative stability of that structure: the more stable structures make the larger contribution.

(c) *The resonance hybrid is more stable than any of the contributing structures.* This increase in stability is called the **resonance energy**. The more nearly equal in stability the contributing structures, the greater the resonance energy.

There can be resonance only between structures that contain the *same number of odd electrons*. We need concern ourselves about this restriction only in dealing with *di*-radicals: molecules that contain *two* unpaired electrons. There cannot be resonance between a diradical structure and a structure with all electrons paired.

6.24 Resonance structure of the allyl radical

In the language of the resonance theory, then, the allyl radical is a resonance hybrid of the two structures, I and II.

$$\underset{\text{I}}{CH_2{=}CH{-}CH_2\cdot} \qquad \underset{\text{II}}{\cdot CH_2{-}CH{=}CH_2}$$

This simply means that the allyl radical does not correspond to either I or II, but rather to a structure intermediate between I and II. Furthermore, since I and II are exactly equivalent, and hence have exactly the same stability, the resonance hybrid is equally related to I and to II; that is, I and II are said to make *equal contributions to the hybrid.*

This does *not* mean that the allyl radical consists of molecules half of which correspond to I and half to II, nor does it mean that an individual molecule changes back and forth between I and II. All molecules are the same; each one has a structure intermediate between I and II.

An analogy to biological hybrids that was suggested by Professor G. W. Wheland of the University of Chicago is helpful. When we refer to a mule as a hybrid of a horse and a donkey, we do not mean that some mules are horses and some mules are donkeys; nor do we mean that an individual mule is a horse part of the time and a donkey part of the time. We mean simply that a mule is an animal that is related to both a horse and a donkey, and that can be conveniently defined in terms of those familiar animals.

An analogy used by Professor John D. Roberts of the California Institute of Technology is even more apt. A medieval European traveler returns home from a journey to India, and describes a rhinoceros as a sort of cross between a dragon and a unicorn—a quite satisfactory description of a real animal in terms of two familiar but entirely imaginary animals.

It must be understood that our drawing of two structures to represent the allyl radical does not imply that either of these structures (or the molecules each would singly represent) has any existence. The two pictures are necessary because of the limitations of our rather crude methods of representing molecules. We draw two pictures because no *single* one would suffice. It is not surprising that certain molecules cannot be represented by one structure of the sort we have employed; on the contrary, the surprising fact is that the crude dot-and-dash representation used by organic chemists has worked out to the extent that it has.

The resonance theory further tells us that the allyl radical does not contain one carbon–carbon single bond and one carbon–carbon double bond (as in I or II), but rather contains two *identical* bonds, each one intermediate between a single and a double bond. This new type of bond—this **hybrid bond**—has been described as a *one-and-a-half bond.* It is said to possess one-half single-bond character and one-half double-bond character.

$$\left[\underset{\text{I}}{CH_2{=}CH{-}CH_2\cdot} \quad \underset{\text{II}}{\cdot CH_2{-}CH{=}CH_2}\right] \quad \textit{equivalent to} \quad \underset{\text{III}}{\underbrace{CH_2 \overset{...}{-} CH \overset{...}{-} CH_2}_{\cdot}}$$

The odd electron is not localized on one carbon or the other but is *delocalized*, being equally distributed over both terminal carbons. We might represent this symmetrical hybrid molecule as in III, where the broken lines represent half bonds.

Problem 6.13 The nitro group, $—NO_2$, is usually represented as

$$-N\overset{\diagup\!\!\!\diagup O}{\underset{\diagdown O}{}}$$

Actual measurement shows that the two nitrogen–oxygen bonds of a nitro compound have exactly the same length. In nitromethane, CH_3NO_2, for example, the two nitrogen–oxygen bond lengths are each 1.21 A, as compared with a usual length of 1.36 A for a nitrogen–oxygen single bond and 1.18 A for a nitrogen–oxygen double bond. What is a better representation of the $—NO_2$ group?

Problem 6.14 The carbonate ion, CO_3^{--}, might be represented as

$$O{=}C\overset{\diagup O^-}{\underset{\diagdown O^-}{}}$$

Actual measurement shows that all the carbon–oxygen bonds in $CaCO_3$ have the same length, 1.31 A, as compared with a usual length of about 1.36 A for a carbon–oxygen single bond and about 1.23 A for a carbon–oxygen double bond. What is a better representation of the CO_3^{--} ion?

6.25 Stability of the allyl radical

A further, most important outcome of the resonance theory is this: *as a resonance hybrid, the allyl radical is more stable* (i.e., *contains less energy*) *than either of the contributing structures.* This additional stability possessed by the molecule is referred to as *resonance energy.* Since these particular contributing structures are exactly equivalent and hence of the same stability, we expect stabilization due to resonance to be large.

Just *how* large is the resonance energy of the allyl radical? To know the exact value, we would have to compare the actual, hybrid allyl radical with a *non-existent* radical of structure I or II—something we cannot do, experimentally. We can, however, estimate the resonance energy by comparing two reactions: dissociation of propane to form a *n*-propyl radical, and dissociation of propylene to form an allyl radical.

$$\underset{\text{Propane}}{CH_3CH_2CH_3} \longrightarrow \underset{n\text{-Propyl radical}}{CH_3CH_2CH_2\cdot} + H\cdot \qquad \Delta H = +\,98\text{ kcal}$$

$$\underset{\text{Propylene}}{CH_2{=}CH{-}CH_3} \longrightarrow \underset{\text{Allyl radical}}{CH_2{=}CH{-}CH_2\cdot} + H\cdot \qquad \Delta H = +88$$

Propane, the *n*-propyl radical, and propylene are each fairly satisfactorily represented by a single structure; the allyl radical, on the other hand, is a resonance hybrid. We see that the energy difference between propylene and the allyl radical is 10 kcal/mole less (98 − 88) than the energy difference between propane and the *n*-propyl radical; we attribute the lower dissociation energy entirely to resonance stabilization of the allyl radical, and estimate the resonance energy to be 10 kcal/mole.

6.26 Orbital picture of the allyl radical

To get a clearer picture of what a resonance hybrid is—and, especially, to understand how resonance stabilization arises—let us consider the bond orbitals in the allyl radical.

Since each carbon is bonded to three other atoms, it uses sp^2 orbitals (as in ethylene, Sec. 5.2). Overlap of these orbitals with each other and with the *s* orbitals of five hydrogen atoms gives the molecular skeleton shown in Fig. 6.5, with all bond angles 120°. In addition, each carbon atom has a *p* orbital which, as we know, consists of two equal lobes, one lying above and the other lying below the plane of the σ bonds; it is occupied by a single electron.

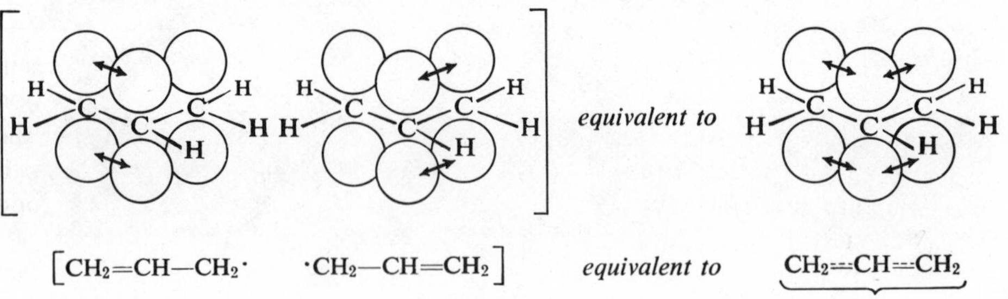

Figure 6.5. Allyl radical. The *p* orbital of the middle carbon overlaps *p* orbitals on both sides to permit delocalization of electrons.

As in the case of ethylene, the *p* orbital of one carbon can overlap the *p* orbital of an adjacent carbon atom, permitting the electrons to pair and a bond to be formed. In this way we would arrive at either of the contributing structures, I or II, with the odd electron occupying the *p* orbital of the remaining carbon atom. But the overlap is not limited to a pair of *p* orbitals as it was in ethylene; the *p* orbital of the middle carbon atom overlaps equally well the *p* orbitals of *both* the carbon atoms to which it is bonded. The result is two continuous π electron clouds, one lying above and one lying below the plane of the atoms.

Since no more than two electrons may occupy the same orbital (Pauli exclusion principle), these π clouds are actually made up of *two* orbitals (See 29.5). One of these, containing two π electrons, encompasses all three carbon atoms; the other, containing the third (odd) π electron, is divided equally between the terminal carbons.

The overlap of the *p* orbitals in both directions, and the resulting participation of each electron in two bonds, is equivalent to our earlier description of the allyl radical as a resonance hybrid of two structures. These two methods of representation, the drawing of several resonance structures and the drawing of an electron cloud, are merely our crude attempts to convey by means of pictures the idea that *a given pair of electrons may serve to bind together more than two nuclei.* It is this ability of π electrons to participate in several bonds, this **delocalization of electrons**, that results in stronger bonds and a more stable molecule. For this reason the term *delocalization energy* is frequently used instead of *resonance energy.*

The covalent bond owes its strength to the fact that an electron is attracted more strongly by two nuclei than by one. In the same way an electron is more strongly attracted by three nuclei than by two.

We saw earlier (Sec. 2.21) that the methyl radical may not be quite flat: that hybridization of carbon may be intermediate between sp^2 and sp^3. For the allyl radical, on the other hand—and for many other free radicals—flatness is clearly required to permit the overlap of p orbitals that leads to stabilization of the radical.

In terms of the conventional valence-bond structures we employ, it is difficult to visualize a single structure that is intermediate between the two structures, I and II. The orbital approach, on the other hand, gives us a rather clear picture of the allyl radical: the density of electrons holding the central carbon to each of the others is intermediate between that of a single bond and that of a double bond.

6.27 Using the resonance theory

The great usefulness, and hence the great value, of the resonance theory lies in the fact that it retains the simple though crude type of structural representation which we have used so far in this book. Particularly helpful is the fact that the stability of a structure can often be roughly estimated from its **reasonableness.** If only one reasonable structure can be drawn for a molecule, the chances are good that this one structure adequately describes the molecule.

The criterion of reasonableness is not so vague as it might appear. The fact that a particular structure seems reasonable to us means that we have previously encountered a compound whose properties are pretty well accounted for by a structure of that type; the structure must, therefore, represent a fairly stable kind of arrangement of atoms and electrons. For example, each of the contributing structures for the allyl radical appears quite reasonable because we have encountered compounds, alkenes and free radicals, that possess the features of this structure.

There are a number of other criteria that we can use to estimate relative stabilities, and hence relative importance, of contributing structures. One of these has to do with (a) *electronegativity and location of charge.*

For example, a convenient way of indicating the polarity (*ionic character*) of the hydrogen–chlorine bond is to represent HCl as a hybrid of structures I and II. We judge that II is appreciably stable and hence makes significant contribution, because in it a negative charge is located on a highly electronegative atom, chlorine.

$$\underset{\text{I}}{\text{H—Cl}} \qquad \underset{\text{II}}{\text{H}^+\text{Cl}^-}$$

On the other hand, we consider methane to be represented adequately by the single structure III.

```
    H
    |
H—C—H
    |
    H
   III
```

Although it is possible to draw additional, ionic structures like IV and V, we judge these to be unstable since in them a negative charge is located on an atom of low

$$\underset{\text{IV}}{\begin{array}{c} \mathrm{H^+} \\ \mathrm{H{-}\bar{C}{-}H} \\ | \\ \mathrm{H} \end{array}} \qquad \underset{\text{V}}{\begin{array}{c} \mathrm{H} \\ | \\ \mathrm{H{-}C^- \; H^+} \\ | \\ \mathrm{H} \end{array}} \quad \text{etc.}$$

electronegativity, carbon. We expect IV and V to make negligible contribution to the hybrid and hence we ignore them.

In later sections we shall use certain other criteria to help us estimate stabilities of possible contributing structures: (b) *number of bonds* (Sec. 8.17); (c) *dispersal of charge* (Sec. 11.19); (d) *complete vs. incomplete octet* (Sec. 11.20); (e) *separation of charge* (Sec. 18.12).

Finally, we shall find certain cases where the overwhelming weight of evidence —bond lengths, dipole moments, reactivity—indicate that an accurate description of a given molecule requires contribution from structures of a sort that may appear quite unreasonable to us (Secs. 6.28 and 8.18); this simply reminds us that, after all, we know very little about the structure of molecules, and must be prepared to change our ideas of what is reasonable to conform with evidence provided by experimental facts.

In the next section, we shall encounter contributing structures that are very strange looking indeed.

Problem 6.15 The ionization potential of the allyl radical is 188 kcal/mole. (a) How does the allyl cation compare in stability with the simple alkyl cations of Sec. 5.18? (b) Is the cation adequately represented by the structure $CH_2{=}CHCH_2^+$? Describe its structure in both valence-bond and orbital terms. (Check your answer in Sec. 8.21.)

Problem 6.16 *Benzene*, C_6H_6, is a flat molecule with all bond angles 120° and all carbon–carbon bonds 1.39 A long. Its heat of hydrogenation (absorption of three moles of hydrogen) is 49.8 kcal/mole, as compared with values of 28.6 for cyclohexene (one mole of hydrogen) and 55.4 for 1,3-cyclohexadiene (two moles of hydrogen). (a) Is benzene adequately represented by the Kekulé formula shown? (b) Suggest a better structure for benzene in both valence-bond and orbital terms. (Check your answer in Secs. 10.7–10.8.)

Cyclohexene

1,3-Cyclohexadiene

Benzene (Kekulé formula)

6.28 Resonance stabilization of alkyl radicals. Hyperconjugation

The relative stabilities of tertiary, secondary, and primary alkyl radicals are accounted for on exactly the same basis as the stability of the allyl radical:

delocalization of electrons, this time through overlap between the p orbital occupied by the odd electron and a σ orbital of the alkyl group (Fig. 6.6). Through this

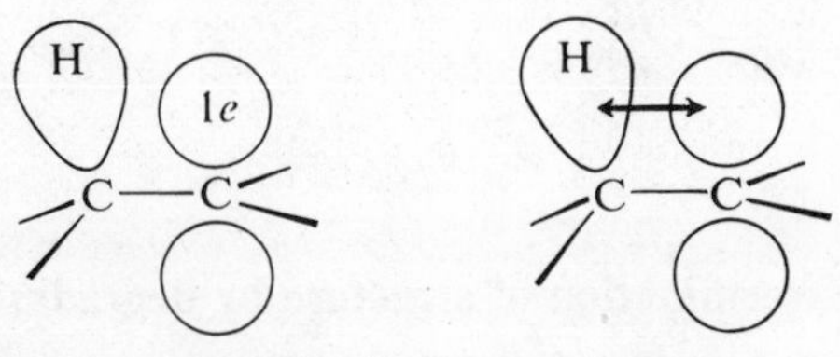

Figure 6.6. Hyperconjugation in an alkyl free radical. (a) Separate σ and p orbitals. (b) Overlapping orbitals.

overlap, individual electrons can, to an extent, help bind together three nuclei, two carbons and one hydrogen. This kind of delocalization, involving σ bond orbitals, is called **hyperconjugation.**

In resonance language, we would say that the ethyl radical, for example, is a hybrid of not only the usual structure, I, but also three additional structures, II,

```
   H  H                H  H             ·H  H              H  H
   |  |                |  |                 |              |  |
H—C—C·           H·   C=C           H—C=C           H—C=C
   |  |                |  |              |  |                 |
   H  H                H  H              H  H             ·H  H
     I                   II                III                IV
```

III, and IV, in which a double bond joins the two carbons, and the odd electron is held by a hydrogen atom.

Individually, each of these "no-bond" resonance structures appears strange but, taken together, they mean that the carbon–hydrogen bond is something less than a single bond, that the carbon–carbon bond has some double bond character, and that the odd electron is partly accommodated by hydrogen atoms. Contribution from these unstable structures is not nearly so important as from, say, the equivalent structures for the allyl radical, and the resulting stabilization is not nearly so large. It is believed, however, to stabilize the ethyl radical to the extent of 6 kcal relative to the methyl radical (104 − 98, Sec. 3.24), for which such resonance is not possible.

If we extend this idea to the isopropyl radical, we find that instead of three hyperconjugation structures we now have six. (*Draw them.*) The larger number of contributing structures means more extensive delocalization of the odd electron, and hence greater stabilization of the radical. In agreement with this expectation, we find that the bond dissociation energy of the isopropyl–hydrogen bond is only 95 kcal, indicating a resonance energy of 9 kcal/mole (104 − 95).

For the *tert*-butyl radical there should be nine such hyperconjugation structures. (*Draw them.*) Here we find a bond dissociation energy of 91 kcal, indicating a resonance stabilization of 13 kcal/mole (104 − 91).

In summary, the relative stabilities of the free radicals we have studied are determined by delocalization of electrons. Delocalization takes place through overlap of the p orbital occupied by the odd electron: overlap with the π cloud of a double bond in the allyl radical, or overlap with σ bonds in alkyl radicals.

Problem 6.17 It has been postulated that the relative stabilities of alkyl cations are determined not only by inductive effects but also by resonance stabilization. How might you account for the following order of stability of cations?

tert-butyl > isopropyl > ethyl > methyl

6.29 Ozonolysis. Determination of structure by degradation

Along with addition and substitution we may consider a third general kind of alkene reaction, **cleavage**: a reaction in which the double bond is completely broken and the alkene molecule converted into two smaller molecules.

The classical reagent for cleaving the carbon–carbon double bond is ozone. **Ozonolysis** (cleavage by ozone) is carried out in two stages: first, addition of ozone to the double bond to form an *ozonide*; and second, hydrolysis of the ozonide to yield the cleavage products.

Ozone gas is passed into a solution of the alkene in some inert solvent like carbon tetrachloride; evaporation of the solvent leaves the ozonide as a viscous oil. This unstable, explosive compound is not purified, but is treated directly with water, generally in the presence of a reducing agent.

In the cleavage products a doubly-bonded oxygen is found attached to each of the originally doubly-bonded carbons:

Ozonolysis

$$-\overset{|}{C}=\overset{|}{C}- \xrightarrow{O_3} \text{Molozonide} \longrightarrow \text{Ozonide} \xrightarrow[Zn]{H_2O} -\overset{|}{C}=O + O=\overset{|}{C}-$$

Alkene Molozonide Ozonide Cleavage products (Aldehydes and ketones)

These compounds containing the C=O group are called *aldehydes* and *ketones*; at this point we need only know that they are compounds that can readily be identified (Sec. 19.17). The function of the reducing agent, which is frequently zinc dust, is to prevent formation of hydrogen peroxide, which would otherwise react with the aldehydes and ketones. (Aldehydes, RCHO, are often converted into acids, RCOOH, for ease of isolation.)

Knowing the number and arrangement of carbon atoms in these aldehydes and ketones, we can work back to the structure of the original alkene. For example, for three of the isomeric hexylenes:

$$\underset{\text{Aldehydes}}{CH_3CH_2CH_2\overset{H}{\overset{|}{C}}=O + O=\overset{H}{\overset{|}{C}}CH_3} \xleftarrow{H_2O/Zn} \xleftarrow{O_3} \underset{\text{2-Hexene}}{CH_3CH_2CH_2CH=CHCH_3}$$

$$\underset{\text{Aldehydes}}{CH_3CH_2\overset{H}{\overset{|}{C}}=O + O=\overset{H}{\overset{|}{C}}CH_2CH_3} \xleftarrow{H_2O/Zn} \xleftarrow{O_3} \underset{\text{3-Hexene}}{CH_3CH_2CH=CHCH_2CH_3}$$

$$\underset{\text{Aldehyde}}{CH_3CH_2\overset{\overset{H}{|}}{C}{=}O} + \underset{\text{Ketone}}{O{=}\overset{\overset{CH_3}{|}}{C}{-}CH_3} \xleftarrow{H_2O/Zn} \xleftarrow{O_3} \underset{\text{2-Methyl-2-pentene}}{CH_3CH_2CH{=}\overset{\overset{CH_3}{|}}{C}{-}CH_3}$$

One general approach to the determination of the structure of an unknown compound is **degradation**, the breaking down of the unknown compound into a number of smaller, more easily identifiable fragments. Ozonolysis is a typical means of degradation.

Another method of degradation that gives essentially the same information—although somewhat less reliable—is vigorous oxidation by permanganate, which is believed to involve formation and cleavage of intermediate glycols (Sec. 6.20).

$$-\overset{|}{C}{=}\overset{|}{C}- \xrightarrow{KMnO_4} \left[-\underset{OH}{\overset{|}{\underset{|}{C}}}-\underset{OH}{\overset{|}{\underset{|}{C}}}-\right] \longrightarrow \text{acids, ketones, } CO_2$$

Carboxylic acids, RCOOH, are obtained instead of aldehydes, RCHO. A terminal $=CH_2$ group is oxidized to CO_2. For example:

$$\underset{\substack{\text{Carboxylic}\\\text{acid}}}{CH_3COOH} + \underset{\text{Ketone}}{O{=}\overset{\overset{CH_3}{|}}{C}{-}CH_3} \xleftarrow{KMnO_4} \underset{\text{2-Methyl-2-butene}}{CH_3CH{=}\overset{\overset{CH_3}{|}}{C}{-}CH_3}$$

$$\underset{\substack{\text{Carboxylic}\\\text{acid}}}{CH_3CH_2CH_2COOH} + \underset{\substack{\text{Carbon}\\\text{dioxide}}}{CO_2} \xleftarrow{KMnO_4} \underset{\text{1-Pentene}}{CH_3CH_2CH_2CH{=}CH_2}$$

Problem 6.18 What products would you expect from each of the dimers of isobutylene (Sec. 6.15) upon cleavage by: (a) ozonolysis, (b) $KMnO_4$?

6.30 Analysis of alkenes

The functional group of an alkene is the carbon–carbon double bond. To characterize an unknown compound as an alkene, therefore, we must show that it undergoes the reactions typical of the carbon–carbon double bond. Since there are so many of these reactions, we might at first assume that this is an easy job. But let us look at the problem more closely.

First of all, which of the many reactions of alkenes do we select? Addition of hydrogen bromide, for example? Hydrogenation? Let us imagine ourselves in the laboratory, working with gases and liquids and solids, with flasks and test tubes and bottles.

We could pass dry hydrogen bromide from a tank through a test tube of an unknown liquid. But what would we see? How could we tell whether or not a reaction takes place? A colorless gas bubbles through a colorless liquid; a different colorless liquid may or may not be formed.

We could attempt to hydrogenate the unknown compound. Here, we might say, we could certainly tell whether or not reaction takes place: a drop in the hydrogen pressure would show us that addition had occurred. This is true, and hydrogenation can be a useful analytical tool. But a catalyst must be prepared, and a fairly elaborate piece of apparatus must be used; the whole operation might take hours.

Whenever possible, *we select for a characterization test a reaction that is rapidly and conveniently carried out, and that gives rise to an easily observed change.* We select a test that requires a few minutes and a few test tubes, a test in which a color appears or disappears, or bubbles of gas are evolved, or a precipitate forms or dissolves.

Experience has shown that an alkene is best characterized, then, by its property of decolorizing both a solution of bromine in carbon tetrachloride (Sec. 6.5) and a cold, dilute, neutral permanganate solution (the Baeyer test, Sec. 6.20). Both tests are easily carried out; in one, a red color disappears, and in the other, a purple color disappears and is replaced by brown manganese dioxide.

$$\underset{\text{Alkene}}{>C=C<} + \underset{\textit{Red}}{Br_2/CCl_4} \longrightarrow \underset{\textit{Colorless}}{-\underset{Br}{\overset{|}{\underset{|}{C}}}-\underset{Br}{\overset{|}{\underset{|}{C}}}-}$$

$$\underset{\text{Alkene}}{>C=C<} + \underset{\textit{Purple}}{MnO_4^-} \longrightarrow \underset{\textit{Brown ppt.}}{MnO_2} + \underset{\textit{Colorless}}{-\underset{OH}{\overset{|}{\underset{|}{C}}}-\underset{OH}{\overset{|}{\underset{|}{C}}}-} \text{ or other products}$$

Granting that we have selected the best tests for the characterization of alkenes, let us go on to another question. We add bromine in carbon tetrachloride to an unknown organic compound, let us say, and the red color disappears. What does this tell us? Only that our unknown is a compound that reacts with bromine. It *may* be an alkene. But it is not enough merely to know that a particular kind of compound reacts with a given reagent; we must also know what *other* kinds of compounds also react with the reagent. In this case, the unknown may equally well be an alkyne. (It may also be any of a number of compounds that undergo rapid *substitution* by bromine; in that case, however, hydrogen bromide would be evolved and could be detected by the cloud it forms when we blow our breath over the test tube.)

In the same way, decolorization of permanganate does not prove that a compound is an alkene, but only that it contains some functional group that can be oxidized by permanganate. The compound *may* be an alkene; but it may instead be an alkyne, an aldehyde, or any of a number of easily oxidized compounds. It may even be a compound that is contaminated with an *impurity* that is oxidized; alcohols, for example, are not oxidized under these conditions, but often contain impurities that *are*. We can usually rule out this by making sure that more than a drop or two of the reagent is decolorized.

By itself, a single characterization test seldom proves that an unknown is one particular kind of compound. It may limit the number of possibilities, so that a final decision can then be made on the basis of additional tests. Or, conversely, if certain possibilities have already been eliminated, a single test may permit a final choice to be made. Thus, the bromine or permanganate test would be sufficient to differentiate an alkene from an alkane, or an alkene from an alkyl halide, or an alkene from an alcohol.

The tests most used in characterizing alkenes, then, are the following: (a) rapid decolorization of bromine in carbon tetrachloride without evolution of HBr, a test also given by alkynes; (b) decolorization of cold, dilute, neutral, aqueous permanganate solution (the Baeyer test), a test also given by alkynes and aldehydes. Also helpful is the solubility of alkenes in cold concentrated sulfuric acid, a test also given by a great many other compounds, including all those containing oxygen (they form soluble oxonium salts) and compounds that are readily sulfonated (Secs. 12.11 and 17.8). Alkanes or alkyl halides are not soluble in cold concentrated sulfuric acid.

Of the compounds we have dealt with so far, alcohols also dissolve in sulfuric acid. Alcohols can be distinguished from alkenes, however, by the fact that alcohols give a negative test with bromine in carbon tetrachloride and a negative Baeyer test—so long as we are not misled by impurities. Primary and secondary alcohols *are* oxidized by chromic anhydride, CrO_3, in aqueous sulfuric acid: within *two seconds*, the clear orange solution turns blue-green and becomes opaque.

$$\underset{1^\circ\ or\ 2^\circ}{ROH} + \underset{\substack{Clear,\\ orange}}{HCrO_4^-} \longrightarrow \textit{Opaque, blue-green}$$

Tertiary alcohols do not give this test; nor do alkenes.

Problem 6.19 Describe simple chemical tests (if any) that would distinguish between: (a) an alkene and an alkane; (b) an alkene and an alkyl halide; (c) an alkene and a secondary alcohol; (d) an alkene, an alkane, an alkyl halide, and a secondary alcohol. Tell exactly what you would *do* and *see*.

Problem 6.20 Assuming the choice to be limited to alkane, alkene, alkyl halide, secondary alcohol, and tertiary alcohol, characterize compounds A, B, C, D, and E on the basis of the following information:

Compound	Qual. elem. anal.	H_2SO_4	Br_2/CCl_4	$KMnO_4$	CrO_3
A	----	Insoluble	−	−	−
B	----	Soluble	−	−	+
C	Cl	Insoluble	−	−	−
D	----	Soluble	+	+	−
E	----	Soluble	−	−	−

Once characterized as an alkene, an unknown may then be identified as a previously reported alkene on the basis of its physical properties, including its infrared spectrum and molecular weight. Proof of structure of a new compound is best accomplished by degradation: cleavage by ozone or permanganate, followed by identification of the fragments formed (Sec. 6.29).

(Spectroscopic analysis of alkenes will be discussed in Secs. 13.15–13.16.)

PROBLEMS

1. Draw a structural formula and give (when you can) an alternative name for:

(a) ethylene bromide
(b) ethyl bromide
(c) bromoethylene
(d) ethylene glycol
(e) propylene glycol
(f) propylene bromohydrin
(g) vinyl bromide
(h) allyl chloride

2. Give structures and names of the products (if any) expected from reaction of isobutylene with:

(a) H_2, Ni
(b) Cl_2
(c) Br_2
(d) I_2
(e) HBr
(f) HBr (peroxides)
(g) HI
(h) HI (peroxides)
(i) H_2SO_4
(j) H_2O, H^+
(k) Br_2, H_2O
(l) Br_2 + NaCl(aq)
(m) H_2SO_4 ($\longrightarrow C_8H_{16}$)
(n) isobutane + HF
(o) cold alkaline $KMnO_4$
(p) hot $KMnO_4$
(q) HCO_2OH
(r) O_3; then Zn, H_2O

3. Which alkene of each pair would you expect to be more reactive toward addition of H_2SO_4?

(a) ethylene or propylene
(b) ethylene or vinyl bromide
(c) propylene or 2-butene
(d) 2-butene or isobutylene
(e) vinyl chloride or 1,2-dichloroethene
(f) 1-pentene or 2-methyl-1-butene
(g) ethylene or $CH_2{=}CHCOOH$
(h) propylene or 3,3,3-trifluoropropene

4. Give structures and names of the principal products expected from addition of HI to:

(a) 2-butene
(b) 2-pentene
(c) 2-methyl-1-butene
(d) 2-methyl-2-butene
(e) 3-methyl-1-butene (2 products)
(f) vinyl bromide
(g) 2,3-dimethyl-1-butene
(h) 2,4,4-trimethyl-2-pentene

5. Draw the structure of 6-methyl-2-heptene. Label each set of hydrogen atoms to show their relative reactivities toward chlorine atoms, using (1) for the most reactive, (2) for the next, etc.

6. Account for the fact that addition of $CBrCl_3$ in the presence of peroxides takes place faster to 2-ethyl-l-hexene than to 1-octene.

7. In methyl alcohol solution (CH_3OH), bromine adds to ethylene to yield not only ethylene bromide but also $Br{-}CH_2CH_2{-}OCH_3$. How can you account for this? Write equations for all steps.

8. As an alternative to the one-step 1,2-hydride shift described in Sec. 5.22, one might instead propose—in view of the reactions we have studied in this chapter—that carbonium ions rearrange by a two-step mechanism, involving the intermediate formation of an alkene:

$$-\underset{H}{\overset{|}{C}}-\underset{\oplus}{\overset{|}{C}}- \longrightarrow -\overset{|}{C}{=}\overset{|}{C}- + H^+ \longrightarrow -\underset{\oplus}{\overset{|}{C}}-\underset{H}{\overset{|}{C}}-$$

When (by a reaction we have not yet taken up) the isobutyl cation was generated in D_2O containing D_3O^+, there was obtained *tert*-butyl alcohol containing *no* deuterium attached to carbon. How does this experiment permit one to rule out the two-step mechanism?

9. In Sec. 6.17 a mechanism was presented for free-radical addition of hydrogen bromide. Equally consistent with the evidence given there is the following alternative mechanism:

(2a) $\quad$ Rad· + HBr $\longrightarrow$ Rad—Br + H·

(3a) $\quad H\cdot + -\overset{|}{C}{=}\overset{|}{C}- \longrightarrow -\underset{H}{\overset{|}{C}}-\underset{\cdot}{\overset{|}{C}}-$

$$\text{(4a)}\quad -\underset{\text{H}}{\overset{|}{\underset{|}{\text{C}}}}-\overset{|}{\underset{\cdot}{\text{C}}}- + \text{HBr} \longrightarrow -\underset{\text{H}}{\overset{|}{\underset{|}{\text{C}}}}-\underset{\text{Br}}{\overset{|}{\underset{|}{\text{C}}}}- + \text{H}\cdot$$

then (3a), (4a), (3a), (4a), etc.

(a) In steps (2a) and (4a) an alkyl radical abstracts bromine instead of hydrogen from hydrogen bromide. On the basis of bond dissociation energies (Table 1.2, p. 21), is this mechanism more or less likely than (2)–(4) on p. 203? Explain.

(b) The esr study (p. 204) showed that the intermediate free radical from a given alkene is the *same* whether HBr or DBr (deuterium bromide) is being added to the double bond. Explain how this evidence permits a definite choice between mechanism (2a)–(4a) and mechanism (2)–(4).

10. (a) Write all steps in the free-radical addition of HBr to propylene. (b) Write all steps that would be involved in the free-radical addition of HCl to propylene.

(c) List ΔH for each reaction in (a) and (b). Assume the following bond dissociation energies: π bond, 68 kcal; 1° R—Br, 69 kcal; 1° R—Cl, 82 kcal; 2° R—H, 95 kcal.

(d) Suggest a possible reason why the peroxide effect is observed for HBr but not for HCl.

11. When isobutylene and chlorine are allowed to react in the dark at 0° in the absence of peroxides, the principal product is not the addition product but methallyl chloride (3-chloro-2-methyl-1-propene). Bubbling oxygen through the reaction mixture produces no change.

This reaction was carried out with labeled isobutylene (1-^{14}C-2-methyl-1-propene, $(CH_3)_2C{=}^{14}CH_2$), and the methallyl chloride contained was collected, purified, and subjected to ozonolysis. Formaldehyde ($H_2C{=}O$) and chloroacetone ($ClCH_2COCH_3$) were obtained; all (97% or more) of the radioactivity was present in the chloroacetone.

(a) Give the structure, including the position of the isotopic label, of the methallyl chloride obtained. (b) Judging from the evidence, is the reaction ionic or free radical? (c) Using only steps with which you are already familiar, outline a mechanism that accounts for the formation of this product. (d) Can you suggest one reason why isobutylene is more prone than 1- or 2-butene to undergo this particular reaction? (e) Under similar conditions, and in the presence of oxygen, 3,3-dimethyl-1-butene yields mostly the addition product, but also a small yield of 4-chloro-2,3-dimethyl-1-butene. In light of your answer to (c) how do you account for the formation of this minor product?

12. How do you account for the following facts: formic acid, HCOOH, contains one carbon–oxygen bond of 1.36 A and another of 1.23 A, yet sodium formate, $HCOO^-Na^+$, contains two equal carbon–oxygen bonds, each of 1.27 A. (Check your answer in Sec. 18.13.)

pages 597–598

$$\text{H}-\text{C}\begin{matrix}\text{//O}\\ \\ \text{\\OH}\end{matrix}$$

Formic acid

13. (a) When 1-octene is allowed to react with N-bromosuccinimide, there is obtained not only 3-bromo-1-octene but also 1-bromo-2-octene. How can you account for this? (b) Propylene, $CH_3CH{=}^{14}CH_2$, labeled with carbon-14 (a radioactive isotope) is converted into allyl bromide by free-radical bromination. What would you predict about the position of the tagged atom (^{14}C) in the product?

14. Give the structure of the alkene that yields on ozonolysis:

(a) $CH_3CH_2CH_2CHO$ and HCHO 1-pentene

(b) $CH_3-\underset{CH_3}{\underset{|}{CH}}-CHO$ and CH_3CHO 4-methy-2-pentene

(c) Only $CH_3-CO-CH_3$ 2,3-dimethyl-2-butene

(d) CH_3CHO and HCHO and $OHC-CH_2-CHO$ 1,4, hexadiene

(e) What would each of these alkenes yield upon cleavage by $KMnO_4$?

15. Describe simple chemical tests that would distinguish between:

(a) 2-chloropentane and *n*-heptane
(b) 2-hexene and *tert*-butyl bromide
(c) isobutane and isobutylene
(d) allyl bromide and 1-hexene
(e) *sec*-butyl alcohol and *n*-heptane
(f) 1-octene and *n*-pentyl alcohol
(g) *tert*-pentyl alcohol and 2,2-dimethylhexane
(h) *n*-propyl alcohol and allyl alcohol ($CH_2{=}CHCH_2OH$)

Tell exactly what you would *do* and *see*. (Qualitative elemental analysis is a simple chemical test; degradation is not.)

16. A hydrocarbon, A, adds one mole of hydrogen in the presence of a platinum catalyst to form *n*-hexane. When A is oxidized vigorously with $KMnO_4$, a single carboxylic acid, containing three carbon atoms, is isolated. Give the structure and name of A. Show your reasoning, including equations for all reactions.

17. Outline all steps in a possible laboratory synthesis of each of the following compounds, using only the organic source given, plus any necessary solvents and inorganic reagents. (See general instructions about synthesis below.)

(a) ethylene from ethane
(b) propylene from propane
(c) ethyl iodide from ethane
(d) 2-bromopropane from propane (*Note:* simple monobromination of propane yields, of course, a mixture of 1-bromopropane and 2-bromopropane, and is therefore not satisfactory for this synthesis. The mixture might, however, be used as an *intermediate.*)
(e) 1,2-dibromopropane from propane
(f) 1,2-dibromobutane from 1-bromobutane
(g) 2-iodobutane from 1-chlorobutane
(h) 2-methylpentane from propylene
(i) 3-methylheptane from *n*-butyl bromide
(j) 1,2-dibromo-2-methylpropane from isobutane
(k) 2-iodobutane from *n*-butyl alcohol
(l) *n*-propyl bromide from isopropyl bromide
(m) propylene chlorohydrin from *n*-propyl iodide
(n) isohexane from $(CH_3)_2C(OH)CH_2CH_2CH_3$
(o) 2,2-dimethylbutane from 3-chloro-2,2-dimethylbutane

About Synthesis

Each synthesis should be the one that gives a reasonably pure product in reasonably good yield.

It is not necessary to complete and balance each equation. Simply draw the structure of the organic compounds, and write on the arrow the necessary reagents and any critical conditions. For example:

$$CH_3CH_2OH \xrightarrow{H^+,\ heat} CH_2{=}CH_2 \xrightarrow{H_2,\ Ni} CH_3CH_3$$

At this stage you may be asked to make a particular compound by a method that would never actually be used for that compound: for example, the synthesis of ethane just above. But if you can work out a way to make ethane from ethyl alcohol, then, when the need arises, you will also know how to make a complicated alkane from a complicated alcohol, and, in fact, how to replace an —OH group by —H in just about any compound you encounter. Furthermore, you will have gained practice in putting together what you have learned about several different kinds of compounds.

Chapter 7 Stereochemistry II. Preparation and Reactions of Stereoisomers

7.1 Stereoisomerism

Stereoisomers, we have learned, are isomers that differ only in the way their atoms are oriented in space. So far, our study has been limited to finding out what the various kinds of stereoisomers are, how to predict their existence, how to name them, and, in a general way, how their properties compare.

In Chap. 4, we learned that stereoisomers exist of the kind called *enantiomers* (mirror-image isomers), that they can be optically active, and that both their existence and their optical activity are the result of the *chirality* of certain molecules, that is, of the non-superimposability of such molecules on their mirror images. We learned how to predict, from a simple examination of molecular structure, whether or not a particular compound can display this kind of isomerism. We learned how to specify the configuration of a particular enantiomer by use of the letters R and S.

We learned about *diastereomers*: stereoisomers that are *not* mirror images. Some of these (Secs. 4.17 and 4.18) were of the kind that contained more than one chiral center. Others (Sec. 5.6) were the kind, *geometric isomers*, that owe their existence to hindered rotation about double bonds.

In Secs. 4.20 and 5.6, we learned that stereoisomers can be classified not only as to whether or not they are mirror images, but also—and quite independently of the other classification—as to how they are interconverted. Altogether, we have: (a) *configurational isomers*, interconverted by inversion (turning-inside-out) at a chiral center; (b) *geometric isomers*, interconverted—in principle—by rotation about a double bond; and (c) *conformational isomers*, interconverted by rotations about single bonds.

The operation required—rotation—is the same for interconversion of geometric and conformational isomers, and it has been suggested that they be called collectively *rotational* (or *torsional*) *isomers*. Geometric isomers are thus double-bond rotational isomers, and conformational isomers are single-bond rotational isomers.

On the other hand, from the very practical standpoint of *isolability*, geometric isomers are more akin to configurational isomers: interconversion requires bond breaking—a π bond in the case of geometric isomers—and hence is always a difficult process. Conformational isomers are interconverted by the (usually) easy process of rotation about single bonds.

For convenience, we laid down (Sec. 4.20) the following "ground rule" for discussions and problems in this book: unless specifically indicated otherwise, *the terms "stereoisomers," "enantiomers," and "diastereomers" will refer only to configurational isomers, including geometric isomers*, and will exclude conformational isomers. The latter will be referred to as "conformational isomers," "conformers," "conformational enantiomers," and "conformational diastereomers."

7.2 Reactions involving stereoisomers

Now let us go on from the *existence* of stereoisomers, and look at their *involvement* in chemical reactions: reactions in which stereoisomers are *formed*, and reactions in which stereoisomers are *consumed*; reactions in which the reagent is of the ordinary (i.e., optically inactive) kind and those in which the reagent is optically active.

We shall take up:

(a) the conversion of an achiral molecule into a chiral molecule, with the generation of a chiral center;

(b) reactions of chiral molecules in which bonds to the chiral center are not broken, and see how such reactions can be used to relate the configuration of one compound to that of another;

(c) reactions of the kind in (b) in which a second chiral center is generated;

(d) reactions of chiral compounds with optically active reagents.

Then we shall examine the stereochemistry of several reactions we have already studied—free-radical halogenation of alkanes, and electrophilic addition of halogens to alkenes—and see how stereochemistry can be used to get information about reaction mechanisms. In doing this, we shall take up:

(e) a reaction of a chiral compound in which a bond to a chiral center is broken;

(f) a reaction of an achiral compound in which two chiral centers are generated at the same time.

7.3 Generation of a chiral center. Synthesis and optical activity

One of the products of chlorination of *n*-butane is the chiral compound, *sec*-butyl chloride. It can exist as two enantiomers, I and II, which are specified

$$\underset{\substack{n\text{-Butane} \\ \textit{Achiral}}}{CH_3CH_2CH_2CH_3} \xrightarrow{Cl_2,\ \text{heat or light}} \underset{\substack{sec\text{-Butyl chloride} \\ \textit{Chiral}}}{CH_3CH_2-\overset{*}{C}H(Cl)-CH_3} + n\text{-Butyl chloride}$$

(Sec. 4.16) as S and R, respectively.

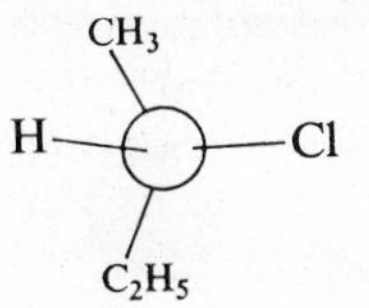

CH_3 H Cl C_2H_5 CH_3 Cl H C_2H_5

I   II

S-(+)-*sec*-butyl chloride   R-(−)-*sec*-butyl chloride

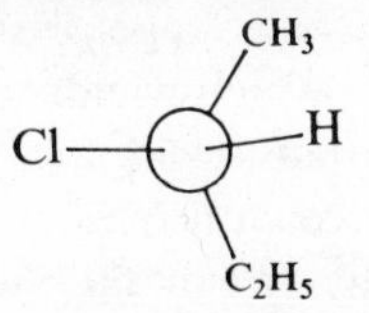

Each enantiomer should, of course, be optically active. Now, if we were to put the *sec*-butyl chloride actually prepared by the chlorination of *n*-butane into a polarimeter, would it rotate the plane of polarized light? The answer is *no*, because prepared as described it would consist of the racemic modification. The next question is: *why is the racemic modification formed?*

In the first step of the reaction, a chlorine atom abstracts hydrogen to yield hydrogen chloride and a *sec*-butyl free radical. The carbon that carries the odd electron in the free radical is sp^2-hybridized (*trigonal*, Sec. 2.21), and hence a part of the molecule is *flat*, the trigonal carbon and the three atoms attached to it lying in the same plane. In the second step, the free radical abstracts chlorine from a chlorine molecule to yield *sec*-butyl chloride. But chlorine may become attached to either face of the flat radical, and, depending upon which face, yield either of two products: R or S (see Fig. 7.1). Since the chance of attachment to one face is exactly the same as for attachment to the other face, the enantiomers are obtained in exactly equal amounts. The product is the racemic modification.

Cl_2 *a*   *b* CH_3 H C_2H_5

a   *b*

CH_3 Cl H C_2H_5 CH_3 H Cl C_2H_5

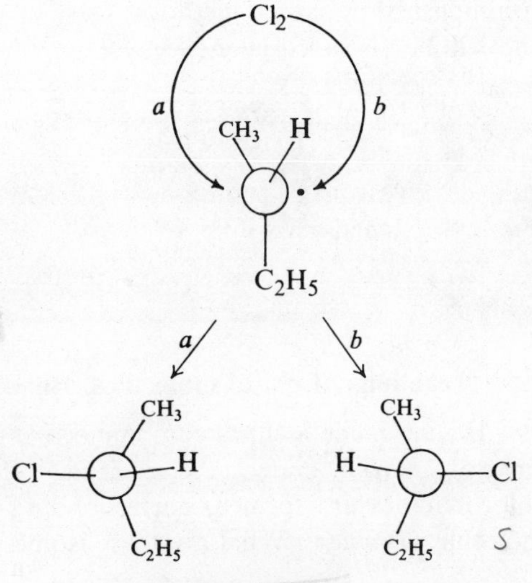

Enantiomers
Formed in equal amounts

Figure 7.1. Generation of a chiral center. Chlorine becomes attached to either face of flat free radical, via (*a*) or (*b*), to give enantiomers, and in equal amounts.

If we were to apply the approach just illustrated to the synthesis of any compound whatsoever—and on the basis of any mechanism, correct or incorrect—we would arrive at the same conclusion: as long as neither the starting material nor the reagent (nor the environment) is optically active, we should obtain an optically inactive product. At some stage of the reaction sequence, there will be two alternative paths, one of which yields one enantiomer and the other the opposite enantiomer. The two paths will always be equivalent, and selection between them *random*. The facts agree with these predictions. **Synthesis of chiral compounds from achiral reactants always yields the racemic modification.** This is simply one aspect of the more general rule: **optically inactive reactants yield optically inactive products.**

Problem 7.1 Show in detail why racemic *sec*-butyl chloride would be obtained if: (a) the *sec*-butyl radical were not flat, but pyramidal; (b) chlorination did not involve a free *sec*-butyl radical at all, but proceeded by a mechanism in which a chlorine atom displaced a hydrogen atom, taking the position on the carbon atom formerly occupied by that hydrogen.

To purify the *sec*-butyl chloride obtained by chlorination of *n*-butane, we would carry out a fractional distillation. But since the enantiomeric *sec*-butyl chlorides have exactly the same boiling point, they cannot be separated, and are collected in the same distillation fraction. If recrystallization is attempted, there can again be no separation since their solubilities in every (optically inactive) solvent are identical. It is easy to see, then, that whenever a racemic modification is *formed* in a reaction, we will *isolate* (by ordinary methods) a racemic modification.

If an ordinary chemical synthesis yields a racemic modification, and if this cannot be separated by our usual methods of distillation, crystallization, etc., how do we know that the product obtained *is* a racemic modification? It is optically inactive; how do we know that it is actually made up of a mixture of two optically active substances? The separation of enantiomers (called *resolution*) can be accomplished by special methods; these involve the use of optically active reagents, and will be discussed later (Sec. 7.9).

Problem 7.2 Isopentane is allowed to undergo free-radical chlorination, and the reaction mixture is separated by careful fractional distillation. (a) How many fractions of formula $C_5H_{11}Cl$ would you expect to collect? (b) Draw structural formulas, stereochemical where pertinent, for the compounds making up each fraction. Specify each enantiomer as R or S. (c) Which if any, of the fractions, as collected, would show optical activity? (d) Account in detail—just as was done in the preceding section—for the optical activity or inactivity of each fraction.

7.4 Reactions of chiral molecules. Bond breaking

Having made a chiral compound, *sec*-butyl chloride, let us see what happens when it, in turn, undergoes free-radical chlorination. A number of isomeric dichlorobutanes are formed, corresponding to attack at various positions in the molecule. (*Problem:* What are these isomers?)

$$\underset{\textit{sec-Butyl chloride}}{CH_3CH_2\text{—}\overset{*}{\underset{\underset{Cl}{|}}{C}}H\text{—}CH_3} \xrightarrow{Cl_2,\ \text{heat or light}} \underset{\text{1,2-Dichlorobutane}}{CH_3CH_2\text{—}\overset{*}{\underset{\underset{Cl}{|}}{C}}H\text{—}CH_2Cl} + \text{other products}$$

Let us take, say, (S)-*sec*-butyl chloride (which, we saw in Sec. 7.3, happens to rotate light to the right), and consider only the part of the reaction that yields 1,2-dichlorobutane. Let us make a model (I) of the starting molecule, using a single ball for $—C_2H_5$ but a separate ball for each atom in $—CH_3$. Following the familiar steps of the mechanism, we remove an —H from $—CH_3$ and replace it with a —Cl. Since we break no bond to the chiral center in either step, the model we arrive at necessarily has configuration II, in which the spatial arrangement

CH_3 / H— —Cl / C_2H_5 $\xrightarrow{Cl\cdot}$ $CH_2\cdot$ / H— —Cl / C_2H_5 $\xrightarrow{Cl_2}$ CH_2Cl / H— —Cl / C_2H_5

I (S)-*sec*-Butyl chloride

II (R)-1,2-Dichlorobutane

about the chiral center is unchanged—or, as we say, *configuration is retained*—with $—CH_2Cl$ now occupying the same relative position that was previously occupied by $—CH_3$. It is an axiom of stereochemistry that molecules, too, behave in just this way, and that *a reaction that does not involve the breaking of a bond to a chiral center proceeds with retention of configuration about that chiral center.*

(If a bond to a chiral center *is* broken in a reaction, we can make no general statement about stereochemistry, except that configuration *can* be—and more than likely *will* be—changed. As discussed in Sec. 7.10, just what happens depends on the mechanism of the particular reaction.)

Problem 7.3 We carry out free-radical chlorination of (S)-*sec*-butyl chloride, and by fractional distillation isolate the various isomeric products. (a) Draw stereochemical formulas of the 1,2-, 2,2-, and 1,3-dichlorobutanes obtained in this way. Give each enantiomer its proper R or S specification. (b) Which of these fractions, as isolated, will be optically active, and which will be optically inactive?

Now, let us see how the axiom about bond breaking is applied in relating the configuration of one chiral compound to that of another.

7.5 Reactions of chiral molecules. Relating configurations

We learned (Sec. 4.14) that the configuration of a particular enantiomer can be determined directly by a special kind of x-ray diffraction, which was first applied in 1951 by Bijvoet to (+)-tartaric acid. But the procedure is difficult and time-consuming, and can be applied only to certain compounds. In spite of this limitation, however, the configurations of hundreds of other compounds are now known, since they had already been related by chemical methods to (+)-tartaric acid. Most of these relationships were established by application of the axiom given above; that is, *the configurational relationship between two optically active compounds can be determined by converting one into the other by reactions that do not involve breaking of a bond to a chiral center.*

Let us take as an example (−)-2-methyl-1-butanol (the enantiomer found in fusel oil) and accept, for the moment, that it has configuration III, which we would specify S. We treat this alcohol with hydrogen chloride and obtain the alkyl chloride, 1-chloro-2-methylbutane. Without knowing the mechanism of this reaction, we can see that the carbon–oxygen bond is the one that is broken. *No bond to the chiral center is broken*, and therefore configuration is retained, with

—CH_2Cl occupying the same relative position in the product that was occupied by —CH_2OH in the reactant. We put the chloride into a tube, place this tube in a polarimeter, and find that the plane of polarized light is rotated to the right;

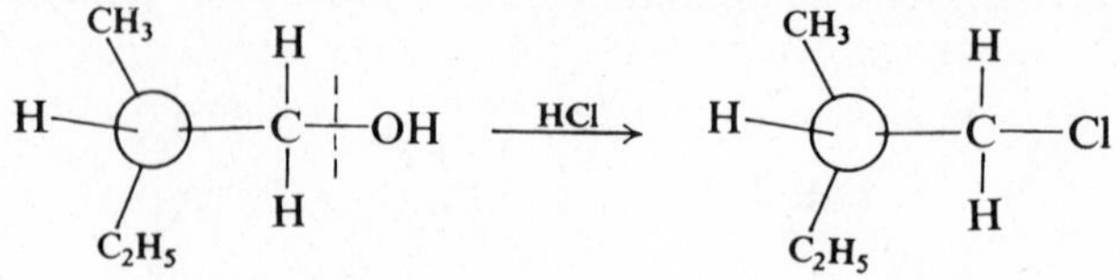

CH_3, H, C_2H_5 / H, C, OH, H $\xrightarrow{HCl}$ CH_3, H, C_2H_5 / H, C, Cl, H

III IV

S-(−)-2-Methyl-1-butanol S-(+)-1-Chloro-2-methylbutane

that is, the product is (+)-1-chloro-2-methylbutane. Since (−)-2-methyl-1-butanol has configuration III, (+)-1-chloro-2-methylbutane must have configuration IV.

Or, we oxidize (−)-2-methyl-1-butanol with potassium permanganate, obtain the acid 2-methylbutanoic acid, and find that this rotates light to the right. Again, no bond to the chiral center is broken, and we assign configuration V to (+)-2-methylbutanoic acid.

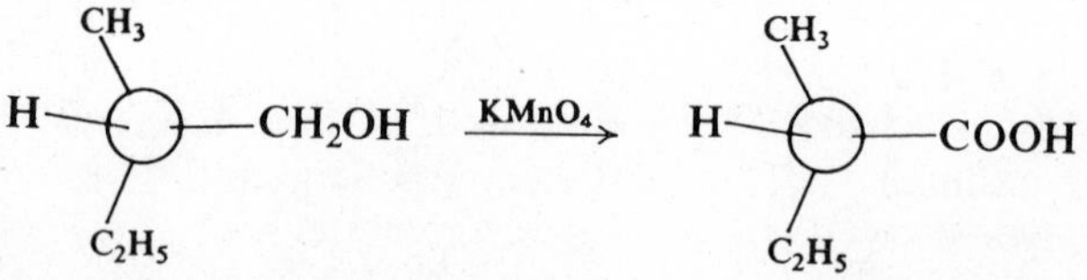

CH_3, H, C_2H_5, CH_2OH $\xrightarrow{KMnO_4}$ CH_3, H, C_2H_5, COOH

III V

(S)-(−)-2-Methyl-1-butanol (S)-(+)-2-Methylbutanoic acid

We can nearly always tell whether or not a bond to a chiral center is broken by simple inspection of the formulas of the reactant and product, as we have done in these cases, and without a knowledge of the reaction mechanism. We must be aware of the possibility, however, that a bond may break and re-form during the course of a reaction without this being evident on the surface. This kind of thing does not happen at random, but in certain specific situations which an organic chemist learns to recognize. Indeed, stereochemistry plays a leading role in this learning process: one of the best ways to detect hidden bond-breaking is so to design the experiment that if such breaking occurs, it must involve a chiral center.

But how do we know in the first place that (−)-2-methyl-1-butanol has configuration III? Its configuration was related in this same manner to that of another compound, and that one to the configuration of still another, and so on, going back ultimately to (+)-tartaric acid and Bijvoet's x-ray analysis.

We say that the (−)-2-methyl-1-butanol, the (+)-chloride, and the (+)-acid have *similar* (or the *same*) configurations. The enantiomers of these compounds, the (+)-alcohol, (−)-chloride, and (−)-acid, form another set of compounds with similar configurations. The (−)-alcohol and, for example, the (−)-chloride are said to have *opposite* configurations. As we shall find, we are usually more interested in knowing whether two compounds have similar or opposite configurations than in knowing what the actual configuration of either compound actually is. That is to say, we are more interested in *relative* configurations than in *absolute* configurations.

In this set of compounds with similar configurations, we notice that two are dextrorotatory and the third is levorotatory. The sign of rotation is important as a means of keeping track of a particular isomer—just as we might use boiling point or refractive index to tell us whether we have *cis*- or *trans*-2-butene, *now that their configurations have been assigned*—but the fact that two compounds happen to have the same sign or opposite sign of rotation means little; they may or may not have similar configurations.

The three compounds all happen to be specified as S, but this is simply because —CH_2Cl and —COOH happen to have the same relative priority as —CH_2OH. If we were to replace the chlorine with deuterium (*Problem:* How could this be done?), the product would be specified R, yet obviously it would have the same configuration as the alcohol, halide, and acid. Indeed, looking back to *sec*-butyl chloride and 1,2-dichlorobutane, we see that the similar configurations I and II *are* specified differently, one S and the other R; here, a group (—CH_3) that has a lower priority than —C_2H_5 is converted into a group (—CH_2Cl) that has a higher priority. We cannot tell whether two compounds have the same or opposite configuration by simply looking at the letters used to specify their configurations; we must work out and compare the absolute configurations indicated by those letters.

Problem 7.4 Which of the following reactions could safely be used to relate configurations?

(a) (+)-$C_6H_5CH(OH)CH_3 + PBr_3 \longrightarrow C_6H_5CHBrCH_3$

(b) (+)-$CH_3CH_2CHClCH_3 + C_6H_6 + AlCl_3 \longrightarrow C_6H_5CH(CH_3)CH_2CH_3$

(c) (−)-$C_6H_5CH(OC_2H_5)CH_2OH + HBr \longrightarrow C_6H_5CH(OC_2H_5)CH_2Br$

(d) (+)-$CH_3CH(OH)CH_2Br + NaCN \longrightarrow CH_3CH(OH)CH_2CN$

(e) (+)-$CH_3CH_2C(=O)—^{18}OCH(CH_3)C_2H_5 + OH^- \longrightarrow CH_3CH_2COO^- + CH_3CH_2CH^{18}OHCH_3$

(f) (−)-$CH_3CH_2CHBrCH_3 + C_2H_5O^-Na^+ \longrightarrow C_2H_5—O—CH(CH_3)CH_2CH_3$

(g) (+)-$CH_3CH_2CHOHCH_3 \xrightarrow{Na} CH_3CH_2CH(ONa)CH_3 \xrightarrow{C_2H_5Br} C_2H_5—O—CH(CH_3)CH_2CH_3$

Problem 7.5 What general conclusion must you draw from each of the following observations? (a) After standing in an aqueous acidic solution, optically active $CH_3CH_2CHOHCH_3$ is found to have lost its optical activity. (b) After standing in solution with potassium iodide, optically active *n*-$C_6H_{13}CHICH_3$ is found to have lost its optical activity. (c) Can you suggest experiments to test your conclusions? (See Sec. 3.29.)

7.6 Optical purity

Reactions in which bonds to chiral centers are not broken can be used to get one more highly important kind of information: the specific rotations of optically pure compounds. For example, the 2-methyl-1-butanol obtained from fusel oil (which happens to have specific rotation −5.756°) is *optically pure*—like most chiral compounds from biological sources—that is, it consists entirely of the one enantiomer, and contains none of its mirror image. When this material is treated with hydrogen chloride, the 1-chloro-2-methylbutane obtained is found to have specific rotation of +1.64°. Since no bond to the chiral center is broken, every molecule of alcohol with configuration III is converted into a molecule of chloride with configuration IV; since the alcohol was optically pure, the chloride of specific rotation +1.64° is also optically pure. Once this *maximum rotation* has been established, anyone can determine the optical purity of a sample of 1-chloro-2-methylbutane in a few moments by simply measuring its specific rotation.

If a sample of the chloride has a rotation of +0.82°, that is, 50% of the maximum, we say that it is *50% optically pure.* We consider the components of the mixture to be (+)-isomer and (±)-isomer (not (+)-isomer and (−)-isomer). (*Problem:* What are the percentages of (+)-isomer and (−)-isomer in this sample?)

Problem 7.6 Predict the specific rotation of the chloride obtained by treatment with hydrogen chloride of 2-methyl-1-butanol of specific rotation +3.12°.

7.7 Reactions of chiral molecules. Generation of a second chiral center

Let us return to the reaction we used as our example in Sec. 7.4, free-radical chlorination of *sec*-butyl chloride, but this time focus our attention on one of the other products, one in which a second chiral center is generated: 2,3-dichlorobutane. This compound, we have seen (Sec. 4.18), exists as three stereoisomers, *meso* and a pair of enantiomers.

$$\underset{\textit{sec}\text{-Butyl chloride}}{\mathrm{CH_3CH_2{-}\underset{Cl}{\overset{*}{C}}H{-}CH_3}} \xrightarrow{\mathrm{Cl_2,\ heat\ or\ light}} \underset{\text{2,3-Dichlorobutane}}{\mathrm{CH_3{-}\underset{Cl}{\overset{*}{C}}H{-}\underset{Cl}{\overset{*}{C}}H{-}CH_3}} + \text{other products}$$

Let us suppose that we take optically active *sec*-butyl chloride (the (S)-isomer, say), carry out the chlorination, and by fractional distillation separate the 2,3-dichlorobutanes from all the other products (the 1,2-isomer, 2,2-isomer, etc.). Which stereoisomers can we expect to have?

Figure 7.2 shows the course of reaction. Three important points are illustrated which apply in all cases where a second chiral center is generated. First, since no bonds to the original chiral center, C–2, are broken, its configuration is retained in all the products. Second, there are two possible configurations about the new chiral center, C–3, and both of these appear; in this particular case, they result from attacks (*a*) and (*b*) on opposite sides of the flat portion of the free radical, giving the diastereomeric S,S and R,S (or *meso*) products. Third, the diastereomeric products will be formed in unequal amounts; in this case because attack (*a*) and attack (*b*) are not equally likely.

In Sec. 7.3 we saw that generation of the first chiral center in a compound yields equal amounts of enantiomers, that is, yields an optically inactive racemic modification. Now we see that generation of a new chiral center in a compound that is already optically active yields an optically active product containing unequal amounts of diastereomers.

Suppose (as is actually the case) that the products from (S)-*sec*-butyl chloride show an S,S:*meso* ratio of 29:71. What would we get from chlorination of (R)-*sec*-butyl chloride? We would get (R,R-) and *meso*-products, and the R,R:*meso* ratio would be exactly 29:71. Whatever factor favors *meso*-product over (S,S)-product will favor *meso*-product over (R,R)-product, and to exactly the same extent.

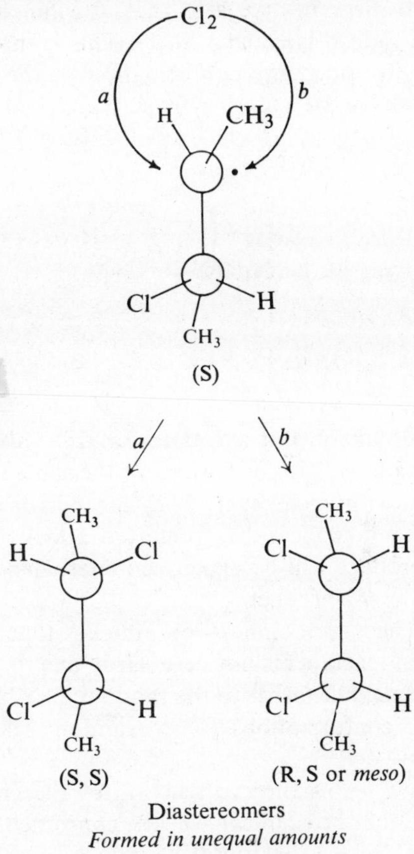

Figure 7.2. Generation of a second chiral center. Configuration at original chiral center unchanged. Chlorine becomes attached via (*a*) or (*b*) to give diastereomers, and in unequal amounts.

(S)

(S, S) (R, S or *meso*)

Diastereomers
Formed in unequal amounts

Finally, what can we expect to get from optically inactive, racemic *sec*-butyl chloride? The (S)-isomer that is present would yield (S,S)- and *meso*-products in the ratio of 29:71; the (R)-isomer would yield (R,R)- and *meso*-products, and in the ratio of 29:71. Since there are exactly equal quantities of (S)- and (R)-reactants, the two sets of products would exactly balance each other, and we would obtain racemic and *meso* products in the ratio of 29:71. Optically inactive reactants yield optically inactive products.

One point requires further discussion. Why are the diastereomeric products formed in unequal amounts? It is because the intermediate 3-chloro-2-butyl radical in Fig. 7.2 already contains a chiral center. The free radical is chiral, and lacks the symmetry that is necessary for attack at the two faces to be equally likely. (Make a model of the radical and assure yourself that this is so.)

In the following section, this point is discussed in more detail.

7.8 Formation of enantiomers and diastereomers: a closer look

To understand better how formation of diastereomers differs from formation of enantiomers, let us contrast the reaction of the chiral 3-chloro-2-butyl radical shown in Fig. 7.2 with the reaction of the achiral *sec*-butyl radical.

In Sec. 7.3, we said that attachment of chlorine to either face of the *sec*-butyl radical is equally likely. This is in effect true, but deserves closer examination. Consider any conformation of the free radical: I, for example. It is clear that attack by chlorine from the top of I and attack from the bottom are *not*

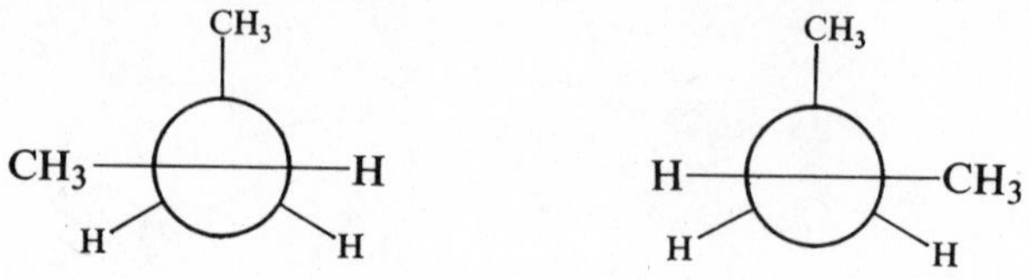

I II

sec-Butyl radical
Achiral

equally likely. But a rotation of 180° about the single bond converts I into II; these are two conformations of the same free radical, and are, of course, in equilibrium with each other. They are mirror images, and hence of equal energy and equal abundance; any preferred attack from, say, the bottom of I to give the (R)-product will be exactly counterbalanced by attack from the bottom of II to give the (S)-product.

The "randomness of attack" that yields the racemic modification from achiral reactants is not necessarily due to the symmetry of any individual reactant molecule, but rather to the random distribution of such molecules between mirror-image conformations (or to random selection between mirror-image transition states).

Now, let us turn to reaction of the chiral 3-chloro-2-butyl radical (Fig. 7.2). Here, the free radical we are concerned with already contains a chiral center, about which it has the (S)-configuration; attack is *not* random on such a radical because mirror-image conformations are not present—they could only come from (R) free radicals, and there are none of those radicals present.

Preferred attack from, say, the bottom of conformation III—a likely preference since this would keep the two chlorine atoms as far apart as possible in the transition state—would yield *meso*-2,3-dichlorobutane. A rotation of 180° about the single bond would convert III into IV. Attack from the bottom of IV would

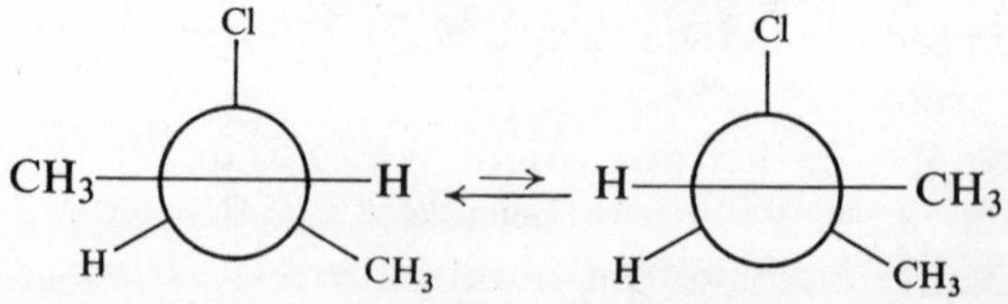

III IV

3-Chloro-2-butyl radical
Chiral

yield the (S,S)-isomer. But III and IV are not mirror images, are not of equal energy, and are not of equal abundance. In particular, because of lesser crowding between the methyl groups, we would expect III to be more stable and hence more abundant than IV, and the *meso* product to predominate over the (S,S)-isomer (as it actually does).

We might have made a different guess about the preferred direction of attack, and even a different estimate about relative stabilities of conformations, but we would still arrive at the same basic conclusion: except by sheer coincidence, the two diastereomers would not be formed in equal amounts.

In this discussion, we have assumed that the relative rates of competing reactions depend on relative populations of the conformations of the reactants. This assumption is correct here, if, as seems likely, reaction of the free radicals with chlorine is easier and faster than the rotation that interconverts conformations.

If, on the other hand, reaction with chlorine were a relatively difficult reaction and much slower than interconversion of conformations, then relative rates would be determined by relative stabilities of the transition states. We would still draw the same general conclusions. In the reaction of the achiral *sec*-butyl radical, the transition states are mirror images and therefore of the same stability, and the rates of formation of the two products would be exactly the same. In the reaction of the chiral 3-chloro-2-butyl radical, the transition states are not mirror-images and therefore not of the same stability, and rates of formation of the two products would be different. (In the latter case, we would even make the same prediction, that the *meso* product would predominate, since the same relationship between methyl groups that would make conformation III more stable would also make the transition state resembling conformation III more stable.)

Problem 7.7 Answer the following questions about the formation of 2,3-dichlorobutane from (R)-*sec*-butyl chloride. (a) Draw conformations (V and VI) of the intermediate radicals that correspond to III and IV above. (b) What is the relationship between V and VI? (c) How will the V:VI ratio compare with the III:IV ratio? (d) Assuming the same preferred direction of attack by chlorine as on III and IV, which stereoisomeric product would be formed from V? From VI? (e) Which product would you expect to predominate? (f) In view of the ratio of products actually obtained from (S)-*sec*-butyl chloride, what ratio of products must be obtained from (R)-*sec*-butyl chloride?

Problem 7.8 Each of the following reactions is carried out, and the products are separated by careful fractional distillation or recrystallization. For each reaction tell how many fractions will be collected. Draw stereochemical formulas of the compound or compounds making up each fraction, and give each its R/S specification. Tell whether each fraction, as collected, will show optical activity or optical inactivity.

(a) monochlorination of (R)-*sec*-butyl chloride at 300°;
(b) monochlorination of racemic *sec*-butyl chloride at 300°;
(c) monochlorination of racemic 1-chloro-2-methylbutane at 300°;
(d) addition of bromine to (S)-3-bromo-1-butene.

7.9 Reactions of chiral molecules with optically active reagents. Resolution

So far in this chapter we have discussed the reactions of chiral compounds only with optically inactive reagents. Now let us turn to reactions with optically *active* reagents, and examine one of their most useful applications: **resolution of a racemic modification**, that is, *the separation of a racemic modification into enantiomers.*

We know (Sec. 7.3) that when optically inactive reactants form a chiral compound, the product is the racemic modification. We know that the enantiomers making up a racemic modification have identical physical properties (except for direction of rotation of polarized light), and hence cannot be separated by the usual methods of fractional distillation or fractional crystallization. Yet throughout this book are frequent references to experiments carried out using optically active compounds like (+)-*sec*-butyl alcohol, (−)-2-bromooctane, (−)-α-phenylethyl chloride, (+)-α-phenylpropionamide. How are such optically active compounds obtained?

Some optically active compounds are obtained from natural sources, since living organisms usually produce only one enantiomer of a pair. Thus only (−)-2-methyl-1-butanol is formed in the yeast fermentation of starches, and only (+)-lactic acid, $CH_3CHOHCOOH$, in the contraction of muscles; only (−)-malic acid, $HOOCCH_2CHOHCOOH$, is obtained from fruit juices, only (−)-quinine from the bark of the cinchona tree. Indeed, we deal with optically active substances to an extent that we may not realize. We eat optically active bread and optically active meat, live in houses, wear clothes, and read books made of optically active cellulose. The proteins that make up our muscles and other tissues, the glycogen in our liver and in our blood, the enzymes and hormones that enable us to grow, and that regulate our bodily processes—all these are optically active. Naturally occurring compounds are optically active because the enzymes that bring about their formation—and often the raw materials from which they are made—are themselves optically active. As to the origin of the optically active enzymes, we can only speculate.

Amino acids, the units from which proteins are made, have been reported present in meteorites, but in such tiny amounts that the speculation has been made that "what appears to be the pitter-patter of heavenly feet is probably instead the print of an earthly thumb." Part of the evidence that the amino acids found in a meteorite by Cyril Ponnamperuma (of NASA) are really extraterrestrial in origin is that they are optically *inactive*—not optically active as earthly contaminants from biological sources would be.

From these naturally occurring compounds, other optically active compounds can be made. We have already seen, for example, how (−)-2-methyl-1-butanol can be converted without loss of configuration into the corresponding chloride or acid (Sec. 7.5); these optically active compounds can, in turn, be converted into many others.

Most optically active compounds are obtained by the resolution of a racemic modification, that is, by a separation of a racemic modification into enantiomers. Most such resolutions are accomplished through the use of reagents that are themselves optically active; these reagents are generally obtained from natural sources.

The majority of resolutions that have been carried out depend upon the reaction of organic bases with organic acids to yield salts. Let us suppose, for example, that we have prepared the racemic acid, (±)-HA. Now, there are isolated from various plants very complicated bases called *alkaloids* (that is, *alkali-like*), among which are cocaine, morphine, strychnine, and quinine. Most alkaloids are produced by plants in only one of two possible enantiomeric forms, and hence they are optically active. Let us take one of these optically active bases, say a levorotatory one, (−)-B, and mix it with our racemic acid (±)-HA. The acid is present in two configurations, but the base is present in only one configuration; there will result, therefore, crystals of two different salts, [(−)-BH^+ (+)-A^-] and [(−)-BH^+-(−)-A^-].

What is the relationship between these two salts? They are not superimposable, since the acid portions are not superimposable. They are not mirror images, since the base portions are not mirror images. The salts are stereoisomers that are not enantiomers, and therefore are *diastereomers*.

$$\left.\begin{array}{l}(+)\text{-HA}\\ (-)\text{-HA}\end{array}\right\} + (-)\text{-B} \longrightarrow \left\{\begin{array}{l}[(-)\text{-BH}^+\ (+)\text{-A}^-] \xrightarrow{H^+} (+)\text{-HA} + (-)\text{-BH}^+\\ [(-)\text{-BH}^+\ (-)\text{-A}^-] \xrightarrow{H^+} (-)\text{-HA} + (-)\text{-BH}^+\end{array}\right.$$

Enantiomers: *in a racemic modification*	Alkaloid *base*	Diastereomers: *separable*	Resolved enantiomers	Alkaloid *as a salt*

These diastereomeric salts have, of course, different physical properties, including solubility in a given solvent. They can therefore be separated by fractional crystallization. Once the two salts are separated, optically active acid can be recovered from each salt by addition of strong mineral acid, which displaces the weaker organic acid. If the salt has been carefully purified by repeated crystallizations to remove all traces of its diastereomer, then the acid obtained from it is *optically pure*. Among the alkaloids commonly used for this purpose are (−)-brucine, (−)-quinine, (−)-strychnine, and (+)-cinchonine.

Resolution of organic bases is carried out by reversing the process just described: using naturally occurring optically active acids, (−)-malic acid, for example. Resolution of alcohols, which we shall find to be of special importance in synthesis, poses a special problem: since alcohols are neither appreciably basic nor acidic, they cannot be resolved by direct formation of salts. Yet they can be resolved by a rather ingenious adaptation of the method we have just described: one attaches to them an acidic "handle," which permits the formation of salts, and then when it is no longer needed can be removed.

Compounds other than organic bases, acids, or alcohols can also be resolved. Although the particular chemistry may differ from the salt formation just described, the principle remains the same: **a racemic modification is converted by an optically active reagent into a mixture of diastereomers which can then be separated.**

7.10 Reactions of chiral molecules. Mechanism of free-radical chlorination

So far, we have discussed only reactions of chiral molecules in which bonds to the chiral center are not broken. What is the stereochemistry of reactions in which the bonds to the chiral center *are* broken? The answer is: *it depends.* It depends on the *mechanism* of the reaction that is taking place; because of this, stereochemistry can often give us information about a reaction that we cannot get in any other way.

For example, stereochemistry played an important part in establishing the mechanism that was the basis of our entire discussion of the halogenation of alkanes (Chap. 3). The chain-propagating steps of this mechanism are:

(2a) $$X\cdot + RH \longrightarrow HX + R\cdot$$

(3a) $$R\cdot + X_2 \longrightarrow RX + X\cdot$$

Until 1940 the existing evidence was just as consistent with the following alternative steps:

(2b) $$X\cdot + RH \longrightarrow RX + H\cdot$$

(3b) $$H\cdot + X_2 \longrightarrow HX + X\cdot$$

To differentiate between these alternative mechanisms, H. C. Brown, M. S. Kharasch, and T. H. Chao, working at the University of Chicago, carried out the photochemical halogenation of optically active S-(+)-1-chloro-2-methylbutane. A number of isomeric products were, of course, formed, corresponding to attack at various positions in the molecule. (*Problem:* What were these products?) They focused their attention on just *one* of these products: 1,2-dichloro-2-methylbutane, resulting from substitution at the chiral center (C–2).

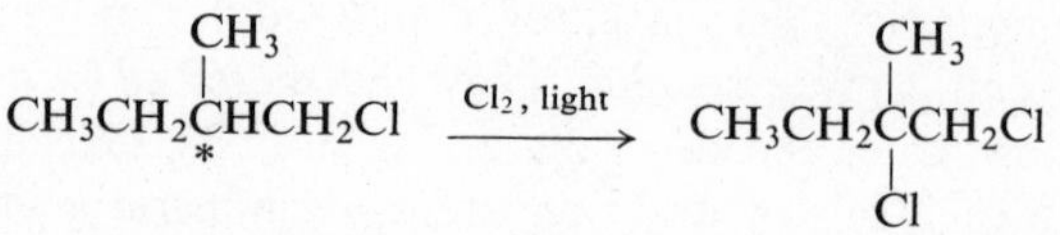

$$CH_3CH_2\overset{CH_3}{\underset{*}{C}}HCH_2Cl \xrightarrow{Cl_2,\ light} CH_3CH_2\overset{CH_3}{\underset{Cl}{C}}CH_2Cl$$

(S)-(+)-1-Chloro-2-methylbutane — *Optically active*

(±)-1,2-Dichloro-2-methylbutane — *Optically inactive*

They had planned the experiment on the following basis. The two mechanisms differed as to whether or not a free alkyl radical is an intermediate. The most likely structure for such a radical, they thought, was *flat*—as, it turns out, it very probably is—and the radical would lose the original chirality. Attachment of chlorine to either face would be equally likely, so that an optically inactive, racemic product would be formed. That is to say, the reaction would take place *with racemization* (see Fig. 7.3).

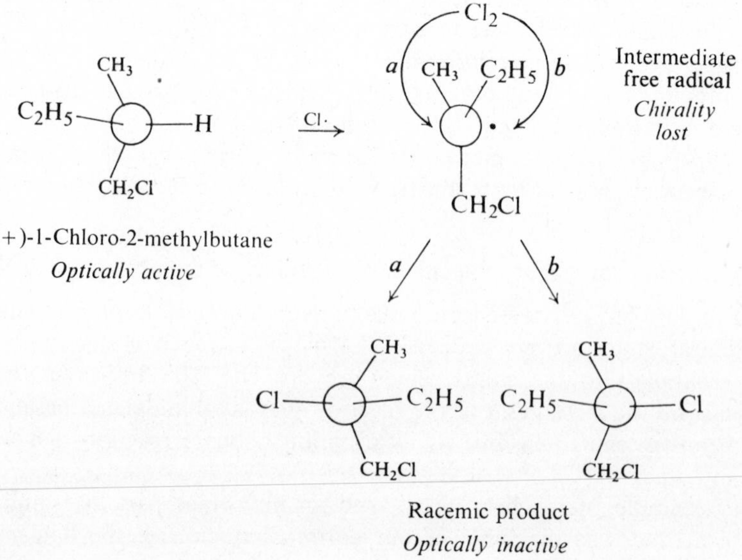

Figure 7.3. Racemization through free-radical formation. Chlorine becomes attached to either face of free radical, via (*a*) or (*b*), to give enantiomers, and in equal amounts.

For the alternative mechanism, in which chlorine would become attached to the molecule while the hydrogen was being displaced, they could make no prediction, except that formation of an optically inactive product would be highly unlikely: there was certainly no reason to expect that *back-side* attack (on the face opposite the hydrogen) would take place to exactly the same extent as *front-side* attack. (In ionic displacements, attack is generally back-side.)

By careful fractional distillation they separated the 1,2-dichloro-2-methylbutane from the reaction mixture, and found it to be *optically inactive*. From this they concluded that the mechanism involving free alkyl radicals, (2a), (3a), is the correct one. This mechanism is accepted without question today, and the work of Brown, Kharasch, and Chao is frequently referred to as evidence of the stereochemical behavior of free radicals, with the original significance of the work exactly reversed.

We can begin to see how stereochemistry provides the organic chemist with one of his most powerful tools for finding out what is going on in a chemical reaction.

Problem 7.9 This work does *not* prove that free radicals are flat. Racemization is consistent with what other structure for free radicals? Explain. (*Hint:* See Sec. 2.21.)

Problem 7.10 Altogether, the free-radical chlorination of (S)-(+)-1-chloro-2-methylbutane gave six fractions of formula $C_5H_{10}Cl_2$. Four fractions were found to be optically active, and two fractions optically inactive. Draw structural formulas for the compounds making up each fraction. Account in detail for optical activity or inactivity in each case.

7.11 Stereoselective and stereospecific reactions. *syn-* and *anti-*Addition

As our second example of the application of stereochemistry to the study of reaction mechanisms, let us take another familiar reaction: addition of halogens to alkenes. In this section we shall look at the stereochemical facts and, in the next, see how these facts can be interpreted.

Addition of bromine to 2-butene yields 2,3-dibromobutane. Two chiral centers are generated in the reaction, and the product, we know, can exist as a *meso* compound and a pair of enantiomers.

$$\underset{\text{2-Butene}}{CH_3CH{=}CHCH_3} + Br_2 \longrightarrow \underset{\text{2,3-Dibromobutane}}{CH_3{-}\overset{*}{\underset{Br}{CH}}{-}\overset{*}{\underset{Br}{CH}}{-}CH_3}$$

The reactant, too, exists as diastereomers: a pair of geometric isomers. If we start with, say, *cis*-2-butene, which of the stereoisomeric products do we get? A mixture of all of them? No. *cis*-2-Butene yields *only* racemic 2,3-dibromobutane; none of the *meso* compound is obtained. *A reaction that yields predominantly one stereoisomer (or one pair of enantiomers) of several diastereomeric possibilities is called a* **stereoselective reaction.**

Now, suppose we start with *trans*-2-butene. Does this, too, yield the racemic dibromide? No. *trans*-2-Butene yields *only meso*-2,3-dibromobutane. *A reaction in which stereochemically different reactants give stereochemically different products is called a* **stereospecific reaction.**

Addition of bromine to alkenes is both stereoselective and stereospecific. We say it is *completely* stereoselective since, from a given alkene, we obtain *only* one diastereomer (or one pair of enantiomers). We say it is stereospecific, since just which stereoisomer we obtain depends upon which stereoisomeric alkene we start with.

In the above definition, *stereochemically different* means, in practice, *diastereomerically different*. The term *stereospecific* is not applied to reactions, like those in Secs. 7.4 and 7.5, in which enantiomerically different reactants give enantiomerically different products.

All stereospecific reactions are necessarily stereoselective, but the reverse is not true. There are reactions from which one particular stereoisomer is the predominant product *regardless* of the stereochemistry of the reactant; there are reactions in which the reactant cannot exist as stereoisomers, but from which one particular stereoisomer is the predominant product. Such reactions are stereoselective but not stereospecific.

To describe stereospecificity in addition reactions, the concepts of *syn*-addition and *anti*-addition are used. These terms are not the names of specific mechanisms. They simply indicate the stereochemical facts: that the product obtained is the one to be expected if the two portions of the reagent were to add to the same face of the alkene (*syn*) or to opposite faces (*anti*).

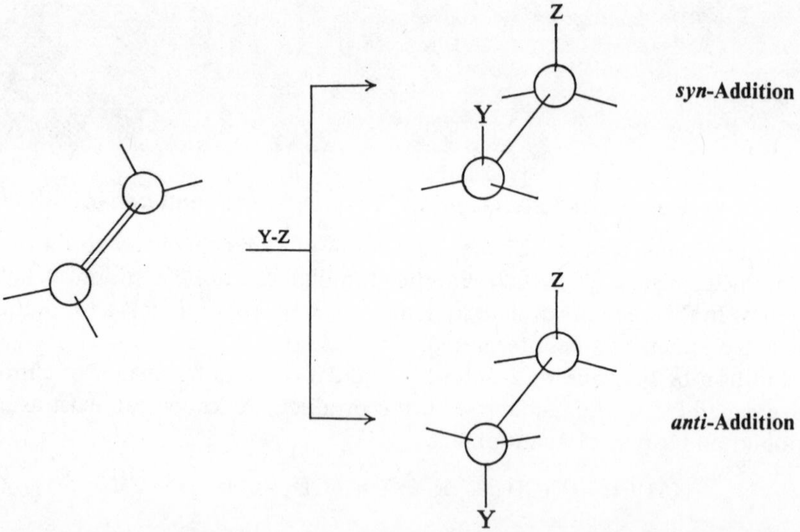

Y-Z

Z
Y
***syn*-Addition**

Z
Y
***anti*-Addition**

Addition of bromine to the 2-butenes involves *anti*-addition. If we start (Fig. 7.4) with *cis*-2-butene, we can attach the bromine atoms to opposite faces of the alkene either as in (*a*) or in (*b*) and thus obtain the enantiomers. Since, whatever the mechanism, (*a*) and (*b*) should be equally likely, we obtain the racemic modification.

Starting with *trans*-2-butene (Fig. 7.5), we can again attach the bromine atoms to opposite faces of the alkene in two ways but, whichever way we choose, we obtain the *meso*-dibromide.

anti-Addition is the general rule for the reaction of bromine or chlorine with simple alkenes. We shall encounter other examples of stereospecific additions, both *anti* and *syn*. We shall find that other reactions besides addition can be

***anti*-Addition**

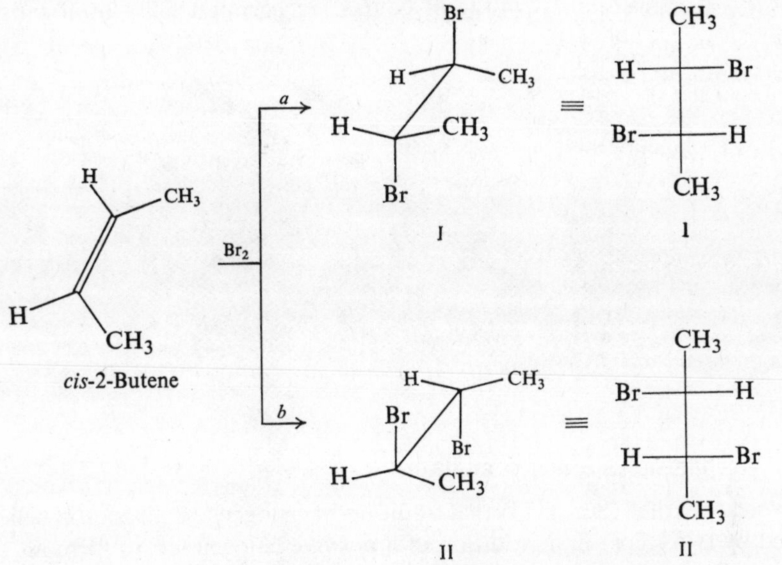

cis-2-Butene

I

II

I *and* II *are enantiomers*

Racemic 2,3-dibromobutane

Figure 7.4. *anti*-Addition to *cis*-2-butene. Attachment as in (*a*) or (*b*) equally likely: gives racemic modification.

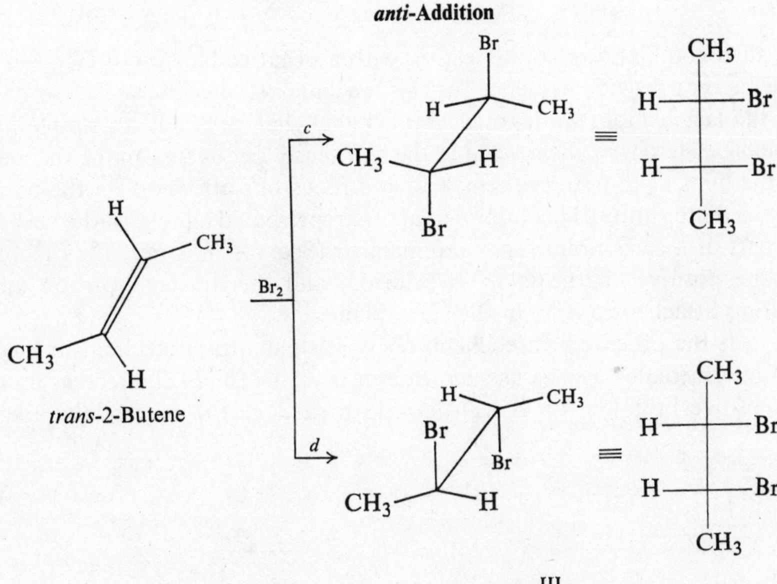

Figure 7.5. *anti*-Addition to *trans*-2-butene. Attachment as in (*c*) or (*d*) gives *meso* product.

stereospecific—and also that some can be non-stereospecific. Whatever the stereochemistry of a reaction, it must, of course, be accounted for by a satisfactory mechanism.

Problem 7.11 On treatment with permanganate, *cis*-2-butene yields a glycol of m.p. 34°, and *trans*-2-butene yields a glycol of m.p. 19°. Both glycols are optically inactive. Handling as described in Sec. 7.9 converts the glycol of m.p. 19° (but not the one of m.p. 34°) into two optically active fractions of equal but opposite rotation.

(a) What is the configuration of the glycol of m.p. 19°? Of m.p. 34°?

(b) Assuming these results are typical (they are), *what is the stereochemistry of hydroxylation with permanganate?*

(c) Treatment of the same alkenes with peroxy acids gives the opposite results: the glycol of m.p. 19° from *cis*-2-butene, and the glycol of m.p. 34° from *trans*-2-butene. *What is the stereochemistry of hydroxylation with peroxy acids?*

7.12 Mechanism of halogen addition

We saw earlier (Sec. 6.13) that addition of halogens to alkenes is believed to proceed by two steps: first, addition of a positive halogen ion to form an organic

(1) 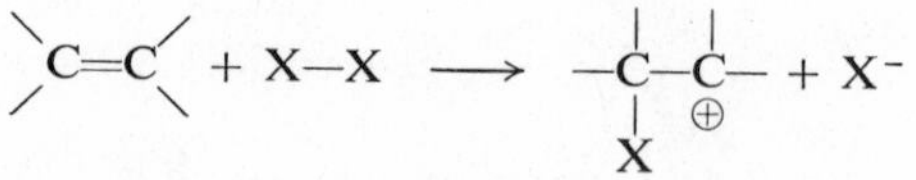 $>C{=}C< + X{-}X \longrightarrow -\underset{X}{C}-\overset{\oplus}{C}- + X^-$

(2) 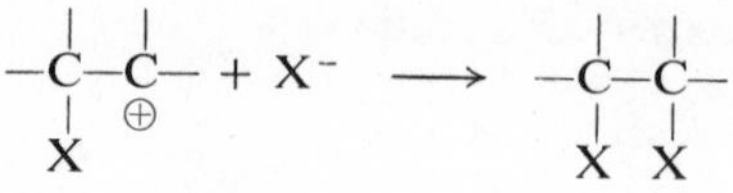 $-\underset{X}{C}-\overset{\oplus}{C}- + X^- \longrightarrow -\underset{X}{C}-\underset{X}{C}-$

cation; then combination of this cation with a negative halide ion. We saw some of the facts that provide evidence for this mechanism.

In the last section, we learned another fact: halogens add to simple alkenes with *complete* stereospecificity, and in the *anti* sense. Let us reexamine the mechanism in the light of this stereochemistry, and focus our attention on the nature of the intermediate cation. This intermediate we represented simply as the carbonium ion. A part of a carbonium ion, we remember (Sec. 5.16), is *flat*: the carbon that carries the positive charge is sp^2-hybridized, and this trigonal carbon and the three atoms attached to it lie in the same plane.

Now, is the observed stereochemistry consistent with a mechanism involving such an intermediate? Let us use addition of bromine to *cis*-2-butene as an example. A positive bromine ion is transferred to, say, the top face of the alkene to

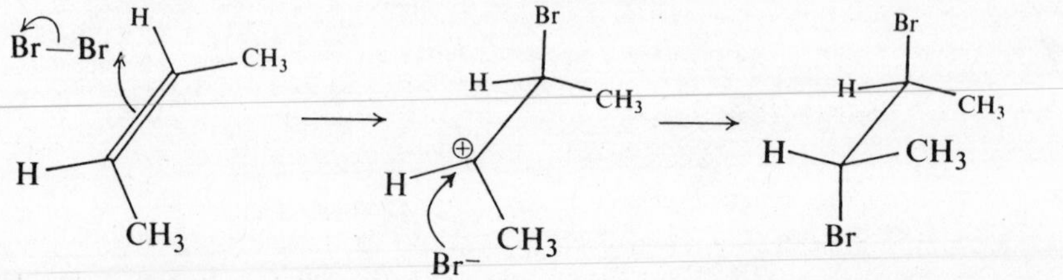

Br—Br H CH3 H CH3 → Br H CH3 ⊕ H CH3 Br⁻ → Br H CH3 H CH3 Br

cis-2-Butene Cation I (S,S)-2,3-Dibromobutane

form the carbonium ion I. Then, a bromide ion attacks the *bottom* face of the positively charged carbon to complete the *anti* addition; attack at this face is preferred, we might say, because it permits the two bromines to be as far apart as possible in the transition state. (We obtain the racemic product: the S,S-dibromide as shown, the R,R-dibromide through attachment of positive bromine to the near end of the alkene molecule.)

But this picture of the reaction is not satisfactory, and for two reasons. First, to account for the *complete* stereospecificity of addition, we must assume that attack at the bottom face of the cation is not just preferred, but is the *only* line of attack: conceivable, but—especially in view of other reactions of carbonium ions (Sec. 14.13)—not likely. Then, even if we accept this exclusively bottom-side attack, we are faced with a second problem. Rotation about the carbon–carbon bond would convert cation I into cation II; bottom–side attack on cation II would

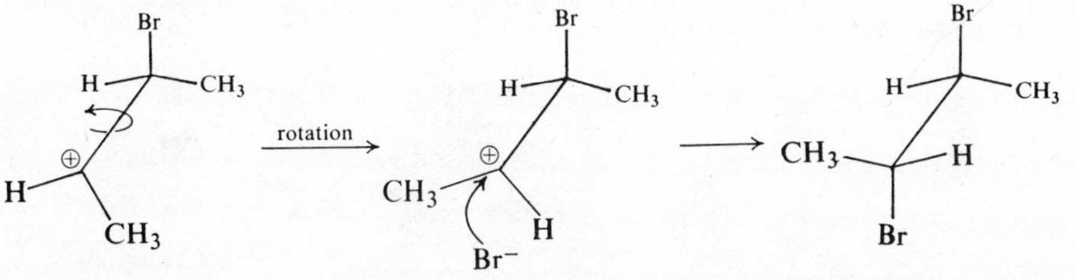

rotation →

Cation I Cation II *meso*-2,3-Dibromobutane

yield not the racemic dibromide but the *meso* dibromide—in effect *syn*-addition, and contrary to fact.

To accommodate the stereochemical facts, then, we would have to make two assumptions about halogen addition: after the carbonium ion is formed, it is attacked by bromide ion (a) before rotation about the single bond can occur, and (b) exclusively from the side away from the halogen already in the cation. Neither of these assumptions is very likely; together, they make the idea of a simple carbonium ion intermediate hard to accept.

In 1937, to account better for the observed stereochemistry, I. Roberts and G. E. Kimball at Columbia University proposed the following mechanism. In step (1) of the addition of bromine, for example, positive bromine attaches itself

(1) Br—Br + >C=C< ⟶ Br^- + cyclic bromonium ion

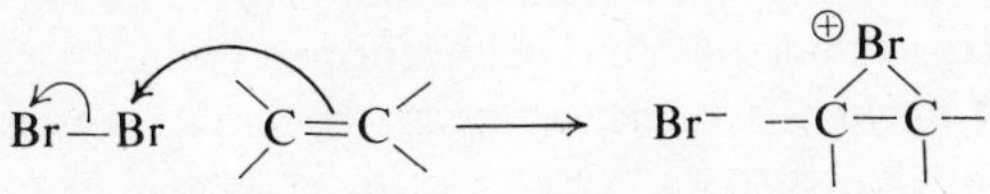

A bromonium ion

(2) bromonium ion + Br^- ⟶ —C(Br)—C(Br)—

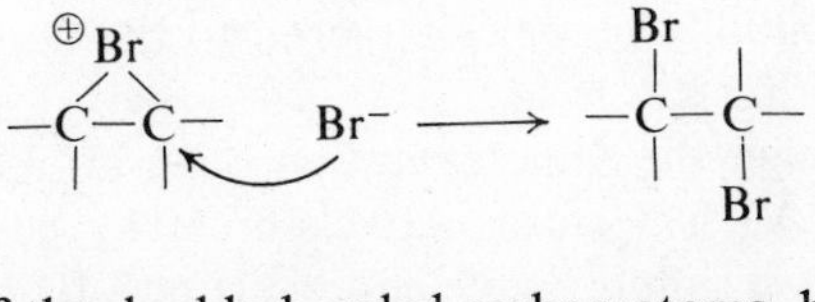

not to just one of the doubly-bonded carbon atoms, but to both, forming a cyclic **bromonium ion**. In step (2), bromide ion attacks this bromonium ion to yield the dibromide.

Now, how does the bromonium ion mechanism account for *anti*-addition? Using models, let us first consider addition of bromine to *cis*-2-butene (Fig. 7.6).

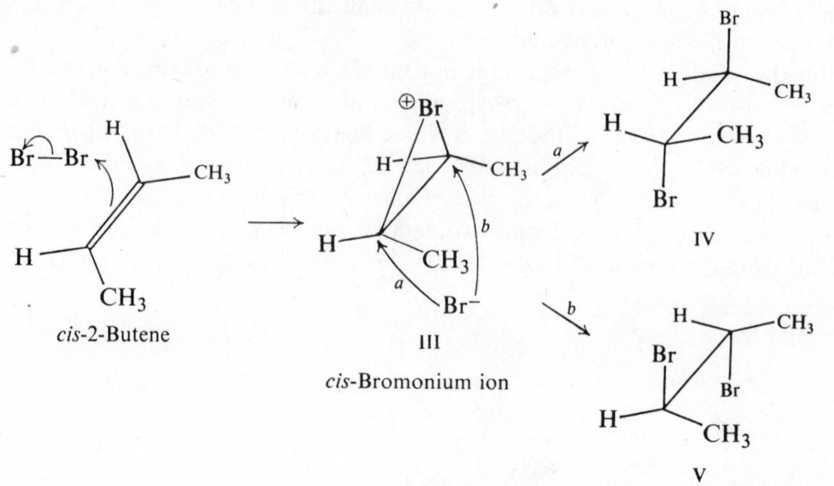

Br—Br H CH$_3$ H CH$_3$ *cis*-2-Butene → ⊕Br H CH$_3$ H CH$_3$ Br$^-$ *a* *b* III *cis*-Bromonium ion

a → Br H CH$_3$ H CH$_3$ Br IV

b → H CH$_3$ Br Br H CH$_3$ V

IV *and* V *are enantiomers*
Racemic 2,3-dibromobutane

Figure 7.6. Addition of bromine to *cis*-2-butene via cyclic bromonium ion. Opposite-side attacks (*a*) and (*b*) equally likely, give enantiomers in equal amounts.

In the first step, positive bromine becomes attached to either the top or bottom face of the alkene. Let us see what we would get if bromine becomes attached to the top face. When this happens, the carbon atoms of the double bond tend to become tetrahedral, and the hydrogens and methyls are displaced downward. The methyl groups are, however, still located across from each other, as they were in the alkene. In this way, bromonium ion III is formed.

Now bromonium ion III is attacked by bromide ion. A new carbon–bromine bond is formed, and an old carbon–bromine bond is broken. This attack occurs on the bottom face of III, so that the bond being formed is on the opposite of carbon from the bond being broken. Attack can occur by path (*a*) to yield structure IV or by path (*b*) to yield structure V. We recognize IV and V as enantiomers. Since attack by either (*a*) or (*b*) is equally likely, the enantiomers are formed in equal amounts, and thus we obtain the racemic modification. The same results are obtained if positive bromine initially becomes attached to the bottom face of *cis*-2-butene. (Show with models that this is so.)

Next, let us carry through the same operation on *trans*-2-butene (Fig. 7.7). This time, bromonium ion VI is formed. Attack on it by path (*c*) yields VII, attack by (*d*) yields VIII. If we simply rotate either VII or VIII about the carbon–carbon bond, we readily recognize the symmetry of the compound. It is *meso*-2,3-dibromo-butane; VII and VIII are identical. The same results are obtained if positive bromine is initially attached to the bottom face of *trans*-2-butene. (Show with models that this is so.)

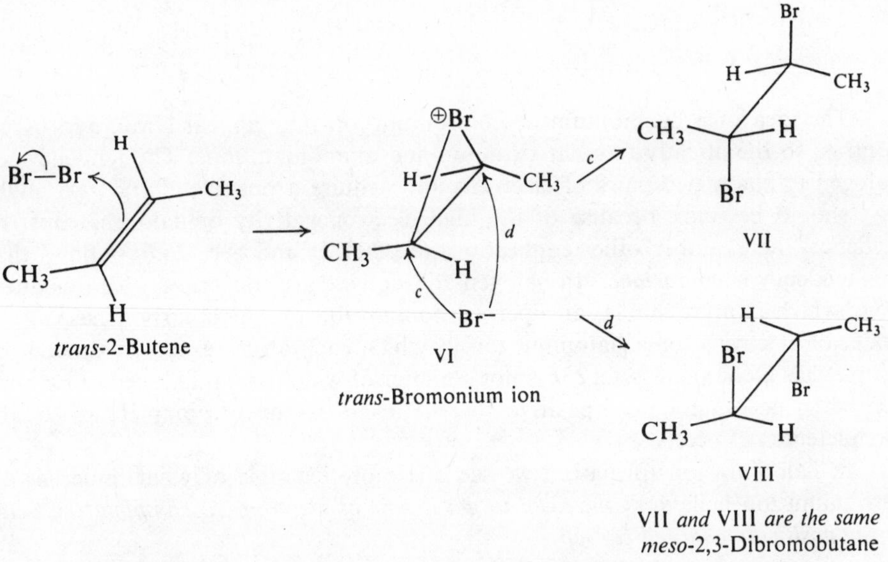

trans-2-Butene → VI *trans*-Bromonium ion → (c) VII, (d) VIII

VII *and* VIII *are the same* *meso*-2,3-Dibromobutane

Figure 7.7. Addition of bromine to *trans*-2-butene via cyclic bromonium ion. Opposite-side attacks (*c*) and (*d*) give same product.

Problem 7.12 (a) What is the relationship between the bromonium ions formed by attachment of positive bromine to the top and bottom faces of *trans*-2-butene? In what proportions are they formed? (b) Answer the same questions for *cis*-2-butene. (c) For *trans*-2-pentene. (d) For *cis*-2-pentene.

Problem 7.13 (a) Predict the products of addition of bromine to *trans*-2-pentene. Is attack by bromide ion by the two paths equally likely? Account for the fact that inactive material is actually obtained. (b) Do the same for *cis*-2-pentene.

The concept of a halonium ion solves both of the problems associated with an open carbonium ion: a halogen bridge prevents rotation about the carbon–carbon bond, and at the same time restricts bromide ion attack exclusively to the opposite face of the cation. This opposite-side approach, we shall find (Sec. 14.10), is *typical* of attack by bases (nucleophiles) on tetrahedral carbon.

That such cyclic intermediates *can* give rise to *anti*-addition is demonstrated by hydroxylation with peroxy acids (Problem 7.11, p. 242): there, analogous intermediates—perfectly respectable compounds called *epoxides* (Chap. 17)—can actually be isolated and studied.

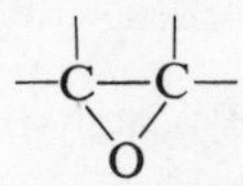

An epoxide

Cyclic halonium ions were first proposed, then, simply as the most reasonable explanation for the observed stereochemistry. Since that time, however, more positive evidence has been discovered. In 1967, Olah (p. 160) prepared cations whose nmr spectra indicate that they are indeed cyclic halonium ions. For example:

$$\underset{\substack{| \quad\; | \\ F \quad Br}}{(CH_3)_2C{-}CHCH_3} + SbF_5 \xrightarrow{\text{liquid } SO_2} \underset{\substack{\diagdown \; \diagup \\ Br^{\oplus}}}{(CH_3)_2C{-}CHCH_3} \quad SbF_6^-$$

The idea of a bromonium or chloronium ion may appear strange to us, in contrast to the already familiar oxonium and ammonium ions. The tendency for halogen to share two pairs of electrons and acquire a positive charge, we might say, should be weak because of the high electronegativity of halogens. But the evidence—here, and in other connections (Sec. 11.21 and Sec. 25.6)—shows that this tendency is *appreciable.* In halogen addition we are concerned with this question: which is more stable, an open carbonium ion in which carbon has only a sextet of electrons, or a halonium ion in which each atom (except hydrogen, of course) has a complete octet? It is not a matter of which atom, halogen or carbon, can better accommodate a positive charge; it is a matter of completeness or incompleteness of octets.

In halonium ion formation we see one more example of what underlies all carbonium ion behavior: *the need to get a pair of electrons to complete the octet of the positively charged carbon.*

There are exceptions to the rule of *anti*-addition of halogens, but exceptions that are quite understandable. If the alkene contains substituents that can strongly stabilize the open carbonium ion—as, for example, in a *benzyl* cation (Sec. 12.19)—then addition proceeds with little or no stereospecificity. Carbon is getting the electrons it needs, but in a different way.

Problem 7.14 Olah treated compounds of the formula $(CH_3)_2CXCF(CH_3)_2$ with SbF_5. He observed the formation of halonium ions when X=Cl, Br, or I, but an open carbonium ion when X=F. How do you account for the difference in behavior of the difluoro compound? (*Hint:* See Sec. 1.15.)

PROBLEMS

1. Each of the following reactions is carried out, and the products are separated by careful fractional distillation or recrystallization. For each reaction tell how many fractions will be collected. Draw stereochemical formulas of the compound or compounds making up each fraction, and give each its R/S specification. Tell whether each fraction, as collected, will show optical activity or optical inactivity.

(a) *n*-pentane + Cl_2 (300°) ⟶ $C_5H_{11}Cl$;
(b) 1-chloropentane + Cl_2 (300°) ⟶ $C_5H_{10}Cl_2$;
(c) (S)-2-chloropentane + Cl_2 (300°) ⟶ $C_5H_{10}Cl_2$;
(d) (R)-2-chloro-2,3-dimethylpentane + Cl_2 (300°) ⟶ $C_7H_{14}Cl_2$;
(e) *meso*-$HOCH_2CHOHCHOHCH_2OH$ + HNO_3 ⟶ $HOCH_2CHOHCHOHCOOH$;
(f) (R)-*sec*-butyl chloride + KOH (alc);
(g) (S)-3-chloro-1-butene + HCl;
(h) racemic $C_6H_5COCHOHC_6H_5$ + H_2, catalyst ⟶ $C_6H_5CHOHCHOHC_6H_5$.

2. In Problem 7.11 we saw that *hydroxylation with permanganate is syn*, and *hydroxylation with peroxy acids is anti*. Keeping in mind that reaction of epoxides (Sec. 17.12) is acid-catalyzed, give a detailed mechanism for hydroxylation with peroxy acids. (Check your answer in Sec. 17.12.)

3. Give the absolute configuration and R/S specification of compounds A–G.

(a) (R)-$HOCH_2CHOHCH{=}CH_2$ + cold alkaline $KMnO_4$ $\longrightarrow$ A (optically active) + B (optically inactive);
(b) (S)-1-chloro-2-methylbutane + Li, then + CuI $\longrightarrow$ C;
(c) C + (S)-1-chloro-2-methylbutane $\longrightarrow$ D;
(d) (R,R)-$HOCH_2CHOHCHOHCH_2OH$ + HBr $\longrightarrow$ E ($HOCH_2CHOHCHOHCH_2Br$);
(e) (R)-3-methyl-2-ethyl-1-pentene + H_2/Ni $\longrightarrow$ F (optically active) + G (optically inactive).

4. An excess of the racemic acid $CH_3CHClCOOH$ is allowed to react with (S)-2-methyl-1-butanol to form the ester, $CH_3CHCl\underset{\underset{O}{\|}}{C}{-}OCH_2CH(CH_3)CH_2CH_3$, and the reaction mixture is carefully distilled. Three fractions are obtained, each of which is optically active. Draw stereochemical formulas of the compound or compounds making up each fraction.

5. Addition of chlorine water to 2-butene yields not only 2,3-dichlorobutane but the chlorohydrin, 3-chloro-2-butanol. *cis*-2-Butene gives only the *threo* chlorohydrin, and *trans*-2-butene gives only the *erythro* chlorohydrin. What is the stereochemistry of chlorohydrin formation, and how do you account for it?

$$\begin{array}{ccc} CH_3 & & CH_3 \\ Cl{-}\!\!\!+\!\!\!{-}H & \qquad & H{-}\!\!\!+\!\!\!{-}Cl \\ H{-}\!\!\!+\!\!\!{-}OH & & H{-}\!\!\!+\!\!\!{-}OH \\ CH_3 & & CH_3 \end{array}$$

and enantiomer — *Threo* and enantiomer — *Erythro*

3-Chloro-2-butanol

6. (a) How do you account for the fact that when allyl bromide is treated with dilute H_2SO_4, there is obtained not only 1-bromo-2-propanol, but also 2-bromo-1-propanol? (b) In contrast, allyl chloride yields only one product, 1-chloro-2-propanol. How do you account for this difference between the chloride and the bromide?

7. (a) Alfred Hassner (University of Colorado) has found iodine azide, IN_3, to add to terminal alkenes with the orientation shown, and with complete stereospecificity

$$RCH{=}CH_2 + IN_3 \longrightarrow R\underset{\underset{N_3}{|}}{C}HCH_2I$$

(*anti*) to the 2-butenes. Suggest a mechanism for this reaction.

(b) In polar solvents like nitromethane, BrN_3 adds with the same orientation and stereospecificity as IN_3. In non-polar solvents like *n*-pentene, however, orientation is reversed, and addition is non-stereospecific. In solvents of intermediate polarity like methylene chloride, mixtures of products are obtained; light or peroxides favor formation of $RCHBrCH_2N_3$; oxygen favors formation of $RCH(N_3)CH_2Br$. Account in detail for these observations.

Chapter 8

Alkynes and Dienes

8.1 Introduction

Alkanes have the general formula C_nH_{2n+2}; alkenes have the general formula C_nH_{2n}. In this chapter we shall take up two kinds of hydrocarbons that have the same general formula, C_nH_{2n-2}: the **alkynes** and the **dienes**. As the formula indicates, they contain an even smaller proportion of hydrogen than the alkenes, and display an even higher degree of unsaturation. In spite of having the same general formula, alkynes and dienes have different functional groups, and hence different properties.

ALKYNES

8.2 Structure of acetylene. The carbon–carbon triple bond

The simplest member of the alkyne family is **acetylene**, C_2H_2. Using the methods we applied to the structure of ethylene (Sec. 5.2), we arrive at a structure in which the carbon atoms share *three* pairs of electrons, that is, are joined by a *triple bond. The* **carbon–carbon triple bond** *is the distinguishing feature of the alkyne structure.*

H:C:::C:H   H—C≡C—H

Acetylene

Again, quantum mechanics tells us a good deal more about acetylene, and about the carbon–carbon triple bond. To form bonds with two other atoms, carbon makes use of two equivalent hybrid orbitals: *sp* orbitals, formed by the mixing of *one s* and *one p* orbital (Sec. 1.9). These *sp* orbitals lie along a straight line that passes through the carbon nucleus; the angle between the two orbitals is thus 180°. This **linear** arrangement (Fig. 1.5) permits the hybrid orbitals to be as far apart as possible. Just as mutual repulsion among orbitals gives four tetrahedral bonds or three trigonal bonds, so it gives two linear bonds.

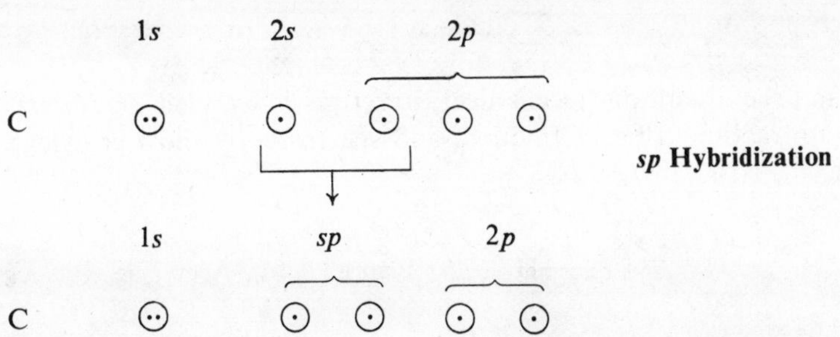

If we arrange the two carbons and the two hydrogens of acetylene to permit maximum overlap of orbitals, we obtain the structure shown in Fig. 8.1.

Figure 8.1. Acetylene molecule: only σ bonds shown.

$$\mathrm{H}\text{——}\mathrm{C}\overset{\sigma}{\text{——}}\mathrm{C}\overset{\sigma}{\text{——}}\mathrm{H}$$

Acetylene is a *linear molecule*, all four atoms lying along a single straight line. Both carbon–hydrogen and carbon–carbon bonds are cylindrically symmetrical about a line joining the nuclei, and are therefore σ bonds.

The molecule is not yet complete, however. In forming the *sp* orbitals already described, each carbon atom has used only one of its three *p* orbitals; it has two remaining *p* orbitals. Each of these consists of two equal lobes, whose axis lies at right angles both to the axis of the other *p* orbital and to the line of the *sp* orbitals; each *p* orbital is occupied by a single electron. But the sum of two perpendicular *p* orbitals is not four spherical lobes, but a single doughnut-shaped cloud (Fig. 8.2). Overlap of the *p* orbitals on one carbon with the *p* orbitals on the other car-

Figure 8.2. Acetylene molecule: two *p* orbitals on one carbon (doughnut-shaped cloud) can overlap two *p* orbitals on other carbon.

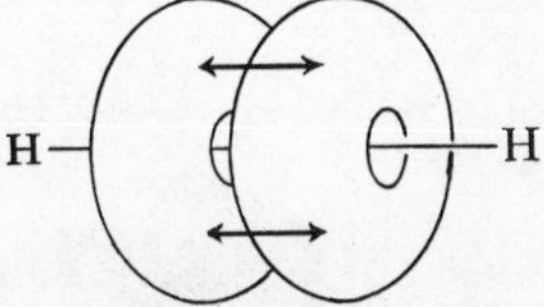

bon permits pairing of electrons. Two π bonds are formed, which together make a single cylindrical sheath about the line joining the nuclei (Fig. 8.3).

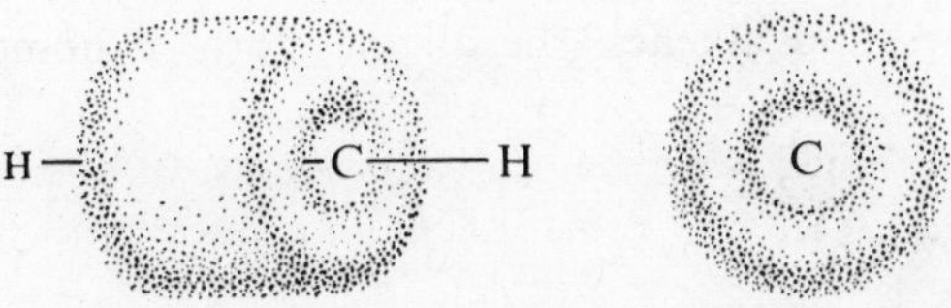

Figure 8.3. Acetylene molecule: carbon–carbon triple bond. π cloud forms cylindrical sheath.

The carbon–carbon "triple bond" is thus made up of one strong σ bond and two weaker π bonds; it has a total strength of 198 kcal. It is stronger than the carbon–carbon double bond of ethylene (163 kcal) or the carbon–carbon single bond of ethane (88 kcal), and therefore is shorter than either.

Again, the quantum mechanical structure is verified by direct evidence. Electron diffraction, x-ray diffraction, and spectroscopy show acetylene (Fig. 8.4)

$$\mathrm{H{-}C{\equiv}C{-}H}\quad (\mathrm{C{\equiv}C}\ 1.21\ \text{A};\ \mathrm{C{-}H}\ 1.08\ \text{A};\ 180^\circ)$$

Figure 8.4. Acetylene molecule: shape and size.

to be a linear molecule. The C—C distance is 1.21 A, as compared with 1.34 A in ethylene and 1.53 A in ethane. As in the case of the double bond, the structure of the triple bond is verified—although this time in a negative way—by the evidence of isomer number. As we can readily see from models, the linearity of the bonding should not permit geometric isomerism; no such isomers have ever been found.

The C—H distance in acetylene is 1.08 A, even shorter than in ethylene (1.103 A); because of their greater *s* character, *sp* orbitals are smaller than sp^2 orbitals, and *sp*-hybridized carbon forms shorter bonds than sp^2-hybridized carbon. The C—H bond dissociation energy in acetylene is not known, but we would expect it to be even greater than in ethylene. Oddly enough, the same *sp* hybridization that almost certainly makes cleavage of the C—H bond to form free radicals (*homolysis*) more difficult, makes cleavage to form ions (*heterolysis*) easier, as we shall see (Sec. 8.10).

$$\mathrm{HC{\equiv}C{:}H \longrightarrow HC{\equiv}C\cdot + H\cdot}$$

Homolysis: *one electron to each fragment*

$$\mathrm{HC{\equiv}C{:}H \longrightarrow HC{\equiv}C{:}^- + H^+}$$

Heterolysis: *both electrons to one fragment*

Problem 8.1 Compare the electronic configurations of CO_2, which is a linear molecule (check your answer to Problem 1.6, p. 25), and H_2O, which has a bond angle of 105°.

8.3 Higher alkynes. Nomenclature

Like the alkanes and alkenes, the alkynes form a homologous series, the increment again being —CH_2—.

The alkynes are named according to two systems. In one, they are considered to be derived from acetylene by replacement of one or both hydrogen atoms by alkyl groups.

$\mathrm{H{-}C{\equiv}C{-}C_2H_5}$	$\mathrm{CH_3{-}C{\equiv}C{-}CH_3}$	$\mathrm{CH_3{-}C{\equiv}C{-}CH(CH_3)_2}$
Ethylacetylene	Dimethylacetylene	Methylisopropylacetylene
1-Butyne	2-Butyne	4-Methyl-2-pentyne

For more complicated alkynes the **IUPAC** names are used. The rules are exactly the same as for the naming of alkenes, except that the ending **-yne** replaces

-ene. The parent structure is the longest continuous chain that contains the triple bond, and the positions both of substituents and of the triple bond are indicated by numbers. The triple bond is given the number of the *first* triply-bonded carbon encountered, starting from the end of the chain nearest the triple bond.

8.4 Physical properties of alkynes

Being compounds of low polarity, the alkynes have physical properties that are essentially the same as those of the alkanes and alkenes. They are insoluble in water but quite soluble in the usual organic solvents of low polarity: ligroin, ether, benzene, carbon tetrachloride. They are less dense than water. Their boiling points (Table 8.1) show the usual increase with increasing carbon number,

Table 8.1 ALKYNES

Name	Formula	M.p., °C	B.p., °C	Density (at 20°)
Acetylene	$HC{\equiv}CH$	− 82	− 75	
Propyne	$HC{\equiv}CCH_3$	−101.5	− 23	
1-Butyne	$HC{\equiv}CCH_2CH_3$	−122	9	
1-Pentyne	$HC{\equiv}C(CH_2)_2CH_3$	− 98	40	0.695
1-Hexyne	$HC{\equiv}C(CH_2)_3CH_3$	−124	72	.719
1-Heptyne	$HC{\equiv}C(CH_2)_4CH_3$	− 80	100	.733
1-Octyne	$HC{\equiv}C(CH_2)_5CH_3$	− 70	126	.747
1-Nonyne	$HC{\equiv}C(CH_2)_6CH_3$	− 65	151	.763
1-Decyne	$HC{\equiv}C(CH_2)_7CH_3$	− 36	182	.770
2-Butyne	$CH_3C{\equiv}CCH_3$	− 24	27	.694
2-Pentyne	$CH_3C{\equiv}CCH_2CH_3$	−101	55	.714
3-Methyl-1-butyne	$HC{\equiv}CCH(CH_3)_2$		29	.665
2-Hexyne	$CH_3C{\equiv}C(CH_2)_2CH_3$	− 92	84	.730
3-Hexyne	$CH_3CH_2C{\equiv}CCH_2CH_3$	− 51	81	.725
3,3-Dimethyl-1-butyne	$HC{\equiv}CC(CH_3)_3$	− 81	38	.669
4-Octyne	$CH_3(CH_2)_2C{\equiv}C(CH_2)_2CH_3$		131	.748
5-Decyne	$CH_3(CH_2)_3C{\equiv}C(CH_2)_3CH_3$		175	.769

and the usual effects of chain-branching; they are very nearly the same as the boiling points of alkanes or alkenes with the same carbon skeletons.

8.5 Industrial source of acetylene

The alkyne of chief industrial importance is the simplest member of the family, **acetylene**. It can be prepared by the action of water on calcium carbide, CaC_2, which itself is prepared by the reaction between calcium oxide and coke at the very high temperatures of the electric furnace. The calcium oxide and coke are in turn obtained from limestone and coal, respectively. Acetylene is thus obtained by a few steps from three abundant, cheap raw materials: water, coal, limestone.

$$\begin{array}{l} \text{Coal} \longrightarrow \text{coke} \\ \text{Limestone} \longrightarrow \text{CaO} \end{array} \Big\} \xrightarrow{2000^\circ} CaC_2 \xrightarrow{H_2O} H{-}C{\equiv}C{-}H$$

An alternative synthesis, based on petroleum, is displacing the carbide process. This involves the controlled, high-temperature partial oxidation of methane.

$$6CH_4 + O_2 \xrightarrow{1500^\circ} 2HC{\equiv}CH + 2CO + 10H_2$$

(The economic feasibility of this process is partly due to use of side-products: carbon monoxide and hydrogen for production of alcohols, and some hydrogen as fuel to maintain the oxidation temperature.)

Enormous quantities of acetylene are consumed each year. Dissolved under pressure in acetone contained in tanks, it is sold to be used as fuel for the oxyacetylene torch. It is the organic starting material for the large-scale synthesis of important organic compounds, including acetic acid and a number of unsaturated compounds that are used to make plastics and synthetic rubber. Many of the synthetic uses of acetylene have grown out of work done in Germany before and during World War II by W. Reppe (at the I. G. Farbenindustrie). Aimed at replacing petroleum (scarce in Germany) by the more abundant coal as the primary organic source, this work has revolutionized the industrial chemistry of acetylene.

8.6 Preparation of alkynes

A carbon–carbon triple bond is formed in the same way as a double bond: elimination of atoms or groups from two adjacent carbons. The groups eliminated

$$-\overset{W}{\underset{Y}{C}}-\overset{X}{\underset{Z}{C}}- \longrightarrow -\overset{W}{C}{=}\overset{X}{C}- \longrightarrow -C{\equiv}C-$$

and the reagents used are essentially the same as in the preparations of alkenes.

PREPARATION OF ALKYNES

1. Dehydrohalogenation of alkyl dihalides. Discussed in Sec. 8.6.

$$\left[-\overset{H}{C}{=}\overset{H}{C}- \xrightarrow{X_2}\right] -\overset{H}{\underset{X}{C}}-\overset{H}{\underset{X}{C}}- \xrightarrow{\text{KOH (alc)}} -\underset{X}{C}{=}\overset{H}{C}- \xrightarrow{NaNH_2} -C{\equiv}C-$$

Example:

$$CH_3CH{=}CH_2 \xrightarrow{Br_2} \underset{\substack{\text{1,2-Dibromopropane} \\ \text{(Propylene bromide)}}}{CH_3\underset{Br}{CH}-\underset{Br}{CH_2}} \xrightarrow{\text{KOH (alc)}} \underset{\text{1-Bromo-1-propene}}{CH_3CH{=}CHBr} \xrightarrow{NaNH_2} \underset{\text{Propyne}}{CH_3C{\equiv}CH}$$

2. **Reaction of sodium acetylides with primary alkyl halides.** Discussed in Sec. 8.12.

$$-C{\equiv}CH \xrightarrow[\text{or Na metal}]{NaNH_2} -C{\equiv}C{:}^-Na^+ + RX \longrightarrow -C{\equiv}C{-}R + NaX$$

(R *must* be *1°*)

Examples:

$$\underset{\text{Sodium acetylide}}{HC{\equiv}C{:}^-Na^+} + \underset{n\text{-Butyl bromide}}{CH_3CH_2CH_2CH_2Br} \longrightarrow \underset{\substack{\text{1-Hexyne} \\ (n\text{-Butylacetylene})}}{HC{\equiv}CCH_2CH_2CH_2CH_3}$$

$$\underset{\text{Sodium } n\text{-pentylacetylide}}{CH_3(CH_2)_4C{\equiv}C{:}^-Na^+} + \underset{n\text{-Pentyl chloride}}{CH_3(CH_2)_3CH_2Cl} \longrightarrow \underset{\text{6-Dodecyne}}{CH_3(CH_2)_4C{\equiv}C(CH_2)_4CH_3}$$

3. **Dehalogenation of tetrahalides.** Discussed in Sec. 8.6.

$$-\underset{X}{\overset{X}{C}}-\underset{X}{\overset{X}{C}}- + 2Zn \longrightarrow -C{\equiv}C- + 2ZnX_2$$

Example:

$$CH_3-\underset{Br}{\overset{Br}{C}}-\underset{Br}{\overset{Br}{C}}H \xrightarrow{Zn} \underset{\text{Propyne}}{CH_3-C{\equiv}CH}$$

Dehydrohalogenation of vicinal dihalides is particularly useful since the dihalides themselves are readily obtained from the corresponding alkenes by addition of halogen. This amounts to conversion—by several steps—of a double bond into a triple bond.

Dehydrohalogenation can generally be carried out in two stages as shown.

$$-\underset{X}{\overset{H}{C}}-\underset{X}{\overset{H}{C}}- \xrightarrow{KOH\ (alc)} \underset{\substack{\text{A vinyl halide} \\ \textit{Very unreactive}}}{-\underset{X}{C}{=}\overset{H}{C}-} \xrightarrow{NaNH_2} -C{\equiv}C-$$

Carried through only the first stage, it is a valuable method for preparing unsaturated halides. The halides thus obtained, with halogen attached directly to doubly-bonded carbon, are called **vinyl halides**, and are very unreactive (Sec. 25.5). Under mild conditions, therefore, dehydrohalogenation stops at the vinyl halide stage; more vigorous conditions—use of a stronger base—are required for alkyne formation.

Reaction of sodium acetylides with alkyl halides permits conversion of smaller alkynes into larger ones. Practically, the reaction is limited to the use of primary halides because of the great tendency for secondary and tertiary halides to undergo a side reaction, elimination; this point will be discussed further (Sec. 8.12) after we have learned something about the nature of acetylides.

Dehalogenation of tetrahalides is severely limited by the fact that these halides are themselves generally prepared from the alkynes. As is the case with the double bond and a dihalide, the triple bond may be protected by conversion into a tetrahalide with subsequent regeneration of the triple bond by treatment with zinc.

8.7 Reactions of alkynes

Just as alkene chemistry is the chemistry of the carbon–carbon double bond, so alkyne chemistry is the chemistry of the carbon–carbon triple bond. Like alkenes, alkynes undergo electrophilic addition, and for the same reason: availability of the loosely held π electrons. For reasons that are not understood, the carbon–carbon triple bond is *less* reactive than the carbon–carbon double bond toward electrophilic reagents.

Reasonably enough, the triple bond is *more* reactive than the double bond toward reagents that are themselves electron-rich. Thus alkynes undergo a set of reactions, *nucleophilic addition*, that are virtually unknown for simple alkenes. Although time does not permit us to go into these particular reactions here, we shall take up nucleophilic addition later in connection with other kinds of compounds (Chaps. 19 and 27).

Besides addition, alkynes undergo certain reactions that are due to the acidity of a hydrogen atom held by triply-bonded carbon.

REACTIONS OF ALKYNES

Addition Reactions

$$-C{\equiv}C- + YZ \longrightarrow -\overset{}{\underset{Y}{C}}{=}\underset{Z}{C}- \xrightarrow{YZ} -\overset{Y}{\underset{Y}{C}}-\overset{Z}{\underset{Z}{C}}-$$

1. **Addition of hydrogen.** Discussed in Sec. 8.9.

$$\underset{\text{Alkyne}}{-C{\equiv}C-} \xrightarrow{2H_2} \underset{\text{Alkane}}{-\overset{H}{\underset{H}{C}}-\overset{H}{\underset{H}{C}}-}$$

$$-C{\equiv}C- \xrightarrow{\text{Na or Li, } NH_3} \underset{\text{(H and H trans)}}{>C{=}C<} \quad \textbf{\textit{Trans}}$$

$$-C{\equiv}C- \xrightarrow[\text{Pd or Ni-B (P-2)}]{H_2} \underset{\text{(H and H cis)}}{>C{=}C<} \quad \textbf{\textit{Cis}}$$

Examples:

$$\underset{\text{2-Butyne}}{CH_3-C{\equiv}C-CH_3} \xrightarrow{2H_2,\ Ni} \underset{n\text{-Butane}}{CH_3CH_2CH_2CH_3}$$

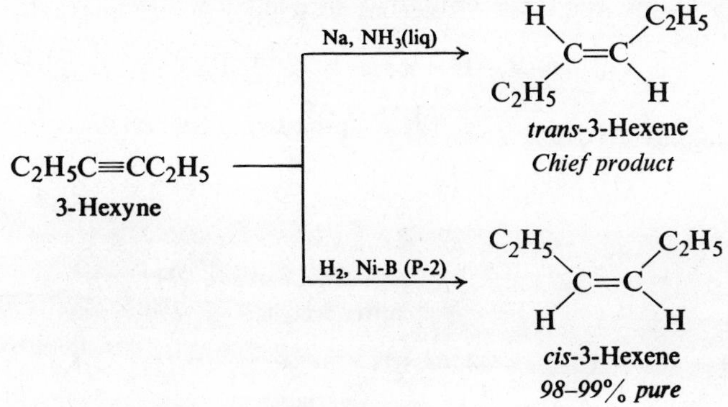

$$C_2H_5C\equiv CC_2H_5 \xrightarrow{Na,\ NH_3(liq)} \underset{\substack{trans\text{-3-Hexene}\\ \textit{Chief product}}}{\overset{H\qquad C_2H_5}{C_2H_5CH{=}CHC_2H_5}}$$

3-Hexyne

$$C_2H_5C\equiv CC_2H_5 \xrightarrow{H_2,\ \text{Ni-B (P-2)}} \underset{\substack{cis\text{-3-Hexene}\\ \textit{98–99\% pure}}}{C_2H_5CH{=}CHC_2H_5}$$

trans-3-Hexene: H and C_2H_5 on opposite sides of C=C. *Chief product*

cis-3-Hexene: both C_2H_5 on the same side of C=C. *98–99% pure*

2. **Addition of halogens.** Discussed in Sec. 8.8.

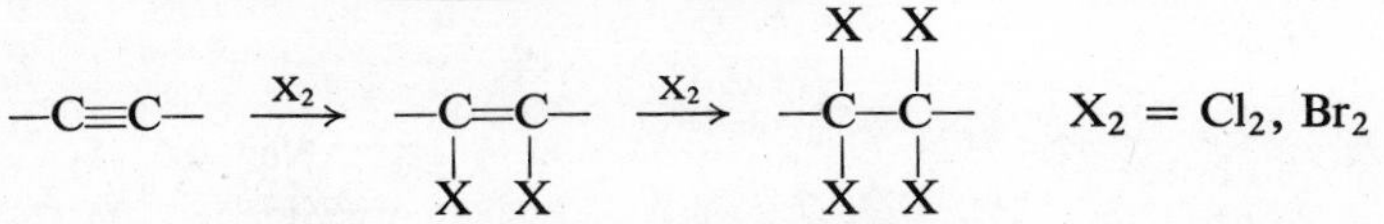

$$-C\equiv C- \xrightarrow{X_2} -\underset{X}{C}=\underset{X}{C}- \xrightarrow{X_2} -\overset{X}{\underset{X}{C}}-\overset{X}{\underset{X}{C}}- \qquad X_2 = Cl_2,\ Br_2$$

Example:

$$CH_3C\equiv CH \xrightarrow{Br_2} CH_3-\underset{Br}{C}=\underset{Br}{C}H \xrightarrow{Br_2} CH_3-\overset{Br}{\underset{Br}{C}}-\overset{Br}{\underset{Br}{C}}H$$

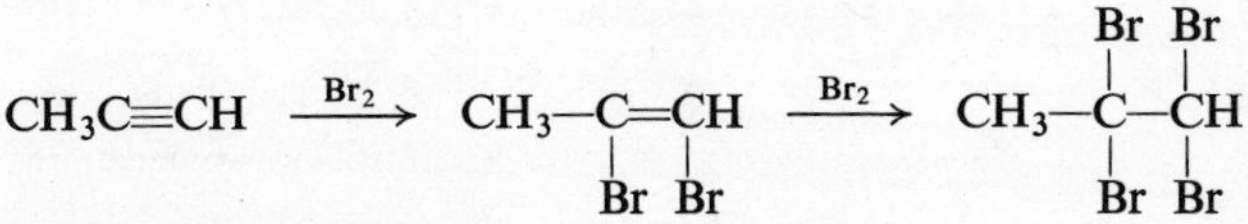

3. **Addition of hydrogen halides.** Discussed in Sec. 8.8.

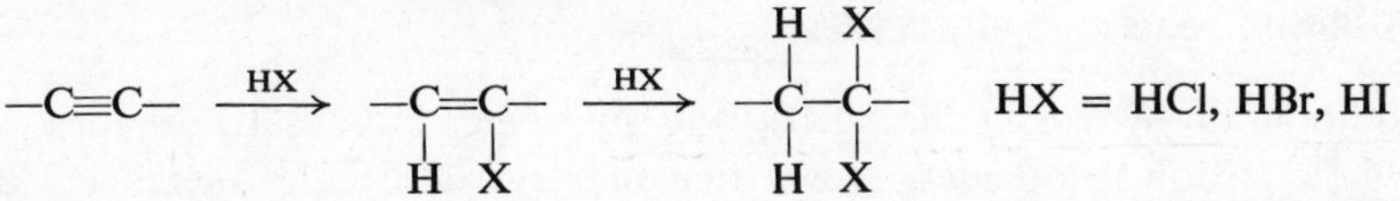

$$-C\equiv C- \xrightarrow{HX} -\underset{H}{C}=\underset{X}{C}- \xrightarrow{HX} -\overset{H}{\underset{H}{C}}-\overset{X}{\underset{X}{C}}- \qquad HX = HCl,\ HBr,\ HI$$

Example:

$$CH_3C\equiv CH \xrightarrow{HCl} CH_3\underset{Cl}{C}=CH_2 \xrightarrow{HI} CH_3-\overset{I}{\underset{Cl}{C}}-CH_3$$

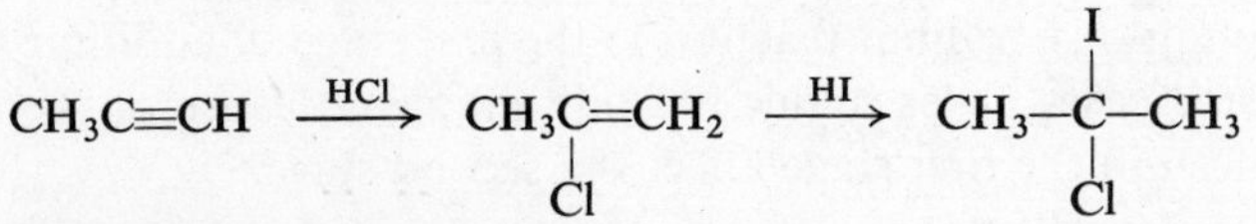

4. **Addition of water. Hydration.** Discussed in Sec. 8.13.

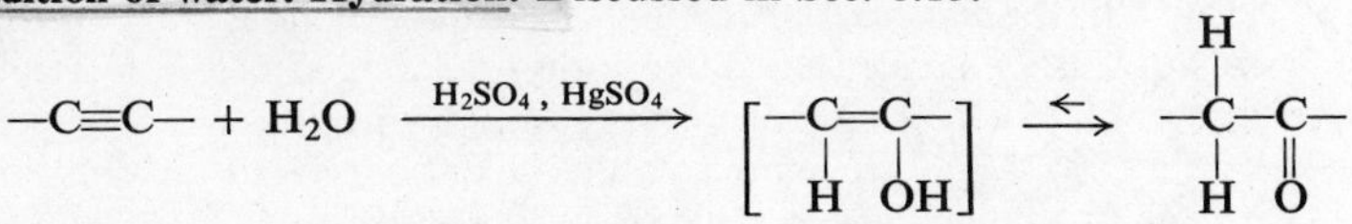

$$-C\equiv C- + H_2O \xrightarrow{H_2SO_4,\ HgSO_4} \left[-\underset{H}{C}=\underset{OH}{C}-\right] \rightleftarrows -\overset{H}{\underset{H}{C}}-\underset{O}{\overset{\|}{C}}-$$

Examples:

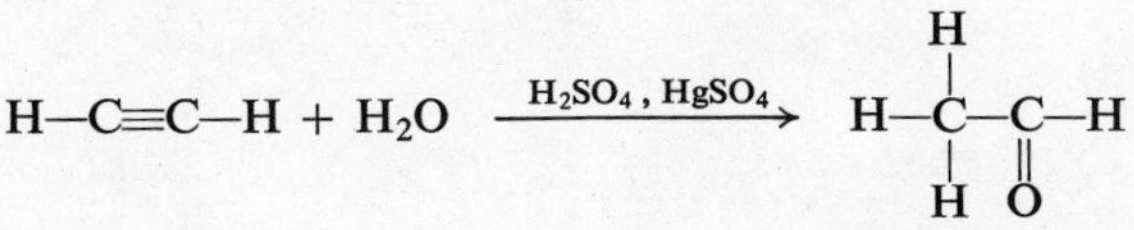

$$H-C\equiv C-H + H_2O \xrightarrow{H_2SO_4,\ HgSO_4} H-\overset{H}{\underset{H}{C}}-\underset{O}{\overset{\|}{C}}-H$$

Acetaldehyde

$$CH_3-C\equiv C-H + H_2O \xrightarrow{H_2SO_4,\ HgSO_4} H-\overset{H}{\underset{H}{C}}-\underset{O}{\overset{\|}{C}}-\overset{H}{\underset{H}{C}}-H$$

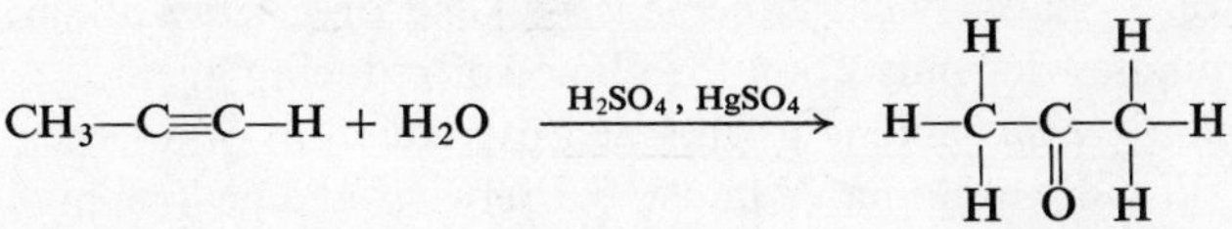

Acetone

Reactions as Acids

$$-C{\equiv}C{-}H + \text{base} \longrightarrow -C{\equiv}C{:}^-$$

5. Formation of heavy metal acetylides. Discussed in Sec. 8.11.

$$-C{\equiv}C{-}H + M^+ \longrightarrow -C{\equiv}C{-}M + H^+$$

Examples:

$$H{-}C{\equiv}C{-}H + 2Ag^+ \xrightarrow{\text{alcohol}} \underset{\text{Silver acetylide}}{Ag{-}C{\equiv}C{-}Ag} + 2H^+$$

$$CH_3C{\equiv}C{-}H + Cu(NH_3)_2^+ \longrightarrow \underset{\text{Cuprous methylacetylide}}{CH_3C{\equiv}C{-}Cu} + NH_4^+ + NH_3$$

Identification of terminal alkynes

6. Formation of alkali metal acetylides. Discussed in Sec. 8.10.

Examples:

$$H{-}C{\equiv}C{-}H + Na \xrightarrow{\text{liq } NH_3} \underset{\text{Sodium acetylide}}{H{-}C{\equiv}C{:}^-Na^+} + \tfrac{1}{2}H_2$$

$$CH_3{-}\underset{\displaystyle CH_3}{\underset{|}{CH}}{-}C{\equiv}C{-}H + NaNH_2 \xrightarrow{\text{ether}} \underset{\text{Sodium isopropylacetylide}}{CH_3{-}\underset{\displaystyle CH_3}{\underset{|}{CH}}{-}C{\equiv}C{:}^-Na^+} + NH_3$$

8.8 Addition reactions of alkynes

Addition of hydrogen, halogens, and hydrogen halides to alkynes is very much like addition to alkenes, except that here *two* molecules of reagent can be consumed for each triple bond. As shown, it is generally possible, by proper selection of conditions, to limit reaction to the first stage of addition, formation of alkenes. In some cases at least, this is made simpler because of the way that the atoms introduced in the first stage affect the second stage.

Problem 8.2 (a) Write the equation for the two-stage addition of bromine to 2-butyne. (b) How will the first two bromine atoms affect the reactivity of the double bond? (c) How will this influence the competition for halogen between 2-butyne and 2,3-dibromo-2-butene? (d) In what proportions would you mix the reagents to help limit reaction to the first stage? (e) Would you bubble 2-butyne into a solution of Br_2 in CCl_4, or drip the bromine solution into a solution of 2-butyne?

8.9 Reduction to alkenes

Reduction of an alkyne to the double-bond stage can—unless the triple bond is at the end of a chain—yield either a *cis*-alkene or a *trans*-alkene. Just which isomer predominates depends upon the choice of reducing agent.

Predominantly *trans*-alkene is obtained by reduction of alkynes with sodium or lithium in liquid ammonia. Almost entirely *cis*-alkene (as high as 98%) is obtained by hydrogenation of alkynes with several different catalysts: a specially prepared palladium called *Lindlar's catalyst*; or a nickel boride called *P-2 catalyst* reported by H. C. Brown (see p. 507) and his son, C. A. Brown.

$$R—C{\equiv}C—R \xrightarrow{\text{Na, NH}_3\text{(liq)}} \underset{\textit{Trans}}{(H)(R)C{=}C(R)(H)}$$

$$R—C{\equiv}C—R \xrightarrow[\text{Pd/C or Ni-B (P-2)}]{H_2} \underset{\textit{Cis}}{(R)(H)C{=}C(R)(H)}$$

Each of these reactions is, then, highly stereoselective. The stereoselectivity in the *cis*-reduction of alkynes is attributed, in a general way, to the attachment of two hydrogens to the same side of an alkyne sitting on the catalyst surface; presumably this same stereochemistry holds for the hydrogenation of terminal alkynes, RC≡CH, which cannot yield *cis*- and *trans*-alkenes.

The mechanism that gives rise to *trans*-reduction is not understood.

Problem 8.3 Most methods of making alkenes (Secs. 5.14 and 5.23) yield predominantly the more stable isomer, usually the *trans*. Outline all steps in the conversion of a mixture of 75% *trans*-2-pentene and 25% *cis*-2-pentene into essentially pure *cis*-2-pentene.

8.10 Acidity of alkynes. Very weak acids

In our earlier consideration of acids (in the Lowry-Brønsted sense, Sec. 1.22), we took *acidity* to be a measure of the tendency of a compound to lose a hydrogen ion. Appreciable acidity is generally shown by compounds in which hydrogen is attached to a rather electronegative atom (e.g., N, O, S, X). The bond holding the hydrogen is polar, and the relatively positive hydrogen can separate as the positive ion; considered from another viewpoint, an electronegative element can better accommodate the pair of electrons left behind. In view of the electronegativity series, $F > O > N > C$, it is not surprising to find that HF is a fairly strong acid, H_2O a comparatively weak one, NH_3 still weaker, and CH_4 so weak that we would not ordinarily consider it an acid at all.

In organic chemistry we are frequently concerned with the acidities of compounds that do not turn litmus red or taste sour, yet have a tendency—even though small—to lose a hydrogen ion.

A triply-bonded carbon acts as though it were an entirely different element—a more electronegative one—from a carbon having only single or double bonds. As a result, hydrogen attached to triply-bonded carbon, as in acetylene or any alkyne with the triple bond at the end of the chain (RC≡C—H), shows appreciable acidity. For example, sodium reacts with acetylene to liberate hydrogen gas and form the compound *sodium acetylide*.

$$HC{\equiv}C—H + Na \longrightarrow \underset{\text{Sodium acetylide}}{HC{\equiv}C{:}^-Na^+} + \tfrac{1}{2}H_2$$

Just how strong an acid is acetylene? Let us compare it with two familiar compounds, ammonia and water.

Sodium metal reacts with ammonia to form sodamide, $NaNH_2$, which is the salt of the weak acid, $H—NH_2$.

$$NH_3 + Na \longrightarrow \underset{\text{Sodamide}}{Na^+NH_2^-} + \tfrac{1}{2}H_2$$

Addition of acetylene to sodamide dissolved in ether yields ammonia and sodium acetylide.

$$\underset{\substack{\text{Stronger}\\\text{acid}}}{HC{\equiv}C{-}H} + \underset{\substack{\text{Stronger}\\\text{base}}}{Na^+NH_2^-} \rightleftharpoons \underset{\substack{\text{Weaker}\\\text{acid}}}{H{-}NH_2} + \underset{\substack{\text{Weaker}\\\text{base}}}{HC{\equiv}C^-Na^+}$$

The weaker acid, $H—NH_2$, is displaced from its salt by the stronger acid, $HC{\equiv}C—H$. In other language, the stronger base, NH_2^-, pulls the hydrogen ion away from the weaker base, $HC{\equiv}C^-$; if NH_2^- holds the hydrogen ion more tightly than $HC{\equiv}C^-$, then $H—NH_2$ must necessarily be a weaker acid than $HC{\equiv}C—H$.

Addition of water to sodium acetylide forms sodium hydroxide and regenerates

$$\underset{\substack{\text{Stronger}\\\text{acid}}}{H{-}OH} + \underset{\substack{\text{Stronger}\\\text{base}}}{HC{\equiv}C^-Na^+} \rightleftharpoons \underset{\substack{\text{Weaker}\\\text{acid}}}{HC{\equiv}C{-}H} + \underset{\substack{\text{Weaker}\\\text{base}}}{Na^+OH^-}$$

acetylene. The weaker acid, $HC{\equiv}C—H$, is displaced from its salt by the stronger acid, $H—OH$.

Thus we see that acetylene is a stronger acid than ammonia, but a weaker acid than water.

Acidity $$H_2O > HC{\equiv}CH > NH_3$$

Other alkynes that have a hydrogen attached to triply-bonded carbon show comparable acidity.

The method we have just described for comparing acidities of acetylene, ammonia, and water is a general one, and has been used to determine relative acidities of a number of extremely weak acids. *One compound is shown to be a stronger acid than another by its ability to displace the second compound from salts.*

$$\underset{\substack{\text{Stronger}\\\text{acid}}}{A{-}H} + B^-M^+ \longrightarrow \underset{\substack{\text{Weaker}\\\text{acid}}}{B{-}H} + A^-M^+$$

How can we account for the fact that hydrogen attached to triply-bonded carbon is especially acidic? How can we account for the fact that acetylene is a stronger acid than, say, ethane? A possible explanation can be found in the electronic configurations of the anions.

If acetylene is a stronger acid than ethane, then the acetylide ion must be a weaker base than the ethide ion, $C_2H_5^-$. In the acetylide anion the unshared pair of electrons occupies an *sp* orbital; in the ethide anion the unshared pair of electrons occupies an sp^3 orbital. The availability of this pair for sharing with acids determines the basicity of the anion. Now, compared with an sp^3 orbital,

$$\underset{\substack{\text{Acetylene}\\ \text{Stronger}\\ \text{acid}}}{HC{\equiv}C{:}H} \rightleftarrows H^+ + \underset{\substack{\text{Acetylide ion}\\ \text{Weaker}\\ \text{base}}}{HC{\equiv}C{:}^-}$$

$$\underset{\substack{\text{Ethane}\\ \text{Weaker}\\ \text{acid}}}{CH_3CH_2{:}H} \longleftarrow H^+ + \underset{\substack{\text{Ethide ion}\\ \text{Stronger}\\ \text{base}}}{CH_3CH_2{:}^-}$$

an *sp* orbital has less *p* character and more *s* character (Sec. 5.4). An electron in a *p* orbital is at some distance from the nucleus and is held relatively loosely; an electron in an *s* orbital, on the other hand, is close to the nucleus and is held more tightly. The acetylide ion is the weaker base since its pair of electrons is held more tightly, in an *sp* orbital.

Problem 8.4 When 1-hexyne was added to a solution of *n*-propylmagnesium bromide, a gas was evolved. The density of the gas showed that it had a molecular weight of 44. When it was bubbled through aqueous $KMnO_4$ or Br_2 in CCl_4, there was no visible change. (a) What was the gas? (b) Write an equation to account for its formation. (c) How could you have predicted such a reaction?

Problem 8.5 What do you suppose the structure of calcium carbide is? Can you suggest another name for it? What is the nature of its reaction with water?

8.11 Formation of heavy metal acetylides

The acidic acetylenes react with certain heavy metal ions, chiefly Ag^+ and Cu^+, to form insoluble acetylides. Formation of a precipitate upon addition of an alkyne to a solution of $AgNO_3$ in alcohol, for example, is an indication of hydrogen attached to triply-bonded carbon. This reaction can be used to differentiate *terminal* alkynes (those with the triple bond at the *end* of the chain) from *non-terminal* alkynes.

$$\underset{\substack{\text{1-Butyne}\\ \textit{A terminal alkyne}}}{CH_3CH_2C{\equiv}C{-}H} \xrightarrow{Ag^+} \underset{\text{Precipitate}}{CH_3CH_2C{\equiv}C{-}Ag} \left[\xrightarrow{HNO_3} \underset{\text{1-Butyne}}{CH_3CH_2C{\equiv}C{-}H} + Ag^+\right]$$

$$\underset{\substack{\text{2-Butyne}\\ \textit{A non-terminal alkyne}}}{CH_3{-}C{\equiv}C{-}CH_3} \xrightarrow{Ag^+} \text{no reaction}$$

If allowed to dry, these heavy metal acetylides are likely to explode. They should be destroyed while still wet by warming with nitric acid; the strong mineral acid regenerates the weak acid, acetylene.

8.12 Reaction of sodium acetylides with alkyl halides. Substitution *vs.* elimination

Sodium acetylides are used in the synthesis of higher alkynes. For example:

$$HC{\equiv}C{:}^{-}Na^{+} + C_2H_5{:}\ddot{\underset{..}{X}}{:} \longrightarrow \underset{\text{1-Butyne}}{HC{\equiv}C{-}C_2H_5} + Na^{+}{:}\ddot{\underset{..}{X}}{:}^{-}$$

$$C_2H_5C{\equiv}C{:}^{-}Na^{+} + CH_3{:}\ddot{\underset{..}{X}}{:} \longrightarrow \underset{\text{2-Pentyne}}{C_2H_5C{\equiv}C{-}CH_3} + Na^{+}{:}\ddot{\underset{..}{X}}{:}^{-}$$

This reaction involves substitution of acetylide ion for halide ion. It results from attack by the acetylide ion on carbon..

$$HC{\equiv}C{:}^{-} + {-}\underset{|}{\overset{X}{\overset{|}{C}}}{-}\underset{H}{\overset{|}{C}}{-} \longrightarrow {-}\underset{HC\equiv C}{\overset{|}{C}}{-}\underset{|}{\overset{|}{C}}{-} + X^{-}$$

Alkyne

Attack on C: *substitution*

Since sodium acetylide is the salt of the extremely weak acid, acetylene, the acetylide ion is an extremely strong base, stronger in fact than hydroxide ion. In our discussion of the synthesis of alkenes from alkyl halides (Sec. 5.13), we saw that the basic hydroxide ion causes elimination by abstracting a hydrogen ion. It is not surprising that the even more basic acetylide ion can also cause elimination.

$${-}\underset{|}{\overset{X}{\overset{|}{C}}}{-}\underset{H}{\overset{|}{C}}{-} + {:}C{\equiv}CH^{-} \longrightarrow X^{-} + \underset{\text{Alkene}}{>C{=}C<} + HC{\equiv}CH$$

Attack on H: *elimination*

The acetylide ion, then, can react with an alkyl halide in two ways: by attack at carbon to give **substitution**, or by attack at hydrogen to give **elimination**. We have seen that the order of reactivity of alkyl halides toward elimination (Sec. 5.14) is 3° > 2° > 1°. In substitution (of the present kind), we shall find (Sec. 14.11) the order of reactivity is just the opposite: 1° > 2° > 3°. It is to be expected, then, that: *where substitution and elimination are competing reactions, the proportion of elimination increases as the structure of an alkyl halide is changed from primary to*

Elimination increases →

RX = 1° 2° 3°

← *Substitution increases*

Elimination (E2) ***vs.*** **Substitution** (S_N2)

secondary to tertiary. Many tertiary halides—fastest at elimination and slowest at substitution—yield exclusively alkenes under these conditions.

When the attacking reagent is a *strong* base like hydroxide or acetylide, that is, when the reagent has a strong affinity for hydrogen ion, elimination is particularly important. Practically speaking, *only primary halides give good yields of the substitution product, the alkyne.* With secondary and tertiary halides, elimination predominates to such an extent that the method is essentially useless. We shall encounter this competition between substitution and elimination again and again in our study of organic chemistry.

Reflecting their ability to form carbon–carbon bonds (Sec. 3.17), copper acetylides, too, are used to couple with organic halides, particularly with the ordinarily unreactive *vinyl* and *aryl* halides (Chap. 25).

8.13 Hydration of alkynes. Tautomerism

Addition of water to acetylene to form *acetaldehyde*, which can then be oxidized to *acetic acid*, is an extremely important industrial process.

From the structure of acetaldehyde, it at first appears that this reaction follows a different pattern from the others, in which two groups attach themselves to the two triply-bonded carbons. Actually, however, the product can be accounted for in a rather simple way.

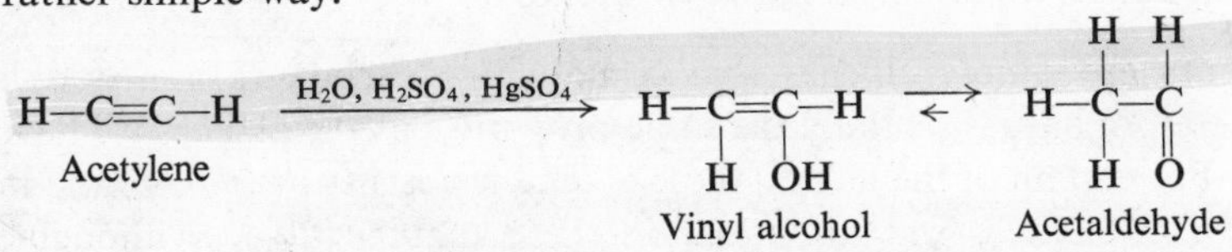

$$\underset{\text{Acetylene}}{H{-}C{\equiv}C{-}H} \xrightarrow{H_2O,\ H_2SO_4,\ HgSO_4} \underset{\text{Vinyl alcohol}}{H{-}\underset{H}{C}{=}\underset{OH}{C}{-}H} \rightleftarrows \underset{\text{Acetaldehyde}}{H{-}\underset{H}{\overset{H}{C}}{-}\underset{O}{\overset{H}{C}}}$$

If hydration of acetylene followed the same pattern as hydration of alkenes, we would expect addition of H— and —OH to the triple bond to yield the structure that we would call *vinyl alcohol*. But all attempts to prepare vinyl alcohol result—like hydration of acetylene—in the formation of acetaldehyde.

A structure with —OH attached to doubly-bonded carbon is called an **enol** (*-ene* for the carbon–carbon double bond, *-ol* for *alcohol*). It is almost always true that when we try to make a compound with the enol structure, we obtain instead a compound with the **keto** structure (one that contains a C=O group).

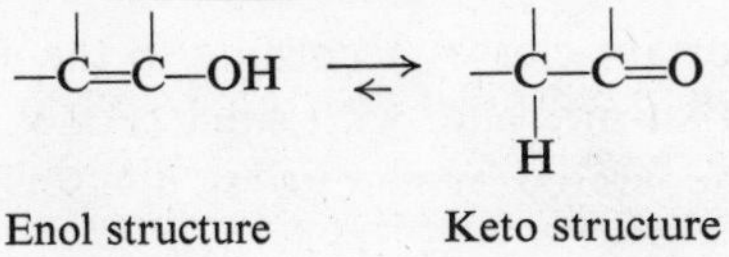

$$\underset{\text{Enol structure}}{-\overset{|}{C}{=}\overset{|}{C}{-}OH} \rightleftarrows \underset{\text{Keto structure}}{-\underset{H}{\overset{|}{C}}{-}\overset{|}{C}{=}O} \qquad \textbf{Keto-enol tautomerism}$$

There is an equilibrium between the two structures, but it generally lies very much in favor of the keto form. Thus, vinyl alcohol is formed initially by hydration of acetylene, but it is rapidly converted into an equilibrium mixture that is almost all acetaldehyde.

Rearrangements of this enol–keto kind take place particularly easily because of the polarity of the —O—H bond. A hydrogen ion separates readily from oxygen to form a hybrid anion; but when a hydrogen ion (most likely a *different* one) returns, it may attach itself either to oxygen or to carbon of the anion. When it returns to oxygen, it may readily come off again; but when it attaches itself to

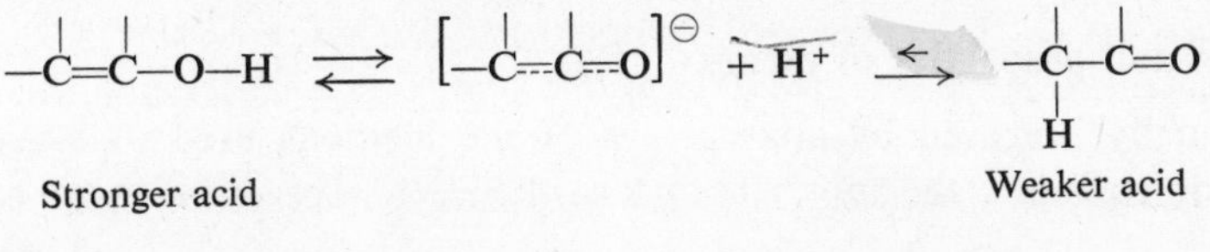

$$\underset{\text{Stronger acid}}{-\overset{|}{C}{=}\overset{|}{C}{-}O{-}H} \rightleftarrows \left[-\overset{|}{C}{\cdots}\overset{|}{C}{\cdots}O\right]^{\ominus} + H^+ \rightleftarrows \underset{\text{Weaker acid}}{-\underset{H}{\overset{|}{C}}{-}\overset{|}{C}{=}O} \qquad \textbf{Keto–enol tautomerism}$$

carbon, it tends to stay there. We recognize this reaction as another example of the conversion of a stronger acid into a weaker acid (Sec. 8.10).

Compounds whose structures differ markedly in arrangement of atoms, but which exist in equilibrium, are called **tautomers**. The most common kind of **tautomerism** involves structures that differ in the point of attachment of *hydrogen*. In these cases, as in **keto–enol tautomerism**, the tautomeric equilibrium generally favors the structure in which hydrogen is bonded to carbon rather than to a more electronegative atom; that is, equilibrium favors the weaker acid.

Problem 8.6 Hydration of propyne yields the ketone *acetone*, CH_3COCH_3, rather than the aldehyde CH_3CH_2CHO. What does this suggest about the orientation of the initial addition?

DIENES

8.14 Structure and nomenclature of dienes

Dienes are simply alkenes that contain two carbon–carbon double bonds. They therefore have essentially the same properties as the alkenes we have already studied. For certain of the dienes, these alkene properties are *modified* in important ways; we shall focus our attention on these modifications. Although we shall consider chiefly *di*enes in this section, what we shall say applies equally well to compounds with more than two double bonds.

Dienes are named by the IUPAC system in the same way as alkenes, except that the ending **-diene** is used, with *two* numbers to indicate the positions of the *two* double bonds. This system is easily extended to compounds containing any number of double bonds.

$CH_2{=}CH{-}CH{=}CH_2$ — 1,3-Butadiene

$CH_2{=}CH{-}CH_2{-}CH{=}CH_2$ — 1,4-Pentadiene

$CH_2{=}CH{-}CH{=}CH{-}CH{=}CH_2$ — 1,3,5-Hexatriene

Dienes are divided into two important classes according to the arrangement of the double bonds. Double bonds that alternate with single bonds are said to be **conjugated**; double bonds that are separated by more than one single bond are said to be **isolated.**

$-\overset{|}{C}{=}\overset{|}{C}{-}\overset{|}{C}{=}\overset{|}{C}-$
Conjugated double bonds

$-\overset{|}{C}{=}\overset{|}{C}{-}\underset{|}{\overset{|}{C}}{-}\overset{|}{C}{=}\overset{|}{C}-$
Isolated double bonds

A third class of dienes, of increasing interest to organic chemists, contain *cumulated* double bonds; these compounds are known as **allenes**:

$-\overset{|}{C}{=}C{=}\overset{|}{C}-$ Cumulated double bonds: allenes

8.15 Preparation and properties of dienes

Dienes are usually prepared by adaptations of the methods used to make simple alkenes. For example, the most important diene, **1,3-butadiene** (used to make synthetic rubber, Sec. 8.25), has been made in this country by a cracking process, and in Germany by dehydration of an alcohol containing two —OH groups:

$$CH_3CH_2CH_2CH_3 \xrightarrow[\text{catalyst}]{\text{heat}} \begin{cases} CH_3CH_2CH{=}CH_2 \text{ (1-Butene)} \xrightarrow[\text{catalyst}]{\text{heat}} \\ CH_3CH{=}CHCH_3 \text{ (2-Butene)} \xrightarrow[\text{catalyst}]{\text{heat}} \end{cases} CH_2{=}CH{-}CH{=}CH_2$$

n-Butane → 1-Butene / 2-Butene → 1,3-Butadiene

$$\underset{OH}{CH_2}CH_2CH_2\underset{OH}{CH_2} \xrightarrow{\text{heat, acid}} CH_2{=}CH{-}CH{=}CH_2$$

1,3-Butadiene

The chemical properties of a diene depend upon the arrangement of its double bonds. Isolated double bonds exert little effect on each other, and hence each reacts as though it were the only double bond in the molecule. Except for the consumption of larger amounts of reagents, then, the chemical properties of the non-conjugated dienes are identical with those of the simple alkenes.

Conjugated dienes differ from simple alkenes in three ways: (a) they are *more stable*, (b) they undergo *1,4-addition*, and (c) toward free radical addition, they are *more reactive*.

8.16 Stability of conjugated dienes

If we look closely at Table 6.1 (p. 183), we find that the heats of hydrogenation of alkenes having similar structures are remarkably constant. For monosubstituted alkenes ($RCH{=}CH_2$) the values are very close to 30 kcal/mole; for disubstituted alkenes ($R_2C{=}CH_2$ or $RCH{=}CHR$), 28 kcal/mole; and for trisubstituted alkenes ($R_2C{=}CHR$), 27 kcal/mole. For a compound containing more than one double bond we might expect a heat of hydrogenation that is the sum of the heats of hydrogenation of the individual double bonds.

For non-conjugated dienes this additive relationship is found to hold. As shown in Table 8.2, 1,4-pentadiene and 1,5-hexadiene, for example, have heats of hydrogenation very close to 2 × 30 kcal, or 60 kcal/mole.

Table 8.2 HEATS OF HYDROGENATION OF DIENES

Diene	ΔH of Hydrogenation, kcal/mole
1,4-Pentadiene	60.8
1,5-Hexadiene	60.5
1,3-Butadiene	57.1
1,3-Pentadiene	54.1
2-Methyl-1,3-butadiene (Isoprene)	53.4
2,3-Dimethyl-1,3-butadiene	53.9
1,2-Propadiene (Allene)	71.3

For conjugated dienes, however, the measured values are slightly lower than expected. For 1,3-butadiene we might expect 2 × 30, or 60 kcal: the actual value, 57 kcal, is 3 kcal lower. In the same way the values for 1,3-pentadiene and 2,3-dimethyl-1,3-butadiene are also below the expected values by 2–4 kcal.

Heats of Hydrogenation

$CH_2{=}CH{-}CH{=}CH_2$
Expected: 30 + 30 = 60 kcal
Observed: 57

$CH_3{-}CH{=}CH{-}CH{=}CH_2$
Expected: 28 + 30 = 58 kcal
Observed: 54

$CH_2{=}C(CH_3){-}C(CH_3){=}CH_2$
Expected: 28 + 28 = 56 kcal
Observed: 54

What do these heats of hydrogenation tell us about the conjugated dienes? Using the approach of Sec. 6.4, let us compare, for example, 1,3-pentadiene (heat of hydrogenation, 54 kcal) and 1,4-pentadiene (heat of hydrogenation, 61 kcal). They both consume two moles of hydrogen and yield the same product, *n*-pentane. If 1,3-pentadiene *evolves* less energy than 1,4-pentadiene, it can only mean that it *contains* less energy; that is to say, the conjugated 1,3-pentadiene is more stable than the non-conjugated 1,4-pentadiene.

In the next three sections we shall see how two different factors have been invoked to account for the relative stabilities of conjugated dienes, and of simple alkenes as well: (a) delocalization of π electrons, and (b) strengthening of σ bonds through changes in hybridization of carbon.

Unusual stability of conjugated dienes is also strongly indicated by the fact that, where possible, they are the preferred diene products of elimination reactions (Sec. 5.14).

Problem 8.7 Predict the major product of dehydrohalogenation of 4-bromo-1-hexene.

Problem 8.8 (a) Predict the heat of hydrogenation of *allene*, $CH_2{=}C{=}CH_2$. (b) The actual value is 71 kcal. What can you say about the stability of a *cumulated* diene?

8.17 Resonance in conjugated dienes

Let us focus our attention on the four key carbon atoms of any conjugated diene system. We ordinarily write the $C_1{-}C_2$ and $C_3{-}C_4$ bonds as double, and the $C_2{-}C_3$ bond as single:

$$-\overset{1}{\underset{|}{C}}{=}\overset{2}{\underset{|}{C}}{-}\overset{3}{\underset{|}{C}}{=}\overset{4}{\underset{|}{C}}-$$

This would correspond to an orbital picture of the molecule (see Fig. 8.5*a*), in which π bonds are formed by overlap of the *p* orbitals of C_1 and C_2, and overlap of the *p* orbitals of C_3 and C_4.

In the allyl radical we saw that resonance resulted from the overlap of the *p* orbital of a carbon atom with *p* orbitals on *both* sides. We might expect that, in the same way, there could be a certain amount of overlap between the p orbitals of C_2 and C_3, as shown in Fig. 8.5b. The resulting delocalization of the π electrons makes the molecule more stable: each pair of electrons attracts—and is attracted by—not just two carbon nuclei, but *four*.

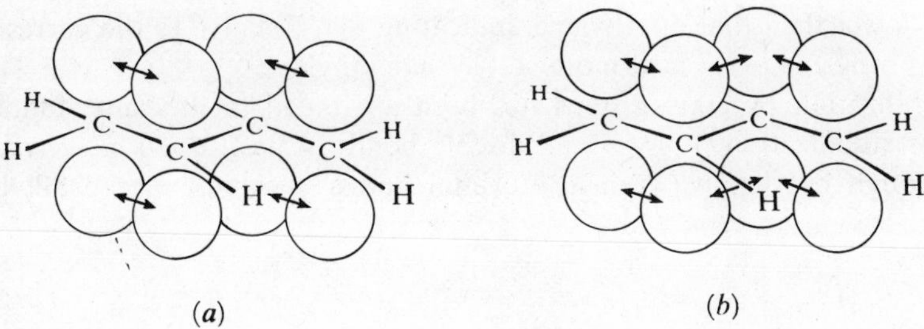

Figure 8.5. Conjugated diene. (a) Overlap of p orbitals to form two double bonds. (b) Overlap of p orbitals to form conjugated system: delocalization of π electrons.

Using the language of conventional valence-bond structures, we say that a conjugated diene is a resonance hybrid of I and II. The dotted line in II represents

$$\underset{\text{I}}{-\overset{1}{\underset{|}{C}}=\overset{2}{\underset{|}{C}}-\overset{3}{\underset{|}{C}}=\overset{4}{\underset{|}{C}}-} \qquad \underset{\text{II}}{-\dot{\underset{|}{C}}-\underset{|}{C}=\underset{|}{C}-\dot{\underset{|}{C}}-}$$

a *formal bond*, and simply means that an electron on C_1 and an electron on C_4 have opposite spins, that is to say, are *paired*.

To the extent that II contributes to the structure, it gives a certain double-bond character to the C_2—C_3 bond and a certain single-bond character to the C_1—C_2 and C_3—C_4 bonds; most important, it makes the molecule more stable than we would expect I (the most stable contributing structure) to be.

Formation of a bond releases energy and stabilizes a system; all other things being equal, the more bonds, the more stable a structure. Consideration of *number of bonds* is one of the criteria (Sec. 6.27) that can be used to estimate relative stability and hence relative importance of a contributing structure. On this basis we would expect II with 10 bonds (the formal bond does not count) to be less stable than I with 11 bonds. The resonance energy for such a hybrid of non-equivalent structures should be less than for a hybrid made up of equivalent structures. The structure of a conjugated diene should resemble I more than II, since the more stable structure I makes the larger contribution to the hybrid.

Consistent with partial double-bond character, the C_2—C_3 bond in 1,3-butadiene is 1.48 A long, as compared with 1.53 A for a pure single bond. The resonance energy of a conjugated diene is only 2–4 kcal/mole, compared with 10 kcal/mole for the allyl radical. (However, for an alternative interpretation, see Sec. 8.19.)

8.18 Resonance in alkenes. Hyperconjugation

Heats of hydrogenation showed us (Sec. 6.4) that alkenes are stabilized not only by conjugation but also by the presence of alkyl groups: *the greater the number of alkyl groups attached to the doubly-bonded carbon atoms, the more stable the alkene.* To take the simplest example, the heat of hydrogenation of propylene is 2.7 kcal lower than that of ethylene, indicating that (relative to the corresponding alkane) propylene is 2.7 kcal more stable than ethylene.

Stabilization by alkyl groups has been attributed to the same fundamental factor as stabilization by a second double bond: *delocalization of electrons*, this time through overlap between a π orbital and a σ orbital of the alkyl group.

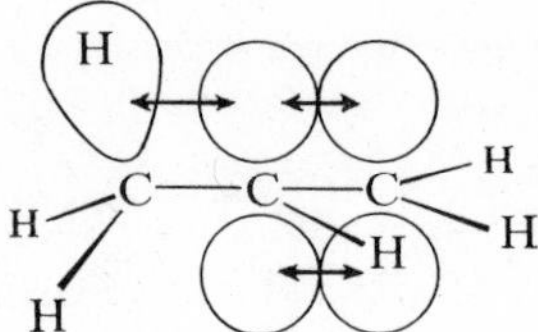

Figure 8.6. Hyperconjugation in an alkene: overlap between σ and π orbitals.

Through this overlap, individual electrons can, to an extent, help bind together four nuclei. Delocalization of this kind, involving σ bond orbitals, we recognize as *hyperconjugation* (Sec. 6.28).

Translated into resonance terminology, such hyperconjugation is represented by contribution from structures like II. (As before, the dotted line in II represents

```
     H  H  H                 H  H  H
     |  |  |                 |  |  |
   H-C--C==C-H             H-C==C--C-H
     |                                :
     H                       H.........:
     3  2  1                 3  2  1
        I                       II
                          and two more
                           equivalent
                           structures
```

a formal bond, indicating that electrons on the two atoms are paired.) Considered by itself, a structure like II is indeed strange, since there is no real bond joining the hydrogen to carbon. This is, however, simply a rough way of indicating that the carbon–hydrogen bond is something *less* than a single bond, that the C_2—C_3 bond has some double-bond character, and that the C_1—C_2 bond has some single-bond character.

Consistent with partial double-bond character, the carbon–carbon "single" bond in propylene is 1.50 A long, as compared to 1.53 A for a pure single bond.

The greater the number of alkyl groups attached to the doubly-bonded carbons, the greater the number of contributing structures like II, the greater the delocalization of electrons, and the more stable the alkene.

Hyperconjugation of the kind described above is called *sacrificial hyperconjugation*, since there is one less real bond in structures like II than in I. In contrast, the kind of hyperconjugation we encountered in connection with free radicals and carbonium ions involves no "sacrifice" of a bond and is called *isovalent hyperconjugation.*

8.19 Stability of dienes and alkenes: an alternative interpretation

We have seen that the carbon–hydrogen bond length decreases as we proceed along the series ethane, ethylene, acetylene, and we attributed this to changes in hybridization of carbon (see Table 8.3). As the *p* character of the bonding orbital decreases, the orbital size decreases, and the bond becomes shorter (Sec. 5.4).

Table 8.3 Carbon–Hydrogen Single Bond Lengths and Hybridization

Compound	Length, A	Hybridization
$CH_3—CH_3$	1.112	sp^3–s
$CH_2{=}CH_2$	1.103	sp^2–s
$HC{\equiv}CH$	1.079	sp–s

The carbon–carbon single-bond length also decreases along an analogous series, ethane, propylene, propyne (Table 8.4). We notice that these differences are bigger than for carbon–hydrogen bonds. Here, the bond-shortening has been attributed to hyperconjugation, as discussed in Sec. 8.18.

Table 8.4 Carbon–Carbon Single Bond Lengths and Hybridization

Compound	Length, A	Hybridization
$CH_3—CH_3$	1.53	sp^3–sp^3
$CH_2{=}CH—CH_3$	1.50	sp^2–sp^3
$HC{\equiv}C—CH_3$	1.46	sp–sp^3

It has been argued, most notably by M. J. S. Dewar of the University of Texas, that there is no need to invoke hyperconjugation in molecules like these, and that the changes in C—C bond length—like the changes in C—H bond length—are due simply to changes in hybridization of carbon.

Furthermore, Dewar has proposed that such shortening of bonds is accompanied by a proportional increase in bond energies (E); that is, shortening a bond makes the molecule more stable. Change in hybridization affects bond lengths more—and hence affects molecular stability more—when carbon–carbon bonds are involved than when carbon–hydrogen bonds are involved. An alkyl substituent stabilizes an alkene, relative to the corresponding alkane, because sp^2 hybridization strengthens a carbon–carbon bond more than a carbon–hydrogen bond.

In a similar way, the unusual stability of conjugated dienes is attributed, not to delocalization of the π electrons, but to the fact that sp^2–sp^2 hybridization makes the $C_2—C_3$ bond short (1.48 A) and strong.

There is little doubt that both factors, delocalization of π electrons and change in σ bonds, are at work. The question is: what is the relative importance of each? The answer may well turn out to be: *both* are important.

In the case of molecules like the allyl radical, where clearly no single structure is acceptable, Dewar has not questioned the importance of π-electron delocalization, although he considers σ-bond stability to play a larger part than has been recognized. He also accepts a more important role for isovalent hyperconjugation—in free radicals and carbonium ions—than for the sacrificial hyperconjugation we have so far discussed.

8.20 Electrophilic addition to conjugated dienes. 1,4-Addition

When 1,4-pentadiene is treated with bromine under conditions (what are they?) that favor formation of the *dihalide*, there is obtained the expected product, 4,5-dibromo-1-pentene. Addition of more bromine yields the 1,2,4,5-tetrabromopentane.

$$CH_2{=}CH{-}CH_2{-}CH{=}CH_2 \xrightarrow{Br_2} \underset{Br\quad\;\; Br}{CH_2{-}CH}{-}CH_2{-}CH{=}CH_2 \xrightarrow{Br_2} \underset{Br\quad\;\; Br}{CH_2{-}CH}{-}CH_2{-}\underset{Br\quad\;\; Br}{CH{-}CH_2}$$

pentane. This is typical of the behavior of dienes containing isolated double bonds: the double bonds react independently, as though they were in different molecules.

When 1,3-butadiene is treated with bromine under similar conditions, there is obtained not only the expected 3,4-dibromo-1-butene, but also 1,4-dibromo-2-butene. Treatment with HCl yields not only 3-chloro-1-butene, but also 1-chloro-2-butene. Hydrogenation yields not only 1-butene but also 2-butene.

$CH_2{=}CH{-}CH{=}CH_2$ (1,3-Butadiene)

$$\xrightarrow{Br_2} \underset{Br\qquad Br}{CH_2{-}CH}{-}CH{=}CH_2 \text{ (1,2-addition)} \quad\text{and}\quad \underset{Br}{CH_2}{-}CH{=}CH{-}\underset{Br}{CH_2} \text{ (1,4-addition)}$$

$$\xrightarrow{HCl} \underset{H\qquad Cl}{CH_2{-}CH}{-}CH{=}CH_2 \text{ (1,2-addition)} \quad\text{and}\quad \underset{H}{CH_2}{-}CH{=}CH{-}\underset{Cl}{CH_2} \text{ (1,4-addition)}$$

$$\xrightarrow[cat.]{H_2} \underset{H\qquad H}{CH_2{-}CH}{-}CH{=}CH_2 \text{ (1,2-addition)} \quad\text{and}\quad \underset{H}{CH_2}{-}CH{=}CH{-}\underset{H}{CH_2} \text{ (1,4-addition)}$$

Study of many conjugated dienes and many reagents shows that such behavior is typical: *in* **additions to conjugated dienes,** *a reagent may attach itself not only to a pair of adjacent carbons* **(1,2-addition)**, *but also to the carbons at the two ends of the conjugated system* **(1,4-addition)**. Very often the 1,4-addition product is the major one.

$$\overset{1}{-C}{=}\overset{2}{C}{-}\overset{3}{C}{=}\overset{4}{C-} \xrightarrow{YZ} \underset{Y\quad\; Z}{-C{-}C}{-}C{=}C{-} \text{ (1,2-addition)} \quad\text{and}\quad \underset{Y}{-C}{-}C{=}C{-}\underset{Z}{C-} \text{ (1,4-addition)}$$

8.21 Allyl cations. Delocalization in carbonium ions

How can we account for the products obtained? We have seen (Secs. 6.10 and 6.11) that electrophilic addition is a two-step process, and that the first step takes place in the way that yields the more stable carbonium ion. Let us apply this principle to the addition, for example, of HCl to 2,4-hexadiene, which yields 4-chloro-2-hexene and 2-chloro-3-hexene:

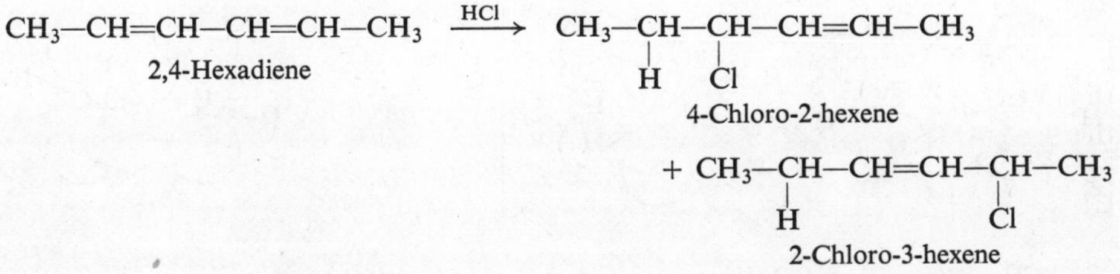

$$\underset{\text{2,4-Hexadiene}}{CH_3{-}CH{=}CH{-}CH{=}CH{-}CH_3} \xrightarrow{HCl} \underset{\text{4-Chloro-2-hexene}}{CH_3{-}\underset{H}{\underset{|}{C}H}{-}\underset{Cl}{\underset{|}{C}H}{-}CH{=}CH{-}CH_3}$$

$$+\ \underset{\text{2-Chloro-3-hexene}}{CH_3{-}\underset{H}{\underset{|}{C}H}{-}CH{=}CH{-}\underset{Cl}{\underset{|}{C}H}{-}CH_3}$$

These products show that hydrogen adds to C–2 to yield carbonium ion I, rather than to C–3 to yield carbonium ion II:

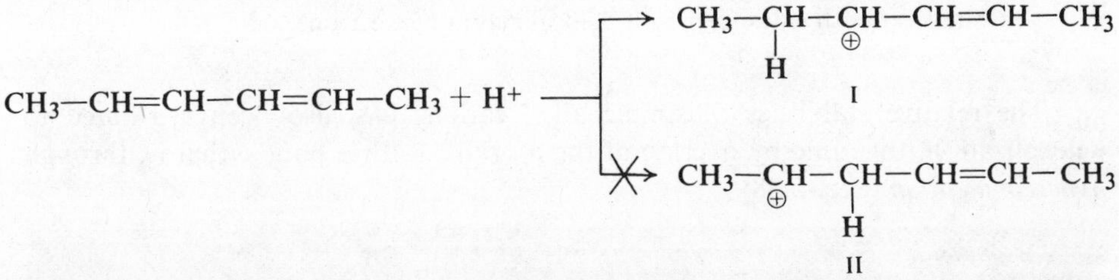

$$CH_3{-}CH{=}CH{-}CH{=}CH{-}CH_3 + H^+ \longrightarrow \underset{\text{I}}{CH_3{-}\underset{H}{\underset{|}{C}H}{-}\underset{\oplus}{C}H{-}CH{=}CH{-}CH_3}$$

$$CH_3{-}CH{=}CH{-}CH{=}CH{-}CH_3 + H^+ \not\longrightarrow \underset{\text{II}}{CH_3{-}\underset{\oplus}{C}H{-}\underset{H}{\underset{|}{C}H}{-}CH{=}CH{-}CH_3}$$

Since both I and II are secondary cations, how can we account for the preference? I is not simply a secondary cation, but is an *allyl* cation as well, since the carbon bearing the positive charge is attached to a doubly-bonded carbon.

Let us look more closely at such cations, using the parent allyl cation, $CH_2{=}CH{-}CH_2^+$, as our example. Bond dissociation energies showed us that allyl radicals are unusually stable, and we attributed this stability to resonance between equivalent structures (Secs. 6.24–6.25). The ionization potential (188 kcal) of the allyl radical enables us to calculate that the allyl cation, too, is unusually stable. Even though we have just drawn its structure as that of a primary cation, it is 24 kcal more stable than the ethyl cation, and just about as stable as the isopropyl cation. We can now expand the sequence of Sec. 5.18.

Stability of carbonium ions

$$3° > \overset{\text{allyl}}{2°} > 1° > CH_3^+$$

Like the allyl radical, the allyl cation is a resonance hybrid of two exactly equivalent structures:

$$\Big[\underset{\text{III}}{CH_2{=}CH{-}CH_2^+} \quad \underset{\text{IV}}{^+CH_2{-}CH{=}CH_2}\Big] \quad \textit{equivalent to} \quad \underset{\oplus}{\underbrace{CH_2{\cdots}CH{\cdots}CH_2}}$$

In either of the contributing structures, there is an empty *p* orbital on the electron-deficient carbon. Overlap of this empty *p* orbital with the π cloud of the double bond results in delocalization of the π electrons: each of them helps to hold together all three carbon nuclei (Fig. 8.7). We see how *flatness* is required to permit the overlap that provides electrons to the electron-deficient carbon and stabilizes the cation.

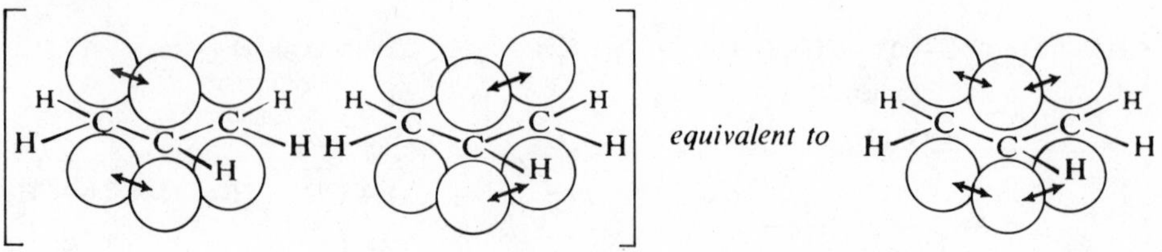

$$\left[CH_2{=}CH{-}CH_2^+ \quad {}^+CH_2{-}CH{=}CH_2\right] \quad \textit{equivalent to} \quad \underbrace{CH_2\text{---}CH\text{---}CH_2}_{\oplus}$$

Figure 8.7 Allyl cation. The *p* orbital of the middle carbon overlaps *p* orbitals on both sides to permit delocalization of electrons.

The relative stabilities of simple alkyl cations has also been attributed to delocalization, this time by overlap of the *p* orbital with σ bonds, that is, through *hyperconjugation* (Sec. 6.28).

Problem 8.9 Draw resonance structures to show how the order of stability of alkyl cations could be accounted for on the basis of hyperconjugation.

The products obtained from addition to conjugated dienes are always consistent with the formation of the most stable intermediate carbonium ion: an allyl cation. This requires the first step to be *addition to one of the ends* of the conjugated system.

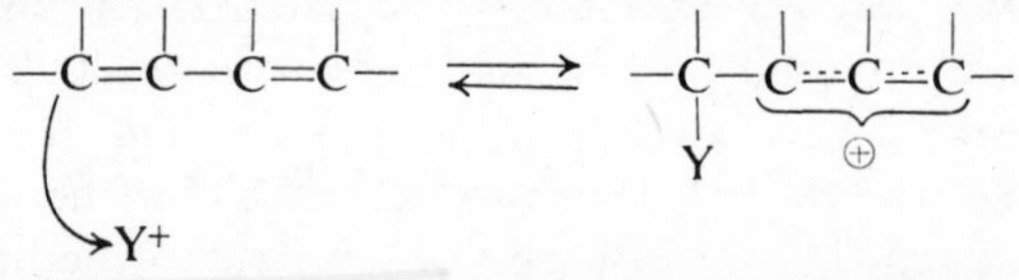

$$-\overset{|}{C}{=}\overset{|}{C}{-}\overset{|}{C}{=}\overset{|}{C}- \;\xrightarrow{Y^+}\; -\underset{Y}{\overset{|}{C}}-\underbrace{\overset{|}{C}\text{---}\overset{|}{C}\text{---}\overset{|}{C}}_{\oplus}-$$

Adds to end of conjugated system An allyl carbonium ion

The first step of addition to 2,4-hexadiene yields, then, not cation I, but the hybrid cation VI in which the charge is spread over two carbons:

$$\left[\underset{\text{I}}{CH_3{-}\underset{H}{CH}{-}\overset{\oplus}{CH}{-}CH{=}CH{-}CH_3} \qquad \underset{\text{V}}{CH_3{-}\underset{H}{CH}{-}CH{=}CH{-}\overset{\oplus}{CH}{-}CH_3}\right]$$

equivalent to

$$\underset{\text{VI}}{CH_3{-}\underset{H}{CH}{-}\underbrace{CH\text{---}CH\text{---}CH}_{\oplus}{-}CH_3}$$

In the second step, the negative chloride ion can attach itself to either of these carbons and thus yield the 1,2- or 1,4-product.

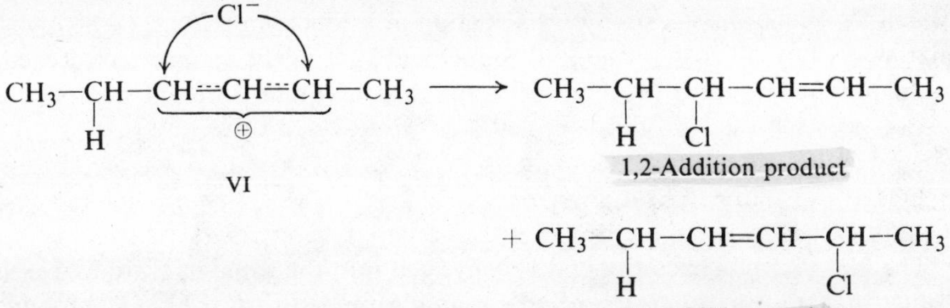

VI

1,2-Addition product

1,4-Addition product

We have not shown *why* 1,4-addition occurs; we have simply shown that it is not unreasonable that it *does* happen. In summary:

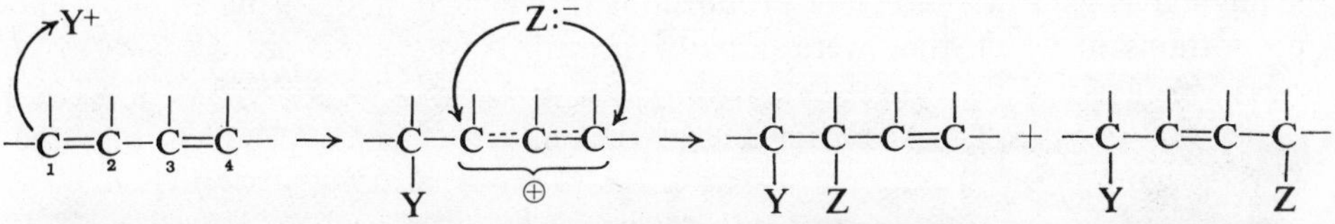

Addition to end of conjugated system — Allyl carbonium ion — 1,2-Addition — 1,4-Addition

Thus the hybrid nature of the allyl cation governs both steps of electrophilic addition to conjugated dienes: the first, through stabilization; the second, by permitting attachment to either of two carbon atoms.

Problem 8.10 Account for the fact that 2-methyl-1,3-butadiene reacts (a) with HCl to yield only 3-chloro-3-methyl-1-butene and 1-chloro-3-methyl-2-butene; (b) with bromine to yield only 3,4-dibromo-3-methyl-1-butene and 1,4-dibromo-2-methyl-2-butene.

8.22 1,2- *vs.* 1,4-Addition. Rate *vs.* equilibrium

A very important principle emerges when we look at the relative amounts of 1,2- and 1,4-addition products obtained.

Addition of HBr to 1,3-butadiene yields both the 1,2- and the 1,4-products; the *proportions* in which they are obtained are markedly affected by the temperature

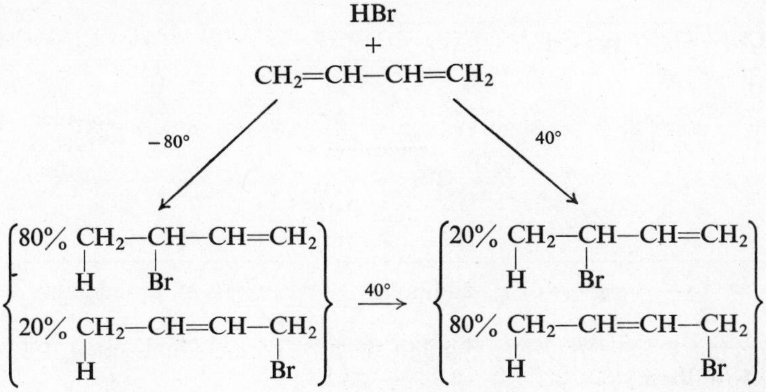

at which the reaction is carried out. Reaction at a low temperature (−80°) yields a mixture containing 20% of the 1,4-product and 80% of the 1,2-product. Reaction at a higher temperature (40°) yields a mixture of quite different composition, 80% 1,4- and 20% 1,2-product. At intermediate temperatures, mixtures of intermediate compositions are obtained. Although each isomer is quite stable at low temperatures, prolonged heating of either the 1,4- or the 1,2-compound yields the same mixture. How are these observations to be interpreted?

The fact that either compound is converted into the same mixture by heating indicates that this mixture is the result of equilibrium between the two compounds. The fact that the 1,4-compound predominates in the equilibrium mixture indicates that it is the more stable of the two.

The fact that more 1,2- than 1,4-product is obtained at −80° indicates that the 1,2-product is formed *faster* than the 1,4-product; since each compound remains unchanged at −80°, the proportions in which they are isolated show the proportions in which they were initially formed. As the reaction temperature is raised, the proportions in which the products are initially formed may remain the same, but there is faster conversion of the initially formed products into the equilibrium mixture.

The proportions of products actually isolated from the low-temperature addition are determined by the **rates** of addition, whereas for the high-temperature addition they are determined by the **equilibrium** between the two isomers.

Let us examine the matter of 1,2- and 1,4-addition more closely by drawing a potential energy curve for the reactions involved (Fig. 8.8). The carbonium ion initially formed reacts to yield the 1,2-product faster than the 1,4-product; consequently, the energy of activation leading to the 1,2-product must be less than

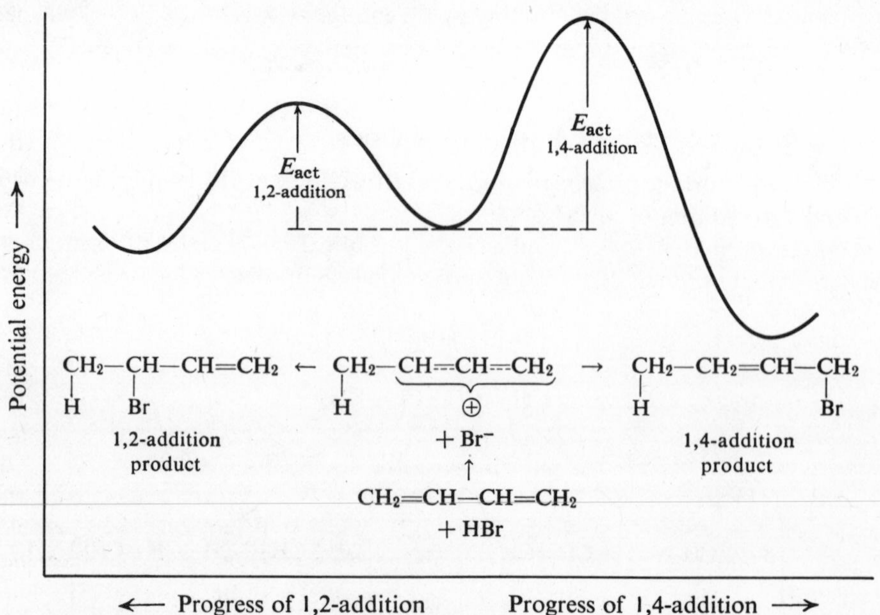

Figure 8.8. Potential energy changes during progress of reaction: 1,2- vs. 1,4-addition.

that leading to the 1,4-product. We represent this by the lower hill leading from the ion to the 1,2-product. More collisions have enough energy to climb the low hill than the high hill, so that the 1,2-compound is formed faster than the 1,4-compound. The 1,4-product, however, is more stable than the 1,2-product, and hence we must place its valley at a lower level than that of the 1,2-product.

We shall see later (Sec. 14.12) that alkyl halides, and particularly allyl halides, can undergo ionization. Now ionization of either bromo compound yields the same carbonium ion; the most likely—and simplest—way in which the 1,2- and 1,4-products reach equilibrium is through this ion.

$$\underset{\text{1,2-}}{\underset{\text{H}\quad\ \text{Br}}{CH_2{-}CH{-}CH{=}CH_2}} \rightleftarrows \underset{\text{H}}{\overset{\oplus}{CH_2{-}CH{\cdots}CH{\cdots}CH_2}} + Br^- \rightleftarrows \underset{\text{1,4-}}{\underset{\text{H}\qquad\qquad\ \text{Br}}{CH_2{-}CH{=}CH{-}CH_2}}$$

$$\uparrow$$

$$CH_2{=}CH{-}CH{=}CH_2 + HBr$$

Ionization of the bromides involves climbing the potential hills back toward this carbonium ion. But there is a higher hill separating the ion from the 1,4-product than from the 1,2-product; consequently, the 1,4-product will ionize more slowly than the 1,2-product. Equilibrium is reached when the rates of the opposing reactions are equal. The 1,2-product is formed rapidly, but ionizes rapidly. The 1,4-product is formed slowly, but ionizes even more slowly; once formed, the 1,4-product tends to persist. At temperatures high enough for equilibrium to be reached —that is, high enough for significantly fast ionization—the more stable 1,4-product predominates.

We have not tried to account for the fact that the 1,2-product is formed faster than the 1,4-product, or for the fact that the 1,4-product is more stable than the 1,2-product (although we notice that this is consistent with our generalization that disubstituted alkenes are more stable than monosubstituted alkenes). We have accepted these facts and have simply tried to show what they mean in terms of energy considerations. Similar relationships have been observed for other dienes and reagents.

These facts illustrate two important points. First, we must be cautious when we interpret product composition in terms of rates of reaction; we must be sure that one product is not converted into the other *after* its formation. Second, the more stable product is by no means *always* formed faster. *On the basis of much evidence*, we have concluded that *generally* the more stable a carbonium ion or free radical, the faster it is formed; a consideration of the transition states for the various reactions has shown (Secs. 3.26, 5.21, and 6.11) that this is reasonable. *We must not, however, extend this principle to other reactions unless the evidence warrants it.*

Problem 8.11 Addition of one mole of bromine to 1,3,5-hexatriene yields only 5,6-dibromo-1,3-hexadiene and 1,6-dibromo-2,4-hexadiene. (a) Are these products consistent with the formation of the most stable intermediate carbonium ion? (b) What other product or products would also be consistent? (c) Actually, which factor appears to be in control, rate or position of equilibrium?

8.23 Free-radical addition to conjugated dienes: orientation

Like other alkenes, conjugated dienes undergo addition not only by electrophilic reagents but also by free radicals. In free-radical addition, conjugated dienes show two special features: they undergo **1,4-addition** as well as 1,2-addition, and they are **much more reactive** than ordinary alkenes. We can account for both features—*orientation* and *reactivity*—by examining the structure of the intermediate free radical.

Let us take, as an example, addition of $BrCCl_3$ to 1,3-butadiene in the presence of a peroxide. As we have seen (Sec. 6.18), the peroxide decomposes (step 1) to yield a free radical, which abstracts bromine from $BrCCl_3$ (step 2) to generate a CCl_3 radical.

(1) $$\text{Peroxide} \longrightarrow \text{Rad}\cdot$$

(2) $$\text{Rad}\cdot + BrCCl_3 \longrightarrow \text{Rad—Br} + \cdot CCl_3$$

The $\cdot CCl_3$ radical thus formed adds to the butadiene (step 3). Addition to one of the *ends* of the conjugated system is the preferred reaction, since this yields a resonance-stabilized allyl free radical.

(3) $$\overset{\cdot CCl_3}{\underset{1}{CH_2}}{=}\underset{2}{CH}{-}\underset{3}{CH}{=}\underset{4}{CH_2} \longrightarrow \begin{bmatrix} Cl_3C{-}CH_2{-}\dot{C}H{-}CH{=}CH_2 \\ Cl_3C{-}CH_2{-}CH{=}CH{-}\dot{C}H_2 \end{bmatrix}$$

Addition to end of conjugated system

equivalent to

$$Cl_3C{-}CH_2{-}\underbrace{CH{\cdots}CH{\cdots}CH_2}_{\cdot}$$

Allylic free radical

The allyl free radical then abstracts bromine from a molecule of $BrCCl_3$ (step 4) to complete the addition, and in doing so forms a new $\cdot CCl_3$ radical which can carry on the chain. In step (4) bromine can become attached to either C–2 or C–4 to yield either the 1,2- or 1,4-product.

(4) $$Cl_3C{-}CH_2{-}\underbrace{CH{\cdots}CH{\cdots}CH_2}_{\cdot} \xrightarrow{BrCCl_3} Cl_3C{-}CH_2{-}\underset{Br}{\underset{|}{CH}}{-}CH{=}CH_2$$

Allylic free radical — 1,2-Addition product

and $$Cl_3C{-}CH_2{-}CH{=}CH{-}CH_2{-}Br$$

1,4-Addition product

8.24 Free-radical addition to conjugated dienes: reactivity

If $BrCCl_3$ is allowed to react with a 50:50 mixture of 1,3-butadiene and a simple alkene like 1-octene, addition occurs almost exclusively to the 1,3-butadiene. Evidently the $\cdot CCl_3$ radical adds much more rapidly to the conjugated diene than to the simple alkene. Similar results have been observed in a great many radical additions.

How can we account for the unusual reactivity of conjugated dienes? In our discussion of halogenation of the simple alkenes (Sec. 3.27), we found that not only orientation but also relative reactivity was related to the stability of the free radical formed in the first step. On this basis alone, we might expect addition to a conjugated diene, which yields a stable allyl free radical, to occur faster than addition to a simple alkene.

On the other hand, we have just seen (Sec. 8.16) that conjugated dienes are more stable than simple alkenes. On this basis alone, we might expect addition to conjugated dienes to occur more slowly than to simple alkenes.

The relative rates of the two reactions depend chiefly upon the E_{act}'s. Stabilization of the incipient allyl free radical lowers the energy level of the transition state; stabilization of the diene lowers the energy of the reactants. Whether the net E_{act} is larger or smaller than for addition to a simple alkene depends upon *which* is stabilized *more* (see Fig. 8.9).

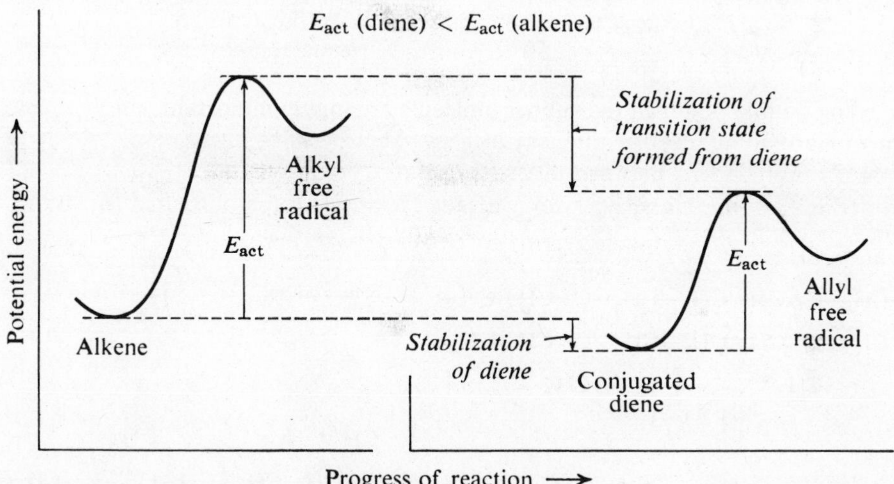

Figure 8.9. Molecular structure and rate of reaction. Transition state from diene stabilized more than diene itself: E_{act} is lowered. (Plots aligned with each other for easy comparison.)

The fact is that conjugated dienes are more reactive than simple alkenes. In the present case, then—and in most cases involving alkenes and free radicals, or alkenes and carbonium ions—the factors stabilizing the transition state are more important than the factors stabilizing the reactant. However, this is *not always* true. (It does not seem to be true, for example, in electrophilic addition to conjugated dienes.)

8.25 Free-radical polymerization of dienes. Rubber and rubber substitutes

Like substituted ethylenes, conjugated dienes, too, undergo free-radical polymerization. From 1,3-butadiene, for example, there is obtained a polymer

$$\underset{\text{1,3-Butadiene}}{CH_2{=}CH{-}CH{=}CH_2} \qquad \underset{\text{Polybutadiene}}{[{-}CH_2{-}CH{=}CH{-}CH_2{-}]_n}$$

whose structure indicates that 1,4-addition occurs predominantly:

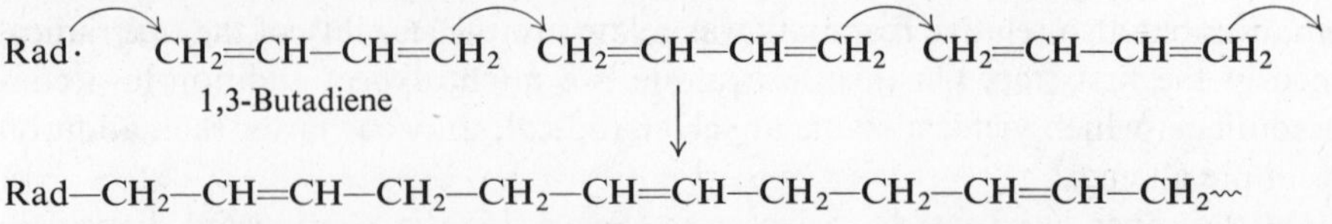

$$\text{Rad}\cdot \quad \underset{\text{1,3-Butadiene}}{CH_2{=}CH{-}CH{=}CH_2} \quad CH_2{=}CH{-}CH{=}CH_2 \quad CH_2{=}CH{-}CH{=}CH_2$$

$$\downarrow$$

$$\text{Rad}{-}CH_2{-}CH{=}CH{-}CH_2{-}CH_2{-}CH{=}CH{-}CH_2{-}CH_2{-}CH{=}CH{-}CH_2{\sim}$$

Such a polymer differs from the polymers of simple alkenes in one very important way: each unit still contains one double bond.

Natural rubber has a structure that strongly resembles these synthetic polydienes. We could consider it to be a polymer of the conjugated diene 2-methyl-1,3-butadiene, **isoprene.**

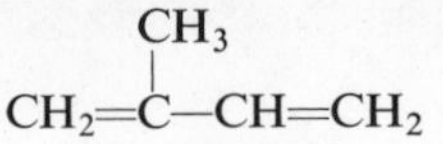

$$\underset{\text{Isoprene}}{CH_2{=}\overset{\overset{CH_3}{|}}{C}{-}CH{=}CH_2} \qquad \underset{\substack{\textit{cis}\text{-Polyisoprene}\\ \text{Natural rubber}}}{\left[{-}CH_2{-}\overset{\overset{CH_3}{|}}{C}{=}CH{-}CH_2{-}\right]_n}$$

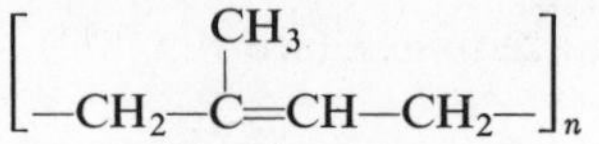

The double bonds in the rubber molecule are highly important, since—apparently by providing reactive allylic hydrogens—they permit *vulcanization*, the formation of sulfur bridges between different chains. These *cross-links* make the rubber harder and stronger, and do away with the tackiness of the untreated rubber.

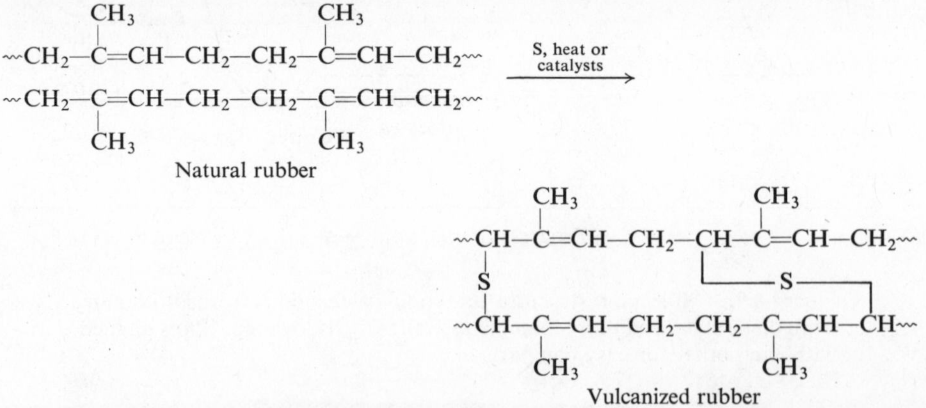

$$\underset{\text{Natural rubber}}{\begin{array}{l} {\sim}CH_2{-}\overset{\overset{CH_3}{|}}{C}{=}CH{-}CH_2{-}CH_2{-}\overset{\overset{CH_3}{|}}{C}{=}CH{-}CH_2{\sim} \\ {\sim}CH_2{-}\underset{\underset{CH_3}{|}}{C}{=}CH{-}CH_2{-}CH_2{-}\underset{\underset{CH_3}{|}}{C}{=}CH{-}CH_2{\sim} \end{array}} \xrightarrow[\text{catalysts}]{\text{S, heat or}}$$

$$\underset{\text{Vulcanized rubber}}{\begin{array}{l} {\sim}CH{-}\overset{\overset{CH_3}{|}}{C}{=}CH{-}CH_2{-}CH{-}\overset{\overset{CH_3}{|}}{C}{=}CH{-}CH_2{\sim} \\ \;\;|\qquad\qquad\qquad\qquad\quad \llcorner\text{—S—}\urcorner \\ \;\;S \\ \;\;| \\ {\sim}CH{-}\underset{\underset{CH_3}{|}}{C}{=}CH{-}CH_2{-}CH_2{-}\underset{\underset{CH_3}{|}}{C}{=}CH{-}CH{\sim} \end{array}}$$

Polymerization of dienes to form substitutes for rubber was the forerunner of the enormous present-day plastics industry. *Polychloroprene* (Neoprene, Duprene) was the first commercially successful rubber substitute in the United States.

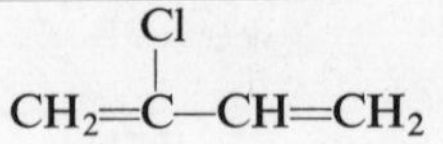

$$\underset{\text{Chloroprene}}{CH_2{=}\overset{\overset{Cl}{|}}{C}{-}CH{=}CH_2} \qquad \underset{\text{Polychloroprene}}{\left[{-}CH_2{-}\overset{\overset{Cl}{|}}{C}{=}CH{-}CH_2{-}\right]_n}$$

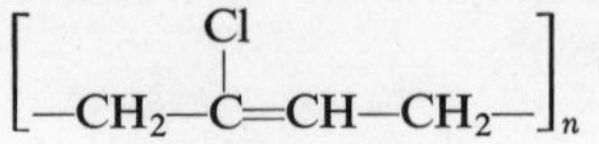

The properties of rubber substitutes—like those of other polymers—are determined, in part, by the nature of the substituent groups. Polychloroprene, for example, is inferior to natural rubber in some properties, but superior in its resistance to oil, gasoline, and other organic solvents.

Polymers of isoprene, too, can be made artificially: they contain the same unsaturated chain and the same substituent (the $—CH_3$ group) as natural rubber. But polyisoprene made by the free-radical process we have been talking about was—in the properties that really matter—a far cry from natural rubber. It differed in *stereochemistry*: natural rubber has the *cis*-configuration at (nearly) every double bond; the artificial material was a mixture of *cis* and *trans*. Not until 1955 could a true synthetic *rubber* be made; what was needed was an entirely new kind of catalyst and an entirely new mechanism of polymerization (Sec. 32.6). With these, it became possible to carry out a stereoselective polymerization of isoprene to a material virtually identical with natural rubber: *cis*-1,4-polyisoprene.

$$\sim CH_2(CH_3)C{=}CH{-}CH_2{-}CH_2{-}(CH_3)C{=}CH{-}CH_2{-}CH_2{-}(CH_3)C{=}CH{-}CH_2 \sim$$

Natural rubber
All cis-configurations

8.26 Isoprene and the isoprene rule

The isoprene unit is one of nature's favorite building blocks. It occurs not only in rubber, but in a wide variety of compounds isolated from plant and animal sources. For example, nearly all the *terpenes* (found in the essential oils of many plants) have carbon skeletons made up of isoprene units joined in a regular, head-to-tail way. Recognition of this fact—the so-called **isoprene rule**—has been of great help in working out structures of terpenes.

$$\text{(ring: } H_2C{-}C(CH_3)_2{-}C{=}C(CH_3){-}CH_2{-}H_2C) \; C{-}CH{=}CH{-}C(CH_3){=}CH{-}CH{=}CH{-}C(CH_3){=}CH{-}CH_2OH$$

Vitamin A

$$CH_2{=}C(CH_3){-}CH_2{-}CH_2{-}CH_2{-}CH(CH_3){-}CH_2{-}CH_2OH$$

Citronellol: *a terpene*
(found in oil of geranium)

$$\text{ring: } C(CH_3){=}CH{-}CH_2{-}C(CH(CH_3)_2){=}CH{-}CH_2$$

γ-Terpinene: *a terpene*
(found in coriander oil)

A fascinating area of research linking organic chemistry and biology is the study of the *biogenesis* of natural products: the detailed sequence of reactions by which a compound is formed in living systems, plant or animal. All the isoprene units in nature, it appears, originate from the same compound, "isopentenyl" pyrophosphate.

$$CH_2{=}\overset{\displaystyle CH_3}{\overset{|}{C}}{-}CH_2{-}CH_2{-}O{-}\overset{\displaystyle O}{\overset{\|}{\underset{\displaystyle O}{\underset{|}{P}}}}{-}O{-}\overset{\displaystyle O}{\overset{\|}{\underset{\displaystyle O}{\underset{|}{P}}}}{-}OH$$

Isopentenyl pyrophosphate

Work done since about 1950 has shown how compounds as seemingly different from rubber as *cholesterol* (p. 507) are built up, step by step, from isoprene units.

O—P—O—P—OH

Isopentenyl pyrophosphate

HO HO

Squalene ⟶ Lanosterol ⟶ Cholesterol

Problem 8.12 (a) Mark off the isoprene units making up the squalene molecule. (b) There is one deviation from the head-to-tail sequence. Where is it? Does its particular location suggest anything to you—in general terms—about the biogenesis of this molecule? (c) What skeletal changes, if any, accompany the conversion of squalene into lanosterol? Of lanosterol into cholesterol?

8.27 Analysis of alkynes and dienes

Alkynes and dienes respond to characterization tests in the same way as alkenes: they decolorize bromine in carbon tetrachloride without evolution of hydrogen bromide, and they decolorize cold, neutral, dilute permanganate; they are not oxidized by chromic anhydride. They are, however, more unsaturated than alkenes. This property can be detected by determination of their molecular formulas (C_nH_{2n-2}) and by a quantitative hydrogenation (two moles of hydrogen are taken up per mole of hydrocarbon).

Proof of structure is best accomplished by the same degradative methods that are used in studying alkenes. Upon ozonolysis alkynes yield carboxylic acids, whereas alkenes yield aldehydes and ketones. For example:

$$\underset{\text{2-Pentyne}}{CH_3CH_2C{\equiv}CCH_3} \xrightarrow{O_3} \xrightarrow{H_2O} \underset{\text{Carboxylic acids}}{CH_3CH_2COOH + HOOCCH_3}$$

Ozonolysis of dienes yields aldehydes and ketones, including double-ended ones containing two C=O groups per molecule. For example:

$$\underset{\substack{|\\ CH_3}}{}\quad CH_2{=}\overset{\overset{CH_3}{|}}{C}{-}CH{=}CH_2 \xrightarrow{O_3} \xrightarrow{H_2O,\ Zn} H{-}\overset{\overset{H}{|}}{C}{=}O + O{=}\overset{\overset{CH_3}{|}}{C}{-}\underset{\underset{H}{|}}{C}{=}O + O{=}\overset{\overset{H}{|}}{C}{-}H$$

A terminal alkyne (RC≡CH) is characterized, and differentiated from isomers, by its conversion into insoluble silver and cuprous acetylides (Sec. 8.11).

(Spectroscopic analysis of alkynes and dienes is discussed in Secs. 13.15–13.16.)

Problem 8.13 Contrast the ozonolysis products of the following isomers: (a) 1-pentyne, (b) 2-pentyne, (c) 3-methyl-1-butyne, (d) 1,3-pentadiene, (e) 1,4-pentadiene, (f) isoprene (2-methyl-1,3-butadiene).

Problem 8.14 Predict the ozonolysis products from polybutadiene, $(C_4H_6)_n$: (a) if 1,2-addition is involved in the polymerization; (b) if 1,4-addition is involved.

Problem 8.15 Ozonolysis of natural rubber yields chiefly (90%) the compound

$$O{=}\overset{\overset{H}{|}}{C}{-}CH_2{-}CH_2{-}\overset{\overset{CH_3}{|}}{C}{=}O$$

What does this tell us about the structure of rubber?

PROBLEMS

1. (a) Draw structures of the seven isomeric alkynes of formula C_6H_{10}. (b) Give the IUPAC and derived name of each. (c) Indicate which ones will react with Ag^+ or $Cu(NH_3)_2^+$. (d) Draw structures of the ozonolysis products expected from each.

2. (a) Draw structures of all isomeric dienes of formula C_6H_{10}, omitting cumulated dienes. (b) Name each one. (c) Indicate which ones are conjugated. (d) Indicate which ones can show geometric isomerism, and draw the isomeric structures. (e) Draw structures of the ozonolysis products expected from each. (f) Which isomers (other than *cis-trans* pairs) could not be distinguished on the basis of (e)?

3. Write equations for all steps in the manufacture of acetylene starting from limestone and coal.

4. Outline all steps in the synthesis of propyne from each of the following compounds, using any needed organic or inorganic reagents. Follow the other directions given on page 224.

(a) 1,2-dibromopropane
(b) propylene
(c) isopropyl bromide
(d) propane
(e) *n*-propyl alcohol
(f) 1,1-dichloropropane
(g) acetylene
(h) 1,1,2,2-tetrabromopropane

5. Outline all steps in the synthesis from acetylene of each of the following compounds, using any needed organic or inorganic reagents.

(a) ethylene
(b) ethane
(c) ethylidene bromide (1,1-dibromoethane)
(d) vinyl chloride
(e) 1,2-dichloroethane
(f) acetaldehyde
(g) propyne
(h) 1-butyne
(i) 2-butyne
(j) *cis*-2-butene
(k) *trans*-2-butene
(l) 1-pentyne
(m) 2-pentyne
(n) 3-hexyne

6. Give structures and names of the organic products expected from the reaction (if any) of 1-butyne with:

(a) 1 mole H_2, Ni
(b) 2 moles H_2, Ni
(c) 1 mole Br_2
(d) 2 moles Br_2
(e) 1 mole HCl
(f) 2 moles HCl
(g) H_2O, H^+, Hg^{++}
(h) Ag^+
(i) product (h) + HNO_3
(j) $NaNH_2$
(k) product (j) + C_2H_5Br
(l) product (j) + *tert*-butyl chloride
(m) C_2H_5MgBr
(n) product (m) + H_2O
(o) O_3, then H_2O
(p) hot $KMnO_4$

7. Answer Problem 6 for 1,3-butadiene instead of 1-butyne.

8. Answer Problem 6 for 1,4-pentadiene instead of 1-butyne.

9. Give structures and names of the products from dehydrohalogenation of each of the following halides. Where more than one product is expected, indicate which will be the major product.

(a) 1-chlorobutane; 2-chlorobutane
(b) 1-chlorobutane; 4-chloro-1-butene
(c) 2-bromo-2-methylbutane; 3-bromo-2-methylbutane
(d) 1-bromo-2-methylbutane; 4-bromo-2-methylbutane
(e) 1-chloro-2,3-dimethylbutane; 2-chloro-2,3-dimethylbutane
(f) 4-chloro-1-butene; 5-chloro-1-pentene

10. Which alkyl halide of each pair in Problem 9 would you expect to undergo dehydrohalogenation faster?

11. Give structures of the chief product or products expected from addition of one mole of HCl to each of the following compounds:

(a) 1,3-butadiene; 1-butene
(b) 1,3-butadiene; 1,4-pentadiene
(c) 1,3-butadiene; 2-methyl-1,3-butadiene
(d) 1,3-butadiene; 1,3-pentadiene

12. Answer Problem 11 for the addition of $BrCCl_3$ in the presence of peroxides (Sec. 6.18) instead of addition of HCl.

13. Which compound of each pair in Problem 12 would you expect to be more reactive toward addition of $BrCCl_3$?

14. (a) The heat of hydrogenation of acetylene (converted into ethane) is 75.0 kcal/mole. Calculate ΔH for hydrogenation of acetylene to ethylene. (b) How does the stability of an alkyne relative to an alkene compare with the stability of an alkene relative to an alkane? (c) Solely on the basis of your answer to (b), would you expect acetylene to be more or less reactive than ethylene toward addition of a free methyl radical, $CH_3\cdot$? (d) Draw the structure of the free radical expected from addition of $CH_3\cdot$ to acetylene: from addition of $CH_3\cdot$ to ethylene. Judging only from the relative stabilities of the radicals being formed, would you expect $CH_3\cdot$ to add to acetylene faster or slower than to ethylene? (e) $CH_3\cdot$ has been found to add more slowly to acetylene than to ethylene. Which factor—reactant stability or radical stability—is more important here?

15. (a) Make a model of *allene*, $CH_2{=}C{=}CH_2$, a cumulated diene. What is the spatial relationship between the pair of hydrogens at one end of the molecule and the pair of hydrogens at the other end? (b) Substituted allenes of the type $RCH{=}C{=}CHR$ have been obtained in optically active form. Is this consistent with the shape of the molecule in (a)? Where are the chiral centers in the substituted allene? (c) Work out the electronic configuration of allene. (*Hint:* How many atoms are attached to the middle carbon? To each of the end carbons?) Does this lead to the same shape of molecule that you worked out in (a) and (b)?

16. A useful method of preparing 1-alkenes involves reaction of Grignard reagents with the unusually reactive halide, allyl bromide:

$$RMgX + BrCH_2CH{=}CH_2 \longrightarrow R{-}CH_2CH{=}CH_2$$

When 1-hexene (b.p. 63.5°) is prepared in this way, it is contaminated with *n*-hexane (b.p. 69°) and 1,5-hexadiene (b.p. 60°); these are difficult to remove because of the closeness of boiling points. The mixture is treated with bromine and the product distilled. There are obtained three fractions: b.p. 68–69°; b.p. 77–78° at 15 mm pressure; and a high-boiling residue.

(a) What does each of these fractions contain? (b) What would you do next to get pure 1-hexene? (c) Show how this procedure could be applied to the separation of *n*-pentane (b.p. 36°) and 1-pentene (b.p. 30°); 1-decene (b.p. 171°) and 5-decyne (b.p. 175°).

17. Outline all steps in a possible laboratory synthesis of each of the following, using alcohols of four carbons or fewer as your only organic source, and any necessary inorganic reagents. (*Remember:* Work backwards.)

(a) *meso*-3,4-dibromohexane;
(b) (2R,3R;2S,3S)-2,3-heptanediol, a racemic modification.

18. Treatment with phosphoric acid converts 2,7-dimethyl-2,6-octadiene into I.

```
  H3C     CH3
     \   /
       C              CH3
     /   \           /
  H2C     CH—C
    |     |          \\
  H2C—CH2             CH2
```

I

1,1-Dimethyl-2-isopropenylcyclopentane

Using reaction steps already familiar to you, suggest a mechanism for this reaction.

19. *Gutta percha* is a non-elastic naturally-occurring polymer used in covering golf balls and underwater cables. It has the same formula, $(C_5H_8)_n$, and yields the same hydrogenation product and the same ozonolysis product (Problem 8.15, page 279) as natural rubber. Using structural formulas, show the most likely structural difference between gutta percha and rubber.

20. Describe simple chemical tests that would distinguish between:

(a) 2-pentyne and *n*-pentane
(b) 1-pentyne and 1-pentene
(c) 1-pentyne and 2-pentyne
(d) 1,3-pentadiene and *n*-pentane
(e) 1,3-pentadiene and 1-pentyne
(f) 2-hexyne and isopropyl alcohol
(g) allyl bromide and 2,3-dimethyl-1,3-butadiene

Tell exactly what you would *do* and *see*.

21. Describe chemical methods (not necessarily simple tests) that would distinguish between:

(a) 2-pentyne and 2-pentene
(b) 1,4-pentadiene and 2-pentene
(c) 1,4-pentadiene and 2-pentyne
(d) 1,4-pentadiene and 1,3-pentadiene

22. On the basis of physical properties, an unknown compound is believed to be one of the following:

n-pentane (b.p. 36°)
2-pentene (b.p. 36°)
1-chloropropene (b.p. 37°)
trimethylethylene (b.p. 39°)
1-pentyne (b.p. 40°)
methylene chloride (b.p. 40°)
3,3-dimethyl-1-butene (b.p. 41°)
1,3-pentadiene (b.p. 42°)

Describe how you would go about finding out which of the possibilities the unknown actually is. Where possible, use simple chemical tests; where necessary, use more elaborate chemical methods like quantitative hydrogenation and cleavage. Tell exactly what you would *do* and *see*.

23. A hydrocarbon of formula C_6H_{10} absorbs only *one* mole of H_2 upon catalytic hydrogenation. Upon ozonolysis the hydrocarbon yields

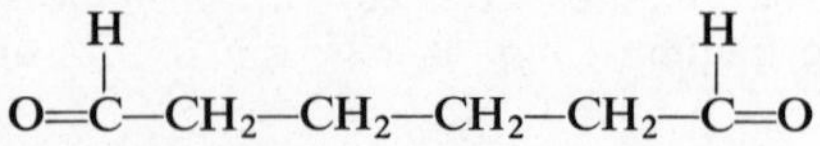

$$O{=}\overset{\displaystyle H}{\overset{|}{C}}{-}CH_2{-}CH_2{-}CH_2{-}CH_2{-}\overset{\displaystyle H}{\overset{|}{C}}{=}O$$

What is the structure of the hydrocarbon? (Check your answer in Sec. 9.17.)

24. A hydrocarbon was found to have a molecular weight of 80–85. A 10.02-mg sample took up 8.40 cc of H_2 gas measured at 0° and 760 mm pressure. Ozonolysis yielded only

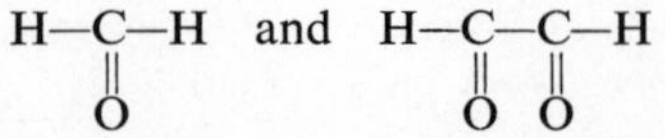

$$H{-}\underset{\displaystyle O}{\underset{\|}{C}}{-}H \quad \text{and} \quad H{-}\underset{\displaystyle O}{\underset{\|}{C}}{-}\underset{\displaystyle O}{\underset{\|}{C}}{-}H$$

What was the hydrocarbon?

25. *Myrcene*, $C_{10}H_{16}$, a terpene isolated from oil of bay, absorbs three moles of hydrogen to form $C_{10}H_{22}$. Upon ozonolysis myrcene yields:

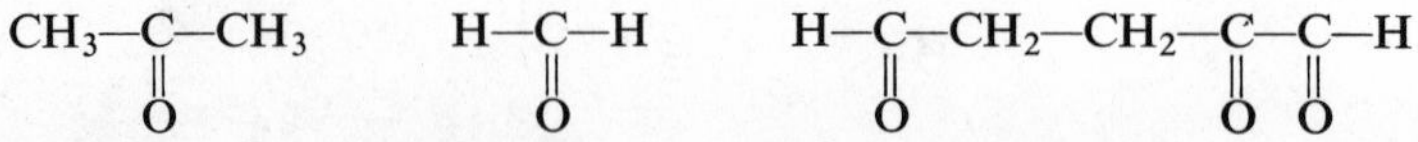

$$CH_3{-}\underset{\displaystyle O}{\underset{\|}{C}}{-}CH_3 \qquad H{-}\underset{\displaystyle O}{\underset{\|}{C}}{-}H \qquad H{-}\underset{\displaystyle O}{\underset{\|}{C}}{-}CH_2{-}CH_2{-}\underset{\displaystyle O}{\underset{\|}{C}}{-}\underset{\displaystyle O}{\underset{\|}{C}}{-}H$$

(a) What structures are consistent with these facts?
(b) On the basis of the isoprene rule (Sec. 8.26), what is the most likely structure for myrcene?

26. *Dihydromyrcene*, $C_{10}H_{18}$, formed from myrcene (Problem 25), absorbs two moles of hydrogen to form $C_{10}H_{22}$. Upon cleavage by $KMnO_4$, dihydromyrcene yields:

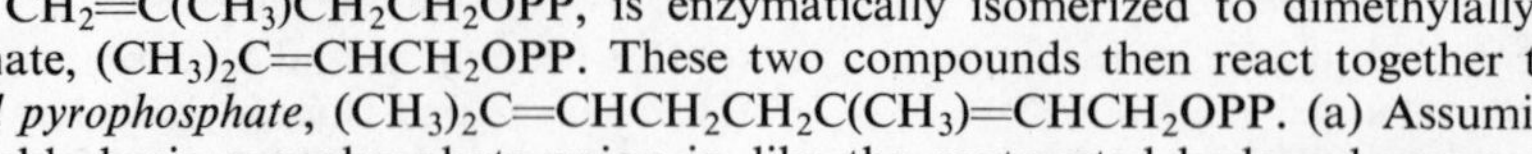

$$CH_3{-}\underset{\displaystyle O}{\underset{\|}{C}}{-}CH_3 \qquad CH_3{-}\underset{\displaystyle O}{\underset{\|}{C}}{-}OH \qquad CH_3{-}\underset{\displaystyle O}{\underset{\|}{C}}{-}CH_2{-}CH_2{-}\underset{\displaystyle O}{\underset{\|}{C}}{-}OH$$

(a) Keeping in mind the isoprene rule, what is the most likely structure for dihydromyrcene? (b) Is it surprising that a compound of this structure is formed by reduction of myrcene?

27. At the beginning of the biogenesis of squalene (Sec. 8.26) isopentenyl pyrophosphate, $CH_2{=}C(CH_3)CH_2CH_2OPP$, is enzymatically isomerized to dimethylallyl pyrophosphate, $(CH_3)_2C{=}CHCH_2OPP$. These two compounds then react together to yield *geranyl pyrophosphate*, $(CH_3)_2C{=}CHCH_2CH_2C(CH_3){=}CHCH_2OPP$. (a) Assuming that the weakly basic pyrophosphate anion is, like the protonated hydroxyl group, a good leaving group,

$$R{-}OPP \longrightarrow R^{\oplus} + OPP^{-}$$

can you suggest a series of familiar steps by which geranyl pyrophosphate might be formed? (b) Geranyl pyrophosphate then reacts with another molecule of isopentenyl pyrophosphate to form *farnesyl pyrophosphate*. What is the structure of farnesyl pyrophosphate? (c) What is the relationship between farnesyl pyrophosphate and squalene? (d) An enzyme system from the rubber plant catalyzes the conversion of isopentenyl pyrophosphate into rubber; dimethylallyl pyrophosphate appears to act as an initiator for the process. Can you suggest a "mechanism" for the formation of natural rubber?

Chapter 9

Alicyclic Hydrocarbons

9.1 Open-chain and cyclic compounds

In the compounds that we have studied in previous chapters, the carbon atoms are attached to one another to form *chains*; these are called **open-chain** compounds. In many compounds, however, the carbon atoms are arranged to form *rings*; these are called **cyclic** compounds.

In this chapter we shall take up the *alicyclic* hydrocarbons (*ali*phatic *cyclic* hydrocarbons). Much of the chemistry of cycloalkanes and cycloalkenes we already know, since it is essentially the chemistry of open-chain alkanes and alkenes. But the cyclic nature of some of these compounds confers very special properties on them. It is because of these special properties that, during the past fifteen years, alicyclic chemistry has become what Professor Lloyd Ferguson, of the California State College at Los Angeles, has called "the playground for organic chemists." It is on some of these special properties that we shall focus our attention.

9.2 Nomenclature

Cyclic aliphatic hydrocarbons are named by prefixing **cyclo-** to the name of the corresponding open-chain hydrocarbon having the same number of carbon atoms as the ring. For example:

H_2C–CH_2–H_2C (three-membered ring)	H_2C—CH_2, H_2C—CH_2 (four-membered ring)	CH=CH, H_2C, H_2C—CH_2 (five-membered ring)
Cyclopropane	**Cyclobutane**	**Cyclopentene**

Substituents on the ring are named, and their positions are indicated by numbers,

Table 9.1 Cyclic Aliphatic Hydrocarbons

Name	M.p., °C	B.p., °C	Density (at 20°C)
Cyclopropane	−127	−33	
Cyclobutane	−80	13	
Cyclopentane	−94	49	0.746
Cyclohexane	6.5	81	.778
Cycloheptane	−12	118	.810
Cyclooctane	14	149	.830
Methylcyclopentane	−142	72	.749
cis-1,2-Dimethylcyclopentane	−62	99	.772
trans-1,2-Dimethylcyclopentane	−120	92	.750
Methylcyclohexane	−126	100	.769
Cyclopentene	−93	46	.774
1,3-Cyclopentadiene	−85	42	.798
Cyclohexene	−104	83	.810
1,3-Cyclohexadiene	−98	80.5	.840
1,4-Cyclohexadiene	−49	87	.847

the lowest combination of numbers being used. In simple cycloalkenes and cycloalkynes the doubly- and triply-bonded carbons are considered to occupy positions 1 and 2. For example:

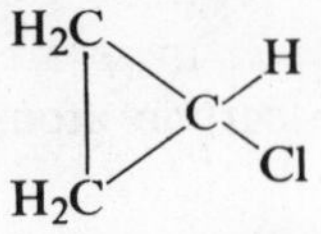

H_2C, H_2C, C(H)(Cl)

Chlorocyclopropane

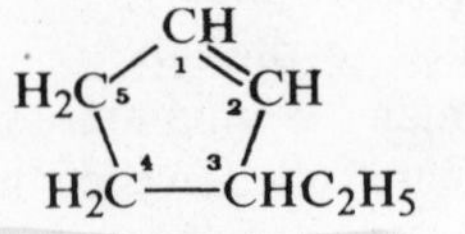

CH (1), CH (2), CHC_2H_5 (3), H_2C (4), H_2C (5)

3-Ethylcyclopentene

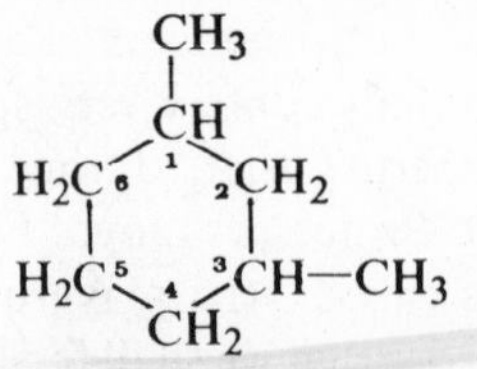

CH_3 on CH (1), CH_2 (2), $CH{-}CH_3$ (3), CH_2 (4), H_2C (5), H_2C (6)

1,3-Dimethylcyclohexane

CH (1), CH (2), CH (3), CH (4), H_2C (5), H_2C (6)

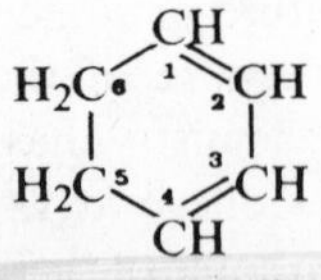

1,3-Cyclohexadiene

For convenience, aliphatic rings are often represented by simple geometric figures: a triangle for cyclopropane, a square for cyclobutane, a pentagon for cyclopentane, a hexagon for cyclohexane, and so on. It is understood that two hydrogens are located at each corner of the figure unless some other group is indicated. For example:

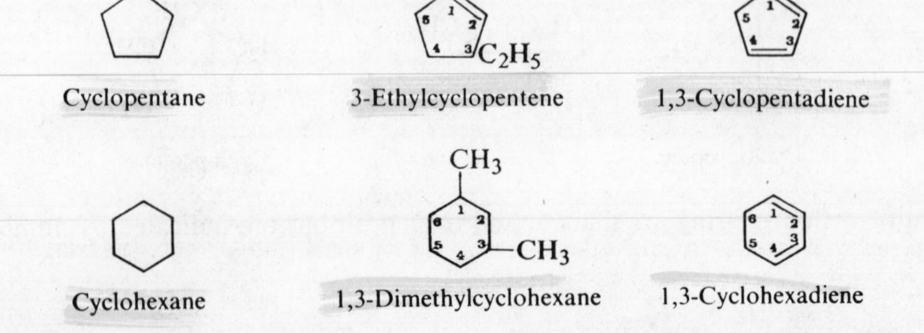

Cyclopentane 3-Ethylcyclopentene 1,3-Cyclopentadiene

Cyclohexane 1,3-Dimethylcyclohexane 1,3-Cyclohexadiene

Polycyclic compounds contain two or more rings that share two or more carbon atoms. We can illustrate the naming system with *norbornane*, whose systematic name is bicyclo[2.2.1]heptane: (a) *heptane*, since it contains a total of *seven* carbon atoms; (b) *bicyclo*, since it contains *two* rings, that is, breaking two carbon–

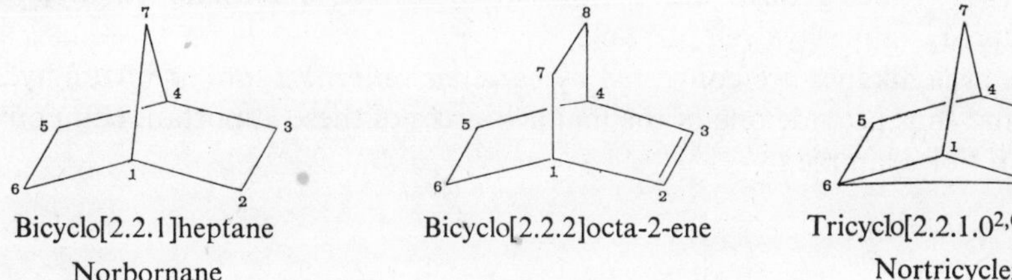

Bicyclo[2.2.1]heptane Norbornane

Bicyclo[2.2.2]octa-2-ene

Tricyclo[2.2.1.$0^{2,6}$]heptane Nortricyclene

carbon bonds converts it into an open-chain compound; (c) [2.2.1], since the number of carbons between bridgeheads (shared carbons) is *two* (C–2 and C–3), *two* (C–5 and C–6), and *one* (C–7).

Polycyclic compounds in a variety of strange and wonderful shapes have been made, and their properties have revealed unexpected facets of organic chemistry. Underlying much of this research there has always been the challenge: *can such a compound be made*?

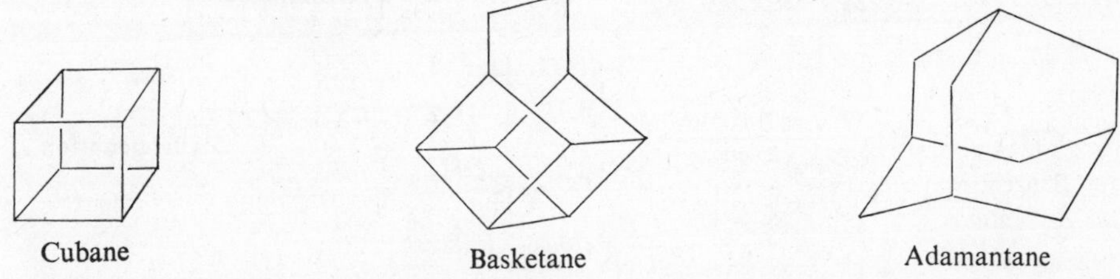

Cubane

Basketane

Adamantane

The ultimate polycyclic aliphatic system is *diamond* which is, of course, not a hydrocarbon at all, but one of the allotropic forms of elemental carbon. In diamond each

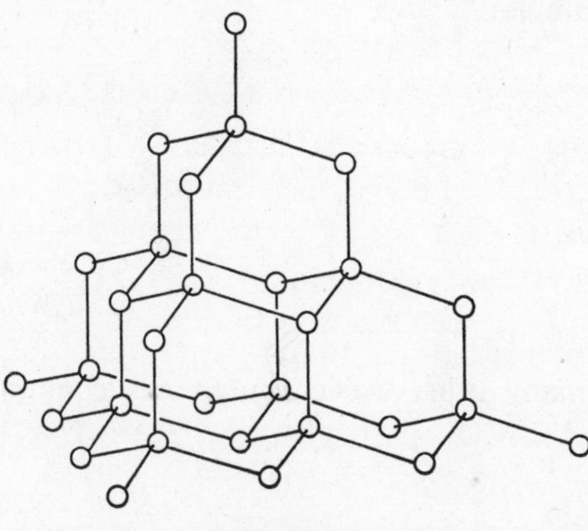

Diamond

carbon atom is attached to four others by tetrahedral bonds of the usual single bond length, 1.54 A. (Note the cyclohexane chairs, Sec. 9.11.)

9.3 Industrial source

We have already mentioned (Sec. 3.13) that petroleum from certain areas (in particular California) is rich in cycloalkanes, known to the petroleum industry as *naphthenes*. Among these are cyclohexane, methylcyclohexane, methylcyclopentane, and 1,2-dimethylcyclopentane.

These cycloalkanes are converted by *catalytic reforming* into aromatic hydrocarbons, and thus provide one of the major sources of these important compounds (Sec. 12.4). For example:

$$\underset{\substack{\textbf{Methylcyclohexane}\\ \textit{Aliphatic}}}{\begin{array}{ccc} & CH_2 & \\ H_2C & & CHCH_3 \\ | & & | \\ H_2C & & CH_2 \\ & CH_2 & \end{array}} \xrightarrow[300\ lb/in.^2]{Mo_2O_3{\cdot}Al_2O_3,\ 560^\circ} \underset{\substack{\textbf{Toluene}\\ \textit{Aromatic}}}{C_6H_5CH_3} + 3H_2 \qquad \textbf{Dehydrogenation}$$

Just as elimination of hydrogen from cyclic aliphatic compounds yields aromatic compounds, so addition of hydrogen to aromatic compounds yields cyclic aliphatic compounds, specifically cyclohexane derivatives. An important example of this is the hydrogenation of benzene to yield pure cyclohexane.

$$\underset{\substack{\text{Benzene}\\ \textit{Aromatic}}}{C_6H_6} + 3H_2 \xrightarrow[25\ atm.]{Ni,\ 150\text{–}250^\circ} \underset{\substack{\text{Cyclohexane}\\ \textit{Aliphatic}}}{\begin{array}{ccc} & CH_2 & \\ H_2C & & CH_2 \\ | & & | \\ H_2C & & CH_2 \\ & CH_2 & \end{array}} \qquad \textbf{Hydrogenation}$$

As we might expect, hydrogenation of substituted benzenes yields substituted cyclohexanes. For example:

$$\underset{\substack{\text{Phenol}\\ \textit{Aromatic}}}{C_6H_5OH} + 3H_2 \xrightarrow[150\ atm.]{Ni,\ 150\text{–}200^\circ} \underset{\substack{\text{Cyclohexanol}\\ \textit{Aliphatic}}}{\begin{array}{ccc} & CH_2 & \\ H_2C & & CHOH \\ | & & | \\ H_2C & & CH_2 \\ & CH_2 & \end{array}}$$

From cyclohexanol many other cyclic compounds containing a six-membered ring can be made.

9.4 Preparation

Preparation of alicyclic hydrocarbons from other aliphatic compounds generally involves two stages: (a) conversion of some open-chain compound or compounds into a compound that contains a ring, a process called *cyclization*; (b) conversion of the cyclic compound thus obtained into the kind of compound that we want: for example, conversion of a cyclic alcohol into a cyclic alkene, or of a cyclic alkene into a cyclic alkane.

Very often, cyclic compounds are made by the *adapting* of a standard method of preparation to the job of closing a ring. For example, we have seen (Sec. 3.17) that the alkyl groups of two alkyl halides can be coupled together through conversion of one halide into an organometallic compound (a lithium dialkylcopper):

$$\begin{array}{llll} CH_3CH_2{-}Cl & \longrightarrow CH_3CH_2{-}M & \searrow & CH_3CH_2 \\ CH_3CH_2{-}Cl & & \nearrow \longrightarrow & CH_3CH_2 \end{array}$$

Ethyl chloride (2 moles) → *n*-Butane

The same method applied to a *di*halide can bring about coupling between two alkyl groups *that are part of the same molecule*:

$$H_2C\begin{cases} CH_2{-}Cl \\ CH_2{-}Cl \end{cases} \xrightarrow{\text{Zn, NaI}} H_2C\begin{cases} CH_2{-}ZnX \\ CH_2{-}X \end{cases} \longrightarrow H_2C\begin{cases} CH_2 \\ | \\ CH_2 \end{cases}$$

1,3-Dichloropropane → Cyclopropane

In this case zinc happens to do a good job. Although this particular method works well only for the preparation of cyclopropane, it illustrates an important principle: the carrying out of what is normally an *intermolecular* (between-molecules) reaction under such circumstances that it becomes an *intramolecular* (within-a-molecule) reaction. As we can see, it involves tying together the ends of a difunctional molecule.

Alicyclic hydrocarbons are prepared from other cyclic compounds (e.g., halides or alcohols) by exactly the same methods that are used for preparing open-chain hydrocarbons from other open-chain compounds.

Problem 9.1 Starting with cyclohexanol (Sec. 9.3), how would you prepare: (a) cyclohexene, (b) 3-bromocyclohexene, (c) 1,3-cyclohexadiene?

Problem 9.2 Bromocyclobutane can be obtained from open-chain compounds. How would you prepare cyclobutane from it?

The most important route to rings of many different sizes is through the important class of reactions called **cycloadditions**: *reactions in which molecules are added together to form rings.* We shall see one example of cycloaddition in Secs. 9.15–9.16, and others later on.

9.5 Reactions

With certain very important and interesting exceptions, alicyclic hydrocarbons undergo the same reactions as their open-chain analogs.

Cycloalkanes undergo chiefly free-radical substitution (compare Sec. 3.19). For example:

$$\text{Cyclopropane} + Cl_2 \xrightarrow{\text{light}} \text{Chlorocyclopropane} + HCl$$

$$\text{Cyclopentane} + Br_2 \xrightarrow{300^\circ} \text{Bromocyclopentane} + HBr$$

Cycloalkenes undergo chiefly addition reactions, both electrophilic and free radical (compare Sec. 6.2); like other alkenes, they can also undergo cleavage and allylic substitution. For example:

$$\text{Cyclohexene} + Br_2 \longrightarrow \text{1,2-Dibromocyclohexane}$$

$$\text{1-Methylcyclopentene} + HI \longrightarrow \text{1-Iodo-1-methylcyclopentane}$$

$$\text{3,5-Dimethylcyclopentene} \xrightarrow{O_3} \xrightarrow{H_2O/Zn} O{=}\overset{H}{\overset{|}{C}}{-}\overset{CH_3}{\overset{|}{CH}}{-}CH_2{-}\overset{CH_3}{\overset{|}{CH}}{-}\overset{H}{\overset{|}{C}}{=}O$$

A dialdehyde

The two smallest cycloalkanes, cyclopropane and cyclobutane, show certain chemical properties that are entirely different from those of the other members of their family. Some of these exceptional properties fit into a pattern and, as we shall see, can be understood in a general way.

The chemistry of bicyclic compounds is even more remarkable, and is right now one of the most intensively studied areas of organic chemistry (Sec. 28.13).

9.6 Reactions of small-ring compounds. Cyclopropane and cyclobutane

Besides the free-radical substitution reactions that are characteristic of cycloalkanes and of alkanes in general, cyclopropane and cyclobutane undergo certain addition reactions. These addition reactions destroy the cyclopropane and cyclobutane ring systems, and yield open-chain products. For example:

Cyclopropane (H_2C–CH_2–H_2C ring):

$\xrightarrow{\text{Ni, } H_2\text{, } 80°}$ $CH_2CH_2CH_2$ with H at each end (H, H) — Propane

$\xrightarrow{Cl_2\text{, } FeCl_3}$ $CH_2CH_2CH_2$ with Cl at each end (Cl, Cl) — 1,3-Dichloropropane

$\xrightarrow{\text{conc. } H_2SO_4}$ $CH_2CH_2CH_2$ with H and OH at the ends — *n*-Propyl alcohol

In each of these reactions a carbon–carbon bond is broken, and the two atoms of the reagent appear at the ends of the propane chain:

Y–Z + cyclopropane (H_2C, CH_2, H_2C) $\longrightarrow$ $CH_2CH_2CH_2$ with Y and Z at the ends

In general, cyclopropane undergoes addition less readily than propylene: chlorination, for example, requires a Lewis acid catalyst to polarize the chlorine molecule (compare Sec. 11.11). Yet the reaction with sulfuric acid and other aqueous protic acids takes place considerably faster for cyclopropane than for propylene. (Odder still, treatment with bromine and $FeBr_3$ yields a grand mixture of bromopropanes.)

Cyclobutane does not undergo most of the ring-opening reactions of cyclopropane; it is hydrogenated, but only under more vigorous conditions than those required for cyclopropane. Thus cyclobutane undergoes addition less readily than cyclopropane and, with some exceptions, cyclopropane less readily than an alkene. The remarkable thing is that these cycloalkanes undergo addition at all.

9.7 Baeyer strain theory

In 1885 Adolf von Baeyer (of the University of Munich) proposed a theory to account for certain aspects of the chemistry of cyclic compounds. The part of his theory dealing with the ring-opening tendencies of cyclopropane and cyclobutane is generally accepted today, although it is dressed in more modern language. Other parts of his theory have been shown to be based on false assumptions, and have been discarded.

Baeyer's argument was essentially the following. In general, when carbon is bonded to four other atoms, the angle between any pair of bonds is the tetrahedral angle 109.5°. But the ring of cyclopropane is a triangle with three angles of 60°, and the ring of cyclobutane is a square with four angles of 90°. In cyclopropane or cyclobutane, therefore, one pair of bonds to each carbon cannot assume the tetrahedral angle, but must be compressed to 60° or 90° to fit the geometry of the ring.

These deviations of bond angles from the "normal" tetrahedral value cause the molecules to be *strained*, and hence to be unstable compared with molecules in which the bond angles are tetrahedral. Cyclopropane and cyclobutane undergo ring-opening reactions since these relieve the strain and yield the more stable open-chain compounds. Because the deviation of the bond angles in cyclopropane (109.5° − 60° = 49.5°) is greater than in cyclobutane (109.5° − 90° = 19.5°), cyclopropane is more highly strained, more unstable, and more prone to undergo ring-opening reactions than is cyclobutane.

The angles of a regular pentagon (108°) are very close to the tetrahedral angle (109.5°), and hence cyclopentane should be virtually free of angle strain. The angles of a regular hexagon (120°) are somewhat larger than the tetrahedral angle, and hence, Baeyer proposed (incorrectly), there should be a certain amount of strain in cyclohexane. Further, he suggested (incorrectly) that as one proceeded to cycloheptane, cyclooctane, etc., the deviation of the bond angles from 109.5° would become progressively larger, and the molecules would become progressively more strained.

Thus Baeyer considered that rings smaller or larger than cyclopentane or cyclohexane were unstable; it was because of this instability that the three- and four-membered rings underwent ring-opening reactions; it was because of this instability that great difficulty had been encountered in the synthesis of the larger rings. How does Baeyer's strain theory agree with the facts?

9.8 Heats of combustion and relative stabilities of the cycloalkanes

We recall (Sec. 2.6) that the heat of combustion is the quantity of heat evolved when one mole of a compound is burned to carbon dioxide and water. Like heats of hydrogenation (Secs. 6.4 and 8.16), heats of combustion can often furnish valuable information about the relative stabilities of organic compounds. Let us see if the heats of combustion of the various cycloalkanes support Baeyer's proposal that rings smaller or larger than cyclopentane and cyclohexane are unstable.

Examination of the data for a great many compounds has shown that the heat of combustion of an aliphatic hydrocarbon agrees rather closely with that calculated by assuming a certain characteristic contribution from each structural unit. For open-chain alkanes each methylene group, —CH_2—, contributes very close to 157.4 kcal/mole to the heat of combustion. Table 9.2 lists the heats of combustion that have been measured for some of the cycloalkanes.

Table 9.2 HEATS OF COMBUSTION OF CYCLOALKANES

Ring size	Heat of combustion per CH_2, kcal/mole	Ring size	Heat of combustion per CH_2, kcal/mole
3	166.6	10	158.6
4	164.0	11	158.4
5	158.7	12	157.6
6	157.4	13	157.8
7	158.3	14	157.4
8	158.6	15	157.5
9	158.8	17	157.2
	Open-chain 157.4		

We notice that for cyclopropane the heat of combustion per —CH_2— group is 9 kcal higher than the open-chain value of 157.4; for cyclobutane it is 7 kcal higher than the open-chain value. Whatever the compound in which it occurs, a —CH_2— group yields the same products on combustion: carbon dioxide and water.

$$-CH_2- + \tfrac{3}{2}O_2 \longrightarrow CO_2 + H_2O + \text{heat}$$

If cyclopropane and cyclobutane evolve more energy per —CH_2— group than an open-chain compound, it can mean only that they *contain* more energy per —CH_2— group. In agreement with the Baeyer angle-strain theory, then, cyclopropane and cyclobutane are less stable than open-chain compounds; it is reasonable to suppose that their tendency to undergo ring-opening reactions is related to this instability.

According to Baeyer, rings larger than cyclopentane and cyclohexane also should be unstable, and hence also should have high heats of combustion; furthermore relative instability—and, with it, heat of combustion—should increase steadily with ring size. However, we see from Table 9.2 that almost exactly the opposite is true. For none of the rings larger than four carbons does the heat of combustion per —CH_2— deviate much from the open-chain value of 157.4. Indeed, one of the biggest deviations is for Baeyer's "most stable" compound, cyclopentane: 1.3 kcal per —CH_2—, or 6.5 kcal for the molecule. Rings containing seven to eleven carbons have about the same value as cyclopentane, and when we reach rings of twelve carbons or more, heats of combustion are indistinguishable from the open-chain values. Contrary to Baeyer's theory, then, none of these rings is appreciably less stable than open-chain compounds, and the larger ones are completely free of strain. Furthermore, once they have been synthesized, these large-ring cycloalkanes show little tendency to undergo the ring-opening reactions characteristic of cyclopropane and cyclobutane.

What is wrong with Baeyer's theory that it does not apply to rings larger than four members? Simply this: the angles that Baeyer used for each ring were based on the assumption that the rings were *flat*. For example, the angles of a regular (flat) hexagon are 120°, the angles for a regular decagon are 144°. But the cyclohexane ring is not a regular hexagon, and the cyclodecane ring is not a regular decagon. These rings are not flat, but are puckered (see Fig. 9.1) so that each bond angle of carbon can be 109.5°.

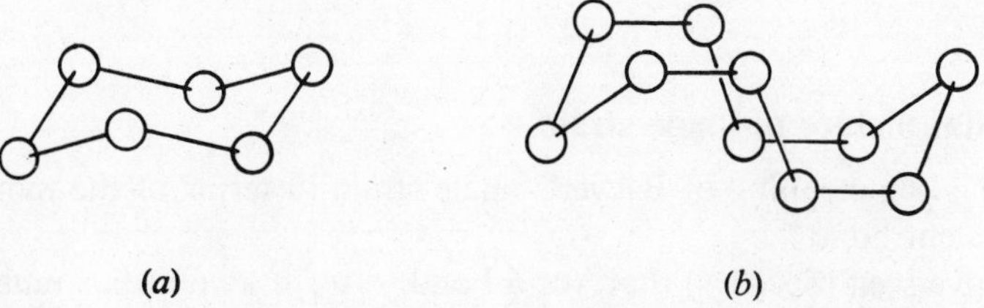

Figure 9.1. Puckered rings. (*a*) Cyclohexane. (*b*) Cyclodecane.

A three-membered ring must be planar, since three points (the three carbon nuclei) define a plane. A four-membered ring need not be planar, but puckering here would increase (angle) strain. A five-membered ring need not be planar, but in this case a planar arrangement would permit the bond angles to have nearly the tetrahedral value. All rings larger than this are puckered. (Actually, as we shall see, cyclobutane and cyclopentane are puckered, too, but this is *in spite of* increased angle strain.)

If large rings are stable, why are they difficult to synthesize? Here we encounter Baeyer's second false assumption. The fact that a compound is difficult to synthesize does not necessarily mean that it is unstable. The closing of a ring requires that two ends of a chain be brought close enough to each other for a bond to form. The larger the ring one wishes to synthesize, the longer must be the chain from which it is made, and the less is the likelihood of the two ends of the chain approaching each other. Under these conditions the end of one chain is more likely to encounter the end of a *different* chain, and thus yield an entirely different product (see Fig. 9.2).

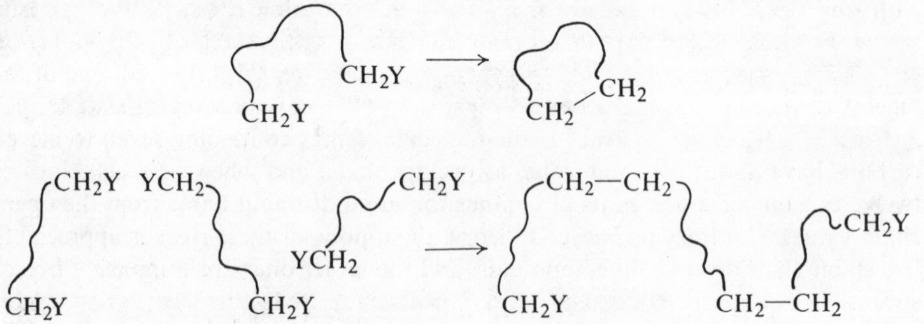

CH_2Y CH_2Y → CH_2—CH_2

CH_2Y YCH_2 CH_2Y CH_2Y YCH_2 CH_2Y → CH_2—CH_2 CH_2Y CH_2Y CH_2—CH_2

Figure 9.2. Ring closure (upper) *vs.* chain lengthening (lower).

The methods that are used successfully to make large rings take this fact into consideration. Reactions are carried out in highly dilute solutions where collisions between two different chains are unlikely; under these conditions the ring-closing reaction, although slow, is the principal one. Five- and six-membered rings are the kind most commonly encountered in organic chemistry because they are large enough to be free of angle strain, and small enough that ring closure is likely.

9.9 Orbital picture of angle strain

What is the meaning of Baeyer's angle strain in terms of the modern picture of the covalent bond?

We have seen (Sec. 1.8) that, for a bond to form, two atoms must be located so that an orbital of one overlaps an orbital of the other. For a given pair of atoms, the greater the overlap of atomic orbitals, the stronger the bond. When carbon is bonded to four other atoms, its bonding orbitals (sp^3 orbitals) are directed to the corners of a tetrahedron; the angle between any pair of orbitals is thus 109.5°. Formation of a bond with another carbon atom involves overlap of one of these sp^3 orbitals with a similar sp^3 orbital of the other carbon atom.

This overlap is most effective, and hence the bond is strongest, when the two atoms are located so that an sp^3 orbital of each atom points toward the other atom. This means that when carbon is bonded to two other carbon atoms the C—C—C bond angle should be 109.5°.

In cyclopropane, however, the C—C—C bond angle cannot be 109.5°, but instead must be 60°. As a result, the carbon atoms cannot be located to permit

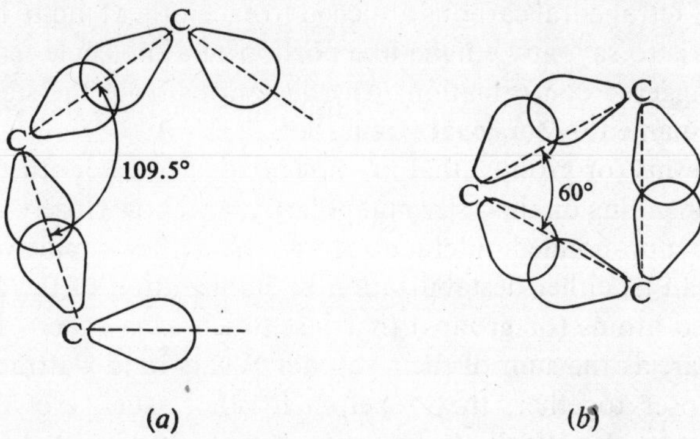

Figure 9.3. Angle strain. (*a*) Maximum overlap permitted for open-chain or large-ring compounds. (*b*) Poor overlap for cyclopropane ring. Bent bonds have much *p* character.

their sp^3 orbitals to point toward each other (see Fig. 9.3). There is less overlap and the bond is weaker than the usual carbon–carbon bond.

The decrease in stability of a cyclic compound attributed to *angle strain* is due to poor overlap of atomic orbitals in the formation of the carbon–carbon bonds.

On the basis of quantum mechanical calculations, C. A. Coulson and W. A. Moffitt (of Oxford University) proposed *bent bonds* between carbon atoms of cyclopropane rings; this idea is supported by electron density maps based on X-ray studies. Carbon uses sp^2 orbitals for carbon–hydrogen bonds (which are short and strong), and orbitals with much *p* character (sp^4 to sp^5) for the carbon–carbon bonds. The high *p* character of these carbon–carbon bonds, and their location—largely outside the ring—seems to underlie much of the unusual chemistry of these rings. The carbon–carbon bond orbitals can overlap orbitals on adjacent atoms; the resulting delocalization is responsible for the effects of cyclopropyl as a substituent. The carbon–carbon bond orbitals provide a site for the attack by acids that is the first step of ring-opening. (Indeed, "edge-protonated" cyclopropanes seem to be key intermediates in many reactions that do not, on the surface, seem to involve cyclopropane rings.)

Ring-opening is *due to* the weakness of the carbon–carbon bonds, but the *way in which it happens* reflects the unusual nature of the bonds; all this stems ultimately from the geometry of the rings and angle strain.

9.10 Factors affecting stability of conformations

To go more deeply into the chemistry of cyclic compounds, we must use conformational analysis (Sec. 4.20). As preparation for that, let us review the factors that determine the stability of a conformation.

Any atom tends to have bond angles that match those of its bonding orbitals: tetrahedral (109.5°) for sp^3-hybridized carbon, for example. Any deviations from the "normal" bond angles are accompanied by **angle strain** (Secs. 9.8–9.9).

Any pair of tetrahedral carbons attached to each other tend to have their bonds staggered. That is to say, any ethane-like portion of a molecule tends, like ethane, to take up a staggered conformation. Any deviations from the staggered arrangement are accompanied by **torsional strain** (Sec. 3.3).

Any two atoms (or groups) that are not bonded to each other can interact in several ways, depending on their size and polarity, and how closely they are brought together. These non-bonded interactions can be either repulsive or attractive, and the result can be either destabilization or stabilization of the conformation.

Non-bonded atoms (or groups) that just touch each other—that is, that are about as far apart as the sum of their van der Waals radii—attract each other. If brought any closer together, they repel each other: such crowding together is accompanied by **van der Waals strain (steric strain)** (Secs. 1.19, 3.5).

Non-bonded atoms (or groups) tend to take positions that result in the most favorable **dipole–dipole interactions**: that is, positions that minimize dipole–dipole repulsions or maximize dipole–dipole attractions. (A particularly powerful attraction results from the special kind of dipole–dipole interaction called the **hydrogen bond** (Sec. 1.19).

All these factors, working together or opposing each other, determine the net stability of a conformation. To figure out what the most stable conformation of a particular molecule should be, one ideally should consider all possible combinations of bond angles, angles of rotation, and even bond lengths, and see which combination results in the lowest energy content. A start in this direction—feasible only by use of computers—has been made, most notably by Professor James B. Hendrickson (of Brandeis University).

Both calculations and experimental measurements show that the final result is a compromise, and that few molecules have the idealized conformations that we assign them and, for convenience, usually work with. For example, probably no tetravalent carbon compound—except one with four identical substituents—has *exactly* tetrahedral bond angles: a molecule accepts a certain amount of angle strain to relieve van der Waals strain or dipole–dipole interaction. In the *gauche* conformer of *n*-butane (Sec. 3.5), the dihedral angle between the methyl groups is not 60°, but almost certainly larger: the molecule accepts some torsional strain to ease van der Waals strain between the methyl groups.

9.11 Conformations of cycloalkanes

Let us look more closely at the matter of puckered rings, starting with cyclohexane, the most important of the cycloalkanes. Let us make a model of the molecule, and examine the conformations that are free of angle strain.

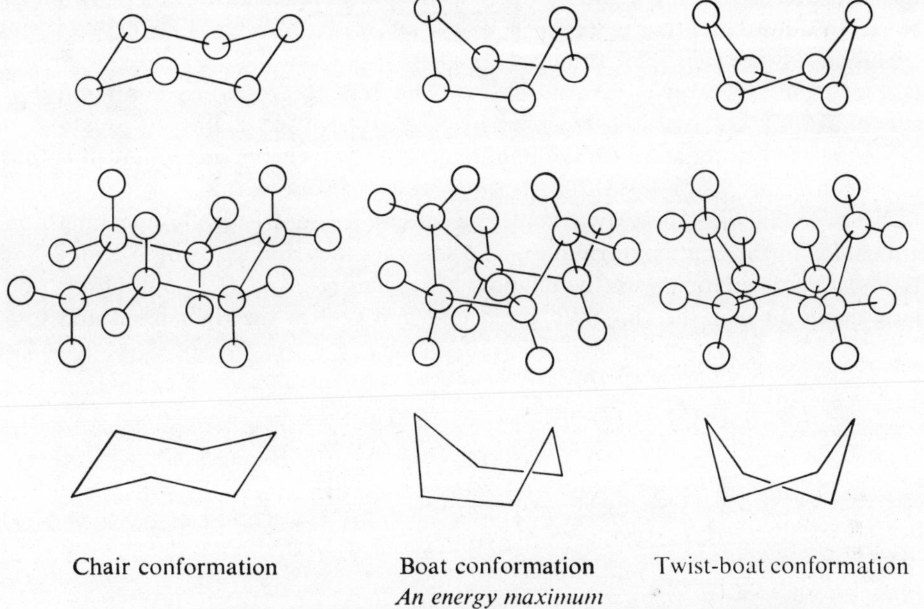

Figure 9.4. Conformations of cyclohexane that are free of angle strain.

First, there is the **chair form** (Fig. 9.4). If we sight along each of the carbon–carbon bonds in turn, we see in every case perfectly staggered bonds:

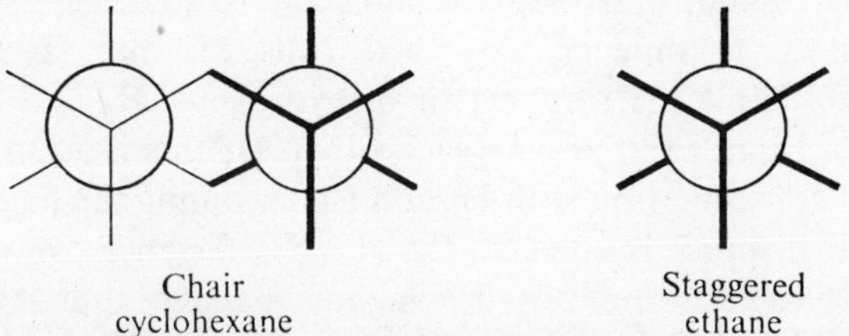

The conformation is thus not only free of angle strain but free of torsional strain as well. It lies at an energy minimum, and is therefore a conformational isomer. *The chair form is the most stable conformation of cyclohexane, and, indeed, of nearly every derivative of cyclohexane.*

Next, let us flip the "left" end of the molecule up (Fig. 9.4) to make the *boat conformation*. (Like all the transformations we shall carry out in this section, this involves only rotations about single bonds; what we are making are indeed conformations.) This is not a very happy arrangement. Sighting along either of two carbon–carbon bonds, we see sets of exactly eclipsed bonds,

Flagpole bonds

Boat cyclohexane

Eclipsed ethane

and hence we expect considerable torsional strain: as much as in *two* ethane molecules. In addition, there is van der Waals strain due to crowding between the "flagpole" hydrogens, which lie only 1.83 A apart, considerably closer than the sum of their van der Waals radii (2.5 A). The boat conformation is a good deal less stable (7.1 kcal/mole, it has been calculated) than the chair conformation. It is believed to lie, not at an energy minimum, but at an energy maximum; it is thus not a conformer, but a transition state between two conformers.

Now, what are these two conformers that lie—energetically speaking—on either side of the boat conformation? To see what they are, let us hold a model of the boat conformation with the flagpole hydrogens (H_a and H_b) pointing up, and look down through the ring. We grasp C–2 and C–3 in the right hand and C–5

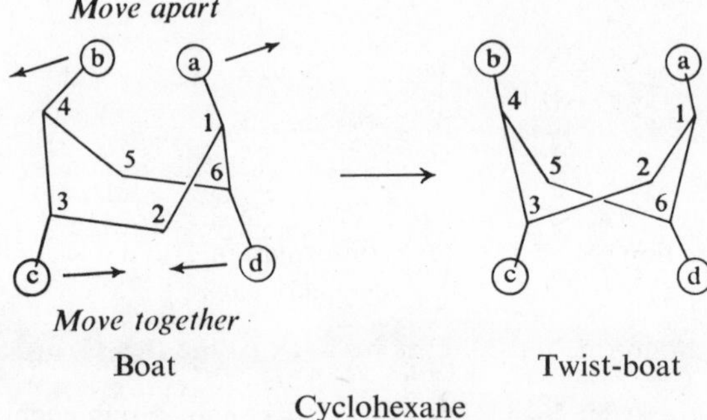

Boat Twist-boat

Cyclohexane

and C–6 in the left hand, and *twist* the molecule so that, say, C–3 and C–6 go *down*, and C–2 and C–5 come *up*. As we do this, H_a and H_b move diagonally apart, and we see (below the ring) a pair of hydrogens, H_c and H_d (on C–3 and C–6, respectively), begin to approach each other. (If this motion is continued, we make a new boat conformation with H_c and H_d becoming the flagpole hydrogens.) When the H_a—H_b distance is equal to the H_c—H_d distance, we stop and examine the molecule. We have minimized the flagpole–flagpole interactions, and at the same time have partly relieved the torsional strain at the C_2—C_3 and C_5—C_6 bonds.

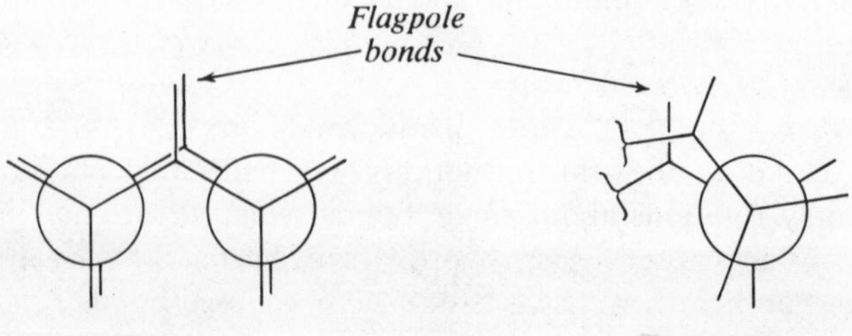

Flagpole bonds

Boat cyclohexane

Twist-boat cyclohexane

This new configuration is the **twist-boat form.** It is a conformer, lying at an energy minimum 5.5 kcal above the chair conformation. The twist-boat conformer is separated from another, enantiomeric twist-boat conformer by an energy barrier 1.6 kcal high, at the top of which is the boat conformation.

Between the chair form and the twist-boat form lies the highest barrier of all: a transition state conformation (the *half-chair*) which, with angle strain and torsional strain, lies about 11 kcal above the chair form.

The overall relationships are summarized in Fig. 9.5. Equilibrium exists between the chair and twist-boat forms, with the more stable chair form being favored—10,000 to 1 at room temperature.

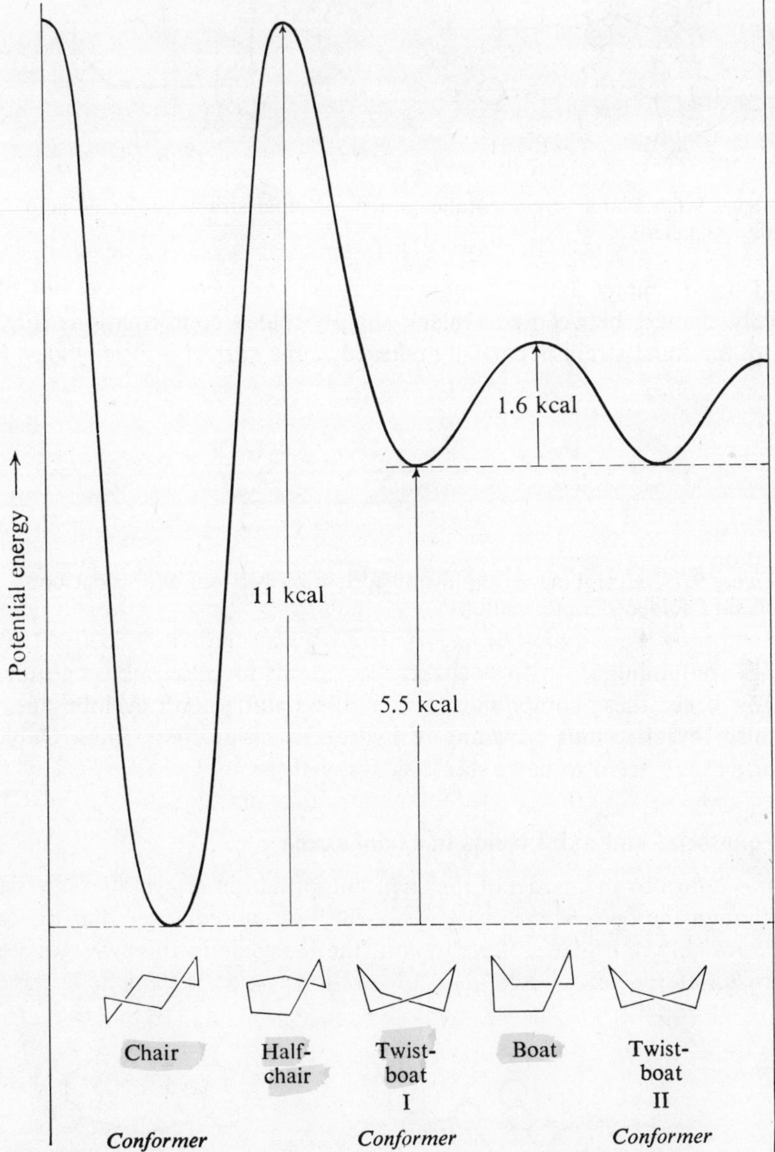

Figure 9.5. Potential energy relationships among conformations of cyclohexane.

If chair cyclohexane is, conformationally speaking, the perfect specimen of a cycloalkane, planar cyclopentane (Fig. 9.6) must certainly be the poorest: there is exact bond eclipsing between every pair of carbons. To (partially) relieve this torsional strain, cyclopentane takes on a slightly puckered conformation, even at the cost of a little angle strain. (See also Problem 10, p. 316.)

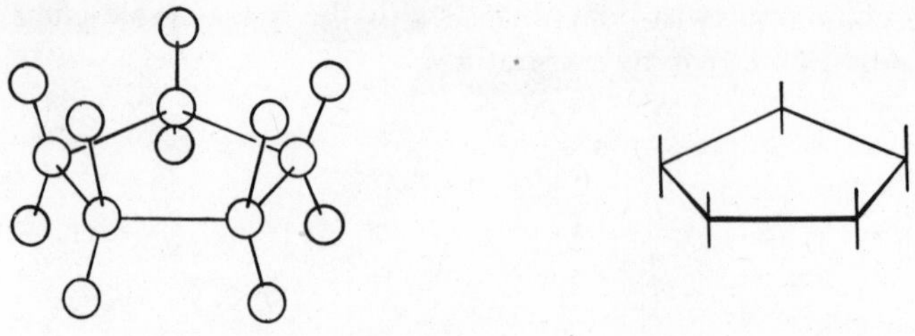

Figure 9.6. Planar cyclopentane: much torsional strain. Molecule actually puckered.

Evidence of many kinds strongly indicates that cyclobutane is not planar, but rapidly changes between equivalent, slightly folded conformations (Fig. 9.7). Here, too, torsional strain is partially relieved at the cost of a little angle strain.

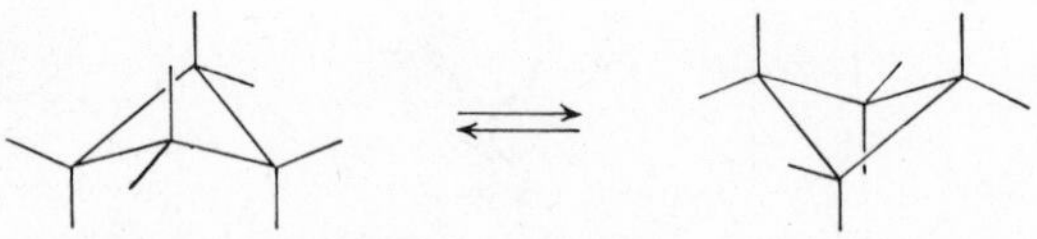

Figure 9.7. Cyclobutane: rapid transformation between equivalent nonplanar "folded" conformations.

Rings containing seven to twelve carbon atoms are also subject to torsional strain, and hence these compounds, too, are less stable than cyclohexane; scale models also reveal serious crowding of hydrogens inside these rings. Only quite large ring systems seem to be as stable as cyclohexane.

9.12 Equatorial and axial bonds in cyclohexane

Let us return to the model of the chair conformation of cyclohexane (see Fig. 9.8). Although the cyclohexane ring is not flat, we can consider that the carbon atoms lie roughly in a plane. If we look at the molecule in this way, we see that the hydrogen atoms occupy two kinds of position: six hydrogens lie in the plane,

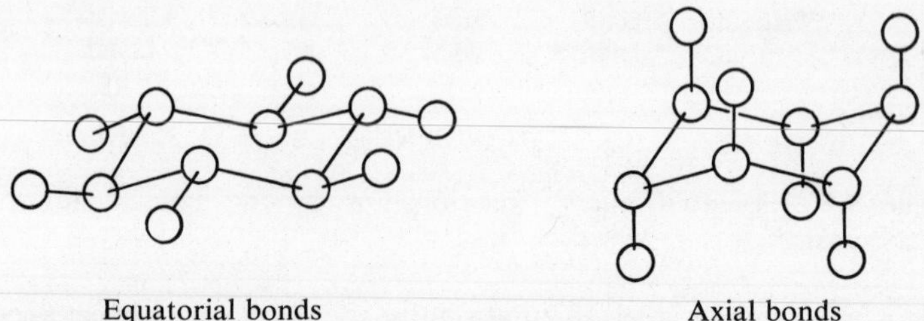

Figure 9.8. Equatorial and axial bonds in cyclohexane.

while six hydrogens lie above or below the plane. The bonds holding the hydrogens that are in the plane of the ring lie in a belt about the "equator" of the ring, and are called **equatorial bonds.** The bonds holding the hydrogen atoms that are above and below the plane are pointed along an axis perpendicular to the plane and are called **axial bonds.** In the chair conformation each carbon atom has one equatorial bond and one axial bond.

Cyclohexane itself, in which only hydrogens are attached to the carbon atoms, is not only free of angle strain and torsional strain, but free of van der Waals strain as well. Hydrogens on adjacent carbons are the same distance apart (2.3 A) as in (staggered) ethane and, if anything, feel mild van der Waals attraction for each other. We notice that the three axial hydrogens on the same side of the molecule are thrown rather closely together, despite the fact that they are attached to alternate carbon atoms; as it happens, however, they are the same favorable distance apart (2.3 A) as the other hydrogens are.

If, now, a hydrogen is replaced by a larger atom or group, crowding occurs. The most severe crowding is among atoms held by the three axial bonds on the same side of the molecule; the resulting interaction is called **1,3-diaxial interaction.** Except for hydrogen, *a given atom or group has more room in an equatorial position than in an axial position.*

As a simple example of the importance of 1,3-diaxial interactions, let us consider methylcyclohexane. In estimating relative stabilities of various conformations of this compound, we must focus our attention on methyl, since it is the largest substituent on the ring and hence the one most subject to crowding. There are two

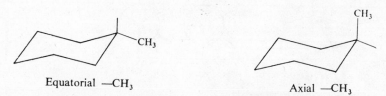

CH3

CH3

Equatorial —CH3

Axial —CH3

Figure 9.9. Chair conformations of methylcyclohexane.

possible chair conformations (see Fig. 9.9), one with $-CH_3$ in an equatorial position, the other with $-CH_3$ in an axial position. As shown in Fig. 9.10, the two axial hydrogens (on C-3 and C-5) approach the axial $-CH_3$ (on C-1) more closely than any hydrogens approach the equatorial $-CH_3$. We would expect

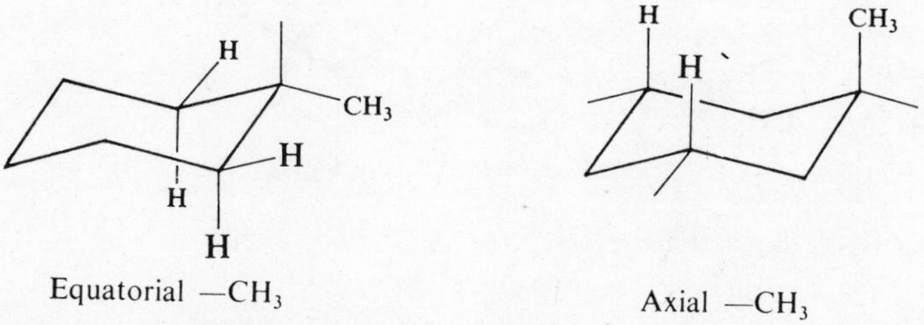

H

CH3

H

H

H

H

H

CH3

H

Equatorial —CH3

Axial —CH3

Figure 9.10. 1,3-Diaxial interaction in methylcyclohexane. Axial $-CH_3$ more crowded than equatorial $-CH_3$.

the equatorial conformation to be the more stable, and it is, by about 1.8 kcal. Most molecules (about 95% at room temperature) exist in the conformation with methyl in the uncrowded equatorial position.

In an equatorial position, we see, $-CH_3$ points *away from* its nearest neighbors: the two hydrogens—one axial, and one equatorial—on the adjacent carbons. This is not true of $-CH_3$ in an axial position, since it is held by a bond that is *parallel to* the bonds holding its nearest neighbors: the two axial hydrogens.

Conformational analysis can account not only for the fact that one conformation is more stable than another, but often—with a fair degree of accuracy—for just *how much* more stable it is. We have attributed the 1.8-kcal energy difference between the two conformations of methylcyclohexane to 1,3-diaxial interactions between a methyl group and *two* hydrogens. If, on that basis, we assign a value of 0.9 kcal/mole to each 1,3-diaxial methyl–hydrogen interaction, we shall find that we can account amazingly well for the energy differences between conformations of a variety of cyclohexanes containing more than one methyl group.

We notice that 0.9 kcal is nearly the same value that we earlier (Sec. 3.5) assigned to a *gauche* interaction in *n*-butane; examination of models shows that this is not just accidental.

Let us make a model of the conformation of methylcyclohexane with axial methyl. If we hold it so that we can sight along the $C_1—C_2$ bond, we see something like this, represented by a Newman projection:

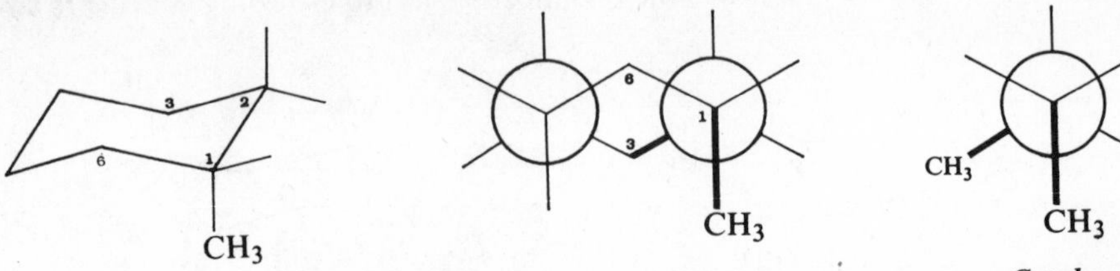

3 2 6 1 CH3 6 1 3 CH3 CH3 CH3

Axial $-CH_3$

Gauche
n-Butane

The methyl group and C–3 of the ring have the same relative locations as the two methyl groups in the *gauche* conformation of *n*-butane (Sec. 3.5). If we now sight along the $C_1—C_6$ bond, we see a similar arrangement but with C–5 taking the place of C–3.

Next, let us make a model of the conformation with equatorial methyl. This time, if we sight along the $C_1—C_2$ bond, we see this:

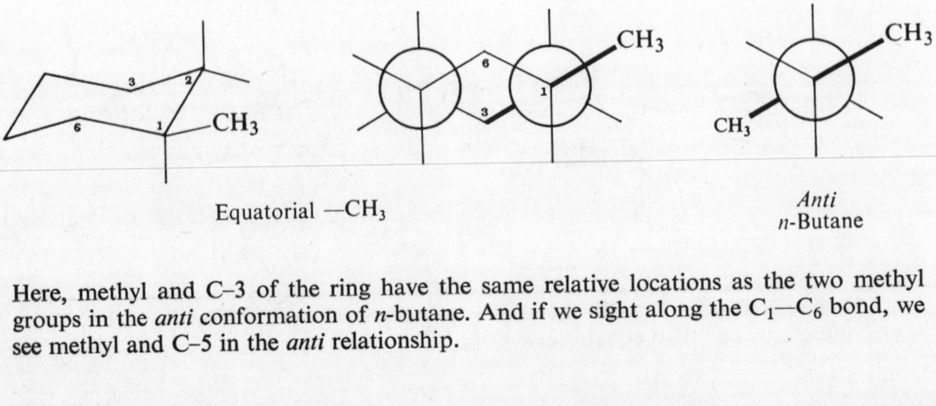

3 2 6 1 CH3 CH3 6 1 3 CH3 CH3

Equatorial $-CH_3$

Anti
n-Butane

Here, methyl and C–3 of the ring have the same relative locations as the two methyl groups in the *anti* conformation of *n*-butane. And if we sight along the $C_1—C_6$ bond, we see methyl and C–5 in the *anti* relationship.

Thus, for each 1,3-diaxial methyl–hydrogen interaction there is a "butane-*gauche*" interaction between the methyl group and a carbon atom of the ring. Of the two approaches, however, looking for 1,3-diaxial interactions is much the easier and has the advantage, when we study substituents other than methyl, of focusing our attention on the sizes of the groups being crowded together.

In general, then, it has been found that (a) chair conformations are more stable than twist conformations, and (b) the most stable chair conformations are those in which the largest groups are in equatorial positions. There are exceptions to both these generalizations (which we shall encounter later in problems), but the exceptions are understandable ones.

Problem 9.3 For other alkylcyclohexanes the difference in energy between equatorial and axial conformations has been found to be: ethyl, 1.9 kcal/mole; isopropyl, 2.1 kcal/mole; and *tert*-butyl, more than 5 kcal/mole. Using models, can you account for the big increase at *tert*-butyl? (*Hint:* Don't forget freedom of rotation about *all* the single bonds.)

9.13 Stereoisomerism of cyclic compounds: *cis*- and *trans*-isomers

Let us turn for the moment from conformational analysis, and look at configurational isomerism in cyclic compounds.

We shall begin with the glycol of cyclopentene, 1,2-cyclopentanediol. Using models, we find that we can arrange the atoms of this molecule as in I, in which both hydroxyls lie below (or above) the plane of the ring, and as in II, in which one hydroxyl lies above and the other lies below the plane of the ring.

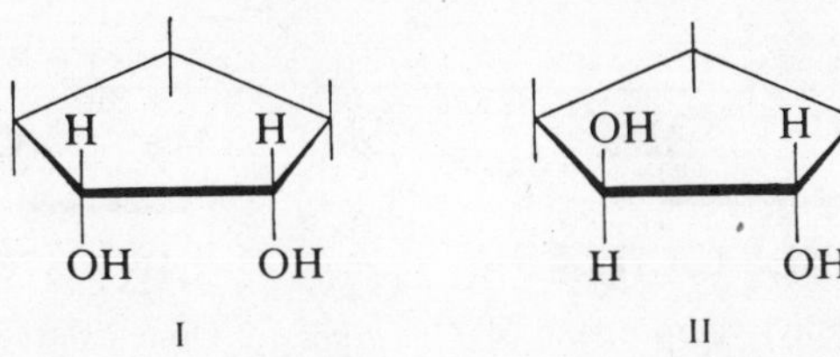

I *cis*-1,2-Cyclopentanediol

II *trans*-1,2-Cyclopentanediol

I and II cannot be superimposed, and hence are isomers. They differ only in the way their atoms are oriented in space, and hence are stereoisomers. No amount of rotation about bonds can interconvert I and II, and hence they are not conformational isomers. They are configurational isomers; they are interconverted only by breaking of bonds, and hence are isolable. They are not mirror images, and hence are diastereomers; they should, therefore, have different physical properties, as the two glycols actually have. Configuration I is designated the *cis*-configuration, and II is designated the *trans*-configuration. (Compare *cis*- and *trans*-alkenes, Sec. 5.6.)

Problem 9.4 You have two bottles labeled "1,2-Cyclopentanediol," one containing a compound of m.p. 30°, the other a compound of m.p. 55°; both compounds are optically inactive. How could you decide, beyond any doubt, which bottle should be labeled "*cis*" and which "*trans*"?

Problem 9.5 (a) Starting from cyclopentanol, outline a synthesis of stereochemically pure *cis*-1,2-cyclopentanediol. (b) Of stereochemically pure *trans*-1,2-cyclopentanediol.

Stereoisomerism of this same sort should be possible for compounds other than glycols, and for rings other than cyclopentane. Some examples of isomers that have been isolated are:

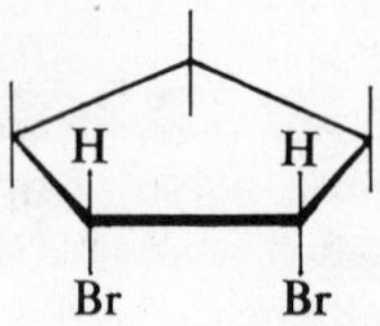

cis-1,2-Dibromocyclopentane

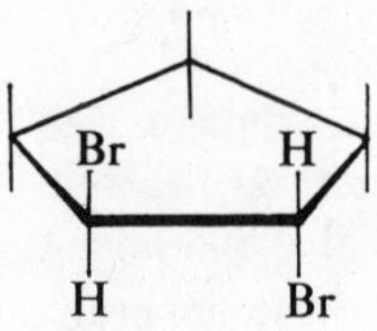

trans-1,2-Dibromocyclopentane

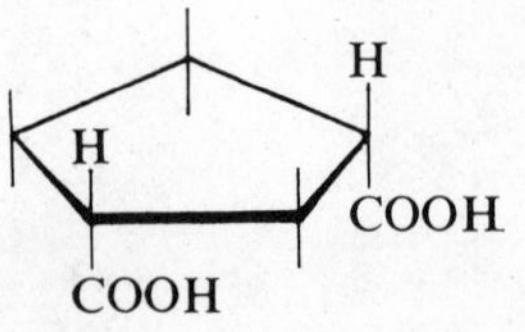

cis-1,3-Cyclopentanedicarboxylic acid

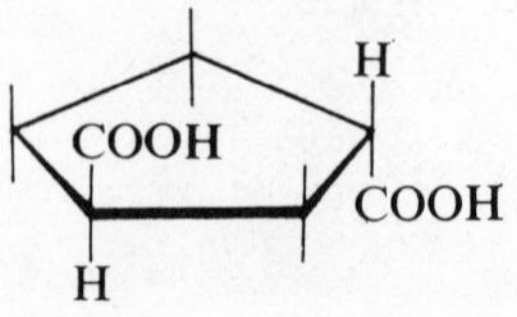

trans-1,3-Cyclopentanedicarboxylic acid

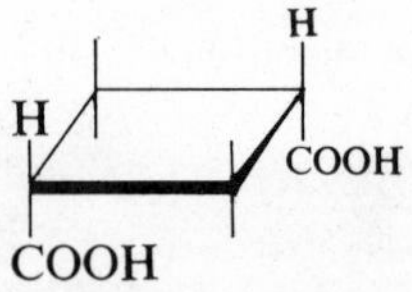

cis-1,3-Cyclobutanedicarboxylic acid

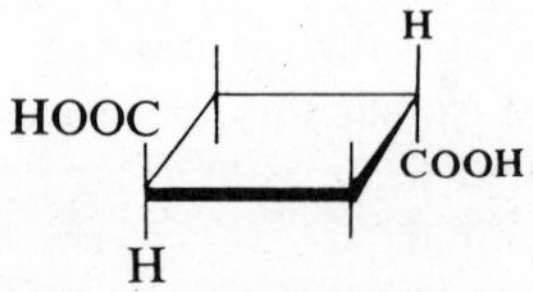

trans-1,3-Cyclobutanedicarboxylic acid

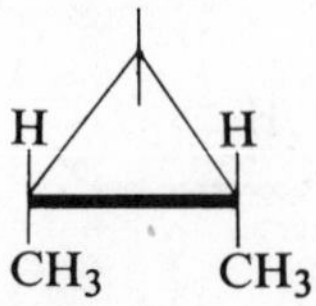

cis-1,2-Dimethylcyclopropane

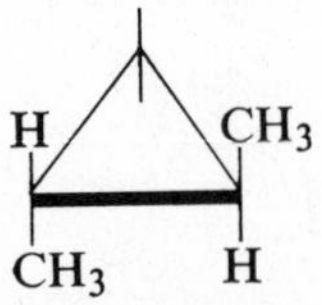

trans-1,2-Dimethylcyclopropane

If we examine models of *cis*- and *trans*-1,2-cyclopentanediol more closely, we find that each compound contains two chiral centers. We know (Sec. 4.18) that compounds containing more than one chiral center are often—but not always—chiral. Are these glycols chiral? As always, to test for possible chirality, we construct a model of the molecule and a model of its mirror image, and see if the two are superimposable. When we do this for the *trans*-glycol, we find that the models

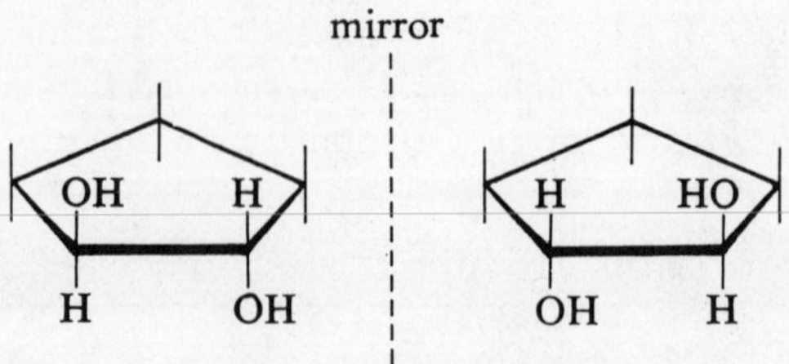

Not superimposable
Enantiomers: resolvable
trans-1,2-Cyclopentanediol

are not superimposable. The *trans* glycol is chiral, and the two models we have constructed therefore correspond to enantiomers. Next, we find that the models are not interconvertible by rotation about single bonds. They therefore represent, not conformational isomers, but configurational isomers; they should be capable of isolation—*resolution*—and, when isolated, each should be optically active.

Next let us look at *cis*-1,2-cyclopentanediol. This, too, contains two chiral centers; is it also chiral? This time we find that a model of the molecule and a model of its mirror image *are* superimposable. In spite of its chiral centers, *cis*-1,2-cyclopentanediol is not chiral; it cannot exist in two enantiomeric forms, and cannot be optically active. It is a *meso* compound.

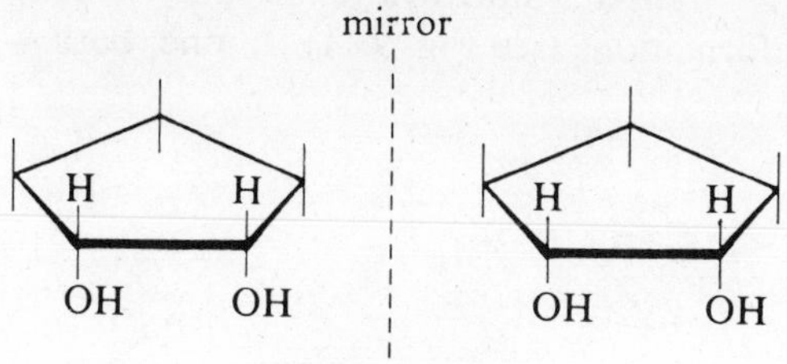

mirror

H H | H H

OH OH | OH OH

Superimposable
A meso compound
cis-1,2-Cyclopentanediol

We might have recognized *cis*-1,2-cyclopentanediol as a *meso* structure on sight from the fact that one half of the molecule is the mirror image of the other half (Sec. 4.18):

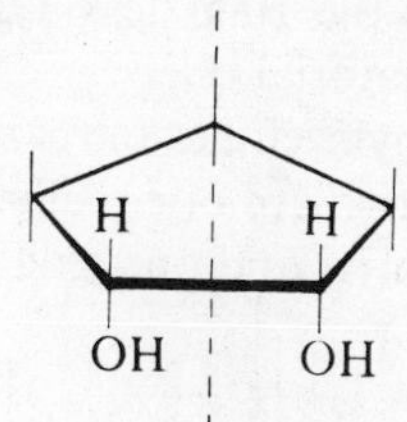

A meso compound
cis-1,2-Cyclopentanediol

Thus, of the two 1,2-cyclopentanediols obtainable from cyclopentene, only one is separable into enantiomers, that is, is *resolvable*; this must necessarily be the *trans*-glycol. The other glycol is a single, inactive, nonresolvable compound, and it must have the *cis* configuration.

What is the relationship between the *meso cis*-glycol and either of the enantiomeric *trans*-glycols? They are *diastereomers*, since they are stereoisomers that are not enantiomers.

Problem 9.6 Five of the eight structures shown at the top of p. 302 are achiral. Which are these?

9.14 Stereoisomerism of cyclic compounds. Conformational analysis

So far, we have described the relative positions of groups in *cis*- and *trans*-isomers in terms of flat rings: both groups are below (or above) the plane of the ring, or one group is above and the other is below the plane of the ring. In view of what we have said about puckering, however, we realize that this is a highly simplified picture even for four- and five-membered rings, and for six-membered rings is quite inaccurate.

Let us apply the methods of conformational analysis to the stereochemistry of cyclohexane derivatives; and, since we are already somewhat familiar with interactions of the methyl group, let us use the dimethylcyclohexanes as our examples.

If we consider only the more stable, chair conformations, we find that a particular molecule of *trans*-1,2-dimethylcyclohexane, to take our first example, can exist in two conformations (see Fig. 9.11). In one, both $—CH_3$ groups are in

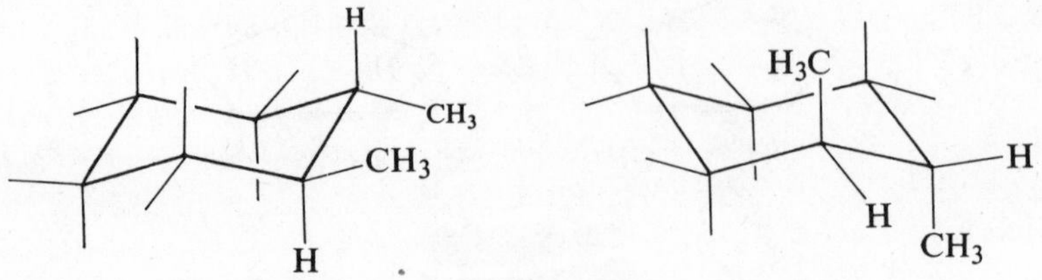

Diequatorial Diaxial

Figure 9.11. Chair conformations of *trans*-1,2-dimethylcyclohexane.

equatorial positions, and in the other, both $—CH_3$ groups are in axial positions. Thus, we see, the two $—CH_3$ groups of the *trans*-isomer are not necessarily on opposite sides of the ring; in fact, because of lesser crowding between $—CH_3$ groups and axial hydrogens of the ring (less 1,3-diaxial interaction), the more stable conformation is the diequatorial one.

A molecule of *cis*-1,2-dimethylcyclohexane can also exist in two conformations (see Fig. 9.12). In this case, the two are of equal stability (they are mirror images) since in each there is one equatorial and one axial $—CH_3$ group.

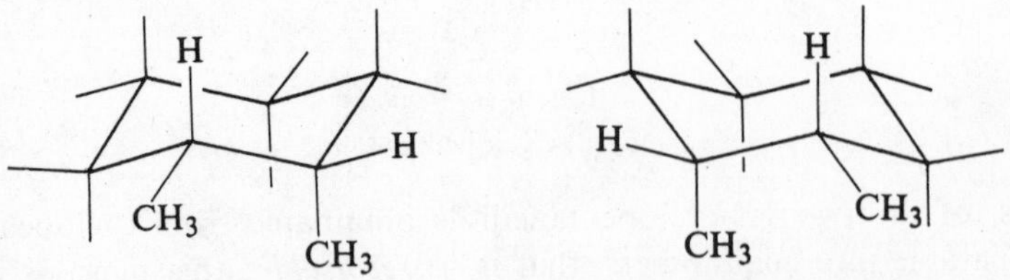

Equatorial-axial Axial-equatorial

Figure 9.12. Chair conformations of *cis*-1,2-dimethylcyclohexane.

In the most stable conformation of *trans*-1,2-dimethylcyclohexane, both $—CH_3$ groups occupy uncrowded equatorial positions. In either conformation of the *cis*-1,2-dimethylcyclohexane, only one $—CH_3$ group can occupy an equatorial position. It is not surprising to find that *trans*-1,2-dimethylcyclohexane is more stable than *cis*-1,2-dimethylcyclohexane.

It is interesting to note that in the most stable conformation (diequatorial) of the *trans*-isomer, the $—CH_3$ groups are exactly the same distance apart as they are in either conformation of the *cis*-isomer. Clearly, it is not repulsion between the $-CH_3$ groups—as one might incorrectly infer from planar representations—that causes the difference in stability between the *trans*- and *cis*-isomers: the cause is 1,3-diaxial interactions (Sec. 9.12).

Now, just *how much* more stable is the *trans*-isomer? In the *cis*-1,2-dimethylcyclohexane there is one axial methyl group, which means *two* 1,3-diaxial methyl–hydrogen interactions: one with each of two hydrogen atoms. (Or, what is equivalent (Sec. 9.12), there are two butane-*gauche* interactions between the methyl groups and carbon atoms of the ring.) In addition, there is one butane-*gauche* interaction between the two methyl groups. On the basis of 0.9 kcal for each 1,3-diaxial methyl–hydrogen interaction or butane-*gauche* interaction, we calculate a total of 2.7 kcal of van der Waals strain for the *cis*-1,2-dimethylcyclohexane. In the (diequatorial) *trans*-isomer there are no 1,3-diaxial methyl–hydrogen interactions, but there is one butane-*gauche* interaction between the methyl groups; this confers 0.9 kcal of van der Waals strain on the molecule. We subtract 0.9 kcal from 2.7 kcal and conclude that the *trans*-isomer should be more stable than the *cis*-isomer by 1.8 kcal/mole, in excellent agreement with the measured value of 1.87 kcal.

Problem 9.7 Compare stabilities of the possible chair conformations of: (a) *cis*-1,2-dimethylcyclohexane; (b) *trans*-1,2-dimethylcyclohexane; (c) *cis*-1,3-dimethylcyclohexane; (d) *trans*-1,3-dimethylcyclohexane; (e) *cis*-1,4-dimethylcyclohexane; (f) *trans*-1,4-dimethylcyclohexane. (g) On the basis of 0.9 kcal/mole per 1,3-diaxial methyl–hydrogen interaction, predict (where you can) the potential energy difference between the members of each pair of conformations.

Problem 9.8 On theoretical grounds, K. S. Pitzer (then at the University of California) calculated that the energy difference between the conformations of *cis*-1,3-dimethylcyclohexane should be about 5.4 kcal, much larger than that between the chair conformations of *trans*-1,2-dimethylcyclohexane or of *trans*-1,4-dimethylcyclohexane. (a) What special factor must Pitzer have recognized in the *cis*-1,3-isomer? (b) Using the 0.9 kcal value where it applies, what value must you assign to the factor you invoked in (a), if you are to arrive at the energy difference of 5.4 kcal for the *cis*-1,3-conformations? (c) The potential energy difference between *cis*- and *trans*-1,1,3,5-tetramethylcyclohexane was then measured by Norman L. Allinger (at Wayne State University) as 3.7 kcal/mole. This measurement was carried out because of its direct bearing on the matter of *cis*-1,3-dimethylcyclohexane. What is the connection between this measurement and parts (a) and (b)? Does Allinger's measurement support Pitzer's calculation?

Problem 9.9 Predict the relative stabilities of the *cis*- and *trans*-isomers of: (a) 1,3-dimethylcyclohexane; (b) 1,4-dimethylcyclohexane. (c) On the basis of 0.9 kcal/mole per 1,3-diaxial methyl–hydrogen interaction or butane-*gauche* interaction, and assuming that each stereoisomer exists exclusively in its more stable conformation, predict the potential energy difference between members of each pair of stereoisomers.

Conformational analysis of cyclohexane derivatives containing several *different* substituents follows along the same lines as that of the dimethylcyclohexanes. We need to keep in mind that, of two groups, the larger one will tend to call the tune. Because of its very large 1,3-diaxial interactions (Problem 9.3, p. 301), the bulky *tert*-butyl group is particularly prone to occupy an equatorial position. If—as is usually the case—other substituents are considerably smaller than *tert*-butyl, the molecule is virtually locked in a single conformation: the one with an equatorial *tert*-butyl group. Consider cyclohexanes I and II containing a 4-*tert*-butyl group *cis* or *trans* to another substituent —G. In each diastereomer, *tert*-butyl holds —G exclusively in the axial or in the equatorial position, yet, because

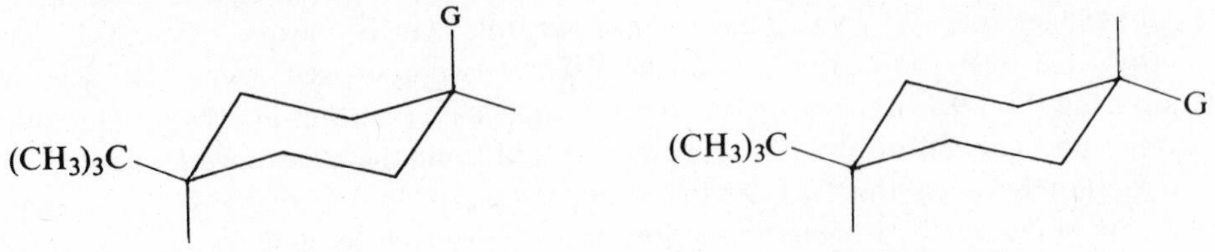

G
$(CH_3)_3C$

I

A *cis*-4-*tert*-butyl substituted cyclohexane

$(CH_3)_3C$ G

II

A *trans*-4-*tert*-butyl substituted cyclohexane

of its distance, exerts little electronic effect on —G. Following a suggestion by Professor Saul Winstein (of the University of California, Los Angeles), *tert*-butyl has been widely used as a holding group, to permit the study of physical and chemical properties associated with a purely axial or purely equatorial substituent.

Problem 9.10 Use the energy differences given in Problem 9.3 (p. 301) to calculate values for the various alkyl–hydrogen 1,3-diaxial interactions, and from these calculate the difference in energy between the two conformations of:

(a) *cis*-4-*tert*-butylmethylcyclohexane;
(b) *trans*-4-*tert*-butylmethylcyclohexane;
(c) *trans*-3-*cis*-4-dimethyl-*tert*-butylcyclohexane.

Now, what can we say about the possible chirality of the 1,2-dimethylcyclohexanes? Let us make a model of *trans*-1,2-dimethylcyclohexane—in the more stable diequatorial conformation, say—and a model of its mirror image. We find

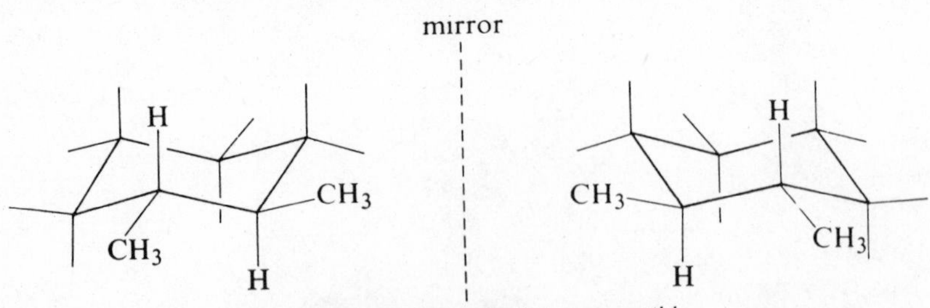

Not superimposable; not interconvertible
trans-1,2-Dimethylcyclohexane
A resolvable racemic modification

they are not superimposable, and therefore are enantiomers. We find that they are not interconvertible, and hence are configurational isomers. (When we flip one of these into the opposite chair conformation, it is converted, not into its mirror image, but into a diaxial conformation.) Thus, *trans*-1,2-dimethylcyclohexane should, in principle, be resolvable into (configurational) enantiomers, each of which should be optically active.

Next, let us make a model of *cis*-1,2-dimethylcyclohexane and a model of its mirror image. We find they are not superimposable, and hence are enantiomers.

mirror

Not superimposable; but interconvertible
cis-1,2-Dimethylcyclohexane
A non-resolvable racemic modification

In contrast to what we have said for the *trans*-compound, however, we find that these models *are* interconvertible by flipping one chair conformation into the other. These are conformational enantiomers and hence, except possibly at low temperatures, should interconvert too rapidly for resolution and measurement of optical activity.

Thus, just as with the *cis*- and *trans*-1,2-cyclopentanediols (Sec. 9.13), we could assign configurations to the *cis*- and *trans*-1,2-dimethylcyclohexanes by finding out which of the two is resolvable. The *cis*-1,2-dimethylcyclohexane is not literally a *meso* compound, but it is a non-resolvable racemic modification, which for most practical purposes amounts to the same thing.

To summarize, then, 1,2-dimethylcyclohexane exists as a pair of (configurational) diastereomers: the *cis*- and *trans*-isomers. The *cis*-isomer exists as a pair of conformational enantiomers. The *trans*-isomer exists as a pair of configurational enantiomers, each of which in turn exists as two conformational diastereomers (axial–axial and equatorial–equatorial).

Because of the ready interconvertibility of chair conformations, it is possible to use planar drawings to predict the configurational stereoisomerism of cyclo-

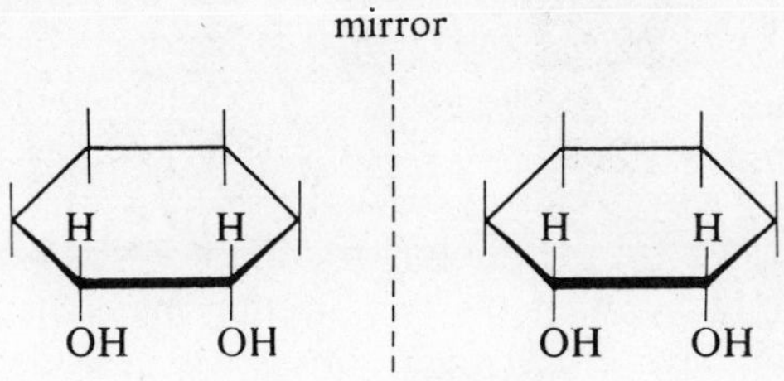

Superimposable
cis-1,2-Cyclohexanediol

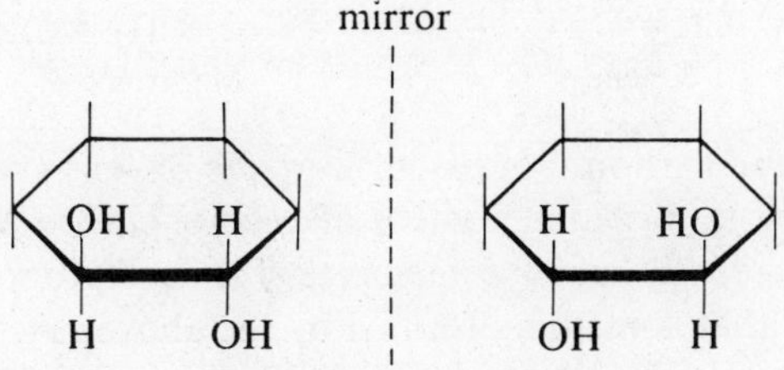

Not superimposable
trans-1,2-Cyclohexanediol

hexane derivatives. To understand the true geometry of such molecules, however, and with it the matter of stability, one must use models and formulas like those in Figs. 9.11 and 9.12.

Problem 9.11 Which of the following compounds are resolvable, and which are non-resolvable? Which are truly *meso* compounds? Use models as well as drawings.

(a) *cis*-1,2-cyclohexanediol
(b) *trans*-1,2-cyclohexanediol
(c) *cis*-1,3-cyclohexanediol
(d) *trans*-1,3-cyclohexanediol
(e) *cis*-1,4-cyclohexanediol
(f) *trans*-1,4-cyclohexanediol

Problem 9.12 Tell which, if any, of the compounds of Problem 9.11 exist as:

(a) a single conformation;
(b) a pair of conformational enantiomers;
(c) a pair of conformational diastereomers;
(d) a pair of (configurational) enantiomers, each of which exists as a single conformation;
(e) a pair of (configurational) enantiomers, each of which exists as a pair of conformational diastereomers;
(f) none of the above answers. (Give the correct answer.)

Problem 9.13 Draw structural formulas for all stereoisomers of the following. Label any *meso* compounds and indicate pairs of enantiomers. Do any (like *cis*-1,2-dimethylcyclohexane) exist as a non-resolvable racemic modification?

(a) *cis*-2-chlorocyclohexanol
(b) *trans*-2-chlorocyclohexanol
(c) *cis*-3-chlorocyclopentanol
(d) *trans*-3-chlorocyclopentanol
(e) *cis*-4-chlorocyclohexanol
(f) *trans*-4-chlorocyclohexanol

9.15 Carbenes. Methylene

The difference between successive members of a homologous series, we have seen, is the CH_2 unit, or *methylene*. But methylene is more than just a building block for the mental construction of compounds; it is an actual molecule, and its chemistry and the chemistry of its derivatives, the **carbenes**, has become one of the most exciting and productive fields of organic research.

Methylene is formed by the photolysis of either *diazomethane*, CH_2N_2, or *ketene*, $CH_2{=}C{=}O$. (Notice that the two starting materials and the two other

$$\underset{\text{Diazomethane}}{CH_2{=}\overset{+}{N}{=}\overset{-}{N}} \xrightarrow{\text{ultraviolet light}} \underset{\text{Methylene}}{CH_2} + N_2$$

$$\underset{\text{Ketene}}{CH_2{=}C{=}O} \xrightarrow{\text{ultraviolet light}} \underset{\text{Methylene}}{CH_2} + CO$$

products, nitrogen and carbon monoxide, are pairs of *isoelectronic* molecules, that is, molecules containing the same number of valence electrons.)

Methylene as a highly reactive molecule was first proposed in the 1930s to account for the fact that something formed by the above reactions was capable of removing certain metal mirrors (compare Problem 16, p. 72). Its existence was definitely established in 1959 by spectroscopic studies.

Figure 9.13. Evidence of early (1944) research on methylene, CH_2, by D. Duck. (As unearthed by Professors P. P. Gaspar and G. S. Hammond of the California Institute of Technology.)

These studies revealed that methylene not only exists but exists in two different forms (different spin states), generally referred to by their spectroscopic designations: *singlet* methylene, in which the unshared electrons are paired:

$$CH_2\!: \qquad \begin{matrix} H \\ H\!:\!\ddot{C}\!: \end{matrix} \qquad H\!-\!\!\!-\!C\!:\ (\text{other } H;\ 103°;\ 1.12\ \text{A})$$

Singlet methylene
Unshared electrons paired

and *triplet* methylene, in which the unshared electrons are *not* paired.

$$\cdot CH_2 \cdot \qquad H\!:\!\dot{\underset{\cdot}{C}}\!:\!H \qquad H\!-\!\dot{\underset{\cdot}{C}}\!-\!H\ (180°;\ 1.03\ \text{A})$$

Triplet methylene
Unshared electrons not paired: a diradical

Triplet methylene is thus a free radical: in fact, it is a *di*radical. As a result of the difference in electronic configuration, the two kinds of molecules differ in shape and in chemical properties. Singlet methylene is the less stable form, and is often the form first generated, in the initial photolysis.

The exact chemical properties observed depend upon which form of methylene is reacting, and this in turn depends upon the experimental conditions. In the liquid phase, the first-formed singlet methylene reacts rapidly with the abundant solvent molecules before it loses energy. In the gas phase—especially in the presence of an inert gas like nitrogen or argon—singlet methylene loses energy through collisions and is converted into triplet methylene, which then reacts.

When methylene is generated in the presence of alkenes, there are obtained cyclopropanes. For example:

$$\underset{\text{2-Butene}}{CH_3CH{=}CHCH_3} + \underset{\text{Diazomethane}}{CH_2N_2} \xrightarrow{\text{light}} \underset{\text{1,2-Dimethylcyclopropane}}{CH_3CH{-}CHCH_3\ (\text{ring with } CH_2)} + N_2$$

This is an example of the most important reaction of methylene and other carbenes: *addition to the carbon–carbon double bond.* Its most striking feature is that it can occur with two different kinds of stereochemistry.

$$>C{=}C< + CH_2 \longrightarrow -\overset{|}{C}{-}\overset{|}{C}-\ (\text{ring with } CH_2)$$ **Addition**

For example, photolysis of diazomethane in liquid *cis*-2-butene gives only *cis*-1,2-dimethylcyclopropane, and in liquid *trans*-2-butene gives only *trans*-1,2-dimethylcyclopropane. Addition here is stereospecific and *syn*. Photolysis of diazomethane in gaseous 2-butene—either *cis* or *trans*—gives *both cis*- and *trans*-1,2-dimethylcyclopropanes. Addition here is non-stereospecific.

There seems to be little doubt that the following interpretation, due to P. S. Skell of Pennsylvania State University, is the correct one.

It is **singlet** methylene that undergoes the *stereospecific addition*. Although neutral, singlet methylene is electron-deficient and hence electrophilic; like other

$$CH_2 + >C{=}C< \longrightarrow \left[-\overset{|}{C}{\cdots}\overset{|}{C}-\ (\cdots CH_2 \cdots)\right] \longrightarrow -\overset{|}{C}{-}\overset{|}{C}-\ (\text{ring with } CH_2)$$

Singlet methylene
Stereospecific
syn-addition

electrophiles, it can find electrons at the carbon–carbon double bond. The stereochemistry strongly indicates simultaneous attachment to both doubly-bonded carbon atoms. (However, on both theoretical and experimental grounds, the transition state is believed to be unsymmetrical: attachment to one carbon has proceeded further than attachment to the other, with the development of considerable positive charge on the second carbon.)

It is **triplet** methylene that undergoes the *non-stereospecific addition*. Triplet

$$CH_2 + >C{=}C< \longrightarrow \underset{\text{I}}{-\overset{|}{C}{-}\overset{|}{\dot{C}}-\ (\text{with } CH_2\cdot \text{ on first C})} \longrightarrow -C{-}C-\ (\text{ring with } CH_2)$$

Triplet methylene
Non-stereospecific
addition

methylene is a diradical, and it adds by a free-radical-like two-step mechanism: actually, addition followed by combination. The intermediate diradical I lasts long enough for rotation to occur about the central carbon–carbon bond, and both *cis* and *trans* products are formed. (*Problem:* Using the approach of Sec. 7.12, assure yourself that this is so.)

Besides addition, methylene undergoes another reaction which, quite literally, belongs in a class by itself: *insertion.*

$$-\overset{|}{\underset{|}{C}}-H + CH_2 \longrightarrow -\overset{|}{\underset{|}{C}}-CH_2-H \qquad \textbf{Insertion}$$

Methylene can *insert itself* into every carbon–hydrogen bond of most kinds of molecules. We cannot take time to say more here about this remarkable reaction, except that when addition is the desired reaction, insertion becomes an annoying side-reaction.

Problem 9.14 In the gas phase, with low alkene concentration and in the presence of an inert gas, addition of methylene to the 2-butenes is, we have seen, nonstereospecific. If, however, there is present in this system a little oxygen, addition becomes almost completely stereospecific (*syn*). Account in detail for the effect of oxygen. (*Hint:* See Sec. 2.14.)

9.16 Substituted carbenes. α-Elimination

A more generally useful way of making cyclopropanes is illustrated by the reaction of 2-butene with chloroform in the presence of potassium *tert*-butoxide (*t*-Bu = *tert*-butyl):

$$\underset{\text{2-Butene}}{CH_3CH{=}CHCH_3} + \underset{\text{Chloroform}}{CHCl_3} \xrightarrow{t\text{-BuO}^-K^+} \underset{\text{3,3-Dichloro-1,2-dimethylcyclopropane}}{CH_3\underset{\diagdown\ \diagup}{CH-CH}CH_3 \atop \overset{C}{\diagup\ \diagdown} \atop Cl\quad Cl} + t\text{-BuOH} + KCl$$

The dichlorocyclopropanes obtained can be reduced to hydrocarbons or hydrolyzed to *ketones*, the starting point for many syntheses (Chap. 19).

Here, too, reaction involves a divalent carbon compound, a derivative of methylene: *dichlorocarbene*, $:CCl_2$. It is generated in two steps, initiated by attack on chloroform by the very strong base, *tert*-butoxide ion, and then adds to the alkene.

(1) $$t\text{-BuO}:^- + H:CCl_3 \rightleftharpoons :CCl_3^- + t\text{-BuO}:H$$

(2) $$:CCl_3^- \longrightarrow \underset{\text{Dichlorocarbene}}{:CCl_2} + Cl^-$$

(3) $$CH_3CH{=}CHCH_3 + :CCl_2 \longrightarrow CH_3\underset{\diagdown\ \diagup}{CH-CH}CH_3 \atop \overset{C}{\diagup\ \diagdown} \atop Cl\quad Cl$$

It is believed that, because of the presence of the halogen atoms, the singlet form, with the electrons paired, is the more stable form of dichlorocarbene, and is the one adding to the double bond. (Stabilization by the halogen atoms is presumably one reason why dihalocarbenes do not generally undergo the insertion reaction that is so characteristic of unsubstituted singlet methylene.)

The addition of dihalocarbenes, like that of singlet methylene, is *stereospecific* and *syn*.

Problem 9.15 (a) Addition of $:CCl_2$ to cyclopentene yields a single compound. What is it? (b) Addition of $:CBrCl$ to cyclopentene yields a mixture of stereoisomers. In light of (a), how do you account for this? What are the isomers likely to be? (*Hint:* Use models.)

In dehydrohalogenation of alkyl halides (Sec. 5.13), we have already encountered a reaction in which hydrogen ion and halide ion are eliminated from a molecule by the action of base; there —H and —X were lost from adjacent carbons, and so the process is called *β-elimination.* In the generation of the methylene shown here, both —H and —X are eliminated from the same carbon, and the process is called *α-elimination.* (Later on, in Sec. 24.12, we shall see some of the evidence for the mechanism of α-elimination shown above.)

$$-\overset{X}{\underset{|}{\overset{|}{C}}}-\underset{H}{\underset{|}{\overset{|}{C}}}- \xrightarrow{\text{base}} \;>C=C<$$ **Beta-elimination**

$$-\underset{X}{\underset{|}{\overset{|}{C}}}-H \xrightarrow{\text{base}} -\overset{|}{C}:$$ **Alpha-elimination**

Problem 9.16 (a) Why does $CHCl_3$ not undergo β-elimination through the action of base? (b) What factor would you expect to make α-elimination from $CHCl_3$ easier than from, say, CH_3Cl?

There are many ways of generating what appear to be carbenes. But in some cases at least, it seems clear that no *free* carbene is actually an intermediate; instead, a *carbenoid* (carbene-like) reagent transfers a carbene unit directly to a double bond. For example, in the extremely useful Simmons-Smith reaction

$$CH_2I_2 + Zn(Cu) \longrightarrow ICH_2ZnI$$

$$>C=C< + ICH_2ZnI \longrightarrow \left[\begin{array}{c} -\overset{|}{C}\text{---}\overset{|}{C}- \\ CH_2 \\ I-Zn\text{---}I \end{array}\right] \longrightarrow -\overset{|}{C}\text{---}\overset{|}{C}- \; (CH_2 \text{ bridge}) + ZnI_2$$

(H. E. Simmons and R. D. Smith of the du Pont Company) the carbenoid is an organozinc compound which delivers methylene stereospecifically (and without competing insertion) to the double bond.

9.17 Analysis of alicyclic hydrocarbons

A cyclopropane readily dissolves in concentrated sulfuric acid, and in this resembles an alkene or alkyne. It can be differentiated from these unsaturated hydrocarbons, however, by the fact that it is not oxidized by cold, dilute, neutral permanganate.

Other alicyclic hydrocarbons have the same kind of properties as their open-chain counterparts, and they are characterized in the same way: cycloalkanes by their general inertness, and cycloalkenes and cycloalkynes by their response to tests for unsaturation (bromine in carbon tetrachloride, and aqueous permanganate). That one is dealing with cyclic hydrocarbons is shown by molecular formulas and by degradation products.

The properties of cyclohexane, for example, show clearly that it is an alkane. However, combustion analysis and molecular weight determination show its molecular formula to be C_6H_{12}. Only a cyclic structure (although not necessarily a six-membered ring) is consistent with both sets of data.

Similarly, the absorption of only one mole of hydrogen shows that cyclohexene contains only one carbon–carbon double bond; yet its molecular formula is C_6H_{10}, which in an open-chain compound would correspond to two carbon–carbon double bonds or one triple bond. Again, only a cyclic structure fits the facts.

Problem 9.17 Compare the molecular formulas of: (a) *n*-hexane and cyclohexane; (b) *n*-pentane and cyclopentane; (c) 1-hexene and cyclohexene; (d) dodecane, *n*-hexylcyclohexane, and cyclohexylcyclohexane. (e) In general, how can you deduce the number of rings in a compound from its molecular formula and degree of unsaturation?

Problem 9.18 What is the molecular formula of: (a) cyclohexane; (b) methylcyclopentane; (c) 1,2-dimethylcyclobutane? (d) Does the molecular formula give any information about the *size* of ring in a compound?

Problem 9.19 The yellow plant pigments α-, β-, and *γ-carotene*, and the red pigment of tomatoes, *lycopene*, are converted into Vitamin A in the liver. All four have the molecular formula $C_{40}H_{56}$. Upon catalytic hydrogenation, α- and β-carotene yield $C_{40}H_{78}$, γ-carotene yields $C_{40}H_{80}$, and lycopene yields $C_{40}H_{82}$. How many rings, if any, are there in each compound?

Cleavage products of cycloalkenes and cycloalkynes also reveal the cyclic structure. Ozonolysis of cyclohexene, for example, does not break the molecule into two aldehydes of lower carbon number, but simply into a single six-carbon compound containing *two* aldehyde groups.

Cyclohexene (H_2C–CH_2–CH=CH–CH_2–H_2C, ring) $\xrightarrow{O_3}$ $\xrightarrow{H_2O,\ Zn}$ OHC–CH_2–CH_2–CH_2–CH_2–CHO

A di-aldehyde

Cyclohexene $\xrightarrow{KMnO_4}$ $HOOC$–CH_2–CH_2–CH_2–CH_2–$COOH$

A di-acid

Problem 9.20 Predict the ozonolysis products of: (a) cyclohexene; (b) 1-methylcyclopentene; (c) 3-methylcyclopentene; (d) 1,3-cyclohexadiene; (e) 1,4-cyclohexadiene.

Problem 9.21 Both cyclohexene and 1,7-octadiene yield the di-aldehyde $OHC(CH_2)_4CHO$ upon ozonolysis. What other facts would enable you to distinguish between the two compounds?

(Analysis of cyclic aliphatic hydrocarbons by spectroscopy will be discussed in Secs. 13.15–13.16.)

PROBLEMS

1. Draw structural formulas of:

(a) methylcyclopentane
(b) 1-methylcyclohexene
(c) 3-methylcyclopentene
(d) *trans*-1,3-dichlorocyclobutane
(e) *cis*-2-bromo-1-methylcyclopentane
(f) cyclohexylcyclohexane
(g) cyclopentylacetylene
(h) 1,1-dimethyl-4-chlorocycloheptane
(i) bicyclo[2.2.1]hepta-2,5-diene
(j) 1-chlorobicyclo[2.2.2]octane

2. Give structures and names of the principal organic products expected from each of the following reactions:

(a) cyclopropane + Cl_2, $FeCl_3$
(b) cyclopropane + Cl_2 (300°)
(c) cyclopropane + conc. H_2SO_4
(d) cyclopentane + Cl_2, $FeCl_3$
(e) cyclopentane + Cl_2 (300°)
(f) cyclopentane + conc. H_2SO_4
(g) cyclopentene + Br_2/CCl_4
(h) cyclopentene + Br_2 (300°)
(i) 1-methylcyclohexene + HCl
(j) 1-methylcyclohexene + Br_2(aq)
(k) 1-methylcyclohexene + HBr (peroxides)
(l) 1,3-cyclohexadiene + HCl
(m) cyclopentanol + H_2SO_4 (heat)
(n) bromocyclohexane + KOH(alc)
(o) cyclopentene + cold $KMnO_4$
(p) cyclopentene + HCO_2OH
(q) cyclopentene + hot $KMnO_4$
(r) chlorocyclopentane + $(C_2H_5)_2CuLi$
(s) 1-methylcyclopentene + cold conc. H_2SO_4
(t) 3-methylcyclopentene + O_3, then H_2O/Zn
(u) cyclohexene + $H_2SO_4 \longrightarrow C_{12}H_{20}$
(v) cyclopentene + $CHCl_3$ + *t*-BuOK
(w) cyclopentene + CH_2I_2 + Zn(Cu)

3. Outline all steps in the laboratory synthesis of each of the following from cyclohexanol.

(a) cyclohexene
(b) cyclohexane
(c) *trans*-1,2-dibromocyclohexane
(d) *cis*-1,2-cyclohexanediol
(e) *trans*-1,2-cyclohexanediol
(f) $OHC(CH_2)_4CHO$
(g) adipic acid, $HOOC(CH_2)_4COOH$
(h) bromocyclohexane
(i) 2-chlorocyclohexanol
(j) 3-bromocyclohexene
(k) 1,3-cyclohexadiene
(l) cyclohexylcyclohexane
(m) *norcarane*, bicyclo[4.1.0]heptane

4. Give structure of all isomers of the following. For cyclohexane derivatives, planar formulas (p. 307) will be sufficient here. Label pairs of enantiomers, and *meso* compounds.

(a) dichlorocyclopropanes
(b) dichlorocyclobutanes
(c) dichlorocyclopentanes
(d) dichlorocyclohexanes
(e) chloro-1,1-dimethylcyclohexanes
(f) 1,3,5-trichlorocyclohexanes

(g) There are a number of stereoisomeric 1,2,3,4,5,6-hexachlorocyclohexanes. Without attempting to draw all of them, give the structure of the most stable isomer, and show its preferred conformation.

5. (a) 2,5-Dimethyl-1,1-cyclopentanedicarboxylic acid (I) can be prepared as two optically inactive substances (A and B) of different m.p. Draw their structures. (b) Upon heating, A yields two 2,5-dimethylcyclopentanecarboxylic acids (II), and B yields only one. Assign structures to A and B.

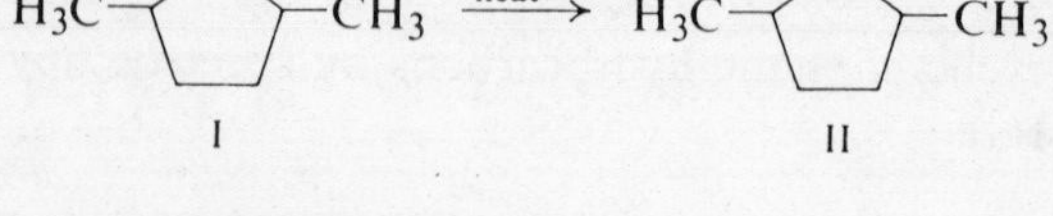

HOOC COOH / H_3C— —CH_3 (I) $\xrightarrow{\text{heat}}$ H COOH / H_3C— —CH_3 (II)

6. (a) The following compounds can be resolved into optically active enantiomers.

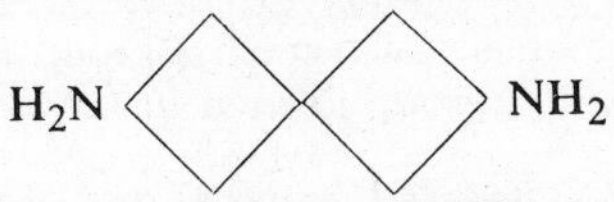

H_2N (spiro structure) NH_2 — 3,3′-Diaminospiro[3.3]heptane

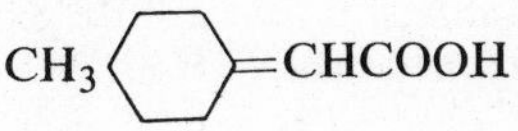

CH_3 (cyclohexane) $=CHCOOH$ — 4-Methylcyclohexylideneacetic acid

Using models and then drawing three-dimensional formulas, account for this. Label the chiral center in each compound.

(b) Addition of bromine to optically active 4-methylcyclohexylideneacetic acid yields two optically active dibromides. Assuming a particular configuration for the starting material, draw stereochemical formulas for the products.

7. (a) *trans*-1,2-Dimethylcyclohexane exists about 99% in the diequatorial conformation. *trans*-1,2-Dibromocyclohexane (or *trans*-1,2-dichlorocyclohexane), on the other hand, exists about equally in the diequatorial and diaxial conformations; furthermore, the fraction of the diaxial conformation decreases with increasing polarity of the solvent. How do you account for the contrast between the dimethyl and dibromo (or dichloro) compounds? (*Hint:* See Problem 11, p. 141.)

(b) If *trans*-3-*cis*-4-dibromo-*tert*-butylcyclohexane is subjected to prolonged heating, it is converted into an equilibrium mixture (about 50:50) of itself and a diastereomer. What is the diastereomer likely to be? How do you account for the approximately equal stability of these two diastereomers? (Here, and in (c), consider the more stable conformation of each diastereomer to be the one with an equatorial *tert*-butyl group.)

(c) There are two more diastereomeric 3,4-dibromo-*tert*-butylcyclohexanes. What are they? How do you account for the fact that neither is present to an appreciable extent in the equilibrium mixture?

8. The compound *decalin*, $C_{10}H_{18}$, consists of two fused cyclohexane rings:

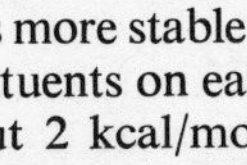

Decalin

(a) Using models, show how there can be two isomeric decalins, *cis* and *trans*. (b) How many different conformations free of angle strain are possible for *cis*-decalin? For *trans*-decalin? (c) Which is the most stable conformation of *cis*-decalin? Of *trans*-decalin? (d) Account for the fact that *trans*-decalin is more stable than *cis*-decalin. (*Hint:* Consider each ring in turn. What are the largest substituents on each ring?) (e) The difference in stability between *cis*- and *trans*-decalin is about 2 kcal/mole; conversion of one into the other takes place only under very vigorous conditions. The chair and twist-boat forms of cyclohexane, on the other hand, differ in stability by about 6 kcal/mole, yet are readily interconverted at room temperature. How do you account for the contrast? Draw energy curves to illustrate your answer.

9. Allinger (p. 305) found the energy difference between *cis*- and *trans*-1,3-di-*tert*-butylcyclohexane to be 5.9 kcal/mole, and considers that this value represents the energy difference between the chair and twist-boat forms of cyclohexane. Defend Allinger's position.

10. It has been suggested that in certain substituted cyclopentanes the ring exists preferentially in the "envelope" form:

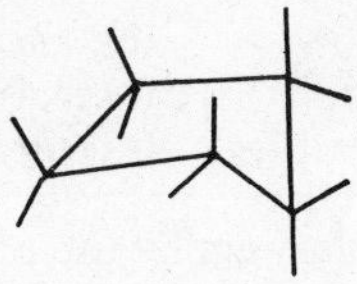

Using models, suggest a possible explanation for each of the following facts:

(a) The attachment of a methyl group to the badly strained cyclopentane ring raises the heat of combustion very little more than attachment of a methyl group to the unstrained cyclohexane ring. (*Hint:* Where is the methyl group located in the "envelope" form?)

(b) Of the 1,2-dimethylcyclopentanes, the *trans*-isomer is more stable than the *cis*. Of the 1,3-dimethylcyclopentanes, on the other hand, the *cis*-isomer is more stable than the *trans*.

(c) The *cis*-isomer of methyl 3-methylcyclobutanecarboxylate

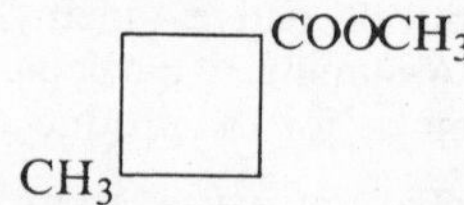

is more stable than the *trans*-isomer

11. Each of the following reactions is carried out, and the products are separated by careful distillation, recrystallization, or chromatography. For each reaction tell how many fractions will be collected. Draw a stereochemical formula of the compound or compounds making up each fraction. Tell whether each fraction, as collected, will be optically active or optically inactive.

(a) (R)-3-hydroxycyclohexene + $KMnO_4 \longrightarrow C_6H_{12}O_3$;
(b) (R)-3-hydroxycyclohexene + $HCO_2OH \longrightarrow C_6H_{12}O_3$;
(c) (S,S)-1,2-dichlorocyclopropane + Cl_2 (300°) $\longrightarrow C_3H_3Cl_3$;
(d) *racemic* 4-methylcyclohexene + Br_2/CCl_4.

12. Outline all steps in a possible laboratory synthesis of each of the following, using alcohols of four carbons or fewer as your only organic source, and any necessary inorganic reagents. (*Remember:* Work backwards.)

(a) *cis*-1,2-di(*n*-propyl)cyclopropane;
(b) racemic *trans*-1-methyl-2-ethyl-3,3-dichlorocyclopropane.

13. Describe simple chemical tests that would distinguish between:

(a) cyclopropane and propane
(b) cyclopropane and propylene
(c) 1,2-dimethylcyclopropane and cyclopentane
(d) cyclobutane and 1-butene
(e) cyclopentane and 1-pentene
(f) cyclopentane and cyclopentene
(g) cyclohexanol and *n*-butylcyclohexane
(h) 1,2-dimethylcyclopentene and cyclopentanol
(i) cyclohexane, cyclohexene, cyclohexanol, and bromocyclohexane

14. How many rings does each of the following contain?
(a) *Camphane*, $C_{10}H_{18}$, a terpene related to camphor, takes up no hydrogen. (*b*) *Cholestane*, $C_{27}H_{48}$, a steroid of the same ring structure as cholesterol, cortisone, and the sex hormones, takes up no hydrogen. (c) *β-Phellandrene*, $C_{10}H_{16}$, a terpene, reacts with bromine to form $C_{10}H_{16}Br_4$. (d) *Ergocalciferol* (so-called "Vitamin D_2"), $C_{28}H_{44}O$, an alcohol, gives $C_{28}H_{52}O$ upon catalytic hydrogenation. (e) How many double bonds does ergocalciferol contain?

15. On the basis of the results of catalytic hydrogenation, how many rings does each of the following aromatic hydrocarbons contain?

(a) *benzene* (C_6H_6) $\longrightarrow C_6H_{12}$
(b) *naphthalene* ($C_{10}H_8$) $\longrightarrow C_{10}H_{18}$
(c) *toluene* (C_7H_8) $\longrightarrow C_7H_{14}$
(d) *anthracene* ($C_{14}H_{10}$) $\longrightarrow C_{14}H_{24}$
(e) *phenanthrene* ($C_{14}H_{10}$) $\longrightarrow C_{14}H_{24}$
(f) 3,4-*benzpyrene* ($C_{20}H_{12}$) $\longrightarrow C_{20}H_{32}$
(g) *chrysene* ($C_{18}H_{12}$) $\longrightarrow C_{18}H_{30}$

(Check your answers by use of the index.)

16. (a) A hydrocarbon of formula $C_{10}H_{16}$ absorbs only one mole of H_2 upon hydrogenation. How many rings does it contain? (b) Upon ozonolysis it yields 1,6-cyclodecanedione (III). What is the hydrocarbon?

$$O{=}C\begin{matrix}(CH_2)_4\\ \\(CH_2)_4\end{matrix}C{=}O$$

III

$$CH_3{-}\underset{\underset{O}{\|}}{C}{-}CH_2{-}CH_2{-}CH\begin{matrix}\nearrow \overset{O}{\overset{\|}{C}}{-}CH_3\\ \searrow CH_2COOH\end{matrix}$$

IV

17. *Limonene*, $C_{10}H_{16}$, a terpene found in orange, lemon, and grapefruit peel, absorbs only two moles of hydrogen, forming *p-menthane*, $C_{10}H_{20}$. Oxidation by permanganate converts limonene into IV. (a) How many rings, if any, are there in limonene? (b) What structures are consistent with the oxidation product? (c) On the basis of the isoprene rule (Sec. 8.26), which structure is most likely for limonene? For *p*-menthane? (d) Addition of one mole of H_2O converts limonene into an alcohol. What are the most likely structures for this alcohol? (e) Addition of two moles of H_2O to limonene yields *terpin hydrate*. What is the most likely structure for terpin hydrate?

18. *α-Terpinene*, $C_{10}H_{16}$, a terpene found in coriander oil, absorbs only two moles of hydrogen, forming *p-menthane*, $C_{10}H_{20}$. Ozonolysis of α-terpinene yields V; permanganate cleavage yields VI.

$$CH_3{-}\underset{\underset{O}{\|}}{C}{-}CH_2{-}CH_2{-}\underset{\underset{O}{\|}}{C}{-}CH(CH_3)_2$$

V

$$HOOC{-}\overset{CH_3}{\underset{OH}{C}}{-}CH_2{-}CH_2{-}\overset{CH(CH_3)_2}{\underset{OH}{C}}{-}COOH$$

VI

(a) How many rings, if any, are there in α-terpinene? (b) On the basis of the cleavage products, V and VI, and the isoprene rule, what is the most likely structure for α-terpinene? (c) How do you account for the presence of the —OH groups in VI?

19. Using only chemistry that you have already encountered, can you suggest a mechanism for the conversion of *nerol* ($C_{10}H_{18}O$) into α-terpineol ($C_{10}H_{18}O$) in the presence of dilute H_2SO_4?

$$CH_3{-}\overset{CH_3}{\overset{|}{C}}{=}CH{-}CH_2{-}CH_2{-}\overset{CH_3}{\overset{|}{C}}{=}CH{-}CH_2OH \xrightarrow{H_2O,\ H^+}$$

Nerol (found in bergamot)

α-Terpineol: ring of C(CH_3)=CH–CH_2–CH–H_2C–H_2C, with CH bearing C(OH)(H_3C)(CH_3)

α-Terpineol

Chapter 10 Benzene
Aromatic Character

10.1 Aliphatic and aromatic compounds

Chemists have found it useful to divide all organic compounds into two broad classes: **aliphatic** compounds and **aromatic** compounds. The original meanings of the words "aliphatic" (*fatty*) and "aromatic" (*fragrant*) no longer have any significance.

Aliphatic compounds are open-chain compounds and those cyclic compounds that resemble the open-chain compounds. The families we have studied so far—alkanes, alkenes, alkynes, and their cyclic analogs—are all members of the aliphatic class.

Aromatic compounds *are benzene and compounds that resemble benzene in chemical behavior.* Aromatic properties are those properties of benzene that distinguish it from aliphatic hydrocarbons. Some compounds that possess aromatic properties have structures that seem to differ considerably from the structure of benzene: actually, however, there is a basic similarity in electronic configuration (Sec. 10.10).

Aliphatic hydrocarbons, as we have seen, undergo chiefly addition and free-radical substitution; addition occurs at multiple bonds, and free-radical substitution occurs at other points along the aliphatic chain. In contrast, we shall find that *aromatic hydrocarbons are characterized by a tendency to undergo ionic substitution.* We shall find this contrast maintained in other families of compounds (i.e., acids, amines, aldehydes, etc.); the hydrocarbon parts of their molecules undergo reactions characteristic of either aliphatic or aromatic hydrocarbons.

It is important not to attach undue weight to the division between aliphatic and aromatic compounds. Although extremely useful, it is often less important than some other classification. For example, the similarities between aliphatic and aromatic acids, or between aliphatic and aromatic amines, are more important than the differences.

10.2 Structure of benzene

It is obvious from our definition of aromatic compounds that any study of their chemistry must begin with a study of benzene. Benzene has been known since 1825; its chemical and physical properties are perhaps better known than those of any other single organic compound. In spite of this, no satisfactory structure for benzene had been advanced until about 1931, and it was ten to fifteen years before this structure was generally used by organic chemists.

The difficulty was not the complexity of the benzene molecule, but rather the limitations of the structural theory as it had so far developed. Since an understanding of the structure of benzene is important both in our study of aromatic compounds and in extending our knowledge of the structural theory, we shall examine in some detail the facts upon which this structure of benzene is built.

10.3 Molecular formula. Isomer number. Kekulé structure

(*a*) *Benzene has the molecular formula* C_6H_6. From its elemental composition and molecular weight, benzene was known to contain six carbon atoms and six hydrogen atoms. The question was: how are these atoms arranged?

In 1858, August Kekulé (of the University of Bonn) had proposed that carbon atoms can join to one another to form *chains*. Then, in 1865, he offered an answer to the question of benzene: these carbon chains can sometimes be closed, to form *rings*.

> "I was sitting writing at my textbook, but the work did not progress; my thoughts were elsewhere. I turned my chair to the fire, and dozed. Again the atoms were gamboling before my eyes. This time the smaller groups kept modestly in the background. My mental eye, rendered more acute by repeated visions of this kind, could now distinguish larger structures of manifold conformations; long rows, sometimes more closely fitted together; all twisting and turning in snake-like motion. But look! What was that? One of the snakes had seized hold of its own tail, and the form whirled mockingly before my eyes. As if by a flash of lightning I woke;... I spent the rest of the night working out the consequences of the hypothesis. Let us learn to dream, gentlemen, and then perhaps we shall learn the truth."— August Kekulé, 1865.

Kekulé's structure of benzene was one that we would represent today as I.

```
        H                      H                      H
        |                      |                      |
        C                      C                      C
     //   \                  /  |  \              //     \
H—C         C—H        H—C     |     C—H     H—C           C=CH2
   |        ||            ||   |    ||          \         /
H—C         C—H        H—C     |     C—H          C=====C
     \   //                  \  |  /            /         \
        C                      C               H           H
        |                      |
        H                      H

        I                      II                    III
  Kekulé formula        "Dewar" formula
```

$$CH_3-C\equiv C-C\equiv C-CH_3 \qquad\qquad CH_2=CH-C\equiv C-CH=CH_2$$

IV V

Other structures are, of course, consistent with the formula C_6H_6: for example, II–V. Of all these, Kekulé's structure was accepted as the most nearly satisfactory;

the evidence was of a kind with which we are already familiar: **isomer number** (Sec. 4.2).

(*b*) *Benzene yields only one monosubstitution product*, C_6H_5Y. Only one bromobenzene, C_6H_5Br, is obtained when one hydrogen atom is replaced by bromine; similarly, only one chlorobenzene, C_6H_5Cl, or one nitrobenzene, $C_6H_5NO_2$, etc., has ever been made. This fact places a severe limitation on the structure of benzene: each hydrogen must be exactly equivalent to every other hydrogen, since the replacement of any one of them yields the same product.

Structure V, for example, must now be rejected, since it would yield two isomeric monobromo derivatives, the 1-bromo and the 2-bromo compounds; all hydrogens are not equivalent in V. Similar reasoning shows us that II and III are likewise unsatisfactory. (How many monosubstitution products would each of these yield?) I and IV, among others, are still possibilities, however.

(*c*) *Benzene yields three isomeric disubstitution products*, $C_6H_4Y_2$ or C_6H_4YZ. Three and only three isomeric dibromobenzenes, $C_6H_4Br_2$, three chloronitrobenzenes, $C_6H_4ClNO_2$, etc., have ever been made. This fact further limits our choice of a structure; for example, IV must now be rejected. (How many disubstitution products would IV yield?)

At first glance, structure I seems to be consistent with this new fact; that is, we can expect three isomeric dibromo derivatives, the 1,2- the 1,3-, and the 1,4-dibromo compounds shown:

1,2-Dibromobenzene   1,3-Dibromobenzene   1,4-Dibromobenzene

Closer examination of structure I shows, however, that *two* 1,2-dibromo isomers (VI and VII), differing in the positions of bromine relative to the double bonds, should be possible:

VI   VII

But Kekulé visualized the benzene molecule as a dynamic thing: ". . . the form whirled mockingly before my eyes . . ." He described it in terms of two structures, VIII and IX, between which the benzene molecule alternates. As a consequence, the two 1,2-dibromobenzenes (VI and VII) would be in rapid equilibrium and hence could not be separated.

```
          H                          H
          |                          |
          C                          C
     H-C⁄   ⧵C-H               H-C⁄⁄   ⧵C-H
       ||    |        ⇌          |     ||
     H-C⧵   ⁄C-H               H-C⧵⧵   ⁄C-H
          C                          C
          |                          |
          H                          H
        VIII                        IX

          Br                         Br
          |                          |
          C                          C
     H-C⁄   ⧵C-Br              H-C⁄⁄   ⧵C-Br
       ||    |        ⇌          |     ||
     H-C⧵   ⁄C-H               H-C⧵⧵   ⁄C-H
          C                          C
          |                          |
          H                          H
         VI                         VII
```

Later, when the idea of tautomerism (Sec. 8.13) became defined, it was assumed that Kekulé's "alternation" essentially amounted to tautomerism.

On the other hand, it is believed by some that Kekulé had intuitively anticipated by some 75 years our present concept of delocalized electrons, and drew two pictures (VIII and IX)—as we shall do, too—as a crude representation of something that neither picture alone satisfactorily represents. Rightly or wrongly, the term "Kekulé structure" has come to mean a (hypothetical) molecule with alternating single and double bonds—just as the term "Dewar benzene" has come to mean a structure (II) that James Dewar devised in 1867 as an example of what benzene was *not*.

10.4 Stability of the benzene ring. Reactions of benzene

Kekulé's structure, then, accounts satisfactorily for facts (*a*), (*b*), and (*c*) in Sec. 10.3. But there are a number of facts that are still not accounted for by this structure; most of these unexplained facts seem related to unusual stability of the benzene ring. The most striking evidence of this stability is found in the chemical reactions of benzene.

(*d*) *Benzene undergoes substitution rather than addition.* Kekulé's structure of benzene is one that we would call "cyclohexatriene." We would expect this cyclohexatriene, like the very similar compounds, cyclohexadiene and cyclohexene, to undergo readily the addition reactions characteristic of the alkene structure. As the examples in Table 10.1 show, this is not the case; under conditions that cause an alkene to undergo rapid addition, benzene reacts either not at all or very slowly.

Table 10.1 CYCLOHEXENE *vs.* BENZENE

Reagent	Cyclohexene gives	Benzene gives
$KMnO_4$ (cold, dilute, aqueous)	Rapid oxidation	No reaction
Br_2/CCl_4 (in the dark)	Rapid addition	No reaction
HI	Rapid addition	No reaction
H_2 + Ni	Rapid hydrogenation at 25°, 20 lb/in.^{2}	Slow hydrogenation at 100–200°, 1500 lb/in.^{2}

In place of addition reactions, benzene readily undergoes a new set of reactions, all involving **substitution.** The most important are shown below.

REACTIONS OF BENZENE

1. Nitration. Discussed in Sec. 11.8.

$$C_6H_6 + HONO_2 \xrightarrow{H_2SO_4} C_6H_5NO_2 + H_2O$$

Nitrobenzene

2. Sulfonation. Discussed in Sec. 11.9.

$$C_6H_6 + HOSO_3H \xrightarrow{SO_3} C_6H_5SO_3H + H_2O$$

Benzenesulfonic acid

3. Halogenation. Discussed in Sec. 11.11.

$$C_6H_6 + Cl_2 \xrightarrow{Fe} C_6H_5Cl + HCl$$

Chlorobenzene

$$C_6H_6 + Br_2 \xrightarrow{Fe} C_6H_5Br + HBr$$

Bromobenzene

4. Friedel-Crafts alkylation. Discussed in Secs. 11.10 and 12.6.

$$C_6H_6 + RCl \xrightarrow{AlCl_3} C_6H_5R + HCl$$

An alkylbenzene

5. Friedel-Crafts acylation. Discussed in Sec. 19.6.

$$C_6H_6 + RCOCl \xrightarrow{AlCl_3} C_6H_5COR + HCl$$

An acyl chloride — A ketone

In each of these reactions an atom or group has been substituted for one of the hydrogen atoms of benzene. The product can itself undergo further substitution of the same kind; the fact that it has retained the characteristic properties of benzene indicates that it has retained the characteristic structure of benzene.

It would appear that benzene resists addition, in which the benzene ring system would be destroyed, whereas it readily undergoes substitution, in which the ring system is preserved.

10.5 Stability of the benzene ring. Heats of hydrogenation and combustion

Besides the above qualitative indications that the benzene ring is more stable than we would expect cyclohexatriene to be, there exist quantitative data which show *how much* more stable.

(*e*) *Heats of hydrogenation and combustion of benzene are lower than expected.* We recall (Sec. 6.3) that heat of hydrogenation is the quantity of heat evolved when one mole of an unsaturated compound is hydrogenated. In most cases the value is about 28–30 kcal for each double bond the compound contains. It is not surprising, then, that cyclohexene has a heat of hydrogenation of 28.6 kcal and cyclohexadiene has one about twice that (55.4 kcal.)

We might reasonably expect cyclohexatriene to have a heat of hydrogenation about three times as large as cyclohexene, that is, about 85.8 kcal. Actually, the value for benzene (49.8 kcal) is *36 kcal less* than this expected amount.

This can be more easily visualized, perhaps, by means of an energy diagram (Fig. 10.1), in which the height of a horizontal line represents the potential energy content of a molecule. The broken lines represent the expected values, based upon three equal steps of 28.6 kcal. The final product, cyclohexane, is the same in all three cases.

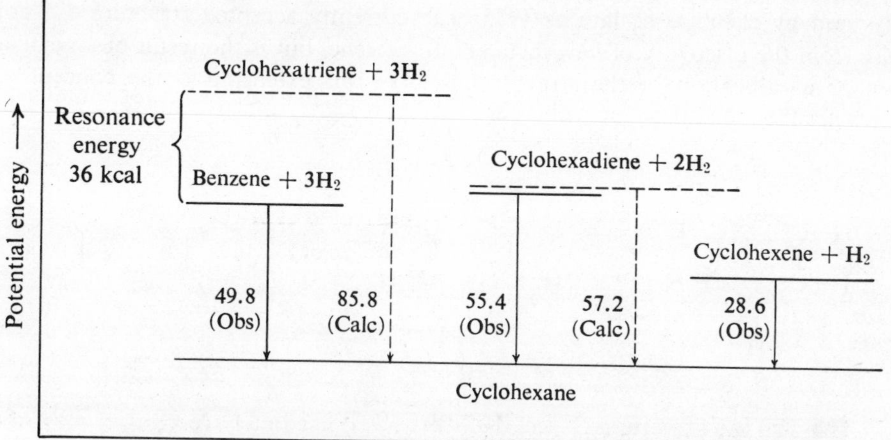

Figure 10.1. Heats of hydrogenation and stability: benzene, cyclohexadiene, and cyclohexene.

The fact that benzene *evolves* 36 kcal less energy than predicted can only mean that benzene *contains* 36 kcal less energy than predicted; in other words, benzene is more stable by 36 kcal than we would have expected cyclohexatriene to be. The heat of combustion of benzene is also lower than that expected, and by about the same amount.

Problem 10.1 From Fig. 10.1 determine the ΔH of the following reactions: (a) benzene + H_2 ⟶ 1,3-cyclohexadiene; (b) 1,3-cyclohexadiene + H_2 ⟶ cyclohexene.

Problem 10.2 For a large number of organic compounds, the heat of combustion actually measured agrees rather closely with that calculated by assuming a certain characteristic contribution from each kind of bond, e.g., 54.0 kcal for each C—H bond, 49.3 kcal for each C—C bond, and 117.4 kcal for each C=C bond (*cis*-1,2-disubstituted). (a) On this basis, what is the calculated heat of combustion for cyclohexatriene? (b) How does this compare with the measured value of 789.1 kcal for benzene?

10.6 Carbon–carbon bond lengths in benzene

(*f*) *All carbon–carbon bonds in benzene are equal and are intermediate in length between single and double bonds.* Carbon–carbon double bonds in a wide variety of compounds are found to be about 1.34 A long. Carbon–carbon single bonds, in which the nuclei are held together by only one pair of electrons, are considerably longer: 1.53 A in ethane, for example, 1.50 A in propylene, 1.48 A in 1,3-butadiene.

If benzene actually possessed three single and three double bonds, as in a Kekulé structure, we would expect to find three short bonds (1.34 A) and three long bonds (1.48 A, probably, as in 1,3-butadiene). Actually, x-ray diffraction studies show that the six carbon–carbon bonds in benzene are equal and have a length of 1.39 A, and are thus intermediate between single and double bonds.

10.7 Resonance structure of benzene

The Kekulé structure of benzene, while admittedly unsatisfactory, was generally used by chemists as late as 1945. The currently accepted structure did not arise from the discovery of new facts about benzene, but is the result of an extension or modification of the structural theory; this extension is the concept of *resonance* (Sec. 6.23).

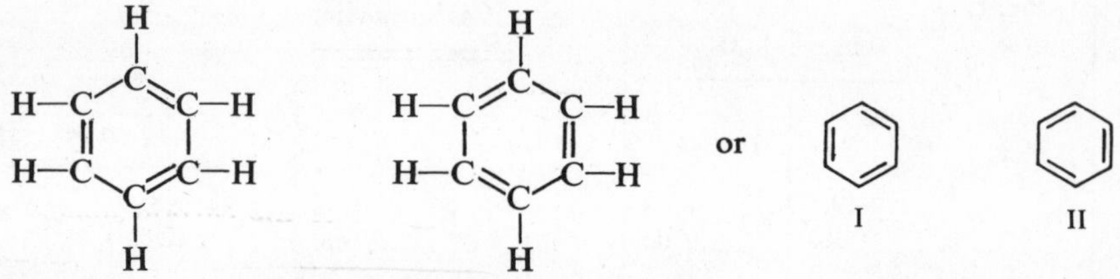

I II or I II

The Kekulé structures I and II, we now immediately recognize, meet the conditions for resonance: *structures that differ only in the arrangement of electrons.* Benzene is a *hybrid* of I and II. Since I and II are exactly equivalent, and hence of exactly the same stability, they make equal contributions to the hybrid. And, also since I and II are exactly equivalent, stabilization due to resonance should be large.

The puzzling aspects of benzene's properties now fall into place. The six bond lengths are identical because the six bonds are identical: they are one-and-a-half bonds and their length, 1.39 A, is intermediate between the lengths of single and double bonds.

When it is realized that all carbon–carbon bonds in benzene are equivalent, there is no longer any difficulty in accounting for the number of isomeric disubstitution products. It is clear that there should be just three, in agreement with experiment:

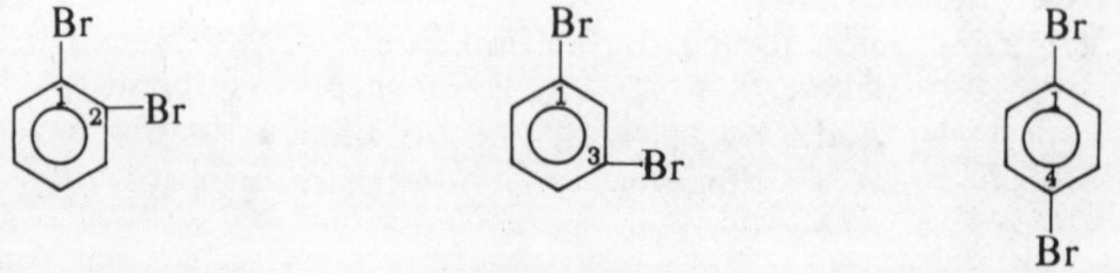

1,2-Dibromobenzene 1,3-Dibromobenzene 1,4-Dibromobenzene

Finally, the "unusual" stability of benzene is not unusual at all: it is what one would expect of a hybrid of equivalent structures. The 36 kcal of energy that benzene does not contain—compared with cyclohexatriene—is resonance energy. It is the 36 kcal of resonance energy that is responsible for the new set of properties we call *aromatic properties*.

Addition reactions convert an alkene into a more stable saturated compound. Hydrogenation of cyclohexene, for example, is accompanied by the evolution of 28.6 kcal; the product lies 28.6 kcal lower than the reactants on the energy scale (Fig. 10.1).

But addition would convert benzene into a *less* stable product by destroying the resonance-stabilized benzene ring system; for example, according to Fig. 10.1 the first stage of hydrogenation of benzene requires 5.6 kcal to convert benzene into the less stable cyclohexadiene. As a consequence, it is easier for reactions of benzene to take an entirely different course, one in which the ring system is retained: *substitution.*

(This is not quite all of the story in so far as stability goes. As we shall see in Sec. 10.10, an additional factor besides resonance is necessary to make benzene what it is.)

10.8 Orbital picture of benzene

A more detailed picture of the benzene molecule is obtained from a consideration of the bond orbitals in this molecule.

Since each carbon is bonded to three other atoms, it uses *sp*² orbitals (as in ethylene, Sec. 5.2). These lie in the same plane, that of the carbon nucleus, and are directed toward the corners of an equilateral triangle. If we arrange the six carbons and six hydrogens of benzene to permit maximum overlap of these orbitals, we obtain the structure shown in Fig. 10.2*a*.

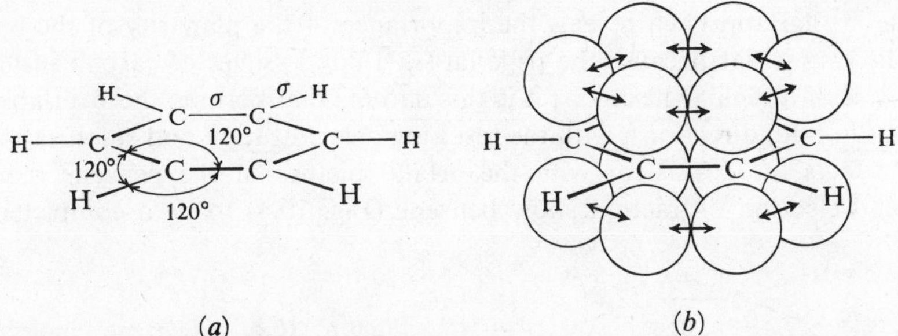

Figure 10.2. Benzene molecule. (*a*) Only σ bonds shown. (*b*) *p* orbitals overlap to form π bonds.

Benzene is a *flat molecule*, with every carbon and every hydrogen lying in the same plane. It is a very *symmetrical molecule*, too, with each carbon atom lying at the angle of a regular hexagon; every bond angle is 120°. Each bond orbital is cylindrically symmetrical about the line joining the atomic nuclei and hence, as before, these bonds are designated as σ bonds.

The molecule is not yet complete, however. There are still six electrons to be accounted for. In addition to the three orbitals already used, each carbon atom has a fourth orbital, a *p* orbital. As we know, this *p* orbital consists of two equal lobes, one lying above and the other lying below the plane of the other three orbitals, that is, above and below the plane of the ring; it is occupied by a single electron.

As in the case of ethylene, the *p* orbital of one carbon can overlap the *p* orbital of an adjacent carbon atom, permitting the electrons to pair and an additional π bond to be formed (see Fig. 10.2*b*). But the overlap here is not limited to a pair of *p* orbitals as it was in ethylene; the *p* orbital of any one carbon atom overlaps equally well the *p* orbitals of *both* carbon atoms to which it is bonded. The result (see Fig. 10.3) is two continuous doughnut-shaped electron clouds, one lying above and the other below the plane of the atoms.

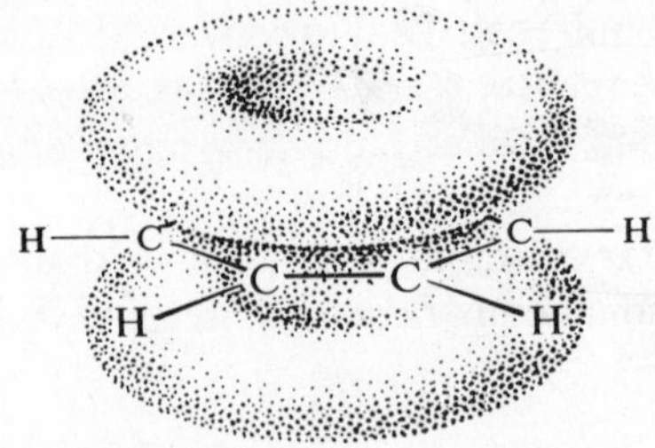

Figure 10.3. Benzene molecule. π clouds above and below plane of ring.

As with the allyl radical, it is the overlap of the *p* orbitals in both directions, and the resulting participation of each electron in several bonds that corresponds to our description of the molecule as a resonance hybrid of two structures. Again it is the *delocalization* of the π electrons—their participation in several bonds—that makes the molecule more stable.

To accommodate six π electrons, there must be *three* orbitals (Sec. 29.5). Their sum is, however, the symmetrical π clouds we have described.

The orbital approach reveals the importance of the planarity of the benzene ring. The ring is flat because the trigonal (sp^2) bond angles of carbon just fit the 120° angles of a regular hexagon; it is this flatness that permits the overlap of the *p* orbitals in both directions, with the resulting delocalization and stabilization.

The facts are consistent with the orbital picture of the benzene molecule. X-ray and electron diffraction show benzene (Fig. 10.4) to be a completely flat,

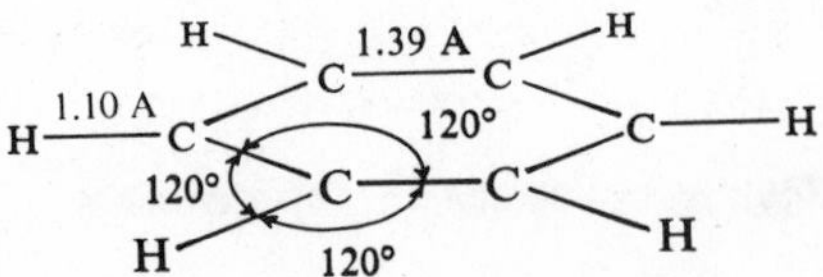

Figure 10.4. Benzene molecule: shape and size.

symmetrical molecule with all carbon–carbon bonds equal, and all bond angles 120°.

As we shall see, the chemical properties of benzene are just what we would expect of this structure. Despite delocalization, the π electrons are nevertheless more loosely held than the σ electrons. The π electrons are thus particularly available to a reagent that is seeking electrons: *the typical reactions of the benzene ring are those in which it serves as a source of electrons for electrophilic (acidic) reagents.* Because of the resonance stabilization of the benzene ring, *these reactions lead to substitution*, in which the aromatic character of the benzene ring is preserved.

Problem 10.3 The carbon–hydrogen bond dissociation energy for benzene (112 kcal) is considerably larger than for cyclohexane. On the basis of the orbital picture of benzene, what is one factor that may be responsible for this? What piece of physical evidence tends to support your answer? (*Hint:* Look at Fig. 10.4 and see Sec. 5.4.)

Problem 10.4 The molecules of *pyridine*, C_5H_5N, are flat, with all bond angles about 120°. All carbon–carbon bonds are 1.39 A long and the two carbon–nitrogen bonds are 1.36 A long. The measured heat of combustion is 23 kcal lower than that calculated by the method of Problem 10.2 on page 323. Pyridine undergoes such substitution reactions as nitration and sulfonation (Sec. 10.4). (a) Is pyridine adequately represented by formula I? (b) Account for the properties of pyridine by both valence-bond and orbital structures. (Check your answer in Sec. 31.6.)

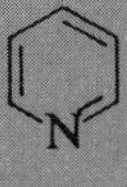

I

Problem 10.5 The compound *borazole*, $B_3N_3H_6$, is shown by electron diffraction to have a flat cyclic structure with alternating boron and nitrogen atoms, and all boron–nitrogen bond lengths the same. (a) How would you represent borazole by valence-bond structures? (b) In terms of orbitals? (c) How many π electrons are there, and which atoms have they "come from"?

10.9 Representation of the benzene ring

For convenience we shall represent the benzene ring by a regular hexagon containing a circle (I); it is understood that a hydrogen atom is attached to each angle of the hexagon unless another atom or group is indicated.

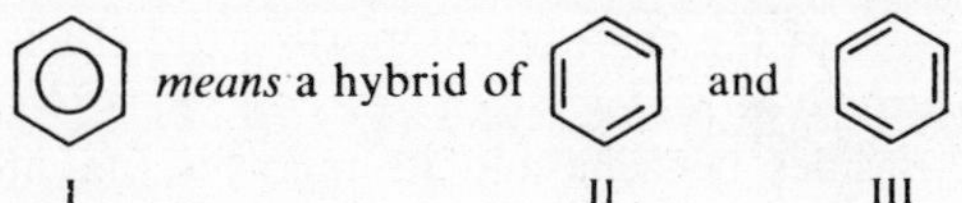

I represents a resonance hybrid of the Kekulé structures II and III. The straight lines stand for the σ bonds joining carbon atoms. The circle stands for the cloud of six delocalized π electrons. (From another viewpoint, the straight lines stand for single bonds, and the circle stands for the extra *half-bonds*.)

I is a particularly useful representation of the benzene ring, since it emphasizes the equivalence of the various carbon–carbon bonds. The presence of the circle distinguishes the benzene ring from the cyclohexane ring, which is often represented today by a plain hexagon.

There is no complete agreement among chemists about how to represent the benzene ring. The student should expect to encounter it most often as one of the Kekulé formulas. The representation adopted in this book has certain advantages, and its use seems to be gaining ground. It is interesting that very much the same representation was advanced as long ago as 1899 by Johannes Thiele (of the University of Munich), who used a broken circle to stand for partial bonds ("partial valences").

10.10 Aromatic character. The Hückel $4n + 2$ rule

We have defined aromatic compounds as those that resemble benzene. But just which properties of benzene must a compound possess before we speak of it as being aromatic? Besides the compounds that contain benzene rings, there are many other substances that are called aromatic; yet some of these superficially bear little resemblance to benzene.

What properties do all aromatic compounds have in common?

From the experimental standpoint, aromatic compounds are compounds whose molecular formulas would lead us to expect a high degree of unsaturation, and yet which are resistant to the addition reactions generally characteristic of unsaturated compounds. Instead of addition reactions, we often find that these aromatic compounds undergo electrophilic substitution reactions like those of benzene. Along with this resistance toward addition—and presumably the cause of it—we find evidence of unusual stability: low heats of hydrogenation and low heats of combustion. Aromatic compounds are cyclic—generally containing five-, six-, or seven-membered rings—and when examined by physical methods, they are found to have flat (or nearly flat) molecules. Their protons show the same sort of *chemical shift* in nmr spectra (Sec. 13.8) as the protons of benzene and its derivatives.

From a theoretical standpoint, to be aromatic a compound must have a molecule that contains *cyclic clouds of delocalized π electrons above and below the plane of the molecule*; furthermore, *the π clouds must contain a total of* **(4n + 2) π electrons**. That is to say, for the particular degree of stability that characterizes an aromatic compound, delocalization alone is not enough. There must be a particular number of π electrons: 2, or 6, or 10, etc. This requirement, called the *4n + 2 rule* or *Hückel rule* (after Erich Hückel, of the Institut für theoretische Physik, Stuttgart), is based on quantum mechanics, and has to do with the filling up of the various orbitals that make up the π cloud (Sec. 29.6). The Hückel rule is strongly supported by the facts.

Let us look at some of the evidence supporting the Hückel rule. Benzene has six π electrons, the *aromatic sextet*; six is, of course, a Hückel number, corresponding to $n = 1$. Besides benzene and its relatives (naphthalene, anthracene, phenanthrene, Chap. 30), we shall encounter a number of heterocyclic compounds (Chap. 31) that are clearly aromatic; these aromatic heterocycles, we shall see, are just the ones that can provide an aromatic sextet.

Or, as further examples, consider these six compounds, for each of which just one contributing structure is shown:

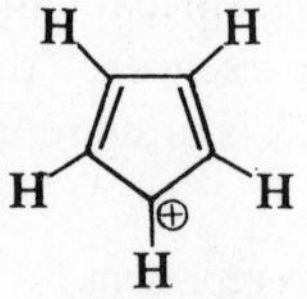

Cyclopentadienyl cation
Four π electrons

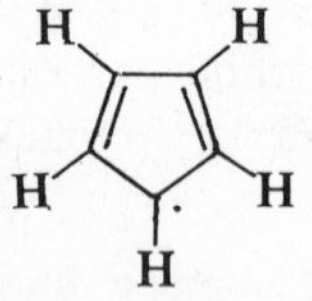

Cyclopentadienyl radical
Five π electrons

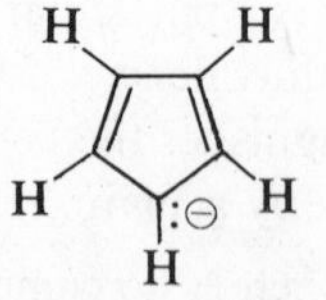

Cyclopentadienyl anion
Six π electrons
Aromatic

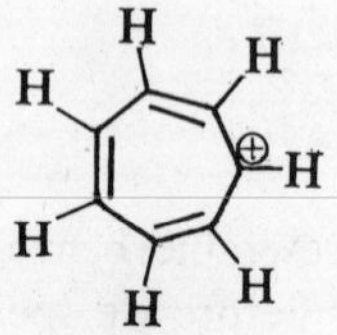

Cycloheptatrienyl cation
(Tropylium ion)
Six π electrons
Aromatic

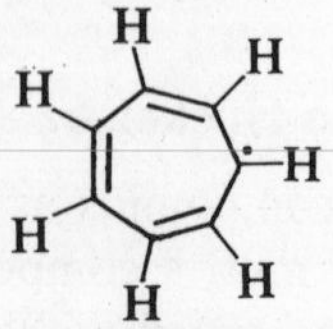

Cycloheptatrienyl radical
Seven π electrons

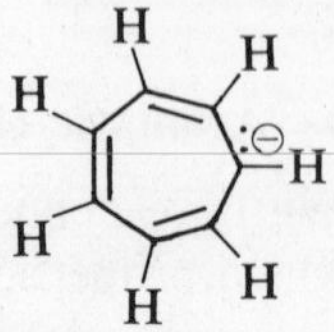

Cycloheptatrienyl anion
Eight π electrons

Each molecule is a hybrid of either five or seven equivalent structures, with the charge or odd electron on each carbon. Yet, of the six compounds, only *two* give evidence of *unusually* high stability: the cyclopentadienyl anion and the cycloheptatrienyl cation (*tropylium ion*).

For a hydrocarbon, cyclopentadiene is an unusually strong acid ($K_a = 10^{-15}$), indicating that loss of a hydrogen ion gives a particularly stable anion. (It is, for example, a much stronger acid than cycloheptatriene, $K_a = 10^{-45}$, despite the fact that the latter gives an anion that is stabilized by seven contributing structures.) Dicyclopentadienyliron (*ferrocene*), $[(C_5H_5)^-]_2Fe^{++}$, is a stable molecule that has been shown to be a "sandwich" of an iron atom between two flat five-membered rings. All carbon–carbon bonds are 1.4 A long. The rings of ferrocene undergo two typically aromatic substitution reactions: sulfonation and the Friedel-Crafts reaction.

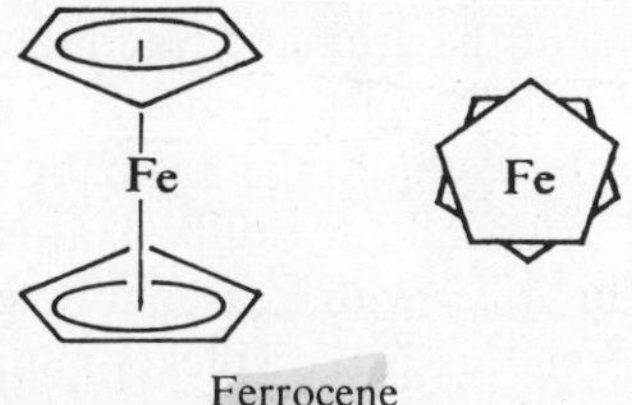

Ferrocene

Of the cycloheptatrienyl derivatives, on the other hand, it is the cation that is unusual. Tropylium bromide, C_7H_7Br, melts above 200°, is soluble in water but insoluble in non-polar solvents, and gives an immediate precipitate of AgBr when treated with silver nitrate. This is strange behavior for an organic bromide, and strongly suggests that, even in the solid, we are dealing with an ionic compound, R^+Br^-, the cation of which is actually a *stable* carbonium ion.

Consider the electronic configuration of the cyclopentadienyl anion (Fig. 10.5). Each carbon, trigonally hybridized, is held by a σ bond to two other carbons

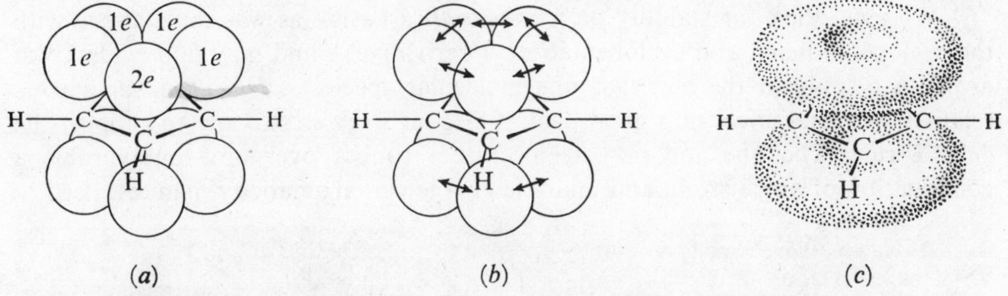

Figure 10.5. Cyclopentadienyl anion. (*a*) Two electrons in *p* orbital of one carbon; one electron in *p* orbital of each of the other carbons. (*b*) Overlap of *p* orbitals to form π bonds. (*c*) π clouds above and below plane of ring; total of six π electrons, the aromatic sextet.

and one hydrogen. The ring is a regular pentagon, whose angles (108°) are not a bad fit for the 120° trigonal angle; any instability due to imperfect overlap (angle strain) is more than made up for by the delocalization that is to follow. Four carbons have one electron each in *p* orbitals; the fifth carbon (the "one" that lost the proton, but actually, of course, indistinguishable from the others) has two electrons. Overlap of the p orbitals gives rise to π clouds containing a total of six electrons, the aromatic sextet.

In a similar way, we arrive at the configuration of the tropylium ion. It is a regular heptagon (angles 128.5°). Six carbons contribute one p electron each, and the seventh contributes only an empty p orbital. Result: the aromatic sextet. The ions are conveniently represented as:

Cyclopentadienyl anion

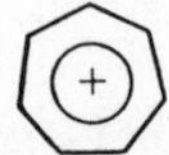

Cycloheptatrienyl cation (Tropylium ion)

Six is the Hückel number most often encountered, and for good reason. To provide p orbitals, the atoms of the aromatic ring must be trigonally (sp^2) hybridized, which means, ideally, bond angles of 120°. To permit the overlap of the p orbitals that gives rise to the π cloud, the aromatic compound must be flat, or nearly so. The number of trigonally hybridized atoms that will fit a flat ring without undue angle strain (i.e., with reasonably good overlap for π bond formation) is five, six, or seven. Six is the Hückel number of π electrons that can be provided—as we have just seen—by these numbers of atoms. (It is surely no coincidence that benzene, our model for aromatic character, is the "perfect" specimen: six carbons to provide six π electrons and to make a hexagon whose angles exactly match the trigonal angle.)

Now, what evidence is there that other Hückel numbers—2, 10, 14, etc.—are also "magic" numbers? We cannot expect aromatic character necessarily to appear here in the form of highly stable compounds comparable to benzene and its derivatives. The rings will be too small or too large to accommodate trigonally hybridized atoms very well, so that any stabilization due to aromaticity may be largely offset by angle strain or poor overlap of p orbitals, or both.

We must look for stability on a *comparative* basis—as was done above with the cyclopentadienyl and cycloheptatrienyl derivatives—and may find evidence of aromaticity only in the fact that one molecular species is *less unstable* than its relatives. The net effect of a great deal of elegant work is strongly to support the $4n + 2$ rule. The question now seems rather to be: over how unfavorable a combination of angle strain and multiple charge can aromaticity manifest itself?

Problem 10.6 Ronald Breslow (of Columbia University) found that treatment of 3-chlorocyclopropene with $SbCl_5$ yields a stable crystalline solid, I, of formula

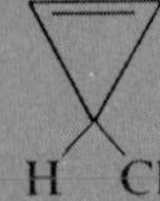

3-Chlorocyclopropene

$C_3H_3SbCl_6$, insoluble in non-polar solvents but soluble in polar solvents like nitromethane, acetonitrile, or sulfur dioxide. The nmr spectrum of I shows three exactly equivalent protons. 3-Chlorocyclopropene reacts with $AgBF_4$ to give AgCl and a solution with an nmr spectrum identical to that of I. Treatment of I with chloride ion regenerates 3-chlorocyclopropene.

Conversion of I into $C_3H_3^+$ by electron impact (Sec. 5.16) requires 235 kcal/mole, as compared with 255 kcal/mole for conversion of allyl chloride into $C_3H_5^+$.

(a) Give in detail the most likely structure of I, and show how this structure accounts for the various observations. (b) Of what theoretical significance are these findings?

Problem 10.7 1,3,5,7-*Cyclooctatetraene*, C_8H_8, has a heat of combustion (compare Problem 10.2, p. 323) of 1095 kcal; it rapidly decolorizes cold aqueous $KMnO_4$ and reacts with Br_2/CCl_4 to yield $C_8H_8Br_8$. (a) How should its structure be represented? (b) Upon what theoretical grounds might one have predicted its structure and properties? (c) Treatment of cyclooctatetraene with potassium metal has been found to yield a stable compound $2K^+C_8H_8^{--}$. Of what significance is the formation of this salt? (d) Using models, suggest a possible shape (or shapes) for cyclooctatetraene. What shape would you predict for the $C_8H_8^{--}$ anion?

10.11 Nomenclature of benzene derivatives

In later chapters we shall consider in detail the chemistry of many of the derivatives of benzene. Nevertheless, for our present discussion of the reactions of the benzene ring it will be helpful for us to learn to name some of the more important of these derivatives.

For many of these derivatives we simply prefix the name of the substituent group to the word *–benzene*, as, for example, in *chlorobenzene*, *bromobenzene*, *iodobenzene*, or *nitrobenzene*. Other derivatives have special names which may

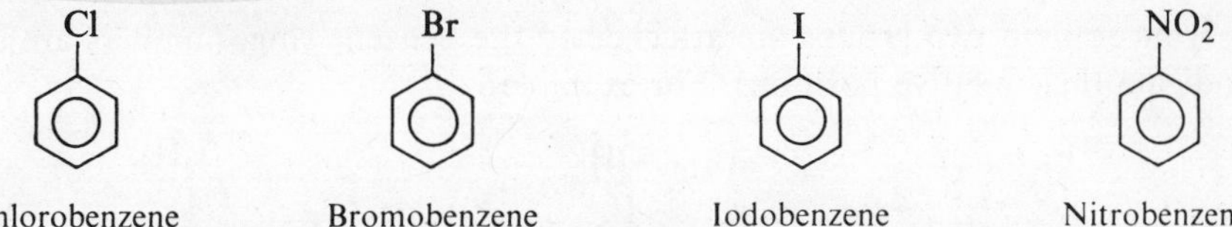

show no resemblance to the name of the attached substituent group. For example, methylbenzene is always known as *toluene*, aminobenzene as *aniline*, hydroxybenzene as *phenol*, and so on. The most important of these special compounds are:

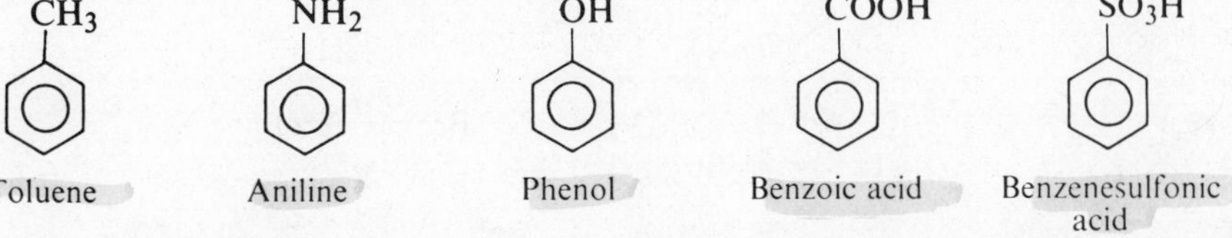

If several groups are attached to the benzene ring, we must not only tell what they are, but also indicate their relative positions. The three possible isomers of a disubstituted benzene are differentiated by the use of the names ***ortho***, ***meta***, and ***para***. For example:

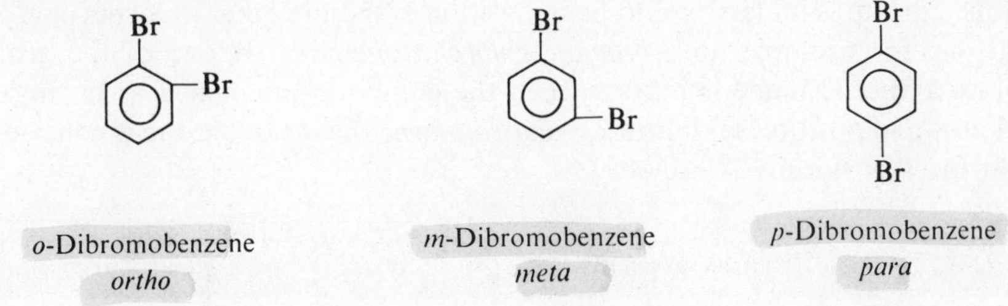

If the two groups are different, and neither is a group that gives a special name to the molecule, we simply name the two groups successively and end the word with *-benzene*, as, for example, *chloronitrobenzene*, *bromoiodobenzene*, etc. If one of the two groups is the kind that gives a special name to the molecule, then the compound is named as a derivative of that special compound, as, for example, *nitrotoluene*, *bromophenol*, etc.

p-Bromoiodobenzene 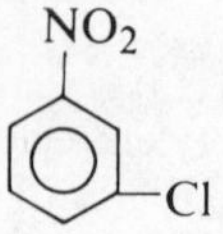   *m*-Chloronitrobenzene 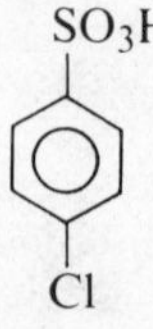   *p*-Chlorobenzenesulfonic acid

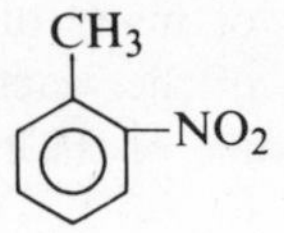

o-Nitrotoluene 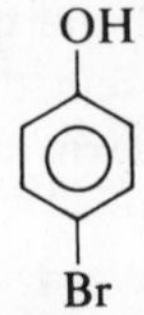   *p*-Bromophenol 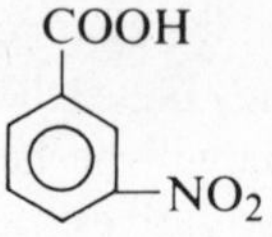   *m*-Nitrobenzoic acid 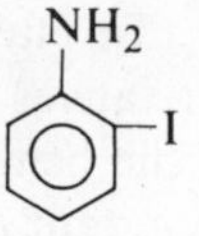   *o*-Iodoaniline

If more than two groups are attached to the benzene ring, numbers are used to indicate their relative positions. For example:

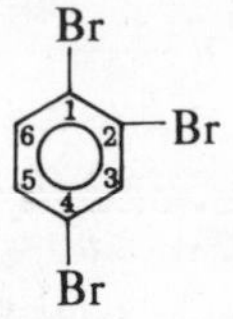

1,2,4-Tribromobenzene 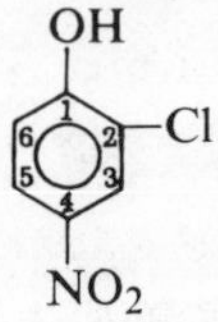   2-Chloro-4-nitrophenol 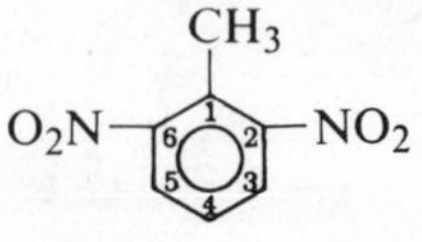   2,6-Dinitrotoluene

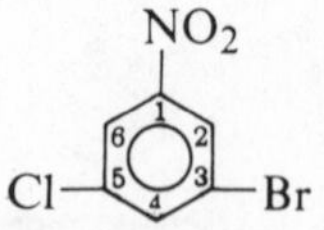

3-Bromo-5-chloronitrobenzene 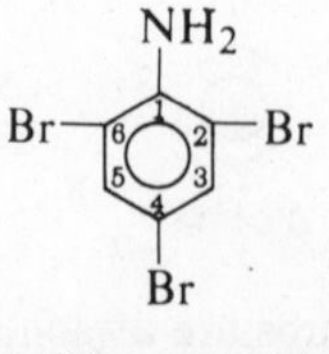   2,4,6-Tribromoaniline

If all the groups are the same, each is given a number, the sequence being the one that gives the lowest combination of numbers; if the groups are different, then the last-named group is understood to be in position 1 and the other numbers conform to that, as, for example, in *3-bromo-5-chloronitrobenzene*. If one of the groups that gives a special name is present, then the compound is named as having the special group in position 1; thus in *2,6-dinitrotoluene* the methyl group is considered to be at the 1-position.

Problem 10.8 You have three bottles containing the three isomeric dibromobenzenes; they have the melting points +87°, +6°, and −7°. By a great deal of work, you prepare six dibromonitrobenzenes ($C_6H_3Br_2NO_2$) and find that, of the six, *one* is related to (derived from or convertible into) the dibromobenzene of m.p. +87°, *two* to the isomer of m.p. +6°, and *three* to the isomer of m.p. −7°.

Label each bottle with the correct name of *ortho*, *meta*, or *para*.

(This work was actually carried out by Wilhelm Körner, of the University of Milan, and was the first example of the **Körner method of absolute orientation.**)

10.12 Quantitative elemental analysis: nitrogen and sulfur

This chapter has dealt with the structure of benzene and with some of its reactions. It is well to remind ourselves again that all this discussion has meaning only because it is based upon solid facts. As we saw earlier (Sec. 2.24), we can discuss the structure and reactions of a compound only when we know its molecular formula and the molecular formulas of its products.

To know a molecular formula we must know what elements are present in the compound, and in what proportions. In Sec. 2.25 we saw how various elements can be detected in an organic compound, and in Sec. 2.26 how the percentage of carbon, hydrogen, and halogen can be measured.

Quantitative analysis for nitrogen is carried out either (a) by the *Dumas method* or (b) by the *Kjeldahl method.* The Kjeldahl method is somewhat more convenient, particularly if many analyses must be carried out; however, it cannot be used for all kinds of nitrogen compounds.

In the Dumas method, the organic compound is passed through a tube containing, first, hot copper oxide and, next, hot copper metal gauze. The copper oxide oxidizes the compound (as in the carbon–hydrogen combustion, Sec. 2.26), converting combined nitrogen into molecular nitrogen. The copper gauze reduces any nitrogen oxides that may be formed, also to molecular nitrogen. The nitrogen gas is collected and its volume is measured. For example, an 8.32-mg sample of *aniline* yields 1.11 cc of nitrogen at 21° and 743 mm pressure (corrected for the vapor pressure of water). We calculate the volume at standard temperature and pressure,

$$\text{vol. } N_2 \text{ at S.T.P.} = 1.11 \times \frac{273}{273 + 21} \times \frac{743}{760} = 1.01 \text{ cc}$$

and, from it, the weight of nitrogen,

$$\text{wt. N} = \frac{1.01}{22400} \times (2 \times 14.01) = 0.00126 \text{ g } \textit{or } 1.26 \text{ mg}$$

and, finally, the percentage of nitrogen in the sample

$$\%\text{N} = \frac{1.26}{8.32} \times 100 = 15.2\%$$

Problem 10.9 Why is the nitrogen in the Dumas analysis collected over 50% aqueous KOH rather than, say, pure water, aqueous NaCl, or mercury?

In the Kjeldahl method, the organic compound is digested with concentrated sulfuric acid, which converts combined nitrogen into ammonium sulfate. The solution is then made alkaline. The ammonia thus liberated is distilled, and its amount is determined by titration with standard acid. For example, the ammonia formed from a 3.51-mg sample of aniline neutralizes 3.69 ml of 0.0103 N acid. For every milliequivalent of acid there is a milliequivalent of ammonia, and a

$$\text{milligram-atoms N} = \text{milliequivalents NH}_3 = \text{milliequivalents acid}$$
$$= 3.69 \times 0.0103 = 0.0380$$

milligram-atom of nitrogen. From this, the weight and, finally, the percentage of nitrogen in the compound can be calculated.

$$\text{wt. N} = \text{milligram-atoms N} \times 14.01 = 0.0380 \times 14.01 = 0.53 \text{ mg}$$

$$\%\text{N} = \frac{0.53}{3.51} \times 100 = 15.1\%$$

Sulfur in an organic compound is converted into sulfate ion by the methods used in halogen analysis (Sec. 2.26): treatment with sodium peroxide or with nitric acid (*Carius method*). This is then converted into barium sulfate, which is weighed.

Problem 10.10 A Dumas nitrogen analysis of a 5.72-mg sample of *p-phenylenediamine* gave 1.31 cc of nitrogen at 20° and 746 mm. The gas was collected over saturated aqueous KOH solution (the vapor pressure of water, 6 mm). Calculate the percentage of nitrogen in the compound.

Problem 10.11 A Kjeldahl nitrogen analysis of a 3.88-mg sample of *ethanolamine* required 5.73 ml of 0.0110 N hydrochloric acid for titration of the ammonia produced. Calculate the percentage of nitrogen in the compound.

Problem 10.12 A Carius sulfur analysis of a 4.81-mg sample of *p-toluenesulfonic acid* gave 6.48 mg of $BaSO_4$. Calculate the percentage of sulfur in the compound.

Problem 10.13 How does each of the above answers compare with the theoretical value calculated from the formula of the compound? (Each compound is listed in the index.)

PROBLEMS

1. Draw structures of:

(a) *p*-dinitrobenzene
(b) *m*-bromonitrobenzene
(c) *o*-chlorobenzoic acid
(d) *m*-nitrotoluene
(e) *p*-bromoaniline
(f) *m*-iodophenol
(g) mesitylene (1,3,5-trimethylbenzene)
(h) 3,5-dinitrobenzenesulfonic acid
(i) 4-chloro-2,3-dinitrotoluene
(j) 2-amino-5-bromo-3-nitrobenzoic acid
(k) *p*-hydroxybenzoic acid
(l) 2,4,6-trinitrophenol (picric acid)

2. Give structures and names of all the possible isomeric:

(a) xylenes (dimethylbenzenes)
(b) aminobenzoic acids ($H_2NC_6H_4COOH$)
(c) trimethylbenzenes
(d) dibromonitrobenzenes
(e) bromochlorotoluenes
(f) trinitrotoluenes

3. (a) How many isomeric monosubstitution products are theoretically possible from each of the following structures of formula C_6H_6? (b) How many disubstitution products? (c) Which structures, if any, would be acceptable for benzene on the basis of isomer number?

$HC{\equiv}C{-}CH_2{-}CH_2{-}C{\equiv}CH$ (I) $HC{\equiv}C{-}CH_2{-}C{\equiv}C{-}CH_3$ (II)

$HC{\equiv}C{-}C{\equiv}C{-}CH_2{-}CH_3$ (III)

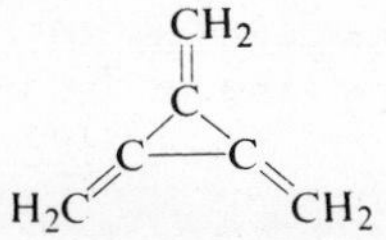

IV: a cyclopropane ring of three C atoms, one bearing $=CH_2$ and the other two each bearing $=CH_2$ ($H_2C{=}$, $=CH_2$)

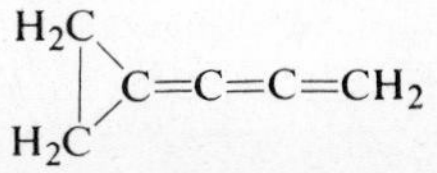

V: H_2C, H_2C (ring) $>C{=}C{=}C{=}CH_2$

4. Give structures and names of all theoretically possible products of the ring mononitration of:

(a) *o*-dichlorobenzene
(b) *m*-dichlorobenzene
(c) *p*-dichlorobenzene
(d) *o*-bromochlorobenzene
(e) *m*-bromochlorobenzene
(f) *p*-bromochlorobenzene
(g) *o*-chloronitrobenzene
(h) *m*-chloronitrobenzene
(i) *p*-chloronitrobenzene
(j) 1,3,5-trimethylbenzene
(k) 4-bromo-1,2-dimethylbenzene
(l) *p*-ethyltoluene

5. Give structures and names of all benzene derivatives that *theoretically* can have the indicated number of isomeric ring-substituted derivatives.

(a) C_8H_{10}: one monobromo derivative
(b) C_8H_{10}: two monobromo derivatives
(c) C_8H_{10}: three monobromo derivatives
(d) C_9H_{12}: one mononitro derivative
(e) C_9H_{12}: two mononitro derivatives
(f) C_9H_{12}: three mononitro derivatives
(g) C_9H_{12}: four mononitro derivatives

6. There are three known tribromobenzenes, of m.p. 44°, 87°, and 120°. Could these isomers be assigned structures by use of the Körner method (Problem 10.8, p. 332)? Justify your answer.

7. For a time the prism formula VI, proposed in 1869 by Albert Ladenburg of Germany, was considered as a possible structure for benzene, on the grounds that it would yield one monosubstitution product and three isomeric disubstitution products.

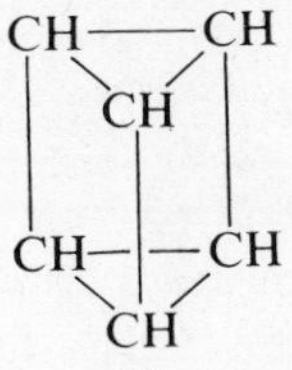

VI: prism structure of six CH units (two triangular CH faces joined by three vertical bonds)

(a) Draw Ladenburg structures of three possible isomeric dibromobenzenes.
(b) On the basis of the Körner method of absolute orientation, label each Ladenburg structure in (a) as *ortho*, *meta*, or *para*.
(c) In light of Chap. 4, can the Ladenburg formula actually pass the test of isomer number?
(Derivatives of Ladenburg "benzene," called *prismanes*, have actually been made.)

8. In 1874 Griess (p. 1077) reported that he had decarboxylated the six known diaminobenzoic acids, $C_6H_3(NH_2)_2COOH$, to the diaminobenzenes. Three acids gave a diamine of m.p. 63°, two acids gave a diamine of m.p. 104°, and one acid gave a diamine of m.p. 142°. Draw the structural formulas for the three isomeric diaminobenzenes and label each with its melting point.

9. For which of the following might you expect aromaticity (geometry permitting)?

(a) The annulenes containing up to 20 carbons. (*Annulenes* are monocyclic compounds of the general formula $[—CH{=}CH—]_n$.)

(b) The monocyclic polyenes C_9H_{10}, $C_9H_9^+$, $C_9H_9^-$.

10. The properties of *pyrrole*, commonly represented by VII,

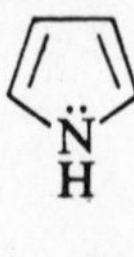

VII

show that it is aromatic. Account for its aromaticity on the basis of orbital theory. (*Hint:* See Sec. 10.10. Check your answer in Sec. 31.2.)

11. When benzene is treated with chlorine under the influence of ultraviolet light, a solid material of m.wt. 291 is formed. Quantitative analysis gives an empirical formula of CHCl. (a) What is the molecular formula of the product? (b) What is a possible structural formula? (c) What kind of reaction has taken place? (d) Is the product aromatic? (e) Actually, the product can be separated into six isomeric compounds, one of which is used as an insecticide (Gammexane or Lindane). How do these isomers differ from each other? (f) Are more than six isomers possible?

12. Can you account for the following order of acidity. (*Hint:* See Sec. 8.10.)

acetylene > benzene > *n*-pentane

Chapter 11 Electrophilic Aromatic Substitution

11.1 Introduction

We have already seen that the characteristic reactions of benzene involve substitution, in which the resonance-stabilized ring system is preserved. What kind of reagents bring about this substitution? What is the mechanism by which these reactions take place?

Above and below the plane of the benzene ring there is a cloud of π electrons. Because of resonance, these π electrons are more involved in holding together carbon nuclei than are the π electrons of a carbon–carbon double bond. Still, in comparison with σ electrons, these π electrons are loosely held and are available to a reagent that is seeking electrons.

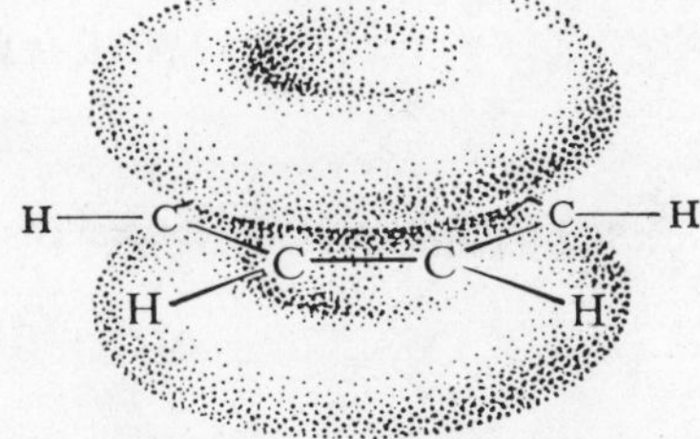

Figure 11.1. Benzene ring: π cloud is source of electrons.

It is not surprising that *in its typical reactions the benzene ring serves as a* ***source of electrons***, that is, as a **base**. The compounds with which it reacts are deficient in electrons, that is, are electrophilic reagents or acids. Just as the typical reactions of the alkenes are electrophilic addition reactions, so *the typical reactions of the benzene ring are* **electrophilic substitution reactions.**

These reactions are characteristic not only of benzene itself, but of the benzene ring wherever it is found—and, indeed, of many aromatic rings, benzenoid and non-benzenoid.

Electrophilic aromatic substitution includes a wide variety of reactions: nitration, halogenation, sulfonation, and Friedel-Crafts reactions, undergone by nearly all aromatic rings; reactions like nitrosation and diazo coupling, undergone only by rings of high reactivity; and reactions like desulfonation, isotopic exchange, and many ring closures which, although apparently unrelated, are found on closer examination to be properly and profitably viewed as reactions of this kind. In synthetic importance electrophilic aromatic substitution is probably unequaled by any other class of organic reactions. It is the initial route of access to nearly all aromatic compounds: it permits the direct introduction of certain substituent groups which can then be converted, by replacement or by transformation, into other substituents, including even additional aromatic rings.

ELECTROPHILIC AROMATIC SUBSTITUTION

Ar = *aryl*, any aromatic group with attachment directly to ring carbon

1. **Nitration.** Discussed in Sec. 11.8.

$$\mathrm{ArH + HONO_2 \xrightarrow{H_2SO_4} ArNO_2 + H_2O}$$

A nitro compound

2. **Sulfonation.** Discussed in Sec. 11.9.

$$\mathrm{ArH + HOSO_3H \xrightarrow{SO_3} ArSO_3H + H_2O}$$

A sulfonic acid

3. **Halogenation.** Discussed in Sec. 11.11.

$$\mathrm{ArH + Cl_2 \xrightarrow{Fe} ArCl + HCl}$$

An aryl chloride

$$\mathrm{ArH + Br_2 \xrightarrow{Fe} ArBr + HBr}$$

An aryl bromide

4. **Friedel-Crafts alkylation.** Discussed in Sec. 11.10.

$$\mathrm{ArH + RCl \xrightarrow{AlCl_3} ArR + HCl}$$

An alkylbenzene

5. **Friedel-Crafts acylation.** Discussed in Sec. 19.6.

$$\mathrm{ArH + RCOCl \xrightarrow{AlCl_3} ArCOR + HCl}$$

An acyl chloride — A ketone

6. **Protonation.** Discussed in Sec. 11.12.

$$\mathrm{ArSO_3H + H^+ \xrightarrow{H_2O} ArH + H_2SO_4} \qquad \textit{Desulfonation}$$

$$\mathrm{ArH + D^+ \longrightarrow ArD + H^+} \qquad \textit{Hydrogen exchange}$$

7. **Thallation.** Discussed in Sec. 11.13.

$$ArH + Tl(OOCCF_3)_3 \xrightarrow{CF_3COOH} ArTl(OOCCF_3)_2 + CF_3COOH$$

Thallium trifluoroacetate — An arylthallium ditrifluoroacetate

8. **Nitrosation.** Discussed in Secs. 23.11 and 24.10.

$$ArH + HONO \longrightarrow ArN{=}O + H_2O$$

A nitroso compound — *Only for highly reactive* ArH

9. **Diazo coupling.** Discussed in Sec. 23.17.

$$ArH + Ar'N_2{}^+X^- \longrightarrow ArN{=}NAr' + HX$$

A diazonium salt — An azo compound — *Only for highly reactive* ArH

10. **Kolbe reaction.** Discussed in Sec. 24.11. *Only for phenols.*

11. **Reimer-Tiemann reaction.** Discussed in Sec. 24.12. *Only for phenols.*

11.2 Effect of substituent groups

Like benzene, toluene undergoes electrophilic aromatic substitution: sulfonation, for example. Although there are three possible monosulfonation products, this reaction actually yields appreciable amounts of only two of them: the *o*- and *p*-isomers.

$$C_6H_5CH_3 \xrightarrow{H_2SO_4,\ SO_3,\ 35^\circ} p\text{-}CH_3C_6H_4SO_3H \text{ and } o\text{-}CH_3C_6H_4SO_3H \text{ and } 6\%\ m\text{-isomer}$$

Toluene — *p*-Toluenesulfonic acid *62%* — *o*-Toluenesulfonic acid *32%*

Benzene and toluene are insoluble in sulfuric acid, whereas the sulfonic acids are readily soluble; completion of reaction is indicated simply by disappearance of the hydrocarbon layer. When shaken with fuming sulfuric acid at room temperature, benzene reacts completely within 20 to 30 minutes, whereas toluene is found to react within only a minute or two.

Studies of nitration, halogenation, and Friedel-Crafts alkylation of toluene give analogous results. In some way the methyl group makes the ring more reactive than unsubstituted benzene, and *directs* the attacking reagent to the *ortho* and *para* positions of the ring.

On the other hand, nitrobenzene, to take a different example, has been found to undergo substitution more slowly than benzene, and to yield chiefly the *meta isomer*.

Like methyl or nitro, any group attached to a benzene ring affects the **reactivity** of the ring and determines the **orientation** of substitution. When an electrophilic reagent attacks an aromatic ring, it is the group already attached to the ring that determines *how readily* the attack occurs and *where* it occurs.

A group that makes the ring more reactive than benzene is called an **activating group**. A group that makes the ring less reactive than benzene is called a **deactivating group.**

A group that causes attack to occur chiefly at positions **ortho** and **para** to it is called an **ortho,para director**. A group that causes attack to occur chiefly at positions **meta** to it is called a **meta director.**

In this chapter we shall examine the methods that are used to measure these effects on reactivity and orientation, the results of these measurements, and a theory that accounts for these results. The theory is, of course, based on the most likely mechanism for electrophilic aromatic substitution; we shall see what this mechanism is, and some of the evidence supporting it. First let us look at the facts.

11.3 Determination of orientation

To determine the effect of a group on orientation is, in principle, quite simple: the compound containing this group attached to benzene is allowed to undergo substitution and the product is analyzed for the proportions of the three isomers. Identification of each isomer as *ortho*, *meta*, or *para* generally involves comparison with an authentic sample of that isomer prepared by some other method from a compound whose structure is known. In the last analysis, of course, all these identifications go back to absolute determinations of the Körner type (Problem 10.8, p. 332).

In this way it has been found that every group can be put into one of two classes: *ortho,para* directors or *meta* directors. Table 11.1 summarizes the orientation of nitration in a number of substituted benzenes. Of the five positions open to attack, three (60%) are *ortho* and *para* to the substituent group, and two (40%) are *meta* to the group; if there were no selectivity in the substitution reaction, we

Table 11.1 ORIENTATION OF NITRATION OF C_6H_5Y

Y	*Ortho*	*Para*	*Ortho* plus *para*	*Meta*
—OH	50–55	45–50	100	trace
—$NHCOCH_3$	19	79	98	2
—CH_3	58	38	96	4
—F	12	88	100	trace
—Cl	30	70	100	trace
—Br	37	62	99	1
—I	38	60	98	2
—NO_2	6.4	0.3	6.7	93.3
—$N(CH_3)_3^+$	0	11	11	89
—CN	—	—	19	81
—COOH	19	1	20	80
—SO_3H	21	7	28	72
—CHO	—	—	28	72

would expect the *ortho* and *para* isomers to make up 60% of the product, and the *meta* isomer to make up 40%. We see that seven of the groups direct 96–100% of nitration to the *ortho* and *para* positions; the other six direct 72–94% to the *meta* positions.

A given group causes the same general kind of orientation—predominantly *ortho,para* or predominantly *meta*—whatever the electrophilic reagent involved. The actual distribution of isomers may vary, however, from reaction to reaction. In Table 11.2, for example, compare the distribution of isomers obtained from toluene by sulfonation or bromination with that obtained by nitration.

Table 11.2 ORIENTATION OF SUBSTITUTION IN TOLUENE

	Ortho	*Meta*	*Para*
Nitration	58	4	38
Sulfonation	32	6	62
Bromination	33	—	67

11.4 Determination of relative reactivity

A group is classified as *activating* if the ring it is attached to is more reactive than benzene, and is classified as *deactivating* if the ring it is attached to is less reactive than benzene. The reactivities of benzene and a substituted benzene are compared in one of the following ways.

The **time required** for reactions to occur under identical conditions can be measured. Thus, as we just saw, toluene is found to react with fuming sulfuric acid in about one-tenth to one-twentieth the time required by benzene. Toluene is more reactive than benzene, and $—CH_3$ is therefore an activating group.

The **severity of conditions** required for comparable reaction to occur within the same period of time can be observed. For example, benzene is nitrated in less than an hour at 60° by a mixture of concentrated sulfuric acid and concentrated nitric acid; comparable nitration of nitrobenzene requires treatment at 90° with fuming nitric acid and concentrated sulfuric acid. Nitrobenzene is evidently less reactive than benzene, and the nitro group, $—NO_2$, is a deactivating group.

For an exact, quantitative comparison under identical reaction conditions, **competitive reactions** can be carried out, in which the compounds to be compared are allowed to compete for a limited amount of a reagent (Sec. 3.22). For example, if equimolar amounts of benzene and toluene are treated with a small amount of nitric acid (in a solvent like nitromethane or acetic acid, which will dissolve both

$$\underset{1}{C_6H_5NO_2} \xleftarrow{C_6H_6} HNO_3 \xrightarrow{C_6H_5CH_3} \underset{25}{CH_3C_6H_4NO_2} \quad o,\ m, \text{ and } p$$

$$\underset{30}{C_6H_5NO_2} \xleftarrow{C_6H_6} HNO_3 \xrightarrow{C_6H_5Cl} \underset{1}{ClC_6H_4NO_2} \quad o,\ m, \text{ and } p$$

organic and inorganic reactants), about 25 times as much nitrotoluene as nitrobenzene is obtained, showing that toluene is 25 times as reactive as benzene. On the other hand, a mixture of benzene and chlorobenzene yields a product in which nitrobenzene exceeds the nitrochlorobenzenes by 30:1, showing that chlorobenzene is only one-thirtieth as reactive as benzene. The chloro group is therefore classified as deactivating, the methyl group as activating. The activation or deactivation caused by some groups is extremely powerful: aniline, $C_6H_5NH_2$, is roughly one million times as reactive as benzene, and nitrobenzene, $C_6H_5NO_2$, is roughly one-millionth as reactive as benzene.

11.5 Classification of substituent groups

The methods described in the last two sections have been used to determine the effects of a great number of groups on electrophilic substitution. As shown in Table 11.3, nearly all groups fall into one of two classes: activating and *ortho,para*-directing, or deactivating and *meta*-directing. The halogens are in a class by themselves, being deactivating but *ortho,para*-directing.

Table 11.3 EFFECT OF GROUPS ON ELECTROPHILIC AROMATIC SUBSTITUTION

Activating: *Ortho,para* Directors	Deactivating: *Meta* Directors
Strongly activating	$—NO_2$
$—NH_2$ ($—NHR$, $—NR_2$)	$—N(CH_3)_3{}^+$
$—OH$	$—CN$
	$—COOH$ ($—COOR$)
Moderately activating	$—SO_3H$
$—OCH_3$ ($—OC_2H_5$, etc.)	$—CHO$, $—COR$
$—NHCOCH_3$	
	Deactivating: *Ortho,para* Directors
Weakly activating	$—F$, $—Cl$, $—Br$, $—I$
$—C_6H_5$	
$—CH_3$ ($—C_2H_5$, etc.)	

Just by knowing the effects summarized in these short lists, we can now predict fairly accurately the course of hundreds of aromatic substitution reactions. We now know, for example, that bromination of nitrobenzene will yield chiefly the *m*-isomer and that the reaction will go more slowly than the bromination of benzene itself; indeed, it will probably require severe conditions to go at all. We now know that nitration of $C_6H_5NHCOCH_3$ (*acetanilide*) will yield chiefly the *o*- and *p*-isomers and will take place more rapidly than nitration of benzene.

Although, as we shall see, it is possible to account for these effects in a reasonable way, it is necessary for the student to memorize the classifications in Table 11.3 so that he may deal rapidly with synthetic problems involving aromatic compounds.

11.6 Orientation in disubstituted benzenes

The presence of two substituents on a ring makes the problem of orientation more complicated, but even here we can frequently make very definite predictions. First of all, the two substituents may be located so that the directive influence of one *reinforces* that of the other; for example, in I, II, and III the orientation clearly must be that indicated by the arrows.

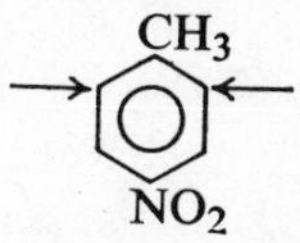

I 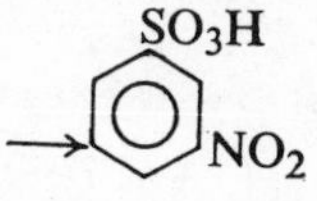 II 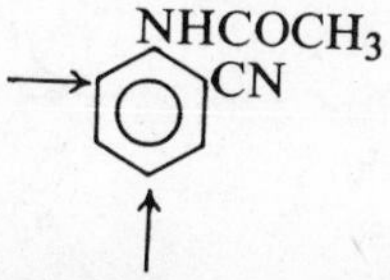 III

On the other hand, when the directive effect of one group *opposes* that of the other, it may be difficult to predict the major product; in such cases complicated mixtures of several products are often obtained.

Even where there are opposing effects, however, it is still possible in certain cases to make predictions in accordance with the following generalizations.

(a) *Strongly activating groups generally win out over deactivating or weakly activating groups.* The differences in directive power in the sequence

$$-NH_2, -OH > -OCH_3, -NHCOCH_3 > -C_6H_5, -CH_3 > \textit{meta} \text{ directors}$$

are great enough to be used in planning feasible syntheses. For example:

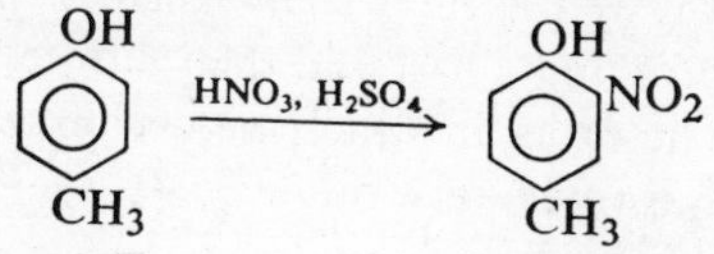

HNO_3, H_2SO_4

Sole product

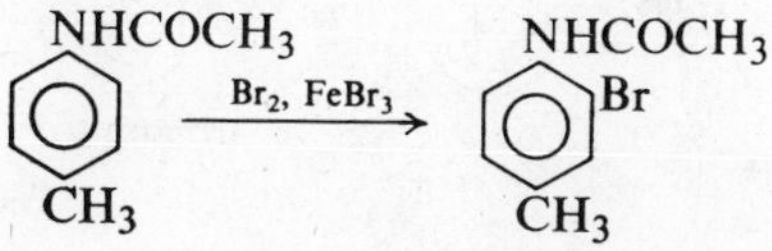

$Br_2, FeBr_3$

Chief product

$Br_2, FeBr_3$

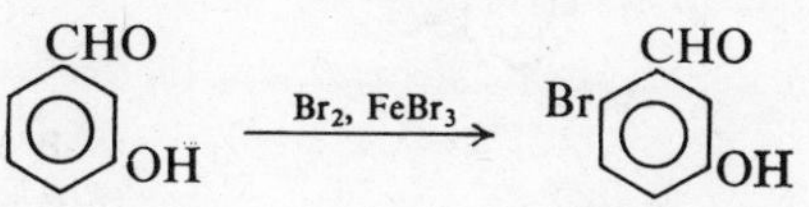

Chief product

There must be, however, a fairly large difference in the effects of the two groups for clear-cut results; otherwise one gets results like these:

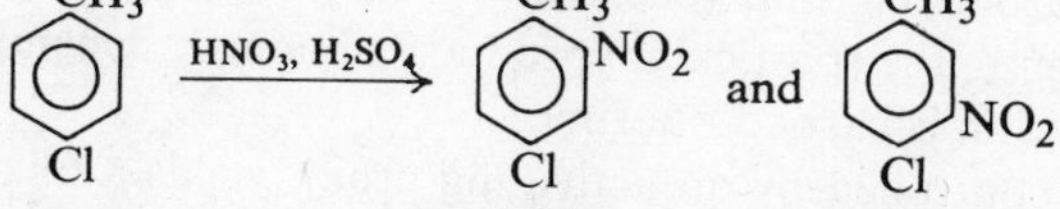

HNO_3, H_2SO_4 and

58% 42%

(b) *There is often little substitution between two groups that are meta to each other.* In many cases it seems as though there just is not enough room between two groups located *meta* to each other for appreciable substitution to occur there, as illustrated by IV and V:

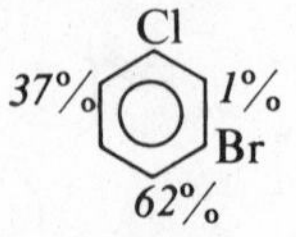

Cl
37% 1%
Br
62%
Nitration
IV

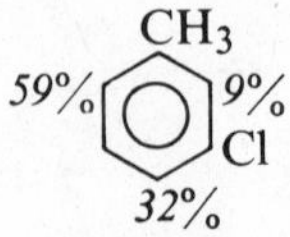

CH_3
59% 9%
Cl
32%
Nitration
V

11.7 Orientation and synthesis

As we discussed earlier (Sec. 3.14), a laboratory synthesis is generally aimed at obtaining a single, pure compound. Whenever possible we should avoid use of a reaction that produces a mixture, since this lowers the yield of the compound we want and causes difficult problems of purification. With this in mind, let us see some of the ways in which we can apply our knowledge of orientation to the synthesis of pure aromatic compounds.

First of all, *we must consider the order in which we introduce these various substituents into the ring*. In the preparation of the bromonitrobenzenes, for example, it is obvious that if we nitrate first and then brominate, we will obtain the *m*-isomer; whereas if we brominate first and then nitrate, we will obtain a mixture of the *o*- and *p*-isomers. The order in which we decide to carry out the two steps, then, depends upon which isomer we want.

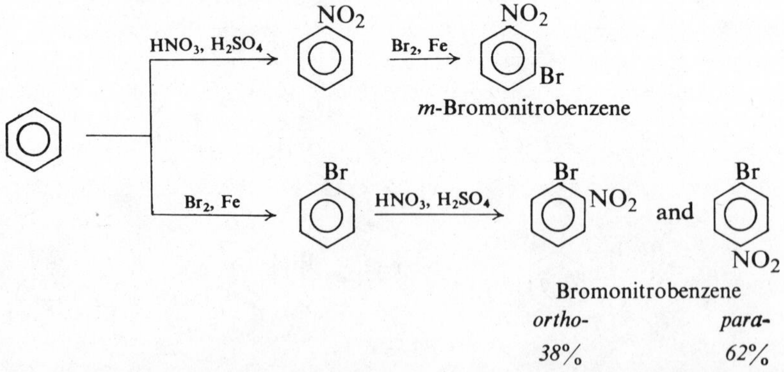

HNO_3, H_2SO_4 → NO_2 → Br_2, Fe → NO_2, Br
m-Bromonitrobenzene

Br_2, Fe → Br → HNO_3, H_2SO_4 → Br, NO_2 and Br, NO_2
Bromonitrobenzene
ortho- 38% *para*- 62%

Next, if our synthesis involves conversion of one group into another, *we must consider the proper time for this conversion*. For example, oxidation of a methyl group yields a carboxyl group (Sec. 12.10). In the preparation of nitrobenzoic acids from toluene, the particular product obtained depends upon whether oxidation or nitration is carried out first.

Substitution controlled by an activating group yields a mixture of *ortho* and *para* isomers; nevertheless, we must often make use of such reactions, as in the examples just shown. It is usually possible to obtain the pure *para* isomer from the mixture by fractional crystallization. As the more symmetrical isomer, it is the less soluble (Sec. 12.3), and crystallizes while the solvent still retains the soluble *ortho* isomer. Some *para* isomer, of course, remains in solution to contaminate the *ortho* isomer, which is therefore difficult to purify. As we shall see, special approaches are often used to prepare *ortho* isomers.

Toluene (CH_3) $\xrightarrow{KMnO_4}$ (COOH) $\xrightarrow{HNO_3,\ H_2SO_4}$ (COOH, NO_2) *m*-Nitrobenzoic acid

Toluene (CH_3) $\xrightarrow{HNO_3,\ H_2SO_4}$ (CH_3, NO_2) and (CH_3, NO_2)

$\downarrow K_2Cr_2O_7$ $\downarrow K_2Cr_2O_7$

(COOH, NO_2) (COOH, NO_2)

o-Nitrobenzoic acid *p*-Nitrobenzoic acid

In the special case of nitro compounds, the difference in boiling points is often large enough that both *ortho* and *para* isomers can be obtained pure by fractional distillation. As a result, many aromatic compounds are best prepared not by direct substitution but by conversion of one group into another, in the last analysis starting from an original nitro compound; we shall take up these methods of conversion later.

A goal of aromatic synthesis is control of orientation: the preparation, at will and from the same substrate, of a pure *ortho*, a pure *meta*, or a pure *para* isomer. Steps toward this goal have been taken very recently by Edward C. Taylor (Princeton University) and Alexander McKillop (University of East Anglia), chiefly through the chemistry of *thallium*: thallium as the cation in organic salts; thallium salts as Lewis acids; arylthallium compounds (Sec. 11.13) as reactive organometallic intermediates. One approach to regiospecific substitution involves complexing—attachment through a Lewis acid–base reaction—of the attacking reagent by some other molecule. Complexing of the reagent by the substituent group prior to reaction tends to favor attack at the *nearest* position: *ortho*. Complexing of the reagent by a bulky molecule tends to favor attack at the *least crowded* position: *para*. If reaction can be carried out so that orientation is governed, not by relative rates of reaction—as it usually is—but by position of equilibrium, then the *most stable* isomer is favored: often the *meta* isomer. We shall see examples of all these ways of controlling orientation.

11.8 Mechanism of nitration

Now that we have seen the effects that substituent groups exert on orientation and reactivity in electrophilic aromatic substitution, let us see how we can account for these effects. The first step in doing this is to examine the mechanism for the reaction. Let us begin with nitration, using benzene as the aromatic substrate.

The commonly accepted mechanism for nitration with a mixture of nitric and sulfuric acids (the widely used "mixed acid" of the organic chemist) involves the following sequence of reactions:

(1) $HONO_2 + 2H_2SO_4 \rightleftharpoons H_3O^+ + 2HSO_4^- + \oplus NO_2$
Nitronium ion

(2) $\oplus NO_2 + C_6H_6 \longrightarrow C_6H_5^{\oplus}(H)(NO_2)$ *Slow*

(3) $C_6H_5^{\oplus}(H)(NO_2) + HSO_4^- \longrightarrow C_6H_5NO_2 + H_2SO_4$ *Fast*

Step (1) generates the **nitronium ion**, $\oplus NO_2$, which is the electrophilic particle that actually attacks the benzene ring. This reaction is simply an acid–base equilibrium in which sulfuric acid serves as the acid and the much weaker nitric acid serves as a base. We may consider that the very strong acid, sulfuric acid, causes nitric acid to ionize in the sense, $HO^- \ldots {}^+NO_2$, rather than in the usual way, $H^+ \ldots {}^-ONO_2$. The nitronium ion is well known, existing in salts such as nitronium perchlorate, $NO_2^+ClO_4^-$, and nitronium fluoborate, $NO_2^+BF_4^-$. Indeed, solutions of these stable nitronium salts in solvents like nitromethane or acetic acid have been found by George Olah (of Case Western Reserve University) to nitrate aromatic compounds smoothly and in high yield at room temperature.

Needing electrons, the nitronium ion finds them particularly available in the π cloud of the benzene ring, and so in step (2) attaches itself to one of the carbon atoms by a covalent bond. This forms the carbonium ion,

$$C_6H_5^{\oplus}(H)(NO_2)$$

often called a *benzenonium ion.*

Just what is the structure of this carbonium ion? We find that we can represent it by three structures (I, II, and III) that differ from each other only in position of double bonds and positive charge. The actual ion must then be a resonance hybrid of these three structures.

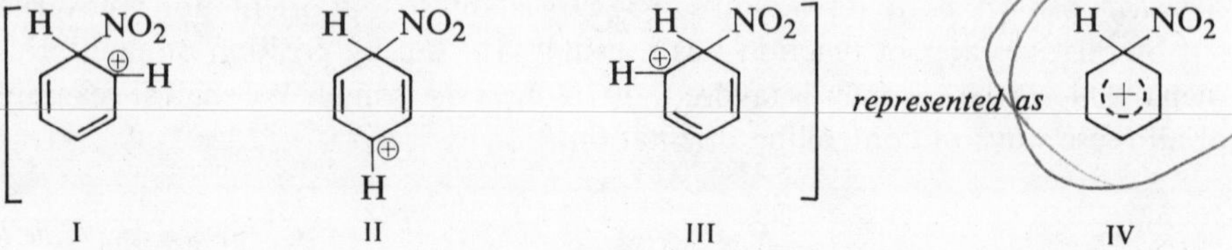

[Structures I, II, III (H, NO_2; $\oplus$, H) *represented as* IV (H, NO_2; (+))]

I II III IV

This means, of course, that the positive charge is not localized on one carbon atom, but is distributed over the molecule, being particularly strong on the carbon atoms *ortho* and *para* to the carbon bearing the $—NO_2$ group. (As we shall see later, this *ortho,para* distribution is significant.) The dispersal of the positive charge over the molecule by resonance makes this ion more stable than an ion with a localized positive charge. It is probably because of this stabilization that the carbonium ion forms at all, in view of the stability of the original benzene itself. Sometimes the hybrid carbonium ion is represented as IV, where the broken line stands for the fractional bonds due to the delocalized π electrons.

Thus far the reaction is like addition to alkenes: an electrophilic particle, attracted by the π electrons, attaches itself to the molecule to form a carbonium ion. But the fate of this carbonium ion is different from the fate of the ion formed from an alkene. Attachment of a basic group to the benzenonium ion to yield the addition product would destroy the aromatic character of the ring. Instead, the basic ion, HSO_4^-, abstracts a hydrogen ion (step 3) to yield the substitution product, which retains the resonance-stabilized ring. Loss of a hydrogen ion, as we have seen, is one of the reactions typical of a carbonium ion (Sec. 5.20); it is the *preferred* reaction in this case.

As with other carbonium ion reactions we have studied, it is the *formation* of the carbonium ion (step 2) that is the more difficult step; once formed, the carbonium ion rapidly loses a hydrogen ion (step 3) to form the products. (We shall see proof of this in Sec. 11.16.)

Electrophilic substitution, then, like electrophilic addition, is a stepwise process involving an intermediate carbonium ion. The two reactions differ, however, in the fate of the carbonium ion. While the mechanism of nitration is, perhaps, better established than the mechanisms for other aromatic substitution reactions, it seems clear that all these reactions follow the same course.

Problem 11.1 Nitration by nitric acid alone is believed to proceed by essentially the same mechanism as nitration in the presence of sulfuric acid. Write an equation for the generation of NO_2^+ from nitric acid alone.

11.9 Mechanism of sulfonation

Sulfonation of many aromatic compounds involves the following steps:

(1) $$2H_2SO_4 \rightleftharpoons H_3O^+ + HSO_4^- + SO_3$$

(2) $$SO_3 + C_6H_6 \rightleftharpoons \overset{\oplus}{C_6H_5}\genfrac{}{}{0pt}{}{\diagup H}{\diagdown SO_3^-}$$ *Slow*

(3) $$\overset{\oplus}{C_6H_5}\genfrac{}{}{0pt}{}{\diagup H}{\diagdown SO_3^-} + HSO_4^- \rightleftharpoons C_6H_5SO_3^- + H_2SO_4$$ *Fast*

(4) $$C_6H_5SO_3^- + H_3O^+ \rightleftharpoons C_6H_5SO_3H + H_2O$$ *Equilibrium far to the left*

Again the first step, which generates the electrophilic sulfur trioxide, is simply an acid–base equilibrium, this time between molecules of sulfuric acid. For sulfonation we commonly use sulfuric acid containing an excess of SO_3; even if this is not done, it appears that SO_3 formed in step (1) can be the electrophile.

$$:\ddot{O}: \quad \ddot{S}:\ddot{O}: \quad :\ddot{O}:$$

In step (2) the electrophilic reagent, SO_3, attaches itself to the benzene ring to form the intermediate carbonium ion. Although sulfur trioxide is not positively charged, it is electron-deficient, and hence an acid, nevertheless.

Step (3) is the loss of a hydrogen ion to form the resonance-stabilized substitution product, this time the anion of benzenesulfonic acid which, being a strong acid, is highly dissociated (step 4).

With some aromatic substrates and at certain acidities, the electrophile may be HSO_3^+ or molecules that can readily transfer SO_3 or HSO_3^+ to the aromatic ring.

Problem 11.2 Write an equation for the formation from H_2SO_4 of each of the following sulfonating electrophiles: (a) $H_3SO_4^+$; (b) HSO_3^+; (c) $H_2S_2O_7$.

11.10 Mechanism of Friedel-Crafts alkylation

In Friedel-Crafts alkylation, the electrophile is typically a carbonium ion. It, too, is formed in an acid-base equilibrium, this time in the Lewis sense:

(1) $$RCl + AlCl_3 \rightleftharpoons AlCl_4^- + R^{\oplus}$$

(2) $$R^{\oplus} + C_6H_6 \rightleftharpoons C_6H_5^{\oplus}(H)(R) \qquad \textit{Slow}$$

(3) $$C_6H_5^{\oplus}(H)(R) + AlCl_4^- \rightleftharpoons C_6H_5R + HCl + AlCl_3 \qquad \textit{Fast}$$

In certain cases, there is no free carbonium ion involved. Instead, the alkyl group is transferred—without a pair of electrons—directly to the aromatic ring from the polar complex, I, between $AlCl_3$ and the alkyl halide:

$$\underset{\text{I}}{Cl_3\overset{\ominus}{Al}—\overset{\oplus}{Cl}—R} + C_6H_6 \longrightarrow C_6H_5^{\oplus}(H)(R) + AlCl_4^- \qquad \textit{Slow}$$

The electrophile is thus either (a) R^+ or (b) a molecule like I that can readily *transfer* R^+ to the aromatic ring. *This duality of mechanism is common in electrophilic aromatic substitution.* In either case, the Lewis acid R^+ is displaced from RCl by the other Lewis acid, $AlCl_3$.

We speak of the Friedel-Crafts reaction as electrophilic substitution and, from the viewpoint of the aromatic ring, it is. But, just as an acid reacts with a base, so an electrophile reacts with a *nucleophile* (nucleus-lover), a molecule which can provide the electrons that the electrophile seeks. From the opposite point of view, then, this reaction involves nucleophilic attack by the aromatic ring on the alkyl group of complex I. The $AlCl_4^-$ ion is a better leaving group than Cl^- would be; the Lewis acid, $AlCl_3$, serves the same purpose here that a Lowry-Brønsted acid does in protonation of an alcohol (Sec. 5.20).

As we shall find out when we take up the Friedel-Crafts reaction as a synthetic tool (Sec. 12.6), the Friedel-Crafts reaction in its widest sense involves reactants other than alkyl halides and Lewis acids other than aluminium chloride: BF_3, $SnCl_4$, HF, and even H^+.

Problem 11.3 How do you account for the fact that benzene in the presence of $AlCl_3$ reacts: (a) with *n*-propyl bromide to give isopropylbenzene; (b) with isobutyl bromide to yield *tert*-butylbenzene; (c) with neopentyl bromide to yield *tert*-pentylbenzene? (d) By which of the alternative mechanisms for the Friedel-Crafts reaction are these products probably formed?

Problem 11.4 Write all steps in the most likely mechanism for the reaction of benzene: (a) with *tert*-butyl alcohol in the presence of H_2SO_4 to yield *tert*-butylbenzene; (b) with propylene in the presence of H_3PO_4 to form isopropylbenzene.

11.11 Mechanism of halogenation

Aromatic halogenation, illustrated for chlorination, involves the following steps.

(1) $$Cl_2 + FeCl_3 \rightleftharpoons \underset{\text{II}}{Cl_3\overset{\ominus}{Fe}\text{—}\overset{\oplus}{Cl}\text{—}Cl}$$

(2) $$\underset{\text{II}}{Cl_3\overset{\ominus}{Fe}\text{—}\overset{\oplus}{Cl}\text{—}Cl} + C_6H_6 \longrightarrow \overset{\oplus}{C_6H_5}\genfrac{}{}{0pt}{}{\diagup H}{\diagdown Cl} + FeCl_4^- \qquad \textit{Slow}$$

(3) $$\overset{\oplus}{C_6H_5}\genfrac{}{}{0pt}{}{\diagup H}{\diagdown Cl} + FeCl_4^- \longrightarrow C_6H_5Cl + HCl + FeCl_3 \qquad \textit{Fast}$$

The key step (2) is the attachment of positive chlorine to the aromatic ring. It seems unlikely, though, that an actually free Cl^+ ion is involved. Instead, ferric chloride combines with Cl_2 to form complex II, from which chlorine is transferred, without its electrons, directly to the ring.

Addition of halogens to alkenes, we have seen (Sec. 6.13), similarly involves attack by positive halogen to form an intermediate carbonium ion. The loosely held π electrons of an alkene make it more reactive, however, and positive halogen is transferred from the halogen molecule itself, X_2, with loss of Cl^-. The less reactive benzene molecule needs the assistance of a Lewis acid; reaction occurs with the loss of the better leaving group, $FeCl_4^-$. Indeed, more highly reactive aromatic compounds, i.e., those whose π electrons are more available, do react with halogens in the absence of any added Lewis acid.

Problem 11.5 Certain activated benzene rings can be chlorinated by hypochlorous acid, HOCl, and this reaction is catalyzed by H^+. In light of the above discussion, can you suggest a possible function of H^+?

Problem 11.6 Aromatic bromination catalyzed by the Lewis acid thallium acetate, $Tl(OOCCH_3)_3$, gives only the *para* isomer. Suggest an explanation for this regiospecificity. (*Hint:* See Sec. 11.7.)

11.12 Desulfonation. Mechanism of protonation

When an aromatic sulfonic acid is heated to 100–175° with aqueous acid, it is converted into sulfuric acid and an aromatic hydrocarbon. This *desulfonation* is the exact reverse of the sulfonation process by which the sulfonic acid was originally made.

$$\underset{\substack{\text{Hydrocarbon}\\ \textit{Volatile}}}{C_6H_6} + H_2SO_4 \overset{H^+}{\rightleftharpoons} \underset{\substack{\text{Sulfonic acid}\\ \textit{Non-volatile}}}{C_6H_5SO_3H} + H_2O$$

By applying the usual equilibrium principles, we can select conditions that will drive the reaction in the direction we want it to go. To sulfonate we use a large excess of concentrated or fuming sulfuric acid; high concentration of sulfonating agent and low concentration of water (or its removal by reaction with SO_3) shift the equilibrium toward sulfonic acid. To desulfonate we use dilute acid and often pass superheated steam through the reaction mixture; high concentration of water and removal of the relatively volatile hydrocarbon by steam distillation shift the equilibrium toward hydrocarbon.

According to the principle of microscopic reversibility (p. 170), the mechanism of desulfonation must be the exact reverse of the mechanism of sulfonation.

(1) $$C_6H_5SO_3^- + H^+ \rightleftharpoons C_6H_5^{\oplus}\langle^{H}_{SO_3^-}$$

(2) $$C_6H_5^{\oplus}\langle^{H}_{SO_3^-} \rightleftharpoons C_6H_6 + SO_3$$

The reaction is simply another example of electrophilic aromatic substitution. The electrophile is the proton, H^+, and the reaction is *protonation* or, more specifically, *protodesulfonation.*

Sulfonation is unusual among electrophilic aromatic substitution reactions in its reversibility. It is also unusual in another way: in sulfonation, ordinary hydrogen (protium) is displaced from an aromatic ring about twice as fast as deuterium. These two facts are related to each other and, as we shall see in Sec. 11.16, give us a more detailed picture of sulfonation and of electrophilic aromatic substitution in general.

Problem 11.7 Predict the product or products of: (a) monobromination of toluene; (b) monobromination of *p*-toluenesulfonic acid followed by treatment with acid and superheated steam. (c) Using the principle of (b), and following the guidelines of Sec. 11.7, outline a synthesis from benzene of *o*-dibromobenzene; of *o*-bromochlorobenzene.

11.13 Thallation

Treatment of aromatic compounds with thallium trifluoroacetate, $Tl(OOCCF_3)_3$, dissolved in trifluoroacetic acid (CF_3COOH) gives rapidly and in high yield arylthallium ditrifluoroacetates, stable crystalline compounds. Reaction is believed by Taylor and McKillop (p. 345) to involve electrophilic attack on the aromatic ring by the (Lewis) acidic thallium.

$$ArH + Tl(OOCCF_3)_3 \xrightarrow{CF_3COOH} ArTl(OOCCF_3)_2 + CF_3COOH$$

Thallium trifluoroacetate — Arylthallium ditrifluoroacetate

Thallium compounds are very poisonous, and must be handled with extreme care.

Although substituent groups affect the reactivity of the aromatic substrate as expected for electrophilic substitution, orientation is unusual in a number of ways, and it is here that much of the usefulness of thallation lies. Thallation is almost exclusively *para* to —R, —Cl, and $—OCH_3$, and this is attributed to the bulk of the electrophile, thallium trifluoroacetate, which seeks out the uncrowded *para* position.

Thallation is almost exclusively *ortho* to certain substituents like —COOH, $—COOCH_3$, and $—CH_2OCH_3$ (even though some of these are normally *meta*-directing), and this is attributed to prior complexing of the electrophile with the substituent; thallium is held at just the right distance for easy intramolecular delivery to the *ortho* position. For example:

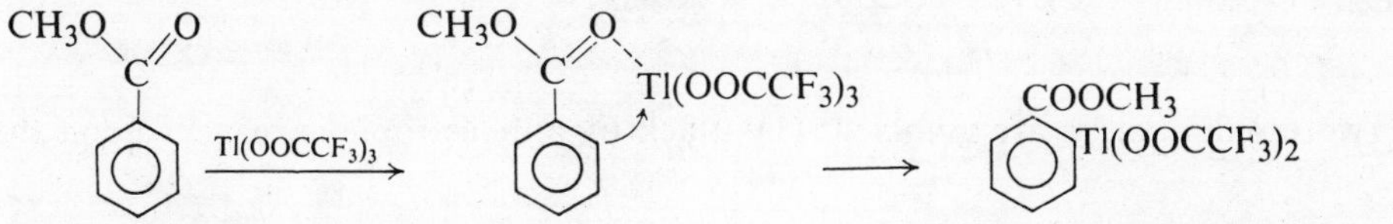

CH_3O—C(=O)—C_6H_5 $\xrightarrow{Tl(OOCCF_3)_3}$ CH_3O—C(=O···$Tl(OOCCF_3)_3$)—C_6H_5 $\longrightarrow$ o-$C_6H_4(COOCH_3)Tl(OOCCF_3)_2$

Methyl benzoate — *o*-Carbomethoxyphenylthallium ditrifluoroacetate *Only product*

(In $C_6H_5CH_2CH_2COOH$, however, it is evidently held too far from the ring, and must *leave* the substituent before attacking the ring intermolecularly—at the *para* position.)

Thallation is *reversible*, and when carried out at a higher temperature (73° instead of room temperature) yields the *more stable* isomer: usually the *meta* (compare 1,2- and 1,4-addition, Sec. 8.22). For example:

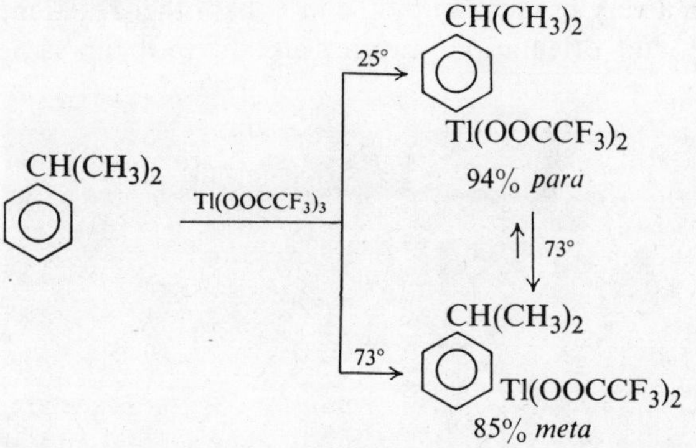

$C_6H_5CH(CH_3)_2$ + $Tl(OOCCF_3)_3$:

- 25° → p-$(CH_3)_2CHC_6H_4Tl(OOCCF_3)_2$, 94% *para* — *Rate-controlled*
- 73° → m-$(CH_3)_2CHC_6H_4Tl(OOCCF_3)_2$, 85% *meta* — *Equilibrium-controlled*

(*para* ⇌ *meta* at 73°)

Now, these arylthallium compounds are useful, not in themselves, but as intermediates in the synthesis of a variety of other aromatic compounds. Thallium can be replaced by other atoms or groups which cannot themselves be introduced directly into the aromatic ring—or at least not with the same regiospecificity. In this way one can prepare phenols (ArOH, Sec. 24.5) and aryl iodides (Sec. 25.3). Direct iodination of most aromatic rings does not work very well, but the process of thallation followed by treatment with iodide ion gives aryl iodides in high yields.

$$\text{ArH} + \text{Tl(OOCCF}_3)_3 \xrightarrow{\text{CF}_3\text{COOH}} \text{ArTl(OOCCF}_3)_2 \xrightarrow{\text{KI}} \text{ArI}$$

11.14 Mechanism of electrophilic aromatic substitution: a summary

Electrophilic aromatic substitution reactions seem, then, to proceed by a single mechanism, whatever the particular reagent involved. This can be summarized for the reagent YZ as follows:

(1) $$C_6H_6 + Y^+ \longrightarrow C_6H_5{\overset{\oplus}{\,}}\!<^{H}_{Y} \qquad \textit{Slow}$$

(2) $$C_6H_5{\overset{\oplus}{\,}}\!<^{H}_{Y} + :Z^- \longrightarrow C_6H_5Y + H{:}Z \qquad \textit{Fast}$$

Two essential steps are involved: (1) attack by an electrophilic reagent upon the ring to form a carbonium ion, $C_6H_5{\overset{\oplus}{\,}}\!<^{H}_{Y}$, and (2) abstraction of a hydrogen ion from this carbonium ion by some base. In each case there is a preliminary acid–base reaction which generates the attacking particle; the actual substitution, however, is contained in these two steps.

Most of the support for this mechanism comes from evidence about the nature of the attacking particle in each of these reactions: evidence, that is, that substitution is *electrophilic*. This evidence, in turn, comes largely from kinetics, augmented by various other observations: the nitrating power of preformed nitronium salts (Sec. 11.8), for example, or carbonium ion-like rearrangements in some Friedel-Crafts alkylations (Problem 11.3 above). The electrophilic nature of these reactions is supported in a very broad way by the fact that other reactions which show the same reactivity and orientation features also fit into the same mechanistic pattern.

Problem 11.8 In each of the following reactions, groups on the ring under attack exert the kinds of effects summarized in Sec. 11.5. Suggest a likely electrophile in each case, and write a likely mechanism.

(a) $$\text{ArH} + \text{R—C(=O)—Cl} \xrightarrow{\text{AlCl}_3} \text{Ar—C(=O)—R}$$

(b) $$\text{ArH} + \text{Ar'N}_2{}^+\text{Cl}^- \longrightarrow \text{Ar—N=N—Ar'}$$

(c) $$\text{ArH} + \text{HONO} \xrightarrow{\text{H}^+} \text{Ar—NO} + \text{H}_2\text{O}$$

Problem 11.9 When phenol is treated with D_2SO_4 in D_2O (deuterium sulfate in heavy water), there is formed phenol containing deuterium instead of hydrogen at positions *ortho* and *para* to the —OH group. Benzene undergoes similar exchange but at a much lower rate; under the same conditions benzenesulfonic acid does not undergo exchange at all. (a) Outline the most probable mechanism for hydrogen–deuterium exchange in aromatic compounds. (b) What is the attacking reagent in each case, and to what general class does this reaction belong?

But this is only part of the mechanism. Granting that substitution is electrophilic, how do we know that it involves *two* steps, as we have shown, and not just *one*? And how do we know that, of the two steps, the first is much slower than the second? To understand the answer to these questions, we must first learn something about *isotope effects.*

11.15 Isotope effects

Different isotopes of the same element have, by definition, the same electronic configuration, and hence similar chemical properties. This similarity is the basis of the isotopic tracer technique (Sec. 3.29): one isotope does pretty much what another will do, but, from its radioactivity or unusual mass, can be traced through a chemical sequence.

Yet different isotopes have, also by definition, different masses, and because of this their chemical properties are *not identical*: the same reactions can occur but at somewhat different rates (or, for reversible reactions, with different positions of equilibrium). *A difference in rate (or position of equilibrium) due to a difference in the isotope present in the reaction system is called an* **isotope effect.**

Theoretical considerations, which we cannot go into, supported by much experimental evidence, lead to the conclusion: *if a particular atom is less tightly bound in the transition state of a reaction than in the reactant, the reaction involving the heavier isotope of that atom will go more slowly*. The hydrogen isotopes have the greatest proportional differences in mass: deuterium (D) is twice as heavy as protium (H), and tritium (T) is three times as heavy. As a result, hydrogen isotope effects are the biggest, the easiest to measure, and—because of the special importance of hydrogen in organic chemistry—the most often studied. (If you doubt the importance of hydrogen, look at the structure of almost any compound in this book.)

One kind of reaction in which an atom is less tightly bound in the transition state than in the reactant is a reaction in which a bond to that atom is being broken. Isotope effects due to the breaking of a bond to the isotopic atom are called *primary isotope effects.* They are in general the biggest effects observed for a particular set of isotopes.

In this book we shall be concerned with **primary hydrogen isotope effects,** which amount to this: *a bond to protium (H) is broken faster than a bond to deuterium (D).* For many reactions of this kind,

$$\text{(1)} \qquad \sim\!\text{C—H} + \text{Z} \xrightarrow{k^{\text{H}}} [\sim\!\text{C---H---Z}] \longrightarrow \sim\!\text{C} + \text{H—Z}$$

$$\text{(2)} \qquad \sim\!\text{C—D} + \text{Z} \xrightarrow{k^{\text{D}}} [\sim\!\text{C---D---Z}] \longrightarrow \sim\!\text{C} + \text{D—Z}$$

in which hydrogen is abstracted as an atom, positive ion, or negative ion, deuterium isotope effects (k^H/k^D) in the range 5 to 8 (at room temperature) have been observed; that is to say, the reaction is 5 to 8 times as fast for ordinary hydrogen as for deuterium. (Tritium isotope effects, k^H/k^T, are about twice as large as deuterium isotope effects.)

These differences in rate can be measured in a variety of ways. In some cases, the rates of the two individual reactions (1) and (2) can be measured directly and the results compared. Usually, however, it is more feasible, as well as more satisfactory, to use our familiar method of competition (Sec. 3.22) in either of two ways.

In *intermolecular* competition, a mixture of labeled and unlabeled reactants compete for a limited amount of reagent; reactions (1) and (2) thus go on in the same mixture, and we measure the relative amounts of H—Z and D—Z produced. (Sometimes, larger amounts of the reagent Z are used, and the relative amounts of the two reactants—ordinary and labeled—left *unconsumed* are measured; the less reactive will have been used up more slowly and will predominate. The relative rates of reaction can be calculated without much difficulty.)

In *intramolecular* competition, a single reactant is used which contains several equivalent positions, some labeled and some not:

```
                 ┌──→ H—Z + C      ──→ Product(D)
C—H              │        ⌇
⌇     Z          │        C—D
C—D ─────────────┤
                 │
                 └──→ D—Z + C—H    ──→ Product(H)
                          ⌇
                          C
```

One can then measure either the relative amounts of H—Z and D—Z, or the relative amounts of the D-containing product formed by reaction (3) and the H-containing product formed by reaction (4).

Problem 11.10 (a) When excess toluene-α-*d* ($C_6H_5CH_2D$) was photochemically monochlorinated at 80° with 0.1 mole of chlorine, there were obtained 0.0212 mole DCl and 0.0868 mole HCl. What is the value of the isotope effect k^H/k^D (*per hydrogen atom*, of course)? (b) What relative amounts of DCl and HCl would you expect to get from $C_6H_5CHD_2$?

11.16 Mechanism of electrophilic aromatic substitution: the two steps

Now that we know what isotope effects are and, in a general way, how they arise, we are ready to see why they are of interest to the organic chemist. Let us return to the questions we asked before: how do we know that electrophilic aromatic substitution involves *two* steps,

(1) $\quad ArH + Y^+ \longrightarrow \overset{\oplus}{Ar}\begin{smallmatrix} H \\ Y \end{smallmatrix}$ **Slow:** *rate-determining*

(2) $\quad \overset{\oplus}{Ar}\begin{smallmatrix} H \\ Y \end{smallmatrix} + :Z \longrightarrow ArY + H:Z$ **Fast**

instead of just *one*,

$$\text{(1a)} \qquad ArH + Y^+ \longrightarrow \left[Ar \begin{matrix} H \\ \\ Y \end{matrix} \right]^+ \longrightarrow ArY + H^+$$

and how do we know that, of these two steps, the first is much slower than the second?

The answer is found in a series of studies begun by Lars Melander (of the Nobel Institute of Chemistry, Stockholm) and extended by many other workers. A variety of aromatic compounds labeled with deuterium or tritium were subjected to nitration, bromination, and Friedel-Crafts alkylation. It was found that in these reactions deuterium or tritium is replaced at the *same* rate as protium; *there is no significant isotope effect.*

We have seen that a carbon–deuterium bond is broken more slowly than a carbon–protium bond, and a carbon–tritium bond more slowly yet. How then, are we to interpret the fact that there is no isotope effect here? If the rates of replacement of the various hydrogen isotopes are the same, it can only mean that the reactions *whose rates we are comparing* do not involve the breaking of a carbon–hydrogen bond.

This interpretation is consistent with our mechanism. The rate of the overall substitution is determined by the slow attachment of the electrophilic reagent to the aromatic ring to form the carbonium ion. Once formed, the carbonium ion rapidly loses hydrogen ion to form the products. Step (1) is thus the *rate-determining step.* Since it does not involve the breaking of a carbon–hydrogen bond, its rate—and hence the rate of the overall reaction—is independent of the particular hydrogen isotope that is present.

If substitution involved a *single* step, as in (1a), this step would necessarily be the rate-determining step and, since it involves breaking of the carbon–hydrogen bond, an isotope effect would be observed. Or, if step (2) of the two-step sequence were slow enough relative to step (1) to affect the overall rate, again we would expect an isotope effect. (Indeed, sulfonation *does* show a small isotope effect and, as we shall see, for just this reason. Even in sulfonation, however, the overall rate is controlled chiefly by step (1).)

Thus the absence of isotope effects establishes not only the two-step nature of electrophilic aromatic substitution, but also the relative speeds of the steps. Attachment of the electrophile to a carbon atom of the ring is the difficult step (see Fig. 11.2); but it is equally difficult whether the carbon carries protium or deuterium. The next step, loss of hydrogen ion, is easy. Although it occurs more slowly for deuterium than for protium, this really makes no difference; slightly faster or slightly slower, its speed has no effect on the overall rate.

Let us look at this matter more closely (Fig. 11.2, insert). Every carbonium ion formed, whether I(H) or I(D), goes on to product, since the energy barrier to

$$\overset{\oplus}{Ar}\begin{matrix} H \\ NO_2 \end{matrix} \qquad\qquad \overset{\oplus}{Ar}\begin{matrix} D \\ NO_2 \end{matrix}$$

I(H) I(D)

the right (ahead of the carbonium ion)—whether slightly higher for deuterium or slightly lower for protium—is still considerably lower than the barrier to the left (behind the carbonium ion). But the barrier behind the carbonium ion is the E_{act} for the *reverse* of step (1). It is this reverse reaction that must be much slower than step (2) if step (1) is to be truly rate-determining (see Sec. 14.12). Summarized in terms of the *rate constants*, *k*, for the various steps, we have:

$$ArH + {}^{+}NO_2 \underset{k_{-1}}{\overset{k_1}{\rightleftharpoons}} \underset{\text{I}}{\overset{\oplus}{Ar}}\!\begin{smallmatrix} H \\ NO_2 \end{smallmatrix} \xrightarrow{k_2} ArNO_2 + H^+ \qquad k_2 \gg k_{-1}$$

(1) and (2) label the two steps.

We can see why nitration and reactions like it are not reversible. In the reverse of nitration, nitrobenzene is protonated (the reverse of reaction 2) to form carbonium ion I; but this is, of course, no different from the ion I formed in the nitration process, and it does the same thing: (re)forms nitrobenzene.

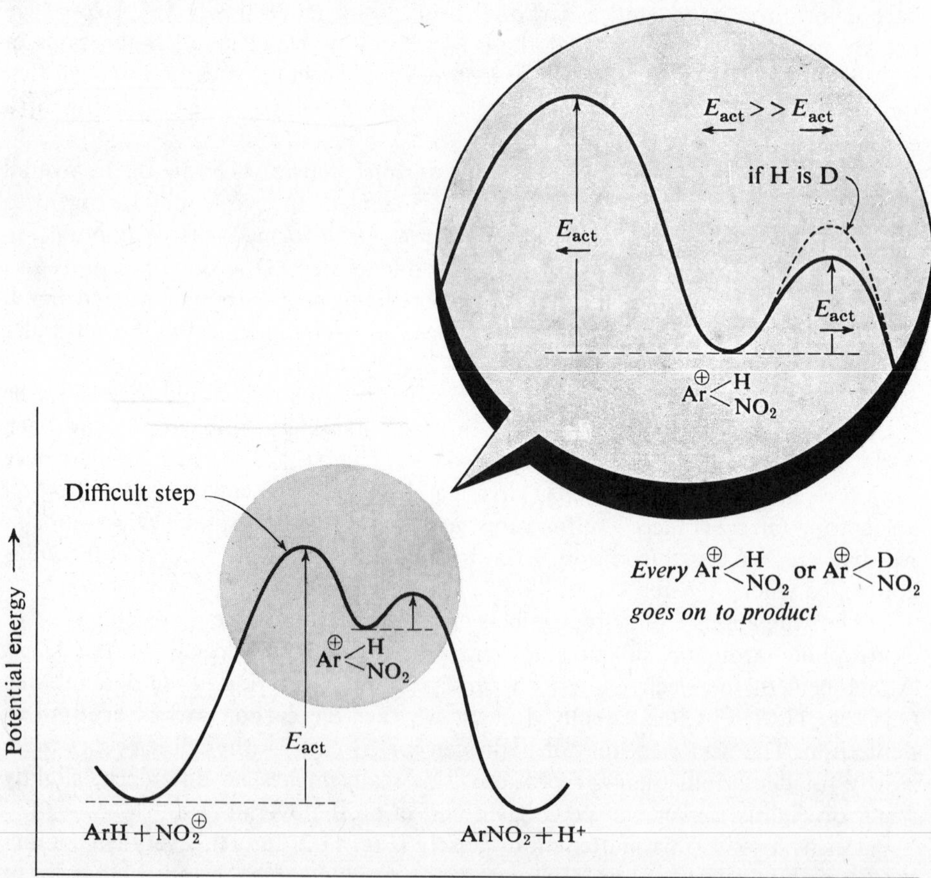

Figure 11.2. Nitration. Formation of carbonium ion is rate-controlling step; occurs equally rapidly whether protium (H) or deuterium (D) at point of attack. All carbonium ions go on to product. There is no isotope effect, and nitration is irreversible.

Unlike most other electrophilic substitution reactions, sulfonation shows a moderate isotope effect: ordinary hydrogen (protium) is displaced from an aromatic ring about twice as fast as deuterium. Does this mean that sulfonation takes place by a different mechanism than nitration, one involving a single step? Almost certainly not.

$$\mathrm{ArH + SO_3} \underset{k_{-1}}{\overset{k_1}{\rightleftharpoons}} \underset{\text{II}}{\overset{\oplus}{\mathrm{Ar}}\begin{matrix}\diagup \mathrm{H} \\ \diagdown \mathrm{SO_3^-}\end{matrix}} \xrightarrow{k_2} \mathrm{ArSO_3^- + H^+} \qquad k_2 \sim k_{-1}$$

(1) (2)

Unlike most other electrophilic substitution reactions, sulfonation is reversible, and this fact gives us our clue. Reversibility means that carbonium ion II can lose SO_3 to form the hydrocarbon. Evidently here reaction (2) is *not* much

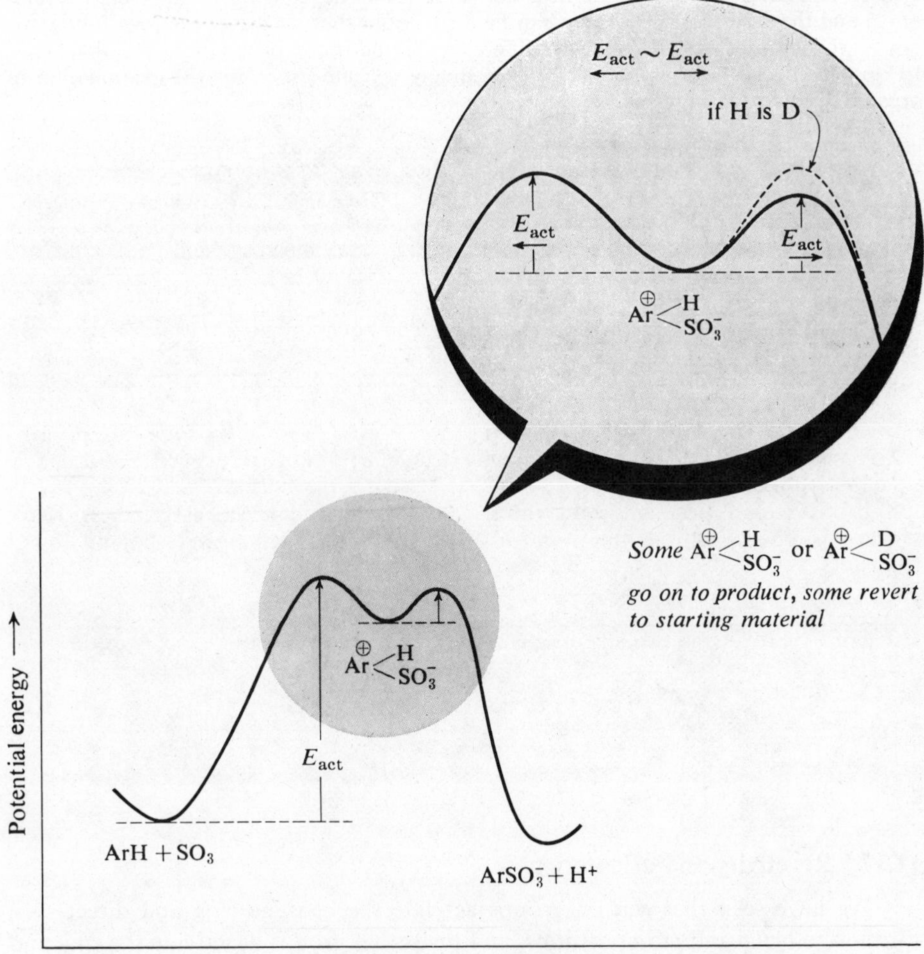

Figure 11.3 Sulfonation. Some carbonium ions go on to product, some revert to starting material. There is an isotope effect, and sulfonation is reversible.

faster than the reverse of reaction (1). In sulfonation, the energy barriers on either side of the carbonium ion II must be roughly the same height; some ions go one way, some go the other (Fig. 11.3). Now, whether the carbonium ion is II(D) or II(H), the barrier to the left (behind it) is the same height. But to climb the barrier to the right (ahead), a carbon–hydrogen bond must be broken, so this barrier is higher for carbonium ion II(D) than for carbonium ion II(H). More deuterated ions than ordinary ions revert to starting material, and so overall sulfonation is slower for the deuterated benzene. Thus, the particular shape of potential energy curve that makes sulfonation reversible also permits an isotope effect to be observed.

By use of especially selected aromatic substrates—highly hindered ones—isotope effects can be detected in other kinds of electrophilic aromatic substitution, even in nitration. In certain reactions the *size* of the isotope can be deliberately varied by changes in experimental conditions—and in a way that shows dependence on the relative rates of (2) and the reverse of (1). There can be little doubt that all these reactions follow the same two-step mechanism, but with differences in the shape of potential energy curves. In isotope effects the chemist has an exceedingly delicate probe for the examination of organic reaction mechanisms.

Problem 11.11 From the reaction of mesitylene (1,3,5-trimethylbenzene) with HF and BF_3, Olah (see p. 346) isolated at low temperatures a bright-yellow solid whose elemental composition corresponds to mesitylene:HF:BF_3 in the ratio 1:1:1. The compound was poorly soluble in organic solvents and, when melted, conducted an electric current; chemical analysis showed the presence of the BF_4^- ion. When heated, the compound evolved BF_3 and regenerated mesitylene.

What is a likely structure for the yellow compound? The isolation of this and related compounds is considered to be strong support for the mechanism of electrophilic aromatic substitution. Why should this be so?

Problem 11.12 Dehydrobromination by $C_2H_5O^-Na^+$ of ordinary isopropyl bromide and of labeled isopropyl bromide, $(CD_3)_2CHBr$, at 25° has been studied, and the rates found to be in the ratio 1.76:0.26. (a) What is the value of the isotope effect? (b) Is this isotope effect consistent with the mechanism for dehydrohalogenation given in Sec. 5.13? (c) With the following two-step mechanism involving a carbonium ion?

$$\text{RBr} \xrightarrow{\text{slow}} \text{R}^+\text{Br}^- \xrightarrow{\text{fast}} \text{alkene}$$

(d) With the following two-step mechanism involving a carbanion?

$$-\underset{\text{H}}{\overset{|}{\underset{|}{\text{C}}}}-\underset{\text{Br}}{\overset{|}{\underset{|}{\text{C}}}}- + {}^-\text{OC}_2\text{H}_5 \xrightarrow{\text{slow}} -\underset{\ominus}{\overset{|}{\underset{\vdots}{\text{C}}}}-\underset{\text{Br}}{\overset{|}{\underset{|}{\text{C}}}}- \xrightarrow{\text{fast}} \text{alkene}$$

11.17 Reactivity and orientation

We have seen that certain groups activate the benzene ring and direct substitution to *ortho* and *para* positions, and that other groups deactivate the ring and (except halogens) direct substitution to *meta* positions. Let us see if we can account for these effects on the basis of principles we have already learned.

First of all, we must remember that reactivity and orientation are both matters of relative rates of reaction. Methyl is said to activate the ring because it makes the ring react *faster* than benzene; it causes *ortho,para* orientation because it makes the *ortho* and *para* positions react *faster* than the *meta* positions.

Now, we know that, whatever the specific reagent involved, the rate of electrophilic aromatic substitution is determined by the same slow step—attack of the electrophile on the ring to form a carbonium ion:

$$C_6H_6 + Y^+ \longrightarrow C_6H_5\overset{\oplus}{}\!\!<^{H}_{Y} \qquad \textbf{Slow:}\ \textit{rate-determining}$$

Any differences in rate of substitution must therefore be due to differences in the rate of this step.

For closely related reactions, a difference in rate of formation of carbonium ions is largely determined by a difference in E_{act}, that is, by a difference in stability of transition states. As with other carbonium ion reactions we have studied, factors that stabilize the ion by dispersing the positive charge should for the same reason stabilize the incipient carbonium ion of the transition state. Here again we expect the more stable carbonium ion to be formed more rapidly. We shall therefore concentrate on the relative stabilities of the carbonium ions.

In electrophilic aromatic substitution the intermediate carbonium ion is a hybrid of structures I, II, and III, in which the positive charge is distributed about the ring, being strongest as the positions *ortho* and *para* to the carbon atom being attacked.

[Structures I, II, III (resonance forms of the carbonium ion with H and Y on the attacked carbon) *represented as* IV]

I II III IV

A group already attached to the benzene ring should affect the stability of the carbonium ion by dispersing or intensifying the positive charge, depending upon its electron-releasing or electron-withdrawing nature. It is evident from the structure of the ion (I–III) that this stabilizing or destabilizing effect should be especially important when the group is attached *ortho* or *para* to the carbon being attacked.

11.18 Theory of reactivity

To compare rates of substitution in benzene, toluene, and nitrobenzene, we compare the structures of the carbonium ions formed from the three compounds:

[Structures: carbonium ions with H, Y; II bearing CH_3; III bearing NO_2]

I II III

By releasing electrons, the methyl group (II) tends to neutralize the positive charge of the ring and so become more positive itself; this dispersal of the charge stabilizes the carbonium ion. In the same way the inductive effect stabilizes the developing positive charge in the transition state and thus leads to a faster reaction.

Transition state: *developing positive charge*

Carbonium ion: *full positive charge*

The $-NO_2$ group, on the other hand, has an electron-withdrawing inductive effect (III); this tends to intensify the positive charge, destabilizes the carbonium ion, and thus causes a slower reaction.

Reactivity in electrophilic aromatic substitution depends, then, upon the tendency of a substituent group to release or withdraw electrons. **A group that releases electrons activates the ring; a group that withdraws electrons deactivates the ring.**

Electrophilic Aromatic Substitution

G *releases electrons: stabilizes carbonium ion, activates*

G = $-NH_2$, $-OH$, $-OCH_3$, $-NHCOCH_3$, $-C_6H_5$, $-CH_3$

G *withdraws electrons: destabilizes carbonium ion, deactivates*

G = $-N(CH_3)_3^+$, $-NO_2$, $-CN$, $-SO_3H$, $-COOH$, $-CHO$, $-COR$, $-X$

Like $-CH_3$, other alkyl groups release electrons, and like $-CH_3$ they activate the ring. For example, *tert*-butylbenzene is 16 times as reactive as benzene toward nitration. Electron release by $-NH_2$ and $-OH$, and by their derivatives $-OCH_3$ and $-NHCOCH_3$, is due not to their inductive effect but to resonance, and is discussed later (Sec. 11.20).

We are already familiar with the electron-withdrawing effect of the halogens (Sec. 6.11). The full-fledged positive charge of the $-N(CH_3)_3^+$ group has, of course, a powerful attraction for electrons. In the other deactivating groups (e.g., $-NO_2$, $-CN$, $-COOH$), the atom next to the ring is attached by a multiple bond to oxygen or nitrogen. These electronegative atoms attract the mobile π electrons, making the atom next to the ring electron-deficient; to make up this deficiency, the atom next to the ring withdraws electrons from the ring.

We might expect replacement of hydrogen in $—CH_3$ by halogen to decrease the electron-releasing tendency of the group, and perhaps to convert it into an electron-withdrawing group. This is found to be the case. Toward nitration,

Activating	Weakly deactivating	Moderately deactivating	Strongly deactivating

toluene is 25 times as reactive as benzene; benzyl chloride is only one-third as reactive as benzene. The $—CH_2Cl$ group is thus weakly deactivating. Further replacement of hydrogen by halogen to yield the $—CHCl_2$ and the $—CCl_3$ groups results in stronger deactivation.

11.19 Theory of orientation

Before we try to account for orientation in electrophilic substitution, let us look more closely at the facts.

An activating group activates all positions of the benzene ring; even the positions *meta* to it are more reactive than any single position in benzene itself. It directs *ortho* and *para* simply because it activates the *ortho* and *para* positions much *more* than it does the *meta*.

A deactivating group deactivates all positions in the ring, even the positions *meta* to it. It directs *meta* simply because it deactivates the *ortho* and *para* positions even *more* than it does the *meta*.

Thus both *ortho,para* orientation and *meta* orientation arise in the same way: **the effect of any group—whether activating or deactivating—is strongest at the *ortho* and *para* positions.**

To see if this is what we would expect, let us compare, for example, the carbonium ions formed by attack at the *para* and *meta* positions of toluene, a compound that contains an activating group. Each of these is a hybrid of three structures, I–III for *para*, IV–VI for *meta*. In one of these six structures, II, the positive charge is located on the carbon atom to which $—CH_3$ is attached. Although $—CH_3$ releases electrons to all positions of the ring, it does so most strongly to the car-

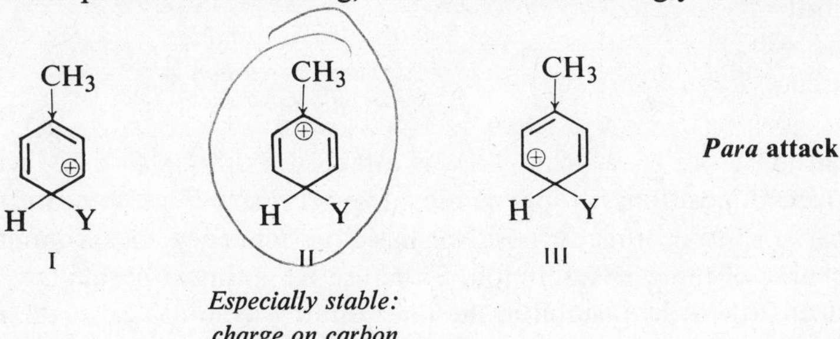

Especially stable: charge on carbon carrying substituent

bon atom nearest it; consequently, structure II is a particularly stable one. Because of contribution from structure II, the hybrid carbonium ion resulting from

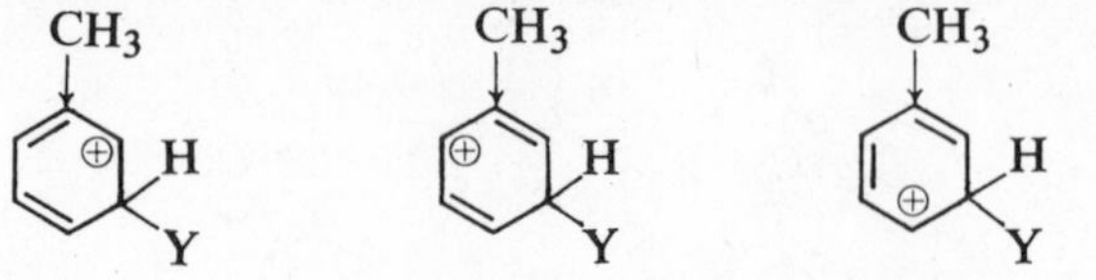

IV V VI *Meta* attack

attack at the *para* position is more stable than the carbonium ion resulting from attack at a *meta* position. *Para* substitution, therefore, occurs faster than *meta* substitution.

In the same way, it can be seen that attack at an *ortho* position (VII–IX)

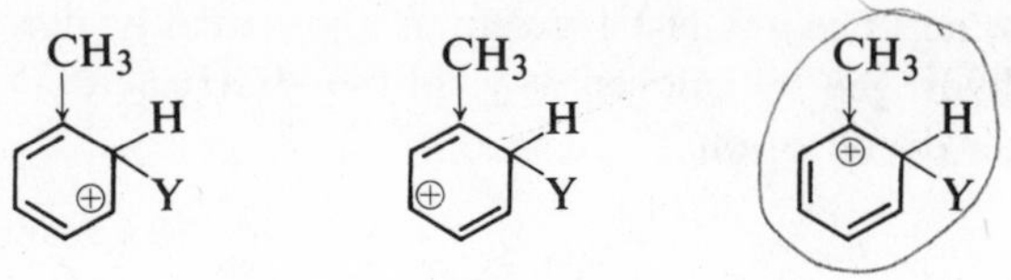

VII VIII IX *Ortho* attack

Especially stable: charge on carbon carrying substituent

also yields a more stable carbonium ion, through contribution from IX, than attack at a *meta* position.

In toluene, *ortho,para* substitution is thus faster than *meta* substitution because electron release by —CH_3 is more effective during attack at the positions *ortho* and *para* to it.

Next, let us compare the carbonium ions formed by attack at the *para* and *meta* positions of nitrobenzene, a compound that contains a deactivating group. Each of these is a hybrid of three structures, X–XII for *para* attack, XIII–XV for *meta* attack. In one of the six structures, XI, the positive charge is located on the

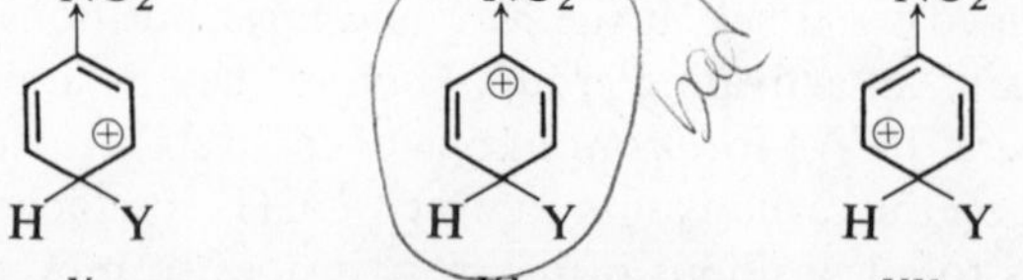

X XI XII *Para* attack

Especially unstable: charge on carbon carrying substituent

carbon atom to which —NO_2 is attached. Although —NO_2 withdraws electrons from all positions, it does so most from the carbon atom nearest it, and hence this carbon atom, already positive, has little tendency to accommodate the positive charge of the carbonium ion. Structure XI is thus a particularly unstable one and does little to help stabilize the ion resulting from attack at the *para* position. The ion for *para* attack is virtually a hybrid of only two structures, X and XII; the positive charge is mainly restricted to only *two* carbon atoms. It is less stable than the ion resulting from attack at a *meta* position, which is a hybrid of three structures, and in which the positive charge is accommodated by *three* carbon atoms. *Para* substitution, therefore, occurs more slowly than *meta* substitution.

XIII XIV XV **Meta attack**

In the same way it can be seen that attack at an *ortho* position (XVI–XVIII) yields a less stable carbonium ion, because of the instability of XVIII, than attack at a *meta* position.

XVI XVII XVIII **Ortho attack**

Especially unstable: charge on carbon carrying substituent

In nitrobenzene, *ortho,para* substitution is thus slower than *meta* substitution because electron withdrawal by —NO_2 is more effective during attack at the positions *ortho* and *para* to it.

Thus we see that both *ortho,para* orientation by activating groups and *meta* orientation by deactivating groups follow logically from the structure of the intermediate carbonium ion. The charge of the carbonium ion is strongest at the positions *ortho* and *para* to the point of attack, and hence a group attached to one of these positions can exert the strongest effect, whether activating or deactivating.

The unusual behavior of the halogens, which direct *ortho* and *para* although deactivating, results from a combination of two opposing factors, and will be taken up in Sec. 11.21.

11.20 Electron release via resonance

We have seen that a substituent group affects both reactivity and orientation in electrophilic aromatic substitution by its tendency to release or withdraw electrons. So far, we have considered electron release and electron withdrawal only as inductive effects, that is, as effects due to the electronegativity of the group concerned.

But certain groups (—NH_2 and —OH, and their derivatives) act as powerful activators toward electrophilic aromatic substitution, even though they contain electronegative atoms and can be shown in other ways to have electron-withdrawing inductive effects. If our approach to the problem is correct, these groups must release electrons in some other way than through their inductive effects; they are believed to do this by a resonance effect. But before we discuss this, let us review a little of what we know about nitrogen and oxygen.

Although electronegative, the nitrogen of the —NH_2 group is basic and tends to share its last pair of electrons and acquire a positive charge. Just as ammonia accepts a hydrogen ion to form the ammonium (NH_4^+) ion, so organic compounds related to ammonia accept hydrogen ions to form substituted ammonium ions.

$$\ddot{N}H_3 + H^+ \longrightarrow NH_4^+ \qquad R\ddot{N}H_2 + H^+ \longrightarrow RNH_3^+$$

$$R_2\ddot{N}H + H^+ \longrightarrow R_2NH_2^+ \qquad R_3\ddot{N} + H^+ \longrightarrow R_3NH^+$$

The —OH group shows similar but weaker basicity; we are already familiar with oxonium ions, ROH_2^+.

$$H_2\ddot{O} + H^+ \longrightarrow H_3O^+ \qquad R\ddot{O}H + H^+ \longrightarrow ROH_2^+$$

The effects of —NH_2 and —OH on electrophilic aromatic substitution can be accounted for by assuming that nitrogen and oxygen can share more than a pair of electrons with the ring and can accommodate a positive charge.

The carbonium ion formed by attack *para* to the —NH_2 group of aniline, for example, is considered to be a hybrid not only of structures I, II, and III, with positive charges located on carbons of the ring, but also of structure IV in which the

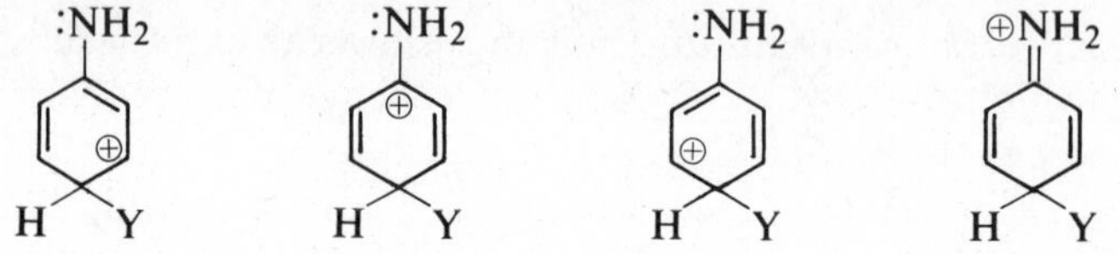

:NH₂ :NH₂ :NH₂ ⊕NH₂

H Y H Y H Y H Y

I II III IV

Para attack

Especially stable: every atom has octet

:NH₂ :NH₂ :NH₂

H Y H Y H Y

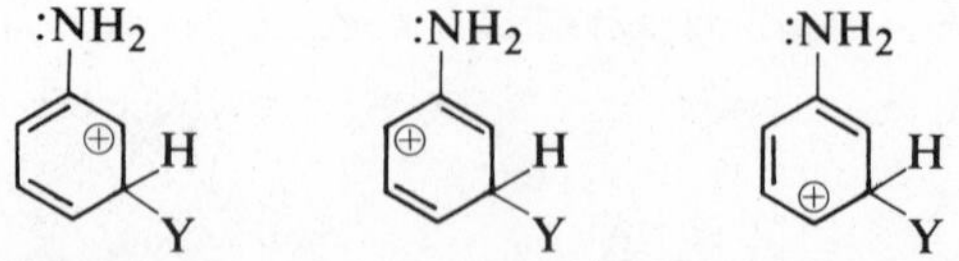

V VI VII

***Meta* attack**

positive charge is carried by nitrogen. Structure IV is especially stable, since in it *every atom* (except hydrogen, of course) *has a complete octet of electrons.* This carbonium ion is much more stable than the one obtained by attack on benzene itself, or the one obtained (V–VII) from attack *meta* to the —NH_2 group of aniline; in neither of these cases is a structure like IV possible. (Compare, for example, the stabilities of the ions NH_4^+ and CH_3^+. Here it is not a matter of which atom, nitrogen or carbon, can better accommodate a positive charge; it is a matter of which atom has a complete octet of electrons.)

Examination of the corresponding structures (VIII–XI) shows that *ortho* attack is much like *para* attack:

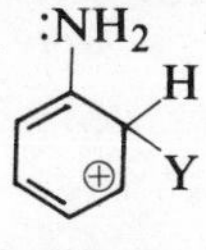

:NH_2 H Y (VIII) :NH_2 H Y (IX) :NH_2 H Y (X) ⊕NH_2 H Y (XI) *Ortho* **attack**

VIII 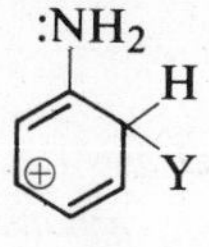 IX 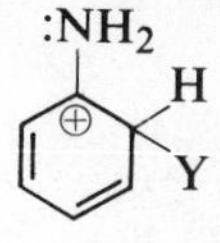 X 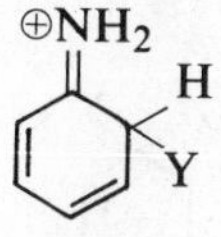 XI

Especially stable:
every atom has octet

Thus substitution in aniline occurs faster than substitution in benzene, and occurs predominantly at the positions *ortho* and *para* to $—NH_2$.

In the same way activation and *ortho,para* orientation by the —OH group is accounted for by contribution of structures like XII and XIII, in which every atom has a complete octet of electrons:

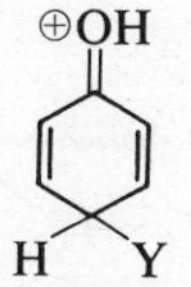

⊕OH H Y ⊕OH H Y

XII
Para attack

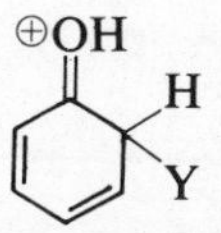

XIII
Ortho attack

The similar effects of the derivatives of $—NH_2$ and —OH are accounted for by similar structures (shown only for *para* attack):

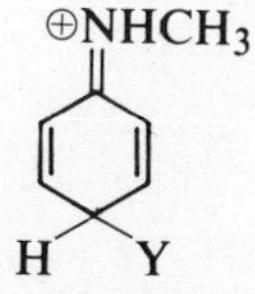

⊕$NHCH_3$ ⊕$N(CH_3)_2$ ⊕$NHCOCH_3$ ⊕OCH_3

H Y H Y H Y H Y

$—NHCH_3$ 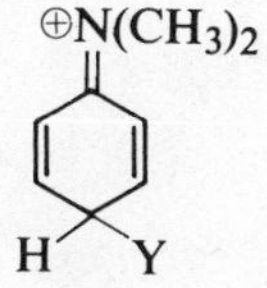 $—N(CH_3)_2$ 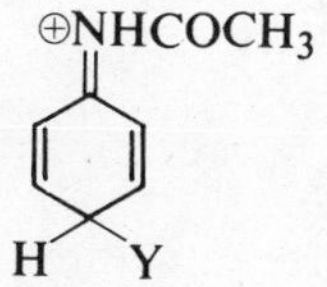 $—NHCOCH_3$ 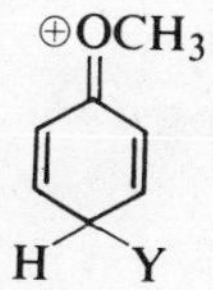 $—OCH_3$

The tendency of oxygen and nitrogen in groups like these to share more than a pair of electrons with an aromatic ring is shown in a number of other ways, which will be discussed later (Sec. 23.2 and Sec. 24.7).

11.21 Effect of halogen on electrophilic aromatic substitution

Halogens are unusual in their effect on electrophilic aromatic substitution: they are deactivating yet *ortho,para*-directing. Deactivation is characteristic of electron withdrawal, whereas *ortho,para* orientation is characteristic of electron release. Can halogen both withdraw and release electrons?

The answer is *yes*. Halogen withdraws electrons through its inductive effect, and releases electrons through its resonance effect. So, presumably, can the $—NH_2$ and —OH groups, but there the much stronger resonance effect greatly outweighs the other. For halogen, the two effects are more evenly balanced, and we observe the operation of both.

Let us first consider **reactivity**. Electrophilic attack on benzene yields car-

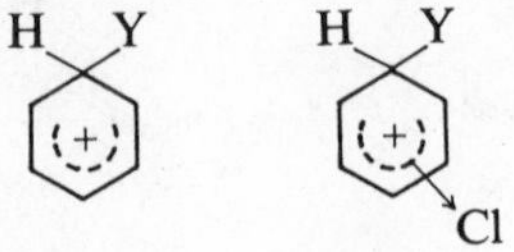

I II

—Cl *withdraws electrons: destabilizes carbonium ion, deactivates ring*

bonium ion I, attack on chlorobenzene yields carbonium ion II. The electron-withdrawing inductive effect of chlorine intensifies the positive charge in carbonium ion II, makes the ion less stable, and causes a slower reaction.

Next, to understand **orientation**, let us compare the structures of the carbonium ions formed by attack at the *para* and *meta* positions of chlorobenzene. Each of

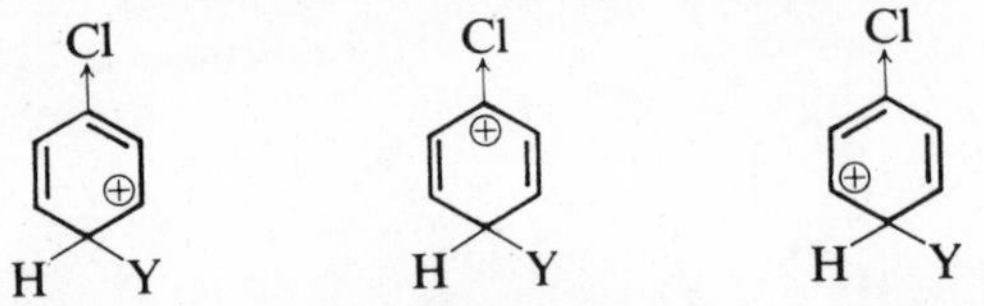

III IV V ***Para* attack**

IV *Especially unstable: charge on carbon bearing substituent*

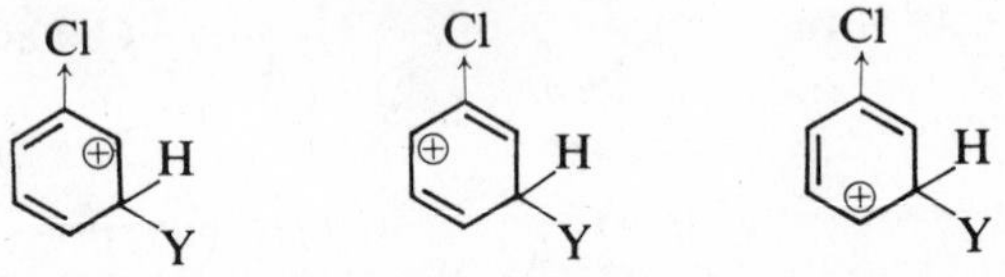

VI VII VIII ***Meta* attack**

these is a hybrid of three structures, III–V for *para*, VI–VIII for *meta*. In one of these six structures, IV, the positive charge is located on the carbon atom to which chlorine is attached. Through its inductive effect chlorine withdraws electrons most from the carbon to which it is joined, and thus makes structure IV especially unstable. As before, we expect IV to make little contribution to the hybrid, which should therefore be less stable than the hybrid ion resulting from attack at the *meta* positions. If only the inductive effect were involved, then, we would expect not only deactivation but also *meta* orientation.

But the existence of halonium ions (Sec. 7.12) has shown us that halogen can share more than a pair of electrons and can accommodate a positive charge. If we apply that idea to the present problem, what do we find? The ion resulting from *para* attack is a hybrid not only of structures III–V, but also of structure IX, in which chlorine bears a positive charge and is joined to the ring by a double bond. This structure should be comparatively stable, since in it every atom (except hydrogen, of course) has a *complete octet of electrons*. (Structure IX is exactly analogous to those proposed to account for activation and *ortho,para* direction by $—NH_2$ and —OH.) No such structure is possible for the ion resulting from *meta* attack. To the extent that structure IX contributes to the hybrid, it makes the ion resulting from *para* attack more stable than the ion resulting from *meta* attack.

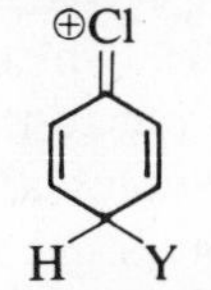

⊕Cl

H Y

IX
Comparatively stable: every atom has octet

***Para* attack**

Although we could not have predicted the relative importance of the two factors—the instability of IV and the stabilization by IX—the result indicates that the contribution from IX is the more important.

In the same way it can be seen that attack at an *ortho* position also yields an ion (X–XIII) that can be stabilized by accommodation of the positive charge by chlorine.

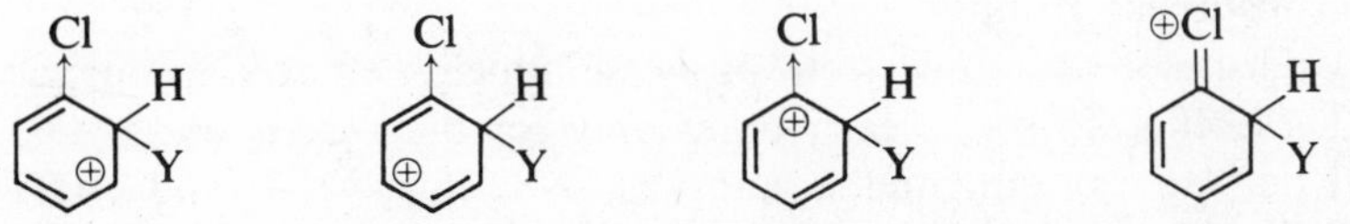

Cl H Y ⊕ — X

Cl H Y ⊕ — XI

Cl H Y ⊕ — XII
Especially unstable: charge on carbon bearing substituent

⊕Cl H Y — XIII
Comparatively stable: every atom has octet

***Ortho* attack**

Through its inductive effect halogen tends to withdraw electrons and thus to destabilize the intermediate carbonium ion. This effect is felt for attack at all positions, but particularly for attack at the positions *ortho* and *para* to the halogen.

Through its resonance effect halogen tends to release electrons and thus to stabilize the intermediate carbonium ion. This electron release is effective only for attack at the positions *ortho* and *para* to the halogen.

The inductive effect is stronger than the resonance effect and causes net electron withdrawal—and hence deactivation—for attack at all positions. The resonance effect tends to oppose the inductive effect for attack at the *ortho* and *para* positions, and hence makes the deactivation less for *ortho,para* attack than for *meta.*

Reactivity is thus controlled by the stronger inductive effect, and orientation is controlled by the resonance effect, which, although weaker, seems to be more selective.

Problem 11.13 Hydrogen iodide adds to vinyl chloride more slowly than to ethylene, and yields 1-chloro-1-iodoethane. (a) Draw the formula of the carbonium ion formed in the initial step of the addition to vinyl chloride. (b) Of addition to ethylene. (c) Judging from the relative rates of reaction, which would appear to be the more stable carbonium ion? (d) Account for the difference in stability.

(e) Draw the formula for the carbonium ion that would be formed if vinyl chloride were to yield 1-chloro-2-iodoethane. (f) Judging from the actual orientation of addition, which carbonium ion from vinyl chloride is the more stable, (a) or (e)? (g) Account for the difference in stability.

(h) Which effect, inductive or resonance, controls reactivity in electrophilic addition to vinyl halides? (i) Which effect controls orientation?

Thus we find that a single structural concept—partial double-bond formation between halogen and carbon—helps to account for unusual chemical properties of such seemingly different compounds as aryl halides and vinyl halides. The structures involving doubly-bonded halogen, which probably make important contribution not only to benzenonium ions but to the parent aryl halides as well (Sec. 25.6), certainly do not seem to meet our usual standard of reasonableness (Sec. 6.27). The sheer weight of evidence forces us to accept the idea that certain carbon–halogen bonds possess double-bond character. If this idea at first appears strange to us, it simply shows how little, after all, we really know about molecular structure.

11.22 Relation to other carbonium ion reactions

In summary, we can say that both reactivity and orientation in electrophilic aromatic substitution are determined by the rates of formation of the intermediate carbonium ions concerned. These rates parallel the stabilities of the carbonium ions, which are determined by the electron-releasing or electron-withdrawing tendencies of the substituent groups.

A group may release or withdraw electrons by an inductive effect, a resonance effect, or both. These effects oppose each other only for the $—NH_2$ and —OH groups (and their derivatives) and for the halogens, —X. For $—NH_2$ and —OH the resonance effect is much the more important; for —X the effects are more evenly matched. It is because of this that the halogens occupy the unusual position of being deactivating groups but *ortho,para* directors.

We have accounted for the facts of electrophilic aromatic substitution in exactly the way that we accounted for the relative ease of dehydration of alcohols, and for reactivity and orientation in electrophilic addition to alkenes: the more stable the carbonium ion, the faster it is formed; the faster the carbonium ion is formed, the faster the reaction goes.

In all this we have estimated the stability of a carbonium ion on the same basis: **the dispersal or concentration of the charge** due to electron release or electron withdrawal by the substituent groups. As we shall see, the approach that has worked so well for elimination, for addition, and for electrophilic aromatic substitution works for still another important class of organic reactions in which a positive charge develops: *nucleophilic aliphatic substitution by the* S_N1 *mechanism* (Sec. 14.14). It works equally well for *nucleophilic aromatic substitution* (Sec. 25.9), in which a negative charge develops. Finally, we shall find that this approach will help us to understand *acidity* or *basicity* of such compounds as carboxylic acids, sulfonic acids, amines, and phenols.

PROBLEMS

1. Give structures and names of the principal products expected from the ring monobromination of each of the following compounds. In each case, tell whether bromination will occur faster or slower than with benzene itself.

(a) acetanilide ($C_6H_5NHCOCH_3$)
(b) iodobenzene
(c) *sec*-butylbenzene
(d) N-methylaniline ($C_6H_5NHCH_3$)
(e) ethyl benzoate ($C_6H_5COOC_2H_5$)
(f) acetophenone ($C_6H_5COCH_3$)

(g) phenetole ($C_6H_5OC_2H_5$)
(h) diphenylmethane ($C_6H_5CH_2C_6H_5$)
(i) benzonitrile (C_6H_5CN)
(j) benzotrifluoride ($C_6H_5CF_3$)
(k) biphenyl ($C_6H_5—C_6H_5$)

2. Give structures and names of the principal organic products expected from mononitration of:

(a) *o*-nitrotoluene
(b) *m*-dibromobenzene
(c) *p*-nitroacetanilide (p-$O_2NC_6H_4NHCOCH_3$)
(d) *m*-dinitrobenzene
(e) *m*-cresol (m-$CH_3C_6H_4OH$)
(f) *o*-cresol
(g) *p*-cresol
(h) *m*-nitrotoluene
(i) *p*-xylene (p-$C_6H_4(CH_3)_2$)
(j) terephthalic acid (p-$C_6H_4(COOH)_2$)
(k) anilinium hydrogen sulfate ($C_6H_5NH_3{}^+HSO_4{}^-$)

3. Give structures and names of the principal organic products expected from the monosulfonation of :

(a) cyclohexylbenzene
(b) nitrobenzene
(c) anisole ($C_6H_5OCH_3$)
(d) benzenesulfonic acid
(e) salicylaldehyde (o-HOC_6H_4CHO)
(f) *m*-nitrophenol
(g) *o*-fluoroanisole
(h) *o*-nitroacetanilide (o-$O_2NC_6H_4NHCOCH_3$)
(i) *o*-xylene
(j) *m*-xylene
(k) *p*-xylene

4. Arrange the following in order of reactivity toward ring nitration, listing by structure the most reactive at the top, the least reactive at the bottom.

(a) benzene, mesitylene (1,3,5-$C_6H_3(CH_3)_3$), toluene, *m*-xylene, *p*-xylene
(b) benzene, bromobenzene, nitrobenzene, toluene
(c) acetanilide ($C_6H_5NHCOCH_3$), acetophenone ($C_6H_5COCH_3$), aniline, benzene
(d) terephthalic acid, toluene, *p*-toluic acid (p-$CH_3C_6H_4COOH$), *p*-xylene
(e) chlorobenzene, *p*-chloronitrobenzene, 2,4-dinitrochlorobenzene
(f) 2,4-dinitrochlorobenzene, 2,4-dinitrophenol
(g) *m*-dinitrobenzene, 2,4-dinitrotoluene

5. Even though 1,3,5-trinitrobenzene (TNB) has more shattering power (more *brisance*) and is no more dangerous to handle, 2,4,6-trinitrotoluene (TNT) has always been the high explosive in more general use. Can you suggest a reason (connected with manufacture) for the popularity of TNT? (Benzene and toluene are both readily available materials; for many years benzene was cheaper.)

6. For each of the following compounds, indicate which ring you would expect to be attacked in nitration, and give structures of the principal products.

(a) 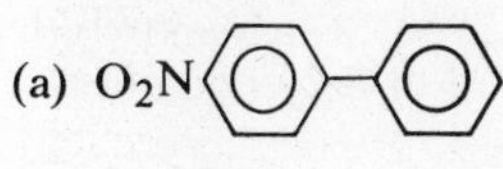

p-Nitrobiphenyl

(b) 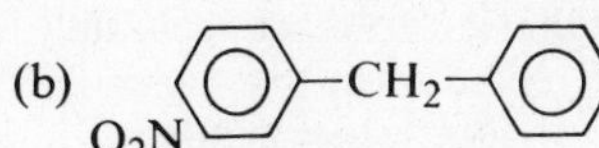
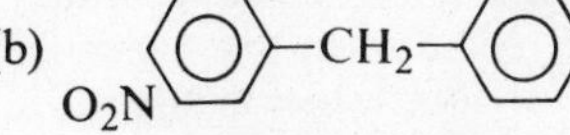

m-Nitrodiphenylmethane

(c) 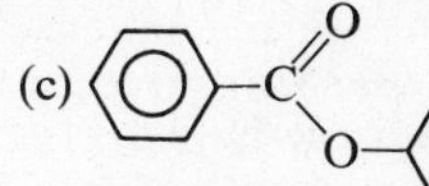

Phenyl benzoate

7. Arrange the compounds of each set in order of reactivity toward electrophilic substitution. Indicate in each set which would yield the highest percentage of *meta* isomer, and which would yield the lowest.

(a) $C_6H_5N(CH_3)_3{}^+$, $C_6H_5CH_2N(CH_3)_3{}^+$, $C_6H_5CH_2CH_2N(CH_3)_3{}^+$, $C_6H_5CH_2CH_2CH_2N(CH_3)_3{}^+$
(b) $C_6H_5NO_2$, $C_6H_5CH_2NO_2$, $C_6H_5CH_2CH_2NO_2$
(c) $C_6H_5CH_3$, $C_6H_5CH_2COOC_2H_5$, $C_6H_5CH(COOC_2H_5)_2$, $C_6H_5C(COOC_2H_5)_3$

8. There is evidence that the phenyl group, C_6H_5–, has an electron-withdrawing inductive effect. Yet each ring of biphenyl, C_6H_5–C_6H_5, is more reactive than benzene

toward electrophilic substitution, and the chief products are *ortho* and *para* isomers. Show how reactivity and orientation can be accounted for on the basis of resonance.

9. When β-phenylethyl alcohol, $C_6H_5CH_2CH_2OH$, is treated with thallium trifluoroacetate followed by potassium iodide, there is obtained predominantly one aryl iodide, the particular isomer depending upon the conditions of thallation: (a) 25°, *ortho*; (b) 75°, *meta*; (c) prior conversion to the ester, $C_6H_5CH_2CH_2OCOCH_3$, then 25°, *para*. Suggest an explanation for each case of regiospecificity.

10. There is evidence that the reaction between HNO_3 and H_2SO_4 to generate $^+NO_2$ (which we have summarized in one equation, Sec. 11.8) actually involves three steps, the second of which is the slowest one and the one that actually produces $^+NO_2$. Can you suggest a reasonable sequence of reactions? (*Hint:* See Sec. 5.20.)

11. Treatment of *sulfanilic acid* (p-$H_2NC_6H_4SO_3H$) with 3 moles of bromine yields 2,4,6-tribromoaniline. Treatment of 4-hydroxy-1,3-benzenedisulfonic acid with nitric acid yields picric acid, 2,4,6-trinitrophenol. (a) Outline the most probable mechanism for the replacement of $—SO_3H$ by —Br and by $—NO_2$. (b) To what general class of organic reactions do those reactions belong?

12. Using only individual steps with which you are already familiar, outline a likely mechanism for the following reaction.

$$C_6H_5C(CH_3)_3 + Br_2(AlBr_3) \longrightarrow C_6H_5Br + HBr + (CH_3)_2C{=}CH_2$$

13. In light of what you have learned in this chapter, predict the major products of each of the following reactions.

(a) $(CH_3)_3\overset{+}{N}CH{=}CH_2 + HI$
(b) $CH_2{=}CHCF_3 + HBr(AlBr_3)$
(c) What is the function of $AlBr_3$ in (b)? Why is it needed here?

14. You are trying to find out whether or not there is an isotope effect in a particular kind of substitution in which the electrophile Y replaces a hydrogen of an aromatic ring. In each of the following cases, tell what you would *do*, and what you would *expect to observe* if there were an isotope effect. (You can quantitatively analyze mixtures of isomers. Your mass spectrometer will tell you what percentage of the hydrogen in a compound is deuterium, but not the location of deuterium in a molecule.)

(a) C_6H_6 and C_6D_6 are allowed to react separately but under identical conditions.
(b) A 50:50 mixture of C_6H_6 and C_6D_6 is allowed to react with a limited amount of the reagent.
(c) Anisole and anisole-4-d are allowed to react separately. (Both your watch and your mass spectrometer are under repair when this particular experiment is carried out.)
(d) Benzene-1,3,5-d_3 (1,3,5-trideuteriobenzene) is allowed to react.

15. Outline all steps in the laboratory synthesis of the following compounds from benzene and/or toluene, using any needed aliphatic or inorganic reagents. (Review the general instructions on p. 224. Assume that a pure *para* isomer can be separated from an *ortho,para* mixture.)

(a) *p*-nitrotoluene
(b) *p*-bromonitrobenzene
(c) *p*-dichlorobenzene
(d) *m*-bromobenzenesulfonic acid
(e) *p*-bromobenzenesulfonic acid
(f) *p*-bromobenzoic acid
(g) *m*-bromobenzoic acid
(h) *o*-iodobenzoic acid
(i) 1,3,5-trinitrobenzene
(j) 2-bromo-4-nitrotoluene
(k) 2-bromo-4-nitrobenzoic acid
(l) 4-bromo-3-nitrobenzoic acid
(m) 3,5-dinitrobenzoic acid
(n) 4-nitro-1,2-dibromobenzene
(o) 2-nitro-1,4-dichlorobenzene
(p) *m*-iodotoluene

16. Outline all steps in the following laboratory syntheses, using any needed aliphatic or inorganic reagents. (Follow the other instructions in Problem 15).

(a) 4-nitro-2,6-dibromoanisole from anisole ($C_6H_5OCH_3$)
(b) 4-bromo-2-nitrobenzoic acid from *o*-nitrotoluene
(c) 2,4,6-tribromoaniline from aniline
(d) 2,4-dinitroacetanilide from acetanilide ($C_6H_5NHCOCH_3$)
(e) 5-nitroisophthalic acid from *m*-xylene
(f) 4-nitroisophthalic acid from *m*-xylene
(g) 2-nitroterephthalic acid from *p*-xylene (two ways)
(h) Which way in (g) is preferable? Why?

Chapter

12

Arenes

12.1 Aliphatic-aromatic hydrocarbons

From our study so far, we know what kind of chemical properties to expect of an aliphatic hydrocarbon, that is, of an alkane, alkene, or alkyne. We know what kind of chemical behavior to expect of the parent aromatic hydrocarbon, benzene. Many important compounds are not just aliphatic or just aromatic, however, but contain both aliphatic and aromatic units; hydrocarbons of this kind are known collectively as **arenes**. *Ethylbenzene*, for example, contains a benzene ring and an aliphatic side chain.

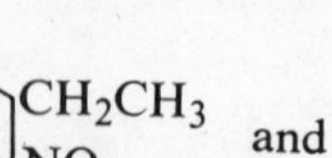

$C_6H_5CH_2CH_3$

Ethylbenzene

What kind of chemical properties might we expect of one of these mixed aliphatic-aromatic hydrocarbons? First, we might expect it to show *two* sets of chemical properties. The ring of ethylbenzene should undergo the electrophilic

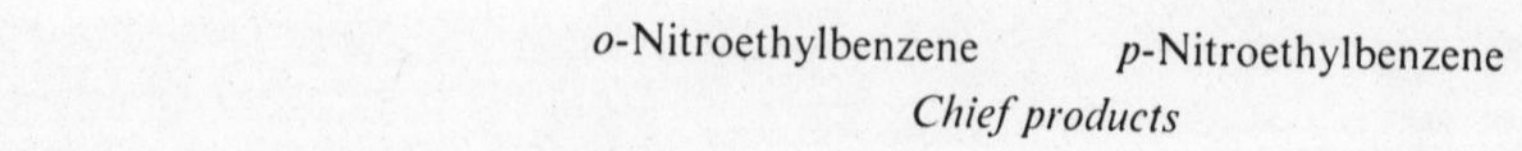

$$C_6H_5CH_2CH_3 \xrightarrow{HNO_3,\ H_2SO_4} o\text{-}O_2NC_6H_4CH_2CH_3 \text{ and } p\text{-}O_2NC_6H_4CH_2CH_3$$

o-Nitroethylbenzene   *p*-Nitroethylbenzene

Chief products

More readily than for benzene

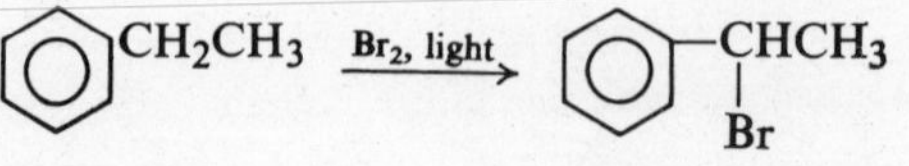

$$C_6H_5CH_2CH_3 \xrightarrow{Br_2,\ light} C_6H_5CHBrCH_3$$

1-Bromo-1-phenylethane

(α-Phenylethyl bromide)

Only product

More readily than for ethane

substitution characteristic of benzene, and the side chain should undergo the free-radical substitution characteristic of ethane. Second, the properties of each portion of the molecule should be modified by the presence of the other portion. The ethyl group should modify the aromatic properties of the ring, and the ring should modify the aliphatic properties of the side chain.

These predictions are correct. Treatment of ethylbenzene with nitric acid and sulfuric acid, for instance, introduces a nitro group into the ring; treatment with bromine in the presence of light introduces a bromine atom into the side chain. But because of the ethyl group, nitration takes place more readily than with benzene itself, and occurs chiefly at the positions *ortho* and *para* to the ethyl group; and because of the ring, bromination takes place more readily than with ethane, and occurs exclusively on the carbon nearer the ring. Thus *each portion of the molecule affects the* **reactivity** *of the other portion and determines the* **orientation** *of attack.*

In the same way we may have a molecule that is part aromatic and part alkene, or part aromatic and part alkyne. Again each portion of such a molecule shows the properties characteristic of its particular structure, although these properties are modified by the other portion of the molecule.

We shall examine most closely the compounds made up of aromatic and alkane units, the **alkylbenzenes**. We shall look much more briefly at the aromatic-alkene compounds (**alkenylbenzenes**) and aromatic-alkyne compounds (**alkynylbenzenes**).

We shall encounter the *benzyl* free radical and the *benzyl* carbonium ion, which pretty much complete our lists of these reactive particles, and shall see how their relative stabilities can be accounted for.

12.2 Structure and nomenclature

The simplest of the alkylbenzenes, methylbenzene, is given the special name of **toluene**. Compounds containing longer side chains are named by prefixing the name of the alkyl group to the word *-benzene*, as, for example, in *ethylbenzene*, *n-propylbenzene*, and *isobutylbenzene*.

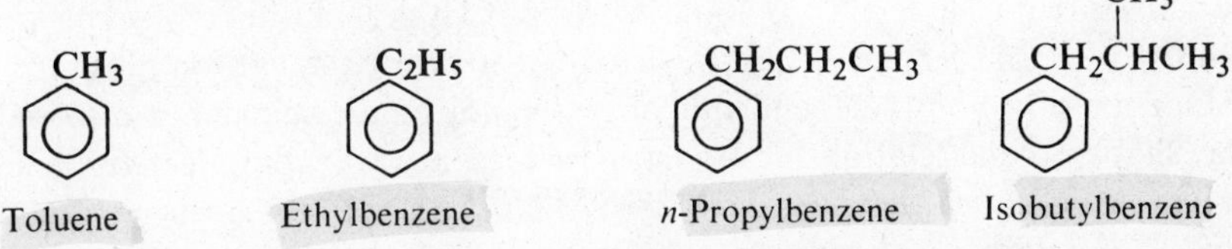

Toluene Ethylbenzene *n*-Propylbenzene Isobutylbenzene

The simplest of the dialkylbenzenes, the dimethylbenzenes, are given the special names of **xylenes**; we have, then, *o-xylene*, *m-xylene*, and *p-xylene*. Dialkylbenzenes containing one methyl group are named as derivatives of toluene, while others are named by prefixing the names of both alkyl groups to the word *-benzene*. A compound containing a very complicated side chain might be named as a

CH_3 CH_3 CH_3 CH_3 CH_3 CH_3

o-Xylene *m*-Xylene *p*-Xylene

p-$CH_3C_6H_4C_2H_5$ — *p*-Ethyltoluene

m-$C_2H_5C_6H_4CH(CH_3)_2$ — *m*-Ethylisopropylbenzene

$CH_3CH(CH_3)CH(C_6H_5)CH_2CH_3$ — 2-Methyl-3-phenylpentane

phenylalkane (C_6H_5 = **phenyl**). Compounds containing more than one benzene ring are nearly always named as derivatives of alkanes.

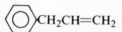

$C_6H_5CH_2C_6H_5$ — Diphenylmethane

$C_6H_5CH_2CH_2C_6H_5$ — 1,2-Diphenylethane

The simplest alkenylbenzene has the special name **styrene**. Others are generally named as substituted alkenes, occasionally as substituted benzenes. Alkynylbenzenes are named as substituted alkynes.

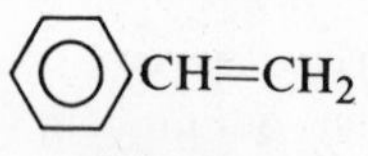

Styrene (Vinylbenzene) (Phenylethylene)

$C_6H_5CH_2CH{=}CH_2$ — Allylbenzene (3-Phenylpropene)

$C_6H_5C(CH_3){=}CHCH_3$ — 2-Phenyl-2-butene

$C_6H_5C{\equiv}CH$ — Phenylacetylene

12.3 Physical properties

As compounds of low polarity, the alkylbenzenes possess physical properties that are essentially the same as those of the hydrocarbons we have already studied. They are insoluble in water, but quite soluble in non-polar solvents like ether, carbon tetrachloride, or ligroin. They are almost always less dense than water. As we can see from Table 12.1, boiling points rise with increasing molecular weight, the boiling point increment being the usual 20–30° for each carbon atom.

Since melting points depend not only on molecular weight but also on molecular shape, their relationship to structure is a very complicated one. One important general relationship does exist, however, between melting point and structure of aromatic compounds: *among isomeric disubstituted benzenes, the para isomer generally melts considerably higher than the other two*. The xylenes, for example, boil within six degrees of one another; yet they differ widely in melting point, the *o*- and *m*-isomers melting at −25° and −48°, and the *p*-isomer melting at +13°. Since dissolution, like melting, involves overcoming the intermolecular forces of the crystal, it is not surprising to find that *generally the para isomer is also the least soluble in a given solvent*.

The higher melting point and lower solubility of a *para* isomer is only a special example of the general effect of molecular symmetry on intracrystalline forces. The more symmetrical a compound, the better it fits into a crystal lattice and hence the higher the melting point and the lower the solubility. *Para* isomers are simply the most symmetrical of disubstituted benzenes. We can see (Table 12.1) that

Table 12.1 ALIPHATIC-AROMATIC HYDROCARBONS

Name	Formula	M.p., °C	B.p., °C	Density (20°C)
Benzene	C_6H_6	5.5	80	0.879
Toluene	$C_6H_5CH_3$	− 95	111	.866
o-Xylene	1,2-$C_6H_4(CH_3)_2$	− 25	144	.880
m-Xylene	1,3-$C_6H_4(CH_3)_2$	− 48	139	.864
p-Xylene	1,4-$C_6H_4(CH_3)_2$	13	138	.861
Hemimellitene	1,2,3-$C_6H_3(CH_3)_3$	− 25	176	.895
Pseudocumene	1,2,4-$C_6H_3(CH_3)_3$	− 44	169	.876
Mesitylene	1,3,5-$C_6H_3(CH_3)_3$	− 45	165	.864
Prehnitene	1,2,3,4-$C_6H_2(CH_3)_4$	− 6.5	205	.902
Isodurene	1,2,3,5-$C_6H_2(CH_3)_4$	− 24	197	
Durene	1,2,4,5-$C_6H_2(CH_3)_4$	80	195	
Pentamethylbenzene	$C_6H(CH_3)_5$	53	231	
Hexamethylbenzene	$C_6(CH_3)_6$	165	264	
Ethylbenzene	$C_6H_5C_2H_5$	− 95	136	.867
n-Propylbenzene	$C_6H_5CH_2CH_2CH_3$	− 99	159	.862
Cumene	$C_6H_5CH(CH_3)_2$	− 96	152	.862
n-Butylbenzene	$C_6H_5(CH_2)_3CH_3$	− 81	183	.860
Isobutylbenzene	$C_6H_5CH_2CH(CH_3)_2$		171	.867
sec-Butylbenzene	$C_6H_5CH(CH_3)C_2H_5$	− 83	173.5	.864
tert-Butylbenzene	$C_6H_5C(CH_3)_3$	− 58	169	.867
p-Cymene	1,4-$CH_3C_6H_4CH(CH_3)_2$	− 70	177	.857
Biphenyl	$C_6H_5C_6H_5$	70	255	
Diphenylmethane	$C_6H_5CH_2C_6H_5$	26	263	
Triphenylmethane	$(C_6H_5)_3CH$	93	360	
1,2-Diphenylethane	$C_6H_5CH_2CH_2C_6H_5$	52	284	
Styrene	$C_6H_5CH{=}CH_2$	− 31	145	.907
trans-Stilbene	*trans*-$C_6H_5CH{=}CHC_6H_5$	124	307	
cis-Stilbene	*cis*-$C_6H_5CH{=}CHC_6H_5$	6		
unsym-Diphenylethylene	$(C_6H_5)_2C{=}CH_2$	9	277	1.02
Triphenylethylene	$(C_6H_5)_2C{=}CHC_6H_5$	73		
Tetraphenylethylene	$(C_6H_5)_2C{=}C(C_6H_5)_2$	227	425	
Phenylacetylene	$C_6H_5C{\equiv}CH$	− 45	142	0.930
Diphenylacetylene	$C_6H_5C{\equiv}CC_6H_5$	62.5	300	

1,2,4,5-tetramethylbenzene melts 85° to 100° higher than the less symmetrical 1,2,3,5- and 1,2,3,4-isomers. A particularly striking example of the effect of symmetry on melting point is that of benzene and toluene. The introduction of a single methyl group into the extremely symmetrical benzene molecule lowers the melting point from 5° to −95°.

12.4 Industrial source of alkylbenzenes

It would be hard to exaggerate the importance to the chemical industry and to our entire economy of the large-scale production of benzene and the alkylbenzenes. Just as the alkanes obtained from petroleum are ultimately the source of nearly all our aliphatic compounds, so benzene and the alkylbenzenes are ultimately the source of nearly all our aromatic compounds. When a chemist wishes to make a complicated aromatic compound, whether in the laboratory or in industry, he does not make a benzene ring; he takes a simpler compound already containing a benzene ring and then adds to it, piece by piece, until he has built the structure he wants.

Just where do the enormous quantities of simple aromatic compounds come from? There are two large reservoirs of organic material, **coal** and **petroleum**, and aromatic compounds are obtained from both. Aromatic compounds are separated as such from coal tar, and are synthesized from the alkanes of petroleum.

By far the larger portion of coal that is mined today is converted into coke, which is needed for the smelting of iron to steel. When coal is heated in the absence of air, it is partly broken down into simpler, volatile compounds which are driven out; the residue is *coke*. The volatile materials consist of *coal gas* and a liquid known as **coal tar**.

From coal tar by distillation there are obtained a number of aromatic compounds. Upon coking, a ton of soft coal may yield about 120 pounds of coal tar. From this 120 pounds the following aromatic compounds can be separated: benzene, 2 pounds; toluene, 0.5 pound; xylenes, 0.1 pound; phenol, 0.5 pound; cresols, 2 pounds; naphthalene, 5 pounds. Two pounds of benzene from a ton of coal does not represent a very high percentage yield, yet so much coal is coked every year that the annual production of benzene from coal tar is enormous.

Still larger quantities of aromatic hydrocarbons are needed, and these are synthesized from alkanes through the process of **catalytic reforming** (Sec. 9.3). This can bring about not only *dehydrogenation*, as in the formation of toluene from methylcyclohexane, but also *cyclization* and *isomerization*, as in the formation of toluene from *n*-heptane or 1,2-dimethylcyclopentane. In an analogous way, benzene is obtained from cyclohexane and methylcyclopentane, as well as from the *hydrodealkylation* of toluene.

Today, petroleum is the *chief* source of the enormous quantities of benzene, toluene, and the xylenes required for chemicals and fuels. Half of the toluene and xylenes are utilized in high-test gasoline where, in a sense, they replace the aliphatic compounds—inferior as fuels—from which they were made. (A considerable fraction even of naphthalene, the major component of coal tar distillate, is now being produced from petroleum hydrocarbons.)

12.5 Preparation of alkylbenzenes

Although a number of the simpler alkylbenzenes are available from industrial sources, the more complicated compounds must be synthesized in one of the ways outlined below.

PREPARATION OF ALKYLBENZENES

1. **Attachment of alkyl group: Friedel-Crafts alkylation.** Discussed in Secs. 12.6–12.8.

$$C_6H_6 + RX \xrightarrow{\text{Lewis acid}} C_6H_5R + HX \quad R \textit{ may rearrange}$$

Lewis acid: $AlCl_3$, BF_3, HF, etc.

Ar-X *cannot be used in place of* R-X

2. Conversion of side chain. Discussed in Sec. 19.10.

$$C_6H_5-\underset{\underset{O}{\|}}{C}-R \xrightarrow[\text{or } N_2H_4\text{, base, heat}]{\text{Zn(Hg), HCl, heat}} C_6H_5CH_2R$$ **Clemmensen or Wolff-Kishner reduction**

A ketone

$$C_6H_5CH{=}CHR \xrightarrow{H_2,\ Ni} C_6H_5CH_2CH_2R$$

Friedel-Crafts alkylation is extremely useful since it permits the direct attachment of an alkyl group to the aromatic ring. There are, however, a number of limitations to its use (Sec. 12.8), including the fact that the alkyl group that becomes attached to the ring is not always the same as the alkyl group of the parent halide; this **rearrangement** of the alkyl group is discussed in Sec. 12.7.

There are frequently available aromatic compounds containing aliphatic side chains that are not simple alkyl groups. An alkylbenzene can be prepared from one of these compounds by converting the side chain into an alkyl group. Although there is an aromatic ring in the molecule, this conversion is essentially the preparation of an alkane from some other aliphatic compound. The methods used are those that we have already learned for the preparation of alkanes: hydrogenation of a carbon–carbon double bond in a side chain, for example. Many problems of the alkylbenzenes are solved by a consideration of simple alkane chemistry.

The most important side-chain conversion involves **reduction of ketones** either by amalgamated zinc and HCl (*Clemmensen reduction*) or by hydrazine and strong base (*Wolff-Kishner reduction*). This method is important because the necessary ketones are readily available through a modification of the Friedel-Crafts reaction that involves acid chlorides (see Sec. 19.6). Unlike alkylation by the Friedel-Crafts reaction, this method does not involve rearrangement.

Problem 12.1 How might you prepare ethylbenzene from: (a) benzene and ethyl alcohol; (b) acetophenone, $C_6H_5COCH_3$; (c) styrene, $C_6H_5CH{=}CH_2$; (d) α-phenylethyl alcohol, $C_6H_5CHOHCH_3$; and (e) β-phenylethyl chloride, $C_6H_5CH_2CH_2Cl$?

Problem 12.2 How might you prepare 2,3-diphenylbutane from α-phenylethyl alcohol, $C_6H_5CHOHCH_3$?

12.6 Friedel-Crafts alkylation

If a small amount of anhydrous aluminum chloride is added to a mixture of benzene and methyl chloride, a vigorous reaction occurs, hydrogen chloride gas is

$$C_6H_6 + CH_3Cl \xrightarrow{AlCl_3} C_6H_5CH_3 + HCl$$

Toluene

evolved, and toluene can be isolated from the reaction mixture. This is the simplest example of the reaction discovered in 1877 at the University of Paris by the French-American team of chemists, Charles Friedel and James Crafts. *Considered in its various modifications, the Friedel-Crafts reaction is by far the most important method for attaching alkyl side chains to aromatic rings.*

Each of the components of the simple example just given can be varied. The alkyl halide may contain an alkyl group more complicated than methyl, and a halogen atom other than chlorine; in some cases alcohols are used or—especially in industry—alkenes. Substituted alkyl halides, like benzyl chloride, $C_6H_5CH_2Cl$, also can be used. Because of the low reactivity of halogen attached to an aromatic ring (Sec. 25.5), aryl halides (Ar—X, e.g., bromo- or chlorobenzene) *cannot* be used in place of alkyl halides.

The aromatic ring to which the side chain becomes attached may be that of benzene itself, certain substituted benzenes (chiefly alkylbenzenes and halobenzenes), or more complicated aromatic ring systems like naphthalene and anthracene (Chap. 30).

In place of aluminum chloride, other Lewis acids can be used, in particular BF_3, HF, and phosphoric acid.

The reaction is carried out by simply mixing together the three components; usually the only problems are those of moderating the reaction by cooling and of trapping the hydrogen halide gas. Since the attachment of an alkyl side chain makes the ring more susceptible to further attack (Sec. 11.5), steps must be taken to limit substitution to *mono*alkylation. As in halogenation of alkanes (Sec. 2.8), this is accomplished by using an *excess* of the hydrocarbon. In this way an alkyl carbonium ion seeking an aromatic ring is more likely to encounter an unsubstituted ring than a substituted one. Frequently the aromatic compound does double duty, serving as solvent as well as reactant.

From polyhalogenated alkanes it is possible to prepare compounds containing more than one aromatic ring:

$$2C_6H_6 + CH_2Cl_2 \xrightarrow{AlCl_3} \underset{\text{Diphenylmethane}}{C_6H_5CH_2C_6H_5} + 2HCl$$

$$2C_6H_6 + ClCH_2CH_2Cl \xrightarrow{AlCl_3} \underset{\text{1,2-Diphenylethane}}{C_6H_5CH_2CH_2C_6H_5} + 2HCl$$

$$3C_6H_6 + CHCl_3 \xrightarrow{AlCl_3} C_6H_5\!-\!\overset{\overset{\displaystyle C_6H_5}{|}}{\underset{\underset{\displaystyle H}{|}}{C}}\!-\!C_6H_5 + 3HCl$$

Triphenylmethane

$$3C_6H_6 + CCl_4 \xrightarrow{AlCl_3} C_6H_5\!-\!\overset{\overset{\displaystyle C_6H_5}{|}}{\underset{\underset{\displaystyle Cl}{|}}{C}}\!-\!C_6H_5 + 3HCl$$

Triphenylchloromethane

12.7 Mechanism of Friedel-Crafts alkylation

In Sec. 11.10 we said that two mechanisms are possible for Friedel-Crafts alkylation. Both involve electrophilic aromatic substitution, but they differ as to the nature of the electrophile.

One mechanism for Friedel-Crafts alkylation involves the following steps,

$$\text{(1)} \qquad \mathrm{RCl + AlCl_3 \rightleftharpoons AlCl_4^- + R^{\oplus}}$$

$$\text{(2)} \qquad \mathrm{R^{\oplus} + C_6H_6 \rightleftharpoons \overset{\oplus}{C_6H_5}\!\left<\begin{matrix}R\\H\end{matrix}\right.}$$

$$\text{(3)} \qquad \mathrm{\overset{\oplus}{C_6H_5}\!\left<\begin{matrix}R\\H\end{matrix}\right. + AlCl_4^- \rightleftharpoons C_6H_5R + HCl + AlCl_3}$$

in which the electrophile is an alkyl carbonium ion. The function of the aluminum chloride is to generate this carbonium ion by abstracting the halogen from the alkyl halide. It is not surprising that other Lewis acids can function in the same way and thus take the place of aluminum chloride:

$$\mathrm{R{:}\ddot{\underset{..}{X}}{:} + \begin{matrix}:\ddot{Cl}:\\ \ddot{Al}{:}\ddot{\underset{..}{Cl}}{:}\\ :\underset{..}{Cl}:\end{matrix} \rightleftharpoons R^{\oplus} + \begin{matrix}:\ddot{Cl}:\\ :\ddot{\underset{..}{X}}{:}\ddot{\underset{..}{Al}}{:}\ddot{\underset{..}{Cl}}{:}\\ :\underset{..}{Cl}:\end{matrix}^{\ominus}}$$

$$\mathrm{R{:}\ddot{\underset{..}{X}}{:} + \begin{matrix}:\ddot{F}:\\ \ddot{B}{:}\ddot{\underset{..}{F}}{:}\\ :\underset{..}{F}:\end{matrix} \rightleftharpoons R^{\oplus} + \begin{matrix}:\ddot{F}:\\ :\ddot{\underset{..}{X}}{:}\ddot{\underset{..}{B}}{:}\ddot{\underset{..}{F}}{:}\\ :\underset{..}{F}:\end{matrix}^{\ominus}}$$

Carbonium ions from alkyl halides

$$\mathrm{R{:}\ddot{\underset{..}{X}}{:} + \begin{matrix}:\ddot{Cl}:\\ \ddot{Fe}{:}\ddot{\underset{..}{Cl}}{:}\\ :\underset{..}{Cl}:\end{matrix} \rightleftharpoons R^{\oplus} + \begin{matrix}:\ddot{Cl}:\\ :\ddot{\underset{..}{X}}{:}\ddot{\underset{..}{Fe}}{:}\ddot{\underset{..}{Cl}}{:}\\ :\underset{..}{Cl}:\end{matrix}^{\ominus}}$$

$$\mathrm{R{:}\ddot{\underset{..}{X}}{:} + H{:}\ddot{\underset{..}{F}}{:} \rightleftharpoons R^{\oplus} + :\ddot{\underset{..}{X}}{:}{-}{-}H{:}\ddot{\underset{..}{F}}{:}^{\ominus}}$$

Judging from the mechanism just described, we might expect the benzene ring to be attacked by carbonium ions generated in other ways: by the action of acid on alcohols (Sec. 5.20) and on alkenes (Sec. 6.10).

$$\mathrm{ROH + H^+ \rightleftharpoons ROH_2^{\oplus} \rightleftharpoons R^{\oplus} + H_2O}$$

$$\mathrm{-\overset{|}{C}{=}\overset{|}{C}- + H^+ \rightleftharpoons -\underset{H}{\overset{|}{\underset{|}{C}}}-\overset{|}{\underset{|}{C}}^{\oplus}}$$

Carbonium ions from alcohols and from alkenes

This expectation is correct: alcohols and alkenes, in the presence of acids, alkylate aromatic rings in what we may consider to be a modification of the Friedel-Crafts reaction.

$$\underset{\textit{tert}\text{-Butyl alcohol}}{C_6H_6 + (CH_3)_3COH} \xrightarrow{H_2SO_4} \underset{\textit{tert}\text{-Butylbenzene}}{C_6H_5{-}C(CH_3)_3}$$

$$\underset{\text{Isobutylene}}{C_6H_6 + (CH_3)_2C{=}CH_2} \xrightarrow{H_2SO_4} \underset{\textit{tert}\text{-Butylbenzene}}{C_6H_5{-}C(CH_3)_3}$$

Also judging from the mechanism, we might expect Friedel-Crafts alkylation to be accompanied by the kind of rearrangement that is characteristic of carbonium ion reactions (Sec. 5.22). This expectation, too, is correct. As the following examples show, alkylbenzenes containing rearranged alkyl groups are not only formed but are sometimes the sole products. In each case, we see that the

$$\underset{n\text{-Propyl chloride}}{C_6H_6 + CH_3CH_2CH_2Cl} \xrightarrow[-18^\circ \text{ to } 80^\circ]{AlCl_3} \underset{\substack{n\text{-Propylbenzene} \\ 35\text{–}31\%}}{C_6H_5CH_2CH_2CH_3} \text{ and } \underset{\substack{\text{Isopropylbenzene} \\ 65\text{–}69\%}}{C_6H_5\overset{\overset{CH_3}{|}}{C}HCH_3}$$

$$\underset{n\text{-Butyl chloride}}{C_6H_6 + CH_3CH_2CH_2CH_2Cl} \xrightarrow[0^\circ]{AlCl_3} \underset{\substack{n\text{-Butylbenzene} \\ 34\%}}{C_6H_5CH_2CH_2CH_2CH_3} \text{ and } \underset{\substack{\textit{sec}\text{-Butylbenzene} \\ 66\%}}{C_6H_5\overset{\overset{CH_3}{|}}{C}HCH_2CH_3}$$

$$\underset{\text{Isobutyl chloride}}{C_6H_6 + CH_3\overset{\overset{CH_3}{|}}{C}HCH_2Cl} \xrightarrow[-18^\circ \text{ to } 80^\circ]{AlCl_3} \underset{\substack{\textit{tert}\text{-Butylbenzene} \\ \textit{Only product}}}{C_6H_5\overset{\overset{CH_3}{|}}{\underset{\underset{CH_3}{|}}{C}}CH_3}$$

$$\underset{\text{Neopentyl alcohol}}{C_6H_6 + CH_3\overset{\overset{CH_3}{|}}{\underset{\underset{CH_3}{|}}{C}}CH_2OH} \xrightarrow[60^\circ]{BF_3} \underset{\substack{\textit{tert}\text{-Pentylbenzene} \\ \textit{Only product}}}{C_6H_5\overset{\overset{CH_3}{|}}{\underset{\underset{CH_3}{|}}{C}}CH_2CH_3}$$

particular kind of rearrangement corresponds to what we would expect if a less stable (1°) carbonium ion were to rearrange by a 1,2-shift to a more stable (2° or 3°) carbonium ion.

We can now make another addition to our list of carbonium ion reactions (Sec. 6.16). **A carbonium ion may:**

(a) eliminate a hydrogen ion to form an alkene;
(b) rearrange to a more stable carbonium ion;
(c) combine with a negative ion or other basic molecule;
(d) add to an alkene to form a larger carbonium ion;
(e) abstract a hydride ion from an alkane;
(f) alkylate an aromatic ring.

A carbonium ion formed by (b) or (d) can subsequently undergo any of the reactions.

In alkylation, as in its other reactions, the carbonium ion gains a pair of electrons to complete the octet of the electron-deficient carbon—this time from the π cloud of an aromatic ring.

Problem 12.3 *tert*-Pentylbenzene is the major product of the reaction of benzene in the presence of BF_3 with each of the following alcohols: (a) 2-methyl-1-butanol, (b) 3-methyl-2-butanol, (c) 3-methyl-1-butanol, and (d) neopentyl alcohol. Account for its formation in each case.

In some of the examples given above, we see that *part* of the product is made up of *unrearranged* alkylbenzenes. Must we conclude that part of the reaction does not go by way of carbonium ions? Not *necessarily*. Attack on an aromatic ring is probably one of the most difficult jobs a carbonium ion is called on to do; that is to say, toward carbonium ions an aromatic ring is a reagent of low reactivity and hence high selectivity. Although there may be present a higher concentration of the more stable, rearranged carbonium ions, the aromatic ring may tend to seek out the scarce unrearranged ions because of their higher reactivity. In some cases, it is quite possible that some of the carbonium ions react with the aromatic ring before they have time to rearrange; the same low stability that makes primary carbonium ions, for example, prone to rearrangement also makes them highly reactive.

On the other hand, there is additional evidence (of a kind we cannot go into here) that makes it very likely that there *is* a second mechanism for Friedel-Crafts alkylation. In this mechanism, the electrophile is not an alkyl carbonium ion, but an acid–base complex of alkyl halide and Lewis acid, from which the alkyl group is transferred *in one step* from halogen to the aromatic ring.

$$\mathrm{Cl{-}\overset{\ominus}{Al}(Cl)_2{-}\overset{\oplus}{Cl}{-}R} + \mathrm{C_6H_6} \longrightarrow \left[\overset{\delta+}{\mathrm{C_6H_5}}\!\begin{matrix}\mathrm{H}\\ \cdots\mathrm{R}\text{---}\overset{\delta-}{\mathrm{ClAlCl_3}}\end{matrix}\right] \longrightarrow \overset{\oplus}{\mathrm{C_6H_5}}\!\begin{matrix}\mathrm{H}\\ \mathrm{R}\end{matrix} + \mathrm{AlCl_4^-}$$

This duality of mechanism does not reflect exceptional behavior, but is usual for electrophilic aromatic substitution. It also fits into the usual pattern for *nucleophilic aliphatic substitution* (Sec. 14.16), which—from the standpoint of the alkyl halide—is the kind of reaction taking place. Furthermore, the particular halides (1° and methyl) which appear to react by this second mechanism are just the ones that would have been *expected* to do so.

12.8 Limitations of Friedel-Crafts alkylation

We have encountered three limitations to the use of Friedel-Crafts alkylation: (a) the danger of polysubstitution; (b) the possibility that the alkyl group will rearrange; and (c) the fact that aryl halides cannot take the place of alkyl halides. Besides these, there are several other limitations.

(d) An aromatic ring less reactive than that of the halobenzenes does not undergo the Friedel-Crafts reaction; evidently the carbonium ion, R^+, is a less powerful electrophile than NO_2^+ and the other electron-deficient reagents that bring about electrophilic aromatic substitution.

Next, (e) aromatic rings containing the $—NH_2$, $—NHR$, or $—NR_2$ group do not undergo Friedel-Crafts alkylation, partly because the strongly basic nitrogen ties up the Lewis acid needed for ionization of the alkyl halide:

$$C_6H_5\ddot{N}H_2 + AlCl_3 \longrightarrow C_6H_5\overset{\ominus AlCl_3}{\underset{\oplus}{\ddot{N}H_2}}$$

I

Problem 12.4 Tying up of the acidic catalyst by the basic nitrogen is not the only factor that prevents alkylation, since even when excess catalyst is used, reaction does not occur. Looking at the structure of the complex (I) shown for aniline, can you suggest another factor? (*Hint:* See Sec. 11.18.)

Despite these numerous limitations, the Friedel-Crafts reaction, in its various modifications (for example, acylation, Sec. 19.6), is an extremely useful synthetic tool.

12.9 Reactions of alkylbenzenes

The most important reactions of the alkylbenzenes are outlined below, with toluene and ethylbenzene as specific examples; essentially the same behavior is shown by compounds bearing other side chains. Except for hydrogenation and oxidation, these reactions involve either **electrophilic substitution in the aromatic ring** or **free-radical substitution in the aliphatic side chain**.

In following sections we shall be mostly concerned with (a) how experimental conditions determine which portion of the molecule—aromatic or aliphatic—is attacked, and (b) how each portion of the molecule modifies the reactions of the other portion.

REACTIONS OF ALKYLBENZENES

1. Hydrogenation.

Example:

$$C_6H_5CH_2CH_3 + 3H_2 \xrightarrow{\text{Ni, Pt, Pd}} C_6H_{11}CH_2CH_3$$

Ethylbenzene → Ethylcyclohexane

2. Oxidation. Discussed in Sec. 12.10.

Example:

$$C_6H_5CH_2CH_3 \xrightarrow[\text{(or } K_2Cr_2O_7\text{, or dil. } HNO_3)]{KMnO_4} C_6H_5COOH \ (+ CO_2)$$

Ethylbenzene → Benzoic acid

3. Substitution in the ring. Electrophilic aromatic substitution. Discussed in Sec. 12.11.

Examples:

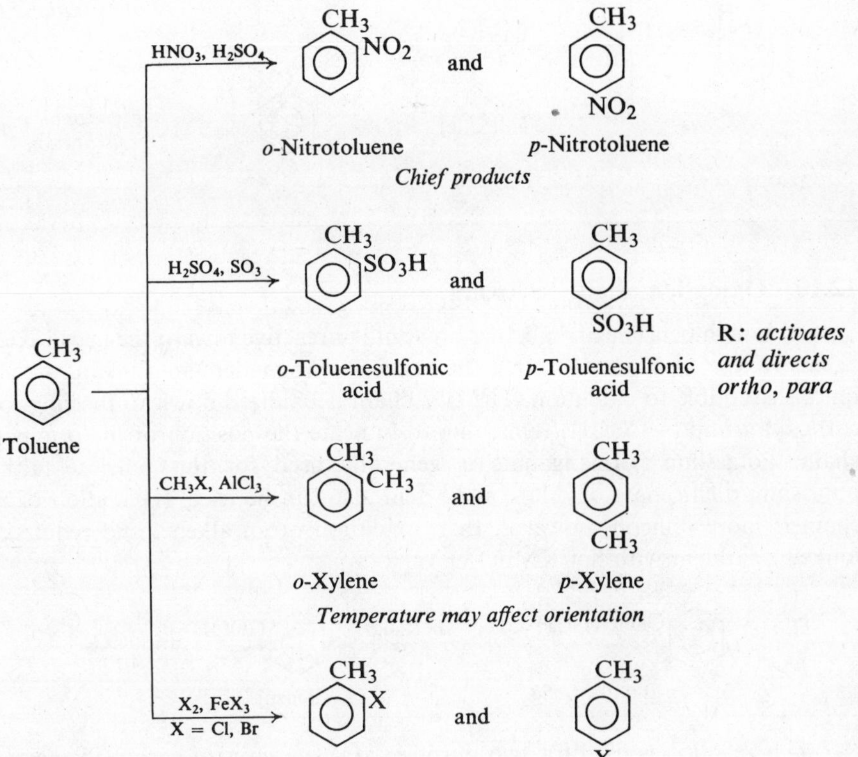

4. Substitution in the side chain. Free-radical halogenation. Discussed in Secs. 12.12–12.14.

Examples:

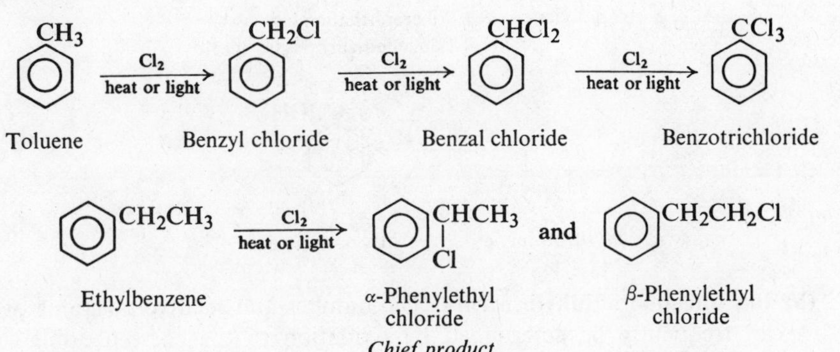

Note: Competition between ring and side chain. Discussed in Sec. 12.12.

Toluene ($C_6H_5CH_3$) + Cl_2:

- heat or light → $C_6H_5CH_2Cl$ — *Free radical substitution*
- $FeCl_3$ → o-chlorotoluene (CH_3, Cl) and p-chlorotoluene (CH_3, Cl) — *Electrophilic substitution*

12.10 Oxidation of alkylbenzenes

Although benzene and alkanes are quite unreactive toward the usual oxidizing agents ($KMnO_4$, $K_2Cr_2O_7$, etc.), the benzene ring renders an aliphatic side chain quite susceptible to oxidation. The side chain is oxidized down to the ring, only a carboxyl group (—COOH) remaining to indicate the position of the original side chain. Potassium permanganate is generally used for this purpose, although potassium dichromate or dilute nitric acid also can be used. (Oxidation of a side chain is more difficult, however, than oxidation of an alkene, and requires prolonged treatment with hot $KMnO_4$.)

$$C_6H_5CH_2CH_2CH_2CH_3 \xrightarrow{\text{hot } KMnO_4} C_6H_5COOH \text{ and } CO_2$$

n-Butylbenzene → Benzoic acid

This reaction is used for two purposes: (a) synthesis of carboxylic acids, and (b) identification of alkylbenzenes.

(a) Synthesis of carboxylic acids. One of the most useful methods of preparing an aromatic carboxylic acid involves oxidation of the proper alkylbenzene. For example:

$$p\text{-}CH_3C_6H_4CH_3 \xrightarrow{KMnO_4} p\text{-}HOOCC_6H_4COOH$$

p-Xylene → Terephthalic acid (1,4-Benzenedicarboxylic acid)

$$p\text{-}CH_3C_6H_4NO_2 \xrightarrow{Cr_2O_7^{--},\ H^+} p\text{-}HOOCC_6H_4NO_2$$

p-Nitrotoluene → *p*-Nitrobenzoic acid

(b) Identification of alkylbenzenes. The number and relative positions of side chains can frequently be determined by oxidation to the corresponding acids.

Suppose, for example, that we are trying to identify an unknown liquid of formula C_8H_{10} and boiling point 137–139° that we have shown in other ways to be an alkylbenzene (Sec. 12.22). Looking in Table 12.1 (p. 375), we find that it could be any one of four compounds: *o*-, *m*-, or *p*-xylene, or ethylbenzene. As shown below, oxidation of each of these possible hydrocarbons yields a different acid, and these acids can readily be distinguished from each other by their melting points or the melting points of derivatives.

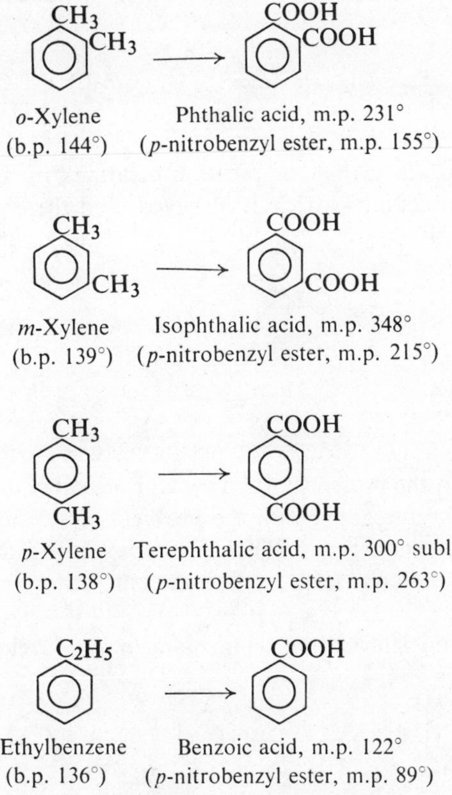

12.11 Electrophilic aromatic substitution in alkylbenzenes

Because of its electron-releasing effect, an alkyl group activates a benzene ring to which it is attached, and directs *ortho* and *para* (Secs. 11.18 and 11.19).

Problem 12.5 Treatment with methyl chloride and $AlCl_3$ at 0° converts toluene chiefly into *o*- and *p*-xylenes; at 80°, however, the chief product is *m*-xylene. Furthermore, either *o*- or *p*-xylene is readily converted into *m*-xylene by treatment with $AlCl_3$ and HCl at 80°.

How do you account for this effect of temperature on orientation? Suggest a role for the HCl.

Problem 12.6 Why is polysubstitution a complicating factor in Friedel-Crafts alkylation but not in aromatic nitration, sulfonation, or halogenation?

12.12 Halogenation of alkylbenzenes: ring *vs.* side chain

Alkylbenzenes clearly offer two main areas to attack by halogens: the ring and the side chain. We can control the position of attack simply by choosing the proper reaction conditions.

Halogenation of alkanes requires conditions under which halogen atoms are formed, that is, high temperature or light. Halogenation of benzene, on the other hand, involves transfer of positive halogen, which is promoted by acid catalysts like ferric chloride.

$$CH_4 + Cl_2 \xrightarrow{\text{heat or light}} CH_3Cl + HCl$$

$$C_6H_6 + Cl_2 \xrightarrow{\text{FeCl}_3\text{, cold}} C_6H_5Cl + HCl$$

We might expect, then, that the position of attack in, say, toluene would be governed by which attacking particle is involved, and therefore by the conditions employed. This is so: if chlorine is bubbled into boiling toluene that is exposed to

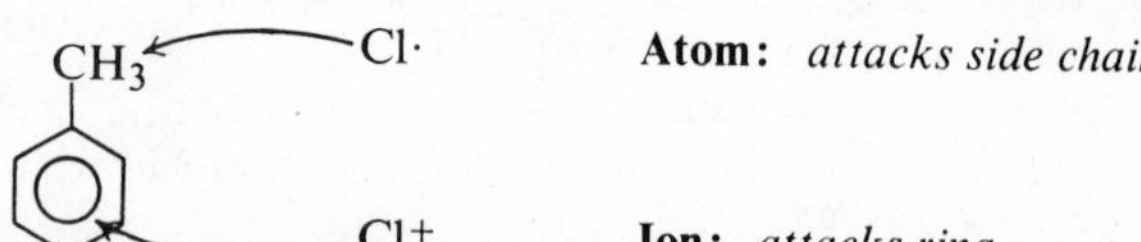

ultraviolet light, substitution occurs almost exclusively in the side chain; in the absence of light and in the presence of ferric chloride, substitution occurs mostly in the ring. (Compare the foregoing with the problem of substitution *vs.* addition in the halogenation of alkenes (Sec. 6.21), where atoms bring about substitution and ions—or, more accurately, molecules that can transfer ions—bring about addition.)

Like nitration and sulfonation, ring halogenation yields chiefly the *o*- and

$$\text{Toluene} \xrightarrow{Cl_2,\ Fe,\ \text{or}\ FeCl_3} o\text{-Chlorotoluene (58\%)} \text{ and } p\text{-Chlorotoluene (42\%)}$$

p-isomers. Similar results are obtained with other alkylbenzenes, and with bromine as well as chlorine.

Side-chain halogenation, like halogenation of alkanes, may yield polyhalogenated products; even when reaction is limited to monohalogenation, it may yield a mixture of isomers.

Side-chain chlorination of toluene can yield successively the mono-, di-, and trichloro compounds. These are known as *benzyl chloride*, *benzal chloride*, and

$$C_6H_5CH_3 \xrightarrow[\text{heat, light}]{Cl_2} C_6H_5CH_2Cl \xrightarrow[\text{heat, light}]{Cl_2} C_6H_5CHCl_2 \xrightarrow[\text{heat, light}]{Cl_2} C_6H_5CCl_3$$

Toluene — Benzyl chloride — Benzal chloride — Benzotrichloride

benzotrichloride; such compounds are important intermediates in the synthesis of alcohols, aldehydes, and acids.

12.13 Side-chain halogenation of alkylbenzenes

Chlorination and bromination of side chains differ from one another in orientation and reactivity in one very significant way. Let us look first at bromination, and then at chlorination.

An alkylbenzene with a side chain more complex than methyl may offer more than one position for attack, and so we must consider the likelihood of obtaining a mixture of isomers. Bromination of ethylbenzene, for example, could theoretically yield two products: 1-bromo-1-phenylethane and 2-bromo-1-phenylethane. Despite

$$C_6H_5CH_2CH_3 \xrightarrow[\text{heat, light}]{Br_2} C_6H_5CHBrCH_3 \quad (\text{not } C_6H_5CH_2CH_2Br)$$

Ethylbenzene → 1-Bromo-1-phenylethane *Only product*; ✗→ 2-Bromo-1-phenylethane

a probability factor that favors 2-bromo-1-phenylethane by 3:2, the *only* product found is 1-bromo-1-phenylethane. Evidently abstraction of the hydrogens attached to the carbon next to the aromatic ring is greatly preferred.

Hydrogen atoms attached to carbon joined directly to an aromatic ring are called **benzylic hydrogens.**

$$C_6H_5-\overset{|}{\underset{H}{C}}-$$

Benzylic hydrogen:
easy to abstract

The relative ease with which benzylic hydrogens are abstracted is shown not only by orientation of bromination but also—and in a more exact way—by comparison of reactivities of different compounds. Competition experiments (Sec. 3.22) show, for example, that at 40° a benzylic hydrogen of toluene is 3.3 times as reactive toward bromine atoms as the tertiary hydrogen of an alkane—and nearly 100 million times as reactive as a hydrogen of methane!

Examination of reactions that involve attack not only by halogen atoms but by other free radicals as well has shown that this is a general rule: benzylic hydrogens are extremely easy to abstract and thus resemble allylic hydrogens. We can now expand the reactivity sequence of Sec. 6.22:

Ease of abstraction of hydrogen atoms allylic, benzylic > 3° > 2° > 1° > CH_4, vinylic

Side-chain halogenation of alkylbenzenes proceeds by the same mechanism as halogenation of alkanes. Bromination of toluene, for example, would include the following steps:

$$C_6H_5CH_3 \xrightarrow{Br\cdot} C_6H_5CH_2\cdot \xrightarrow{Br_2} C_6H_5CH_2Br$$

Toluene — Benzyl radical — Benzyl bromide

The fact that benzylic hydrogens are unusually easy to abstract means that benzyl radicals are unusually easy to form.

Ease of formation of free radicals allyl, benzyl $> 3° > 2° > 1° > CH_3\cdot$, vinyl

Again we ask the question: are these findings in accord with our rule that *the more stable the radical, the more rapidly it is formed*? Is the rapidly formed benzyl radical relatively stable?

The bond dissociation energies in Table 1.2 (p. 21) show that only 85 kcal is needed for formation of benzyl radicals from a mole of toluene, as compared with 91 kcal for formation of *tert*-butyl radicals and 88 kcal for formation of allyl radicals. Relative to the hydrocarbon from which each is formed, then, a benzyl radical contains less energy and is more stable than a *tert*-butyl radical.

We can now expand the sequence of radical stabilities (Sec. 6.22). Relative to the hydrocarbon from which each is formed, the relative stability of free radicals is:

Stability of free radicals allyl, benzyl $> 3° > 2° > 1° > CH_3\cdot$, vinyl

Orientation of chlorination shows that chlorine atoms, like bromine atoms, preferentially attack benzylic hydrogen; but, as we see, the preference is less marked:

$$C_6H_5CH_2CH_3 \xrightarrow[\text{heat, light}]{Cl_2} C_6H_5CH(Cl)CH_3 \text{ and } C_6H_5CH_2CH_2Cl$$

Ethylbenzene — 1-Chloro-1-phenylethane, *Major product, 91%* — 2-Chloro-1-phenylethane, 9%

Furthermore, competition experiments show that, under conditions where 3°, 2°, and 1° hydrogens show relative reactivities of 5.0:3.8:1.0, the relative rate per benzylic hydrogen of toluene is only 1.3. As in its attack on alkanes (Sec. 3.28), the more reactive chlorine atom is less selective than the bromine atom: less selective between hydrogens in a single molecule, and less selective between hydrogens in different molecules.

In the attack by the comparatively unreactive bromine atom, we have said (Sec. 2.23), the transition state is reached late in the reaction process: the carbon–hydrogen bond is largely broken, and the organic group has acquired a great deal of free-radical character. The factors that stabilize the benzyl free radical stabilize the incipient benzyl free radical in the transition state.

In contrast, in the attack by the highly reactive chlorine atom, the transition state is reached early in the reaction process: the carbon–hydrogen bond is only slightly broken, and the organic group has acquired little free-radical character. The factors that stabilize the benzyl radical have little effect on this transition state.

Just why benzylic hydrogens are *less* reactive toward chlorine atoms than even secondary hydrogens is not understood. It has been attributed to *polar factors* (Sec. 32.4), but this hypothesis has been questioned.

12.14 Resonance stabilization of the benzyl radical

How are we to account for the stability of the benzyl radical? Bond dissociation energies indicate that 19 kcal/mole less energy (104 − 85) is needed to form the benzyl radical from toluene than to form the methyl radical from methane.

$$\underset{\text{Toluene}}{C_6H_5CH_3} \longrightarrow \underset{\text{Benzyl radical}}{C_6H_5CH_2\cdot} + H\cdot \qquad \Delta H = +85 \text{ kcal}$$

As we did for the allyl radical (Sec. 6.24), let us examine the structures involved. Toluene contains the benzene ring and is therefore a hybrid of the two Kekulé structures, I and II:

CH_3 CH_3

I II

Similarly, the benzyl radical is a hybrid of the two Kekulé structures, III and IV:

$CH_2\cdot$ $CH_2\cdot$

III IV

This resonance causes stabilization, that is, lowers the energy content. However, resonance involving Kekulé structures presumably stabilizes both molecule and radical to the same extent, and hence does not affect the *difference* in their energy contents. If there were no other factors involved, then we might reasonably expect the bond dissociation energy for a benzylic hydrogen to be about the same as that of a methane hydrogen (see Fig. 12.1).

Considering further, however, we find that we can draw three additional structures for the radical: V, VI, and VII. In these structures there is a double bond between the side chain and the ring, and the odd electron is located on the carbon atoms *ortho* and *para* to the side chain. Drawing these pictures is, of

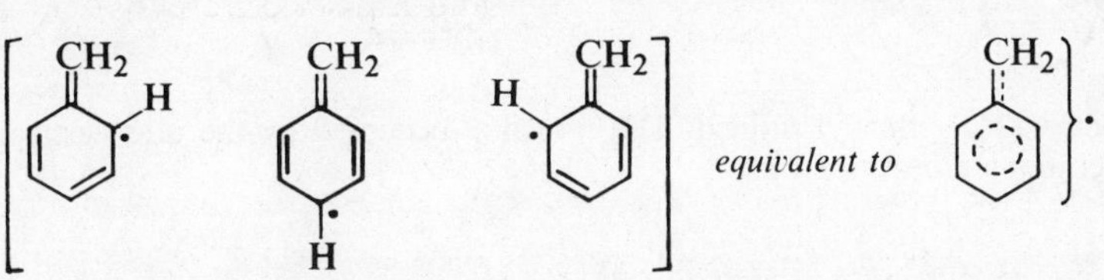

CH_2 H CH_2 H CH_2 H *equivalent to* CH_2

V VI VII

course, our way of indicating that the odd electron is not localized on the side chain but is *delocalized*, being distributed about the ring. We cannot draw comparable structures for the toluene molecule.

Contribution from the three structures, V–VII, stabilizes the radical in a way that is not possible for the molecule. Resonance thus lowers the energy content of the benzyl radical more than it lowers the energy content of toluene. This extra stabilization of the radical evidently amounts to 19 kcal/mole (Fig. 12.1).

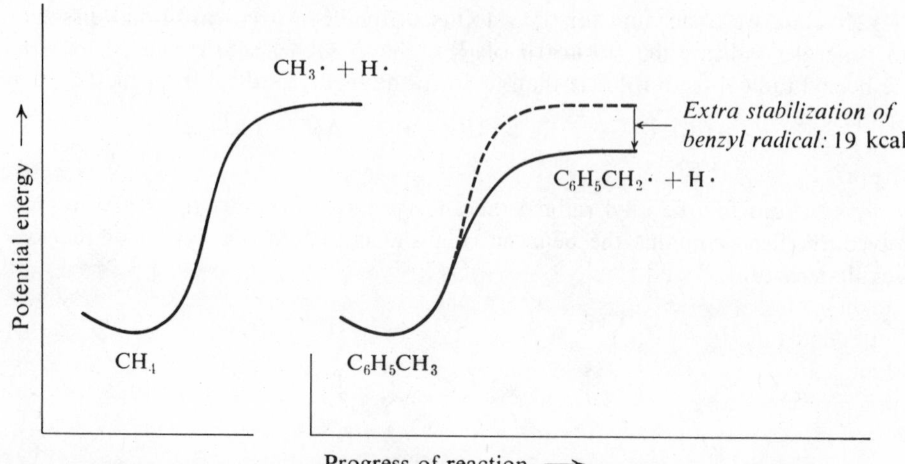

Figure 12.1. Molecular structure and rate of reaction. Resonance-stabilized benzyl radical formed faster than methyl radical. (Plots aligned with each other for easy comparison.)

We say, then, that the benzyl radical is *stabilized by resonance*. When we use this expression, we must always bear in mind that we actually mean that the benzyl radical is stabilized by resonance *to a greater extent than* the hydrocarbon from which it is formed.

In terms of orbitals, delocalization results from overlap of the *p* orbital occupied by the odd electron with the π cloud of the ring.

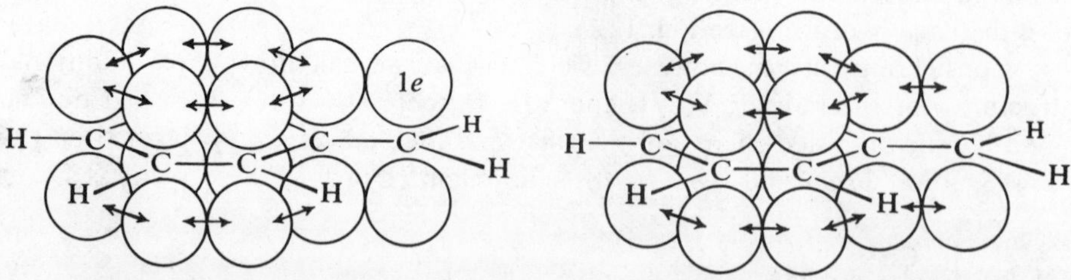

Figure 12.2. Benzyl radical. The *p* orbital occupied by the odd electron overlaps π cloud of ring.

Problem 12.7 It is believed that the side-chain hydrogens of the benzyl radical lie in the same plane as the ring. Why should they?

Problem 12.8 The strength of the bond holding side-chain hydrogen in *m*-xylene is the same as in toluene; in *o*- and *p*-xylene it is 3–4 kcal lower. How do you account for these differences?

12.15 Triphenylmethyl: a stable free radical

We have said that benzyl and allyl free radicals are stabilized by resonance; but we must realize, of course, that they are stable only in comparison with simple alkyl radicals like methyl or ethyl. Benzyl and allyl free radicals are extremely reactive, unstable particles, whose fleeting existence (a few thousandths of a second) has been proposed simply because it is the best way to account for certain experimental observations. We do not find bottles on the laboratory shelf labeled "benzyl radicals" or "allyl radicals." Is there, then, any direct evidence for the existence of free radicals?

In 1900 a remarkable paper appeared in the *Journal of the American Chemical Society* and in the *Berichte der deutschen chemischen Gesellschaft*; its author was the young Russian-born chemist Moses Gomberg, who was at that time an instructor at the University of Michigan. Gomberg was interested in completely phenylated alkanes. He had prepared tetraphenylmethane (a synthesis a number of eminent chemists had previously attempted, but unsuccessfully), and he had now set himself the task of synthesizing hexaphenylethane. Having available triphenylchloromethane (Sec. 12.6), he went about the job in just the way we might today: he tried to couple together two triphenylmethyl groups by use of a metal (Sec. 9.4). Since sodium did not work very well, he used instead finely divided silver, mercury, or, best of all, zinc dust. He allowed a benzene solution of triphenylchloromethane to stand over one of these metals, and then filtered the solution free of the metal halide. When the benzene was evaporated, there was left behind a white crystalline solid which after recrystallization melted at 185°; this he thought was hexaphenylethane.

$$(C_6H_5)_3C{-}Cl + Zn + Cl{-}C(C_6H_5)_3 \longrightarrow \text{Hexaphenylethane} + ZnCl_2$$

Triphenylchloromethane
2 moles

Hexaphenylethane: *Expected product*

As a chemist always does with a new compound, Gomberg analyzed his product for its carbon and hydrogen content. To his surprise, the analysis showed 88% carbon and 6% hydrogen, a total of only 94%. Thinking that combustion had not been complete, he carried out the analysis again, this time more carefully and under more vigorous conditions; he obtained the same results as before. Repeated analysis of samples prepared from both triphenylchloromethane and triphenylbromomethane, and purified by recrystallization from a variety of solvents, finally convinced him that he had prepared not a hydrocarbon—not hexaphenylethane—but a compound containing 6% of some other element, probably oxygen.

Oxygen could have come from impure metals; but extremely pure samples of metals, carefully freed of oxygen, gave the same results.

Oxygen could have come from the air, although he could not see how molecular oxygen could react at room temperature with a hydrocarbon. He carried out the reaction again, this time under an atmosphere of carbon dioxide. When he filtered the solution (also under carbon dioxide) and evaporated the solvent, there was left behind not his compound of m.p. 185° but an entirely different substance, much more soluble in benzene than his first product, and having a much lower melting point. This new substance was eventually purified, and on analysis it gave the correct composition for hexaphenylethane: 93.8% carbon, 6.2% hydrogen.

Dissolved in benzene, the new substance gave a yellow solution. When a small amount of air was admitted to the container, the yellow color disappeared, and then after a few minutes reappeared. When more oxygen was admitted, the same thing happened: disappearance of the color and slow reappearance. Finally the color disappeared for good; evaporation of the solvent yielded the original compound of m.p. 185°.

Not only oxygen but also halogens were rapidly absorbed by ice-cold solutions of this substance; even solutions of normally unreactive iodine were instantly decolorized.

The compound of m.p. 185° was the peroxide,

$$(C_6H_5)_3C{-}O{-}O{-}C(C_6H_5)_3$$

as Gomberg showed by preparing it in an entirely different way. The products of the halogen reactions were the triphenylhalomethanes, $(C_6H_5)_3C{-}X$.

If this new substance he had made was indeed hexaphenylethane, it was behaving very strangely. Cleavage of a carbon–carbon bond by such mild reagents as oxygen and iodine was unknown to organic chemists.

$$(C_6H_5)_3C{-}Cl + Zn \longrightarrow (C_6H_5)_3C\cdot$$

Triphenylchloromethane Triphenylmethyl

$$(C_6H_5)_3C\cdot + \cdot C(C_6H_5)_3 \rightleftharpoons \text{Dimer}$$

Triphenylmethyl *Yellow* Dimer *Colorless*

"The experimental evidence presented above forces me to the conclusion that we have to deal here with a free radical, triphenylmethyl, $(C_6H_5)_3C$. On this assumption alone do the results described above become intelligible and receive an adequate explanation." Gomberg was proposing that he had prepared a *stable* free radical.

It was nearly ten years before Gomberg's proposal was generally accepted. It now seems clear that what happens is the following: the metal abstracts a chlorine atom from triphenylchloromethane to form the free radical triphenylmethyl; two of these radicals then combine to form a dimeric hydrocarbon. But the carbon–carbon bond in the dimer is a very weak one, and even at room temperature can break to regenerate the radicals. Thus an equilibrium exists between the free radicals and the hydrocarbon. Although this equilibrium tends to favor the hydrocarbon, any solution of the dimer contains an appreciable concentration of free triphenylmethyl radicals. The fraction of material existing as free radicals is about 2% in a 1 *M* solution, 10% in a 0.01 *M* solution, and nearly 100% in very dilute solutions. We could quite correctly label a bottle containing a dilute solution of this substance as "triphenylmethyl radicals."

Triphenylmethyl is yellow; both the dimer and the peroxide are colorless. A solution of the dimer is yellow because of the triphenylmethyl present in the equilibrium mixture. When oxygen is admitted, the triphenylmethyl rapidly reacts to form the peroxide, and the yellow color disappears. More dimer dissociates to restore equilibrium and the yellow color reappears. Only when all the dimer–triphenylmethyl mixture is converted into the peroxide does the yellow color fail to appear. In a similar way it is triphenylmethyl that reacts with iodine.

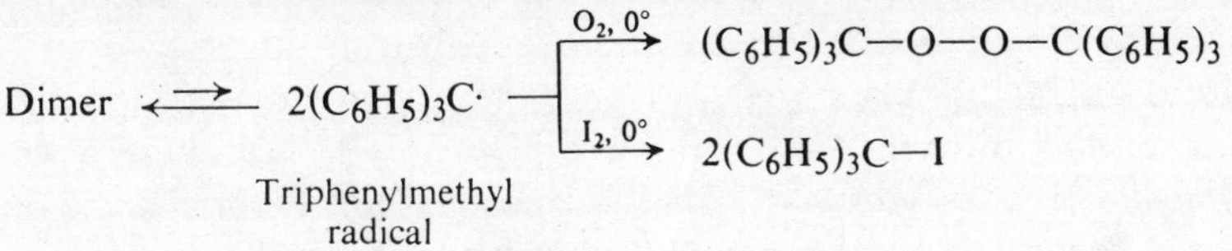

$$\text{Dimer} \rightleftarrows \underset{\text{Triphenylmethyl radical}}{2(C_6H_5)_3C\cdot} \begin{cases} \xrightarrow{O_2,\ 0^\circ} (C_6H_5)_3C{-}O{-}O{-}C(C_6H_5)_3 \\ \xrightarrow{I_2,\ 0^\circ} 2(C_6H_5)_3C{-}I \end{cases}$$

Thus the dimer undergoes its surprising reactions by first dissociating into triphenylmethyl, which, although unusually stable for a free radical, is nevertheless an exceedingly reactive particle.

Now, what *is* this dimer? For nearly 70 years it was believed to be hexaphenylethane. It—and dozens of analogs—were studied exhaustively, and the equilibria between them and triarylmethyl radicals were interpreted on the basis of the hexaarylethane structure. Then, in 1968, the dimer was shown to have the structure I.

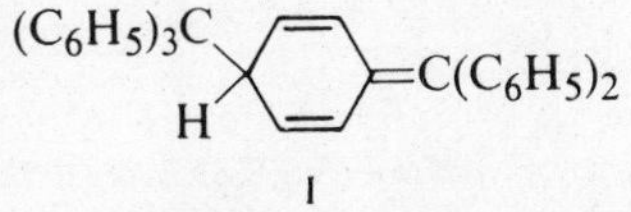

$(C_6H_5)_3C$, H [cyclohexadiene ring] $=C(C_6H_5)_2$

I

Gomberg's original task is still unaccomplished: hexaphenylethane, it seems, has never been made.

The basic significance of Gomberg's work remains unchanged. Many dimers have been prepared, and the existence of free triarylmethyl radicals has been substantiated in a number of ways; indeed, certain of these compounds seem to exist entirely as the free radical even in the solid state. The most convincing evidence for the free-radical nature of these substances lies in properties that arise directly from the odd electron that characterizes a free radical. Two electrons that occupy the same orbital and thus make up a pair have opposite spins (Sec. 1.6); the magnetic moments corresponding to their spins exactly cancel each other. But, by definition (Sec. 2.12), the odd electron of a free radical is not paired, and hence the effect of its spin is not canceled. This spin gives to the free radical a net magnetic moment. This magnetic moment reveals itself in two ways: (a) the compound is *paramagnetic*; that is, unlike most matter, it is attracted by a magnetic field; and (b) the compound gives a characteristic *paramagnetic resonance absorption* spectrum (or *electron spin resonance* spectrum, Sec. 13.14) which depends upon the orientation of the spin of an unpaired electron in a changing external magnetic field. This latter property permits the detection not only of stable free radicals but of low concentrations of short-lived radical intermediates in chemical reactions, and can even give information about their structure. (See, for example, Sec. 6.17).

The remarkable dissociation to form free radicals is the result of two factors. First, triphenylmethyl radicals are unusually stable because of resonance of the sort we have proposed for the benzyl radical. Here, of course, there are an even larger number of structures (36 of them) that stabilize the radical but not the hydrocarbon; the odd electron is highly delocalized, being distributed over three aromatic rings.

Second, crowding among the large aromatic rings tends to stretch and weaken the carbon–carbon bond joining the triphenylmethyl groups in the dimer. Once the radicals are formed, the bulky groups make it difficult for the carbon atoms to approach each other closely enough for bond formation: *so* difficult, in fact, that hexaphenylethane is not formed at all, but instead dimer I—even with the sacrifice of aromaticity of one ring. Even so, there is crowding in the dimer, and the total effect is to lower the dissociation energy to only 11 kcal/mole, as compared with a dissociation energy of 80–90 kcal for most carbon–carbon single bonds.

It would be hard to overestimate the importance of Gomberg's contribution to the field of free radicals and to organic chemistry as a whole. Although triphenylmethyl was isolable only because it was *not a typical* free radical, its chemical properties showed what kind of behavior to expect of free radicals *in general*; most important of all, it proved that such things as free radicals could exist.

Problem 12.9 The ΔH for dissociation of the dimer I has been measured as 11 kcal/mole, the E_{act} as 19 kcal/mole. (a) Draw the potential energy curve for the reaction. (b) What is the energy of activation for the reverse reaction, combination of triphenylmethyl radicals? (c) How do you account for this unusual fact? (Compare Sec. 2.17.)

Problem 12.10 When 1.5 g of "diphenyltetra(*o*-tolyl)ethane" is dissolved in 50 g of benzene, the freezing point of the solvent is lowered 0.5° (the cryoscopic constant for benzene is 5°). Interpret these results.

12.16 Preparation of alkenylbenzenes. Conjugation with ring

An aromatic hydrocarbon with a side chain containing a double bond can be prepared by essentially the same methods as simple alkenes (Secs. 5.12 and 5.19). In general, these methods involve elimination of atoms or groups from two adjacent carbons. The presence of the aromatic ring in the molecule may affect the orientation of elimination and the ease with which it takes place.

On an industrial scale, the elimination generally involves *dehydrogenation*. For example, **styrene**, the most important of these compounds—and perhaps the most important synthetic aromatic compound—can be prepared by simply heating ethylbenzene to about 600° in the presence of a catalyst. The ethylbenzene, in

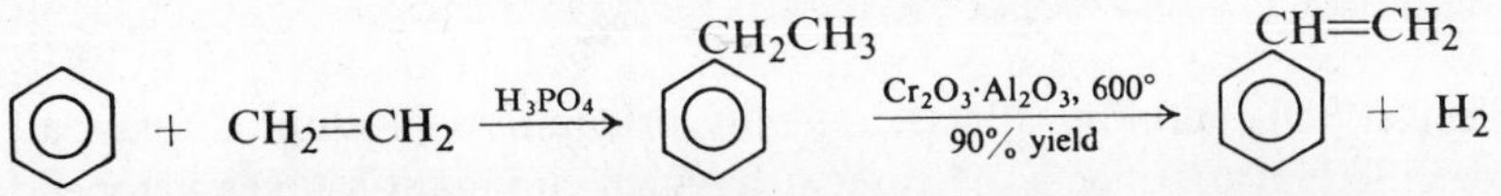

$$C_6H_6 + CH_2{=}CH_2 \xrightarrow{H_3PO_4} C_6H_5CH_2CH_3 \xrightarrow[90\%\ \text{yield}]{Cr_2O_3\cdot Al_2O_3,\ 600°} C_6H_5CH{=}CH_2 + H_2$$

Ethylbenzene Styrene

turn, is prepared by a Friedel-Crafts reaction between two simple hydrocarbons, benzene and ethylene.

In the laboratory, however, we are most likely to use dehydrohalogenation or dehydration.

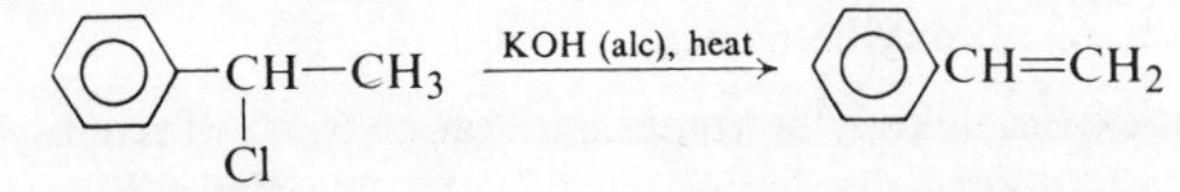

$$C_6H_5\text{—}CH(Cl)\text{—}CH_3 \xrightarrow{\text{KOH (alc), heat}} C_6H_5CH{=}CH_2$$

1-Phenyl-1-chloroethane Styrene

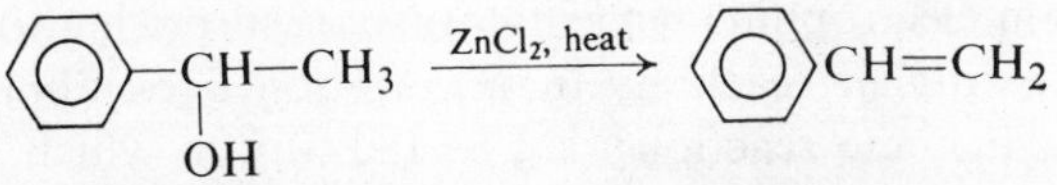

$$C_6H_5\text{—}CH(OH)\text{—}CH_3 \xrightarrow{ZnCl_2,\ \text{heat}} C_6H_5CH{=}CH_2$$

1-Phenylethanol Styrene

Dehydrohalogenation of 1-phenyl-2-chloropropane, or dehydration of 1-phenyl-2-propanol, could yield two products: 1-phenylpropene or 3-phenylpropene. Actually, only the first of these products is obtained. We saw earlier (Secs. 5.14 and 5.23) that where isomeric alkenes can be formed by elimination, the

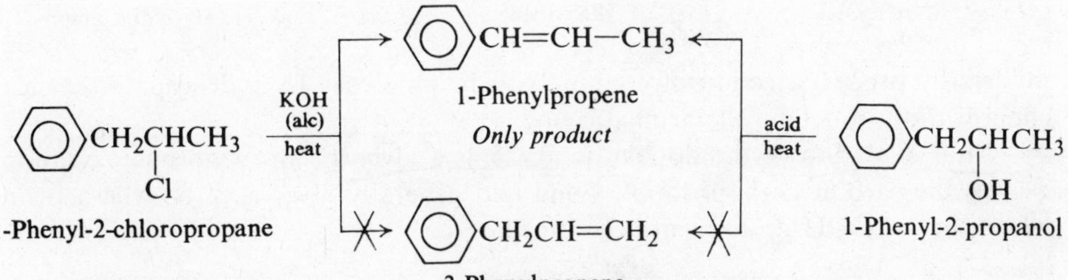

$$C_6H_5CH_2CH(Cl)CH_3 \xrightarrow[\text{heat}]{\text{KOH (alc)}} C_6H_5CH{=}CH\text{—}CH_3 \xleftarrow[\text{heat}]{\text{acid}} C_6H_5CH_2CH(OH)CH_3$$

1-Phenyl-2-chloropropane → 1-Phenylpropene (*Only product*) ← 1-Phenyl-2-propanol

$$C_6H_5CH_2CH(Cl)CH_3 \not\rightarrow C_6H_5CH_2CH{=}CH_2 \not\leftarrow C_6H_5CH_2CH(OH)CH_3$$

3-Phenylpropene

preferred product is the more stable alkene. This seems to be the case here, too. That 1-phenylpropene is much more stable than its isomer is shown by the fact that 3-phenylpropene is rapidly converted into 1-phenylpropene by treatment with hot alkali.

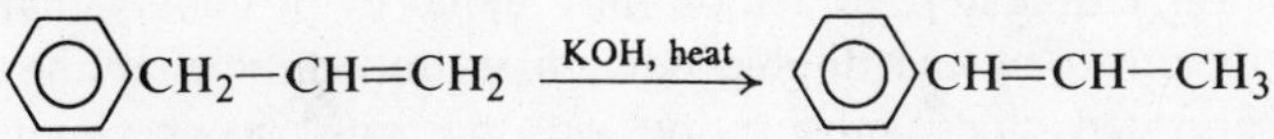

$$C_6H_5CH_2\text{—}CH{=}CH_2 \xrightarrow{\text{KOH, heat}} C_6H_5CH{=}CH\text{—}CH_3$$

3-Phenylpropene (Allylbenzene) 1-Phenylpropene

A double bond that is separated from a benzene ring by one single bond is said to be *conjugated with the ring*. Such conjugation confers unusual stability on

$$C_6H_5-\overset{|}{C}=\overset{|}{C}-$$

Double bond conjugated with ring: *unusually stable system*

a molecule. This stability affects not only orientation of elimination, but, as we shall see (Sec. 21.6), affects the ease with which elimination takes place.

> **Problem 12.11** Account for the stability of alkenes like styrene on the basis of: (a) delocalization of π electrons, showing both resonance structures and orbital overlap; and (b) change in hybridization.

12.17 Reactions of alkenylbenzenes

As we might expect, alkenylbenzenes undergo two sets of reactions: **substitution in the ring**, and **addition to the double bond in the side chain**. Since both ring and double bond are good sources of electrons, there may be competition between the two sites for certain electrophilic reagents; it is not surprising that, in general, the double bond shows higher reactivity than the resonance-stabilized benzene ring. Our main interest in these reactions will be the way in which the aromatic ring affects the reactions of the double bond.

Although both the benzene ring and the carbon–carbon double bond can be hydrogenated catalytically, the conditions required for the double bond are much

$$\underset{\text{Styrene}}{C_6H_5CH=CH_2} \xrightarrow[\text{75 minutes}]{H_2,\ Ni,\ 20^\circ,\ 2\text{–}3\ atm} \underset{\text{Ethylbenzene}}{C_6H_5CH_2CH_3} \xrightarrow[\text{100 minutes}]{H_2,\ Ni,\ 125^\circ,\ 110\ atm} \underset{\text{Ethylcyclohexane}}{C_6H_{11}CH_2CH_3}$$

milder; by proper selection of conditions it is quite easy to hydrogenate the side chain without touching the aromatic ring.

Mild oxidation of the double bond yields a glycol; more vigorous oxidation cleaves the carbon–carbon double bond and generally gives a carboxylic acid in which the —COOH group is attached to the ring.

$$\underset{\text{Styrene}}{C_6H_5CH=CH_2} \xrightarrow[HCO_2H]{H_2O_2} \underset{\text{A glycol}}{C_6H_5\underset{OH}{\underset{|}{CH}}-\underset{OH}{\underset{|}{CH_2}}} \xrightarrow{KMnO_4} \underset{\text{Benzoic acid}}{C_6H_5COOH}$$

Both double bond and ring react with halogens by ionic mechanisms that have essentially the same first step: attack on the π cloud by positively charged halogen. Halogen is consumed by the double bond first, and only after the side chain is completely saturated does substitution on the ring occur. Ring-halogenated alkenylbenzenes must be prepared, therefore, by generation of the double bond after halogen is already present on the ring. For example:

$$C_6H_5C_2H_5 \xrightarrow{Cl_2,\ FeCl_3} p\text{-}ClC_6H_4C_2H_5 \xrightarrow{Cl_2,\ heat} p\text{-}ClC_6H_4CHClCH_3 \xrightarrow{KOH} p\text{-}ClC_6H_4CH{=}CH_2$$

p-Chlorostyrene

In a similar way, alkenylbenzenes undergo the other addition reactions characteristic of the carbon–carbon double bond. Let us look further at the reactions of *conjugated* alkenylbenzenes, and the way in which the ring affects *orientation* and *reactivity*.

12.18 Addition to conjugated alkenylbenzenes: orientation. Stability of the benzyl cation

Addition of an unsymmetrical reagent to a double bond may in general yield two different products. In our discussion of alkenes (Secs. 6.11 and 6.17), we found that usually one of the products predominates, and that we can predict which it will be in a fairly simple way: *in either electrophilic or free-radical addition, the first step takes place in the way that yields the more stable particle*, carbonium ion in one kind of reaction, free radical in the other kind. Does this rule apply to reactions of alkenylbenzenes?

The effect of the benzene ring on orientation can be well illustrated by a single example, addition of HBr to 1-phenylpropene. In the absence of peroxides, bromine becomes attached to the carbon adjacent to the ring; in the presence of peroxides, bromine becomes attached to the carbon once removed from the ring. According to the mechanisms proposed for these two reactions, these products are formed as follows:

$$C_6H_5CH{=}CHCH_3 \xrightarrow{HBr} C_6H_5\overset{\oplus}{C}HCH_2CH_3 \xrightarrow{Br^-} C_6H_5CH(Br)CH_2CH_3 \qquad \textbf{No peroxides}$$

A benzyl cation

$$C_6H_5CH{=}CHCH_3 \xrightarrow{Br\cdot} C_6H_5\dot{C}HCH(Br)CH_3 \xrightarrow{HBr} C_6H_5CH_2CH(Br)CH_3 \qquad \textbf{Peroxides present}$$

A benzyl free radical

The first step of each of these reactions takes place in the way that yields the *benzyl* cation or the *benzyl* free radical rather than the alternative secondary cation or secondary free radical. Is this consistent with our rule that the more stable particle is formed faster?

Consideration of bond dissociation energies has already shown us that a benzyl free radical is an extremely stable one. We have accounted for this stability on the basis of resonance involving the benzene ring (Sec. 12.14).

What can we say about a benzyl cation? From the ionization potential (179 kcal) of the benzyl free radical, we can calculate (Sec. 5.18) that the benzyl cation is 69 kcal more stable than the methyl cation, and just about as stable as the *tert*-butyl cation. We can now expand our sequence of Sec. 8.21 to include the benzyl cation:

Stability of carbonium ions $\quad \begin{matrix}\text{benzyl}\\3°\end{matrix} > \begin{matrix}\text{allyl}\\2°\end{matrix} > 1° > CH_3^+$

The stability of a benzyl cation—relative to the compounds from which it is made—is also accounted for by resonance involving the benzene ring. Both the carbonium ion and the compound from which it is made are hybrids of Kekulé structures. In addition, the carbonium ion can be represented by three other structures, I, II, and III, in which the positive charge is located on the *ortho* and *para* carbon atoms. Whether we consider this as resonance stabilization or simply

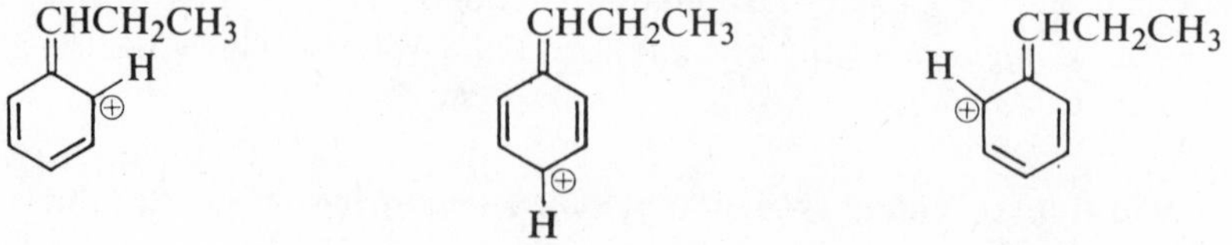

$CHCH_2CH_3$ H ⊕ — I

$CHCH_2CH_3$ ⊕ H — II

$CHCH_2CH_3$ H ⊕ — III

as dispersal of charge, contribution from these structures stabilizes the carbonium ion.

The orbital picture of the benzyl cation is similar to that of the benzyl free radical (Sec. 12.14) except that the p orbital that overlaps the π cloud is an *empty* one. The p orbital contributes no electrons, but permits further delocalization of the π electrons to include the carbon nucleus of the side chain.

Problem 12.12 How do you account for the following facts? (a) Triphenylchloromethane is completely ionized in certain solvents (e.g., liquid SO_2); (b) triphenylcarbinol, $(C_6H_5)_3COH$, dissolves in concentrated H_2SO_4 to give a solution that has the same intense yellow color as triphenylchloromethane solutions. (*Note:* This yellow color is different from that of solutions of triphenylmethyl.)

Problem 12.13 In light of Problem 12.12, can you suggest a possible reason, besides steric hindrance, why the reaction of CCl_4 with benzene stops at triphenylchloromethane? (See Sec. 12.6.)

12.19 Addition to conjugated alkenylbenzenes: reactivity

On the basis of the stability of the particle being formed, we might expect addition to a conjugated alkenylbenzene, which yields a stable *benzyl* cation or free radical, to occur faster than addition to a simple alkene.

On the other hand, we have seen (Sec. 12.17) that conjugated alkenylbenzenes are more stable than simple alkenes. On this basis alone, we might expect addition to conjugated alkenylbenzenes to occur more slowly than to simple alkenes.

The situation is exactly analogous to the one discussed for addition to conjugated dienes (Sec. 8.24). Both *reactant* and *transition state* are stabilized by resonance; whether reaction is faster or slower than for simple alkenes depends upon *which* is stabilized *more* (see Fig. 8.9, p. 275).

The fact is that conjugated alkenylbenzenes are much more reactive than simple alkenes toward both ionic and free-radical addition. Here again—as in *most* cases of this sort—resonance stabilization of the transition state leading to a carbonium ion or free radical is more important than resonance stabilization of the reactant. We must realize, however, that this is *not always* true.

Problem 12.14 Draw a potential energy diagram similar to Fig. 8.9 (p. 275) to summarize what has been said in this section.

Problem 12.15 Suggest one reason why tetraphenylethylene does not react with bromine in carbon tetrachloride.

12.20 Alkynylbenzenes

The preparations and properties of the alkynylbenzenes are just what we might expect from our knowledge of benzene and the alkynes.

Problem 12.16 Outline all steps in the conversion of: (a) ethylbenzene into phenylacetylene; (b) *trans*-1-phenylpropene into *cis*-1-phenylpropene.

12.21 Analysis of alkylbenzenes

Aromatic hydrocarbons with saturated side chains are distinguished from alkenes by their failure to decolorize bromine in carbon tetrachloride (without evolution of hydrogen bromide) and by their failure to decolorize cold, dilute, neutral permanganate solutions. (Oxidation of the side chains requires more vigorous conditions; see Sec. 12.10.)

They are distinguished from alkanes by the readiness with which they are sulfonated by—and thus dissolve in—cold fuming sulfuric acid (see Sec. 11.4).

They are distinguished from alcohols and other oxygen-containing compounds by their failure to dissolve immediately in cold concentrated sulfuric acid, and from primary and secondary alcohols by their failure to give a positive chromic anhydride test (Sec. 6.30).

Upon treatment with chloroform and aluminum chloride, alkylbenzenes give orange to red colors. These colors are due to triarylmethyl cations, Ar_3C^+, which are probably produced by a Friedel-Crafts reaction followed by a transfer of hydride ion (Sec. 6.16):

$$ArH \xrightarrow{CHCl_3,\ AlCl_3} ArCHCl_2 \xrightarrow{ArH,\ AlCl_3} Ar_2CHCl \xrightarrow{ArH,\ AlCl_3} Ar_3CH$$

$$Ar_2CHCl \xrightarrow{AlCl_3} Ar_2CH^+AlCl_4^- \quad + \quad Ar_3CH \longrightarrow Ar_2CH_2 + \underset{\text{Orange to red color}}{Ar_3C^+\ AlCl_4^-}$$

This test is given by any aromatic compound that can undergo the Friedel-Crafts reaction, with the particular color produced being characteristic of the aromatic system involved: orange to red from halobenzenes, blue from *naphthalene*, purple from *phenanthrene*, green from *anthracene* (Chap. 30).

Problem 12.17 Describe simple chemical tests (if any) that would distinguish between: (a) *n*-propylbenzene and *o*-chlorotoluene; (b) benzene and toluene; (c) *m*-chlorotoluene and *m*-dichlorobenzene; (d) bromobenzene and bromocyclohexane; (e) bromobenzene and 3-bromo-1-hexene; (f) ethylbenzene and benzyl alcohol ($C_6H_5CH_2OH$). Tell exactly what you would *do* and *see*.

The number and orientation of side chains in an alkylbenzene is shown by the carboxylic acid produced on vigorous oxidation (Sec. 12.10).

Problem 12.18 On the basis of characterization tests and physical properties, an unknown compound of b.p. 182° is believed to be either *m*-diethylbenzene or *n*-butylbenzene. How could you distinguish between the two possibilities?

(Analysis of alkylbenzenes by spectroscopic methods will be discussed in Secs. 13.15–13.16.)

12.22 Analysis of alkenyl- and alkynylbenzenes

Aromatic hydrocarbons with unsaturated side chains undergo the reactions characteristic of aromatic rings and of the carbon–carbon double or triple bond. (Their analysis by spectroscopic methods is discussed in Secs. 13.15–13.16.)

Problem 12.19 Predict the response of allylbenzene to the following test reagents: (a) cold concentrated sulfuric acid; (b) Br_2 in CCl_4; (c) cold, dilute, neutral permanganate; (d) $CHCl_3$ and $AlCl_3$; (e) CrO_3 and H_2SO_4.

Problem 12.20 Describe simple chemical tests (if any) that would distinguish between: (a) styrene and ethylbenzene; (b) styrene and phenylacetylene; (c) allylbenzene and 1-nonene; (d) allylbenzene and allyl alcohol ($CH_2{=}CH{-}CH_2OH$). Tell exactly what you would *do* and *see*.

PROBLEMS

1. Draw the structure of:

(a) *m*-xylene
(b) mesitylene
(c) *o*-ethyltoluene
(d) *p*-di-*tert*-butylbenzene
(e) cyclohexylbenzene
(f) 3-phenylpentane
(g) isopropylbenzene (cumene)
(h) *trans*-stilbene
(i) 1,4-diphenyl-1,3-butadiene
(j) *p*-dibenzylbenzene
(k) *m*-bromostyrene
(l) diphenylacetylene

2. Outline all steps in the synthesis of ethylbenzene from each of the following compounds, using any needed aliphatic or inorganic reagents.

(a) benzene
(b) styrene
(c) phenylacetylene
(d) α-phenylethyl alcohol ($C_6H_5CHOHCH_3$)
(e) β-phenylethyl alcohol ($C_6H_5CH_2CH_2OH$)
(f) 1-chloro-1-phenylethane
(g) 2-chloro-1-phenylethane
(h) *p*-bromoethylbenzene
(i) acetophenone ($C_6H_5C(=O)CH_3$)

3. Give structures and names of the principal organic products expected from reaction (if any) of *n*-propylbenzene with each of the following. Where more than one product is to be expected, indicate which will predominate.

(a) H_2, Ni, room temperature, low pressure
(b) H_2, Ni, 200°, 100 atm.
(c) cold dilute $KMnO_4$
(d) hot $KMnO_4$
(e) $K_2Cr_2O_7$, H_2SO_4, heat
(f) boiling NaOH(aq)
(g) boiling HCl(aq)
(h) HNO_3, H_2SO_4
(i) H_2SO_4, SO_3
(j) Tl $(OOCCF_3)_3$
(k) Cl_2, Fe
(l) Br_2, Fe
(m) I_2, Fe
(n) Br_2, heat, light
(o) CH_3Cl, $AlCl_3$, 0°
(p) $C_6H_5CH_2Cl$, $AlCl_3$, 0° (*Note:* A benzyl halide is *not* an aryl halide.)
(q) C_6H_5Cl, $AlCl_3$, 80°
(r) isobutylene, HF
(s) *tert*-butyl alcohol, H_2SO_4
(t) cyclohexene, HF

4. Give structures and names of the principal organic products expected from reaction (if any) of *trans*-1-phenyl-1-propene with:

(a) H_2, Ni, room temperature, low pressure
(b) H_2, Ni, 200°, 100 atm.
(c) Br_2 in CCl_4
(d) excess Br_2, Fe
(e) HCl
(f) HBr
(g) HBr (peroxides)
(h) cold conc. H_2SO_4
(i) Br_2, H_2O
(j) cold dilute $KMnO_4$
(k) hot $KMnO_4$
(l) HCO_2OH
(m) O_3, then H_2O/Zn
(n) Br_2, 300°
(o) $CHBr_3$, *t*-BuOK
(p) product (c), KOH(alc)

5. Give structures and names of the principal organic products expected from each of the following reactions:

(a) benzene + cyclohexene + HF
(b) phenylacetylene + alcoholic $AgNO_3$
(c) *m*-nitrobenzyl chloride + $K_2Cr_2O_7$ + H_2SO_4 + heat
(d) allylbenzene + HCl
(e) *p*-chlorotoluene + hot $KMnO_4$
(f) *eugenol* ($C_{10}H_{12}O_2$, 2-methoxy-4-allylphenol) + hot KOH
 $\longrightarrow$ *isoeugenol* ($C_{10}H_{12}O_2$)
(g) benzyl chloride + Mg + dry ether
(h) product of (g) + H_2O
(i) *p*-xylene + Br_2 + Fe
(j) 1-phenyl-1,3-butadiene + one mole H_2 + Ni, 2 atm., 30°
(k) *trans*-stilbene + O_3, then H_2O/Zn
(l) 1,3-diphenylpropyne + H_2, Pd $\longrightarrow C_{15}H_{14}$
(m) 1,3-diphenylpropyne + Li, NH_3(liq) $\longrightarrow C_{15}H_{14}$
(n) *p*-$CH_3OC_6H_4CH{=}CHC_6H_5$ + HBr

6. Treatment of benzyl alcohol ($C_6H_5CH_2OH$) with cold concentrated H_2SO_4 yields a high-boiling resinous material. What is a likely structure for this material, and how is it probably formed?

7. Label each set of hydrogens in each of the following compounds in order of expected ease of abstraction by bromine atoms. Use (1) for the most reactive, (2) for the next, etc.

(a) 1-phenyl-2-hexene
(b) 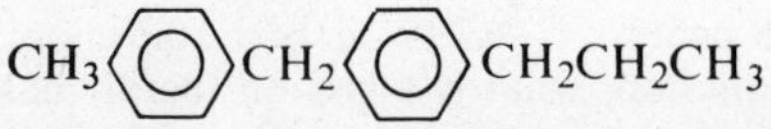 CH_3⟨◯⟩CH_2⟨◯⟩$CH_2CH_2CH_3$
(c) 1,2,4-trimethylbenzene (*Hint:* See Problem 12.8, p. 390.)
(d) What final monobromination product or products would abstraction of each kind of hydrogen in (a) lead to?

8. Give structures and names of the products expected from dehydrohalogenation of each of the following. Where more than one product can be formed, predict the major product.

(a) 1-chloro-1-phenylbutane
(b) 1-chloro-2-phenylbutane
(c) 2-chloro-2-phenylbutane
(d) 2-chloro-1-phenylbutane
(e) 3-chloro-2-phenylbutane

9. Answer Problem 8 for dehydration of the alcohol corresponding to each of the halides given. (*Hint:* Do not forget Sec. 5.22.)

10. Arrange in order of ease of dehydration: (a) the alcohols of Problem 9; (b) $C_6H_5CH_2CH_2OH$, $C_6H_5CHOHCH_3$, $(C_6H_5)_2C(OH)CH_3$.

11. Arrange the compounds of each set in order of reactivity toward the indicated reaction.

(a) addition of HCl: styrene, *p*-chlorostyrene, *p*-methylstyrene
(b) dehydration: α-phenylethyl alcohol ($C_6H_5CHOHCH_3$), α-(*p*-nitrophenyl)ethyl alcohol, α-(*p*-aminophenyl)ethyl alcohol.

12. (a) Draw structures of all possible products of addition of one mole of Br_2 to 1-phenyl-1,3-butadiene. (b) Which of these possible products are consistent with the intermediate formation of the most stable carbonium ion? (c) Actually, only 1-phenyl-3,4-dibromo-1-butene is obtained. What is the most likely explanation of this fact?

13. (a) The heats of hydrogenation of the stereoisomeric stilbenes (1,2-diphenylethenes) are: *cis*-, 26.3 kcal; *trans*-, 20.6 kcal. Which isomer is the more stable? (b) *cis*-Stilbene is converted into *trans*-stilbene (but not vice versa) either (i) by action of a very small amount of Br_2 in the presence of light, or (ii) by action of a very small amount of HBr (but not HCl) in the presence of peroxides. What is the agent that probably brings about the conversion? Can you suggest a way in which the conversion might take place? (c) Why is *trans*-stilbene not converted into *cis*-stilbene?

14. One mole of triphenylcarbinol lowers the freezing point of 1000 g of 100% sulfuric acid twice as much as one mole of methanol. How do you account for this?

15. Can you account for the order of acidity: triphenylmethane > diphenylmethane > toluene > *n*-pentane (*Hint:* See Sec. 12.18.)

16. When a mixture of toluene and $CBrCl_3$ was irradiated with ultraviolet light, there were obtained, in almost exactly equimolar amounts, benzyl bromide and $CHCl_3$. (a) Show in detail all steps in the most likely mechanism for this reaction. (b) There were also obtained, in small amounts, HBr and C_2Cl_6; the ratio of $CHCl_3$ to HBr was 20:1. How do you account for the formation of HBr? Of C_2Cl_6? What, specifically, does the 20:1 ratio tell you about the reaction?

17. When the product of the HF-catalyzed reaction of benzene with 1-dodecene, previously reported to be pure 2-phenyldodecane, was analyzed by gas chromatography, five evenly-spaced peaks of about the same size were observed, indicating the presence of five components, probably closely related in structure. What five compounds most likely make up this mixture, and how could you have anticipated their formation?

18. The bond dissociation energy for the central C—C bond of hexacyclopropylethane is only 45 kcal/mole. Besides steric interaction, what is a second factor that may contribute to the weakness of this bond? (*Hint:* See Sec. 9.9.)

19. On theoretical grounds it is believed that a primary isotope effect is greatest if bond breaking and bond making have proceeded to an equal extent in the transition state. (a) In free-radical halogenation of the side chain of toluene, k^H/k^D is about 2 in chlorination and about 5 in bromination. There are two possible interpretations of this. What are they? (b) In light of Sec. 2.23, which interpretation is the more likely?

20. The three xylenes are obtained as a mixture from the distillation of coal tar; further separation by distillation is difficult because of the closeness of their boiling points (see Table 12.1, p. 375), and so a variety of chemical methods have been used. In each case below tell which isomer you would expect to react preferentially, and why.

(a) An old method: treatment of the mixture at room temperature with 80% sulfuric acid.

(b) Another old method: sulfonation of all three xylenes, and then treatment of the sulfonic acids with dilute aqueous acid.

(c) A current method: extraction of one isomer into a BF_3/HF layer.
(d) A proposed method:

$$C_6H_5C(CH_3)_2{}^-Na^+ + \text{xylenes} \rightleftarrows C_6H_5CH(CH_3)_2 + A + \text{two xylenes}$$

(*Hint* to part (d): See Secs. 8.10, 5.17, and 11.18.)

21. Upon ionic addition of bromine, *cis*-1-phenyl-1-propene gives a mixture of 17% *erythro* dibromide and 83% *threo*; *trans*-1-phenyl-1-propene gives 88% *erythro*, 12% *threo*; and *trans*-1-(*p*-methoxyphenyl)propene gives 63% *erythro*, 37% *threo*.

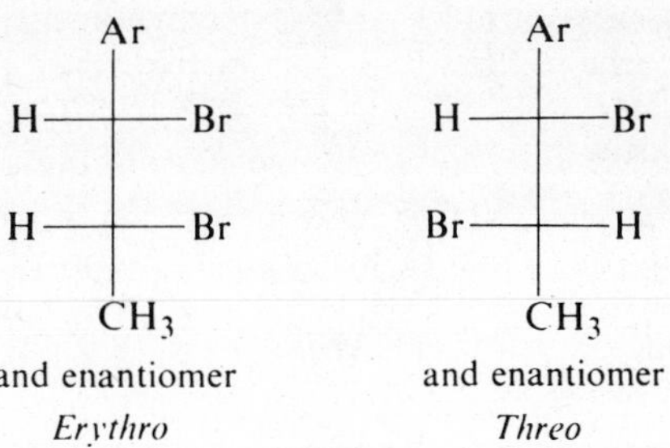

How do these results compare with those obtained with the 2-butenes (Sec. 7.11)? Suggest a possible explanation for the difference. What is the effect of the *p*-methoxy group, and how might you account for this?

22. Outline all steps in a possible laboratory synthesis of each of the following compounds from benzene and/or toluene, using any necessary aliphatic or inorganic reagents. Follow instructions on p. 224. Assume a pure *para* isomer can be separated from an *ortho,para* mixture.

(a) ethylbenzene
(b) styrene
(c) phenylacetylene
(d) isopropylbenzene
(e) 2-phenylpropene
(f) 3-phenylpropene (allylbenzene)
(g) 1-phenylpropyne (two ways)
(h) *trans*-1-phenylpropene
(i) *cis*-1-phenylpropene
(j) *p*-*tert*-butyltoluene
(k) *p*-nitrostyrene
(l) *p*-bromobenzyl bromide
(m) *p*-nitrobenzal bromide
(n) *p*-bromobenzoic acid
(o) *m*-bromobenzoic acid
(p) 1,2-diphenylethane
(q) *p*-nitrodiphenylmethane (p-$O_2NC_6H_4CH_2C_6H_5$) (*Hint:* See Problem 3(p).)

23. Describe simple chemical tests that would distinguish between:

(a) benzene and cyclohexane
(b) benzene and 1-hexene
(c) toluene and *n*-heptane
(d) cyclohexylbenzene and 1-phenylcyclohexene
(e) benzyl alcohol ($C_6H_5CH_2OH$) and *n*-pentylbenzene
(f) cinnamyl alcohol ($C_6H_5CH{=}CHCH_2OH$) and 3-phenyl-1-propanol ($C_6H_5CH_2CH_2CH_2OH$)
(g) chlorobenzene and ethylbenzene
(h) nitrobenzene and *m*-dibromobenzene

24. Describe chemical methods (not necessarily simple tests) that would enable you to distinguish between the compounds of each of the following sets. (For example, make use of Table 18.1, page 580.

(a) 1-phenylpropene, 2-phenylpropene, 3-phenylpropene (allylbenzene)
(b) all alkylbenzenes of formula C_9H_{12}
(c) *m*-chlorotoluene and benzyl chloride
(d) *p*-divinylbenzene (p-$C_6H_4(CH{=}CH_2)_2$) and 1-phenyl-1,3-butadiene
(e) $C_6H_5CHClCH_3$, p-$CH_3C_6H_4CH_2Cl$, and p-$ClC_6H_4C_2H_5$

25. An unknown compound is believed to be one of the following. Describe how you would go about finding out which of the possibilities the unknown actually is. Where possible, use simple chemical tests; where necessary, use more elaborate chemical methods like quantitative hydrogenation, cleavage, etc. Where necessary, make use of Table 18.1, page 580.

	b.p.		b.p.
bromobenzene	156°	*p*-chlorotoluene	162°
3-phenylpropene	157	*o*-ethyltoluene	162
m-ethyltoluene	158	*p*-ethyltoluene	163
n-propylbenzene	159	mesitylene	165
o-chlorotoluene	159	2-phenylpropene	165
m-chlorotoluene	162		

26. The compound *indene*, C_9H_8, found in coal tar, rapidly decolorizes Br_2/CCl_4 and dilute $KMnO_4$. Only one mole of hydrogen is absorbed readily to form *indane*, C_9H_{10}. More vigorous hydrogenation yields a compound of formula C_9H_{16}. Vigorous oxidation of indene yields phthalic acid. What is the structure of indene? Of indane? (*Hint:* See Problem 9.17, p. 313.)

27. A solution of 0.01 mole *tert*-butyl peroxide (p. 114) in excess ethylbenzene was irradiated with ultraviolet light for several hours. Gas chromatographic analysis of the product showed the presence of nearly 0.02 mole of *tert*-butyl alcohol. Evaporation of the alcohol and unreacted ethylbenzene left a solid residue which was separated by chromatography into just two products: X (1 g) and Y (1 g). X and Y each had the empirical formula C_8H_9 and m.w. 210; each was inert toward cold dilute $KMnO_4$ and toward Br_2/CCl_4.

When isopropylbenzene was substituted for ethylbenzene in the above reaction, exactly similar results were obtained, except that the single compound Z (2.2 g) was obtained instead of X and Y. Z had the empirical formula C_9H_{11}, m.w. 238, and was inert toward cold, dilute $KMnO_4$ and toward Br_2/CCl_4.

What are the most likely structures for X, Y, and Z, and what is the most likely mechanism by which they are formed?

Chapter 13 Spectroscopy and Structure

13.1 Determination of structure: spectroscopic methods

Near the beginning of our study (Sec. 3.32), we outlined the general steps an organic chemist takes when he is confronted with an unknown compound and sets out to find the answer to the question: *what is it?* We have seen, in more detail, some of the ways in which he carries out the various steps: determination of molecular weight and molecular formula; detection of the presence—or absence—of certain functional groups; degradation to simpler compounds; conversion into derivatives; synthesis by an unambiguous route.

At every stage of structure determination—from the isolation and purification of the unknown substance to its final comparison with an authentic sample—the use of instruments has, since World War II, revolutionized organic chemical practice. Instruments not only help an organic chemist to do what he does *faster* but, more important, let him do what could not be done *at all* before: to analyze complicated mixtures of closely related compounds; to describe the structure of molecules in detail never imagined before; to detect, identify, and measure the concentration of short-lived intermediates whose very existence was, not so long ago, only speculation.

By now, we are familiar with some of the features of the organic chemical landscape; so long as we do not wander too far from home, we can find our way about without becoming lost. We are ready to learn a little about how to interpret the kind of information these modern instruments give, so that they can help us to see more clearly the new things we shall meet, and to recognize them more readily when we encounter them again. The instruments most directly concerned with our primary interest, molecular structure, are the *spectrometers*—measurers of spectra. Of the various spectra, we shall actually work with only two: *infrared* (*ir*) *and nuclear magnetic resonance* (*nmr*), since they are the workhorses of the organic chemical laboratory today; of these, we shall spend most of our time with nmr.

We shall look very briefly at three other kinds of spectra: *mass*, *ultraviolet* (*uv*), and *electron spin resonance* (*esr*).

In all this, we must constantly keep in mind that what we learn at this stage must be greatly simplified. There are many exceptions to the generalizations we shall learn; there are many pitfalls into which we can stumble. Our ability to apply spectroscopic methods to the determination of organic structure is limited by our understanding of organic chemistry as a whole—and in this we are, of course, only beginners. But so long as we are aware of the dangers of a little learning, and are willing to make mistakes and profit from them, it is worthwhile for us to become beginners in this area of organic chemistry, too.

Let us look first at the mass spectrum, and then at the others, which, as we shall see, are all parts—different ranges of wavelengths—of a single spectrum: that of electromagnetic radiation.

13.2 The mass spectrum

In the mass spectrometer, molecules are bombarded with a beam of energetic electrons. The molecules are ionized and broken up into many fragments, some of which are positive ions. Each kind of ion has a particular ratio of mass to charge, or *m/e value*. For most ions, the charge is 1, so that *m/e* is simply the mass of the ion. Thus, for neopentane:

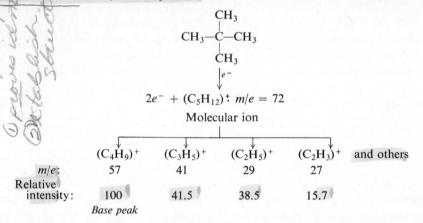

$$\begin{array}{c} CH_3 \\ | \\ CH_3\text{—}C\text{—}CH_3 \\ | \\ CH_3 \end{array}$$

$$\downarrow e^-$$

$$2e^- + (C_5H_{12})^{\dot{+}} \quad m/e = 72$$

Molecular ion

	$(C_4H_9)^+$	$(C_3H_5)^+$	$(C_2H_5)^+$	$(C_2H_3)^+$	and others
m/e:	57	41	29	27	
Relative intensity:	100	41.5	38.5	15.7	
	Base peak				

The set of ions is analyzed in such a way that a signal is obtained for each value of *m/e* that is represented; the intensity of each signal reflects the relative abundance of the ion producing the signal. The largest peak is called the *base peak*; its intensity is taken as 100, and the intensities of the other peaks are expressed relative to it. A plot—or even a list—showing the relative intensities of signals at the various *m/e* values is called a *mass spectrum*, and is highly characteristic of a particular compound. Compare, for example, the spectra of two isomers shown in Fig. 13.1.

Mass spectra can be used in two general ways: (a) to prove the identity of two compounds, and (b) to help establish the structure of a new compound.

Two compounds are shown to be identical by the fact that they have identical physical properties: melting point, boiling point, density, refractive index, etc. The greater the number of physical properties measured, the stronger the evidence.

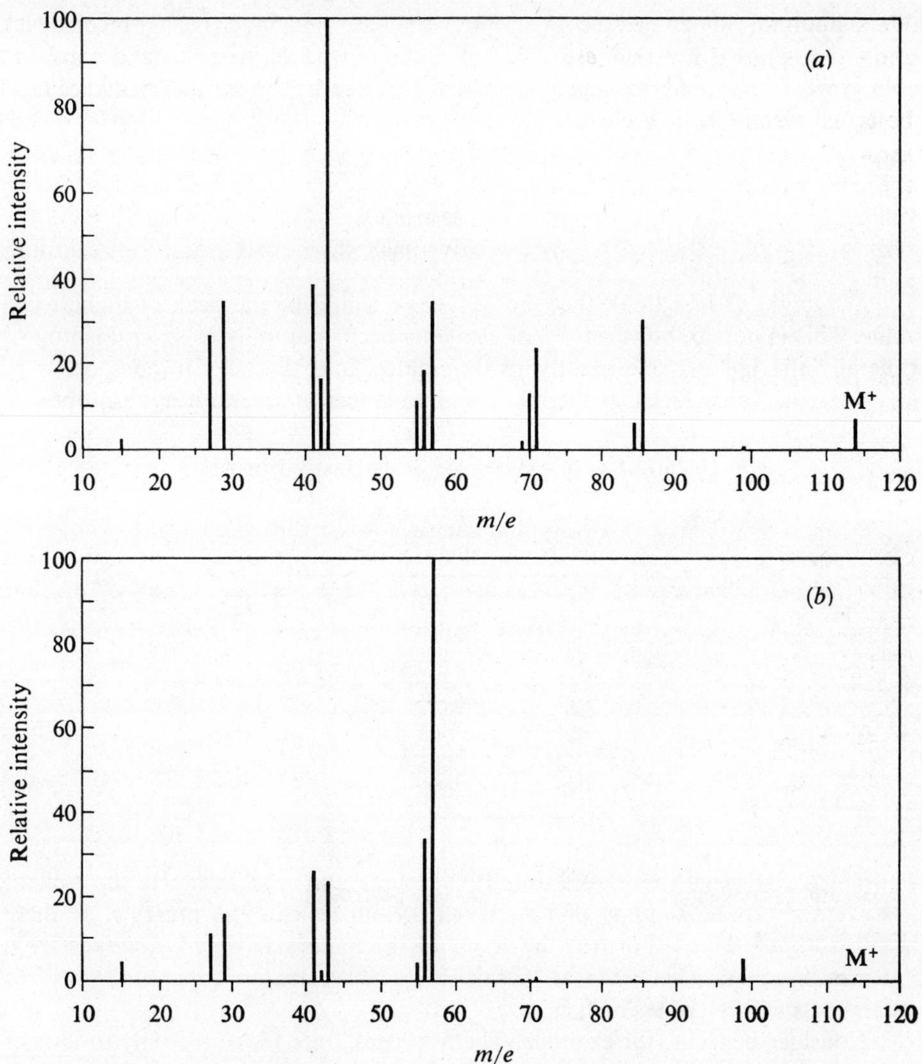

Figure 13.1. Mass spectra of two isomeric alkanes. (*a*) *n*-Octane; (*b*) 2,2,4-trimethylpentane.

Now, a single mass spectrum amounts to dozens of physical properties, since it shows the relative abundances of dozens of different fragments. If we measure the mass spectrum of an unknown compound and find it to be identical with the spectrum of a previously reported compound of known structure, then we can conclude that--almost beyond the shadow of a doubt—the two compounds are identical.

The mass spectrum helps to establish the structure of a *new* compound in several different ways: it can give an exact molecular weight; it can give a molecular formula—or at least narrow the possibilities to a very few; and it can indicate the presence in a molecule of certain structural units.

If one electron is removed from the parent molecule, there is produced the *molecular ion* (or *parent ion*), whose *m/e* value is, of course, the molecular weight of the compound. Sometimes the M^+ peak is the base peak, and is easily recognized; often, though, it is not the base peak—it may even be very small—and considerable work is required to locate it. Once identified, it gives the most accurate molecular weight obtainable.

$$M + e^- \longrightarrow \underset{\substack{\text{Molecular ion} \\ \text{(Parent ion)} \\ m/e\,=\,\text{mol. wt.}}}{M^+} + 2e^-$$

We might at first think that the M^+ peak would be the peak of highest m/e value. This is not so, however. Most elements occur naturally as several isotopes; generally the lightest one greatly predominates, and the heavier ones occur to lesser extent. Table 13.1 lists the relative abundances of several heavy isotopes.

Table 13.1 ABUNDANCE OF SOME HEAVY ISOTOPES

Heavy isotope	Abundance relative to isotope of lowest atomic weight
^2H	0.015%
^{13}C	1.11
^{15}N	0.37
^{18}O	0.20
^{33}S	0.78
^{34}S	4.4
^{37}Cl	32.5
^{81}Br	98.0

The molecular weight that one usually measures and works with is the sum of the average atomic weights of the elements, and reflects the presence of these heavy isotopes. This is not true, however, of the molecular weight obtained from the mass spectrum; here, the M^+ peak is due to molecules containing only the commonest isotope of each element.

Consider benzene, for example. The M^+ peak, m/e 78, is due only to ions of formula $C_6H_6^+$. There is a peak at m/e 79, the M + 1 peak, which is due to $C_5{}^{13}CH_6^+$ and $C_6H_5D^+$. There is an M + 2 peak at m/e 80, due to $C_4{}^{13}C_2H_6^+$, $C_5{}^{13}CH_5D^+$, and $C_6H_4D_2^+$. Now, because of the low natural abundance of most heavy isotopes, these isotopic peaks are generally much less intense than the M^+ peak; just how much less intense depends upon which elements they are due to. In the case of benzene, the M + 1 and M + 2 peaks are, respectively, 6.75% and 0.18% as intense as the M^+ peak. (Table 13.1 shows us, however, that a monochloro compound would have an M + 2 peak about one-third as intense as the M^+ peak, and a monobromo compound would have M and M + 2 peaks of about equal intensity.)

It is these isotopic peaks that make it possible for us to determine the molecular formula of the compound. Knowing the relative natural abundances of isotopes, one can calculate for any molecular formula the relative intensity to be expected for each isotopic peak: M + 1, M + 2, etc. The results of such calculations are available in tables. Consider, for example, a compound for which M^+ is 44.

The compound might be (among other less likely possibilities) N_2O, CO_2, C_2H_4O, or C_3H_8. By use of Table 13.2, we clearly could pick out the most likely formula from the mass spectral data.

Table 13.2 CALCULATED INTENSITIES OF ISOTOPIC PEAKS

	M	M + 1	M + 2
N_2O	100	0.80	0.20
CO_2	100	1.16	0.40
C_2H_4O	100	1.91	0.01
C_3H_8	100	3.37	0.04

Finally, study of compounds of known structure is beginning to reveal the factors that determine which fragments a particular structure is likely to break into. In this we can find much that is familiar to us: the preferential formation of carbonium ions that we recognize as being relatively stable ones; elimination of small, stable molecules like water, ammonia, and carbon monoxide. Under the energetic conditions, extensive rearrangement can occur, complicating the interpretation; but here, too, patterns are emerging. The *direction* of rearrangement is, as we would expect, toward more stable ions. As this knowledge accumulates, the process is reversed: from the kind of fragmentation an unknown compound gives, its structure is deduced.

Problem 13.1 (a) Referring to the neopentane fragmentation (p. 406), what is a likely structure for $C_4H_9^+$; $C_3H_5^+$; $C_2H_5^+$; $C_2H_3^+$? (b) Write a balanced equation for the formation of $C_4H_9^+$ from the molecular ion $C_5H_{12}^{+\cdot}$.

13.3 The electromagnetic spectrum

We are already familiar with various kinds of electromagnetic radiation: light—visible, ultraviolet, infrared—x-rays, radio and radar waves. These are simply different parts of a broad spectrum that stretches from gamma rays, whose wavelengths are measured in fractions of an Angstrom unit, to radio waves, whose wavelengths are measured in meters or even kilometers. All these waves have the same velocity, 3×10^{10} centimeters per second. Their frequency is related to the wavelength by the expression

$$\nu = c/\lambda$$

where ν = frequency, in Hz (*Hertz*, cycles/sec)
λ = wavelength, in cm
c = velocity, 3×10^{10} cm/sec

The shorter the wavelength, the higher the frequency.

When a beam of electromagnetic radiation is passed through a substance, the radiation can be either absorbed or transmitted, depending upon its frequency and the structure of the molecules it encounters. Electromagnetic radiation is energy, and hence when a molecule absorbs radiation, it gains energy. Just how much energy it gains depends upon the frequency of the radiation: the higher the frequency (the shorter the wavelength), the greater the gain in energy.

$$\Delta E = h\nu$$

where ΔE = gain in energy, in ergs
h = Planck's constant, 6.5×10^{-27} erg-sec
ν = frequency, in Hz

The energy gained by the molecule in this way may bring about increased vibration or rotation of the atoms, or may raise electrons to higher energy levels. The particular frequency of radiation that a given molecule can absorb depends upon the changes in vibrations or rotations or electronic states that are permitted to a molecule of that structure. The spectrum of a compound is a plot that shows how much electromagnetic radiation is absorbed (or transmitted) at each frequency. It can be highly characteristic of the compound's structure.

13.4 The infrared spectrum

Of all the properties of an organic compound, the one that, by itself, gives the most information about the compound's structure is its infrared spectrum.

A molecule is constantly vibrating: its bonds *stretch* (and contract), and *bend* with respect to each other. Changes in vibrations of a molecule are caused by absorption of infrared light; light lying beyond (lower frequency, longer wavelength, less energy) the red end of the visible spectrum.

A particular part of the infrared spectrum is referred to either by its wavelength or—and this is considered preferable—by its frequency. Wavelength is expressed in microns, μ ($1\ \mu = 10^{-4}$ cm or 10^4 A). Frequency is expressed, not in Hertz, but in *wavenumbers*, cm^{-1}, often called *reciprocal centimeters*; the wavenumber is simply the number of waves per centimeter, and is equal to the reciprocal of the wavelength in centimeters.

Like the mass spectrum, an infrared spectrum is a highly characteristic property of an organic compound—see, for example, the spectra in Fig. 13.2, p. 411—and can be used both to establish the identity of two compounds and to reveal the structure of a new compound.

Two substances that have identical infrared spectra are, in effect, identical in thousands of different physical properties—the absorption of light at thousands of different frequencies—and must almost certainly be the same compound. (One region of the infrared spectrum is called, appropriately, the *fingerprint* region.)

The infrared spectrum helps to reveal the structure of a new compound by telling us what groups are present in—or absent from—the molecule. A particular group of atoms gives rise to *characteristic absorption bands*; that is to say, a particular group absorbs light of certain frequencies that are much the same from compound to compound. For example, the —OH group of alcohols absorbs strongly at 3200–3600 cm^{-1}; the C=O group of ketones at 1710 cm^{-1}; the —C≡N group, at 2250 cm^{-1}; the $—CH_3$ group at 1450 and 1375 cm^{-1}.

Interpretation of an infrared spectrum is not a simple matter. Bands may be obscured by the overlapping of other bands. Overtones (harmonics) may appear at just twice the frequency of the fundamental band. The absorption band of a

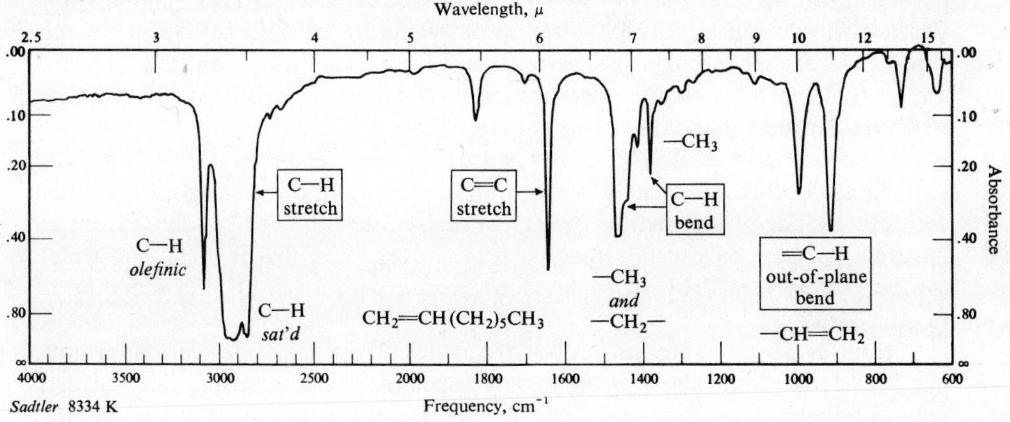

A-111

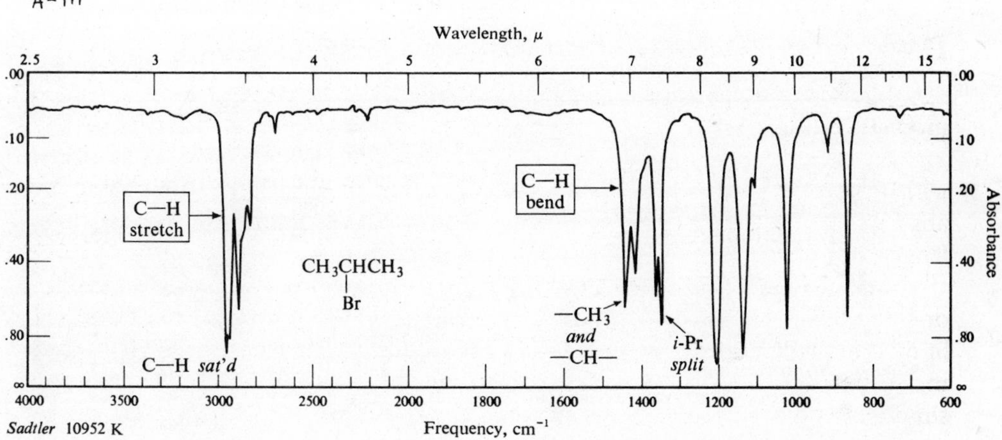

A-112

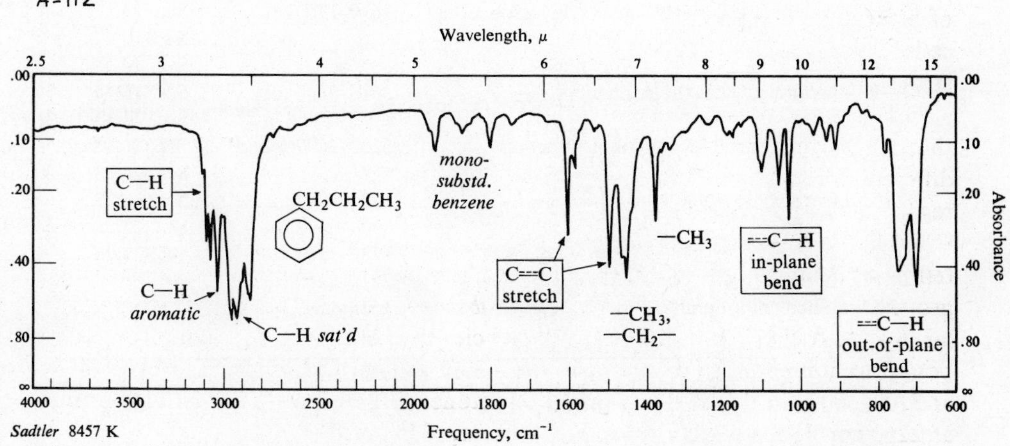

Figure 13.2. Infrared spectra. (*a*) 1-Octene; (*b*) isopropyl bromide; (*c*) *n*-butylbenzene.

particular group may be *shifted* by various structural features—conjugation, electron withdrawal by a neighboring substituent, angle strain or van der Waals strain, hydrogen bonding—and be mistaken for a band of an entirely different group. (On the other hand, recognized for what they are, such shifts *reveal* the structural features that cause them.)

In our work we shall have modest aims: to learn to recognize a few of the more striking absorption bands, and to gain a little practice in correlating infrared data with other kinds of information. We must realize that we shall be taking from an infrared spectrum only a tiny fraction of the information that is there, and which can be gotten from it by an experienced person with a broad understanding of organic structure.

Table 13.3 lists infrared absorption frequencies characteristic of various groups. We shall look more closely at the infrared spectra of hydrocarbons in Sec. 13.15 and, in following chapters, at the infrared spectra of other families of compounds.

Table 13.3 CHARACTERISTIC INFRARED ABSORPTION FREQUENCIES[a]

Bond	Compound type	Frequency range, cm^{-1}	Reference
C—H	Alkanes	2850–2960	Sec. 13.15
		1350–1470	
C—H	Alkenes	3020–3080 (*m*)	Sec. 13.15
		675–1000	
C—H	Aromatic rings	3000–3100 (*m*)	Sec. 13.15
		675–870	
C—H	Alkynes	3300	Sec. 13.15
C=C	Alkenes	1640–1680 (*v*)	Sec. 13.15
C≡C	Alkynes	2100–2260 (*v*)	Sec. 13.15
C⋯C	Aromatic rings	1500, 1600 (*v*)	Sec. 13.15
C—O	Alcohols, ethers, carboxylic acids, esters	1080–1300	Sec. 16.13
			Sec. 17.17
			Sec. 18.22
			Sec. 20.25
C=O	Aldehydes, ketones, carboxylic acids, esters	1690–1760	Sec. 19.17
			Sec. 18.22
			Sec. 20.25
O—H	Monomeric alcohols, phenols	3610–3640 (*v*)	Sec. 16.13
			Sec. 24.13
	Hydrogen-bonded alcohols, phenols	3200–3600(*broad*)	Sec. 16.13
			Sec. 24.13
	Carboxylic acids	2500–3000 (*broad*)	Sec. 18.22
N—H	Amines	3300–3500 (*m*)	Sec. 23.20
C—N	Amines	1180–1360	Sec. 23.20
C≡N	Nitriles	2210–2260 (*v*)	
$—NO_2$	Nitro compounds	1515–1560	
		1345–1385	

[a] All bands strong unless marked: *m*, moderate; *w*, weak; *v*, variable.

13.5 The ultraviolet spectrum

Light of wavelength between about 400 nm and 750 nm is visible. (A *nanometer*, nm, is 10^{-7}cm, and equals one mμ.) Just beyond the red end of the visible spectrum (λ greater than 750 nm) lies the infrared region. Just beyond the violet end of the visible spectrum (λ less than 400 nm) lies the ultraviolet region.

The ultraviolet spectrometers commonly used measure absorption of light in the visible and "near" ultraviolet region, that is, in the 200–750 nm range. This light is of higher frequency (and greater energy) than infrared light and, when it is absorbed by a molecule, the changes it produces are, naturally, ones that require greater energy: changes in electronic states.

In a transition to a higher electronic level, a molecule can go *from* any of a number of sub-levels—corresponding to various vibrational and rotational states—*to* any of a number of sub-levels; as a result, ultraviolet absorption bands are broad. Where an infrared spectrum shows many sharp peaks, a typical ultraviolet spectrum shows only a few broad humps. One can conveniently describe such a spectrum in terms of the *position* of the top of the hump (λ_{max}) and the *intensity* of that absorption (ϵ_{max}, the extinction coefficient).

When we speak of a molecule as being raised to a higher electronic level, we mean that an electron has been changed from one orbital to another orbital of higher energy. This electron can be of any of the kinds we have encountered: a σ electron, a π electron, or an n electron (a non-bonding electron—that is, one of an unshared pair). A σ electron is held tightly, and a good deal of energy is required to excite it: energy corresponding to ultraviolet light of short wavelength, in a region—"far" ultraviolet—outside the range of the usual spectrometer. It is chiefly excitations of the comparatively loosely held n and π electrons that appear in the (near) ultraviolet spectrum, and, of these, only jumps to the lower—more stable—excited states.

The electronic transitions of most concern to the organic chemist are: (a) $n \rightarrow \pi^*$, in which the electron of an unshared pair goes to an unstable (*antibonding*) π orbital, as, for example,

$$\mathrm{>C{=}\ddot{O}:} \longrightarrow \mathrm{>C\dot{=}\dot{O}:} \qquad n \longrightarrow \pi^*$$

and (b) $\pi \rightarrow \pi^*$, in which an electron goes from a stable (*bonding*) π orbital to an unstable π orbital, as, for example,

$$\mathrm{>C{=}\ddot{O}:} \longrightarrow \mathrm{>\dot{C}{-}\ddot{O}:} \qquad \pi \longrightarrow \pi^*$$

A $\pi \rightarrow \pi^*$ transition can occur for even a simple alkene, like ethylene, but absorption occurs in the far ultraviolet. Conjugation of double bonds, however, lowers the energy required for the transition, and absorption moves to longer wavelengths, where it can be more conveniently measured. If there are enough double bonds in conjugation, absorption will move into the visible region, and the compound will be colored. β-Carotene, for example, is a yellow pigment found in carrots and green leaves, and is a precursor of vitamin A; it contains eleven carbon–carbon double bonds in conjugation, and owes its color to absorption at the violet end of the visible spectrum (λ_{max} 451 nm).

How does conjugation bring about this effect? We have seen (Sec. 8.17) that 1,3-butadiene, for example, is stabilized by contribution from structures involving formal bonds. Stabilization is not very great, however, since such structures—and additional, ionic structures—are not very stable and make only small contribution to the hybrid. Similar structures contribute to an excited state of butadiene, too, but here, because of the instability of the molecule, they make much larger contribution. Resonance stabilizes the excited state *more* than it stabilizes the ground state, and thus reduces the difference between them.

In contrast to the infrared spectrum, the ultraviolet spectrum is not used primarily to show the presence of individual functional groups, but rather to show relationships between functional groups, chiefly conjugation: conjugation between two or more carbon–carbon double (or triple) bonds; between carbon–carbon and carbon–oxygen double bonds; between double bonds and an aromatic ring; and even the presence of an aromatic ring itself. It can, in addition, reveal the number and location of substituents attached to the carbons of the conjugated system.

Problem 13.2 In Problem 9.19, page 313, you calculated the number of rings in β-carotene. Taking into account also the molecular formula, the number of double bonds, conjugation, its natural occurrence, and its conversion into vitamin A (p. 277), what possible structure for β-carotene occurs to you?

Problem 13.3 Compounds A, B, and C have the formula C_5H_8, and on hydrogenation all yield *n*-pentane. Their ultraviolet spectra show the following values of λ_{max}: A, 176 nm; B, 211 nm; C, 215 nm. (1-Pentene has λ_{max} 178 nm.) (a) What is a likely structure for A? For B and C? (b) What kind of information might enable you to assign specific structures to B and C?

13.6 The nuclear magnetic resonance (nmr) spectrum

Like electrons, the nuclei of certain atoms are considered to *spin*. The spinning of these charged particles—the circulation of charge—generates a *magnetic moment* along the axis of spin, so that these nuclei act like tiny bar magnets. One such nucleus—and the one we shall be mostly concerned with—is the *proton*, the nucleus of ordinary hydrogen, ^{1}H.

Now, if a proton is placed in an external magnetic field, its magnetic moment, according to quantum mechanics, can be aligned in either of two ways: *with* or *against* the external field. Alignment with the field is the more stable, and energy must be absorbed to "flip" the tiny proton magnet over to the less stable alignment, against the field.

Just how much energy is needed to flip the proton over depends, as we might expect, on the strength of the external field: the stronger the field, the greater the tendency to remain lined up with it, and the higher the frequency (*Remember:* $\Delta E = h\nu$) of the radiation needed to do the job.

$$\nu = \frac{\gamma H_0}{2\pi}$$

where ν = frequency, in Hz

H_0 = strength of the magnetic field, in gauss

γ = a nuclear constant, the *gyromagnetic ratio*, 26,750 for the proton

In a field of 14,092 gauss, for example, the energy required corresponds to electromagnetic radiation of frequency 60 MHz (60 megahertz or 60 million cycles per second): radiation in the radiofrequency range, and of much lower energy (lower frequency, longer wavelength) than even infrared light.

In principle, we could place a substance in a magnetic field of constant strength, and then obtain a spectrum in the same way we obtain an infrared or an ultraviolet spectrum: pass radiation of steadily changing frequency through the substance, and observe the frequency at which radiation is absorbed. In practice, however, it has been found more convenient to keep the radiation frequency constant, and to vary the strength of the magnetic field; at some value of the field strength the energy required to flip the proton matches the energy of the radiation, absorption occurs, and a signal is observed. Such a spectrum is called a *nuclear magnetic resonance* (nmr) *spectrum* (Fig. 13.3).

Since the nucleus involved is the proton, the spectrum is sometimes called a *pmr* (proton magnetic resonance) spectrum, to differentiate it from spectra involving such nuclei as ^{13}C (called *cmr* spectra) or ^{19}F.

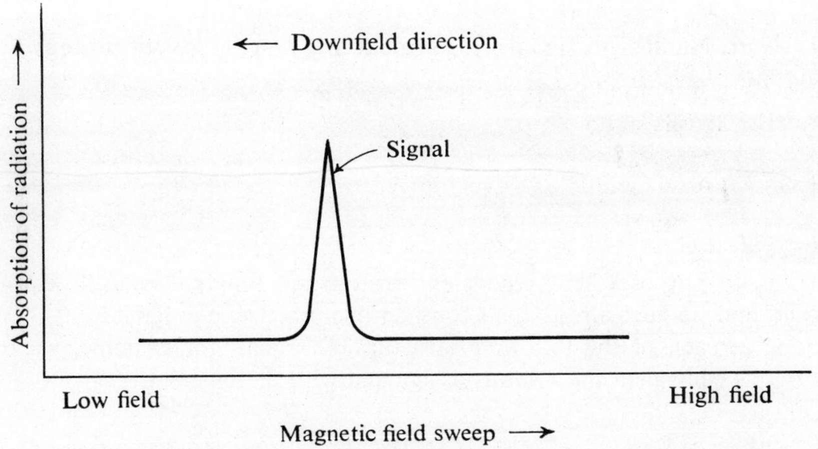

Figure 13.3. The nmr spectrum.

Now, if the situation were as simple as we have so far described it, all the protons in an organic molecule would absorb at exactly the same field strength, and the spectrum would consist of a single signal that would tell us little about the structure of the molecule. But the frequency at which a proton absorbs depends on the magnetic field which that proton *feels*, and this *effective* field strength is not exactly the same as the *applied* field strength. The effective field strength at each proton depends on the environment of that proton—on, among other things, the electron density at the proton, and the presence of other, nearby protons. Each proton—or, more precisely, each set of equivalent protons—will have a slightly different environment from every other set of protons, and hence will require a slightly *different applied* field strength to produce the *same effective* field strength: the particular field strength at which absorption takes place.

At a given radiofrequency, then, *all protons absorb at the same effective field strength, but they absorb at different applied field strengths.* It is this applied field strength that is measured, and against which the absorption is plotted.

The result is a spectrum showing many absorption peaks, whose relative positions, reflecting as they do differences in environment of protons, can give almost unbelievably detailed information about molecular structure.

In the following sections, we shall look at various aspects of the nmr spectrum:

(a) the *number of signals*, which tells us how many different "kinds" of protons there are in a molecule;

(b) the *positions of the signals*, which tell us something about the electronic environment of each kind of proton;

(c) the *intensities of the signals*, which tell us how many protons of each kind there are; and

(d) the *splitting of a signal* into several peaks, which tells us about the environment of a proton with respect to other, nearby protons.

13.7 Nmr. Number of signals. Equivalent and non-equivalent protons

In a given molecule, protons with the same environment absorb at the same (applied) field strength; protons with different environments absorb at different (applied) field strengths. A set of protons with the same environment are said to be *equivalent*; the number of signals in the nmr spectrum tells us, therefore, how many sets of equivalent protons—how many "kinds" of protons—a molecule contains.

For our purposes here, equivalent protons are simply chemically equivalent protons, and we have already had considerable practice in judging what these are. Looking at each of the following structural formulas, for example, we readily pick out as equivalent the protons designated with the same letter:

$CH_3—CH_2—Cl$	$CH_3—CHCl—CH_3$	$CH_3—CH_2—CH_2—Cl$
a *b*	*a* *b* *a*	*a* *b* *c*
2 nmr signals	*2 nmr signals*	*3 nmr signals*
Ethyl chloride	Isopropyl chloride	*n*-Propyl chloride

Realizing that, to be chemically equivalent, protons must also be *stereo*chemically equivalent, we find we can readily analyze the following formulas, too:

$(CH_3)_a(CH_3)_a$C=CH_bH_b	$(CH_3)_a$(Br)C=CH_bH_c	H_a(Cl)C=CH_bH_c	Cyclopropane with CH_3 (*a*), H (*b*), H (*c*), H (*c*), H (*d*), H (*d*)
2 nmr signals	*3 nmr signals*	*3 nmr signals*	*4 nmr signals*
Isobutylene	2-Bromopropene	Vinyl chloride	Methylcyclopropane

1,2-Dichloropropane (optically active or optically inactive) gives four nmr signals, and it takes only a little work with models or stereochemical formulas to see that this should indeed be so.

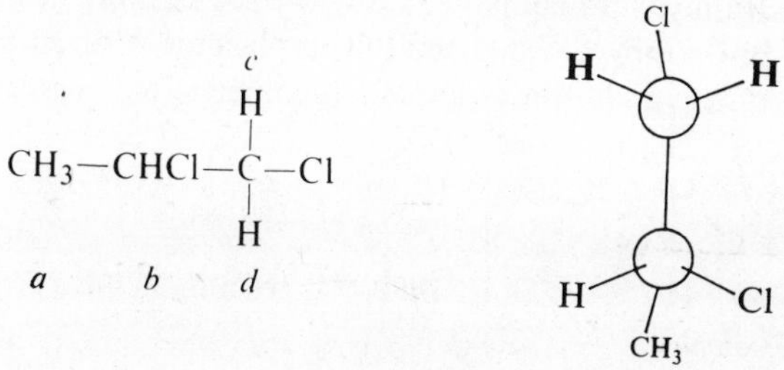

$$\underset{a}{CH_3}-\underset{b}{CHCl}-\overset{\overset{c}{H}}{\underset{\underset{d}{H}}{C}}-Cl$$

4 nmr signals
1,2-Dichloropropane

The environments of the two protons on C–1 are *not* the same (and no amount of rotation about single bonds will make them so); the protons are not equivalent, and will absorb at different field strengths.

We can tell from a formula which protons are in different environments and hence should give different signals. We cannot always tell—particularly with stereochemically different protons—just *how* different these environments are; they may not be different enough for the signals to be noticeably separated, and we may see *fewer* signals than we predict.

Now, just how did we arrive at the conclusions of the last few paragraphs? Most of us—perhaps without realizing it—judge the equivalence of protons by following the approach of isomer number (Sec. 4.2). This is certainly the easiest way to do it. We imagine each proton in turn to be replaced by some other atom Z. If replacement of either of two protons by Z would yield the same product—or enantiomeric products—then the two protons are chemically equivalent. We ignore the existence of conformational isomers and, as we shall see in Sec. 13.13, this is just what we should do.

Take, for example, ethyl chloride. Replacement of a methyl proton would give $CH_2Z{-}CH_2Cl$; replacement of a methylene proton would give $CH_3{-}CHZCl$. These are, of course, different products, and we easily recognize the methyl protons as being non-equivalent to the methylene protons.

The product $CH_2Z{-}CH_2Cl$ is the same regardless of *which one* of the three methyl protons is replaced. The (average) environment of the three protons is identical, and hence we expect one nmr signal for all three.

Replacement of either of the two methylene protons would give one of a pair of enantiomers:

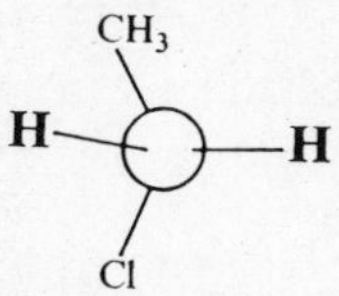

Enantiotopic protons
Ethyl chloride

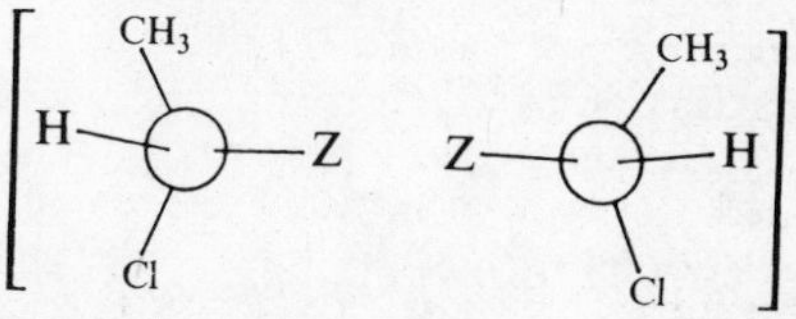

Such pairs of protons are called **enantiotopic protons.** The environments of these two protons are mirror images of each other; in an achiral medium, these protons behave *as if* they were equivalent, and we see one nmr signal for the pair.

Turning to 2-bromopropene, we see that replacement of either of the vinylic protons gives one of a pair of diastereomers (geometric isomers, in this case):

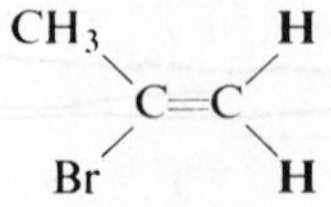

$$\begin{matrix} CH_3 & & H \\ & C{=}C & \\ Br & & H \end{matrix} \qquad \left[\begin{matrix} CH_3 & & Z \\ & C{=}C & \\ Br & & H \end{matrix} \qquad \begin{matrix} CH_3 & & H \\ & C{=}C & \\ Br & & Z \end{matrix}\right]$$

Diastereotopic protons
2-Bromopropene

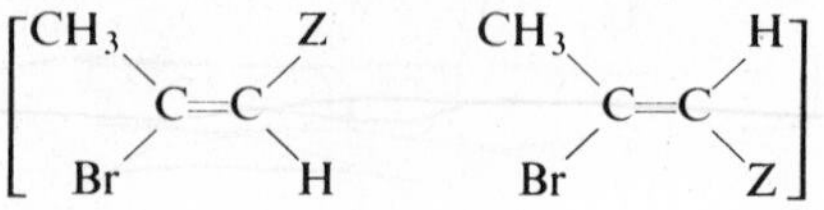

Such pairs of protons are called **diastereotopic protons**. The environments of these two protons are neither identical nor mirror images of each other; these protons are non-equivalent, and we expect an nmr signal from each one.

Similarly, in 1,2-dichloropropane the two protons on C–1 are diastereotopic, non-equivalent, and give separate nmr signals.

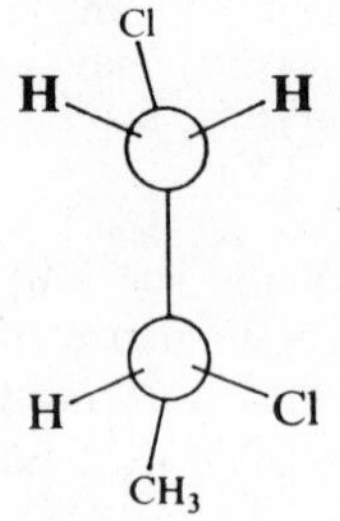

Cl; **H**, **H**; H, Cl; CH_3 [Newman projection]

Diastereotopic protons
1,2-Dichloropropane

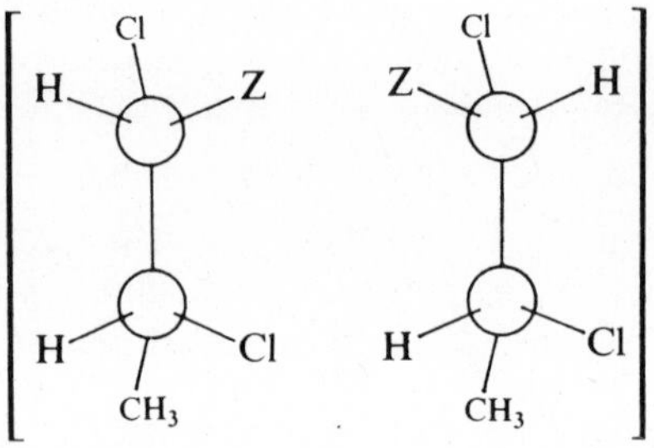

[Cl; H, Z; H, Cl; CH_3 — Cl; Z, H; H, Cl; CH_3]

In Sec. 13.13, we shall take a closer look at equivalence. The guidelines we have laid down here, however—based on rapid rotation about single bonds—hold for most spectra taken under ordinary conditions, specifically, at room temperature.

Problem 13.4 Draw the structural formula of each of the following compounds (disregarding enantiomerism), and label all sets of equivalent protons. How many nmr signals would you expect to see from each?

(a) the two isomers of formula $C_2H_4Cl_2$
(b) the four isomers of $C_3H_6Br_2$
(c) ethylbenzene and *p*-xylene
(d) mesitylene, *p*-ethyltoluene, isopropylbenzene
(e) CH_3CH_2OH and CH_3OCH_3
(f) $CH_3CH_2OCH_2CH_3$, $CH_3OCH_2CH_2CH_3$, $CH_3OCH(CH_3)_2$, $CH_3CH_2CH_2CH_2OH$
(g) $\begin{matrix} CH_2{-}CH_2 \\ | \quad\; | \\ CH_2{-}O \end{matrix}$, $CH_3{-}CH{-}CH_2$ (epoxide ring through O) (*Hint:* Make Models.)

(h) $CH_3CH_2\underset{\underset{O}{\|}}{C}{-}H$, $CH_3\underset{\underset{O}{\|}}{C}CH_3$, and $CH_2{=}CHCH_2OH$

Problem 13.5 Three isomeric dimethylcyclopropanes give, respectively, 2, 3, and 4 nmr signals. Draw a stereoisomeric formula for the isomer giving rise to each number of signals.

Problem 13.6 How many nmr signals would you expect from cyclohexane? Why?

13.8 Nmr. Positions of signals. Chemical shift

Just as the number of signals in an nmr spectrum tells us how many kinds of protons a molecule contains, so the *positions of the signals* help to tell us *what kinds* of protons they are: aromatic, aliphatic, primary, secondary, tertiary; benzylic, vinylic, acetylenic; adjacent to halogen or to other atoms or groups. These different kinds of protons have different electronic environments, and it is the electronic environment that determines just where in the spectrum a proton absorbs.

When a molecule is placed in a magnetic field—as it is when one determines an nmr spectrum—its electrons are caused to circulate and, in circulating, they generate secondary magnetic fields: *induced* magnetic fields.

Circulation of electrons *about the proton itself* generates a field aligned in such a way that—at the proton—it opposes the applied field. The field felt by the proton is thus diminished, and the proton is said to be **shielded**.

Circulation of electrons—specifically, π electrons—*about nearby nuclei* generates a field that can either oppose or reinforce the applied field at the proton, depending on the proton's location (Fig. 13.4). If the induced field opposes the

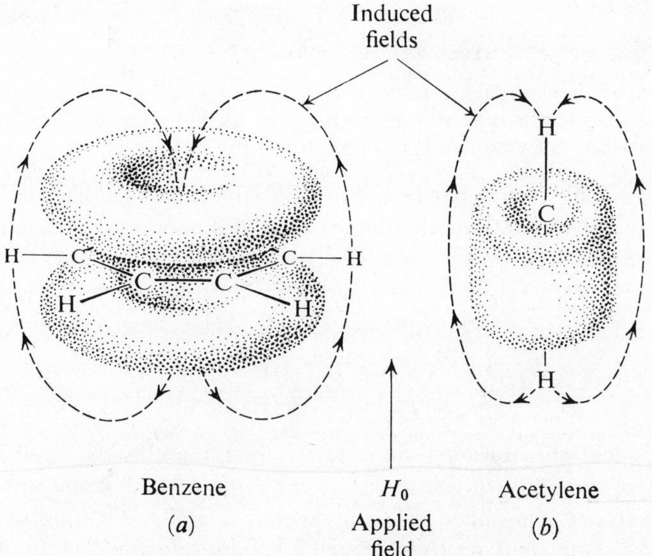

Figure 13.4. Induced field (*a*) reinforces applied field at the aromatic protons, and (*b*) opposes applied field at the acetylenic protons. Aromatic protons are deshielded; acetylenic protons are shielded.

applied field, the proton is shielded, as before. If the induced field reinforces the applied field, then the field felt by the proton is augmented, and the proton is said to be **deshielded.**

Compared with a naked proton, a shielded proton requires a higher applied field strength—and a deshielded proton requires a lower applied field strength—to provide the particular effective field strength at which absorption occurs. Shielding thus shifts the absorption upfield, and deshielding shifts the absorption downfield. Such shifts in the position of nmr absorptions, arising from shielding and deshielding by electrons, are called **chemical shifts.**

How are the direction and magnitude—the *value*—of a particular chemical shift to be measured and expressed?

The unit in which a chemical shift is most conveniently expressed is parts per million (ppm) of the total applied magnetic field. Since shielding and deshielding arise from *induced* secondary fields, the magnitude of a chemical shift is proportional to the strength of the applied field—or, what is equivalent, proportional to the radiofrequency the field must match. If, however, it is expressed as a *fraction* of the applied field—that is, if the observed shift is divided by the particular radiofrequency used—then a chemical shift has a constant value that is independent of the radiofrequency and the magnetic field that the nmr spectrometer employs.

The **reference point** from which chemical shifts are measured is, for practical reasons, not the signal from a naked proton, but the signal from an actual compound: usually tetramethylsilane, $(CH_3)_4Si$. Because of the low electronegativity of silicon, the shielding of the protons in the silane is greater than in most other organic molecules; as a result, most nmr signals appear in the same direction from the tetramethylsilane signal: *downfield.*

The most commonly used scale is the δ (*delta*) scale. The position of the tetramethylsilane signal is taken as 0.0 ppm. Most chemical shifts have δ values between 0 and 10 (minus 10, actually). A *small* δ value represents a *small* downfield shift, and a *large* δ value represents a *large* downfield shift.

One commonly encounters another scale: the τ (*tau*) scale, on which the tetramethylsilane signal is taken as 10.0 ppm. Most τ values lie between 0 and 10. The two scales are related by the expression $\tau = 10 - \delta$.

An nmr signal from a particular proton appears at a different field strength than the signal from tetramethylsilane. This difference—the chemical shift—is measured not in gauss, as we might expect, but in the equivalent frequency units (*Remember:* $\nu = \gamma H_0/2\pi$), and it is divided by the frequency of the spectrometer used. Thus, for a spectrometer operating at 60 MHz, that is, at 60×10^6 Hz:

$$\delta = \frac{\text{observed shift (Hz)} \times 10^6}{60 \times 10^6 \text{ (Hz)}}$$

The chemical shift for a proton is determined, then, by the electronic environment of the proton. In a given molecule, protons with different environments—non-equivalent protons—have different chemical shifts. Protons with the same environment—equivalent protons—have the same chemical shift. (So, too, do protons with mirror-image environments—enantiotopic protons.) We have already seen what the equivalence of protons means in terms of molecular structure.

Table 13.4 CHARACTERISTIC PROTON CHEMICAL SHIFTS

Type of proton		Chemical shift, ppm
		δ
Cyclopropane		0.2
Primary	RCH_3	0.9
Secondary	R_2CH_2	1.3
Tertiary	R_3CH	1.5
Vinylic	C=C—H	4.6–5.9
Acetylenic	C≡C—H	2–3
Aromatic	Ar—H	6–8.5
Benzylic	Ar—C—H	2.2–3
Allylic	$C{=}C{-}CH_3$	1.7
Fluorides	HC—F	4–4.5
Chlorides	HC—Cl	3–4
Bromides	HC—Br	2.5–4
Iodides	HC—I	2–4
Alcohols	HC—OH	3.4–4
Ethers	HC—OR	3.3–4
Esters	RCOO—CH	3.7–4.1
Esters	HC—COOR	2–2.2
Acids	HC—COOH	2–2.6
Carbonyl compounds	HC—C=O	2–2.7
Aldehydic	RCHO	9–10
Hydroxylic	ROH	1–5.5
Phenolic	ArOH	4–12
Enolic	C=C—OH	15–17
Carboxylic	RCOOH	10.5–12
Amino	RNH_2	1–5

Furthermore, it has been found that a proton with a particular environment shows much the same chemical shift, whatever the molecule it happens to be part of. Take, for example, our familiar classes of hydrogens: primary, secondary, and tertiary. In the absence of other nearby substituents, absorption occurs at about these values:

RCH_3	δ 0.9
R_2CH_2	δ 1.3
R_3CH	δ 1.5

All these protons, in turn, differ widely from aromatic protons which, because of the powerful deshielding due to the circulation of the π electrons (see Fig. 13.4, p. 419), absorb far downfield:

Ar—H δ 6–8.5

Attachment of chlorine to the carbon bearing the proton causes a downfield shift. If the chlorine is attached to the carbon once removed from the carbon bearing the proton, there is again a downward shift, but this time much weaker.

$CH_3{-}Cl$	δ 3.0	$CH_3{-}C{-}Cl$	δ 1.5
$R{-}CH_2{-}Cl$	δ 3.4	$R{-}CH_2{-}C{-}Cl$	δ 1.7
$R_2CH{-}Cl$	δ 4.0	$R_2CH{-}C{-}Cl$	δ 1.6

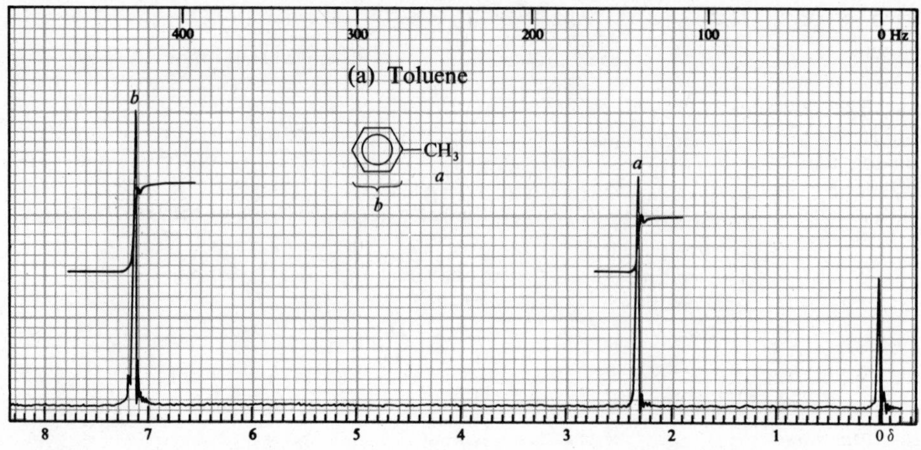

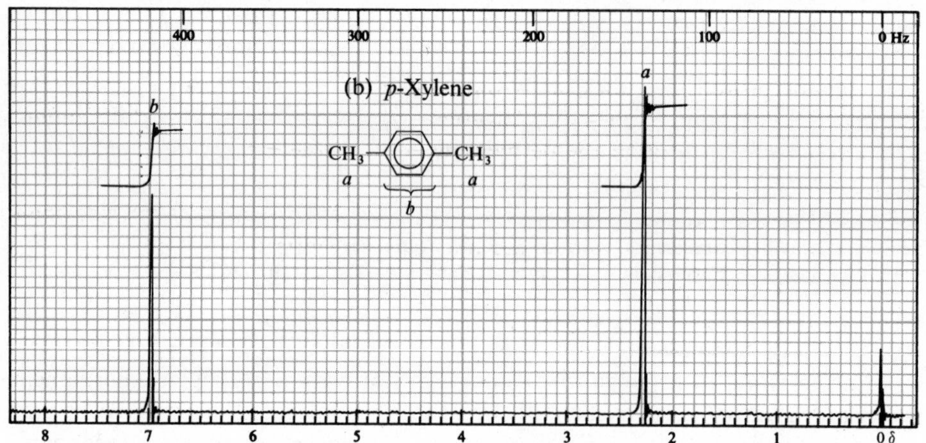

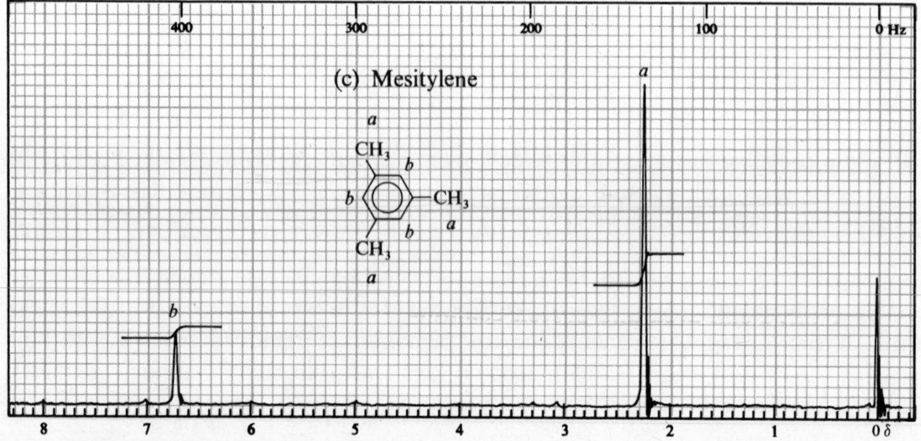

Figure 13.5. Nmr spectra: chemical shift. (a) Toluene; (b) *p*-xylene; (c) mesitylene.

Two chlorines cause a greater downfield shift. Other halogens show similar effects.

The downfield shift caused by chlorine is what we might have expected from its inductive effect: electron withdrawal lowers the electron density in the vicinity of the proton and thus causes deshielding. The effect of a substituent on the chemical shift is unquestionably the net result of many factors; yet we shall often observe chemical shifts which strongly suggest that an inductive effect is at least one of the factors at work.

Table 13.4 lists chemical shifts for protons in a variety of environments.

The nmr spectra (Fig. 13.5, p. 422) of the alkylbenzenes *toluene*, *p-xylene*, and *mesitylene* illustrate the points we have just made. In each spectrum there are two signals: one for the side-chain protons, and one for the ring protons. (Here, as in some—though not most—aromatic compounds, the *ortho*, *meta*, and *para* protons have nearly the same chemical shifts.)

In each spectrum, the ring protons show the low-field absorption we have said is characteristic of aromatic protons. Absorption is not only at low field, but at nearly the *same* field strength for the three compounds: at δ 7.17, 7.05, and 6.78. (These values are not *exactly* the same, however, since the environments of the aromatic protons are not exactly the same in the three compounds.)

In each compound, side-chain protons—benzylic protons—are close enough to the ring to feel a little of the deshielding effect of the π electrons (Fig. 13.4, p. 419), and hence absorb somewhat downfield from ordinary alkyl protons: at δ 2.32, 2.30, and 2.25. In all three compounds, the environment of the side-chain protons is almost identical, and so are the chemical shifts.

The similarity in structure among these three alkylbenzenes is thus reflected in the similarity of their nmr spectra. There is, however, a major difference in their structures—a difference in *numbers* of aromatic and side-chain protons—and, as we shall see in the next section, this is reflected in a major difference in nmr spectra.

The chemical shift is fundamental to the nmr spectrum since, by separating the absorption peaks due to the various protons of a molecule, it reveals all the other features of the spectrum. The *numerical values* of chemical shifts, although significant, do not have the overriding importance that absorption frequencies have in the infrared spectrum. In our work with nmr, we shall escape much of the uncertainty that accompanies the beginner's attempts to identify precisely infrared absorption bands; at the same time, we have a greater *variety* of concepts to learn about—but these, at our present level, we may find more satisfying and intellectually more stimulating.

Problem 13.7 What is a possible explanation for the following differences in chemical shift for aromatic protons? Benzene δ 7.37; toluene δ 7.17; *p*-xylene δ 7.05; mesitylene δ 6.78.

13.9 Nmr. Peak area and proton counting

Let us look again at the nmr spectra (Fig. 13.5, p. 422) of toluene, *p*-xylene, and mesitylene, and this time focus our attention, not on the positions of the signals, but on their relative *intensities*, as indicated by the sizes of the absorption peaks.

Judging roughly from the peak heights, we see that the (high-field) peak for side-chain protons is smaller than the (low-field) peak for aromatic protons in the case of toluene, somewhat larger in the case of *p*-xylene, and considerably larger in the case of mesitylene. More exact comparison, based on the *areas under the peaks*, shows that the peaks for side-chain and aromatic protons have sizes in the ratio 3:5 for toluene; 3:2 (or 6:4) for *p*-xylene; and 3:1 (or 9:3) for mesitylene.

This illustrates a general quality of all nmr spectra. *The area under an nmr signal is directly proportional to the number of protons giving rise to the signal.*

It is not surprising that this is so. The absorption of every quantum of energy is due to exactly the same thing: the flipping over of a proton in the same effective magnetic field. The more protons flipping, the more the energy absorbed, and the greater is the area under the absorption peak.

Areas under nmr signals are measured by an electronic integrator, and are usually given on the spectrum chart in the form of a stepped curve; heights of steps are proportional to peak areas. Nmr chart paper is cross-hatched, and we can conveniently estimate step heights by simply counting squares. We arrive at a set of numbers that are in the same ratio as the numbers of different kinds of protons. We convert this set of numbers into a set of smallest whole numbers just as we did in calculating empirical formulas (Sec. 2.27). The number of protons

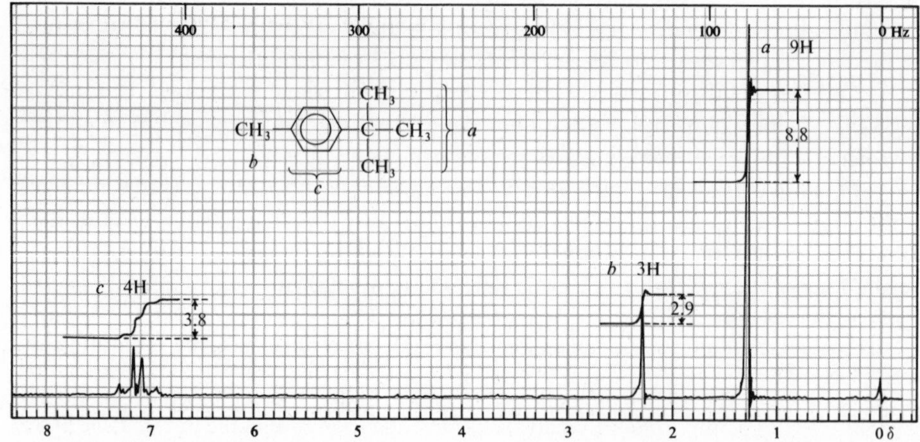

Figure 13.6. Nmr spectrum of *p-tert*-butyltoluene. Proton counting.
The ratio of step heights $a:b:c$ is

$$8.8:2.9:3.8 = 3.0:1.0:1.3 = 9.0:3.0:3.9$$

Alternatively, since the molecular formula $C_{11}H_{16}$ is known,

$$\frac{16 \text{ H}}{15.5 \text{ units}} = 1.03 \text{ H per unit}$$

$$a = 1.03 \times 8.8 = 9.1 \qquad b = 1.03 \times 2.9 = 3.0 \qquad c = 1.03 \times 3.8 = 3.9$$

Either way, we find: *a*, 9H; *b*, 3H; *c*, 4H.

The 4H of *c* (δ 7.1) are in the aromatic range, suggesting a disubstituted benzene $—C_6H_4—$. The 3H of *b* (δ 2.28) have a shift expected for benzylic protons, giving $CH_3—C_6H_4—$. There is left C_4H_9 which, in view of the 9H of *a* (δ 1.28) must be $—C(CH_3)_3$; since these are once removed from the ring their shift is nearly normal for an alkyl group. The compound is *tert*-butyltoluene (actually, as shown by the absorption pattern of the aromatic protons, the *p*-isomer).

giving rise to each signal is equal to the whole number for that signal—or to some multiple of it. See, for example, Fig. 13.6.

We take any shortcuts we can. If we know the molecular formula and hence the total number of protons, we can calculate from the combined step heights the number of squares per proton. If we suspect a particular structural feature that gives a characteristic signal—an aldehydic (—CHO) or carboxylic (—COOH) proton, say, which gives a far-downfield peak—we can use this step height as a starting point.

Working the following problems will give us some idea of the tremendous help "proton counting" by nmr can be in assigning a structure to a compound.

Problem 13.8 Go back to Problem 13.4 (p. 418), where you predicted the number of nmr signals from several compounds. Tell, where you can, the relative positions of the signals (that is, their sequence as one moves downfield) and, roughly, the δ value expected for each. For each signal tell the number of protons giving rise to it.

Problem 13.9 Give a structure or structures consistent with each of the nmr spectra shown in Fig. 13.7 (p. 426).

13.10 Nmr. Splitting of signals. Spin–spin coupling

An nmr spectrum, we have said, shows a signal for each kind of proton in a molecule; the few spectra we have examined so far bears this out. If we look much further, however, we soon find that most spectra are—or *appear* to be—much more complicated than this. Figure 13.8 (p. 427), for example, shows the nmr spectra for three compounds,

$CH_2Br—CHBr_2$	$CH_3—CHBr_2$	$CH_3—CH_2Br$
1,1,2-Tribromoethane	1,1-Dibromoethane	Ethyl bromide

each of which contains only two kinds of protons; yet, instead of two peaks, these spectra show *five*, *six*, and *seven* peaks, respectively

What does this multiplicity of peaks mean? How does it arise, and what can it tell us about molecular structure?

The answer is that we are observing the *splitting* of nmr signals caused by spin–spin coupling. The signal we expect from each set of equivalent protons is appearing, not as a single peak, but as a *group* of peaks. Splitting reflects the environment of the absorbing protons: not with respect to electrons, but with respect to other, nearby protons. It is as though we were permitted to sit on a proton and look about in all directions: we can *see* and *count* the protons attached to the carbon atoms next to our own carbon atom and, sometimes, even see protons still farther away.

Let us take the case of adjacent carbon atoms carrying, respectively, a pair of secondary protons and a tertiary proton, and consider first the absorption by one of the secondary protons:

$$—\overset{|}{C}H—CH_2—$$

The magnetic field that a secondary proton feels at a particular instant is slightly increased or slightly decreased by the spin of the neighboring tertiary proton: *increased* if the tertiary proton happens at that instant to be aligned *with* the applied

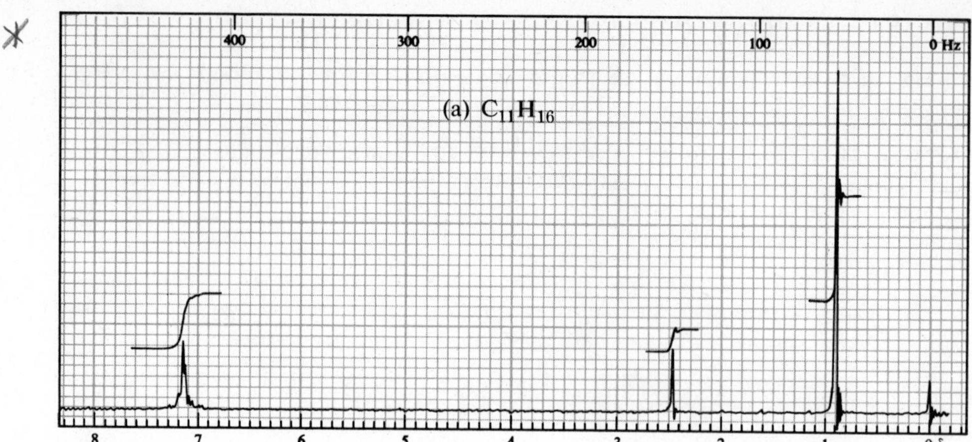

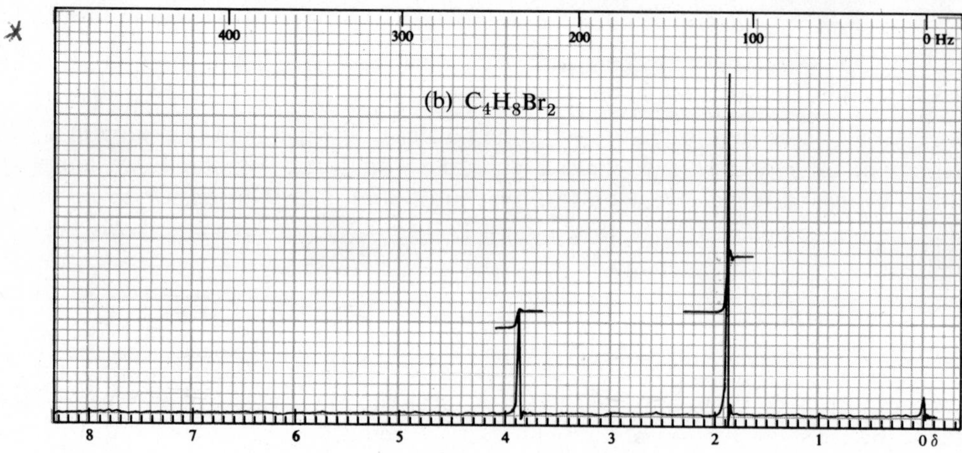

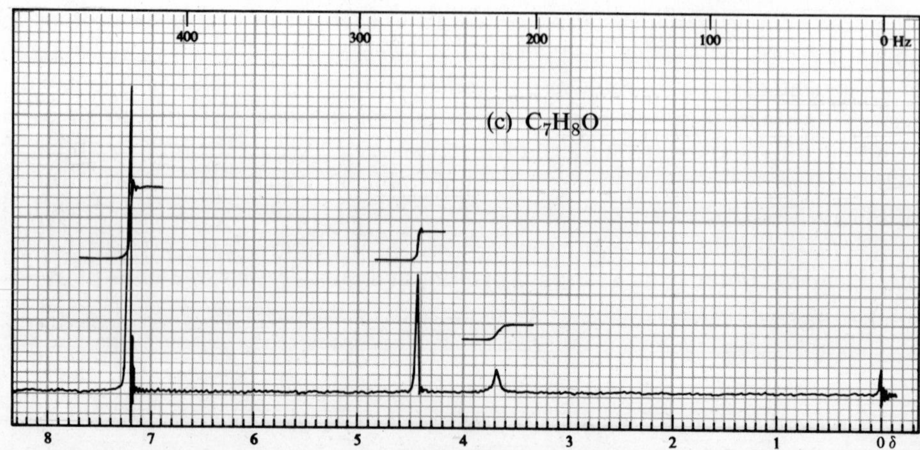

Figure 13.7. Nmr spectra for Problem 13.9, p. 425.

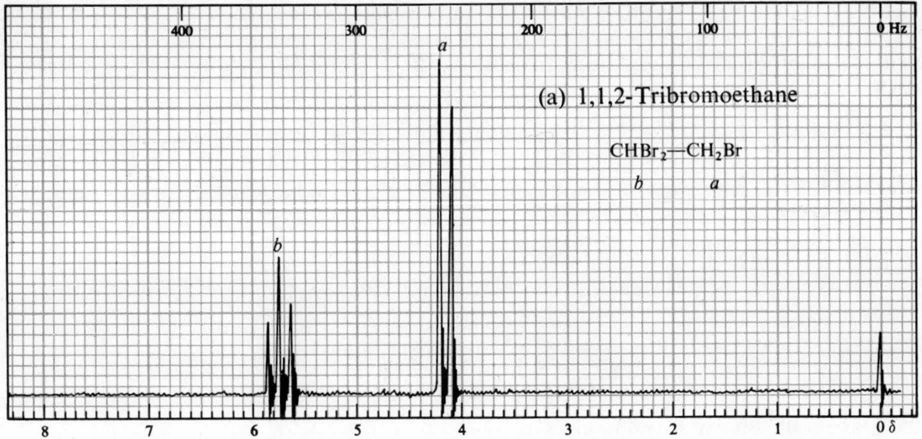

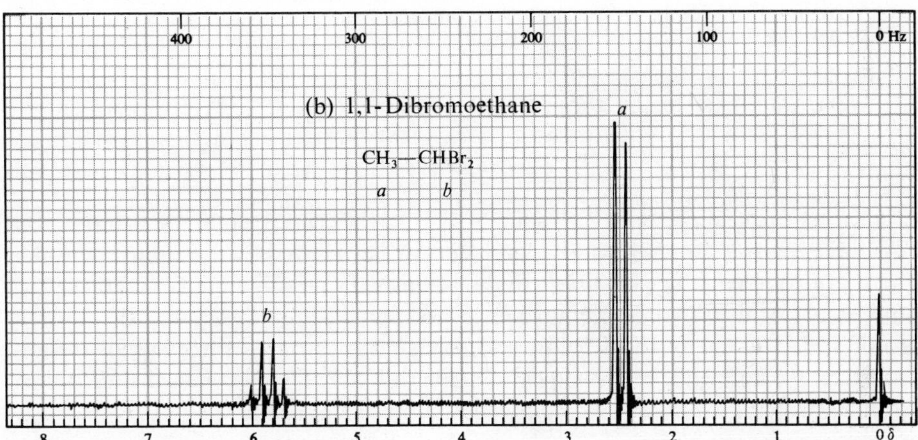

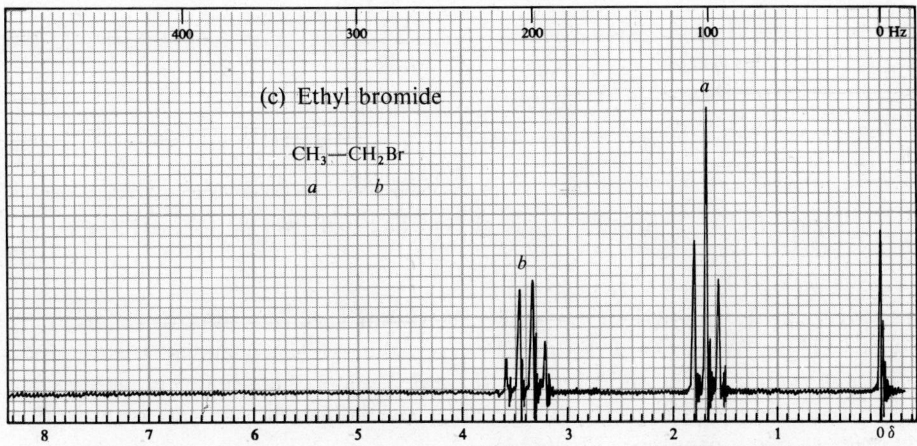

Figure 13.8. Nmr spectra: splitting of signals. (a) 1,1,2-Tribromoethane; (b) 1,1-dibromoethane; (c) ethyl bromide.

field; or *decreased* if the tertiary proton happens to be aligned *against* the applied field.

For half the molecules, then, absorption by a secondary proton is shifted slightly downfield, and for the other half of the molecules the absorption is shifted slightly upfield. The signal is split into *two* peaks: a *doublet*, with equal peak intensities (Fig. 13.9).

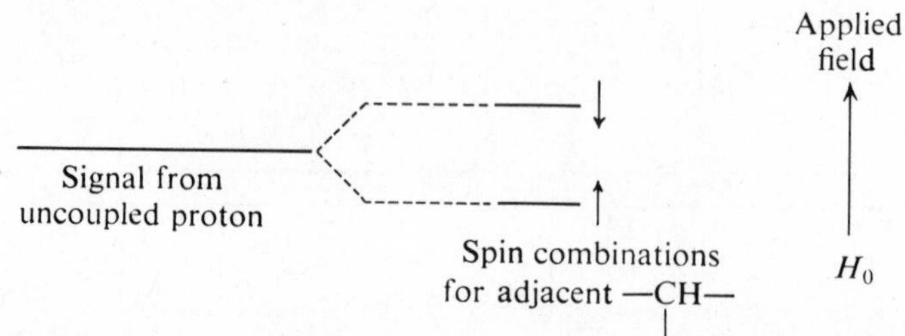

Figure 13.9. Spin–spin coupling. Coupling with one proton gives a 1:1 doublet.

Next, what can we say about the absorption by the tertiary proton?

$$-\overset{|}{C}H-CH_2-$$

It is, in its turn, affected by the spin of the neighboring secondary protons. But now there are *two* protons whose alignments in the applied field we must consider. There are four equally probable combinations of spin alignments for these two protons, of which two are equivalent. At any instant, therefore, the tertiary proton feels any one of three fields, and its signal is split into three equally spaced peaks: a *triplet*, with relative peak intensities 1:2:1, reflecting the combined (double) probability of the two equivalent combinations (Fig. 13.10).

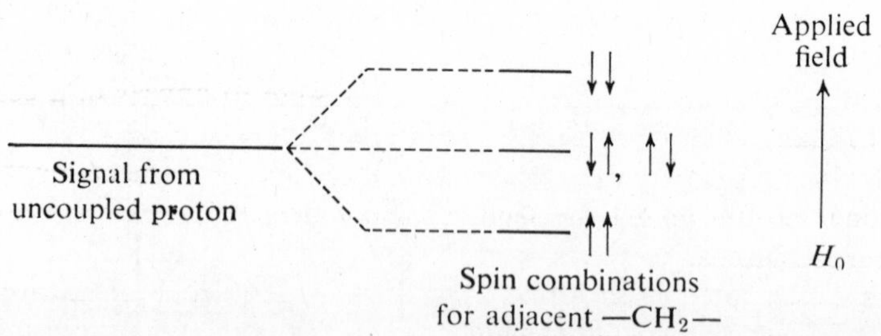

Figure 13.10. Spin–spin coupling. Coupling with two protons gives a 1:2:1 triplet.

Figure 13.11 (p. 429) shows an idealized nmr spectrum due to the grouping —CH—CH_2—. We see a 1:1 doublet (from the —CH_2—) and a 1:2:1 triplet (from the —CH—). The total area (both peaks) under the doublet is *twice* as big as the total area (all three peaks) of the triplet, since the doublet is due to absorption by twice as many protons as the triplet.

A little measuring shows us that the separation of peaks (the *coupling constant*,

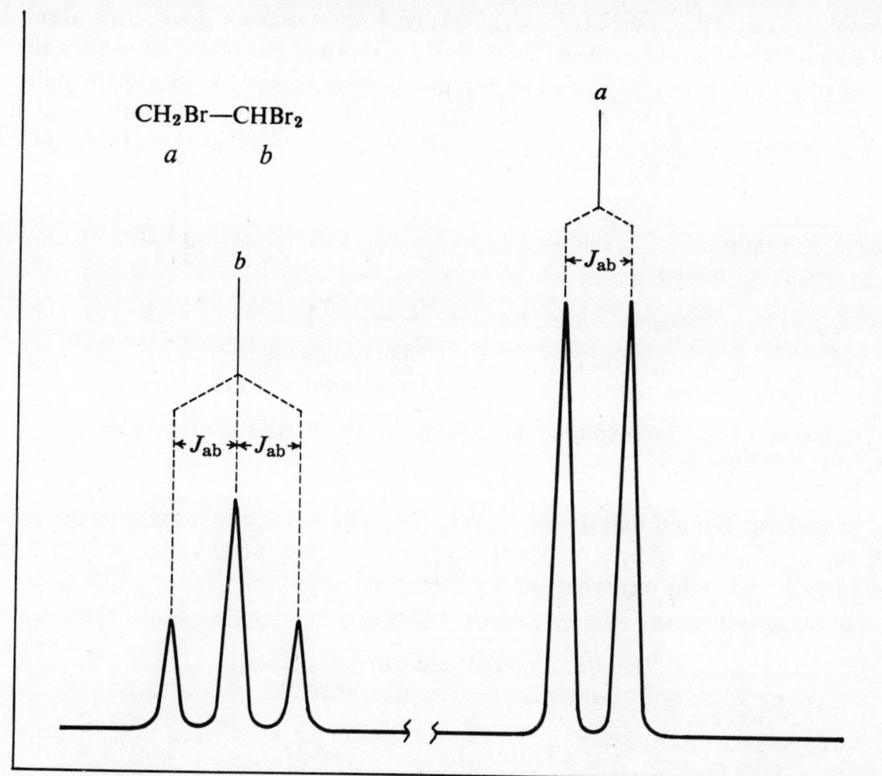

Figure 13.11. Spin–spin splitting. Signal *a* is split into a doublet by coupling with one proton; signal *b* is split into a triplet by two protons. Spacings in both sets the same (J_{ab}).

J, Sec. 13.11) in the doublet is exactly the same as the separation of peaks in the triplet. (Spin–spin coupling is a *reciprocal* affair, and the effect of the secondary protons on the tertiary proton must be identical with the effect of the tertiary proton on the secondary protons.) Even if they were to appear in a complicated spectrum of many absorption peaks, the identical peak separations would tell us that this doublet and triplet were related: that the (two) protons giving the doublet and the (one) proton giving the triplet are coupled, and hence are attached to adjacent carbon atoms.

We have seen that an nmr signal is split into a doublet by one nearby proton, and into a triplet by two (equivalent) nearby protons. What splitting can we expect more than two protons to produce? In Fig. 13.12 (p. 430), we see that three equivalent protons split a signal into four peaks—a quartet—with the intensity pattern 1:3:3:1.

It can be shown that, in general, *a set of n equivalent protons will split an nmr signal into n + 1 peaks.*

If we turn once more to Fig. 13.8 (p. 427), we no longer find these spectra so confusing. We now see not just five or six or seven peaks, but instead a doublet and a triplet, or a doublet and a quartet, or a triplet and a quartet. We recognize each of these multiplets from the even spacings within it, and from its symmetrical intensity pattern (1:1, or 1:2:1, or 1:3:3:1). Each spectrum does show absorption by just two kinds of protons; but clearly it shows a great deal more than that.

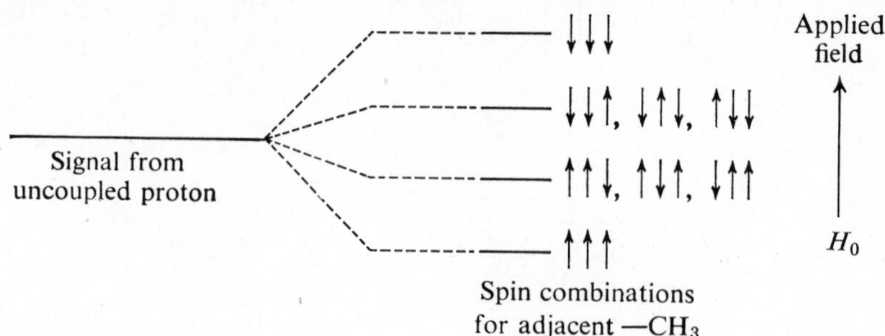

Figure 13.12. Spin–spin coupling. Coupling with three protons gives a 1:3:3:1 quartet.

If we keep in mind that the peak area reflects the number of *absorbing* protons, and the multiplicity of splittings reflects the number of *neighboring* protons, we find in each spectrum just what we would expect.

In the spectrum of $CHBr_2$—CH_2Br we see

Downfield triplet	and	*Upfield doublet*
Area: 1		*Area: 2*
—C(H)—CH_2—		—C(H)—CH_2—

In the spectrum of CH_3—$CHBr_2$ we see

Downfield quartet	and	*Upfield doublet*
Area: 1		*Area: 3*
—C(H)—CH_3		—C(H)—CH_3

and in the spectrum of CH_3—CH_2Br we see

Downfield quartet	and	*Upfield triplet*
Area: 2		*Area: 3*
—CH_2—CH_3		—CH_2—CH_3

We see chemical shifts that are consistent with the deshielding effect of halogens: in each spectrum, the protons on the carbon carrying the greater number of halogens absorb farther downfield (larger δ).

In each spectrum, we see that the spacing of the peaks within one multiplet is the same as within the other, so that even in a spectrum with many other peaks, we could pick out these two multiplets as being coupled.

Finally, we see a feature that we have not yet discussed: the various multiplets do not show quite the symmetry we have attributed to them. In spectrum (a), we see

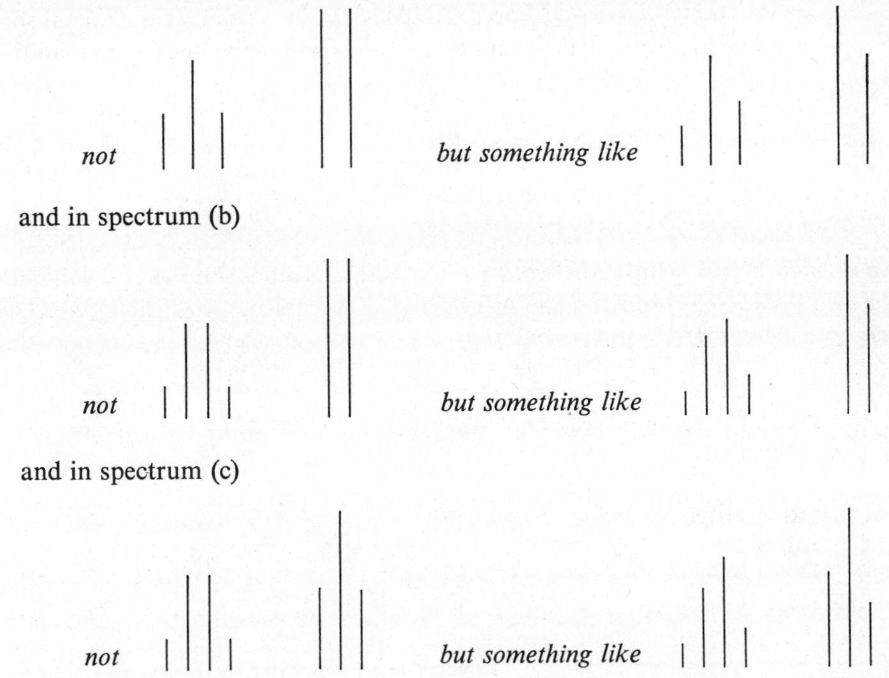

In each case, the inner peaks—the peaks nearer the other, coupled multiplets—are larger than the outer peaks.

Perfectly symmetrical multiplets are to be expected only when the separation between multiplets is very large relative to the separation within multiplets—that is, when the chemical shift is much larger than the coupling constant (Sec. 13.11). The patterns we see here are very commonly observed, and are helpful in matching up multiplets: we know in which direction—upfield or downfield—to look for the second multiplet.

We have not yet answered a very basic question: just which protons in a molecule can be coupled? *We may expect to observe spin–spin splitting only between non-equivalent neighboring protons.* By "non-equivalent" protons we mean protons with different chemical shifts, as we have already discussed (Sec. 13.8). By "neighboring" protons we mean most commonly protons on *adjacent* carbons, as in the examples we have just looked at (Fig. 13.8, p. 427); sometimes protons further removed from each other may also be coupled, particularly if π bonds intervene. (If protons on the *same* carbon are non-equivalent—as they sometimes are—they may show coupling.)

We do *not* observe splitting due to coupling between the protons making up the same $—CH_3$ group, since they are equivalent. We do *not* observe splitting due to coupling between the protons on C-1 and C-2 of 1,2-dichloroethane

$$\underset{\text{Cl}}{\underset{|}{CH_2}}—\underset{\text{Cl}}{\underset{|}{CH_2}}$$

1,2-Dichloroethane

No splitting

since, although on different carbons, they, too, are equivalent.

In the spectrum of 1,2-dibromo-2-methylpropane,

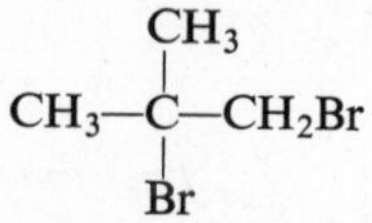

$$\begin{array}{c} CH_3 \\ | \\ CH_3\text{—}C\text{—}CH_2Br \\ | \\ Br \end{array}$$

1,2-Dibromo-2-methylpropane

No splitting

we do *not* observe splitting between the six methyl protons, on the one hand, and the two $—CH_2—$ protons on the other hand. They are non-equivalent, and give rise to different nmr signals, but they are not on adjacent carbons, and their spins do not (noticeably) affect each other. The nmr spectrum contains two singlets, with a peak area ratio of 3:1 (or 6:2). For the same reason, we do *not* observe splitting due to coupling between ring and side-chain protons in alkylbenzenes (Fig. 13.5, p. 422).

We do *not* observe splitting between the two vinyl protons of isobutylene

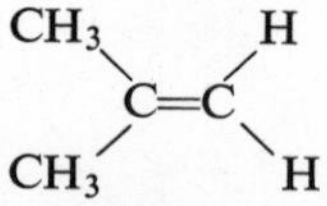

$$(CH_3)_2C{=}CH_2$$

Isobutylene

No splitting

since they are equivalent. On the other hand, we may observe splitting between the two vinyl protons on the same carbon if, as in 2-bromopropene, they are non-equivalent.

$$\begin{array}{ccc} CH_3 & & H_a \\ & C{=}C & \\ Br & & H_b \end{array}$$

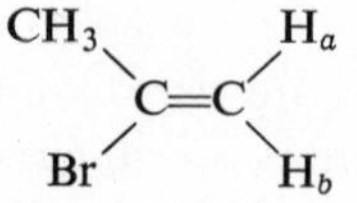

2-Bromopropene

The fluorine (^{19}F) nucleus has magnetic properties of the same kind as the proton. It gives rise to nmr spectra, although at a quite different frequency-field strength combination than the proton. Fluorine nuclei can be coupled not only with each other, but also with protons. *Absorption by* fluorine does not appear in the proton nmr spectrum—it is far off the scale—but the *splitting by* fluorine of proton signals can be seen. The signal for the two protons of 1,2-dichloro-1,1-difluoroethane, for example,

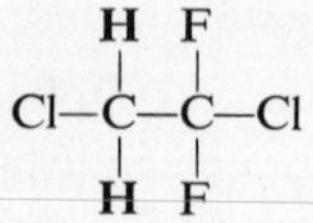

$$\begin{array}{c} H\ \ F \\ |\ \ \ | \\ Cl\text{—}C\text{—}C\text{—}Cl \\ |\ \ \ | \\ H\ \ F \end{array}$$

appears as a 1:2:1 triplet with peak spacings of 11 Hz. (What would you expect to see in the fluorine nmr spectrum?)

Figures 13.13 and 13.14, p. 433, and Fig. 13.15, p. 434, illustrate some of the kinds of splitting we are likely to encounter in nmr spectra.

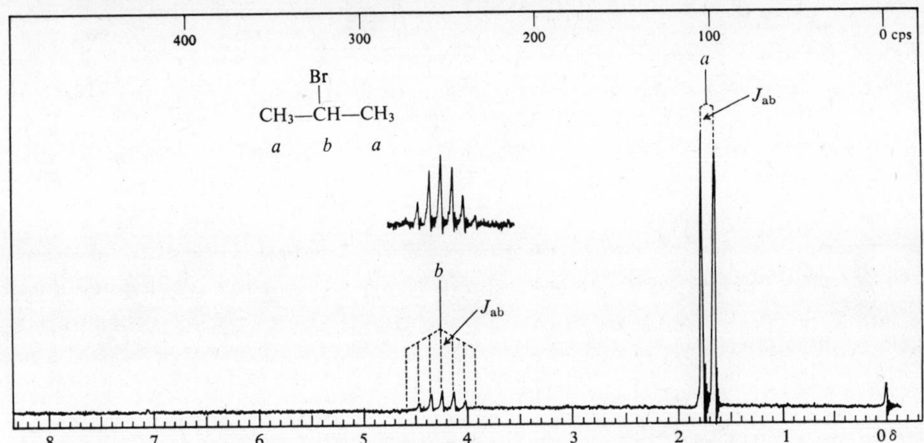

Figure 13.13. Nmr spectrum of isopropyl bromide. Absorption by the six methyl protons H_a appears upfield, split into a doublet by the single adjacent proton H_b. Absorption by the lone proton H_b appears downfield (the inductive effect of bromine) split into a septet by the six adjacent protons—with the small outside peaks typically hard to see.

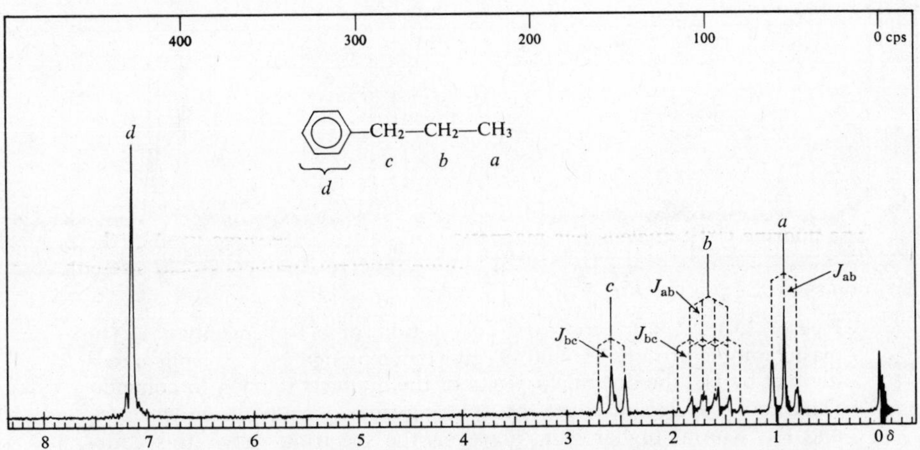

Figure 13.14. Nmr spectrum of *n*-propylbenzene. Moving downfield, we see the expected sequence of signals: *a*, primary (3H); *b*, secondary (2H); *c*, benzylic (2H); and *d*, aromatic (5H). Signals *a* and *c* are each split into a triplet by the two secondary protons H_b. The five protons adjacent to the secondary protons—three on one side and two on the other—are, of course, not equivalent; but the coupling constants, J_{ab} and J_{bc}, are nearly the same, and signal *b* appears as a sextet (5 + 1 peaks). The coupling constants are not *exactly* the same, however, as shown by the broadening of the six peaks.

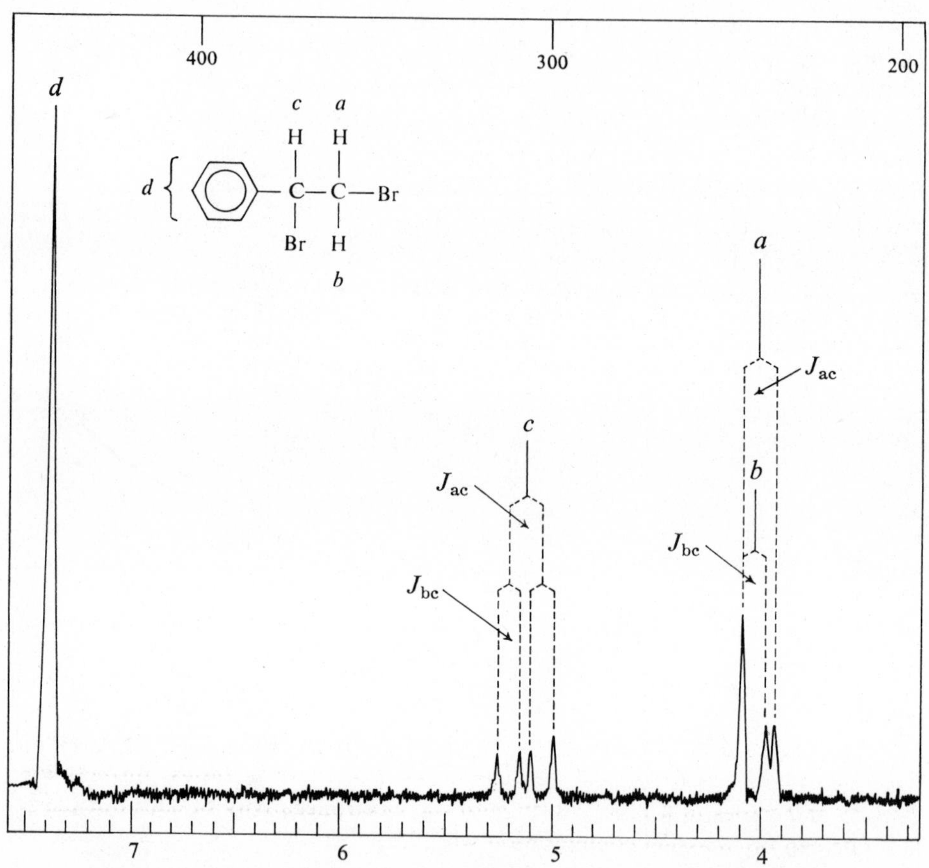

Figure 13.15. Nmr spectrum of 1,2-dibromo-1-phenylethane. The diastereotopic protons H_a and H_b give different signals, each split into a doublet by H_c; the downfield peaks of the doublets happen to coincide. (The above spectrum shows no splitting due to coupling between H_a and H_b. Run at higher gain, however, the spectrum shows this coupling: each doublet is split into a quartet.)

The four-line pattern of *c* is due to successive splittings by H_a and H_b. (If J_{ac} and J_{bc} were equal—as they would have to be if, for example, H_a and H_b were equivalent—the middle peaks of *c* would merge to give the familiar 1:2:1 triplet.)

13.11 Nmr. Coupling constants

The distance between peaks in a multiplet is a measure of the effectiveness of spin–spin coupling, and is called the **coupling constant**, J. Coupling (unlike chemical shift) is not a matter of induced magnetic fields. The value of the coupling constant —as measured, in Hz—remains the same, whatever the applied magnetic field (that is, whatever the radiofrequency used). In this respect, of course, spin–spin splitting differs from chemical shift, and, when necessary, the two can be distinguished on this basis: the spectrum is run at a second, different radiofrequency; when measured in Hz, peak separations due to splitting remain constant, whereas peak separations due to chemical shifts change. (When divided by the radiofrequency and thus converted into ppm, the numerical value of the chemical shift would, of course, remain constant.)

As we can see from the following summary, the size of a coupling constant depends markedly on the structural relationships between the coupled protons.

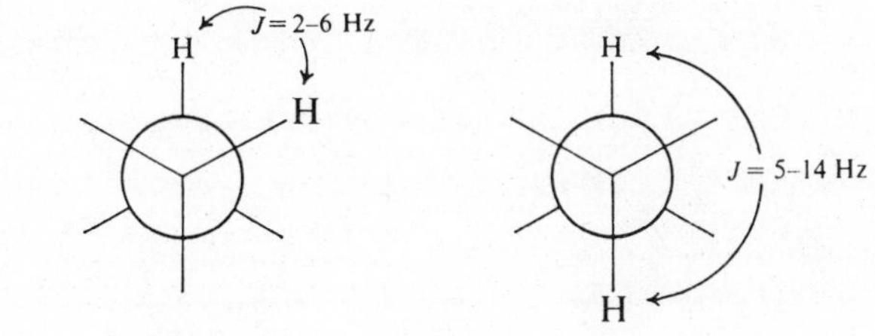

$J = 2\text{–}6$ Hz

Gauche

$J = 5\text{–}14$ Hz

Anti

Vicinal protons
J varies with dihedral angle

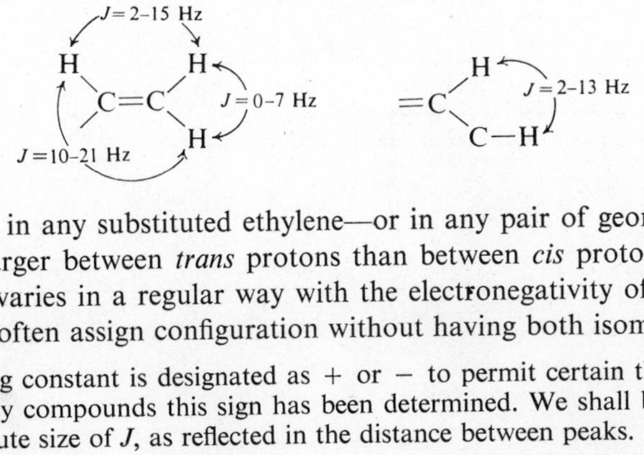

$J = 2\text{–}15$ Hz

$J = 0\text{–}7$ Hz

$J = 10\text{–}21$ Hz

$J = 2\text{–}13$ Hz

Vinylic protons

For example, in any substituted ethylene—or in any pair of geometric isomers—J is always larger between *trans* protons than between *cis* protons; furthermore, the size of J varies in a regular way with the electronegativity of substituents, so that one can often assign configuration without having both isomers in hand.

A coupling constant is designated as + or − to permit certain theoretical correlations; for many compounds this sign has been determined. We shall be concerned only with the absolute size of J, as reflected in the distance between peaks.

Although we shall not work very much with the *values* of coupling constants, we should realize that, to an experienced person, they can often be the most important feature of an nmr spectrum: the feature that gives exactly the kind of information about molecular structure that is being looked for.

Problem 13.10 Go back to Problem 13.8 (p. 425), and tell, where you can, the kind of splitting expected for each signal.

Problem 13.11 In Problem 13.9 (p. 425) you analyzed some nmr spectra. Does the absence of splitting in these spectra now lead you to change any of your answers?

Problem 13.12 Give a structure or structures consistent with each of the nmr spectra shown in Fig. 13.16 (p. 436).

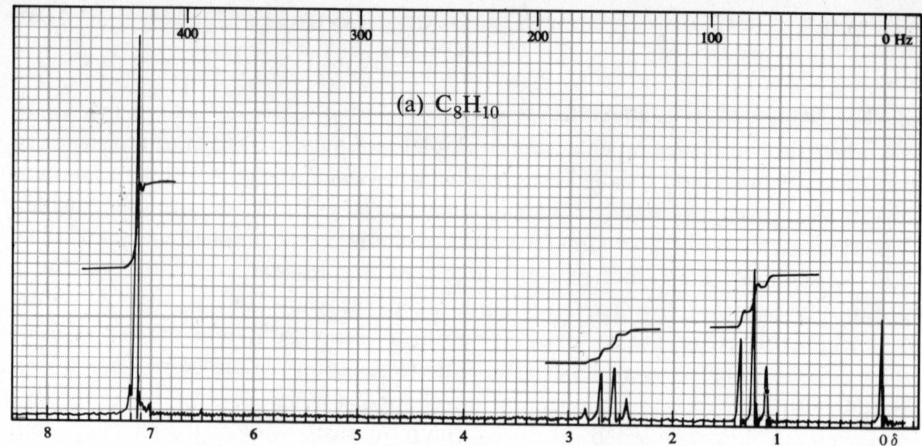

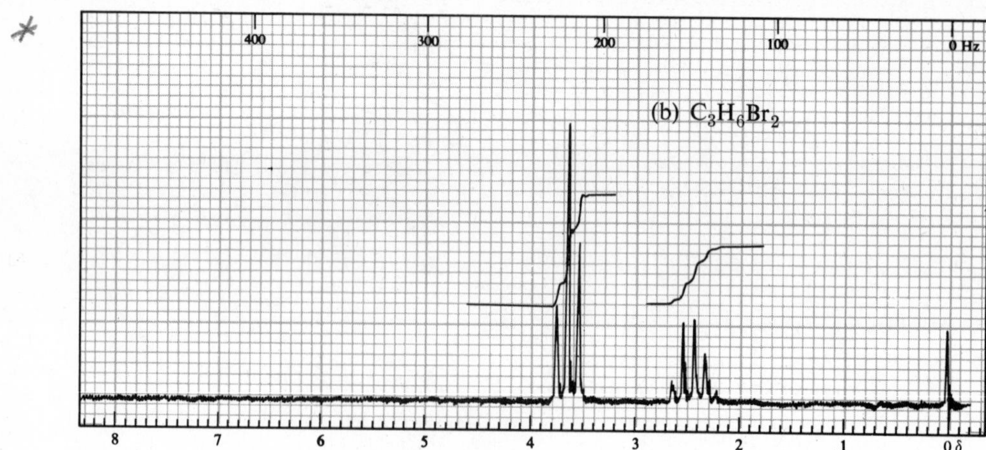

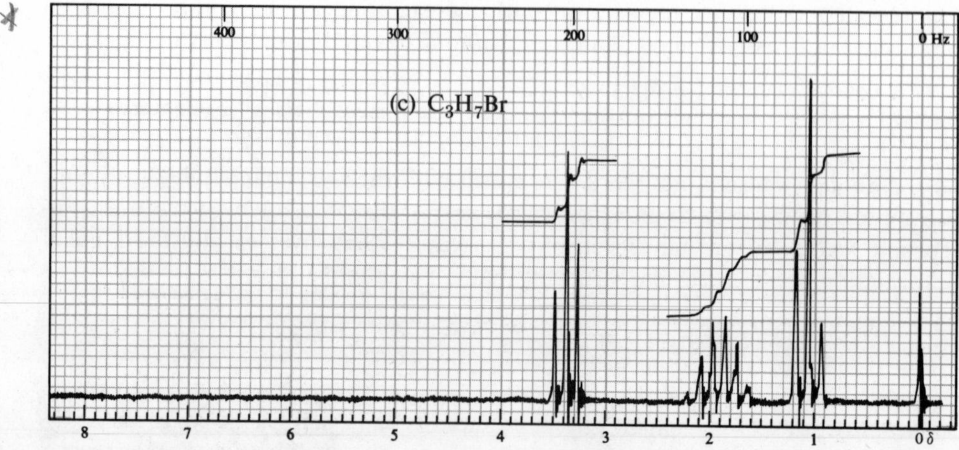

Figure 13.16. Nmr spectra for Problem 13.12, p. 435.

13.12 Nmr. Complicated spectra. Deuterium labeling

Most nmr spectra that the organic chemist is likely to encounter are considerably more complicated than the ones given in this book. How are these analyzed?

First of all, many spectra showing a large number of peaks can be completely analyzed by the same general methods we shall use here. It just takes practice.

Then again, in many cases complete analysis is not necessary for the job at hand. Evidence of other kinds may already have limited the number of possible structures, and all that is required of the nmr spectrum is that it let us choose among these. Sometimes all that we need to know is how many kinds of protons there are—or, perhaps, how many kinds and how many of each kind. Sometimes

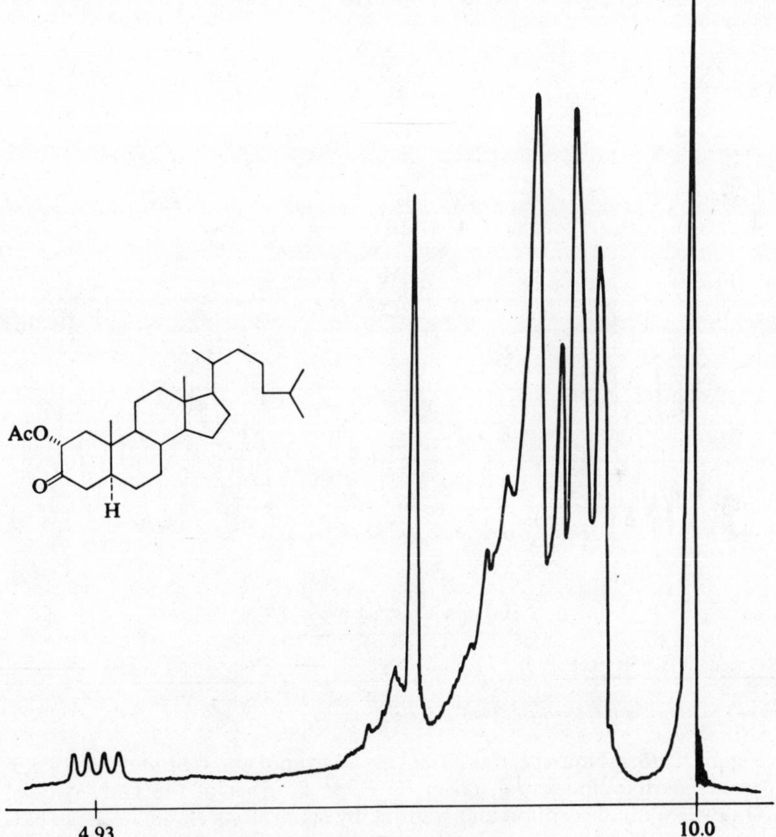

Courtesy of *The Journal of the American Chemical Society*

Figure 13.17. Nmr spectrum of 2-α-acetoxycholestane-3-one, taken by K. L. Williamson and W. S. Johnson of the University of Wisconsin and Stanford University. The four downfield peaks are due to the proton on C–2, whose signal is split successively by the axial proton and the equatorial proton on C–1.

only one structural feature is still in doubt—for example, does the molecule contain two methyl groups or one ethyl group?—and the answer is given in a set of peaks standing clear from the general confusion. (See, for example, Fig. 13.17, above.)

Instrumental techniques are available, and others are being rapidly developed,

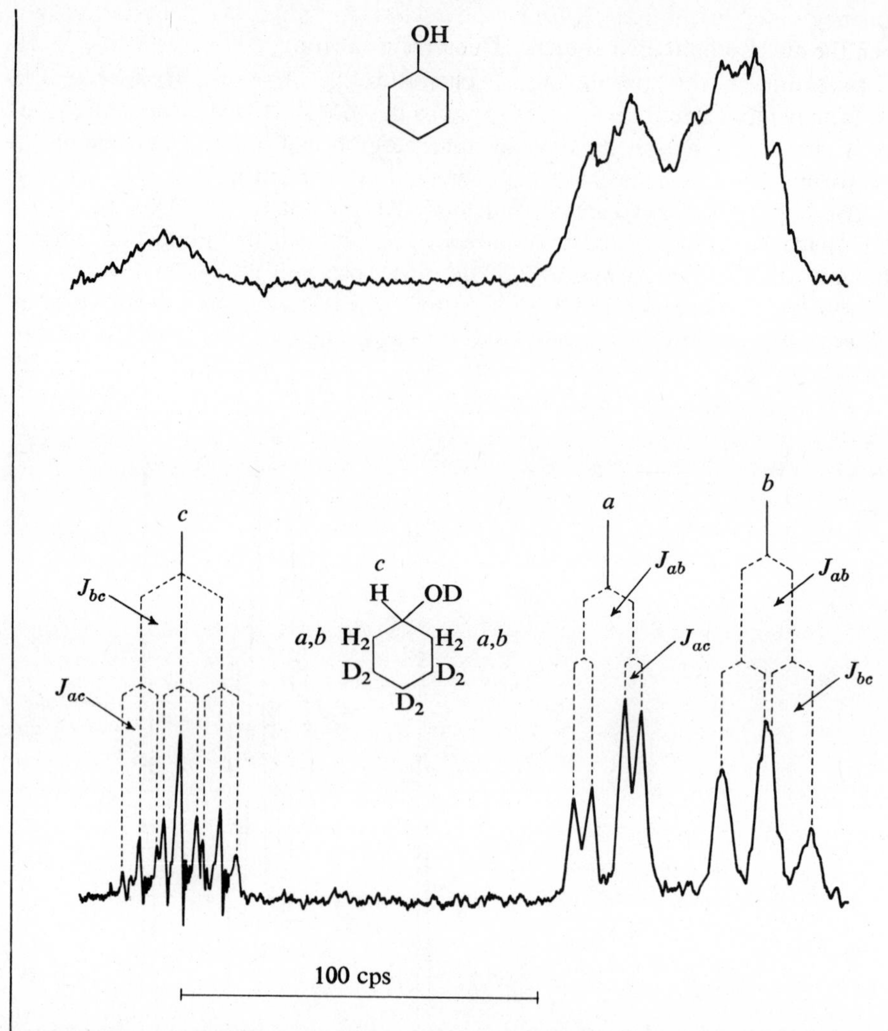

Courtesy of *The Journal of the American Chemical Society*

Figure 13.18. Nmr spectra of (*top*) cyclohexanol and (*bottom*) 3,3,4,4,5,5-hexadeuteriocyclohexanol, taken by F. A. L. Anet of the University of Ottawa. With absorption and splitting by six protons eliminated, the pattern due to the five remaining protons can be analyzed.

The diastereotopic sets of protons, H_a and H_b, give different signals. Signal *a* is split successively into doublets by H_b (only *one* H_b splits each H_a) and by H_c. Signal *b* is split similarly by H_a and H_c. Downfield signal *c* is split successively into triplets by H_a (both protons) and H_b (both protons).

to help in the analysis of complicated spectra, and to simplify the spectra actually measured. By the method of *double resonance* (or *double irradiation*), for example, the spins of two sets of protons can be *decoupled*, and a simpler spectrum obtained.

The molecule is irradiated with two radiofrequency beams: the usual one, whose absorption is being measured; and a second, much stronger beam, whose frequency differs from that of the first in such a way that the following happens. When the field strength is reached at which the proton we are interested in absorbs and gives a signal, the splitting protons are absorbing the other, very strong radiation. These splitting protons are "stirred up" and flip over very rapidly—so rapidly that the signalling proton sees them, not in the various combinations of spin alignments (Sec. 13.10), but in a single *average* alignment. The spins are decoupled, and the signal appears as a single, unsplit peak.

A particularly elegant way to simplify an nmr spectrum—and one that is easily understood by an organic chemist—is the use of *deuterium labeling.*

Because a deuteron has a much smaller magnetic moment than a proton, it absorbs at a much higher field and so gives no signal in the proton nmr spectrum. Furthermore, its coupling with a proton is weak and it ordinarily broadens, but does not split, a proton's signal; even this effect can be eliminated by double irradiation.

As a result, then, the replacement of a proton by a deuteron removes from an nmr spectrum both the signal from that proton and the splitting by it of signals of other protons; it is as though there were no hydrogen at all at that position in the molecule. For example:

CH_3—CH_2—	CH_2D—CH_2—	CH_3—CHD—
Triplet Quartet	Triplet Triplet	Doublet Quartet
3H *2H*	*2H* *2H*	*3H* *1H*

One can use deuterium labeling to find out which signal is produced by which proton or protons: one observes the disappearance of a particular signal when a proton in a known location is replaced by deuterium. One can use deuterium labeling to simplify a complicated spectrum so that a certain set of signals can be seen more clearly: see, for example, Fig. 13.18, p. 438. (This figure also illustrates a point made at the beginning of this section: the formidable looking nine-peak multiplet is analyzed without too much difficulty.)

13.13 Equivalence of protons: a closer look

We have seen that equivalence—or non-equivalence—of protons is fundamental to the nmr spectrum, since it affects both the number of signals and their splitting. Let us look more closely at equivalence, and see how it is affected by the rate at which certain molecular changes occur:

(a) *rotations about single bonds,* as in the interconversion between conformations of substituted ethanes or cyclohexanes;

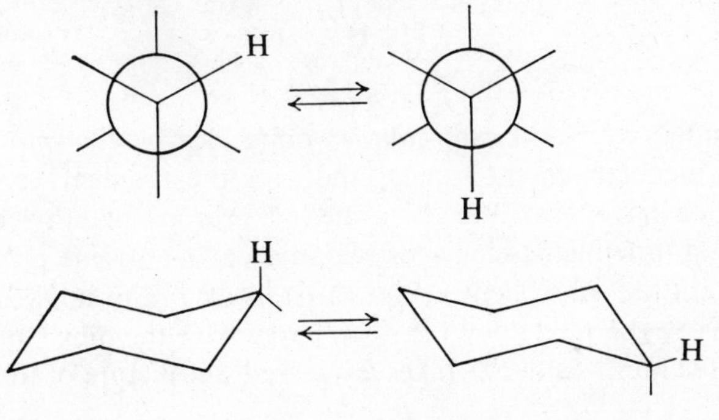

H ⇌ H

H ⇌ H

(b) *inversion of molecules*, that is, the turning inside out of pyramidal molecules like amines (Sec. 22.6);

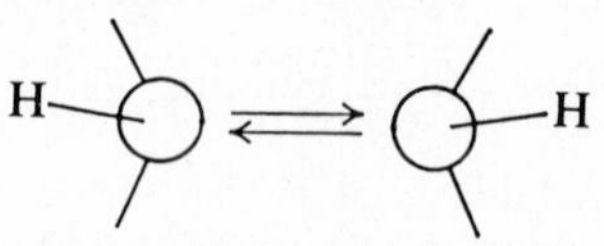

$$\text{H—N} \rightleftharpoons \text{N—H}$$

(c) *proton exchange*, as, for example, of alcohols (Sec. 16.13).

$$\text{R*O—H* + RO—H} \rightleftharpoons \text{R*O—H + RO—H*}$$

Each of these molecular changes can change the environment—both electronic and protonic—of a given proton, and hence can affect both its chemical shift and its coupling with other protons. The basic question that arises is whether or not the nmr spectrometer sees the proton in *each* environment or in an *average of all* of them. The answer is, in short, that it can often see the proton in either way, depending upon the temperature, and in this ability lies much of the usefulness of nmr spectroscopy.

In comparing it with other spectrometers, Professor John D. Roberts of the California Institute of Technology has likened the nmr spectrometer to a camera with a relatively long shutter time—that is, to a "slow" camera. Such a camera photographs the spokes of a wheel in different ways depending upon the speed with which the wheel spins: as sharp, individual spokes if spinning is slow; as blurred spokes if spinning is faster; and as a single circular smear if spinning is faster yet. In the same way, if the molecular change is relatively fast, the nmr spectrometer sees a proton in its average environment—a smeared-out picture; if the molecular process is slow, the spectrometer sees the proton in each of its environments.

In this section we shall examine the effects of rotations about single bonds on the nmr spectrum, and in later sections the effects of the other molecular changes.

Let us return to ethyl chloride (Sec. 13.7), and focus our attention on the methyl protons. If, at any instant, we could look at an individual molecule, we would almost certainly see it in conformation I. One of the methyl protons is *anti* to

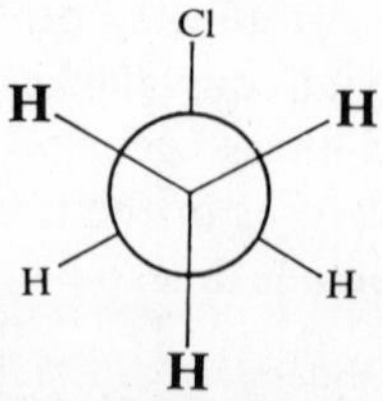

Cl
H H
H H
H
I

the chlorine and two protons are *gauche*; quite clearly, the *anti* proton is in a different environment from the others, and—for the moment—is not equivalent to them. Yet, we have seen, the three methyl protons of ethyl chloride give a single nmr signal (a triplet, because of the adjacent methylene group), and hence must be magnetically equivalent. How can this be? The answer is, of course, that rotation about the single bond is—compared with the nmr "shutter speed"—a fast process; the nmr "camera" takes a smeared-out picture of the three protons.

Each proton is seen in an *average* environment, which is exactly the same as the average environment of each of the other two: one-third *anti*, and two-thirds *gauche*.

There are three conformations of ethyl chloride, II, III, and IV, identical except that a different individual proton occupies the *anti* position. Being of equal stability, the three conformations are exactly equally populated: one-third of the molecules in each. In one of these conformations a given proton is *anti* to chlorine, and in two it is *gauche*.

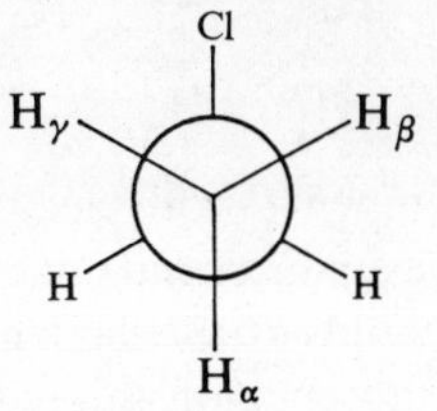

II 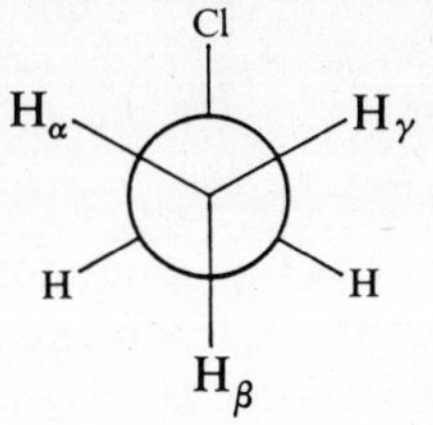 III 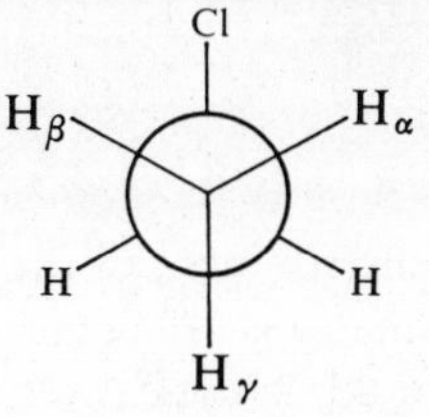 IV

1,1,2-Trichloroethane, to take another example, presents a somewhat different conformational picture, but the net result is the same: identical average environments and hence equivalence for the two methylene protons.

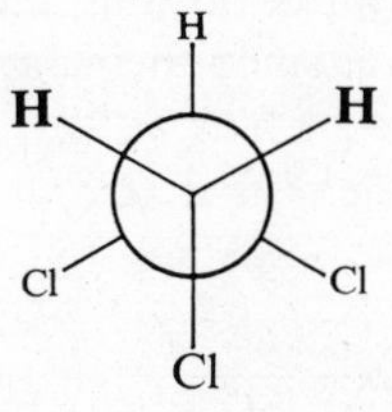

V 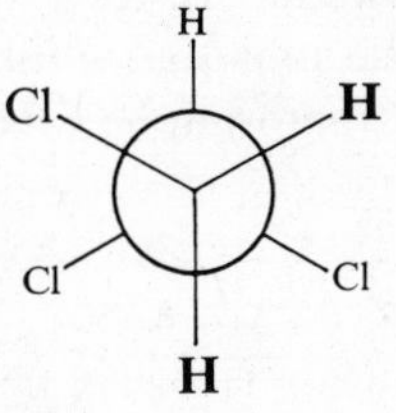 VI 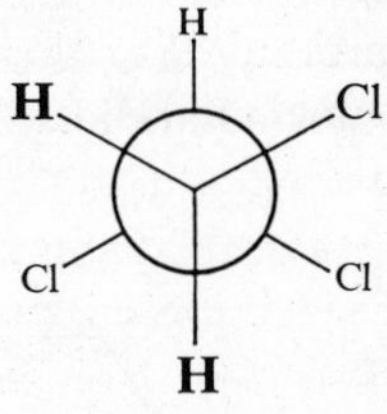 VII

The environments of the two protons are the same in V. The environments are different for the two in VI and VII, but average out the same because of the equal populations of these enantiomeric conformations. (Here, however, we cannot say just *what* the average environment is, unless we know the ratio of V to the racemic modification (VI plus VII).)

With diastereotopic protons, on the other hand, the situation is different: diastereotopic protons are non-equivalent and no rotation will change this. We decided (Sec. 13.7) that the two C–1 protons of 1,2-dichloropropane, $CH_3CHClCH_2Cl$, are diastereotopic, since replacement of either one by an atom Z would yield diastereomers:

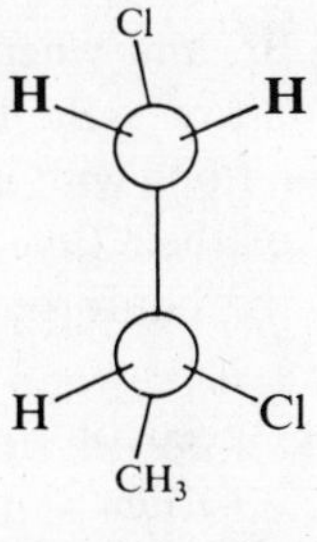

1,2-Dichloropropane

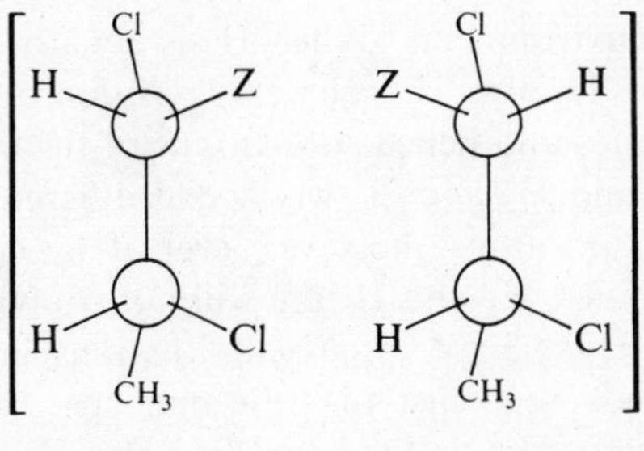

Rotation cannot interconvert the diastereomers, nor can it make the protons, H_a and H_b, equivalent. In none of the conformations (VIII, IX, or X)

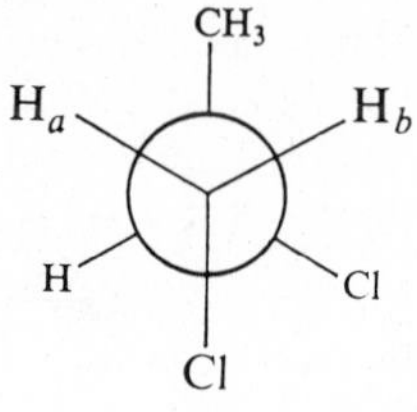

VIII

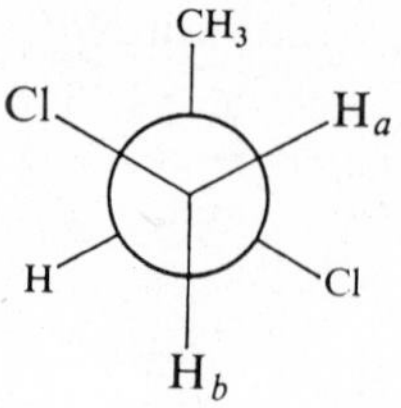

IX

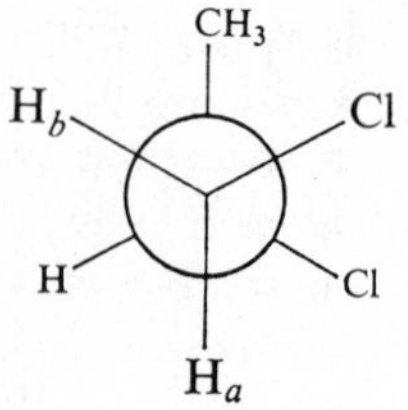

X

is the environment of the two protons the same; nor is there a pair of mirror-image conformations to balance out their environments. (This holds true whether the compound is optically active or inactive; the presence or absence of an enantiomeric molecule has no effect on the environment of a proton in any individual molecule.) These diastereotopic protons give different signals, couple with the proton on C–2 (with different coupling constants), and couple with each other.

Cyclohexane presents an exactly analogous situation, since the transformation of one chair form into another involves rotations about single bonds. In any chair conformation there are two kinds of protons: six equatorial protons and six axial protons. Yet there is a single nmr signal for all twelve, since their *average* environments are identical: half equatorial, half axial.

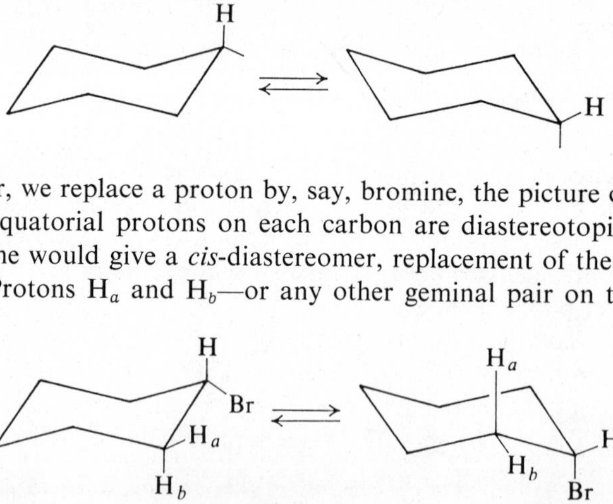

If, however, we replace a proton by, say, bromine, the picture changes. Now, the axial and equatorial protons on each carbon are diastereotopic protons: replacement of one would give a *cis*-diastereomer, replacement of the other a *trans*-diastereomer. Protons H_a and H_b—or any other geminal pair on the ring—have

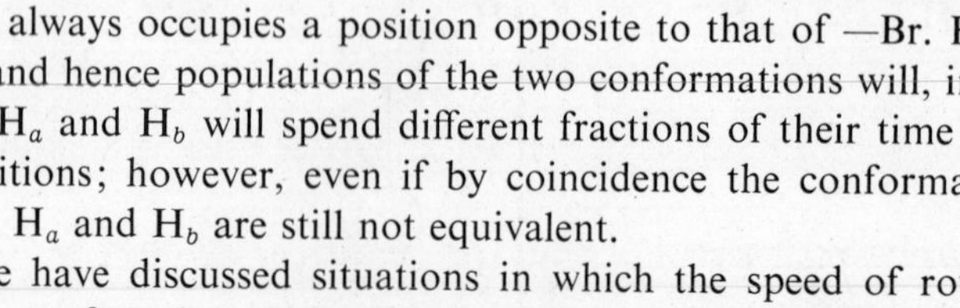

different environments. When H_a is equatorial, so is —Br, and when H_a is axial, so is —Br; H_b always occupies a position opposite to that of —Br. Furthermore, the stabilities and hence populations of the two conformations will, in general, be different, and H_a and H_b will spend different fractions of their time in axial and equatorial positions; however, even if by coincidence the conformations are of equal stability, H_a and H_b are still not equivalent.

So far, we have discussed situations in which the speed of rotation about single bonds is so fast that the nmr spectrometer sees protons in their average environment. This is the *usual* situation. It is this situation in which our earlier test for equivalence would work: if replacement of either of two protons by Z would give the same (or enantiomeric) products, the protons are equivalent. We ignore conformers in judging the identity of two products.

Now, if—by lowering the temperature—we could sufficiently slow down rotations about single bonds, we would expect an nmr spectrum that reflects the "instantaneous" environments of protons in each conformation. *This is exactly what happens.* As cyclohexane, for example, is cooled down, the single sharp peak observed at room temperature is seen to broaden and then, at about $-70°$, to split into two peaks, which at $-100°$ are clearly separated: one peak is due to axial protons, and the other peak is due to equatorial protons.

This does *not* mean that the molecule is frozen into a single conformation; it still flips back and forth between two (equivalent) chair conformations; a given proton is axial one moment and equatorial the next. It is just that now the time between interconversions is long enough that we "photograph" the molecule, not as a blur but sharply as one conformation or the other.

By study of the broadening of the peak, or of the coalescence of the two peaks, it is possible to estimate the E_{act} for rotation. Indeed, it was by this method that the barrier of 11 kcal/mole (Sec. 9.11) was calculated.

Problem 13.13 The fluorine nmr spectrum (Sec. 13.10) of 1,2-difluorotetrachloroethane, $CFCl_2CFCl_2$, shows a single peak at room temperature, but at $-120°$ shows two peaks (singlets) of unequal area. Interpret each spectrum, and account for the difference. What is the significance of the unequal areas of the peaks in the low-temperature spectrum? Why is there no splitting in either spectrum?

Problem 13.14 At room temperature, the fluorine nmr spectrum of CF_2BrCBr_2CN (3,3-difluoro-2,2,3-tribromopropanenitrile) shows a single sharp peak. As the temperature is lowered this peak broadens and, at $-98°$, is split into two doublets (equal spacing) and a singlet. The combined area of the doublets is considerably larger than—more than twice as large as—the area of the singlet. Interpret each spectrum, and account for the relative peak areas in the low-temperature spectrum.

13.14 The electron spin resonance (esr) spectrum

Let us consider a free radical placed in a magnetic field and subjected to electromagnetic radiation; and let us focus our attention, not on the nuclei, but on the odd, unpaired electron. This electron spins and thus generates a magnetic moment, which can be lined up with or against the external magnetic field. Energy is required to change the spin state of the electron, from alignment with the field to the less stable alignment, against the field. This energy is provided by absorption of radiation of the proper frequency. An absorption spectrum is produced, which is called an *electron spin resonance* (*esr*) *spectrum* or an *electron paramagnetic resonance* (*epr*) *spectrum*.

The esr spectrum is thus analogous to the nmr spectrum. An electron has, however, a much larger magnetic moment than the nucleus of a proton, and more energy is required to reverse the spin. In a field of 3200 gauss, for example, where nmr absorption would occur at about 14 MHz, esr absorption occurs at a much higher frequency: 9000 MHz, in the *microwave* region.

Like nmr signals, esr signals show splitting, and from exactly the same cause, coupling with the spins of certain nearby nuclei: for example, protons near carbon atoms that carry—or help to carry—the odd electron. For this reason, esr spectroscopy can be used not only to detect the presence of free radicals and to measure their concentration, but also to give evidence about their structure: what free radicals they are, and how the odd electron is spread over the molecule.

Problem 13.15 Although all electrons spin, only molecules containing unpaired electrons—only free radicals—give esr spectra. Why is this? (*Hint:* Consider the possibility (a) that one electron of a pair has its spin reversed, or (b) that both electrons of a pair have their spins reversed.)

Problem 13.16 In each of the following cases, tell what free radical is responsible for the esr spectrum, and show how the observed splitting arises. (a) X-irradiation of methyl iodide at low temperatures: a four-line signal. (b) γ-irradiation at 77°K of propane and of *n*-butane: symmetrical signals of, respectively, 8 lines and 7 lines. (c) Triphenylmethyl chloride + zinc: a very complex signal.

13.15 Spectroscopic analysis of hydrocarbons. Infrared spectra

In this first encounter with infrared spectra, we shall see absorption bands due to vibrations of carbon–hydrogen and carbon–carbon bonds: bands that will constantly reappear in all the spectra we meet, since along with their various functional groups, compounds of all kinds contain carbon and hydrogen. We must expect to find these spectra complicated and, at first, confusing. Our aim is to learn to pick out of the confusion those bands that are most characteristic of certain structural features.

Let us look first at the various kinds of vibration, and see how the positions of the bands associated with them vary with structure.

Bands due to *carbon–carbon stretching* may appear at about 1500 and 1600 cm^{-1} for aromatic bonds, at 1650 cm^{-1} for double bonds (shifted to about 1600 cm^{-1} by conjugation), and at 2100 cm^{-1} for triple bonds. These bands, however, are often unreliable. (They may disappear entirely for fairly symmetrically substituted alkynes and alkenes, because the vibrations do not cause the change in dipole moment that is essential for infrared absorption.) More generally useful bands are due to the various carbon–hydrogen vibrations.

Absorption due to *carbon–hydrogen stretching*, which occurs at the high-frequency end of the spectrum, is characteristic of the hybridization of the carbon holding the hydrogen: at 2800–3000 cm^{-1} for tetrahedral carbon; at 3000–3100 cm^{-1} for trigonal carbon (alkenes and aromatic rings); and at 3300 cm^{-1} for digonal carbon (alkynes).

Absorption due to various kinds of *carbon–hydrogen bending*, which occurs at lower frequencies, can also be characteristic of structure. Methyl and methylene groups absorb at about 1430–1470 cm^{-1}; for methyl, there is another band, quite characteristic, at 1375 cm^{-1}. The isopropyl "split" is characteristic: a doublet, with equal intensity of the two peaks, at 1370 and 1385 cm^{-1} (confirmed by a band at 1170 cm^{-1}). *tert*-Butyl gives an unsymmetrical doublet: 1370 cm^{-1} (*strong*) and 1395 cm^{-1} (*moderate*).

Carbon–hydrogen bending in alkenes and aromatic rings is both in-plane and out-of-plane, and of these the latter kind is more useful. For **alkenes**, out-of-plane bending gives strong bands in the 800–1000 cm^{-1} region, the exact location depending upon the nature and number of substituents, and the stereochemistry:

$RCH{=}CH_2$	910–920 cm^{-1} 990–1000	*cis*-$RCH{=}CHR$	675–730 cm^{-1} (*variable*)
$R_2C{=}CH_2$	880–900	*trans*-$RCH{=}CHR$	965–975

For **aromatic rings**, out-of-plane C—H bending gives strong absorption in the 675–870 cm^{-1} region, the exact frequency depending upon the number and location of substituents; for many compounds absorption occurs at:

monosubstituted	690–710 cm^{-1} 730–770	*m*-disubstituted	690–710 cm^{-1} 750–810
o-disubstituted	735–770	*p*-disubstituted	810–840

Now, what do we look for in the infrared spectrum of a hydrocarbon? To begin with, we can rather readily tell whether the compound is aromatic or purely aliphatic. The spectra in Fig. 13.2 (p. 411) show the contrast that is typical: aliphatic absorption is strongest at higher frequency and is essentially missing below 900 cm^{-1}; aromatic absorption is strong at lower frequencies (C—H out-of-plane bending) between 650 and 900 cm^{-1}. In addition, an aromatic ring will show C—H stretching at 3000–3100 cm^{-1}; often, there is carbon–carbon stretching at 1500 and 1600 cm^{-1} and C—H in-plane bending in the 1000–1100 cm^{-1} region.

An alkene shows C—H stretching at 3000–3100 cm^{-1} and, most characteristically, strong out-of-plane C—H bending between 800–1000 cm^{-1}, as discussed above.

A terminal alkyne, $RC{\equiv}CH$, is characterized by its C—H stretching band, a strong and sharp band at 3300 cm^{-1}, and by carbon–carbon stretching at 2100 cm^{-1}. A disubstituted alkyne, on the other hand, does not show the 3300 cm^{-1} band and, if the two groups are fairly similar, the 2100 cm^{-1} band may be missing, too.

Some of these characteristic bands are labeled in the spectra of Fig. 13.2, page 411.

Problem 13.17 What is a likely structure for a hydrocarbon of formula C_6H_{12} that shows strong absorption at 2920 and 2840 cm^{-1}, and at 1450 cm^{-1}; none above 2920 cm^{-1}; and below 1450 cm^{-1} none until about 1250 cm^{-1}?

13.16 Spectroscopic analysis of hydrocarbons. Nmr

The application of nmr spectroscopy to hydrocarbons needs no special discussion beyond that already given in Secs. 13.6–13.11. For hydrocarbons as for other kinds of compounds, we shall find that where the infrared spectrum helps to tell us what *kind* of compound we are dealing with, the nmr spectrum will help to tell us *what* compound.

About Analyzing Spectra

In problems you will be given the molecular formula of a compound and asked to deduce its structure from its spectroscopic properties: sometimes from its infrared or nmr spectrum alone, sometimes from both. The compound will generally be a simple one, and you may need to look only at a few features of the spectra to find the answer. To confirm your answer, however, and to gain experience, see how much information you can get from the spectra: try to identify as many infrared bands as you can, to assign all nmr signals to specific protons, and to analyze the various spin–spin splittings. Above all, look at as many spectra as you can find: in the laboratory, in other books, in catalogs of spectra in the library.

PROBLEMS

1. Give a structure or structures consistent with each of the following sets of nmr data.

(a) $C_3H_3Cl_5$
a triplet, δ 4.52, 1H
b doublet, δ 6.07, 2H

(b) $C_3H_5Cl_3$
a singlet, δ 2.20, 3H
b singlet, δ 4.02, 2H

(c) C_4H_9Br
a doublet, δ 1.04, 6H
b multiplet, δ 1.95, 1H
c doublet, δ 3.33, 2H

(d) $C_{10}H_{14}$
a singlet, δ 1.30, 9H
b singlet, δ 7.28, 5H

(e) $C_{10}H_{14}$
a doublet, δ 0.88, 6H
b multiplet, δ 1.86, 1H
c doublet, δ 2.45, 2H
d singlet, δ 7.12, 5H

(f) C_9H_{10}
a quintet, δ 2.04, 2H
b triplet, δ 2.91, 4H
c singlet, δ 7.17, 4H

(g) $C_{10}H_{13}Cl$
a singlet, δ 1.57, 6H
b singlet, δ 3.07, 2H
c singlet, δ 7.27, 5H

(h) $C_{10}H_{12}$
a multiplet, δ 0.65, 2H
b multiplet, δ 0.81, 2H
c singlet, δ 1.37, 3H
d singlet, δ 7.17, 5H

(i) $C_9H_{11}Br$
a quintet, δ 2.15, 2H
b triplet, δ 2.75, 2H
c triplet, δ 3.38, 2H
d singlet, δ 7.22, 5H

(j) $C_3H_5ClF_2$
a triplet, δ 1.75, 3H
b triplet, δ 3.63, 2H

2. Identify the stereoisomeric 1,3-dibromo-1,3-dimethylcyclobutanes on the basis of their nmr spectra.

Isomer X: singlet, δ 2.13, 6H
singlet, δ 3.21, 4H

Isomer Y: singlet, δ 1.88, 6H
doublet, δ 2.84, 2H
doublet, δ 3.54, 2H
doublets have equal spacing

3. When mesitylene (nmr spectrum, Fig. 13.5, p. 422) is treated with HF and SbF_5 in liquid SO_2 solution, the following peaks, all singlets, are observed in the nmr spectrum; δ 2.8, 6H; δ 2.9, 3H; δ 4.6, 2H; and δ 7.7, 2H. To what compound is the spectrum due? Assign all peaks in the spectrum.

Of what general significance to chemical theory is such an observation?

4. (a) On catalytic hydrogenation, compound A, C_5H_8, gave *cis*-1,2-dimethylcyclopropane. On this basis, three isomeric structures were considered possible for A. What were they? (b) Absence of infrared absorption at 890 cm^{-1} made one of the structures unlikely. Which one was it? (c) The nmr spectrum of A showed signals at δ 2.22 and δ 1.04 with intensity ratio 3:1. Which of the three structures in (a) is consistent with this? (d) The base peak in the mass spectrum was found at *m/e* 67. What ion was this peak probably due to, and how do you account for its abundance? (e) Compound A was synthesized in one step from open-chain compounds. How do you think this was done?

5. X-ray analysis shows that the [18]annulene (Problem 9, p. 336, $n = 9$) is planar. The nmr spectrum shows two broad bands: τ 1.1 and τ 11.8, peak area ratio 2:1. (a) Are these properties consistent with aromaticity? Explain. (b) Would you have predicted aromaticity for this compound? Explain. (*Hint:* Carefully draw a structural formula for the compound, keeping in mind bond angles and showing all hydrogen atoms.)

6. Hydrocarbon B, C_6H_6, gave an nmr spectrum with two signals: δ 6.55 and δ 3.84, peak area ratio 2:1. When warmed in pyridine for three hours, B was quantitatively converted into benzene.

Mild hydrogenation of B yielded C, whose spectra showed the following: mass spectrum, mol. wt. 82; infrared spectrum, no double bonds; nmr spectrum, one broad peak at δ 2.34.

(a) How many rings are there in C? (See Problem 9.17, p. 313.) (b) How many rings are there (probably) in B? How many double bonds in B? (c) Can you suggest a structure for B? for C?

(d) In the nmr spectrum of B, the upfield signal was a quintet, and the downfield signal was a triplet. How must you account for these splittings?

7. The five known 1,2,3,4,5,6-hexachlorocyclohexanes can be described in terms of the equatorial (e) or axial (a) disposition of successive chlorines: eeeeee, eeeeea, eeeeaa, eeaeea, eeeaaa. Their nmr spectra have been measured.

Which of these would give: (a) only one peak (two isomers); (b) two peaks, 5H:1H (one isomer); (c) two peaks, 4H:2H (two isomers)?

(d) Which one of the isomers in (a) would you expect to show no change in nmr spectrum at low temperature? Which one would show a split into two peaks? Predict the relative peak areas for the latter case.

8. (a) Although the nmr spectrum of *trans*-4-*tert*-butyl-1-bromocyclohexane is complicated, the signal from one proton stands clear (δ 3.83), downfield from the rest. Which proton is this, and why? (b) The *cis*-isomer shows a corresponding peak, but at δ 4.63. Assuming that the *tert*-butyl group exerts no direct magnetic effect, to what do you attribute the difference in chemical shifts between the two spectra? These data are typical, and are the basis of a generalization relating conformation and chemical shift. What is that generalization?

9. The nmr spectrum of bromocyclohexane shows a downfield peak (1H) at δ 4.16. This signal is a single peak at room temperature, but at $-75°$ separates into two peaks of *unequal* area (but totalling *one* proton): δ 3.97 and δ 4.64 in the ratio 4.6:1.0. How do you account for the separation of peaks? On the basis of your generalization of the previous problem, which conformation of the molecule predominates, and (at $-75°$) what percentage of molecules does it account for?

10. Give a structure or structures consistent with each of the infrared spectra in Fig. 13.19, page 448.

11. Give a structure or structures consistent with each of the nmr spectra in Fig. 13.20, page 449.

12. Give a structure or structures consistent with each of the nmr spectra in Fig. 13.21, page 450.

13. Give a structure or structures of the compound D, whose infrared and nmr spectra are shown in Fig. 13.22, p. 451.

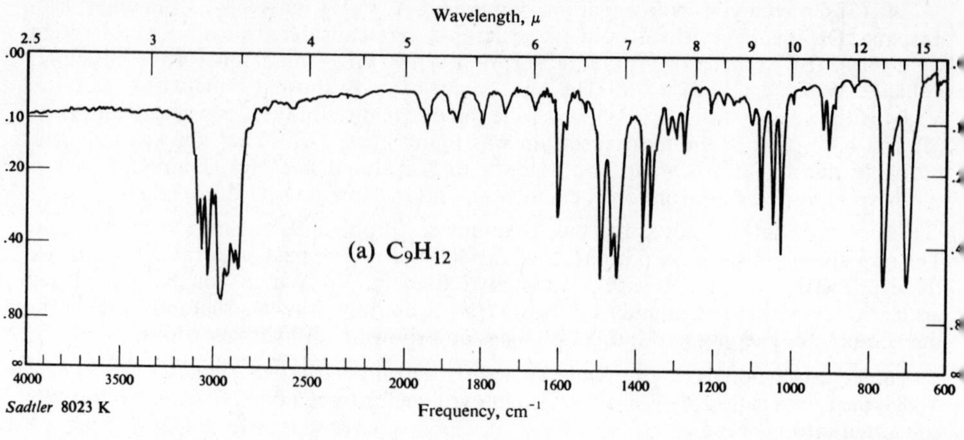

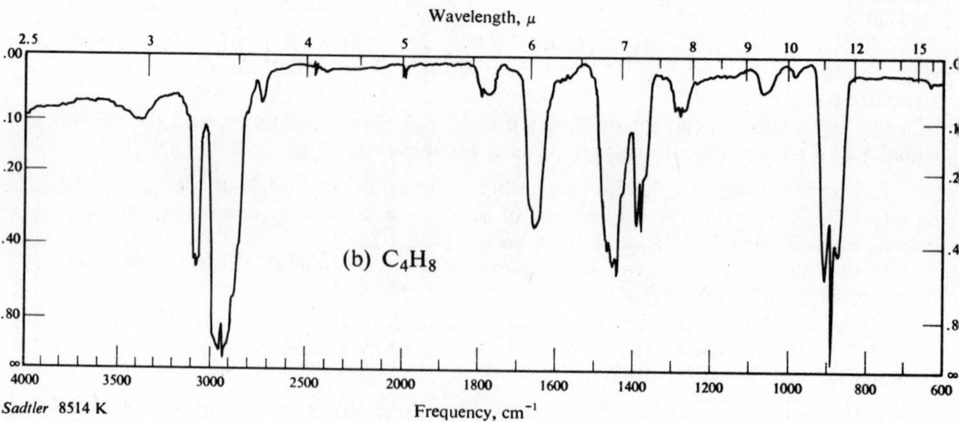

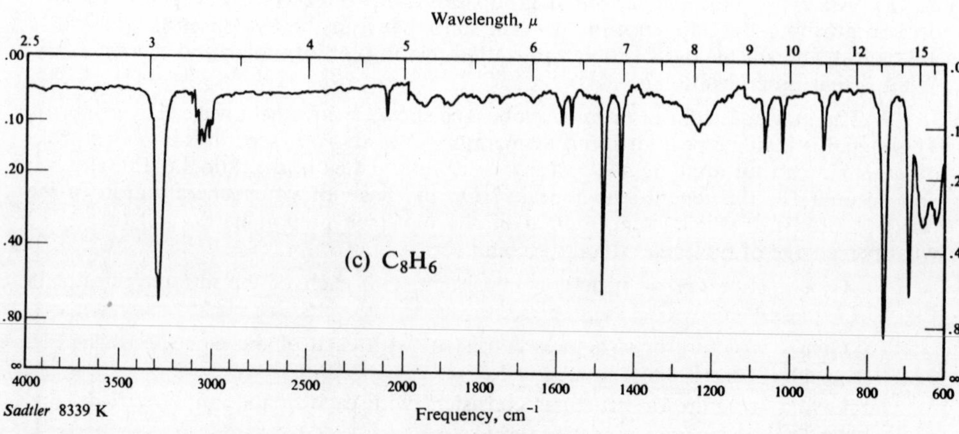

Figure 13.19. Infrared spectra for Problem 10, p. 447.

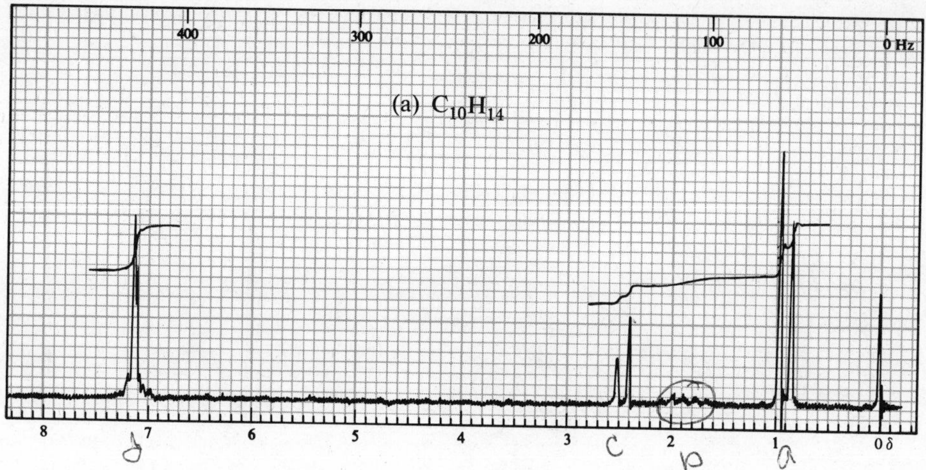

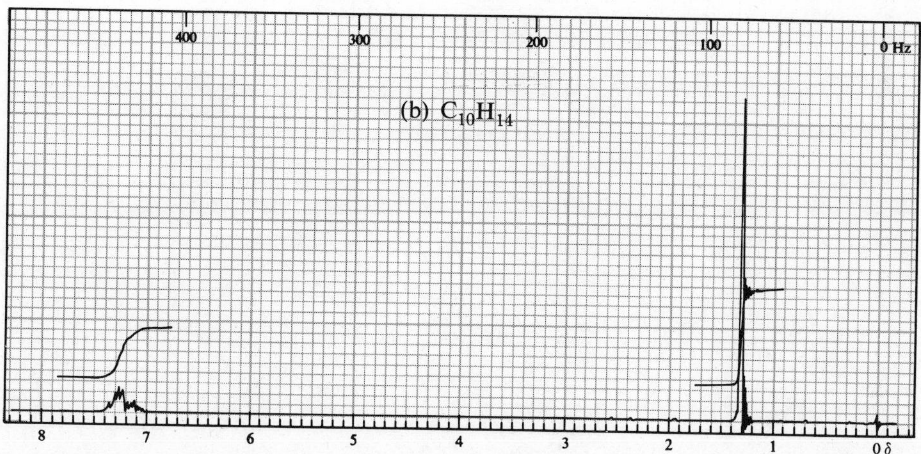

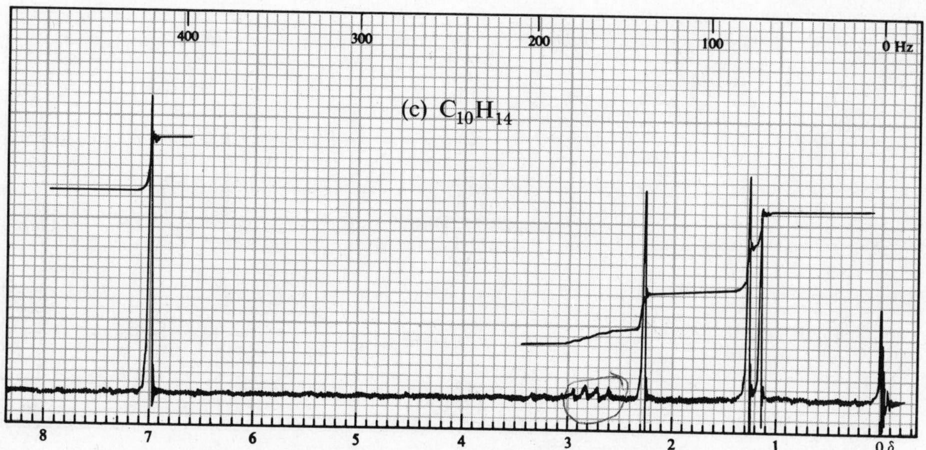

Figure 13.20. Nmr spectra for Problem 11, p. 447.

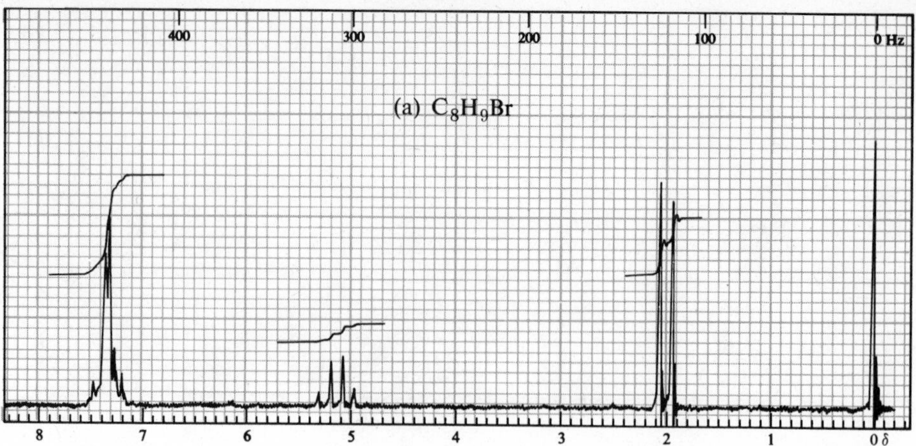

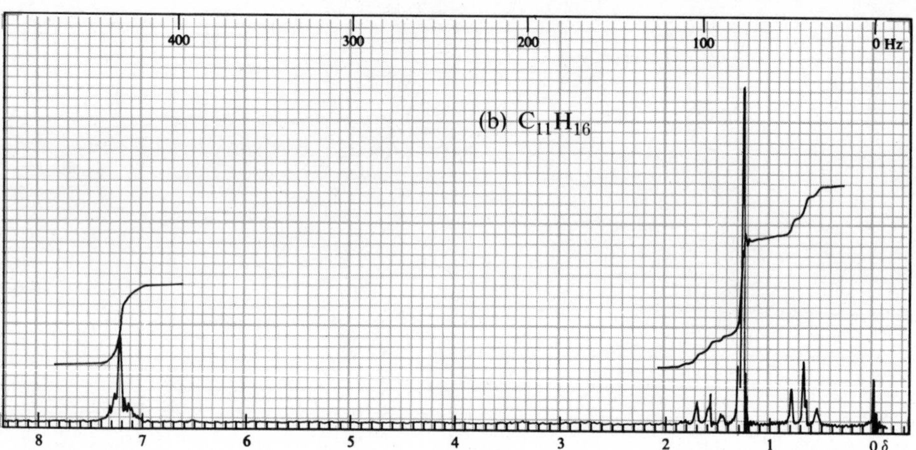

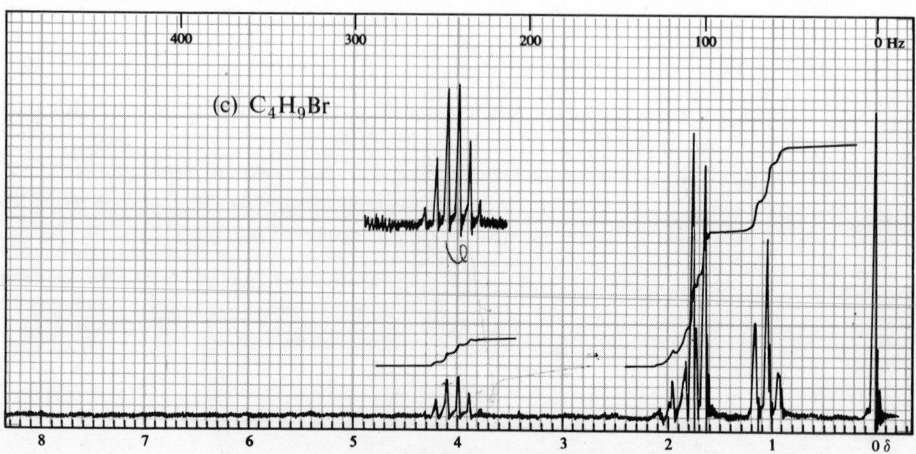

Figure 13.21. Nmr spectra for Problem 12, p. 447.

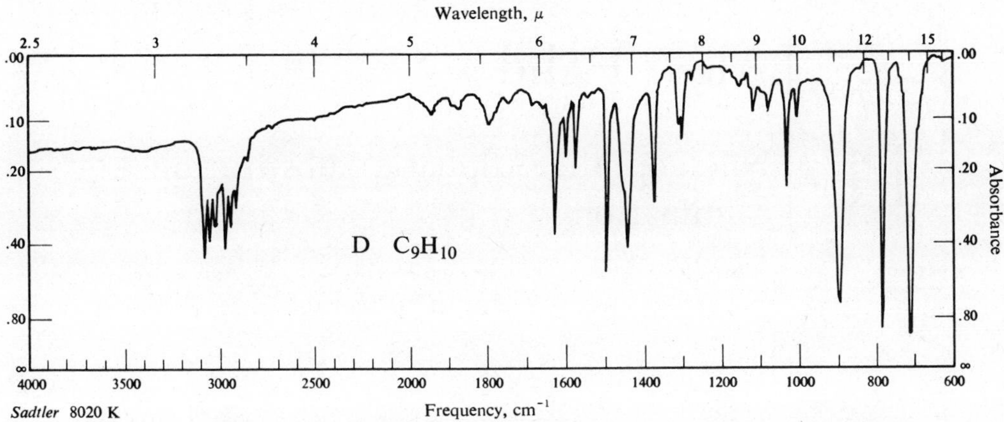

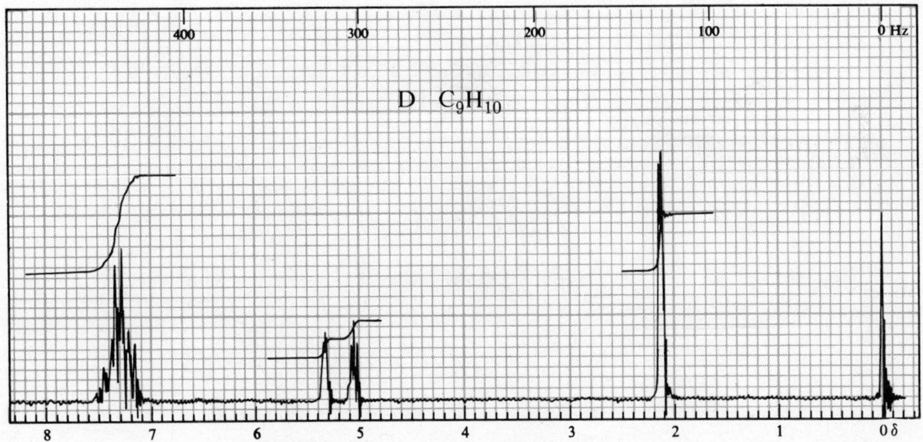

Figure 13.22. Infrared and nmr spectra for compound D, Problem 13, p. 447.

Chapter 14 Alkyl Halides

Nucleophilic Aliphatic Substitution Elimination

14.1 Structure and nomenclature

We shall consider as alkyl halides all compounds of the general formula R—X, where R is any simple alkyl or substituted alkyl group. For example:

$$CH_3-\underset{Cl}{\overset{CH_3}{C}}-CH_3$$

tert-Butyl chloride
2-Chloro-2-methylpropane

$$H_2C{=}CH-CH_2Br$$

Allyl bromide
3-Bromo-1-propene

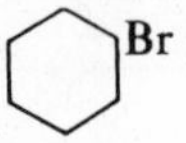

Cyclohexyl bromide

$$C_6H_5CH_2Cl$$

Benzyl chloride

$$O_2NC_6H_4CH_2Br$$

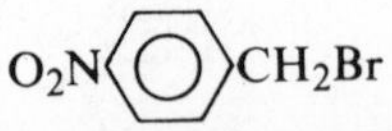

p-Nitrobenzyl bromide

$$CH_2{=}CHCl$$

Vinyl chloride

$$\underset{Br}{CH_2}-\underset{Br}{CH_2}$$

Ethylene bromide
1,2-Dibromoethane

$$\underset{OH}{CH_2}-\underset{Br}{CH_2}$$

Ethylene bromohydrin
2-Bromoethanol

Substituted alkyl halides undergo, of course, the reactions characteristic of their other functional groups—nitration of benzyl chloride, oxidation of ethylene bromohydrin, addition to allyl bromide—but as halides they react very much like ethyl or isopropyl or *tert*-butyl halides.

Compounds in which the halogen atom is attached directly to an aromatic ring (*aryl halides*, e.g., bromobenzene) differ so much from the alkyl halides in their preparations and properties that they will be taken up in a separate chapter (Chap. 25). For the present we need to know that—in the kinds of reaction typical of alkyl halides—*most aryl halides are extremely unreactive.*

As we know from our previous acquaintance with these compounds, alkyl halides are given both common names and IUPAC names.

14.2 Physical properties

Because of greater molecular weight, haloalkanes have considerably higher boiling points than alkanes of the same number of carbons. For a given alkyl group, the boiling point increases with increasing atomic weight of the halogen, so that a fluoride is the lowest boiling, an iodide the highest boiling.

Table 14.1 ALKYL HALIDES

Name	Chloride		Bromide		Iodide	
	B.p., °C	Density at 20°C	B.p., °C	Density at 20°C	B.p., °C	Density at 20°C
Methyl	− 24		5		43	2.279
Ethyl	12.5		38	1.440	72	1.933
n-Propyl	47	.890	71	1.335	102	1.747
n-Butyl	78.5	.884	102	1.276	130	1.617
n-Pentyl	108	.883	130	1.223	157	1.517
n-Hexyl	134	.882	156	1.173	180	1.441
n-Heptyl	160	.880	180		204	1.401
n-Octyl	185	.879	202		225.5	
Isopropyl	36.5	.859	60	1.310	89.5	1.705
Isobutyl	69	.875	91	1.261	120	1.605
sec-Butyl	68	.871	91	1.258	119	1.595
tert-Butyl	51	.840	73	1.222	100*d*	
Cyclohexyl	142.5	1.000	165			
Vinyl (Haloethene)	− 14		16		56	
Allyl (3-Halopropene)	45	.938	71	1.398	103	
Crotyl (1-Halo-2-butene)	84				132	
Methylvinylcarbinyl (3-Halo-1-butene)	64					
Propargyl (3-Halopropyne)	65		90	1.520	115	
Benzyl	179	1.102	201		93^{10}	
α-Phenylethyl	92^{15}		85^{10}			
β-Phenylethyl	92^{20}		92^{11}		127^{19}	
Diphenylmethyl	173^{19}		184^{20}			
Triphenylmethyl	310		230^{15}			
Dihalomethane	40	1.336	99	2.49	180*d*	3.325
Trihalomethane	61	1.489	151	2.89	*subl.*	4.008
Tetrahalomethane	77	1.595	189.5	3.42	*subl.*	4.32
1,1-Dihaloethane	57	1.174	110	2.056	179	2.84
1,2-Dihaloethane	84	1.257	132	2.180	*d*	2.13
Trihaloethylene	87		164	2.708		
Tetrahaloethylene	121				*subl.*	
Benzal halide	205		140^{20}			
Benzotrihalide	221	1.38				

In spite of their polarity, alkyl halides are insoluble in water, probably because of their inability to form hydrogen bonds. They are soluble in the typical organic solvents.

Iodo, bromo, and polychloro compounds are more dense than water.

14.3 Industrial source

On an industrial scale alkyl halides—chiefly the chlorides because of the cheapness of chlorine—are most often prepared by direct halogenation of hydrocarbons at the high temperatures needed for these free-radical reactions (Secs. 3.19, 6.21, and 12.12–12.13). Even though mixtures containing isomers and compounds of different halogen content are generally obtained, these reactions are useful industrially since often a mixture can be used as such or separated into its components by distillation.

Certain important halides are prepared by methods similar to those used in the laboratory; thus, for vinyl chloride:

$$\underset{\text{Acetylene}}{HC{\equiv}CH} \xrightarrow{HCl,\ HgCl_2} \underset{\text{Vinyl chloride}}{CH_2{=}CHCl}$$

$$\underset{\text{Ethylene}}{CH_2{=}CH_2} \xrightarrow{Cl_2} \underset{\quad Cl \qquad Cl}{CH_2{-}CH_2} \xrightarrow{500^\circ} \underset{\text{Vinyl chloride}}{CH_2{=}CHCl}$$

Many fluorine compounds are not prepared by direct fluorination, but rather by replacement of chlorine, using inorganic fluorides:

$$CH_3Cl + Hg_2F_2 \longrightarrow \underset{\substack{\text{Methyl fluoride} \\ \text{(b.p. } -79^\circ)}}{CH_3F} + Hg_2Cl_2$$

$$CCl_4 + SbF_3 \longrightarrow \underset{\substack{\text{Dichlorodifluoromethane} \\ \text{(Freon-12)} \\ \text{(b.p. } -28^\circ)}}{CCl_2F_2}$$

The increasingly important polyfluorides known as *fluorocarbons* are prepared by replacement of hydrogen using inorganic fluorides:

$$\underset{n\text{-Heptane}}{C_7H_{16}} + 32CoF_3 \longrightarrow \underset{\substack{\text{Perfluoroheptane} \\ (per = \text{fully substituted}) \\ \text{(b.p. } 84^\circ)}}{C_7F_{16}} + 16HF + 32CoF_2 \qquad (2CoF_2 + F_2 \longrightarrow 2CoF_3)$$

Cobalt(III) fluoride, CoF_3, is a convenient fluorinating agent.

14.4 Preparation

In the laboratory alkyl halides are most often prepared by the methods outlined below.

PREPARATION OF ALKYL HALIDES

1. From alcohols. Discussed in Secs. 16.4–16.5.

$$R{-}OH \xrightarrow{HX \text{ or } PX_3} R{-}X$$

Examples:

$$\underset{n\text{-Propyl alcohol}}{CH_3CH_2CH_2OH} \xrightarrow[\text{or } NaBr,\ H_2SO_4,\ \text{heat}]{\text{conc. HBr}} \underset{n\text{-Propyl bromide}}{CH_3CH_2CH_2Br}$$

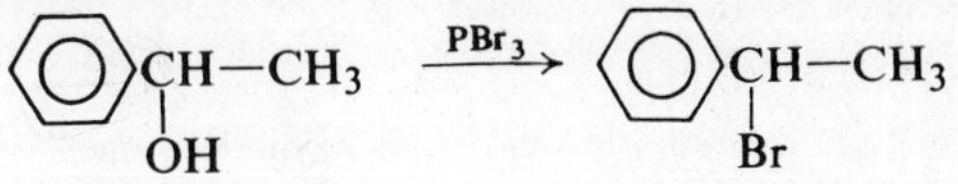

$$C_6H_5{-}\underset{OH}{CH}{-}CH_3 \xrightarrow{PBr_3} C_6H_5{-}\underset{Br}{CH}{-}CH_3$$

1-Phenylethanol → 1-Bromo-1-phenylethane

α-Phenylethyl alcohol → α-Phenylethyl bromide

$$\underset{\text{Ethyl alcohol}}{CH_3CH_2OH} \xrightarrow{P + I_2} \underset{\text{Ethyl iodide}}{CH_3CH_2I}$$

2. Halogenation of certain hydrocarbons. Discussed in Secs. 3.19, 6.21, 12.12–12.14.

$$R{-}H \xrightarrow{X_2} R{-}X + HX$$

Examples:

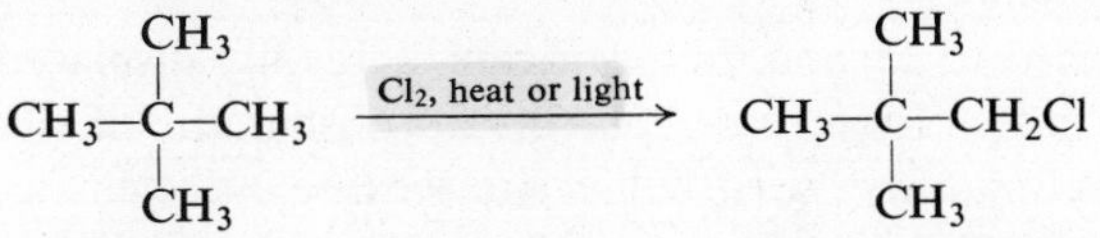

$$\underset{\text{Neopentane}}{CH_3{-}C(CH_3)_2{-}CH_3} \xrightarrow{Cl_2,\ \text{heat or light}} \underset{\text{Neopentyl chloride}}{CH_3{-}C(CH_3)_2{-}CH_2Cl}$$

$$\underset{\text{Toluene}}{C_6H_5CH_3} \xrightarrow{Br_2,\ \text{reflux, light}} \underset{\text{Benzyl bromide}}{C_6H_5CH_2Br}$$

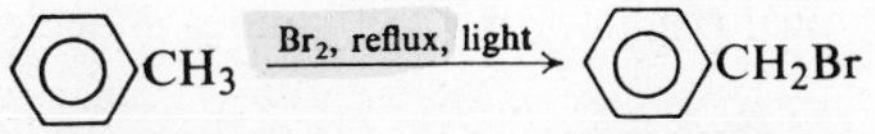

3. Addition of hydrogen halides to alkenes. Discussed in Secs. 6.6–6.7.

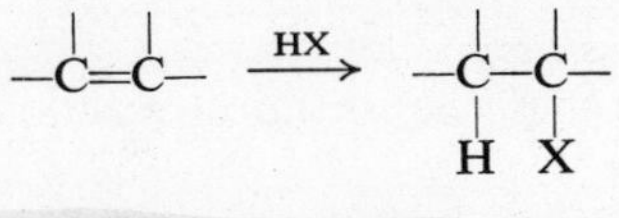

$$-\overset{|}{C}{=}\overset{|}{C}- \xrightarrow{HX} -\underset{H}{\overset{|}{C}}{-}\underset{X}{\overset{|}{C}}-$$

4. Addition of halogens to alkenes and alkynes

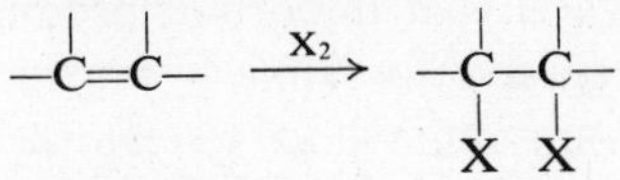

$$-\overset{|}{C}{=}\overset{|}{C}- \xrightarrow{X_2} -\underset{X}{\overset{|}{C}}{-}\underset{X}{\overset{|}{C}}-$$ Discussed in Sec. 6.5.

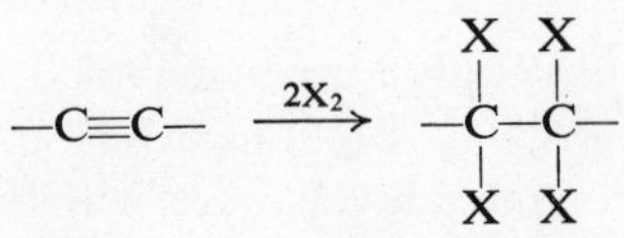

$$-C{\equiv}C- \xrightarrow{2X_2} -\underset{X}{\overset{X}{C}}{-}\underset{X}{\overset{X}{C}}-$$ Discussed in Sec. 8.8.

5. Halide exchange. Discussed in Sec. 14.4.

$$R{-}X + I^- \xrightarrow{\text{acetone}} R{-}I + X^-$$

Alkyl halides are nearly always prepared from alcohols, which are available commercially (Sec. 15.5) or are readily synthesized (Secs. 15.7 and 16.9–16.10). Although certain alcohols tend to undergo rearrangement (Sec. 16.4) during replacement of —OH by —X, this tendency can be minimized by use of phosphorus halides.

Certain halides are best prepared by direct halogenation. The most important of these preparations involve substitution of —X for the unusually reactive allylic or benzylic hydrogens.

$$-\underset{\text{H}}{\underset{|}{\overset{|}{\text{C}}}}-\overset{|}{\text{C}}=\overset{|}{\text{C}}- \longrightarrow -\underset{\text{X}}{\underset{|}{\overset{|}{\text{C}}}}-\overset{|}{\text{C}}=\overset{|}{\text{C}}- \qquad C_6H_5-\underset{\text{H}}{\underset{|}{\overset{|}{\text{C}}}}- \longrightarrow C_6H_5-\underset{\text{X}}{\underset{|}{\overset{|}{\text{C}}}}-$$

Allylic hydrogen — An allyl halide — Benzylic hydrogen — A benzyl halide

An alkyl iodide is often prepared from the corresponding bromide or chloride by treatment with a solution of sodium iodide in acetone; the less soluble bromide or sodium chloride precipitates from solution and can be removed by filtration.

14.5 Reactions

A halide ion is an extremely weak base. Its reluctance to share its electrons is shown by its great tendency to release a hydrogen ion, that is, by the high acidity of the hydrogen halides.

When attached to carbon, halogen can be readily displaced as halide ion by other, stronger bases. These bases possess an unshared pair of electrons and are seeking a relatively positive site, that is, are seeking a nucleus with which to share their electrons.

Basic, electron-rich reagents are called **nucleophilic reagents** (from the Greek, *nucleus-loving*). The typical reaction of alkyl halides is **nucleophilic substitution**:

$$\text{R:X} + \underset{\text{A nucleophilic reagent}}{\text{:Z}} \longrightarrow \text{R:Z} + \underset{\text{Leaving group}}{\text{:X}^-} \qquad \textbf{Nucleophilic substitution}$$

To describe the ease of displacement of the weakly basic halide ions, we refer to them as good *leaving groups.*

(Aryl and vinyl halides undergo these substitution reactions with extreme difficulty, Sec. 25.5.)

Alkyl halides react with a large number of nucleophilic reagents, both inorganic and organic, to yield a wide variety of important products. As we shall see, these reagents include not only negative ions like hydroxide, alkoxide, and cyanide, but also neutral bases like ammonia and water; their characteristic feature is an *unshared pair of electrons.*

As a synthetic tool, nucleophilic substitution involving alkyl halides is one of the three or four most useful classes of organic reactions. Much of the importance of alcohols is due to their ready conversion into alkyl halides, with their good leaving groups.

A large number of nucleophilic substitutions are listed below to give an idea of the versatility of alkyl halides; many will be left to later chapters for detailed discussion.

As we already know (Secs. 5.12 and 8.12), alkyl halides undergo not only substitution but also **elimination**, a reaction that is important in the synthesis of alkenes. Both elimination and substitution are brought about by basic reagents, and hence there must always be *competition* between the two reactions. We shall be interested to see how this competition is affected by such factors as the structure of the halide or the particular nucleophilic reagent used.

We shall look rather closely at both nucleophilic substitution and elimination reactions of the alkyl halides, for they provide a particularly good illustration of the effect of structure on reactivity, and of the methods that may be used to determine mechanisms of reactions.

REACTIONS OF ALKYL HALIDES

1. Nucleophilic substitution

$R{:}X + {:}Z$	$\longrightarrow$	$R{:}Z \quad + {:}X^-$	
$R{:}X + {:}OH^-$	$\longrightarrow$	$R{:}OH + {:}X^-$	Alcohol
$+ H_2O$	$\longrightarrow$	$R{:}OH$	Alcohol
$+ {:}OR'^-$	$\longrightarrow$	$R{:}OR'$	Ether (Williamson synthesis, Sec. 17.5)
$+ {^-}{:}C{\equiv}CR'$	$\longrightarrow$	$R{:}C{\equiv}CR'$	Alkyne (Sec. 8.12)
$+ \overset{\delta-}{R'}—\overset{\delta+}{M}$	$\longrightarrow$	$R{:}R'$	Alkane (coupling, Sec. 3.17)
$+ {:}I^-$	$\longrightarrow$	$R{:}I$	Alkyl iodide
$+ {:}CN^-$	$\longrightarrow$	$R{:}CN$	Nitrile (Sec. 18.8)
$+ R'COO{:}^-$	$\longrightarrow$	$R'COO{:}R$	Ester
$+ {:}NH_3$	$\longrightarrow$	$R{:}NH_2$	Primary amine (Sec. 22.10)
$+ {:}NH_2R'$	$\longrightarrow$	$R{:}NHR'$	Secondary amine (Sec. 22.13)
$+ {:}NHR'R''$	$\longrightarrow$	$R{:}NR'R''$	Tertiary amine (Sec. 22.13)
$+ {:}P(C_6H_5)_3$	$\longrightarrow$	$[R{:}P(C_6H_5)_3]^+X^-$	Phosphonium salt (Sec. 21.10)
$+ {:}SH^-$	$\longrightarrow$	$R{:}SH$	Thiol (mercaptan)
$+ {:}SR'^-$	$\longrightarrow$	$R{:}SR'$	Thioether (sulfide)
$+ ArH + AlCl_3$	$\longrightarrow$	ArR	Alkylbenzene (Friedel-Crafts reaction, Sec. 12.6)
$+ [CH(COOC_2H_5)_2]^-$	$\longrightarrow$	$R{:}CH(COOC_2H_5)_2$	(Malonic ester synthesis, Sec. 26.2)
$+ [CH_3COCHCOOC_2H_5]^-$	$\longrightarrow$	$CH_3CO\underset{\ddot{R}}{C}HCOOC_2H_5$	(Acetoacetic ester synthesis, Sec. 26.3)

2. Dehydrohalogenation: elimination. Discussed in Secs. 5.12–5.14 and 14.18–14.23.

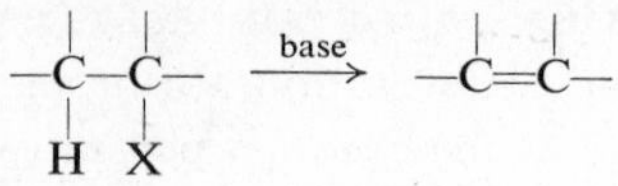

$$-\underset{H}{\overset{|}{C}}-\underset{X}{\overset{|}{C}}- \xrightarrow{\text{base}} -\overset{|}{C}=\overset{|}{C}-$$

3. Preparation of Grignard reagent. Discussed in Secs. 3.16 and 15.12.

$$RX + Mg \xrightarrow{\text{dry ether}} RMgX$$

4. Reduction. Discussed in Sec. 3.15.

$$RX + M + H^+ \longrightarrow RH + M^+ + X^-$$

Examples:

$$(CH_3)_3CCl \xrightarrow{Mg} (CH_3)_3CMgCl \xrightarrow{D_2O} (CH_3)_3CD$$

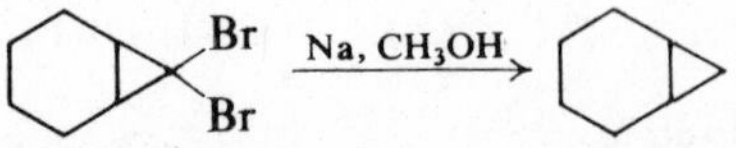

Br, Br — Na, CH_3OH →

7,7-Dibromonorcarane   Norcarane

14.6 Alkyl sulfonates

In following sections, we shall discuss the mechanisms of nucleophilic aliphatic substitution and of elimination using alkyl halides as our examples. But we should realize that these reactions take place in exactly the same ways with a variety of other compounds: compounds which, like alkyl halides, contain *good leaving groups*.

Of these other compounds, alkyl esters of sulfonic acids, $ArSO_2OR$, are most commonly used in place of alkyl halides: usually in the study of reaction mechanisms, but also in synthesis. As the anions of strong acids, sulfonate anions are weak bases and hence are good leaving groups in either nucleophilic substitution or elimination:

$$\underset{\text{Alkyl sulfonate}}{ArSO_2OR} + :Z \longrightarrow R:Z + \underset{\substack{\text{Weak base:} \\ \textit{good leaving group}}}{ArSO_3^-}$$ **Nucleophilic substitution**

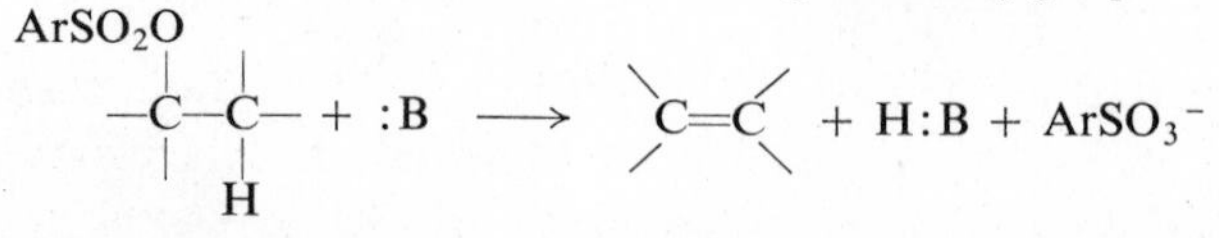

$$\underset{\text{Alkyl sulfonate}}{-\overset{ArSO_2O}{\overset{|}{C}}-\underset{H}{\overset{|}{C}}-} + :B \longrightarrow \;\rangle C=C\langle\; + H:B + ArSO_3^-$$ **Elimination**

Most commonly used are esters of *p*-toluenesulfonic acid: the *p*-toluenesulfonates. The name of the *p*-toluenesulfonyl group is often shortened to *tosyl* (Ts); *p*-toluenesulfonyl chloride thus becomes *tosyl chloride* (TsCl), and *p*-toluenesulfonates become *tosylates* (TsOR).

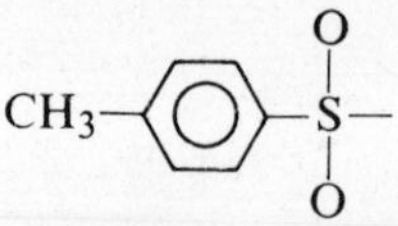

$CH_3-C_6H_4-SO_2-$ $Br-C_6H_4-SO_2-$ CH_3-SO_2-

Tosyl or *Ts* 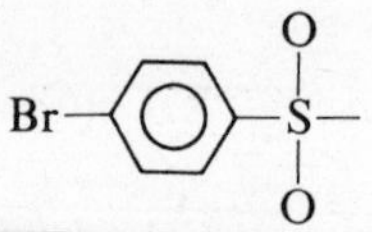   *Brosyl* or *Bs* 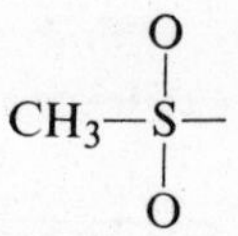   *Mesyl* or *Ms*

Like alkyl halides, alkyl sulfonates are prepared from alcohols but, as we shall see in Sec. 16.7, the two syntheses differ in one very important way.

14.7 Rate of reaction: effect of concentration. Kinetics

Before we discuss nucleophilic substitution involving alkyl halides, let us return briefly to the matter of what determines the rate of a reaction.

We have seen (Sec. 2.18) that the rate of a chemical reaction can be expressed as a product of three factors:

$$\text{rate} = \begin{matrix}\text{collision}\\ \text{frequency}\end{matrix} \times \begin{matrix}\text{energy}\\ \text{factor}\end{matrix} \times \begin{matrix}\text{probability}\\ \text{factor}\end{matrix}$$

So far, we have used this relationship to understand problems of orientation and relative reactivity; in doing this we have compared rates of *different* reactions. When the conditions that we can control (temperature, concentration) are kept the same, closely related reactions proceed at different rates chiefly because they have different energy factors, that is to say, different E_{act}'s. We have been able to account surprisingly well for many differences in E_{act}'s by using structural theory to estimate stabilities of the transition states.

It is also useful to study an *individual* reaction to see how its rate is affected by deliberate changes in experimental conditions. We can determine E_{act}, for example, if we measure the rate at different temperatures (Sec. 2.18). But perhaps the most valuable information about a reaction is obtained by studying the effect of *changes in concentration* on its rate.

How does a change in concentration of reactants affect the rate of a reaction at a constant temperature? An increase in concentration cannot alter the fraction of collisions that have sufficient energy, or the fraction of collisions that have the proper orientation; it can serve only to increase the total number of collisions. If more molecules are crowded into the same space, they will collide more often and the reaction will go faster. Collision frequency, and hence rate, depends in a very exact way upon concentration.

The field of chemistry that deals with rates of reaction, and in particular with dependence of rates on concentration, is called **kinetics**. Let us see what kinetics can tell us about nucleophilic aliphatic substitution.

14.8 Kinetics of nucleophilic aliphatic substitution. Second-order and first-order reactions

Let us take a specific example, the reaction of methyl bromide with sodium hydroxide to yield methanol:

$$CH_3Br + OH^- \longrightarrow CH_3OH + Br^-$$

This reaction would probably be carried out in aqueous ethanol, in which both reactants are soluble.

If the reaction results from collision between a hydroxide ion and a methyl bromide molecule, we would expect the rate to depend upon the concentration of both these reactants. If either OH^- concentration, $[OH^-]$, or CH_3Br concentration, $[CH_3Br]$, is doubled, the collision frequency should be doubled and the reaction rate doubled. If either concentration is cut in half, the collision frequency, and consequently the rate, should be halved.

This is found to be so. We say that the rate of reaction depends upon both $[OH^-]$ and $[CH_3Br]$, and we indicate this by the expression

$$\text{rate} = k[CH_3Br][OH^-]$$

If concentrations are expressed in, say, moles per liter, then k is the number which, multiplied by these concentrations, tells us how many moles of methanol are formed in each liter during each second. At a given temperature and for a given solvent, k always has the same value and is characteristic of this particular reaction; k is called the **rate constant**. For example, for the reaction between methyl bromide and hydroxide ion in a mixture of 80% ethanol and 20% water at 55°, the value of k is 0.0214 liters per mole per second.

What we have just seen is, of course, not surprising; we all know that an increase in concentration causes an increase in rate. But now let us look at the corresponding reaction between *tert*-butyl bromide and hydroxide ion:

$$CH_3-\overset{\overset{CH_3}{|}}{\underset{\underset{Br}{|}}{C}}-CH_3 + OH^- \longrightarrow CH_3-\overset{\overset{CH_3}{|}}{\underset{\underset{OH}{|}}{C}}-CH_3 + Br^-$$

As before, if we double [RBr] the rate doubles; if we cut [RBr] in half the rate is halved. But if we double $[OH^-]$, or if we cut $[OH^-]$ in half, there is no change in the rate. *The rate of reaction is independent of* $[OH^-]$.

The rate of reaction of *tert*-butyl bromide depends only upon [RBr]. This is indicated by the expression

$$\text{rate} = k[RBr]$$

For the reaction of *tert*-butyl bromide in 80% alcohol at 55°, the rate constant is 0.010 per second. This means that of every mole of *tert*-butyl bromide present, 0.010 mole reacts each second, whatever the $[OH^-]$.

The methyl bromide reaction is said to follow **second-order kinetics**, since its rate is dependent upon the concentrations of *two* substances. The *tert*-butyl bromide reaction is said to follow **first-order kinetics**; its rate depends upon the concentration of only *one* substance.

How are we to account for this difference in kinetic order? How are we to account for the puzzling fact that the rate of the *tert*-butyl bromide reaction is independent of $[OH^-]$?

To account for such differences in kinetic order, as well as for many other observations, it has been proposed that *nucleophilic substitution can proceed by two different mechanisms.* In the following sections we shall see what these two mechanisms are believed to be, the facts on which they are based, and how they account for the facts.

Recognition of the duality of mechanism for nucleophilic aliphatic substitution, formulation of the mechanisms themselves, and analysis of the factors influencing competition between them—all are largely due to Sir Christopher Ingold (of University College, London) and the people who worked with him; and this is only a fraction of their total contribution to the theory of organic chemistry.

14.9 The S_N2 reaction: mechanism and kinetics

The reaction between methyl bromide and hydroxide ion to yield methanol follows second-order kinetics; that is, the rate depends upon the concentrations of both reactants:

$$CH_3Br + OH^- \longrightarrow CH_3OH + Br^-$$

$$\text{rate} = k[CH_3Br][OH^-]$$

The simplest way to account for the kinetics is to assume that reaction requires a collision between a hydroxide ion and a methyl bromide molecule. On the basis of evidence we shall shortly discuss, it is known that in its attack the hydroxide ion stays as far away as possible from the bromine; that is to say, it attacks the molecule from the rear.

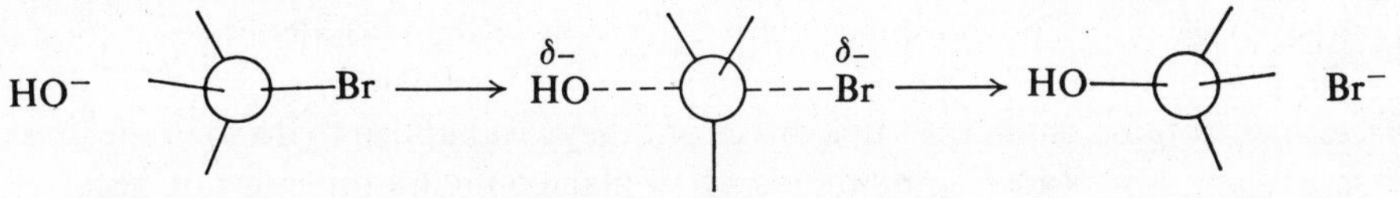

$$HO^- \quad \text{C—Br} \longrightarrow \overset{\delta-}{HO}\text{---C---}\overset{\delta-}{Br} \longrightarrow HO\text{—C} \quad Br^-$$

Figure 14.1. The S_N2 reaction: complete inversion of configuration. Nucleophilic reagent attacks back side.

The reaction is believed to take place as shown in Fig. 14.1. When hydroxide ion collides with a methyl bromide molecule at the face most remote from the bromine, and when such a collision has sufficient energy, a C—OH bond forms and the C—Br bond breaks, liberating the bromide ion.

The transition state can be pictured as a structure in which carbon is partially bonded to both —OH and —Br; the C—OH bond is not completely formed, the C—Br bond is not yet completely broken. Hydroxide has a diminished negative charge, since it has begun to share its electrons with carbon. Bromine has developed a partial negative charge, since it has partly removed a pair of electrons from carbon. At the same time, of course, ion–dipole bonds between hydroxide ion and solvent are being broken and ion–dipole bonds between bromide ion and solvent are being formed.

The —OH and —Br are located as far apart as possible; the three hydrogens and the carbon lie in a single plane, all bond angles being 120°. The C—H bonds are thus arranged like the spokes of a wheel, with the C—OH and the C—Br bonds lying along the axle.

This is the mechanism that is called S_N2: *substitution nucleophilic bimolecular*. The term *bi*molecular is used here since the rate-determining step involves collision of *two* particles.

What evidence is there that alkyl halides can react in this manner? First of all, as we have just seen, the mechanism is consistent with the kinetics of a reaction like the one between methyl bromide and hydroxide ion. In general, **an S_N2 reaction follows second-order kinetics.** Let us look at some of the other evidence.

14.10 The S_N2 reaction: stereochemistry

Both 2-bromooctane and 2-octanol are chiral; that is, they have molecules that are not superimposable on their mirror images. Consequently, these compounds can exist as enantiomers, and can show optical activity. Optically active 2-octanol has been obtained by resolution of the racemic modification (Sec. 7.9), and from it optically active 2-bromooctane has been made.

The following configurations have been assigned (Sec. 7.5):

C_6H_{13} / H— —Br / CH_3 — (−)-2-Bromooctane, [α] = −34.6°

C_6H_{13} / H— —OH / CH_3 — (−)-2-Octanol, [α] = −9.9°

We notice that the (−)-bromide and the (−)-alcohol have similar configurations; that is, —OH occupies the same relative position in the (−)-alcohol as —Br does in the (−)-bromide. As we know, compounds of similar configuration do not *necessarily* rotate light in the same direction; they just happen to do so in the present case. (As we also know, compounds of similar configuration are not necessarily given the same specification of R and S (Sec. 7.5); it just happens that both are R in this case.)

When (−)-2-bromooctane is allowed to react with sodium hydroxide under conditions where second-order kinetics are followed, there is obtained (+)-2-octanol.

C_6H_{13} / H— —Br / CH_3 $\xrightarrow[S_N2]{NaOH}$ C_6H_{13} / HO— —H / CH_3

(−)-2-Bromooctane $\xrightarrow[S_N2]{NaOH}$ (+)-2-Octanol

[α] = −34.6° [α] = +9.9°

optical purity 100% optical purity 100%

We see that the —OH group has not taken the position previously occupied by —Br; the alcohol obtained has a configuration *opposite* to that of the bromide. *A reaction that yields a product whose configuration is opposite to that of the reactant is said to proceed with* **inversion of configuration**.

(In this particular case, inversion of configuration happens to be accompanied by a change in specification, from R to S, but this is not always true. We cannot tell whether a reaction proceeds with inversion or retention of configuration simply by looking at the letters used to specify the reactant and product; we must work out and compare the absolute configurations indicated by those letters.)

Now the question arises: does a reaction like this proceed with *complete inversion*? That is to say, is the configuration of *every* molecule inverted? The answer is *yes*. **An S_N2 reaction proceeds with complete stereochemical inversion.**

To answer a question like this, we must know the optical purity both of the reactant that we start with, and of the product that we obtain: in this case, of 2-bromooctane and 2-octanol. To know these we must, in turn, know the maximum rotation of the bromide and of the alcohol; that is, we must know the rotation of an optically pure sample of each.

Suppose, for example, that we know the rotation of optically pure 2-bromooctane to be 34.6° and that of optically pure 2-octanol to be 9.9°. If, then, a sample of optically pure bromide were found to yield optically pure alcohol, we would know that the reaction had proceeded with complete inversion. Or—and this is much more practicable—if a sample of the halide of rotation, say, $-28.7°$ (83% optically pure) were found to yield alcohol of rotation $+8.22°$ (83% optically pure), we would draw exactly the same conclusion.

In developing the ideas of S_N1 and S_N2 reactions, Ingold (p. 460) studied the reaction of optically active 2-bromooctane and obtained results which—when corrected for a small contribution from the S_N1 reaction, and for the effect of bromide ion generated during the reaction (see Problem 14, p. 488)—led him to conclude that the S_N2 reaction proceeds, within limits of experimental error, with complete inversion.

The particular value that Ingold used for the rotation of optically pure 2-bromooctane has been questioned, but the basic idea of complete inversion in S_N2 reactions is established beyond question: by study of systems other than alkyl halides and by elegant work involving radioactivity and optical activity (Problem 13, p. 488).

It was to account for inversion of configuration that back-side attack was first proposed for substitution of the S_N2 kind. As —OH becomes attached to carbon, three bonds are forced apart until they reach the planar "spoke" arrangement of the transition state; then, as bromide is expelled, they move on to a tetrahedral arrangement *opposite* to the original one. This process has often been likened to the turning-inside-out of an umbrella in a gale.

$$HO^- \quad H\!-\!\underset{CH_3}{\overset{C_6H_{13}}{C}}\!-\!Br \longrightarrow \overset{\delta-}{HO}\text{---}\underset{CH_3}{\overset{C_6H_{13}\ H}{C}}\text{---}\overset{\delta-}{Br} \longrightarrow HO\!-\!\underset{CH_3}{\overset{C_6H_{13}}{C}}\!-\!H \quad Br^-$$

S_N2: *complete inversion*

The stereochemistry of the 2-bromooctane reaction indicates back-side attack in accordance with the S_N2 mechanism; studies of other optically active compounds, under conditions where the reactions follow second-order kinetics, show similar results. It is not possible to study the stereochemistry of most halides, since they are not optically active; however, there seems no reason to doubt that they, too, undergo back-side attack.

The S_N2 mechanism is supported, then, by stereochemical evidence. Indeed, the relationship between mechanism and stereochemistry is so well established that in the absence of other evidence complete inversion is taken to indicate an S_N2 reaction.

We see once more how stereochemistry can give us a kind of information about a reaction that we cannot get by any other means.

Inversion of configuration is the general rule for reactions occurring at chiral centers, being much commoner than retention of configuration. Oddly enough, it is the very prevalence of inversion that made its detection difficult. Paul Walden (at the Polytechnicum in Riga, Latvia) discovered the phenomenon of inversion in 1896 when he encountered one of the exceptional reactions in which inversion does *not* take place.

Problem 14.1 (a) What product would be formed if the reaction of *cis*-4-bromocyclohexanol with OH^- proceeded with inversion? (b) Without inversion? (c) Is it always necessary to use optically active compounds to study the stereochemistry of substitution reactions?

14.11 The S_N2 reaction: reactivity

In what way would we expect changes in structure of the alkyl group to affect reactivity in an S_N2 substitution? In contrast to the free-radical and carbonium ion reactions we have studied, this time the structure of the transition state is *not* intermediate between the structures of the reactant and product; this time we cannot simply assume that factors stabilizing the product will also stabilize the transition state.

First of all, let us compare transition state and reactants with regard to electron distribution. In the transition state, there is a partly formed bond between carbon and hydroxide ion and a partly broken bond between carbon and halide ion: hydroxide ion has brought electrons to carbon, and halide ion has taken electrons away. Unless one of the two processes, bond-making or bond-breaking, has gone much further than the other, the net charge on carbon is not greatly different from what it was at the start of the reaction. Electron withdrawal or electron release by substituents should affect stability of transition state and reactant in much the same way, and therefore should have little influence on reaction rate.

To understand how structure does influence the rate, let us compare transition state and reactants with regard to *shape*, starting with the methyl bromide reaction. The carbon in reactant and product is tetrahedral, whereas carbon in the transition state is bonded to five atoms. As indicated before, the C—H bonds are arranged like the spokes of a wheel, with the C—OH and C—Br bonds lying along the axle (Fig. 14.2).

What would be the effect of replacing the hydrogens successively by methyl

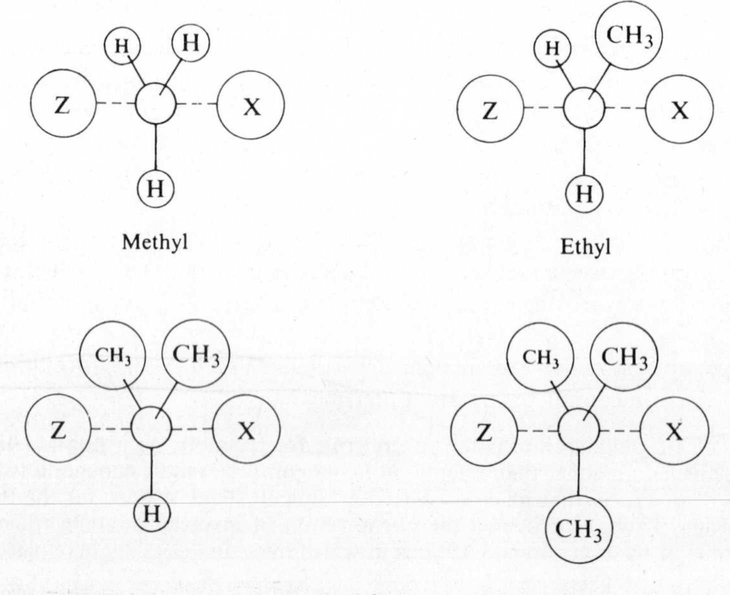

Figure 14.2. Steric factor in the S_N2 reaction. Crowding raises energy of transition state and slows down reaction.

groups? That is, how will the transition state differ as we go from methyl bromide through ethyl bromide and isopropyl bromide to *tert*-butyl bromide? As hydrogen atoms are replaced by the larger methyl groups, there is increased crowding about the carbon; this is particularly severe in the transition state, where the methyls are thrown close to both —OH and —Br (Fig. 14.2). Non-bonded interaction raises the energy of the crowded transition state more than the energy of the roomier reactant; E_{act} is higher and reaction is slower.

In agreement with this prediction, *differences in rate between two S_N2 reactions seem to be due chiefly to* **steric factors**, and not to electronic factors; that is to say, differences in rate are related to the *bulk* of the substituents and not to their ability to withdraw or release electrons. As the number of substituents attached to the carbon bearing the halogen is increased, the reactivity toward S_N2 substitution decreases. These substituents may be aliphatic, or aromatic, or both, as shown in the following two sequences:

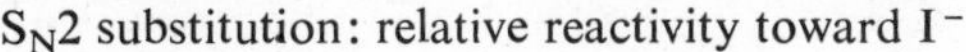

S_N2 substitution: relative reactivity toward I^-

CH_3Br	>	CH_3CH_2Br	>	$(CH_3)_2CHBr$	>	$(CH_3)_3CBr$
Methyl		Ethyl		Isopropyl		*tert*-Butyl
150		1		0.01		0.001

$C_6H_5CH_2Br$	>	$C_6H_5CH(CH_3)Br$	>	$C_6H_5C(CH_3)_2Br$
Benzyl		α-Phenylethyl		β-Phenylisopropyl

(To give an idea of how large these differences may be, the relative rates for a particular S_N2 reaction, substitution by iodide ion, are indicated below the formulas in the first sequence.)

In S_N2 reactions the order of reactivity of RX is CH_3X > 1° > 2° > 3°.

In cases where steric factors are kept constant, electronic effects on S_N2 reactions can be observed; however, these effects are found to be comparatively *small.* Some S_N2 reactions are speeded up slightly by electron release, and others are speeded up slightly by electron withdrawal, but it is not usually possible to predict which will be the case simply from the structures involved.

Problem 14.2 (a) Draw the structures of ethyl, *n*-propyl, isobutyl, and neopentyl bromides. These structures can be considered methyl bromide with one of its hydrogens replaced by various alkyl groups (GCH_2Br). What is the group G in each case?

(b) The relative rates of reaction (with ethoxide ion) are roughly: methyl bromide, 100; ethyl bromide, 6; *n*-propyl bromide, 2; isobutyl bromide, 0.2; neopentyl bromide, 0.00002. What is the effect of the *size* of the group G attached to carbon bearing the halogen? How does this compare with the effect of changing the *number* of groups?

Thus we see that the S_N2 mechanism is supported by three lines of evidence: kinetics, stereochemistry, and effect of structure on reactivity.

Now let us turn to the other mechanism by which nucleophilic aliphatic substitution can take place.

14.12 The S_N1 reaction: mechanism and kinetics. Rate-determining step

The reaction between *tert*-butyl bromide and hydroxide ion to yield *tert*-butyl alcohol follows first-order kinetics; that is, the rate depends upon the concentration of only one reactant, *tert*-butyl bromide.

$$CH_3-\underset{Br}{\overset{CH_3}{C}}-CH_3 + OH^- \longrightarrow CH_3-\underset{OH}{\overset{CH_3}{C}}-CH_3 + Br^-$$

$$\text{rate} = k\,[RBr]$$

How are we to interpret the fact that the rate is independent of $[OH^-]$? If the rate of reaction does not depend upon $[OH^-]$, it can only mean that the reaction *whose rate we are measuring* does not involve OH^-.

These observations are quite consistent with the following mechanism.

(1) $$CH_3-\underset{Br}{\overset{CH_3}{C}}-CH_3 \longrightarrow CH_3-\underset{\oplus}{\overset{CH_3}{C}}-CH_3 + Br^- \qquad \textbf{Slow}$$

S_N1

(2) $$CH_3-\underset{\oplus}{\overset{CH_3}{C}}-CH_3 + OH^- \longrightarrow CH_3-\underset{OH}{\overset{CH_3}{C}}-CH_3 \qquad \textbf{Fast}$$

tert-Butyl bromide slowly dissociates (step 1) into bromide ions and *tert*-butyl cations. The carbonium ions then combine rapidly (step 2) with hydroxide ions to yield *tert*-butyl alcohol.

The rate of the overall reaction is determined by the slow breaking of the C—Br bond to form the carbonium ion; once formed, the carbonium ion reacts rapidly to form the product. *A single step whose rate determines the overall rate of a stepwise reaction is called a* **rate-determining step**. It is not surprising that the rate-determining step here is the one that involves the *breaking* of a bond, an energy-demanding process. The required energy is supplied by formation of many ion–dipole bonds between the two kinds of ion and the solvent. (Although each of these is weak, altogether in reactions like these they supply 110–150 kcal/mole!)

This is the mechanism that is called $\mathbf{S_N1}$: *substitution nucleophilic unimolecular*. The term *uni*molecular is used here since the rate-determining step involves only *one* molecule (disregarding the many necessary solvent molecules).

What evidence is there that alkyl halides can react by this mechanism? As we have just seen, the mechanism is consistent with the first-order kinetics of a reaction like the one between *tert*-butyl bromide and hydroxide ion. In general, **an S_N1 reaction follows first-order kinetics**. The rate of the entire reaction is determined by how fast the alkyl halide ionizes, and hence depends only upon the concentration of alkyl halide.

In the following sections, we shall look at some of the other evidence.

Let us see what we mean by rate-determining step in a reaction like this,

(1) $$A \underset{k_{-1}}{\overset{k_1}{\rightleftharpoons}} R + B$$

(2) $$R + C \xrightarrow{k_2} \text{product}$$

where R is a reactive intermediate (carbonium ion, free radical) whose concentration is maintained at some low *steady state* throughout the reaction. The exact kinetics expression for the formation of the product is

$$\text{rate} = \frac{k_1[\text{A}]}{1 + \dfrac{k_{-1}[\text{B}]}{k_2[\text{C}]}} \tag{3}$$

Without going into the derivation of this equation, let us see what it means.

The term $k_1[\text{A}]$ is in the numerator and the term $k_2[\text{C}]$ is in the denominator of the denominator; the bigger they are, the faster the rate. This is reasonable, since $k_1[\text{A}]$ is the rate of step (1) and $k_2[\text{C}]$ contributes to the rate of step (2). The term $k_{-1}[\text{B}]$ is in the denominator; the bigger it is, the slower the rate. This, too, is understandable, since it contributes to the rate of the reverse of step (1).

Now if $k_2[\text{C}]$ happens to be *much larger* than $k_{-1}[\text{B}]$, the term $k_{-1}[\text{B}]/k_2[\text{C}]$ is very small—insignificant relative to 1—and drops out. Under these conditions we get our familiar rate expression for first-order kinetics:

$$\text{rate} = k_1[\text{A}]$$

But if $k_2[\text{C}]$ is much larger than $k_{-1}[\text{B}]$, it must mean that *step* (2) *is much faster than the reverse of step* (1). This is the *real* requirement for step (1) to be rate-determining If we return to Sec. 11.16, we see that the absence of an isotope effect in nitration was accounted for on just this basis.

Does this mean that, contrary to what was said before, step (1)—in the forward direction—need not be slower than step (2)? Step (1) must still be a slow step, for otherwise the reactive intermediate would be formed faster than it could be consumed, and its concentration would build up—contrary to the nature of the reactive intermediate, and a condition different from the one for which the kinetics expression (3) holds.

Problem 14.3 When iodine is added to a benzene solution of dimer I (Sec. 12.15), the color of the iodine gradually fades, at a rate that depends upon [dimer] but *is independent of* $[I_2]$. When a benzene solution of the dimer is shaken under an atmosphere of NO gas, the pressure of the gas gradually drops, at a rate that depends upon [dimer] but *is independent of* the NO pressure. The rate constants for the two reactions are *identical*. Account for these results.

14.13 The S_N1 reaction: stereochemistry

We have proposed that, under the conditions we have described, methyl bromide reacts with hydroxide ion by the S_N2 mechanism, and that *tert*-butyl bromide reacts by the S_N1 mechanism. Since *sec*-alkyl bromides are intermediate in structure between these two halides, it is not surprising to find that they can react by either or both mechanisms.

An increase in $[OH^-]$ speeds up the second-order reaction but has no effect on the first-order reaction. At high $[OH^-]$, therefore, the second-order reaction is so much the faster that *sec*-alkyl bromides react almost entirely by the S_N2 mechanism. The behavior of optically active 2-bromooctane in an S_N2 reaction has been studied (Sec. 14.10) by use of high $[OH^-]$.

In the same way, a decrease in $[OH^-]$ slows down the second-order reaction, but has no effect on the first-order reaction. The behavior of optically active 2-bromooctane in an S_N1 reaction has been studied by use of low $[OH^-]$.

Problem 14.4 In 80% ethanol at 55°, isopropyl bromide reacts with hydroxide ion according to the following kinetic equation, where the rate is expressed as moles per liter per second:

$$\text{rate} = 4.7 \times 10^{-5}[\text{RX}][\text{OH}^-] + 0.24 \times 10^{-5}[\text{RX}]$$

What percentage of the isopropyl bromide reacts by the S_N2 mechanism when $[OH^-]$ is: (a) 0.001 molar, (b) 0.01 molar, (c) 0.1 molar, (d) 1.0 molar, (e) 5.0 molar?

When (−)-2-bromooctane is converted into the alcohol under conditions (low $[OH^-]$) where first-order kinetics are followed, there is obtained (+)-2-octanol.

$$(-)\text{-}C_6H_{13}CHBrCH_3 \xrightarrow[S_N1]{OH^-,\ H_2O} (+)\text{-}C_6H_{13}CHOHCH_3$$

Lower optical purity

The product has the opposite configuration from the starting materials, as in the S_N2 reaction, but this time there is a loss in optical purity. Optically pure bromide yields alcohol that is only about two-thirds optically pure. Optically pure starting material contains only the one enantiomer, whereas the product clearly must contain both. The product is thus a mixture of the inverted compound and the racemic modification, and we say that the reaction has proceeded with **partial racemization.** How can we account for these stereochemical results?

The carbonium ion, we saw in Sec. 5.16, has a *flat* structure: carbon is bonded to three other atoms, and for this bonding uses sp^2 orbitals. Let us see how this shape affects the stereochemical course of the reaction.

In the first step the optically active 2-bromooctane ionizes to form bromide ion and the flat 2-octyl cation. The nucleophilic reagent OH^- (or very possibly H_2O) then attaches itself to the carbonium ion. But it may attach itself to either

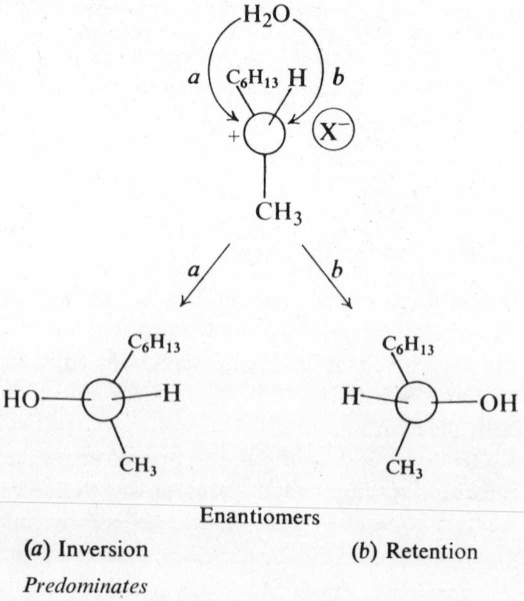

H_2O

a C₆H₁₃ H b

X^-

CH_3

HO—C(C₆H₁₃)(CH₃)—H   H—C(C₆H₁₃)(CH₃)—OH

Enantiomers

(*a*) Inversion   (*b*) Retention

Predominates

Figure 14.3. The S_N1 reaction: racemization plus inversion. Nucleophilic reagent attacks both (*a*) back side and (*b*) front side of carbonium ion. Back-side attack predominates.

face of this flat ion, and depending upon which face, yields one or the other of two products (see Fig. 14.3).

If the attack were purely random, we would expect equal amounts of the two isomers; that is to say, we would expect only the racemic modification. But the product is *not completely* racemized, for the inverted product exceeds its enantiomer. How do we account for this? The simplest explanation is that attack by the nucleophilic reagent occurs before the departing halide ion has completely left the neighborhood of the carbonium ion; to a certain extent the departing ion thus *shields* the front side of the ion from attack. As a result, back-side attack is somewhat preferred.

Racemization in an S_N1 reaction arises, then, from the *loss of configuration* in the intermediate carbonium ion. In some cases racemization may be almost complete; hydrolysis of α-phenylethyl chloride, for example, proceeds with 87% racemization and 13% inversion:

$$\underset{[\alpha] = -34^\circ}{C_6H_5CHClCH_3} \xrightarrow[S_N1]{OH^-,\ H_2O} \underset{[\alpha] = +1.7^\circ}{C_6H_5CHOHCH_3}$$

$$\text{Optical purity} = \frac{-34}{-109} \times 100 = 31\% \qquad \text{Optical purity} = \frac{+1.7}{+42.3} \times 100 = 4\%$$

In contrast to an S_N2 reaction, which proceeds with complete inversion, **an S_N1 reaction proceeds with racemization.**

(We shall look more closely at the stereochemistry—and, indeed, all aspects—of the S_N1 reaction in Sec. 14.17.)

Problem 14.5 Suppose that, under S_N1 conditions, 2-bromooctane of specific rotation −20.8° was found to yield 2-octanol of specific rotation +3.96°. Using the rotations for optically pure samples given on p. 462, calculate: (a) the optical purity of reactant and of product; (b) the percentage of racemization and of inversion accompanying the reaction; (c) the percentage of front-side and of back-side attack on the carbonium ion.

14.14 The S_N1 reaction: reactivity

The rate-determining step of an S_N1 reaction is the formation of a carbonium ion. Judging from our previous experience, therefore, we expect the reactivity of an alkyl halide to depend chiefly upon *how stable a carbonium ion it can form.*

Our expectation is correct: the order of reactivity of alkyl halides in S_N1 reactions is the same as the order of stability of carbonium ions.

In S_N1 reactions the order of reactivity of RX is allyl, benzyl > 3° > 2° > 1° > CH_3X.

As the positive charge develops on the carbon atom in the transition state, it is dispersed by the same factors—inductive effect and resonance—that stabilize the full-fledged carbonium ion.

$$RX \longrightarrow \left[\overset{\delta_+}{R}\text{---}\overset{\delta_-}{X}\right] \longrightarrow R^+ + X^-$$

Transition state — R *has partial positive charge*

Products — R *has full positive charge*

The following example gives some idea of how much the rate of an S_N1 reaction can be changed by changes in structure:

$$RBr + H_2O \xrightarrow{\text{formic acid}} ROH + HBr$$

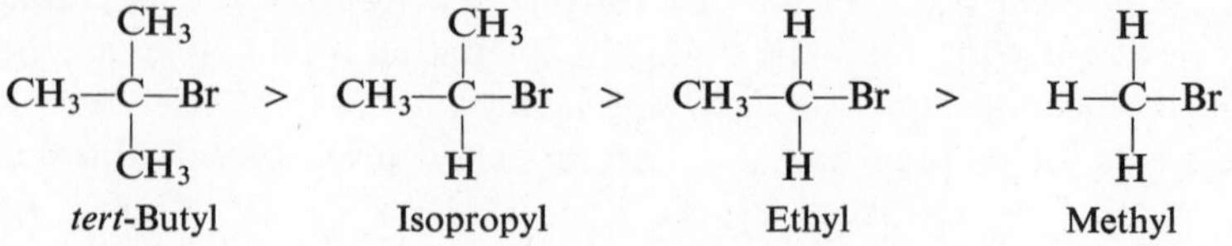

	$CH_3-\overset{CH_3}{\underset{CH_3}{C}}-Br$	>	$CH_3-\overset{CH_3}{\underset{H}{C}}-Br$	>	$CH_3-\overset{H}{\underset{H}{C}}-Br$	>	$H-\overset{H}{\underset{H}{C}}-Br$
	tert-Butyl		Isopropyl		Ethyl		Methyl
Relative rate:	100 million		45		1.7		1.0

(Formic acid is used here as an even better ionizing solvent than water.)

The rate of an S_N2 reaction, we saw, is affected largely by steric factors, that is, by the bulk of the substituents. In contrast, the rate of an S_N1 reaction is affected largely by **electronic factors**, that is, by the tendency of substituents to release or withdraw electrons.

Problem 14.6 Neopentyl halides are notoriously slow in nucleophilic substitution, whatever the experimental conditions. How can you account for this?

14.15 The S_N1 reaction: rearrangement

If the S_N1 reaction involves intermediate carbonium ions, we might expect it to show one of the characteristic features of carbonium ion reactions: *rearrangement*. In an S_N2 reaction, on the other hand, the halide ion does not leave until the nucleophilic reagent has become attached; there is no free intermediate particle and hence we would expect no rearrangement. These expectations are correct.

The following example illustrates this point. We shall see (Sec. 16.5) that the neopentyl cation is particularly prone to rearrange to the more stable *tert*-pentyl cation. Neopentyl bromide reacts (slowly) with ethoxide ion by an S_N2 mechanism to yield neopentyl ethyl ether; it reacts (slowly) with ethyl alcohol by an S_N1 reaction to yield almost entirely rearranged products.

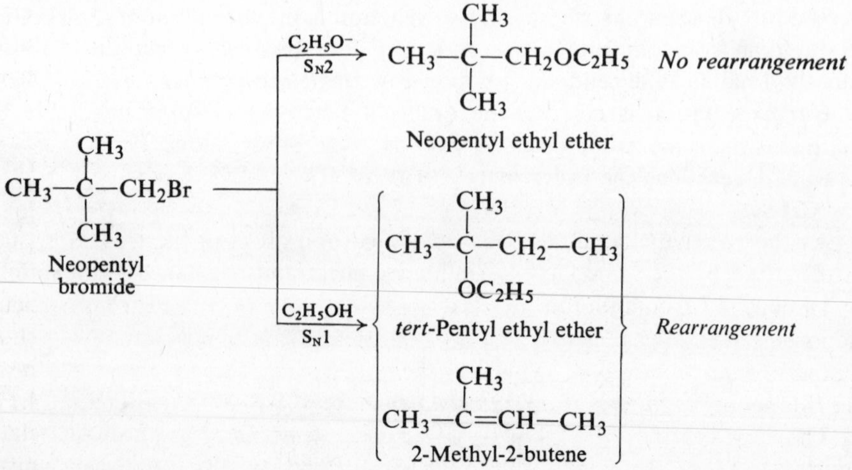

$$CH_3-\overset{CH_3}{\underset{CH_3}{C}}-CH_2Br \xrightarrow[S_N2]{C_2H_5O^-} CH_3-\overset{CH_3}{\underset{CH_3}{C}}-CH_2OC_2H_5 \quad \textit{No rearrangement}$$

Neopentyl bromide → Neopentyl ethyl ether

$$CH_3-\overset{CH_3}{\underset{CH_3}{C}}-CH_2Br \xrightarrow[S_N1]{C_2H_5OH} \left\{ CH_3-\overset{CH_3}{\underset{OC_2H_5}{C}}-CH_2-CH_3, \quad CH_3-\overset{CH_3}{C}=CH-CH_3 \right\} \quad \textit{Rearrangement}$$

Neopentyl bromide → *tert*-Pentyl ethyl ether; 2-Methyl-2-butene

Because of the strong correlation between rearrangement and formation of carbonium ions, in the absence of other information rearrangement is often taken as an indication of an S_N1 mechanism.

We notice that the S_N1 reaction is accompanied by much elimination; expulsion of a proton to yield an alkene is, of course, typical behavior of a carbonium ion.

14.16 S_N2 *vs.* S_N1

The strength of the evidence for the two mechanisms, S_N1 and S_N2, lies in its consistency. Nucleophilic substitutions that follow first-order kinetics also show racemization and rearrangement, and the reactivity sequence 3° > 2° > 1° > CH_3X. Reactions that follow second-order kinetics show complete stereochemical inversion and no rearrangement, and follow the reactivity sequence CH_3X > 1° > 2° > 3°. (The few exceptions to these generalizations are understandable exceptions; see Problem 16.5, p. 525.)

$$RX = \overset{\xleftarrow{S_N2 \text{ increases}}}{\underset{\xrightarrow[S_N1 \text{ increases}]{}}{CH_3X \quad 1^\circ \quad 2^\circ \quad 3^\circ}}$$

S_N2 *vs.* S_N1

Because there are two opposing reactivity sequences, we seldom encounter *either* of them in a pure form, but find instead a sequence that is a combination of the two. Most typically for halides, as we go along the series CH_3, 1°, 2°, 3°, reactivity passes through a *minimum*, usually at 2°:

$$\begin{array}{ccccccc} CH_3X & > & 1^\circ & > & 2^\circ & < & 3^\circ \\ S_N2 & & S_N2 & & \text{Mixed} & & S_N1 \end{array}$$

Reactivity by the S_N2 mechanism decreases from CH_3 to 1°, and at 2° is so low that the S_N1 reaction begins to contribute significantly; reactivity, now by S_N1, rises sharply to 3°. The change in mechanism at 2° is confirmed by kinetics and other evidence.

The occurrence of a minimum or maximum in a property—reactivity, acidity, antibacterial activity, etc.—as one proceeds along a logical series always suggests the working of opposing factors. (See, for example, the effect of acidity on certain carbonyl reactions, Sec. 19.14.) In the case of nucleophilic aliphatic substitution, a minimum of the kind we have just encountered is highly characteristic of a change in molecularity of reaction.

Problem 14.7 In 80% ethanol at 55°, the second-order rate constant for the reaction of ethyl bromide with hydroxide ion is 0.0017 liters/mole/sec. Making use of this rate constant and those in Sec. 14.8 and Problem 14.4, calculate the relative rates of hydrolysis in 0.1N hydroxide for methyl, ethyl, isopropyl, and *tert*-butyl bromides.

Despite the predisposition of a particular class of halide toward a particular reaction mechanism, we can to a certain extent control the reaction by our choice of experimental conditions. (This was done, for example, to obtain the nearly pure S_N2 sequence in Sec. 14.11 and the nearly pure S_N1 sequence in Sec. 14.14.)

The very way in which changes in experimental conditions affect the relative importance of the two mechanisms provides additional evidence for the mechanisms.

We have already seen an example of this: high **concentration of the nucleophilic reagent** favors the S_N2 reaction; low concentration favors the S_N1 reaction.

The **nature of the nucleophilic reagent** also plays an important role: for example, neopentyl bromide reacts with ethoxide ion by the S_N2 mechanism and with ethyl alcohol by the S_N1 mechanism. The strongly nucleophilic (strongly basic) ethoxide ion pushes halogen from the molecule, whereas the weakly nucleophilic ethanol waits to be invited in.

Finally, the **polarity of the solvent** can often determine the mechanism by which reaction occurs. Ionization of an alkyl halide is possible only because most of the energy needed to reach the transition state is supplied by formation of dipole–dipole bonds between the solvent and the polar transition state. The more polar the solvent, the stronger the solvation forces and the faster the ionization.

$$\underset{\text{Reactant}}{R{-}X} \longrightarrow \underset{\text{Transition state}}{\left[\overset{\delta_+}{R}\cdots\overset{\delta_-}{X}\right]} \longrightarrow \underset{\text{Products}}{R^+ + X^-}$$

More polar than reactant: stabilized more by solvation

Changing the solvent, say, from 80% ethanol to the much more polar water should speed up ionization and hence the rate of the S_N1 reaction. What effect will this have on the S_N2 reaction? Here we do not have a transition state that is more polar than the reactants; in fact, since the negative charge is dispersed over —OH and —X, this transition state is *less* strongly solvated than the reactants.

$$\underset{\text{Reactants}}{HO^- + R{-}X} \longrightarrow \underset{\text{Transition state}}{\left[\overset{\delta_-}{HO}\cdots R\cdots\overset{\delta_-}{X}\right]} \longrightarrow \underset{\text{Products}}{HO{-}R + X^-}$$

Reactants: *Concentrated charge: stabilized more than transition state by solvation*

Transition state: *Dispersed charge*

Increasing the polarity of the solvent slows down the S_N2 reaction slightly. Other things being equal, the more polar the solvent, the more likely it is that an alkyl halide will react by the S_N1 mechanism. (For a closer look at this matter, see Sec. 18.11.)

These mechanisms give us some idea of the kind of behavior to expect from a halide of a particular structure: its reactivity under a given set of conditions, the likelihood of racemization or of rearrangement, the extent of elimination. They tell us how to change the experimental conditions—concentration, solvent, the nucleophilic reagent—to achieve the results we want: to speed up reaction, to avoid racemization or rearrangement, to minimize elimination.

Problem 14.8 Benzyl bromide reacts with H_2O in formic acid solution to yield benzyl alcohol; the rate is independent of $[H_2O]$. Under the same conditions *p*-methylbenzyl bromide reacts 58 times as fast.

Benzyl bromide reacts with ethoxide ion in dry alcohol to yield benzyl ethyl ether ($C_6H_5CH_2OC_2H_5$); the rate depends upon both [RBr] and $[OC_2H_5^-]$. Under the same conditions *p*-methylbenzyl bromide reacts 1.5 times as fast.

Interpret these results. What do they illustrate concerning the effect of: (a) polarity of solvent, (b) nucleophilic power of the reagent, and (c) electron release by substituents?

Problem 14.9 The rate of reaction of 3-chloro-1-butene with ethoxide ion in ethyl alcohol depends upon both [RCl] and $[OC_2H_5^-]$; the product is 3-ethoxy-1-butene, $CH_3CH(OC_2H_5)CH{=}CH_2$. The reaction of 3-chloro-1-butene with ethyl alcohol alone, on the other hand, yields not only 3-ethoxy-1-butene but also 1-ethoxy-2-butene, $CH_3CH{=}CHCH_2OC_2H_5$. How do you account for these results? (*Hint:* See Sec. 8.21.)

Problem 14.10 Predict the effect of increasing solvent polarity on the rate of: (a) the S_N2 attack by ammonia on an alkyl halide:

$$RX + NH_3 \longrightarrow RNH_3^+ + X^-$$

(b) the S_N1 reaction of an alkyldimethylsulfonium ion with the solvent:

$$RS(CH_3)_2^+ \longrightarrow (CH_3)_2S + R^+$$

$$R^+ \xrightarrow{H_2O,\ C_2H_5OH} ROH + ROC_2H_5$$

14.17 Solvolysis

Let us turn briefly to the special case of nucleophilic aliphatic substitution in which the solvent is the nucleophile: *solvolysis*. In its various aspects, solvolysis is—and has been for many years—the most intensively studied reaction in organic chemistry. Yet it is the reaction about which there is probably the most intense disagreement.

$$R{:}X + \underset{\text{solvent}}{:S} \longrightarrow R{:}S + X{:}^-$$

There is no added strong nucleophile and so, for many compounds, solvolysis falls into the category we have called S_N1: that is, reaction proceeds by two—or more—steps, with the intermediate formation of a carbonium ion. It is this intermediate that lies at the center of the problem: its nature, how it is formed, and how it reacts. In studying solvolysis one is studying all S_N1 reactions and, in many ways, all reactions involving intermediate carbonium ions.

Perhaps the biggest question to be answered is: just what is the role played by the solvent? Does it, at one extreme, simply cluster about the carbonium ion and the anion—and the transition state leading to their formation—aiding in heterolysis through formation of ion–dipole bonds? Or, at the other extreme, does a single solvent molecule act as a nucleophile and help push the leaving group out of the molecule? Kinetics cannot be used to give a direct answer to this question since the concentration of the solvent does not change during the course of reaction. (We can, of course, study the kinetics of hydrolysis of alkyl halides in, say, formic acid solution—but this is certainly not the same reaction as hydrolysis in water.)

It seems clear that the solvent can give *nucleophilic assistance* to solvolysis. How strong this assistance is depends upon the nucleophilic power of the solvent, and upon how badly the assistance is needed. Water and alcohols, for example, are strongly nucleophilic solvents, acetic acid is weaker, formic acid is weaker yet, and trifluoroacetic acid is very weak. Formation of tertiary cations is relatively easy and needs little nucleophilic assistance; in any case, crowding would discourage such assistance. Reactivity of tertiary substrates depends little on nucleophilic power of the solvent and chiefly on its polarity. Formation of secondary cations needs much nucleophilic assistance; reactivity depends on both nucleophilic power and polarity of the solvent. With most primary substrates, reaction is probably straight-forward S_N2: a single step with solvent acting as nucleophile.

Let us concentrate, then, on secondary alkyl substrates. Just what is meant by the term *nucleophilic assistance*? First of all, it differs from the S_N2 kind of attack in this way: it leads to the formation, not of product, but of the intermediate carbonium ion. Next, it differs from general "solvation" in this way: a single solvent molecule is involved, not a cluster. This solvent molecule attacks the substrate at the back side and, acting as a nucleophile, helps to push the leaving group out the front side. There is formed a carbonium ion—or at least something with a great deal of cation character. Clinging to its back side is the solvent molecule and to the front side, the leaving group. Each may be bonded to carbon through overlap of a lobe of the empty *p* orbital. The geometry is similar to that in the S_N2 transition state, but this is an *intermediate*, and corresponds to an energy minimum in a progress-of-reaction plot. If the leaving group is an anion, and if the solvent is of only moderate polarity, bonding between cation and anion may be chiefly electrostatic, and one speaks of an *ion pair*.

This cationoid intermediate—this "carbonium ion"—now reacts. Ultimately it reacts with a solvent molecule—with formation of a full-fledged bond—to yield product. If, at the time of reaction, the leaving group is still bonded to the front side—or is still lurking there—reaction with solvent occurs at the back side. If, on the other hand, the carbonium ion has lasted long enough for the leaving group to be exchanged for a second solvent molecule—thus forming a symmetrical intermediate—reaction with solvent is equally likely at front or back side. Solvolysis can occur with complete inversion or with inversion plus varying amounts of racemization.

Elegant work by Saul Winstein (of the University of California, Los Angeles) revealed the detailed behavior of ion pairs that are intermediates in certain cases of solvolysis: *tight* (or *intimate*) ion pairs, the cation of which is free enough to pivot about and lose configuration, and yet is held tightly enough that recombination to the covalently bonded compound is the favored process; *loose* (or *solvent-separated*) ion pairs, the cation of which is susceptible to attack by outside nucleophiles.

It has been suggested that there is a continuous spectrum of mechanisms for nucleophilic substitution ranging from the idealized S_N1 reaction (called **Lim**, for limiting) at one end, to the idealized S_N2 reaction (called **N**) at the other. On progress-of-reaction plots, the energy minimum for the carbonium ion becomes shallower and shallower as we move away from the S_N1 end; at the S_N2 end the minimum has disappeared, and we have a single maximum.

This viewpoint may well be correct. But the differences in stability between the various classes of carbonium ions are great enough that, by and large, reactions fall into three separated groups: (a) for primary substrates, S_N2; (b) for tertiary substrates, S_N1, with an intermediate that approximates our idea of a simple (solvated) carbonium ion; and (c) for secondary substrates, a two-step reaction that is S_N1 to the extent that there is a cationoid intermediate, but one formed with nucleophilic assistance and still encumbered with nucleophile (solvent) and leaving group.

14.18 Elimination: E2 and E1

We encountered dehydrohalogenation in Sec. 5.12 as one of the best methods of preparing alkenes. At that time we said that the mechanism involves a single step: base pulls a hydrogen ion away from carbon, and simultaneously a halide ion separates—aided, of course, by solvation.

E2

Bimolecular elimination

$$\underset{\underset{\curvearrowleft :B}{H}}{-\overset{X}{\overset{|}{C}}-\overset{|}{\underset{|}{C}}-} \longrightarrow X^- + \;>C{=}C<\; + H{:}B$$

But we have just learned that alkyl halides, particularly tertiary ones, can dissociate into halide ions and carbonium ions; and we already know (Sec. 5.20) that a carbonium ion can lose a hydrogen ion to a base to form an alkene.

E1

Unimolecular elimination

$$(1)\quad -\overset{X}{\overset{|}{\underset{|}{C}}}-\overset{|}{\underset{\underset{H}{|}}{C}}- \longrightarrow X^- + -\overset{\oplus}{\underset{|}{C}}-\overset{|}{\underset{\underset{H}{|}}{C}}- \qquad \textit{Slow}$$

A carbonium ion

$$(2)\quad -\overset{\oplus}{\underset{|}{C}}-\overset{|}{\underset{\underset{\underset{\curvearrowleft :B}{H}}{|}}{C}}- \longrightarrow \;>C{=}C<\; + H{:}B \qquad \textit{Fast}$$

Are there, then, *two mechanisms* for dehydrohalogenation of alkyl halides? The answer is: *yes*, and they bear the same relationship to each other as do the S_N2 and S_N1 mechanisms for substitution. They are the **E2 mechanism** (*elimination, bimolecular*), which involves two molecules in the rate-determining step, and the **E1 mechanism** (*elimination, unimolecular*), which involves one molecule in the rate-determining step.

The order of reactivity of alkyl halides toward E2 or E1 elimination is the same:

Reactivity toward E2 or E1 elimination $3° > 2° > 1°$

This sequence reflects, for the E2 reaction, the relative stabilities of the alkenes being formed (Sec. 5.14), and for the E1 reaction, the stabilities of the carbonium ions being formed in the first (slow) step.

But, as we might expect, reactions following the two mechanisms differ in kinetics: second-order for E2, and first-order for E1. At the base concentrations ordinarily used for dehydrohalogenation, the E2 mechanism—whose rate depends upon base concentration—is the principal reaction path. The E1 mechanism is generally encountered only with tertiary halides and in solutions of low base concentration. Using the difference in kinetics as our point of departure, let us look at the evidence for each of the mechanisms.

14.19 Evidence for the E1 mechanism

What is the evidence for the **E1 mechanism**? The elimination reactions that

(a) *follow first-order kinetics*

also:

(b) show *the same effect of structure on reactivity* as in S_N1 reactions; and
(c) where the structure permits, *are accompanied by rearrangement.*

The fact that the rate is independent of the base concentration is interpreted as it was in Sec. 14.12 for the S_N1 reaction; indeed, we see that the rate-determining step in E1 and S_N1 reactions is *exactly the same.* It follows that the order of reactivity of halides should be the same as in S_N1 reactions—and it is.

Finally, first-order elimination is accompanied by the same kind of rearrangement that we expect for a reaction proceeding by way of carbonium ions. The 2-methyl-2-butene formed from neopentyl bromide (Sec. 14.15) is clearly the product of E1 elimination. Indeed, the reaction in which we first encountered rearrangement, dehydration of alcohols, is simply E1 elimination involving the protonated alcohol (Sec. 5.22).

14.20 Evidence for the E2 mechanism

What is the evidence for the **E2 mechanism**? The elimination reactions that

(a) *follow second-order kinetics*

also:

(b) *are not accompanied by rearrangements*;
(c) show *a large deuterium isotope effect*;
(d) *do not undergo hydrogen–deuterium exchange*; and
(e) show *a large element effect.*

Under conditions where reactions follow second-order kinetics, dehydrobromination of ordinary isopropyl bromide by sodium ethoxide takes place *seven times as fast* as that of the labeled compound, $(CD_3)_2CHBr$. An isotope effect of this size, we have seen (Sec. 11.15), reveals the breaking of a carbon–hydrogen bond in the transition state of the rate-determining step.

Facts (a), (b), and (c) are, of course, exactly what we would expect for the E2 mechanism. The rate-determining step (the *only* step) involves reaction between a molecule of alkyl halide and a molecule of base, in which a carbon–hydrogen bond is broken, and in which there is no opportunity for rearrangement. In particular, these three facts rule out a carbonium ion (E1) mechanism for second-order elimination.

There is, however, another reasonable mechanism that we must consider: *the carbanion mechanism*, which has as its first step the abstraction of a hydrogen ion to form a negatively charged particle called a *carbanion*. This mechanism, like the E2, is consistent with facts (a), (b), and (c).

$$(1)\quad -\overset{\overset{\displaystyle X}{|}}{\underset{|}{C}}-\overset{|}{\underset{\underset{\displaystyle H\nwarrow :B}{|}}{C}}- \;\rightleftarrows\; -\overset{\overset{\displaystyle X}{|}}{\underset{|}{C}}-\overset{|}{\underset{\ddot{}\ominus}{C}}- \;+\; H{:}B$$

A carbanion

Elimination via carbanion

Seldom observed

$$(2)\quad -\overset{\overset{\displaystyle X}{|}}{\underset{|}{C}}-\overset{|}{\underset{\ddot{}\ominus}{C}}- \;\longrightarrow\; X^- + \;\rangle C{=}C\langle$$

In an attempt to distinguish between these two possibilities, dehydrohalogenation of β-phenylethyl bromide, $C_6H_5CH_2CH_2Br$, was carried out in labeled ethanol, C_2H_5OD, with ethoxide ion as base. If carbanions were formed, *some* of them might be expected to recombine with a hydrogen ion to regenerate the starting material:

$$\underset{\text{H}\quad\text{Br}}{C_6H_5-CH-CH_2} \;\underset{C_2H_5OH}{\overset{C_2H_5O^-}{\rightleftarrows}}\; \underset{\ominus\quad\;\;\text{Br}}{C_6H_5-\ddot{C}H-CH_2} \;\underset{C_2H_5OD}{\overset{C_2H_5O^-}{\leftrightarrows}}\; \underset{\text{D}\quad\text{Br}}{C_6H_5-CH-CH_2}$$

Unlabeled halide
Starting material

Labeled halide
Exchange product

$$\downarrow$$

$$C_6H_5-CH{=}CH_2 + Br^-$$

Elimination product

But in this recombination the carbanion would be almost certain to regain a *deuteron*, not a proton, since nearly all the molecules of alcohol are C_2H_5OD, not C_2H_5OH.

When this reaction was interrupted, and unconsumed β-phenylethyl bromide recovered, it was found by mass spectrometric analysis to contain *no deuterium*. Similar experiments with other systems have given similar results: in typical second-order elimination reactions there is *no hydrogen–deuterium exchange*.

Fact (d) shows, then, that if carbanions are involved in these elimination reactions they lose halide ion much more rapidly than they combine with hydrogen ions. (That is, k_2 would have to be much larger than k_{-1}.) But if this were so, the first step (1) would be rate-determining—just as in an S_N1 reaction, or in aromatic nitration—and the rate of the second step (2) would have no effect on the overall rate of reaction.

Now, is this true? How can we tell whether or not the rate at which halide ion is lost affects the rate of elimination? We might consider looking for an isotope effect, as was done in aromatic nitration. But here that would be a very difficult job. We are not dealing with loss of hydrogen, whose isotopes differ two- or three-fold in mass; we are dealing with loss of elements like chlorine, whose isotopes differ by only a few percent, with correspondingly small differences in the ease with which bonds are broken.

It has been pointed out by Joseph Bunnett (of the University of California, Santa Cruz) that evidence against the carbanion mechanism is available in the *element effect*.

In S_N1 and S_N2 displacements the reactivity of alkyl halides follows the sequence

$$R—I > R—Br > R—Cl > R—F$$

with the ease of breaking the carbon–halogen bond depending upon its strength (see Table 1.2, p. 21). The differences in rate here are quite large: alkyl bromides, for example, react 25 to 50 times as fast as alkyl chlorides. These element effects are, in fact, much larger than the isotope effects observed for the breaking of bonds to protium and deuterium—as, indeed, they *should* be, considering the much greater differences in bond strength.

Now, in these elimination reactions, the reactivity of alkyl halides follows the same sequence as for substitution—and with element effects of just about the same size. Clearly, the rate of breaking the carbon–halogen bond *does* affect the overall rate of reaction. On this evidence, if carbanions were formed, they would find step (2) difficult and would revert to starting material many times before finally losing halide ion. But such reversible carbanion formation has been ruled out by the absence of isotopic exchange.

Thus, only the E2 mechanism fits all the facts.

14.21 Orientation of elimination. The variable E2 transition state

Where elimination can produce a mixture of isomers, which one predominates? We saw earlier (Sec. 5.14) that in dehydrohalogenation the more stable alkene is formed faster: *sec*-butyl bromide, for example, yields more 2-butene than 1-butene. We attributed this orientation to the *alkene character* of the transition

$$\underset{\substack{|\\ Br \\ \textit{sec}\text{-Butyl bromide}}}{CH_3CH_2CHCH_3} \xrightarrow{\text{alc. KOH}} \underset{\substack{\text{2-Butene}\\ \textit{81\%}}}{CH_3CH{=}CHCH_3} + \underset{\substack{\text{1-Butene}\\ \textit{19\%}}}{CH_3CH_2CH{=}CH_2}$$

state: the double bond is partly formed in the transition state, and factors that stabilize an alkene stabilize this incipient alkene.

Predominant formation of the more stable isomer is called **Saytzeff orientation** after the Russian chemist Alexander Saytzeff (University of Kazan), who in 1875 first formulated a "rule" for orientation in dehydrohalogenation.

Problem 14.11 Like Markovnikov, Saytzeff stated his rule in terms, not of product stability, but of numbers of hydrogens on carbon atoms. (a) Suggest a wording for this original Saytzeff rule. (b) Predict the major product of dehydrohalogenation of 2-bromo-1-phenylbutane on the basis of the original rule. (c) On the basis of the modern rule.

But orientation in E2 elimination is not always Saytzeff, particularly when compounds other than alkyl halides are involved. To see the various factors at work here, let us take, as a simple example, elimination from the 2-hexyl halides brought about by the strong base sodium methoxide. The iodide, bromide, and chloride do react with Saytzeff orientation, but the fluoride gives predominantly the less substituted alkene, 1-hexene. Such orientation is called **Hofmann orientation**, since it was first observed by Hofmann (in elimination from quaternary ammonium salts, Sec. 23.5). Furthermore, we can see that there is a steady increase in the fraction of 1-hexene along the series I, Br, Cl, F.

$$CH_3CH_2CH_2CH_2\underset{\displaystyle X}{\underset{|}{C}}HCH_3 \xrightarrow[E2]{CH_3ONa/CH_3OH} \text{2-hexene and 1-hexene}$$

	2-hexene	1-hexene
X = I	81%	19%
Br	72	28
Cl	67	33
F	30	70

Such observations are best understood in terms of what Bunnett (p. 478) has called the *variable transition state* theory of E2 elimination. We are speaking, remember, of a one-step elimination; both the C—H and C—X bonds are being broken in the same transition state. But there is a whole spectrum of E2 transition states which differ in the relative *extent* to which the two bonds are broken.

Variable E2 Transition State

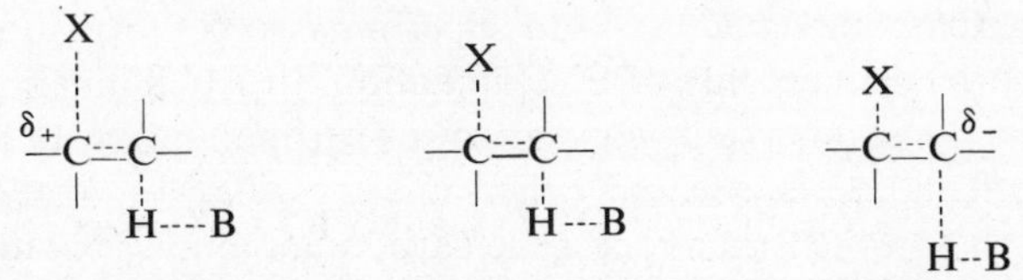

Carbonium-like: X----$\overset{\delta+}{C}$==C—, H----B

Central: X---C==C—, H---B

Carbanion-like: X--C==$\overset{\delta-}{C}$—, H--B

Carbonium-like — Central (*Much alkene character*) — Carbanion-like

At the center of the spectrum is the transition state we have described before for elimination from alkyl halides: both C—H and C—X bonds are broken to a considerable extent, the transition state has considerable alkene character, and orientation is Saytzeff.

But, if breaking of the C—H bond greatly exceeds breaking of the C—X bond, there is little alkene character to the transition state, but instead the development of negative charge on the carbon losing the proton. In this case, the transition state has *carbanion character*, and its stability is controlled as we might expect, by dispersal or intensification of the negative charge: electron-withdrawing groups stabilize, and electron-releasing groups destabilize. At one end of the spectrum, then, we have the carbanion-like transition state.

At the other end of the spectrum is the transition state in which C—X bond-breaking greatly exceeds C—H bond-breaking. Positive charge develops on the carbon losing the leaving group, giving carbonium ion character to the transition state. Alkene character is diminished, and we might expect orientation to be less strongly Saytzeff.

Consider elimination from the 2-hexyl halides. With the iodide, there is considerable breaking of both bonds in the transition state, much alkene character, and preferred formation of the more stable alkene: Saytzeff orientation. As we go along the series I, Br, Cl, F, the C—X bond becomes stronger, and the extent to which it is broken in the transition state decreases. At the same time, the electron-withdrawing effect of X increases, favoring the development of negative charge.

With the fluoride, we have predominant C—H bond-breaking, with little alkene character but considerable carbanion character to the transition state. A primary hydrogen is preferentially abstracted by base, since that permits the negative charge to develop on a primary carbon, to which there is attached only one electron-releasing alkyl group. Orientation is Hofmann.

Bunnett believes that C—F bond-breaking lags behind C—H bond-breaking chiefly because of the strength of the C—F bond. Ingold (p. 460), who was the first to suggest carbanion character as the underlying cause of Hofmann orientation, believed that electron withdrawal by fluorine is the major factor.

In E2 elimination with bases like KOH and CH_3ONa, most alkyl halides give Saytzeff orientation. Certain other compounds (quaternary ammonium salts, Sec. 23.5, for example) give Hofmann orientation. Alkyl sulfonates fall in between. With each kind of compound, orientation is affected—sometimes drastically—by the choice of base and solvent, and by stereochemistry. (The percentage of 1-hexene from 2-hexyl chloride, for example, jumps from 33% in CH_3ONa/CH_3OH to 91% in *t*-BuOK/*t*-BuOH, evidently for steric reasons.) In all this, we should remember that orientation is a matter of relative stabilities of competing transition states; these stabilities are determined by electronic factors—alkene character and carbanion character—with superimposed conformational factors.

So far, we have spoken only of E2 elimination. In E1 elimination, orientation is determined in the second step: conversion of carbonium ion to alkene. As we might expect, orientation is essentially the same regardless of what leaving group has departed earlier in the formation of the carbonium ion. Orientation is strongly Saytzeff, reflecting much alkene character in the transition state.

Problem 14.12 2-Phenylethyl bromide undergoes E2 elimination about 10 times as fast as 1-phenylethyl bromide, even though they both yield the same alkene. Suggest a possible explanation for this.

14.22 Stereochemistry of elimination

Dehydrohalogenation of 1-bromo-1,2-diphenylpropane gives, as we would expect, 1,2-diphenylpropene. But the halide contains two chiral centers, and we

$$C_6H_5\overset{*}{C}H(Br)-\overset{*}{C}H(CH_3)-C_6H_5 \longrightarrow C_6H_5CH{=}C(CH_3)C_6H_5$$

1-Bromo-1,2-diphenylpropane   1,2-Diphenylpropene

can easily show that it can exist as two pairs of enantiomers; each pair is diastereomeric with the other pair. On E2 elimination, one pair of enantiomers yields *only* the *cis*-alkene, and the other pair yields *only* the *trans*-alkene. The reaction is completely *stereospecific* (Sec. 7.11).

As this example and many others show, the bimolecular reaction of alkyl halides involves *anti*-elimination: in the transition state the hydrogen and the leaving group are located as far apart as possible, in the *anti* relationship (Sec. 3.3) as opposed to *gauche* or *eclipsed* (see Fig. 14.4, p. 482).

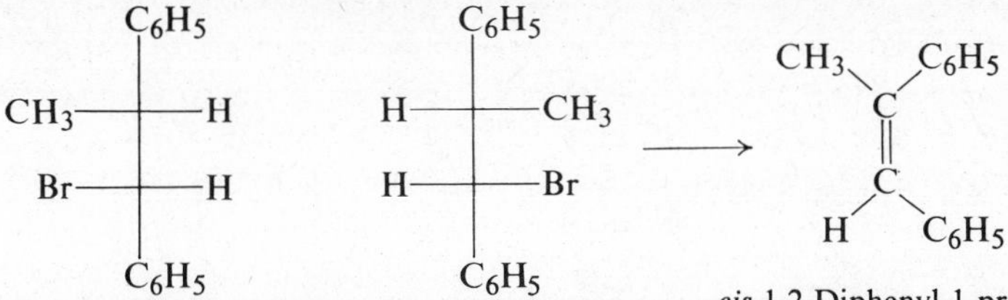

C_6H_5 / CH_3—H / Br—H / C_6H_5 I (1R,2R)

C_6H_5 / H—CH_3 / H—Br / C_6H_5 II (1S,2S)

1-Bromo-1,2-diphenylpropane

→ $CH_3(C_6H_5)C{=}C(H)(C_6H_5)$

cis-1,2-Diphenyl-1-propene

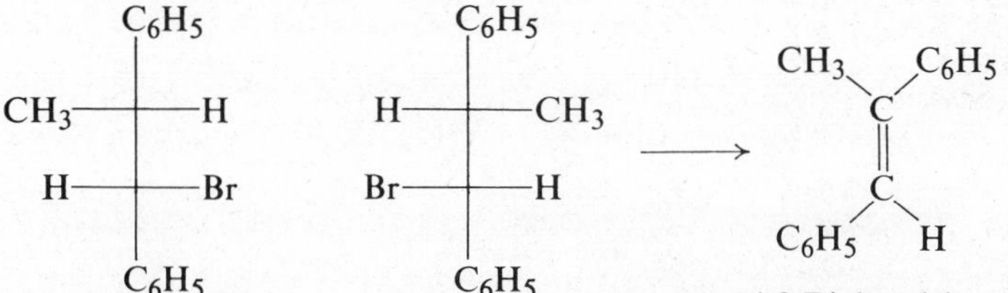

C_6H_5 / CH_3—H / H—Br / C_6H_5 III (1S,2R)

C_6H_5 / H—CH_3 / Br—H / C_6H_5 IV (1R,2S)

1-Bromo-1,2-diphenylpropane

→ $CH_3(C_6H_5)C{=}C(C_6H_5)(H)$

trans-1,2-Diphenyl-1-propene

Thus, diastereomer I or its enantiomer, II, gives the *cis*-alkene:

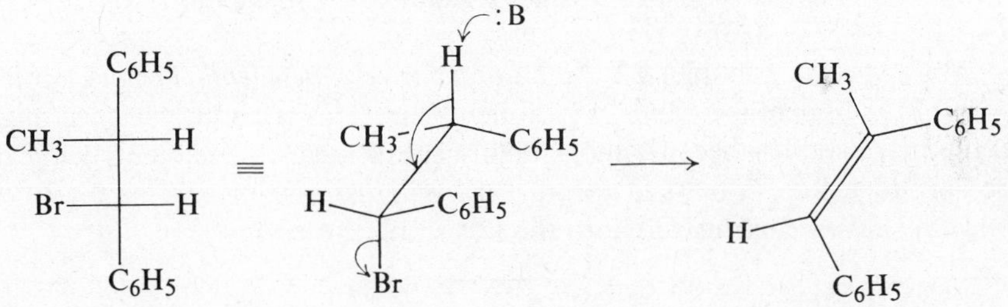

C_6H_5 / CH_3—H / Br—H / C_6H_5 ≡ (:B, H, CH_3, C_6H_5, H, C_6H_5, Br) → CH_3, C_6H_5, H, C_6H_5

I

cis-1,2-Diphenyl-1-propene

and diastereomer III (or its enantiomer, IV) gives the *trans*-alkene:

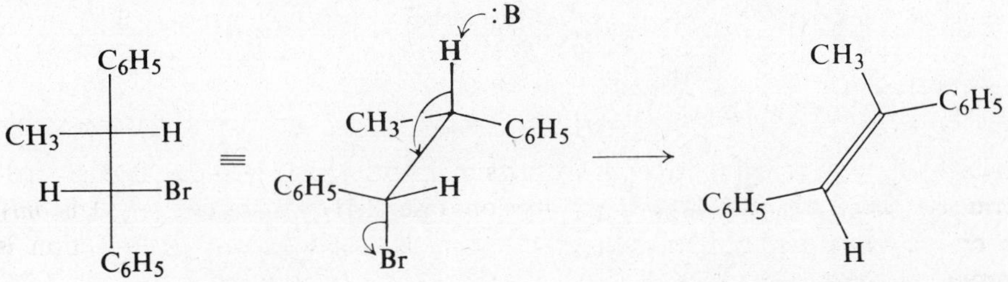

C_6H_5 / CH_3—H / H—Br / C_6H_5 ≡ (:B, H, CH_3, C_6H_5, C_6H_5, H, Br) → CH_3, C_6H_5, C_6H_5, H

III

trans-1,2-Diphenyl-1-propene

The preference for *anti*-elimination from halides can be very strong. To see this is so, we must turn from open-chain compounds to cyclic compounds. In

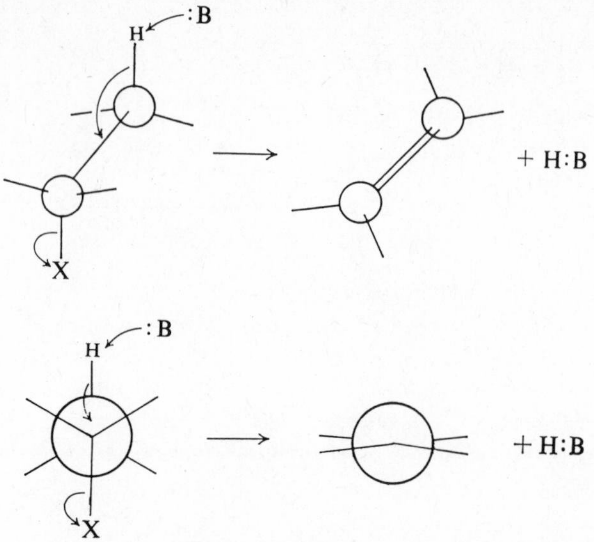

Figure 14.4. The E2 reaction of alkyl halides: *anti*-elimination. Hydrogen and the leaving group, —X, are as far apart as possible, in the *anti* relationship.

cyclohexane rings, 1,2-substituents can take up the *anti* conformation only by occupying axial positions; this, in turn, is possible only if they are *trans* to each other (see Fig. 14.5, p. 483).

To take a specific example: E2 elimination converts *neomenthyl chloride* into a mixture of 75% 3-menthene and 25% 2-menthene. This is about what we might expect, the more stable—because more highly substituted—3-menthene being the preferred product. But, in marked contrast, E2 elimination converts the diastereomeric *menthyl chloride* exclusively into the less stable 2-menthene.

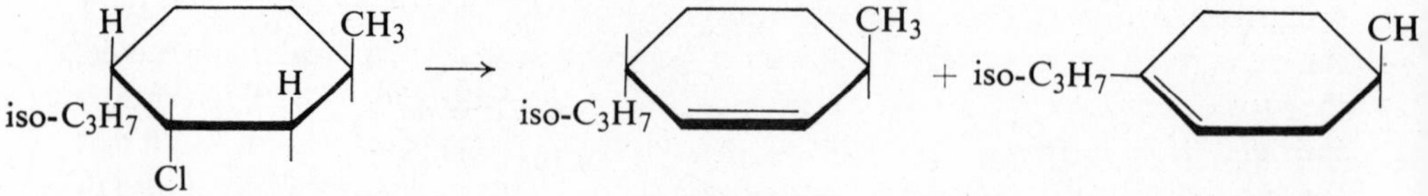

Neomenthyl chloride

2-Menthene
25%

3-Menthene
75%

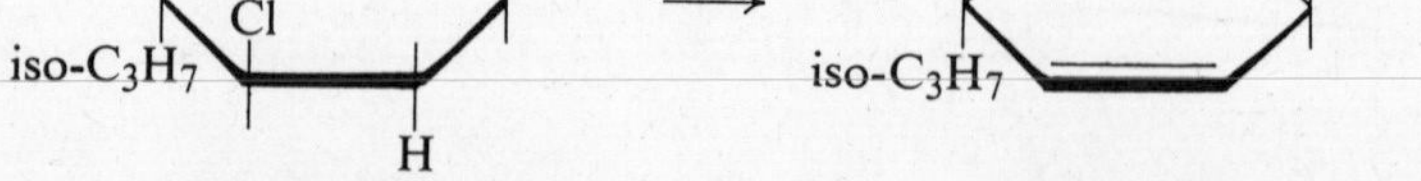

Menthyl chloride

2-Menthene
Only product

How are we to account for these differences in behavior? In neomenthyl chloride there is a hydrogen on either side of the chlorine which is *trans* to the

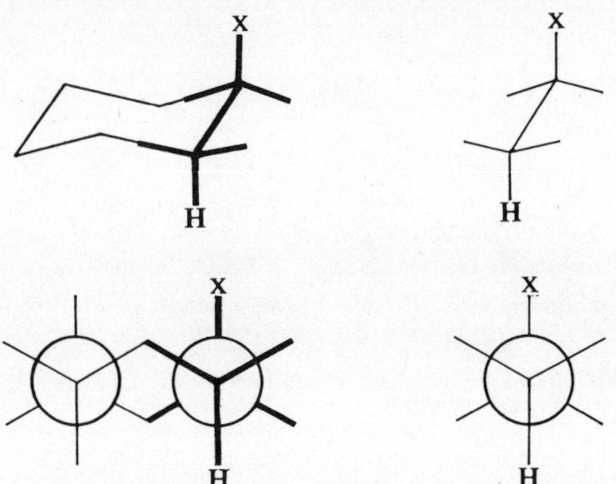

Figure 14.5. Only *trans*-1,2-substituents can assume the *anti* relationship.

chlorine, and which can take up a conformation *anti* to it. Either hydrogen *can* be eliminated, and the ratio of products is determined in the usual way, by the relative stabilities of the alkenes being formed. In menthyl chloride, on the other hand, only one hydrogen is *trans* to the chlorine, and it is the only one that is eliminated, despite the fact that this yields the less stable alkene.

In recent years it has become clear that E2 reactions can also proceed by *syn*-elimination: in the transition state the hydrogen and leaving group are in the *eclipsed* (or *gauche*) relationship. Although uncommon for alkyl halides, *syn*-elimination is often observed for quaternary ammonium salts and sometimes for alkyl sulfonates. On electronic grounds, the most stable transition states seem to be those in which the hydrogen and leaving group are *periplanar* (in the same plane) to permit overlap of incipient *p* orbitals in the partially-formed double bond. Of the two periplanar eliminations, the *anti* is probably easier than the *syn*—other things being equal. But various factors may throw the stereochemistry one way or the other. Conformational effects enter in, and the degree of carbanion character; the stereochemistry is affected by the strength of the base and by its bulk and by the bulk of the leaving group. Ring systems present special situations: it is difficult for *cis*-1,2-substituents to become *syn*-periplanar in cyclohexanes, but easy in cyclopentanes.

Problem 14.13 When treated with *t*-BuOK in DMSO, diastereomer V (and its

```
      CH3                    CH3
       |                      |
  H ---+--- D           D ---+--- H
       |                      |
  H ---+--- OTs         H ---+--- OTs
       |                      |
      CH3                    CH3
       V                      VI
```

enantiomer) gave *cis*-2-butene without loss of deuterium and *trans*-2-butene with loss of deuterium; diastereomer VI (and its enantiomer) gave *trans*-2-butene without loss of deuterium. How do you account for these findings? What is the stereochemistry of elimination here?

Problem 14.14 Of the various isomeric 1,2,3,4,5,6-hexachlorocyclohexanes, one isomer undergoes dehydrohalogenation by base much more slowly than the others. Which isomer is probably the unreactive one, and why is it unreactive?

Problem 14.15 Suggest an explanation for the fact that dehydrohalogenation of *sec*-butyl chloride yields both *cis*- and *trans*-2-butene, but mostly (6:1) the *trans*-isomer. (Assume only *anti* elimination.)

Problem 14.16 How do you account for the fact that when heated in ethanol, in the absence of added base, menthyl chloride yields both 3-menthene (68%) and 2-menthene (32%)?

Problem 14.17 Using models, suggest explanations for the following.

(a) On E2 elimination with *t*-BuOK/*t*-BuOH, both *cis*- and *trans*-2-phenylcyclopentyl tosylates give 1-phenylcyclopentene as the only alkene; the *cis* isomer reacts 14 times as fast as the *trans*.

(b) On E2 elimination with *n*-$C_5H_{11}ONa$/*n*-$C_5H_{11}OH$ to give 2-chloronorbornene, VIII reacts about 100 times as fast as its diastereomer, VII.

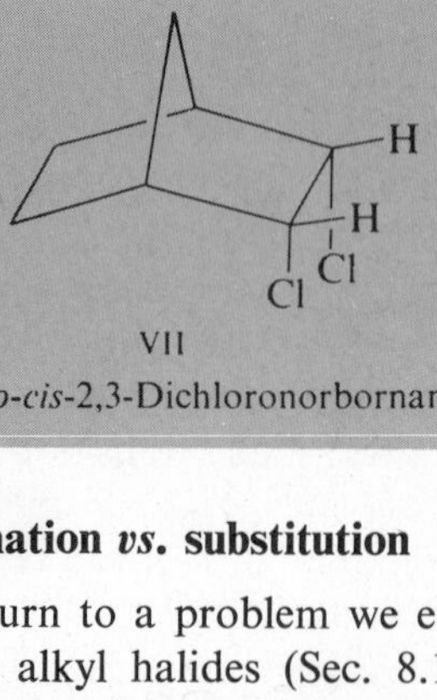

VII

endo-cis-2,3-Dichloronorbornane

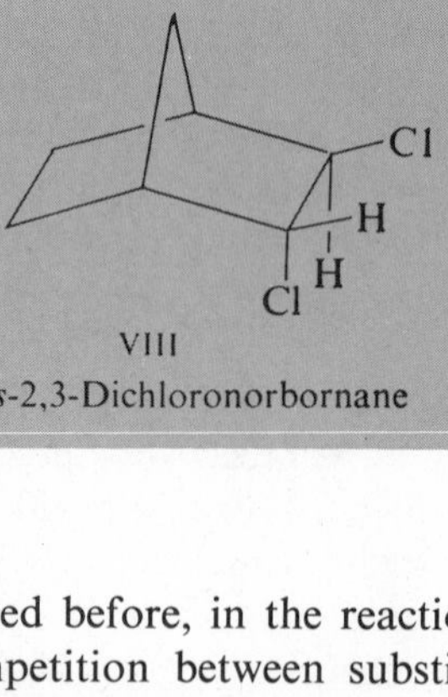

VIII

trans-2,3-Dichloronorbornane

14.23 Elimination *vs.* substitution

Let us return to a problem we encountered before, in the reaction between acetylides and alkyl halides (Sec. 8.12): competition between substitution and elimination. Both reactions result from attack by the same nucleophilic reagent: attack at carbon causes substitution, attack at hydrogen causes elimination.

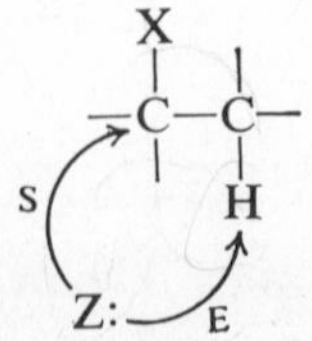

E2 *vs.* S_N2

We can see more clearly now why reaction with acetylides to form alkynes is limited in practice to *primary* halides. Under the conditions of the reaction—a solvent of low polarity (liquid ammonia or ether) and a powerful nucleophilic reagent (acetylide ion)—we would expect substitution, that is, alkyne formation, to take place by an S_N2 mechanism. Primary halides should therefore form alkynes fastest, tertiary halides the slowest.

On the other hand, the speed with which an alkyl halide undergoes elimination depends chiefly (Sec. 5.14) upon the stability of the alkene formed; tertiary halides, which necessarily yield highly branched (more stable) alkenes, undergo elimination fastest.

Primary halides, then, undergo substitution fastest and elimination slowest; tertiary halides undergo substitution slowest and elimination fastest. It is not surprising that the yields of alkynes are good for primary halides and very bad for tertiary halides.

The same considerations hold for the reactions of alkyl halides with other nucleophiles. *Where substitution and elimination are competing reactions, the proportion of elimination increases as the structure of an alkyl halide is changed from primary to secondary to tertiary.* Many tertiary halides yield exclusively alkenes under these conditions.

$$\text{RX} = \overset{\xrightarrow{\text{Elimination increases}}}{\underset{\xleftarrow[\text{Substitution increases}]{}}{1^\circ \qquad 2^\circ \qquad 3^\circ}}$$

Elimination (E2)
vs.
Substitution (S_N2)

Like acetylide ion, hydroxide ion is a *strong* base; that is, it has a strong affinity for hydrogen ion. The preparation of alcohols from alkyl halides gives good yields with primary halides, somewhat poorer yields with secondary halides; it is essentially worthless for the preparation of tertiary alcohols.

Tertiary alcohols are best prepared under conditions that favor the S_N1 reaction: solvent of high polarity, and reagent of low nucleophilic power. This is accomplished by simply boiling with water, which serves both as solvent and nucleophilic reagent. Yet even here the yields of alcohol are not high; considerable elimination occurs since the intermediate is a *tertiary* carbonium ion, which can easily expel a hydrogen ion to yield a relatively stable alkene.

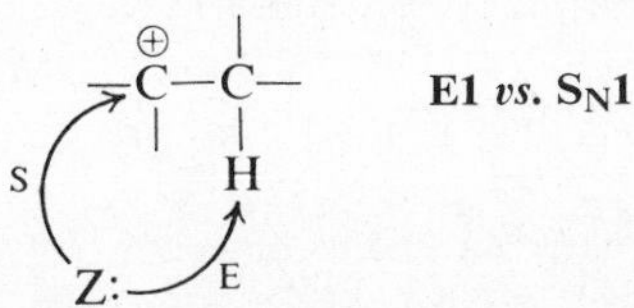

E1 *vs.* S_N1

When we want the product of a substitution reaction, elimination is a nuisance to be avoided. But when we want an alkene from an alkyl halide, elimination is what we are trying to bring about. To do this, we use a solvent of low polarity, and a high concentration of a strong base: concentrated alcoholic potassium hydroxide.

Problem 14.18 Which compound of each of the following sets would you expect to give the higher yield of substitution product under conditions for bimolecular reaction?

(a) ethyl bromide or β-phenylethyl bromide;
(b) α-phenylethyl bromide or β-phenylethyl bromide;
(c) isobutyl bromide or *n*-butyl bromide;
(d) isobutyl bromide or *tert*-butyl bromide.

Problem 14.19 Suggest an explanation for each of the following facts.

(a) Dehydrohalogenation of isopropyl bromide, which requires several hours of refluxing in alcoholic KOH, is brought about in less than a minute at room temperature by t-BuO^-K^+ in DMSO. (*Hint:* See Sec. 1.21.)

(b) The reaction of *tert*-butyl chloride in water to yield (chiefly) *tert*-butyl alcohol is not appreciably affected by dissolved sodium fluoride; in DMSO, however, sodium fluoride brings about rapid formation of isobutylene.

14.24 Analysis of alkyl halides

Simple alkyl halides respond to the common characterization tests in the same manner as alkanes: they are insoluble in cold concentrated sulfuric acid; they are inert to bromine in carbon tetrachloride, to aqueous permanganate, and to chromic anhydride. They are readily distinguished from alkanes, however, by qualitative analysis (Sec. 2.25), which shows the presence of halogen.

In many cases, the presence of halogen can be detected without a sodium fusion or Schöniger oxidation. An unknown is warmed for a few minutes with alcoholic silver nitrate (the alcohol dissolves both the ionic reagent and the organic compound); halogen is indicated by formation of a precipitate that is insoluble in dilute nitric acid.

As in almost all reactions of organic halides, reactivity toward alcoholic silver nitrate follows the sequence RI > RBr > RCl. For a given halogen atom, reactivity decreases in the order 3° > 2° > 1°, the sequence typical of carbonium ion formation; allyl and benzyl halides are highly reactive. Other evidence (stereochemistry, rearrangements) suggests that this reaction is of the S_N1 type. Silver ion is believed to dispose reaction toward this mechanism (rather than the S_N2) by *pulling* halide away from the alkyl group.

$$R{:}X + Ag^+ \longrightarrow R^+ + Ag^+X^-$$

(Vinyl and aryl halides do not react, Sec. 25.5.)

As mentioned earlier (Sec. 14.1), substituted alkyl halides also undergo the reactions characteristic of their other functional groups.

Problem 14.20 Describe simple chemical tests (if any) that would distinguish between: (a) ethylene bromohydrin and ethylene bromide; (b) 4-chloro-1-butene and *n*-butyl chloride; (c) bromocyclohexane and bromobenzene; (d) 1-chloro-2-methyl-2-propanol and 1,2-dichloro-2-methylpropane. Tell exactly what you would *do* and *see*.

14.25 Spectroscopic analysis of alkyl halides

For the spectroscopic analysis of alkyl halides, see the general discussion in Chapter 13, in which many alkyl halides were used as examples.

PROBLEMS

1. Outline the synthesis of ethyl bromide from: (a) ethane, (b) ethylene, (c) ethanol. Which method would one most probably use in the laboratory?

2. Which methods of Problem 1 would be used to prepare pure samples of:

(a) ethyl chloride
(b) ethyl fluoride
(c) ethyl iodide
(d) *n*-propyl bromide
(e) isopropyl bromide
(f) benzyl chloride
(g) α-phenylethyl chloride
(h) cyclohexyl bromide

3. Outline the synthesis of the following compounds from isopropyl alcohol:

(a) isopropyl bromide
(b) allyl chloride
(c) 1-chloro-2-propanol
(d) 1,2-dibromopropane
(e) 2,2-dibromopropane
(f) 2-bromopropene
(g) 1-bromopropene
(h) 1,3-dichloro-2-propanol
(i) 2,3-dibromo-1-propanol
(j) 2,2-dichloro-1-methylcyclopropane

4. Outline all steps in a possible laboratory synthesis of each of the following from cyclohexanol and any necessary aliphatic, aromatic, or inorganic reagents.

(a) bromocyclohexane
(b) iodocyclohexane
(c) *trans*-1,2-dibromocyclohexane
(d) 3-bromocyclohexene
(e) *trans*-2-chlorocyclohexanol
(f) norcarane (see p. 458)

5. Outline all steps in a possible laboratory synthesis of each of the following, using benzene, toluene, and any needed aliphatic or inorganic reagents.

(a) *p*-bromobenzyl chloride
(b) triphenylchloromethane
(c) allyl iodide
(d) benzal bromide
(e) *m*-nitrobenzotrichloride
(f) 1,2-dichloro-1-phenylethane
(g) phenylacetylene
(h) phenylcyclopropane

6. Give the structures and names of the chief organic products expected from the reaction (if any) of *n*-butyl bromide with:

(a) NaOH(aq)
(b) KOH(alc)
(c) cold conc. H_2SO_4
(d) Zn, H^+
(e) Li, then CuI, ethyl bromide
(f) Mg, ether
(g) product (f) + D_2O
(h) H_2, Pt
(i) dilute neutral $KMnO_4$
(j) NaI in acetone
(k) benzene, $AlCl_3$
(l) $CH_3C{\equiv}C^-Na^+$
(m) HgF_2
(n) Br_2/CCl_4

7. Referring when necessary to the list on page 457, give structures of the chief organic products expected from the reaction of *n*-butyl bromide with:

(a) NH_3
(b) $C_6H_5NH_2$
(c) NaCN
(d) $NaOC_2H_5$
(e) CH_3COOAg
(f) $NaSCH_3$

8. Write equations for the most likely side reactions in the conversion of *n*-butyl bromide into:

(a) 1-butanol by aqueous NaOH
(b) methyl *n*-butyl ether by CH_3ONa
(c) 1-butene by alcoholic KOH
(d) 1-hexyne by sodium acetylide

Will each of these side reactions be more or less important if *tert*-butyl bromide is used instead of *n*-butyl bromide?

9. Arrange the compounds of each set in order of reactivity toward S_N2 displacement:

(a) 2-bromo-2-methylbutane, 1-bromopentane, 2-bromopentane
(b) 1-bromo-3-methylbutane, 2-bromo-2-methylbutane, 3-bromo-2-methylbutane
(c) 1-bromobutane, 1-bromo-2,2-dimethylpropane, 1-bromo-2-methylbutane, 1-bromo-3-methylbutane

10. Arrange the compounds of each set in order of reactivity toward S_N1 displacement:

(a) the compounds of Problem 9(a)
(b) the compounds of Problem 9(b)
(c) benzyl chloride, *p*-chlorobenzyl chloride, *p*-methoxybenzyl chloride, *p*-methylbenzyl chloride, *p*-nitrobenzyl chloride
(d) benzyl bromide, α-phenylethyl bromide, β-phenylethyl bromide

11. Arrange the compounds in each set in order of ease of dehydrohalogenation by concentrated alcoholic KOH:

(a) compounds of Problem 9(a)
(b) compounds of Problem 9(b)
(c) 2-bromo-1-phenylpropane and 3-bromo-1-phenylpropane
(d) 5-bromo-1,3-cyclohexadiene, bromocyclohexane, 3-bromocyclohexene
(e) *cis*- and *trans*-2-bromomethylcyclohexane

12. Consider, as an example, the reaction between an alkyl halide and NaOH in a mixture of water and ethanol. In a table, with one column for S_N2 and another for S_N1, compare the two mechanisms with regard to:

(a) stereochemistry
(b) kinetic order
(c) occurrence of rearrangements
(d) relative rates of CH_3X, C_2H_5X, iso-C_3H_7X, *tert*-C_4H_9X
(e) relative rates of RCl, RBr, and RI
(f) effect on rate of a rise in temperature
(g) effect on rate of doubling [RX]
(h) effect on rate of doubling $[OH^-]$
(i) effect on rate of increasing the water content of the solvent
(j) effect on rate of increasing the alcohol content of the solvent

13. When optically active 2-iodooctane was allowed to stand in acetone solution containing $Na^{131}I$ (radioactive iodide), the alkyl halide was observed to lose optical activity and to exchange its ordinary iodine for radioactive iodine. The rate of each of these reactions depended on both [RI] and $[I^-]$, but racemization was exactly *twice* as fast as isotopic exchange. This experiment, reported in 1935 by E. D. Hughes (of University College, London), is considered to have established the stereochemistry of the S_N2 reaction: that each molecule undergoing substitution suffers inversion of configuration. Show exactly how this conclusion is justified. (*Hint:* Take one molecule of alkyl halide at a time, and consider what happens when it undergoes substitution.)

14. Optically active *sec*-butyl alcohol retains its activity indefinitely in contact with aqueous base, but is rapidly converted into optically inactive (racemic) *sec*-butyl alcohol by dilute sulfuric acid. How do you account for these facts? Suggest a detailed mechanism or mechanisms for the racemization by dilute acid.

15. When neomenthyl chloride undergoes E2 elimination, 2-menthane makes up one-fourth of the reaction product (Sec. 14.22). Since menthyl chloride can yield *only* 2-menthene, we might expect it to react at one-fourth of the rate of neomenthyl chloride. Actually, however, it reacts only 1/200 as fast as neomenthyl chloride: that is, only 1/50 as fast as we would have expected. How do you account for this unusually slow elimination from menthyl chloride? (*Hint:* Use models.)

16. When either *cis*- or *trans*-1-phenylpropene was treated with chlorine in CCl_4, and the reaction product was separated by preparative-scale gas chromatography, two fractions, A and B, of formula $C_9H_{10}Cl_2$, were obtained. On treatment with potassium *tert*-butoxide in *tert*-butyl alcohol, each fraction gave 1-chloro-1-phenyl-1-propene. Nmr showed that —Cl and —CH_3 were *trans* in the product from A, *cis* in the product from B.

(a) Give structural formulas for A and B. (b) In this particular system, is addition of halogen stereospecific? Discuss.

17. *cis*-4-*tert*-Butylcyclohexyl tosylate reacts rapidly with NaOEt in EtOH to yield 4-*tert*-butylcyclohexene; the rate of reaction is proportional to the concentration of both tosylate and ethoxide ion. Under the same conditions, *trans*-4-*tert*-butylcyclohexyl tosylate reacts slowly to yield the alkene (plus 4-*tert*-butylcyclohexyl ethyl ether); the rate of reaction depends only on the concentration of the tosylate.

How do you account for these observations?

18. (a) It has been proposed that the conversion of vicinal dihalides into alkenes by the action of iodide ion can proceed by either a one-step mechanism (i) or a three-step mechanism (ii).

(i) $-\overset{|}{\underset{Br}{C}}-\overset{|}{\underset{Br}{C}}- \xrightarrow{I^-} -\overset{|}{C}=\overset{|}{C}- + IBr + Br^-$

(ii) $-\overset{|}{\underset{Br}{C}}-\overset{|}{\underset{Br}{C}}- \xrightarrow{I^-} -\overset{|}{\underset{I}{C}}-\overset{|}{\underset{Br}{C}}- \xrightarrow{I^-} -\overset{|}{C}\overset{\underset{|}{I}-I}{\text{——}}\overset{|}{C}- \longrightarrow -\overset{|}{C}=\overset{|}{C}-$

Show the details, particularly the expected stereochemistry, of each step of each mechanism.

(b) The following stereochemical observations have been made:

meso-1,2-dibromo-1,2-dideuterioethane (CHDBrCHDBr) + I^- $\longrightarrow$ *only cis*-CHD=CHD

meso-2,3-dibromobutane + I^- $\longrightarrow$ *only trans*-2-butene
racemic 2,3-dibromobutane + I^- $\longrightarrow$ *only cis*-2-butene

On the basis of the observed stereochemistry, which mechanism is most probably followed by each halide? Explain in detail. How do you account for the difference in behavior between the halides?

(c) When 1-bromocyclohexene (ordinary bromine) is allowed to react with radioactive Br_2, and the resulting tribromide is treated with iodide ion, there is obtained 1-bromocyclohexene that contains less than 0.3% of radioactive bromine.

19. On treatment with the aromatic base pyridine (Sec. 31.8), racemic 1,2-dibromo-1,2-diphenylethane loses HBr to yield *trans*-1-bromo-1,2-diphenylethene; in contrast, the *meso* dibromide loses Br_2 to yield *trans*-1,2-diphenylethene. (a) Suggest a mechanism for the reaction of each stereoisomer. (b) How do you account for the difference in their behavior?

20. Treatment of neopentyl chloride with the strong base sodamide ($NaNH_2$) yields a hydrocarbon of formula C_5H_{10}, which readily dissolves in concentrated sulfuric acid, but is not oxidized by cold, dilute, neutral permanganate. Its nmr spectrum shows absorption at δ 0.20 and δ 1.05 with peak area ratio 2:3. When the same reaction is carried out using the labeled alkyl halide, $(CH_3)_3CCD_2Cl$, the product obtained has its M^+ peak at m/e 71. What is a likely structure for the hydrocarbon, and how is it probably formed? Is the result of the labeling experiment consistent with your mechanism? (*Hint:* See Sec. 9.16.)

21. (a) In the liquid form, *tert*-butyl fluoride and isopropyl fluoride gave the following nmr spectra.

tert-butyl fluoride: doublet, δ 1.30, J = 20 Hz
isopropyl fluoride: two doublets, δ 1.23, 6H, J = 23 Hz and 4 Hz
two multiplets, δ 4.64, 1H, J = 48 Hz and 4 Hz

How do you account for each of these spectra? (*Hint:* See Sec. 13.10.)

(b) When the alkyl fluorides were dissolved in liquid SbF_5, the following nmr spectra were obtained.

from *tert*-butyl fluoride: singlet, δ 4.35
from isopropyl fluoride: doublet, δ 5.06, 6H, J = 4 Hz
multiplet, δ 13.5, 1H, J = 4 Hz

To what molecule is each of these spectra due? (*Hint:* What does the disappearance of just half the peaks observed in part (a) suggest?) Is the very large downfield shift what you might have expected for molecules like these? Of what fundamental significance to organic theory are these observations?

22. When methallyl chloride, $CH_2{=}C(CH_3)CH_2Cl$, was treated with sodamide in tetrahydrofuran solution, there was obtained a hydrocarbon, C_4H_6, which gave the following nmr spectrum:

a doublet, δ 0.83, 2H, $J = 2$ Hz
b doublet, δ 2.13, 3H, $J = 1$ Hz
c multiplet, δ 6.40, 1H

(a) What is a likely structure for this hydrocarbon, and by what mechanism was it probably formed? (b) What product would you expect to obtain by the same reaction from allyl chloride?

23. Hydrocarbon C has been prepared in two different ways:

(i) Cl—◇—Br + Na ⟶ C

1-Chloro-3-bromocyclobutane

(ii) $CH_2{=}CH{-}CH_2{-}CHN_2 \xrightarrow{\text{light}} C$

Allyldiazomethane

Mass spectrometry shows a molecular weight of 54 for C. (What is its molecular formula?) On gas chromatography, C was found to have a different retention time from cyclobutene, butadiene, or methylenecyclopropane. C was stable at 180° (unlike cyclobutene), but was converted into butadiene at 225°. The nmr spectrum of C showed: *a*, singlet δ 0.45, 2H; *b*, multiplet, δ 1.34, 2H; *c*, multiplet, δ 1.44, 2H.

(a) What single structure for C is consistent with all these facts? (*Hint:* In analyzing the nmr spectrum, take stereochemistry into consideration.) (b) By what familiar reaction is C formed in (i)? in (ii)?

24. Describe simple chemical tests that would serve to distinguish between:

(a) allyl chloride and *n*-propyl chloride
(b) allyl chloride and benzyl chloride
(c) ethylene chlorohydrin, ethylene chloride, and ethylene glycol
(d) cyclohexanol, cyclohexyl bromide, and cyclohexene
(e) *tert*-butyl alcohol, *tert*-butyl chloride, and 1-octene
(f) benzyl chloride and *p*-chlorotoluene.

Tell exactly what you would *do* and *see*.

25. A liquid of boiling point 39–41° was insoluble in water, dilute acids or bases, or concentrated H_2SO_4. It did not react with Br_2/CCl_4 or dilute $KMnO_4$. It was subjected to sodium fusion, and the resulting solution was filtered, acidified with nitric acid, and boiled. Addition of $AgNO_3$ gave a precipitate.

(a) On the basis of Table 14.1, what compound or compounds might this have been? (b) Several milliliters of CCl_4 were added to a portion of the acidified solution from the fusion, and the mixture was shaken with chlorine water. A violet color appeared in the CCl_4 layer. Which compound or compounds of (a) are still possible? (c) How would each of the other possibilities have responded in (b)?

26. An unknown compound is believed to be one of the following. Describe how you would go about finding out which of the possibilities the unknown actually is. Where possible, use simple chemical tests; where necessary, use more elaborate chemical methods like quantitative hydrogenation, cleavage, etc. Where necessary, make use of Table 18.1. p. 580.

(a)	b.p., °C		b.p., °C
n-decane	174	*p*-cymene (*p*-isopropyltoluene)	177
4-methylcyclohexanol	174	limonene (see Problem 17, page 317)	178
1,3-dichloro-2-propanol	176	*n*-heptyl bromide	180

(b)

1-phenyl-1-propene	177	*n*-hexyl iodide	180
benzyl chloride	179	cyclohexylcarbinol	182
2-octanol	179		

(c)

m-diethylbenzene	182	*n*-octyl chloride	185
n-butylbenzene	183	*trans*-decalin (see Problem 8, p. 315)	186
2-ethyl-1-hexanol	184		

Chapter 15 Alcohols I. Preparation and Physical Properties

15.1 Structure

Alcohols are compounds of the general formula ROH, where R is any alkyl or substituted alkyl group. The group may be primary, secondary, or tertiary; it may be open-chain or cyclic; it may contain a double bond, a halogen atom, or an aromatic ring. For example:

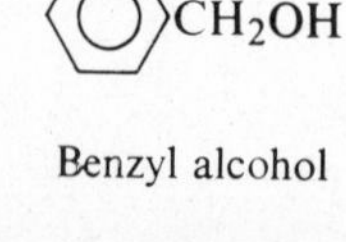

$$CH_3-\underset{OH}{\overset{CH_3}{C}}-CH_3$$
tert-Butyl alcohol

$$H_2C{=}CH-CH_2OH$$
Allyl alcohol

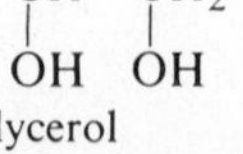

$$C_6H_{11}OH$$
Cyclohexanol

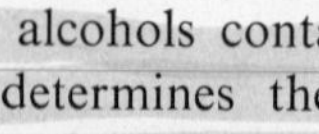

$$C_6H_5CH_2OH$$
Benzyl alcohol

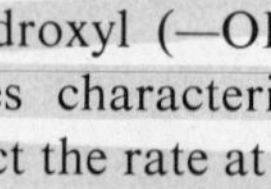

$$\underset{Cl}{CH_2}-\underset{OH}{CH_2}$$
Ethylene chlorohydrin
(β-Chloroethyl alcohol)

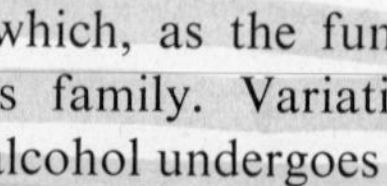

$$\underset{OH}{CH_2}-\underset{OH}{CH}-\underset{OH}{CH_2}$$
Glycerol

All alcohols contain the hydroxyl (—OH) group, which, as the functional group, determines the properties characteristic of this family. Variations in structure of the R group may affect the rate at which the alcohol undergoes certain reactions, and even, in a few cases, may affect the kind of reaction.

Compounds in which the hydroxyl group is attached directly to an aromatic ring are not alcohols; they are *phenols*, and differ so markedly from the alcohols that we shall consider them in a separate chapter.

15.2 Classification

We classify a carbon atom as *primary*, *secondary*, or *tertiary* according to the number of other carbon atoms attached to it (Sec. 3.11). An alcohol is classified according to the kind of carbon that bears the —OH group:

$$\underset{\substack{\text{Primary}\\(1°)}}{\text{R}-\overset{\text{H}}{\underset{\text{H}}{\text{C}}}-\text{OH}} \qquad \underset{\substack{\text{Secondary}\\(2°)}}{\text{R}-\overset{\text{R}}{\underset{\text{H}}{\text{C}}}-\text{OH}} \qquad \underset{\substack{\text{Tertiary}\\(3°)}}{\text{R}-\overset{\text{R}}{\underset{\text{R}}{\text{C}}}-\text{OH}}$$

One reaction, oxidation, which directly involves the hydrogen atoms attached to the carbon bearing the —OH group, takes an entirely different course for each class of alcohol. Usually, however, alcohols of different classes differ only in *rate* or *mechanism* of reaction, and in a way consistent with their structures. Certain substituents may affect reactivity in such a way as to make an alcohol of one class resemble the members of a different class; benzyl alcohol, for example, though formally a primary alcohol, often acts like a tertiary alcohol. We shall find that these variations, too, are consistent with the structures involved.

15.3 Nomenclature

Alcohols are named by three different systems. For the simpler alcohols the **common names**, which we have already encountered (Sec. 5.19), are most often used. These consist simply of the name of the alkyl group followed by the word *alcohol*. For example:

$$\underset{\text{Ethyl alcohol}}{CH_3CH_2OH} \qquad \underset{\text{Isopropyl alcohol}}{CH_3\underset{\text{OH}}{CH}CH_3} \qquad \underset{\text{Isobutyl alcohol}}{CH_3\overset{CH_3}{CH}CH_2OH}$$

$$\underset{\textit{tert}\text{-Pentyl alcohol}}{CH_3CH_2-\overset{CH_3}{\underset{\text{OH}}{C}}-CH_3} \qquad \underset{p\text{-Nitrobenzyl alcohol}}{O_2N-C_6H_4-CH_2OH} \qquad \underset{\alpha\text{-Phenylethyl alcohol}}{C_6H_5-\overset{\alpha}{\underset{\text{OH}}{C}H}\overset{\beta}{C}H_3}$$

We should notice that similar names do not always mean the same classification; for example, isopropyl alcohol is a secondary alcohol, whereas isobutyl alcohol is a primary alcohol.

It is sometimes convenient to name alcohols by the **carbinol** system. According to this system, alcohols are considered to be derived from *methyl alcohol*, CH_3OH, by the replacement of one or more hydrogen atoms by other groups. We simply name the groups attached to the carbon bearing the —OH and then add the suffix *-carbinol* to include the C—OH portion:

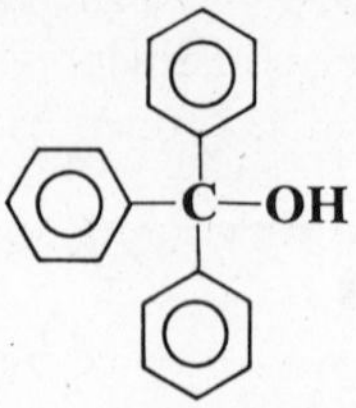

$(C_6H_5)_3C-OH$
Triphenylcarbinol

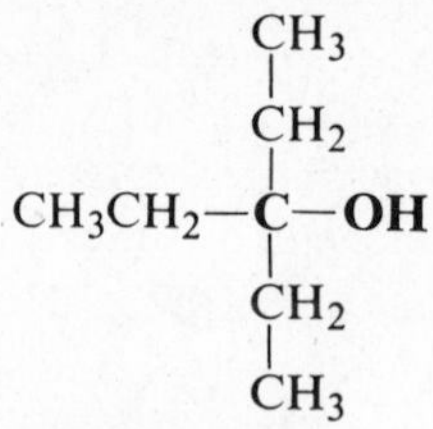

$CH_3CH_2-C(CH_2CH_3)(CH_2CH_3)-OH$
Triethylcarbinol

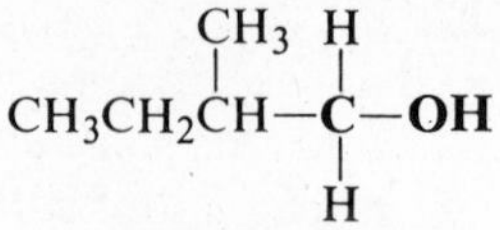

$CH_3CH_2CH(CH_3)-CH_2-OH$
sec-Butylcarbinol

Finally, there is the most versatile system, the **IUPAC**. The rules are:

(1) Select as the parent structure the longest continuous carbon chain *that contains the —OH group*; then consider the compound to have been derived from this structure by replacement of hydrogen by various groups. The parent structure is known as *ethanol, propanol, butanol*, etc., depending upon the number of carbon atoms; each name is derived by replacing the terminal *-e* of the corresponding alkane name by **-ol**.

(2) Indicate by a number the position of the —OH group in the parent chain, generally using the lowest possible number for this purpose.

(3) Indicate by numbers the positions of other groups attached to the parent chain.

CH_3OH
Methanol

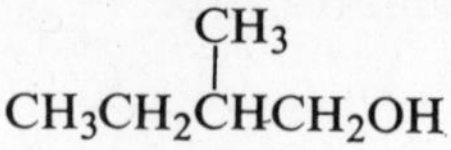

$CH_3CH_2CH(CH_3)CH_2OH$
2-Methyl-1-butanol

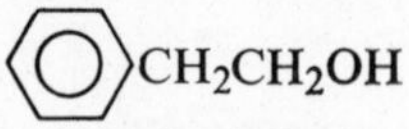

$C_6H_5CH_2CH_2OH$
2-Phenylethanol

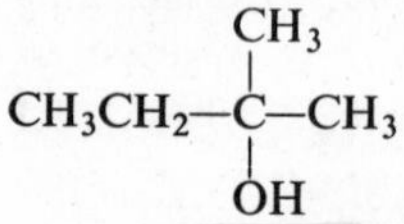

$CH_3CH_2-C(CH_3)(OH)-CH_3$
2-Methyl-2-butanol

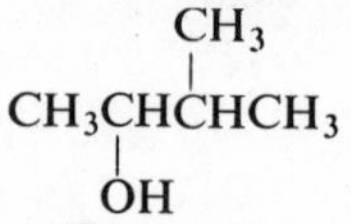

$CH_3CH(OH)CH(CH_3)CH_3$
3-Methyl-2-butanol

$ClCH_2CH_2OH$
2-Chloroethanol

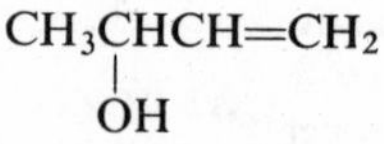

$CH_3CH(OH)CH{=}CH_2$
3-Buten-2-ol

Alcohols containing two hydroxyl groups are called *glycols*. They have both common names and IUPAC names.

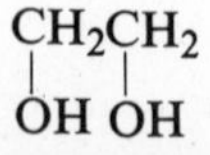

$CH_2(OH)CH_2(OH)$
Ethylene glycol
1,2-Ethanediol

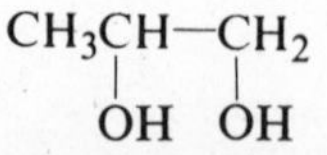

$CH_3CH(OH)-CH_2(OH)$
Propylene glycol
1,2-Propanediol

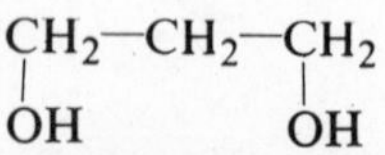

$CH_2(OH)-CH_2-CH_2(OH)$
Trimethylene glycol
1,3-Propanediol

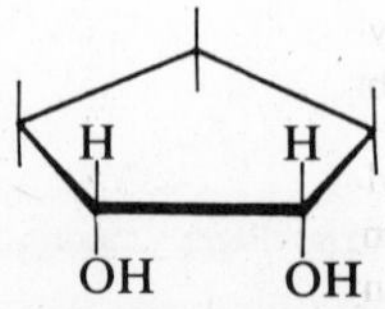

(cyclopentane ring with H, H and OH, OH on adjacent carbons)
cis-1,2-Cyclopentanediol

15.4 Physical properties

The compounds we have studied so far, the various hydrocarbons, are non-polar or nearly so, and have the physical properties that we might expect of such compounds: the relatively low melting points and boiling points that are characteristic of molecules with weak intermolecular forces; solubility in non-polar solvents and insolubility in polar solvents like water.

Alcohols, in contrast, contain the very polar —OH group. In particular, this group contains hydrogen attached to the very electronegative element, oxygen, and therefore permits hydrogen bonding (Sec. 1.19). The physical properties (Table 15.1) show the effects of this hydrogen bonding.

```
          R
          |
R—O---H—O
   |
   H
```

Table 15.1 ALCOHOLS

Name	Formula	M.p., °C	B.p., °C	Density at 20°C	Solub., g/100 g H_2O
Methyl	CH_3OH	− 97	64.5	0.793	∞
Ethyl	CH_3CH_2OH	−115	78.3	.789	∞
n-Propyl	$CH_3CH_2CH_2OH$	−126	97	.804	∞
n-Butyl	$CH_3(CH_2)_2CH_2OH$	− 90	118	.810	7.9
n-Pentyl	$CH_3(CH_2)_3CH_2OH$	− 78.5	138	.817	2.3
n-Hexyl	$CH_3(CH_2)_4CH_2OH$	− 52	156.5	.819	0.6
n-Heptyl	$CH_3(CH_2)_5CH_2OH$	− 34	176	.822	0.2
n-Octyl	$CH_3(CH_2)_6CH_2OH$	− 15	195	.825	0.05
n-Decyl	$CH_3(CH_2)_8CH_2OH$	6	228	.829	
n-Dodecyl	$CH_3(CH_2)_{10}CH_2OH$	24			
n-Tetradecyl	$CH_3(CH_2)_{12}CH_2OH$	38			
n-Hexadecyl	$CH_3(CH_2)_{14}CH_2OH$	49			
n-Octadecyl	$CH_3(CH_2)_{16}CH_2OH$	58.5			
Isopropyl	$CH_3CHOHCH_3$	− 86	82.5	.789	∞
Isobutyl	$(CH_3)_2CHCH_2OH$	−108	108	.802	10.0
sec-Butyl	$CH_3CH_2CHOHCH_3$	−114	99.5	.806	12.5
tert-Butyl	$(CH_3)_3COH$	25.5	83	.789	∞
Isopentyl	$(CH_3)_2CHCH_2CH_2OH$	−117	132	.813	2
active-Amyl	$(-)$-$CH_3CH_2CH(CH_3)CH_2OH$		128	.816	3.6
tert-Pentyl	$CH_3CH_2C(OH)(CH_3)_2$	− 12	102	.809	12.5
Cyclopentanol	*cyclo*-C_5H_9OH		140	.949	
Cyclohexanol	*cyclo*-$C_6H_{11}OH$	24	161.5	.962	
Allyl	$CH_2{=}CHCH_2OH$	−129	97	.855	∞
Crotyl	$CH_3CH{=}CHCH_2OH$		118	.853	16.6
Methylvinyl-carbinol	$CH_2{=}CHCHOHCH_3$		97		
Benzyl	$C_6H_5CH_2OH$	− 15	205	1.046	4
α-Phenylethyl	$C_6H_5CHOHCH_3$		205	1.013	
β-Phenylethyl	$C_6H_5CH_2CH_2OH$	− 27	221	1.02	1.6
Diphenylcarbinol (Benzhydrol)	$(C_6H_5)_2CHOH$	69	298		0.05
Triphenylcarbinol	$(C_6H_5)_3COH$	162.5			
Cinnamyl	$C_6H_5CH{=}CHCH_2OH$	33	257.5		
Ethylene glycol	CH_2OHCH_2OH	− 16	197	1.113	
Propylene glycol	$CH_3CHOHCH_2OH$		187	1.040	
1,3-Propanediol	$HOCH_2CH_2CH_2OH$		215	1.060	
Glycerol	$HOCH_2CHOHCH_2OH$	18	290	1.261	
Pentaerythritol	$C(CH_2OH)_4$	260			6

Let us look first at **boiling points**. Among hydrocarbons the factors that determine boiling point seem to be chiefly molecular weight and shape; this is to be expected of molecules that are held together chiefly by van der Waals forces. Alcohols, too, show increase in boiling point with increasing carbon number, and decrease in boiling point with branching. But the unusual thing about alcohols is that they boil so *high*: as Table 15.2 shows, much higher than hydrocarbons of the same molecular weight, and higher, even, than many other compounds of considerable polarity. How are we to account for this?

Table 15.2 STRUCTURE AND BOILING POINT

Name	Structure	Mol. Wt.	Dipole Moment, D	B.p., °C
n-Pentane	$CH_3CH_2CH_2CH_2CH_3$	72	0	36
Ethyl ether	$CH_3CH_2—O—CH_2CH_3$	74	1.18	35
n-Propyl chloride	$CH_3CH_2CH_2Cl$	79	2.10	47
n-Butyraldehyde	$CH_3CH_2CH_2CHO$	72	2.72	76
n-Butyl alcohol	$CH_3CH_2CH_2CH_2OH$	74	1.63	118

The answer is, of course, that alcohols, like water, are *associated liquids*: their abnormally high boiling points are due to the greater energy needed to break the hydrogen bonds that hold the molecules together. Although ethers and aldehydes contain oxygen, they contain hydrogen that is bonded only to carbon; these hydrogens are not positive enough to bond appreciably with oxygen.

Infrared spectroscopy (Sec. 13.4) has played a key role in the study of hydrogen bonding. In dilute solution in a non-polar solvent like carbon tetrachloride (or in the gas phase), where association between molecules is minimal, ethanol, for example, shows an O—H stretching band at 3640 cm^{-1}. As the concentration of ethanol is increased, this band is gradually replaced by a broader band at 3350 cm^{-1}. The bonding of hydrogen to the second oxygen weakens the O—H bond, and lowers the energy and hence the frequency of vibration.

Problem 15.1 The infrared spectrum of *cis*-1,2-cyclopentanediol has an O—H stretching band at a lower frequency than for a free —OH group, and this band does not disappear even at high dilution. *trans*-1,2-Cyclopentanediol shows no such band. Can you suggest a possible explanation?

Problem 15.2 It has been suggested that there is weak hydrogen bonding: (a) between chloroform molecules; (b) between HCN molecules. How would you account for this? (*Hint:* See Sec. 8.10.)

The **solubility** behavior of alcohols also reflects their ability to form hydrogen bonds. In sharp contrast to hydrocarbons, the lower alcohols are miscible with water. Since alcohol molecules are held together by the same sort of intermolecular forces as water molecules, there can be mixing of the two kinds of molecules: the energy required to break a hydrogen bond between two water molecules or two alcohol molecules is provided by formation of a hydrogen bond between a water molecule and an alcohol molecule.

This is true, however, only for the lower alcohols, where the —OH group constitutes a large portion of the molecule. A long aliphatic chain with a small

—OH group at one end is mostly alkane, and its physical properties show this. The change in solubility with carbon number is a gradual one: the first three primary alcohols are miscible with water; *n*-butyl alcohol is soluble to the extent of 8 g per 100 g water; *n*-pentyl, 2 g; *n*-hexyl, 1 g; and the higher alcohols still less. For practical purposes we consider that the borderline between solubility and insolubility in water occurs at about four to five carbon atoms for normal primary alcohols.

Polyhydroxy alcohols provide more than one site per molecule for hydrogen bonding, and their properties reflect this. The simplest glycol, ethylene glycol, boils at 197°. The lower glycols are miscible with water, and those containing as many as seven carbon atoms show appreciable solubility in water. (Ethylene glycol owes its use as an antifreeze—e.g. Prestone—to its high boiling point, low freezing point, and high solubility in water.)

Problem 15.3 The disaccharide *sucrose*, $C_{12}H_{22}O_{11}$, is a big molecule and yet (it is ordinary table sugar) is extremely soluble in water. What might you guess about its structure? (Check your answer on p. 1119.)

Problem 15.4 How do you account for the fact that, although ethyl ether has a much lower boiling point than *n*-butyl alcohol, it has the same solubility (8 g per 100 g) in water?

15.5 Industrial source

If an organic chemist were allowed to choose ten aliphatic compounds with which to be stranded on a desert island, he would almost certainly pick alcohols. From them he could make nearly every other kind of aliphatic compound: alkenes, alkyl halides, ethers, aldehydes, ketones, acids, esters, and a host of others. From the alkyl halides, he could make Grignard reagents, and from the reaction between these and the aldehydes and ketones obtain more complicated alcohols and so on. Our stranded chemist would use his alcohols not only as raw materials but frequently as the solvents in which reactions are carried out and from which products are recrystallized.

For alcohols to be such important starting materials in aliphatic chemistry, they must be not only versatile in their reactions but also available in large amounts and at low prices. There are two principal ways to get the simple alcohols that are the backbone of aliphatic organic synthesis: by **hydration of alkenes** obtained from the cracking of petroleum, and by **fermentation of carbohydrates**. In addition to these two chief methods, there are some others that have more limited application. (See Fig. 15.1.)

(a) Hydration of alkenes. We have already seen (Sec. 3.31) that alkenes containing up to four or five carbon atoms can be separated from the mixture obtained from the cracking of petroleum. We have also seen (Secs. 6.8 and 6.9) that alkenes are readily converted into alcohols either by direct addition of water, or by addition of sulfuric acid followed by hydrolysis. By this process there can

$$CH_3{-}CH{=}CH_2 + H_2SO_4 \longrightarrow \underset{\displaystyle OSO_3H}{CH_3{-}\underset{|}{C}H{-}CH_3} \xrightarrow{H_2O} \underset{\displaystyle OH}{CH_3{-}\underset{|}{C}H{-}CH_3}$$

Isopropyl alcohol (2°)

$$CH_3-\overset{\displaystyle CH_3}{\overset{|}{C}}=CH_2 + H_2O \xrightarrow{H^+} CH_3-\underset{\underset{\displaystyle {}^+OH_2}{|}}{\overset{\overset{\displaystyle CH_3}{|}}{C}}-CH_3 \xrightarrow{-H^+} CH_3-\underset{\underset{\displaystyle OH}{|}}{\overset{\overset{\displaystyle CH_3}{|}}{C}}-CH_3$$

tert-Butyl alcohol (3°)

be obtained only those alcohols whose formation is consistent with the application of Markovnikov's rule: for example, isopropyl but not *n*-propyl, *sec*-butyl but not *n*-butyl, *tert*-butyl but not isobutyl. Thus the *only* primary alcohol obtainable in this way is ethyl alcohol.

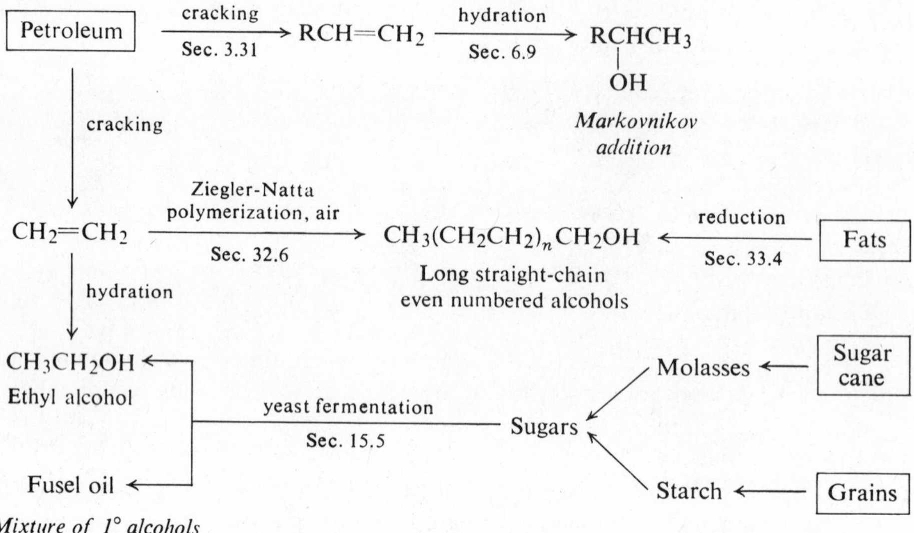

Figure 15.1. Industrial sources of alcohols.

(b) Fermentation of carbohydrates. Fermentation of sugars by yeast, the oldest synthetic chemical process used by man, is still of enormous importance for the preparation of **ethyl alcohol** and certain other alcohols. The sugars come from a variety of sources, mostly molasses from sugar cane, or starch obtained from various grains; the name "grain alcohol" has been given to ethyl alcohol for this reason.

When starch is the starting material, there is obtained, in addition to ethyl alcohol, a smaller amount of *fusel oil* (German: *Fusel*, inferior liquor), a mixture of primary alcohols: mostly isopentyl alcohol with smaller amounts of *n*-propyl alcohol, isobutyl alcohol, and 2-methyl-1-butanol, known as *active amyl alcohol* (*amyl* = *pentyl*).

Problem 15.5 The isopentyl and active amyl alcohols are formed by enzymatic transformation of the amino acids *leucine* and *isoleucine*, which come from hydrolysis of protein material in the starch.

$$\underset{\text{Leucine}}{(CH_3)_2CHCH_2CH(NH_3^+)COO^-} \qquad \underset{\text{Isoleucine}}{CH_3CH_2CH(CH_3)CH(NH_3^+)COO^-}$$

(a) Which amino acid gives which alcohol? (b) Although both amino acids are optically active, and the transformation processes are analogous, only one gives an alcohol that is optically active. Why is this?

15.6 Ethyl alcohol

Ethyl alcohol is not only the oldest synthetic organic chemical used by man, but it is also one of the most important.

In industry ethyl alcohol is widely used as a solvent for lacquers, varnishes, perfumes, and flavorings; as a medium for chemical reactions; and in recrystallizations. In addition, it is an important raw material for synthesis; after we have learned more about the reactions of alcohols (Chap. 16), we can better appreciate the role played by the leading member of the family. For these industrial purposes ethyl alcohol is prepared both by hydration of ethylene and by fermentation of sugar from molasses (or sometimes starch); thus its ultimate source is petroleum, sugar cane, and various grains.

Ethyl alcohol is the alcohol of "alcoholic" beverages. For this purpose it is prepared by fermentation of sugar from a truly amazing variety of vegetable sources. The particular beverage obtained depends upon what is fermented (rye or corn, grapes or elderberries, cactus pulp or dandelions), how it is fermented (whether carbon dioxide is allowed to escape or is bottled up, for example), and what is done after fermentation (whether or not it is distilled). The special flavor of a beverage is not due to the ethyl alcohol but to other substances either characteristic of the particular source, or deliberately added.

Medically, ethyl alcohol is classified as a *hypnotic* (sleep producer); it is less toxic than other alcohols. (Methanol, for example, is quite *poisonous*: drinking it, breathing it for prolonged periods, or allowing it to remain long on the skin can lead to blindness or death.)

Because of its unique position as both a highly taxed beverage and an important industrial chemical, ethyl alcohol poses a special problem: it must be made available to the chemical industry in a form that is unfit to drink. This problem is solved by addition of a *denaturant*, a substance that makes it unpalatable or even poisonous. Two of the eighty-odd legal denaturants, for example, are methanol and high-test gasoline. When necessary, pure undenatured ethyl alcohol is available for chemical purposes, but its use is strictly controlled by the Federal Government.

Except for alcoholic beverages, nearly all the ethyl alcohol used is a mixture of 95% alcohol and 5% water, known simply as *95% alcohol*. What is so special about the concentration of 95%? Whatever the method of preparation, ethyl alcohol is obtained first mixed with water; this mixture is then concentrated by fractional distillation. But it happens that the component of lowest boiling point is not ethyl alcohol (b.p. 78.3°) but a *binary azeotrope* containing 95% alcohol and 5% water (b.p. 78.15°). As an azeotrope, it of course gives a vapor of the same composition, and hence cannot be further concentrated by distillation no matter how efficient the fractionating column used.

Pure ethyl alcohol is known as *absolute alcohol*. Although more expensive than 95% alcohol, it is available for use when specifically required. It is obtained by taking advantage of the existence of another azeotrope, this time a *ternary* one of b.p. 64.9°: 7.5% water, 18.5% ethyl alcohol, and 74% benzene.

Problem 15.6 Describe exactly what will happen if one distills a mixture of 150 g of 95% alcohol and 74 g of benzene.

For certain special purposes (Secs. 26.2 and 26.3) even the slight trace of water found in commercial absolute alcohol must be removed. This can be accomplished by treatment of the alcohol with metallic magnesium; water is converted into insoluble $Mg(OH)_2$, from which the alcohol is then distilled.

15.7 Preparation of alcohols

Most of the simple alcohols and a few of the complicated ones are available from the industrial sources described in Sec. 15.5. Other alcohols must be prepared by one of the methods outlined below.

PREPARATION OF ALCOHOLS

1. Oxymercuration-demercuration. Discussed in Sec. 15.8.

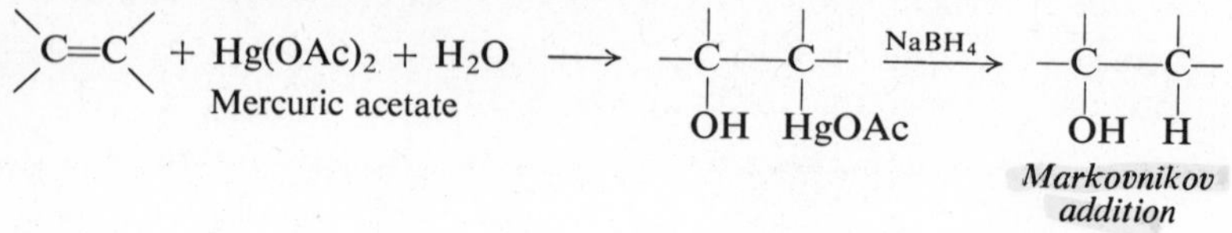

$$\text{>C=C<} + \underset{\text{Mercuric acetate}}{Hg(OAc)_2} + H_2O \longrightarrow \text{—}\underset{OH}{\overset{|}{C}}\text{—}\underset{HgOAc}{\overset{|}{C}}\text{—} \xrightarrow{NaBH_4} \text{—}\underset{OH}{\overset{|}{C}}\text{—}\underset{H}{\overset{|}{C}}\text{—}$$

Markovnikov addition

Examples:

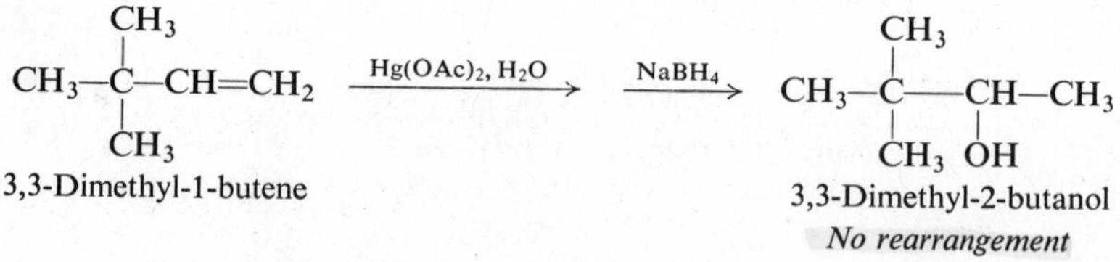

$$\underset{\text{3,3-Dimethyl-1-butene}}{CH_3\text{—}\overset{CH_3}{\underset{CH_3}{C}}\text{—}CH\text{=}CH_2} \xrightarrow{Hg(OAc)_2,\ H_2O} \xrightarrow{NaBH_4} \underset{\text{3,3-Dimethyl-2-butanol}}{CH_3\text{—}\overset{CH_3}{\underset{CH_3}{C}}\text{—}\underset{OH}{CH}\text{—}CH_3}$$

No rearrangement

Norbornene $\xrightarrow{Hg(OAc)_2,\ H_2O}$ $\xrightarrow{NaBH_4}$ *exo*-Norborneol (OH)

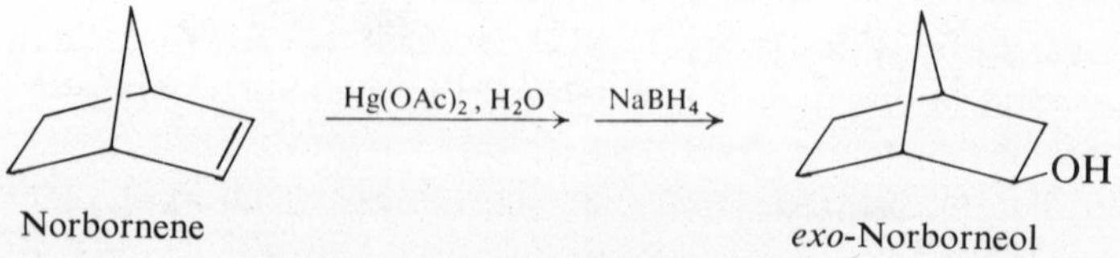

2. Hydroboration-oxidation. Discussed in Secs. 15.9–15.11.

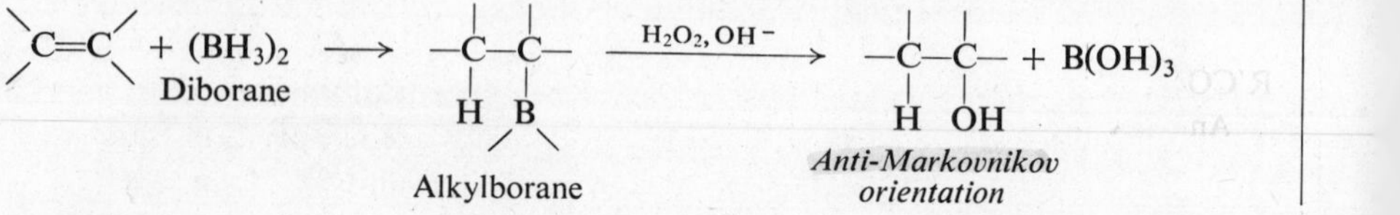

$$\text{>C=C<} + \underset{\text{Diborane}}{(BH_3)_2} \longrightarrow \underset{\text{Alkylborane}}{\text{—}\underset{H}{\overset{|}{C}}\text{—}\underset{B<}{\overset{|}{C}}\text{—}} \xrightarrow{H_2O_2,\ OH^-} \text{—}\underset{H}{\overset{|}{C}}\text{—}\underset{OH}{\overset{|}{C}}\text{—} + B(OH)_3$$

Anti-Markovnikov orientation

Examples:

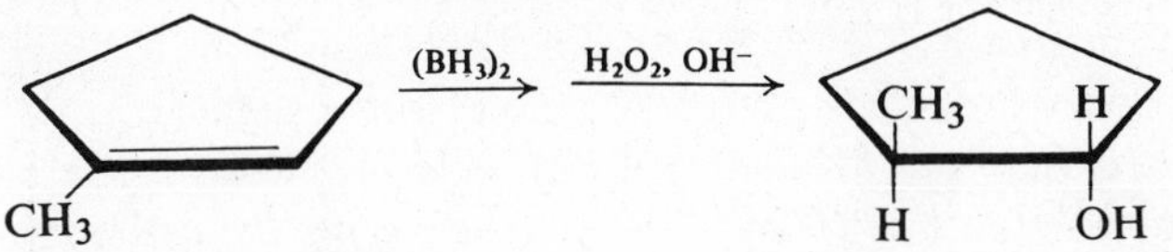

$$\text{1-Methylcyclopentene} \xrightarrow{(BH_3)_2} \xrightarrow{H_2O_2,\ OH^-} \textit{trans}\text{-2-Methyl-1-cyclopentanol}$$

syn-Addition

1-Methylcyclopentene

trans-2-Methyl-1-cyclopentanol

$$CH_3-C(CH_3)_2-CH{=}CH_2 \xrightarrow{(BH_3)_2} \xrightarrow{H_2O_2,\ OH^-} CH_3-C(CH_3)_2-CH_2-CH_2OH$$

No rearrangement

3,3-Dimethyl-1-butene → 3,3-Dimethyl-1-butanol

3. **Grignard synthesis.** Discussed in Secs. 15.12–15.15.

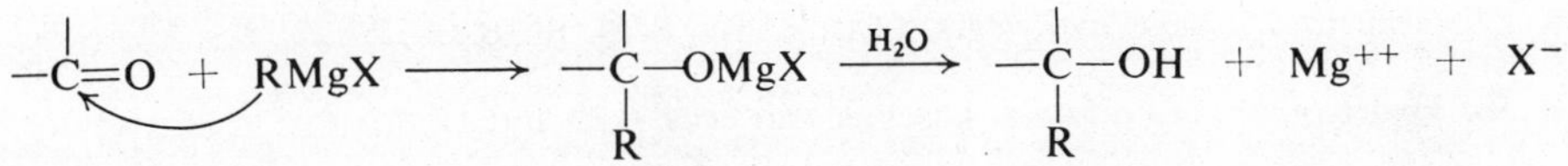

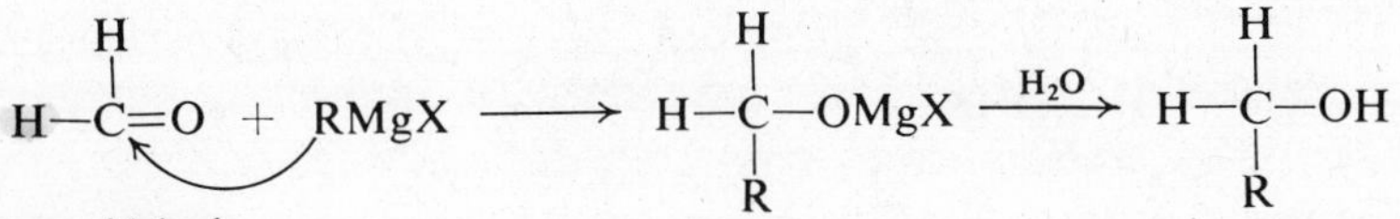

$$-\overset{|}{C}{=}O + RMgX \longrightarrow -\overset{|}{\underset{R}{C}}-OMgX \xrightarrow{H_2O} -\overset{|}{\underset{R}{C}}-OH + Mg^{++} + X^-$$

$$H-\overset{H}{\overset{|}{C}}{=}O + RMgX \longrightarrow H-\overset{H}{\underset{R}{C}}-OMgX \xrightarrow{H_2O} H-\overset{H}{\underset{R}{C}}-OH$$

Formaldehyde — **1° alcohol**

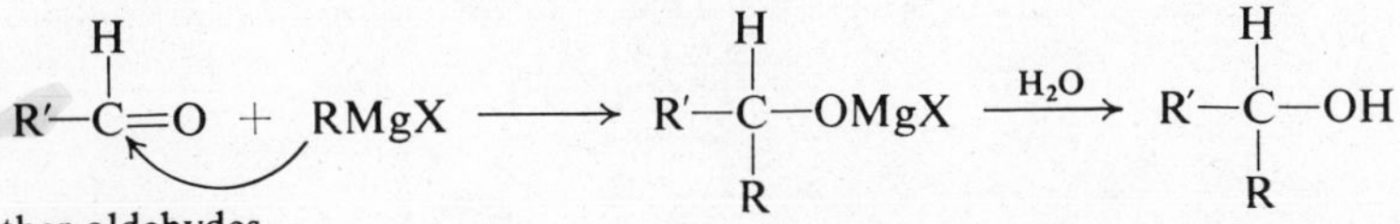

$$R'-\overset{H}{\overset{|}{C}}{=}O + RMgX \longrightarrow R'-\overset{H}{\underset{R}{C}}-OMgX \xrightarrow{H_2O} R'-\overset{H}{\underset{R}{C}}-OH$$

Other aldehydes — **2° alcohol**

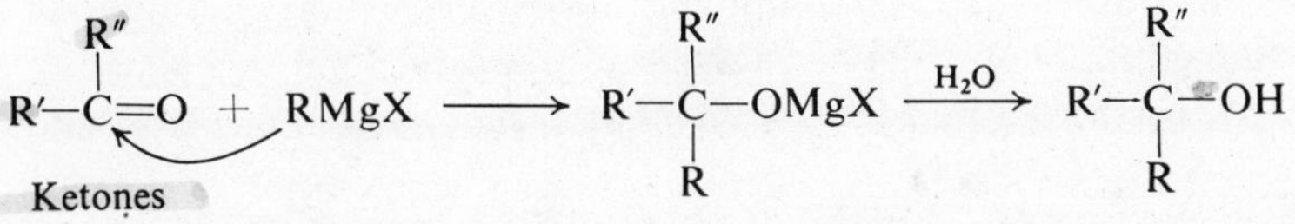

$$R'-\overset{R''}{\overset{|}{C}}{=}O + RMgX \longrightarrow R'-\overset{R''}{\underset{R}{C}}-OMgX \xrightarrow{H_2O} R'-\overset{R''}{\underset{R}{C}}-OH$$

Ketones — **3° alcohol**

$$H_2C\overset{O}{—}CH_2 + RMgX \longrightarrow RCH_2CH_2OMgX \xrightarrow{H_2O} RCH_2CH_2OH$$

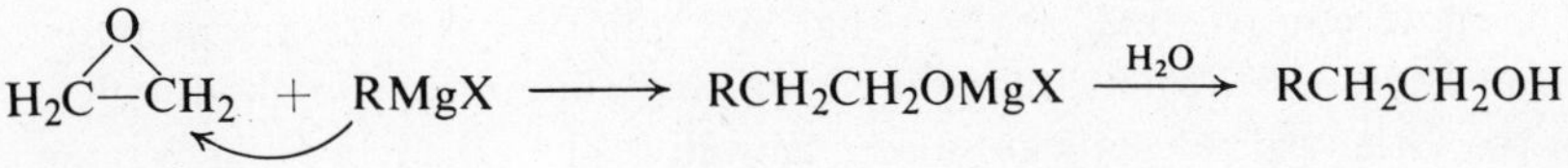

Ethylene oxide — **1° alcohol:** *two carbons added*

$$R'COOC_2H_5 + 2RMgX \longrightarrow R'-\overset{R}{\underset{R}{C}}-OMgX \xrightarrow{H_2O} R'-\overset{R}{\underset{R}{C}}-OH$$

An ester — **3° alcohol**

Discussed in Sec. 20.21.

4. **Hydrolysis of alkyl halides.** Discussed in Sec. 15.7.

$$R{-}X + OH^- \text{ (or } H_2O) \longrightarrow R{-}OH + X^- \text{ (or HX)}$$

Examples:

$$C_6H_5CH_2Cl \xrightarrow{\text{aqueous NaOH}} C_6H_5CH_2OH$$

Benzyl chloride → Benzyl alcohol

$$CH_2{=}CH_2 \xrightarrow{Cl_2,\ H_2O} CH_2Cl{-}CH_2OH \xrightarrow{Na_2CO_3,\ H_2O} CH_2OH{-}CH_2OH$$

Ethylene → Ethylene chlorohydrin → Ethylene glycol

5. **Aldol condensation.** Discussed in Sec. 21.7.

6. **Reduction of carbonyl compounds.** Discussed in Sec. 19.10.

7. **Reduction of acids and esters.** Discussed in Secs. 18.18 and 20.22.

8. **Hydroxylation of alkenes.** Discussed in Secs. 6.20 and 17.12.

$$-\overset{|}{C}{=}\overset{|}{C}- \xrightarrow{KMnO_4} -\overset{|}{\underset{OH}{C}}-\overset{|}{\underset{OH}{C}}- \qquad \textit{Syn}\text{-hydroxylation}$$

$$-\overset{|}{C}{=}\overset{|}{C}- \xrightarrow{RCO_2OH} -\overset{|}{C}\underset{O}{\diagdown\diagup}\overset{|}{C}- \xrightarrow{H_2O,\ H^+} -\overset{|}{\underset{OH}{C}}-\overset{OH}{\underset{|}{C}}- \qquad \textit{Anti}\text{-hydroxylation}$$

We can follow either of two approaches to the synthesis of alcohols—or, for that matter, of most other kinds of compounds. (a) We can retain the original carbon skeleton, and simply convert one functional group into another until we arrive at an alcohol; or (b) we can generate a new, bigger carbon skeleton and at the same time produce an alcohol.

By far the most important method of preparing alcohols is the **Grignard synthesis**. This is an example of the second approach, since it leads to the formation of carbon–carbon bonds. In the laboratory a chemist is chiefly concerned with preparing the more complicated alcohols that he cannot buy; these are prepared by the Grignard synthesis from rather simple starting materials. The alkyl halides from which the Grignard reagents are made, as well as the aldehydes and ketones themselves, are most conveniently prepared from alcohols; thus the method ultimately involves the synthesis of alcohols from less complicated alcohols.

Alcohols can be conveniently made from compounds containing carbon–carbon double bonds in two ways; by **oxymercuration-demercuration** and by **hydroboration-oxidation.** Both amount to addition of water to the double bond, but with *opposite orientation*—Markovnikov and anti-Markovnikov—and hence the two methods neatly complement each other.

Hydrolysis of alkyl halides is severely limited as a method of synthesizing alcohols, since alcohols are usually more available than the corresponding halides; indeed, the best general preparation of halides is from alcohols. The synthesis of benzyl alcohol from toluene, however, is an example of a useful application of this method.

$$\underset{\text{Toluene}}{C_6H_5CH_3} \xrightarrow{Cl_2,\ \text{heat, light}} \underset{\text{Benzyl chloride}}{C_6H_5CH_2Cl} \xrightarrow{\text{aqueous NaOH}} \underset{\text{Benzyl alcohol}}{C_6H_5CH_2OH}$$

For those halides that can undergo elimination, the formation of alkene must always be considered a possible side reaction.

Problem 15.7 Give structures of compounds A through D in the following industrially important synthesis.

$$CH_3CH{=}CH_2 + Cl_2 \xrightarrow{600^\circ} A\ (C_3H_5Cl)$$

$$A + H_2O,\ OH^- \longrightarrow B\ (C_3H_6O)$$

$$B + Cl_2,\ H_2O \longrightarrow C\ (C_3H_7O_2Cl)$$

$$C + H_2O,\ OH^- \longrightarrow D\ (C_3H_8O_3)$$

15.8 Oxymercuration-demercuration

Alkenes react with mercuric acetate in the presence of water to give hydroxymercurial compounds which on reduction yield alcohols.

Oxymercuration **Demercuration**

$$\underset{\text{Alkene}}{>C{=}C<} + \underset{\text{Water}}{H_2O} + \underset{\text{Mercuric acetate}}{Hg(OAc)_2} \longrightarrow -\underset{OH}{\overset{|}{\underset{|}{C}}}-\underset{HgOAc}{\overset{|}{\underset{|}{C}}}- \xrightarrow{NaBH_4} \underset{\text{Alcohol}}{-\underset{OH}{\overset{|}{\underset{|}{C}}}-\underset{H}{\overset{|}{\underset{|}{C}}}-}$$

$$-OAc = CH_3COO-$$

The first stage, *oxymercuration*, involves addition to the carbon–carbon double bond of —OH and —HgOAc. Then, in *demercuration*, the —HgOAc is replaced by —H. The reaction sequence amounts to hydration of the alkene, but is much more widely applicable than direct hydration.

The two-stage process of oxymercuration-demercuration is fast and convenient, takes place under mild conditions, and gives excellent yields—often over 90%. The alkene is added at room temperature to an aqueous solution of mercuric acetate diluted with the solvent tetrahydrofuran. Reaction is generally complete within minutes. The organomercurial compound is not isolated but is simply reduced *in situ* by sodium borohydride, $NaBH_4$. (The mercury is recovered as a ball of elemental mercury.)

Oxymercuration-demercuration is highly regiospecific, and gives alcohols corresponding to *Markovnikov* addition of water to the carbon–carbon double bond. For example:

$$\underset{\text{1-Hexene}}{CH_3(CH_2)_3CH{=}CH_2} \xrightarrow{Hg(OAc)_2,\ H_2O} \xrightarrow{NaBH_4} \underset{\text{2-Hexanol}}{CH_3(CH_2)_3\underset{OH}{\underset{|}{C}}HCH_3}$$

$$\underset{\text{2-Methyl-1-butene}}{CH_3CH_2\overset{CH_3}{\overset{|}{C}}{=}CH_2} \xrightarrow{Hg(OAc)_2,\ H_2O} \xrightarrow{NaBH_4} \underset{\textit{tert}\text{-Pentyl alcohol}}{CH_3CH_2\overset{CH_3}{\overset{|}{\underset{OH}{\underset{|}{C}}}}CH_3}$$

$$\text{1-Methylcyclopentene} \xrightarrow{Hg(OAc)_2,\ H_2O} \xrightarrow{NaBH_4} \text{1-Methylcyclopentanol}$$

$$\underset{\text{3,3-Dimethyl-1-butene}}{CH_3{-}\overset{CH_3}{\overset{|}{\underset{CH_3}{\underset{|}{C}}}}{-}CH{=}CH_2} \xrightarrow{Hg(OAc)_2,\ H_2O} \xrightarrow{NaBH_4} \underset{\text{3,3-Dimethyl-2-butanol}}{CH_3{-}\overset{CH_3}{\overset{|}{\underset{CH_3}{\underset{|}{C}}}}{-}\underset{OH}{\underset{|}{C}}H{-}CH_3}$$

Oxymercuration involves electrophilic addition to the carbon–carbon double bond, with the mercuric ion acting as electrophile. The absence of rearrangement and the high degree of stereospecificity (typically *anti*)—*in the oxymercuration step*—argues against an open carbonium ion as intermediate. Instead, it has been proposed, there is formed a cyclic *mercurinium ion*, analogous to the bromonium

$$\left[-\overset{|}{C}{-}{-}\overset{|}{C}- \atop \diagdown\ Hg\ \diagup \right]^{++}$$

and chloronium ions involved in the addition of halogens. In 1971, Olah (p. 160) reported spectroscopic evidence for the preparation of stable solutions of such mercurinium ions.

The mercurinium ion is attacked by the nucleophilic solvent—water, in the present case—to yield the addition product. This attack is back-side (unless prevented by some structural feature) and the net result is *anti* addition, as in the addition of halogens (Sec. 7.12). Attack is thus of the S_N2 type; yet the orientation of addition shows that the nucleophile becomes attached to the more highly substituted carbon—as though there were a free carbonium ion intermediate. As we shall see (Sec. 17.15), the transition state in reactions of such unstable three-membered rings has much S_N1 character.

Although the demercuration reaction is not really understood, free radicals have been proposed as intermediates. Whatever the mechanism, demercuration is generally not stereospecific and can, in certain special cases, be accompanied by rearrangement.

Despite the stereospecificity of the first stage, then, the overall process is not, in general, stereospecific. Rearrangements *can* occur, but are not common. The reaction of 3,3-dimethyl-1-butene illustrates the absence of the rearrangements that are typical of intermediate carbonium ions.

Mercuration can be carried out in different solvents to yield products other than alcohols. This use of *solvomercuration* as a general synthetic tool is due largely to H. C. Brown (p. 507).

Problem 15.8 Predict the product of the reaction of styrene with mercuric acetate in methanol solution, followed by reduction with $NaBH_4$.

15.9 Hydroboration-oxidation

With the reagent *diborane*, $(BH_3)_2$, alkenes undergo *hydroboration* to yield alkylboranes, R_3B, which on oxidation give alcohols. For example:

$$\underset{\text{Diborane}}{(BH_3)_2} \xrightarrow{H_2C{=}CH_2} CH_3CH_2BH_2 \xrightarrow{H_2C{=}CH_2} (CH_3CH_2)_2BH \xrightarrow{H_2C{=}CH_2} \underset{\text{Triethylboron}}{(CH_3CH_2)_3B}$$

$$\underset{\text{Triethylboron}}{(CH_3CH_2)_3B} + \underset{\text{Hydrogen peroxide}}{3H_2O_2} \xrightarrow{OH^-} \underset{\text{Ethyl alcohol}}{3CH_3CH_2OH} + \underset{\text{Boric acid}}{B(OH)_3}$$

The reaction procedure is simple and convenient, the yields are exceedingly high, and, as we shall see, the products are ones difficult to obtain from alkenes in any other way.

Diborane is the dimer of the hypothetical BH_3 (*borane*) and, in the reactions that concern us, acts much as though it were BH_3. Indeed, in tetrahydrofuran, one of the solvents used for hydroboration, the reagent exists as the monomer, in the form of an acid–base complex with the solvent.

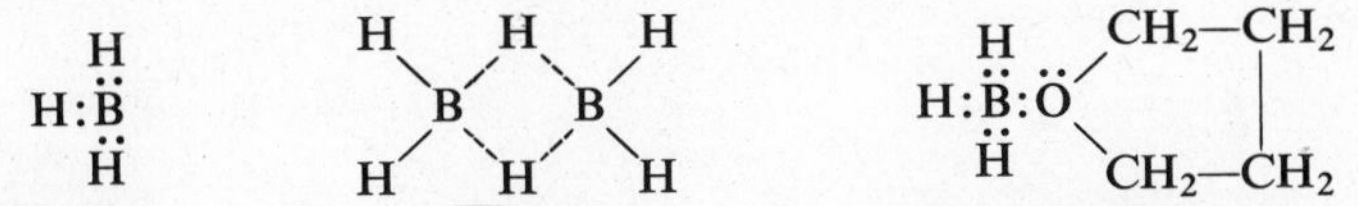

H:B(H)(H) — Borane

$H_2B(\mu\text{-}H)_2BH_2$ — Diborane

$H_3B{:}O(CH_2CH_2)_2$ — Borane-tetrahydrofuran complex

Hydroboration involves addition to the double bond of BH_3 (or, in following stages, BH_2R and BHR_2), with hydrogen becoming attached to one doubly-bonded carbon, and boron to the other. The alkylborane can then undergo oxidation, in which the boron is replaced by —OH (by a mechanism we shall encounter in Sec. 28.6).

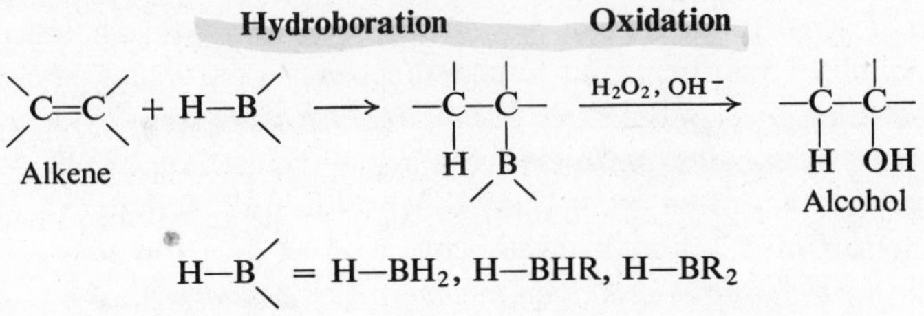

Hydroboration **Oxidation**

$$\underset{\text{Alkene}}{>C{=}C<} + H{-}B< \longrightarrow -\underset{H}{\overset{|}{C}}-\underset{B<}{\overset{|}{C}}- \xrightarrow{H_2O_2,\ OH^-} \underset{\text{Alcohol}}{-\underset{H}{\overset{|}{C}}-\underset{OH}{\overset{|}{C}}-}$$

$$H{-}B< \;=\; H{-}BH_2,\ H{-}BHR,\ H{-}BR_2$$

Thus, the two-stage reaction process of hydroboration-oxidation permits, in effect, the addition to the carbon–carbon double bond of the elements of H—OH.

Reaction is carried out in an ether, commonly tetrahydrofuran or "diglyme" (*di*ethylene *gly*col *me*thyl ether, $CH_3OCH_2CH_2OCH_2CH_2OCH_3$). Diborane is commercially available in tetrahydrofuran solution. The alkylboranes are not isolated, but are simply treated *in situ* with alkaline hydrogen peroxide.

15.10 Orientation and stereochemistry of hydroboration

Hydroboration-oxidation, then, converts alkenes into alcohols. Addition is highly regiospecific; the preferred product here, however, is exactly *opposite* to the one formed by oxymercuration-demercuration or by direct acid-catalyzed hydration. For example:

$$\underset{\text{Propylene}}{CH_3CH{=}CH_2} \xrightarrow{(BH_3)_2} \xrightarrow{H_2O_2,\ OH^-} \underset{\substack{n\text{-Propyl alcohol} \\ (1°)}}{CH_3CH_2CH_2OH}$$

$$\underset{\text{1-Butene}}{CH_3CH_2CH{=}CH_2} \xrightarrow{(BH_3)_2} \xrightarrow{H_2O_2,\ OH^-} \underset{\substack{n\text{-Butyl alcohol} \\ (1°)}}{CH_3CH_2CH_2CH_2OH}$$

$$\underset{\text{Isobutylene}}{CH_3{-}\overset{\overset{CH_3}{|}}{C}{=}CH_2} \xrightarrow{(BH_3)_2} \xrightarrow{H_2O_2,\ OH^-} \underset{\substack{\text{Isobutyl alcohol} \\ (1°)}}{CH_3{-}\overset{\overset{CH_3}{|}}{CH}{-}CH_2OH}$$

$$\underset{\text{2-Methyl-2-butene}}{CH_3{-}CH{=}\overset{\overset{CH_3}{|}}{C}{-}CH_3} \xrightarrow{(BH_3)_2} \xrightarrow{H_2O_2,\ OH^-} \underset{\substack{\text{3-Methyl-2-butanol} \\ (2°)}}{CH_3{-}\underset{\underset{OH}{|}}{CH}{-}\overset{\overset{CH_3}{|}}{CH}{-}CH_3}$$

$$\underset{\text{3,3-Dimethyl-1-butene}}{CH_3{-}\overset{\overset{CH_3}{|}}{\underset{\underset{CH_3}{|}}{C}}{-}CH{=}CH_2} \xrightarrow{(BH_3)_2} \xrightarrow{H_2O_2,\ OH^-} \underset{\substack{\text{3,3-Dimethyl-1-butanol} \\ (1°)}}{CH_3{-}\overset{\overset{CH_3}{|}}{\underset{\underset{CH_3}{|}}{C}}{-}CH_2{-}CH_2OH}$$

The hydroboration-oxidation process gives products corresponding to **anti-Markovnikov** *addition of water to the carbon–carbon double bond.*

The reaction of 3,3-dimethyl-1-butene illustrates a particular advantage of the method. *Rearrangement does not occur in hydroboration*—evidently because carbonium ions are not intermediates—and hence the method can be used without the complications that often accompany other addition reactions.

The reaction of 1,2-dimethylcyclopentene illustrates the stereochemistry of the synthesis: *hydroboration-oxidation involves overall* ***syn addition***.

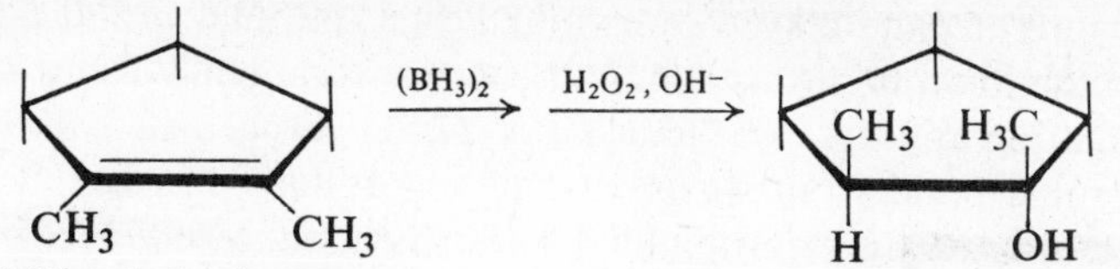

$\xrightarrow{(BH_3)_2}$ $\xrightarrow{H_2O_2,\ OH^-}$

1,2-Dimethylcyclopentene → *cis*-1,2-Dimethylcyclopentanol

Through a combination of features of which we take up only three—orientation, stereochemistry, and freedom from rearrangements—hydroboration-oxidation gains its great synthetic utility: it gives a set of alcohols not obtainable from alkenes by other methods and, through these alcohols (Sec. 16.10), provides a convenient route to corresponding members of many chemical families.

We catch here a brief glimpse of just one of the many applications of hydroboration to organic synthesis that have been discovered by H. C. Brown (of Purdue University). Although generally recognized as an outstanding organic chemist, Professor Brown was originally trained as an inorganic chemist, in the laboratory of H. I. Schlesinger at the University of Chicago. It was in this laboratory—in the course of a search for volatile uranium compounds, during World War II—that lithium aluminum hydride and sodium borohydride (Sec. 19.10) were first made and their reducing properties first observed; and it was here that Brown's interest in borohydrides originated.

The examples we have used to show the fundamentals of hydroboration-oxidation have been, necessarily, simple ones. In practice, synthesis generally involves more complicated molecules, but the principles remain the same. For example:

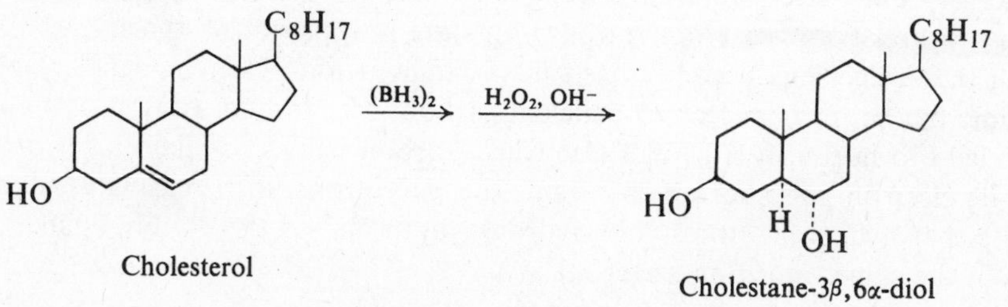

$\xrightarrow{(BH_3)_2}$ $\xrightarrow{H_2O_2,\ OH^-}$

Cholesterol → Cholestane-3β,6α-diol

$$C_8H_{17} = -CH(CH_3)CH_2CH_2CH_2CH(CH_3)_2$$

Problem 15.9 Predict the products of hydroboration-oxidation of: (a) *cis*-2-phenyl-2-butene; (b) *trans*-2-phenyl-2-butene; (c) 1-methylcyclohexene.

Problem 15.10 The stereochemistry of hydroboration-oxidation is the *net* result of the stereochemistry of the two steps, and is consistent with either of two combinations of stereochemistry for the individual steps. What are these two combinations?

15.11 Mechanism of hydroboration

Much of the usefulness of hydroboration-oxidation lies in the "unusual" orientation of the hydration. The —OH simply takes the position occupied by boron in the intermediate alkylborane, and hence the final product reflects the orientation of the hydroboration step. Is this orientation really "unusual"?

The orientation appears to be unusual because hydrogen adds to the opposite end of the double bond from where it adds in ordinary electrophilic addition. But the fundamental idea in electrophilic addition is that the *electrophilic* part of the reagent—the *acidic* part—becomes attached, using the π electrons, in such a way that the carbon being deprived of the π electrons is the one best able to stand the deprivation. Thus, with propylene as an example:

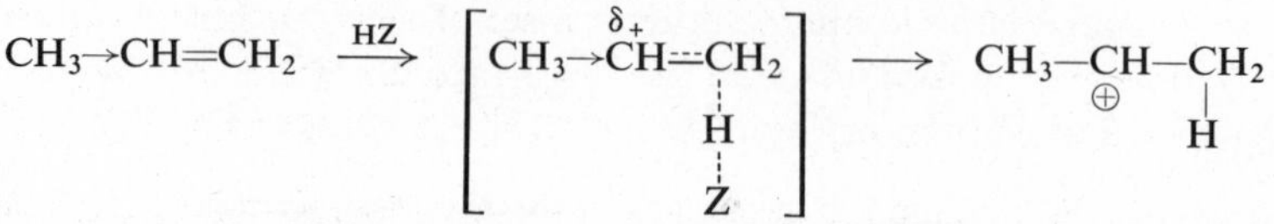

$$CH_3{\rightarrow}CH{=}CH_2 \xrightarrow{HZ} \left[\begin{array}{c} CH_3{\rightarrow}\overset{\delta_+}{C}H{\overset{\dots}{=}}CH_2 \\ \qquad\quad \vdots \\ \qquad\quad H \\ \qquad\quad \vdots \\ \qquad\quad Z \end{array}\right] \longrightarrow \begin{array}{c} CH_3{-}\underset{\oplus}{C}H{-}CH_2 \\ \qquad\quad | \\ \qquad\quad H \end{array}$$

Now, what is the center of acidity in BH_3? Clearly, *boron*, with only six electrons. It is not at all surprising that boron should seek out the π electrons of the double bond and begin to attach itself to carbon. In doing this, it attaches itself in such a way that the positive charge can develop on the carbon best able to accommodate it. Thus:

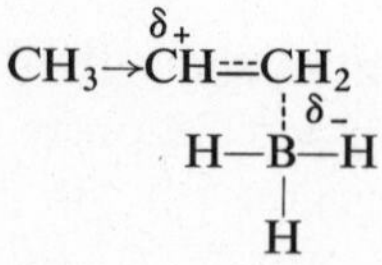

$$\begin{array}{c} CH_3{\rightarrow}\overset{\delta_+}{C}H{\overset{\dots}{=}}CH_2 \\ \qquad\quad \vdots\ \delta_- \\ \qquad H{-}B{-}H \\ \qquad\quad | \\ \qquad\quad H \end{array}$$

Unlike ordinary electrophilic addition, however, the reaction does not proceed to give a carbonium ion. As the transition state is approached, the carbon that is losing the π electrons becomes itself increasingly acidic: electron-deficient boron is acidic but so, too, is electron-deficient carbon. Not too far away is a hydrogen atom held to boron by a pair of electrons. Carbon begins to take that hydrogen, with its electron pair; boron, as it gains the π electrons, is increasingly willing to release that hydrogen. Boron and hydrogen both add to the doubly-bonded carbons in the same transition state:

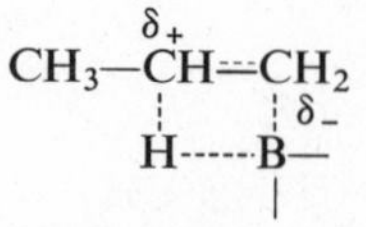

$$\begin{array}{c} CH_3{-}\overset{\delta_+}{C}H{\overset{\dots}{=}}CH_2 \\ \qquad \vdots \qquad \vdots\ \delta_- \\ \qquad H{\cdots\cdots}B{-} \\ \qquad\qquad | \end{array}$$

Transition state for hydroboration

In view of the basic nature of alkenes and the acidic nature of BH_3, the principal driving force of the reaction is almost certainly *attachment of boron to carbon*. In the transition state attachment of boron to C–1 has proceeded to a greater extent than attachment of hydrogen to C–2. Thus loss of (π) electrons by C–2 to the C_1—B bond exceeds its gain of electrons from hydrogen, and so C–2, the carbon that can best accommodate the charge, has become positive.

On theoretical grounds (Chap. 29) it has been postulated that the step we have described must follow a preliminary step in which boron attaches itself to both carbon atoms, or perhaps to the π electrons.

Thus orientation of addition in hydroboration is controlled in fundamentally the same way as in two-step electrophilic addition. Hydrogen becomes attached to opposite ends of the double bond in the two reactions because it adds without electrons in one case (as a *proton*, an acid), and with electrons in the other case (as a *hydride ion*, a base).

Because of the Lowry-Brønsted treatment of acids and bases, we tend to think of hydrogen chiefly in its proton character. Actually, its hydride character has considerably more *reality*. Solid lithium hydride, for example, has an ionic crystalline lattice made up of Li^+ and H^-; by contrast, a naked unsolvated proton is not encountered by the organic chemist.

We are already familiar with the facile transfer of hydride from carbon to carbon: within a single molecule (hydride shift in rearrangements), and between molecules (abstraction by carbonium ion, Sec. 6.16). Later on we shall encounter a set of remarkably versatile reducing agents (hydrides like *lithium aluminum hydride*, $LiAlH_4$, and *sodium borohydride*, $NaBH_4$) that function by transfer of hydride ion to organic molecules.

Problem 15.11 Identify the acids and bases (Lewis or Lowry-Brønsted) in each of the following reactions:

(a) $Li^+H^- + H_2O \longrightarrow H_2 + Li^+OH^-$

(b) $(C_2H_5)_3B + NH_3 \longrightarrow (C_2H_5)_3\overset{-}{B}{:}\overset{+}{N}H_3$

(c) $(BH_3)_2 + 2(CH_3)_3N \longrightarrow 2H_3\overset{-}{B}{:}\overset{+}{N}(CH_3)_3$

(d) $2Li^+H^- + (BH_3)_2 \longrightarrow 2Li^+BH_4^-$

Problem 15.12 In light of the mechanism, what stereochemistry would you expect for the hydroboration step? On this basis, which of the two combinations in Problem 15.10 (p. 507) would be the correct one? What would the stereochemistry of the oxidation step be?

(Actually, the stereochemistry, worked out in a way we cannot go into here, is part of the basis for the mechanism, and not the other way around.)

15.12 Grignard synthesis of alcohols

The Grignard reagent, we recall, has the formula RMgX, and is prepared by the reaction of metallic magnesium with the appropriate organic halide (Sec. 3.16). This halide can be alkyl (1°, 2°, 3°), allylic, aralkyl (e.g., benzyl), or aryl (phenyl

$$RX + Mg \xrightarrow{\text{anhydrous ether}} \underset{\text{A Grignard reagent}}{RMgX}$$

or substituted phenyl). The halogen may be —Cl, —Br or —I. (Arylmagnesium *chlorides* must be made in the cyclic ether tetrahydrofuran instead of ethyl ether.)

One of the most important uses of the Grignard reagent is its reaction with aldehydes and ketones to yield alcohols. Aldehydes and ketones have the general formulas:

$$\underset{\text{An aldehyde}}{R-\overset{\overset{H}{|}}{C}=O} \qquad \underset{\text{A ketone}}{R-\overset{\overset{R}{|}}{C}=O}$$

The functional group of both is the **carbonyl group**, $-\overset{|}{C}=O$, and, as we shall see later (Chap. 19), aldehydes and ketones resemble each other closely in most of their reactions. Like the carbon–carbon double bond, the carbonyl group is unsaturated, and like the carbon–carbon bond, it undergoes addition. One of its typical reactions is addition of the Grignard reagent.

Since the electrons of the carbonyl double bond hold together atoms of quite different electronegativity, we would not expect the electrons to be equally shared; in particular, the mobile π cloud should be pulled strongly toward the more electronegative atom, oxygen. Whatever the mechanism involved, addition of an unsymmetrical reagent is oriented so that the nucleophilic (basic) portion attaches itself to carbon, and the electrophilic (acidic) portion attaches itself to oxygen.

The carbon–magnesium bond of the Grignard reagent is a highly polar bond, carbon being negative relative to electropositive magnesium. It is not surprising, then, that in the addition to carbonyl compounds, the organic group becomes attached to carbon and magnesium to oxygen. The product is the magnesium

$$\overset{\delta+}{C}=\overset{\delta-}{O} + \overset{\delta-}{R}\!:\!\overset{\delta+}{MgX} \longrightarrow -\underset{R}{\overset{|}{C}}-OMgX \xrightarrow{H_2O} -\underset{R}{\overset{|}{C}}-OH + Mg(OH)X$$

An alcohol

$$Mg(OH)X \xrightarrow{H^+} Mg^{++} + X^- + H_2O$$

salt of the weakly acidic alcohol and is easily converted into the alcohol itself by the addition of the stronger acid, water. Since the Mg(OH)X thus formed is a gelatinous material difficult to handle, dilute mineral acid (HCl, H_2SO_4) is commonly used instead of water, so that water-soluble magnesium salts are formed.

15.13 Products of the Grignard synthesis

The class of alcohol that is obtained from a Grignard synthesis depends upon the type of carbonyl compound used: *formaldehyde*, HCHO, *yields primary alcohols; other aldehydes*, RCHO, *yield secondary alcohols; and ketones*, R_2CO, *yield tertiary alcohols.*

$$H-\overset{H}{\overset{|}{C}}=O + RMgX \longrightarrow H-\underset{R}{\overset{H}{C}}-OMgX \xrightarrow{H_2O} H-\underset{R}{\overset{H}{C}}-OH$$

Formaldehyde — 1° alcohol

$$R'-\overset{H}{\overset{|}{C}}=O + RMgX \longrightarrow R'-\underset{R}{\overset{H}{C}}-OMgX \xrightarrow{H_2O} R'-\underset{R}{\overset{H}{C}}-OH$$

Higher aldehydes — 2° alcohol

$$R'-\overset{R''}{\overset{|}{C}}=O + RMgX \longrightarrow R'-\underset{R}{\overset{R''}{C}}-OMgX \xrightarrow{H_2O} R'-\underset{R}{\overset{R''}{C}}-OH$$

Ketones — 3° alcohol

This relationship arises directly from our definitions of aldehydes and ketones, and our definitions of primary, secondary, and tertiary alcohols. The number of hydrogens attached to the carbonyl carbon defines the carbonyl compound as formaldehyde, higher aldehyde, or ketone. The carbonyl carbon is the one that finally bears the —OH group in the product; here the number of hydrogens defines the alcohol as primary, secondary, or tertiary. For example:

$$\underset{\substack{sec\text{-Butylmagnesium}\\ \text{bromide}}}{\underset{\substack{|\\ MgBr}}{CH_3CH_2CHCH_3}} + \underset{\text{Formaldehyde}}{H{-}\overset{\substack{H\\ |}}{C}{=}O} \longrightarrow CH_3CH_2\overset{\substack{CH_3\\ |}}{C}HCH_2OMgBr \xrightarrow{H_2O} \underset{\substack{\text{A 1° alcohol}\\ sec\text{-Butylcarbinol}\\ \text{(2-Methyl-1-butanol)}}}{CH_3CH_2\overset{\substack{CH_3\\ |}}{C}HCH_2OH}$$

$$\underset{\substack{\text{Phenylmagnesium}\\ \text{bromide}}}{C_6H_5MgBr} + \underset{\text{Acetaldehyde}}{CH_3{-}\overset{\substack{H\\ |}}{C}{=}O} \longrightarrow C_6H_5\overset{\substack{CH_3\\ |}}{C}HOMgBr \xrightarrow{H_2O} \underset{\substack{\text{A 2° alcohol}\\ \text{Phenylmethylcarbinol}\\ \text{(1-Phenylethanol)}}}{C_6H_5\overset{\substack{CH_3\\ |}}{C}HOH}$$

$$\underset{\substack{n\text{-Butylmagnesium}\\ \text{bromide}}}{n\text{-}C_4H_9MgBr} + \underset{\text{Acetone}}{CH_3{-}\overset{\substack{CH_3\\ |}}{C}{=}O} \longrightarrow n\text{-}C_4H_9{-}\underset{\substack{|\\ CH_3}}{\overset{\substack{CH_3\\ |}}{C}}{-}OMgBr \xrightarrow{H_2O} \underset{\substack{\text{A 3° alcohol}\\ n\text{-Butyldimethylcarbinol}\\ \text{(2-Methyl-2-hexanol)}}}{n\text{-}C_4H_9{-}\underset{\substack{|\\ CH_3}}{\overset{\substack{CH_3\\ |}}{C}}{-}OH}$$

A related synthesis utilizes *ethylene oxide* (Sec. 17.14) to make *primary alcohols containing two more carbons* than the Grignard reagent. Here, too, the organic

$$\underset{\text{Ethylene oxide}}{H_2C\overset{O}{—}CH_2} + RMgX \longrightarrow RCH_2CH_2OMgX \xrightarrow{H_2O} \underset{\substack{\text{A 1° alcohol:}\\ \textit{two carbons added}}}{RCH_2CH_2OH}$$

group becomes attached to carbon and magnesium to oxygen, this time with the breaking of a carbon–oxygen σ bond in the highly strained three-membered ring (Sec. 9.9). For example:

$$\underset{\substack{\text{Phenylmagnesium}\\ \text{bromide}}}{C_6H_5MgBr} + \underset{\substack{\text{Ethylene}\\ \text{oxide}}}{H_2C\overset{O}{—}CH_2} \longrightarrow C_6H_5CH_2CH_2OMgBr \xrightarrow{H_2O} \underset{\substack{\beta\text{-Phenylethyl alcohol}\\ \text{(2-Phenylethanol)}}}{C_6H_5CH_2CH_2OH}$$

15.14 Planning a Grignard synthesis

How do we decide which Grignard reagent and which carbonyl compound to use in preparing a particular alcohol? We have only to look at the structure of the alcohol we want. Of the groups attached to the carbon bearing the —OH group, one must come from the Grignard reagent, the other two (including any hydrogens) must come from the carbonyl compound.

Most alcohols can be obtained from more than one combination of reagents; we usually choose the combination that is most readily available. Consider, for example, the synthesis of 2-phenyl-2-hexanol:

$$CH_3CH_2CH_2CH_2-C(CH_3)(OH)-C_6H_5 \longleftarrow CH_3CH_2CH_2CH_2MgBr + C_6H_5-C(CH_3)=O$$

2-Phenyl-2-hexanol — *n*-Butylmagnesium bromide — Acetophenone

$$CH_3CH_2CH_2CH_2-C(CH_3)(OH)-C_6H_5 \longleftarrow CH_3CH_2CH_2CH_2-C(CH_3)=O + BrMg-C_6H_5$$

2-Phenyl-2-hexanol — Methyl *n*-butyl ketone — Phenylmagnesium bromide

As shown, we could make this either from the four-carbon Grignard reagent and the aromatic ketone, or from the phenyl Grignard reagent and the six-carbon aliphatic ketone. As we shall know when we have studied aldehydes and ketones (Chap. 19), the first route uses the more readily available carbonyl compound and is the one actually used to make this alcohol.

15.15 Limitations of the Grignard synthesis

The very reactivity that makes a Grignard reagent so useful strictly limits how we may use it. We must keep this reactivity in mind when we plan the experimental conditions of the synthesis, when we select the halide that is to become the Grignard reagent, and when we select the compound with which it is to react.

In our first encounter with the Grignard reagent (Sec. 3.16), we allowed it to react with water to form an alkane; the stronger acid, water, displaced the extremely weak acid, the alkane, from its salt. In the same way, *any* compound containing hydrogen attached to an electronegative element—oxygen, nitrogen, sulfur, or even triply-bonded carbon—is acidic enough to decompose a Grignard reagent. A Grignard reagent reacts rapidly with oxygen and carbon dioxide, and with nearly every organic compound containing a carbon–oxygen or carbon–nitrogen multiple bond.

How does all this affect our reaction between a Grignard reagent and, say, an aldehyde? First of all, alkyl halide, aldehyde, and the ether used as solvent must be scrupulously dried and freed of the alcohol from which each was very probably made; a Grignard reagent will not even form in the presence of water. Our apparatus must be completely dry before we start. We must protect the reaction system from the water vapor, oxygen, and carbon dioxide of the air: water vapor can be kept out by use of calcium chloride tubes, and oxygen and carbon dioxide can be swept out of the system with dry nitrogen. Having done all this we may hope to obtain a good yield of product—providing we have properly chosen the halide and the aldehyde.

We cannot prepare a Grignard reagent from a compound (e.g., $HOCH_2CH_2Br$) that contains, in addition to halogen, some group (e.g., —OH) that will react with a Grignard reagent; if this were tried, as fast as a molecule of Grignard reagent formed it would react with the active group (—OH) in another molecule to yield an undesired product ($HOCH_2CH_2$—H).

We must be particularly watchful in the preparation of an arylmagnesium halide, in view of the wide variety of substituents that might be present on the benzene ring. Carboxyl (—COOH), hydroxyl (—OH), amino (—NH_2), and —SO_3H all contain hydrogen attached to oxygen or nitrogen, and therefore are so acidic that they will decompose a Grignard reagent. We have just learned that a Grignard reagent adds to the carbonyl group (C=O), and we shall learn that it adds similarly to —COOR and —C≡N groups. The nitro (—NO_2) group oxidizes a Grignard reagent. It turns out that only a comparatively few groups may be present in the halide molecule from which we prepare a Grignard reagent; among these are —R, —Ar, —OR, and —Cl (of an aryl chloride).

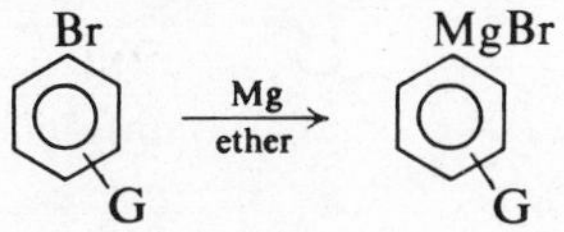

$$\text{Br-C}_6\text{H}_4\text{-G} \xrightarrow[\text{ether}]{\text{Mg}} \text{BrMg-C}_6\text{H}_4\text{-G}$$

G *may not be:*

—COOH	—C=O
—OH	—COOR
—NH_2	—C≡N
—SO_3H	—NO_2

and many others

G *may be:*

—R	—OR
—Ar(aryl)	—Cl

By the same token, the aldehyde (or other compound) with which a Grignard reagent is to react may not contain other groups that are reactive toward a Grignard reagent. For example, a Grignard reagent would be decomposed before it could add to the carbonyl group of:

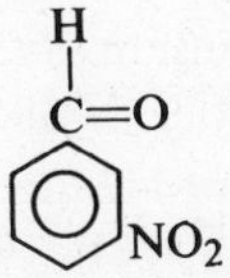

m-Nitrobenzaldehyde (m-O_2N-C_6H_4-CHO) *p*-Aminoacetophenone 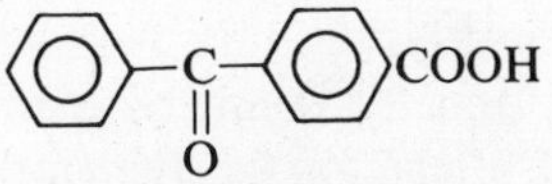 (p-H_2N-C_6H_4-CO-CH_3) *p*-Benzoylbenzoic acid (C_6H_5-CO-C_6H_4-COOH)

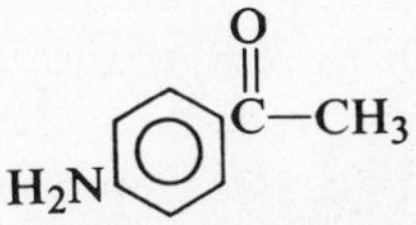

These may seem like severe limitations, and they are. Nevertheless, the number of acceptable combinations is so great that the Grignard reagent is one of our most valuable synthetic tools. The kind of precautions described here must be taken in any kind of organic synthesis: we must not restrict our attention to the group we happen to be interested in, but must look for possible interference by other functional groups.

15.16 Steroids

Cholesterol (p. 507), notorious as the substance deposited on the walls of arteries and as the chief constituent of gallstones, is the kind of alcohol called a *sterol.* Sterols belong, in turn, to the class of compounds called **steroids**: compounds of the general formula

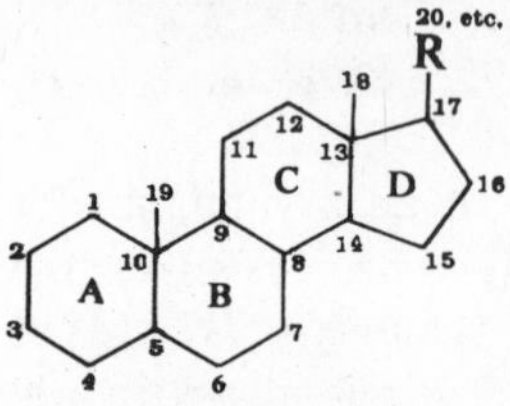

A steroid

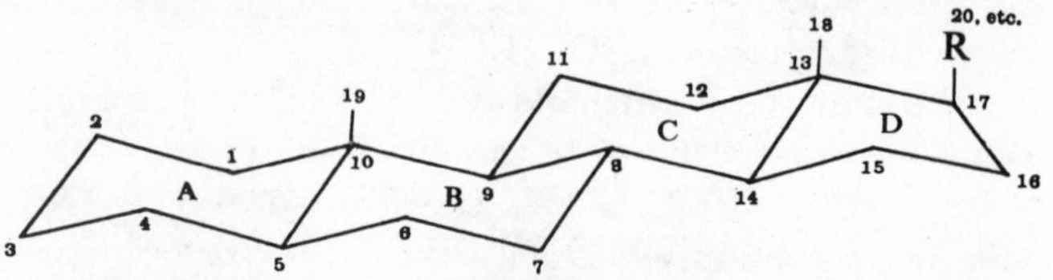

The rings are (generally) aliphatic. Lines like the vertical ones attached to the 10- and 13-positions represent *angular methyl* groups. Commonly, in cholesterol, for example,

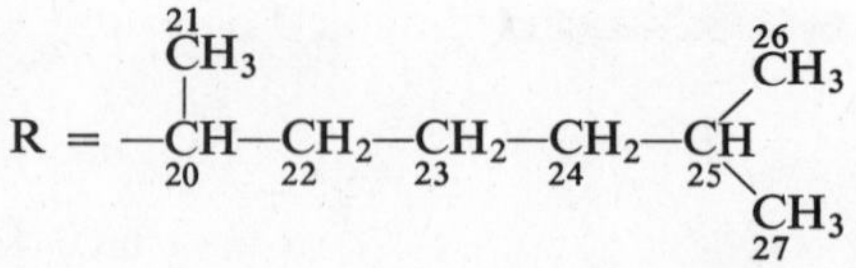

$$R = -\overset{21}{\underset{20}{CH}}(CH_3)-\underset{22}{CH_2}-\underset{23}{CH_2}-\underset{24}{CH_2}-\underset{25}{CH}(\overset{26}{CH_3})(\underset{27}{CH_3})$$

Stereochemistry is indicated by solid lines (β-bonds, coming *out* of the plane of the paper) and dotted lines (α-bonds, going *behind* the plane of the paper).

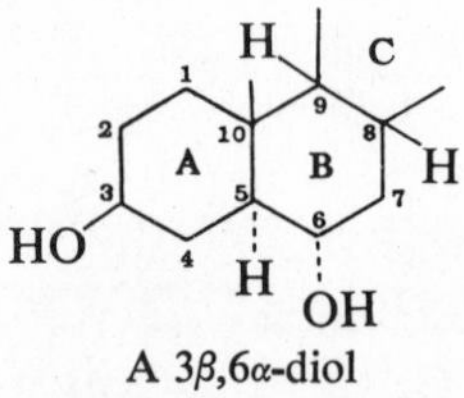

A 3β,6α-diol

I

Thus in I the —H and —OH at the 5- and 6-positions are *cis* to each other, but *trans* to the 3–OH and to the angular methyl at the 10-position. Fusion of the rings to each other can be *cis* or *trans*, thus increasing the complications of the stereochemistry.

Finally, in any rigid cyclic system like this, conformational effects are marked, and often completely control the course of reaction.

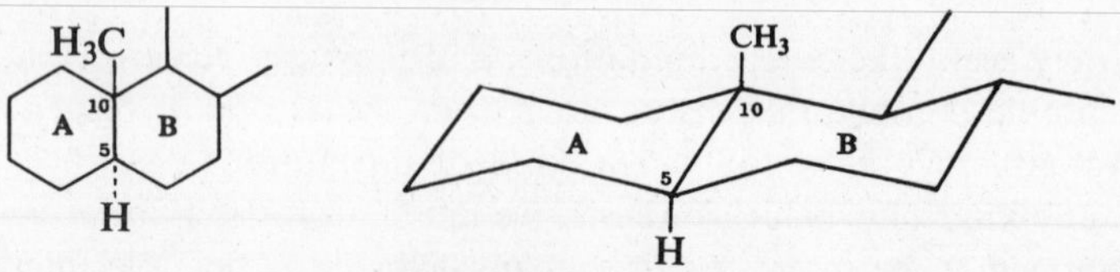

trans-Fusion

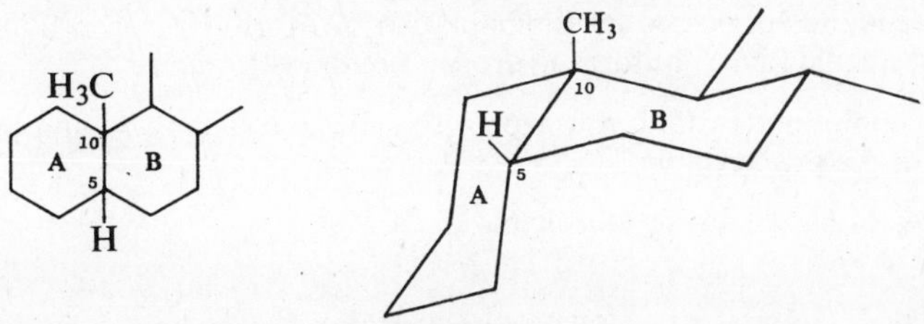

cis-Fusion

Steroids include sex hormones and adrenal cortical hormones (*cortisone* is one), cardiac glycosides, and bile acids. Because of their biological importance—and, undoubtedly, because of the fascinating complexity of the chemistry—the study of steroids has been, and is now, one of the most active areas of organic chemical research.

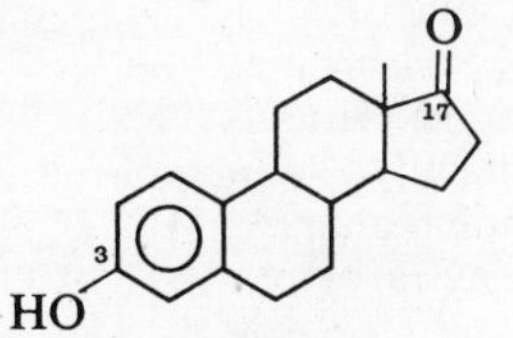

Estrone

An *estrogen*, or female sex hormone

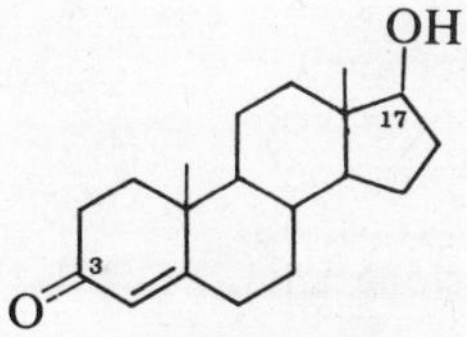

Testosterone

An *androgen*, or male sex hormone

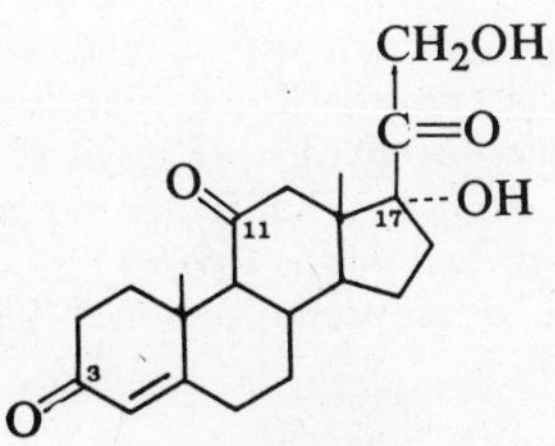

Cortisone

An adrenocortical hormone

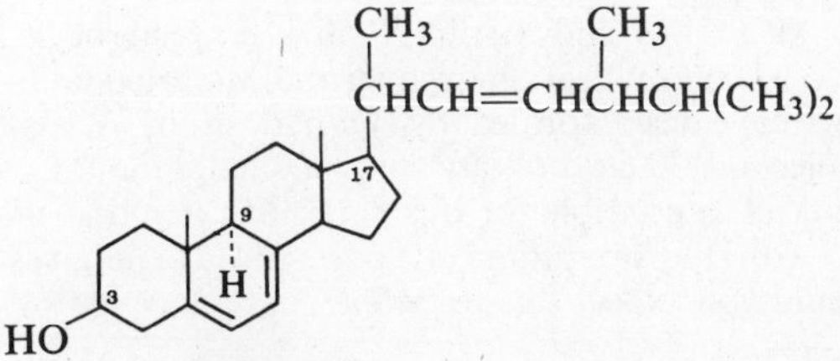

Ergosterol

A precursor of Vitamin D

PROBLEMS

1. (a) Ignoring enantiomerism, draw the structures of the eight isomeric pentyl alcohols, $C_5H_{11}OH$. (b) Name each by the IUPAC system and by the carbinol system. (c) Label each as primary, secondary, or tertiary. (d) Which one is isopentyl alcohol? *n*-Pentyl alcohol? *tert*-Pentyl alcohol? (e) Give the structure of a primary, a secondary, and a tertiary alcohol of the formula $C_6H_{13}OH$. (f) Give the structure of a primary, a secondary, and a tertiary *cyclic* alcohol of the formula C_5H_9OH.

2. Without referring to tables, arrange the following compounds in order of decreasing boiling point: (a) 3-hexanol; (b) *n*-hexane; (c) dimethyl-*n*-propylcarbinol; (d) *n*-octyl alcohol; (e) *n*-hexyl alcohol.

3. Looking at the beginning of each chapter for the structure involved, tell which families of compounds discussed in this book can: (a) form hydrogen bonds with other molecules of the same kind; (b) form hydrogen bonds with water.

4. Which compound would you expect to have the higher boiling point? (Check your answers in the proper tables.)

(a) *p*-cresol (p-$CH_3C_6H_4OH$) or anisole ($C_6H_5OCH_3$)

(b) methyl acetate, $CH_3C(=O)OCH_3$, or propionic acid, $CH_3CH_2C(=O)OH$

(c) propionic acid or *n*-pentyl alcohol.

5. Write equations to show how isopropyl alcohol might be prepared: (a) from an alkene; (b) from an alkyl halide; (c) by a Grignard reaction. (d) Which method is used industrially? Why?

6. Give structures of the Grignard reagent and the aldehyde or ketone that would react to yield each of the following alcohols. If more than one combination of reactants is possible, show each of the combinations.

(a)–(h) each of the isomeric pentyl alcohols of Problem 1(a)
(i) 1-phenyl-1-propanol
(j) 2-phenyl-2-propanol
(k) 1-phenyl-2-propanol
(l) 3-phenyl-1-propanol
(m) 1-methylcyclohexanol
(n) cyclohexylcarbinol
(o) 1-cyclohexylethanol
(p) 2,4-dimethyl-3-pentanol
(q) 1-(*p*-tolyl)ethanol, p-$CH_3C_6H_4CHOHCH_3$
(r) triphenylcarbinol, $(C_6H_5)_3COH$

7. For many 2-substituted ethanols, GCH_2CH_2OH, the *gauche* conformation is more stable than the *anti*:

$$G = -OH, -NH_2, -F, -Cl, -Br, -OCH_3, -NHCH_3, -N(CH_3)_2, \text{ and } -NO_2.$$

How might this be accounted for?

8. (a) As shown on p. 507, cholesterol is converted into cholestane-3β,6α-diol through *syn*-hydration by hydroboration-oxidation. What stereoisomeric product could also have been formed by *syn*-hydration? Actually, the reaction gives a 78% yield of cholestane-3β,6α-diol, and only a small amount of its stereoisomer. What factor do you think is responsible for this particular stereospecificity? (*Hint:* See pp. 514–515.)

(b) Hydroboration of androst-9(11)-ene gives 90% of a single stereoisomer. Which would you expect this to be?

Androst-9(11)-ene

9. (a) Using models and then drawing formulas, show the possible chair conformations for *cis*-1,3-cyclohexanediol. (b) On the basis solely of 1,3-interaction, which would you expect to be the more stable conformation? (c) Infrared evidence indicates intramolecular hydrogen bonding in *cis*-1,3-cyclohexanediol. Just how would the infrared spectrum show this? Which conformation in (a) is indicated by this evidence, and what is the source of its stability?

10. The infrared spectrum of the stereoisomer of 2,5-di-*tert*-butyl-1,4-cyclohexanediol in which all four substituents are *cis* to each other shows the presence of an intramolecular hydrogen bond. In what conformation does the molecule exist? (*Hint:* Use models.)

11. (a) What are the two diastereomeric products that could be formed by *anti*-addition of bromine to cholesterol? to 2-cholestene? (b) Actually, one product greatly predominates in each case, as shown:

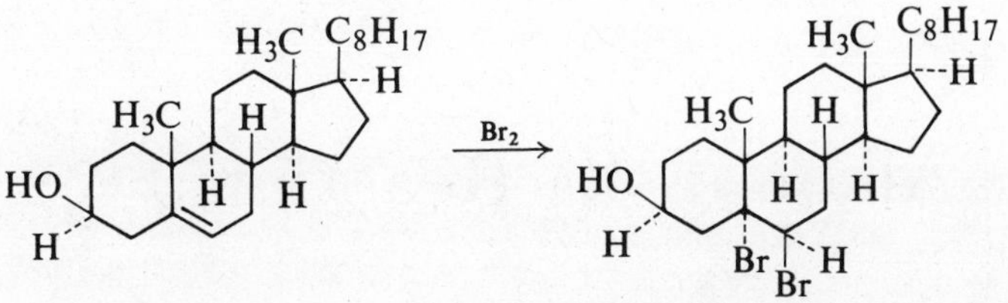

Cholesterol $\xrightarrow{Br_2}$ 5α,6β-Dibromo-3β-hydroxycholestane

85% yield

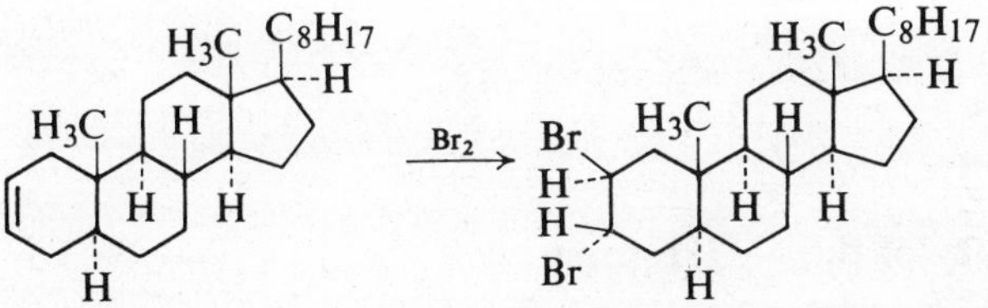

2-Cholestene $\xrightarrow{Br_2}$ 2β,3α-Dibromocholestane

70% yield

How do you account for the observed stereochemistry? (It is *not* a matter of relative stability of the diastereomers.) (*Hint:* Consider carefully the stereochemical possibilities at each step of the mechanism.)

12. On treatment with a variety of reagents (water, acetylide ion), borate esters of the kind shown are converted into alkenes:

$$(RO)_2BCH_2CH_2Br \longrightarrow CH_2{=}CH_2$$

The *cis* and *trans* esters (I) were prepared, and their configurations were assigned by nmr. Each ester was treated with bromine, and the resulting dibromide was treated with water. *cis*-I gave only *trans*-II as the final product, and *trans*-I gave only *cis*-II.

$CH_3CH{=}C(CH_3)B(OR)_2$ — I (*cis* or *trans*)

$CH_3CH{=}CBrCH_3$ — II (*cis* or *trans*)

Making use of what you know about the addition of bromine to alkenes, what do you conclude about the stereochemistry of this elimination reaction? Show the most likely mechanism for the elimination, including the part played by water (or acetylide ion).

Chapter 16 Alcohols II. Reactions

16.1 Chemistry of the —OH group

The chemical properties of an alcohol, ROH, are determined by its functional group, —OH, the hydroxyl group. When we have learned the chemistry of the alcohols, we shall have learned much of the chemistry of the hydroxyl group in whatever compound it may occur; we shall know, in part at least, what to expect of hydroxyhalides, hydroxyacids, hydroxyaldehydes, etc.

Reactions of an alcohol can involve the breaking of either of two bonds: the C---OH bond, with removal of the —OH group; or the O---H bond, with removal of —H. Either kind of reaction can involve substitution, in which a group replaces the —OH or —H, or elimination, in which a double bond is formed.

Differences in the structure of R cause differences in reactivity, and in a few cases even profoundly alter the course of the reaction. We shall see what some of these effects of structure on reactivity are, and how they can be accounted for.

16.2 Reactions

Some of the more important reactions of alcohols are listed below, and are discussed in following sections.

REACTIONS OF ALCOHOLS

C---OH BOND CLEAVAGE

R$\vdots$OH

1. Reaction with hydrogen halides. Discussed in Secs. 16.4–16.5.

$$R{-}OH + HX \longrightarrow RX + H_2O \qquad R\ \textit{may rearrange}$$

Reactivity of HX: HI > HBr > HCl

Reactivity of ROH: allyl, benzyl > 3° > 2° > 1°

Examples:

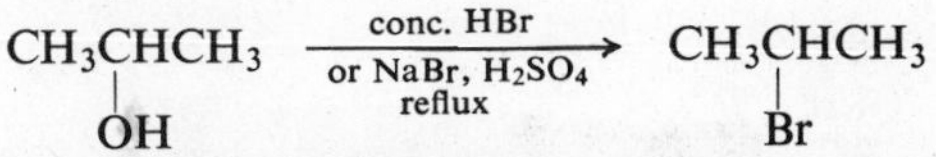

$$\underset{\text{Isopropyl alcohol}}{CH_3\underset{\underset{OH}{|}}{C}HCH_3} \xrightarrow[\text{or NaBr, }H_2SO_4\text{ reflux}]{\text{conc. HBr}} \underset{\text{Isopropyl bromide}}{CH_3\underset{\underset{Br}{|}}{C}HCH_3}$$

$$\underset{n\text{-Pentyl alcohol}}{CH_3CH_2CH_2CH_2CH_2OH} \xrightarrow[\text{heat}]{HCl,\ ZnCl_2} \underset{n\text{-Pentyl chloride}}{CH_3CH_2CH_2CH_2CH_2Cl}$$

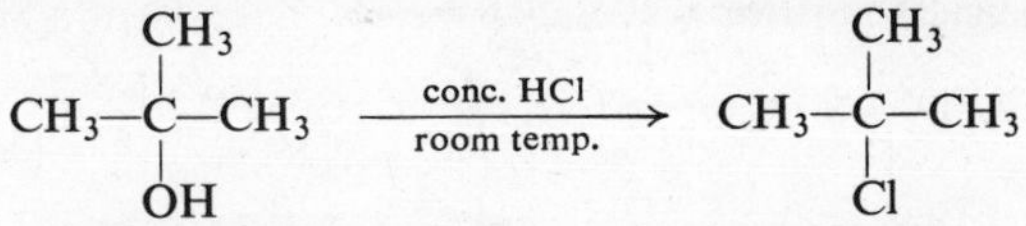

$$\underset{tert\text{-Butyl alcohol}}{CH_3-\overset{\overset{CH_3}{|}}{\underset{\underset{OH}{|}}{C}}-CH_3} \xrightarrow[\text{room temp.}]{\text{conc. HCl}} \underset{tert\text{-Butyl chloride}}{CH_3-\overset{\overset{CH_3}{|}}{\underset{\underset{Cl}{|}}{C}}-CH_3}$$

2. Reaction with phosphorus trihalides. Discussed in Sec. 16.10.

$$R{-}OH + PX_3 \longrightarrow RX + H_3PO_3$$
$$(PX_3 = PBr_3,\ PI_3)$$

Examples:

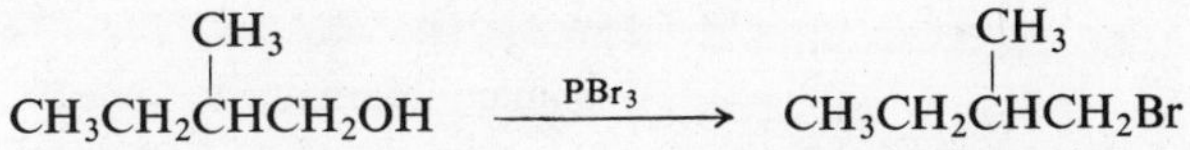

$$\underset{\text{2-Methyl-1-butanol}}{CH_3CH_2\overset{\overset{CH_3}{|}}{C}HCH_2OH} \xrightarrow{PBr_3} \underset{\text{2-Methyl-1-bromobutane}}{CH_3CH_2\overset{\overset{CH_3}{|}}{C}HCH_2Br}$$

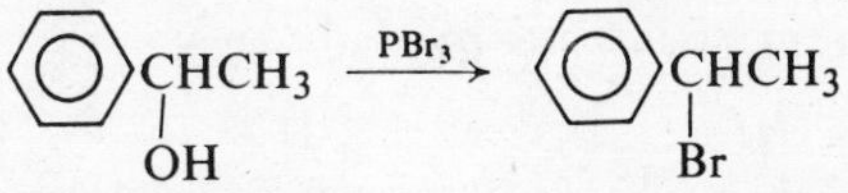

$$\underset{\text{1-Phenylethanol}}{C_6H_5\underset{\underset{OH}{|}}{C}HCH_3} \xrightarrow{PBr_3} \underset{\text{1-Bromo-1-phenylethane}}{C_6H_5\underset{\underset{Br}{|}}{C}HCH_3}$$

$$\underset{\text{Ethyl alcohol}}{CH_3CH_2OH} \xrightarrow{P+I_2} \underset{\text{Ethyl iodide}}{CH_3CH_2I}$$

3. Dehydration. Discussed in Secs. 5.19–5.23, and 16.3.

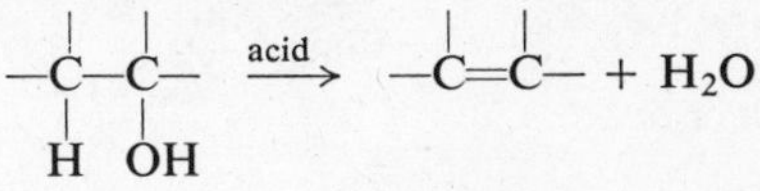

$$-\underset{\underset{H}{|}}{\overset{|}{C}}-\underset{\underset{OH}{|}}{\overset{|}{C}}- \xrightarrow{\text{acid}} -\overset{|}{C}{=}\overset{|}{C}- + H_2O \qquad \textit{Rearrangement may occur}$$

Reactivity of ROH: 3° > 2° > 1°

Examples:

$$\underset{n\text{-Butyl alcohol}}{CH_3CH_2CH_2CH_2OH} \xrightarrow{H_2SO_4,\text{ heat}} \underset{\substack{\text{2-Butene}\\ \textit{Major product}}}{CH_3CH{=}CHCH_3} \text{ and } \underset{\text{1-Butene}}{CH_3CH_2CH{=}CH_2}$$

$$\underset{\text{Cyclohexanol}}{C_6H_{11}OH} \xrightarrow{Al_2O_3,\ 250°} \underset{\text{Cyclohexene}}{C_6H_{10}}$$

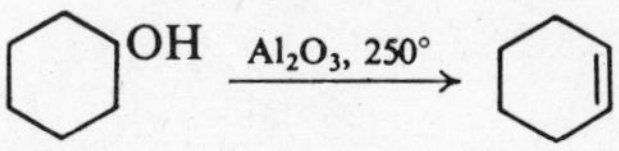

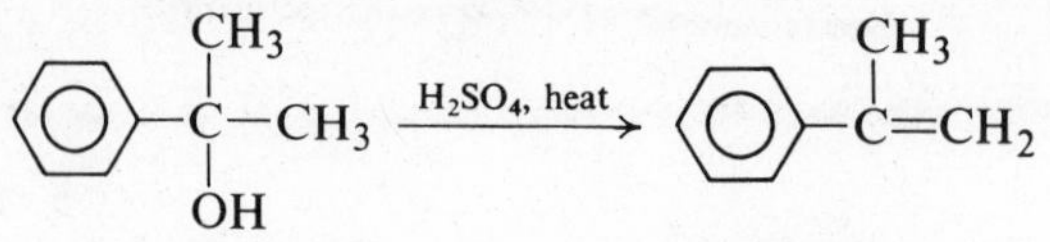

$$\text{C}_6\text{H}_5\text{—C(CH}_3)_2\text{—OH} \xrightarrow{H_2SO_4,\ \text{heat}} \text{C}_6\text{H}_5\text{—C(CH}_3)\text{=CH}_2$$

2-Phenyl-2-propanol → 2-Phenylpropene

O--H BOND CLEAVAGE

RO⋮H

4. Reaction as acids: reaction with active metals. Discussed in Sec. 16.6.

$$RO\text{—}H + M \longrightarrow RO^-M^+ + \tfrac{1}{2}H_2 \qquad M = \text{Na, K, Mg, Al, etc.}$$

Reactivity of ROH: $CH_3OH > 1° > 2° > 3°$

Examples:

$$CH_3CH_2OH \xrightarrow{Na} CH_3CH_2O^-Na^+ + \tfrac{1}{2}H_2$$

Sodium ethoxide

$$CH_3\text{—}CH(CH_3)\text{—}OH \xrightarrow{Al} (CH_3\text{—}CH(CH_3)\text{—}O)_3Al$$

Aluminium isopropoxide

$$CH_3\text{—}C(CH_3)_2\text{—}OH \xrightarrow{K} CH_3\text{—}C(CH_3)_2\text{—}O^-K^+$$

tert-Butyl alcohol → Potassium *tert*-butoxide

5. Ester formation. Discussed in Secs. 16.7 and 18.16.

Examples:

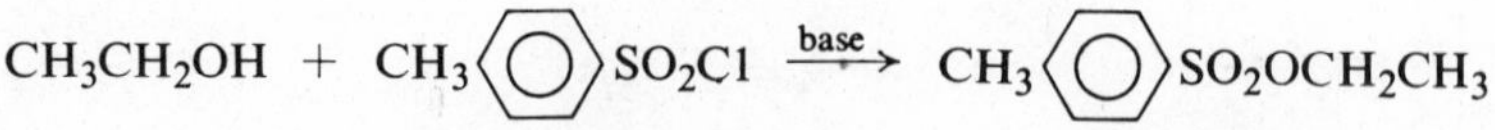

$$CH_3CH_2OH + CH_3C_6H_4SO_2Cl \xrightarrow{base} CH_3C_6H_4SO_2OCH_2CH_3$$

Tosyl chloride (*p*-Toluenesulfonyl chloride) → Ethyl tosylate

$$CH_3CH_2OH + CH_3C(=O)OH \xrightarrow{H^+} CH_3C(=O)OC_2H_5 + H_2O$$

Acetic acid → Ethyl acetate

6. Oxidation. Discussed in Sec. 16.8.

Primary: $R\text{—}CH_2OH \xrightarrow{K_2Cr_2O_7} R\text{—}CH\text{=}O$ (An aldehyde) $\xrightarrow{KMnO_4 \text{ or } K_2Cr_2O_7} R\text{—}COOH$ (A carboxylic acid)

$R\text{—}CH_2OH \xrightarrow{KMnO_4} R\text{—}COOH$

Secondary:

$$R_2CHOH \xrightarrow{K_2Cr_2O_7 \text{ or } CrO_3} R_2C{=}O$$

A ketone

Tertiary:

$$R_3C{-}OH \xrightarrow{\text{neut. } KMnO_4} \text{no reaction}$$

Examples:

$$\underset{\substack{n\text{-Propyl alcohol} \\ (1^\circ)}}{CH_3CH_2CH_2OH} \xrightarrow{K_2Cr_2O_7} \underset{\text{Propionaldehyde}}{CH_3CH_2CH{=}O} + H_2$$

$$\underset{\substack{\text{2-Methyl-1-butanol} \\ (1^\circ)}}{CH_3CH_2CH(CH_3)CH_2OH} \xrightarrow{KMnO_4} \underset{\text{2-Methylbutanoic acid}}{CH_3CH_2CH(CH_3)COOH}$$

$$\underset{\substack{\text{Cyclohexanol} \\ (2^\circ)}}{C_6H_{11}OH} \xrightarrow{K_2Cr_2O_7} \underset{\text{Cyclohexanone}}{C_6H_{10}{=}O}$$

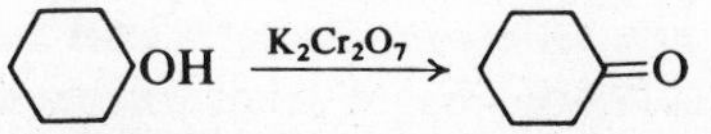

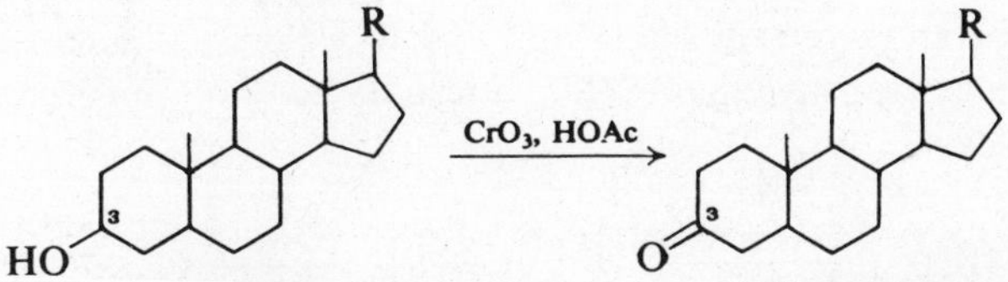

$$\underset{\substack{\text{3-Cholestanol} \\ (2^\circ)}}{\text{HO-(steroid)-R}} \xrightarrow{CrO_3,\ HOAc} \underset{\text{3-Cholestanone}}{\text{O=(steroid)-R}}$$

We can see that alcohols undergo many kinds of reactions, to yield many kinds of products. Because of the availability of alcohols, each of these reactions is one of the best ways to make the particular kind of product. After we have learned a little more about the reactions themselves, we shall look at some of the ways in which they can be applied to synthetic problems.

16.3 Dehydration

We discussed the dehydration of alcohols at some length earlier (Secs. 5.19–5.23). It might be well, however, to summarize what we know about this reaction at our present level of sophistication.

(a) Mechanism. According to the commonly accepted mechanism, we remember, dehydration involves (1) formation of the protonated alcohol, ROH_2^+, (2) its slow dissociation into a carbonium ion, and (3) fast expulsion of a hydrogen ion from the carbonium ion to form an alkene. Acid is required to convert the alcohol into the protonated alcohol, which dissociates—by loss of the weakly basic water molecule—much more easily than the alcohol itself.

$$\underset{\text{Alcohol}}{\underset{\text{H}\ \ \text{OH}}{-\text{C}-\text{C}-}} \underset{}{\overset{\text{H}^+}{\rightleftharpoons}} \underset{\text{Protonated alcohol}}{\underset{\text{H}\ {}^+\text{OH}_2}{-\text{C}-\text{C}-}} \overset{-\text{H}_2\text{O}}{\rightleftharpoons} \underset{\text{Carbonium ion}}{\underset{\text{H}\qquad}{-\text{C}-\overset{}{\underset{\oplus}{\text{C}}}-}} \overset{-\text{H}^+}{\rightleftharpoons} \underset{\text{Alkene}}{-\text{C}{=}\text{C}-}$$

We recognize this mechanism as an example of E1 elimination with the protonated alcohol as substrate. We can account, in a general way, for the contrast between alcohols and alkyl halides, which mostly undergo elimination by the E2 mechanism. Since the alcohol must be protonated to provide a reasonably good leaving group, H_2O, dehydration requires an acidic medium. But for E2 elimination we need a fairly strong base to attack the substrate without waiting for it to dissociate into carbonium ions. A strong base and an acidic medium are, of course, incompatible: any base much stronger than the alcohol itself would become protonated at the expense of the alcohol.

(b) Reactivity. We know that the rate of elimination depends greatly upon the rate of formation of the carbonium ion, which in turn depends upon its stability.

We know how to estimate the stability of a carbonium ion, on the basis of inductive effects and resonance. Because of the electron-releasing inductive effect of alkyl groups, stability and hence rate of formation of the simple alkyl cations follows the sequence 3° > 2° > 1°.

We know that because of resonance stabilization (Sec. 12.19) the benzyl cation should be an extremely stable ion, and so we are not surprised to find that an alcohol such as 1-phenylethanol (like a tertiary alcohol) undergoes dehydration extremely rapidly.

$$\underset{\text{1-Phenylethanol}}{C_6H_5\underset{\text{OH}}{\underset{|}{\text{CHCH}_3}}} \xrightarrow{\text{acid}} \underset{\text{A benzyl cation}}{C_6H_5\underset{\oplus}{\text{CHCH}_3}} \xrightarrow{-\text{H}^+} \underset{\text{Styrene}}{C_6H_5\text{CH}{=}\text{CH}_2}$$

(c) Orientation. We know that expulsion of the hydrogen ion takes place in such a way as to favor the formation of the more stable alkene. We can estimate the relative stability of an alkene on the basis of the number of alkyl groups attached to the doubly-bonded carbons, and on the basis of conjugation with a benzene ring or with another carbon–carbon double bond. It is understandable, then, that *sec*-butyl alcohol yields chiefly 2-butene, and 1-phenyl-2-propanol yields only 1-phenylpropene.

$$\underset{\textit{sec-}\text{Butyl alcohol}}{\text{CH}_3\text{CH}_2\underset{\text{OH}}{\underset{|}{\text{CHCH}_3}}} \xrightarrow{\text{acid}} \underset{\substack{\text{2-Butene} \\ \textit{Chief product}}}{\text{CH}_3\text{CH}{=}\text{CHCH}_3}$$

$$\underset{\text{1-Phenyl-2-propanol}}{C_6H_5\text{CH}_2\underset{\text{OH}}{\underset{|}{\text{CHCH}_3}}} \xrightarrow{\text{acid}} \underset{\substack{\text{1-Phenylpropene} \\ \textit{Only product}}}{C_6H_5\text{CH}{=}\text{CHCH}_3}$$

(d) Rearrangement. Finally, we know that a carbonium ion can rearrange, and that this rearrangement seems to occur whenever a 1,2-shift of hydrogen or alkyl group can form a more stable carbonium ion.

In all this we must not lose sight of the fact that the rates of formation of carbonium ions and of alkenes depend chiefly upon the stabilities of the transition states leading to their formation. A more stable carbonium ion is formed faster because the factors—inductive effects and resonance—that disperse the charge of a carbonium ion tend also to disperse the developing positive charge of an incipient carbonium ion in the transition state. In the same way, the factors that stabilize an alkene—conjugation of hyperconjugation, or perhaps change in hybridization—tend to stabilize the developing double bond in the transition state.

16.4 Reaction with hydrogen halides: facts

Alcohols react readily with hydrogen halides to yield alkyl halides and water. The reaction is carried out either by passing the dry hydrogen halide gas into the alcohol, or by heating the alcohol with the concentrated aqueous acid. Sometimes hydrogen bromide is generated in the presence of the alcohol by reaction between sulfuric acid and sodium bromide.

The least reactive of the hydrogen halides, HCl, requires the presence of zinc chloride for reaction with primary and secondary alcohols; on the other hand, the very reactive *tert*-butyl alcohol is converted to the chloride by simply being shaken with concentrated hydrochloric acid at room temperature. For example:

$$\text{C}_6\text{H}_{11}\text{OH} \xrightarrow{\text{dry HBr}} \text{C}_6\text{H}_{11}\text{Br}$$

Cyclohexanol → Cyclohexyl bromide

$$CH_3CH_2CH_2CH_2OH \xrightarrow[\text{reflux}]{\text{NaBr, } H_2SO_4} CH_3CH_2CH_2CH_2Br$$

n-Butyl alcohol → *n*-Butyl bromide

$$CH_3CH_2CH_2OH \xrightarrow[\text{heat}]{\text{HCl + ZnCl}_2} CH_3CH_2CH_2Cl$$

n-Propyl alcohol → *n*-Propyl chloride

$$(CH_3)_3C\text{—}OH \xrightarrow[\text{room temp.}]{\text{conc. HCl}} (CH_3)_3C\text{—}Cl$$

tert-Butyl alcohol → *tert*-Butyl chloride

Let us list some of the facts that are known about the reaction between alcohols and hydrogen halides.

(a) The reaction is catalyzed by acids. Even though the aqueous hydrogen halides are themselves strong acids, the presence of additional sulfuric acid speeds up the formation of halides.

Problem 16.1 How do you account for the catalysis by $ZnCl_2$ of the HCl reaction? (*Hint:* $ZnCl_2$ is sometimes used as a (weak) Friedel-Crafts catalyst.)

(b) Rearrangement of the alkyl group occurs, except with most primary alcohols. The alkyl group in the halide does not always have the same structure as the alkyl group in the parent alcohol. For example:

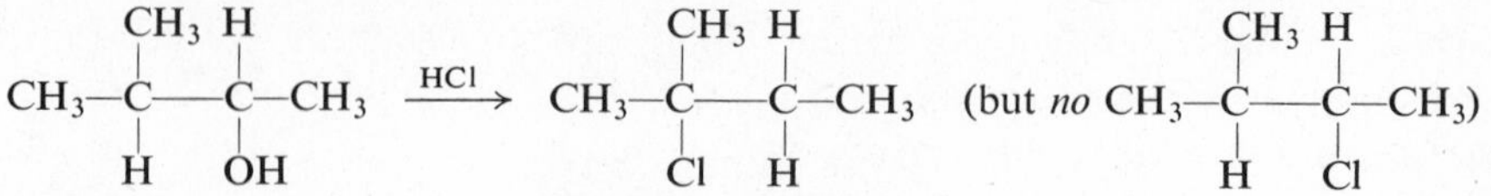

$$\underset{\text{3-Methyl-2-butanol}}{CH_3-\overset{CH_3}{\underset{H}{C}}-\overset{H}{\underset{OH}{C}}-CH_3} \xrightarrow{HCl} \underset{\substack{\text{2-Chloro-2-methylbutane}\\(\textit{tert}\text{-Pentyl chloride})}}{CH_3-\overset{CH_3}{\underset{Cl}{C}}-\overset{H}{\underset{H}{C}}-CH_3} \quad (\text{but } \textit{no}\ CH_3-\overset{CH_3}{\underset{H}{C}}-\overset{H}{\underset{Cl}{C}}-CH_3)$$

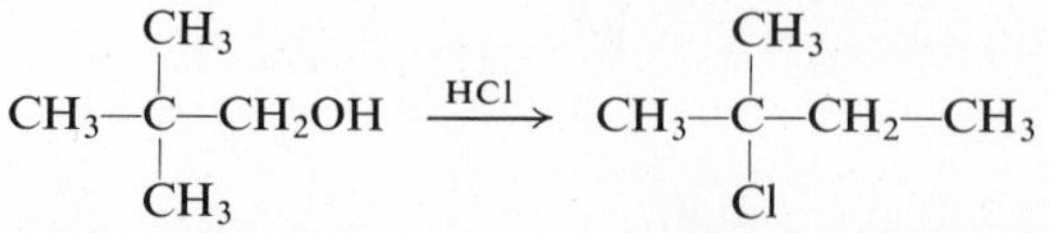

$$\underset{\text{Neopentyl alcohol}}{CH_3-\overset{CH_3}{\underset{CH_3}{C}}-CH_2OH} \xrightarrow{HCl} \underset{\textit{tert}\text{-Pentyl chloride}}{CH_3-\overset{CH_3}{\underset{Cl}{C}}-CH_2-CH_3}$$

We see that the halogen does not always become attached to the carbon that originally held the hydroxyl (the first example); even the carbon skeleton may be different from that of the starting material (the second example).

On the other hand, as shown on p. 523 for *n*-propyl and *n*-butyl alcohols, most primary alcohols give high yields of primary halides *without* rearrangement.

(c) The order of reactivity of alcohols toward HX is allyl, benzyl $> 3° > 2° > 1° < CH_3$. Reactivity decreases through most of the series (and this order is the basis of the *Lucas test*, Sec. 16.11), passes through a *minimum* at 1°, and rises again at CH_3.

16.5 Reaction with hydrogen halides: mechanism

What do the facts that we have just listed suggest to us about the mechanism of reaction between alcohols and hydrogen halides?

Catalysis by acid suggests that here, as in dehydration, the protonated alcohol ROH_2^+ is involved. The occurrence of *rearrangement* suggests that carbonium ions are intermediates—although *not* with primary alcohols. The idea of carbonium ions is strongly supported by the *order of reactivity* of alcohols, which parallels the stability of carbonium ions—*except* for methyl.

On the basis of this evidence, we formulate the following mechanism. The

(1) $$ROH + HX \rightleftharpoons ROH_2^+ + X^-$$

(2) $$ROH_2^+ \rightleftharpoons R^+ + H_2O$$

(3) $$R^+ + X^- \longrightarrow RX$$

S_N1: *all except methanol and most 1° alcohols*

alcohol accepts (step 1) the hydrogen ion to form the protonated alcohol, which dissociates (step 2) into water and a carbonium ion; the carbonium ion then combines (step 3) with a halide ion (not necessarily the one from step 1) to form the alkyl halide.

Looking at the mechanism we have written, we recognize the reaction for what it is: *nucleophilic substitution*, with the protonated alcohol as substrate and halide ion as the nucleophile. Once the reaction type is recognized, the other pieces of evidence fall into place.

The particular set of equations written above is, of course, the S_N1 mechanism for substitution. Primary alcohols do not undergo rearrangement simply because they do not react by this mechanism. Instead, they react by the alternative S_N2 mechanism:

$$X^- + ROH_2^+ \longrightarrow \left[\overset{\delta_-}{X}\text{---}R\text{---}\overset{\delta_+}{OH_2}\right] \longrightarrow X{-}R + H_2O$$

S_N2:
most 1° alcohols and methanol

What we see here is another example of that characteristic of nucleophilic substitution: a shift in the molecularity of reaction, in this particular case between 2° and 1°. This shift is confirmed by the fact that reactivity passes through a minimum at 1° and rises again at methyl. Because of poor accommodation of the positive charge, formation of primary carbonium ions is very slow; so slow in this instance that the unimolecular reaction is replaced by the relatively unhindered bimolecular attack. The bimolecular reaction is even faster for the still less hindered methanol.

Thus alcohols, like halides, undergo substitution by both S_N1 and S_N2 mechanisms; but alcohols lean more toward the unimolecular mechanism. We encountered the same situation in elimination (Sec. 16.3), and the explanation here is essentially the same: we cannot have a strong nucleophile—a strong *base*—present in the acidic medium required for protonation of the alcohol.

Neopentyl alcohol reacts with almost complete rearrangement, showing that, although primary, it follows the carbonium ion mechanism. This unusual behavior is easily explained. Although neopentyl is a primary group, it is a very bulky one and, as we have seen (Problem 14.2, p. 465), compounds containing this group undergo S_N2 reactions very slowly. Formation of the neopentyl cation from neopentyl alcohol is slow, but is nevertheless much faster than the alternative bimolecular reaction.

Problem 16.2 Because of the great tendency of the neopentyl cation to rearrange, neopentyl chloride cannot be prepared from the alcohol. How might neopentyl chloride be prepared?

Problem 16.3 Predict the relative rates at which the following alcohols will react with aqueous HBr:

(a) benzyl alcohol, *p*-methylbenzyl alcohol, *p*-nitrobenzyl alcohol;
(b) benzyl alcohol, α-phenylethyl alcohol, β-phenylethyl alcohol.

Problem 16.4 When allowed to react with aqueous HBr, 3-buten-2-ol ($CH_3CHOHCH{=}CH_2$) yields not only 3-bromo-1-butene ($CH_3CHBrCH{=}CH_2$) but also 1-bromo-2-butene ($CH_3CH{=}CHCH_2Br$). (a) How do you account for these results? (*Hint:* See Sec. 8.21.) (b) Predict the product of the reaction between HBr and 2-buten-1-ol ($CH_3CH{=}CHCH_2OH$). (c) How does this "rearrangement" differ from those described in the last section?

Problem 16.5 (a) Write the steps in the reaction of an alcohol with HCl by the S_N1 mechanism. (b) What is the rate-determining step? (c) The rate of reaction depends upon the concentration of what substance? (d) The concentration of this substance depends in turn upon the concentrations of what other compounds? (e) Will the rate depend *only* on [ROH]? Does an S_N1 reaction always follow first-order kinetics?

16.6 Alcohols as acids

We have seen that an alcohol, acting as a base, can accept a hydrogen ion to form the protonated alcohol, ROH_2^+. Let us now turn to reactions in which an alcohol, acting as an acid, loses a hydrogen ion to form the alkoxide ion, RO^-.

Since an alcohol contains hydrogen bonded to the very electronegative element oxygen, we would expect it to show appreciable acidity. The polarity of the O—H bond should facilitate the separation of the relatively positive hydrogen as the ion; viewed differently, electronegative oxygen should readily accommodate the negative charge of the electrons left behind.

The acidity of alcohols is shown by their reaction with active metals to form hydrogen gas, and by their ability to displace the weakly acidic hydrocarbons from their salts (e.g., Grignard reagents):

$$\mathrm{ROH + Na \longrightarrow RO^-Na^+ + \tfrac{1}{2}H_2}$$

$$\underset{\text{Stronger acid}}{\mathrm{ROH}} + \mathrm{R'MgX} \longrightarrow \underset{\text{Weaker acid}}{\mathrm{R'H}} + \mathrm{Mg(OR)X}$$

With the possible exception of methanol, they are weaker acids than water, but stronger acids than acetylene or ammonia:

$$\underset{\text{Stronger base}}{\mathrm{RO^-Na^+}} + \underset{\text{Stronger acid}}{\mathrm{H{-}OH}} \longrightarrow \underset{\text{Weaker base}}{\mathrm{Na^+OH^-}} + \underset{\text{Weaker acid}}{\mathrm{RO{-}H}}$$

$$\underset{\text{Stronger base}}{\mathrm{HC{\equiv}C^-Na^+}} + \underset{\text{Stronger acid}}{\mathrm{RO{-}H}} \longrightarrow \underset{\text{Weaker base}}{\mathrm{RO^-Na^+}} + \underset{\text{Weaker acid}}{\mathrm{HC{\equiv}C{-}H}}$$

As before, these relative acidities are determined by displacement (Sec. 8.10). We may expand our series of acidities and basicities, then, to the following:

Relative acidities: $H_2O > ROH > HC{\equiv}CH > NH_3 > RH$

Relative basicities: $OH^- < OR^- < HC{\equiv}C^- < NH_2^- < R^-$

Not only does the alkyl group make an alcohol less acidic than water, but the *bigger* the alkyl group, the less acidic the alcohol: methanol is the strongest acid and tertiary alcohols are the weakest.

This acid-weakening effect of alkyl groups is *not* an electronic effect, as was once believed, with electron release destabilizing the anion and making it a stronger base. In the gas phase, the relative acidities of various alcohols and of alcohols and water are reversed; evidently, the easily polarized alkyl groups are helping to accommodate the negative charge, just as they help to accommodate the positive charge in carbonium ions (Secs. 5.18 and 11.18). Alcohols *are* weaker acids than water *in solution*—which is where we are normally concerned with acidity—and this is a solvation effect; a bulky group interferes with the ion–dipole interactions that stabilize the anion.

Since an alcohol is a weaker acid than water, an alkoxide is not prepared from the reaction of the alcohol with sodium hydroxide, but is prepared instead by reaction of the alcohol with the active metal itself.

As we shall see, the alkoxides are extremely useful reagents; they are used as powerful bases (stronger than hydroxide) and to introduce the —OR group into a molecule.

Problem 16.6 Which would you expect to be the stronger acid: (a) β-chloroethyl alcohol or ethyl alcohol? (b) *p*-Nitrobenzyl alcohol or benzyl alcohol? (c) *n*-Propyl alcohol or glycerol, $HOCH_2CHOHCH_2OH$?

Problem 16.7 Sodium metal was added to *tert*-butyl alcohol and allowed to react. When the metal was consumed, ethyl bromide was added to the resulting mixture. Work-up of the reaction mixture yielded a compound of formula $C_6H_{14}O$.

In a similar experiment, sodium metal was allowed to react with ethanol. When *tert*-butyl bromide was added, a gas was evolved, and work-up of the remaining mixture gave ethanol as the only organic material.

(a) Write equations for all reactions. (b) What familiar reaction type is involved in each case? (c) Why did the reactions take different courses?

16.7 Formation of alkyl sulfonates

Sulfonyl chlorides (the acid chlorides of sulfonic acids) are prepared by the action of phosphorus pentachloride or thionyl chloride on sulfonic acids or their salts:

$$\underset{(\text{or } ArSO_3Na)}{ArSO_2OH} + PCl_5 \xrightarrow{\text{heat}} \underset{\substack{\text{A sulfonyl}\\ \text{chloride}}}{ArSO_2Cl} + POCl_3 + \underset{(\text{or NaCl})}{HCl}$$

Alcohols react with these sulfonyl chlorides to form esters, *alkyl sulfonates*:

$$ArSO_2Cl + ROH \xrightarrow[\text{or pyridine}]{\text{aqueous } OH^-} \underset{\text{An alkyl sulfonate}}{ArSO_2OR} + Cl^- + H_2O$$

We have already seen (Sec. 14.6) that the weak basicity of the sulfonate anion, $ArSO_3^-$, makes it a good leaving group, and as a result alkyl sulfonates undergo nucleophilic substitution and elimination in much the same manner as alkyl halides.

Alkyl sulfonates offer a very real advantage over alkyl halides in reactions where stereochemistry is important; this advantage lies, not in the reactions of alkyl sulfonates, but in their *preparation*. Whether we use an alkyl halide or sulfonate, and whether we let it undergo substitution or elimination, our starting point for the study is almost certainly the alcohol. The sulfonate *must* be prepared from the alcohol; the halide nearly always *will* be. It is at the alcohol stage that any resolution will be carried out, or any diastereomers separated; the alcohol is then converted into the halide or sulfonate, the reaction we are studying is carried out, and the products are examined.

Now, any preparation of a halide from an alcohol must involve breaking of the carbon–oxygen bond, and hence is accompanied by the likelihood of stereo-

$$R \vdots O{-}H \xrightarrow{HX \text{ or } PX_3} R{-}X$$

chemical inversion and the possibility of racemization. Preparation of a sulfonate, on the other hand, does not involve the breaking of the carbon–oxygen bond, and hence proceeds with complete retention; when we carry out a reaction with this sulfonate, we know exactly what we are starting with.

$$\mathrm{R{-}O{+}H} \;+\; \mathrm{Cl{-}\overset{\overset{\displaystyle O}{|}}{\underset{\underset{\displaystyle O}{|}}{S}}{-}Ar} \longrightarrow \mathrm{R{-}O{-}\overset{\overset{\displaystyle O}{|}}{\underset{\underset{\displaystyle O}{|}}{S}}{-}Ar}$$

Problem 16.8 Outline all steps in the synthesis of *sec*-butyl tosylate, starting with benzene, toluene, and any necessary aliphatic and inorganic reagents.

Problem 16.9 You prepare *sec*-butyl tosylate from alcohol of $[\alpha]$ $+6.9°$. On hydrolysis with aqueous base, this ester gives *sec*-butyl alcohol of $[\alpha]$ $-6.9°$. Without knowing the configuration or optical purity of the starting alcohol, what (if anything) can you say about the stereochemistry of the hydrolysis step?

16.8 Oxidation of alcohols

The compound that is formed by oxidation of an alcohol depends upon the number of hydrogens attached to the carbon bearing the —OH group, that is, upon whether the alcohol is primary, secondary, or tertiary. We have already encountered these products—aldehydes, ketones, and carboxylic acids—and should recognize them from their structures, even though we have not yet discussed much of their chemistry. They are important compounds, and their preparation by the oxidation of alcohols is of great value in organic synthesis (Secs. 16.9 and 16.10).

The number of oxidizing agents available to the organic chemist is growing at a tremendous rate. As with all synthetic methods, emphasis is on the development of highly *selective* reagents, which will operate on only one functional group in a complex molecule, and leave the other functional groups untouched. Of the many reagents that can be used to oxidize alcohols, we can consider only the most common ones, those containing Mn(VII) and Cr(VI).

Primary alcohols can be oxidized to carboxylic acids, RCOOH, usually by heating with aqueous $KMnO_4$. When reaction is complete, the aqueous solution of the soluble potassium salt of the carboxylic acid is filtered from MnO_2, and the acid is liberated by the addition of a stronger mineral acid.

$$\underset{\text{1° alcohol}}{RCH_2OH} + \underset{\textit{Purple}}{KMnO_4} \longrightarrow \underset{\textit{Sol. in }H_2O}{RCOO^-K^+} + \underset{\textit{Brown}}{MnO_2} + KOH$$

$$\downarrow H^+$$

$$\underset{\substack{\text{A carboxylic acid}\\ \textit{Insol. in }H_2O}}{RCOOH}$$

Primary alcohols can be oxidized to aldehydes, RCHO, by the use of $K_2Cr_2O_7$. Since, as we shall see (Sec. 19.9), aldehydes are themselves readily oxidized to acids, the aldehyde must be removed from the reaction mixture by special techniques before it is oxidized further.

$$\underset{1^\circ\text{ alcohol}}{RCH_2OH} + \underset{\textit{Orange-red}}{Cr_2O_7^{--}} \longrightarrow \underset{\text{An aldehyde}}{R-\overset{H}{\overset{|}{C}}=O} + \underset{\textit{Green}}{Cr^{+++}}$$

$$\downarrow K_2Cr_2O_7$$

$$\underset{\text{A carboxylic acid}}{R-C(=O)OH}$$

Secondary alcohols are oxidized to ketones, R_2CO, by chromic acid in a form selected for the job at hand: aqueous $K_2Cr_2O_7$, CrO_3 in glacial acetic acid, CrO_3

$$\underset{\text{A 2° alcohol}}{R-\overset{R'}{\overset{|}{C}}HOH} \xrightarrow{K_2Cr_2O_7 \text{ or } CrO_3} \underset{\text{A ketone}}{R-\overset{R'}{\overset{|}{C}}=O}$$

in pyridine, etc. Hot permanganate also oxidizes secondary alcohols; it is seldom used for the synthesis of ketones, however, since oxidation tends to go past the ketone stage, with breaking of carbon–carbon bonds.

With no hydrogen attached to the carbinol carbon, **tertiary alcohols are not oxidized at all under alkaline conditions**. If acid is present, they are rapidly dehydrated to alkenes, which are then oxidized.

Let us look briefly at the mechanism of just one oxidation reaction, to see the kind of thing that is involved here. Oxidation of secondary alcohols by Cr(VI) is believed to involve (1) formation of a chromate ester, which (2) loses a proton and an

(1) $$R_2CHOH + \underset{Cr(VI)}{HCrO_4^-} + H^+ \longrightarrow R_2CHOCrO_3H + H_2O$$

(2) $$R-\overset{R}{\overset{|}{\underset{\underset{H_2O}{H}}{C}}}-O-CrO_3H \longrightarrow R-\overset{R}{\overset{|}{C}}=O + H_3O^+ + \underset{Cr(IV)}{HCrO_3^-}$$

(3) $$R_2CHOH + Cr(IV) \longrightarrow R_2\dot{C}OH + Cr(III)$$

(4) $$R_2\dot{C}OH + Cr(VI) \longrightarrow R_2C=O + Cr(V)$$

(5) $$R_2CHOH + Cr(V) \xrightarrow{\text{via an ester}} R_2C=O + Cr(III)$$

$HCrO_3^-$ ion to form the ketone. It is possible that the proton is lost to an oxygen of the ester group in a cyclic mechanism (2a). Additional alcohol is then oxidized, evidently by reactions (3)–(5), with chromium finally reaching the Cr(III) state.

(2a) $$R-\overset{R}{\overset{|}{\underset{H}{C}}}-O-Cr(=O)(=O)OH \longrightarrow R_2C=O + \underset{Cr(IV)}{H_2CrO_3}$$

The difficult step in all this is breaking the carbon–hydrogen bond; this is made possible by the synchronous departure of $HCrO_3^-$, in what is really an E2 elimination—but here with the formation of a carbon–oxygen double bond.

In connection with analysis, we shall encounter two reagents used to oxidize alcohols of special kinds: (a) *hypohalite* (Sec. 16.11), and (b) *periodic acid* (Sec. 16.12).

16.9 Synthesis of alcohols

Let us try to get a broader picture of the synthesis of complicated alcohols. We learned (Sec. 15.12) that they are most often prepared by the reaction of Grignard reagents with aldehydes or ketones. In this chapter we have learned that aldehydes and ketones, as well as the alkyl halides from which the Grignard reagents are made, are themselves most often prepared from alcohols. Finally, we know that the simple alcohols are among our most readily available compounds. We have available to us, then, a synthetic route leading from simple alcohols to more complicated ones.

alcohol $\longrightarrow$ **alkyl halide** $\longrightarrow$ **Grignard reagent**

alcohol $\longrightarrow$ **aldehyde or ketone**

(together) $\longrightarrow$ **more complicated alcohol**

As a simple example, consider conversion of the two-carbon ethyl alcohol into the four-carbon *sec*-butyl alcohol:

$$\underset{\text{Ethyl alcohol}}{CH_3CH_2OH} \xrightarrow{HBr} CH_3CH_2Br \xrightarrow{Mg} CH_3CH_2MgBr$$

$$\underset{\text{Ethyl alcohol}}{CH_3CH_2OH} \xrightarrow{K_2Cr_2O_7} \underset{\text{Acetaldehyde}}{CH_3-\overset{H}{\overset{|}{C}}=O}$$

$$CH_3CH_2MgBr + CH_3CHO \longrightarrow CH_3CH_2-\underset{OMgBr}{\underset{|}{C}HCH_3} \xrightarrow{H_2O,\ H^+} \underset{\textit{sec-}\text{Butyl alcohol}}{CH_3CH_2\underset{OH}{\underset{|}{C}HCH_3}}$$

Using the *sec*-butyl alcohol thus obtained, we could prepare even larger alcohols:

$$\underset{\textit{sec-}\text{Butyl alcohol}}{CH_3CH_2\overset{CH_3}{\overset{|}{C}}HOH} \xrightarrow{PBr_3} CH_3CH_2\overset{CH_3}{\overset{|}{C}}HBr \xrightarrow{Mg} CH_3CH_2\overset{CH_3}{\overset{|}{C}}HMgBr \xrightarrow{CH_3CHO} \underset{\text{3-Methyl-2-pentanol}}{CH_3CH_2\overset{CH_3}{\overset{|}{C}}H-\underset{OH}{\underset{|}{C}HCH_3}}$$

$$\underset{\textit{sec-}\text{Butyl alcohol}}{CH_3CH_2\overset{CH_3}{\overset{|}{C}}HOH} \xrightarrow{K_2Cr_2O_7} CH_3CH_2\overset{CH_3}{\overset{|}{C}}=O \xrightarrow{C_2H_5MgBr} \underset{\text{3-Methyl-3-pentanol}}{CH_3CH_2\overset{CH_3}{\overset{|}{\underset{OH}{\underset{|}{C}}}}-CH_2CH_3}$$

By combining our knowledge of alcohols with what we know about alkylbenzenes and aromatic substitution, we can extend our syntheses to include aromatic alcohols. For example:

$$C_6H_6 \xrightarrow{Br_2,\ Fe} C_6H_5Br \xrightarrow{Mg} C_6H_5MgBr$$

Phenylmagnesium bromide

$$CH_3CH_2OH \xrightarrow{K_2Cr_2O_7} CH_3{-}\underset{}{\overset{H}{\overset{|}{C}}}{=}O$$

$$C_6H_5MgBr + CH_3CHO \longrightarrow C_6H_5{-}\underset{OH}{\underset{|}{\overset{H}{\overset{|}{C}}}}{-}CH_3$$

1-Phenylethanol

$$C_6H_5CH_3 \xrightarrow{Cl_2,\ heat} C_6H_5CH_2Cl \xrightarrow{Mg} C_6H_5CH_2MgCl$$

$$CH_3\underset{OH}{\underset{|}{C}}HCH_3 \xrightarrow{K_2Cr_2O_7} CH_3{-}\underset{O}{\underset{\|}{C}}{-}CH_3$$

$$C_6H_5CH_2MgCl + CH_3COCH_3 \longrightarrow C_6H_5CH_2{-}\underset{OH}{\underset{|}{\overset{CH_3}{\overset{|}{C}}}}{-}CH_3$$

1-Phenyl-2-methyl-2-propanol

Granting that we know the chemistry of the individual steps, how do we go about planning a route to these more complicated alcohols? In almost every organic synthesis it is best to **work backward** from the compound we want. There are relatively few ways to make a complicated alcohol; there are relatively few ways to make the Grignard reagent or the aldehyde or ketone; and so on back to our ultimate starting materials. On the other hand, alcohols can undergo so many different reactions that, if we go at the problem the other way around, we find a bewildering number of paths, few of which take us where we want to go.

Let us suppose (and this is quite reasonable) that we have available all alcohols of four carbons or fewer, and that we want to make, say, 2-methyl-2-hexanol. Let us set down the structure and see what we need to make it.

$$CH_3CH_2CH_2CH_2{-}\underset{OH}{\underset{|}{\overset{CH_3}{\overset{|}{C}}}}{-}CH_3$$

2-Methyl-2-hexanol

Since it is a tertiary alcohol, we must use a Grignard reagent and a ketone. But which Grignard reagent? And which ketone? Using the same approach as before (Sec. 15.14), we see that there are two possibilities:

$$CH_3CH_2CH_2CH_2 \vdots {-}\underset{OH}{\underset{|}{\overset{CH_3}{\overset{|}{C}}}}{-}CH_3 \longleftarrow CH_3CH_2CH_2CH_2MgBr + \underset{O}{\underset{\|}{\overset{CH_3}{\overset{|}{C}}}}{-}CH_3$$

2-Methyl-2-hexanol   *n*-Butylmagnesium bromide   Acetone

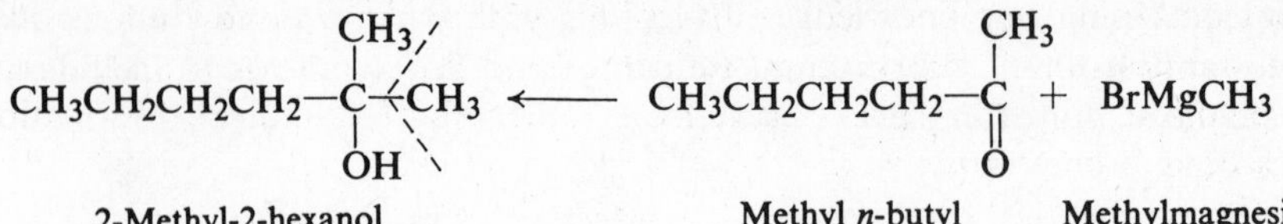

$$CH_3CH_2CH_2CH_2-\underset{OH}{\overset{CH_3}{C}}-CH_3 \longleftarrow CH_3CH_2CH_2CH_2-\underset{O}{\overset{CH_3}{C}} + BrMgCH_3$$

2-Methyl-2-hexanol | Methyl *n*-butyl ketone | Methylmagnesium bromide

Of these two possibilities we would select the one involving the four-carbon Grignard reagent and the three-carbon ketone; now how are we to make *them*? The Grignard reagent can be made only from the corresponding alkyl halide, *n*-butyl bromide, and that in turn most likely from an alcohol, *n*-butyl alcohol. Acetone requires, of course, isopropyl alcohol. Putting together the entire synthesis, we have the following sequence:

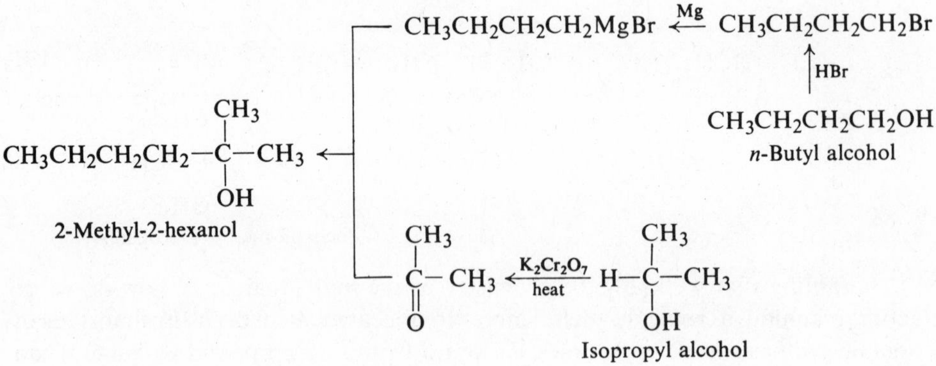

$$CH_3CH_2CH_2CH_2MgBr \xleftarrow{Mg} CH_3CH_2CH_2CH_2Br \xleftarrow{HBr} CH_3CH_2CH_2CH_2OH$$

n-Butyl alcohol

$$CH_3\underset{O}{\overset{\|}{C}}-CH_3 \xleftarrow[\text{heat}]{K_2Cr_2O_7} H-\underset{OH}{\overset{CH_3}{C}}-CH_3$$

Isopropyl alcohol

$$\longrightarrow CH_3CH_2CH_2CH_2-\underset{OH}{\overset{CH_3}{C}}-CH_3$$

2-Methyl-2-hexanol

Let us consider that in addition to our alcohols of four carbons or fewer we have available benzene and toluene, another reasonable assumption, and that we wish to make, say, 1-phenyl-3-methyl-2-butanol. Again we set down the structure of the desired alcohol and work backward to the starting materials. For a

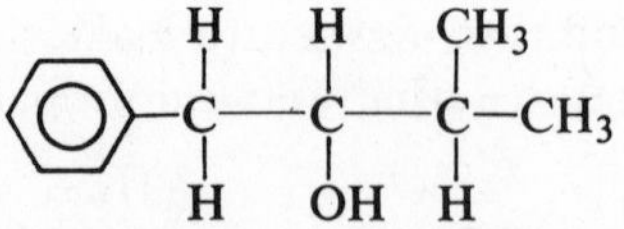

$$C_6H_5-\underset{H}{\overset{H}{C}}-\underset{OH}{\overset{H}{C}}-\underset{H}{\overset{CH_3}{C}}-CH_3$$

1-Phenyl-3-methyl-2-butanol

secondary alcohol, a Grignard reagent and an aldehyde are indicated, and again there are two choices: we may consider the molecule to be put together between (a) C–1 and C–2 or (b) C–2 and C–3. Of the two possibilities we select the first,

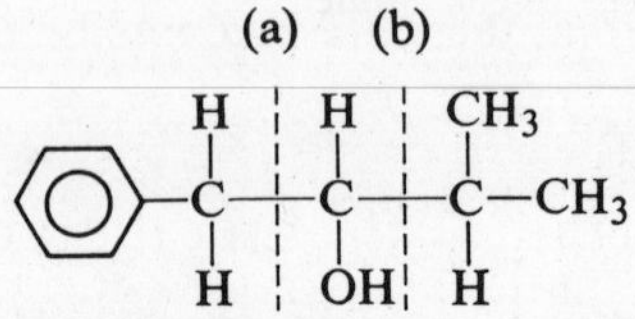

$$C_6H_5-\underset{H}{\overset{H}{C}}\overset{(a)}{\vdots}\underset{OH}{\overset{H}{C}}\overset{(b)}{\vdots}\underset{H}{\overset{CH_3}{C}}-CH_3$$

since this requires a compound with only one carbon attached to the benzene ring, which we have available in toluene. We need, then, a four-carbon aldehyde and benzylmagnesium chloride. The aldehyde can readily be made from isobutyl alcohol, but how about benzylmagnesium chloride? This is, of course, made from benzyl chloride, which in turn is made from toluene by free-radical chlorination. Our synthesis is complete:

$$C_6H_5CH_3 \xrightarrow{Cl_2,\ heat} C_6H_5CH_2Cl \xrightarrow{Mg} C_6H_5CH_2MgCl$$

Toluene → Benzylmagnesium chloride

$$HOCH_2{-}CH(CH_3){-}CH_3 \xrightarrow{K_2Cr_2O_7} O{=}CH{-}CH(CH_3){-}CH_3$$

Isobutyl alcohol

$$C_6H_5CH_2MgCl + O{=}CH{-}CH(CH_3){-}CH_3 \longrightarrow C_6H_5CH_2{-}CH(OH){-}CH(CH_3){-}CH_3$$

1-Phenyl-3-methyl-2-butanol

Now that we know how to make complicated alcohols from simple ones, what can we use them for?

16.10 Syntheses using alcohols

The alcohols that we have learned to make can be converted into other kinds of compounds having the same carbon skeleton; from complicated alcohols we can make complicated aldehydes, ketones, acids, halides, alkenes, alkynes, alkanes, etc.

Alkyl halides are prepared from alcohols by use of hydrogen halides or phosphorus halides. Phosphorus halides are often preferred because they tend less to bring about rearrangement (Sec. 16.4).

Alkenes are prepared from alcohols either by direct dehydration or by dehydrohalogenation of intermediate alkyl halides; to avoid rearrangement we often select dehydrohalogenation of halides even though this route involves an extra step. (Or, sometimes better, we use elimination from alkyl sulfonates.)

$$\text{alcohol} \xrightarrow{PBr_3 \text{ or } PI_3} \text{alkyl halide} \xrightarrow{KOH} \text{alkene}$$

$$\text{alcohol} \xrightarrow{TsCl} \text{alkyl tosylate} \xrightarrow{base} \text{alkene}$$

$$\text{alcohol} \xrightarrow{acid} \text{dehydration to alkene}$$

$$\text{alcohol} \xrightarrow{HX} \text{alkyl halide}$$

(last two: *Rearrangements possible*)

Alkanes, we learned (Sec. 3.15), are best prepared from the corresponding alkenes by hydrogenation, so that now we have a route from complicated alcohols to complicated alkanes.

Complicated aldehydes and ketones are made by oxidizing complicated alcohols. By reaction with Grignard reagents these aldehydes and ketones can be converted into even more complicated alcohols, and so on.

Given the time, necessary inorganic reagents, and the single alcohol ethanol, our chemical Crusoe of Sec. 15.5 could synthesize all the aliphatic compounds that have ever been made—and for that matter the aromatic ones, too.

In planning the synthesis of these other kinds of compounds, we again follow our system of working backward. We try to limit the synthesis to as few steps as possible, but nevertheless do not sacrifice purity for time. For example, where rearrangement is likely to occur we prepare an alkene in two steps via the halide rather than by the single step of dehydration.

Assuming again that we have available alcohols of four carbons or fewer, benzene, and toluene, let us take as an example 3-methyl-1-butene. It could be

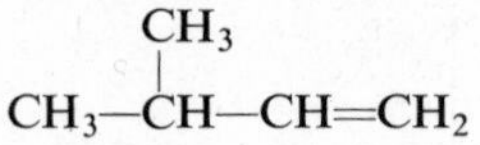

$$CH_3-\underset{\displaystyle |}{\overset{\displaystyle CH_3}{\overset{|}{C}}}H-CH=CH_2$$

3-Methyl-1-butene

prepared by dehydrohalogenation of an alkyl halide of the same carbon skeleton, or by dehydration of an alcohol. If the halogen or hydroxyl group were attached to C–2, we would obtain some of the desired product, but much more of its isomer, 2-methyl-2-butene:

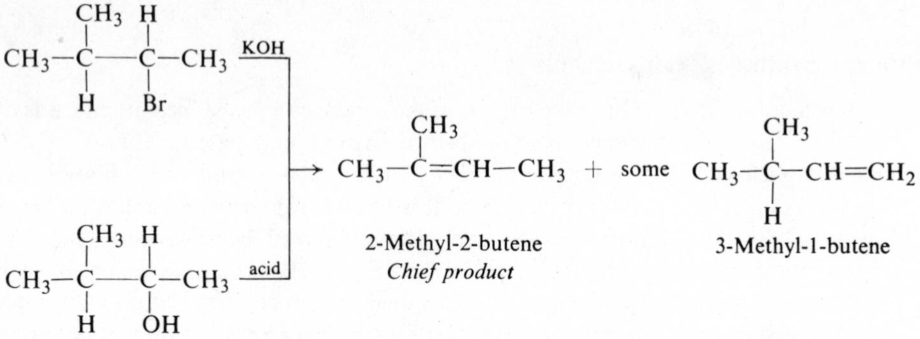

$$CH_3-\underset{H}{\overset{CH_3}{C}}-\underset{Br}{\overset{H}{C}}-CH_3 \xrightarrow{KOH}$$

$$CH_3-\underset{H}{\overset{CH_3}{C}}-\underset{OH}{\overset{H}{C}}-CH_3 \xrightarrow{acid}$$

$$\longrightarrow CH_3-\overset{CH_3}{C}=CH-CH_3 \quad + \text{ some } \quad CH_3-\underset{H}{\overset{CH_3}{C}}-CH=CH_2$$

2-Methyl-2-butene *Chief product*   3-Methyl-1-butene

We would select, then, the compound with the functional group attached to C–1. Even so, if we were to use the alcohol, there would be extensive rearrangement to yield, again, the more stable 2-methyl-2-butene:

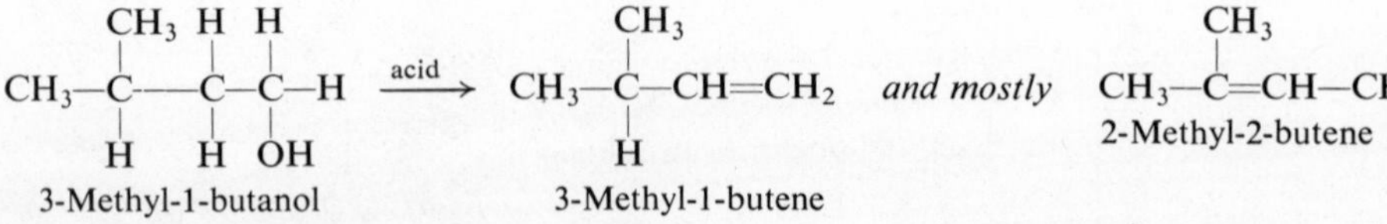

$$CH_3-\underset{H}{\overset{CH_3}{C}}-\underset{H}{\overset{H}{C}}-\underset{OH}{\overset{H}{C}}-H \xrightarrow{acid} CH_3-\underset{H}{\overset{CH_3}{C}}-CH=CH_2 \quad \textit{and mostly} \quad CH_3-\overset{CH_3}{C}=CH-CH_3$$

3-Methyl-1-butanol   3-Methyl-1-butene   2-Methyl-2-butene

Only dehydrohalogenation of 1-bromo-3-methylbutane would yield the desired product in pure form:

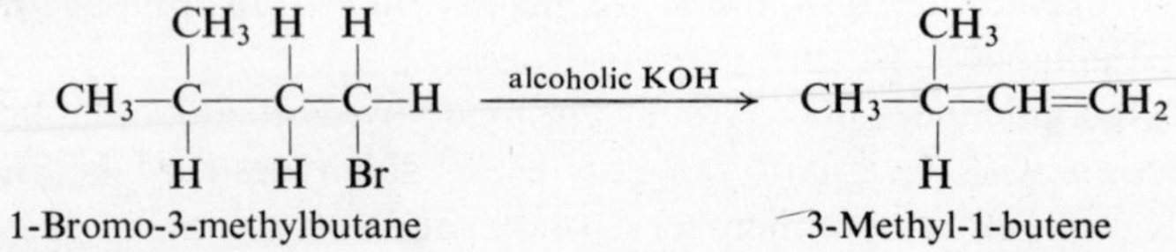

$$CH_3-\underset{H}{\overset{CH_3}{C}}-\underset{H}{\overset{H}{C}}-\underset{Br}{\overset{H}{C}}-H \xrightarrow{\text{alcoholic KOH}} CH_3-\underset{H}{\overset{CH_3}{C}}-CH=CH_2$$

1-Bromo-3-methylbutane   3-Methyl-1-butene

How do we prepare the necessary alkyl halide? Certainly not by bromination of an alkane, since even if we could make the proper alkane in some way, bromination would occur almost entirely at the tertiary position to give the wrong product. (Chlorination would give the proper chloride—but as a minor component of a grand mixture.) As usual, then, we would prepare the halide from the corresponding alcohol, in this case 3-methyl-1-butanol. Since this is a primary alcohol (without branching near the —OH group), and hence does not form the halide via the carbonium ion, rearrangement is not likely; we might use, then, either hydrogen bromide or PBr_3.

$$CH_3-\underset{}{\overset{CH_3}{\overset{|}{C}H}}-CH_2-CH_2Br \xleftarrow{PBr_3} CH_3-\overset{CH_3}{\overset{|}{C}H}-CH_2-CH_2OH$$

3-Methyl-1-butanol

Now, how do we make 3-methyl-1-butanol? It is a primary alcohol and contains one carbon more than our largest available alcohol; therefore we would use the reaction of a Grignard reagent with formaldehyde. The necessary Grignard reagent is isobutylmagnesium bromide, which we could have prepared from

$$\underset{\text{3-Methyl-1-butanol}}{CH_3-\overset{CH_3}{\overset{|}{C}H}-CH_2-CH_2OH} \longleftarrow \begin{cases} \underset{\text{Formaldehyde}}{H-\overset{H}{\overset{|}{C}}=O} \\ \underset{\text{Isobutylmagnesium bromide}}{CH_3-\overset{CH_3}{\overset{|}{C}H}-CH_2MgBr} \end{cases}$$

isobutyl bromide, and that in turn from isobutyl alcohol. The formaldehyde is made by oxidation of methanol. The entire sequence, from which we could expect to obtain quite pure 3-methyl-1-butene, is the following:

$$\underset{\text{3-Methyl-1-butene}}{CH_3-\overset{CH_3}{\overset{|}{C}H}-CH=CH_2} \xleftarrow{KOH} CH_3-\overset{CH_3}{\overset{|}{C}H}-CH_2-CH_2Br \xleftarrow{PBr_3} \underset{\text{3-Methyl-1-butanol}}{CH_3-\overset{CH_3}{\overset{|}{C}H}-CH_2-CH_2OH}$$

$$\underset{\text{3-Methyl-1-butanol}}{CH_3-\overset{CH_3}{\overset{|}{C}H}-CH_2-CH_2OH} \longleftarrow \begin{cases} H-\overset{H}{\overset{|}{C}}=O \xleftarrow{K_2Cr_2O_7} \underset{\text{Methanol}}{CH_3OH} \\ CH_3-\overset{CH_3}{\overset{|}{C}H}-CH_2MgBr \xleftarrow{Mg} CH_3-\overset{CH_3}{\overset{|}{C}H}-CH_2Br \end{cases}$$

$$CH_3-\overset{CH_3}{\overset{|}{C}H}-CH_2Br \xleftarrow{PBr_3} \underset{\text{Isobutyl alcohol}}{CH_3-\overset{CH_3}{\overset{|}{C}H}-CH_2OH}$$

16.11 Analysis of alcohols. Characterization. Iodoform test

Alcohols dissolve in cold concentrated sulfuric acid. This property they share with alkenes, amines, practically all compounds containing oxygen, and easily sulfonated compounds. (Alcohols, like other oxygen-containing compounds, form oxonium salts, which dissolve in the highly polar sulfuric acid.)

Alcohols are not oxidized by cold, dilute, neutral permanganate (although primary and secondary alcohols are, of course, oxidized by permanganate under more vigorous conditions). However, as we have seen (Sec. 6.30), alcohols often contain impurities that *are* oxidized under these conditions, and so the permanganate test must be interpreted with caution.

Alcohols do not decolorize bromine in carbon tetrachloride. This property serves to distinguish them from alkenes and alkynes.

Alcohols are further distinguished from alkenes and alkynes—and, indeed, from nearly every other kind of compound—by their oxidation by chromic anhydride, CrO_3, in aqueous sulfuric acid: within *two seconds*, the clear orange solution turns blue-green and becomes opaque.

$$\underset{1^\circ \text{ or } 2^\circ}{\text{ROH}} + \underset{\substack{\textit{Clear,}\\ \textit{orange}}}{\text{HCrO}_4^-} \longrightarrow \textit{Opaque, blue-green}$$

Tertiary alcohols do not give this test. Aldehydes do, but are easily differentiated in other ways (Sec. 19.17).

Reaction of alcohols with sodium metal, with the evolution of hydrogen gas, is of some use in characterization; a *wet* compound of any kind, of course, will do the same thing, until the water is used up.

The presence of the —OH group in a molecule is often indicated by the formation of an ester upon treatment with an acid chloride or anhydride (Sec. 18.16). Some esters are sweet-smelling; others are solids with sharp melting points, and can be derivatives in identifications. (If the molecular formulas of starting material and product are determined, it is possible to calculate *how many* —OH groups are present.)

Problem 16.10 Make a table to show the response of each kind of compound we have studied so far toward the following reagents: (a) cold concentrated H_2SO_4; (b) cold, dilute, neutral $KMnO_4$; (c) Br_2 in CCl_4; (d) CrO_3 in H_2SO_4; (e) cold fuming sulfuric acid; (f) $CHCl_3$ and $AlCl_3$; (g) sodium metal.

Whether an alcohol is primary, secondary, or tertiary is shown by the **Lucas test**, which is based upon the difference in reactivity of the three classes toward hydrogen halides (Sec. 16.4). Alcohols (of not more than six carbons) are soluble in the *Lucas reagent*, a mixture of concentrated hydrochloric acid and zinc chloride. (Why are they more soluble in this than in water?) The corresponding alkyl chlorides are insoluble. Formation of a chloride from an alcohol is indicated by the cloudiness that appears when the chloride separates from the solution; hence, the time required for cloudiness to appear is a measure of the reactivity of the alcohol.

A tertiary alcohol reacts immediately with the Lucas reagent, and a secondary alcohol reacts within five minutes; a primary alcohol does not react appreciably at room temperature. As we have seen, benzyl alcohol and allyl alcohol react as rapidly as tertiary alcohols with the Lucas reagent; allyl chloride, however, is soluble in the reagent. (Why?)

Whether or not an alcohol contains one particular structural unit is shown by the **iodoform test**. The alcohol is treated with iodine and sodium hydroxide (sodium hypoiodite, NaOI); an alcohol of the structure

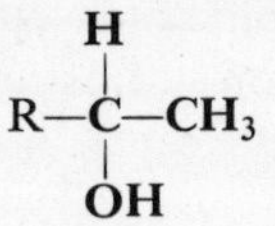

$$\mathrm{R{-}\underset{\displaystyle OH}{\overset{\displaystyle H}{C}}{-}CH_3} \quad \textit{where R is H or an alkyl or aryl group}$$

yields a yellow precipitate of iodoform (CHI_3, m.p. 119°). For example:

Gives positive iodoform test	**Gives negative iodoform test**
$\mathrm{CH_3{-}\underset{\displaystyle OH}{\overset{\displaystyle H}{C}}{-}H}$	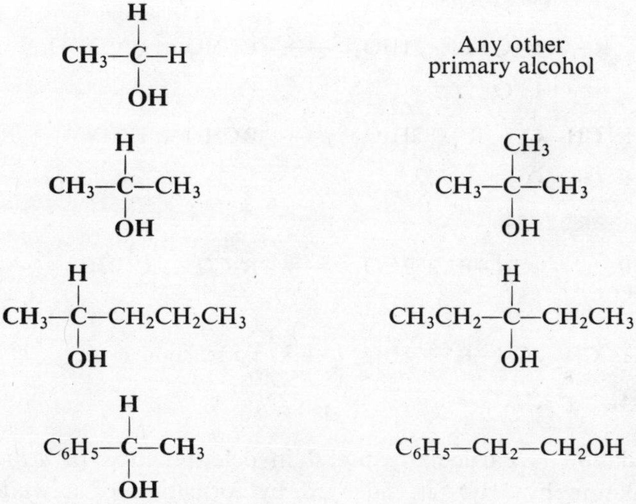 Any other primary alcohol
$\mathrm{CH_3{-}\underset{\displaystyle OH}{\overset{\displaystyle H}{C}}{-}CH_3}$	$\mathrm{CH_3{-}\underset{\displaystyle OH}{\overset{\displaystyle CH_3}{C}}{-}CH_3}$
$\mathrm{CH_3{-}\underset{\displaystyle OH}{\overset{\displaystyle H}{C}}{-}CH_2CH_2CH_3}$	$\mathrm{CH_3CH_2{-}\underset{\displaystyle OH}{\overset{\displaystyle H}{C}}{-}CH_2CH_3}$
$\mathrm{C_6H_5{-}\underset{\displaystyle OH}{\overset{\displaystyle H}{C}}{-}CH_3}$	$\mathrm{C_6H_5{-}CH_2{-}CH_2OH}$

The reaction involves oxidation, halogenation, and cleavage.

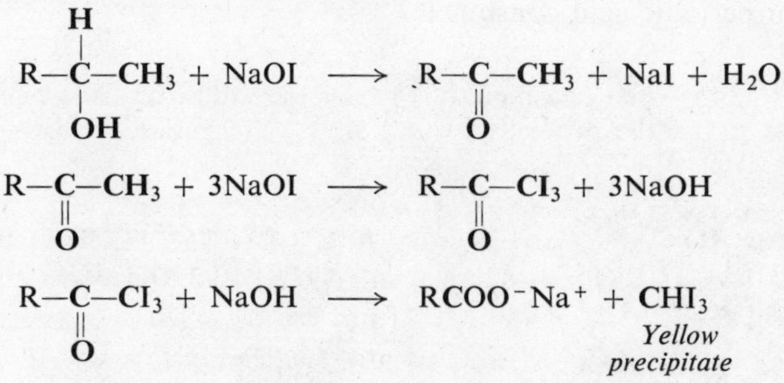

$$\mathrm{R{-}\underset{\displaystyle OH}{\overset{\displaystyle H}{C}}{-}CH_3 + NaOI \longrightarrow R{-}\underset{\displaystyle O}{\underset{\displaystyle \|}{C}}{-}CH_3 + NaI + H_2O}$$

$$\mathrm{R{-}\underset{\displaystyle O}{\underset{\displaystyle \|}{C}}{-}CH_3 + 3NaOI \longrightarrow R{-}\underset{\displaystyle O}{\underset{\displaystyle \|}{C}}{-}CI_3 + 3NaOH}$$

$$\mathrm{R{-}\underset{\displaystyle O}{\underset{\displaystyle \|}{C}}{-}CI_3 + NaOH \longrightarrow RCOO^-Na^+ + \underset{\textit{Yellow precipitate}}{CHI_3}}$$

As would be expected from the equations, a compound of structure

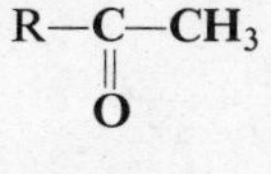

$$\mathrm{R{-}\underset{\displaystyle O}{\underset{\displaystyle \|}{C}}{-}CH_3} \quad \textit{where R is H or an alkyl or aryl group}$$

also gives a positive test (Sec. 19.17).

In certain special cases this reaction is used not as a test, but to synthesize the carboxylic acid, RCOOH. Here, hypobromite or the cheaper hypochlorite would probably be used.

16.12 Analysis of glycols. Periodic acid oxidation

Upon treatment with periodic acid, HIO_4, compounds containing two or more —OH or =O groups attached to *adjacent* carbon atoms undergo oxidation with cleavage of carbon–carbon bonds. For example:

$$\text{R—}\underset{\text{OH}}{\text{CH}}\text{—}\underset{\text{OH}}{\text{CH}}\text{—R}' + \text{HIO}_4 \longrightarrow \text{RCHO} + \text{R}'\text{CHO} \quad (+\ \text{HIO}_3)$$

$$\text{R—}\underset{\text{O}}{\underset{\|}{\text{C}}}\text{—}\underset{\text{O}}{\underset{\|}{\text{C}}}\text{—R}' + \text{HIO}_4 \longrightarrow \text{RCOOH} + \text{R}'\text{COOH}$$

$$\text{R—}\underset{\text{OH}}{\text{CH}}\text{—}\underset{\text{O}}{\underset{\|}{\text{C}}}\text{—R}' + \text{HIO}_4 \longrightarrow \text{RCHO} + \text{R}'\text{COOH}$$

$$\text{R—}\underset{\text{OH}}{\text{CH}}\text{—}\underset{\text{OH}}{\text{CH}}\text{—}\underset{\text{OH}}{\text{CH}}\text{—R}' + 2\text{HIO}_4 \longrightarrow \text{RCHO} + \text{HCOOH} + \text{R}'\text{CHO}$$

$$\text{R—}\overset{\text{R}}{\underset{\text{OH}}{\text{C}}}\text{——}\underset{\text{OH}}{\text{CH}}\text{—R}' + \text{HIO}_4 \longrightarrow \text{R}_2\text{CO} + \text{R}'\text{CHO}$$

$$\text{R—}\underset{\text{OH}}{\text{CH}}\text{—CH}_2\text{—}\underset{\text{OH}}{\text{CH}}\text{—R}' + \text{HIO}_4 \longrightarrow \text{no reaction}$$

The oxidation is particularly useful in determination of structure. Qualitatively, oxidation by HIO_4 is indicated by formation of a white precipitate ($AgIO_3$) upon addition of silver nitrate. Since the reaction is usually quantitative, valuable information is given by the nature and amounts of the products, and by the quantity of periodic acid consumed.

Problem 16.11 When one mole of each of the following compounds is treated with HIO_4, what will the products be, and how many moles of HIO_4 will be consumed?

(a) $CH_3CHOHCH_2OH$
(b) $CH_3CHOHCHO$
(c) $CH_2OHCHOHCH_2OCH_3$
(d) $CH_2OHCH(OCH_3)CH_2OH$
(e) *cis*-1,2-cyclopentanediol
(f) $CH_2OH(CHOH)_3CHO$
(g) $CH_2OH(CHOH)_3CH_2OH$

Problem 16.12 Assign a structure to each of the following compounds:

$$\begin{aligned}
&\text{A} + \text{one mole HIO}_4 \longrightarrow \text{CH}_3\text{COCH}_3 + \text{HCHO}\\
&\text{B} + \text{one mole HIO}_4 \longrightarrow \text{OHC(CH}_2)_4\text{CHO}\\
&\text{C} + \text{one mole HIO}_4 \longrightarrow \text{HOOC(CH}_2)_4\text{CHO}\\
&\text{D} + \text{one mole HIO}_4 \longrightarrow 2\text{HOOC—CHO}\\
&\text{E} + 3\text{HIO}_4 \longrightarrow 2\text{HCOOH} + 2\text{HCHO}\\
&\text{F} + 3\text{HIO}_4 \longrightarrow 2\text{HCOOH} + \text{HCHO} + \text{CO}_2\\
&\text{G} + 5\text{HIO}_4 \longrightarrow 5\text{HCOOH} + \text{HCHO}
\end{aligned}$$

16.13 Spectroscopic analysis of alcohols

Infrared. In the infrared spectrum of a hydrogen-bonded alcohol—and this is the kind that we commonly see—the most conspicuous feature is a strong, broad band in the 3200–3600 cm^{-1} region due to O—H stretching (see Fig. 16.1).

O—H stretching, *strong, broad*

Alcohols, ROH (or phenols, ArOH) 3200–3600 cm^{-1}

(A monomeric alcohol, as discussed in Sec. 15.4, gives a sharp, variable band at 3610–3640 cm^{-1}.)

Another strong, broad band, due to C—O stretching, appears in the 1000–1200 cm^{-1} region, the exact frequency depending on the nature of the alcohol:

C—O stretching, *strong, broad*

1° ROH	about 1050 cm^{-1}	3° ROH	about 1150 cm^{-1}
2° ROH	about 1100 cm^{-1}	ArOH	about 1230 cm^{-1}

(Compare the locations of this band in the spectra of Fig. 16.1.)

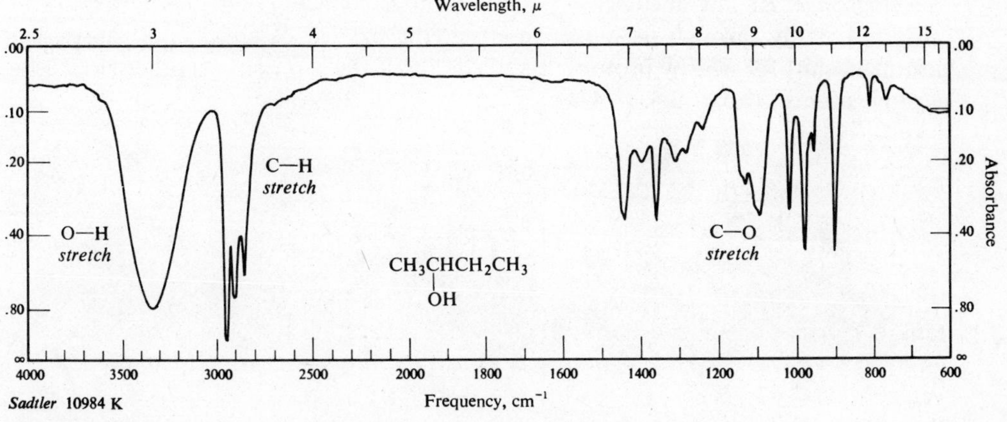

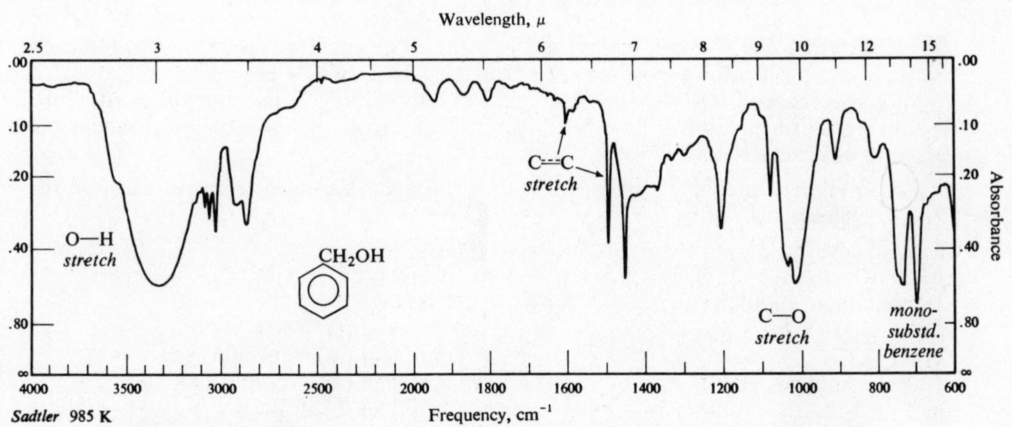

Figure 16.1. Infrared spectra of (a) *sec*-butyl alcohol and (b) benzyl alcohol.

Phenols (ArOH) also show both these bands, but the C—O stretching appears at somewhat higher frequencies. Ethers show C—O stretching, but the O—H band is absent. Carboxylic acids and esters show C—O stretching, but give absorption characteristic of the carbonyl group, C=O, as well. (For a comparison of certain oxygen compounds, see Table 20.3, p. 689.)

Nmr. Nmr absorption by a hydroxylic proton (O—H) is shifted downfield by hydrogen bonding. The chemical shift that is observed depends, therefore, on the degree of hydrogen bonding, which in turn depends on temperature, concentration, and the nature of the solvent (Sec. 15.4). As a result, the signal can appear anywhere in the range δ 1–5. It may be hidden among the peaks due to alkyl protons, although its presence there is often revealed through proton counting.

A hydroxyl proton ordinarily gives rise to a singlet in the nmr spectrum: its signal is not split by nearby protons, nor does it split their signals. Proton exchange between two (identical) molecules of alcohol

$$R^*\text{—O—}H^* + R\text{—O—}H \rightleftharpoons R^*\text{—O—}H + R\text{—O—}H^*$$

is so fast that the proton—now in one molecule and in the next instant in another—cannot see nearby protons in their various combinations of spin alignments, but in a single *average* alignment.

Presumably through its inductive effect, the oxygen of an alcohol causes a downfield shift for nearby protons: a shift of about the same size as other electronegative atoms (Table 13.4, p. 421).

Problem 16.13 Can you suggest a procedure that might move a hidden O—H peak into the open? (*Hint:* See Sec. 15.4.)

Problem 16.14 (a) Very dry, pure samples of alcohols show spin–spin splitting of the O—H signals. What splitting would you expect for a primary alcohol? a secondary alcohol? a tertiary alcohol? (b) This splitting disappears on the addition of a trace of acid or base. Write equations to show just how proton exchange would be speeded up by an acid (H:B); by a base (:B).

PROBLEMS

1. Refer to the isomeric pentyl alcohols of Problem 1(a), p. 515. (a) Indicate which (if any) will give a positive iodoform test. (b) Describe how each will respond to the Lucas reagent. (c) Describe how each will respond to chromic anhydride. (d) Outline all steps in a possible synthesis of each, starting from alcohols of four carbons or less, and using any necessary inorganic reagents.

2. Give structures and names of the chief products expected from the reaction (if any) of cyclohexanol with:

(a) cold conc. H_2SO_4
(b) H_2SO_4, heat
(c) cold dilute $KMnO_4$
(d) CrO_3, H_2SO_4
(e) Br_2/CCl_4
(f) conc. aqueous HBr
(g) $P + I_2$
(h) Na
(i) CH_3COOH, H^+
(j) H_2, Ni
(k) CH_3MgBr
(l) NaOH(aq)
(m) product (f) + Mg
(n) product (m) + product (d)
(o) product (b) + cold alk. $KMnO_4$
(p) product (b) + Br_2/CCl_4
(q) product (b) + C_6H_6, HF
(r) product (b) + H_2, Ni

(s) product (q) + HNO_3/H_2SO_4
(t) product (b) + N-bromosuccinimide
(u) product (b) + $CHCl_3$ + *t*-BuOK
(v) product (d) + C_6H_5MgBr
(w) tosyl chloride, OH^-
(x) product (w) + *t*-BuOK

3. Outline all steps in a possible laboratory synthesis of each of the following compounds from *n*-butyl alcohol, using any necessary inorganic reagents. Follow the general instructions on p. 224.

(a) *n*-butyl bromide
(b) 1-butene
(c) *n*-butyl hydrogen sulfate
(d) potassium *n*-butoxide
(e) *n*-butyraldehyde, $CH_3CH_2CH_2CHO$
(f) *n*-butyric acid, $CH_3CH_2CH_2COOH$
(g) *n*-butane
(h) 1,2-dibromobutane
(i) 1-chloro-2-butanol
(j) 1-butyne
(k) ethylcyclopropane
(l) 1,2-butanediol
(m) *n*-octane
(n) 3-octyne
(o) *cis*-3-octene
(p) *trans*-3-octene
(q) 4-octanol
(r) 4-octanone, $CH_3CH_2CH_2CH_2C(=O)CH_2CH_2CH_3$
(s) 5-(*n*-propyl)-5-nonanol
(t) *n*-butyl *n*-butyrate, $CH_3CH_2CH_2C(=O){-}OCH_2CH_2CH_2CH_3$

4. Give structures and (where possible) names of the principal organic products of the following:
(a) benzyl alcohol + Mg
(b) isobutyl alcohol + benzoic acid + H^+
(c) ethylene bromide + excess NaOH(aq)
(d) *n*-butyl alcohol + H_2, Pt
(e) crotyl alcohol ($CH_3CH{=}CHCH_2OH$) + Br_2/H_2O
(f) $CH_3OH + C_2H_5MgBr$
(g) *p*-bromobenzyl bromide + NaOH(aq)
(h) *tert*-butyl alcohol + C_6H_6 + H_2SO_4

5. In Great Britain during the past few years, thousands of motorists have been (politely) stopped by the police and asked to blow into a "breathalyser": a glass tube containing silica gel impregnated with certain chemicals, and leading into a plastic bag. If, for more than half the length of the tube, the original yellow color turns green, the motorist looks very unhappy and often turns red. What chemicals are impregnated on the silica gel, why does the tube turn green, and why does the motorist turn red?

6. Arrange the alcohols of each set in order of reactivity toward aqueous HBr:
(a) the isomeric pentyl alcohols of Problem 1(a), p. 515. (*Note:* It may be necessary to list these in groups of about the same reactivity.)
(b) 1-phenyl-1-propanol, 3-phenyl-1-propanol, 1-phenyl-2-propanol
(c) benzyl alcohol, *p*-cyanobenzyl alcohol, *p*-hydroxylbenzyl alcohol
(d) 2-buten-1-ol, 3-buten-1-ol
(e) cyclopentylcarbinol, 1-methylcyclopentanol, *trans*-2-methylcyclopentanol
(f) benzyl alcohol, diphenylcarbinol, methanol, triphenylcarbinol

hydride shifts

7. Outline the sequence of steps that best accounts for the following facts.

(a) 3-methyl-1-butene + HCl yields both 3-chloro-2-methylbutane and 2-chloro-2-methylbutane.

(b) Either 2-pentanol or 3-pentanol + HCl yields both 2-chloropentane and 3-chloropentane.

(c) 2,2,4-trimethyl-3-pentanol $\xrightarrow{Al_2O_3,\ heat}$ 2,4,4-trimethyl-2-pentene + 2,4,4-trimethyl-1-pentene + 2,3,4-trimethyl-2-pentene + 2,3,4-trimethyl-1-pentene + 3-methyl-2-isopropyl-1-butene + 3,3,4-trimethyl-1-pentene.

(d) 2,2-dimethylcyclohexanol $\xrightarrow{H^+}$ 1,2-dimethylcyclohexene + 1-isopropylcyclopentene. (*Hint:* Use models.)

(e) cyclobutyldiethylcarbinol $\xrightarrow{H^+}$ 1,2-diethylcyclopentene

(f) 1-methylcyclopentyl ring bearing CH_3 and $CHCH_3$(OH) on the same carbon $\xrightarrow{H^+}$ 1,2-dimethylcyclohexene

8. Outline all steps in a possible laboratory synthesis of each of the following compounds from cyclohexanol and any necessary aliphatic, aromatic, or inorganic reagents.

(a) cyclohexanone ($C_6H_{10}O$)
(b) bromocyclohexane
(c) 1-methylcyclohexanol
(d) 1-methylcyclohexene
(e) *trans*-2-methylcyclohexanol
(f) cyclohexylmethylcarbinol
(g) *trans*-1,2-dibromocyclohexane
(h) cyclohexylcarbinol
(i) 1-bromo-1-phenylcyclohexane
(j) cyclohexanecarboxylic acid
(k) adipic acid, $HOOC(CH_2)_4COOH$
(l) norcarane (see p. 458)

9. Outline all steps in a possible laboratory synthesis of each of the following compounds from benzene, toluene, and alcohols of four carbons or fewer.

(a) 2,3-dimethyl-2-butanol
(b) 2-phenyl-2-propanol
(c) 2-phenylpropene
(d) 2-methyl-1-butene
(e) isopentane
(f) 1,2-dibromo-2-methylbutane
(g) 3-hexanol
(h) 3-hexanone (I)
(i) 4-ethyl-4-heptanol
(j) 2-bromo-2-methylhexane
(k) methylacetylene
(l) *trans*-1,2-dimethylcyclopropane
(m) 1-chloro-1-phenylethane (α-phenylethyl chloride)
(n) *sec*-butylbenzene
(o) methyl isopropyl ketone (II)
(p) 2-methylhexane
(q) benzyl methyl ketone (III)
(r) 2,2-dimethylhexane
(s) 2-bromo-1-phenylpropane
(t) 3-heptyne
(u) ethyl propionate (IV)

$$CH_3CH_2CH_2\underset{\underset{\displaystyle O}{\|}}{C}CH_2CH_3 \qquad \text{I}$$

$$CH_3\underset{\underset{\displaystyle O}{\|}}{C}CH(CH_3)_2 \qquad \text{II}$$

$$C_6H_5CH_2\underset{\underset{\displaystyle O}{\|}}{C}CH_3 \qquad \text{III}$$

$$CH_3CH_2\underset{\underset{\displaystyle O}{\|}}{C}{-}OCH_2CH_3 \qquad \text{IV}$$

10. Compounds "labeled" at various positions by isotopic atoms are useful in determining reaction mechanisms and in following the fate of compounds in biological systems. Outline a possible synthesis of each of the following labeled compounds using $^{14}CH_3OH$ as the source of ^{14}C, and D_2O as the source of deuterium.

(a) 2-methyl-1-propanol-1-^{14}C, $(CH_3)_2CH^{14}CH_2OH$
(b) 2-methyl-1-propanol-2-^{14}C, $(CH_3)_2{}^{14}CHCH_2OH$
(c) 2-methyl-1-propanol-3-^{14}C, $^{14}CH_3CH(CH_3)CH_2OH$
(d) propene-1-^{14}C, $CH_3CH{=}^{14}CH_2$
(e) propene-2-^{14}C, $CH_3{}^{14}CH{=}CH_2$
(f) propene-3-^{14}C, $^{14}CH_3CH{=}CH_2$
(g) C_6H_5D
(h) $C_6H_5CH_2D$
(i) *p*-$DC_6H_4CH_3$
(j) $CH_3CH_2CHD^{14}CH_3$

11. When *trans*-2-methylcyclopentanol is treated with tosyl chloride and the product with potassium *tert*-butoxide, the only alkene obtained is 3-methylcyclopentene. (a) What is the stereochemistry of this reaction? (b) This is the final step of a general synthetic route to 3-alkylcyclopentenes, starting from cyclopentanone. Outline all steps in this

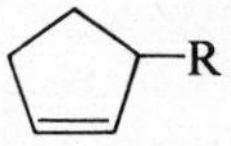

R

route, carefully choosing your reagents in each step. (c) What advantage does this sequence have over an analogous one involving an intermediate halide instead of a tosylate?

12. Making use of any necessary organic or inorganic reagents, outline all steps in the conversion of:

(a) androst-9(11)-ene (p. 516) into the saturated 11-keto derivative.

(b)

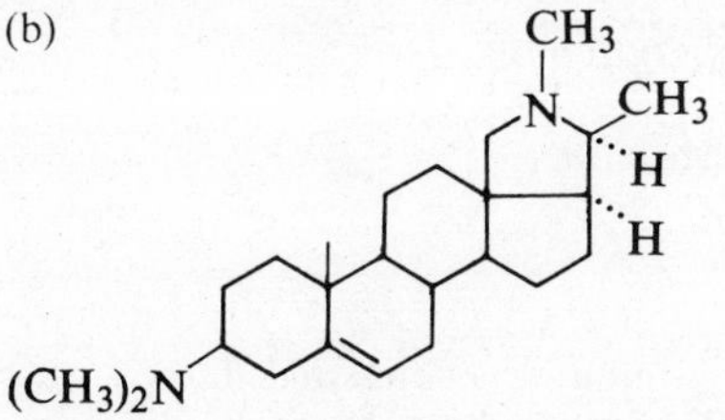

Conesine
(3β-dimethylaminocon-5-enine)
An alkaloid

into

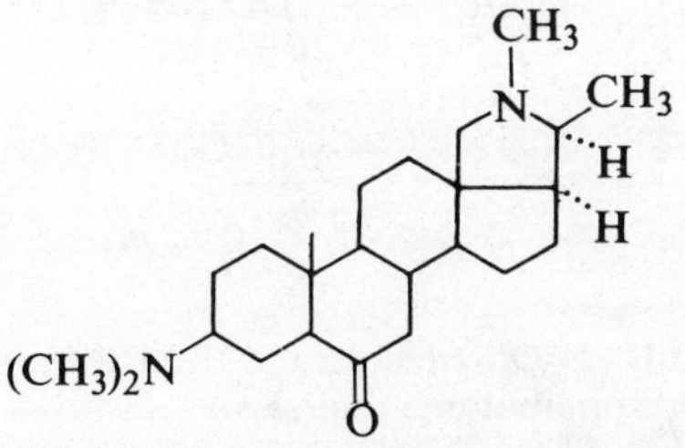

3β-Dimethylaminoconanin-6-one

(c)

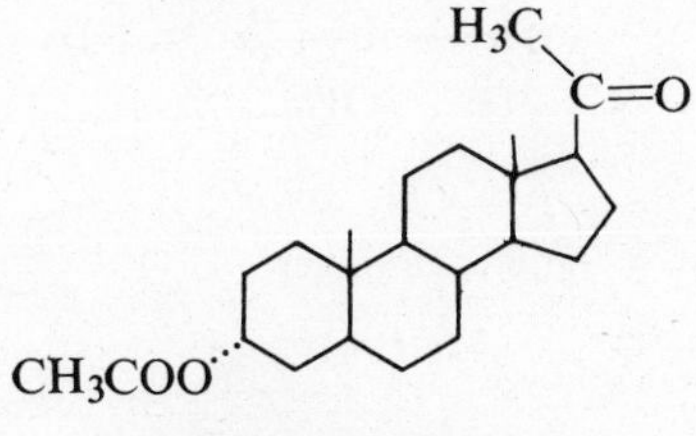

5α-Pregnane-3α-ol-20-one
(acetate ester)

into

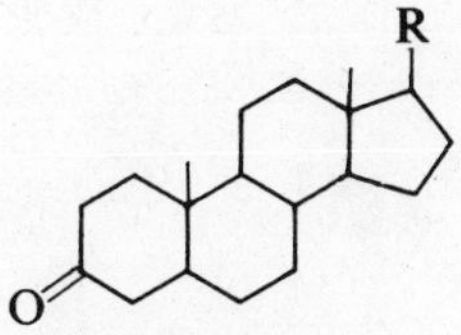

3-Cholestanone

where $R = -CH(CH_3)CH_2CH_2CH_2CH(CH_3)_2$

(*Hint:* $CH_3COOCH_3 + H_2O \xrightarrow{OH^-,\ heat} CH_3COO^- + CH_3OH$.)

13. Assign structures to the compounds A through HH.

(a) ethylene + $Cl_2(aq) \longrightarrow$ A (C_2H_5OCl)
A + $NaHCO_3(aq) \longrightarrow$ B ($C_2H_6O_2$)

(b) ethylene + $Cl_2(aq) \longrightarrow$ A (C_2H_5OCl)
A + $HNO_3 \longrightarrow$ C ($C_2H_3O_2Cl$)
C + $H_2O \longrightarrow$ D ($C_2H_4O_3$)

(c) E + $6HIO_4 \longrightarrow$ 6HCOOH

(d) F ($C_{18}H_{34}O_2$) + $HCO_2OH \longrightarrow$ G ($C_{18}H_{36}O_4$)
G + $HIO_4 \longrightarrow CH_3(CH_2)_7CHO + OHC(CH_2)_7COOH$

(e) allyl alcohol + $Br_2/CCl_4 \longrightarrow$ H ($C_3H_6OBr_2$)
H + $HNO_3 \longrightarrow$ I ($C_3H_4O_2Br_2$)
I + Zn $\longrightarrow$ J ($C_3H_4O_2$)

(f) 1,2,3-tribromopropane + KOH(alc) $\longrightarrow$ K ($C_3H_4Br_2$)
K + NaOH(aq) $\longrightarrow$ L (C_3H_5OBr)
L + KOH(alc) $\longrightarrow$ M (C_3H_4O)
(g) 2,2-dichloropropane + NaOH(aq) $\longrightarrow$ [N ($C_3H_8O_2$)] $\longrightarrow$ O (C_3H_6O)
(h) propyne + Cl_2(aq) $\longrightarrow$ [P ($C_3H_6O_2Cl_2$)] $\longrightarrow$ Q ($C_3H_4OCl_2$)
Q + Cl_2(aq) $\longrightarrow$ R ($C_3H_3OCl_3$)
R + NaOH(aq) $\longrightarrow$ $CHCl_3$ + S ($C_2H_3O_2Na$)
(i) cyclohexene + $KMnO_4$ $\longrightarrow$ T ($C_6H_{12}O_2$)
T + CH_3COOH, H^+ $\longrightarrow$ U ($C_{10}H_{16}O_4$)
(j) V ($C_3H_8O_3$) + CH_3COOH, H^+ $\longrightarrow$ W ($C_9H_{14}O_6$)
(k) cyclohexanol + $K_2Cr_2O_7$, H^+ $\longrightarrow$ X ($C_6H_{10}O$)
X + *m*-$CH_3C_6H_4MgBr$, followed by H_2O $\longrightarrow$ Y ($C_{13}H_{18}O$)
Y + heat $\longrightarrow$ Z ($C_{13}H_{16}$)
Z + Ni (300°) $\longrightarrow$ AA ($C_{13}H_{12}$)
(l) (R)-(+)-1-bromo-2,4-dimethylpentane + Mg $\longrightarrow$ BB
BB + $(CH_3)_2CHCH_2CHO$, then H_2O $\longrightarrow$ CC ($C_{12}H_{26}O$), *a mixture*
CC + CrO_3 $\longrightarrow$ DD ($C_{12}H_{24}O$)
DD + CH_3MgBr, then H_2O $\longrightarrow$ EE ($C_{13}H_{28}O$), *a mixture*
EE + I_2, heat $\longrightarrow$ FF ($C_{13}H_{26}$), *a mixture*
FF + H_2, Ni $\longrightarrow$ GG ($C_{13}H_{28}$) + HH ($C_{13}H_{28}$)
Optically active *Optically inactive*

14. (a) On treatment with HBr, *threo*-3-bromo-2-butanol is converted into racemic 2,3-dibromobutane, and *erythro*-3-bromo-2-butanol is converted into *meso*-2,3-dibromobutane. What appears to be the stereochemistry of the reaction? Does it proceed with inversion or retention of configuration?

	CH_3	
Br	—	H
H	—	OH
	CH_3	

and enantiomer
Threo

	CH_3	
H	—	Br
H	—	OH
	CH_3	

and enantiomer
Erythro

3-Bromo-2-butanol

(b) When optically active *threo*-3-bromo-2-butanol is treated with HBr, *racemic* 2,3-dibromobutane is obtained. Now what is the stereochemistry of the reaction? Can you think of a mechanism that accounts for this stereochemistry?

(c) These observations, reported in 1939 by Saul Winstein (p. 474) and Howard J. Lucas (of The California Institute of Technology), are the first of many described as "neighboring group effects." Does this term help you find an answer to (b)?

(d) On treatment with aqueous HBr, both *cis*- and *trans*-2-bromocyclohexanol are converted into the same product. In light of (b), what would you expect this product to be?

15. Interpret the following observations. (a) When dissolved in HSO_3F–SbF_5–SO_2 at −60°, methanol gave the following nmr spectrum: *a*, triplet, δ 4.7, 3H; *b*, quartet, δ 9.4, 2H. Under the same conditions, isobutyl alcohol gave: *a*, doublet, δ 1.1, 6H; *b*, multiplet, δ 2.3, 1H; *c*, two overlapping triplets, δ 4.7, 2H; *d*, triplet, δ 9.4, 2H.

(b) Warming to +50° had no effect on the methanol solution. At −30°, however, the isobutyl alcohol spectrum slowly disappeared, to be replaced by a single peak at δ 4.35. (c) Even at −60°, *tert*-butyl alcohol dissolved in HSO_3F–SbF_5–SO_2 to give immediately a single peak at δ 4.35.

16. Tricyclopropylcarbinol (R_3COH, R = cyclopropyl) gives a complex nmr spectrum in the region δ 0.2–1.1, and is transparent in the near ultraviolet. A solution of the alcohol in concentrated H_2SO_4 has the following properties:

(i) A freezing-point lowering corresponding to four particles for each molecule dissolved;
(ii) intense ultraviolet absorption (λ_{max} 270 nm, ϵ_{max} 22,000);
(iii) an nmr spectrum with one peak, a singlet, δ 2.26.

When the solution is diluted and neutralized, the original alcohol is recovered.

(a) What substance is formed in sulfuric acid solution? Show how its formation accounts for each of the facts (i)–(iii). How do you account for the evident stability of this substance? (*Hint:* See Secs. 9.9 and 12.18.)
(b) A solution of 2-cyclopropyl-2-propanol in strong acid gives the following nmr spectrum:

a singlet, δ 2.60, 3H
b singlet, δ 3.14, 3H
c multiplet, δ 3.5–4, 5H

A similar solution of 2-cyclopropyl-1,1,1-trideuterio-2-propanol gives a similar spectrum except that *a* and *b* are each reduced to one-half their former area.

What general conclusion about the relative locations of the two methyl groups must you make? Can you suggest a specific geometry for the molecule that is consistent not only with this spectrum but also with your answer to part (a)? (*Hint:* Use models.)

17. By use of Table 16.1 tell which alcohol or alcohols each of the following is likely to be. Tell what further steps you would take to identify it or to confirm your identification. (Below, Ar = α-naphthyl, Sec. 30.2.)

$$\underset{\text{An isocyanate}}{\text{ArNCO}} + \text{ROH} \longrightarrow \underset{\text{A urethane}}{\text{ArNHCOOR}}$$

II: b.p. 115–7°; Lucas test, secondary; 3,5-dinitrobenzoate, m.p. 95–6°
JJ: b.p. 128–30°; negative halogen test; Lucas test, primary
KK: b.p. 128–31°; positive iodoform test
LL: b.p. 115–8°; 3,5-dinitrobenzoate, m.p. 60–1°
MM: b.p. 117–9°; α-naphthylurethane, m.p. 69–71°

Table 16.1 DERIVATIVES OF SOME ALCOHOLS

Alcohol	B.p., °C	α-Naphthylurethane M.p., °C	3,5-Dinitrobenzoate M.p., °C
3-Methyl-2-butanol	114	112	76
3-Pentanol	116	71	97
n-Butyl alcohol	118	71	64
2-Pentanol	119	76	61
1-Chloro-2-propanol	127	—	83
2-Methyl-1-butanol	128	97	62
Ethylene chlorohydrin	129	101	92
4-Methyl-2-pentanol	131	88	65
3-Methyl-1-butanol	132	67	62
2-Chloro-1-propanol	132	—	76

18. Describe simple chemical tests that would serve to distinguish between:

(a) *n*-butyl alcohol and *n*-octane
(b) *n*-butyl alcohol and 1-octene
(c) *n*-butyl alcohol and *n*-pentyl bromide
(d) *n*-butyl alcohol and 3-buten-1-ol
(e) 3-buten-1-ol and 2-buten-1-ol

(f) 3-pentanol and 1-pentanol
(g) 3-pentanol and 2-pentanol
(h) 3-phenyl-1-propanol and cinnamyl alcohol (3-phenyl-2-propen-1-ol)
(i) 1,2-propanediol and 1,3-propanediol
(j) *n*-butyl alcohol and *tert*-pentyl alcohol
(k) *p*-bromobenzyl alcohol and *p*-ethylbenzyl alcohol
(l) α-phenylethyl alcohol and β-phenylethyl alcohol

Tell exactly what you would *do* and *see*.

19. Although it is a secondary alcohol, 1-chloro-2-propanol behaves like a primary alcohol in the Lucas test. Can you suggest a reason for this behavior?

20. (a) Compound NN of formula $C_9H_{12}O$ responded to a series of tests as follows:

(1) Na $\longrightarrow$ slow formation of gas bubbles
(2) acetic anhydride $\longrightarrow$ pleasant smelling product
(3) CrO_3/H_2SO_4 $\longrightarrow$ opaque blue-green *immediately*
(4) hot $KMnO_4$ $\longrightarrow$ benzoic acid
(5) Br_2/CCl_4 $\longrightarrow$ no decolorization
(6) I_2 + NaOH $\longrightarrow$ yellow solid
(7) rotated plane-polarized light

What was NN? Write equations for all the above reactions.

(b) Compound OO, an isomer of NN, also was found to be optically active. It showed the same behavior as NN except for test (6). From the careful oxidation of OO by $KMnO_4$ there was isolated an acid of formula $C_9H_{10}O_2$. What was OO?

21. Identify each of the following isomers of formula $C_{20}H_{18}O$:

Isomer PP (m.p. 88°)
a singlet, δ 2.23, 1H
b doublet, δ 3.92, 1H, J = 7 Hz
c doublet, δ 4.98, 1H, J = 7 Hz
d singlet, δ 6.81, 10 H
e singlet, δ 6.99, 5H

Isomer QQ (m.p. 88°)
a singlet, δ 2.14, 1H
b singlet, δ 3.55, 2H
c broad peak, δ 7.25, 15H

What single simple chemical test would distinguish between these two isomers?

22. Give a structure or structures consistent with each of the nmr spectra in Fig. 16.2, p. 548.

23. Give a structure or structures consistent with each of the nmr spectra in Fig. 16.3, p. 549.

24. Upon hydrogenation, compound RR (C_4H_8O) is converted into SS ($C_4H_{10}O$). On the basis of their infrared spectra (Fig. 16.4, p. 550) give the structural formulas of RR and SS.

25. Give a structure or structures for the compound TT, whose infrared and nmr spectra are shown in Fig. 16.5, p. 550, and Fig. 16.6, p. 551.

26. *Geraniol*, $C_{10}H_{18}O$, a terpene found in rose oil, gives the infrared and nmr spectra shown in Fig. 16.7 (p. 551). In the next problem, chemical evidence is given from which its structure can be deduced; before working that problem, however, let us see how much information we can get from the spectra alone.

(a) Examine the infrared spectrum. Is geraniol aliphatic or aromatic? What functional group is clearly present? From the molecular formula, what other groupings must also be present in the molecule? Is their presence confirmed by the infrared spectrum?

(b) In the nmr spectrum, assign the number of protons to each signal on the basis of the integration curve. From the chemical shift values, and keeping in mind the infrared information, what kind of proton probably gives rise to each signal?

(c) When geraniol is shaken with D_2O, the peak at δ 3.32 disappears. Why?

(d) Write down likely groupings in the molecule. How many (if any) methyl groups are there? Methylene groups? Vinylic or allylic protons?

(e) What relationships among these groupings are suggested by chemical shift values, splittings, etc.?

(f) Draw a structure or structures consistent with the spectra. Taking into account the source of geraniol, are any of these more likely than others?

27. *Geraniol*, $C_{10}H_{18}O$, a terpene found in rose oil, adds two moles of bromine to form a tetrabromide, $C_{10}H_{18}OBr_4$. It can be oxidized to a ten-carbon aldehyde or to a ten-carbon carboxylic acid. Upon vigorous oxidation, geraniol yields:

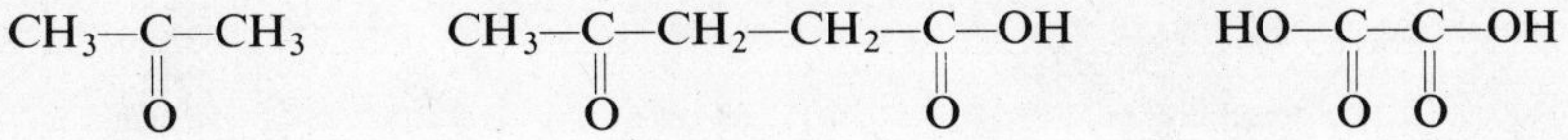

$$CH_3{-}\underset{\underset{O}{\|}}{C}{-}CH_3 \qquad CH_3{-}\underset{\underset{O}{\|}}{C}{-}CH_2{-}CH_2{-}\underset{\underset{O}{\|}}{C}{-}OH \qquad HO{-}\underset{\underset{O}{\|}}{C}{-}\underset{\underset{O}{\|}}{C}{-}OH$$

(a) Keeping in mind the isoprene rule (Sec. 8.26), what is the most likely structure for geraniol? (b) Nerol (Problem 19, p. 317) can be converted into the same saturated alcohol as geraniol, and yields the same oxidation products as geraniol, yet has different physical properties. What is the most probable structural relationship between geraniol and nerol? (c) Like nerol, geraniol is converted by sulfuric acid into α-terpineol (Problem 19, p. 317), but much more slowly than nerol. On this basis, what structures might you assign to nerol and geraniol? (*Hint:* Use models.)

28. Upon treatment with HBr, both geraniol (preceding problem) and *linalool* (from oil of lavender, bergamot, coriander) yield the same bromide, of formula $C_{10}H_{17}Br$. How do you account for this fact?

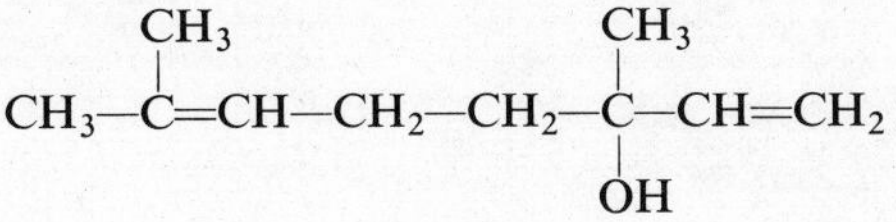

$$CH_3{-}\overset{\overset{CH_3}{|}}{C}{=}CH{-}CH_2{-}CH_2{-}\underset{\underset{OH}{|}}{\overset{\overset{CH_3}{|}}{C}}{-}CH{=}CH_2$$

Linalool

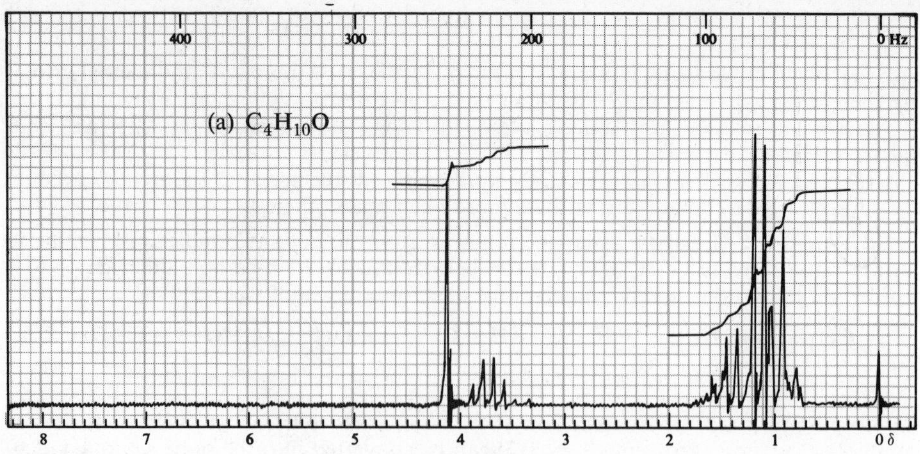

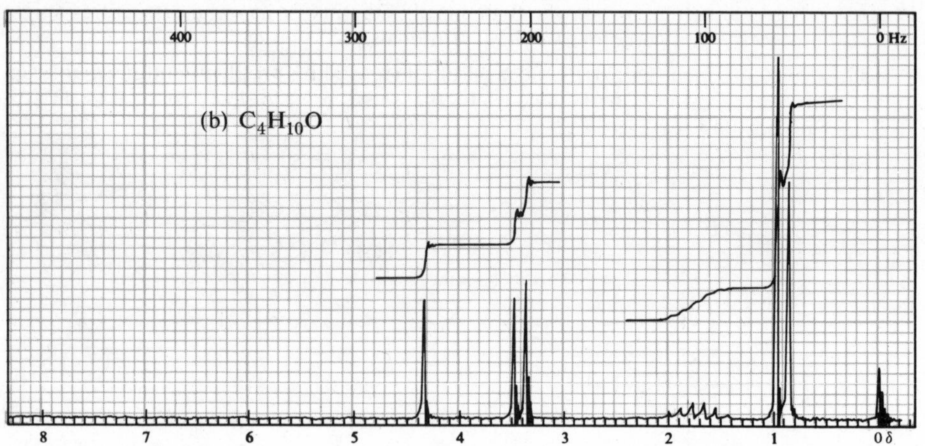

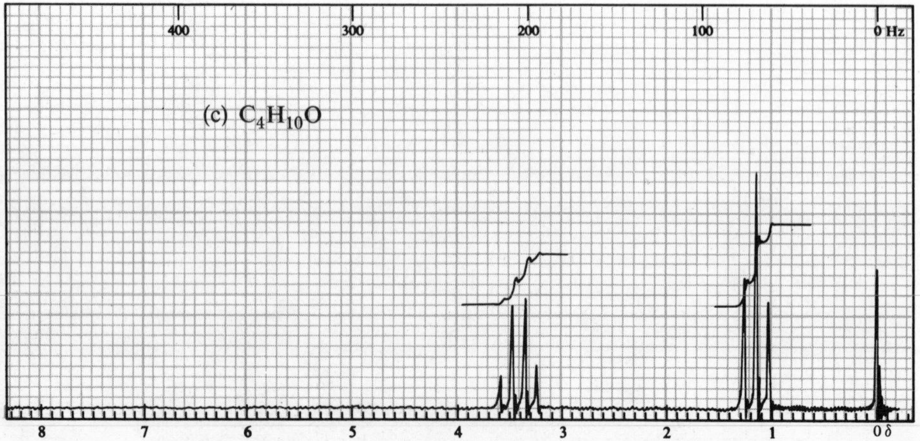

Figure 16.2. Nmr spectra for Problem 22, p. 546.

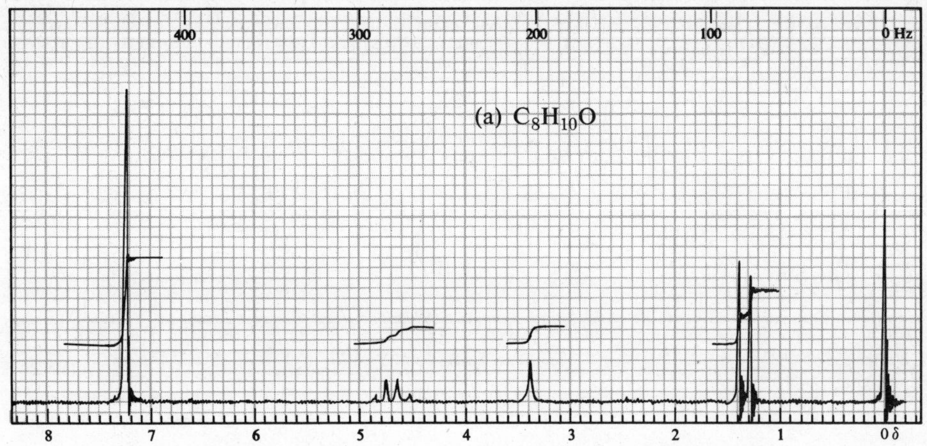

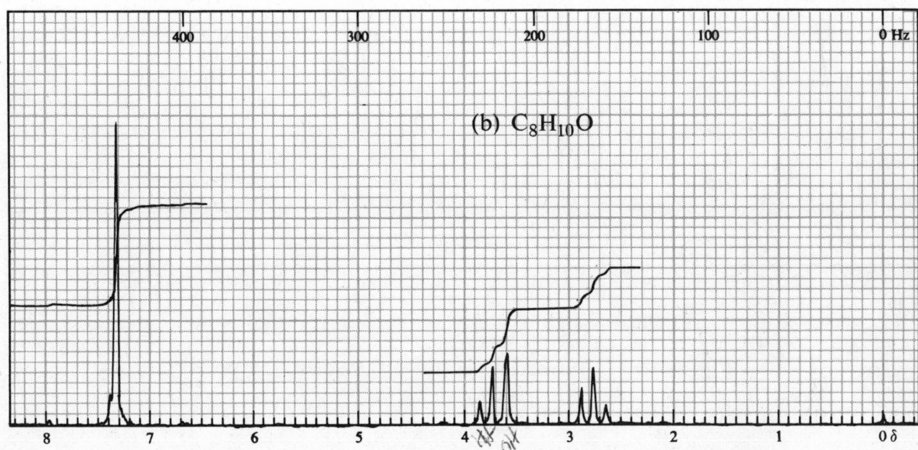

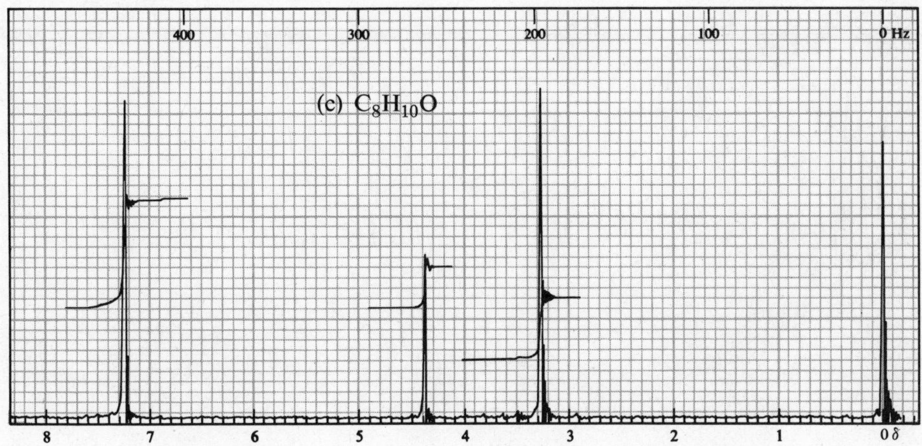

Figure 16.3. Nmr spectra for Problem 23, p. 546.

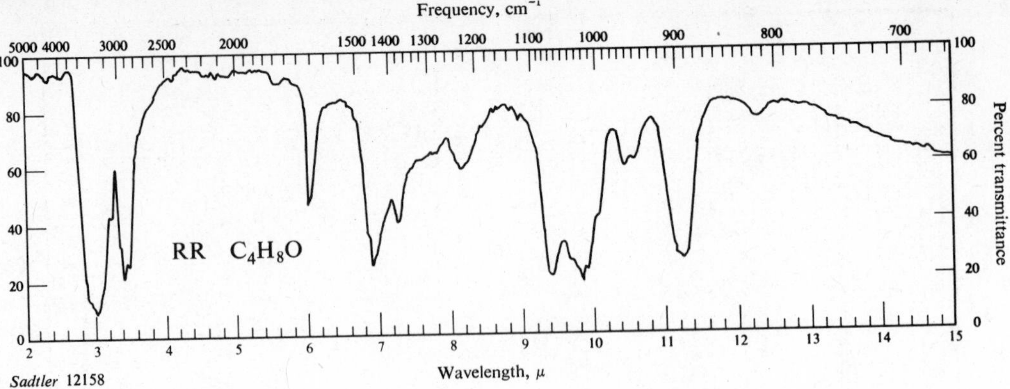

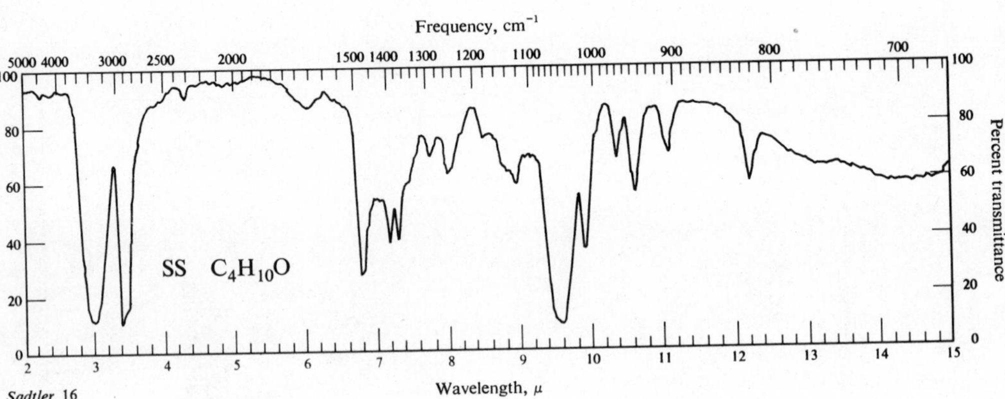

Figure 16.4. Infrared spectra for Problem 24, p. 546.

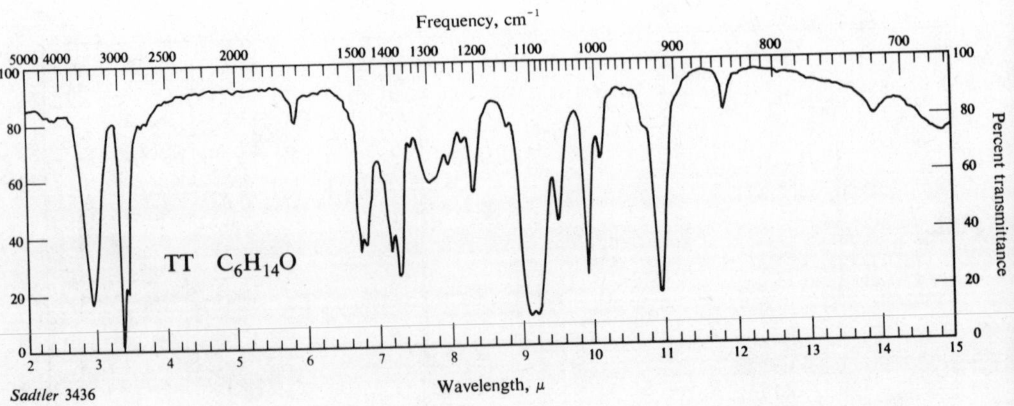

Figure 16.5. Infrared spectrum for Problem 25, p. 546.

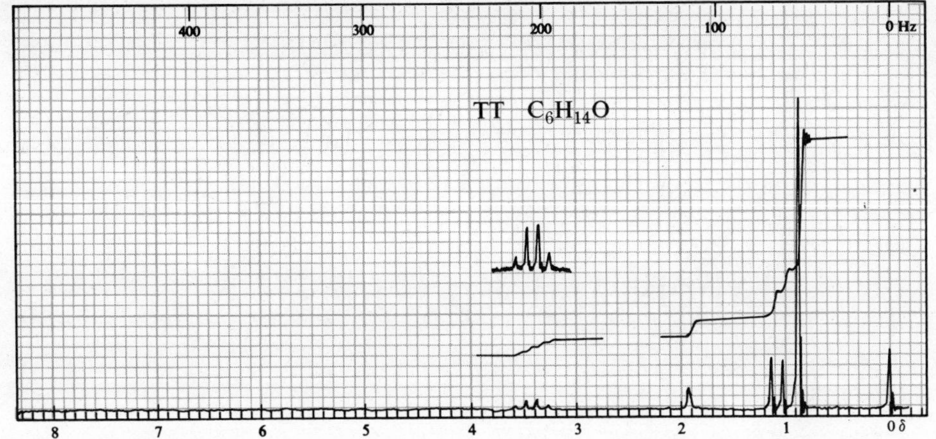

Figure 16.6. Nmr spectrum for Problem 25, p. 546.

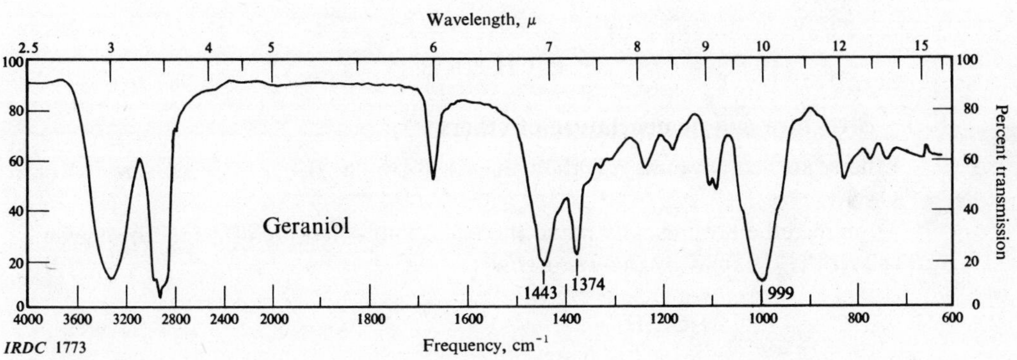

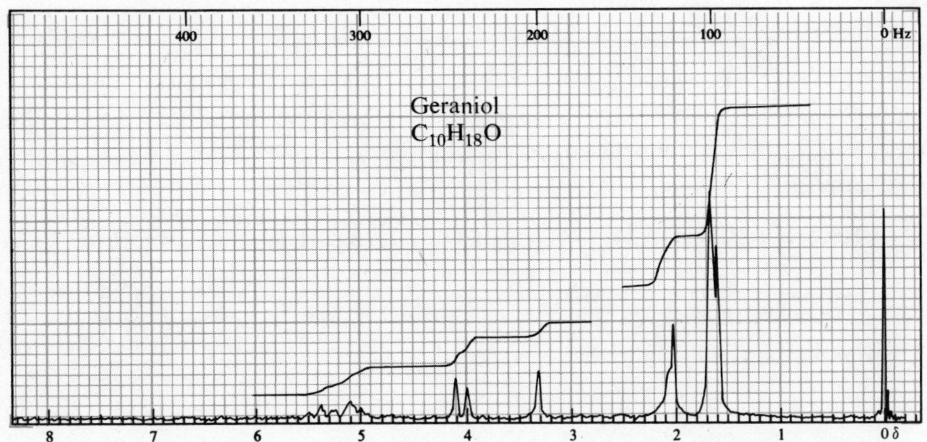

Figure 16.7. Infrared and nmr spectra for Problem 26, p. 546.

Chapter 17 Ethers and Epoxides

ETHERS

17.1 Structure and nomenclature of ethers

Ethers are compounds of the general formula R—O—R, Ar—O—R, or Ar—O—Ar.

To name ethers we usually name the two groups that are attached to oxygen, and follow these names by the word *ether*:

$C_2H_5OC_2H_5$

Ethyl ether

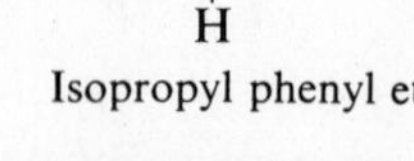

$C_6H_5-O-C_6H_5$

Phenyl ether

$CH_3-O-C(CH_3)_2-CH_3$

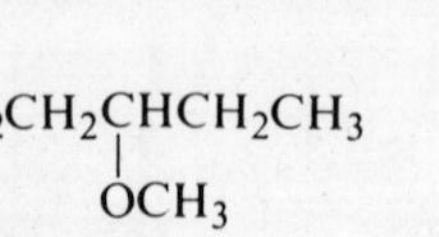

Methyl *tert*-butyl ether

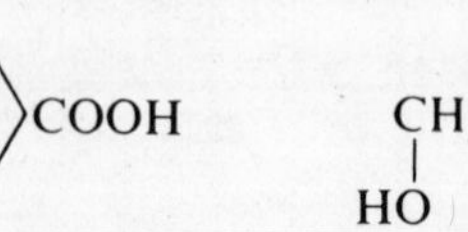

$CH_3-CH(CH_3)-O-C_6H_5$

Isopropyl phenyl ether

If one group has no simple name, the compound may be named as an *alkoxy* derivative:

$CH_3CH_2CH_2CH(OCH_3)CH_2CH_3$

3-Methoxyhexane

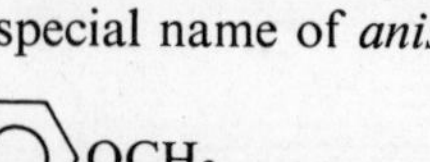

$C_2H_5O-C_6H_4-COOH$

p-Ethoxybenzoic acid

$HOCH_2CH_2OC_2H_5$

2-Ethoxyethanol

The simplest aryl alkyl ether has the special name of *anisole*.

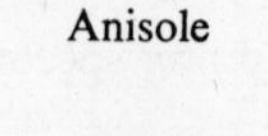

$C_6H_5OCH_3$

Anisole

If the two groups are identical, the ether is said to be *symmetrical* (e.g., *ethyl ether*, *phenyl ether*), if different, *unsymmetrical* (e.g., *methyl tert-butyl ether*, *anisole*).

17.2 Physical properties of ethers

Since the C—O—C bond angle is not 180°, the dipole moments of the two C—O bonds do not cancel each other; consequently, ethers possess a small net dipole moment (e.g., 1.18 D for ethyl ether).

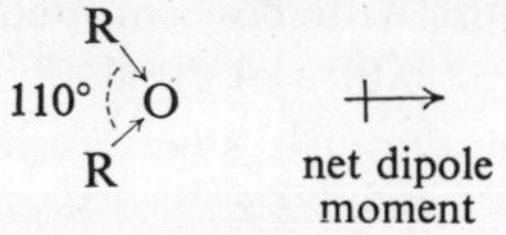

R
110° O
R

net dipole moment

This weak polarity does not appreciably affect the boiling points of ethers, which are about the same as those of alkanes having comparable molecular weights, and much lower than those of isomeric alcohols. Compare, for example, the boiling points of *n*-heptane (98°), methyl *n*-pentyl ether (100°), and *n*-hexyl alcohol (157°). The hydrogen bonding that holds alcohol molecules strongly together is not possible for ethers, since they contain hydrogen bonded only to carbon (Sec. 15.4).

On the other hand, ethers show a solubility in water comparable to that of the alcohols, both ethyl ether and *n*-butyl alcohol, for example, being soluble to the extent of about 8 g per 100 g of water. We attributed the water solubility of the lower alcohols to hydrogen bonding between water molecules and alcohol molecules; presumably the water solubility of ether arises in the same way.

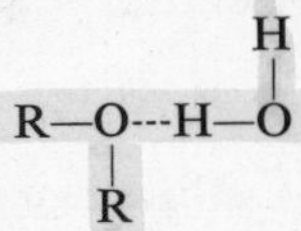

Table 17.1 ETHERS

Name	M.p., °C	B.p., °C	Name	M.p., °C	B.p., °C
Methyl ether	−140	−24	Anisole	−37	154
Ethyl ether	−116	34.6	Phenetole (Ethyl phenyl ether)	−33	172
n-Propyl ether	−122	91	Phenyl ether	27	259
Isopropyl ether	−60	69	1,4-Dioxane	11	101
n-Butyl ether	−95	142	Tetrahydrofuran	−108	66
Vinyl ether		35			
Allyl ether		94			

```
          H
          |
R—O---H—O
  |
  R
```

17.3 Industrial sources of ethers. Dehydration of alcohols

A number of symmetrical ethers containing the lower alkyl groups are prepared on a large scale, chiefly for use as solvents. The most important of these is

ethyl ether, the familiar anesthetic and the solvent we use in extractions and in the preparation of Grignard reagents; others include isopropyl ether and *n*-butyl ether.

These ethers are prepared by reactions of the corresponding alcohols with sulfuric acid. Since a molecule of water is lost for every pair of alcohol molecules, the reaction is a kind of *dehydration*. Dehydration to ethers rather than to alkenes

$$2\text{R—O—H} \xrightarrow{\text{H}_2\text{SO}_4\text{, heat}} \text{R—O—R} + \text{H}_2\text{O}$$

is controlled by the choice of reaction conditions. For example, ethylene is prepared by heating ethyl alcohol with concentrated sulfuric acid to 180°; ethyl ether is prepared by heating a mixture of ethyl alcohol and concentrated sulfuric acid to 140°, alcohol being continuously added to keep it in excess.

Dehydration is generally limited to the preparation of symmetrical ethers, because, as we might expect, a combination of two alcohols usually yields a mixture of three ethers.

Ether formation by dehydration is an example of nucleophilic substitution, with the protonated alcohol as substrate and a second molecule of alcohol as nucleophile.

Problem 17.1 (a) Give all steps of a likely mechanism for the dehydration of an alcohol to an ether. (b) Is this the only possibility? Give all steps of an alternative mechanism. (*Hint:* See Sec. 14.16.) (c) Dehydration of *n*-butyl alcohol gives *n*-butyl ether. Which of your alternatives appears to be operating here?

Problem 17.2 In ether formation by dehydration, as in other cases of substitution, there is a competing elimination reaction. What is this reaction, and what products does it yield? For what alcohols would elimination be most important?

Problem 17.3 (a) Upon treatment with sulfuric acid, a mixture of ethyl and *n*-propyl alcohols yields a mixture of three ethers. What are they? (b) On the other hand, a mixture of *tert*-butyl alcohol and ethyl alcohol gives a good yield of a single ether. What ether is this likely to be? How do you account for the good yield?

On standing in contact with air, most aliphatic ethers are converted slowly into unstable peroxides. Although present in only low concentrations, these peroxides are very dangerous, since they can cause violent explosions during the distillations that normally follow extractions with ether.

The presence of peroxides is indicated by formation of a red color when the ether is shaken with an aqueous solution of ferrous ammonium sulfate and potassium thiocyanate; the peroxide oxidizes ferrous ion to ferric ion, which reacts with thiocyanate ion to give the characteristic blood-red color of the complex.

$$\text{peroxide} + \text{Fe}^{++} \longrightarrow \text{Fe}^{+++} \xrightarrow{\text{SCN}^-} \underset{\textit{Red}}{\text{Fe(SCN)}_n{}^{-(3-n)}} \qquad (n = 1 \text{ to } 6)$$

Peroxides can be removed from ethers in a number of ways, including washing with solutions of ferrous ion (which reduces peroxides), or distillation from concentrated H_2SO_4 (which oxidizes peroxides).

For use in the preparation of Grignard reagents, the ether (usually ethyl) must be free of traces of water and alcohol. This so-called **absolute ether** can be prepared by distillation of ordinary ether from concentrated H_2SO_4 (which removes not only water and alcohol but also peroxides), and subsequent storing over metallic sodium. There is available today commercial anhydrous ether of such high quality that only the treatment with sodium is needed to make it ready for the Grignard reaction.

It is hard to overemphasize the hazards met in using ethyl ether, even when it is free of peroxides: it is highly volatile, and the flammability of its vapors makes explosions and fires ever-present dangers unless proper precautions are observed.

17.4 Preparation of ethers

The following methods are generally used for the laboratory preparation of ethers. (The Williamson synthesis is used for the preparation of aryl alkyl ethers industrially, as well.)

PREPARATION OF ETHERS

1. **Williamson synthesis.** Discussed in Sec. 17.5.

$$RX + \begin{matrix} R'O^-Na^+ \\ \text{or} \\ ArO^-Na^+ \end{matrix} \longrightarrow \begin{matrix} ROR' \\ \text{or} \\ ROAr \end{matrix}$$

Yield from RX: $CH_3 > 1° > 2° (> 3°)$

Examples:

$$\left[(CH_3)_2CHOH \xrightarrow{Na}\right] \underset{\text{Sodium isopropoxide}}{(CH_3)_2CHO^-Na^+} + \underset{n\text{-Propyl bromide}}{CH_3CH_2CH_2Br} \longrightarrow \underset{n\text{-Propyl isopropyl ether}}{CH_3CH_2CH_2OCH(CH_3)_2}$$

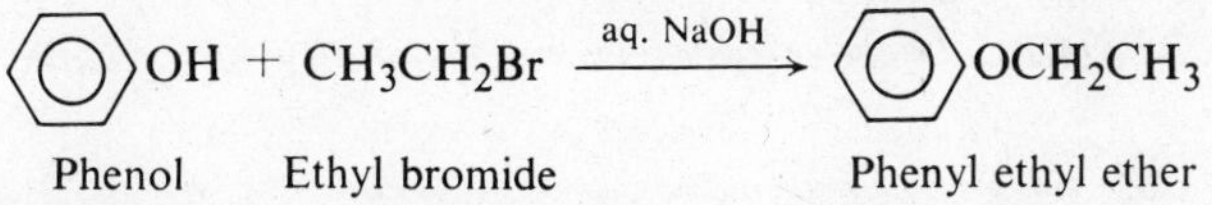

$$\underset{\text{Phenol}}{C_6H_5OH} + \underset{\text{Ethyl bromide}}{CH_3CH_2Br} \xrightarrow{\text{aq. NaOH}} \underset{\text{Phenyl ethyl ether}}{C_6H_5OCH_2CH_3}$$

2. **Alkoxymercuration-demercuration.** Discussed in Sec. 17.6.

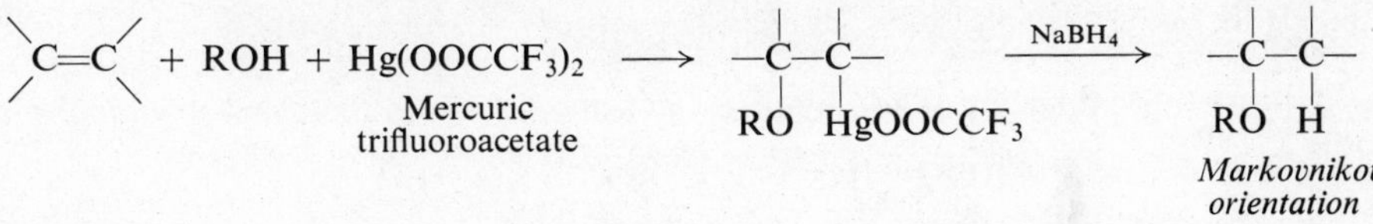

$$\text{>C=C<} + ROH + \underset{\text{Mercuric trifluoroacetate}}{Hg(OOCCF_3)_2} \longrightarrow \underset{RO\quad HgOOCCF_3}{-\overset{|}{\underset{|}{C}}-\overset{|}{\underset{|}{C}}-} \xrightarrow{NaBH_4} \underset{RO\quad H}{-\overset{|}{\underset{|}{C}}-\overset{|}{\underset{|}{C}}-}$$

Markovnikov orientation

Example:

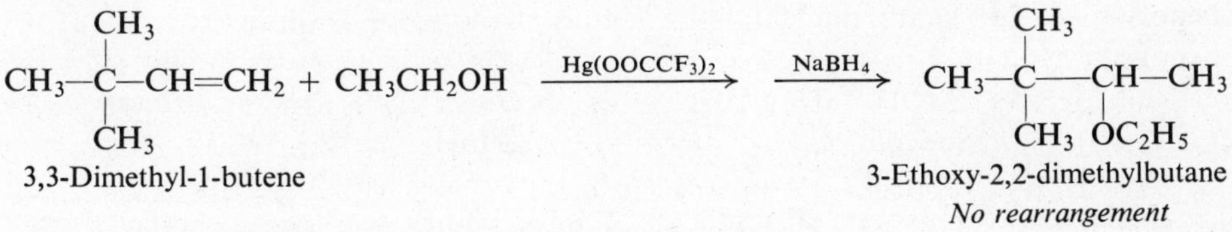

$$\underset{\text{3,3-Dimethyl-1-butene}}{CH_3-C(CH_3)_2-CH=CH_2} + CH_3CH_2OH \xrightarrow{Hg(OOCCF_3)_2} \xrightarrow{NaBH_4} \underset{\text{3-Ethoxy-2,2-dimethylbutane}}{CH_3-C(CH_3)_2-CH(OC_2H_5)-CH_3}$$

No rearrangement

17.5 Preparation of ethers. Williamson synthesis

In the laboratory, the Williamson synthesis of ethers is important because of its versatility: it can be used to make unsymmetrical ethers as well as symmetrical ethers, and aryl alkyl ethers as well as dialkyl ethers.

In the Williamson synthesis an alkyl halide (or substituted alkyl halide) is allowed to react with a sodium alkoxide or a sodium phenoxide:

$$R{-}X + Na^{+}\,{}^{-}O{-}R' \longrightarrow R{-}O{-}R' + Na^{+}X^{-}$$

$$R{-}X + Na^{+}\,{}^{-}O{-}Ar \longrightarrow R{-}O{-}Ar + Na^{+}X^{-}$$

For the preparation of methyl aryl ethers, *methyl sulfate*, $(CH_3)_2SO_4$, is frequently used instead of the more expensive methyl halides.

$$CH_3Br + Na^{+}\,{}^{-}O{-}C(CH_3)_3 \longrightarrow CH_3{-}O{-}C(CH_3)_3$$

Sodium *tert*-butoxide — Methyl *tert*-butyl ether

$$C_6H_5OH + CH_3OSO_2OCH_3 \xrightarrow{\text{aq. NaOH}} C_6H_5OCH_3 + CH_3OSO_3^{-}\,Na^{+}$$

Phenol — Methyl sulfate — Anisole

$$C_6H_5CH_2Br + HOC_6H_5 \xrightarrow{\text{aq. NaOH}} C_6H_5CH_2OC_6H_5$$

Benzyl bromide — Phenol — Benzyl phenyl ether

The Williamson synthesis involves nucleophilic substitution of alkoxide ion or phenoxide ion for halide ion; it is strictly analogous to the preparation of alcohols by treatment of alkyl halides with aqueous hydroxide (Sec. 15.7). *Aryl halides cannot in general be used, because of their low reactivity toward nucleophilic substitution.*

Problem 17.4 (a) On what basis could you have predicted that methyl sulfate would be a good methylating agent in reactions like those presented above? (*Hint:* What is the *leaving group*? See Sec. 14.6.) (b) Can you suggest another class of compounds that might serve in place of alkyl halides in the Williamson synthesis?

Sodium alkoxides are made by direct action of sodium metal on dry alcohols:

$$ROH + Na \longrightarrow RO^{-}Na^{+} + \tfrac{1}{2}H_2$$

An alkoxide

Sodium phenoxides, on the other hand, because of the appreciable acidity of phenols (Sec. 24.7), are made by the action of aqueous sodium hydroxide on phenols:

$$ArOH + Na^{+}\,{}^{-}OH \longrightarrow ArO^{-}Na^{+} + H_2O$$

Stronger acid — A phenoxide — Weaker acid

If we wish to make an unsymmetrical dialkyl ether, we have a choice of two combinations of reagents; one of these is nearly always better than the other.

In the preparation of ethyl *tert*-butyl ether, for example, the following combinations are conceivable:

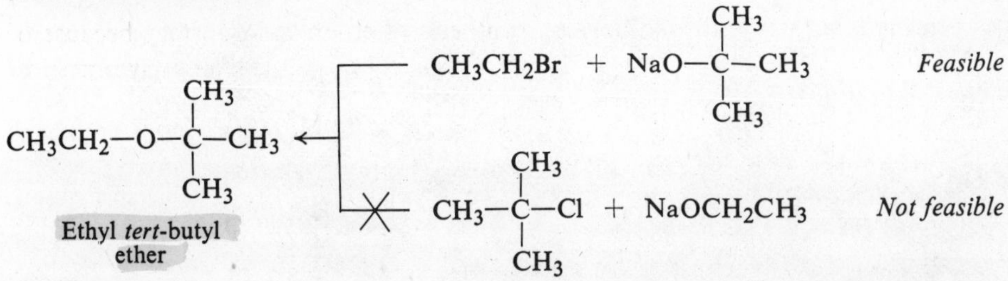

$$CH_3CH_2{-}O{-}C(CH_3)_2{-}CH_3 \longleftarrow \begin{cases} CH_3CH_2Br + NaO{-}C(CH_3)_2{-}CH_3 & \textit{Feasible} \\ CH_3{-}C(CH_3)_2{-}Cl + NaOCH_2CH_3 \;(\times) & \textit{Not feasible} \end{cases}$$

Ethyl *tert*-butyl ether

Which do we choose? As always, we must consider the danger of elimination competing with the desired substitution; elimination should be particularly serious here because of the strong basicity of the alkoxide reagent. We therefore reject the use of the tertiary halide, which we expect to yield mostly—or all—elimination product; we must use the other combination. The disadvantage of the slow

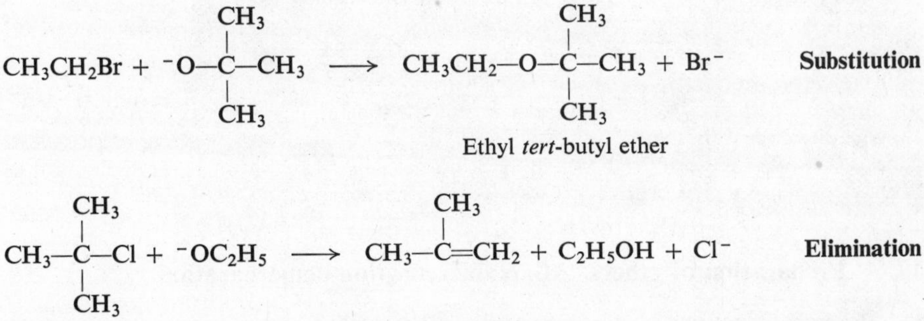

$$CH_3CH_2Br + {}^-O{-}C(CH_3)_2{-}CH_3 \longrightarrow CH_3CH_2{-}O{-}C(CH_3)_2{-}CH_3 + Br^- \qquad \textbf{Substitution}$$

Ethyl *tert*-butyl ether

$$CH_3{-}C(CH_3)_2{-}Cl + {}^-OC_2H_5 \longrightarrow CH_3{-}C(CH_3){=}CH_2 + C_2H_5OH + Cl^- \qquad \textbf{Elimination}$$

reaction between sodium and *tert*-butyl alcohol (Sec. 16.6) in the preparation of the alkoxide is more than offset by the tendency of the primary halide to undergo substitution rather than elimination. In planning a Williamson synthesis of a dialkyl ether, we must always keep in mind that the tendency for alkyl halides to undergo dehydrohalogenation is 3° > 2° > 1°.

For the preparation of an aryl alkyl ether there are again two combinations to be considered; here, one combination can usually be rejected out of hand. *n*-Propyl phenyl ether, for example, can be prepared only from the alkyl halide and the phenoxide, since the aryl halide is quite unreactive toward alkoxides.

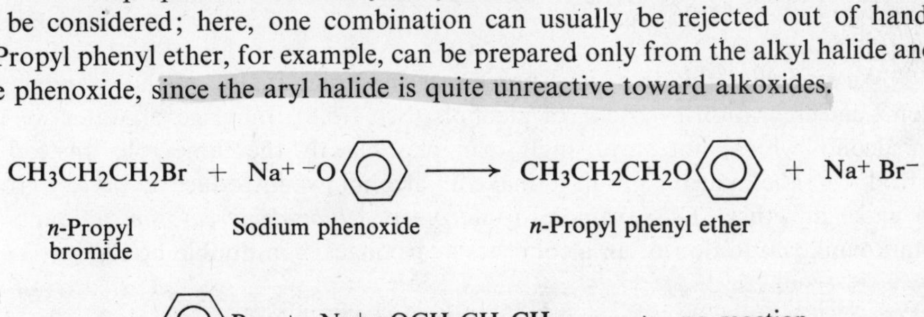

$$CH_3CH_2CH_2Br + Na^+ \, {}^-OC_6H_5 \longrightarrow CH_3CH_2CH_2OC_6H_5 + Na^+ \, Br^-$$

n-Propyl bromide; Sodium phenoxide; *n*-Propyl phenyl ether

$$C_6H_5Br + Na^+ \, {}^-OCH_2CH_2CH_3 \longrightarrow \text{no reaction}$$

Bromobenzene; Sodium *n*-propoxide

Since alkoxides and phenoxides are prepared from the corresponding alcohols and phenols, and since alkyl halides are commonly prepared from the alcohols, the Williamson method ultimately involves the synthesis of an ether from two hydroxy compounds.

Problem 17.5 Outline the synthesis, from alcohols and/or phenols, of:

(a) ethyl *tert*-butyl ether
(b) *n*-propyl phenyl ether
(c) isobutyl *sec*-butyl ether
(d) cyclohexyl methyl ether

Problem 17.6 Outline the synthesis of phenyl *p*-nitrobenzyl ether from any of these starting materials: toluene, bromobenzene, phenol. (*Caution:* Double-check the nitration stage.)

Problem 17.7 When optically active 2-octanol of specific rotation −8.24° is converted into its sodium salt, and the salt is then treated with ethyl bromide, there is obtained the optically active ether, 2-ethoxyoctane, with specific rotation −14.6°. Making use of the configuration and maximum rotation of 2-octanol given on p. 462, what, if anything, can you say about: (a) the configuration of (−)-2-ethoxyoctane? (b) the maximum rotation of 2-ethoxyoctane?

Problem 17.8 (*Work this after Problem 17.7.*) When (−)-2-bromooctane of specific rotation −30.3° is treated with ethoxide ion in ethyl alcohol, there is obtained 2-ethoxyoctane of specific rotation +15.3°. Using the configuration and maximum rotation of the bromide given on p. 462, answer the following questions. (a) Does this reaction involve complete retention of configuration, complete inversion, or inversion plus racemization? (b) By what mechanism does this reaction appear to proceed? (c) In view of the reagent and solvent, is this the mechanism you would have expected to operate? (d) What mechanism do you suppose is involved in the alternative synthesis (Problem 17.7) of 2-ethoxyoctane from the salt of 2-octanol and ethyl bromide? (e) Why, then, do the products of the two syntheses have *opposite* rotations?

17.6 Preparation of ethers. Alkoxymercuration-demercuration

Alkenes react with mercuric trifluoroacetate in the presence of an alcohol to give alkoxymercurial compounds which on reduction yield ethers.

Alkoxymercuration **Demercuration**

$$\underset{\text{Alkene}}{>C{=}C<} + \underset{\text{Alcohol}}{ROH} + \underset{\substack{\text{Mercuric}\\\text{trifluoroacetate}}}{Hg(OOCCF_3)_2} \longrightarrow -\underset{OR}{\overset{|}{\underset{|}{C}}}-\underset{HgOOCCF_3}{\overset{|}{\underset{|}{C}}}- \xrightarrow{NaBH_4} \underset{\text{Ether}}{-\underset{OR}{\overset{|}{\underset{|}{C}}}-\underset{H}{\overset{|}{\underset{|}{C}}}-}$$

We recognize this two-stage process as the exact analog of the oxymercuration—demercuration synthesis of alcohols (Sec. 15.8). In place of water we use an alcohol which, not surprisingly, can play exactly the same role. Instead of introducing the hydroxy group to make an alcohol, we introduce an *alkoxy* group to make an ether. This example of *solvomercuration-demercuration* amounts to Markovnikov addition of an alcohol to a carbon–carbon double bond.

Problem 17.9 Write all steps of a likely mechanism for alkoxymercuration.

Alkoxymercuration-demercuration has all the advantages we saw for its counterpart: speed, convenience, high yield, and the virtual absence of rearrangement. Compared with the Williamson synthesis, it has one tremendous advantage: there is no competing elimination reaction. As a result, it can be used for the synthesis of nearly every kind of alkyl ether except—evidently for steric reasons—di-*tert*-alkyl ethers. For example:

$$\underset{\text{3,3-Dimethyl-1-butene}}{CH_3-\overset{\overset{CH_3}{|}}{\underset{\underset{CH_3}{|}}{C}}-CH{=}CH_2} + (CH_3)_2CHOH \xrightarrow{Hg(OOCCF_3)_2} \xrightarrow{NaBH_4} \underset{\text{3-Isopropoxy-2,2-dimethylbutane}}{CH_3-\overset{\overset{CH_3}{|}}{\underset{\underset{CH_3}{|}}{C}}-\underset{\underset{OCH(CH_3)_2}{|}}{CH}-CH_3}$$

$$\underset{\text{Styrene}}{C_6H_5CH{=}CH_2} + (CH_3)_3COH \xrightarrow{Hg(OOCCF_3)_2} \xrightarrow{NaBH_4} \underset{\alpha\text{-Phenylethyl } tert\text{-butyl ether}}{C_6H_5\underset{\underset{OC(CH_3)_3}{|}}{CH}CH_3}$$

We notice that, instead of the mercuric acetate which was used in the preparation of alcohols, here mercuric *trifluoro*acetate is used. With a bulky alcohol—secondary or tertiary—as solvent, the trifluoroacetate is required for a good yield of ether.

Problem 17.10 In the presence of a secondary or tertiary alcohol, mercuric acetate adds to alkenes to give much—or even chiefly—organic acetate instead of ether

$$-\underset{\underset{OAc}{|}}{\overset{|}{C}}-\underset{\underset{HgOAc}{|}}{\overset{|}{C}}- \quad \textit{instead of} \quad -\underset{\underset{OR}{|}}{\overset{|}{C}}-\underset{\underset{HgOAc}{|}}{\overset{|}{C}}-$$

as the product. How do you account for the advantage of using mercuric trifluoroacetate? (*Hint:* Trifluoroacetic acid is a much stronger acid than acetic.)

Problem 17.11 Starting with any alcohols, outline all steps in the synthesis of each of the following ethers, using the Williamson synthesis or alkoxymercuration-demercuration, whichever you think is best suited for the particular job.

(a) *n*-hexyl isopropyl ether
(b) 2-hexyl isopropyl ether
(c) cyclohexyl *tert*-butyl ether
(d) cyclohexyl ether

17.7 Reactions of ethers. Cleavage by acids

Ethers are comparatively unreactive compounds. The ether linkage is quite stable toward bases, oxidizing agents, and reducing agents. In so far as the ether linkage itself is concerned, ethers undergo just one kind of reaction, **cleavage by acids**:

$$R{-}O{-}R' + HX \longrightarrow R{-}X + R'{-}OH \xrightarrow{HX} R'{-}X$$

$$Ar{-}O{-}R + HX \longrightarrow R{-}X + Ar{-}OH$$

Reactivity of HX: HI > HBr > HCl

Cleavage takes place only under quite vigorous conditions: concentrated acids (usually HI or HBr) and high temperatures.

An alkyl ether yields initially an alkyl halide and an alcohol; the alcohol may react further to form a second mole of alkyl halide. Because of the low reactivity at the bond between oxygen and an aromatic ring, an aryl alkyl ether undergoes cleavage of the alkyl–oxygen bond and yields a phenol and an alkyl halide. For example:

$$\underset{\text{Isopropyl ether}}{CH_3-\overset{CH_3}{\overset{|}{C}}H-O-\overset{CH_3}{\overset{|}{C}}H-CH_3} \xrightarrow[130\text{–}140^\circ]{48\%\ HBr} \underset{\text{Isopropyl bromide}}{2CH_3-\overset{CH_3}{\overset{|}{C}}H-Br}$$

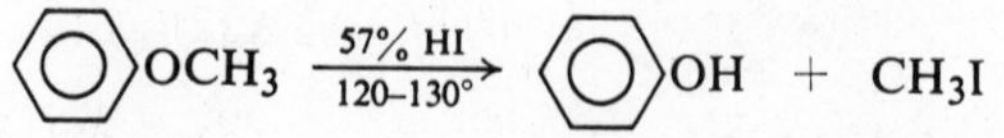

$$\underset{\text{Anisole}}{C_6H_5OCH_3} \xrightarrow[120\text{–}130^\circ]{57\%\ HI} \underset{\text{Phenol}}{C_6H_5OH} + \underset{\text{Methyl iodide}}{CH_3I}$$

Cleavage involves nucleophilic attack by halide ion on the protonated ether, with displacement of the weakly basic alcohol molecule:

$$R\ddot{O}R' + HX \rightleftharpoons R\overset{H\oplus}{\ddot{O}}R' + X^- \xrightarrow[S_N2]{S_N1 \text{ or}} RX + \underset{\substack{\text{Weak base:}\\ \textit{good leaving group}}}{R'OH}$$

Such a reaction occurs much more readily than displacement of the strongly basic alkoxide ion from the neutral ether.

$$ROR' + X^- \not\longrightarrow RX + \underset{\substack{\text{Strong base:}\\ \textit{poor leaving group}}}{R'O^-}$$

Reaction of a protonated ether with halide ion, like the corresponding reaction of a protonated alcohol, can proceed by either an S_N1 or S_N2 mechanism, depending upon conditions and the structure of the ether. As we might expect, a primary

S_N1

(1) $$R\overset{H}{\overset{|}{O}}R'^{+} \xrightarrow{\text{slow}} R^+ + HOR'$$

(2) $$R^+ + X^- \xrightarrow{\text{fast}} R\text{—}X$$

S_N2

$$R\overset{H}{\overset{|}{O}}R'^{+} + X^- \longrightarrow \left[\overset{\delta-}{X}\text{---}R\text{---}\underset{\delta+}{\overset{H}{\overset{|}{O}}}R'\right] \longrightarrow RX + HOR'$$

alkyl group tends to undergo S_N2 displacement, whereas a tertiary alkyl group tends to undergo S_N1 displacement.

Problem 17.12 Cleavage of optically active methyl *sec*-butyl ether by anhydrous HBr yields chiefly methyl bromide and *sec*-butyl alcohol; the *sec*-butyl alcohol has the same configuration and optical purity as the starting material. How do you interpret these results?

17.8 Electrophilic substitution in aromatic ethers

The alkoxy group, —OR, was listed (Sec. 11.5) as *ortho,para*-directing toward electrophilic aromatic substitution, and moderately activating. It is a much stronger activator than —R, but much weaker than —OH.

The carbonium ions resulting from *ortho* and *para* attack were considered (Sec. 11.20) to be stabilized by contribution from structures I and II. These structures

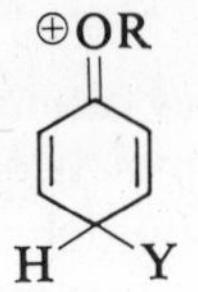

⊕OR ⊕OR H Y H Y

I 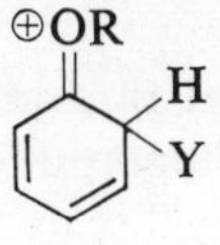 II

are especially stable ones, since in them every atom (except hydrogen, of course) has a complete octet of electrons.

The ability of the oxygen to share more than a pair of electrons with the ring and to accommodate a positive charge is consistent with the basic character of ethers.

Problem 17.13 Predict the principal products of: (a) bromination of *p*-methylanisole; (b) nitration of *m*-nitroanisole; (c) nitration of benzyl phenyl ether.

17.9 Cyclic ethers

In their preparation and properties, most cyclic ethers are just like the ethers we have already studied: the chemistry of the ether linkage is essentially the same whether it forms part of an open chain or part of an aliphatic ring.

Problem 17.14 *1,4-Dioxane* is prepared industrially (for use as a water-soluble solvent) by dehydration of an alcohol. What alcohol is used?

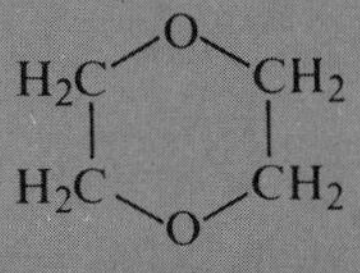

O H_2C CH_2 H_2C CH_2 O — 1,4-Dioxane

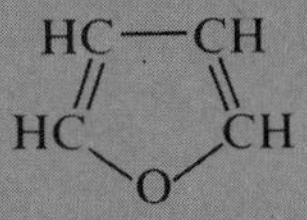

HC—CH HC CH O — Furan

H_2C—CH_2 H_2C CH_2 O — Tetrahydrofuran

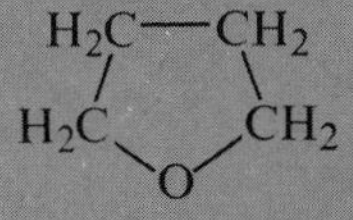

Problem 17.15 The unsaturated cyclic ether *furan* can readily be made from substances isolated from oat hulls and corncobs; one of its important uses involves its conversion into (a) *tetrahydrofuran*, and (b) 1,4-dichlorobutane. Using your knowledge of alkene chemistry and ether chemistry, show how these conversions can be carried out.

Cyclic ethers of one class deserve special attention because of their unusual reactivity; these compounds are the *epoxides*.

EPOXIDES

17.10 Preparation of epoxides

Epoxides are compounds containing the three-membered ring:

$$-\overset{|}{C}-\overset{|}{C}- \text{ (bridged by } O\text{)}$$

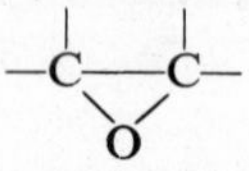

Epoxide ring
(Oxirane ring)

They are ethers, but the three-membered ring gives them unusual properties.

By far the most important epoxide is the simplest one, ethylene oxide. It is prepared on an industrial scale by catalytic oxidation of ethylene by air.

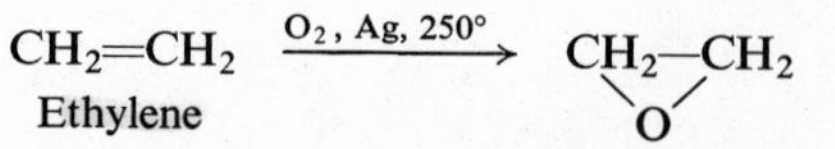

$$\underset{\text{Ethylene}}{CH_2{=}CH_2} \xrightarrow{O_2,\ Ag,\ 250°} \underset{\text{Ethylene oxide}}{CH_2{-}CH_2 \text{ (bridged by } O)}$$

Other epoxides are prepared by the following methods.

PREPARATION OF EPOXIDES

1. From halohydrins. Discussed in Sec. 17.10.

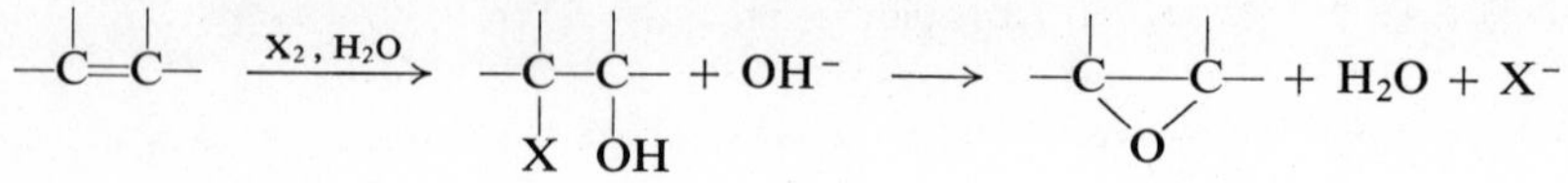

$$-\overset{|}{C}{=}\overset{|}{C}- \xrightarrow{X_2,\ H_2O} -\underset{X}{\overset{|}{C}}-\underset{OH}{\overset{|}{C}}- + OH^- \longrightarrow -\overset{|}{C}-\overset{|}{C}- \text{ (bridged by } O) + H_2O + X^-$$

Example:

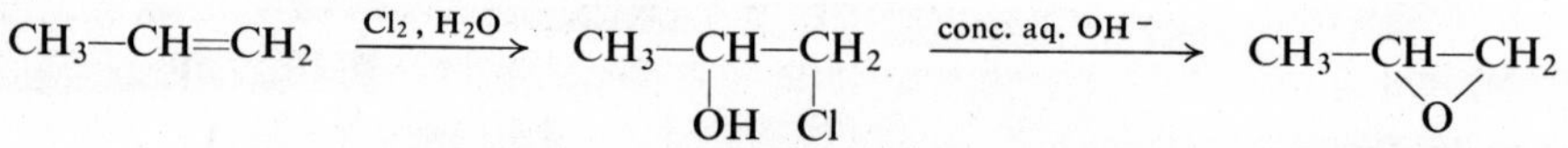

$$CH_3{-}CH{=}CH_2 \xrightarrow{Cl_2,\ H_2O} \underset{\text{Propylene chlorohydrin}}{CH_3{-}\underset{OH}{CH}{-}\underset{Cl}{CH_2}} \xrightarrow{\text{conc. aq. } OH^-} \underset{\text{Propylene oxide}}{CH_3{-}CH{-}CH_2 \text{ (bridged by } O)}$$

2. Peroxidation of carbon–carbon double bonds. Discussed in Sec. 17.10.

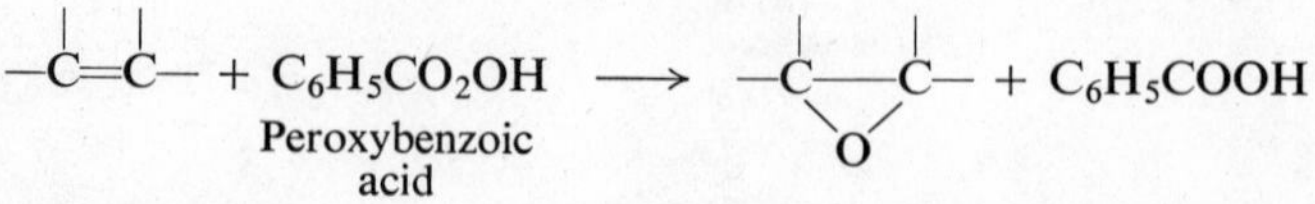

$$-\overset{|}{C}{=}\overset{|}{C}- + \underset{\text{Peroxybenzoic acid}}{C_6H_5CO_2OH} \longrightarrow -\overset{|}{C}-\overset{|}{C}- \text{ (bridged by } O) + C_6H_5COOH$$

Examples:

$$\underset{\text{Styrene}}{C_6H_5{-}CH{=}CH_2} \xrightarrow{\text{peroxybenzoic acid}} \underset{\text{Styrene oxide}}{C_6H_5{-}CH{-}CH_2 \text{ (bridged by } O)}$$

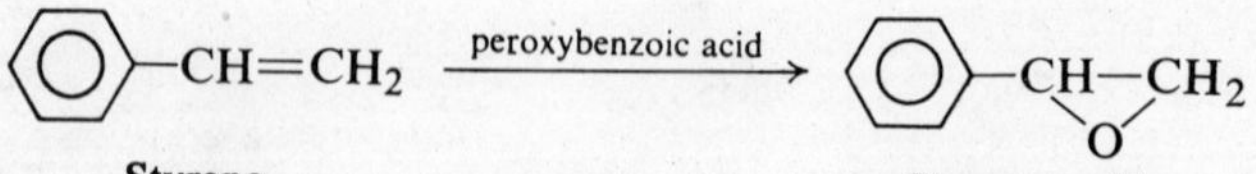

$$\text{Cyclohexene} \xrightarrow{\text{peroxybenzoic acid}} \text{Cyclohexene oxide}$$

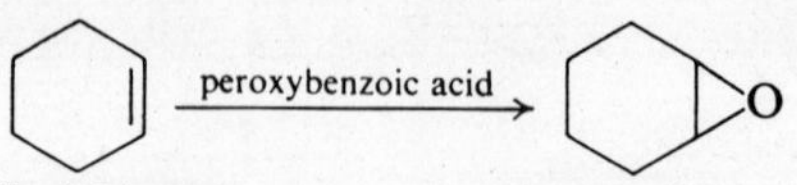

The conversion of halohydrins into epoxides by the action of base is simply an adaptation of the Williamson synthesis (Sec. 17.5); a cyclic compound is obtained because both alcohol and halide happen to be part of the same molecule. In the presence of hydroxide ion a small proportion of the alcohol exists as alkoxide; this alkoxide displaces halide ion from another portion of the same molecule to yield the cyclic ether.

(1) $$CH_2(Br){-}CH_2(OH) + OH^- \rightleftarrows H_2O + CH_2(Br){-}CH_2(O^{\ominus})$$

(2) $$CH_2(Br){-}CH_2(:O^{\ominus}) \longrightarrow \left[H{-}C(H)(\overset{\delta-}{Br})\cdots C(H){-}H \text{ with } \overset{\delta-}{O} \right] \longrightarrow \underset{\diagdown O \diagup}{CH_2{-}CH_2} + Br^-$$

Since halohydrins are nearly always prepared from alkenes by addition of halogen and water to the carbon–carbon double bond (Sec. 6.14), this method amounts to the conversion of an alkene into an epoxide.

Alternatively, the carbon–carbon double bond may be oxidized directly to the epoxide group by peroxybenzoic acid:

$$C_6H_5{-}C(=O){-}O{-}O{-}H$$

Peroxybenzoic acid

When allowed to stand in ether or chloroform solution, the peroxy acid and the unsaturated compound—which need not be a simple alkene—react to yield benzoic acid and the epoxide. For example:

$$\text{Cyclopentene} + C_6H_5C(=O){-}O{-}OH \longrightarrow \text{Cyclopentene oxide} + C_6H_5C(=O){-}OH$$

Cyclopentene, Peroxybenzoic acid, Cyclopentene oxide, Benzoic acid

$$C_6H_5{-}CH{=}CH{-}CH_2OH + C_6H_5C(=O){-}O{-}OH \longrightarrow C_6H_5{-}\underset{\diagdown O \diagup}{CH{-}CH}{-}CH_2OH + C_6H_5C(=O){-}OH$$

3-Phenyl-2-propen-1-ol
Cinnamyl alcohol

17.11 Reactions of epoxides

Epoxides owe their importance to their high reactivity, which is due to the ease of opening of the highly strained three-membered ring. The bond angles of the ring, which average 60°, are considerably less than the normal tetrahedral carbon angle of 109.5°, or the divalent oxygen angle of 110° for open-chain ethers (Sec. 17.2). Since the atoms cannot be located to permit maximum overlap of orbitals (Sec. 9.9), the bonds are weaker than in an ordinary ether, and the molecule is less stable.

Epoxides undergo acid-catalyzed reactions with extreme ease, and—unlike ordinary ethers—can even be cleaved by bases. Some of the important reactions are outlined below.

REACTIONS OF EPOXIDES

anti

1. Acid-catalyzed cleavage. Discussed in Sec. 17.12.

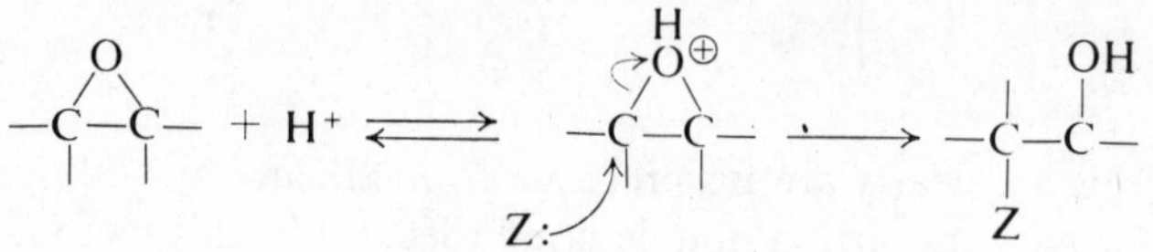

$$-\underset{|}{C}\overset{O}{\text{—}}\underset{|}{C}- + H^+ \rightleftharpoons -\underset{|}{C}\overset{\overset{H}{O^{\oplus}}}{\text{—}}\underset{|}{C}- \xrightarrow{Z:} -\underset{\underset{Z}{|}}{C}-\overset{\overset{OH}{|}}{C}-$$

Examples:

$$H_2O + CH_2\text{—}CH_2\ (\text{O}) \xrightarrow{H^+} \underset{OH}{CH_2}\text{—}\underset{OH}{CH_2}$$

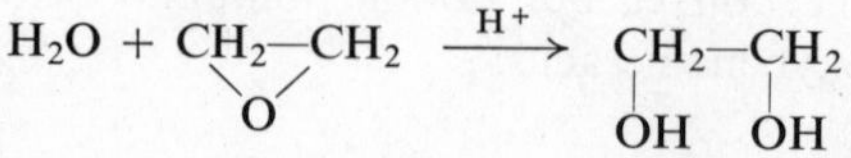

Ethylene glycol
(1,2-Ethanediol)

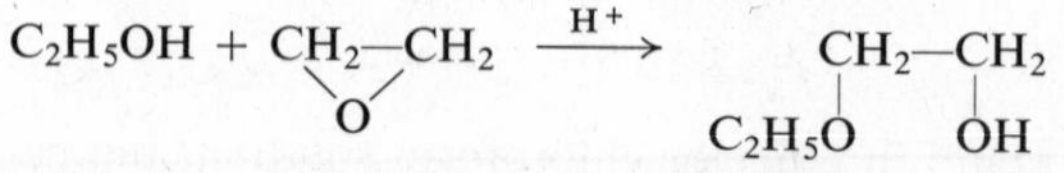

$$C_2H_5OH + CH_2\text{—}CH_2\ (\text{O}) \xrightarrow{H^+} \underset{C_2H_5O}{CH_2}\text{—}\underset{OH}{CH_2}$$

2-Ethoxyethanol

$$C_6H_5OH + CH_2\text{—}CH_2\ (\text{O}) \xrightarrow{H^+} C_6H_5OCH_2CH_2OH$$

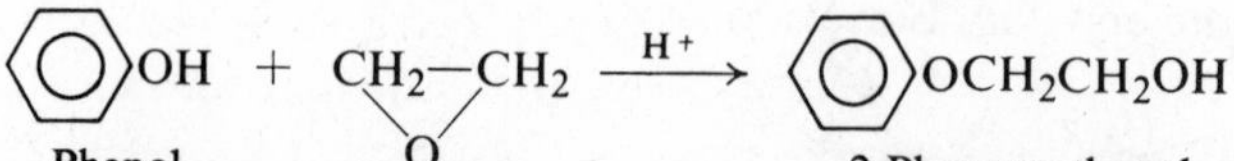

Phenol — 2-Phenoxyethanol

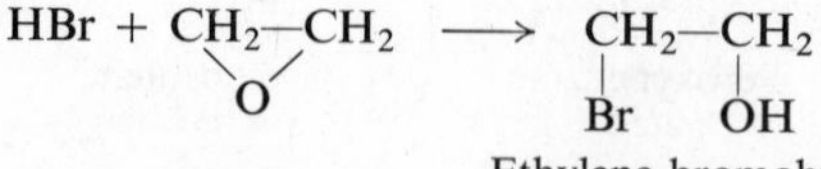

$$HBr + CH_2\text{—}CH_2\ (\text{O}) \longrightarrow \underset{Br}{CH_2}\text{—}\underset{OH}{CH_2}$$

Ethylene bromohydrin
(2-Bromoethanol)

2. Base-catalyzed cleavage. Discussed in Sec. 17.13.

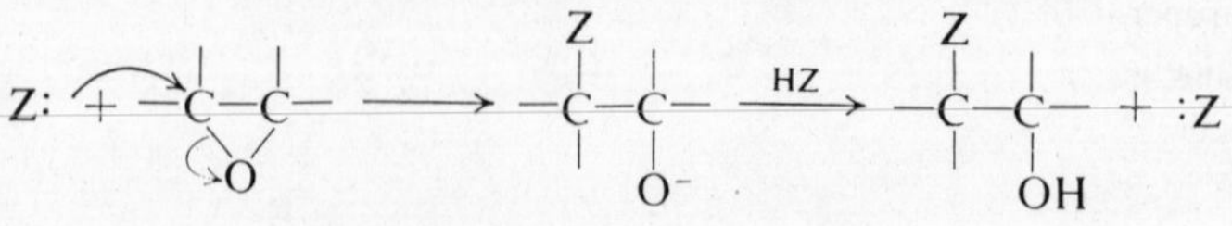

$$Z: + -\underset{|}{C}\overset{}{\text{—}}\underset{|}{C}-\ (\text{O}) \longrightarrow -\overset{\overset{Z}{|}}{\underset{|}{C}}-\underset{\underset{O^-}{|}}{C}- \xrightarrow{HZ} -\overset{\overset{Z}{|}}{\underset{|}{C}}-\underset{\underset{OH}{|}}{C}- + :Z$$

Examples:

$$C_2H_5O^-Na^+ + CH_2\text{—}CH_2\ (\text{O}) \longrightarrow C_2H_5OCH_2CH_2OH$$

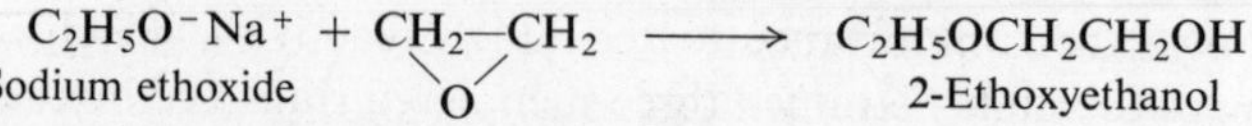

Sodium ethoxide — 2-Ethoxyethanol

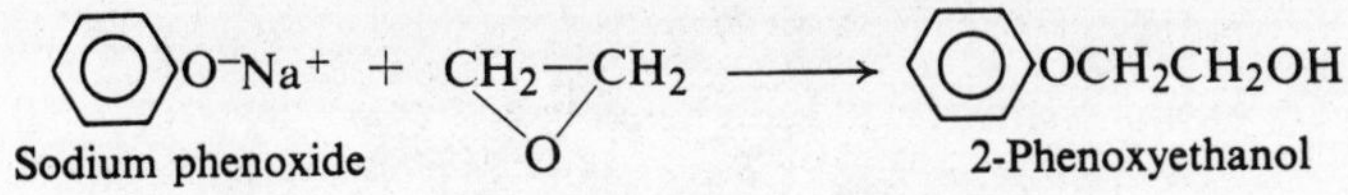

$$C_6H_5O^-Na^+ + CH_2{-}CH_2(O) \longrightarrow C_6H_5OCH_2CH_2OH$$

Sodium phenoxide — 2-Phenoxyethanol

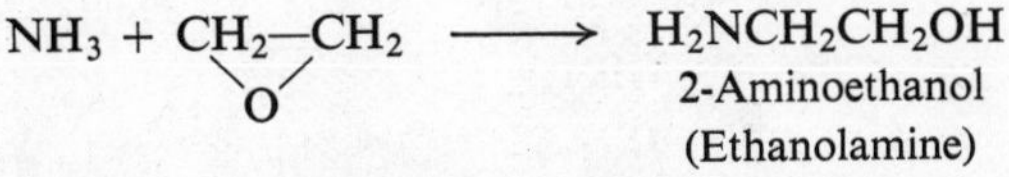

$$NH_3 + CH_2{-}CH_2(O) \longrightarrow H_2NCH_2CH_2OH$$

2-Aminoethanol (Ethanolamine)

3. Reaction with Grignard reagents. Discussed in Sec. 17.14.

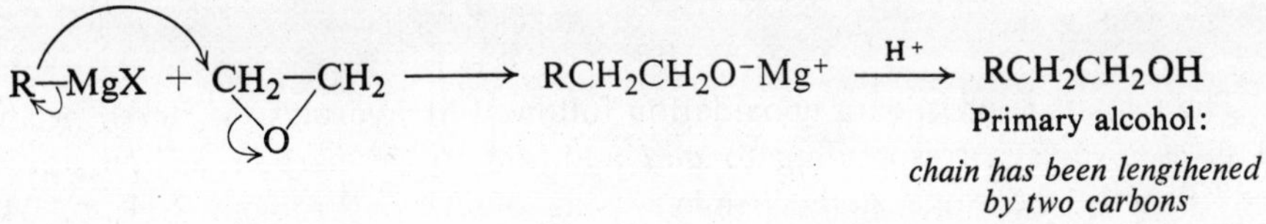

$$R{-}MgX + CH_2{-}CH_2(O) \longrightarrow RCH_2CH_2O^-Mg^+ \xrightarrow{H^+} RCH_2CH_2OH$$

Primary alcohol: *chain has been lengthened by two carbons*

Examples:

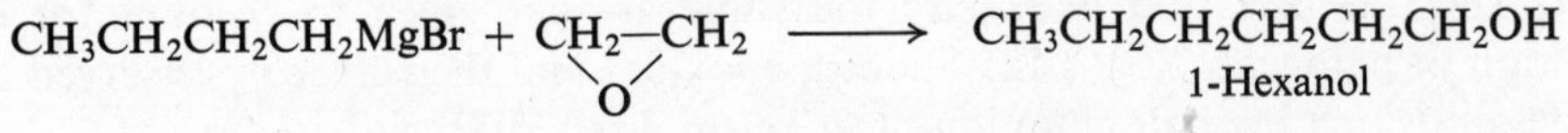

$$CH_3CH_2CH_2CH_2MgBr + CH_2{-}CH_2(O) \longrightarrow CH_3CH_2CH_2CH_2CH_2CH_2OH$$

1-Hexanol

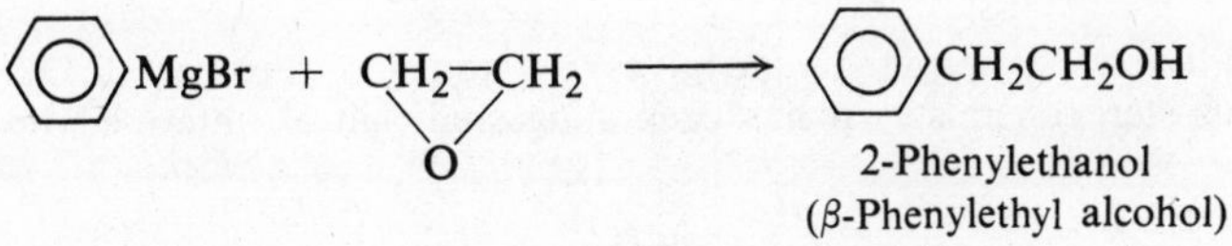

$$C_6H_5MgBr + CH_2{-}CH_2(O) \longrightarrow C_6H_5CH_2CH_2OH$$

2-Phenylethanol (β-Phenylethyl alcohol)

17.12 Acid-catalyzed cleavage of epoxides. *anti*-Hydroxylation

Like other ethers, an epoxide is converted by acid into the protonated epoxide, which can then undergo attack by any of a number of nucleophilic reagents.

An important feature of the reactions of epoxides is the formation of compounds that contain *two* functional groups. Thus, reaction with water yields a glycol; reaction with an alcohol yields a compound that is both ether and alcohol.

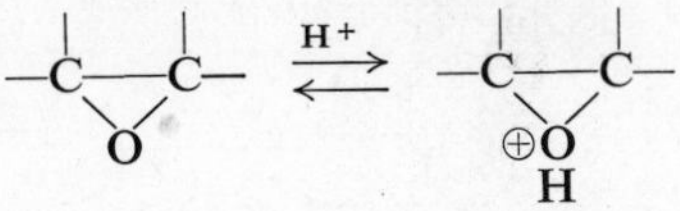

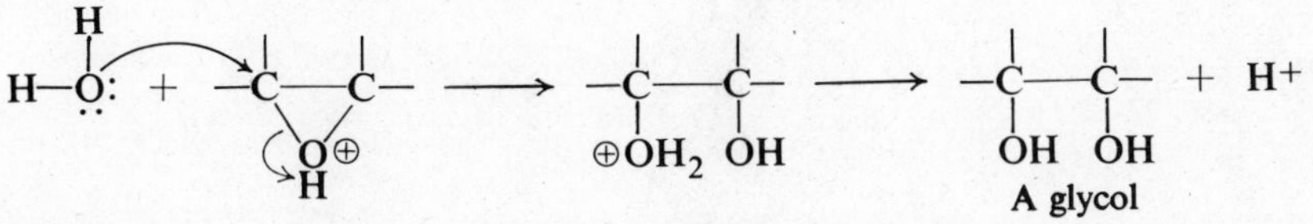

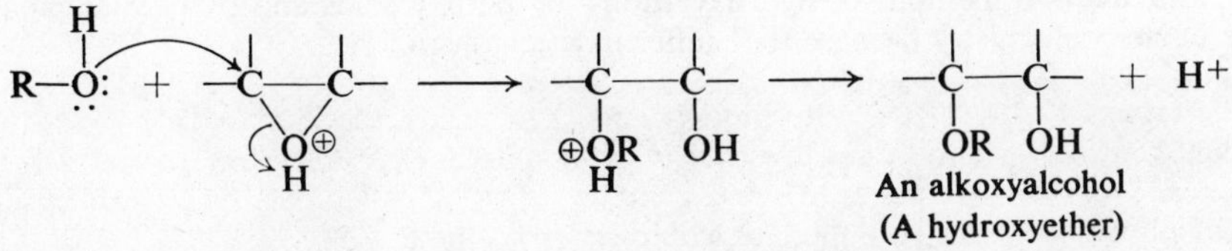

$$-\overset{|}{C}-\overset{|}{C}-(O) \overset{H^+}{\rightleftharpoons} -\overset{|}{C}-\overset{|}{C}-(\overset{\oplus}{O}H)$$

$$H_2\ddot{O}: + -\overset{|}{C}-\overset{|}{C}-(\overset{\oplus}{O}H) \longrightarrow -\overset{|}{C}(\oplus OH_2)-\overset{|}{C}(OH)- \longrightarrow -\overset{|}{C}(OH)-\overset{|}{C}(OH)- + H^+$$

A glycol

$$R\ddot{O}H: + -\overset{|}{C}-\overset{|}{C}-(\overset{\oplus}{O}H) \longrightarrow -\overset{|}{C}(\oplus ORH)-\overset{|}{C}(OH)- \longrightarrow -\overset{|}{C}(OR)-\overset{|}{C}(OH)- + H^+$$

An alkoxyalcohol (A hydroxyether)

Problem 17.16 The following compounds are commercially available for use as water-soluble solvents. How could each be made?

(a) $CH_3CH_2—O—CH_2CH_2—O—CH_2CH_2—OH$ Carbitol
(b) $C_6H_5—O—CH_2CH_2—O—CH_2CH_2—OH$ Phenyl carbitol
(c) $HO—CH_2CH_2—O—CH_2CH_2—OH$ Diethylene glycol
(d) $HO—CH_2CH_2—O—CH_2CH_2—O—CH_2CH_2—OH$ Triethylene glycol

Problem 17.17 Show in detail (including structures and transition states) the steps in the acid-catalyzed hydrolysis of ethylene oxide by an S_N1 mechanism; by an S_N2 mechanism.

The two-stage process of epoxidation followed by hydrolysis is stereospecific, and gives glycols corresponding to *anti* addition to the carbon–carbon double bond. Exactly the same stereochemistry was observed (Problem 7.11, p. 242) for hydroxylation of alkenes by peroxyformic acid—and for good reason: an epoxide is formed there, too, but is rapidly cleaved in the acidic medium, formic acid. The interpretation is exactly the same as that given to account for *anti* addition of halogens (Sec. 7.12); indeed, epoxides and their hydrolysis served as a model on which the halonium ion mechanism was patterned.

Hydroxylation with permanganate gives *syn*-addition (Problem 7.11, p. 242). To account for this stereochemistry it has been suggested that an intermediate like I is involved:

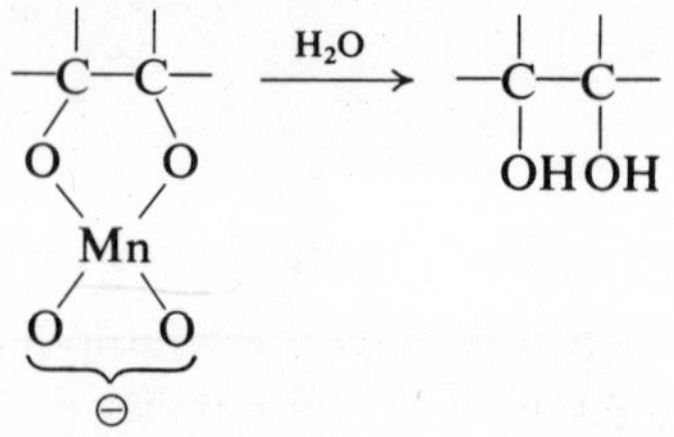

$$\underset{\text{I}}{\text{—C—C— (O—Mn(O)(O)—O, cyclic; charge } \ominus)} \xrightarrow{H_2O} \text{—C(OH)—C(OH)—}$$

Hydrolysis of such an intermediate would yield the *cis*-glycol. This mechanism is supported by the fact that osmium tetroxide, OsO_4, which also yields the *cis*-glycol, actually forms stable intermediates of structure II.

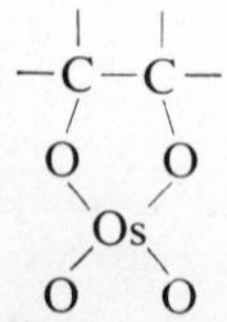

$$\underset{\text{II}}{\text{—C—C— (O—Os(O)(O)—O, cyclic)}}$$

Thus, the two methods of hydroxylation—by peroxy acids and by permanganate—differ in stereochemistry because they differ in mechanism.

Problem 17.18 Using both models and drawings of the kind in Sec. 7.12, show all steps in the formation and hydrolysis of the epoxide of: (a) cyclopentene; (b) *cis*-2-butene; (c) *trans*-2-butene; (d) *cis*-2-pentene; (e) *trans*-2-pentene. (f) Which (if any) of the above products, as obtained, would be optically active?

17.13 Base-catalyzed cleavage of epoxides

Unlike ordinary ethers, epoxides can be cleaved under alkaline conditions. Here it is the epoxide itself, not the protonated epoxide, that undergoes nucleophilic attack. The lower reactivity of the non-protonated epoxide is compensated for by the more basic, more strongly nucleophilic reagent: alkoxide, phenoxide, ammonia, etc.

Let us look, for example, at the reaction of ethylene oxide with phenol. Acid catalyzes reaction by converting the epoxide into the highly reactive protonated epoxide. Base catalyzes reaction by converting the phenol into the more strongly nucleophilic phenoxide ion.

$$C_6H_5OH + \underset{\overset{\oplus}{O}H}{CH_2{-}CH_2} \longrightarrow C_6H_5\overset{H}{\underset{\oplus}{O}}CH_2CH_2OH \xrightarrow{-H^+} C_6H_5OCH_2CH_2OH$$

Weakly nucleophilic reagent; Protonated epoxide, *Highly reactive*

$$C_6H_5O^- + \underset{O}{CH_2{-}CH_2} \longrightarrow C_6H_5OCH_2CH_2O^- \xrightarrow{H^+} C_6H_5OCH_2CH_2OH$$

Strongly nucleophilic reagent; Non-protonated epoxide

Problem 17.19 Write equations for the reaction of ethylene oxide with (a) methanol in the presence of a little H_2SO_4; (b) methanol in the presence of a little $CH_3O^-Na^+$; (c) aniline.

Problem 17.20 Using the reaction between phenol and ethylene oxide as an example, show why it is not feasible to bring about reaction between the protonated epoxide and the highly nucleophilic reagent phenoxide ion. (*Hint:* Consider what would happen if one started with a solution of sodium phenoxide and ethylene oxide and added acid to it.)

Problem 17.21 Poly(oxypropylene)glycols,

$$HO{-}\overset{CH_3}{\overset{|}{C}H}{-}CH_2{-}O\left[CH_2\overset{CH_3}{\overset{|}{C}H}{-}O\right]_n{-}CH_2\overset{CH_3}{\overset{|}{C}HOH}$$

which are used in the manufacture of polyurethane foam rubber, are formed by the action of base (e.g., hydroxide ion) on propylene oxide in the presence of propylene glycol as an initiator. Write all steps in a likely mechanism for their formation.

17.14 Reaction of ethylene oxide with Grignard reagents

Reaction of Grignard reagents with ethylene oxide is an important method of preparing primary alcohols since the product contains two carbons more than the alkyl or aryl group of the Grignard reagent. As in reaction with the carbonyl group (Sec. 15.12), we see the nucleophilic (basic) alkyl or aryl group of the

Grignard reagent attach itself to the relatively positive carbon and the electrophilic (acidic) magnesium attach itself to the relatively negative oxygen. Use of higher epoxides is complicated by rearrangements and formation of mixtures.

$$R{-}MgX + \underset{\diagdown O \diagup}{CH_2{-}CH_2} \longrightarrow RCH_2CH_2O^-MgX^+ \xrightarrow{H^+} RCH_2CH_2OH$$

17.15 Orientation of cleavage of epoxides

There are two carbon atoms in an epoxide ring and, in principle, either one can suffer nucleophilic attack. In a symmetrical epoxide like ethylene oxide, the two carbons are equivalent, and attack occurs randomly at both. But in an unsymmetrical epoxide, the carbons are *not* equivalent, and the product we obtain depends upon which one is preferentially attacked. Just what is the orientation of cleavage of epoxides, and how does one account for it?

The preferred point of attack, it turns out, depends chiefly on whether the reaction is acid-catalyzed or base-catalyzed. Consider, for example, two reactions of isobutylene oxide:

$$CH_3{-}\overset{\overset{CH_3}{|}}{C}{-}CH_2\ (\text{epoxide, }{-}O{-}) + H_2{}^{18}O \xrightarrow{H^+} CH_3{-}\overset{\overset{CH_3}{|}}{\underset{\underset{^{18}OH}{|}}{C}}{-}CH_2OH$$

$$CH_3{-}\overset{\overset{CH_3}{|}}{C}{-}CH_2\ (\text{epoxide, }{-}O{-}) + CH_3OH \xrightarrow{CH_3ONa} CH_3{-}\overset{\overset{CH_3}{|}}{\underset{\underset{OH}{|}}{C}}{-}CH_2OCH_3$$

Here, as in general, the nucleophile attacks the more substituted carbon in acid-catalyzed cleavage, and the less substituted carbon in base-catalyzed cleavage.

Our first thought is that two different mechanisms are involved here, S_N1 and S_N2. But the evidence indicates pretty clearly that both are of the S_N2 type: cleavage of the carbon–oxygen bond and attack by the nucleophile occur in a single step. (There is not only stereochemical evidence—complete inversion—but also evidence of several kinds that we cannot go into here.) How, then, are we to account for the difference in orientation—in particular, for S_N2 attack at the *more hindered* position in acid-catalyzed cleavage?

In an S_N2 reaction, we said earlier (Sec. 14.11), carbon loses electrons to the leaving group and gains electrons from the nucleophile, and as a result does not become appreciably positive or negative in the transition state; electronic factors are unimportant, and steric factors control reactivity. But in acid-catalyzed cleavage of an epoxide, the carbon–oxygen bond, already weak because of the angle strain of the three-membered ring, is further weakened by protonation: the leaving group is a very good one, a weakly basic alcohol hydroxyl. The nucleophile, on the other hand, is a poor one (water, alcohol, phenol). Although there are both bond-breaking and bond-making in the transition state, bond-breaking has proceeded further than bond-making; the leaving group has taken electrons away to a much greater extent than the nucleophile has brought them up, and the carbon has acquired a considerable positive charge.

Crowding, on the other hand, is relatively unimportant, because both leaving group and nucleophile are far away. Stability of the transition state is determined chiefly by electronic factors, not steric factors. We speak of such a reaction as having considerable S_N1 *character*. Attack occurs not at the less hindered carbon, but at *the carbon that can best accommodate the positive charge.*

Acid-catalyzed S_N2 cleavage

Z: + —C—C— (with O⊕—H bridge) ⟶ [Z···C(δ+)—C— ··· Oδ+—H] ⟶ Z—C—C—OH

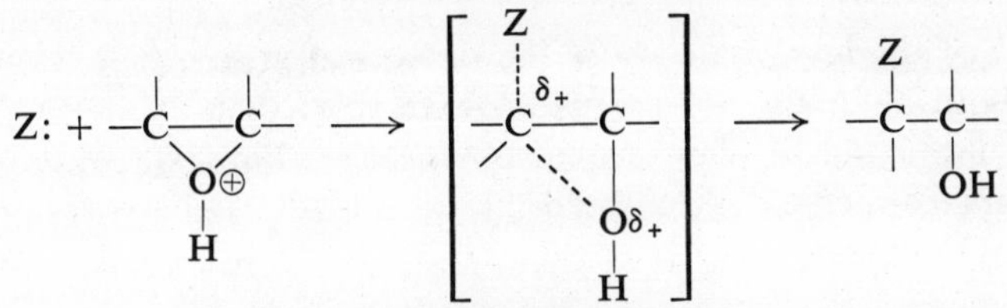

Bond-breaking exceeds bond-making: positive charge on carbon

In base-catalyzed cleavage, the leaving group is a poorer one—a strongly basic alkoxide oxygen—and the nucleophile is a good one (hydroxide, alkoxide, phenoxide). Bond-breaking and bond-making are more nearly balanced, and reactivity is controlled in the more usual way, by steric factors. *Attack occurs at the less hindered carbon.*

Base-catalyzed S_N2 cleavage

Z: + —C—C— (with O bridge) ⟶ [Z···C—C— ··· Oδ−] ⟶ Z—C—C—O−

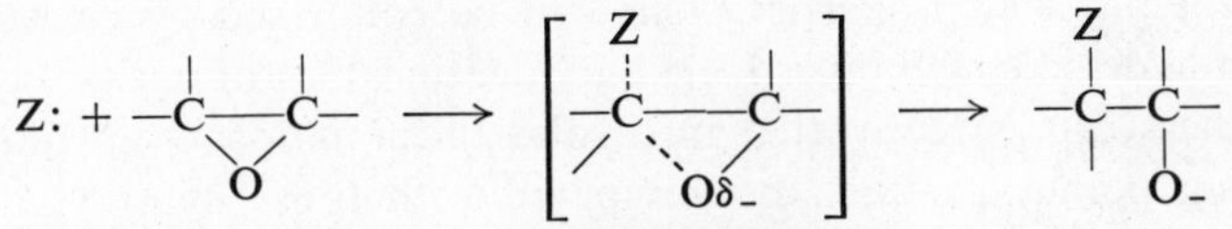

Bond-making balances bond-breaking: no particular charge on carbon

Problem 17.22 Predict the chief product of each of the following reactions:

(a) styrene oxide + dry HCl
(b) styrene oxide + CH_3OH + a little CH_3ONa
(c) propylene oxide + aniline
(d) trimethylethylene oxide + HCl

One further point. We have encountered the two-step addition of unsymmetrical reagents in which the first step is attack by positive halogen; formation of halohydrins (Sec. 6.14), and ionic addition of IN_3 and BrN_3 (Problem 7, p. 247). The orientation is what would be expected if a carbonium ion were the intermediate. Propylene chlorohydrin, for example, is $CH_3CHOHCH_2Cl$; IN_3 adds to terminal alkenes to yield $RCH(N_3)CH_2I$. Yet the exclusively *anti* stereochemistry

(Problems 5 and 7, p. 247) indicates that the intermediate is not an open cation but a *halonium ion*; cleavage of this ring must involve attack by the nucleophile (H_2O or N_3^-) at the more hindered carbon. This is not really surprising, in view of what we have just said about epoxides. The halonium ion ring is even less stable than that of a protonated epoxide; cleavage has much S_N1 character, and takes place at the carbon atom that can best accommodate the positive charge. (Consider, too, the orientation of solvomercuration, in which the intermediate is a cyclic *mercurinium ion*.)

17.16 Analysis of ethers

Because of the low reactivity of the functional group, the chemical behavior of ethers—both aliphatic and aromatic—resembles that of the hydrocarbons to which they are related. They are distinguished from hydrocarbons, however, by their solubility in cold concentrated sulfuric acid through formation of oxonium salts.

Problem 17.23 Because of their highly reactive benzene rings, aryl ethers may decolorize bromine in carbon tetrachloride. How could this behavior be distinguished from the usual unsaturation test? (*Hint:* See Sec. 6.30.)

Problem 17.24 Expand the table you made in Problem 16.10, p. 536, to include ethers.

Problem 17.25 Describe simple chemical tests (if any) that would distinguish between an aliphatic ether and (a) an alkane; (b) an alkene; (c) an alkyne; (d) an alkyl halide; (e) a primary or secondary alcohol; (f) a tertiary alcohol; (g) an alkyl aryl ether

Identification as a previously reported ether is accomplished through the usual comparison of physical properties. This can be confirmed by cleavage with hot concentrated hydriodic acid (Sec. 17.7) and identification of one or both products. Aromatic ethers can be converted into solid bromination or nitration products whose melting points can then be compared with those of previously reported derivatives.

Proof of structure of a new ether would involve cleavage by hydriodic acid and identification of the products formed. Cleavage is used quantitatively in the **Zeisel method** to show the number of alkoxyl groups in an alkyl aryl ether.

Problem 17.26 How many methoxyl groups per molecule of papaverine would be indicated by the following results of a Zeisel analysis?

Treatment of *papaverine* ($C_{20}H_{21}O_4N$, one of the opium alkaloids) with hot concentrated hydriodic acid yields CH_3I, indicating the presence of the methoxyl group $-OCH_3$. When 4.24 mg of papaverine is treated with hydriodic acid and the CH_3I thus formed is passed into alcoholic silver nitrate, 11.62 mg of silver iodide is obtained.

17.17 Spectroscopic analysis of ethers

Infrared. The infrared spectrum of an ether does not, of course, show the O—H band characteristic of alcohols; but the strong band due to C—O stretching is still present, in the 1060–1300 cm^{-1} range, and is the striking feature of the spectrum. (See Fig. 17.1).

C—O stretching, *strong, broad*

Alkyl ethers 1060–1150 cm^{-1}

Aryl and vinyl ethers 1200–1275 cm^{-1} (and, weaker, at 1020–1075 cm^{-1})

Carboxylic acids and esters show C—O stretching, but show carbonyl absorption as well. (For a comparison of certain oxygen compounds, see Table 20.3, p. 689.)

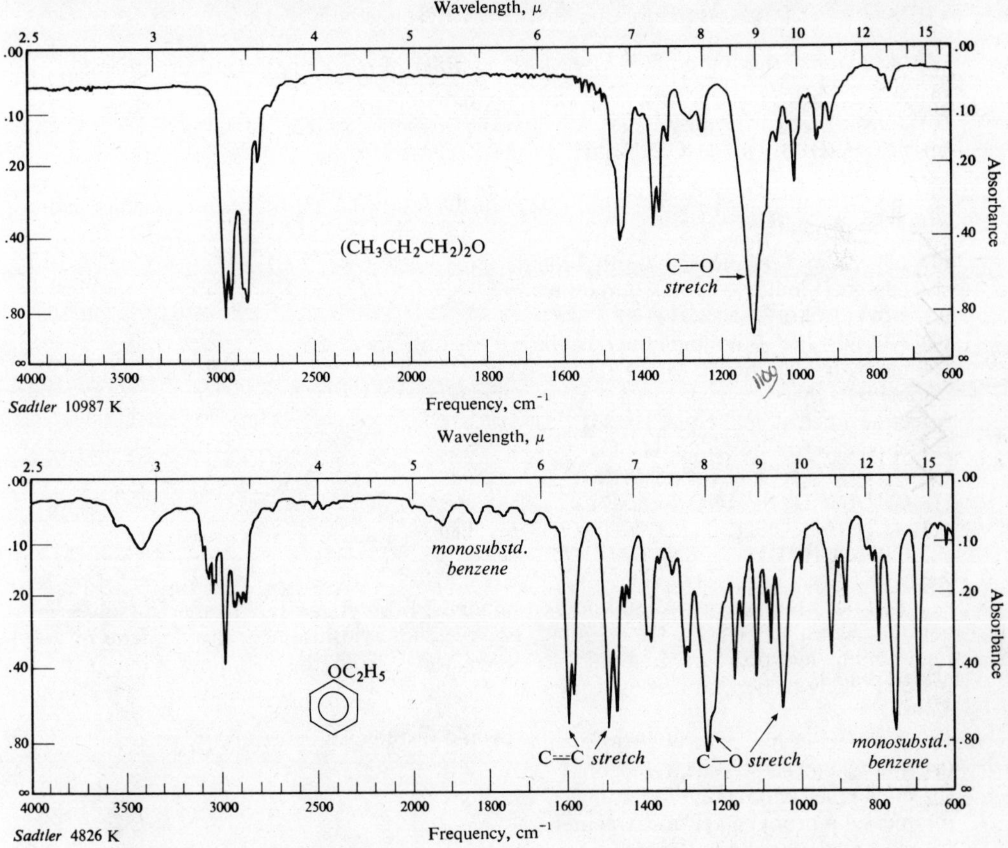

Figure 17.1. Infrared spectra of (*a*) *n*-propyl ether and (*b*) phenetole.

PROBLEMS

1. Write structural formulas for:

(a) methyl ether
(b) isopropyl ether
(c) methyl *n*-butyl ether
(d) isobutyl *tert*-butyl ether
(e) 3-methoxyhexane
(f) vinyl ether
(g) allyl ether
(h) β-chloroethyl ether
(i) anisole
(j) phenetole
(k) phenyl ether
(l) cyclohexene oxide
(m) *p*-nitrobenzyl *n*-propyl ether
(n) 1,2-epoxypentane

2. Name the following structures:

(a) $(CH_3)_2CHCH_2—O—CH_2CH(CH_3)_2$
(b) $CH_3—O—CH(CH_3)_2$
(c) $(CH_3)_3C—O—CH_2CH_3$
(d) $CH_3CH_2CH_2CH(OCH_3)CH_2CH_2CH_3$
(e) *p*-$BrC_6H_4OC_2H_5$
(f) *o*-$O_2NC_6H_4CH_2OC_6H_5$
(g) 2,4-$Br_2C_6H_3OCH_3$

3. Outline a possible laboratory synthesis of each of the following compounds from alcohols and phenols:

(a) methyl *tert*-butyl ether
(b) phenetole ($C_6H_5OC_2H_5$)
(c) *n*-butyl cyclohexyl ether
(d) *p*-tolyl benzyl ether
(e) isopropyl isobutyl ether
(f) isopropyl *tert*-butyl ether
(g) resorcinol dimethyl ether (1,3-dimethoxybenzene)

4. Arrange the compounds of each set in order of reactivity toward bromine:

(a) anisole, benzene, chlorobenzene, nitrobenzene, phenol
(b) anisole, *m*-hydroxyanisole, *o*-methylanisole, *m*-methylanisole
(c) *p*-$C_6H_4(OH)_2$, *p*-$CH_3OC_6H_4OH$, *p*-$C_6H_4(OCH_3)_2$

5. Write a balanced equation for each of the following. (If no reaction occurs, indicate "no reaction.")

(a) potassium *tert*-butoxide + ethyl iodide
(b) *tert*-butyl iodide + potassium ethoxide
(c) ethyl alcohol + H_2SO_4 (140°)
(d) *n*-butyl ether + boiling aqueous NaOH
(e) methyl ethyl ether + excess HI (hot)
(f) methyl ether + Na
(g) ethyl ether + cold conc. H_2SO_4
(h) ethyl ether + hot conc. H_2SO_4
(i) $C_6H_5OC_2H_5$ + hot conc. HBr
(j) $C_6H_5OC_2H_5 + HNO_3, H_2SO_4$
(k) *p*-$CH_3C_6H_4OCH_3 + KMnO_4 + KOH$ + heat
(l) $C_6H_5OCH_2C_6H_5 + Br_2$, Fe

6. Like other oxygen-containing compounds, *n*-butyl *tert*-butyl ether dissolves in cold concentrated H_2SO_4. On standing, however, an acid-insoluble layer, made up of high-boiling hydrocarbon material, slowly separates from the solution. What is this material likely to be, and how is it formed?

7. Describe simple chemical tests that would distinguish between:

(a) *n*-butyl ether and *n*-pentyl alcohol
(b) ethyl ether and methyl iodide
(c) methyl *n*-propyl ether and 1-pentene
(d) isopropyl ether and allyl ether
(e) anisole and toluene
(f) vinyl ether and ethyl ether
(g) *n*-butyl *tert*-butyl ether and *n*-octane

Tell exactly what you would *do* and *see*.

8. An unknown compound is believed to be one of the following. Describe how you would go about finding out which of the possibilities the unknown actually is. Where possible, use simple chemical tests; where necessary, use more elaborate chemical methods like quantitative hydrogenation, cleavage, etc. Make use of any needed tables of physical constants.

(a) *n*-propyl ether (b.p. 91°) and 2-methylhexane (b.p. 91°)
(b) benzyl ethyl ether (b.p. 188°) and allyl phenyl ether (b.p. 192°)
(c) methyl *p*-tolyl ether (b.p. 176°) and methyl *m*-tolyl ether (b.p. 177°)

(d) ethyl *n*-propyl ether (b.p. 64°), 1-hexene (b.p. 64°), and methanol (b.p. 65°)
(e) anisole (b.p. 154°), bromobenzene (b.p. 156°), *o*-chlorotoluene (b.p. 159°), *n*-propylbenzene (b.p. 159°), and cyclohexanol (b.p. 162°)
(f) ethyl ether (b.p. 35°), *n*-pentane (b.p. 36°), and isoprene (b.p. 34°)
(g) methyl *o*-tolyl ether (b.p. 171°), phenetole (b.p. 172°), and isopentyl ether (b.p. 173°)

9. Three compounds, A, B, and C, have the formula C_8H_9OBr. They are insoluble in water, but are soluble in cold concentrated H_2SO_4. B is the only one of the three that gives a precipitate when treated with $AgNO_3$. The three compounds are unaffected by dilute $KMnO_4$ and Br_2/CCl_4. Further investigation of their chemical properties leads to the following results:

oxidation by hot alkaline $KMnO_4$:

A $\longrightarrow$ D ($C_8H_7O_3Br$), an acid
B $\longrightarrow$ E ($C_8H_8O_3$), an acid
C $\longrightarrow$ unaffected

treatment with hot conc. HBr:

A $\longrightarrow$ F (C_7H_7OBr)
B $\longrightarrow$ G (C_7H_7OBr)
C $\longrightarrow$ H (C_6H_5OBr), identified as *o*-bromophenol
E $\longrightarrow$ I ($C_7H_6O_3$), identified as salicylic acid, *o*-HOC_6H_4COOH

p-hydroxybenzoic acid $\xrightarrow{(CH_3)_2SO_4,\ NaOH}$ $\xrightarrow{HCl}$ J ($C_8H_8O_3$)
J + Br_2 + Fe $\longrightarrow$ D

What are the probable structures of A, B, and C? Of compounds D through J? Write equations for all reactions involved.

10. Before doing the chemical work described in the preceding problem, we could quickly have learned a good deal about the structure of A, B, and C from examination of their nmr spectra. What would you expect to see in the nmr spectrum of each compound? Give approximate chemical shift values, splittings, and relative peak areas.

11. Give the structures and names of the products you would expect from the reaction of ethylene oxide with:

(a) H_2O, H^+
(b) H_2O, OH^-
(c) C_2H_5OH, H^+
(d) product of (c), H^+
(e) $HOCH_2CH_2OH$, H^+
(f) product of (e), H^+
(g) anhydrous HBr
(h) HCN
(i) HCOOH
(j) C_6H_5MgBr
(k) NH_3
(l) diethylamine ($C_2H_5NHC_2H_5$)
(m) phenol, H^+
(n) phenol, OH^-
(o) $HC{\equiv}C^-Na^+$

12. Propylene oxide can be converted into propylene glycol by the action of either dilute acid or dilute base. When optically active propylene oxide is used, the glycol obtained from acidic hydrolysis has a rotation opposite to that obtained from alkaline hydrolysis. What is the most likely interpretation of these facts?

13. In Sec. 17.10 a mechanism is proposed for the conversion of ethylene bromohydrin into ethylene oxide in the presence of base. (a) To what general class does this reaction belong? (b) Using models, show the likely steric course of this reaction. (c) Can you suggest a reason why sodium hydroxide readily converts *trans*-2-chlorocyclohexanol into cyclohexene oxide, but converts the *cis*-isomer into entirely different products? (d) Account for the fact that addition of chlorine and water to oleic acid (*cis*-9-octadecenoic acid) followed by treatment with base gives the same epoxide (same stereoisomer) as does treatment of oleic acid with a peroxy acid.

14. (a) Draw formulas for all the stereoisomers of I.

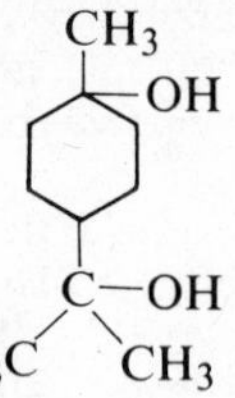

CH_3 —OH (ring) —C—OH, H_3C CH_3

I

(b) Indicate which isomers, when separated from all others, will be optically active, and which will be optically inactive. (c) One of these stereoisomers is very readily converted into an ether, $C_{10}H_{18}O$. Which isomer is this, and what is the structure of the ether?

15. Give the structures (including configurations where pertinent) of compounds K through Y:

(a) $CH_2{=}CH_2 + Cl_2/H_2O \longrightarrow$ K (C_2H_5OCl)
K + H_2SO_4 + heat $\longrightarrow$ L ($C_4H_8OCl_2$)
L + alc. KOH $\longrightarrow$ M (C_4H_6O)

(b) $ClCH_2CH{-}CH_2$ (epoxide, O bridging) + $CH_3OH + H_2SO_4 \longrightarrow$ N ($C_4H_9O_2Cl$)
N + NaOCl $\longrightarrow$ $CHCl_3$ + O ($C_3H_6O_3$)
N + NaOH(aq) $\longrightarrow$ P ($C_4H_8O_2$)

(c) $ClCH_2CH_2CH_2OH$ + KOH $\longrightarrow$ Q (C_3H_6O)

(d) benzene + ethylene oxide + BF_3 $\longrightarrow$ R ($C_8H_{10}O$)

(e) $CH_2{=}CHCH_2CH_2CH_2OH + Hg(OAc)_2 + H_2O$, then $NaBH_4$ $\longrightarrow$ S ($C_5H_{10}O$)

(f) methyl vinyl ether + dil. H_2SO_4 $\longrightarrow$ T (C_2H_4O)

(g) cyclohexene oxide + anhydrous HCl $\longrightarrow$ U ($C_6H_{11}OCl$)

(h) 1-methylcyclohexene + HCO_2OH $\longrightarrow$ V ($C_7H_{14}O_2$)

(i) racemic 3,4-epoxy-1-butene + cold alkaline $KMnO_4$, then dilute acid $\longrightarrow$ W ($C_4H_{10}O_4$)

(j) *cis*-2-butene + Cl_2/H_2O, then OH^-, then dilute acid $\longrightarrow$ X ($C_4H_{10}O_2$)

(k) *trans*-2-butene treated as in (j) $\longrightarrow$ Y ($C_4H_{10}O_2$)

16. Give a structure or structures for the compound whose infrared spectrum is shown in Fig. 17.2 (p. 575). If you find more than one structure consistent with the spectrum, could you decide among the possibilities on the basis of the nmr spectrum? Tell what you would expect to see in each case.

17. Give a structure or structures for the compound Z, whose infrared and nmr spectra are shown in Fig. 17.3 (p. 575).

18. Give a structure or structures consistent with each nmr spectrum shown in Fig. 17.4 (p. 576).

19. Give the structures of compounds AA, BB, and CC on the basis of their infrared spectra (Fig. 17.5, p. 577) and their nmr spectra (Fig. 17.6, p. 578).

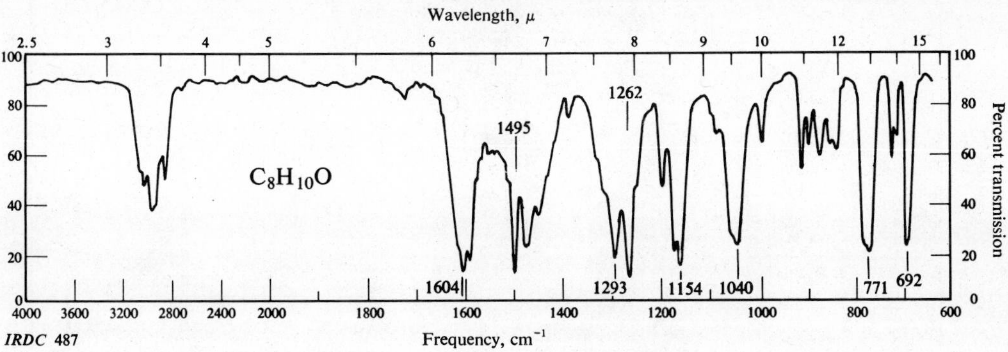

Figure 17.2. Infrared spectrum for Problem 16, p. 574.

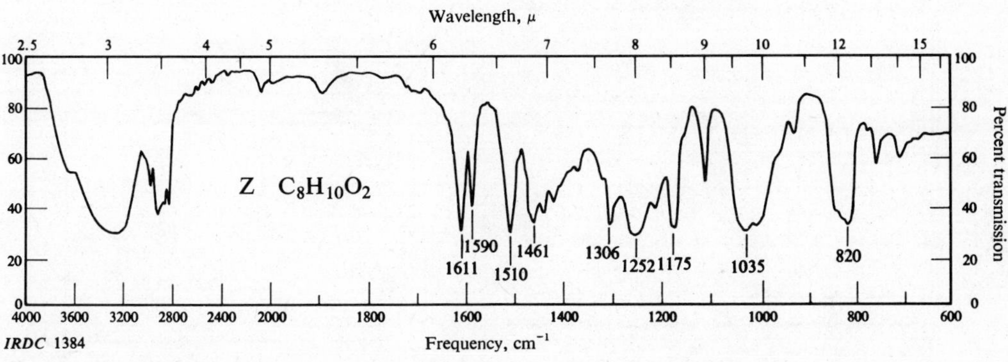

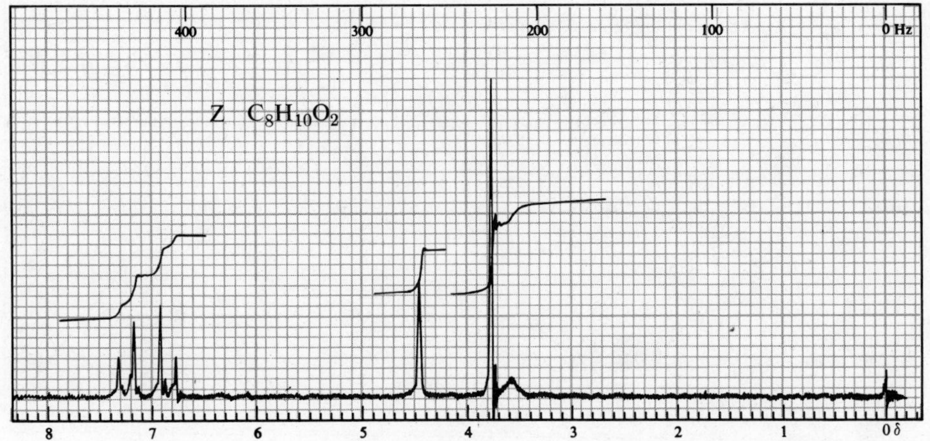

Figure 17.3. Infrared and nmr spectra for Problem 17, p. 574.

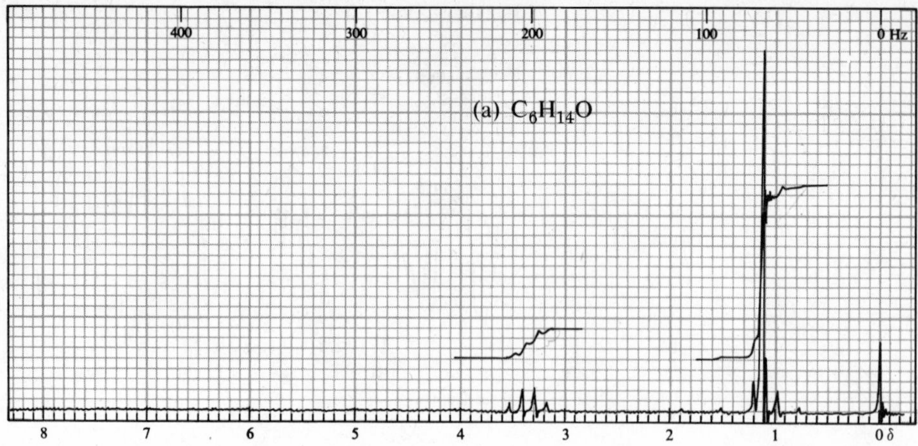

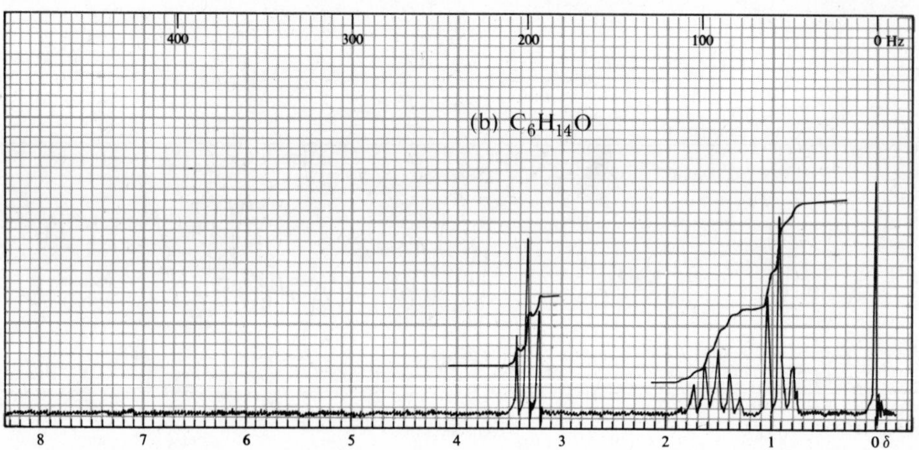

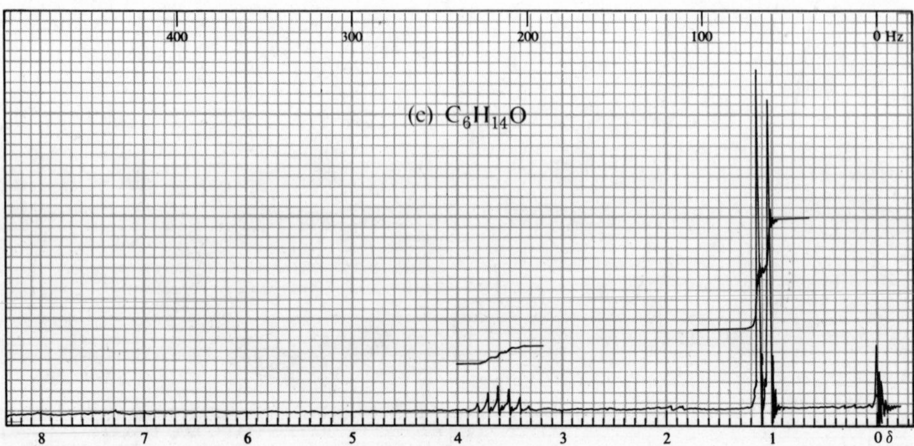

Figure 17.4. Nmr spectra for Problem 18, p. 574.

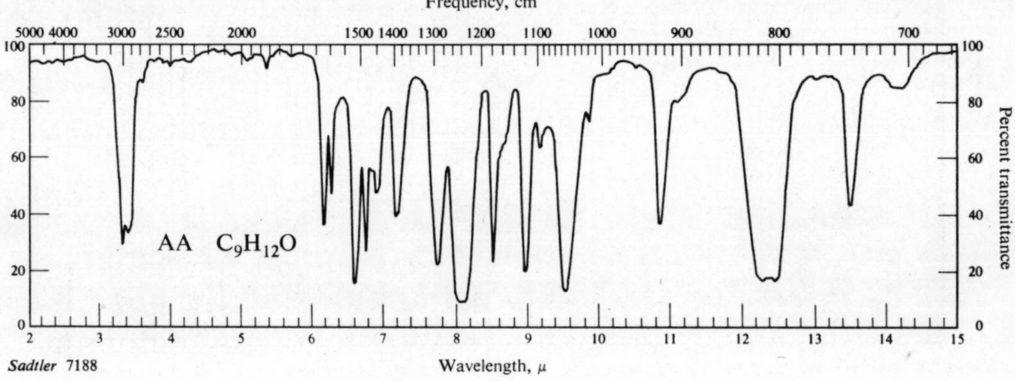

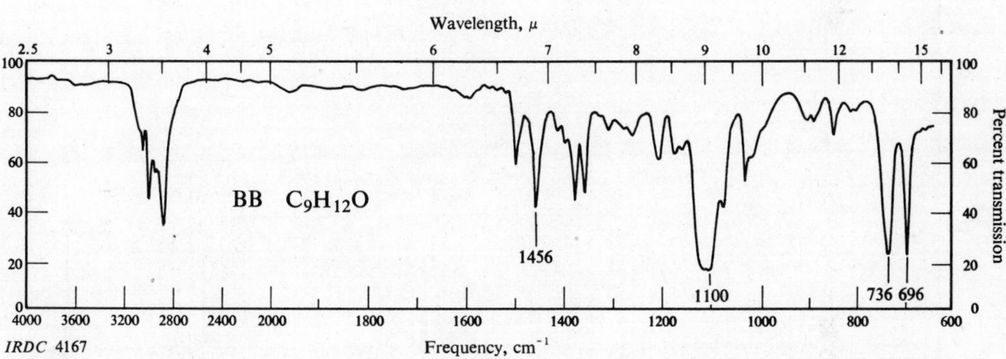

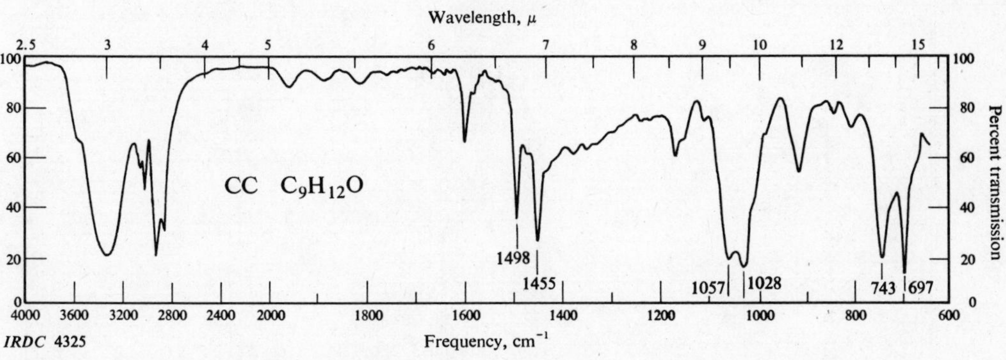

Figure 17.5. Infrared spectra for Problem 19, p. 574.

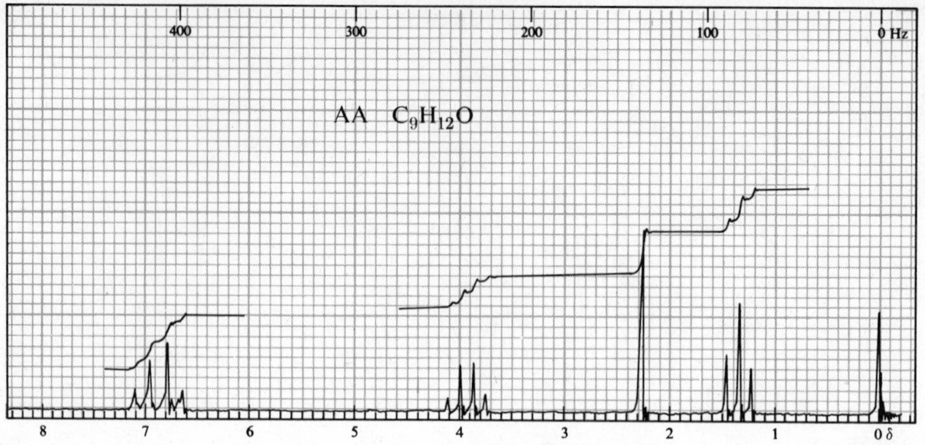

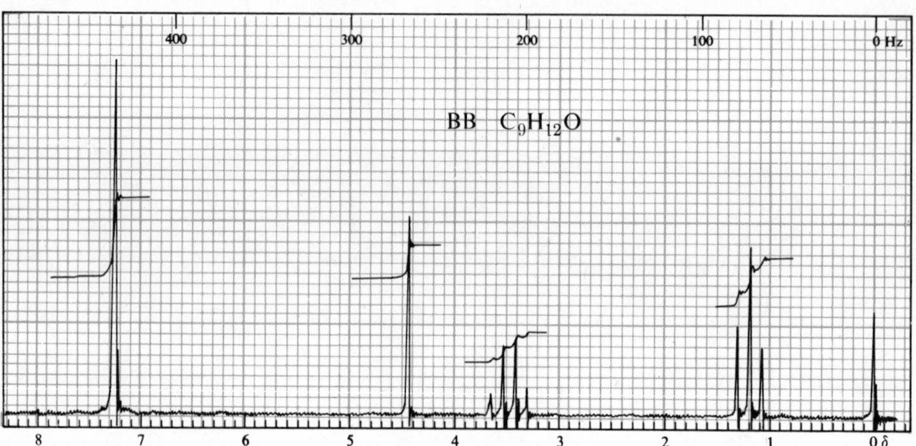

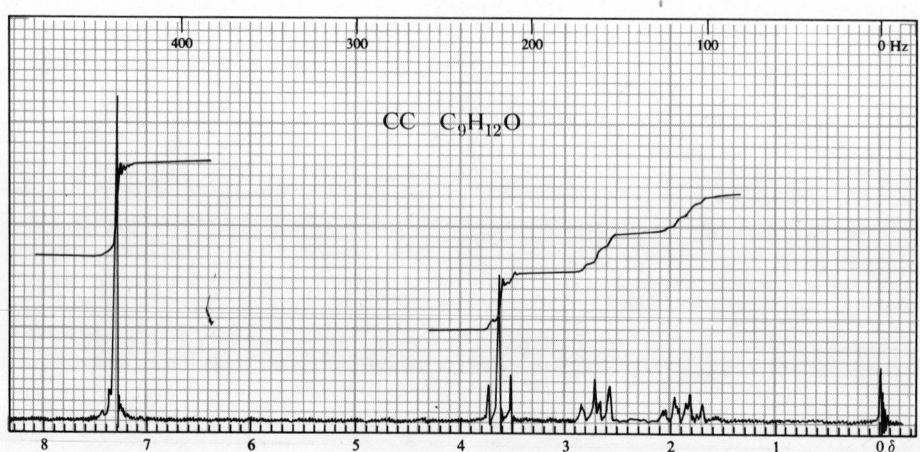

Figure 17.6. Nmr spectra for Problem 19, p. 574.

Chapter 18

Carboxylic Acids

18.1 Structure

Of the organic compounds that show appreciable acidity, by far the most important are the carboxylic acids. These compounds contain the **carboxyl group**

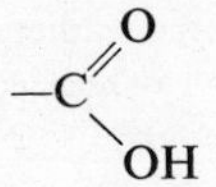

$$-C\begin{matrix}\nearrow O \\ \searrow OH\end{matrix}$$

attached to either an alkyl group (RCOOH) or an aryl group (ArCOOH). For example:

$HCOOH$	CH_3COOH	$CH_3(CH_2)_{10}COOH$	$CH_3(CH_2)_7CH{=}CH(CH_2)_7COOH$
Formic acid	Acetic acid	Lauric acid	Oleic acid
Methanoic acid	Ethanoic acid	Dodecanoic acid	*cis*-9-Octadecenoic acid

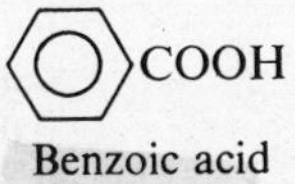

C_6H_5COOH	$p\text{-}O_2NC_6H_4COOH$	$C_6H_5CH_2COOH$
Benzoic acid	*p*-Nitrobenzoic acid	Phenylacetic acid

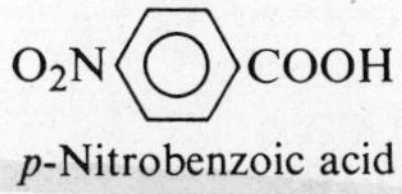

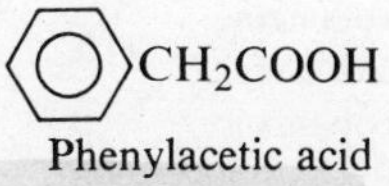

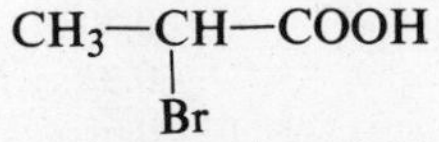

$CH_3-CH(Br)-COOH$	$C_6H_{11}COOH$	$CH_2{=}CHCOOH$
α-Bromopropionic acid	Cyclohexanecarboxylic acid	Acrylic acid
2-Bromopropanoic acid		Propenoic acid

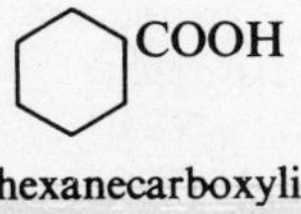

Whether the group is aliphatic or aromatic, saturated or unsaturated, substituted or unsubstituted, the properties of the carboxyl group are essentially the same.

18.2 Nomenclature

The aliphatic carboxylic acids have been known for a long time, and as a result have common names that refer to their sources rather than to their chemical structures. The **common names** of the more important acids are shown in Table 18.1. *Formic acid*, for example, adds the sting to the bite of an ant (Latin: *formica*, ant); *butyric acid* gives rancid butter its typical smell (Latin: *butyrum*, butter);

Table 18.1 CARBOXYLIC ACIDS

Name	Formula	M.p., °C	B.p., °C	Solub., g/100 g H_2O
Formic	HCOOH	8	100.5	∞
Acetic	CH_3COOH	16.6	118	∞
Propionic	CH_3CH_2COOH	−22	141	∞
Butyric	$CH_3(CH_2)_2COOH$	− 6	164	∞
Valeric	$CH_3(CH_2)_3COOH$	−34	187	3.7
Caproic	$CH_3(CH_2)_4COOH$	− 3	205	1.0
Caprylic	$CH_3(CH_2)_6COOH$	16	239	0.7
Capric	$CH_3(CH_2)_8COOH$	31	269	0.2
Lauric	$CH_3(CH_2)_{10}COOH$	44	225_{100}	i.
Myristic	$CH_3(CH_2)_{12}COOH$	54	251_{100}	i.
Palmitic	$CH_3(CH_2)_{14}COOH$	63	269_{100}	i.
Stearic	$CH_3(CH_2)_{16}COOH$	70	287_{100}	i.
Oleic	*cis*-9-Octadecenoic	16	223_{10}	i.
Linoleic	*cis,cis*-9,12-Octadecadienoic	− 5	230_{16}	i.
Linolenic	*cis,cis,cis*-9,12,15-Octadecatrienoic	−11	232_{17}	i.
Cyclohexanecarboxylic	*cyclo*-$C_6H_{11}COOH$	31	233	0.20
Phenylacetic	$C_6H_5CH_2COOH$	77	266	1.66
Benzoic	C_6H_5COOH	122	250	0.34
o-Toluic	*o*-$CH_3C_6H_4COOH$	106	259	0.12
m-Toluic	*m*-$CH_3C_6H_4COOH$	112	263	0.10
p-Toluic	*p*-$CH_3C_6H_4COOH$	180	275	0.03
o-Chlorobenzoic	*o*-ClC_6H_4COOH	141		0.22
m-Chlorobenzoic	*m*-ClC_6H_4COOH	154		0.04
p-Chlorobenzoic	*p*-ClC_6H_4COOH	242		0.009
o-Bromobenzoic	*o*-BrC_6H_4COOH	148		0.18
m-Bromobenzoic	*m*-BrC_6H_4COOH	156		0.04
p-Bromobenzoic	*p*-BrC_6H_4COOH	254		0.006
o-Nitrobenzoic	*o*-$O_2NC_6H_4COOH$	147		0.75
m-Nitrobenzoic	*m*-$O_2NC_6H_4COOH$	141		0.34
p-Nitrobenzoic	*p*-$O_2NC_6H_4COOH$	242		0.03
Phthalic	*o*-$C_6H_4(COOH)_2$	231		0.70
Isophthalic	*m*-$C_6H_4(COOH)_2$	348		0.01
Terephthalic	*p*-$C_6H_4(COOH)_2$	300 *subl.*		0.002
Salicylic	*o*-HOC_6H_4COOH	159		0.22
p-Hydroxybenzoic	*p*-HOC_6H_4COOH	213		0.65
Anthranilic	*o*-$H_2NC_6H_4COOH$	146		0.52
m-Aminobenzoic	*m*-$H_2NC_6H_4COOH$	179		0.77
p-Aminobenzoic	*p*-$H_2NC_6H_4COOH$	187		0.3
o-Methoxybenzoic	*o*-$CH_3OC_6H_4COOH$	101		0.5
m-Methoxybenzoic	*m*-$CH_3OC_6H_4COOH$	110		
p-Methoxybenzoic (Anisic)	*p*-$CH_3OC_6H_4COOH$	184		0.04

and *caproic*, *caprylic*, and *capric acids* are all found in goat fat (Latin: *caper*, goat).

Branched-chain acids and substituted acids are named as derivatives of the straight-chain acids. To indicate the position of attachment, the Greek letters, α-, β-, γ-, δ-, etc., are used; the α-carbon is the one bearing the carboxyl group.

$$\overset{\delta}{C}-\overset{\gamma}{C}-\overset{\beta}{C}-\overset{\alpha}{C}-COOH$$ *Used in common names*

For example:

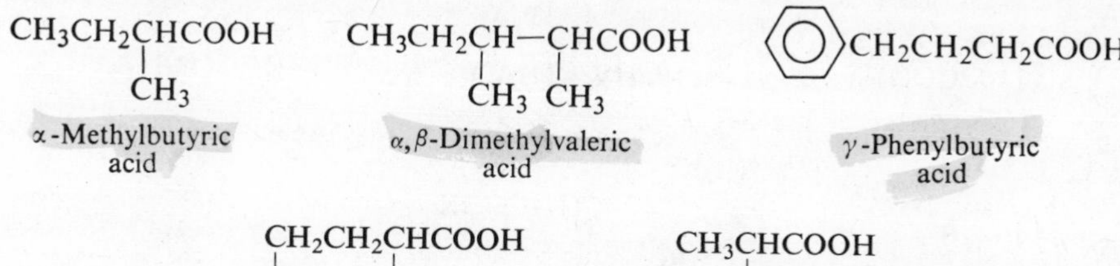

α-Methylbutyric acid   α,β-Dimethylvaleric acid   γ-Phenylbutyric acid

$CH_2CH_2CHCOOH$ (with Cl on the first carbon and CH_3 on the third)
γ-Chloro-α-methylbutyric acid

$CH_3CHCOOH$ (with OH on the second carbon)
α-Hydroxypropionic acid
Lactic acid

Generally the parent acid is taken as the one of longest carbon chain, although some compounds are named as derivatives of acetic acid.

Aromatic acids, ArCOOH, are usually named as derivatives of the parent acid, **benzoic acid**, C_6H_5COOH. The methylbenzoic acids are given the special name of *toluic acids*.

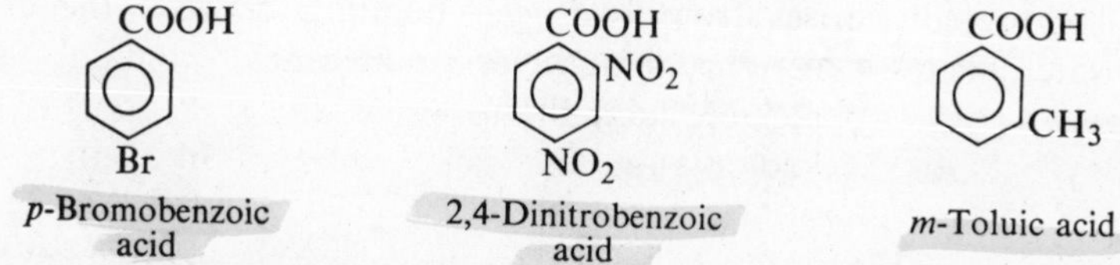

p-Bromobenzoic acid   2,4-Dinitrobenzoic acid   *m*-Toluic acid

The **IUPAC names** follow the usual pattern. The longest chain carrying the carboxyl group is considered the parent structure, and is named by replacing the –*e* of the corresponding alkane with –**oic acid**. For example:

$CH_3CH_2CH_2CH_2COOH$
Pentanoic acid

$CH_3CH_2CHCOOH$ (with CH_3 on the third carbon)
2-Methylbutanoic acid

$C_6H_5CH_2CH_2COOH$
3-Phenylpropanoic acid

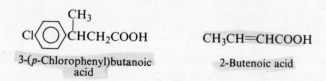

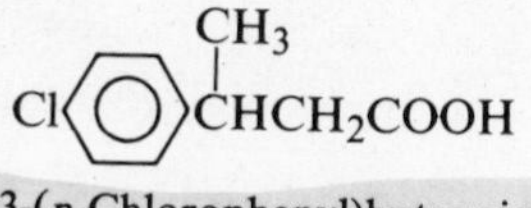

3-(*p*-Chlorophenyl)butanoic acid

$CH_3CH{=}CHCOOH$
2-Butenoic acid

The position of a substituent is indicated as usual by a number. We should notice

$$\overset{5}{C}-\overset{4}{C}-\overset{3}{C}-\overset{2}{C}-\overset{1}{C}OOH \qquad \textit{Used in IUPAC names}$$

that the carboxyl carbon is always considered as C–1, and hence C–2 corresponds to α of the common names, C–3 to β, and so on. (*Caution:* Do not mix Greek letters with IUPAC names, or Arabic numerals with common names.)

The name of a **salt** of a carboxylic acid consists of the name of the cation (*sodium, potassium, ammonium,* etc.) followed by the name of the acid with the ending *–ic acid* changed to **–ate**. For example:

C_6H_5COONa	$(CH_3COO)_2Ca$	$HCOONH_4$
Sodium benzoate	Calcium acetate	Ammonium formate

$$\underset{\large Br}{CH_2}-\underset{\large Br}{CH}-COOK$$

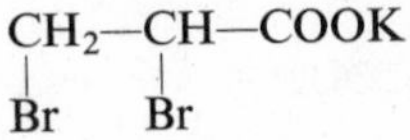

Potassium α,β-dibromopropionate
(Potassium 2,3-dibromopropanoate)

18.3 Physical properties

As we would expect from their structure, carboxylic acid molecules are polar, and like alcohol molecules can form hydrogen bonds with each other and with other kinds of molecules. The aliphatic acids therefore show very much the same solubility behavior as the alcohols: the first four are miscible with water, the five-carbon acid is partly soluble, and the higher acids are virtually insoluble. Water solubility undoubtedly arises from hydrogen bonding between the carboxylic acid and water. The simplest aromatic acid, benzoic acid, contains too many carbon atoms to show appreciable solubility in water.

Carboxylic acids are soluble in less polar solvents like ether, alcohol, benzene, etc.

We can see from Table 18.1 that as a class the carboxylic acids are even higher boiling than alcohols. For example, propionic acid (b.p. 141°) boils more than twenty degrees higher than the alcohol of comparable molecular weight, *n*-butyl alcohol (b.p. 118°). These very high boiling points are due to the fact that a pair of carboxylic acid molecules are held together not by one but by two hydrogen bonds:

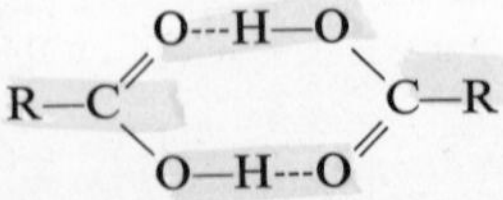

$$R-C\begin{matrix}\nearrow\!\!\!\!= O\text{---}H-O\searrow \\ \searrow O-H\text{---}O\,\!\!=\!\!\!\nearrow\end{matrix}C-R$$

Problem 18.1 At 110° and 454 mm pressure, 0.11 g acetic acid vapor occupies 63.7 cc; at 156° and 458 mm, 0.081 g occupies 66.4 cc. Calculate the molecular weight of acetic acid in the vapor phase at each temperature. How do you interpret these results?

The odors of the lower aliphatic acids progress from the sharp, irritating odors of formic and acetic acids to the distinctly unpleasant odors of butyric, valeric, and caproic acids; the higher acids have little odor because of their low volatility.

18.4 Salts of carboxylic acids

Although much weaker than the strong mineral acids (sulfuric, hydrochloric, nitric), the carboxylic acids are tremendously more acidic than the very weak organic acids (alcohols, acetylene) we have so far studied; they are much stronger acids than water. Aqueous hydroxides therefore readily convert carboxylic acids into their salts; aqueous mineral acids readily convert the salts back into the carboxylic acids. Since we can do little with carboxylic acids without encountering

$$\underset{\text{Acid}}{RCOOH} \underset{H^+}{\overset{OH^-}{\rightleftharpoons}} \underset{\text{Salt}}{RCOO^-}$$

this conversion to and from their salts, it is worthwhile for us to examine the properties of these salts.

Salts of carboxylic acid—like all salts—are crystalline non-volatile solids made up of positive and negative ions; their properties are what we would expect of such structures. The strong electrostatic forces holding the ions in the crystal lattice can be overcome only by heating to a high temperature, or by a very polar solvent. The temperature required for melting is so high that before it can be reached carbon–carbon bonds break and the molecule decomposes, generally in the neighborhood of 300–400°. A decomposition point is seldom useful for the identification of a compound, since it usually reflects the rate of heating rather than the identity of the compound.

The alkali metal salts of carboxylic acids (sodium, potassium, ammonium) are soluble in water but insoluble in non-polar solvents; most of the heavy metal salts (iron, silver, copper, etc.) are insoluble in water.

Thus we see that, except for the acids of four carbons or less, which are soluble both in water and in organic solvents, *carboxylic acids and their alkali metal salts show exactly opposite solubility behavior*. Because of the ready interconversion of acids and their salts, this difference in solubility behavior may be used in two important ways: for *identification* and for *separation*.

A water-insoluble organic compound that dissolves in cold dilute aqueous sodium hydroxide must be either a carboxylic acid or one of the few other kinds of organic compounds more acidic than water; that it is indeed a carboxylic acid can then be shown in other ways.

$$\underset{\substack{\text{Stronger acid} \\ \textit{Insoluble in } H_2O}}{RCOOH} + NaOH \longrightarrow \underset{\substack{\textit{Soluble in} \\ H_2O}}{RCOONa} + \underset{\substack{\text{Weaker} \\ \text{acid}}}{H_2O}$$

Instead of sodium hydroxide, we can use aqueous sodium bicarbonate; even if the unknown is water-soluble, its acidity is shown by the evolution of bubbles of CO_2.

$$\underset{\textit{Insoluble in } H_2O}{RCOOH} + NaHCO_3 \longrightarrow \underset{\textit{Soluble in } H_2O}{RCOONa} + H_2O + CO_2\uparrow$$

We can separate a carboxylic acid from non-acidic compounds by taking advantage of its solubility and their insolubility in aqueous base; once the separation has been accomplished, we can regenerate the acid by acidification of the aqueous solution. If we are dealing with solids, we simply stir the mixture with aqueous base and then filter the solution from insoluble, non-acidic materials; addition of mineral acid to the filtrate precipitates the carboxylic acid, which can be collected on a filter. If we are dealing with liquids, we shake the mixture with aqueous base in a separatory funnel and separate the aqueous layer from the insoluble organic layer; addition of acid to the aqueous layer again liberates the carboxylic acid, which can then be separated from the water. For completeness of separation and ease of handling, we often add a water-insoluble solvent like ether to the acidified mixture. The carboxylic acid is extracted from the water by the ether, in which it is more soluble; the volatile ether is readily removed by distillation from the comparatively high-boiling acid.

For example, an aldehyde prepared by the oxidation of a primary alcohol (Sec. 16.8) may very well be contaminated with the carboxylic acid; this acid can be simply washed out with dilute aqueous base. The carboxylic acid prepared by oxidation of an alkylbenzene (Sec. 12.10) may very well be contaminated with unreacted starting material; the carboxylic acid can be taken into solution by aqueous base, separated from the insoluble hydrocarbon, and regenerated by addition of mineral acid.

Since separations of this kind are more clear-cut and less wasteful of material, they are preferred wherever possible over recrystallization or distillation.

18.5 Industrial source

Acetic acid, by far the most important of all carboxylic acids, is prepared by air oxidation of acetaldehyde, which is readily available from the hydration of acetylene (Sec. 8.13), or the dehydrogenation of ethanol.

$$\left.\begin{array}{l}\underset{\text{Acetylene}}{HC{\equiv}CH} \xrightarrow{H_2O,\ H_2SO_4,\ HgSO_4} \\[1em] \underset{\text{Ethanol}}{CH_3CH_2OH} \xrightarrow{Cu,\ 250\text{–}300^\circ}\end{array}\right] \longrightarrow \underset{\text{Acetaldehyde}}{CH_3\overset{\overset{\large H}{|}}{C}{=}O} \xrightarrow{O_2,\ Mn^{++}} \underset{\text{Acetic acid}}{CH_3COOH}$$

Large amounts of acetic acid are also produced as the dilute aqueous solution known as *vinegar*. Here, too, the acetic acid is prepared by air oxidation; the compound that is oxidized is ethyl alcohol, and the catalysts are bacterial (*Acetobacter*) enzymes.

The most important sources of aliphatic carboxylic acids are the animal and vegetable **fats** (Secs. 33.2–33.4). From fats there can be obtained, in purity of over 90%, straight-chain carboxylic acids of even carbon number ranging from six to eighteen carbon atoms. These acids can be converted into the corresponding alcohols (Sec. 18.18), which can then be used, in the ways we have already studied (Sec. 16.10), to make a great number of other compounds containing long, straight-chain units.

The most important of the aromatic carboxylic acids, **benzoic acid** and the

phthalic acids, are prepared on an industrial scale by a reaction we have already encountered: oxidation of alkylbenzenes (Sec. 12.10). The toluene and xylenes required are readily available from coal tar and, by catalytic reforming of aliphatic hydrocarbons (Sec. 12.4), from petroleum; another precursor of phthalic acid (the *ortho* isomer) is the aromatic hydrocarbon *naphthalene*, also found in coal tar. Cheap oxidizing agents like chlorine or even air (in the presence of catalysts) are used.

Coal tar or petroleum (catalytic reforming) → Toluene ($C_6H_5CH_3$) $\xrightarrow[\text{heat}]{Cl_2}$ Benzotrichloride ($C_6H_5CCl_3$) $\xrightarrow{H_2O,\ OH^-}$ Benzoic acid (C_6H_5COOH)

Coal tar or petroleum (catalytic reforming) → *o*-Xylene ($C_6H_4(CH_3)_2$) or Naphthalene $\xrightarrow{O_2,\ V_2O_5}$ Phthalic acid ($C_6H_4(COOH)_2$) $\xrightarrow{\text{heat, catalyst, } -CO_2}$ Benzoic acid

Problem 18.2 In the presence of peroxides, carboxylic acids (or esters) react with 1-alkenes to yield more complicated acids. For example:

$$n\text{-}C_4H_9CH{=}CH_2 + CH_3CH_2CH_2COOH \xrightarrow{\text{peroxides}} n\text{-}C_4H_9CH_2CH_2CH(C_2H_5)COOH$$

1-Hexene; *n*-Butyric acid; 2-Ethyloctanoic acid (*70% yield*)

(a) Outline all steps in a likely mechanism for this reaction. (*Hint:* See Sec. 6.18.) Predict the products of similar reactions between: (b) 1-octene and propionic acid; (c) 1-decene and isobutyric acid; and (d) 1-octene and ethyl malonate, $CH_2(COOC_2H_5)_2$.

Problem 18.3 (a) Carbon monoxide converts a sulfuric acid solution of each of the following into 2,2-dimethylbutanoic acid: 2-methyl-2-butene, *tert*-pentyl alcohol, neopentyl alcohol. Suggest a likely mechanism for this method of synthesizing carboxylic acids. (b) *n*-Butyl alcohol and *sec*-butyl alcohol give the same product. What would you expect it to be?

18.6 Preparation

The straight-chain aliphatic acids up to C_6, and those of even carbon number up to C_{18}, are commercially available, as are the simple aromatic acids. Other carboxylic acids can be prepared by the methods outlined below.

PREPARATION OF CARBOXYLIC ACIDS

1. **Oxidation of primary alcohols.** Discussed in Sec. 16.8.

$$RCH_2OH \xrightarrow{KMnO_4} RCOOH$$

Examples:

$$\underset{\text{2-Methyl-1-butanol}}{CH_3CH_2\overset{\overset{CH_3}{|}}{C}HCH_2OH} \xrightarrow{KMnO_4} \underset{\text{2-Methylbutanoic acid}}{CH_3CH_2\overset{\overset{CH_3}{|}}{C}HCOOH}$$

$$\underset{\text{Isobutyl alcohol}}{CH_3\overset{\overset{CH_3}{|}}{C}HCH_2OH} \xrightarrow{KMnO_4} \underset{\text{Isobutyric acid}}{CH_3\overset{\overset{CH_3}{|}}{C}HCOOH}$$

2. Oxidation of alkylbenzenes. Discussed in Sec. 12.10.

$$Ar{-}R \xrightarrow{KMnO_4 \text{ or } K_2Cr_2O_7} Ar{-}COOH$$

Examples:

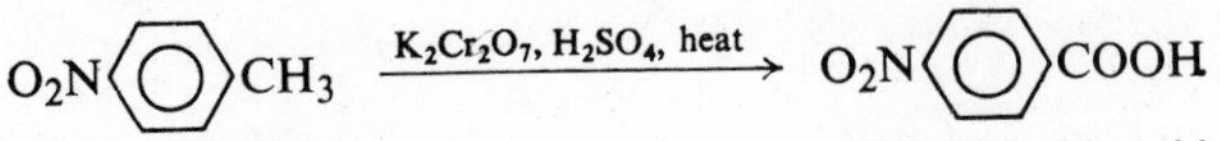

$$\underset{p\text{-Nitrotoluene}}{O_2N{-}C_6H_4{-}CH_3} \xrightarrow{K_2Cr_2O_7,\ H_2SO_4,\ \text{heat}} \underset{p\text{-Nitrobenzoic acid}}{O_2N{-}C_6H_4{-}COOH}$$

$$\underset{o\text{-Bromotoluene}}{o\text{-}BrC_6H_4CH_3} \xrightarrow[\text{heat}]{KMnO_4,\ OH^-} \underset{o\text{-Bromobenzoic acid}}{o\text{-}BrC_6H_4COOH}$$

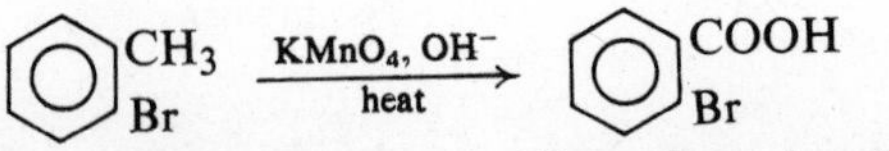

3. Carbonation of Grignard reagents. Discussed in Sec. 18.7.

$$\underset{\text{(or ArX)}}{RX} \xrightarrow{Mg} RMgX \xrightarrow{CO_2} RCOOMgX \xrightarrow{H^+} \underset{\text{(or ArCOOH)}}{RCOOH}$$

Examples:

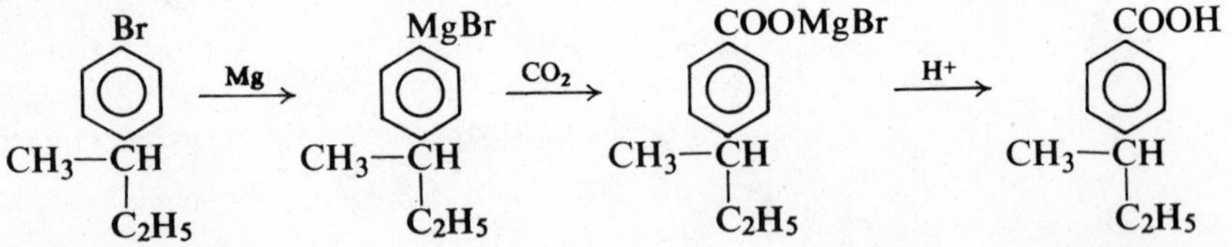

$$\underset{p\text{-Bromo-}sec\text{-butylbenzene}}{p\text{-}Br{-}C_6H_4{-}CH(CH_3)C_2H_5} \xrightarrow{Mg} p\text{-}BrMg{-}C_6H_4{-}CH(CH_3)C_2H_5 \xrightarrow{CO_2} p\text{-}BrMgOOC{-}C_6H_4{-}CH(CH_3)C_2H_5 \xrightarrow{H^+} \underset{p\text{-}sec\text{-Butylbenzoic acid}}{p\text{-}HOOC{-}C_6H_4{-}CH(CH_3)C_2H_5}$$

$$\underset{tert\text{-Pentyl chloride}}{C_2H_5{-}\overset{\overset{CH_3}{|}}{\underset{\underset{CH_3}{|}}{C}}{-}Cl} \xrightarrow{Mg} C_2H_5{-}\overset{\overset{CH_3}{|}}{\underset{\underset{CH_3}{|}}{C}}{-}MgCl \xrightarrow{CO_2} \xrightarrow{H^+} \underset{\text{Ethyldimethylacetic acid (2,2-Dimethylbutanoic acid)}}{C_2H_5{-}\overset{\overset{CH_3}{|}}{\underset{\underset{CH_3}{|}}{C}}{-}COOH}$$

4. Hydrolysis of nitriles. Discussed in Sec. 18.8.

$$\begin{matrix} R{-}C{\equiv}N \\ \text{or} \\ Ar{-}C{\equiv}N \end{matrix} + H_2O \xrightarrow{\text{acid or base}} \begin{matrix} R{-}COOH \\ \text{or} \\ Ar{-}COOH \end{matrix} + NH_3$$

Examples:

$$C_6H_5CH_2Cl \xrightarrow{NaCN} C_6H_5CH_2CN \xrightarrow{70\%\ H_2SO_4,\ reflux} C_6H_5CH_2COOH + NH_4^+$$

Benzyl chloride — Phenylacetonitrile — Phenylacetic acid

$$n\text{-}C_4H_9Br \xrightarrow{NaCN} n\text{-}C_4H_9CN \xrightarrow{aq.\ alc.\ NaOH,\ reflux} n\text{-}C_4H_9COO^- + NH_3 \xrightarrow{H^+} n\text{-}C_4H_9COOH + NH_4^+$$

n-Butyl bromide — *n*-Valeronitrile (Pentanenitrile) — *n*-Valeric acid (Pentanoic acid)

$$\text{Diazonium salt} \longrightarrow o\text{-}CH_3C_6H_4CN \xrightarrow{75\%\ H_2SO_4,\ 150\text{–}160^\circ} o\text{-}CH_3C_6H_4COOH + NH_4^+$$

o-Tolunitrile — *o*-Toluic acid

5. **Malonic ester synthesis.** Discussed in Sec. 26.2.

6. **Special methods for phenolic acids.** Discussed in Sec. 24.11.

All the methods listed are important; our choice is governed by the availability of starting materials.

Oxidation is the most direct and is generally used when possible, some lower aliphatic acids being made from the available alcohols, and substituted aromatic acids from substituted toluenes.

The **Grignard synthesis** and the **nitrile synthesis** have the special advantage of increasing the length of a carbon chain, and thus extending the range of available materials. In the aliphatic series both Grignard reagents and nitriles are prepared from halides, which in turn are usually prepared from alcohols. The syntheses thus amount to the preparation of acids from alcohols containing one less carbon atom.

$$RCH_2OH \xrightarrow{KMnO_4} RCOOH \quad \textit{Same carbon number}$$

Higher carbon number

$$RCH_2OH \xrightarrow{PBr_3} RCH_2Br \xrightarrow{Mg} RCH_2MgBr \xrightarrow{CO_2} \xrightarrow{H^+} RCH_2COOH$$

$$RCH_2Br \xrightarrow{CN^-} RCH_2CN \xrightarrow{H_2O} RCH_2COOH$$

Problem 18.4 What carboxylic acid can be prepared from *p*-bromotoluene: (a) by direct oxidation? (b) by free-radical chlorination followed by the nitrile synthesis?

Aromatic nitriles generally cannot be prepared from the unreactive aryl halides (Sec. 25.5). Instead they are made from diazonium salts by a reaction we shall discuss later (Sec. 23.13). Diazonium salts are prepared from aromatic amines, which in turn are prepared from nitro compounds. Thus the carboxyl group eventually occupies the position on the ring where a nitro group was originally introduced by direct nitration (Sec. 11.8).

$$\underset{}{ArH} \longrightarrow \underset{\text{Nitro compound}}{ArNO_2} \longrightarrow \underset{\text{Amine}}{ArNH_2} \longrightarrow \underset{\text{Diazonium ion}}{ArN_2^+} \longrightarrow \underset{\text{Nitrile}}{ArC{\equiv}N} \longrightarrow \underset{\text{Acid}}{ArCOOH}$$

For the preparation of quite complicated acids, the most versatile method of all is used, the *malonic ester synthesis* (Sec. 26.2).

18.7 Grignard synthesis

The Grignard synthesis of a carboxylic acid is carried out by bubbling gaseous CO_2 into the ether solution of the Grignard reagent, or by pouring the Grignard reagent on crushed Dry Ice (solid CO_2); in the latter method Dry Ice serves not only as reagent but also as cooling agent.

The Grignard reagent adds to the carbon–oxygen double bond just as in the reaction with aldehydes and ketones (Sec. 15.12). The product is the magnesium salt of the carboxylic acid, from which the free acid is liberated by treatment with mineral acid.

$$R{-}MgX + O{=}C{=}O \longrightarrow RCOO^-MgX^+ \xrightarrow{H^+} RCOOH + Mg^{++} + X^-$$

The Grignard reagent can be prepared from primary, secondary, tertiary, or aromatic halides; the method is limited only by the presence of other reactive groups in the molecule (Sec. 15.15). The following syntheses illustrate the application of this method:

$$\underset{\textit{tert}\text{-Butyl alcohol}}{(CH_3)_3C{-}OH} \xrightarrow{HCl} \underset{\textit{tert}\text{-Butyl chloride}}{(CH_3)_3C{-}Cl} \xrightarrow{Mg} (CH_3)_3C{-}MgCl \xrightarrow{CO_2} \xrightarrow{H^+} \underset{\text{Trimethylacetic acid}}{(CH_3)_3C{-}COOH}$$

$$\underset{\text{Mesitylene}}{C_6H_3(CH_3)_3} \xrightarrow{Br_2} \underset{\text{Bromomesitylene}}{2,4,6\text{-}(CH_3)_3C_6H_2Br} \xrightarrow{Mg} 2,4,6\text{-}(CH_3)_3C_6H_2MgBr \xrightarrow{CO_2} \xrightarrow{H^+} \underset{\text{Mesitoic acid (2,4,6-Trimethylbenzoic acid)}}{2,4,6\text{-}(CH_3)_3C_6H_2COOH}$$

18.8 Nitrile synthesis

Aliphatic nitriles are prepared by treatment of alkyl halides with sodium cyanide in a solvent that will dissolve both reactants; in dimethyl sulfoxide, reaction occurs rapidly and exothermically at room temperature. The resulting nitrile is then hydrolyzed to the acid by boiling aqueous alkali or acid.

$$RX + CN^- \longrightarrow RC{\equiv}N + X^-$$

$$RC{\equiv}N + H_2O \longrightarrow \begin{cases} \xrightarrow{H^+} RCOOH + NH_4^+ \\ \xrightarrow{OH^-} RCOO^- + NH_3 \end{cases}$$

The reaction of an alkyl halide with cyanide ion involves nucleophilic substitution (Sec. 14.5). The fact that HCN is a very weak acid tells us that cyanide ion is a strong base; as we might expect, this strongly basic ion can abstract hydrogen ion and thus cause elimination as well as substitution. Indeed, with

$$\underset{n\text{-Butyl bromide}}{CH_3CH_2CH_2CH_2Br} + CN^- \longrightarrow \underset{\text{Valeronitrile}}{CH_3CH_2CH_2CH_2CN} \qquad \text{1° halide: } \textit{substitution}$$

$$\underset{tert\text{-Butyl bromide}}{CH_3-\underset{CH_3}{\overset{CH_3}{C}}-Br} + CN^- \longrightarrow \underset{\text{Isobutylene}}{CH_3-\overset{CH_3}{C}{=}CH_2} + HCN \qquad \text{3° halide: } \textit{elimination}$$

tertiary halides elimination is the principal reaction; even with secondary halides the yield of substitution product is poor. Here again we find a nucleophilic substitution reaction that is of synthetic importance *only when primary halides are used.*

As already mentioned, aromatic nitriles are made, not from the unreactive aryl halides, but from diazonium salts (Sec. 23.13).

Although nitriles are sometimes named as *cyanides* or as *cyano* compounds, they generally take their names from the acids they yield upon hydrolysis. They are named by dropping *–ic acid* from the common name of the acid and adding **-nitrile**; usually for euphony an "o" is inserted between the root and the ending (e.g., *acetonitrile*). In the IUPAC system they are named by adding *–nitrile* to the name of the parent hydrocarbon (e.g., *ethanenitrile*). For example:

$CH_3C{\equiv}N$	$CH_3(CH_2)_3C{\equiv}N$	$C_6H_5C{\equiv}N$	$CH_3C_6H_4C{\equiv}N$
Acetonitrile (Ethanenitrile)	*n*-Valeronitrile (Pentanenitrile)	Benzonitrile	*p*-Tolunitrile

18.9 Reactions

The characteristic chemical behavior of carboxylic acids is, of course, determined by their functional group, **carboxyl**, —COOH. This group is made up of a carbonyl group (C=O) and a hydroxyl group (—OH). As we shall see, it is the —OH that actually undergoes nearly every reaction—loss of H^+, or replacement by another group—but *it does so in a way that is possible only because of the effect of the* C=O.

The rest of the molecule undergoes reactions characteristic of its structure; it may be aliphatic or aromatic, saturated or unsaturated, and may contain a variety of other functional groups.

REACTIONS OF CARBOXYLIC ACIDS

1. Acidity. Salt formation. Discussed in Secs. 18.4, 18.10–18.14.

$$RCOOH \rightleftharpoons RCOO^- + H^+$$

Examples:

$$\underset{\text{Acetic acid}}{2CH_3COOH} + Zn \longrightarrow \underset{\text{Zinc acetate}}{(CH_3COO^-)_2Zn^{++}} + H_2$$

$$\underset{\text{Lauric acid}}{CH_3(CH_2)_{10}COOH} + NaOH \longrightarrow \underset{\text{Sodium laurate}}{CH_3(CH_2)_{10}COO^-Na^+} + H_2O$$

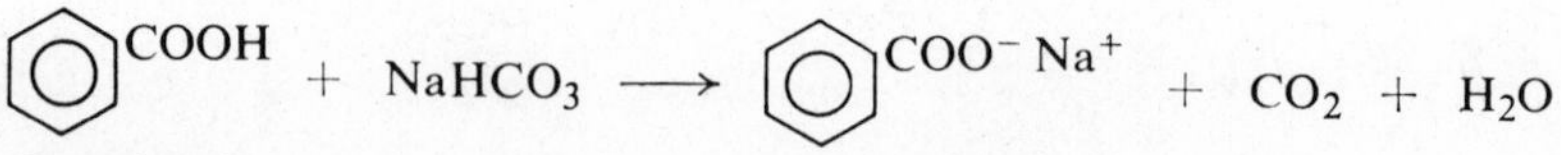

$$\underset{\text{Benzoic acid}}{C_6H_5COOH} + NaHCO_3 \longrightarrow \underset{\text{Sodium benzoate}}{C_6H_5COO^-Na^+} + CO_2 + H_2O$$

2. Conversion into functional derivatives

$$R-C(=O)-OH \longrightarrow R-C(=O)-Z \qquad (Z = -Cl,\ -OR',\ -NH_2)$$

(a) Conversion into acid chlorides. Discussed in Sec. 18.15.

$$R-C(=O)-OH + \left\{\begin{matrix} SOCl_2 \\ PCl_3 \\ PCl_5 \end{matrix}\right\} \longrightarrow \underset{\text{Acid chloride}}{R-C(=O)-Cl}$$

Examples:

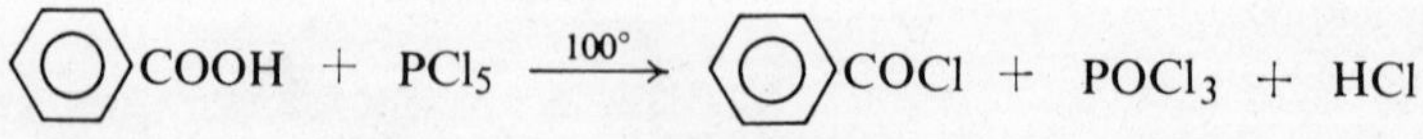

$$\underset{\text{Benzoic acid}}{C_6H_5COOH} + PCl_5 \xrightarrow{100^\circ} \underset{\text{Benzoyl chloride}}{C_6H_5COCl} + POCl_3 + HCl$$

$$\underset{\text{Stearic acid}}{n\text{-}C_{17}H_{35}COOH} + \underset{\text{Thionyl chloride}}{SOCl_2} \xrightarrow{\text{reflux}} \underset{\text{Stearoyl chloride}}{n\text{-}C_{17}H_{35}COCl} + SO_2 + HCl$$

$$\underset{\text{Acetic acid}}{3CH_3COOH} + PCl_3 \xrightarrow{50^\circ} \underset{\text{Acetyl chloride}}{3CH_3COCl} + H_3PO_3$$

(b) Conversion into esters. Discussed in Secs. 18.16 and 20.15.

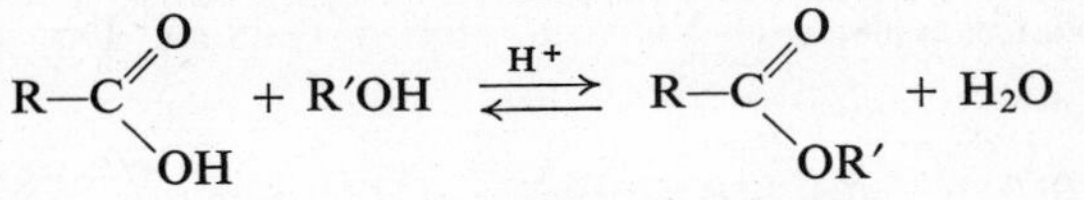

$$R{-}C(=O){-}OH + R'OH \overset{H^+}{\rightleftharpoons} R{-}C(=O){-}OR' + H_2O$$

An ester

Reactivity of R′OH: 1° > 2° (> 3°)

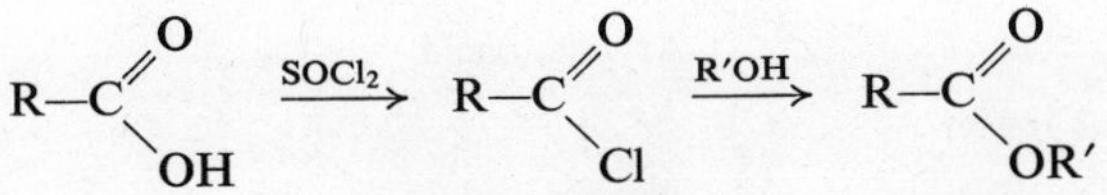

$$R{-}C(=O){-}OH \xrightarrow{SOCl_2} R{-}C(=O){-}Cl \xrightarrow{R'OH} R{-}C(=O){-}OR'$$

An acid chloride → An ester

Examples:

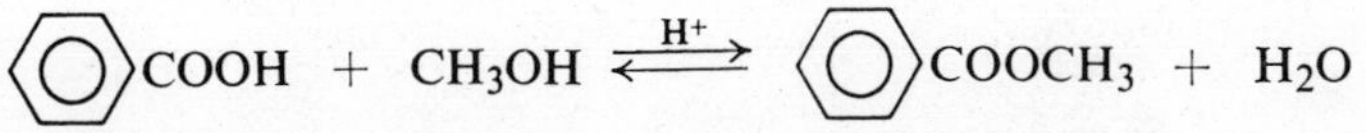

$$C_6H_5COOH + CH_3OH \overset{H^+}{\rightleftharpoons} C_6H_5COOCH_3 + H_2O$$

Benzoic acid, Methanol → Methyl benzoate

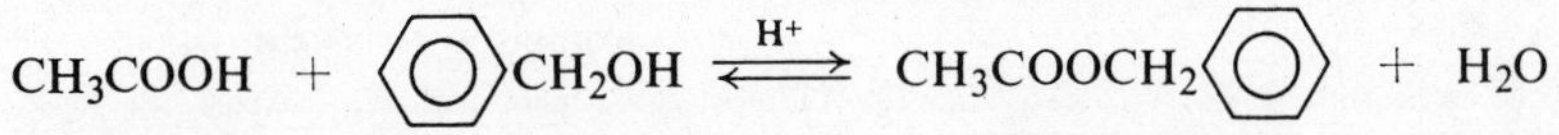

$$CH_3COOH + C_6H_5CH_2OH \overset{H^+}{\rightleftharpoons} CH_3COOCH_2C_6H_5 + H_2O$$

Acetic acid, Benzyl alcohol → Benzyl acetate

$$(CH_3)_3CCOOH \xrightarrow{SOCl_2} (CH_3)_3CCOCl \xrightarrow{C_2H_5OH} (CH_3)_3CCOOC_2H_5$$

Trimethylacetic acid → Ethyl trimethylacetate

(c) Conversion into amides. Discussed in Sec. 18.17.

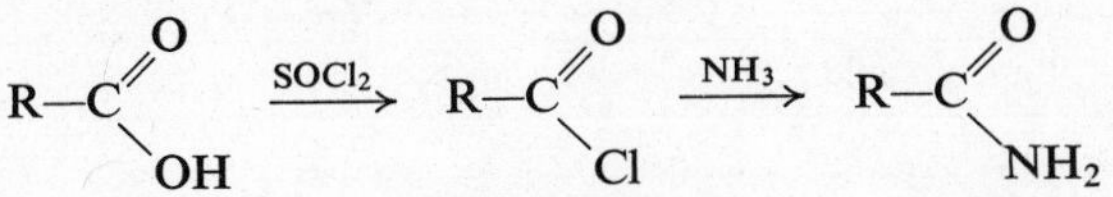

$$R{-}C(=O){-}OH \xrightarrow{SOCl_2} R{-}C(=O){-}Cl \xrightarrow{NH_3} R{-}C(=O){-}NH_2$$

An acid chloride → An amide

Example:

$$C_6H_5CH_2COOH \xrightarrow{SOCl_2} C_6H_5CH_2COCl \xrightarrow{NH_3} C_6H_5CH_2CONH_2$$

Phenylacetic acid → Phenylacetyl chloride → Phenylacetamide

3. Reduction. Discussed in Sec. 18.18.

$$RCOOH \xrightarrow{LiAlH_4} RCH_2OH$$

1° alcohol

Also reduced via esters (Sec. 20.22)

Examples:

$$4(CH_3)_3CCOOH + 3LiAlH_4 \xrightarrow{ether} [(CH_3)_3CCH_2O]_4AlLi + 2LiAlO_2 + 4H_2 \xrightarrow{H^+} (CH_3)_3CCH_2OH$$

Trimethylacetic acid → Neopentyl alcohol (2,2-Dimethyl-1-propanol)

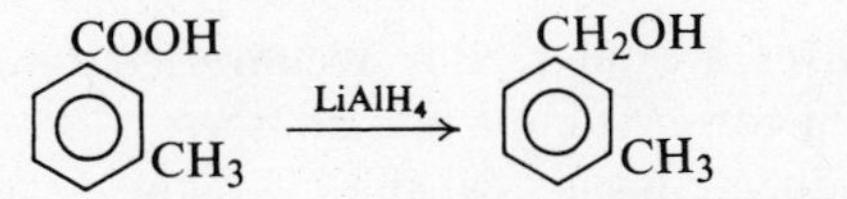

$$m\text{-}CH_3C_6H_4COOH \xrightarrow{LiAlH_4} m\text{-}CH_3C_6H_4CH_2OH$$

m-Toluic acid → *m*-Methylbenzyl alcohol

4. Substitution in alkyl or aryl group

(a) Alpha-halogenation of aliphatic acids. Hell-Volhard-Zelinsky reaction. Discussed in Sec. 18.19.

$$\mathrm{RCH_2COOH + X_2 \xrightarrow{P} \underset{\underset{X}{|}}{RCHCOOH} + HX} \qquad X_2 = Cl_2, Br_2$$

An α-haloacid

Examples:

$$\underset{\text{Acetic acid}}{\mathrm{CH_3COOH}} \xrightarrow{Cl_2, P} \underset{\text{Chloroacetic acid}}{\mathrm{ClCH_2COOH}} \xrightarrow{Cl_2, P} \underset{\text{Dichloroacetic acid}}{\mathrm{Cl_2CHCOOH}} \xrightarrow{Cl_2, P} \underset{\text{Trichloroacetic acid}}{\mathrm{Cl_3CCOOH}}$$

$$\underset{\text{Isovaleric acid}}{\mathrm{CH_3\overset{\overset{CH_3}{|}}{C}HCH_2COOH}} \xrightarrow{Br_2, P} \underset{\alpha\text{-Bromoisovaleric acid}}{\mathrm{CH_3\overset{\overset{CH_3}{|}}{C}H\underset{\underset{Br}{|}}{C}HCOOH}}$$

(b) Ring substitution in aromatic acids. Discussed in Secs. 11.5 and 11.17.

—COOH: deactivates, and directs *meta* in electrophilic substitution.

Example:

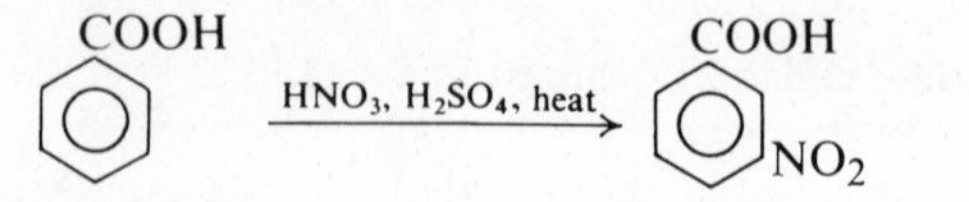

$$\underset{\text{Benzoic acid}}{\mathrm{C_6H_5COOH}} \xrightarrow{HNO_3,\ H_2SO_4,\ \text{heat}} \underset{m\text{-Nitrobenzoic acid}}{m\text{-}\mathrm{O_2NC_6H_4COOH}}$$

The most characteristic property of the carboxylic acids is the one that gives them their name: **acidity.** Their tendency to give up a hydrogen ion is such that in aqueous solution a measurable equilibrium exists between acid and ions; they are thus much more acidic than any other class of organic compounds we have studied so far.

$$\mathrm{RCOOH + H_2O \rightleftharpoons RCOO^- + H_3O^+}$$

The OH of an acid can be replaced by a number of groups—Cl, OR′, NH_2—to yield compounds known as *acid chlorides*, *esters*, and *amides*. These compounds are called **functional derivatives** of acids; they all contain the **acyl group**:

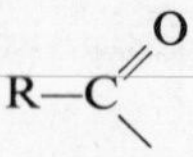

$$\mathrm{R{-}C{\overset{\displaystyle O}{/\!\!/}}\ \diagdown}$$

The functional derivatives are all readily reconverted into the acid by simple hydrolysis, and are often converted one into another.

One of the few reducing agents capable of reducing an acid directly to an alcohol is *lithium aluminum hydride*, $LiAlH_4$.

The hydrocarbon portion of an aliphatic acid can undergo the free-radical halogenation characteristic of alkanes, but because of the random nature of the substitution it is seldom used. The presence of a small amount of phosphorus, however, causes halogenation (by an ionic mechanism) to take place *exclusively at the alpha position*. This reaction is known as the **Hell-Volhard-Zelinsky reaction,** and it is of great value in synthesis.

An aromatic ring bearing a carboxyl group undergoes the aromatic electrophilic substitution reactions expected of a ring carrying a deactivating, *meta*-directing group. Deactivation is so strong that the Friedel-Crafts reaction does not take place. We have already accounted for this effect of the —COOH group on the basis of its strong electron-withdrawing tendencies (Sec. 11.18).

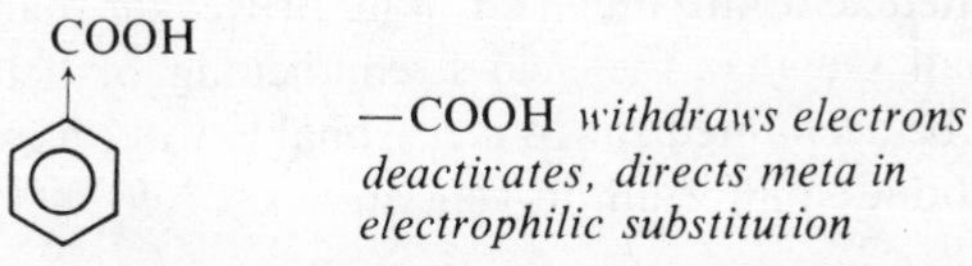

COOH
—COOH *withdraws electrons: deactivates, directs meta in electrophilic substitution*

Decarboxylation—elimination of the —COOH group as CO_2—is of limited importance for aromatic acids, and highly important for certain substituted aliphatic acids: malonic acids (Sec. 26.2) and β-keto acids (Sec. 26.3). It is worthless for most simple aliphatic acids, yielding a complicated mixture of hydrocarbons.

18.10 Ionization of carboxylic acids. Acidity constant

In aqueous solution a carboxylic acid exists in equilibrium with the carboxylate anion and the hydrogen ion (actually, of course, the hydronium ion, H_3O^+).

$$\text{RCOOH} + \text{H}_2\text{O} \rightleftarrows \text{RCOO}^- + \text{H}_3\text{O}^+$$

As for any equilibrium, the concentrations of the components are related by the expression

$$K_a = \frac{[\text{RCOO}^-][\text{H}_3\text{O}^+]}{[\text{RCOOH}]}$$

(Since the concentration of water, the solvent, remains essentially constant, this term is usually omitted.) The equilibrium constant is called here the **acidity constant, K_a** (*a for acidity*).

Every carboxylic acid has its characteristic K_a, which indicates how strong an acid it is. Since the acidity constant is the ratio of ionized to unionized material, the larger the K_a the greater the extent of the ionization (under a given set of conditions) and the stronger the acid. We use the K_a's, then, to compare in an exact way the strengths of different acids.

We see in Table 18.2 (p. 600) that unsubstituted aliphatic and aromatic acids have K_a's of about 10^{-4} to 10^{-5} (0.0001 to 0.00001). This means that they are weakly acidic, with only a slight tendency to release protons.

By the same token, carboxylate anions are moderately basic, with an appreciable tendency to combine with protons. They react with water to increase the concentration of hydroxide ions, a reaction often referred to as *hydrolysis*. As

$$RCOO^- + H_2O \rightleftharpoons RCOOH + OH^-$$

a result aqueous solutions of carboxylate salts are slightly alkaline. (The basicity of an aqueous solution of a carboxylate salt is due chiefly, of course, to the carboxylate anions, not to the comparatively few hydroxide ions they happen to generate.)

We may now expand the series of relative acidities and basicities:

Relative acidities: $RCOOH > HOH > ROH > HC{\equiv}CH > NH_3 > RH$

Relative basicities: $RCOO^- < HO^- < RO^- < HC{\equiv}C^- < NH_2^- < R^-$

Certain substituted acids are much stronger or weaker than a typical acid like CH_3COOH. We shall see that the acid-strengthening or acid-weakening effect of a substituent can be accounted for in a reasonable way; however, we must first learn a little more about equilibrium in general.

18.11 Equilibrium

So far we have dealt very little with the problem of equilibrium. Under the conditions employed, most of our reactions have been essentially irreversible; that is, they have been one-way reactions. With a few exceptions—1,4-addition, for example (Sec. 8.22)—the products obtained, and their relative yields, have been determined by how fast reactions go and not by how nearly to completion they proceed before equilibrium is reached. Consequently, we have been concerned with the relationship between structure and rate; now we shall turn to the relationship between structure and equilibrium.

Let us consider the reversible reaction between A and B to form C and D. The

$$A + B \rightleftharpoons C + D$$

yield of C and D does not depend upon how fast A and B react, but rather upon how completely they have reacted when equilibrium is reached.

The concentrations of the various components are related by the familiar expression,

$$K_{eq} = \frac{[C][D]}{[A][B]}$$

in which K_{eq} is the equilibrium constant. The more nearly a reaction has proceeded to completion when it reaches equilibrium, the larger is [C][D] compared with [A][B], and hence the larger the K_{eq}. The value of K_{eq} is therefore a measure of the tendency of the reaction to go to completion.

The value of K_{eq} is determined by the change in *free energy*, G, on proceeding from reactants to products (Fig. 18.1). The exact relationship is given by the expression,

$$\Delta G^\circ = -2.303RT \log K_{eq}$$

where ΔG° is the *standard free energy change*.

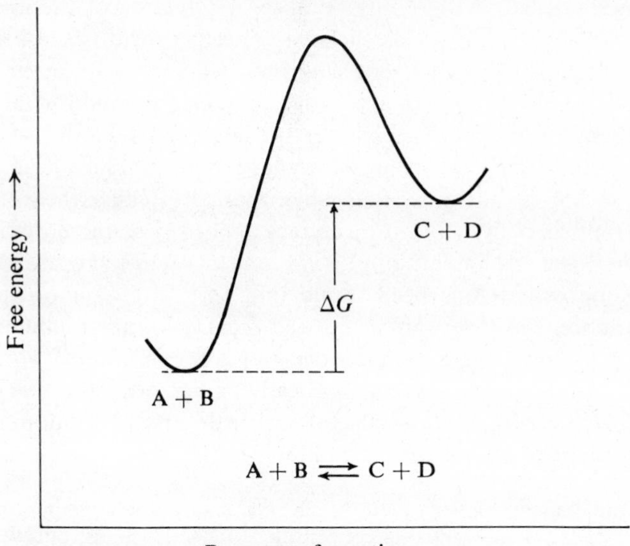

Figure 18.1. Free energy curve for a reversible reaction.

Free energy change is related to our familiar quantity ΔH (precisely $\Delta H°$, which is only slightly different) by the expression,

$$\Delta G° = \Delta H - T\Delta S°$$

where $\Delta S°$ is the *standard entropy change*. Entropy corresponds roughly, to the *randomness* of the system. To the extent that $T\Delta S°$ contributes to $\Delta G°$, equilibrium tends to shift toward the side in which fewer restrictions are placed on the positions of atoms and molecules. ("Die Energie der Welt ist constant. Die Entropie der Welt strebt einem Maximum zu." *Clausius, 1865.*)

Under the same experimental conditions two reversible reactions have K_{eq}'s of different sizes because of a difference in $\Delta G°$. In attempting to understand the effect of structure on position of equilibrium, we shall estimate differences in relative stabilities of reactants and products. Now, what we estimate in this way are not differences in free energy change but differences in potential energy change. It turns out that very often these differences are *proportional to* differences in $\Delta G°$. So long as we compare closely related compounds, the predictions we make by this approach are generally good ones.

These predictions are good ones despite the fact that the free energy changes on which they depend are made up to varying degrees of ΔH and $\Delta S°$. For example, *p*-nitrobenzoic acid is a stronger acid than benzoic acid. We attribute this (Sec. 18.14) to stabilization of the *p*-nitrobenzoate anion (relative to the benzoate anion) through dispersal of charge by the electron-withdrawing nitro group. Yet, in this case, the greater acidity is due about as much to a more favorable $\Delta S°$ as to a more favorable ΔH. How can our simple "stabilization by dispersal of charge" account for an effect that involves the randomness of a system?

Stabilization *is* involved, but it appears partly in the $\Delta S°$ for this reason.

Ionization of an acid is possible only because of solvation of the ions produced: the many ion–dipole bonds provide the energy needed for dissociation. But solvation requires that molecules of solvent leave their relatively unordered arrangement to cluster in some ordered fashion about the ions. This is good for the ΔH but bad for the $\Delta S°$. Now, because of its greater intrinsic stability, the *p*-nitrobenzoate anion does not *need* as many solvent molecules to help stabilize it as the benzoate anion does. The $\Delta S°$ is thus more favorable. We can visualize the *p*-nitrobenzoate ion accepting only as many solvent molecules as it has to, and stopping when the gain in stability (decrease in enthalpy) is no longer worth the cost in entropy.

(In the same way, it has been found that very often a more polar solvent speeds up a reaction—as, for example, an S_N1 reaction of alkyl halides (Sec. 14.16) —not so much by lowering E_{act} as by bringing about a more favorable entropy of activation. A more polar solvent is already rather ordered, and its clustering about the ionizing molecule amounts to very little loss of randomness—indeed, it may even amount to an *increase* in randomness.)

By the organic chemist's approach we can make *very* good predictions indeed. We can not only account for, say, the relative acidities of a set of acids, but we can correlate these acidities *quantitatively* with the relative acidities of another set of acids, or even with the relative rates of a set of reactions. These relationships are summarized in the Hammett equation (named for Louis P. Hammett of Columbia University),

$$\log \frac{K}{K_0} = \rho\sigma \qquad \text{or} \qquad \log \frac{k}{k_0} = \rho\sigma$$

where K or k refers to the reaction of a *m*- or *p*-substituted phenyl compound (say, ionization of a substituted benzoic acid) and K_0 or k_0 refers to the same reaction of the unsubstituted compound (say, ionization of benzoic acid).

The *substituent constant* (σ, *sigma*) is a number (+ or −) indicating the relative electron-withdrawing or electron-releasing effect of a particular substituent. The *reaction constant* (ρ, *rho*) is a number (+ or −) indicating the relative *need* of a particular reaction for electron withdrawal or electron release.

A vast amount of research has shown that the Hammett relationship holds for *hundreds of sets of reactions*. (Ionization of 40-odd *p*-substituted benzoic acids, for example, is *one* set.) By use of just two tables—one of σ constants and one of ρ constants—we can calculate the relative K_{eq}'s or relative rates for thousands of individual reactions. For example, from the σ value for *m*-NO_2 (+0.710) and the ρ value for ionization of benzoic acids in water at 25° (+1.000), we can calculate that K_a for *m*-nitrobenzoic acid is 5.13 times as big as the K_a for benzoic acid. Using the same σ value, and the ρ value for acid-catalyzed hydrolysis of benzamides in 60% ethanol at 80° (−0.298), we can calculate that *m*-nitrobenzamide will be hydrolyzed only 0.615 as fast as benzamide.

The Hammett relationship is called a *linear free energy relationship* since it is based on—and reveals—the fact that a linear relationship exists between free energy change and the effect exerted by a substituent. Other linear free energy relationships are known, which take into account steric as well as electronic effects, and which apply to *ortho* substituted phenyl compounds as well as *meta* and *para*, and to aliphatic as well as aromatic compounds. Together they make up what is perhaps the greatest accomplishment of physical-organic chemistry.

In dealing with rates, we compare the stability of the reactants with the stability of the transition state. In dealing with equilibria, we shall compare the stability of the reactants with the stability of the products. For closely related reactions, we are justified in assuming that the more stable the products relative to the reactants, the further reaction proceeds toward completion.

18.12 Acidity of carboxylic acids

Let us see how the acidity of carboxylic acids is related to structure. In doing this we shall assume that acidity is determined chiefly by the difference in stability between the acid and its anion.

First, and most important, there is the fact that carboxylic acids are acids at all. How can we account for the fact that the —OH of a carboxylic acid tends to release a hydrogen ion so much more readily than the —OH of, say, an alcohol? Let us examine the structures of the reactants and products in these two cases.

We see that the alcohol and alkoxide ion are each represented satisfactorily by a single structure. However, we can draw two reasonable structures (I and II) for the carboxylic acid and two reasonable structures (III and IV) for the carboxylate anion. Both acid and anion are resonance hybrids. But is resonance equally

$$R{-}O{-}H \rightleftharpoons H^+ + R{-}O^-$$

$$\left[\underset{\text{I}}{R{-}C(=O){-}OH} \quad \underset{\text{II}}{R{-}C(-O^-){=}\overset{+}{O}H} \right] \rightleftharpoons H^+ + \left[\underset{\text{III}}{R{-}C(=O){-}O^-} \quad \underset{\text{IV}}{R{-}C(-O^-){=}O} \right]$$

Non-equivalent: *resonance less important*

Equivalent: *resonance more important*

important in the two cases? By the principles of Sec. 6.27 we know that resonance is much more important between the exactly equivalent structures III and IV than between the non-equivalent structures I and II. As a result, although both acid and anion are stabilized by resonance, stabilization is far greater for the anion than for the acid (see Fig. 18.2). Equilibrium is shifted in the direction of increased ionization, and K_a is increased.

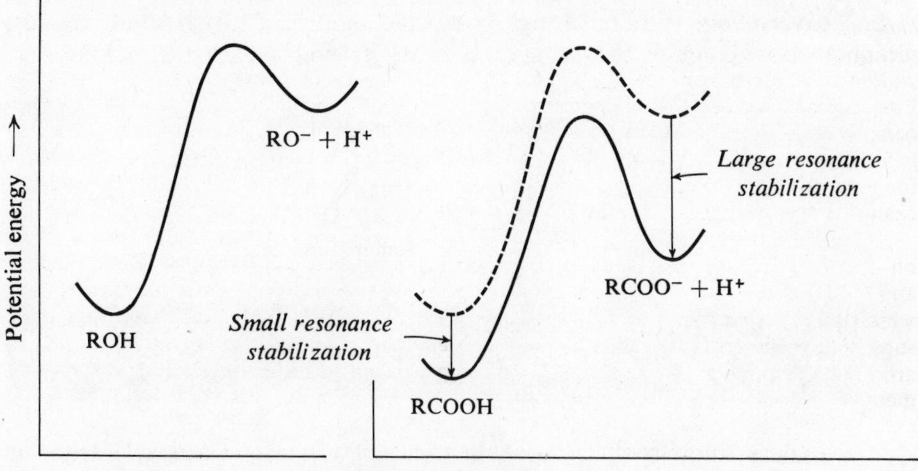

Figure 18.2. Molecular structure and position of equilibrium. Carboxylic acid yields resonance-stabilized anion; is stronger acid than alcohol. (Plots aligned with each other for easy comparison.)

Strictly speaking, resonance is less important for the acid because the contributing structures are of *different stability*, whereas the equivalent structures for the ion must necessarily be of *equal stability*. In structure II two atoms of similar electronegativity carry opposite charges; since energy must be supplied to separate opposite charges, II should contain more energy and hence be less stable than I. Consideration of *separation of charge* is one of the rules of thumb (Sec. 6.27) that can be used to estimate relative stability and hence relative importance of a contributing structure.

The acidity of a carboxylic acid is thus due to the powerful resonance stabilization of its anion. *This stabilization and the resulting acidity are possible only because of the presence of the carbonyl group.*

18.13 Structure of carboxylate ions

According to the resonance theory, then, a carboxylate ion is a hybrid of two structures which, being of equal stability, contribute equally. Carbon is joined to each oxygen by a "one-and-one-half" bond. The negative charge is evenly distributed over both oxygen atoms.

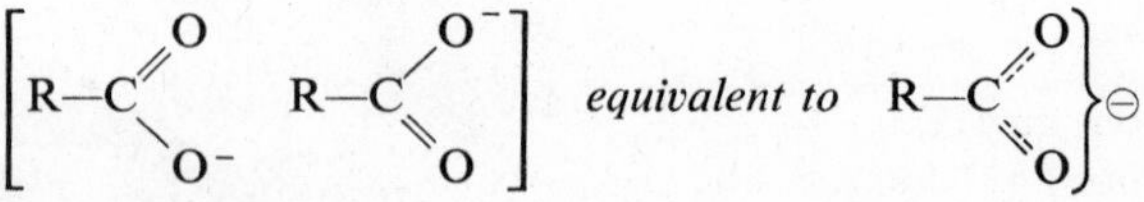

$$\left[R-C\begin{matrix} \diagup\!\!\diagup O \\ \diagdown O^- \end{matrix} \quad R-C\begin{matrix} \diagup O^- \\ \diagdown\!\!\diagdown O \end{matrix} \right] \quad \textit{equivalent to} \quad R-C\left.\begin{matrix} O \\ O \end{matrix}\right\}^{\ominus}$$

That the anion is indeed a resonance hybrid is supported by the evidence of bond length. Formic acid, for example, contains a carbon–oxygen double bond and a carbon–oxygen single bond; we would expect these bonds to have different lengths. Sodium formate, on the other hand, if it is a resonance hybrid, ought to contain two equivalent carbon–oxygen bonds; we would expect these to have the same length, intermediate between double and single bonds. X-ray and electron diffraction show that these expectations are correct. Formic acid contains one carbon–oxygen bond of 1.36 A (single bond) and another of 1.23 A (double bond); sodium formate contains two equal carbon–oxygen bonds, each 1.27 A long.

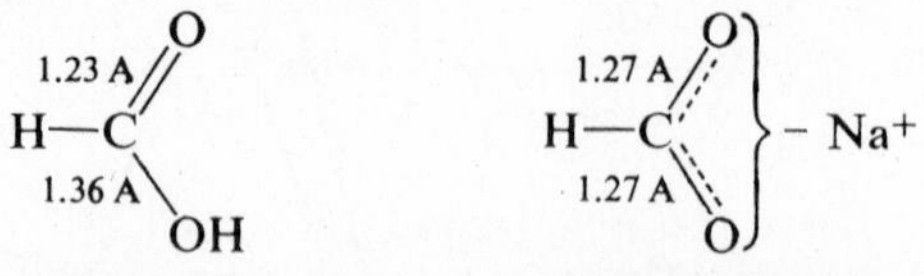

$$H-C\begin{matrix} \overset{1.23\,A}{\diagup\!\!\diagup} O \\ \underset{1.36\,A}{\diagdown} OH \end{matrix} \qquad\qquad H-C\left.\begin{matrix} \overset{1.27\,A}{\diagup} O \\ \underset{1.27\,A}{\diagdown} O \end{matrix}\right\}^{-} Na^+$$

Formic acid Sodium formate

Problem 18.5 How do you account for the fact that the three carbon–oxygen bonds in $CaCO_3$ have the same length, and that this length (1.31 A) is greater than that found in sodium formate?

What does this resonance mean in terms of orbitals? Carboxyl carbon is joined to the three other atoms by σ bonds (Fig. 18.3); since these bonds utilize sp^2 orbitals (Sec. 5.2), they lie in a plane and are 120° apart. The remaining p orbital of the carbon overlaps equally well p orbitals from *both* of the oxygens, to form hybrid bonds (compare benzene, Sec. 10.8). In this way the electrons are bound not just to one or two nuclei but to *three* nuclei (one carbon and two oxygens); they are therefore held more tightly, the bonds are stronger, and the

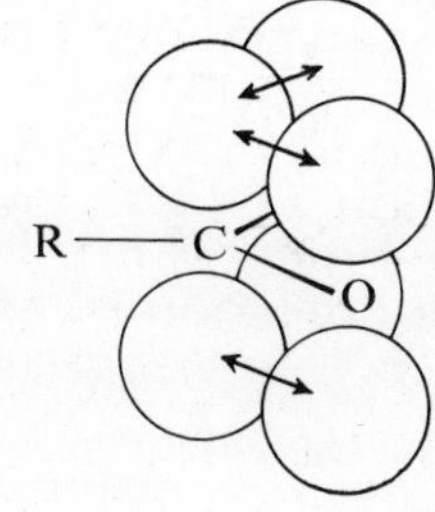

Figure 18.3. Carboxylate ion. Overlap of *p* orbitals in both directions: delocalization of π electrons, and dispersal of charge.

anion is more stable. This participation of electrons in more than one bond, this smearing-out or delocalization of the electron cloud, is what is meant by representing the anion as a resonance hybrid of two structures.

Problem 18.6 How do you account for the fact that the α-hydrogens of an aldehyde (say, *n*-butyraldehyde) are much more acidic than any other hydrogens in the molecule? (Check your answer in Sec. 21.1.)

$$\overset{\gamma}{C}H_3\overset{\beta}{C}H_2\overset{\alpha}{C}H_2\overset{H}{C}{=}O$$

n-Butyraldehyde

18.14 Effect of substituents on acidity

Next, let us see how changes in the structure of the group bearing the —COOH affect the acidity. Any factor that stabilizes the anion more than it stabilizes the acid should increase the acidity; any factor that makes the anion less stable should decrease acidity. From what we have learned about carbonium ions, we know what we might reasonably expect. Electron-withdrawing substituents should disperse the negative charge, stabilize the anion, and thus increase acidity. Electron-releasing substituents should intensify the negative charge, destabilize the anion, and thus decrease acidity.

Acid Strength

$$G{\leftarrow}C\left\{\begin{matrix}O\\ \\ O\end{matrix}\right\}^{\ominus} \qquad\qquad G{\rightarrow}C\left\{\begin{matrix}O\\ \\ O\end{matrix}\right\}^{\ominus}$$

G withdraws electrons: *stabilizes anion, strengthens acid*

G releases electrons: *destabilizes anion, weakens acid*

The K_a's listed in Table 18.2 are in agreement with this prediction.

Looking first at the aliphatic acids, we see that the electron-withdrawing halogens strengthen acids: chloroacetic acid is 100 times as strong as acetic acid, dichloroacetic acid is still stronger, and trichloroacetic acid is more than 10,000 times as strong as the unsubstituted acid. The other halogens exert similar effects.

Table 18.2 ACIDITY CONSTANTS OF CARBOXYLIC ACIDS

	K_a		K_a
HCOOH	17.7×10^{-5}	$CH_3CHClCH_2COOH$	8.9×10^{-5}
CH_3COOH	1.75 "	$ClCH_2CH_2CH_2COOH$	2.96 "
$ClCH_2COOH$	136 "	FCH_2COOH	260 "
$Cl_2CHCOOH$	5530 "	$BrCH_2COOH$	125 "
Cl_3CCOOH	23200 "	ICH_2COOH	67 "
$CH_3CH_2CH_2COOH$	1.52 "	$C_6H_5CH_2COOH$	4.9 "
$CH_3CH_2CHClCOOH$	139 "	$p\text{-}O_2NC_6H_4CH_2COOH$	14.1 "

ACIDITY CONSTANTS OF SUBSTITUTED BENZOIC ACIDS

K_a of benzoic acid $= 6.3 \times 10^{-5}$

	K_a		K_a		K_a
p-NO_2	36×10^{-5}	*m*-NO_2	32×10^{-5}	*o*-NO_2	670×10^{-5}
p-Cl	10.3 "	*m*-Cl	15.1 "	*o*-Cl	120 "
p-CH_3	4.2 "	*m*-CH_3	5.4 "	*o*-CH_3	12.4 "
p-OCH_3	3.3 "	*m*-OCH_3	8.2 "	*o*-OCH_3	8.2 "
p-OH	2.6 "	*m*-OH	8.3 "	*o*-OH	105 "
p-NH_2	1.4 "	*m*-NH_2	1.9 "	*o*-NH_2	1.6 "

Problem 18.7 (a) What do the K_a's of the monohaloacetic acids tell us about the relative strengths of the inductive effects of the different halogens? (b) On the basis of Table 18.2, what kind of inductive effect does the phenyl group, $—C_6H_5$, appear to have?

α-Chlorobutyric acid is about as strong as chloroacetic acid. As the chlorine is moved away from the —COOH, however, its effect rapidly dwindles: β-chlorobutyric acid is only six times as strong as butyric acid, and γ-chlorobutyric acid is only twice as strong. It is typical of inductive effects that they decrease rapidly with distance, and are seldom important when acting through more than four atoms.

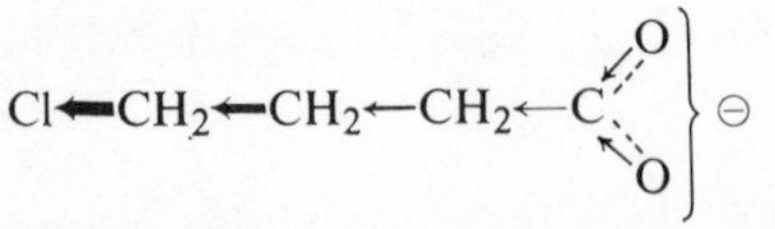

$$\left[Cl \leftarrow CH_2 \leftarrow CH_2 \leftarrow CH_2 \leftarrow C\begin{matrix} O \\ O \end{matrix} \right]^{\ominus}$$

Inductive effect: *decreases with distance*

The aromatic acids are similarly affected by substituents: $—CH_3$, —OH, and $—NH_2$ make benzoic acid weaker, and —Cl and $—NO_2$ make benzoic acid stronger. We recognize the acid-weakening groups as the ones that activate the ring toward electrophilic substitution (and deactivate toward nucleophilic substitution). The acid-strengthening groups are the ones that deactivate toward electrophilic substitution (and activate toward nucleophilic substitution). Furthermore, the groups that have the largest effects on reactivity—whether activating or deactivating—have the largest effects on acidity.

The —OH and $—OCH_3$ groups display both kinds of effect we have attributed to them (Sec. 11.20): from the *meta* position, an electron-withdrawing acid-strengthening inductive effect; and from the *para* position, an electron-releasing acid-weakening resonance effect (which at this position outweighs the inductive effect). Compare the two effects exerted by halogen (Sec. 11.21).

ortho-Substituted aromatic acids do not fit into the pattern set by their *meta* and *para* isomers, and by aliphatic acids. Nearly all *ortho* substituents exert an effect of the same kind—acid-strengthening—whether they are electron-withdrawing or electron-releasing, and the effect is unusually large. (Compare, for example, the effects of *o*-NO_2 and *o*-CH_3, of *o*-NO_2 and *m*- or *p*-NO_2.) This *ortho* effect undoubtedly has to do with the *nearness* of the groups involved, but is more than just steric hindrance arising from their bulk.

Thus we see that the same concepts—inductive effect and resonance—that we found so useful in dealing with rates of reaction are also useful in dealing with equilibria. By using these concepts to estimate the stabilities of anions, we are able to predict the relative strengths of acids; in this way we can account not only for the effect of substituents on the acid strength of carboxylic acids but also for the very fact that the compounds are acids.

Problem 18.8 There is evidence that certain groups like *p*-methoxy weaken the acidity of benzoic acids not so much by destabilizing the anion as by stabilizing the acid. Draw structures to show the kind of resonance that might be involved. Why would you expect such resonance to be more important for the acid than for the anion?

18.15 Conversion into acid chlorides

A carboxylic acid is perhaps more often converted into the acid chloride than into any other of its functional derivatives. From the highly reactive acid chloride there can then be obtained many other kinds of compounds, including esters and amides (Sec. 20.8).

An acid chloride is prepared by substitution of —Cl for the —OH of a carboxylic acid. Three reagents are commonly used for this purpose: *thionyl chloride*, $SOCl_2$; *phosphorus trichloride*, PCl_3; and *phosphorus pentachloride*, PCl_5. (Of what inorganic acids might we consider these reagents to be the acid chlorides?) For example:

$$C_6H_5\text{—}C(=O)\text{—}OH + SOCl_2 \xrightarrow{\text{reflux}} C_6H_5\text{—}C(=O)\text{—}Cl + SO_2 + HCl$$

Benzoic acid → Benzoyl chloride

$$3,5\text{-}(O_2N)_2C_6H_3COOH + PCl_5 \xrightarrow{\text{heat}} 3,5\text{-}(O_2N)_2C_6H_3COCl + POCl_3 + HCl$$

3,5-Dinitrobenzoic acid → 3,5-Dinitrobenzoyl chloride

Thionyl chloride is particularly convenient, since the products formed besides the acid chloride are gases and thus easily separated from the acid chloride; any excess of the low-boiling thionyl chloride (79°) is easily removed by distillation.

18.16 Conversion into esters

Acids are frequently converted into their esters via the acid chlorides:

$$\underset{\text{Acid}}{\text{RCOOH}} \xrightarrow{\text{SOCl}_2\text{, etc.}} \underset{\text{Acid chloride}}{\text{RCOCl}} \xrightarrow{\text{R'OH}} \underset{\text{Ester}}{\text{RCOOR'}}$$

A carboxylic acid is converted directly into an ester when heated with an alcohol in the presence of a little mineral acid, usually concentrated sulfuric acid or dry hydrogen chloride. This reaction is reversible, and generally reaches equilibrium when there are appreciable quantities of both reactants and products present.

$$\underset{\text{Acid}}{\text{RCOOH}} + \underset{\text{Alcohol}}{\text{R'OH}} \overset{\text{H}^+}{\rightleftharpoons} \underset{\text{Ester}}{\text{RCOOR'}} + \text{H}_2\text{O}$$

For example, when we allow one mole of acetic acid and one mole of ethyl alcohol to react in the presence of a little sulfuric acid until equilibrium is reached (after several hours), we obtain a mixture of about two-thirds mole each of ester and water, and one-third mole each of acid and alcohol. We obtain this same equilibrium mixture, of course, if we start with one mole of ester and one mole of water, again in the presence of sulfuric acid. *The same catalyst, hydrogen ion, that catalyzes the forward reaction, esterification, necessarily catalyzes the reverse reaction, hydrolysis.*

This reversibility is a disadvantage in the preparation of an ester directly from an acid; the preference for the acid chloride route is due to the fact that both steps—preparation of acid chloride from acid, and preparation of ester from acid chloride—are essentially irreversible and go to completion.

Direct esterification, however, has the advantage of being a single-step synthesis; it can often be made useful by application of our knowledge of equilibria. If either the acid or the alcohol is cheap and readily available, it can be used in large excess to shift the equilibrium toward the products and thus to increase the yield of ester. For example, it is worthwhile to use eight moles of cheap ethyl alcohol to convert one mole of valuable γ-phenylbutyric acid more completely into the ester:

$$\underset{\substack{\gamma\text{-Phenylbutyric acid} \\ \textit{1 mole}}}{\text{C}_6\text{H}_5\text{CH}_2\text{CH}_2\text{CH}_2\text{C(=O)OH}} + \underset{\substack{\text{Ethyl alcohol} \\ \textit{8 moles}}}{\text{C}_2\text{H}_5\text{OH}} \xrightleftharpoons{\text{H}_2\text{SO}_4\text{, reflux}} \underset{\substack{\text{Ethyl } \gamma\text{-phenylbutyrate} \\ \textit{85–88\% yield}}}{\text{C}_6\text{H}_5\text{CH}_2\text{CH}_2\text{CH}_2\text{C(=O)OC}_2\text{H}_5} + \text{H}_2\text{O}$$

Sometimes the equilibrium is shifted by removing one of the products. An elegant way of doing this is illustrated by the preparation of ethyl adipate. The dicarboxylic acid adipic acid, an excess of ethyl alcohol, and toluene are heated with a little sulfuric acid under a distillation column. The lowest boiling component (b.p. 75°) of the reaction mixture is an azeotrope of water, ethyl alcohol, and toluene (compare Sec. 15.6); consequently, as fast as water is formed it is removed as the azeotrope by distillation. In this way a 95–97% yield of ester is obtained:

$$\underset{\substack{\text{Adipic acid}\\ \textit{Non-volatile}}}{HOOC(CH_2)_4COOH} + \underset{\substack{\text{Ethyl alcohol}\\ \textit{B.p. 78°}}}{2C_2H_5OH} \underset{\longleftarrow}{\xrightarrow{\text{toluene (b.p. 111°), } H_2SO_4}} \underset{\substack{\text{Ethyl adipate}\\ \textit{B.p. 245°}}}{C_2H_5OOC(CH_2)_4COOC_2H_5} + \underset{\substack{\textit{Removed as}\\ \textit{azeotrope, b.p. 75°}}}{2H_2O}$$

The equilibrium is particularly unfavorable when phenols (ArOH) are used instead of alcohols; yet, if water is removed during the reaction, phenolic esters (RCOOAr) are obtained in high yield.

The presence of bulky groups near the site of reaction, whether in the alcohol or in the acid, slows down esterification (as well as its reverse, hydrolysis). This

Reactivity in esterification

$CH_3OH > 1° > 2° (> 3°)$

$HCOOH > CH_3COOH > RCH_2COOH > R_2CHCOOH > R_3CCOOH$

steric hindrance can be so marked that special methods are required to prepare esters of tertiary alcohols or esters of acids like 2,4,6-trimethylbenzoic acid (mesitoic acid).

The mechanism of esterification is necessarily the exact reverse of the mechanism of hydrolysis of esters. We shall discuss both mechanisms when we take up the chemistry of esters (Sec. 20.18) after we have learned a little more about the carbonyl group.

Problem 18.9 (a) In the formation of an acid chloride, which bond of a carboxylic acid is broken, C—OH or CO—H? (b) When labeled methanol, $CH_3{}^{18}OH$, was allowed to react with ordinary benzoic acid, the methyl benzoate produced was found to be enriched in ^{18}O, whereas the water formed contained only ordinary oxygen. In this esterification, which bond of the carboxylic acid is broken, C—OH or CO—H? Which bond of the alcohol?

18.17 Conversion into amides

Amides are compounds in which the —OH of the carboxylic acid has been

$$\underset{\text{Acid}}{RCOOH} \longrightarrow \underset{\text{Acid chloride}}{RCOCl} \xrightarrow{NH_3} \underset{\text{Amide}}{R{-}C(=O)NH_2}$$

replaced by $-NH_2$. These are generally prepared by reaction of ammonia with acid chlorides.

18.18 Reduction of acids to alcohols

Conversion of alcohols into acids (Sec. 18.6) is important because, in general, alcohols are more available than acids. This is not always true, however; long straight-chain acids from fats are more available than are the corresponding alcohols, and here the reverse process becomes important: reduction of acids to alcohols.

Lithium aluminum hydride, $LiAlH_4$, is one of the few reagents that can reduce an acid to an alcohol; the initial product is an alkoxide from which the alcohol is liberated by hydrolysis:

$$4RCOOH + 3LiAlH_4 \longrightarrow 4H_2 + 2LiAlO_2 + (RCH_2O)_4AlLi \xrightarrow{H_2O} \underset{1°\ \text{alcohol}}{4RCH_2OH}$$

Because of the excellent yields it gives, $LiAlH_4$ is widely used in the laboratory for the reduction of not only acids but many other classes of compounds. Since it is somewhat expensive, it can be used in industry only for the reduction of small amounts of valuable raw materials, as in the synthesis of certain drugs and hormones.

As an alternative to direct reduction, acids are often converted into alcohols by a two-step process: esterification, and reduction of the ester. Esters can be reduced in a number of ways (Sec. 20.22) that are adaptable to both laboratory and industry.

We have seen (Sec. 18.5) that in the carboxylic acids obtained from fats we have available long straight-chain units for use in organic synthesis. Reduction of these acids to alcohols (either directly or as esters) is a fundamental step in the utilization of these raw materials, since from the alcohols, as we know, a host of other compounds can be prepared (Sec. 16.10). Although only acids of even carbon number are available, it is possible, of course, to increase the chain length and thus prepare compounds of odd carbon number. (For an alternative source of alcohols both of even and odd carbon number, see Sec. 32.6.)

Problem 18.10 Outline the synthesis from lauric acid (n-$C_{11}H_{23}COOH$, dodecanoic acid) of the following compounds: (a) 1-bromododecane; (b) tridecanoic acid (C_{13} acid); (c) 1-tetradecanol; (d) 1-dodecene; (e) dodecane; (f) 1-dodecyne; (g) methyl n-decyl ketone; (h) 2-dodecanol; (i) undecanoic acid; (j) 2-tetradecanol; (k) 2-methyl-2-tetradecanol.

18.19 Halogenation of aliphatic acids. Substituted acids

In the presence of a small amount of phosphorus, aliphatic carboxylic acids react smoothly with chlorine or bromine to yield a compound in which α-hydrogen has been replaced by halogen. This is the **Hell-Volhard-Zelinsky reaction.** Because of its specificity—*only alpha halogenation*—and the readiness with which it takes place, it is of considerable importance in synthesis.

$$CH_3COOH \xrightarrow{Cl_2,\ P} ClCH_2COOH \xrightarrow{Cl_2,\ P} Cl_2CHCOOH \xrightarrow{Cl_2,\ P} Cl_3CCOOH$$

$$CH_3CH_2COOH \xrightarrow{Br_2,\ P} CH_3CHBrCOOH \xrightarrow{Br_2,\ P} CH_3CBr_2COOH$$

$$\downarrow Br_2,\ P$$

no further substitution

The function of the phosphorus is ultimately to convert a little of the acid into acid halide. In this form (for reasons we cannot go into here) each molecule of acid sooner or later undergoes α-halogenation.

$$P + X_2 \longrightarrow PX_3$$

$$RCH_2COOH + PX_3 \longrightarrow RCH_2COX$$

$$RCH_2COX + X_2 \longrightarrow \underset{\displaystyle X}{\underset{|}{RCHCOX}} + HX$$

$$\underset{\displaystyle X}{\underset{|}{RCHCOX}} + RCH_2COOH \rightleftarrows \underset{\substack{\displaystyle X \\ \alpha\text{-Haloacid}}}{\underset{|}{RCHCOOH}} + RCH_2COX$$

The halogen of these halogenated acids undergoes *nucleophilic displacement* and *elimination* much as it does in the simpler alkyl halides (Secs. 14.5 and 5.12). Halogenation is therefore the first step in the conversion of a carboxylic acid into many important substituted carboxylic acids:

$$\underset{\substack{\displaystyle Br \\ \text{An } \alpha\text{-halogenated acid}}}{\underset{|}{RCHCOOH}} + \text{large excess of } NH_3 \longrightarrow \underset{\substack{\displaystyle NH_2 \\ \text{An } \alpha\text{-amino acid}}}{\underset{|}{RCHCOOH}}$$

$$\underset{\displaystyle Br}{\underset{|}{RCHCOOH}} + NaOH \longrightarrow \underset{\displaystyle OH}{\underset{|}{RCHCOONa}} \xrightarrow{H^+} \underset{\substack{\displaystyle OH \\ \text{An } \alpha\text{-hydroxy acid}}}{\underset{|}{RCHCOOH}}$$

$$RCH_2\underset{\displaystyle Br}{\underset{|}{C}}HCOOH + KOH\ (alc) \longrightarrow RCH{=}CHCOO^- \xrightarrow{H^+} \underset{\text{An } \alpha,\beta\text{-unsaturated acid}}{RCH{=}CHCOOH}$$

These new substituents can, in turn, undergo *their* characteristic reactions.

Problem 18.11 Predict the product of each of the following reactions:

(a) $CH_2{=}CHCOOH + H_2/Ni$
(b) *trans*-$CH_3CH{=}CHCOOH + Br_2/CCl_4$
(c) $C_6H_5CH(OH)CH_2COOH + H^+$, heat $\longrightarrow C_9H_8O_2$
(d) *o*-$HOOCC_6H_4CH_2OH + H^+$, heat $\longrightarrow C_8H_6O_2$

18.20 Dicarboxylic acids

If the substituent is a second carboxyl group, we have a *dicarboxylic acid.* For example:

$HOOCCH_2COOH$	$HOOCCH_2CH_2COOH$	$HOOCCH_2CH_2CH_2CH_2COOH$
Malonic acid	Succinic acid	Adipic acid
Propanedioic acid	Butanedioic acid	Hexanedioic acid

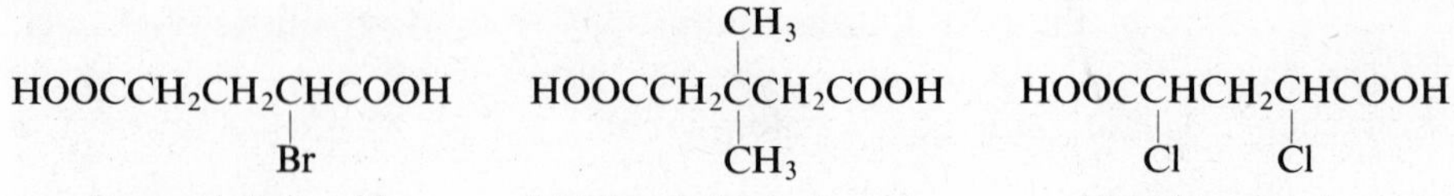

$$\underset{\text{Br}}{\text{HOOCCH}_2\text{CH}_2\overset{|}{\text{C}}\text{HCOOH}}$$
α-Bromoglutaric acid
2-Bromopentanedioic acid

$$\text{HOOCCH}_2\overset{\text{CH}_3}{\underset{\text{CH}_3}{\text{C}}}\text{CH}_2\text{COOH}$$
β,β-Dimethylglutaric acid
3,3-Dimethylpentanedioic acid

$$\text{HOOC}\underset{\text{Cl}}{\text{C}}\text{HCH}_2\underset{\text{Cl}}{\text{C}}\text{HCOOH}$$
α,α'-Dichloroglutaric acid
2,4-Dichloro-pentanedioic acid

We have already encountered the benzenedicarboxylic acids, the *phthalic acids* (Sec. 12.10).

Table 18.3 Dicarboxylic Acids

Name	Formula	M.p., °C	Solub., g/100 g H_2O at 20°	K_1	K_2
Oxalic	HOOC—COOH	189	9	5400×10^{-5}	5.2×10^{-5}
Malonic	$HOOCCH_2COOH$	136	74	140	0.20
Succinic	$HOOC(CH_2)_2COOH$	185	6	6.4	0.23
Glutaric	$HOOC(CH_2)_3COOH$	98	64	4.5	0.38
Adipic	$HOOC(CH_2)_4COOH$	151	2	3.7	0.39
Maleic	*cis*-HOOCCH=CHCOOH	130.5	79	1000	0.055
Fumaric	*trans*-HOOCCH=CHCOOH	302	0.7	96	4.1
Phthalic	1,2-$C_6H_4(COOH)_2$	231	0.7	110	0.4
Isophthalic	1,3-$C_6H_4(COOH)_2$	348.5	0.01	24	2.5
Terephthalic	1,4-$C_6H_4(COOH)_2$	300 *subl*	0.002	29	3.5

Most dicarboxylic acids are prepared by adaptation of methods used to prepare monocarboxylic acids. Where hydrolysis of a nitrile yields a monocarboxylic acid, hydrolysis of a dinitrile or a cyanocarboxylic acid yields a dicarboxylic acid; where oxidation of a methylbenzene yields a benzoic acid, oxidation of a dimethylbenzene yields a phthalic acid. For example:

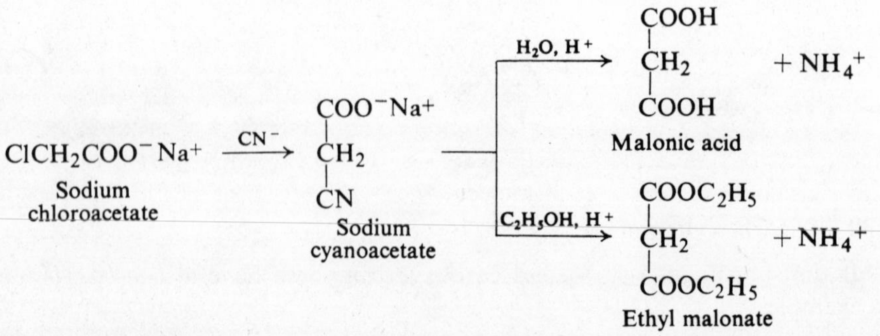

$$\underset{\text{Sodium chloroacetate}}{\text{ClCH}_2\text{COO}^-\text{Na}^+} \xrightarrow{\text{CN}^-} \underset{\text{Sodium cyanoacetate}}{\text{CN—CH}_2\text{—COO}^-\text{Na}^+} \begin{cases} \xrightarrow{\text{H}_2\text{O, H}^+} \underset{\text{Malonic acid}}{\text{COOH—CH}_2\text{—COOH}} + \text{NH}_4^+ \\ \xrightarrow{\text{C}_2\text{H}_5\text{OH, H}^+} \underset{\text{Ethyl malonate}}{\text{COOC}_2\text{H}_5\text{—CH}_2\text{—COOC}_2\text{H}_5} + \text{NH}_4^+ \end{cases}$$

Problem 18.12 Why is chloroacetic acid converted into its salt before treatment with cyanide in the above preparation?

Problem 18.13 Outline a synthesis of: (a) pentanedioic acid from 1,3-propanediol (available from a fermentation of glycerol); (b) nonanedioic acid from *cis*-9-octadecenoic acid (oleic acid, obtained from fats); (c) succinic acid from 1,4-butynediol (available from acetylene and formaldehyde).

In general, dicarboxylic acids show the same chemical behavior as monocarboxylic acids. It is possible to prepare compounds in which only one of the carboxyl groups has been converted into a derivative; it is possible to prepare compounds in which the two carboxyl groups have been converted into different derivatives.

Problem 18.14 Predict the products of the following reactions:

(a) adipic acid (146 g) + 95% ethanol (146 g) + benzene + conc. H_2SO_4, 100°
(b) adipic acid (146 g) + 95% ethanol (50 g) + benzene + conc. H_2SO_4, 100°
(c) succinic acid + $LiAlH_4$
(d) pentanedioic acid + 1 mole Br_2, P
(e) terephthalic acid + excess $SOCl_2$
(f) maleic acid (*cis*-butenedioic acid) + Br_2/CCl_4

As with other acids containing more than one ionizable hydrogen (H_2SO_4, H_2CO_3, H_3PO_4, etc.), ionization of the second carboxyl group occurs less readily than ionization of the first (compare K_1's with K_2's in Table 18.3). More energy

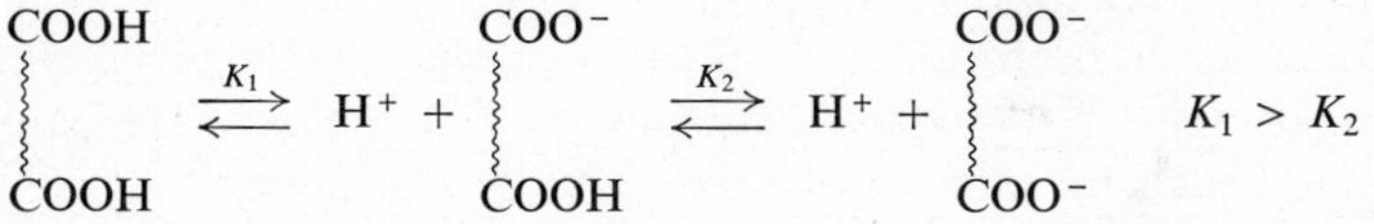

$$\begin{matrix} COOH \\ \wr \\ COOH \end{matrix} \underset{}{\overset{K_1}{\rightleftharpoons}} H^+ + \begin{matrix} COO^- \\ \wr \\ COOH \end{matrix} \underset{}{\overset{K_2}{\rightleftharpoons}} H^+ + \begin{matrix} COO^- \\ \wr \\ COO^- \end{matrix} \qquad K_1 > K_2$$

is required to separate a positive hydrogen ion from the doubly charged anion than from the singly charged anion.

Problem 18.15 Compare the acidity (first ionization) of oxalic acid with that of formic acid; of malonic acid with that of acetic acid. How do you account for these differences?

Problem 18.16 Arrange oxalic, malonic, succinic, and glutaric acids in order of acidity (first ionization). How do you account for this order?

In addition to the reactions typical of any carboxylic acid, we shall find, some of these dicarboxylic acids undergo reactions that are possible only because there are two carboxyl groups in each molecule, and because these carboxyl groups are located in a particular way with respect to each other.

Problem 18.17 Give a likely structure for the product of each of the following reactions:

(a) oxalic acid + ethylene glycol $\longrightarrow$ $C_4H_4O_4$
(b) succinic acid + heat $\longrightarrow$ $C_4H_4O_3$
(c) terephthalic acid + ethylene glycol $\longrightarrow$ $(C_{10}H_8O_4)_n$, the polymer Dacron

18.21 Analysis of carboxylic acids. Neutralization equivalent

Carboxylic acids are recognized through their acidity. They dissolve in aqueous sodium hydroxide and in aqueous sodium bicarbonate. The reaction with bicarbonate releases bubbles of carbon dioxide (see Sec. 18.4).

(Phenols, Sec. 24.7, are more acidic than water, but—with certain exceptions—are considerably weaker than carboxylic acids; they dissolve in aqueous sodium hydroxide, but *not* in aqueous sodium bicarbonate. Sulfonic acids are even more acidic than carboxylic acids, but they contain sulfur, which can be detected by elemental analysis.)

Once characterized as a carboxylic acid, an unknown is identified as a particular acid on the usual basis of its physical properties and the physical properties of derivatives. The derivatives commonly used are *amides* (Secs. 20.11 and 23.6) and *esters* (Sec. 20.15).

Problem 18.18 Expand the table you made in Problem 17.24, p. 570, to include the kinds of compounds and tests we have taken up since then.

Particularly useful both in identification of previously studied acids and in proof of structure of new ones is the **neutralization equivalent**: *the equivalent weight of the acid as determined by titration with standard base.* A weighed sample of the acid is dissolved in water or aqueous alcohol, and the volume of standard base needed to neutralize the solution is measured. For example, a 0.224-g sample of an unknown acid (m.p. 139–140°) required 13.6 ml of 0.104 N sodium hydroxide solution for neutralization (to a phenolphthalein end point). Since each 1000 ml of the base contains 0.104 equivalents, and since the number of equivalents of base required equals the number of equivalents of acid present,

$$\frac{13.6}{1000} \times 0.104 \text{ equivalents of acid} = 0.224 \text{ g}$$

and

$$1 \text{ equivalent of acid} = 0.224 \times \frac{1000}{13.6} \times \frac{1}{0.104} = 158 \text{ g}$$

Problem 18.19 Which of the following compounds might the above acid be: (a) *o*-chlorobenzoic acid (m.p. 141°) or (b) 2,6-dichlorobenzoic acid (m.p. 139°)?

Problem 18.20 A 0.187-g sample of an acid (b.p. 203–205°) required 18.7 ml of 0.0972 N NaOH for neutralization. (a) What is the neutralization equivalent? (b) Which of the following acids might it be: *n*-caproic acid (b.p. 205°), methoxyacetic acid (b.p. 203°), or ethoxyacetic acid (b.p. 206°)?

Problem 18.21 (a) How many equivalents of base would be neutralized by one mole of phthalic acid? What is the neutralization equivalent of phthalic acid? (b) What is the relation between neutralization equivalent and the number of acidic hydrogens per molecule of acid? (c) What is the neutralization equivalent of 1,3,5-benzenetricarboxvlic acid? Of mellitic acid, $C_6(COOH)_6$?

A metal salt of a carboxylic acid is recognized through these facts: (a) it leaves a residue when strongly heated (*ignition test*); (b) it decomposes at a fairly high temperature, instead of melting; and (c) it is converted into a carboxylic acid upon treatment with dilute mineral acid.

Problem 18.22 The residue left upon ignition of a sodium salt of a carboxylic acid was white, soluble in water, turned moist litmus blue, and reacted with dilute hydrochloric acid with the formation of bubbles. What was its probable chemical composition?

18.22 Spectroscopic analysis of carboxylic acids

Infrared. The carboxyl group is made up of a carbonyl group (C=O) and a hydroxyl group (OH), and the infrared spectrum of carboxylic acids reflects both these structural units. For hydrogen-bonded (dimeric) acids, O—H stretching gives a strong, broad band in the 2500–3000 cm^{-1} range (see Fig. 18.4, below).

O—H stretching, *strong, broad*

—COOH and enols	2500–3000 cm^{-1}
ROH and ArOH	3200–3600 cm^{-1}

With acids we encounter, for the first time, absorption due to stretching of the carbonyl group. This strong band appears in a region that is usually free of other

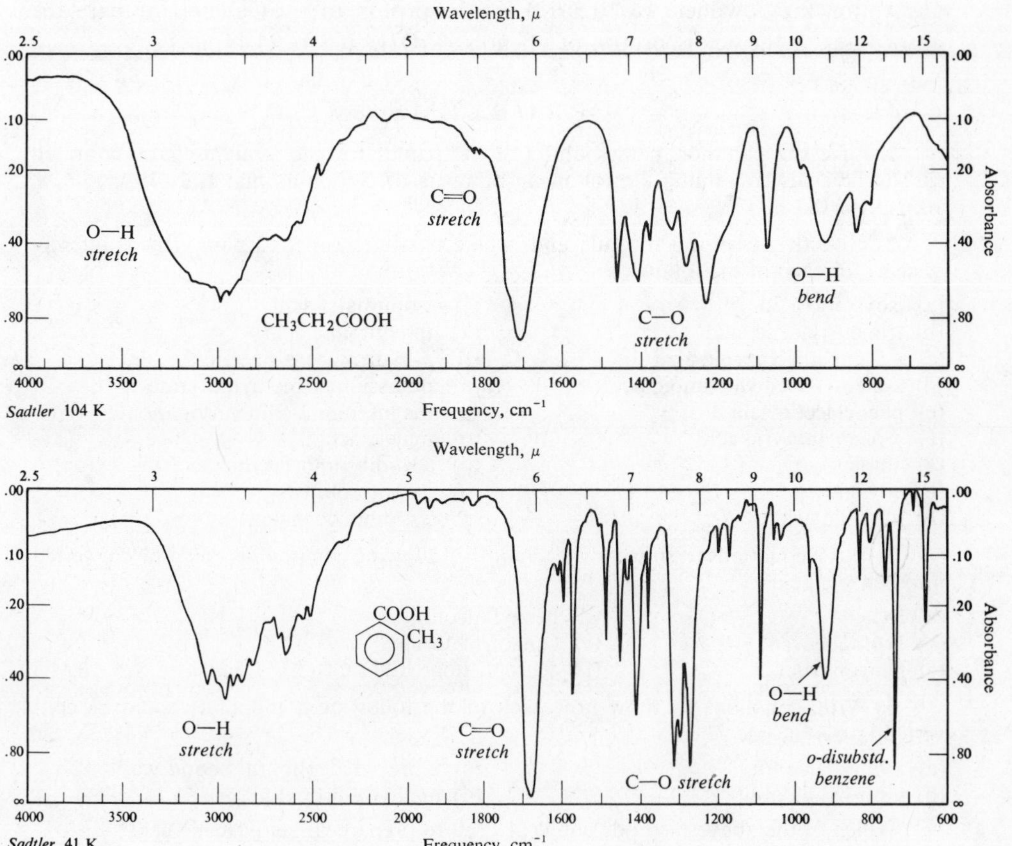

Figure 18.4. Infrared spectra of (*a*) propionic acid and (*b*) *o*-toluic acid.

strong absorption, and by its exact frequency gives much information about structure. For (hydrogen-bonded) acids, the C=O band is at about 1700 cm^{-1}.

C=O stretching, *strong*

$R-C(=O)-OH$	1700–1725 cm^{-1}	$-C=C-C(=O)-OH$	1680–1700 cm^{-1}
$Ar-C(=O)-OH$	1680–1700 cm^{-1}	$-C(OH)=CH-C(=O)-$ (OH·······O) (enols)	1540–1640 cm^{-1}

Acids also show a C—O stretching band at about 1250 cm^{-1} (compare alcohols, Sec. 16.13, and ethers, Sec. 17.17), and bands for O—H bending near 1400 cm^{-1} and 920 cm^{-1} (*broad*).

Enols, too, show both O—H and C=O absorption; these can be distinguished by the particular frequency of the C=O band. Aldehydes, ketones, and esters show carbonyl absorption, but the O—H band is missing. (For a comparison of certain oxygen compounds, see Table 20.3, p. 689.)

Nmr. The outstanding feature of the nmr spectrum of a carboxylic acid is the absorption far downfield (δ 10.5–12) by the proton of —COOH. (Compare the absorption by the acidic proton of phenols, ArOH, in Sec. 24.14.)

PROBLEMS

1. Give the common names and IUPAC names for the straight-chain saturated carboxylic acids containing the following numbers of carbon atoms: 1, 2, 3, 4, 5, 6, 8, 10, 12, 16, 18.

2. Give the structural formula and, where possible, a second name (by a different system) for each of the following:

(a) isovaleric acid
(b) trimethylacetic acid
(c) α,β-dimethylcaproic acid
(d) 2-methyl-4-ethyloctanoic acid
(e) phenylacetic acid
(f) γ-phenylbutyric acid
(g) adipic acid
(h) *p*-toluic acid
(i) phthalic acid
(j) isophthalic acid
(k) terephthalic acid
(l) *p*-hydroxybenzoic acid
(m) potassium α-methylbutyrate
(n) magnesium 2-chloropropanoate
(o) maleic acid
(p) α,α'-dibromosuccinic acid
(q) isobutyronitrile
(r) 2,4-dinitrobenzonitrile

3. Write equations to show how each of the following compounds could be converted into benzoic acid:

(a) toluene
(b) bromobenzene
(c) benzonitrile
(d) benzyl alcohol
(e) benzotrichloride
(f) acetophenone, $C_6H_5COCH_3$ (*Hint:* See Sec. 16.11.)

4. Write equations to show how each of the following compounds could be converted into *n*-butyric acid:

(a) *n*-butyl alcohol
(b) *n*-propyl alcohol
(c) *n*-propyl alcohol (a second way)
(d) methyl *n*-propyl ketone

Which of the above methods could be used to prepare trimethylacetic acid?

5. Write equations to show how tetrahydrofuran could be converted into: (a) succinic acid; (b) glutaric acid; (c) adipic acid.

6. Write equations to show the reaction (if any) of benzoic acid with:

(a) KOH
(b) Al
(c) CaO
(d) Na_2CO_3
(e) $NH_3(aq)$
(f) H_2, Ni, 20°, 1 atm.
(g) $LiAlH_4$
(h) hot $KMnO_4$
(i) PCl_5
(j) PCl_3
(k) $SOCl_2$
(l) Br_2/Fe
(m) $Br_2 + P$
(n) HNO_3/H_2SO_4
(o) fuming sulfuric acid
(p) CH_3Cl, $AlCl_3$
(q) $Tl(OOCCF_3)_3$
(r) *n*-propyl alcohol, H^+

7. Answer Problem 6 for *n*-valeric acid.

8. Write equations to show how isobutyric acid could be converted into each of the following, using any needed reagents.

(a) ethyl isobutyrate
(b) isobutyryl chloride
(c) isobutyramide
(d) magnesium isobutyrate
(e) isobutyl alcohol

9. Write equations to show all steps in the conversion of benzoic acid into:

(a) sodium benzoate
(b) benzoyl chloride
(c) benzamide
(d) benzene
(e) *n*-propyl benzoate
(f) *p*-tolyl benzoate
(g) *m*-bromophenyl benzoate
(h) benzyl alcohol

10. Write equations to show how phenylacetic acid could be converted into each of the following, using any needed reagents.

(a) sodium phenylacetate
(b) ethyl phenylacetate
(c) phenylacetyl chloride
(d) phenylacetamide
(e) *p*-bromophenylacetic acid
(f) *p*-nitrophenylacetic acid
(g) β-phenylethyl alcohol
(h) α-bromophenylacetic acid
(i) α-aminophenylacetic acid
(j) α-hydroxyphenylacetic acid
(k) phenylmalonic acid, $C_6H_5CH(COOH)_2$

11. Complete the following, giving the structures and names of the principal organic products.

(a) $C_6H_5CH{=}CHCOOH + KMnO_4 + OH^- +$ heat
(b) $p\text{-}CH_3C_6H_4COOH + HNO_3 + H_2SO_4$
(c) succinic acid + $LiAlH_4$, followed by H^+
(d) $C_6H_5COOH + C_6H_5CH_2OH + H^+$
(e) product (d) + $HNO_3 + H_2SO_4$
(f) $C_6H_5CH_2COOH + Tl(OOCCF_3)_3$
(g) cyclo-$C_6H_{11}MgBr + CO_2$, followed by H_2SO_4
(h) product (g) + $C_2H_5OH + H^+$
(i) product (g) + $SOCl_2$ + heat
(j) $m\text{-}CH_3C_6H_4OCH_3 + KMnO_4 + OH^-$
(k) mesitylene + $K_2Cr_2O_7 + H_2SO_4$
(l) isobutyric acid + isobutyl alcohol + H^+
(m) salicylic acid ($o\text{-}HOC_6H_4COOH$) + Br_2, Fe
(n) sodium acetate + *p*-nitrobenzyl bromide (What would you predict?)
(o) linolenic acid + excess H_2, Ni
(p) oleic acid + $KMnO_4$, heat
(q) linoleic acid + O_3, then H_2O, Zn
(r) benzoic acid ($C_7H_6O_2$) + H_2, Ni, heat, pressure $\longrightarrow C_7H_{12}O_2$
(s) benzoic acid + ethylene glycol + $H^+ \longrightarrow C_{16}H_{14}O_4$
(t) phthalic acid + ethyl alcohol + $H^+ \longrightarrow C_{12}H_{14}O_4$
(u) oleic acid + Br_2/CCl_4
(v) product (u) + KOH (alcoholic)
(w) oleic acid + HCO_2OH

12. Outline a possible laboratory synthesis of the following labeled compounds, using $Ba^{14}CO_3$ or $^{14}CH_3OH$ as the source of ^{14}C.

(a) $CH_3CH_2CH_2{}^{14}COOH$
(b) $CH_3CH_2{}^{14}CH_2COOH$
(c) $CH_3{}^{14}CH_2CH_2COOH$
(d) $^{14}CH_3CH_2CH_2COOH$

13. Outline all steps in a possible laboratory synthesis of each of the following compounds from toluene and any needed aliphatic and inorganic reagents.

(a) benzoic acid
(b) phenylacetic acid
(c) *p*-toluic acid
(d) *m*-chlorobenzoic acid
(e) *p*-chlorobenzoic acid
(f) *p*-bromophenylacetic acid
(g) α-chlorophenylacetic acid

14. Outline a possible laboratory synthesis of each of the following compounds from benzene, toluene, and alcohols of four carbons or fewer, using any needed inorganic reagents.

(a) ethyl α-methylbutyrate
(b) 3,5-dinitrobenzoyl chloride
(c) α-amino-*p*-bromophenylacetic acid
(d) α-hydroxypropionic acid
(e) p-$HO_3SC_6H_4COOH$
(f) 2-pentenoic acid
(g) *p*-toluamide
(h) *n*-hexyl benzoate
(i) 3-bromo-4-methylbenzoic acid
(j) α-methylphenylacetic acid
(k) 2-bromo-4-nitrobenzoic acid
(l) 1,2,4-benzenetricarboxylic acid

15. Without referring to tables, arrange the compounds of each set in order of acidity:

(a) butanoic acid, 2-bromobutanoic acid, 3-bromobutanoic acid, 4-bromobutanoic acid
(b) benzoic acid, *p*-chlorobenzoic acid, 2,4-dichlorobenzoic acid, 2,4,6-trichlorobenzoic acid
(c) benzoic acid, *p*-nitrobenzoic acid, *p*-toluic acid
(d) α-chlorophenylacetic acid, *p*-chlorophenylacetic acid, phenylacetic acid, α-phenylpropionic acid
(e) *p*-nitrobenzoic acid, *p*-nitrophenylacetic acid, β-(*p*-nitrophenyl)propionic acid
(f) acetic acid, acetylene, ammonia, ethane, ethanol, sulfuric acid, water
(g) acetic acid, malonic acid, succinic acid

16. Arrange the monosodium salts of the acids in Problem 15(f) in order of basicity.

17. The two water-insoluble solids, benzoic acid and *o*-chlorobenzoic acid, can be separated by treatment with an aqueous solution of sodium formate. What reaction takes place? (*Hint:* Look at the K_a's in Table 18.2.)

18. Arrange the compounds of each set in order of reactivity in the indicated reaction:

(a) esterification by benzoic acid: *sec*-butyl alcohol, methanol, *tert*-pentyl alcohol, *n*-propyl alcohol
(b) esterification by ethyl alcohol: benzoic acid, 2,6-dimethylbenzoic acid, *o*-toluic acid
(c) esterification by methanol: acetic acid, formic acid, isobutyric acid, propionic acid, trimethylacetic acid

19. Give stereochemical formulas of compounds A–F:

(a) racemic β-bromobutyric acid + one mole Br_2, P $\longrightarrow$ A + B
(b) fumaric acid + HCO_2OH $\longrightarrow$ C ($C_4H_6O_6$)
(c) 1,4-cyclohexadiene + $CHBr_3/t$-BuOK $\longrightarrow$ D ($C_7H_8Br_2$)
D + $KMnO_4$ $\longrightarrow$ E ($C_7H_8Br_2O_4$)
E + H_2, Ni(base) $\longrightarrow$ F ($C_7H_{10}O_4$)

20. Give structures of compounds G through J:

$$\text{acetylene} + CH_3MgBr \longrightarrow G + CH_4$$
$$G + CO_2 \longrightarrow H \xrightarrow{H^+} I\ (C_3H_2O_2)$$
$$I \xrightarrow{H_2O,\ H_2SO_4,\ HgSO_4} J\ (C_3H_4O_3)$$
$$J + KMnO_4 \longrightarrow CH_2(COOH)_2$$

21. Describe simple chemical tests (other than color change of an indicator) that would serve to distinguish between:

(a) propionic acid and *n*-pentyl alcohol
(b) isovaleric acid and *n*-octane
(c) ethyl *n*-butyrate and isobutyric acid
(d) propionyl chloride and propionic acid
(e) *p*-aminobenzoic acid and benzamide
(f) $C_6H_5CH{=}CHCOOH$ and $C_6H_5CH{=}CHCH_3$

Tell exactly what you would *do* and *see*.

22. Compare benzoic acid and sodium benzoate with respect to:

(a) volatility
(b) melting point
(c) solubility in water and (d) in ether
(e) degree of ionization of solid
(f) degree of ionization in water
(g) acidity and basicity

Does this comparison hold generally for acids and their salts?

23. Tell how you would separate by chemical means the following mixtures, recovering each component in reasonably pure form:

(a) caproic acid and ethyl caproate
(b) *n*-butyl ether and *n*-butyric acid
(c) isobutyric acid and 1-hexanol
(d) sodium benzoate and triphenylcarbinol

Tell exactly what you would *do* and *see*.

24. An unknown compound is believed to be one of the following. Describe how you would go about finding out which of the possibilities the unknown actually is. Where possible, use simple chemical tests; where necessary, use more elaborate chemical methods like quantitative hydrogenation, cleavage, neutralization equivalent, etc. Make use of any needed tables of physical constants.

(a) acrylic acid ($CH_2{=}CHCOOH$, b.p. 142°) and propionic acid (b.p. 141°)
(b) mandelic acid ($C_6H_5CHOHCOOH$, m.p. 120°) and benzoic acid (m.p. 122°)
(c) *o*-chlorobenzoic acid (m.p. 141°), mesotartaric acid (m.p. 140°), *m*-nitrobenzoic acid (m.p. 141°), and suberic acid ($HOOC(CH_2)_6COOH$, m.p. 144°)
(d) chloroacetic acid (b.p. 189°), α-chloropropionic acid (b.p. 186°), dichloroacetic acid (b.p. 194°), and *n*-valeric acid (b.p. 187°)
(e) 3-nitrophthalic acid (m.p. 220°) and 2,4,6-trinitrobenzoic acid (m.p. 220°)
(f) *p*-chlorobenzoic acid (m.p. 242°), *p*-nitrobenzoic acid (m.p. 242°), *o*-nitrocinnamic acid (o-$O_2NC_6H_4CH{=}CHCOOH$, m.p. 240°)
(g) The following compounds, all of which boil within a few degrees of each other:

o-chloroanisole
β-chlorostyrene
p-cresyl ethyl ether
cis-decalin (see Problem 8, p. 315)
2,4-dichlorotoluene
isodurene
linalool (see Problem 28, p. 547)
4-methylpentanoic acid
α-phenylethyl chloride
o-toluidine (o-$CH_3C_6H_4NH_2$)

25. By use of Table 18.4 tell which acid or acids each of the following is likely to be. Tell what further steps you would take to identify it or to confirm your identification.

K: m.p. 155–7°; positive halogen test; *p*-nitrobenzyl ester, m.p. 104–6°; neutralization equivalent, 158 ± 2
L: m.p. 152–4°; negative tests for halogen and nitrogen
M: m.p. 153–5°; positive chlorine test; neutralization equivalent, 188 ± 4
N: m.p. 72–3°; anilide, m.p. 117–8°; amide, m.p. 155–7°
O: m.p. 79–80°; amide, m.p. 97–9°
P: m.p. 78–80°; negative tests for halogen and nitrogen; positive test with CrO_3/H_2SO_4

Table 18.4 DERIVATIVES OF SOME CARBOXYLIC ACIDS

	Acid M.p., °C	Amide M.p., °C	Anilide M.p., °C	*p*-Nitrobenzyl ester M.p., °C
trans-Crotonic ($CH_3CH{=}CHCOOH$)	72	161	118	67
Phenylacetic	77	156	118	65
Arachidic (n-$C_{19}H_{39}COOH$)	77	108	92	—
α-Hydroxyisobutyric	79	98	136	80
Glycolic ($HOCH_2COOH$)	80	120	97	107
β-Iodopropionic	82	101	—	—
Iodoacetic	83	95	143	—
Adipic ($HOOC(CH_2)_4COOH$)	151	220	241	106
p-Nitrophenylacetic	153	198	198	—
2,5-Dichlorobenzoic	153	155	—	—
m-Chlorobenzoic	154	134	122	107
2,4,6-Trimethylbenzoic	155	—	—	188
m-Bromobenzoic	156	155	136	105
p-Chlorophenoxyacetic	158	133	125	—
Salicylic (o-HOC_6H_4COOH)	159	142	136	98

26. An unknown acid was believed to be either *o*-nitrobenzoic acid (m.p. 147°) or anthranilic acid (m.p. 146°). A 0.201-g sample neutralized 12.4 ml of 0.098 N NaOH. Which acid was it?

27. Carboxylic acid Q contained only carbon, hydrogen, and oxygen, and had a neutralization equivalent of 149 ± 3. Vigorous oxidation by $KMnO_4$ converted Q into R, m.p. 345–50°, neutralization equivalent 84 ± 2.

When Q was heated strongly with soda lime a liquid S of b.p. 135–7° distilled. Vigorous oxidation by $KMnO_4$ converted S into T, m.p. 121–2°, neutralization equivalent 123 ± 2.

U, an isomer of Q, gave upon oxidation V, m.p. 375–80°, neutralization equivalent 70 ± 2.

What were compounds Q through V? (Make use of any needed tables of physical constants.)

28. *Tropic acid* (obtained from the alkaloid atropine, found in deadly nightshade, *Atropa belladona*), $C_9H_{10}O_3$, gives a positive CrO_3/H_2SO_4 test and is oxidized by hot $KMnO_4$ to benzoic acid. Tropic acid is converted by the following sequence of reactions into *hydratropic acid*:

$$\text{tropic acid} \xrightarrow{HBr} C_9H_9O_2Br \xrightarrow{OH^-} C_9H_8O_2 \text{ (atropic acid)}$$

$$\text{atropic acid} \xrightarrow{H_2,Ni} \text{hydratropic acid } (C_9H_{10}O_2)$$

(a) What structure or structures are possible at this point for hydratropic acid? For tropic acid?

(b) When α-phenylethyl chloride is treated with magnesium in ether, the resulting solution poured over dry ice, and the mixture then acidified, there is obtained an acid whose amide has the same melting point as the amide of hydratropic acid. A mixed melting point determination shows no depression. Now what is the structure of hydratropic acid? Of tropic acid?

29. Give a structure or structures consistent with each of the following sets of nmr data:

(a) $C_3H_5ClO_2$
a doublet, δ 1.73, 3H
b quartet, δ 4.47, 1H
c singlet, δ 11.22, 1H

(b) $C_3H_5ClO_2$
a singlet, δ 3.81, 3H
b singlet, δ 4.08, 2H

(c) $C_4H_7BrO_2$
a triplet, δ 1.30, 3H
b singlet, δ 3.77, 2H
c quartet, δ 4.23, 2H

(d) $C_4H_7BrO_2$
a triplet, δ 1.08, 3H
b quintet, δ 2.07, 2H
c triplet, δ 4.23, 1H
d singlet, δ 10.97, 1H

CH3-CH2CH(Br)CO2H

(e) $C_4H_8O_3$
a triplet, δ 1.27, 3H
b quartet, δ 3.66, 2H
c singlet, δ 4.13, 2H
d singlet, δ 10.95, 1H

CH3CH2OCH2CO2H

30. Which (if any) of the following compounds could give rise to each of the infrared spectra shown in Fig. 18.5 (p. 616)?

n-butyric acid
crotonic acid ($CH_3CH{=}CHCOOH$)
malic acid ($HOOCCHOHCH_2COOH$)
benzoic acid
p-nitrobenzoic acid
mandelic acid ($C_6H_5CHOHCOOH$)
p-nitrobenzyl alcohol

a– CH3-CH(Cl)-CO2H

α-chloropropionic acid

b. Cl-CH2-C(=O)-OCH3

methyl chloroacetate

c. BrCH2C(=O)-O-CH2CH3

ethyl bromoacetate

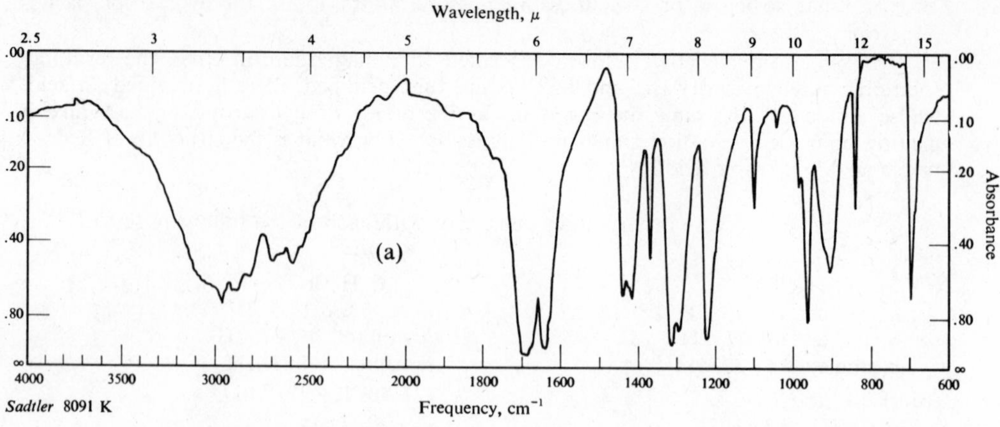

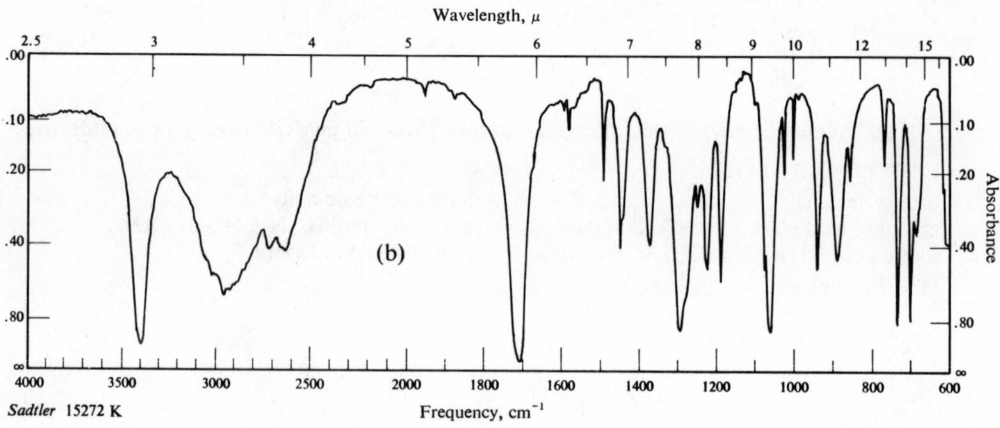

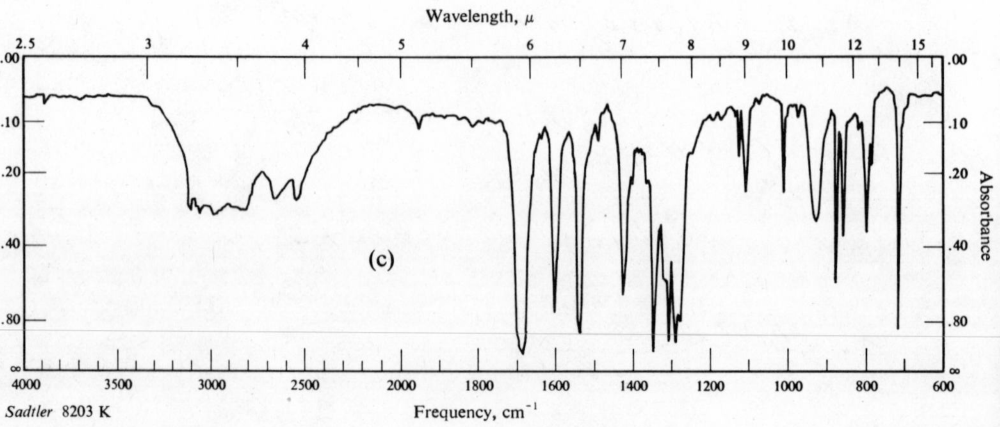

Figure 18.5. Infrared spectra for Problem 30, p. 615.

Chapter 19 Aldehydes and Ketones

Nucleophilic Addition

19.1 Structure

Aldehydes are compounds of the general formula RCHO; ketones are compounds of the general formula RR′CO. The groups R and R′ may be aliphatic or aromatic.

$$\underset{\text{An aldehyde}}{\text{H(R)C=O}} \qquad \underset{\text{A ketone}}{\text{R}'\text{(R)C=O}}$$

Both aldehydes and ketones contain the carbonyl group, C=O, and are often referred to collectively as **carbonyl compounds.** *It is the carbonyl group that largely determines the chemistry of aldehydes and ketones.*

It is not surprising to find that aldehydes and ketones resemble each other closely in most of their properties. However, there is a hydrogen atom attached to the carbonyl group of aldehydes, and there are two organic groups attached to the carbonyl group of ketones. This difference in structure affects their properties in two ways: (a) aldehydes are quite easily oxidized, whereas ketones are oxidized only with difficulty; (b) aldehydes are usually more reactive than ketones toward nucleophilic addition, the characteristic reaction of carbonyl compounds.

Let us examine the structure of the carbonyl group. Carbonyl carbon is joined to three other atoms by σ bonds; since these bonds utilize sp^2 orbitals (Sec. 1.10), they lie in a plane, and are 120° apart. The remaining p orbital of the carbon overlaps a p orbital of oxygen to form a π bond; carbon and oxygen are thus

$$\text{R}'\text{(R)}\overset{\delta_+}{\text{C}}=\overset{\delta_-}{\text{O}} \quad (120°)$$

joined by a double bond. The part of the molecule immediately surrounding carbonyl carbon is *flat*; oxygen, carbonyl carbon, and the two atoms directly attached to carbonyl carbon lie in a plane.

The electrons of the carbonyl double bond hold together atoms of quite different electronegativity, and hence the electrons are not equally shared; in particular, the mobile π cloud is pulled strongly toward the more electronegative atom, oxygen.

The facts are consistent with the orbital picture of the carbonyl group. Electron diffraction and spectroscopic studies of aldehydes and ketones show that carbon, oxygen, and the two other atoms attached to carbonyl carbon lie in a plane; the three bond angles of carbon are very close to 120°. The large dipole moments (2.3–2.8 D) of aldehydes and ketones indicate that the electrons of the carbonyl group are quite unequally shared. We shall see how the physical and chemical properties of aldehydes and ketones are determined by the structure of the carbonyl group.

19.2 Nomenclature

The common names of aldehydes are derived from the names of the corresponding carboxylic acids by replacing *–ic acid* by *–aldehyde.*

The IUPAC names of aldehydes follow the usual pattern. The longest chain carrying the —CHO group is considered the parent structure and is named by replacing the *–e* of the corresponding alkane by *–al.* The position of a substituent is indicated by a number, the carbonyl carbon always being considered as C–1. Here, as with the carboxylic acids, we notice that C–2 of the IUPAC name corresponds to *alpha* of the common name.

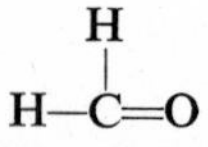

$H{-}C(H){=}O$ — Formaldehyde, Methanal

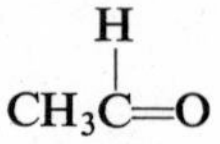

$CH_3C(H){=}O$ — Acetaldehyde, Ethanal

$CH_3CH_2C(H){=}O$ — Propionaldehyde, Propanal

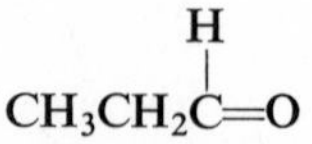

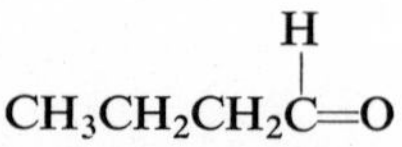

$CH_3CH_2CH_2C(H){=}O$ — *n*-Butyraldehyde, Butanal

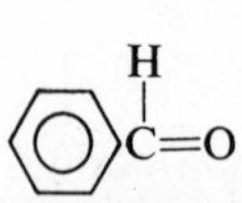

$C_6H_5C(H){=}O$ — Benzaldehyde

$O_2NC_6H_4C(H){=}O$ — *p*-Nitrobenzaldehyde

$CH_3C_6H_4C(H){=}O$ — *p*-Tolualdehyde

$o\text{-}HOC_6H_4C(H){=}O$ — Salicylaldehyde (*o*-Hydroxybenzaldehyde)

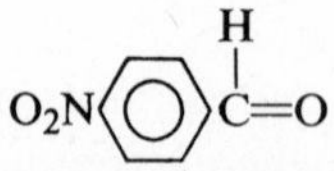

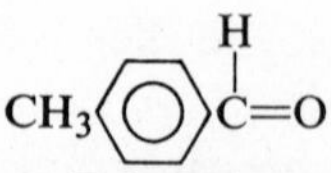

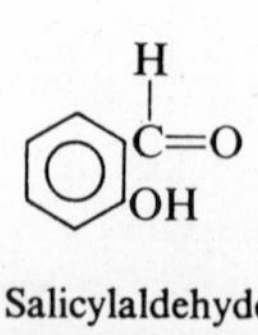

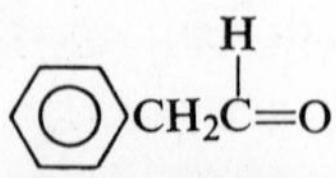

$C_6H_5CH_2C(H){=}O$ — Phenylacetaldehyde (Phenylethanal)

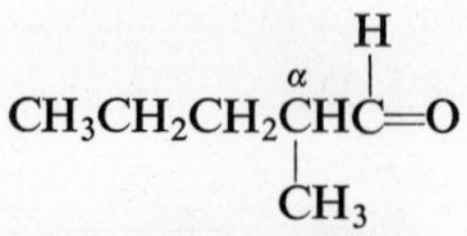

$CH_3CH_2CH_2\overset{\alpha}{C}H(CH_3)C(H){=}O$ — α-Methylvaleraldehyde, 2-Methylpentanal

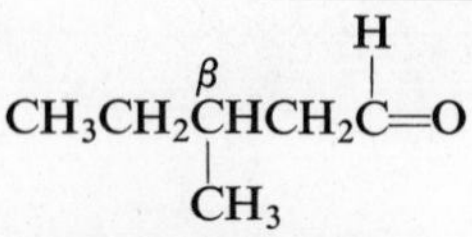

$CH_3CH_2\overset{\beta}{C}H(CH_3)CH_2C(H){=}O$ — β-Methylvaleraldehyde, 3-Methylpentanal

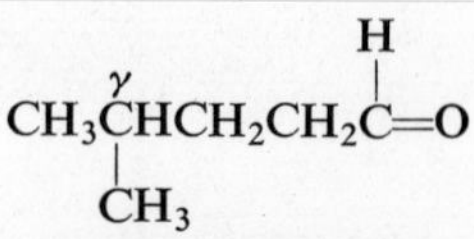

$CH_3\overset{\gamma}{C}H(CH_3)CH_2CH_2C(H){=}O$ — Isocaproaldehyde, γ-Methylvaleraldehyde, 4-Methylpentanal

The simplest aliphatic ketone has the common name of *acetone*. For most other aliphatic ketones we name the two groups that are attached to carbonyl carbon, and follow these names by the word *ketone*. A ketone in which the carbonyl group is attached to a benzene ring is named as a *–phenone*, as illustrated below.

According to the IUPAC system, the longest chain carrying the carbonyl group is considered the parent structure, and is named by replacing the *–e* of the corresponding alkane with *–one*. The positions of various groups are indicated by numbers, the carbonyl carbon being given the lowest possible number.

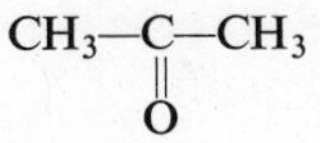

$CH_3{-}C({=}O){-}CH_3$
Acetone
Propanone

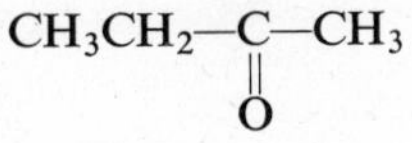

$CH_3CH_2{-}C({=}O){-}CH_3$
Methyl ethyl ketone
Butanone

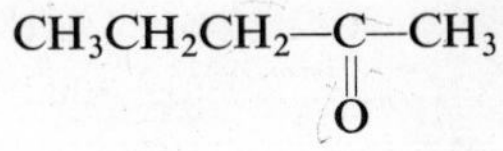

$CH_3CH_2CH_2{-}C({=}O){-}CH_3$
Methyl *n*-propyl ketone
2-Pentanone

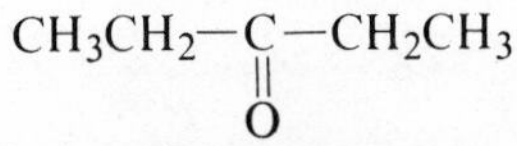

$CH_3CH_2{-}C({=}O){-}CH_2CH_3$
Ethyl ketone
3-Pentanone

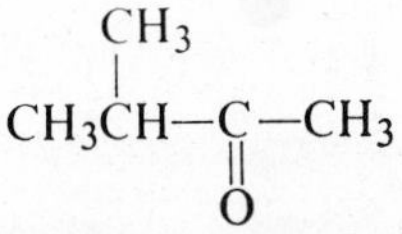

$CH_3CH(CH_3){-}C({=}O){-}CH_3$
Methyl isopropyl ketone
3-Methyl-2-butanone

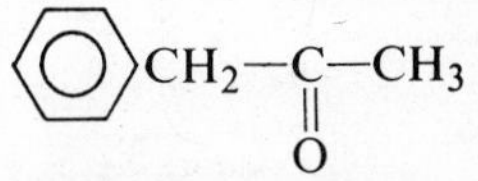

$C_6H_5CH_2{-}C({=}O){-}CH_3$
Benzyl methyl ketone
1-Phenyl-2-propanone

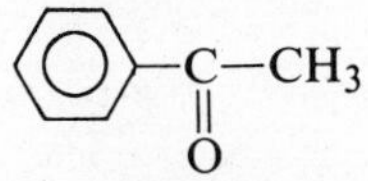

$C_6H_5{-}C({=}O){-}CH_3$
Acetophenone

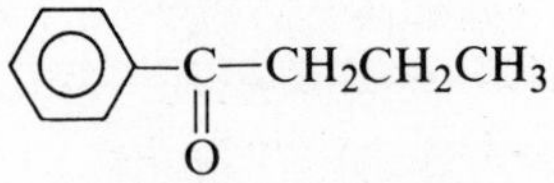

$C_6H_5{-}C({=}O){-}CH_2CH_2CH_3$
n-Butyrophenone

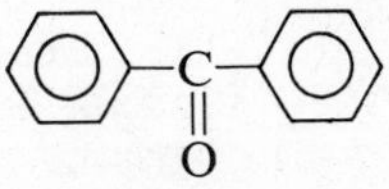

$C_6H_5{-}C({=}O){-}C_6H_5$
Benzophenone

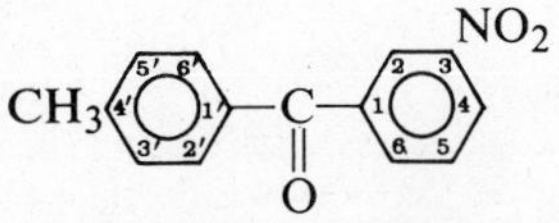

$CH_3C_6H_4{-}C({=}O){-}C_6H_4NO_2$ (ring positions 1′–6′ and 1–6; CH_3 at 4′, NO_2 at 3)
3-Nitro-4′-methylbenzophenone

19.3 Physical properties

The polar carbonyl group makes aldehydes and ketones polar compounds, and hence they have higher boiling points than non-polar compounds of comparable molecular weight. By themselves, they are not capable of intermolecular hydrogen bonding since they contain hydrogen bonded only to carbon; as a result they have lower boiling points than comparable alcohols or carboxylic acids. For example, compare *n*-butyraldehyde (b.p. 76°) and methyl ethyl ketone (b.p. 80°) with *n*-pentane (b.p. 36°) and ethyl ether (b.p. 35°) on the one hand, and with *n*-butyl alcohol (b.p. 118°) and propionic acid (b.p. 141°) on the other.

The lower aldehydes and ketones are appreciably soluble in water, presumably because of hydrogen bonding between solute and solvent molecules; borderline solubility is reached at about five carbons. Aldehydes and ketones are soluble in the usual organic solvents.

Table 19.1 ALDEHYDES AND KETONES

	M.p., °C	B.p., °C	Solub., g/100 g H_2O
Formaldehyde	− 92	− 21	v.sol.
Acetaldehyde	−121	20	∞
Propionaldehyde	− 81	49	16
n-Butyraldehyde	− 99	76	7
n-Valeraldehyde	− 91	103	sl.s
Caproaldehyde		131	sl.s
Heptaldehyde	− 42	155	0.1
Phenylacetaldehyde		194	sl.s
Benzaldehyde	− 26	178	0.3
o-Tolualdehyde		196	
m-Tolualdehyde		199	
p-Tolualdehyde		205	
Salicyaldehyde (*o*-Hydroxybenzaldehyde)	2	197	1.7
p-Hydroxybenzaldehyde	116		1.4
Anisaldehyde	3	248	0.2
Vanillin	82	285	1
Piperonal	37	263	0.2
Acetone	− 94	56	∞
Methyl ethyl ketone	− 86	80	26
2-Pentanone	− 78	102	6.3
3-Pentanone	− 41	101	5
2-Hexanone	− 35	150	2.0
3-Hexanone		124	sl.s
Methyl isobutyl ketone	− 85	119	1.9
Acetophenone	21	202	
Propiophenone	21	218	
n-Butyrophenone	11	232	
Benzophenone	48	306	

Formaldehyde is a gas (b.p. −21°), and is handled either as an aqueous solution (*Formalin*), or as one of its solid polymers: *paraformaldehyde* $(CH_2O)_n$, or *trioxane*, $(CH_2O)_3$. When dry formaldehyde is desired, as, for example, for reaction with a Grignard reagent, it is obtained by heating paraformaldehyde or trioxane.

$$\sim CH_2OCH_2OCH_2O \sim$$

Paraformaldehyde

$$\begin{array}{ccc} & O & \\ H_2C & & CH_2 \\ | & & | \\ O & & O \\ & CH_2 & \end{array}$$

Trioxane

Acetaldehyde (b.p. 20°) is often generated from its higher-boiling trimer by heating the trimer with acid:

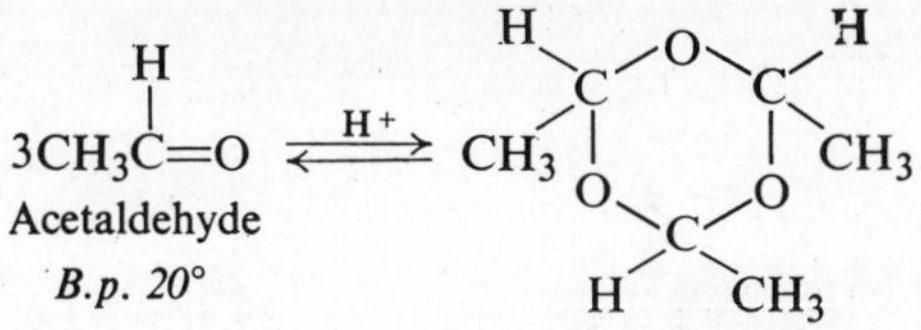

$$3CH_3\overset{H}{\overset{|}{C}}{=}O \underset{}{\overset{H^+}{\rightleftharpoons}} \text{Paraldehyde (cyclic trimer: } (CH_3CHO)_3\text{)}$$

Acetaldehyde
B.p. 20°

Paraldehyde
B.p. 125°

19.4 Preparation

A few of the many laboratory methods of preparing aldehydes and ketones are outlined below; most of these are already familiar to us. Some of the methods involve oxidation or reduction in which an alcohol, hydrocarbon, or acid chloride is converted into an aldehyde or ketone of the same carbon number. Other methods involve the formation of new carbon–carbon bonds, and yield aldehydes or ketones of higher carbon number than the starting materials.

Industrial preparation is generally patterned after these laboratory methods, but with use of cheaper reagents: alcohols are oxidized catalytically with air, or by dehydrogenation over hot copper.

PREPARATION OF ALDEHYDES

1. Oxidation of primary alcohols. Discussed in Secs. 16.8 and 19.5.

$$\underset{\text{1° Alcohol}}{RCH_2OH} \xrightarrow{K_2Cr_2O_7} \underset{\text{Aldehyde}}{R{-}\overset{H}{\overset{|}{C}}{=}O}$$

Example:

$$\underset{\substack{n\text{-Butyl alcohol}\\ \text{(1-Butanol)}\\ B.p.\ 118°}}{CH_3CH_2CH_2CH_2OH} \xrightarrow{K_2Cr_2O_7,\ H_2SO_4,\ \text{warm}} \underset{\substack{n\text{-Butyraldehyde}\\ \text{(Butanal)}\\ B.p.\ 76°}}{CH_3CH_2CH_2CHO}$$

2. Oxidation of methylbenzenes. Discussed in Sec. 19.5.

$$ArCH_3 \xrightarrow{Cl_2,\ \text{heat}} ArCHCl_2 \xrightarrow{H_2O} ArCHO$$

$$ArCH_3 \xrightarrow{CrO_3,\ \text{acetic anhydride}} ArCH(OOCCH_3)_2 \xrightarrow{H_2O} ArCHO$$

Examples:

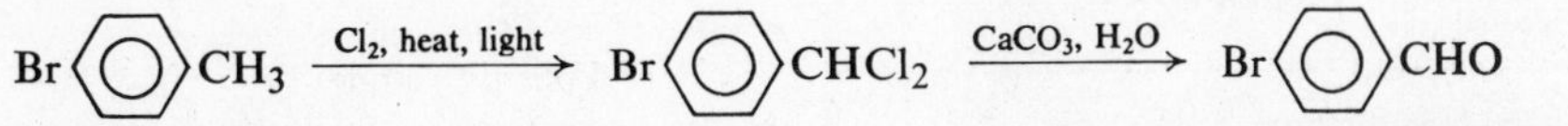

$$\underset{p\text{-Bromotoluene}}{Br{-}C_6H_4{-}CH_3} \xrightarrow{Cl_2,\ \text{heat, light}} Br{-}C_6H_4{-}CHCl_2 \xrightarrow{CaCO_3,\ H_2O} \underset{p\text{-Bromobenzaldehyde}}{Br{-}C_6H_4{-}CHO}$$

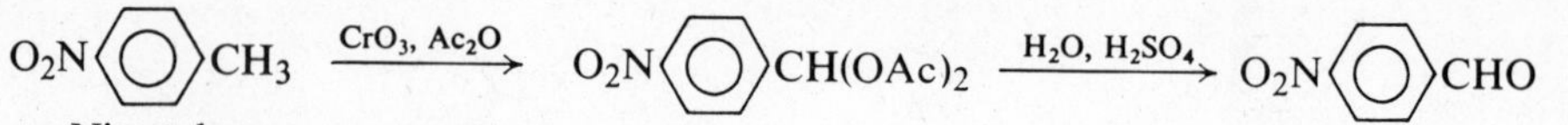

$$O_2N\langle\bigcirc\rangle CH_3 \xrightarrow{CrO_3,\ Ac_2O} O_2N\langle\bigcirc\rangle CH(OAc)_2 \xrightarrow{H_2O,\ H_2SO_4} O_2N\langle\bigcirc\rangle CHO$$

p-Nitrotoluene ... *p*-Nitrobenzaldehyde

3. Reduction of acid chlorides. Discussed in Sec. 19.4.

$$\underset{\text{Acid chloride}}{RCOCl \text{ or } ArCOCl} \xrightarrow{LiAlH(OBu\text{-}t)_3} \underset{\text{Aldehyde}}{RCHO \text{ or } ArCHO}$$

Examples:

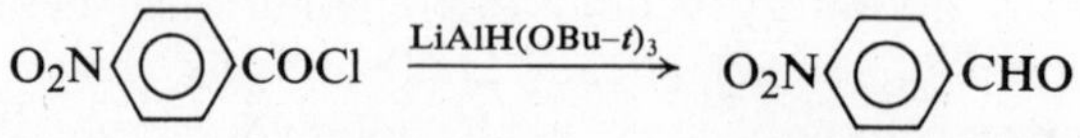

$$\underset{p\text{-Nitrobenzoyl chloride}}{O_2N\langle\bigcirc\rangle COCl} \xrightarrow{LiAlH(OBu\text{-}t)_3} \underset{p\text{-Nitrobenzaldehyde}}{O_2N\langle\bigcirc\rangle CHO}$$

4. Reimer-Tiemann reaction. Phenolic aldehydes. Discussed in Sec. 24.12.

PREPARATION OF KETONES

1. Oxidation of secondary alcohols. Discussed in Sec. 16.8.

$$\underset{2^\circ\text{ Alcohol}}{RCHOHR'} \xrightarrow{CrO_3 \text{ or } K_2Cr_2O_7} \underset{\text{Ketone}}{R{-}\underset{\underset{O}{\|}}{C}{-}R'}$$

Example:

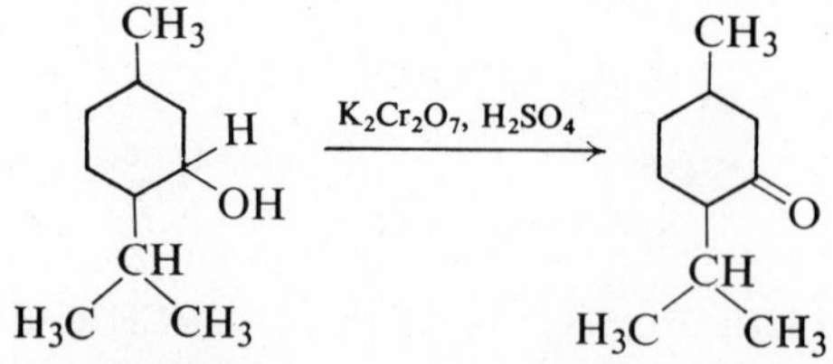

$$\underset{(-)\text{-Menthol}}{\text{menthol}} \xrightarrow{K_2Cr_2O_7,\ H_2SO_4} \underset{(-)\text{-Menthone}}{\text{menthone}}$$

2. Friedel-Crafts acylation. Discussed in Sec. 19.6.

$$\underset{\text{Acid chloride}}{R{-}C(=O)Cl} + ArH \xrightarrow[\text{or other Lewis acid}]{AlCl_3} \underset{\text{Ketone}}{R{-}\underset{\underset{O}{\|}}{C}{-}Ar} + HCl$$

Examples:

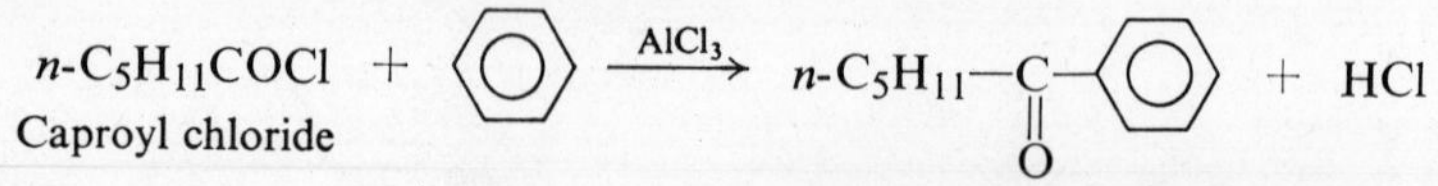

$$\underset{\text{Caproyl chloride}}{n\text{-}C_5H_{11}COCl} + \langle\bigcirc\rangle \xrightarrow{AlCl_3} n\text{-}C_5H_{11}{-}\underset{\underset{O}{\|}}{C}{-}\langle\bigcirc\rangle + HCl$$

n-Pentyl phenyl ketone

No rearrangement of n-pentyl group

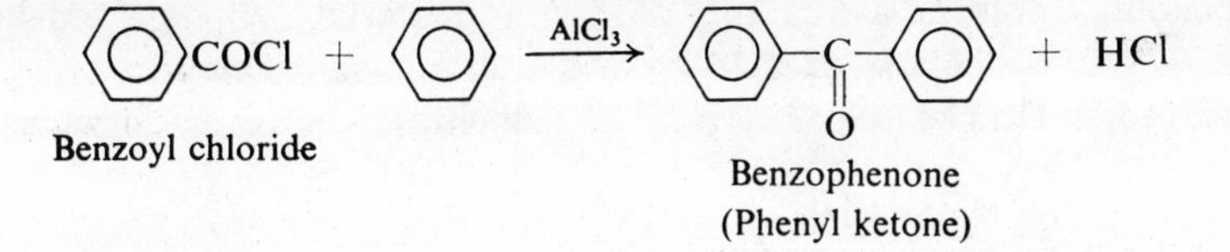

$$C_6H_5COCl + C_6H_6 \xrightarrow{AlCl_3} C_6H_5-\overset{O}{\overset{\|}{C}}-C_6H_5 + HCl$$

Benzoyl chloride — Benzophenone (Phenyl ketone)

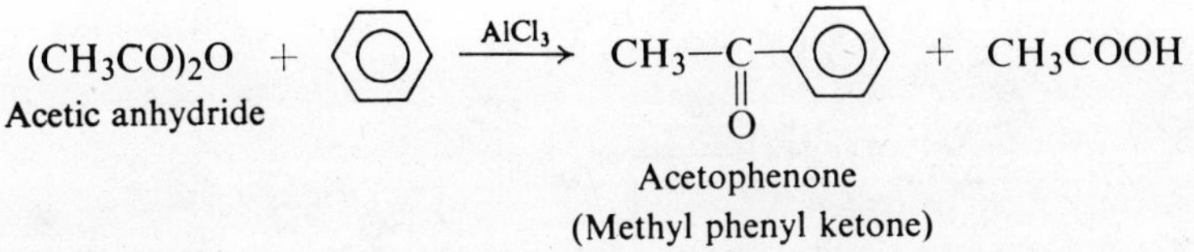

$$(CH_3CO)_2O + C_6H_6 \xrightarrow{AlCl_3} CH_3-\overset{O}{\overset{\|}{C}}-C_6H_5 + CH_3COOH$$

Acetic anhydride — Acetophenone (Methyl phenyl ketone)

3. Reaction of acid chlorides with organocadmium compounds. Discussed in Sec. 19.7.

$$R'MgX \xrightarrow{CdCl_2} R'_2Cd$$

$$R'_2Cd + RCOCl \text{ or } ArCOCl \longrightarrow R-\overset{O}{\overset{\|}{C}}-R' \text{ or } Ar-\overset{O}{\overset{\|}{C}}-R'$$

R′ *must be aryl or primary alkyl*

Examples:

$$CH_3CH_2CH_2CH_2MgBr \xrightarrow{CdCl_2} (CH_3CH_2CH_2CH_2)_2Cd$$

Di-*n*-butylcadmium

$$(CH_3CH_2CH_2CH_2)_2Cd + 2CH_3\overset{CH_3}{\overset{|}{C}}HCOCl \longrightarrow 2CH_3CH_2CH_2CH_2\underset{O}{\underset{\|}{C}}\overset{CH_3}{\overset{|}{C}}HCH_3$$

Isobutyryl chloride

n-Butyl isopropyl ketone (2-Methyl-3-heptanone)

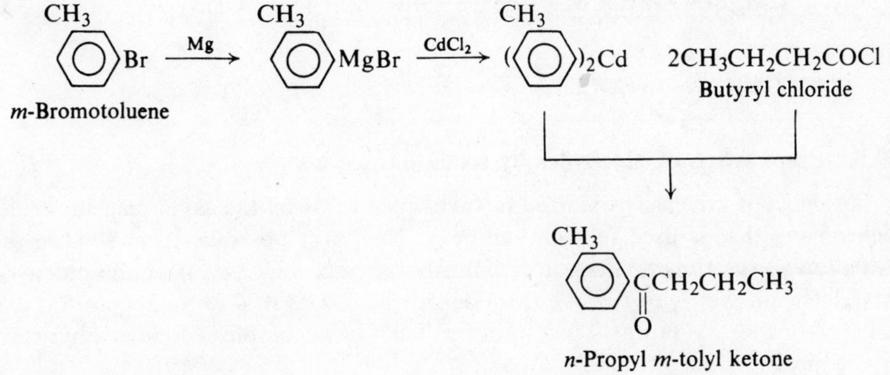

$$m\text{-}CH_3C_6H_4Br \xrightarrow{Mg} m\text{-}CH_3C_6H_4MgBr \xrightarrow{CdCl_2} (m\text{-}CH_3C_6H_4)_2Cd$$

m-Bromotoluene

$$(m\text{-}CH_3C_6H_4)_2Cd + 2CH_3CH_2CH_2COCl \longrightarrow m\text{-}CH_3C_6H_4\underset{O}{\underset{\|}{C}}CH_2CH_2CH_3$$

Butyryl chloride

n-Propyl *m*-tolyl ketone

4. Acetoacetic ester synthesis. Discussed in Sec. 26.3.

Depending upon the availability of starting materials, **aliphatic aldehydes** can be prepared from alcohols or acid chlorides of the same carbon skeleton, and **aromatic aldehydes** can be prepared from methylbenzenes or aromatic acid chlorides.

$$\left.\begin{array}{r} RCH_2OH \\ RCOOH \longrightarrow RCOCl \end{array}\right] \longrightarrow RCHO$$

Preparation of aliphatic aldehydes

$$\left.\begin{array}{r} ArCH_3 \\ ArCOOH \longrightarrow ArCOCl \end{array}\right] \longrightarrow ArCHO$$

Preparation of aromatic aldehydes

There are, in addition, a number of methods by which the aldehyde group is introduced into an aromatic ring: for example, the Reimer-Tiemann synthesis of phenolic aldehydes (Sec. 24.12).

Aliphatic ketones are readily prepared from the corresponding secondary alcohols, if these are available. More complicated aliphatic ketones can be prepared by the reaction of acid chlorides with organocadmium compounds. A

$$\left.\begin{array}{r} RR'CHOH \\ RCOOH \longrightarrow RCOCl \xrightarrow{R'_2Cd} \end{array}\right] \longrightarrow R-\underset{\underset{O}{\|}}{C}-R'$$

Preparation of aliphatic ketones

particularly useful method for making complicated aliphatic ketones, the acetoacetic ester synthesis, will be discussed later (Sec. 26.3). **Aromatic ketones** containing a carbonyl group attached directly to an aromatic ring are conveniently prepared by Friedel-Crafts acylation (Sec. 19.6).

$$\left.\begin{array}{r} ArH \xrightarrow{RCOCl\ (Ar'COCl),\ AlCl_3} \\ ArBr \longrightarrow ArMgBr \longrightarrow Ar_2Cd \xrightarrow{RCOCl\ (Ar'COCl)} \\ ArCOOH \longrightarrow ArCOCl \xrightarrow{R_2Cd\ (Ar'_2Cd)} \end{array}\right] \longrightarrow Ar\underset{\underset{O}{\|}}{C}R \quad (Ar\underset{\underset{O}{\|}}{C}Ar')$$

Preparation of aromatic ketones

19.5 Preparation of aldehydes by oxidation methods

Aldehydes are easily oxidized to carboxylic acids by the same reagent, acidic dichromate, that is used in their synthesis. How is it possible, then, to stop the oxidation of a primary alcohol or a methylbenzene (Sec. 19.4) at the aldehyde stage? The answer is to remove the aldehyde as fast as it is formed, before it can undergo further oxidation. This "removal" can be accomplished either physically or chemically.

An aldehyde always has a lower boiling point than the alcohol from which it is formed. (Why?) Acetaldehyde, for example, has a boiling point of 20°; ethyl alcohol has a boiling point of 78°. When a solution of dichromate and sulfuric acid is dripped into boiling ethyl alcohol, acetaldehyde is formed in a medium whose temperature is some 60 degrees above its boiling point; before it can undergo appreciable oxidation, it escapes from the reaction medium. Reaction is carried out under a fractionating column that allows aldehyde to pass but returns alcohol to the reaction vessel.

In the case of methylbenzenes, oxidation of the side chain can be interrupted by trapping the aldehyde in the form of a non-oxidizable derivative, the *gem*-diacetate (Latin: *Gemini*, twins), which is isolated and then hydrolyzed.

$$\mathrm{ArCH_3} \xrightarrow{\text{acetic anhydride}} \mathrm{ArCH(O\overset{\displaystyle O}{\overset{\|}{C}}CH_3)_2} \xrightarrow{\text{hydrolysis}} \mathrm{ArCHO}$$

A *gem*-diacetate
Not oxidized

Problem 19.1 A *gem*-diacetate is the ester of what "alcohol"?

Problem 19.2 Optically active alcohols in which the chiral center carries the —OH undergo racemization in acidic solutions. (Why?) Give a detailed experimental procedure (including apparatus) for studying the stereochemistry of acidic hydrolysis of *sec*-butyl benzoate that would prevent racemization of the alcohol subsequent to hydrolysis. *sec*-Butyl benzoate has a boiling point of 234°; an azeotrope of 68% *sec*-butyl alcohol and 32% water has a boiling point of 88.5°.

19.6 Preparation of ketones by Friedel-Crafts acylation

One of the most important modifications of the Friedel-Crafts reaction involves the use of acid chlorides rather than alkyl halides. An acyl group, RCO—, becomes attached to the aromatic ring, thus forming a ketone; the process is called **acylation**. As usual for the Friedel-Crafts reaction (Sec. 12.8), the aromatic ring undergoing substitution must be at least as reactive as that of a halobenzene; catalysis by aluminum chloride or another Lewis acid is required.

$$\mathrm{ArH} + \mathrm{R{-}C}\begin{smallmatrix}\nearrow O\\ \searrow Cl\end{smallmatrix} \xrightarrow{\mathrm{AlCl_3}} \mathrm{Ar{-}\underset{\underset{\displaystyle O}{\|}}{C}{-}R} + \mathrm{HCl}$$

A ketone

The most likely mechanism for Friedel-Crafts acylation is analogous to the carbonium ion mechanism for Friedel-Crafts alkylation (Sec. 11.10), and involves the following steps:

(1) $$\mathrm{RCOCl} + \mathrm{AlCl_3} \longrightarrow \mathrm{RC{\equiv}\overset{\oplus}{O}} + \mathrm{AlCl_4^-}$$

(2) $$\mathrm{ArH} + \mathrm{RC{\equiv}\overset{\oplus}{O}} \longrightarrow \overset{\oplus}{\mathrm{Ar}}\begin{smallmatrix}\nearrow H\\ \searrow COR\end{smallmatrix}$$

(3) $$\overset{\oplus}{\mathrm{Ar}}\begin{smallmatrix}\nearrow H\\ \searrow COR\end{smallmatrix} + \mathrm{AlCl_4^-} \longrightarrow \mathrm{Ar{-}\underset{\underset{\displaystyle O}{\|}}{C}{-}R} + \mathrm{HCl} + \mathrm{AlCl_3}$$

This fits the pattern of electrophilic aromatic substitution, the attacking reagent this time being the **acylium ion**, $R{-}C{\equiv}\overset{\oplus}{O}$. The acylium ion is considerably more stable than ordinary carbonium ions since in it every atom has an octet of electrons.

Alternatively, it may be that the electrophile is a complex between acid chloride and Lewis acid:

$$R{-}C(=\overset{+}{O}{-}\overset{-}{A}lCl_3){-}Cl$$

In this case, from the standpoint of the acid chloride, reaction is acid-catalyzed nucleophilic acyl substitution, of the kind discussed in Sec. 20.4, with the aromatic ring acting as the nucleophile.

In planning the synthesis of diaryl ketones, ArCOAr′, it is particularly important to select the right combination of ArCOCl and Ar′H. In the preparation of *m*-nitrobenzophenone, for example, the nitro group can be present in the acid chloride but not in the ring undergoing substitution, since as a strongly deactivating group it prevents the Friedel-Crafts reaction (Sec. 12.8).

Benzene + *m*-Nitrobenzoyl chloride $\xrightarrow{AlCl_3}$ *m*-Nitrobenzophenone

m-Nitrobenzoic acid $\xrightarrow{PCl_5}$ *m*-Nitrobenzoyl chloride

Toluene $\xrightarrow{KMnO_4}$ Benzoic acid $\xrightarrow{HNO_3,\ H_2SO_4}$ *m*-Nitrobenzoic acid

Nitrobenzene + Benzoyl chloride $\xrightarrow{AlCl_3}$ No reaction

Friedel-Crafts acylation is one of the most important methods of preparing ketones in which the carbonyl group is attached to an aromatic ring. Once formed, these ketones may be converted into secondary alcohols by reduction, into tertiary alcohols by reaction with Grignard reagents, and into many other important classes of compounds, as we shall see.

Of particular importance is the conversion of the acyl group into an alkyl group. This can be accomplished by the **Clemmensen reduction** (amalgamated zinc and concentrated hydrochloric acid), or the **Wolff-Kishner reduction** (hydrazine and base). For example:

$$C_6H_5C(=O)(CH_2)_4CH_3 \xrightarrow{Zn(Hg),\ HCl} C_6H_5CH_2(CH_2)_4CH_3$$

n-Pentyl phenyl ketone → *n*-Hexylbenzene

$$m\text{-}H_3CC_6H_4C(=O)CH_2CH_2CH_3 \xrightarrow{NH_2NH_2,\ OH^-,\ 200°} m\text{-}H_3CC_6H_4CH_2CH_2CH_2CH_3$$

n-Propyl *m*-tolyl ketone → *m*-(*n*-Butyl)toluene

A straight-chain alkyl group longer than ethyl generally cannot be attached in good yield to an aromatic ring by Friedel-Crafts alkylation because of rearrangement (Sec. 12.7). Such a group is readily introduced, however, in two steps: (1) formation of a ketone by Friedel-Crafts acylation (or by the reaction of an organocadmium compound with an acyl chloride, described in the following section); (2) Clemmensen or Wolff-Kishner reduction of the ketone.

19.7 Preparation of ketones by use of organocadmium compounds

Grignard reagents react with dry cadmium chloride to yield the corresponding organocadmium compounds, which react with acid chlorides to yield ketones:

$$2R'MgX + CdCl_2 \longrightarrow R'_2Cd + 2MgXCl$$

R′ must be aryl or primary alkyl

$$R'_2Cd + 2RCOCl \longrightarrow 2R{-}C(=O){-}R' + CdCl_2$$

A ketone

Here, as in its other reactions (Sec. 20.7), the acid chloride is undergoing nucleophilic substitution, the nucleophile being the basic alkyl or aryl group of the organometallic compound.

Only organocadmium compounds containing aryl or primary alkyl groups are stable enough for use. In spite of this limitation, the method is one of the most valuable for the synthesis of ketones.

Grignard reagents themselves react readily with acid chlorides, but the products are usually tertiary alcohols; these presumably result from reaction of initially formed ketones with more Grignard reagent. (If tertiary alcohols are desired, they are better prepared from esters than from acid chlorides, Sec. 20.21.) Organocadmium compounds, being less reactive, do not react with ketones.

The comparatively low reactivity of organocadmium compounds not only makes the synthesis of ketones possible, but in addition widens the applicability of the method. Organocadmium compounds do not react with many of the functional groups with which the Grignard reagent does react: $-NO_2$, $-CN$, $-CO-$, $-COOR$, for example. Consequently, the presence of one of these groups in the acid chloride molecule does not interfere with the synthesis of a ketone (compare with Sec. 15.15). For example:

$$2\,O_2N\text{-}C_6H_4\text{-}COCl + (CH_3)_2Cd \longrightarrow 2\,O_2N\text{-}C_6H_4\text{-}\underset{\underset{O}{\|}}{C}\text{-}CH_3 + CdCl_2$$

p-Nitrobenzoyl chloride — Dimethylcadmium — *p*-Nitroacetophenone (Methyl *p*-nitrophenyl ketone)

$$CH_3O\underset{\underset{O}{\|}}{C}CH_2CH_2\underset{\underset{O}{\|}}{C}Cl + [(CH_3)_2CHCH_2CH_2]_2Cd \longrightarrow$$

Diisopentylcadmium

$$CH_3O\underset{\underset{O}{\|}}{C}CH_2CH_2\underset{\underset{O}{\|}}{C}CH_2CH_2CH(CH_3)_2$$

Methyl 4-oxo-7-methyloctanoate (A γ-keto ester)

Problem 19.3 Would it be feasible to make *p*-nitroacetophenone via the reaction between di(*p*-nitrophenyl)cadmium, $(p\text{-}O_2NC_6H_4)_2Cd$, and acetyl chloride?

19.8 Reactions. Nucleophilic addition

The carbonyl group, C=O, governs the chemistry of aldehydes and ketones. It does this in two ways: (a) by providing a site for nucleophilic addition, and (b) by increasing the acidity of the hydrogen atoms attached to the *alpha* carbon. Both these effects are quite consistent with the structure of the carbonyl group and, in fact, are due to the same thing: the ability of oxygen to accommodate a negative charge.

In this section, we shall examine the carbonyl group as a site for nucleophilic addition; in Sec. 21.1, we shall see how the acid-strengthening effect arises.

The carbonyl group contains a carbon–oxygen double bond; since the mobile π electrons are pulled strongly toward oxygen, carbonyl carbon is electron-deficient and carbonyl oxygen is electron-rich. Because it is flat, this part of the molecule is open to relatively unhindered attack from above or below, in a direction perpendicular to the plane of the group. It is not surprising that this accessible, polarized group is highly reactive.

What kind of reagents will attack such a group? Since the important step in these reactions is the formation of a bond to the electron-deficient (acidic) carbonyl carbon, the carbonyl group is most susceptible to attack by electron-rich, nucleophilic reagents, that is, by bases. **The typical reaction of aldehydes and ketones is nucleophilic addition.**

Nucleophilic addition

$$R'R\text{C}{=}O + {:}Z \longrightarrow \left[R'R\overset{\overset{Z}{\vdots}}{C}{\cdots}O^{\delta-}\right] \longrightarrow R'R\overset{\overset{Z}{|}}{C}\text{-}O^- \xrightarrow{H_2O} R'R\overset{\overset{Z}{|}}{C}\text{-}OH$$

Reactant	Transition state	Product
Trigonal	*Becoming tetrahedral*	*Tetrahedral*
	Partial negative charge on oxygen	*Negative charge on oxygen*

As might be expected, we can get a much truer picture of the reactivity of the carbonyl group by looking at the transition state for attack by a nucleophile. In the reactant, carbon is trigonal. In the transition state, carbon has begun to acquire the tetrahedral configuration it will have in the product; the attached groups are thus being brought closer together. We might expect moderate steric hindrance in this reaction; that is, larger groups (R and R′) will tend to resist crowding more than smaller groups. But the transition state is a relatively roomy one compared, say, with the transition state for an S_N2 reaction, with its pentavalent carbon; it is this comparative uncrowdedness that we are really referring to when we say that the carbonyl group is "accessible" to attack.

In the transition state, oxygen has started to acquire the electrons—and the negative charge—that it will have in the product. *It is the tendency of oxygen to acquire electrons—its ability to carry a negative charge—that is the real cause of the reactivity of the carbonyl group toward nucleophiles.* (The polarity of the carbonyl group is not the *cause* of the reactivity; it is simply another *manifestation* of the electronegativity of oxygen.)

Aldehydes generally undergo nucleophilic addition more readily than ketones. This difference in reactivity is consistent with the transition states involved, and seems to be due to a combination of electronic and steric factors. A ketone contains a second alkyl or aryl group where an aldehyde contains a hydrogen atom. A second alkyl or aryl group of a ketone is larger than the hydrogen of an aldehyde, and resists more strongly the crowding together in the transition state. An alkyl group releases electrons, and thus destabilizes the transition state by intensifying the negative charge developing on oxygen.

We might have expected an aryl group, with its electron-withdrawing inductive effect (Problem 18.7, p. 600), to stabilize the transition state and thus speed up reaction; however, it seems to stabilize the *reactant* even more, by resonance (contribution by I), and thus causes net deactivation.

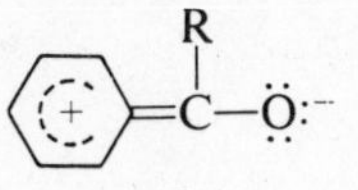

$$\underset{\text{I}}{(\text{C}_6\text{H}_5)^{+}\!=\!\overset{\text{R}}{\overset{|}{\text{C}}}\!-\!\ddot{\text{O}}\text{:}^-}$$

If acid is present, hydrogen ion becomes attached to carbonyl oxygen. This prior protonation lowers the E_{act} for nucleophilic attack, since it permits oxygen to

Acid-catalyzed nucleophilic addition

$$\begin{matrix}\text{R}'\\ \text{R}\end{matrix}\!\!>\!\text{C=O} \underset{}{\overset{\text{H}^+}{\rightleftharpoons}} \begin{matrix}\text{R}'\\ \text{R}\end{matrix}\!\!>\!\text{C=}\overset{\oplus}{\text{O}}\text{H} \xrightarrow{:\text{Z}} \left[\text{R}'\!-\!\overset{\text{Z}}{\overset{\vdots}{\text{C}}}\!\cdots\!\overset{\delta+}{\text{OH}}\right] \longrightarrow \text{R}'\!-\!\overset{\text{Z}}{\overset{|}{\text{C}}}\!\begin{matrix}\\ \text{R} \quad \text{OH}\end{matrix}$$

Undergoes nucleophilic attack more readily

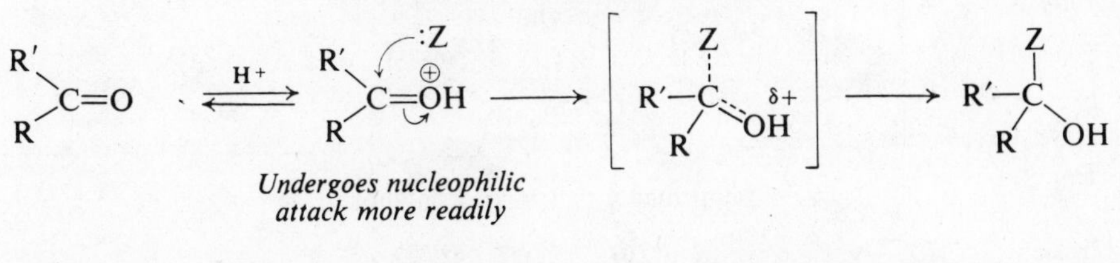

acquire the π electrons without having to accept a negative charge. Thus nucleophilic addition to aldehydes and ketones can be catalyzed by acids (sometimes, by *Lewis* acids).

REACTIONS OF ALDEHYDES AND KETONES

1. Oxidation. Discussed in Sec. 19.9.

(a) Aldehydes

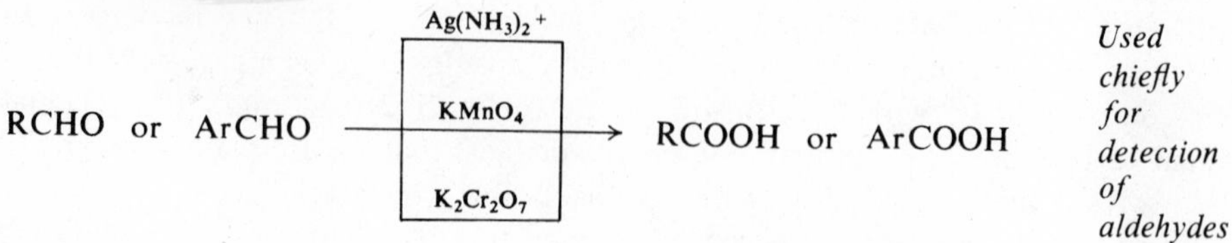

Examples:

$$CH_3CHO + 2Ag(NH_3)_2^+ + 3OH^- \longrightarrow 2Ag + CH_3COO^- + 4NH_3 + 2H_2O$$ *Tollens' test*

(*Colorless solution*; 2Ag: *Silver mirror*)

(b) Methyl ketones

$$R{-}\underset{\underset{O}{\|}}{C}{-}CH_3 \text{ or } Ar{-}\underset{\underset{O}{\|}}{C}{-}CH_3 \xrightarrow{OX^-} RCOO^- \text{ or } ArCOO^- + CHX_3$$ *Haloform reaction*

Examples:

$$C_2H_5{-}\underset{\underset{O}{\|}}{C}{-}CH_3 + 3OI^- \longrightarrow C_2H_5COO^- + CHI_3 + 2OH^-$$

(CHI_3: Iodoform, *Yellow; m.p. 119°*)

$$CH_3\overset{CH_3}{\overset{|}{C}}{=}CH\underset{\underset{O}{\|}}{C}CH_3 \xrightarrow[60^\circ]{KOCl} CHCl_3 + CH_3\overset{CH_3}{\overset{|}{C}}{=}CHCOOK \xrightarrow{H_2SO_4} CH_3\overset{CH_3}{\overset{|}{C}}{=}CHCOOH$$

Mesityl oxide (4-Methyl-3-penten-2-one) → 3-Methyl-2-butenoic acid

2. Reduction

(a) Reduction to alcohols. Discussed in Sec. 19.10.

$$>C{=}O \xrightarrow[\text{LiAlH}_4 \text{ or NaBH}_4\text{; then } H^+]{H_2 + \text{Ni, Pt, or Pd}} -\underset{\underset{H}{|}}{\overset{|}{C}}-OH$$

Examples:

$$\text{Cyclopentanone} \xrightarrow{H_2,Ni} \text{Cyclopentanol}$$

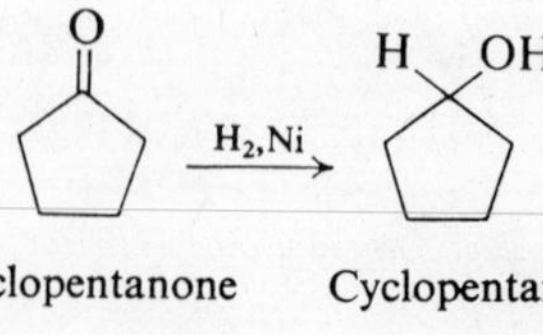

Cyclopentanone Cyclopentanol

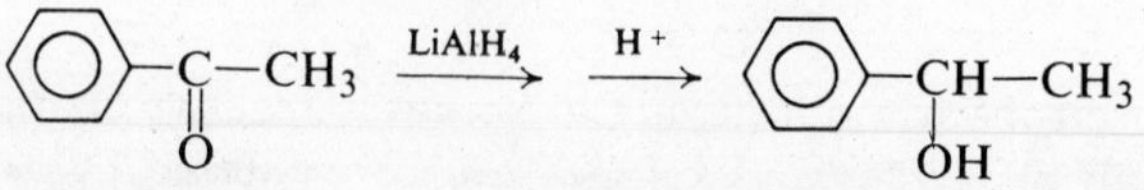

$$C_6H_5{-}\underset{\underset{O}{\|}}{C}{-}CH_3 \xrightarrow{LiAlH_4} \xrightarrow{H^+} C_6H_5{-}\underset{\underset{OH}{|}}{CH}{-}CH_3$$

Acetophenone α-Phenylethyl alcohol

(b) Reduction to hydrocarbons. Discussed in Sec. 19.10.

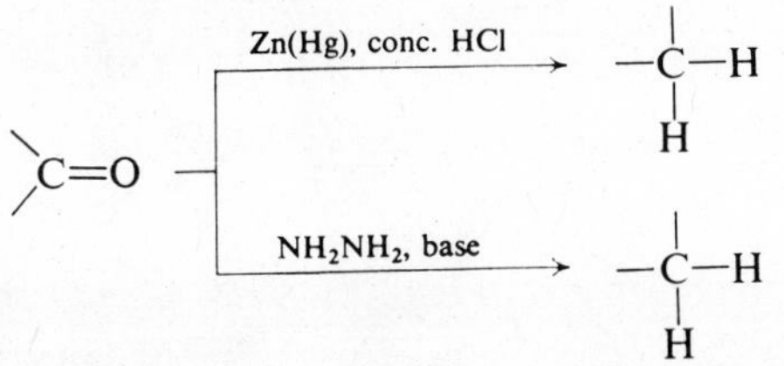

$$>C{=}O \xrightarrow{\text{Zn(Hg), conc. HCl}} -\overset{|}{\underset{H}{C}}-H$$

Clemmensen reduction
for compounds sensitive to base

$$>C{=}O \xrightarrow{\text{NH}_2\text{NH}_2\text{, base}} -\overset{|}{\underset{H}{C}}-H$$

Wolff-Kishner reduction
for compounds sensitive to acid

Examples:

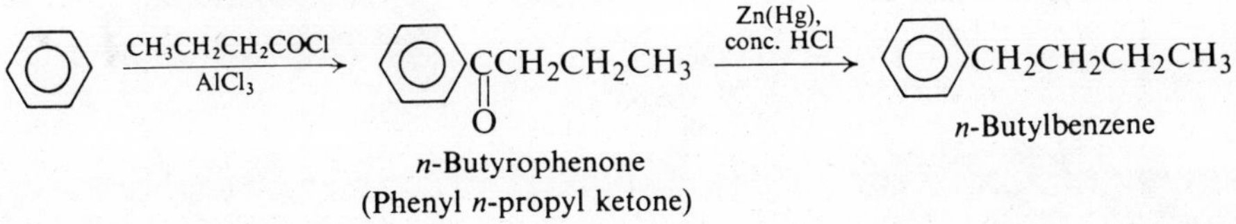

$$C_6H_6 \xrightarrow[\text{AlCl}_3]{CH_3CH_2CH_2COCl} C_6H_5\underset{O}{\underset{\|}{C}}CH_2CH_2CH_3 \xrightarrow[\text{conc. HCl}]{\text{Zn(Hg),}} C_6H_5CH_2CH_2CH_2CH_3$$

n-Butyrophenone
(Phenyl *n*-propyl ketone)

n-Butylbenzene

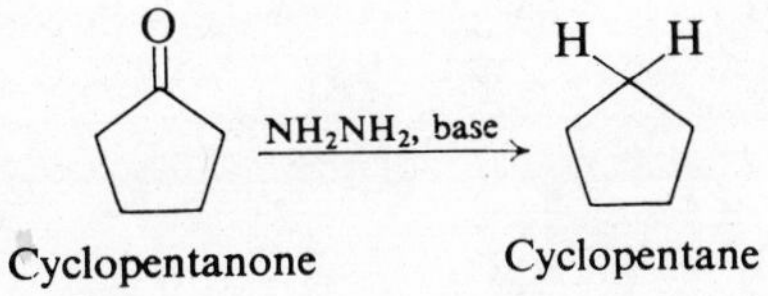

$$\text{Cyclopentanone} \xrightarrow{\text{NH}_2\text{NH}_2\text{, base}} \text{Cyclopentane}$$

Cyclopentanone Cyclopentane

(c) Reductive amination. Discussed in Sec. 22.11.

3. Addition of Grignard reagents. Discussed in Secs. 15.12–15.15 and 19.11.

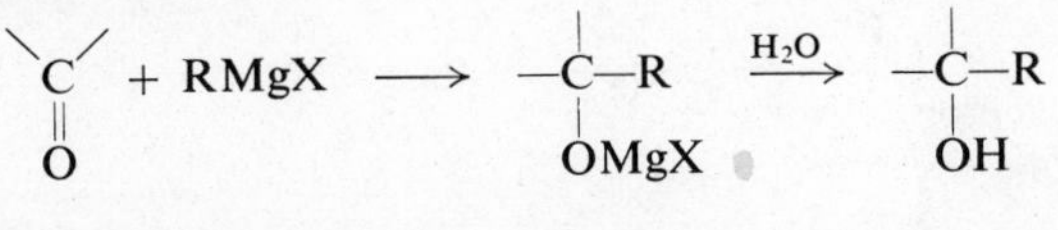

$$>\underset{O}{\underset{\|}{C}}< + RMgX \longrightarrow -\overset{|}{\underset{OMgX}{\underset{|}{C}}}-R \xrightarrow{H_2O} -\overset{|}{\underset{OH}{\underset{|}{C}}}-R$$

4. Addition of cyanide. Cyanohydrin formation. Discussed in Sec. 19.12.

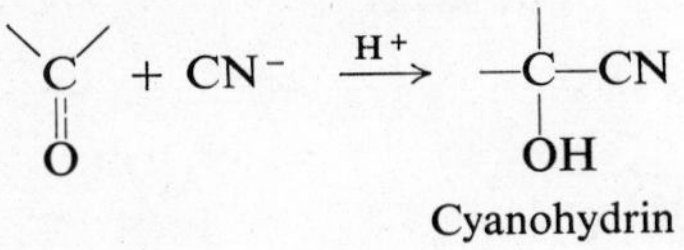

$$>\underset{O}{\underset{\|}{C}}< + CN^- \xrightarrow{H^+} -\overset{|}{\underset{OH}{\underset{|}{C}}}-CN$$

Cyanohydrin

Examples:

$$C_6H_5-\overset{H}{\overset{|}{C}}{=}O \xrightarrow[\text{NaHSO}_3]{\text{NaCN}} C_6H_5-\overset{H}{\overset{|}{\underset{OH}{\underset{|}{C}}}}-CN \xrightarrow{H_2O,\ HCl} C_6H_5-\overset{H}{\overset{|}{\underset{OH}{\underset{|}{C}}}}-COOH$$

Benzaldehyde Mandelonitrile Mandelic acid

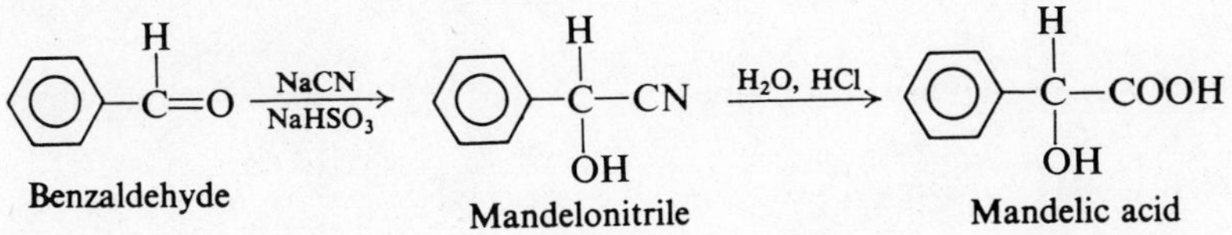

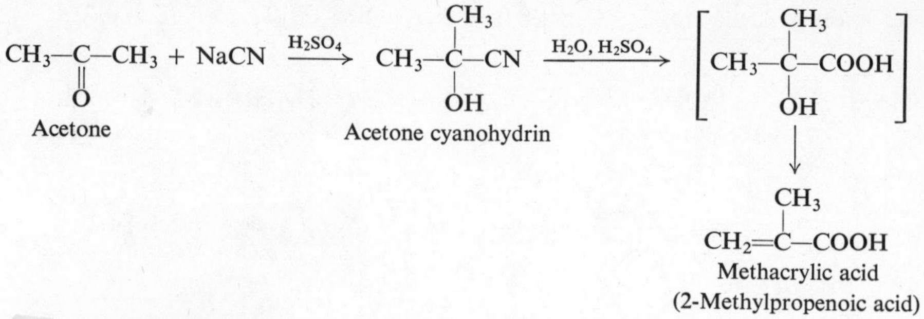

$$CH_3-\underset{\underset{O}{\|}}{C}-CH_3 + NaCN \xrightarrow{H_2SO_4} CH_3-\underset{OH}{\overset{CH_3}{C}}-CN \xrightarrow{H_2O,\ H_2SO_4} \left[CH_3-\underset{OH}{\overset{CH_3}{C}}-COOH \right] \longrightarrow CH_2=\overset{CH_3}{C}-COOH$$

Acetone — Acetone cyanohydrin — Methacrylic acid (2-Methylpropenoic acid)

5. Addition of bisulfite. Discussed in Sec. 19.13.

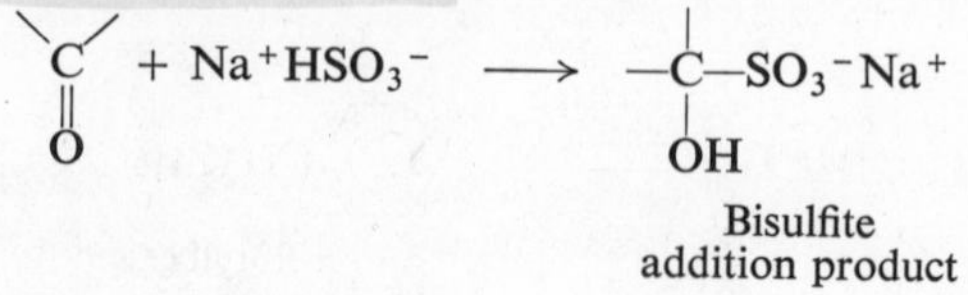

$$\underset{O}{\overset{\diagdown\ \diagup}{\underset{\|}{C}}} + Na^+HSO_3^- \longrightarrow -\underset{OH}{\overset{|}{C}}-SO_3^-Na^+$$

Bisulfite addition product

Used in purification
Not for hindered ketones

Examples:

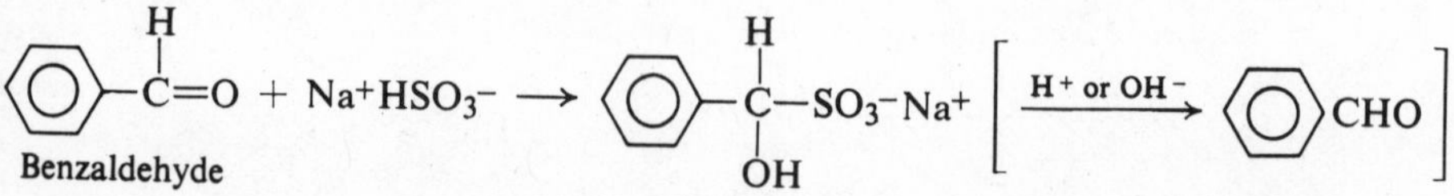

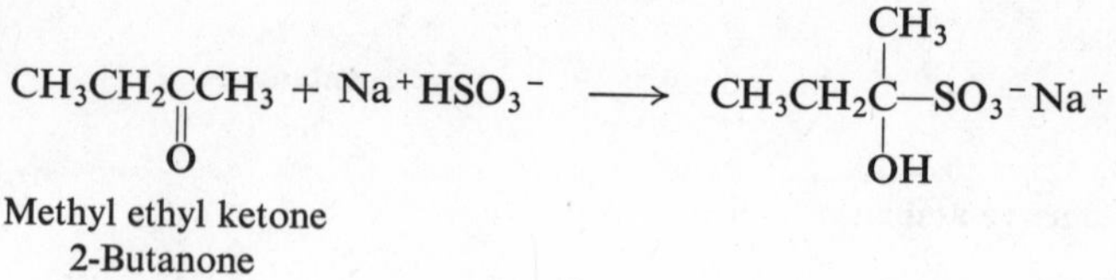

$$C_6H_5-\overset{H}{C}=O + Na^+HSO_3^- \longrightarrow C_6H_5-\underset{OH}{\overset{H}{C}}-SO_3^-Na^+ \quad \left[\xrightarrow{H^+ \text{ or } OH^-} C_6H_5CHO \right]$$

Benzaldehyde

$$CH_3CH_2\underset{\underset{O}{\|}}{C}CH_3 + Na^+HSO_3^- \longrightarrow CH_3CH_2\underset{OH}{\overset{CH_3}{C}}-SO_3^-Na^+$$

Methyl ethyl ketone
2-Butanone

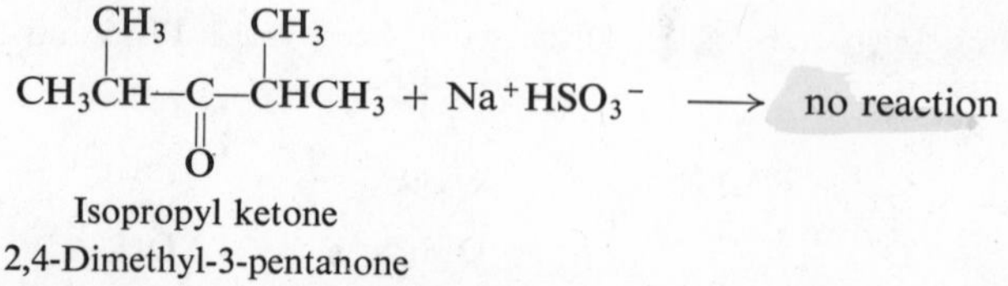

$$CH_3\overset{CH_3}{CH}-\underset{\underset{O}{\|}}{C}-\overset{CH_3}{CH}CH_3 + Na^+HSO_3^- \longrightarrow \text{no reaction}$$

Isopropyl ketone
2,4-Dimethyl-3-pentanone

6. Addition of derivatives of ammonia. Discussed in Sec. 19.14.

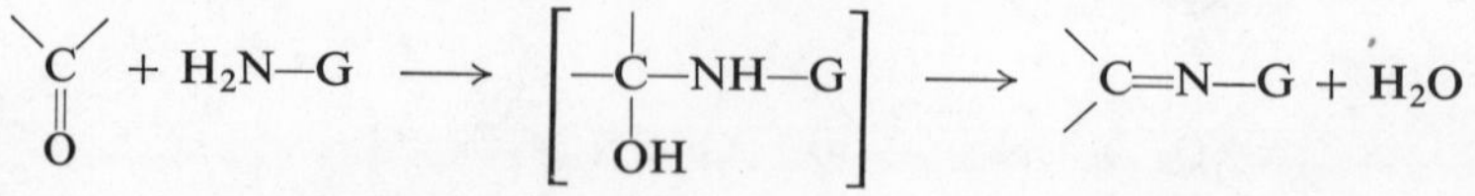

$$\underset{O}{\overset{\diagdown\ \diagup}{\underset{\|}{C}}} + H_2N-G \longrightarrow \left[-\underset{OH}{\overset{|}{C}}-NH-G \right] \longrightarrow \,\rangle C=N-G + H_2O$$

Used for identification

H_2N-G		Product	
H_2N-OH	Hydroxylamine	$\rangle C=NOH$	Oxime
H_2N-NH_2	Hydrazine	$\rangle C=NNH_2$	Hydrazone
$H_2N-NHC_6H_5$	Phenylhydrazine	$\rangle C=NNHC_6H_5$	Phenylhydrazone
$H_2N-NHCONH_2$	Semicarbazide	$\rangle C=NNHCONH_2$	Semicarbazone

Examples:

$$CH_3\overset{H}{\overset{|}{C}}{=}O + H_2N{-}OH \xrightarrow{H^+} CH_3\overset{H}{\overset{|}{C}}{=}NOH + H_2O$$

Acetaldehyde Hydroxylamine Acetaldoxime

$$C_6H_5{-}\overset{H}{\overset{|}{C}}{=}O + H_2N{-}NHC_6H_5 \xrightarrow{H^+} C_6H_5{-}\overset{H}{\overset{|}{C}}{=}NNHC_6H_5 + H_2O$$

Benzaldehyde Phenylhydrazine Benzaldehyde phenylhydrazone

7. Addition of alcohols. Acetal formation. Discussed in Sec. 19.15.

$$\underset{\underset{O}{\|}}{\diagdown C \diagup} + 2ROH \underset{}{\overset{H^+}{\rightleftharpoons}} {-}\overset{|}{\underset{\underset{OR}{|}}{C}}{-}OR + H_2O$$

An acetal

Example:

$$CH_3{-}\overset{H}{\overset{|}{C}}{=}O + 2C_2H_5OH \overset{HCl}{\rightleftharpoons} CH_3{-}\overset{H}{\overset{|}{\underset{\underset{OC_2H_5}{|}}{C}}}{-}OC_2H_5 + H_2O$$

Acetaldehyde

Acetal (Acetaldehyde diethyl acetal)

8. Cannizzaro reaction. Discussed in Sec. 19.16.

$$2\ {-}\overset{H}{\overset{|}{C}}{=}O \xrightarrow{\text{strong base}} {-}COO^- + {-}CH_2OH$$

An aldehyde with no α-hydrogens Acid salt Alcohol

Examples:

$$2\ HCHO \xrightarrow{\text{50\% NaOH, room temperature}} HCOO^- + CH_3OH$$

Formaldehyde Formate ion Methanol

$$2\ m\text{-}ClC_6H_4CHO \xrightarrow{\text{50\% KOH}} m\text{-}ClC_6H_4COO^- + m\text{-}ClC_6H_4CH_2OH$$

m-Chlorobenzaldehyde *m*-Chlorobenzoate ion *m*-Chlorobenzyl alcohol

$$3,4\text{-}(OCH_3)_2C_6H_3CHO + HCHO \xrightarrow{\text{50\% NaOH, 65°}} 3,4\text{-}(OCH_3)_2C_6H_3CH_2OH + HCOO^-$$

Crossed Cannizzaro reaction

Veratraldehyde 3,4-Dimethoxybenzaldehyde

3,4-Dimethoxybenzyl alcohol

9. Halogenation of ketones. Discussed in Secs. 21.3–21.4.

$$-\overset{\overset{\displaystyle O}{\|}}{C}-\underset{\underset{\displaystyle H}{|}}{\overset{|}{C}}- + X_2 \xrightarrow{\text{acid or base}} -\overset{\overset{\displaystyle O}{\|}}{C}-\underset{\underset{\displaystyle X}{|}}{\overset{|}{C}}- + HX \qquad \textit{α-Halogenation}$$

$$X_2 = Cl_2,\ Br_2,\ I_2$$

10. Addition of carbanions.

(a) **Aldol condensation.** Discussed in Secs. 21.5–21.8.

(b) **Reactions related to aldol condensation.** Discussed in Sec. 21.9.

(c) **Wittig reaction.** Discussed in Sec. 21.10.

(d) **Reformatsky reaction.** Discussed in Sec. 21.13.

19.9 Oxidation

Aldehydes are easily oxidized to carboxylic acids; ketones are not. Oxidation is the reaction in which aldehydes differ most from ketones, and this difference stems directly from their difference in structure: by definition, an aldehyde has a hydrogen atom attached to the carbonyl carbon, and a ketone has not. Regardless of exact mechanism, this hydrogen is abstracted in oxidation, either as a proton or an atom, and the analogous reaction for a ketone—abstraction of an alkyl or aryl group—does not take place.

Oxidation by chromic acid, for example, seems to involve a rate-determining step analogous to that for oxidation of secondary alcohols (Sec. 16.8): elimination (again possibly by a cyclic mechanism) from an intermediate chromate ester.

$$\underset{H}{\overset{R}{>}}C=O + \underset{\text{Cr(VI)}}{HCrO_4^-} + H^+ \rightleftharpoons R-\underset{\underset{\displaystyle H}{|}}{\overset{\overset{\displaystyle OH}{|}}{C}}-O-CrO_3H \longrightarrow$$

$$R-\overset{\overset{\displaystyle OH}{|}}{C}=O + H^+ + \underset{\text{Cr(IV)}}{HCrO_3^-}$$

The intermediate is the chromate ester of the aldehyde *hydrate*, $RCH(OH)_2$; it seems likely that the ester is formed *from* the hydrate, which exists in equilibrium with the aldehyde. In that case, what we are dealing with is essentially oxidation of a special kind of alcohol—a *gem*-diol.

Aldehydes are oxidized not only by the same reagents that oxidize primary and secondary alcohols—permanganate and dichromate—but also by the very mild oxidizing agent silver ion. Oxidation by silver ion requires an alkaline medium; to prevent precipitation of the insoluble silver oxide, a complexing agent is added: ammonia.

Tollens' reagent contains the silver ammonia ion, $Ag(NH_3)_2^+$. Oxidation of the aldehyde is accompanied by reduction of silver ion to free silver (in the form of a *mirror* under the proper conditions).

$$RCHO + Ag(NH_3)_2^+ \longrightarrow RCOO^- + Ag$$

Colorless solution   *Silver mirror*

(Oxidation by complexed cupric ion is a characteristic of certain substituted carbonyl compounds, and will be taken up with *carbohydrates* in Sec. 34.6.)

Oxidation by Tollens' reagent is useful chiefly for detecting aldehydes, and in particular for differentiating them from ketones (see Sec. 19.17). The reaction is of value in synthesis in those cases where aldehydes are more readily available than the corresponding acids: in particular, for the synthesis of unsaturated acids from the unsaturated aldehydes obtained from the aldol condensation (Sec. 21.6), where advantage is taken of the fact that Tollens' reagent does not attack carbon–carbon double bonds.

$$\overset{\beta}{R\text{CH}}{=}\overset{\alpha}{\text{CH}}{-}\overset{\text{H}}{\overset{|}{\text{C}}}{=}\text{O} \xrightarrow{\text{Tollens' reagent}} \overset{\beta}{R\text{CH}}{=}\overset{\alpha}{\text{CH}}{-}\text{COOH}$$

α,β-Unsaturated aldehyde α,β-Unsaturated acid

Oxidation of ketones requires breaking of carbon–carbon bonds, and (except for the haloform reaction) takes place only under vigorous conditions. Cleavage involves the double bond of the *enol* form (Sec. 8.13) and, where the structure

$$-\underset{\text{H}}{\underset{|}{\overset{|}{\text{C}}}}-\overset{\text{OH}}{\overset{|}{\text{C}}}{=}\overset{|}{\text{C}}- \rightleftarrows -\underset{\text{H}}{\underset{|}{\overset{|}{\text{C}}}}-\overset{\text{O}}{\overset{\|}{\text{C}}}-\underset{\text{H}}{\underset{|}{\overset{|}{\text{C}}}}- \rightleftarrows -\overset{|}{\text{C}}{=}\overset{\text{OH}}{\overset{|}{\text{C}}}-\underset{\text{H}}{\underset{|}{\overset{|}{\text{C}}}}-$$

Enol   Ketone   Enol

permits, occurs on either side of the carbonyl group; in general, then, mixtures of carboxylic acids are obtained (see Sec. 6.29).

Problem 19.4 Predict the product(s) of vigorous oxidation of: (a) 3-hexanone; (b) cyclohexanone.

Methyl ketones are oxidized smoothly by means of hypohalite in the haloform reaction (Sec. 16.11). Besides being commonly used to detect these ketones (Sec. 19.17), this reaction is often useful in synthesis, hypohalite having the special advantage of not attacking carbon–carbon double bonds. For example:

$$C_6H_5-\overset{\text{H}}{\overset{|}{\text{C}}}{=}\overset{\text{CH}_3}{\overset{|}{\text{C}}}-\underset{\text{O}}{\underset{\|}{\text{C}}}-\text{CH}_3 \xrightarrow{\text{KOCl}} C_6H_5-\overset{\text{H}}{\overset{|}{\text{C}}}{=}\overset{\text{CH}_3}{\overset{|}{\text{C}}}-\text{COOH} + \text{CHCl}_3$$

Available by aldol condensation (Sec. 21.8) α-Methylcinnamic acid

19.10 Reduction

Aldehydes can be reduced to primary alcohols, and ketones to secondary alcohols, either by catalytic hydrogenation or by use of chemical reducing agents like lithium aluminum hydride, $LiAlH_4$. Such reduction is useful for the preparation of certain alcohols that are less available than the corresponding carbonyl compounds, in particular carbonyl compounds that can be obtained by the aldol condensation (Sec. 21.7). For example:

$$\text{Cyclopentanone} \xrightarrow{LiAlH_4} \xrightarrow{H^+} \text{Cyclopentanol}$$

Cyclopentanone Cyclopentanol

$$CH_3CH{=}CHCHO \xrightarrow{H_2,\ Ni} CH_3CH_2CH_2CH_2OH$$

Crotonaldehyde *n*-Butyl alcohol

From aldol condensation of acetaldehyde

$$C_6H_5CH{=}CHCHO \xrightarrow{NaBH_4} \xrightarrow{H^+} C_6H_5CH{=}CHCH_2OH$$

Cinnamaldehyde Cinnamyl alcohol

From aldol condensation of benzaldehyde and acetaldehyde
(Sec. 21.8)

Sodium borohydride, $NaBH_4$, does not reduce carbon–carbon double bonds, not even those conjugated with carbonyl groups, and is thus useful for the reduction of such unsaturated carbonyl compounds to unsaturated alcohols.

Aldehydes and ketones can be reduced to hydrocarbons by the action (a) of amalgamated zinc and concentrated hydrochloric acid, the **Clemmensen reduction**; or (b) of hydrazine, NH_2NH_2, and a strong base like KOH or potassium *tert*-butoxide, the **Wolff-Kishner reduction**. These are particularly important when applied to the alkyl aryl ketones obtained from Friedel-Crafts acylation, since this reaction sequence permits, indirectly, the attachment of straight alkyl chains to the benzene ring. For example:

$$\text{Resorcinol} \xrightarrow{CH_3(CH_2)_4COOH,\ ZnCl_2} C_6H_3(OH)_2CO(CH_2)_4CH_3 \xrightarrow{Zn(Hg),\ HCl} C_6H_3(OH)_2CH_2(CH_2)_4CH_3$$

Resorcinol — $CO(CH_2)_4CH_3$ — $CH_2(CH_2)_4CH_3$

4-*n*-Hexylresorcinol
Used as an antiseptic

A special sort of oxidation and reduction, the *Cannizzaro reaction*, will be discussed in Sec. 19.16.

Let us look a little more closely at reduction by metal hydrides. Alcohols are formed from carbonyl compounds, smoothly and in high yield, by the action of such compounds as lithium aluminum hydride, $LiAlH_4$. Here again, we see

$$4R_2C{=}O + LiAlH_4 \longrightarrow (R_2CHO)_4AlLi \xrightarrow{H_2O} 4R_2CHOH + LiOH + Al(OH)_3$$

nucleophilic addition: this time the nucleophile is hydrogen transferred with a pair of electrons—as a hydride ion, H:⁻—from the metal to carbonyl carbon:

$$\rangle C{=}O + H{-}AlH_3^- \longrightarrow -\overset{H}{\underset{|}{\overset{|}{C}}}-\overset{-}{O}AlH_3 \xrightarrow{3\ \rangle C=O} (-\overset{|}{\underset{|}{C}}-O)_4Al^-$$

19.11 Addition of Grignard reagents

The addition of Grignard reagents to aldehydes and ketones has already been discussed as one of the most important methods of preparing complicated alcohols (Secs. 15.12–15.15).

The organic group, transferred *with a pair of electrons* from magnesium to carbonyl carbon, is a powerful nucleophile.

$$\rangle C{=}O + R{-}MgX \longrightarrow -\overset{R}{\underset{|}{\overset{|}{C}}}-\overset{-}{O}\overset{+}{M}gX$$

19.12 Addition of cyanide

The elements of HCN add to the carbonyl group of aldehydes and ketones to yield compounds known as **cyanohydrins**:

$$\rangle C{=}O + CN^- \xrightarrow{H^+} -\underset{OH}{\underset{|}{\overset{|}{C}}}-CN$$

A cyanohydrin

The reaction is often carried out by adding mineral acid to a mixture of the carbonyl compound and aqueous sodium cyanide. In a useful modification, cyanide is added to the bisulfite addition product (Sec. 19.13) of the carbonyl compound, the bisulfite ion serving as the necessary acid:

$$-\underset{OH}{\underset{|}{\overset{|}{C}}}-SO_3^-Na^+ \rightleftarrows \rangle C{=}O + Na^+HSO_3^- \xrightarrow{CN^-} -\underset{OH}{\underset{|}{\overset{|}{C}}}-CN + SO_3^{--} + Na^+$$

Addition appears to involve nucleophilic attack on carbonyl carbon by the strongly basic cyanide ion; subsequently (or possibly simultaneously) oxygen accepts a hydrogen ion to form the cyanohydrin product:

$$\underset{\text{O}}{\overset{\diagdown\ \diagup}{\underset{\|}{\text{C}}}} \xrightarrow{:\text{CN}^-} -\underset{\text{O}^-}{\overset{|}{\underset{|}{\text{C}}}}-\text{CN} \xrightarrow{\text{H}^+} -\underset{\text{OH}}{\overset{|}{\underset{|}{\text{C}}}}-\text{CN}$$

Nucleophilic reagent (:CN⁻) — Cyanohydrin

Although it is the elements of HCN that become attached to the carbonyl group, a highly acidic medium—in which the concentration of un-ionized HCN is highest—actually retards reaction. This is to be expected, since the very weak acid HCN is a poor source of cyanide ion.

Cyanohydrins are nitriles, and their principal use is based on the fact that, like other nitriles, they undergo hydrolysis; in this case the products are α-hydroxy-acids or unsaturated acids. For example:

$$m\text{-}O_2NC_6H_4-\overset{\text{H}}{\overset{|}{\text{C}}}=\text{O} \xrightarrow{\text{CN}^-,\ \text{H}^+} m\text{-}O_2NC_6H_4-\underset{\text{OH}}{\underset{|}{\overset{\text{H}}{\overset{|}{\text{C}}}}}-\text{CN} \xrightarrow{\text{HCl, heat}} m\text{-}O_2NC_6H_4-\underset{\text{OH}}{\underset{|}{\overset{\text{H}}{\overset{|}{\text{C}}}}}-\text{COOH}$$

m-Nitrobenzaldehyde → *m*-Nitromandelic acid

$$CH_3CH_2-\overset{CH_3}{\overset{|}{C}}=O \xrightarrow{CN^-,\ H^+} CH_3CH_2-\underset{OH}{\underset{|}{\overset{CH_3}{\overset{|}{C}}}}-CN \xrightarrow{H_2SO_4,\ \text{heat}} \left[CH_3CH_2-\underset{OH}{\underset{|}{\overset{CH_3}{\overset{|}{C}}}}-COOH\right] \longrightarrow CH_3CH=\overset{CH_3}{\overset{|}{C}}-COOH$$

Methyl ethyl ketone
2-Butanone

2-Methyl-2-butenoic acid

Problem 19.5 Each of the following is converted into the cyanohydrin, and the products are separated by careful fractional distillation or crystallization. For each reaction tell how many fractions will be collected, and whether each fraction, as collected, will be optically active or inactive, resolvable or non-resolvable.

(a) Acetaldehyde; (b) benzaldehyde; (c) acetone;
(d) R-(+)-glyceraldehyde, $CH_2OHCHOHCHO$; (e) (±)-glyceraldehyde.
(f) How would your answer to each of the above be changed if each mixture were subjected to hydrolysis to hydroxy acids before fractionation?

19.13 Addition of bisulfite

Sodium bisulfite adds to most aldehydes and to many ketones (especially methyl ketones) to form bisulfite addition products:

$$\underset{\text{O}}{\overset{\diagdown\ \diagup}{\underset{\|}{\text{C}}}} + Na^+HSO_3^- \rightleftharpoons -\underset{\text{OH}}{\overset{|}{\underset{|}{\text{C}}}}-SO_3^-Na^+$$

A bisulfite addition product

The reaction is carried out by mixing the aldehyde or ketone with a concentrated aqueous solution of sodium bisulfite; the product separates as a crystalline solid. Ketones containing bulky groups usually fail to react with bisulfite, presumably for steric reasons.

Addition involves nucleophilic attack by bisulfite ion on carbonyl carbon, followed by attachment of a hydrogen ion to carbonyl oxygen:

$$\underset{\text{Nucleophilic reagent}}{\overset{\displaystyle >C=O}{:SO_3H^-}} \rightleftharpoons -\overset{|}{\underset{O^-}{C}}-SO_3^- \xrightleftharpoons{H^+} -\overset{|}{\underset{OH}{C}}-SO_3^-$$

Like other carbonyl addition reactions, this one is reversible. Addition of acid or base destroys the bisulfite ion in equilibrium with the addition product, and regenerates the carbonyl compound.

$$-\overset{|}{\underset{OH}{C}}-SO_3^-Na^+ \rightleftharpoons >C=O + HSO_3^- \begin{cases} \xrightarrow{H^+} SO_2 + H_2O \\ \xrightarrow{OH^-} SO_3^{--} + H_2O \end{cases}$$

Bisulfite addition products are generally prepared for the purpose of separating a carbonyl compound from non-carbonyl compounds. The carbonyl compound can be purified by conversion into its bisulfite addition product, separation of the crystalline addition product from the non-carbonyl impurities, and subsequent regeneration of the carbonyl compound. A non-carbonyl compound can be freed of carbonyl impurities by washing it with aqueous sodium bisulfite; any contaminating aldehyde or ketone is converted into its bisulfite addition product which, being somewhat soluble in water, dissolves in the aqueous layer.

Problem 19.6 Suggest a practical situation that might arise in the laboratory in which you would need to (a) separate an aldehyde from undesired non-carbonyl materials; (b) remove an aldehyde that is contaminating a non-carbonyl compound. Describe how you could carry out the separations, telling exactly what you would do and see.

19.14 Addition of derivatives of ammonia

Certain compounds related to ammonia add to the carbonyl group to form derivatives that are important chiefly for the characterization and identification of aldehydes and ketones (Sec. 19.17). The products contain a carbon–nitrogen double bond resulting from elimination of a molecule of water from the initial addition products. Some of these reagents and their products are:

$$\underset{\text{O}}{\overset{\diagdown\ \diagup}{\underset{\|}{\text{C}}}} + :\text{NH}_2\text{OH} \xrightarrow{H^+} \left[\begin{array}{c}|\\ -\text{C}-\text{NHOH}\\ |\\ \text{OH}\end{array}\right] \longrightarrow \;\rangle\text{C}{=}\text{NOH} + H_2O$$

Hydroxylamine — Oxime

$$\underset{\text{O}}{\overset{\diagdown\ \diagup}{\underset{\|}{\text{C}}}} + :\text{NH}_2\text{NHC}_6\text{H}_5 \xrightarrow{H^+} \left[\begin{array}{c}|\\ -\text{C}-\text{NHNHC}_6\text{H}_5\\ |\\ \text{OH}\end{array}\right] \longrightarrow \;\rangle\text{C}{=}\text{NNHC}_6\text{H}_5 + H_2O$$

Phenylhydrazine — Phenylhydrazone

$$\underset{\text{O}}{\overset{\diagdown\ \diagup}{\underset{\|}{\text{C}}}} + :\text{NH}_2\text{NHCONH}_2 \xrightarrow{H^+} \left[\begin{array}{c}|\\ -\text{C}-\text{NHNHCONH}_2\\ |\\ \text{OH}\end{array}\right] \longrightarrow \;\rangle\text{C}{=}\text{NNHCONH}_2 + H_2O$$

Semicarbazide — Semicarbazone

Like ammonia, these derivatives of ammonia are basic, and therefore react with acids to form salts: hydroxylamine hydrochloride, $HONH_3^+Cl^-$; phenylhydrazine hydrochloride, $C_6H_5NHNH_3^+Cl^-$; and semicarbazide hydrochloride, $NH_2CONHNH_3^+Cl^-$. The salts are less easily oxidized by air than the free bases, and it is in this form that the reagents are best preserved and handled. When needed, the basic reagents are liberated from their salts in the presence of the carbonyl compound by addition of a base, usually sodium acetate.

$$C_6H_5NHNH_3^+Cl^- + CH_3COO^-Na^+ \rightleftarrows C_6H_5NHNH_2 + CH_3COOH + Na^+Cl^-$$

Phenylhydrazine hydrochloride	Sodium acetate	Phenylhydrazine	Acetic acid
Stronger acid	*Stronger base*	*Weaker base*	*Weaker acid*

It is often necessary to adjust the reaction medium to just the right acidity. Addition involves nucleophilic attack by the basic nitrogen compound on carbonyl carbon. Protonation of carbonyl oxygen makes carbonyl carbon more susceptible to nucleophilic attack; in so far as the carbonyl compound is concerned, then, addition will be favored by high acidity. But the ammonia derivative, $H_2N{-}G$, can also undergo protonation to form the ion, $^+H_3N{-}G$, which lacks unshared electrons and is no longer nucleophilic; in so far as the nitrogen compound is concerned, then, addition is favored by low acidity. The conditions under which

$$\underset{\text{O}}{\overset{\diagdown\ \diagup}{\underset{\|}{\text{C}}}} \overset{H^+}{\rightleftarrows} \underset{\oplus\text{OH}}{\overset{\diagdown\ \diagup}{\underset{\|}{\text{C}}}} \longrightarrow \left[\begin{array}{c} \quad\;\text{H}\\ |\quad |\oplus\\ -\text{C}-\text{N}-\text{G}\\ |\quad |\\ \text{HO}\;\;\text{H}\end{array}\right] \longrightarrow \;\rangle\text{C}{=}\text{N}{-}\text{G} + H_2O + H^+$$

$$H_2\ddot{N}{-}G \overset{H^+}{\rightleftarrows} {}^+H_3N{-}G$$

Free base:	Salt:
nucleophilic	*not nucleophilic*

addition proceeds most rapidly are thus the result of a compromise: the solution must be acidic enough for an appreciable fraction of the carbonyl compound to be protonated, but not so acidic that the concentration of the free nitrogen compound is too low. The exact conditions used depend upon the basicity of the reagent, and upon the reactivity of the carbonyl compound.

Problem 19.7 Semicarbazide (1 mole) is added to a mixture of cyclohexanone (1 mole) and benzaldehyde (1 mole). If the product is isolated immediately, it consists almost entirely of the semicarbazone of cyclohexanone; if the product is isolated after several hours, it consists almost entirely of the semicarbazone of benzaldehyde. How do you account for these observations? (*Hint:* See Sec. 8.22.)

19.15 Addition of alcohols. Acetal formation

Alcohols add to the carbonyl group of aldehydes in the presence of anhydrous acids to yield **acetals**:

$$\underset{\text{Aldehyde}}{R'\!-\!\overset{\displaystyle H}{\overset{|}{C}}\!=\!O} + \underset{\text{Alcohol}}{2ROH} \xrightleftharpoons{\text{dry HCl}} \underset{\text{Acetal}}{R'\!-\!\overset{\displaystyle H}{\overset{|}{\underset{\underset{\displaystyle OR}{|}}{C}}}\!-\!OR} + H_2O$$

The reaction is carried out by allowing the aldehyde to stand with an excess of the anhydrous alcohol and a little anhydrous acid, usually hydrogen chloride. In the preparation of ethyl acetals the water is often removed as it is formed by means of the azeotrope of water, benzene, and ethyl alcohol (b.p. 64.9°, Sec. 15.6). (Simple *ketals* are usually difficult to prepare by reaction of ketones with alcohols, and are made in other ways.)

$$\underset{\text{Benzaldehyde}}{C_6H_5\!-\!\overset{\displaystyle H}{\overset{|}{C}}\!=\!O} + \underset{\text{Ethyl alcohol}}{2C_2H_5OH} \xrightleftharpoons{\text{dry HCl}} \underset{\text{Diethyl acetal of benzaldehyde}}{C_6H_5\!-\!\overset{\displaystyle H}{\overset{|}{\underset{\underset{\displaystyle OC_2H_5}{|}}{C}}}\!-\!OC_2H_5} + H_2O$$

There is good evidence that in alcoholic solution an aldehyde exists in equilibrium with a compound called a **hemiacetal**:

$$R'\!-\!\overset{\displaystyle H}{\overset{|}{C}}\!=\!O + ROH \xrightleftharpoons{H^+} \underset{\text{A hemiacetal}}{R'\!-\!\overset{\displaystyle H}{\overset{|}{\underset{\underset{\displaystyle OH}{|}}{C}}}\!-\!OR}$$

A hemiacetal is formed by the addition of the nucleophilic alcohol molecule to the carbonyl group; it is both an ether and an alcohol. With a few exceptions, hemiacetals are too unstable to be isolated.

In the presence of acid the hemiacetal, acting as an alcohol, reacts with more of the solvent alcohol to form the acetal, an ether:

$$\underset{\substack{\text{Hemiacetal}\\\text{(An alcohol)}}}{\text{R}'\text{—}\underset{\text{OH}}{\overset{\text{H}}{\underset{|}{\overset{|}{\text{C}}}}}\text{—OR}} + \text{ROH} \overset{\text{H}^+}{\rightleftharpoons} \underset{\substack{\text{Acetal}\\\text{(An ether)}}}{\text{R}'\text{—}\underset{\text{OR}}{\overset{\text{H}}{\underset{|}{\overset{|}{\text{C}}}}}\text{—OR}} + \text{H}_2\text{O}$$

The reaction involves the formation (step 1) of the ion I, which then combines (step 2) with a molecule of alcohol to yield the protonated acetal. As we can see,

$$(1)\qquad \underset{\text{Hemiacetal}}{\text{R}'\text{—}\underset{\text{OH}}{\overset{\text{H}}{\underset{|}{\overset{|}{\text{C}}}}}\text{—OR}} + \text{H}^+ \rightleftharpoons \text{R}'\text{—}\underset{\oplus\text{OH}_2}{\overset{\text{H}}{\underset{|}{\overset{|}{\text{C}}}}}\text{—OR} \rightleftharpoons \underset{\text{I}}{\text{R}'\text{—}\overset{\text{H}}{\overset{|}{\text{C}}}\text{=}\overset{\oplus}{\text{O}}\text{R}} + \text{H}_2\text{O}$$

$$(2)\qquad \underset{\text{I}}{\text{R}'\text{—}\overset{\text{H}}{\overset{|}{\text{C}}}\text{=}\overset{\oplus}{\text{O}}\text{R}} + \text{ROH} \rightleftharpoons \text{R}'\text{—}\underset{\substack{\oplus\text{OR}\\|\\\text{H}}}{\overset{\text{H}}{\underset{|}{\overset{|}{\text{C}}}}}\text{—OR} \rightleftharpoons \underset{\text{Acetal}}{\text{R}'\text{—}\underset{\text{OR}}{\overset{\text{H}}{\underset{|}{\overset{|}{\text{C}}}}}\text{—OR}} + \text{H}^+$$

this mechanism is strictly analogous to the S_N1 route we have previously encountered (Sec. 17.3) for the formation of ethers.

Acetal formation thus involves (a) nucleophilic addition to a carbonyl group, and (b) ether formation via a carbonium ion.

Acetals have the structure of ethers and, like ethers, are cleaved by acids and are stable toward bases. Acetals differ from ethers, however, in the extreme *ease* with which they undergo acidic cleavage; they are rapidly converted even at room

$$\underset{\text{Acetal}}{\text{R}'\text{—}\underset{\text{OR}}{\overset{\text{H}}{\underset{|}{\overset{|}{\text{C}}}}}\text{—OR}} + \text{H}_2\text{O} \xrightarrow[\text{fast}]{\text{H}^+} \underset{\text{Aldehyde}}{\text{R}'\text{—}\overset{\text{H}}{\overset{|}{\text{C}}}\text{=O}} + \underset{\text{Alcohol}}{2\text{ROH}}$$

temperature into the aldehyde and alcohol by dilute mineral acids. The mechanism of hydrolysis is exactly the reverse of that by which acetals are formed.

Problem 19.8 Account for the fact that anhydrous acids bring about formation of acetals whereas aqueous acids bring about hydrolysis of acetals.

The heart of the chemistry of acetals is the "carbonium" ion,

$$\left[\underset{\text{Ia}}{\text{R—}\underset{\oplus}{\overset{\text{H}}{\overset{|}{\text{C}}}}\text{—OR}} \qquad \underset{\text{Ib}}{\text{R—}\overset{\text{H}}{\overset{|}{\text{C}}}\text{=}\underset{\oplus}{\text{O}}\text{R}}\right]$$

Especially stable: every atom has octet

which is a hybrid of structures Ia and Ib. Contribution from Ib, in which every atom has an octet of electrons, makes this ion considerably more stable than ordinary carbonium ions. (Indeed, Ib *alone* may pretty well represent the ion, in which case it is not a carbonium ion at all but an *oxonium* ion.)

Now, generation of this ion is the rate-determining step both in formation of acetals (reading to the right in equation 1) and in their hydrolysis (reading to the left in equation 2). The same factor—the providing of electrons by oxygen—that stabilizes the ion also stabilizes the transition state leading to its formation. Generation of the ion is speeded up, and along with it the entire process: formation or hydrolysis of the acetal.

(Oddly enough, oxygen causes activation in *nucleophilic* substitution here in precisely the same way it activates aromatic ethers toward *electrophilic* substitution (Sec. 17.8); the common feature is, of course, development of a positive charge in the transition state of the rate-determining step.)

We shall find the chemistry of hemiacetals and acetals to be fundamental to the study of carbohydrates (Chaps. 34 and 35).

Problem 19.9 (a) The following reaction is an example of what familiar synthesis?

$$C_6H_4(OH)_2 + CH_2I_2 \xrightarrow{\text{aqueous NaOH}} C_6H_4{<}^{O}_{O}{>}CH_2$$

II

(b) To what family of compounds does II belong? (c) What will II yield upon treatment with acid? With base?

Problem 19.10 Suggest a convenient chemical method for separating unreacted benzaldehyde from benzaldehyde diethyl acetal. (Compare Problem 19.6, p. 639.)

Problem 19.11 *Glyceraldehyde*, $CH_2OHCHOHCHO$, is commonly made from the acetal of acrolein, $CH_2{=}CH{-}CHO$. Show how this could be done. Why is acrolein itself not used?

Problem 19.12 How do you account for the following differences in ease of hydrolysis?

(a) $RC(OR')_3 \gg RCH(OR')_2 \gg RCH_2OR'$
An ortho ester; An acetal; An ether

(b) $R_2C(OR')_2 > RCH(OR')_2 > H_2C(OR')_2$
A ketal; An acetal; A formal

Problem 19.13 The simplest way to prepare an aldehyde, $RCH^{18}O$, labeled at the carbonyl oxygen, is to allow an ordinary aldehyde to stand in $H_2{}^{18}O$ in the presence of a little acid. Suggest a detailed mechanism for this oxygen exchange.

19.16 Cannizzaro reaction

In the presence of concentrated alkali, aldehydes containing no α-hydrogens undergo self-oxidation-and-reduction to yield a mixture of an alcohol and a salt of a carboxylic acid. This reaction, known as the **Cannizzaro reaction**, is generally brought about by allowing the aldehyde to stand at room temperature with concentrated aqueous or alcoholic hydroxide. (Under these conditions an aldehyde containing α-hydrogens would undergo aldol condensation faster, Sec. 21.5.)

$$\underset{\text{Formaldehyde}}{2HCHO} \xrightarrow{50\%\ NaOH} \underset{\text{Methanol}}{CH_3OH} + \underset{\text{Sodium formate}}{HCOO^-Na^+}$$

$$\underset{p\text{-Nitrobenzaldehyde}}{O_2N\text{-}C_6H_4\text{-}CHO} \xrightarrow{35\%\ NaOH} \underset{p\text{-Nitrobenzyl alcohol}}{O_2N\text{-}C_6H_4\text{-}CH_2OH} + \underset{\text{Sodium } p\text{-nitrobenzoate}}{O_2N\text{-}C_6H_4\text{-}COO^-\ Na^+}$$

In general, a mixture of two aldehydes undergoes a Cannizzaro reaction to yield all possible products. If one of the aldehydes is formaldehyde, however, reaction yields almost exclusively sodium formate and the alcohol corresponding to the other aldehyde:

$$ArCHO + HCHO \xrightarrow{\text{conc. NaOH}} ArCH_2OH + HCOO^-Na^+$$

The high tendency for formaldehyde to undergo oxidation makes this **crossed Cannizzaro reaction** a useful synthetic tool. For example:

$$\underset{\substack{\text{Anisaldehyde}\\(p\text{-Methoxybenzaldehyde})}}{p\text{-}CH_3O\text{-}C_6H_4\text{-}CHO} + HCHO \xrightarrow{\text{conc. NaOH}} \underset{p\text{-Methoxybenzyl alcohol}}{p\text{-}CH_3O\text{-}C_6H_4\text{-}CH_2OH} + HCOO^-Na^+$$

Evidence, chiefly from kinetics and experiments with isotopically labeled compounds, indicates that even this seemingly different reaction follows the familiar pattern for carbonyl compounds: nucleophilic addition. Two successive additions

(1) $$Ar\text{—}\overset{H}{\overset{|}{C}}\text{=O} + OH^- \rightleftharpoons \underset{\text{I}}{Ar\text{—}\overset{H}{\overset{|}{C}}\text{—O}^-}\ (\text{with OH on C})$$

(2) $$Ar\text{—}\overset{H}{\overset{|}{C}}\text{=O} + \underset{\text{I}}{Ar\text{—}\overset{H}{\overset{|}{\underset{OH}{\underset{|}{C}}}}\text{—O}^-} \longrightarrow Ar\text{—}\overset{H}{\overset{|}{\underset{H}{\underset{|}{C}}}}\text{—O}^- + Ar\text{—}\underset{OH}{\underset{|}{C}}\text{=O}$$

$$Ar\text{—}CH_2\text{—O}^- \xrightarrow{+H^+} ArCH_2OH \qquad Ar\text{—}C(=O)OH \xrightarrow{-H^+} ArCOO^-$$

are involved: addition of hydroxide ion (step 1) to give intermediate I; and addition of a hydride ion from I (step 2) to a second molecule of aldehyde. The presence of the negative charge on I aids in the loss of hydride ion.

Problem 19.14 In the case of some aldehydes there is evidence that intermediate II is the hydride donor in the Cannizzaro reactions. (a) How would II be formed from I?

$$\begin{array}{c} \text{H} \\ | \\ \text{R—C—O}^- \\ | \\ \text{O}^- \\ \text{II} \end{array}$$

(b) Why would you expect II to be a better hydride donor than I? (*Hint:* What is one product of the hydride transfer from II?)

Problem 19.15 Suggest an experiment to prove that a hydride transfer of the kind shown in step (2) is actually involved, that is, that hydrogen is transferred from I and not from the solvent.

Problem 19.16 From examination of the mechanism, can you suggest one factor that would tend to make a crossed Cannizzaro reaction involving formaldehyde take place in the particular way it does?

Problem 19.17 Phenylglyoxal, C_6H_5COCHO, is converted by aqueous sodium hydroxide into sodium mandelate, $C_6H_5CHOHCOONa$. Suggest a likely mechanism for this conversion.

Problem 19.18 In the **benzilic acid rearrangement**, the diketone *benzil* is converted by sodium hydroxide into the salt of *benzilic acid.*

$$\underset{\text{Benzil}}{C_6H_5COCOC_6H_5} \xrightarrow{OH^-} (C_6H_5)_2C(OH)COO^- \xrightarrow{H^+} \underset{\text{Benzilic acid}}{(C_6H_5)_2C(OH)COOH}$$

If sodium methoxide is used instead of sodium hydroxide, the ester $(C_6H_5)_2C(OH)COOCH_3$ is obtained. Suggest a possible mechanism for this rearrangement.

19.17 Analysis of aldehydes and ketones

Aldehydes and ketones are characterized through the addition to the carbonyl group of nucleophilic reagents, especially derivatives of ammonia (Sec. 19.14). An aldehyde or ketone will, for example, react with 2,4-dinitrophenylhydrazine to form an insoluble yellow or red solid.

Aldehydes are characterized, and in particular are differentiated from ketones, through their ease of oxidation: aldehydes give a positive test with Tollens' reagent (Sec. 19.9); ketones do not. A positive Tollens' test is also given by a few other kinds of easily oxidized compounds, e.g., certain phenols and amines; these compounds do not, however, give positive tests with 2,4-dinitrophenylhydrazine.

Aldehydes are also, of course, oxidized by many other oxidizing agents: by cold, dilute, neutral $KMnO_4$ and by CrO_3 in H_2SO_4 (Sec. 6.30).

A highly sensitive test for aldehydes is the *Schiff test*. An aldehyde reacts with the fuchsin-aldehyde reagent to form a characteristic magenta color.

Aliphatic aldehydes and ketones having α-hydrogen react with Br_2 in CCl_4. This reaction is generally too slow to be confused with a test for unsaturation, and moreover it liberates HBr.

Aldehydes and ketones are generally identified through the melting points of derivatives like 2,4-dinitrophenylhydrazones, oximes, and semicarbazones (Sec. 19.14).

Methyl ketones are characterized through the iodoform test (see Sec. 16.11).

Problem 19.19 Make a table to summarize the behavior of each class of compound we have studied toward each of the oxidizing agents we have studied.

Problem 19.20 A convenient test for aldehydes and most ketones depends upon the fact that a carbonyl compound generally causes a change in color when it is added to a solution of hydroxylamine hydrochloride and an acid-base indicator. What is the basis of this test?

Problem 19.21 Expand the table you made in Problem 18.18, p. 608, to include aldehydes and ketones, and, in particular, emphasize oxidizing agents.

19.18 Spectroscopic analysis of aldehydes and ketones

Infrared. Infrared spectroscopy is by far the best way to detect the presence of a carbonyl group in a molecule. The strong band due to C=O stretching appears at about 1700 cm^{-1}, where it is seldom obscured by other strong absorptions; it is one of the most useful bands in the infrared spectrum, and is often the first one looked for (see Fig. 19.1).

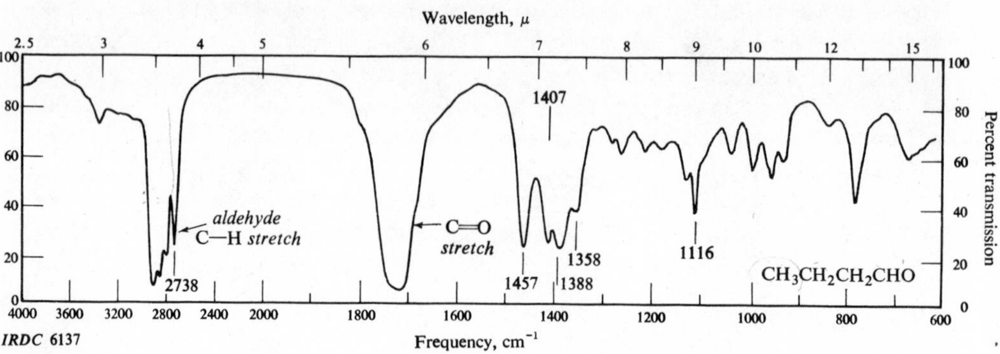

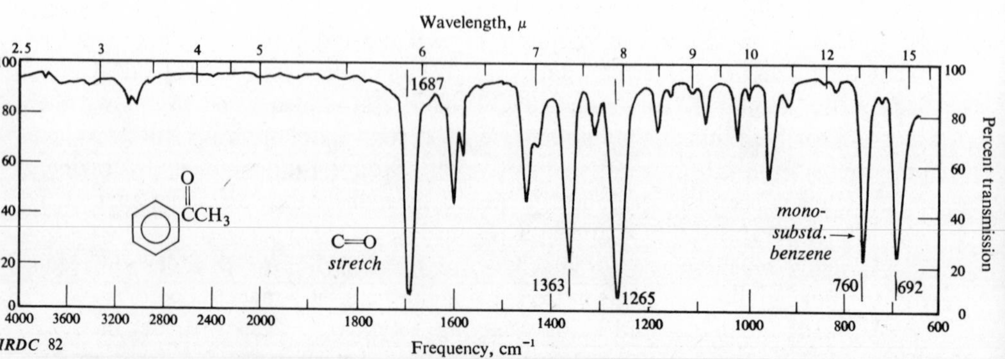

Figure 19.1. Infrared spectra of (*a*) *n*-butyraldehyde and (*b*) acetophenone.

The carbonyl band is given not only by aldehydes and ketones, but also by carboxylic acids and their derivatives. Once identified as arising from an aldehyde or ketone (see below), its exact frequency can give a great deal of information about the structure of the molecule.

C=O stretching, *strong*

RCHO 1725 cm^{-1}	R_2CO 1710 cm^{-1}	Cyclobutanones 1780 cm^{-1}
ArCHO 1700 cm^{-1}	ArCOR 1690 cm^{-1}	Cyclopentanones 1740 cm^{-1}
—C=C—CHO 1685 cm^{-1}	—C=C—C=O 1675 cm^{-1}	—C=C—C— 1540–1640 cm^{-1} (OH-----O) (enols)

The —CHO group of an aldehyde has a characteristic C—H stretching band near 2720 cm^{-1}; this, in conjunction with the carbonyl band, is fairly certain evidence for an aldehyde (see Fig. 19.1).

Carboxylic acids (Sec. 18.22) and esters (Sec. 20.25) also show carbonyl absorption, and in the same general region as aldehydes and ketones. Acids, however, also show the broad O—H band. Esters usually show the carbonyl band at somewhat higher frequencies than ketones of the same general structure; furthermore, esters show characteristic C—O stretching bands. (For a comparison of certain oxygen compounds, see Table 20.3, p. 689.)

Nmr. The proton of an aldehyde group, —CHO, absorbs far downfield, at δ 9–10. Coupling of this proton with adjacent protons has a small constant (J 1–3 Hz), and the fine splitting is often seen superimposed on other splittings.

Ultraviolet. The ultraviolet spectrum can tell a good deal about the structure of carbonyl compounds: particularly, as we might expect from our earlier discussion (Sec. 13.5), about conjugation of the carbonyl group with a carbon–carbon double bond.

Saturated aldehydes and ketones absorb weakly in the near ultraviolet. Conjugation moves this weak band (the R band) to longer wavelengths (why?) and, more important, moves a very intense band (the K band) from the far ultraviolet to the near ultraviolet.

—C=O	—C=C—C=O	
λ_{max} 270–300 nm	λ_{max} 300–350 nm	λ_{max} 215–250 nm
ϵ_{max} 10–20	ϵ_{max} 10–20	ϵ_{max} 10,000–20,000

The exact position of this K band gives information about the number and location of substituents in the conjugated system.

PROBLEMS

1. Neglecting enantiomerism, give structural formulas, common names, and IUPAC names for:

(a) the seven carbonyl compounds of formula $C_5H_{10}O$
(b) the five carbonyl compounds of formula C_8H_8O that contain a benzene ring

2. Give the structural formula of:

(a) acetone
(b) benzaldehyde
(c) methyl isobutyl ketone
(d) trimethylacetaldehyde
(e) acetophenone
(f) cinnamaldehyde
(g) 4-methylpentanal
(h) phenylacetaldehyde
(i) benzophenone
(j) α,γ-dimethylcaproaldehyde
(k) 3-methyl-2-pentanone
(l) 2-butenal
(m) 4-methyl-3-penten-2-one (mesityl oxide)
(n) 1,3-diphenyl-2-propen-1-one (benzalacetophenone)
(o) 3-hydroxypentanal
(p) benzyl phenyl ketone
(q) salicyaldehyde
(r) *p,p'*-dihydroxybenzophenone
(s) *m*-tolualdehyde

3. Write balanced equations, naming all organic products, for the reaction (if any) of phenylacetaldehyde with:

(a) Tollens' reagent
(b) CrO_3/H_2SO_4
(c) cold dilute $KMnO_4$
(d) $KMnO_4$, H^+, heat
(e) H_2, Ni, 20 lb/in^2, 30°
(f) $LiAlH_4$
(g) $NaBH_4$
(h) C_6H_5MgBr, then H_2O
(i) isopropylmagnesium chloride, then H_2O
(j) $NaHSO_3$
(k) CN^-; H^+
(l) hydroxylamine
(m) phenylhydrazine
(n) 2,4-dinitrophenylhydrazine
(o) semicarbazide
(p) ethyl alcohol, dry HCl(g)

4. Answer Problem 3 for cyclohexanone.

5. Write balanced equations, naming all organic products, for the reaction (if any) of benzaldehyde with:

(a) conc. NaOH
(b) formaldehyde, conc. NaOH
(c) CN^-, H^+
(d) product (c) + H_2O, H^+, heat
(e) CH_3MgI, then H_2O
(f) product (e) + H^+, heat
(g) $(CH_3)_2{}^{14}CHMgBr$, then H_2O
(h) $H_2{}^{18}O$, H^+

6. Write equations for all steps in the synthesis of the following from propionaldehyde, using any other needed reagents:

(a) *n*-propyl alcohol
(b) propionic acid
(c) α-hydroxybutyric acid
(d) *sec*-butyl alcohol
(e) 1-phenyl-1-propanol
(f) methyl ethyl ketone
(g) *n*-propyl propionate
(h) 2-methyl-3-pentanol

7. Write equations for all steps in the synthesis of the following from acetophenone, using any other needed reagents:

(a) ethylbenzene
(b) benzoic acid
(c) α-phenylethyl alcohol
(d) 2-phenyl-2-butanol
(e) diphenylmethylcarbinol
(f) α-hydroxy-α-phenylpropionic acid

8. Outline all steps in a possible laboratory synthesis of each of the following from benzene, toluene, and alcohols of four carbons or fewer, using any needed inorganic reagents:

(a) isobutyraldehyde
(b) phenylacetaldehyde
(c) *p*-bromobenzaldehyde
(d) methyl ethyl ketone
(e) 2,4-dinitrobenzaldehyde
(f) *p*-nitrobenzophenone
(g) 2-methyl-3-pentanone
(h) benzyl methyl ketone
(i) *m*-nitrobenzophenone
(j) *n*-propyl *p*-tolyl ketone
(k) α-methylbutyraldehyde
(l) *n*-butyl isobutyl ketone
(m) *p*-nitroacetophenone
(n) 3-nitro-4'-methylbenzophenone
(o) *p*-nitropropiophenone

9. Outline all steps in a possible laboratory synthesis of each of the following from benzene, toluene, and alcohols of four carbons or fewer, using any needed inorganic reagents:

(a) *n*-butylbenzene
(b) α-hydroxy-*n*-valeric acid
(c) 2-methylheptane
(d) 2,3,5-trimethyl-3-hexanol
(e) *p*-nitro-α-hydroxyphenylacetic acid
(f) 1,2-diphenyl-2-propanol
(g) ethylphenyl-*p*-bromophenylcarbinol
(h) 3-methyl-2-butenoic acid

10. (a) What are A, B, and C?

$C_6H_5C(CH_3)_2CH_2COOH + PCl_3 \longrightarrow$ A ($C_{11}H_{13}OCl$)
A + $AlCl_3/CS_2 \longrightarrow$ B ($C_{11}H_{12}O$)
B + N_2H_4, OH^-, heat, high-boiling solvent $\longrightarrow$ C ($C_{11}H_{14}$)

C gave the following nmr spectrum:

a singlet, δ 1.22, 6H
b triplet, δ 1.85, 2H, $J = 7$ Hz
c triplet, δ 2.83, 2H, $J = 7$ Hz
d singlet, δ 7.02, 4H

(b) C was also formed by treatment of the alcohol D ($C_{11}H_{16}O$) with concentrated sulfuric acid. What is the structure of D?

11. In the oxidation of an alcohol RCH_2OH to an aldehyde by chromic acid, the chief side-reaction is formation, not of the carboxylic acid, but of the ester $RCOOCH_2R$. Experiment has shown that a mixture of isobutyl alcohol and isobutyraldehyde is oxidized much faster than either compound alone. Suggest a possible explanation for these facts. (*Hint:* See Sec. 19.9.)

12. Give stereochemical formulas for compounds E–J.

R-(+)-*glyceraldehyde* ($CH_2OHCHOHCHO$) + CN^-, $H^+ \longrightarrow$ E + F
(both E and F have the formula $C_4H_7O_3N$)
E + F + OH^-, H_2O, heat; then $H^+ \longrightarrow$ G + H (both $C_4H_8O_5$)
G + $HNO_3 \longrightarrow$ I ($C_4H_6O_4$), *optically active*
H + $HNO_3 \longrightarrow$ J ($C_4H_6O_4$), *optically inactive*

13. (a) *cis*-1,2-Cyclopentanediol reacts with acetone in the presence of dry HCl to yield compound K, $C_8H_{14}O_2$, which is resistant to boiling alkali, but which is readily converted into the starting materials by aqueous acids. What is the most likely structure of K? To what class of compounds does it belong?

(b) *trans*-1,2-Cyclopentanediol does not form an analogous compound. How do you account for this fact?

14. The oxygen exchange described in Problem 19.13 (p. 643) can be carried out by use of hydroxide ion instead of hydrogen ion as catalyst. Suggest a detailed mechanism for exchange under these conditions. (*Hint:* See Sec. 19.16.)

15. Vinyl alkyl ethers, RCH=CHOR′, are very rapidly hydrolyzed by dilute aqueous acid to form the alcohol R′OH and the aldehyde RCH_2CHO. Hydrolysis in $H_2{}^{18}O$ gives alcohol R′OH containing only ordinary oxygen. Outline all steps in the most likely mechanism for the hydrolysis. Show how this mechanism accounts not only for the results of the tracer experiment, but also for the extreme ease with which hydrolysis takes place.

16. (a) Optically active 2-octyl brosylate was found to react with pure water to yield 2-octanol with complete inversion of configuration. With mixtures of water and the "inert" solvent dioxane (p. 561), however, inversion was accompanied by racemization, the extent of racemization increasing with the concentration of dioxane. From this and other evidence, R. A. Sneen (of Purdue University) has proposed that inverted alcohol is formed through (S_N2) attack by water, and that retained alcohol is formed via an initial attack by dioxane.

Show in detail how nucleophilic attack by dioxane could ultimately lead to the formation of alcohol with retention of configuration.

(b) In the mixed solvent methanol and acetone (*no* water present), 2-octyl brosylate was found to yield not only the 2-octyl methyl ether, but also some *2-octanol.* When the same reaction was carried out in the presence of the base pyridine (to neutralize the sulfonic acid formed), no 2-octanol was obtained; there was obtained instead, in impure form, a substance whose infrared spectrum showed no absorption in the carbonyl region, but which reacted with an acidic solution of 2,4-dinitrophenylhydrazine to yield the 2,4-dinitrophenylhydrazone of acetone. Sneen has proposed that the 2-octanol was formed by a series of reactions initiated by nucleophilic attack on 2-octyl brosylate by acetone.

Outline all steps in mechanism for the formation of 2-octanol under these conditions. What compound is probably responsible for the formation of the 2,4-dinitrophenylhydrazone? How do you account for the effect of added base?

17. On treatment with bromine, certain diarylcarbinols (I) are converted into a 50:50 mixture of aryl bromide (II) and aldehyde (III).

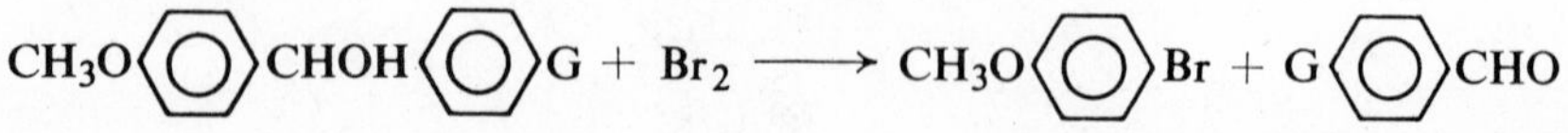

$$CH_3O\langle\bigcirc\rangle CHOH\langle\bigcirc\rangle G + Br_2 \longrightarrow CH_3O\langle\bigcirc\rangle Br + G\langle\bigcirc\rangle CHO$$

I II III

Whether G is $-NO_2$, $-H$, $-Br$, or $-CH_3$, bromine appears *only* in the ring containing the $-OCH_3$ group. The rate of reaction is affected moderately by the nature of G, decreasing along the series: $G = -CH_3 > -H > -Br > -NO_2$. The rate of reaction is slowed down by the presence of added bromide ion.

Outline all steps in the most likely mechanism for this reaction. Show how your mechanism accounts for each of the above facts.

18. A naïve graduate student needed a quantity of benzhydrol, $(C_6H_5)_2CHOH$, and decided to prepare it by the reaction between phenymagnesium bromide and benzaldehyde. He prepared a mole of the Grignard reagent. To insure a good yield, he then added, not one, but *two* moles of the aldehyde. On working up the reaction mixture, he was at first gratified to find he had obtained a good yield of a crystalline product, but his hopes were dashed when closer examination revealed that he had made, not benzhydrol, but the ketone benzophenone. Bewildered, the student made the first of many trips to his research director's office.

He returned shortly, red-faced, to the laboratory, carried out the reaction again using equimolar amounts of the reactants, and obtained a good yield of the compound he wanted.

What had gone wrong in his first attempt? How had his generosity with benzaldehyde betrayed him? (*Hint:* See Sec. 19.16. Examine the structure of the initial addition product.) (In Problem 20, p. 724, we shall follow his further adventures.)

19. (a) Give structural formulas of compounds L and M, and of *isoeugenol* and *vanillin.*

eugenol (below) + KOH, 225° $\longrightarrow$ isoeugenol ($C_{10}H_{12}O_2$)

isoeugenol + $(CH_3CO)_2O$ $\longrightarrow$ L ($C_{12}H_{14}O_3$) (See Sec. 20.10)

L + $K_2Cr_2O_7$, H_2SO_4, 75° $\longrightarrow$ M ($C_{10}H_{10}O_4$)

M + HSO_3^-, H_2O, boil $\longrightarrow$ vanillin ($C_8H_8O_3$)

(b) Account for the conversion of eugenol into isoeugenol.

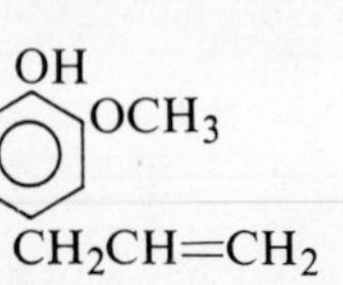

OH, OCH_3, $CH_2CH{=}CH_2$ — Eugenol

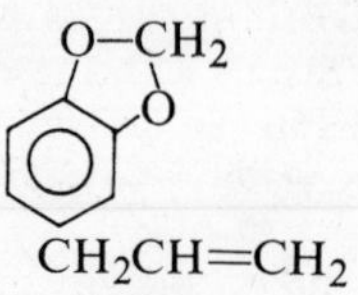

O–CH_2–O, $CH_2CH{=}CH_2$ — Safrole

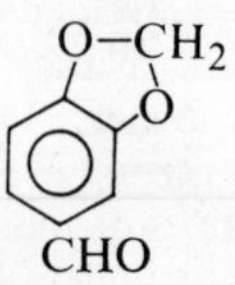

O–CH_2–O, CHO — Piperonal

(c) Suggest a way to convert *safrole* into *piperonal* (above).

20. Suggest a mechanism for the following reaction.

$$(CH_3)_2C{=}CHCH_2CH_2C(CH_3){=}CHCHO + H_3O^+ \longrightarrow$$

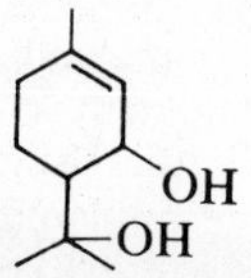

OH
OH
3,8-Carvomenthenediol

The ring-closing step can be considered as either nucleophilic addition or electrophilic addition depending on one's point of view. Show how this is so, identifying both the electrophile and the nucleophile.

21. The trimer of trichloroacetaldehyde (compare *paraldehyde*, p. 621) exists in two forms, N and O, which give the following nmr data.

N: singlet, δ 4.28
O: two singlets, δ 4.63 and δ 5.50, peak area ratio 2:1

Show in as much detail as you can the structure of each of these.

22. How do you account for the difference in behavior between diastereomers IV and V? (*Hint:* Draw Newman projections. What are the bulkiest groups?)

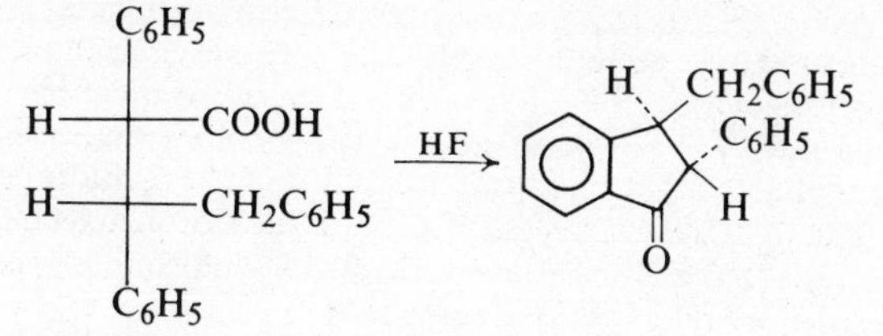

C_6H_5
H—COOH
H—$CH_2C_6H_5$
C_6H_5
IV

HF →

H $CH_2C_6H_5$
C_6H_5
H
O

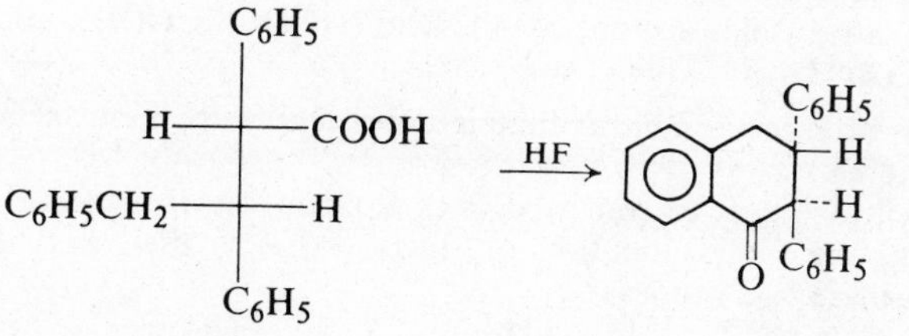

C_6H_5
H—COOH
$C_6H_5CH_2$—H
C_6H_5
V

HF →

C_6H_5
H
H
O C_6H_5

23. The acetal (VI) of glycerol and benzaldehyde has been found to exist in two configurations. (a) Draw them. (b) One of these exists preferentially in a conformation

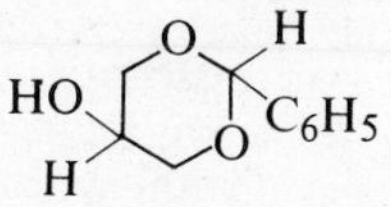

O H
HO C_6H_5
O
H
VI

in which the phenyl group occupies an axial position. Which configuration is this, and what counterbalances the unfavorable steric factor?

24. Describe a simple chemical test that would serve to distinguish between:

(a) *n*-valeraldehyde and ethyl ketone
(b) phenylacetaldehyde and benzyl alcohol
(c) cyclohexanone and methyl *n*-caproate
(d) 2-pentanone and 3-pentanone
(e) propionaldehyde and ethyl ether
(f) diethyl acetal and *n*-valeraldehyde
(g) diethyl acetal and *n*-propyl ether
(h) methyl *m*-tolyl ketone and propiophenone
(i) 2-pentanone and 2-pentanol
(j) paraldehyde and isobutyl ether
(k) dioxane and trioxane

Tell exactly what you would *do* and *see*.

25. An unknown compound is believed to be one of the following, all of which boil within a few degrees of each other. Describe how you would go about finding out which of the possibilities the unknown actually is. Where possible use simple chemical tests; where necessary use more elaborate chemical methods such as quantitative hydrogenation, cleavage, neutralization equivalent, saponification equivalent, etc. Make use of any needed tables of physical constants.

(a) phenylacetaldehyde
m-tolualdehyde
o-tolualdehyde
acetophenone
p-tolualdehyde

(b) methyl β-phenylethyl ketone
cyclohexylbenzene
benzyl *n*-butyrate
γ-phenylpropyl alcohol
n-caprylic acid

(c) isophorone (3,5,5-trimethyl-2-cyclohexen-1-one)
n-dodecane
benzyl *n*-butyl ether
ethyl benzoate
m-cresyl acetate
n-nonyl alcohol

(d) *p*-chloroacetophenone
methyl *o*-chlorobenzoate
p-chlorobenzyl chloride
m-chloronitrobenzene

26. *Citral*, $C_{10}H_{16}O$, is a terpene that is the major constituent of lemongrass oil. It reacts with hydroxylamine to yield a compound of formula $C_{10}H_{17}ON$, and with Tollens' reagent to give a silver mirror and a compound of formula $C_{10}H_{16}O_2$. Upon vigorous oxidation citral yields acetone, oxalic acid (HOOC—COOH), and levulinic acid ($CH_3COCH_2CH_2COOH$).

(a) Propose a structure for citral that is consistent with these facts and with the isoprene rule (Sec. 8.26.)

(b) Actually citral seems to consist of two isomers, citral *a* (*geranial*) and citral *b* (*neral*), which yield the same oxidation products. What is the most likely structural difference between these two isomers?

(c) Citral *a* is obtained by mild oxidation of geraniol (Problem 27, p. 547); citral *b* is obtained in a similar way from nerol. On this basis assign structures to citral *a* and citral *b*.

27. (+)-*Carvotanacetone*, $C_{10}H_{16}O$, is a terpene found in thuja oil. It reacts with hydroxylamide and semicarbazide to form crystalline derivatives. It gives negative tests with Tollens' reagent, but rapidly decolorizes cold dilute $KMnO_4$.

Carvotanacetone can be reduced successively to *carvomenthone*, $C_{10}H_{18}O$, and *carvomenthol*, $C_{10}H_{20}O$. Carvomenthone reacts with hydroxylamine but not with cold dilute $KMnO_4$. Carvomenthol does not react with hydroxylamine or cold dilute $KMnO_4$, but gives a positive test with CrO_3/H_2SO_4.

One set of investigators found that oxidation of carvotanacetone gave isopropylsuccinic acid and pyruvic acid, $CH_3COCOOH$; another set of investigators isolated acetic acid and β-isopropylglutaric acid.

$$\underset{\displaystyle \text{Isopropylsuccinic acid}}{\begin{array}{c}HOOCCHCH_2COOH\\ |\\ CH(CH_3)_2\end{array}} \qquad \underset{\displaystyle \beta\text{-Isopropylglutaric acid}}{\begin{array}{c}HOOCCH_2CHCH_2COOH\\ |\\ CH(CH_3)_2\end{array}}$$

What single structure for carvotanacetone is consistent with all these facts?

28. Which (if any) of the following compounds could give rise to each of the infrared spectra shown in Fig. 19.2 (p. 654)?

isobutyraldehyde	ethyl vinyl ether
2-butanone	cyclopropylcarbinol
tetrahydrofuran	3-buten-2-ol

29. Give a structure or structures consistent with each of the nmr spectra in Fig. 19.3 (p. 655).

30. Give the structures of compounds P, Q, and R on the basis of their infrared spectra (Fig. 19.4, p. 656) and their nmr spectra (Fig. 19.5, p. 657).

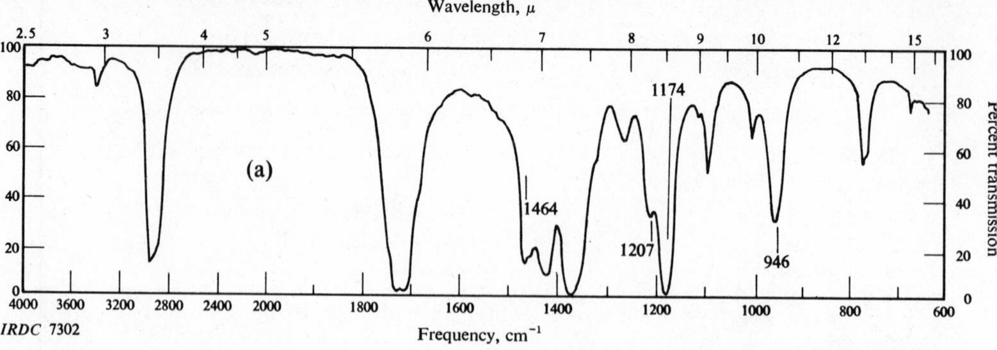

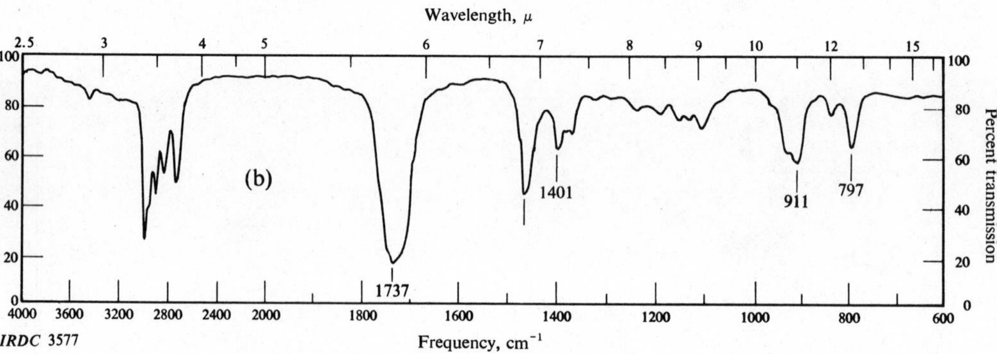

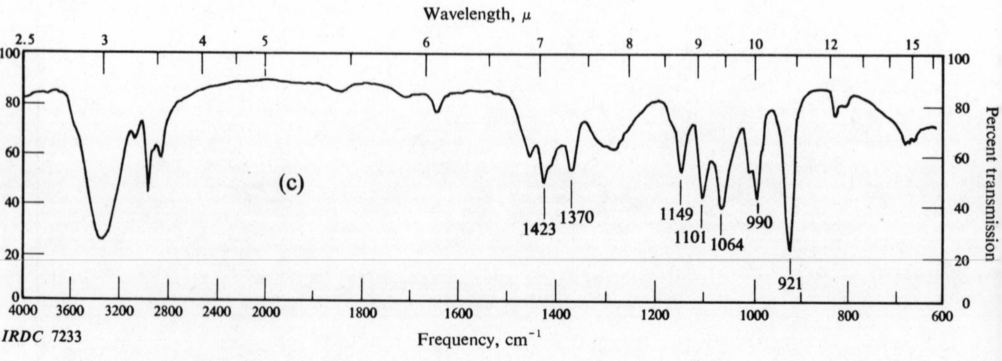

Figure 19.2. Infrared spectra for Problem 28, p. 653.

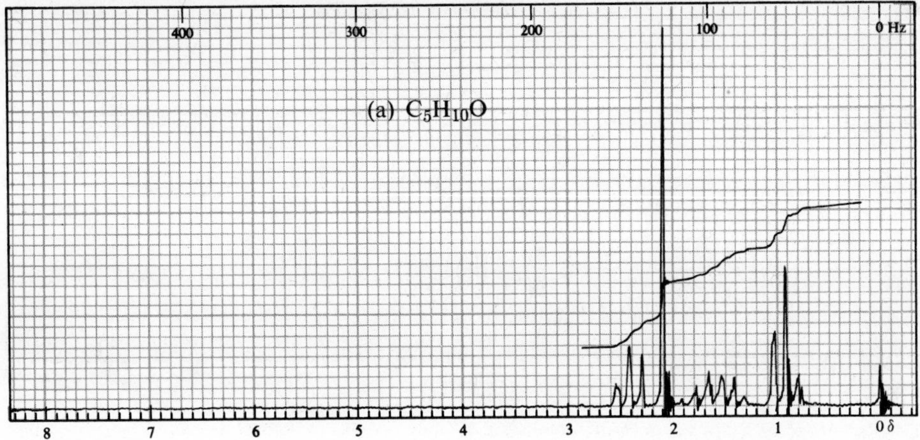

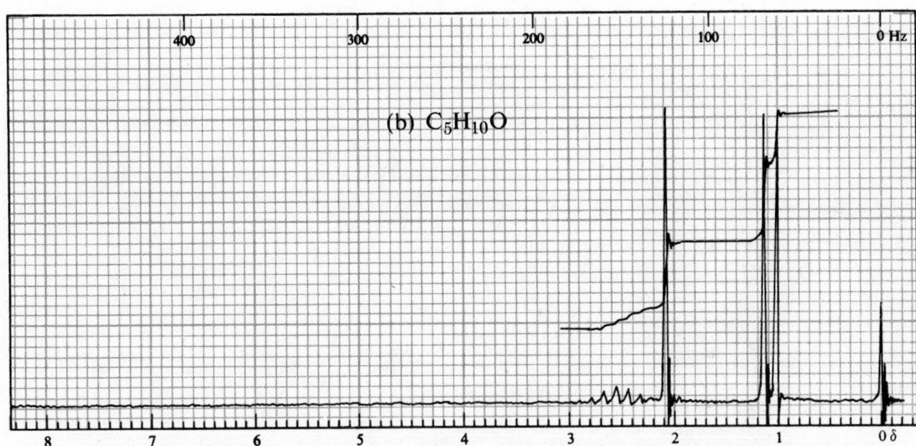

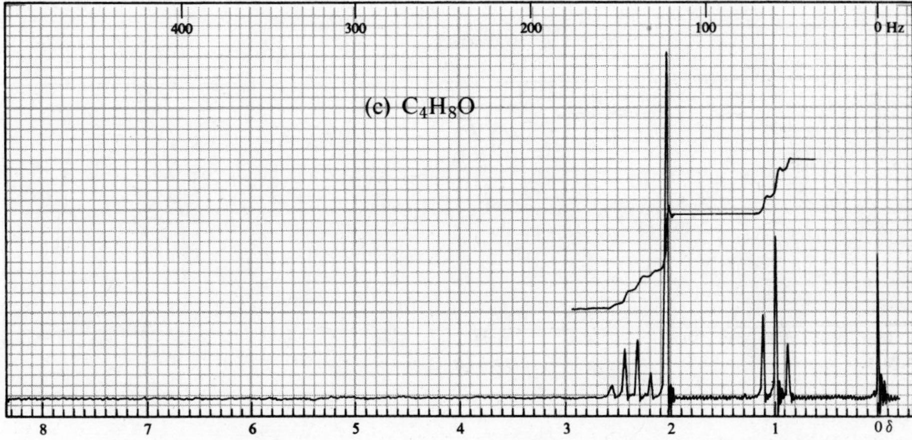

Figure 19.3. Nmr spectra for Problem 29, p. 653.

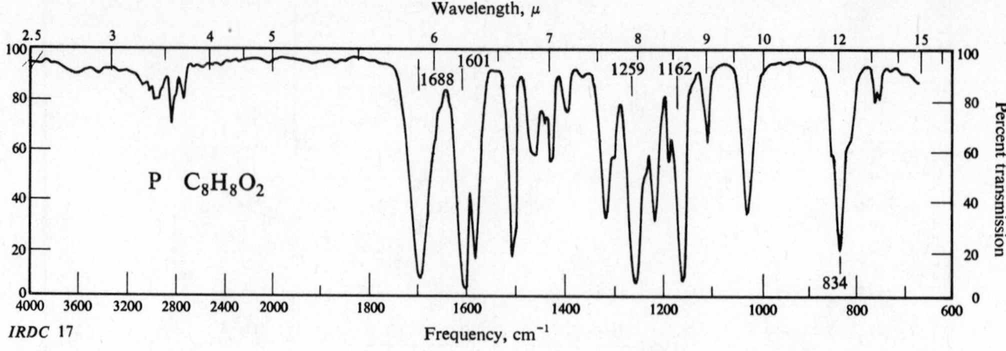

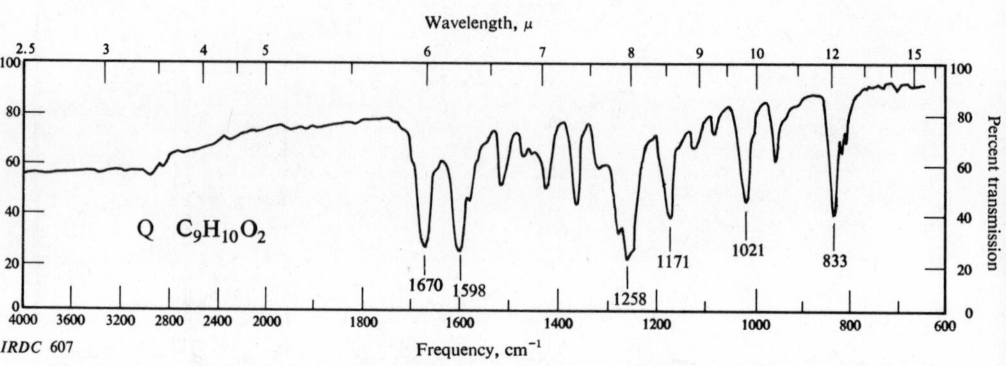

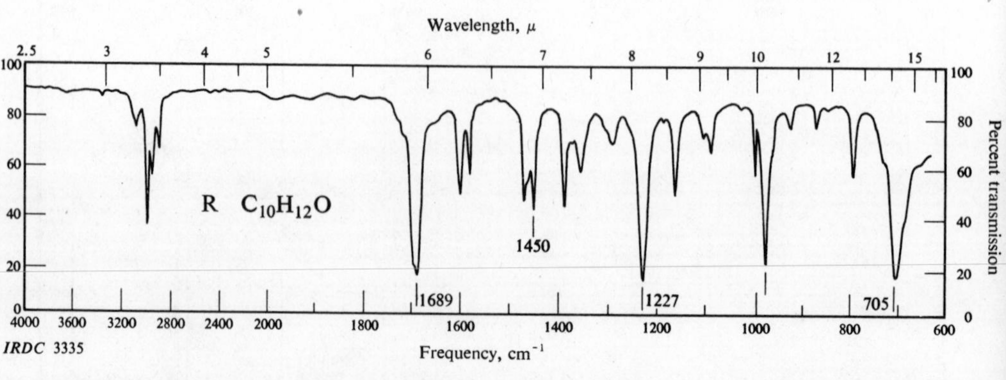

Figure 19.4. Infrared spectra for Problem 30, p. 653.

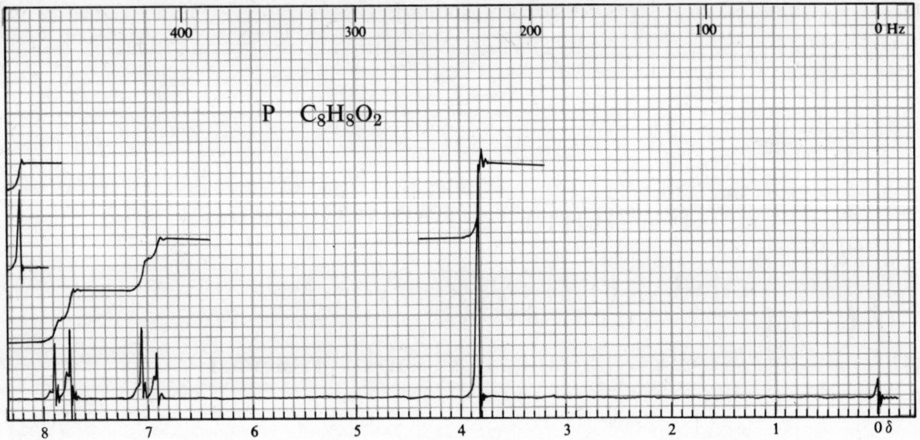

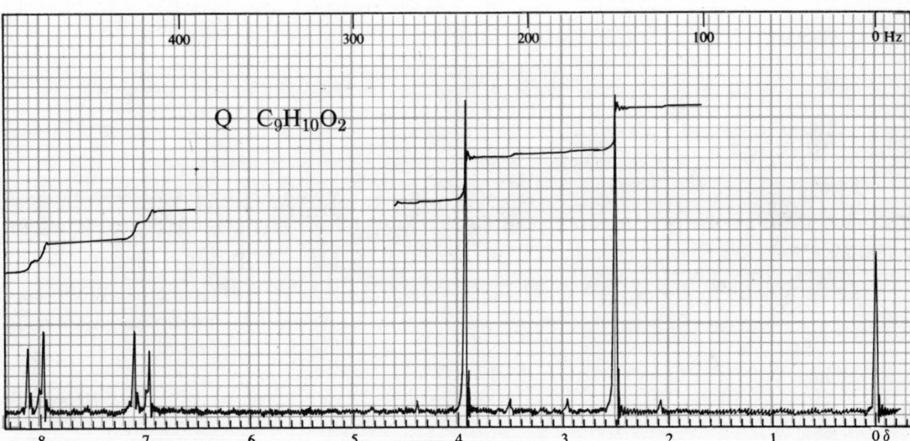

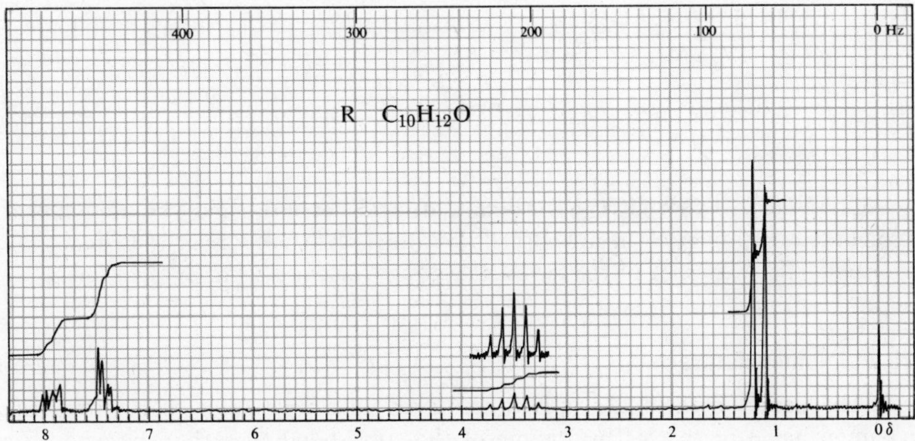

Figure 19.5. Nmr spectra for Problem 30, p. 653.

Chapter 20 Functional Derivatives of Carboxylic Acids
Nucleophilic Acyl Substitution

20.1 Structure

Closely related to the carboxylic acids and to each other are a number of chemical families known as **functional derivatives of carboxylic acids**: *acid chlorides, anhydrides, amides,* and *esters.* These derivatives are compounds in which the —OH of a carboxyl group has been replaced by —Cl, —OOCR, —NH_2, or —OR′.

$$\underset{\text{Acid chloride}}{R{-}\overset{O}{\overset{\|}{C}}{-}Cl} \qquad \underset{\text{Anhydride}}{R{-}\overset{O}{\overset{\|}{C}}{-}O{-}\overset{O}{\overset{\|}{C}}{-}R} \qquad \underset{\text{Amide}}{R{-}\overset{O}{\overset{\|}{C}}{-}NH_2} \qquad \underset{\text{Ester}}{R{-}\overset{O}{\overset{\|}{C}}{-}OR'}$$

R *may be alkyl or aryl*

They all contain the **acyl group**:

$$\underset{\text{Acyl group}}{R{-}\overset{O}{\overset{\|}{C}}{-}}$$

Like the acid to which it is related, an acid derivative may be aliphatic or aromatic, substituted or unsubstituted; whatever the structure of the rest of the molecule, the properties of the functional group remain essentially the same.

20.2 Nomenclature

The names of acid derivatives are taken in simple ways from either the common name or the IUPAC name of the corresponding carboxylic acid. For example:

Structure	Structure	Change:
$CH_3-C(=O)OH$ Acetic acid Ethanoic acid	$C_6H_5-C(=O)OH$ Benzoic acid	
$CH_3-C(=O)Cl$ Acetyl chloride Ethanoyl chloride	$C_6H_5-C(=O)Cl$ Benzoyl chloride	*–ic acid* to *–yl chloride*
$CH_3-C(=O)-O-C(=O)-CH_3$ Acetic anhydride Ethanoic anhydride	$C_6H_5-C(=O)-O-C(=O)-C_6H_5$ Benzoic anhydride	*acid* to *anhydride*
$CH_3-C(=O)NH_2$ Acetamide Ethanamide	$C_6H_5-C(=O)NH_2$ Benzamide	*–ic acid* of common name (or *–oic acid* of IUPAC name) to *–amide*
$CH_3-C(=O)OC_2H_5$ Ethyl acetate Ethyl ethanoate	$C_6H_5-C(=O)OC_2H_5$ Ethyl benzoate	*–ic acid* to *–ate*, preceded by name of alcohol or phenol group

20.3 Physical properties

The presence of the C=O group makes the acid derivatives polar compounds. Acid chlorides and anhydrides (Table 20.1) and esters (Table 20.2, p. 674) have boiling points that are about the same as those of aldehydes or ketones of comparable molecular weight (see Sec. 15.4). Amides (Table 20.1) have quite high boiling points because they are capable of strong intermolecular hydrogen bonding.

$$\cdots O{=}C(R)-N(H)-H\cdots O{=}C(R)-N(H)-H\cdots$$

The border line for solubility in water ranges from three to five carbons for the esters to five or six carbons for the amides. The acid derivatives are soluble in the usual organic solvents.

Volatile esters have pleasant, rather characteristic odors; they are often used in the preparation of perfumes and artificial flavorings. Acid chlorides have sharp, irritating odors, at least partly due to their ready hydrolysis to HCl and carboxylic acids.

Table 20.1 ACID CHLORIDES, ANHYDRIDES, AND AMIDES

Name	M.p., °C	B.p., °C	Name	M.p., °C	B.p., °C
Acetyl chloride	−112	51	Succinic anhydride	120	
Propionyl chloride	− 94	80	Maleic anhydride	60	
n-Butyryl chloride	− 89	102			
n-Valeryl chloride	−110	128	Formamide	3	200*d*
Stearoyl chloride	23	215^{15}	Acetamide	82	221
Benzoyl chloride	− 1	197	Propionamide	79	213
p-Nitrobenzoyl chloride	72	154^{15}	*n*-Butyramide	116	216
			n-Valeramide	106	232
3,5-Dinitrobenzoyl chloride	74	196^{12}	Stearamide	109	251^{12}
			Benzamide	130	290
Acetic anhydride	− 73	140	Succinimide	126	
Phthalic anhydride	131	284	Phthalimide	238	

20.4 Nucleophilic acyl substitution. Role of the carbonyl group

Before we take up each kind of acid derivative separately, it will be helpful to outline certain general patterns into which we can then fit the rather numerous individual facts.

Each derivative is nearly always prepared—directly or indirectly—from the corresponding carboxylic acid, and can be readily converted back into the carboxylic acid by simple hydrolysis. Much of the chemistry of acid derivatives involves their conversion one into another, and into the parent acid. In addition, each derivative has certain characteristic reactions of its own.

The derivatives of carboxylic acids, like the acids themselves, contain the carbonyl group, C=O. This group is retained in the products of most reactions undergone by these compounds, and does not suffer any permanent changes itself. But by its presence in the molecule it determines the characteristic reactivity of these compounds, and is the key to the understanding of their chemistry.

Here, too, as in aldehydes and ketones, the carbonyl group performs two functions: (a) it provides a site for nucleophilic attack, and (b) it increases the acidity of hydrogens attached to the *alpha* carbon.

(We shall discuss reactions resulting from the acidity of α-hydrogens in Secs. 21.11–21.12 and 26.1–26.3.)

Acyl compounds—carboxylic acids and their derivatives—typically undergo **nucleophilic substitution** in which —OH, —Cl, —OOCR, $—NH_2$, or —OR′ is replaced by some other basic group. Substitution takes place much more readily than at a saturated carbon atom; indeed, many of these substitutions do not usually take place at all in the absence of the carbonyl group, as, for example, replacement of $—NH_2$ by —OH.

$$R-C(=O)-W + :Z \longrightarrow R-C(O^-)(W)-Z \longrightarrow R-C(=O)-Z + :W$$

$$-W = -OH, -Cl, -OOCR, -NH_2, -OR'$$

To account for the properties of acyl compounds, let us turn to the carbonyl group. We have encountered this group in our study of aldehydes and ketones (Secs. 19.1 and 19.8), and we know what it is like and what in general to expect of it.

Carbonyl carbon is joined to three other atoms by σ bonds; since these bonds utilize sp^2 orbitals (Sec. 1.10), they lie in a plane and are 120° apart. The remaining p orbital of the carbon overlaps a p orbital of oxygen to form a π bond; carbon and oxygen are thus joined by a double bond. The part of the molecule immediately surrounding carbonyl carbon is *flat*; oxygen, carbonyl carbon, and the two atoms directly attached to carbonyl carbon lie in a plane:

$$\begin{matrix} W \\ R \end{matrix} > \overset{\delta+}{C} = \overset{\delta-}{O} \quad (120°)$$

We saw before that both electronic and steric factors make the carbonyl group particularly susceptible to nucleophilic attack at the carbonyl carbon: (a) the tendency of oxygen to acquire electrons even at the expense of gaining a negative charge; and (b) the relatively unhindered transition state leading from the trigonal reactant to the tetrahedral intermediate. These factors make acyl compounds, too, susceptible to nucleophilic attack.

It is in the second step of the reaction that acyl compounds differ from aldehydes and ketones. The tetrahedral intermediate from an aldehyde or ketone gains a proton, and the result is *addition*. The tetrahedral intermediate from an acyl

$$R-C(=O)-R' + :Z \longrightarrow R-C(O^-)(R')-Z \xrightarrow{H^+} R-C(OH)(R')-Z$$

Aldehyde or ketone
Addition

$$R-C(=O)-W + :Z \longrightarrow R-C(O^-)(W)-Z \longrightarrow R-C(=O)-Z + :W$$

Acyl compound
Substitution

compound ejects the :W group, returning to a trigonal compound, and thus the result is *substitution*.

We can see why the two classes of compounds differ as they do. The ease with which :W is lost depends upon its basicity: the weaker the base, the better the leaving group. For acid chlorides, acid anhydrides, esters, and amides, :W is, respectively: the very weak base Cl^-; the moderately weak base $RCOO^-$; and the strong bases $R'O^-$ and NH_2^-. But for an aldehyde or ketone to undergo substitution, the leaving group would have to be hydride ion ($:H^-$) or alkide ion

(:R^-) which, as we know, are the strongest bases of all. (Witness the low acidity of H_2 and RH.) And so with aldehydes and ketones addition almost always takes place instead.

Problem 20.1 Suggest a likely mechanism for each of the following reactions, and account for the behavior shown:

(a) The last step in the haloform reaction (Sec. 16.11),

$$OH^- + R\text{—}\underset{\underset{O}{\|}}{C}\text{—}CX_3 \xrightarrow{H_2O} RCOO^- + CHX_3$$

(b) The reaction of *o*-fluorobenzophenone with amide ion,

$$C_6H_4F\text{—}\overset{\overset{O}{\|}}{C}\text{—}C_6H_5 \xrightarrow[NH_3]{NH_2^-} C_6H_5F + H_2N\text{—}\underset{\underset{O}{\|}}{C}\text{—}C_6H_5$$

Thus, nucleophilic acyl substitution proceeds by two steps, with the intermediate formation of a tetrahedral compound. Generally, the overall rate is affected by the rate of both steps, but the *first* step is the more important. The first step, formation of the tetrahedral intermediate, is affected by the same factors

Nucleophilic acyl substitution

$$\underset{W}{\overset{R}{>}}C{=}O \;+\; :Z \longrightarrow \left[\,R\text{—}\overset{\overset{Z}{\vdots}}{\underset{W}{C}}\cdots O^{\delta-}\right] \longrightarrow R\text{—}\overset{\overset{Z}{|}}{\underset{W}{C}}\text{—}O^- \longrightarrow \underset{Z}{\overset{R}{>}}C{=}O + :W$$

Reactant	Transition state	Intermediate	Product	Leaving group
Trigonal	*Becoming tetrahedral*	*Tetrahedral*	*Trigonal*	*Weaker base leaves more readily*
	Partial negative charge on oxygen	*Negative charge on oxygen*		

as in addition to aldehydes and ketones (Sec. 19.8): it is favored by electron withdrawal, which stabilizes the developing negative charge; and it is hindered by the presence of bulky groups, which become crowded together in the transition state. The second step depends, as we have seen, on the basicity of the leaving group, :W.

If acid is present, H^+ becomes attached to carbonyl oxygen, thus making the

Acid-catalyzed nucleophilic acyl substitution

$$\underset{W}{\overset{R}{>}}C{=}O \;+\; H^+ \rightleftharpoons \underset{W}{\overset{R}{>}}C{=}\overset{\oplus}{O}H \;\xrightarrow{:Z}\; \left[\,R\text{—}\overset{\overset{Z}{\vdots}}{\underset{W}{C}}\cdots \overset{\delta+}{O}H\right] \longrightarrow R\text{—}\overset{\overset{Z}{|}}{\underset{W}{C}}\text{—}OH \longrightarrow \underset{Z}{\overset{R}{>}}C{=}O + H:W$$

Undergoes nucleophilic attack more readily

carbonyl group even more susceptible to the nucleophilic attack; oxygen can now acquire the π electrons without having to accept a negative charge.

It is understandable that acid derivatives are hydrolyzed more readily in either alkaline or acidic solution than in neutral solution: alkaline solutions provide hydroxide ion, which acts as a strongly nucleophilic reagent; acid solutions provide hydrogen ion, which attaches itself to carbonyl oxygen and thus renders the molecule vulnerable to attack by the weakly nucleophilic reagent, water.

Alkaline hydrolysis

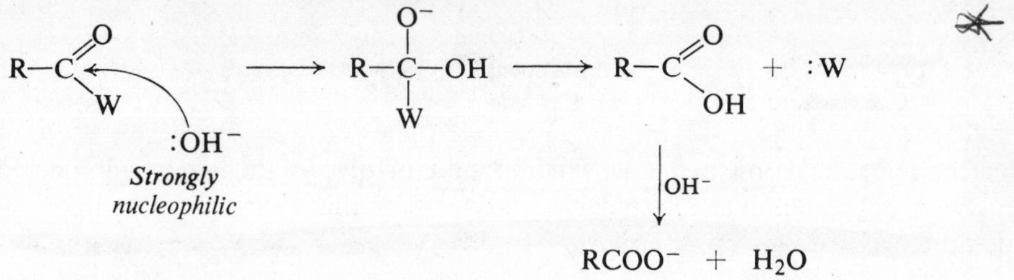

$$\text{R—C(=O)W} + :\text{OH}^- \longrightarrow \text{R—C(O}^-\text{)(OH)—W} \longrightarrow \text{R—C(=O)OH} + :\text{W}$$

$:OH^-$ *Strongly nucleophilic*

$$\text{R—C(=O)OH} \xrightarrow{\text{OH}^-} \text{RCOO}^- + \text{H}_2\text{O}$$

Acidic hydrolysis

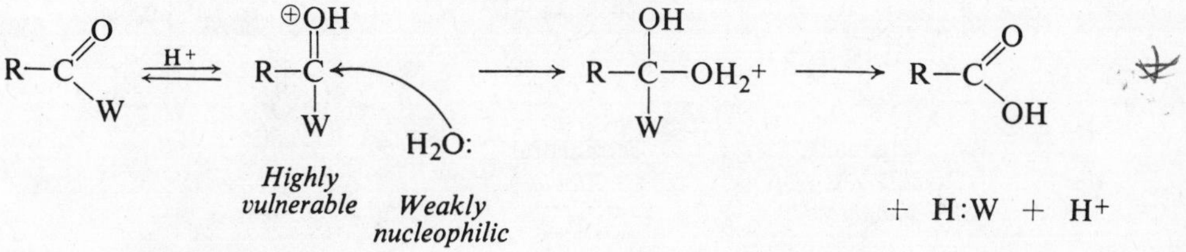

$$\text{R—C(=O)W} \underset{}{\overset{\text{H}^+}{\rightleftharpoons}} \text{R—C(=}\overset{\oplus}{\text{O}}\text{H)W} \xrightarrow{\text{H}_2\text{O:}} \text{R—C(OH)(W)—OH}_2^+ \longrightarrow \text{R—C(=O)OH} + \text{H:W} + \text{H}^+$$

Highly vulnerable (protonated acid derivative); $H_2O:$ *Weakly nucleophilic*

20.5 Nucleophilic substitution: alkyl *vs.* acyl

As we have said, nucleophilic substitution takes place much more readily at an acyl carbon than at saturated carbon. Thus, toward nucleophilic attack acid chlorides are more reactive than alkyl chlorides, amides are more reactive than amines (RNH_2), and esters are more reactive than ethers.

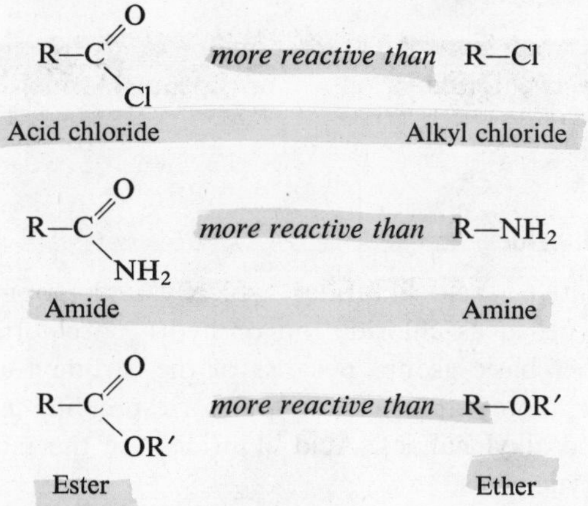

Reactivity in nucleophilic displacement

$$\underset{\text{Acid chloride}}{\text{R—C(=O)Cl}} \quad \textit{more reactive than} \quad \underset{\text{Alkyl chloride}}{\text{R—Cl}}$$

$$\underset{\text{Amide}}{\text{R—C(=O)NH}_2} \quad \textit{more reactive than} \quad \underset{\text{Amine}}{\text{R—NH}_2}$$

$$\underset{\text{Ester}}{\text{R—C(=O)OR}'} \quad \textit{more reactive than} \quad \underset{\text{Ether}}{\text{R—OR}'}$$

It is, of course, the carbonyl group that makes acyl compounds more reactive than alkyl compounds. Nucleophilic attack (S_N2) on a tetrahedral alkyl carbon involves a badly crowded transition state containing pentavalent carbon; a bond must be partly broken to permit the attachment of the nucleophile:

Alkyl nucleophilic substitution

Z: + C—W $\xrightarrow{S_N2}$ Z---C---W ⟶ Z—C + :W

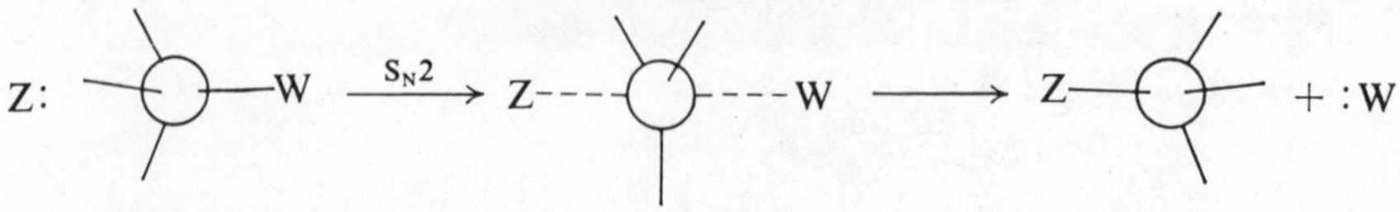

Tetrahedral C
Attack hindered

Pentavalent C
Unstable

Nucleophilic attack on a flat acyl compound involves a relatively unhindered transition state leading to a tetrahedral intermediate that is actually a compound; since the carbonyl group is unsaturated, attachment of the nucleophile requires

Acyl nucleophilic substitution

Z: + R(W)C=O ⟶ Z—C(R)(W)—O⁻ ⟶ R(Z)C=O + :W

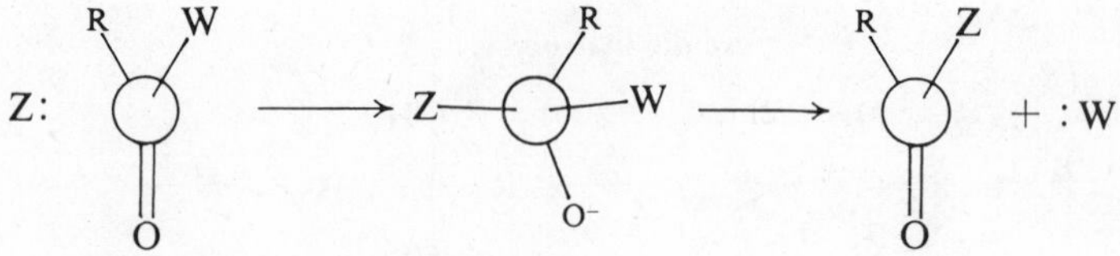

Trigonal C
Attack relatively unhindered

Tetrahedral C
Stable

breaking only of the weak π bond, and places a negative charge on an atom quite willing to accept it, oxygen.

ACID CHLORIDES

20.6 Preparation of acid chlorides

Acid chlorides are prepared from the corresponding acids by reaction with thionyl chloride, phosphorus trichloride, or phosphorus pentachloride, as discussed in Sec. 18.15.

20.7 Reactions of acid chlorides

Like other acid derivatives, acid chlorides typically undergo nucleophilic substitution. Chlorine is expelled as chloride ion or hydrogen chloride, and its place is taken by some other basic group. Because of the carbonyl group these reactions take place much more rapidly than the corresponding nucleophilic substitution reactions of the alkyl halides. Acid chlorides are the most reactive of the derivatives of carboxylic acids.

REACTIONS OF ACID CHLORIDES

1. Conversion into acids and derivatives. Discussed in Sec. 20.8.

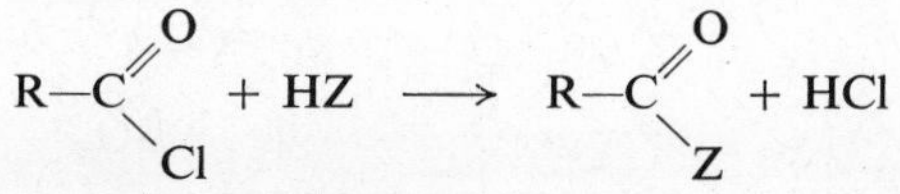

$$R{-}C(=O){-}Cl + HZ \longrightarrow R{-}C(=O){-}Z + HCl$$

(a) Conversion into acids. Hydrolysis.

$$\underset{\text{}}{RCOCl} + H_2O \longrightarrow \underset{\text{An acid}}{RCOOH} + HCl$$

Example:

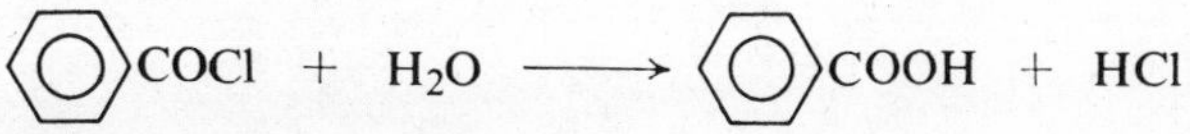

$$\underset{\text{Benzoyl chloride}}{C_6H_5COCl} + H_2O \longrightarrow \underset{\text{Benzoic acid}}{C_6H_5COOH} + HCl$$

(b) Conversion into amides. Ammonolysis

$$RCOCl + 2NH_3 \longrightarrow \underset{\text{An amide}}{RCONH_2} + NH_4Cl$$

Example:

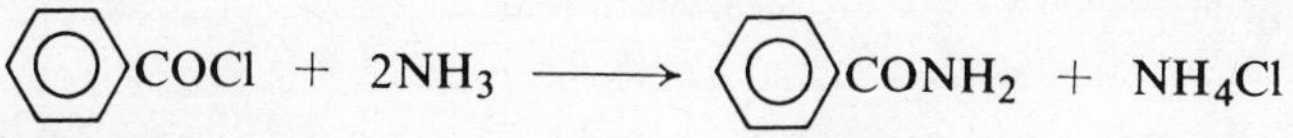

$$\underset{\text{Benzoyl chloride}}{C_6H_5COCl} + 2NH_3 \longrightarrow \underset{\text{Benzamide}}{C_6H_5CONH_2} + NH_4Cl$$

(c) Conversion into esters. Alcoholysis

$$RCOCl + R'OH \longrightarrow \underset{\text{An ester}}{RCOOR'} + HCl$$

Example:

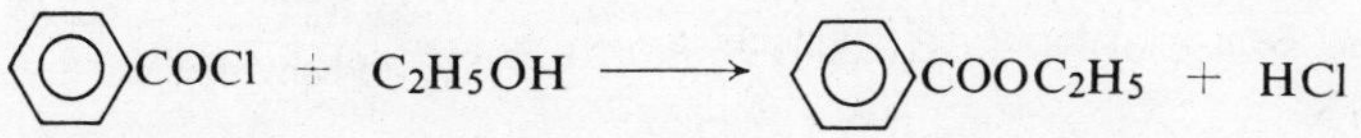

$$\underset{\text{Benzoyl chloride}}{C_6H_5COCl} + \underset{\text{Ethyl alcohol}}{C_2H_5OH} \longrightarrow \underset{\text{Ethyl benzoate}}{C_6H_5COOC_2H_5} + HCl$$

2. Formation of ketones. Friedel-Crafts acylation. Discussed in Sec. 19.6.

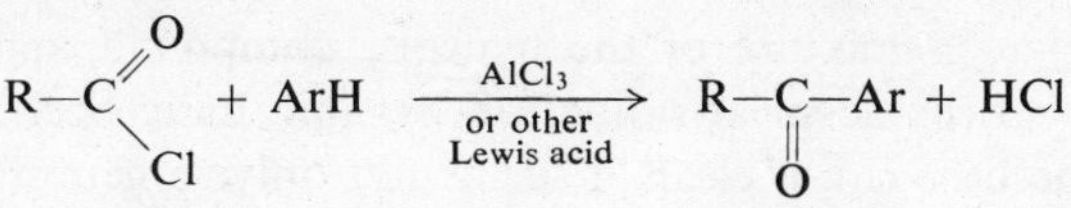

$$R{-}C(=O){-}Cl + ArH \xrightarrow[\text{or other Lewis acid}]{AlCl_3} \underset{\text{A ketone}}{R{-}\overset{}{\underset{\|}{C}}{-}Ar} + HCl$$

(A ketone: $R{-}CO{-}Ar$)

3. Formation of ketones. Reaction with organocadmium compounds. Discussed in Sec. 19.7.

$$RMgX \xrightarrow{CdCl_2} R'_2Cd \;+\; RCOCl \text{ or } ArCOCl \longrightarrow R{-}\underset{\|}{\overset{}{C}}{-}R' \text{ (with } =O\text{) or } Ar{-}C({=}O){-}R'$$

Ketone

R′ must be aryl or primary alkyl

4. Formation of aldehydes by reduction. Discussed in Sec. 19.4.

$$RCOCl \text{ or } ArCOCl \xrightarrow{LiAlH(OBu\text{-}t)_3} RCHO \text{ or } ArCHO$$

Aldehyde

20.8 Conversion of acid chlorides into acid derivatives

In the laboratory, amides and esters are usually prepared from the acid chloride rather than from the acid itself. Both the preparation of the acid chloride and its reactions with ammonia or an alcohol are rapid, essentially irreversible reactions. It is more convenient to carry out these two steps than the single slow, reversible reaction with the acid. For example:

$$n\text{-}C_{17}H_{35}COOH \xrightarrow[\text{heat}]{SOCl_2} n\text{-}C_{17}H_{35}COCl \xrightarrow[\text{cold}]{NH_3} n\text{-}C_{17}H_{35}CONH_2$$

Stearic acid — Stearoyl chloride — Stearamide

$$3,5\text{-}(O_2N)_2C_6H_3COOH \xrightarrow[\text{heat}]{PCl_5} 3,5\text{-}(O_2N)_2C_6H_3COCl \xrightarrow[\text{pyridine}]{n\text{-}C_3H_7OH} 3,5\text{-}(O_2N)_2C_6H_3COOC_3H_7$$

3,5-Dinitrobenzoic acid — 3,5-Dinitrobenzoyl chloride — *n*-Propyl 3,5-dinitrobenzoate

$$C_6H_5C({=}O)Cl + HOC_6H_5 \xrightarrow{\text{aqueous NaOH}} C_6H_5C({=}O)OC_6H_5$$

Benzoyl chloride — Phenol — Phenyl benzoate

Aromatic acid chlorides (ArCOCl) are considerably less reactive than the aliphatic acid chlorides. With cold water, for example, acetyl chloride reacts almost explosively, whereas benzoyl chloride reacts only very slowly. The reaction of aromatic acid chlorides with an alcohol or a phenol is often carried out using the **Schotten-Baumann** technique: the acid chloride is added in portions (followed by vigorous shaking) to a mixture of the hydroxy compound and a base, usually aqueous sodium hydroxide or pyridine (an organic base, Sec. 31.11). Although the function of the base is not clear, it seems not only to neutralize the hydrogen chloride that would otherwise be liberated, but also to catalyze the reaction.

ACID ANHYDRIDES

20.9 Preparation of acid anhydrides

Only one monocarboxylic acid anhydride is encountered very often; however, this one, **acetic anhydride**, is immensely important. It is prepared by the reaction of acetic acid with **ketene**, $CH_2{=}C{=}O$, which itself is prepared by high-temperature dehydration of acetic acid.

$$CH_3COOH \xrightarrow[700^\circ]{AlPO_4} H_2O + \underset{\text{Ketene}}{CH_2{=}C{=}O} \xrightarrow{CH_3COOH} \underset{\text{Acetic anhydride}}{(CH_3CO)_2O}$$

Ketene is an extremely reactive, interesting compound, which we have already encountered as a source of *methylene* (Sec. 9.15). It is made in the laboratory

$$CH_3COCH_3 \xrightarrow{700\text{–}750^\circ} CH_4 + \underset{\text{Ketene}}{CH_2{=}C{=}O}$$

by pyrolysis of acetone, and ordinarily used as soon as it is made.

In contrast to monocarboxylic acids, certain *di*carboxylic acids yield anhydrides on simple heating: in those cases where a five- or six-membered ring is produced. For example:

$$\text{Succinic acid } (HOOC{-}CH_2{-}CH_2{-}COOH) \xrightarrow{\text{heat}} \underset{\text{Succinic anhydride}}{\text{(cyclic } CH_2CH_2(CO)_2O)} + \underset{\text{Water}}{H_2O}$$

$$C_6H_4(COOH)_2 \xrightarrow{200^\circ} \underset{\text{Phthalic anhydride}}{C_6H_4(CO)_2O} + \underset{\text{Water}}{H_2O}$$

Ring size is crucial: with adipic acid, for example, anhydride formation would produce a seven-membered ring, and does not take place. Instead, carbon dioxide is lost and cyclopentanone, a ketone with a five-membered ring, is formed.

$$HOOC{-}CH_2CH_2CH_2CH_2{-}COOH \xrightarrow{Ba(OH)_2,\ 285\text{–}295^\circ} \underset{\text{Cyclopentanone}}{(CH_2)_4C{=}O} + \underset{\text{Carbon dioxide}}{CO_2} + \underset{\text{Water}}{H_2O}$$

Problem 20.2 Cyclic anhydrides can be formed from only the *cis*-1,2-cyclopentanedicarboxylic acid, but from both the *cis*- and *trans*-1,2-cyclohexanedicarboxylic acids. How do you account for this?

Problem 20.3 *Maleic acid* ($C_4H_4O_4$, m.p. 130°, highly soluble in water, heat of combustion 327 kcal) and *fumaric acid* ($C_4H_4O_4$, m.p. 302°, insoluble in water, heat of combustion 320 kcal) are both dicarboxylic acids; they both decolorize Br_2 in CCl_4 and aqueous $KMnO_4$; on hydrogenation both yield succinic acid. When heated (maleic acid at 100°, fumaric acid at 250–300°), both acids yield the same anhydride, which is converted by cold water into maleic acid. Interpret these facts.

20.10 Reactions of acid anhydrides

Acid anhydrides undergo the same reactions as acid chlorides, but a little more slowly; where acid chlorides yield a molecule of HCl, anhydrides yield a molecule of carboxylic acid.

Compounds containing the acetyl group are often prepared from acetic anhydride; it is cheap, readily available, less volatile and more easily handled than acetyl chloride, and it does not form corrosive hydrogen chloride. It is widely used industrially for the esterification of the polyhydroxy compounds known as *carbohydrates*, especially cellulose (Chap. 35).

REACTIONS OF ACID ANHYDRIDES

1. Conversion into acids and acid derivatives. Discussed in Sec. 20.10.

$$(RCO)_2O + HZ \longrightarrow RCOZ + RCOOH$$

(a) Conversion into acids. Hydrolysis

Example:

$$\underset{\text{Acetic anhydride}}{(CH_3CO)_2O} + H_2O \longrightarrow \underset{\text{Acetic acid}}{2CH_3COOH}$$

(b) Conversion into amides. Ammonolysis

Examples:

$$\underset{\text{Acetic anhydride}}{(CH_3CO)_2O} + 2NH_3 \longrightarrow \underset{\text{Acetamide}}{CH_3CONH_2} + \underset{\text{Ammonium acetate}}{CH_3COO^-NH_4^+}$$

$$\underset{\text{Succinic anhydride}}{\begin{array}{c} H_2C-C(=O) \\ | \quad\quad \backslash O \\ H_2C-C(=O) \end{array}} + 2NH_3 \longrightarrow \underset{\text{Ammonium succinamate}}{\begin{array}{l} CH_2CONH_2 \\ | \\ CH_2COONH_4 \end{array}} \xrightarrow{H^+} \underset{\text{Succinamic acid}}{\begin{array}{l} CH_2CONH_2 \\ | \\ CH_2COOH \end{array}}$$

(c) Conversion into esters. Alcoholysis

Examples:

$$(CH_3CO)_2O + CH_3OH \longrightarrow CH_3COOCH_3 + CH_3COOH$$

Acetic anhydride — Methyl acetate (An ester) — Acetic acid

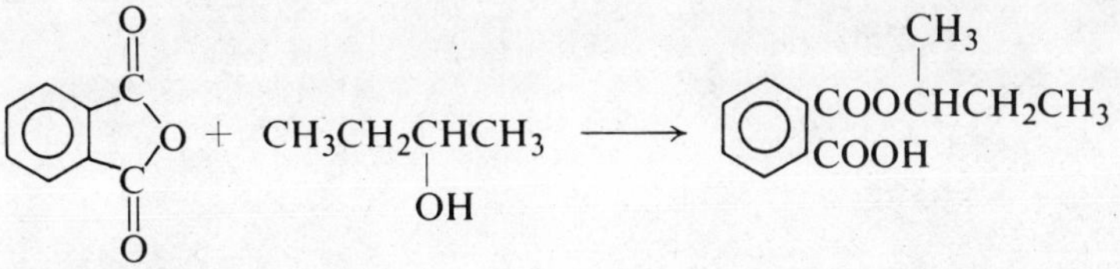

$$\text{Phthalic anhydride} + CH_3CH_2CH(OH)CH_3 \longrightarrow o\text{-}C_6H_4(COOCH(CH_3)CH_2CH_3)(COOH)$$

Phthalic anhydride — *sec*-Butyl alcohol — *sec*-Butyl hydrogen phthalate

2. Formation of ketones. Friedel-Crafts acylation. Discussed in Sec. 19.6.

$$(RCO)_2O + ArH \xrightarrow[\text{or other Lewis acid}]{AlCl_3} R{-}\underset{\substack{\|\\ O}}{C}{-}Ar + RCOOH$$

A ketone

Examples:

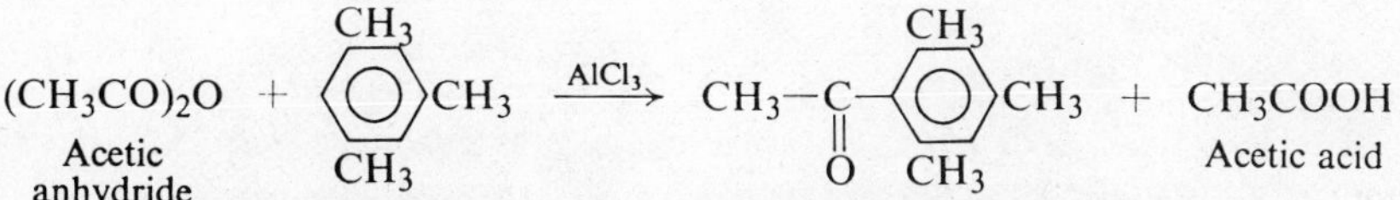

$$(CH_3CO)_2O + 1,3,5\text{-}C_6H_3(CH_3)_3 \xrightarrow{AlCl_3} CH_3{-}\underset{\substack{\|\\ O}}{C}{-}C_6H_2(CH_3)_3 + CH_3COOH$$

Acetic anhydride — Mesitylene — Methyl mesityl ketone — Acetic acid

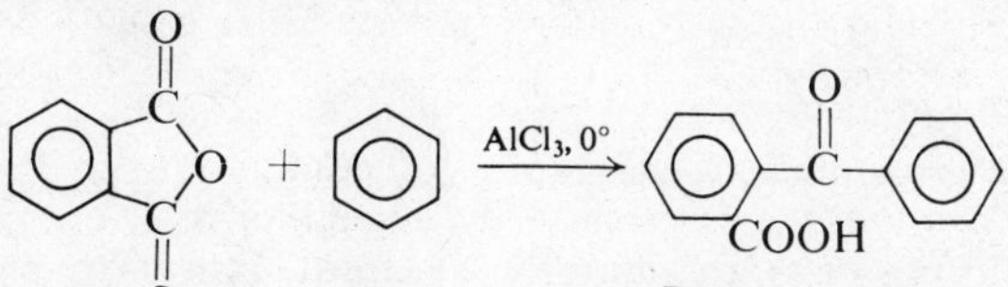

$$\text{Phthalic anhydride} + C_6H_6 \xrightarrow{AlCl_3,\ 0^\circ} o\text{-}C_6H_4(COOH){-}\underset{\substack{\|\\ O}}{C}{-}C_6H_5$$

Phthalic anhydride — *o*-Benzoylbenzoic acid

Only "half" of the anhydride appears in the acyl product; the other "half" forms a carboxylic acid. A cyclic anhydride, we see, undergoes exactly the same reactions as any other anhydride. However, since both "halves" of the anhydride are attached to each other by carbon–carbon bonds, the acyl compound and the carboxylic acid formed will have to be part of the same molecule. Cyclic anhydrides can thus be used to make compounds containing both the acyl group and the carboxyl group: compounds that are, for example, both acids and amides, both acids and esters, etc. These difunctional compounds are of great value in further synthesis.

Problem 20.4 Give structural formulas for compounds A through G.

$$\text{Benzene} + \text{succinic anhydride} \xrightarrow{AlCl_3} \text{A } (C_{10}H_{10}O_3)$$

$$\text{A} + \text{Zn(Hg)} \xrightarrow{HCl} \text{B } (C_{10}H_{12}O_2)$$

$$\text{B} + SOCl_2 \longrightarrow \text{C } (C_{10}H_{11}OCl)$$

$$\text{C} \xrightarrow{AlCl_3} \text{D } (C_{10}H_{10}O)$$

$$\text{D} + H_2 \xrightarrow{Pt} \text{E } (C_{10}H_{12}O)$$

$$\text{E} + H_2SO_4 \xrightarrow{\text{heat}} \text{F } (C_{10}H_{10})$$

$$\text{F} \xrightarrow{\text{Pt, heat}} \text{G } (C_{10}H_8) + H_2$$

Problem 20.5 (a) What product will be obtained if D of the preceding problem is treated with C_6H_5MgBr and then water? (b) What will you finally get if the product from (a) replaces E in the preceding problem?

Problem 20.6 When heated with acid (e.g., concentrated H_2SO_4), *o*-benzoylbenzoic acid yields a product of formula $C_{14}H_8O_2$. What is the structure of this product? What general type of reaction has taken place?

Problem 20.7 Predict the products of the following reactions:
(a) toluene + phthalic anhydride + $AlCl_3$
(b) the product from (a) + conc. H_2SO_4 + heat

Problem 20.8 (a) The two 1,3-cyclobutanedicarboxylic acids (p. 302) have been assigned configurations on the basis of the fact that one can be converted into an anhydride and the other cannot. Which configuration would you assign to the one that can form the anhydride, and why? (b) The method of (a) cannot be used to assign configurations to the 1,2-cyclohexanedicarboxylic acids, since *both* give anhydrides. Why is this? (c) Could the method of (a) be used to assign configurations to the 1,3-cyclohexanedicarboxylic acids?

Problem 20.9 Alcohols are the class of compounds most commonly resolved (Sec. 7.9), despite the fact that they are not acidic enough or basic enough to form (stable) salts. Outline all steps in a procedure for the resolution of *sec*-butyl alcohol, using as resolving agent the base (−)-B.

AMIDES

20.11 Preparation of amides

In the laboratory amides are prepared by the reaction of ammonia with acid chlorides or, when available, acid anhydrides (Secs. 20.8 and 20.10). In industry they are often made by heating the ammonium salts of carboxylic acids.

20.12 Reactions of amides

An amide is hydrolyzed when heated with aqueous acids or aqueous bases. The products are ammonia and the carboxylic acid, although one product or the other is obtained in the form of a salt, depending upon the acidity or basicity of the medium.

Another reaction of importance, the Hoffmann degradation of amides, will be discussed later (Sec. 22.12).

REACTIONS OF AMIDES

1. Hydrolysis. Discussed in Sec. 20.13.

$$RCONH_2 + H_2O \xrightarrow{H^+} RCOOH + NH_4^+$$

$$RCONH_2 + H_2O \xrightarrow{OH^-} RCOO^- + NH_3$$

Examples:

$$C_6H_5CONH_2 + H_2SO_4 + H_2O \longrightarrow C_6H_5COOH + NH_4^+HSO_4^-$$

Benzamide → Benzoic acid

$$CH_3CH_2CH_2CONH_2 + NaOH + H_2O \longrightarrow CH_3CH_2CH_2COO^-Na^+ + NH_3$$

Butyramide → Sodium butyrate

2. Conversion into imides. Discussed in Sec. 20.14.

3. Hofmann degradation of amides. Discussed in Secs. 22.12 and 28.2–28.5.

$$RCONH_2 \text{ or } ArCONH_2 \xrightarrow{OBr^-} RNH_2 \text{ or } ArNH_2 + CO_3^{--}$$

Amide → 1° amine

20.13 Hydrolysis of amides

Hydrolysis of amides is typical of the reactions of carboxylic acid derivatives. It involves nucleophilic substitution, in which the —NH_2 group is replaced by —OH. Under acidic conditions hydrolysis involves attack by water on the protonated amide:

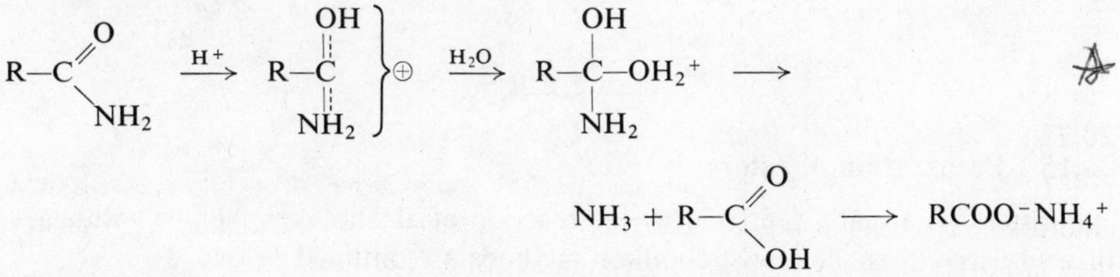

$$R{-}C(=O){-}NH_2 \xrightarrow{H^+} \left[R{-}C(OH){-}NH_2\right]^{\oplus} \xrightarrow{H_2O} R{-}C(OH)(NH_2){-}OH_2^+ \longrightarrow$$

$$NH_3 + R{-}C(=O){-}OH \longrightarrow RCOO^-NH_4^+$$

Under alkaline conditions hydrolysis involves attack by the strongly nucleophilic hydroxide ion on the amide itself:

$$R{-}C(=O){-}NH_2 \xrightarrow{OH^-} R{-}C(O^-)(OH){-}NH_2 \longrightarrow RCOO^- + NH_3$$

20.14 Imides

Like other anhydrides, cyclic anhydrides react with ammonia to yield amides; in this case the product contains both $-CONH_2$ and $-COOH$ groups. If this acid–amide is heated, a molecule of water is lost, a ring forms, and a product is obtained in which two acyl groups have become attached to nitrogen; compounds of this sort are called **imides**. Phthalic anhydride gives *phthalamic acid* and *phthalimide*:

$$\text{Phthalic anhydride} \xrightarrow{2NH_3} C_6H_4(CONH_2)(COO^-NH_4^+) \text{ (Ammonium phthalamate)} \xrightarrow{H^+} C_6H_4(CONH_2)(COOH) \text{ (Phthalamic acid)}$$

Phthalic anhydride $\xrightarrow{NH_3,\ heat}$ Phthalimide; Ammonium phthalamate $\xrightarrow{heat,\ 300°}$ Phthalimide; Phthalamic acid $\xrightarrow{heat}$ Phthalimide

Phthalimide

Problem 20.10 Outline all steps in the synthesis of *succinimide* from succinic acid.

Problem 20.11 Account for the following sequence of acidities. (*Hint:* See Sec. 18.12.)

	K_a
Ammonia	10^{-33}
Benzamide	10^{-14} to 10^{-15}
Phthalimide	5×10^{-9}

ESTERS

20.15 Preparation of esters

Esters are usually prepared by the reaction of alcohols or phenols with acids or acid derivatives. The most common methods are outlined below.

PREPARATION OF ESTERS

1. From acids. Discussed in Secs. 18.16 and 20.18.

$$\underset{\substack{\text{Carboxylic acid} \\ R\ \textit{may be alkyl or aryl}}}{RCOOH} + \underset{\substack{\text{Alcohol} \\ R'\ \textit{is usually alkyl}}}{R'OH} \underset{}{\overset{H^+}{\rightleftharpoons}} \underset{\text{Ester}}{RCOOR'} + H_2O$$

Reactivity of R′OH: $1° > 2° (> 3°)$

Examples:

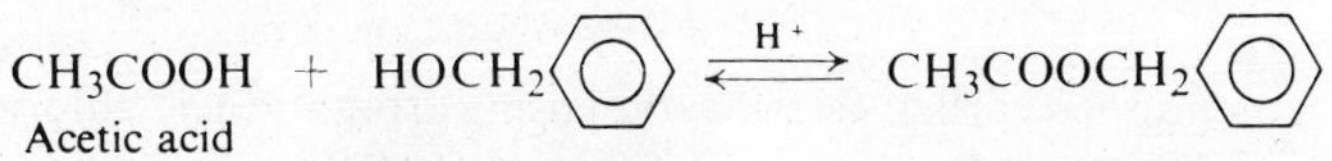

$$\underset{\text{Acetic acid}}{CH_3COOH} + \underset{\text{Benzyl alcohol}}{HOCH_2C_6H_5} \overset{H^+}{\rightleftharpoons} \underset{\text{Benzyl acetate}}{CH_3COOCH_2C_6H_5}$$

$$\underset{\text{Benzoic acid}}{C_6H_5COOH} + \underset{\text{Isobutyl alcohol}}{HOCH_2CH(CH_3)CH_3} \overset{H^+}{\rightleftharpoons} \underset{\text{Isobutyl benzoate}}{C_6H_5COOCH_2CH(CH_3)CH_3}$$

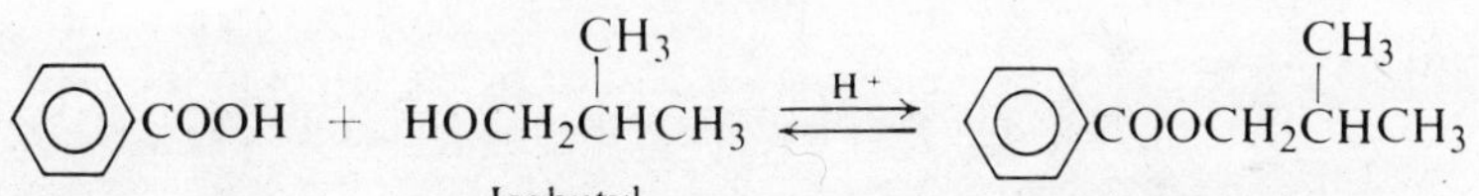

2. From acid chlorides or anhydrides. Discussed in Secs. 20.8 and 20.10.

$$RCOCl + R'OH\ (\text{or ArOH}) \longrightarrow RCOOR'\ (\text{or RCOOAr}) + HCl$$

$$(RCO)_2O + R'OH\ (\text{or ArOH}) \longrightarrow RCOOR'\ (\text{or RCOOAr}) + RCOOH$$

Examples:

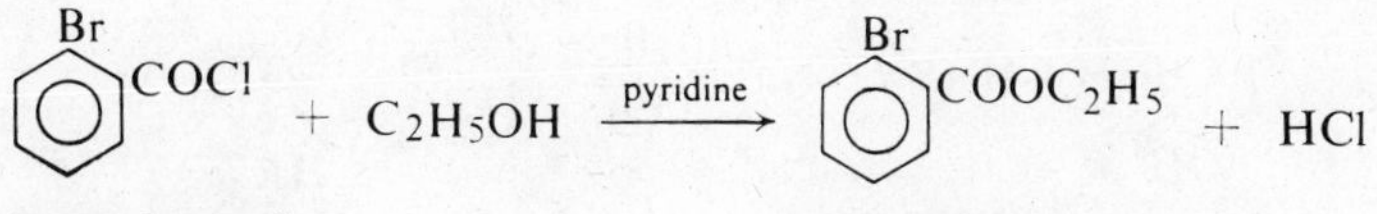

$$\underset{o\text{-Bromobenzoyl chloride}}{o\text{-}BrC_6H_4COCl} + C_2H_5OH \xrightarrow{\text{pyridine}} \underset{\text{Ethyl } o\text{-bromobenzoate}}{o\text{-}BrC_6H_4COOC_2H_5} + HCl$$

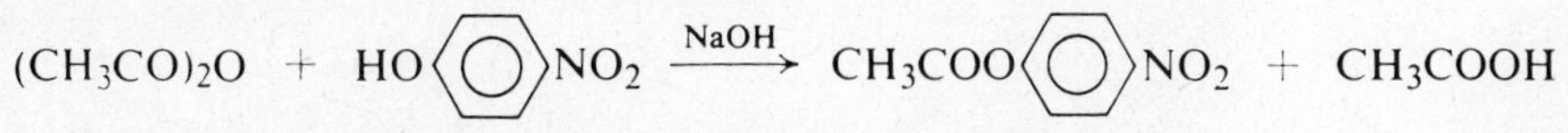

$$\underset{\text{Acetic anhydride}}{(CH_3CO)_2O} + \underset{p\text{-Nitrophenol}}{HOC_6H_4NO_2} \xrightarrow{\text{NaOH}} \underset{p\text{-Nitrophenyl acetate}}{CH_3COOC_6H_4NO_2} + CH_3COOH$$

3. From esters. Transesterification. Discussed in Sec. 20.20.

The direct reaction of alcohols or phenols with acids involves an equilibrium and—especially in the case of phenols—requires effort to drive to completion (see Sec. 18.16). In the laboratory, reaction with an acid chloride or anhydride is more commonly used.

The effect of the structure of the alcohol and of the acid on ease of esterification has already been discussed (Sec. 18.16).

Table 20.2 ESTERS OF CARBOXYLIC ACIDS

Name	M.p., °C	B.p., °C	Name	M.p., °C	B.p., °C
Methyl acetate	−98	57.5	Ethyl formate	−80	54
Ethyl acetate	−84	77	Ethyl acetate	−84	77
n-Propyl acetate	−92	102	Ethyl propionate	−74	99
n-Butyl acetate	−77	126	Ethyl *n*-butyrate	−93	121
n-Pentyl acetate		148	Ethyl *n*-valerate	−91	146
Isopentyl acetate	−78	142	Ethyl stearate	34	215^{15}
Benzyl acetate	−51	214	Ethyl phenylacetate		226
Phenyl acetate		196	Ethyl benzoate	−35	213

As was mentioned earlier, esterification using aromatic acid chlorides, ArCOCl, is often carried out in the presence of base (the Schotten-Baumann technique, Sec. 20.8).

Problem 20.12 When benzoic acid is esterified by methanol in the presence of a little sulfuric acid, the final reaction mixture contains five substances: benzoic acid, methanol, water, methyl benzoate, sulfuric acid. Outline a procedure for the separation of the pure ester.

A hydroxy acid is both alcohol and acid. In those cases where a five- or six-membered ring can be formed, *intramolecular* esterification occurs. Thus, a γ- or δ-hydroxy acid loses water spontaneously to yield a cyclic ester known as a **lactone**. Treatment with base (actually hydrolysis of an ester) rapidly opens the

$$\underset{\text{OH}}{\text{RCH}}CH_2CH_2COO^-Na^+ \underset{OH^-}{\overset{H^+}{\rightleftharpoons}} \text{(ring: } H_2C{-}C(=O){-}O{-}CHR{-}H_2C\text{)}$$

Salt of a
γ-hydroxy acid

A γ-lactone
A cyclic ester: five-membered ring

$$\underset{\text{OH}}{\text{RCH}}CH_2CH_2CH_2COO^-Na^+ \underset{OH^-}{\overset{H^+}{\rightleftharpoons}} \text{(ring: } H_2C{-}C(=O){-}O{-}CHR{-}H_2C{-}H_2C\text{)}$$

Salt of a
δ-hydroxy acid

A δ-lactone
A cyclic ester: six-membered ring

lactone ring to give the open-chain salt. We shall encounter lactones again in our study of carbohydrates (Sec. 34.8).

Problem 20.13 Suggest a likely structure for the product formed by heating each of these acids. (a) *Lactic acid*, $CH_3CHOHCOOH$, gives *lactide*, $C_6H_8O_4$. (b) 10-Hydroxydecanoic acid gives a material of high molecular weight (1000–9000).

20.16 Reactions of esters

Esters undergo the nucleophilic substitution that is typical of carboxylic acid derivatives. Attack occurs at the electron-deficient carbonyl carbon, and results in the replacement of the —OR′ group by —OH, —OR″, or $—NH_2$:

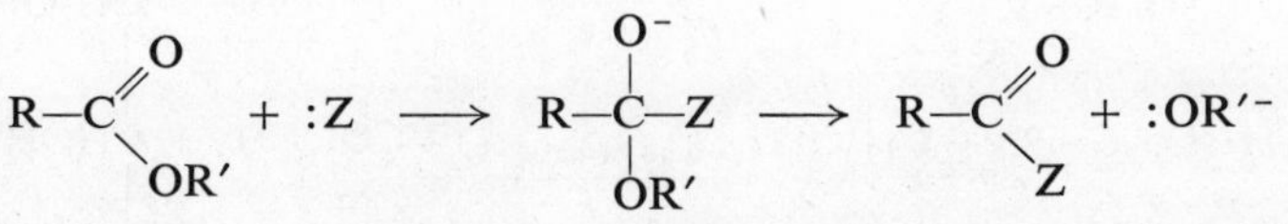

$$R-C(=O)-OR' + :Z \longrightarrow R-C(O^-)(Z)-OR' \longrightarrow R-C(=O)-Z + :OR'^-$$

$$:Z = :OH^-, \quad :OR''^-, \quad :NH_3$$

These reactions are sometimes carried out in the presence of acid. In these acid-catalyzed reactions, H^+ attaches itself to the oxygen of the carbonyl group, and thus renders carbonyl carbon even more susceptible to nucleophilic attack.

$$R-C(=O)-OR' + H^+ \rightleftharpoons \left\{R-C(OH)(OR')\right\}^{\oplus} \quad \leftarrow :Z$$

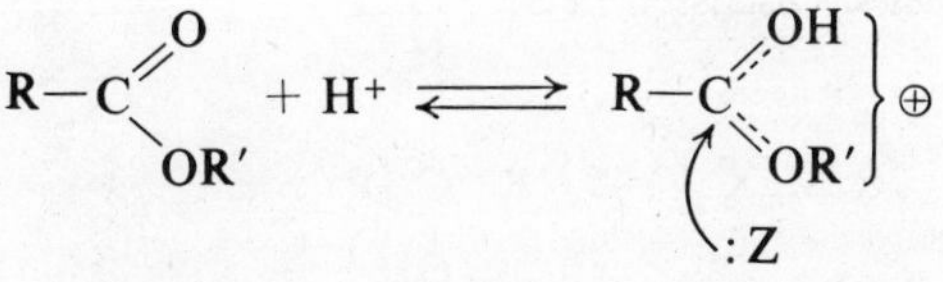

Acid catalysis: *makes carbon more susceptible to nucleophilic attack*

REACTIONS OF ESTERS

1. Conversion into acids and acid derivatives.

(a) Conversion into acids. Hydrolysis. Discussed in Secs. 20.17 and 20.18.

$$RCOOR' + H_2O \xrightarrow{H^+} RCOOH + R'OH$$

$$RCOOR' + H_2O \xrightarrow{OH^-} RCOO^- + R'OH$$

Example:

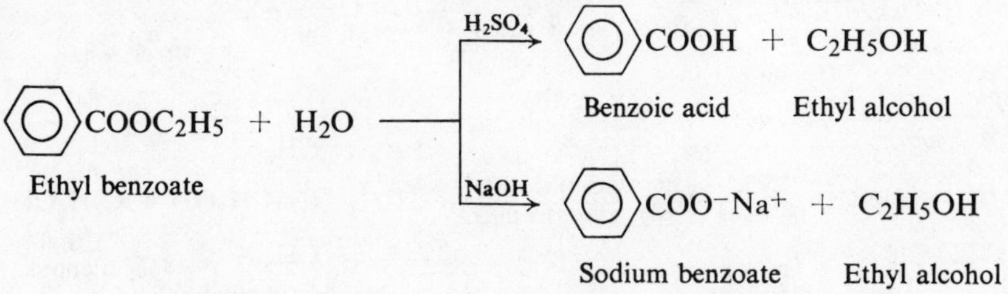

$$C_6H_5COOC_2H_5 + H_2O \xrightarrow{H_2SO_4} C_6H_5COOH + C_2H_5OH$$

Ethyl benzoate → Benzoic acid, Ethyl alcohol

$$C_6H_5COOC_2H_5 + H_2O \xrightarrow{NaOH} C_6H_5COO^-Na^+ + C_2H_5OH$$

Ethyl benzoate → Sodium benzoate, Ethyl alcohol

(b) Conversion into amides. Ammonolysis. Discussed in Sec. 20.19.

$$RCOOR' + NH_3 \longrightarrow RCONH_2 + R'OH$$

Example:

$$CH_3COOC_2H_5 + NH_3 \longrightarrow CH_3CONH_2 + C_2H_5OH$$

Ethyl acetate → Acetamide, Ethyl alcohol

(c) Conversion into esters. Transesterification. Alcoholysis. Discussed in Sec. 20.20.

$$\text{RCOOR}' + \text{R}''\text{OH} \xrightleftharpoons{\text{acid or base}} \text{RCOOR}'' + \text{R}'\text{OH}$$

Example:

$$\begin{array}{l} CH_2\text{—O—}\overset{\displaystyle O}{\overset{\|}{C}}\text{—R} \\ | \\ CH\text{—O—}\overset{\displaystyle O}{\overset{\|}{C}}\text{—R}' \\ | \\ CH_2\text{—O—}\overset{\displaystyle O}{\overset{\|}{C}}\text{—R}'' \end{array} + CH_3OH \xrightarrow{\text{acid or base}} \begin{array}{c} RCOOCH_3 \\ + \\ R'COOCH_3 \\ + \\ R''COOCH_3 \end{array} + \begin{array}{c} CH_2OH \\ | \\ CHOH \\ | \\ CH_2OH \end{array}$$

A glyceride (A fat) — Mixture of methyl esters — Glycerol

2. Reaction with Grignard reagents. Discussed in Sec. 20.21.

$$\text{RCOOR}' + 2\text{R}''\text{MgX} \longrightarrow \text{R—}\underset{\underset{\displaystyle OH}{|}}{\overset{\overset{\displaystyle R''}{|}}{C}}\text{—R}''$$

Tertiary alcohol

Example:

$$CH_3\overset{\overset{\displaystyle CH_3}{|}}{C}HCOOC_2H_5 + 2CH_3MgI \longrightarrow CH_3\overset{\overset{\displaystyle CH_3}{|}}{C}H\text{—}\underset{\underset{\displaystyle OH}{|}}{\overset{\overset{\displaystyle CH_3}{|}}{C}}\text{—}CH_3$$

Ethyl isobutyrate — Methylmagnesium iodide *2 moles* — 2,3-Dimethyl-2-butanol

3. Reduction to alcohols. Discussed in Sec. 20.22.

(a) Catalytic hydrogenation. Hydrogenolysis

$$\text{RCOOR}' + 2H_2 \xrightarrow[\text{3000–6000 lb/in.}^2]{CuO.CuCr_2O_4,\ 250°} \underset{1°\ \text{alcohol}}{RCH_2OH} + R'OH$$

Example:

$$CH_3\text{—}\underset{\underset{\displaystyle CH_3}{|}}{\overset{\overset{\displaystyle CH_3}{|}}{C}}\text{—}COOC_2H_5 + 2H_2 \xrightarrow[\text{250°, 3300 lb/in.}^2]{CuO.CuCr_2O_4} CH_3\text{—}\underset{\underset{\displaystyle CH_3}{|}}{\overset{\overset{\displaystyle CH_3}{|}}{C}}\text{—}CH_2OH + C_2H_5OH$$

Ethyl trimethylacetate (Ethyl 2,2-dimethylpropanoate) — Neopentyl alcohol (2,2-Dimethylpropanol) — Ethyl alcohol

(b) Chemical reduction

$$4\text{RCOOR}' + 2\text{LiAlH}_4 \xrightarrow{\text{anhyd. ether}} \left\{\begin{array}{c} LiAl(OCH_2R)_4 \\ + \\ LiAl(OR')_4 \end{array}\right\} \xrightarrow{H^+} \left\{\begin{array}{c} RCH_2OH \\ + \\ R'OH \end{array}\right\}$$

Example:

$$CH_3(CH_2)_7CH{=}CH(CH_2)_7COOCH_3 \xrightarrow{LiAlH_4} CH_3(CH_2)_7CH{=}CH(CH_2)_7CH_2OH$$

Methyl oleate (Methyl *cis*-9-octadecenoate) — Oleyl alcohol (*cis*-9-Octadecen-1-ol)

4. Reaction with carbanions. Claisen condensation. Discussed in Secs. 21.11 and 21.12.

$$-C(=O)OR' + -\overset{|}{\underset{H}{C}}-C(=O)OR' \xrightarrow{^-OC_2H_5} -\underset{O}{\overset{}{C}}-\overset{|}{\underset{|}{C}}-C(=O)OR'$$

A β-keto ester

20.17 Alkaline hydrolysis of esters

A carboxylic ester is hydrolyzed to a carboxylic acid and an alcohol or phenol when heated with aqueous acid or aqueous base. Under alkaline conditions, of course, the carboxylic acid is obtained as its salt, from which it can be liberated by addition of mineral acid.

Base promotes hydrolysis of esters by providing the strongly nucleophilic reagent OH^-. This reaction is essentially irreversible, since a resonance-stabilized

$$R-C(=O)OR' + OH^- \longrightarrow R-\overset{O^-}{\underset{OH}{C}}-OR' \longrightarrow R-C\left\{\begin{matrix}O\\O\end{matrix}\right\}^{\ominus} + R'OH$$

Ester — Hydroxide — Salt — Alcohol

carboxylate anion (Sec. 18.13) shows little tendency to react with an alcohol.

Let us look at the various aspects of the mechanism we have written, and see what evidence there is for each of them.

First, reaction involves attack on the ester by hydroxide ion. This is consistent with the **kinetics**, which is second-order, with the rate depending on both ester concentration and hydroxide concentration.

Next, hydroxide attacks at the carbonyl carbon and displaces alkoxide ion. That is to say, reaction involves cleavage of the bond between oxygen and the acyl group, RCO$\vdots$OR′. For this there are two lines of evidence, the first being the **stereochemistry**.

Let us consider, for example, the formation and subsequent hydrolysis of an ester of optically active *sec*-butyl alcohol. Reaction of (+)-*sec*-butyl alcohol with benzoyl chloride must involve cleavage of the hydrogen–oxygen bond and hence cannot change the configuration about the chiral center (see Sec. 7.4). If hydrolysis of this ester involves cleavage of the bond between oxygen and the *sec*-butyl group, we would expect almost certainly inversion (or inversion plus racemization if the reaction goes by an S_N1 type of mechanism):

$$C_6H_5\overset{O}{\overset{\|}{C}}{-}Cl \quad H{-}O{-}C(C_2H_5)(CH_3)H \longrightarrow C_6H_5\overset{O}{\overset{\|}{C}}{-}O{-}C(C_2H_5)(CH_3)H \xrightarrow{OH^-} C_6H_5COO^- + H{-}C(C_2H_5)(CH_3){-}OH$$

(+)-*sec*-Butyl alcohol — Cleavage between oxygen and alkyl group: *inversion* — (−)-*sec*-Butyl alcohol

If, on the other hand, the bond between oxygen and the *sec*-butyl group remains intact during hydrolysis, then we would expect to obtain *sec*-butyl alcohol of the same configuration as the starting material:

$$C_6H_5\overset{O}{\overset{\|}{C}}{-}Cl \quad H{-}O{-}C(C_2H_5)(CH_3)H \longrightarrow C_6H_5\overset{O}{\overset{\|}{C}}{-}O{-}C(C_2H_5)(CH_3)H \xrightarrow{OH^-} C_6H_5COO^- + HO{-}C(C_2H_5)(CH_3){-}H$$

(+)-*sec*-Butyl alcohol — Cleavage between oxygen and acyl group: *retention* — (+)-*sec*-Butyl alcohol

When *sec*-butyl alcohol of rotation +13.8° was actually converted into the benzoate and the benzoate was hydrolyzed in alkali, there was obtained *sec*-butyl alcohol of rotation +13.8°. This complete retention of configuration strongly indicates that bond cleavage occurs between oxygen and the acyl group.

Tracer studies have confirmed the kind of bond cleavage indicated by the stereochemical evidence. When ethyl propionate labeled with ^{18}O was hydrolyzed by base in ordinary water, the ethanol produced was found to be enriched in ^{18}O; the propionic acid contained only the ordinary amount of ^{18}O:

$$CH_3CH_2{-}C(=O){-}^{18}OC_2H_5 + OH^- \longrightarrow CH_3CH_2{-}C(=O){-}OH + C_2H_5{}^{18}OH$$

The alcohol group retained the oxygen that it held in the ester; cleavage occurred between oxygen and the acyl group.

The study of a number of other hydrolyses by both tracer and stereochemical methods has shown that cleavage between oxygen and the acyl group is the usual one in ester hydrolysis. This behavior indicates that the preferred point of nucleophilic attack is the carbonyl carbon rather than the alkyl carbon; this is, of course, what we might have expected in view of the generally greater reactivity of carbonyl carbon (Sec. 20.5).

Finally, according to the mechanism, attack by hydroxide ion on carbonyl carbon does not displace alkoxide ion in one step,

$$OH^- + R{-}C(=O){-}OR' \longrightarrow \left[HO^{\delta-}{\cdots}\overset{O}{\overset{\|}{\underset{R}{\underset{|}{C}}}}{\cdots}OR'^{\delta-} \right] \longrightarrow HO{-}C(=O){-}R + R'O^- \quad \text{Does not happen}$$

Transition state

$$HO{-}C(=O){-}R \xrightarrow{OH^-} {}^-OOCR$$

but rather in *two steps* with the intermediate formation of a tetrahedral compound. These alternative mechanisms were considered more or less equally likely until 1950 when elegant work on **isotopic exchange** was reported by Myron Bender (now at Northwestern University).

Bender carried out the alkaline hydrolysis of carbonyl-labeled ethyl benzoate, $C_6H_5C^{18}OOC_2H_5$, in ordinary water, and focused his attention, not on the product, but on the *reactant.* He interrupted the reaction after various periods of time, and isolated the unconsumed ester and analyzed it for ^{18}O content. He found that in the alkaline solution the ester was undergoing not only hydrolysis but also *exchange of its* ^{18}O *for ordinary oxygen from the solvent.*

$$\underset{\substack{\text{Labeled ester}\\ \textit{Starting material}}}{R{-}C(={}^{18}O){-}OC_2H_5} + OH^- \rightleftharpoons \underset{\text{I}}{R{-}C({}^{18}O^-)(OH){-}OC_2H_5} \longrightarrow R{-}C(={}^{18}O){-}OH + OC_2H_5^-$$

$$\text{I} \underset{}{\overset{H_2O}{\rightleftharpoons}} \underset{\text{II}}{R{-}C({}^{18}OH)(OH){-}OC_2H_5} \overset{H_2O}{\rightleftharpoons} \text{III}$$

$$R{-}C(={}^{18}O){-}OH \longrightarrow \underset{\text{Hydrolysis products}}{R{-}C({}^{18}O)(O)^{\ominus} + HOC_2H_5}$$

$$\underset{\substack{\text{Unlabeled ester}\\ \textit{Exchange product}}}{R{-}C(=O){-}OC_2H_5} + {}^{18}OH^- \rightleftharpoons \underset{\text{III}}{R{-}C({}^{18}OH)(O^-){-}OC_2H_5} \longrightarrow R{-}C(=O){-}{}^{18}OH + OC_2H_5^- \longrightarrow \text{Hydrolysis products}$$

Oxygen exchange is not consistent with the one-step mechanism, which provides no way for it to happen. Oxygen exchange is consistent with a two-step mechanism in which intermediate I is not only formed, but partly reverts into starting material and partly is converted (probably via the neutral species II) into III—an intermediate that is equivalent to I except for the position of the label. If all this is so, the "reversion" of intermediate III into "starting material" yields ester that has lost its ^{18}O.

Bender's work does not *prove* the mechanism we have outlined. Conceivably, oxygen exchange—and hence the tetrahedral intermediate—simply represent a blind-alley down which ester molecules venture but which does not lead to hydrolysis. Such coincidence is unlikely, however, particularly in light of certain kinetic relationships between oxygen exchange and hydrolysis.

Similar experiments have indicated the reversible formation of tetrahedral intermediates in hydrolysis of other esters, amides, anhydrides, and acid chlorides, and are the basis of the general mechanism we have shown for nucleophilic acyl substitution.

Exchange experiments are also the basis of our estimate of the relative importance of the two steps: differences in rate of hydrolysis of acyl derivatives depend chiefly on how fast intermediates are formed, and also on what fraction of the intermediate goes on to product. As we have said, the rate of formation of the intermediate is affected by both electronic and steric factors: in the transition state, a negative charge is developing and carbon is changing from trigonal toward tetrahedral.

Even in those cases where oxygen exchange cannot be detected, we cannot rule out the possibility of an intermediate; it may simply be that it goes on to hydrolysis products much faster than it does anything else.

Problem 20.14 The relative rates of alkaline hydrolysis of ethyl *p*-substituted benzoates, *p*-$GC_6H_4COOC_2H_5$, are:

$$\begin{array}{ccccccccc} G = NO_2 & > & Cl & > & H & > & CH_3 & > & OCH_3 \\ 110 & & 4 & & 1 & & 0.5 & & 0.2 \end{array}$$

(a) How do you account for this order of reactivity? (b) What kind of effect, activating or deactivating, would you expect from *p*-Br? from *p*-NH_2? from *p*-$C(CH_3)_3$? (c) Predict the order of reactivity toward alkaline hydrolysis of: *p*-aminophenyl acetate, *p*-methylphenyl acetate, *p*-nitrophenyl acetate, phenyl acetate.

Problem 20.15 The relative rates of alkaline hydrolysis of alkyl acetates, CH_3COOR, are:

$$\begin{array}{ccccccc} R = CH_3 & > & C_2H_5 & > & (CH_3)_2CH & > & (CH_3)_3C \\ 1 & & 0.6 & & 0.15 & & 0.008 \end{array}$$

(a) What two factors might be at work here? (b) Predict the order of reactivity toward alkaline hydrolysis of: methyl acetate, methyl formate, methyl isobutyrate, methyl propionate, and methyl trimethylacetate.

Problem 20.16 Exchange experiments show that the fraction of the tetrahedral intermediate that goes on to products follows the sequence:

acid chloride > acid anhydride > ester > amide

What is one factor that is probably at work here?

20.18 Acidic hydrolysis of esters

Hydrolysis of esters is promoted not only by base but also by acid. Acidic hydrolysis, as we have seen (Sec. 18.16), is reversible,

$$RCOOR' + H_2O \underset{H^+}{\overset{H^+}{\rightleftharpoons}} RCOOH + R'OH$$

and hence the mechanism for hydrolysis is also—taken in the opposite direction—

the mechanism for esterification. Any evidence about one reaction must apply to both.

The mechanism for acid-catalyzed hydrolysis and esterification is contained in the following equilibria:

$$\underset{\text{Ester}}{\overset{\substack{H^+\\+\\O\\\|}}{R-C-OR'}} \rightleftharpoons \overset{\substack{H_2O\\\text{Water}\\+\\OH^{\oplus}\\\|}}{R-C-OR'} \rightleftharpoons \underset{\substack{|\\\oplus OH_2\\\text{I}}}{\overset{\substack{OH\\|}}{R-C-OR'}} \rightleftharpoons$$

$$\underset{\substack{|\quad H\\OH\\\text{II}}}{\overset{\substack{OH\\|\quad\oplus}}{R-C-OR'}} \rightleftharpoons \left\{\underset{\substack{\|\\OH}}{\overset{\substack{OH\\\|}}{R-C}}\right\}^{\oplus} + \underset{\text{Alcohol}}{R'OH} \rightleftharpoons \underset{\text{Acid}}{\overset{\substack{O\\\|}}{R-C-OH}} + H^+$$

Mineral acid speeds up both processes by protonating carbonyl oxygen and thus rendering carbonyl carbon more susceptible to nucleophilic attack (Sec. 20.4). In hydrolysis, the nucleophile is a water molecule and the leaving group is an alcohol; in esterification, the roles are exactly reversed.

As in alkaline hydrolysis, there is almost certainly a tetrahedral intermediate —or, rather, several of them. The existence of more than one intermediate is required by, among other things, the reversible nature of the reaction. Looking only at hydrolysis, intermediate II is *likely*, since it permits separation of the weakly basic alcohol molecule instead of the strongly basic alkoxide ion; but consideration of esterification shows that II almost certainly *must* be involved, since it is the product of attack by alcohol on the protonated acid.

The evidence for the mechanism is much the same as in alkaline hydrolysis. The position of cleavage, RCO$\vdash$OR′ and RCO$\vdash$OH, has been shown by ^{18}O studies of both hydrolysis and esterification. The existence of the tetrahedral intermediates was demonstrated, as in the alkaline reaction, by ^{18}O exchange between the carbonyl oxygen of the ester and the solvent.

Problem 20.17 Write the steps to account for exchange between $RC^{18}OOR'$ and H_2O in acidic solution. There is reason to believe that a key intermediate here is identical with one in alkaline hydrolysis. What might this intermediate be?

Problem 20.18 Account for the fact (Sec. 18.16) that the presence of bulky substituents in either the alcohol group or the acid group slows down both esterification and hydrolysis.

Problem 20.19 Acidic hydrolysis of *tert*-butyl acetate in water enriched in ^{18}O has been found to yield *tert*-butyl alcohol enriched in ^{18}O and acetic acid containing ordinary oxygen. Acidic hydrolysis of the acetate of optically active 3,7-dimethyl-3-octanol has been found to yield alcohol of much lower optical purity than the starting alcohol, and having the opposite sign of rotation. (a) How do you interpret these two sets of results? (b) Is it surprising that these particular esters should show this kind of behavior?

20.19 Ammonolysis of esters

Treatment of an ester with ammonia, generally in ethyl alcohol solution, yields the amide. This reaction involves nucleophilic attack by a base, ammonia, on the electron-deficient carbon; the alkoxy group, —OR′, is replaced by $-NH_2$. For example:

$$\underset{\text{Ethyl acetate}}{CH_3-C(=O)-OC_2H_5} + NH_3 \longrightarrow \underset{\text{Acetamide}}{CH_3-C(=O)-NH_2} + C_2H_5OH$$

20.20 Transesterification

In the esterification of an acid, an alcohol acts as a nucleophilic reagent; in hydrolysis of an ester, an alcohol is displaced by a nucleophilic reagent. Knowing this, we are not surprised to find that one alcohol is capable of displacing another alcohol from an ester. This *alcoholysis* (cleavage by an alcohol) of an ester is called **transesterification.**

$$RCOOR' + R''OH \underset{}{\overset{H^+ \text{ or } OR''^-}{\rightleftharpoons}} RCOOR'' + R'OH$$

Transesterification is catalyzed by acid (H_2SO_4 or dry HCl) or base (usually alkoxide ion). The mechanisms of these two reactions are exactly analogous to those we have already studied. For acid-catalyzed transesterification:

$$\underset{\text{Ester A}}{R-C(=O)-OR'} + H^+ \rightleftharpoons R-C(\overset{\oplus}{=}OH)-OR' \xrightleftharpoons{R''OH \text{ (Alcohol B)}} R-C(OH)(OR')-\overset{\oplus}{O}HR'' \rightleftharpoons$$

$$R-C(OH)(OR'')-\overset{\oplus}{O}HR' \rightleftharpoons \left\{R-C(OH)(OR'')\right\}^{\oplus} + \underset{\text{Alcohol A}}{R'OH} \rightleftharpoons \underset{\text{Ester B}}{R-C(=O)-OR''} + H^+$$

For base-catalyzed transesterification:

$$\underset{\text{Ester A}}{R-C(=O)-OR'} + \underset{\text{Alkoxide B}}{^-OR''} \rightleftharpoons R-C(O^-)(OR'')-OR' \rightleftharpoons \underset{\text{Ester B}}{R-C(=O)-OR''} + \underset{\text{Alkoxide A}}{^-OR'}$$

Transesterification is an equilibrium reaction. To shift the equilibrium to the right, it is necessary to use a large excess of the alcohol whose ester we wish to make, or else to remove one of the products from the reaction mixture. The second approach is the better one when feasible, since in this way the reaction can be driven to completion.

20.21 Reaction of esters with Grignard reagents

The reaction of carboxylic esters with Grignard reagents is an excellent method for preparing tertiary alcohols. As in the reaction with aldehydes and ketones (Sec. 19.11), the nucleophilic (basic) alkyl or aryl group of the Grignard reagent attaches itself to the electron-deficient carbonyl carbon. Expulsion of the alkoxide group would yield a ketone, and in certain special cases ketones are indeed isolated from this reaction. However, as we know, ketones themselves readily react with Grignard reagents to yield tertiary alcohols (Sec. 15.13); in the present case the products obtained correspond to the addition of the Grignard reagent to such a ketone:

$$\underset{\text{Ester}}{R-C(=O)OR'} \xrightarrow{R''MgX} \left[R-\underset{\underset{O}{\|}}{C}-R''\right] + R'OMgX \xrightarrow{R''MgX} R-\underset{OMgX}{\overset{R''}{\overset{|}{\underset{|}{C}}}}-R'' \xrightarrow{H_2O} \underset{3^\circ \text{ alcohol}}{R-\underset{OH}{\overset{R''}{\overset{|}{\underset{|}{C}}}}-R''}$$

Two of the three groups attached to the carbon bearing the hydroxyl group in the alcohol come from the Grignard reagent and hence must be identical; this, of course, places limits upon the alcohols that can be prepared by this method. But, where applicable, reaction of a Grignard reagent with an ester is preferred to reaction with a ketone because esters are generally more accessible.

Problem 20.20 Starting from valeric acid, and using any needed reagents, outline the synthesis of 3-ethyl-3-heptanol via the reaction of a Grignard reagent with: (a) a ketone; (b) an ester.

Problem 20.21 (a) Esters of which acid would yield *secondary* alcohols on reaction with Grignard reagents? (b) Starting from alcohols of four carbons or fewer, outline all steps in the synthesis of 4-heptanol.

20.22 Reduction of esters

Like many organic compounds, esters can be reduced in two ways: (a) by catalytic hydrogenation using molecular hydrogen, or (b) by chemical reduction. In either case, the ester is cleaved to yield (in addition to the alcohol or phenol from which it was derived) a primary alcohol corresponding to the acid portion of the ester.

$$\underset{\text{Ester}}{RCOOR'} \xrightarrow{\text{reduction}} \underset{1^\circ \text{ alcohol}}{RCH_2OH} + R'OH$$

Hydrogenolysis (cleavage by hydrogen) of an ester requires more severe conditions than simple hydrogenation of (addition of hydrogen to) a carbon–

carbon double bond. High pressures and elevated temperatures are required: the catalyst used most often is a mixture of oxides known as *copper chromite*, of approximately the composition $CuO.CuCr_2O_4$. For example:

$$CH_3(CH_2)_{10}COOCH_3 \xrightarrow[150°,\ 5000\ lb/in.^2]{H_2,\ CuO.CuCr_2O_4} CH_3(CH_2)_{10}CH_2OH + CH_3OH$$

Methyl laurate (Methyl dodecanoate) → Lauryl alcohol (1-Dodecanol)

Chemical reduction is carried out by use of sodium metal and alcohol, or more usually by use of lithium aluminium hydride. For example:

$$CH_3(CH_2)_{14}COOC_2H_5 \xrightarrow{LiAlH_4} CH_3(CH_2)_{14}CH_2OH$$

Ethyl palmitate (Ethyl hexadecanoate) → 1-Hexadecanol

Problem 20.22 Predict the products of the hydrogenolysis of *n*-butyl oleate over copper chromite.

20.23 Functional derivatives of carbonic acid

Much of the chemistry of the functional derivatives of carbonic acid is already quite familiar to us through our study of carboxylic acids. The first step in dealing with one of these compounds is to recognize just how it is related to the parent acid. Since carbonic acid is bifunctional, each of its derivatives, too, contains two functional groups; these groups can be the same or different. For example:

$[HO{-}C({=}O){-}OH]$	$Cl{-}C({=}O){-}Cl$	$H_2N{-}C({=}O){-}NH_2$	$C_2H_5O{-}C({=}O){-}OC_2H_5$
Carbonic acid	Phosgene (Carbonyl chloride)	Urea (Carbamide)	Ethyl carbonate
Acid	*Acid chloride*	*Amide*	*Ester*

$C_2H_5O{-}C({=}O){-}Cl$	$H_2N{-}C{\equiv}N$	$H_2N{-}C({=}O){-}OC_2H_5$
Ethyl chlorocarbonate	Cyanamide	Urethane (Ethyl carbamate)
Acid chloride–ester	*Amide–nitrile*	*Ester–amide*

We use these functional relationships to carbonic acid simply for convenience. Many of these compounds could just as well be considered as derivatives of other acids, and, indeed, are often so named. For example:

$[H_2N{-}C({=}O){-}OH]$	$H_2N{-}C({=}O){-}NH_2$	$H_2N{-}C({=}O){-}OC_2H_5$
Carbamic acid	Carbamide	Ethyl carbamate
Acid	*Amide*	*Ester*

$[HO{-}C{\equiv}N]$	$H_2N{-}C{\equiv}N$
Cyanic acid	Cyanamide
Acid	*Amide*

In general, a derivative of carbonic acid containing an —OH group is unstable, and decomposes to carbon dioxide. For example:

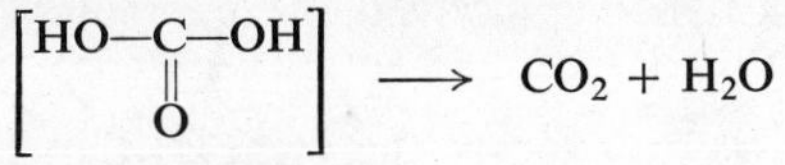

$$\left[\mathrm{HO{-}\overset{\displaystyle O}{\overset{\|}{C}}{-}OH}\right] \longrightarrow CO_2 + H_2O$$

Carbonic acid

$$\left[\mathrm{RO{-}\overset{\displaystyle O}{\overset{\|}{C}}{-}OH}\right] \longrightarrow CO_2 + ROH$$

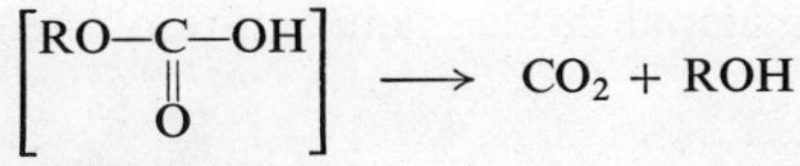

Alkyl hydrogen carbonate

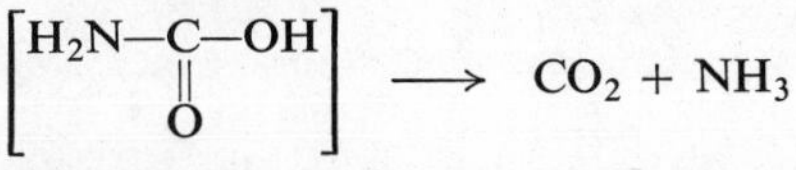

$$\left[\mathrm{H_2N{-}\overset{\displaystyle O}{\overset{\|}{C}}{-}OH}\right] \longrightarrow CO_2 + NH_3$$

Carbamic acid

$$\left[\mathrm{Cl{-}\overset{\displaystyle O}{\overset{\|}{C}}{-}OH}\right] \longrightarrow CO_2 + HCl$$

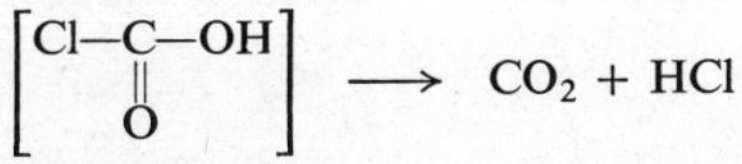

Chlorocarbonic acid

Most derivatives of carbonic acid are made from one of three industrially available compounds: phosgene, urea, or cyanamide.

Phosgene, $COCl_2$, a highly poisonous gas, is manufactured by the reaction between carbon monoxide and chlorine.

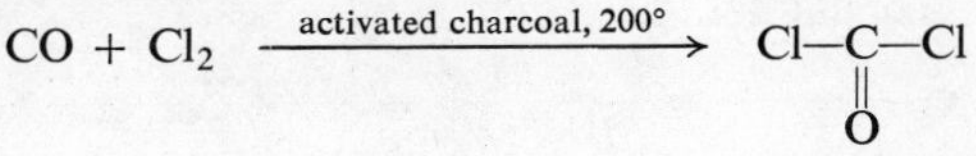

$$CO + Cl_2 \xrightarrow{\text{activated charcoal, 200°}} \mathrm{Cl{-}\overset{\displaystyle O}{\overset{\|}{C}}{-}Cl}$$

Phosgene

It undergoes the usual reactions of an acid chloride.

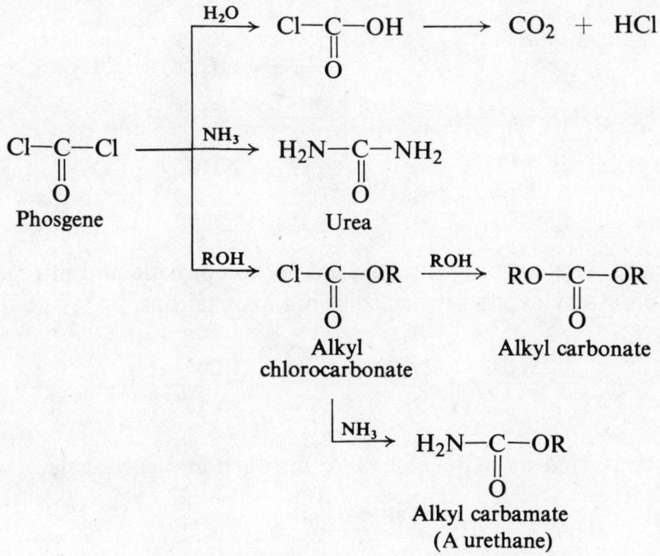

$$\mathrm{Cl{-}\overset{\displaystyle O}{\overset{\|}{C}}{-}Cl}\ (\text{Phosgene}) \xrightarrow{H_2O} \mathrm{Cl{-}\overset{\displaystyle O}{\overset{\|}{C}}{-}OH} \longrightarrow CO_2 + HCl$$

$$\mathrm{Cl{-}\overset{\displaystyle O}{\overset{\|}{C}}{-}Cl} \xrightarrow{NH_3} \mathrm{H_2N{-}\overset{\displaystyle O}{\overset{\|}{C}}{-}NH_2}\ (\text{Urea})$$

$$\mathrm{Cl{-}\overset{\displaystyle O}{\overset{\|}{C}}{-}Cl} \xrightarrow{ROH} \mathrm{Cl{-}\overset{\displaystyle O}{\overset{\|}{C}}{-}OR}\ (\text{Alkyl chlorocarbonate}) \xrightarrow{ROH} \mathrm{RO{-}\overset{\displaystyle O}{\overset{\|}{C}}{-}OR}\ (\text{Alkyl carbonate})$$

$$\mathrm{Cl{-}\overset{\displaystyle O}{\overset{\|}{C}}{-}OR} \xrightarrow{NH_3} \mathrm{H_2N{-}\overset{\displaystyle O}{\overset{\|}{C}}{-}OR}$$

Alkyl carbamate (A urethane)

Problem 20.23 Suggest a possible synthesis of (a) 2-pentylurethane, $H_2NCOO-CH(CH_3)(n-C_3H_7)$, used as a hypnotic; (b) benzyl chlorocarbonate (*carbobenzoxy chloride*), $C_6H_5CH_2OCOCl$, used in the synthesis of peptides (Sec. 36.10).

Urea, H_2NCONH_2, is excreted in the urine as the chief nitrogen-containing end product of protein metabolism. It is synthesized on a large scale for use as a fertilizer and as a raw material in the manufacture of urea–formaldehyde plastics and of drugs.

$$CO_2 + 2NH_3 \rightleftharpoons \underset{\text{Ammonium carbamate}}{H_2NCOONH_4} \xrightleftharpoons{\text{heat, pressure}} \underset{\text{Urea}}{H_2N-\underset{\underset{O}{\|}}{C}-NH_2}$$

Urea is weakly basic, forming salts with strong acids. The fact that it is a stronger base than ordinary amides is attributed to resonance stabilization of the cation:

$$H_2N-\underset{\underset{O}{\|}}{C}-NH_2 + H^+ \rightleftharpoons \left[H_2N-\underset{\underset{\oplus OH}{\|}}{C}-NH_2 \quad H_2\overset{\oplus}{N}=\underset{\underset{OH}{|}}{C}-NH_2 \quad H_2N-\underset{\underset{OH}{|}}{C}=\overset{\oplus}{N}H_2 \right]$$

$$\textit{equivalent to} \quad \left. H_2N\text{---}\underset{\underset{OH}{\vdots}}{C}\text{---}NH_2 \right\}\oplus$$

Problem 20.24 Account for the fact that *guanidine*, $(H_2N)_2C=NH$, is *strongly* basic.

Urea undergoes hydrolysis in the presence of acids, bases, or the enzyme *urease* (isolable from jack beans; generated by many bacteria, such as *Micrococcus ureae*).

$$\underset{\text{Urea}}{H_2N-\underset{\underset{O}{\|}}{C}-NH_2} \xrightarrow{H_2O} \begin{cases} \xrightarrow{H^+} NH_4^+ + CO_2 \\ \xrightarrow{OH^-} NH_3 + CO_3^{--} \\ \xrightarrow{\text{urease}} NH_3 + CO_2 \end{cases}$$

Urea reacts with nitrous acid to yield carbon dioxide and nitrogen; this is a useful way to destroy excess nitrous acid in diazotizations.

$$H_2N-\underset{\underset{O}{\|}}{C}-NH_2 \xrightarrow{HONO} CO_2 + N_2$$

Urea is converted by hypohalites into nitrogen and carbonate.

$$H_2N-\underset{\underset{O}{\|}}{C}-NH_2 \xrightarrow{Br_2\ OH^-} N_2 + CO_3^{--} + Br^-$$

Treatment of urea with acid chlorides or anhydrides yields **ureides**. Of special

$$H_2N-\underset{\underset{O}{\|}}{C}-NH_2 + CH_3COCl \longrightarrow CH_3CONH-\underset{\underset{O}{\|}}{C}-NH_2$$

Acetylurea
A ureide

importance are the cyclic ureides formed by reaction with malonic esters; these are known as **barbiturates** and are important hypnotics (sleep-producers). For example:

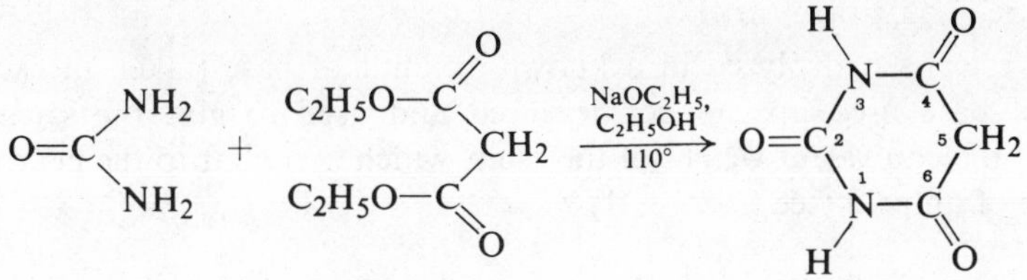

$$O{=}C(NH_2)_2 + (C_2H_5O-CO)_2CH_2 \xrightarrow[110^\circ]{NaOC_2H_5,\ C_2H_5OH} \text{barbituric acid (N1-H, C2=O, N3-H, C4=O, C5H}_2\text{, C6=O ring)}$$

Urea — Ethyl malonate — Barbituric acid (Malonylurea)

Cyanamide, $H_2N-C{\equiv}N$, is obtained in the form of its calcium salt by the high-temperature reaction between calcium carbide and nitrogen. This reaction is

$$\underset{\text{Calcium carbide}}{CaC_2} + N_2 \xrightarrow{1000^\circ} \underset{\text{Calcium cyanamide}}{CaNCN} + C$$

important as a method of nitrogen fixation; calcium cyanamide is used as a fertilizer, releasing ammonia by the action of water.

Problem 20.25 Give the electronic structure of the cyanamide anion, $(NCN)^{--}$. Discuss its molecular shape, bond lengths, and location of charge.

Problem 20.26 Give equations for the individual steps probably involved in the conversion of calcium cyanamide into ammonia in the presence of water. What other product or products will be formed in this process? Label each step with the name of the fundamental reaction type to which it belongs.

Problem 20.27 Cyanamide reacts with water in the presence of acid or base to yield urea; with methanol in the presence of acid to yield methylisourea, $H_2NC({=}NH)OCH_3$; with hydrogen sulfide to yield *thiourea*, $H_2NC({=}S)NH_2$; and with ammonia to yield *guanidine*, $H_2NC({=}NH)NH_2$. (a) What functional group of cyanamide is involved in each of these reactions? (b) To what general class of reaction do these belong? (c) Show the most probable mechanisms for these reactions, pointing out the function of acid or base wherever involved.

20.24 Analysis of carboxylic acid derivatives. Saponification equivalent

Functional derivatives of carboxylic acids are recognized by their hydrolysis—under more or less vigorous conditions—to carboxylic acids. Just *which kind* of derivative it is is indicated by the other products of the hydrolysis.

Problem 20.28 Which kind (or kinds) of acid derivative: (a) rapidly forms a white precipitate (insoluble in HNO_3) upon treatment with alcoholic silver nitrate?

(b) reacts with boiling aqueous NaOH to liberate a gas that turns moist litmus paper blue? (c) reacts immediately with cold NaOH to liberate a gas that turns moist litmus blue? (d) yields *only* a carboxylic acid upon hydrolysis? (e) yields an alcohol when heated with acid or base?

Identification or proof of structure of an acid derivative involves the identification or proof of structure of the carboxylic acid formed upon hydrolysis (Sec. 18.21). In the case of an ester, the alcohol that is obtained is also identified (Sec. 16.11). (In the case of a substituted amide, Sec. 23.6, the amine obtained is identified, Sec. 23.19.)

If an ester is hydrolyzed in a known amount of base (taken in excess), the amount of base used up can be measured and used to give the **saponification equivalent**: the equivalent weight of the ester, which is similar to the neutralization equivalent of an acid (see Sec. 18.21).

$$\underset{\textit{one equivalent}}{RCOOR'} + \underset{\textit{one equivalent}}{OH^-} \longrightarrow RCOO^- + R'OH$$

Problem 20.29 (a) What is the saponification equivalent of *n*-propyl acetate? (b) There are eight other simple aliphatic esters that have the same saponification equivalent. What are they? (c) In contrast, how many simple aliphatic acids have this equivalent weight? (d) Is saponification equivalent as helpful in identification as neutralization equivalent?

Problem 20.30 (a) How many equivalents of base would be used up by one mole of methyl phthalate, o-$C_6H_4(COOCH_3)_2$? What is the saponification equivalent of methyl phthalate? (b) What is the relation between saponification equivalent and the number of ester groups per molecule? (c) What is the saponification equivalent of glyceryl stearate (tristearoylglyerol)?

20.25 Spectroscopic analysis of carboxylic acid derivatives

Infrared. The infrared spectrum of an acyl compound shows the strong band in the neighborhood of 1700 cm^{-1} that we have come to expect of C=O stretching (see Fig. 20.1).

The exact frequency depends on the family the compound belongs to (see Table 20.3, p. 689) and, for a member of a particular family, on its exact structure. For esters, for example:

C=O stretching, *strong*

RCOOR 1740 cm^{-1}	ArCOOR 1715–1730 cm^{-1}	RCOOAr 1770 cm^{-1}
	or	or
	$-\overset{|}{C}=\overset{|}{C}-COOR$	$RCOO\overset{|}{C}=\overset{|}{C}-$

Esters are distinguished from acids by the absence of the O—H band. They are distinguished from ketones by two strong C—O stretching bands in the 1050–1300 cm^{-1} region; the exact position of these bands, too, depends on the ester's structure.

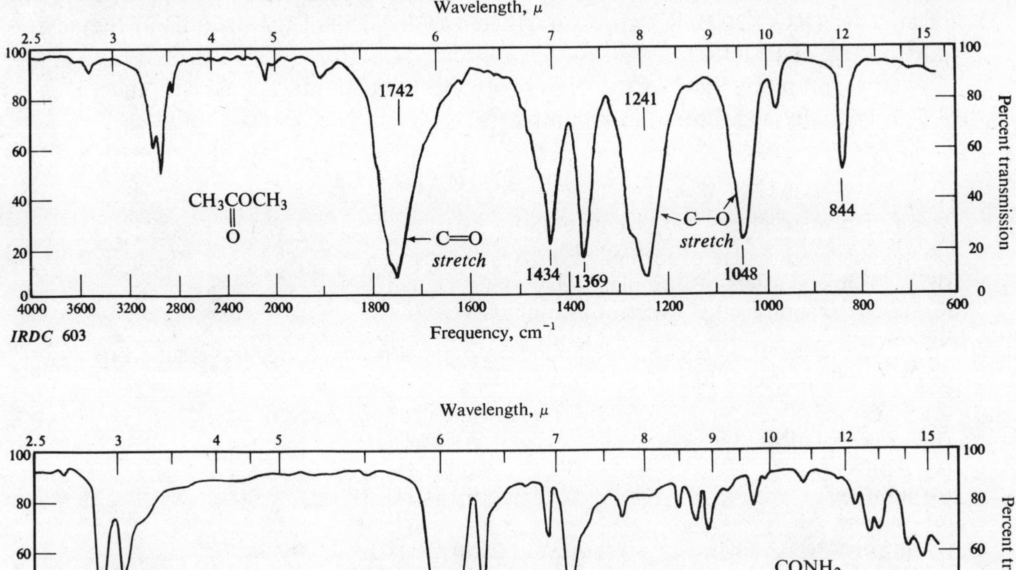

Figure 20.1. Infrared spectra of (*a*) methyl acetate and (*b*) benzamide.

Besides the carbonyl band, amides ($RCONH_2$) show absorption due to N—H stretching in the 3050–3550 cm^{-1} region (the number of bands and their location depending on the degree of hydrogen bonding), and absorption due to N—H bending in the 1600–1640 cm^{-1} region.

Table 20.3 INFRARED ABSORPTION BY SOME OXYGEN COMPOUNDS

Compound	O—H	C—O	C=O
Alcohols	3200–3600 cm^{-1}	1000–1200 cm^{-1}	—
Phenols	3200–3600	1140–1230	—
Ethers, aliphatic	—	1060–1150	—
Ethers, aromatic	—	1200–1275 1020–1075	—
Aldehydes, ketones	—	—	1675–1725 cm^{-1}
Carboxylic acids	2500–3000	1250	1680–1725
Esters	—	1050–1300 (*two bands*)	1715–1740
Acid chlorides	—	—	1750–1810
Amides ($RCONH_2$)	(N—H 3050–3550)	—	1650–1690

Nmr. As we can see in Table 13.4 (p. 421), the protons in the alkyl portion of an ester ($RCOOCH_2R'$) absorb farther downfield than the protons in the acyl portion (RCH_2COOR').

Absorption by the —CO—NH protons of an amide appears in the range δ 5–8, typically as a broad, low hump.

PROBLEMS

1. Draw structures and give names of:

(a) nine isomeric esters of formula $C_5H_{10}O_2$
(b) six isomeric esters of formula $C_8H_8O_2$
(c) three isomeric methyl esters of formula $C_7H_{12}O_4$

2. Write balanced equations, naming all organic products, for the reaction (if any) of *n*-butyryl chloride with:

(a) H_2O
(b) isopropyl alcohol
(c) *p*-nitrophenol
(d) ammonia
(e) toluene, $AlCl_3$
(f) nitrobenzene, $AlCl_3$
(g) $NaHCO_3$ (aq)
(h) alcoholic $AgNO_3$
(i) CH_3NH_2
(j) $(CH_3)_2NH$
(k) $(CH_3)_3N$
(l) $C_6H_5NH_2$
(m) $(C_6H_5)_2Cd$
(n) C_6H_5MgBr

(Check your answers to (i) through (l) in Sec. 23.6.)

3. Answer Problem 2, parts (a) through (l) for acetic anhydride.

4. Write equations to show the reaction (if any) of succinic anhydride with:

(a) hot aqueous NaOH
(b) aqueous ammonia
(c) aqueous ammonia, then cold dilute HCl
(d) aqueous ammonia, then strong heat
(e) benzyl alcohol
(f) toluene, $AlCl_3$, heat

5. Write balanced equations, naming all organic products, for the reaction (if any) of phenylacetamide with:

(a) hot HCl (aq)
(b) hot NaOH (aq)

6. Answer Problem 5 for phenylacetonitrile.

7. Write balanced equations, naming all organic products, for the reaction (if any) of methyl *n*-butyrate with:

(a) hot H_2SO_4 (aq)
(b) hot KOH (aq)
(c) isopropyl alcohol + H_2SO_4
(d) benzyl alcohol + $C_6H_5CH_2ONa$
(e) ammonia
(f) phenylmagnesium bromide
(g) isobutylmagnesium bromide
(h) $LiAlH_4$, then acid

8. Outline the synthesis of each of the following labeled compounds, using $H_2{}^{18}O$ as the source of ^{18}O.

(a) $C_6H_5{-}\overset{\displaystyle O}{\overset{\|}{C}}{-}{}^{18}OCH_3$ (b) $C_6H_5{-}\overset{\displaystyle {}^{18}O}{\overset{\|}{C}}{-}OCH_3$ (c) $C_6H_5{-}\overset{\displaystyle {}^{18}O}{\overset{\|}{C}}{-}{}^{18}OCH_3$

Predict the product obtained from each upon alkaline hydrolysis in ordinary H_2O.

9. Outline the synthesis of each of the following labeled compounds, using $^{14}CO_2$ or $^{14}CH_3OH$ and $H_2{}^{18}O$ as the source of the "tagged" atoms.

(a) $CH_3CH_2{}^{14}COCH_3$
(b) $CH_3CH_2CO^{14}CH_3$
(c) $CH_3{}^{14}CH_2COCH_3$
(d) $^{14}CH_3CH_2COCH_3$
(e) $C_6H_5{}^{14}CH_2CH_3$
(f) $C_6H_5CH_2{}^{14}CH_3$
(g) $CH_3CH_2C^{18}OCH_3$

10. Predict the product of the reaction of γ-butyrolactone with (a) ammonia, (b) $LiAlH_4$, (c) $C_2H_5OH + H_2SO_4$.

11. When *sec*-butyl alcohol of rotation $+13.8°$ was treated with tosyl chloride, and the resulting tosylate was allowed to react with sodium benzoate, there was obtained *sec*-butyl benzoate. Alkaline hydrolysis of this ester gave *sec*-butyl alcohol of rotation $-13.4°$. In which step must inversion have taken place? How do you account for this?

12. Account for the following observations. (*Hint:* See Sec. 14.13, and Problem 14.9 on p. 473.)

$C_6H_5-CH(OCOR)-CH=CH-CH_3$ (*optically active*)

- $\xrightarrow{\text{5N NaOH}}$ $C_6H_5-CH(OH)-CH=CH-CH_3$ (*complete retention*)
- $\xrightarrow{\text{dil. NaOH}}$ $C_6H_5-CH=CH-CH(OH)-CH_3$ (*inactive*)

$C_6H_5-CH=CH-CH(OCOR)-CH_3$ (*optically active*)

- $\xrightarrow{\text{dil. NaOH}}$ $C_6H_5-CH=CH-CH(OH)-CH_3$ (*inactive*)
- $\xrightarrow{\text{5N NaOH}}$ $C_6H_5-CH=CH-CH(OH)-CH_3$ (*complete retention*)

13. An unknown compound is believed to be one of the following, all of which boil within a few degrees of each other. Describe how you would go about finding out which of the possibilities the unknown actually is. Where possible use simple chemical tests; where necessary use more elaborate chemical methods like quantitative hydrogenation, cleavage, neutralization equivalent, saponification equivalent, etc. Make use of any needed tables of physical constants.

benzyl acetate	methyl *o*-toluate
ethyl benzoate	methyl *m*-toluate
isopropyl benzoate	methyl *p*-toluate
methyl phenylacetate	

14. Describe simple chemical tests that would serve to distinguish between:

(a) propionic acid and methyl acetate
(b) *n*-butyryl chloride and *n*-butyl chloride
(c) *p*-nitrobenzamide and ethyl *p*-nitrobenzoate
(d) glyceryl tristearate and glyceryl trioleate
(e) benzonitrile and nitrobenzene
(f) acetic anhydride and *n*-butyl alcohol
(g) glyceryl monopalmitate and glyceryl tripalmitate
(h) ammonium benzoate and benzamide
(i) *p*-bromobenzoic acid and benzoyl bromide

Tell exactly what you would *do* and *see*.

15. Tell how you would separate by chemical means the following mixtures, recovering each component in reasonably pure form: (a) benzoic acid and ethyl benzoate; (b) *n*-valeronitrile and *n*-valeric acid; (c) ammonium benzoate and benzamide. Tell exactly what you would do and see.

16. Carboxyl groups are often masked by reaction with **dihydropyran**, which yields esters that are stable toward base but easily hydrolyzed by dilute aqueous acids. Account in detail both for the formation of these esters and for their ease of hydrolysis. (*Hint*: See Sec. 19.15.)

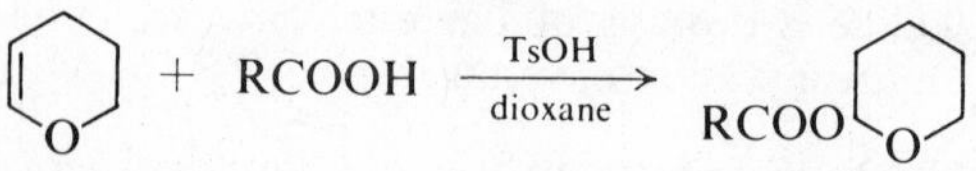

Dihydropyran (DHP) + RCOOH —(TsOH, dioxane)→ Tetrahydropyranyl ester (RCOOTHP)

17. Treatment of 2,4-pentanedione with KCN and acetic acid, followed by hydrolysis, gives two products, A and B. Both A and B are dicarboxylic acids of formula $C_7H_{12}O_6$. A melts at 98°. When heated, B gives first a lactonic acid ($C_7H_{10}O_5$, m.p. 90°) and finally a dilactone ($C_7H_8O_4$, m.p. 105°). (a) What structure must B have that permits ready formation of both a monolactone and a dilactone? (b) What is the structure of A? (*Hint:* Use models.)

18. Give the structures (including configurations where pertinent) of compounds C through O.

(a) Urea (H_2NCONH_2) + hot dilute NaOH ⟶ C + NH_3
(b) Phosgene ($COCl_2$) + 1 mole C_2H_5OH, then + NH_3 ⟶ D ($C_3H_7O_2N$)
(c) bromobenzene + Mg, ether ⟶ E (C_6H_5MgBr)
E + ethylene oxide, followed by H^+ ⟶ F ($C_8H_{10}O$)
F + PBr_3 ⟶ G (C_8H_9Br)
G + NaCN ⟶ H (C_9H_9N)
H + H_2SO_4, H_2O, heat ⟶ I ($C_9H_{10}O_2$)
I + $SOCl_2$ ⟶ J (C_9H_9OCl)
J + anhydrous HF ⟶ K (C_9H_8O)
K + H_2, catalyst ⟶ L ($C_9H_{10}O$)
L + H_2SO_4, warm ⟶ M (C_9H_8)
(d) *trans*-2-methylcyclohexanol + acetyl chloride ⟶ N
N + NaOH (aq) + heat ⟶ O + sodium acetate

19. *Progesterone* is a hormone, secreted by the corpus luteum, that is involved in the control of pregnancy. Its structure was established, in part, by the following synthesis from the steroid *stigmasterol*, obtained from soybean oil.

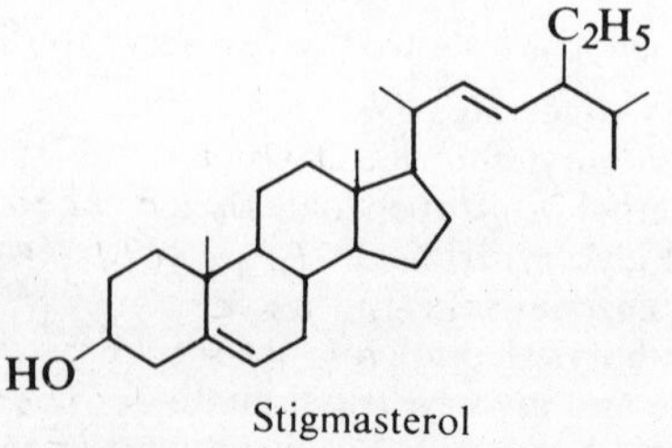

Stigmasterol

Stigmasterol ($C_{29}H_{48}O$) + $(CH_3CO)_2O$ ⟶ P ($C_{31}H_{50}O_2$)
P + Br_2 ⟶ Q ($C_{31}H_{50}O_2Br_2$)
Q + O_3, then Ag_2O ⟶ R ($C_{24}H_{36}O_4Br_2$)
R + Zn/CH_3COOH ⟶ S ($C_{24}H_{36}O_4$)
S + C_2H_5OH, H^+ ⟶ T ($C_{26}H_{40}O_4$)
T + C_6H_5MgBr, then H_2O ⟶ U ($C_{36}H_{46}O_3$)

U + acid, warm $\longrightarrow$ V ($C_{36}H_{44}O_2$)
V + Br_2; then CrO_3, H^+ $\longrightarrow$ W ($C_{23}H_{34}O_3Br_2$)
W + Zn/CH_3COOH $\longrightarrow$ X ($C_{23}H_{34}O_3$)
X + H_2O, H^+, heat $\longrightarrow$ Y ($C_{21}H_{32}O_2$), *pregnenolone*
Y + Br_2; then CrO_3, H^+ $\longrightarrow$ Z ($C_{21}H_{30}O_2Br_2$)
Z + Zn/CH_3COOH $\longrightarrow$ *progesterone* ($C_{21}H_{30}O_2$)

(a) Give structures for progesterone and the intermediates P–Z.

(b) Progesterone shows strong absorption in the near ultraviolet: λ_{max} 240 nm, ϵ_{max} 17,600. On this basis, what is the structure for progesterone?

20. On the basis of the following evidence assign structures to: (a) Compounds AA to DD, isomers of formula $C_3H_8O_2$; (b) compounds EE to MM, isomers of formula $C_3H_6O_2$. (*Note:* α-Hydroxy ketones, —CHOH—CO—, give positive tests with Tollens' reagent and with Fehling's and Benedict's solutions (p. 1075), but negative Schiff's tests.

	NaHCO₃	*Acetic anhydride*	*Tollens'*	*Schiff's*	*HIO₄*
(a) AA	—	$C_7H_{12}O_4$	—	—	—
BB	—	$C_7H_{12}O_4$	—	—	+
CC	—	$C_5H_{10}O_3$	—	—	—
DD	—	—	—[1]	—[1]	—
(b) EE	—	$C_5H_8O_3$	+	+	+
FF	—	$C_5H_8O_3$	+	—	+
GG	—	$C_5H_8O_3$	+	+	—
HH	CO_2	—	—	—	—
II	—[2]	—	+	—	—
JJ	—	—	—	—	—
KK	—	$C_7H_{10}O_4$	—	—	+
LL	—	—	—[1]	—[1]	—[1]
MM	—	$C_5H_8O_3$	—	—	—[1]

[1] After treatment with dilute acid, solution gives positive test.
[2] After treatment with NaOH, solution gives positive iodoform test.

21. 2,5-Dimethyl-1,1-cyclopentanedicarboxylic acid can be prepared as a mixture of two optically inactive substances of different physical properties, NN and OO. When each is heated and the reaction mixture worked up by fractional crystallization, NN yields a single product, PP, of formula $C_8H_{14}O_2$, and OO yields two products, QQ and RR, both of formula $C_8H_{14}O_2$.

(a) Give stereochemical formulas for NN, OO, PP, QQ, and RR. (b) Describe another method by which you could assign configurations to NN and OO.

22. (a) (−)-*Erythrose*, $C_4H_8O_4$, gives tests with Tollens' reagent and Benedict's solution (p. 1075), and is oxidized by bromine water to an optically active acid, $C_4H_8O_5$. Treatment with acetic anhydride yields $C_{10}H_{14}O_7$. Erythrose consumes three moles of HIO_4 and yields three moles of formic acid and one mole of formaldehyde. Oxidation of erythrose by nitric acid yields an *optically inactive* compound of formula $C_4H_6O_6$.

(−)-*Threose*, an isomer of erythrose, shows similar chemical behavior except that nitric acid oxidation yields an *optically active* compound of formula $C_4H_6O_6$.

On the basis of this evidence what structure or structures are possible for (−)-erythrose? For (−)-threose?

(b) When R-glyceraldehyde, $CH_2OHCHOHCHO$, is treated with cyanide and the resulting product is hydrolyzed, two monocarboxylic acids are formed (see Problem 12, p. 649). These acids are identical with the acids obtained by oxidation with bromine water of (−)-threose and (−)-erythrose.

Assign a single structure to (−)-erythrose and to (−)-threose.

23. Which (if any) of the following compounds could give rise to each of the infrared spectra shown in Fig. 20.2 (p. 695)?

ethyl acetate
ethyl acrylate ($CH_2{=}CHCOOC_2H_5$)
isobutyric acid
methacrylic acid [$CH_2{=}C(CH_3)COOH$]
methacrylamide [$CH_2{=}C(CH_3)CONH_2$]
phenylacetamide

24. Give a structure or structures consistent with each of the nmr spectra shown in Fig. 20.3 (p. 696).

25. Give the structures of compounds SS, TT, and UU on the basis of their infrared spectra (Fig. 20.4, p. 697) and their nmr spectra (Fig. 20.5, p. 698).

26. Give a structure or structures consistent with the nmr spectrum shown in Fig. 20.6 (p. 699).

27. Give the structure of compound VV on the basis of its infrared and nmr spectra shown in Fig. 20.7 (p. 699).

28. Give a structure or structures consistent with each of the nmr spectra shown in Fig. 20.8 (p. 700).

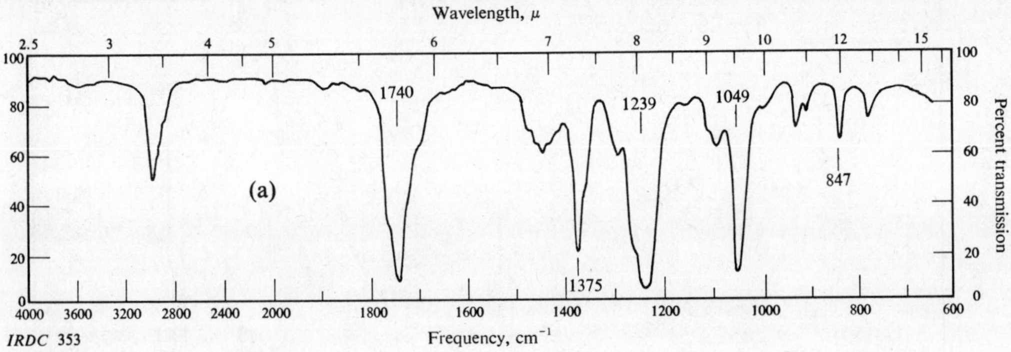

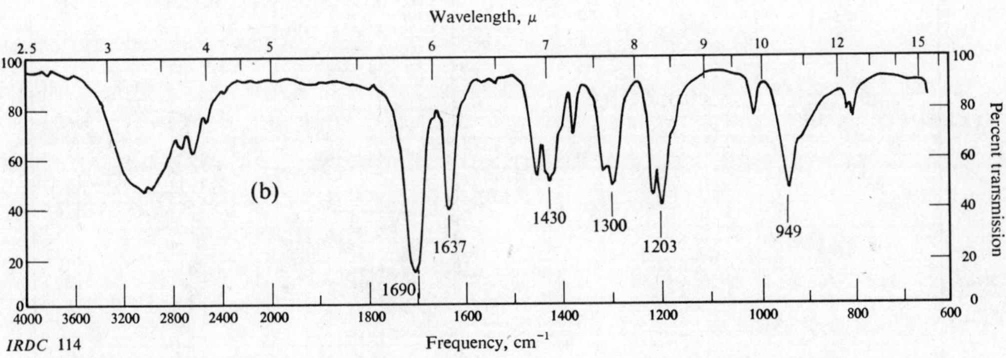

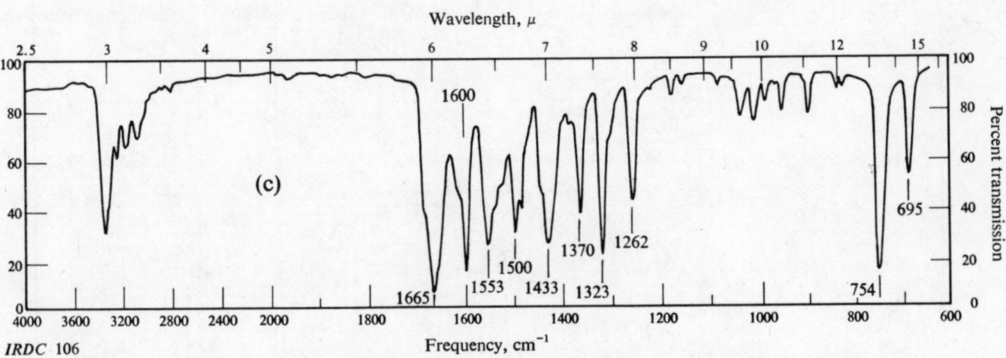

Figure 20.2. Infrared spectra for Problem 23, p. 694.

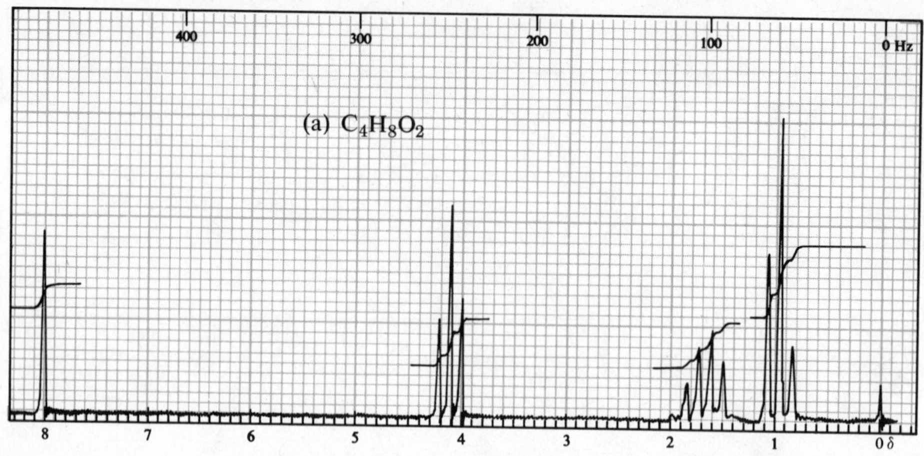

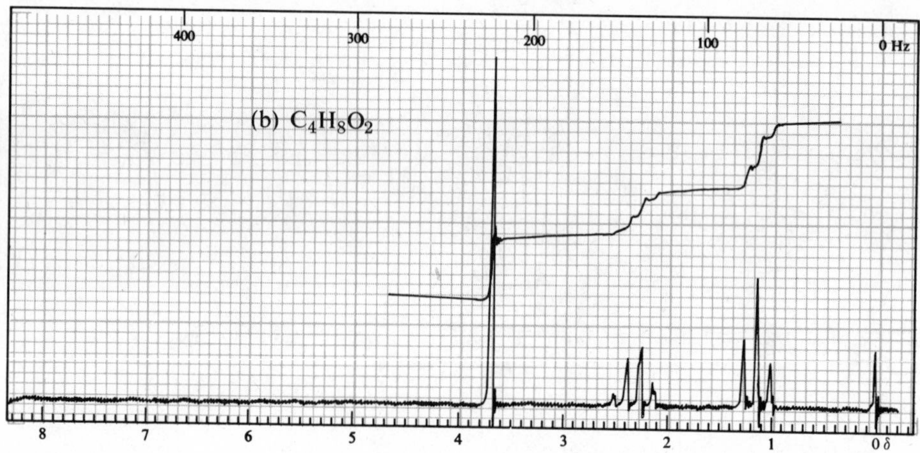

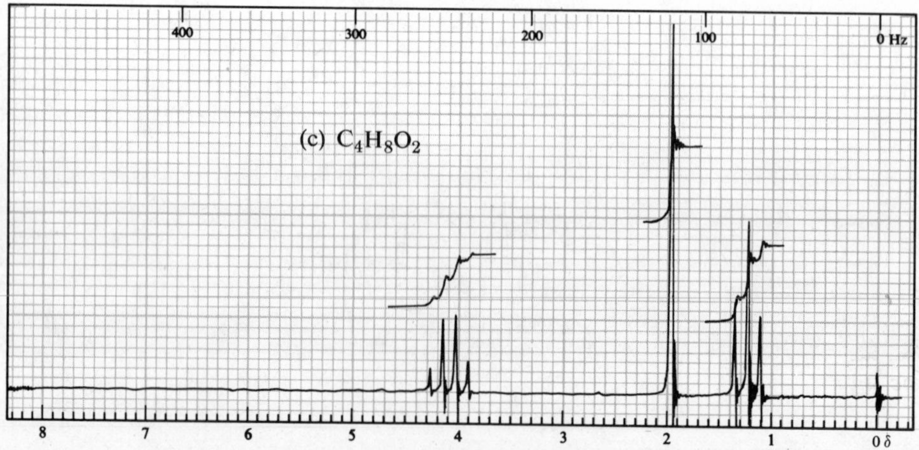

Figure 20.3. Nmr spectra for Problem 24, p. 694.

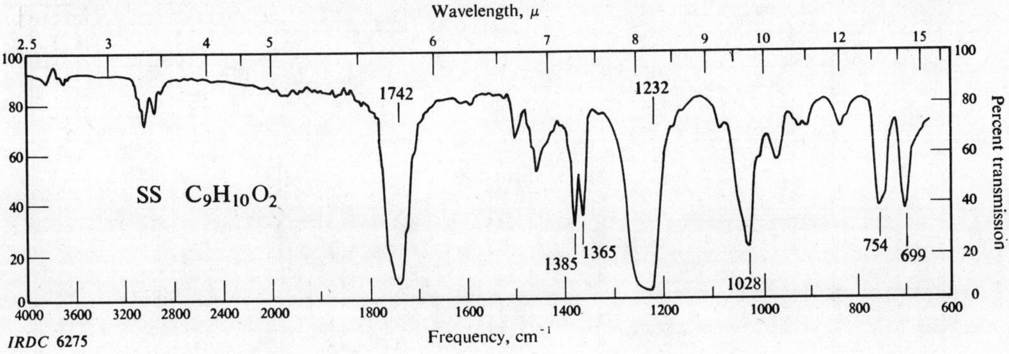

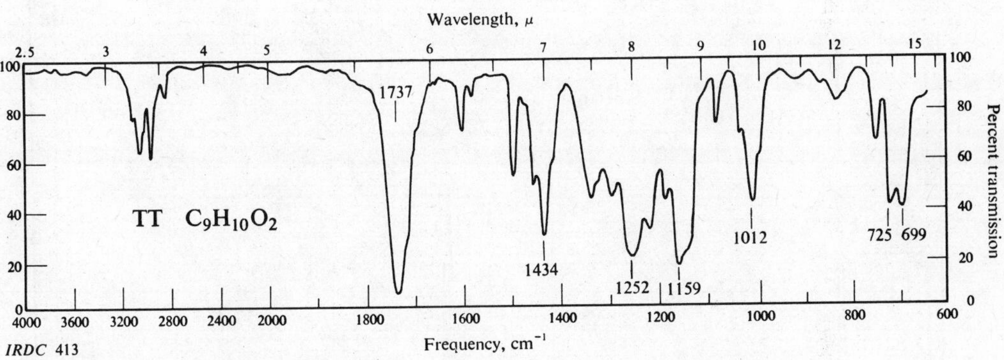

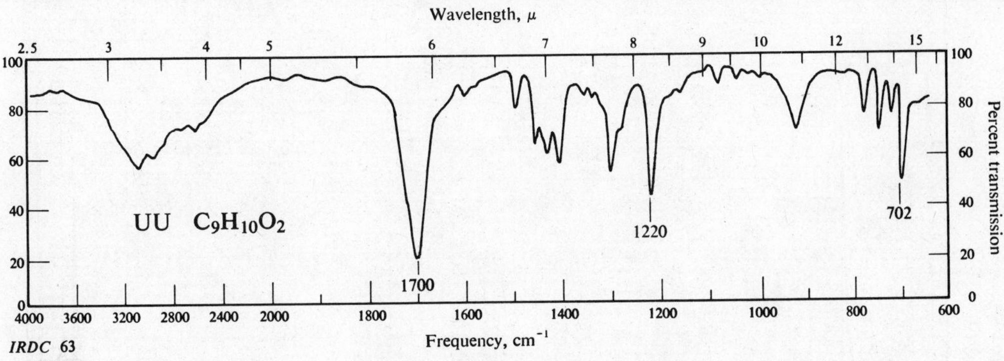

Figure 20.4. Infrared spectra for Problem 25, p. 694.

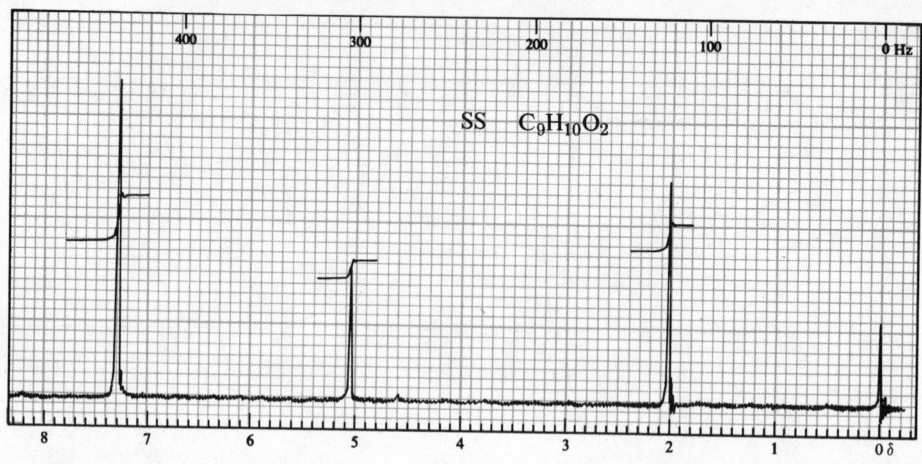

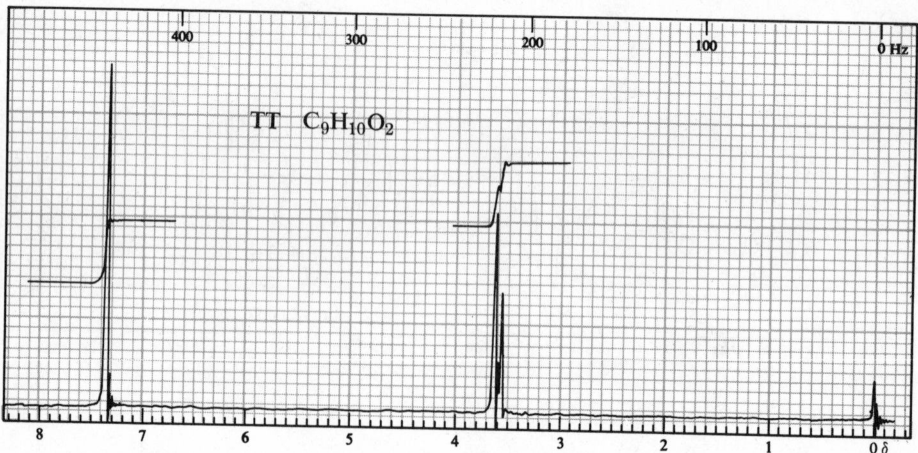

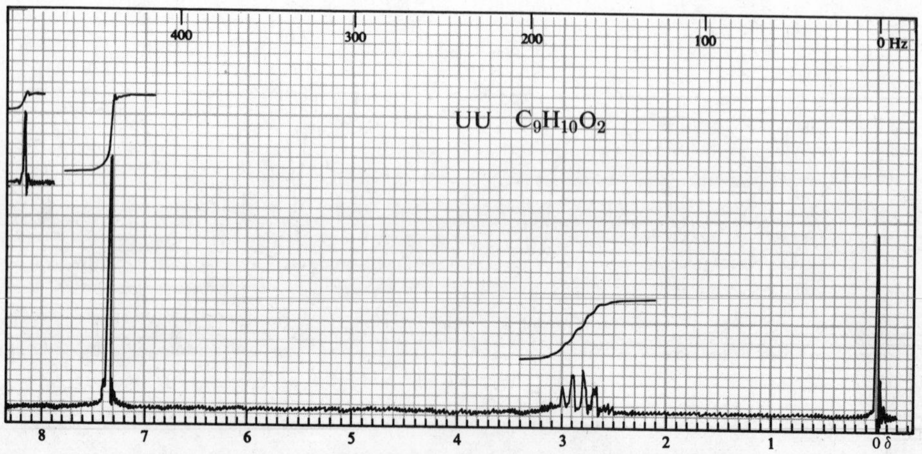

Figure 20.5. Nmr spectra for Problem 25, p. 694.

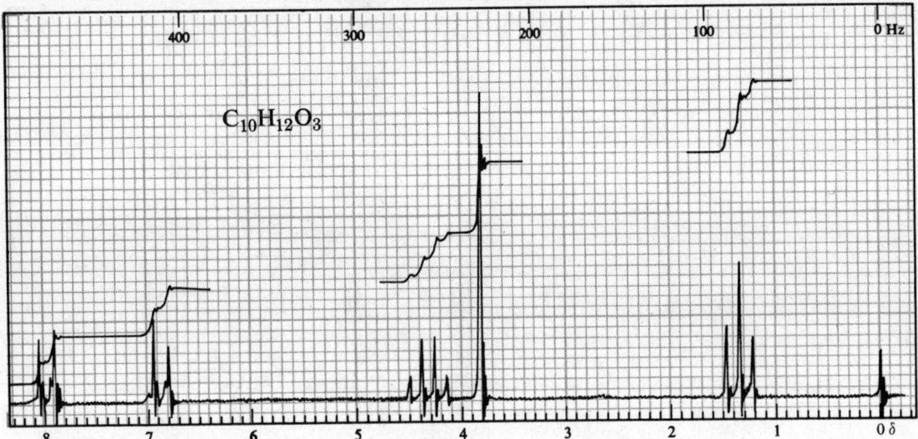

Figure 20.6. Nmr spectrum for Problem 26, p. 694.

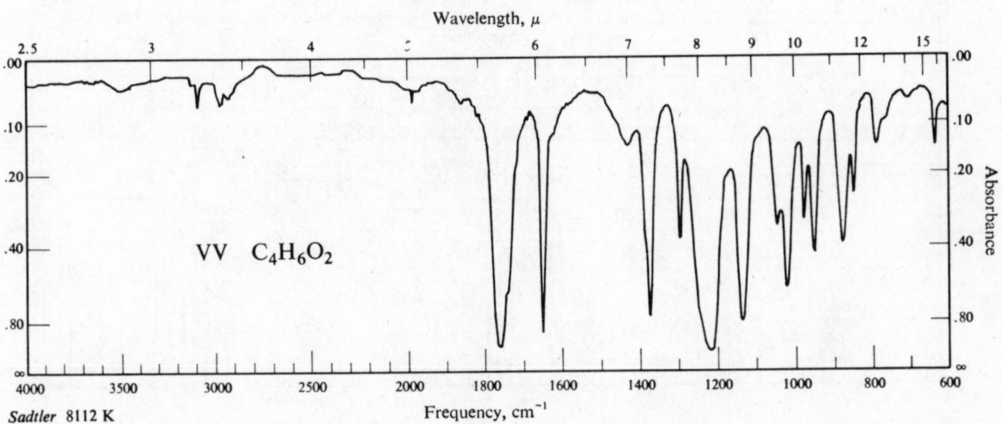

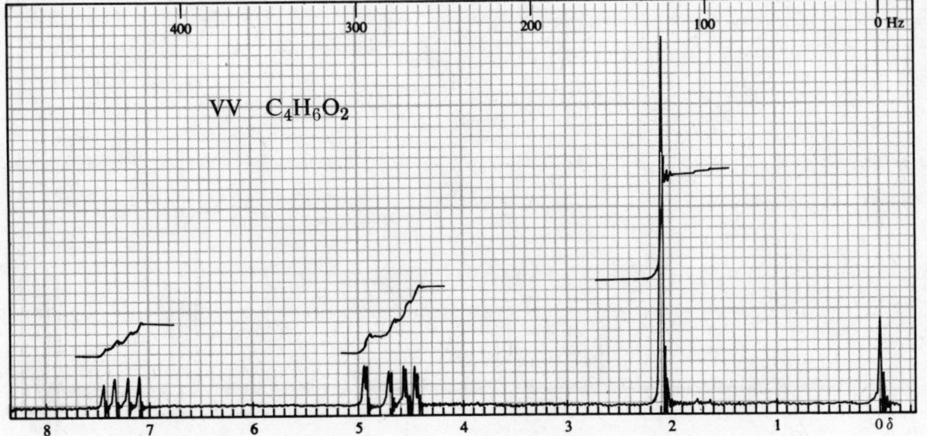

Figure 20.7. Infrared and nmr spectra for Problem 27, p. 594.

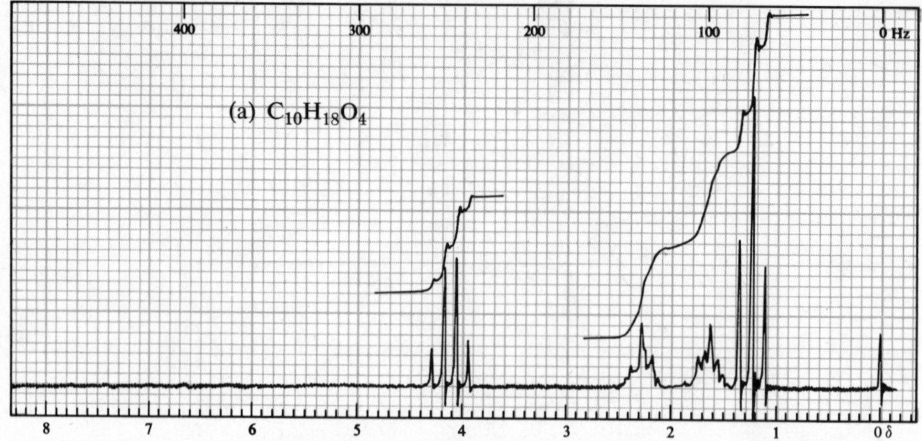

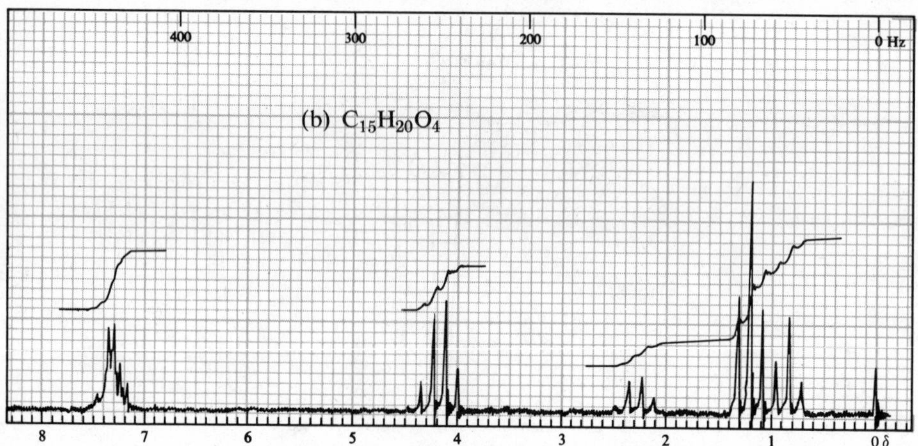

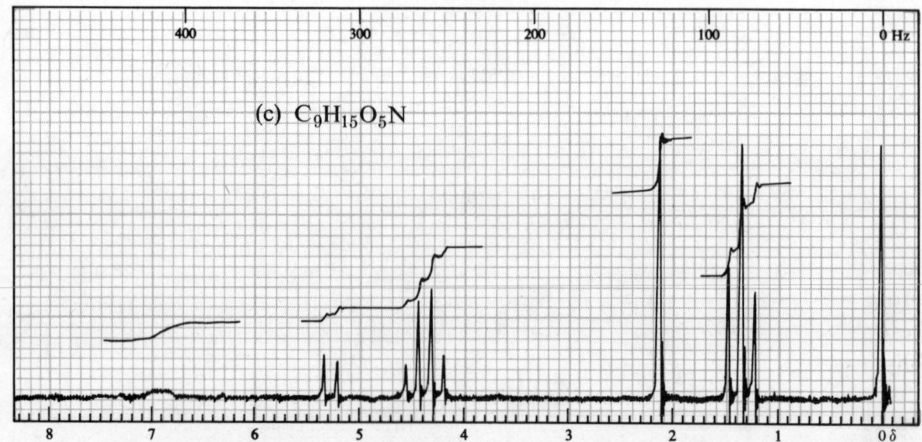

Figure 20.8. Nmr spectra for Problem 28, p. 694.

Chapter 21 Carbanions I

Aldol and Claisen Condensations

21.1 Acidity of α-hydrogens

In our introduction to aldehydes and ketones, we learned that it is the carbonyl group that largely determines the chemistry of aldehydes and ketones. At that time, we saw in part how the carbonyl group does this: by providing a site at which nucleophilic addition can take place. Now we are ready to learn another part of the story: how the carbonyl group strengthens the acidity of the hydrogen atoms attached to the α-carbon and, by doing this, gives rise to a whole set of chemical reactions.

Ionization of an α-hydrogen,

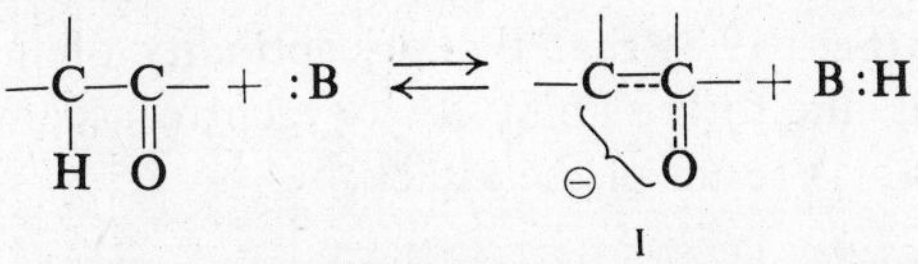

$$-\underset{\mathrm{H}}{\overset{|}{\mathrm{C}}}-\underset{\mathrm{O}}{\overset{}{\mathrm{C}}}- \; + \; :\mathrm{B} \;\rightleftarrows\; -\overset{|}{\mathrm{C}}\!\overset{---}{}\!\overset{|}{\mathrm{C}}- \; + \; \mathrm{B}:\mathrm{H}$$

(C=O in the reactant; in the product I the negative charge ⊖ is delocalized over C---C---O)

I

yields a carbanion I that is a resonance hybrid of two structures II and III,

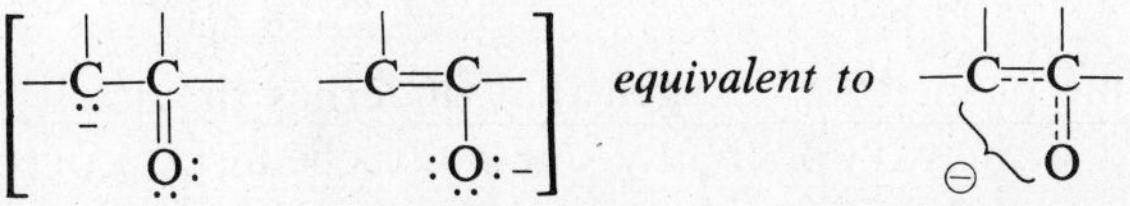

$$\left[\; -\overset{|}{\underset{\cdot\cdot}{\mathrm{C}}}^{-}-\underset{\mathrm{\ddot{O}:}}{\overset{|}{\mathrm{C}}}- \qquad -\overset{|}{\mathrm{C}}=\underset{:\mathrm{\ddot{O}}:^{-}}{\overset{|}{\mathrm{C}}}- \;\right] \quad \textit{equivalent to} \quad -\overset{|}{\mathrm{C}}\text{---}\overset{|}{\mathrm{C}}- \;(\ominus \text{ over C---C---O})$$

II   III   I

resonance that is possible only through participation by the carbonyl group. Resonance of this kind is *not* possible for carbanions formed by ionization of β-hydrogens, γ-hydrogens, etc., from saturated carbonyl compounds.

Problem 21.1 Which structure, II or III, would you expect to make the larger contribution to the carbanion I? Why?

Problem 21.2 Account for the fact that the diketone acetylacetone (2,4-pentanedione) is about as acidic as phenol, and much more acidic than, say, acetone. Which hydrogens are the most acidic?

Problem 21.3 How do you account for the following order of acidity?

$$(C_6H_5)_3CH > (C_6H_5)_2CH_2 > C_6H_5CH_3 > CH_4$$

The carbonyl group thus affects the acidity of α-hydrogens in just the way it affects the acidity of carboxylic acids: by helping to accommodate the negative charge of the anion.

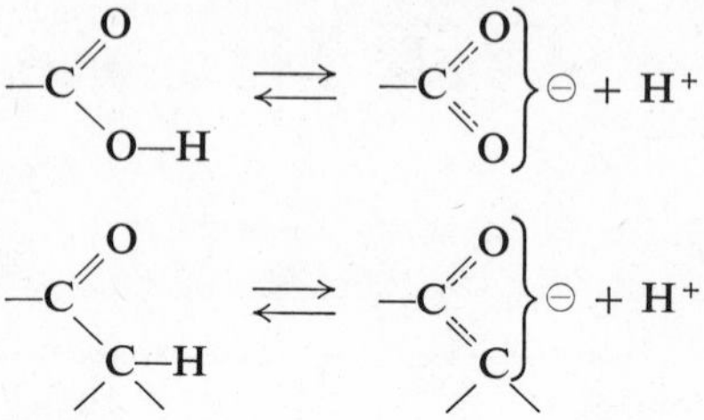

$$-C(=O)-O-H \rightleftharpoons -C\left\{\begin{matrix}O\\O\end{matrix}\right\}^{\ominus} + H^+$$

$$-C(=O)-\overset{|}{C}-H \rightleftharpoons -C\left\{\begin{matrix}O\\C\end{matrix}\right\}^{\ominus} + H^+$$

I

Resonance in I involves structures (II and III) of quite different stabilities, and hence is much less important than the resonance involving equivalent structures in a carboxylate ion. Compared with the hydrogen of a —COOH group, the α-hydrogen atoms of an aldehyde or ketone are very weakly acidic; the important thing is that they are considerably more acidic than hydrogen atoms anywhere else in the molecule, and that they are acidic enough for *significant*—even though very low—concentrations of carbanions to be generated.

We shall use the term *carbanion* to describe ions like I since *part* of the charge is carried by carbon, even though the stability that gives these ions their importance is due to the very fact that most of the charge is *not* carried by carbon but by oxygen.

We saw before (Sec. 19.8) that the susceptibility of the carbonyl group to nucleophilic attack is due to the ability of oxygen to accommodate the negative charge that develops as a result of the attack,

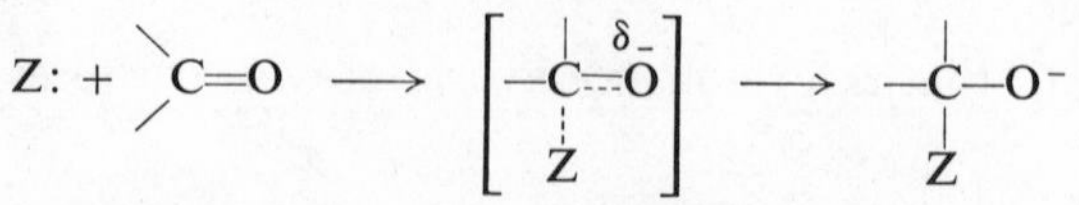

$$Z: + \;\rangle C{=}O \longrightarrow \left[-\overset{|}{\underset{\vdots}{\underset{Z}{C}}}\overset{\delta_-}{\cdots}O\right] \longrightarrow -\overset{|}{\underset{|}{\underset{Z}{C}}}-O^-$$

precisely the same property of oxygen that underlies the acidity of α-hydrogens. We have started with two apparently unrelated chemical properties of carbonyl compounds and have traced them to a common origin—an indication of the simplicity underlying the seeming confusion of organic chemistry.

Problem 21.4 In the reaction of aqueous NaCN with an α,β-unsaturated ketone like

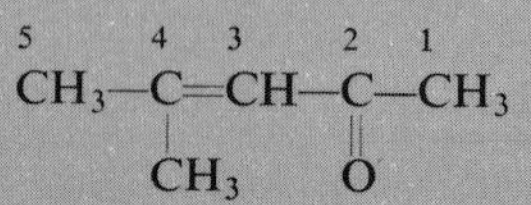

$$\overset{5}{CH_3}-\underset{\underset{CH_3}{|}}{\overset{4}{C}}{=}\overset{3}{CH}-\underset{\underset{O}{||}}{\overset{2}{C}}-\overset{1}{CH_3}$$

CN^- adds, not to C-2, but to C-4. (a) How do you account for this behavior?

(b) What product would you expect to isolate from the reaction mixture? (*Hint:* See Secs. 19.12 and 8.20.) (Check your answers in Sec. 27.5.)

21.2 Reactions involving carbanions

The carbonyl group occurs in compounds other than aldehydes and ketones—in esters, for example—and, wherever it is, it makes any α-hydrogens acidic and thus aids in formation of carbanions. Since these α-hydrogens are only weakly acidic, however, the carbanions are highly basic, exceedingly reactive particles. In their reactions they behave as we would expect: as *nucleophiles*. As nucleophiles, carbanions can attack carbon and, in doing so, form carbon–carbon bonds. *From the standpoint of synthesis, acid-strengthening by carbonyl groups is probably the most important structural effect in organic chemistry.*

We shall take up first the behavior of ketones toward the halogens, and see evidence that carbanions do indeed exist; at the same time, we shall see an elegant example of the application of kinetics, stereochemistry, and isotopic tracers to the understanding of reaction mechanisms. And while we are at it, we shall see something of the role that keto–enol tautomerism plays in the chemistry of carbonyl compounds.

Next, we shall turn to reactions in which the carbonyl group plays *both* its roles: the *aldol condensation*, in which a carbanion generated from one molecule of aldehyde or ketone adds, as a nucleophile, to the carbonyl group of a second molecule; and the *Claisen condensation*, in which a carbanion generated from one molecule of ester attacks the carbonyl group of a second molecule, with acyl substitution as the final result.

REACTIONS INVOLVING CARBANIONS

1. Halogenation of ketones. Discussed in Secs. 21.3–21.4.

$$-\underset{\text{H}}{\overset{|}{\text{C}}}-\underset{\text{O}}{\overset{\|}{\text{C}}}- + X_2 \xrightarrow{\text{H}^+ \text{ or OH}^-} -\underset{\text{X}}{\overset{|}{\text{C}}}-\underset{\text{O}}{\overset{\|}{\text{C}}}- + \text{HX} \qquad X_2 = Cl_2, Br_2, I_2$$

Ketone $\qquad$ α-Halo ketone

Examples:

$$\text{Cyclohexanone} + Br_2 \xrightarrow{H^+} \text{2-Bromocyclohexanone} + HBr$$

Cyclohexanone $\qquad$ 2-Bromocyclohexanone

$$CH_3-\underset{CH_3}{\overset{CH_3}{\underset{|}{\overset{|}{C}}}}-\underset{O}{\overset{}{\underset{\|}{C}}}-CH_3 + I_2 + OH^- \longrightarrow \left[CH_3-\underset{CH_3}{\overset{CH_3}{\underset{|}{\overset{|}{C}}}}-\underset{O}{\underset{\|}{C}}-CI_3\right] \xrightarrow{OH^-} CH_3-\underset{CH_3}{\overset{CH_3}{\underset{|}{\overset{|}{C}}}}-COO^- + CHI_3$$

Methyl *tert*-butyl ketone (3,3-Dimethyl-2-butanone) $\qquad$ Trimethylacetate ion $\qquad$ Iodoform

2. Nucleophilic addition to carbonyl compounds.

(a) Aldol condensation. Discussed in Secs. 21.5–21.8.

$$\underset{\text{O}}{\overset{\diagdown\ \diagup}{\text{C}}} + -\overset{|}{\underset{\text{H}}{\text{C}}}-\overset{|}{\text{C}}=\text{O} \xrightarrow{\text{base or acid}} -\overset{|}{\underset{\text{OH}}{\text{C}}}-\overset{|}{\underset{|}{\text{C}}}-\overset{|}{\text{C}}=\text{O}$$

An aldol
(A β-hydroxy carbonyl compound)

Examples:

$$\underset{\substack{\text{Acetaldehyde}\\ \textit{2 moles}}}{\text{CH}_3\overset{\text{H}}{\text{C}}=\text{O} + \text{CH}_2\overset{\text{H}}{\text{C}}=\text{O}} \text{ (H on } \text{CH}_2\text{)} \xrightarrow{\text{OH}^-} \underset{\substack{\text{Acetaldol}\\ \text{(3-Hydroxybutanal)}}}{\text{CH}_3\overset{\text{H}}{\underset{\text{OH}}{\text{C}}}-\text{CH}_2\overset{\text{H}}{\text{C}}=\text{O}} \xrightarrow[\text{heat}]{\text{NaHSO}_4,} \underset{\substack{\text{Crotonaldehyde}\\ \text{(2-Butenal)}}}{\text{CH}_3-\overset{\text{H}}{\text{C}}=\overset{\text{H}}{\text{C}}-\overset{\text{H}}{\text{C}}=\text{O}} + \text{H}_2\text{O}$$

$$\underset{\substack{\text{Acetone}\\ \textit{2 moles}}}{\text{CH}_3\overset{\text{CH}_3}{\text{C}}=\text{O} + \text{CH}_2\overset{\text{CH}_3}{\text{C}}=\text{O}} \text{ (H on } \text{CH}_2\text{)} \xrightarrow{\text{OH}^-} \underset{\text{Diacetone alcohol}}{\text{CH}_3\overset{\text{CH}_3}{\underset{\text{OH}}{\text{C}}}-\text{CH}_2\underset{\text{O}}{\underset{\|}{\text{C}}}\text{CH}_3} \xrightarrow[\text{heat}]{\text{NaHSO}_4,} \underset{\substack{\text{Mesityl oxide}\\ \text{(4-Methyl-3-penten-2-one)}}}{\text{CH}_3\overset{\text{CH}_3}{\text{C}}=\text{CH}\underset{\text{O}}{\underset{\|}{\text{C}}}\text{CH}_3} + \text{H}_2\text{O}$$

$$\underset{\text{Benzaldehyde}}{\text{C}_6\text{H}_5-\overset{\text{H}}{\text{C}}=\text{O}} + \underset{\text{Acetaldehyde}}{\underset{\text{H}}{\text{CH}_2}\overset{\text{H}}{\text{C}}=\text{O}} \xrightarrow{\text{OH}^-} \left[\text{C}_6\text{H}_5-\overset{\text{H}}{\underset{\text{OH}}{\text{C}}}-\text{CH}_2\overset{\text{H}}{\text{C}}=\text{O}\right] \rightarrow \underset{\substack{\text{Cinnamaldehyde}\\ \text{(3-Phenyl-2-propenal)}}}{\text{C}_6\text{H}_5-\overset{\text{H}}{\text{C}}=\overset{\text{H}}{\text{C}}-\overset{\text{H}}{\text{C}}=\text{O}} + \text{H}_2\text{O}$$

$$\underset{\text{Benzaldehyde}}{\text{C}_6\text{H}_5-\overset{\text{H}}{\text{C}}=\text{O}} + \underset{\text{Acetone}}{\underset{\text{H}}{\text{CH}_2}\underset{\text{O}}{\underset{\|}{\text{C}}}\text{CH}_3} \xrightarrow{\text{OH}^-} \left[\text{C}_6\text{H}_5-\overset{\text{H}}{\underset{\text{OH}}{\text{C}}}-\text{CH}_2\underset{\text{O}}{\underset{\|}{\text{C}}}\text{CH}_3\right] \rightarrow \text{C}_6\text{H}_5-\overset{\text{H}}{\text{C}}=\overset{\text{H}}{\text{C}}-\underset{\text{O}}{\underset{\|}{\text{C}}}-\text{CH}_3$$

Benzalacetone
(*Benzal* is $C_6H_5CH=$)
(4-Phenyl-3-buten-2-one)

$$\underset{\text{Benzaldehyde}}{\text{C}_6\text{H}_5-\overset{\text{H}}{\text{C}}=\text{O}} + \underset{\text{Acetophenone}}{\underset{\text{H}}{\text{CH}_2}-\underset{\text{O}}{\underset{\|}{\text{C}}}-\text{C}_6\text{H}_5} \xrightarrow{\text{OH}^-} \underset{\substack{\text{Benzalacetophenone}\\ \text{(1,3-Diphenyl-2-propen-1-one)}}}{\text{C}_6\text{H}_5-\overset{\text{H}}{\text{C}}=\overset{\text{H}}{\text{C}}-\underset{\text{O}}{\underset{\|}{\text{C}}}-\text{C}_6\text{H}_5} + \text{H}_2\text{O}$$

(b) Reactions related to aldol condensation. Discussed in Sec. 21.9.

(c) Addition of Grignard reagents. Discussed in Sec. 19.11.

(d) Addition of organozinc compounds. Reformatsky reaction. Discussed in Sec. 21.13.

(e) **Wittig reaction.** Discussed in Sec. 21.10.

$$\underset{\text{O}}{\overset{\diagdown\diagup}{\underset{\|}{\text{C}}}} + \underset{\text{An ylide}}{\text{Ph}_3\text{P}=\overset{\overset{\text{R}'}{|}}{\text{C}}-\text{R}} \longrightarrow \underset{\text{A betaine}}{-\underset{\underset{\ominus\text{O}}{|}}{\overset{|}{\text{C}}}-\underset{\underset{\oplus\text{PPh}_3}{|}}{\overset{\overset{\text{R}'}{|}}{\text{C}}}-\text{R}} \longrightarrow -\overset{|}{\text{C}}=\overset{\overset{\text{R}'}{|}}{\text{C}}-\text{R} + \text{Ph}_3\text{PO}$$

Examples:

$$\underset{\text{Cinnamaldehyde}}{C_6H_5CH{=}CHCHO} + \underset{\text{Methylenetriphenylphosphorane}}{Ph_3P{=}CH_2} \longrightarrow C_6H_5CH{=}CH-\underset{\underset{\ominus O}{|}}{\overset{\overset{H}{|}}{C}}-\underset{\underset{\oplus PPh_3}{|}}{\overset{\overset{H}{|}}{CH}} \longrightarrow \underset{\substack{\text{1-Phenyl-1,3-butadiene}\\(69\%)}}{C_6H_5CH{=}CH-CH{=}CH_2}$$

$$\underset{\text{Cyclohexanone}}{C_6H_{10}{=}O} + Ph_3P{=}CH_2 \longrightarrow C_6H_{10}(O^{\ominus})(CH_2\overset{\oplus}{P}Ph_3) \longrightarrow \underset{\substack{\text{Methylenecyclohexane}\\(48\%)}}{C_6H_{10}{=}CH_2} + Ph_3PO$$

3. Nucleophilic acyl substitution.

(a) **Claisen condensation.** Discussed in Secs. 21.11–21.12.

$$-\overset{\overset{O}{/\!/}}{C}\diagdown_{OR'} + -\underset{\underset{H}{|}}{\overset{|}{C}}-\overset{\overset{O}{/\!/}}{C}\diagdown_{OR'} \xrightarrow{^-OC_2H_5} \underset{\text{A }\beta\text{-keto ester}}{-\underset{\underset{O}{\|}}{C}-\overset{|}{\underset{|}{C}}-\overset{\overset{O}{/\!/}}{C}\diagdown_{OR'}}$$

Examples:

$$\underset{\substack{\text{Ethyl acetate}\\ \textit{2 moles}}}{CH_3\overset{\overset{O}{/\!/}}{C}\diagdown_{OC_2H_5} + \underset{\underset{H}{|}}{CH_2}\overset{\overset{O}{/\!/}}{C}\diagdown_{OC_2H_5}} \xrightarrow{^-OC_2H_5} \underset{\text{Ethyl acetoacetate}}{CH_3\underset{\underset{O}{\|}}{C}CH_2\overset{\overset{O}{/\!/}}{C}\diagdown_{OC_2H_5}}$$

$$\underset{\text{Ethyl benzoate}}{C_6H_5COOC_2H_5} + \underset{\text{Ethyl acetate}}{CH_3COOC_2H_5} \xrightarrow{^-OC_2H_5} \underset{\text{Ethyl benzoylacetate}}{C_6H_5-\underset{\underset{O}{\|}}{C}-CH_2COOC_2H_5} + C_2H_5OH$$

(b) **Acylation of organocadmium compounds.** Discussed in Sec. 19.7.

4. Nucleophilic aliphatic substitution.

(a) **Coupling of alkyl halides with organometallic compounds.** Discussed in Sec. 3.17.

(b) Synthesis of acetylides. Discussed in Sec. 8.12.

(c) Alkylation of malonic ester and acetoacetic ester. Discussed in Secs. 26.2–26.3.

5. Addition to α,β-unsaturated carbonyl compounds. Michael addition. Discussed in Sec. 27.7.

21.3 Base-promoted halogenation of ketones

Acetone reacts with bromine to form bromoacetone; the reaction is accelerated by bases (e.g., hydroxide ion, acetate ion, etc.). Study of the kinetics shows that

$$\underset{\text{Acetone}}{CH_3COCH_3} + Br_2 + :B \longrightarrow \underset{\text{Bromoacetone}}{CH_3COCH_2Br} + Br^- + H:B$$

the rate of reaction depends upon the concentration of acetone, [acetone], and of base, [:B], but is *independent of bromine concentration*:

$$\text{rate} = k\,[\text{acetone}][:B]$$

We have encountered this kind of situation before (Sec. 14.12) and know, in a general way, what it must mean: if the rate of reaction does not depend upon $[Br_2]$, it can only mean that the reaction *whose rate we are measuring* does not involve Br_2.

The kinetics is quite consistent with the following mechanism. The base

(1) $$CH_3\underset{\underset{O}{\|}}{C}CH_3 + :B \rightleftharpoons H:B + \underset{\text{I}}{CH_3\underset{\underset{O^{\ominus}}{\|}}{C}{=}CH_2}$$ **Slow:** *rate-determining*

(2) $$\underset{\text{I}}{CH_3\underset{\underset{O^{\ominus}}{\|}}{C}{=}CH_2} + Br_2 \longrightarrow CH_3\underset{\underset{O}{\|}}{C}CH_2Br + Br^-$$ **Fast**

slowly abstracts a proton (step 1) from acetone to form carbanion I, which then reacts rapidly with bromine (step 2) to yield bromoacetone. Step (1), generation of the carbanion, is the rate-determining step, since its rate determines the overall rate of the reaction sequence. As fast as carbanions are generated, they are snapped up by bromine molecules.

Strong support for this interpretation comes from the kinetics of iodination. Here, too, the rate of reaction depends upon [acetone] and [:B] but is independent of $[I_2]$. Furthermore, and most significant, at a given [acetone] and [:B], bromination and iodination *proceed at identical rates*. That is to say, in the rate expression

$$\text{rate} = k\,[\text{acetone}][:B]$$

the value of k is the same regardless of which halogen is involved. It *should* be, of course, according to the proposed mechanism, since in both cases it is the rate constant for the same reaction, abstraction of a proton from the ketone.

Study of the bromination of acetone, done by A. Lapworth (of the University of

Manchester) in 1904, showed for the first time how kinetics could be used to reveal the mechanism of an organic reaction. The carbanion mechanism has since been confirmed not only by the iodination work, but also by studies of stereochemistry and isotopic exchange.

Problem 21.5 Show in detail exactly how each of the following facts provides evidence for the carbanion mechanism of base-promoted halogenation of ketones.

(a) In basic solution, (+)-phenyl *sec*-butyl ketone undergoes racemization; the rate constant for loss of optical activity is identical with the rate constant for bromination of this ketone.

(b) Ketone II undergoes racemization in basic solution, but ketone III does not.

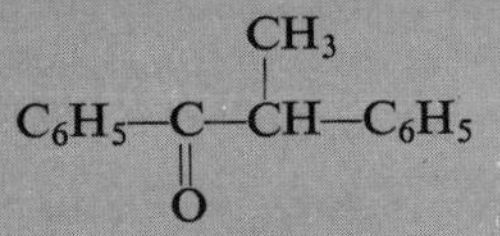

$$\underset{\text{II}}{C_6H_5-\underset{\underset{O}{\|}}{C}-\overset{\overset{CH_3}{|}}{C}H-C_6H_5} \qquad \underset{\text{III}}{C_6H_5-\underset{\underset{O}{\|}}{C}-\overset{\overset{CH_3}{|}}{\underset{\underset{C_4H_9\text{-}n}{|}}{C}}-C_6H_5}$$

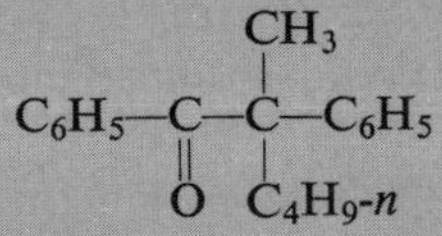

(c) When (+)-phenyl *sec*-butyl ketone is allowed to stand in D_2O containing OD^-, it not only undergoes racemization, but also becomes labeled with deuterium at the α-position; the rate constants for racemization and hydrogen exchange are identical.

Problem 21.6 (a) Suggest a mechanism for the base-catalyzed racemization of the optically active ester, ethyl mandelate, $C_6H_5CHOHCOOC_2H_5$. (b) How do you account for the fact that optically active mandelic acid undergoes racemization in base *much more slowly* than the ester? (*Hint:* See Sec. 18.20.) (c) What would you predict about the rate of base-catalyzed racemization of α-methylmandelic acid, $C_6H_5C(CH_3)(OH)COOH$?

Problem 21.7 Suppose, as an alternative to the carbanion mechanism, that hydrogen exchange and racemization were both to arise by some kind of direct displacement of one hydrogen (H) by another (D) with inversion of configuration. What relationship would you then expect between the rates of racemization and exchange? (*Hint:* Take one molecule at a time, and see what happens when H is replaced by D with inversion.)

21.4 Acid-catalyzed halogenation of ketones. Enolization

Acids, like bases, speed up the halogenation of ketones. Acids are not, however, consumed, and hence we may properly speak of acid-*catalyzed* halogenation (as contrasted to base-*promoted* halogenation). Although the reaction is not,

$$\underset{\text{Acetone}}{CH_3COCH_3} + Br_2 \xrightarrow{\text{acid}} \underset{\text{Bromoacetone}}{CH_3COCH_2Br} + HBr$$

strictly speaking, a part of carbanion chemistry, this is perhaps the best place to take it up, since it shows a striking parallel in every aspect to the base-promoted reaction we have just left.

Here, too, the kinetics show the rate of halogenation to be independent of halogen concentration, but dependent upon ketone concentration and, this time, acid concentration. Here, too, we find the remarkable identity of rate constants for apparently different reactions: for bromination and iodination of acetone, and exchange of its hydrogens for deuterium; for iodination and racemization of phenyl *sec*-butyl ketone.

The interpretation, too, is essentially the same as the one we saw before: *preceding* the step that involves halogen, there is a rate-determining reaction that can lead not only to halogenation but also to racemization and to hydrogen exchange.

The rate-determining reaction here is the formation of the *enol*, which involves two steps: rapid, reversible protonation (step 1) of the carbonyl oxygen, followed by the slow loss of an α-hydrogen (step 2).

(1) $CH_3{-}\underset{\underset{O}{\|}}{C}{-}CH_3 + H{:}B \rightleftharpoons CH_3{-}\underset{\underset{+OH}{\|}}{C}{-}CH_3 + {:}B$ **Fast**

(2) $CH_3{-}\underset{\underset{+OH}{\|}}{C}{-}CH_3 + {:}B \longrightarrow CH_3{-}\underset{\underset{OH}{|}}{C}{=}CH_2 + H{:}B$ **Slow**

Enol

(3) $CH_3{-}\underset{\underset{OH}{|}}{C}{=}CH_2 + X_2 \longrightarrow CH_3{-}\underset{\underset{+OH}{\|}}{C}{-}CH_2X + X^-$ **Fast**

I

(4) $CH_3{-}\underset{\underset{+OH}{\|}}{C}{-}CH_2X + {:}B \rightleftharpoons CH_3{-}\underset{\underset{O}{\|}}{C}{-}CH_2X + H{:}B$ **Fast**

Once formed, the enol reacts rapidly with halogen (step 3). We might have expected the unsaturated enol to undergo addition and, indeed, the reaction starts out exactly as though this were going to happen: positive halogen attaches itself to form a cation. As usual (Sec. 6.11), attachment occurs in the way that yields the more stable cation.

The ion formed in this case, I, is an exceedingly stable one, owing its stability to the fact that it is hardly a "carbonium" ion at all, since oxygen can carry the charge and still have an octet of electrons. The ion is, actually, a protonated ketone; loss of the proton yields the product, bromoacetone.

We may find it odd, considering that we call this reaction "acid-catalyzed," that the rate-determining step (2) is really the same as in the base-promoted reaction: abstraction of an α-hydrogen by a base—here, by the conjugate base of the catalyzing acid. Actually, what we see here must always hold true: a reaction that is truly *catalyzed* by acid or base is catalyzed by *both acid and base*. In our case, transfer of the proton from the acid H:B to carbonyl oxygen (step 1) makes the ketone more reactive and hence speeds up enolization. But, if this is truly catalysis, the acid must not be *consumed*. Regeneration of the acid H:B requires that the conjugate base :B get a proton from somewhere; it takes it from the α-carbon (step 2), and thus completes the enolization. Both acid and base speed up the rate-determining step (2): base directly, as one of the reactants, and acid indirectly, by increasing the concentration of the other reactant, the protonated ketone. Using a strong mineral acid in aqueous solution, we would not be aware of the role played by the base; the acid is H_3O^+ and the conjugate base, H_2O, is the solvent.

Problem 21.8 Show in detail how the enolization mechanism accounts for the following facts: (a) the rate constants for acid-catalyzed hydrogen–deuterium exchange and bromination of acetone are identical; (b) the rate constants for acid-catalyzed racemization and iodination of phenyl *sec*-butyl ketone are identical.

Problem 21.9 (a) In the acid-catalyzed dehydration of alcohols (Sec. 5.20), what is the base involved? (b) In the base-catalyzed racemization and hydrogen exchange of phenyl *sec*-butyl ketone (Problem 21.5, p. 707), what is the acid involved?

21.5 Aldol condensation

Under the influence of dilute base or dilute acid, two molecules of an aldehyde or a ketone may combine to form a β-hydroxyaldehyde or β-hydroxyketone. This reaction is called the **aldol condensation**. In every case the product results from addition of one molecule of aldehyde (or ketone) to a second molecule in such a way that the α-carbon of the first becomes attached to the carbonyl carbon of the second. For example:

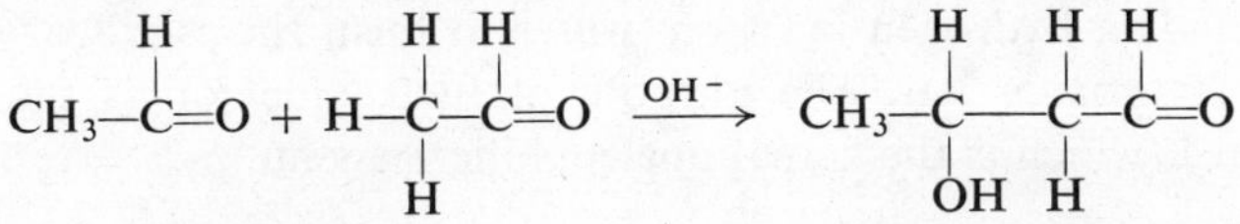

$$\mathrm{CH_3{-}\overset{H}{\overset{|}{C}}{=}O + H{-}\underset{H}{\underset{|}{\overset{H}{\overset{|}{C}}}}{-}\overset{H}{\overset{|}{C}}{=}O \xrightarrow{OH^-} CH_3{-}\underset{OH}{\underset{|}{\overset{H}{\overset{|}{C}}}}{-}\underset{H}{\underset{|}{\overset{H}{\overset{|}{C}}}}{-}\overset{H}{\overset{|}{C}}{=}O}$$

Acetaldehyde *2 moles* → Aldol (β-Hydroxybutyraldehyde) (3-Hydroxybutanal)

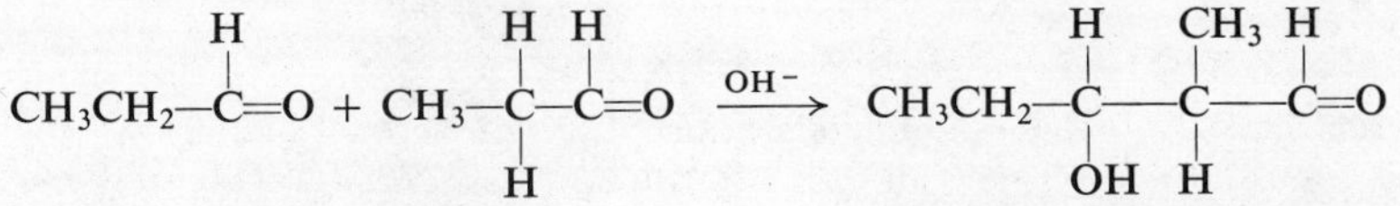

$$\mathrm{CH_3CH_2{-}\overset{H}{\overset{|}{C}}{=}O + CH_3{-}\underset{H}{\underset{|}{\overset{H}{\overset{|}{C}}}}{-}\overset{H}{\overset{|}{C}}{=}O \xrightarrow{OH^-} CH_3CH_2{-}\underset{OH}{\underset{|}{\overset{H}{\overset{|}{C}}}}{-}\underset{H}{\underset{|}{\overset{CH_3}{\overset{|}{C}}}}{-}\overset{H}{\overset{|}{C}}{=}O}$$

Propionaldehyde *2 moles* → β-Hydroxy-α-methylvaleraldehyde (3-Hydroxy-2-methylpentanal)

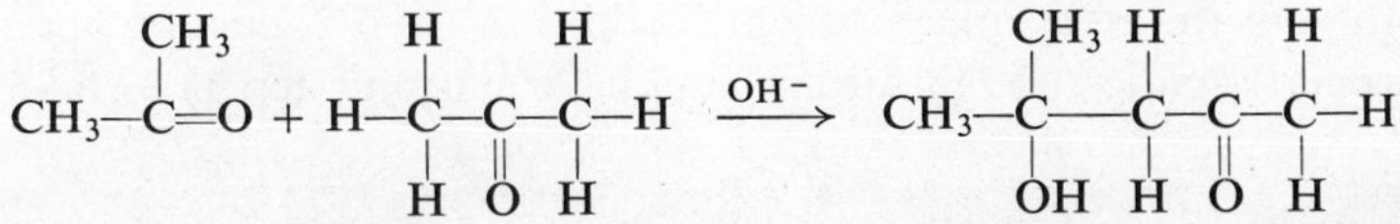

$$\mathrm{CH_3{-}\overset{CH_3}{\overset{|}{C}}{=}O + H{-}\underset{H}{\underset{|}{\overset{H}{\overset{|}{C}}}}{-}\underset{O}{\underset{\|}{C}}{-}\underset{H}{\underset{|}{\overset{H}{\overset{|}{C}}}}{-}H \xrightarrow{OH^-} CH_3{-}\underset{OH}{\underset{|}{\overset{CH_3}{\overset{|}{C}}}}{-}\underset{H}{\underset{|}{\overset{H}{\overset{|}{C}}}}{-}\underset{O}{\underset{\|}{C}}{-}\underset{H}{\underset{|}{\overset{H}{\overset{|}{C}}}}{-}H}$$

Acetone *2 moles* → 4-Hydroxy-4-methyl-2-pentanone (Diacetone alcohol)

If the aldehyde or ketone does not contain an α-hydrogen, a simple aldol condensation cannot take place. For example:

No α-hydrogen atoms

$$\left.\begin{array}{r}\mathrm{ArCHO}\\ \mathrm{HCHO}\\ \mathrm{(CH_3)_3CCHO}\\ \mathrm{ArCOAr}\\ \mathrm{ArCOCR_3}\end{array}\right\} \xrightarrow{\text{dilute OH}^-} \text{no reaction}$$

(In concentrated base, however, these may undergo the Cannizzaro reaction, Sec. 19.16.)

The generally accepted mechanism for the base-catalyzed condensation involves the following steps, acetaldehyde being used as an example. Hydroxide ion

(1) $$\mathrm{CH_3CHO} + \underset{\text{Basic catalyst}}{\mathrm{OH^-}} \rightleftarrows \mathrm{H_2O} + \underset{\text{I}}{[\mathrm{CH_2CHO}]^-}$$

$$\text{(2)} \qquad CH_3\!-\!\overset{\displaystyle H}{\overset{|}{C}}\!=\!O + \underset{\substack{\text{I}\\ \textit{Nucleophilic}\\ \textit{reagent}}}{[CH_2CHO]^-} \rightleftharpoons CH_3\!-\!\underset{\substack{|\\ O^-\\ \text{II}}}{\overset{\displaystyle H}{\overset{|}{C}}}\!-\!CH_2CHO$$

$$\text{(3)} \qquad CH_3\!-\!\underset{\substack{|\\ O^-\\ \text{II}}}{\overset{\displaystyle H}{\overset{|}{C}}}\!-\!CH_2CHO + H_2O \rightleftharpoons CH_3\!-\!\underset{\substack{|\\ OH\\ \text{III}}}{\overset{\displaystyle H}{\overset{|}{C}}}\!-\!CH_2CHO + OH^-$$

abstracts (step 1) a hydrogen ion from the α-carbon of the aldehyde to form carbanion I, which attacks (step 2) carbonyl carbon to form ion II. II (an alkoxide) abstracts (step 3) a hydrogen ion from water to form the β-hydroxyaldehyde III, regenerating hydroxide ion. The purpose of hydroxide ion is thus to produce the carbanion I, which is the actual nucleophilic reagent.

Problem 21.10 Illustrate these steps for:

(a) propionaldehyde
(b) acetone
(c) acetophenone
(d) cyclohexanone
(e) phenylacetaldehyde

Problem 21.11 The aldol condensation of unsymmetrical ketones (methyl ethyl ketone, for example) is usually of little value in synthesis. Why do you think this is so?

The carbonyl group plays two roles in the aldol condensation. It not only provides the unsaturated linkage at which addition (step 2) occurs, but also makes the α-hydrogens acidic enough for carbanion formation (step 1) to take place.

Problem 21.12 In *acid-catalyzed aldol condensations*, acid is believed to perform two functions: to catalyze conversion of carbonyl compound into the enol form, and to provide protonated carbonyl compound with which the enol can react. The reaction that then takes place can, depending upon one's point of view, be regarded either as acid-catalyzed nucleophilic addition to a carbonyl group, or as electrophilic addition to an alkene. On this basis, write all steps in the mechanism of acid-catalyzed aldol condensation of acetaldehyde. In the actual *condensation* step, identify the nucleophile and the electrophile.

Problem 21.13 (a) When acetaldehyde at fairly high concentration was allowed to undergo base-catalyzed aldol condensation in heavy water (D_2O), the product was found to contain almost no deuterium bound to carbon. This finding has been taken as one piece of evidence that the slow step in this aldol condensation is formation of the carbanion. How would you justify this conclusion? (b) The kinetics also supports this conclusion. What kinetics would you expect if this were the case? (*Remember:* Two molecules of acetaldehyde are involved in aldol condensation.) (c) When the experiment in part (a) was carried out at low acetaldehyde concentration, the product was found to contain considerable deuterium bound to carbon. How do you account for this? (*Hint:* See Sec. 14.20.) (d) In contrast to acetaldehyde, acetone was found to undergo base-catalyzed hydrogen–deuterium exchange much faster than aldol condensation. What is one important factor contributing to this difference in behavior?

Problem 21.14 In alkaline solution, 4-methyl-4-hydroxy-2-pentanone is partly converted into acetone. What does this reaction amount to? Show all steps in the most likely mechanism. (*Hint:* See Problem 5.8, p. 170.)

21.6 Dehydration of aldol products

The β-hydroxyaldehydes and β-hydroxyketones obtained from aldol condensations are very easily dehydrated; the major products have the carbon–carbon double bond between the α- and β-carbon atoms. For example:

$$CH_3-\underset{OH}{\overset{H}{C}}-\underset{H}{\overset{H}{C}}-\overset{H}{C}=O \xrightarrow{\text{dil. HCl, warm}} CH_3-\overset{H}{C}=\overset{H}{C}-\overset{H}{C}=O + H_2O$$

Aldol — Crotonaldehyde (2-Butenal)

$$CH_3-\underset{OH}{\overset{CH_3}{C}}-\underset{H}{\overset{H}{C}}-\underset{O}{\underset{\|}{C}}-CH_3 \xrightarrow{\text{I}_2\text{ (a Lewis acid), distill}} CH_3-\overset{CH_3}{C}=\overset{H}{C}-\underset{O}{\underset{\|}{C}}-CH_3 + H_2O$$

Diacetone alcohol (4-Hydroxy-4-methyl-2-pentanone) — Mesityl oxide (4-Methyl-3-penten-2-one)

Both the ease and the orientation of elimination are related to the fact that the alkene obtained is a particularly stable one, since the carbon–carbon double bond is conjugated with the carbon–oxygen double bond of the carbonyl group (compare Sec. 8.16).

Problem 21.15 Draw resonance structures to account for the unusual stability of an α,β-unsaturated aldehyde or ketone. What is the significance of these structures in terms of orbitals? (See Sec. 8.17.)

As we know, an alkene in which the carbon–carbon double bond is conjugated with an aromatic ring is particularly stable (Sec. 12.17); in those cases where elimination of water from the aldol product can form such a conjugated alkene, the unsaturated aldehyde or ketone is the product actually isolated from the reaction. For example:

$$C_6H_5-\underset{O}{\underset{\|}{C}}-CH_3 + CH_3-\underset{O}{\underset{\|}{C}}-C_6H_5 \xrightarrow{NaOC_2H_5} \left[C_6H_5-\underset{OH}{\overset{CH_3}{C}}-\underset{H}{\overset{H}{C}}-\underset{O}{\underset{\|}{C}}-C_6H_5\right] \xrightarrow{-H_2O} C_6H_5-\overset{CH_3}{C}=\overset{H}{C}-\underset{O}{\underset{\|}{C}}-C_6H_5$$

Acetophenone
2 moles

1,3-Diphenyl-2-buten-1-one

21.7 Use of aldol condensation in synthesis

Catalytic hydrogenation of α,β-unsaturated aldehydes and ketones yields saturated alcohols, addition of hydrogen occurring both at carbon–carbon and at carbon–oxygen double bonds. It is for the purpose of ultimately preparing saturated alcohols that the aldol condensation is often carried out. For example, *n*-butyl alcohol and 2-ethyl-1-hexanol are both prepared on an industrial scale in this way:

$$\underset{\text{Acetaldehyde}}{2CH_3CHO} \xrightarrow{OH^-} \underset{\text{Aldol}}{CH_3CHOHCH_2CHO} \xrightarrow{-H_2O} \underset{\substack{\text{Crotonaldehyde}\\ \text{(2-Butenal)}}}{CH_3CH{=}CHCHO} \xrightarrow{H_2,\ Ni} \underset{n\text{-Butyl alcohol}}{CH_3CH_2CH_2CH_2OH}$$

$$\underset{n\text{-Butyl alcohol}}{CH_3CH_2CH_2CH_2OH} \xrightarrow{Cu,\ 250°} \underset{n\text{-Butyraldehyde}}{CH_3CH_2CH_2CHO} \xrightarrow{OH^-} CH_3CH_2CH_2CHOH\underset{C_2H_5}{\underset{|}{C}}HCHO$$

$$\xrightarrow{-H_2O} CH_3CH_2CH_2CH{=}\underset{C_2H_5}{\underset{|}{C}}CHO \xrightarrow{H_2,\ Ni} \underset{\text{2-Ethyl-1-hexanol}}{CH_3CH_2CH_2CH_2\underset{C_2H_5}{\underset{|}{C}}HCH_2OH}$$

Unsaturated alcohols can be prepared if a reagent is selected that reduces only the carbonyl group and leaves the carbon–carbon double bond untouched; one such reagent is sodium borohydride, $NaBH_4$.

$$\underset{\substack{\alpha,\beta\text{-Unsaturated carbonyl}\\ \text{compound}}}{R\overset{\beta}{C}H{=}\overset{\alpha}{C}H{-}\underset{\underset{O}{\|}}{C}{-}R'} \xrightarrow{NaBH_4} \xrightarrow{H^+} \underset{\text{Unsaturated alcohol}}{RCH{=}CH{-}\underset{\underset{OH}{|}}{C}H{-}R'}$$

Problem 21.16 Outline the synthesis of the following alcohols starting from alcohols of smaller carbon number:

(a) 2-methyl-1-pentanol
(b) 4-methyl-2-pentanol
(c) 2-cyclohexylcyclohexanol
(d) 2,4-diphenyl-1-butanol
(e) 1,3-diphenyl-2-buten-1-ol

Problem 21.17 The insect repellent "6-12" (2-ethyl-1,3-hexanediol) is produced by the same chemical company that produces *n*-butyl alcohol and 2-ethyl-1-hexanol; suggest a method for its synthesis. How could you synthesize 2-methyl-2,4-pentanediol?

21.8 Crossed aldol condensation

An aldol condensation between two different carbonyl compounds—a so-called **crossed aldol condensation**—is not always feasible in the laboratory, since a mixture of the four possible products may be obtained. On a commercial scale, however, such a synthesis may be worthwhile if the mixture can be separated and the components marketed.

Problem 21.18 *n*-Butyl alcohol, *n*-hexyl alcohol, 2-ethyl-1-hexanol, and 2-ethyl-1-butanol are marketed by the same chemical concern; how might they be prepared from cheap, readily available compounds?

Under certain conditions, a good yield of a single product can be obtained from a crossed aldol condensation: (a) one reactant contains no α-hydrogens and therefore is incapable of condensing with itself (e.g., aromatic aldehydes or formaldehyde); (b) this reactant is mixed with the catalyst; and then (c) a carbonyl

$$C_6H_5CHO \text{ (Benzaldehyde)} \xrightarrow{OH^-} \begin{cases} \xrightarrow{CH_3CHO,\ 20^\circ} C_6H_5\text{–}CH\!=\!CH\text{–}CH\!=\!O & \text{Cinnamaldehyde} \\ \xrightarrow{CH_3COCH_3,\ 100^\circ} C_6H_5\text{–}CH\!=\!CH\text{–}CO\text{–}CH_3 & \text{Benzalacetone} \\ \xrightarrow{CH_3COC_6H_5,\ 20^\circ} C_6H_5\text{–}CH\!=\!CH\text{–}CO\text{–}C_6H_5 & \text{Benzalacetophenone} \end{cases}$$

Crossed aldol condensations

compound that contains α-hydrogens is added slowly to this mixture. There is thus present at any time only a very low concentration of the ionizable carbonyl compound, and the carbanion it forms reacts almost exclusively with the other carbonyl compound, which is present in large excess.

Problem 21.19 Outline the synthesis of each of the following from benzene or toluene and any readily available alcohols:

(a) 4-phenyl-2-butanol
(b) 1,3-diphenyl-1-propanol
(c) 1,3-diphenylpropane
(d) 2,3-diphenyl-1-propanol
(e) 1,5-diphenyl-1,4-pentadien-3-one (dibenzalacetone)

Problem 21.20 (a) What prediction can you make about the acidity of the γ-hydrogens of α,β-unsaturated carbonyl compounds,

$$\overset{\gamma}{\underset{H}{-\overset{|}{C}}}-\overset{\beta}{\overset{|}{C}}=\overset{\alpha}{\overset{|}{C}}-\overset{|}{C}=O$$

as, for example, in crotonaldehyde? (b) In view of your answer to (a), suggest a way to synthesize 5-phenyl-2,4-pentadienal, $C_6H_5CH{=}CH{-}CH{=}CH{-}CHO$.

21.9 Reactions related to the aldol condensation

There are a large number of condensations that are closely related to the aldol condensation. Each of these reactions has its own name—*Perkin, Knoevenagel, Doebner, Claisen, Dieckmann*, for example—and at first glance each may seem quite different from the others. Closer examination shows, however, that like the aldol condensation each of these involves attack by a carbanion on a carbonyl group. In each case the carbanion is generated in very much the same way: the abstraction by base of a hydrogen ion *alpha* to a carbonyl group. Different bases may be used—sodium hydroxide, sodium ethoxide, sodium acetate, amines—and the carbonyl group to which the hydrogen is *alpha* may vary—aldehyde, ketone, anhydride, ester—but the chemistry is essentially the same as that of the aldol condensation. We shall take up a few of these condensations in the following problems and in following sections; in doing this, we must not lose sight of the fundamental resemblance of each of them to the aldol condensation.

Problem 21.21 Esters can be condensed with aromatic aldehydes in the presence of alkoxides; thus benzaldehyde and ethyl acetate, in the presence of sodium ethoxide, give ethyl cinnamate, $C_6H_5CH{=}CHCOOC_2H_5$. Show all steps in the most likely mechanism for this condensation.

Problem 21.22 Account for the following reactions:

(a) $C_6H_5CHO + CH_3NO_2 \xrightarrow{KOH} C_6H_5CH{=}CHNO_2 + H_2O$

(b) $C_6H_5CHO + C_6H_5CH_2CN \xrightarrow{NaOC_2H_5} C_6H_5CH{=}C(C_6H_5){-}CN + H_2O$

(c) $C_6H_5CHO + CH_3C_6H_3(NO_2)_2 \xrightarrow{2^\circ\ amine} C_6H_5CH{=}CHC_6H_3(NO_2)_2 + H_2O$ (2,4-dinitrotoluene → 2,4-dinitrostilbene; NO_2 groups ortho and para to the side chain)

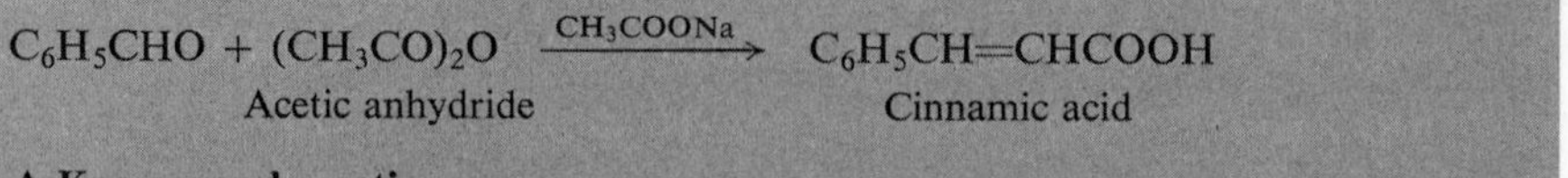

(d) $CH_3CHO + NaC{\equiv}CH \xrightarrow{NH_3(l)} CH_3CH(ONa)C{\equiv}CH \xrightarrow{NH_4Cl} CH_3CH(OH)C{\equiv}CH$

(e) **A Perkin condensation**:

$$\underset{\text{Acetic anhydride}}{C_6H_5CHO + (CH_3CO)_2O} \xrightarrow{CH_3COONa} \underset{\text{Cinnamic acid}}{C_6H_5CH{=}CHCOOH}$$

(f) **A Knoevenagel reaction**:

$$C_6H_5CHO + CH_2(COOC_2H_5)_2 \xrightarrow{2^\circ\ amine} C_6H_5CH{=}C(COOC_2H_5)_2$$

(g) **A Cope reaction**:

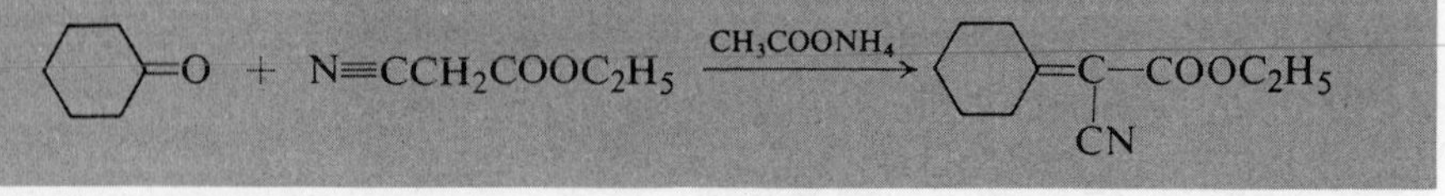

$$\text{cyclohexanone} + N{\equiv}CCH_2COOC_2H_5 \xrightarrow{CH_3COONH_4} C_6H_{10}{=}C(CN){-}COOC_2H_5$$

21.10 The Wittig reaction

In 1954, Georg Wittig (then at the University of Tübingen) reported a method of synthesizing alkenes from carbonyl compounds, which amounts to the replacement of carbonyl oxygen, =O, by the group =CRR′. The heart of the synthesis

$$\gt C{=}O + Ph_3P{=}\underset{\text{An ylide}}{\overset{R'}{\overset{|}{C}}}{-}R \longrightarrow \underset{\text{A betaine}}{-\underset{{}^-O}{\overset{|}{\underset{|}{C}}}-\underset{\overset{+}{P}Ph_3}{\overset{R'}{\overset{|}{\underset{|}{C}}}}-R} \longrightarrow -\overset{|}{C}{=}\overset{R'}{\overset{|}{C}}{-}R + \underset{\text{Triphenylphosphine oxide}}{Ph_3PO}$$

is the nucleophilic attack on carbonyl carbon by an *ylide* to form a *betaine* which—often spontaneously—undergoes elimination to yield the product. For example:

$$\underset{\text{Benzophenone}}{(C_6H_5)_2C{=}O} + \underset{\text{Methylenetriphenylphosphorane}}{Ph_3P{=}CH_2} \longrightarrow (C_6H_5)_2\underset{{}^-O}{\underset{|}{C}}-\underset{\overset{+}{P}Ph_3}{\underset{|}{CH_2}} \longrightarrow \underset{\text{1,1-Diphenylethene}}{(C_6H_5)_2C{=}CH_2}$$

$$\underset{\text{Benzaldehyde}}{C_6H_5CHO} + C_6H_5CH{=}CH{-}CH{=}PPh_3 \longrightarrow C_6H_5\underset{{}^-O}{\underset{|}{CH}}-\underset{\overset{+}{P}Ph_3}{\underset{|}{CH}}-CH{=}CHC_6H_5 \longrightarrow$$

$$\underset{\text{1,4-Diphenyl-1,3-butadiene}}{C_6H_5CH{=}CH{-}CH{=}CHC_6H_5}$$

The reaction is carried out under mild conditions, and the position of the carbon–carbon double bond is not in doubt. Carbonyl compounds may contain a wide variety of substituents, and so may the ylide. (Indeed, in its broadest form, the Wittig reaction involves reactants other than carbonyl compounds, and may lead to products other than substituted alkenes.)

The phosphorus ylides have hybrid structures, and it is the negative charge on

$$\left[Ph_3P{=}\overset{R'}{\overset{|}{C}}{-}R \qquad Ph_3\overset{}{\underset{+}{P}}{-}\underset{-}{\overset{R'}{\overset{|}{C}}}{-}R \right]$$

carbon—the carbanion character of ylides—that is responsible for their characteristic reactions: in this case, nucleophilic attack on carbonyl carbon.

The preparation of ylides is a two-stage process, each stage of which belongs to a familiar reaction type: nucleophilic attack on an alkyl halide, and abstraction of a proton by a base.

$$R{-}\overset{R'}{\overset{|}{C}}HX + \underset{\text{Triphenylphosphine}}{Ph_3P} \longrightarrow \underset{\text{A phosphonium salt}}{Ph_3\overset{+}{P}{-}\overset{R'}{\overset{|}{C}}H{-}R\ X^-} \xrightarrow{\text{base}} \underset{\text{An ylide}}{Ph_3P{=}\overset{R'}{\overset{|}{C}}{-}R} + \text{base:H}$$

Many different bases have been used—chiefly alkoxides and organometallics—and in a variety of solvents. For example:

$$CH_3Br + Ph_3P \longrightarrow Ph_3\overset{+}{P}{-}CH_3\,Br^- \xrightarrow[\text{THF}]{C_6H_5Li} Ph_3P{=}CH_2 + C_6H_6 + LiBr$$

$$CH_2{=}CHCH_2Cl + Ph_3P \longrightarrow Ph_3\overset{+}{P}{-}CH_2CH{=}CH_2\ Cl^- \xrightarrow[DMF]{NaOEt} Ph_3P{=}CHCH{=}CH_2$$

Problem 21.23 What side reactions would you expect to encounter in the preparation of an ylide like $Ph_3P{=}C(CH_3)CH_2CH_3$?

Problem 21.24 Give the structure of an ylide and a carbonyl compound from which each of the following could be made.

(a) $CH_3CH_2CH_2CH{=}C(CH_3)CH_2CH_3$
(b) $C_6H_5C(CH_3){=}CHCH_2C_6H_5$
(c) $C_6H_5CH{=}CHC_6H_5$

(d) cyclopentylidene$=CHCH_3$

(e) 1,4-diphenyl-1,3-butadiene (an alternative to the set of reagents used on p. 715)
(f) $CH_2{=}CHCH{=}C(CH_3)COOCH_3$

Problem 21.25 Outline all steps in a possible laboratory synthesis of each ylide and each carbonyl compound in the preceding problem, starting from benzene, toluene, alcohols of four carbons or fewer, acetic anhydride, triphenylphosphine, and cyclopentanol, and using any needed inorganic reagents.

Problem 21.26 Give the structures of compounds A–C.

$C_6H_5OCH_2Cl + Ph_3P$, then t-BuOK $\longrightarrow$ A ($C_{25}H_{21}OP$)
A + methyl ethyl ketone $\longrightarrow$ Ph_3PO + B ($C_{11}H_{14}O$)
B + dilute aqueous acid $\longrightarrow$ C ($C_5H_{10}O$)

The above sequence offers a general route to what class of compounds?

Problem 21.27 Give the structures of compounds D–F.

(a) $C_6H_5COCH_2CH_2CH_2CH_2Br + Ph_3P$, then NaOEt $\longrightarrow$ D ($C_{11}H_{12}$)
(b) $BrCH_2CH_2CH_2Br + Ph_3P$, then base $\longrightarrow$ E ($C_{39}H_{34}P_2$)
E + o-$C_6H_4(CHO)_2$ $\longrightarrow$ F ($C_{11}H_{10}$)

Problem 21.28 Give the structures of compounds G and H, and account for the stereochemistry of each step.

trans-2-octene + $C_6H_5CO_2OH$ $\longrightarrow$ G ($C_8H_{16}O$)
G + Ph_2PLi, then CH_3I $\longrightarrow$ H ($C_{21}H_{29}OP$)
H $\longrightarrow$ *cis*-2-octene

21.11 Claisen condensation. Formation of β-keto esters

An α-hydrogen in an ester, like an α-hydrogen in an aldehyde or ketone, is weakly acidic, and for the same reason. through resonance, the carbonyl group helps accommodate the negative charge of the carbanion. Let us look at an exceedingly important reaction of esters that depends upon the acidity of α-hydrogens. It is—for esters—the exact counterpart of the aldol condensation; reaction takes a different turn at the end, but a turn that is typical of the chemistry of acyl compounds.

When ethyl acetate is treated with sodium ethoxide, and the resulting mixture is acidified, there is obtained ethyl β-ketobutyrate (ethyl 3-oxobutanoate), generally known as **ethyl acetoacetate** or **acetoacetic ester**:

$$\underset{\substack{\text{Ethyl acetate}\\ \textit{2 moles}}}{2CH_3COOC_2H_5} + \underset{\substack{\text{Sodium}\\ \text{ethoxide}}}{Na^{+\,-}OC_2H_5} \xrightarrow{C_2H_5OH} \underset{\text{Sodioacetoacetic ester}}{CH_3COCHCOOC_2H_5{}^-Na^+} + 2C_2H_5OH$$

$$\downarrow H^+$$

$$\underset{\substack{\text{Ethyl acetoacetate}\\ \text{Acetoacetic ester}\\ \textit{A β-keto ester}}}{CH_3\overset{\beta}{C}OCH_2COOC_2H_5}$$

Ethyl acetoacetate is the ester of a *β-keto acid*; its preparation illustrates the reaction known as the **Claisen condensation.**

The generally accepted mechanism for the Claisen condensation (shown here for ethyl acetate) is:

(1) $$CH_3COOC_2H_5 + {}^-OC_2H_5 \rightleftharpoons C_2H_5OH + \underset{\text{I}}{{}^-CH_2COOC_2H_5}$$

(2) $$CH_3-\overset{\overset{O}{\|}}{C}-OC_2H_5 + \underset{\text{I}}{{}^-CH_2COOC_2H_5} \rightleftharpoons CH_3-\overset{\overset{O^-}{|}}{\underset{\underset{OC_2H_5}{|}}{C}}-CH_2COOC_2H_5$$

$$\updownarrow$$

$$CH_3\overset{\overset{O}{\|}}{C}CH_2COOC_2H_5 + {}^-OC_2H_5$$

(3) $$\underset{\text{Stronger acid}}{CH_3\overset{\overset{O}{\|}}{C}CH_2COOC_2H_5} + {}^-OC_2H_5 \rightleftharpoons CH_3COCHCOOC_2H_5{}^- + \underset{\text{Weaker acid}}{C_2H_5OH}$$

Ethoxide ion abstracts (step 1) a hydrogen ion from the α-carbon of the ester to form carbanion I. The powerfully nucleophilic carbanion I attacks (step 2) the carbonyl carbon of a second molecule of ester to displace ethoxide ion and yield the keto ester.

Like the aldol condensation and related reactions, the Claisen condensation involves nucleophilic attack by a carbanion on an electron-deficient carbonyl carbon. *In the aldol condensation, nucleophilic attack leads to addition, the typical reaction of aldehydes and ketones; in the Claisen condensation, nucleophilic attack leads to substitution, the typical reaction of acyl compounds* (Sec. 20.4).

When reaction is complete there is present, not acetoacetic ester, but its sodium salt, *sodioacetoacetic ester*. The α-hydrogens of acetoacetic ester are located *alpha* to *two* carbonyl groups, and hence ionization yields a particularly stable carbanion in which two carbonyl groups help accommodate the charge. As a result acetoacetic ester is a much stronger acid than ordinary esters or other compounds containing a single carbonyl group. It is considerably stronger than ethyl alcohol, and hence it reacts (step 3) with ethoxide ion to form ethyl alcohol and the anion

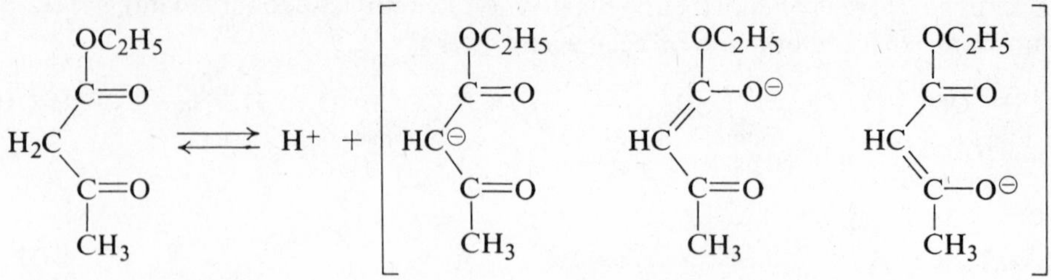

$$\underset{\text{Acetoacetic ester}}{\begin{matrix} OC_2H_5 \\ | \\ C{=}O \\ / \\ H_2C \\ \backslash \\ C{=}O \\ | \\ CH_3 \end{matrix}} \rightleftharpoons H^+ + \left[\begin{matrix} OC_2H_5 \\ | \\ C{=}O \\ / \\ HC^{\ominus} \\ \backslash \\ C{=}O \\ | \\ CH_3 \end{matrix} \quad \begin{matrix} OC_2H_5 \\ | \\ C{-}O^{\ominus} \\ /\!/ \\ HC \\ \backslash \\ C{=}O \\ | \\ CH_3 \end{matrix} \quad \begin{matrix} OC_2H_5 \\ | \\ C{=}O \\ / \\ HC \\ \backslash\!\backslash \\ C{-}O^{\ominus} \\ | \\ CH_3 \end{matrix} \right]$$

equivalent to

$$\begin{matrix} OC_2H_5 \\ | \\ C{\overset{...}{=}}O \\ HC \\ C{\overset{...}{=}}O \\ | \\ CH_3 \end{matrix} \Big\} \ominus$$

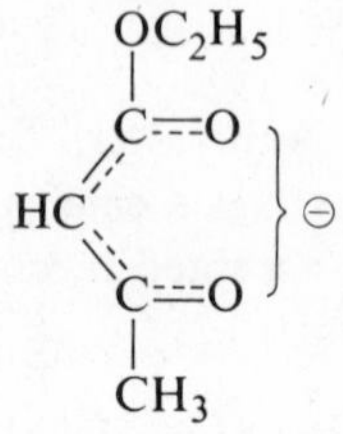

of sodioacetoacetic ester. Formation of the salt of acetoacetic ester is essential to the success of the reaction; of the various equilibria involved in the reaction, only (3) is favorable to the product we want.

Problem 21.29 Better yields are obtained if the Claisen condensation is carried out in ether with alcohol-free sodium ethoxide as catalyst instead of in ethyl alcohol solution. How do you account for this?

As we might expect, the Claisen condensation of more complicated esters yields the products resulting from ionization of an α-hydrogen of the ester; as a result, it is always the α-carbon of one molecule that becomes attached to the carbonyl carbon of another. For example:

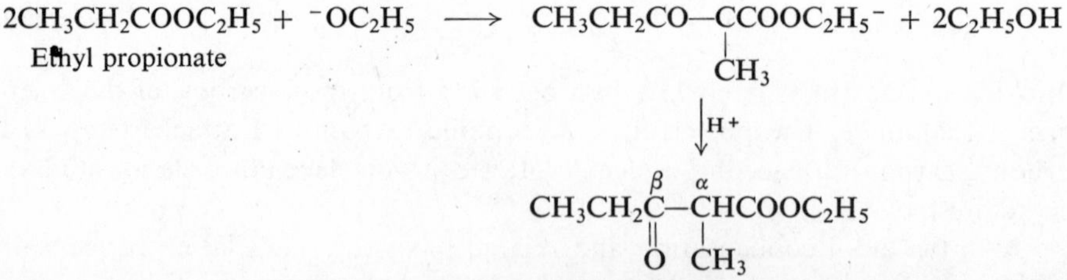

$$\underset{\text{Ethyl propionate}}{2CH_3CH_2COOC_2H_5} + {}^-OC_2H_5 \longrightarrow CH_3CH_2CO{-}\underset{\underset{CH_3}{|}}{C}COOC_2H_5{}^- + 2C_2H_5OH$$

$$\downarrow H^+$$

$$CH_3CH_2\overset{\beta}{\underset{\underset{O}{\|}}{C}}{-}\overset{\alpha}{\underset{\underset{CH_3}{|}}{C}}HCOOC_2H_5$$

Ethyl 3-oxo-2-methylpentanoate
Ethyl α-methyl-β-ketovalerate
A β-keto ester

Problem 21.30 Sodium ethoxide converts ethyl adipate into 2-carbethoxycyclopentanone (II). This is an example of the **Dieckmann condensation.**

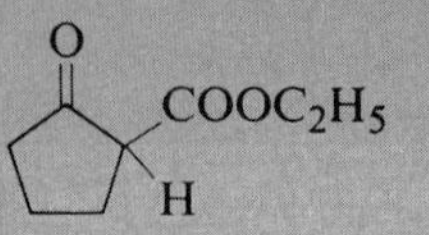

[Structure: cyclopentanone ring with C=O; adjacent carbon bearing $COOC_2H_5$ and H]

II

(a) How do you account for formation of II? (b) What product would you expect from the action of sodium ethoxide on ethyl pimelate (ethyl heptanedioate)? (c) Would you expect similar behavior from ethyl glutarate or ethyl succinate? Actually, ethyl succinate reacts with sodium ethoxide to yield a compound of formula $C_{12}H_{16}O_6$ containing a six-membered ring. What is the likely structure for this last product?

21.12 Crossed Claisen condensation

Like a crossed aldol condensation (Sec. 21.8), a **crossed Claisen condensation** is generally feasible only when one of the reactants has no α-hydrogens and thus is incapable of undergoing self-condensation. For example:

$$C_6H_5COOC_2H_5 + CH_3COOC_2H_5 \xrightarrow{^-OC_2H_5} C_6H_5\text{—}\underset{\underset{O}{\|}}{C}\text{—}CH_2COOC_2H_5 + C_2H_5OH$$

Ethyl benzoate; Ethyl acetate; Ethyl benzoylacetate

$$HCOOC_2H_5 + CH_3COOC_2H_5 \xrightarrow{^-OC_2H_5} H\text{—}\underset{\underset{O}{\|}}{C}\text{—}CH_2COOC_2H_5 + C_2H_5OH$$

Ethyl formate; Ethyl acetate; Ethyl formylacetate (known only as the Na salt)

$$\begin{array}{l}COOC_2H_5\\ |\\ COOC_2H_5\end{array} + CH_3COOC_2H_5 \xrightarrow{^-OC_2H_5} C_2H_5OOC\text{—}\underset{\underset{O}{\|}}{C}\text{—}CH_2COOC_2H_5 + C_2H_5OH$$

Ethyl oxalate; Ethyl acetate; Ethyl oxaloacetate

$$C_2H_5O\text{—}\underset{\underset{O}{\|}}{C}\text{—}OC_2H_5 + C_6H_5CH_2COOC_2H_5 \xrightarrow{^-OC_2H_5} C_2H_5O\text{—}\underset{\underset{O}{\|}}{C}\text{—}\overset{\overset{C_6H_5}{|}}{C}HCOOC_2H_5 + C_2H_5OH$$

Ethyl carbonate; Ethyl phenylacetate; Ethyl phenylmalonate, Phenylmalonic ester

Problem 21.31 In what order should the reactants be mixed in each of the above crossed Claisen condensations? (*Hint:* See Sec. 21.8.)

Problem 21.32 Ketones (but not aldehydes) undergo a crossed Claisen condensation with esters. For example:

$$CH_3COOC_2H_5 + CH_3COCH_3 \xrightarrow{NaOC_2H_5} CH_3COCH_2COCH_3 + C_2H_5OH$$

Ethyl acetate; Acetone; Acetylacetone

(a) Outline all steps in the most likely mechanism for this reaction. (b) Predict the principal products expected from the reaction in the presence of sodium ethoxide of ethyl propionate and acetone; (c) of ethyl benzoate and acetophenone; (d) of ethyl oxalate and cyclohexanone.

Problem 21.33 Outline the synthesis from simple esters of:

(a) ethyl α-phenylbenzoylacetate, $C_6H_5COCH(C_6H_5)COOC_2H_5$
(b) ethyl 2,3-dioxo-1,4-cyclopentanedicarboxylate (I). (*Hint:* Use ethyl oxalate as one ester.)

(c) ethyl 1,3-dioxo-2-indanecarboxylate (II)

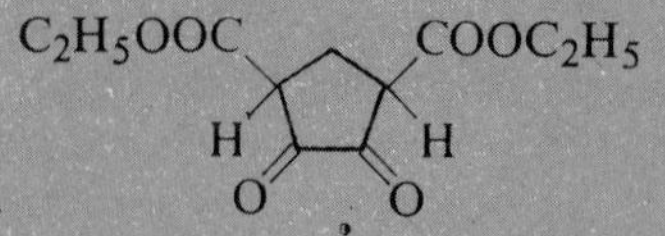

$$\text{I: cyclopentane ring with } C_2H_5OOC\text{, H and } COOC_2H_5\text{, H on adjacent carbons, and two adjacent C=O}$$

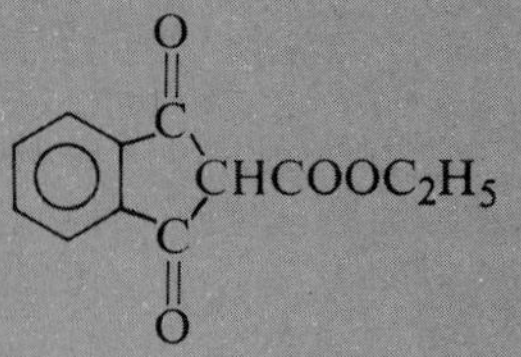

$$\text{II: benzene ring fused to } \overset{O}{\overset{\|}{C}}-CHCOOC_2H_5-\underset{O}{\underset{\|}{C}}$$

21.13 Reformatsky reaction. Preparation of β-hydroxy esters

In the Claisen condensation, we have just seen, carbanions are generated from esters through abstraction of an α-hydrogen by base. But we are familiar with another way of generating carbanions—or rather, groups with considerable carbanion character: through formation of organometallic compounds. This approach, too, plays a part in the chemistry of esters.

If an α-bromo ester is treated with metallic zinc in the presence of an aldehyde or ketone, there is obtained a β-hydroxy ester. This reaction, known as the **Reformatsky reaction**, is the most important method of preparing β-hydroxy acids and their derivatives. For example:

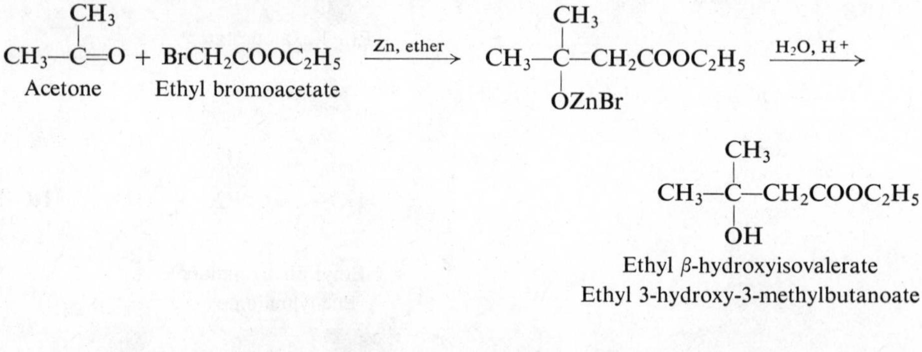

$$\underset{\text{Acetone}}{CH_3-\overset{CH_3}{\overset{|}{C}}=O} + \underset{\text{Ethyl bromoacetate}}{BrCH_2COOC_2H_5} \xrightarrow{\text{Zn, ether}} CH_3-\overset{CH_3}{\overset{|}{\underset{OZnBr}{\underset{|}{C}}}}-CH_2COOC_2H_5 \xrightarrow{H_2O,\ H^+}$$

$$\underset{\substack{\text{Ethyl } \beta\text{-hydroxyisovalerate} \\ \text{Ethyl 3-hydroxy-3-methylbutanoate}}}{CH_3-\overset{CH_3}{\overset{|}{\underset{OH}{\underset{|}{C}}}}-CH_2COOC_2H_5}$$

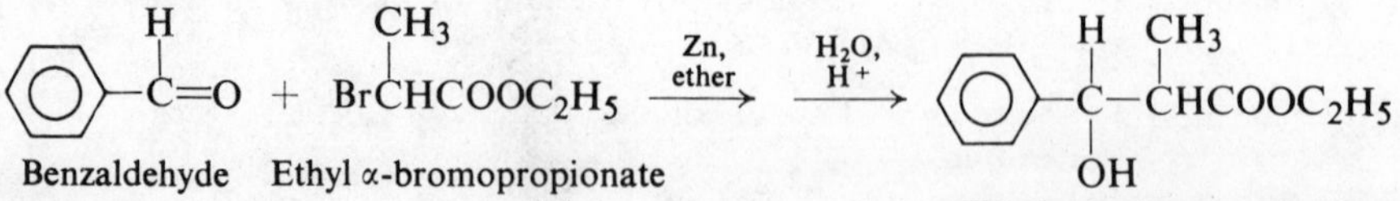

$$\underset{\text{Benzaldehyde}}{C_6H_5-\overset{H}{\overset{|}{C}}=O} + \underset{\text{Ethyl } \alpha\text{-bromopropionate}}{Br\overset{CH_3}{\overset{|}{C}}HCOOC_2H_5} \xrightarrow{\text{Zn, ether}} \xrightarrow{H_2O,\ H^+} \underset{\substack{\text{Ethyl } \beta\text{-hydroxy-}\beta\text{-phenyl-} \\ \alpha\text{-methylpropionate}}}{C_6H_5-\overset{H}{\overset{|}{\underset{OH}{\underset{|}{C}}}}-\overset{CH_3}{\overset{|}{C}}HCOOC_2H_5}$$

The α-bromo ester and zinc react in absolute ether to yield an intermediate organozinc compound, which then adds to the carbonyl group of the aldehyde or ketone. The formation and subsequent reaction of the organozinc compound is similar to the formation and reaction of a Grignard reagent. Zinc is used in place of magnesium simply because the organozinc compounds are less reactive than Grignard reagents; they do not react with the ester function but only with the aldehyde or ketone.

$$\underset{\text{Ethyl bromoacetate}}{BrCH_2COOC_2H_5} \xrightarrow{Zn} BrZnCH_2COOC_2H_5$$

$$BrZnCH_2COOC_2H_5 + \underset{\text{Acetone}}{CH_3-\overset{\displaystyle CH_3}{\overset{|}{C}}=O} \longrightarrow CH_3-\overset{\displaystyle CH_3}{\overset{|}{C}}(-OZnBr)-CH_2COOC_2H_5 \xrightarrow{H^+} \underset{\text{Ethyl }\beta\text{-hydroxyisovalerate}}{CH_3-\overset{\displaystyle CH_3}{\overset{|}{C}}(-OH)-CH_2COOC_2H_5}$$

The Reformatsky reaction takes place only with esters containing bromine in the *alpha* position, and hence necessarily yields *beta*-hydroxy esters. By the proper

$$R-\overset{\displaystyle R'}{\overset{|}{C}}=O + BrC(R'')(H)COOC_2H_5 \qquad \text{R, R}', \text{R}'' \textit{ may be } \text{H, alkyl, } \textit{or} \text{ aryl}$$

$$\xrightarrow{Zn} \xrightarrow{H^+} R-C(R')(HO)-C(R'')(H)-COOC_2H_5 \xrightarrow{-H_2O} R-C(R')=C(R'')-COOC_2H_5 \xrightarrow{H_2,\ Ni} R-C(R')(H)-C(R'')(H)-COOC_2H_5 \xrightarrow{\text{hydrolysis}} R-C(R')(H)-C(R'')(H)-COOH$$

selection of ester and carbonyl compound, a wide variety of rather complicated β-hydroxy carboxylic acids can be prepared.

Like β-hydroxyaldehydes and -ketones, β-hydroxyesters and -acids are readily dehydrated. The unsaturated compounds thus obtained (chiefly α,β-unsaturated) can be hydrogenated to saturated carboxylic acids. Extended in this way, the Reformatsky reaction is a useful general method for preparing carboxylic acids, paralleling the aldol route to alcohols.

In planning the synthesis of a carboxylic acid by the Reformatsky reaction, our problem is to select the proper starting materials; to do this, we have only to look at the structure of the product we want. For example:

Acid wanted:	*Requires:*	*Starting materials:*
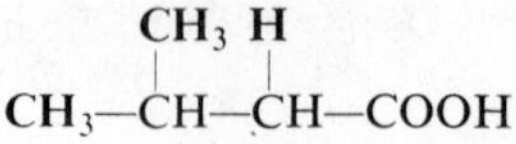 $CH_3-CH(CH_3)-CH(H)-COOH$	R = CH_3-; R′ = CH_3-; R″ = H—	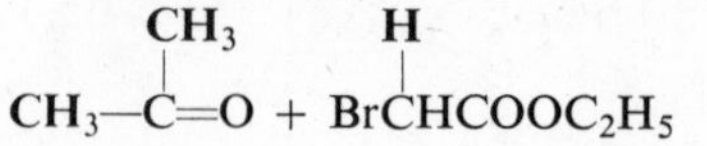 $CH_3-C(CH_3)=O + BrCH(H)COOC_2H_5$
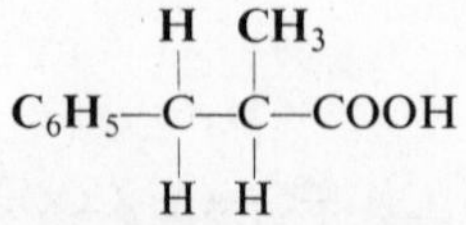 $C_6H_5-CH_2-CH(CH_3)-COOH$	R = C_6H_5-; R′ = H—; R″ = CH_3-	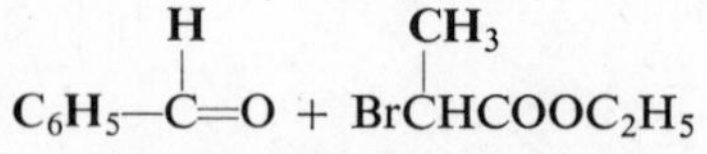 $C_6H_5-C(H)=O + BrCH(CH_3)COOC_2H_5$

Problem 21.34 Outline the synthesis of the following acids via the Reformatsky reaction:

(a) *n*-valeric acid; (b) α,γ-dimethylvaleric acid; (c) cinnamic acid; (d) α-methyl-β-phenylpropionic acid.

Problem 21.35 Outline the synthesis of the following, starting from benzaldehyde and ethyl bromoacetate:

(a) $C_6H_5CH_2CH_2COOH$
(b) $C_6H_5CH_2CH_2CHO$
(c) $C_6H_5CH_2CH_2CH_2CH_2COOH$

Problem 21.36 Give structures of compounds A, B, and C:

ethyl oxalate + ethyl acetate + sodium ethoxide $\longrightarrow$ A ($C_8H_{12}O_5$)
A + ethyl bromoacetate + Zn, then H_2O $\longrightarrow$ B ($C_{12}H_{20}O_7$)
B + OH^- + heat, then H^+ $\longrightarrow$ C ($C_6H_8O_7$), *citric acid*

PROBLEMS

1. Write balanced equations, naming all organic products, for the reaction (if any) of phenylacetaldehyde with:

(a) dilute NaOH
(b) dilute HCl
(c) aqueous Na_2CO_3
(d) Br_2/CCl_4
(e) $Ph_3P{=}CH_2$

2. Answer Problem 1 for cyclohexanone.

3. Write balanced equations, naming all organic products, for the reaction (if any) of benzaldehyde with:

(a) dilute NaOH
(b) conc. NaOH
(c) acetaldehyde, dilute NaOH
(d) propionaldehyde, dilute NaOH
(e) acetone, dilute NaOH
(f) product (e), dilute NaOH
(g) acetophenone, NaOH
(h) acetic anhydride, sodium acetate, heat
(i) ethyl acetate, sodium ethoxide
(j) ethyl phenylacetate, sodium ethoxide
(k) formaldehyde, conc. NaOH
(l) crotonaldehyde, NaOH
(m) $Ph_3P{=}CHCH{=}CH_2$
(n) $Ph_3P{=}CH(OC_6H_5)$
(o) product (n), dilute acid

4. Write equations for all steps in the synthesis of the following from propionaldehyde, using any other needed reagents:

(a) α-methyl-β-hydroxyvaleraldehyde
(b) 2-methyl-1-pentanol
(c) 2-methyl-2-pentenal
(d) 2-methyl-2-penten-1-ol
(e) 2-methyl-1,3-pentanediol
(f) α-methylvaleric acid
(g) 2-methyl-3-phenylpropenal
(h) CH_3CD_2CHO
(i) $CH_3CH_2CH^{18}O$
(j) 2-methyl-3-hexene

5. Write equations for all steps in the synthesis of the following from acetophenone, using any other needed reagents:

(a) benzoic acid
(b) 1,3-diphenyl-2-buten-1-one
(c) 1,3-diphenyl-1-butanol
(d) 1,3-diphenyl-2-buten-1-ol
(e) 1,3-diphenyl-2-propen-1-one
(f) α-phenylpropionaldehyde (*Hint:* See Problem 21.26.)

6. Give the structures of the principal products expected from the reaction in the presence of sodium ethoxide of:

(a) ethyl *n*-butyrate
(b) ethyl phenylacetate
(c) ethyl isovalerate
(d) ethyl formate and ethyl propionate
(e) ethyl oxalate and ethyl succinate
(f) ethyl benzoate and ethyl phenylacetate
(g) ethyl propionate and cyclohexanone
(h) ethyl phenylacetate and acetophenone
(i) ethyl carbonate and acetophenone

7. Sodium ethoxide is added to a mixture of ethyl acetate and ethyl propionate (a) Give the structures of the products expected. (b) Would this reaction be a good method of synthesizing any one of these?

8. Outline all steps in a possible synthesis of each of the following via the Claisen condensation, using any needed reagents:

(a) $C_6H_5COCH(CH_3)COOC_2H_5$
(b) $C_6H_5CH_2COCH(C_6H_5)COOC_2H_5$
(c) $C_2H_5OOCCOCH(CH_3)COOC_2H_5$
(d) $C_6H_5CH(CHO)COOC_2H_5$
(e) $(CH_3)_2CHCOCH_2COCH_3$
(f) $C_6H_5COCH_2COCH_3$
(g) 2-benzoylcyclohexanone
(h) $C_2H_5OOCCH(CHO)CH_2COOC_2H_5$

9. The cinnamic acid obtained by the Perkin condensation is the more stable *trans*-isomer. Suggest a method of preparing *cis*-cinnamic acid. (*Hint:* See Sec. 8.9.)

10. Outline all steps in a possible laboratory synthesis of each of the following from benzene, toluene, acetic anhydride, triphenylphosphine, and alcohols of four carbons or fewer, using any needed inorganic reagents:

(a) 4-methyl-4-hydroxy-2-pentanone
(b) 4-methyl-2-pentanol
(c) crotonaldehyde, $CH_3CH{=}CHCHO$
(d) cinnamyl alcohol, $C_6H_5CH{=}CHCH_2OH$
(e) *p*-nitrocinnamaldehyde
(f) 1,3-butanediol
(g) 3-methyl-2-butenoic acid (via aldol condensation)
(h) 3-methyl-2-butenoic acid (a second way)
(i) 3-methyl-1-pentyn-3-ol (*Oblivon*, a hypnotic)
(j) 1-phenyl-1,3,5-hexatriene
(k) 1,6-diphenyl-1,3,5-hexatriene
(l) 2,3-dimethyl-2-pentenoic acid
(m) 3-hydroxy-4-phenylbutanoic acid
(n) α,α-dimethylcaproic acid
(o) indanone (I)
(p) racemic *erythro*-2,3-dihydroxy-3-phenylpropanoic acid (II and its enantiomer)

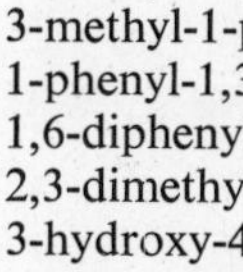

O

I

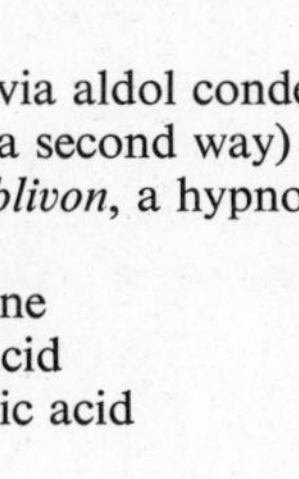

COOH
H—OH
H—OH
C_6H_5

II

$CH_3-HC-CH-COOH$
O CH_2
C
O

γ-Methylparaconic acid

11. How do you account for the formation of γ-methylparaconic acid (previous page) from the reaction of acetaldehyde with succinic acid?

12. Considerable quantities of acetone are consumed in the manufacture of methyl isobutyl ketone (MIBK). How do you think the synthesis of MIBK is accomplished?

13. Methyl ethyl ketone can be made to undergo the Claisen condensation with a given ester to yield either of two products, depending upon experimental conditions. (a) What are these two products? (b) How could you tell quickly and simply which product you had obtained? (*Note:* Use ethyl benzoate as the ester.)

14. The acetylenic ester $CH_3—C{\equiv}C—COOC_2H_5$ can be converted into ethyl acetoacetate. (a) How? (b) Outline a synthesis of the acetylenic ester from acetylene and any needed reagents.

15. The compound *pentaerythritol*, $C(CH_2OH)_4$, used in making explosives, is obtained from the reaction of acetaldehyde and formaldehyde in the presence of calcium hydroxide. Outline the probable steps in this synthesis.

16. The labeled alkene, 1,3,3-trideuteriocyclohexene, needed for a particular stereochemical study, was prepared from cyclohexanone. Outline all steps in such a synthesis.

17. (a) The haloform test (Sec. 16.11) depends upon the fact that three hydrogens on the same carbon atom are successively replaced by halogen. Using acetone as an example, show why the carbon that suffers the initial substitution should be the preferred site of further substitution. (*Hint:* See Sec. 18.14.)

(b) The haloform test also depends upon the ease with which the trihalomethyl ketone produced in (a) is cleaved by base. What is the most likely mechanism for this cleavage? What factor makes such a reaction possible in this particular case?

18. Upon treatment with dilute NaOH, β-methylcrotonaldehyde, $(CH_3)_2C{=}CHCHO$, yields a product of formula $C_{10}H_{14}O$, called *dehydrocitral*. What is a likely structure for this product, and how is it formed? (*Hint:* See *citral*, Problem 26, p. 652.)

19. As part of the total synthesis of vitamin D_3, compound III was converted into IV by a number of stages, two of which involved use of the Wittig reaction. Show how this conversion might have been carried out.

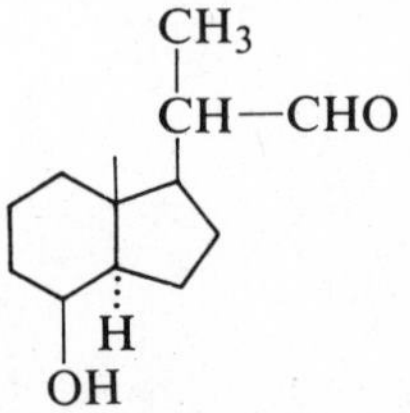

CH3
CH—CHO
H
OH

III

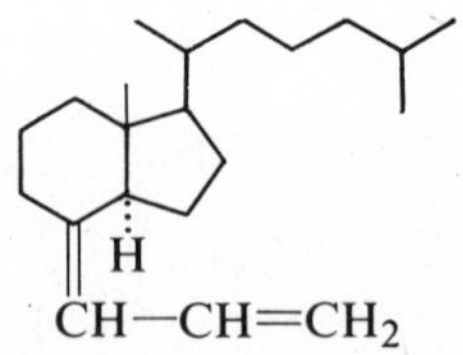

H
CH—CH=CH2

IV

20. Meanwhile, back at the laboratory, our naïve graduate student (Problem 18, p. 650) had need of the hydroxy ester $(CH_3)_2C(OH)CH_2COOC_2H_5$. Turning once again to the Grignard reaction, he prepared methylmagnesium iodide and to it he added acetoacetic ester. Everything went well; indeed, even without the application of heat, the reaction mixture bubbled merrily. Working carefully and with great skill, he isolated an excellent yield of the starting material, acetoacetic ester. He poured this down the sink and fled, sobbing, to his research director's office, where he begged for a new research problem.

What reaction had taken place? What was the bubbling due to? (In Problem 12, p. 881, we shall see how he made out with his new research problem.)

21. In contrast to simple carbonyl compounds, 1,3-dicarbonyl compounds—like acetoacetic ester or 2,4-pentanedione (acetylacetone)—exist to an appreciable extent in the enol form.

(a) Pure samples of keto and enol forms of acetoacetic ester have been isolated. Each retained its identity for weeks if acids and bases were carefully excluded. Write equations to show exactly how keto–enol interconversion is speeded up by a base. By an acid.

(b) Draw the structure of the enol form of, say, 2,4-pentanedione. Can you suggest one factor that would tend to stabilize the enol form of such a compound?

(c) Although the enol form of acetoacetic ester is an alcohol, it does *not* have a higher boiling point than the keto form. (Actually, it boils somewhat *lower*.) Can you suggest a second factor that would tend to stabilize the enol form of a 1,3-dicarbonyl compound?

22. Draw the structures (stereochemical where pertinent) of products A and B.

(a) [cyclopentane ring bearing H_3C and $COOC_2H_5$ on one carbon; adjacent carbon bearing H and CH_2COCH_3] + NaOEt, then H_2O $\longrightarrow$ A ($C_{10}H_{14}O_2$), *highly enolic*

(b) methyl ethyl ketone + ethyl oxalate + NaOEt $\longrightarrow$ B ($C_6H_6O_3$)

23. (a) Fig. 21.1(*a*) (below) shows the nmr spectrum of a solution of acetylacetone, $CH_3COCH_2COCH_3$, in chloroform. Besides the peaks shown, there is a small hump, *e*, near δ 15 of about the same area as the peak *d* at δ 5.5. How do you interpret this spectrum? What *quantitative* conclusion can you draw?

(b) Fig. 21.1(*b*) (p. 726) shows the nmr spectrum of benzoylacetone, $C_6H_5COCH_2COCH_3$. There is an additional peak, *d*, near δ 16 of about the same area as the peak *b* at δ 6.1. How do you interpret this spectrum? How do you account for the difference between it and the spectrum in (*a*)?

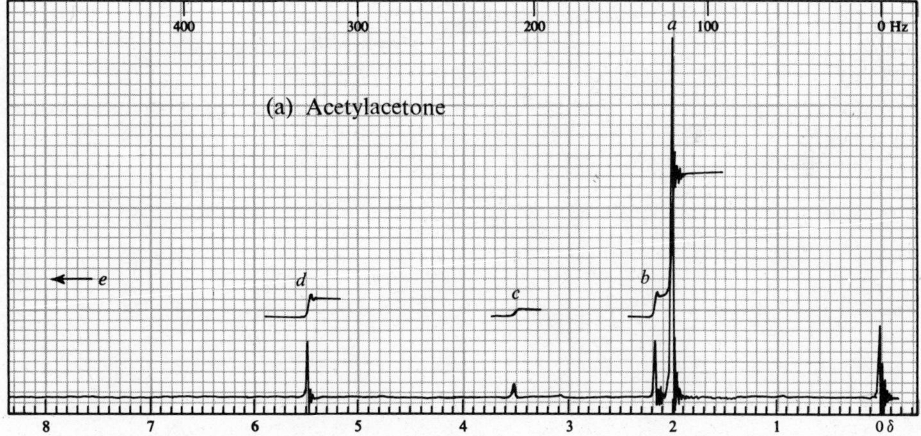

Figure 21.1(*a*). Nmr spectrum of acetylacetone.

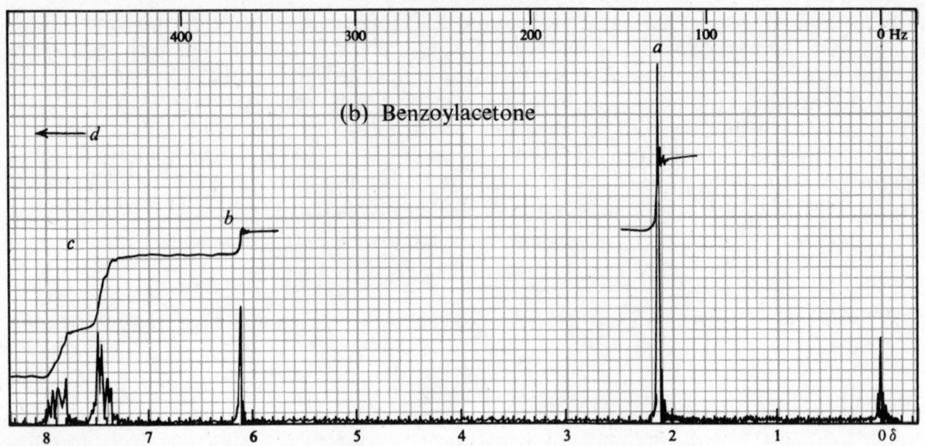

Figure 21.1(*b*). Nmr spectrum of benzoylacetone.

Chapter 22 Amines I. Preparation and Physical Properties

22.1 Structure

Nearly all the organic compounds that we have studied so far are bases, although very weak ones. Much of the chemistry of alcohols, ethers, esters, and even of alkenes and aromatic hydrocarbons is understandable in terms of the basicity of these compounds.

Of the organic compounds that show appreciable basicity (for example, those strong enough to turn litmus blue), by far the most important are the **amines.** An amine has the general formula RNH_2, R_2NH, or R_3N, where R is any alkyl or aryl group. For example:

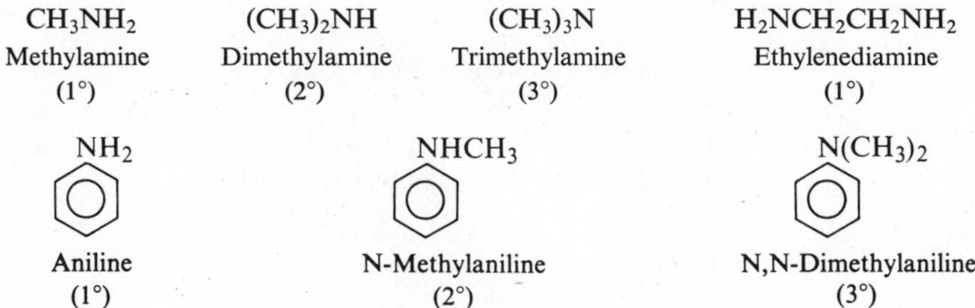

22.2 Classification

Amines are classified as **primary**, **secondary**, or **tertiary**, according to the number of groups attached to the nitrogen atom.

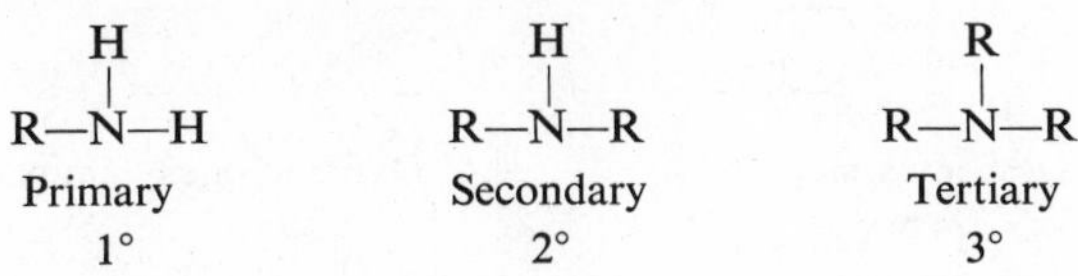

In their fundamental properties—*basicity* and the accompanying *nucleophilicity*—amines of different classes are very much the same. In many of their reactions, however, the final products depend upon the number of hydrogen atoms attached to the nitrogen atom, and hence are different for amines of different classes.

22.3 Nomenclature

Aliphatic amines are named by naming the alkyl group or groups attached to nitrogen, and following these by the word *-amine*. More complicated ones are often named by prefixing *amino–* (or *N-methylamino–*, *N,N-diethylamino–*, etc.) to the name of the parent chain. For example:

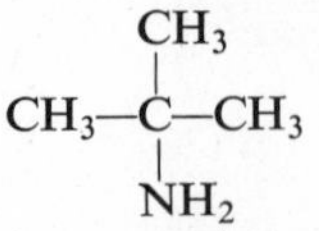

$CH_3-C(CH_3)(NH_2)-CH_3$
tert-Butylamine
(1°)

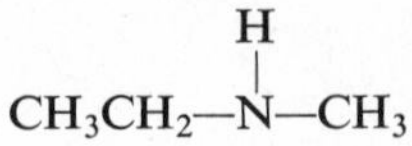

$CH_3CH_2-NH-CH_3$
Methylethylamine
(2°)

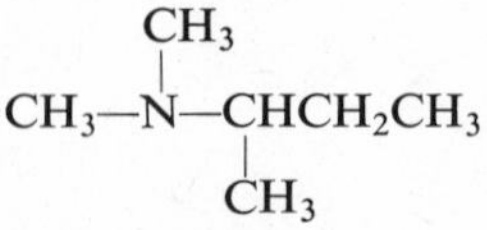

$CH_3-N(CH_3)-CH(CH_3)CH_2CH_3$
Dimethyl-*sec*-butylamine
(3°)

$H_2NCH_2CH_2CH_2COOH$
γ-Aminobutyric acid
(1°)

$H_2NCH_2CH_2OH$
2-Aminoethanol
(Ethanolamine)
(1°)

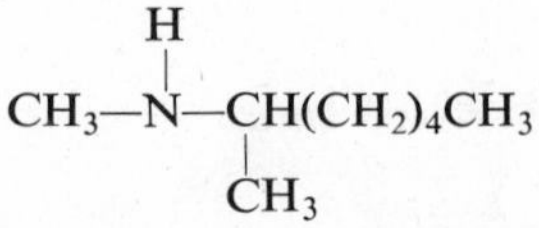

$CH_3-NH-CH(CH_3)(CH_2)_4CH_3$
2-(N-Methylamino)heptane
(2°)

Aromatic amines—those in which nitrogen is attached directly to an aromatic ring—are generally named as derivatives of the simplest aromatic amine, **aniline**. An aminotoluene is given the special name of *toluidine*. For example:

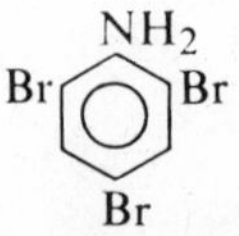

$2,4,6\text{-}Br_3C_6H_2NH_2$
2,4,6-Tribromoaniline
(1°)

$C_6H_5N(CH_3)C_2H_5$
N-Methyl-N-ethylaniline
(3°)

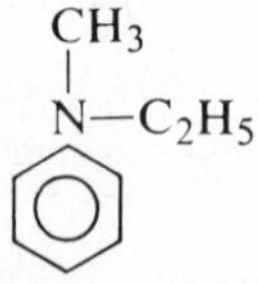

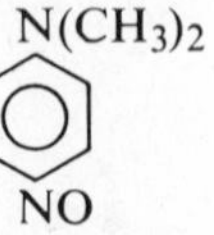

$p\text{-}ON-C_6H_4-N(CH_3)_2$
p-Nitroso-N,N-dimethylaniline
(3°)

$p\text{-}CH_3C_6H_4NH_2$
p-Toluidine
(1°)

$C_6H_5-NH-C_6H_5$
Diphenylamine
(2°)

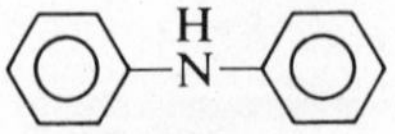

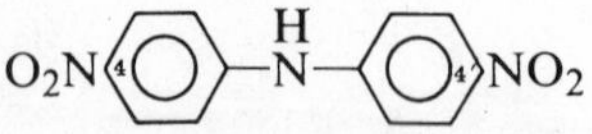

$O_2N-C_6H_4-NH-C_6H_4-NO_2$ (4,4')
4,4′-Dinitrodiphenylamine
(2°)

Salts of amines are generally named by replacing *–amine* by *–ammonium* (or *–aniline* by *–anilinium*), and adding the name of the anion (*chloride*, *nitrate*, *sulfate*, etc.). For example:

$(C_2H_5NH_3^+)_2SO_4^{--}$	$(CH_3)_3NH^+NO_3^-$	$C_6H_5NH_3^+Cl^-$
Ethylammonium sulfate	Trimethylammonium nitrate	Anilinium chloride

22.4 Physical properties of amines

Like ammonia, amines are polar compounds and, except for tertiary amines, can form intermolecular hydrogen bonds. Amines have higher boiling points

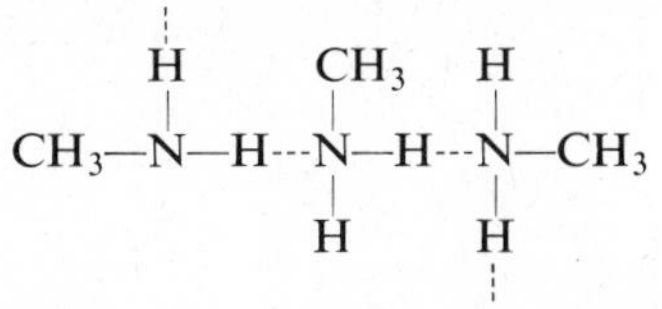

```
      ⋮
      H      CH3   H
      |      |     |
CH3—N—H···N—H···N—CH3
             |     |
             H     H
                   ⋮
```

than non-polar compounds of the same molecular weight, but lower boiling points than alcohols or carboxylic acids.

Amines of all three classes are capable of forming hydrogen bonds with water. As a result, smaller amines are quite soluble in water, with borderline solubility

Table 22.1 AMINES

Name	M.p., °C	B.p., °C	Solub., g/100 g H_2O	K_b
Methylamine	− 92	− 7.5	v.sol.	4.5×10^{-4}
Dimethylamine	− 96	7.5	v.sol.	5.4
Trimethylamine	−117	3	91	0.6
Ethylamine	− 80	17	∞	5.1
Diethylamine	− 39	55	v.sol.	10.0
Triethylamine	−115	89	14	5.6
n-Propylamine	− 83	49	∞	4.1
Di-*n*-propylamine	− 63	110	s.sol.	10
Tri-*n*-propylamine	− 93	157	s.sol.	4.5
Isopropylamine	−101	34	∞	4
n-Butylamine	− 50	78	v.sol.	4.8
Isobutylamine	− 85	68	∞	3
sec-Butylamine	−104	63	∞	4
tert-Butylamine	− 67	46	∞	5
Cyclohexylamine		134	s.sol.	5
Benzylamine		185	∞	0.2
α-Phenylethylamine		187	4.2	
β-Phenylethylamine		195	s.	
Ethylenediamine	8	117	s.	
Tetramethylenediamine [$H_2N(CH_2)_4NH_2$]	27	158	v.sol.	0.85
Hexamethylenediamine	39	196	v.sol.	5
Tetramethylammonium hydroxide	63	135*d*	220	strong base

Table 22.1 AMINES (*continued*)

Name	M.p., °C	B.p,, °C	Solub., g/100 g H_2O	K_b
Aniline	− 6	184	3.7	4.2×10^{-10}
Methylaniline	− 57	196	v.sl.sol.	7.1
Dimethylaniline	3	194	1.4	11.7
Diphenylamine	53	302	i.	0.0006
Triphenylamine	127	365	i.	
o-Toluidine	− 28	200	1.7	2.6
m-Toluidine	− 30	203	s.sol.	5
p-Toluidine	44	200	0.7	12
o-Anisidine (*o*-$CH_3OC_6H_4NH_2$)	5	225	s.sol.	3
m-Anisidine		251	s.sol.	2
p-Anisidine	57	244	v.sl.sol.	20
o-Chloroaniline	− 2	209	i.	0.05
m-Chloroaniline	− 10	236		0.3
p-Chloroaniline	70	232		1
o-Bromoaniline	32	229	s.sol.	0.03
m-Bromoaniline	19	251	v.sl.sol.	0.4
p-Bromoaniline	66	*d*	i.	0.7
o-Nitroaniline	71	284	0.1	0.00006
m-Nitroaniline	114	307*d*	0.1	0.029
p-Nitroaniline	148	332	0.05	0.001
2,4-Dinitroaniline	187		s.sol.	
2,4,6-Trinitroaniline (picramide)	188		0.1	
o-Phenylenediamine [*o*-$C_6H_4(NH_2)_2$]	104	252	3	3
m-Phenylenediamine	63	287	25	10
p-Phenylenediamine	142	267	3.8	140
Benzidine	127	401	0.05	9
p-Aminobenzoic acid	187		0.3	0.023
Sulfanilic acid	288*d*		1	0.17
Sulfanilamide	163		0.4	

Name	Formula	M.p., °C
Acetanilide	$C_6H_5NHCOCH_3$	114
Benzanilide	$C_6H_5NHCOC_6H_5$	163
Aceto-*o*-toluidide	*o*-$CH_3C_6H_4NHCOCH_3$	110
Aceto-*m*-toluidide	*m*-$CH_3C_6H_4NHCOCH_3$	66
Aceto-*p*-toluidide	*p*-$CH_3C_6H_4NHCOCH_3$	147
o-Nitroacetanilide	*o*-$O_2NC_6H_4NHCOCH_3$	93
m-Nitroacetanilide	*m*-$O_2NC_6H_4NHCOCH_3$	154
p-Nitroacetanilide	*p*-$O_2NC_6H_4NHCOCH_3$	216

being reached at about six carbon atoms. Amines are soluble in less polar solvents like ether, alcohol, benzene, etc. The methylamines and ethylamines smell very much like ammonia; the higher alkylamines have decidedly "fishy" odors.

Aromatic amines are generally very toxic; they are readily absorbed through the skin, often with fatal results.

Aromatic amines are very easily oxidized by air, and although most are colorless when pure, they are often encountered discolored by oxidation products.

22.5 Salts of amines

Aliphatic amines are about as basic as ammonia; aromatic amines are considerably less basic. Although amines are much weaker bases than hydroxide ion or ethoxide ion, they are much stronger bases than alcohols, ethers, esters, etc.; they are much stronger bases than water. Aqueous mineral acids or carboxylic acids readily convert amines into their salts; aqueous hydroxide ion readily converts the salts back into the free amines. As with the carboxylic acids, we can

$$\left.\begin{matrix} \text{RNH}_2 \\ \text{1° amine} \\ \text{R}_2\text{NH} \\ \text{2° amine} \\ \text{R}_3\text{N} \\ \text{3° amine} \end{matrix}\right\} \underset{\text{OH}^-}{\overset{\text{H}^+}{\rightleftarrows}} \left\{\begin{matrix} \text{RNH}_3^+ \\ \text{salt} \\ \text{R}_2\text{NH}_2^+ \\ \text{salt} \\ \text{R}_3\text{NH}^+ \\ \text{salt} \end{matrix}\right.$$

Insoluble in water *Soluble in water*

do little with amines without encountering this conversion into and from their salts; it is therefore worthwhile to look at the properties of these salts.

In Sec. 18.4 we contrasted physical properties of carboxylic acids with those of their salts; amines and their salts show the same contrast. Amine salts are typical ionic compounds. They are non-volatile solids, and when heated generally decompose before the high temperature required for melting is reached. The halides, nitrates, and sulfates are soluble in water but are insoluble in non-polar solvents.

The difference in solubility behavior between amines and their salts can be used both to detect amines and to separate them from non-basic compounds. A water-insoluble organic compound that dissolves in cold, dilute aqueous hydrochloric acid must be appreciably basic, which means almost certainly that it is an amine. An amine can be separated from non-basic compounds by its solubility in acid; once separated, the amine can be regenerated by making the aqueous solution alkaline. (See Sec. 18.4 for a comparable situation for carboxylic acids.)

Problem 22.1 Describe exactly how you would go about separating a mixture of the three water-insoluble liquids, aniline (b.p. 184°), *n*-butylbenzene (b.p. 183°), and *n*-valeric acid (b.p. 187°), recovering each compound pure and in essentially quantitative yield. Do the same for a mixture of the three water-insoluble solids, *p*-toluidine, *o*-bromobenzoic acid, and *p*-nitroanisole.

22.6 Stereochemistry of nitrogen

So far in our study of organic chemistry, we have devoted considerable time to the spatial arrangement of atoms and groups attached to carbon atoms, that is, to the stereochemistry of carbon. Now let us look briefly at the stereochemistry of nitrogen.

Amines are simply ammonia in which one or more hydrogen atoms have been replaced by organic groups. Nitrogen uses sp^3 orbitals, which are directed to the corners of a tetrahedron. Three of these orbitals overlap *s* orbitals of hydrogen or carbon; the fourth contains an unshared pair of electrons (see Fig. 1.11, p. 18). Amines, then, are like ammonia, pyramidal, and with very nearly the same bond angles (108° in trimethylamine, for example).

From an examination of models, we can see that a molecule in which nitrogen carries three different groups is not superimposable on its mirror image; it is chiral and should exist in two enantiomeric forms (I and II) each of which—

R, R′, R″ (I) ⇌ R, R′, R″ (II)

separated from the other—might be expected to show optical activity.

But such enantiomers have not yet been isolated—for simple amines—and spectroscopic studies have shown why: the energy barrier between the two pyramidal arrangements about nitrogen is ordinarily so low that they are rapidly interconverted. Just as rapid rotation about carbon–carbon single bonds prevents isolation of conformational enantiomers (Sec. 4.20), so rapid *inversion* about nitrogen prevents isolation of enantiomers like I and II. Evidently, an unshared pair of electrons of nitrogen cannot ordinarily serve as a fourth group to maintain configuration.

Next, let us consider the quaternary ammonium salts, compounds in which four alkyl groups are attached to nitrogen. Here all four sp^3 orbitals are used to form bonds, and quaternary nitrogen is tetrahedral. Quaternary ammonium salts in which nitrogen holds four different groups have been found to exist as *configurational* enantiomers, capable of showing optical activity: methylallylphenylbenzylammonium iodide, for example.

Problem 22.2 At room temperature, the nmr spectrum of 1-ethylaziridine (III) shows the triplet-quartet of an ethyl group, and two other signals of equal peak area. When the temperature is raised to 120°, the latter two signals merge into a single signal. How do you interpret these observations?

H_2C, H_2C ring with N—C_2H_5 (III) H_2C, HC(CH_3) ring with N—Cl (IV)

Problem 22.3 Account for the following, drawing all pertinent stereochemical formulas. (a) 1-Chloro-2-methylaziridine (IV, above) was prepared in two isomeric forms separable at 25° by ordinary gas chromatography. (b) The reaction of $(C_6H_5)_2C{=}NCH_3$ with R-(+)-2-phenylperoxypropionic acid gave a product, $C_{14}H_{13}ON$, with [α] +12.5°, which showed no loss of optical activity up to (at least) 90°.

Problem 22.4 Racemization in certain free-radical and carbonium ion reactions has been attributed (Secs. 7.10 and 14.13) to loss of configuration in a flat intermediate. Account for the fact that the formation of alkyl carbanions, R:⁻—which are believed to be *pyramidal*—can also lead to racemization.

22.7 Industrial source

Some of the simplest and most important amines are prepared on an industrial scale by processes that are not practicable as laboratory methods.

The most important of all amines, **aniline**, is prepared in several ways: (a) reduction of nitrobenzene by the cheap reagents, iron and dilute hydrochloric acid (or by catalytic hydrogenation, Sec. 22.9); (b) treatment of chlorobenzene with

$$C_6H_5NO_2 \xrightarrow{\text{Fe, 30\% HCl, heat}} C_6H_5NH_3^+Cl^- \xrightarrow{Na_2CO_3} C_6H_5NH_2$$

Nitrobenzene — Anilinium chloride — Aniline

$$C_6H_5Cl \xrightarrow{NH_3,\ Cu_2O,\ 200^\circ,\ 900\ \text{lb/in.}^2} C_6H_5NH_2$$

Chlorobenzene — Aniline

ammonia at high temperatures and high pressures in the presence of a catalyst. Process (b), we shall see (Chap. 25), involves nucleophilic aromatic substitution.

Methylamine, dimethylamine, and trimethylamine are synthesized on an industrial scale from methanol and ammonia:

$$NH_3 \xrightarrow[Al_2O_3,\ 450^\circ]{CH_3OH} CH_3NH_2 \xrightarrow[Al_2O_3,\ 450^\circ]{CH_3OH} (CH_3)_2NH \xrightarrow[Al_2O_3,\ 450^\circ]{CH_3OH} (CH_3)_3N$$

Ammonia — Methylamine — Dimethylamine — Trimethylamine

Alkyl halides are used to make some higher alkylamines, just as in the laboratory (Sec. 22.10). The acids obtained from fats (Sec. 33.4) can be converted into long-chain 1-aminoalkanes of even carbon number via reduction of nitriles (Sec. 22.8).

$$RCOOH \xrightarrow{NH_3,\ \text{heat}} RCONH_2 \xrightarrow{\text{heat}} RC{\equiv}N \xrightarrow{H_2,\ \text{cat.}} RCH_2NH_2$$

Acid — Amide — Nitrile — Amine

22.8 Preparation

Some of the many methods that are used to prepare amines in the laboratory are outlined on the following pages.

PREPARATION OF AMINES

1. Reduction of nitro compounds. Discussed in Sec. 22.9.

$$\underset{\text{Nitro compound}}{ArNO_2 \text{ or } RNO_2} \xrightarrow{\text{metal, } H^+\text{; or } H_2\text{, catalyst}} \underset{1^\circ \text{ amine}}{ArNH_2 \text{ or } RNH_2}$$

Chiefly for aromatic amines

Examples:

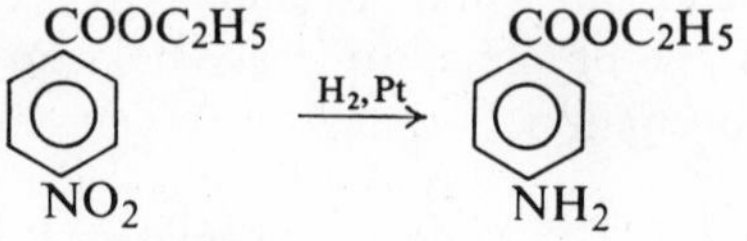

$$p\text{-}NO_2C_6H_4COOC_2H_5 \xrightarrow{H_2, Pt} p\text{-}NH_2C_6H_4COOC_2H_5$$

Ethyl *p*-nitrobenzoate → Ethyl *p*-aminobenzoate

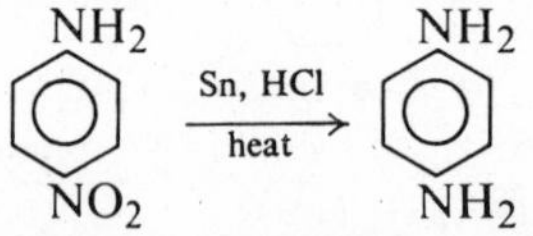

$$p\text{-}NO_2C_6H_4NH_2 \xrightarrow[\text{heat}]{\text{Sn, HCl}} p\text{-}NH_2C_6H_4NH_2$$

p-Nitroaniline → *p*-Phenylenediamine

$$CH_3CH_2CH_2NO_2 \xrightarrow{\text{Fe, HCl}} CH_3CH_2CH_2NH_2$$

1-Nitropropane → *n*-Propylamine

2. Reaction of halides with ammonia or amines. Discussed in Secs. 22.10 and 22.13.

$$NH_3 \xrightarrow{RX} RNH_2 \xrightarrow{RX} R_2NH \xrightarrow{RX} R_3N \xrightarrow{RX} R_4N^+X^-$$

1° amine, 2° amine, 3° amine, Quaternary ammonium salt (4°)

RX must be alkyl, or aryl with electron-withdrawing substituents

Examples:

$$CH_3COOH \xrightarrow[P]{Cl_2} CH_2ClCOOH \xrightarrow{NH_3} CH_2(NH_2)COO^-NH_4^+ \xrightarrow{H^+} CH_2(NH_2)COOH \text{ (or } CH_2(^+NH_3)COO^-\text{)}$$

Acetic acid → Chloroacetic acid → ... → Aminoacetic acid (Glycine; an amino acid) (1°)

$$C_2H_5Cl \xrightarrow{NH_3} C_2H_5NH_2 \xrightarrow{CH_3Cl} C_2H_5\text{—}NH\text{—}CH_3$$

Ethyl chloride → Ethylamine (1°) → Methylethylamine (2°)

$$C_6H_5CH_2Cl \xrightarrow{NH_3} C_6H_5CH_2NH_2 \xrightarrow{2CH_3Cl} C_6H_5CH_2\text{—}N(CH_3)\text{—}CH_3$$

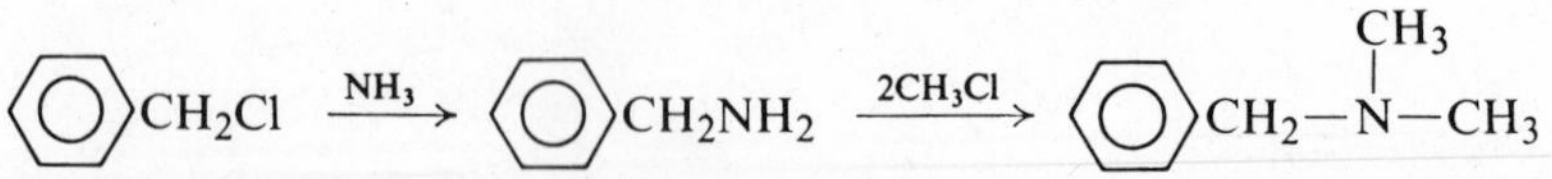

Benzyl chloride → Benzylamine (1°) → Benzyldimethylamine (3°)

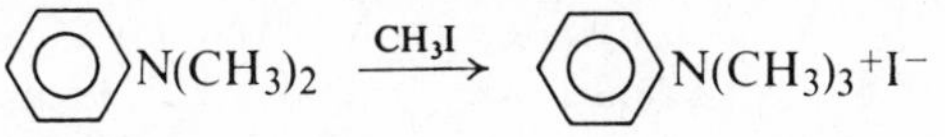

$$C_6H_5N(CH_3)_2 \xrightarrow{CH_3I} C_6H_5N(CH_3)_3^+I^-$$

N,N-Dimethylaniline (3°) → Phenyltrimethylammonium iodide (4°)

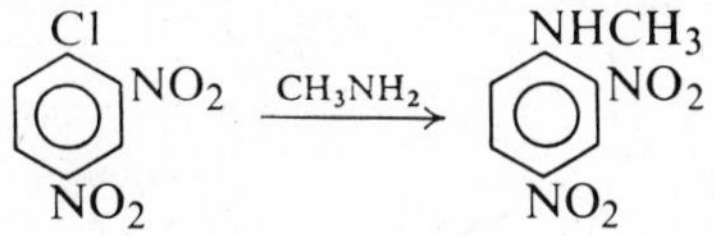

$$2,4\text{-}(NO_2)_2C_6H_3Cl \xrightarrow{CH_3NH_2} 2,4\text{-}(NO_2)_2C_6H_3NHCH_3$$

2,4-Dinitrochlorobenzene → N-Methyl-2,4-dinitroaniline (2°)

3. **Reductive amination.** Discussed in Sec. 22.11.

$$>C{=}O + NH_3 \xrightarrow[\text{or } NaBH_3CN]{H_2,\ Ni} >CH{-}NH_2 \quad 1° \text{ amine}$$

$$+ RNH_2 \xrightarrow[\text{or } NaBH_3CN]{H_2,\ Ni} >CH{-}NHR \quad 2° \text{ amine}$$

$$+ R_2NH \xrightarrow[\text{or } NaBH_3CN]{H_2,\ Ni} >CH{-}NR_2 \quad 3° \text{ amine}$$

Examples:

$$CH_3{-}\underset{\underset{O}{\|}}{C}{-}CH_3 + NH_3 + H_2 \xrightarrow{Ni} CH_3{-}\underset{\underset{NH_2}{|}}{CH}{-}CH_3$$

Acetone → Isopropylamine (1°)

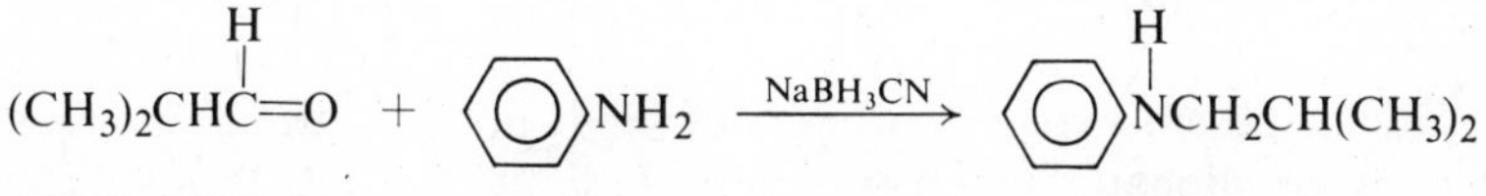

$$(CH_3)_2CH\overset{\overset{H}{|}}{C}{=}O + C_6H_5NH_2 \xrightarrow{NaBH_3CN} C_6H_5\overset{\overset{H}{|}}{N}CH_2CH(CH_3)_2$$

Isobutyraldehyde + Aniline (1°) → N-Isobutylaniline (2°)

$$CH_3\overset{\overset{H}{|}}{C}{=}O + (CH_3)_2NH + H_2 \xrightarrow{Ni} CH_3CH_2{-}\overset{\overset{CH_3}{|}}{N}{-}CH_3$$

Acetaldehyde + Dimethylamine (2°) → Dimethylethylamine (3°)

4. **Reduction of nitriles.** Discussed in Sec. 22.8.

$$RC{\equiv}N \xrightarrow{2H_2,\ \text{catalyst}} RCH_2NH_2$$

Nitrile → 1° amine

Examples:

$$C_6H_5CH_2Cl \xrightarrow{NaCN} C_6H_5CH_2CN \xrightarrow{H_2,\ Ni,\ 140°} C_6H_5CH_2CH_2NH_2$$

Benzyl chloride → Phenylacetonitrile (Benzyl cyanide) → β-Phenylethylamine (1°)

$$\underset{\text{1,4-Dichlorobutane}}{ClCH_2CH_2CH_2CH_2Cl} \xrightarrow{NaCN} \underset{\text{Adiponitrile}}{NC(CH_2)_4CN} \xrightarrow{H_2,\ Ni} \underset{\substack{\text{Hexamethylenediamine}\\ \text{(1,6-Diaminohexane)}\\ (1°)}}{H_2NCH_2(CH_2)_4CH_2NH_2}$$

5. Hofmann degradation of amides. Discussed in Secs. 22.12 and 28.2–28.5.

$$\underset{\text{Amide}}{RCONH_2 \text{ or } ArCONH_2} \xrightarrow{OBr^-} \underset{1°\text{ amine}}{RNH_2 \text{ or } ArNH_2} + CO_3^{--}$$

Examples:

$$\underset{\substack{\text{Caproamide}\\ \text{(Hexanamide)}}}{CH_3(CH_2)_4CONH_2} \xrightarrow{KOBr} \underset{n\text{-Pentylamine}}{CH_3(CH_2)_4NH_2}$$

$$\underset{m\text{-Bromobenzamide}}{m\text{-}BrC_6H_4CONH_2} \xrightarrow{KOBr} \underset{m\text{-Bromoaniline}}{m\text{-}BrC_6H_4NH_2}$$

Reduction of aromatic nitro compounds is by far the most useful method of preparing amines, since it uses readily available starting materials, and yields the most important kind of amines, *primary aromatic amines*. These amines can be converted into aromatic diazonium salts, which are among the most versatile class of organic compounds known (see Secs. 23.11–23.17). The sequence

$$\text{nitro compound} \longrightarrow \text{amine} \longrightarrow \text{diazonium salt}$$

provides the best possible route to dozens of kinds of aromatic compounds.

Reduction of aliphatic nitro compounds is limited by the availability of the starting materials.

Ammonolysis of halides is usually limited to the aliphatic series, because of the generally low reactivity of aryl halides toward nucleophilic substitution. (However, see Chap. 25.) Ammonolysis has the disadvantage of yielding a mixture of different classes of amines. It is important to us as one of the most general methods of introducing the amino ($-NH_2$) group into molecules of all kinds; it can be used, for example, to convert bromoacids into amino acids. The exactly analogous reaction of halides with amines permits the preparation of every class of amine (as well as quaternary ammonium salts, $R_4N^+X^-$).

Reductive amination, the catalytic or chemical reduction of aldehydes (RCHO) and ketones (R_2CO) in the presence of ammonia or an amine, accomplishes much the same purpose as the reaction of halides. It too can be used to prepare any class of amine, and has certain advantages over the halide reaction. The formation of mixtures is more readily controlled in reductive amination than in ammonolysis of halides. Reductive amination of ketones yields amines containing a *sec*-alkyl group; these amines are difficult to prepare by ammonolysis because of the tendency of *sec*-alkyl halides to undergo elimination rather than substitution.

Synthesis via **reduction of nitriles** has the special feature of *increasing the length of a carbon chain*, producing a primary amine that has one more carbon atom than the alkyl halide from which the nitrile was made. The **Hofmann degradation of amides** has the feature of *decreasing the length of a carbon chain* by one carbon atom; it is also of interest as an example of an important class of reactions involving rearrangement.

$$RCH_2OH \xrightarrow{KMnO_4} RCOOH \xrightarrow{SOCl_2} RCOCl \xrightarrow{NH_3} RCONH_2 \xrightarrow{OBr^-} RNH_2 \quad \textit{Lower carbon number}$$

$$RCH_2OH \xrightarrow{PBr_3} RCH_2Br \xrightarrow{NaCN} RCH_2C{\equiv}N \xrightarrow{H_2,\ Ni} RCH_2CH_2NH_2 \quad \textit{Higher carbon number}$$

$$RCH_2Br \xrightarrow{NH_3} RCH_2NH_2 \quad \textit{Same carbon number}$$

$$RCH_2OH \xrightarrow{K_2Cr_2O_7} RCH{=}O \xrightarrow{NH_3,\ H_2,\ Ni} RCH_2NH_2 \quad \textit{Same carbon number}$$

Problem 22.5 Show how *n*-pentylamine can be synthesized from available materials by the four routes just outlined.

22.9 Reduction of nitro compounds

Like many organic compounds, nitro compounds can be reduced in two general ways: (a) by catalytic hydrogenation using molecular hydrogen, or (b) by chemical reduction, usually by a metal and acid.

Hydrogenation of a nitro compound to an amine takes place smoothly when a solution of the nitro compound in alcohol is shaken with finely divided nickel or platinum under hydrogen gas. For example:

$$o\text{-}CH_3CONHC_6H_4NO_2 \xrightarrow{H_2,\ Pt} o\text{-}CH_3CONHC_6H_4NH_2$$

o-Nitroacetanilide → *o*-Aminoacetanilide

This method cannot be used when the molecule also contains some other easily hydrogenated group, such as a carbon–carbon double bond.

Chemical reduction in the laboratory is most often carried out by adding hydrochloric acid to a mixture of the nitro compound and a metal, usually granulated tin. In the acidic solution, the amine is obtained as its salt; the free amine is liberated by the addition of base, and is steam-distilled from the reaction

$$p\text{-}CH_3C_6H_4NO_2 \xrightarrow[\text{heat}]{\text{Sn, HCl}} (p\text{-}CH_3C_6H_4NH_3^+)_2SnCl_6^{--} \xrightarrow{OH^-} p\text{-}CH_3C_6H_4NH_2 + SnO_3^{--}$$

p-Nitrotoluene *p*-Toluidine

mixture. The crude amine is generally contaminated with some unreduced nitro compound, from which it can be separated by taking advantage of the basic properties of the amine; the amine is soluble in aqueous mineral acid, and the nitro compound is not.

Reduction of nitro compounds to amines is an essential step in what is probably the most important synthetic route in aromatic chemistry. Nitro compounds are readily prepared by direct nitration; when a mixture of *o*- and *p*-isomers is obtained, it can generally be separated to yield the pure isomers. The primary aromatic amines obtained by the reduction of these nitro compounds are readily converted into diazonium salts; the diazonium group, in turn, can be replaced by a large number of other groups (Sec. 23.11). In most cases this sequence is the best method of introducing these other groups into the aromatic ring. In addition, diazonium salts can be used to prepare the extremely important class of compounds, the *azo dyes*.

$$ArH \longrightarrow ArNO_2 \longrightarrow ArNH_2 \longrightarrow ArN_2^+ \longrightarrow \begin{cases} ArX \\ ArOH \\ ArCN \\ \text{azo dyes} \end{cases}$$

22.10 Ammonolysis of halides

Many organic halogen compounds are converted into amines by treatment with aqueous or alcoholic solutions of ammonia. The reaction is generally carried out either by allowing the reactants to stand together at room temperature or by heating them under pressure. Displacement of halogen by NH_3 yields the amine salt, from which the free amine can be liberated by treatment with hydroxide ion.

$$RX + NH_3 \longrightarrow RNH_3^+X^-$$

$$RNH_3^+X^- + OH^- \longrightarrow RNH_2 + H_2O + X^-$$

Ammonolysis of halides belongs to the class of reactions that we have called nucleophilic substitution. The organic halide is attacked by the nucleophilic ammonia molecule in the same way that it is attacked by hydroxide ion, alkoxide ion, cyanide ion, acetylide ion, and water:

$$H_3N: + R{-}X \longrightarrow \left[\overset{\delta_+}{H_3N}\text{---}R\text{---}\overset{\delta_-}{X}\right] \longrightarrow H_3\overset{+}{N}{-}R + X^-$$

Like these other nucleophilic substitution reactions, ammonolysis is limited chiefly to alkyl halides or substituted alkyl halides. As with other reactions of this kind, elimination tends to compete (Sec. 14.23) with substitution: ammonia can attack hydrogen to form alkene as well as attack carbon to form amine. Ammonolysis thus gives the highest yields with primary halides (where substitution predominates) and is virtually worthless with tertiary halides (where elimination predominates).

$$CH_3CH_2CH_2CH_2Br \xrightarrow{NH_3} CH_3CH_2CH_2CH_2NH_3^+Br^- \quad \textit{Substitution}$$

$$CH_3-\underset{Br}{\overset{CH_3}{C}}-CH_3 \xrightarrow{NH_3} CH_3-\overset{CH_3}{C}=CH_2 + NH_4Br \quad \textit{Elimination}$$

Because of their generally low reactivity, aryl halides are converted into amines only (a) if the ring carries $-NO_2$ groups, or other strongly electron-withdrawing groups, at positions *ortho* and *para* to the halogen, or (b) if a high temperature or a strongly basic reagent is used (Chap. 25).

Some examples of the application of ammonolysis to synthesis are:

$$C_6H_5CH_3 \xrightarrow[heat]{Cl_2} C_6H_5CH_2Cl \xrightarrow{NH_3} C_6H_5CH_2NH_2$$

Toluene — Benzyl chloride — Benzylamine

$$\underset{\text{Propionic acid}}{CH_3CH_2COOH} \xrightarrow[P]{Br_2} \underset{\alpha\text{-Bromopropionic acid}}{CH_3\underset{Br}{CH}COOH} \xrightarrow{NH_3} \underset{\text{Alanine}\ (\alpha\text{-Aminopropionic acid})}{CH_3\underset{NH_2}{CH}COOH}$$

$$\underset{\text{Ethylene}}{CH_2=CH_2} \xrightarrow{Cl_2} \underset{\text{Ethylene chloride}}{ClCH_2CH_2Cl} \xrightarrow{2NH_3} \underset{\text{Ethylenediamine}}{H_2NCH_2CH_2NH_2}$$

A serious disadvantage to the synthesis of amines by ammonolysis is the formation of more than one class of amine. The primary amine salt, formed by

$$RX + NH_3 \longrightarrow \underset{1^\circ\ \text{amine salt}}{RNH_3^+X^-}$$

the initial substitution, reacts with the reagent ammonia to yield the ammonium salt and the free primary amine; the following equilibrium thus exists:

$$RNH_3^+ + NH_3 \rightleftharpoons \underset{1^\circ\ \text{amine}}{RNH_2} + NH_4^+$$

The free primary amine, like the ammonia from which it was made, is a nucleophilic reagent; it too can attack the alkyl halide, to yield the salt of a secondary amine:

$$\underset{1^\circ\ \text{amine}}{RNH_2} + RX \longrightarrow R_2NH_2^+X^- \underset{}{\overset{NH_3}{\rightleftharpoons}} \underset{2^\circ\ \text{amine}}{R_2NH}$$

The secondary amine, which is in equilibrium with its salt, can in turn attack the alkyl halide to form the salt of a tertiary amine:

$$\underset{2^\circ\text{ amine}}{R_2NH} + RX \longrightarrow R_3NH^+X^- \overset{NH_3}{\rightleftharpoons} \underset{3^\circ\text{ amine}}{R_3N}$$

Finally, the tertiary amine can attack the alkyl halide to form a compound of the formula $R_4N^+X^-$, called a *quaternary ammonium salt* (discussed in Sec. 23.5):

$$\underset{3^\circ\text{ amine}}{R_3N} + RX \longrightarrow \underset{\substack{\text{Quaternary ammonium salt}\\(4^\circ)}}{R_4N^+X^-}$$

The presence of a large excess of ammonia lessens the importance of these last reactions and increases the yield of primary amine; under these conditions, a molecule of alkyl halide is more likely to encounter, and be attacked by, one of the numerous ammonia molecules rather than one of the relatively few amine molecules. At best, the yield of primary amine is always cut down by the formation of the higher classes of amines. Except in the special case of methylamine, the primary amine can be separated from these by-products by distillation.

22.11 Reductive amination

Many aldehydes (RCHO) and ketones (R_2CO) are converted into amines by **reductive amination**: reduction in the presence of ammonia. Reduction can be accomplished catalytically or by use of sodium cyanohydridoborate, $NaBH_3CN$. Reaction involves reduction of an intermediate compound (an *imine*, RCH=NH or R_2C=NH) that contains a carbon–nitrogen double bond.

$$\underset{\text{An aldehyde}}{R\text{—}\overset{\overset{\displaystyle H}{|}}{C}\text{=}O} + NH_3 \longrightarrow \left[\underset{\text{An imine}}{R\text{—}\overset{\overset{\displaystyle H}{|}}{C}\text{=}NH}\right] \xrightarrow[\text{or } NaBH_3CN]{H_2,\ Ni} \underset{\text{A 1° amine}}{R\text{—}\overset{\overset{\displaystyle H}{|}}{\underset{\underset{\displaystyle H}{|}}{C}}\text{—}NH_2}$$

$$\underset{\text{A ketone}}{R\text{—}\overset{\overset{\displaystyle R'}{|}}{C}\text{=}O} + NH_3 \longrightarrow \left[\underset{\text{An imine}}{R\text{—}\overset{\overset{\displaystyle R'}{|}}{C}\text{=}NH}\right] \xrightarrow[\text{or } NaBH_3CN]{H_2,\ Ni} \underset{\text{A 1° amine}}{R\text{—}\overset{\overset{\displaystyle R'}{|}}{\underset{\underset{\displaystyle H}{|}}{C}}\text{—}NH_2}$$

Reductive amination has been used successfully with a wide variety of aldehydes and ketones, both aliphatic and aromatic. For example:

$$\underset{\substack{\text{Heptaldehyde}\\\text{(Heptanal)}}}{CH_3(CH_2)_5CHO} \xrightarrow{NH_3,\ H_2,\ Ni} \underset{\substack{n\text{-Heptylamine}\\\text{(1-Aminoheptane)}}}{CH_3(CH_2)_5CH_2NH_2}$$

$$\underset{\text{Benzaldehyde}}{C_6H_5CHO} \xrightarrow{NH_3,\ H_2,\ Ni} \underset{\text{Benzylamine}}{C_6H_5CH_2NH_2}$$

$$CH_3(CH_2)_2\underset{\underset{O}{\|}}{C}CH_3 \xrightarrow{NH_3,\ H_2,\ Ni} CH_3(CH_2)_2\underset{\underset{NH_2}{|}}{C}HCH_3$$

2-Pentanone (Methyl *n*-propyl ketone) → 2-Aminopentane

$$C_6H_5-\underset{\underset{O}{\|}}{C}-CH_3 \xrightarrow{NH_3,\ NaBH_3CN} C_6H_5-\underset{\underset{NH_2}{|}}{C}H-CH_3$$

Acetophenone (Methyl phenyl ketone) → α-Phenylethylamine

Reductive amination of ketones yields amines containing a *sec*-alkyl group; such amines are difficult to obtain by ammonolysis because of the tendency for *sec*-alkyl halides to undergo elimination. For example, cyclohexanone is converted into cyclohexylamine in good yield, whereas ammonolysis of bromocyclohexane yields only cyclohexene.

$$\text{Cyclohexanol} \xrightarrow{K_2Cr_2O_7,\ H_2SO_4} \text{Cyclohexanone} \xrightarrow{NH_3,\ H_2,\ Ni} \text{Cyclohexylamine}$$

$$\text{Cyclohexanol} \xrightarrow{PBr_3} \text{Bromocyclohexane} \xrightarrow{NH_3} \text{Cyclohexene}$$

During reductive amination the aldehyde or ketone can react not only with ammonia but also with the primary amine that has already been formed, and thus yield a certain amount of secondary amine. The tendency for the reaction to go

$$\underset{\text{Aldehyde}}{R-\overset{\overset{H}{|}}{C}=O} + \underset{1^\circ\ \text{amine}}{H_2N-CH_2R} \longrightarrow \left[\underset{\text{Imine}}{R-\overset{\overset{H}{|}}{C}=N-CH_2R}\right] \xrightarrow{\text{reduction}} \underset{2^\circ\ \text{amine}}{RCH_2-\overset{\overset{H}{|}}{N}-CH_2R}$$

beyond the desired stage can be fairly well limited by the proportions of reactants employed and is seldom a serious handicap.

22.12 Hofmann degradation of amides

As a method of synthesis of amines, the Hofmann degradation of amides has the special feature of yielding a product containing one less carbon than the starting material. As we can see, reaction involves migration of a group from carbonyl

$$\underset{\text{An amide}}{R-C(=O)NH_2} \xrightarrow{OBr^-} \underset{\text{A }1^\circ\text{ amine}}{R-NH_2} + CO_3^{--}$$

carbon to the adjacent nitrogen atom, and thus is an example of a *molecular rearrangement.* We shall return to the Hofmann degradation (Secs. 28.2–28.5) and discuss its mechanism in detail.

Problem 22.6 Using a different method in each case, show how the following amines could be prepared from *toluene* and any aliphatic reagents:

(a) $C_6H_5CH_2NH_2$

(b) $p\text{-}CH_3C_6H_4CH(NH_2)CH_3$

(c) $C_6H_5CH_2CH_2NH_2$

(d) $p\text{-}CH_3C_6H_4NH_2$

(e) $C_6H_5NH_2$

22.13 Synthesis of secondary and tertiary amines

So far we have been chiefly concerned with the synthesis of primary amines. Secondary and tertiary amines are prepared by adaptations of one of the processes already described: ammonolysis of halides or reductive amination. For example:

$$\underset{\substack{n\text{-Butylamine}\\(1^\circ)}}{CH_3CH_2CH_2CH_2NH_2} + \underset{\text{Ethyl bromide}}{CH_3CH_2Br} \longrightarrow \underset{\substack{\text{Ethyl-}n\text{-butylamine}\\(2^\circ)}}{CH_3CH_2CH_2CH_2\text{—}NH\text{—}CH_2CH_3}$$

$$\underset{\substack{\text{Butanone}\\\text{(Methyl ethyl ketone)}}}{CH_3CH_2C(=O)CH_3} + \underset{\substack{\text{Methylamine}\\(1^\circ)}}{CH_3NH_2} \xrightarrow{H_2,\ Ni} \underset{\substack{\text{Methyl-}sec\text{-butylamine}\\(2^\circ)}}{CH_3CH_2CH(NHCH_3)CH_3}$$

$$\underset{\substack{\text{Aniline}\\(1^\circ)}}{C_6H_5NH_2} \xrightarrow{CH_3Cl} \underset{\substack{\text{N-Methylaniline}\\(2^\circ)}}{C_6H_5NHCH_3} \xrightarrow{CH_3Cl} \underset{\substack{\text{N,N-Dimethylaniline}\\(3^\circ)}}{C_6H_5N(CH_3)_2}$$

$$\underset{\substack{\text{Ethyl-}n\text{-butylamine}\\(2^\circ)}}{CH_3CH_2CH_2CH_2\text{—}NH\text{—}CH_2CH_3} + \underset{\substack{\text{Methyl}\\\text{bromide}}}{CH_3Br} \longrightarrow \underset{\substack{\text{Methylethyl-}n\text{-butylamine}\\(3^\circ)}}{CH_3CH_2CH_2CH_2\text{—}N(CH_3)\text{—}CH_2CH_3}$$

Where ammonia has been used to produce a primary amine, a primary amine can be used to produce a secondary amine, or a secondary amine can be used to produce a tertiary amine. In each of these syntheses there is a tendency for reaction to proceed beyond the first stage and to yield an amine of a higher class than the one that is wanted.

PROBLEMS

1. Draw structures, give names, and classify as primary, secondary, or tertiary:

(a) the eight isomeric amines of formula $C_4H_{11}N$
(b) the five isomeric amines of formula C_7H_9N that contain a benzene ring

2. Give the structural formulas of the following compounds:

(a) *sec*-butylamine
(b) *o*-toluidine
(c) anilinium chloride
(d) diethylamine
(e) *p*-aminobenzoic acid
(f) benzylamine
(g) isopropylammonium benzoate
(h) *o*-phenylenediamine
(i) N,N-dimethylaniline
(j) ethanolamine (2-aminoethanol)
(k) β-phenylethylamine
(l) N,N-dimethylaminocyclohexane
(m) diphenylamine
(n) 2,4-dimethylaniline
(o) tetra-*n*-butylammonium iodide
(p) *p*-anisidine

3. Show how *n*-propylamine could be prepared from each of the following:

(a) *n*-propyl bromide
(b) *n*-propyl alcohol
(c) propionaldehyde
(d) 1-nitropropane
(e) propionitrile
(f) *n*-butyramide
(g) *n*-butyl alcohol
(h) ethyl alcohol

Which of these methods can be applied to the preparation of aniline? Of benzylamine?

4. Outline all steps in a possible laboratory synthesis of each of the following compounds from benzene, toluene, and alcohols of four carbons or less, using any needed inorganic reagents.

(a) isopropylamine
(b) *n*-pentylamine
(c) *p*-toluidine
(d) ethylisopropylamine
(e) α-phenylethylamine
(f) β-phenylethylamine
(g) *m*-chloroaniline
(h) *p*-aminobenzoic acid
(i) 3-aminoheptane
(j) N-ethylaniline
(k) 2,4-dinitroaniline
(l) the drug *benzedrine* (2-amino-1-phenylpropane)
(m) *p*-nitrobenzylamine
(n) 2-amino-1-phenylethanol

5. Outline all steps in a possible laboratory synthesis from palmitic acid, $n\text{-}C_{15}H_{31}COOH$, of:

(a) $n\text{-}C_{16}H_{33}NH_2$
(b) $n\text{-}C_{17}H_{35}NH_2$
(c) $n\text{-}C_{15}H_{31}NH_2$
(d) $n\text{-}C_{15}H_{31}CH(NH_2)\text{-}n\text{-}C_{16}H_{33}$

6. On the basis of the following synthesis give the structures of *putrescine* and *cadaverine*, found in rotting flesh:

(a) ethylene bromide $\xrightarrow{KCN}$ $C_4H_4N_2$ $\xrightarrow{Na,\ C_2H_5OH}$ putrescine ($C_4H_{12}N_2$)

(b) $Br(CH_2)_5Br$ $\xrightarrow{NH_3}$ cadaverine ($C_5H_{14}N_2$)

7. One of the raw materials for the manufacture of Nylon 66 is *hexamethylenediamine*, $NH_2(CH_2)_6NH_2$. Much of this amine is made by a process that begins with the 1,4-addition of chlorine to 1,3-butadiene. What do you think might be the subsequent steps in this process?

8. Outline all steps in a possible synthesis of β-alanine (β-aminopropionic acid) from succinic anhydride.

9. Using models and then drawing formulas, show the stereoisomeric forms in which each of the following compounds can exist. Tell which stereoisomers when separated from all others would be optically active and which would be optically inactive.

(a) α-phenylethylamine
(b) N-methyl-N-ethylaniline
(c) methylethyl-*n*-propylphenylammonium bromide

(d)

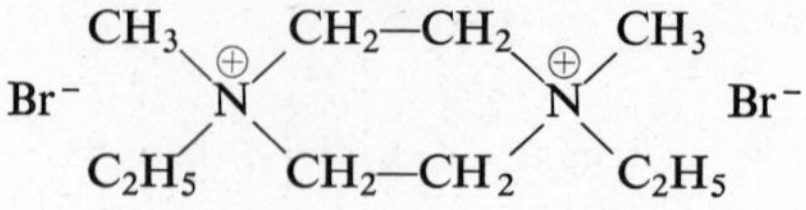

```
      CH3  ⊕  CH2—CH2  ⊕  CH3
Br⁻      N               N        Br⁻
      C2H5    CH2—CH2     C2H5
```

(e)

```
   H      CH2—CH2  ⊕  CH2—CH2     H
      C              N              C              Br⁻
   C6H5   CH2—CH2     CH2—CH2     COOC2H5
```

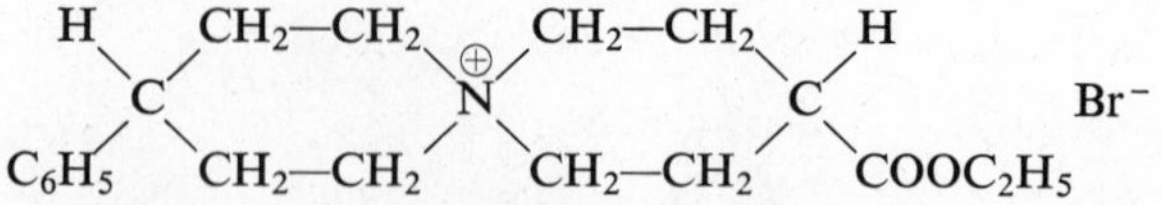

(f) methylethylphenylamine oxide, $(CH_3)(C_2H_5)(C_6H_5)N{-}O$

10. Two geometric isomers of benzaldoxime, $C_6H_5CH{=}NOH$, are known. (a) Draw their structures, showing the geometry of the molecules. (b) Show how this geometry results from their electronic configurations. (c) Would you predict geometric isomerism for benzophenoneoxime, $(C_6H_5)_2C{=}NOH$? For acetophenoneoxime, $C_6H_5C(CH_3){=}NOH$? For azobenzene, $C_6H_5N{=}NC_6H_5$?

11. (a) Give structural formulas of compounds A through D.

$$\text{phthalimide (Sec. 20.14)} + \text{KOH (alc.)} \longrightarrow \text{A } (C_8H_4O_2NK)$$

$$\text{A} + CH_3CH_2CH_2Br\text{, heat} \longrightarrow \text{B } (C_{11}H_{11}O_2N)$$

$$\text{B} + H_2O\text{, } OH^-\text{, heat} \longrightarrow \text{C } (C_3H_9N) + \text{D}$$

(b) This sequence illustrates the **Gabriel synthesis**. What class of compounds does it produce? What particular advantage does it have over alternative methods for the production of these compounds? On what special property of phthalimide does the synthesis depend?

Chapter 23 Amines II. Reactions

23.1 Reactions

Like ammonia, the three classes of amines contain nitrogen that bears an unshared pair of electrons; as a result, amines closely resemble ammonia in chemical properties. The tendency of nitrogen to share this pair of electrons underlies the entire chemical behavior of amines: their basicity, their action as nucleophiles, and the unusually high reactivity of aromatic rings bearing amino or substituted amino groups.

REACTIONS OF AMINES

1. Basicity. Salt formation. Discussed in Secs. 22.5 and 23.2–23.4.

$$RNH_2 + H^+ \rightleftharpoons RNH_3^+$$

$$R_2NH + H^+ \rightleftharpoons R_2NH_2^+$$

$$R_3N + H^+ \rightleftharpoons R_3NH^+$$

Examples:

$$C_6H_5NH_2 + HCl \rightleftharpoons C_6H_5NH_3^+Cl^-$$

Aniline → Anilinium chloride (Aniline hydrochloride)

$$(CH_3)_2NH + HNO_3 \rightleftharpoons (CH_3)_2NH_2^+NO_3^-$$

Dimethylamine → Dimethylammonium nitrate

$$C_6H_5N(CH_3)_2 + CH_3COOH \rightleftharpoons C_6H_5\overset{H}{N}(CH_3)_2^+ \; ^-OOCCH_3$$

N,N-Dimethylaniline → N,N-Dimethylanilinium acetate

2. Alkylation. Discussed in Secs. 22.13 and 23.5.

$$RNH_2 \xrightarrow{RX} R_2NH \xrightarrow{RX} R_3N \xrightarrow{RX} R_4N^+X^-$$

$$ArNH_2 \xrightarrow{RX} ArNHR \xrightarrow{RX} ArNR_2 \xrightarrow{RX} ArNR_3{}^+X^-$$

Examples:

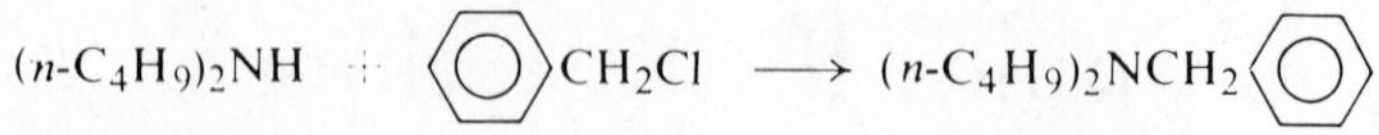

$$(n\text{-}C_4H_9)_2NH + C_6H_5CH_2Cl \longrightarrow (n\text{-}C_4H_9)_2NCH_2C_6H_5$$

Di-*n*-butylamine (2°) — Benzyl chloride — Benzyldi(*n*-butyl)amine (3°)

$$n\text{-}C_3H_7NH_2 \xrightarrow{CH_3I} n\text{-}C_3H_7N(H)CH_3 \xrightarrow{CH_3I} n\text{-}C_3H_7N(CH_3)CH_3 \xrightarrow{CH_3I} n\text{-}C_3H_7N(CH_3)_2CH_3{}^+I^-$$

n-Propylamine (1°) — *n*-Propylmethylamine (2°) — *n*-Propyldimethylamine (3°) — *n*-Propyltrimethylammonium iodide (4°)

3. Conversion into amides. Discussed in Sec. 23.6.

Primary: RNH_2
- $\xrightarrow{R'COCl}$ $R'CONHR$ An N-substituted amide
- $\xrightarrow{ArSO_2Cl}$ $ArSO_2NHR$ An N-substituted sulfonamide

Secondary: R_2NH
- $\xrightarrow{R'COCl}$ $R'CONR_2$ An N,N-disubstituted amide
- $\xrightarrow{ArSO_2Cl}$ $ArSO_2NR_2$ An N,N-disubstituted sulfonamide

Tertiary: R_3N
- $\xrightarrow{R'COCl}$ No reaction
- $\xrightarrow{ArSO_2Cl}$ No reaction under conditions of Hinsberg test (*but see* Sec. 23.18).

Examples:

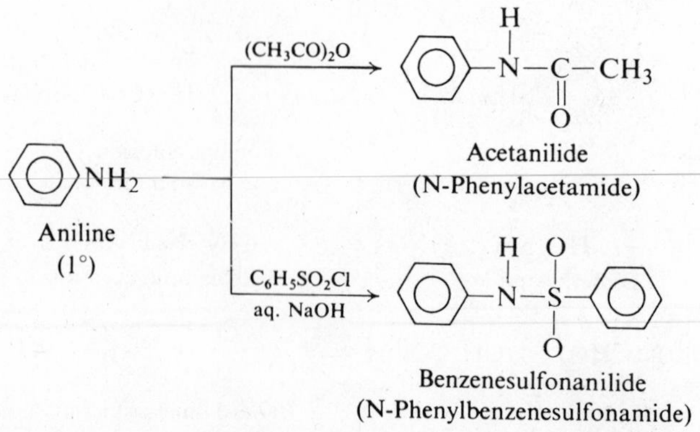

$$C_6H_5NH_2 \xrightarrow{(CH_3CO)_2O} C_6H_5\text{-}N(H)\text{-}C(=O)\text{-}CH_3$$

Aniline (1°) — Acetanilide (N-Phenylacetamide)

$$C_6H_5NH_2 \xrightarrow[\text{aq. NaOH}]{C_6H_5SO_2Cl} C_6H_5\text{-}N(H)\text{-}S(=O)_2\text{-}C_6H_5$$

Benzenesulfonanilide (N-Phenylbenzenesulfonamide)

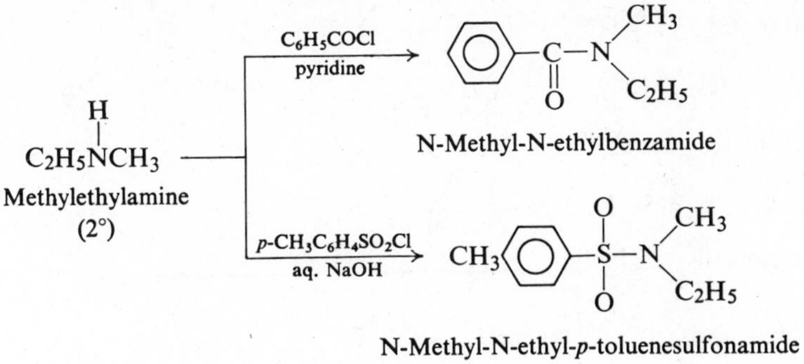

$C_2H_5NHCH_3$ (Methylethylamine, 2°)

$\xrightarrow[\text{pyridine}]{C_6H_5COCl}$ $C_6H_5CON(CH_3)C_2H_5$

N-Methyl-N-ethylbenzamide

$\xrightarrow[\text{aq. NaOH}]{p\text{-}CH_3C_6H_4SO_2Cl}$ $p\text{-}CH_3C_6H_4SO_2N(CH_3)C_2H_5$

N-Methyl-N-ethyl-*p*-toluenesulfonamide

4. **Ring substitution in aromatic amines.** Discussed in Secs. 23.7, 23.10, and 23.17.

$-NH_2$, $-NHR$, $-NR_2$: Activate powerfully, and direct *ortho,para* in electrophilic aromatic substitution

$-NHCOR$: Less powerful activator than $-NH_2$

Examples:

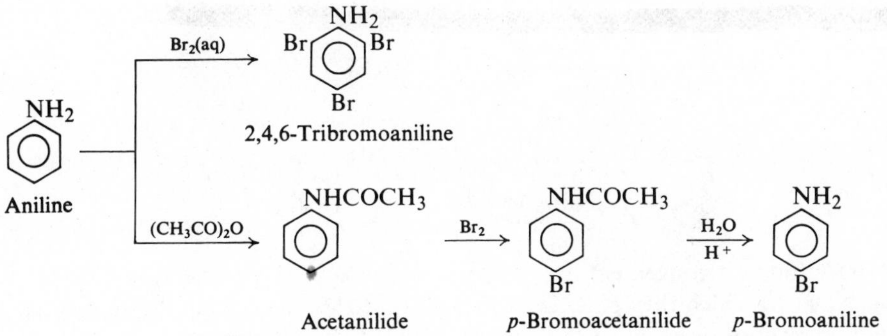

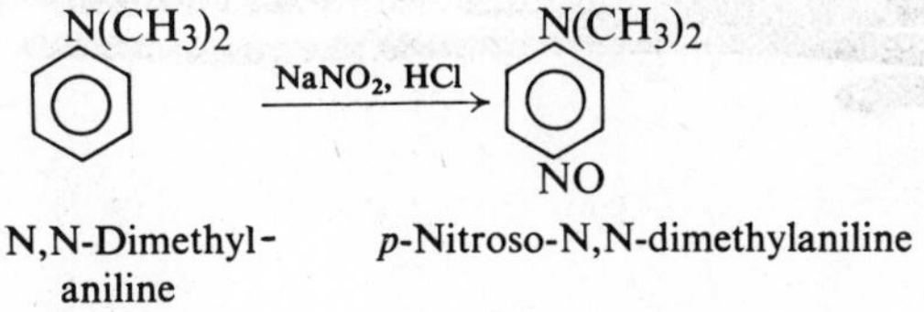

Aniline $\xrightarrow{Br_2(aq)}$ 2,4,6-Tribromoaniline

Aniline $\xrightarrow{(CH_3CO)_2O}$ Acetanilide $\xrightarrow{Br_2}$ *p*-Bromoacetanilide $\xrightarrow[H^+]{H_2O}$ *p*-Bromoaniline

N,N-Dimethylaniline $\xrightarrow{NaNO_2,\ HCl}$ *p*-Nitroso-N,N-dimethylaniline

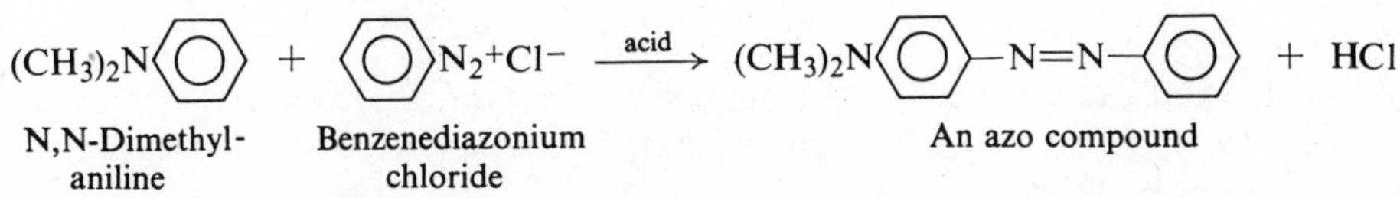

$(CH_3)_2NC_6H_5$ + $C_6H_5N_2^+Cl^-$ $\xrightarrow{\text{acid}}$ $(CH_3)_2NC_6H_4-N{=}N-C_6H_5$ + HCl

N,N-Dimethylaniline + Benzenediazonium chloride → An azo compound

5. **Hofmann elimination from quaternary ammonium salts.** Discussed in Sec. 23.5.

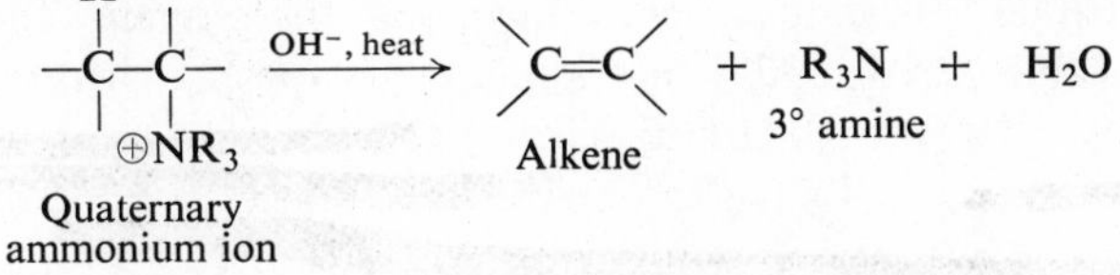

$$-\overset{H}{\underset{|}{C}}-\underset{\oplus NR_3}{C}- \xrightarrow{OH^-,\ heat} >C{=}C< + R_3N + H_2O$$

Quaternary ammonium ion → Alkene + R_3N (3° amine) + H_2O

6. Reactions with nitrous acid. Discussed in Secs. 23.10–23.11.

Primary aromatic: $ArNH_2 \xrightarrow{HONO} Ar{-}N{\equiv}N^+$ Diazonium salt

Primary aliphatic: $RNH_2 \xrightarrow{HONO} [R{-}N{\equiv}N^+] \xrightarrow{H_2O} N_2$ + mixture of alcohols and alkenes

Secondary aromatic or aliphatic: $ArNHR$ or $R_2NH \xrightarrow{HONO} Ar{-}N(R){-}N{=}O$ or $R_2N{-}N{=}O$ N-Nitrosoamine

Tertiary aromatic: $C_6H_5NR_2 \xrightarrow{HONO} O{=}N{-}C_6H_4{-}NR_2$ *p*-Nitroso compound

23.2 Basicity of amines. Basicity constant

Like ammonia, amines are converted into their salts by aqueous mineral acids and are liberated from their salts by aqueous hydroxides. Like ammonia, therefore, amines are more basic than water and less basic than hydroxide ion:

$$\underset{\text{Stronger base}}{RNH_2} + H_3O^+ \longrightarrow RNH_3^+ + \underset{\text{Weaker base}}{H_2O}$$

$$RNH_3^+ + \underset{\text{Stronger base}}{OH^-} \longrightarrow \underset{\text{Weaker base}}{RNH_2} + H_2O$$

We found it convenient to compare acidities of carboxylic acids by measuring the extent to which they give up hydrogen ion to water; the equilibrium constant for this reaction was called the acidity constant, K_a. In the same way, it is convenient to compare basicities of amines by measuring the extent to which they accept hydrogen ion from water; the equilibrium constant for this reaction is called a **basicity constant**, K_b.

$$RNH_2 + H_2O \rightleftharpoons RNH_3^+ + OH^-$$

$$K_b = \frac{[RNH_3^+][OH^-]}{[RNH_2]}$$

(As in the analogous expression for an acidity constant, the concentration of the solvent, water, is omitted.) Each amine has its characteristic K_b; the larger the K_b, the stronger the base.

We must not lose sight of the fact that the principal base in an aqueous solution of an amine (or of ammonia, for that matter) is the *amine* itself, not hydroxide ion. Measurement of $[OH^-]$ is simply a convenient way to compare basicities.

We see in Table 22.1 (p. 729) that aliphatic amines of all three classes have K_b's of about 10^{-3} to 10^{-4} (0.001 to 0.0001); they are thus somewhat stronger bases than ammonia ($K_b = 1.8 \times 10^{-5}$). Aromatic amines, on the other hand, are considerably weaker bases than ammonia, having K_b's of 10^{-9} or less. Substituents on the ring have a marked effect on the basicity of aromatic amines, *p*-nitroaniline, for example, being only 1/4000 as basic as aniline (Table 23.1).

Table 23.1 BASICITY CONSTANTS OF SUBSTITUTED ANILINES

K_b of aniline = 4.2×10^{-10}

	K_b		K_b		K_b
p-NH_2	140×10^{-10}	*m*-NH_2	10×10^{-10}	*o*-NH_2	3×10^{-10}
p-OCH_3	20	*m*-OCH_3	2	*o*-OCH_3	3
p-CH_3	12	*m*-CH_3	5	*o*-CH_3	2.6
p-Cl	1	*m*-Cl	.3	*o*-Cl	.05
p-NO_2	.001	*m*-NO_2	.029	*o*-NO_2	.00006

23.3 Structure and basicity

Let us see how basicity of amines is related to structure. We shall handle basicity just as we handled acidity: we shall compare the stabilities of amines with the stabilities of their ions; the more stable the ion relative to the amine from which it is formed, the more basic the amine.

First of all, amines are more basic than alcohols, ethers, esters, etc., for the same reason that ammonia is more basic than water: nitrogen is less electronegative than oxygen, and can better accommodate the positive charge of the ion.

An aliphatic amine is more basic than ammonia because the electron-releasing alkyl groups tend to disperse the positive charge of the substituted ammonium ion, and therefore stabilize it in a way that is not possible for the unsubstituted ammonium ion. Thus an *ammonium* ion is stabilized by electron release in the same way as a *carbonium* ion (Sec. 5.17). From another point of view, we can consider that an alkyl group pushes electrons toward nitrogen, and thus makes the fourth pair more available for sharing with an acid. (The differences in basicity among primary, secondary, and tertiary aliphatic amines are due to a combination of solvation and electronic factors.)

$$R\rightarrow\overset{\displaystyle H}{\underset{\displaystyle H}{\overset{|}{\underset{|}{N}}}}: + H^+ \rightleftarrows R\rightarrow\overset{\displaystyle H}{\underset{\displaystyle H}{\overset{|}{\underset{|}{N}}}}-H^+$$

R *releases electrons: makes unshared pair more available* — R *releases electrons: stabilizes ion, increases basicity*

How can we account for the fact that aromatic amines are weaker bases than ammonia? Let us compare the structures of aniline and the anilinium ion with the structures of ammonia and the ammonium ion. We see that ammonia and the ammonium ion are each represented satisfactorily by a single structure:

$$\overset{\displaystyle H}{H:\underset{\cdot\cdot}{\overset{\cdot\cdot}{N}}:H} \qquad\qquad \overset{\displaystyle H}{\underset{\displaystyle H}{H:\overset{\cdot\cdot}{\underset{\cdot\cdot}{N}}:H^+}}$$

Ammonia — Ammonium ion

Aniline and anilinium ion contain the benzene ring and therefore are hybrids of the Kekulé structures I and II, and III and IV. This resonance presumably stabilizes

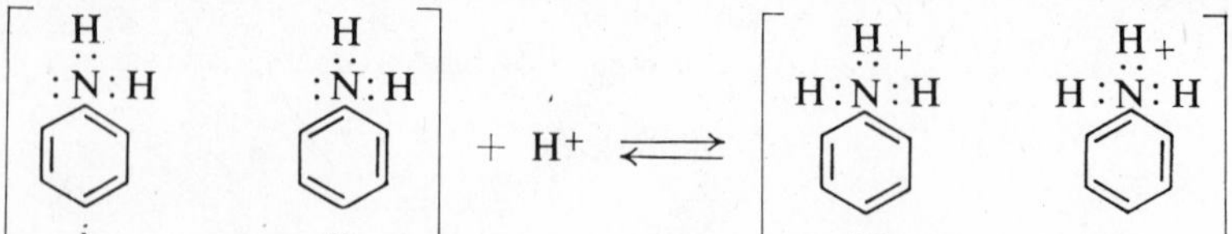

$$\left[\begin{array}{cc} \text{I} & \text{II} \end{array}\right] + H^+ \rightleftharpoons \left[\begin{array}{cc} \text{III} & \text{IV} \end{array}\right]$$

Aniline Anilinium ion

both amine and ion to the same extent. It lowers the energy content of each by the same number of kcal/mole, and hence does not affect the *difference* in their energy contents, that is, does not affect ΔG of ionization. If there were no other factors involved, then, we might expect the basicity of aniline to be about the same as the basicity of ammonia.

However, there are additional structures to be considered. To account for the powerful activating effect of the $-NH_2$ group on electrophilic aromatic substitution (Sec. 11.20), we considered that the intermediate carbonium ion is stabilized by structures in which there is a double bond between nitrogen and the ring; contribution from these structures is simply a way of indicating the tendency for nitrogen to share its fourth pair of electrons and to accept a positive charge. It is generally believed that the $-NH_2$ group tends to share electrons with the ring, not only in the carbonium ion which is the intermediate in electrophilic aromatic substitution, but also in the aniline molecule itself.

Thus aniline is a hybrid not only of structures I and II but also of structures V, VI, and VII. We cannot draw comparable structures for the anilinium ion.

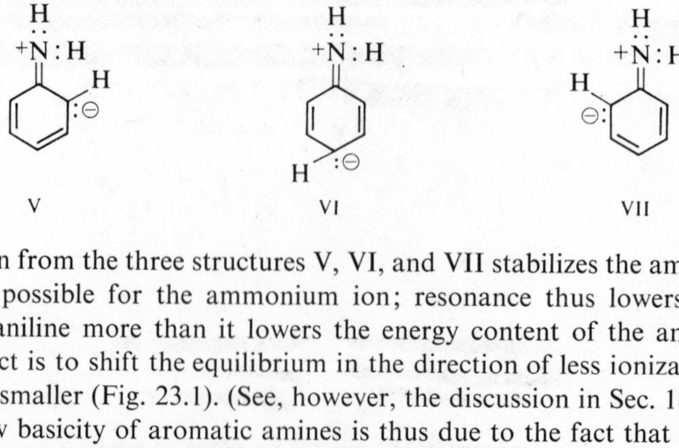

V VI VII

Contribution from the three structures V, VI, and VII stabilizes the amine in a way that is not possible for the ammonium ion; resonance thus lowers the energy content of aniline more than it lowers the energy content of the anilinium ion. The net effect is to shift the equilibrium in the direction of less ionization, that is, to make K_b smaller (Fig. 23.1). (See, however, the discussion in Sec. 18.11.)

The low basicity of aromatic amines is thus due to the fact that the amine is stabilized by resonance to a greater extent than is the ion.

From another point of view, we can say that aniline is a weaker base than ammonia because the fourth pair of electrons is partly shared with the ring and is thus less available for sharing with a hydrogen ion. The tendency (through resonance) for the $-NH_2$ group to release electrons to the aromatic ring makes the ring more reactive toward electrophilic attack; at the same time this tendency necessarily makes the amine less basic. Similar considerations apply to other aromatic amines.

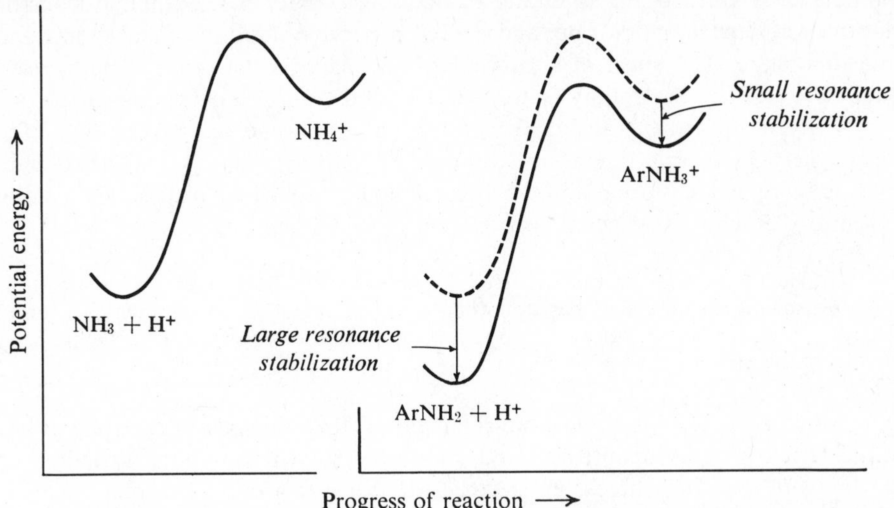

Figure 23.1. Molecular structure and position of equilibrium. Resonance-stabilized aromatic amine is weaker base than ammonia. (Plots aligned with each other for easy comparison.)

23.4 Effect of substituents on basicity of aromatic amines

How is the basicity of an aromatic amine affected by substituents on the ring?

In Table 23.1 (p. 749) we see that an electron-releasing substituent like $—CH_3$ increases the basicity of aniline, and an electron-withdrawing substituent like —X or $—NO_2$ decreases the basicity. These effects are understandable. Electron release tends to disperse the positive charge of the anilinium ion, and thus stabilizes the ion relative to the amine. Electron withdrawal tends to intensify the positive charge of the anilinium ion, and thus destabilizes the ion relative to the amine.

Basicity of Aromatic Amines

NH_2–C_6H_4–G $+ H^+ \rightleftharpoons$ $^+NH_3$–C_6H_4–G

G releases electrons: stabilizes cation, increases basicity

G = $—NH_2$, $—OCH_3$, $—CH_3$

NH_2–C_6H_4–G $+ H^+ \rightleftharpoons$ $^+NH_3$–C_6H_4–G

G withdraws electrons: destabilizes cation, decreases basicity

G = $—NH_3^+$, $—NO_2$, $—SO_3^-$, —COOH, —X

We notice that the base-strengthening substituents are the ones that activate an aromatic ring toward electrophilic substitution; the base-weakening substituents are the ones that deactivate an aromatic ring toward electrophilic substitution (see Sec. 11.5). Basicity depends upon position of equilibrium, and hence on relative stabilities of reactants and products. Reactivity in electrophilic aromatic substitution depends upon rate, and hence on relative stabilities of reactants and transition state. The effect of a particular substituent is the same in both cases, however, since the controlling factor is accommodation of a positive charge.

A given substituent affects the basicity of an amine and the acidity of a carboxylic acid in opposite ways (compare Sec. 18.14). This is to be expected, since basicity depends upon ability to accommodate a positive charge, and acidity depends upon ability to accommodate a negative charge.

Once again we see the operation of the **ortho effect** (Sec. 18.14). Even electron-releasing substituents weaken basicity when they are *ortho* to the amino group, and electron-withdrawing substituents do so to a much greater extent from the *ortho* position than from the *meta* or *para* position.

From another point of view, we can consider that an electron-releasing group pushes electrons toward nitrogen and makes the fourth pair more available for sharing with an acid, whereas an electron-withdrawing group helps pull electrons away from nitrogen and thus makes the fourth pair less available for sharing.

Problem 23.1 (a) Besides destabilizing the anilinium ion, how else might a nitro group affect basicity? (*Hint:* See structures V–VII on p. 750.) (b) Why does the nitro group exert a larger base-weakening effect from the *para* position than from the nearer *meta* position?

Problem 23.2 Draw the structural formula of the product expected (if any) from the reaction of trimethylamine and BF_3.

23.5 Quaternary ammonium salts. Exhaustive methylation. Hofmann elimination

Like ammonia, an amine can react with an alkyl halide; the product is an amine of the next higher class. The alkyl halide undergoes nucleophilic substitution, with the basic amine serving as the nucleophilic reagent. We see that one of

$$\underset{1^\circ}{RNH_2} \xrightarrow{RX} \underset{2^\circ}{R_2NH} \xrightarrow{RX} \underset{3^\circ}{R_3N} \xrightarrow{RX} \underset{4^\circ}{R_4N^+X^-}$$

the hydrogens attached to nitrogen has been replaced by an alkyl group; the reaction is therefore often referred to as *alkylation of amines.* The amine can be aliphatic or aromatic, primary, secondary, or tertiary; the halide is generally an alkyl halide.

We have already encountered alkylation of amines as a side reaction in the preparation of primary amines by the ammonolysis of halides (Sec. 22.10), and as a method of synthesis of secondary and tertiary amines (Sec. 22.13). Let us look at one further aspect of this reaction, the formation of quaternary ammonium salts.

Quaternary ammonium salts are the products of the final stage of alkylation of nitrogen. They have the formula $R_4N^+X^-$. Four organic groups are covalently bonded to nitrogen, and the positive charge of this ion is balanced by some negative ion. When the salt of a primary, secondary, or tertiary amine is treated with hydroxide ion, nitrogen gives up a hydrogen ion and the free amine is liberated. The quaternary ammonium ion, having no proton to give up, is not affected by hydroxide ion.

$$\mathrm{R{:}\overset{R}{\underset{R}{\ddot{N}}}{:}R^+X^-} \xrightarrow{Ag_2O} \mathrm{R{:}\overset{R}{\underset{R}{\ddot{N}}}{:}R^+OH^-} + \mathrm{AgX}$$

Quaternary ammonium salt — Quaternary ammonium hydroxide — *Insoluble*

When a solution of a quaternary ammonium halide is treated with silver oxide, silver halide precipitates. When the mixture is filtered and the filtrate is evaporated to dryness, there is obtained a solid which is free of halogen. An aqueous solution of this substance is strongly alkaline, and is comparable to a solution of sodium hydroxide or potassium hydroxide. A compound of this sort is called a **quaternary ammonium hydroxide**. It has the structure $R_4N^+OH^-$. Its aqueous solution is basic for the same reason that solutions of sodium or potassium hydroxide are basic: the solution contains hydroxide ions.

When a quaternary ammonium hydroxide is heated strongly (to 125° or higher), it decomposes to yield water, a tertiary amine, and an alkene. Trimethyl-*n*-propylammonium hydroxide, for example, yields trimethylamine and propylene:

$$\mathrm{CH_3{-}\overset{CH_3}{\underset{CH_3}{N^+}}{-}CH_2CH_2CH_3\ OH^-} \xrightarrow{heat} \mathrm{CH_3{-}\overset{CH_3}{\underset{CH_3}{N}}} + \mathrm{CH_2{=}CHCH_3} + \mathrm{H_2O}$$

Trimethyl-*n*-propylammonium hydroxide — Trimethylamine — Propylene

This reaction, called the **Hofmann elimination**, is quite analogous to the dehydrohalogenation of an alkyl halide (Sec. 14.18). Most commonly, reaction is E2: hydroxide ion abstracts a proton from carbon; a molecule of tertiary amine is expelled, and the double bond is generated. Bases other than hydroxide ion can be used.

$$\mathrm{-\overset{R_3N^{\oplus}}{\underset{|}{\overset{|}{C}}}-\underset{\underset{OH^-}{H}}{\overset{|}{C}}-} \longrightarrow \mathrm{>C{=}C<} + \mathrm{R_3N{:}} + \mathrm{H_2O}$$

E1 elimination from quaternary ammonium ions is also known. Competing with either E2 or E1 elimination there is, as usual, substitution: either S_N2 or S_N1. (*Problem:* What products would you expect from substitution?)

Orientation in the E2 reaction is typically strongly Hofmann (Sec. 14.21)—not surprisingly, since it was for this reaction that Hofmann formulated his rule. For example:

$$\underset{\text{2-Pentyltrimethyl-ammonium ion}}{\underset{\oplus N(CH_3)_3}{CH_3CH_2CH_2\overset{|}{C}HCH_3}} \xrightarrow[C_2H_5OH]{C_2H_5ONa} \underset{\substack{\text{1-Pentene} \\ 96\%}}{CH_3CH_2CH_2CH{=}CH_2} + \underset{\substack{\text{2-Pentene} \\ 4\%}}{CH_3CH_2CH{=}CHCH_3}$$

The transition state has considerable carbanion character, at least partly because powerful electron withdrawal by the positively charged nitrogen favors development of negative charge. There is preferential abstraction of a proton from the carbon that can best accommodate the partial negative charge: in the example given, from the primary carbon rather than the secondary.

Sulfonium ions, R_3S^+, react similarly to quaternary ammonium ions.

The stereochemistry of Hofmann elimination is commonly *anti*, but less so than was formerly believed. *Syn* elimination is important for certain cyclic compounds, and can be made important even for open-chain compounds by the proper choice of base and solvent. Quaternary ammonium ions are more prone to *syn* elimination than alkyl halides and sulfonates. Electronically, *anti* formation of the double bond is favored in eliminations; but when the alkene character of the transition state is slight—as here—other factors come into play: conformational factors, it has been postulated.

Problem 23.3 Predict the major products of E2 elimination from: (a) 2-methyl-3-pentyltrimethylammonium ion; (b) diethyldi-*n*-propylammonium ion; (c) dimethylethyl(2-chloroethyl)ammonium ion; (d) dimethylethyl-*n*-propylammonium ion.

Problem 23.4 When dimethyl-*tert*-pentylsulfonium ethoxide is heated in ethanol, the alkene obtained is chiefly (86%) 2-methyl-1-butene; when the corresponding sulfonium iodide is heated in ethanol, the alkene obtained is chiefly (86%) 2-methyl-2-butene.

(a) How do you account for the difference in products? (b) From the sulfonium iodide reaction there is also obtained considerable material identified as an ether. What ether would you expect it to be, and how is it formed? (c) What ether would you expect to obtain from the sulfonium ethoxide reaction?

The formation of quaternary ammonium salts, followed by an elimination of the kind just described, is very useful in the determination of the structures of certain complicated nitrogen-containing compounds. The compound, which may be a primary, secondary, or tertiary amine, is converted into the quaternary ammonium hydroxide by treatment with excess methyl iodide and silver oxide. The number of methyl groups taken up by nitrogen depends upon the class of the amine; a primary amine will take up three methyl groups, a secondary amine will take up two, and a tertiary amine only one. This process is known as **exhaustive methylation of amines**.

When heated, a quaternary ammonium hydroxide undergoes elimination to an alkene and a tertiary amine. From the structures of these products it is often possible to deduce the structure of the original amine. As a simple example, contrast the products (I and II) obtained from the following isomeric cyclic amines:

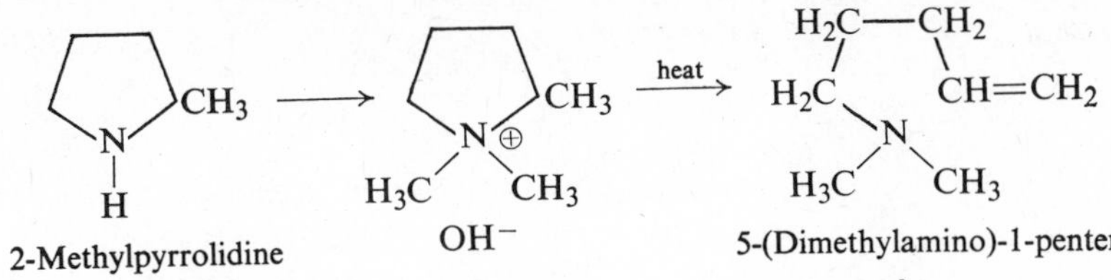

2-Methylpyrrolidine $\longrightarrow$ 1,1,2-trimethylpyrrolidinium OH^- $\xrightarrow{\text{heat}}$ $CH_2=CH-CH_2-CH_2-CH_2-N(CH_3)_2$

5-(Dimethylamino)-1-pentene

I

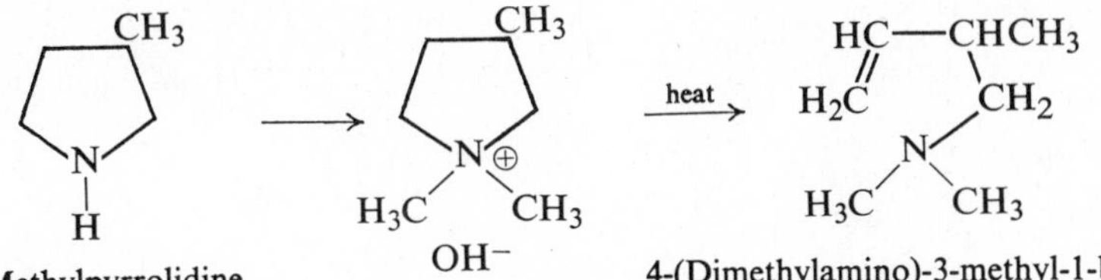

3-Methylpyrrolidine $\longrightarrow$ 1,1,3-trimethylpyrrolidinium OH^- $\xrightarrow{\text{heat}}$ $CH_2=CH-CH(CH_3)-CH_2-N(CH_3)_2$

4-(Dimethylamino)-3-methyl-1-butene

II

Problem 23.5 (a) What products would be expected from the hydrogenation of I and II? (b) How could you prepare an authentic sample of each of these expected hydrogenation products?

Problem 23.6 What products would be expected if I and II were subjected to exhaustive methylation and elimination?

23.6 Conversion of amines into substituted amides

We have learned (Sec. 20.11) that ammonia reacts with acid chlorides of carboxylic acids to yield amides, compounds in which —Cl has been replaced by

$$NH_3 + R-C(=O)Cl \longrightarrow R-C(=O)NH_2$$

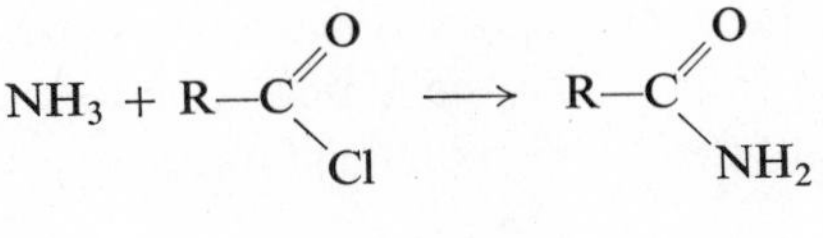

the $-NH_2$ group. Not surprisingly, acid chlorides of sulfonic acids react similarly.

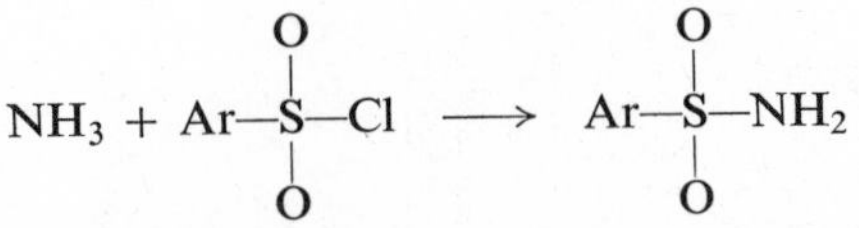

$$NH_3 + Ar-SO_2-Cl \longrightarrow Ar-SO_2-NH_2$$

A sulfonyl chloride A sulfonamide

In these reactions ammonia serves as a nucleophilic reagent, attacking the carbonyl carbon or sulfur and displacing chloride ion. In the process nitrogen loses a proton to a second molecule of ammonia or another base.

In a similar way primary and secondary amines can react with acid chlorides to form **substituted amides**, compounds in which —Cl has been replaced by the —NHR or $-NR_2$ group:

Primary: RNH_2 $\xrightarrow{R'COCl}$ $R'CONHR$ An N-substituted amide

$\xrightarrow{ArSO_2Cl}$ $ArSO_2NHR$ An N-substituted sulfonamide

Secondary: R_2NH $\xrightarrow{R'COCl}$ $R'CONR_2$ An N,N-disubstituted amide

$\xrightarrow{ArSO_2Cl}$ $ArSO_2NR_2$ An N,N-disubstituted sulfonamide

Tertiary: R_3N $\xrightarrow{R'COCl}$ No reaction

$\xrightarrow{ArSO_2Cl}$ No reaction under conditions of Hinsberg test (*but see* Sec. 23.18).

Tertiary amines, although basic, fail to yield amides, presumably because they cannot lose a proton (to stabilize the product) after attaching themselves to carbon or to sulfur. Here is a reaction which requires not only that amines be basic, but also that they possess a hydrogen atom attached to nitrogen. (However, see Sec. 23.19.)

Substituted amides are generally named as derivatives of the unsubstituted amides. For example:

$CH_3C(=O)NHC_2H_5$
N-Ethylacetamide

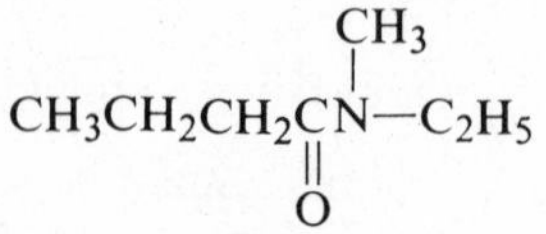

$CH_3CH_2CH_2C(=O)N(CH_3)-C_2H_5$
N-Methyl-N-ethylbutyramide

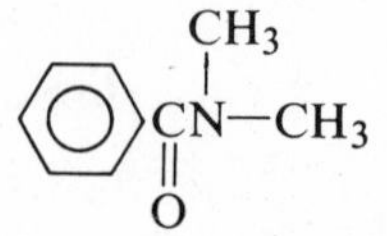

$C_6H_5C(=O)N(CH_3)-CH_3$
N,N-Dimethylbenzamide

In many cases, and particularly where aromatic amines are involved, we are more interested in the amine from which the amide is derived than in the acyl group. In these cases the substituted amide is named as an acyl derivative of the amine. For example:

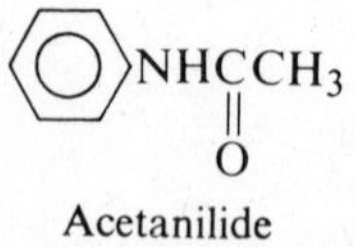

$C_6H_5NHC(=O)CH_3$
Acetanilide

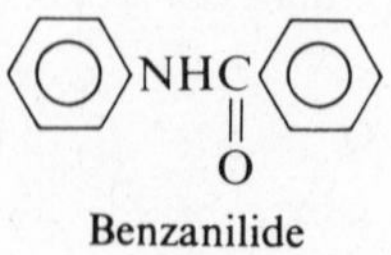

$C_6H_5NHC(=O)C_6H_5$
Benzanilide

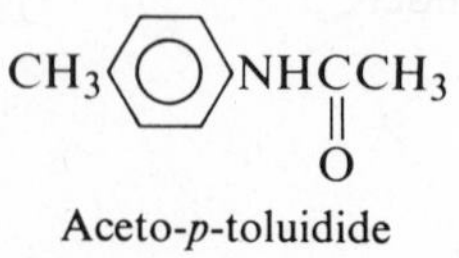

$CH_3C_6H_4NHC(=O)CH_3$
Aceto-*p*-toluidide

Substituted amides of aromatic carboxylic acids or of sulfonic acids are prepared by the Schotten-Baumann technique: the acid chloride is added to the amine in the presence of a base, either aqueous sodium hydroxide or pyridine. For example:

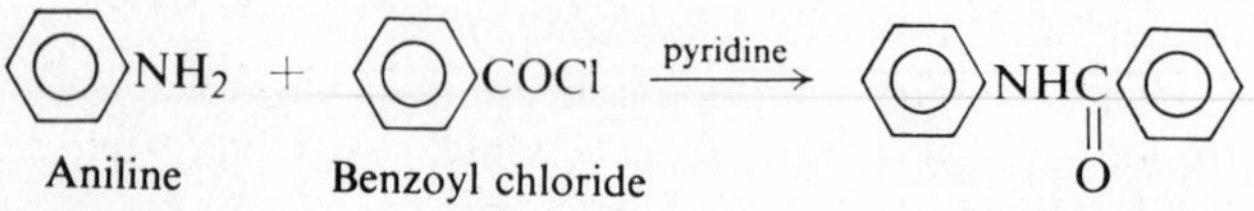

$$C_6H_5NH_2 + C_6H_5COCl \xrightarrow{\text{pyridine}} C_6H_5NHC(=O)C_6H_5$$

Aniline — Benzoyl chloride — Benzanilide

$$(n\text{-}C_4H_9)_2NH + C_6H_5SO_2Cl \xrightarrow{NaOH} C_6H_5SO_2N(C_4H_9)_2$$

Di-*n*-butylamine Benzenesulfonyl chloride N,N-Di-*n*-butylbenzenesulfonamide

Acetylation is generally carried out using acetic anhydride rather than acetyl chloride. For example:

$$o\text{-}CH_3C_6H_4NH_2 + (CH_3CO)_2O \xrightarrow{CH_3COONa} o\text{-}CH_3C_6H_4NHCOCH_3 + CH_3COOH$$

o-Toluidine Acetic anhydride Aceto-*o*-toluidide

Like simple amides, substituted amides undergo hydrolysis; the products are the acid and the amine, although one or the other is obtained as its salt, depending upon the acidity or alkalinity of the medium.

$$C_6H_5CON(CH_3)C_6H_5 + NaOH \xrightarrow{heat} C_6H_5COO^-Na^+ + C_6H_5NHCH_3$$

N-Methylbenzanilide Sodium benzoate N-Methylaniline

$$p\text{-}BrC_6H_4NHCOCH_3 + H_2O + HCl \xrightarrow{heat} CH_3COOH + p\text{-}BrC_6H_4NH_3^+Cl^-$$

p-Bromoacetanilide Acetic acid *p*-Bromoanilinium chloride

Sulfonamides are hydrolyzed more slowly than amides of carboxylic acids; examination of the structures involved shows us what probably underlies this difference. Nucleophilic attack on a trigonal acyl carbon (Sec. 20.4) is relatively unhindered; it involves the temporary attachment of a fourth group, the nucleophilic reagent. Nucleophilic attack on tetrahedral sulfonyl sulfur is relatively hindered; it involves the temporary attachment of a *fifth* group. The tetrahedral

$$R{-}C({=}O){-}W + :Z \longrightarrow R{-}C(O^-)(W){-}Z$$

Trigonal C
Attack relatively unhindered

Tetrahedral C
Stable octet

Acyl nucleophilic substitution

$$Ar{-}SO_2{-}W + :Z \longrightarrow \left[Z{-}S(Ar)(O)(O){-}W\right]^-$$

Tetrahedral S
Attack hindered

Pentavalent S
Unstable decet

Sulfonyl nucleophilic substitution

carbon of the acyl intermediate makes use of the permitted octet of electrons; although sulfur may be able to use more than eight electrons in covalent bonding, this is a less stable system than the octet. Thus both steric and electronic factors tend to make sulfonyl compounds less reactive than acyl compounds.

There is a further contrast between the amides of the two kinds of acids. The substituted amide from a primary amine still has a hydrogen attached to nitrogen, and as a result is *acidic*: in the case of a sulfonamide, this acidity is appreciable, and much greater than for the amide of a carboxylic acid. A monosubstituted sulfonamide is less acidic than a carboxylic acid, but about the same as a phenol (Sec. 24.7); it reacts with aqueous hydroxides to form salts.

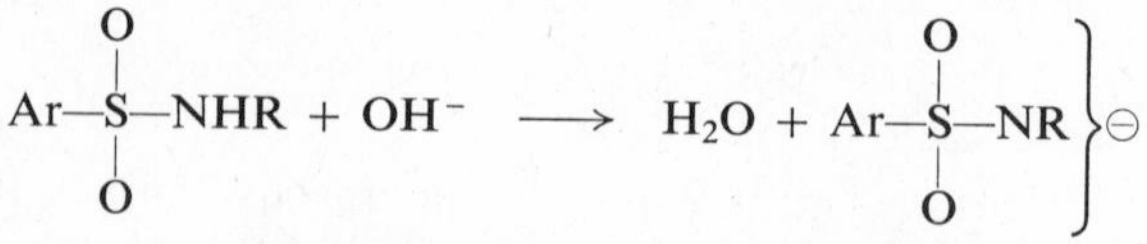

$$Ar{-}\underset{O}{\overset{O}{S}}{-}NHR + OH^- \longrightarrow H_2O + \left\{Ar{-}\underset{O}{\overset{O}{S}}{-}NR\right\}^{\ominus}$$

This difference in acidity, too, is understandable. A sulfonic acid is more acidic than a carboxylic acid because the negative charge of the anion is dispersed over three oxygens instead of just two. In the same way, a sulfonamide is more acidic than the amide of a carboxylic acid because the negative charge is dispersed over two oxygens plus nitrogen instead of over just one oxygen plus nitrogen.

Problem 23.7 (a) Although amides of carboxylic acids are very weakly acidic ($K_a = 10^{-14}$ to 10^{-15}), they are still enormously more acidic than ammonia ($K_a = 10^{-33}$) or amines, RNH_2. Account in detail for this.

(b) Diacetamide, $(CH_3CO)_2NH$, is much more acidic ($K_a = 10^{-11}$) than acetamide ($K_a = 8.3 \times 10^{-16}$), and roughly comparable to benzenesulfonamide ($K_a = 10^{-10}$). How can you account for this?

Problem 23.8 In contrast to carboxylic esters, we know, alkyl sulfonates undergo nucleophilic attack at alkyl carbon. What *two* factors are responsible for this difference

$$R{-}C(=O){-}OR' \quad \text{(Z: attacks C)} \qquad\qquad Ar{-}\underset{O}{\overset{O}{S}}{-}O{-}R \quad \text{(Z: attacks R)}$$

in behavior? (*Hint:* See Sec. 14.6.)

The conversion of an amine into a sulfonamide is used in determining the class of the amine; this is discussed in the section on analysis (Sec. 23.18).

23.7 Ring substitution in aromatic amines

We have already seen that the $-NH_2$, $-NHR$, and $-NR_2$ groups act as powerful activators and *ortho,para* directors in electrophilic aromatic substitution. These effects were accounted for by assuming that the intermediate carbonium ion is stabilized by structures like I and II in which nitrogen bears a positive charge

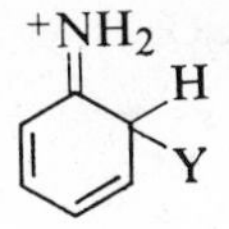

$^+NH_2$ H Y

I

$^+NH_2$ H Y

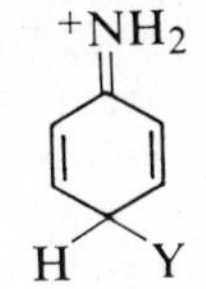

II

and is joined to the ring by a double bond. Such structures are especially stable since in them every atom (except hydrogen) has a complete octet of electrons; indeed, structure I or II *by itself* must pretty well represent the intermediate.

In such structures nitrogen shares more than one pair of electrons with the ring, and hence carries the charge of the "carbonium ion." Thus the basicity of nitrogen accounts for one more characteristic of aromatic amines.

The acetamido group, $—NHCOCH_3$, is also activating and *ortho,para*-directing, but less powerfully so than a free amino group. Electron withdrawal by oxygen of the carbonyl group makes the nitrogen of an amide a much poorer source of electrons than the nitrogen of an amine. Electrons are less available for sharing with a hydrogen ion, and therefore amides are much weaker bases than amines: amides of carboxylic acids do not dissolve in dilute aqueous acids. Electrons are less available for sharing with an aromatic ring, and therefore an acetamido group activates an aromatic ring less strongly than an amino group.

More precisely, electron withdrawal by carbonyl oxygen destabilizes a positive charge on nitrogen, whether this charge is acquired by *protonation* or by *electrophilic attack on the ring.*

(We have seen (Sec. 11.5) that the $—NR_3^+$ group is a powerful deactivator and *meta* director. In a quaternary ammonium salt, nitrogen no longer has electrons to share with the ring; on the contrary, the full-fledged positive charge on nitrogen makes the group strongly electron-attracting.)

In electrophilic substitution, the chief problem encountered with aromatic amines is that they are *too* reactive. In halogenation, substitution tends to occur at every available *ortho* or *para* position. For example:

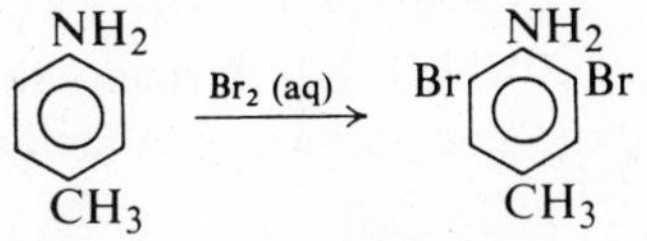

NH_2 / CH_3 $\xrightarrow{Br_2\ (aq)}$ NH_2 / Br Br / CH_3

p-Toluidine 3,5-Dibromo-4-aminotoluene

Nitric acid not only nitrates, but oxidizes the highly reactive ring as well, with loss of much material as tar. Furthermore, in the strongly acidic nitration medium, the amine is converted into the anilinium ion; substitution is thus controlled not by the $—NH_2$ group but by the $—NH_3^+$ group which, because of its positive charge, directs much of the substitution to the *meta* position.

There is, fortunately, a simple way out of these difficulties. We *protect* the amino group: we acetylate the amine, then carry out the substitution, and finally hydrolyze the amide to the desired substituted amine. For example:

$$p\text{-Toluidine} \xrightarrow{(CH_3CO)_2O} \text{Aceto-}p\text{-toluidide} \xrightarrow[CH_3COOH]{Br_2} \xrightarrow[H^+]{H_2O} \text{3-Bromo-4-aminotoluene}$$

NH2 / CH3 — p-Toluidine; NHCOCH3 / CH3 — Aceto-p-toluidide; NHCOCH3, Br / CH3; NH2, Br / CH3 — 3-Bromo-4-aminotoluene

$$\text{Acetanilide} \xrightarrow{HNO_3,\ H_2SO_4,\ 15°} p\text{-Nitroacetanilide} \xrightarrow[\text{heat}]{H_2O,\ H^+} p\text{-Nitroaniline}$$

NHCOCH3 — Acetanilide; NHCOCH3 / NO2 — p-Nitroacetanilide; NH2 / NO2 — p-Nitroaniline

Problem 23.9 Nitration of un-acetylated aniline yields a mixture of about two-thirds *meta* and one-third *para* product. Since almost all the aniline is in the form of the anilinium ion, how do you account for the fact that even more *meta* product is not obtained?

23.8 Sulfonation of aromatic amines. Dipolar ions

Aniline is usually sulfonated by "baking" the salt, anilinium hydrogen sulfate, at 180–200°; the chief product is the *p*-isomer. In this case we cannot discuss orientation on our usual basis of which isomer is formed *faster*. Sulfonation is

$$\text{Aniline} \xrightarrow{H_2SO_4} \text{Anilinium hydrogen sulfate} \xrightarrow{180\text{–}200°} \text{Sulfanilic acid}$$

NH2 — Aniline; $NH_3{}^+HSO_4{}^-$ — Anilinium hydrogen sulfate; $NH_3{}^+$ / $SO_3{}^-$ — Sulfanilic acid

known to be reversible, and the *p*-isomer is known to be the most stable isomer; it may well be that the product obtained, the *p*-isomer, is determined by the position of an equilibrium and not by relative rates of formation (see Sec. 8.22 and Sec. 12.11). It also seems likely that, in some cases at least, sulfonation of amines proceeds by a mechanism that is entirely different from ordinary aromatic substitution.

Whatever the mechanism by which it is formed, the chief product of this reaction is *p*-aminobenzenesulfonic acid, known as **sulfanilic acid**; it is an important and interesting compound.

First of all, its properties are not those we would expect of a compound containing an amino group and a sulfonic acid group. Both aromatic amines and aromatic sulfonic acids have low melting points; benzenesulfonic acid, for example, melts at 66°, and aniline at $-6°$. Yet sulfanilic acid has such a high melting point that on being heated it decomposes (at 280–300°) before its melting point can be reached. Sulfonic acids are generally very soluble in water; indeed, we have seen that the sulfonic acid group is often introduced into a molecule to make it water-soluble. Yet sulfanilic acid is not only insoluble in organic solvents, but also nearly insoluble in water. Amines dissolve in aqueous mineral acids because of their conversion into water-soluble salts. Sulfanilic acid is soluble in aqueous bases but insoluble in aqueous acids.

These properties of sulfanilic acid are understandable when we realize that sulfanilic acid actually has the structure I which contains the $—NH_3^+$ and $—SO_3^-$ groups. Sulfanilic acid is a salt, but of a rather special kind, called a **dipolar ion**

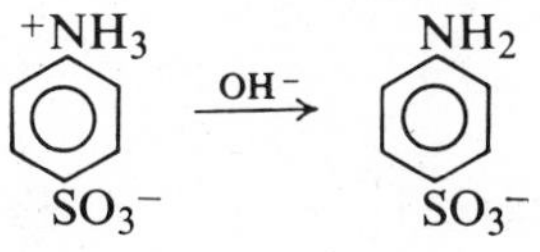

$$\underset{\substack{\text{I} \\ \textit{Insoluble in water}}}{^{+}NH_3—C_6H_4—SO_3^-} \xrightarrow{OH^-} \underset{\substack{\text{II} \\ \textit{Soluble in water}}}{NH_2—C_6H_4—SO_3^-}$$

(sometimes called a *zwitterion*, from the German, *Zwitter*, hermaphrodite). It is the product of reaction between an acidic group and a basic group that are part of the same molecule. The hydrogen ion is attached to nitrogen rather than oxygen simply because the $—NH_2$ group is a stronger base than the $—SO_3^-$ group. A high melting point and insolubility in organic solvents are properties we would expect of a salt. Insolubility in water is not surprising, since many salts are insoluble in water. In alkaline solution, the strongly basic hydroxide ion pulls hydrogen ion away from the weakly basic $—NH_2$ group to yield the *p*-aminobenzenesulfonate ion (II), which, like most sodium salts, is soluble in water. In aqueous acid, however, the sulfanilic acid structure is not changed, and therefore the compound remains insoluble; sulfonic acids are strong acids and their anions (very weak bases) show little tendency to accept hydrogen ion from H_3O^+.

We can expect to encounter dipolar ions whenever we have a molecule containing both an amino group and an acid group, providing the amine is more basic than the anion of the acid.

Problem 23.10 *p*-Aminobenzoic acid is not a dipolar ion, whereas glycine (aminoacetic acid) is a dipolar ion. How can you account for this?

23.9 Sulfanilamide. The sulfa drugs

The amide of sulfanilic acid (*sulfanilamide*) and certain related substituted amides are of considerable medical importance as the *sulfa drugs*. Although they have been supplanted to a wide extent by the antibiotics (such as penicillin, terramycin, chloromycetin, and aureomycin), the sulfa drugs still have their medical uses, and make up a considerable portion of the output of the pharmaceutical industry.

Sulfonamides are prepared by the reaction of a sulfonyl chloride with ammonia or an amine. The presence in a sulfonic acid molecule of an amino group, however, poses a special problem: if sulfanilic acid were converted to the acid chloride, the sulfonyl group of one molecule could attack the amino group of another to form an amide linkage. This problem is solved by protecting the amino group through acetylation prior to the preparation of the sulfonyl chloride. Sulfanilamide and related compounds are generally prepared in the following way:

$$\underset{\text{Aniline}}{C_6H_5NH_2} \xrightarrow{(CH_3CO)_2O} \underset{\text{Acetanilide}}{C_6H_5NHCOCH_3} \xrightarrow{ClSO_3H} \underset{p\text{-Acetamidobenzenesulfonyl chloride}}{p\text{-}CH_3CONHC_6H_4SO_2Cl}$$

$$p\text{-}CH_3CONHC_6H_4SO_2Cl \xrightarrow{NH_3} p\text{-}CH_3CONHC_6H_4SO_2NH_2 \xrightarrow[H^+]{H_2O} \underset{\text{Sulfanilamide}}{p\text{-}H_2NC_6H_4SO_2NH_2}$$

$$p\text{-}CH_3CONHC_6H_4SO_2Cl \xrightarrow{RNH_2} p\text{-}CH_3CONHC_6H_4SO_2NHR \xrightarrow[H^+]{H_2O} \underset{\text{Substituted sulfanilamide}}{p\text{-}H_2NC_6H_4SO_2NHR}$$

The selective removal of the acetyl group in the final step is consistent with the general observation that amides of carboxylic acids are more easily hydrolyzed than amides of sulfonic acids.

$$CH_3\text{—}\overset{O}{\overset{\|}{C}}\text{—}NH\text{—}C_6H_4\text{—}SO_2\text{—}NH_2 \xrightarrow[\text{heat}]{\text{dilute HCl}} \underset{\text{Sulfanilamide}}{H_2N\text{—}C_6H_4\text{—}SO_2\text{—}NH_2}$$

Hydrolysis occurs here

The antibacterial activity—and toxicity—of a sulfanilamide stems from a rather simple fact: enzymes in the bacteria (and in the patients) confuse it for *p*-aminobenzoic acid, which is an essential metabolite. In what is known as *metabolite antagonism*, the sulfanilamide competes with *p*-aminobenzoic acid for reactive

p-Aminobenzoic acid   Substituted sulfanilamide

sites on the enzymes; deprived of the essential metabolite, the organism fails to reproduce, and dies.

Just how good a drug the sulfanilamide is depends upon the nature of the group R attached to amido nitrogen. This group must confer just the right degree of acidity to the amido hydrogen (Sec. 23.6), but acidity is clearly only one of the factors involved. Of the hundreds of such compounds that have been synthesized, only a half dozen or so have had the proper combination of high antibacterial activity and low toxicity to human beings that is necessary for an effective drug; in nearly all these effective compounds the group R contains a heterocyclic ring (Chap. 31).

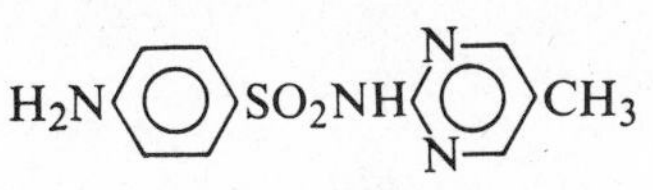

$H_2N\langle\bigcirc\rangle SO_2NH\langle$pyrimidine$\rangle CH_3$ HOOCCH$_2$CH$_2$CONH$\langle\bigcirc\rangle SO_2NH\langle$thiazole$\rangle$

Sulfamerazine 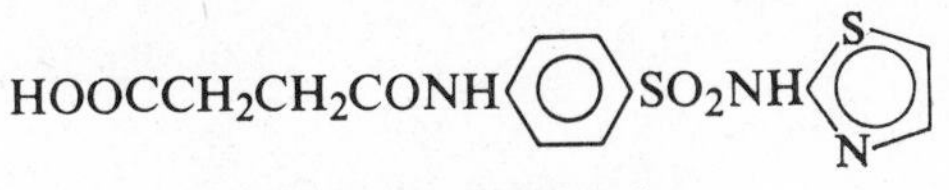 Succinoylsulfathiazole

23.10 Reactions of amines with nitrous acid

Each class of amine yields a different kind of product in its reaction with nitrous acid, HONO. This unstable reagent is generated in the presence of the amine by the action of mineral acid on sodium nitrite.

Primary aromatic amines react with nitrous acid to yield *diazonium salts*; this is one of the most important reactions in organic chemistry. Following sections are devoted to the preparation and properties of aromatic diazonium salts.

$$ArNH_2 + NaNO_2 + 2HX \xrightarrow{\text{cold}} ArN_2^+X^- + NaX + 2H_2O$$

1° aromatic amine A diazonium salt

Primary aliphatic amines also react with nitrous acid to yield diazonium salts; but since aliphatic diazonium salts are quite unstable and break down to yield a complicated mixture of organic products (see Problem 23.11, below), this reaction is of little synthetic value. The fact that nitrogen is evolved quantitatively is of some

$$RNH_2 + NaNO_2 + HX \longrightarrow [RN_2^+X^-] \xrightarrow{H_2O} N_2 + \text{mixture of alcohols and alkenes}$$

1° aliphatic amine *Unstable*

importance in analysis, however, particularly of amino acids and proteins.

Problem 23.11 The reaction of *n*-butylamine with sodium nitrite and hydrochloric acid yields nitrogen and the following mixture: *n*-butyl alcohol, 25%; *sec*-butyl alcohol, 13%; 1-butene and 2-butene, 37%; *n*-butyl chloride, 5%; *sec*-butyl chloride, 3%. (a) What is the most likely intermediate common to all of these products? (b) Outline reactions that account for the various products.

Problem 23.12 Predict the organic products of the reaction of: (a) isobutylamine with nitrous acid; (b) neopentylamine with nitrous acid.

Secondary amines, both aliphatic and aromatic, react with nitrous acid to yield N-nitrosoamines.

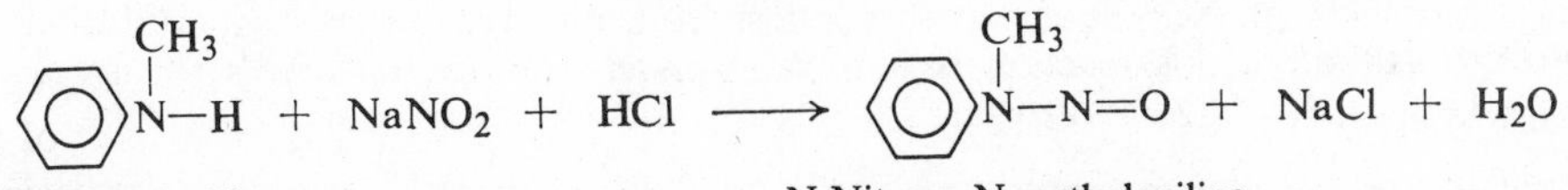

$$C_6H_5N(CH_3)\text{—H} + NaNO_2 + HCl \longrightarrow C_6H_5N(CH_3)\text{—N=O} + NaCl + H_2O$$

N-Methylaniline N-Nitroso-N-methylaniline

Tertiary aromatic amines undergo ring substitution, to yield compounds in which a nitroso group, —N=O, is joined to carbon; thus N,N-dimethylaniline yields chiefly *p*-nitroso-N,N-dimethylaniline.

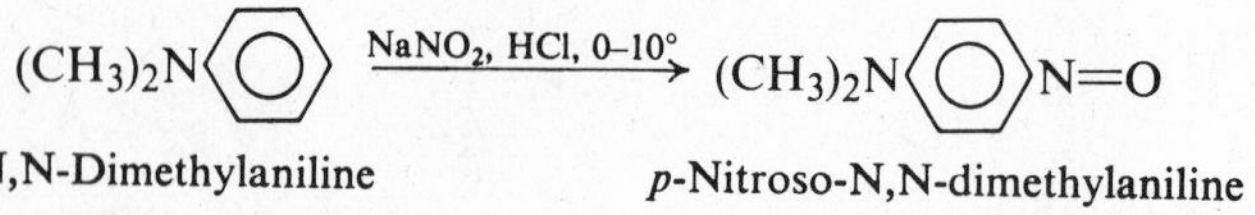

$$(CH_3)_2N\text{-}C_6H_5 \xrightarrow{NaNO_2,\ HCl,\ 0\text{–}10°} (CH_3)_2N\text{-}C_6H_4\text{-}N{=}O$$

N,N-Dimethylaniline — *p*-Nitroso-N,N-dimethylaniline

Ring nitrosation is an electrophilic aromatic substitution reaction, in which the attacking reagent is either the *nitrosonium ion*, ^+NO, or some species (like $H_2\overset{+}{O}$—NO or NOCl) that can easily transfer ^+NO to the ring. The nitrosonium ion is very weakly electrophilic compared with the reagents involved in nitration, sulfonation, halogenation, and the Friedel-Crafts reaction; nitrosation ordinarily occurs only in rings bearing the powerfully activating dialkylamino (—NR_2) or hydroxy (—OH) group.

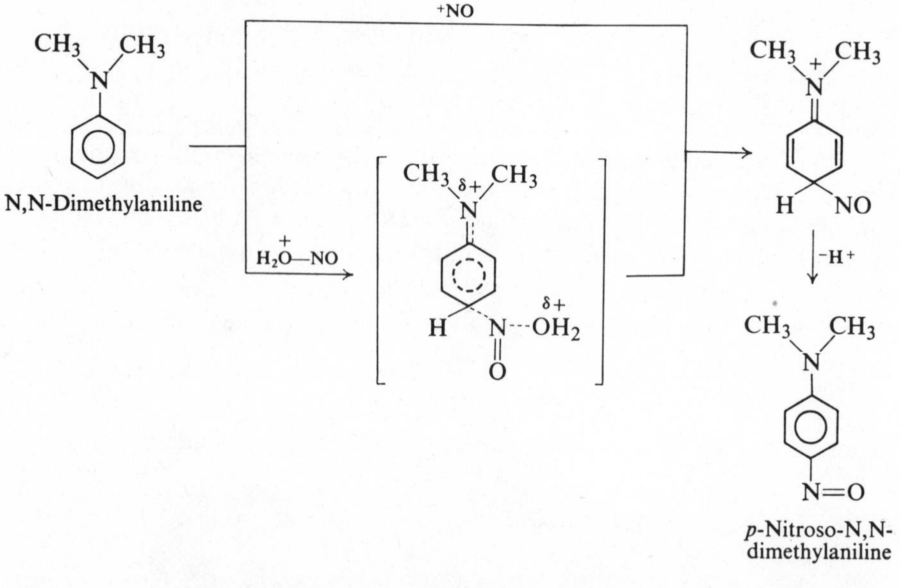

^+NO

N,N-Dimethylaniline

$H_2\overset{+}{O}$—NO

$-H^+$

p-Nitroso-N,N-dimethylaniline

Despite the differences in final product, the reaction of nitrous acid with all these amines involves the same initial step: *electrophilic attack by* ^+NO *with displacement of* H^+. This attack occurs at the position of highest electron availability in primary and secondary amines: at nitrogen. Tertiary aromatic amines are attacked at the highly reactive ring.

Tertiary aliphatic amines (and, to an extent, tertiary aromatic amines, too, particularly if the *para* position is blocked) react with nitrous acid to yield an N-nitroso derivative of a *secondary* amine; the group that is lost from nitrogen appears as an aldehyde or ketone. Although this reaction is not really understood, it too seems to involve the initial attack by ^+NO on nitrogen.

Problem 23.13 (a) Write equations to show how the molecule $H_2\overset{+}{O}$—NO is formed in the nitrosating mixture. (b) Why can this transfer ^+NO to the ring more easily than HONO can? (c) Write equations to show how NOCl can be formed from $NaNO_2$ and aqueous hydrochloric acid. (d) Why is NOCl a better nitrosating agent than HONO?

Problem 23.14 (a) Which, if either, of the following seems likely? (i) The ring of N-methylaniline is much less reactive toward electrophilic attack than the ring of N,N-dimethylaniline. (ii) Nitrogen of N-methylaniline is much more reactive toward electrophilic attack than nitrogen of N,N-dimethylaniline.

(b) How do you account for the fact that the two amines give different products with nitrous acid?

23.11 Diazonium salts. Preparation and reactions

When a primary aromatic amine, dissolved or suspended in cold aqueous mineral acid, is treated with sodium nitrite, there is formed a diazonium salt.

$$\underset{\text{1° aromatic amine}}{ArNH_2} + NaNO_2 + 2HX \xrightarrow{\text{cold}} \underset{\text{A diazonium salt}}{ArN{\equiv}N{:}^+X^-} + NaX + 2H_2O$$

Since diazonium salts slowly decompose even at ice-bath temperatures, the solution is used immediately after preparation.

The large number of reactions undergone by diazonium salts may be divided into two classes: **replacement**, in which nitrogen is lost as N_2, and some other atom or group becomes attached to the ring in its place; and **coupling**, in which the nitrogen is retained in the product.

REACTIONS OF DIAZONIUM SALTS

1. Replacement of nitrogen

$$ArN_2^+ + :Z \longrightarrow ArZ + N_2$$

(a) **Replacement by —Cl, —Br, and —CN. Sandmeyer reaction.** Discussed in Secs. 23.12–23.13.

$$ArN_2^+ \xrightarrow{CuCl} ArCl + N_2$$

$$ArN_2^+ \xrightarrow{CuBr} ArBr + N_2$$

$$ArN_2^+ \xrightarrow{CuCN} ArCN + N_2$$

Examples:

$$\underset{o\text{-Toluidine}}{o\text{-}CH_3C_6H_4NH_2} \xrightarrow{NaNO_2,\ HCl} \underset{o\text{-Toluenediazonium chloride}}{o\text{-}CH_3C_6H_4N_2^+Cl^-} \xrightarrow{CuCl} \underset{o\text{-Chlorotoluene}}{o\text{-}CH_3C_6H_4Cl} + N_2$$

$$\underset{o\text{-Toluidine}}{o\text{-}CH_3C_6H_4NH_2} \xrightarrow{NaNO_2,\ H_2SO_4} o\text{-}CH_3C_6H_4N_2^+HSO_4^- \xrightarrow{CuBr} \underset{o\text{-Bromotoluene}}{o\text{-}CH_3C_6H_4Br} + N_2$$

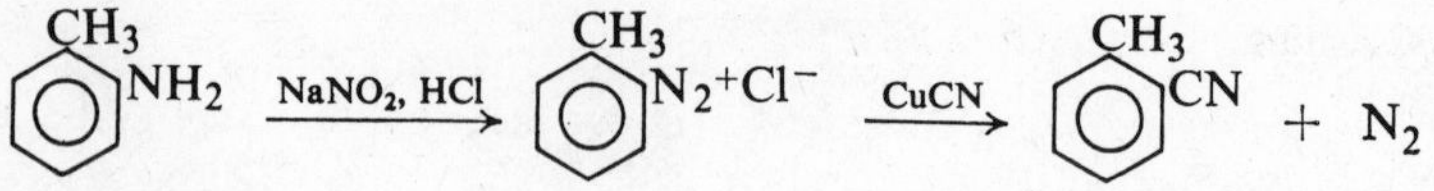

$$o\text{-}CH_3C_6H_4NH_2 \xrightarrow{NaNO_2,\ HCl} o\text{-}CH_3C_6H_4N_2^+Cl^- \xrightarrow{CuCN} o\text{-}CH_3C_6H_4CN + N_2$$

o-Toluidine → *o*-Tolunitrile

(b) Replacement by —I. Discussed in Sec. 23.12.

$$ArN_2^+ + I^- \longrightarrow ArI + N_2$$

Example:

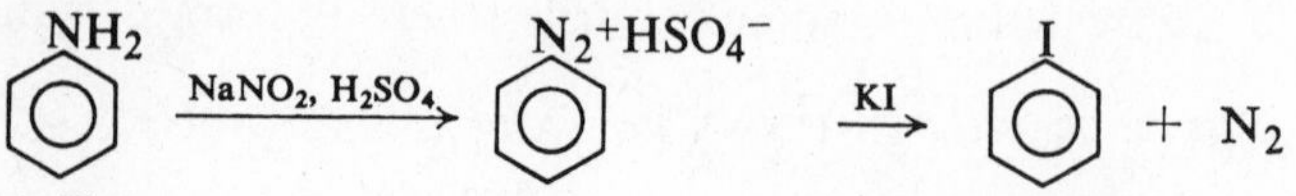

$$C_6H_5NH_2 \xrightarrow{NaNO_2,\ H_2SO_4} C_6H_5N_2^+HSO_4^- \xrightarrow{KI} C_6H_5I + N_2$$

Aniline → Iodobenzene

(c) Replacement by —F. Discussed in Sec. 23.12.

$$ArN_2^+BF_4^- \xrightarrow{heat} ArF + N_2 + BF_3$$

Example:

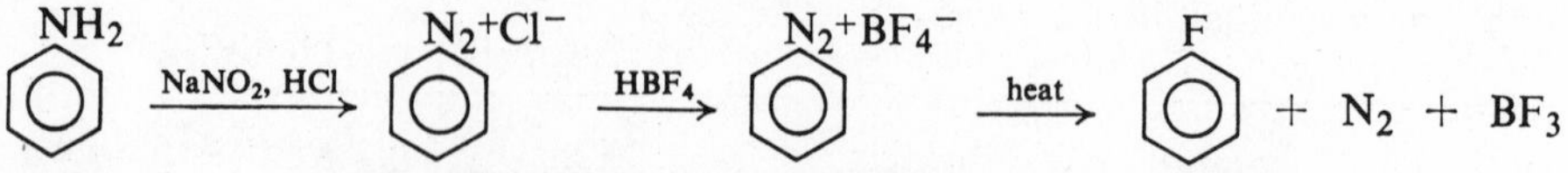

$$C_6H_5NH_2 \xrightarrow{NaNO_2,\ HCl} C_6H_5N_2^+Cl^- \xrightarrow{HBF_4} C_6H_5N_2^+BF_4^- \xrightarrow{heat} C_6H_5F + N_2 + BF_3$$

Aniline → Benzenediazonium chloride → Benzenediazonium fluoborate (*Isolated as crystalline salt*) → Fluorobenzene

(d) Replacement by —OH. Discussed in Sec. 23.14.

$$ArN_2^+ + H_2O \xrightarrow{H^+} ArOH + N_2$$

A phenol

Example:

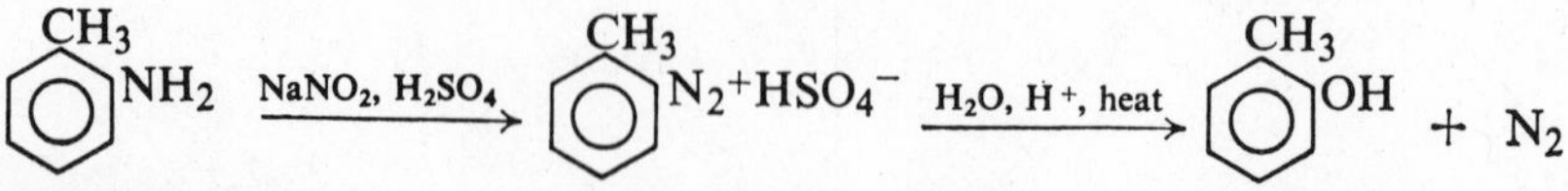

$$o\text{-}CH_3C_6H_4NH_2 \xrightarrow{NaNO_2,\ H_2SO_4} o\text{-}CH_3C_6H_4N_2^+HSO_4^- \xrightarrow{H_2O,\ H^+,\ heat} o\text{-}CH_3C_6H_4OH + N_2$$

o-Toluidine → *o*-Cresol

(e) Replacement by —H. Discussed in Sec. 23.15.

$$ArN_2^+ + H_3PO_2 \xrightarrow{H_2O} ArH + H_3PO_3 + N_2$$

Example:

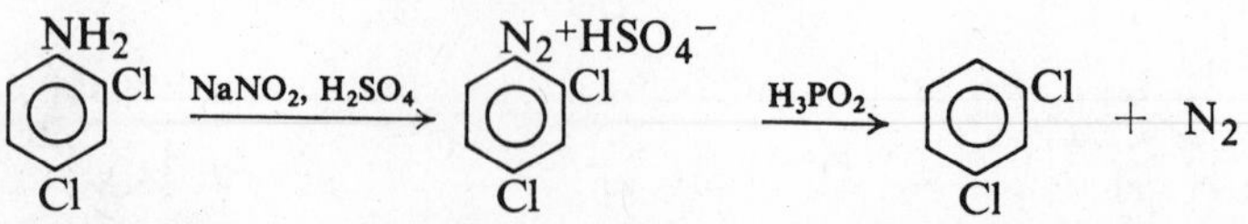

$$2,4\text{-}Cl_2C_6H_3NH_2 \xrightarrow{NaNO_2,\ H_2SO_4} 2,4\text{-}Cl_2C_6H_3N_2^+HSO_4^- \xrightarrow{H_3PO_2} m\text{-}C_6H_4Cl_2 + N_2$$

2,4-Dichloroaniline → *m*-Dichlorobenzene

2. Coupling. Discussed in Sec. 23.17.

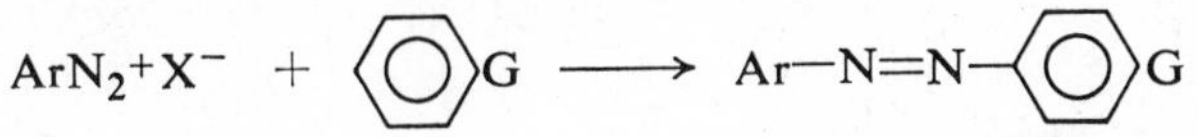

$$ArN_2^+X^- + C_6H_5G \longrightarrow Ar{-}N{=}N{-}C_6H_4G$$

An azo compound

G *must be a strongly electron-releasing group:* OH, NR_2, NHR, NH_2

Example:

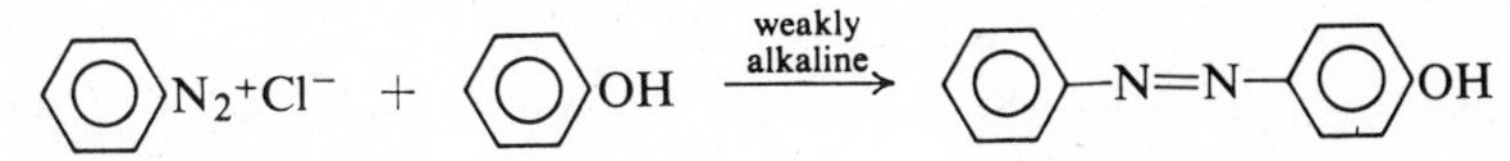

$$C_6H_5N_2^+Cl^- + C_6H_5OH \xrightarrow{\text{weakly alkaline}} C_6H_5{-}N{=}N{-}C_6H_4OH$$

Benzenediazonium chloride — Phenol — *p*-Hydroxyazobenzene *p*-(Phenylazo)phenol

Replacement of the diazonium group is the best general way of introducing F, Cl, Br, I, CN, OH, and H into an aromatic ring. Diazonium salts are valuable in synthesis not only because they react to form so many classes of compounds, but also because they can be prepared from nearly all primary aromatic amines. There are few groups whose presence in the molecule interferes with diazotization; in this respect, diazonium salts are quite different from Grignard reagents (Sec. 15.15). The amines from which diazonium compounds are prepared are readily obtained from the corresponding nitro compounds, which are prepared by direct nitration. Diazonium salts are thus the most important link in the sequence:

$$ArH \longrightarrow ArNO_2 \longrightarrow ArNH_2 \longrightarrow ArN_2^+ \longrightarrow \begin{cases} Ar{-}F \\ Ar{-}Cl \\ Ar{-}Br \\ Ar{-}I \\ Ar{-}CN \longrightarrow Ar{-}COOH \\ Ar{-}OH \\ Ar{-}H \end{cases}$$

In addition to the atoms and groups just listed, there are dozens of other groups that can be attached to an aromatic ring by replacement of the diazonium nitrogen, as, for example, —Ar, $—NO_2$, —OR, —SH, —SR, —NCS, —NCO, $—PO_3H_2$, $—AsO_3H_2$, $—SbO_3H_2$; the best way to introduce most of these groups is via diazotization.

The coupling of diazonium salts with aromatic phenols and amines yields *azo compounds*, which are of tremendous importance to the dye industry.

23.12 Diazonium salts. Replacement by halogen. Sandmeyer reaction

Replacement of the diazonium group by —Cl or —Br is carried out by mixing the solution of freshly prepared diazonium salt with cuprous chloride or cuprous bromide. At room temperature, or occasionally at elevated temperatures, nitrogen is steadily evolved, and after several hours the aryl chloride or aryl bromide can be isolated from the reaction mixture. This procedure, using cuprous halides, is generally referred to as the **Sandmeyer reaction.**

$$ArN_2^+X^- \xrightarrow{CuX} ArX + N_2$$

Sometimes the synthesis is carried out by a modification known as the *Gattermann reaction*, in which copper powder and hydrogen halide are used in place of the cuprous halide.

Replacement of the diazonium group by —I does not require the use of a cuprous halide or copper; the diazonium salt and potassium iodide are simply mixed together and allowed to react.

$$ArN_2^+X^- + I^- \longrightarrow ArI + N_2 + X^-$$

Replacement of the diazonium group by —F is carried out in a somewhat different way. Addition of fluoboric acid, HBF_4, to the solution of diazonium salt causes the precipitation of the diazonium fluoborate, $ArN_2^+BF_4^-$, which can be collected on a filter, washed, and dried. The diazonium fluoborates are unusual among diazonium salts in being fairly stable compounds. On being heated, the dry diazonium fluoborate decomposes to yield the aryl fluoride, boron trifluoride,

$$ArN_2^+X^- \xrightarrow{HBF_4} ArN_2^+BF_4^- \xrightarrow{heat} ArF + BF_3 + N_2$$

and nitrogen. An analogous procedure involves the diazonium hexafluorophosphate, $ArN_2^+PF_6^-$.

The advantages of the synthesis of aryl halides from diazonium salts will be discussed in detail in Sec. 25.3. Aryl fluorides and iodides cannot generally be prepared by direct halogenation. Aryl chlorides and bromides can be prepared by direct halogenation, but, when a mixture of *o*- and *p*-isomers is obtained, it is difficult to isolate the pure compounds because of their similarity in boiling point. Diazonium salts ultimately go back to nitro compounds, which are usually obtainable in pure form.

23.13 Diazonium salts. Replacement by —CN. Synthesis of carboxylic acids

Replacement of the diazonium group by —CN is carried out by allowing the diazonium salt to react with cuprous cyanide. To prevent loss of cyanide as HCN, the diazonium solution is neutralized with sodium carbonate before being mixed with the cuprous cyanide.

$$ArN_2^+X^- \xrightarrow{CuCN} ArCN + N_2$$

Hydrolysis of nitriles yields carboxylic acids. The synthesis of nitriles from diazonium salts thus provides us with an excellent route from nitro compounds to carboxylic acids. For example:

$$\underset{\substack{p\text{-Toluic}\\ \text{acid}}}{p\text{-}CH_3C_6H_4COOH} \xleftarrow[H^+]{H_2O} \underset{p\text{-Tolunitrile}}{p\text{-}CH_3C_6H_4CN} \xleftarrow{CuCN} \underset{\substack{p\text{-Toluenediazonium}\\ \text{chloride}}}{p\text{-}CH_3C_6H_4N_2^+Cl^-} \xleftarrow[HCl]{NaNO_2} \underset{p\text{-Toluidine}}{p\text{-}CH_3C_6H_4NH_2} \xleftarrow[HCl]{Fe} \underset{p\text{-Nitrotoluene}}{p\text{-}CH_3C_6H_4NO_2} \xleftarrow[H_2SO_4]{HNO_3} \underset{\text{Toluene}}{C_6H_5CH_3}$$

This way of making aromatic carboxylic acids is more generally useful than either carbonation of a Grignard reagent or oxidation of side chains. We have just seen that pure bromo compounds, which are needed to prepare the Grignard reagent, are themselves most often prepared via diazonium salts; furthermore, there are many groups that interfere with the preparation and use of the Grignard reagent (Sec. 15.15). The nitro group can generally be introduced into a molecule more readily than an alkyl side chain; furthermore, conversion of a side chain into a carboxyl group cannot be carried out on molecules that contain other groups sensitive to oxidation.

23.14 Diazonium salts. Replacement by —OH. Synthesis of phenols

Diazonium salts react with water to yield phenols. This reaction takes place

$$ArN_2^+X^- + H_2O \longrightarrow ArOH + N_2 + H^+$$

slowly in the ice-cold solutions of diazonium salts, and is the reason diazonium salts are used immediately upon preparation; at elevated temperatures it can be made the chief reaction of diazonium salts.

As we shall see, phenols can couple with diazonium salts to form azo compounds (Sec. 23.17); the more acidic the solution, however, the more slowly this coupling occurs. To minimize coupling during the synthesis of a phenol, therefore —coupling, that is, between phenol that has been formed and diazonium ion that has not yet reacted—the diazonium solution is added slowly to a large volume of boiling dilute sulfuric acid.

This is the best general way to make the important class of compounds, the phenols.

23.15 Diazonium salts. Replacement by —H

Replacement of the diazonium group by —H can be brought about by a number of reducing agents; perhaps the most useful of these is *hypophosphorus acid*, H_3PO_2. The diazonium salt is simply allowed to stand in the presence of the hypophosphorous acid; nitrogen is lost, and hypophosphorous acid is oxidized to phosphorous acid:

$$ArN_2^+X^- + H_3PO_2 + H_2O \longrightarrow ArH + N_2 + H_3PO_3 + HX$$

An especially elegant way of carrying out this replacement is to use hypophosphorous acid as the diazotizing acid. The amine is dissolved in hypophosphorous acid, and sodium nitrite is added; the diazonium salt is reduced as fast as it is formed.

This reaction of diazonium salts provides a method of removing an $—NH_2$ or $—NO_2$ group from an aromatic ring. This process can be extremely useful in synthesis, as is shown in some of the examples in the following section.

23.16 Syntheses using diazonium salts

Let us look at a few examples of how diazonium salts can be used in organic synthesis.

To begin with, we might consider some rather simple compounds, the three isomeric bromotoluenes. The best synthesis of each employs diazotization, but not for the same purpose in the three cases. The *o*- and *p*-bromotoluenes are prepared from the corresponding *o*- and *p*-nitrotoluenes:

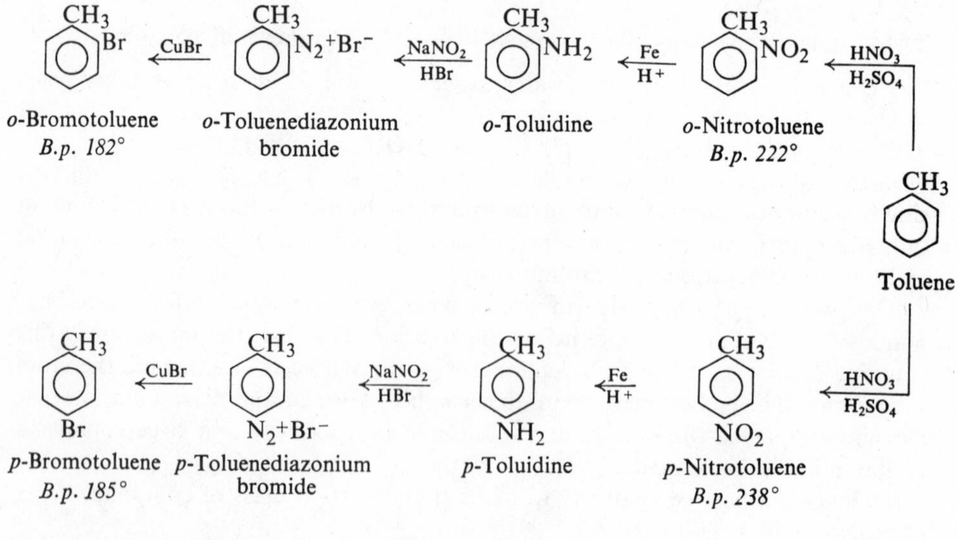

Toluene $\xrightarrow{HNO_3,\ H_2SO_4}$ *o*-Nitrotoluene (B.p. 222°) $\xrightarrow{Fe,\ H^+}$ *o*-Toluidine $\xrightarrow{NaNO_2,\ HBr}$ *o*-Toluenediazonium bromide $\xrightarrow{CuBr}$ *o*-Bromotoluene (B.p. 182°)

Toluene $\xrightarrow{HNO_3,\ H_2SO_4}$ *p*-Nitrotoluene (B.p. 238°) $\xrightarrow{Fe,\ H^+}$ *p*-Toluidine $\xrightarrow{NaNO_2,\ HBr}$ *p*-Toluenediazonium bromide $\xrightarrow{CuBr}$ *p*-Bromotoluene (B.p. 185°)

The advantage of these many-step syntheses over direct bromination is, as we have seen, that a pure product is obtained. Separation of the *o*- and *p*-bromotoluenes obtained by direct bromination is not feasible.

Synthesis of *m*-bromotoluene is a more complicated matter. The problem here is one of preparing a compound in which two *ortho,para*-directing groups are situated *meta* to each other. Bromination of toluene or methylation of bromobenzene would not yield the correct isomer. *m*-Bromotoluene is obtained by the following sequence of reactions:

$$\text{Toluene} \xrightarrow[H_2SO_4]{HNO_3} p\text{-Nitrotoluene} \xrightarrow[H^+]{Fe} p\text{-Toluidine} \xrightarrow{(CH_3CO)_2O} \text{Aceto-}p\text{-toluidide} \xrightarrow{Br_2} CH_3C_6H_3(Br)NHCOCH_3 \xrightarrow[H^+]{H_2O} CH_3C_6H_3(Br)NH_2 \xrightarrow[HCl]{NaNO_2} CH_3C_6H_3(Br)N_2^+Cl^- \text{ (Diazonium salt)} \xrightarrow{H_3PO_2} m\text{-Bromotoluene}$$

The key to the synthesis is the introduction of a group that is a much stronger *ortho,para* director than $—CH_3$, and that can be easily removed after it has done its job of directing bromine to the correct position. Such a group is the $—NHCOCH_3$ group: it is introduced into the *para* position of toluene via nitration, reduction, and acetylation; it is readily removed by hydrolysis, diazotization, and reduction.

Problem 23.15 Outline the synthesis from benzene or toluene of the following compounds: *m*-nitrotoluene, *m*-iodotoluene, 3,5-dibromotoluene, 1,3,5-tribromobenzene, the three toluic acids ($CH_3C_6H_4COOH$), the three methylphenols (cresols).

In the synthesis of *m*-bromotoluene, advantage was taken of the fact that the diazonium group is prepared from a group that is strongly *ortho,para*-directing. Ultimately, however, the diazonium group is prepared from the $—NO_2$ group, which is a strongly *meta*-directing group. Advantage can be taken of this fact, too, as in the preparation of *m*-bromophenol:

$$\text{Nitrobenzene} \xrightarrow[Fe]{Br_2} m\text{-Bromonitrobenzene} \xrightarrow[HCl]{Sn} m\text{-Bromoaniline} \xrightarrow[HCl]{NaNO_2} m\text{-Bromobenzenediazonium chloride} \xrightarrow[H^+]{H_2O} m\text{-Bromophenol}$$

Here again there is the problem of preparing a compound with two *ortho,para* directors situated *meta* to each other. Bromination at the nitro stage gives the necessary *meta* orientation.

Problem 23.16 Outline the synthesis from benzene or toluene of the following compounds: *m*-dibromobenzene, *m*-bromoiodobenzene.

As a final example, let us consider the preparation of 1,2,3-tribromobenzene:

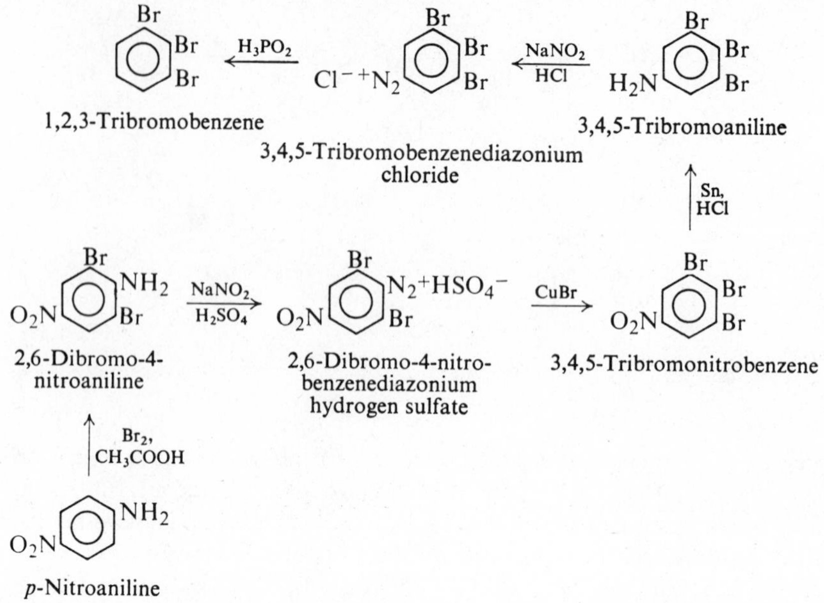

1,2,3-Tribromobenzene $\xleftarrow{H_3PO_2}$ 3,4,5-Tribromobenzenediazonium chloride ($Cl^{-}\,{}^{+}N_2$) $\xleftarrow[HCl]{NaNO_2}$ 3,4,5-Tribromoaniline (H_2N)

$\uparrow$ Sn, HCl

2,6-Dibromo-4-nitroaniline $\xrightarrow[H_2SO_4]{NaNO_2}$ 2,6-Dibromo-4-nitrobenzenediazonium hydrogen sulfate ($N_2{}^{+}HSO_4{}^{-}$) $\xrightarrow{CuBr}$ 3,4,5-Tribromonitrobenzene

$\uparrow$ Br_2, CH_3COOH

p-Nitroaniline

In this synthesis advantage is taken of the fact that the $—NO_2$ group is a *meta* director, that the $—NH_2$ group is an *ortho,para* director, and that each of them can be converted into a diazonium group. One diazonium group is replaced by —Br, the other by —H.

Problem 23.17 Outline the synthesis from benzene or toluene of the following compounds: 2,6-dibromotoluene, 3,5-dibromonitrobenzene.

23.17 Coupling of diazonium salts. Synthesis of azo compounds

Under the proper conditions, diazonium salts react with certain aromatic compounds to yield products of the general formula Ar—N=N—Ar′, called **azo compounds**. In this reaction, known as **coupling**, the nitrogen of the diazonium group is retained in the product, in contrast to the replacement reactions we have studied up to this point, in which nitrogen is lost.

$$ArN_2{}^{+} + Ar'H \longrightarrow \underset{\text{An azo compound}}{Ar{-}N{=}N{-}Ar'} + H^{+}$$

The aromatic ring (Ar′H) undergoing attack by the diazonium ion must, in general, contain a powerfully electron-releasing group, generally —OH, $—NR_2$, —NHR, or $—NH_2$. Substitution usually occurs *para* to the activating group. Typically, coupling with phenols is carried out in mildly alkaline solution, and with amines in mildly acidic solution.

Activation by electron-releasing groups, as well as the evidence of kinetics studies, indicates that coupling is electrophilic aromatic substitution in which the diazonium ion is the attacking reagent:

$$\text{G--C}_6\text{H}_5 + \text{ArN}_2^+ \longrightarrow [\text{G--C}_6\text{H}_5(\text{H})(\text{N=N--Ar})]^{(+)} \longrightarrow \text{G--C}_6\text{H}_4\text{--N=N--Ar} + \text{H}^+$$

It is significant that the aromatic compounds which undergo coupling are also the ones which undergo nitrosation. Like the nitrosonium ion, ^+NO, the diazonium ion, ArN_2^+, is evidently very weakly electrophilic, and is capable of attacking only very reactive rings.

Problem 23.18 Benzenediazonium chloride couples with phenol, but not with the less reactive anisole. 2,4-Dinitrobenzenediazonium chloride, however, couples with anisole; 2,4,6-trinitrobenzenediazonium chloride even couples with the hydrocarbon mesitylene (1,3,5-trimethylbenzene). (a) How can you account for these differences in behavior? (b) Would you expect *p*-toluenediazonium chloride to be more or less reactive as a coupling reagent than benzenediazonium chloride?

In the laboratory we find that coupling involves more than merely mixing together a diazonium salt and a phenol or amine. Competing with any other reaction of diazonium salts is the reaction with water to yield a phenol. If coupling proceeds slowly because of unfavorable conditions, phenol formation may very well become the major reaction. Furthermore, the phenol formed from the diazonium salt can itself undergo coupling; even a relatively small amount of this undesired coupling product could contaminate the desired material—usually a dye whose color should be as pure as possible—to such an extent that the product would be worthless. Conditions under which coupling proceeds as rapidly as possible must therefore be selected.

It is most important that the coupling medium be adjusted to the right degree of acidity or alkalinity. This is accomplished by addition of the proper amount of hydroxide or salts like sodium acetate or sodium carbonate. It will be well to examine this matter in some detail, since it illustrates a problem that is frequently encountered in organic chemical practice.

The electrophilic reagent is the diazonium ion, ArN_2^+. In the presence of hydroxide ion, the diazonium ion exists in equilibrium with an un-ionized compound, Ar—N=N—OH, and salts (Ar—N=N—O$^-$Na$^+$) derived from it:

$$\underset{\textit{Couples}}{\text{Ar—N}\equiv\text{N}^+\text{OH}^-} \underset{\text{H}^+}{\overset{\text{NaOH}}{\rightleftharpoons}} \underset{\textit{Does not couple}}{\text{Ar—N=N—OH}} \underset{\text{H}^+}{\overset{\text{NaOH}}{\rightleftharpoons}} \underset{\textit{Does not couple}}{\text{Ar—N=N—O}^-\text{Na}^+}$$

For our purpose we need only know that hydroxide tends to convert diazonium ion, which couples, into compounds which do not couple. In so far as the electrophilic reagent is concerned, then, coupling will be favored by a low concentration of hydroxide ion, that is, by high acidity.

But what is the effect of high acidity on the amine or phenol with which the diazonium salt is reacting? Acid converts an amine into its ion, which, because of the positive charge, is relatively unreactive toward electrophilic aromatic substitution: much too unreactive to be attacked by the weakly electrophilic diazonium ion. The higher the acidity, the higher the proportion of amine that exists as its ion, and the lower the rate of coupling.

$$C_6H_5NH_2 \underset{OH^-}{\overset{H^+}{\rightleftharpoons}} C_6H_5NH_3^+$$

Couples *Does not couple*

An analogous situation exists for a phenol. A phenol is appreciably acidic; in aqueous solutions it exists in equilibrium with phenoxide ion:

$$C_6H_5O^- \underset{OH^-}{\overset{H^+}{\rightleftharpoons}} C_6H_5OH$$

Couples rapidly *Couples slowly*

The fully developed negative charge makes —O^- much more powerfully electron-releasing than —OH; the phenoxide ion is therefore much more reactive than the un-ionized phenol toward electrophilic aromatic substitution. The higher the acidity of the medium, the higher the proportion of phenol that is un-ionized, and the lower the rate of coupling. In so far as the amine or phenol is concerned, then, coupling is favored by low acidity.

The conditions under which coupling proceeds most rapidly are the result of a compromise. The solution must not be so alkaline that the concentration of diazonium ion is too low; it must not be so acidic that the concentration of free amine or phenoxide ion is too low. It turns out that amines couple fastest in mildly acidic solutions, and phenols couple fastest in mildly alkaline solutions.

Problem 23.19 Suggest a reason for the use of *excess* mineral acid in the diazotization process.

Problem 23.20 (a) Coupling of diazonium salts with primary or secondary aromatic amines (but not with tertiary aromatic amines) is complicated by a side reaction that yields an isomer of the azo compound. Judging from the reaction of secondary aromatic amines with nitrous acid (Sec. 23.10), suggest a possible structure for this by-product.

(b) Upon treatment with mineral acid, this by-product regenerates the original reactants which recombine to form the azo compound. What do you think is the function of the acid in this regeneration? (*Hint:* See Problem 5.8, p. 170.)

Azo compounds are the first compounds we have encountered that as a class are strongly colored. They can be intensely yellow, orange, red, blue, or even green, depending upon the exact structure of the molecule. Because of their color, the azo compounds are of tremendous importance as dyes; about half of the dyes in industrial use today are azo dyes. Some of the acid–base indicators with which the student is already familiar are azo compounds.

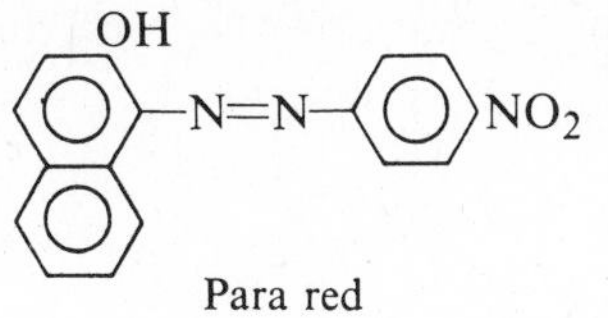

OH

$$\text{(1-naphthyl, 2-OH)}-N{=}N-C_6H_4-NO_2 \qquad Na^{+}\,{}^{-}O_3S-C_6H_4-N{=}N-C_6H_4-N(CH_3)_2$$

Para red
A red dye

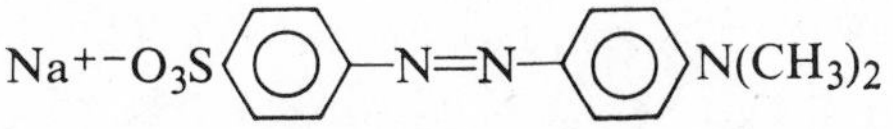

Methyl orange
An acid-base indicator:
red in acid, *yellow* in base

Problem 23.21 An azo compound is cleaved at the azo linkage by stannous chloride, $SnCl_2$, to form two amines. (a) What is the structure of the azo compound that is cleaved to 3-bromo-4-aminotoluene and 2-methyl-4-aminophenol? (b) Outline a synthesis of this azo compound, starting with benzene and toluene.

Problem 23.22 Show how *p*-amino-N,N-dimethylaniline can be made via an azo compound.

23.18 Analysis of amines. Hinsberg test

Amines are characterized chiefly through their basicity. A water-insoluble compound that dissolves in cold dilute hydrochloric acid—or a water-soluble compound (not a salt, Sec. 18.21) whose aqueous solution turns litmus blue—must almost certainly be an amine (Secs. 22.5 and 23.2). Elemental analysis shows the presence of nitrogen.

Whether an amine is primary, secondary, or tertiary is best shown by the **Hinsberg test**. The amine is shaken with benzenesulfonyl chloride in the presence of aqueous *potassium* hydroxide (Sec. 23.6). Primary and secondary amines form substituted sulfonamides; tertiary amines do not—*if* the test is carried out properly.

The monosubstituted sulfonamide from a primary amine has an acidic hydrogen attached to nitrogen. Reaction with potassium hydroxide converts this amide into a soluble salt which, *if the amine contained fewer than eight carbons*, is at least partly soluble. Acidification of this solution regenerates the insoluble amide.

The disubstituted sulfonamide from a secondary amine has no acidic hydrogen and remains insoluble in the alkaline reaction mixture.

What do we observe when we treat an amine with benzenesulfonyl chloride and excess potassium hydroxide? A *primary amine* yields a clear solution, from which, upon acidification, an insoluble material separates. A *secondary amine* yields an insoluble compound, which is unaffected by acid. A *tertiary amine* yields an insoluble compound (the unreacted amine itself) which dissolves upon acidification of the mixture.

$$\underset{\text{1° Amine}}{RNH_2} + C_6H_5SO_2Cl \xrightarrow{OH^-} [C_6H_5SO_2NHR] \xrightarrow{KOH} \underset{\textit{Clear solution}}{C_6H_5SO_2NR^-K^+} \xrightarrow{H^+} \underset{\textit{Insoluble}}{C_6H_5SO_2NHR}$$

$$\underset{2° \text{ Amine}}{R_2NH} + C_6H_5SO_2Cl \xrightarrow{OH^-} \underset{Insoluble}{C_6H_5SO_2NR_2} \xrightarrow{\text{KOH or } H^+} \text{No reaction}$$

$$\underset{3° \text{ Amine}}{R_3N} + C_6H_5SO_2Cl \xrightarrow{OH^-} \underset{Insoluble}{R_3N} \xrightarrow{HCl} \underset{Clear\ solution}{R_3NH^+Cl^-}$$

Like all experiments, the Hinsberg test must be done *carefully* and interpreted *thoughtfully*. Among other things, misleading side-reactions can occur if the proportions of reagents are incorrect, or if the temperature is too high or the time of reaction too long. Tertiary amines evidently *react*—after all, they are just as nucleophilic as other amines; but the initial product (I) has no acidic proton to

$$C_6H_5SO_2Cl + R_3N \longrightarrow \underset{I}{C_6H_5SO_2NR_3{}^+Cl^-} \xrightarrow{OH^-} C_6H_5SO_3{}^- + R_3N + Cl^-$$

lose, and ordinarily is hydrolyzed to regenerate the amine.

Problem 23.23 In non-aqueous medium, the product $C_6H_5SO_2N(CH_3)_3{}^+Cl^-$ can actually be isolated from the reaction of benzenesulfonyl chloride with one equivalent of trimethylamine. When *two* equivalents of the amine are used, there is formed, slowly, $C_6H_5SO_2N(CH_3)_2$ and $(CH_3)_4N^+Cl^-$. (a) Give all steps in a likely mechanism for this latter reaction. What fundamental type of reaction is probably involved?

(b) If, in carrying out the Hinsberg test, the reaction mixture is heated or allowed to stand, many tertiary amines give precipitates. What are these precipitates likely to be? What incorrect conclusion about the unknown amine are you likely to draw?

Problem 23.24 The sulfonamides of big primary amines are only partially soluble in aqueous KOH. (a) In the Hinsberg test, what incorrect conclusion might you draw about such an amine? (b) How might you modify the procedure to avoid this mistake?

Behavior toward nitrous acid (Sec. 23.10) is of some use in determining the class of an amine. In particular, the behavior of primary aromatic amines is quite characteristic: treatment with nitrous acid converts them into diazonium salts, which yield highly colored azo compounds upon treatment with β-naphthol (a phenol, see Sec. 23.17).

Among the numerous derivatives useful in identifying amines are: amides (e.g., acetamides, benzamides, or sulfonamides) for primary and secondary amines; quaternary ammonium salts (e.g., those from benzyl chloride or methyl iodide) for tertiary amines.

We have already discussed proof of structure by use of exhaustive methylation and elimination (Sec. 23.5).

23.19 Analysis of substituted amides

A substituted amide of a carboxylic acid is characterized by the presence of nitrogen, insolubility in dilute acid and dilute base, and hydrolysis to a carboxylic acid and an amine. It is generally identified through identification of its hydrolysis products (Secs. 18.21 and 23.18).

23.20 Spectroscopic analysis of amines and substituted amides

Infrared. The number and positions of absorption bands depend on the class to which the amine belongs (see Fig. 23.2).

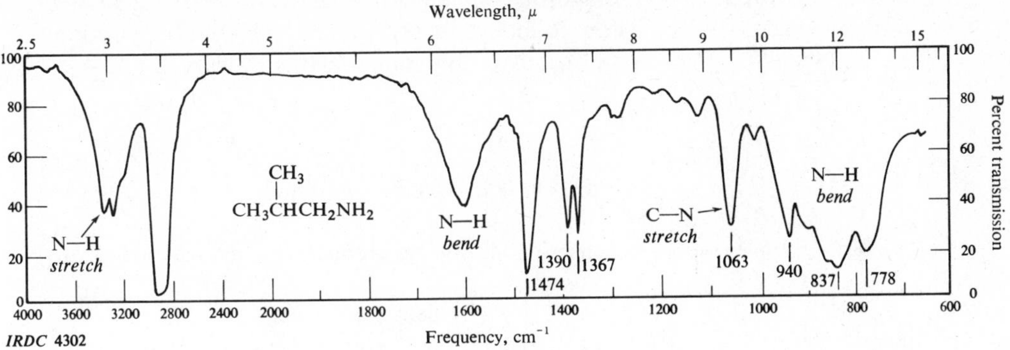

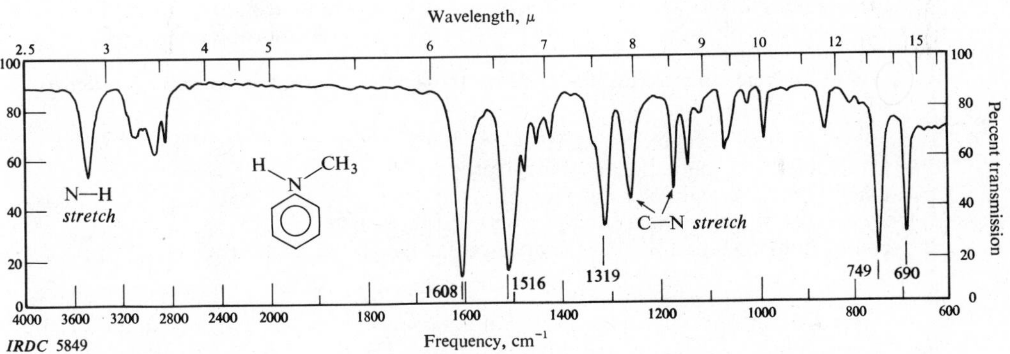

Figure 23.2. Infrared spectra of (*a*) isobutylamine and (*b*) N-methylaniline.

An amide, substituted or unsubstituted, shows the C=O band in the 1640–1690 cm^{-1} region. In addition, if it contains a free N—H group, it will show N—H stretching at 3050–3550 cm^{-1}, and —NH bending at 1600–1640 cm^{-1} (RCONH$_2$) or 1530–1570 cm^{-1} (RCONHR′).

N—H stretching 3200–3500 cm^{-1}

1° Amines	2° Amines	3° Amines
Often two bands	*One band*	*No band*

N—H bending

1° Amines Strong bands 650–900 cm^{-1} (*broad*) and 1560–1650 cm^{-1}

C—N stretching

Aliphatic 1030–1230 cm^{-1} (*weak*)	Aromatic 1180–1360 cm^{-1} (*strong*)
(3°: *usually a doublet*)	*Two bands*

Nmr. Absorption by N—H protons of amines falls in the range δ 1–5, where it is often detected only by proton counting. Absorption by —CO—NH— protons of amides (Sec. 20.25) appears as a broad, low hump farther downfield (δ 5–8).

PROBLEMS

1. Write complete equations, naming all organic products, for the reaction (if any) of *n*-butylamine with:

(a) dilute HCl
(b) dilute H_2SO_4
(c) acetic acid
(d) dilute NaOH
(e) acetic anhydride
(f) isobutyryl chloride
(g) *p*-nitrobenzoyl chloride + pyridine
(h) benzenesulfonyl chloride + KOH (aq)
(i) ethyl bromide
(j) benzyl bromide
(k) bromobenzene
(l) excess methyl iodide, then Ag_2O
(m) product (l) + strong heat
(n) $CH_3COCH_3 + H_2 + Ni$
(o) HONO ($NaNO_2$ + HCl)
(p) phthalic anhydride
(q) sodium chloroacetate
(r) 2,4,6-trinitrochlorobenzene

2. Without referring to tables, arrange the compounds of each set in order of basicity:

(a) ammonia, aniline, cyclohexylamine
(b) ethylamine, 2-aminoethanol, 3-amino-1-propanol
(c) aniline, *p*-methoxyaniline, *p*-nitroaniline
(d) benzylamine, *m*-chlorobenzylamine, *m*-ethylbenzylamine
(e) *p*-chloro-N-methylaniline, 2,4-dichloro-N-methylaniline, 2,4,6-trichloro-N-methylaniline

3. Which is the more strongly basic, an aqueous solution of trimethylamine or an aqueous solution of tetramethylammonium hydroxide? Why? (*Hint:* What is the principal base in each solution?)

4. Compare the behavior of the three amines, aniline, N-methylaniline, and N,N-dimethylaniline, toward each of the following reagents:

(a) dilute HCl
(b) $NaNO_2$ + HCl (aq)
(c) methyl iodide
(d) benzenesulfonyl chloride + KOH (aq)
(e) acetic anhydride
(f) benzoyl chloride + pyridine
(g) bromine water

5. Answer Problem 4 for ethylamine, diethylamine, and triethylamine.

6. Give structures and names of the principal organic products expected from the action (if any) of sodium nitrite and hydrochloric acid on:

(a) *p*-toluidine
(b) N,N-diethylaniline
(c) *n*-propylamine
(d) sulfanilic acid
(e) N-methylaniline
(f) 2-amino-3-methylbutane
(g) benzidine (4,4′-diaminobiphenyl)
(h) benzylamine

7. Write equations for the reaction of *p*-nitrobenzenediazonium sulfate with:

(a) *m*-phenylenediamine
(b) hot dilute H_2SO_4
(c) HBr + Cu
(d) *p*-cresol
(e) KI
(f) CuCl
(g) CuCN
(h) HBF_4, then heat
(i) H_3PO_2

8. Give the reagents and any special conditions necessary to convert *p*-toluenediazonium chloride into:

(a) toluene
(b) *p*-cresol, *p*-$CH_3C_6H_4OH$
(c) *p*-chlorotoluene
(d) *p*-bromotoluene
(e) *p*-iodotoluene
(f) *p*-fluorotoluene
(g) *p*-tolunitrile, *p*-$CH_3C_6H_4CN$
(h) 4-methyl-4′-(N,N-dimethylamino)azobenzene
(i) 2,4-dihydroxy-4′-methylazobenzene

9. Write balanced equations, naming all organic products, for the following reactions:

(a) *n*-butyryl chloride + methylamine
(b) acetic anhydride + N-methylaniline
(c) tetra-*n*-propylammonium hydroxide + heat
(d) isovaleryl chloride + diethylamine
(e) tetramethylammonium hydroxide + heat
(f) trimethylamine + acetic acid
(g) N,N-dimethylacetamide + boiling dilute HCl
(h) benzanilide + boiling aqueous NaOH
(i) methyl formate + aniline
(j) excess methylamine + phosgene ($COCl_2$)
(k) *m*-$O_2NC_6H_4NHCH_3$ + $NaNO_2$ + H_2SO_4
(l) aniline + Br_2 (aq) in excess
(m) *m*-toluidine + Br_2 (aq) in excess
(n) *p*-toluidine + Br_2 (aq) in excess
(o) *p*-toluidine + $NaNO_2$ + HCl
(p) $C_6H_5NHCOCH_3$ + HNO_3 + H_2SO_4
(q) *p*-$CH_3C_6H_4NHCOCH_3$ + HNO_3 + H_2SO_4
(r) *p*-$C_2H_5C_6H_4NH_2$ + large excess of CH_3I
(s) benzanilide + Br_2 + Fe

10. Outline all steps in a possible laboratory synthesis of each of the following compounds from benzene, toluene, and alcohols of four carbons or fewer, using any needed inorganic reagents.

(a) 4-amino-2-bromotoluene
(b) 4-amino-3-bromotoluene
(c) *p*-aminobenzenesulfonanilide (*p*-$H_2NC_6H_4SO_2NHC_6H_5$)
(d) monoacetyl *p*-phenylenediamine (*p*-aminoacetanilide)
(e) *p*-nitroso-N,N-diethylaniline
(f) 4-amino-3-nitrobenzoic acid
(g) 2,6-dibromo-4-isopropylaniline
(h) *p*-aminobenzylamine
(i) N-nitroso-N-isopropylaniline
(j) N-ethyl-N-methyl-*n*-valeramide
(k) *n*-hexylamine
(l) 1-amino-1-phenylbutane
(m) aminoacetamide
(n) hippuric acid ($C_6H_5CONHCH_2COOH$)

11. Outline all steps in a possible laboratory synthesis from benzene, toluene, and any needed inorganic reagents of:

(a) the six isomeric dibromotoluenes, $CH_3C_6H_3Br_2$. (*Note:* One may be more difficult to make than any of the others.)
(b) the three isomeric chlorobenzoic acids, each one free of the others
(c) the three isomeric bromofluorobenzenes

Review the instructions on page 224. Assume that an *ortho,para* mixture of isomeric nitro compounds can be separated by distillation (see Sec. 11.7).

12. Outline all steps in a possible laboratory synthesis of each of the following compounds from benzene and toluene and any needed aliphatic and inorganic reagents.

(a) *p*-fluorotoluene
(b) *m*-fluorotoluene
(c) *p*-iodobenzoic acid
(d) *m*-bromoaniline
(e) 3-bromo-4-methylbenzoic acid
(f) 2-bromo-4-methylbenzoic acid
(g) *m*-ethylphenol
(h) 3,5-dibromoaniline
(i) 3-bromo-4-iodotoluene
(j) 2-amino-4-methylphenol
(k) 2,6-dibromoiodobenzene
(l) 4-iodo-3-nitrotoluene
(m) *p*-hydroxyphenylacetic acid
(n) 2-bromo-4-chlorotoluene

13. When adipic acid (hexanedioic acid) and hexamethylenediamine (1,6-diaminohexane) are mixed, a salt is obtained. On heating, this salt is converted into Nylon 66, a high-molecular-weight compound of formula $(C_{12}H_{22}O_2N_2)_n$. (a) Draw the structural formula for Nylon 66. To what class of compounds does it belong? (b) Write an equation for the chemistry involved when a drop of hydrochloric acid makes a hole in a Nylon 66 stocking.

14. Account for the following reactions, making clear the role played by tosyl chloride.

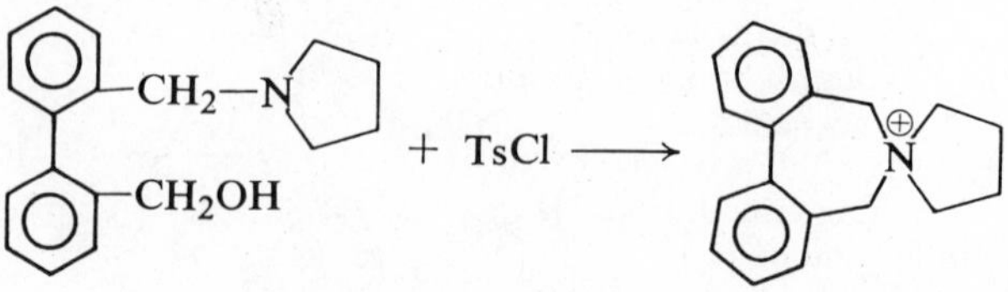

$$\text{(biphenyl: }2\text{-}CH_2\text{—N(pyrrolidine)}, 2'\text{-}CH_2OH) + TsCl \longrightarrow \text{(cyclic quaternary ammonium ion, }\overset{\oplus}{N}\text{ spiro-pyrrolidinium)}$$

15. If halide ion is present during hydrolysis of benzenediazonium ion or *p*-nitrobenzenediazonium ion, there is obtained not only the phenol, but also the aryl halide: the higher the halide ion concentration, the greater the proportion of aryl halide obtained. The presence of halide ion has no effect on the rate of decomposition of benzenediazonium ion, but speeds up decomposition of the *p*-nitrobenzenediazonium ion.

(a) Suggest a mechanism or mechanisms to account for these facts. (b) What factor is responsible for the unusually high reactivity of diazonium ions in this reaction—and, indeed, in most of their reactions? (*Hint:* See Sec. 14.5.)

16. Describe simple chemical tests (other than color reactions with indicators) that would serve to distinguish between:

(a) N-methylaniline and *o*-toluidine
(b) aniline and cyclohexylamine
(c) $n\text{-}C_4H_9NH_2$ and $(n\text{-}C_4H_9)_2NH$
(d) $(n\text{-}C_4H_9)_2NH$ and $(n\text{-}C_4H_9)_3N$
(e) $(CH_3)_3NHCl$ and $(CH_3)_4NCl$
(f) $C_6H_5NH_3Cl$ and $o\text{-}ClC_6H_4NH_2$
(g) $(C_2H_5)_2NCH_2CH_2OH$ and $(C_2H_5)_4NOH$
(h) aniline and acetanilide
(i) $(C_6H_5NH_3)_2SO_4$ and $p\text{-}H_3\overset{+}{N}C_6H_4SO_3^-$
(j) $ClCH_2CH_2NH_2$ and $CH_3CH_2NH_3Cl$
(k) 2,4,6-trinitroaniline and aniline
(l) $C_6H_5NHSO_2C_6H_5$ and $C_6H_5NH_3HSO_4$

Tell exactly what you would *do* and *see*.

17. Describe simple chemical methods for the separation of the following mixtures, recovering each component in essentially pure form:

(a) triethylamine and *n*-heptane
(b) aniline and anisole
(c) stearamide and octadecylamine
(d) $o\text{-}O_2NC_6H_4NH_2$ and $p\text{-}H_3\overset{+}{N}C_6H_4SO_3^-$
(e) $C_6H_5NHCH_3$ and $C_6H_5N(CH_3)_2$
(f) *n*-caproic acid, tri-*n*-propylamine, and cyclohexane
(g) *o*-nitrotoluene and *o*-toluidine
(h) *p*-ethylaniline and propionanilide

Tell exactly what you would *do* and *see*.

18. The compounds in each of the following sets boil (or melt) within a few degrees of each other. Describe simple chemical tests that would serve to distinguish among the members of each set.

(a) aniline, benzylamine, and N,N-dimethylbenzylamine
(b) *o*-chloroacetanilide and 2,4-diaminochlorobenzene
(c) N-ethylbenzylamine, N-ethyl-N-methylaniline, β-phenylethylamine, and *o*-toluidine
(d) acetanilide and ethyl oxamate ($C_2H_5OOCCONH_2$)

(e) benzonitrile, N,N-dimethylaniline, and formamide
(f) N,N-dimethyl-*m*-toluidine, nitrobenzene, and *m*-tolunitrile
(g) N-(*sec*-butyl)benzenesulfonamide
p-chloroaniline — *o*-nitroaniline
N,N-dibenzylaniline — *p*-nitrobenzyl chloride
2,4-dinitroaniline — *p*-toluenesulfonyl chloride
N-ethyl-N-(*p*-tolyl)-*p*-toluenesulfonamide

Tell exactly what you would *do* and *see*.

19. An unknown amine is believed to be one of those in Table 23.2. Describe how you would go about finding out which of the possibilities the unknown actually is. Where possible use simple chemical tests.

Table 23.2 DERIVATIVES OF SOME AMINES

Amine	B.p., °C	Benzene-sulfonamide M.p., °C	Acetamide M.p., °C	Benzamide M.p., °C	*p*-Toluene-sulfonamide M.p., °C
m-Toluidine	203	95	66	125	114
N-Ethylaniline	205		54	60	87
N-Methyl-*m*-toluidine	206		66		
N,N-Diethyl-*o*-toluidine	206				
N-Methyl-*o*-toluidine	207		55	66	120
N-Methyl-*p*-toluidine	207	64	83	53	60
N,N-Dimethyl-*o*-chloroaniline	207				
o-Chloroaniline	209	129	87	99	105

20. *Choline*, a constituent of *phospholipids* (fat-like phosphate esters of great physiological importance), has the formula $C_5H_{15}O_2N$. It dissolves readily in water to form a strongly basic solution. It can be prepared by the reaction of ethylene oxide with trimethylamine in the presence of water.

(a) What is a likely structure for choline? (b) What is a likely structure for its acetyl derivative, *acetylcholine*, $C_7H_{17}O_3N$, important in nerve action?

21. *Novocaine*, a local anesthetic, is a compound of formula $C_{13}H_{20}O_2N_2$. It is insoluble in water and dilute NaOH, but soluble in dilute HCl. Upon treatment with $NaNO_2$ and HCl and then with β-naphthol, a highly colored solid is formed.

When Novocaine is boiled with aqueous NaOH, it slowly dissolves. The alkaline solution is shaken with ether and the layers are separated.

Acidification of the aqueous layer causes the precipitation of a white solid A; continued addition of acid causes A to redissolve. Upon isolation A is found to have a melting point of 185–6° and the formula $C_7H_7O_2N$.

Evaporation of the ether layer leaves a liquid B of formula $C_6H_{15}ON$. B dissolves in water to give a solution that turns litmus blue. Treatment of B with acetic anhydride gives C, $C_8H_{17}O_2N$, which is insoluble in water and dilute base, but soluble in dilute HCl.

B is found to be identical with the compound formed by the action of diethylamine on ethylene oxide.

(a) What is the structure of Novocaine? (b) Outline all steps in a complete synthesis of Novocaine from toluene and readily available aliphatic and inorganic reagents.

22. A solid compound D, of formula $C_{15}H_{15}ON$, was insoluble in water, dilute HCl, or dilute NaOH. After prolonged heating of D with aqueous NaOH, a liquid, E, was observed floating on the surface of the alkaline mixture. E did not solidify upon cooling to room temperature; it was steam-distilled and separated. Acidification of the alkaline mixture with hydrochloric acid caused precipitation of a white solid, F.

Compound E was soluble in dilute HCl, and reacted with benzenesulfonyl chloride and excess KOH to give a base-insoluble solid, G.

Compound F, m.p. 180°, was soluble in aqueous $NaHCO_3$, and contained no nitrogen.

What were compounds D, E, F, and G?

23. Give the structures of compounds H through Q:

CH_3N =O $\xrightarrow{\text{reduction}}$ H($C_9H_{17}ON$, an alcohol)

H + heat $\longrightarrow$ I ($C_9H_{15}N$)
I + CH_3I, then Ag_2O $\longrightarrow$ J ($C_{10}H_{19}ON$)
J + heat $\longrightarrow$ K ($C_{10}H_{17}N$)
K + CH_3I, then Ag_2O $\longrightarrow$ L ($C_{11}H_{21}ON$)
L + heat $\longrightarrow$ M (C_8H_{10})
M + Br_2 $\longrightarrow$ N ($C_8H_{10}Br_2$)
N + $(CH_3)_2NH$ $\longrightarrow$ O ($C_{12}H_{22}N_2$)
O + CH_3I, then Ag_2O $\longrightarrow$ P ($C_{14}H_{30}O_2N_2$)
P + heat $\longrightarrow$ Q (C_8H_8)

24. *Pantothenic acid,* $C_9H_{17}O_5N$, occurs in Coenzyme A (p. 1173), essential to metabolism of carbohydrates and fats. It reacts with dilute NaOH to give $C_9H_{16}O_5NNa$, with ethyl alcohol to give $C_{11}H_{21}O_5N$, and with hot NaOH to give compound V (see below) and β-aminopropionic acid. Its nitrogen is non-basic. Pantothenic acid has been synthesized as follows:

isobutyraldehyde + formaldehyde + K_2CO_3 $\longrightarrow$ R ($C_5H_{10}O_2$)
R + $NaHSO_3$, then KCN $\longrightarrow$ S ($C_6H_{11}O_2N$)
S + H_2O, H^+, heat $\longrightarrow$ [T ($C_6H_{12}O_4$)] $\longrightarrow$ U ($C_6H_{10}O_3$)
U + NaOH(aq), warm $\longrightarrow$ V ($C_6H_{11}O_4Na$)
U + sodium β-aminopropionate, then H^+ $\longrightarrow$ pantothenic acid ($C_9H_{17}O_5N$)

What is the structure of pantothenic acid?

25. An unknown compound W contained chlorine and nitrogen. It dissolved readily in water to give a solution that turned litmus red. Titration of W with standard base gave a neutralization equivalent of 131 $\pm$ 2.

When a sample of W was treated with aqueous NaOH a liquid separated; it contained nitrogen but not chlorine. Treatment of the liquid with nitrous acid followed by β-naphthol gave a red precipitate.

What was W? Write equations for all reactions.

26. Which (if any) of the following compounds could give rise to each of the infrared spectra shown in Fig. 23.3 (p. 783)?

n-butylamine	*o*-anisidine
diethylamine	*m*-anisidine
N-methylformamide	aniline
N,N-dimethylformamide	N,N-dimethyl-*o*-toluidine
2-(dimethylamino)ethanol	acetanilide

27. Give a structure or structures consistent with each of the nmr spectra shown in Fig. 23.4 (p. 784).

28. Give the structures of compounds X, Y, and Z on the basis of their infrared spectra (Fig. 23.5, p. 785) and their nmr spectra (Fig. 23.6, p. 786).

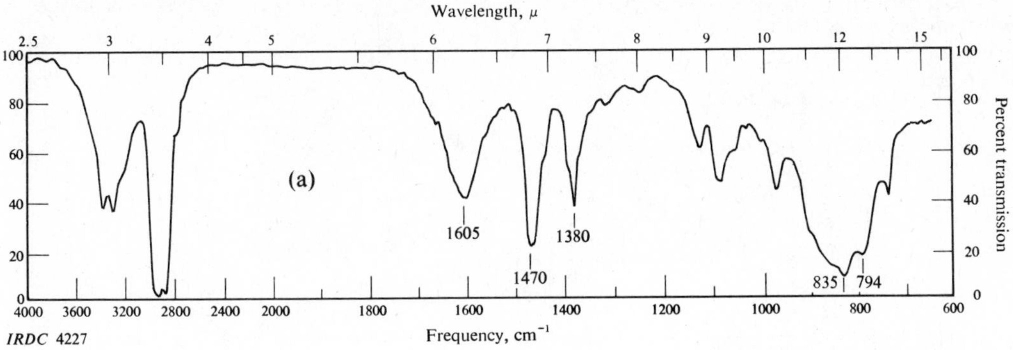

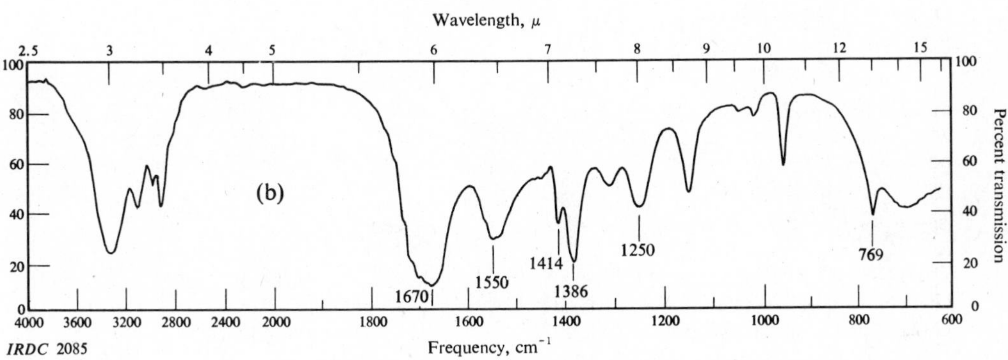

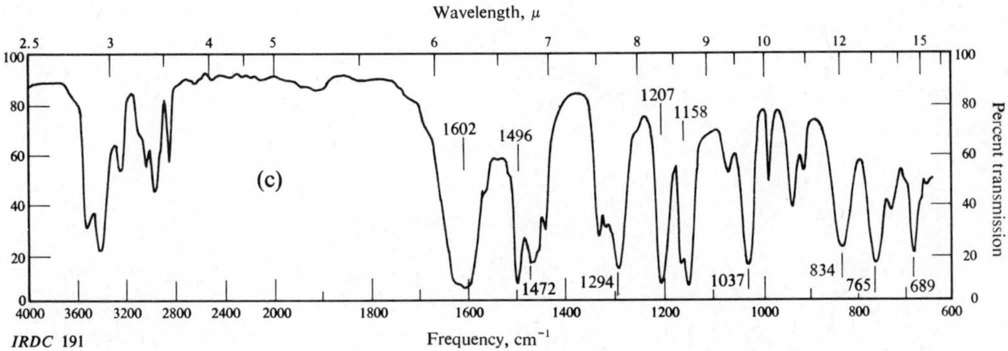

Figure 23.3. Infrared spectra for Problem 26, p. 782.

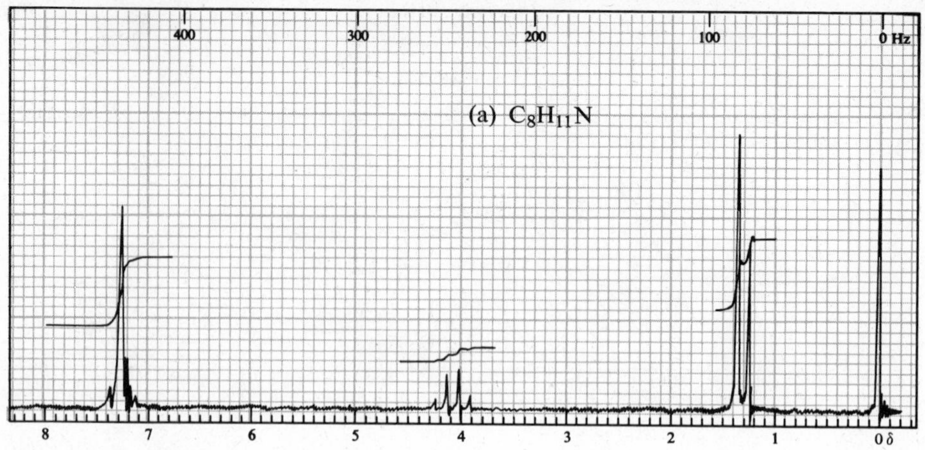

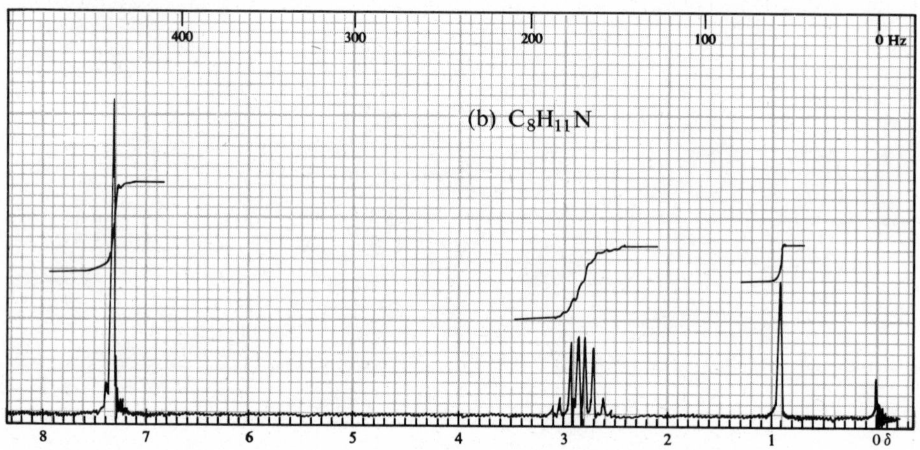

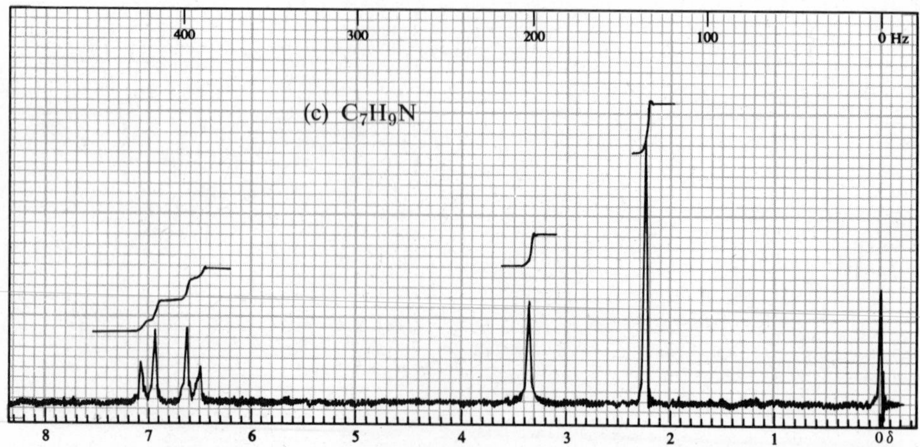

Figure 23.4. Nmr spectra for Problem 27, p. 782.

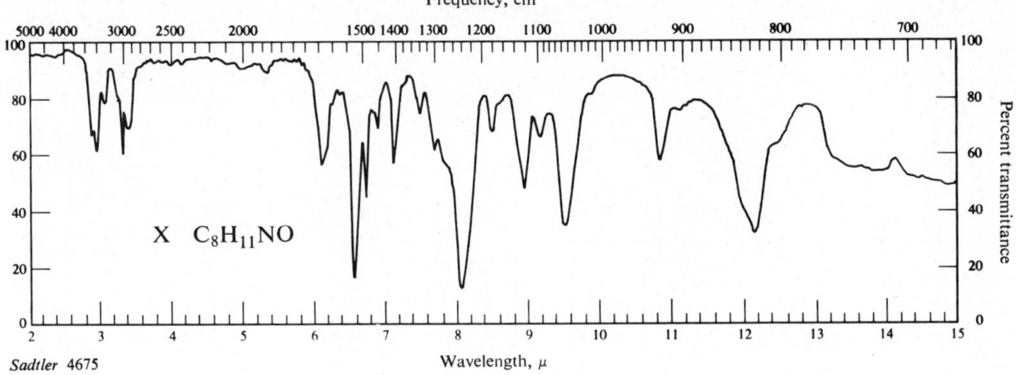

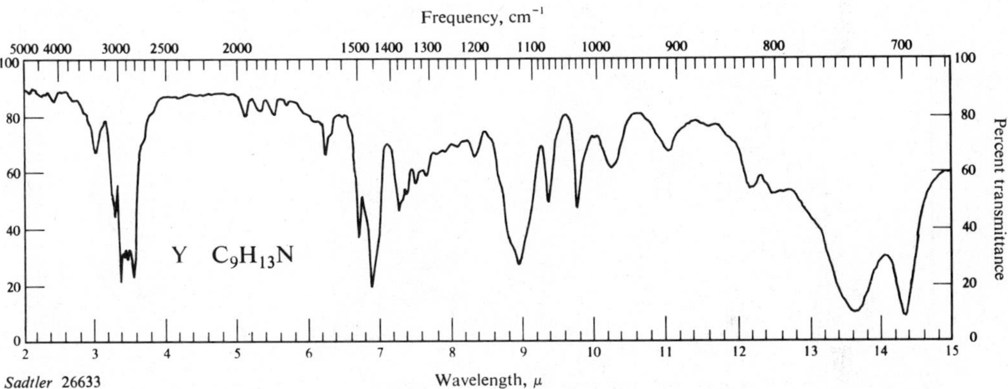

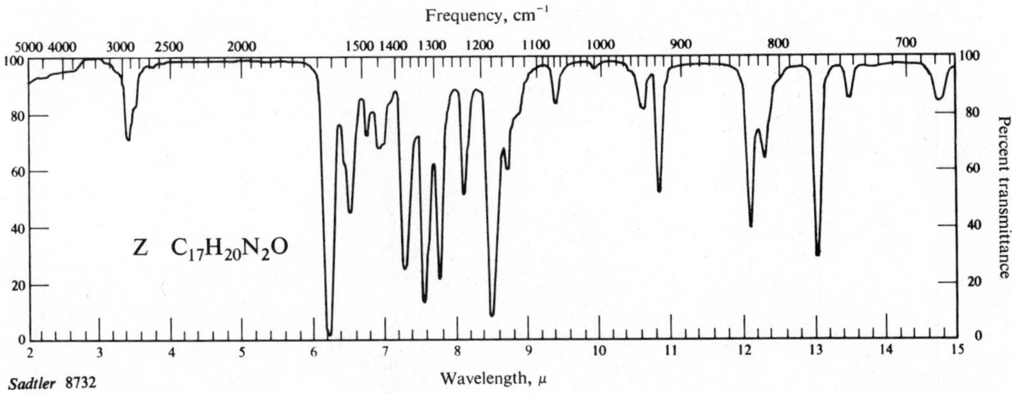

Figure 23.5. Infrared spectra for Problem 28, p. 782.

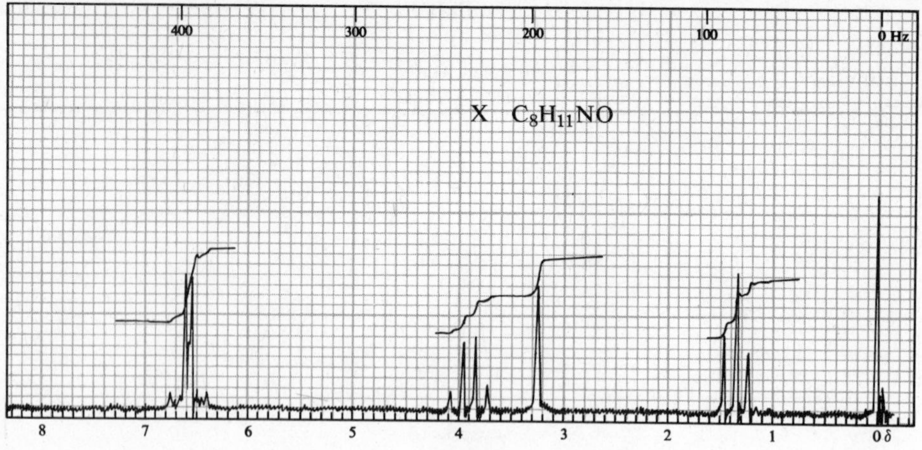

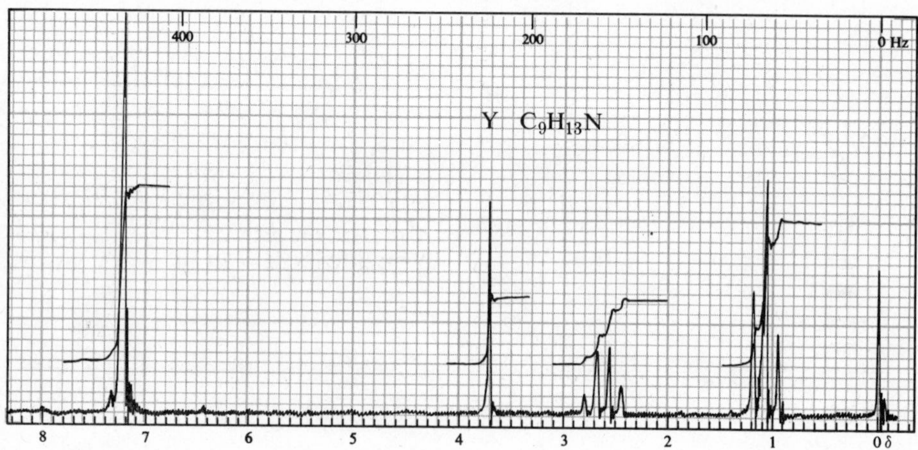

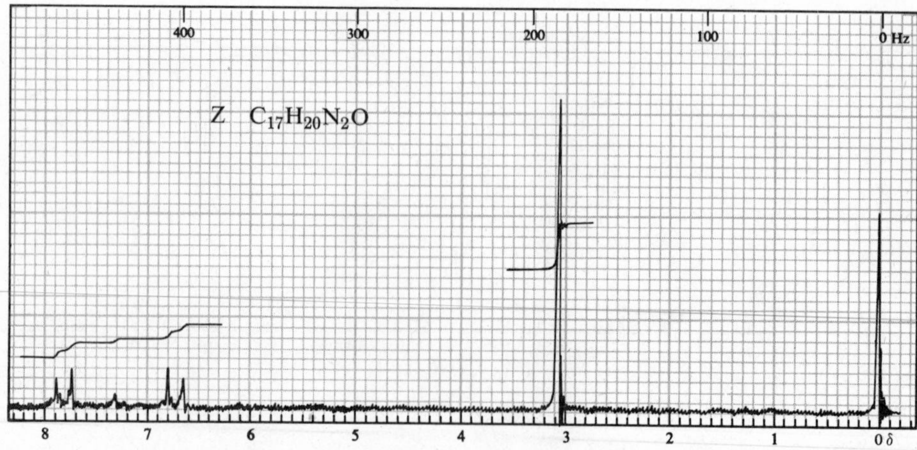

Figure 23.6. Nmr spectra for Problem 28, p. 782.

Chapter

24 Phenols

24.1 Structure and nomenclature

Phenols are compounds of the general formula ArOH, where Ar is phenyl, substituted phenyl, or one of the other aryl groups we shall study later (e.g., naphthyl, Chap. 30). *Phenols differ from alcohols in having the* —OH *group attached directly to an aromatic ring.*

Phenols are generally named as derivatives of the simplest member of the family, **phenol**. The methylphenols are given the special name of *cresols*. Occasionally phenols are named as *hydroxy–* compounds.

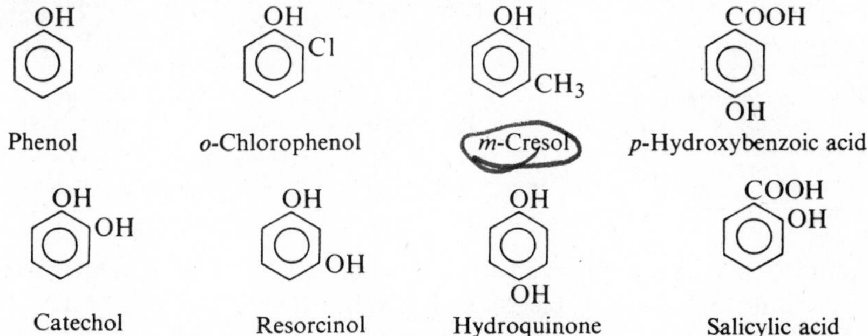

Both phenols and alcohols contain the —OH group, and as a result the two families resemble each other to a limited extent. We have already seen, for example, that both alcohols and phenols can be converted into ethers and esters. In most of their properties, however, and in their preparations, the two kinds of compound differ so greatly that they well deserve to be classified as different families.

24.2 Physical properties

The simplest phenols are liquids or low-melting solids; because of hydrogen bonding, they have quite high boiling points. Phenol itself is somewhat soluble in water (9 g per 100 g of water), presumably because of hydrogen bonding with the water; most other phenols are essentially insoluble in water. Unless some group capable of producing color is present, phenols themselves are colorless. However, like aromatic amines, they are easily oxidized; unless carefully purified, many phenols are colored by oxidation products.

Table 24.1 PHENOLS

Name	M.p., °C	B.p., °C	Solub., g/100 g H_2O at 25°	K_a
Phenol	41	182	9.3	1.1×10^{-10}
o-Cresol	31	191	2.5	0.63
m-Cresol	11	201	2.6	0.98
p-Cresol	35	202	2.3	0.67
o-Fluorophenol	16	152		15
m-Fluorophenol	14	178		5.2
p-Fluorophenol	48	185		1.1
o-Chlorophenol	9	173	2.8	77
m-Chlorophenol	33	214	2.6	16
p-Chlorophenol	43	220	2.7	6.3
o-Bromophenol	5	194		41
m-Bromophenol	33	236		14
p-Bromophenol	64	236	1.4	5.6
o-Iodophenol	43			34
m-Iodophenol	40			13
p-Iodophenol	94			6.3
o-Aminophenol	174		1.7^{0}	2.0
m-Aminophenol	123		2.6	69
p-Aminophenol	186		1.1^{0}	
o-Nitrophenol	45	217	0.2	600
m-Nitrophenol	96		1.4	50
p-Nitrophenol	114		1.7	690
2,4-Dinitrophenol	113		0.6	1000000
2,4,6-Trinitrophenol (picric acid)	122		1.4	very large
Catechol	104	246	45	1
Resorcinol	110	281	123	3
Hydroquinone	173	286	8	2

An important point emerges from a comparison of the physical properties of the isomeric nitrophenols (Table 24.2). We notice that *o*-nitrophenol has a much lower boiling point and much lower solubility in water than its isomers; it is the only one of the three that is readily steam-distillable. How can these differences be accounted for?

Table 24.2 PROPERTIES OF THE NITROPHENOLS

	B.p., °C at 70 mm	Solub., g/100 g H_2O	
o-Nitrophenol	100	0.2	Volatile in steam
m-Nitrophenol	194	1.35	Non-volatile in steam
p-Nitrophenol	*dec.*	1.69	Non-volatile in steam

Let us consider first the *m*- and *p*-isomers. They have very high boiling points because of intermolecular hydrogen bonding:

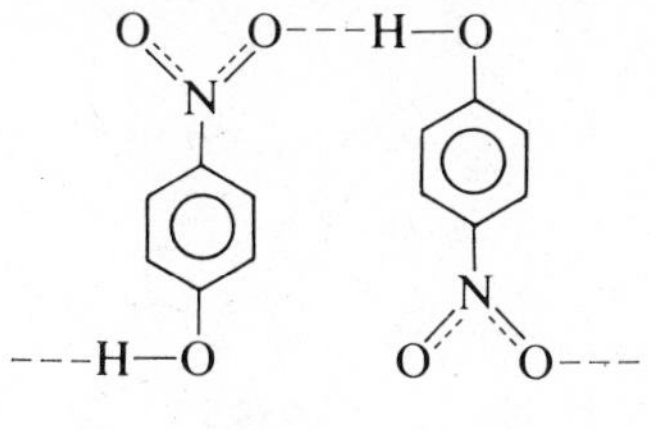

Intermolecular hydrogen bonding

Their solubility in water is due to hydrogen bonding with water molecules:

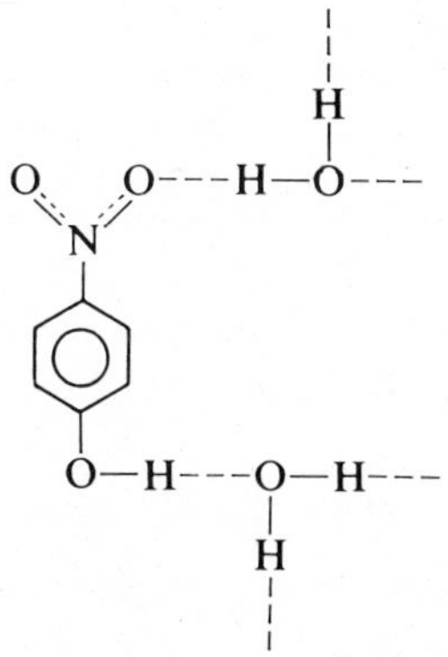

Steam distillation depends upon a substance having an appreciable vapor pressure at the boiling point of water; by lowering the vapor pressure, intermolecular hydrogen bonding inhibits steam distillation of the *m*- and *p*-isomers.

What is the situation for the *o*-isomer? Examination of models shows that the $—NO_2$ and —OH groups are located exactly right for the formation of a

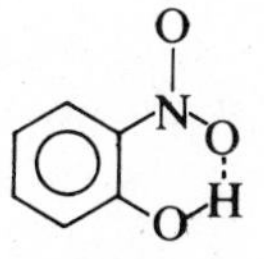

o-Nitrophenol

Intramolecular hydrogen bonding: chelation

hydrogen bond *within a single molecule.* This **intramolecular hydrogen bonding** takes the place of *inter*molecular hydrogen bonding with other phenol molecules and with water molecules; therefore *o*-nitrophenol does not have the low volatility of an associated liquid, nor does it have the solubility characteristic of a compound that forms hydrogen bonds with water.

The holding of a hydrogen or metal atom between two atoms of a single molecule is called **chelation** (Greek: *chele*, claw). See, for example, *chlorophyll* (p. 1004) and *heme* (p. 1152).

Intramolecular hydrogen bonding seems to occur whenever the structure of a compound permits; we shall encounter other examples of its effect on physical properties.

Problem 24.1 Interpret the following observations. The O—H bands (Sec. 15.4) for the isomeric nitrophenols in solid form (KBr pellets) and in $CHCl_3$ solution are:

	KBr	$CHCl_3$
o-	3200 cm^{-1}	3200 cm^{-1}
m-	3330	3520
p-	3325	3530

Problem 24.2 In which of the following compounds would you expect intramolecular hydrogen bonding to occur: *o*-nitroaniline, *o*-cresol, *o*-hydroxybenzoic acid (salicylic acid), *o*-hydroxybenzaldehyde (salicylaldehyde), *o*-fluorophenol, *o*-hydroxybenzonitrile.

24.3 Salts of phenols

Phenols are fairly acidic compounds, and in this respect differ markedly from alcohols, which are even more weakly acidic than water. Aqueous hydroxides convert phenols into their salts; aqueous mineral acids convert the salts back into the free phenols. As we might expect, phenols and their salts have opposite solubility properties, the salts being soluble in water and insoluble in organic solvents.

$$\text{ArOH} \underset{H^+}{\overset{OH^-}{\rightleftharpoons}} \text{ArO}^-$$

A phenol	A phenoxide ion
(acid)	(salt)
Insoluble in water	*Soluble in water*

Most phenols have K_a's in the neighborhood of 10^{-10}, and are thus considerably weaker acids than the carboxylic acids (K_a's about 10^{-5}). Most phenols are weaker than carbonic acid, and hence, unlike carboxylic acids, do not dissolve in aqueous bicarbonate solutions. Indeed, phenols are conveniently liberated from their salts by the action of carbonic acid.

$$CO_2 + H_2O \rightleftharpoons \underset{\text{Stronger acid}}{H_2CO_3} + \underset{\textit{Soluble in water}}{ArO^-Na^+} \longrightarrow \underset{\substack{\text{Weaker acid} \\ \textit{Insoluble in water}}}{ArOH} + Na^+HCO_3^-$$

The acid strength of phenols and the solubility of their salts in water are useful both in analysis and in separations. A water-insoluble substance that dissolves in aqueous hydroxide but not in aqueous bicarbonate must be more acidic than water, but less acidic than a carboxylic acid; most compounds in this range of acidity are phenols. A phenol can be separated from non-acidic compounds by means of its solubility in base; it can be separated from carboxylic acids by means of its insolubility in bicarbonate.

Problem 24.3 Outline the separation by chemical methods of a mixture of *p*-cresol, *p*-toluic acid, *p*-toluidine, and *p*-nitrotoluene. Describe exactly what you would *do* and *see*.

24.4 Industrial source

Most phenols are made industrially by the same methods that are used in the laboratory; these are described in Sec. 24.5. There are, however, special ways of obtaining certain of these compounds on a commercial scale, including the most important one, phenol. In quantity produced, phenol ranks near the top of the list of synthetic aromatic compounds. Its principal use is in the manufacture of the phenol–formaldehyde polymers (Sec. 32.7).

A certain amount of phenol, as well as the cresols, is obtained from coal tar (Sec. 12.4). Most of it (probably over 90%) is synthesized. One of the synthetic processes used is the fusion of sodium benzenesulfonate with alkali (Sec. 30.12); another is the Dow process, in which chlorobenzene is allowed to react with aqueous sodium hydroxide at a temperature of about 360°. Like the synthesis of aniline from chlorobenzene (Sec. 22.7), this second reaction involves nucleophilic substitution under conditions that are not generally employed in the laboratory (Sec. 25.4).

$$C_6H_5Cl \xrightarrow[\text{4500 lb/in.}^2]{\text{NaOH, 360°}} C_6H_5O^-Na^+ \xrightarrow{\text{HCl}} C_6H_5OH$$

Chlorobenzene — Sodium phenoxide — Phenol

An increasingly important process for the synthesis of phenol starts with *cumene*, isopropylbenzene. Cumene is converted by air oxidation into cumene hydroperoxide, which is converted by aqueous acid into phenol and acetone.

$$C_6H_5-CH(CH_3)_2 \xrightarrow{O_2} C_6H_5-C(CH_3)_2-OOH \xrightarrow{H_2O,\ H^+} C_6H_5OH + CH_3-C(=O)-CH_3$$

Cumene — Cumene hydroperoxide — Phenol — Acetone

(The mechanism of this reaction is discussed in Sec. 28.6.)

Problem 24.4 Outline a synthesis of cumene from cheap, readily available hydrocarbons.

Certain phenols and their ethers are isolated from the *essential oils* of various plants (so called because they contain the *essence*—odor or flavor—of the plants). A few of these are:

$$\text{4-}(CH_2CH{=}CH_2)\text{-2-}(OCH_3)C_6H_3OH$$
Eugenol
Oil of cloves

$$\text{4-}(CH{=}CHCH_3)\text{-2-}(OCH_3)C_6H_3OH$$
Isoeugenol
Oil of nutmeg

$$\text{4-}(CH{=}CHCH_3)C_6H_4OCH_3$$
Anethole
Oil of aniseed

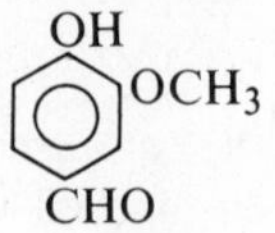

Vanillin (OH, OCH_3, CHO)
Oil of vanilla bean

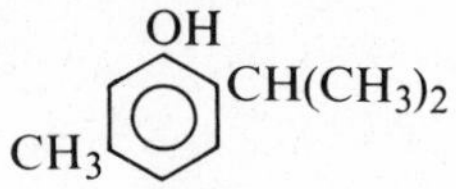

Thymol (OH, $CH(CH_3)_2$, CH_3)
Oil of thyme and mint

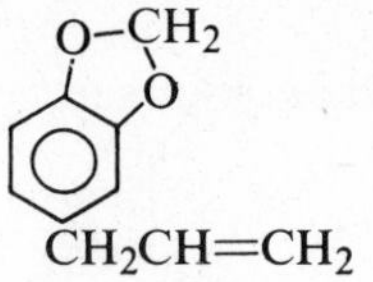

Safrole ($O-CH_2-O$, $CH_2CH{=}CH_2$)
Oil of sassafras

24.5 Preparation

In the laboratory, phenols are generally prepared by one of the three methods outlined below.

PREPARATION OF PHENOLS

1. Hydrolysis of diazonium salts. Discussed in Sec. 23.14.

$$ArN_2^+ + H_2O \longrightarrow ArOH + H^+ + N_2$$

Example:

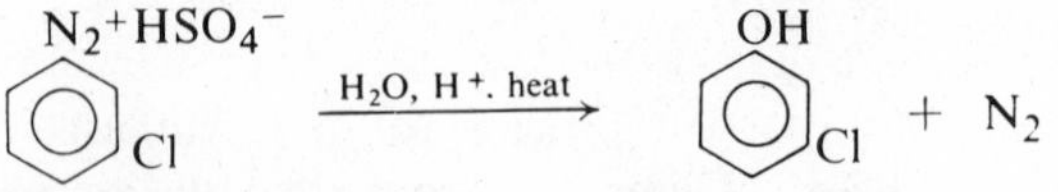

$$m\text{-}ClC_6H_4N_2^+HSO_4^- \xrightarrow{H_2O,\ H^+,\ \text{heat}} m\text{-}ClC_6H_4OH + N_2$$

m-Chlorobenzenediazonium hydrogen sulfate → *m*-Chlorophenol

2. Oxidation of arylthallium compounds. Discussed in Sec. 24.5.

$$ArTl(OOCCF_3)_2 \xrightarrow[Ph_3P]{Pb(OAc)_4} ArOOCCF_3 \xrightarrow[\text{heat}]{H_2O,\ OH^-} ArO^- \xrightarrow{H^+} ArOH$$

Arythallium trifluoroacetate; Aryl trifluoroacetate

$$ArH \xrightarrow{Tl(OOCCF_3)_3} ArTl(OOCCF_3)_2$$

Example:

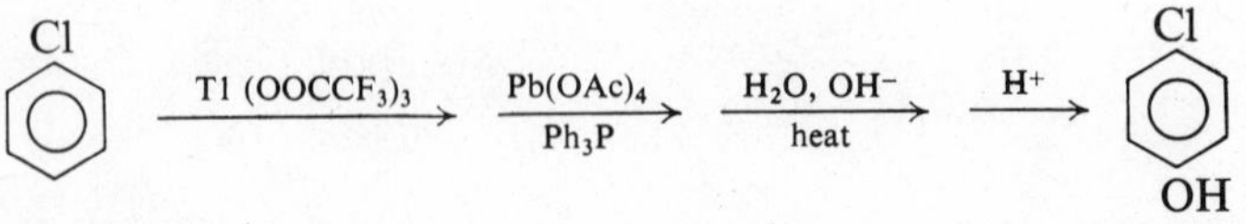

$$C_6H_5Cl \xrightarrow{Tl\ (OOCCF_3)_3} \xrightarrow[Ph_3P]{Pb(OAc)_4} \xrightarrow[\text{heat}]{H_2O,\ OH^-} \xrightarrow{H^+} p\text{-}ClC_6H_4OH$$

Chlorobenzene → *p*-Chlorophenol *Only isomer*

3. Alkali fusion of sulfonates. Discussed in Sec. 30.12.

Hydrolysis of diazonium salts is a highly versatile method of making phenols. It is the last step in a synthetic route that generally begins with nitration (Secs. 23.11 and 23.14).

Much simpler and more direct is a recently developed route via thallation. An arylthallium compound is oxidized by lead tetraacetate (in the presence of triphenylphosphine, Ph_3P) to the phenolic ester of trifluoroacetic acid, which on hydrolysis yields the phenol. The entire sequence, including thallation, can be carried out without isolation of intermediates. Although the full scope of the method has not yet been reported, it has two advantages over the diazonium route: (a) the speed and high yield made possible by the fewer steps; and (b) orientation control in the thallation step. (Review Secs. 11.7 and 11.13.)

Of limited use is the hydrolysis of aryl halides containing strongly electron-withdrawing groups *ortho* and *para* to the halogen (Sec. 25.9); 2,4-dinitrophenol and 2,4,6-trinitrophenol (*picric acid*) are produced in this way on a large scale:

$$\text{2,4-Dinitrochlorobenzene (Cl, NO}_2\text{, NO}_2) \xrightarrow{\text{NaOH}} \text{Sodium 2,4-dinitrophenoxide (ONa)} \xrightarrow{H^+} \text{2,4-Dinitrophenol (OH)} \xrightarrow[H_2SO_4]{HNO_3} \text{2,4,6-Trinitrophenol (}O_2N\text{, OH, NO}_2\text{, NO}_2)$$

2,4-Dinitrochlorobenzene — Sodium 2,4-dinitrophenoxide — 2,4-Dinitrophenol — 2,4,6-Trinitrophenol, Picric acid

Problem 24.5 Outline all steps in the synthesis *from toluene* of: (a) *p*-cresol via diazotization; (b) *p*-cresol via thallation; (c) and (d) *m*-cresol via each route. (*Hint:* See Secs. 23.16 and 11.13.)

24.6 Reactions

Aside from acidity, the most striking chemical property of a phenol is the extremely high reactivity of its ring toward electrophilic substitution. Even in ring substitution, acidity plays an important part; ionization of a phenol yields the $—O^-$ group, which, because of its full-fledged negative charge, is even more strongly electron-releasing than the —OH group.

Phenols undergo not only those electrophilic substitution reactions that are typical of most aromatic compounds, but also many others that are possible only because of the unusual reactivity of the ring. We shall have time to take up only a few of these reactions.

REACTIONS OF PHENOLS

1. **Acidity. Salt formation.** Discussed in Secs. 24.3 and 24.7.

$$ArOH + H_2O \rightleftharpoons ArO^- + H_3O^+$$

Example:

$$C_6H_5OH + NaOH \longrightarrow C_6H_5O^-Na^+ + H_2O$$

Phenol — Sodium phenoxide

2. **Ether formation. Williamson synthesis.** Discussed in Secs. 17.5 and 24.8.

$$ArO^- + RX \longrightarrow ArOR + X^-$$

Examples:

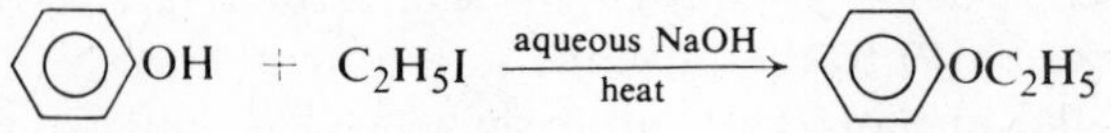

$$C_6H_5OH + C_2H_5I \xrightarrow[\text{heat}]{\text{aqueous NaOH}} C_6H_5OC_2H_5$$

Phenol — Ethyl iodide — Phenyl ethyl ether (Phenetole)

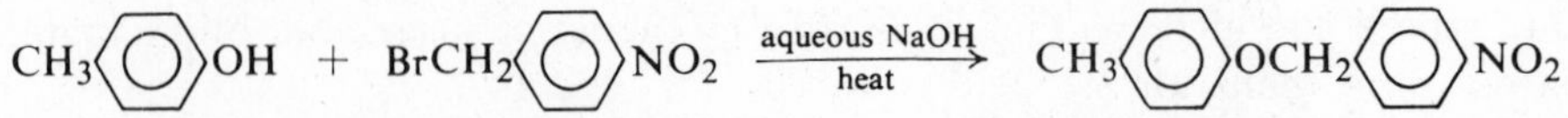

$$p\text{-}CH_3C_6H_4OH + p\text{-}BrCH_2C_6H_4NO_2 \xrightarrow[\text{heat}]{\text{aqueous NaOH}} p\text{-}CH_3C_6H_4OCH_2C_6H_4NO_2\text{-}p$$

p-Cresol — p-Nitrobenzyl bromide — p-Tolyl p-nitrobenzyl ether

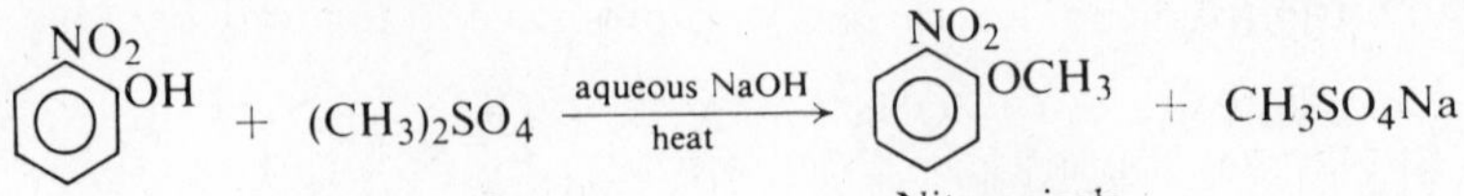

$$o\text{-}NO_2C_6H_4OH + (CH_3)_2SO_4 \xrightarrow[\text{heat}]{\text{aqueous NaOH}} o\text{-}NO_2C_6H_4OCH_3 + CH_3SO_4Na$$

o-Nitrophenol — Methyl sulfate — o-Nitroanisole (o-Nitrophenyl methyl ether)

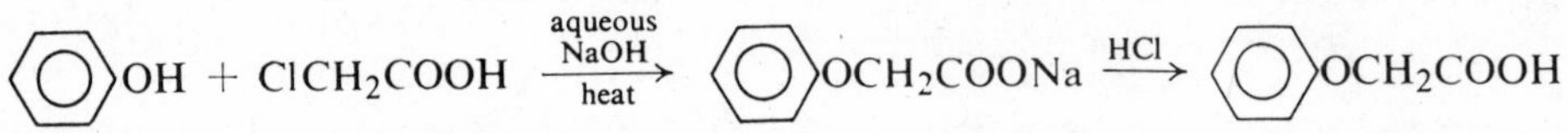

$$C_6H_5OH + ClCH_2COOH \xrightarrow[\text{heat}]{\text{aqueous NaOH}} C_6H_5OCH_2COONa \xrightarrow{\text{HCl}} C_6H_5OCH_2COOH$$

Phenol — Chloroacetic acid — Phenoxyacetic acid

3. Ester formation. Discussed in Secs. 20.8, 20.15, and 24.9.

$$ArOH \xrightarrow{RCOCl} RCOOAr$$

$$ArOH \xrightarrow{Ar'SO_2Cl} Ar'SO_2OAr$$

Examples:

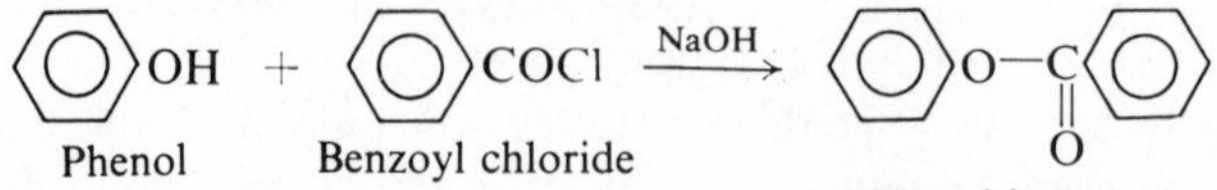

$$C_6H_5OH + C_6H_5COCl \xrightarrow{\text{NaOH}} C_6H_5O\text{—}C(=O)\text{—}C_6H_5$$

Phenol — Benzoyl chloride — Phenyl benzoate

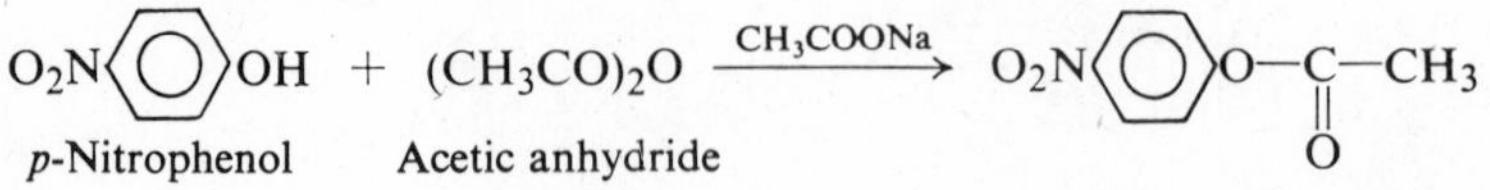

$$p\text{-}O_2NC_6H_4OH + (CH_3CO)_2O \xrightarrow{CH_3COONa} p\text{-}O_2NC_6H_4O\text{—}C(=O)\text{—}CH_3$$

p-Nitrophenol — Acetic anhydride — p-Nitrophenyl acetate

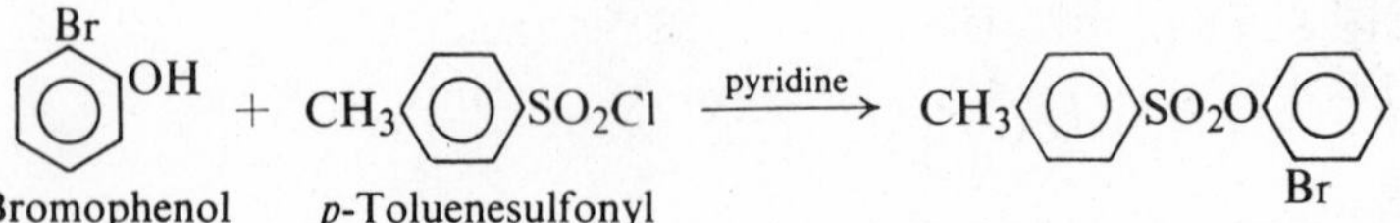

$$o\text{-}BrC_6H_4OH + p\text{-}CH_3C_6H_4SO_2Cl \xrightarrow{\text{pyridine}} p\text{-}CH_3C_6H_4SO_2OC_6H_4Br\text{-}o$$

o-Bromophenol — p-Toluenesulfonyl chloride — o-Bromophenyl p-toluenesulfonate

4. Ring substitution. Discussed in Sec. 24.10.

—OH, —O⁻ } Activate powerfully, and direct *ortho,para* in electrophilic aromatic substitution.

—OR: Less powerful activator than —OH.

(a) **Nitration.** Discussed in Sec. 24.10.

Example:

Phenol $\xrightarrow{\text{dilute } HNO_3,\ 20°}$ *o*-Nitrophenol and *p*-Nitrophenol

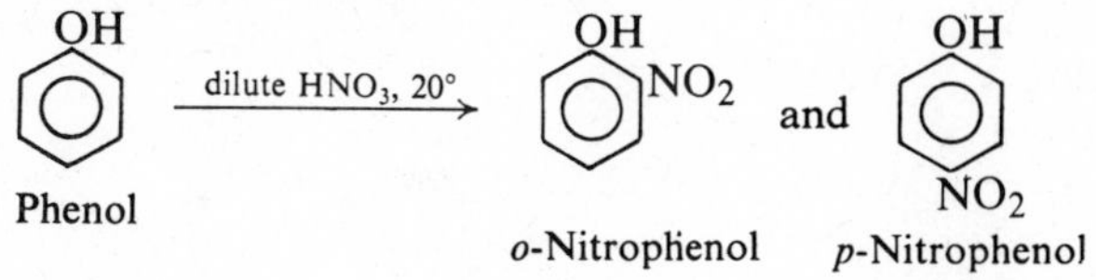

(b) **Sulfonation.** Discussed in Sec. 24.10.

Example:

Phenol $\xrightarrow{H_2SO_4}$ (15–20°) *o*-Phenolsulfonic acid; (100°) *p*-Phenolsulfonic acid

o-Phenolsulfonic acid $\xrightarrow{H_2SO_4,\ 100°}$ *p*-Phenolsulfonic acid

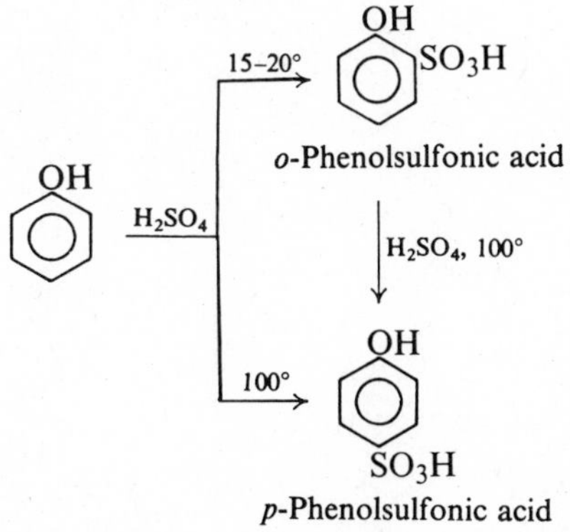

(c) **Halogenation.** Discussed in Sec. 24.10.

Examples:

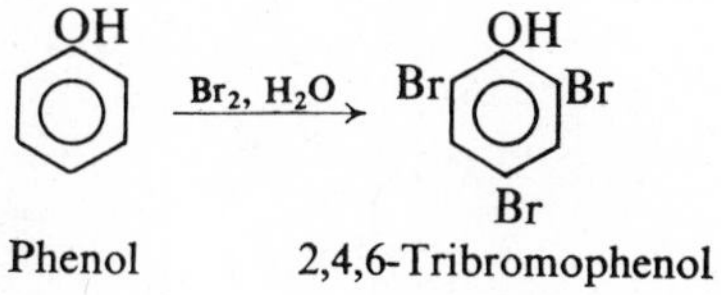

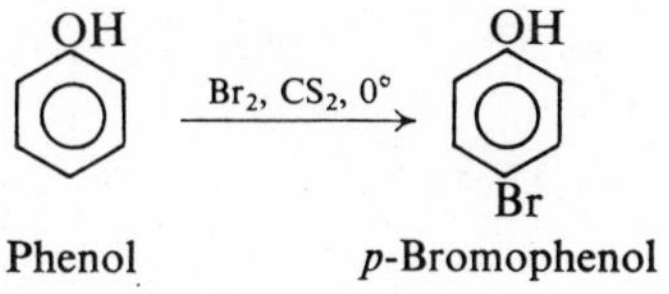

Phenol $\xrightarrow{Br_2,\ H_2O}$ 2,4,6-Tribromophenol

Phenol $\xrightarrow{Br_2,\ CS_2,\ 0°}$ *p*-Bromophenol

(d) **Friedel-Crafts alkylation.** Discussed in Sec. 24.10.

Example:

Phenol + $CH_3-\underset{Cl}{\overset{CH_3}{C}}-CH_3$ (*tert*-Butyl chloride) $\xrightarrow{HF}$ *p-tert*-Butylphenol

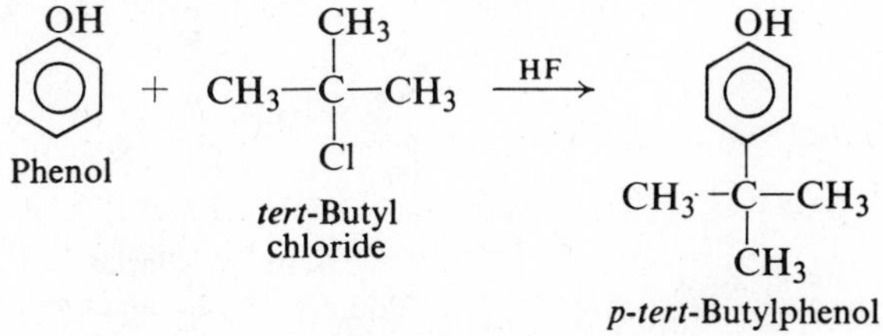

(e) **Friedel-Crafts acylation. Fries rearrangement.** Discussed in Secs. 24.9 and 24.10.

Examples:

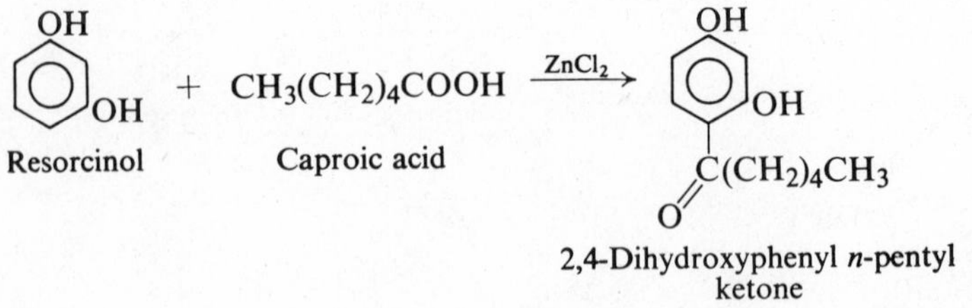

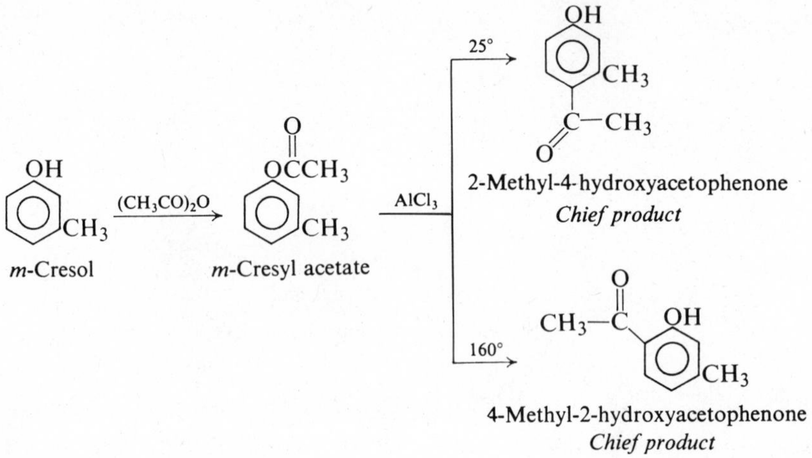

$$\text{Resorcinol} + CH_3(CH_2)_4COOH \xrightarrow{ZnCl_2} \text{2,4-Dihydroxyphenyl } n\text{-pentyl ketone}$$

Resorcinol — Caproic acid — 2,4-Dihydroxyphenyl *n*-pentyl ketone (ring with OH, OH and $C(=O)(CH_2)_4CH_3$)

$$m\text{-Cresol} \xrightarrow{(CH_3CO)_2O} m\text{-Cresyl acetate } (OCOCH_3,\ CH_3) \xrightarrow{AlCl_3}$$

- $25°$: 2-Methyl-4-hydroxyacetophenone (OH, CH_3, $C(=O)-CH_3$) — *Chief product*
- $160°$: 4-Methyl-2-hydroxyacetophenone ($CH_3-C(=O)$, OH, CH_3) — *Chief product*

(f) **Nitrosation.** Discussed in Sec. 24.10.

Example:

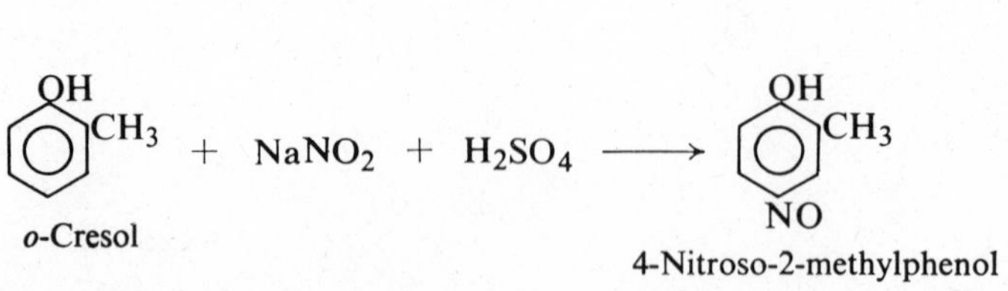

$$o\text{-Cresol} + NaNO_2 + H_2SO_4 \longrightarrow \text{4-Nitroso-2-methylphenol}$$

o-Cresol (OH, CH_3) — 4-Nitroso-2-methylphenol (OH, CH_3, NO)

(g) **Coupling with diazonium salts.** Discussed in Secs. 23.17 and 24.10.

(h) **Carbonation. Kolbe reaction.** Discussed in Sec. 24.11.

Example:

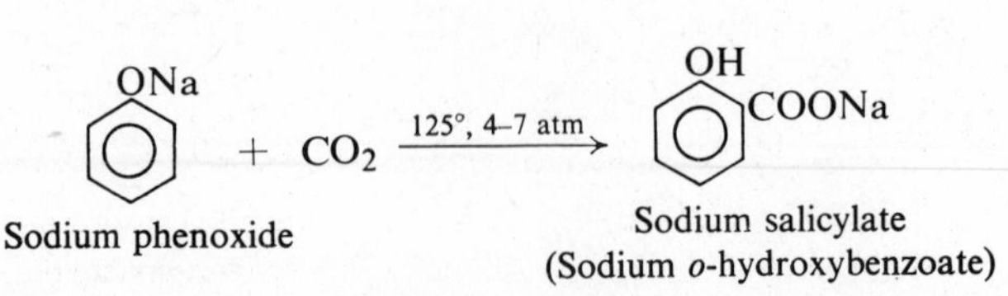

$$\text{Sodium phenoxide (ONa)} + CO_2 \xrightarrow{125°,\ 4\text{–}7\ \text{atm}} \text{Sodium salicylate (OH, COONa)}$$

Sodium phenoxide — Sodium salicylate (Sodium *o*-hydroxybenzoate)

(i) **Aldehyde formation. Reimer-Tiemann reaction.** Discussed in Sec. 24.12.

Example:

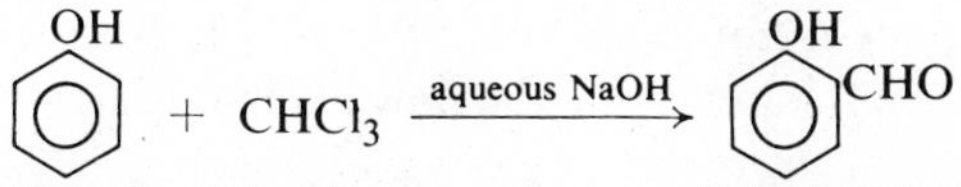

$$\text{Phenol (C}_6\text{H}_5\text{OH)} + \underset{\text{Chloroform}}{CHCl_3} \xrightarrow{\text{aqueous NaOH}} \underset{\substack{\text{Salicylaldehyde}\\(o\text{-Hydroxybenzaldehyde})}}{o\text{-HOC}_6\text{H}_4\text{CHO}}$$

(j) **Reaction with formaldehyde.** Discussed in Sec. 32.7.

24.7 Acidity of phenols

Phenols are converted into their salts by aqueous hydroxides, but not by aqueous bicarbonates. The salts are converted into the free phenols by aqueous mineral acids, carboxylic acids, or carbonic acid.

$$\underset{\substack{\text{Stronger}\\\text{acid}}}{ArOH} + OH^- \longrightarrow ArO^- + \underset{\substack{\text{Weaker}\\\text{acid}}}{H_2O}$$

$$ArO^- + \underset{\substack{\text{Stronger}\\\text{acid}}}{H_2CO_3} \longrightarrow \underset{\substack{\text{Weaker}\\\text{acid}}}{ArOH} + HCO_3^-$$

Phenols must therefore be considerably stronger acids than water, but considerably weaker acids than the carboxylic acids. Table 24.1 (p. 788) shows that this is indeed so: most phenols have K_a's of about 10^{-10}, whereas carboxylic acids have K_a's of about 10^{-5}.

Although weaker than carboxylic acids, phenols are tremendously more acidic than alcohols, which have K_a's in the neighborhood of 10^{-16} to 10^{-18} How does it happen that an —OH attached to an aromatic ring is so much more acidic than an —OH attached to an alkyl group? The answer is to be found in an examination of the structures involved. As usual we shall assume that differences in acidity are due to differences in stabilities of reactants and products (Sec. 18.12).

Let us examine the structures of reactants and products in the ionization of an alcohol and of phenol. We see that the alcohol and the alkoxide ion are each represented satisfactorily by a single structure. Phenol and the phenoxide ion

$$\underset{\text{Alcohol}}{R\text{—}\ddot{\underset{\cdot\cdot}{O}}\text{:H}} \rightleftharpoons H^+ + \underset{\text{Alkoxide ion}}{R\text{—}\ddot{\underset{\cdot\cdot}{O}}\text{:}^-}$$

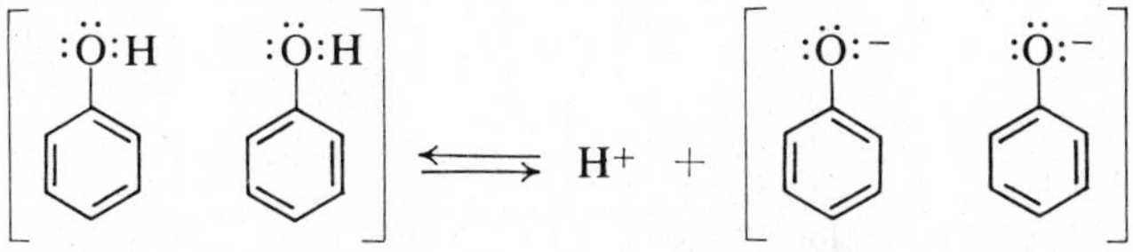

$$\underset{\text{Phenol}}{\left[\ \underset{\text{I}}{C_6H_5\ddot{\underset{\cdot\cdot}{O}}\text{:H}} \quad \underset{\text{II}}{C_6H_5\ddot{\underset{\cdot\cdot}{O}}\text{:H}}\ \right]} \rightleftharpoons H^+ + \underset{\text{Phenoxide ion}}{\left[\ \underset{\text{III}}{C_6H_5\ddot{\underset{\cdot\cdot}{O}}\text{:}^-} \quad \underset{\text{IV}}{C_6H_5\ddot{\underset{\cdot\cdot}{O}}\text{:}^-}\ \right]}$$

contain a benzene ring and therefore must be hybrids of the Kekulé structures I and II, and III and IV. This resonance presumably stabilizes both molecule and ion to the same extent. It lowers the energy content of each by the same number of kcal/mole, and hence does not affect the *difference* in their energy contents. If there were no other factors involved, then, we might expect the acidity of a phenol to be about the same as the acidity of an alcohol.

However, there are additional structures to be considered. Being basic, oxygen can share more than a pair of electrons with the ring; this is indicated by contribution from structures V–VII for phenol, and VIII–X for the phenoxide ion.

V VI VII
Phenol

VIII IX X
Phenoxide ion

Now, are these two sets of structures equally important? Structures V-VII for phenol carry both positive and negative charges; structures VIII–X for phenoxide ion carry only a negative charge. Since energy must be supplied to separate opposite charges, the structures for the phenol should contain more energy and hence be less stable than the structures for phenoxide ion. (We have already encountered the effect of *separation of charge* on stability in Sec. 18.12.) The net effect of resonance is therefore to stabilize the phenoxide ion to a greater extent than the phenol, and thus to shift the equilibrium toward ionization and make K_a larger than for an alcohol (Fig. 24.1).

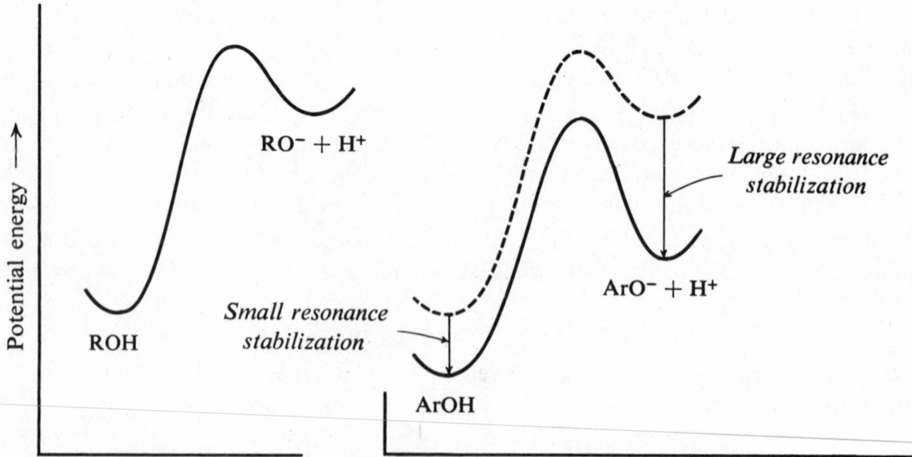

Figure 24.1. Molecular structure and position of equilibrium. Phenol yields resonance-stabilized anion; is stronger acid than alcohol. (Plots aligned with each other for easy comparison.)

We have seen (Sec. 23.3) that aromatic amines are weaker bases than aliphatic amines, since resonance stabilizes the free amine to a greater extent than it does the ion. Here we have exactly the opposite situation, phenols being stronger acids than their aliphatic counterparts, the alcohols, because resonance stabilizes the ion to a greater extent than it does the free phenol. (Actually, of course, resonance with the ring exerts the *same* effect in both cases; it stabilizes—and thus weakens—the base: amine or phenoxide ion.)

In Table 24.1 (p. 788) we see that electron-attracting substituents like —X or $—NO_2$ increase the acidity of phenols, and electron-releasing substituents like $—CH_3$ decrease acidity. Thus substituents affect acidity of phenols in the same way that they affect acidity of carboxylic acids (Sec. 18.14); it is, of course, opposite to the way these groups affect basicity of amines (Sec. 23.4). Electron-attracting substituents tend to disperse the negative charge of the phenoxide ion, whereas electron-releasing substituents tend to intensify the charge.

Problem 24.6 How do you account for the fact that, unlike most phenols, 2,4-dinitrophenol and 2,4,6-trinitrophenol are soluble in aqueous sodium bicarbonate?

We can see that a group attached to an aromatic ring affects *position of equilibrium* in reversible reactions in the same way that it affects *rate* in irreversible reactions. An electron-releasing group favors reactions in which the ring becomes more positive, as in electrophilic substitution or in the conversion of an amine into its salt. An electron-withdrawing group favors reactions in which the ring becomes more negative, as in nucleophilic substitution (Chap. 25) or in the conversion of a phenol or an acid into its salt.

24.8 Formation of ethers. Williamson synthesis

As already discussed (Sec. 17.5), phenols are converted into ethers by reaction in alkaline solution with alkyl halides; methyl ethers can also be prepared by reaction with methyl sulfate. In alkaline solutions a phenol exists as the phenoxide ion which, acting as a nucleophilic reagent, attacks the halide (or the sulfate) and displaces halide ion (or sulfate ion).

$$\text{ArOH} \xrightarrow{\text{OH}^-} \text{ArO}^- \begin{cases} \xrightarrow{\text{RX}} \text{Ar—O—R} + \text{X}^- \\ \xrightarrow{(\text{CH}_3)_2\text{SO}_4} \text{Ar—O—CH}_3 + \text{CH}_3\text{OSO}_3^- \end{cases}$$

Certain ethers can be prepared by the reaction of unusually active aryl halides with sodium alkoxides. For example:

$$\text{2,4-}(NO_2)_2C_6H_3Cl + Na^{+}\,{}^{-}OCH_3 \longrightarrow \text{2,4-}(NO_2)_2C_6H_3OCH_3$$

2,4-Dinitrochlorobenzene

2,4-Dinitroanisole
(2,4-Dinitrophenyl methyl ether)

While alkoxy groups are activating and *ortho,para*-directing in electrophilic aromatic substitution, they are considerably less so than the —OH group. As a result, ethers do not generally undergo those reactions (Secs. 24.10–24.12) which require the especially high reactivity of phenols: coupling, Kolbe reaction, Reimer-Tiemann reaction, etc. This difference in reactivity is probably due to the fact that, unlike a phenol, an ether cannot ionize to form the extremely reactive phenoxide ion.

As a consequence of the lower reactivity of the ring, an aromatic ether is less sensitive to oxidation than a phenol. For example:

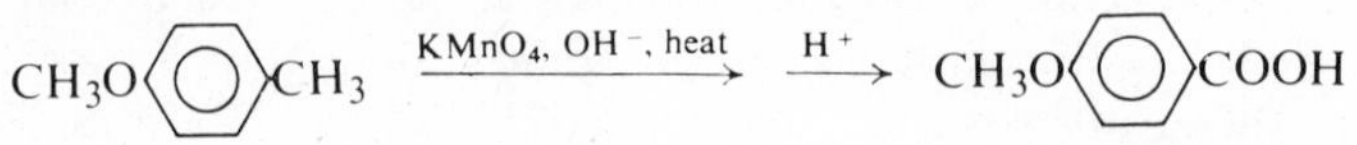

$$CH_3O\langle\bigcirc\rangle CH_3 \xrightarrow{KMnO_4,\ OH^-,\ heat} \xrightarrow{H^+} CH_3O\langle\bigcirc\rangle COOH$$

p-Methylanisole — Anisic acid

We have already discussed the cleavage of ethers by acids (Sec. 17.7). Cleavage of methyl aryl ethers by concentrated hydriodic acid is the basis of an important analytical procedure (the *Zeisel procedure*, Sec. 17.16).

Problem 24.7 2,4-Dichlorophenoxyacetic acid is the important weed-killer known as 2,4-D. Outline the synthesis of this compound starting from benzene or toluene and acetic acid.

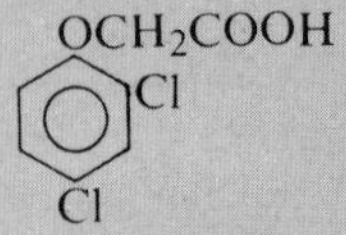

OCH_2COOH (ring with Cl at 2- and 4-positions)

2,4-Dichlorophenoxyacetic acid
(2,4-D)

Problem 24.8 The *n*-propyl ether of 2-amino-4-nitrophenol is one of the sweetest compounds ever prepared, being about 5000 times as sweet as the common sugar sucrose. It can be made from the dinitro compound by reduction with ammonium bisulfide. Outline the synthesis of this material starting from benzene or toluene and any aliphatic reagents.

24.9 Ester formation. Fries rearrangement

Phenols are usually converted into their esters by the action of acids, acid chlorides, or anhydrides as discussed in Secs. 18.16, 20.8, and 20.15.

Problem 24.9 Predict the products of the reaction between phenyl benzoate and one mole of bromine in the presence of iron.

When esters of phenols are heated with aluminum chloride, the acyl group migrates from the phenolic oxygen to an *ortho* or *para* position of the ring, thus yielding a ketone. This reaction, called the *Fries rearrangement*, is often used instead of direct acylation for the synthesis of phenolic ketones. For example:

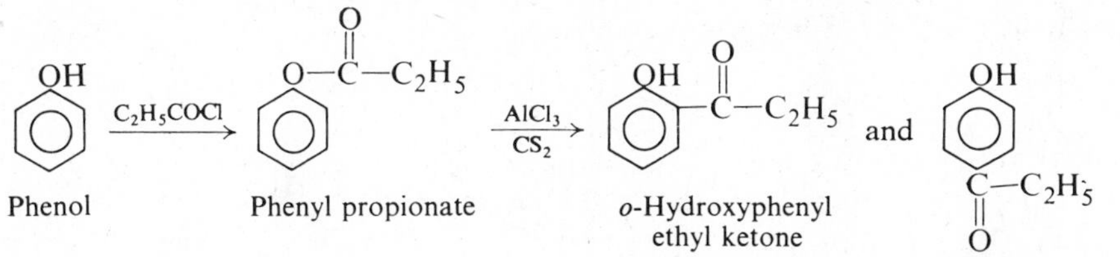

Phenol $\xrightarrow{C_2H_5COCl}$ Phenyl propionate $\xrightarrow[CS_2]{AlCl_3}$ *o*-Hydroxyphenyl ethyl ketone (*o*-Hydroxypropiophenone) *Volatile in steam* and *p*-Hydroxyphenyl ethyl ketone (*p*-Hydroxypropiophenone) *Non-volatile in steam*

In at least some cases, rearrangement appears to involve generation of an acylium ion, RCO^+, which then attacks the ring as in ordinary Friedel-Crafts acylation.

Problem 24.10 A mixture of *o*- and *p*-isomers obtained by the Fries rearrangement can often be separated by steam distillation, only the *o*-isomer distilling. How do you account for this?

Problem 24.11 4-*n*-Hexylresorcinol is used in certain antiseptics. Outline its preparation starting with resorcinol and any aliphatic reagents.

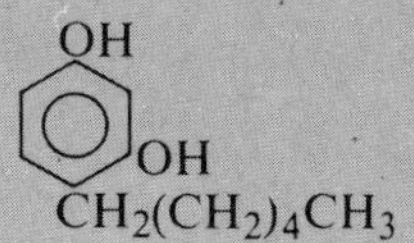

4-*n*-Hexylresorcinol

24.10 Ring substitution

Like the amino group, the phenolic group powerfully activates aromatic rings toward electrophilic substitution, and in essentially the same way. The intermediates are hardly carbonium ions at all, but rather oxonium ions (like I and II), in which every atom (except hydrogen) has a complete octet of electrons;

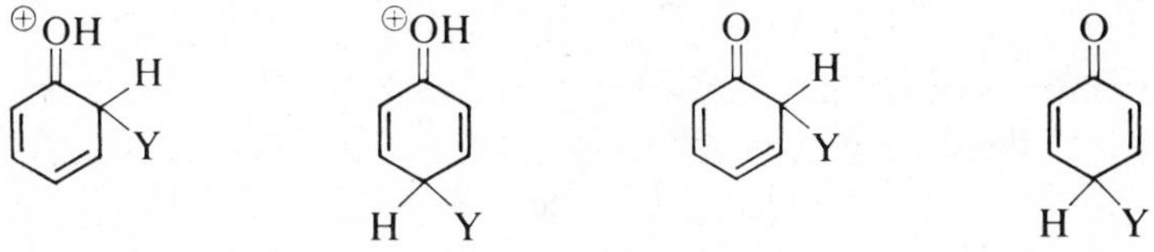

I II III IV

they are formed tremendously faster than the carbonium ions derived from benzene itself. Attack on a phenoxide ion yields an even more stable—and even more rapidly formed—intermediate, an unsaturated ketone (like III and IV).

With phenols, as with amines, special precautions must often be taken to prevent polysubstitution and oxidation.

Treatment of phenols with aqueous solutions of bromine results in replacement of every hydrogen *ortho* or *para* to the —OH group, and may even cause displacement of certain other groups. For example:

$$o\text{-Cresol} + 2Br_2(aq) \longrightarrow \text{4,6-Dibromo-2-methylphenol} + 2HBr$$

$$p\text{-Phenolsulfonic acid} + 3Br_2(aq) \longrightarrow \text{2,4,6-Tribromophenol} + 3HBr + H_2SO_4$$

If halogenation is carried out in a solvent of low polarity, such as chloroform, carbon tetrachloride, or carbon disulfide, reaction can be limited to monohalogenation. For example:

$$\text{Phenol} \xrightarrow{Br_2,\ CS_2,\ 0^\circ} p\text{-Bromophenol (Chief product)} \text{ and } o\text{-Bromophenol}$$

Phenol is converted by concentrated nitric acid into 2,4,6-trinitrophenol (*picric acid*), but the nitration is accompanied by considerable oxidation. To

$$\text{Phenol} \xrightarrow{HNO_3} \text{2,4,6-Trinitrophenol (Picric acid)}$$

obtain mononitrophenols, it is necessary to use dilute nitric acid at a low temperature; even then the yield is poor. (The isomeric products are readily separated by

$$\text{Phenol} \xrightarrow{\text{dilute } HNO_3,\ 20^\circ} o\text{-Nitrophenol (40\% yield)} \text{ and } p\text{-Nitrophenol (13\% yield)}$$

steam distillation. *Why?*)

Problem 24.12 Picric acid can be prepared by treatment of 2,4-phenoldisulfonic acid with nitric acid. (a) Show in detail the mechanism by which this happens. (b) What advantage does this method of synthesis have over the direct nitration of phenol?

Alkylphenols can be prepared by Friedel-Crafts alkylation of phenols, but the yields are often poor.

Although phenolic ketones can be made by direct acylation of phenols, they are more often prepared in two steps by means of the Fries rearrangement (Sec. 24.9).

Problem 24.13 The product of sulfonation of phenol depends upon the temperature of reaction: chiefly *ortho* at 15–20°, chiefly *para* at 100°. Once formed, *o*-phenolsulfonic acid is converted into the *p*-isomer by sulfuric acid at 100°. How do you account for these facts? (*Hint:* See Sec. 8.22.)

In addition, phenols undergo a number of other reactions that also involve electrophilic substitution, and that are possible only because of the especially high reactivity of the ring.

Nitrous acid converts phenols into nitrosophenols:

$$\text{Phenol (C}_6\text{H}_5\text{OH)} \xrightarrow{NaNO_2,\ H_2SO_4,\ 7\text{-}8^\circ} p\text{-}HOC_6H_4NO$$

Phenol

p-Nitrosophenol
80% yield

Phenols are one of the few classes of compounds reactive enough to undergo attack by the weakly electrophilic nitrosonium ion, ^+NO.

Problem 24.14 The —NO group is readily oxidized to the —NO_2 group by nitric acid. Suggest a better way to synthesize *p*-nitrophenol than the one given earlier in this section.

As we have seen, the ring of a phenol is reactive enough to undergo attack by diazonium salts, with the formation of azo compounds. This reaction is discussed in detail in Sec. 23.17.

24.11 Kolbe reaction. Synthesis of phenolic acids

Treatment of the salt of a phenol with carbon dioxide brings about substitution of the carboxyl group, —COOH, for hydrogen of the ring. This reaction is known as the **Kolbe reaction**; its most important application is in the conversion of phenol itself into *o*-hydroxybenzoic acid, known as *salicylic acid.* Although some *p*-hydroxybenzoic acid is formed as well, the separation of the two isomers can be

ONa + $O^{\delta-}$=$C^{\delta+}$=$O^{\delta-}$ —125°, 4–7 atm→ OH, COONa (Sodium salicylate, *Chief product*) —H^+→ OH, COOH (Salicylic acid)

carried out readily by steam distillation, the *o*-isomer being the more volatile. (Why?)

It seems likely that CO_2 attaches itself initially to phenoxide oxygen rather than to the ring. In any case, the final product almost certainly results from electrophilic attack by electron-deficient carbon on the highly reactive ring.

Problem 24.15 *Aspirin* is acetylsalicylic acid (*o*-acetoxybenzoic acid, *o*-CH_3COO-C_6H_4COOH); *oil of wintergreen* is the ester, methyl salicylate. Outline the synthesis of these two compounds from phenol.

24.12 Reimer-Tiemann reaction. Synthesis of phenolic aldehydes. Dichlorocarbene

Treatment of a phenol with chloroform and aqueous hydroxide introduces an aldehyde group, —CHO, into the aromatic ring, generally *ortho* to the —OH. This reaction is known as the **Reimer-Tiemann reaction**. For example:

Phenol (OH) —$CHCl_3$, aq. NaOH, 70°→ [O^-, $CHCl_2$] ⟶ O^-, CHO —HCl→ OH, CHO (Salicylaldehyde, *Chief product*)

A substituted benzal chloride is initially formed, but is hydrolyzed by the alkaline reaction medium.

The Reimer-Tiemann reaction involves electrophilic substitution on the highly reactive phenoxide ring. The electrophilic reagent is dichlorocarbene, $:CCl_2$, generated from chloroform by the action of base. Although electrically neutral, dichlorocarbene contains a carbon atom with only a sextet of electrons and hence is strongly electrophilic.

$$OH^- + CHCl_3 \rightleftharpoons H_2O + {}^-:CCl_3 \longrightarrow Cl^- + :CCl_2$$

Chloroform / Dichlorocarbene

O^- (phenoxide) + $:CCl_2$ (*Electrophilic reagent*) ⟶ [O=, H, $CCl_2^{\ominus}$] ⟶ O^-, $CHCl_2$

We encountered dichlorocarbene earlier (Sec. 9.16) as a species adding to carbon–carbon double bonds. There, as here, it is considered to be formed from chloroform by the action of a strong base.

The formation of dichlorocarbene by the sequence

(1) $CHCl_3 + OH^- \rightleftharpoons CCl_3^- + H_2O$

(2) $CCl_3^- \rightleftharpoons Cl^- + :CCl_2$

$\xrightarrow{\text{fast}}$ products (addition to alkenes, Reimer-Tiemann reaction, hydrolysis, etc.)

is indicated by many lines of evidence, due mostly to elegant work by Jack Hine of the Ohio State University.

Problem 24.16 What bearing does each of the following facts have on the mechanism above? Be specific.

(a) $CHCl_3$ undergoes alkaline hydrolysis much more rapidly than CCl_4 or CH_2Cl_2.

(b) Hydrolysis of ordinary chloroform is carried out in D_2O in the presence of OD^-. When the reaction is interrupted, and unconsumed chloroform is recovered, it is found to contain deuterium. (*Hint:* See Sec. 20.17.)

(c) The presence of added Cl^- *slows down* alkaline hydrolysis of $CHCl_3$.

(d) When alkaline hydrolysis of $CHCl_3$ in the presence of I^- is interrupted, there is recovered not only $CHCl_3$ but also $CHCl_2I$. (In the absence of base, $CHCl_3$ does not react with I^-.)

(e) In the presence of base, $CHCl_3$ reacts with acetone to give 1,1,1-trichloro-2-methyl-2-propanol.

24.13 Analysis of phenols

The most characteristic property of phenols is their particular degree of acidity. Most of them (Secs. 24.3 and 24.7) are stronger acids than water but weaker acids than carbonic acid. Thus, a water-insoluble compound that dissolves in aqueous sodium hydroxide but *not* in aqueous sodium bicarbonate is most likely a phenol.

Many (but not all) phenols form colored complexes (ranging from green through blue and violet to red) with ferric chloride. (This test is also given by *enols*.)

Phenols are often identified through bromination products and certain esters and ethers.

Problem 24.17 Phenols are often identified as their aryloxyacetic acids, $ArOCH_2COOH$. Suggest a reagent and a procedure for the preparation of these derivatives. (*Hint:* See Sec. 24.8.) Aside from melting point, what other property of the aryloxyacetic acids would be useful in identifying phenols? (*Hint:* See Sec. 18.21.)

24.14 Spectroscopic analysis of phenols

Infrared. As can be seen in Fig. 24.2 (p. 806), phenols show a strong, broad band due to O—H stretching in the same region, 3200–3600 cm^{-1}, as alcohols.

O—H stretching, *strong, broad*

Phenols (or alcohols), 3200–3600 cm^{-1}

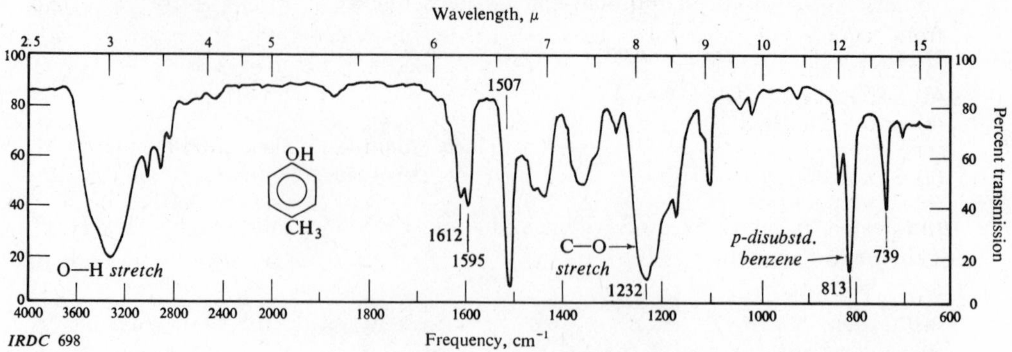

Figure 24.2. Infrared spectrum of *p*-cresol.

Phenols differ from alcohols, however, in the position of the C—O stretching band (compare Sec. 16.13).

C—O stretching, *strong, broad*

Phenols, about 1230 cm^{-1} Alcohols, 1050–1200 cm^{-1}

Phenolic ethers do not, of course, show the O—H band, but do show C—O stretching.

C—O stretching, *strong, broad*

Aryl and vinyl ethers, 1200–1275 cm^{-1}, and weaker, 1020–1075 cm^{-1}

Alkyl ethers, 1060–1150 cm^{-1}

(For a comparison of certain oxygen compounds, see Table 20.3, p. 689.)

Nmr. Absorption by the O—H proton of a phenol, like that of an alcohol (Sec. 16.13), is affected by the degree of hydrogen bonding, and hence by the temperature, concentration, and nature of the solvent. The signal may appear anywhere in the range δ 4–7, or, if there is intramolecular hydrogen bonding, still lower: δ 6–12.

PROBLEMS

1. Write structural formulas for:

(a) 2,4-dinitrophenol
(b m-cresol
(c) hydroquinone
(d) resorcinol
(e) 4-*n*-hexylresorcinol
(f) catechol
(g) picric acid
(h) phenyl acetate
(i) anisole
(j) salicylic acid
(k) ethyl salicylate

2. Give the reagents and any critical conditions necessary to prepare phenol from:

(a) aniline
(b) benzene
(c) chlorobenzene
(d) cumene (isopropylbenzene)

3. Outline the steps in a possible industrial synthesis of:

(a) catechol from *guaiacol*, o-$CH_3OC_6H_4OH$, found in beech-wood tar
(b) catechol from phenol
(c) resorcinol from benzene
(d) picric acid from chlorobenzene
(e) *veratrole*, o-$C_6H_4(OCH_3)_2$, from catechol

4. Outline a possible laboratory synthesis of each of the following compounds from benzene and/or toluene, using any needed aliphatic and inorganic reagents.

(a)–(c) the three cresols
(d) *p*-iodophenol
(e) *m*-bromophenol
(f) *o*-bromophenol
(g) 3-bromo-4-methylphenol
(h) 2-bromo-4-methylphenol
(i) 2-bromo-5-methylphenol
(j) 5-bromo-2-methylphenol
(k) 2,4-dinitrophenol
(l) *p*-isopropylphenol
(m) 2,6-dibromo-4-isopropylphenol
(n) 2-hydroxy-5-methylbenzaldehyde
(o) *o*-methoxybenzyl alcohol

5. Give structures and names of the principal organic products of the reaction (if any) of *o*-cresol with:

(a) aqueous NaOH
(b) aqueous $NaHCO_3$
(c) hot conc. HBr
(d) methyl sulfate, aqueous NaOH
(e) benzyl bromide, aqueous NaOH
(f) bromobenzene, aqueous NaOH
(g) 2,4-dinitrochlorobenzene, aqueous NaOH
(h) acetic acid, H_2SO_4
(i) acetic anhydride
(j) phthalic anhydride
(k) *p*-nitrobenzoyl chloride, pyridine
(l) benzenesulfonyl chloride, aqueous NaOH
(m) product (i) + $AlCl_3$
(n) thionyl chloride
(o) ferric chloride solution
(p) H_2, Ni, 200°, 20 atm.
(q) cold dilute HNO_3
(r) H_2SO_4, 15°
(s) H_2SO_4, 100°
(t) bromine water
(u) Br_2, CS_2
(v) $NaNO_2$, dilute H_2SO_4
(w) product (v) + HNO_3
(x) *p*-nitrobenzenediazonium chloride
(y) CO_2, NaOH, 125°, 5 atm.
(z) $CHCl_3$, aqueous NaOH, 70°

6. Answer Problem 5 for anisole.

7. Answer Problem 5, parts (a) through (o), for benzyl alcohol.

8. Without referring to tables, arrange the compounds of each set in order of acidity:

(a) benzenesulfonic acid, benzoic acid, benzyl alcohol, phenol
(b) carbonic acid, phenol, sulfuric acid, water
(c) *m*-bromophenol, *m*-cresol, *m*-nitrophenol, phenol
(d) *p*-chlorophenol, 2,4-dichlorophenol, 2,4,6-trichlorophenol

9. Describe simple chemical tests that would serve to distinguish between:

(a) phenol and *o*-xylene
(b) *p*-ethylphenol, *p*-methylanisole, and *p*-methylbenzyl alcohol
(c) 2,5-dimethylphenol, phenyl benzoate, *m*-toluic acid
(d) anisole and *o*-toluidine
(e) acetylsalicylic acid, ethyl acetylsalicylate, ethyl salicylate, and salicylic acid
(f) *m*-dinitrobenzene, *m*-nitroaniline, *m*-nitrobenzoic acid, and *m*-nitrophenol

Tell exactly what you would *do* and *see*.

10. Describe simple chemical methods for the separation of the compounds of Problem 9, parts (a), (c), (d), and (f), recovering each component in essentially pure form.

11. Outline all steps in a possible laboratory synthesis of each of the following compounds starting from the aromatic source given, and using any needed aliphatic and inorganic reagents:

(a) 2,4-diaminophenol (Amidol, used as a photographic developer) from chlorobenzene
(b) 4-amino-1,2-dimethoxybenzene from catechol
(c) 2-nitro-1,3-dihydroxybenzene from resorcinol (*Hint:* See Problem 11.7, p. 350.)
(d) 2,4,6-trimethylphenol from mesitylene
(e) *p-tert*-butylphenol from phenol
(f) 4-(*p*-hydroxyphenyl)-2,2,4-trimethylpentane from phenol
(g) 2-phenoxy-1-bromoethane from phenol (*Hint:* Together with $C_6H_5OCH_2CH_2OC_6H_5$.)
(h) phenyl vinyl ether from phenol

(i) What will phenyl vinyl ether give when heated with acid?
(j) 2,6-dinitro-4-*tert*-butyl-3-methylanisole (synthetic musk) from *m*-cresol
(k) 5-methyl-1,3-dihydroxybenzene (*orcinol*, the parent compound of the litmus dyes) from toluene

12. Outline a possible synthesis of each of the following from benzene, toluene, or any of the natural products shown in Sec. 24.4, using any other needed reagents.
(a) *caffeic acid*, from coffee beans
(b) *tyramine*, found in ergot (*Hint:* See Problem 21.22a, p. 714.)
(c) *noradrenaline*, an adrenal hormone

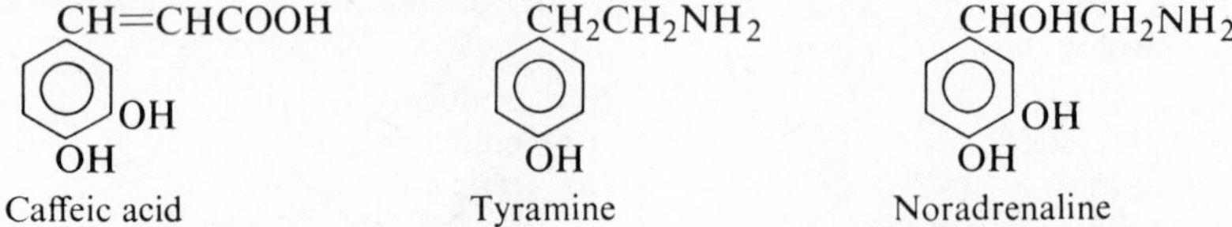

Caffeic acid Tyramine Noradrenaline

13. The reaction between benzyl chloride and sodium phenoxide follows second-order kinetics in a variety of solvents; the nature of the products, however, varies considerably. (a) In dimethylformamide, dioxane, or tetrahydrofuran, reaction yields only benzyl phenyl ether. Show in detail the mechanism of this reaction. To what general class does it belong? (b) In aqueous solution, the yield of ether is cut in half, and there is obtained, in addition, *o*- and *p*-benzylphenol. Show in detail the mechanism by which the latter products are formed. To what general class (or classes) does the reaction belong? (c) What is a possible explanation for the difference between (a) and (b)? (*Hint:* See Sec. 1.21.) (d) In methanol or ethanol, reaction occurs as in (a); in liquid phenol or 2,2,2-trifluoroethanol, reaction is as in (b). How can you account for these differences?

14. When *phloroglucinol*, 1,3,5-trihydroxybenzene, is dissolved in concentrated $HClO_4$, its nmr spectrum shows two peaks of equal area at δ 6.12 and δ 4.15. Similar solutions of 1,3,5-trimethoxybenzene and 1,3,5-triethoxybenzene show similar nmr peaks. On dilution, the original compounds are recovered unchanged. Solutions of these compounds in D_2SO_4 also show these peaks, but on standing the peaks gradually disappear.

How do you account for these observations? What is formed in the acidic solutions? What would you expect to recover from the solution of 1,3,5-trimethoxybenzene in D_2SO_4?

15. When the terpene *citral* is allowed to react in the presence of dilute acid with *olivetol*, there is obtained a mixture of products containing I, the racemic form of one of the physiologically active components of hashish (marijuana). (C_5H_{11} is *n*-pentyl.) Show all steps in a likely mechanism for the formation of I.

$(CH_3)_2C{=}CHCH_2CH_2C(CH_3){=}CHCHO$ + [Olivetol: benzene ring with OH, HO, C_5H_{11}]

Citral Olivetol

↓

[Structure I: OH, O, C_5H_{11}]

I

Δ^1-3,4-*trans*-Tetrahydrocannabinol

16. Give structures of all compounds below:

(a) *p*-nitrophenol + C_2H_5Br + NaOH (aq) ⟶ A ($C_8H_9O_3N$)
A + Sn + HCl ⟶ B ($C_8H_{11}ON$)
B + $NaNO_2$ + HCl, then phenol ⟶ C ($C_{14}H_{14}O_2N_2$)
C + ethyl sulfate + NaOH (aq) ⟶ D ($C_{16}H_{18}O_2N_2$)
D + $SnCl_2$ ⟶ E ($C_8H_{11}ON$)
E + acetyl chloride ⟶ *phenacetin* ($C_{10}H_{13}O_2N$), an analgesic ("pain-killer") and antipyretic ("fever-killer")

(b) β-(*o*-hydroxyphenyl)ethyl alcohol + HBr ⟶ F (C_8H_9OBr)
F + KOH ⟶ *coumarane* (C_8H_8O), insoluble in NaOH

(c) phenol + $ClCH_2COOH$ + NaOH (aq), then HCl ⟶ G ($C_8H_8O_3$)
G + $SOCl_2$ ⟶ H ($C_8H_7O_2Cl$)
H + $AlCl_3$ ⟶ *3-cumaranone* ($C_8H_6O_2$)

(d) *p*-cymene (*p*-isopropyltoluene) + conc. H_2SO_4 ⟶ I + J (both $C_{10}H_{14}O_3S$)
I + KOH + heat, then H^+ ⟶ *carvacrol* ($C_{10}H_{14}O$), found in some essential oils
J + KOH + heat, then H^+ ⟶ *thymol* ($C_{10}H_{14}O$), from oil of thyme
I + HNO_3 ⟶ K ($C_8H_8O_5S$)
p-toluic acid + fuming sulfuric acid ⟶ K

(e) anethole (p. 791) + HBr ⟶ L ($C_{10}H_{13}OBr$)
L + Mg ⟶ M ($C_{20}H_{26}O_2$)
M + HBr, heat ⟶ *hexestrol* ($C_{18}H_{22}O_2$), a synthetic estrogen (female sex hormone)

17. The adrenal hormone (−)-*adrenaline* was the first hormone isolated and the first synthesized. Its structure was proved by the following synthesis:

catechol + $ClCH_2COCl$ $\xrightarrow{POCl_3}$ N ($C_8H_7O_3Cl$)
N + CH_3NH_2 ⟶ O ($C_9H_{11}O_3N$)
O + H_2, Pd ⟶ (±)-adrenaline ($C_9H_{13}O_3N$)
N + NaOI, then H^+ ⟶ 3,4-dihydroxybenzoic acid

What is the structure of adrenaline?

18. (−)-*Phellandral*, $C_{10}H_{16}O$, is a terpene found in eucalyptus oils. It is oxidized by Tollens' reagent to (−)-phellandric acid, $C_{10}H_{16}O_2$, which readily absorbs only one mole of hydrogen, yielding dihydrophellandric acid, $C_{10}H_{18}O_2$. (±)-Phellandral has been synthesized as follows:

isopropylbenzene + H_2SO_4 + SO_3 ⟶ P ($C_9H_{12}O_3S$)
P + KOH, fuse ⟶ Q ($C_9H_{12}O$)
Q + H_2, Ni ⟶ R ($C_9H_{18}O$)
R + $K_2Cr_2O_7$, H_2SO_4 ⟶ S ($C_9H_{16}O$)
S + KCN + H^+ ⟶ T ($C_{10}H_{17}ON$)
T + acetic anhydride ⟶ U ($C_{12}H_{19}O_2N$)
U + heat (600°) ⟶ V ($C_{10}H_{15}N$) + CH_3COOH
V + H_2SO_4 + H_2O ⟶ W ($C_{10}H_{16}O_2$)
W + $SOCl_2$ ⟶ X ($C_{10}H_{15}OCl$)
X $\xrightarrow{\text{reduction}}$ (±)-phellandral

(a) What is the most likely structure of phellandral? (b) Why is synthetic phellandral optically inactive? At what stage in the synthesis does inactivity of this sort first appear? (c) Dihydrophellandric acid is actually a mixture of two optically inactive isomers. Give the structures of these isomers and account for their optical inactivity.

19. Compound Y, C_7H_8O, is insoluble in water, dilute HCl, and aqueous $NaHCO_3$; it dissolves in dilute NaOH. When Y is treated with bromine water it is converted rapidly into a compound of formula $C_7H_5OBr_3$. What is the structure of Y?

20. Two isomeric compounds, Z and AA, are isolated from oil of bay leaf; both are found to have the formula $C_{10}H_{12}O$. Both are insoluble in water, dilute acid, and dilute base. Both give positive tests with dilute $KMnO_4$ and Br_2/CCl_4. Upon vigorous oxidation, both yield anisic acid, *p*-$CH_3OC_6H_4COOH$.

(a) At this point what structures are possible for Z and AA?
(b) Catalytic hydrogenation converts Z and AA into the same compound, $C_{10}H_{14}O$. Now what structures are possible for Z and AA?
(c) Describe chemical procedures (other than synthesis) by which you could assign structures to Z and AA.
(d) Compound Z can be synthesized as follows:

$$p\text{-bromoanisole} + \text{Mg} + \text{ether, then allyl bromide} \longrightarrow \text{Z}$$

What is the structure of Z?
(e) Z is converted into AA when heated strongly with concentrated base. What is the most likely structure for AA?
(f) Suggest a synthetic sequence starting with *p*-bromoanisole that would independently confirm the structure assigned to AA.

21. Compound BB ($C_{10}H_{12}O_3$) was insoluble in water, dilute HCl, and dilute aqueous $NaHCO_3$; it was soluble in dilute NaOH. A solution of BB in dilute NaOH was boiled, and the distillate was collected in a solution of NaOI, where a yellow precipitate formed.

The alkaline residue in the distillation flask was acidified with dilute H_2SO_4; a solid, CC, precipitated. When this mixture was boiled, CC steam-distilled and was collected. CC was found to have the formula $C_7H_6O_3$; it dissolved in aqueous $NaHCO_3$ with evolution of a gas.

(a) Give structures and names for BB and CC. (b) Write complete equations for all the above reactions.

22. *Chavibetol*, $C_{10}H_{12}O_2$, is found in betel-nut leaves. It is soluble in aqueous NaOH but not in aqueous $NaHCO_3$.

Treatment of chavibetol (a) with methyl sulfate and aqueous NaOH gives compound DD, $C_{11}H_{14}O_2$; (b) with hot hydriodic acid gives methyl iodide; (c) with hot concentrated base gives compound EE, $C_{10}H_{12}O_2$.

Compound DD is insoluble in aqueous NaOH, and readily decolorizes dilute $KMnO_4$ and Br_2/CCl_4. Treatment of DD with hot concentrated base gives FF, $C_{11}H_{14}O_2$.

Ozonolysis of EE gives a compound that is isomeric with vanillin (p. 792).

Ozonolysis of FF gives a compound that is identical with the one obtained from the treatment of vanillin with methyl sulfate.

What is the structure of chavibetol?

23. *Piperine*, $C_{17}H_{19}O_3N$, is an alkaloid found in black pepper. It is insoluble in water, dilute acid, and dilute base. When heated with aqueous alkali, it yields *piperic acid*, $C_{12}H_{10}O_4$, and the cyclic secondary amine *piperidine* (see Sec. 31.12), $C_5H_{11}N$.

Piperic acid is insoluble in water, but soluble in aqueous NaOH and aqueous $NaHCO_3$. Titration gives an equivalent weight of 215 ± 6. It reacts readily with Br_2/CCl_4, without evolution of HBr, to yield a compound of formula $C_{12}H_{10}O_4Br_4$. Careful oxidation of piperic acid yields *piperonylic acid*, $C_8H_6O_4$, and *tartaric acid*, HOOCCHOHCHOHCOOH.

When piperonylic acid is heated with aqueous HCl at 200° it yields formaldehyde and *protocatechuic acid*, 3,4-dihydroxybenzoic acid.

(a) What kind of compound is piperine? (b) What is the structure of piperonylic acid? Of piperic acid? Of piperine?

(c) Does the following synthesis confirm your structure?

catechol + $CHCl_3$ + NaOH $\longrightarrow$ GG ($C_7H_6O_3$)
GG + CH_2I_2 + NaOH $\longrightarrow$ HH ($C_8H_6O_3$)
HH + CH_3CHO + NaOH $\longrightarrow$ II ($C_{10}H_8O_3$)
II + acetic anhydride + sodium acetate $\longrightarrow$ piperic acid ($C_{12}H_{10}O_4$)
piperic acid + PCl_5 $\longrightarrow$ JJ ($C_{12}H_9O_3Cl$)
JJ + piperidine $\longrightarrow$ piperine

24. *Hordinene*, $C_{10}H_{15}ON$, is an alkaloid found in germinating barley. It is soluble in dilute HCl and in dilute NaOH; it reprecipitates from the alkaline solution when CO_2 is bubbled in. It reacts with benzenesulfonyl chloride to yield a product KK that is soluble in dilute acids.

When hordinene is treated with methyl sulfate and base, a product, LL, is formed. When LL is oxidized by alkaline $KMnO_4$, there is obtained anisic acid, p-$CH_3OC_6H_4$-COOH. When LL is heated strongly there is obtained p-methoxystyrene.

(a) What structure or structures are consistent with this evidence? (b) Outline a synthesis or syntheses that would prove the structure of hordinene.

25. The structure of the terpene *α-terpineol* (found in oils of cardamom and marjoram) was proved in part by the following synthesis:

p-toluic acid + fuming sulfuric acid $\longrightarrow$ MM ($C_8H_8O_5S$)
MM + KOH $\xrightarrow{\text{fusion}}$ NN ($C_8H_8O_3$)
NN + Na, alcohol $\longrightarrow$ OO ($C_8H_{14}O_3$)
OO + HBr $\longrightarrow$ PP ($C_8H_{13}O_2Br$)
PP + base, heat $\longrightarrow$ QQ ($C_8H_{12}O_2$)
QQ + C_2H_5OH, HCl $\longrightarrow$ RR ($C_{10}H_{16}O_2$)
RR + CH_3MgI, then H_2O $\longrightarrow$ α-terpineol ($C_{10}H_{18}O$)

What is the most likely structure for α-terpineol?

26. *Coniferyl alcohol*, $C_{10}H_{12}O_3$, is obtained from the sap of conifers. It is soluble in aqueous NaOH but not in aqueous $NaHCO_3$.

Treatment of coniferyl alcohol (a) with benzoyl chloride and pyridine gives compound SS, $C_{24}H_{20}O_5$; (b) with cold HBr gives $C_{10}H_{11}O_2Br$; (c) with hot hydriodic acid gives a volatile compound identified as methyl iodide; (d) with methyl iodide and aqueous base gives compound TT, $C_{11}H_{14}O_3$.

Both SS and TT are insoluble in dilute NaOH, and rapidly decolorize dilute $KMnO_4$ and Br_2/CCl_4.

Ozonolysis of coniferyl alcohol gives vanillin.

What is the structure of coniferyl alcohol?

Write equations for all the above reactions.

27. When α-(p-tolyloxy)isobutyric acid (prepared from p-cresol) is treated with Br_2, there is obtained UU.

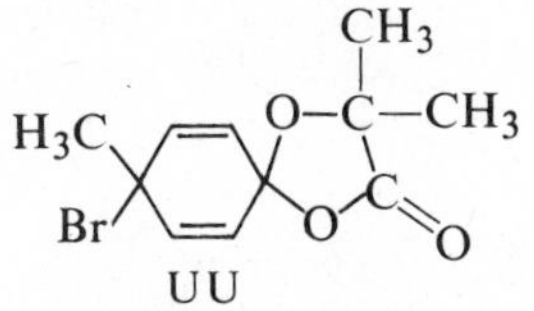

UU

(a) To what class of compounds does UU belong? Suggest a mechanism for its formation.

(b) Give structural formulas for compounds VV, WW, and XX.

UU + $AgNO_3$, CH_3OH $\longrightarrow$ VV ($C_{12}H_{16}O_4$)
VV + H_2, Rh $\longrightarrow$ WW ($C_{12}H_{20}O_4$)
WW + H_2O, OH^- $\longrightarrow$ XX ($C_8H_{14}O$)

(c) The reactions outlined in (b) can be varied. Of what general synthetic utility do you think this general process might be?

28. Compounds AAA–FFF are phenols or related compounds whose structures are given in Problem 19, p. 650, or Sec. 24.4. Assign a structure to each one on the basis of infrared and/or nmr spectra shown as follows.

AAA, BBB, and CCC: infrared spectra in Fig. 24.3 (p. 812)
nmr spectra in Fig. 24.4 (p. 813)
DDD: nmr spectrum in Fig. 24.5 (p. 814)
EEE and FFF: infrared spectra in Fig. 24.6 (p. 814)

(*Hint:* After you have worked out some of the structures, compare infrared spectra.)

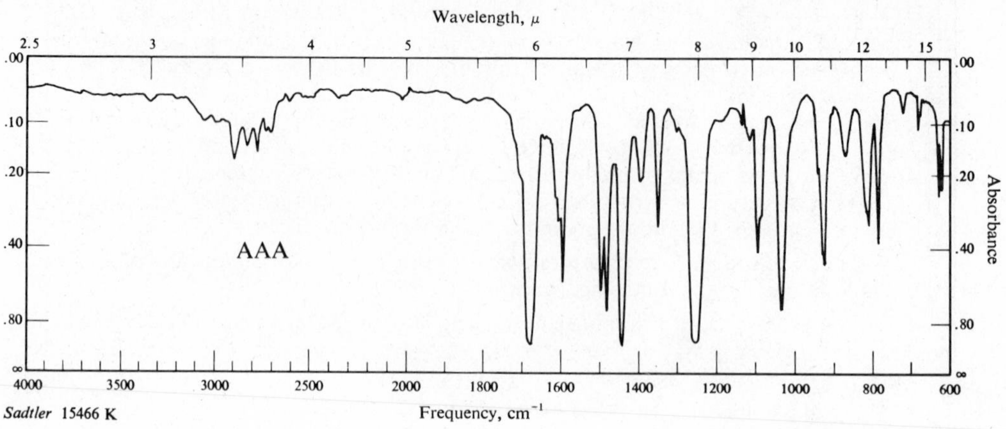

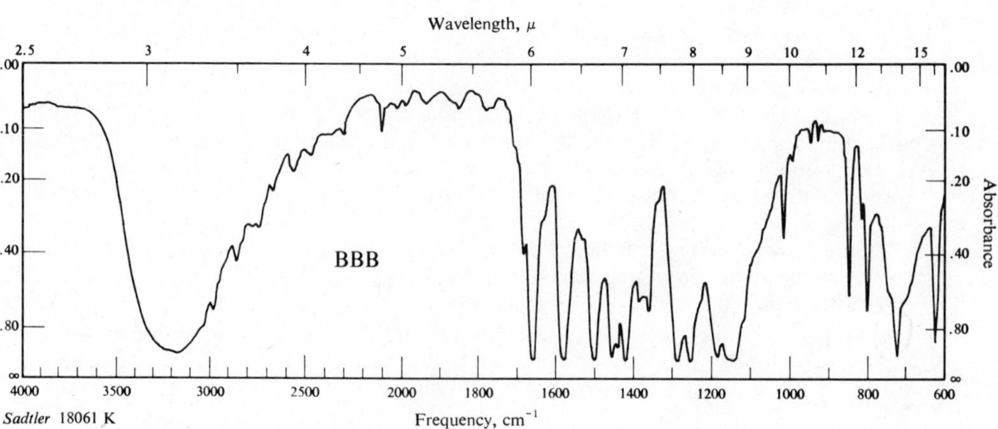

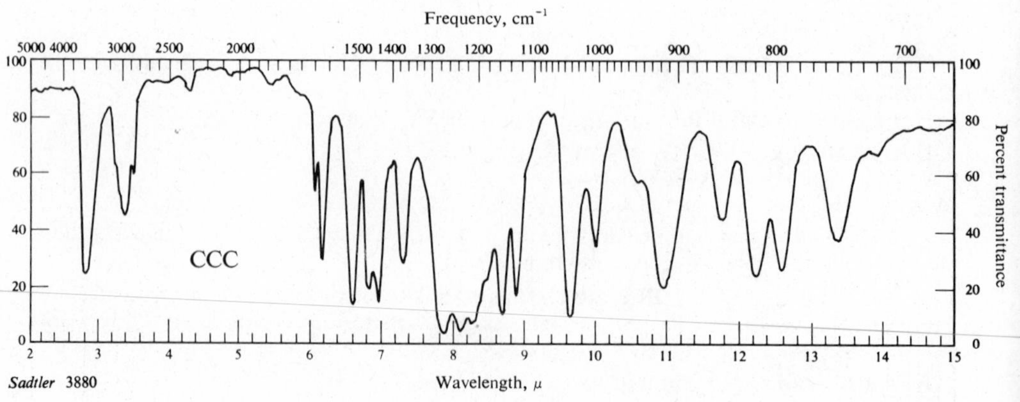

Figure 24.3. Infrared spectra for Problem 28, p. 811.

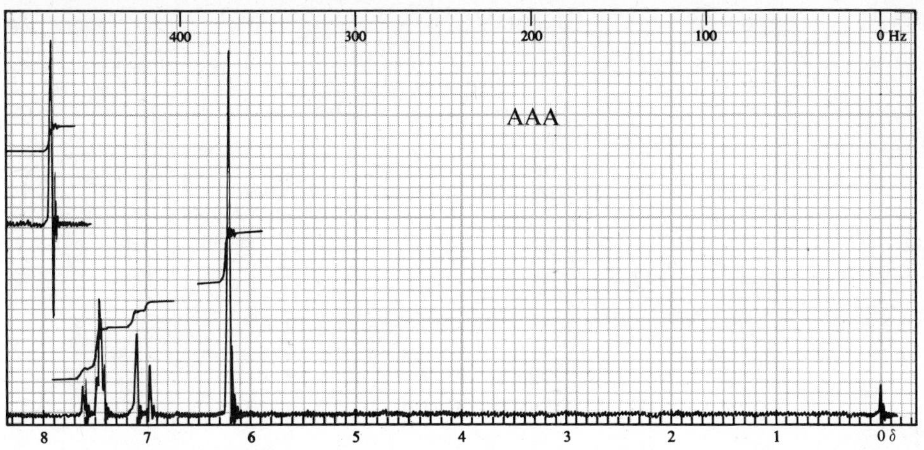

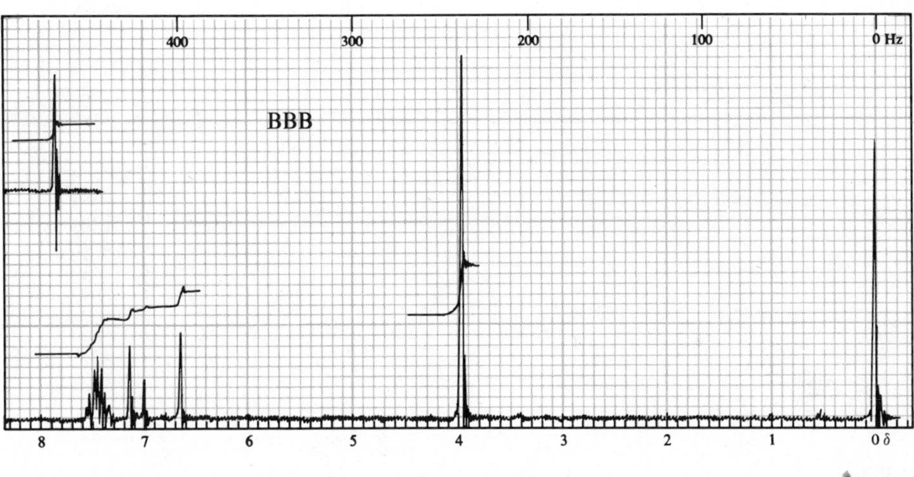

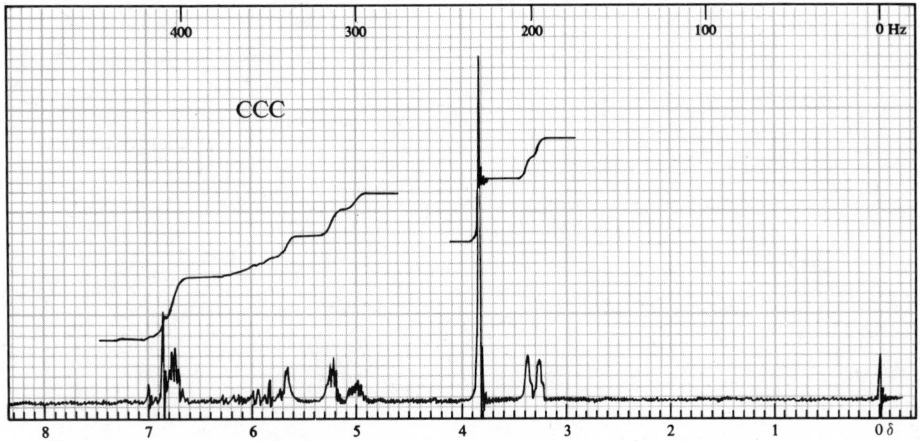

Figure 24.4. Nmr spectra for Problem 28, p. 811.

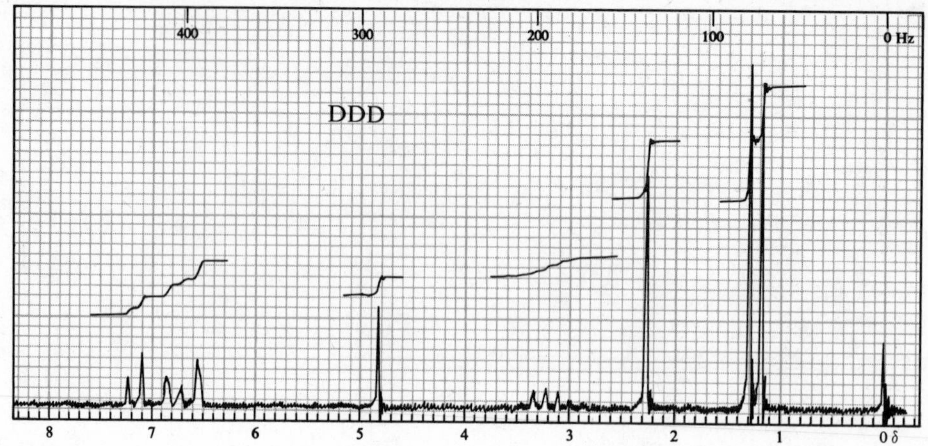

Figure 24.5. Nmr spectrum for Problem 28, p. 811.

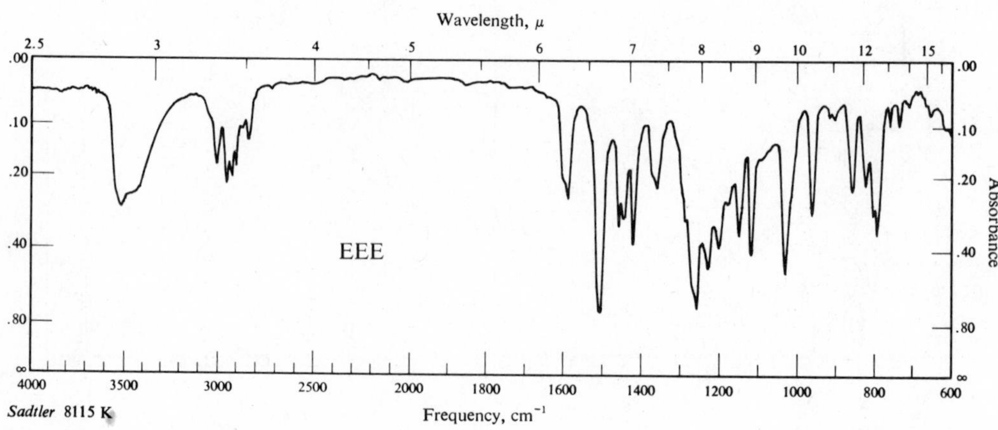

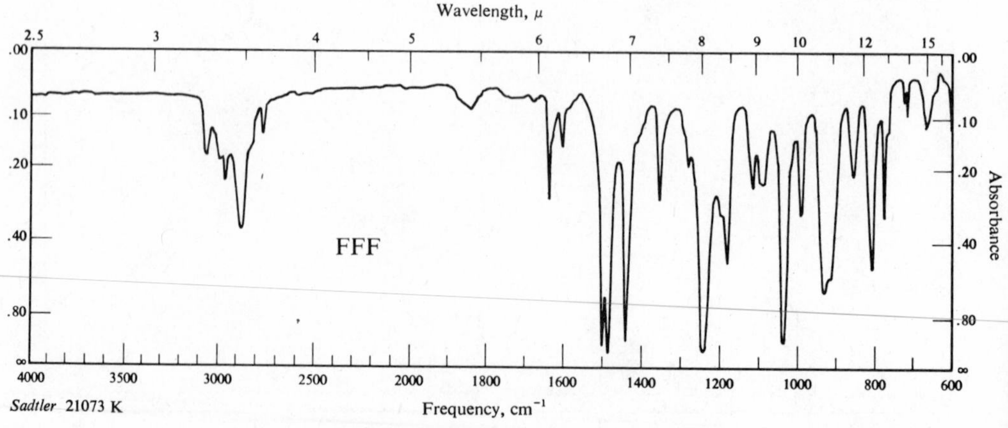

Figure 24.6. Infrared spectra for Problem 28, p. 811.

PART II

Special Topics

Chapter 25 Aryl Halides

Nucleophilic Aromatic Substitution

25.1 Structure

Aryl halides are compounds containing halogen attached directly to an aromatic ring. They have the general formula ArX, where Ar is phenyl, substituted phenyl, or one of the other aryl groups that we shall study (e.g., naphthyl, Chap. 30):

Bromobenzene

Cl
NO$_2$
m-Chloronitrobenzene

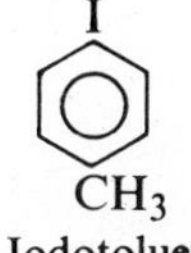

p-Iodotoluene

COOH
Cl
o-Chlorobenzoic acid

An aryl halide is not just any halogen compound containing an aromatic ring. Benzyl chloride, for example, is not an aryl halide, for halogen is not attached to the aromatic ring; in structure and properties it is simply a substituted alkyl halide and was studied with the compounds it closely resembles (Chap. 14).

We take up the aryl halides in a separate chapter because they differ so much from the alkyl halides in their preparation and properties. Aryl halides as a class are comparatively unreactive toward the nucleophilic substitution reactions so characteristic of the alkyl halides. The presence of certain other groups on the aromatic ring, however, greatly increases the reactivity of aryl halides; in the absence of such groups, reaction can still be brought about by very basic reagents or high temperatures. We shall find that **nucleophilic aromatic substitution** can follow two very different paths: the *bimolecular displacement mechanism*, for activated aryl halides; and the *elimination-addition mechanism*, which involves the remarkable intermediate called *benzyne*.

It will be useful to compare aryl halides with certain other halides that are not aromatic at all: *vinyl halides*, compounds in which halogen is attached directly

$$-\overset{|}{C}=\overset{|}{C}-X$$

A vinyl halide

to a doubly-bonded carbon.

Vinyl halides, we have already seen, show an interesting parallel to aryl halides. Each kind of compound contains another functional group besides halogen: aryl halides contain a ring, which undergoes electrophilic substitution; vinyl halides contain a carbon–carbon double bond, which undergoes electrophilic addition. In each of these reactions, halogen exerts an anomalous influence on reactivity and orientation. In electrophilic substitution, halogen deactivates, yet directs *ortho,para* (Sec. 11.21); in electrophilic addition, halogen deactivates, yet causes Markovnikov orientation (Problem 11.13, p. 367). In both cases we attributed the influence of halogen to the working of opposing factors. Through its inductive effect, halogen withdraws electrons and deactivates the entire molecule toward electrophilic attack. Through its resonance effect, halogen releases electrons and tends to activate—but only toward attack *at certain positions.*

Problem 25.1 Drawing all pertinent structures, account in detail for the fact that: (a) nitration of chlorobenzene is slower than that of benzene, yet occurs predominantly *ortho,para*; (b) addition of hydrogen iodide to vinyl chloride is slower than to ethylene, yet yields predominantly 1-chloro-1-iodoethane.

The parallel between aryl and vinyl halides goes further: both are unreactive toward nucleophilic substitution and, as we shall see, for basically the same reason. Moreover, this low reactivity is caused—partly, at least—by the same structural feature that is responsible for their anomalous influence on electrophilic attack: partial double-bond character of the carbon–halogen bond.

We must keep in mind that aryl halides are of "low reactivity" only with respect to certain sets of familiar reactions typical of the more widely studied alkyl halides. Before 1953, aryl halides appeared to undergo essentially only one reaction—and that one, rather poorly. It is becoming increasingly evident that aryl halides are actually capable of doing many different things; as with the "unreactive" alkanes (Sec. 3.18), it is only necessary to provide the proper conditions—and to have the ingenuity to *observe* what is going on. Of these reactions, we shall have time to take up only two. But we should be aware that there *are* others: free-radical reactions, for example, and what Joseph Bunnett (p. 478) has named the *base-catalyzed halogen dance* (Problem 23, p. 845).

25.2 Physical properties

Unless modified by the presence of some other functional group, the physical properties of the aryl halides are much like those of the corresponding alkyl halides. Chlorobenzene and bromobenzene, for example, have boiling points very nearly the same as those of *n*-hexyl chloride and *n*-hexyl bromide; like the alkyl halides, the aryl halides are insoluble in water and soluble in organic solvents.

Table 25.1 ARYL HALIDES

	M.p., °C	B.p., °C	*Ortho* M.p., °C	*Ortho* B.p., °C	*Meta* M.p., °C	*Meta* B.p., °C	*Para* M.p., °C	*Para* B.p., °C
Fluorobenzene	− 45	85						
Chlorobenzene	− 45	132						
Bromobenzene	− 31	156						
Iodobenzene	− 31	189						
Fluorotoluene				115	−111	115		116
Chlorotoluene			− 34	159	− 48	162	8	162
Bromotoluene			− 26	182	− 40	184	28	185
Iodotoluene				206		211	35	211
Difluorobenzene			− 34	92	− 59	83	− 13	89
Dichlorobenzene			− 17	180	− 24	173	52	175
Dibromobenzene			6	221	− 7	217	87	219
Diiodobenzene			27	287	35	285	129	285
Nitrochlorobenzene			32	245	48	236	83	239
2,4-Dinitro-chlorobenzene	53	315						
2,4,6-Trinitro-chlorobenzene (picryl chloride)	83							
Vinyl chloride	−160	− 14						
Vinyl bromide	−138	16						

The physical constants listed in Table 25.1 illustrate very well a point previously made (Sec. 12.3) about the boiling points and melting points of *ortho*, *meta*, and *para* isomers. The isomeric dihalobenzenes, for example, have very nearly the same boiling points: between 173° and 180° for the dichlorobenzenes, 217° to 221° for the dibromobenzenes, and 285° to 287° for the diiodobenzenes. Yet the melting points of these same compounds show a considerable spread; in each case, the *para* isomer has a melting point that is some 70–100 degrees higher than the *ortho* or *meta* isomer. The physical constants of the halotoluenes show a similar relationship.

Here again we see that, having the most symmetrical structure, the *para* isomer fits better into a crystalline lattice and has the highest melting point. We can see how it is that a reaction product containing both *ortho* and *para* isomers frequently deposits crystals of only the *para* isomer upon cooling. Because of the strong intracrystalline forces, the higher melting *para* isomer also is less soluble in a given solvent than the *ortho* isomer, so that purification of the *para* isomer is often possible by recrystallization. The *ortho* isomer that remains in solution is generally heavily contaminated with the *para* isomer, and is difficult to purify.

25.3 Preparation

Aryl halides are most often prepared in the laboratory by the methods outlined below, and on an industrial scale by adaptations of these methods.

PREPARATION OF ARYL HALIDES

1. From diazonium salts. Discussed in Secs. 23.12 and 25.3.

$$ArH \xrightarrow[H_2SO_4]{HNO_3} ArNO_2 \xrightarrow{redn.} ArNH_2 \xrightarrow[0^\circ]{HONO} ArN_2^+ \text{ (Diazonium salt)} \longrightarrow \begin{cases} \xrightarrow{BF_4^-} ArF \\ \xrightarrow{CuCl} ArCl \\ \xrightarrow{CuBr} ArBr \\ \xrightarrow{I^-} ArI \end{cases} + N_2$$

Example:

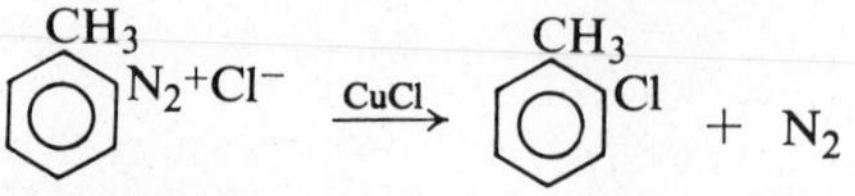

$$o\text{-}CH_3C_6H_4N_2^+Cl^- \xrightarrow{CuCl} o\text{-}CH_3C_6H_4Cl + N_2$$

o-Toluenediazonium chloride → *o*-Chlorotoluene

2. Halogenation. Discussed in Secs. 11.11 and 12.12.

$$ArH + X_2 \xrightarrow{\text{Lewis acid}} ArX + HX$$

$$X_2 = Cl_2, Br_2$$

$$\text{Lewis acid} = FeCl_3, AlCl_3, Tl(OAc)_3, \text{etc.}$$

Examples:

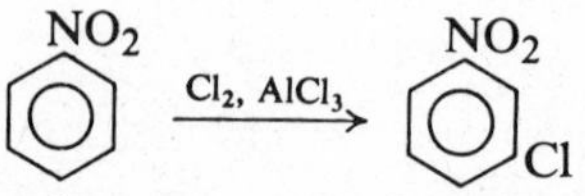

$$C_6H_5NO_2 \xrightarrow{Cl_2, AlCl_3} m\text{-}ClC_6H_4NO_2$$

Nitrobenzene → *m*-Chloronitrobenzene

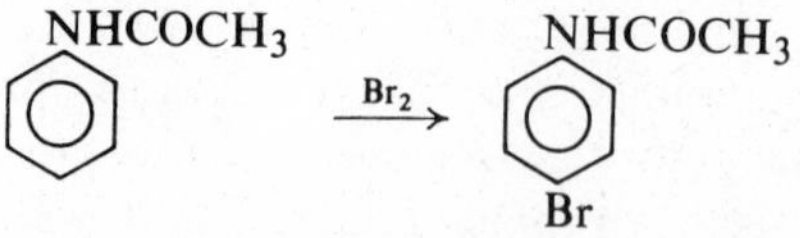

$$C_6H_5NHCOCH_3 \xrightarrow{Br_2} p\text{-}BrC_6H_4NHCOCH_3$$

Acetanilide → *p*-Bromoacetanilide *Major product*

3. From arylthallium compounds. Discussed in Sec. 25.3.

$$[ArH + Tl(OOCCF_3)_3 \longrightarrow] \; ArTl(OOCCF_3)_2 + KI \longrightarrow ArI$$

Arylthallium trifluoroacetate; *For iodides only*

Examples:

$$C_6H_5CH_3 \xrightarrow{Tl(OOCCF_3)_3} \xrightarrow{KI} p\text{-}IC_6H_4CH_3$$

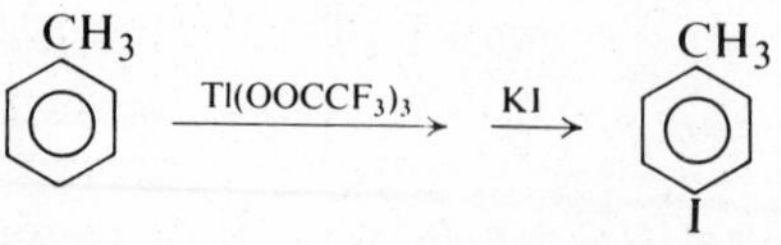

Toluene → *p*-Iodotoluene

COOH

Benzoic acid

$\xrightarrow{Tl(OOCCF_3)_3}$ $\xrightarrow{KI}$

COOH, I

o-Iodobenzoic acid

These methods, we notice, differ considerably from the methods of preparing alkyl halides. (a) Direct halogenation of the aromatic ring is more useful than direct halogenation of alkanes; although mixtures may be obtained (e.g., *ortho* + *para*), attack is not nearly so random as in the free-radical halogenation of aliphatic hydrocarbons. Furthermore, by use of bulky thallium acetate (Sec. 11.7) as the Lewis acid, one can direct bromination *exclusively* to the *para* position. (b) Alkyl halides are most often prepared from the corresponding alcohols; aryl halides are not prepared from the phenols. Instead, aryl halides are most commonly prepared by replacement of the nitrogen of a **diazonium salt**; as the sequence above shows, this ultimately comes from a nitro group which was itself introduced directly into the ring. *From the standpoint of synthesis, then, the nitro compounds bear much the same relationship to aryl halides that alcohols do to alkyl halides.* (These reactions of diazonium salts have been discussed in detail in Secs. 23.11–23.12.)

The preparation of aryl halides from diazonium salts is more important than direct halogenation for several reasons. First of all, fluorides and iodides, which can seldom be prepared by direct halogenation, can be obtained from the diazonium salts. Second, where direct halogenation yields a mixture of *ortho* and *para* isomers, the *ortho* isomer, at least, is difficult to obtain pure. On the other hand, the *ortho* and *para* isomers of the corresponding nitro compounds, from which the diazonium salts ultimately come, can often be separated by fractional distillation (Sec. 11.7). For example, the *o*- and *p*-bromotoluenes boil only three degrees apart: 182° and 185°. The corresponding *o*- and *p*-nitrotoluenes, however, boil sixteen degrees apart: 222° and 238°.

Aryl *iodides* can be prepared by simple treatment of arylthallium compounds with iodide. As in the synthesis of phenols (Sec. 24.5) the thallation route has the advantages of speed, high yield, and orientation control (see Secs. 11.7 and 11.13).

Problem 25.2 Using a different approach in each case, outline all steps in the synthesis of the following from toluene: (a) *p*-bromotoluene; (b) *p*-iodotoluene; (c) *m*-bromotoluene; (d) *m*-iodotoluene; (e) *o*-bromotoluene.

25.4 Reactions

The typical reaction of alkyl halides, we have seen (Sec. 14.5), is nucleophilic substitution. Halogen is displaced as halide ion by such bases as OH^-, OR^-, NH_3, CN^-, etc., to yield alcohols, ethers, amines, nitriles, etc. Even Friedel-Crafts alkylation is, from the standpoint of the alkyl halide, nucleophilic substitution by the basic aromatic ring.

$$R{:}X + {:}Z \longrightarrow R{:}Z + {:}X^-$$

$$Z = OH^-, OR^-, NH_3, CN^-, \text{etc.}$$

It is typical of **aryl halides** *that they undergo nucleophilic substitution only with extreme difficulty.* Except for certain industrial processes where very severe conditions are feasible, one does not ordinarily prepare phenols (ArOH), ethers (ArOR), amines ($ArNH_2$), or nitriles (ArCN) by nucleophilic attack on aryl halides. We cannot use aryl halides as we use alkyl halides in the Friedel-Crafts reaction.

However, aryl halides do undergo nucleophilic substitution readily if the aromatic ring contains, in addition to halogen, certain other properly placed groups: electron-withdrawing groups like $—NO_2$, —NO, or —CN, located *ortho* or *para* to halogen. For aryl halides having this special kind of structure, nucleophilic substitution proceeds readily and can be used for synthetic purposes.

The reactions of unactivated aryl halides with strong bases or at high temperatures, which proceed via benzyne, are finding increasing synthetic importance. The Dow process, which has been used for many years in the manufacture of phenol (Sec. 24.4), turns out to be what Bunnett (p. 478) calls "benzyne chemistry on the tonnage scale!"

The aromatic ring to which halogen is attached can, of course, undergo the typical electrophilic aromatic substitution reactions: nitration, sulfonation, halogenation, Friedel-Crafts alkylation. Like any substituent, halogen affects the reactivity and orientation in these reactions. As we have seen (Sec. 11.5), halogen is unusual in being deactivating, yet *ortho,para*-directing.

REACTIONS OF ARYL HALIDES

1. Formation of Grignard reagent. Limitations are discussed in Sec. 15.15.

$$ArBr + Mg \xrightarrow{\text{dry ether}} ArMgBr$$

$$ArCl + Mg \xrightarrow{\text{tetrahydrofuran}} ArMgCl$$

2. Substitution in the ring. Electrophilic aromatic substitution. Discussed in Sec. 11.21.

X: Deactivates and directs *ortho,para*
in electrophilic aromatic substitution.

3. Nucleophilic aromatic substitution. Bimolecular displacement. Discussed in Secs. 25.7–25.13.

$$Ar{:}X + {:}Z \longrightarrow Ar{:}Z + {:}X^-$$

Ar *must contain strongly electron-withdrawing groups ortho and/or para to* —X.

Examples:

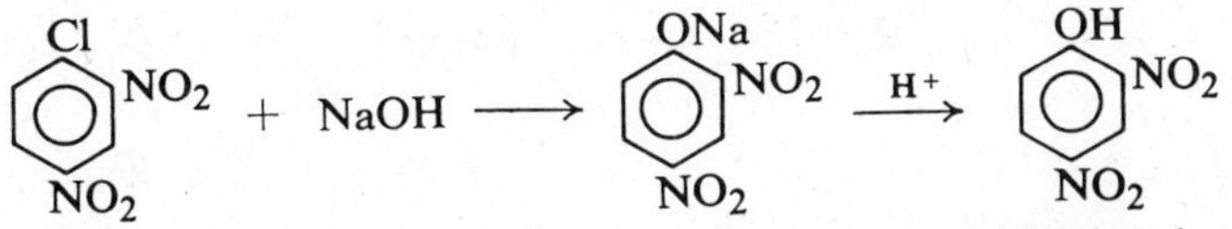

2,4-Dinitrochlorobenzene + NaOH ⟶ (2,4-dinitrophenoxide, ONa) $\xrightarrow{H^+}$ 2,4-Dinitrophenol

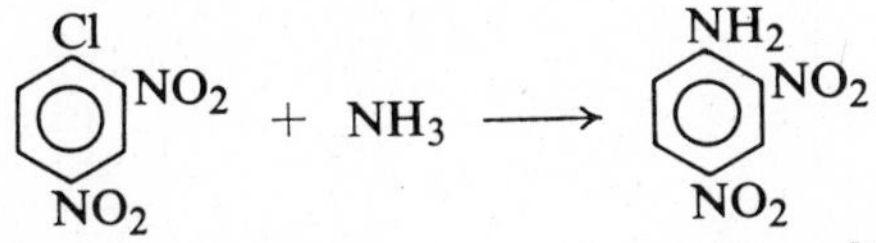

2,4-Dinitrochlorobenzene + NH_3 ⟶ 2,4-Dinitroaniline

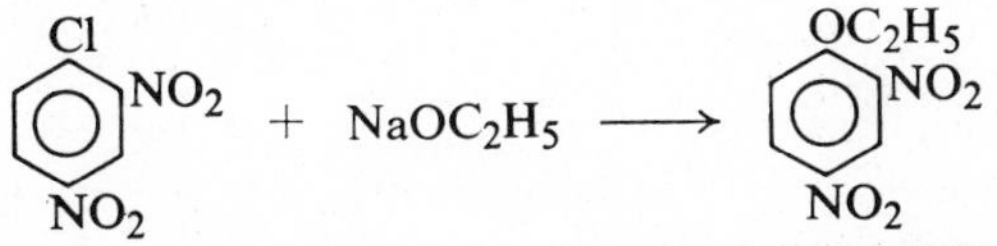

2,4-Dinitrochlorobenzene + $NaOC_2H_5$ ⟶ 2,4-Dinitrophenyl ethyl ether

4. Nucleophilic aromatic substitution. Elimination-addition. Discussed in Sec. 25.14.

$$Ar{:}X + {:}Z \longrightarrow Ar{:}Z + {:}X^-$$

:Z = Strong base

Ring not activated toward bimolecular displacement

Examples:

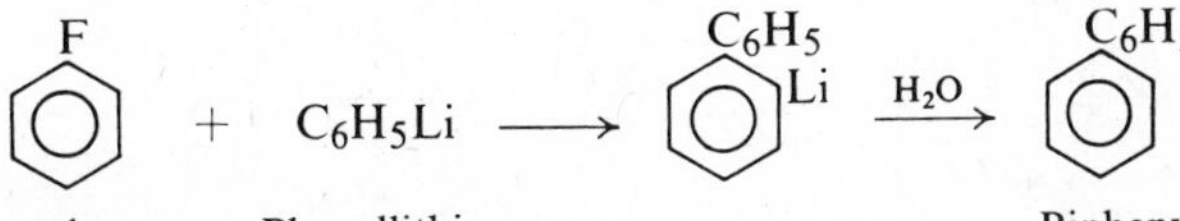

Fluorobenzene + Phenyllithium (C_6H_5Li) ⟶ (C_6H_5, Li) $\xrightarrow{H_2O}$ Biphenyl

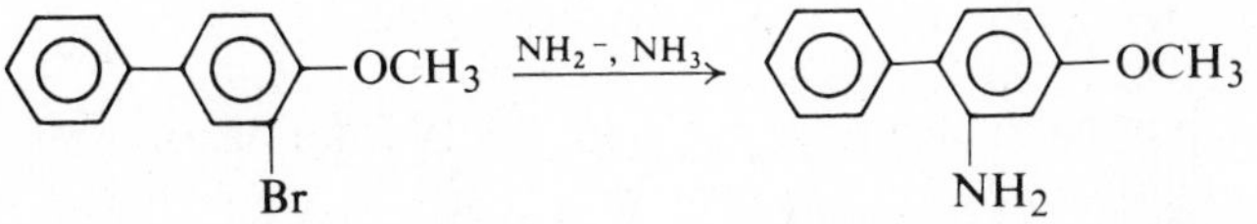

3-Bromo-4-methoxybiphenyl $\xrightarrow{NH_2^-,\ NH_3}$ 2-Amino-4-methoxybiphenyl

25.5 Low reactivity of aryl and vinyl halides

We have seen (Sec. 14.24) that an alkyl halide is conveniently detected by the precipitation of insoluble silver halide when it is warmed with alcoholic silver nitrate. The reaction occurs nearly instantaneously with tertiary, allyl, and benzyl bromides, and within five minutes or so with primary and secondary bromides. Compounds containing halogen joined directly to an aromatic ring or to a doubly-bonded carbon, however, do not yield silver halide under these conditions. Bromobenzene or vinyl bromide can be heated with alcoholic $AgNO_3$ for days without the slightest trace of AgBr being detected. In a similar way, attempts to convert aryl or vinyl halides into phenols (or alcohols), ethers, amines, or nitriles by treatment with the usual nucleophilic reagents are also unsuccessful; aryl or vinyl halides cannot be used in place of alkyl halides in the Friedel-Crafts reaction.

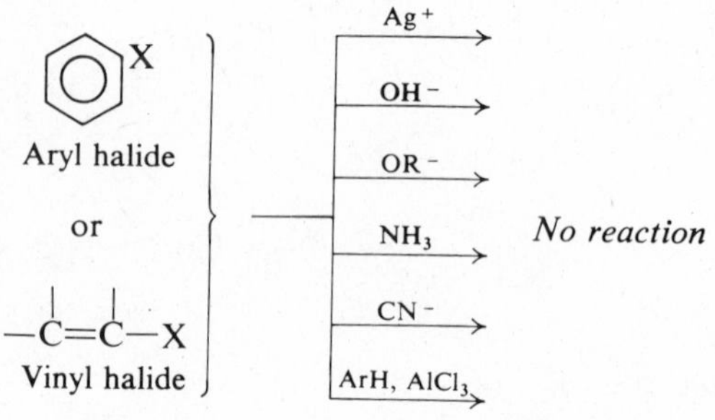

$$\left.\begin{array}{c}\text{Aryl halide (C}_6\text{H}_5\text{X)} \\ \text{or} \\ -\overset{|}{\text{C}}=\overset{|}{\text{C}}-\text{X} \\ \text{Vinyl halide}\end{array}\right\} \xrightarrow{\text{Ag}^+,\ \text{OH}^-,\ \text{OR}^-,\ \text{NH}_3,\ \text{CN}^-,\ \text{ArH, AlCl}_3} \textit{No reaction}$$

How can the low reactivity of these halides be accounted for? To find possible answers, let us look at their structures.

25.6 Structure of aryl and vinyl halides

The low reactivity of aryl and vinyl halides toward displacement has, like the stabilities of alkenes and dienes (Secs. 8.17–8.19), been attributed to two different factors: (a) delocalization of electrons by resonance; and (b) differences in (σ) bond energies due to differences in hybridization of carbon.

Let us look first at the resonance interpretation.

Chlorobenzene is considered to be a hybrid of not only the two Kekulé structures, I and II, but also of three structures, III, IV, and V, in which chlorine is

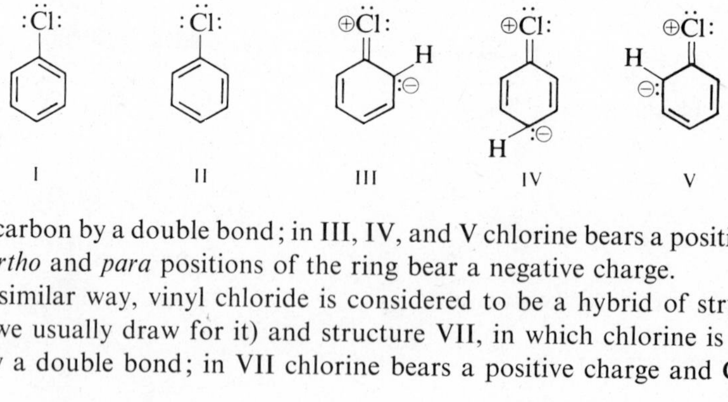

I II III IV V

joined to carbon by a double bond; in III, IV, and V chlorine bears a positive charge and the *ortho* and *para* positions of the ring bear a negative charge.

In a similar way, vinyl chloride is considered to be a hybrid of structure VI (the one we usually draw for it) and structure VII, in which chlorine is joined to carbon by a double bond; in VII chlorine bears a positive charge and C–2 bears

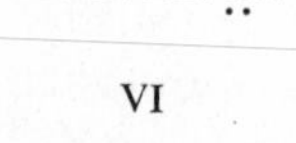

$$\underset{\text{VI}}{\text{H}:\overset{\text{H}}{\underset{2}{\text{C}}}::\overset{\text{H}}{\underset{1}{\text{C}}}:\ddot{\text{Cl}}:} \qquad \underset{\text{VII}}{\text{H}:\overset{\text{H}}{\underset{\ominus}{\ddot{\text{C}}}}:\overset{\text{H}}{\text{C}}::\underset{\oplus}{\ddot{\text{Cl}}}:}$$

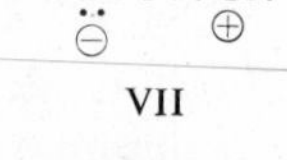

a negative charge. Other aryl and vinyl halides are considered to have structures exactly analogous to these.

Contribution from III, IV, and V, and from VII stabilizes the chlorobenzene and vinyl chloride molecules, and gives double-bond character to the carbon–

chlorine bond. Carbon and chlorine are thus held together by something more than a single pair of electrons, and the carbon–chlorine bond is stronger than if it were a pure single bond. The low reactivity of these halides toward nucleophilic substitution is due (partly, at least) to resonance stabilization of the halides (by a factor that in this case does not stabilize the transition state to the same extent); this stabilization increases the E_{act} for displacement, and thus slows down reaction. For aryl halides, another factor—which may well be the most important one—is stabilization of the molecule by resonance involving the Kekulé structures.

The alternative interpretation is simple. In alkyl halides the carbon holding halogen is sp^3-hybridized. In aryl and vinyl halides, carbon is sp^2-hybridized; the bond to halogen is shorter and stronger, and the molecule is more stable (see Sec. 5.4).

What evidence is there to support either interpretation, other than the fact that it would account for *the low reactivity of aryl and vinyl halides*?

The carbon–halogen bonds of aryl and vinyl halides are unusually short. In chlorobenzene and vinyl chloride the C—Cl bond length is only 1.69 A, as compared with a length of 1.77–1.80 A in a large number of alkyl chlorides (Table 25.2). In bromobenzene and vinyl bromide the C—Br bond length is only 1.86 A, as compared with a length of 1.91–1.92 A in alkyl bromides.

Now, as we have seen (Sec. 5.2), a double bond is shorter than a single bond joining the same pair of atoms; if the carbon–halogen bond in aryl and vinyl halides has double-bond character, it should be shorter than the carbon–halogen bond in alkyl halides. Alternatively, a bond formed by overlap of an sp^2 orbital should be shorter than the corresponding bond involving an sp^3 orbital.

Dipole moments of aryl and vinyl halides are unusually small. Organic halogen compounds are polar molecules; displacement of electrons toward the more electronegative element makes halogen relatively negative and carbon relatively positive. Table 25.2 shows that the dipole moments of a number of alkyl chlorides and bromides range from 2.02 D to 2.15 D. The mobile π electrons of the benzene ring and of the carbon–carbon double bond should be particularly easy to displace; hence we might have expected aryl and vinyl halides to have even larger dipole moments than alkyl halides.

However, we see that this is not the case. Chlorobenzene and bromobenzene have dipole moments of only 1.7 D, and vinyl chloride and vinyl bromide have dipole moments of only 1.4 D. This is consistent with the resonance picture of these molecules. In the structures that contain doubly-bonded halogen (III, IV,

Table 25.2 BOND LENGTHS AND DIPOLE MOMENTS OF HALIDES

	Bond Lengths, A		Dipole Moments, D	
	C—Cl	C—Br	R—Cl	R—Br
CH_3—X	1.77	1.91	—	—
C_2H_5—X	1.77	1.91	2.05	2.02
n-C_3H_7—X	—	—	2.10	2.15
n-C_4H_9—X	—	—	2.09	2.15
$(CH_3)_3C$—X	1.80	1.92	2.13	—
CH_2=CH—X	1.69	1.86	1.44	1.41
C_6H_5—X	1.69	1.86	1.73	1.71

V, and VII) there is a positive charge on halogen and a negative charge on carbon; to the extent that these structures contribute to the hybrids, they tend to oppose the usual displacement of electrons toward halogen. Although there is still a net displacement of electrons toward halogen in aryl halides and in vinyl halides, it is less than in other organic halides.

Alternatively, sp^2-hybridized carbon is, in effect, a more electronegative atom than an sp^3-hybridized carbon (see Sec. 8.10), and is less willing to release electrons to chlorine.

As was discussed in Secs. 11.21 and 25.1, contribution from structures in which halogen is doubly bonded and bears a positive charge accounts for *the way halogen affects the reactions of the benzene ring or of the carbon–carbon double bond to which it is joined.*

The counterargument is that this simply indicates that resonance of this kind can occur—but not how important it is in the halide molecules.

Finally, the *existence of cyclic halonium ions* (Sec. 7.12) certainly shows that halogen *can* share more than a pair of electrons.

It is hard to believe that the stability of these molecules is not affected by the particular kind of hybridization; on the other hand, it seems clear that there is resonance involving halogen and the π electrons. The question, once more, is one of their relative importance. As in the case of alkenes and dienes, it is probable that *both* are important.

As we shall see, in the rate-determining step of nucleophilic aromatic substitution a nucleophile attaches itself to the carbon bearing halogen; this carbon becomes tetrahedral, and the ring acquires a negative charge. Such a reaction is made more difficult by the fact that it destroys the aromaticity of the ring and disrupts the resonance between ring and halogen; and, if Dewar is correct (Sec. 8.19), because energy is required to change the hybridization of carbon from sp^2 to sp^3

Problem 25.3 In Sec. 25.3 we learned that, unlike alkyl halides, aryl halides are not readily prepared from the corresponding hydroxy compounds. How might you account for this contrast between alcohols and phenols? (*Hint:* See Sec. 24.7.)

25.7 Nucleophilic aromatic substitution: bimolecular displacement

We have seen that the aryl halides are characterized by very low reactivity toward the nucleophilic reagents like OH^-, OR^-, NH_3, and CN^- that play such an important part in the chemistry of the alkyl halides. Consequently, nucleophilic aromatic substitution is much less important in synthesis than either nucleophilic aliphatic substitution or electrophilic aromatic substitution.

However, the presence of certain groups at certain positions of the ring markedly activates the halogen of aryl halides toward displacement. We shall have a look at some of these activation effects, and then try to account for them on the basis of the chemical principles we have learned. We shall find a remarkable parallel between the two kinds of aromatic substitution, electrophilic and nucleophilic, with respect both to mechanism and to the ways in which substituent groups affect reactivity and orientation.

Chlorobenzene is converted into phenol by aqueous sodium hydroxide only at temperatures over 300°. The presence of a nitro group *ortho* or *para* to the chlorine greatly increases its reactivity: *o*- or *p*-chloronitrobenzene is converted into the nitrophenol by treatment with aqueous sodium hydroxide at 160°. A nitro group *meta* to the chlorine, on the other hand, has practically no effect on reactivity. As the number of *ortho* and *para* nitro groups on the ring is increased, the reactivity increases: the phenol is obtained from 2,4-dinitrochlorobenzene by treatment with hot aqueous sodium carbonate, and from 2,4,6-trinitrochlorobenzene by simple treatment with water.

Chlorobenzene $\xrightarrow{\text{6–8\% NaOH, 350°, 4500 lb/in.}^2}$ Phenol

p-Chloronitrobenzene $\xrightarrow{\text{15\% NaOH, 160°}}$ *p*-Nitrophenol

2,4-Dinitrochlorobenzene $\xrightarrow{\text{boiling aq. } Na_2CO_3\text{, 130°}}$ 2,4-Dinitrophenol

2,4,6-Trinitrochlorobenzene $\xrightarrow{H_2O\text{, warm}}$ 2,4,6-Trinitrophenol (Picric acid)

Similar effects are observed when other nucleophilic reagents are used. Ammonia or sodium methoxide, for example, reacts with chloro- or bromobenzene only under very vigorous conditions. For example:

Chlorobenzene $\xrightarrow{NH_3\text{, } Cu_2O\text{, 200°, 900 lb/in.}^2}$ Aniline

Yet if the ring contains a nitro group—or preferably two or three of them—*ortho* or *para* to the halogen, reaction proceeds quite readily. For example:

2,4-Dinitrochlorobenzene $\xrightarrow{NH_3\text{, 170°}}$ 2,4-Dinitroaniline

$$\text{2,4,6-Trinitrochlorobenzene (Cl, } O_2N, NO_2, NO_2) \xrightarrow{NaOCH_3,\ 20^\circ} \text{2,4,6-Trinitroanisole (}OCH_3, O_2N, NO_2, NO_2)$$

2,4,6-Trinitrochlorobenzene 2,4,6-Trinitroanisole

Like $—NO_2$, certain other groups have been found to activate halogen located *ortho* or *para* to them: $—N(CH_3)_3^+$, $—CN$, $—SO_3H$, $—COOH$, $—CHO$, $—COR$. This is a familiar list. All these are electron-withdrawing groups, which are deactivating and *meta*-directing toward *electrophilic* substitution (see Table 11.3, p. 342).

Although our concern here is primarily with displacement of halogen, it is important to know that these electron-withdrawing substituents activate many groups other than halogen toward nucleophilic substitution. (Hydrogen is generally not displaced from the aromatic ring, since this would require the separation of the very strongly basic hydride ion, $:H^-$.)

Problem 25.4 When *p*-nitroso-N,N-dimethylaniline is heated with aqueous KOH, dimethylamine is evolved; this reaction is sometimes used to prepare pure dimethylamine, free from methylamine and trimethylamine. (a) What are the other products of the reaction? (b) To what class of organic reactions does this belong? (c) Upon what property of the nitroso group does this reaction depend? (d) Outline all steps in the preparation of pure diethylamine starting from nitrobenzene and ethyl alcohol.

Problem 25.5 How do you account for the following observations?

(a) Although most ethers are inert toward bases, 2,4-dinitroanisole is readily cleaved to methanol and 2,4-dinitrophenol when refluxed with dilute aqueous NaOH.

(b) Although amides can be hydrolyzed by either aqueous acid or aqueous alkali, hydrolysis of *p*-nitroacetanilide is best carried out in acidic solution.

(c) Treatment of *o*-chloronitrobenzene by aqueous sodium sulfite yields sodium *o*-nitrobenzenesulfonate. Give the structure of the reagent involved. How does this reagent compare with the one in ordinary sulfonations?

(d) Would you expect the method of (c) to be a general one for preparation of sulfonic acids? Could it be used, for example, to prepare benzenesulfonic acid?

(e) Washing crude *m*-dinitrobenzene with aqueous sodium sulfite removes contaminating *o*- and *p*-dinitrobenzene.

If electron-withdrawing groups activate toward nucleophilic substitution, we might expect electron-releasing groups to *deactivate*. This is found to be so. Furthermore, the degree of deactivation depends upon how strongly they release electrons: $—NH_2$ and $—OH$ deactivate strongly; $—OR$, moderately; and $—R$, weakly.

In nucleophilic as in electrophilic aromatic substitution, then, a substituent group affects reactivity by its ability to attract or release electrons; in nucleophilic as in electrophilic aromatic substitution, a substituent group exerts its effect chiefly at the position *ortho* and *para* to it. The kind of effect that each group exerts, however, is exactly opposite to the kind of effect it exerts in electrophilic aromatic substitution. *In* **nucleophilic aromatic substitution** *electron withdrawal causes activation, and electron release causes deactivation.*

To account for these effects, we must look at the mechanism for the kind of nucleophilic aromatic substitution we have been talking about.

25.8 Bimolecular displacement mechanism for nucleophilic aromatic substitution

The bimolecular displacement mechanism for nucleophilic aromatic substitution (shown here for chlorobenzene) is:

(1) $C_6H_5Cl + :Z \longrightarrow \overset{\ominus}{C_6H_5}(Cl)(Z)$ (I) **Slow**

(2) $\overset{\ominus}{C_6H_5}(Cl)(Z)$ (I) $\longrightarrow C_6H_5Z + :Cl^-$ **Fast**

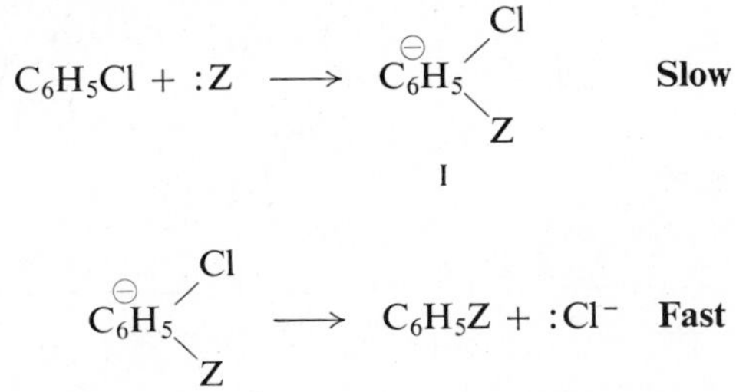

Bimolecular displacement

There are two essential steps: attack of a nucleophilic reagent upon the ring to form a carbanion (I), and the expulsion of halide ion from this carbanion to yield the product.

The intermediate carbanion (I) is a hybrid of II, III, and IV; this hybrid is sometimes represented by the single structure V:

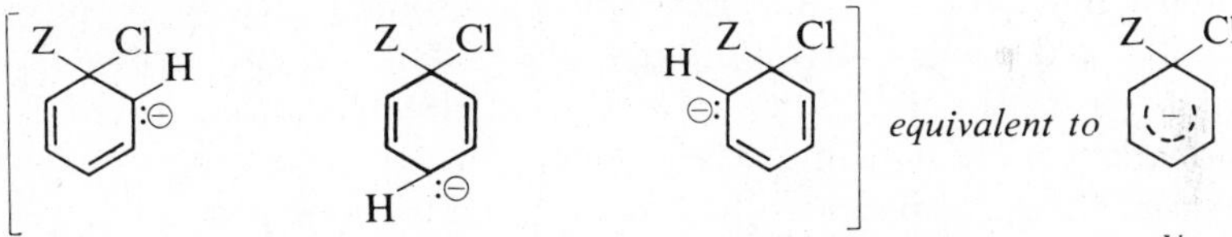

II III IV *equivalent to* V

In nucleophilic aliphatic substitution (S_N2), the intermediate in which carbon is bonded to both the attacking group and the displaced group is considered to be a transition state; a structure (VI) containing carbon bonded to five atoms must be

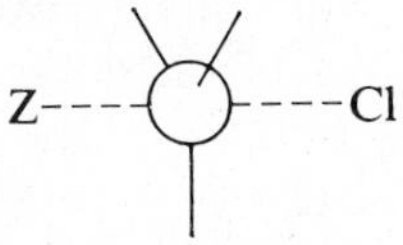

Z---C---Cl

VI
Aliphatic S_N2
Pentavalent carbon, transition state

V
Aromatic
Tetrahedral carbon, compound

Intermediates in nucleophilic substitution

unstable and so corresponds to the top of an energy hill (Fig. 25.1). In nucleophilic aromatic substitution, on the other hand, the intermediate is an actual compound; a structure (V) containing tetrahedral carbon and having the negative charge distributed about the ring is comparatively stable, and corresponds to an energy valley (Fig. 25.2).

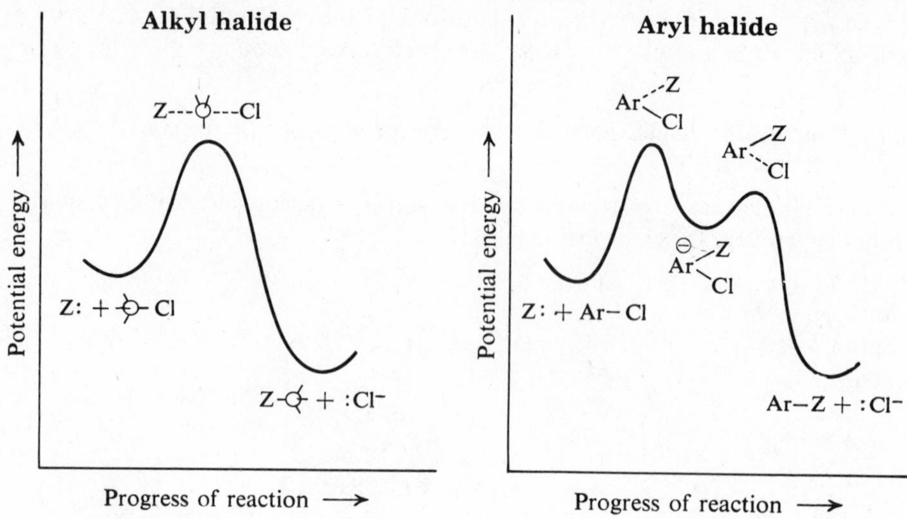

Figure 25.1. Energy curve for nucleophilic aliphatic (S_N2) substitution. One-step reaction: intermediate is a transition state.

Figure 25.2. Energy curve for nucleophilic aromatic substitution. Two-step reaction: intermediate is a compound.

25.9 Reactivity in nucleophilic aromatic substitution

For reactions involving an intermediate carbonium ion, we have seen that the overall rate depends only on the rate of formation of the carbonium ion. In nucleophilic aromatic substitution an analogous situation seems to exist: the first step, formation of the carbanion, largely determines the overall rate of reaction; once formed, the carbanion rapidly reacts to yield the final product.

For closely related reactions, we might expect a difference in rate of formation of carbanions to be largely determined by a difference in E_{act}, that is, by a difference in stability of the transition states. Factors that stabilize the carbanion by dispersing the charge should for the same reason stabilize the incipient carbanion of the transition state. Just as the more stable carbonium ion is formed more rapidly, so, we expect, the more stable carbanion should be formed more rapidly. We shall therefore concentrate our attention on the relative stabilities of the intermediate carbanions.

Cl :Z ⟶ [Z Cl (δ−)] ⟶ Z Cl (−)

Transition state: *developing negative charge*

Carbanion: *full negative charge*

To compare the rates of substitution in chlorobenzene itself, a chlorobenzene containing an electron-withdrawing group, and a chlorobenzene containing an electron-releasing group, we compare the structures of carbanions I, II, and III.

Z Cl / Z Cl / Z Cl

G / G

I 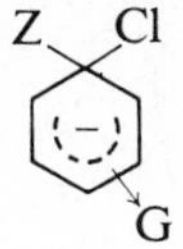 II 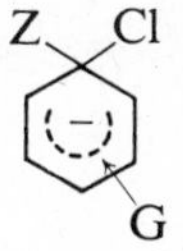 III

A group that withdraws electrons (II) tends to neutralize the negative charge of the ring and so to become more negative itself; this dispersal of the charge stabilizes the carbanion. In the same way, electron withdrawal stabilizes the transition state with its developing negative charge, and thus speeds up reaction. A group that releases electrons (III) tends to intensify the negative charge, destabilizes the carbanion (and the transition state), and thus slows down reaction.

Nucleophilic Aromatic Substitution

Cl, G + :Z ⟶ Z Cl, G

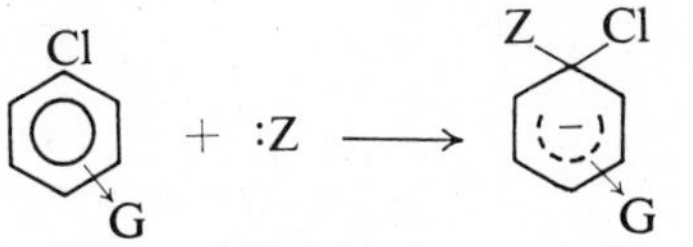

G *withdraws electrons: stabilizes carbanion, activates*

G = $-N(CH_3)_3^+$, $-NO_2$, $-CN$, $-SO_3H$, $-COOH$, $-CHO$, $-COR$, $-X$

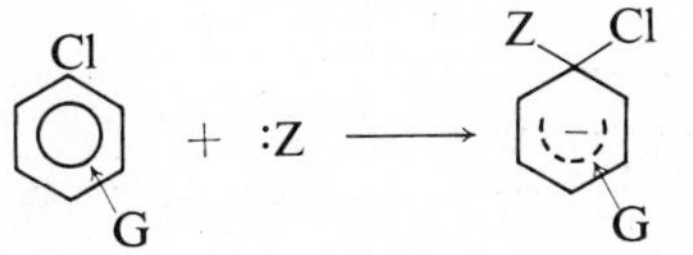

Cl, G + :Z ⟶ Z Cl, G

G *releases electrons: destabilizes carbanion, deactivates*

G = $-NH_2$, $-OH$, $-OR$, $-R$

It is clear, then, why a given substituent group affects nucleophilic and electrophilic aromatic substitution in opposite ways: it affects the stability of negatively and positively charged ions in opposite ways.

25.10 Orientation in nucleophilic aromatic substitution

To see why it is that a group activates the positions *ortho* and *para* to it most strongly, let us compare, for example, the carbanions formed from *p*-chloronitrobenzene and *m*-chloronitrobenzene. Each of these is a hybrid of three structures, I–III for *para* attack, IV–VI for *meta* attack. In one of these six structures, II,

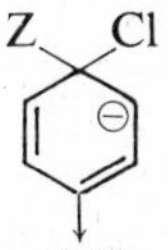

Z Cl / Z Cl / Z Cl

NO_2 / NO_2 / NO_2

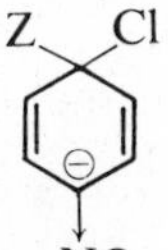

I II III

***Para* attack**

Especially stable: charge on carbon bearing substituent

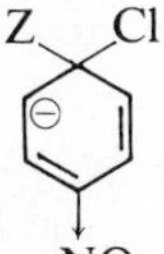

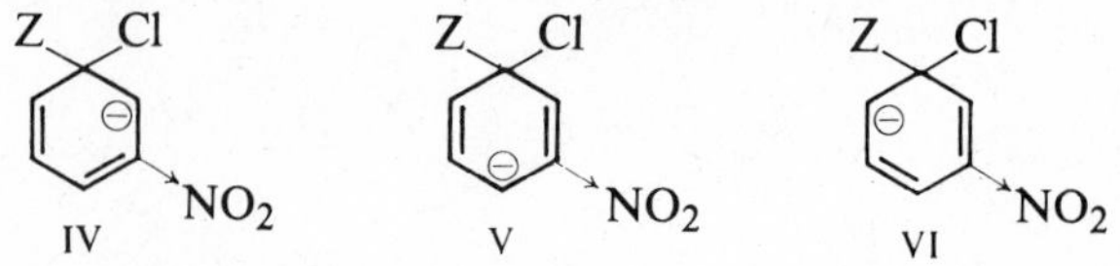

IV V VI *Meta* attack

the negative charge is located on the carbon atom to which $—NO_2$ is attached. Although $—NO_2$ attracts electrons from all positions of the ring, it does so most from the carbon atom nearest it; consequently, structure II is a particularly stable one. Because of contribution from structure II, the hybrid carbanion resulting from attack on *p*-chloronitrobenzene is more stable than the carbanion resulting from attack on *m*-chloronitrobenzene. The *para* isomer therefore reacts faster than the *meta* isomer.

In the same way, it can be seen that attack on *o*-chloronitrobenzene (VII–IX) also yields a more stable carbanion, because of contribution from IX, than attack on *m*-chloronitrobenzene.

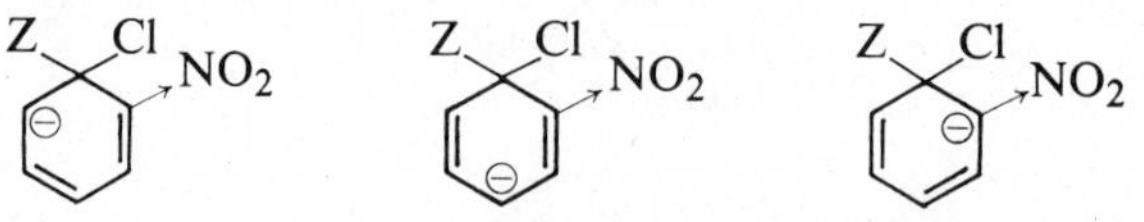

VII VIII IX *Ortho* attack

Especially stable: charge on carbon bearing substituent

By considerations similar to those of Sec. 11.19, we can see that deactivation by an electron-releasing group should also be strongest when it is *ortho* or *para* to the halogen.

Nucleophilic and electrophilic aromatic substitution are similar, then, in that a group exerts its strongest influence—whether activating or deactivating—at the positions *ortho* and *para* to it. This similarity is due to a similarity in the intermediate ions: in both cases the charge of the intermediate ion—whether negative or positive—is strongest at the positions *ortho* and *para* to the point of attack, and hence a group attached to one of these positions can exert the strongest influence.

25.11 Electron withdrawal by resonance

The activation by $—NO_2$ and other electron-attracting groups can be accounted for, as we have seen, simply on the basis of inductive effects. However, it is generally believed that certain of these groups withdraw electrons by resonance as well. Let us see what kind of structures are involved.

The intermediate carbanions formed by nucleophilic attack on *o*- and *p*-chloronitrobenzene are considered to be hybrids not only of structures with negative charges carried by carbons of the ring (as shown in the last section), but also of structures I and II in which the negative charge is carried by oxygen of the $—NO_2$ group. Being highly electronegative, oxygen readily accommodates a negative charge, and hence I and II should be especially stable structures. The carbanions to which these structures contribute are therefore much more stable than the ones

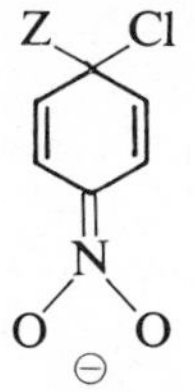

I

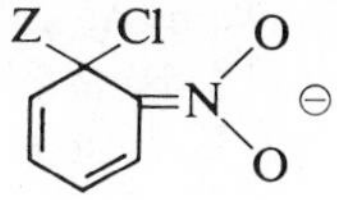

II

formed by attack on chlorobenzene itself or on *m*-chloronitrobenzene, for which structures like I and II are not possible. Thus resonance involving the $—NO_2$ group strengthens the activation toward nucleophilic substitution caused by the inductive effect.

The activating effect of a number of other electron-attracting groups is considered to arise, in part, from the contribution of similar structures (shown only for *para* isomers) to the intermediate carbanions.

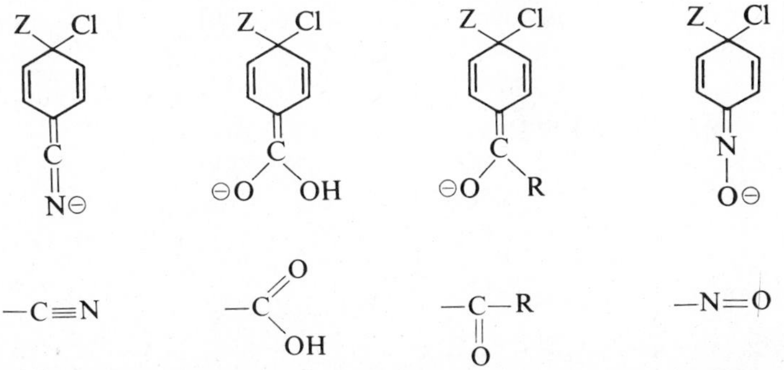

—C≡N —COOH —CO—R —N=O

Problem 25.6 There is evidence to suggest that the nitroso group, $—\ddot{N}=\ddot{O}:$, activates *ortho* and *para* positions toward *both* nucleophilic and electrophilic aromatic substitution; the group apparently can either withdraw or release electrons upon demand by the attacking reagent. Show how this might be accounted for. (*Hint:* See Sec. 11.20.)

25.12 Evidence for the two steps in bimolecular displacement

Our interpretation of reactivity and orientation in nucleophilic aromatic substitution has been based on one all-important assumption that we have not yet justified: *displacement involves two steps, of which the first one is much slower than the second.*

(1) 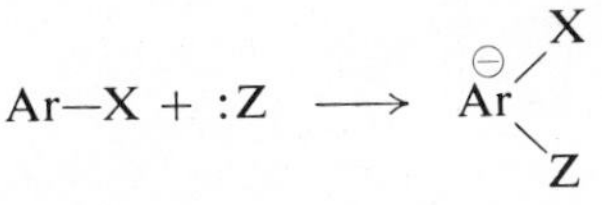 $$\text{Ar—X} + :\text{Z} \longrightarrow \overset{\ominus}{\text{Ar}}\!\begin{smallmatrix}\text{X}\\ \text{Z}\end{smallmatrix} \qquad \textit{Slow}$$

(2) 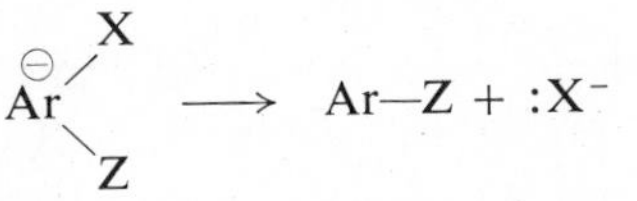 $$\overset{\ominus}{\text{Ar}}\!\begin{smallmatrix}\text{X}\\ \text{Z}\end{smallmatrix} \longrightarrow \text{Ar—Z} + :\text{X}^- \qquad \textit{Fast}$$

The problem here reminds us of the analogous problem in electrophilic aromatic substitution (Sec. 11.16). There the answer was found in the absence of an isotope effect: although carbon–deuterium bonds are broken more slowly than carbon–hydrogen bonds, deuterium and hydrogen were found to be displaced at the same rate. Reactivity is determined by the rate of a reaction that does not involve the breaking of a carbon–hydrogen bond.

But in nucleophilic aromatic substitution, we are dealing with displacement, not of hydrogen, but of elements like the halogens; as was discussed in connection with dehydrohalogenation, any isotope effects would be small, and hard to measure.

The answer came from Joseph Bunnett (p. 478), who is responsible for much of what we understand about nucleophilic aromatic substitution. It was while studying this reaction that he first conceived the idea of *element effect* (Sec. 14.20), and showed how it gave evidence for the two-step mechanism.

In S_N1 and S_N2 displacement, we recall, the reactivity of alkyl halides follows the sequence

$$\text{R—I} > \text{R—Br} > \text{R—Cl} > \text{R—F}$$

The ease of breaking the carbon–halogen bond depends upon its strength, and the resulting differences in rate are quite large.

Yet, in nucleophilic *aromatic* substitution, there is often very little difference in reactivity among the various halides and, more often than not, the fluoride—containing the carbon–halogen bond hardest to break—is the *most* reactive. If reactivity is independent of the strength of the carbon–halogen bond, we can only conclude that the reaction *whose rate we are observing* does not involve breaking

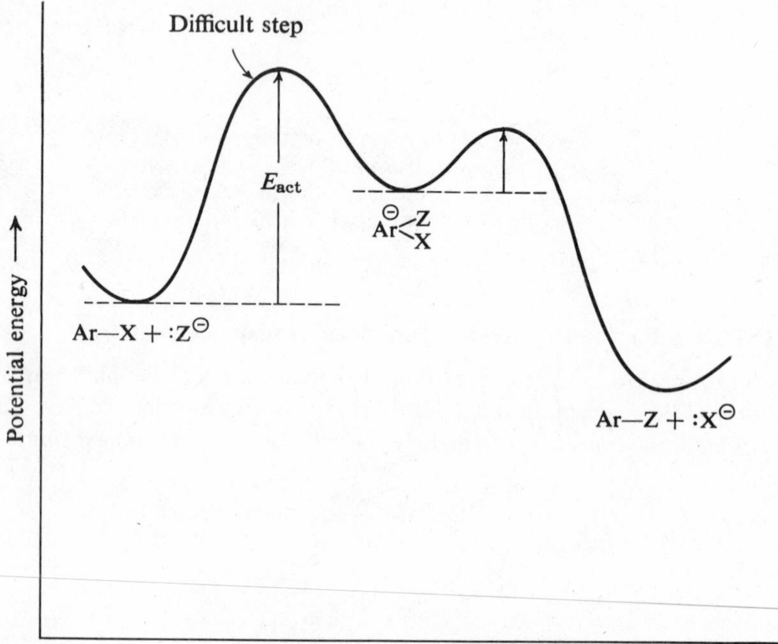

Figure 25.3. Potential energy changes during course of reaction: nucleophilic aromatic substitution. Formation of carbanion is rate-controlling step; strength of C—X bond does not affect over-all rate.

of the carbon–halogen bond. In nucleophilic aromatic substitution, as in electrophilic aromatic substitution, the rate of reaction is determined by the rate of attachment of the attacking particle to the ring (Fig. 25.3).

The *faster* reaction of aryl fluorides is attributed to the very strong inductive effect of fluorine; by withdrawing electrons it stabilizes the transition state of the first step of a reaction that will ultimately lead to its displacement.

Problem 25.7 When 2,4,6-trinitroanisole is treated with sodium ethoxide, a product of formula $C_9H_{10}O_8N_3^-Na^+$ is formed. A product of the same formula is formed by the treatment of trinitrophenetole by sodium methoxide. When treated with acid, both products give the same mixture of trinitroanisole and trinitrophenetole. What structure (or structures) would you assign to these products?

25.13 Nucleophilic substitution: aliphatic and aromatic

We can see a regular progression in the three kinds of nucleophilic substitution that we have studied so far. The departing group leaves the molecule *before* the entering group becomes attached in an S_N1 reaction, *at the same time* in an S_N2 reaction, and *after* in nucleophilic aromatic substitution. A *positive charge* thus develops on carbon during an S_N1 reaction, *no particular charge* during an S_N2 reaction, and a *negative charge* during nucleophilic aromatic substitution. As a result, an S_N1 reaction is favored by *electron release*, an S_N2 reaction is relatively *insensitive to electronic factors*, and nucleophilic aromatic substitution is favored by *electron withdrawal*.

$$\text{R—X} \longrightarrow \underset{\text{X}^-}{\overset{}{\text{R}^+ \atop +}} \xrightarrow{:\text{Z}} \text{R—Z}$$

S_N1
Positive charge develops on carbon

$$\text{R—X} \xrightarrow{:\text{Z}} \left[\text{Z}\cdots\text{R}\cdots\overset{\delta-}{\text{X}}\right] \longrightarrow \text{R—Z} + \text{X}^-$$

S_N2
Little charge develops on carbon

$$\text{C}_6\text{H}_5\text{X} \xrightarrow{:\text{Z}} [\text{C}_6\text{H}_5(\text{Z})(\text{X})]^- \longrightarrow \text{C}_6\text{H}_5\text{Z} + \text{X}^-$$

Nucleophilic aromatic
Negative charge develops on carbon

25.14 Elimination-addition mechanism for nucleophilic aromatic substitution. Benzyne

We have seen that electron-withdrawing groups activate aryl halides toward nucleophilic substitution. In the absence of such activation, substitution can be *made* to take place, by use of very strong bases, for example. But when this is done, substitution does not take place by the mechanism we have just discussed (the so-called *bimolecular mechanism*), but by an entirely different mechanism: the *benzyne* (or *elimination-addition*) *mechanism*. Let us first see what this mechanism is, and then examine some of the evidence for it.

When an aryl halide like chlorobenzene is treated with the very strongly basic amide ion, NH_2^-, in liquid ammonia, it is converted into aniline. This is not the simple displacement that, on the surface, it appears to be. Instead, the reaction involves two stages: *elimination* and then *addition*. The intermediate is the molecule called *benzyne* (or *dehydrobenzene*).

$$C_6H_5X \xrightarrow[NH_3]{NH_2^-} \text{Benzyne} \xrightarrow[NH_3]{NH_2^-} C_6H_5NH_2$$

Aryl halide — Benzyne — Aniline

Benzyne has the structure shown in Fig. 25.4, in which an additional bond is formed between two carbons (the one originally holding the halogen and the one

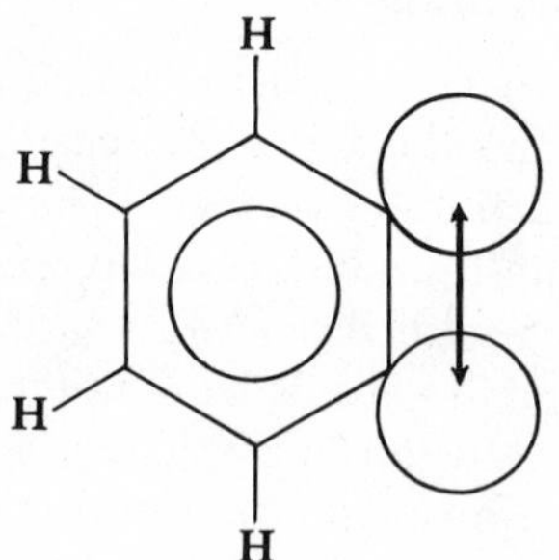

Figure 25.4. Benzyne molecule. Sideways overlap of sp^2 orbitals forms π bond out of plane of aromatic π cloud.

originally holding the hydrogen) by sideways overlap of sp^2 orbitals. This new bond orbital lies along the side of the ring, and has little interaction with the π cloud lying above and below the ring. The sideways overlap is not very good, the new bond is a weak one, and benzyne is a highly reactive molecule.

The elimination stage, in which benzyne is formed, involves two steps: abstraction of a hydrogen ion (step 1) by the amide ion to form ammonia and carbanion I, which then loses halide ion (step 2) to form benzyne.

(1) $C_6H_5X + NH_2^- \longrightarrow C_6H_4X^{:-} \text{ (I)} + NH_3$

Elimination

(2) $C_6H_4X^{:-} \text{ (I)} \longrightarrow \text{Benzyne} + X^-$

The addition stage, in which benzyne is consumed, may also involve two steps: attachment of the amide ion (step 3) to form carbanion II, which then reacts with an acid, ammonia, to abstract a hydrogen ion (step 4). It may be that step (3) and step (4) are concerted, and addition involves a single step; if this is so, the transition state is probably one in which attachment of nitrogen has proceeded to a greater extent than attachment of hydrogen, so that it has considerable carbanion

(3) Benzyne + NH_2^- ⟶ II

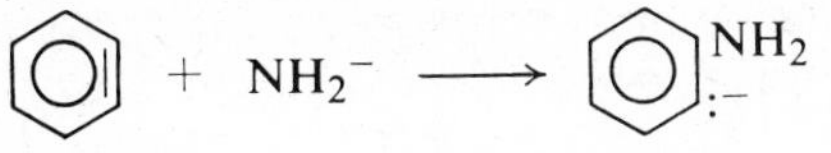

Addition

(4) 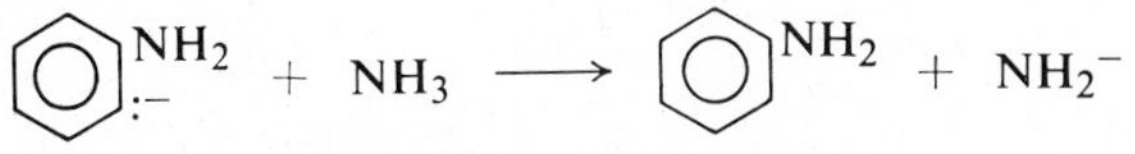 II + NH_3 ⟶ Aniline + NH_2^-

character. (This is analogous to hydroboration (Sec. 15.11), in which the transition state has considerable carbonium ion character.)

Let us look at the facts on which the above mechanism is based.

(a) *Fact*. Labeled chlorobenzene in which ^{14}C held the chlorine atom was allowed to react with amide ion. In *half* the aniline obtained the amino group was held by ^{14}C and in *half* it was held by an adjacent carbon.

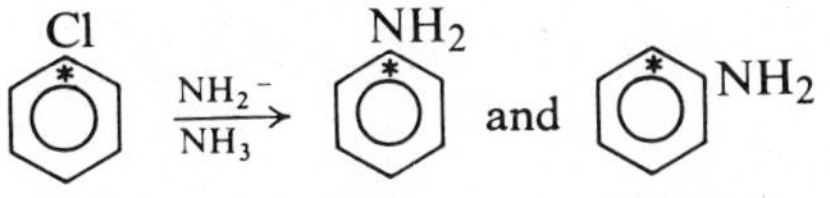

Cl —NH_2^-/NH_3→ NH_2 (47%) and NH_2 (53%)

Interpretation. In benzyne the labeled carbon and the ones next to it become equivalent, and NH_2^- adds randomly (except for a small isotope effect) to one or the other.

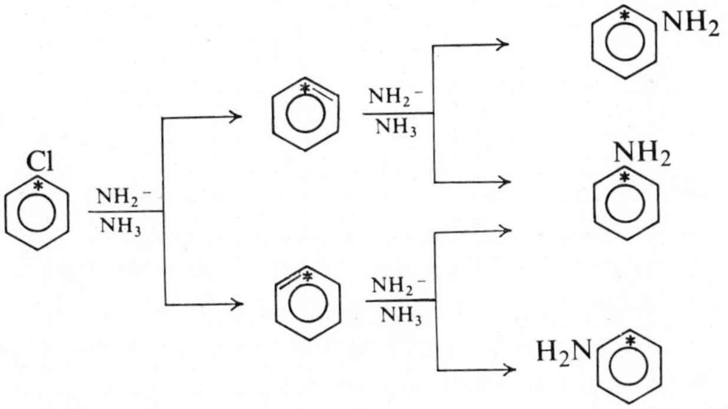

Cl —NH_2^-/NH_3→ benzynes —NH_2^-/NH_3→ NH_2, NH_2, H_2N

Although foreshadowed by certain earlier observations, this experiment, reported in 1953 by John D. Roberts of the California Institute of Technology, marks the real beginning of benzyne chemistry.

(b) *Fact*. Compounds containing two groups *ortho* to halogen, like 2-bromo-3-methylanisole, do not react at all.

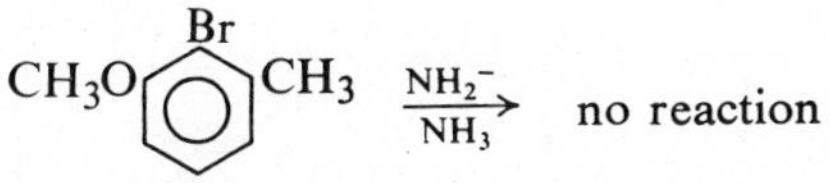

CH_3O, Br, CH_3 —NH_2^-/NH_3→ no reaction

Interpretation. With no *ortho* hydrogen to be lost, benzyne cannot form.

(c) *Fact*. When a 50:50 mixture of bromobenzene and *o*-deuteriobromobenzene is allowed to react with a limited amount of amide ion, recovered unreacted material contains more of the deuteriobromobenzene than bromobenzene; the deuterated compound is less reactive and is consumed more slowly.

$$C_6H_4(Br)(H) \xrightarrow[NH_3]{NH_2^-} \text{aniline} \xleftarrow[NH_3]{NH_2^-} C_6H_4(Br)(D)$$

o-Deuteriobromobenzene
Reacts more slowly: more left unconsumed

Interpretation. This isotope effect (Sec. 11.15) shows not only that the *ortho* hydrogen is involved, but that it is involved in a rate-determining step. Deuterium is abstracted more slowly in the first step (equation 1, p. 836), and the whole reaction sequence is slowed down.

$$C_6H_4(Br)(H) \xrightarrow{NH_2^-} C_6H_4(Br)(:^-) \xleftarrow{NH_2^-} C_6H_4(Br)(D)$$

Bond to D broken more slowly

(d) *Fact. o*-Deuteriofluorobenzene is converted into aniline only very slowly, but loses its deuterium rapidly to yield ordinary fluorobenzene.

$$C_6H_4(F)(D) \xrightarrow[NH_3]{NH_2^-} C_6H_4(F)(H)$$

Interpretation. Abstraction of hydrogen (step 1) takes place, but before the very strong carbon–fluorine bond can break, the carbanion reacts with the acid—which is almost *all* NH_3 with only a trace of NH_2D—to regenerate fluorobenzene, but without its deuterium.

In the case of *o*-deuterio*bromo*benzene, on the other hand, breaking of the weaker carbon–bromide bond (step 2) is much faster than the protonation by ammonia (reverse of step 1): as fast as a carbanion is formed, it loses bromide ion. In this case, isotopic exchange is not important. (It may even be that here steps (1) and (2) are concerted.)

(1) $$C_6H_5X + NH_2^- \underset{k_{-1}}{\overset{k_1}{\rightleftharpoons}} C_6H_4(X)(:^-) + NH_3$$

(2) $$C_6H_4(X)(:^-) \xrightarrow{k_2} C_6H_4 \text{ (benzyne)} + X^-$$

$$\text{for } X = F, \quad k_{-1} \gg k_2$$

$$X = Br, \quad k_2 \gg k_{-1}$$

(e) *Fact.* Both *m*-bromoanisole and *o*-bromoanisole yield the same product: *m*-anisidine (*m*-aminoanisole).

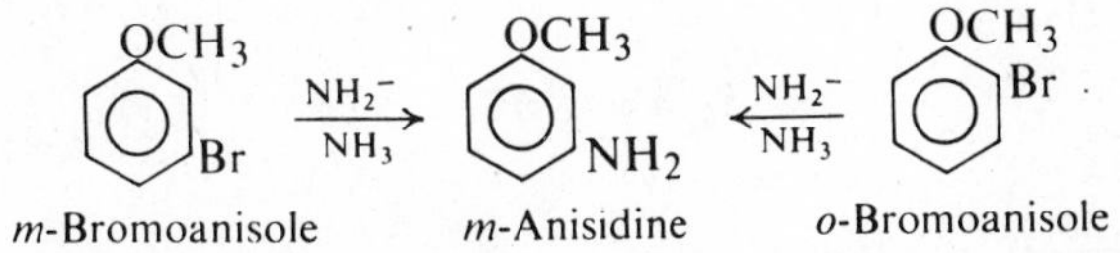

OCH_3, Br — $\xrightarrow[NH_3]{NH_2^-}$ — OCH_3, NH_2 — $\xleftarrow[NH_3]{NH_2^-}$ — OCH_3, Br

m-Bromoanisole *m*-Anisidine *o*-Bromoanisole

Interpretation. They yield the same product because they form the same intermediate benzyne.

Which benzyne is this, and how is it that it yields *m*-anisidine? To deal with orientation—both in the elimination stage and the addition stage—we must remember that a methoxyl group has an electron-withdrawing inductive effect. Since the electrons in carbanions like I and II (pp. 836–837) are out of the plane of the π cloud, there is no question of resonance interaction; only the inductive effect, working along the σ bonds (or perhaps through space), is operative.

o-Bromoanisole yields the benzyne shown (III, 2,3-dehydroanisole) because it has to. *m*-Bromoanisole yields III because, in the first step, the negative charge

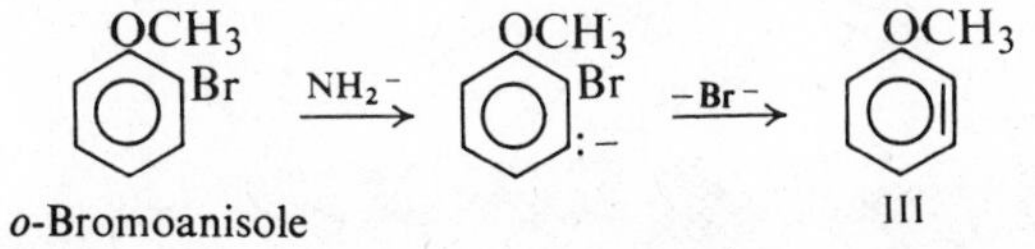

OCH_3, Br $\xrightarrow{NH_2^-}$ OCH_3, Br, :− $\xrightarrow{-Br^-}$ OCH_3

o-Bromoanisole III

appears preferentially on the carbon that can best accommodate it: the carbon next to the electron-withdrawing group. Whatever its source, III yields *m*-anisidine

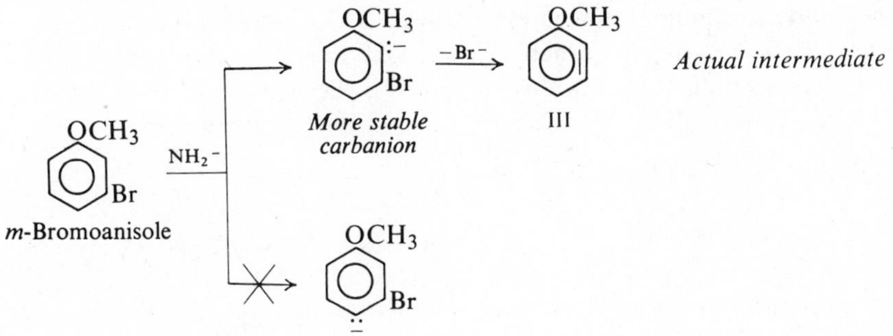

m-Bromoanisole (OCH_3, Br) $\xrightarrow{NH_2^-}$ → OCH_3, :−, Br (*More stable carbanion*) $\xrightarrow{-Br^-}$ OCH_3 III *Actual intermediate*

↛ OCH_3, Br, :−

for the same reason: addition of NH_2^- occurs in such a way that the negative charge appears on the carbon next to methoxyl.

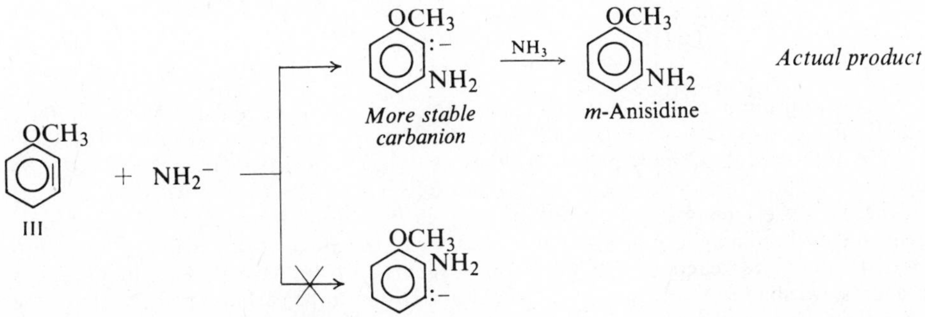

III (OCH_3) + NH_2^- → OCH_3, :−, NH_2 (*More stable carbanion*) $\xrightarrow{NH_3}$ OCH_3, NH_2 *m*-Anisidine *Actual product*

↛ OCH_3, NH_2, :−

Another common way to generate benzyne involves use of organolithium compounds. For example:

$$\text{o-}CH_3OC_6H_4F + \overset{\delta-}{C_6H_5}\text{—}\overset{\delta+}{Li}; \text{ then } H_2O \longrightarrow \text{m-}CH_3OC_6H_4C_6H_5$$

Here benzyne formation involves abstraction of a proton (reaction 5) by the base $C_6H_5^-$ to form a carbanion which loses fluoride ion (reaction 6) to give benzyne.

(5)
$$\underset{\text{Stronger acid}}{\text{o-}CH_3OC_6H_4F} + \underset{\text{Stronger base}}{\overset{\delta-}{C_6H_5}\text{—}\overset{\delta+}{Li}} \longrightarrow \underset{\text{Weaker base}}{2\text{-}CH_3O\text{-}3\text{-}F\text{-}C_6H_3\overset{\delta-}{}\text{—}Li^{\delta+}} + \underset{\text{Weaker acid}}{C_6H_6}$$

(6)
$$2\text{-}CH_3O\text{-}3\text{-}F\text{-}C_6H_3\overset{\delta-}{}\text{—}Li^{\delta+} \longrightarrow 3\text{-methoxybenzyne} + Li^+F^-$$

Problem 25.8 Account for the relative strengths of these acids and bases.

Addition of phenyllithium (reaction 7) to the benzyne gives the organolithium compound IV. From one point of view, this is the same reaction sequence observed for the amide ion–ammonia reaction (above), but it stops at the carbanion stage for want of strong acid. (Alternatively, the Lewis acid Li^+ has completed the sequence.) Addition of water—in this company, a very strong acid—yields (reaction 8) the final product. (The strong acid H^+ has displaced the weaker acid Li^+.)

(7)
$$3\text{-methoxybenzyne} + \overset{\delta-}{C_6H_5}\text{—}\overset{\delta+}{Li} \longrightarrow \underset{\text{IV}}{2\text{-}CH_3O\text{-}C_6H_3(\overset{\delta-}{}\text{—}Li^{\delta+})\text{-}3\text{-}C_6H_5}$$

(8)
$$2\text{-}CH_3O\text{-}C_6H_3(\overset{\delta-}{}\text{—}Li^{\delta+})\text{-}3\text{-}C_6H_5 + H_2O \longrightarrow \text{m-}CH_3OC_6H_4C_6H_5 + Li^+OH^-$$

Organolithium compounds, RLi, resemble Grignard reagents, RMgX, in their reactions. As in Grignard reagents (Sec. 3.16), the carbon–metal bond can probably best be described as a highly polar covalent bond or, in another manner of speaking, as a bond with much *ionic character* (a resonance hybrid of R—M and R^-M^+). Because of the greater electropositivity of lithium, the carbon–lithium bond is even more ionic than the carbon–magnesium bond and, partly as a result of this, organolithium compounds are more reactive than Grignard reagents. As we have done with Grignard reagents, we shall for convenience focus our attention on the carbanion character of the organic group in discussing these reactions as acid–base chemistry. In the reactions involving $K^+NH_2^-$ we indicated free carbanions as intermediates, although even here the attractive forces—whatever they are—between carbon and potassium may be of great importance.

Problem 25.9 Account for the following facts: (a) treatment of the reaction mixture in reaction (8) with carbon dioxide instead of water gives V; (b) treatment of

V: benzene ring bearing OCH_3, $COOH$, C_6H_5

VI: benzene ring bearing OCH_3, $C(OH)(C_6H_5)_2$, C_6H_5

the reaction mixture in reaction (8) with benzophenone gives VI; (c) benzyne can be generated by treatment of *o*-bromofluorobenzene with magnesium metal.

25.15 Analysis of aryl halides

Aryl halides show much the same response to characterization tests as the hydrocarbons from which they are derived: insolubility in cold concentrated sulfuric acid; inertness toward bromine in carbon tetrachloride and toward permanganate solutions; formation of orange to red colors when treated with chloroform and aluminum chloride; dissolution in cold fuming sulfuric acid, but at a slower rate than that of benzene.

Aryl halides are distinguished from aromatic hydrocarbons by the presence of halogen, as shown by elemental analysis. Aryl halides are distinguished from most alkyl halides by their inertness toward silver nitrate; in this respect they resemble vinyl halides (Sec. 25.5).

Any other functional groups that may be present in the molecule undergo their characteristic reactions.

Problem 25.10 Describe simple chemical tests (if any) that will distinguish between: (a) bromobenzene and *n*-hexyl bromide; (b) *p*-bromotoluene and benzyl bromide; (c) chlorobenzene and 1-chloro-1-hexene; (d) α-(*p*-bromophenyl)ethyl alcohol (*p*-$BrC_6H_4CHOHCH_3$) and *p*-bromo-*n*-hexylbenzene; (e) α-(*p*-chlorophenyl)ethyl alcohol and β-(*p*-chlorophenyl)ethyl alcohol (*p*-$ClC_6H_4CH_2CH_2OH$). Tell exactly what you would *do* and *see*.

Problem 25.11 Outline a procedure for distinguishing by chemical means (not necessarily simple tests) between: (a) *p*-bromoethylbenzene and 4-bromo-1,3-dimethylbenzene; (b) *o*-chloropropenylbenzene (*o*-$ClC_6H_4CH{=}CHCH_3$) and *o*-chloroallylbenzene (*o*-$ClC_6H_4CH_2CH{=}CH_2$).

PROBLEMS

1. Give structures and names of the principal organic products of the reaction (if any) of each of the following reagents with bromobenzene:

(a) Mg, ether
(b) boiling 10% aqueous NaOH
(c) boiling alcoholic KOH
(d) sodium acetylide
(e) sodium ethoxide
(f) NH_3, 100°
(g) boiling aqueous NaCN
(h) HNO_3, H_2SO_4
(i) fuming sulfuric acid
(j) Cl_2, Fe

(k) I_2, Fe
(l) C_6H_6, $AlCl_3$
(m) CH_3CH_2Cl, $AlCl_3$
(n) cold dilute $KMnO_4$
(o) hot $KMnO_4$

2. Answer Problem 1 for *n*-butyl bromide.

3. Answer Problem 1, parts (b), (e), (f), and (g) for 2,4-dinitrobromobenzene.

4. Outline a laboratory method for the conversion of bromobenzene into each of the following, using any needed aliphatic and inorganic reagents.

(a) benzene
(b) *p*-bromonitrobenzene
(c) *p*-bromochlorobenzene
(d) *p*-bromobenzenesulfonic acid
(e) 1,2,4-tribromobenzene
(f) *p*-bromotoluene
(g) benzyl alcohol
(h) α-phenylethyl alcohol
(i) 2-phenyl-2-propanol
(j) 2,4-dinitrophenol
(k) allylbenzene (*Hint:* See Problem 16, p. 281.)
(l) benzoic acid
(m) aniline

5. Give the structure and name of the product expected when phenylmagnesium bromide is treated with each of the following compounds and then with water:

(a) H_2O
(b) HBr (dry)
(c) C_2H_5OH
(d) allyl bromide
(e) HCHO
(f) CH_3CHO
(g) C_6H_5CHO
(h) *p*-$CH_3C_6H_4CHO$
(i) CH_3COCH_3
(j) cyclohexanone
(k) 3,3-dimethylcyclohexanone
(l) $C_6H_5COCH_3$
(m) $C_6H_5COC_6H_5$
(n) $(-)$-$C_6H_5COCH(CH_3)C_2H_5$
(o) acetylene

Which products (if any) would be single compounds? Which (if any) would be racemic modifications? Which (if any) would be optically active as isolated?

6. Arrange the compounds in each set in order of reactivity toward the indicated reagent. Give the structure and name of the product expected from the compound you select as the most reactive in each set.

(a) NaOH: chlorobenzene, *m*-chloronitrobenzene, *o*-chloronitrobenzene, 2,4-dinitrochlorobenzene, 2,4,6-trinitrochlorobenzene
(b) HNO_3/H_2SO_4: benzene, chlorobenzene, nitrobenzene, toluene
(c) alcoholic $AgNO_3$: 1-bromo-1-butene, 3-bromo-1-butene, 4-bromo-1-butene
(d) fuming sulfuric acid: bromobenzene, *p*-bromotoluene, *p*-dibromobenzene, toluene
(e) KCN: benzyl chloride, chlorobenzene, ethyl chloride
(f) alcoholic $AgNO_3$: 2-bromo-1-phenylethene, α-phenylethyl bromide, β-phenylethyl bromide

7. In the preparation of 2,4-dinitrochlorobenzene from chlorobenzene, the excess nitric acid and sulfuric acid must be washed from the product. Which would you select for this purpose: aqueous sodium hydroxide or aqueous sodium bicarbonate? Why?

8. Give structures and names of the principal organic products expected from each of the following reactions:

(a) 2,3-dibromopropene + NaOH(aq)
(b) *p*-bromobenzyl bromide + NH_3(aq)
(c) *p*-chlorotoluene + hot $KMnO_4$
(d) *m*-bromostyrene + Br_2/CCl_4
(e) 3,4-dichloronitrobenzene + 1 mole $NaOCH_3$
(f) *p*-bromochlorobenzene + Mg, ethyl ether
(g) *p*-bromobenzyl alcohol + cold dilute $KMnO_4$
(h) *p*-bromobenzyl alcohol + conc. HBr
(i) α-(*o*-chlorophenyl)ethyl bromide + KOH(alc)

(j) *p*-bromotoluene + 1 mole Cl_2, heat, light
(k) *o*-bromobenzotrifluoride + $NaNH_2/NH_3$
(l) *o*-bromoanisole + $K^{+} {}^{-}NEt_2/Et_2NH$

9. Outline all steps in a possible laboratory synthesis of each of the following compounds from benzene and/or toluene, using any needed aliphatic or inorganic reagents:

(a) *m*-chloronitrobenzene
(b) *p*-chloronitrobenzene
(c) *m*-bromobenzoic acid
(d) *p*-bromobenzoic acid
(e) *m*-chlorobenzotrichloride
(f) 3,4-dibromonitrobenzene
(g) *p*-bromobenzal chloride
(h) 2,4-dinitroaniline
(i) *p*-bromostyrene
(j) 2,4-dibromobenzoic acid
(k) *m*-iodotoluene
(l) *p*-bromobenzenesulfonic acid
(m) *p*-chlorobenzyl alcohol
(n) 2-(*p*-tolyl)propane

10. Halogen located at the 2- or 4-position of the aromatic heterocyclic compound *pyridine* (Sec. 31.6) is fairly reactive toward nucleophilic displacement. For example:

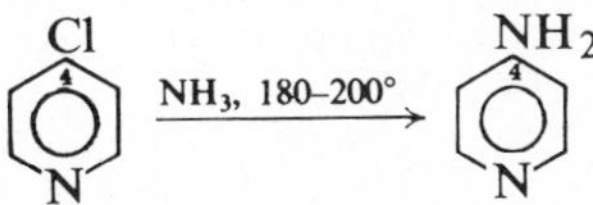

4-Chloropyridine 4-Aminopyridine

How do you account for the reactivity of these compounds? (Check your answer in Sec. 31.10.)

11. The insecticide called DDT, 1,1,1-trichloro-2,2-bis-(*p*-chlorophenyl(ethane, $(p\text{-ClC}_6H_4)_2CHCCl_3$, is manufactured by the reaction between chlorobenzene and trichloroacetaldehyde in the presence of sulfuric acid. Outline the series of steps by which this synthesis most probably takes place; make sure you show the function of the H_2SO_4. Label each step according to its fundamental reaction type.

12. In the Dow process for the manufacture of phenol, two by-products are diphenyl ether and *p*-phenylphenol. It has been suggested that these two compounds are formed via the same intermediate. How might this happen?

13. In KNH_2/NH_3, protium-deuterium exchange takes place at the following relative rates:

$$\begin{array}{ccccccc} o\text{-}C_6H_4DF & > & m\text{-}C_6H_4DF & > & p\text{-}C_6H_4DF & > & C_6H_5D \\ 4{,}000{,}000 & & 4{,}000 & & 200 & & 1 \end{array}$$

How do you account for this sequence of reactivity?

14. Reduction of 2,6-dibromobenzenediazonium chloride, which would be expected to give *m*-dibromobenzene, actually yields chiefly *m*-bromochlorobenzene. How do you account for this?

15. The reaction of 2,4-dinitrofluorobenzene with N-methylaniline to give N-methyl-2,4-dinitrodiphenylamine is catalyzed by weak bases like acetate ion. The reaction of the corresponding bromo compound is faster, and is not catalyzed by bases. How do you account for these observations? (*Hint:* Examine in detail every step of the mechanism.)

16. (a) The labeled ether $2,4\text{-}(NO_2)_2C_6H_3{}^{18}OC_6H_5$ reacts more slowly than the unlabeled ether with the secondary amine piperidine (Sec. 31.12). How do you account for this?

(b) The isotope effect in part (a) becomes weaker as the piperidine concentration is raised. Account in detail for this observation. (*Hint:* See the preceding problem.)

17. The rate of reaction between *p*-fluoronitrobenzene and azide ion (N_3^-) is affected markedly by the nature of the solvent. How do you account for the following relative rates: in methanol, 1; in formamide, 5.6; in N-methylformamide, 15.7; in dimethylformamide, 2.4×10^4.

18. The dry diazonium salt I was subjected to a flash discharge, and an especially

[Structure I: benzene ring with N_2^+ and COO^- ortho substituents]

I

adapted mass spectrometer scanned the spectrum of the products at rapid intervals after the flash. After about 50 microseconds there appeared simultaneously masses 28, 44, and 76. As time passed (about 250 microseconds) mass 76 gradually disappeared and a peak at mass 152 approached maximum intensity.

(a) What are the peaks at 28, 44, and 76 due to? What happens as time passes, and what is the substance of mass 152? (b) From what compound was the diazonium salt I prepared?

19. When a trace of KNH_2 is added to a solution of chlorobenzene and potassium triphenylmethide, $(C_6H_5)_3C^-K^+$, in liquid ammonia, a rapid reaction takes place to yield a product of formula $C_{25}H_{20}$. What is the product? What is the role of KNH_2, and why is it needed?

20. How do you account for each of the following observations?

(a) When *p*-iodotoluene is treated with aqueous NaOH at 340°, there is obtained a mixture of *p*-cresol (51%) and *m*-cresol (49%). At 250°, reaction is, of course, slower, and yields only *p*-cresol.

(b) When diazotized 4-nitro-2-aminobenzoic acid is heated in *tert*-butyl alcohol, there is obtained carbon dioxide, nitrogen, and a mixture of *m*- and *p*-nitrophenyl *tert*-butyl ethers.

(c) When *o*-chlorobenzoic acid is treated with $NaNH_2/NH_3$ in the presence of acetonitrile (CH_3CN) there is obtained a 70% yield of *m*-$HOOCC_6H_4CH_2CN$ and 10–20% of a 1:2 mixture of *o*- and *m*-aminobenzoic acids.

21. When either II or III is treated with $KN(C_2H_5)_2/HN(C_2H_5)_2$, there is obtained in

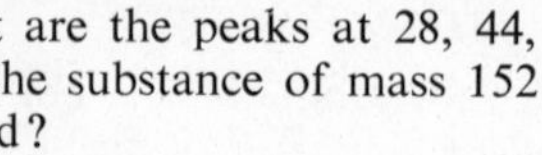

[Structure II: benzene ring with $CH_2CH_2NHCH_3$ and ortho Cl] [Structure III: benzene ring with $CH_2CH_2NHCH_3$ and meta Cl]

II 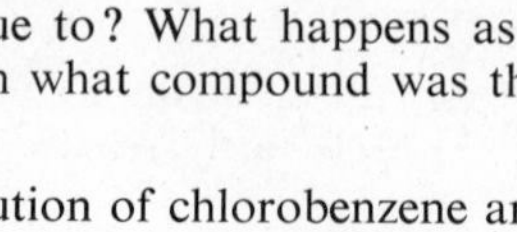 III

good yield the same product, of formula $C_9H_{11}N$. What is the product, and how is it formed?

22. An unknown compound is believed to be one of the following. Describe how you would go about finding out which of the possibilities the unknown actually is. Where possible use simple chemical tests; where necessary use more elaborate chemical methods like quantitative hydrogenation, cleavage, etc. Where necessary, make use of Table 18.1, p. 580.

(a) $C_6H_5CH{=}CHBr$ (b.p. 221°), *o*-$C_6H_4Br_2$ (b.p. 221°), $BrCH_2(CH_2)_3CH_2Br$ (b.p. 224°)
(b) *o*-$CH_3C_6H_4Br$ (b.p. 182°), *m*-$CH_3C_6H_4Br$ (b.p. 184°), *p*-$CH_3C_6H_4Br$ (b.p. 185°)
(c) *o*-$ClC_6H_4C_2H_5$ (b.p. 178°), $C_6H_5CH_2Cl$ (b.p. 179°), *o*-$C_6H_4Cl_2$ (b.p. 180°)
(d) $ClCH_2CH_2OH$ (b.p. 129°), 4-octyne (b.p. 131°), isopentyl alcohol (b.p. 132°), C_6H_5Cl (b.p. 132°), ethylcyclohexane (b.p. 132°), 1-chlorohexane (b.p. 134°)
(e) [benzene ring with $CHBrCH_2Br$] (m.p. 73°); [benzene ring with CH_3, Br, Br, CH_3] (m.p. 74°); [benzene ring with CH_2Br and ortho CH_2Br] (m.p. 76°)

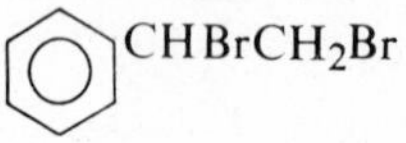

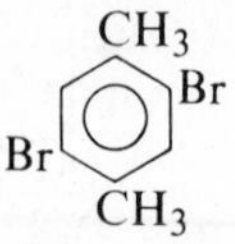

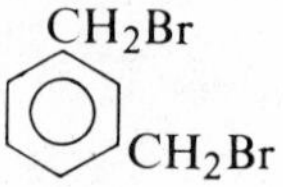

23. In studying the *base-catalyzed halogen dance*, Bunnett has made the following observations. When IV is treated with C_6H_5NHK/NH_3, it is isomerized to V. There is

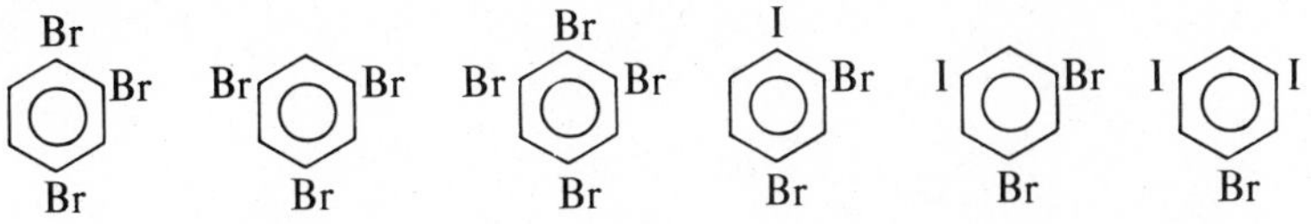

IV: 1,2,4-tribromobenzene; V: 1,3,5-tribromobenzene; VI: 1,2,3,5-tetrabromobenzene; VII; VIII; IX

found, in addition, VI, *m*- and *p*-dibromobenzenes, and unconsumed IV. Similar treatment of VII gives chiefly VIII, along with IX, IV, and V. When IV labeled at the 1-position with radioactive bromine is allowed to react, the recovered IV had the label statistically distributed among all three positions.

(a) Bunnett first considered a mechanism involving intermediate benzynes. Show how you could account for the above observations on this basis.

(b) When IV is allowed to isomerize in the presence of much KI, no iodobromobenzenes are found. On this and other grounds, Bunnett rejected the benzyne mechanism. Explain.

(c) From the isomerization of IV, some unconsumed IV is *always* obtained. Yet the reaction of V gives IV *only if* there is present a small amount of VI to start with. (This is a *real* effect; highly purified materials give the same results.) In the presence of a little VI, the same mixture (about 50:50) of IV and V is formed whether one starts with IV or with V.

Suggest a complete mechanism for the base-catalyzed halogen dance, and show how it accounts for all the facts. It may help to go at the problem in this way. First, start with V and the base, in the presence of VI, and show how IV can be formed. Show how, under the same conditions, V can be formed from IV.

Next, start with *only* IV and base, and show how all the products are formed (V, VI, *m*- and *p*-dibromobenzenes), and account for the scrambling of the bromine label.

Finally, the hardest part: why must VI be added to bring about isomerization of V but not the isomerization of IV? (*Hint:* Simply write for V equations analogous to those you have written for IV, and keep in mind Problem 13, p. 843.)

Chapter 26 Carbanions II

Malonic Ester and Acetoacetic Ester Syntheses

26.1 Carbanions in organic synthesis

We have already seen something of the importance to organic synthesis of the formation of carbon–carbon bonds: it enables us to make big molecules out of little ones. In this process a key role is played by negatively charged carbon. Such *nucleophilic carbon* attacks carbon holding a good leaving group—in alkyl halides or sulfonates, usually—or carbonyl or acyl carbon. Through nucleophilic substitution or nucleophilic addition, a new carbon–carbon bond is formed.

Nucleophilic carbon is of two general kinds. (a) There are the carbanion-like groups in organometallic compounds, usually generated through reaction of an organic halide with a metal: Grignard and organocadmium reagents, for example; the lithium dialkylcopper reagents used in the Corey-House synthesis of hydrocarbons; the organozinc compounds that are intermediates in the Reformatsky reaction. (b) There are the more nearly full-fledged carbanions generated through abstraction of α-hydrogens by base, as in the aldol and Claisen condensations and their relatives.

The difference between these two kinds of carbon is one of degree, not kind. There is interaction—just how much depending on the metal and the solvent—even between electropositive ions like sodium or potassium or lithium and the anion from carbonyl compounds. These intermediates, too, could be called organometallic compounds; the bonding is simply more ionic than that in, say, a Grignard reagent.

In this chapter we shall continue with our study of carbanion chemistry, with emphasis on the attachment of alkyl groups to the α-carbons of carbonyl and acyl compounds. Such *alkylation* reactions owe their great importance to the special nature of the carbonyl group, and in two ways. First, the carbonyl group makes α-hydrogens acidic, so that alkylation can take place. Next, the products obtained still contain the carbonyl group and hence are highly reactive; they are ideal intermediates for *further* molecule-building.

Of the very many alkylation methods that have been developed, we can look at only a few: first, two classics of organic synthesis, the *malonic ester synthesis* and the *acetoacetic ester synthesis*; and then, several newer methods. In doing this we shall be concerned not only with learning a bit more about how to make new molecules from old ones, but also with seeing the variety of ways in which carbanion chemistry is involved.

26.2 Malonic ester synthesis of carboxylic acids

One of the most valuable methods of preparing carboxylic acids makes use of ethyl malonate (*malonic ester*), $CH_2(COOC_2H_5)_2$, and is called the **malonic ester synthesis**. This synthesis depends upon (a) the high acidity of the α-hydrogens of malonic ester, and (b) the extreme ease with which malonic acid and substituted malonic acids undergo decarboxylation. (As we shall see, this combination of properties is more than a happy accident, and can be traced to a single underlying cause.)

Like acetoacetic ester (Sec. 21.11), and for exactly the same reason, malonic ester contains α-hydrogens that are particularly acidic: they are *alpha* to *two* carbonyl groups. When treated with sodium ethoxide in absolute alcohol, malonic ester is converted largely into its salt, *sodiomalonic ester*:

$$\underset{\text{Stronger acid}}{CH_2(COOC_2H_5)_2} + Na^{+}\,{}^{-}OC_2H_5 \rightleftarrows \underset{\text{Sodiomalonic ester}}{CH(COOC_2H_5)_2{}^{-}Na^{+}} + \underset{\text{Weaker acid}}{HOC_2H_5}$$

Reaction of this salt with an alkyl halide yields a substituted malonic ester, an *ethyl alkylmalonate*, often called an *alkylmalonic ester*:

$$CH(COOC_2H_5)_2{}^{-}Na^{+} + RX \longrightarrow \underset{\substack{\text{Ethyl alkylmalonate} \\ \text{Alkylmalonic ester}}}{RCH(COOC_2H_5)_2} + Na^{+}X^{-}$$

This reaction involves nucleophilic attack on the alkyl halide by the carbanion, $CH(COOC_2H_5)_2{}^{-}$, and, as we might expect, gives highest yields with primary alkyl halides, lower yields with secondary alkyl halides, and is worthless for tertiary alkyl halides and for aryl halides.

The alkylmalonic ester still contains one ionizable hydrogen, and on treatment with sodium ethoxide it, too, can be converted into its salt; this salt can react with an alkyl halide—which may be the same as, or different from, the first alkyl halide—to yield a dialkylmalonic ester:

$$RCH(COOC_2H_5)_2 + Na^{+}\,{}^{-}OC_2H_5 \rightleftarrows RC(COOC_2H_5)_2{}^{-}Na^{+} + C_2H_5OH$$

$$\downarrow R'X$$

$$\underset{\text{Dialkylmalonic ester}}{RR'C(COOC_2H_5)_2} + Na^{+}X^{-}$$

The acidity of malonic ester thus permits the preparation of substituted malonic esters containing one or two alkyl groups. How can these substituted malonic esters be used to make carboxylic acids? When heated above its melting point, malonic acid readily loses carbon dioxide to form acetic acid; in a similar way substituted malonic acids readily lose carbon dioxide to form substituted acetic acids. The monoalkyl- and dialkylmalonic esters we have prepared are readily converted into monocarboxylic acids by hydrolysis, acidification, and heat:

$$\underset{\text{A monoalkylmalonic ester}}{RCH(COOC_2H_5)_2} \xrightarrow{H_2O,\ OH^-,\ heat} RCH(COO^-)_2 \xrightarrow{H^+} RCH(COOH)_2 \xrightarrow{heat,\ 140^\circ} \underset{\text{A monosubstituted acetic acid}}{RCH_2COOH} + CO_2$$

$$\underset{\text{A dialkylmalonic ester}}{RR'C(COOC_2H_5)_2} \xrightarrow{H_2O,\ OH^-,\ heat} RR'C(COO^-)_2 \xrightarrow{H^+} RR'C(COOH)_2 \xrightarrow{heat,\ 140^\circ} \underset{\text{A disubstituted acetic acid}}{RR'CHCOOH} + CO_2$$

A malonic ester synthesis yields an acetic acid in which one or two hydrogens have been replaced by alkyl groups.

In planning a malonic ester synthesis, our problem is to select the proper alkyl halide or halides; to do this, we have only to look at the structure of the acid we want. Isocaproic acid, for example, $(CH_3)_2CHCH_2CH_2COOH$, can be considered as acetic acid in which one hydrogen has been replaced by an isobutyl group. To prepare this acid by the malonic ester synthesis, we would have to use isobutyl bromide as the alkylating agent:

$$\underset{\text{Isobutyl bromide}}{CH_3\overset{CH_3}{C}HCH_2Br} + Na^+\,{}^-CH(COOC_2H_5)_2 \longrightarrow \underset{\text{Ethyl isobutylmalonate (Isobutylmalonic ester)}}{CH_3\overset{CH_3}{C}HCH_2CH(COOC_2H_5)_2}$$

(The sodiomalonic ester is obtained from $\underset{\text{Malonic ester}}{CH_2(COOC_2H_5)_2}$ with $Na^+\,{}^-OC_2H_5$.)

$$CH_3\overset{CH_3}{C}HCH_2CH(COOC_2H_5)_2 \xrightarrow{H_2O,\ OH^-,\ heat} CH_3\overset{CH_3}{C}HCH_2CH(COO^-)_2 \xrightarrow{H^+} CH_3\overset{CH_3}{C}HCH_2CH(COOH)_2 \xrightarrow{heat,\ -CO_2} \underset{\text{Isocaproic acid}}{CH_3\overset{CH_3}{C}HCH_2CH_2COOH}$$

An isomer of isocaproic acid, α-methylvaleric acid, $CH_3CH_2CH_2CH(CH_3)COOH$, can be considered as acetic acid in which one hydrogen has been replaced by a

$$\underset{\alpha\text{-Methylvaleric acid}}{CH_3CH_2CH_2\underset{\displaystyle CH_3}{\underset{|}{C}}HCOOH} \xleftarrow{\text{heat, } -CO_2} CH_3CH_2CH_2\overset{\displaystyle COOH}{\overset{|}{\underset{\displaystyle CH_3}{\underset{|}{C}}}}COOH \xleftarrow{H^+} CH_3CH_2CH_2\overset{\displaystyle COO^-}{\overset{|}{\underset{\displaystyle CH_3}{\underset{|}{C}}}}COO^-$$

$$\uparrow H_2O,\ OH^-,\ \text{heat}$$

$$\underset{\text{Methyl bromide}}{CH_3Br} + Na^+CH_3CH_2CH_2C(COOC_2H_5)_2^- \longrightarrow CH_3CH_2CH_2\overset{\displaystyle COOC_2H_5}{\overset{|}{\underset{\displaystyle CH_3}{\underset{|}{C}}}}COOC_2H_5$$

Ethyl methyl-*n*-propylmalonate
Methyl-*n*-propylmalonic ester

$$\uparrow Na^+\ ^-OC_2H_5$$

$$CH_3CH_2CH_2CH(COOC_2H_5)_2$$

$$\uparrow$$

$$\underset{n\text{-Propyl bromide}}{CH_3CH_2CH_2Br} + Na^+CH(COOC_2H_5)_2^-$$

$$\uparrow Na^+\ ^-OC_2H_5$$

$$\underset{\text{Malonic ester}}{CH_2(COOC_2H_5)_2}$$

n-propyl group and a second hydrogen has been replaced by a methyl group; we must therefore use two alkyl halides, *n*-propyl bromide and methyl bromide.

The basic malonic ester synthesis we have outlined can be modified. Often one can advantageously use: different bases as, for example, potassium *tert*-butoxide; alkyl sulfonates instead of halides; polar aprotic solvents like DMSO or DMF (Sec. 1.21).

In place of simple alkyl halides, certain other halogen-containing compounds may be used, in particular the readily available α-bromo esters (why can α-bromo-*acids* not be used?), which yield substituted succinic acids by the malonic ester synthesis. For example:

$$\underset{\alpha\text{-Methylsuccinic acid}}{HOOC\overset{CH_3}{CH}CH_2COOH} \xleftarrow{\text{heat, } -CO_2} HOOC\overset{CH_3}{CH}CH(COOH)_2 \xleftarrow{H^+} {}^-OOC\overset{CH_3}{CH}CH(COO^-)_2$$

$$\uparrow H_2O,\ OH^-,\ \text{heat}$$

$$\underset{\text{Ethyl }\alpha\text{-bromopropionate}}{CH_3\underset{Br}{CH}COOC_2H_5} + Na^+CH(COOC_2H_5)_2^- \longrightarrow C_2H_5OOC\overset{CH_3}{CH}CH(COOC_2H_5)_2$$

$$\uparrow Na^+\ ^-OC_2H_5$$

$$\underset{\text{Malonic ester}}{CH_2(COOC_2H_5)_2}$$

Problem 26.1 Outline the synthesis of the following compounds from malonic ester and alcohols of four carbons or less:

(a) the isomeric acids, *n*-valeric, isovaleric, and α-methylbutyric. (Why can the malonic ester synthesis not be used for the preparation of trimethylacetic acid?)
(b) *leucine* (α-aminoisocaproic acid)
(c) *isoleucine* (α-amino-β-methylvaleric acid)

Problem 26.2 *Adipic acid* is obtained from a malonic ester synthesis in which the first step is addition of one mole of ethylene bromide to a large excess of sodiomalonic ester in alcohol. *Cyclopropanecarboxylic acid* is the final product of a malonic ester synthesis in which the first step is addition of one mole of sodiomalonic ester to two moles of ethylene bromide followed by addition of one mole of sodium ethoxide.

$HOOCCH_2CH_2CH_2CH_2COOH$

Adipic acid

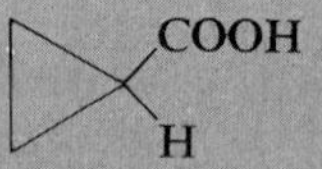

Cyclopropane-carboxylic acid

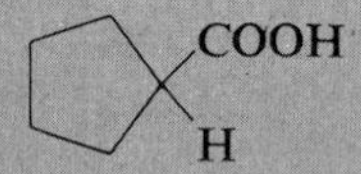

Cyclopentane-carboxylic acid

(a) Account for the difference in the products obtained in the two syntheses. (b) Tell exactly how you would go about synthesizing *cyclopentanecarboxylic acid.*

Problem 26.3 (a) Malonic ester reacts with benzaldehyde in the presence of piperidine (a secondary amine, Sec. 31.12) to yield a product of formula $C_{14}H_{16}O_4$. What is this compound, and how is it formed? (This is an example of the **Knoevenagel reaction**. Check your answer in Problem 21.22 (f), p. 714.) (b) What compound would be obtained if the product of (a) were subjected to the sequence of hydrolysis, acidification, and heating? (c) What is another way to synthesize the product of (b)?

Problem 26.4 (a) Cyclohexanone reacts with cyanoacetic ester (ethyl cyanoacetate, $N{\equiv}CCH_2COOC_2H_5$) in the presence of ammonium acetate to yield a product of formula $C_{11}H_{15}O_2N$. What is this compound, and how is it formed? (This is an example of the **Cope reaction**. Check your answer in Problem 21.22 (g), p. 714.) (b) What compound would be formed from the product of (a) by the sequence of hydrolysis, acidification, and heating?

Problem 26.5 In an example of the **Michael condensation,** malonic ester reacts with ethyl 2-butenoate in the presence of sodium ethoxide to yield A, of formula $C_{13}H_{22}O_6$. The sequence of hydrolysis, acidification, and heating converts A into 3-methylpentanedioic acid. What is A, and how is it formed? (*Hint:* See Sec. 8.20. Check your answer in Sec. 27.7.)

26.3 Acetoacetic ester synthesis of ketones

One of the most valuable methods of preparing ketones makes use of ethyl acetoacetate (*acetoacetic ester*), $CH_3COCH_2COOC_2H_5$, and is called the **acetoacetic ester synthesis of ketones**. This synthesis closely parallels the malonic ester synthesis of carboxylic acids.

Acetoacetic ester is converted by sodium ethoxide into the sodioacetoacetic ester, which is then allowed to react with an alkyl halide to form an alkylacetoacetic ester (an ethyl alkylacetoacetate), $CH_3COCHRCOOC_2H_5$; if desired, the alkylation can be repeated to yield a dialkylacetoacetic ester, $CH_3COCRR'COOC_2H_5$. All alkylations are conducted in absolute alcohol.

When hydrolyzed by dilute aqueous alkali (or by acid), these monoalkyl- or dialkylacetoacetic esters yield the corresponding acids, $CH_3COCHRCOOH$ or $CH_3COCRR'COOH$, which undergo decarboxylation to form ketones, CH_3COCH_2R or $CH_3COCHRR'$. This loss of carbon dioxide occurs even more readily than from malonic acid, and may even take place before acidification of the hydrolysis mixture.

$$CH_3COCH_2COOC_2H_5$$
Acetoacetic ester

$$\downarrow {}^{-}OC_2H_5$$

$$CH_3COCHCOOC_2H_5^-$$

$$\downarrow RX$$

$$\underset{\text{Monoalkylacetoacetic ester}}{CH_3CO\underset{R}{\underset{|}{C}}HCOOC_2H_5} \xrightarrow{OH^-} CH_3CO\underset{R}{\underset{|}{C}}HCOO^- \xrightarrow{H_2O \text{ or } H^+} CH_3CO\underset{R}{\underset{|}{C}}HCOOH \xrightarrow{-CO_2} \underset{\text{A monosubstituted acetone}}{CH_3COCH_2R}$$

$$\downarrow {}^{-}OC_2H_5$$

$$CH_3COCRCOOC_2H_5^-$$

$$\downarrow R'X$$

$$\underset{\text{Dialkylacetoacetic ester}}{CH_3CO\overset{R'}{\underset{R}{C}}COOC_2H_5} \xrightarrow{OH^-} CH_3CO\overset{R'}{\underset{R}{C}}COO^- \xrightarrow{H_2O \text{ or } H^+} CH_3CO\overset{R'}{\underset{R}{C}}COOH \xrightarrow{-CO_2} \underset{\text{A disubstituted acetone}}{CH_3COCHRR'}$$

The acetoacetic ester synthesis of ketones yields an acetone molecule in which one or two hydrogens have been replaced by alkyl groups.

In planning an acetoacetic ester synthesis, as in planning a malonic ester synthesis, our problem is to select the proper alkyl halide or halides. To do this, we have only to look at the structure of the ketone we want. For example, 5-methyl-2-hexanone can be considered as acetone in which one hydrogen has been replaced by an isobutyl group. In order to prepare this ketone by the acetoacetic ester synthesis, we would have to use isobutyl bromide as the alkylating agent:

$$\underset{\text{Isobutyl bromide}}{CH_3\overset{CH_3}{CH}CH_2Br} + Na^+CH_3COCHCOOC_2H_5^- \longrightarrow \underset{\text{Ethyl }\alpha\text{-isobutylacetoacetate}}{CH_3\overset{CH_3}{CH}CH_2\overset{COOC_2H_5}{CH}\underset{O}{\underset{\|}{C}}CH_3}$$

$$Na^+CH_3COCHCOOC_2H_5^- \xleftarrow{Na^+\ {}^-OC_2H_5} \underset{\text{Ethyl acetoacetate}}{CH_3COCH_2COOC_2H_5}$$

$$\xrightarrow{OH^-} CH_3\overset{CH_3}{CH}CH_2\overset{COO^-}{CH}\underset{O}{\underset{\|}{C}}CH_3 \xrightarrow{H_2O \text{ or } H^+} CH_3\overset{CH_3}{CH}CH_2\overset{COOH}{CH}\underset{O}{\underset{\|}{C}}CH_3 \xrightarrow{-CO_2} \underset{\text{5-Methyl-2-hexanone}}{CH_3\overset{CH_3}{CH}CH_2CH_2\underset{O}{\underset{\|}{C}}CH_3}$$

The isomeric ketone 3-methyl-2-hexanone can be considered as acetone in which one hydrogen has been replaced by a *n*-propyl group and a second hydrogen (on the same carbon) has been replaced by a methyl group; we must therefore use two alkyl halides, *n*-propyl bromide and methyl bromide:

$$\underset{\text{3-Methyl-2-hexanone}}{CH_3CH_2CH_2\underset{CH_3}{CH}—\underset{O}{\overset{\|}{C}}CH_3} \xleftarrow{-CO_2} CH_3CH_2CH_2\overset{COOH}{\underset{CH_3}{C}}—\underset{O}{\overset{\|}{C}}CH_3 \xleftarrow{H_2O \text{ or } H^+} CH_3CH_2CH_2\overset{COO^-}{\underset{CH_3}{C}}—\underset{O}{\overset{\|}{C}}CH_3$$

$$CH_3CH_2CH_2\overset{COO^-}{\underset{CH_3}{C}}—\underset{O}{\overset{\|}{C}}CH_3 \xleftarrow{OH^-} CH_3CH_2CH_2\overset{COOC_2H_5}{\underset{CH_3}{C}}—\underset{O}{\overset{\|}{C}}CH_3$$

$$\underset{\text{Methyl bromide}}{CH_3Br} + Na^+CH_3CH_2CH_2\overset{COOC_2H_5}{C}COCH_3^- \longrightarrow \underset{\text{Ethyl }\alpha\text{-methyl-}\alpha\text{-}n\text{-propylacetoacetate}}{CH_3CH_2CH_2\overset{COOC_2H_5}{\underset{CH_3}{C}}—\underset{O}{\overset{\|}{C}}CH_3}$$

$$CH_3CH_2CH_2\overset{COOC_2H_5}{CH}\underset{O}{\overset{\|}{C}}CH_3 \xrightarrow{Na^+\,^-OC_2H_5} Na^+CH_3CH_2CH_2\overset{COOC_2H_5}{C}COCH_3^-$$

$$\underset{n\text{-Propyl bromide}}{CH_3CH_2CH_2Br} + Na^+CH_3COCHCOOC_2H_5^- \longrightarrow CH_3CH_2CH_2\overset{COOC_2H_5}{CH}\underset{O}{\overset{\|}{C}}CH_3$$

$$\underset{\text{Ethyl acetoacetate}}{CH_3COCH_2COOC_2H_5} \xrightarrow{Na^+\,^-OC_2H_5} Na^+CH_3COCHCOOC_2H_5^-$$

Like the malonic ester synthesis, this synthesis, too, can be modified by changes in the base, solvent, and alkylating agent.

Problem 26.6 To what general class does the reaction between sodioacetoacetic ester and an alkyl halide belong? Predict the relative yields using primary, secondary, and tertiary halides. Can aryl halides be used?

Problem 26.7 (a) Predict the product of the acetoacetic ester synthesis in which ethyl bromoacetate (why not bromoacetic *acid*?) is used as the halide. To what general class of compounds does this product belong? (b) Predict the product of the acetoacetic ester synthesis in which benzoyl chloride is used as the halide; in which chloroacetone is used as the halide. To what general classes of compounds do these products belong?

Problem 26.8 Outline the synthesis of the following compounds from acetoacetic ester, benzene, and alcohols of four carbons or less:

(a)–(c) the isomeric ketones:
methyl *n*-butyl ketone (2-hexanone)
methyl isobutyl ketone (4-methyl-2-pentanone)
methyl *sec*-butyl ketone (3-methyl-2-pentanone)

(d) Why can the acetoacetic ester synthesis not be used for the preparation of methyl *tert*-butyl ketone?

(e) 2,4-pentanedione (*acetylacetone*)

(f) 2,5-hexanedione (*acetonylacetone*)

(g) 1-phenyl-1,4-pentanedione

Problem 26.9 The best general preparation of **α-keto acids** is illustrated by the sequence:

$$\text{ethyl propionate} + \text{ethyl oxalate} \xrightarrow{NaOC_2H_5} \text{A } (C_9H_{14}O_5)$$

$$\text{A} + \text{dil. } H_2SO_4 \xrightarrow{\text{boil}} CO_2 + 2C_2H_5OH + CH_3CH_2\underset{\underset{O}{\|}}{C}COOH \ (\alpha\text{-ketobutyric acid})$$

What familiar reactions are involved? What is the structure of A?

Problem 26.10 Outline the synthesis from simple esters of:

(a) α-ketoisocaproic acid
(b) α-keto-β-phenylpropionic acid
(c) α-ketoglutaric acid
(d) *leucine* (α-aminoisocaproic acid). (*Hint:* See Sec. 22.11.)
(e) *glutamic acid* (α-aminoglutaric acid)

26.4 Decarboxylation of β-keto acids and malonic acids

The acetoacetic ester synthesis thus depends on (a) the high acidity of the α-hydrogens of β-keto esters, and (b) the extreme ease with which β-keto acids undergo decarboxylation. These properties are exactly parallel to those on which the malonic ester synthesis depends.

We have seen that the higher acidity of the α-hydrogens is due to the ability of the keto group to help accommodate the negative charge of the acetoacetic ester anion. The ease of decarboxylation is, in part, due to *exactly the same factor.* (So, too, is the occurrence of the Claisen condensation, by which the acetoacetic ester is made in the first place.)

Decarboxylation of β-keto acids involves both the free acid and the carboxylate ion. Loss of carbon dioxide from the anion

$$CH_3-\underset{\underset{O}{\|}}{C}-CH_2-COO^- \longrightarrow CO_2 + \underset{\text{I}}{CH_3-\underset{\underset{O}{\|}}{C}\!=\!CH_2}\ \ominus$$

yields the carbanion I. This carbanion is formed faster than the simple carbanion (R:$^-$) that would be formed from a simple carboxylate ion ($RCOO^-$) because it is more stable. It is more stable, of course, due to the accommodation of the negative charge by the keto group.

Problem 26.11 Decarboxylation of malonic acid involves both the free acid and the monoanion, but not the doubly-charged anion. (a) Account for the ease of decarboxylation of the monoanion. Which end loses carbon dioxide? (b) How do you account for the lack of reactivity of the doubly-charged anion? (*Hint:* See Sec. 18.20.)

Problem 26.12 In contrast to most carboxylic acids (benzoic acid, say) 2,4,6-trinitrobenzoic acid is decarboxylated extremely easily: by simply boiling it in aqueous acid. How do you account for this?

Decarboxylation of free acetoacetic acid involves transfer of the acidic hydrogen to the keto group, either *prior to* (as shown here) or *simultaneously with*

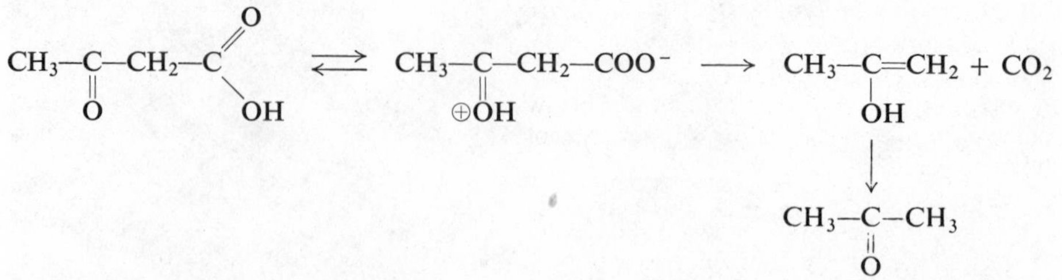

$$CH_3{-}\underset{\underset{O}{\|}}{C}{-}CH_2{-}C\begin{matrix}\nearrow O \\ \searrow OH\end{matrix} \rightleftharpoons CH_3{-}\underset{\underset{\oplus OH}{\|}}{C}{-}CH_2{-}COO^- \longrightarrow CH_3{-}\underset{\underset{OH}{|}}{C}{=}CH_2 + CO_2$$

$$\downarrow$$

$$CH_3{-}\underset{\underset{O}{\|}}{C}{-}CH_3$$

loss of carbon dioxide. We are quite familiar with the function of protonation to reduce the basicity of a leaving group.

Problem 26.13 When dimethylacetoacetic acid is decarboxylated in the presence of iodine or bromine, there is obtained an iododimethylacetone or a bromodimethylacetone (3-halo-3-methyl-2-butanone), although under these conditions neither iodine nor bromine reacts significantly with the dimethylacetone. What bearing does this experiment have on the mechanism of decarboxylation?

Problem 26.14 Suggest a mechanism for the decarboxylation of free malonic acid.

Problem 26.15 Account for the comparative ease with which phenylpropiolic acid, $C_6H_5C{\equiv}CCOOH$, undergoes decarboxylation in alkaline solution.

26.5 Direct and indirect alkylation of esters and ketones

By the malonic ester and acetoacetic ester we make α-substituted acids and α-substituted ketones. But why not do the job *directly*? Why not convert simple acids (or esters) and ketones into their carbanions, and allow these to react with alkyl halides? There are a number of obstacles: (a) self-condensation—aldol condensation, for example, of ketones; (b) polyalkylation; and (c) for unsymmetrical ketones, alkylation at *both* α-carbons, or at the wrong one. Consider self-condensation. A carbanion can be generated from, say, a simple ketone; but competing with attack on an alkyl halide is attack at the carbonyl carbon of another ketone molecule. What is needed is a base-solvent combination that can convert the ketone *rapidly* and essentially *completely* into the carbanion before appreciable self-condensation can occur. Steps toward solving this problem have been taken, and there are available methods—so far, of limited applicability—for the direct alkylation of acids and ketones.

A tremendous amount of work has gone into the development of alternatives to direct alkylation. Another group is introduced temporarily to do one or more of these things: increase the acidity of the α-hydrogens, prevent self-condensation, and direct alkylation to a specific position. The malonic ester and acetoacetic ester syntheses are, of course, typical of this approach. In the acetoacetic ester synthesis, for example, the carbethoxy group, —COOEt, enhances the acidity of α-hydrogens, but only those on one particular α-carbon, so that alkylation will take place there. Then, when alkylation is over, the carbethoxy group is easily removed by hydrolysis and decarboxylation.

In the biosynthesis of fats (Sec. 37.6), long-chain carboxylic acids are made via a series of what are basically malonic ester syntheses. Although in this case reactions are catalyzed by enzymes, the system still finds it worthwhile to consume carbon dioxide to make a malonyl compound, then form a new carbon–carbon bond, and finally eject the carbon dioxide.

To get some idea of the way problems like these are being approached, let us look at just a few of the other alternatives to direct alkylation.

26.6 Synthesis of acids and esters via 2-oxazolines

Reaction of a carboxylic acid with 2-amino-2-methyl-1-propanol yields a heterocyclic compound called a *2-oxazoline* (I). From this compound the acid can be regenerated, in the form of its ethyl ester, by ethanolysis.

$$RCH_2COOH + HOCH_2C(CH_3)_2NH_2 \longrightarrow \underset{\text{I}}{RCH_2\text{-oxazoline}(4,4\text{-}(CH_3)_2)} \xrightarrow[H_2SO_4]{EtOH} RCH_2COOEt$$

2-Amino-2-methyl-1-propanol — A 2-alkyl-4,4-dimethyl-2-oxazoline

Using this way to protect the carboxyl group, A. I. Meyers (Colorado State University) has recently opened an elegant route to alkylated acetic acids—or, by modification along Reformatsky lines, to β-hydroxy esters.

$$RCH_2COOH \xrightarrow{(CH_3)_2C(NH_2)CH_2OH} \underset{\text{I}}{RCH_2\text{-oxazoline}} \xrightarrow{n\text{-BuLi}} \underset{\text{II}}{RCH^-Li^+\text{-oxazoline}}$$

$$\xrightarrow{R'X} RR'CH\text{-oxazoline} \xrightarrow[H_2SO_4]{EtOH} RR'CHCOOH$$

Treatment of the 2-oxazoline with the strong base, *n*-butyllithium, yields the lithio derivative II. This, like sodiomalonic ester, can be alkylated and, if desired, re-alkylated—up to a total of *two* substituents on the α-carbon. Ethanolysis of the new 2-oxazoline yields the substituted ester.

The synthesis depends on: (a) the ease of formation and hydrolysis of 2-oxazolines; (b) the fact that the α-hydrogens retain their acidity in the oxazoline (*Why?*); and (c) the inertness of the 2-oxazoline ring toward the lithio derivative. (The ring is inert toward the Grignard reagent as well, and can be used to protect the carboxyl group in a wide variety of syntheses.)

Problem 26.16 Using the Meyers oxazoline method, outline all steps in the synthesis of: (a) *n*-butyric acid from acetic acid; (b) isobutyric acid from acetic acid; (c) isobutyric acid from propionic acid; (d) β-phenylpropionic acid from acetic acid.

Problem 26.17 (a) Give structural formulas of compounds A and B.

Oxazoline I (R = H) + *n*-BuLi, then $CH_3(CH_2)_5CHO \longrightarrow$ A

A + EtOH, $H_2SO_4 \longrightarrow$ B ($C_{11}H_{22}O_3$)

(b) Outline all steps in the synthesis of ethyl 3-(*n*-propyl)-3-hydroxyhexanoate. (c) Of ethyl 2-ethyl-3-phenyl-3-hydroxypropanoate.

Problem 26.18 (a) Give structural formulas of compounds C–E.

4-hydroxycyclohexanecarboxylic acid + $(CH_3)_2C(NH_2)CH_2OH \longrightarrow$ C ($C_{11}H_{19}O_2N$)

C + CrO_3/pyridine $\longrightarrow$ D ($C_{11}H_{17}O_2N$)

D + C_6H_5MgBr, then C_2H_5OH, $H_2SO_4 \longrightarrow$ E ($C_{15}H_{18}O_2$)

(b) Using benzene, toluene, and any needed aliphatic and inorganic reagents, how would you make $C_6H_5COCH_2CH_2COOH$? (*Hint:* See Sec. 20.10.) (c) Now, how would you make $C_6H_5C(C_2H_5){=}CHCH_2COOH$? (d) Outline a possible synthesis of *p*-$CH_3CH_2CHOHC_6H_4COOC_2H_5$. (e) Of $C_6H_5CHOHC_6H_4COOC_2H_5$-*p*.

26.7 Organoborane synthesis of acids and ketones

Hydroboration of alkenes yields alkylboranes, and these, we have seen (Sec. 15.9), can be converted through oxidation into alcohols. But oxidation is only one of many reactions undergone by alkylboranes. Since the discovery of hydroboration in 1957, H. C. Brown and his co-workers (p. 507) have shown that alkylboranes are perhaps the most versatile class of organic reagents known.

In the presence of base, alkylboranes react with bromoacetone to yield alkylacetones, and with ethyl bromoacetate to yield ethyl alkylacetates.

$$\underset{\text{Bromoacetone}}{R_3B + BrCH_2COCH_3} \xrightarrow{\text{base}} \underset{\text{An alkylacetone}}{RCH_2COCH_3}$$

$$\underset{\text{Ethyl bromoacetate}}{R_3B + BrCH_2COOEt} \xrightarrow{\text{base}} \underset{\text{An ethyl alkylacetate}}{RCH_2COOEt}$$

The following mechanism has been postulated, illustrated for reaction with bromoacetone. Base abstracts (1) a proton—one that is *alpha* both to the carbonyl group and to bromine—to give the carbanion I. Being a strong base, carbanion I

$$\text{(1)} \qquad \text{Base:} + CH_2BrCOCH_3 \rightleftharpoons \underset{\text{I}}{^-CHBrCOCH_3} + \text{Base:H}$$

$$\text{(2)} \qquad R_3B + \underset{\text{I}}{^-CHBrCOCH_3} \longrightarrow \underset{\text{II}}{R_3\overset{\ominus}{B}{:}CHBrCOCH_3}$$

$$\text{(3)} \qquad \underset{\text{II}}{R{-}\overset{\ominus}{\underset{R}{\overset{R}{B}}}{-}\underset{Br}{CH}COCH_3} \longrightarrow \underset{\text{III}}{R{-}\underset{R}{\overset{}{B}}{-}\overset{R}{CH}COCH_3} + Br^-$$

$$\text{(4)} \qquad \underset{\text{III}}{R_2B\text{—}\overset{\overset{\textstyle R}{|}}{C}HCOCH_3} + \text{Base:H} \longrightarrow RCH_2COCH_3 + R_2B\text{:Base}$$

combines (2) with the (Lewis) acidic alkylborane to give II. Intermediate II now rearranges (3) with loss of halide ion to form III. Finally, III undergoes (4) protonolysis (a Lowry-Brønsted acid-base reaction this time) to yield the alkylated ketone.

The key step is (3), in which a new carbon–carbon bond is formed. In II, boron carries a negative charge. Made mobile by this negative charge, and attracted by the adjacent carbon holding a good leaving group, an alkyl group migrates to this carbon—taking its electrons along—and displaces the weakly basic halide ion.

We have, then, three acid-base reactions and a 1,2-alkyl shift: all familiar reaction types. Step (1) involves formation of a carbanion; step (3) involves intramolecular nucleophilic (S_N2) attack by a carbanion-like alkyl group; and step (4) involves attachment of a proton to a carbanion or a carbanion-like moiety.

Protonolysis of alkylboranes is much more difficult than protonolysis of, say, Grignard reagents. The course of reaction (4) is evidently not equilibrium-controlled, but rate-controlled: it is not the stronger base, $R\!:^-$, that gets the proton, but instead the resonance-stabilized carbanion $[RCHCOCH_3]^-$.

Problem 26.19 Trialkylboranes are inert to water, but are particularly prone to protonolysis by carboxylic acids. Can you suggest a specific mechanism for protonolysis of R_3B by a carboxylic acid?

As a synthetic route, this organoborane synthesis parallels the acetoacetic ester and malonic ester syntheses. An acetone unit is furnished by acetoacetic ester or, here, by bromoacetone; an acetic acid unit is furnished by malonic ester or, here, by bromoacetic ester. In these syntheses, bromine plays the same part that the —COOEt group did: by increasing the acidity of certain α-hydrogens, it determines *where* in the molecule reaction will take place; it is easily lost from the molecule when its job is done. Unlike the loss of —COOEt, the departure of —Br is an integral part of the alkylation process.

Consistently high yields depend on the proper selection of reagents. In general, the best base is the bulky potassium 2,6-di-*tert*-butylphenoxide. The best alkylating agent is B-alkyl-9-borabicyclo[3.3.1]nonane, or "B-alkyl-9-BBN," available via successive hydroborations of alkenes:

$$\text{1,5-Cyclooctadiene} \xrightarrow{(BH_3)_2} \text{9-Borabicyclo[3.3.1]nonane (9-BBN) (B—H)} \xrightarrow{RCH{=}CH_2} \text{B-Alkyl-9-borabicyclo[3.3.1]nonane (B—}CH_2CH_2R\text{)}$$

1,5-Cyclooctadiene | 9-Borabicyclo[3.3.1]nonane (9-BBN) *As dimer* | B-Alkyl-9-borabicyclo[3.3.1]nonane (B-Alkyl-9-BBN)

The overall sequence thus amounts to the conversion of alkenes into ketones and esters. For example:

$$(CH_3)_2C{=}CH_2 \xrightarrow{\text{9-BBN}} \text{B}{-}CH_2CH(CH_3)_2 \text{ (9-BBN)} \xrightarrow{BrCH_2COCH_3} (CH_3)_2CHCH_2{-}CH_2COCH_3$$

Isobutylene B-Isobutyl-BBN 5-Methyl-2-hexanone

$$\text{Cyclopentene} \xrightarrow{\text{9-BBN}} \text{B-Cyclopentyl-9-BBN} \xrightarrow[\text{base}]{BrCH_2COOEt} C_5H_9{-}CH_2COOEt$$

Cyclopentene Ethyl cyclopentylacetate

Besides bromoacetone, other bromomethyl ketones ($BrCH_2COR$) can be used if they are available. Bromination is best carried out with cupric bromide as the reagent, and on ketones in which R contains no α-hydrogens to compete with those on methyl: acetophenone, for example, or methyl *tert*-butyl ketone.

Problem 26.20 Using 9-BBN plus any alkenes and unhalogenated acids or ketones, outline all steps in the synthesis of: (a) 2-heptanone; (b) 4-methylpentanoic acid; (c) 4-methyl-2-hexanone; (d) 1-cyclohexyl-2-propanone; (e) ethyl (*trans*-2-methylcyclopentyl)acetate; (f) 1-phenyl-4-methyl-1-pentanone; (g) 1-cyclopentyl-3,3-dimethyl-2-butanone.

26.8 Alkylation of carbonyl compounds via enamines

As we might expect, amines react with carbonyl compounds by nucleophilic addition. If the amine is *primary*, the initial addition product undergoes dehydration (compare Sec. 19.14) to form a compound containing a carbon–nitrogen

$$\gt C{=}O + H_2NR' \longrightarrow -\overset{|}{\underset{OH}{\underset{|}{C}}}-NHR' \longrightarrow \gt C{=}NR'$$

A 1° amine An imine

double bond, an *imine*. Elimination occurs with this orientation even if the carbonyl compound contains an α-hydrogen: that is, the preferred product is the

$$-\underset{H}{\underset{|}{\overset{|}{C}}}-\underset{OH}{\underset{|}{\overset{|}{C}}}-\underset{H}{\underset{|}{N}}-R'$$

$$-\overset{|}{C}{=}\overset{|}{C}-\underset{H}{\underset{|}{N}}-R' \rightleftarrows -\underset{H}{\underset{|}{\overset{|}{C}}}-\overset{|}{C}{=}N-R' \qquad \textbf{Imine-enamine tautomerism}$$

Enamine Imine *More stable form*

imine rather than the *enamine* (*ene* for the carbon–carbon double bond, *amine* for the amino group). If some enamine should be formed initially it rapidly tautomerizes into the more stable imino form.

The system is strictly analogous to the keto–enol one (Secs. 8.13 and 21.4). The proton is acidic, and therefore separates fairly readily from the hybrid anion; it can return to either carbon or nitrogen, but when it returns to carbon, it tends to stay there. Equilibrium favors formation of the weaker acid.

Now, a secondary amine, too, can react with a carbonyl compound, and to yield the same kind of initial product. But here there is no hydrogen left on nitrogen; if dehydration is to occur, it must be in the other direction, to form a carbon–carbon double bond. A stable enamine is the product.

$$\underset{\text{H}}{\overset{|}{-\text{C}}}-\overset{|}{\text{C}}=\text{O} + \underset{\text{A 2° amine}}{\text{R}'_2\text{NH}} \longrightarrow \underset{\text{H}}{\overset{|}{-\text{C}}}-\underset{\text{OH}}{\overset{|}{\text{C}}}-\overset{\text{R}'}{\text{N}}-\text{R}' \longrightarrow \underset{\text{An enamine}}{-\overset{|}{\text{C}}=\overset{|}{\text{C}}-\overset{\text{R}'}{\text{N}}-\text{R}'}$$

In 1954 Gilbert Stork (of Columbia University) showed how enamines could be used in the alkylation and acylation of aldehydes and ketones, and in the years since then enamines have been intensively studied and used in organic synthesis in a wide variety of ways. All we can do here is to try to understand a little of the basic chemistry underlying the use of enamines.

The usefulness of enamines stems from the fact that they contain *nucleophilic carbon*. The electrons responsible for this nucleophilicity are, in the final analysis, the (formally) unshared pair on nitrogen; but they are available for nucleophilic attack by carbon of the enamine. Thus, in alkylation:

$$-\overset{|}{\text{C}}=\overset{|}{\text{C}}-\ddot{\text{N}}\text{R}'_2 \quad \text{X}-\text{R} \longrightarrow \underset{\text{An iminium ion}}{\underset{\text{R}}{\overset{|}{-\text{C}}}-\overset{|}{\text{C}}=\overset{\oplus}{\text{N}}\text{R}'_2} + \text{X}^-$$

The product of alkylation is an iminium ion, which is readily hydrolyzed to regenerate the carbonyl group. The overall process, then, is:

$$\underset{\text{Ketone}}{\underset{\text{H}}{\overset{|}{-\text{C}}}-\overset{|}{\text{C}}=\text{O}} \xrightarrow{\text{R}'_2\text{NH}} \underset{\text{Enamine}}{-\overset{|}{\text{C}}=\overset{|}{\text{C}}-\text{NR}'_2} \xrightarrow{\text{RX}} \underset{\text{Iminium ion}}{\underset{\text{R}}{\overset{|}{-\text{C}}}-\overset{|}{\text{C}}=\overset{\oplus}{\text{N}}\text{R}'_2} \xrightarrow{\text{H}_2\text{O, H}^+} \underset{\text{Alkylated ketone}}{\underset{\text{R}}{\overset{|}{-\text{C}}}-\overset{|}{\text{C}}=\text{O}}$$

(In enamines the nitrogen too, is nucleophilic, but attack there, which yields quaternary *ammo*nium ions, is generally an unwanted side-reaction. Heating often converts N-alkylated compounds into the desired C-alkylated products.)

Nitrogen in enamines plays the same role it does in the chemistry of aromatic amines—not surprisingly, when we realize that enamines are, after all, *vinyl amines*. (Remember the similarities between vinyl and aryl halides.) For example, bromination of aniline involves, we say, electrophilic attack by bromine on the aromatic ring; but from the opposite, and equally valid, point of view, it involves nucleophilic attack on bromine by carbons of the ring—with nitrogen furnishing the electrons.

Commonly used secondary amines are the heterocyclic compounds *pyrrolidine* and *morpholine*:

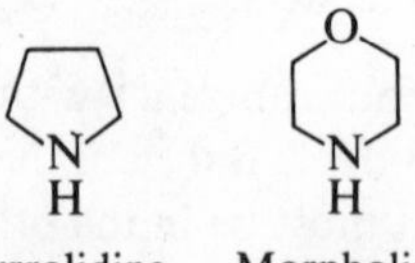

Pyrrolidine Morpholine

Best yields are obtained with reactive halides like benzyl and allyl halides, α-halo esters, and α-halo ketones. For example:

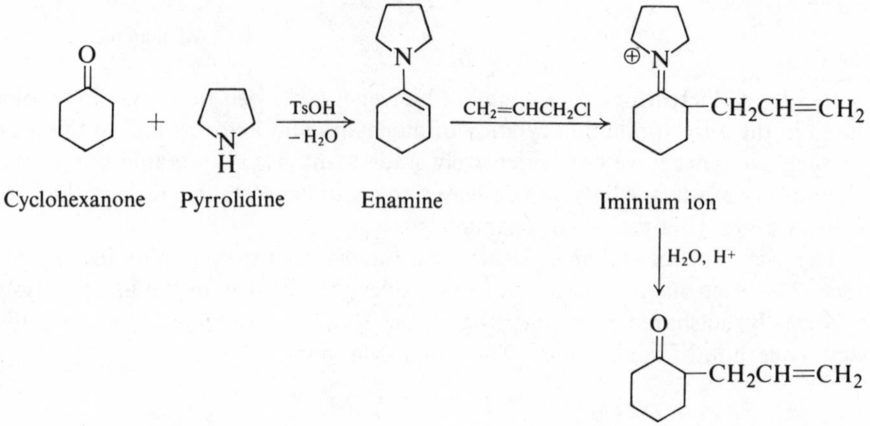

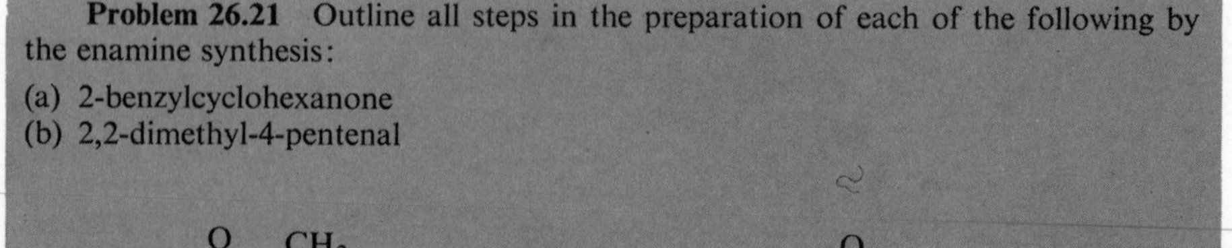

Cyclohexanone + Pyrrolidine $\xrightarrow[-H_2O]{TsOH}$ Enamine $\xrightarrow{CH_2{=}CHCH_2Cl}$ Iminium ion (2-$CH_2CH{=}CH_2$) $\xrightarrow{H_2O,\ H^+}$ 2-$CH_2CH{=}CH_2$-cyclohexanone

β-Tetralone $\xrightarrow{\text{Pyrrolidine}}$ Enamine (NR_2) $\xrightarrow{BrCH_2COOEt}$ $\xrightarrow[H^+]{H_2O}$ 1-CH_2COOEt-2-tetralone

Problem 26.21 Outline all steps in the preparation of each of the following by the enamine synthesis:

(a) 2-benzylcyclohexanone
(b) 2,2-dimethyl-4-pentenal
(c) 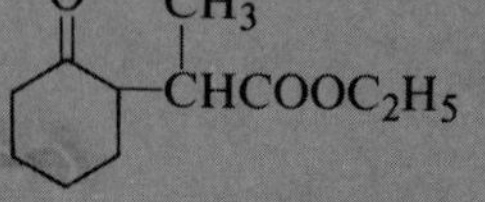 2-($CHCOOC_2H_5$ with CH_3)cyclohexanone
(d) 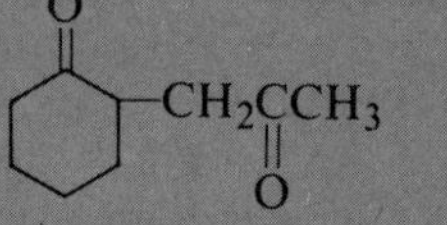 2-(CH_2CCH_3, C=O)cyclohexanone
(e) 2-(2,4-dinitrophenyl)cyclohexanone
(f) 2,2-dimethyl-3-oxobutanal, $CH_3COC(CH_3)_2CHO$

Problem 26.22 Give structural formulas of compounds A–F.

(a) cyclopentanone + morpholine, then TsOH $\longrightarrow$ A ($C_9H_{15}ON$)
A + C_6H_5CHO, then H_2O, H^+ $\longrightarrow$ B ($C_{12}H_{12}O$)
(b) isobutyraldehyde + *tert*-butylamine $\longrightarrow$ C ($C_8H_{17}N$)
C + C_2H_5MgBr $\longrightarrow$ D ($C_8H_{16}NMgBr$) + E
D + $C_6H_5CH_2Cl$, then H_2O, H^+ $\longrightarrow$ F ($C_{11}H_{14}O$)

PROBLEMS

1. Outline the synthesis of each of the following from malonic ester and any other reagents:

(a) *n*-caproic acid
(b) isobutyric acid
(c) β-methylbutyric acid
(d) α,β-dimethylbutyric acid
(e) 2-ethylbutanoic acid
(f) dibenzylacetic acid
(g) α,β-dimethylsuccinic acid
(h) glutaric acid
(i) cyclobutanecarboxylic acid

2. Outline the synthesis of each of the following from acetoacetic ester and any other needed reagents. Do (j)–(m) after Problem 11, below.

(a) methyl ethyl ketone
(b) 3-ethyl-2-pentanone
(c) 3-ethyl-2-hexanone
(d) 5-methyl-2-heptanone
(e) 3,6-dimethyl-2-heptanone
(f) 4-oxo-2-methylpentanoic acid
(g) γ-hydroxy-*n*-valeric acid
(h) 3-methyl-2-hexanol
(i) 2,5-dimethylheptane
(j) β-methylcaproic acid
(k) β-methylbutyric acid
(l) methylsuccinic acid
(m) 2,5-hexanediol

3. What product would you expect from the hydrolysis by dilute alkali of 2-carbethoxycyclopentanone (see Problem 21.30, p. 718)? Suggest a method of synthesis of 2-methylcyclopentanone.

4. Give structures of compounds A through J:

(a) 1,3-dibromopropane + 2 moles sodiomalonic ester $\longrightarrow$ A ($C_{17}H_{28}O_8$)
A + 2 moles sodium ethoxide, then CH_2I_2 $\longrightarrow$ B ($C_{18}H_{28}O_8$)
B + OH^-, heat; then H^+; then heat $\longrightarrow$ C ($C_8H_{12}O_4$)
(b) ethylene bromide + 2 moles sodiomalonic ester $\longrightarrow$ D ($C_{16}H_{26}O_8$)
D + 2 moles sodium ethoxide, then 1 mole ethylene bromide $\longrightarrow$ E ($C_{18}H_{28}O_8$)
E + OH^-, heat; then H^+; then heat $\longrightarrow$ F ($C_8H_{12}O_4$)
(c) 2 moles sodiomalonic ester + I_2 $\longrightarrow$ G ($C_{14}H_{22}O_8$) + 2NaI
G + OH^-, heat; then H^+; then heat $\longrightarrow$ H ($C_4H_6O_4$)
(d) D + 2 moles sodium ethoxide, then I_2 $\longrightarrow$ I ($C_{16}H_{24}O_8$)
I + OH^-, heat; then H^+; then heat $\longrightarrow$ J ($C_6H_8O_4$)
(e) Suggest a possible synthesis for 1,3-cyclopentanedicarboxylic acid; for 1,2-cyclopentanedicarboxylic acid; for 1,1-cyclopentanedicarboxylic acid.

5. Give structures of compounds K through O:

allyl bromide + Mg $\longrightarrow$ K (C_6H_{10})
K + HBr $\longrightarrow$ L ($C_6H_{12}Br_2$)
sodiomalonic ester + excess L $\longrightarrow$ M ($C_{13}H_{23}O_4Br$)
M + sodium ethoxide $\longrightarrow$ N ($C_{13}H_{22}O_4$)
N + OH^-, heat; then H^+; then heat $\longrightarrow$ O ($C_8H_{14}O_2$)

6. When sodium trichloroacetate is heated in diglyme solution with alkenes, there are formed 1,1-dichlorocyclopropanes. How do you account for this?

7. (a) How could you synthesize 2,7-octanedione? (*Hint:* See Problem 26.2, p. 850). (b) Actually, the expected ketone reacts further to give

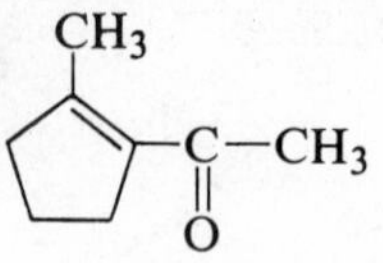

$$\text{1-acetyl-2-methylcyclopentene: cyclopentene ring bearing } CH_3 \text{ and } -\overset{\displaystyle O}{\overset{\|}{C}}-CH_3$$

How does this last reaction occur? To what general types does it belong? (c) How could you synthesize 2,6-heptanedione? (d) What would happen to this ketone under the conditions of (b)?

8. Outline all steps in a possible synthesis of each of the following from simple esters:

(a) 1,2-cyclopentanedione (*Hint:* See Problem 21.33, p. 719–720.)
(b) $CH_3CH_2CH_2COCOOC_2H_5$ (*Hint:* See Problem 26.9, p. 853.)

9. Outline the synthesis from readily available compounds of the following hypnotics (see Sec. 20.23):

(a) 5,5-diethylbarbituric acid (Barbital, Veronal; long-acting)
(b) 5-allyl-5-(2-pentyl)barbituric acid (Seconal; short-acting)
(c) 5-ethyl-5-isopentylbarbituric acid (Amytal; intermediate length of action)

10. (a) Contrast the structures of barbituric acid and Veronal (5,5-diethylbarbituric acid). (b) Account for the appreciable acidity ($K_a = 10^{-8}$) of Veronal.

11. When treated with *concentrated* alkali, acetoacetic ester is converted into two moles of sodium acetate. (a) Outline all steps in a likely mechanism for this reaction. (*Hint:* See Sec. 21.11 and Problem 5.8, p. 170.) (b) Substituted acetoacetic esters also undergo this reaction. Outline the steps in a general synthetic route from acetoacetic ester to carboxylic acids. (c) Outline the steps in the synthesis of 2-hexanone via acetoacetic ester. What acids will be formed as by-products? Outline a procedure for purification of the desired ketone. (Remember that the alkylation is carried out in alcohol; that NaBr is formed; that aqueous base is used for hydrolysis; and that ethyl alcohol is a product of the hydrolysis.)

12. (a) Suggest a mechanism for the alkaline cleavage of β-diketones, as, for example:

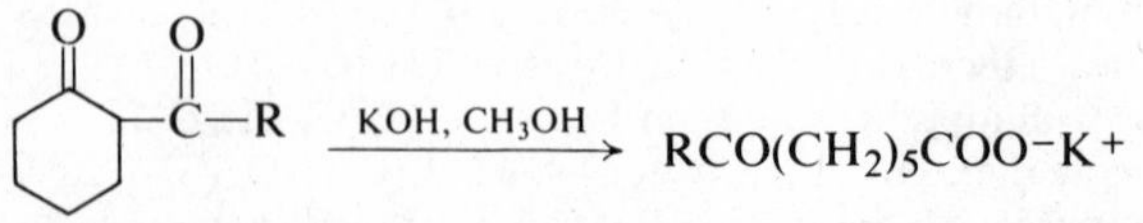

$$\text{2-(RCO)cyclohexanone} \xrightarrow{\text{KOH, } CH_3OH} RCO(CH_2)_5COO^-K^+$$

(b) Starting from cyclohexanone, and using any other needed reagents, outline all steps in a possible synthesis of 7-phenylheptanoic acid. (c) Of pentadecanedioic acid, $HOOC(CH_2)_{13}COOH$.

13. Give structures of compounds P through S:

heptanal (heptaldehyde) + ethyl bromoacetate + Zn, then $H_2O \longrightarrow$ P ($C_{11}H_{22}O_3$)
P + CrO_3 in glacial acetic acid $\longrightarrow$ Q ($C_{11}H_{20}O_3$)
Q + sodium ethoxide, then benzyl chloride $\longrightarrow$ R ($C_{18}H_{26}O_3$)
R + OH^-, heat; then H^+, warm $\longrightarrow$ S ($C_{15}H_{22}O$)

14. Treatment of 1,5-cyclooctadiene with diborane gives a material, T, which is oxidized by alkaline H_2O_2 to a mixture of 72% *cis*-1,5-cyclooctanediol and 28% *cis*-1,4-cyclooctanediol. If T is refluxed for an hour in THF solution (or simply distilled), there is obtained a white crystalline solid, U, which is oxidized to 99%-pure *cis*-1,5-cyclooctanediol.

(a) What is T? What is U? (b) Account for the conversion of T into U.

15. On treatment with concentrated KOH, 2,6-dichlorobenzaldehyde is converted into 1,3-dichlorobenzene and potassium formate. The kinetics shows that the aldehyde and two moles of hydroxide ion are in equilibrium with a reactive intermediate that (ultimately) yields product. (a) Outline a likely mechanism that is consistent with these facts. (*Hint:* See Sec. 19.16.) (b) How do you account for the difference in behavior between this aldehyde and most aromatic aldehydes under these conditions?

16. Give structural formulas of compounds V and W, and tell *exactly* how each is formed:

γ-butyrolactone + CH_3ONa $\longrightarrow$ V ($C_8H_{10}O_3$)
V + conc. HCl $\longrightarrow$ W ($C_7H_{12}OCl_2$)
W + aq. NaOH $\longrightarrow$ dicyclopropyl ketone

17. The structure of *nerolidol*, $C_{15}H_{26}O$, a terpene found in oil of neroli, was established by the following synthesis:

geranyl chloride (RCl) + sodioacetoacetic ester $\longrightarrow$ X ($RC_6H_9O_3$)
X + $Ba(OH)_2$, then H^+, warm $\longrightarrow$ Y (RC_3H_5O)
Y + NaC≡CH, then H_2O $\longrightarrow$ Z (RC_5H_7O)
Z $\xrightarrow{\text{reduction}}$ AA (RC_5H_9O), nerolidol

(a) Give the structure of nerolidol, using R for the geranyl group.
(b) Referring to Problem 27, p. 547, what is the complete structure of nerolidol?

18. The structure of *menthone*, $C_{10}H_{18}O$, a terpene found in peppermint oil, was first established by synthesis in the following way:

ethyl β-methylpimelate + sodium ethoxide, then H_2O $\longrightarrow$ BB ($C_{10}H_{16}O_3$)
BB + sodium ethoxide, then isopropyl iodide $\longrightarrow$ CC ($C_{13}H_{22}O_3$)
CC + OH^-, heat; then H^+; then heat $\longrightarrow$ menthone

(a) What structures for menthone are consistent with this synthesis? (b) On the basis of the isoprene rule (Sec. 8.26) which structure is the more likely? (c) On vigorous reduction menthone yields *p-menthane*, 4-isopropyl-1-methylcyclohexane. Now what structure or structures are most likely for menthone?

19. The structure of *camphoronic acid* (a degradation product of the terpene camphor) was established by the following synthesis:

sodioacetoacetic ester + CH_3I $\longrightarrow$ DD $\xrightarrow{NaOC_2H_5}$ $\xrightarrow{CH_3I}$ EE ($C_8H_{14}O_3$)
EE + ethyl bromoacetate + Zn, then H_2O $\longrightarrow$ FF ($C_{12}H_{22}O_5$)
FF + PCl_5, then KCN $\longrightarrow$ GG ($C_{13}H_{21}O_4N$)
GG + H_2O, H^+, heat $\longrightarrow$ camphoronic acid ($C_9H_{14}O_6$)

What is the structure of camphoronic acid?

20. Two of the oxidation products of the terpene α-terpineol are *terebic acid* and *terpenylic acid.* Their structures were first established by the following synthesis:

ethyl chloroacetate + sodioacetoacetic ester $\longrightarrow$ HH ($C_{10}H_{16}O_5$)
HH + one mole CH_3MgI, then H_2O $\longrightarrow$ II ($C_{11}H_{20}O_5$)
II + OH^-, H_2O, heat, then H^+ $\longrightarrow$ [JJ ($C_7H_{12}O_5$)] $\longrightarrow$ terebic acid ($C_7H_{10}O_4$)
HH + sodium ethoxide, then ethyl chloroacetate $\longrightarrow$ KK ($C_{14}H_{22}O_7$)
KK + OH^-, then H^+, warm $\longrightarrow$ LL ($C_7H_{10}O_5$)
LL + ethyl alcohol, H^+ $\longrightarrow$ MM ($C_{11}H_{18}O_5$)

MM + one mole CH_3MgI, then H_2O $\longrightarrow$ NN ($C_{12}H_{22}O_5$)
NN + OH^-, H_2O, heat, then H^+ $\longrightarrow$ [OO ($C_8H_{14}O_5$)] $\longrightarrow$ terpenylic acid ($C_8H_{12}O_4$)

What is the structure of terebic acid? Of terpenylic acid?

21. Isopentenyl pyrophosphate, the precursor of isoprene units in nature (Sec. 8.26 and Problem 27, p. 282), is formed enzymatically from the pyrophosphate of *mevalonic acid* by the action of ATP (adenosine triphosphate) and Mn^{++} ion.

$$\underset{\text{Mevalonic acid 5-pyrophosphate}}{CH_3-\overset{\overset{CH_2-OPP}{|}\atop{CH_2}}{\underset{CH_2COOH}{C}}-OH} \xrightarrow[Mn^{++}]{ATP} \underset{\text{Mevalonic acid 5-pyrophosphate 3-phosphate}}{CH_3-\overset{\overset{CH_2-OPP}{|}\atop{CH_2}}{\underset{CH_2COOH}{C}}-OPO_3H_2} \longrightarrow \underset{\text{Isopentenyl pyrophosphate}}{CH_3-\overset{\overset{CH_2-OPP}{|}\atop{CH_2}}{C}=CH_2}$$

It is believed that the function of ATP is to phosphorylate mevalonic acid pyrophosphate at the 3-position.

Just what happens in the last step of this conversion? Why should the 3-phosphate undergo this reaction more easily than the 3-hydroxy compound?

Chapter 27

α,β-Unsaturated Carbonyl Compounds

Conjugate Addition

27.1 Structure and properties

In general, a compound that contains both a carbon–carbon double bond and a carbon–oxygen double bond has properties that are characteristic of both functional groups. At the carbon–carbon double bond an unsaturated ester or unsaturated ketone undergoes electrophilic addition of acids and halogens, hydrogenation, hydroxylation, and cleavage; at the carbonyl group it undergoes the nucleophilic substitution typical of an ester or the nucleophilic addition typical of a ketone.

Problem 27.1 What will be the products of the following reactions?

(a) $CH_3CH{=}CHCOOH + H_2 + Pt$
(b) $CH_3CH{=}CHCOOC_2H_5 + OH^- + H_2O$ + heat
(c) $C_6H_5CH{=}CHCOCH_3 + I_2 + OH^-$
(d) $CH_3CH{=}CHCHO + C_6H_5NHNH_2$ + acid catalyst
(e) $CH_3CH{=}CHCHO + Ag(NH_3)_2^+$
(f) $C_6H_5CH{=}CHCOC_6H_5 + O_3$, followed by $Zn + H_2O$
(g) $CH_3CH{=}CHCHO$ + excess H_2 + Ni, heat, pressure
(h) *trans*-$HOOCCH{=}CHCOOH + Br_2/CCl_4$
(i) *trans*-$HOOCCH{=}CHCOOH$ + cold alkaline $KMnO_4$

Problem 27.2 What are A, B, and C, given the following facts?

(a) Cinnamaldehyde ($C_6H_5CH{=}CHCHO$) + H_2 + Ni, at low temperatures and pressures ⟶ A.
(b) Cinnamaldehyde + H_2 + Ni, at high temperatures and pressures ⟶ B.
(c) Cinnamaldehyde + $NaBH_4$, followed by H^+ ⟶ C.

	A	B	C
$KMnO_4$ test	positive	negative	positive
Br_2/CCl_4 test	negative	negative	positive
Tollens' test	positive	negative	negative
$NaHSO_3$ test	positive	negative	negative

In the α,β-unsaturated carbonyl compounds, the carbon–carbon double bond and the carbon–oxygen double bond are separated by just one carbon–carbon single bond; that is, the double bonds are *conjugated.* Because of this conjugation,

$$\overset{\beta}{-\underset{|}{C}}=\overset{\alpha}{\underset{|}{C}}-\underset{|}{C}=O$$

α,β-Unsaturated carbonyl compound
Conjugated system

such compounds possess not only the properties of the individual functional groups, but certain other properties besides. In this chapter we shall concentrate on the α,β-unsaturated compounds, and on the special reactions characteristic of the conjugated system.

Table 27.1 α,β-UNSATURATED CARBONYL COMPOUNDS

Name	Formula	M.p., °C	B.p., °C
Acrolein	$CH_2{=}CHCHO$	−88	52
Crotonaldehyde	$CH_3CH{=}CHCHO$	−69	104
Cinnamaldehyde	$C_6H_5CH{=}CHCHO$	−7	254
Mesityl oxide	$(CH_3)_2C{=}CHCOCH_3$	42	131
Benzalacetone	$C_6H_5CH{=}CHCOCH_3$	42	261
Dibenzalacetone	$C_6H_5CH{=}CHCOCH{=}CHC_6H_5$	113	
Benzalacetophenone (Chalcone)	$C_6H_5CH{=}CHCOC_6H_5$	62	348
Dypnone	$C_6H_5C(CH_3){=}CHCOC_6H_5$		150–51
Acrylic acid	$CH_2{=}CHCOOH$	12	142
Crotonic acid	*trans*-$CH_3CH{=}CHCOOH$	72	189
Isocrotonic acid	*cis*-$CH_3CH{=}CHCOOH$	16	172*d*
Methacrylic acid	$CH_2{=}C(CH_3)COOH$	16	162
Sorbic acid	$CH_3CH{=}CHCH{=}CHCOOH$	134	
Cinnamic acid	*trans*-$C_6H_5CH{=}CHCOOH$	137	300
Maleic acid	*cis*-$HOOCCH{=}CHCOOH$	130.5	
Fumaric acid	*trans*-$HOOCCH{=}CHCOOH$	302	
Maleic anhydride		60	202
Methyl acrylate	$CH_2{=}CHCOOCH_3$		80
Methyl methacrylate	$CH_2{=}C(CH_3)COOCH_3$		101
Ethyl cinnamate	$C_6H_5CH{=}CHCOOC_2H_5$	12	271
Acrylonitrile	$CH_2{=}CH{-}C{\equiv}N$	−82	79

Table 27.1 lists some of the more important of these compounds. Many have common names which the student must expect to encounter. For example:

$CH_2{=}CH{-}CHO$	$CH_2{=}CH{-}COOH$	$CH_2{=}CH{-}C{\equiv}N$	$CH_2{=}\underset{}{\overset{CH_3}{\overset{\vert}{C}}}{-}COOH$
Acrolein	Acrylic acid	Acrylonitrile	Methacrylic acid
Propenal	Propenoic acid	Propenenitrile	2-Methylpropenoic acid

$CH_3CH{=}CHCHO$
Crotonaldehyde
2-Butenal

$C_6H_5CH{=}CHCHO$
Cinnamaldehyde
3-Phenylpropenal

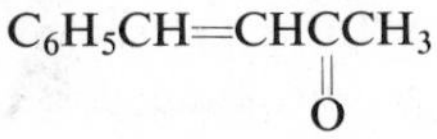

$C_6H_5CH{=}CH\overset{\displaystyle O}{\overset{\|}{C}}CH_3$ (C=O shown below)
Benzalacetone
4-Phenyl-3-buten-2-one

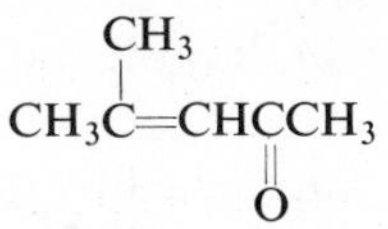

$CH_3\overset{CH_3}{\overset{|}{C}}{=}CHC(=O)CH_3$
Mesityl oxide
4-Methyl-3-penten-2-one

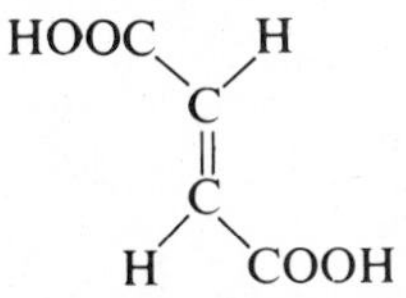

$HOOC(H)C{=}C(H)COOH$ (trans)
Fumaric acid
trans-Butenedioic acid

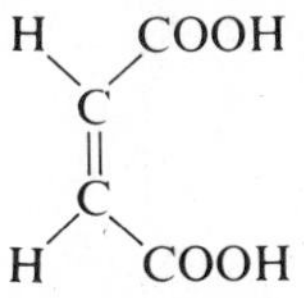

$H(COOH)C{=}C(H)COOH$ (cis)
Maleic acid
cis-Butenedioic acid

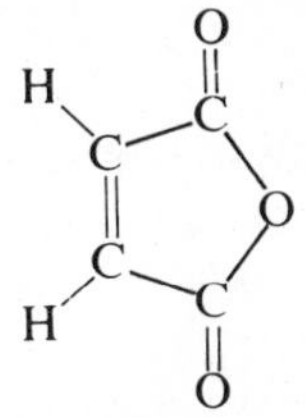

Maleic anhydride
cis-Butenedioic anhydride

27.2 Preparation

There are several general ways to make compounds of this kind: the **aldol condensation**, to make unsaturated aldehydes and ketones; **dehydrohalogenation of α-halo acids** and the **Perkin condensation**, to make unsaturated acids. Besides these, there are certain methods useful only for making single compounds.

All these methods make use of chemistry with which we are already familiar: the fundamental chemistry of alkenes and carbonyl compounds.

Problem 27.3 Outline a possible synthesis of:
(a) crotonaldehyde from acetylene
(b) cinnamaldehyde from compounds of lower carbon number
(c) cinnamic acid from compounds of lower carbon number
(d) 4-methyl-2-pentenoic acid via a malonic ester synthesis

Problem 27.4 The following compounds are of great industrial importance for the manufacture of polymers: acrylonitrile (for Orlon), methyl acrylate (for Acryloid), methyl methacrylate (for Lucite and Plexiglas). Outline a possible industrial synthesis of: (a) acrylonitrile from ethylene; (b) methyl acrylate from ethylene; (c) methyl methacrylate from acetone and methanol.

(d) Polymerization of these compounds is similar to that of ethylene, vinyl chloride, etc. (Sec. 6.19). Draw a structural formula for each of the polymers.

Problem 27.5 Acrolein, $CH_2{=}CHCHO$, is prepared by heating glycerol with sodium hydrogen sulfate, $NaHSO_4$. (a) Outline the likely steps in this synthesis, which involves acid-catalyzed dehydration and keto–enol tautomerization. (*Hint:* Which —OH is easier to eliminate, a primary or a secondary?) (b) How could acrolein be converted into acrylic acid?

27.3 Interaction of functional groups

We have seen (Sec. 6.11) that, with regard to electrophilic addition, a carbon–carbon double bond is activated by an electron-releasing substituent and deactivated by an electron-withdrawing substituent. The carbon–carbon double bond serves as a source of electrons for the electrophilic reagent; the availability of its electrons is determined by the groups attached to it. More specifically, an electron-releasing substituent stabilizes the transition state leading to the initial carbonium ion by dispersing the developing positive charge; an electron-withdrawing substituent destabilizes the transition state by intensifying the positive charge.

Electrophilic Addition

$$-\overset{|}{C}=\overset{|}{C}-G + Y^+ \longrightarrow \left[\begin{array}{c} -\overset{|}{C}\text{---}\overset{|}{C}-G \\ \vdots \quad \delta_+ \\ Y\,\delta_+ \end{array} \right] \longrightarrow -\overset{|}{\underset{Y}{C}}-\overset{|}{\underset{\oplus}{C}}-G$$

G *releases electrons: activates*
G *withdraws electrons: deactivates*

The C=O, —COOH, —COOR, and —CN groups are powerful electron-withdrawing groups, and therefore would be expected to deactivate a carbon–carbon double bond toward electrophilic addition. This is found to be true: α,β-unsaturated ketones, acids, esters, and nitriles are in general less reactive than simple alkenes toward reagents like bromine and the hydrogen halides.

But this powerful electron withdrawal, which deactivates a carbon–carbon double bond toward reagents seeking electrons, at the same time *activates* toward reagents that are electron-rich. As a result, the carbon–carbon double bond of an α,β-unsaturated ketone, acid, ester, or nitrile is susceptible to nucleophilic attack, and undergoes a set of reactions, **nucleophilic addition**, that is uncommon for the simple alkenes.

27.4 Electrophilic addition

The presence of the carbonyl group not only lowers the **reactivity** of the carbon–carbon double bond toward electrophilic addition, but also controls the **orientation** of the addition.

In general, it is observed that addition of an unsymmetrical reagent to an α,β-unsaturated carbonyl compound takes place in such a way that hydrogen becomes attached to the α-carbon and the negative group becomes attached to the β-carbon. For example:

$$\underset{\text{Acrolein}}{CH_2=CH-CHO} + HCl(g) \xrightarrow{-10^\circ} \underset{\beta\text{-Chloropropionaldehyde}}{\underset{Cl}{\underset{|}{CH_2}}-\underset{H}{\underset{|}{CH}}-CHO}$$

$$\underset{\text{Acrylic acid}}{CH_2=CH-COOH} + H_2O \xrightarrow{H_2SO_4,\ 100^\circ} \underset{\beta\text{-Hydroxypropionic acid}}{\underset{OH}{\underset{|}{CH_2}}-\underset{H}{\underset{|}{CH}}-COOH}$$

$$\underset{\text{Crotonic acid}}{CH_3-CH=CH-COOH} + HBr(g) \xrightarrow{20^\circ} \underset{\beta\text{-Bromobutyric acid}}{CH_3-\underset{Br}{\underset{|}{CH}}-\underset{H}{\underset{|}{CH}}-COOH}$$

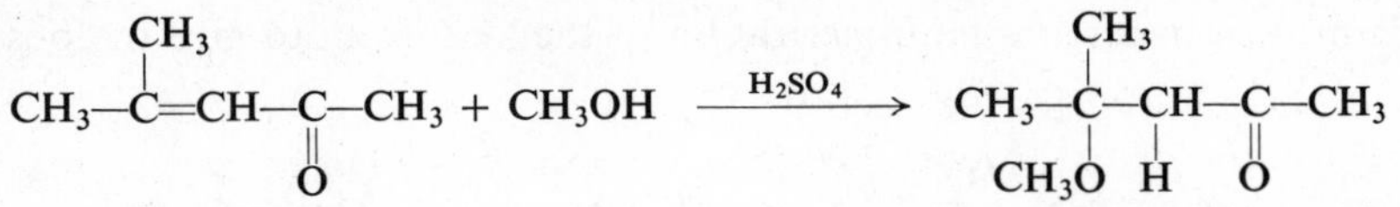

$$\underset{\text{Mesityl oxide}}{CH_3-\underset{CH_3}{\overset{|}{C}}=CH-\underset{\underset{O}{\|}}{C}-CH_3} + CH_3OH \xrightarrow{H_2SO_4} \underset{\text{4-Methoxy-4-methyl-2-pentanone}}{CH_3-\overset{CH_3}{\underset{CH_3O}{C}}-\underset{H}{CH}-\underset{\underset{O}{\|}}{C}-CH_3}$$

Electrophilic addition to simple alkenes takes place in such a way as to form the most stable intermediate carbonium ion. Addition to α,β-unsaturated carbonyl compounds, too, is consistent with this principle; to see that this is so, however, we must look at the conjugated system as a whole. As in the case of conjugated dienes (Sec. 8.20), addition to an *end* of the conjugated system is preferred, since this yields (step 1) a resonance-stabilized carbonium ion. Addition to the carbonyl oxygen end would yield carbonium ion I; addition to the β-carbon end would yield carbonium ion II.

(1) 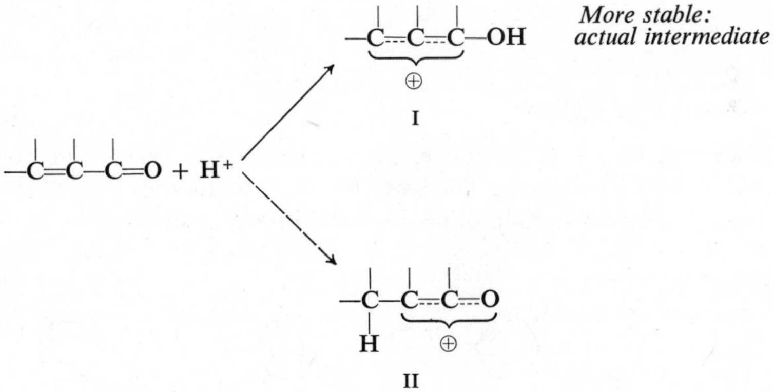 $-C=C-C=O + H^+ \longrightarrow \underset{\text{I}}{\underbrace{-C{\cdots}C{\cdots}C}_{\oplus}-OH}$ *More stable: actual intermediate*

$\longrightarrow \underset{\text{II}}{-\underset{H}{C}-\underbrace{C{\cdots}C{\cdots}O}_{\oplus}}$

Of the two, I is the more stable, since the positive charge is carried by carbon atoms alone, rather than partly by the highly electronegative oxygen atom.

In the second step of addition, a negative ion or basic molecule attaches itself either to the carbonyl carbon or to the β-carbon of the hybrid ion I.

(2) 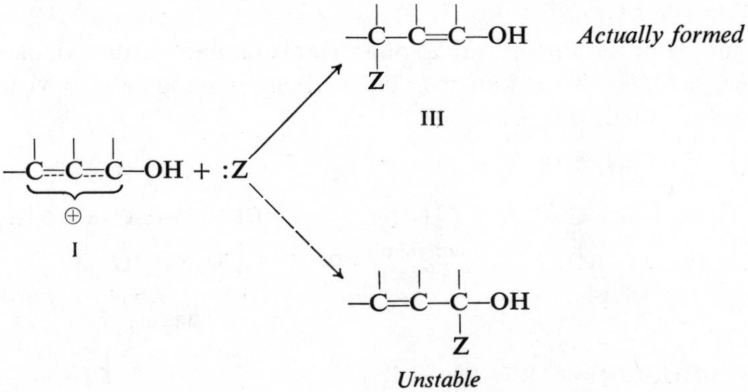 $\underset{\text{I}}{\underbrace{-C{\cdots}C{\cdots}C}_{\oplus}-OH} + :Z \longrightarrow \underset{\text{III}}{-\underset{Z}{C}-C=C-OH}$ *Actually formed*

$\longrightarrow -C=C-\underset{Z}{C}-OH$ *Unstable*

Of the two possibilities, only addition to the β-carbon yields a stable product (III), which is simply the enol form of the saturated carbonyl compound. The enol form then undergoes tautomerization to the keto form to give the observed product (IV).

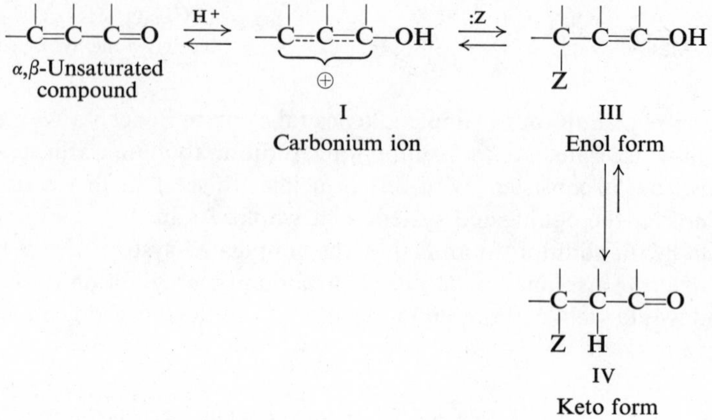

$$-\overset{|}{C}=\overset{|}{C}-\overset{|}{C}=O \underset{}{\overset{H^+}{\rightleftarrows}} \underbrace{-\overset{|}{C}\text{---}\overset{|}{C}\text{---}\overset{|}{C}}_{\oplus}-OH \overset{:Z}{\rightleftarrows} -\underset{Z}{\underset{|}{\overset{|}{C}}}-\overset{|}{C}=\overset{|}{C}-OH$$

α,β-Unsaturated compound; I Carbonium ion; III Enol form

$$\updownarrow$$

$$-\underset{Z}{\underset{|}{\overset{|}{C}}}-\underset{H}{\underset{|}{\overset{|}{C}}}-\overset{|}{C}=O$$

IV Keto form

27.5 Nucleophilic addition

Aqueous sodium cyanide converts α,β-unsaturated carbonyl compounds into β-cyano carbonyl compounds. The reaction amounts to addition of the elements of HCN to the carbon–carbon double bond. For example:

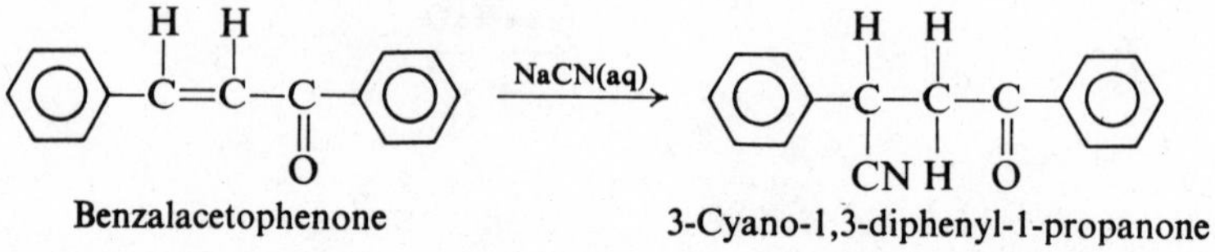

$$C_6H_5-\overset{H}{\overset{|}{C}}=\overset{H}{\overset{|}{C}}-\underset{O}{\underset{\|}{C}}-C_6H_5 \xrightarrow{NaCN(aq)} C_6H_5-\overset{H}{\overset{|}{\underset{CN}{\underset{|}{C}}}}-\overset{H}{\overset{|}{\underset{H}{\underset{|}{C}}}}-\underset{O}{\underset{\|}{C}}-C_6H_5$$

Benzalacetophenone → 3-Cyano-1,3-diphenyl-1-propanone

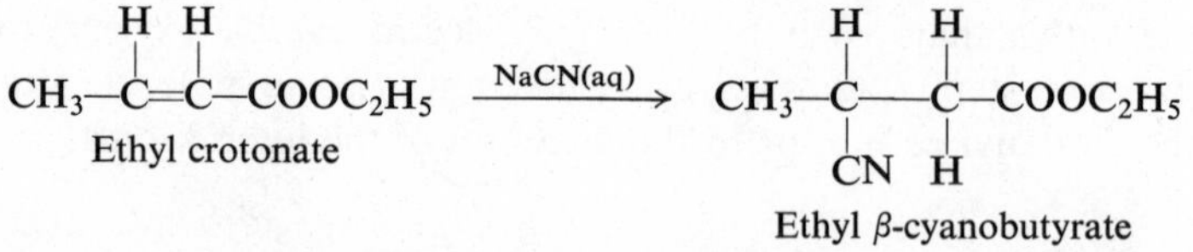

$$CH_3-\overset{H}{\overset{|}{C}}=\overset{H}{\overset{|}{C}}-COOC_2H_5 \xrightarrow{NaCN(aq)} CH_3-\overset{H}{\overset{|}{\underset{CN}{\underset{|}{C}}}}-\overset{H}{\overset{|}{\underset{H}{\underset{|}{C}}}}-COOC_2H_5$$

Ethyl crotonate → Ethyl β-cyanobutyrate

Ammonia or certain derivatives of ammonia (amines, hydroxylamine, phenylhydrazine, etc.) add to α,β-unsaturated carbonyl compounds to yield β-amino carbonyl compounds. For example:

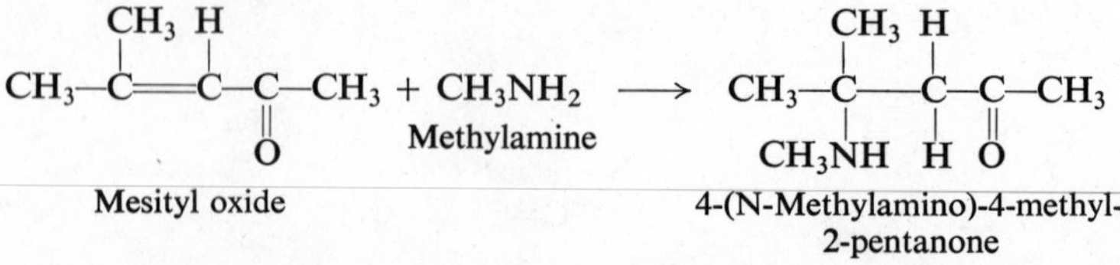

$$CH_3-\overset{CH_3}{\overset{|}{C}}=\overset{H}{\overset{|}{C}}-\underset{O}{\underset{\|}{C}}-CH_3 + CH_3NH_2 \longrightarrow CH_3-\overset{CH_3}{\overset{|}{\underset{CH_3NH}{\underset{|}{C}}}}-\overset{H}{\overset{|}{\underset{H}{\underset{|}{C}}}}-\underset{O}{\underset{\|}{C}}-CH_3$$

Mesityl oxide + Methylamine → 4-(N-Methylamino)-4-methyl-2-pentanone

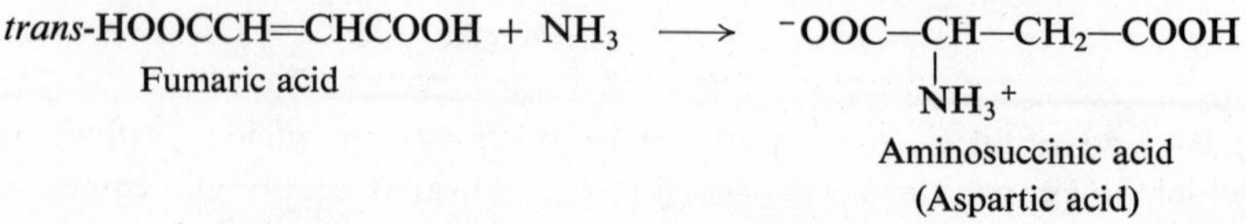

$$trans\text{-}HOOCCH=CHCOOH + NH_3 \longrightarrow {}^-OOC-\underset{NH_3^+}{\underset{|}{CH}}-CH_2-COOH$$

Fumaric acid → Aminosuccinic acid (Aspartic acid)

$$\mathrm{C_6H_5{-}\overset{H}{\underset{}{C}}{=}\overset{H}{C}{-}COOH} + \mathrm{NH_2OH} \longrightarrow \mathrm{C_6H_5{-}\overset{H}{\underset{NHOH}{C}}{-}\overset{H}{\underset{H}{C}}{-}COOH}$$

Cinnamic acid   Hydroxylamine   3-(N-Hydroxylamino)-3-phenylpropanoic acid

These reactions are believed to take place by the following mechanism:

(1) $$-\overset{|}{C}{=}\overset{|}{C}{-}\overset{|}{C}{=}O + :Z \longrightarrow -\underset{Z}{\overset{|}{C}}{-}\underbrace{\overset{|}{C}{\cdots}\overset{|}{C}{\cdots}O}_{\ominus}$$

I

(2) $$-\underset{Z}{\overset{|}{C}}{-}\underbrace{\overset{|}{C}{\cdots}\overset{|}{C}{\cdots}O}_{\ominus} + H^+ \longrightarrow -\underset{Z}{\overset{|}{C}}{-}\overset{|}{C}{=}\overset{|}{C}{-}OH \;\text{(Enol)} \rightleftharpoons -\underset{Z}{\overset{|}{C}}{-}\underset{H}{\overset{|}{C}}{-}\overset{|}{C}{=}O \;\text{(Keto)}$$

I   Enol   Keto

The nucleophilic reagent adds (step 1) to the carbon–carbon double bond to yield the hybrid anion I, which then accepts (step 2) a hydrogen ion from the solvent to yield the final product. This hydrogen ion can add either to the α-carbon or to oxygen, and thus yield either the keto or the enol form of the product; in either case the same equilibrium mixture, chiefly keto, is finally obtained.

In the examples we have just seen, the nucleophilic reagent, :Z, is either the strongly basic anion, $:CN^-$, or a neutral base like ammonia and its derivatives, $:NH_2$—G. These are the same reagents which, we have seen, add to the carbonyl group of simple aldehydes and ketones. (Indeed, nucleophilic reagents rarely add to the carbon–carbon double bond of α,β-unsaturated *aldehydes*, but rather to the highly reactive carbonyl group.)

These nucleophilic reagents add to the conjugated system in such a way as to form the most stable intermediate anion. The most stable anion is I, which is the hybrid of II and III.

$$-\underset{Z}{\overset{|}{C}}{-}\underbrace{\overset{|}{C}{\cdots}\overset{|}{C}{\cdots}O}_{\ominus} \quad \textit{equivalent to} \quad \left[-\underset{Z}{\overset{|}{C}}{-}\underset{\ominus}{\overset{|}{C}}{-}\overset{|}{C}{=}O \quad -\underset{Z}{\overset{|}{C}}{-}\overset{|}{C}{=}\overset{|}{C}{-}O^{\ominus} \right]$$

I   II   III

As usual, initial addition occurs to an *end* of the conjugated system, and in this case to the particular end (β-carbon) *that enables the electronegative element oxygen to accommodate the negative charge.*

The tendency for α,β-unsaturated carbonyl compounds to undergo nucleophilic addition is thus due not simply to the electron-withdrawing ability of the carbonyl group, but to the existence of the conjugated system that permits formation of the resonance-stabilized anion I. The importance in synthesis of α,β-unsaturated aldehydes, ketones, acids, esters, and nitriles is due to the fact that they provide such a conjugated system.

Problem 27.6 Draw structures of the anion expected from nucleophilic addition to each of the other positions in the conjugated system, and compare its stability with that of I.

Problem 27.7 Treatment of crotonic acid, $CH_3CH{=}CHCOOH$, with phenylhydrazine yields compound IV.

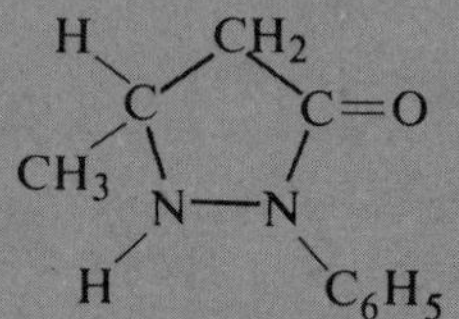

H CH₂ C C=O CH₃ N—N H C₆H₅

IV

To what simple class of compounds does IV belong? How can you account for its formation? (*Hint:* See Sec. 20.11.)

Problem 27.8 Treatment of acrylonitrile, $CH_2{=}CHCN$, with ammonia yields a mixture of two products: β-aminopropionitrile, $H_2NCH_2CH_2CN$, and di(β-cyanoethyl)amine, $NCCH_2CH_2NHCH_2CH_2CN$. How do you account for their formation?

Problem 27.9 Treatment of ethyl acrylate, $CH_2{=}CHCOOC_2H_5$, with methylamine yields $CH_3N(CH_2CH_2COOC_2H_5)_2$. How do you account for its formation?

27.6 Comparison of nucleophilic and electrophilic addition

We can see that nucleophilic addition is closely analogous to electrophilic addition: (a) addition proceeds in two steps; (b) the first and controlling step is the formation of an intermediate ion; (c) both orientation of addition and reactivity are determined by the stability of the intermediate ion, or, more exactly, by the stability of the transition state leading to its formation; (d) this stability depends upon dispersal of the charge.

The difference between nucleophilic and electrophilic addition is, of course, that the intermediate ions have opposite charges: negative in nucleophilic addition, positive in electrophilic addition. As a result, the effects of substituents are exactly opposite. Where an electron-withdrawing group deactivates a carbon–carbon double bond toward electrophilic addition, it activates toward nucleophilic addition. An electron-withdrawing group stabilizes the transition state leading to the formation of an intermediate anion in nucleophilic addition by helping to disperse the developing negative charge:

Nucleophilic addition

$$-\overset{|}{C}=\overset{|}{C}\rightarrow G + :Z \longrightarrow \left[-\overset{|}{\underset{\vdots}{C}}\overset{\delta-}{=}\overset{|}{C}\rightarrow G \atop \ddot{Z} \right] \longrightarrow -\overset{|}{\underset{Z}{C}}-\overset{|}{\underset{\ominus}{\ddot{C}}}\rightarrow G$$

G *withdraws electrons: activates*

Addition to an α,β-unsaturated carbonyl compound can be understood best in terms of an attack on the entire conjugated system. To yield the most stable intermediate ion, this attack must occur at an end of the conjugated system. A nucleophilic reagent attacks at the β-carbon to form an ion in which the negative charge is partly accommodated by the electronegative atom oxygen; an electrophilic reagent attacks oxygen to form a carbonium ion in which the positive charge is accommodated by carbon.

$$-\overset{|}{C}=\overset{|}{C}-\overset{|}{C}=O \xrightarrow{H^+} \underbrace{-\overset{|}{C}\text{---}\overset{|}{C}\text{---}\overset{|}{C}}_{\oplus}-OH \qquad \textit{Electrophilic attack}$$

$$-\overset{|}{C}=\overset{|}{C}-\overset{|}{C}=O \xrightarrow{:Z} -\overset{|}{\underset{Z}{C}}-\underbrace{\overset{|}{C}\text{---}\overset{|}{C}\text{---}O}_{\ominus} \qquad \textit{Nucleophilic attack}$$

27.7 The Michael addition

Of special importance in synthesis is the nucleophilic addition of carbanions to α,β-unsaturated carbonyl compounds known as the **Michael addition**. Like the reactions of carbanions that we studied in the previous chapter, it results in formation of carbon–carbon bonds. For example:

$$\underset{\text{Benzalacetophenone}}{C_6H_5-\overset{H}{\overset{|}{C}}=\overset{H}{\overset{|}{C}}-\underset{O}{\underset{\|}{C}}-C_6H_5} + \underset{\text{Ethyl malonate}}{CH_2(COOC_2H_5)_2} \xrightarrow{\text{piperidine}} C_6H_5-\overset{H}{\overset{|}{\underset{CH(COOC_2H_5)_2}{\underset{|}{C}}}}-\overset{H}{\overset{|}{\underset{H}{\underset{|}{C}}}}-\underset{O}{\underset{\|}{C}}-C_6H_5$$

$$\underset{\text{Ethyl cinnamate}}{C_6H_5-\overset{H}{\overset{|}{C}}=\overset{H}{\overset{|}{C}}-COOC_2H_5} + \underset{\text{Ethyl malonate}}{CH_2(COOC_2H_5)_2} \xrightarrow{OC_2H_5^-} C_6H_5-\overset{H}{\overset{|}{\underset{CH(COOC_2H_5)_2}{\underset{|}{C}}}}-\overset{H}{\overset{|}{\underset{H}{\underset{|}{C}}}}-COOC_2H_5$$

$$\underset{\text{Ethyl crotonate}}{CH_3-\overset{H}{\overset{|}{C}}=\overset{H}{\overset{|}{C}}-COOC_2H_5} + \underset{\text{Ethyl methylmalonate}}{CH_3-CH(COOC_2H_5)_2} \xrightarrow{OC_2H_5^-} CH_3-\overset{H}{\overset{|}{\underset{\underset{CH_3}{\underset{|}{C(COOC_2H_5)_2}}}{\underset{|}{C}}}}-\overset{H}{\overset{|}{\underset{H}{\underset{|}{C}}}}-COOC_2H_5$$

$$\underset{\text{Ethyl }\alpha\text{-methylacrylate}}{H{-}\overset{H}{\overset{|}{C}}{=}\overset{CH_3}{\overset{|}{C}}{-}COOC_2H_5} + \underset{\text{Ethyl cyanoacetate}}{\overset{COOC_2H_5}{\overset{|}{\underset{\underset{CN}{|}}{CH_2}}}} \xrightarrow{OC_2H_5^-} H{-}\overset{H}{\overset{|}{\underset{\underset{\underset{\underset{CN}{|}}{CHCOOC_2H_5}}{|}}{C}}}{-}\overset{CH_3}{\overset{|}{\underset{\underset{H}{|}}{C}}}{-}COOC_2H_5$$

The Michael addition is believed to proceed by the following mechanism (shown for malonic ester):

(1) $$CH_2(COOC_2H_5)_2 + :Base \longrightarrow H{:}Base^+ + CH(COOC_2H_5)_2^-$$

(2) $$-\overset{|}{C}{=}\overset{|}{C}{-}\overset{|}{C}{=}O + \underset{\textit{Nucleophilic reagent}}{CH(COOC_2H_5)_2^-} \longrightarrow -\overset{|}{\underset{\underset{CH(COOC_2H_5)_2}{|}}{C}}{-}\underbrace{\overset{|}{C}{\overset{\cdots}{=}}\overset{|}{C}{\overset{\cdots}{=}}O}_{\ominus}$$

(3) $$-\overset{|}{\underset{\underset{CH(COOC_2H_5)_2}{|}}{C}}{-}\underbrace{\overset{|}{C}{\overset{\cdots}{=}}\overset{|}{C}{\overset{\cdots}{=}}O}_{\ominus} + H{:}Base^+ \longrightarrow -\overset{|}{\underset{\underset{CH(COOC_2H_5)_2}{|}}{C}}{-}\overset{|}{\underset{\underset{H}{|}}{C}}{-}\overset{|}{C}{=}O + :Base$$

The function of the base is to abstract (step 1) a hydrogen ion from malonic ester and thus generate a carbanion which, acting as a nucleophilic reagent, then attacks (step 2) the conjugated system in the usual manner.

In general, the compound from which the carbanion is generated must be a fairly acidic substance, so that an appreciable concentration of the carbanion can be obtained. Such a compound is usually one that contains a $-CH_2-$ or $-CH-$ group flanked by two electron-withdrawing groups which can help accommodate the negative charge of the anion. In place of ethyl malonate, compounds like ethyl cyanoacetate and ethyl acetoacetate can be used.

$$\underset{\text{Ethyl malonate}}{H_2C\begin{cases}C{=}O(OC_2H_5)\\ C{=}O(OC_2H_5)\end{cases}} + :Base \rightleftharpoons H{:}Base^+ + \left\{HC\begin{matrix}\overset{\cdots}{C}{\overset{\cdots}{=}}O(OC_2H_5)\\ \overset{\cdots}{C}{\overset{\cdots}{=}}O(OC_2H_5)\end{matrix}\right\}^{\ominus}$$

$$\underset{\text{Ethyl cyanoacetate}}{H_2C\begin{cases}C{=}O(OC_2H_5)\\ C{\equiv}N\end{cases}} + :Base \rightleftharpoons H{:}Base^+ + \left\{HC\begin{matrix}\overset{\cdots}{C}{\overset{\cdots}{=}}O(OC_2H_5)\\ \overset{\cdots}{C}{\overset{\cdots}{\equiv}}N\end{matrix}\right\}^{\ominus}$$

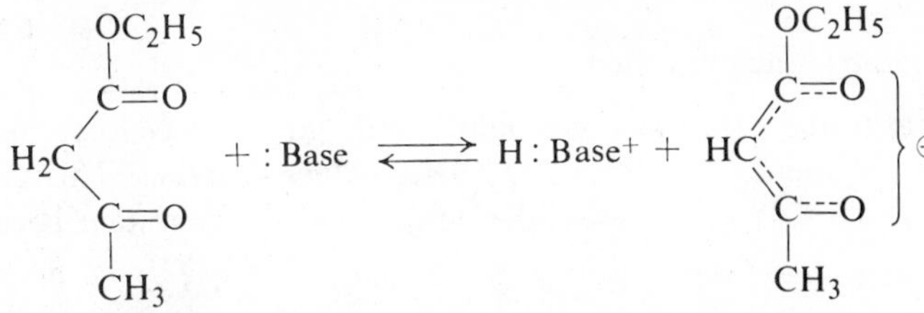

$$\underset{\text{Ethyl acetoacetate}}{H_2C(COOC_2H_5)(COCH_3)} + :\text{Base} \rightleftharpoons H:\text{Base}^+ + \left[HC(\text{---}C(=O)OC_2H_5)(\text{---}C(=O)CH_3)\right]^{\ominus}$$

Problem 27.10 Predict the products of the following Michael additions:

(a) ethyl crotonate + malonic ester $\longrightarrow$ A $\xrightarrow{OH^-}$ $\xrightarrow{H^+}$ $\xrightarrow{\text{heat}}$ B

(b) ethyl acrylate + ethyl acetoacetate $\longrightarrow$ C $\xrightarrow{H_2O,\ H^+}$ D

(c) methyl vinyl ketone + malonic ester $\longrightarrow$ E

(d) benzalacetophenone + acetophenone $\longrightarrow$ F

(e) acrylonitrile + allyl cyanide $\longrightarrow$ G $\xrightarrow{H_2O,\ H^+}$ H + $2NH_4^+$

(f) $C_2H_5OOC—C{\equiv}C—COOC_2H_5$ (1 mole) + ethyl acetoacetate (1 mole) $\longrightarrow$ I

(g) I $\xrightarrow{\text{strong } OH^-,\ H_2O}$ $\xrightarrow{H^+}$ J + CH_3COOH

Problem 27.11 Formaldehyde and malonic ester react in the presence of ethoxide ion to give K, $C_8H_{12}O_4$. (a) What is the structure of K? (*Hint:* See Problem 26.3, p. 850.) (b) How can K be converted into L, $(C_2H_5OOC)_2CHCH_2CH(COOC_2H_5)_2$? (c) What would you get if L were subjected to hydrolysis, acidification, and heat?

Problem 27.12 Show how a Michael addition followed by an aldol condensation can transform a mixture of methyl vinyl ketone and cyclohexanone into $\Delta^{1,9}$-octalone.

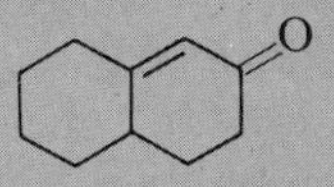

$\Delta^{1,9}$-Octalone

Problem 27.13 When mesityl oxide, $(CH_3)_2C{=}CHCOCH_3$, is treated with ethyl malonate in the presence of sodium ethoxide, compound M is obtained. (a) Outline the steps in its formation. (b) How could M be turned into 5,5-dimethyl-1,3-cyclohexanedione?

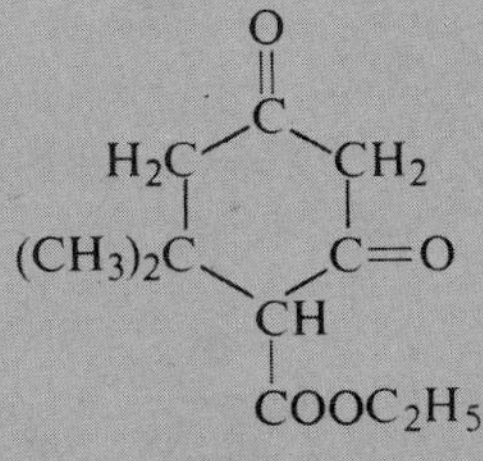

M

Problem 27.14 In the presence of piperidine (a secondary amine, Sec. 31.12), 1,3-cyclopentadiene and benzal-*p*-bromoacetophenone yield N. Outline the steps in its formation.

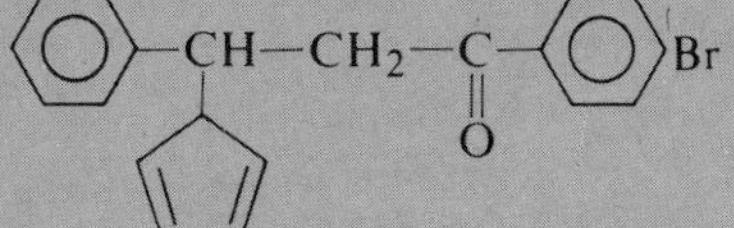

N

27.8 The Diels-Alder reaction

α,β-Unsaturated carbonyl compounds undergo an exceedingly useful reaction with conjugated dienes, known as the **Diels-Alder reaction.** This is an addition reaction in which C–1 and C–4 of the conjugated diene system become attached

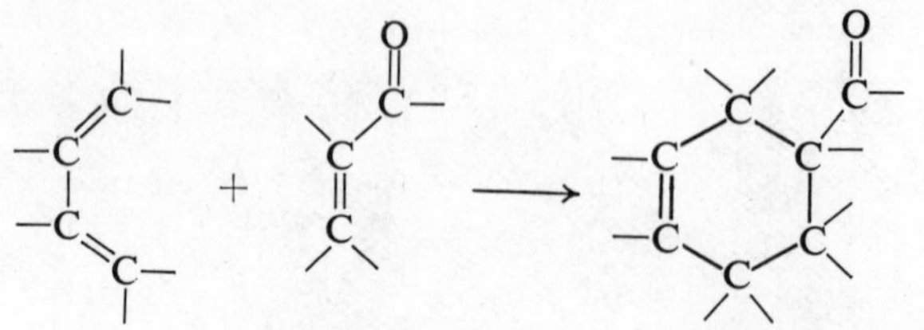

Diene + Dienophile (Greek: diene-loving) ⟶ Adduct *Six-membered ring*

to the doubly-bonded carbons of the unsaturated carbonyl compound to form a six-membered ring. A concerted, single-step mechanism is almost certainly involved; both new carbon–carbon bonds are partly formed in the same transition state, although not necessarily to the same extent. The Diels-Alder reaction is the most important example of *cycloaddition*, which is discussed further in Sec. 29.9. Since reaction involves a system of 4 π electrons (the diene) and a system of 2 π electrons (the dienophile), it is known as a [4 + 2] cycloaddition.

The Diels-Alder reaction is useful not only because a ring is generated, but also because it takes place so readily for a wide variety of reactants. Reaction is favored by electron-withdrawing substituents in the dienophile, but even simple alkenes can react. Reaction often takes place with the evolution of heat when the reactants are simply mixed together. A few examples of the Diels-Alder reaction are:

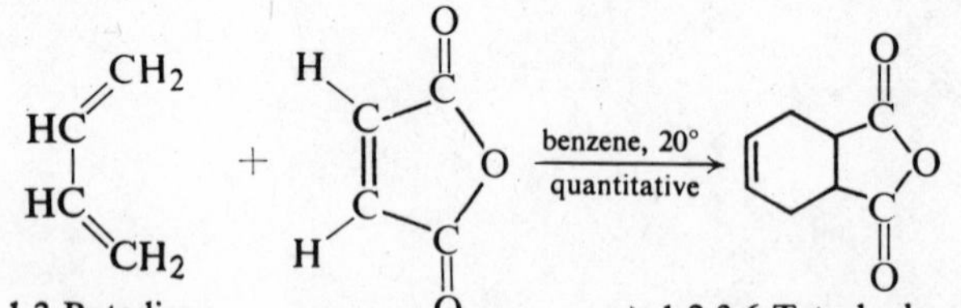

1,3-Butadiene + Maleic anhydride $\xrightarrow[\text{quantitative}]{\text{benzene, 20°}}$ *cis*-1,2,3,6-Tetrahydrophthalic anhydride

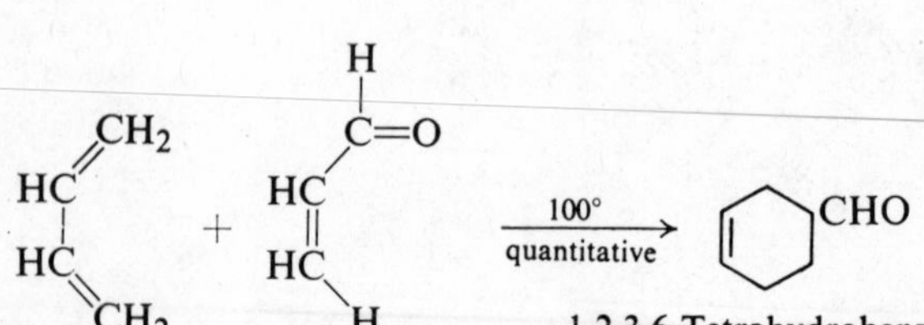

1,3-Butadiene + Acrolein $\xrightarrow[\text{quantitative}]{100°}$ 1,2,3,6-Tetrahydrobenzaldehyde

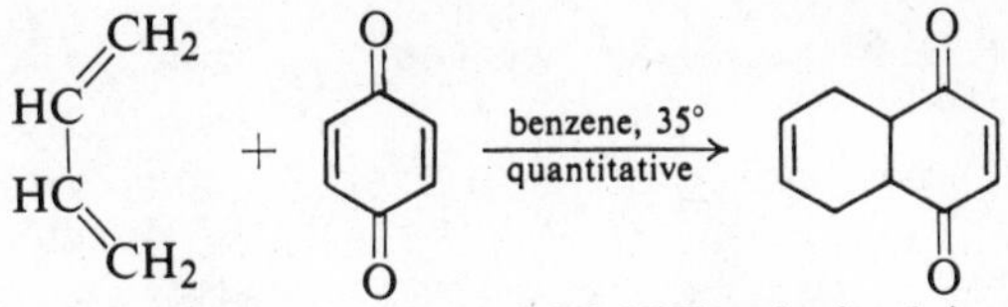

$CH_2{=}CH{-}CH{=}CH_2$ + p-benzoquinone $\xrightarrow[\text{quantitative}]{\text{benzene, 35°}}$ 5,8,9,10-tetrahydro-1,4-naphthoquinone

1,3-Butadiene

p-Benzoquinone

5,8,9,10-Tetrahydro-1,4-naphthoquinone

$\downarrow$ 1,3-butadiene, 100°

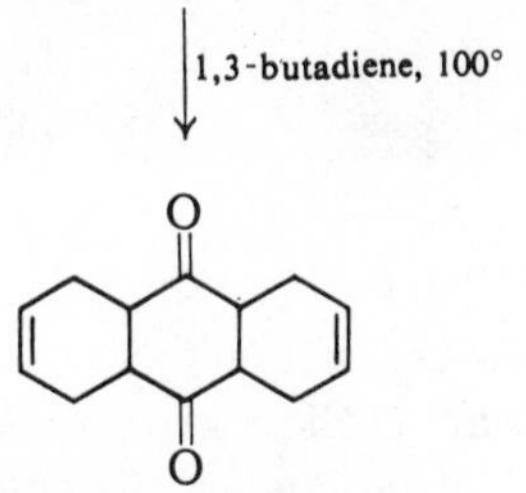

1,4,5,8,11,12,13,14-Octahydro-9,10-anthraquinone

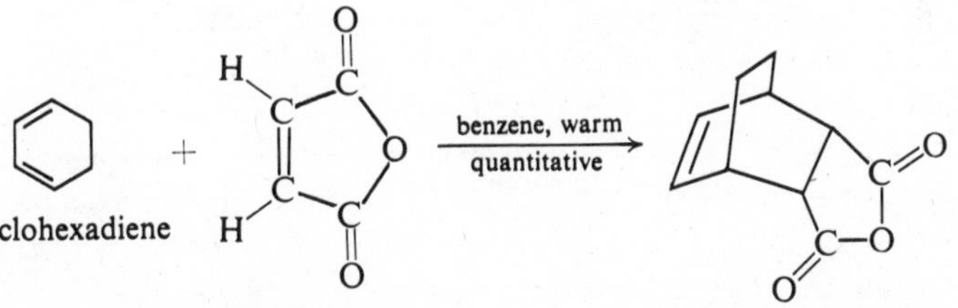

1,3-Cyclohexadiene + Maleic anhydride $\xrightarrow[\text{quantitative}]{\text{benzene, warm}}$ adduct

1,3-Cyclohexadiene

Maleic anhydride

Problem 27.15 From what reactants could each of the following compounds be synthesized?

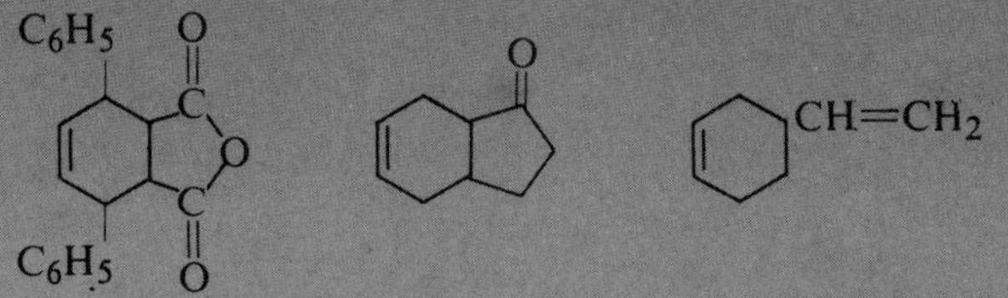

C_6H_5 C_6H_5 CH=CH$_2$

Problem 27.16 (a) In one synthesis of the hormone *cortisone* (by Lewis Sarett of Merck, Sharp and Dohme), the initial step was the formation of I by a Diels-Alder reaction. What were the starting materials?

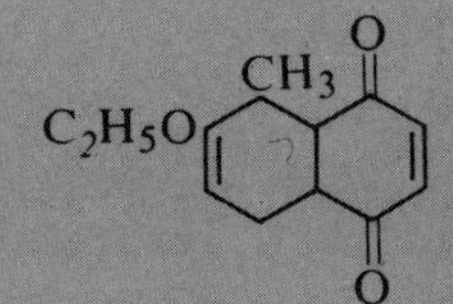

CH_3 C_2H_5O

I

(b) In another synthesis of cortisone (by R. B. Woodward, p. 938), the initial step was the formation of II by a Diels-Alder reaction. What were the starting materials?

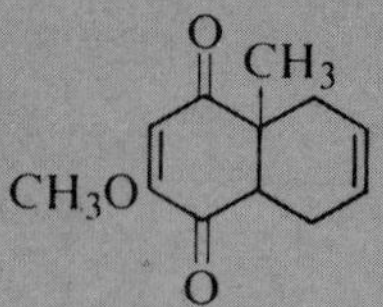

II

27.9 Quinones

α,β-Unsaturated ketones of a rather special kind are given the name of **quinones**: these are cyclic diketones of such a structure that they are converted by reduction into hydroquinones, phenols containing two —OH groups. For example:

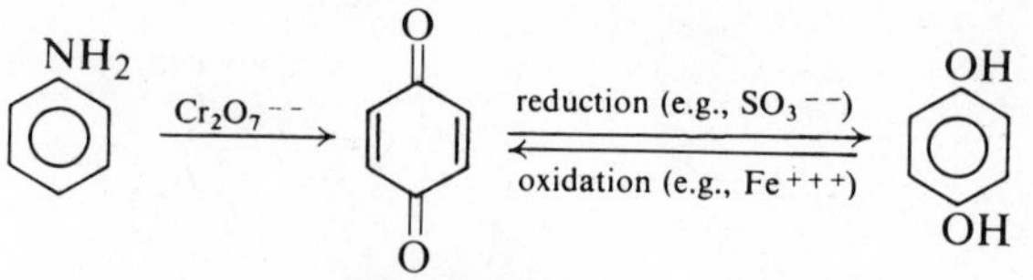

NH_2 — $Cr_2O_7^{--}$ → *p*-Benzoquinone (Quinone) *Yellow* ⇌ (reduction (e.g., SO_3^{--}) / oxidation (e.g., Fe^{+++})) Hydroquinone

Because they are highly conjugated, quinones are colored; *p*-benzoquinone, for example, is yellow.

Also because they are highly conjugated, quinones are rather closely balanced, energetically, against the corresponding hydroquinones. The ready interconversion provides a convenient oxidation–reduction system that has been studied intensively. Many properties of quinones result from the tendency to form the aromatic hydroquinone system.

Quinones—some related to more complicated aromatic systems (Chap. 30)—have been isolated from biological sources (molds, fungi, higher plants). In many cases they seem to take part in oxidation–reduction cycles essential to the living organism.

Problem 27.17 When *p*-benzoquinone is treated with HCl, there is obtained 2-chlorohydroquinone. It has been suggested that this product arises via an initial 1,4-addition. Show how this might be so.

Problem 27.18 (a) Hydroquinone is used in photographic developers to aid in the conversion of silver ion into free silver. What property of hydroquinone is being taken advantage of here?

(b) *p*-Benzoquinone can be used to convert iodide ion into iodine. What property of the quinone is being taken advantage of here?

Problem 27.19 How do you account for the fact that the treatment of phenol with nitrous acid yields the mono-oxime of *p*-benzoquinone?

PROBLEMS

1. Outline all steps in a possible laboratory synthesis of each of the unsaturated carbonyl compounds in Table 27.1, p. 866, using any readily available monofunctional compounds: simple alcohols, aldehydes, ketones, acids, esters, and hydrocarbons.

2. Give the structures of the organic products expected from the reaction of benzalacetone, $C_6H_5CH{=}CHCOCH_3$, with each of the following:

(a) H_2, Ni
(b) $NaBH_4$
(c) NaOI
(d) O_3, then Zn, H_2O
(e) Br_2
(f) HCl
(g) HBr
(h) H_2O, H^+
(i) CH_3OH, H^+
(j) NaCN (aq)
(k) CH_3NH_2
(l) aniline
(m) NH_3
(n) NH_2OH
(o) benzaldehyde, base
(p) ethyl malonate, base
(q) ethyl cyanoacetate, base
(r) ethyl methylmalonate, base
(s) ethyl acetoacetate, base
(t) 1,3-butadiene
(u) 1,3-cyclohexadiene
(v) 1,3-cyclopentadiene

3. In the presence of base the following pairs of reagents undergo Michael addition. Give the structures of the expected products.

(a) benzalacetophenone + ethyl cyanoacetate
(b) ethyl cinnamate + ethyl cyanoacetate
(c) ethyl fumarate + ethyl malonate
(d) ethyl acetylenedicarboxylate + ethyl malonate
(e) mesityl oxide + ethyl malonate
(f) mesityl oxide + ethyl acetoacetate
(g) ethyl crotonate + ethyl methylmalonate
(h) formaldehyde + 2 moles ethyl malonate
(i) acetaldehyde + 2 moles ethyl acetoacetate
(j) methyl acrylate + nitromethane
(k) 2 moles ethyl crotonate + nitromethane
(l) 3 moles acrylonitrile + nitromethane
(m) 1 mole acrylonitrile + $CHCl_3$

4. Give the structures of the compounds expected from the hydrolysis and decarboxylation of the products obtained in Problem 3, parts (a) through (i).

5. Depending upon reaction conditions, dibenzalacetone and ethyl malonate can be made to yield any of three products by Michael addition.

dibenzalacetone + 2 moles ethyl malonate $\longrightarrow$ A (no unsaturation)
dibenzalacetone + 1 mole ethyl malonate $\longrightarrow$ B (one carbon–carbon double bond)
dibenzalacetone + 1 mole ethyl malonate $\longrightarrow$ C (no unsaturation)

What are A, B, and C?

6. Give the structure of the product of the Diels-Alder reaction between:

(a) maleic anhydride and isoprene
(b) maleic anhydride and 1,1′-bicyclohexenyl (I)
(c) maleic anhydride and 1-vinyl-1-cyclohexene
(d) 1,3-butadiene and methyl vinyl ketone
(e) 1,3-butadiene and crotonaldehyde
(f) 2 moles 1,3-butadiene and dibenzalacetone
(g) 1,3-butadiene and β-nitrostyrene ($C_6H_5CH{=}CHNO_2$)
(h) 1,3-butadiene and 1,4-naphthoquinone (II)
(i) *p*-benzoquinone and 1,3-cyclohexadiene
(j) *p*-benzoquinone and 1,1′-bicyclohexenyl (I)
(k) *p*-benzoquinone and 2 moles 1,3-cyclohexadiene

(l) *p*-benzoquinone and 2 moles 1,1′-bicyclohexenyl (I)
(m) 1,3-cyclopentadiene and acrylonitrile
(n) 1,3-cyclohexadiene and acrolein

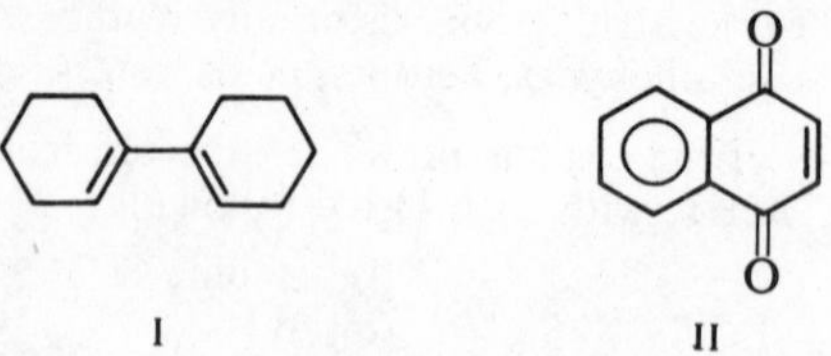

I II

7. From what reactants could the following be synthesized by the Diels-Alder reaction?

(a) $CH_2{=}CH$

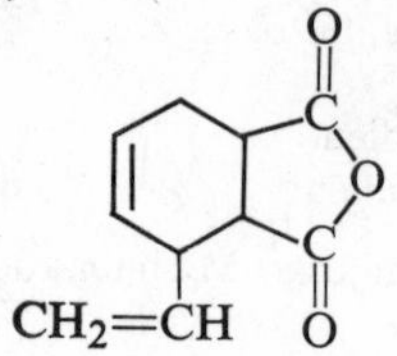

(b) CH_3, CH_3

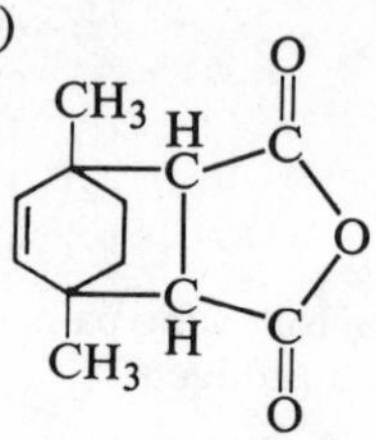

(c) $COCH_3$, C_6H_5

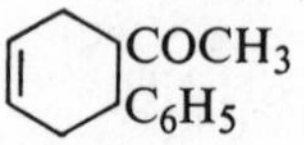

(d) COOH, COOH

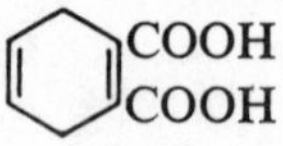

(e)

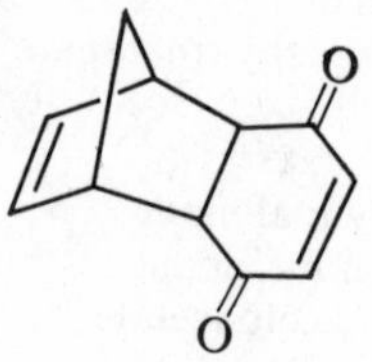

(f)

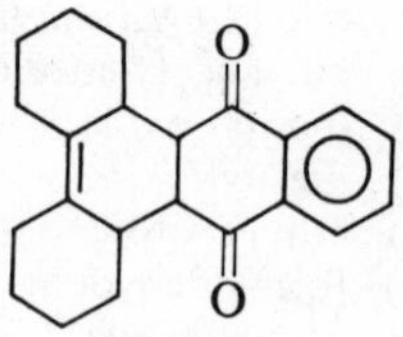

(g) CHO, CH_3

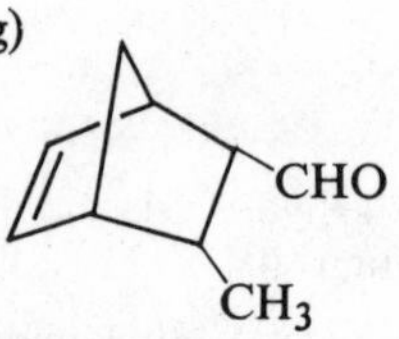

(h) $COCH_3$

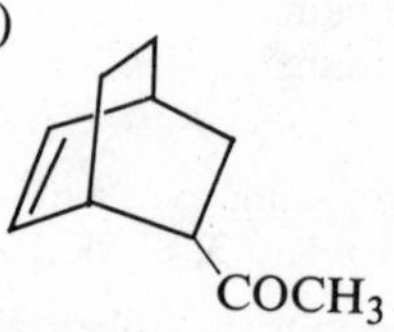

(i)

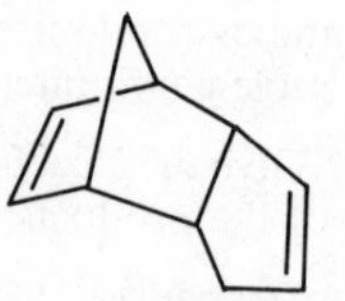

8. The following observations illustrate one aspect of the stereochemistry of the Diels-Alder reaction:

maleic anhydride + 1,3-butadiene ⟶ D ($C_8H_8O_3$)
D + H_2O, heat ⟶ E ($C_8H_{10}O_4$)
E + H_2, Ni ⟶ F ($C_8H_{12}O_4$), m.p. 192°
fumaryl chloride (*trans*-ClOCCH=CHCOCl) + 1,3-butadiene ⟶ G ($C_8H_8O_2Cl_2$)
G + H_2O, heat ⟶ H ($C_8H_{10}O_4$)
H + H_2, Ni ⟶ I ($C_8H_{12}O_4$), m.p. 215°
I can be resolved; F cannot be resolved.

Does the Diels-Alder reaction involve a *syn*-addition or an *anti*-addition?

9. On the basis of your answer to Problem 8, give the stereochemical formulas of the products expected from each of the following reactions. Label meso compounds and racemic modifications.

(a) crotonaldehyde (*trans*-2-butenal) + 1,3-butadiene
(b) *p*-benzoquinone + 1,3-butadiene
(c) maleic anhydride + 1,3-butadiene, followed by cold alkaline $KMnO_4$
(d) maleic anhydride + 1,3-butadiene, followed by hot $KMnO_4$ ⟶ $C_8H_{10}O_8$

10. Account for the following observations:

(a) Dehydration of 3-hydroxy-2,2-dimethylpropanoic acid yields 2-methyl-2-butenoic acid.

(b) $C_2H_5OOC{-}COOC_2H_5$ (Ethyl oxalate) + $CH_3CH{=}CHCOOC_2H_5$ (Ethyl crotonate) $\xrightarrow{OC_2H_5^-}$ $C_2H_5OOC{-}C({=}O){-}CH_2CH{=}CHCOOC_2H_5$

(c) $CH_2{=}CH{-}\overset{+}{P}Ph_3\ Br^-$ + salicylaldehyde + a little base $\longrightarrow$ 2H-chromene + Ph_3PO

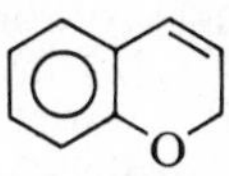

(d) $CH_3CH{=}CHCOOC_2H_5 + Ph_3P{=}CH_2 \longrightarrow CH_3{-}CH{-}CH{-}COOC_2H_5 + Ph_3P$ (the two CH carbons bridged by CH_2)

(e) o-bromofluorobenzene + furan + Li $\longrightarrow$ 1,4-dihydronaphthalene-1,4-endoxide

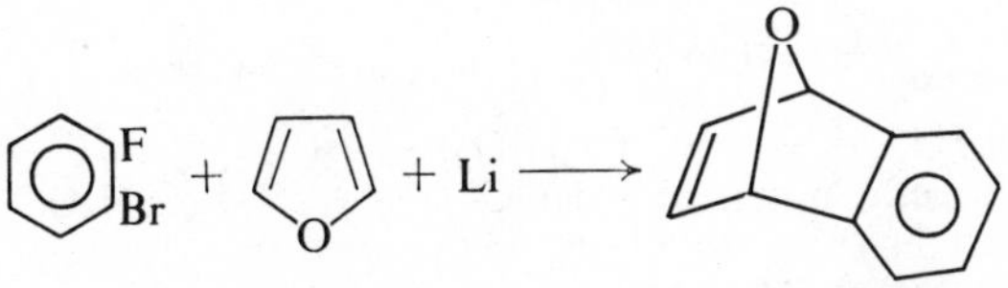

11. When *citral* (Problem 26, p. 652) is refluxed with aqueous potassium carbonate, acetaldehyde distills from the mixture and 6-methyl-5-hepten-2-one is obtained in high yield. Show all steps in a likely mechanism. (*Hint:* See Sec. 21.5.)

12. In connection with his new research problem, our naïve graduate student (Problem 18, p. 650, and Problem 20, p. 724) needed a quantity of the unsaturated alcohol $C_6H_5CH{=}CHC(OH)(CH_3)(C_2H_5)$. He added a slight excess of benzalacetone, $C_6H_5CH{=}CHCOCH_3$, to a solution of ethylmagnesium bromide, and, by use of a color test, found that the Grignard reagent had been consumed. He worked up the reaction mixture in the usual way with dilute acid. Having learned a little (but not much) from his earlier sad experiences, he tested the product with iodine and sodium hydroxide; when a copious precipitate of iodoform appeared, he concluded that he had simply recovered his starting material.

He threw his product into the waste crock, carefully and methodically destroyed his glassware, burned his laboratory coat, left school, and went into politics, where he did quite well; his career in Washington was marred only, in the opinion of some, by his blind antagonism toward all appropriations for scientific research and his frequent attacks —alternately vitriolic and caustic—on the French.

What had he thrown into the waste crock? How had it been formed?

13. Treatment of $CF_3(C_6H_5)C{=}CF_2$ with EtONa/EtOH yields chiefly $CF_3(C_6H_5)C{=}CF(OEt)$. Similar treatment of $CF_2Cl(C_6H_5)C{=}CF_2$ yields $EtOCF_2(C_6H_5)C{=}CF_2$. The rates of the two reactions are almost identical. It has been suggested that both reactions proceed by the same mechanism.

Show all steps in a mechanism that is consistent with the nature of these reactants, and that accounts for the similarity in rate despite the difference in final product.

14. Give structures of compounds J through QQ:

(a) glycerol + $NaHSO_4$, heat $\longrightarrow$ J (C_3H_4O)
J + ethyl alcohol + HCl $\longrightarrow$ K ($C_7H_{15}O_2Cl$)
K + NaOH, heat $\longrightarrow$ L ($C_7H_{14}O_2$)
L + cold neutral $KMnO_4$ $\longrightarrow$ M ($C_7H_{16}O_4$)
M + dilute H_2SO_4 $\longrightarrow$ N ($C_3H_6O_3$) + ethyl alcohol

(b) $C_2H_5OOC{-}C{\equiv}C{-}COOC_2H_5$ + sodiomalonic ester $\longrightarrow$ O ($C_{15}H_{22}O_8$)
O + OH^-, heat; then H^+; then heat $\longrightarrow$ P ($C_6H_6O_6$), *aconitic acid*, found in sugar cane and beetroot

(c) ethyl fumarate + sodiomalonic ester $\longrightarrow$ Q ($C_{15}H_{24}O_8$)
Q + OH^-, heat; then H^+; then heat $\longrightarrow$ R ($C_6H_8O_6$), *tricarballylic acid*

(d) benzil ($C_6H_5COCOC_6H_5$) + benzyl ketone ($C_6H_5CH_2COCH_2C_6H_5$) + base $\longrightarrow$ S ($C_{29}H_{20}O$), "tetracyclone"
S + maleic anhydride $\longrightarrow$ T ($C_{33}H_{22}O_4$)
T + heat $\longrightarrow$ CO + H_2 + U ($C_{32}H_{20}O_3$)

(e) S + $C_6H_5C{\equiv}CH$ $\longrightarrow$ V ($C_{37}H_{26}O$)
V + heat $\longrightarrow$ CO + W ($C_{36}H_{26}$)

(f) acetone + $BrMgC{\equiv}COC_2H_5$, then H_2O $\longrightarrow$ X ($C_7H_{12}O_2$)
X + H_2, Pd/$CaCO_3$ $\longrightarrow$ Y ($C_7H_{14}O_2$)
Y + H^+, warm $\longrightarrow$ Z (C_5H_8O), β-methylcrotonaldehyde

(g) ethyl 3-methyl-2-butenoate + ethyl cyanoacetate + base $\longrightarrow$ AA ($C_{12}H_{19}O_4N$)
AA + OH^-, heat; then H^+; then heat $\longrightarrow$ BB ($C_7H_{12}O_4$)

(h) mesityl oxide + ethyl malonate + base $\longrightarrow$ CC ($C_{13}H_{22}O_5$)
CC + NaOBr, OH^-, heat; then H^+ $\longrightarrow$ $CHBr_3$ + BB ($C_7H_{12}O_4$)

(i) $CH_3C{\equiv}CNa$ + acetaldehyde $\longrightarrow$ DD (C_5H_8O)
DD + $K_2Cr_2O_7$, H_2SO_4 $\longrightarrow$ EE (C_5H_6O)

(j) 3-pentyn-2-one + H_2O, Hg^{++}, H^+ $\longrightarrow$ FF ($C_5H_8O_2$)

(k) mesityl oxide + NaOCl, then H^+ $\longrightarrow$ GG ($C_5H_8O_2$)

(l) methallyl chloride (3-chloro-2-methylpropene) + HOCl $\longrightarrow$ HH ($C_4H_8OCl_2$)
HH + KCN $\longrightarrow$ II ($C_6H_8ON_2$)
II + H_2SO_4, H_2O, heat $\longrightarrow$ JJ ($C_6H_8O_4$)

(m) ethyl adipate + NaOEt $\longrightarrow$ KK ($C_8H_{12}O_3$)
KK + methyl vinyl ketone + base $\xrightarrow{\text{Michael}}$ LL ($C_{12}H_{18}O_4$)
LL + base $\xrightarrow{\text{aldol}}$ MM ($C_{12}H_{16}O_3$)

(n) hexachloro-1,3-cyclopentadiene + CH_3OH + KOH $\longrightarrow$ NN ($C_7H_6Cl_4O_2$)
NN + $CH_2{=}CH_2$, heat, pressure $\longrightarrow$ OO ($C_9H_{10}Cl_4O_2$)
OO + Na + *t*-BuOH $\longrightarrow$ PP ($C_9H_{14}O_2$)
PP + dilute acid $\longrightarrow$ QQ (C_7H_8O), *7-ketonorbornene*

15. *Spermine*, $H_2NCH_2CH_2CH_2NHCH_2CH_2CH_2CH_2NHCH_2CH_2CH_2NH_2$, found in seminal fluid, has been synthesized from acrylonitrile and 1,4-diaminobutane (putrescine). Show how this was probably done.

16. Outline all steps in each of the following syntheses:

(a) $HOOC{-}CH{=}CH{-}CH{=}CH{-}COOH$ from adipic acid
(b) $HC{\equiv}C{-}CHO$ from acrolein (*Hint:* See Problem 14(a) above.)
(c) $CH_3COCH{=}CH_2$ from acetone and formaldehyde
(d) $CH_3COCH{=}CH_2$ from vinylacetylene
(e) β-phenylglutaric acid from benzaldehyde and aliphatic reagents
(f) phenylsuccinic acid from benzaldehyde and aliphatic reagents
(g) 4-phenyl-2,6-heptanedione from benzaldehyde and aliphatic reagents (*Hint:* See Problem 3(f), above.)

17. Treatment of ethyl acetoacetate with acetaldehyde in the presence of the base piperidine was found to give a product of formula $C_{14}H_{22}O_6$. Controversy arose about its structure: did it have open-chain structure III or cyclic structure IV, each formed by combinations of aldol and Michael condensations?

$C_2H_5OOC{-}CH(COCH_3){-}CH(CH_3){-}CH(COCH_3){-}COOC_2H_5$

III

(cyclohexanone ring bearing HO, CH_3; $COOC_2H_5$; CH_3; $COOC_2H_5$; =O)

IV

(a) Show just how each possible product could have been formed.
(b) Then the nmr spectrum of the compound was found to be the following:

a complex, δ 0.95–1.10, 3H
b singlet, δ 1.28, 3H
c triplet, centered at δ 1.28, 3H
d triplet, centered at δ 1.32, 3H
e singlet, δ 2.5, 2H
f broad singlet, δ 3.5, 1H
g complex, δ 2–4, total of 3H
h quartet, δ 4.25, 2H
i quartet, δ 4.30, 2H

Which structure is the correct one? Assign all peaks in the spectrum. Describe the spectrum you would expect from the other possibility.

18. Give the likely structures for UU and VV.

1,3-butadiene + propiolic acid (HC≡CCOOH) $\longrightarrow$ RR ($C_7H_8O_2$)
RR + 1 mole $LiAlH_4$ $\longrightarrow$ SS ($C_7H_{10}O$)
SS + methyl chlorocarbonate (CH_3OCOCl) $\longrightarrow$ TT ($C_9H_{12}O_3$)
TT + heat (short time) $\longrightarrow$ toluene + UU (C_7H_8)
UU + tetracyanoethylene $\longrightarrow$ VV ($C_{13}H_8N_4$)

Compound UU is not toluene or 1,3,5-cycloheptatriene; on standing at room temperature it is converted fairly rapidly into toluene. Compound UU gives the following spectral data. Ultraviolet: λ_{max} 303 nm, ϵ_{max} 4400. Infrared: strong bands at 3020, 2900, 1595, 1400, 864, 692, and 645 cm^{-1}; medium bands at 2850, 1152, and 790 cm^{-1}.

19. Give structures of compounds WW through YY, and account for their formation:

cyclopentanone + pyrrolidine, then acid $\longrightarrow$ WW ($C_9H_{15}N$)
WW + CH_2=$CHCOOCH_3$ $\longrightarrow$ XX ($C_{13}H_{21}O_2N$)
XX + H_2O, H^+, heat $\longrightarrow$ YY ($C_9H_{14}O_3$)

20. Irradiation by ultraviolet light of 2,2,4,4-tetramethyl-1,3-cyclobutanedione (V) produces tetramethylethylene and two moles of carbon monoxide. When the irradiation is carried out in furan (VI), there is obtained a product believed to have the structure VII.

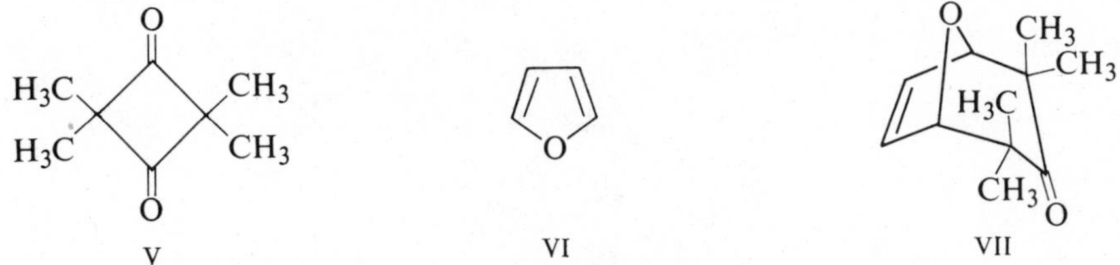

(a) Chief support for structure VII comes from elemental analysis, mol. wt. determination, and nmr data:

a singlet, δ 0.85, 6H
b singlet, δ 1.25, 6H
c singlet, δ 4.32, 2H
d singlet, δ 6.32, 2H

Show how the nmr data support the proposed structure. Why should there be two singlets of 6H each instead of one peak of 12H?

(b) It is proposed that, in the formation of tetramethylethylene, one mole of carbon monoxide is lost at a time. Draw electronic structures to show all steps in such a two-stage mechanism. How does the formation of VII support such a mechanism?

21. β-Lactones cannot be made from β-hydroxyacids. The β-lactone VIII was obtained, however, by treatment of sodium maleate (or sodium fumarate) with bromine water.

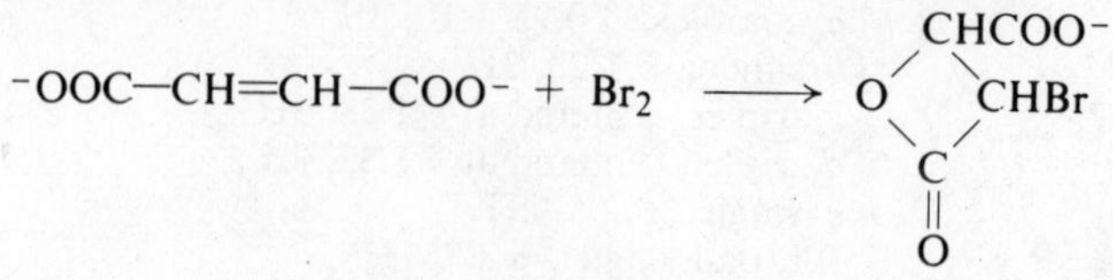

$$^{-}OOC{-}CH{=}CH{-}COO^{-} + Br_2 \longrightarrow \text{VIII (four-membered ring: O, CHCOO}^{-}\text{, CHBr, C=O)}$$

VIII

This experiment, reported in 1937 by P. D. Bartlett and D. S. Tarbell (of Harvard University), was an important step in the establishment of the mechanism of addition of halogens to carbon–carbon double bonds. Why is this so? How do you account for the formation of the β-lactone?

22. When the sodium salt of diazocyclopentadiene-2-carboxylic acid (IX) is heated above 140°, N_2 and CO_2 are involved. If IX is heated in solution with tetracyclone (X),

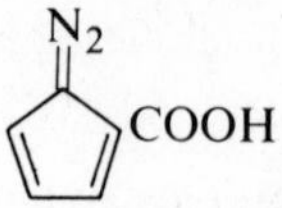

IX 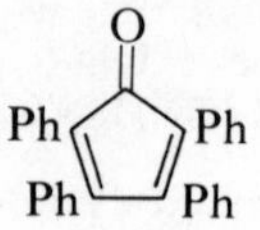 X 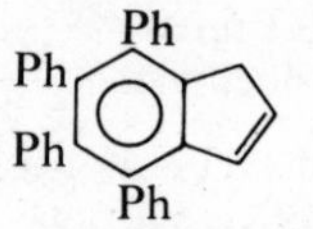 XI

CO is evolved as well, and 4,5,6,7-tetraphenylindene (XI) is obtained. Show all steps in a likely mechanism for the formation of XI. (*Hint*: See Problem 10(e) above.) Of what special theoretical interest are these findings?

Chapter
28 Rearrangements and Neighboring Group Effects
Nonclassical Ions

28.1 Rearrangements and neighboring group effects: intramolecular nucleophilic attack

Carbonium ions, we know, can rearrange through migration of an organic group or a hydrogen atom, with its pair of electrons, to the electron-deficient

$$-\overset{R}{\underset{|}{\overset{|}{C}}}-\overset{\oplus}{\underset{|}{C}}- \longrightarrow -\overset{\oplus}{\underset{|}{C}}-\overset{R}{\underset{|}{\overset{|}{C}}}-$$

carbon. Indeed, when carbonium ions were first postulated as reactive intermediates (p. 160), it was to account for rearrangements of a particular kind. Such rearrangements still provide the best single clue that we are dealing with a carbonium ion reaction.

The driving force behind all carbonium ion reactions is the need to provide electrons to the electron-deficient carbon. When an electron-deficient carbon is generated, a near-by group may help to relieve this deficiency. It may, of course, remain in place and release electrons through the molecular framework, inductively or by resonance. Or—and this is what we are concerned with here—it may actually *carry the electrons* to where they are needed. Other atoms besides carbon can be electron-deficient—in particular, nitrogen and oxygen—and they, too, can get electrons through rearrangement. The most important class of molecular rearrangements is that involving *1,2-shifts to electron-deficient atoms.* It is the kind of rearrangement that we shall deal with in this chapter.

An electron-deficient carbon is most commonly generated by the departure of a leaving group which takes the bonding electrons with it. The migrating group is, of course, a nucleophile, and so a rearrangement of this sort amounts to *intramolecular nucleophilic substitution.* Now, as we have seen, nucleophilic substitution can be of two kinds, S_N2 and S_N1. Exactly the same possibilities exist for a rearrangement: it can be S_N2-like, with the migrating group helping to push out the leaving group in a single-step reaction; or it can be S_N1-like, with the migrating

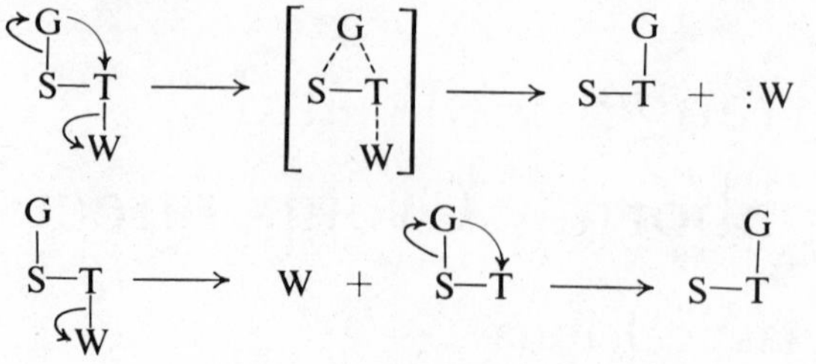

S_N2-like migration

S_N1-like migration

G = migrating group
S = migration source
T = migration terminus

group waiting for the departure of the leaving group before it moves. This matter of *timing* of bond-breaking and bond-making is, as we shall see, of major concern in the study of rearrangements.

The term *anchimeric assistance* (Gr., *anchi* + *meros*, adjacent parts) is often used to describe the help given by a migrating group in the expelling of a leaving group.

In a rearrangement, a near-by group carries electrons to an electron-deficient atom, and then *stays there.* But sometimes, it happens, a group brings electrons and then *goes back to where it came from.* This gives rise to what are called **neighboring group effects**: intramolecular effects exerted on a reaction through direct participation—that is, through movement to within bonding distance—by a group near the reaction center.

Neighboring group effects involve the same basic process as rearrangement. Indeed, in many cases there *is* rearrangement, but it is *hidden.* What we see on the surface may be this:

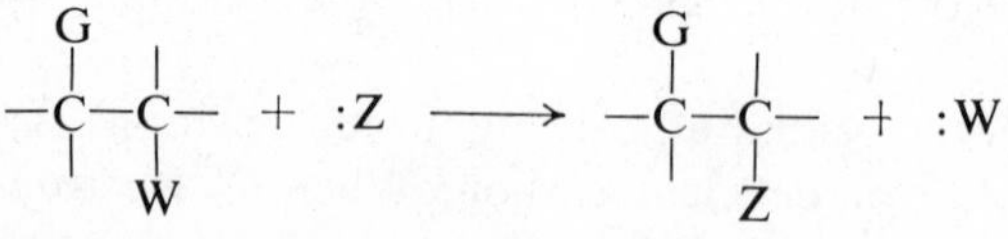

But what is actually happening may be this:

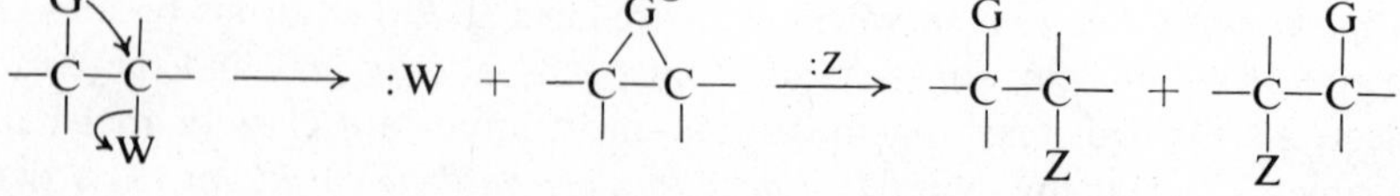

I

The neighboring group, acting as an internal nucleophile, attacks carbon at the reaction center; the leaving group is lost, and there is formed a *bridged intermediate* (I), usually a cation. This undergoes attack by an external nucleophile to yield the product. The overall stereochemistry is determined by the way in which the bridged ion is formed and the way in which it reacts, and typically differs from the stereochemistry observed for simple attack by an external nucleophile. If a neighboring group helps to push out the leaving group—that is, gives anchimeric assistance—it may accelerate the reaction, sometimes tremendously. Thus, neighboring group participation is most often revealed by a *special kind of stereochemistry* or by an *unusually fast rate of reaction.*

We have, of course, encountered internal nucleophilic attack before. In the preparation of epoxides by action of base on halohydrins (Sec. 17.10), the bridged intermediate—the epoxide—happens to be stable in the reaction medium, persists, and is isolated.

If a neighboring group is to form a bridged cation, it must have electrons to form the extra bond. These may be *unshared pairs* on atoms like sulfur, nitrogen, oxygen, or bromine; π *electrons* of a double bond or aromatic ring; or even, in some cases, σ *electrons.*

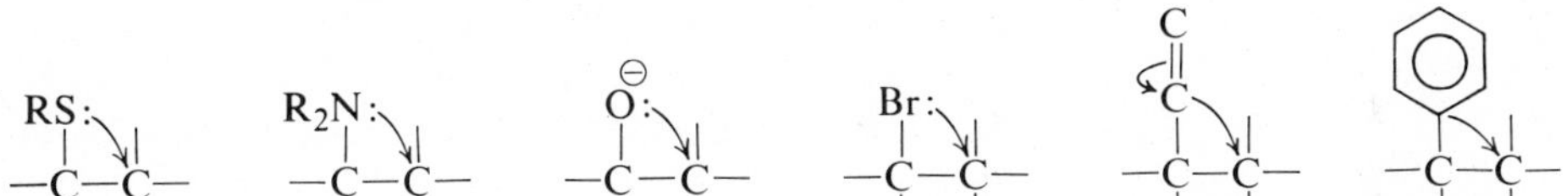

In making its nucleophilic attack, a neighboring group competes with outside molecules that are often intrinsically much stronger nucleophiles. Yet the evidence clearly shows that the neighboring group enjoys—for its nucleophilic power—a tremendous advantage over these outside nucleophiles. Why is this? The answer is quite simple: *because it is there.*

The neighboring group is there, in the same molecule, poised in the proper position for attack. It does not have to wait until its path happens to cross that of the substrate; its "effective concentration" is extremely high. It does not have to give up precious freedom of motion (translational entropy) when it becomes locked into a transition state. Between it and the reaction center there are no tightly clinging solvent molecules that must be stripped away as reaction takes place. Finally, the electronic reorganization—changes in overlap—that accompanies reaction undoubtedly happens more easily in this cyclic system.

Enzymes function by accelerating, very specifically, rates of the organic reactions involved in life processes. They evidently do this by bringing reactants together into exactly the right positions for reaction to occur. Underlying much enzyme activity, it appears, are what amount to neighboring group effects.

Problem 28.1 Draw the structure of the bridged intermediate (I, above) expected if each of the following were to act as a neighboring group. To what class of compounds does each intermediate belong?

(a) $-N(CH_3)_2$
(b) $-SCH_3$
(c) $-OH$
(d) $-O^-$
(e) $-Br$
(f) $-C_6H_5$
(g) $-C_6H_4OCH_3$-*p*
(h) $-C_6H_4O^-$-*p*
(i) $-CH{=}CHR$

28.2 Hofmann rearrangement. Migration to electron-deficient nitrogen

Let us begin with a reaction that we encountered earlier as a method of synthesis of amines: the Hofmann degradation of amides. Whatever the mechanism

$$\underset{\text{An amide}}{R{-}C({=}O)NH_2} \xrightarrow{OBr^-} \underset{\text{A 1° amine}}{R{-}NH_2} + CO_3^{--}$$

of the reaction, it is clear that rearrangement occurs, since the group joined to carbonyl carbon in the amide is found joined to nitrogen in the product.

The reaction is believed to proceed by the following steps:

(1) $$R{-}C({=}O){-}\ddot{N}H_2 + OBr^- \longrightarrow R{-}C({=}O){-}\ddot{N}H{-}Br + OH^-$$

(2) $$R{-}C({=}O){-}\ddot{N}H{-}Br + OH^- \longrightarrow R{-}C({=}O){-}\ddot{N}^{\ominus}{-}Br + H_2O$$

(3) $$R{-}C({=}O){-}\ddot{N}^{\ominus}{-}Br \longrightarrow R{-}C({=}O){-}\ddot{N}: + Br^-$$

(4) $$R{-}C({=}O){-}\ddot{N}: \longrightarrow R{-}\ddot{N}{=}C{=}O$$

Steps (3) and (4): *Simultaneous*

(5) $$R{-}\ddot{N}{=}C{=}O + 2OH^- \xrightarrow{H_2O} R{-}\ddot{N}H_2 + CO_3^{--}$$

Step (1) is the halogenation of an amide. This is a known reaction, an N-haloamide being isolated if no base is present. Furthermore, if the N-haloamide isolated in this way is then treated with base, it is converted into the amine.

Step (2) is the abstraction of a hydrogen ion by hydroxide ion. This is reasonable behavior for hydroxide ion, especially since the presence of the electron-withdrawing bromine increases the acidity of the amide. Unstable salts have actually been isolated in certain of these reactions.

Step (3) involves the separation of a halide ion, which leaves behind an electron-deficient nitrogen atom.

In Step (4) the actual rearrangement occurs. Steps (3) and (4) are generally

(3,4) $$R{-}C({=}O){-}\ddot{N}^{\ominus}{-}Br \longrightarrow R{-}N{=}C{=}O + Br^-$$

believed to occur simultaneously, the attachment of R to nitrogen helping to push out halide ion. That is, migration is S_N2-like, and provides anchimeric assistance.

Step (5) is the hydrolysis of an isocyanate (R—N=C=O) to form an amine and carbonate ion. This is a known reaction of isocyanates. If the Hofmann degradation is carried out in the absence of water, an isocyanate can actually be isolated.

Like the rearrangement of carbonium ions that we have already encountered (Sec. 5.22), the Hofmann rearrangement involves a 1,2-shift. In the rearrangement of carbonium ions a group migrates with its electrons to an electron-deficient carbon; in the present reaction the group migrates with its electrons to an electron-deficient *nitrogen*. We consider nitrogen to be electron-deficient even though it probably loses electrons—to bromide ion—*while* migration takes place, rather than before.

The strongest support for the mechanism just outlined is the fact that many of the proposed intermediates have been isolated, and that these intermediates have been shown to yield the products of the Hofmann degradation. The mechanism is also supported by the fact that analogous mechanisms account satisfactorily for observations made on a large number of related rearrangements. Furthermore, the actual rearrangement step fits the broad pattern of 1,2-shifts to electron-deficient atoms.

In addition to evidence indicating what the various steps in the Hofmann degradation are, there is also evidence that gives us a rather intimate view of just how the rearrangement step takes place. In following sections we shall see what some of that evidence is. We shall be interested in this not just for what it tells us about the Hofmann degradation, but because it will give us an idea of the kind of thing that can be done in studying rearrangements of many kinds.

Problem 28.2 Reaction of acid chlorides with sodium azide, NaN_3, yields *acyl azides*, $RCON_3$. When heated, these undergo the *Curtius rearrangement* to amines, RNH_2, or, in a non-hydroxylic solvent, to isocyanates, RNCO. Using the structure

$$R-\overset{\overset{\displaystyle O}{\|}}{C}-\overset{\ominus}{N}-\overset{\oplus}{N}\equiv N$$

for the azide, suggest a mechanism for the rearrangement. (*Hint:* Write balanced equations.)

28.3 Hofmann rearrangement. Intramolecular or intermolecular?

One of the first questions asked in the study of a rearrangement is this: Is the rearrangement *intra*molecular or *inter*molecular? That is, does the migrating group move from one atom to another atom within the same molecule, or does it move from one molecule to another?

In the mechanism outlined above, the Hofmann rearrangement is shown as intramolecular. How do we know that this is so? To answer this question, T. J. Prosser and E. L. Eliel (of the University of Notre Dame) carried out degradation of a mixture of *m*-deuteriobenzamide and benzamide-^{15}N. When they analyzed the product with the mass spectrometer, they found only *m*-deuterioaniline and aniline-^{15}N. There was *none* of the mixture of cross-products that would have been formed if a phenyl group from one molecule had become attached to the nitrogen of another. The results of this elegant double labeling experiment thus show beyond doubt that the Hofmann rearrangement is intramolecular.

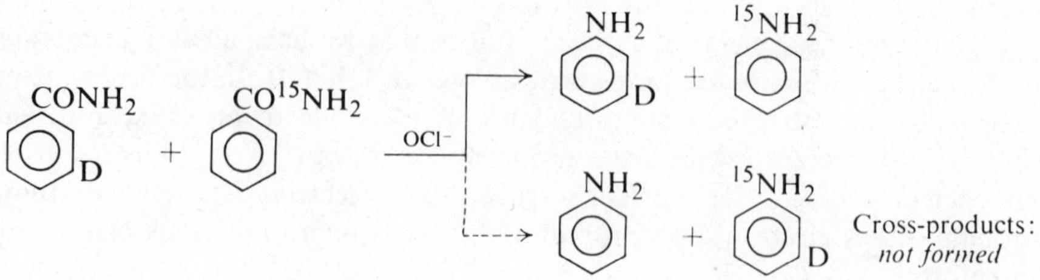

$CONH_2$ (D-substituted benzene) + $CO^{15}NH_2$ (benzene) $\xrightarrow{OCl^-}$ NH_2 (D-substituted benzene) + $^{15}NH_2$ (benzene)

$\dashrightarrow$ NH_2 (benzene) + $^{15}NH_2$ (D-substituted benzene)

Cross-products: *not formed*

28.4 Hofmann rearrangement. Stereochemistry at the migrating group

When optically active α-phenylpropionamide undergoes the Hofmann degradation, α-phenylethylamine of the same configuration and of essentially the same optical purity is obtained:

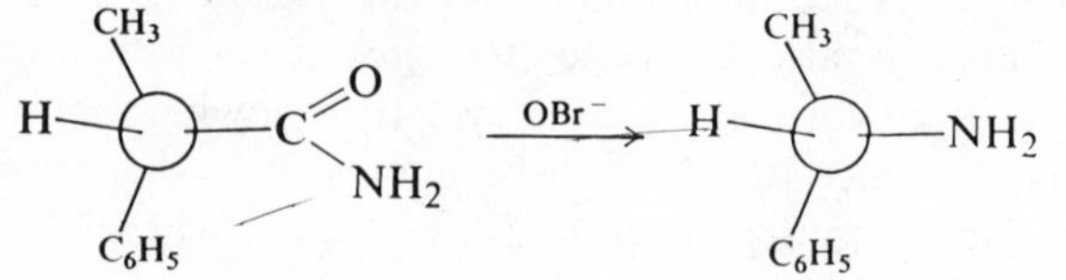

H, CH_3, C_6H_5 — C(=O)NH_2 $\xrightarrow{OBr^-}$ H, CH_3, C_6H_5 — NH_2

(+)-α-Phenylpropionamide (−)-α-Phenylethylamine

Retention of configuration

Rearrangement proceeds *with complete retention of configuration* about the chiral center of the migrating group.

These results tell us two things. First, nitrogen takes the same relative position on the chiral carbon that was originally occupied by the carbonyl carbon. Second, the chiral carbon does not break away from the carbonyl carbon until it has started to attach itself to nitrogen. If the group were actually to become free during its migration, we would expect considerable loss of configuration and hence a partially racemic product. (If the group were to become free—*really* free—we would expect reaction to be, in part, intermolecular, also contrary to fact.)

We may picture the migrating group as moving from carbon to nitrogen via a transition state, I, in which carbon is pentavalent:

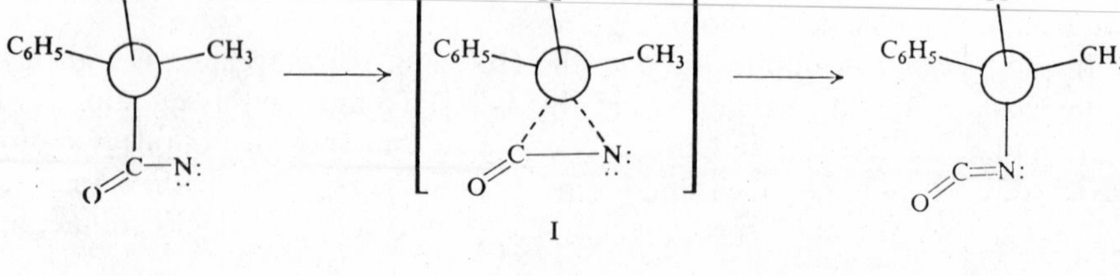

H, C_6H_5, CH_3 — C(=O)—N: $\longrightarrow$ [H, C_6H_5, CH_3 — bridging C(=O)···N:] $\longrightarrow$ H, C_6H_5, CH_3 — N=C=O

I

The migrating group *steps* from atom to atom; it does *not* jump.

There is much evidence to suggest that the stereochemistry of all 1,2-shifts has this common feature: *complete retention of configuration in the migrating group.*

Problem 28.3 Many years before the Hofmann degradation of optically active α-phenylpropionamide was studied, the following observations were made: when the cyclopentane derivative II, in which the —COOH and —$CONH_2$ groups are *cis* to each other, was treated with hypobromite, compound III was obtained; compound III could be converted by heat into the amide IV (called a *lactam*). What do these results show about the mechanism of the rearrangement? (*Use models.*)

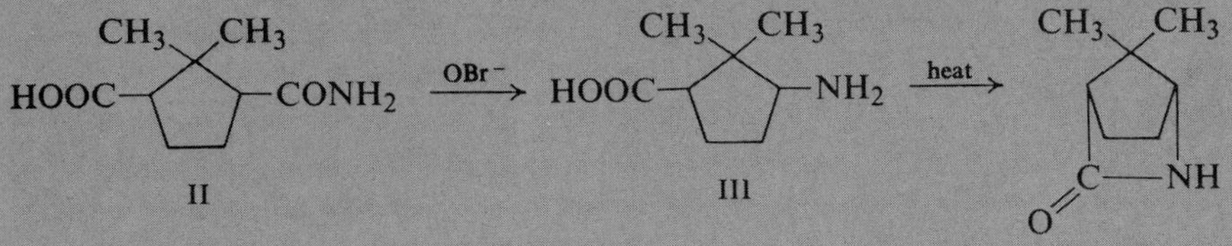

CH_3 CH_3 HOOC—(cyclopentane)—$CONH_2$ (II) $\xrightarrow{OBr^-}$ CH_3 CH_3 HOOC—(cyclopentane)—NH_2 (III) $\xrightarrow{heat}$ CH_3 CH_3 (bicyclic lactam, O=C—NH) (IV)

28.5 Hofmann rearrangement. Timing of the steps

We said that steps (3) and (4) of the mechanism are believed to be simultaneous, that is, that loss of bromide ion and migration occur in the same step:

(3,4) $$R{-}C({=}O){-}\ddot{N}^{\ominus}{-}Br \longrightarrow R{-}N{=}C{=}O + Br^-$$

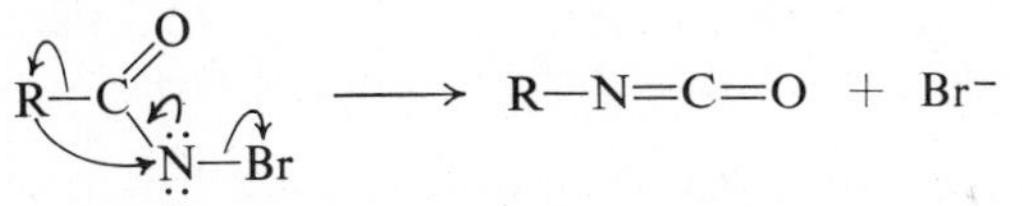

One reason for believing this is simply the anticipated difficulty of forming a highly unstable intermediate in which an electronegative element like nitrogen has only a sextet of electrons. Such a particle should be even less stable than primary carbocations, and those, we know, are seldom formed; reaction takes the easier, S_N2-like path. Another reason is the effect of structure on rate of reaction. Let us examine this second reason.

When the migrating group is aryl, the rate of the Hofmann degradation is increased by the presence of electron-releasing substituents in the aromatic ring; thus substituted benzamides show the following order of reactivity:

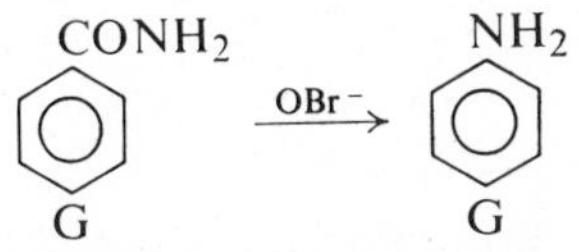

$$G{-}C_6H_4{-}CONH_2 \xrightarrow{OBr^-} G{-}C_6H_4{-}NH_2$$

G: $-OCH_3 > -CH_3 > -H > -Cl > -NO_2$

Now, how could electron release speed up Hofmann degradation? One way could be through its effect on the rate of migration. Migration of an alkyl group must involve a transition state containing pentavalent carbon, like I in the preceding section. Migration of an aryl group, on the other hand, takes place via a structure like V. This structure is a familiar one; from the standpoint of the migrating aryl group, rearrangement is simply electrophilic aromatic substitution, with the electron-deficient atom—nitrogen, in this case—acting as the attacking reagent. In at least some rearrangements, as we shall see, there is evidence that structures

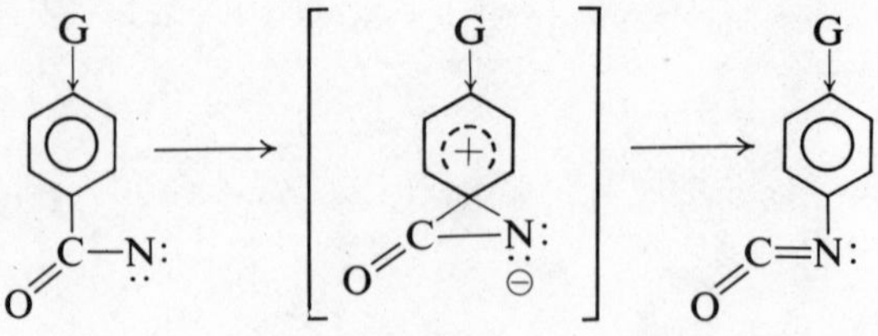

V

like V are actual intermediate compounds, as in the ordinary kind of electrophilic aromatic substitution (Sec. 11.16). Electron-releasing groups disperse the developing charge on the aromatic ring and thus speed up formation of V. Viewed in this way, substituents affect the rate of rearrangement—the *migratory aptitude*—of an aryl group in exactly the same way as they affect the rate of aromatic nitration, halogenation, or sulfonation. (As we shall see, however, conformational effects can sometimes completely outweigh these electronic effects.)

There is another way in which electron release might be speeding up reaction: by speeding up formation of the electron-deficient species in equation (3). But the observed effect is a strong one, and more consistent with the development of the positive charge *in the ring itself*, as during rearrangement.

We should be clear about what the question is here. It is *not* whether some groups migrate faster than others—there is little doubt about that—but whether the rate of rearrangement affects the overall rate—the *measured* rate—of the Hofmann degradation.

It is likely, then, that electron-releasing substituents speed up Hofmann degradation by speeding up rearrangement. Now, under what conditions can this happen? Consider the sequence (3) and (4). Loss of bromide ion (3) could be fast

(3) $\mathrm{R{-}C({=}O){-}\ddot{N}^{\ominus}{-}Br \longrightarrow R{-}C({=}O){-}\ddot{N} + Br^-}$

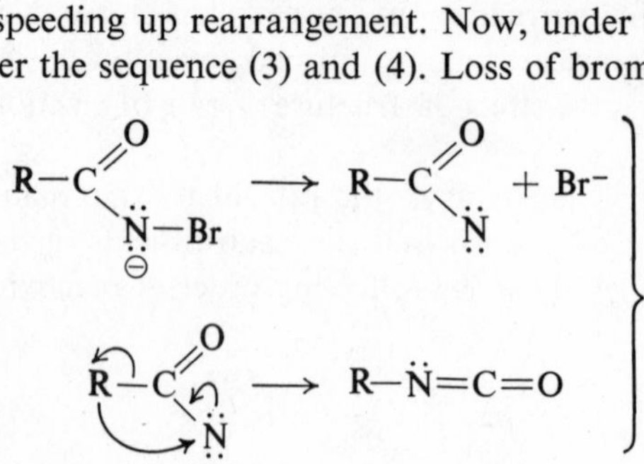

(4) $\mathrm{R{-}C({=}O){-}\ddot{N} \longrightarrow R{-}\ddot{N}{=}C{=}O}$

Simultaneous

and reversible, followed by slow rearrangement (4). Rearrangement would be rate-determining, as required, but in that case something else would not fit. The reverse of (3) is combination of the particle ArCON with bromide ion; if this were taking place, so should combination of ArCON with the solvent, water—more abundant and more nucleophilic—to form the *hydroxyamic acid* ArCONHOH. But hydroxamic acids are *not* formed in the Hofmann degradation.

If ArCON were indeed an intermediate, then, it would have to be undergoing rearrangement as fast as it is formed; that is, (4) would have to be fast compared with (3). But in that case, the overall rate would be independent of the rate of rearrangement, contrary to fact.

We are left with the concerted mechanism (3,4). Attachment of the migrating group helps to push out bromide ion, and overall rate *does* depend on the rate of rearrangement. As the amount of anchimeric assistance varies, so does the observed rate of reaction.

At the migrating group, we said, rearrangement amounts to electrophilic substitution. But at the electron-deficient nitrogen, rearrangement amounts to *nucleophilic* substitution: the migrating group (with its electrons) is a nucleophile, and bromide ion is the leaving group. The sequence (3) and (4) corresponds to an S_N1 mechanism; the concerted reaction (3,4) corresponds to a S_N2 mechanism. Dependence of overall rate on the nature of the nucleophile is consistent with the S_N2-like mechanism, but not with the S_N1-like mechanism.

28.6 Rearrangement of hydroperoxides. Migration to electron-deficient oxygen

In Sec. 24.4 we encountered the synthesis of phenol via cumene hydroperoxide:

$$C_6H_5CH(CH_3)_2 \xrightarrow{O_2} C_6H_5{-}\underset{CH_3}{\overset{CH_3}{\underset{|}{\overset{|}{C}}}}{-}O{-}OH \xrightarrow{H_2O,\ H^+} C_6H_5OH + CH_3COCH_3$$

Isopropylbenzene (Cumene) → Cumene hydroperoxide → Phenol + Acetone

The phenyl group is joined to carbon in the hydroperoxide and to oxygen in phenol: clearly, rearrangement takes place. This time, it involves a 1,2-shift to electron-deficient *oxygen*.

(1) $$CH_3{-}\underset{CH_3}{\overset{C_6H_5}{\underset{|}{\overset{|}{C}}}}{-}O{-}OH + H^+ \rightleftharpoons CH_3{-}\underset{CH_3}{\overset{C_6H_5}{\underset{|}{\overset{|}{C}}}}{-}O{-}\overset{+}{O}H_2$$

Cumene hydroperoxide
I

(2) $$CH_3{-}\underset{CH_3}{\overset{C_6H_5}{\underset{|}{\overset{|}{C}}}}{-}O{-}\overset{+}{O}H_2 \longrightarrow CH_3{-}\underset{CH_3}{\overset{C_6H_5}{\underset{|}{\overset{|}{C}}}}{-}O^+ + H_2O$$

(3) $$CH_3{-}\underset{CH_3}{\overset{C_6H_5}{\underset{|}{\overset{|}{C}}}}{-}O^+ \longrightarrow CH_3{-}\underset{CH_3}{\underset{|}{C}}{=}\overset{+}{O}{-}C_6H_5$$

II

Steps (2) and (3): *Simultaneous*

(4) $CH_3-\underset{CH_3}{\underset{|}{C}}=\overset{+}{O}-C_6H_5 + H_2O \longrightarrow CH_3-\underset{CH_3}{\underset{|}{\overset{+OH_2}{\overset{|}{C}}}}-O-C_6H_5 \rightleftharpoons CH_3-\underset{CH_3}{\underset{|}{\overset{OH}{\overset{|}{C}}}}-O-C_6H_5 + H^+$

II → III

(5) $CH_3-\underset{CH_3}{\underset{|}{\overset{OH}{\overset{|}{C}}}}-O-C_6H_5 \xrightarrow{H^+} CH_3-\underset{CH_3}{\underset{|}{C}}=O + HO-C_6H_5$

III → Acetone + Phenol

Acid converts (step 1) the peroxide I into the protonated peroxide, which loses (step 2) a molecule of water to form an intermediate in which oxygen bears only six electrons. A 1,2-shift of the phenyl group from carbon to electron-deficient oxygen yields (step 3) the "carbonium" ion II, which reacts with water to yield (step 4) the hydroxy compound III. Compound III is a hemi-acetal (Sec. 19.15) which breaks down (step 5) to give phenol and acetone.

Every step of the reaction involves chemistry with which we are already quite familiar: protonation of a hydroxy compound with subsequent ionization to leave an electron-deficient particle; a 1,2-shift to an electron-deficient atom; reaction of a carbonium ion with water to yield a hydroxy compound; decomposition of a hemi-acetal. In studying organic chemistry we encounter many new things; but much of what seems new is found to fit into old familiar patterns of behavior.

It is very probable that steps (2) and (3) are simultaneous, the migrating phenyl group helping to push out (2,3) the molecule of water. This concerted

(2,3) $CH_3-\underset{CH_3}{\underset{|}{\overset{Ph}{\overset{|}{C}}}}-O-OH_2^+ \longrightarrow CH_3-\underset{CH_3}{\underset{|}{C}}=\overset{+}{O}-Ph + H_2O$

mechanism is supported by the same line of reasoning that we applied to the Hofmann rearrangement. (a) A highly unstable intermediate containing oxygen with only a sextet of electrons should be very difficult to form. (b) There is evidence that, if there *is* such an intermediate, it must undergo rearrangement as fast as it is formed; that is, if (2) and (3) are separate steps, (3) must be fast compared with (2). (c) The rate of overall reaction is speeded up by electron-releasing substituents in migrating aryl groups, and in a way that resembles, *quantitatively*, the effect of these groups on ordinary electrophilic aromatic substitution. Almost certainly, then, substituents affect the overall rate of reaction by affecting the rate of migration, and hence migration must take place in the rate-determining step. This rules out the possibility of a fast (3), and leaves us with the concerted reaction (2,3).

Problem 28.4 When α-phenylethyl hydroperoxide, $C_6H_5CH(CH_3)O{-}OH$, undergoes acid-catalyzed rearrangement in $H_2{}^{18}O$, recovered unrearranged hydroperoxide is found to contain *no* oxygen-18. Taken with the other evidence, what does this finding tell us about the mechanism of reaction?

28.7 Rearrangement of hydroperoxides. Migratory aptitude

The rearrangement of hydroperoxides lets us see something that the Hofmann rearrangement could not: the preferential migration of one group rather than another. That is, we can observe the relative speeds of migration—the relative migratory aptitudes—of two groups, not as a difference in rate of reaction, but as a difference in the product obtained. In cumene hydroperoxide, for example, any one of three groups could migrate: phenyl and two methyls. If, instead of phenyl,

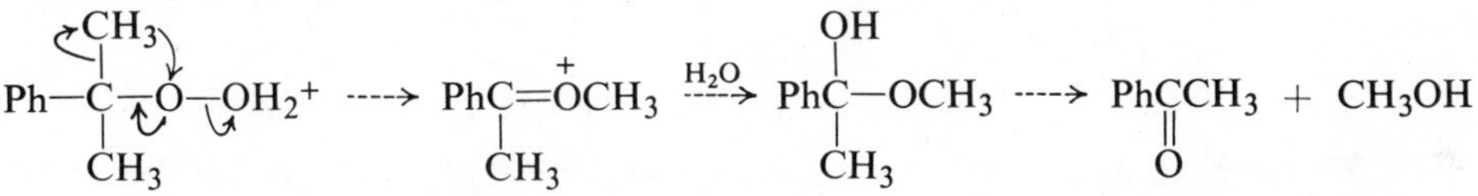

$$\text{Ph}-\underset{\text{CH}_3}{\overset{\text{CH}_3}{\text{C}}}-\text{O}-\text{OH}_2^+ \dashrightarrow \underset{\text{CH}_3}{\text{PhC}}=\overset{+}{\text{O}}\text{CH}_3 \xrightarrow{\text{H}_2\text{O}} \text{Ph}\underset{\text{CH}_3}{\overset{\text{OH}}{\text{C}}}-\text{OCH}_3 \dashrightarrow \text{Ph}\underset{\text{O}}{\overset{}{\text{C}}}\text{CH}_3 + \text{CH}_3\text{OH}$$

Acetophenone Methanol
Not obtained

methyl were to migrate, reaction would be expected to yield methanol and acetophenone. Actually, phenol and acetone are formed quantitatively, showing that a phenyl group migrates much faster than a methyl.

It is generally true in 1,2-shifts that aryl groups have greater migratory aptitudes than alkyl groups. We can see why this should be so. Migration of an alkyl group must involve a transition state containing pentavalent carbon (IV). Migration

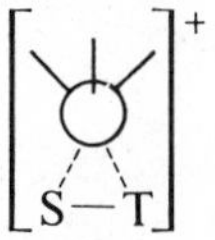

IV
Alkyl migration:
pentavalent carbon

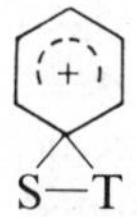

V
Aryl migration:
benzenonium ion

of an aryl group, on the other hand, takes place via a structure of the benzenonium ion type (V); transition state or actual intermediate, V clearly offers an easier path for migration than does IV.

The hydroperoxide may contain several aryl groups and, if they are different, we can observe competition in migration between them, too. As was observed in

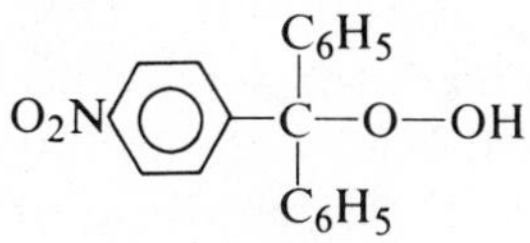

$$O_2N{-}C_6H_4{-}C(C_6H_5)_2{-}O{-}OH \qquad\qquad O_2N{-}C_6H_4{-}\underset{\text{O}}{\overset{}{\text{C}}}{-}C_6H_5$$

VI 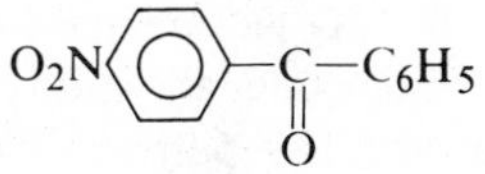 VII the rate study, the relative migratory aptitude of an aryl group is raised by electron-releasing substituents, and lowered by electron-withdrawing substituents. For example, when *p*-nitrotriphenylmethyl hydroperoxide (VI) is treated with acid, it yields exclusively phenol and *p*-nitrobenzophenone (VII); as we would have expected, phenyl migrates in preference to *p*-nitrophenyl.

Problem 28.5 When *p*-methylbenzyl hydroperoxide, *p*-$CH_3C_6H_4CH_2O—OH$, is treated with acid, there are obtained *p*-methylbenzaldehyde (61%) and *p*-cresol (38%). (a) How do you account for the formation of each of these? What other products must have been formed? (b) What do the relative yields of the aromatic products show?

Problem 28.6 Treatment of aliphatic hydroperoxides, $RCH_2O—OH$ and $R_2CHO—OH$, with aqueous acid yields aldehydes and ketones as the only organic products. What conclusion do you draw about migratory aptitudes?

28.8 Pinacol rearrangement. Migration to electron-deficient carbon

Upon treatment with mineral acids, 2,3-dimethyl-2,3-butanediol (often called *pinacol*) is converted into methyl *tert*-butyl ketone (often called *pinacolone*). The

$$CH_3—\underset{OH}{\overset{CH_3}{C}}—\underset{OH}{\overset{CH_3}{C}}—CH_3 \xrightarrow{H^+} CH_3—\underset{O}{\overset{\|}{C}}—\underset{CH_3}{\overset{CH_3}{C}}—CH_3 + H_2O$$

Pinacol
2,3-Dimethyl-2,3-butanediol

Pinacolone
Methyl *tert*-butyl ketone
3,3-Dimethyl-2-butanone

glycol undergoes dehydration, and in such a way that rearrangement of the carbon skeleton occurs. Other glycols undergo analogous reactions, which are known collectively as **pinacol rearrangements.**

The pinacol rearrangement is believed to involve two important steps: (1) loss of water from the protonated glycol to form a carbonium ion; and (2) rearrangement of the carbonium ion by a 1,2-shift to yield the protonated ketone.

$$(1)\quad R—\underset{OH}{\overset{R}{C}}—\underset{OH}{\overset{R}{C}}—R \overset{H^+}{\rightleftarrows} R—\underset{OH}{\overset{R}{C}}—\underset{^+OH_2}{\overset{R}{C}}—R \longrightarrow H_2O + R—\underset{OH}{\overset{R}{C}}—\underset{\oplus}{\overset{R}{C}}—R$$

$$(2)\quad R—\underset{OH}{\overset{R}{C}}—\underset{\oplus}{\overset{R}{C}}—R \longrightarrow R—\underset{\oplus OH}{\overset{\|}{C}}—\underset{R}{\overset{R}{C}}—R \rightleftarrows H^+ + R—\underset{O}{\overset{\|}{C}}—\underset{R}{\overset{R}{C}}—R$$

Both steps in this reaction are already familiar to us: formation of a carbonium ion from an alcohol under the influence of acid, followed by a 1,2-shift to the electron-deficient atom. The pattern is also familiar: rearrangement of a cation to a more stable cation, in this case to the protonated ketone. The driving force is the usual one behind carbonium ion reactions: the need to provide the electron-deficient carbon with electrons. The special feature of the pinacol rearrangement is the presence in the molecule of the second oxygen atom; it is this oxygen atom, with its unshared pairs, that ultimately provides the needed electrons.

Problem 28.7 Account for the products of the following reactions:

(a) 1,1,2-triphenyl-2-amino-1-propanol $\xrightarrow{\text{HONO}}$ 1,2,2-triphenyl-1-propanone
(*Hint:* See Problem 23.11, p. 763.)

(b) 2-phenyl-1-iodo-2-propanol + Ag^+ $\longrightarrow$ benzyl methyl ketone

When the groups attached to the carbon atoms bearing —OH differ from one another, the pinacol rearrangement can conceivably give rise to more than one compound. The product actually obtained is determined (a) by which —OH group is lost in step (1), and then (b) by which group migrates in step (2) to the electron-deficient carbon thus formed. For example, let us consider the rearrangement of 1-phenyl-1,2-propanediol. The structure of the product actually obtained, methyl benzyl ketone, indicates that the benzyl cation (I) is formed in preference to the secondary cation (II), and that —H migrates in preference to $—CH_3$.

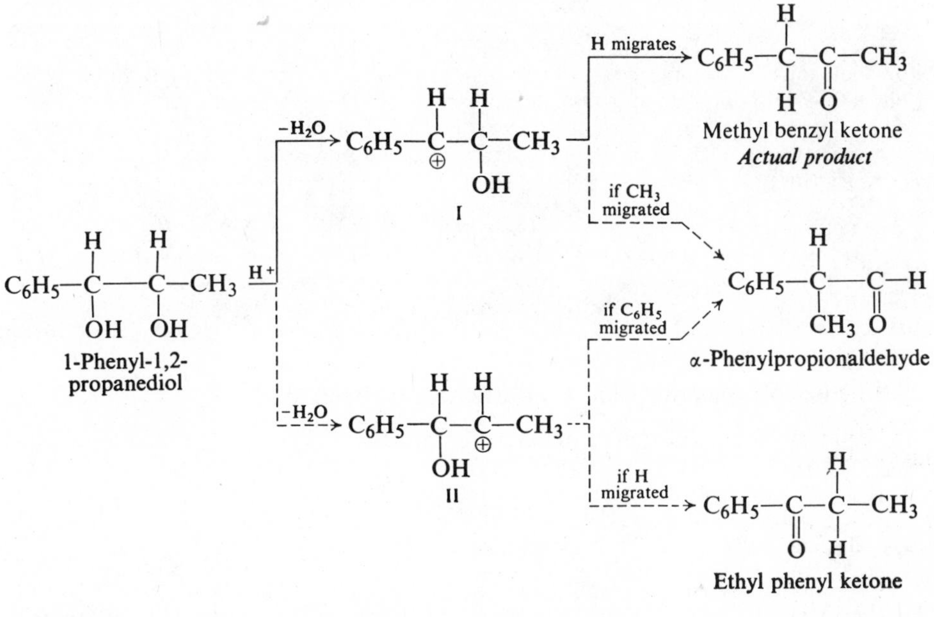

$$C_6H_5-\underset{OH}{\overset{H}{C}}-\underset{OH}{\overset{H}{C}}-CH_3 \xrightarrow{H^+,\ -H_2O} C_6H_5-\underset{\oplus}{\overset{H}{C}}-\underset{OH}{\overset{H}{C}}-CH_3 \xrightarrow{\text{H migrates}} C_6H_5-\underset{H}{\overset{H}{C}}-\underset{O}{\overset{\|}{C}}-CH_3$$

1-Phenyl-1,2-propanediol; I; Methyl benzyl ketone *Actual product*

I $\xrightarrow{\text{if } CH_3 \text{ migrated}}$ $C_6H_5-\underset{CH_3}{\overset{H}{C}}-\underset{O}{\overset{\|}{C}}-H$ $\xleftarrow{\text{if } C_6H_5 \text{ migrated}}$ II

α-Phenylpropionaldehyde

$$C_6H_5-\underset{OH}{\overset{H}{C}}-\underset{OH}{\overset{H}{C}}-CH_3 \xrightarrow{H^+,\ -H_2O} C_6H_5-\underset{OH}{\overset{H}{C}}-\underset{\oplus}{\overset{H}{C}}-CH_3 \xrightarrow{\text{if H migrated}} C_6H_5-\underset{O}{\overset{\|}{C}}-\underset{H}{\overset{H}{C}}-CH_3$$

II; Ethyl phenyl ketone

Study of a large number of pinacol rearrangements has shown that usually the product is the one expected if, first, ionization occurs to yield the more stable carbonium ion, and then, once the preferred ionization has taken place, migration takes place according to the sequence —Ar > —R. (We have already seen how it is that an aryl group migrates faster than an alkyl.) Hydrogen can migrate, too, but we cannot predict its relative migratory aptitude. Hydrogen may migrate in preference to —R or —Ar, but this is not always the case; indeed, it sometimes happens that with a given pinacol either —H or —R can migrate, depending upon experimental conditions.

Among aryl groups, relative migratory aptitude depends—other things being equal—on the ability of the ring to accommodate a positive charge. However, as we shall see in the next section, strong stereochemical factors can be involved, and may outweigh these electronic factors.

Problem 28.8 For the rearrangement of each of the following glycols show which carbonium ion you would expect to be the more stable, and then the rearrangement that this carbonium ion would most likely undergo:

(a) 1,2-propanediol
(b) 2-methyl-1,2-propanediol
(c) 1-phenyl-1,2-ethanediol
(d) 1,1-diphenyl-1,2-ethanediol
(e) 1-phenyl-1,2-propanediol
(f) 1,1-diphenyl-2,2-dimethyl-1,2-ethanediol
(g) 1,1,2-triphenyl-2-methyl-1,2-ethanediol
(h) 2-methyl-3-ethyl-2,3-pentanediol
(i) 1,1-bis(*p*-methoxyphenyl)-2,2-diphenyl-1,2-ethanediol

We have depicted the pinacol rearrangement as a two-step process with an actual carbonium ion as intermediate. There is good evidence that this is so, at least when a tertiary or benzylic cation can be formed. Evidently the stability of the incipient cation in the transition state permits (S_N1-like) loss of water without anchimeric assistance from the migrating group. This is, we note, in contrast to what happens in migration to electron-deficient nitrogen or oxygen.

Problem 28.9 The following reactions have all been found to yield a mixture of pinacol and pinacolone, and *in the same proportions*: treatment of 3-amino-2,3-dimethyl-2-butanol with nitrous acid; treatment of 3-chloro-2,3-dimethyl-2-butanol with aqueous silver ion; and acid-catalyzed hydrolysis of the epoxide of 2,3-dimethyl-2-butene. What does this finding indicate about the mechanism of the pinacol rearrangement?

Problem 28.10 When pinacol was treated with acid in $H_2{}^{18}O$ solution, recovered unrearranged pinacol was found to contain oxygen-18. Studies showed that oxygen exchange took place two to three times as fast as rearrangement. What bearing does this fact have on the mechanism of rearrangement?

28.9 Pinacolic deamination. Conformational effects

Primary aliphatic amines react with nitrous acid to form diazonium salts (Sec. 23.10).

$$\underset{1^\circ\text{ aliphatic amine}}{RNH_2} + NaNO_2 + HX \longrightarrow RN_2{}^+X^- \longrightarrow R^+X^- + N_2$$

$$R^+X^- \longrightarrow \text{products}$$

Unlike their aromatic counterparts, however, these diazonium ions are extremely unstable, and lose nitrogen rapidly to give products that strongly suggest intermediate formation of carbonium ions (Problem 23.11, p. 763).

If such an amino group is located *alpha* to a hydroxyl group, then treatment with nitrous acid causes a reaction closely related to the pinacol rearrangement, *pinacolic deamination*:

$$\mathrm{R_2C(OH)\!-\!CR(NH_2)R} \xrightarrow{\text{HONO}} \mathrm{R_2C(OH)\!-\!\overset{\oplus}{C}R_2} \longrightarrow \mathrm{RC(\!=\!\overset{\oplus}{O}H)\!-\!CR_3} \longrightarrow \mathrm{RC(\!=\!O)\!-\!CR_3}$$

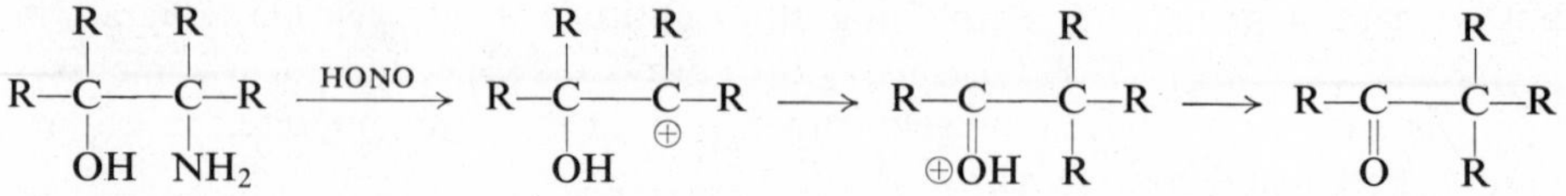

This system permits many studies not possible with pinacols, since here the electron-deficiency is generated at a pre-determined position: at the carbon that held the amino group.

Problem 28.11 Give the structure of the carbonium ion generated: (a) by action of acid on 1,1-diphenyl-1,2-propanediol; (b) by action of nitrous acid on 1,1-diphenyl-2-amino-1-propanol.

Let us examine the stereochemistry of pinacolic deamination in some detail. In this we shall see the operation of a factor we have not yet encountered in rearrangements: *conformational effects*. More important, we shall get some idea of the methods used to attack problems like this.

When optically active 2-amino-1,1-diphenyl-1-propanol is treated with nitrous acid, there is obtained 1,2-diphenyl-1-propanone of inverted configuration but lower optical purity than the starting material. Reaction has taken place with

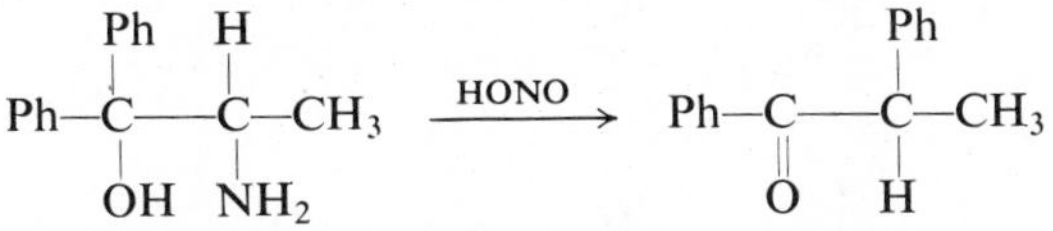

Inversion (77%) plus racemization (23%)
equivalent to
inversion (88.5%) plus retention (11.5%)

racemization plus inversion: stereochemistry typical of S_N1 reactions, and consistent with the idea of an open carbonium ion as intermediate.

In a series of elegant experiments, Clair Collins (of Oak Ridge National Laboratory) has given us intimate details about the reaction: the intermediacy of open carbonium ions, their approximate life-time, and the conformational factors that affect their chemistry. Collins, too, carried out deamination of optically active 2-amino-1,1-diphenyl-1-propanol, but his starting material was labeled stereospecifically (I) with carbon-14 in one of the phenyl groups. He resolved

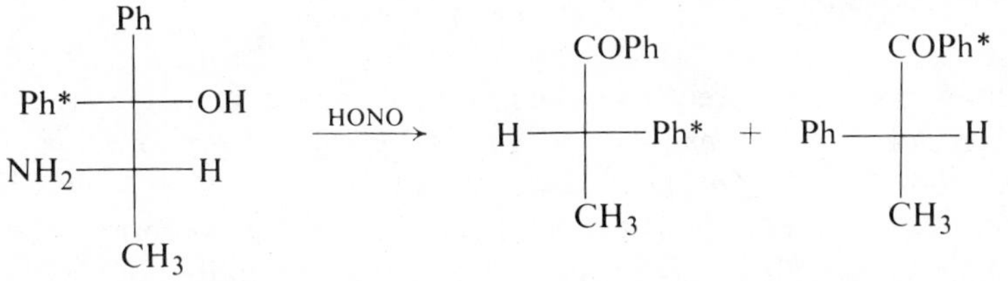

I

1,1-Diphenyl-2-amino-1-propanol

Inversion: *migration of Ph** (88%)

Retention: *migration of Ph* (12%)

the products and, by degradation studies, determined the location of the radioactive label in each. The inverted product had been formed exclusively by migration of the labeled group, Ph*; the product of retained configuration was formed exclusively by migration of the unlabeled group, Ph. (The 12% of retention observed by Collins agrees, of course, quite well with the results of the earlier simple stereochemical study.)

On the basis of these results, Collins pictured the reaction as taking place as shown in Fig. 28.1. Three conclusions were drawn. (a) *An open carbonium ion is formed.* If, instead, migration of phenyl were concerted with loss of N_2, attack

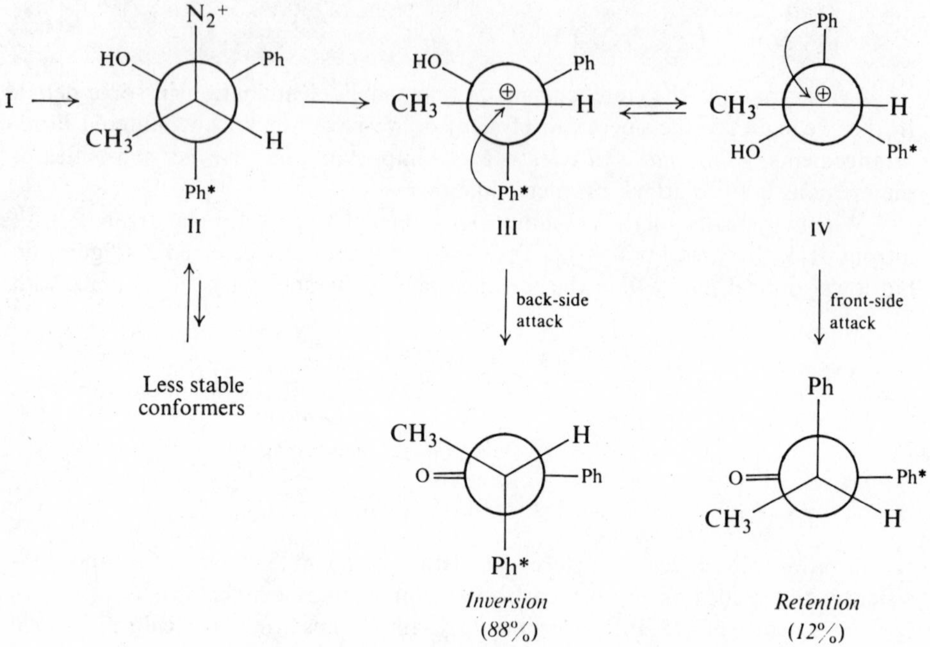

Figure 28.1. Pinacolic deamination of optically active labeled 2-amino-1,1-diphenyl-1-propanol.The most abundant conformer, II, of the diazonium ion yields cation III. Cation III does two things: (a) rearranges by back-side migration of Ph*, and (b) rotates, in the easiest way possible, to form cation IV, which rearranges by front-side migration of Ph.

would have been exclusively back-side, with complete inversion. (b) *The carbonium ion does not last long enough for very much rotation to occur about the central carbon–carbon bond.* If, instead, the carbonium ion were long-lived, there would have been equilibration between the equally stable conformations III and IV, leading to complete racemization and equal migration of Ph and Ph*. (c) *Conformational effects largely determine the course of rearrangement.* The most stable and hence most abundant conformation of the diazonium ion is II, in which the bulky phenyl groups flank tiny hydrogen. Nitrogen is lost to yield carbonium ion III. This first-formed cation is the species that undergoes most of the rearrangement, and in the way consistent with its conformation: back-side migration of Ph*. Some cations last long enough for partial rotation (involving eclipsing only of the small groups $—CH_3$ and —OH) to conformation IV, which rearranges by front-side migration of Ph.

The course of rearrangement is thus determined largely by the conformation of the first-formed ion and, to a lesser extent, of the ion most easily formed from it by limited rotation. These conformations reflect, in turn, the most stable conformation of the parent diazonium ion.

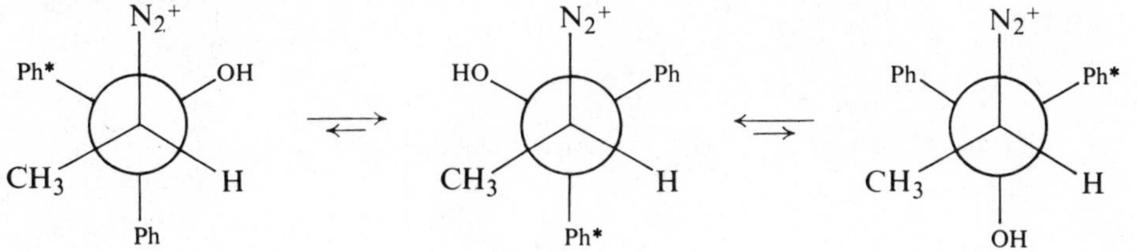

II

Most abundant conformer

Furthermore, we can see that rearrangement of either cation III or cation IV involves a transition state in which the bulky non-migrating groups—methyl and one phenyl—are on opposite sides of the molecule: a so-called *trans* transition state. In contrast, front-side migration of Ph* would require, first, formation of the less stable cation V (either from a less abundant conformation of the diazonium

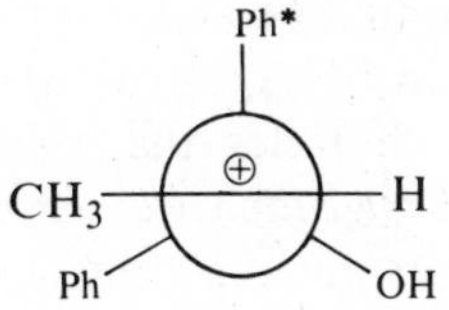

V

CH_3 and Ph crowded together:
cation slow to form,
slow to rearrange

ion or, by rotation, from cation III); and then, its reaction via a crowded *cis* transition state. Both these processes are slow, and their combination does not happen to a measurable extent.

The fine print on page 235 described two extreme situations for the reaction of different conformers. (a) If reaction of the conformers is much faster than the rotation that interconverts them, then the ratio of products obtained reflects the relative populations of the conformers. (b) If reaction of the conformers is much slower than their interconversion, then the ratio of products reflects the relative stabilities of the transition states involved. It was pointed out that, whichever situation exists, we will in general make the same rough prediction about products, since a particular spacial relationship will affect conformer stabilities and transition state stabilities in much the same way.

In pinacolic deamination we have a rather special situation, where reaction and rotation are of roughly comparable speeds, and hence both the populations and the reactivities of conformers affect the product ratio. Most interesting of all, perhaps, is the ingenuity that Collins used to show that this is so.

Conformational factors can determine more than the stereochemistry of rearrangement. In light of Collins' findings, let us examine work done earlier by D. Y. Curtin (of the University of Illinois) with 2-amino-1-anisyl-1-phenyl-1-propanol. This resembles Collins' labeled compound (p. 899), except that an anisyl group (*p*-methoxyphenyl group) takes the place of one of the phenyls. Here, the competition in migration is between a phenyl and an anisyl, instead of between labeled and unlabeled phenyl groups.

Curtin prepared both diastereomeric forms, VI and VII, each as a racemic

Ph
An—|—OH
H_2N—|—H
CH_3
VI (and enantiomer)

$\xrightarrow{\text{HONO}}$

$$\text{Ph—}\underset{\underset{\text{O}}{\|}}{\text{C}}\text{—}\overset{\overset{\text{An}}{|}}{\text{CH}}\text{—CH}_3 + \text{An—}\underset{\underset{\text{O}}{\|}}{\text{C}}\text{—}\overset{\overset{\text{Ph}}{|}}{\text{CH}}\text{—CH}_3$$

Anisyl migration (*94%*) — Phenyl migration (*6%*)

Ph
HO—|—An
H_2N—|—H
CH_3
VII (and enantiomer)

$\xrightarrow{\text{HONO}}$

$$\text{An—}\underset{\underset{\text{O}}{\|}}{\text{C}}\text{—}\overset{\overset{\text{Ph}}{|}}{\text{CH}}\text{—CH}_3 + \text{Ph—}\underset{\underset{\text{O}}{\|}}{\text{C}}\text{—}\overset{\overset{\text{An}}{|}}{\text{CH}}\text{—CH}_3$$

Phenyl migration (*88%*) — Anisyl migration (*12%*)

modification. In the deamination of VI, migration of anisyl was found to exceed that of phenyl, 94:6. This, we might say, is to be expected: with its electron-releasing methoxyl group, anisyl migrates much faster than phenyl. But in the deamination of the diastereomer VII, *phenyl migration was found to exceed that of anisyl, 88:12.* Clearly, migratory aptitude is *not* the controlling factor in the reaction of VII—nor then, most probably, in the reaction of VI, either.

The most reasonable interpretation of Curtin's work is outlined in Fig. 28.2. This assumes a situation exactly analogous to that indicated by Collins' work, a reasonable assumption since anisyl and phenyl are of the same bulk. Whether phenyl or anisyl migrates predominantly depends on which group is in the proper location in the first-formed carbonium ion, and this again depends on the most stable conformation of the parent diazonium ion. The minor product in each case is due to front-side migration of the aryl group brought into position by the easiest rotation of the carbonium ion.

In the case of diastereomer VII, for example, phenyl is in position to migrate in cation VII*b*, and does so. Competing with this migration is rotation about the single bond to form cation VII*c*, which reacts by anisyl migration. We notice that the percentage of back-side attack by phenyl is the same (88%) as in the original stereochemical study of 2-amino-1,1-diphenyl-1-propanol (p. 899). This *should* be so, since the same competition is involved in both cases: phenyl migration *vs.* rotation about a bond that is sterically the same. (Indeed, it was the quantitative similarity of results in the two studies that gave Collins his first clue as to what might be involved in such reactions, and led to his labeling experiment.)

In these particular reactions, then, just which group migrates is controlled, not electronically by intrinsic migratory aptitude, but sterically by conformational factors. This does *not* negate the idea of migratory aptitude. Groups *do* differ in their tendencies to migrate, and in some cases the effects of such differences can be very great. What we see here is simply that conformational factors can, sometimes, outweigh migratory aptitudes.

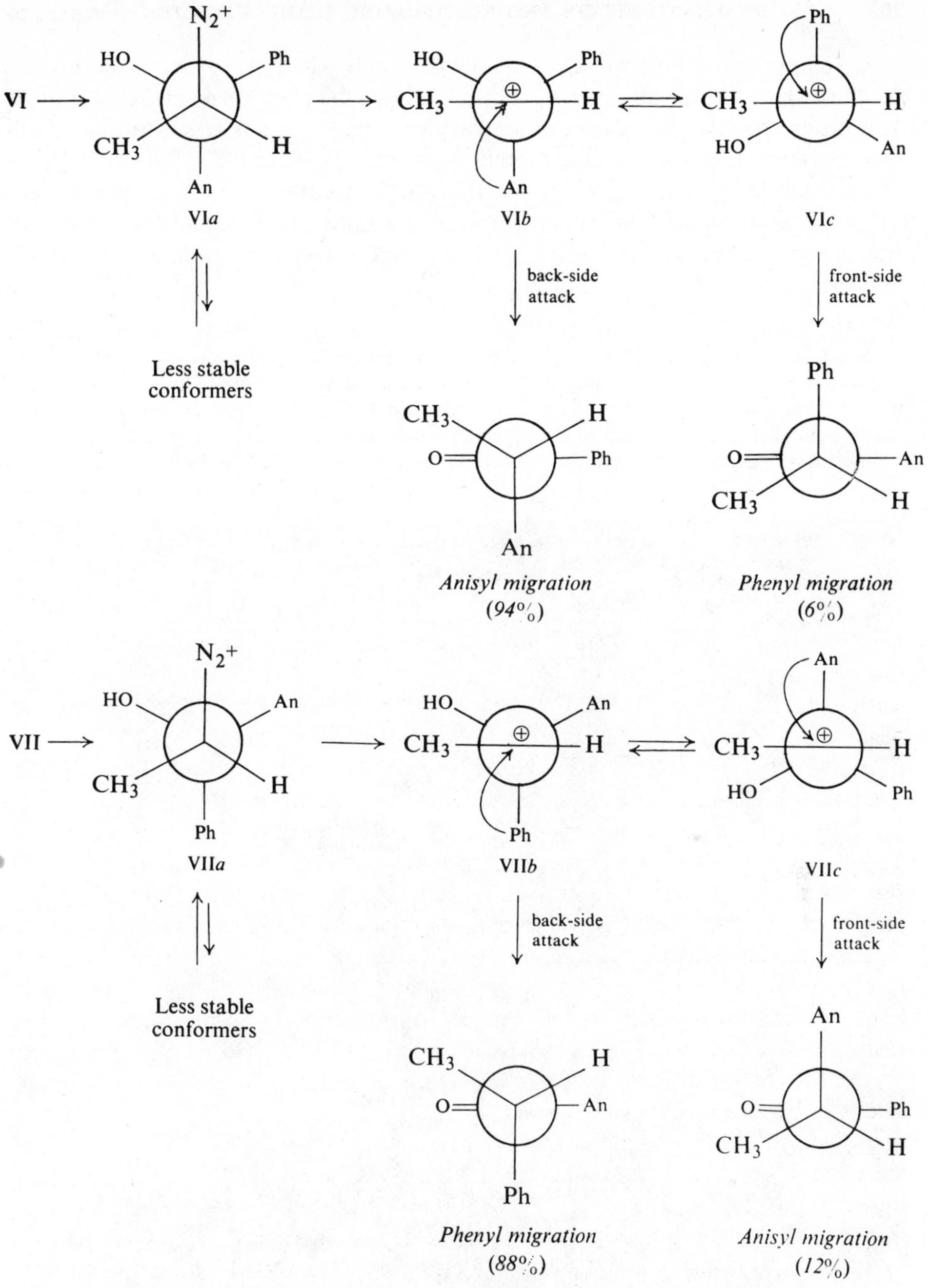

Figure 28.2. Pinacolic deamination of diastereomeric 2-amino-1-anisyl-1-phenyl-1-propanols. In each case the most abundant conformer, VI*a* or VII*a*, of the diazonium ion yields a cation in which an aryl group is in position for back-side migration via a *trans* transition state: anisyl in VI*b*, phenyl in VII*b*. Such rearrangement predominates. Some of each first-formed cation is converted through rotation into another cation, in which the other aryl group is in position for front-side migration via a *trans* transition state: phenyl in VI*c*, anisyl in VII*c*. Such rearrangement gives the minor product.

Another point: here we have reactions not only where steric factors are powerful, but where electronic factors are weak. From the standpoint of the migrating aryl group, remember, rearrangement is electrophilic aromatic substitution with the migration terminus as electrophilic reagent. The reagent, a full-fledged carbonium ion in this case, is highly reactive and hence not very selective; it prefers to attack anisyl rather than phenyl and, other things being equal, would do so. But the preference is not a strong one, and here is less important than the steric factors.

The strength of these electronic factors depends on how badly they are needed. In S_N2-like rearrangements, where the migrating group is needed to help push out the leaving group, differences in migratory tendencies are very great. (That is, the migration terminus is an unreactive reagent and hence is highly selective.) Indeed, as we shall see in Sec. 28.12, the strength of the effects of substituents in migrating aryl groups can be used to measure the relative importance of S_N1-like and S_N2-like rearrangements.

Problem 28.12 When Collins (p. 899) prepared, in optically active form, diastereomers of 1-amino-1-phenyl-2-*p*-tolyl-2-propanol (VIII and IX, Ar = *p*-tolyl) and

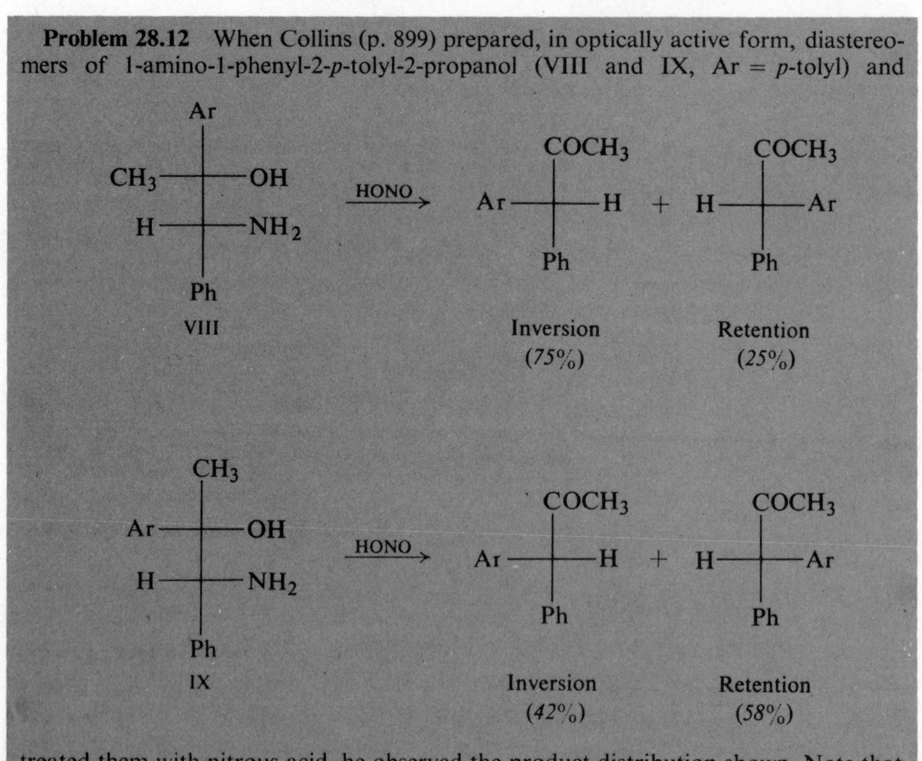

treated them with nitrous acid, he observed the product distribution shown. Note that IX rearranges with *predominant retention*. When Ar = *p*-methoxyphenyl, essentially the same results were obtained. Account in detail for these findings.

28.10 Neighboring group effects: stereochemistry

When treated with concentrated hydrobromic acid, the bromohydrin 3-bromo-2-butanol is converted into 2,3-dibromobutane. This, we say, involves nothing out of the ordinary; it is simply nucleophilic attack (S_N1 or S_N2) by bromide ion on the protonated alcohol. But in 1939 Saul Winstein (p. 474) and Howard J. Lucas (of the California Institute of Technology) described the stereochemistry of this reaction and, in doing this, opened the door to a whole new concept in organic chemistry: the *neighboring group effect.*

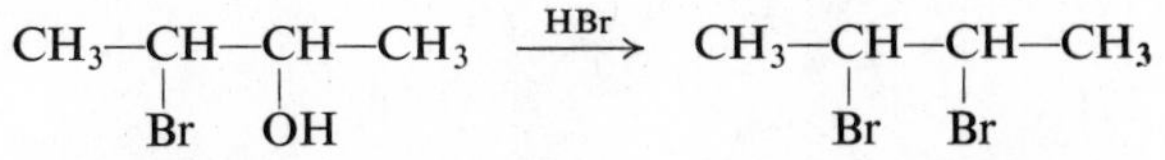

$$\underset{\text{3-Bromo-2-butanol}}{CH_3-\underset{Br}{\underset{|}{CH}}-\underset{OH}{\underset{|}{CH}}-CH_3} \xrightarrow{HBr} \underset{\text{2,3-Dibromobutane}}{CH_3-\underset{Br}{\underset{|}{CH}}-\underset{Br}{\underset{|}{CH}}-CH_3}$$

First, Winstein and Lucas found (Fig. 28.3) that (racemic) *erythro* bromohydrin yields only the *meso* dibromide, and (racemic) *threo* bromohydrin yields

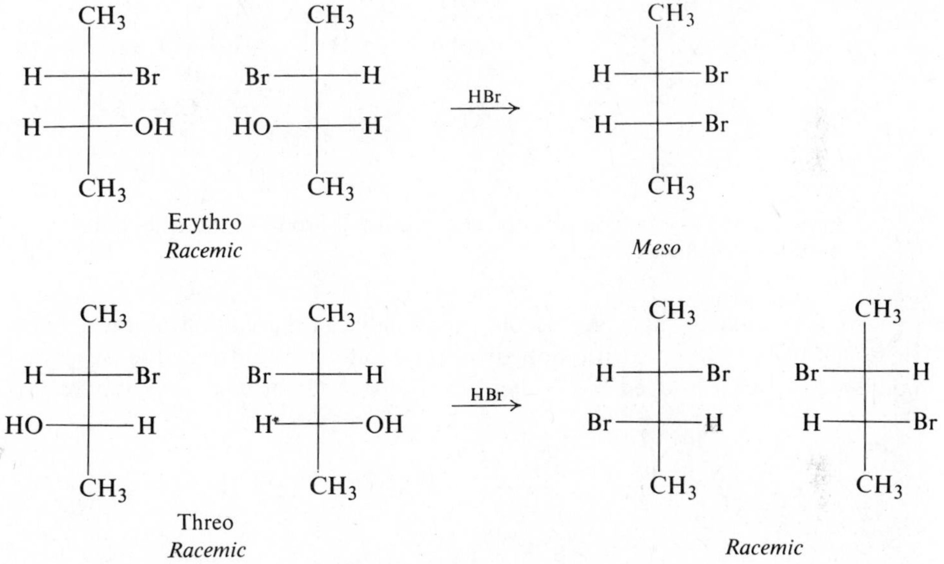

$$\begin{array}{c} CH_3 \\ H-\!\!\!-\!\!\!+\!\!\!-\!\!\!-Br \\ H-\!\!\!-\!\!\!+\!\!\!-\!\!\!-OH \\ CH_3 \end{array} \quad \begin{array}{c} CH_3 \\ Br-\!\!\!-\!\!\!+\!\!\!-\!\!\!-H \\ HO-\!\!\!-\!\!\!+\!\!\!-\!\!\!-H \\ CH_3 \end{array} \quad \xrightarrow{HBr} \quad \begin{array}{c} CH_3 \\ H-\!\!\!-\!\!\!+\!\!\!-\!\!\!-Br \\ H-\!\!\!-\!\!\!+\!\!\!-\!\!\!-Br \\ CH_3 \end{array}$$

Erythro
Racemic

Meso

$$\begin{array}{c} CH_3 \\ H-\!\!\!-\!\!\!+\!\!\!-\!\!\!-Br \\ HO-\!\!\!-\!\!\!+\!\!\!-\!\!\!-H \\ CH_3 \end{array} \quad \begin{array}{c} CH_3 \\ Br-\!\!\!-\!\!\!+\!\!\!-\!\!\!-H \\ H-\!\!\!-\!\!\!+\!\!\!-\!\!\!-OH \\ CH_3 \end{array} \quad \xrightarrow{HBr} \quad \begin{array}{c} CH_3 \\ H-\!\!\!-\!\!\!+\!\!\!-\!\!\!-Br \\ Br-\!\!\!-\!\!\!+\!\!\!-\!\!\!-H \\ CH_3 \end{array} \quad \begin{array}{c} CH_3 \\ Br-\!\!\!-\!\!\!+\!\!\!-\!\!\!-H \\ H-\!\!\!-\!\!\!+\!\!\!-\!\!\!-Br \\ CH_3 \end{array}$$

Threo
Racemic

Racemic

Figure 28.3. Conversion of racemic 3-bromo-2-butanols into 2,3-dibromobutanes.

only the *racemic* dibromide. Apparently, then, reaction proceeds with complete retention of configuration—unusual for nucleophilic substitution. But something even more unusual was still to come.

They carried out the same reaction again but this time used *optically active* starting materials (Fig. 28.4). From optically active *erythro* bromohydrin they obtained, of course, optically inactive product: the *meso* dibromide. But *optically active threo bromohydrin also yielded optically inactive product: the racemic dibromide.*

In one of the products (I) from the *threo* bromohydrin, there is retention of configuration. But in the other product (II), there is inversion, not only at the carbon that held the hydroxyl group, but also at the carbon that held bromine—a carbon that, on the surface, is not even involved in the reaction. How is one to account for the fact that exactly half the molecules react with complete retention, and the other half with this strange double inversion?

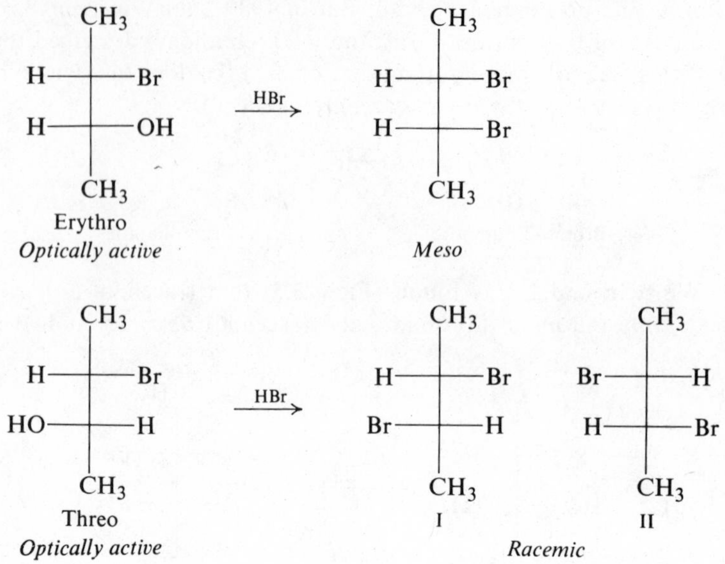

Erythro (H–C–Br, H–C–OH between two CH_3) *Optically active* $\xrightarrow{HBr}$ *Meso* (H–C–Br, H–C–Br between two CH_3)

Threo (H–C–Br, HO–C–H between two CH_3) *Optically active* $\xrightarrow{HBr}$ I (H–C–Br, Br–C–H) + II (Br–C–H, H–C–Br) *Racemic*

Figure 28.4. Conversion of optically active 3-bromo-2-butanols into 2,3-dibromobutanes.

Winstein and Lucas gave the only reasonable interpretation of these facts. In step (1) the protonated bromohydrin loses water to yield, not the open carbonium ion, but a bridged bromonium ion. In step (2) bromide ion attacks this

(1) 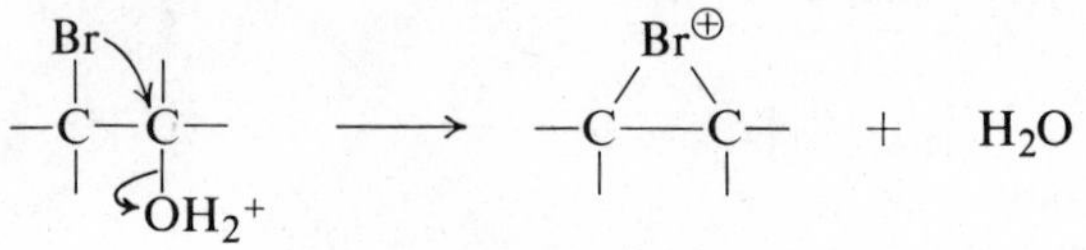 $-\overset{Br}{\overset{|}{C}}-\underset{OH_2^+}{\underset{|}{C}}- \longrightarrow -C\overset{Br^{\oplus}}{—}C- + H_2O$

A bromonium ion

(2) 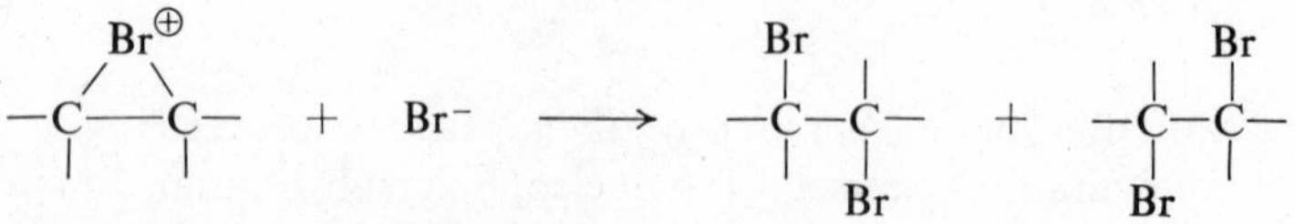 $-C\overset{Br^{\oplus}}{—}C- + Br^- \longrightarrow -\overset{Br}{\overset{|}{C}}-\underset{Br}{\underset{|}{C}}- + -\underset{Br}{\underset{|}{C}}-\overset{Br}{\overset{|}{C}}-$

bromonium ion to give the dibromide. But it can attack the bromonium ion *at either of two carbon atoms*: attack at one gives the product with retention at both chiral centers; attack at the other gives the product with inversion about both centers. Figure 28.5 depicts the reaction of the optically active *threo* bromohydrin.

The bromonium ion has the same structure as that proposed two years earlier by Roberts and Kimball (Sec. 7.12) as an intermediate in the addition of bromine to alkenes. Here it is formed in a different way, but its reaction is the same, and so is the final product.

Reaction consists of two successive nucleophilic substitutions. In the first one the nucleophile is the neighboring bromine; in the second, it is bromide ion from outside the molecule. Both substitutions are pictured as being S_N2-like;

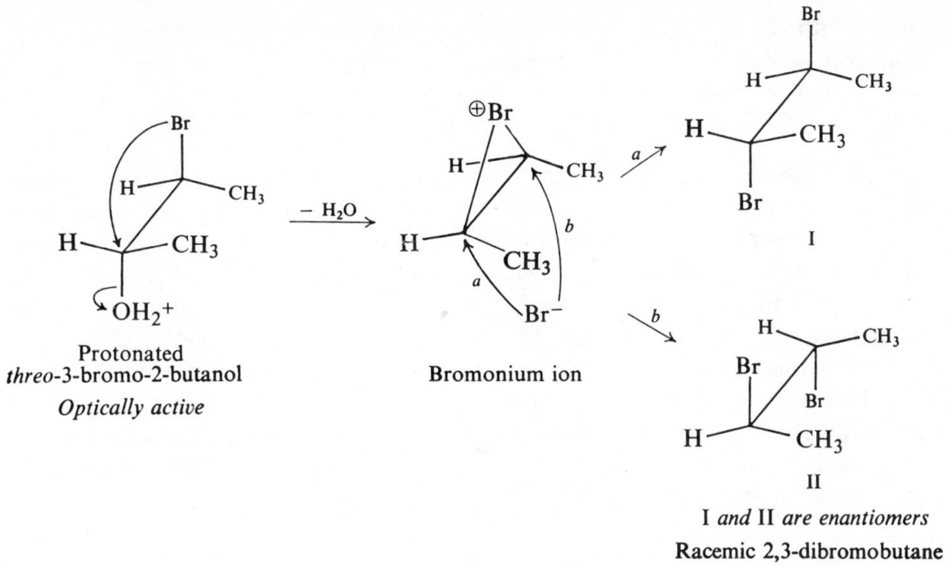

Protonated *threo*-3-bromo-2-butanol
Optically active

Bromonium ion

I and II are enantiomers
Racemic 2,3-dibromobutane

Figure 28.5. Conversion of optically active *threo*-3-bromo-2-butanol into racemic 2,3-dibromobutane via cyclic bromonium ion. Opposite-side attacks *a* and *b* equally likely, give enantiomers in equal amounts.

that is, single-step processes with attachment of the nucleophile and loss of the leaving group taking place in the same transition state. This is consistent with the complete stereospecificity: an open carbonium ion in either (1) or (2) might be expected to result in the formation of a mixture of diastereomers. (As we shall see, there is additional evidence indicating that a neighboring bromine is likely to provide assistance in step (1).)

The basic process is, we see, the same as in rearrangements: intramolecular (1,2) nucleophilic attack. Indeed, there *is* rearrangement here; in half the molecules formed, the bromine has migrated from one carbon to the next.

Problem 28.13 Drawing structures like those in Fig. 28.5, show the stereochemical course of reaction of optically active *erythro*-3-bromo-2-butanol with hydrogen bromide.

Problem 28.14 Actually, the door opened by Winstein and Lucas (Sec. 28.10) was already ajar. In 1937, E. D. Hughes, Ingold (p. 460), and their co-workers reported that, in contrast to the neutral acid or its ester, sodium α-bromopropionate undergoes hydrolysis with *retention* of configuration.

$$CH_3CHBrCOO^-Na^+ \xrightarrow[H_2O]{OH^-} CH_3CHOHCOO^-Na^+$$

Sodium α-bromopropionate — Sodium lactate

Give a likely interpretation of these findings.

28.11 Neighboring group effects: rate of reaction. Anchimeric assistance

Like other alkyl halides, mustard gas (β,β′-dichlorodiethyl sulfide) undergoes hydrolysis. But this hydrolysis is unusual in several ways: (a) the kinetics is first-order, with the rate independent of added base; and (b) it is *enormously* faster than hydrolysis of ordinary primary alkyl chlorides.

$$ClCH_2CH_2—S—CH_2CH_2Cl \xrightarrow{H_2O} ClCH_2CH_2—S—CH_2CH_2OH$$

We have encountered this kind of kinetics before in S_N1 reactions and know, in a general way, what it must mean: in the rate-determining step, the substrate is reacting unimolecularly to form an intermediate, which then reacts rapidly with solvent or other nucleophile. But what is this intermediate? It can hardly be the carbonium ion. A primary cation is highly unstable and hard to form, so that primary alkyl chlorides ordinarily react by S_N2 reactions instead; and here we have electron-withdrawing sulfur further to destabilize a carbonium ion.

This is another example of a neighboring group effect, one that shows itself not in stereochemistry but in *rate of reaction*. Sulfur helps to push out chloride ion, forming a cyclic *sulfonium ion* in the process. As fast as it is formed, this intermediate reacts with water to yield the product.

$$RS{:}\text{—}CH_2\text{—}CH_2Cl \xrightarrow[-Cl^-]{k_1} \underset{\text{A sulfonium ion}}{CH_2\text{—}\overset{R\,|\,S^{\oplus}}{}\text{—}CH_2} \xrightarrow[H_2O]{k_2} RSCH_2CH_2OH$$

$$k_2 \gg k_1$$

Reaction thus involves formation of a cation, but not a highly unstable carbonium ion with its electron-deficient carbon; instead, it is a cation in which every atom has an octet of electrons. Open-chain sulfonium ions, R_3S^+, are well-known, stable molecules; here, because of angle strain, the sulfonium ion is less stable and highly reactive—but still enormously more stable and easier to form than a carbonium ion.

The first, rate-determining step is unimolecular, but it is S_N2-like. As with other primary halides, a nucleophile is needed to help push out the leaving group. Here the nucleophile happens to be part of the same molecule. Sulfur has unshared electrons it is willing to share, and hence is highly nucleophilic. Most important, *it is there*: poised in just the right position for attack. The result is an enormous increase in rate.

There is much additional evidence to support the postulate that the effect of neighboring sulfur is due to anchimeric assistance. Cyclohexyl chloride undergoes solvolysis in ethanol-water to yield a mixture of alcohol and ether. As usual for secondary alkyl substrates, reaction is S_N1 with nucleophilic assistance from the solvent (see Sec. 14.17). A C_6H_5S— group on the adjacent carbon can speed

$$C_6H_{10}(G)Cl \xrightarrow{H_2O/EtOH} C_6H_{10}(G)OH + C_6H_{10}(G)OEt$$

Relative rates of reaction

G: *trans*-C_6H_5S $\gg$ H $>$ *cis*-C_6H_5S
70,000 1.00 0.16

up reaction powerfully—*but only if it is trans to chlorine*. The *cis* substituted chloride actually reacts more slowly than the unsubstituted compound.

The *trans* sulfide group evidently gives strong anchimeric assistance. Why cannot the *cis* sulfide? The answer is found in the examination of molecular models. Like other nucleophiles, a neighboring group attacks carbon at the side away from the leaving group. In an open-chain compound like mustard gas—or like either diastereomer of 3-bromo-2-butanol—rotation about a carbon–carbon bond can bring the neighboring group into the proper position for back-side attack: *anti* to the leaving group (Fig. 28.6*a*). But in cyclohexane derivatives, 1,2-substituents are *anti* to each other only when they both occupy axial positions—possible only for *trans* substituents (Fig. 28.6*b*). Hence, only the *trans* chloride shows the

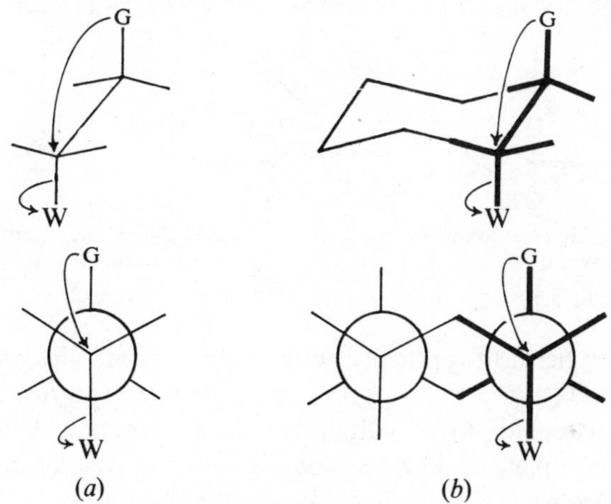

Figure 28.6. Anchimeric assistance. (*a*) *Anti* relationship between neighboring group and leaving group required for back-side attack. (*b*) In cyclohexane derivatives, only *trans*-1,2-substituents can assume *anti* relationship.

neighboring group effect, anchimeric assistance from sulfur. The *cis* isomer reacts without anchimeric assistance; through its electron-withdrawing inductive effect, sulfur slows down formation of the carbonium ion, and thus the rate of reaction.

Let us look at another example of solvolysis. A very commonly studied system is one in which the solvent is acetic acid (HOAc) and the substrates are alkyl esters of sulfonic acids: ROTs, alkyl tosylates (alkyl *p*-toluenesulfonates); ROBs, alkyl

$$\underset{\text{Alkyl tosylate}}{\text{R—OTs}} \xrightarrow[\text{KOAc}]{\text{HOAc}} \underset{\text{Alkyl acetate}}{\text{R—OAc}}$$

brosylates (alkyl *p*-bromobenzenesulfonates); etc. Loss of the weakly basic sulfonate anion, with more or less nucleophilic assistance from the solvent, generates a cation—as part of an ion pair—which combines with the solvent to yield the acetate.

When 2-acetoxycyclohexyl tosylate is heated in acetic acid there is obtained, as expected, the diacetate of 1,2-cyclohexanediol. The reactant exists as diastereomers, and just what happens—and how fast it happens—depends upon which

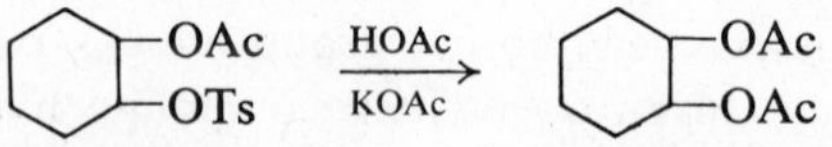

OAc, OTs $\xrightarrow[\text{KOAc}]{\text{HOAc}}$ OAc, OAc

2-Acetoxycyclohexyl tosylate — 1,2-Cyclohexanediol diacetate

diastereomer we start with. The *cis* tosylate yields chiefly the *trans* diacetate. Reaction takes the usual course for nucleophilic substitution, predominant inversion. But the *trans* tosylate also yields *trans* diacetate. Here, apparently, reaction takes place with *retention*, unusual for nucleophilic substitution, and in contrast to what is observed for the *cis* isomer. Two pieces of evidence show us clearly

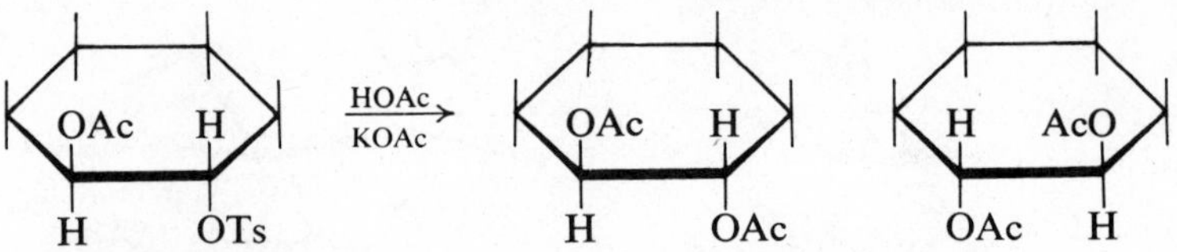

OAc H / H OTs $\xrightarrow[\text{KOAc}]{\text{HOAc}}$ OAc H / H OAc + H AcO / OAc H

(+)-*trans*-2-Acetoxycyclohexyl tosylate
Optically active

trans-1,2-Cyclohexanediol diacetate
Racemic

what is happening here: (a) optically active *trans* tosylate yields *optically inactive trans* diacetate; and (b) the *trans* tosylate reacts *800 times as fast as the cis isomer.*

The apparent retention of configuration in the reaction of the *trans* tosylate is a neighboring group effect. The neighboring group is acetoxy, containing oxygen with unshared electrons. Through back-side nucleophilic attack, acetoxy helps to push out the tosylate anion (1) and, in doing this, inverts the configuration at the

(1) CH_3–C(=O)–O–C–C–OTs ⟶ [CH_3–δ_+C…O…C–C…$OTs^{\delta-}$] ⟶ CH_3–⊕C(O)(O)–C–C– + OTs^-

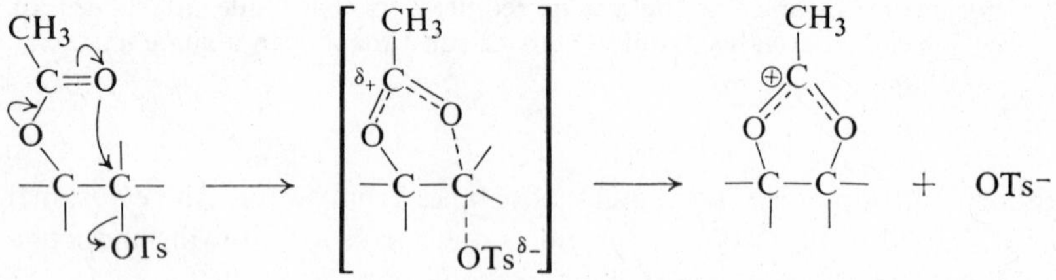

An acetoxonium ion

(2) CH_3–⊕C(O)(O)–C–C– + HOAc $\xrightarrow{-H^+}$ CH_3–C(=O)–O–C–C–OAc + CH_3–C(=O)–O–C–C–AcO

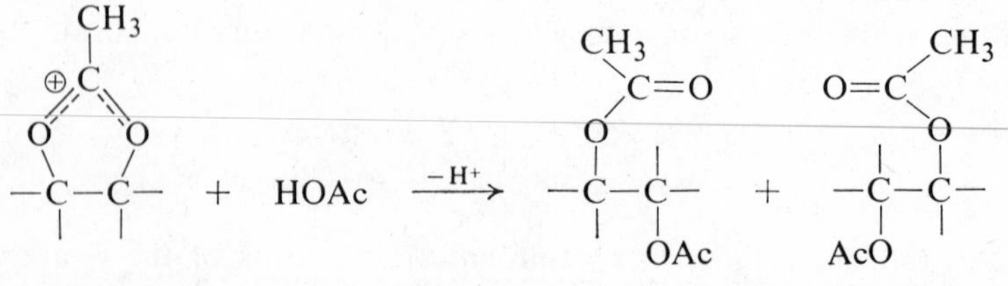

carbon under attack. There is formed an *acetoxonium ion*. This symmetrical intermediate undergoes nucleophilic attack (2) by the solvent at either of two carbons—

again with inversion—and yields the product. The result: in half the molecules, retention at both carbons; in the other half, inversion at both carbons.

The *cis* tosylate cannot assume the diaxial conformation needed for back-side attack by acetoxy, and there is no neighboring group effect. Stereochemistry is normal, and reaction is much slower than for the *trans* tosylate.

Compared with unsubstituted cyclohexyl tosylate, the 2-acetoxycyclohexyl tosylates show the following relative reactivities toward acetolysis:

Cyclohexyl tosylate	>	*trans*-2-Acetoxycyclohexyl tosylate	>	*cis*-2-Acetoxycyclohexyl tosylate
1.00		0.30		0.00045

Reaction of the *cis* tosylate is much slower than that of cyclohexyl tosylate, and this we can readily understand: powerful electron-withdrawal by acetoxy slows down formation of the carbonium ion in the S_N1 process. Reaction of the *trans* tosylate, although much faster than that of its diastereomer, is still somewhat slower than that of cyclohexyl tosylate. But should not the anchimerically assisted reaction be much *faster* than the unassisted reaction of the unsubstituted tosylate? The answer is, *not necessarily*. We must not forget the electronic effect of the acetoxy substituent. Although S_N2-like, attack by acetoxy has considerable S_N1 character (see Sec. 17.15); deactivation by electron withdrawal tends to offset activation by anchimeric assistance. The *cis* tosylate is electronically similar to the *trans*, and is a much better standard by which to measure anchimeric assistance. (This point will be discussed further in the next section.)

In Sec. 17.15 we said that the orientation of opening of strained rings like halonium ions and protonated epoxides indicates considerable S_N1 character in the transition state. But if ring-*opening* has S_N1 character so, according to the principle of microscopic reversibility, must ring-*closing*.

Problem 28.15 Of what structures is the acetoxonium ion a hybrid? To what does it owe its stability, relative to a carbonium ion?

Problem 28.16 How do you account for the following relative rates of acetolysis of 2-substituted cyclohexyl brosylates? In which cases is there evidence of a neighboring group effect?

(2-G-cyclohexyl-OBs) $\xrightarrow[\text{KOAc}]{\text{HOAc}}$ (2-G-cyclohexyl-OAc)

G	Relative rates *cis*	*trans*
Cl	1.6	5.9
Br	1.5	1250
I		2.2×10^8
H	1.2×10^4	

28.12 Neighboring group effects. Neighboring aryl

In 1949, at the University of California at Los Angeles, Donald J. Cram published the first of a series of papers on the effects of neighboring aryl groups, and set off a controversy that only recently, after twenty years, shows signs of being resolved. Let us look at just one example of the kind of thing he discovered.

Solvolysis of 3-phenyl-2-butyl tosylate in acetic acid yields the acetate. The tosylate contains two chiral centers, and exists as two racemic modifications; so,

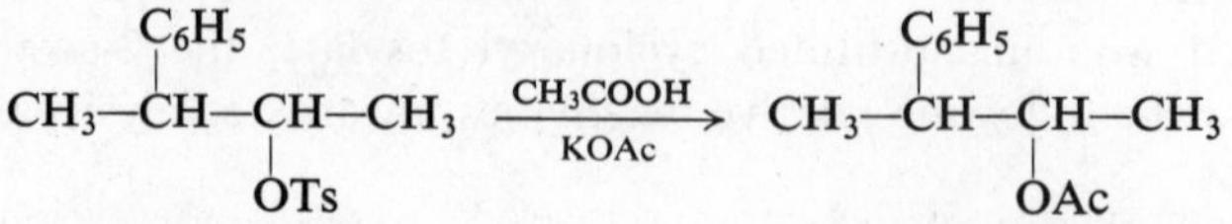

$$\underset{\text{3-Phenyl-2-butyl tosylate}}{CH_3-\overset{C_6H_5}{\overset{|}{C}H}-\underset{OTs}{\underset{|}{C}H}-CH_3} \xrightarrow[KOAc]{CH_3COOH} \underset{\text{3-Phenyl-2-butyl acetate}}{CH_3-\overset{C_6H_5}{\overset{|}{C}H}-\underset{OAc}{\underset{|}{C}H}-CH_3}$$

too, does the acetate. Solvolysis is completely stereospecific and proceeds, it at first appears, with *retention* of configuration: racemic *erythro* tosylate gives only racemic *erythro* acetate, and racemic *threo* tosylate gives only racemic *threo* acetate (Fig. 28.7). When, however, optically active *threo* tosylate is used, it is found to yield

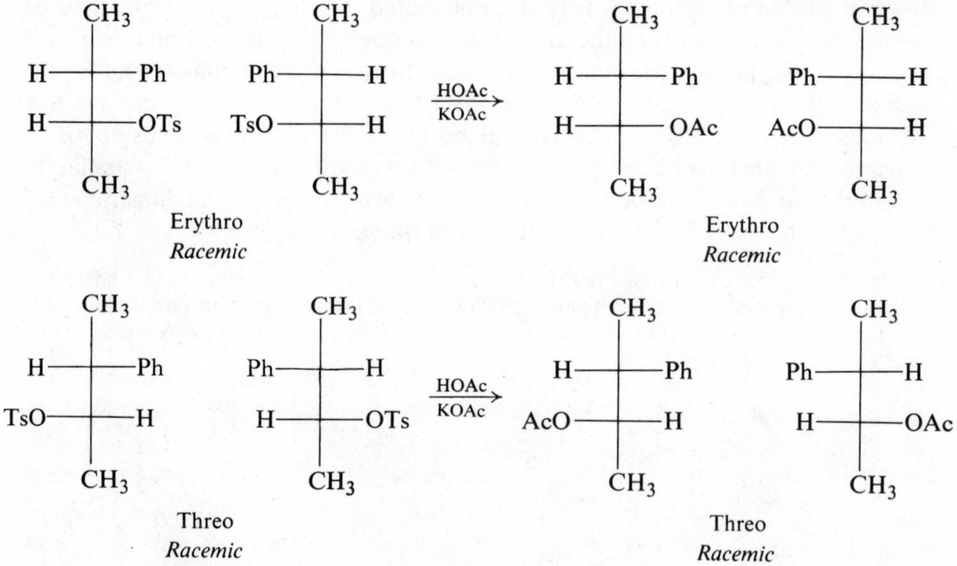

CH3 / H—Ph / H—OTs / CH3 ; CH3 / Ph—H / TsO—H / CH3 —(HOAc, KOAc)→ CH3 / H—Ph / H—OAc / CH3 ; CH3 / Ph—H / AcO—H / CH3

Erythro *Racemic* → Erythro *Racemic*

CH3 / H—Ph / TsO—H / CH3 ; CH3 / Ph—H / H—OTs / CH3 —(HOAc, KOAc)→ CH3 / H—Ph / AcO—H / CH3 ; CH3 / Ph—H / H—OAc / CH3

Threo *Racemic* → Threo *Racemic*

Figure 28.7. Acetolysis of racemic 3-phenyl-2-butyl tosylates.

optically *inactive* product, racemic *threo* acetate. We see here the same pattern as in Sec. 28.10: retention at both carbons in half the molecules of the product, inversion at both carbons in the other half (Fig. 28.8).

Cram interpreted these results in the following way. The neighboring phenyl group, with its π electrons, helps to push out (1) the tosylate anion. There is formed

(1) 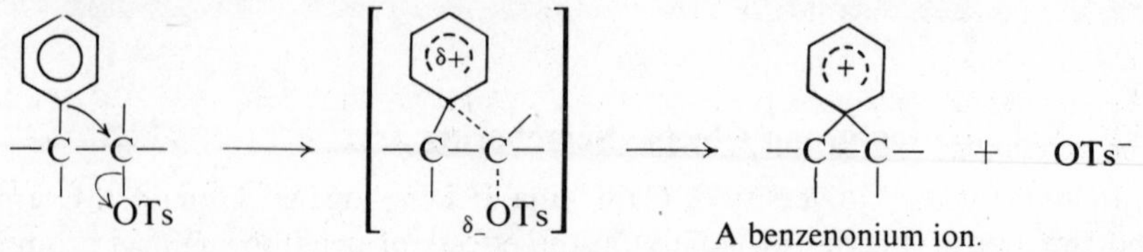 $-\overset{Ph}{\overset{|}{\underset{|}{C}}}-\overset{|}{\underset{OTs}{C}}- \longrightarrow \left[-\overset{(\delta+)}{C}\cdots C- \;\; {}_{\delta-}OTs \right] \longrightarrow -\overset{(+)}{C-C}- + OTs^-$

A benzenonium ion.

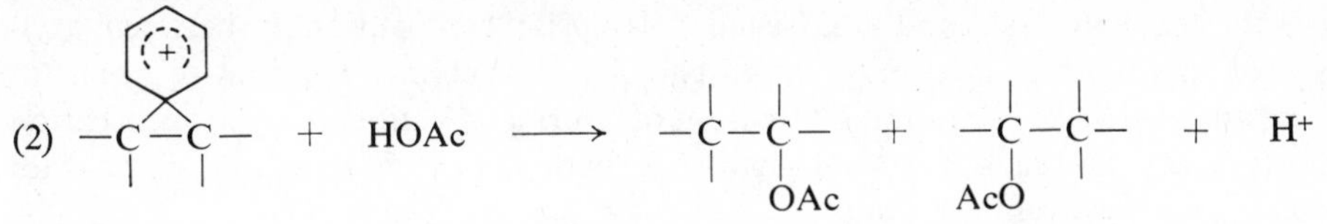

(2) —C—C— (bridged by benzenonium ring, (+)) + HOAc ⟶ —C—C(OAc)— + —C(AcO)—C— + H^+

an intermediate *bridged* ion—a benzenonium ion. This undergoes nucleophilic attack (2) by acetic acid at either of the two equivalent carbons to yield the product.

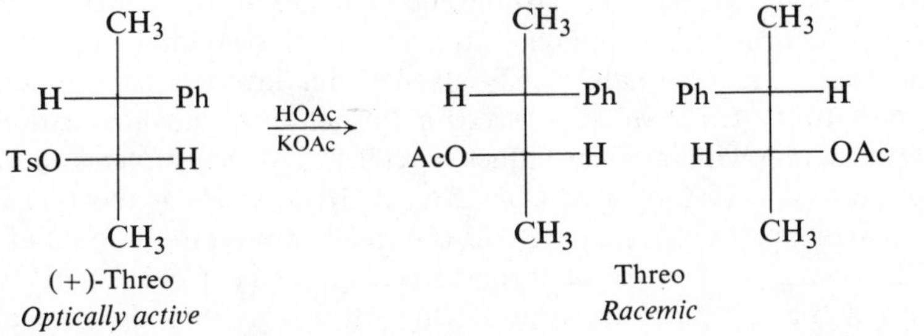

CH_3		CH_3	CH_3
H—Ph	$\xrightarrow[KOAc]{HOAc}$	H—Ph	Ph—H
TsO—H		AcO—H	H—OAc
CH_3		CH_3	CH_3
(+)-Threo *Optically active*		Threo *Racemic*	

Figure 28.8. Acetolysis of optically active *threo*-3-phenyl-2-butyl tosylate.

Problem 28.17 (a) Drawing structures of the kind in Fig. 28.5 (p. 907), show how Cram's mechanism accounts for the conversion of optically active *threo*-3-phenyl-2-butyl tosylate into racemic acetate. (b) In contrast, optically active *erythro* tosylate yields optically *active erythro* acetate. Show that this, too, fits Cram's interpretation of the reaction.

In the controversy that developed, the point under attack was not so much the existence of the intermediate bridged ion—although this was questioned, too—as its mode of formation. The 3-phenyl-2-butyl tosylates undergo solvolysis at much the same rate as does unsubstituted *sec*-butyl tosylate: formolysis a little faster, acetolysis a little slower. Yet, as depicted by Cram, phenyl gives anchimeric assistance to the reaction. Why, then, is there no rate acceleration?

Several alternatives were proposed: one, that participation by phenyl in expulsion of tosylate occurs, but is weak; another, that bridging occurs, not in the rate-determining step, but rapidly, following formation of an open cation. H. C. Brown (p. 507) suggested that—for unsubstituted phenyl, at least—the intermediate is not a bridged ion at all, but a pair of rapidly equilibrating open carbonium ions; phenyl, now on one carbon and now on the other, blocks back-side attack by the solvent and thus gives rise to the observed stereochemistry.

By 1971, a generally accepted picture of these reactions had begun to emerge, based on work by a number of investigators, prominent among them Paul Schleyer at Princeton University. The big stumbling-block had been the widely held idea that secondary cations are formed, like tertiary cations, with little assistance from the solvent (Sec. 14.17). Using as standards certain special secondary substrates whose structure *prevents* solvent assistance, Schleyer showed that ordinary secondary substrates do indeed react with much solvent assistance.

Cram's original proposal seems to be essentially correct: aryl *can* give anchimeric assistance through formation of bridged ions. Competition is *not* between aryl-assisted solvolysis and unassisted solvolysis; competition is between aryl-assisted solvolysis and solvent-assisted solvolysis. Anchimeric assistance need not cause anchimeric acceleration. Formation of a bridged cation and an open cation may proceed at much the same rate, one with aryl assistance, the other with equally strong solvent assistance.

On the assumption of these two competing processes, successful quantitative correlations have been made among data of various kinds: rate of reaction, stereochemistry, scrambling of isotopic labels, and Hammett constants (Sec. 18.11) to represent the relative electronic effects of various substituents in aromatic rings. If neighboring aryl contains strongly electron-withdrawing substituents, reaction products are normal—chiefly alkenes plus inverted ester—and the rate of solvolysis is what one would expect for formation of an open cation slowed down by electron-withdrawing inductive effects. As substituents become increasingly electron-releasing (p-Cl, m-CH_3, p-CH_3, p-CH_3O) the rate increases *more* than expected if only inductive effects were operating; the amount of "extra" speed matches the amount of abnormal stereochemistry. Consider, for example, acetolysis of 3-aryl-2-butyl brosylates. One calculates from the rate data that m-tolyl assists in 73% of reaction; 68% of the product is found to have retained configuration. For p-methyl, calculated 87%, found 88%; for p-methoxyphenyl, calculated 99%, found 100%.

How much anchimeric assistance there is, then, depends on how nucleophilic the neighboring group is. It also depends on how badly anchimeric assistance is *needed.* The more nucleophilic the solvent, the more assistance *it* gives, and the less the neighboring group participates. Or, if the open cation is a relatively stable one—tertiary or benzylic—it may need little assistance of any kind, either from the solvent or from the neighboring group.

In summary, an incipient cation can get electrons in three different ways: (a) from a substituent, through an inductive effect or resonance; (b) from the solvent; (c) from a neighboring group.

In all this, H. C. Brown played a role familiar to him: that of gad-fly—the organic chemist's conscience—forcing careful examination of ideas that had been accepted perhaps too readily because of their neatness. The turning point in this part of the great debate was marked by the joint publication of a paper by Brown and Schleyer setting forth essentially the interpretation we have just given.

In 1970, Olah (p. 160) prepared a molecule whose carbon-13 nmr spectrum (cmr) was consistent with a bridged benzenonium ion, and *not* with a pair of equilibrating open cations.

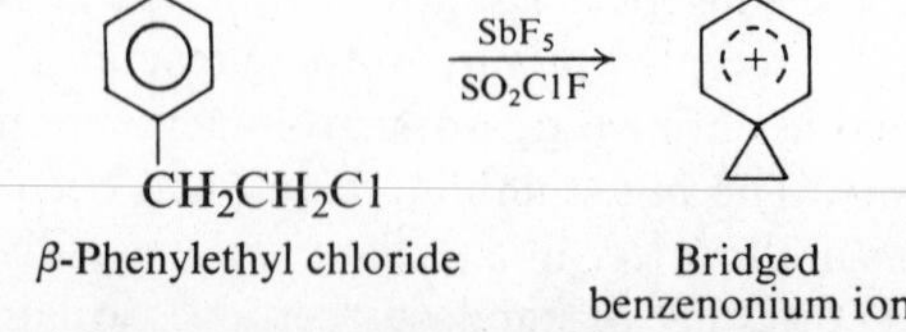

β-Phenylethyl chloride   Bridged benzenonium ion

Problem 28.18 Quenching of Olah's solution with water gave a 3:1 mixture of β-phenylethyl alcohol and α-phenylethyl alcohol. The spectrum showed the presence not only of the bridged cation but, in lesser amounts, of an open cation. What is a likely structure for the open cation, and how is it formed?

28.13 Neighboring group effects: nonclassical ions

The rearrangement of carbonium ions was first postulated, by Meerwein (p. 160) in 1922, to account for the conversion of camphene hydrochloride into isobornyl chloride. Oddly enough, this chemical landmark is the most poorly

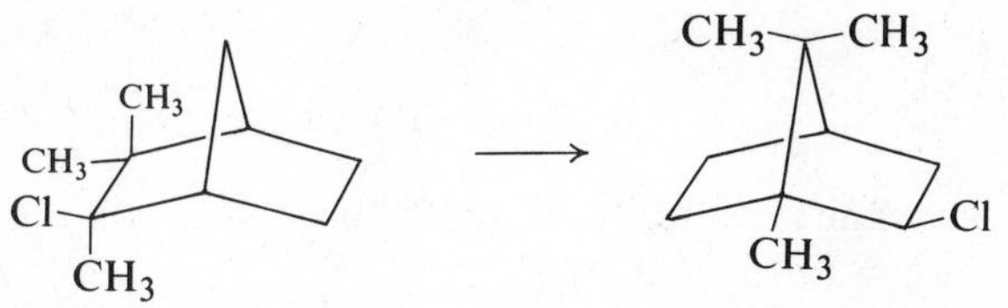

Camphene hydrochloride

Isobornyl chloride

understood of all such rearrangements. With various modifications in structure, this bicyclic system has been for over 20 years the object of closer scrutiny than any other in organic chemistry.

We can see, in a general way, how this particular rearrangement could take place. Camphene hydrochloride loses chloride ion to form cation I, which rearranges by a 1,2-alkyl shift to form cation II. Using models, and keeping careful

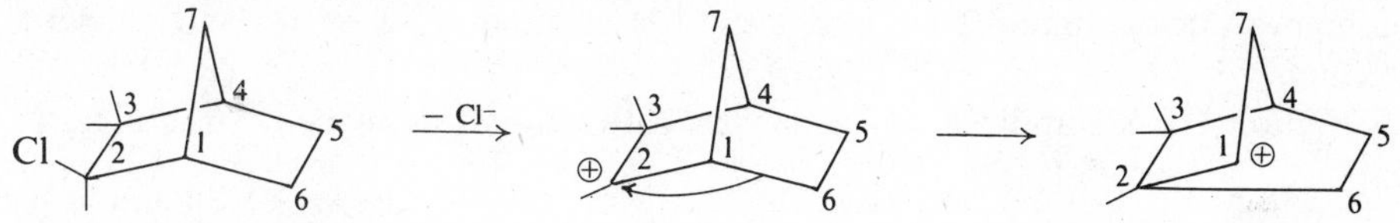

Camphene hydrochloride I II

track of the various carbon atoms, we find that cation II need only combine with a chloride ion to yield isobornyl chloride.

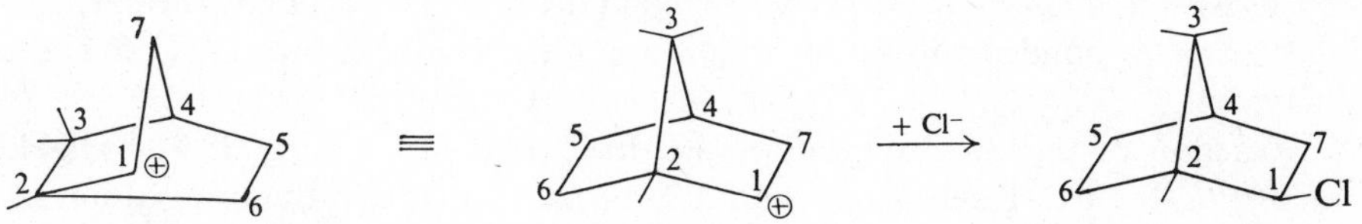

II II Isobornyl chloride

We have accounted for the observed change in carbon skeleton, but we have not answered two questions that have plagued the organic chemist for a generation. Why is only the *exo* chloride, isobornyl chloride, obtained, and none of its *endo* isomer, bornyl chloride? Why does camphene hydrochloride undergo solvolysis thousands of times as fast as, say, *tert*-butyl chloride? To see the kind of answers that have been given, let us turn to a simpler but basically similar system.

In 1949 Winstein reported these findings. On acetolysis, the diastereomeric *exo*- and *endo*-norbornyl brosylates both yield *exo*-norbornyl acetate:

OBs, and enantiomer — $\xrightarrow[\text{KOAc}]{\text{HOAc}}$ — OAc, and enantiomer — $\xleftarrow[\text{KOAc}]{\text{HOAc}}$ — OBs, and enantiomer

exo-Norbornyl brosylate — *exo*-Norbornyl acetate — *endo*-Norbornyl brosylate

If the starting brosylate is optically active, the product is still the optically inactive racemic modification. For example:

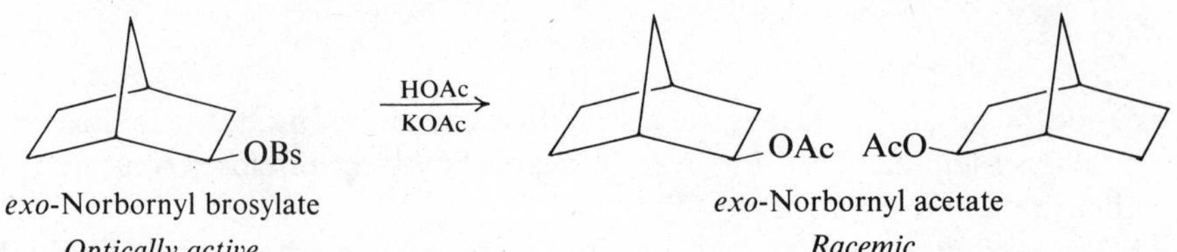

exo-Norbornyl brosylate
Optically active

exo-Norbornyl acetate
Racemic

Finally, *exo*-norbornyl brosylate reacts 350 times as fast as the *endo* brosylate.

Winstein interpreted the behavior of these compounds in the following way (Fig. 28.9). Loss of brosylate anion yields (1) the bridged cation III, which undergoes nucleophilic attack by solvent (2) at either C–2 or C–1 to yield the product.

Cation III is stabilized by resonance between two equivalent structures, IV

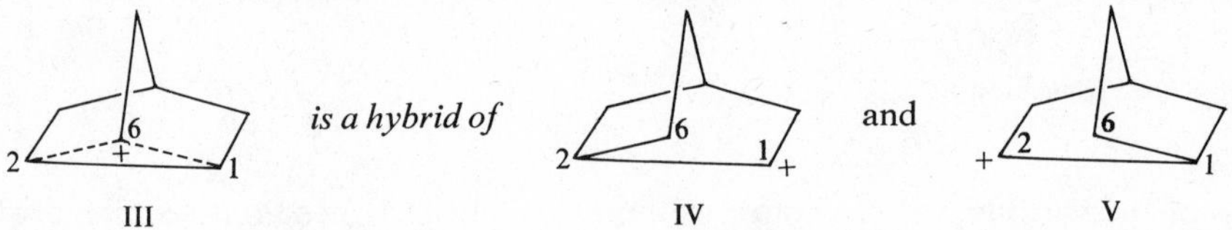

III — IV — V

and V, each corresponding to an open cation. The charge is divided between two carbons (C–1 and C–2) each of which—held in the proper position by the particular ring system—is bonded to C–6 by a half-bond. The bridging carbon (C–6) is thus pentavalent.

Reaction of the *exo* brosylate is S_N2-like, as shown in Fig. 28.9: back-side attack by C–6 on C–1 helps to push out brosylate, and yields the bridged ion in a single step. The geometry of the *endo* brosylate does not permit such back-side attack, and consequently it undergoes an S_N1-like reaction: slow formation of the open cation followed by rapid conversion into the bridged ion.

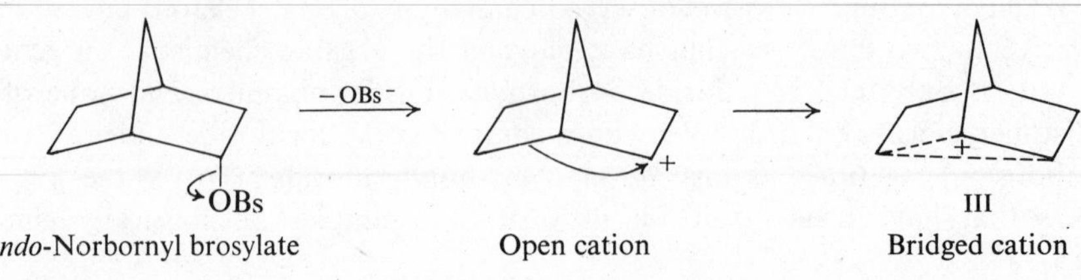

endo-Norbornyl brosylate — Open cation — III Bridged cation

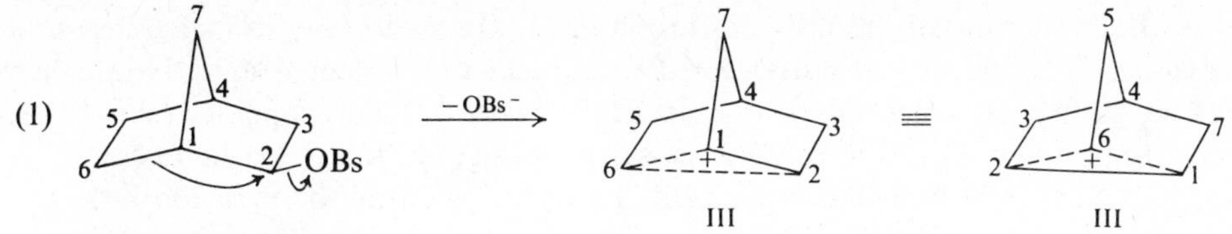

exo-Norbornyl brosylate
Optically active

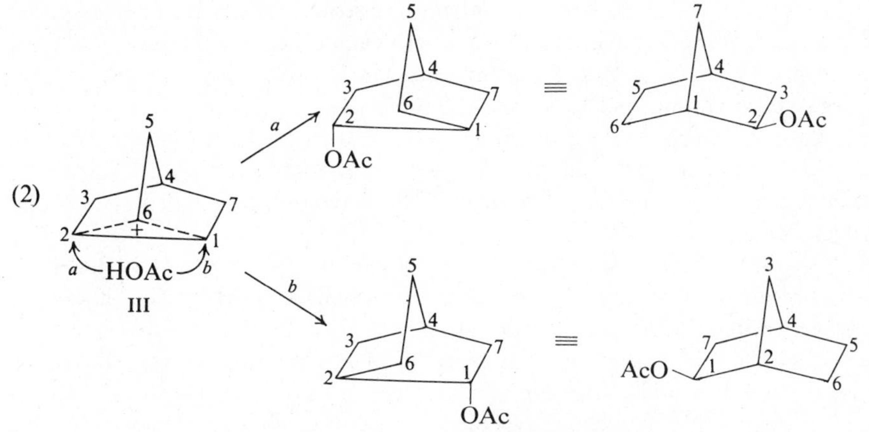

exo-Norbornyl acetate
Racemic

Figure 28.9. Conversion of optically active *exo*-norbornyl brosylate into racemic *exo*-norbornyl acetate via nonclassical ion. Brosylate anion is lost with anchimeric assistance from C–6, to give bridged cation III. Cation III undergoes back-side attack at either C–2 (path *a*) or C–1 (path *b*). Attacks *a* and *b* are equally likely, and give racemic product.

The two diastereomers yield the same product, racemic *exo* acetate, because they react via the same intermediate. But only the *exo* brosylate reacts with anchimeric assistance, and hence it reacts at the faster rate.

What Winstein was proposing was that *saturated* carbon using σ electrons could act as a neighboring group, to give anchimeric assistance to the expulsion of a leaving group, and to form an intermediate bridged cation containing pentavalent carbon. Bridged ions of this kind, with delocalized bonding σ electrons, have become known as *nonclassical ions*.

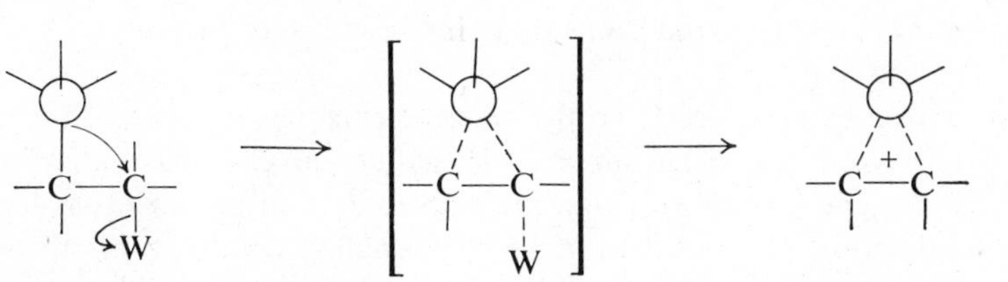

A nonclassical bridged ion

Interpretation of the behavior of the norbornyl and many related systems on the basis of nonclassical ions seemed to be generally accepted until 1962, when H. C. Brown declared, "But the Emperor is naked!" Brown's point was not that the idea of nonclassical ions was necessarily wrong, but that it was *not necessarily right*. It had been accepted too readily, he thought, on the basis of too little evidence, and needed closer examination.

Brown suggested alternative interpretations. The norbornyl cation, for example, might not be a bridged ion but a pair of equilibrating open carbonium ions. That is to say, IV and V are not contributing structures to a resonance hybrid, but two distinct compounds in equilibrium with each other. Each ion can combine with solvent: IV at C–1, V at C–2. Substitution is exclusively *exo* because the *endo* face of each cation lies in a fold of the molecule, and is screened from attack. Differences in rate, too, are attributed to steric factors. It is not that the *exo* substrate reacts unusually fast, but that the *endo* substrate reacts unusually *slowly*, due to steric hindrance to the departure of the leaving group with its cluster of solvent molecules.

To test these alternative hypotheses, a tremendous amount of work has been done, by Brown and by others. For example, camphene hydrochloride is known to undergo ethanolysis 6000 times as fast as *tert*-butyl chloride, and this had been attributed to anchimeric assistance with formation of a bridged ion. Brown pointed out that the wrong standard for comparison had been chosen. He showed that a number of substituted (3°) cyclopentyl chlorides (examine the structure of camphene hydrochloride closely) *also* react much faster than *tert*-butyl chloride. He attributed these fast reactions—including that of camphene hydrochloride—to *relief of steric strain*. On ionization, chloride ion is lost and the methyl group on the sp^2-hybridized carbon moves into the plane of the ring: four non-bonded interactions thus disappear, two for chlorine and two for methyl. For certain systems at least, it became clear that one need not invoke a nonclassical ion to account for the facts.

In 1970, Olah reported that he had prepared a stable norbornyl cation in SbF_5–SO_2. From its pmr, cmr, and Raman spectra, he concluded that it has, indeed, the nonclassical structure with delocalization of σ electrons. The 2-phenylnorbornyl cation, on the other hand, has the classical structure; this benzylic

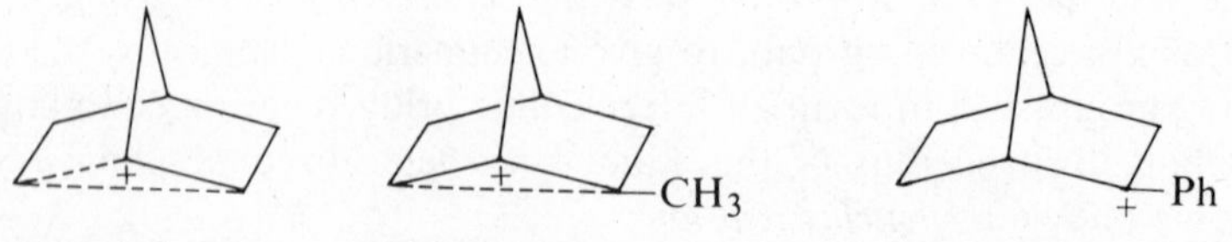

Norbornyl cation	2-Methylnorbornyl cation	2-Phenylnorbornyl cation
Bridged ion	*Some bridging*	*Open ion*

cation, stabilized by electrons from the benzene ring, has no need of bridging. The tertiary 2-methylnorbornyl cation is intermediate in character: there is *partial* σ delocalization and hence bridging, but weaker than in the unsubstituted cation. (Interestingly enough, delocalization in the 2-methyl cation seems to come, not from the C_6—C_7 bond, but from the C_6—H bond; Olah pictures the back lobe of the carbon–hydrogen bond overlapping the p orbital of C_2.)

Thus, it seems, there *are* such things as nonclassical cations. What is still to be settled is just how much they are involved in the chemistry of ordinary solvolytic reactions.

Problem 28.19 (a) Show how a nonclassical ion intermediate could account for both the stereospecificity and the unusually fast rate (if it *is* unusually fast) of rearrangement of camphene hydrochloride into isobornyl chloride. (b) How do you account for the fact that optically active product is formed here, in contrast to what is obtained from solvolysis of norbornyl compounds?

PROBLEMS

1. Give a detailed interpretation of each of the following observations.

(a) $CH_3CH_2CH_2CD_2NH_2 \xrightarrow[H_2O]{HONO}$ 1-butanol + 2-butanol

The 2-butanol is $CH_3CH_2CHOHCHD_2$ + $CH_3CHOHCH_2CHD_2$
76.8% *23.2%*

(b) $CH_3CH_2CD_2CH_2NH_2 \xrightarrow[H_2O]{HONO}$ $CH_3CH_2CD_2CH_2OH$ + $CH_3CH_2CDOHCH_2D$
68% *23.9%*

+ $CH_3CHOHCHDCH_2D$ + ($CH_3CHDCHOHCH_2D$ + $CH_3CDOHCH_2CH_2D$)
7.7% *0.4%*

2. Treatment of 1-methyl-1-cyclohexyl hydroperoxide with acid gives a product of formula $C_7H_{14}O_2$, which gives positive tests with CrO_3/H_2SO_4, 2,4-dinitrophenylhydrazine, and NaOI. What is a likely structure for this compound, and how is it formed?

3. (a) Describe simple chemical tests that would serve to distinguish among the possible products of rearrangement of 1-phenyl-1,2-propanediol shown on page 897. Tell exactly what you would do and see. (b) Alternatively, you could use the nmr spectrum. Tell exactly what you would expect to see in the spectrum of each possible product.

4. In the presence of base, acyl derivatives of hydroxamic acids undergo the **Lossen rearrangement** to yield isocyanates or amines.

$$R{-}\overset{O}{\overset{\|}{C}}{-}\underset{H}{\underset{|}{N}}{-}O{-}\overset{O}{\overset{\|}{C}}{-}R' \xrightarrow{\text{base}} RNCO \xrightarrow{H_2O} RNH_2$$

(a) Write a detailed mechanism for the rearrangement.

(b) Study of a series of compounds in which R and R′ were *m*- and *p*-substituted phenyl groups showed that reaction is speeded up by electron-releasing substituents in R and by electron-withdrawing substituents in R′. How do you account for these effects?

5. (a) Show all steps in the mechanisms probably involved in the following transformation. (*Hint:* Don't forget Sec. 21.5.)

3-Hydroperoxycyclohexene $\xrightarrow{H^+}$ Adipaldehyde (*6%*) + Cyclopentene-1-carboxaldehyde (*39%*)

(b) An important difference in migratory aptitude is illustrated here. What is it?

6. Benzophenone oxime, $C_{13}H_{11}ON$, m.p. 141°, like other oximes, is soluble in aqueous NaOH and gives a color with ferric chloride. When heated with acids it is transformed into a solid A, $C_{13}H_{11}ON$, m.p. 163°, which is insoluble in aqueous NaOH and in aqueous HCl.

After prolonged heating of A with aqueous NaOH, a liquid B separates and is collected by steam distillation. Acidification of the aqueous residue causes precipitation of a white solid C, m.p. 120–1°.

Compound B, b.p. 184°, is soluble in dilute HCl. When this acidic solution is chilled and then treated successively with $NaNO_2$ and β-naphthol, a red solid is formed. B reacts with acetic anhydride to give a compound that melts at 112.5–114°.

(a) What is the structure of A? (b) The transformation of benzophenone oxime into A illustrates a reaction to which the name **Beckmann** is attached. To what general class of reactions must this transformation belong? (c) Suggest a likely series of steps, each one basically familiar, for this transformation? (*Hint:* See Secs. 16.5, 6.10, and 8.13.)

(d) Besides acids like sulfuric, other compounds "catalyze" this reaction. How might PCl_5 do the job? Tosyl chloride?

(e) What product or products corresponding to A would you expect from a similar transformation of acetone oxime; of acetophenone oxime; of *p*-nitrobenzophenone oxime; of methyl *n*-propyl ketoxime? (f) How would you go about identifying each of the products in (e)?

7. Urea is converted by hypohalites into nitrogen and carbonate. Given the fact

$$H_2N{-}\overset{\displaystyle O}{\overset{\|}{C}}{-}NH_2 \xrightarrow{Br_2,\ OH^-} N_2 + CO_3^{--} + Br^-$$

that hydrazine, $H_2N{-}NH_2$, is oxidized to nitrogen by hypohalite, show that this reaction of urea is simply an example of the Hofmann degradation of amides.

8. Treatment of triarylcarbinols, Ar_3COH, with acidic hydrogen peroxide yields a 50:50 mixture of ketone, ArCOAr, and phenol, ArOH. (a) Show all steps in a likely mechanism for this reaction. (b) Predict the major products obtained from *p*-methoxytriphenylcarbinol, $p\text{-}CH_3OC_6H_4(C_6H_5)_2COH$. From *p*-chlorotriphenylcarbinol.

9. (a) Upon treatment with acid I (R = C_2H_5) yields II and III. Show all steps in these transformations.

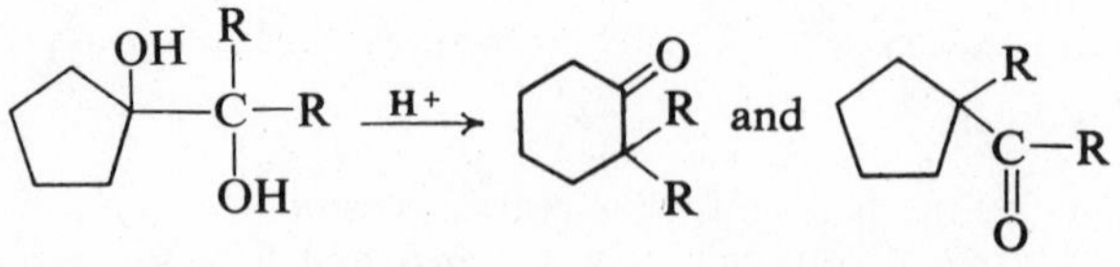

I: 1-(C(OH)R₂)-cyclopentanol (OH; R–C–R; OH) $\xrightarrow{H^+}$ II: 2,2-dialkylcyclohexanone (O; R; R) and III: cyclopentane bearing R and C(=O)–R

I II III

(b) Account for the fact that when R = C_6H_5, I yields only II.

(c) Show the most likely steps in the following transformation:

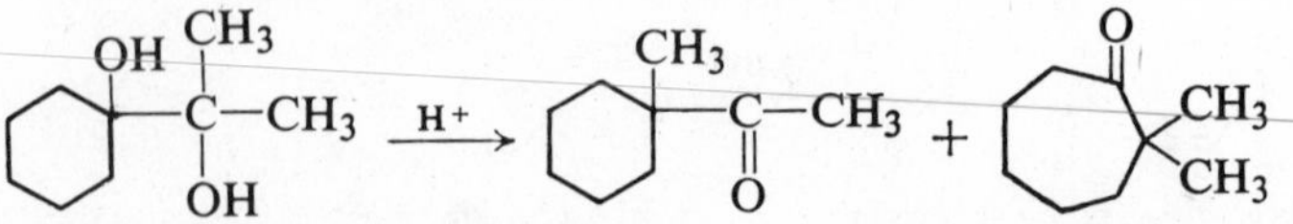

1-(C(OH)(CH₃)₂)-cyclohexanol (OH; CH₃; C–CH₃; OH) $\xrightarrow{H^+}$ cyclohexane bearing CH₃ and C(=O)–CH₃ + 2,2-dimethylcycloheptanone (O; CH₃; CH₃)

(d) Predict the products of the pinacol rearrangement of 2,3-diphenyl-2,3-butanediol; of 3-phenyl-1,2-propanediol. Describe a simple chemical test that would show whether your prediction was correct or incorrect.

10. When dissolved in HSO_3F–SbF_5–SO_2, the glycol 1,3-propanediol is rapidly converted into propionaldehyde. Write all steps in a likely mechanism for this reaction.

11. In the oxidation stage of hydroboration-oxidation, alkylboranes are converted into alkyl borates, which are hydrolyzed to alcohols. It has been suggested that the formation of the borates involves the reagent HOO^-.

$$H_2O_2 + OH^- \overset{(1)}{\rightleftharpoons} HOO^- + H_2O$$

$$\underset{\text{Trialkylborane}}{R_3B} + 3HOO^- \xrightarrow{(2)} \underset{\text{Alkyl borate}}{(RO)_3B} \xrightarrow{(3)} 3ROH$$

(a) Show all steps in a possible mechanism for step (2), the formation of the borate.

(b) What did you conclude (Problem 15.10, p. 507) was the likely stereochemistry of the oxidation stage of hydroboration-oxidation? Is your mechanism in (a) consistent with this stereochemistry?

12. Account in detail for each of the following sets of observations:

(a) On treatment with aqueous HBr, both *cis*- and *trans*-2-bromocyclohexanol are converted into *trans*-1,2-dibromocyclohexane.

(b) Treatment of *either* epoxide IV *or* epoxide V with aqueous OH^- gives the same product VI.

IV: $CH_3—CH—CH—CH_2Br$ (epoxide O bridging C2–C3)

V: $CH_3—CH(Br)—CH—CH_2$ (epoxide O bridging C3–C4)

VI: $CH_3—CH—CH—CH_2OH$ (epoxide O bridging C2–C3)

(c) $[C_6H_5COC(CH_3)_2]^-Na^+ + H_2C—CH—CH_2Cl$ (epoxide O bridging C1–C2) $\longrightarrow$ bicyclic product bearing H_3C, H_3C, C_6H_5 and two ring O atoms

(d) The relative rates of formolysis of *p*-$GC_6H_4CH_2CH_2OTs$ for various G's are: H 2.1, OCH_3 160, O^- 10^8.

(e) $CH_3CHClCH_2NEt_2 \xrightarrow{OH^-} CH_3CH(NEt_2)CH_2OH$

(f) *Either* $CH_3CHOHCH_2SEt$ *or* $CH_3CH(SEt)CH_2OH \xrightarrow{HCl} CH_3CHClCH_2SEt$

13. (a) In acetic acid solution nitrous acid converts 3-phenyl-2-butylamine into a mixture of acetates. Examination of these products shows that in the reaction of the

(+)-Erythro: CH_3 (top); H—C—Ph; H—C—NH_2; CH_3 (bottom)

(+)-Threo: CH_3 (top); H—C—Ph; NH_2—C—H; CH_3 (bottom)

3-Phenyl-2-butylamine

(+)-*erythro* amine phenyl migration exceeds methyl migration 8:1, whereas in the reaction of the (+)-*threo* amine methyl migration exceeds phenyl migration 1.5:1. Suggest a likely explanation.

(b) In contrast, solvolysis of the corresponding tosylates (Sec. 28.12) gives acetates indicating *no* methyl migration for either diastereomer. How do you account for this difference between the two systems?

14. Spectroscopic and thin layer chromatographic analysis has shown that, even when not found in the final product, epoxides are present during the reaction of such pinacols as 1,1,2,2-tetraphenyl-1,2-ethanediol. It seems most likely that epoxides represent a blind alley down which many molecules stray before pinacolone is finally formed. (a) How are these epoxides probably formed? (b) What probably happens to them in the reaction medium?

15. Labeled $ArCH_2{}^{14}CH_2NH_2$ was treated with HONO, and the $ArCH_2CH_2OH$ obtained was oxidized to ArCOOH. The fraction of the original radioactivity found in the ArCOOH depended on the nature of Ar: p-$NO_2C_6H_4$ 8%, C_6H_5 27%, p-$CH_3OC_6H_4$ 45%. How do you account for these findings?

16. Collins (p. 899) prepared 1,1,2-triphenylethyl acetate triply labeled with ^{14}C (indicated as C*) and studied reactions (1)–(3) in ordinary acetic acid. The equilibrium

(1) $Ph_2CH-\underset{\displaystyle OAc}{\underset{|}{C^*HPh}} \xrightarrow{HOAc} \underset{50\%}{Ph_2CH-\underset{\displaystyle OAc}{\underset{|}{C^*HPh}}} + \underset{50\%}{Ph_2C^*H-\underset{\displaystyle OAc}{\underset{|}{CHPh}}}$

(2) $Ph_2CH-\underset{\displaystyle OAc}{\underset{|}{CHPh^*}} \xrightarrow{HOAc} \underset{33\%}{Ph_2CH-\underset{\displaystyle OAc}{\underset{|}{CHPh^*}}} + \underset{67\%}{Ph^*PhCH-\underset{\displaystyle OAc}{\underset{|}{CHPh}}}$

(3) $Ph_2CH-\underset{\displaystyle OAc^*}{\underset{|}{CHPh}} \xrightarrow{HOAc} Ph_2CH-\underset{\displaystyle OAc}{\underset{|}{CHPh}}$

product of (1) and (2) had the indicated distribution of labels. The rate of acetate exchange (3) was found to be *identical* with the rates of (1) and (2).

Collins concluded that bridged ions are *not* involved in this particular system.

(a) Explain in detail how his conclusion is justified. Show just what probably *does* happen. (Among other things: what results would be expected if a bridged ion *were* involved?)

(b) Why might this system be expected to differ from, say, the 3-phenyl-2-butyl one?

17. Account in detail for each of the following sets of observations.

(a) Compound VII reacts with acetic acid 1200 times as fast as does ethyl tosylate,

$$\underset{\text{VII}}{(CH_3)_2C{=}CHCH_2CH_2OTs} \xrightarrow{HOAc}$$

$$\underset{\text{VIII}}{(CH_3)_2C{=}CHCH_2CH_2OAc} + \underset{\text{IX}}{\begin{matrix} H_2C \\ | \\ H_2C \end{matrix}\!\!>\!CHC(CH_3){=}CH_2}$$

and yields not only VIII but also IX. When the labeled compound VIIa is used, product VIII consists of equal amounts of VIIIa and VIIIb.

$$\underset{\text{VIIa}}{(CH_3)_2C{=}CHCH_2CD_2OTs} \qquad \underset{\text{VIIIa}}{(CH_3)_2C{=}CHCH_2CD_2OAc}$$

$$\underset{\text{VIIIb}}{(CH_3)_2C{=}CHCD_2CH_2OAc}$$

(b) The cyclopentene derivative X (ONs = *p*-nitrobenzenesulfonate) undergoes solvolysis in acetic acid 95 times as fast as the analogous saturated compound (XI), and gives *exo*-norbornyl acetate (XII).

NsO

X

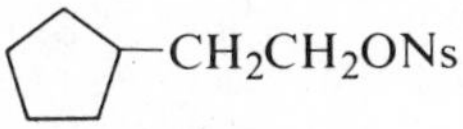

$-CH_2CH_2ONs$

XI

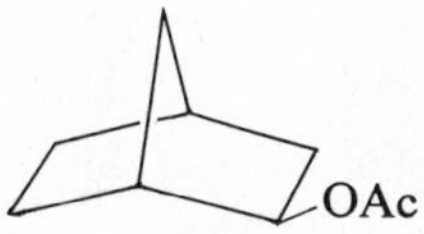

OAc

XII

(c) *anti*-7-Norbornylene tosylate (XIII) reacts with acetic acid 10^{11} times as fast

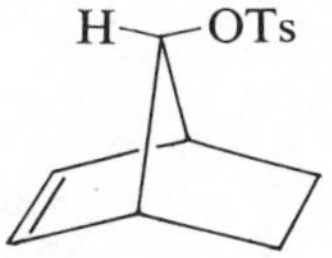

H OTs

XIII

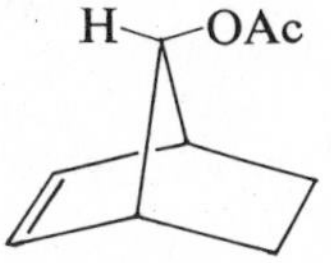

H OAc

XIV

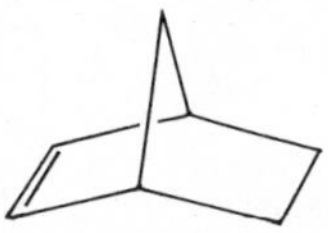

XV

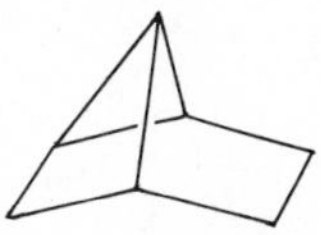

XVI

as the saturated analog, and yields *anti*-7-norbornylene acetate (XIV) with *retention* of configuration. Solvolysis of XIII in the presence of $NaBH_4$ gives XV and XVI.

18. (a) We saw (Sec. 28.13) that optically active *exo*-norbornyl brosylate reacts with acetic acid to give optically inactive *exo*-norbornyl acetate. The related brosylate XVII similarly reacts to give XVIII; yet in this case optically active brosylate yields optically

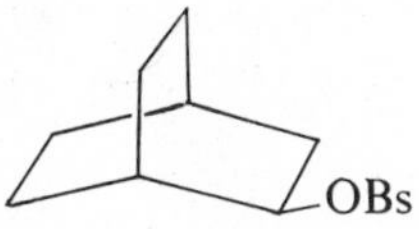

OBs

XVII

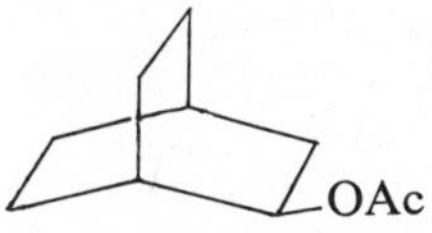

OAc

XVIII

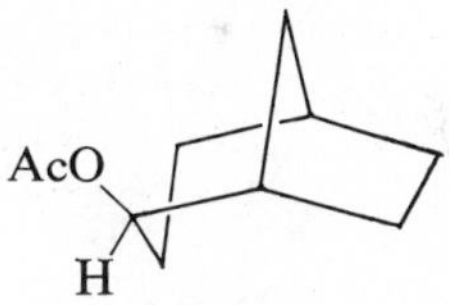

AcO

H

XIX

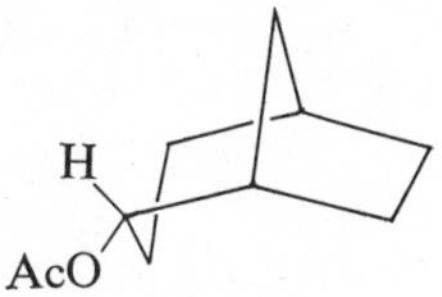

H

AcO

XX

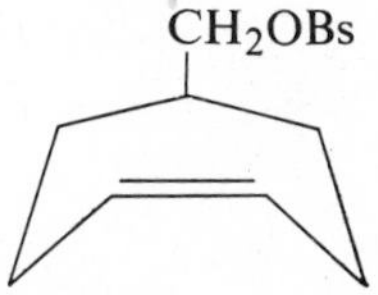

CH_2OBs

XXI

active acetate. Oddly enough, the complete racemization in the norbornyl reaction and the complete retention here are taken as evidence of the same fundamental behavior. On what common basis can you account for all of the above observations? (*Hint:* See also part (b).)

(b) Brosylate XVII also yields XIX, but no XX. When XVII is optically active, so is the XIX that is obtained. How do these facts fit into your answer to (a)?

(c) Brosylate XXI reacts with acetic acid 30 times as fast as the corresponding saturated compound does, and yields (optically inactive) XX, but no XIX. How do you account for these observations?

19. (a) Treatment of XXII with $NaOCH_3$ gives product XXIII; treatment of XXII with R_2NH gives the corresponding product XXIV. Show all steps in the most likely mechanism for these rearrangements.

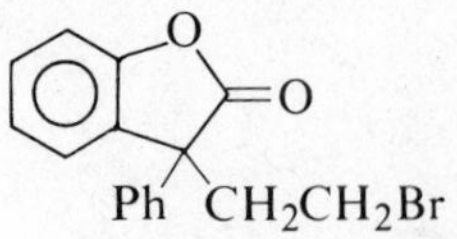

XXII

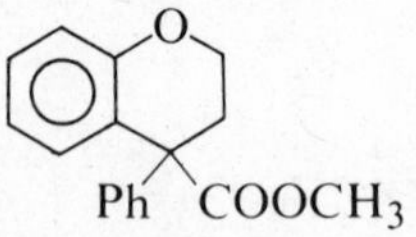

XXIII

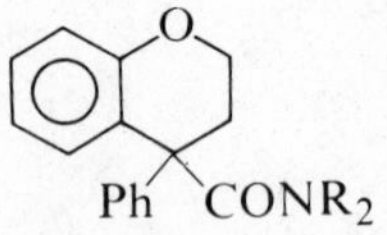

XXIV

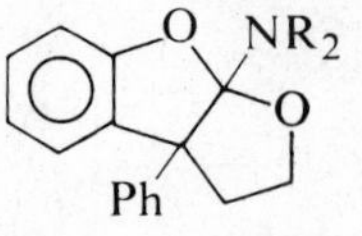

XXV

(b) From the reaction of XXII with R_2NH, there is also obtained XXV. How is XXV probably formed? Of what general significance is its isolation?

Chapter 29 Molecular Orbitals. Orbital Symmetry

29.1 Molecular orbital theory

The structure of molecules is best understood through quantum mechanics. Exact quantum mechanical calculations are enormously complicated, and so various methods of approximation have been worked out to simplify the mathematics. The method that is often the most useful for the organic chemist is based on the concept of *molecular orbitals*: orbitals that are centered, not about individual nuclei, but about all the nuclei in the molecule.

What are the various molecular orbitals of a molecule like? What is their order of stability? How are electrons distributed among them? These are things we must know if we are to understand the relative stability of molecules: why certain molecules are aromatic, for example. These are things we must know if we are to understand the course of many chemical reactions: their stereochemistry, for example, and how easy or difficult they are to bring about; indeed, whether or not they will occur at all.

We cannot learn here how to make quantum mechanical calculations, but we can see what the results of some of these calculations are, and learn a little about how to use them.

In this chapter, then, we shall learn what is meant by the *phase* of an orbital, and what *bonding* and *antibonding* orbitals are. We shall see, in a non-mathematical way, what lies behind the Hückel $4n + 2$ rule for aromaticity. And finally, we shall take a brief look at a recent—and absolutely fundamental—development in chemical theory: the application of the concept of *orbital symmetry* to the understanding of organic reactions.

29.2 Wave equations. Phase

In our first description of atomic and molecular structure, we said that electrons show properties not only of particles but also of waves. We must examine a little more closely the wave character of electrons, and see how this is involved in chemical bonding. First, let us look at some properties of waves in general.

Let us consider the *standing waves* (or *stationary waves*) generated by the vibration of a string secured at both ends: the wave generated by, say, the plucking of a guitar string (Fig. 29.1). As we proceed horizontally along the string from

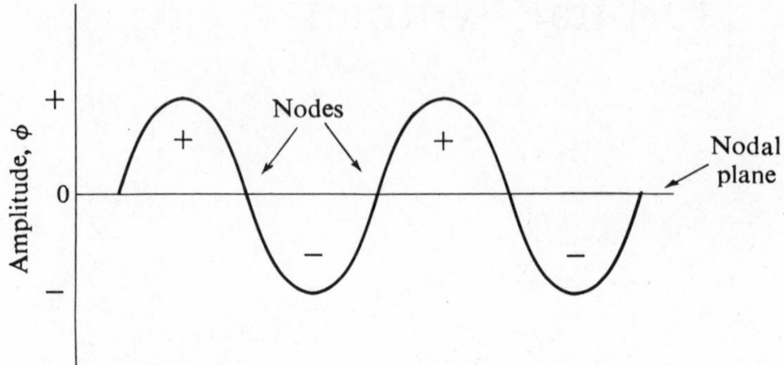

Figure 29.1. Standing waves. Plus and minus signs show relative phases.

left to right, we find that the vertical displacement—the *amplitude* of the wave—increases in one direction, passes through a maximum, decreases to zero, and then increases in the opposite direction. The places where the amplitude is zero are called *nodes*. In Fig. 29.1 they lie in a plane—the *nodal plane*—perpendicular to the plane of the paper. Displacement upward and displacement downward correspond to opposite *phases* of the wave. To distinguish between phases, we arbitrarily assign algebraic signs to the amplitude: plus for, say, displacement upward, and minus for displacement downward. If we were to superimpose similar waves on one another exactly *out of phase*—that is, with the crests of one lined up with the troughs of the other—they would cancel each other; that is to say, the sum of their amplitudes, + and −, would be zero.

The differential equation that describes the wave is a *wave equation*. Solution of this equation gives the amplitude, ϕ, as a function, $f(x)$, of the distance, x, along the wave. Such a function is a *wave function*.

Now, electron waves are described by a wave equation of the same general form as that for string waves. The wave functions that are acceptable solutions to this equation again give the amplitude, ϕ, this time as a function, not of a single coordinate, but of the three coordinates necessary to describe motion in three dimensions. It is these electron wave functions that we call *orbitals*.

Any wave equation has a *set* of solutions—an infinity of them, actually—each corresponding to a different energy level. The *quantum* thus comes naturally out of the mathematics.

Like a string wave, an electron wave can have nodes, where the amplitude is zero. On opposite sides of a node the amplitude has opposite signs, that is, the wave is of opposite phases. Of special interest to us is the fact that between the two lobes of a *p* orbital lies a nodal plane, perpendicular to the axis of the orbital

(Fig. 29.2). The two lobes are of opposite phase, and this is often indicated by + and − signs.

As used here, the signs do not have anything to do with charge. They simply indicate that the amplitude is of opposite algebraic sign in the two lobes. To avoid

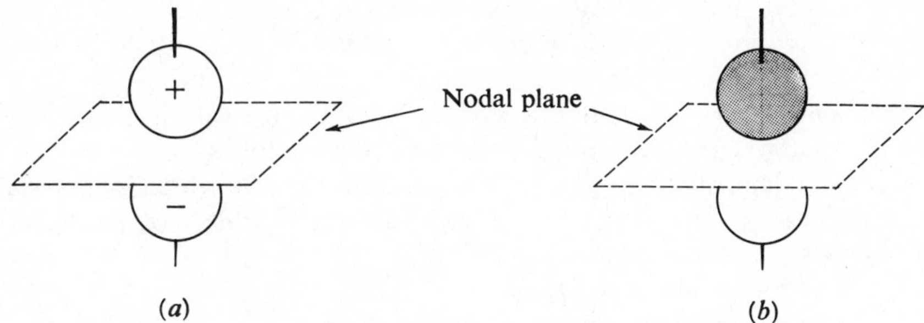

Figure 29.2. The *p* orbital. The two lobes are of opposite phase, indicated either (*a*) by plus and minus signs or (*b*) by shading.

confusion, we shall show lobes as shaded and unshaded. Two shaded lobes are of the same phase, both plus or both minus—it does not matter which. Similarly, two unshaded lobes are of the same phase; a shaded lobe and an unshaded lobe are of opposite phase.

The amplitude or wave function, ϕ, *is* the orbital. As is generally true for waves, however, it is the square of the amplitude, ϕ^2, that has physical meaning. For electron waves, ϕ^2 represents the probability of finding an electron at any particular place. The fuzzy balls or simple spheres we draw to show the "shapes" of orbitals are crude representations of the space within which ϕ^2 has a particular value—the space within which the electron spends, say, 95% of its time. Whether ϕ is positive or negative, ϕ^2 is of course positive; this makes sense, since probability cannot be negative. The usual practice is to draw the lobes of a *p* orbital to represent ϕ^2; if + or − signs are added, or one lobe is shaded and the other unshaded, this is to show the relative signs of ϕ.

29.3 Molecular orbitals. LCAO method

As chemists, we picture molecules as collections of atoms held together by bonds. We consider the bonds to arise from the overlap of an atomic orbital of one atom with an atomic orbital of another atom. A new orbital is formed, which is occupied by a pair of electrons of opposite spin. Each electron is attracted by both positive nuclei, and the increase in electrostatic attraction gives the bond its strength, that is, stabilizes the molecule relative to the isolated atoms.

This highly successful qualitative model parallels the most convenient quantum mechanical approach to molecular orbitals: **the method of linear combination of atomic orbitals (LCAO).** We have assumed that the shapes and dispositions of bond orbitals are related in a simple way to the shapes and dispositions of atomic orbitals. The LCAO method makes the same assumption *mathematically*: to calculate an approximate molecular orbital, ψ, one uses a *linear combination* (that is, a combination through addition or subtraction) of atomic orbitals.

$$\psi = \phi_A + \phi_B$$

where ψ is the molecular orbital
ϕ_A is atomic orbital A
ϕ_B is atomic orbital B

The rationale for this assumption is simple: when the electron is near atom A, ψ resembles ϕ_A; when the electron is near atom B, ψ resembles ϕ_B.

Now this combination is *effective*—that is, the molecular orbital is appreciably more stable than the atomic orbitals—only if the atomic orbitals ϕ_A and ϕ_B:

(a) overlap to a considerable extent;
(b) are of comparable energy; and
(c) have the same symmetry about the bond axis.

These requirements can be justified mathematically. Qualitatively, we can say this: if there is not considerable overlap, the energy of ψ is equal to either that of ϕ_A or that of ϕ_B; if the energy of ϕ_A and ϕ_B are quite different, the energy of ψ is essentially that of the more stable atomic orbital. In either case, there is no significant stabilization, and no bond formation.

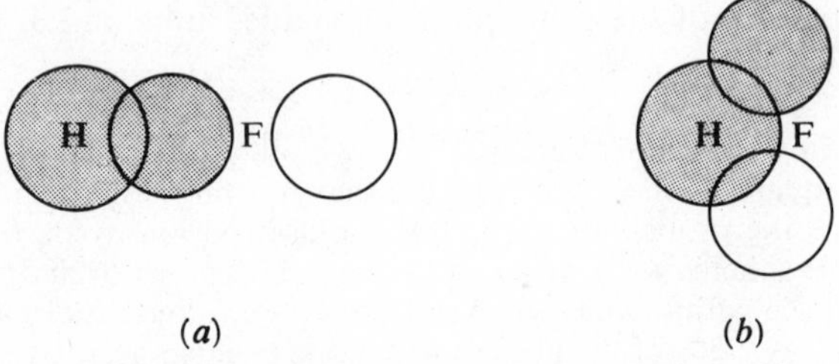

Figure 29.3. The hydrogen fluoride molecule: dependence of overlap on orbital symmetry. (*a*) Overlap of lobes of same phase leads to bonding. (*b*) Positive overlap and negative overlap cancel each other.

When we speak of the symmetry of orbitals, we are referring to the relative phases of lobes, and their disposition in space. To see what is meant by requirement (c), that the overlapping orbitals have the same symmetry, let us look at one example: hydrogen fluoride. This molecule can be pictured as resulting from overlap of the *s* orbital of hydrogen with a *p* orbital of fluorine. In Fig. 29.3*a*, we use the $2p_x$ orbital, where the x coordinate is taken as the H—F axis. The shaded *s* orbital overlaps the shaded lobe of the *p* orbital, and a bond forms. If, however, we were to use the $2p_z$ (or $2p_y$) orbital as in Fig. 29.3*b*, overlap of *both* lobes—plus and minus—would occur and cancel each other. That is, the positive overlap integral would be exactly canceled by the negative overlap integral; the net effect would be *no overlap*, and no bond formation. The dependence of overlap on phase is fundamental to chemical bonding.

29.4 Bonding and antibonding orbitals

Quantum mechanics shows that linear combination of two functions gives, not one, but *two* combinations and hence *two* molecular orbitals: a *bonding* orbital, more stable than the component atomic orbitals; and an *antibonding* orbital, less stable than the component orbitals.

$$\psi_{+} = \phi_{A} + \phi_{B}$$ Bonding orbital: *stabilizes molecule*

$$\psi_{-} = \phi_{A} - \phi_{B}$$ Antibonding orbital: *destabilizes molecule*

Two *s* orbitals, for example, can be added,

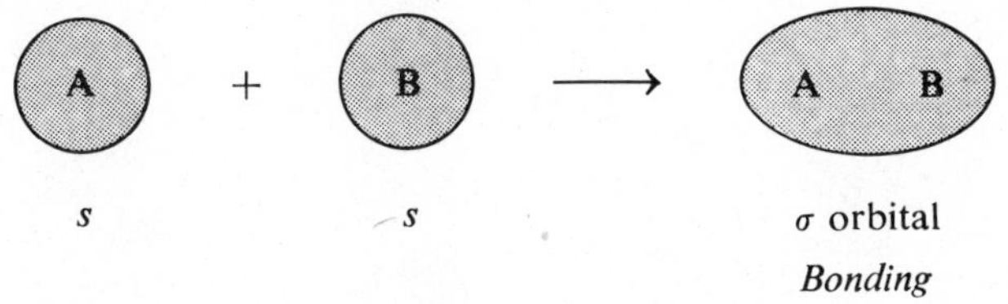

A + B ⟶ A B

s *s* σ orbital

Bonding

or subtracted.

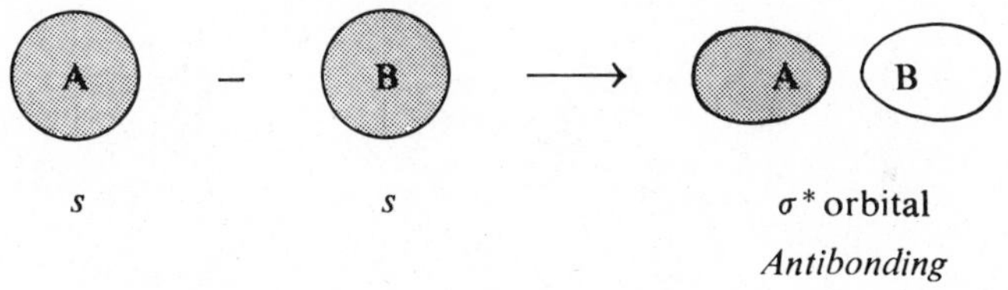

A − B ⟶ A B

s *s* σ* orbital

Antibonding

We can see, in a general way, why there must be two combinations. There can be as many as two electrons in each component atomic orbital, making a total of four electrons; two molecular orbitals are required to accommodate them.

Figure 29.4 shows schematically the shapes of the molecular orbitals, bonding and antibonding, that result from overlap of various kinds of atomic orbitals. We recognize the bonding orbitals, σ and π, although until now we have not shown the two lobes of a π orbital as being of opposite phase. An antibonding orbital, we see, has a nodal plane perpendicular to the bond axis, and cutting between the atomic nuclei. The antibonding sigma orbital, σ*, thus consists of two lobes, of opposite phase. The antibonding pi orbital, π*, consists of four lobes.

In a bonding orbital, electrons are concentrated in the region between the nuclei, where they can be attracted by both nuclei. The increase in electrostatic attraction lowers the energy of the system. In an antibonding orbital, by contrast, electrons are *not* concentrated between the nuclei; electron charge is zero in the nodal plane. Electrons spend most of their time farther from a nucleus than in the separated atoms. There is a decrease in electrostatic attraction, and an increase in repulsion between the nuclei. The energy of the system is higher than that of the separated atoms. *Where electrons in a bonding orbital tend to hold the atoms together, electrons in an antibonding orbital tend to force the atoms apart.*

It may at first seem strange that electrons in certain orbitals can actually weaken the bonding. Should not *any* electrostatic attraction, even if less than optimum, be better than none? We must remember that it is the bond dissociation energy we are concerned with. We are not comparing the electrostatic attraction in an antibonding orbital with no electrostatic attraction; we are comparing it with the stronger electrostatic attraction in the separated atoms.

There are, in addition, orbitals of a third kind, *non-bonding orbitals*. As the name indicates, electrons in these orbitals—unshared pairs, for example—neither strengthen nor weaken the bonding between atoms.

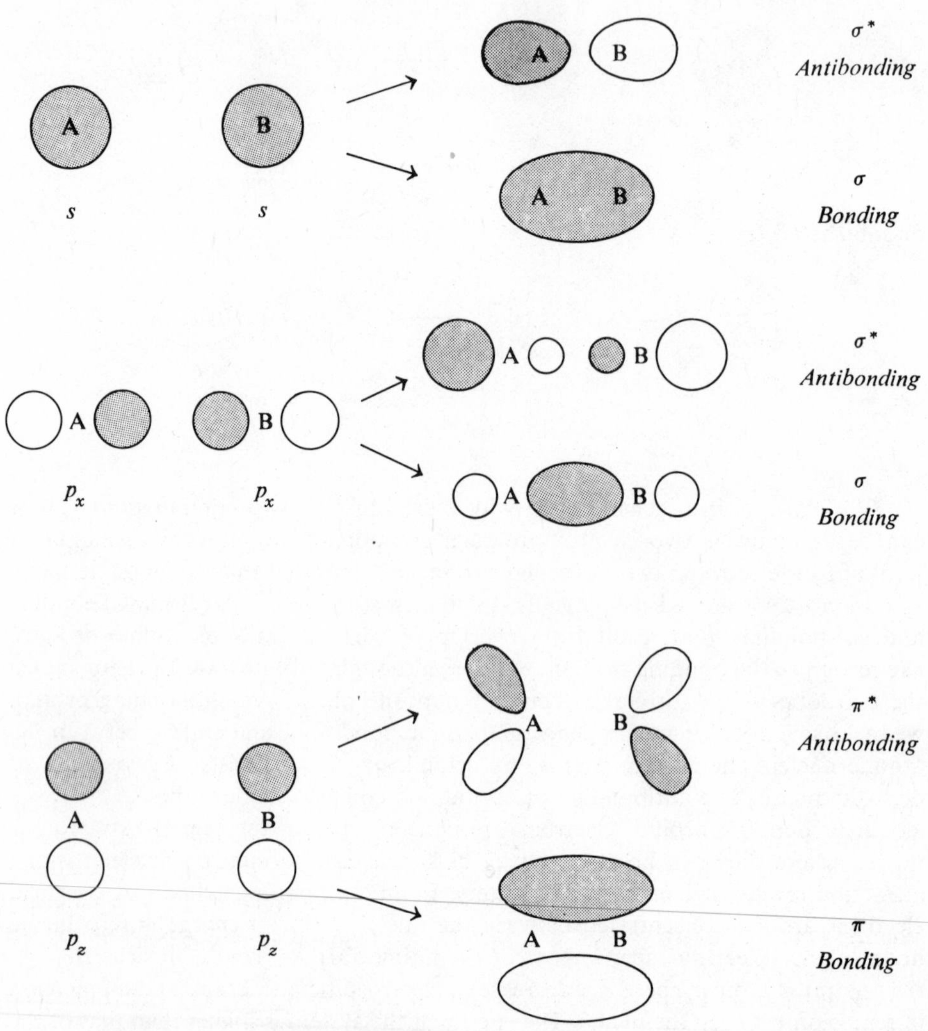

Figure 29.4. Bonding and antibonding orbitals.

29.5 Electronic configurations of some molecules

Let us look at the electronic configurations of some familiar molecules. The shapes and relative stabilities of the various molecular orbitals are calculated by quantum mechanics, and we shall simply use the results of these calculations. We picture the nuclei in place, with the molecular orbitals mapped out about them, and we feed electrons into the orbitals. In doing this we follow the same rules that we followed in arriving at the electronic configurations of atoms. There can be only two electrons—and of opposite spin—in each orbital, with orbitals of lower energy being filled up first. If there are orbitals of equal energy, each gets an electron before any one of them gets a pair of electrons. We shall limit our attention to orbitals containing π electrons, since these electrons will be the ones of chief interest to us.

For the π electrons of ethylene (Fig. 29.5), there are two molecular orbitals since there are two linear combinations of the two component p orbitals. The broken line in the figure indicates the non-bonding energy level; below it lies the bonding orbital, π, and above it lies the antibonding orbital, π^*.

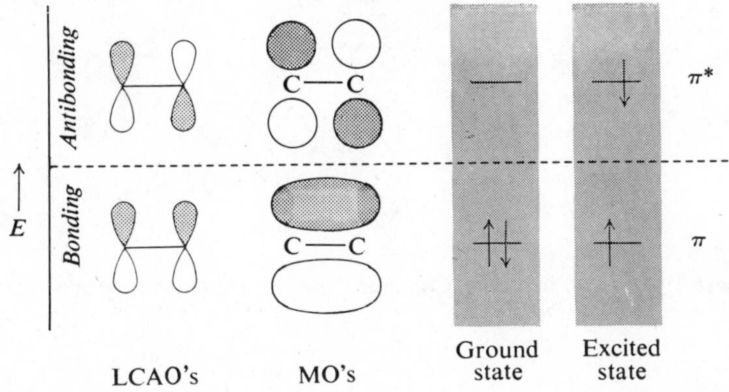

Figure 29.5. Ethylene. Configuration of π electrons in ground state and excited state.

Normally, a molecule exists in the state of lowest energy, the *ground state*. But, as we have seen (Sec. 13.5), absorption of light of the right frequency (in the ultraviolet region) raises a molecule to an *excited state*, a state of higher energy. In the ground state of ethylene, we see, both π electrons are in the π orbital; this configuration is specified as π^2, where the superscript tells the number of electrons in that orbital. In the excited state one electron is in the π orbital and the other—still of opposite spin—is in the π^* orbital; this configuration, $\pi\pi^*$, is naturally the less stable since only one electron helps to hold the atoms together, while the other tends to force them apart.

For 1,3-butadiene, with four component p orbitals, there are four molecular orbitals for π electrons (Fig. 29.6). The ground state has the configuration $\psi_1^2\psi_2^2$; that is, there are two electrons in each of the bonding orbitals, ψ_1 and ψ_2. The higher of these, ψ_2, resembles two isolated π orbitals, although it is of somewhat lower

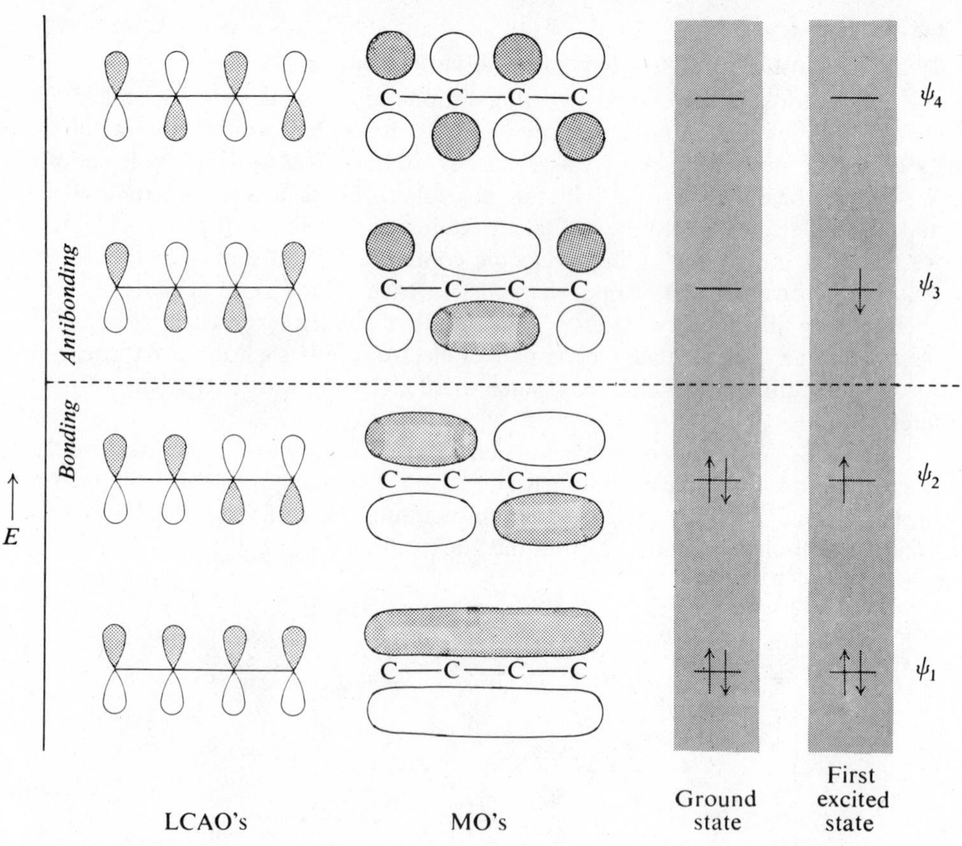

Figure 29.6. 1,3-Butadiene. Configuration of π electrons in ground state and first excited state.

energy. Orbital ψ_1 encompasses all four carbons; this delocalization provides the net stabilization of the conjugated system. Absorption of light of the right frequency raises one electron to ψ_3.

$$\underset{\text{Ground state}}{\psi_1^2\psi_2^2} \xrightarrow{h\nu} \underset{\text{Lowest excited state}}{\psi_1^2\psi_2\psi_3}$$

Next, let us look at the allyl system: cation, free radical, and anion. Regardless

$$\underbrace{CH_2\text{---}CH\text{---}CH_2}_{\oplus} \qquad \underbrace{CH_2\text{---}CH\text{---}CH_2}_{\cdot} \qquad \underbrace{CH_2\text{---}CH\text{---}CH_2}_{\ominus}$$

Allyl cation — Allyl free radical — Allyl anion

of the number of π electrons, there are three component p orbitals, one on each carbon, and they give rise to three molecular orbitals, ψ_1, ψ_2, and ψ_3. As shown in Fig. 29.7, ψ_1 is bonding and ψ_3 is antibonding. Orbital ψ_2 encompasses only the end carbons (there is a node at the middle carbon) and is of the same energy as an isolated p orbital; it is therefore non-bonding.

The allyl cation has π electrons only in the bonding orbital. The free radical has one electron in the non-bonding orbital as well, and the anion has two in the non-bonding orbital. The bonding orbital ψ_1 encompasses all three carbons, and

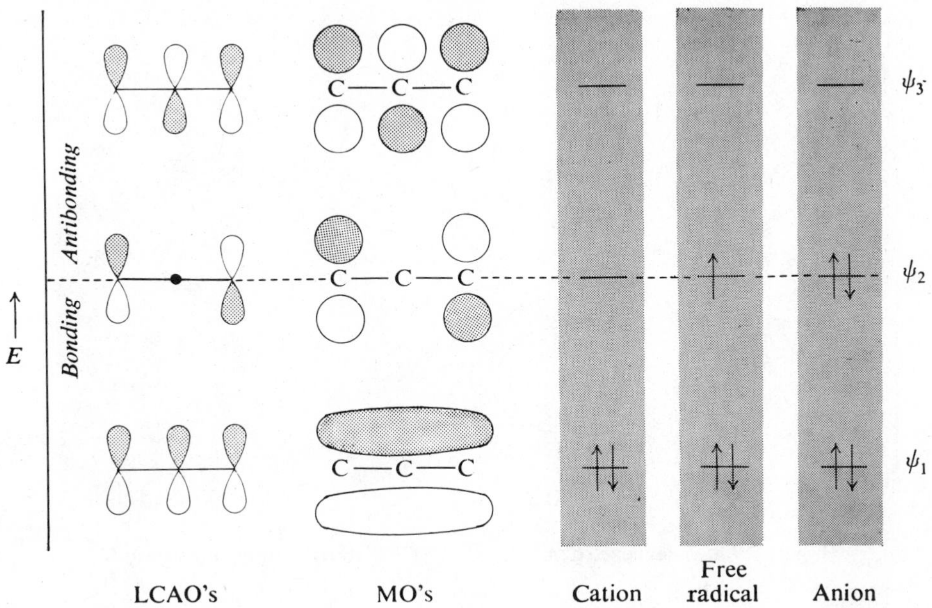

Figure 29.7. Allyl system. Configuration of π electrons in cation, free radical, and anion.

is more stable than a localized π orbital involving only two carbons; it is this delocalization that gives allylic particles their special stability. We see the symmetry we have attributed to allylic particles on the basis of the resonance theory; the two ends of each of these molecules are equivalent.

Finally, let us look at benzene. There are six combinations of the six component p orbitals, and hence six molecular orbitals. Of these, we shall consider only these combinations, which correspond to the three most stable molecular

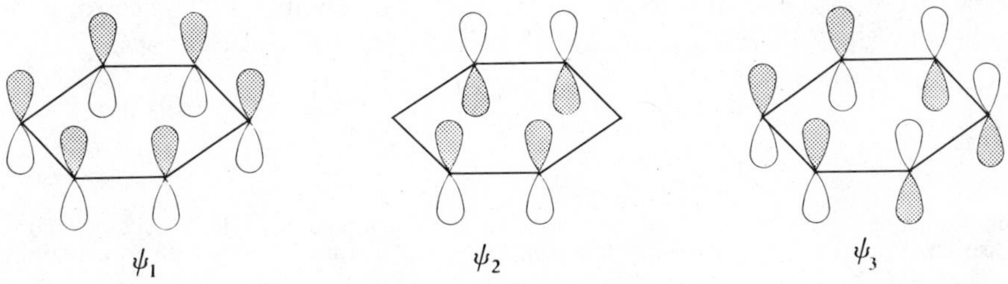

Benzene: first three LCAO's

orbitals, all bonding orbitals (Fig. 29.8). Each contains a pair of electrons. The lowest orbital, ψ_1, encompasses all six carbons. Orbitals ψ_2 and ψ_3 are of different shape, but equal energy; together they provide—as does ψ_1—equal electron density

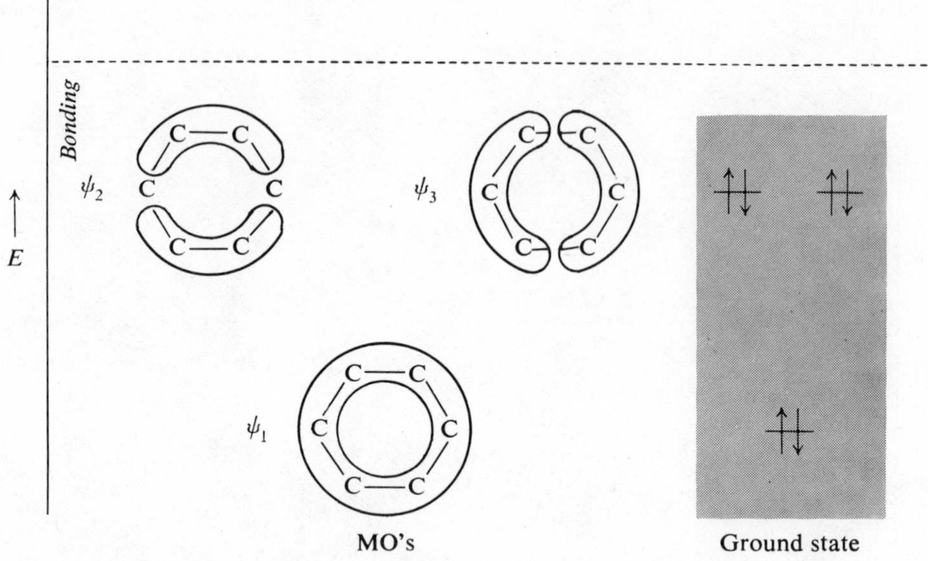

Figure 29.8. Benzene. Configuration of π electrons in ground state.

at all six carbons. The net result, then, is a highly symmetrical molecule with considerable delocalization of π electrons. But this is only part of the story; in the next section we shall look more closely at just what makes benzene such a special kind of molecule.

29.6 Aromatic character. The Hückel 4*n* + 2 rule

In Chap. 10 we discussed the structure of aromatic compounds. An aromatic molecule is flat, with cyclic clouds of delocalized π electrons above and below the plane of the molecule. We have just seen, for benzene, the molecular orbitals that permit this delocalization. But delocalization alone is not enough. For that special degree of stability we call *aromaticity*, the number of π electrons must conform to **Hückel's rule**: *there must be a total of* **(4*n* + 2) π electrons.**

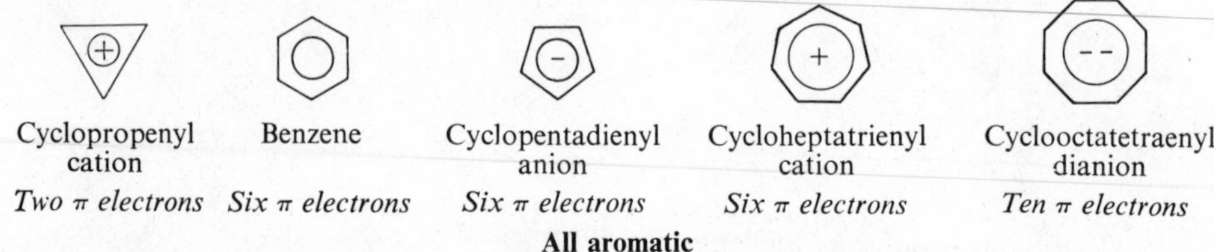

In Sec. 10.10, we saw evidence of special stability associated with the "magic" numbers of 2, 6, and 10 π electrons, that is, with systems where n is 0, 1, and 2 respectively. Problem 5 (p. 447) described the nmr spectrum of cyclooctadecanonaene, which contains 18 π electrons (n is 4). Twelve protons lie outside the ring,

Cyclooctadecanonaene
Eighteen π electrons
Aromatic

are deshielded, and absorb downfield; but, because of the particular geometry of the large flat molecule, six protons lie *inside* the ring, are shielded (see Fig. 13.4, p. 419), and absorb upfield. The spectrum is unusual, but exactly what we would expect if this molecule were aromatic.

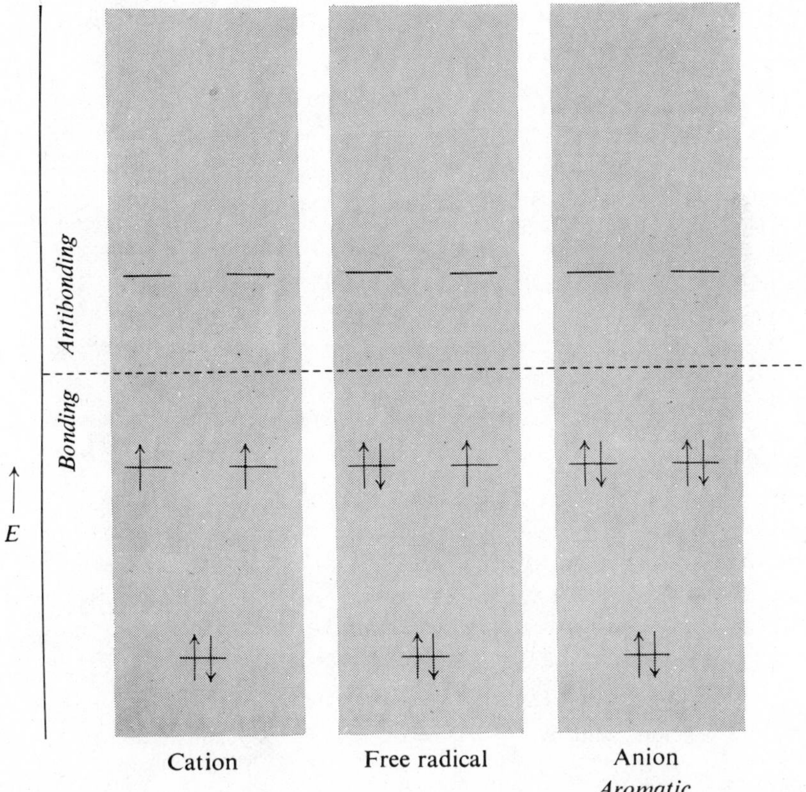

Figure 29.9. Cyclopentadienyl system. Configuration of π electrons in cation, free radical, and anion.

Hückel (p. 328) was a pioneer in the field of molecular orbital theory. He developed the LCAO method in its simplest form, yet "Hückel molecular orbitals" have proved enormously successful in dealing with organic molecules. Hückel proposed the $4n + 2$ rule in 1931. It has been tested in many ways since then, and it *works*. Now, what is the theoretical basis for this rule?

Let us begin with the cyclopentadienyl system. Five sp^2-hybridized carbons have five component p orbitals, which give rise to five molecular orbitals (Fig. 29.9, p. 935). At the lowest energy level there is a single molecular orbital. Above this, the orbitals appear as *degenerate* pairs, that is, pairs of orbitals of equal energy. The lowest degenerate pair are bonding, the higher ones are antibonding.

The cyclopentadienyl cation has four electrons. Two of these go into the lower orbital. Of the other two electrons, one goes into each orbital of the lower degenerate pair. The cyclopentadienyl free radical has one more electron, which fills one orbital of the pair. The anion has still another electron, and with this we fill the remaining orbital of the pair. The six π electrons of the cyclopentadienyl anion are *just enough to fill all the bonding orbitals*. Fewer than six leaves bonding orbitals unfilled; more than six, and electrons would have to go into antibonding orbitals. Six π electrons gives maximum bonding and hence maximum stability.

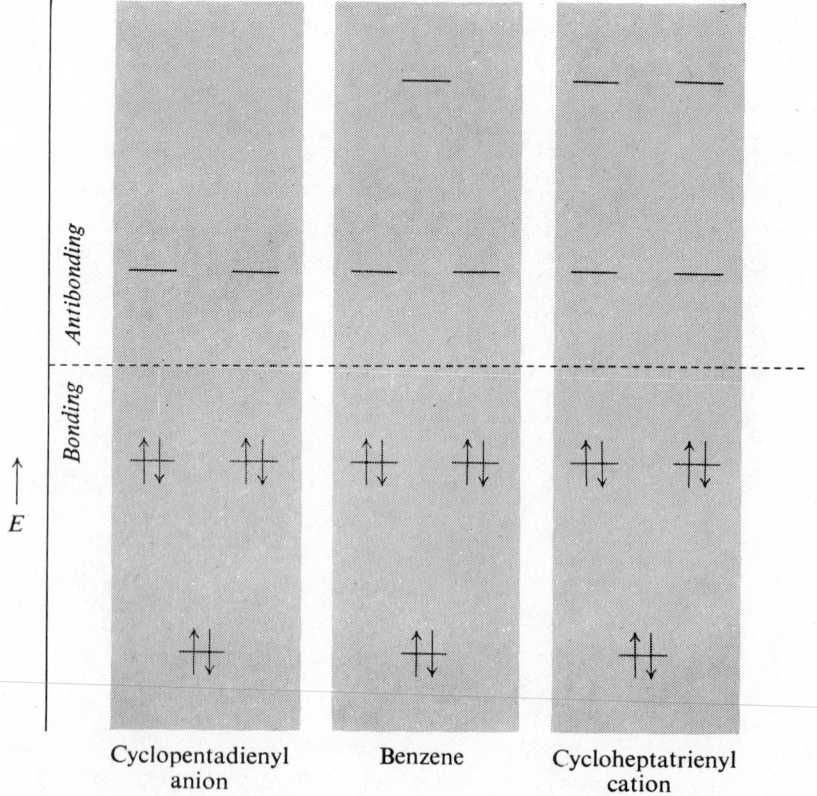

Figure 29.10. Aromatic compounds with 6 π electrons. Configuration of π electrons in cyclopentadienyl anion, benzene, and cycloheptatrienyl cation.

Figure 29.10 shows the molecular orbitals for rings containing five, six, and seven sp^2-hybridized carbons. We see the same pattern for all of them: a single orbital at the lowest level, and above it a series of degenerate pairs. It takes $(4n + 2)$ π electrons to *fill* a set of these bonding orbitals: 2 electrons for the lowest orbital, and 4 for each of n degenerate pairs. Such an electron configuration has been likened to the rare gas configuration of an atom, with its closed shell. It is the filling of these molecular orbital shells that makes these molecules aromatic.

In Problem 10.6 (p. 330) we saw that the cyclopropenyl cation is unusually stable: 20 kcal/mole more stable even than the allyl cation. In contrast, the cyclo-

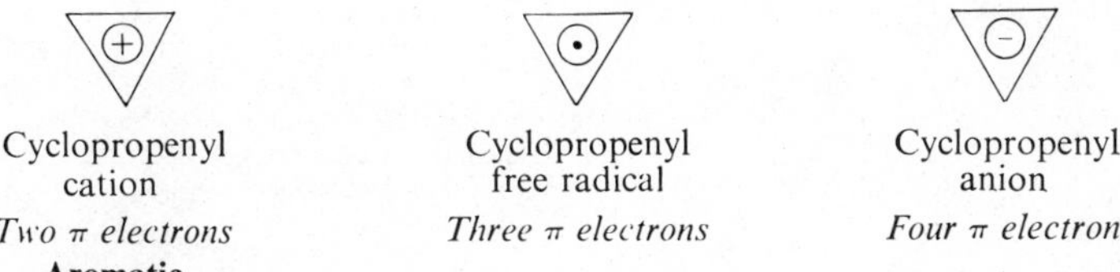

Cyclopropenyl cation	Cyclopropenyl free radical	Cyclopropenyl anion
Two π electrons	*Three π electrons*	*Four π electrons*
Aromatic		

propenyl free radical and anion are *not* unusually stable; indeed, the anion seems to be particularly *unstable*. The cation has the Hückel number of two π electrons (n is zero) and is aromatic. Here, too, aromaticity results from the filling up of a molecular orbital shell (Fig. 29.11).

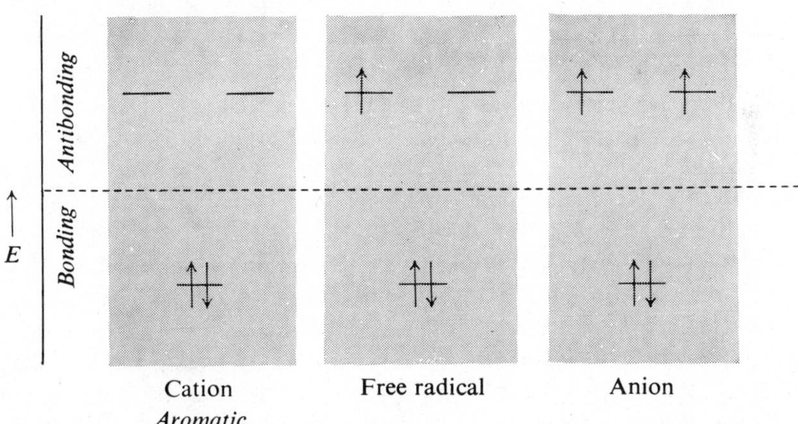

Figure 29.11. Cyclopropenyl system. Configuration of π electrons in cation, free radical, and anion.

In the allyl system (Fig. 29.7) the third and fourth electrons go into a non-bonding orbital, whereas here they go into antibonding orbitals. As a result, the cyclopropenyl free radical and anion are less stable than their open-chain counterparts. For the cyclopropenyl anion in particular, with two electrons in antibonding orbitals, simple calculations indicate no net stabilization due to delocalization, that is, zero resonance energy. Some calculations indicate that the molecule is actually less stable than if there were no conjugation at all. Such cyclic molecules, in which delocalization actually leads to *de*stabilization, are not just non-aromatic; they are *anti*aromatic.

Problem 29.1 When 3,4-dichloro-1,2,3,4-tetramethylcyclobutene was dissolved at $-78°$ in SbF_5–SO_2, the solution obtained gave three nmr peaks, at δ 2.07, δ 2.20, and δ 2.68, in the ratio 1:1:2. As the solution stood, these peaks slowly disappeared and were replaced by a single peak at δ 3.68. What compound is each spectrum probably due to? Of what theoretical significance are these findings?

Problem 29.2 (a) Cyclopropenones (I) have been made, and found to have rather unusual properties.

```
R—C═══C—R
    \ /
     C          R = phenyl or n-propyl
     ‖
     O
     I
```

They have very high dipole moments: about 5 D, compared with about 3 D for benzophenone or acetone. They are highly basic for ketones, reacting with perchloric acid to yield salts of formula $(R_2C_3OH)^+ClO_4^-$. What factor may be responsible for these unusual properties?

(b) Diphenylcyclopropenone was allowed to react with phenylmagnesium bromide, and the reaction mixture was hydrolyzed with perchloric acid. There was obtained, not a tertiary alcohol, but a salt of formula $[(C_6H_5)_3C_3]^+ClO_4^-$. Account for the formation of this salt.

(c) The synthesis of the cyclopropenones involved the addition to alkynes of CCl_2, which was generated from $Cl_3CCOONa$. Show all steps in the most likely mechanism for the formation of CCl_2. (*Hint:* See Sec. 9.16.)

29.7 Orbital symmetry and the chemical reaction

A chemical reaction involves the crossing of an energy barrier. In crossing this barrier, the reacting molecules seek the easiest path: a low path, to avoid climbing any higher than is necessary; and a broad path, to avoid undue restrictions on the arrangement of atoms. As reaction proceeds, there is a change in bonding among the atoms, from the bonding in the reactants to the bonding in the products. Bonding is a stabilizing factor; the stronger the bonding, the more stable the system. If a reaction is to follow the easiest path, it must take place in the way that *maintains maximum bonding during the reaction process*. Now, bonding, as we visualize it, results from overlap of orbitals. Overlap requires that portions of different orbitals occupy the same space, and that they be *of the same phase*.

This line of reasoning seems perfectly straightforward. Yet the central idea, that the course of reaction can be controlled by orbital symmetry, was a revolutionary one, and represents one of the really giant steps forward in chemical theory. A number of people took part in the development of this concept: K. Fukui in Japan, H. C. Longuet-Higgins in England. But organic chemists became aware of the power of this approach chiefly through a series of papers published in 1965 by R. B. Woodward and Roald Hoffmann working at Harvard University.

Very often in organic chemistry, theory lags behind experiment; many facts are accumulated, and a theory is proposed to account for them. This is a perfectly respectable process, and extremely valuable. But with orbital symmetry, just the reverse has been true. The theory lay in the mathematics, and what was needed was the spark of genius to see the applicability to chemical reactions. Facts were sparse, and Woodward and Hoffmann made *predictions*, which have since been borne out by experiment. All this is the more convincing because these predictions were of the kind called "risky": that is, the events predicted seemed unlikely on any grounds other than the theory being tested.

Orbital symmetry effects are observed in *concerted* reactions, that is, in reactions where several bonds are being made or broken simultaneously. Woodward and Hoffmann formulated "rules," and described certain reaction paths as *symmetry-allowed* and others as *symmetry-forbidden. All of this applies only to concerted reactions*, and refers to the relative ease with which they take place. A "symmetry-forbidden" reaction is simply one for which the concerted mechanism is very difficult, so difficult that, if reaction is to occur at all, it will probably do so in a different way: by a different concerted path that is symmetry-allowed; or, if there is none, by a stepwise, non-concerted mechanism. In the following brief discussion, and in the problems based on it, we have not the space to give the evidence indicating that each reaction is indeed concerted; but there must *be* such evidence, and gathering it is often the hardest job the investigator has to do.

Nor have we space here for a full, rigorous treatment of concerted reactions, which considers the correlation of symmetry between all the molecular orbitals of the products. We shall focus our attention on certain key orbitals, which contain the "valence" electrons of the molecules. Even this simplified approach, we shall find, is tremendously powerful; it is highly graphic, and in some cases gives information that the more detailed treatment does not.

29.8 Electrocyclic reactions

Under the influence of heat or light, a conjugated polyene can undergo isomerization to form a cyclic compound with a single bond between the terminal carbons of the original conjugated system; one double bond disappears, and the remaining double bonds shift their positions. For example, 1,3,5-hexatrienes yield 1,3-cyclohexadienes:

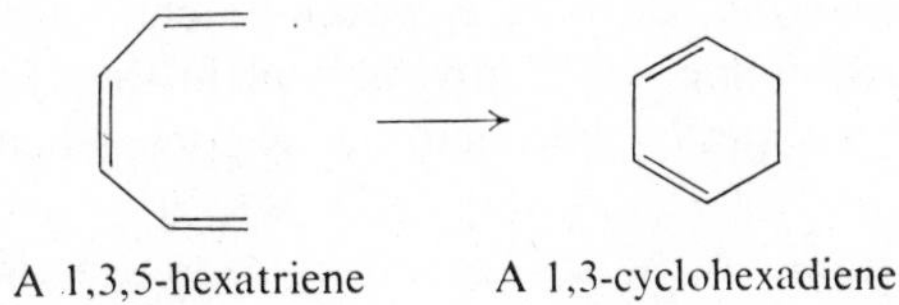

A 1,3,5-hexatriene A 1,3-cyclohexadiene

The reverse process can also take place: a single bond is broken and a cyclic compound yields an open-chain polyene. Cyclobutenes, for example, are converted into butadienes:

A cyclobutene A 1,3-butadiene

Such interconversions are called **electrocyclic reactions.**

It is the stereochemistry of electrocyclic reactions that is of chief interest to us. To observe this, we must have suitably substituted molecules. Let us consider first the interconversion of 3,4-dimethylcyclobutene and 2,4-hexadiene (Fig. 29.12). The cyclobutene exists as *cis* and *trans* isomers. The hexadiene exists in three forms: *cis,cis*; *cis,trans*; and *trans,trans*. As we can see, the *cis* cyclobutene yields only

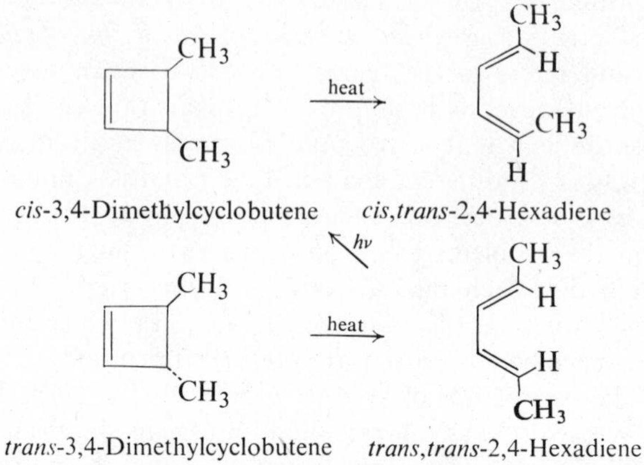

CH_3 CH_3 heat CH_3 H CH_3 H

cis-3,4-Dimethylcyclobutene *cis,trans*-2,4-Hexadiene

$h\nu$

CH_3 CH_3 heat CH_3 H H CH_3

trans-3,4-Dimethylcyclobutene *trans,trans*-2,4-Hexadiene

Figure 29.12. Interconversions of 3,4-dimethylcyclobutenes and 2,4-hexadienes.

one of the three isomeric dienes; the *trans* cyclobutene yields a different isomer. Reaction is thus *completely stereospecific*. Furthermore, photochemical cyclization of the *trans,trans* diene gives a different cyclobutene than the one from which the diene is formed by the thermal (heat-promoted) ring-opening.

The interconversions of the corresponding dimethylcyclohexadienes and the 2,4,6-octatrienes are also stereospecific (Fig. 29.13). Here, too, thermal and photochemical reactions differ in stereochemistry. If we examine the structures closely, we see something else: the stereochemistry of the triene-cyclohexadiene interconversions is *opposite to* that of the diene-cyclobutene interconversions. For the thermal reactions, for example, *cis* methyl groups in the cyclobutene become *cis* and *trans* in the diene; *cis* methyl groups in the cyclohexadiene are *trans* and *trans* in the related triene.

Electrocyclic reactions, then, are completely stereospecific. The exact stereochemistry depends upon two things: (a) the number of double bonds in the polyene, and (b) whether reaction is thermal or photochemical. It is one of the triumphs of the orbital symmetry approach that it can account for all these facts; indeed, most of the examples known today were *predicted* by Woodward and Hoffmann before the facts were known.

It is easier to examine these interconversions from the standpoint of cyclization; according to the principle of microscopic reversibility, whatever applies to this reaction applies equally well to the reverse process, ring-opening. In cyclization, two π electrons of the polyene form the new σ bond of the cycloalkene. But which two electrons? We focus our attention on the *highest occupied molecular orbital* (*HOMO*) *of the polyene*. Electrons in this orbital are the "valence" elec-

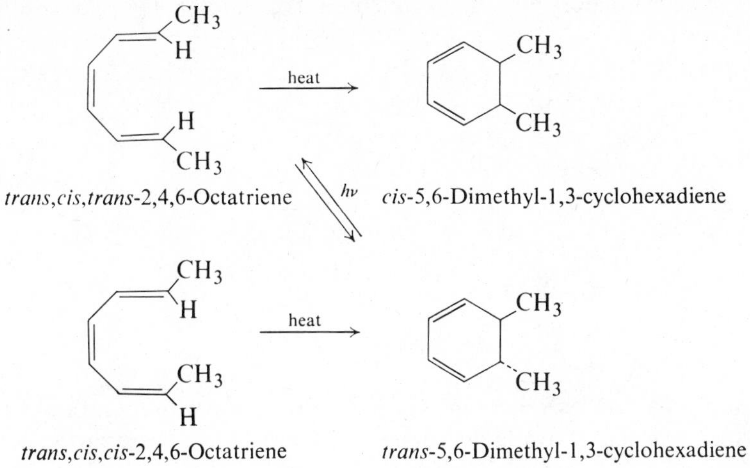

Figure 29.13. Interconversions of 2,4,6-octatrienes and 5,6-dimethyl-1,3-cyclohexadienes.

trons of the molecule; they are the least tightly held, and the most easily pushed about during reaction.

Let us begin with the thermal cyclization of a disubstituted butadiene, RCH=CH—CH=CHR. As we have already seen (Fig. 29.6, p. 932), the highest occupied molecular orbital of a conjugated diene is ψ_2. It is the electrons in this

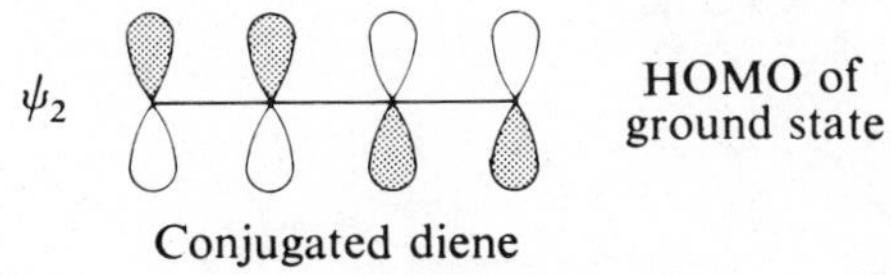

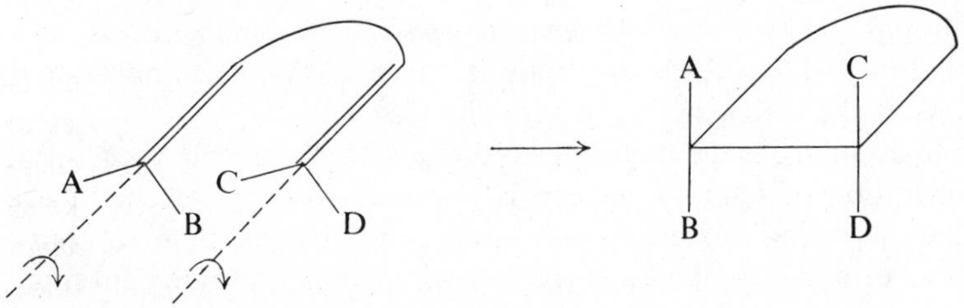

orbital that will form the bond that closes the ring. Bond formation requires overlap, in this case overlap of lobes on C–1 and C–4 of the diene: the front carbons in Fig. 29.14. We see that to bring these lobes into position for overlap, there must be rotation about two bonds, C_1—C_2 and C_3—C_4. This rotation can take place in two different ways: there can be **conrotatory** motion, in which the bonds rotate in the same direction, or there can be **disrotatory** motion, in which the bonds rotate in opposite directions.

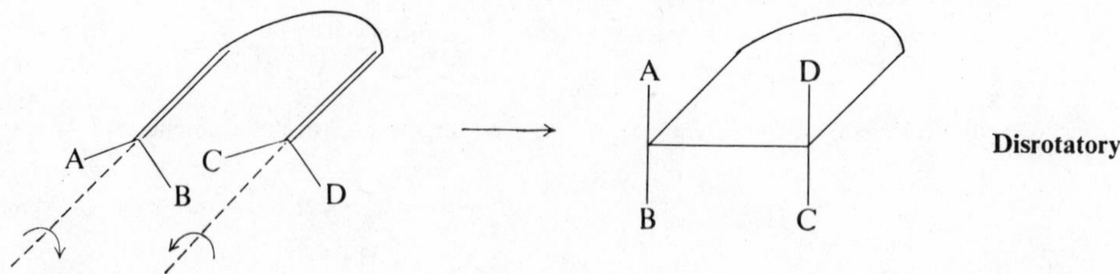

Now, in this case, as we see in Fig. 29.14, conrotatory motion brings together lobes of the *same phase*; overlap occurs and a bond forms. Disrotatory motion, on the other hand, brings together lobes of *opposite phase*; here interaction is antibonding, and repulsive. As Fig. 29.15 on opposite page shows, it is conrotatory motion that produces the stereochemistry actually observed.

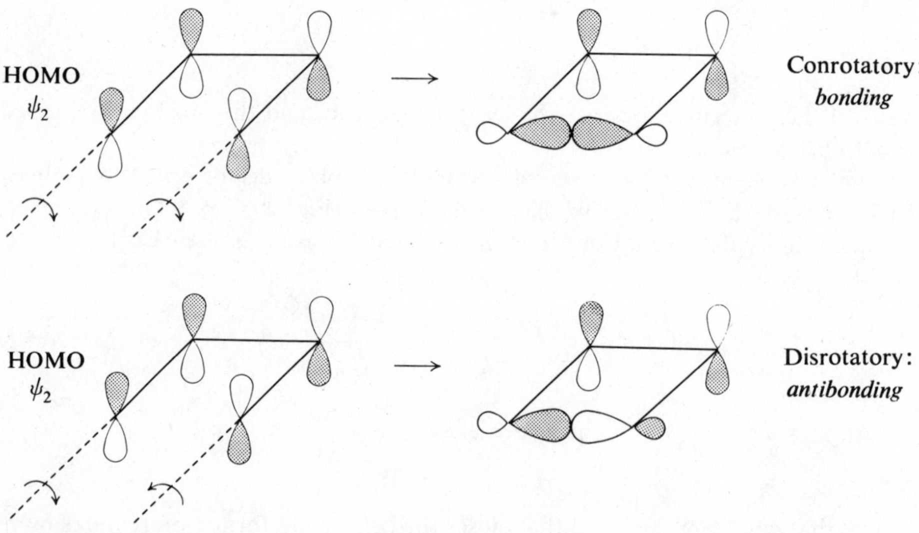

Figure 29.14. Thermal cyclization of a 1,3-butadiene to a cyclobutene. Conrotatory motion leads to bonding. Disrotatory motion leads to antibonding.

How are we to account for the opposite stereochemistry in the photochemical reaction? On absorption of light, butadiene is converted into the excited state shown in Fig. 29.6, in which one electron from ψ_2 has been raised to ψ_3. Now the highest occupied orbital is ψ_3, and it is the electron here that we are concerned

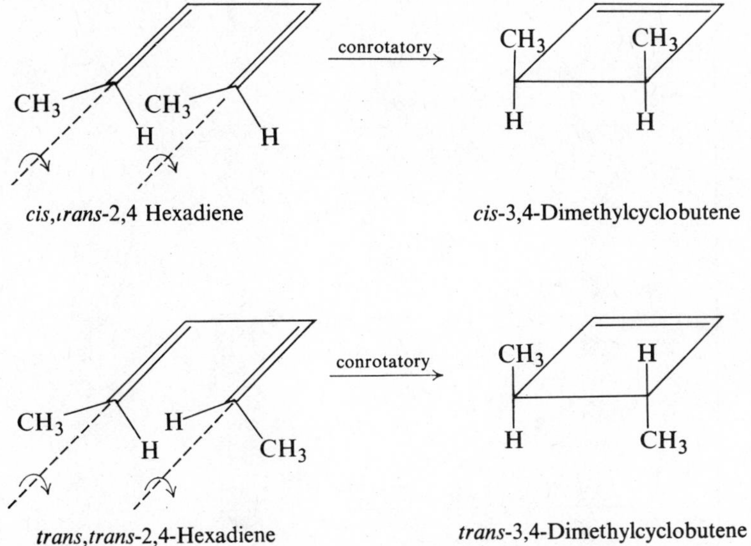

Figure 29.15. Thermal cyclization of substituted butadienes. Observed stereochemistry indicates conrotatory motion.

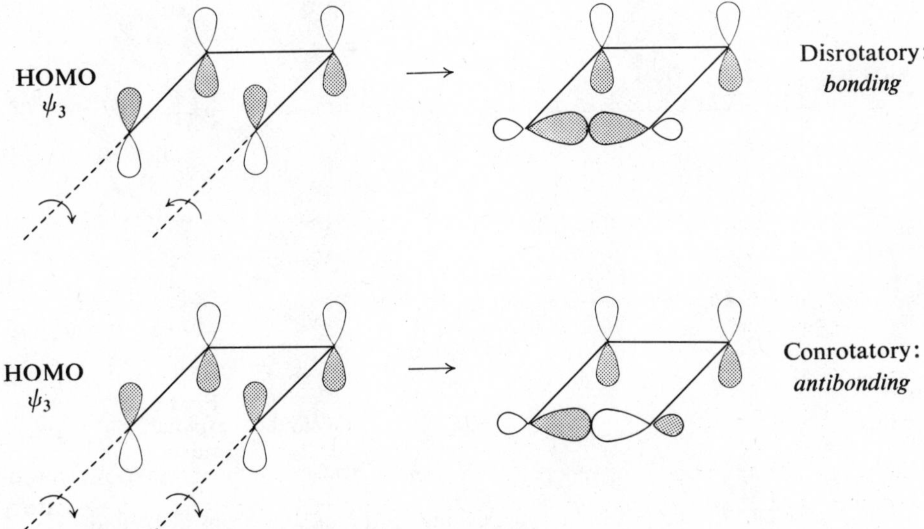

Figure 29.16. Photochemical cyclization of a 1,3-butadiene to a cyclobutene. Disrotatory motion leads to bonding. Conrotatory motion leads to antibonding.

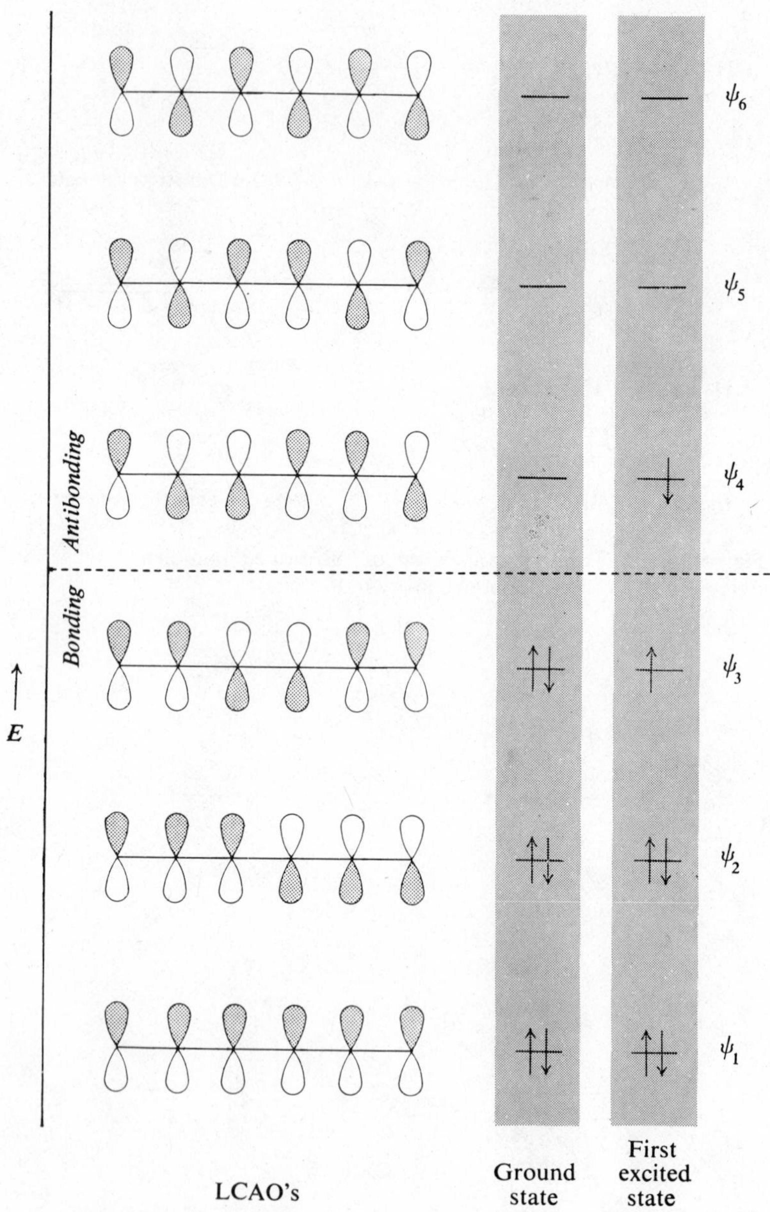

Figure 29.17. A 1,3,5-hexatriene. Configuration of π electrons in ground state and first excited state.

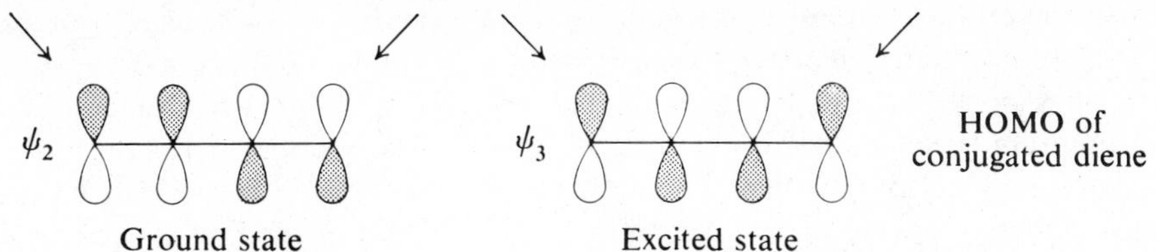

with. But in ψ_3 the relative symmetry of the terminal carbons is opposite to that in ψ_2. Now it is the *disrotatory* motion that brings together lobes of the same phase, and the stereochemistry is reversed (Fig. 29.16).

Next, let us look at the thermal cyclization of a disubstituted hexatriene, RCH=CH—CH=CH—CH=CHR, whose electronic configuration is shown in Fig. 29.17. The HOMO for the ground state of the hexatriene is ψ_3. If we compare this with the HOMO for the ground state of butadiene (ψ_2 in Fig. 29.6), we see that the relative symmetry about the terminal carbons is opposite in the two cases.

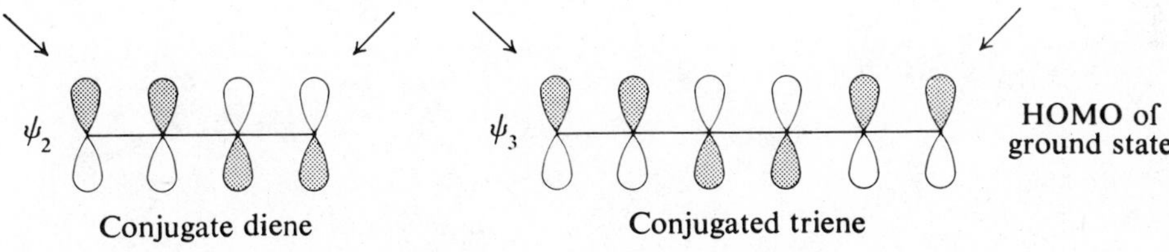

For ground state hexatriene it is disrotatory motion that leads to bonding and, as shown in Fig. 29.18, gives rise to the observed stereochemistry.

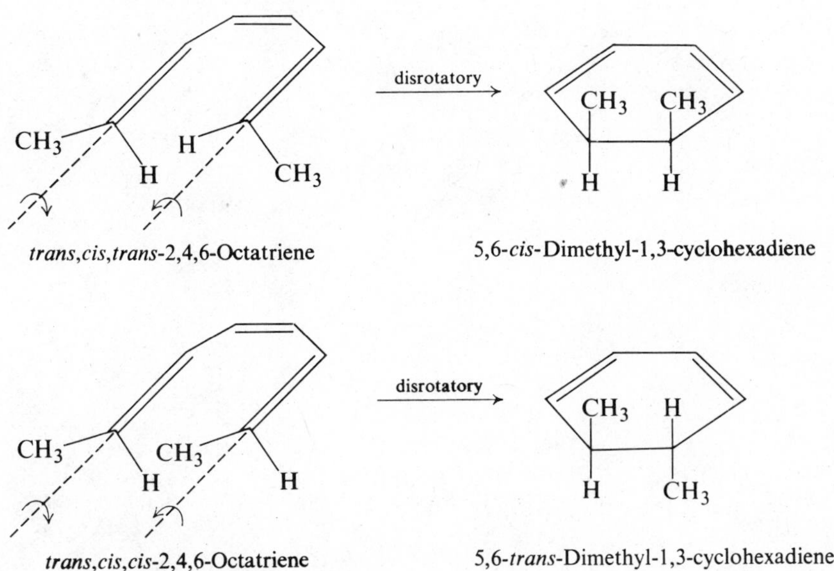

Figure 29.18. Thermal cyclization of substituted hexatrienes. Observed stereochemistry indicates disrotatory motion.

In the excited state of hexatriene, ψ_4 is the HOMO, and once again we see a reversal of symmetry: here, conrotatory motion is the favored process.

What we see here is part of a regular pattern (Table 29.1) that emerges from the quantum mechanics. As the number of pairs of π electrons in the polyene increases, the relative symmetry about the terminal carbons in the HOMO alternates regularly. Furthermore, symmetry in the HOMO of the first excited state is always opposite to that in the ground state.

Table 29.1 WOODWARD-HOFFMANN RULES FOR ELECTROCYCLIC REACTIONS

Number of π electrons	Reaction	Motion
$4n$	thermal	conrotatory
$4n$	photochemical	disrotatory
$4n + 2$	thermal	disrotatory
$4n + 2$	photochemical	conrotatory

Problem 29.3 Thermal ring closure of three stereoisomeric 2,4,6,8-decatetraenes (I, II, and III) has been found to be in agreement with the Woodward-Hoffmann rules.

I II III

Two of these stereoisomers give one dimethylcyclooctatriene, and the third stereoisomer gives a different dimethylcyclooctatriene. (a) Which decatetraenes give which cyclooctatrienes? (b) Predict the product of photochemical ring closure of each.

Problem 29.4 The commonly observed conversion of cyclopropyl cations into allyl cations is considered to be an example of an electrocyclic reaction. (a) What is

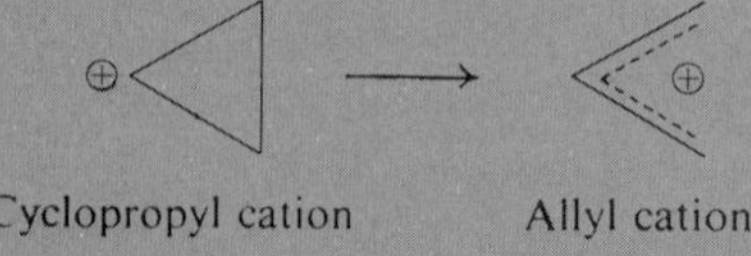

Cyclopropyl cation Allyl cation

the HOMO of the allyl cation? How many π electrons has it? (b) Where does this reaction fit in Table 29.1? Would you expect conrotatory or disrotatory motion? (c) What prediction would you make about interconversion of allyl and cyclopropyl *anions*? (d) About the interconversion of pentadienyl cations and cyclopentenyl cations?

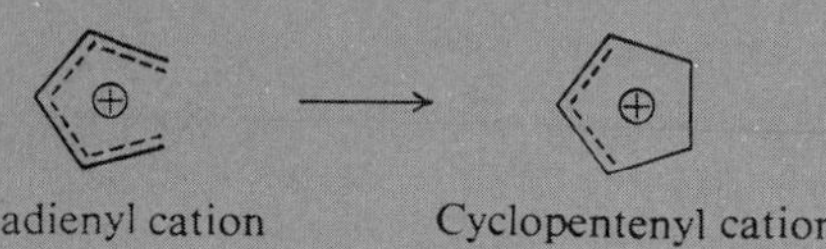

Pentadienyl cation Cyclopentenyl cation

Problem 29.5 Each of the following reactions involves one or more concerted steps that take place in accordance with the Woodward-Hoffmann rules. In each case, show exactly what is happening.

(a) 7-Cyano-7-trifluoromethylnorcaradiene ⇌ 7-Cyano-7-trifluoromethyltropylidene

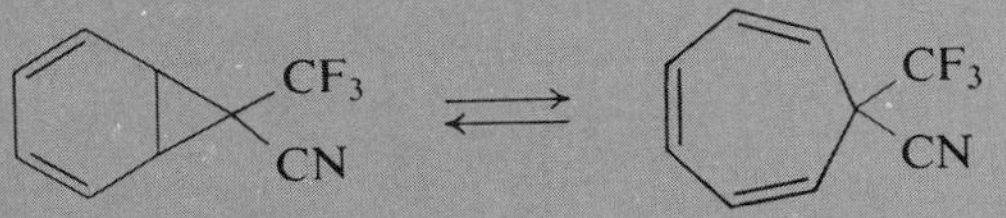

(b) (25°)

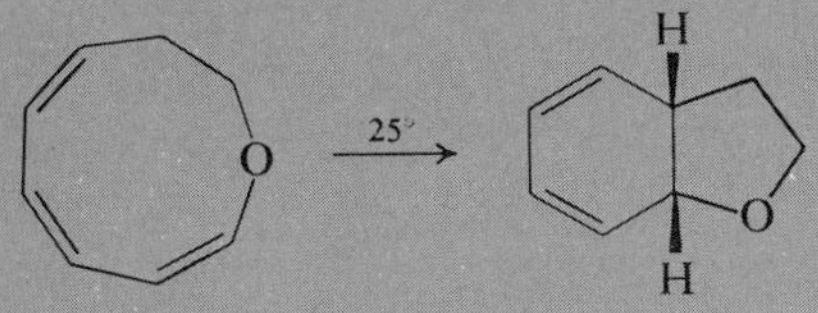

(c) 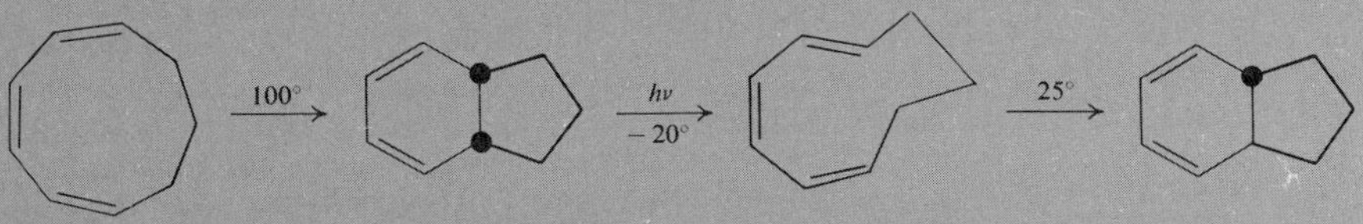 *cis,cis,cis*-Cyclonona-1,3,5-triene —100°→ *cis*-Bicyclo[4.3.0]nona-2,4-diene —$h\nu$, −20°→ *cis,cis,trans*-Cyclonona-1,3,5-triene —25°→ *trans*-Bicyclo[4.3.0]nona-2,4-diene

(d) *cis*-Bicyclo[6.2.0]deca-2,9-diene —140°→ *trans,cis,cis*-Cyclodeca-1,3,5-triene —220°→ *trans*-Bicyclo[4.4.0]deca-2,4-diene

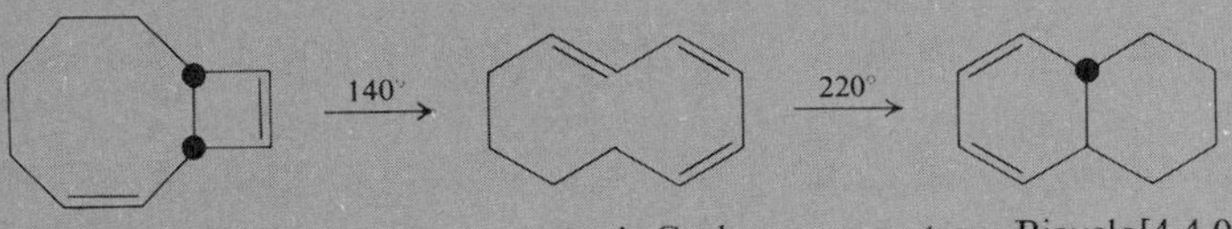

(e) *endo*-2-Chloro-*exo*-2-bromobicyclo[3.1.0]hexane + Ag^+ —H_2O→ 2-Bromo-2-cyclohexenol + AgCl

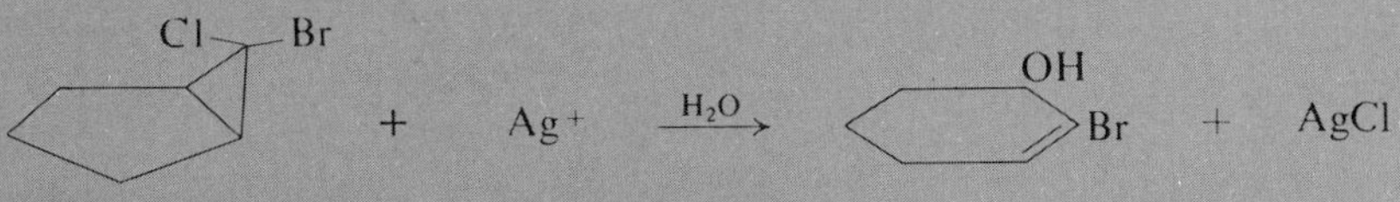

(f) —H_3PO_4→

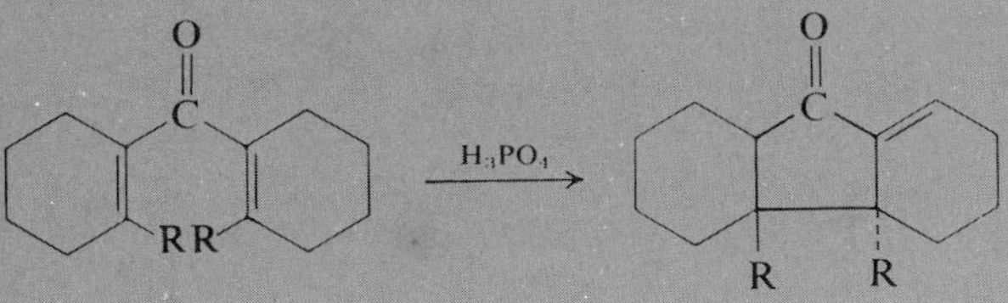

Problem 29.6 Stereoisomers IV and V are easily interconverted by heating. After 51 days at 124°—during which time, it was calculated, 2.6×10^6 interconversions took

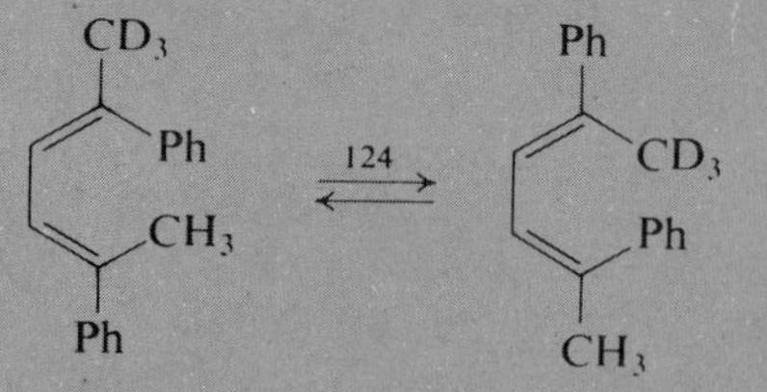

CD_3 Ph CH_3 Ph
IV

124 ⇄

Ph CD_3 Ph CH_3
V

place—*only* IV and V were found to be present; there was *none* of their stereoisomers VI and VII. Propose a mechanism for the interconversion that would account for this remarkable stereospecificity.

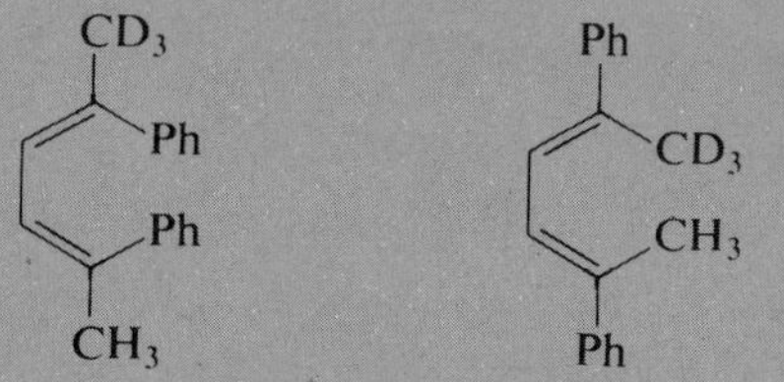

CD_3 Ph Ph CH_3
VI

Ph CD_3 CH_3 Ph
VII

29.9 Cycloaddition reactions

In Sec. 27.8, we encountered the Diels-Alder reaction, in which a conjugated diene and a substituted alkene—the dienophile—react to form a cyclohexene.

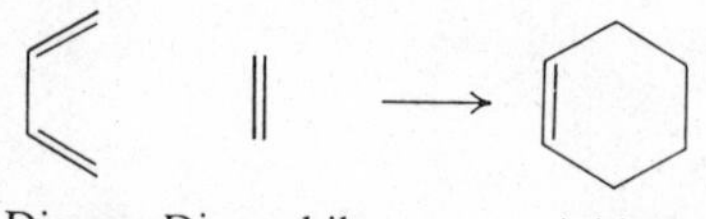

Diene Dienophile → Adduct

Diels-Alder reaction:
a [4 + 2] cycloaddition

This is an example of **cycloaddition**, a reaction in which two unsaturated molecules combine to form a cyclic compound, with π electrons being used to form two new σ bonds. The Diels-Alder reaction is a [4 + 2] cycloaddition, since it involves a system of 4 π electrons and a system of 2 π electrons.

Reaction takes place very easily, often spontaneously, and at most requires moderate application of heat.

There are several aspects to the stereochemistry of the Diels-Alder reaction. (a) First, we have taken for granted—correctly—that the diene must be in the

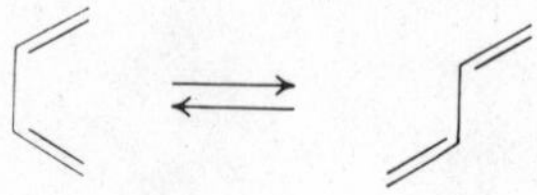

s-cis ⇄ *s-trans*

Required for Diels-Alder reaction

conformation (*s-cis*) that permits the ends of the conjugated system to *reach* the doubly-bonded carbons of the dienophile.

(b) Next, with respect to the alkene (dienophile) addition is clear-cut *syn* (Problem 8, p. 880); this stereospecificity is part of the evidence that the Diels-Alder

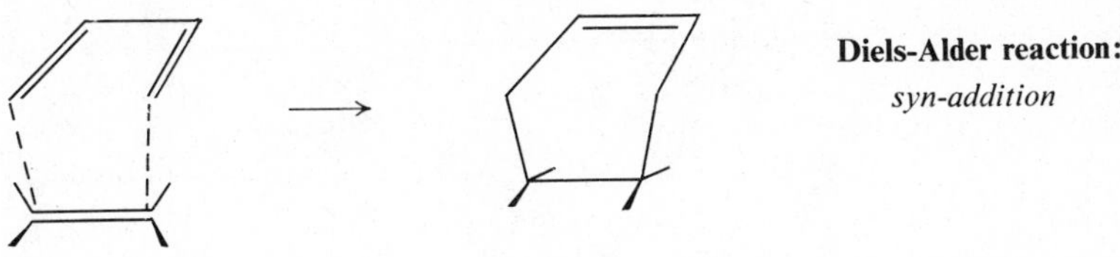

reaction is, indeed, a concerted one, that is, that both new bonds are formed in the same transition state.

(c) Finally, the Diels-Alder reaction takes place in the *endo*, rather than *exo*, sense. That is to say, any other unsaturated groups in the dienophile (for example, —CO—O—CO— in maleic anhydride) tend to lie *near* the developing double bond in the diene moiety (Fig. 29.19). For the *endo* preference to be *seen*, of course, the diene must be suitably substituted.

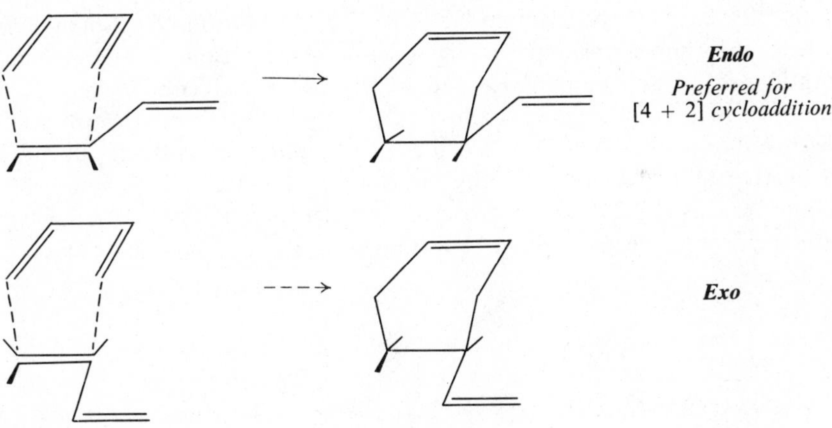

Figure 29.19. Stereochemistry of the Diels-Alder reaction, illustrated for the reaction between two moles of 1,3-butadiene.

Now are there such reactions as [2 + 2] cycloadditions? Can, say, two molecules of ethylene combine to form cyclobutane? The answer is: yes, but not easily under thermal conditions. Under vigorous conditions cycloaddition may occur, but step-wise—via diradicals—and not in a concerted fashion. Photochemical [2 + 2] cycloadditions, on the other hand, are very common. (Although some of these, too, may be stepwise reactions, many are clearly concerted.)

Of thermal cycloadditions, then, [4 + 2] is easy and [2 + 2] is difficult. Of [2 + 2] cycloadditions, the thermal reaction is difficult and the photochemical reaction is easy. How are we to account for these contrasts?

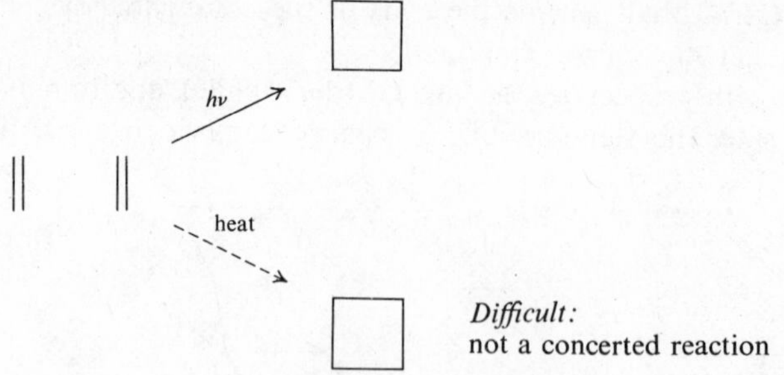

In cycloaddition, two new σ bonds are formed by use of π electrons of the reactants. The concerted reaction results from overlap of orbitals of one molecule with orbitals of the other. As before, it is on electrons in the HOMO that we focus attention. But which orbital does the HOMO overlap? Each new orbital in the product can contain only two electrons. The HOMO of each reactant already contains two electrons, so it must overlap an *empty* orbital of the other reactant; it picks the most stable of these, the lowest unoccupied molecular orbital (LUMO). In the transition state of cycloaddition, then, *stabilization comes chiefly from overlap between the HOMO of one reactant and the LUMO of the other.*

On this basis, let us examine the [4 + 2] cycloaddition of 1,3-butadiene and ethylene, the simplest example of the Diels-Alder reaction. The electronic configurations of these compounds—and of dienes and alkenes in general—have been given in Fig. 29.5 (p. 931) and Fig. 29.6 (p. 932). There are two combinations: overlap of the HOMO of butadiene (ψ_2) with the LUMO of ethylene (π^*); and overlap of the HOMO of ethylene (π) with the LUMO of butadiene (ψ_3). In either case, as Fig. 29.20 shows, overlap brings together lobes of the same phase. There is a flow of electrons from HOMO to LUMO, and bonding occurs.

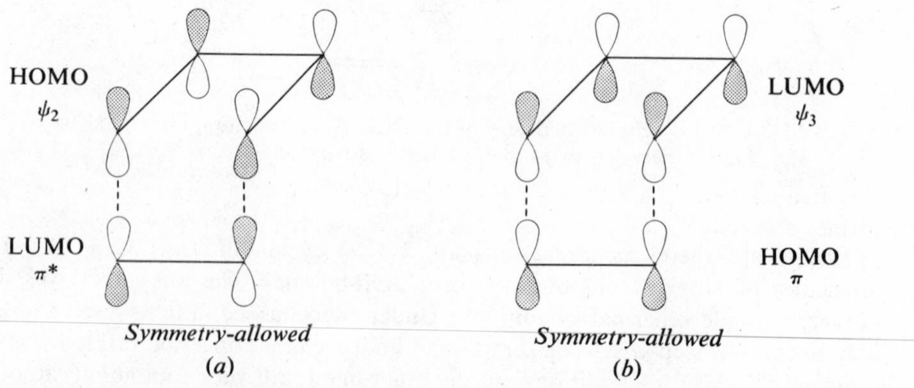

Figure 29.20. Symmetry-allowed thermal [4 + 2] cycloaddition: 1,3-butadiene and ethylene. Overlap of (*a*) HOMO of 1,3-butadiene and LUMO of ethylene, and (*b*) HOMO of ethylene and LUMO of 1,3-butadiene.

Now, consider a thermal [2 + 2] cyclization, dimerization of ethylene. This would involve overlap of the HOMO, π, of one molecule with the LUMO, π^*, of the other. But π and π^* are of opposite symmetry, and, as Fig. 29.21 shows, lobes of opposite phase would approach each other. Interaction is antibonding and repulsive, and concerted reaction does not occur.

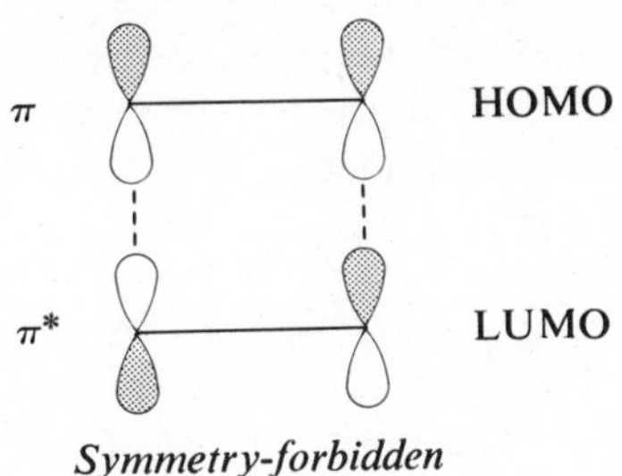

Figure 29.21. Symmetry-forbidden thermal [2 + 2] cycloaddition: two molecules of ethylene. Interaction is antibonding.

Photochemical [2 + 2] cycloadditions are symmetry-allowed. Here we have (Fig. 29.22) overlap of the HOMO (π^*) of an excited molecule with the LUMO (also π^*) of a ground-state molecule.

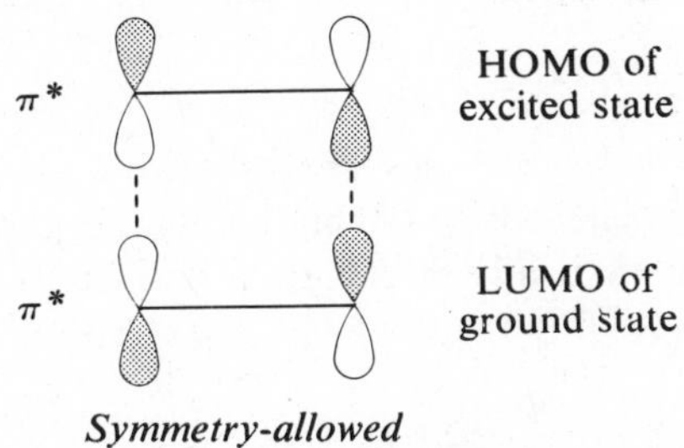

Figure 29.22. Symmetry-allowed photochemical [2 + 2] cycloaddition: two molecules of ethylene, one excited and one in ground-state. Interaction is bonding.

If, in a concerted reaction of this kind, both bonds to a component are being formed (or broken) on the same face, the process is said to be *suprafacial.* If the bonds are being formed (or broken) on opposite faces, the process is *antarafacial.*

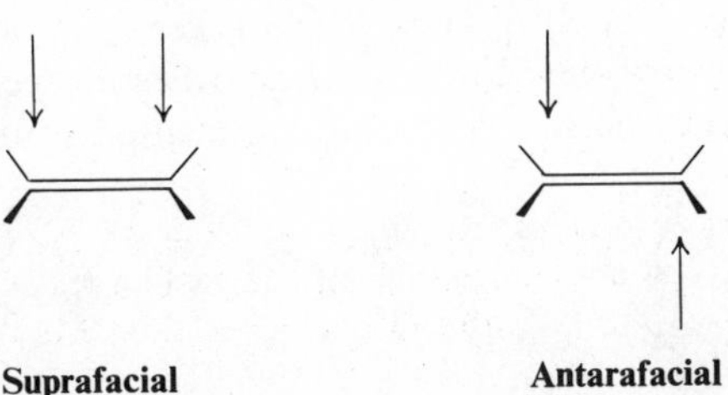

These terms resemble the familiar ones *syn* and *anti*, but with this difference. *Syn* and *anti* describe the net stereochemistry of a reaction. We have seen *anti* addition, for example, as the overall result of a two-step mechanism. *Suprafacial* and *antarafacial*, in contrast, refer to actual processes: the simultaneous making (or breaking) of two bonds on the same face or opposite faces of a component.

So far, our discussion of cycloaddition has assumed that reaction is suprafacial with respect to both components. For [4 + 2] cycloadditions, the stereochemistry shows that this is indeed the case. Now, as far as orbital symmetry is concerned, thermal [2 + 2] cycloaddition *could* occur if it were suprafacial with respect to one component and antarafacial with respect to the other (Fig. 29.23).

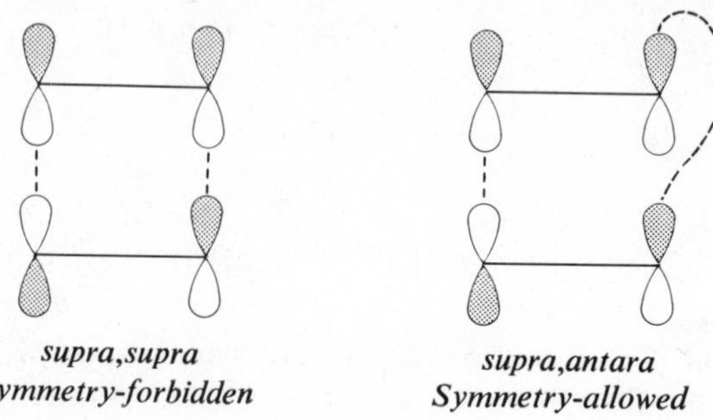

Figure 29.23. [2 + 2] Cycloaddition. *Supra,supra*: geometrically possible, but symmetry-forbidden. *Supra,antara*: symmetry-allowed, but geometrically difficult.

Almost certainly, such a *supra,antara* process is impossible here on geometric grounds. But if the ring being formed is big enough, both *supra,supra* and *supra, antara* processes are geometrically possible; in that case orbital symmetry determines, not *whether* cycloaddition occurs, but *how* it occurs (Table 29.2).

Table 29.2 WOODWARD-HOFFMANN RULES FOR [$i + j$] CYCLOADDITIONS

$i + j$	Thermal	Photochemical
$4n$	supra-antara antara-supra	supra-supra antara-antara
$4n + 2$	supra-supra antara-antara	supra-antara antara-supra

Cycloadditions are reversible. These *cycloreversions* (for example, the *retro*-Diels-Alder reaction) follow the same symmetry rules as cycloadditions—as they must, of course, since they occur via the same transition states.

Problem 29.7 Give structural formulas for the products expected from each of the following reactions. Tell *why* you expect the particular products.

(a) *trans,trans*-2,4-hexadiene + ethylene
(b) *trans*-1,3-pentadiene + maleic anhydride

(c) *trans,trans*-1,4-diphenyl-1,3-butadiene + maleic anhydride

(d) *cis*-2-butene $\xrightarrow{h\nu}$ A + B

(e) *trans*-2-butene $\xrightarrow{h\nu}$ A + C

(f) *cis*-2-butene + *trans*-2-butene $\xrightarrow{h\nu}$ A + B + C + D

Problem 29.8 On standing, cyclopentadiene spontaneously forms *dicyclopentadiene* (I), from which it can be regenerated by heating under a fractionating column.

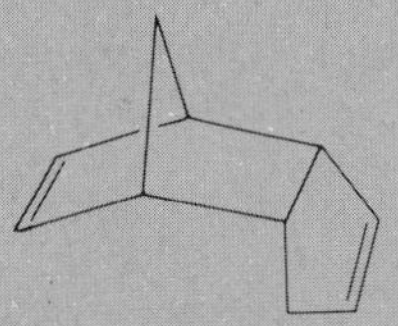

I
Dicyclopentadiene

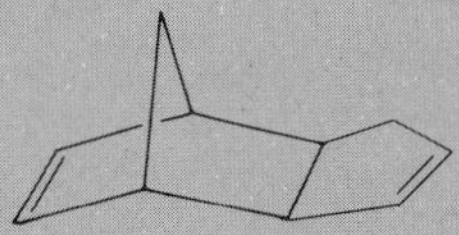

II

(a) What reaction has taken place in the formation of dicyclopentadiene? In the regeneration of cyclopentadiene? (b) On what basis could you have predicted that dicyclopentadiene would have the structure I rather than the structure II?

Problem 29.9 Each of the following reactions is believed to be concerted. Tell what kind of reaction is involved in each case, and what significance it bears on orbital symmetry theory.

(a) $COOC_2H_5$ N + $(CN)_2C{=}C(CN)_2$ ⟶ $(CN)_2$ $(CN)_2$ N $COOC_2H_5$

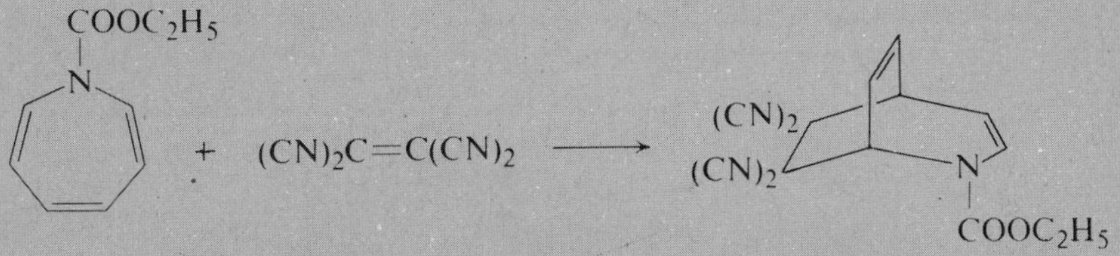

(b) $\xrightarrow{h\nu}$

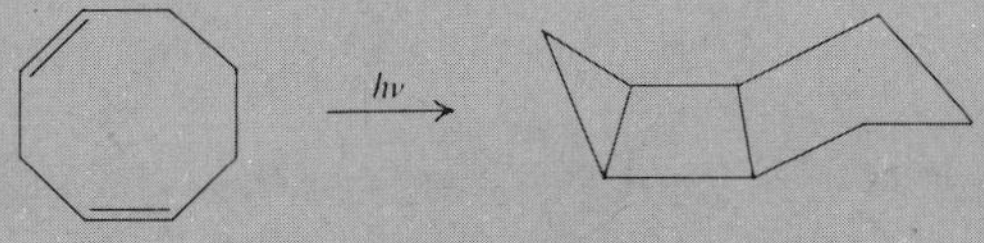

(c) O + ⟶ O

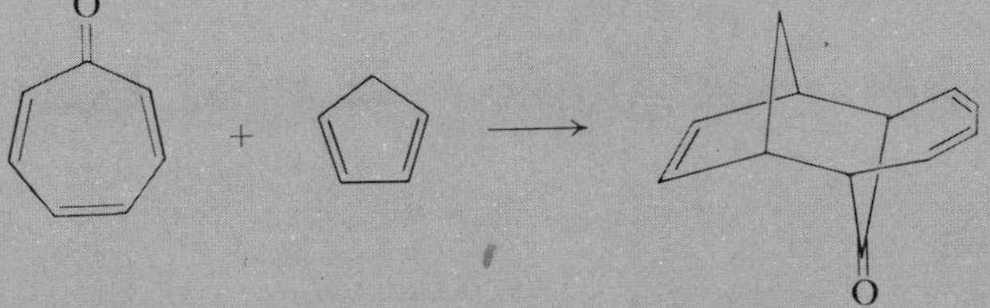

(d) + $CH_3OOCC{\equiv}CCOOCH_3$ ⟶ $COOCH_3$ $COOCH_3$

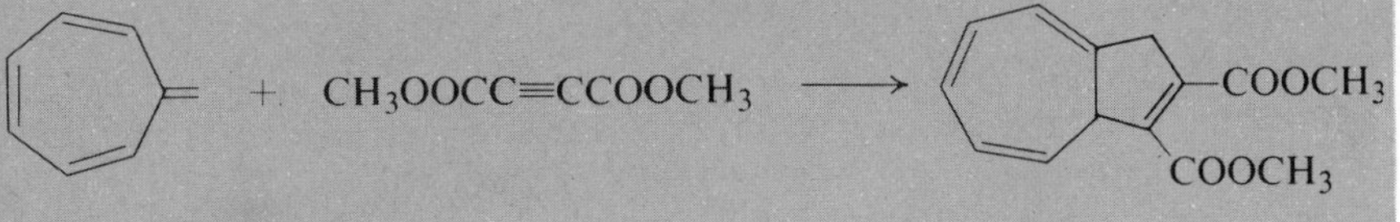

(e) [bi(cycloheptatrienylidene)] + $(CN)_2C{=}C(CN)_2$ ⟶ [adduct with H, H, NC, NC, CN, CN]

29.10 Sigmatropic reactions

A concerted reaction of the type,

$$\overset{G}{\underset{|}{}}C{-}(C{=}C)_n \longrightarrow (C{=}C)_n{-}\overset{G}{\underset{|}{}}C$$

in which a group migrates with its σ bond within a π framework—an ene or a polyene—is called a **sigmatropic reaction.**

The migration is accompanied by a shift in π bonds. For example:

G(1)–C1–C2=C3 ⟶ [G bridging C⋯C⋯C] ⟶ C=C–C(–G) **[1,3] Sigmatropic reaction**

G(1)–C1–C2=C3–C4=C5 ⟶ [G bridging six-membered transition state] ⟶ C=C–C=C–C(–G) **[1,5] Sigmatropic reaction**

In the designations [1,3] and [1,5] the "3" and "5" refer to the number of the carbon to which group G is migrating (the migration terminus). The "1" does *not* refer to the migration source; instead, it specifies that in both reactant and product bonding is to the same atom (number 1) in the migrating group. The important *Cope rearrangement* of hexa-1,5-dienes, for example, is a

[C1–C2=C3 / C1–C2=C3 hexadiene] ⟶ [six-membered transition state] ⟶ [rearranged diene] **[3,3] Sigmatropic reaction**

A 1,5-hexadiene

[3,3] sigmatropic reaction, in which there is a change in position of attachment in G as well as in the π framework—indeed, G itself is a π framework.

In the transition state of a sigmatropic reaction, the migrating group is bonded to both the migration source and the migration terminus; it is the nature of this transition state that we are concerned with. In Sec. 1.8, for convenience we considered bonding in the H_2 molecule to arise from overlap between orbitals on two hydrogen atoms. In the same way, and simply *for convenience*, we consider bonding in the transition state for sigmatropic reactions to arise from overlap between an orbital of an atom or free radical (G) and an orbital of an allylic free radical (the π framework).

This does *not* mean that rearrangement actually involves the separation and reattachment of a free radical. Such a stepwise reaction would not be a concerted one, and hence is not the kind of reaction we are dealing with here. Indeed, a stepwise reaction would be a (high-energy) alternative open to a system if a (concerted) sigmatropic rearrangement were symmetry-forbidden.

In the transition state, there is overlap between the HOMO of one component and the HOMO of the other. Each HOMO is singly occupied, and together they provide a pair of electrons.

The HOMO of an allylic radical depends on the number of carbons in the π framework. The migrating group is passed from one end of the allylic radical to the other, and so it is the end carbons that we are concerned with. We see that

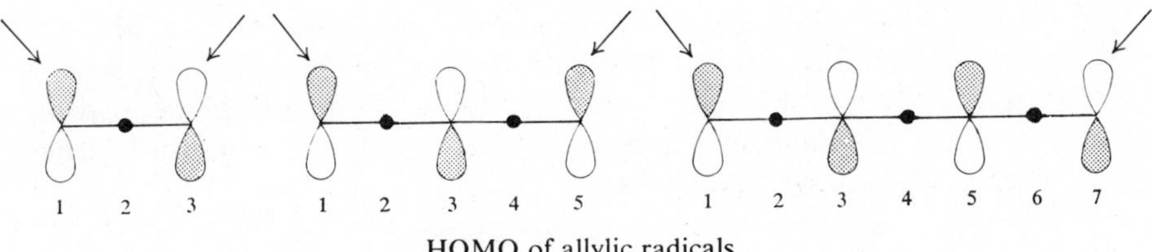

HOMO of allylic radicals

the symmetry at these end carbons alternates regularly as we pass from C–3 to C–5 to C–7, and so on. The HOMO of the migrating group depends, as we shall see, on the nature of the group.

Let us consider first the simplest case: **migration of hydrogen.** Stereochemically, this shift can be suprafacial or antarafacial:

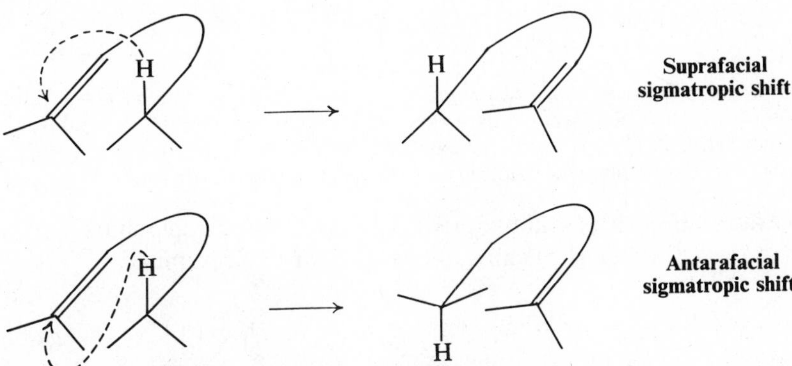

In the transition state, a *three-center bond* is required, and this must involve overlap between the *s* orbital of the hydrogen and lobes of *p* orbitals of the two terminal carbons. Whether a suprafacial or antarafacial shift is allowed depends upon the symmetry of these terminal orbitals:

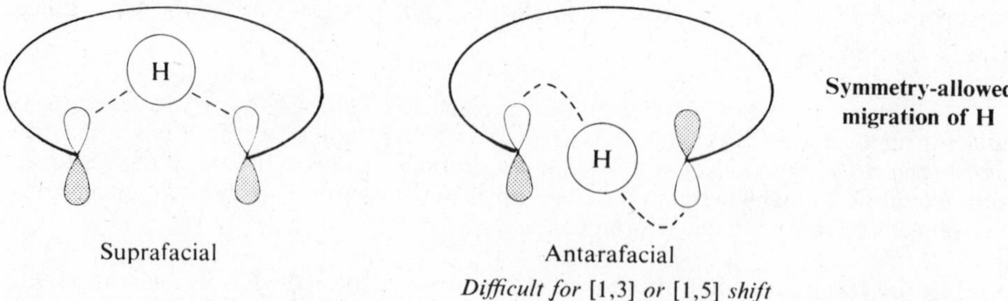

Suprafacial

Antarafacial
Difficult for [1,3] *or* [1,5] *shift*

Symmetry-allowed migration of H

Whether a sigmatropic rearrangement actually takes place, though, depends not only on the symmetry requirements but also on the *geometry* of the system. In particular, [1,3] and [1,5] *antara* shifts should be extremely difficult, since they would require the π framework to be twisted far from the planarity that it requires for delocalization of electrons.

Practically, then, [1,3] and [1,5] sigmatropic reactions seem to be limited to *supra* shifts. A [1,3] *supra* shift of hydrogen is symmetry-forbidden; since the *s* orbital of hydrogen would have to overlap *p* lobes of opposite phase, hydrogen cannot be bonded simultaneously to both carbons. A [1,5] *supra* shift of hydrogen, on the other hand, is symmetry-allowed.

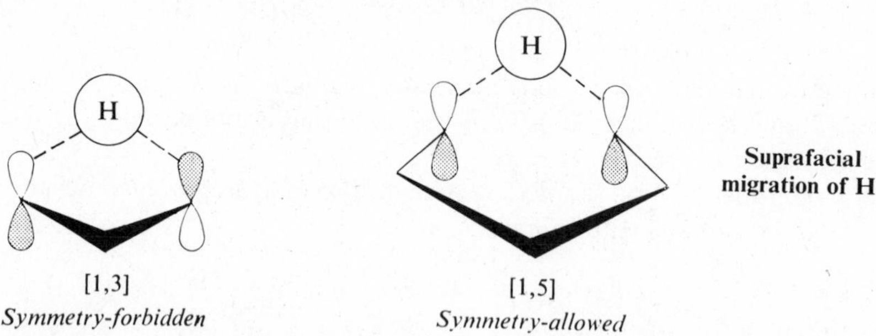

[1,3]
Symmetry-forbidden

[1,5]
Symmetry-allowed

Suprafacial migration of H

For larger π frameworks, both *supra* and *antara* shifts should be possible on geometric grounds, and here we would expect the stereochemistry to depend simply on orbital symmetry. A [1,7]-H shift, for example, should be *antara*, a [1,9]-H shift, *supra*, and so on. For photochemical reactions, as before, predictions are exactly reversed.

The facts agree with the above predictions: [1,3] sigmatropic shifts of hydrogen are not known, whereas [1,5] shifts are well known. For example:

CH_3 … $=CD_2$ $\xrightarrow[\text{heat}]{[1,5]\text{-H}}$ $=CH_2$ … CHD_2

The preference for [1,5]-H shifts over [1,3]-H shifts has been demonstrated many times. For example, the heating of 3-deuterioindene (I) causes scrambling of the label to *all three* non-aromatic positions. Let us examine this reaction.

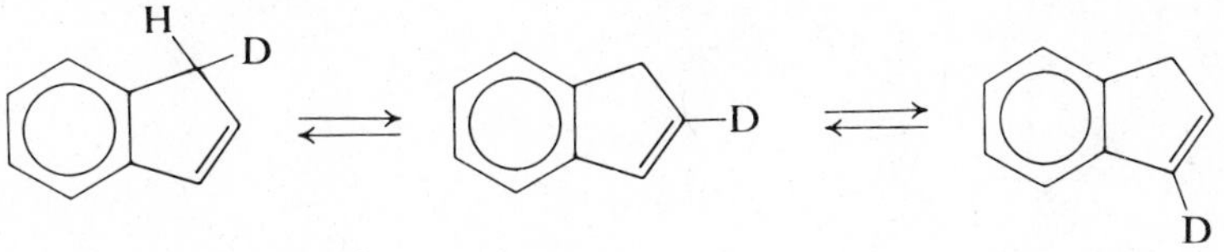

I II III

We cannot account for the formation of II on the basis of [1,3] shifts: migration of D would regenerate I; migration of H would yield only III.

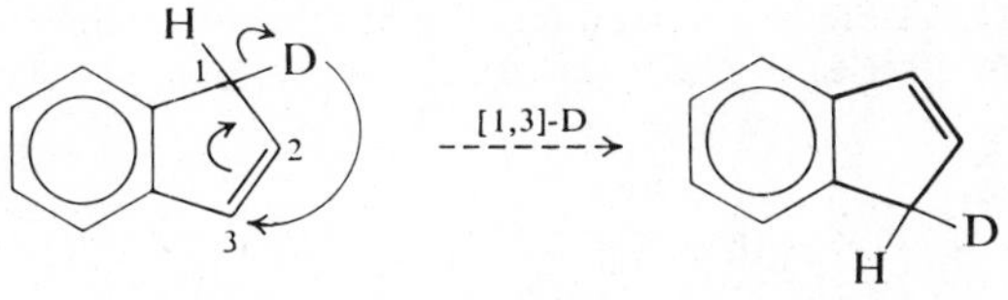

[1,3]-D

I I

[1,3]-H

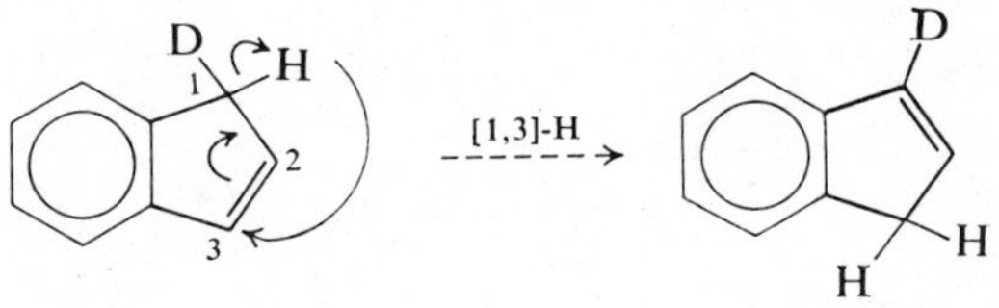

I III

But if we include the *p* orbitals of the benzene ring, and count along the edge of this ring, we see that a [1,5] shift of D would yield the unstable non-aromatic

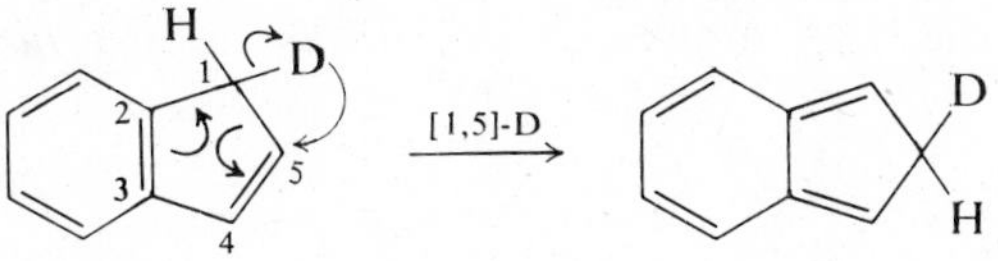

[1,5]-D

I IV*a*

intermediate IV*a*. This, in turn, can transfer H or D by [1,5] shifts to yield all the observed products (see Fig. 29.24).

So far we have discussed only migration of hydrogen, which is necessarily limited to the overlap of an *s* orbital. Now let us turn to **migration of carbon.** Here, we have two possible kinds of bonding to the migrating group. One of these is similar to what we have just described for migration of hydrogen: bonding of both ends of the π framework to the same lobe on carbon. Depending on the symmetry of the π framework, the symmetry-allowed migration may be suprafacial or antarafacial.

I ⇌ IVa ([1,5]-D) ⇌ II ([1,5]-H); I ⇌ IVb ([1,5]-H) ⇌ III

Figure 29.24. Deuterium scrambling in indene via unstable intermediates IV*a* and IV*b*: a series of [1,5] hydrogen shifts.

With carbon, a new aspect appears: the stereochemistry in the migrating group. Bonding through the same lobe on carbon means attachment to the same face of the atom, that is to say, *retention of configuration in the migrating group.*

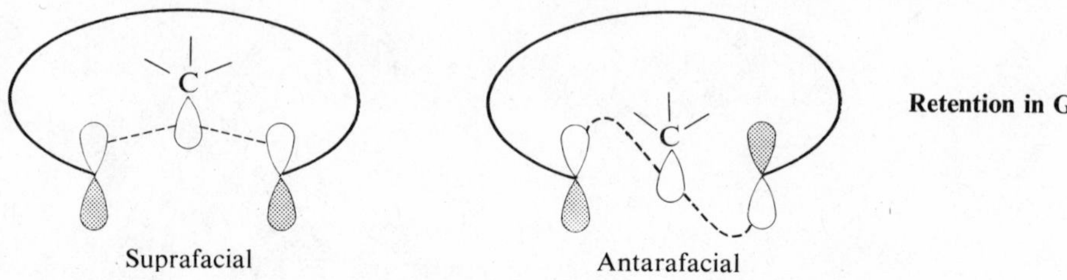

But there is a second possibility for carbon: bonding to the two ends of the π framework through different lobes of a p orbital. These lobes are on opposite faces of carbon—exactly as in an S_N2 reaction—and there is *inversion of configuration in the migrating group.*

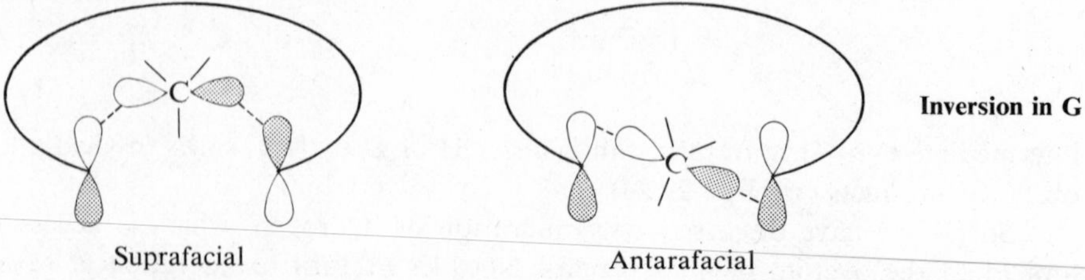

For [1,3] and [1,5] shifts, the geometry pretty effectively prevents antarafacial migration. Limiting ourselves, then, to suprafacial migrations, we make these predictions: [1,3] migration with inversion; [1,5] migration with retention. *These predictions have been borne out by experiment.*

In 1968, Jerome Berson (of the University of Wisconsin) reported that the deuterium-labeled bicyclo[3.2.0]heptene V is converted stereospecifically into the

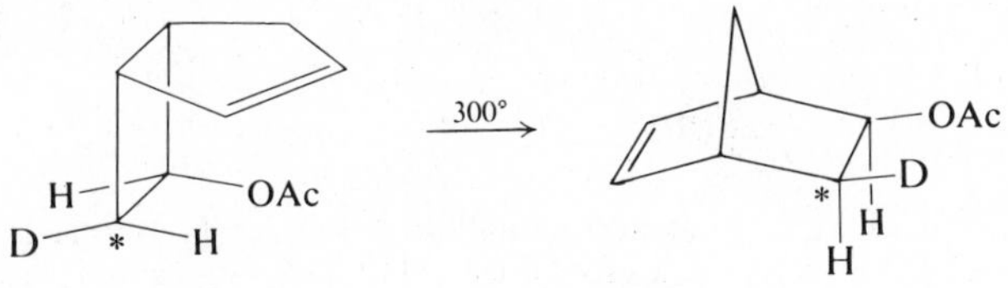

exo-norbornene VI. As Fig. 29.25 shows, this reaction proceeds by a [1,3] migration and with *complete inversion* of configuration in the migrating group.

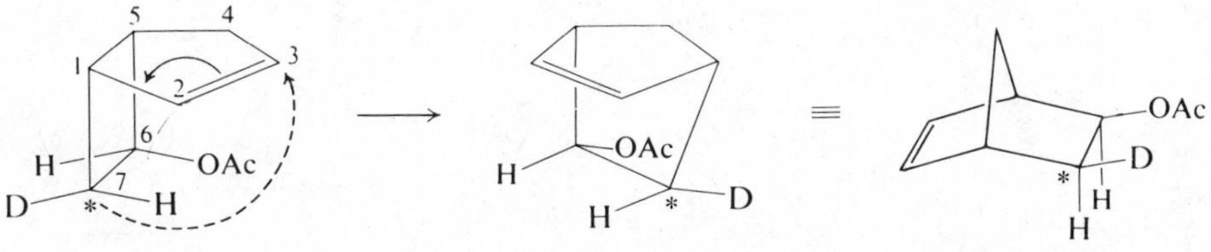

Figure 29.25. The deuterium-labeled bicyclo[3.2.0]heptene V rearranges via a [1,3]-C shift to the norbornene VI. There is *inversion of configuration* at C–7: from R to S. (Or, using C–6 as our standard, we see that H eclipses OAc in V, and D eclipses OAc in VI.)

In 1970, H. Kloosterziel (of the University of Technology, Eindhoven, The Netherlands) reported a study of the rearrangement of the diastereomeric 6,9-dimethylspiro[4.4]nona-1,3-dienes (*cis*-VII and *trans*-VII) to the dimethylbicyclo-[4.3.0]nonadienes VIII, IX, and X. These reactions are completely stereospecific.

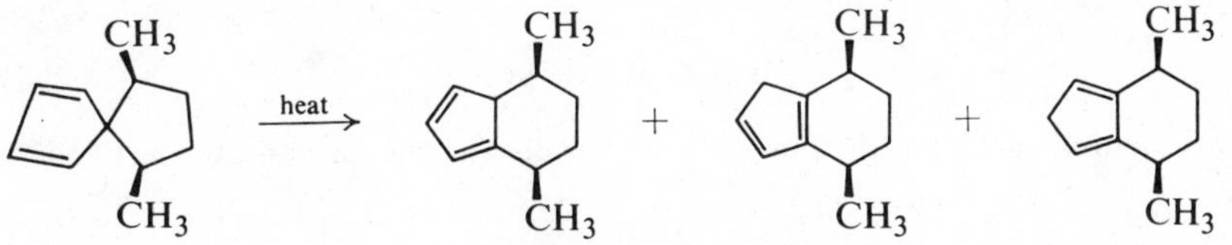

cis-VII → (heat) *cis*-VIII + *cis*-IX (*Chief product*) + *cis*-X

As Fig. 29.26 shows, they proceed by [1,5] migrations and with *complete retention* of configuration in the migrating group.

To predict a different stereochemistry between [1,3] and [1,5] migrations, and in particular to predict *inversion* in the [1,3] shift—certainly not the easier path on geometric grounds—is certainly "risky". The fulfillment of such predictions demonstrates both the validity and the power of the underlying theory.

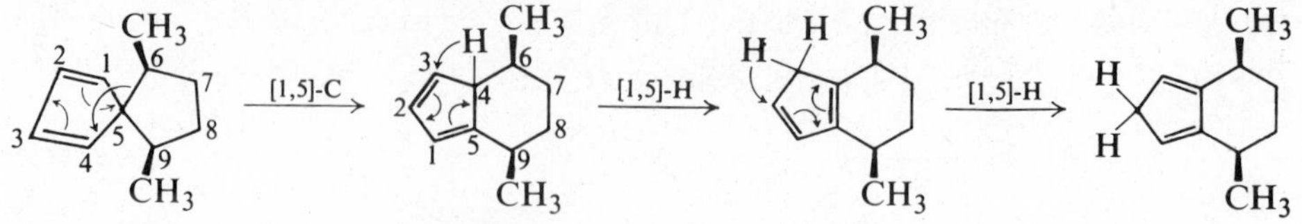

cis-VII → [1,5]-C → cis-VIII → [1,5]-H → cis-IX → [1,5]-H → cis-X

Figure 29.26. Rearrangement of *cis*-6,9-dimethylspiro[4.4]nona-1,3-diene. Migration of C–6 from C–5 to C–4 is a [1,5]-C shift. (We count 5, 1, 2, 3, 4.) Configuration at C–6 is *retained*, as shown by its relationship to configuration at C–9. Successive [1,5]-H shifts then yield the other products.

Problem 29.10 In each of the following, the high stereospecificity or regiospecificity provides confirmation of predictions based on orbital symmetry. Show how this is so. (*Use models.*)

(a) → heat → +

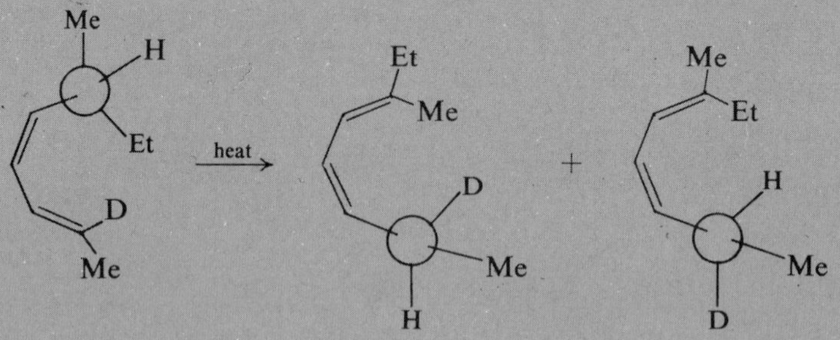

(b) When 1,3,5-cyclooctatriene labeled with deuterium at the 7 and 8 positions was heated, it gave products labeled only at the 3, 4, 7, and 8 positions.

(c) → 120° →

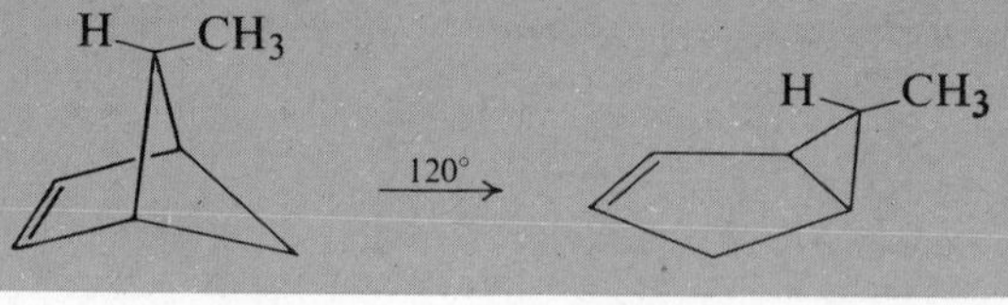

PROBLEMS

1. Tropolone (I, $C_7H_7O_2$) has a flat molecule with all carbon–carbon bonds of the

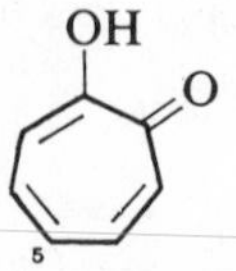

Tropolone

I

same length (1.40 A). The measured heat of combustion is 20 kcal lower than that calculated by the method of Problem 10.2 (p. 323). Its dipole moment is 3.71 D; that of 5-bromotropolone is 2.07 D.

Tropolone undergoes the Reimer-Tiemann reaction, couples with diazonium ions, and is nitrated by dilute nitric acid. It gives a green color with ferric chloride, and does not react with 2,4-dinitrophenylhydrazine. Tropolone is both acidic ($K_a = 10^{-7}$) and weakly basic, forming a hydrochloride in ether.

(a) What class of compounds does tropolone resemble? Is it adequately represented by formula I? (b) Using both valence-bond and orbital structures, account for the properties of tropolone.

(c) In what direction is the dipole moment of tropolone? Is this consistent with the structure you have proposed?

(d) The infrared spectrum of tropolone shows a broad band at about 3150 cm^{-1} that changes only slightly upon dilution. What does this tell you about the structure of tropolone?

2. Each transformation shown below is believed to involve a concerted reaction. In each case show just what is happening.

(a) 110° → 250° →

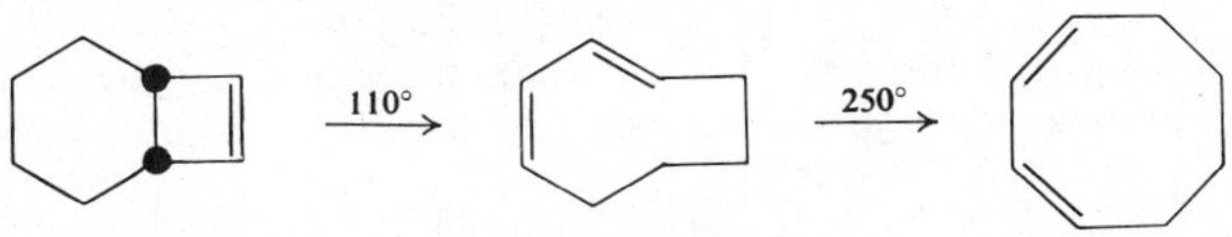

cis-Bicyclo[4.2.0]-octa-7-ene *cis,trans*-Cyclo-octa-1,3-diene *cis,cis*-Cyclo-octa-1,3-diene

(b) heat → →

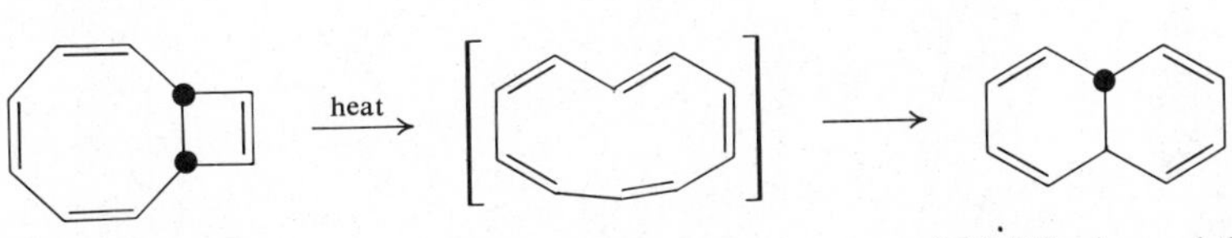

cis-Bicyclo[6.2.0]-deca-2,4,6,9-tetraene *cis,cis,cis,cis,trans*-Deca-1,3,5,7,9-pentaene *trans*-9,10-Dihydronaphthalene

(c) CH_3 CH_3 CH_3 $h\nu$ → CH_3 H CH_3 CH_3 + CH_3 CH_3 CH_3

$h\nu$ (⇅)

CH_3 H CH_3 CH_3

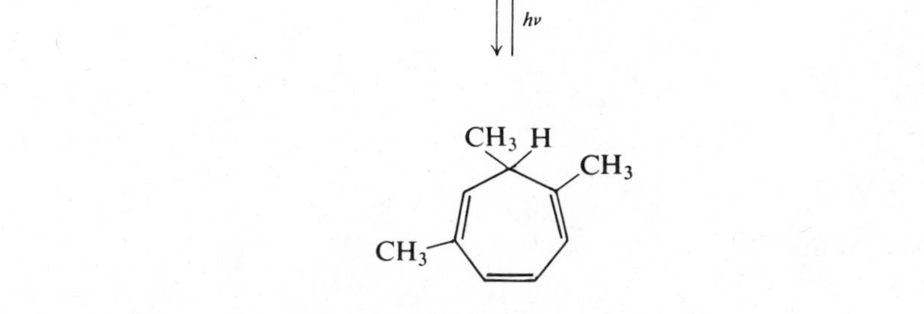

(d) $h\nu$ → →

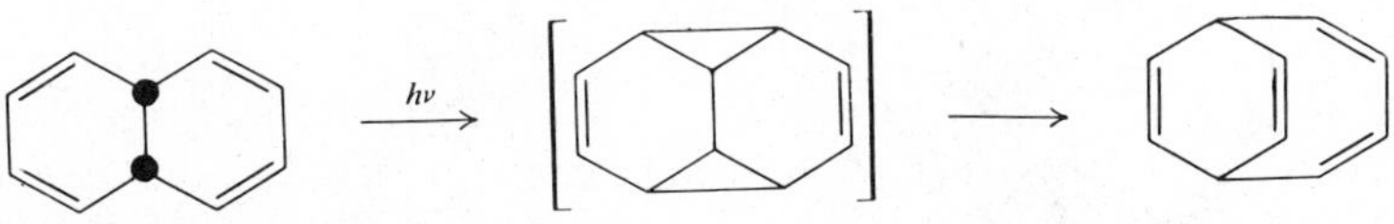

cis-9,10-Dihydronaphthalene

(e) [cyclopentadiene] + $CH_2{=}C(CH_3)CH_2I$ + $AgOOCCCl_3$ ⟶

[bicyclic product]—CH_3 + [bicyclic product]$=CH_2$

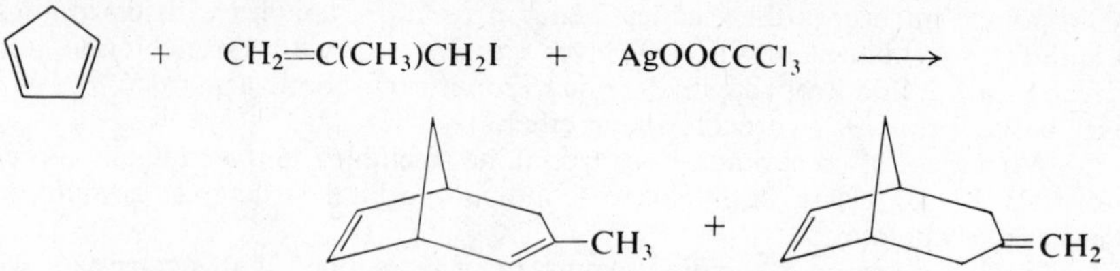

(f) CH_3—[structure with CH_3, CH_3, D, D] $\overset{300°}{\rightleftarrows}$ [structure with CH_3, CH_3, CH_3, D, D] $\overset{300°}{\rightleftarrows}$ D—[structure with CH_3, CH_3, CH_3, D]

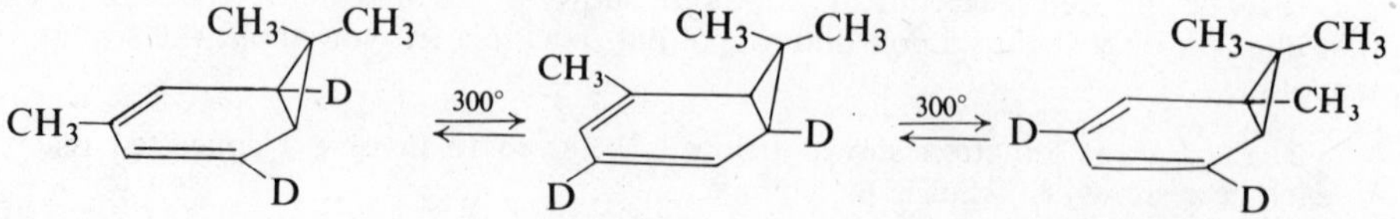

3. Each of the following transformations is believed to proceed by the indicated sequence of concerted reactions. Show just what each step involves, and give structures of compounds A–J.

(a) Electrocyclic closure; electrocyclic closure.

[structure with H, R, R, H] $\xrightarrow{\text{heat}}$ [A] $\xrightarrow{\text{heat}}$ [structure with H, R, H, R]

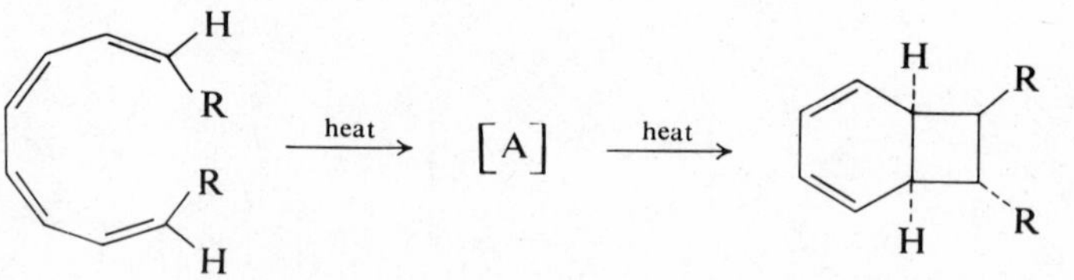

(b) [1,5]-H shift; electrocyclic opening.

[structure] $\xrightarrow{200°}$ B $\xrightarrow{260°}$ C

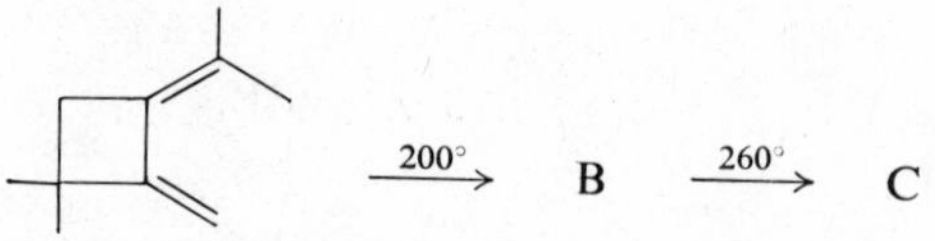

(c) Electrocyclic opening; electrocyclic closure. Final products are not interconvertible at 170°; be sure you account for *both* of them.

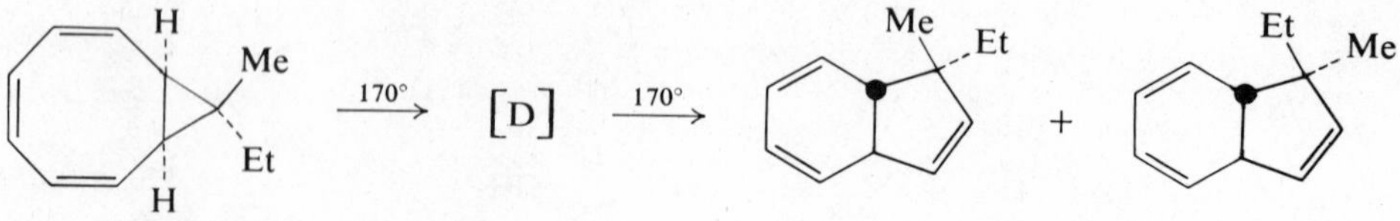

[structure with H, Me, Et, H] $\xrightarrow{170°}$ [D] $\xrightarrow{170°}$ [structure with Me, Et] + [structure with Et, Me]

(d) Three electrocyclic closures.

[structure] $\xrightarrow{h\nu}$ E $\xleftarrow{175°}$ F $\xrightarrow{h\nu}$ G

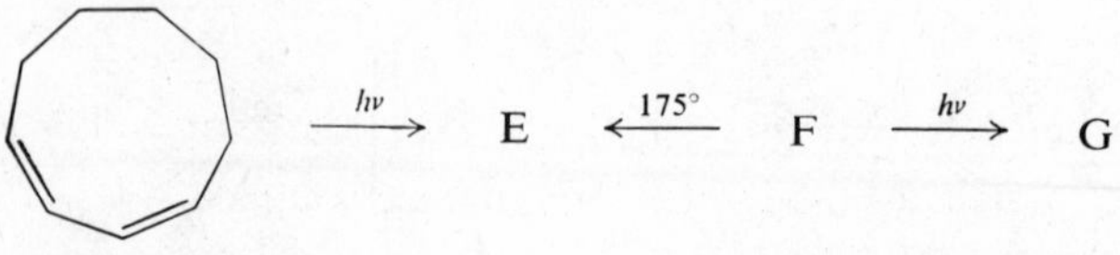

cis,cis-Cyclonona-1,3-diene

(e) A series of *supra* H shifts.

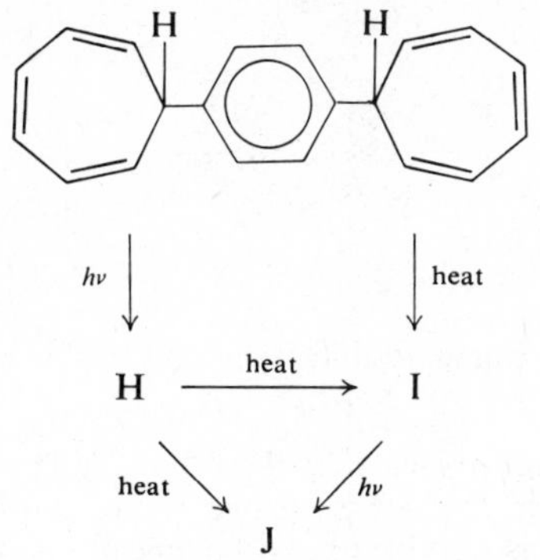

$$\text{H} \xrightarrow{\text{heat}} \text{I}$$

(starting compound: $h\nu$ → H; heat → I; H —heat→ J; I —$h\nu$→ J)

4. Account for the difference in conditions required to bring about the following transformations:

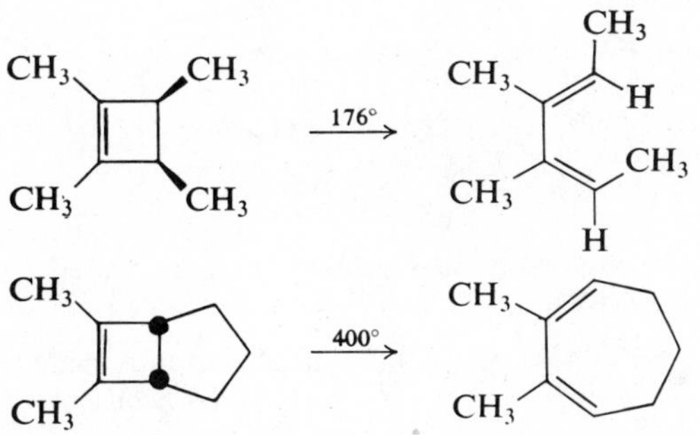

(tetramethylcyclobutene $\xrightarrow{176°}$ dimethylhexadiene; dimethylbicyclo compound $\xrightarrow{400°}$ dimethylcycloheptadiene)

5. Give stereochemical structures of K and L, and tell exactly what process is taking place in each reaction.

cis,cis,cis-Cycloocta-1,3,5-triene $\xrightarrow{80\text{–}100°}$ K (C_8H_{10})

K + $CH_3OOCC{\equiv}CCOOCH_3$ ⟶ L ($C_{14}H_{16}O_4$)

L $\xrightarrow{\text{heat}}$ cyclobutene + dimethyl phthalate

6. (a) The familiar rearrangement of a carbonium ion by a 1,2-alkyl shift is, as we have described it (Sec. 5.22), a concerted reaction. Its ease certainly suggests that it is symmetry-allowed. Discuss the reaction from the standpoint of orbital symmetry. What stereochemistry would you predict in the migrating group?

(b) There is evidence that concerted 1,4-alkyl shifts of the kind

$$\underset{\text{R}}{\text{C}}\text{—}\underbrace{\text{C}{=\!=}\text{C}{=\!=}\text{C}}_{\oplus} \longrightarrow \underbrace{\text{C}{=\!=}\text{C}{=\!=}\text{C}}_{\oplus}\text{—}\underset{\text{R}}{\text{C}}$$

can occur. What stereochemistry would you predict in the migrating group?

7. Discuss the direct, concerted, non-catalytic addition of H_2 to an alkene from the standpoint of orbital symmetry.

8. The deuterium scrambling between II and III has been accounted for on the basis of intramolecular Diels-Alder and *retro*-Diels-Alder reactions. Show how this might oc-

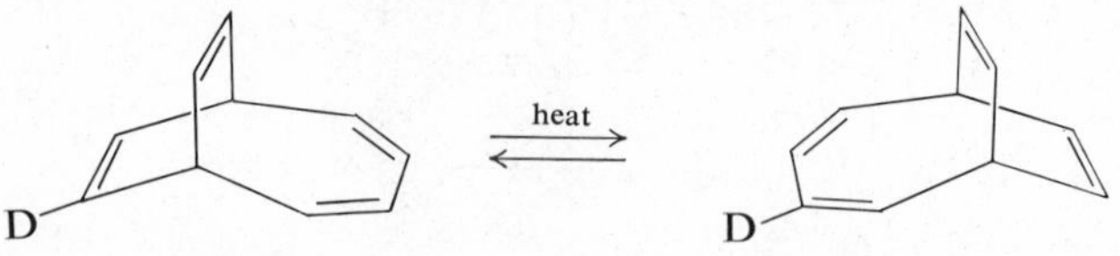

II III

cur. (*Hint:* Look for an intermediate that is symmetrical except for the presence of deuterium.)

9. Suggest an explanation for each of the following facts.

(a) When the diazonium salt IV is treated with *trans,trans*-2,4-hexadiene, N_2 and CO_2 are evolved, and there is obtained stereochemically pure V. (*Hint:* See Problem 18, p. 844.)

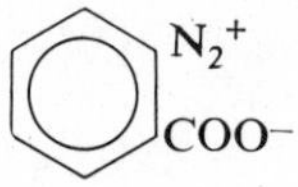

IV

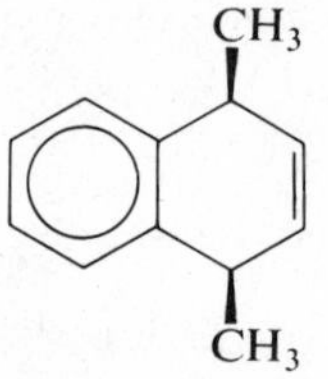

cis-V

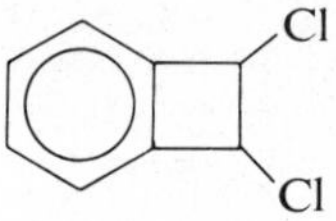

VI

(b) In contrast, decomposition of IV in *either cis*- or *trans*-1,2-dichloroethene yields a mixture of *cis*- and *trans*-VI.

10. For each of the following reactions suggest an intermediate that would account for the formation of the product. *Show exact stereochemistry.* (For a hint, see Fig. 29.24, p. 958.)

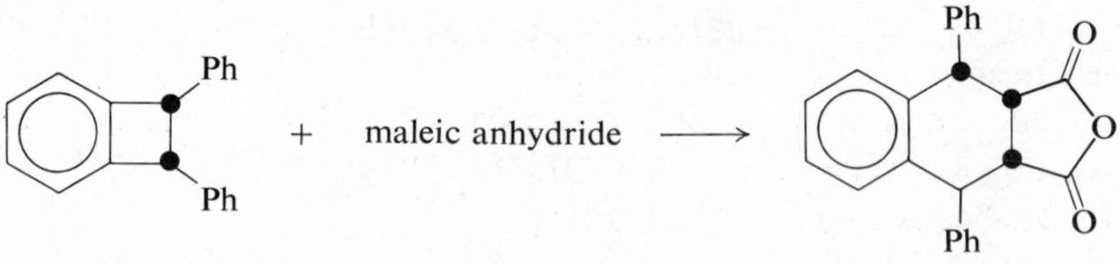

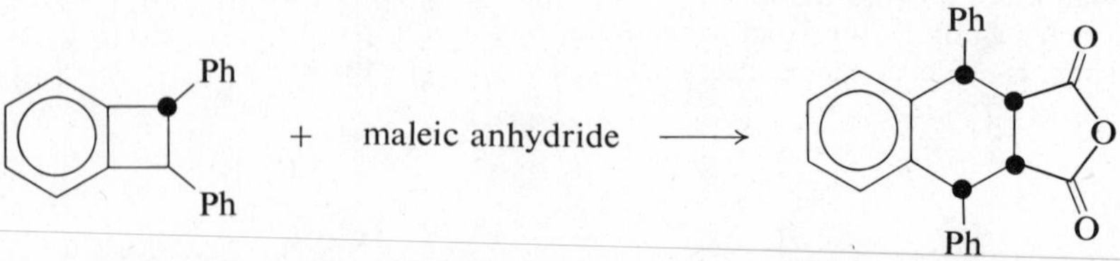

11. (a) The diastereomeric 6,9-dimethylspiro[4.4]nona-1,3-dienes (p. 959) were synthesized by reaction of cyclopentadiene with diastereomeric 2,5-dibromohexanes in the presence of sodium amide. Which 2,5-dibromohexane would you expect to yield each spirane?

(b) The stereochemistry of the spiranes obtained was shown by comparison of their nmr spectra, specifically, of the peaks due to the olefinic hydrogens. Explain.

12. (a) Berson synthesized the stereospecifically labeled compound V (p. 959) by the following sequence. Give structures for compounds M–R.

+ B_2D_6, then H_2O_2 $\longrightarrow$ M + N (both C_7H_9DO)

M + N $\xrightarrow{\text{oxidation}}$ O + P (both C_7H_7DO), *separated*

O + $LiAl(OBu\text{-}t)_3H$, then $(CH_3CO)_2O$ $\longrightarrow$ Q + R (both $C_8H_{11}O_2$), *separated*

Q is compound V on p. 959.

(b) Berson's study of the rearrangement of V to VI (p. 959) was complicated by the tendency of VI, once formed, to decompose into cyclopentadiene and vinyl acetate. What kind of reaction is this decomposition?

13. (a) Woodward and Hoffmann have suggested that the *endo* preference in Diels-Alder reactions is a "secondary" effect of orbital symmetry, and there is experimental evidence to support this suggestion. Using the dimerization of butadiene (Fig. 29.19, p. 949) as example, show how these secondary effects could arise. (*Hint:* Draw the orbitals involved and examine the structures closely.)

(b) In contrast, [6 + 4] cycloaddition was predicted to take place in the *exo* sense. This has been confirmed by experiment. Using the reaction of *cis*-1,3,5-hexatriene with 1,3-butadiene as example, show how this prediction could have been made.

14. (a) The sex attractant of the male boll weevil has been synthesized by the following sequence. Give stereochemical structures for compounds S–Y.

ethylene + 3-methyl-2-cyclohexenone $\xrightarrow{h\nu}$ S ($C_9H_{14}O$)

S $\xrightarrow{\text{bromination}}$ T ($C_9H_{13}OBr$)

T + CO_3^{--} $\longrightarrow$ U ($C_9H_{12}O$)

U + CH_3Li $\longrightarrow$ V ($C_{10}H_{16}O$), *a single stereoisomer* (*Hint:* Examine structure of U.)

V + IO_4^-/OsO_4 $\longrightarrow$ W ($C_9H_{14}O_3$), *a carboxylic acid*

W + excess $Ph_3P{=}CH_2$ $\longrightarrow$ X ($C_{10}H_{16}O_2$) $\xrightarrow{NaAl(OR)_2H_2}$ Y ($C_{10}H_{18}O$), *the sex attractant*

(b) The stereochemistry of the sex attractant was confirmed by the following reaction. Give a stereochemical formula for Z, and show how this confirms the stereochemistry.

Y + $Hg(OAc)_2$, then $NaBH_4$ $\longrightarrow$ Z ($C_{10}H_{18}O$)

15. (a) Although "Dewar benzene," VII, is less stable by 60 kcal than its isomer benzene, its conversion into benzene is surprisingly slow, with an E_{act} of about 37 kcal. It has a half-life at room temperature of two days; at 90° complete conversion into benzene takes one-half hour.

The high E_{act} for conversion of VII into benzene is attributed to the fact that the reaction is symmetry-forbidden. Explain.

VII VIII

(b) In Problem 18, p. 883, we outlined the synthesis of VIII. Although much less stable than its aromatic isomer toluene, this compound is surprisingly long-lived. Here, too, the conversion is considered to be symmetry-forbidden. Explain.

16. (a) In the skin of animals exposed to sunlight, 7-dehydrocholesterol is converted

C_8H_{17}

HO

7-Dehydrocholesterol

$$C_8H_{17} = -\underset{}{\overset{CH_3}{CH}}CH_2CH_2CH_2CH(CH_3)_2$$

into the hormone *cholecalciferol*, the so-called "vitamin" D_3 that plays a vital role in the development of bones. In the laboratory, the following sequence was observed:

7-Dehydrocholesterol $\underset{}{\overset{h\nu}{\rightleftarrows}}$ Pre-cholecalciferol $\underset{}{\overset{\text{warm}}{\rightleftarrows}}$ Cholecalciferol

What processes are actually taking place in these two reactions? Show details.

(b) An exactly analogous reaction sequence is used to convert the plant steroid ergosterol (p. 515) into *ergocalciferol*, the "vitamin" D_2 that is added to milk:

$$\text{ergosterol} \xrightarrow{h\nu} \text{pre-ergocalciferol} \xrightarrow{\text{warm}} \text{ergocalciferol}$$

What is the structure of pre-ergocalciferol? Of ergocalciferol?

(c) On heating at 190°, pre-ergocalciferol is converted into IX and X, stereoisomers of ergosterol. What reaction is taking place, and what are the structures of IX and X?

(d) Still another stereoisomer of ergosterol, XI, can be converted by ultraviolet light into pre-ergocalciferol. What must XI be?

17. On photolysis at room temperature, *trans*-XII was converted into *cis*-XII. When *trans*-XII was photolyzed at −190°, however, no *cis*-XII could be detected in the

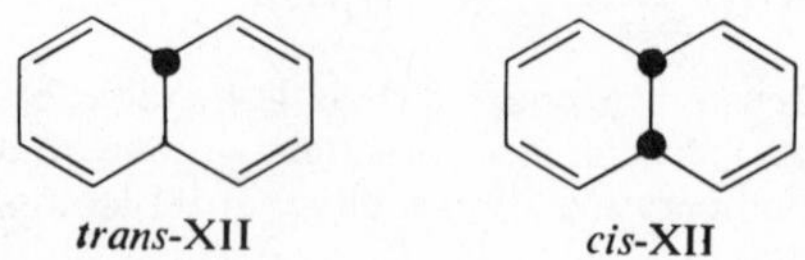

trans-XII *cis*-XII

reaction mixture. When *trans*-XII was photolyzed at −190°, allowed to warm to room temperature, and then cooled again to −190°, *cis*-XII was obtained. If, instead, the low-temperature photolysis mixture was reduced at −190°, cyclodecane was formed; reduction of the room-temperature photolysis mixture gave only a trace of cyclodecane.

On the basis of these and other facts, E. E. van Tamelen (of Stanford University) proposed a two-step mechanism, consistent with orbital symmetry theory, for the conversion of *trans*-XII into *cis*-XII.

(a) Suggest a mechanism for the transformation. Show how it accounts for the facts.

(b) The intermediate proposed by van Tamelen—never isolated and never before identified—is of considerable theoretical interest. Why? What conclusion do you draw about this compound from the facts?

Chapter 30 Polynuclear Aromatic Compounds

30.1 Fused-ring aromatic compounds

Two aromatic rings that share a pair of carbon atoms are said to be *fused*. In this chapter we shall study the chemistry of the simplest and most important of the fused-ring hydrocarbons, **naphthalene**, $C_{10}H_8$, and look briefly at two others of formula $C_{14}H_{10}$, **anthracene** and **phenanthrene**.

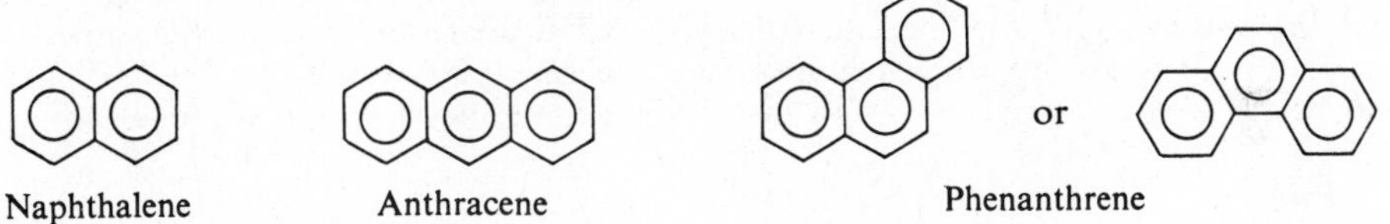

Naphthalene Anthracene Phenanthrene

Table 30.1 POLYNUCLEAR AROMATIC COMPOUNDS

Name	M.p., °C	B.p., °C	Name	M.p., °C	B.p., °C
Naphthalene	80	218	1-Naphthalenesulfonic acid	90	
1,4-Dihydronaphthalene	25	212	2-Naphthalenesulfonic acid	91	
Tetralin	− 30	208	1-Naphthol	96	280
cis-Decalin	− 43	194	2-Naphthol	122	286
trans-Decalin	− 31	185	1,4-Naphthoquinone	125	
1-Methylnaphthalene	− 22	241	Anthracene	217	354
2-Methylnaphthalene	38	240	9,10-Anthraquinone	286	380
1-Bromonaphthalene	6	281	Phenanthrene	101	340
2-Bromonaphthalene	59	281	9,10-Phenanthrenequinone	207	
1-Chloronaphthalene		263	Chrysene	255	
2-Chloronaphthalene	46	265	Pyrene	150	
1-Nitronaphthalene	62	304	1,2-Benzanthracene	160	
2-Nitronaphthalene	79		1,2,5,6-Dibenzanthracene	262	
1-Naphthylamine	50	301	Methylcholanthrene	180	
2-Naphthylamine	113	294			

All three of these hydrocarbons are obtained from coal tar, naphthalene being the most abundant (5%) of all constituents of coal tar.

If diamond (p. 285) is the ultimate polycyclic aliphatic system, then the other allotropic form of elemental carbon, *graphite*, might be considered the ultimate in fused-ring aromatic systems. X-ray analysis shows that the carbon atoms are arranged in layers. Each layer is a continuous network of planar, hexagonal rings; the carbon atoms within a layer are held together by strong, covalent bonds 1.42 A long (only slightly longer than those in benzene, 1.39 A). The different layers, 3.4 A apart, are held to each other by comparatively weak forces. The lubricating properties of graphite (its "greasy" feel) may be due to slipping of layers (with adsorbed gas molecules between) over one another.

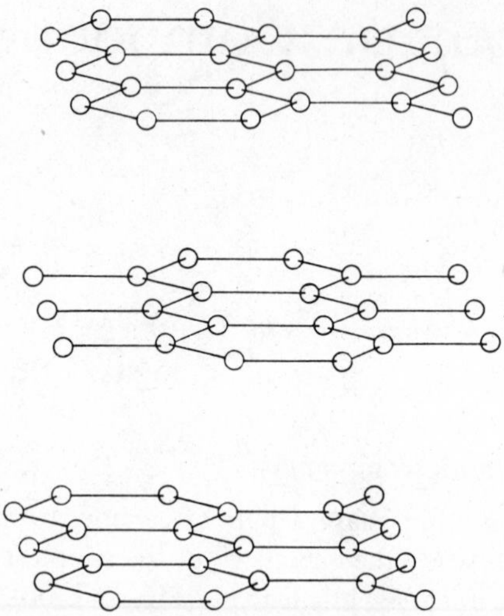

NAPHTHALENE

30.2 Nomenclature of naphthalene derivatives

Positions in the naphthalene ring system are designated as in I. Two isomeric

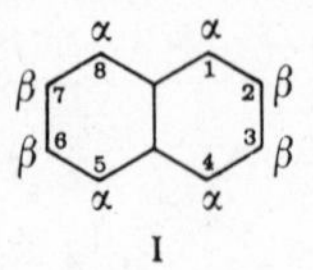

I

monosubstituted naphthalenes are differentiated by the prefixes 1- and 2-, or α- and β-. The arrangement of groups in more highly substituted naphthalenes is indicated by numbers. For example:

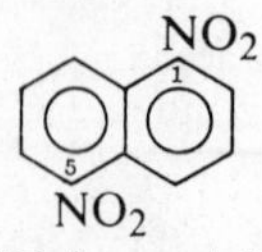

1,5-Dinitronaphthalene

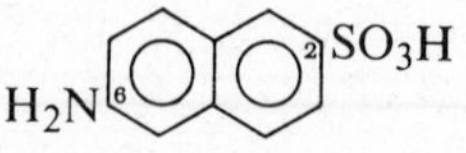

6-Amino-2-naphthalenesulfonic acid

2-Naphthol
β-Naphthol

2,4-Dinitro-1-naphthylamine

Problem 30.1 How many different mononitronaphthalenes are possible? Dinitronaphthalenes? Nitronaphthylamines?

30.3 Structure of naphthalene

Naphthalene is classified as aromatic because its properties resemble those of benzene (see Sec. 10.10). Its molecular formula, $C_{10}H_8$, might lead one to expect a high degree of unsaturation; yet naphthalene is resistant (although less so than benzene) to the addition reactions characteristic of unsaturated compounds. Instead, the typical reactions of naphthalene are electrophilic substitution reactions, in which hydrogen is displaced as hydrogen ion and the naphthalene ring system is preserved. Like benzene, naphthalene is unusually stable: its heat of combustion is 61 kcal lower than that calculated on the assumption that it is aliphatic (see Problem 10.2, p. 323).

From the experimental standpoint, then, naphthalene is classified as aromatic on the basis of its properties. From a theoretical standpoint, naphthalene has the structure required of an aromatic compound: it contains flat six-membered rings, and consideration of atomic orbitals shows that the structure can provide π clouds containing six electrons—the *aromatic sextet* (Fig. 30.1). Ten carbons lie at the

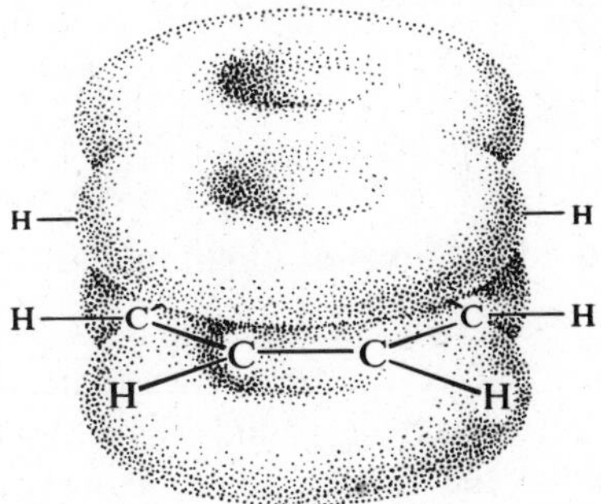

Figure 30.1. Naphthalene molecule. π clouds above and below plane of rings.

corners of two fused hexagons. Each carbon is attached to three other atoms by σ bonds; since these σ bonds result from the overlap of trigonal sp^2 orbitals, all carbon and hydrogen atoms lie in a single plane. Above and below this plane there is a cloud of π electrons formed by the overlap of p orbitals and shaped like a figure 8. We can consider this cloud as two partially overlapping sextets that have a pair of π electrons in common.

In terms of valence bonds, naphthalene is considered to be a resonance hybrid of the three structures I, II, and III. Its resonance energy, as shown by the heat of combustion, is 61 kcal/mole.

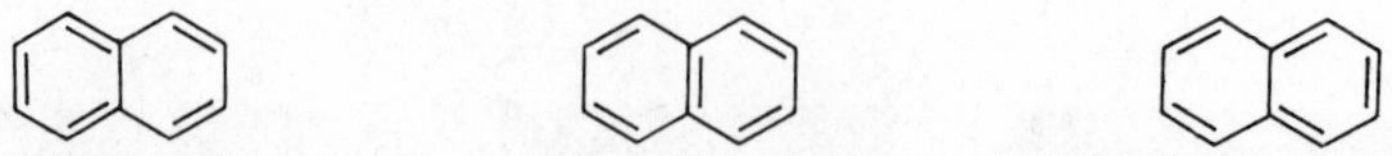

I II III

X-ray analysis shows that, in contrast to benzene, all carbon–carbon bonds in naphthalene are not the same; in particular, the C_1—C_2 bond is considerably shorter (1.365 A) than the C_2—C_3 bond (1.404 A). Examination of structures I, II, and III shows us that this difference in bond lengths is to be expected. The C_1—C_2 bond is double in two structures and single in only one; the C_2—C_3 bond is single in two structures and double in only one. We would therefore expect the C_1—C_2 bond to have more double-bond character than single, and the C_2—C_3 bond to have more single-bond character than double.

For convenience, we shall represent naphthalene as the single structure IV,

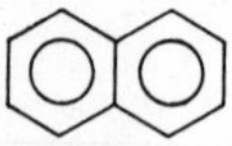

IV

in which the circles stand for partially overlapping aromatic sextets.

Although representation IV suggests a greater symmetry for naphthalene than exists, it has the advantage of emphasizing the aromatic nature of the system.

30.4 Reactions of naphthalene

Like benzene, naphthalene typically undergoes electrophilic substitution; this is one of the properties that entitle it to the designation of "aromatic." An electrophilic reagent finds the π cloud a source of available electrons, and attaches itself to the ring to form an intermediate carbonium ion; to restore the stable aromatic system, the carbonium ion then gives up a proton.

Naphthalene undergoes oxidation or reduction more readily than benzene, but only to the stage where a substituted benzene is formed; further oxidation or reduction requires more vigorous conditions. Naphthalene is stabilized by resonance to the extent of 61 kcal/mole; benzene is stabilized to the extent of 36 kcal/mole. When the aromatic character of one ring of naphthalene is destroyed, only 25 kcal of resonance energy is sacrificed; in the next stage, 36 kcal has to be sacrificed.

REACTIONS OF NAPHTHALENE

1. Oxidation. Discussed in Sec. 30.5.

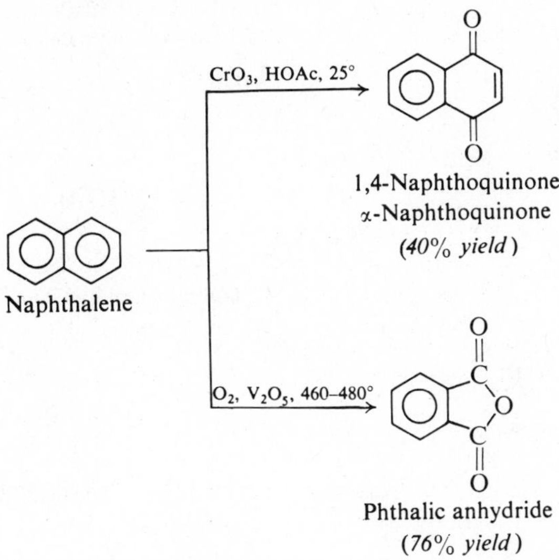

2. Reduction. Discussed in Sec. 30.6.

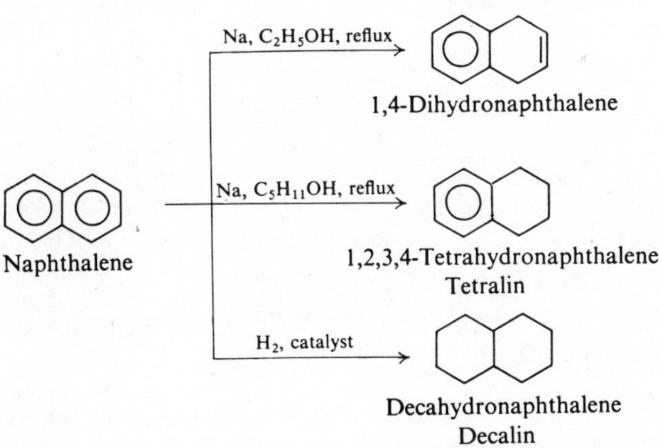

3. Electrophilic substitution. Discussed in Secs. 30.8–30.13.

(a) Nitration. Discussed in Sec. 30.8.

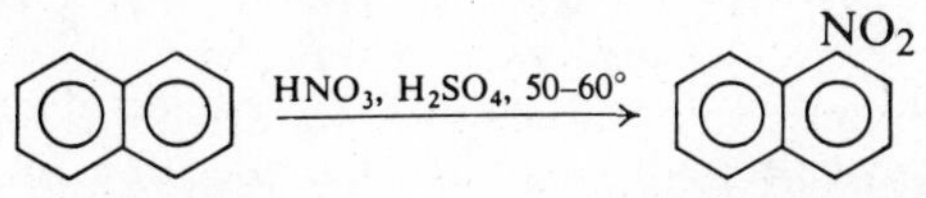

Naphthalene $\xrightarrow{HNO_3,\ H_2SO_4,\ 50\text{–}60^\circ}$ 1-Nitronaphthalene (NO_2)

1-Nitronaphthalene
α-Nitronaphthalene
(*90–95% yield*)

(b) Halogenation. Discussed in Sec. 30.8.

Naphthalene $\xrightarrow[\text{No Lewis acid needed}]{Br_2,\ CCl_4,\ \text{reflux}}$ 1-Bromonaphthalene
α-Bromonaphthalene
(75% yield)

(c) Sulfonation. Discussed in Sec. 30.11.

Naphthalene $\xrightarrow{\text{conc. } H_2SO_4,\ 80°}$ 1-Naphthalenesulfonic acid
α-Naphthalenesulfonic acid

1-Naphthalenesulfonic acid $\xrightarrow{\text{conc. } H_2SO_4,\ 160°}$ 2-Naphthalenesulfonic acid

Naphthalene $\xrightarrow{\text{conc. } H_2SO_4,\ 160°}$ 2-Naphthalenesulfonic acid
β-Naphthalenesulfonic acid

(d) Friedel-Crafts acylation. Discussed in Sec. 30.10.

Naphthalene $\xrightarrow{CH_3COCl,\ AlCl_3}$

$\xrightarrow{\text{solvent: } C_2H_2Cl_4}$ 1-Acetonaphthalene
1-Naphthyl methyl ketone
(93% yield)

$\xrightarrow{\text{solvent: } C_6H_5NO_2}$ 2-Acetonaphthalene
2-Naphthyl methyl ketone
(90% yield)

30.5 Oxidation of naphthalene

Oxidation of naphthalene by oxygen in the presence of vanadium pentoxide destroys one ring and yields phthalic anhydride. Because of the availability of naphthalene from coal tar, and the large demand for phthalic anhydride (for example, see Secs. 30.18 and 32.7), this is an important industrial process.

Oxidation of certain naphthalene derivatives destroys the aromatic character of one ring in a somewhat different way, and yields diketo compounds known as *quinones* (Sec. 27.9). For example:

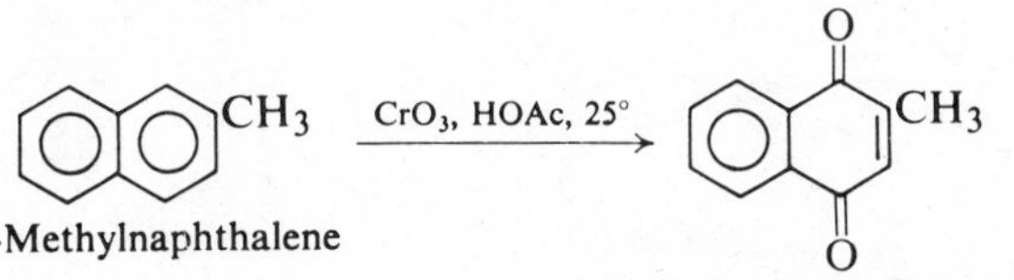

CH3 $\xrightarrow{CrO_3,\ HOAc,\ 25°}$ O CH3 O

2-Methylnaphthalene

2-Methyl-1,4-naphthoquinone
(*70% yield*)

Because of this tendency to form quinones, it is not always feasible to prepare naphthalenecarboxylic acids as we do benzoic acids, by oxidation of methyl side chains.

Problem 30.2 Show how 1- and 2-naphthalenecarboxylic acids (α- and β-*naphthoic acids*) can be obtained from naphthalene by way of the corresponding acetonaphthalenes.

30.6 Reduction of naphthalene

In contrast to benzene, naphthalene can be reduced by chemical reducing agents. It is converted by sodium and ethanol into 1,4-dihydronaphthalene, and by sodium and isopentyl alcohol into 1,2,3,4-tetrahydronaphthalene (*tetralin*). The temperature at which each of these sodium reductions is carried out is the boiling point of the alcohol used; at the higher temperature permitted by isopentyl alcohol (b.p. 132°), reduction proceeds further than with the lower boiling ethyl alcohol (b.p. 78°).

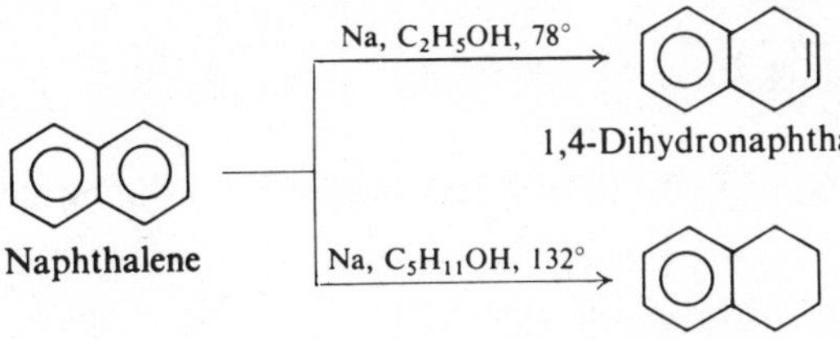

$\xrightarrow{Na,\ C_2H_5OH,\ 78°}$

1,4-Dihydronaphthalene

Naphthalene

$\xrightarrow{Na,\ C_5H_{11}OH,\ 132°}$

1,2,3,4-Tetrahydronaphthalene
Tetralin

The tetrahydronaphthalene is simply a dialkyl derivative of benzene. As with other benzene derivatives, the aromatic ring that remains is reduced only by vigorous catalytic hydrogenation.

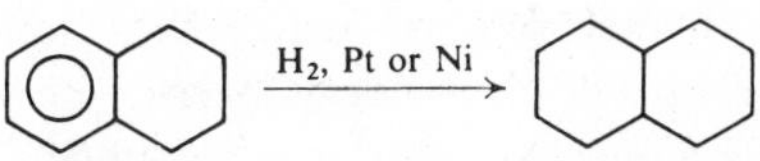

$\xrightarrow{H_2,\ Pt\ or\ Ni}$

Tetralin Decalin

Problem 30.3 *Decalin* exists in two stereoisomeric forms, *cis*-decalin (b.p. 194°) and *trans*-decalin (b.p. 185°).

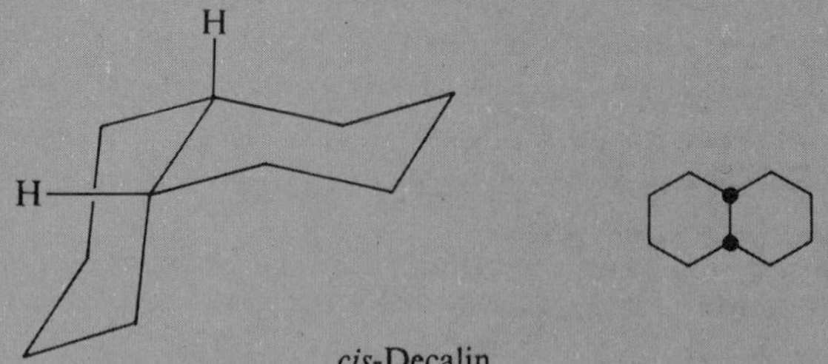

cis-Decalin

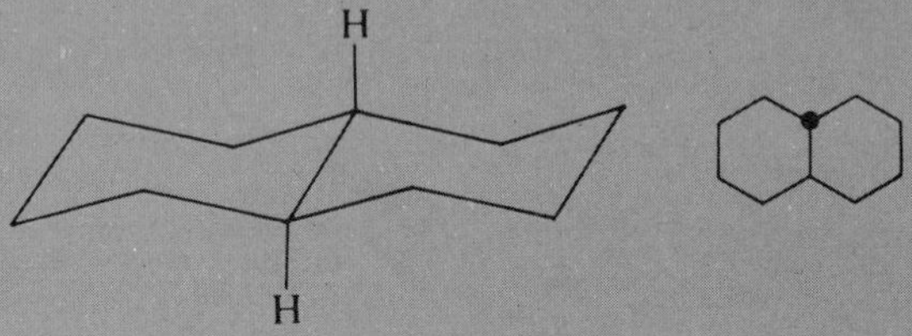

trans-Decalin

(a) Build models of these compounds and see that they differ from one another. Locate in the models the pair of hydrogen atoms, attached to the fused carbons, that are *cis* or *trans* to each other.

(b) In *trans*-decalin is one ring attached to the other by two equatorial bonds, by two axial bonds, or by one axial bond and one equatorial bond? In *cis*-decalin? Remembering (Sec. 9.12) that an equatorial position gives more room than an axial position for a bulky group, predict which should be the more stable isomer, *cis*- or *trans*-decalin.

(c) Account for the following facts: rapid hydrogenation of tetralin over a platinum black catalyst at low temperatures yields *cis*-decalin, while slow hydrogenation of tetralin over nickel at high temperatures yields *trans*-decalin. Compare this with 1,2- and 1,4-addition to conjugated dienes (Sec. 8.22), Friedel-Crafts alkylation of toluene (Sec. 12.11), sulfonation of phenol (Problem 24.13, p. 803), and sulfonation of naphthalene (Sec. 30.11).

30.7 Dehydrogenation of hydroaromatic compounds. Aromatization

Compounds like 1,4-dihydronaphthalene, tetralin, and decalin, which contain the carbon skeleton of an aromatic system but too many hydrogen atoms for aromaticity, are called *hydroaromatic compounds*. They are sometimes prepared, as we have seen, by partial or complete hydrogenation of an aromatic system.

More commonly, however, the process is reversed, and hydroaromatic compounds are converted into aromatic compounds. Such a process is called **aromatization**.

One of the best methods of aromatization is **catalytic dehydrogenation,** accomplished by heating the hydroaromatic compound with a catalyst like platinum, palladium, or nickel. We recognize these as the catalysts used for hydrogenation; since they lower the energy barrier between hydrogenated and dehydrogenated compounds, they speed up reaction in *both* directions (see Sec. 6.3). The position of the equilibrium is determined by other factors: hydrogenation is favored by an excess of hydrogen under pressure; dehydrogenation is favored by sweeping away the hydrogen in a stream of inert gas. For example:

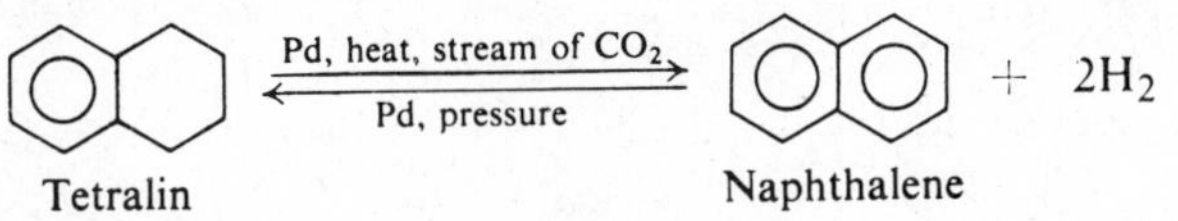

Tetralin $\underset{\text{Pd, pressure}}{\overset{\text{Pd, heat, stream of } CO_2}{\rightleftharpoons}}$ Naphthalene $+ \ 2H_2$

In an elegant modification of dehydrogenation, hydrogen is *transferred* from the hydroaromatic compound to a compound that readily accepts hydrogen. For example:

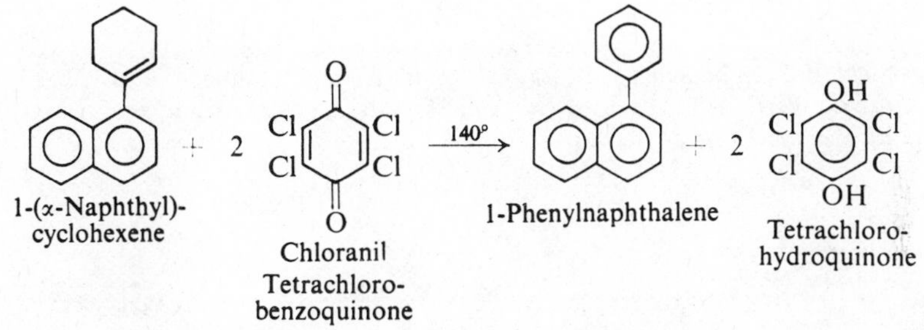

1-(α-Naphthyl)-cyclohexene + 2 Chloranil (Tetrachlorobenzoquinone) $\xrightarrow{140°}$ 1-Phenylnaphthalene + 2 Tetrachlorohydroquinone

The tendency to form the stable aromatic system is so strong that, when necessary, groups can be eliminated: for example, a methyl group located at the point of fusion between two rings, a so-called *angular methyl group* (Sec. 15.16).

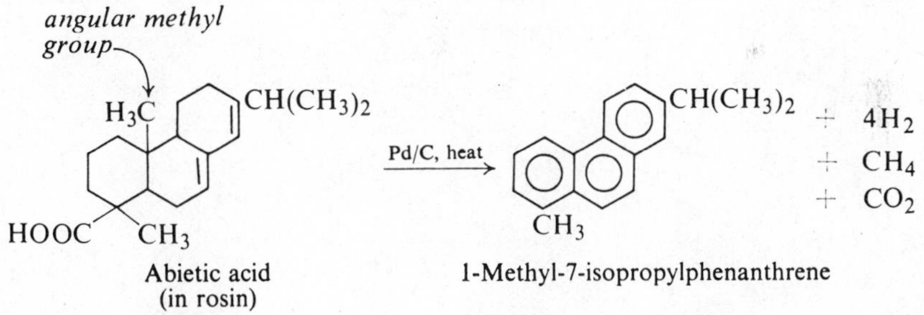

angular methyl group → H_3C

Abietic acid (in rosin) [HOOC, CH_3, $CH(CH_3)_2$] $\xrightarrow{\text{Pd/C, heat}}$ 1-Methyl-7-isopropylphenanthrene [CH_3, $CH(CH_3)_2$] $+ \ 4H_2 + CH_4 + CO_2$

Aromatization has also been accomplished by heating hydroaromatic compounds with selenium, sulfur, or organic disulfides, RSSR. Here hydrogen is eliminated as H_2Se, H_2S, or RSH.

Problem 30.4 In a convenient laboratory preparation of dry hydrogen bromide, Br_2 is dripped into boiling tetralin; the vapors react to form naphthalene and four moles of hydrogen bromide. Account, step by step, for the formation of these products. What familiar reactions are involved in this aromatization?

Aromatization is important in both *synthesis* and *analysis*. Many polynuclear aromatic compounds are made from open-chain compounds by ring closure; the last step in such a synthesis is aromatization (see, for example, Secs. 30.14, 30.19, and 31.13). Many naturally occurring substances are hydroaromatic;

conversion into identifiable aromatic compounds gives important information about their structures. For example:

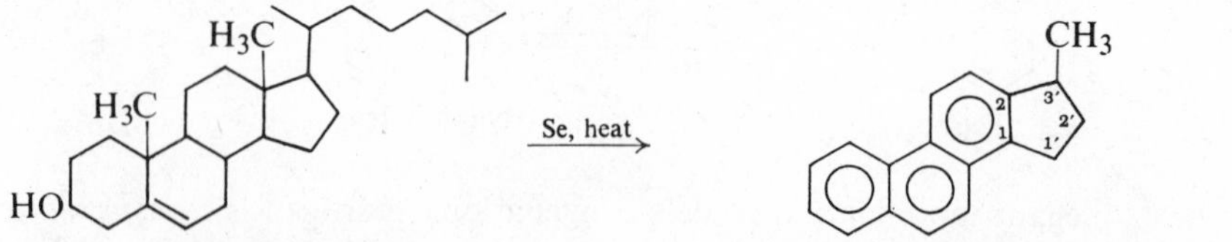

Cholesterol: a steroid
Occurs in all animal tissues

Se, heat →

3′-Methyl-1,2-cyclopentenophenanthrene
(Diels' hydrocarbon)

Problem 30.5 *Cadinene*, $C_{15}H_{24}$, is found in oil of cubebs. Dehydrogenation with sulfur converts cadinene into *cadalene*, $C_{15}H_{18}$, which can be synthesized from *carvone* by the following sequence:

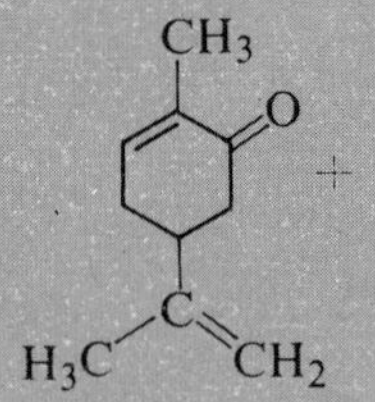

Carvone + $BrCH_2COOC_2H_5$ + Zn ⟶ A ($C_{14}H_{22}O_3$)

A + acid $\xrightarrow{\text{heat}}$ [B] $\xrightarrow{\text{isomerization}}$ C ($C_{12}H_{16}O_2$), *a benzene derivative*

C + C_2H_5OH + H_2SO_4 ⟶ D ($C_{14}H_{20}O_2$)

D + Na + alcohol ⟶ E ($C_{12}H_{18}O$) $\xrightarrow{\text{HBr}}$ F ($C_{12}H_{17}Br$)

F + $CH_3C(COOC_2H_5)_2^-Na^+$ ⟶ G ($C_{20}H_{30}O_4$)

G + H_2SO_4 $\xrightarrow{\text{heat}}$ H ($C_{15}H_{22}O_2$) $\xrightarrow{SOCl_2}$ I ($C_{15}H_{21}OCl$)

I + $AlCl_3$ ⟶ J ($C_{15}H_{20}O$) $\xrightarrow{H_2,\ Ni}$ K ($C_{15}H_{22}O$)

K + sulfur $\xrightarrow{\text{strong heating}}$ cadalene

(a) What is the structure and systematic name of cadalene? (b) What is a likely carbon skeleton for cadinene?

30.8 Nitration and halogenation of naphthalene

Nitration and halogenation of naphthalene occur almost exclusively in the 1-position. Chlorination or bromination takes place so readily that a Lewis acid is not required for catalysis.

As we would expect, introduction of these groups opens the way to the preparation of a series of *alpha*-substituted naphthalenes: from 1-nitronaphthalene via the amine and diazonium salts, and from 1-bromonaphthalene via the Grignard reagent.

Synthesis of α-substituted naphthalenes

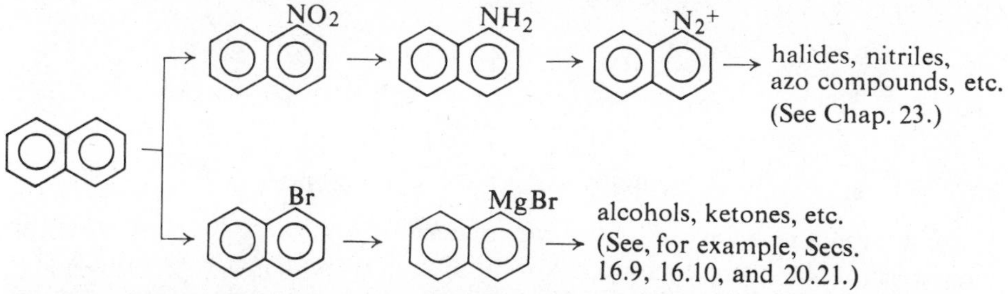

Problem 30.6 Starting with 1-nitronaphthalene, and using any inorganic or aliphatic reagents, prepare:

(a) 1-naphthylamine
(b) α-iodonaphthalene
(c) α-naphthonitrile
(d) α-naphthoic acid (1-naphthalenecarboxylic acid)
(e) α-naphthoyl chloride
(f) 1-naphthyl ethyl ketone
(g) 1-(aminomethyl)naphthalene, $C_{10}H_7CH_2NH_2$
(h) 1-(*n*-propyl)naphthalene
(i) α-naphthaldehyde
(j) (1-naphthyl)methanol
(k) 1-chloromethylnaphthalene
(l) (1-naphthyl)acetic acid
(m) N-(1-naphthyl)acetamide

Problem 30.7 Starting with 1-bromonaphthalene, and using any inorganic or aliphatic reagents, prepare:

(a) 1-naphthylmagnesium bromide
(b) α-naphthoic acid (1-naphthalenecarboxylic acid)
(c) 2-(1-naphthyl)-2-propanol (dimethyl-1-naphthylcarbinol)
(d) 1-isopropylnaphthalene
(e) 1-naphthylcarbinol ($1\text{-}C_{10}H_7CH_2OH$)
(f) methyl-1-naphthylcarbinol (1-(1-naphthyl)ethanol)
(g) 2-(1-naphthyl)ethanol

Problem 30.8 (a) When 1-chloronaphthalene is treated with sodium amide, $Na^+NH_2^-$, in the secondary amine *piperidine* (Sec. 31.12), there is obtained not only I but also II,

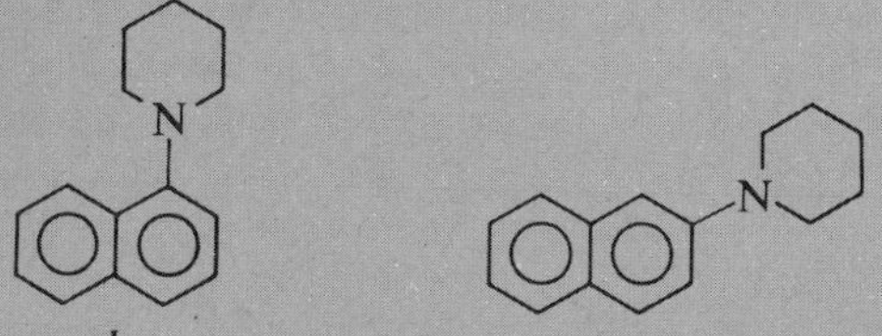

in the ratio of 1:2. Similar treatment of 1-bromo- or 1-iodonaphthalene yields the same products *and in the same* 1:2 *ratio*. Show all steps in a mechanism that accounts for these observations. Can you suggest possible factors that might tend to favor II over I?

(b) Under the conditions of part (a), 1-fluoro-2-methylnaphthalene reacts to

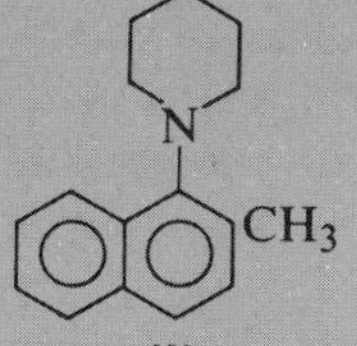

yield III. By what mechanism must this reaction proceed?

(c) Under the conditions of part (a), 1-fluoronaphthalene yields I and II, but in the ratio of 3:2. How do you account for this different ratio of products? What two factors make the fluoronaphthalene behave differently from the other halonaphthalenes?

30.9 Orientation of electrophilic substitution in naphthalene

Nitration and halogenation of naphthalene take place almost exclusively in the α-position. Is this orientation of substitution what we might have expected?

In our study of electrophilic substitution in the benzene ring (Chap. 11), we found that we could account for the observed orientation on the following basis: (a) the controlling step is the attachment of an electrophilic reagent to the aromatic ring to form an intermediate carbonium ion; and (b) this attachment takes place in such a way as to yield the most stable intermediate carbonium ion. Let us see if this approach can be applied to the nitration of naphthalene.

Attack by nitronium ion at the α-position of naphthalene yields an intermediate carbonium ion that is a hybrid of structures I and II in which the positive charge is accommodated by the ring under attack, and several structures like III in which the charge is accommodated by the other ring.

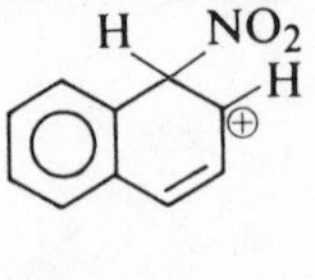

I — *More stable: aromatic sextet preserved*

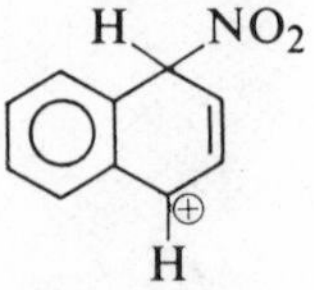

II — *More stable: aromatic sextet preserved*

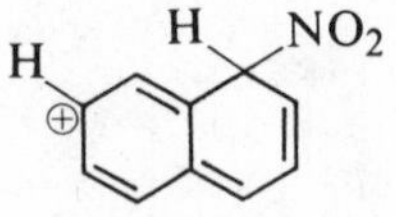

III — *Less stable: aromatic sextet disrupted*

Alpha attack

Attack at the β-position yields an intermediate carbonium ion that is a hybrid of IV and V in which the positive charge is accommodated by the ring under attack, and several structures like VI in which the positive charge is accommodated by the other ring.

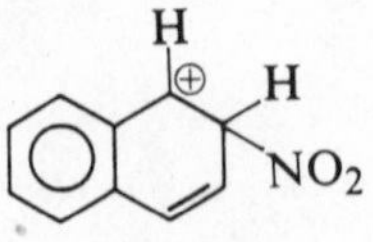

IV — *More stable: aromatic sextet preserved*

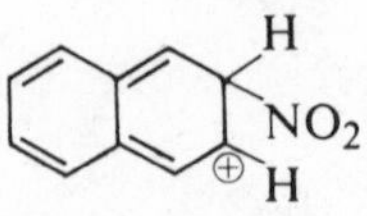

V — *Less stable: aromatic sextet disrupted*

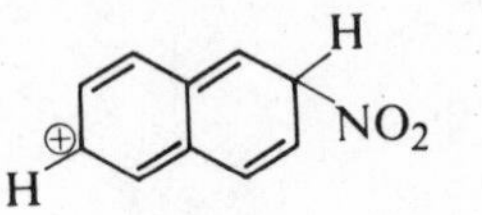

VI — *Less stable: aromatic sextet disrupted*

Beta attack

In structures I, II, and IV, the aromatic sextet is preserved in the ring that is not under attack; these structures thus retain the full resonance stabilization of one benzene ring (36 kcal/mole). In structures like III, V, and VI, on the other hand, the aromatic sextet is disrupted in both rings, with a large sacrifice of resonance stabilization. Clearly, structures like I, II, and IV are much the more stable.

But there are two of these stable contributing structures (I and II) for attack at the α-position and only one (IV) for attack at the β-position. On this basis we would expect the carbonium ion resulting from attack at the α-position (and also the transition state leading to that ion) to be much more stable than the carbonium ion (and the corresponding transition state) resulting from attack at the β-position, and that nitration would therefore occur much more rapidly at the α-position.

Throughout our study of polynuclear hydrocarbons, we shall find that the matter of orientation is generally understandable on the basis of this principle: of the large number of structures contributing to the intermediate carbonium ion, the important ones are those that require the smallest sacrifice of resonance stabilization. Indeed, we shall find that this principle accounts for orientation not only in electrophilic substitution but also in oxidation, reduction, and addition.

30.10 Friedel-Crafts acylation of naphthalene

Naphthalene can be acetylated by acetyl chloride in the presence of aluminum chloride. The orientation of substitution is determined by the particular solvent used: predominantly *alpha* in carbon disulfide or solvents like tetrachloroethane, predominantly *beta* in nitrobenzene. (The effect of nitrobenzene has been attributed to its forming a complex with the acid chloride and aluminum chloride which, because of its bulkiness, attacks the roomier *beta* position.)

Naphthalene $\xrightarrow{CH_3COCl,\ AlCl_3}$

solvent: $C_2H_2Cl_4$ → $COCH_3$ (1-position): 1-Acetonaphthalene, Methyl α-naphthyl ketone

solvent: $C_6H_5NO_2$ → $COCH_3$ (2-position): 2-Acetonaphthalene, Methyl β-naphthyl ketone

Thus acetylation (as well as sulfonation, Sec. 30.11) affords access to the *beta* series of naphthalene derivatives. Treatment of 2-acetonaphthalene with hypohalite, for example, provides the best route to β-naphthoic acid.

2-Acetonaphthalene (Methyl β-naphthyl ketone), $COCH_3$ $\xrightarrow{NaOCl,\ 60–70°}$ β-Naphthoic acid, $COOH$ (*88% yield*) + $CHCl_3$

Acylation of naphthalene by succinic anhydride yields a mixture of *alpha* and *beta* products. These are separable, however, and both are of importance in the synthesis of higher ring systems (see Sec. 30.19).

Naphthalene + Succinic anhydride $\xrightarrow{AlCl_3,\ C_6H_5NO_2}$

4-(1-Naphthyl)-4-oxobutanoic acid
β-(1-Naphthoyl)propionic acid ($COCH_2CH_2COOH$ at the 1-position)

4-(2-Naphthyl)-4-oxobutanoic acid
β-(2-Naphthoyl)propionic acid ($COCH_2CH_2COOH$ at the 2-position)

Friedel-Crafts alkylation of naphthalene is of little use, probably for a combination of reasons: the high reactivity of naphthalene which causes side reactions and polyalkylations, and the availability of alkylnaphthalenes via acylation or ring closure (Sec. 30.14).

Problem 30.9 The position of the —COOH in β-naphthoic acid was shown by vigorous oxidation and identification of the product. What was this product? What product would have been obtained from α-naphthoic acid?

Problem 30.10 Outline the synthesis of the following compounds via an initial acylation:

(a) 2-ethylnaphthalene
(b) methylethyl-2-naphthylcarbinol (2-(2-naphthyl)-2-butanol)
(c) 2-(*sec*-butyl)naphthalene
(d) 1-(2-naphthyl)ethanol
(e) γ-(2-naphthyl)butyric acid
(f) 4-(2-naphthyl)-1-butanol
(g) 5-(2-naphthyl)-2-methyl-2-pentanol
(h) 2-isohexylnaphthalene
(i) 1-amino-1-(2-naphthyl)ethane
(j) β-vinylnaphthalene

30.11 Sulfonation of naphthalene

Sulfonation of naphthalene at 80° yields chiefly 1-naphthalenesulfonic acid; sulfonation at 160° or higher yields chiefly 2-naphthalenesulfonic acid. When 1-naphthalenesulfonic acid is heated in sulfuric acid at 160°, it is largely converted into the 2-isomer. These facts become understandable when we recall that sulfonation is readily reversible (Sec. 11.12).

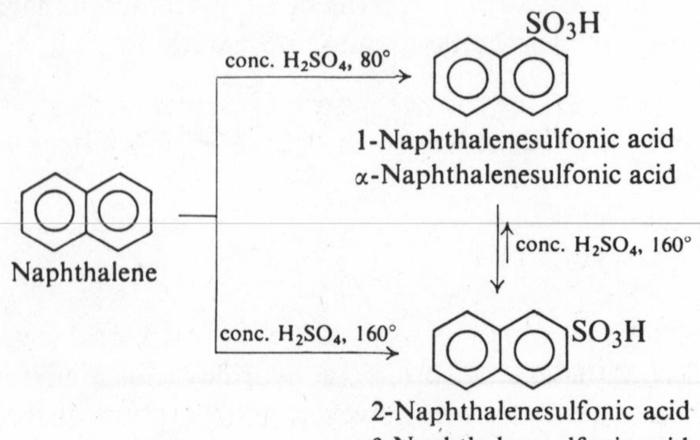

Sulfonation, like nitration and halogenation, occurs more rapidly at the α-position, since this involves the more stable intermediate carbonium ion. But, for the same reason, attack by hydrogen ion, with subsequent desulfonation, also occurs more readily at the α-position. Sulfonation at the β-position occurs more slowly but, once formed, the β-sulfonic acid tends to resist desulfonation. At low temperatures desulfonation is slow and we isolate the product that is formed faster, the *alpha* naphthalenesulfonic acid. At higher temperatures, desulfonation becomes important, equilibrium is more readily established, and we isolate the product that is more stable, the *beta* naphthalenesulfonic acid.

SO_3H | $\xleftarrow[H^+]{SO_3}$ | $\xrightarrow[H^+]{SO_3}$ | SO_3H

α-Isomer
Formed rapidly; desulfonated rapidly

β-Isomer
Formed slowly; desulfonated slowly

We see here a situation exactly analogous to one we have encountered several times before: in 1,2- and 1,4-addition to conjugated dienes (Sec. 8.22), in Friedel-Crafts alkylation of toluene (Sec. 12.11), and in sulfonation of phenols (Problem 24.13, p. 803). At low temperatures the controlling factor is *rate of reaction*, at high temperatures, *position of equilibrium*.

Sulfonation is of special importance in the chemistry of naphthalene because it gives access to the *beta*-substituted naphthalenes, as shown in the next section.

Problem 30.11 (a) Show all steps in the sulfonation and desulfonation of naphthalene. (b) Draw a potential energy curve for the reactions involved. (Compare your answer with Fig. 8.8, p. 272.)

30.12 Naphthols

Like the phenols we have already studied, naphthols can be prepared from the corresponding sulfonic acids by fusion with alkali. Naphthols can also be made

$SO_3^-Na^+$ $\xrightarrow{NaOH,\ H_2O,\ 300°}$ O^-Na^+ $\xrightarrow{dil.\ H_2SO_4}$ OH

Sodium 2-naphthalenesulfonate

Sodium 2-naphthoxide

2-Naphthol
β-Naphthol

from the naphthylamines by direct hydrolysis under acidic conditions. (This reaction, which does not work in the benzene series, is superior to hydrolysis of diazonium salts.)

NH_2 $\xrightarrow{dil.\ H_2SO_4,\ 200°,\ 14\ atm.}$ OH + NH_3

1-Naphthylamine

1-Naphthol
α-Naphthol
(95% yield)

The α-substituted naphthalenes, like substituted benzenes, are most commonly prepared by a sequence of reactions that ultimately goes back to a nitro compound (Sec. 30.8). Preparation of β-substituted naphthalenes, on the other hand, cannot start with the nitro compound, since nitration does not take place in the β-position. The route to β-naphthylamine, and through it to the versatile diazonium salts, lies through β-naphthol. β-Naphthol is made from the β-sulfonic acid; it is converted into β-naphthylamine when heated under pressure with ammonia and ammonium sulfite (the **Bucherer reaction**, not useful in the benzene series except in rare cases).

Synthesis of β-substituted naphthalenes

Naphthalene → 2-Naphthalenesulfonic acid (SO_3H) → 2-Naphthol (OH) —[NH_3, $(NH_4)_2SO_3$, heat, pressure]→ 2-Naphthylamine (NH_2)

2-Naphthylamine → 2-Naphthalenediazonium salt (N_2^+) → Halides, nitriles, azo compounds, etc. (See Chap. 23)

Naphthols undergo the usual reactions of phenols. Coupling with diazonium salts is particularly important in dye manufacture (see Sec. 23.17); the orientation of this substitution is discussed in the following section.

Problem 30.12 Starting from naphthalene, and using any readily available reagents, prepare the following compounds:

(a) 2-bromonaphthalene
(b) 2-fluoronaphthalene
(c) β-naphthonitrile
(d) β-naphthoic acid
(e) β-naphthaldehyde
(f) 3-(2-naphthyl)propenoic acid

Problem 30.13 Diazonium salts can be converted into nitro compounds by treatment with sodium nitrite, usually in the presence of a catalyst. Suggest a method for preparing 2-nitronaphthalene.

30.13 Orientation of electrophilic substitution in naphthalene derivatives

We have seen that naphthalene undergoes nitration and halogenation chiefly at the α-position, and sulfonation and Friedel-Crafts acylation at either the α- or β-position, depending upon conditions. Now, to what position will a *second* substituent attach itself, and how is the orientation influenced by the group already present?

Orientation of substitution in the naphthalene series is more complicated than in the benzene series. An entering group may attach itself either to the ring that already carries the first substituent, or to the other ring; there are seven different positions open to attack, in contrast to only three positions in a monosubstituted benzene.

The major products of further substitution in a monosubstituted naphthalene can usually be predicted by the following rules. As we shall see, these rules are reasonable ones in light of structural theory and our understanding of electrophilic aromatic substitution.

(a) An activating group (electron-releasing group) tends to direct further substitution into the same ring. An activating group in position 1 directs further substitution to position 4 (and, to a lesser extent, to position 2). An activating group in position 2 directs further substitution to position 1.

(b) A deactivating group (electron-withdrawing group) tends to direct further substitution into the other ring: at an α-position in nitration or halogenation, or at an α- or β-position (depending upon temperature) in sulfonation.

For example:

1-Naphthol + $C_6H_5N_2^+Cl^-$ $\xrightarrow{\text{NaOH, 0–10°}}$ 4-Phenylazo-1-naphthol (OH; $N{=}N{-}C_6H_5$)

1-Naphthol + HNO_3 $\xrightarrow{H_2SO_4,\ 20°}$ 2,4-Dinitro-1-naphthol (OH; NO_2; NO_2)

2-Naphthol + $C_6H_5N_2^+Cl^-$ $\xrightarrow{\text{NaOH, 0–5°}}$ 1-Phenylazo-2-naphthol ($N{=}N{-}C_6H_5$; OH)

1-Nitronaphthalene (NO_2) + HNO_3 $\xrightarrow{H_2SO_4,\ 0°}$ 1,5-Dinitronaphthalene (NO_2; NO_2) and 1,8-Dinitronaphthalene (NO_2 NO_2) *Chief product*

2-Methylnaphthalene (CH_3) + Br_2 $\xrightarrow{\text{dark}}$ 1-Bromo-2-methylnaphthalene (Br; CH_3)

These rules do not always hold in sulfonation, because the reaction is reversible and at high temperatures tends to take place at a β-position. However, the observed products can usually be accounted for if this feature of sulfonation is kept in mind.

Problem 30.14 Predict the orientation in each of the following reactions, giving structural formulas and names for the predicted products:

(a) 1-methylnaphthalene + Br_2
(b) 1-methylnaphthalene + HNO_3 + H_2SO_4
(c) 1-methylnaphthalene + CH_3COCl + $AlCl_3$
(d) the same as (a), (b), and (c) for 2-methylnaphthalene
(e) 2-nitronaphthalene + Br_2
(f) 2-methoxynaphthalene + Br_2

Problem 30.15 How do you account for the following observed orientations?

(a) 2-methoxynaphthalene + CH_3COCl + $AlCl_3$ + CS_2 $\longrightarrow$ 1-aceto compound
(b) 2-methoxynaphthalene + CH_3COCl + $AlCl_3$ + $C_6H_5NO_2$ $\longrightarrow$ 6-aceto compound
(c) 2-methylnaphthalene + H_2SO_4 above 100° $\longrightarrow$ 6-sulfonic acid
(d) 2,6-dimethylnaphthalene + H_2SO_4 at 40° $\longrightarrow$ 8-sulfonic acid
(e) 2,6-dimethylnaphthalene + H_2SO_4 + 140° $\longrightarrow$ 3-sulfonic acid
(f) 2-naphthalenesulfonic acid + HNO_3 + H_2SO_4 $\longrightarrow$ 5-nitro and 8-nitro compounds

Problem 30.16 Give the steps for the synthesis of each of the following from naphthalene and any needed reagents:

(a) 4-nitro-1-naphthylamine
(b) 1,4-dinitronaphthalene
(*Hint:* See Problem 30.13, p. 982.)
(c) 2,4-dinitro-1-naphthylamine
(d) 1,3-dinitronaphthalene
(e) 1,2-dinitronaphthalene
(f) 4-amino-1-naphthalenesulfonic acid (*naphthionic acid*)
(g) 8-amino-1-naphthalenesulfonic acid
(h) 5-amino-2-naphthalenesulfonic acid
(i) 8-amino-2-naphthalenesulfonic acid

We have seen (Sec. 30.9) that orientation in naphthalene can be accounted for on the same basis as orientation in substituted benzenes: formation of the more stable intermediate carbonium ion. In judging the relative stabilities of these naphthalene carbonium ions, we have considered that those in which an aromatic sextet is preserved are by far the more stable and hence the more important. Let us see if we can account for orientation in substituted naphthalenes in the same way.

The structures preserving an aromatic sextet are those in which the positive charge is carried by the ring under attack; it is in this ring, therefore, that the charge chiefly develops. Consequently, attack occurs most readily on whichever ring can best accommodate the positive charge: the ring that carries an electron-releasing (activating) group or the ring that does *not* carry an electron-withdrawing (deactivating) group. (We have arrived at the quite reasonable conclusion that a substituent exerts its greatest effect—activating or deactivating—on the ring to which it is attached.)

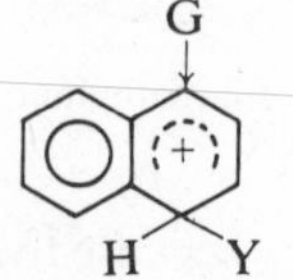

G *is electron-releasing:*
activating,
attack in same ring

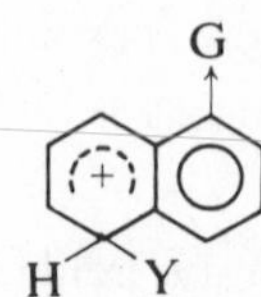

G *is electron-withdrawing:*
deactivating,
attack in other ring

An electron-releasing group located at position 1 can best help accommodate the positive charge if attack occurs at position 4 (or position 2), through the contribution of structures like I and II.

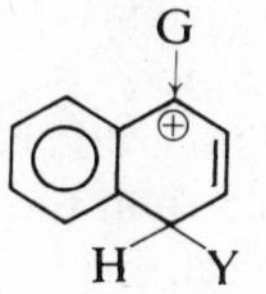

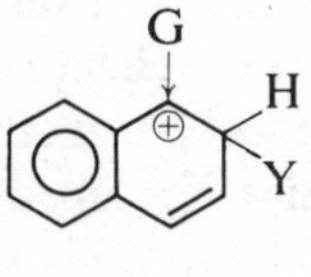

G is electron-releasing: when on position 1, it directs attack to positions 4 or 2

This is true whether the group releases electrons by an inductive effect or by a resonance effect. For example:

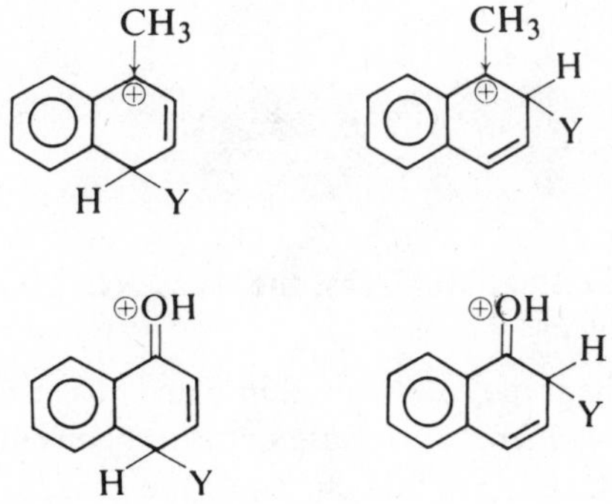

An electron-releasing group located at position 2 could help accommodate the positive charge if attack occurred at position 1 (through structures like III), or if attack occurred at position 3 (through structures like IV).

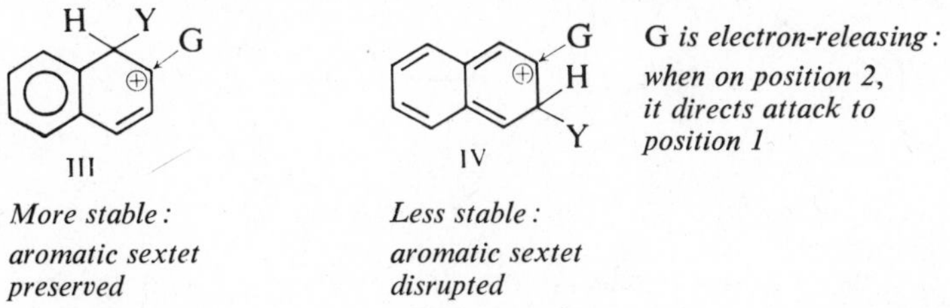
H
Y
G
III
G
H
Y
IV
G is electron-releasing: when on position 2, it directs attack to position 1

More stable: aromatic sextet preserved

Less stable: aromatic sextet disrupted

However, we can see that only the structures like III preserve an aromatic sextet; these are much more stable than the structures like IV, and are the important ones. It is not surprising, therefore, that substitution occurs almost entirely at position 1.

30.14 Synthesis of naphthalene derivatives by ring closure. The Haworth synthesis

Derivatives of benzene, we have seen, are almost always prepared from a compound that already contains the benzene ring: benzene itself or some simple substituted benzene. One seldom generates the benzene ring in the course of a synthesis.

While compounds containing other aromatic ring systems, too, are often prepared from the parent hydrocarbon, there are important exceptions: syntheses in which the ring system, or part of it, is actually generated. Such syntheses usually involve two stages: **ring closure** (or **cyclization**) and **aromatization.**

As an example, let us look at just one method used to make certain naphthalene derivatives: the **Haworth synthesis** (developed by R. D. Haworth at the University of Durham, England). Figure 30.2 (p. 987) shows the basic scheme, which would yield naphthalene itself (not, of course, actually prepared in this way).

All the steps are familiar ones. The reaction in which the second ring is formed is simply Friedel-Crafts acylation that happens to involve two parts of the same molecule. Like many methods of ring closure, this one does not involve a new reaction, but merely an adaptation of an old one.

Problem 30.17 Why is ring closure possible after the first Clemmensen reduction but not before?

To obtain substituted naphthalenes, the basic scheme can be modified in any or all of the following ways:

(a) A substituted benzene can be used in place of benzene and a β-substituted naphthalene obtained. Toluene or anisole or bromobenzene, for example, undergoes the initial Friedel-Crafts reaction chiefly at the *para* position; when the ring is closed, the substituent originally on the benzene ring must occupy a β-position in naphthalene.

G=—R,—X,—OCH_3 → HOC(=O)... → → → → β-Substituted naphthalene

(b) The intermediate cyclic ketone (an α-tetralone) can be treated with a Grignard reagent, and an alkyl (or aryl) group introduced into an α-position.

α-Tetralone —RMgX→ (R, OH) —acid, heat, $-H_2O$→ (R) —Pd, heat→ α-Alkylnaphthalene

(c) The original keto acid (in the form of its ester) can be treated with a Grignard reagent, and an alkyl (or aryl) group introduced into an α-position.

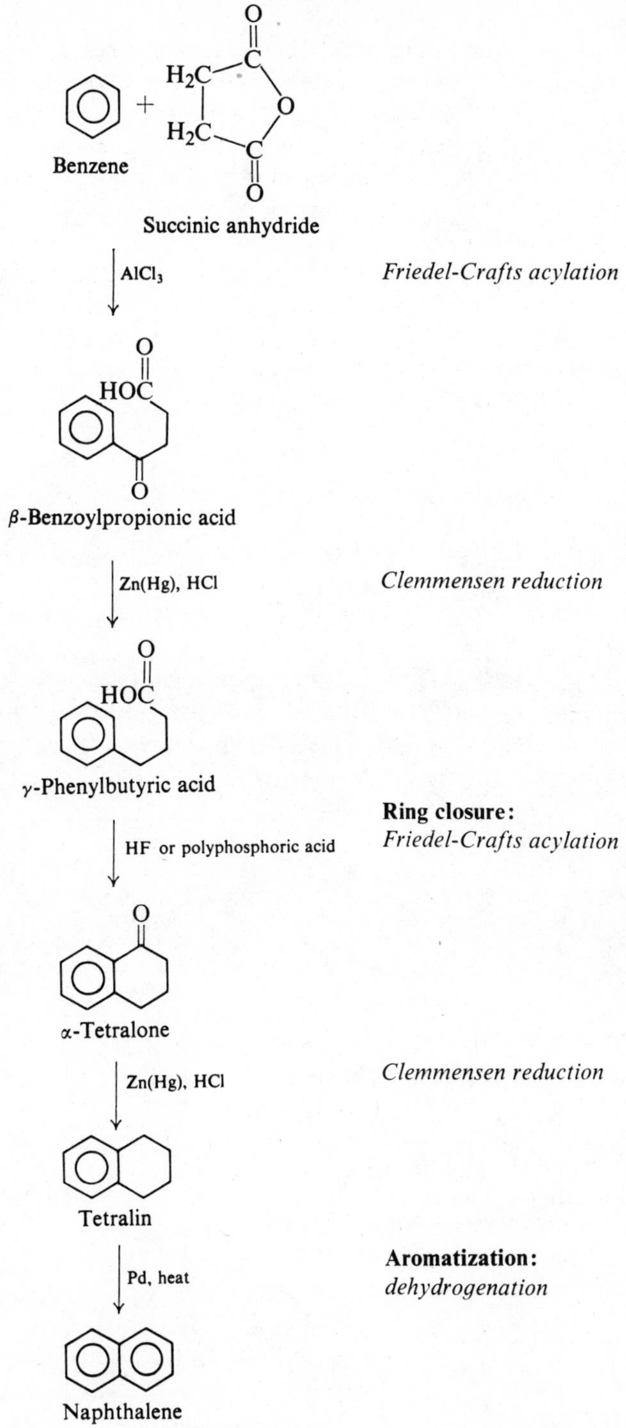

Benzene + Succinic anhydride

$\xrightarrow{AlCl_3}$ *Friedel-Crafts acylation*

β-Benzoylpropionic acid

$\xrightarrow{Zn(Hg),\ HCl}$ *Clemmensen reduction*

γ-Phenylbutyric acid

$\xrightarrow{HF \text{ or polyphosphoric acid}}$ **Ring closure:** *Friedel-Crafts acylation*

α-Tetralone

$\xrightarrow{Zn(Hg),\ HCl}$ *Clemmensen reduction*

Tetralin

$\xrightarrow{Pd,\ heat}$ **Aromatization:** *dehydrogenation*

Naphthalene

Figure 30.2. Haworth synthesis of naphthalene derivatives.

The success of this reaction depends upon the fact that a ketone reacts much faster than an ester with a Grignard reagent.

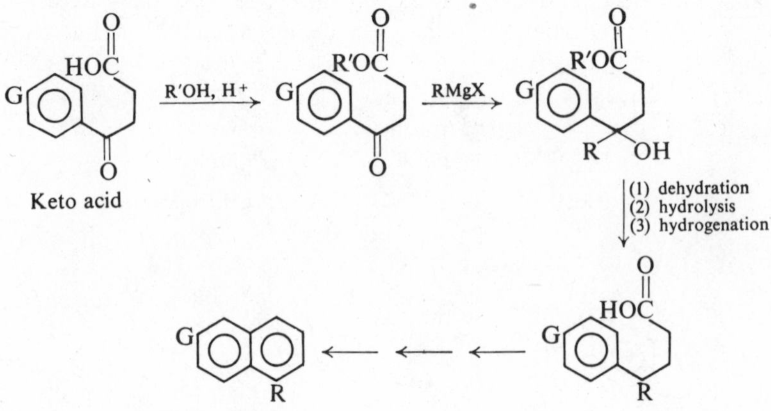

Keto acid $\xrightarrow{R'OH,\ H^+}$ $\xrightarrow{RMgX}$

(1) dehydration
(2) hydrolysis
(3) hydrogenation

$\longleftarrow \longleftarrow \longleftarrow$

1,6-Disubstituted naphthalene

By proper combinations of these modifications, a wide variety of substituted naphthalenes can be prepared.

Problem 30.18 Outline all steps in the synthesis of the following compounds, starting from benzene and using any necessary aliphatic and inorganic reagents:

(a) 2-methylnaphthalene
(b) 1-methylnaphthalene
(c) 1,4-dimethylnaphthalene
(d) 1,7-dimethylnaphthalene
(e) 1,6-dimethylnaphthalene
(f) 1,4,6-trimethylnaphthalene
(g) 1-ethyl-4-methylnaphthalene
(h) 7-bromo-1-ethylnaphthalene
(i) 1-phenylnaphthalene

Problem 30.19 Outline the Haworth sequence of reactions, starting with naphthalene and succinic anhydride. What is the final hydrocarbon or hydrocarbons? (Remember the orientation rules for naphthalene.) Check your answer in Sec. 30.19.

ANTHRACENE AND PHENANTHRENE

30.15 Nomenclature of anthracene and phenanthrene derivatives

The positions in anthracene and phenanthrene are designated by numbers as shown:

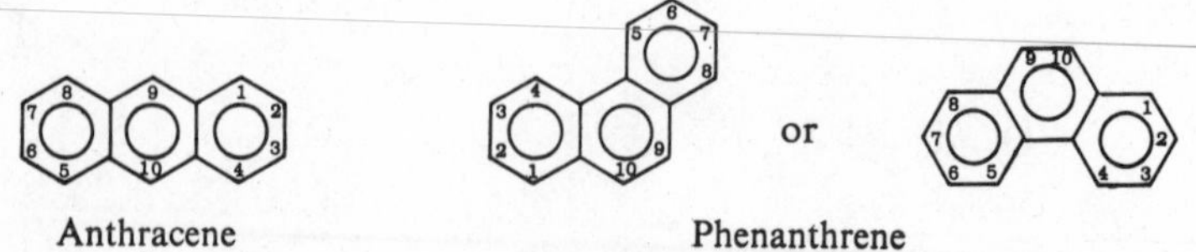

Anthracene Phenanthrene

Examples are found in the various reactions that follow.

30.16 Structure of anthracene and phenanthrene

Like naphthalene, anthracene and phenanthrene are classified as aromatic on the basis of their properties. Consideration of atomic orbitals follows the same pattern as for naphthalene, and leads to the same kind of picture: a flat structure with partially overlapping π clouds lying above and below the plane of the molecule.

In terms of valence bonds, anthracene is considered to be a hybrid of structures I–IV,

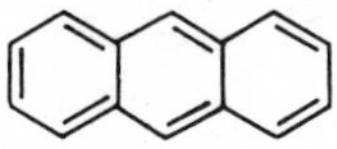

I 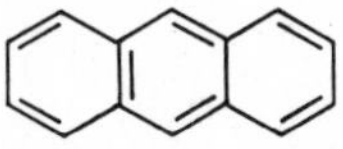 II 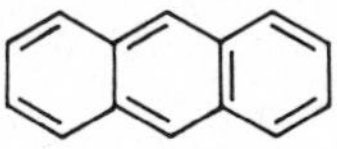 III 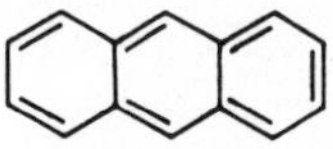 IV

Anthracene

and phenanthrene, a hybrid of structures V–IX. Heats of combustion indicate

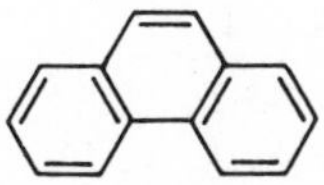

V 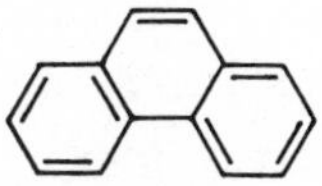 VI 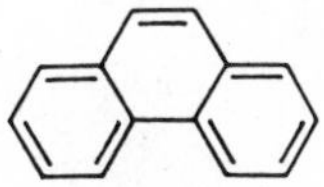 VII

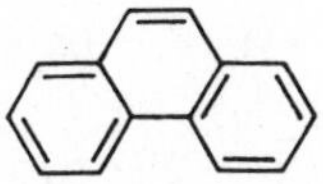

VIII 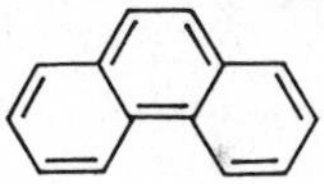 IX

Phenanthrene

that anthracene has a resonance energy of 84 kcal/mole, and that phenanthrene has a resonance energy of 92 kcal/mole.

For convenience we shall represent anthracene as the single structure X, and phenanthrene as XI, in which the circles can be thought of as representing partially overlapping aromatic sextets.

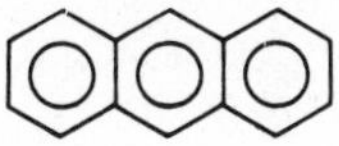

X
Anthracene

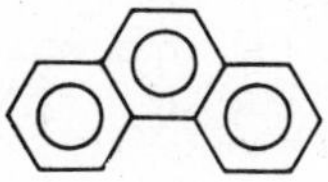

XI
Phenanthrene

30.17 Reactions of anthracene and phenanthrene

Anthracene and phenanthrene are even less resistant toward oxidation or reduction than naphthalene. Both hydrocarbons are oxidized to the 9,10-quinones and reduced to the 9,10-dihydro compounds. Both the orientation of these reactions and the comparative ease with which they take place are understandable on the basis of the structures involved. Attack at the 9- and 10-positions leaves two benzene rings intact; thus there is a sacrifice of only 12 kcal of resonance energy (84 − 2 × 36) for anthracene, and 20 kcal (92 − 2 × 36) for phenanthrene.

Anthracene → ($K_2Cr_2O_7$, H_2SO_4) → 9,10-Anthraquinone

Anthracene → (Na, C_2H_5OH, reflux) → 9,10-Dihydroanthracene

Phenanthrene → ($K_2Cr_2O_7$, H_2SO_4) → 9,10-Phenanthrenequinone

Phenanthrene → (Na, $C_5H_{11}OH$, reflux) → 9,10-Dihydrophenanthrene

(In the case of phenanthrene, the two remaining rings are conjugated; to the extent that this conjugation stabilizes the product—estimated at anywhere from 0 to 8 kcal/mole—the sacrifice is even less than 20 kcal.)

Problem 30.20 How much resonance energy would be sacrificed by oxidation or reduction of one of the outer rings of anthracene? Of phenanthrene?

Both anthracene and phenanthrene undergo electrophilic substitution. With a few exceptions, however, these reactions are of little value in synthesis because of the formation of mixtures and polysubstitution products. Derivatives of these two hydrocarbons are usually obtained in other ways: by electrophilic substitution in 9,10-anthraquinone or 9,10-dihydrophenanthrene, for example, or by ring closure methods (Secs. 30.18 and 30.19).

Bromination of anthracene or phenanthrene takes place at the 9-position. (9-Bromophenanthrene is a useful intermediate for the preparation of certain 9-substituted phenanthrenes.) In both cases, especially for anthracene, there is a tendency for addition to take place with the formation of the 9,10-dibromo-9,10-dihydro derivatives.

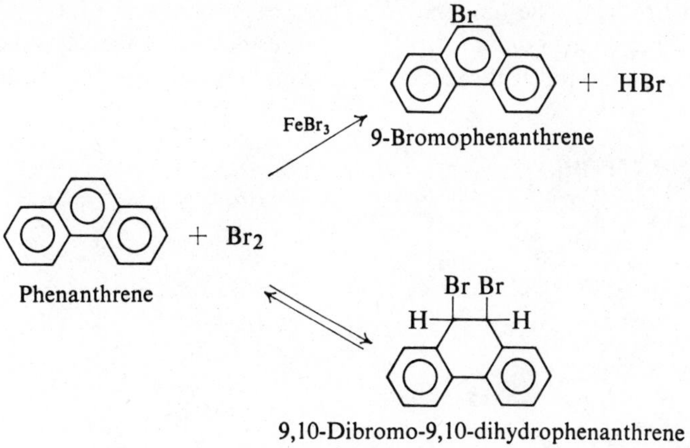

Br

+ HBr

$FeBr_3$

9-Bromophenanthrene

+ Br_2

Phenanthrene

Br Br

H H

9,10-Dibromo-9,10-dihydrophenanthrene

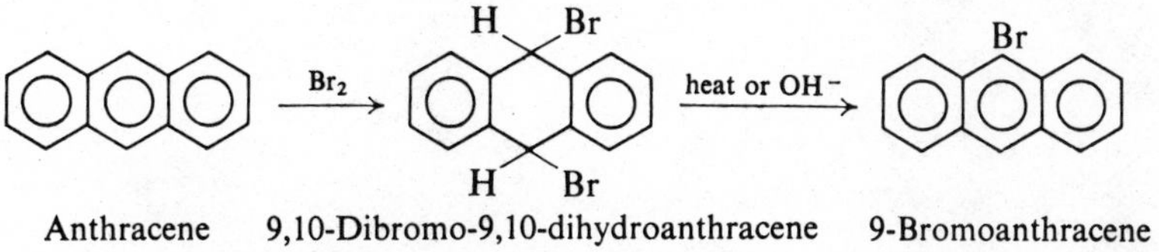

H Br

Br

Br_2

heat or OH^-

H Br

Anthracene 9,10-Dibromo-9,10-dihydroanthracene 9-Bromoanthracene

This reactivity of the 9- and 10-positions toward electrophilic attack is understandable, whether reaction eventually leads to substitution or addition. The carbonium ion initially formed is the most stable one, I or II, in which aromatic

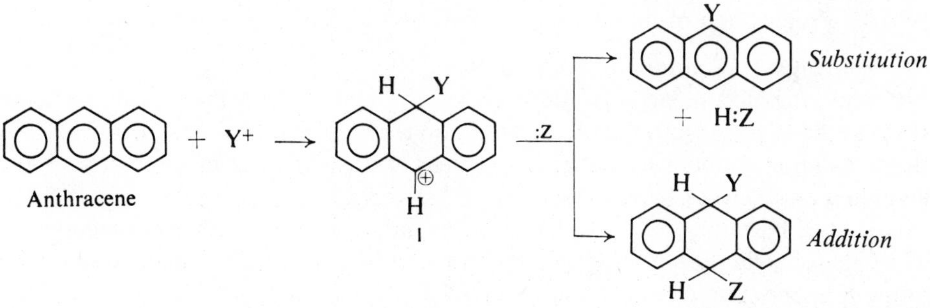

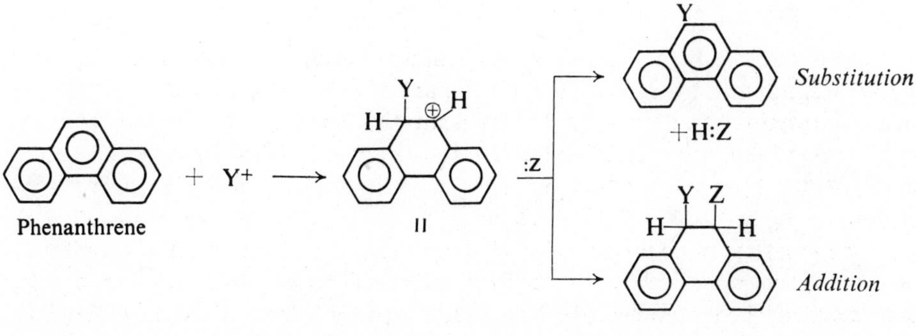

Y

Substitution

H Y

+ H:Z

+ Y^+

:Z

⊕

H Y

Anthracene

H

I

Addition

H Z

Y

Substitution

Y ⊕ H

H

+H:Z

+ Y^+

:Z

Y Z

Phenanthrene

II

H H

Addition

sextets are preserved in two of the three rings. This carbonium ion can then either (a) give up a proton to yield the substitution product, or (b) accept a base to yield the addition product. The tendency for these compounds to undergo addition is undoubtedly due to the comparatively small sacrifice in resonance energy that this entails (12 kcal/mole for anthracene, 20 kcal/mole or less for phenanthrene).

Problem 30.21 Nitric acid converts anthracene into any of a number of products, III–VI, depending upon the exact conditions. How could each be accounted for?

(a) Nitric acid and acetic acid yields III
(b) Nitric acid and ethyl alcohol yields IV
(c) Excess nitric acid yields V
(d) Nitric acid and acetic anhydride yields 9-nitroanthracene (VI)

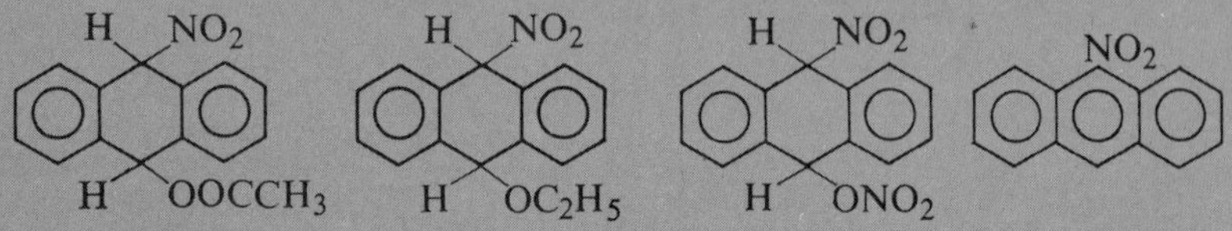

H NO_2 / H $OOCCH_3$ — III; H NO_2 / H OC_2H_5 — IV; H NO_2 / H ONO_2 — V; NO_2 — VI

Problem 30.22 Account for the following observations: (a) Upon treatment with hydrogen and nickel, 9,10-dihydroanthracene yields 1,2,3,4-tetrahydroanthracene. (b) In contrast to bromination, sulfonation of anthracene yields the 1-sulfonic acid.

30.18 Preparation of anthracene derivatives by ring closure. Anthraquinones

Derivatives of anthracene are seldom prepared from anthracene itself, but rather by ring-closure methods. As in the case of naphthalene, the most important method of ring closure involves adaptation of Friedel-Crafts acylation. The products initially obtained are **anthraquinones**, which can be converted into corresponding anthracenes by reduction with zinc and alkali. This last step is seldom carried out, since the quinones are by far the more important class of compounds.

The following reaction sequence shows the basic scheme. (Large amounts of anthraquinones are manufactured for the dye industry in this way.)

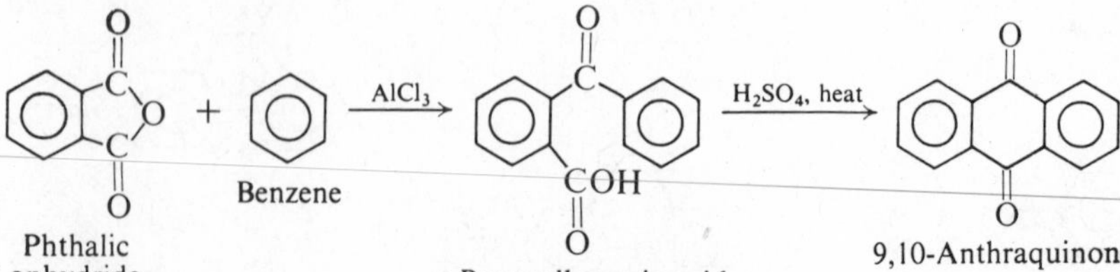

Phthalic anhydride + Benzene $\xrightarrow{AlCl_3}$ o-Benzoylbenzoic acid $\xrightarrow{H_2SO_4,\ heat}$ 9,10-Anthraquinone

The basic scheme can be modified in a number of ways.

(a) A monosubstituted benzene can be used in place of benzene, and a

2-substituted anthraquinone obtained. (The initial acylation goes chiefly *para*. If the *para* position is blocked, *ortho* acylation is possible.) For example:

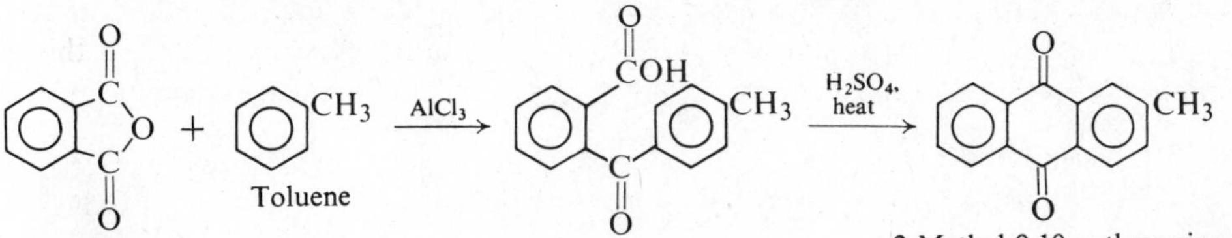

Phthalic anhydride + Toluene $\xrightarrow{AlCl_3}$ *o*-(*p*-Toluyl)benzoic acid $\xrightarrow{H_2SO_4,\ heat}$ 2-Methyl-9,10-anthraquinone

(b) A polynuclear compound can be used in place of benzene, and a product having more than three rings obtained. For example:

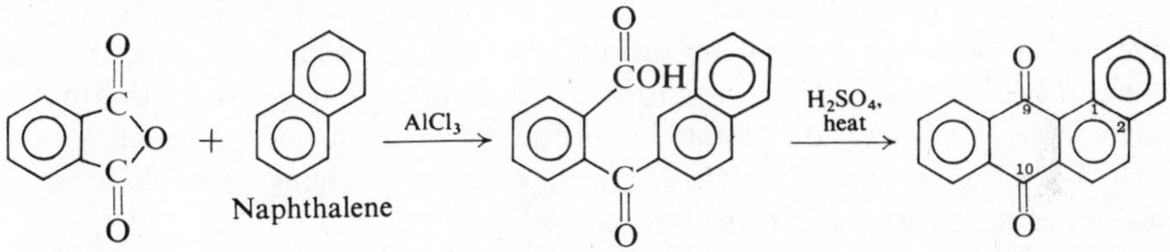

Phthalic anhydride + Naphthalene $\xrightarrow{AlCl_3}$ *o*-(2-Naphthoyl)benzoic acid $\xrightarrow{H_2SO_4,\ heat}$ 1,2-Benz-9,10-anthraquinone

(c) The intermediate *o*-aroylbenzoic acid can be reduced before ring closure, and 9-substituted anthracenes obtained via Grignard reactions.

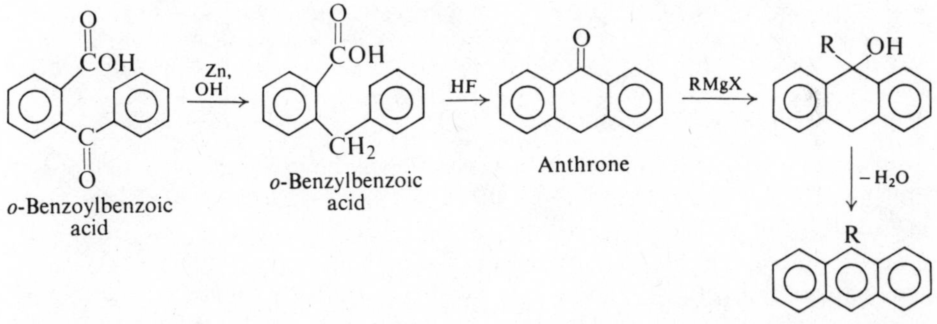

o-Benzoylbenzoic acid $\xrightarrow{Zn,\ OH^-}$ *o*-Benzylbenzoic acid $\xrightarrow{HF}$ Anthrone $\xrightarrow{RMgX}$ (R, OH) $\xrightarrow{-H_2O}$ 9-Alkylanthracene

Anthraquinoid dyes are of enormous technological importance, and much work has been done in devising syntheses of large ring systems embodying the quinone structure. Several examples of anthraquinoid dyes are:

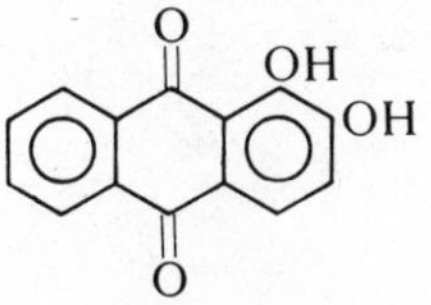

Alizarin

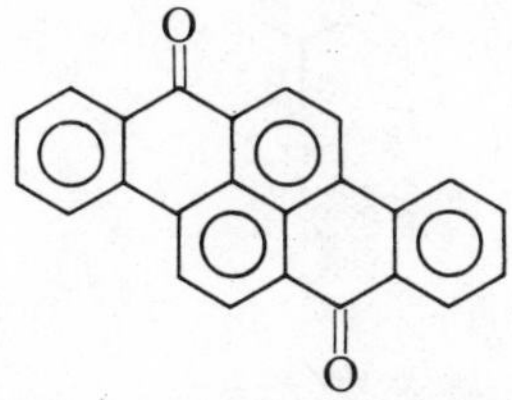

Indanthrene Golden Yellow GK

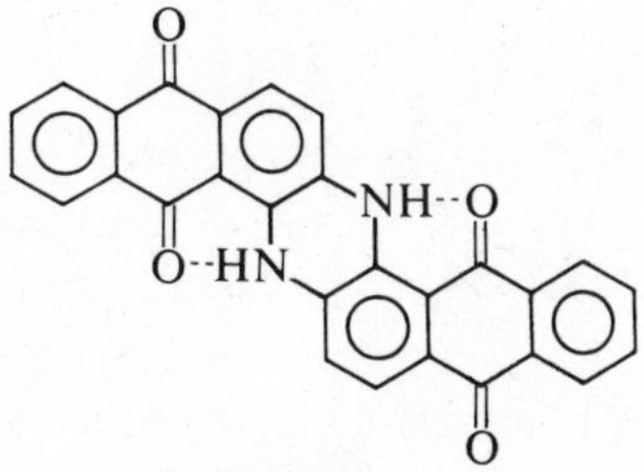

Indanthrone

Problem 30.23 Outline the synthesis of the following, starting from compounds having fewer rings:

(a) 1,4-dimethylanthraquinone
(b) 1,2-dimethylanthraquinone
(c) 1,3-dimethylanthraquinone
(d) 2,9-dimethylanthracene
(e) 9-methyl-1,2-benzanthracene (a potent cancer-producing hydrocarbon)

Problem 30.24 What anthraquinone or anthraquinones would be expected from a sequence starting with 3-nitrophthalic anhydride and (a) benzene, (b) toluene?

30.19 Preparation of phenanthrene derivatives by ring closure

Starting from naphthalene instead of benzene, the Haworth succinic anhydride synthesis (Sec. 30.14) provides an excellent route to substituted phenanthrenes.

The basic scheme is outlined in Fig. 30.3. Naphthalene is acylated by succinic anhydride at both the 1- and 2-positions; the two products are separable, and either can be converted into phenanthrene. We notice that γ-(2-naphthyl)butyric acid undergoes ring closure at the 1-position to yield phenanthrene rather than at the 3-position to yield anthracene; the electron-releasing side chain at the 2-position directs further substitution to the 1-position (Sec. 30.13).

Substituted phenanthrenes are obtained by modifying the basic scheme in the ways already described for the Haworth method (Sec. 30.14).

Problem 30.25 Apply the Haworth method to the synthesis of the following, starting from naphthalene or a monosubstituted naphthalene:

(a) 9-methylphenanthrene
(b) 4-methylphenanthrene
(c) 1-methylphenanthrene
(d) 1,9-dimethylphenanthrene
(e) 4,9-dimethylphenanthrene
(f) 1,4-dimethylphenanthrene
(g) 1,4,9-trimethylphenanthrene
(h) 2-methoxyphenanthrene (*Hint:* See Problem 30.15, p. 984.)

Problem 30.26 Give structural formulas for all intermediates in the following synthesis of 2-methylphenanthrene. Tell what kind of reaction each step involves.

$$\text{Naphthalene} + CH_3CH_2COCl + AlCl_3 \xrightarrow{C_6H_5NO_2} \text{A } (C_{13}H_{12}O)$$

$$\text{A} + Br_2 \longrightarrow \text{B } (C_{13}H_{11}OBr)$$

$$\text{B} + CH(COOC_2H_5)_2{}^-Na^+ \longrightarrow \text{C } (C_{20}H_{22}O_5)$$

$$\text{C} \xrightarrow{\text{aq. KOH, heat}} \text{D} \xrightarrow{HCl} \text{E} \xrightarrow{\text{heat}} \text{F } (C_{15}H_{14}O_3) + CO_2$$

$$\text{F} + Zn(Hg) + HCl \longrightarrow \text{G } (C_{15}H_{16}O_2)$$

$$\text{G} \xrightarrow{\text{polyphosphoric acid}} \text{H } (C_{15}H_{14}O)$$

$$\text{H} + Zn(Hg) + HCl \longrightarrow \text{I } (C_{15}H_{16})$$

$$\text{I} \xrightarrow{\text{Pd, heat}} \text{2-methylphenanthrene}$$

Problem 30.27 Follow the instructions for Problem 30.26 for the following synthesis of phenanthrene (the **Bogert-Cook synthesis**).

$$\beta\text{-Phenylethyl bromide} + Mg \longrightarrow \text{A } (C_8H_9MgBr)$$

$$\text{A} + \text{cyclohexanone} \longrightarrow \text{B} \xrightarrow{H_2O} \text{C } (C_{14}H_{20}O)$$

$$\text{C} \xrightarrow{H_2SO_4} \text{D } (C_{14}H_{18})$$

$$\text{D} \xrightarrow{H_2SO_4} \text{E } (C_{14}H_{18})$$

$$\text{E} \xrightarrow{\text{Se, heat}} \text{phenanthrene}$$

How could β-phenylethyl bromide be made from benzene?

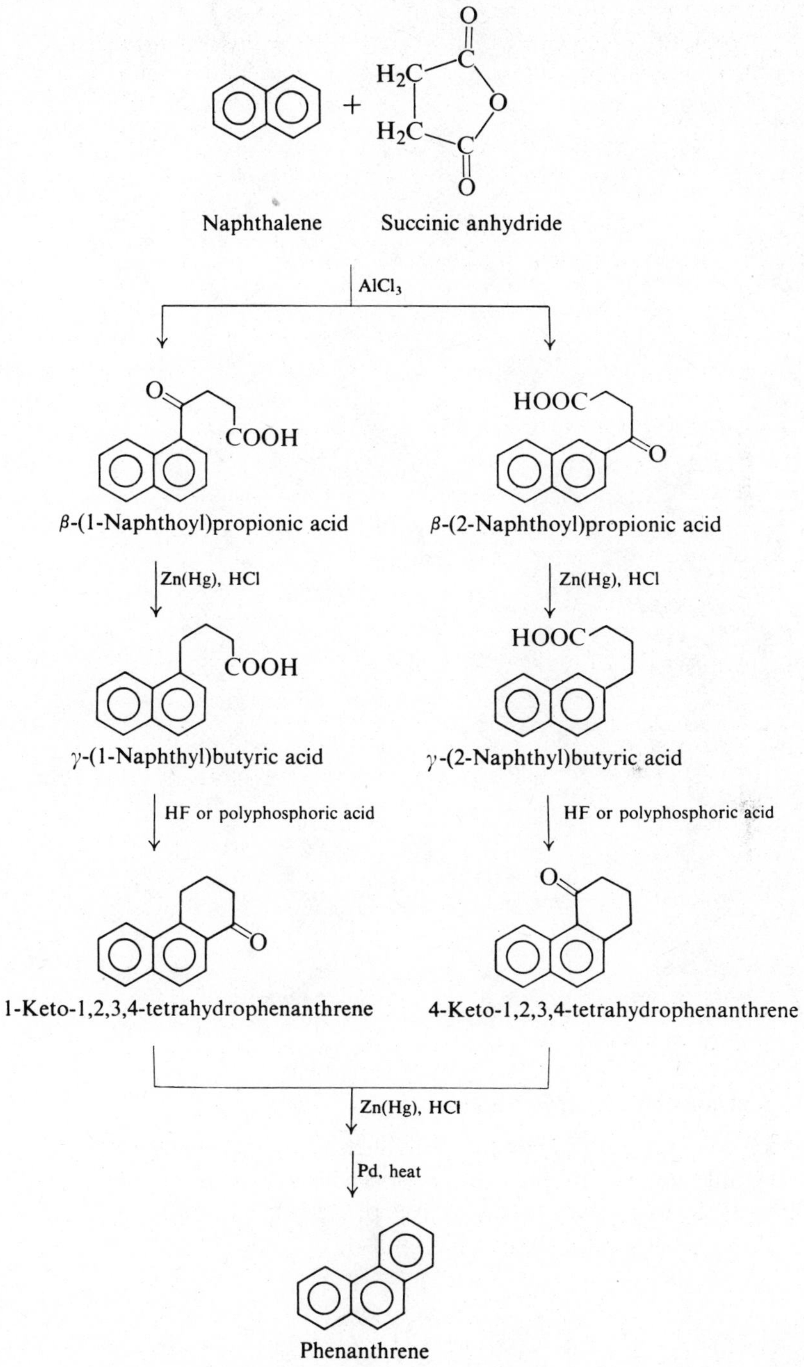

Naphthalene + Succinic anhydride

$\downarrow$ $AlCl_3$

β-(1-Naphthoyl)propionic acid β-(2-Naphthoyl)propionic acid

$\downarrow$ Zn(Hg), HCl

γ-(1-Naphthyl)butyric acid γ-(2-Naphthyl)butyric acid

$\downarrow$ HF or polyphosphoric acid

1-Keto-1,2,3,4-tetrahydrophenanthrene 4-Keto-1,2,3,4-tetrahydrophenanthrene

$\downarrow$ Zn(Hg), HCl

$\downarrow$ Pd, heat

Phenanthrene

Figure 30.3. Haworth synthesis of phenanthrene derivatives.

Problem 30.28 Follow the instructions for Problem 30.26 for the following synthesis of phenanthrene (the **Bardhan-Sengupta synthesis**).

Potassium + ethyl 2-keto-1-cyclohexanecarboxylate $\longrightarrow$ A ($C_9H_{13}O_3K$)

A + β-phenylethyl bromide $\longrightarrow$ B ($C_{17}H_{22}O_3$)

B $\xrightarrow{\text{aq. KOH, heat}}$ C $\xrightarrow{\text{HCl}}$ D ($C_{14}H_{18}O$)

D + Na + moist ether $\longrightarrow$ E ($C_{14}H_{20}O$)

E $\xrightarrow{P_2O_5}$ [F ($C_{14}H_{18}$)] $\xrightarrow{P_2O_5}$ G ($C_{14}H_{18}$)

G $\xrightarrow{\text{Se, heat}}$ phenanthrene

Problem 30.29 Follow the instructions for Problem 30.26 for the following synthesis of *pyrene*:

4-Keto-1,2,3,4-tetrahydrophenanthrene ($C_{14}H_{12}O$)

+ $BrCH_2COOC_2H_5$ + Zn $\xrightarrow{\text{ether}}$ A $\xrightarrow{H_2O,\ H^+}$ B ($C_{18}H_{20}O_3$)

B + acid + heat $\longrightarrow$ C ($C_{18}H_{18}O_2$)

C + aq. NaOH + heat $\longrightarrow$ D $\xrightarrow{\text{HCl}}$ E ($C_{16}H_{14}O_2$)

E $\xrightarrow{\text{HF}}$ F ($C_{16}H_{12}O$)

F + Zn(Hg) + HCl $\longrightarrow$ G ($C_{16}H_{14}$)

G $\xrightarrow{\text{Pd, heat}}$ pyrene ($C_{16}H_{10}$)

How could you make the starting material?

Problem 30.30 Outline a possible synthesis of *chrysene* by the Bogert-Cook method (Problem 30.27, p. 944), starting from naphthalene and using any aliphatic or inorganic reagents. (*Hint:* See Problem 30.7(g), p. 977.)

Chrysene

Problem 30.31 Outline an alternative synthesis of chrysene by the Bogert-Cook method, starting from benzene and using any aliphatic or inorganic reagents.

30.20 Carcinogenic hydrocarbons

Much of the interest in complex polynuclear hydrocarbons has arisen because a considerable number of them have cancer-producing properties. Some of the most powerful carcinogens are derivatives of 1,2-benzanthracene:

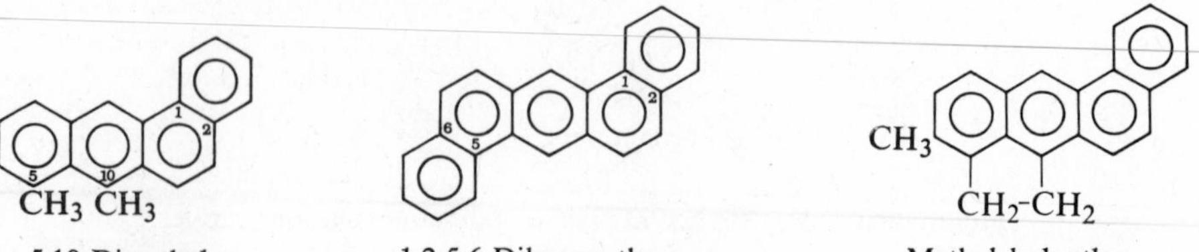

5,10-Dimethyl-1,2-benzanthracene　　1,2,5,6-Dibenzanthracene　　Methylcholanthrene

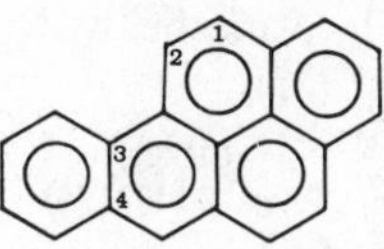

3,4-Benzpyrene

The relationship between carcinogenic activity and chemical properties is far from clear, but the possibility of uncovering this relationship has inspired a tremendous amount of research in the fields of synthesis and of structure and reactivity.

PROBLEMS

1. Give the structures and names of the principal products of the reaction (if any) of naphthalene with:

(a) CrO_3, CH_3COOH
(b) O_2, V_2O_5
(c) Na, C_2H_5OH
(d) Na, $C_5H_{11}OH$
(e) H_2, Ni
(f) HNO_3, H_2SO_4
(g) Br_2
(h) conc. H_2SO_4, 80°
(i) conc. H_2SO_4, 160°
(j) CH_3COCl, $AlCl_3$, CS_2
(k) CH_3COCl, $AlCl_3$, $C_6H_5NO_2$
(l) succinic anhydride, $AlCl_3$, $C_6H_5NO_2$

2. Give the structures and names of the principal products of the reaction of HNO_3/H_2SO_4 with:

(a) 1-methylnaphthalene
(b) 2-methylnaphthalene
(c) 1-nitronaphthalene
(d) 2-nitronaphthalene
(e) 1-naphthalenesulfonic acid
(f) 2-naphthalenesulfonic acid
(g) N-(1-naphthyl)acetamide
(h) N-(2-naphthyl)acetamide
(i) α-naphthol
(j) β-naphthol
(k) anthracene

3. When 2-methylnaphthalene is nitrated, three isomeric mononitro derivatives are obtained. Upon vigorous oxidation one of these yields 3-nitro-1,2,4-benzenetricarboxylic acid, and the other two both yield 3-nitrophthalic acid. Give the names and structures of the original three isomeric nitro compounds.

4. Outline all steps in a possible synthesis of each of the following from naphthalene, using any needed organic and inorganic reagents:

(a) α-naphthol
(b) β-naphthol
(c) α-naphthylamine
(d) β-naphthylamine
(e) 1-iodonaphthalene
(f) 2-iodonaphthalene
(g) 1-nitronaphthalene
(h) 2-nitronaphthalene
(i) α-naphthoic acid
(j) β-naphthoic acid
(k) 4-(1-naphthyl)butanoic acid
(l) α-naphthaldehyde
(m) β-naphthaldehyde
(n) 1-phenylazo-2-naphthol
(o) 1-amino-2-naphthol (*Hint:* Use product of (n).)
(p) 4-amino-1-naphthol
(q) 1-bromo-2-methoxynaphthalene
(r) 1,5-diaminonaphthalene
(s) 4,8-dibromo-1,5-diiodonaphthalene
(t) 5-nitro-2-naphthalenesulfonic acid
(u) 1,2-diaminonaphthalene
(v) 1,3-diaminonaphthalene
(w) *o*-aminobenzoic acid
(x) phenanthrene
(y) 9,10-anthraquinone
(z) anthracene

5. Naphthalene was transformed into another hydrocarbon by the following sequence of reactions:

naphthalene + Na, $C_5H_{11}OH$ $\longrightarrow$ A ($C_{10}H_{12}$)
A + succinic anhydride, $AlCl_3$ $\longrightarrow$ B ($C_{14}H_{16}O_3$)
B + Zn(Hg) + HCl $\longrightarrow$ C ($C_{14}H_{18}O_2$)
C + anhydrous HF $\longrightarrow$ D ($C_{14}H_{16}O$)
D + Zn(Hg) + HCl $\longrightarrow$ E ($C_{14}H_{18}$)
E + Pd/C + heat $\longrightarrow$ F ($C_{14}H_{10}$, m.p. 100–101°) + $4H_2$
What was F?

6. Outline all steps in a possible synthesis of each of the following from hydrocarbons containing fewer rings:

(a) 6-methoxy-4-phenyl-1-methylnaphthalene
(b) 1,2-benzanthracene
(c) 9-phenylanthracene
(d) 1-phenylphenanthrene
(e) 1,9-diphenylphenanthrene

7. Acylation of phenanthrene by succinic anhydride takes place at the 2- and 3-positions. The sequence of reduction, ring closure, and aromatization converts the 2-isomer into G and H, and converts the 3-isomer into G.

What is the structure and name of G? Of H?

8. When 4-phenyl-3-butenoic acid is refluxed there is formed a product, $C_{10}H_8O$, which is soluble in aqueous NaOH but not in aqueous $NaHCO_3$, and which reacts with benzenediazonium chloride to yield a red-orange solid. What is the product, and by what series of steps is it probably formed?

9. Anthracene reacts readily with maleic anhydride to give I, $C_{18}H_{12}O_3$, which can be hydrolyzed to J, a dicarboxylic acid of formula $C_{18}H_{14}O_4$. (a) What reaction do you think is involved in the formation of I? (b) What is the most probable structure of I? Of J?

Anthracene reacts with methyl fumarate to give a product that on hydrolysis yields K, a dicarboxylic acid of formula $C_{18}H_{14}O_4$. (c) Compare the structures of J and K. (*Hint:* See Problem 8, p. 880.)

Anthracene reacts with *p*-benzoquinone to yield L, $C_{20}H_{14}O_2$. In acid, L undergoes rearrangement to a hydroquinone M, $C_{20}H_{14}O_2$. Oxidation of M gives a new quinone N, $C_{20}H_{12}O_2$. Reductive amination of N gives a diamine O, $C_{20}H_{16}N_2$. Deamination of O by the usual method gives the hydrocarbon *triptycene*, $C_{20}H_{14}$. (d) What is a likely structure for triptycene?

10. Reduction of aromatic rings by the action of Li metal in ammonia generally gives 1,4-addition and yields a dihydro compound. Thus from naphthalene, $C_{10}H_8$, one can obtain $C_{10}H_{10}$. (a) Draw the structure of this dihydro compound.

Similar reduction is possible for 2-methoxynaphthalene (methyl 2-naphthyl ether). (b) Draw the structure of this dihydro compound. (c) If this dihydro ether is cleaved by acid, what is the structure of the initial product? (d) What further change will this initial product undoubtedly undergo, and what will be the final product?

11. Reduction of naphthalene by Li metal in $C_2H_5NH_2$ gives a 52% yield of 1,2,3,4,5,6,7,8-octahydronaphthalene. (a) What will this compound yield upon ozonolysis?

Treatment of the ozonolysis product ($C_{10}H_{16}O_2$) with base yields an unsaturated ketone ($C_{10}H_{14}O$). (b) What is its structure? (c) Show how this ketone can be transformed into *azulene*, $C_{10}H_8$, a blue hydrocarbon that is isomeric with naphthalene.

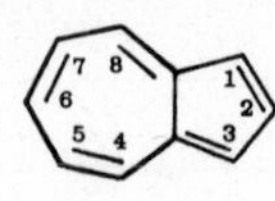

I

Azulene

12. (a) Azulene (preceding problem) is a planar molecule, and has a heat of combustion about 40 kcal/mole lower than that calculated by the method of Problem 10.2 (p. 323). It couples with diazonium salts and undergoes nitration and Friedel-Crafts acylation. Using both valence-bond and orbital structures, account for these properties of azulene. What might be a better representation of azulene than the formula I?

(b) The dipole moment of azulene is 1.08 D; that of 1-chloroazulene is 2.69 D. What is the direction of the dipole of azulene? Is this consistent with the structure you arrived at in (a)?

13. (a) In CF_3COOH solution, azulene gives the following nmr spectrum:

a singlet, δ 4.4, 2H
b doublet, δ 7.8, 1H
c doublet, δ 8.1, 1H
d multiplet, δ 9, 5H

and in CF_3COOD solution, the following spectrum:

a singlet, δ 8.1, 1H
b multiplet, δ 9, 5H

What compound gives rise to the spectrum in CF_3COOH? in CF_3COOD? Identify all nmr signals.

(b) In light of your structure for azulene (preceding problem), how do you account for what happens in CF_3COOH solution? What would you expect to obtain on neutralization of this solution?

(c) Show in detail just how the compound giving rise to the spectrum observed in CF_3COOD must have been formed. What would you expect to obtain on neutralization of this solution?

(d) At which position or positions in azulene would you expect nitration, Friedel-Crafts acylation, and diazonium coupling to occur?

14. Azulene reacts with *n*-butyllithium to yield, after hydrolysis and dehydrogenation, an *n*-butylazulene, and similarly with sodamide to yield an aminoazulene. To what class of reactions do these substitutions belong? In which ring would you expect such substitution to have occurred? At which position?

15. The structure of *eudalene*, $C_{14}H_{16}$, a degradation product of eudesmol (a terpene found in eucalyptus oil), was first established by the following synthesis:

p-isopropylbenzaldehyde + ethyl bromoacetate, Zn; then H_2O $\longrightarrow$ P ($C_{14}H_{20}O_3$)
P + acid, heat $\longrightarrow$ Q ($C_{14}H_{18}O_2$)
Q + Na, ethyl alcohol $\longrightarrow$ R ($C_{12}H_{18}O$)
R $\xrightarrow{HBr}$ $\xrightarrow{KCN}$ $\xrightarrow{H_2O,\ H^+}$ $\xrightarrow{SOCl_2}$ S ($C_{13}H_{17}OCl$)
S + $AlCl_3$, warm $\longrightarrow$ T ($C_{13}H_{16}O$)
T + CH_3MgBr, then H_2O $\longrightarrow$ U ($C_{14}H_{20}O$)
U + acid, heat $\longrightarrow$ V ($C_{14}H_{18}$)
V + sulfur, heat $\longrightarrow$ eudalene ($C_{14}H_{16}$)

What is the structure and systematic name of eudalene?

16. Many polynuclear aromatic compounds do not contain fused ring systems, e.g., biphenyl and triphenylmethane. Give structures and names of compounds W through II, formed in the following syntheses of such polynuclear compounds.

(a) 1,2,4,5-tetrachlorobenzene + H_2O, heat $\longrightarrow$ W ($C_6H_3OCl_3$)
W + HCHO + H_2SO_4 $\longrightarrow$ [X, $C_7H_5O_2Cl_3$] $\longrightarrow$ Y ($C_{13}H_6O_2Cl_6$), "Hexachlorophene," soluble in base

(b) *m*-bromotoluene + Mg, ether $\longrightarrow$ Z (C_7H_7MgBr)
Z + 4-methylcyclohexanone, then H_2O $\longrightarrow$ AA ($C_{14}H_{20}O$)
AA + H^+, heat $\longrightarrow$ BB ($C_{14}H_{18}$)
BB + Pd/C, heat $\longrightarrow$ CC ($C_{14}H_{14}$)

(c) ethyl benzoate + C_6H_5MgBr, then H_2O $\longrightarrow$ DD ($C_{19}H_{16}O$)
DD + conc. HBr $\longrightarrow$ EE ($C_{19}H_{15}Br$)
EE + Ag $\longrightarrow$ FF ($C_{38}H_{30}$)
(d) $(C_6H_5)_3COH$ + $C_6H_5NH_2$ + acid $\longrightarrow$ GG ($C_{25}H_{21}N$)
GG + $NaNO_2$ + HCl; then H_3PO_2 $\longrightarrow$ HH ($C_{25}H_{20}$)
(e) $C_6H_5COCH_3$ + acid + heat $\longrightarrow$ II ($C_{24}H_{18}$) (*Hint:* Acids catalyze aldol condensations.)

17. When 1-nitro-2-aminonaphthalene is treated with sodium nitrite and HCl, and then with warm water, there is obtained not only 1-nitro-2-naphthol, but also 1-chloro-2-naphthol. How do you account for the formation of the chloronaphthol? Consider carefully the stage at which chlorine is introduced into the molecule.

18. Treatment of phenanthrene with diazomethane yields a product JJ for which mass spectrometry indicates a molecular weight of 192. The infrared spectrum of JJ resembles that of 9,10-dihydrophenanthrene; its nmr spectrum shows two signals of one proton each at δ -0.12 and δ 1.48.

(a) What is a likely structure for JJ, and how is it probably formed? How do you account for the formation of JJ rather than one of its isomers?

(b) When a solution of JJ in *n*-pentane was irradiated with ultraviolet light, there were obtained phenanthrene, 2-methylpentane, 3-methylpentane, and *n*-hexane; the alkanes were obtained in the ratio 34:17:49. What happened in this reaction? What is the driving force?

(c) The irradiation of JJ in cyclohexene gave four products of formula C_7H_{12}. What would you expect these products to be?

(d) What would you expect to obtain from the irradiation of JJ in *cis*-4-methyl-2-pentene? In *trans*-4-methyl-2-pentene?

19. When *dihydropentalene* is treated with a little more than two moles of *n*-butyllithium, a stable white crystalline material KK is obtained. In contrast to the rather complicated nmr spectrum of dihydropentalene, the nmr spectrum of KK is simple:

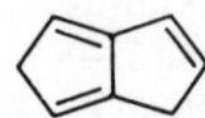

Dihydropentalene

a doublet, δ 4.98, J = 3 cps
b triplet, δ 5.73, J = 3 cps
peak area ratio a:b = 2:1

What is a likely structure for KK? Of what theoretical significance is its formation and stability?

20. (a) When *either* 1-chloronaphthalene *or* 2-chloronaphthalene is treated with lithium piperidide and piperidine (Sec. 31.12) dissolved in ether, the *same* mixture of products is obtained: I and II of Problem 30.8 (p. 977) in the ratio 31:69. Show all steps in a mechanism that accounts for these observations. In particular, show why 2-chloronaphthalene yields the same mixture as 1-chloronaphthalene.

(b) Under the conditions of (a), 1-bromonaphthalene and 1-iodonaphthalene give I and II in the same ratio as 1-chloronaphthalene does. With 1-fluoronaphthalene, however, the ratio of products depends on the concentration of piperidine. At high piperidine concentration, I makes up as much as 84% of the product; at low piperidine concentrations, the product ratio levels off at the 31:69 value.

Account in detail for these facts. Tell what is happening to change the product ratio, why the ratio is affected by piperidine concentration, and why the fluoride should behave differently from the other halides.

21. Give structural formulas for LL through UU. Account in detail for the properties of compound UU.

3,5-dibromo-4-methylanisole + CuCN ⟶ LL ($C_{10}H_8ON_2$)
LL + KOH, then CH_3OH, H^+ ⟶ MM ($C_{12}H_{14}O_5$)
MM + $LiAlH_4$ ⟶ NN ($C_{10}H_{14}O_3$)
NN + PBr_3 ⟶ OO ($C_{10}H_{12}OBr_2$)
OO + Na ⟶ PP ($C_{20}H_{24}O_2$)
PP + CrO_3 ⟶ QQ ($C_{18}H_{18}O_2$), a pale yellow solid (*Hint:* A carbon–carbon bond is formed.)
QQ + 2NaOH ⟶ RR ($C_{18}H_{16}O_2Na_2$), soluble in water
RR + O_2 ⟶ SS ($C_{18}H_{14}O_2$), a yellow solid
SS + $LiAlH_4$ ⟶ TT ($C_{18}H_{18}$)
TT + 2,3-dichloro-5,6-dicyanoquinone ("D.D.Q.") ⟶ UU ($C_{18}H_{16}$)

Compound UU undergoes nitration, bromination, and Friedel-Crafts acylation. X-ray analysis shows that (except for the two methyl groups) UU is flat or nearly flat. Ten carbon–carbon bonds are between 1.386 A and 1.401 A long. The nmr spectrum shows peaks for 10H downfield, and for 6H *far upfield*:

a singlet, δ −4.25 (τ 14.25), 6H
b triplet, δ 8.11, 2H
c doublet, δ 8.62, 4H
d singlet, δ 8.67, 4H

Chapter 31 Heterocyclic Compounds

31.1 Heterocyclic systems

A **heterocyclic compound** is one that contains a ring made up of more than one kind of atom.

In most of the cyclic compounds that we have studied so far—benzene, naphthalene, cyclohexanol, cyclopentadiene—the rings are made up only of carbon atoms; such compounds are called *homocyclic* or *alicyclic* compounds. But there are also rings containing, in addition to carbon, other kinds of atoms, most commonly nitrogen, oxygen, or sulfur. For example:

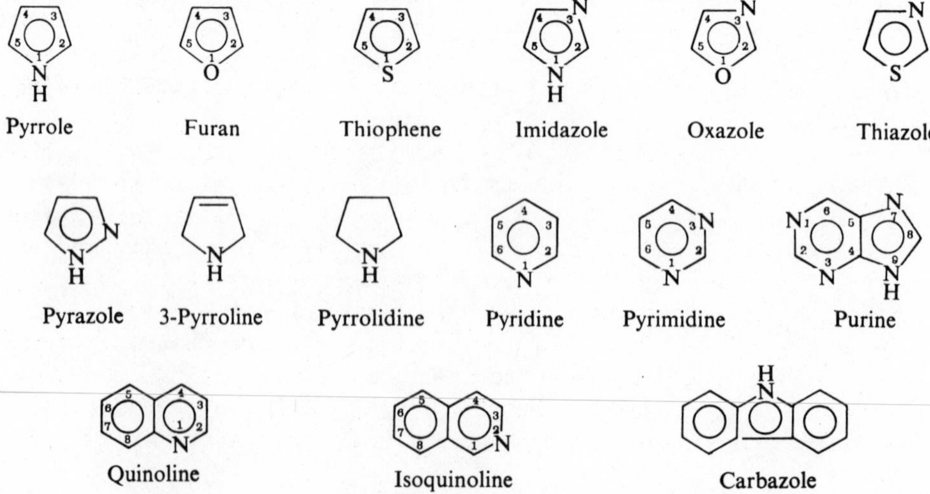

We notice that, in the numbering of ring positions, hetero atoms are generally given the lowest possible numbers.

Table 31.1 HETEROCYCLIC COMPOUNDS

Name	M.p., °C	B.p., °C	Name	M.p., °C	B.p., °C
Furan	− 30	32	Pyridine	− 42	115
Tetrahydrofuran	−108	66	α-Picoline	− 64	128
Furfuryl alcohol		171	β-Picoline		143
Furfural	− 36	162	γ-Picoline		144
Furoic acid	134		Piperidine	− 9	106
Pyrrole		130	Picolinic acid	137	
Pyrrolidine		88	Nicotinic acid	237	
Thiophene	− 40	84	Isonicotinic acid	317	
			Indole	53	254
			Quinoline	− 19	238
			Isoquinoline	23	243

Actually, of course, we have already encountered numerous heterocyclic compounds: *cyclic anhydrides* (Sec. 20.9) and *cyclic imides* (Sec. 20.14), for example: *lactones* (Sec. 20.15) and *lactams* (Problem 28.3, p. 891); *cyclic acetals* of dihydroxy alcohols (Problem 23, p. 651); the solvents *dioxane* and *tetrahydrofuran* (Sec. 17.9). In all these, the chemistry is essentially that of their open-chain analogs.

We have encountered three-membered heterocyclic rings which, because of ring strain, are highly reactive: *epoxides* (Secs. 17.10–17.15) and *aziridines* (Sec. 22.6); the fleeting but important intermediates, cyclic *halonium ions* (Secs. 7.12 and 28.10) and cyclic *sulfonium ions* (Sec. 28.11).

Heterocyclic intermediates are being used more and more in synthesis as *protecting groups*, readily generated and, when their job is done, readily removed. We have seen two examples of this: the temporary incorporation of the carboxyl group into a *2-oxazoline* ring (Sec. 26.6), and the temporary formation of *tetrahydropyranyl (THP) esters*, resistant toward alkali but extremely easily cleaved by acid (Problem 16, p. 692).

In the biological world, as we shall see in the final chapters of this book, heterocyclic compounds are everywhere. Carbohydrates are heterocyclic; so are chlorophyll and hemin, which make leaves green and blood red and bring life to plants and animals. Heterocycles form the sites of reaction in many enzymes and coenzymes. Heredity comes down, ultimately, to the particular sequence of attachment of a half-dozen heterocyclic rings to the long chains of nucleic acids.

In this chapter we can take up only a very few of the many different heterocyclic systems, and look only briefly at them. Among the most important and most interesting heterocycles are the ones that possess aromatic properties; we shall focus our attention on a few of these, and in particular upon their aromatic properties.

We can get some idea of the importance—as well as complexity—of heterocyclic systems from the following examples. Some others are *hemin* (p. 1152), *nicotinamide adenine dinucleotide* (p. 1153), and *oxytocin* (p. 1143).

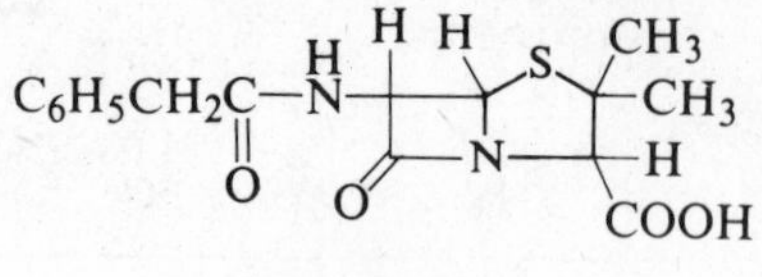

Penicillin G
Antibiotic

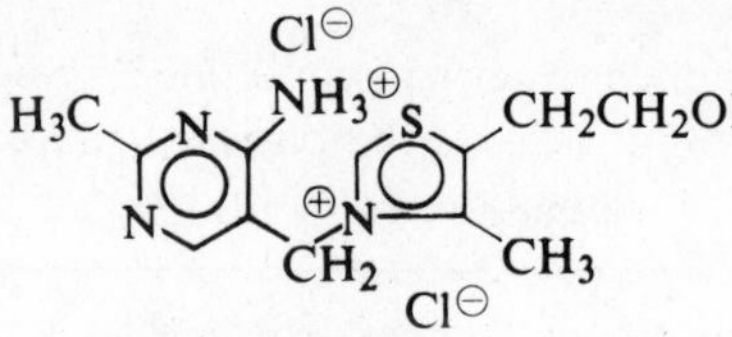

Thiamine
Vitamin B_1
Anti-beriberi factor

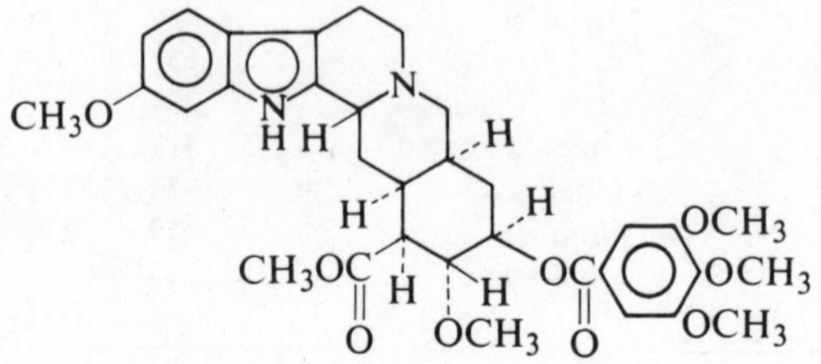

Reserpine
A tranquilizing drug

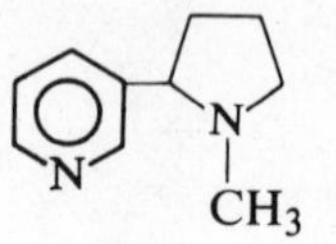

Nicotine
A tobacco alkaloid

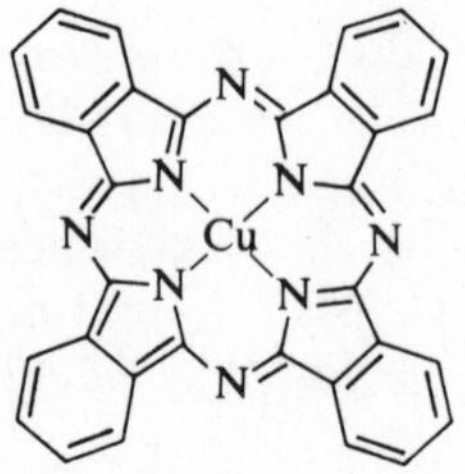

Copper phthalocyanine
A blue pigment

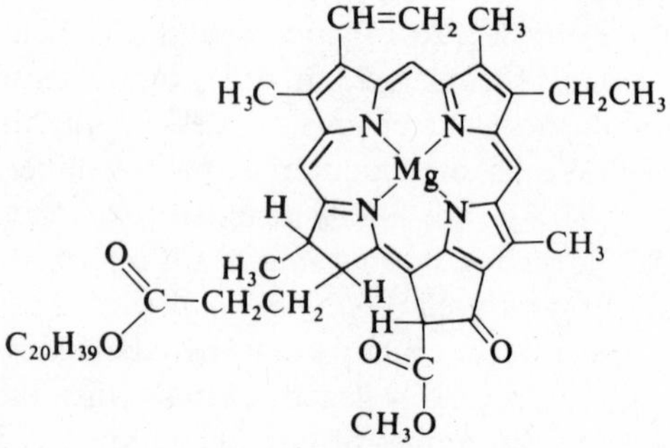

Chlorophyll a
Green plant pigment: catalyst for photosynthesis

FIVE-MEMBERED RINGS

31.2 Structure of pyrrole, furan, and thiophene

The simplest of the five-membered heterocyclic compounds are **pyrrole, furan**, and **thiophene**, each of which contains a single hetero atom.

Judging from the commonly used structures I, II, and III, we might expect each of these compounds to have the properties of a conjugated diene and of an amine, an ether, or a sulfide (thioether). Except for a certain tendency to undergo addi-

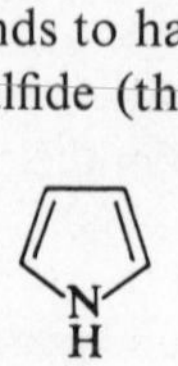

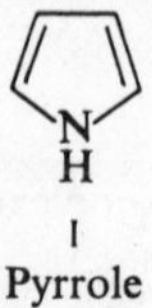

I
Pyrrole

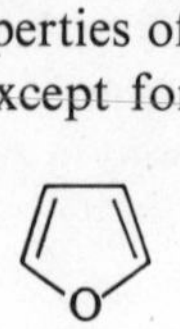

II
Furan

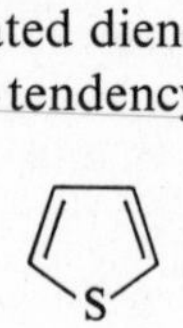

III
Thiophene

tion reactions, however, these heterocycles do not have the expected properties: thiophene does not undergo the oxidation typical of a sulfide, for example; pyrrole does not possess the basic properties typical of amines.

Instead, these heterocycles and their derivatives most commonly undergo electrophilic substitution: nitration, sulfonation, halogenation, Friedel-Crafts acylation, even the Reimer-Tiemann reaction and coupling with diazonium salts. Heats of combustion indicate resonance stabilization to the extent of 22–28 kcal/mole; somewhat less than the resonance energy of benzene (36 kcal/mole), but much greater than that of most conjugated dienes (about 3 kcal/mole). On the basis of these properties, pyrrole, furan, and thiophene must be considered *aromatic*. Clearly, formulas I, II, and III do not adequately represent the structures of these compounds.

Let us look at the orbital picture of one of these molecules, pyrrole. Each atom of the ring, whether carbon or nitrogen, is held by a σ bond to three other atoms. In forming these bonds, the atom uses three sp^2 orbitals, which lie in a plane and are 120° apart. After contributing one electron to each σ bond, each carbon atom of the ring has left *one* electron and the nitrogen atom has left *two* electrons; these electrons occupy p orbitals. Overlap of the p orbitals gives rise to π clouds, one above and one below the plane of the ring; the π clouds contain a total of six electrons, the *aromatic sextet* (Fig. 31.1).

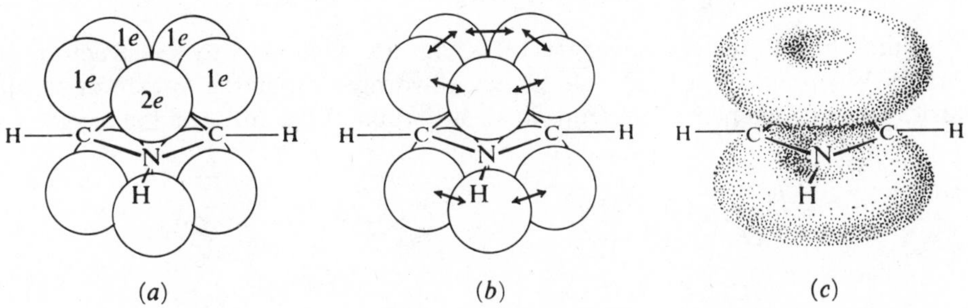

Figure 31.1. Pyrrole molecule. (*a*) Two electrons in p orbital of nitrogen; one electron in p orbital of each carbon. (*b*) Overlap of p orbitals to form π bonds. (*c*) Clouds above and below plane of ring; total of six π electrons, the aromatic sextet.

Delocalization of the π electrons stabilizes the ring. As a result, pyrrole has an abnormally low heat of combustion; it tends to undergo reactions in which the stabilized ring is retained, that is, to undergo substitution.

Nitrogen's extra pair of electrons, which is responsible for the usual basicity of nitrogen compounds, is involved in the π cloud, and is not available for sharing with acids. In contrast to most amines, therefore, pyrrole is an extremely weak base ($K_b \sim 2.5 \times 10^{-14}$). By the same token, there is a high electron density in the ring, which causes pyrrole to be extremely reactive toward electrophilic substitution: it undergoes reactions like nitrosation and coupling with diazonium salts which are characteristic of only the most reactive benzene derivatives, phenols and amines.

It thus appears that pyrrole is better represented by IV,

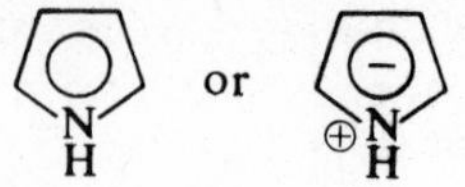

IV
Pyrrole

in which the circle represents the aromatic sextet.

What does IV mean in terms of conventional valence-bond structures? Pyrrole can be considered a hybrid of structures V–IX. Donation of electrons to the ring by nitrogen

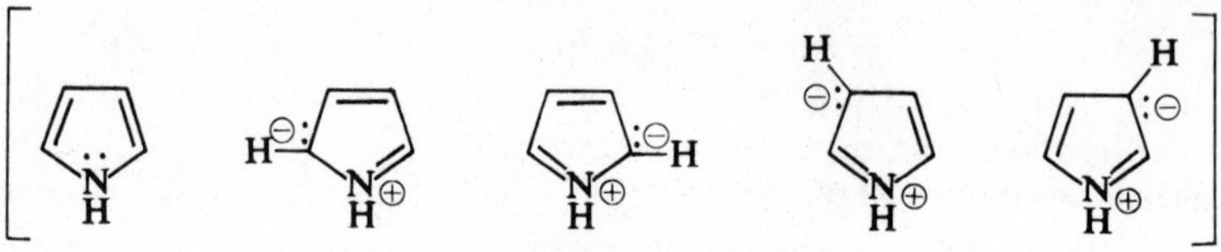

V VI VII VIII IX
Pyrrole

is indicated by the ionic structures in which nitrogen bears a positive charge and the carbon atoms of the ring bear a negative charge.

Furan and thiophene have structures that are analogous to the structure of pyrrole. Where nitrogen in pyrrole carries a hydrogen atom, the oxygen or sulfur carries an unshared pair of electrons in an sp^2 orbital. Like nitrogen, the oxygen or

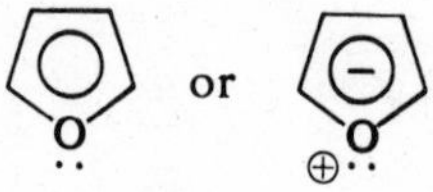

Furan 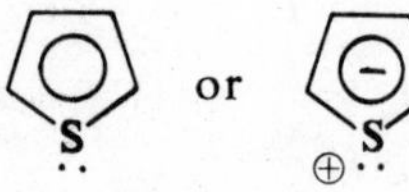 Thiophene

sulfur atom provides two electrons for the π cloud; as a result these compounds, too, behave like extremely reactive benzene derivatives.

31.3 Source of pyrrole, furan, and thiophene

Pyrrole and thiophene are found in small amounts in coal tar. During the fractional distillation of coal tar, thiophene (b.p. 84°) is collected along with the benzene (b.p. 80°); as a result ordinary benzene contains about 0.5% of thiophene, and must be specially treated if *thiophene-free benzene* is desired.

Thiophene can be synthesized on an industrial scale by the high-temperature reaction between *n*-butane and sulfur.

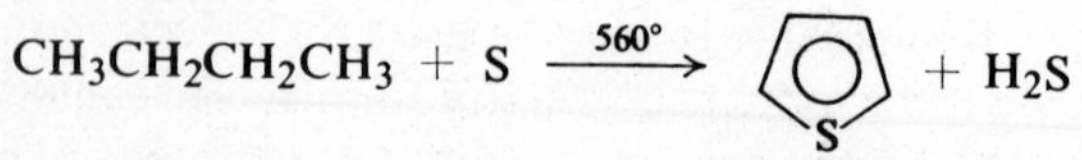

$$CH_3CH_2CH_2CH_3 + S \xrightarrow{560^\circ} \text{Thiophene} + H_2S$$

n-Butane Thiophene

Pyrrole can be synthesized in a number of ways. For example:

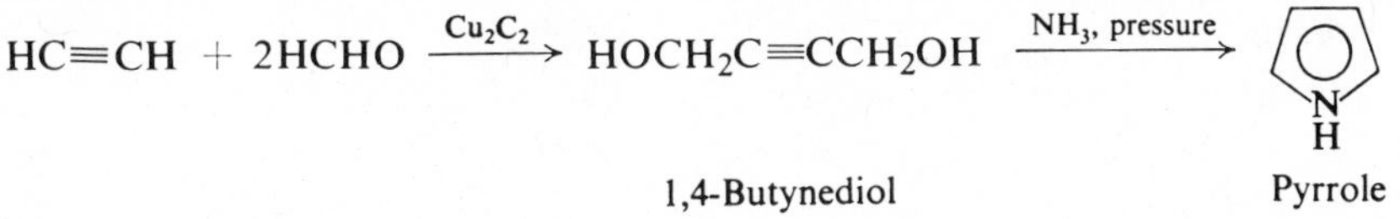

$$HC{\equiv}CH + 2HCHO \xrightarrow{Cu_2C_2} HOCH_2C{\equiv}CCH_2OH \xrightarrow{NH_3,\ pressure} \text{Pyrrole}$$

1,4-Butynediol Pyrrole

The pyrrole ring is the basic unit of the *porphyrin* system, which occurs, for example, in chlorophyll (p. 1004) and in hemoglobin (p. 1152).

Furan is most readily prepared by decarbonylation (elimination of carbon monoxide) of **furfural** (furfuraldehyde), which in turn is made by the treatment of oat hulls, corncobs, or rice hulls with hot hydrochloric acid. In the latter reaction pentosans (polypentosides) are hydrolyzed to pentoses, which then undergo dehydration and cyclization to form furfural.

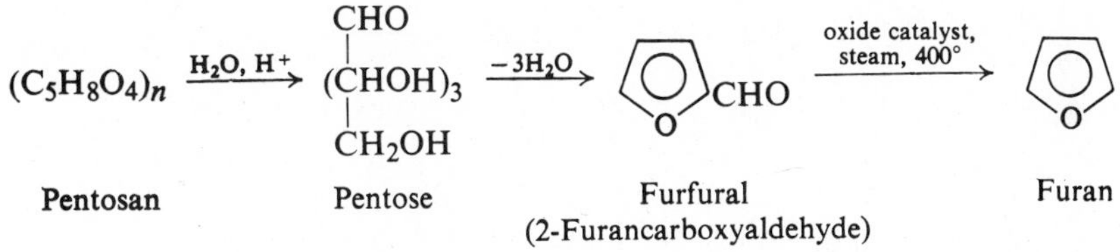

$$(C_5H_8O_4)_n \xrightarrow{H_2O,\ H^+} CHO{-}(CHOH)_3{-}CH_2OH \xrightarrow{-3H_2O} \text{Furfural} \xrightarrow{\text{oxide catalyst, steam, } 400^\circ} \text{Furan}$$

Pentosan Pentose Furfural (2-Furancarboxyaldehyde) Furan

Certain substituted pyrroles, furans, and thiophenes can be prepared from the parent heterocycles by substitution (see Sec. 31.4); most, however, are prepared from open-chain compounds by ring closure. For example:

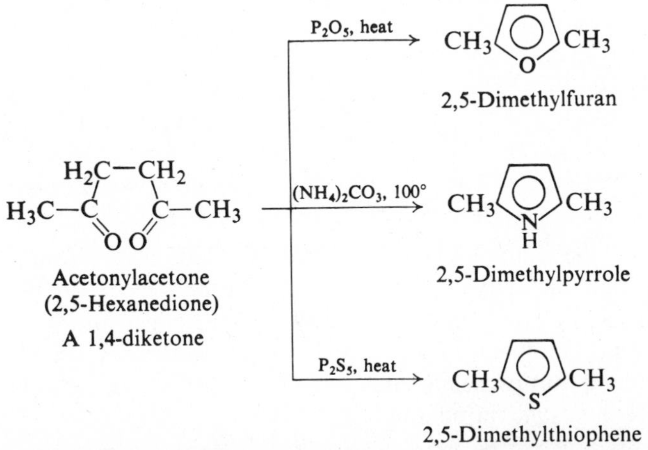

$$H_3C{-}CO{-}CH_2{-}CH_2{-}CO{-}CH_3 \xrightarrow{P_2O_5,\ heat} \text{2,5-Dimethylfuran}$$

$$H_3C{-}CO{-}CH_2{-}CH_2{-}CO{-}CH_3 \xrightarrow{(NH_4)_2CO_3,\ 100^\circ} \text{2,5-Dimethylpyrrole}$$

$$H_3C{-}CO{-}CH_2{-}CH_2{-}CO{-}CH_3 \xrightarrow{P_2S_5,\ heat} \text{2,5-Dimethylthiophene}$$

Acetonylacetone (2,5-Hexanedione) A 1,4-diketone

2,5-Dimethylfuran

2,5-Dimethylpyrrole

2,5-Dimethylthiophene

Problem 31.1 Give structural formulas for all intermediates in the following synthesis of acetonylacetone (2,5-hexanedione):

ethyl acetoacetate + $NaOC_2H_5$ ⟶ A ($C_6H_9O_3Na$)
A + I_2 ⟶ B ($C_{12}H_{18}O_6$) + NaI
B + dilute acid + heat ⟶ 2,5-hexanedione + carbon dioxide + ethanol

Problem 31.2 Outline a synthesis of 2,5-diphenylfuran, starting from ethyl benzoate and ethyl acetate.

31.4 Electrophilic substitution in pyrrole, furan, and thiophene. Reactivity and orientation

Like other aromatic compounds, these five-membered heterocycles undergo nitration, halogenation, sulfonation, and Friedel-Crafts acylation. They are much more reactive than benzene, and resemble the most reactive benzene derivatives (amines and phenols) in undergoing such reactions as the Reimer-Tiemann reaction, nitrosation, and coupling with diazonium salts.

Reaction takes place predominantly at the 2-position. For example:

Furan + pyridine : SO_3 ⟶ 2-Furansulfonic acid (SO_3H)

Furan + $(CH_3CO)_2O$ + $(C_2H_5)_2O : BF_3$ (Boron trifluoride etherate) $\xrightarrow{0°}$ 2-Acetylfuran ($COCH_3$)

Thiophene + C_6H_5COCl + $SnCl_4$ ⟶ 2-Benzoylthiophene (COC_6H_5)

Pyrrole + $C_6H_5N{\equiv}N^+Cl^-$ ⟶ 2-(Phenylazo)pyrrole ($N{=}NC_6H_5$)

Pyrrole + $CHCl_3$ + KOH ⟶ 2-Pyrrolecarboxaldehyde (CHO)
(Low yield)

In some of the examples we notice modifications in the usual electrophilic reagents. The high reactivity of these rings makes it possible to use milder reagents in many cases, as, for example, the weak Lewis acid stannic chloride in the Friedel-Crafts acylation of thiophene. The sensitivity to protic acids of furan (which undergoes ring opening) and pyrrole (which undergoes polymerization) makes it necessary to modify the usual sulfonating agent.

Problem 31.3 Furan undergoes ring opening upon treatment with sulfuric acid; it reacts almost explosively with halogens. Account for the fact that 2-furoic acid, however, can be sulfonated (in the 5-position) by treatment with fuming sulfuric acid, and brominated (in the 5-position) by treatment with bromine at 100°.

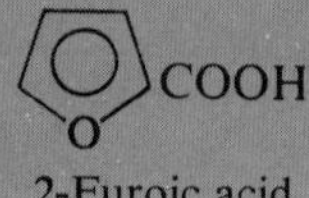

2-Furoic acid

Problem 31.4 Upon treatment with formaldehyde and acid, ethyl 2,4-dimethyl-3-pyrrolecarboxylate is converted into a compound of formula $C_{19}H_{26}O_4N_2$. What is the most likely structure for this product? How is it formed? (*Hint:* See Sec. 32.7.)

Problem 31.5 Predict the products from the treatment of furfural (2-furan-carboxaldehyde) with concentrated aqueous NaOH.

Problem 31.6 Sulfur trioxide dissolves in the tertiary amine pyridine to form a salt:

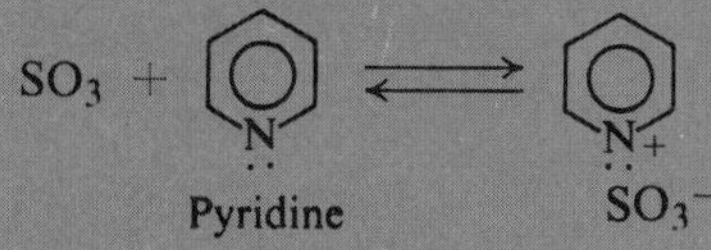

SO_3 + Pyridine ⇌ $C_5H_5N^+—SO_3^-$

Pyridine

Show all steps in the most likely mechanism for the sulfonation of an aromatic compound by this reagent.

In our study of electrophilic aromatic substitution (Sec. 11.19 and Sec. 30.9), we found that we could account for orientation on the following basis: the controlling step is the attachment of the electrophilic reagent to the aromatic ring, which takes place in such a way as to yield the most stable intermediate carbonium ion. Let us apply this approach to the reactions of pyrrole.

Attack at position 3 yields a carbonium ion that is a hybrid of structures I and II. Attack at position 2 yields a carbonium ion that is a hybrid not only of structures III and IV (analogous to I and II) but also of structure V; the extra stabilization conferred by V makes this ion the more stable one.

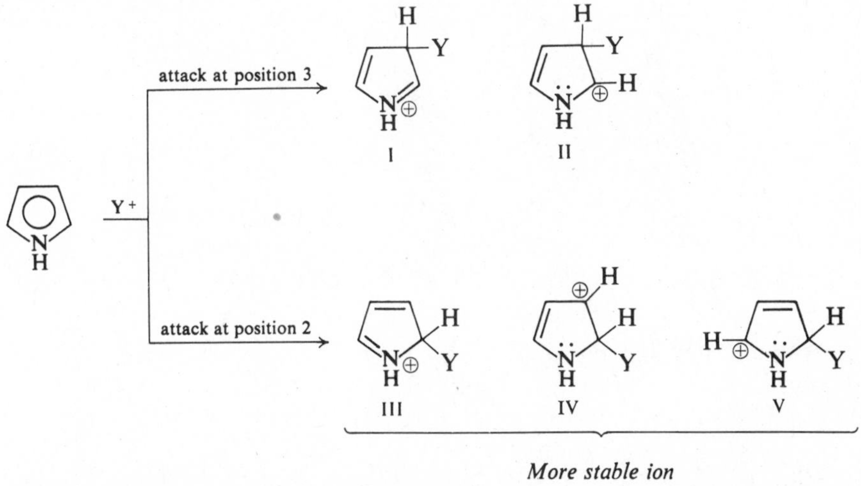

Pyrrole + Y^+ → attack at position 3 → I, II

attack at position 2 → III, IV, V

More stable ion

Viewed differently, attack at position 2 is faster because the developing positive charge is accommodated by *three* atoms of the ring instead of by only two.

Pyrrole is highly reactive, compared with benzene, because of contribution from the relatively stable structure III. In III *every atom has an octet of electrons*; nitrogen accommodates the positive charge simply by *sharing* four pairs of electrons. It is no accident that pyrrole resembles aniline in reactivity: both owe their high reactivity to the ability of nitrogen to share four pairs of electrons.

Orientation of substitution in furan and thiophene, as well as their high reactivity, can be accounted for in a similar way.

Problem 31.7 The heterocycle *indole*, commonly represented as formula VI, is found in coal tar and in orange blossoms.

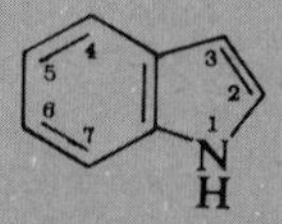

VI
Indole

It undergoes electrophilic substitution, chiefly at position 3. Account (a) for the aromatic properties of indole, and (b) for the orientation in electrophilic substitution. (*Hint:* See Sec. 30.9.)

31.5 Saturated five-membered heterocycles

Catalytic hydrogenation converts pyrrole and furan into the corresponding saturated heterocycles, *pyrrolidine* and *tetrahydrofuran*. Since thiophene poisons most catalysts, *tetrahydrothiophene* is synthesized instead from open-chain compounds.

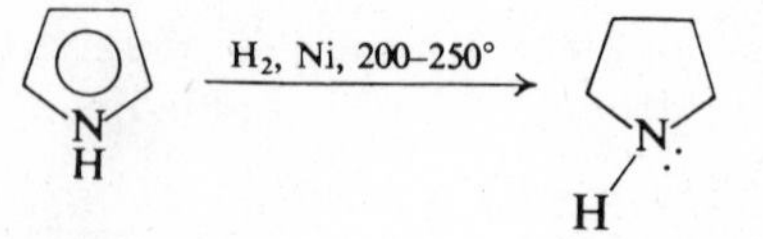

$\xrightarrow{H_2,\ Ni,\ 200\text{–}250°}$

Pyrrole ($K_b \sim 10^{-14}$) → Pyrrolidine ($K_b \sim 10^{-3}$)

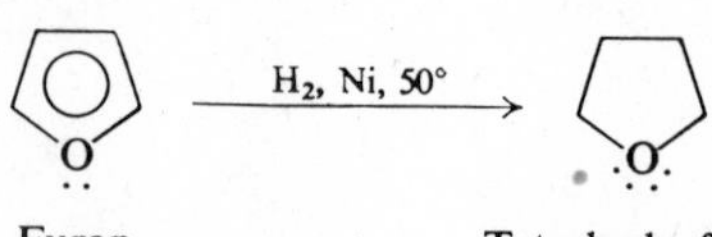

$\xrightarrow{H_2,\ Ni,\ 50°}$

Furan → Tetrahydrofuran

$BrCH_2CH_2CH_2CH_2Br + Na_2S \xrightarrow{heat}$ Tetrahydrothiophene

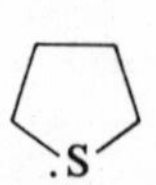

Saturation of these rings destroys the aromatic structure and, with it, the aromatic properties. Each of the saturated heterocycles has the properties we would expect of it: the properties of a secondary aliphatic amine, an aliphatic ether, or an aliphatic sulfide. With nitrogen's extra pair of electrons now available for sharing with acids, pyrrolidine ($K_b \sim 10^{-3}$) has the normal basicity of an aliphatic amine. Hydrogenation of pyrrole increases the base strength by a factor of 10^{11} (100 billion); clearly a fundamental change in structure has taken place.

Tetrahydrofuran is an important solvent, used, for example, in reductions with lithium aluminum hydride, in the preparation of arylmagnesium chlorides (Sec.

25.4), and in hydroborations. Oxidation of tetrahydrothiophene yields *tetramethylene sulfone* (or *sulfolane*), also used as a solvent (Sec. 1.21).

Tetramethylene sulfone
(Sulfolane)

We have encountered pyrrolidine as a secondary amine commonly used in making enamines (Sec. 26.8). The pyrrolidine ring occurs naturally in a number of alkaloids (Sec. 7.9), providing the basicity that gives these compounds their name (*alkali-like*).

Problem 31.8 An older process for the synthesis of both the adipic acid and the hexamethylenediamine needed in the manufacture of Nylon 66 (Sec. 32.7) started with tetrahydrofuran. Using only familiar chemical reactions, suggest possible steps in their synthesis.

Problem 31.9 Predict the products of the treatment of pyrrolidine with:

(a) aqueous HCl
(b) aqueous NaOH
(c) acetic anhydride
(d) benzenesulfonyl chloride + aqueous NaOH
(e) methyl iodide, followed by aqueous NaOH
(f) repeated treatment with methyl iodide, followed by Ag_2O and then strong heating

Problem 31.10 The alkaloid *hygrine* is found in the coca plant. Suggest a structure for it on the basis of the following evidence:

Hygrine ($C_8H_{15}ON$) is insoluble in aqueous NaOH but soluble in aqueous HCl. It does not react with benzenesulfonyl chloride. It reacts with phenylhydrazine to yield a phenylhydrazone. It reacts with NaOI to yield a yellow precipitate and a carboxylic acid ($C_7H_{13}O_2N$). Vigorous oxidation by CrO_3 converts hygrine into *hygrinic acid* ($C_6H_{11}O_2N$).

Hygrinic acid can be synthesized as follows:

$BrCH_2CH_2CH_2Br + CH(COOC_2H_5)_2^-Na^+ \longrightarrow A\ (C_{10}H_{17}O_4Br)$

$A + Br_2 \longrightarrow B\ (C_{10}H_{16}O_4Br_2)$

$B + CH_3NH_2 \longrightarrow C\ (C_{11}H_{19}O_4N)$

$C + aq.\ Ba(OH)_2 + heat \longrightarrow D \xrightarrow{HCl} E \xrightarrow{heat}$ hygrinic acid + CO_2

SIX-MEMBERED RINGS

31.6 Structure of pyridine

Of the six-membered aromatic heterocycles, we shall take up only one, **pyridine**.

Pyridine is classified as aromatic on the basis of its properties. It is flat, with bond angles of 120°; the four carbon–carbon bonds are of the same length, and so are the two carbon–nitrogen bonds. It resists addition and undergoes electrophilic substitution. Its heat of combustion indicates a resonance energy of 23 kcal/mole.

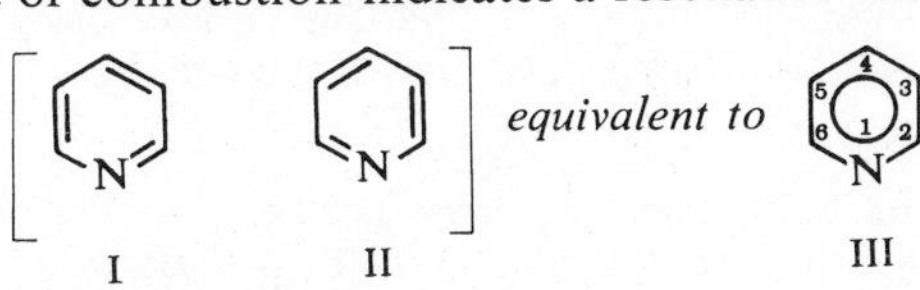

Pyridine can be considered a hybrid of the Kekulé structures I and II. We shall represent it as structure III, in which the circle represents the aromatic sextet.

In electronic configuration, the nitrogen of pyridine is considerably different from the nitrogen of pyrrole. In pyridine the nitrogen atom, like each of the carbon atoms, is bonded to other members of the ring by the use of sp^2 orbitals, and provides one electron for the π cloud. The third sp^2 orbital of each carbon atom is used to form a bond to hydrogen; the third sp^2 orbital of nitrogen simply contains a pair of electrons, which are available for sharing with acids (Fig. 31.2).

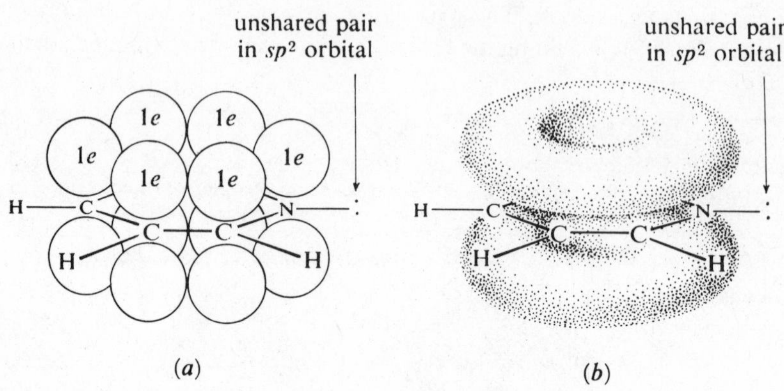

Figure 31.2. Pyridine molecule. (*a*) One electron in each *p* orbital; two electrons in sp^2 orbital of nitrogen. (*b*) The *p* orbitals overlap to form π clouds above and below plane of ring; two unshared electrons still in sp^2 orbital of nitrogen.

Because of this electronic configuration, the nitrogen atom makes pyridine a much stronger base than pyrrole, and affects the reactivity of the ring in a quite different way, as we shall see.

31.7 Source of pyridine compounds

Pyridine is found in coal tar. Along with it are found a number of methylpyridines, the most important of which are the monomethyl compounds, known as *picolines*.

Oxidation of the picolines yields the pyridinecarboxylic acids.

$$C_5H_4N\text{-}CH_3 \xrightarrow{KMnO_4} C_5H_4N\text{-}COOH$$

Picoline (2-, 3-, or 4-) → Pyridinecarboxylic acid (2-, 3-, or 4-)

The 3-isomer (*nicotinic acid* or *niacin*) is a vitamin. The 4-isomer (*isonicotinic acid*) has been used, in the form of its hydrazide, in the treatment of tuberculosis.

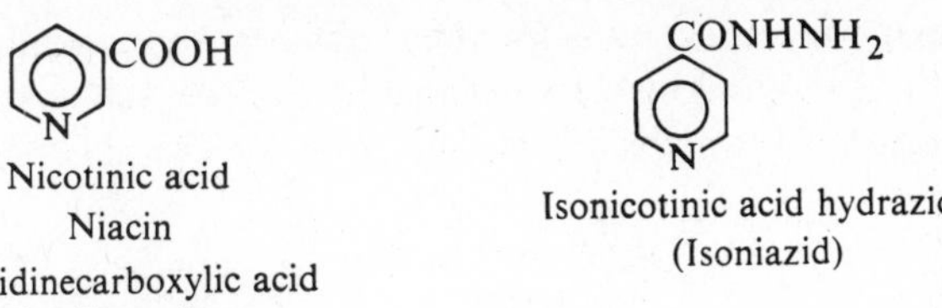

Nicotinic acid
Niacin
3-Pyridinecarboxylic acid
Anti-pellagra factor

Isonicotinic acid hydrazide
(Isoniazid)

The increasing demand for certain pyridine derivatives has led to the development of syntheses involving ring closure. For example:

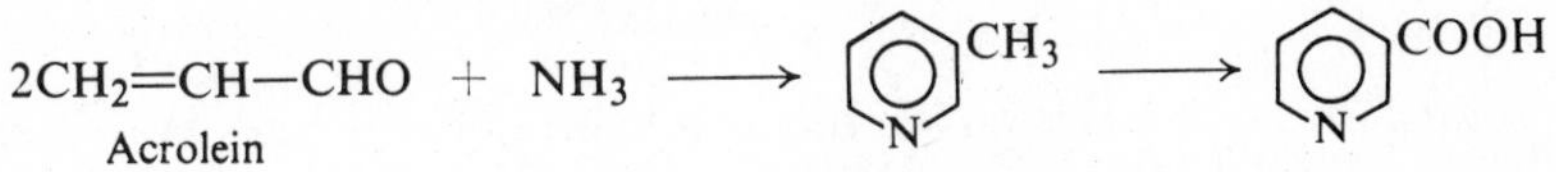

$$2CH_2{=}CH{-}CHO + NH_3 \longrightarrow \text{3-Methylpyridine} \longrightarrow \text{Nicotinic acid}$$

Acrolein; 3-Methylpyridine, β-Picoline; Nicotinic acid

31.8 Reactions of pyridine

The chemical properties of pyridine are those we would expect on the basis of its structure. The ring undergoes the substitution, both electrophilic and nucleophilic, typical of aromatic rings; our interest will lie chiefly in the way the nitrogen atom affects these reactions.

There is another set of reactions in which pyridine acts as a base or nucleophile; these reactions involve nitrogen directly and are due to its unshared pair of electrons.

31.9 Electrophilic substitution in pyridine

Toward electrophilic substitution pyridine resembles a highly deactivated benzene derivative. It undergoes nitration, sulfonation, and halogenation only under very vigorous conditions, and does not undergo the Friedel-Crafts reaction at all.

Substitution occurs chiefly at the 3- (or β-) position.

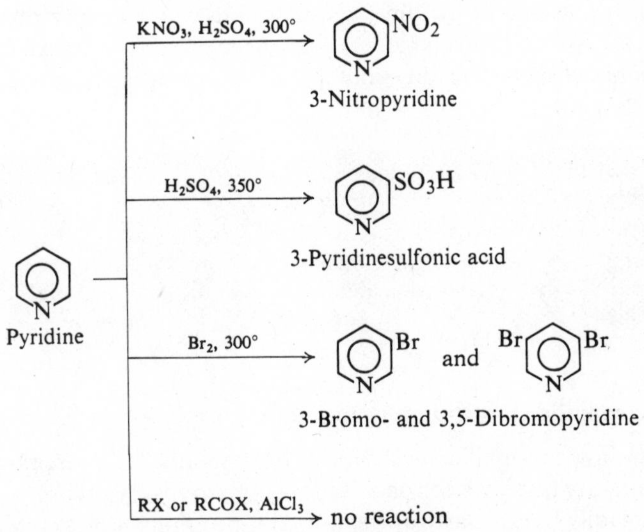

Pyridine:

- $\xrightarrow{KNO_3,\ H_2SO_4,\ 300°}$ 3-Nitropyridine (NO_2)
- $\xrightarrow{H_2SO_4,\ 350°}$ 3-Pyridinesulfonic acid (SO_3H)
- $\xrightarrow{Br_2,\ 300°}$ 3-Bromo- and 3,5-Dibromopyridine
- $\xrightarrow{RX \text{ or } RCOX,\ AlCl_3}$ no reaction

Let us see if we can account for the reactivity and orientation on our usual basis of stability of the intermediate carbonium ion. Attack at the 4-position yields a carbonium ion that is a hybrid of structures I, II, and III;

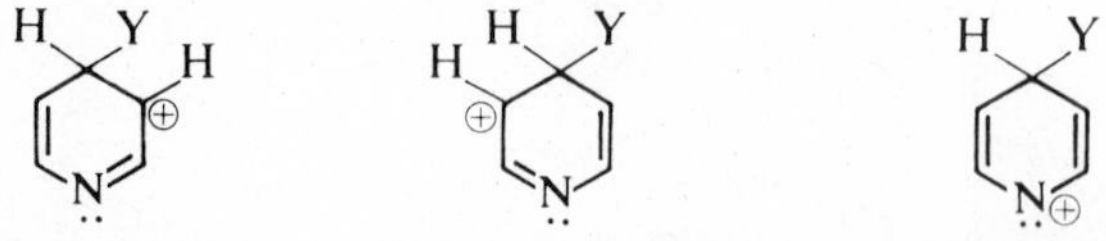

I II III

Especially unstable: nitrogen has sextet

Electrophilic attack at 4-position

Attack at the 3-position yields an ion that is a hybrid of structures IV, V, and VI.

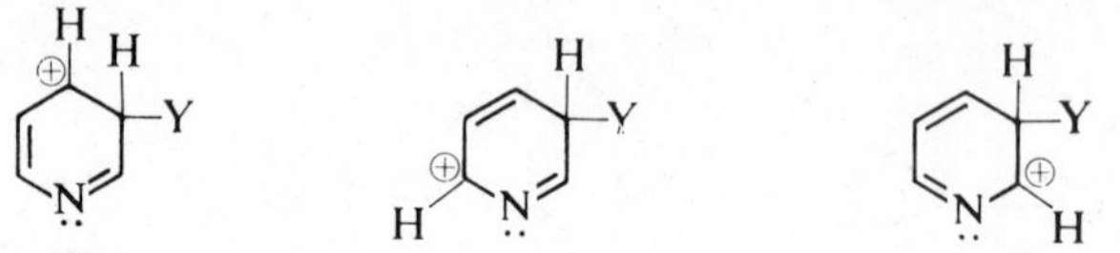

IV V VI

Electrophilic attack at 3-position

(Attack at the 2-position resembles attack at the 4-position just as *ortho* attack resembles *para* attack in the benzene series.)

All these structures are less stable than the corresponding ones for attack on benzene, because of electron withdrawal by the nitrogen atom. As a result, pyridine undergoes substitution more slowly than benzene.

Of these structures, III is *especially* unstable, since in it the electronegative nitrogen atom has only a sextet of electrons. As a result, attack at the 4-position (or 2-position) is especially slow, and substitution occurs predominantly at the 3-position.

It is important to see the difference between substitution in pyridine and substitution in pyrrole. In the case of pyrrole, a structure in which nitrogen bears a positive charge (see Sec. 31.4) is especially stable since every atom has an octet of electrons; nitrogen accommodates the positive charge simply by sharing four pairs of electrons. In the case of pyridine, a structure in which nitrogen bears a positive charge (III) is especially unstable since nitrogen has only a sextet of electrons; nitrogen *shares* electrons readily, but as an electronegative atom it resists the *removal* of electrons.

Problem 31.11 2-Aminopyridine can be nitrated or sulfonated under much milder conditions than pyridine itself; substitution occurs chiefly at the 5-position. Account for these facts.

Problem 31.12 Because of the difficulty of nitrating pyridine, 3-aminopyridine is most conveniently made via nicotinic acid. Outline the synthesis of 3-aminopyridine from β-picoline.

31.10 Nucleophilic substitution in pyridine

Here, as in electrophilic substitution, the pyridine ring resembles a benzene ring that contains strongly electron-withdrawing groups. Nucleophilic substitution takes place readily, particularly at the 2- and 4-positions. For example:

2-Bromopyridine $\xrightarrow{NH_3,\ 180–200°}$ 2-Aminopyridine

4-Chloropyridine $\xrightarrow{NH_3,\ 180–200°}$ 4-Aminopyridine

The reactivity of pyridine toward nucleophilic substitution is so great that even the powerfully basic hydride ion, $:H^-$, can be displaced. Two important examples of this reaction are amination by sodium amide (**Chichibabin reaction**), and alkylation or arylation by organolithium compounds.

Pyridine + $Na^+NH_2^-$ (Sodium amide) $\xrightarrow{heat}$ [intermediate, Na^+] $\xrightarrow{NH_3}$ 2-Aminopyridine + $Na^+NH_2^-$ + H:H

2-Aminopyridine ⇅ Sodium salt of 2-aminopyridine (NH^-Na^+) + NH_3

Pyridine + $\overset{\delta-}{C_6H_5}—\overset{\delta+}{Li}$ (Phenyllithium) $\xrightarrow{heat}$ [intermediate, Li^+] ⟶ 2-Phenylpyridine + Li:H

As we have seen (Sec. 25.8), nucleophilic aromatic substitution can take place by a mechanism that is quite analogous to the mechanism for electrophilic substitution. Reaction proceeds by two steps; the rate of the first step, formation of a charged particle, determines the rate of the overall reaction. In electrophilic substitution, the intermediate is positively charged; in nucleophilic substitution, the intermediate is negatively charged. The ability of the ring to accommodate the charge determines the stability of the intermediate and of the transition state leading to it, and hence determines the rate of the reaction.

Nucleophilic attack at the 4-position yields a carbanion that is a hybrid of structures I, II, and III:

I II III

Especially stable: negative charge on nitrogen

Nucleophilic attack at 4-position

Attack at the 3-position yields a carbanion that is a hybrid of structures IV, V, and VI:

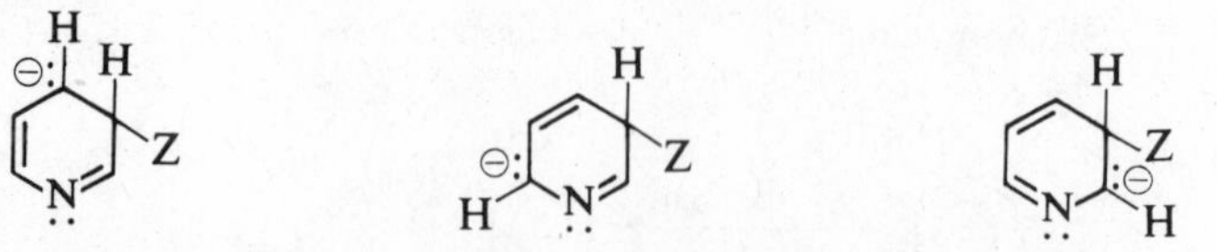

Nucleophilic attack at 3-position

IV V VI

(As before, attack at the 2-position resembles attack at the 4-position.)

All these structures are more stable than the corresponding ones for attack on a benzene derivative, because of electron withdrawal by the nitrogen atom. Structure III is *especially* stable, since the negative charge is located on the atom that can best accommodate it, the electronegative nitrogen atom. It is reasonable, therefore, that nucleophilic substitution occurs more rapidly on the pyridine ring than on the benzene ring, and more rapidly at the 2- and 4-positions than at the 3-position.

The same electronegativity of nitrogen that makes pyridine unreactive toward electrophilic substitution makes pyridine highly reactive toward nucleophilic substitution.

31.11 Basicity of pyridine

Pyridine is a base with $K_b = 2.3 \times 10^{-9}$. It is thus much stronger than pyrrole ($K_b \sim 2.5 \times 10^{-14}$) but much weaker than aliphatic amines ($K_b \sim 10^{-4}$).

Pyridine has a pair of electrons (in an sp^2 orbital) that is available for sharing with acids; pyrrole has not, and can accept an acid only at the expense of the aromatic character of the ring.

The fact that pyridine is a weaker base than aliphatic amines is more difficult to account for, but at least it fits into a pattern. Let us turn for a moment to the basicity of the carbon analogs of amines, the carbanions, and use the approach of Sec. 8.10.

Benzene is a stronger acid than an alkane, as shown by its ability to displace an alkane from its salts; this, of course, means that the phenyl anion, $C_6H_5^-$, is a weaker base than an alkyl anion, R^-.

$$\underset{\text{Stronger base}}{R:^-Na^+} + \underset{\text{Stronger acid}}{C_6H_5:H} \rightleftharpoons \underset{\text{Weaker acid}}{R:H} + \underset{\text{Weaker base}}{C_6H_5:^-Na^+}$$

In the same way, acetylene is a stronger acid than benzene, and the acetylide ion is a weaker base than the phenyl anion.

$$\underset{\text{Stronger base}}{C_6H_5:^-Na^+} + \underset{\text{Stronger acid}}{HC{\equiv}C:H} \rightleftharpoons \underset{\text{Weaker acid}}{C_6H_5:H} + \underset{\text{Weaker base}}{HC{\equiv}C:^-Na^+}$$

Thus we have the following sequences of acidity of hydrocarbons and basicity of their anions:

Relative acidity: $HC{\equiv}C:H > C_6H_5:H > R:H$

Relative basicity: $HC{\equiv}C:^- < C_6H_5:^- < R:^-$

A possible explanation for these sequences can be found in the electronic configuration of the carbanions. In the alkyl, phenyl, and acetylide anions, the unshared pair of electrons occupies respectively an sp^3, an sp^2, and an sp orbital. The availability of this pair for sharing with acids determines the basicity of the particular anion. As we proceed along the series sp^3, sp^2, sp, the p character of the orbital decreases and the s character increases. Now, an electron in a p orbital is at some distance from the nucleus and is held relatively loosely; an electron in an s orbital, on the other hand, is close to the nucleus and is held more tightly. Of the three anions, the alkyl ion is the strongest base since its pair of electrons is held most loosely, in an sp^3 orbital. The acetylide ion is the weakest base since its pair of electrons is held most tightly, in an sp orbital.

Pyridine bears the same relationship to an aliphatic amine as the phenyl anion bears to an alkyl anion. The pair of electrons that gives pyridine its basicity occupies an sp^2 orbital; it is held more tightly and is less available for sharing with acids than the pair of electrons of an aliphatic amine, which occupies an sp^3 orbital.

Problem 31.13 Predict the relative basicities of amines (RCH_2NH_2), imines ($RCH{=}NH$), and nitriles ($RC{\equiv}N$).

Pyridine is widely used in organic chemistry as a water-soluble base, as, for example, in the Schotten-Baumann acylation procedure (Sec. 20.8).

Problem 31.14 Ethyl bromosuccinate is converted into the unsaturated ester ethyl fumarate by the action of pyridine. What is the function of the pyridine? What advantage does it have here over the usual alcoholic KOH?

Like other amines, pyridine has nucleophilic properties, and reacts with alkyl halides to form quaternary ammonium salts.

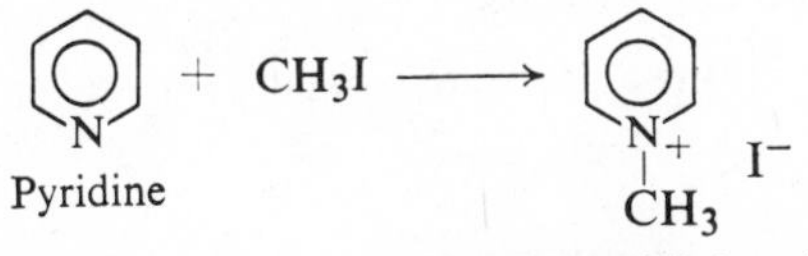

Pyridine + CH_3I ⟶ N-Methylpyridinium iodide (Pyridine methiodide)

N
Pyridine

N^+ I^-
CH_3

N-Methylpyridinium iodide
(Pyridine methiodide)

Problem 31.15 Like any other tertiary amine, pyridine can be converted (by peroxybenzoic acid) into its N-oxide.

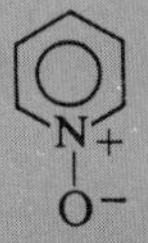

N^+
O^-

Pyridine N-oxide

In contrast to pyridine itself, pyridine N-oxide readily undergoes nitration, chiefly in the 4-position. How do you account for this reactivity and orientation?

Problem 31.16 Pyridine N-oxides not only are reactive toward electrophilic substitution, but also seem to be reactive toward nucleophilic substitution, particularly at the 2- and 4-positions. For example, treatment of 4-nitropyridine N-oxide with hydrobromic acid gives 4-bromopyridine N-oxide. How do you account for this reactivity and orientation?

Problem 31.17 The oxygen of pyridine N-oxide is readily removed by treatment with PCl_3. Suggest a practical route to 4-nitropyridine. To 4-bromopyridine.

31.12 Reduction of pyridine

Catalytic hydrogenation of pyridine yields the aliphatic heterocyclic compound **piperidine**, $C_5H_{11}N$.

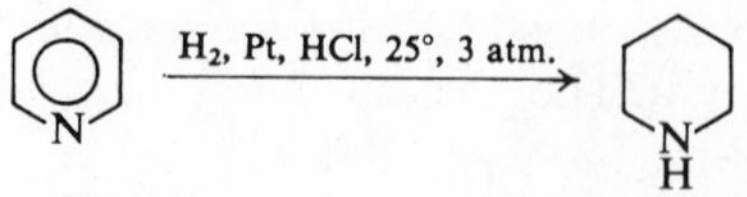

$$\text{Pyridine } (K_b = 2.3 \times 10^{-9}) \xrightarrow{H_2,\ Pt,\ HCl,\ 25°,\ 3\ atm.} \text{Piperidine } (K_b = 2 \times 10^{-3})$$

Piperidine ($K_b = 2 \times 10^{-3}$) has the usual basicity of a secondary aliphatic amine. Like pyridine, it is often used as a basic catalyst in such reactions as the Knoevenagel reaction (Problem 21.22 (f), p. 714) or Michael addition (Sec. 27.7).

Like the pyrrolidine ring, the piperidine and pyridine rings are found in a number of alkaloids, including *nicotine*, *strychnine*, *cocaine*, and *reserpine* (see p. 1004).

Problem 31.18 Why can piperidine not be used in place of pyridine in the Schotten-Baumann procedure?

FUSED RINGS

31.13 Quinoline. The Skraup synthesis

Quinoline, C_9H_7N, contains a benzene ring and a pyridine ring fused as shown in I.

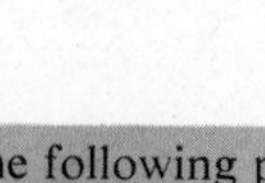

I

Quinoline

($K_b = 3 \times 10^{-10}$)

In general, its properties are the ones we would expect from what we have learned about pyridine and naphthalene.

Problem 31.19 Account for the following properties of quinoline:

(a) Treatment with nitric and sulfuric acids gives 5- and 8-nitroquinolines; treatment with fuming sulfuric acid gives 5- and 8-quinolinesulfonic acids.
(b) Oxidation by $KMnO_4$ gives 2,3-pyridinedicarboxylic acid (*quinolinic acid*).
(c) Treatment with sodamide gives 2-aminoquinoline; treatment with alkyllithium compounds gives 2-alkylquinolines.

Problem 31.20 *8-Hydroxyquinoline* (8-quinolinol) is a reagent in inorganic analysis. Suggest a method of synthesizing it.

Quinoline is found in coal tar. Although certain derivatives of quinoline can be made from quinoline itself by substitution, most are prepared from benzene derivatives by ring closure.

Perhaps the most generally useful method for preparing substituted quinolines is the **Skraup synthesis**. In the simplest example, quinoline itself is obtained from the reaction of aniline with glycerol, concentrated sulfuric acid, nitrobenzene, and ferrous sulfate.

$$\underset{\text{Aniline}}{C_6H_5NH_2} + \underset{\text{Glycerol}}{CH_2OH{-}CHOH{-}CH_2OH} + \underset{\text{Nitrobenzene}}{C_6H_5NO_2} \xrightarrow{H_2SO_4,\ FeSO_4,\ \text{heat}} \text{Quinoline} + C_6H_5NH_2 + H_2O$$

The following steps seem to be involved:

(1) Dehydration of glycerol by hot sulfuric acid to yield the unsaturated aldehyde acrolein:

$$\underset{\text{Glycerol}}{\underset{OH}{CH_2}{-}\underset{OH}{CH}{-}\underset{OH}{CH_2}} \xrightarrow{H_2SO_4,\ \text{heat}} \underset{\text{Acrolein}}{CH_2{=}CH{-}CHO} + 2H_2O$$

(2) Nucleophilic addition of aniline to acrolein to yield β-(phenylamino)propionaldehyde:

$$\underset{\text{Aniline}}{C_6H_5NH_2} + \underset{\text{Acrolein}}{H{-}C(=O){-}CH{=}CH_2} \longrightarrow \underset{\beta\text{-(Phenylamino)propionaldehyde}}{C_6H_5NH{-}CH_2{-}CH_2{-}C(=O)H}$$

(3) Electrophilic attack on the aromatic ring by the electron-deficient carbonyl carbon of the protonated aldehyde (this is the actual ring-closing step):

$$C_6H_5NH{-}CH_2{-}CH_2{-}CHO + H^+ \rightleftarrows C_6H_5NH{-}CH_2{-}CH_2{-}CH{=}\overset{\oplus}{O}H \longrightarrow \text{(4-hydroxy-1,2,3,4-tetrahydroquinoline)} + H^+$$

$$\downarrow$$

$$\underset{\text{1,2-Dihydroquinoline}}{\text{1,2-dihydroquinoline}} + H_2O$$

(4) Oxidation by nitrobenzene resulting in the aromatization of the newly formed ring:

$$3\ \text{1,2-Dihydroquinoline} + C_6H_5NO_2 \xrightarrow{H^+} 3\ \text{Quinoline} + C_6H_5NH_2 + 2H_2O$$

1,2-Dihydroquinoline   Quinoline

Ferrous sulfate in some way moderates the otherwise very vigorous reaction.

Thus we see that what at first appears to be a complicated reaction is actually a sequence of simple steps involving familiar, fundamental types of reactions: acid-catalyzed dehydration, nucleophilic addition to an α,β-unsaturated carbonyl compound, electrophilic aromatic substitution, and oxidation.

The components of the basic synthesis can be modified to yield a wide variety of quinoline derivatives. For example:

aniline + crotonaldehyde ⟶ 2-methylquinoline (quinaldine)

3-nitro-4-aminoanisole + glycerol ⟶ 6-methoxy-8-nitroquinoline

2-aminonaphthalene + glycerol ⟶ 5,6-Benzoquinoline (1-Azaphenanthrene)

Nitrobenzene is often replaced as oxidizing agent by arsenic acid, H_3AsO_4, which usually gives a less violent reaction; vanadium pentoxide is sometimes added as a catalyst. Sulfuric acid can be replaced by phosphoric acid or other acids.

Problem 31.21 Show all steps in the Skraup syntheses mentioned above.

Problem 31.22 The dehydration of glycerol to yield acrolein involves acid-catalyzed dehydration and keto–enol tautomerization. Outline the possible steps in the dehydration. (*Hint:* Which —OH is easier to eliminate, a primary or a secondary?)

Problem 31.23 What is the product of the application of the Skraup synthesis to (a) *o*-nitroaniline, (b) *o*-aminophenol, (c) *o*-phenylenediamine, (d) *m*-phenylenediamine, (e) *p*-toluidine?

Problem 31.24 Outline the synthesis of 6-bromoquinoline. Of 8-methylquinoline.

Problem 31.25 In the **Doebner-von Miller** modification of the Skraup synthesis, aldehydes, ketones, or mixtures of aldehydes and ketones replace the glycerol. If acetaldehyde is used, for example, the product from aniline is 2-methylquinoline (*quinaldine*). (a) Account for its formation. (b) Predict the product if methyl vinyl ketone were used. (c) If a mixture of benzaldehyde and pyruvic acid, $CH_3COCOOH$, were used.

Problem 31.26 Account for the formation of 2,4-dimethylquinoline from aniline and acetylacetone (2,4-pentanedione) by the Doebner-von Miller synthesis. (*Hint:* See Problem 21, p. 724.)

31.14 Isoquinoline. The Bischler-Napieralski synthesis

Isoquinoline, C_9H_7N, contains a benzene ring and a pyridine ring fused as shown in I:

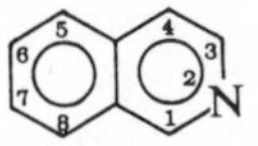

I
Isoquinoline
($K_b = 1.1 \times 10^{-9}$)

Isoquinoline, like quinoline, has the properties we would expect from what we know about pyridine and naphthalene.

Problem 31.27 Account for the following properties of isoquinoline. (*Hint:* Review orientation in β-substituted naphthalenes, Sec. 30.13.)

(a) Nitration gives 5-nitroisoquinoline.

(b) Treatment with potassium amide, KNH_2, gives 1-aminoisoquinoline, and treatment with alkyllithium compounds gives 1-alkylisoquinoline; the 3-substituted products are not obtained.

(c) 1-Methylisoquinoline reacts with benzaldehyde to yield compound II, whereas 3-methylisoquinoline undergoes no reaction. (*Hint:* See Problem 21.22 (c), p. 714.)

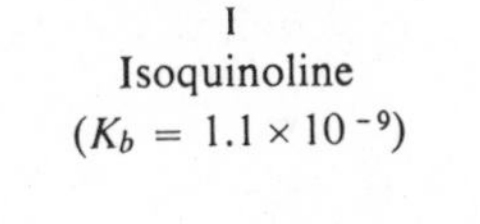

$CH{=}CHC_6H_5$
II

An important method for making derivatives of isoquinoline is the **Bischler-Napieralski synthesis**. Acyl derivatives of β-phenylethylamine are cyclized by treatment with acids (often P_2O_5) to yield dihydroisoquinolines, which can then be aromatized.

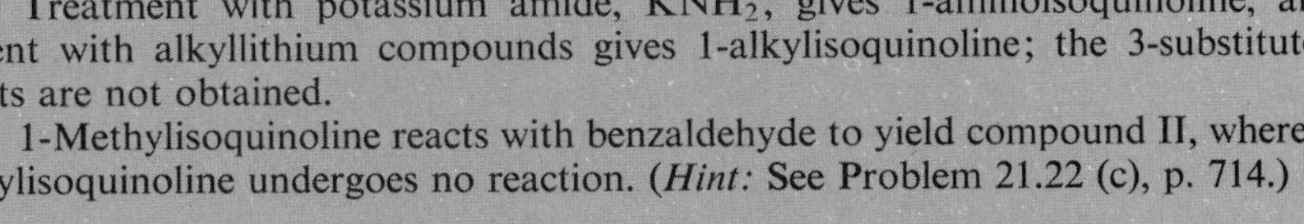

N-(2-phenylethyl)acetamide $\xrightarrow{P_2O_5,\ heat}$ 1-Methyl-3,4-dihydroisoquinoline $\xrightarrow{Pd,\ heat}$ 1-Methylisoquinoline + H_2

Problem 31.28 To what general class of reactions does the ring closure belong? What is the function of the acid? (Check your answers in Sec. 32.7.)

Problem 31.29 Outline the synthesis of N-(2-phenylethyl)acetamide from toluene and aliphatic and inorganic reagents.

PROBLEMS

1. Give structures and names of the principal products from the reaction (if any) of pyridine with:

(a) Br_2, 300°
(b) H_2SO_4, 350°
(c) acetyl chloride, $AlCl_3$
(d) KNO_3, H_2SO_4, 300°
(e) $NaNH_2$, heat
(f) C_6H_5Li
(g) dilute HCl
(h) dilute NaOH
(i) acetic anhydride
(j) benzenesulfonyl chloride
(k) ethyl bromide
(l) benzyl chloride
(m) peroxybenzoic acid
(n) peroxybenzoic acid, then HNO_3, H_2SO_4
(o) H_2, Pt

2. Give structures and names of the principal products from each of the following reactions:

(a) thiophene + conc. H_2SO_4
(b) thiophene + acetic anhydride, $ZnCl_2$
(c) thiophene + acetyl chloride, $TiCl_4$
(d) thiophene + fuming nitric acid in acetic anhydride
(e) product of (d) + Sn, HCl
(f) thiophene + one mole Br_2
(g) product of (f) + Mg; then CO_2; then H^+
(h) pyrrole + pyridine:SO_3
(i) pyrrole + diazotized sulfanilic acid
(j) product of (i) + $SnCl_2$
(k) pyrrole + H_2, Ni $\longrightarrow$ C_4H_9N
(l) furfural + acetone + base
(m) quinoline + HNO_3/H_2SO_4
(n) quinoline N-oxide + HNO_3/H_2SO_4
(o) isoquinoline + *n*-butyllithium

3. Pyrrole can be reduced by zinc and acetic acid to a *pyrroline*, C_4H_7N. (a) What structures are possible for this pyrroline?

(b) On the basis of the following evidence which structure must the pyrroline have?

pyrroline + O_3; then H_2O; then H_2O_2 $\longrightarrow$ A ($C_4H_7O_4N$)
chloroacetic acid + NH_3 $\longrightarrow$ B ($C_2H_5O_2N$)
B + chloroacetic acid $\longrightarrow$ A

4. Furan and its derivatives are sensitive to protic acids. The following reactions illustrate what happens.

2,5-dimethylfuran + dilute H_2SO_4 $\longrightarrow$ C ($C_6H_{10}O_2$)
C + NaOI $\longrightarrow$ succinic acid

(a) What is C? (b) Outline a likely series of steps for its formation from 2,5-dimethylfuran.

5. Pyrrole reacts with formaldehyde in hot pyridine to yield a mixture of products from which there can be isolated a small amount of a compound of formula $(C_5H_5N)_4$. Suggest a possible structure for this compound. (*Hint:* See Sec. 32.7 and p. 1004.)

6. There are three isomeric pyridinecarboxylic acids, $(C_5H_4N)COOH$: D, m.p. 137°; E, m.p. 234–7°; and F, m.p. 317°. Their structures were proved as follows:

quinoline + $KMnO_4$, OH^- $\longrightarrow$ a diacid ($C_7H_5O_4N$) $\xrightarrow{\text{heat}}$ E, m.p. 234–7°
isoquinoline + $KMnO_4$, OH^- $\longrightarrow$ a diacid ($C_7H_5O_4N$) $\xrightarrow{\text{heat}}$ E, m.p. 234–7° and F, m.p. 317°

What structures should be assigned to D, E, and F?

7. (a) What structures are possible for G?

m-toluidine + glycerol $\xrightarrow{\text{Skraup}}$ G ($C_{10}H_9N$)

(b) On the basis of the following evidence which structure must G actually have?

2,3-diaminotoluene + glycerol $\xrightarrow{\text{Skraup}}$ H ($C_{10}H_{10}N_2$)
H + $NaNO_2$, HCl; then H_3PO_2 $\longrightarrow$ G

8. Outline all steps in a possible synthesis of each of the following from benzene, toluene, and any needed aliphatic and inorganic reagents:

(a) 1-phenylisoquinoline
(b) 1-benzylisoquinoline
(c) 1,5-dimethylisoquinoline
(d) 6-nitroquinoline
(e) 2-methyl-6-quinolinecarboxylic acid
(f) 1,8-diazaphenanthrene (*Hint:* Use the Skraup synthesis twice.)

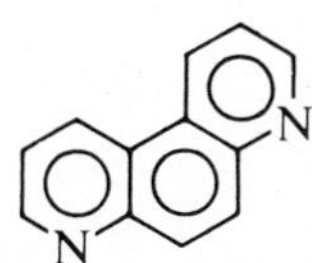

1,8-Diazaphenanthrene

9. Outline all steps in each of the following syntheses, using any other needed reagents:

(a) β-cyanopyridine from β-picoline
(b) 2-methylpiperidine from pyridine
(c) 5-aminoquinoline from quinoline
(d) ethyl 5-nitro-2-furoate from furfural
(e) furylacrylic acid, 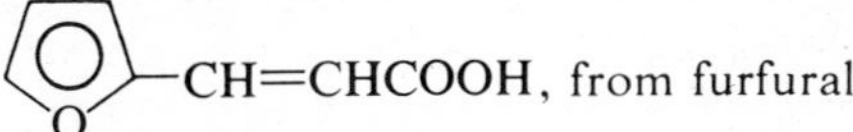 CH=CHCOOH, from furfural

(f) 1,2,5-trichloropentane from furfural
(g) 3-indolecarboxaldehyde from indole

10. Give the structures of compounds I through JJ formed in the following syntheses of heterocyclic systems.

(a) ethyl malonate + urea, base, heat $\longrightarrow$ I ($C_4H_4O_3N_2$), a *pyrimidine* (1,3-diazine)
(b) 2,5-hexanedione + $H_2N—NH_2$ $\longrightarrow$ J ($C_6H_{10}N_2$)
J + air $\longrightarrow$ K ($C_6H_8N_2$), a *pyridazine* (1,2-diazine)
(c) 2,4-pentanedione + $H_2N—NH_2$ $\longrightarrow$ L ($C_5H_8N_2$), a *pyrazole*
(d) 2,3-butanedione + *o*-$C_6H_4(NH_2)_2$ $\longrightarrow$ M ($C_{10}H_{10}N_2$), a *quinoxaline*
(e) ethylene glycol + phosgene $\longrightarrow$ N ($C_3H_4O_3$), a *1,3-dioxolanone*
(f) *o*-aminobenzoic acid + chloroacetic acid $\longrightarrow$ O ($C_9H_9O_4N$)
O + base, strong heat $\longrightarrow$ P (C_8H_7ON), *indoxyl*, an intermediate in the synthesis of indigo
(g) aminoacetone $\longrightarrow$ Q ($C_6H_{10}N_2$)
Q + air $\longrightarrow$ R ($C_6H_8N_2$), a *pyrazine* (1,4-diazine)
(h) ethylenediamine + ethyl carbonate $\longrightarrow$ S ($C_3H_6ON_2$), an *imidazolidone*
(i) *o*-$C_6H_4(NH_2)_2$ + acetic acid, strong heat $\longrightarrow$ T ($C_8H_8N_2$), a *benzimidazole*
(j) ethyl *o*-aminobenzoate + malonic ester $\longrightarrow$ U ($C_{14}H_{17}O_5N$), insoluble in dilute acid
U $\xrightarrow{NaOC_2H_5}$ V ($C_{12}H_{11}O_4N$)
V + acid, warm $\longrightarrow$ W ($C_9H_7O_2N$), a *quinoline*
(k) repeat (j) starting with ethyl 3-amino-2-pyridinecarboxylate $\longrightarrow$ a *1,5-diazanaphthalene*
(l) benzalacetophenone + KCN + acetic acid $\longrightarrow$ X ($C_{16}H_{13}ON$)
X + CH_3OH, H^+, H_2O $\longrightarrow$ Y ($C_{17}H_{16}O_3$) + NH_4^+
Y + phenylhydrazine $\longrightarrow$ Z ($C_{22}H_{18}ON_2$), a *dihydro-1,2-diazine*

(m) acrylic acid + $H_2N—NH_2$ ⟶ AA ($C_3H_8O_2N_2$) ⟶ BB ($C_3H_6ON_2$), a *pyrazolidone*

(n) *o*-$C_6H_4(NH_2)_2$ + glycerol $\xrightarrow{\text{Skraup}}$ CC ($C_{12}H_8N_2$), a *4,5-diazaphenanthrene*

(o) di(*o*-nitrophenyl)acetylene + Br_2 ⟶ DD ($C_{14}H_8O_4N_2Br_2$)

DD + Sn, HCl ⟶ EE ($C_{14}H_{12}N_2Br_2$)

EE $\xrightarrow{\text{warm}}$ [FF ($C_{14}H_{11}N_2Br$)] ⟶ GG ($C_{14}H_{10}N_2$), which contains four fused aromatic rings

(p) *m*-$ClC_6H_4CH_2CH_2CH_2NHCH_3$ + C_6H_5Li ⟶ HH ($C_{10}H_{13}N$), a *tetrahydroquinoline*

(q) *o*-$ClC_6H_4NHCOC_6H_5$ + KNH_2/NH_3 ⟶ II ($C_{13}H_9ON$), a *benzoxazole*

(r) *trans*-I + base ⟶ JJ ($C_{13}H_{15}ON$), an *oxazoline*

OTs

$NHCOC_6H_5$

I

(s) How do you account for the fact that *cis*-I undergoes reaction (r) much more slowly than *trans*-I?

11. The structure of *papaverine*, $C_{20}H_{21}O_4N$, one of the opium alkaloids, has been established by the following synthesis:

3,4-dimethoxybenzyl chloride + KCN ⟶ KK ($C_{10}H_{11}O_2N$)
KK + hydrogen, Ni ⟶ LL ($C_{10}H_{15}O_2N$)
KK + aqueous acid, heat ⟶ MM $\xrightarrow{PCl_5}$ NN ($C_{10}H_{11}O_3Cl$)
LL + NN ⟶ OO ($C_{20}H_{25}O_5N$)
OO + P_2O_5, heat ⟶ PP ($C_{20}H_{23}O_4N$)
PP + Pd, 200° ⟶ papaverine

12. *Plasmochin* (also called *Pamaquine*), a drug effective against malaria, has been synthesized as follows:

ethylene oxide + diethylamine ⟶ QQ ($C_6H_{15}ON$)
QQ + $SOCl_2$ ⟶ RR ($C_6H_{14}NCl$)
RR + sodioacetoacetic ester ⟶ SS ($C_{12}H_{23}O_3N$)
SS + dilute H_2SO_4, warm ⟶ TT ($C_9H_{19}ON$) + CO_2 + C_2H_5OH
TT + H_2, Ni ⟶ UU ($C_9H_{21}ON$)
UU + conc. HBr ⟶ VV ($C_9H_{20}NBr$)
4-amino-3-nitroanisole + glycerol $\xrightarrow{\text{Skraup}}$ WW ($C_{10}H_8O_3N_2$)
WW + Sn + HCl ⟶ XX ($C_{10}H_{10}ON_2$)
VV + XX ⟶ Plasmochin ($C_{19}H_{29}ON_3$)

What is the most likely structure of Plasmochin?

13. (−)-*Nicotine*, the alkaloid in tobacco, can be synthesized in the following way:

nicotinic acid + $SOCl_2$, heat ⟶ nicotinoyl chloride (C_6H_4ONCl)
nicotinoyl chloride + $C_2H_5OCH_2CH_2CH_2CdCl$ ⟶ YY ($C_{11}H_{15}O_2N$)
YY + NH_3, H_2, catalyst ⟶ ZZ ($C_{11}H_{18}ON_2$)
ZZ + HBr + strong heat ⟶ AAA ($C_9H_{12}N_2$) + ethyl bromide
AAA + CH_3I, NaOH ⟶ (±)-nicotine ($C_{10}H_{14}N_2$)
(±)-nicotine + (+)-tartaric acid ⟶ BBB and CCC (both $C_{14}H_{20}O_6N_2$)
BBB + NaOH ⟶ (−)-nicotine + sodium tartrate

What is the structure of (±)-nicotine? Write equations for all the above reactions.

14. The red and blue colors of many flowers and fruits are due to the *anthocyanins*, glycosides of pyrylium salts. The parent structure of the pyrylium salts is *flavylium chloride*, which can be synthesized as follows:

salicylaldehyde + acetophenone $\xrightarrow{\text{aldol}}$ DDD ($C_{15}H_{12}O_2$)
DDD + HCl $\longrightarrow$ flavylium chloride, a salt containing three aromatic rings

Flavylium chloride

(a) What is the structure of DDD? (b) Outline a likely series of steps leading from DDD to flavylium chloride. (c) Account for the aromatic character of the fused ring system.

15. *Tropinic acid*, $C_8H_{13}O_4N$, is a degradation product of atropine, an alkaloid of the deadly nightshade, *Atropa belladonna*. It has a neutralization equivalent of 94 ± 1. It does not react with benzenesulfonyl chloride, cold dilute $KMnO_4$, or Br_2/CCl_4. Exhaustive methylation gives the following results:

tropinic acid + CH_3I $\longrightarrow$ EEE ($C_9H_{16}O_4NI$)
EEE + Ag_2O, then strong heat $\longrightarrow$ FFF ($C_9H_{15}O_4N$)
FFF + CH_3I $\longrightarrow$ GGG ($C_{10}H_{18}O_4NI$)
GGG + Ag_2O, then strong heat $\longrightarrow$ HHH ($C_7H_8O_4$) + $(CH_3)_3N$ + H_2O
HHH + H_2, Ni $\longrightarrow$ heptanedioic acid (pimelic acid)

(a) What structures are likely for tropinic acid?
(b) Tropinic acid is formed by oxidation with CrO_3 of *tropinone*, whose structure has been shown by synthesis to be

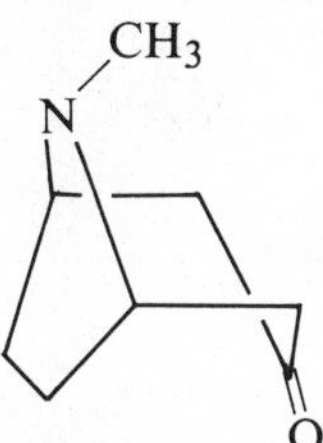

Tropinone

Now what is the most likely structure for tropinic acid?

16. *Tropilidene*, 1,3,5-cycloheptatriene, has been made from tropinone (Problem 15). Show how this might have been done. (*Hint:* See Problem 23, p. 782.)

17. Reduction of tropinone (Problem 15) gives *tropine* and *pseudotropine*, both $C_8H_{15}ON$. When heated with base, tropine is converted into pseudotropine. Give likely structures for tropine and pseudotropine, and explain your answer.

18. *Arecaidine*, $C_7H_{11}O_2N$, an alkaloid of betel nut, has been synthesized in the following way:

ethyl acrylate + NH_3 $\xrightarrow{\text{Michael}}$ III ($C_5H_{11}O_2N$)
III + ethyl acrylate $\xrightarrow{\text{Michael}}$ JJJ ($C_{10}H_{19}O_4N$)
JJJ + sodium ethoxide $\xrightarrow{\text{Dieckmann}}$ KKK ($C_8H_{13}O_3N$)
KKK + benzoyl chloride $\longrightarrow$ LLL ($C_{15}H_{17}O_4N$)
LLL + H_2, Ni $\longrightarrow$ MMM ($C_{15}H_{19}O_4N$)
MMM + acid, heat $\longrightarrow$ NNN ($C_6H_9O_2N$), *guvacine*, another betel nut alkaloid + C_6H_5COOH + C_2H_5OH
NNN + CH_3I $\longrightarrow$ arecaidine ($C_7H_{11}O_2N$)

(a) What is the most likely structure of arecaidine? Of guvacine?
(b) What will guvacine give upon dehydrogenation?

19. Give the structures of compounds OOO through UUU. (*Hint:* Sec. 32.7.)

thiophene + 3-hexanone + H_2SO_4 ⟶ OOO ($C_{14}H_{18}S_2$)
OOO + $(CH_3CO)_2O$ + $HClO_4$ ⟶ PPP ($C_{16}H_{20}OS_2$)
PPP + N_2H_4 + KOH + heat ⟶ QQQ ($C_{16}H_{22}S_2$)
QQQ + $C_6H_5N(CH_3)CHO$ ⟶ RRR ($C_{17}H_{22}OS_2$), an aldehyde
RRR + Ag_2O ⟶ SSS ($C_{17}H_{22}O_2S_2$)
SSS *was resolved*
(+)-SSS + Cu, quinoline, heat ⟶ CO_2 + (+)-TTT ($C_{16}H_{22}S_2$)
(+)-TTT + H_2/Ni ⟶ UUU ($C_{16}H_{34}$), *optically inactive*

What is the significance of the optical inactivity of UUU?

20. (a) Account for the aromatic properties of the imidazole ring.

(b) Arrange the nitrogen atoms of *histamine* (the substance responsible for many allergenic reactions) in order of their expected basicity, and account for your answer.

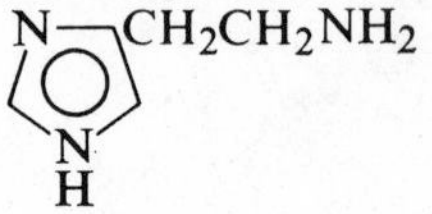

N—$CH_2CH_2NH_2$
N
H

Histamine

21. When heated in solution, 2-pyridinecarboxylic acid (II) loses carbon dioxide and forms pyridine. The rate of this decarboxylation is slowed down by addition of either acid or base. When decarboxylation is carried out in the presence of a ketone, R_2CO, there is obtained not only pyridine but also the tertiary alcohol III. The N-methyl derivative (IV) is decarboxylated much faster than II.

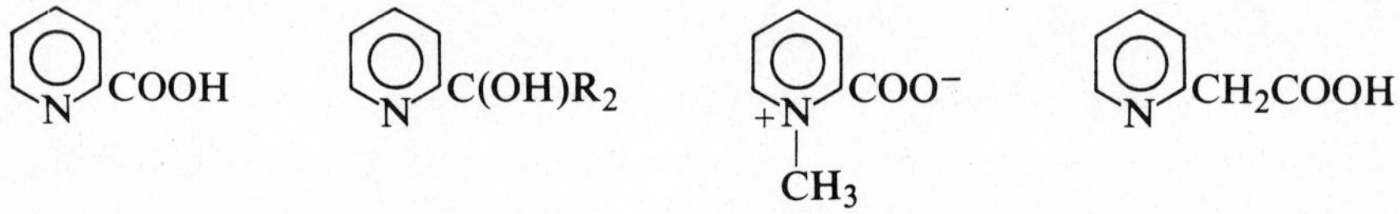

II	III	IV	V
N, COOH	N, $C(OH)R_2$	$+N$, COO^-, CH_3	N, CH_2COOH

(a) Show all steps in the most likely mechanism for decarboxylation of II. Show how this mechanism is consistent with each of the above facts.

(b) In the decarboxylation of the isomeric pyridinecarboxylic acids (II and its isomers), the order of reactivity is:

$$2 > 3 > 4$$

In the decarboxylation of the isomeric pyridineacetic acids (V and its isomers), on the other hand, the order of reactivity is:

$$2 \text{ or } 4 > 3$$

How do you account for each order of reactivity? Why is there a difference between the two sets of acids? (The same mechanism seems to be involved in both cases.)

Chapter 32 Macromolecules. Polymers and Polymerization

32.1 Macromolecules

So far, our study of organic chemistry has dealt mainly with rather small molecules, containing perhaps as many as 50 to 75 atoms. But there also exist enormous molecules called *macromolecules*, which contain hundreds of thousands of atoms. Some of these are naturally occurring, and make up classes of compounds that are, quite literally, vital: the *polysaccharides* starch and cellulose, which provide us with food, clothing, and shelter (Chap. 35); *proteins*, which constitute much of the animal body, hold it together, and run it (Chap. 36); and *nucleic acids*, which control heredity on the molecular level (Chap. 37).

Macromolecules can be man-made, too. The first syntheses were aimed at making substitutes for the natural macromolecules, rubber and silk; but a vast technology has grown up that now produces hundreds of substances that have no natural counterparts. Synthetic macromolecular compounds include: **elastomers**, which have the particular kind of elasticity characteristic of rubber; **fibers**, long, thin, and threadlike, with the great strength *along the fiber* that characterizes cotton, wool, and silk; and **plastics**, which can be extruded as sheets or pipes, painted on surfaces, or molded to form countless objects. We wear these man-made materials, eat and drink from them, sleep between them, sit and stand on them; turn knobs, pull switches, and grasp handles made of them; with their help we hear sounds and see sights remote from us in time and space; we live in houses and move about in vehicles that are increasingly made of them.

We sometimes deplore the resistance to the elements of these seemingly all too immortal materials, and fear that civilization may some day be buried beneath a pile of plastic debris—plastic cigar tips have been found floating in the Sargasso Sea—but with them we can do things never before possible. By use of plastics, blind people can be made to see, and cripples to walk; heart valves can be repaired and arteries patched; damaged tracheas, larynxes, and ureters can be replaced, and some day, perhaps, entire hearts. These materials protect us against heat and cold, electric shock and fire, rust and decay. As tailor-made solvents, they may soon be used to extract fresh water from the sea. Surely the ingenuity that has produced these substances can devise ways of disposing of the waste they create: the problem is not one of technology, but of sociology and, ultimately, of politics.

In this chapter, we shall be first—and chiefly—concerned with the chemical reactions by which macromolecules are formed, and the structures that these reactions produce. Then, we shall see how these structures lead to the properties on which the use of the macromolecules depend: why rubber is elastic, for example, and why nylon is a strong fiber. In later chapters, we shall take up the natural macromolecules—polysaccharides, proteins, and nucleic acids—and study them in much the same way.

In all this, we must remember that what makes macromolecules special is, of course, their great size. This great size permits a certain complexity of structure, not just on the molecular level, but on a *secondary* level that involves the disposition of molecules with respect to each other. Are the molecules stretched out neatly alongside one another, or coiled up independently? What forces act between different molecules? What happens to a collection of giant molecules when it is heated, or cooled, or stretched? As we shall see, the answers to questions like these are found ultimately in structure as we have known it: the nature of functional groups and substituents, their sequence in the molecule, and their arrangement in space.

32.2 Polymers and polymerization

Macromolecules, both natural and man-made, owe their great size to the fact that they are *polymers* (Greek: many parts); that is, each one is made up of a great many simpler units—identical to each other or at least chemically similar—joined together in a regular way. They are formed by a process we touched on earlier: **polymerization**, the *joining together of many small molecules to form very large molecules*. The simple compounds from which polymers are made are called *monomers*.

Polymers are formed in two general ways.

(a) In **chain-reaction polymerization**, there is a series of reactions each of which consumes a reactive particle and produces another, similar particle; each individual reaction thus depends upon the previous one. The reactive particles can be free radicals, cations, or anions. A typical example is the polymerization of ethylene

$$\text{Rad}\cdot + CH_2{=}CH_2 \longrightarrow \text{RadCH}_2\text{CH}_2\cdot \xrightarrow{CH_2=CH_2} \text{RadCH}_2\text{CH}_2\text{CH}_2\text{CH}_2\cdot \longrightarrow \textit{etc.}$$

(Sec. 6.19). Here the chain-carrying particles are free radicals, each of which adds to a monomer molecule to form a new, bigger free radical.

(b) In **step-reaction polymerization**, there is a series of reactions each of which is essentially independent of the preceding one; a polymer is formed simply because the monomer happens to undergo reaction at more than one functional group. A glycol, for example, reacts with a dicarboxylic acid to form an ester; but each moiety of the simple ester still contains a group that can react to generate another ester linkage and hence a larger molecule, which itself can react further, and so on.

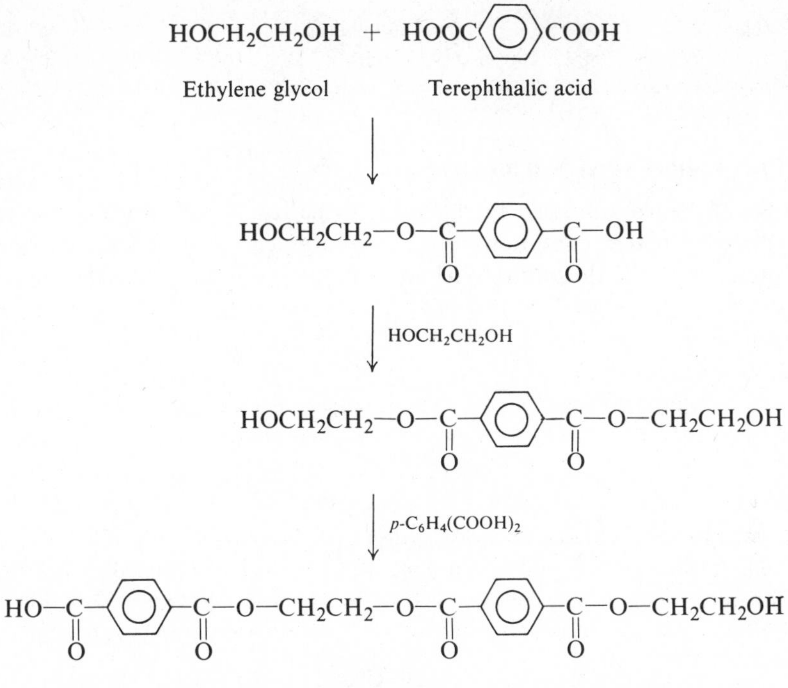

$$HOCH_2CH_2OH + HOOC\text{-}C_6H_4\text{-}COOH$$

Ethylene glycol Terephthalic acid

$$\downarrow$$

$$HOCH_2CH_2\text{—O—}\underset{\underset{O}{\|}}{C}\text{—}C_6H_4\text{—}\underset{\underset{O}{\|}}{C}\text{—OH}$$

$$\downarrow HOCH_2CH_2OH$$

$$HOCH_2CH_2\text{—O—}\underset{\underset{O}{\|}}{C}\text{—}C_6H_4\text{—}\underset{\underset{O}{\|}}{C}\text{—O—}CH_2CH_2OH$$

$$\downarrow p\text{-}C_6H_4(COOH)_2$$

$$HO\text{—}\underset{\underset{O}{\|}}{C}\text{—}C_6H_4\text{—}\underset{\underset{O}{\|}}{C}\text{—O—}CH_2CH_2\text{—O—}\underset{\underset{O}{\|}}{C}\text{—}C_6H_4\text{—}\underset{\underset{O}{\|}}{C}\text{—O—}CH_2CH_2OH$$

There is an alternative, somewhat less meaningful system of classification: *addition* polymerization, in which molecules of monomer are simply added together; and *condensation polymerization*, in which monomer molecules combine with loss of some simple molecules like water. As it happens, the two systems almost exactly coincide; nearly all cases of chain-reaction polymerization involve addition polymerization; nearly all cases of step-reaction polymerization involve condensation polymerization. Indeed, some chemists use the term "addition polymerization" to *mean* polymerization via chain reactions.

Let us look first at chain-reaction polymerization, starting with the kind that involves free radicals.

Problem 32.1 Examine the structure of each of the following synthetic polymers. Tell what class of compound it belongs to and give structures of the most likely monomers.

(a) nylon 6,6 (fibers), $\sim\underset{\underset{O}{\|}}{C}(CH_2)_4\underset{\underset{O}{\|}}{C}NH(CH_2)_6NH\underset{\underset{O}{\|}}{C}(CH_2)_4\underset{\underset{O}{\|}}{C}NH(CH_2)_6NH\sim$

(b) nylon 6 (fibers), $\sim\underset{\underset{O}{\|}}{C}CH_2(CH_2)_4NH\underset{\underset{O}{\|}}{C}CH_2(CH_2)_4NH\sim$

(c) Carbowax (water-soluble wax), $\sim OCH_2CH_2OCH_2CH_2OCH_2CH_2\sim$

(d) Neoprene (oil-resistant elastomer), $\sim CH_2\underset{\underset{Cl}{|}}{C}{=}CHCH_2CH_2\underset{\underset{Cl}{|}}{C}{=}CHCH_2\sim$

(e) Saran (packaging film, seat covers), $\sim CH_2CCl_2CH_2CCl_2\sim$

Problem 32.2 Answer the questions of Problem 32.1 for each of the following kinds of natural macromolecules: (a) a protein, p. 1151; (b) a nucleic acid, p. 1178; (c) starch (amylose), p. 1121; (d) cellulose, p. 1126.

32.3 Free-radical vinyl polymerization

In Sec. 6.19 we discussed briefly the polymerization of ethylene and substituted ethylenes under conditions where free radicals are generated—typically in the presence of small amounts of an initiator, such as a peroxide. Reaction

$$n\text{-}\underset{\text{Vinyl monomer}}{CH_2{=}\underset{\displaystyle G}{\underset{|}{CH}}} \xrightarrow{\text{initiator}} \sim CH_2\underset{G}{\underset{|}{CH}}CH_2\underset{G}{\underset{|}{CH}}CH_2\underset{G}{\underset{|}{CH}}CH_2\underset{G}{\underset{|}{CH}}\sim \quad \text{or } (-CH_2\underset{G}{\underset{|}{CH}}-)_n$$

Polymer

occurs at the doubly-bonded carbons—the vinyl groups—and is called *vinyl polymerization*. A wide variety of unsaturated monomers may be used, to yield polymers with different *pendant groups* (G) attached to the polymer backbone. For example:

$$CH_2{=}\underset{Cl}{\underset{|}{CH}} \longrightarrow \sim CH_2-\underset{Cl}{\underset{|}{CH}}-CH_2-\underset{Cl}{\underset{|}{CH}}-CH_2-\underset{Cl}{\underset{|}{CH}}\sim$$

Vinyl chloride → Poly(vinyl chloride) (PVC)

$$CH_2{=}\underset{CN}{\underset{|}{CH}} \longrightarrow \sim CH_2-\underset{CN}{\underset{|}{CH}}-CH_2-\underset{CN}{\underset{|}{CH}}-CH_2-\underset{CN}{\underset{|}{CH}}\sim$$

Acrylonitrile → Polyacrylonitrile (Orlon)

$$CH_2{=}\underset{C_6H_5}{\underset{|}{CH}} \longrightarrow \sim CH_2-\underset{C_6H_5}{\underset{|}{CH}}-CH_2-\underset{C_6H_5}{\underset{|}{CH}}-CH_2-\underset{C_6H_5}{\underset{|}{CH}}\sim$$

Styrene → Polystyrene

$$CH_2{=}\overset{CH_3}{\overset{|}{\underset{COOCH_3}{\underset{|}{C}}}} \longrightarrow \sim CH_2-\overset{CH_3}{\overset{|}{\underset{COOCH_3}{\underset{|}{C}}}}-CH_2-\overset{CH_3}{\overset{|}{\underset{COOCH_3}{\underset{|}{C}}}}-CH_2-\overset{CH_3}{\overset{|}{\underset{COOCH_3}{\underset{|}{C}}}}\sim$$

Methyl methacrylate → Poly(methyl methacrylate) (Plexiglas, Lucite)

Polymerization involves addition of free radicals to the double bond of the monomer: addition, first, of the free radical generated from the initiator, and then of the growing polymer molecule. This is, of course, an example of chain-reaction polymerization.

$$(1) \qquad \text{Peroxide} \longrightarrow \text{Rad}\cdot$$

$$(2) \qquad \text{Rad}\cdot + CH_2{=}\underset{G}{\underset{|}{CH}} \longrightarrow \text{Rad}CH_2-\underset{G}{\underset{|}{CH}}\cdot$$

Steps (1) and (2): **Chain-initiating steps**

Chain-propagating steps:

(3) $\text{RadCH}_2\text{—}\underset{\text{G}}{\underset{|}{\text{CH}}}\cdot + \text{CH}_2\text{═}\underset{\text{G}}{\underset{|}{\text{CH}}} \longrightarrow \text{RadCH}_2\text{—}\underset{\text{G}}{\underset{|}{\text{CH}}}\text{—CH}_2\text{—}\underset{\text{G}}{\underset{|}{\text{CH}}}\cdot$

then steps like (3) *repeated, until finally:*

Chain-terminating steps:

(4) $2\text{Rad(CH}_2\underset{\text{G}}{\underset{|}{\text{CH}}})_n\text{CH}_2\underset{\text{G}}{\underset{|}{\text{CH}}}\cdot$

$\downarrow$ combination

$\text{Rad(CH}_2\underset{\text{G}}{\underset{|}{\text{CH}}})_n\text{CH}_2\underset{\text{G}}{\underset{|}{\text{CH}}}\text{—}\underset{\text{G}}{\underset{|}{\text{CH}}}\text{CH}_2(\underset{\text{G}}{\underset{|}{\text{CH}}}\text{CH}_2)_n\text{Rad}$

or

(5) $2\text{Rad(CH}_2\underset{\text{G}}{\underset{|}{\text{CH}}})_n\text{CH}_2\underset{\text{G}}{\underset{|}{\text{CH}}}\cdot$

$\downarrow$ disproportionation

$\text{Rad(CH}_2\underset{\text{G}}{\underset{|}{\text{CH}}})_n\text{CH}_2\underset{\text{G}}{\underset{|}{\text{CH}_2}} + \text{Rad(CH}_2\underset{\text{G}}{\underset{|}{\text{CH}}})_n\text{CH═}\underset{\text{G}}{\underset{|}{\text{CH}}}$

In each step the consumption of a free radical is accompanied by the formation of a new, bigger free radical. Eventually, the reaction chain is terminated by steps that consume but do not form free radicals: *combination* or *disproportionation* of two free radicals.

Problem 32.3 Free-radical polymerization of 1,3-butadiene gives molecules containing both the following units,

$$\sim\text{CH}_2\text{CH═CHCH}_2\sim \quad \text{and} \quad \sim\text{CH}_2\underset{\text{CH═CH}_2}{\underset{|}{\text{CH}}}\sim$$

the exact proportions depending on the temperature. Account in detail for the formation of the two different units.

Problem 32.4 Polystyrene formed with isotopically labeled AIBN as initiator

$$(\text{CH}_3)_2\underset{\text{CN}}{\underset{|}{\text{C}}}\text{—N═N—}\underset{\text{CN}}{\underset{|}{\text{C}}}(\text{CH}_3)_2 \longrightarrow \text{N}_2 + 2(\text{CH}_3)_2\underset{\text{CN}}{\underset{|}{\text{C}}}\cdot$$

Azoisobutyronitrile
(AIBN)

was found to contain *two* initiator fragments per molecule. What termination reaction is indicated by this finding?

Added compounds can modify the polymerization process drastically. For example, in the presence of carbon tetrachloride, styrene undergoes polymerization at the same rate as in its absence, but the polystyrene obtained has a lower average molecular weight; furthermore, it contains small amounts of chlorine.

This is an example of **chain-transfer**, the termination of one polymerization chain (7) with the simultaneous initiation of another (8).

(6) $\sim CH_2-\underset{Ph}{CH}\cdot + CH_2=\underset{Ph}{CH} \xrightarrow{\text{polymerization}} \sim CH_2-\underset{Ph}{CH}-CH_2-\underset{Ph}{CH}\cdot$

(7) $\sim CH_2-\underset{Ph}{CH}\cdot + CCl_4 \xrightarrow{\text{chain-transfer}} \sim CH_2-\underset{Ph}{CH}-Cl + \cdot CCl_3$

(8) $Cl_3C\cdot + CH_2=\underset{Ph}{CH} \longrightarrow Cl_3C-CH_2-\underset{Ph}{CH}\cdot \xrightarrow{\text{styrene}} \text{polymer}$

Ordinarily a growing polystyrene radical adds (6) to styrene monomer to continue the reaction chain. Every so often, however, it abstracts an atom from the chain-transfer agent (7) to end the original polymerization chain and generate a new particle ($CCl_3\cdot$ in this case) that initiates a new polymerization chain (8). Since one reaction chain is replaced by another, the rate of polymerization is unaffected. Since the average number of chain-propagating steps in each reaction chain is reduced, the average molecular weight of the polymer is lowered. A transfer agent thus competes with the monomer for the growing radicals. The ratio of rate constants for (7) and (6), $k_{transfer}/k_{polymerization}$, is called the *transfer constant*; it is a measure of how effective the transfer agent is at lowering the molecular weight of the polymer.

Problem 32.5 For polymerization of styrene at 60°, the following chain-transfer constants have been measured. Account for the relative effectiveness of the members of each sequence.

(a) benzene 0.018, *tert*-butylbenzene 0.04, toluene 0.125, ethylbenzene 0.67, isopropylbenzene 0.86;
(b) *n*-heptane 0.42, 2-heptene 2.7;
(c) CCl_4 90, CBr_4 13,600.

An added compound may react with the growing free radical to generate a new free radical that is not reactive enough to add to monomer; a reaction chain is terminated but no new one is begun. Such a compound is, of course, an **inhibitor** (Sec. 2.14). Many amines, phenols, and quinones act as inhibitors. Although their exact mode of action is not understood, it seems clear that they are converted into free radicals that do not add to monomer; instead, they may combine or disproportionate, or combine with another growing radical to halt a second reaction chain.

$$\sim \underset{G}{CH_2}\cdot + \text{Inhibitor} \longrightarrow \sim \underset{G}{CH_2}CH_2 + \text{Inhibitor}\cdot$$

Unreactive: cannot initiate new chain

Since even traces of certain impurities, acting as chain-transfer agents or inhibitors, can interfere with the polymerization process, the monomers used are among the purest organic chemicals produced.

In an extreme case—if the alkene is of low reactivity and the transfer agent of high reactivity—chain transfer is so effective that there is *no* polymerization. Then we observe simply addition of the "transfer agent" to the double bond, a reaction we encountered in Sec. 6.18. For example:

$$n\text{-}C_6H_{13}CH{=}CH_2 + CBr_4 \xrightarrow{\text{peroxides}} n\text{-}C_6H_{13}\underset{\underset{\textstyle Br}{|}}{C}HCH_2CBr_3$$

Problem 32.6 (a) Chain transfer can cause *branching* of a polymer molecule. Show how this could happen. What is the chain-transfer agent? (b) Rather short branches (4 or 5 carbons) are attributed to "back-biting." What do you think is meant by this term? Show the chemical reactions probably involved.

32.4 Copolymerization

So far, we have discussed only polymerization of a single monomeric compound to form a *homopolymer*, a polymer made up—except, of course, at the two ends of the long molecule—of identical units.

Now, if a mixture of two (or more) monomers is allowed to undergo polymerization, there is obtained a **copolymer**: a polymer that contains two (or more) kinds of monomeric units in the same molecule. For example:

$$\underset{\text{Styrene}}{CH_2{=}\underset{\underset{\textstyle Ph}{|}}{C}H} + \underset{\text{Methyl methacrylate}}{CH_2{=}\overset{\overset{\textstyle CH_3}{|}}{\underset{\underset{\textstyle COOCH_3}{|}}{C}}} \xrightarrow{\text{initiator}} \underset{\text{Polystyrene(co-methyl methacrylate)}}{\sim CH_2{-}\underset{\underset{\textstyle Ph}{|}}{C}H{-}CH_2{-}\overset{\overset{\textstyle CH_3}{|}}{\underset{\underset{\textstyle COOCH_3}{|}}{C}}\sim}$$

Through copolymerization there can be made materials with different properties than those of either homopolymer, and thus another dimension is added to the technology. Consider, for example, styrene. Polymerized alone, it gives a good electric insulator that is molded into parts for radios, television sets, and automobiles. Copolymerization with butadiene (30%) adds toughness; with acrylonitrile (20–30%) increases resistance to impact and to hydrocarbons; with maleic anhydride yields a material that, on hydrolysis, is water-soluble, and is used as a dispersant and sizing agent. The copolymer in which butadiene predominates (75% butadiene, 25% styrene) is an elastomer, and since World War II has been the principal rubber substitute manufactured in the United States.

Let us look more closely at the copolymerization process. Consider free radical vinyl polymerization of two monomers, M_1 and M_2. In each step the growing free

$$M_1M_2M_2M_1M_2M_1M_1\cdot \begin{cases} \xrightarrow{M_1} M_1M_2M_2M_1M_2M_1M_1M_1\cdot \\ \xrightarrow{M_2} M_1M_2M_2M_1M_2M_1M_1M_2\cdot \end{cases}$$

radical can react with either monomer to continue the reaction chain. What are the factors that determine *which* monomer it preferentially reacts with?

First, of course, there are the relative *concentrations* of the two monomers; the higher the concentration of a particular monomer, the greater its chance of being incorporated into the chain, and the more abundant its units are in the final product.

Next, there are the relative *reactivities* of the monomers toward free radical addition; in general, the more reactive the monomer, the greater its chance of being incorporated into the polymer. We know that the reactivity of a carbon–carbon double bond toward free radical addition is affected by the stability of the new free radical being formed; factors that tend to stabilize the free radical product tend to stabilize the incipient free radical in the transition state, so that the more stable free radical tends to be formed faster. Now, stability of a free radical depends upon accommodation of the odd electron. The group G stabilizes the radical

$$\sim\!M\cdot + CH_2{=}\underset{\displaystyle G}{\underset{|}{CH}} \longrightarrow \left[\sim\!\overset{\delta\cdot}{M}\cdots CH_2\overset{\cdots}{=}\underset{\displaystyle G}{\underset{|}{\overset{\delta\cdot}{CH}}}\right] \longrightarrow \sim\!M{-}CH_2{-}\underset{\displaystyle G}{\underset{|}{CH}}\cdot$$

by delocalization: the phenyl group in styrene, through formation of a benzylic radical; the vinyl group of 1,3-butadiene, through formation of an allylic radical; the $-COOCH_3$ group of methyl methacrylate, through formation of a radical in which acyl oxygen helps carry the odd electron. (*Problem:* Draw resonance structures to show how this last effect could arise.)

(We notice that the above discussion does not take into account the nature of the attacking radical, and hence would predict the same relative reactivities for a pair of alkenes toward all free radicals. We shall return to this point later.)

Now, let us see what kind of copolymer we would expect to get on the basis of what we have said so far. In the copolymerization of styrene (M_1) and butadiene (M_2), for example, reaction can proceed via either of two growing radicals: one ending in a styrene unit ($\sim\!M_1\cdot$), or one ending in a butadiene unit ($\sim\!M_2\cdot$). Either radical can add to either monomer, to form a copolymer with styrene and butadiene units distributed *randomly* along the molecule:

$$\sim\!M_1M_2M_2M_1M_2M_1M_1\!\sim \qquad \textbf{Random copolymer}$$

With these particular monomers, copolymerization is in fact random. Now, toward either free radical type, it happens, butadiene is about 1.4 times as reactive as styrene, so that, if monomer concentrations were equal, butadiene units would tend to predominate in the product. Furthermore, since butadiene is consumed faster, the relative concentrations of monomers would change as reaction goes on, and so would the composition of the polymer being produced. These effects can be compensated for by adjusting the ratio of monomers fed into the reaction vessel; indeed, by control of the feed ratio, copolymers of any desired composition can be made.

Random copolymerization, of the kind observed for styrene and butadiene, is actually rather rare. In general, copolymerization shows, to a greater or lesser extent, a tendency to *alternation* of monomer units. An extreme case is that of

$$\sim\!M_1M_2M_1M_2M_1M_2M_1M_2\!\sim \qquad \textbf{Alternating copolymer}$$

stilbene (1,2-diphenylethene) and maleic anhydride, which copolymerize with absolutely regular alternation of units; regardless of the feed ratio, a 50:50 copolymer is obtained.

How are we to account for this tendency toward alternation? It must mean that a growing radical ending in one unit tends to add to the *opposite* monomer.

$$\sim\!M_1\cdot \xrightarrow{M_2} \sim\!M_1M_2\cdot \xrightarrow{M_1} \sim\!M_1M_2M_1\cdot \xrightarrow{M_2} \textit{etc.}$$

Clearly, the relative reactivity of a monomer *does* depend upon the nature of the radical that is attacking it. Maleic anhydride is much more reactive than stilbene toward radicals ending in a stilbene unit, and stilbene is much more reactive than maleic anhydride toward the other kind of radical. (Indeed, these two compounds, individually, undergo self-polymerization only with extreme difficulty.) A more modest—and more typical—tendency toward alternation is shown by styrene and methyl methacrylate. Here, toward either radical ($\sim\!M_1\cdot$) the "opposite" monomer (M_2) is about twice as reactive as the "same" monomer (M_1).

The alternating tendency in copolymerization was established on a quantitative basis by Frank R. Mayo (of the Stanford Research Institute) and Cheves Walling (of the University of Utah) while working in the laboratories of the U.S. Rubber Company. Their work was fundamental to the development of free radical chemistry: it showed clearly for the first time the dependence of reactivity on the nature of the attacking free radical, and led directly to the concept of *polar factors*, working not only in copolymerization and other additions of free radicals, but in free radical reactions of all kinds.

Basically, Mayo and Walling's interpretation was the following. Although free radicals are neutral, they have certain tendencies to gain or lose electrons, and hence they partake of the character of electrophilic or nucleophilic reagents. The transition states for their reactions can be polar, with the radical moiety acquiring a partial negative or positive charge at the expense of the substrate—the alkene, in the case of addition. In copolymerization, a substituent generally exerts the same polar effect—electron-withdrawing or electron-releasing—on a free radical as on the alkene (monomer) from which the free radical was derived. Electron-withdrawal makes a free radical electrophilic, but makes an alkene less able to supply the electrons which that radical is seeking. An electrophilic radical will, then, preferentially add to a monomer containing an electron-releasing group. In a similar way, a nucleophilic radical, containing an electron-releasing substituent, will seek out a monomer containing an electron-withdrawing substituent.

Styrene and methyl methacrylate tend to alternate because their substituents are of opposite polarity: in methacrylate the $-COOCH_3$ group tends to withdraw electrons; in styrene the phenyl group tends (via resonance) to release electrons. The transition states for addition to the opposite monomers are thus stabilized:

$$\sim\!CH_2-\underset{\displaystyle Ph}{\underset{|}{CH}}\cdot + CH_2=\overset{\displaystyle CH_3}{\overset{|}{\underset{\displaystyle COOCH_3}{\underset{|}{C}}}} \longrightarrow \left[\sim\!CH_2-\overset{\delta+}{\underset{\displaystyle Ph}{\underset{|}{CH}}}\cdots CH_2\overset{\cdots}{=}\overset{\displaystyle CH_3}{\overset{|}{\underset{\displaystyle COOCH_3}{\underset{|}{C}}}}{}^{\delta-}\right]$$

$$\sim\!CH_2-\overset{\displaystyle CH_3}{\overset{|}{\underset{\displaystyle COOCH_3}{\underset{|}{C}}}}\cdot + CH_2=\underset{\displaystyle Ph}{\underset{|}{CH}} \longrightarrow \left[\sim\!\overset{\delta-}{CH_2}-\overset{\displaystyle CH_3}{\overset{|}{\underset{\displaystyle COOCH_3}{\underset{|}{C}}}}\cdots CH_2\overset{\cdots}{=}\overset{\delta+}{\underset{\displaystyle Ph}{\underset{|}{CH}}}\right]$$

Perhaps the most convincing evidence for the play of polar forces comes from copolymerization of a series of ring-substituted styrenes; here relative reactivities toward a variety of monomers not only fall into a pattern consistent with the familiar electronic effects of the substituents, but show the same quantitative relationships (the Hammett sigma-rho relationship, Sec. 18.11) as do *ionic* reactions: dissociation of carboxylic acids, for example, or hydrolysis of esters.

The concept of polar transition states in free-radical reactions has recently been questioned, at least for reactions in which hydrogen is abstracted—halogenation, for example. Here, it has been suggested, electron-withdrawing or electron-releasing groups affect reactivity simply by strengthening or weakening the bonds holding hydrogen in the substrate.

If we define polar effects on free-radical reactions as effects due to electron-withdrawal or electron-release—rather than to accommodation of the odd electron—then there is no doubt about their existence; it is the *interpretation* of such effects that is open to question.

We must realize that polar effects are *superimposed on* effects due to delocalization of the odd electron. Styrene and butadiene, for example, are highly reactive toward any radical since the transition state contains an incipient benzylic or allylic free radical. This high reactivity is modified—enhanced or lowered—by the demands of the particular attacking radical.

Problem 32.7 (a) Draw structures to account for the strong alternating tendency in copolymerization of butadiene (M_1) and acrylonitrile (M_2). (b) Toward $\sim\!\sim M_1\cdot$ acrylonitrile is 2.5 times as reactive as butadiene, but toward $\sim\!\sim M_2\cdot$ butadiene is *20 times as reactive* as acrylonitrile. How do you account for this contrast?

Copolymers can be made not just from two different monomers but from three, four, or even more. They can be made not only by free-radical chain reactions, but by any of the polymerization methods we shall take up: ionic, coordination, or step-reaction. The monomer units may be distributed in various ways, depending on the technique used. As we have seen, they may alternate along a chain, either randomly or with varying degrees of regularity. In *block copolymers*, sections made up of one monomer alternate with sections of another:

$$\sim\!\sim M_1M_1M_1M_1M_1M_2M_2M_2M_2\sim\!\sim \qquad \textbf{Block copolymer}$$

In *graft copolymers*, a branch of one kind is grafted to a chain of another kind:

$$\begin{array}{l}\sim\!\sim M_1M_1M_1M_1M_1M_1M_1\sim\!\sim \\ \qquad\;\; | \\ \qquad M_2M_2M_2M_2\end{array} \qquad \textbf{Graft copolymer}$$

Problem 32.8 Graft copolymers can be made by each of the following processes. Show the chemistry most likely involved, and the structure of the product. (a) Polybutadiene is treated with styrene in the presence of a free-radical initiator. (b) Poly(vinyl chloride) is treated with methyl methacrylate in the presence of benzoyl peroxide, $(C_6H_5COO)_2$.

32.5 Ionic polymerization. Living polymers

Chain-reaction polymerization can proceed with ions instead of free radicals as the chain-carrying particles: either cations or anions, depending on the kind of initiator that is used.

Cationic polymerization

$$\underset{\text{An acid}}{Y}\ \ CH_2{=}\underset{\underset{G}{|}}{CH} \longrightarrow Y{:}CH_2{-}\underset{\underset{G}{|}}{\overset{\oplus}{C}H}$$

A carbonium ion

$$Y{:}CH_2{-}\underset{\underset{G}{|}}{\overset{\oplus}{C}H}\ \ CH_2{=}\underset{\underset{G}{|}}{CH} \longrightarrow Y{:}CH_2{-}\underset{\underset{G}{|}}{CH}{-}CH_2{-}\underset{\underset{G}{|}}{\overset{\oplus}{C}H} \longrightarrow \textit{etc.}$$

Anionic polymerization

$$\underset{\text{A base}}{Z{:}}\ \ CH_2{=}\underset{\underset{G}{|}}{CH} \longrightarrow Z{:}CH_2{-}\underset{\underset{G}{|}}{\overset{\ominus}{C}H{:}}$$

A carbanion

$$Z{:}CH_2{-}\underset{\underset{G}{|}}{\overset{\ominus}{C}H{:}}\ \ CH_2{=}\underset{\underset{G}{|}}{CH} \longrightarrow Z{:}CH_2{-}\underset{\underset{G}{|}}{CH}{-}CH_2{-}\underset{\underset{G}{|}}{\overset{\ominus}{C}H{:}} \longrightarrow \textit{etc.}$$

Cationic polymerization is initiated by *acids*. Isobutylene, for example, undergoes cationic polymerization to a tacky material used in adhesives. Copolymerization with a little isoprene gives *butyl rubber*, used to make automobile innertubes and tire liners. A variety of acids can be used; sulfuric acid; $AlCl_3$ or BF_3 plus a trace of water. We recognize this process as an extension of the dimerization discussed in Sec. 6.15.

$$H^+\ \ CH_2{=}\underset{CH_3}{\overset{CH_3}{C}}\ \ CH_2{=}\underset{CH_3}{\overset{CH_3}{C}}\ \ CH_2{=}\underset{CH_3}{\overset{CH_3}{C}} \longrightarrow CH_3{-}\underset{CH_3}{\overset{CH_3}{C}}{-}CH_2{-}\underset{CH_3}{\overset{CH_3}{C}}{-}CH_2{-}\underset{CH_3}{\overset{CH_3}{C}}\sim$$

Anionic polymerization, as we might expect, is initiated by *bases*: $Li^+NH_2^-$, for example, or organometallic compounds like *n*-butyllithium. For example:

$$NH_2^-K^+ + CH_2{=}\underset{\underset{Ph}{|}}{CH_2} \longrightarrow NH_2{-}CH_2{-}\underset{\underset{Ph}{|}}{CH^-}K^+ \xrightarrow{CH_2=CHPh}$$

$$NH_2{-}CH_2{-}\underset{\underset{Ph}{|}}{CH}{-}CH_2{-}\underset{\underset{Ph}{|}}{CH^-}K^+ \longrightarrow \textit{etc.}$$

$$n\text{-BuLi} + CH_2{=}\underset{\displaystyle COOCH_3}{\overset{\displaystyle CH_3}{\overset{|}{\underset{|}{C}}}} \longrightarrow n\text{-Bu—}CH_2\text{—}\underset{\displaystyle CH_3}{\overset{\displaystyle CH_3}{\overset{|}{\underset{|}{C}}}}{}^{-}Li^{+} \longrightarrow etc.$$

Active metals like Na or Li can be used; here the initiation becomes a little more complicated, as in the polymerization of styrene by the action of sodium metal and naphthalene. A sodium atom transfers an electron (1) to naphthalene to form a *radical-anion*, which then donates the electron to styrene (2) to form the styrene

(1) $$\text{Na} + \text{naphthalene} \rightleftarrows \text{Na}^{+}\ \text{naphthalene}^{\overset{\cdot}{-}}$$

Naphthalene radical-anion

(2) $$\text{naphthalene}^{\overset{\cdot}{-}} + \underset{\displaystyle Ph}{\underset{|}{CH}}{=}CH_2 \longrightarrow \text{naphthalene} + \underset{\displaystyle Ph}{\underset{|}{CH}}CH_2{}^{\overset{\cdot}{-}}$$

Styrene radical-anion
Formed by one-electron transfer

(3) $$2\,\underset{\displaystyle Ph}{\underset{|}{CH}}CH_2{}^{\overset{\cdot}{-}} \longrightarrow {}^{-}\underset{\displaystyle Ph}{\underset{|}{CH}}\text{—}CH_2\text{—}CH_2\text{—}\underset{\displaystyle Ph}{\underset{|}{CH}}{}^{-}$$

A dianion

radical-anion. Like many other free-radicals, these dimerize (3). The resulting dianion is the true initiator, and begins to grow *at both ends*:

$${}^{-}\underset{\displaystyle Ph}{\underset{|}{CH}}CH_2\text{—}CH_2\underset{\displaystyle Ph}{\underset{|}{CH}}{}^{-}$$

$$\Big\downarrow\ CH_2{=}CHPh$$

$${}^{-}\underset{\displaystyle Ph}{\underset{|}{CH}}\text{—}CH_2\text{~}\underset{\displaystyle Ph}{\underset{|}{CH}}\text{—}CH_2\text{—}\underset{\displaystyle Ph}{\underset{|}{CH}}\text{—}CH_2\text{—}CH_2\text{—}\underset{\displaystyle Ph}{\underset{|}{CH}}\text{—}CH_2\text{—}\underset{\displaystyle Ph}{\underset{|}{CH}}\text{~}CH_2\text{—}\underset{\displaystyle Ph}{\underset{|}{CH}}{}^{-}$$

Anionic polymerization is not limited to the vinyl kind, involving addition to carbon–carbon double bonds. Ethylene oxide, for example, is converted by a small amount of base into a high-molecular-weight polyether.

$$CH_3O^{-} + \underbrace{CH_2\text{—}CH_2}_{\diagdown O \diagup} \longrightarrow CH_3O\text{—}CH_2\text{—}CH_2O^{-}$$

$$CH_3O\text{—}CH_2\text{—}CH_2O^{-} + \underbrace{CH_2\text{—}CH_2}_{\diagdown O \diagup} \longrightarrow CH_3O\text{—}CH_2\text{—}CH_2\text{—}O\text{—}CH_2\text{—}CH_2O^{-}$$

Problem 32.9 The presence of methanol during the polymerization of ethylene oxide by sodium methoxide tends to lower the molecular weight of the product. (a) How do you think it does this? What process is this an example of? (b) What product will be obtained in the presence of *much* methanol?

In 1956, Michael Szwarc (of the State University of New York at Syracuse) reported the following observations. When a sample of styrene was treated with a little sodium naphthalene initiator, rapid polymerization took place and was complete within a few seconds. When more styrene was added, it, too, underwent polymerization; viscosity measurements showed that the molecular weight of the polystyrene was now *higher than before*. If, instead of the second batch of styrene, there was added butadiene, polymerization again took place; the product was found to contain no (homo)polystyrene—all styrene units were now part of a (block) copolymer with butadiene. How are these results to be interpreted?

So far, we have not mentioned chain-terminating steps for ionic polymerization. Such steps do exist. In cationic polymerization, for example, the growing carbonium ion can undergo either of two familiar reactions: ejection of a proton to form an alkene, or combination with an anion. But, particularly in the case of anionic polymerization, termination often involves reaction with an impurity or some other molecule not a necessary part of the polymerization system. Under carefully controlled conditions, there are *no* termination steps—or at least none that happen very fast. Reaction stops when all monomer is consumed, but the reaction mixture contains what Szwarc has named *living* polymer molecules. When these are "fed" additional monomer—either styrene or butadiene, for example—they continue to grow. They are not immortal, however, but can be "killed" by addition of a compound that reacts with carbanions: water, say. The generation of living polymers is of immense practical importance; it provides the best route to block copolymers, and permits the introduction of a variety of terminal groups.

Problem 32.10 Draw the structure of the product expected from the killing of living polystyrene by each of the following reagents: (a) water; (b) carbon dioxide, then water; (c) a small amount of ethylene oxide, then water; (d) a large amount of ethylene oxide, then water.

32.6 Coordination polymerization

When we speak of organic ions as chain-carriers, we realize, of course, that each of these must be balanced by an ion of opposite charge, a *counterion*. A growing carbanion, for example, has more or less closely associated with it a metallic cation like Li^+ or Na^+. Ion pairs—or even higher aggregates—can play important parts in polymerization. If the bonding between the reactive center and the metal is appreciably covalent, the process is called *coordination polymerization*. The growing organic chain is not a full-fledged anion, but its reactivity is due to its *anion-like character*.

Until 1953, almost all vinyl polymerization of commercial importance was of the free-radical type. Since that time, however, ionic polymerization, chiefly in the form of coordination polymerization, has revolutionized the field. Following discoveries by Karl Ziegler (of the Max Planck Institute for Coal Research) and by Guilio Natta (of the Polytechnic Institute of Milan)—who jointly received the Nobel Prize in 1963 for this work—catalysts have been developed that permit control of the polymerization process to a degree never before possible.

These Ziegler-Natta catalysts are complexes of transition metal halides with organometallic compounds: typically, triethylaluminum–titanium trichloride.

Reaction involves nucleophilic addition to the carbon–carbon double bond in the monomer, with the carbanion-like organic group of the growing organometallic compound as nucleophile. The transition metal may play a further role in complexing with the π electrons of the monomer and thus holding it at the reaction site. Polymerization thus amounts to insertion of alkene molecules into the bond between metal and the growing alkyl group. For example, in the formation of polyethylene:

$$M-CH_2CH_3 \xrightarrow{CH_2=CH_2} M-CH_2CH_2CH_2CH_3 \xrightarrow{CH_2=CH_2} M-CH_2CH_2CH_2CH_2CH_2CH_3 \xrightarrow{CH_2=CH_2} etc.$$

Polymerization with Ziegler-Natta catalysts has two important advantages over free-radical polymerization: (a) it gives *linear* polymer molecules; and (b) it permits *stereochemical control.*

Polyethylene made by the free-radical process has a highly branched structure due to chain-transfer of a special kind, in which the transfer agent *is a polymer molecule*: a hydrogen atom is abstracted from somewhere along the polymer chain,

$$\sim CH_2CH(H)CH_2CH_2\sim \xrightarrow{\sim CH_2CH_2\cdot} \sim CH_2CH_2-H + \sim CH_2\dot{C}HCH_2CH_2\sim \xrightarrow{CH_2=CH_2} \sim CH_2CH(CH_2\dot{C}H_2)CH_2CH_2\sim \longrightarrow etc.$$

and a branch grows at the point of attack. In contrast, polyethylene made by the coordination process is virtually unbranched. These unbranched molecules fit together well, and the polymer is said to have a *high degree of crystallinity*; as a result, it has a higher melting point and higher density than the older (*low density*) polyethylene, and is mechanically much stronger. (We shall look at the crystallinity of polymers and its effect on their properties in Sec. 32.8.)

A second, far-reaching development in coordination polymerization is *stereochemical control.* Propylene, for example, could polymerize to any of three different arrangements (Fig. 32.1): *isotactic*, with all methyl groups on one side of an extended chain; *syndiotactic*, with methyl groups alternating regularly from side to side; and *atactic*, with methyl groups distributed at random. By proper choice of experimental conditions—catalyst, temperature, solvent—each of these stereoisomeric polymers has been made. Atactic polypropylene is a soft, elastic, rubbery material. Both isotactic and syndiotactic polypropylenes are highly crystalline: regularity of structure permits their molecules to fit together well. Over a billion pounds of isotactic polypropylene is produced every year, to be molded or extruded as sheets, pipes, and filaments; it is on its way to becoming one of the principal synthetic fibers.

Coordination catalysts also permit stereochemical control about the carbon–carbon double bond. By their use, isoprene has been polymerized to a material virtually identical with natural rubber: *cis*-1,4-polyisoprene. (See Sec. 8.25.)

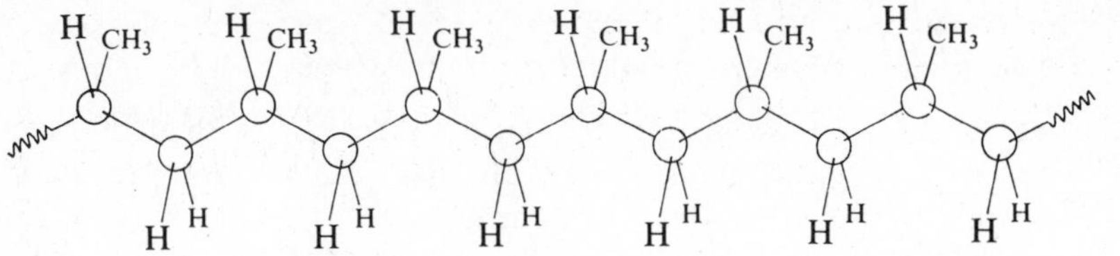

(a)

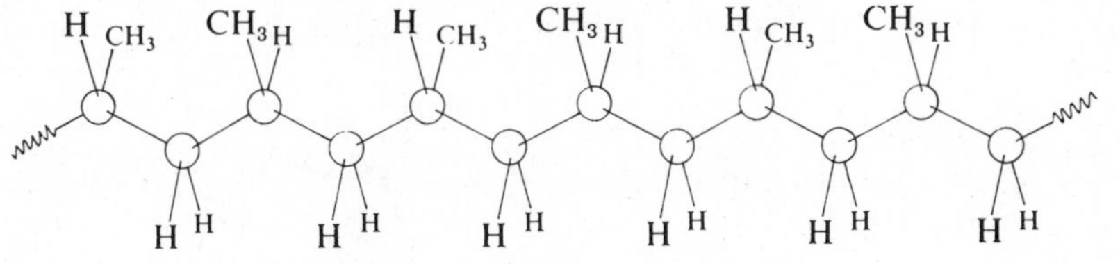

(b)

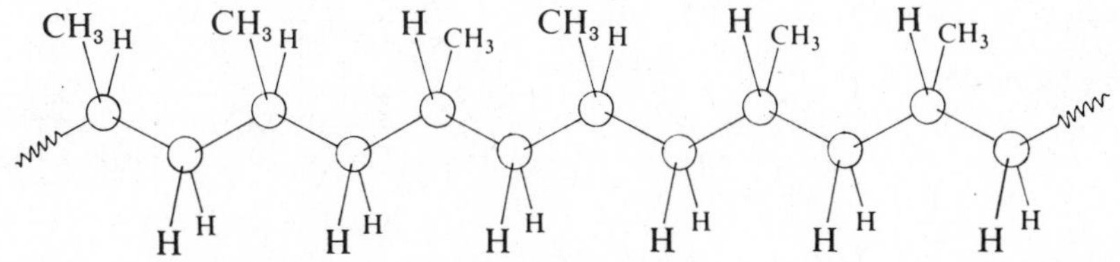

(c)

Figure 32.1. Polypropylene. (*a*) Isotactic. (*b*) Syndiotactic. (*c*) Atactic.

The Ziegler-Natta polymerization of ethylene can be adapted to make molecules of only modest size (C_6–C_{20}) and containing certain functional groups. If, for example, the metal-alkyls initially obtained are heated (in the presence of ethylene and a nickel catalyst), the hydrocarbon groups are displaced as straight-chain 1-alkenes of even carbon number. Large quantities of such alkenes in the C_{12}–C_{20} range are

$$M{-}(CH_2CH_2)_nCH_2CH_3 \xrightarrow[\text{heat}]{CH_2{=}CH_2,\ Ni} CH_2{=}CH{-}(CH_2CH_2)_{n-1}CH_2CH_3$$

consumed in the manufacture of detergents (Sec. 33.5). Alternatively, the metal-alkyls can be oxidized by air to give straight-chain primary alcohols:

$$M{-}(CH_2CH_2)_nCH_2CH_3 \xrightarrow[30\text{–}95^\circ]{\text{air}} M{-}O(CH_2CH_2)_nCH_2CH_3 \xrightarrow[40^\circ]{H_2O,\ H_2SO_4} HO(CH_2CH_2)_nCH_2CH_3$$

"A chemist setting out to build a giant molecule is in the same position as an architect designing a building. He has a number of building blocks of certain shapes and sizes, and his task is to put them together in a structure to serve a particular purpose. . . . What makes high polymer chemistry still more exciting just now is that almost overnight, within the last few years, there have come discoveries of new ways to put the building blocks together—discoveries which promise a great harvest of materials that have never existed on the earth." (Giulio Natta, *Scientific American*, September, 1957, p. 98.)

32.7 Step-reaction polymerization

Carboxylic acids react with amines to yield amides, and with alcohols to form esters. When an acid that contains more than one —COOH group reacts with an amine that contains more than one $—NH_2$ group, or with an alcohol that contains more than one —OH group, then the products are *polyamides* and *polyesters.* For example:

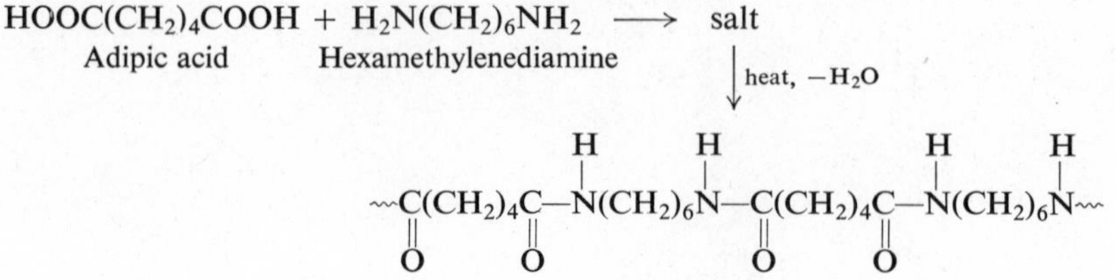

$$\underset{\text{Adipic acid}}{HOOC(CH_2)_4COOH} + \underset{\text{Hexamethylenediamine}}{H_2N(CH_2)_6NH_2} \longrightarrow \text{salt} \xrightarrow{\text{heat, } -H_2O}$$

$$\sim C(=O)(CH_2)_4C(=O)-N(H)(CH_2)_6N(H)-C(=O)(CH_2)_4C(=O)-N(H)(CH_2)_6N(H)\sim$$

Nylon 66
A polyamide

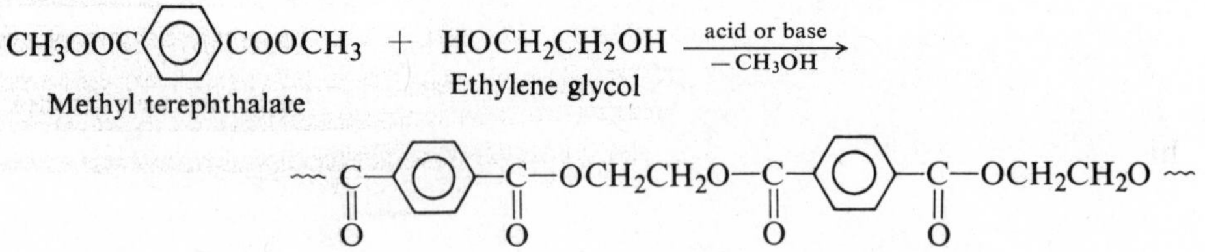

$$\underset{\text{Methyl terephthalate}}{CH_3OOC-C_6H_4-COOCH_3} + \underset{\text{Ethylene glycol}}{HOCH_2CH_2OH} \xrightarrow[-CH_3OH]{\text{acid or base}}$$

$$\sim C(=O)-C_6H_4-C(=O)-OCH_2CH_2O-C(=O)-C_6H_4-C(=O)-OCH_2CH_2O\sim$$

Dacron
A polyester

$$\underset{\text{Phthalic anhydride}}{C_6H_4(CO)_2O} + \underset{\text{Glycerol}}{CH_2(OH)-CH(OH)-CH_2(OH)} \xrightarrow{-H_2O} \text{Glyptal (an alkyd resin)}$$

A polyester

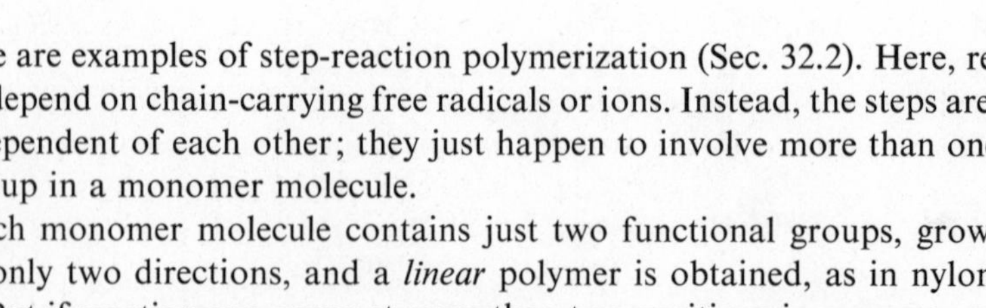

These are examples of step-reaction polymerization (Sec. 32.2). Here, reaction does not depend on chain-carrying free radicals or ions. Instead, the steps are essentially independent of each other; they just happen to involve more than one functional group in a monomer molecule.

If each monomer molecule contains just two functional groups, growth can occur in only two directions, and a *linear* polymer is obtained, as in nylon 66 or Dacron. But if reaction can occur at more than two positions in a monomer, there is formed a highly cross-linked *space network* polymer, as in Glyptal, an *alkyd resin.* Dacron and Glyptal are both polyesters, but their structures are quite different and, as we shall see, so are their uses.

Problem 32.11 Work out a possible structure for an alkyd resin formed from phthalic anhydride and glycerol, considering the following points: (a) In the first stage a linear polyester is formed. (Which hydroxyl groups are esterified more rapidly, primary or secondary?) (b) In the second stage these linear molecules are cross-linked to give a rather rigid network.

Step-reaction polymerization can involve a wide variety of functional groups and a wide variety of reaction types. Among the oldest of the synthetic polymers, and still extremely important, are those resulting from reaction between phenols and formaldehyde: the *phenol-formaldehyde resins* (Bakelite and related polymers). When phenol is treated with formaldehyde in the presence of alkali or acid, there is obtained a high molecular weight substance in which many phenol rings are held together by $—CH_2—$ groups:

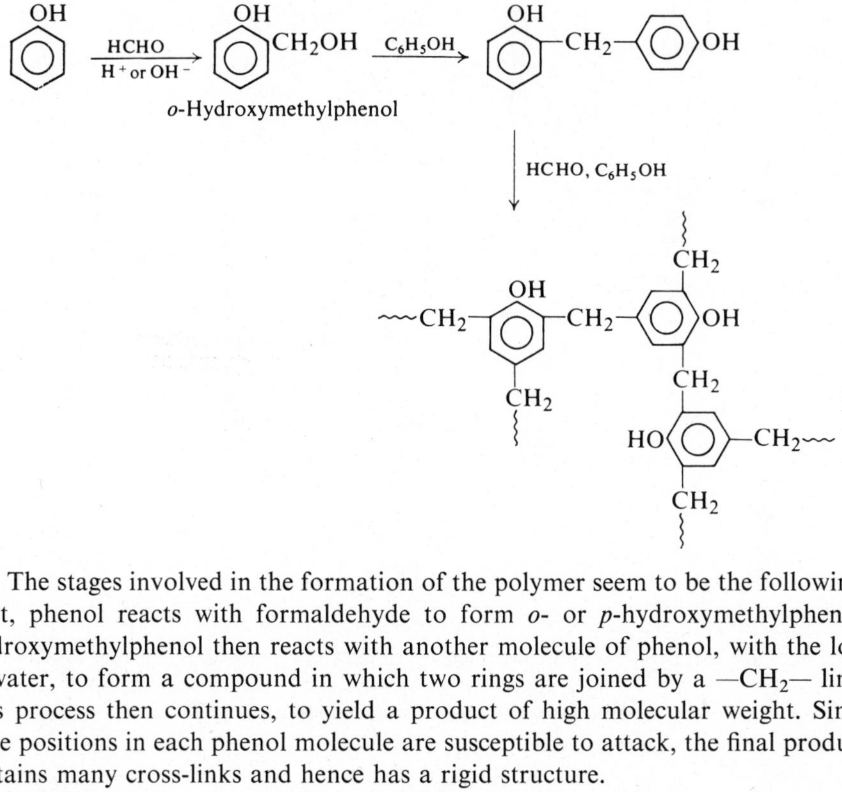

$$C_6H_5OH \xrightarrow[H^+ \text{ or } OH^-]{HCHO} o\text{-}HOC_6H_4CH_2OH \xrightarrow{C_6H_5OH} HOC_6H_4—CH_2—C_6H_4OH$$

o-Hydroxymethylphenol

$$\downarrow \; HCHO,\ C_6H_5OH$$

$$\sim CH_2—C_6H_2(OH)(CH_2\sim)—CH_2—C_6H_2(OH)(CH_2\sim)—CH_2—C_6H_2(OH)(CH_2\sim)—CH_2\sim$$

The stages involved in the formation of the polymer seem to be the following. First, phenol reacts with formaldehyde to form *o*- or *p*-hydroxymethylphenol. Hydroxymethylphenol then reacts with another molecule of phenol, with the loss of water, to form a compound in which two rings are joined by a $—CH_2—$ link. This process then continues, to yield a product of high molecular weight. Since three positions in each phenol molecule are susceptible to attack, the final product contains many cross-links and hence has a rigid structure.

The first stage can be viewed as both electrophilic substitution on the ring by the electron-deficient carbon of formaldehyde, and nucleophilic addition of the aromatic ring to the carbonyl group. Base catalyzes reaction by converting phenol into the more reactive (more nucleophilic) phenoxide ion. Acid catalyzes reaction by protonating formaldehyde and increasing the electron deficiency of the carbonyl carbon.

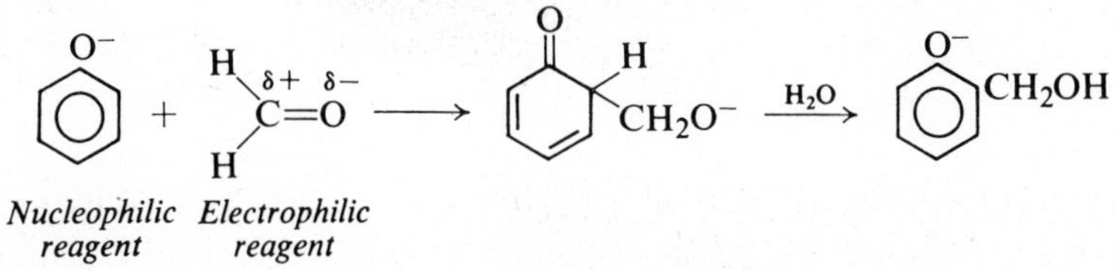

$$C_6H_5O^- + \overset{\delta+}{H_2C}=\overset{\delta-}{O} \longrightarrow \text{(cyclohexadienone with } H,\ CH_2O^-) \xrightarrow{H_2O} o\text{-}{}^-OC_6H_4CH_2OH$$

Basic catalysis

Nucleophilic reagent *Electrophilic reagent*

$$\text{C}_6\text{H}_5\text{OH} + \text{H}_2\text{C}{=}\overset{+}{\text{O}}\text{H} \longrightarrow \text{(ring-protonated intermediate: }{=}\overset{+}{\text{O}}\text{H, H, CH}_2\text{OH)} \xrightarrow{\text{H}_2\text{O}} o\text{-HOC}_6\text{H}_4\text{CH}_2\text{OH}$$

Acidic catalysis

Nucleophilic reagent *Electrophilic reagent*

Urea reacts with formaldehyde to form the *urea-formaldehyde resins*, highly important in molded plastics. Here, too, a space-network polymer is formed.

$$\underset{\text{Formaldehyde}}{\text{H—}\overset{\text{H}}{\overset{|}{\text{C}}}\text{=O}} + \underset{\text{Urea}}{\text{H}_2\text{N—}\underset{\text{O}}{\underset{\|}{\text{C}}}\text{—NH}_2} \longrightarrow \underset{\text{Methylolurea}}{\text{HOCH}_2\text{—}\overset{\text{H}}{\overset{|}{\text{N}}}\text{—}\underset{\text{O}}{\underset{\|}{\text{C}}}\text{—NH}_2}$$

$$\text{HOCH}_2\text{—}\overset{\text{H}}{\overset{|}{\text{N}}}\text{—}\underset{\text{O}}{\underset{\|}{\text{C}}}\text{—NH}_2 + \text{O=}\overset{\text{H}}{\overset{|}{\text{C}}}\text{—H} \longrightarrow \underset{\text{Dimethylolurea}}{\text{HOCH}_2\text{—}\overset{\text{H}}{\overset{|}{\text{N}}}\text{—}\underset{\text{O}}{\underset{\|}{\text{C}}}\text{—}\overset{\text{H}}{\overset{|}{\text{N}}}\text{—CH}_2\text{OH}}$$

$$\text{HOCH}_2\text{—}\overset{\text{H}}{\overset{|}{\text{N}}}\text{—}\underset{\text{O}}{\underset{\|}{\text{C}}}\text{—}\overset{\text{H}}{\overset{|}{\text{N}}}\text{—CH}_2\text{OH} \xrightarrow{\text{HCHO, urea}} \text{(space-network polymer: ∼CH}_2\text{—N∼, CH}_2\text{, ∼N—CH}_2\text{—N—C(=O)—N∼, O=C, CH}_2\text{, ∼N—CH}_2\text{—N—C(=O)—N∼, H}_2\text{C, ∼N—CH}_2\text{∼)}$$

Organic *isocyanates*, RNCO, undergo reactions of the following kinds (compare Sec. 20.23), all of which are used, in one way or another, in the synthesis of

$$\underset{\text{An isocyanate}}{\text{RN=C=O}} + \text{R'OH} \longrightarrow \underset{\text{A urethane}}{\text{RNH—}\underset{\text{O}}{\underset{\|}{\text{C}}}\text{—OR'}}$$

$$\text{RN=C=O} + \text{R'NH}_2 \longrightarrow \underset{\text{A substituted urea}}{\text{RNH—}\underset{\text{O}}{\underset{\|}{\text{C}}}\text{—NHR'}}$$

$$\text{RN=C=O} + \text{H}_2\text{O} \longrightarrow \underset{\substack{\text{A carbamic acid}\\ \textit{Unstable}}}{\left[\text{RNH—}\underset{\text{O}}{\underset{\|}{\text{C}}}\text{—OH}\right]} \longrightarrow \text{RNH}_2 + \text{CO}_2$$

$$\text{RNH}_2 \xrightarrow{\text{RNCO}} \text{RNH—}\underset{\text{O}}{\underset{\|}{\text{C}}}\text{—NHR}$$

polymers. Reaction of *di*hydroxy alcohols with *di*isocyanates gives the important polyurethanes.

Problem 32.12 Give the structure of the polymer expected from the reaction of ethylene glycol and 2,4-tolylene diisocyanate, 2,4-$(OCN)_2C_6H_3CH_3$.

32.8 Structure and properties of macromolecules

The characteristic thing about macromolecules, we have said, is their great size. This size has little effect on chemical properties. A functional group reacts much as we would expect, whether it is in a big or little molecule: an ester is hydrolyzed, an epoxide undergoes ring-opening, an allylic hydrogen is susceptible to abstraction by free radicals.

Problem 32.13 Describe reagents and conditions—if any—that would be expected to cleave the natural polymers of Problem 32.2 (p. 1030) into monomers.

Problem 32.14 When poly(vinyl acetate) is treated with methanol (b.p. 65°) in the presence of a little sulfuric acid, a substance of b.p. 57° distills from the mixture, and a new polymer is left behind. (a) What reaction has taken place? What is the structure of the new polymer? Why must it be prepared in this indirect manner? (b) When this new polymer is treated with *n*-butyraldehyde in the presence of a little phosphoric acid, a third polymer is formed, Butvar, which is used in making safety glass. What reaction has taken place here, and what is the structure of Butvar?

It is in their physical properties that macromolecules differ from ordinary molecules, and it is on these that their special functions depend. To begin with, let us look at the property of *crystallinity*. In a crystalline solid, we know, the structural units—molecules, in the case of a non-ionic compound—are arranged in a very regular, symmetrical way, with a geometric pattern repeated over and over. If a long molecule is to fit into such a pattern, it cannot be looped and coiled into a random conformation, but must be extended in a regular zig-zag (see Fig. 32.2). This lack of randomness corresponds to an unfavorable entropy for the system (Sec. 18.11). On the other hand, the regularity and close fitting of the molecules in a crystal permits operation of strong intermolecular forces—hydrogen bonding, dipole–dipole attractions, van der Waals forces—which result in a favorable enthalpy (heat content). As we shall see, this tug-of-war between entropy and enthalpy is a key factor in determining the use to which a macromolecule can be put.

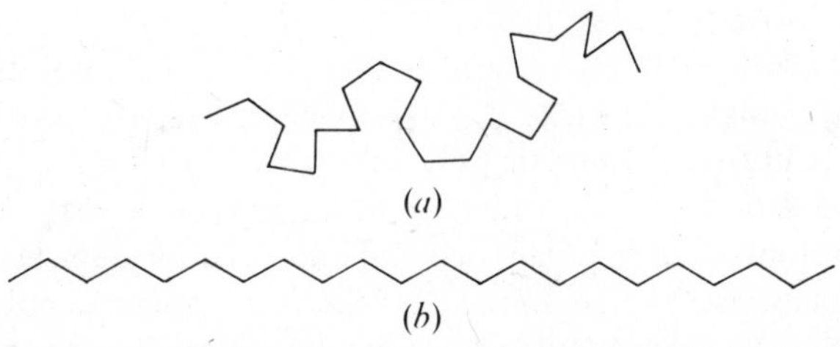

Figure 32.2. Long chain (*a*) in a random conformation, and (*b*) extended.

Now, in general, a high polymer does not exist entirely in crystalline form—not even a polymer whose regularity of molecular structure might be expected to permit this. The problem is the size of the molecule. As solidification begins, the viscosity of the material rises and the polymer molecules find it difficult to move about and arrange their long chains in the regular pattern needed for crystal formation. Chains become entangled; a change in shape of a chain must involve rotation about single bonds, and this becomes difficult because of hindrance to the swinging about of pendant groups. Polymers, then, form solids made up of regions of crystallinity, called *crystallites*, embedded in amorphous material. We speak of the *degree of crystallinity* of a polymer to mean the extent to which it is composed of crystallites.

Problem 32.15 Although both polymers are prepared by free-radical processes, poly(vinyl chloride) is amorphous and poly(vinylidene chloride) (Saran) is highly crystalline. How do you account for the difference? (Vinylidene chloride is 1,1-dichloroethene.)

Let us examine the various uses of polymers, and see how these depend on their structure—molecular and intermolecular.

Fibers are long, thin, threadlike bits of material that are characterized by great tensile (pulling) strength *in the direction of the fiber*. The natural fibers—cotton, wool, silk—are typical. Fibers are twisted into threads, which can then be woven into cloth, or embedded in plastic material to impart strength. The tensile strength can be enormous, some synthetic fibers rivalling—on a weight basis—steel.

The gross characteristics of fibers are reflected on the molecular level—the molecules, too, are long, thin, and threadlike. Furthermore, and most essential, they lie stretched out alongside each other, *lined up in the direction of the fiber*. The strength of the fiber resides, ultimately, in the strength of the chemical bonds of the polymer chains. The lining-up is brought about by *drawing*—stretching—the polymeric material. Once lined up, the molecules stay that way; the tendency to return to random looping and coiling is overcome by strong intermolecular attractions. In a fiber, enthalpy wins out over entropy. This high degree of molecular orientation is usually—although not always—accompanied by appreciable crystallinity.

The key requirements of a fiber are, then, a molecular shape—linear—that permits side-by-side alignment, and strong intermolecular forces to maintain this alignment. In addition, the intermolecular forces prevent "slipping" of one molecule past another. Now, what are these intermolecular forces?

The principal synthetic fibers are polyamides (the nylons), polyesters (Dacron, Terylene, Vycron), polyacrylonitrile ("acrylic fibers," Orlon, Acrilan), polyurethanes (Spandex, Vycra), and isotactic polypropylene. In nylon and polyurethanes, molecular chains are held to each other by hydrogen bonds (Fig. 32.3). In polyesters and polyacrylonitrile, the polar carbonyl and cyano groups lead to powerful dipole–dipole attractions. The stereoregular chains of isotactic polypropylene fit together so well that van der Waals forces are strong enough to maintain alignment.

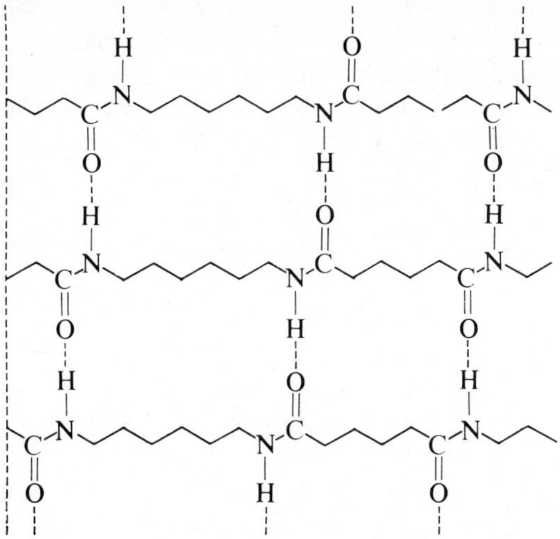

Figure 32.3. Hydrogen bonding in crystallites of nylon 66.

An **elastomer** possesses the high degree of elasticity that is characteristic of rubber: it can be greatly deformed—stretched to eight times its original length, for example—and yet return to its original shape. Here, as in fibers, the molecules are long and thin; as in fibers, they become lined up when the material is stretched. The big difference is this: when the stretching force is removed, the molecular chains of an elastomer do not remain extended and aligned, but return to their original random conformations favored by entropy. They do not remain aligned because the intermolecular forces necessary to hold them that way are weaker than in a fiber. In general, elastomers do not contain highly polar groups or sites for hydrogen bonding; the extended chains do not fit together well enough for van der Waals forces to do the job. In an elastomer entropy beats enthalpy.

One further requirement: the long chains of an elastomer must be connected to each other by occasional cross-links: enough of them to prevent slipping of molecules past one another; not so many as to deprive the chains of the flexibility that is needed for ready extension and return to randomness.

Natural rubber illustrates these structural requirements of an elastomer: long, flexible chains; weak intermolecular forces; and occasional cross-linking. Rubber is *cis*-1,4-polyisoprene. With no highly polar substituents, intermolecular attraction is largely limited to van der Waals forces. But these are weak because of

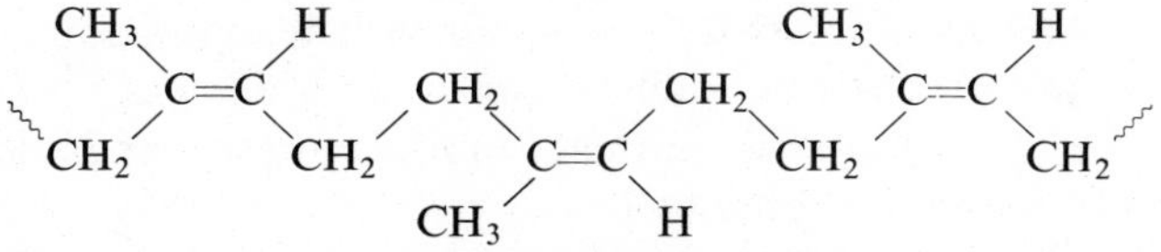

Natural rubber
All cis-configurations

the all-*cis* configuration about the double bond. Figure 32.4 compares the extended chains of rubber with those of its *trans* stereoisomer. As we can see, the *trans* configuration permits highly regular zig-zags that fit together well; the *cis* configuration does not. The all-*trans* stereoisomer occurs naturally as *gutta percha*; it is highly crystalline and non-elastic.

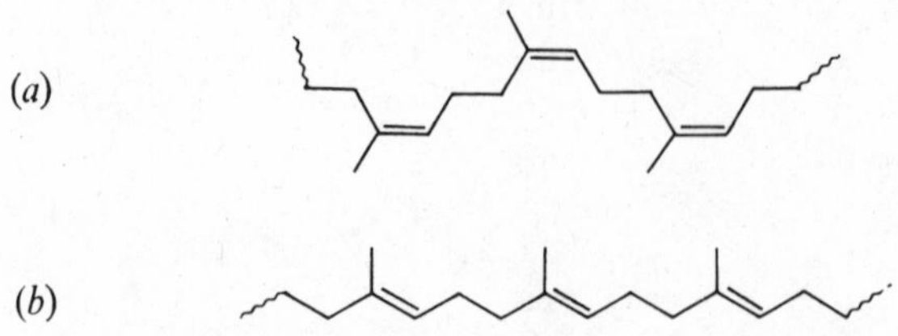

Figure 32.4. Extended chains of (*a*) natural rubber, *cis*-1,4-polyisoprene, and of (*b*) gutta percha, its *trans* stereoisomer.

Cross-linking in rubber, as we have seen (Sec. 8.25), is brought about by *vulcanizing*—heating with sulfur—which causes formation of sulfur bridges between molecules. This reaction involves reactive allylic positions, and thus depends on the double bond in the polymer.

Chief among the synthetic elastomers is SBR, a copolymer of butadiene (75%) and styrene (25%) produced under free-radical conditions; it competes with natural rubber in the main use of elastomers, the making of automobile tires. All-*cis* polybutadiene and polyisoprene can be made by Ziegler-Natta polymerization.

An elastomer that is entirely or mostly polydiene is, of course, highly unsaturated. All that is required of an elastomer, however, is enough unsaturation to permit cross-linking. In making butyl rubber (Sec. 32.5), for example, only 5% of isoprene is copolymerized with isobutylene.

Problem 32.16 (a) A versatile elastomer is obtained by Ziegler-Natta copolymerization of ethylene and propylene in the presence of a little diene, followed by vulcanization. How does the use of ethylene *and* propylene—instead of just one or the other—help to give the polymer elasticity?

(b) A similar copolymer can be made without the diene. This is cured by heating, not with sulfur, but with benzoyl peroxide. Why is this? What is the nature of the cross-links generated here?

Although enormous quantities of man-made fibers and elastomers are produced each year, the major consumption of synthetic polymers is as **plastics**, materials used in the form of sheets, pipes, films, and, most important of all, molded objects: toys and bottles; knobs, handles, and switches; dishes, fountain pens, toothbrushes; valves, gears, bearings; cases for radios and television sets; boats, automobile bodies, and even houses.

The molecular structure of plastics is of two general kinds: long molecules, either linear or branched; and space-network molecules.

The *linear* and *branched* polymers may be more or less crystalline, and include some of the materials also used as fibers: nylon, for example. They include the various polyalkenes we have mentioned: polyethylene, poly(vinyl chloride), polystyrene, etc. On heating, these polymers soften, and for this reason are called *thermoplastic*. It is in this softened state that they can be molded or extruded.

Space-network polymers (or *resins*) are highly cross-linked to form a rigid but irregular three-dimensional structure, as in phenol-formaldehyde or urea-formaldehyde resins. A sample of such material is essentially one gigantic molecule; heating does not soften it, since softening would require breaking of covalent bonds. Indeed, heating may cause formation of additional cross-links and thus make the material harder; for this reason, these polymers are called *thermosetting* polymers. This continuation of the polymerization process through heating is often coupled with the shaping of the product.

Certain linear, thermoplastic polymers are, like the space-network polymers, amorphous—and for basically the same reason. On cooling, their molecules form a rigid but irregular three-dimensional structure; they are held there, not by covalent cross-links, but by powerful dipole–dipole forces which lock the molecules into position before they can shake down into the regular arrangement required of a crystal. These materials are called *glasses*; poly(methyl methacrylate)—Plexiglas, Lucite—is the commonest one. Like ordinary (inorganic) glass, they lack crystalline planes for reflecting light, and are transparent. Like ordinary glass—and like the space-network polymers—they are brittle; when struck, these molecules cannot "give" with the blow through the sliding of crystalline planes over one another; they either resist—or break.

The rest of this book is devoted to organic compounds of biological importance. Many of these are macromolecules. We shall find that, just as the technological function of a macromolecule—fiber, elastomer, plastic—depends on its structure, so does the biological function: to hold the organism together, to nourish it, to control it, to allow it to reproduce itself.

PROBLEMS

1. Account for the fact that, whatever the mechanism—free-radical, cationic, anionic—vinyl polymerization gives products with almost exclusively "head-to-tail" arrangement of units.

2. Like other oxygen-containing compounds, alcohols dissolve in cold concentrated H_2SO_4 (Sec. 6.30). In the case of some secondary and tertiary alcohols, dissolution is followed by the gradual separation of an insoluble liquid of high boiling point. How do you account for this behavior?

3. Isobutylene does not give the kinds of stereoisomeric polymers (isotactic, etc.) that propylene does. Why not? What can you say about 1-butene?

4. Formaldehyde is polymerized by the action of a strong base like sodium methoxide. Suggest a mechanism for the process, and a structure for the polymer. To what general class of organic reactions does this polymerization belong?

5. A simple process for recycling polyurethanes has been developed by the Ford Motor Company. Can you suggest a way to accomplish this? What products would you expect to obtain?

6. Suggest an explanation for the following order of reactivity toward the addition of $BrCCl_3$ in the presence of peroxides: $C_6H_5CH{=}CH_2$ over 100, 1-octene 1.0, $C_6H_5CH_2CH{=}CH_2$ 0.7, $ClCH_2CH{=}CH_2$ 0.5, $Cl_3CCH_2CH{=}CH_2$ 0.3.

7. Account for each of the following observations. (a) In the presence of peroxides, CCl_4 reacts with 1-octene, $RCH{=}CH_2$, to give not only the 1:1 adduct, $RCHClCH_2CCl_3$, but also the 2:1 adduct, $RCHClCH_2CH(R)CH_2CCl_3$. (b) In contrast, CBr_4 adds to the 1-octene to give *only* the 1:1 product. (c) Styrene reacts with peroxides in the presence of CCl_4 to give only polymer.

8. Outline all steps in a possible synthesis from non-polymeric starting materials of each of the following polymers.

(a) Elastic fibers, used in girdles and bathing suits (Spandex, Lycra).

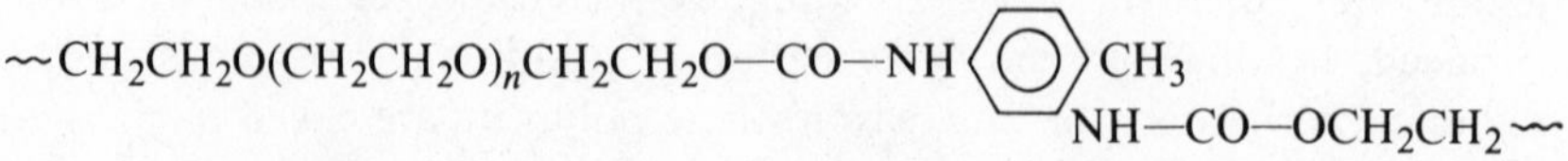

$$\sim CH_2CH_2O(CH_2CH_2O)_nCH_2CH_2O{-}CO{-}NH{-}C_6H_3(CH_3){-}NH{-}CO{-}OCH_2CH_2\sim$$

(b) A polyester resin, used in making pipe, boats, automobile bodies, etc.

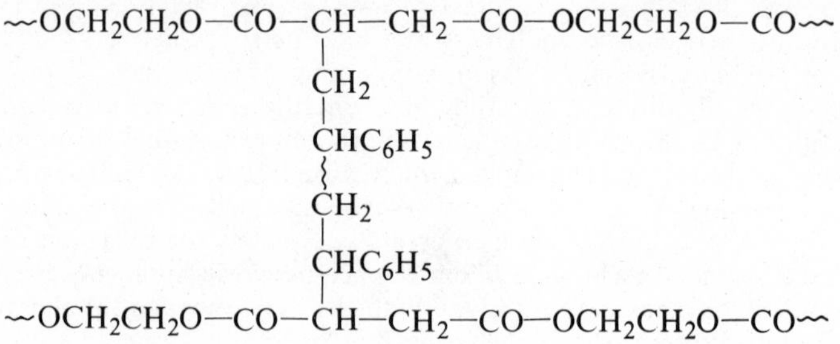

$$\sim OCH_2CH_2O{-}CO{-}CH{-}CH_2{-}CO{-}OCH_2CH_2O{-}CO\sim$$
$$|$$
$$CH_2$$
$$|$$
$$CHC_6H_5$$
$$\wr$$
$$CH_2$$
$$|$$
$$CHC_6H_5$$
$$|$$
$$\sim OCH_2CH_2O{-}CO{-}CH{-}CH_2{-}CO{-}OCH_2CH_2O{-}CO\sim$$

(c) A surface-active polymer.

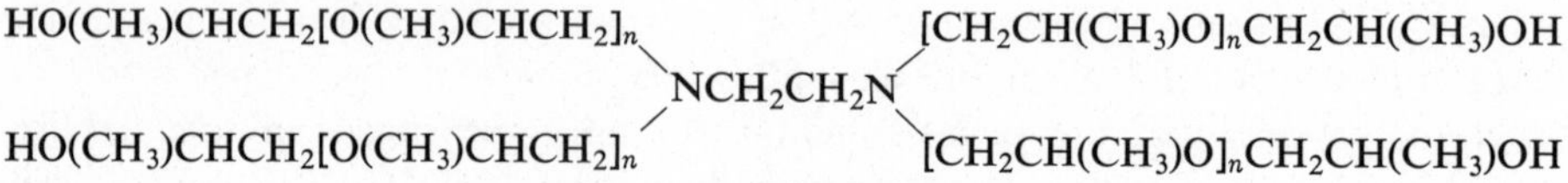

$$[HO(CH_3)CHCH_2[O(CH_3)CHCH_2]_n]_2NCH_2CH_2N[[CH_2CH(CH_3)O]_nCH_2CH(CH_3)OH]_2$$

(d) 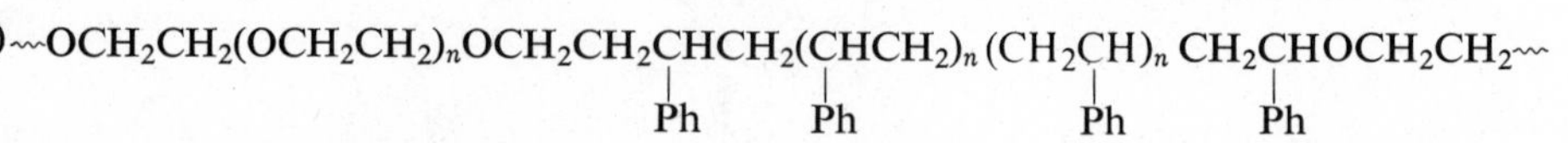 $\sim OCH_2CH_2(OCH_2CH_2)_nOCH_2CH_2CH(Ph)CH_2(CH(Ph)CH_2)_n(CH_2CH(Ph))_nCH_2CH(Ph)OCH_2CH_2\sim$

(e)

$$\wr$$
$$CH_3CCOOCH_3$$
$$|$$
$$CH_2$$
$$|$$
$$CH_3CCOOCH_3$$
$$|$$
$$CH_2$$
$$|$$
$$CH_3CCOOCH_3$$
$$|$$
$$CH_2$$
$$|$$
$$\sim CH_2CH(OAc)CH_2CH(Cl)CH_2C(OAc)(Cl)CH_2CHCH_2CH(Cl)CH_2CH(OAc)\sim$$

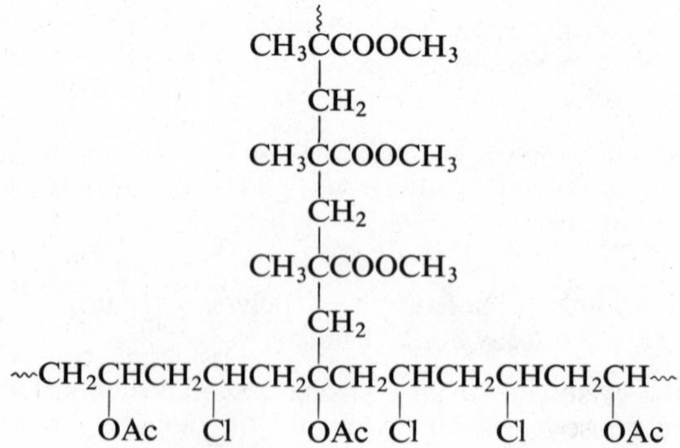

9. Treatment of β-propiolactone with base gives a polymer. Give a likely structure for this polymer, and show a likely mechanism for the process. Is this an example of chain-reaction or step-reaction polymerization?

10. When styrene is treated with KNH_2 in liquid ammonia, the product is a dead polymer that contains one $-NH_2$ group per molecule and no unsaturation. Suggest a termination step for the process.

11. When poly(vinyl acetate) was hydrolyzed, and the product treated with periodic acid and then re-acetylated, there was obtained poly(vinyl acetate) of lower molecular weight than the starting material. What does this indicate about the structure of the original polymer? About the polymerization process?

12. (a) What is the structure of nylon 6, made by alkaline polymerization of caprolactam?

Caprolactam

(b) Suggest a mechanism for the process. Is polymerization of the chain-reaction or step-reaction type?

13. In the *Beckmann rearrangement* (Problem 6, p. 919) oximes are converted into amides by the action of acids. For example:

$$(C_6H_5)_2C{=}NOH \xrightarrow{\text{acid}} C_6H_5C(=O)NHC_6H_5$$

Benzophenone oxime → Benzanilide

Caprolactam (preceding problem) can be made by the Beckmann rearrangement. With what ketone must the process start?

14. Fibers of very high tensile strength ("high-modulus fibers") have been made by reactions like the one between terephthalic acid and *p*-phenylenediamine, *p*-$C_6H_4(NH_2)_2$. Of key importance is the isomer composition of the monomers: the more exclusively *para*, the higher the melting point and the lower the solubility of the polymer, and the stronger the fibers. How do you account for this effect?

15. Evidence of many kinds shows that the metal–carbon bond in compounds like *n*-butyllithium is covalent, although highly polar. Yet living polystyrene solutions, which are colored, have virtually identical spectra whether the metal involved is sodium, potassium, cesium, or lithium. Can you suggest an explanation for this?

16. (a) When the alkane 2,4,6,8-tetramethylnonane was synthesized by an unambiguous method (Problem 13 (l), p. 544), there was obtained a product which was separated by gas chromatography into two components, A and B. The two components had identical mol. wt. and elemental composition, but different m.p., b.p., and infrared and nmr spectra. Looking at the structure of the expected product, what are these two components?

(b) When the same synthesis was carried out starting with an optically active reactant, compound B was obtained in optically active form, but A was still inactive. What is the structure of A? Of B?

(c) The nmr and infrared spectra of A and B were compared with the spectra of isotactic and syndiotactic polypropylenes (Fig. 32.1, p. 1041). With regard to their spectra, A showed a marked resemblance to one of the polymers, and B showed a marked resemblance to the other. It was concluded that the results "confirm the structures originally assigned [by Natta, p. 1039] for the two crystalline polymers of propylene." Which polymer did A resemble? Which polymer did B resemble?

17. Material similar to foam rubber can be made by the following sequence:

adipic acid + excess ethylene glycol ⟶ C
C + excess *p*-OCN—C_6H_4—C_6H_4—NCO-*p* ⟶ D
D + limited H_2O ⟶ E

Write equations for all steps, and show structures for C, D, and E. Be sure to account for the cross-linking in the final polymer, and its *foamy* character. (*Remember:* A foam is a dispersion of a gas in a solid.)

18. In the presence of benzoyl peroxide, allyl acetate gives poor yields of polymer of low molecular weight. The deuterium-labeled ester, $CH_2{=}CHCD_2OAc$, polymerizes 2 to 3 times as fast as the ordinary ester, and gives polymer of about twice the molecular weight. How do you account for these facts?

19. Linseed oil and tung oil, important constituents of paints, are esters (Sec. 33.6) derived from acids that contain two or three double bonds per molecule: 9,12-octadecadienoic acid, for example. On exposure to air, paint forms a tough protective film; oddly enough, after the initial rapid evaporation of solvent, this "drying" of paint is accompanied by a *gain* in weight. What kind of process do you think is involved? Be as specific as you can be.

20. To use an epoxy cement, one mixes the fluid "cement" with the "hardener," applies the mixture to the surfaces being glued together, brings them into contact, and waits for hardening to occur. The fluid cement is a low-molecular-weight polymer prepared by the following reaction:

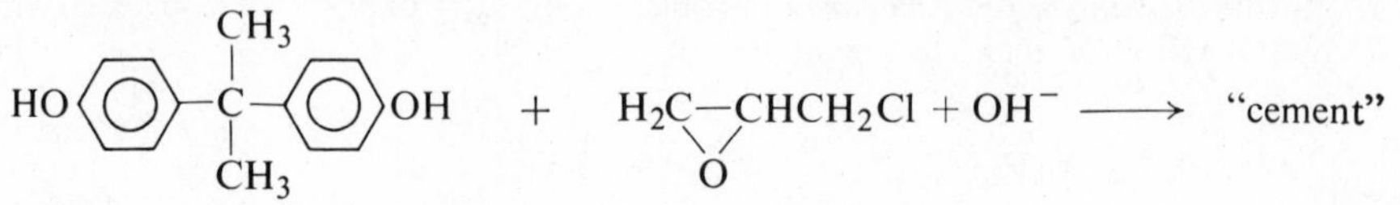

$$HO-C_6H_4-C(CH_3)_2-C_6H_4-OH \quad + \quad \underset{\text{O}}{H_2C{-}CHCH_2Cl} + OH^- \longrightarrow \text{"cement"}$$

2,2-Bis(*p*-hydroxyphenyl)propane ("Bisphenol A") — Epichlorohydrin, *Excess* — *Contains no chlorine*

The hardener can be any of a number of things: $NH_2CH_2CH_2NHCH_2CH_2NH_2$, diethylenetriamine, for example.

(a) What is the structure of the fluid cement, and how is it formed? What is the purpose of using *excess* epichlorohydrin? (b) What happens during hardening? What is the structure of the final epoxy resin? (c) Suggest a method of making bisphenol A, starting from phenol.

21. Poly(methyl methacrylate) was prepared in two different ways: polymer F, with initiation by benzoyl peroxide at 100°; polymer G, with initiation by *n*-butyllithium at $-62°$. Their nmr spectra were, with considerable simplification, as follows:

F *a* singlet, δ 1.10
b singlet, δ 2.0
c singlet, δ 3.58
approximate area ratios, $a{:}b{:}c = 3{:}2{:}3$

G *a* singlet, δ 1.33
b doublet, δ 1.7
c doublet, δ 2.4
d singlet, δ 3.58
approximate area ratios, $a{:}b{:}c{:}d = 3{:}1{:}1{:}3$

Account in detail for the difference in spectra. What, essentially, is polymer F? Polymer G?

PART III

Biomolecules

Chapter 33 Fats

33.1 The organic chemistry of biomolecules

The study of biology at the molecular level is called biochemistry. It is a branch of biology, but it is equally a branch of organic chemistry. Most of the molecules involved, the *biomolecules*, are bigger and more complicated than the ones we have so far studied, and their environment—a living organism—is a far cry from the stark simplicity of the reaction mixture of the organic chemist. But the physical and chemical properties of these compounds depend on molecular structure in exactly the same way as do the properties of other organic compounds.

The detailed chemistry of biological processes is vast and complicated, and is beyond the scope of this book; indeed, the study of biochemistry must be *built upon* a study of the fundamentals of organic chemistry. We can, however, attempt to close the gap between the subject "organic chemistry" and the subject "biochemistry."

In the remaining chapters of this book, we shall take up the principal classes of biomolecules: fats, carbohydrates, proteins, and nucleic acids. Our chief concern will be with their structures—since structure is fundamental to everything else—and with the methods used to determine these structures. Because biomolecules are big ones, we shall encounter structure on several levels: first, of course, the *sequence of functional groups* and the *configuration* at any chiral centers or double bonds; then, *conformation*, with loops, coils, and zig-zags on a grander scale than anything we have seen yet; finally, the arrangement of *collections of molecules*, and even of collections of these collections. We shall see remarkable effects due to our familiar intermolecular forces: operating between biomolecules; between biomolecules—or *parts* of them—and the solvent; between different parts of the same biomolecule.

We shall study the chemical properties of these compounds observed in the test tube, since these properties must lie behind the reactions they undergo in living organisms. In doing this, we shall reinforce our knowledge of basic organic chemistry by applying it to these more complex substances. Finally, mostly in the last chapter, we shall look—very briefly—at a few biochemical processes, just to catch a glimpse of the ways in which molecular structure determines biological behavior.

33.2 Occurrence and composition of fats

Biochemists have found it convenient to define one set of biomolecules, the *lipids*, as substances, insoluble in water, that can be extracted from cells by organic solvents of low polarity like ether or chloroform. This is a catch-all sort of definition, and lipids include compounds of many different kinds: steroids (Sec. 15.16), for example, and terpenes (Sec. 8.26). Of the lipids, we shall take up only the *fats* and certain closely related compounds. These are not the only important lipids—indeed, every compound in an organism seems to play an important role, if only as an unavoidable waste product of metabolism—but they are the most abundant.

Fats are the main constituents of the storage fat cells in animals and plants, and are one of the important food reserves of the organism. We can extract these animal and vegetable fats—liquid fats are often referred to as *oils*—and obtain such substances as corn oil, coconut oil, cottonseed oil, palm oil, tallow, bacon grease, and butter.

Chemically, fats are carboxylic esters derived from the single alcohol, glycerol, $HOCH_2CHOHCH_2OH$, and are known as *glycerides*. More specifically, they are *triacylglycerols*. As Table 33.1 shows, each fat is made up of glycerides derived

$$
\begin{array}{l}
CH_2{-}O{-}\underset{\displaystyle \overset{\|}{O}}{C}{-}R \\
| \\
CH{-}O{-}\underset{\displaystyle \overset{\|}{O}}{C}{-}R' \\
| \\
CH_2{-}O{-}\underset{\displaystyle \overset{\|}{O}}{C}{-}R''
\end{array}
$$

A triacylglycerol
(A glyceride)

from many different carboxylic acids. The proportions of the various acids vary from fat to fat; each fat has its characteristic composition, which does not differ very much from sample to sample.

With only a few exceptions, the fatty acids are all straight-chain compounds, ranging from three to eighteen carbons; except for the C_3 and C_5 compounds, only acids containing an even number of carbons are present in substantial amounts. As we shall see in Sec. 37.6, these even numbers are a natural result of the biosynthesis of fats: the molecules are built up two carbons at a time from acetate units, in steps that closely resemble the malonic ester synthesis of the organic chemist (Sec. 26.2).

Problem 33.1 *n*-Heptadecane is the principal *n*-alkane found both in a 50 million-year-old shale and in the blue-green algae, primitive organisms still existing. When blue-green algae were grown on a medium containing stearic-18-^{14}C acid, essentially all the radioactivity that was not left in unconsumed stearic acid was found in *n*-hepta-

Table 33.1 Fatty Acid Composition of Fats and Oils

Fat or Oil	Saturated Acids, %							Unsaturated Acids, %					
								Enoic				Dienoic	Trienoic
	C_8	C_{10}	C_{12}	C_{14}	C_{16}	C_{18}	$>C_{18}$	$<C_{16}$	C_{16}	C_{18}	$>C_{18}$	C_{18}	C_{18}
Beef tallow			0.2	2–3	25–30	21–26	0.4–1	0.5	2–3	39–42	0.3	2	
Butter	1–2[a]	2–3	1–4	8–13	25–32	8–13	0.4–2	1–2	2–5	22–29	0.2–1.5	3	
Coconut	5–9	4–10	44–51	13–18	7–10	1–4				5–8	0–1	1–3	
Corn				0–2	8–10	1–4			1–2	30–50	0–2	34–56	
Cottonseed				0–3	17–23	1–3				23–44	0–1	34–55	
Lard				1	25–30	12–16		0.2	2–5	41–51	2–3	3–8	
Olive			0–1	0–2	7–20	1–3	0–1		1–3	53–86	0–3	4–22	
Palm				1–6	32–47	1–6				40–52		2–11	
Palm kernel	2–4	3–7	45–52	14–19	6–9	1–3	1–2		0–1	10–18		1–2	
Peanut				0.5	6–11	3–6	5–10		1–2	39–66		17–38	
Soybean				0.3	7–11	2–5	1–3		0–1	22–34		50–60	2–10
											C_{20} $>C_{20}$		
Cod liver				2–6	7–14	0–1		0–2	10–20	25–31	25–32 10–20		
Linseed				0.2	5–9	4–7	0.5–1			9–29		8–29[b]	45–67[c]
Tung										4–13		8–15	78–82[d]

[a] 3–4% C_4, 1–2% C_6.

[b] Linoleic acid, *cis,cis*-9,12-octadecadienoic acid.

[c] Linolenic acid, *cis,cis,cis*-9,12,15-octadecatrienoic acid.

[d] Eleostearic acid, *cis,trans,trans*-9,11,13-octadecatrienoic acid, and 3–6% saturated acids.

decane. By what kind of chemical reaction is the hydrocarbon evidently produced? Of what geological significance is this finding?

Problem 33.2 (a) Acetate is not the only building block for the long chains of lipids. From a 50 million-year-old shale (see Problem 33.1)—as well as from modern organisms—there has been isolated 3,7,11,15-tetramethylhexadecanoic acid,

COOH

3,7,11,15-Tetramethylhexadecanoic acid

What familiar structural unit occurs here?

(b) The long side chain of chlorophyll (p. 1004) is derived from the alcohol *phytol*, which is cis-7(R),11(R)-3,7,11,15-tetramethyl-2-hexadecen-1-ol. The acid in (a)

CH_2OH

cis-7(R)-11(R)-3,7,11,15-Tetramethyl-2-hexadecen-1-ol
Phytol

was found to be a mixture of two diastereomers: the 3(S),7(R),11(R) and 3(R),7(R),-11(R). Of what biogenetic significance is this finding?

Besides saturated acids, there are unsaturated acids containing one or more double bonds per molecule. The most common of these acids are:

$CH_3(CH_2)_7CH{=}CH(CH_2)_7COOH$
Oleic acid
(*cis*-isomer)

$CH_3(CH_2)_4CH{=}CHCH_2CH{=}CH(CH_2)_7COOH$
Linoleic acid
(*cis,cis*-isomer)

$CH_3CH_2CH{=}CHCH_2CH{=}CHCH_2CH{=}CH(CH_2)_7COOH$
Linolenic acid
(*cis,cis,cis*-isomer)

The configuration about these double bonds is almost invariably *cis*, rather than the more stable *trans*.

Unsaturation *with this particular stereochemistry* has an effect that is seemingly trivial but is actually (Sec. 33.8) of vital biological significance: it lowers the melting point. In the solid phase, the molecules of a fat fit together as best they can; the closer they fit, the stronger the intermolecular forces, and the higher the melting point. Saturated acid chains are extended in a linear fashion—with, of course, the zig-zag due to the tetrahedral bond angles—and fit together rather well. *trans*-Unsaturated acid chains can be similarly extended to linear conformations that match saturated chains rather well (Fig. 33.1). But *cis*-unsaturated acid chains have a *bend* at the double bond, and fit each other—and saturated chains—badly. The net result is that *cis* unsaturation lowers the melting point of fat.

While we synthesize fats in our own bodies, we also eat fats synthesized in plants and other animals; they are one of the three main classes of foods, the others being carbohydrates (Chap. 35) and proteins (Chap. 36). Fats are used in enormous amounts as raw materials for many industrial processes; let us look at some of these before we turn our attention to some close relatives of the fats.

(*a*) COOH

(*b*) COOH

(*c*)

Figure 33.1. Extended chains of fatty acids: (*a*) saturated, (*b*) *trans*-unsaturated, (*c*) *cis*-unsaturated. Note bend in (*c*).

33.3 Hydrolysis of fats. Soap. Micelles

The making of soap is one of the oldest of chemical syntheses. (It is not nearly so old, of course, as the production of ethyl alcohol; man's desire for cleanliness is much newer than his desire for intoxication.) When the German tribesmen of Caesar's time boiled goat tallow with potash leached from the ashes of wood fires, they were carrying out the same chemical reaction as the one carried out on a tremendous scale by modern soap manufacturers: *hydrolysis of glycerides.* Hydrolysis yields salts of the carboxylic acids, and glycerol, $CH_2OHCHOHCH_2OH$.

$$\begin{array}{l} CH_2\text{—O—}\underset{\underset{O}{\|}}{C}\text{—R} \\ | \\ CH\text{—O—}\underset{\underset{O}{\|}}{C}\text{—R}' \\ | \\ CH_2\text{—O—}\underset{\underset{O}{\|}}{C}\text{—R}'' \end{array} \xrightarrow{NaOH} \begin{array}{c} CH_2OH \\ | \\ CHOH \\ | \\ CH_2OH \end{array} + \left\{ \begin{array}{l} RCOO^-Na^+ \\ R'COO^-Na^+ \\ R''COO^-Na^+ \end{array} \right\}$$

A glyceride (A fat) — Glycerol — Soap

Ordinary soap today is simply a mixture of sodium salts of long-chain fatty acids. It is a mixture because the fat from which it is made is a mixture, and for washing our hands or our clothes a mixture is just as good as a single pure salt. Soap may vary in composition and method of processing: if made from olive oil, it is *Castile soap*; alcohol can be added to make it transparent; air can be beaten in to make it float; perfumes, dyes, and germicides can be added; if a potassium salt (instead of a sodium salt), it is *soft soap*. Chemically, however, soap remains pretty much the same, and does its job in the same way.

We might at first expect these salts to be water-soluble—and, indeed, one can prepare what are called "soap solutions." But these are not true solutions, in which solute molecules swim about, separately and on their own. Instead, soap is dispersed in spherical clusters called **micelles**, each of which may contain hundreds of soap molecules. A soap molecule has a polar end, —COO^-Na^+, and a non-polar end, the long carbon chain of 12 to 18 carbons. The polar end is water-soluble, and is said to be *hydrophilic* (water-loving). The non-polar end is water-insoluble, and is said to be *hydrophobic* (water-fearing); it is, of course, soluble in non-polar solvents. Molecules like these are called *amphipathic*: they have both polar and non-polar ends and, in addition, are big enough for each end to display its own solubility behavior. In line with the rule of "like dissolves like," each non-polar end seeks a non-polar environment; in this situation, the only such environment about is the non-polar ends of other soap molecules, which therefore huddle together in the center of the micelle (Fig. 33.2). The polar ends project outward into the polar

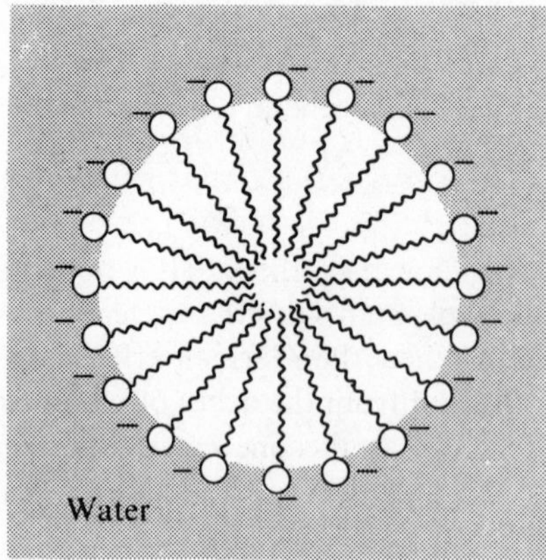

Figure 33.2. Soap micelle. Non-polar hydrocarbon chains "dissolve" in each other. Polar —COO^- groups dissolve in water. Similarly charged micelles repel each other.

solvent, water. Negatively charged carboxylate groups stud the surface of the micelle, and it is surrounded by an ionic atmosphere. Repulsion between similar charges keeps the micelles dispersed.

Now, how does a soap clean? The problem in cleansing is the fat and grease that make up and contain the dirt. Water alone cannot dissolve these hydrophobic substances; oil droplets in contact with water tend to coalesce so that there is a water layer and an oil layer. But the presence of soap changes this. The non-polar ends of soap molecules dissolve in the oil droplet, leaving the carboxylate ends projecting into the surrounding water layer. Repulsion between similar charges keeps the oil droplets from coalescing; a stable emulsion of oil and water forms, and can be removed from the surface being cleaned. As we shall see, this emulsifying, and hence cleansing, property is not limited to carboxylate salts, but is possessed by other amphipathic molecules (Sec. 33.5).

Hard water contains calcium and magnesium salts, which react with soap to form insoluble calcium and magnesium carboxylates (the "ring" in the bathtub).

33.4 Fats as sources of pure acids and alcohols

Treatment of the sodium soaps with mineral acid (or hydrolysis of fats under acidic conditions) liberates a mixture of the free carboxylic acids. In recent years, fractional distillation of these mixtures has been developed on a commercial scale to furnish individual carboxylic acids of over 90% purity.

Fats are sometimes converted by transesterification into the methyl esters of carboxylic acids; the glycerides are allowed to react with methanol in the presence of a basic or acidic catalyst. The mixture of methyl esters can be separated by

$$\begin{array}{l} CH_2—O—\underset{\underset{O}{\|}}{C}—R \\ | \\ CH—O—\underset{\underset{O}{\|}}{C}—R' \\ | \\ CH_2—O—\underset{\underset{O}{\|}}{C}—R'' \end{array} + \underset{\text{Methanol}}{CH_3OH} \xrightarrow{\text{base}} \begin{array}{l} CH_2OH \\ | \\ CHOH \\ | \\ CH_2OH \end{array} + \left\{ \begin{array}{l} RCOOCH_3 \\ R'COOCH_3 \\ R''COOCH_3 \end{array} \right\}$$

A glyceride; Glycerol; Mixture of methyl esters

fractional distillation into individual esters, which can then be hydrolyzed to individual carboxylic acids of high purity. Fats are thus the source of straight-chain acids of even carbon number ranging from six to eighteen carbons.

Alternatively, these methyl esters, either pure or as mixtures, can be catalytically reduced to straight-chain primary alcohols of even carbon number, and from these can be derived a host of compounds (as in Problem 18.10, p. 604). Fats thus provide us with long straight-chain units to use in organic synthesis.

33.5 Detergents

Of the straight-chain primary alcohols obtained from fats—or in other ways (Sec. 32.6)—the C_8 and C_{10} members are used in the production of high-boiling esters used as *plasticizers* (e.g., octyl phthalate). The C_{12} to C_{18} alcohols are used in enormous quantities in the manufacture of *detergents* (cleansing agents).

Although the synthetic detergents vary considerably in their chemical structure, the molecules of all of them have one common feature, a feature they share with ordinary soap: they are amphipathic, and have a large non-polar hydrocarbon end that is oil-soluble, and a polar end that is water-soluble. The C_{12} to C_{18} alcohols are converted into the salts of alkyl hydrogen sulfates. For example:

$$\underset{\text{Lauryl alcohol}}{n\text{-}C_{11}H_{23}CH_2OH} \xrightarrow{H_2SO_4} \underset{\text{Lauryl hydrogen sulfate}}{n\text{-}C_{11}H_{23}CH_2OSO_3H} \xrightarrow{NaOH} \underset{\text{Sodium lauryl sulfate}}{n\text{-}C_{11}H_{23}CH_2OSO_3^-Na^+}$$

For these, the non-polar end is the long chain, and the polar end is the $—OSO_3^-Na^+$.

Treatment of alcohols with ethylene oxide (Sec. 17.13) yields a *non-ionic* detergent:

$$\underset{\text{Lauryl alcohol}}{CH_3(CH_2)_{10}CH_2OH} + 8\underset{\text{Ethylene oxide}}{CH_2—CH_2 \atop \diagdown O \diagup} \xrightarrow{\text{base}} \underset{\text{An ethoxylate}}{CH_3(CH_2)_{10}CH_2(OCH_2CH_2)_8OH}$$

Hydrogen-bonding to the numerous oxygen atoms makes the polyether end of the molecule water-soluble. Alternatively, the ethoxylates can be converted into sulfates and used in the form of the sodium salts.

Perhaps the most widely used detergents are sodium salts of alkylbenzenesulfonic acids. A long-chain alkyl group is attached to a benzene ring by the action

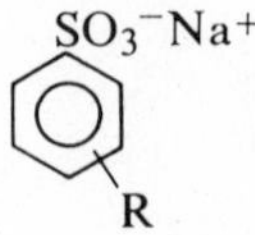

of a Friedel-Crafts catalyst and an alkyl halide, an alkene, or an alcohol. Sulfonation and neutralization yields the detergent.

Formerly, polypropylene was commonly used in the synthesis of these alkylbenzenesulfonates; but the highly-branched side chain it yields blocks the rapid biological degradation of the detergent residues in sewage discharge and septic tanks. Since about 1965 in this country, such "hard" detergents have been replaced by "soft" (biodegradable) detergents: alkyl sulfates, ethoxylates and their sulfates; and alkylbenzenesulfonates in which the phenyl group is randomly attached to the various secondary positions of a long straight chain (C_{12}–C_{18} range). (See Problem 17, p. 403.) The side chains of these "linear" alkylbenzenesulfonates are derived from straight-chain 1-alkenes (Sec. 32.6), or chlorinated straight-chain alkanes separated (by use of molecular sieves) from kerosene.

These detergents act in essentially the same way as soap does. They are used because they have certain advantages. For example, the sulfates and sulfonates retain their efficiency in hard water, since the corresponding calcium and magnesium salts are soluble. Being salts of strong acids, they yield neutral solutions, in contrast to the soaps, which, being salts of weak acids, yield slightly alkaline solutions (Sec. 18.10).

33.6 Unsaturated fats. Hardening of oils. Drying oils

We have seen that fats contain, in varying proportions, glycerides of unsaturated carboxylic acids. We have also seen that, other things being equal, unsaturation in a fat tends to lower its melting point and thus tends to make it a liquid at room temperature. In the United States the long-established use of lard and butter for cooking purposes has led to a prejudice against the use of the cheaper, equally nutritious oils. Hydrogenation of some of the double bonds in such cheap fats as cottonseed oil, corn oil, and soybean oil converts these liquids into solids having a consistency comparable to that of lard or butter. This *hardening* of oils is the basis of an important industry that produces cooking fats (for example, Crisco, Spry) and oleomargarine. Hydrogenation of the carbon–carbon double bonds takes place under such mild conditions (Ni catalyst, 175–190°, 20–40 lb/in.^{2}) that hydrogenolysis of the ester linkage does not occur.

Hydrogenation not only changes the physical properties of a fat, but also—and this is even more important—changes the chemical properties: a hydrogenated fat becomes *rancid* much less readily than does a non-hydrogenated fat. Rancidity is due to the presence of volatile, bad-smelling acids and aldehydes. These compounds result (in part, at least) from attack by oxygen at reactive allylic positions in the fat molecules; hydrogenation slows down the development of rancidity presumably by decreasing the number of double bonds and hence the number of allylic positions.

(In the presence of hydrogenation catalysts, unsaturated compounds undergo not only hydrogenation but also isomerization—shift of double bonds, or stereochemical transformations—which also affects physical and chemical properties.)

Linseed oil and tung oil have special importance because of their high content of glycerides derived from acids that contain two or three double bonds. They are known as **drying oils** and are important constituents of paints and varnishes. The "drying" of paint does not involve merely evaporation of a solvent (turpentine, etc.), but rather a chemical reaction in which a tough organic film is formed. Aside from the color due to the pigments present, protection of a surface by this film is the chief purpose of paint. The film is formed by a polymerization of the unsaturated oils that is brought about by oxygen. The polymerization process and the structure of the polymer are extremely complicated and are not well understood. The process seems to involve, in part, free-radical attack at reactive allylic hydrogens, free-radical chain-reaction polymerization similar to that previously described (Secs. 6.19 and 32.3), and cross-linking by oxygen analogous to that by sulfur in vulcanized rubber (Sec. 8.25).

Problem 33.3 In paints, tung oil "dries" faster than linseed oil. Suggest a reason why. (See Table 33.1.)

33.7 Phosphoglycerides. Phosphate esters

So far, we have talked only about glycerides in which all three ester linkages are to acyl groups, that is, triacylglycerols. There also occur lipids of another kind, phosphoglycerides, which contain only two acyl groups and, in place of the third, a *phosphate* group. The parent structure is *diacylglycerol phosphate*, or *phosphatidic acid.*

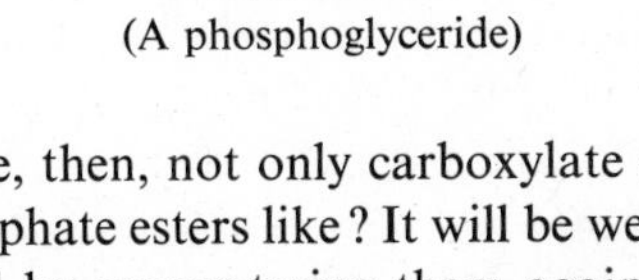

```
R'—C—O—CH2
   ‖     |
   O     |
         |
R''—C—O—CH
    ‖    |          O
    O    |          |
        CH2—O—P—OH
                |
                OH
```

Phosphatidic acid
(A phosphoglyceride)

Phosphoglycerides are, then, not only carboxylate esters but phosphate esters as well. Just what are phosphate esters like? It will be well for us to learn something about them since we shall be encountering them again and again: phospholipids make up the membranes of cells (Sec. 33.8); adenosine triphosphate lies at the heart of the energy system of organisms, and it does its job by converting hosts of other compounds into phosphate esters (Sec. 37.3); nucleic acids, which control heredity, are polyesters of phosphoric acid.

To begin with, phosphates come in various kinds. Phosphoric acid contains three hydroxy groups and can form esters in which one, two, or three of these have been replaced by alkoxy groups. Phosphoric acid is highly acidic, and so are the

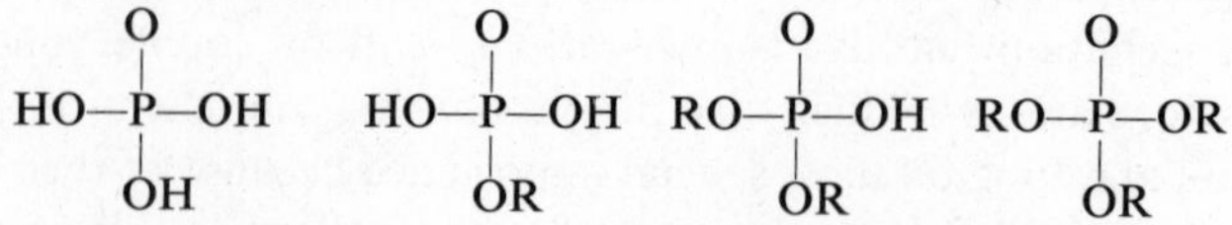

$$\underset{\text{Phosphoric acid}}{\text{HO—}\overset{\text{O}}{\overset{\|}{\text{P}}}\text{—OH} \atop \text{OH}} \qquad \text{HO—}\overset{\text{O}}{\overset{\|}{\text{P}}}\text{—OH} \atop \text{OR} \quad \text{RO—}\overset{\text{O}}{\overset{\|}{\text{P}}}\text{—OH} \atop \text{OR} \quad \text{RO—}\overset{\text{O}}{\overset{\|}{\text{P}}}\text{—OR} \atop \text{OR}$$

Phosphoric acid — Phosphate esters

monoalkyl and dialkyl esters; in aqueous solution they tend to exist as anions, the exact extent of ionization depending, of course, upon the acidity of the medium. For example:

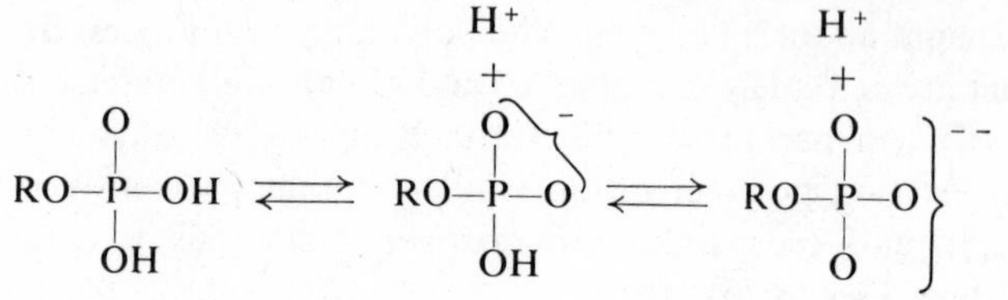

$$\text{RO—P(=O)(OH)—OH} \rightleftarrows \text{H}^+ + \text{RO—P(=O)(OH)—O}^- \rightleftarrows \text{H}^+ + \left\{\text{RO—P(O)(O)—O}\right\}^{--}$$

Like other esters, phosphates undergo hydrolysis to the parent acid and alcohol. Here, the acidity of —OH attached to phosphorus has several effects. In the first place, since acidic phosphate esters can undergo ionization, there may be many species present in the hydrolysis solution. A monoalkyl ester, for example, could exist as dianion, monoanion, neutral ester, and protonated ester; any or all of these could conceivably be undergoing hydrolysis. Actually, the situation is not quite that complicated. From the dissociation constants of these acidic esters, one can calculate the fraction of ester in each form in a given solution. The dependence of rate on acidity of the solution often shows which species is the principal reactant.

In carboxylates, we remember, attack generally occurs at acyl carbon, and in sulfonates, at alkyl carbon, with a resulting difference in point of cleavage. In

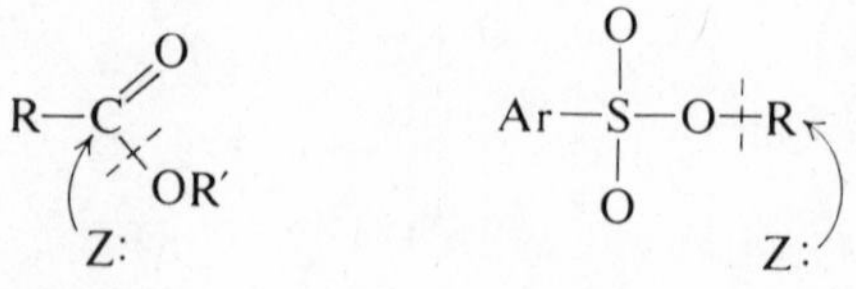

$$\text{R—C(=O)}\,\vert\,\text{OR}' \;\;(\text{Z:} \to \text{C}) \qquad \text{Ar—S(=O)(=O)—O}\,\vert\,\text{R} \;\;(\text{Z:} \to \text{R})$$

hydrolytic behavior, phosphates are intermediate between carboxylates and sulfonates. Cleavage can occur at either position, depending on the nature of the alcohol group.

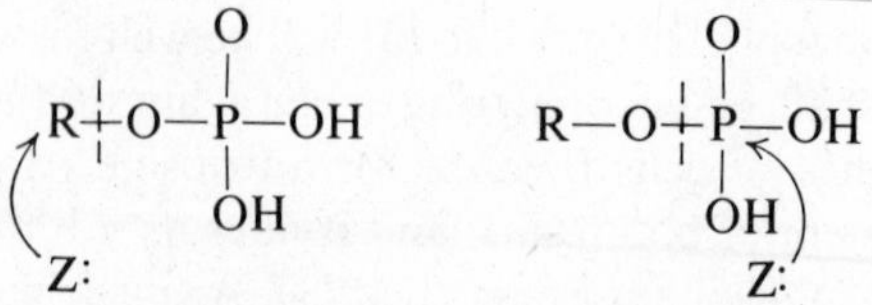

$$\text{R}\,\vert\,\text{O—P(=O)(OH)—OH} \;\;(\text{Z:} \to \text{R}) \qquad \text{R—O}\,\vert\,\text{P(=O)(OH)—OH} \;\;(\text{Z:} \to \text{P})$$

C—O cleavage — P—O cleavage

Here again the acidity of phosphoric acids comes in. Cleavage of the alkyl–oxygen bond in carboxylates is difficult because the carboxylate anion is strongly basic and a poor leaving group; in sulfonates such cleavage is favored because the weakly basic sulfonate anion is a very good leaving group. Phosphoric acid is intermediate in acidity between carboxylic and sulfonic acid; as a result, the phosphate anion is a better leaving group than carboxylate but a poorer one than sulfonate. In these esters, phosphorus is bonded to four groups; but it can accept more—witness stable pentacovalent compounds like PCl_5—and nucleophilic attack at phosphorus competes with attack at alkyl carbon.

In acidic solution, phosphate esters are readily cleaved to phosphoric acid. In alkaline solution, however, only trialkyl phosphates, $(RO)_3PO$, are hydrolyzed, and only one alkoxy group is removed. Monoalkyl and dialkyl esters, $ROPO(OH)_2$ and $(RO)_2PO(OH)$, are inert to alkali, even on long treatment. This may seem unusual behavior, but it has a perfectly rational explanation. The monoalkyl and dialkyl esters contain acidic —OH groups on phosphorus, and in alkaline solution exist as anions; repulsion between like charges prevents attack on these anions by hydroxide ion.

In most phospholipids, phosphate is of the kind

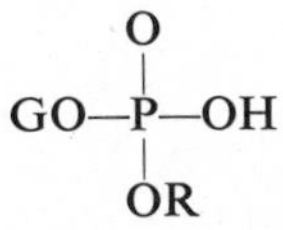

```
      O
      |
GO—P—OH
      |
      OR
```

in which G is the glyceryl group—with its two carboxylates—and R is derived from some other alcohol, ROH, most often *ethanolamine*, $HOCH_2CH_2NH_2$, or *choline*, $HOCH_2CH_2N(CH_3)_3^+$. Since the remaining —OH on phosphorus is

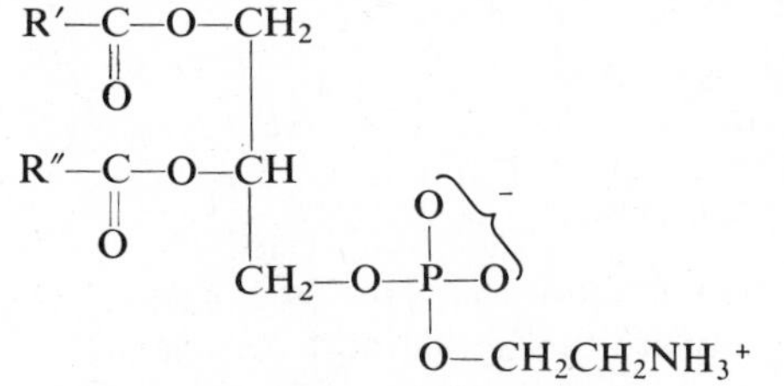

```
R′—C—O—CH2                              R′—C—O—CH2
   ||    |                                 ||    |
   O     |                                 O     |
R″—C—O—CH                               R″—C—O—CH
   ||    |      O  -                       ||    |      O  -
   O     |      |                          O     |      |
        CH2—O—P—O                                CH2—O—P—O
                |                                        |
                O—CH2CH2NH3+                             O—CH2CH2N(CH3)3+
```

Phosphatidyl ethanolamine
(Ethanolamine phosphoglyceride)

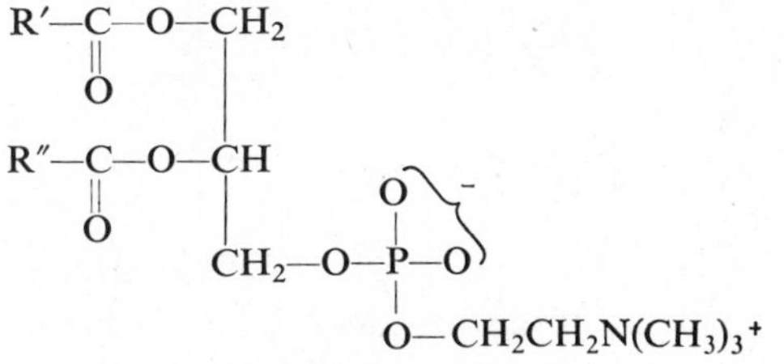

Phosphatidyl choline
(Choline phosphoglyceride)

highly acidic, the ester exists mostly in the ionic form. Furthermore, since the alcohol ROH usually contains an amino group, the phosphate unit carries both positive and negative charges, and the phospholipid is—at this end—a *dipolar ion.* On hydrolysis, these phosphates generally undergo cleavage between phosphorus and oxygen, P⫶O—R.

Problem 33.4 Consider hydrolysis of $(RO)_2PO(OH)$ by aqueous hydroxide, and grant that for electrostatic reasons attack by OH^- cannot occur. Even so, why does not attack by the nucleophile water lead to hydrolysis? After all, water *is* the successful nucleophile in acidic hydrolysis. (*Hint:* See Sec. 20.18.)

33.8 Phospholipids and cell membranes

The fats are found, we said, in storage fat cells of plants and animals. Their function rests on their chemical properties: through oxidation, they are consumed to help provide energy for the life processes.

The phospholipids, on the other hand, are found in the membranes of cells—all cells—and are a basic structural element of living organisms. This vital function depends, in a fascinating way, on their physical properties.

Phosphoglyceride molecules are amphipathic, and in this respect differ from fats—but resemble soaps and detergents. The hydrophobic part is, again, the long fatty acid chains. The hydrophilic part is the dipolar ionic end: the substituted phosphate group with its positive and negative charges. In aqueous solution, as we would expect, phosphoglycerides form micelles. In certain situations, however—at an aperture between two aqueous solutions, for example—they tend to form bilayers: two rows of molecules are lined up, back to back, with their polar ends projecting into water on the two surfaces of the bilayer (Fig. 33.3). Although the

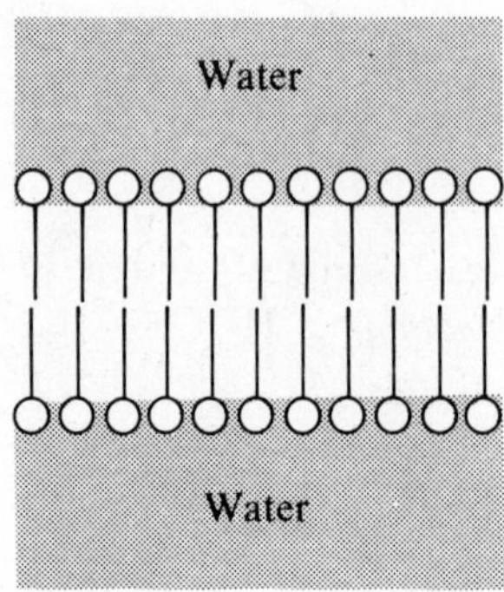

Figure 33.3. A phospholipid bilayer. Hydrophobic fatty chains held together by van der Waals forces. Hydrophilic ends dissolve in water.

polar groups are needed to hold molecules in position, the bulk of the bilayer is made up of the fatty acid chains. Non-polar molecules can therefore dissolve in this mostly hydrocarbon wall and pass through it, but it is an effective barrier to polar molecules and ions.

It is in the form of bilayers that phosphoglycerides are believed to exist in cell membranes. They constitute walls that not only enclose the cell but also very selectively control the passage, in and out, of the various substances—nutrients, waste products, hormones, etc.—even from a solution of low concentration to a solution of high concentration. Now, many of these substances that enter and leave the cells are highly polar molecules like carbohydrates and amino acids, or ions like sodium and potassium. How can these molecules pass through cell membranes when they cannot pass through simple bilayers? And how can permeability be so highly selective?

The answer to both these questions seems to involve the proteins that are also found in cell membrane: embedded in the bilayer, and even extending clear through it. Proteins, as we shall see in Chap. 36, are very long chain amides, polymers of twenty-odd different amino acids. Protein chains can be looped and coiled in a variety of ways; the conformation that is favored for a particular protein molecule depends on the exact sequence of amino acids along its chain.

It has been suggested that transport through membranes happens in the following way. A protein molecule, coiled up to turn its hydrophobic parts outward, is dissolved in the bilayer, forming a part of the cell wall. A molecule approaches: a potassium ion, say. If the particular protein is the one designed to handle potassium ion, it receives the ion into its polar interior. Hidden in this hydrophobic wrapping, the ion is smuggled through the bilayer and released on the other side.

Now, if the transport protein is to do its job, it must be free to move within the membrane. The molecules of the bilayer, while necessarily aligned, must not be locked into a rigid crystalline lattice—as they would be if all the fatty acid chains were saturated. Actually, some of the chains in the membrane phospholipids are unsaturated and these, with their *cis* stereochemistry and the accompanying bend (Fig. 33.1), disrupt the alignment enough to make the membrane semiliquid at physiological temperatures.

Here, we have had a glimpse of just one complex biological process. Yet we can begin to see how the understanding of biology rests on basic chemical concepts: van der Waals forces and ion–dipole bonds; polarity and solubility; melting point and molecular shape; configuration and conformation; and, ultimately, the sequence of atoms in molecular chains.

Problem 33.5 The degree of unsaturation of the membrane lipids in the legs of reindeer is higher in cells near the hooves than in cells near the body. What survival value does this unsaturation gradient have?

PROBLEMS

1. From saponification of cerebrosides, lipids found in the membranes of brain and nerve cells, there is obtained *nervonic acid.* This acid rapidly decolorizes dilute $KMnO_4$ and Br_2/CCl_4 solutions. Hydrogenation in the presence of nickel yields tetracosanoic acid, n-$C_{23}H_{47}COOH$. Vigorous oxidation of nervonic acid yields one acid of neutralization equivalent 156 ± 3 and another acid of neutralization equivalent 137 ± 2. What structure or structures are possible for nervonic acid?

2. When peanut oil is heated very briefly with a little sodium methoxide, its properties are changed dramatically—it becomes so viscous it can hardly be poured—yet saponification yields the same mixture of fatty acids as did the untreated oil. What has probably happened?

3. On oxidation with O_2, methyl oleate (methyl 9-*cis*-octadecenoate) was found to yield a mixture of hydroperoxides of formula $C_{19}H_{36}O_4$. In these, the —OOH group was found attached not only to C–8 and C–11 but also to C–9 and C–10. What is the probable structure of these last two hydroperoxides? How did they arise? Show all steps in a likely mechanism for the reaction.

4. Although alkaline hydrolysis of monoalkyl or monoaryl phosphates is ordinarily very difficult, 2,4-dinitrophenyl phosphate, 2,4-$(NO_2)_2C_6H_3OPO_3H_2$, does react with aqueous base, and with cleavage at the phosphorus–oxygen bond. Suggest an explanation for this.

5. *Spermaceti* (a wax from the head of the sperm whale) resembles high-molecular weight hydrocarbons in physical properties and inertness toward Br_2/CCl_4 and $KMnO_4$; on qualitative analysis it gives positive tests only for carbon and hydrogen. However, its infrared spectrum shows the presence of an ester group, and quantitative analysis gives the empirical formula $C_{16}H_{32}O$.

A solution of the wax and KOH in ethanol is refluxed for a long time. Titration of an aliquot shows that one equivalent of base has been consumed for every 475 ± 10 grams of wax. Water and ether are added to the cooled reaction mixture, and the aqueous and ethereal layers are separated. Acidification of the aqueous layer yields a solid A, m.p. 62–3°, neutralization equivalent 260 ± 5. Evaporation of the ether layer yields a solid B, m.p. 48–9°. (a) What is a likely structure of spermaceti? (b) Reduction by $LiAlH_4$ of either spermaceti or A gives B as the only product. Does this confirm the structure you gave in (a)?

6. As the acidity of the solution is increased, the rate of hydrolysis of monoalkyl phosphates, $ROPO(OH)_2$, rises from essentially zero in alkaline solution, and *passes through a maximum* at the point (moderate acidity, pH about 4) where the concentration of monoanion, $ROPO(OH)(O^-)$, is greatest. Cleavage is at the phosphorus–oxygen bond.

(a) Can you suggest a mechanism or mechanisms that might account for the fact that this species is more reactive than either the dianion, $ROPO(O^-)_2$, or the neutral ester?

(b) At still higher acidity, the rate rises again and continues to rise. To what is the high reactivity now due?

7. On the basis of the following synthesis, give the structure of *vaccenic acid.*

n-hexyl chloride + sodium acetylide $\longrightarrow$ C (C_8H_{14})
C + Na, NH_3; then $I(CH_2)_9Cl$ $\longrightarrow$ D ($C_{17}H_{31}Cl$)
D + KCN $\longrightarrow$ E ($C_{18}H_{31}N$)
E + OH^-, heat; then H^+ $\longrightarrow$ F ($C_{18}H_{32}O_2$)
F + H_2, Pd $\longrightarrow$ *vaccenic acid* ($C_{18}H_{34}O_2$)

8. From the lipids of *Corynebacterium diphtherium* there is obtained *corynomycolenic acid.* Its structure was confirmed by the following synthesis.

n-$C_{13}H_{27}CH_2Br$ + sodiomalonic ester $\longrightarrow$ G ($C_{21}H_{40}O_4$)
G + exactly one mole alc. KOH $\longrightarrow$ H ($C_{19}H_{36}O_4$)
H + dihydropyran (Problem 16, p. 692) $\longrightarrow$ I ($C_{24}H_{44}O_5$)
cis-9-hexadecenoic acid + $SOCl_2$ $\longrightarrow$ J ($C_{16}H_{29}OCl$)
I + Na, then J $\longrightarrow$ K ($C_{40}H_{72}O_6$)
K + dilute acid $\longrightarrow$ L ($C_{34}H_{64}O_3$)
L + $NaBH_4$ $\longrightarrow$ M ($C_{34}H_{66}O_3$)
M + OH^-, heat; then H^+ $\longrightarrow$ (±)-*corynomycolenic acid* ($C_{32}H_{62}O_3$)

What is the structure of corynomycolenic acid?

9. From saponification of the fatty capsule of the tubercle bacillus, there is obtained *tuberculostearic acid.* Its structure was established by the following synthesis.

2-decanol + PBr_3 $\longrightarrow$ N ($C_{10}H_{21}Br$)
N + sodiomalonic ester; then OH^-, heat; then H^+; then heat $\longrightarrow$ O ($C_{12}H_{24}O_2$)
O + $SOCl_2$ $\longrightarrow$ P $\xrightarrow{C_2H_5OH}$ Q ($C_{14}H_{28}O_2$)
Q + $LiAlH_4$ $\longrightarrow$ R ($C_{12}H_{26}O$)
R + PBr_3 $\longrightarrow$ S ($C_{12}H_{25}Br$)
S + Mg; $CdCl_2$; then $C_2H_5OOC(CH_2)_5COCl$ $\longrightarrow$ T ($C_{21}H_{40}O_3$)
T + Zn, HCl $\longrightarrow$ U ($C_{21}H_{42}O_2$)
U + OH^-, heat; then H^+ $\longrightarrow$ tuberculostearic acid ($C_{19}H_{38}O_2$)

What is the structure of tuberculostearic acid?

10. Besides tuberculostearic acid (preceding problem), the capsule of the tubercle bacillus yields *C_{27}-phthienoic acid,* which on injection into animals causes the lesions typical of tuberculosis. On the basis of the following data, assign a structure to this acid.

C_{27}-phthienoic acid ($C_{27}H_{52}O_2$) + O_3; then Zn, H_2O $\longrightarrow$ $CH_3COCOOH$
C_{27}-phthienoic acid + $KMnO_4$ $\longrightarrow$ acid V ($C_{24}H_{48}O_2$)
methyl ester of V + $2C_6H_5MgBr$; then H_2O $\longrightarrow$ W ($C_{36}H_{58}O$)
W + H^+, heat $\longrightarrow$ X ($C_{36}H_{56}$)
X + CrO_3 $\longrightarrow$ $(C_6H_5)_2CO$ + ketone Y ($C_{23}H_{46}O$)
Y + I_2, NaOH $\longrightarrow$ CHI_3
V + Br_2, P $\longrightarrow$ Z $\xrightarrow{\text{alc. KOH}}$ acid AA ($C_{24}H_{46}O_2$)
AA + $KMnO_4$ $\longrightarrow$ among other things, BB ($C_{20}H_{40}O$)
Compound BB was shown to be identical with a sample of $CH_3(CH_2)_{17}COCH_3$.

Caution: $KMnO_4$ is a vigorous reagent, and not all the cleavage occurs at the double bond. Compare the number of carbons in AA and BB.

11. On the basis of the following nmr spectra, assign likely structures to the isomeric fatty acids, CC and DD, of formula $C_{17}H_{35}COOH$.

Isomer CC
a triplet, δ 0.8, 3H
b broad band, δ 1.35, 30H
c triplet, δ 2.3, 2H
d singlet, δ 12.0, 1H

Isomer DD
a triplet, δ 0.8, 3H
b doublet, δ 1.15, 3H
c broad band, δ 1.35, 28H
d multiplet, δ 2.2, 1H
e singlet, δ 12.05, 1H

12. *Juvenile hormones* take part in the delicate balance of hormonal activity that controls development of insects. Applied artificially, they prevent maturing, and thus offer a highly specific way to control insect population.

The structure of the juvenile hormone of the moth *Hyalophora cecropia* was confirmed by the following synthesis. (At each stage where geometric isomers were obtained, these were separated and the desired one—(*Z*) or (*E*)—was selected on the basis of its nmr spectrum.)

2-butanone + $[(CH_3O)_2P(O)CHCOOCH_3]^-Na^+$ (See Sec. 26.2) $\longrightarrow$ [EE ($C_9H_{18}O_6PNa$)]

[EE] $\longrightarrow$ $(CH_3)_2PO_4^-$ + FF ((*Z*)-$C_7H_{12}O_2$) (See Sec. 21.10)
FF + $LiAlH_4$ $\longrightarrow$ GG ($C_6H_{12}O$) $\xrightarrow{PBr_3}$ HH ($C_6H_{11}Br$)
HH + $[CH_3CH_2COCHCOOC_2H_5]^-Na^+$ $\longrightarrow$ II ($C_{13}H_{22}O_3$)
II + OH^-, heat; then H^+; then heat $\longrightarrow$ JJ ($C_{10}H_{18}O$)
JJ + $[(CH_3O)_2P(O)CHCOOCH_3]^-Na^+$ $\longrightarrow$ [KK] $\longrightarrow$ LL ((*E*)-$C_{13}H_{22}O_2$)
LL + $LiAlH_4$ $\longrightarrow$ MM $\xrightarrow{PBr_3}$ NN ($C_{12}H_{21}Br$)
NN + sodioacetoacetic ester $\longrightarrow$ OO ($C_{18}H_{30}O_3$)
OO + OH^-, heat; then H^+; then heat $\longrightarrow$ PP ($C_{15}H_{26}O$)
PP + $[(CH_3O)_2P(O)CHCOOCH_3]^-Na^+$ $\longrightarrow$ [QQ] $\longrightarrow$ RR ((*E*)-$C_{18}H_{30}O_2$)
RR + *m*-$ClC_6H_4CO_2OH$ $\longrightarrow$ SS (*racemic*-$C_{18}H_{30}O_3$)

SS was a mixture of positional isomers, corresponding to attack by perbenzoic acid at various double bonds in RR. Of these, one isomer (a *racemic* modification) was found to be identical, in physical and biological properties, to the natural juvenile hormone. This isomer was the one resulting from reaction at the double bond first introduced into the molecule.

What is the structure of the juvenile hormone of *Hyalophora cecropia*? Account for the fact that the synthesis yields a racemic modification.

Chapter 34 Carbohydrates I. Monosaccharides

34.1 Introduction

In the leaf of a plant, the simple compounds carbon dioxide and water are combined to form the sugar (+)-**glucose**. This process, known as *photosynthesis*, requires catalysis by the green coloring matter *chlorophyll*, and requires energy in the form of light. Thousands of (+)-glucose molecules can then be combined to form the much larger molecules of **cellulose**, which constitutes the supporting framework of the plant. (+)-Glucose molecules can also be combined, in a somewhat different way, to form the large molecules of **starch**, which is then stored in the seeds to serve as food for a new, growing plant.

When eaten by an animal, the starch—and in the case of certain animals also the cellulose—is broken down into the original (+)-glucose units. These can be carried by the bloodstream to the liver to be recombined into **glycogen**, or animal starch; when the need arises, the glycogen can be broken down once more into (+)-glucose. (+)-Glucose is carried by the bloodstream to the tissues, where it is oxidized, ultimately to carbon dioxide and water, with the release of the energy originally supplied as sunlight. Some of the (+)-glucose is converted into fats; some reacts with nitrogen-containing compounds to form amino acids, which in turn are combined to form the proteins that make up a large part of the animal body.

(+)-Glucose, cellulose, starch, and glycogen all belong to the class of organic compounds known as **carbohydrates**. Carbohydrates are the ultimate source of most of our food: we eat starch-containing grain, or feed it to animals to be converted into meat and fat which we then eat. We clothe ourselves with cellulose in the form of cotton and linen, rayon and cellulose acetate. We build houses and furniture from cellulose in the form of wood. Thus carbohydrates quite literally provide us with the necessities of life: food, clothing, and shelter.

Basic necessities aside, our present civilization depends to a surprising degree upon cellulose, particularly as *paper*: the books and newspapers we read, the letters we write, the bills we pay and the money and checks with which we pay them; marriage licenses, drivers' licenses, birth certificates, mortgages; paper in the form of bags and boxes, sheets and rolls.

The study of carbohydrates is one of the most exciting fields of organic chemistry. It extends from the tremendously complicated problem of understanding the process of photosynthesis to the equally difficult problem of unraveling the tangled steps in the enzyme-catalyzed reconversion of (+)-glucose into carbon dioxide and water. Between these two biochemical problems there lie the more traditional problems of the organic chemist: determination of the structure and properties of the carbohydrates, and the study of their conversion into other organic compounds.

In this book we shall learn something of the fundamental chemical properties of the carbohydrates, knowledge that is basic to any further study of these compounds.

34.2 Definition and classification

Carbohydrates are polyhydroxy aldehydes, polyhydroxy ketones, or compounds that can be hydrolyzed to them. A carbohydrate that cannot be hydrolyzed to simpler compounds is called a **monosaccharide**. A carbohydrate that can be hydrolyzed to two monosaccharide molecules is called a **disaccharide**. A carbohydrate that can be hydrolyzed to many monosaccharide molecules is called a **polysaccharide**.

A monosaccharide may be further classified. If it contains an aldehyde group, it is known as an **aldose**; if it contains a keto group, it is known as a **ketose**. Depending upon the number of carbon atoms it contains, a monosaccharide is known as a **triose**, **tetrose**, **pentose**, **hexose**, and so on. An **aldohexose**, for example, is a six-carbon monosaccharide containing an aldehyde group; a **ketopentose** is a five-carbon monosaccharide containing a keto group. Most naturally occurring monosaccharides are pentoses or hexoses.

Carbohydrates that reduce Fehling's (or Benedict's) or Tollens' reagent (p. 1075) are known as **reducing sugars**. All monosaccharides, whether aldose or ketose, are reducing sugars. Most disaccharides are reducing sugars; sucrose (common table sugar) is a notable exception, for it is a non-reducing sugar.

34.3 (+)-Glucose: an aldohexose

Because it is the unit of which starch, cellulose, and glycogen are made up, and because of its special role in biological processes, **(+)-glucose** is by far the most abundant monosaccharide—there are probably more (+)-glucose units in nature than any other organic group—and by far the most important monosaccharide.

Most of what we need to know about monosaccharides we can learn from the study of just this one compound, and indeed from the study of just one aspect: its structure, and how that structure was arrived at. In learning about the structure of (+)-glucose, we shall at the same time learn about its properties, since it is from these properties that the structure has been deduced. (+)-Glucose is typical monosaccharide, so that in learning about its structure and properties, we shall be learning about the structure and properties of the other members of this family.

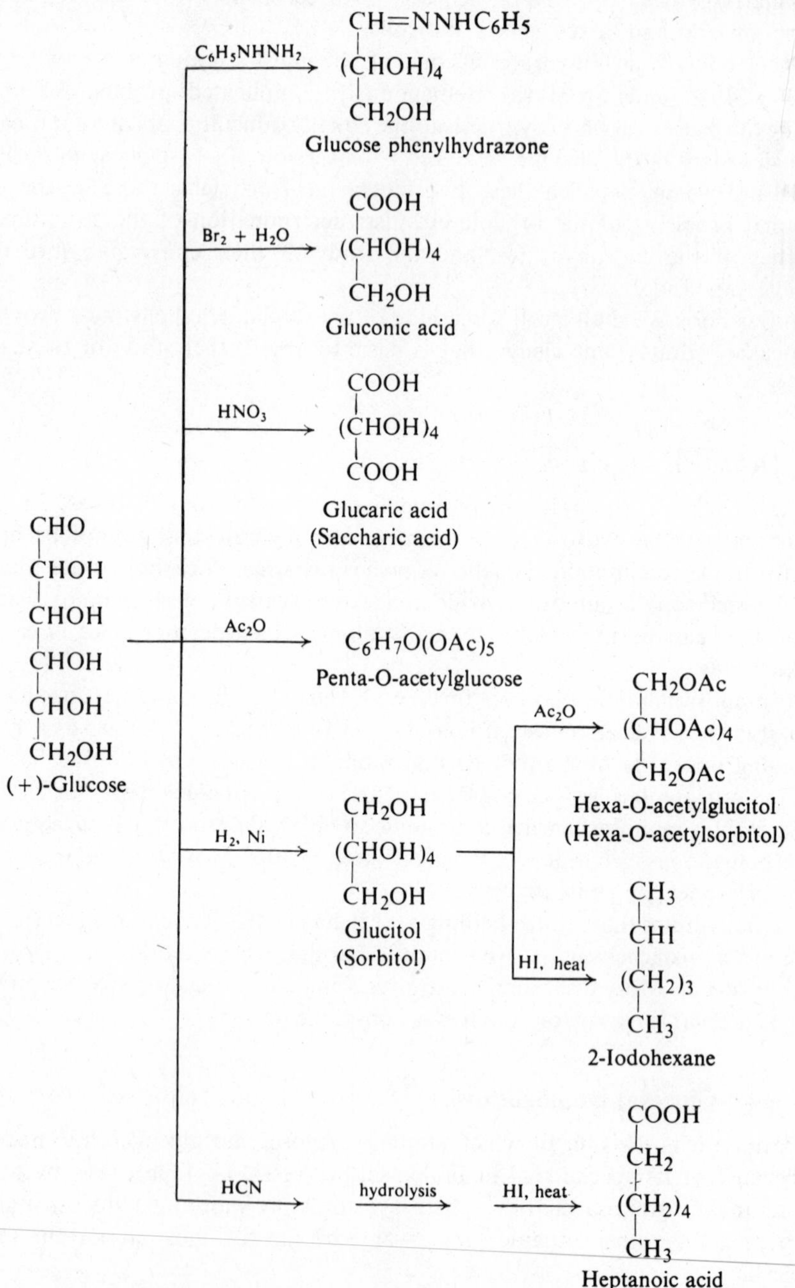

Figure 34.1. (+)-Glucose as an aldohexose.

(+)-Glucose has the molecular formula $C_6H_{12}O_6$, as shown by elemental analysis and molecular weight determination. In Fig. 34.1 is summarized other evidence about its structure: evidence consistent with the idea that (+)-glucose is a six-carbon, straight-chain, pentahydroxy aldehyde, that is, that (+)-glucose is an aldohexose. But this is only the beginning. There are, as we shall see, 16 possible aldohexoses, all stereoisomers of each other, and we want to know which one (+)-glucose is. Beyond this, there is the fact that (+)-glucose exists in *alpha* and *beta* forms, indicating still further stereochemical possibilities that are not accommodated by the simple picture of a pentahydroxy aldehyde. Finally, we must pinpoint the predominant conformation in which the compound exists. All this is the structure of (+)-glucose and, when we have arrived at it, we shall see the features that make it the very special molecule that it is.

Problem 34.1 Assume that you start knowing only the molecular formula of (+)-glucose. You carry out each of the reactions of Fig. 34.1, and study each of the products obtained: characterize the product as to family; determine its molecular weight and, if any, its neutralization equivalent. You identify 2-iodohexane and heptanoic acid by comparison with authentic samples.

(a) For each product, tell what you would actually observe. (b) Take each piece of evidence in turn, and tell what it shows about the structure of (+)-glucose.

34.4 (—)-Fructose: a 2-ketohexose

The most important ketose is **(—)-fructose**, which occurs widely in fruits and, combined with glucose, in the disaccharide *sucrose* (common table sugar).

The following sequence shows that (−)-fructose is a ketone rather than an aldehyde, and gives the position of the keto group in the chain:

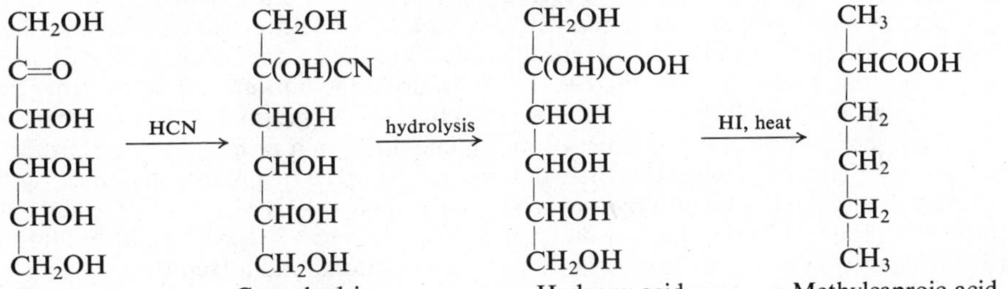

$$
\begin{array}{c}
CH_2OH \\ | \\ C{=}O \\ | \\ CHOH \\ | \\ CHOH \\ | \\ CHOH \\ | \\ CH_2OH \\ \text{Fructose}
\end{array}
\xrightarrow{\text{HCN}}
\begin{array}{c}
CH_2OH \\ | \\ C(OH)CN \\ | \\ CHOH \\ | \\ CHOH \\ | \\ CHOH \\ | \\ CH_2OH \\ \text{Cyanohydrin} \\ \text{(two diastereomers)}
\end{array}
\xrightarrow{\text{hydrolysis}}
\begin{array}{c}
CH_2OH \\ | \\ C(OH)COOH \\ | \\ CHOH \\ | \\ CHOH \\ | \\ CHOH \\ | \\ CH_2OH \\ \text{Hydroxy acid} \\ \text{(two diastereomers)}
\end{array}
\xrightarrow{\text{HI, heat}}
\begin{array}{c}
CH_3 \\ | \\ CHCOOH \\ | \\ CH_2 \\ | \\ CH_2 \\ | \\ CH_2 \\ | \\ CH_3 \\ \alpha\text{-Methylcaproic acid} \\ \text{(racemic modification)}
\end{array}
$$

Fructose is thus a 2-ketohexose.

34.5 Stereoisomers of (+)-glucose. Nomenclature of aldose derivatives

If we examine the structural formula we have drawn for glucose, we see that it contains four chiral centers (marked by asterisks):

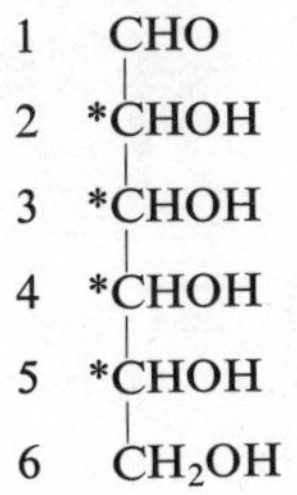

$$
\begin{array}{rl}
1 & CHO \\ & | \\ 2 & {}^*CHOH \\ & | \\ 3 & {}^*CHOH \\ & | \\ 4 & {}^*CHOH \\ & | \\ 5 & {}^*CHOH \\ & | \\ 6 & CH_2OH
\end{array}
$$

Each of the possible stereoisomers is commonly represented by a "cross" formula, as, for example, in I. As always in formulas of this kind, it is understood that

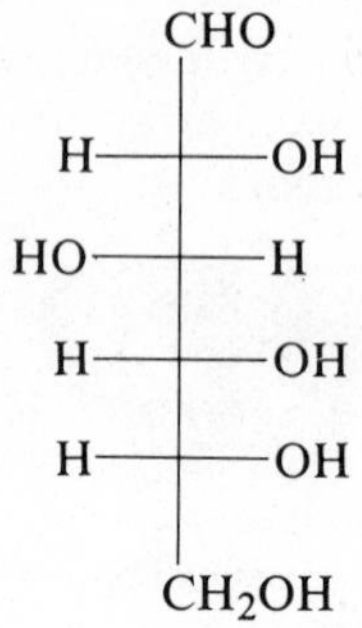

CHO
H—OH
HO—H
H—OH
H—OH
CH_2OH
I

stands for

CHO
H OH
HO H
H OH
H OH
CH_2OH

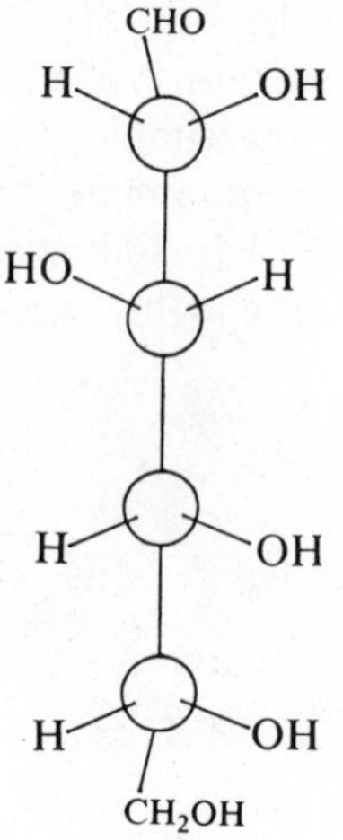

horizontal lines represent bonds coming *toward us* out of the plane of the paper, and *vertical* lines represent bonds going *away from us* behind the plane of the paper.

Only molecular models can show us what is really meant by formulas like I. A correct model of one of these stereoisomers is difficult to build unless we follow certain rules first clearly stated by the great carbohydrate chemist Emil Fischer:

(1) Construct a chain of carbon atoms with a —CHO group at one end and a —CH_2OH group at the other. (2) Hold the —CHO group in one hand and let the rest of the chain hang down. (3) Take the —CH_2OH group at the bottom end in the other hand and bring it up *behind* the chain until it touches the —CHO group. (4) Now one hand can hold both groups firmly and the rest of the chain will form a rather rigid ring projecting *toward you.* (This is the object of the whole operation up to this point: to impart rigidity to an otherwise flexible chain.) By this procedure you have —CHO above —CH_2OH as in formula I, and both these groups directed *away from you.* (5) Finally, still holding the ring as described above, look in turn at each carbon atom, and attach the —OH or —H to the right or to the left just as it appears in the "cross" formula. In each case, these groups will be directed *toward you.*

The dissimilarity of the two ends of an aldohexose molecule prevents the existence of *meso* compounds (Sec. 4.18), and hence we expect that there should be 2^4 or 16 stereoisomers—eight pairs of enantiomers. All 16 of these possible stereoisomers are now known, through either synthesis in the laboratory or isolation from natural sources; only three—(+)-glucose, (+)-mannose, (+)-galactose—are found in abundance.

Problem 34.2 Draw a "cross" formula of one enantiomer of each of these eight pairs, placing —CHO at the top, —CH_2OH at the bottom, and —OH on the right on the lowest chiral center (C-5).

Of these 16 isomers, only one is the (+)-glucose that we have described as the most abundant monosaccharide. A second isomer is (−)-glucose, the enantiomer of the naturally occurring compound. The other 14 isomers are all diastereomers of (+)-glucose, and are given names of their own, for example, *mannose, galactose, gulose*, etc. As we might expect, these other aldohexoses undergo the same set of reactions that we have described for glucose. Although as diastereomers they undergo these reactions at different rates and yield different individual compounds, the chemistry is essentially the same.

The products obtained from these other aldohexoses are generally given names that correspond to the names of the products obtained from glucose. This principle is illustrated in Table 34.1 for the aldohexose (+)-mannose, which occurs naturally in many plants (the name is derived from the Biblical word *manna*).

Table 34.1 NAMES OF ALDOSE DERIVATIVES

Type of Compound	Type Name	Examples of Specific Names	
Monosaccharide $HOCH_2(CHOH)_nCHO$	Aldose	*Glucose*	*Mannose*
Monocarboxylic acid $HOCH_2(CHOH)_nCOOH$	Aldonic acid	*Gluconic acid*	*Mannonic acid*
Dicarboxylic acid $HOOC(CHOH)_nCOOH$	Aldaric acid	*Glucaric acid* (*Saccharic acid*)	*Mannaric acid* (*Mannosaccharic acid*)
Polyhydroxy alcohol $HOCH_2(CHOH)_nCH_2OH$	Alditol	*Glucitol* (*Sorbitol*)	*Mannitol*
Aldehydo acid $HOOC(CHOH)_nCHO$	Uronic acid	*Glucuronic acid*	*Mannuronic acid*

The structural formula we have drawn to represent (+)-glucose so far could actually represent any of the 16 aldohexoses. Only when we have specified the configuration about each of the chiral centers will we have the structural formula that applies only to (+)-glucose itself. Before we can discuss the brilliant way in which the configuration of (+)-glucose was worked out, we must first learn a little more about the chemistry of monosaccharides.

Problem 34.3 (a) How many chiral centers are there in (−)-fructose? (b) How many stereoisomeric 2-ketohexoses should there be? (c) Draw a "cross" formula of one enantiomer of each pair, placing C=O near the top, and —OH on the right on the lowest chiral center (C–5).

34.6 Oxidation. Effect of alkali

Aldoses can be oxidized in four important ways: (a) by Fehling's or Tollens' reagent; (b) by bromine water; (c) by nitric acid; and (d) by periodic acid, HIO_4.

Aldoses reduce **Tollens' reagent**, as we would expect aldehydes to do. They also reduce **Fehling's solution**, an alkaline solution of cupric ion complexed with tartrate ion (or **Benedict's solution**, in which complexing is with citrate ion); the deep-blue color of the solution is discharged, and red cuprous oxide precipitates. These reactions are less useful, however, than we might at first have expected.

In the first place, they cannot be used to differentiate aldoses from ketoses.

Ketoses, too, reduce Fehling's and Tollens' reagents; this behavior is characteristic of α-hydroxy ketones.

In the second place, oxidation by Fehling's or Tollens' reagent cannot be used for the preparation of aldonic acids (monocarboxylic acids) from aldoses. Both Fehling's and Tollens' reagents are alkaline reagents, and the treatment of sugars with alkali can cause extensive isomerization and even decomposition of the chain. Alkali exerts this effect, in part at least, by establishing an equilibrium between the monosaccharide and an enediol structure.

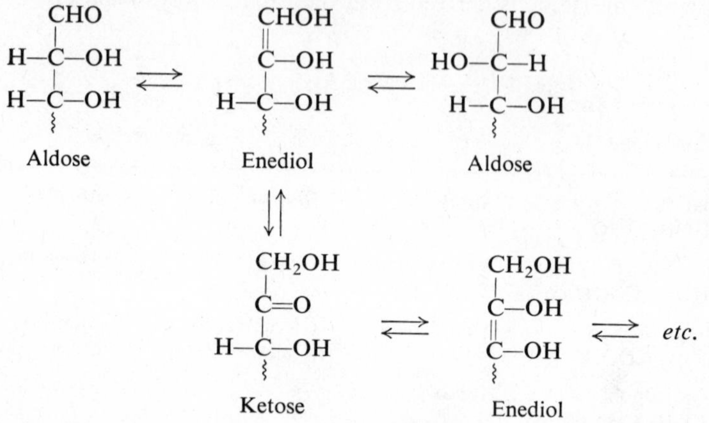

$$\begin{array}{ccccc}
\begin{array}{c} \text{CHO} \\ | \\ \text{H—C—OH} \\ | \\ \text{H—C—OH} \\ \wr \\ \text{Aldose} \end{array} & \rightleftharpoons & \begin{array}{c} \text{CHOH} \\ \| \\ \text{C—OH} \\ | \\ \text{H—C—OH} \\ \wr \\ \text{Enediol} \end{array} & \rightleftharpoons & \begin{array}{c} \text{CHO} \\ | \\ \text{HO—C—H} \\ | \\ \text{H—C—OH} \\ \wr \\ \text{Aldose} \end{array}
\end{array}$$

$$\downarrow\uparrow$$

$$\begin{array}{ccccc}
\begin{array}{c} \text{CH}_2\text{OH} \\ | \\ \text{C=O} \\ | \\ \text{H—C—OH} \\ \wr \\ \text{Ketose} \end{array} & \rightleftharpoons & \begin{array}{c} \text{CH}_2\text{OH} \\ | \\ \text{C—OH} \\ \| \\ \text{C—OH} \\ \wr \\ \text{Enediol} \end{array} & \rightleftharpoons & \textit{etc.}
\end{array}$$

Bromine water oxidizes aldoses, but not ketoses; as an acidic reagent it does not cause isomerization of the molecule. It can therefore be used to differentiate an aldose from a ketose, and is the reagent chosen to synthesize the *aldonic acid* (monocarboxylic acid) from an aldose.

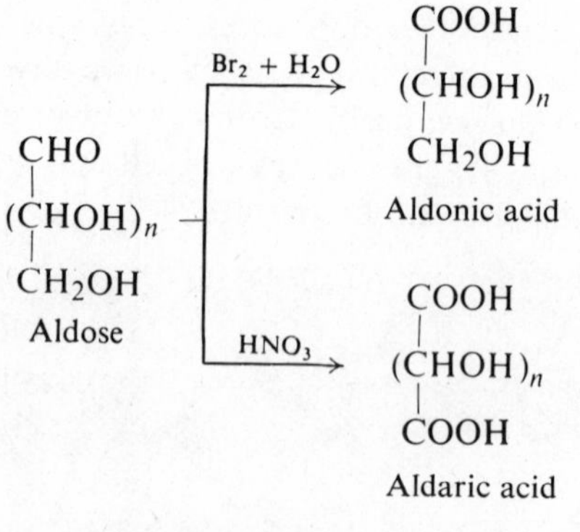

$$\begin{array}{c} \text{CHO} \\ | \\ (\text{CHOH})_n \\ | \\ \text{CH}_2\text{OH} \\ \text{Aldose} \end{array}
\begin{array}{l}
\xrightarrow{Br_2 + H_2O} \begin{array}{c} \text{COOH} \\ | \\ (\text{CHOH})_n \\ | \\ \text{CH}_2\text{OH} \\ \text{Aldonic acid} \end{array} \\[2ex]
\xrightarrow{HNO_3} \begin{array}{c} \text{COOH} \\ | \\ (\text{CHOH})_n \\ | \\ \text{COOH} \\ \text{Aldaric acid} \end{array}
\end{array}$$

Treatment of an aldose with the more vigorous oxidizing agent **nitric acid** brings about oxidation not only of the —CHO group but also of the $—CH_2OH$ group, and leads to the formation of the *aldaric acid* (dicarboxylic acid).

Like other compounds that contain two or more —OH or =O groups on *adjacent* carbon atoms, carbohydrates undergo oxidative cleavage by **periodic acid**, HIO_4 (Sec. 16.12). This reaction, introduced in 1928 by L. Malaprade (at the University of Nancy, France), is one of the most useful tools in modern research on carbohydrate structure.

Problem 34.4 Treatment of (+)-glucose with HIO_4 gives results that confirm its aldohexose structure. What products should be formed, and how much HIO_4 should be consumed?

Problem 34.5 Identify each of the following glucose derivatives:

$$A + 4HIO_4 \longrightarrow 3HCOOH + HCHO + OHC{-}COOH$$
$$B + 5HIO_4 \longrightarrow 4HCOOH + 2HCHO$$
$$C + 3HIO_4 \longrightarrow 2HCOOH + 2OHC{-}COOH$$
$$D + 4HIO_4 \longrightarrow 4HCOOH + OHC{-}COOH$$

34.7 Osazone formation. Epimers

As aldehydes, aldoses react with phenylhydrazine to form phenylhydrazones. If an excess of phenylhydrazine is used, the reaction proceeds further to yield products known as **osazones**, which contain two phenylhydrazine residues per molecule; a third molecule of the reagent is turned into aniline and ammonia. (Just how the —OH group is oxidized is not quite clear.)

$$\underset{\text{Aldose}}{\begin{array}{l}CHO\\ |\\ CHOH\\ \wr\end{array}} \xrightarrow{3C_6H_5NHNH_2} \underset{\text{Osazone}}{\begin{array}{l}CH{=}NNHC_6H_5\\ |\\ C{=}NNHC_6H_5\\ \wr\end{array}} + C_6H_5NH_2 + NH_3$$

Osazone formation is not limited to carbohydrates, but is typical of α-hydroxy aldehydes and α-hydroxy ketones in general (e.g., *benzoin*, $C_6H_5CHOHCOC_6H_5$).

Removal of the phenylhydrazine groups yields dicarbonyl compounds known as **osones**. For example:

$$\underset{\text{Osazone}}{\begin{array}{l}CH{=}NNHC_6H_5\\ |\\ C{=}NNHC_6H_5\\ \wr\end{array}} \xrightarrow{C_6H_5CHO,\ H^+} \underset{\text{Osone}}{\begin{array}{l}CHO\\ |\\ C{=}O\\ \wr\end{array}} + \underset{\text{Benzaldehyde phenylhydrazone}}{2C_6H_5CH{=}NNHC_6H_5}$$

Problem 34.6 Aldehydes are more easily reduced than ketones. On this basis what product would you expect from the reduction of glucosone by zinc and acetic acid? Outline a sequence of reactions by which an aldose can be turned into a 2-ketose.

In 1858 Peter Griess (in time taken from his duties in an English brewery) discovered diazonium salts. In 1875 Emil Fischer (at the University of Munich) found that reduction of benzenediazonium chloride by sulfur dioxide yields phenylhydrazine. Nine years later, in 1884, Fischer reported that the phenylhydrazine he had discovered could be used as a powerful tool in the study of carbohydrates.

One of the difficulties of working with carbohydrates is their tendency to form sirups; these are fine for pouring on pancakes at breakfast, but hard to work with in the laboratory. Treatment with phenylhydrazine converts carbohydrates into solid osazones, which are readily isolated and purified, and can be identified by their characteristic crystalline forms.

Fischer found osazone formation to be useful not only in identifying carbohydrates, but also—and this was much more important—in determining their configurations. For example, the two diastereomeric aldohexoses (+)-glucose and

(+)-mannose yield the same osazone. Osazone formation destroys the configuration about C–2 of an aldose, but does not affect the configuration of the rest of the molecule.

1	CHO		1	HC=NNHC$_6$H$_5$		1	CHO
2	H—C—OH		2	C=NNHC$_6$H$_5$		2	HO—C—H
3	——	$\xrightarrow{3C_6H_5NHNH_2}$	3	——	$\xleftarrow{3C_6H_5NHNH_2}$	3	——
4	——		4	——		4	——
5	——		5	——		5	——
6	CH$_2$OH		6	CH$_2$OH		6	CH$_2$OH

Epimers give the same osazone

It therefore follows that (+)-glucose and (+)-mannose differ only in configuration about C–2, and have the same configuration about C–3, C–4, and C–5. We can see that whenever the configuration of either of these compounds is established, the configuration of the other is immediately known through this osazone relationship. *A pair of diastereomeric aldoses that differ only in configuration about C–2 are called* **epimers**. One way in which a pair of aldoses can be identified as epimers is through the formation of the same osazone.

Problem 34.7 When the ketohexose (−)-fructose is treated with phenylhydrazine, it yields an osazone that is identical with the one prepared from either (+)-glucose or (+)-mannose. How is the configuration of (−)-fructose related to those of (+)-glucose and (+)-mannose?

34.8 Lengthening the carbon chain of aldoses. The Kiliani-Fischer synthesis

In the next few sections we shall examine some of the ways in which an aldose can be converted into a different aldose. These conversions can be used not only to synthesize new carbohydrates, but also, as we shall see, to help determine their configurations.

First, let us look at a method for converting an aldose into another aldose containing one more carbon atom, that is, at a method for lengthening the carbon chain. In 1886, Heinrich Kiliani (at the Technische Hochschule in Munich) showed that an aldose can be converted into two aldonic acids of the next higher carbon number by addition of HCN and hydrolysis of the resulting cyanohydrins. In 1890, Fischer reported that reduction of an aldonic acid (in the form of its lactone, Sec. 20.15) can be controlled to yield the corresponding aldose. In Fig. 34.2, the entire **Kiliani-Fischer synthesis** is illustrated for the conversion of an aldopentose into two aldohexoses.

Addition of cyanide to the aldopentose generates a new chiral center, about which there are two possible configurations. As a result, two diastereomeric cyanohydrins are obtained, which yield diastereomeric carboxylic acids (aldonic acids) and finally diastereomeric aldoses.

The situation is strictly analogous to that in Sec. 7.7. Using models, we can see that the particular configuration obtained here depends upon which face of the carbonyl group is attacked by cyanide ion. Since the aldehyde is already chiral, attack at the two faces is not equally likely. Both possible diastereomeric products are formed, and in unequal amounts.

Since a six-carbon aldonic acid contains —OH groups in the γ- and δ-positions, we would expect it to form a lactone under acidic conditions (Sec. 20.15). This occurs, the γ-lactone generally being the more stable product. It is the lactone that is actually reduced to an aldose in the last step of a Kiliani-Fischer synthesis.

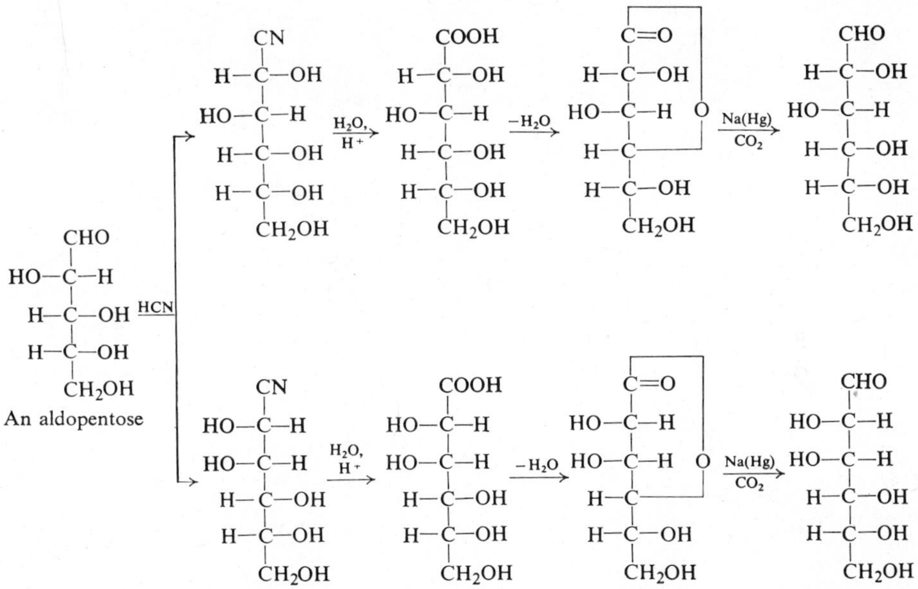

An aldopentose —HCN→ Diastereomeric cyanohydrins —H_2O, H^+→ Diastereomeric aldonic acids (*Mixture usually separated at this point*) —$-H_2O$→ Diastereomeric aldonolactones —Na(Hg), CO_2→ Diastereomeric aldohexoses (*Epimers*)

Figure 34.2. An example of the Kiliani-Fischer synthesis.

The pair of aldoses obtained from the sequence differ only in configuration about C–2, and hence are epimers. A pair of aldoses can be recognized as epimers not only by their conversion into the same osazone (Sec. 34.7), but also by their formation in the same Kiliani-Fischer synthesis.

Like other diastereomers, these epimers differ in physical properties and therefore are separable. However, since carbohydrates are difficult to purify, it is usually more convenient to separate the diastereomeric products at the acid stage, where crystalline salts are easily formed, so that a single pure lactone can be reduced to a single pure aldose.

Problem 34.8 As reducing agent, Fischer used sodium amalgam and acid. Today, lactones are reduced to aldoses by the addition of $NaBH_4$ to an aqueous solution of lactone. If, however, lactone is added to the $NaBH_4$, another product, not the aldose, is obtained. What do you think this other product is? Why is the order of mixing of reagents crucial?

Problem 34.9 (a) Using cross formulas to show configuration, outline all steps in a Kiliani-Fischer synthesis, starting with the aldotriose R-(+)-glyceraldehyde, $CH_2OHCHOHCHO$. How many aldotetroses would be expected? (b) Give configurations of the aldopentoses expected from each of these aldotetroses by a Kiliani-Fischer synthesis; of the aldohexoses expected from each of these aldopentoses.

(c) Make a "family tree" showing configurations of these aldoses hypothetically descended from R-(+)-glyceraldehyde. If the —CHO is placed at the top in each case, what configurational feature is the same in all these formulas? Why?

Problem 34.10 (a) Give the configuration of the dicarboxylic acid (aldaric acid) that would be obtained from each of the tetroses in Problem 34.9 by nitric acid oxidation. (b) Assume that you have actually carried out the chemistry in part (a). In what simple way could you assign configuration to each of your tetroses?

34.9 Shortening the carbon chain of aldoses. The Ruff degradation

There are a number of ways in which an aldose can be converted into another aldose of one less carbon atom. One of these methods for shortening the carbon chain is the **Ruff degradation.** An aldose is oxidized by bromine water to the aldonic acid; oxidation of the calcium salt of this acid by hydrogen peroxide in the presence of ferric salts yields carbonate ion and an aldose of one less carbon atom (see Fig. 34.3).

An aldohexose		An aldonic acid		A calcium aldonate	
CHO		COOH		$COO^-)_2Ca^{++}$	
H—C—OH		H—C—OH		H—C—OH	
HO—C—H	$\xrightarrow{Br_2+H_2O}$	HO—C—H	$\xrightarrow{CaCO_3}$	HO—C—H	$\xrightarrow{H_2O_2,\ Fe^{+++}}$
H—C—OH		H—C—OH		H—C—OH	
H—C—OH		H—C—OH		H—C—OH	
CH_2OH		CH_2OH		CH_2OH	

An aldopentose
CHO
HO—C—H + CO_3^{--}
H—C—OH
H—C—OH
CH_2OH

Figure 34.3. An example of the Ruff degradation.

34.10 Conversion of an aldose into its epimer

In the presence of a tertiary amine, in particular pyridine (Sec. 31.6), an equilibrium is established between an aldonic acid and its epimer. This reaction is the basis of the best method for converting an aldose into its epimer, since the only configuration affected is that at C–2. The aldose is oxidized by bromine water to the aldonic acid, which is then treated with pyridine. From the equilibrium mixture thus formed, the epimeric aldonic acid is separated, and reduced (in the form of its lactone) to the epimeric aldose. See, for example, Fig. 34.4.

34.11 Configuration of (+)-glucose. The Fischer proof

Let us turn back to the year 1888. Only a few monosaccharides were known, among them (+)-glucose, (−)-fructose, (+)-arabinose. (+)-Mannose had just

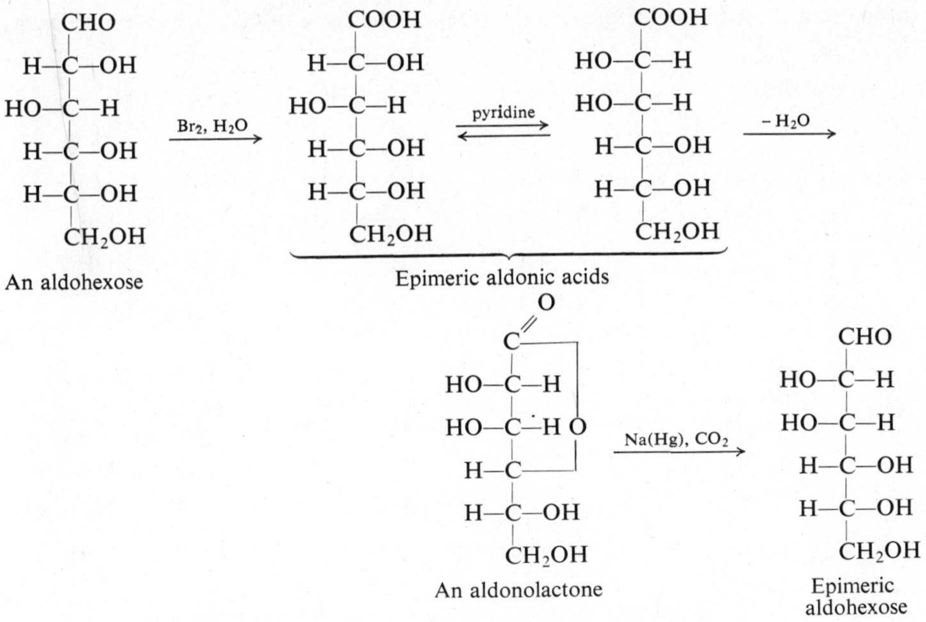

$$
\begin{array}{c}
CHO \\
H-C-OH \\
HO-C-H \\
H-C-OH \\
H-C-OH \\
CH_2OH
\end{array}
\xrightarrow{Br_2,\ H_2O}
\begin{array}{c}
COOH \\
H-C-OH \\
HO-C-H \\
H-C-OH \\
H-C-OH \\
CH_2OH
\end{array}
\overset{\text{pyridine}}{\rightleftarrows}
\begin{array}{c}
COOH \\
HO-C-H \\
HO-C-H \\
H-C-OH \\
H-C-OH \\
CH_2OH
\end{array}
\xrightarrow{-H_2O}
$$

An aldohexose — Epimeric aldonic acids

$$
\begin{array}{c}
\overset{O}{\overset{\|}{C}} \text{———┐} \\
HO-C-H \quad | \\
HO-C-H \quad O \\
H-C \text{———┘} \\
H-C-OH \\
CH_2OH
\end{array}
\xrightarrow{Na(Hg),\ CO_2}
\begin{array}{c}
CHO \\
HO-C-H \\
HO-C-H \\
H-C-OH \\
H-C-OH \\
CH_2OH
\end{array}
$$

An aldonolactone — Epimeric aldohexose

Figure 34.4. Conversion of an aldose into its epimer.

been synthesized. It was known that (+)-glucose was an aldohexose and that (+)-arabinose was an aldopentose. Emil Fischer had discovered (1884) that phenylhydrazine could convert carbohydrates into osazones. The Kiliani cyanohydrin method for lengthening the chain was just two years old.

It was known that aldoses could be reduced to alditols, and could be oxidized to the monocarboxylic aldonic acids and to the dicarboxylic aldaric acids. A theory of stereoisomerism and optical activity had been proposed (1874) by van't Hoff and Le Bel. Methods for separating stereoisomers were known and optical activity could be measured. The concepts of racemic modifications, *meso* compounds, and epimers were well established.

(+)-Glucose was known to be an aldohexose; but as an aldohexose it could have any one of 16 possible configurations. The question was: *which* configuration did it have? In 1888, Emil Fischer (at the University of Würzburg) set out to find the answer to that question, and in 1891 announced the completion of a most remarkable piece of chemical research, for which he received the Nobel Prize in 1902. Let us follow Fischer's steps to the configuration of (+)-glucose. Although somewhat modified, the following arguments are essentially those of Fischer.

The 16 possible configurations consist of eight pairs of enantiomers. Since methods of determining absolute configuration were not then available, Fischer realized that he could at best limit the configuration of (+)-glucose to a pair of enantiomeric configurations; he would not be able to tell which one of the pair was the correct absolute configuration.

To simplify the problem, Fischer therefore rejected eight of the possible configurations, arbitrarily retaining only those (I–VIII) in which C–5 carried the —OH on the right (with the understanding that —H and —OH project toward the observer). He realized that any argument that led to the selection of one of these formulas applied with equal force to the mirror image of that formula. (As it turned out, his arbitrary choice of an —OH on the right of C–5 in (+)-glucose was the correct one.)

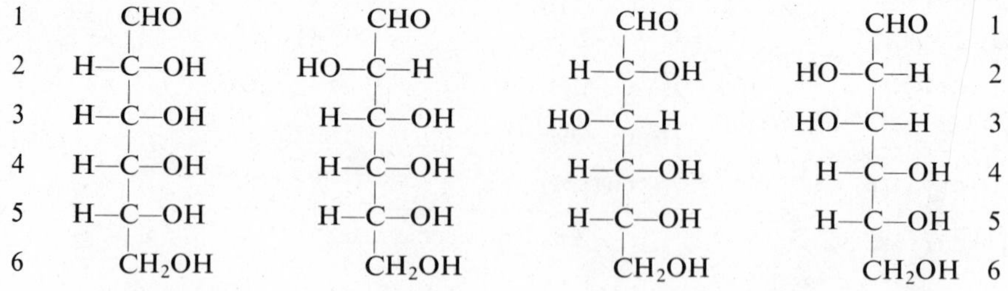

1	CHO	CHO	CHO	CHO	1
2	H—C—OH	HO—C—H	H—C—OH	HO—C—H	2
3	H—C—OH	H—C—OH	HO—C—H	HO—C—H	3
4	H—C—OH	H—C—OH	H—C—OH	H—C—OH	4
5	H—C—OH	H—C—OH	H—C—OH	H—C—OH	5
6	CH_2OH	CH_2OH	CH_2OH	CH_2OH	6
	I	II	III	IV	

1	CHO	CHO	CHO	CHO	1
2	H—C—OH	HO—C—H	H—C—OH	HO—C—H	2
3	H—C—OH	H—C—OH	HO—C—H	HO—C—H	3
4	HO—C—H	HO—C—H	HO—C—H	HO—C—H	4
5	H—C—OH	H—C—OH	H—C—OH	H—C—OH	5
6	CH_2OH	CH_2OH	CH_2OH	CH_2OH	6
	V	VI	VII	VIII	

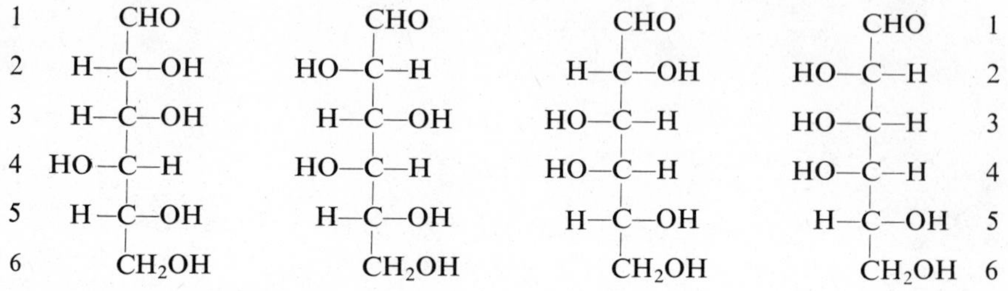

Since his proof depended in part on the relationship between (+)-glucose and the aldopentose (−)-arabinose, Fischer also had to consider the configurations of the five-carbon aldoses. Of the eight possible configurations, he retained only four, IX–XII, again those in which the bottom chiral center carried the —OH on the right.

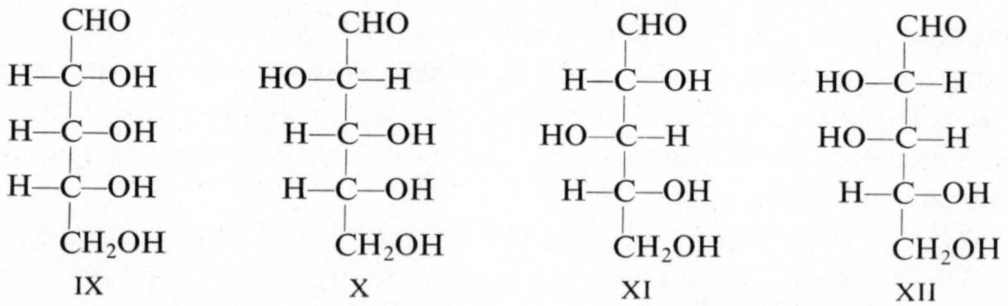

CHO	CHO	CHO	CHO
H—C—OH	HO—C—H	H—C—OH	HO—C—H
H—C—OH	H—C—OH	HO—C—H	HO—C—H
H—C—OH	H—C—OH	H—C—OH	H—C—OH
CH_2OH	CH_2OH	CH_2OH	CH_2OH
IX	X	XI	XII

The line of argument is as follows:

(1) Upon oxidation by nitric acid, (—)-arabinose yields an optically active dicarboxylic acid. Since the —OH on the lowest chiral center is arbitrarily placed on the right, this fact means that the —OH on the uppermost chiral center is on the left (as in X or XII),

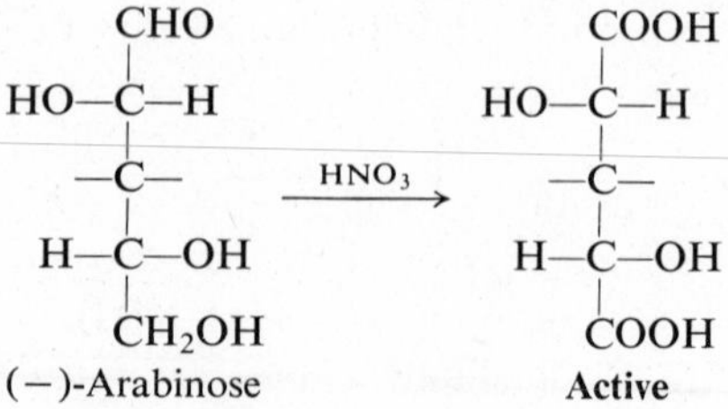

CHO		COOH
HO—C—H		HO—C—H
—C—	$\xrightarrow{HNO_3}$	—C—
H—C—OH		H—C—OH
CH_2OH		COOH
(−)-Arabinose *Partial formula* X or XII		**Active**

for if it were on the right (as in IX or XI), the diacid would necessarily be an inactive *meso* acid.

```
    CHO                    COOH              CHO                    COOH
     |                      |                 |                      |
  H—C—OH                 H—C—OH            H—C—OH                 H—C—OH
     |          HNO3        |                 |          HNO3        |
  H—C—OH      ------>    H—C—OH           HO—C—H       ------>   HO—C—H
     |                      |                 |                      |
  H—C—OH                 H—C—OH            H—C—OH                 H—C—OH
     |                      |                 |                      |
    CH2OH                  COOH              CH2OH                  COOH
    IX                   Inactive             XI                  Inactive
                      A meso compound                          A meso compound
```

(2) (—)-Arabinose is converted by the Kiliani-Fischer synthesis into (+)-glucose and (+)-mannose. (+)-Glucose and (+)-mannose therefore are epimers, differing only in configuration about C–2, and have the same configuration about C–3, C–4, and C–5 as does (−)-arabinose. (+)-Glucose and (+)-mannose must be III and IV, or VII and VIII.

```
                      1      CHO                     CHO    1
                                |                       |
     CHO              2   H—C—OH               HO—C—H      2
      |                         |                       |
  HO—C—H              3  HO—C—H                HO—C—H      3
      |       ---->             |         and           |
    —C—               4       —C—                     —C—   4
      |                         |                       |
   H—C—OH             5   H—C—OH                H—C—OH     5
      |                         |                       |
    CH2OH             6       CH2OH                   CH2OH 6

(−)-Arabinose              (+)-Glucose and (+)-Mannose: epimers
Partial formula                    Partial formulas
X or XII                      III and IV, or VII and VIII
```

(3) Upon oxidation by nitric acid, both (+)-glucose and (+)-mannose yield dicarboxylic acids that are optically active. This means that the —OH on C–4 is on the right, as in III and IV,

```
1      CHO                     COOH        1      CHO                     COOH
        |                        |                 |                        |
2  H—C—OH                  H—C—OH          2 HO—C—H                  HO—C—H
        |                        |                 |                        |
3 HO—C—H        HNO3      HO—C—H           3 HO—C—H       HNO3       HO—C—H
        |      ------>           |                 |     ------>            |
4  H—C—OH                  H—C—OH          4  H—C—OH                  H—C—OH
        |                        |                 |                        |
5  H—C—OH                  H—C—OH          5  H—C—OH                  H—C—OH
        |                        |                 |                        |
6     CH2OH                   COOH         6     CH2OH                   COOH
       III                   Active                IV                   Active
```

for if it were on the left, as in VII and VIII, *one* of the aldaric acids would necessarily be an inactive *meso* acid.

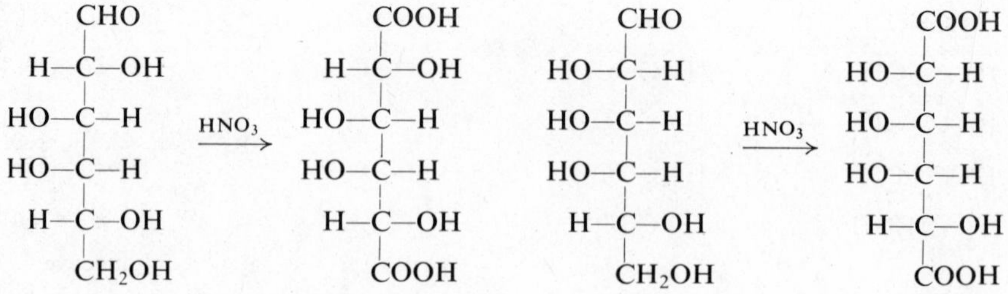

$$
\begin{array}{c}
\text{CHO} \\
\text{H—C—OH} \\
\text{HO—C—H} \\
\text{HO—C—H} \\
\text{H—C—OH} \\
\text{CH}_2\text{OH} \\
\text{VII}
\end{array}
\xrightarrow{\text{HNO}_3}
\begin{array}{c}
\text{COOH} \\
\text{H—C—OH} \\
\text{HO—C—H} \\
\text{HO—C—H} \\
\text{H—C—OH} \\
\text{COOH} \\
\textbf{Inactive} \\
\textit{A meso compound}
\end{array}
\qquad
\begin{array}{c}
\text{CHO} \\
\text{HO—C—H} \\
\text{HO—C—H} \\
\text{HO—C—H} \\
\text{H—C—OH} \\
\text{CH}_2\text{OH} \\
\text{VIII}
\end{array}
\xrightarrow{\text{HNO}_3}
\begin{array}{c}
\text{COOH} \\
\text{HO—C—H} \\
\text{HO—C—H} \\
\text{HO—C—H} \\
\text{H—C—OH} \\
\text{COOH} \\
\textbf{Active}
\end{array}
$$

(−)-Arabinose must also have that same —OH on the right, and hence has configuration X.

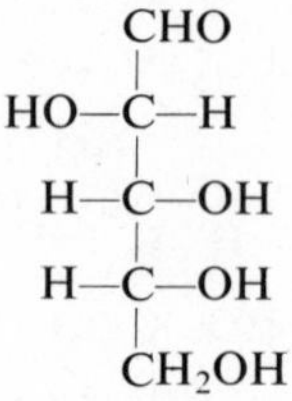

$$
\begin{array}{c}
\text{CHO} \\
\text{HO—C—H} \\
\text{H—C—OH} \\
\text{H—C—OH} \\
\text{CH}_2\text{OH} \\
\text{X} \\
(-)\text{-Arabinose}
\end{array}
$$

(+)-Glucose and (+)-mannose have configurations III and IV, but one question remains: which compound has which configuration? One more step is needed.

(4) Oxidation of another hexose, (+)-gulose, yields the same dicarboxylic acid, (+)-glucaric acid, as does oxidation of (+)-glucose. (The gulose was synthesized for this purpose by Fischer.) If we examine the two possible configurations for (+)-glucaric acid, IIIa and IVa, we see that only IIIa can be derived from two different hexoses: from III and the enantiomer of V.

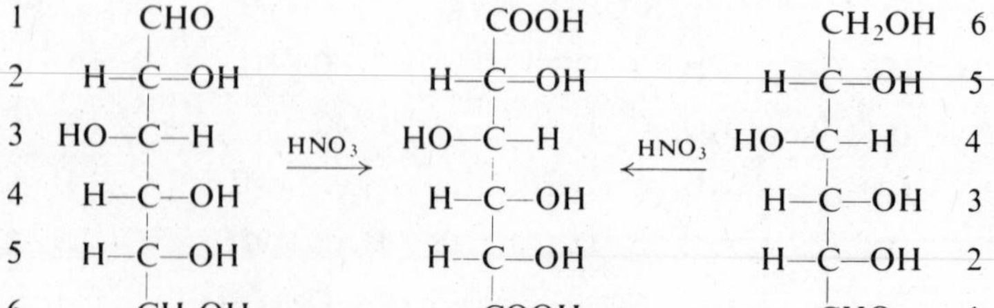

$$
\begin{array}{rc}
1 & \text{CHO} \\
2 & \text{H—C—OH} \\
3 & \text{HO—C—H} \\
4 & \text{H—C—OH} \\
5 & \text{H—C—OH} \\
6 & \text{CH}_2\text{OH} \\
 & \text{III}
\end{array}
\xrightarrow{\text{HNO}_3}
\begin{array}{c}
\text{COOH} \\
\text{H—C—OH} \\
\text{HO—C—H} \\
\text{H—C—OH} \\
\text{H—C—OH} \\
\text{COOH} \\
\text{IIIa}
\end{array}
\xleftarrow{\text{HNO}_3}
\begin{array}{cl}
\text{CH}_2\text{OH} & 6 \\
\text{H—C—OH} & 5 \\
\text{HO—C—H} & 4 \\
\text{H—C—OH} & 3 \\
\text{H—C—OH} & 2 \\
\text{CHO} & 1 \\
\text{Enantiomer of V} &
\end{array}
$$

The acid IVa can be derived from just one hexose: from IV.

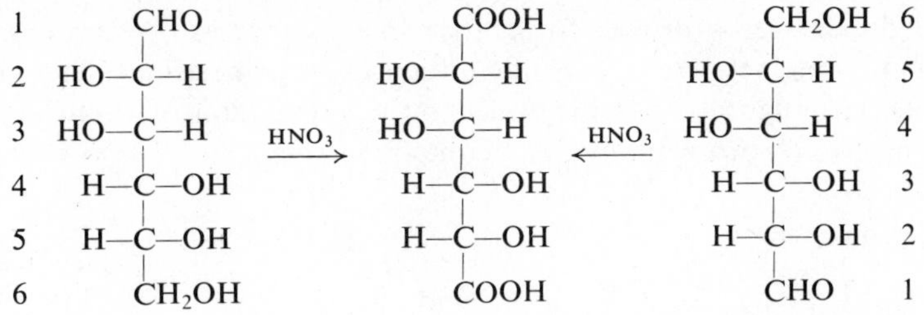

$$
\begin{array}{cc}
1 & \text{CHO} \\
2 & \text{HO—C—H} \\
3 & \text{HO—C—H} \\
4 & \text{H—C—OH} \\
5 & \text{H—C—OH} \\
6 & \text{CH}_2\text{OH} \\
 & \text{IV}
\end{array}
\xrightarrow{\text{HNO}_3}
\begin{array}{c}
\text{COOH} \\
\text{HO—C—H} \\
\text{HO—C—H} \\
\text{H—C—OH} \\
\text{H—C—OH} \\
\text{COOH} \\
\text{IVa}
\end{array}
\xleftarrow{\text{HNO}_3}
\begin{array}{cc}
\text{CH}_2\text{OH} & 6 \\
\text{HO—C—H} & 5 \\
\text{HO—C—H} & 4 \\
\text{H—C—OH} & 3 \\
\text{H—C—OH} & 2 \\
\text{CHO} & 1 \\
\text{IV (rotated 180°)} &
\end{array}
$$

It follows that (+)-glucaric acid has configuration IIIa, and therefore that (+)-glucose has configuration III.

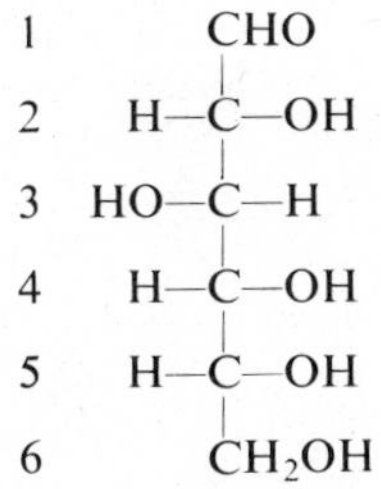

$$
\begin{array}{cc}
1 & \text{CHO} \\
2 & \text{H—C—OH} \\
3 & \text{HO—C—H} \\
4 & \text{H—C—OH} \\
5 & \text{H—C—OH} \\
6 & \text{CH}_2\text{OH} \\
 & \text{III} \\
 & \text{(+)-Glucose}
\end{array}
$$

(+)-Mannose, of course, has configuration IV, and (−)-gulose (the enantiomer of the one used by Fischer) has configuration V.

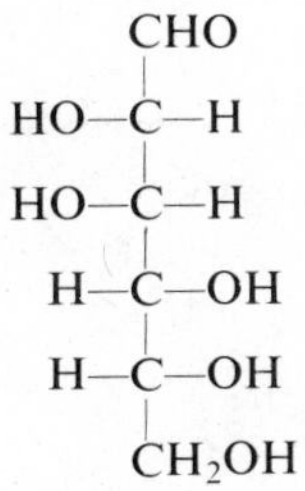

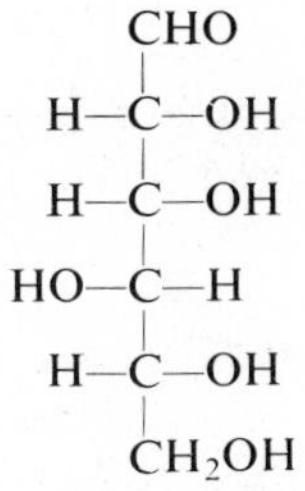

$$
\begin{array}{c}
\text{CHO} \\
\text{HO—C—H} \\
\text{HO—C—H} \\
\text{H—C—OH} \\
\text{H—C—OH} \\
\text{CH}_2\text{OH} \\
\text{IV} \\
\text{(+)-Mannose}
\end{array}
\qquad\qquad
\begin{array}{c}
\text{CHO} \\
\text{H—C—OH} \\
\text{H—C—OH} \\
\text{HO—C—H} \\
\text{H—C—OH} \\
\text{CH}_2\text{OH} \\
\text{V} \\
\text{(−)-Gulose}
\end{array}
$$

34.12 Configurations of aldoses

Today all possible aldoses (and ketoses) of six carbons or less, and many of more than six carbons, are known; most of these do not occur naturally and have been synthesized. The configurations of all these have been determined by application of the same principles that Fischer used to establish the configuration of (+)-glucose; indeed, twelve of the sixteen aldohexoses were worked out by Fischer and his students.

So far in our discussion, we have seen how configurations III, IV, V, and X of the previous section were assigned to (+)-glucose, (+)-mannose, (−)-gulose, and (−)-arabinose, respectively. Let us see how configurations have been assigned to some other monosaccharides.

The aldopentose **(−)-ribose** forms the same osazone as (−)-arabinose. Since (−)-arabinose was shown to have configuration X, (−)-ribose must have configuration IX. This configuration is confirmed by the reduction of (−)-ribose to the optically inactive (*meso*) pentahydroxy compound *ribitol.*

$$
\begin{array}{c}
\mathrm{CHO} \\
\mathrm{HO{-}C{-}H} \\
\mathrm{H{-}C{-}OH} \\
\mathrm{H{-}C{-}OH} \\
\mathrm{CH_2OH} \\
\mathrm{X} \\
\text{(−)-Arabinose}
\end{array}
\xrightarrow{C_6H_5NHNH_2}
\begin{array}{c}
\mathrm{CH{=}NNHC_6H_5} \\
\mathrm{C{=}NNHC_6H_5} \\
\mathrm{H{-}C{-}OH} \\
\mathrm{H{-}C{-}OH} \\
\mathrm{CH_2OH} \\
\\
\text{Osazone}
\end{array}
\xleftarrow{C_6H_5NHNH_2}
\begin{array}{c}
\mathrm{CHO} \\
\mathrm{H{-}C{-}OH} \\
\mathrm{H{-}C{-}OH} \\
\mathrm{H{-}C{-}OH} \\
\mathrm{CH_2OH} \\
\mathrm{IX} \\
\text{(−)-Ribose}
\end{array}
$$

$$
\downarrow H_2,\ Ni
$$

$$
\begin{array}{c}
\mathrm{CH_2OH} \\
\mathrm{H{-}C{-}OH} \\
\mathrm{H{-}C{-}OH} \\
\mathrm{H{-}C{-}OH} \\
\mathrm{CH_2OH} \\
\text{Ribitol} \\
\textit{A meso compound} \\
\textbf{Inactive}
\end{array}
$$

The two remaining aldopentoses, **(+)-xylose** and **(−)-lyxose**, must have the configurations XI and XII. Oxidation by nitric acid converts (+)-xylose into an

$$
\begin{array}{c}
\mathrm{CHO} \\
\mathrm{H{-}C{-}OH} \\
\mathrm{HO{-}C{-}H} \\
\mathrm{H{-}C{-}OH} \\
\mathrm{CH_2OH} \\
\mathrm{XI} \\
\text{(+)-Xylose}
\end{array}
\xrightarrow{HNO_3}
\begin{array}{c}
\mathrm{COOH} \\
\mathrm{H{-}C{-}OH} \\
\mathrm{HO{-}C{-}H} \\
\mathrm{H{-}C{-}OH} \\
\mathrm{COOH} \\
\text{Xylaric acid} \\
\textit{A meso compound} \\
\textbf{Inactive}
\end{array}
\qquad
\begin{array}{c}
\mathrm{CHO} \\
\mathrm{HO{-}C{-}H} \\
\mathrm{HO{-}C{-}H} \\
\mathrm{H{-}C{-}OH} \\
\mathrm{CH_2OH} \\
\mathrm{XII} \\
\text{(−)-Lyxose}
\end{array}
\ \text{would give}\
\begin{array}{c}
\mathrm{COOH} \\
\mathrm{HO{-}C{-}H} \\
\mathrm{HO{-}C{-}H} \\
\mathrm{H{-}C{-}OH} \\
\mathrm{COOH} \\
\\
\textbf{Active}
\end{array}
$$

optically inactive (*meso*) aldaric acid. (+)-Xylose must therefore be XI and (−)-lyxose must be XII

Degradation of (−)-arabinose yields the tetrose **(−)-erythrose**, which therefore has configuration XIII. In agreement with this configuration, (−)-erythrose is found to yield *meso*tartaric acid upon oxidation by nitric acid.

```
     CHO
      |
  HO—C—H
      |                                 CHO                       COOH
   H—C—OH   Ruff degradation →     H—C—OH      HNO3 →       H—C—OH
      |                                 |                          |
   H—C—OH                          H—C—OH                    H—C—OH
      |                                 |                          |
     CH2OH                             CH2OH                      COOH
(−)-Arabinose                          XIII                Mesotartaric acid
                                   (−)-Erythrose               Inactive
```

Degradation of (+)-xylose by the Ruff method yields the tetrose **(−)-threose**, which must therefore have configuration XIV. This is confirmed by oxidation of (−)-threose to optically active (−)-tartaric acid.

```
     CHO
      |
   H—C—OH                               CHO                       COOH
      |                                 |                          |
  HO—C—H   Ruff degradation →     HO—C—H      HNO3 →       HO—C—H
      |                                 |                          |
   H—C—OH                          H—C—OH                    H—C—OH
      |                                 |                          |
     CH2OH                             CH2OH                      COOH
                                        XIV
 (+)-Xylose                        (−)-Threose             (−)-Tartaric acid
                                                                Active
```

Problem 34.11 Assign a name to I, II, VI, VII, and VIII (p. 1082) on the basis of the following evidence and the configurations already assigned:

(a) The aldohexoses **(+)-galactose** and **(+)-talose** yield the same osazone. Degradation of (+)-galactose yields (−)-lyxose. Oxidation of (+)-galactose by nitric acid yields an inactive *meso* acid, *galactaric acid* (also called *mucic acid*).

(b) (−)-Ribose is converted by the Kiliani-Fischer synthesis into the two aldohexoses **(+)-allose** and **(+)-altrose**. Oxidation of (+)-altrose yields optically active (+)-*altraric acid*. Reduction of (+)-allose to a hexahydroxy alcohol yields optically inactive *allitol*.

(c) The aldohexose **(−)-idose** yields the same osazone as (−)-gulose.

Problem 34.12 Go back to the "family tree" you constructed in Problem 34.9, p. 1079, and assign names to all structures.

Problem 34.13 What is the configuration of the 2-ketohexose (−)-fructose? (See Problem 34.7, p. 1078.)

Problem 34.14 Give the configurations of (−)-glucose, (−)-mannose, and (+)-fructose.

34.13 Optical families. D and L

Before we can explore further the structure of (+)-glucose and its relatives, we must examine a topic of stereochemistry we have not yet touched on: use of the prefixes D and L.

Most applications of stereochemistry, as we have already seen, are based upon the *relative* configurations of different compounds, not upon their absolute configurations. We are chiefly interested in whether the configurations of a reactant and its product are the same or different, not in what either configuration actually is. In the days before any absolute configurations had been determined, there was the problem not only of determining the relative configurations of various optically active compounds, but also of indicating these relationships once they had been established. This was a particularly pressing problem with the carbohydrates.

The compound **glyceraldehyde**, $CH_2OHCHOHCHO$, was selected as a standard of reference, because it is the simplest carbohydrate—an aldotriose—capable of optical isomerism. Its configuration could be related to those of the carbohydrates, and because of its highly reactive functional groups, it could be converted into, and thus related to, many other kinds of organic compounds. (+)-Glyceraldehyde was arbitrarily assigned configuration I, and was designated D-glyceraldehyde; (−)-glyceraldehyde was assigned configuration II and was designated L-glyceraldehyde. Configurations were assigned to the glyceraldehydes purely for

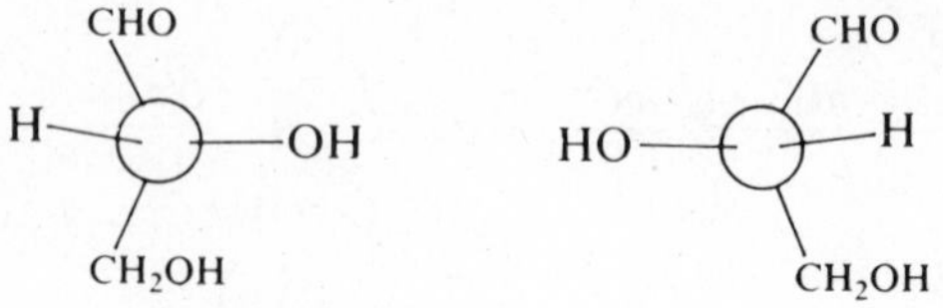

CHO
H— —OH
CH_2OH
I
D-Glyceraldehyde

CHO
HO— —H
CH_2OH
II
L-Glyceraldehyde

convenience; the particular assignment had a 50:50 chance of being correct, and, as it has turned out, the configuration chosen actually is the correct absolute configuration.

Other compounds could be related configurationally to one or the other of the glyceraldehydes by means of reactions that did not involve breaking bonds to a chiral center (Sec. 7.5). On the basis of the *assumed* configuration of the glyceraldehyde, these related compounds could be assigned configurations, too. As it has turned out, these configurations are the correct absolute ones; in any case, for

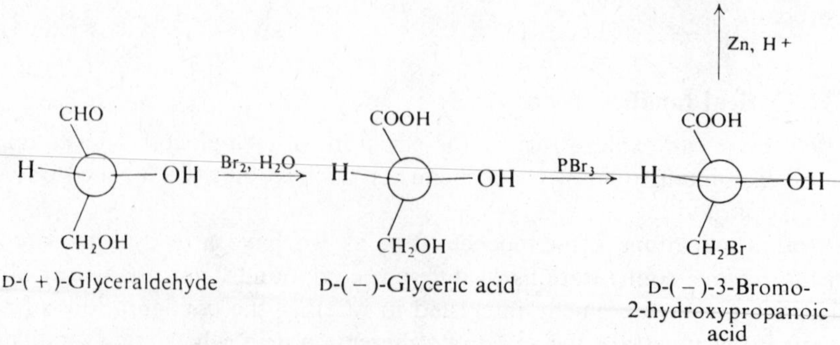

D-(−)-Lactic acid
COOH
H— —OH
CH_3

↑ Zn, H+

CHO
H— —OH
CH_2OH
D-(+)-Glyceraldehyde

$\xrightarrow{Br_2, H_2O}$

COOH
H— —OH
CH_2OH
D-(−)-Glyceric acid

$\xrightarrow{PBr_3}$

COOH
H— —OH
CH_2Br
D-(−)-3-Bromo-2-hydroxypropanoic acid

Figure 34.5. Relating configurations to glyceraldehyde.

many years they served as a convenient way of indicating structural relationships. See, for example, Fig. 34.5.

To indicate the relationship thus established, compounds related to D-glyceraldehyde are given the designation D, and compounds related to L-glyceraldehyde are given the designation L. The symbols D and L (pronounced "dee" and "ell") thus refer to configuration, not to sign of rotation, so that we have, for example, D-(−)-glyceric acid and L-(+)-lactic acid. (One frequently encounters the prefixes *d* and *l*, pronounced "dextro" and "levo," but their meaning is not always clear. Today they usually refer to direction of rotation; in some of the older literature they refer to optical family. It was because of this confusion that D and L were introduced.)

Unfortunately, the use of the designations D and L is not unambiguous. In relating glyceraldehyde to lactic acid, for example, we might envision carrying out a sequence of steps in which the $-CH_2OH$ rather than the $-CHO$ group is converted into the $-COOH$ group:

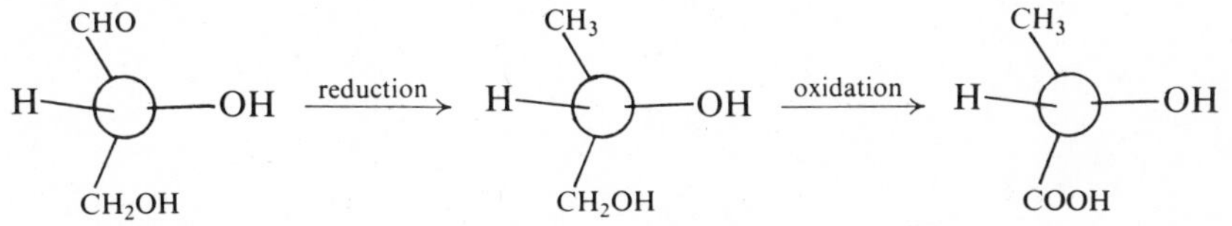

$$\begin{array}{c} CHO \\ H-\!\!\!\bigcirc\!\!\!-OH \\ CH_2OH \end{array} \xrightarrow{\text{reduction}} \begin{array}{c} CH_3 \\ H-\!\!\!\bigcirc\!\!\!-OH \\ CH_2OH \end{array} \xrightarrow{\text{oxidation}} \begin{array}{c} CH_3 \\ H-\!\!\!\bigcirc\!\!\!-OH \\ COOH \end{array}$$

(+)-Glyceraldehyde (+)-1,2-Propanediol (+)-Lactic acid

By this series of reactions, (+)-glyceraldehyde would yield (+)-lactic acid; by the previous sequence, (+)-glyceraldehyde yields (−)-lactic acid. It would appear that, depending upon the particular sequence used, we could designate either of the lactic acids as D-lactic acid; the first sequence is the more direct, and by convention is the accepted one. We should notice that, whatever the ambiguity associated with the use of D and L, there is no ambiguity about the configurational relationship; we arrive at the proper configurations for (+)- and (−)-lactic acids whichever route we use.

The prefixes R and S enable us to specify unambiguously the absolute configuration of a compound, because their use does not depend on a relationship to any other compound. But, by the same token, the letters R and S do not immediately reveal configurational relationships between two compounds; we have to work out and compare the configurations in each case.

The designations D and L, on the other hand, tell us nothing of the configuration of the compound unless we know the route by which the configurational relationship has been established. However, in the case of the carbohydrates (and the amino acids, Chap. 36), there are certain conventions about this which make these designations extremely useful.

Problem 34.15 Which specification, R or S, would you give to the following? (a) D-(+)-glyceraldehyde; (b) D-(−)-glyceric acid; (c) D-(−)-3-bromo-2-hydroxypropionic acid; (d) D-(−)-lactic acid.

Problem 34.16 The transformation of L-(+)-lactic acid into (+)-2-butanol was accomplished by the following sequence of reactions:

$$\text{L-(+)-lactic acid} \xrightarrow{C_2H_5OH,\ H_2SO_4} A \xrightarrow{Na,\ C_2H_5OH} B \xrightarrow{HBr} C$$

$$C \xrightarrow{KCN} D \xrightarrow{H_2O,\ HCl,\ heat} E \xrightarrow{CH_3OH,\ HCl} F \xrightarrow{Na,\ CH_3COOH} G$$

$$G \xrightarrow{HI} H \xrightarrow{H_2,\ Pd} \text{(+)-2-butanol}$$

What is the absolute configuration of (+)-2-butanol?

34.14 Tartaric acid

Tartaric acid, HOOCCHOHCHOHCOOH, has played a key role in the development of stereochemistry, and particularly the stereochemistry of the carbohydrates. In 1848 Louis Pasteur, using a hand lens and a pair of tweezers, laboriously separated a quantity of the sodium ammonium salt of racemic tartaric acid into two piles of mirror-image crystals and, in thus carrying out the first resolution of a racemic modification, was led to the discovery of enantiomerism. Almost exactly 100 years later, in 1951, Bijvoet, using x-ray diffraction—and also laboriously—determined the actual arrangement in space of the atoms of the sodium rubidium salt of (+)-tartaric acid, and thus made the first determination of the absolute configuration of an optically active substance.

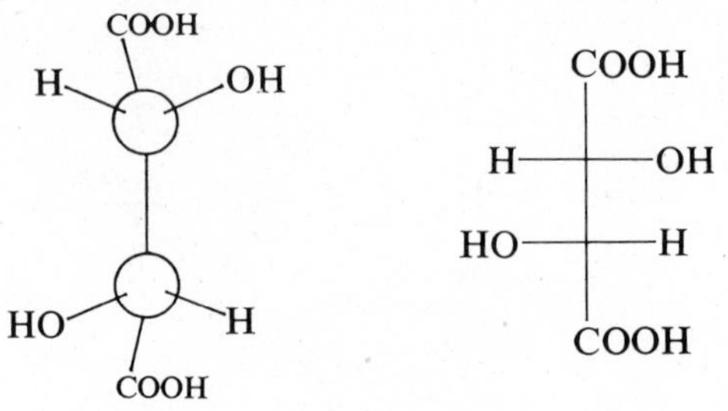

COOH; H, OH; HO, H; COOH

COOH
H—OH
HO—H
COOH

(+)-Tartaric acid

As we shall see in the next section, tartaric acid is the stereochemical link between the carbohydrates and our standard of reference, glyceraldehyde. In 1917, the configurational relationship between glyceraldehyde and tartaric acid was worked out. When the reaction sequence outlined in Fig. 34.6 was carried out starting with D-glyceraldehyde, two products were obtained, one inactive and

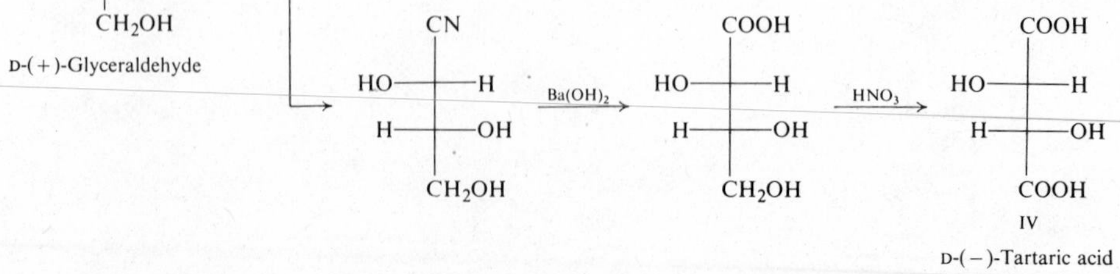

CHO / H—OH / CH_2OH — D-(+)-Glyceraldehyde —HCN→

CN / H—OH / H—OH / CH_2OH —$Ba(OH)_2$→ COOH / H—OH / H—OH / CH_2OH —HNO_3→ COOH / H—OH / H—OH / COOH — III *meso*-Tartaric acid

CN / HO—H / H—OH / CH_2OH —$Ba(OH)_2$→ COOH / HO—H / H—OH / CH_2OH —HNO_3→ COOH / HO—H / H—OH / COOH — IV D-(−)-Tartaric acid

Figure 34.6. Configurational relationship between glyceraldehyde and tartaric acid.

one which rotated the plane of polarized light to the left. The inactive product was, of course, mesotartaric acid, III. The active (−)-tartaric acid thus obtained was assigned configuration IV; since it is related to D-glyceraldehyde, we designate it D-(−)-tartaric acid.

On the basis of the assumed configuration of D-(+)-glyceraldehyde, then, L-(+)-tartaric acid, the enantiomer of D-(−)-tartaric acid, would have configuration V, the mirror image of IV. When Bijvoet determined the absolute configuration

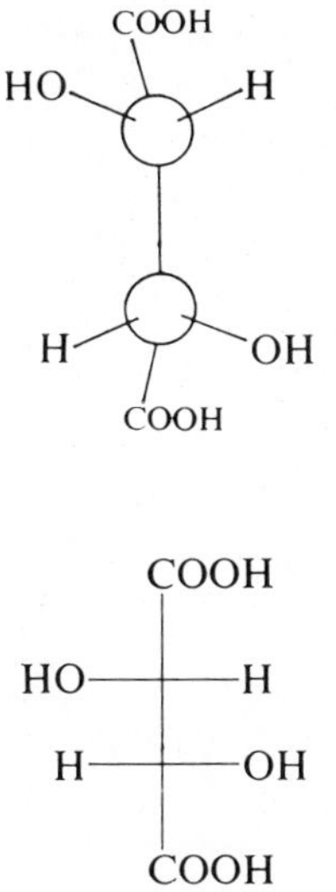

IV
D-(−)-Tartaric acid

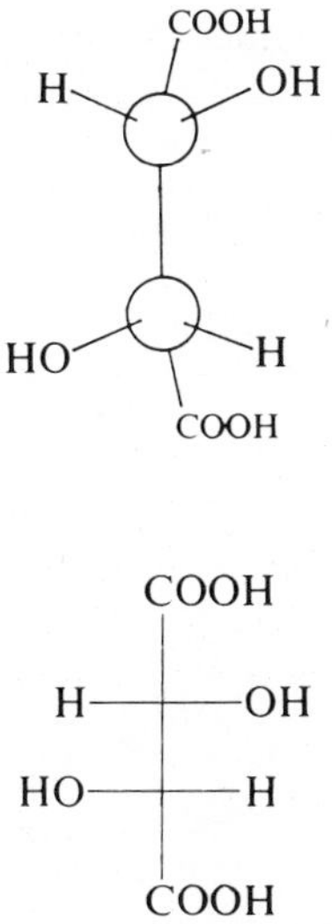

V
L-(+)-Tartaric acid

of (+)-tartaric acid, he found that it actually has the configuration that had been previously assumed. The assumed configurations of the glyceraldehydes, and hence the assumed configurations of all compounds related to them, were indeed the correct ones.

The designation of even the tartaric acids is subject to ambiguity. In this book, we have treated the tartaric acids as one does carbohydrates: by considering —CHO of glyceraldehyde as the position from which the chain is lengthened, via the cyanohydrin reaction. Some chemists, on the other hand, view the tartaric acids as one does the amino acids (Sec. 36.5) and, considering —COOH to be derived from —CHO of glyceraldehyde, designate (−)-tartaric acid as L, and (+)-tartaric acid as D.

Regardless of which convention one follows, this fact remains: (−)- and (+)-tartaric acid—and (+)- and (−)-glyceraldehyde—have the absolute configurations shown on p. 1088 and above.

Problem 34.17 Give the specification by the R/S system of: (a) (−)-tartaric acid; (b) (+)-tartaric acid; (c) mesotartaric acid.

Problem 34.18 (a) From the sequence of Fig. 34.6 the ratio of products III:IV is about 1:3. Why would you have expected to obtain III and IV in unequal amounts?

(b) Outline the same sequence starting from L-(−)-glyceraldehyde. Label each product with its name, showing its rotation and D/L designation. In what ratio will these products be obtained?

(c) Outline the same sequence starting from racemic (±)-glyceraldehyde. How do you account for the fact that only inactive material is obtained in spite of the unequal amounts of diastereomeric products formed from each of the enantiomeric glyceraldehydes?

34.15 Families of aldoses. Absolute configuration

The evidence on which Fischer assigned a configuration to (+)-glucose leads to either of the enantiomeric structures I and II. Fischer, we have seen, arbitrarily selected I, in which the lowest chiral center carries —OH on the right.

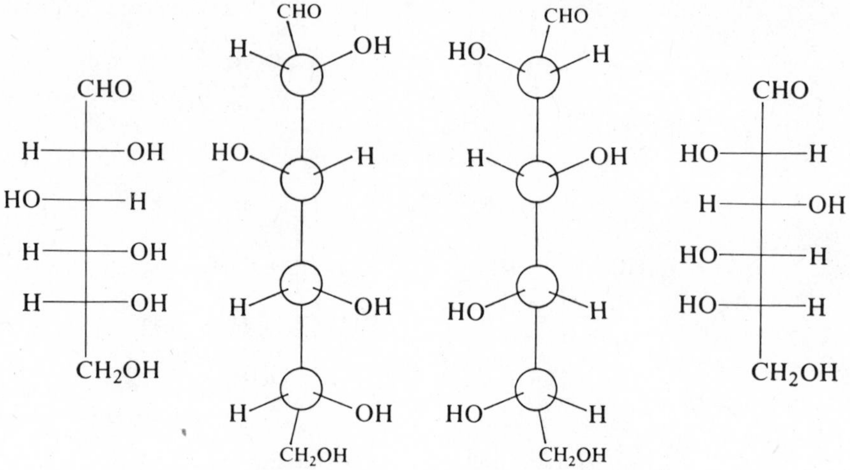

CHO
H—OH
HO—H
H—OH
H—OH
CH_2OH

I
D-(+)-Glucose

CHO
HO—H
H—OH
HO—H
HO—H
CH_2OH

II
L-(−)-Glucose

We recognize I as the enantiomer that would hypothetically be derived from D-(+)-glyceraldehyde by a series of Kiliani-Fischer syntheses, the chiral center of (+)-glyceraldehyde being retained as the *lowest* chiral center of the aldoses derived from it. (See Problem 34.9, p. 1079.) That (+)-glucose is related to D-(+)-glyceraldehyde has been established by a number of reaction sequences, one of which is shown in Fig. 34.7. On this basis, then, structure I becomes D-(+)-glucose, and structure II becomes L-(−)-glucose.

In 1906 the American chemist M. A. Rosanoff (then an instructor at New York University) proposed glyceraldehyde as the standard to which the configurations of carbohydrates should be related. Eleven years later experiment showed that it is the *dextrorotatory* (+)-glyceraldehyde that is related to (+)-glucose. On that basis, (+)-glyceraldehyde was then given the designation D and was assigned a configuration to conform with the one arbitrarily assigned to (+)-glucose by Fischer. Although rejected by Fischer, the Rosanoff convention became universally accepted.

Regardless of the direction in which they rotate polarized light, all monosaccharides are designated as D or L on the basis of the configuration about the lowest chiral center, the carbonyl group being at the top: D if the —OH is on the

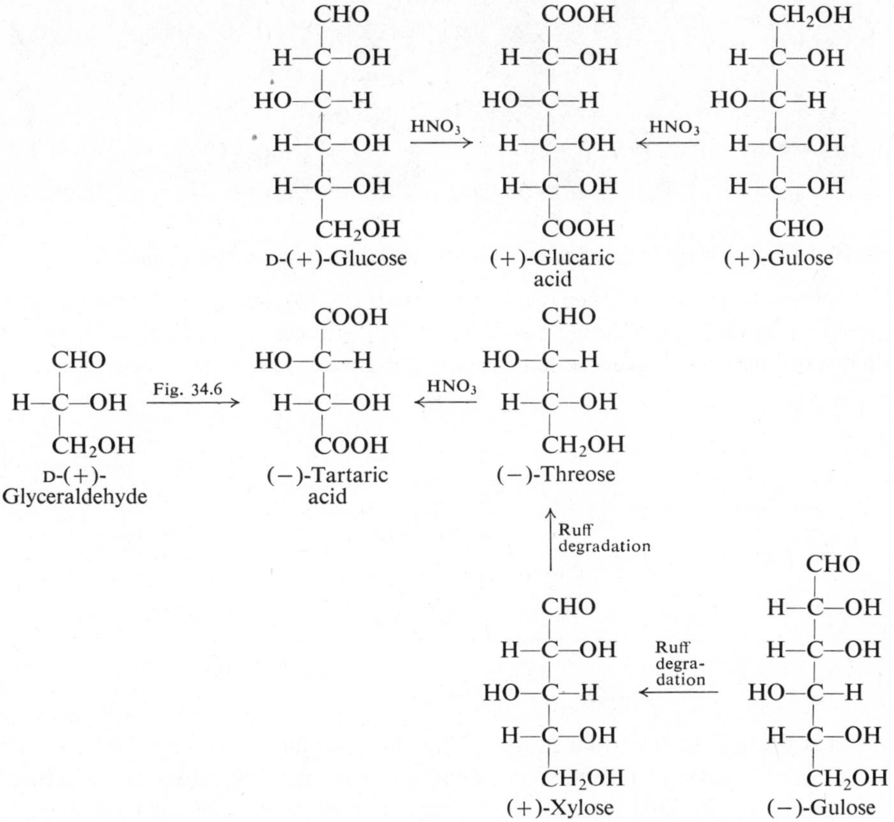

CHO
H—C—OH
HO—C—H
H—C—OH
H—C—OH
CH_2OH
D-(+)-Glucose

$\xrightarrow{HNO_3}$

COOH
H—C—OH
HO—C—H
H—C—OH
H—C—OH
COOH
(+)-Glucaric acid

$\xleftarrow{HNO_3}$

CH_2OH
H—C—OH
HO—C—H
H—C—OH
H—C—OH
CHO
(+)-Gulose

CHO
H—C—OH
CH_2OH
D-(+)-Glyceraldehyde

$\xrightarrow{\text{Fig. 34.6}}$

COOH
HO—C—H
H—C—OH
COOH
(−)-Tartaric acid

$\xleftarrow{HNO_3}$

CHO
HO—C—H
H—C—OH
CH_2OH
(−)-Threose

↑ Ruff degradation

CHO
H—C—OH
HO—C—H
H—C—OH
CH_2OH
(+)-Xylose

$\xleftarrow{\text{Ruff degradation}}$

CHO
H—C—OH
H—C—OH
HO—C—H
H—C—OH
CH_2OH
(−)-Gulose

Figure 34.7. Relating (+)-glucose to D-(+)-glyceraldehyde.

right, L if the —OH is on the left. (As always, it is understood that —H and —OH project toward us from the plane of the paper.) (+)-Mannose and (−)-arabinose, for example, are both assigned to the D-family on the basis of their relationship to D-(+)-glucose, and, through it, to D-(+)-glyceraldehyde.

Until 1949, these configurations were accepted on a purely empirical basis; they were a convenient way to show configurational relationships among the various carbohydrates, and between them and other organic compounds. But so far as anyone knew, the configurations of these compounds might actually have been the mirror images of those assigned; the lowest chiral center in the D-series of monosaccharides might have carried —OH on the left. As we have seen, however, when Bijvoet determined the absolute configuration of (+)-tartaric acid by x-ray analysis in 1949, he found that it actually has the configuration that had been up to then merely assumed. The arbitrary choice that Emil Fischer made in 1891 was the correct one; the configuration he assigned to (+)-glucose—and, through it, to every carbohydrate—is the correct absolute configuration.

Problem 34.19 The (+)-gulose that played such an important part in the proof of configuration of D-(+)-glucose was synthesized by Fischer via the following sequence:

$$\text{D-(+)-glucose} \xrightarrow{HNO_3} \text{(+)-glucaric acid} \xrightarrow{-H_2O} \text{A and B (lactones, separated)}$$

$$\text{A} \xrightarrow{Na(Hg)} \text{C (aldonic acid)} \xrightarrow{-H_2O} \text{D (lactone)} \xrightarrow{Na(Hg),\ acid} \text{D-(+)-glucose}$$

$$\text{B} \xrightarrow{Na(Hg)} \text{E (aldonic acid)} \xrightarrow{-H_2O} \text{F (lactone)} \xrightarrow{Na(Hg),\ acid} \text{(+)-gulose}$$

Give the structures of A through F. What is the configuration of (+)-gulose? Is it a member of the D-family or of the L-family? Why?

34.16 Cyclic structure of D-(+)-glucose. Formation of glucosides

We have seen evidence indicating that D-(+)-glucose is a pentahydroxy aldehyde. We have seen how its configuration has been established. It might seem, therefore, that D-(+)-glucose had been definitely proved to have structure I.

```
        CHO
        |
     H—C—OH
        |
    HO—C—H
        |
     H—C—OH
        |
     H—C—OH
        |
        CH2OH
```

I

D-(+)-Glucose

But during the time that much of the work we have just described was going on, certain facts were accumulating that were inconsistent with this structure of D-(+)-glucose. By 1895 it had become clear that the picture of D-(+)-glucose as a pentahydroxy aldehyde had to be modified.

Among the facts that had still to be accounted for were the following:

(a) **D-(+)-Glucose fails to undergo certain reactions typical of aldehydes.** Although it is readily oxidized, it gives a negative Schiff test and does not form a bisulfite addition product.

(b) **D-(+)-Glucose exists in two isomeric forms which undergo mutarotation.** When crystals of ordinary D-(+)-glucose of m.p. 146° are dissolved in water, the specific rotation gradually drops from an initial +112° to +52.7°. On the other hand, when crystals of D-(+)-glucose of m.p. 150° (obtained by crystallization at temperatures above 98°) are dissolved in water, the specific rotation gradually rises from an initial +19° to +52.7°. The form with the higher positive rotation is called **α-D-(+)-glucose** and that with lower rotation **β-D-(+)-glucose**. The change in rotation of each of these to the equilibrium value is called **mutarotation.**

(c) **D-(+)-Glucose forms two isomeric methyl D-glucosides.** Aldehydes, we remember, react with alcohols in the presence of anhydrous HCl to form acetals (Sec. 19.15). If the alcohol is, say, methanol, the acetal contains two methyl groups:

```
   H                         H                         H
   |      CH3OH, H+          |       CH3OH, H+         |
  —C=O   <========>        —C—OCH3  <========>       —C—OCH3
                             |                         |
                             OH                        OCH3
Aldehyde                 Hemiacetal                  Acetal
```

When D-(+)-glucose is treated with methanol and HCl, the product, **methyl D-glucoside**, contains only one —CH_3 group; yet it has properties resembling those of a full acetal. It does not spontaneously revert to aldehyde and alcohol on contact with water, but requires hydrolysis by aqueous acids.

Furthermore, not just one but two of these monomethyl derivatives of D-(+)-glucose are known, one with m.p. 165° and specific rotation +158°, and the other with m.p. 107° and specific rotation −33°. The isomer of higher positive rotation is called **methyl α-D-glucoside**, and the other is called **methyl-β-D-glucoside.** These glucosides do not undergo mutarotation, and do not reduce Tollens' or Fehling's reagent.

To fit facts like these, ideas about the structure of D-(+)-glucose had to be changed. In 1895, as a result of work by many chemists, including Tollens, Fischer, and Tanret, there emerged a picture of D-(+)-glucose as a *cyclic* structure. In 1926 the ring size was corrected, and in recent years the preferred conformation has been elucidated.

D-(+)-Glucose has the cyclic structure represented crudely by IIa and IIIa, more accurately by IIb and IIIb, and best of all by IIc and IIIc (Fig. 34.8).

Glucose anomers: Hemiacetals

Reducing sugars

Mutarotate

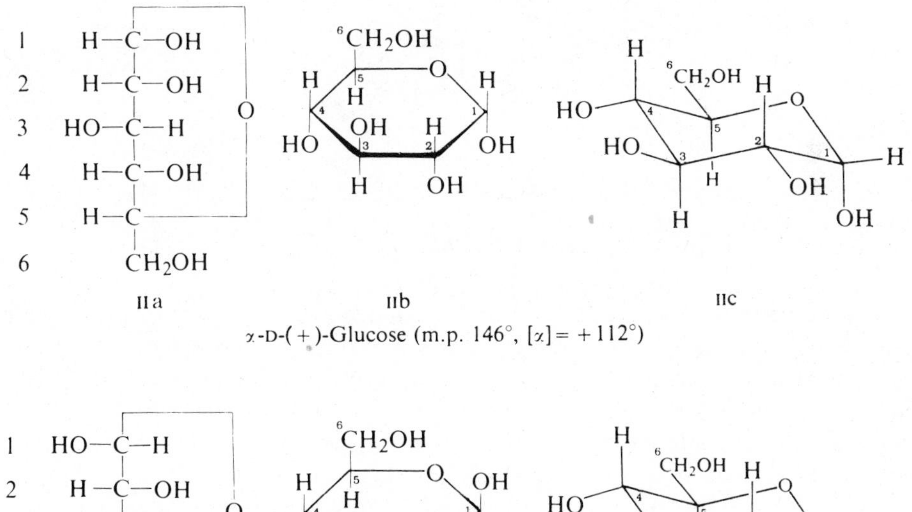

IIa IIb IIc

α-D-(+)-Glucose (m.p. 146°, [α] = +112°)

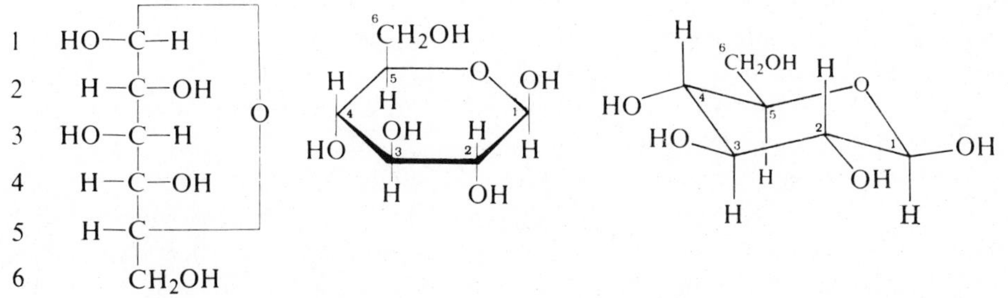

IIIa IIIb IIIc

β-D-(+)-Glucose (m.p. 150°, [α] = +19°)

Figure 34.8. Cyclic structures of D-(+)-glucose.

D-(+)-Glucose is the hemiacetal corresponding to reaction between the aldehyde group and the C–5 hydroxyl group of the open-chain structure (I). It has a cyclic structure simply because aldehyde and alcohol are part of the same molecule.

There are two isomeric forms of D-(+)-glucose because this cyclic structure has one more chiral center than Fischer's original open-chain structure (I). α-D-(+)-Glucose and β-D-(+)-glucose are diastereomers, differing in configuration about C–1. Such a pair of diastereomers are called **anomers**.

As hemiacetals, α- and β-D-(+)-glucose are readily hydrolyzed by water. In aqueous solution either anomer is converted—via the open-chain form—into an equilibrium mixture containing both cyclic isomers. Thus mutarotation results from the ready opening and closing of the hemiacetal ring (Fig. 34.9).

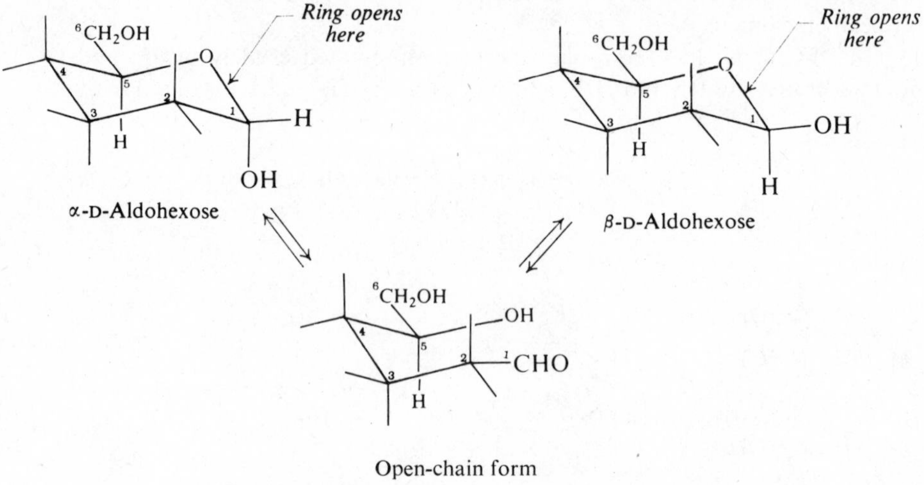

Figure 34.9. Mutarotation.

The typical aldehyde reactions of D-(+)-glucose—osazone formation, and perhaps reduction of Tollens' and Fehling's reagents—are presumably due to a small amount of open-chain compound, which is replenished as fast as it is consumed. The concentration of this open-chain structure is, however, too low (less than 0.5%) for certain easily reversible aldehyde reactions like bisulfite addition and the Schiff test.

The isomeric forms of methyl D-glucoside are anomers and have the cyclic structures IV and V (Fig. 34.10).

Although formed from only one mole of methanol, they are nevertheless full acetals, the other mole of alcohol being D-(+)-glucose itself through the C–5 hydroxyl group. The glucosides do not undergo mutarotation since, being acetals, they are fairly stable in aqueous solution. On being heated with aqueous acids, they undergo hydrolysis to yield the original hemiacetals (II and III). Toward bases glycosides, like acetals generally, are stable. Since they are not readily hydrolyzed to the open-chain aldehyde by the alkali in Tollens' or Fehling's reagent, glucosides are non-reducing sugars.

Glucoside anomers: Acetals

Non-reducing sugars

Do not mutarotate

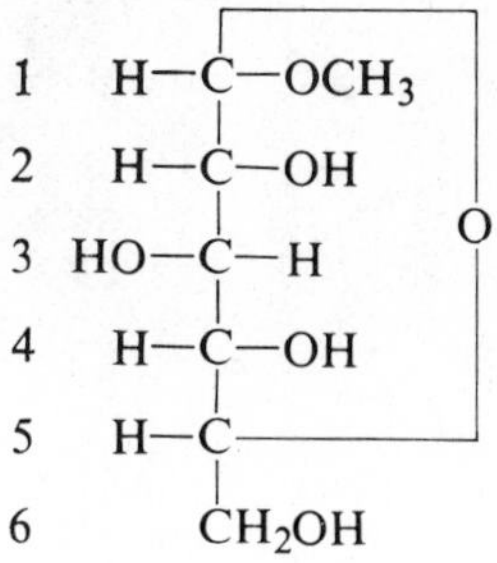

IVa 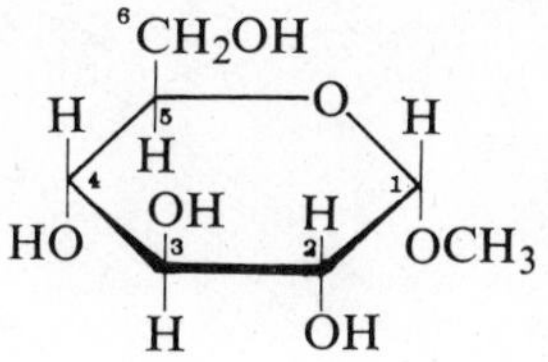 IVb 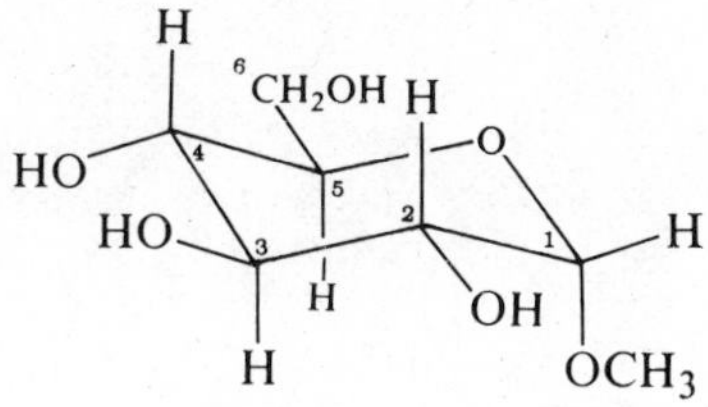 IVc

Methyl α-D-glucoside (m.p. 165°, [α] = +158°)

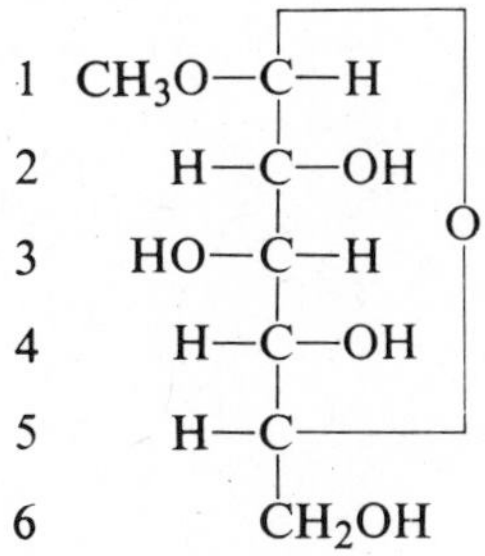

Va 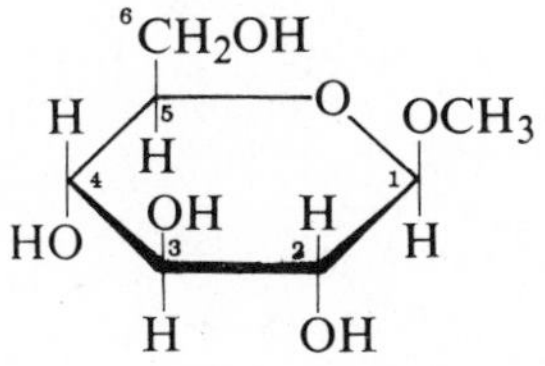 Vb 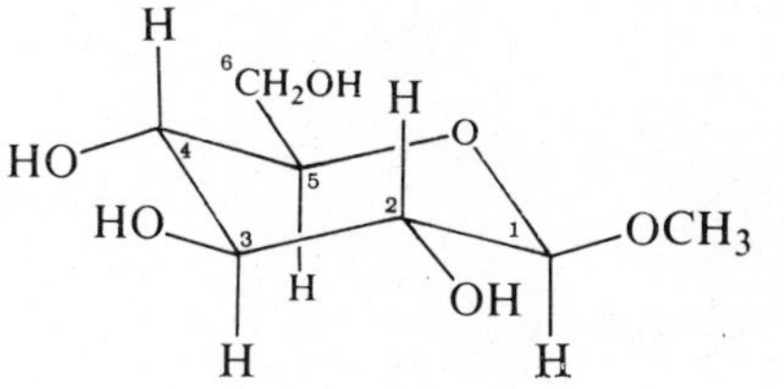 Vc

Methyl β-D-glucoside (m.p. 107°, [α] = −33°)

Figure 34.10. Cyclic structures of methyl D-glucosides.

Like D-(+)-glucose, other monosaccharides exist in anomeric forms capable of mutarotation, and react with alcohols to yield anomeric **glycosides.**

We have represented the cyclic structures of D-glucose and methyl D-glucoside in several different ways: β-D-glucose, for example, by IIIa, IIIb, and IIIc. At this point we should convince ourselves that all three representations correspond to the same structure, and that the configurations about C–2, C–3, C–4, and C–5 are the same as in the open-chain structure worked out by Fischer. These relationships are best seen by use of models.

We can convert the open-chain model of D-glucose into a cyclic model by joining oxygen of the C–5 —OH to the aldehyde carbon C–1. Whether we end up with the α- or β-structure depends upon which face of the flat carbonyl group we join the C–5 oxygen to. IIb and IIIb represent this ring lying on its side, so that groups that were on the right in the vertical model are directed downward, and groups that were on the left in the vertical model are directed upward. (Note particularly that the —CH_2OH group points *upward.*) In the more accurate representations IIc and IIIc, the disposition of these groups is modified by puckering of the six-membered ring, which will be discussed furtner in Sec. 34.20.

Problem 34.20 (a) From the values for the specific rotations of aqueous solutions of pure α- and β-D-(+)-glucose, and for the solution after mutarotation, calculate the relative amounts of α- and of β-forms at equilibrium (assuming a negligible amount of open-chain form).

(b) From examination of structures IIc and IIIc, suggest a reason for the greater proportion of one isomer. (*Hint:* See Sec. 9.14.)

Problem 34.21 From what you learned in Secs. 19.8 and 19.15, suggest a mechanism for the acid-catalyzed mutarotation of D-(+)-glucose.

Problem 34.22 (+)-Glucose reacts with acetic anhydride to give two isomeric pentaacetyl derivatives neither of which reduces Fehling's or Tollens' reagent. Account for these facts.

34.17 Configuration about C–1

Knowledge that aldoses and their glycosides have cyclic structures immediately raises the question: what is the configuration about C–1 in each of these anomeric structures?

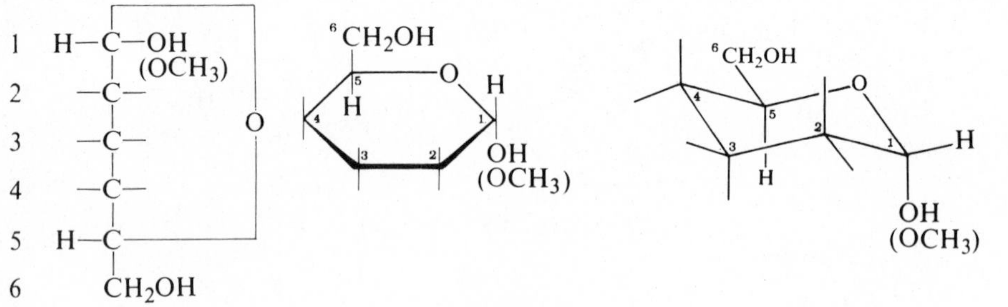

α-D-Anomers

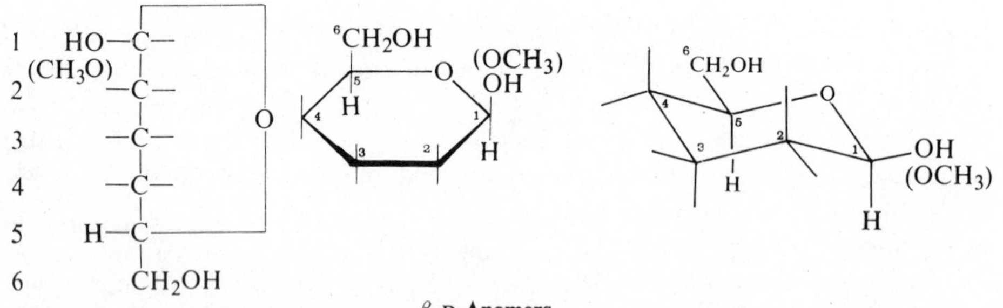

β-D-Anomers

Figure 34.11. Configuration of anomers of aldohexoses.

In 1909 C. S. Hudson (of the U.S. Public Health Service) made the following proposal. *In the* D*-series the more dextrorotatory member of an α,β-pair of anomers is to be named α-*D*-, the other being named β-*D. *In the* L *series the more levorotatory member of such a pair is given the name α-*L *and the other β-*L. Thus the enantiomer of α-D-(+)-glucose is α-L-(−)-glucose.

Furthermore, *the* —OH *or* —OCH_3 *group on* C–1 *is on the right in an α-*D*-anomer and on the left in a β-*D*-anomer*, as shown for Fig. 34.11 for aldohexoses. (Notice that "on the right" means "down" in the cyclic structure.)

Hudson's proposals have been adopted generally. Although they were originally based upon certain apparent but unproved relationships between configuration and optical rotation, all the evidence indicates that the assigned configurations are the correct ones. For example:

α-D-Glucose and methyl α-D-glucoside have the same configuration, as do β-D-glucose and methyl β-D-glucoside. *Evidence:* enzymatic hydrolysis of methyl α-D-glucoside liberates initially the more highly rotating α-D-glucose, and hydrolysis of methyl β-D-glucoside liberates initially β-D-glucose.

The configuration about C–1 is the same in the methyl α-glycosides of all the D-aldohexoses. *Evidence:* they all yield the same compound upon oxidation by HIO_4.

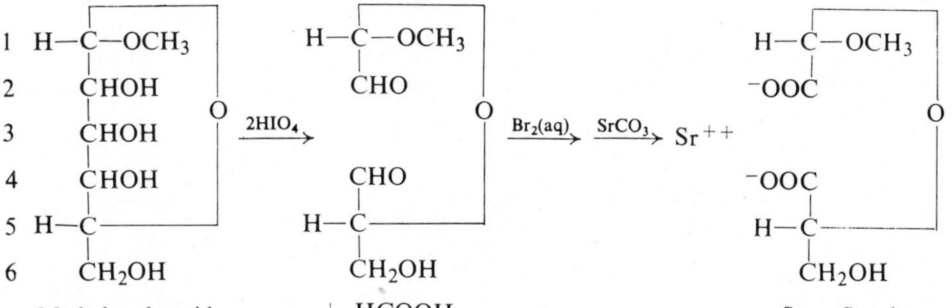

```
1  H—C—OCH3  ┐            H—C—OCH3  ┐                                H—C—OCH3  ┐
2    CHOH    │              CHO     │                                -OOC      │
3    CHOH    O  --2HIO4-->          O  --Br2(aq)--> --SrCO3--> Sr++            O
4    CHOH    │              CHO     │                                -OOC      │
5  H—C———————┘            H—C———————┘                                H—C———————┘
6    CH2OH                  CH2OH                                      CH2OH
```

Methyl α-glycoside of any D-aldohexose + HCOOH Same Sr salt

Oxidation destroys the chiral centers at C–2, C–3, and C–4, but configuration is preserved about C–1 and C–5. Configuration about C–5 is the same for all members of the D-family. The same products can be obtained from all these glycosides *only* if they also have the same configuration about C–1.

The C–1 —OH is on the right in the α-D-series and on the left in the β-D-series. *Evidence:* results of x-ray analysis.

Problem 34.23 (a) What products would be formed from the strontium salts shown above by treatment with dilute HCl?

(b) An oxidation of this sort was used to confirm the configurational relationship between (+)-glucose and (+)-glyceraldehyde. How was this done?

34.18 Methylation

Before we can go on to the next aspect of the structure of D-(+)-glucose, determination of ring size, we must first learn a little more about the methylation of carbohydrates.

As we know, treatment of D-(+)-glucose with methanol and dry hydrogen chloride yields the methyl D-glucosides:

Acetal formation

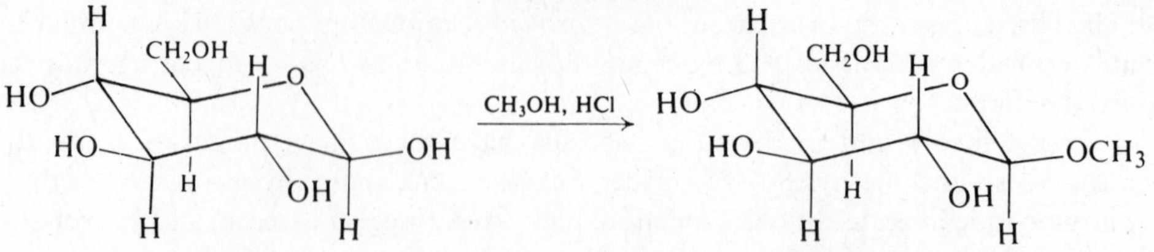

$\xrightarrow{CH_3OH,\ HCl}$

and α-anomer
β-D-(+)-Glucose
Reducing sugar

and α-anomer
Methyl β-D-glucoside
Non-reducing sugar

In this reaction, an aldehyde (or more exactly, its hemiacetal) is converted into an acetal in the usual manner.

Treatment of a methyl D-glucoside with methyl sulfate and sodium hydroxide brings about methylation of the four remaining —OH groups, and yields a methyl tetra-O-methyl-D-glucoside:

Ether formation

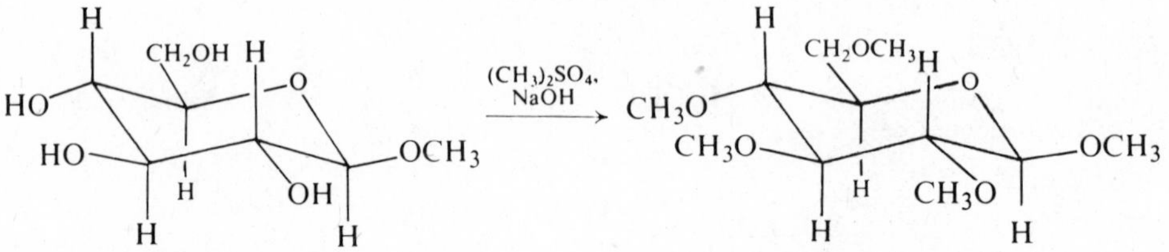

$\xrightarrow{(CH_3)_2SO_4,\ NaOH}$

Methyl β-D-glucoside
Non-reducing sugar

Methyl β-2,3,4,6-tetra-O-methyl-D-glucoside
Non-reducing sugar

In this reaction, ether linkages are formed by a modification of the Williamson synthesis that is possible here because of the comparatively high acidity of these —OH groups. (Why are these —OH groups more acidic than those of an ordinary alcohol?)

There is now an $—OCH_3$ group attached to every carbon in the carbohydrate except the one joined to C–1 through the acetal linkage; if the six-membered ring structure is correct, there is an $—OCH_3$ group on every carbon except C–5.

Treatment of the methyl tetra-O-methyl-D-glucoside with dilute hydrochloric acid removes only one of these $—OCH_3$ groups, and yields a tetra-O-methyl-D-glucose (Fig. 34.12). Only the reactive acetal linkage is hydrolyzed under these mild conditions; the other four $—OCH_3$ groups, held by ordinary ether linkages, remain intact.

What we have just described for D-(+)-glucose is typical of the methylation of any monosaccharide. A fully methylated carbohydrate contains both acetal linkages and ordinary ether linkages; these are formed in different ways and are hydrolyzed under different conditions.

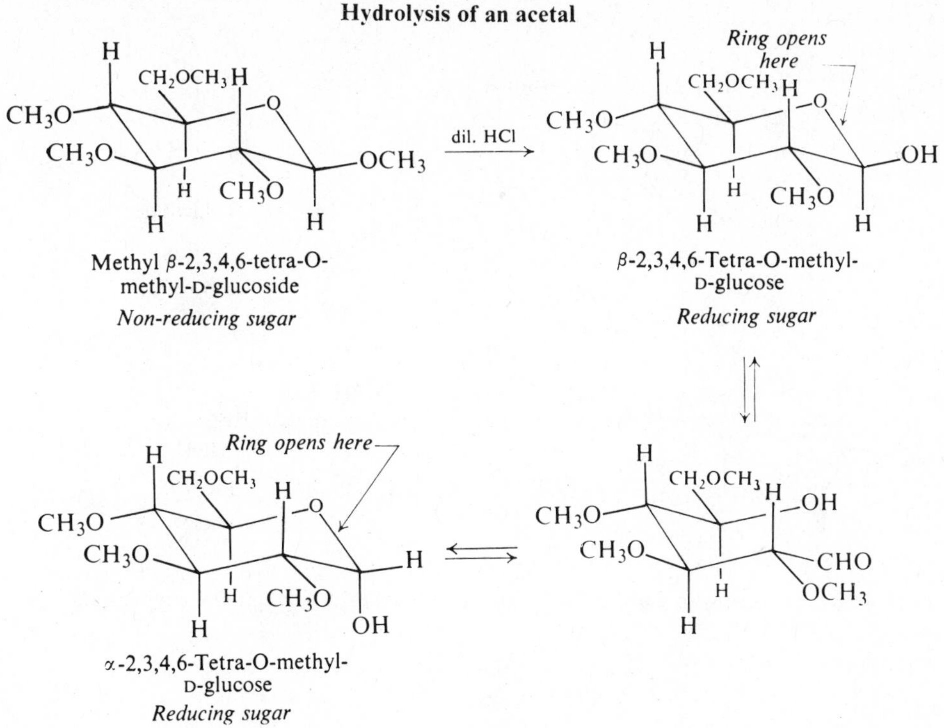

Hydrolysis of an acetal

Methyl β-2,3,4,6-tetra-O-methyl-D-glucoside
Non-reducing sugar

dil. HCl

Ring opens here

β-2,3,4,6-Tetra-O-methyl-D-glucose
Reducing sugar

Ring opens here

α-2,3,4,6-Tetra-O-methyl-D-glucose
Reducing sugar

Figure 34.12. Hydrolysis of a methyl glucoside.

34.19 Determination of ring size

In the cyclic structures that we have used so far for α- and β-D-(+)-glucose and the glucosides, oxygen has been shown as joining together C–1 and C–5; that is, these compounds are represented as containing a six-membered ring. But other ring sizes are possible, in particular, a five-membered ring, one in which C–1 is joined to C–4. What is the evidence that these compounds actually contain a six-membered ring?

When methyl β-D-glucoside is treated with methyl sulfate and sodium hydroxide, and the product is hydrolyzed by dilute hydrochloric acid, there is obtained a tetra-O-methyl-D-glucose. This compound is a cyclic hemiacetal which, in solution, exists in equilibrium with a little of the open-chain form (Fig. 34.13, p. 1102).

This open-chain tetra-O-methyl-D-glucose contains an aldehyde group and four —OCH_3 groups. It also contains a free, unmethylated —OH group at whichever carbon was originally involved in the acetal ring—on C–5, if the six-membered ring is correct. *Determination of ring size becomes a matter of finding out which carbon carries the free —OH group.*

What would we expect to happen if the tetra-O-methyl-D-glucose were vigorously oxidized by nitric acid? The —CHO and the free —OH group should be oxidized to yield a keto acid. But, from what we know about ketones (Sec. 19.9), we would not expect oxidation to stop here: the keto acid should be cleaved on one side or the other of the carbonyl group.

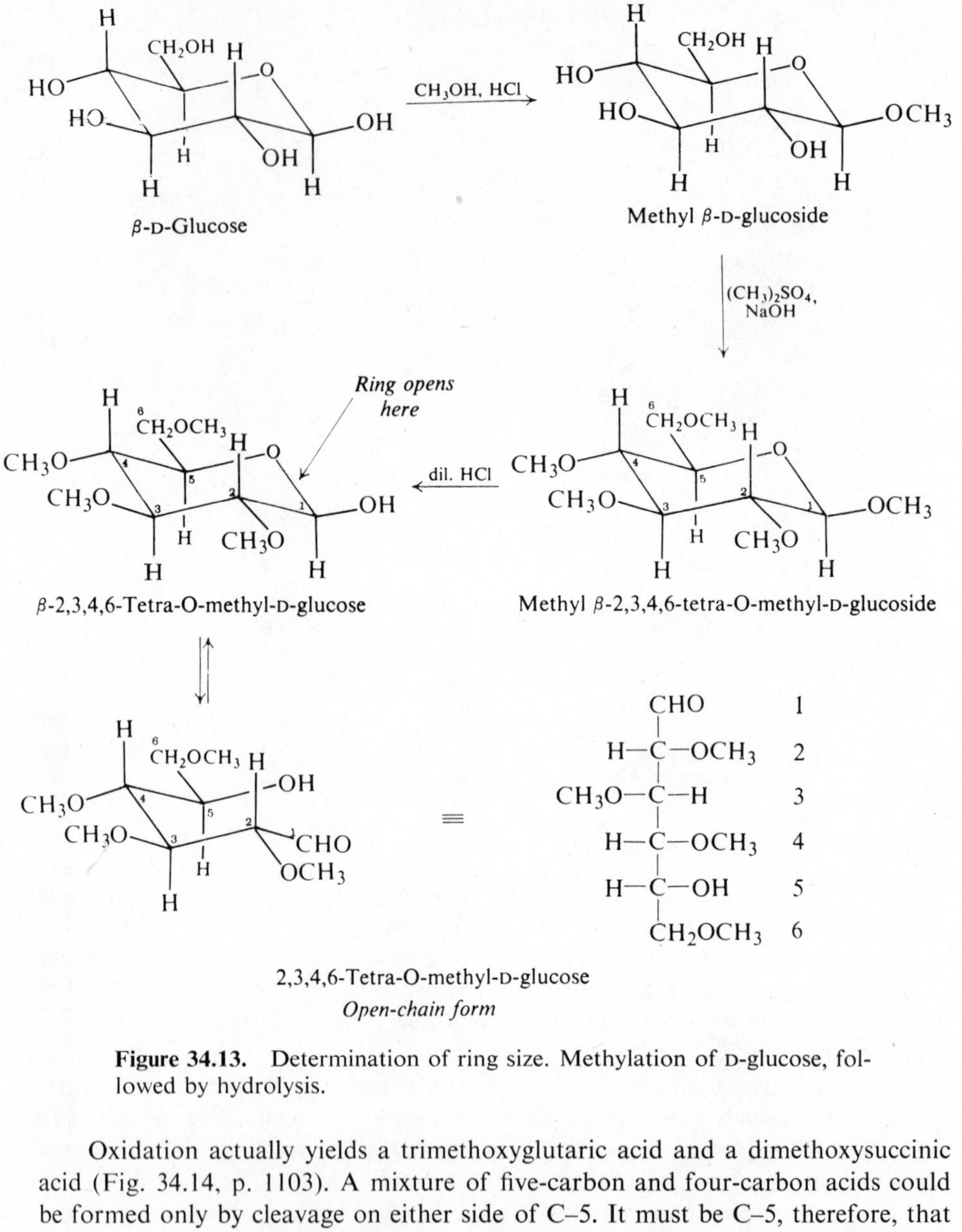

β-D-Glucose $\xrightarrow{CH_3OH,\ HCl}$ Methyl β-D-glucoside

$\downarrow (CH_3)_2SO_4$, NaOH

Ring opens here

β-2,3,4,6-Tetra-O-methyl-D-glucose $\xleftarrow{\text{dil. HCl}}$ Methyl β-2,3,4,6-tetra-O-methyl-D-glucoside

$\rightleftharpoons$

2,3,4,6-Tetra-O-methyl-D-glucose
Open-chain form

CHO	1
H—C—OCH_3	2
CH_3O—C—H	3
H—C—OCH_3	4
H—C—OH	5
CH_2OCH_3	6

Figure 34.13. Determination of ring size. Methylation of D-glucose, followed by hydrolysis.

Oxidation actually yields a trimethoxyglutaric acid and a dimethoxysuccinic acid (Fig. 34.14, p. 1103). A mixture of five-carbon and four-carbon acids could be formed only by cleavage on either side of C–5. It must be C–5, therefore, that carries the carbonyl oxygen of the intermediate keto acid, C–5 that carries the free —OH group in the tetra-O-methyl-D-glucose, C–5 that is involved in the acetal ring of the original glucoside. Methyl β-D-glucoside must contain a six-membered ring.

By the method just described, and largely through the work of Nobel Prize winner Sir W. N. Haworth (of the University of Birmingham, England), it has been established that the six-membered ring is the common one in the glycosides of aldohexoses. Evidence of other kinds (enzymatic hydrolysis, x-ray analysis) indicates that the *free* aldohexoses, too, contain six-membered rings.

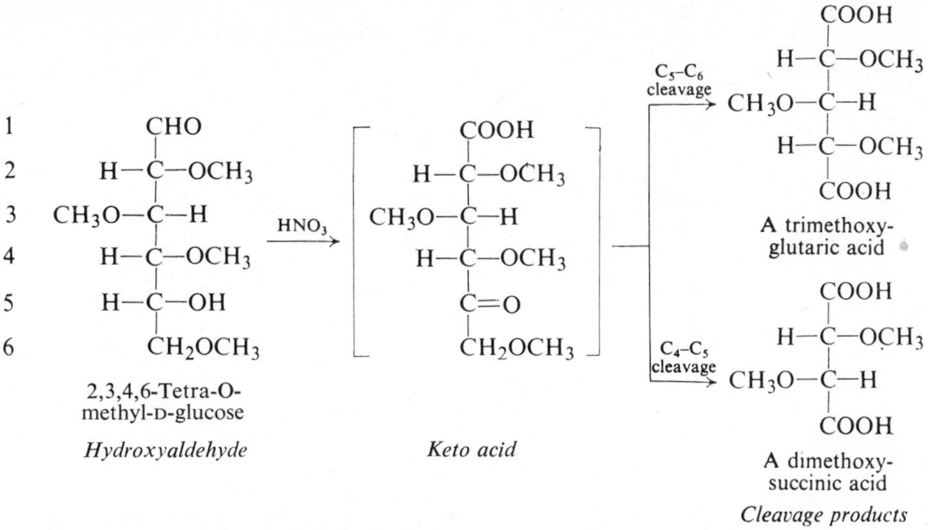

```
                                                                   C5-C6          COOH
                                                                   cleavage   H—C—OCH3
                                                                 ┌─────────→ CH3O—C—H
1        CHO                        COOH                         │             H—C—OCH3
2     H—C—OCH3                   H—C—OCH3                        │               COOH
3  CH3O—C—H       HNO3        CH3O—C—H                           │         A trimethoxy-
4     H—C—OCH3   ───────→        H—C—OCH3     ───────────────────┤         glutaric acid
5     H—C—OH                        C=O                          │
6        CH2OCH3                    CH2OCH3                      │                COOH
                                                                 │  C4-C5     H—C—OCH3
                                                                 └─────────→ CH3O—C—H
                                                                   cleavage       COOH
```

2,3,4,6-Tetra-O-methyl-D-glucose

Hydroxyaldehyde

Keto acid

A trimethoxyglutaric acid

A dimethoxysuccinic acid

Cleavage products

Figure 34.14. Oxidation of 2,3,4,6-tetra-O-methyl-D-glucose.

Problem 34.24 The products of HIO_4 oxidation of the methyl α-glycosides of the D-aldohexoses are shown in Sec. 34.17. What products would have been obtained if these glycosides had contained five-membered rings?

Problem 34.25 When either methyl α-L-arabinoside or methyl β-D-xyloside is methylated, hydrolyzed, and then oxidized by nitric acid, there is obtained a trimethoxyglutaric acid. (a) What ring size is indicated for these aldopentosides? (b) Predict the products of HIO_4 oxidation of each of these aldopentosides.

Problem 34.26 When crystalline methyl α-D-fructoside is methylated, hydrolyzed, oxidized by $KMnO_4$ and then nitric acid, there is obtained a trimethoxyglutaric acid. (a) What ring size is indicated for this 2-ketohexoside? (b) How does this acid compare with the one obtained from methyl α-L-arabinoside?

Problem 34.27 The crystalline methyl α- and β-D-glycosides we have discussed are usually prepared using methanolic HCl at 120°. When D-(+)-glucose is methylated *at room temperature*, there is obtained a liquid methyl D-glucoside. When this so-called "γ"-glucoside is methylated, hydrolyzed, and oxidized by nitric acid, there is obtained a dimethoxysuccinic acid. (a) What ring size is indicated for this "γ"-glucoside? (b) Should the dimethoxysuccinic acid be optically active or inactive? What is its absolute configuration? (c) When the liquid "γ"-glycoside obtained from D-(−)-fructose is methylated, hydrolyzed, and oxidized by nitric acid, there is also obtained a dimethoxysuccinic acid. How does this acid compare with the one in (b)?

If the name of a carbohydrate is exactly to define a particular structure, it must indicate ring size. Following a suggestion made by Haworth, carbohydrates are named to show their relationship to one of the heterocycles *pyran* or *furan*.

Pyran   Furan

A glycose containing a six-membered ring is thus a **pyranose** and its glycosides are **pyranosides**. A glycose containing a five-membered ring is a **furanose** and its glycosides are **furanosides**. For example:

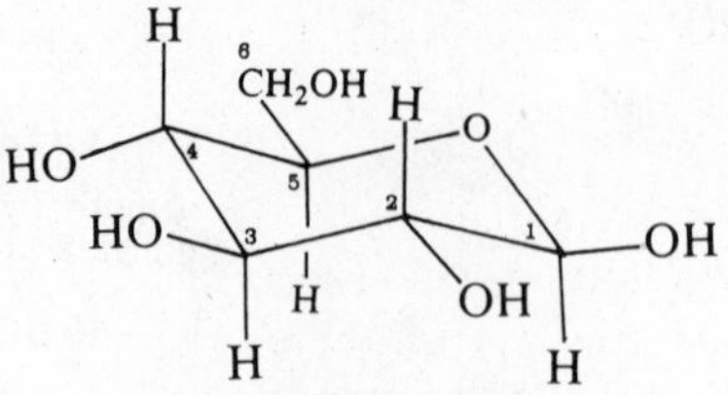

β-D-Glucopyranose

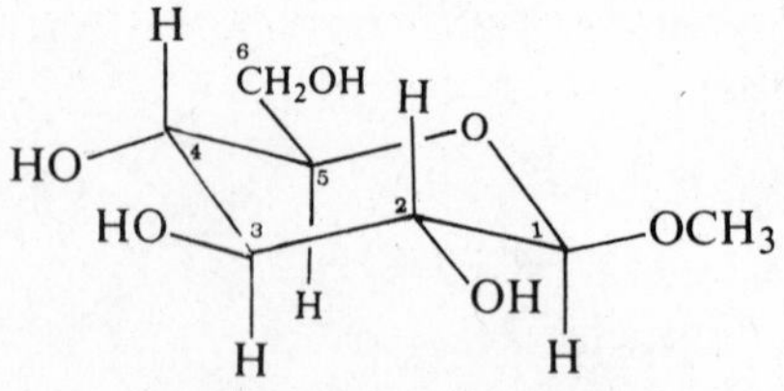

Methyl β-D-glucopyranoside

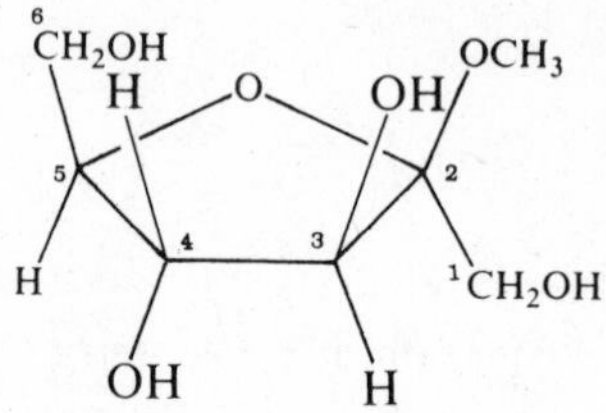

Methyl β-D-fructofuranoside

34.20 Conformation

We have followed the unraveling of the structure of D-(+)-glucose, and with it structures of the other monosaccharides, to the final working out of the ring size in 1926. Left to be discussed is one aspect whose importance has been realized only since about 1950: **conformation.**

D-(+)-Glucose contains the six-membered, pyranose ring. Since the C—O—C bond angle (111°) is very nearly equal to the tetrahedral angle (109.5°), the pyranose ring should be quite similar to the cyclohexane ring (Sec. 9.14). It should be puckered and, to minimize torsional and van der Waals strain, should exist in chair conformations in preference to twist-boat conformations. X-ray analysis shows this reasoning to be correct.

But there are *two* chair conformations possible for a D-(+)-glucopyranose anomer: I and II for β-D-(+)-glucopyranose, for example.

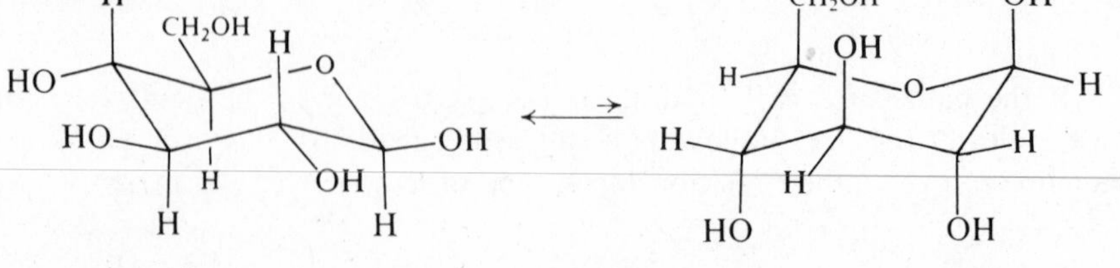

I

More stable:
all bulky groups equatorial

II

Less stable:
all bulky groups axial

β-D-(+)-Glucopyranose

Which of these is the more stable one, the one in which the molecules spend most of the time? For β-D-(+)-glucopyranose, the answer seems clear: I, in which all bulky substituents (—CH_2OH and —OH) occupy roomy equatorial positions, should certainly be much more stable than II, in which all bulky groups are crowded into axial positions. Again, x-ray analysis shows this reasoning to be correct.

What can we say about α-D-(+)-glucose and the other aldohexoses? This problem has been largely worked out by R. E. Reeves (then at the U.S. Southern Regional Research Laboratory) through study of copper complexes.

In general, the more stable conformation is the one in which the bulkiest group, —CH_2OH, occupies an equatorial position. For example:

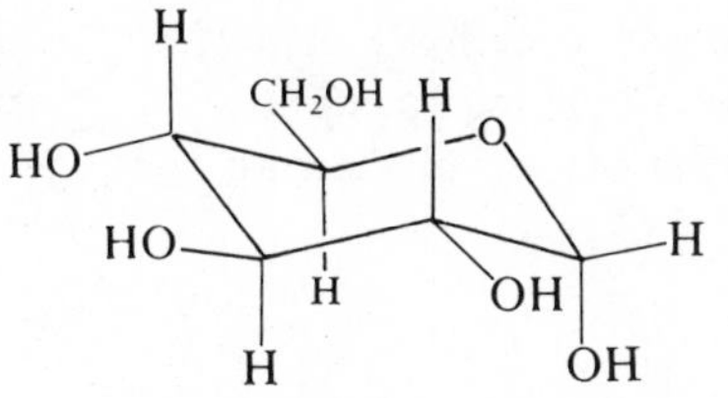

III
α-D-Glucopyranose
Stable conformation

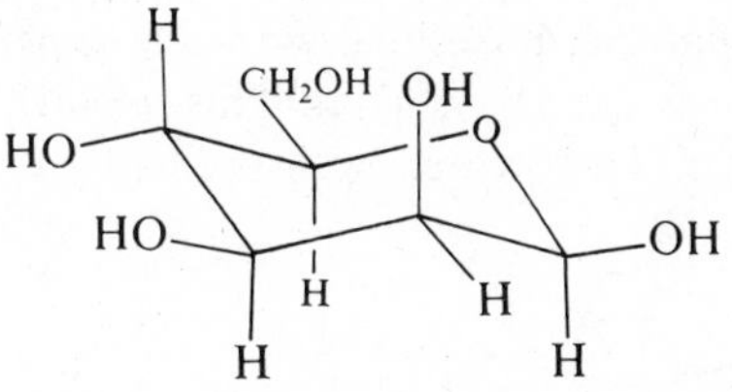

IV
β-D-Mannopyranose
Stable conformation

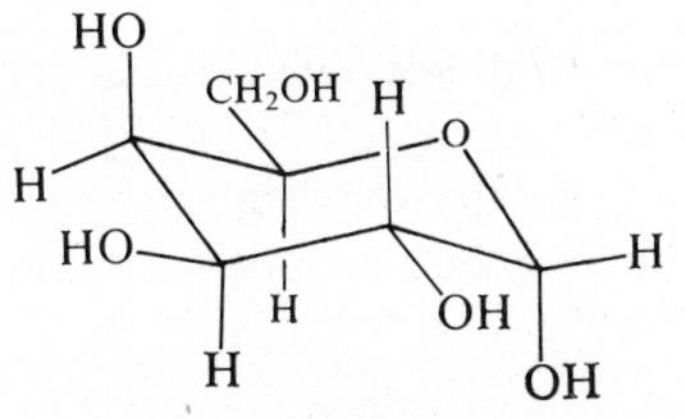

V
α-D-Galactopyranose
Stable conformation

In an extreme case, to permit many —OH groups to take up equatorial positions, the —CH_2OH group may be forced into an axial position. For example:

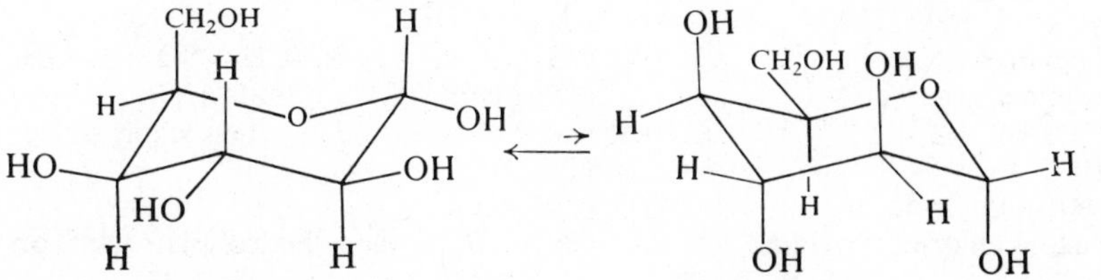

More stable:
4 equatorial OH's,
1 axial —CH_2OH

Less stable:
4 axial OH's,
1 equatorial —CH_2OH

α-D-Idopyranose

We notice that *of all* D*-aldohexoses it is* β-D-(+)*-glucose that can assume a conformation in which every bulky group occupies an equatorial position.* It is hardly accidental that β-D-(+)-glucose is the most widely occurring organic group in nature.

In drawing structural formulas or making models for the aldohexoses, a convenient point of reference is β-D-(+)-glucose. We draw the ring as shown in I—with C–1 down, C–4 up, and oxygen at the right-hand back corner—and place all —OH groups and the —CH_2OH group in equatorial positions. We draw the structures of other D-family aldohexoses merely by taking into account their differences from I. Thus α-D-(+)-glucose (III) differs in configuration at C–1; β-D-mannose (IV) differs in configuration at C–2; α-D-galactose (V) differs at C–1 and C–4. L-Family compounds are, of course, mirror images of these.

In methylated and acetylated pyranoses, too, bulky groups tend to occupy equatorial positions, with one general exception: a methoxy or acetoxy group on C–1 tends to be axial. This *anomeric effect* is attributed to repulsion between the dipoles associated with the C–1 oxygen and the oxygen of the ring.

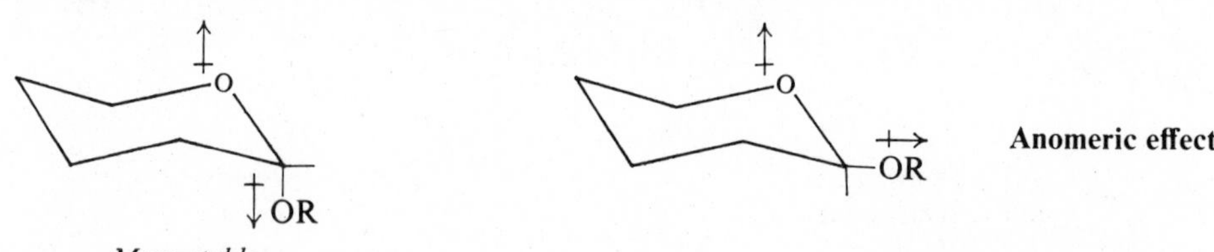

As we would expect for dipole–dipole interactions, the anomeric effect weakens as the polarity of the solvent increases (Sec. 9.10). For free sugars dissolved in water, the anomeric effect is usually outweighed by other factors; D-glucose, for example, exists predominantly as the β-anomer, with the —OH on C–1 equatorial.

Problem 34.28 Draw the conformation you predict to be the most stable for:

(a) β-D-allopyranose
(b) β-D-gulopyranose
(c) β-D-xylopyranose
(d) α-D-arabinopyranose
(e) β-L-(−)-glucopyranose
(f) β-D-(−)-fructopyranose

PROBLEMS

1. Give structures and, where possible, names of the principal products of the reaction (if any) of D-(+)-galactose with:

(a) hydroxylamine
(b) phenylhydrazine
(c) bromine water
(d) HNO_3
(e) HIO_4
(f) acetic anhydride
(g) benzoyl chloride, pyridine
(h) CH_3OH, HCl
(i) CH_3OH, HCl; then $(CH_3)_2SO_4$, NaOH
(j) reagents of (i), then dilute HCl
(k) reagents of (i) and (j), then vigorous oxidation
(l) H_2, Ni
(m) $NaBH_4$
(n) CN^-, H^+; then hydrolysis; then one mole $NaBH_4$
(o) H_2, Ni; then oxidation to monocarboxylic acid
(p) Br_2(aq); then pyridine; then H^+; then Na(Hg), CO_2
(q) phenylhydrazine; then benzaldehyde, H^+
(r) reagents of (q), then reduction to monocarbonyl compound

(s) $Br_2(aq)$; then $CaCO_3$; then H_2O_2, Fe^{+++}
(t) reagents of (i), then NaOH
(u) CH_3OH, HCl; then HIO_4
(v) reagents of (u); then $Br_2(aq)$; then dilute HCl

2. Write equations to show how D-(+)-glucose could be converted into:

(a) methyl β-D-glucoside
(b) methyl β-2,3,4,6-tetra-O-methyl-D-glucoside
(c) 2,3,4,6-tetra-O-methyl-D-glucose
(d) D-mannose
(e) L-gulose
(f) D-arabinose
(g) mesotartaric acid
(h) hexa-O-acetyl-D-glucitol
(i) D-fructose
(j)

```
      CHO
 HO───┼───H
  H───┼───OH
 HO───┼───H
  H───┼───OH
  H───┼───OH
      CH2OH
```

3. Besides D-fructose, there are three D-2-ketohexoses: D-*psicose*, D-*sorbose*, and D-*tagatose*. (a) Draw the possible configurations for these three ketoses. (b) Given the configurations of all aldohexoses, tell how you could assign definite configurations to the ketoses.

4. Draw stereochemical formulas for products A through O, and tell what aldoses E, E′, F, H, I, I′, N, and O are related to.

(a) $ClCH_2CHO + BrMgC{\equiv}CMgBr + OHCCH_2Cl \longrightarrow$ A ($C_6H_8O_2Cl_2$), mainly *meso*
meso-A + KOH $\longrightarrow$ B ($C_6H_6O_2$), a diepoxide
B + H_2O, OH^- $\longrightarrow$ C ($C_6H_{10}O_4$)
C + H_2, Pd/$CaCO_3$ $\longrightarrow$ D ($C_6H_{12}O_4$)
D + cold dilute $KMnO_4$ $\longrightarrow$ E and E′ (both $C_6H_{14}O_6$)
D + peroxyformic acid $\longrightarrow$ F ($C_6H_{14}O_6$)
C + Na, NH_3 $\longrightarrow$ G ($C_6H_{12}O_4$)
G + cold dilute $KMnO_4$ $\longrightarrow$ H ($C_6H_{14}O_6$)
G + peroxyformic acid $\longrightarrow$ I and I′ (both $C_6H_{14}O_6$)

(b) *trans*-2-penten-4-yn-1-ol + HCO_2OH $\longrightarrow$ J ($C_5H_8O_3$), 4-pentyn-1,2,3-triol
J + acetic anhydride, then Pd/$CaCO_3$ + H_2 $\longrightarrow$ K ($C_{11}H_{16}O_6$)
K + HOBr $\longrightarrow$ L and M (both $C_{11}H_{17}O_7Br$)
L + hydrolysis $\longrightarrow$ N ($C_5H_{12}O_5$)
M + hydrolysis $\longrightarrow$ O ($C_5H_{12}O_5$), a racemic modification

(c) Starting with 2-butyn-1,4-diol, outline a synthesis of erythritol; of DL-threitol.

(d) 2-Butyn-1,4-diol (above) is made by reaction under pressure of acetylene with formaldehyde. What kind of reaction is this?

5. When borneol (ROH) is fed to a dog, this toxic substance is excreted as compound P, $C_6H_9O_6$—OR, where R stands for the bornyl group. Compound P does not reduce Benedict's solution. It reacts with aqueous $NaHCO_3$ with the liberation of a gas. Treatment of P with aqueous acid yields borneol (ROH) and D-glucuronic acid (Table 34.1), which is oxidized by bromine water to D-glucaric acid.

(a) What is the structure of P?

(b) Hydrolysis of the polysaccharide *pectin* (from fruits and berries) gives chiefly D-galacturonic acid; hydrolysis of the polysaccharide *algin* (from seaweed) yields D-mannuronic acid. Give the structures of these uronic acids.

(c) There are two uronic acids related to D-fructose. Draw their structures. Give the name and family of the aldonic acids formed from each "fructuronic acid" by reduction of the carbonyl group.

(d) What compound would you expect from the treatment of D-glucosone with bromine water?

6. The rate of oxidation of reducing sugars by cupric ion is found to be proportional to sugar and $[OH^-]$, and to be independent of $[Cu^{++}]$. What does the kinetics suggest about the mechanism of oxidation?

7. Upon oxidation by HIO_4 the methyl glycoside Q yields the same product (shown on p. 1099) as that obtained from methyl α-glycosides of the D-aldohexoses; however, it consumes only one mole of HIO_4 and yields *no* formic acid.

(a) How many carbon atoms are there in Q, and what is the ring size? (b) For which carbon atoms do you know the configuration? (c) When Q is methylated, hydrolyzed, and then vigorously oxidized, the dicarboxylic acid obtained is the di-O-methyl ether of (−)-tartaric acid. What is the complete structure and configuration of Q?

8. *Salicin*, $C_{13}H_{18}O_7$, found in willow (*Salix*, whence the name *salicylic*), is hydrolyzed by emulsin to D-glucose and saligenin, $C_7H_8O_2$. Salicin does not reduce Tollens' reagent. Oxidation of salicin by nitric acid yields a compound that can be hydrolyzed to D-glucose and salicylaldehyde.

Methylation of salicin gives pentamethylsalicin, which on hydrolysis gives 2,3,4,6-tetra-O-methyl-D-glucose.

What is the structure of salicin?

9. The optically inactive carbohydrate *bio-inonose*, $C_6H_{10}O_6$, reduces Benedict's solution, but does not react with bromine water. It is reduced to R and S, of formula $C_6H_{12}O_6$. Compounds R and S are oxidized by HIO_4 to six moles of HCOOH, and react with acetic anhydride to yield products of formula $C_{18}H_{24}O_{12}$. Vigorous oxidation of bio-inonose yields DL-idaric acid (the dicarboxylic acid from idose) as the only six-carbon fragment.

What is the structure of bio-inonose? Of R and S?

10. Much of what is known about photosynthesis has been learned by determining the fate of radioactive carbon dioxide, $^{14}CO_2$. The ^{14}C was found in many products, including glucose, fructose, and sucrose. To measure the radioactivity of each carbon atom in a particular molecule, degradations to one-carbon fragments were carried out.

Tell which position or positions in the molecule each of the following one-carbon products came from.

Show how the activity of the carbon atom in every position could be figured out.

(a) glucose $\xrightarrow{\text{Ruff degradation}}$ CO_2 + arabinose $\xrightarrow{\text{Ruff degradation}}$ CO_2

glucose + HIO_4 ⟶ HCHO

glucose + CH_3OH, HCl; then HIO_4 ⟶ HCOOH

glucose $\xrightarrow{\text{Lactobacillus casei}}$ 2 lactic acid (carboxyls are C–3 and C–4)

↓ $KMnO_4$

CO_2 + CH_3CHO $\xrightarrow{\text{NaOI}}$ CHI_3 + HCOOH

(b) ribulose (a 2-ketopentose) + HIO_4 ⟶ $HOCH_2COOH$ + 2HCOOH + HCHO

ribulose + H_2, Pt; then HIO_4 ⟶ 2HCHO + 3HCOOH

ribulose + $C_6H_5NHNH_2$ ⟶ ribosazone

ribosazone + HIO_4 ⟶ HCHO + HCOOH + $HC{=}NNHC_6H_5$ / $C{=}NNHC_6H_5$ / CHO (a three-carbon chain: HC=NNHC$_6$H$_5$–C(=NNHC$_6$H$_5$)–CHO)

11. *Nucleic acids*, the compounds that control heredity on the molecular level, are polymers composed of nucleotide units. The structures of nucleotides have been determined in the following way, as illustrated for *adenylic acid*, a nucleotide isolated from yeast cells.

Hydrolysis of adenylic acid yields one molecule each of a heterocyclic base, a sugar

T, and phosphoric acid. The base is called *adenine*, and will be represented as R_2NH. Adenylic acid has the formula $R_2N—C_5H_8O_3—OPO_3H_2$.

The sugar T is levorotatory and has the formula $C_5H_{10}O_5$; it reduces Tollens' reagent and Benedict's solution. T is oxidized by bromine water to optically active $C_5H_{10}O_6$, and by nitric acid to optically inactive $C_5H_8O_7$. T forms an osazone that is identical with the osazone obtained from another pentose, (−)-U. Degradation of (−)-U, followed by oxidation by nitric acid, yields optically inactive $C_4H_6O_6$.

(a) What is T?

Careful acidic hydrolysis of adenylic acid yields adenine and a phosphate of T, $C_5H_9O_4—OPO_3H_2$. Reduction of the phosphate with H_2/Pt yields optically inactive V, $C_5H_{11}O_4—OPO_3H_2$. Hydrolysis of V yields optically inactive W, $C_5H_{12}O_5$, which reacts with acetic anhydride to yield optically inactive X, $C_{15}H_{22}O_{10}$.

(b) What is the structure of the phosphate of T?

Adenylic acid does not reduce Tollens' reagent or Benedict's solution. When hydrolyzed by aqueous ammonia, adenylic acid yields phosphoric acid and the nucleoside *adenosine*. Treatment of adenosine with methyl sulfate and NaOH, followed by acidic hydrolysis, yields Y, a methylation product of T. Compound Y has the formula $C_8H_{16}O_5$. Vigorous oxidation of Y yields 2,3-di-O-methylmesotartaric acid and no larger fragments.

Synthesis of adenosine shows that a nitrogen atom of adenine is joined to a carbon atom in T; synthesis also shows that T has the β-configuration.

(c) Give the structure of adenylic acid, using R_2NH for the adenine unit.

(Check your answers in Fig. 37.5, p. 1179.)

12. Give structural formulas for compounds Z through II. Tell what each piece of information—(a), (b), (c), etc.—shows about the structures of Z and AA.

(a) D-glucose + CH_3COCH_3, H_2SO_4 $\longrightarrow$ Z ($C_{12}H_{20}O_6$) + AA ($C_9H_{16}O_6$)
Z or AA + H_2O, OH^- $\longrightarrow$ no reaction
Z $\xrightarrow{H_2O,\ H^+}$ AA $\xrightarrow{H_2O,\ H^+}$ D-glucose + CH_3COCH_3
To what class of compounds do Z and AA belong?

(b) Z or AA + Benedict's solution $\longrightarrow$ no reaction

(c) Z + $(CH_3)_2SO_4$, NaOH $\longrightarrow$ BB ($C_{13}H_{22}O_6$)
BB + H_2O, H^+ $\longrightarrow$ CC ($C_7H_{14}O_6$)
CC + $C_6H_5NHNH_2$ $\longrightarrow$ DD (an osazone)

(d) AA + $(CH_3)_2SO_4$, NaOH $\longrightarrow$ EE ($C_{12}H_{22}O_6$)
EE + H_2O, H^+ $\longrightarrow$ FF ($C_9H_{18}O_6$)
FF + $C_6H_5NHNH_2$ $\longrightarrow$ GG (an osazone)

(e) FF + $(CH_3)_2SO_4$, NaOH $\longrightarrow$ 2,3,5,6-tetra-O-methyl-D-glucofuranose

(f) CC + HNO_3 $\longrightarrow$ HH ($C_6H_{10}O_7$)

(g) CC + HCN, then H_2O, H^+ $\longrightarrow$ II (a δ-lactone)

13. When either D-glyceraldehyde or dihydroxyacetone, $HOCH_2COCH_2OH$, is treated with base, there is obtained a mixture of the following compounds:

```
     CH2OH              CH2OH               CH2OH
     |                  |                   |
     C=O                C=O                 C=O
     |                  |                   |
  HO—C—H              H—C—OH              H—C—OH
     |                  |                   |
   H—C—OH            HO—C—H          HOCH2—C—OH
     |                  |                   |
   H—C—OH             H—C—OH                CH2OH
     |                  |
     CH2OH              CH2OH          and enantiomer
  D-Fructose         D-Sorbose       D,L-Dendroketose
```

Suggest a possible mechanism for this reaction. (*Hints:* See Sec. 34.6. Count the carbons in reactants and products, and consider the reagent used.)

14. In dilute acid, hydrolysis of D-glucose-1-phosphate differs from ordinary alkyl esters of its type ($ROPO_3H_2$) in two ways: it is abnormally fast, and it takes place with cleavage of the carbon–oxygen bond. Can you suggest an explanation for its unusual behavior?

15. In Chap. 13, we learned about certain relationships between nmr spectra and the conformations of six-membered rings: in Problems 8 and 9 (p. 447), that a given proton absorbs farther downfield when in an equatorial position than when in an axial position; in Sec. 13.11, that the coupling constant, J, between *anti* protons (axial,axial) is bigger than between *gauche* protons (axial,equatorial or equatorial,equatorial). It was in the study of carbohydrates that those relationships were first recognized, chiefly by R. U. Lemieux (p. 1119).

(a) In the nmr spectra of aldopyranoses and their derivatives, the signal from one proton is found at lower fields than any of the others. Which proton is this, and why?

(b) In the nmr spectra of the two anomers of D-tetra-O-acetylxylopyranose the downfield peak appears as follows:

Anomer JJ: doublet, δ 5.39, $J = 6$ Hz
Anomer KK: doublet, δ 6.03, $J = 3$ Hz

Identify JJ and KK; that is, tell which is the α-anomer, and which is the β-anomer. Explain your answer.

(c) Answer (b) for the anomers of D-tetra-O-acetylribopyranose:

Anomer LL: doublet, δ 5.72, $J = 5$ Hz
Anomer MM: doublet, δ 5.82, $J = 2$ Hz

(d) Consider two pairs of anomers: NN and OO, and PP and QQ. One pair are the D-penta-O-acetylglucopyranoses, and the other pair are the D-penta-O-acetylmannopyranoses.

Anomer NN: doublet, δ 5.97, $J = 3$ Hz
Anomer OO: doublet, δ 5.68, $J = 3$ Hz
Anomer PP: doublet, δ 5.54, $J = 8$ Hz
Anomer QQ: doublet, δ 5.99, $J = 3$ Hz

Identify NN, OO, PP, and QQ. Explain your answer.

16. The rare sugar (−)-*mycarose* occurs as part of the molecules of several antibiotics. Using the following evidence, work out the structure and configuration of mycarose.

(i) lactone of $CH_3CH(OH)CH{=}C(CH_3)CH_2COOH \xrightarrow{syn\text{-hydroxylation}} RR\ (C_7H_{12}O_4)$
$RR + KBH_4 \longrightarrow$ (±)-mycarose

(ii) In the nmr spectrum of (−)-mycarose and several derivatives, the coupling constant between C_4–H and C_5–H is 9.5–9.7 Hz.

(iii) methyl mycaroside + $HIO_4 \longrightarrow SS\ (C_8H_{14}O_4)$
SS + cold $KMnO_4 \longrightarrow TT\ (C_8H_{14}O_5)$
$TT \xrightarrow{\text{hydrolysis}}$ L-lactic acid

(a) Disregarding stereochemistry, what is the structure of mycarose?

(b) What are the relative configurations about C–3 and C–4? About C–4 and C–5?

(c) What is the absolute configuration at C–5?

(d) What is the absolute configuration of (−)-mycarose? To which family, D or L, does it belong? In what conformation does it preferentially exist?

(e) (−)-Mycarose can be converted into two methyl mycarosides. In the nmr spectrum of one of these, the downfield peak appears as a triplet with $J = 2.4$ Hz. Which anomer, α or β, is this one likely to be? What would you expect to see in the nmr spectrum of the other anomer?

(f) In the nmr spectrum of free (−)-mycarose, the downfield peak (1H) appears as two doublets with $J = 9.5$ and 2.5 Hz. Which anomer of mycarose, α or β, does this appear to be?

17. How do you account for the following facts? (a) In an equilibrium mixture of methyl α-D-glucoside and methyl β-D-glucoside, the α-anomer predominates. (b) In the more stable conformation of *trans*-2,5-dichloro-1,4-dioxane, both chlorines occupy axial positions.

18. From study of the nmr spectra of many compounds, Lemieux (p. 1119) found that the protons of axial acetoxy groups ($—OOCCH_3$) generally absorb at lower field than those of equatorial acetoxy groups.

(a) Draw the two chair conformations of tetra-O-acetyl-β-L-arabinopyranose. On steric grounds, which would you expect to be the more stable? Taking into account the anomeric effect, which would you expect to be the more stable?

(b) In the nmr spectrum of this compound, absorption by the acetoxy protons appears upfield as two equal peaks, at δ 1.92 and δ 2.04. How do you account for the equal sizes of these peaks? What, if anything, does this tell about the relative abundances of the two anomers?

(c) When the acetoxy group on C–1 is replaced by the deuteriated group $—OOCCD_3$, the total area of the upfield peaks is decreased, of course, from 12H to 9H. The ratio of peak areas δ 1.92:δ 2.04 is now 1.46:1.00. Which conformation predominates, and by how much? Is the predominant conformation the one you predicted to be the more stable?

Chapter 35 Carbohydrates II. Disaccharides and Polysaccharides

35.1 Disaccharides

Disaccharides are carbohydrates that are made up of two monosaccharide units. On hydrolysis a molecule of disaccharide yields two molecules of monosaccharide.

We shall study four disaccharides: **(+)-maltose** (malt sugar), **(+)-cellobiose**, **(+)-lactose** (milk sugar), and **(+)-sucrose** (cane or beet sugar). As with the monosaccharides, we shall focus our attention on the structure of these molecules: on which monosaccharides make up the disaccharide, and how they are attached to each other. In doing this, we shall also learn something about the properties of these disaccharides.

35.2 (+)-Maltose

(+)-Maltose can be obtained, among other products, by partial hydrolysis of starch in aqueous acid. (+)-Maltose is also formed in one stage of the fermentation of starch to ethyl alcohol; here hydrolysis is catalyzed by the enzyme *diastase*, which is present in malt (sprouted barley).

Let us look at some of the facts from which the structure of (+)-maltose has been deduced.

(+)-Maltose has the molecular formula $C_{12}H_{22}O_{11}$. It reduces Tollens' and Fehling's reagents and hence is a reducing sugar. It reacts with phenylhydrazine to yield an osazone, $C_{12}H_{20}O_9(=NNHC_6H_5)_2$. It is oxidized by bromine water to a monocarboxylic acid, $(C_{11}H_{21}O_{10})COOH$, *maltobionic acid.* (+)-Maltose exists in *alpha* ($[\alpha] = +168°$) and *beta* ($[\alpha] = +112°$) forms which undergo mutarotation in solution (equilibrium $[\alpha] = +136°$).

All these facts indicate the same thing: (+)-maltose contains a carbonyl group that exists in the reactive hemiacetal form as in the monosaccharides we have studied. It contains only one such "free" carbonyl group, however, since

(a) the osazone contains only two phenylhydrazine residues, and (b) oxidation by bromine water yields only a *mono*carboxylic acid.

When hydrolyzed in aqueous acid, or when treated with the enzyme *maltase* (from yeast), (+)-maltose is completely converted into D-(+)-glucose. This indicates that (+)-maltose ($C_{12}H_{22}O_{11}$) is made up of two D-(+)-glucose units joined together in some manner with the loss of one molecule of water:

$$2C_6H_{12}O_6 - H_2O = C_{12}H_{22}O_{11}$$

Hydrolysis by acid to give a new reducing group (two reducing D-(+)-glucose molecules in place of one (+)-maltose molecule) is characteristic of glycosides; hydrolysis by the enzyme maltase is characteristic of *alpha*-glucosides. A glycoside is an acetal formed by interaction of an alcohol with a carbonyl group of a carbohydrate (Sec. 34.16); in this case the alcohol concerned can only be a second molecule of D-(+)-glucose. We conclude that (+)-maltose contains two D-(+)-glucose units, joined by an *alpha*-glucoside linkage between the carbonyl group of one D-(+)-glucose unit and an —OH group of the other.

Two questions remain: which —OH group is involved, and what are the sizes of the rings in the two D-(+)-glucose units? Answers to both these questions are given by the sequence of oxidation, methylation, and hydrolysis shown in Fig. 35.1.

Oxidation by bromine water converts (+)-maltose into the monocarboxylic acid D-maltobionic acid. Treatment of this acid with methyl sulfate and sodium hydroxide yields octa-O-methyl-D-maltobionic acid. Upon hydrolysis in acidic solution, the methylated acid yields two products, 2,3,5,6-tetra-O-methyl-D-gluconic acid and 2,3,4,6-tetra-O-methyl-D-glucose.

These facts indicate that (+)-maltose has structure I, which is given the name 4-O-(α-D-glucopyranosyl)-D-glucopyranose. It is the —OH group on C-4 that serves as the alcohol in the glucoside formation; both halves of the molecule contain the six-membered, pyranose ring.

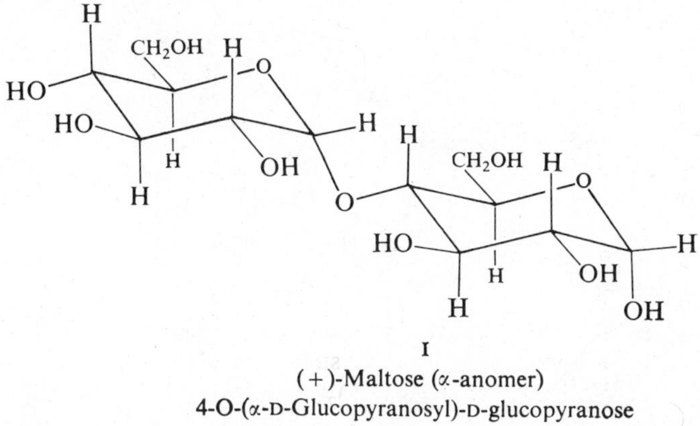

I

(+)-Maltose (α-anomer)

4-O-(α-D-Glucopyranosyl)-D-glucopyranose

Let us see how we arrive at structure I from the experimental facts.

First of all, the initial oxidation labels (with a —COOH group) the D-glucose unit that contains the "free" aldehyde group. Next, methylation labels (as —OCH_3) every free —OH group. Finally, upon hydrolysis, the absence of a methoxyl group shows which —OH groups were *not* free.

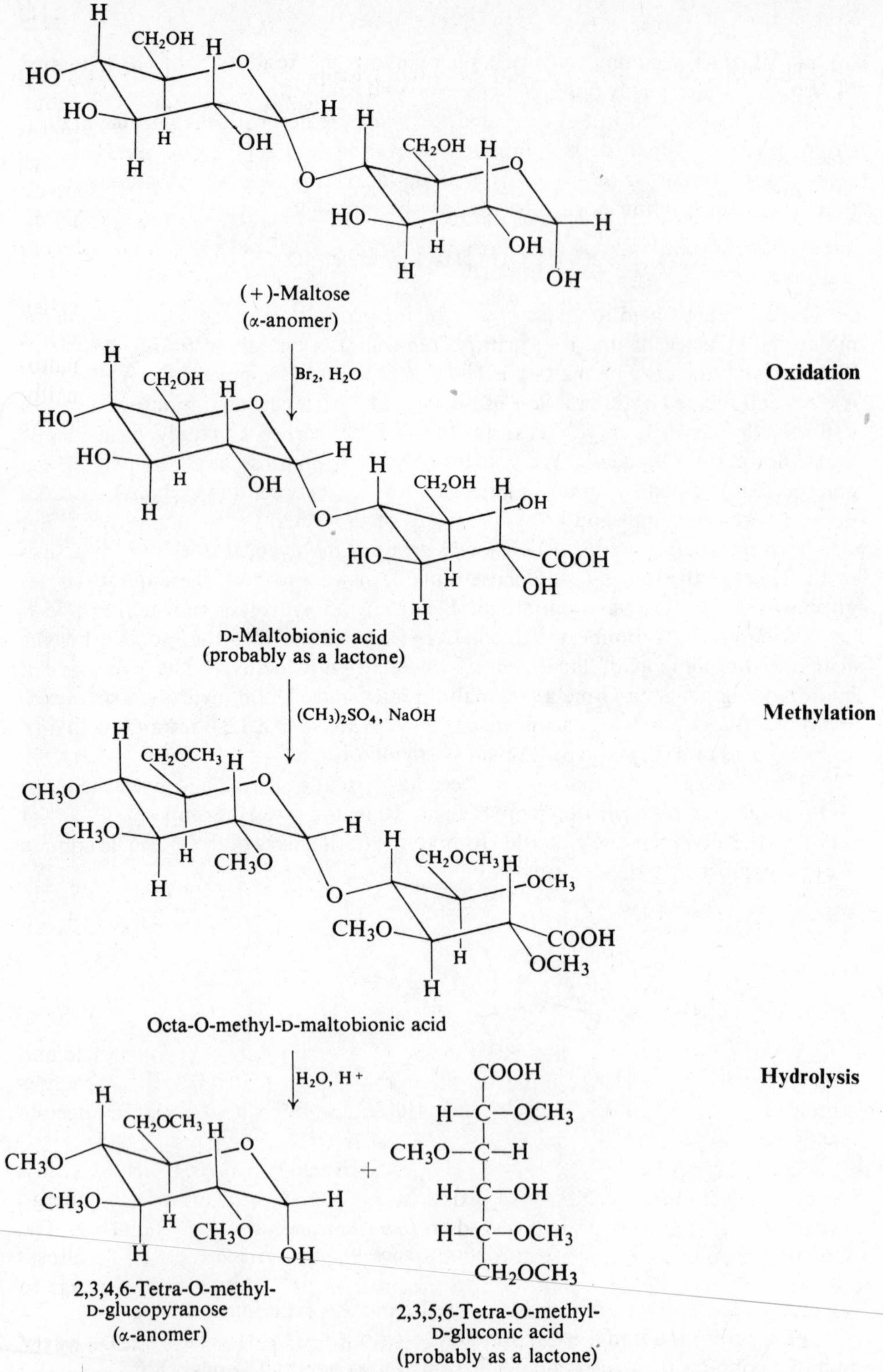

(+)-Maltose
(α-anomer)

Br_2, H_2O — **Oxidation**

D-Maltobionic acid
(probably as a lactone)

$(CH_3)_2SO_4$, NaOH — **Methylation**

Octa-O-methyl-D-maltobionic acid

H_2O, H^+ — **Hydrolysis**

2,3,4,6-Tetra-O-methyl-D-glucopyranose
(α-anomer)

+

COOH
H—C—OCH_3
CH_3O—C—H
H—C—OH
H—C—OCH_3
CH_2OCH_3

2,3,5,6-Tetra-O-methyl-D-gluconic acid
(probably as a lactone)

Figure 35.1. Sequence of oxidation, methylation, and hydrolysis shows that (+)-maltose is 4-O-(α-D-glucopyranosyl)-D-glucopyranose.

The oxidized product, 2,3,5,6-tetra-O-methyl-D-gluconic acid, must have arisen from the reducing (oxidizable) D-glucose unit. The presence of a free —OH group at C–4 shows that this position was not available for methylation at the maltobionic acid stage; hence it is the —OH on C–4 that is tied up in the glucoside linkage of maltobionic acid and of (+)-maltose itself. This leaves only the —OH group on C–5 to be involved in the ring of the reducing (oxidizable) unit in the original disaccharide. On the basis of these facts, therefore, we designate one D-(+)-glucose unit as a 4-O-substituted-D-glucopyranose.

The unoxidized product, 2,3,4,6-tetra-O-methyl-D-glucose, must have arisen from the non-reducing (non-oxidizable) D-glucose unit. The presence of the free —OH group at C–5 indicates that this position escaped methylation at the maltobionic acid stage; hence it is the —OH on C–5 that is tied up as a ring in maltobionic acid and in (+)-maltose itself. On the basis of these facts, therefore, we designate the second D-(+)-glucose unit as an α-D-glucopyranosyl group.

Problem 35.1 Formula I shows the structure of only the α-form of (+)-maltose. What is the structure of the β-(+)-maltose that in solution is in equilibrium with I?

Problem 35.2 The position of the free —OH group in 2,3,4,6-tetra-O-methyl-D-glucose was shown by the products of oxidative cleavage, as described in Sec. 34.19. What products would be expected from oxidative cleavage of 2,3,5,6-tetra-O-methyl-D-gluconic acid?

Problem 35.3 What products would be obtained if (+)-maltose itself were subjected to methylation and hydrolysis? What would this tell us about the structure of (+)-maltose? What uncertainty would remain in the (+)-maltose structure? Why was it necessary to oxidize (+)-maltose first before methylation?

Problem 35.4 When (+)-maltose is subjected to two successive one-carbon degradations, there is obtained a disaccharide that reduces Tollens' and Fehling's reagents but does not form an osazone. What products would be expected from the acidic hydrolysis of this disaccharide? What would these facts indicate about the structure of (+)-maltose?

35.3 (+)-Cellobiose

When cellulose (cotton fibers) is treated for several days with sulfuric acid and acetic anhydride, a combination of acetylation and hydrolysis takes place; there is obtained the octaacetate of (+)-cellobiose. Alkaline hydrolysis of the octaacetate yields (+)-cellobiose itself.

Like (+)-maltose, (+)-cellobiose has the molecular formula $C_{12}H_{22}O_{11}$, is a reducing sugar, forms an osazone, exists in *alpha* and *beta* forms that undergo mutarotation, and can be hydrolyzed to two molecules of D-(+)-glucose. The sequence of oxidation, methylation, and hydrolysis (as described for (+)-maltose) shows that (+)-cellobiose contains two pyranose rings and a glucoside linkage to an —OH group on C–4.

(+)-Cellobiose differs from (+)-maltose in one respect: it is hydrolyzed by the enzyme *emulsin* (from bitter almonds), not by maltase. Since emulsin is known to hydrolyze only β-glucoside linkages, we can conclude that the structure of (+)-cellobiose differs from that of (+)-maltose in only one respect: the D-glucose units are joined by a *beta* linkage rather than by an *alpha* linkage. (+)-Cellobiose is therefore 4-O-(β-D-glucopyranosyl)-D-glucopyranose.

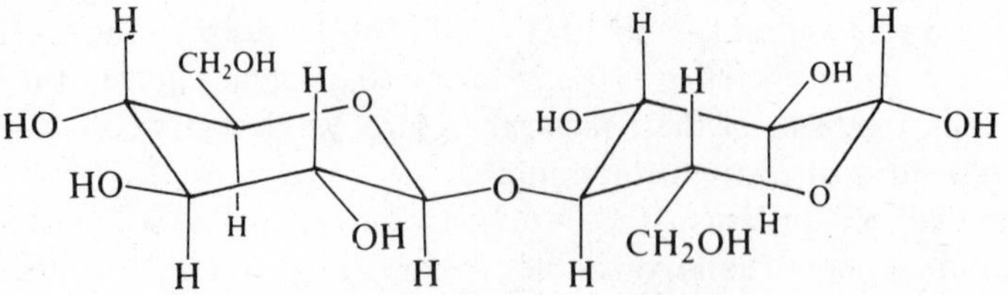

(+)-Cellobiose (β-anomer)
4-O-(β-D-Glucopyranosyl)-D-glucopyranose

Although the D-glucose unit on the right in the formula of (+)-cellobiose may look different from the D-glucose unit on the left, this is only because it has been turned over to permit a reasonable bond angle at the glycosidic oxygen atom.

Problem 35.5 Why is *alkaline* hydrolysis of cellobiose octaacetate (better named octa-O-acetylcellobiose) to (+)-cellobiose preferred over acidic hydrolysis?

Problem 35.6 Write equations for the sequence of oxidation, methylation, and hydrolysis as applied to (+)-cellobiose.

35.4 (+)-Lactose

(+)-Lactose makes up about 5% of human milk and of cow's milk. It is obtained commercially as a by-product of cheese manufacture, being found in the *whey*, the aqueous solution that remains after the milk proteins have been coagulated. Milk *sours* when lactose is converted into lactic acid (sour, like all acids) by bacterial action (e.g., by *Lactobacillus bulgaricus*).

(+)-Lactose has the molecular formula $C_{12}H_{22}O_{11}$, is a reducing sugar, forms an osazone, and exists in *alpha* and *beta* forms which undergo mutarotation. Acidic hydrolysis or treatment with emulsin (which splits β-linkages only) converts (+)-lactose into equal amounts of D-(+)-glucose and D-(+)-galactose. (+)-Lactose is evidently a β-glycoside formed by the union of a molecule of D-(+)-glucose and a molecule of D-(+)-galactose.

The question next arises: which is the reducing monosaccharide unit and which the non-reducing unit? Is (+)-lactose a glucoside or a galactoside? Hydrolysis of lactosazone yields D-(+)-galactose and D-glucosazone; hydrolysis of *lactobionic acid* (monocarboxylic acid) yields D-gluconic acid and D-(+)-galactose (see Fig. 35.2). Clearly, it is the D-(+)-glucose unit that contains the "free" aldehyde group and undergoes osazone formation or oxidation to the acid. (+)-Lactose is thus a substituted D-glucose in which a D-galactosyl unit is attached to one of the oxygens; it is a galactoside, not a glucoside.

The sequence of oxidation, methylation, and hydrolysis gives results analogous to those obtained with (+)-maltose and (+)-cellobiose: the glycoside linkage involves an —OH group on C–4, and both units exist in the six-membered, pyranose form. (+)-Lactose is therefore 4-O-(β-D-galactopyranosyl)-D-glucopyranose.

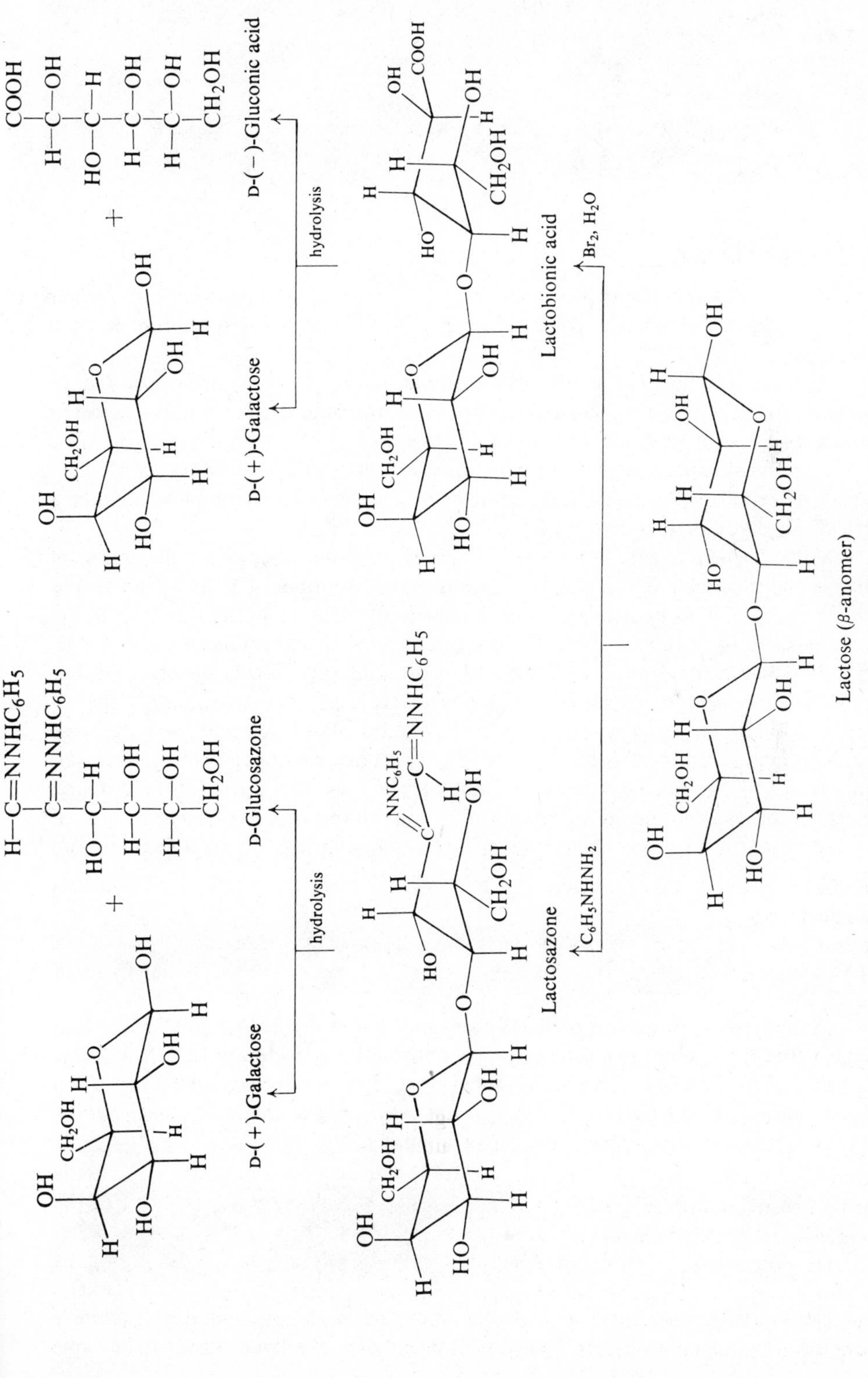

Figure 35.2. Hydrolysis of (+)-lactose derivatives. Shows that glucose is the reducing unit. (+)-Lactose is 4-O-(β-D-galactopyranosyl)-D-glucopyranose.

Problem 35.7 (a) Write equations for the sequence of oxidation, methylation, and hydrolysis as applied to (+)-lactose.

(b) What compounds would be expected from oxidative cleavage of the final products of (a)?

Problem 35.8 What products would be expected if (+)-lactose were subjected to two successive one-carbon degradations followed by acidic hydrolysis?

35.5 (+)-Sucrose

(+)-Sucrose is our common table sugar, obtained from sugar cane and sugar beets. Of organic chemicals, it is the one produced in the largest amount in pure form.

(+)-Sucrose has the molecular formula $C_{12}H_{22}O_{11}$. It does not reduce Tollens' or Fehling's reagent. It is a non-reducing sugar, and in this respect it differs from the other disaccharides we have studied. Moreover, (+)-sucrose does not form an osazone, does not exist in anomeric forms, and does not show mutarotation in solution. All these facts indicate that (+)-sucrose does not contain a "free" aldehyde or ketone group.

When (+)-sucrose is hydrolyzed by dilute aqueous acid, or by the action of the enzyme *invertase* (from yeast), it yields equal amounts of D-(+)-glucose and D-(−)-fructose. This hydrolysis is accompanied by a change in the sign of rotation from positive to negative; it is therefore often called the *inversion* of (+)-sucrose, and the levorotatory mixture of D-(+)-glucose and D-(−)-fructose obtained has been called *invert sugar*. (Honey is mostly invert sugar; the bees supply the invertase.) While (+)-sucrose has a specific rotation of +66.5° and D-(+)-glucose has a specific rotation of +52.7°, D-(−)-fructose has a large negative specific rotation of −92.4°, giving a net negative value for the specific rotation of the mixture. (Because of their opposite rotations and their importance as components of (+)-sucrose, D-(+)-glucose and D-(−)-fructose are commonly called **dextrose** and **levulose.**)

Problem 35.9 How do you account for the experimentally observed $[\alpha] = -19.9°$ for invert sugar?

(+)-Sucrose is made up of a D-glucose unit and a D-fructose unit; since there is no "free" carbonyl group, it must be both a D-glucoside and a D-fructoside. The two hexose units are evidently joined by a glycoside linkage between C–1 of glucose and C–2 of fructose, for only in this way can the single link between the two units effectively block *both* carbonyl functions.

Problem 35.10 What would be the molecular formula of (+)-sucrose if C–1 of glucose were attached to, say, C–4 of fructose, and C–2 of fructose were joined to C–4 of glucose? Would this be a reducing or non-reducing sugar?

Determination of the stereochemistry of the D-glucoside and D-fructoside linkages is complicated by the fact that both linkages are hydrolyzed at the same time. The weight of evidence, including the results of x-ray studies and finally the synthesis of (+)-sucrose (1953), leads to the conclusion that (+)-sucrose is a *beta* D-fructoside and an *alpha* D-glucoside. (The synthesis of sucrose, by R. U. Lemieux of the Prairie Regional Laboratory, Saskatoon, Saskatchewan, has been described as "the Mount Everest of organic chemistry.")

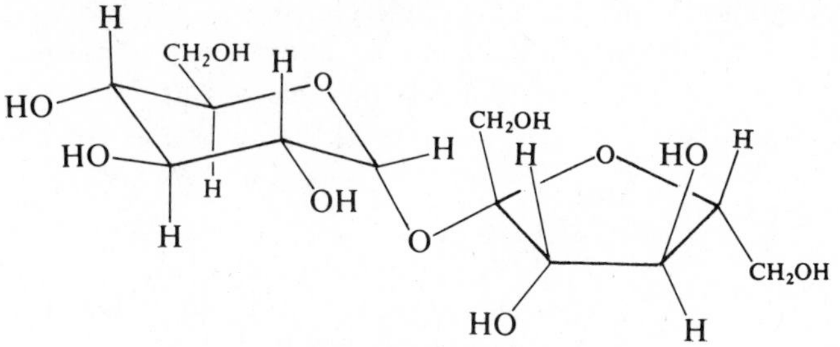

(+)-Sucrose
α-D-Glucopyranosyl β-D-fructofuranoside
β-D-Fructofuranosyl α-D-glucopyranoside
(no anomers; *non-mutarotating*)

Problem 35.11 When (+)-sucrose is hydrolyzed enzymatically, the D-glucose initially obtained mutarotates *downward* to +52.7°. What does this fact indicate about the structure of (+)-sucrose?

Methylation and hydrolysis show that (+)-sucrose contains a D-glucopyranose unit and a D-fructofuranose unit. (The unexpected occurrence of the relatively rare five-membered, furanose ring caused no end of difficulties in both structure proof and synthesis of (+)-sucrose.) (+)-Sucrose is named equally well as either α-D-glucopyranosyl β-D-fructofuranoside or β-D-fructofuranosyl α-D-glucopyranoside.

Problem 35.12 (a) Write equations for the sequence of methylation and hydrolysis as applied to (+)-sucrose.

(b) What compounds would be expected from oxidative cleavage of the final products of (a)?

35.6 Polysaccharides

Polysaccharides are compounds made up of many—hundreds or even thousands—monosaccharide units per molecule. As in disaccharides, these units are held together by glycoside linkages, which can be broken by hydrolysis.

Polysaccharides are naturally occurring polymers, which can be considered as derived from aldoses or ketoses by condensation polymerization. A polysaccharide derived from hexoses, for example, has the general formula $(C_6H_{10}O_5)_n$. This formula, of course, tells us very little about the structure of the polysaccharide. We need to know what the monosaccharide units are and how many there are in each molecule; how they are joined to each other; and whether the huge molecules thus formed are straight-chained or branched, looped or coiled.

By far the most important polysaccharides are **cellulose** and **starch.** Both are produced in plants from carbon dioxide and water by the process of photosynthesis, and both, as it happens, are made up of D-(+)-glucose units. Cellulose is the chief structural material of plants, giving the plants rigidity and form. It is probably the most widespread organic material known. Starch makes up the reserve food supply of plants and occurs chiefly in seeds. It is more water-soluble than cellulose, more easily hydrolyzed, and hence more readily digested.

Both cellulose and starch are, of course, enormously important to us. Generally speaking, we use them in very much the same way as the plant does. We use cellulose for its structural properties: as wood for houses, as cotton or rayon for clothing, as paper for communication and packaging. We use starch as a food: potatoes, corn, wheat, rice, cassava, etc.

35.7 Starch

Starch occurs as granules whose size and shape are characteristic of the plant from which the starch is obtained. When intact, starch granules are insoluble in cold water; if the outer membrane has been broken by grinding, the granules swell in cold water and form a gel. When the intact granule is treated with warm water, a soluble portion of the starch diffuses through the granule wall; in hot water the granules swell to such an extent that they burst.

In general, starch contains about 20% of a water-soluble fraction called **amylose**, and 80% of a water-insoluble fraction called **amylopectin**. These two fractions appear to correspond to different carbohydrates of high molecular weight and formula $(C_6H_{10}O_5)_n$. Upon treatment with acid or under the influence of enzymes, the components of starch are hydrolyzed progressively to dextrin (a mixture of low molecular weight polysaccharides), (+)-maltose, and finally D-(+)-glucose. (A mixture of all these is found in corn sirup, for example.) Both amylose and amylopectin are made up of D-(+)-glucose units, but differ in molecular size and shape.

35.8 Structure of amylose. End group analysis

(+)-Maltose is the only disaccharide that is obtained by hydrolysis of amylose, and D-(+)-glucose is the only monosaccharide. To account for this, it has been proposed that amylose is made up of chains of many D-(+)-glucose units, each unit joined by an *alpha* glycoside linkage to C–4 of the next one.

We could conceive of a structure for amylose in which α- and β-linkages regularly alternate. However, a compound of such a structure would be expected to yield (+)-cellobiose as well as (+)-maltose unless hydrolysis of the β-linkages occurred much faster than hydrolysis of the α-linkages. Since hydrolysis of the β-linkage in (+)-cellobiose is actually slower than hydrolysis of the α-linkage in (+)-maltose, such a structure seems unlikely.

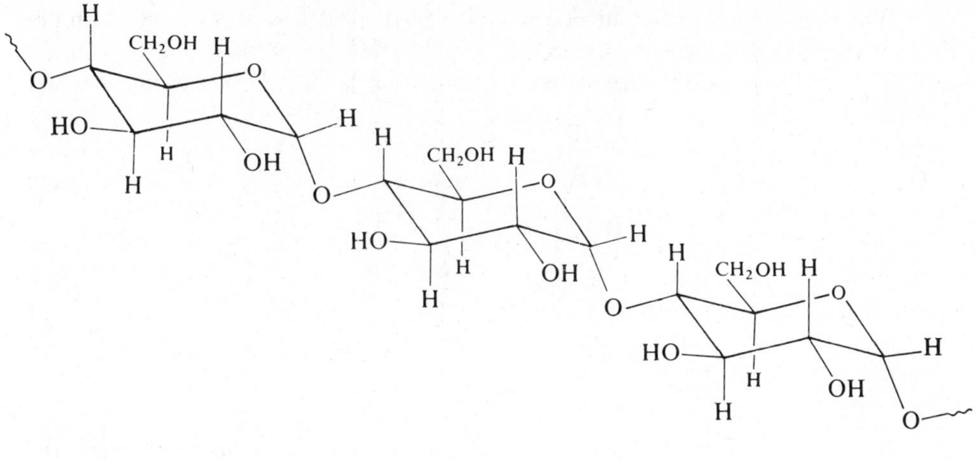

Amylose
(chair conformations assumed)

How many of these α-D-(+)-glucose units are there per molecule of amylose, and what are the shapes of these large molecules? These are difficult questions, and attempts to find the answers have made use of chemical and enzymatic methods, and of physical methods like x-ray analysis, electron microscopy, osmotic pressure and viscosity measurements, and behavior in an ultracentrifuge.

Valuable information about molecular size and shape has been obtained by the combination of methylation and hydrolysis that was so effective in studying the structures of disaccharides. D-(+)-Glucose, a monosaccharide, contains five free —OH groups and forms a pentamethyl derivative, methyl tetra-O-methyl-D-glucopyranoside. When two D-(+)-glucose units are joined together, as in (+)-maltose, each unit contains four free —OH groups; an octamethyl derivative is formed. If each D-(+)-glucose unit in amylose is joined to two others, it contains only three free —OH groups; methylation of amylose should therefore yield a compound containing only three $—OCH_3$ groups per glucose unit. What are the facts?

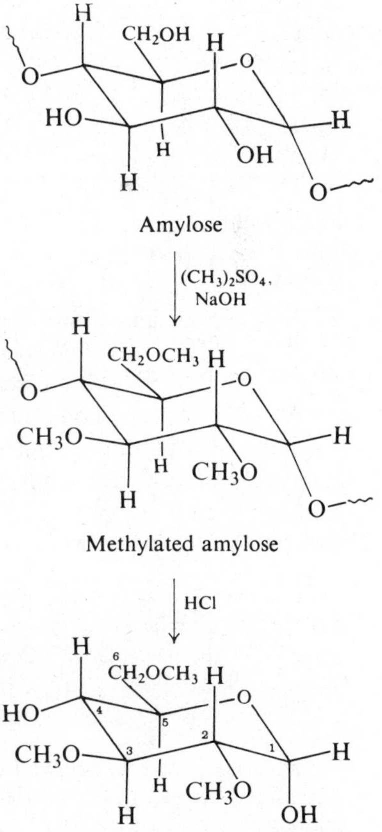

Amylose

$\xrightarrow{(CH_3)_2SO_4,\ NaOH}$

Methylated amylose

$\xrightarrow{HCl}$

2,3,6-Tri-O-methyl-D-glucose
(α-anomer)

When amylose is methylated and hydrolyzed there is obtained, as expected, 2,3,6-tri-O-methyl-D-glucose. But there is also obtained a little bit of 2,3,4,6-tetra-O-methyl-D-glucose, amounting to about 0.2–0.4% of the total product.

H 6 CH_2OCH_3 H O CH_3O CH_3O H H CH_3O H OH

2,3,4,6-Tetra-O-methyl-D-glucose
(α-anomer)

Consideration of the structure of amylose shows that this, too, is to be expected, and an important principle emerges: that of **end group analysis** (Fig. 35.3).

Each D-glucose unit in amylose is attached to two other D-glucose units, one through C–1 and the other through C–4, with C–5 in every unit tied up in the pyranose ring. As a result, free —OH groups at C–2, C–3, and C–6 are available for methylation. But this is not the case for *every* D-glucose unit. Unless the amylose chain is cyclic, it must have two ends. At one end there should be a D-glucose unit that contains a "free" aldehyde group. At the other end there should be a D-glucose unit that has a free —OH on C–4. This last D-glucose unit should undergo methylation at *four* —OH groups, and on hydrolysis should give a molecule of 2,3,4,6-tetra-O-methyl-D-glucose.

Thus each molecule of completely methylated amylose that is hydrolyzed should yield one molecule of 2,3,4,6-tetra-O-methyl-D-glucose; from the number of molecules of tri-O-methyl-D-glucose formed *along with* each molecule of the tetramethyl compound, we can calculate the length of the amylose chain.

Here we see an example of the use of end group analysis to determine chain length. A methylation that yields 0.25% of tetra-O-methyl-D-glucose shows that for every end group (with a free —OH on C–4) there are about 400 chain units.

But physical methods suggest that the chains are even longer than this. Molecular weights range from 150,000 to 600,000, indicating 1000 to 4000 glucose units per molecule. Evidently some degradation of the chain occurs during the methylation step; hydrolysis of only a few glycoside linkages in the alkaline medium would break the chain into much shorter fragments.

Problem 35.13 Consider an amylose chain of 4000 glucose units. At how many places must cleavage occur to lower the average length to 2000 units? To 1000? To 400? What percentage of the total number of glycoside links are hydrolyzed in each case?

Amylose, then, is believed to be made up of long chains, each containing 1000 or more D-glucose units joined together by α-linkages as in (+)-maltose; there is little or no branching of the chain.

Amylose is the fraction of starch that gives the intense blue color with iodine. X-ray analysis shows that the chains are coiled in the form of a helix (like a spiral

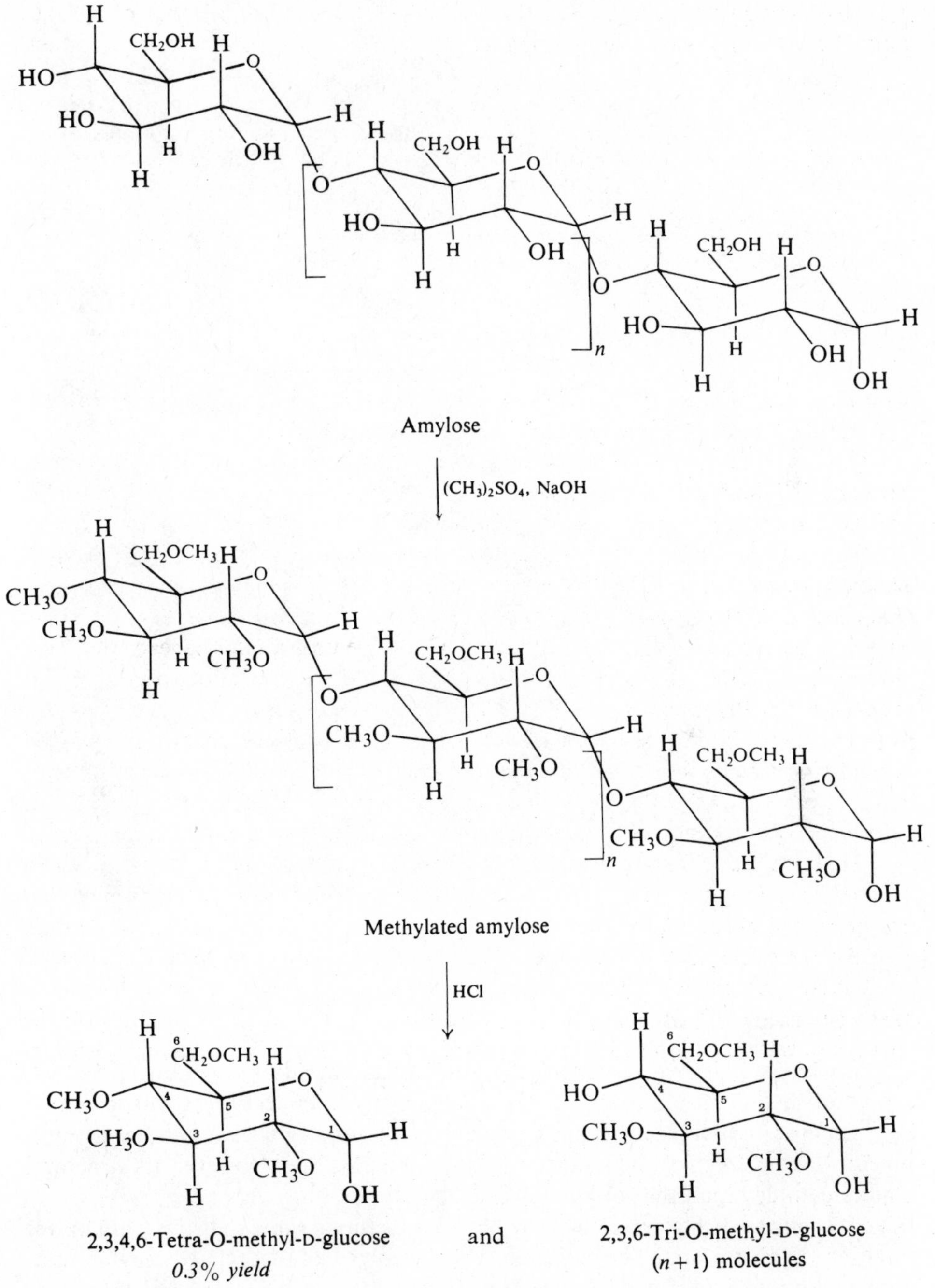

Figure 35.3. End group analysis. Hydrolysis of methylated amylose. End unit of long molecule gives 2,3,4,6-tetra-O-methyl-D-glucose; other units give 2,3,6-tri-O-methyl-D-glucose.

staircase), inside which is just enough space to accommodate an iodine molecule; the blue color is due to entrapped iodine molecules.

Problem 35.14 On the basis of certain evidence, it has been suggested that the rings of amylose have a twist-boat conformation, rather than the usual chair conformation. (a) What feature would tend to make any chair conformation unstable? (b) Suggest a twist-boat conformation that would avoid this difficulty. (*Hint:* What are the largest groups attached to a ring in amylose?)

Problem 35.15 When one mole of a disaccharide like (+)-maltose is treated with periodic acid (under conditions that minimize hydrolysis of the glycoside link), three moles of formic acid (and one of formaldehyde) are obtained.

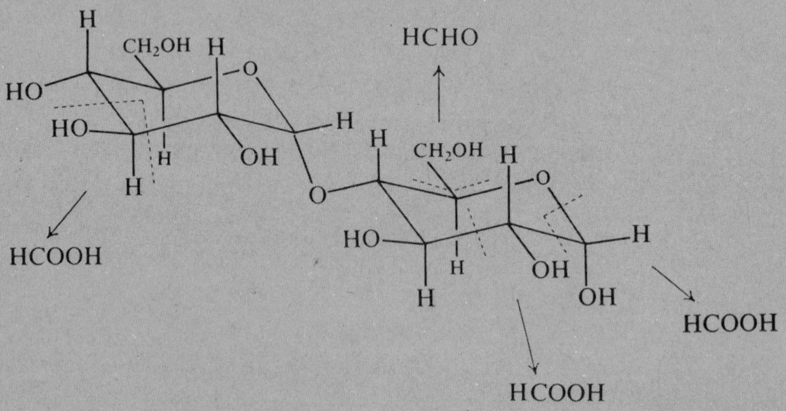

(a) Show what would happen to amylose (see formula on p. 1123) when treated with HIO_4. (b) How could this reaction be used to determine chain length? (c) Oxidation by HIO_4 of 540 mg of amylose (from the sago plant) yielded 0.0102 millimoles of HCOOH. What is the chain length of this amylose?

35.9 Structure of amylopectin

Amylopectin is hydrolyzed to the single disaccharide (+)-maltose; the sequence of methylation and hydrolysis yields chiefly 2,3,6-tri-O-methyl-D-glucose. Like amylose, amylopectin is made up of chains of D-glucose units, each unit joined by an *alpha* glycoside linkage to C–4 of the next one. However, its structure is more complex than that of amylose.

Molecular weights determined by physical methods show that there are up to a million D-glucose units per molecule. Yet hydrolysis of methylated amylopectin gives as high as 5% of 2,3,4,6-tetra-O-methyl-D-glucose, indicating only 20 units per chain. How can these facts be reconciled by the same structure?

The answer is found in the following fact: along with the trimethyl and tetramethyl compounds, hydrolysis yields 2,3-di-O-methyl-D-glucose and in an amount nearly equal to that of the tetramethyl derivative.

Methylated amylopectin

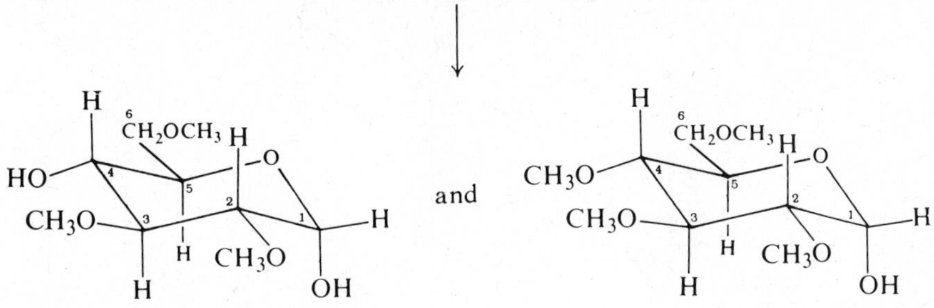

↓

2,3,6-Tri-O-methyl-D-glucose
~90%

and

2,3,4,6-Tetra-O-methyl-D-glucose
~5%

and

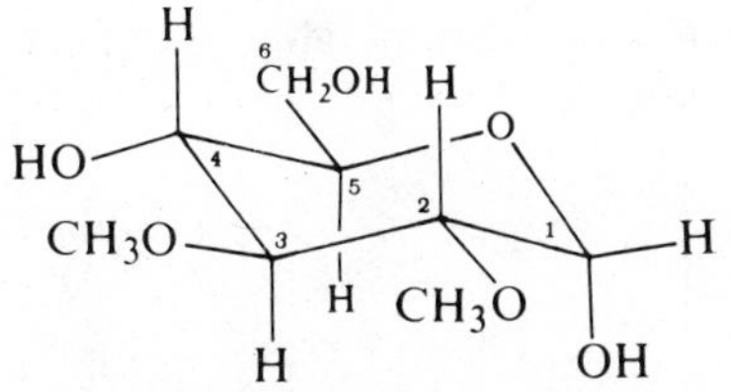

2,3-Di-O-methyl-D-glucose
~5%

Amylopectin has a highly branched structure consisting of several hundred short chains of about 20–25 D-glucose units each. One end of each of these chains is joined through C–1 to a C–6 on the next chain.

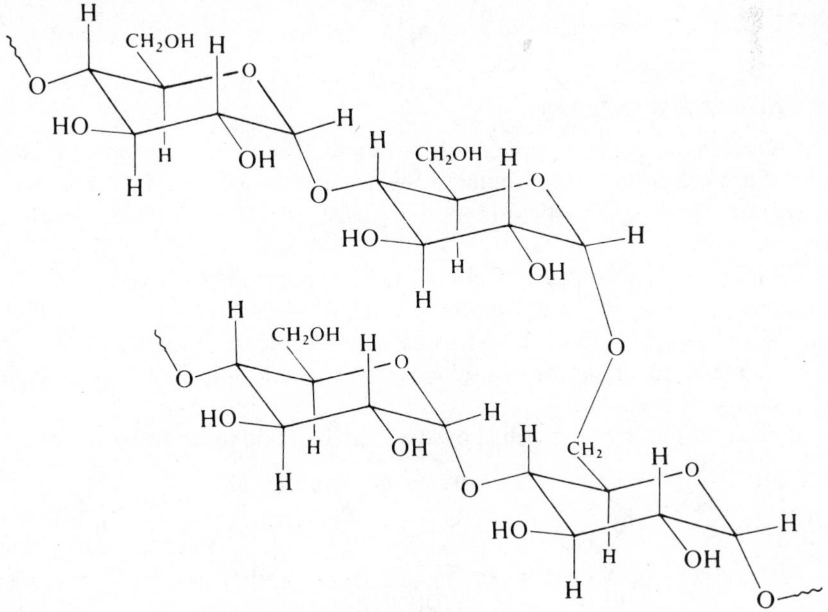

Amylopectin
(chair conformations assumed)

Schematically the amylopectin molecule is believed to be something like this:

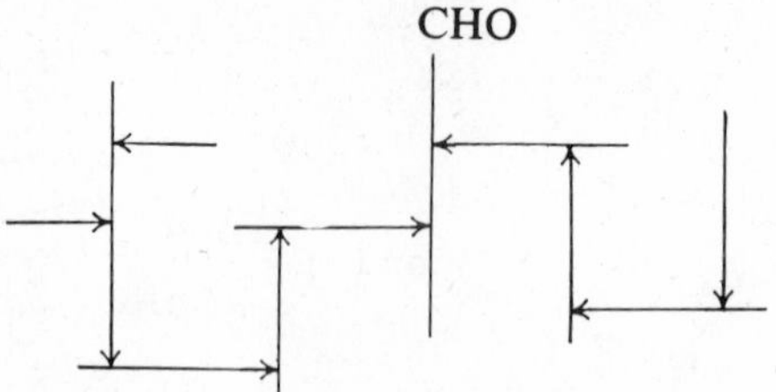

CHO

Amylopectin

Glycogen, the form in which carbohydrate is stored in animals to be released upon metabolic demand, has a structure very similar to that of amylopectin, except that the molecules appear to be more highly branched, and to have shorter chains (12–18 D-glucose units each).

Problem 35.16 Polysaccharides known as *dextrans* have been used as substitutes for blood plasma in transfusions; they are made by the action of certain bacteria on (+)-sucrose. Interpret the following properties of a dextran: Complete hydrolysis by acid yields only D-(+)-glucose. Partial hydrolysis yields only one disaccharide and only one trisaccharide, which contain only α-glycoside linkages. Upon methylation and hydrolysis, there is obtained chiefly 2,3,4-tri-O-methyl-D-glucose, together with smaller amounts of 2,4-di-O-methyl-D-glucose and 2,3,4,6-tetra-O-methyl-D-glucose.

Problem 35.17 Polysaccharides called *xylans* are found along with cellulose in wood and straw. Interpret the following properties of a sample of xylan: Its large negative rotation suggests β-linkages. Complete hydrolysis by acids yields only D-(+)-xylose. Upon methylation and hydrolysis, there is obtained chiefly 2,3-di-O-methyl-D-xylose, together with smaller amounts of 2,3,4-tri-O-methyl-D-xylose and 2-O-methyl-D-xylose.

35.10 Structure of cellulose

Cellulose is the chief component of wood and plant fibers; cotton, for instance, is nearly pure cellulose. It is insoluble in water and tasteless; it is a non-reducing carbohydrate. These properties, in part at least, are due to its extremely high molecular weight.

Cellulose has the formula $(C_6H_{10}O_5)_n$. Complete hydrolysis by acid yields D-(+)-glucose as the only monosaccharide. Hydrolysis of completely methylated cellulose gives a high yield of 2,3,6-tri-O-methyl-D-glucose. Like starch, therefore, cellulose is made up of chains of D-glucose units, each unit joined by a glycoside linkage to C-4 of the next.

Cellulose differs from starch, however, in the configuration of the glycoside

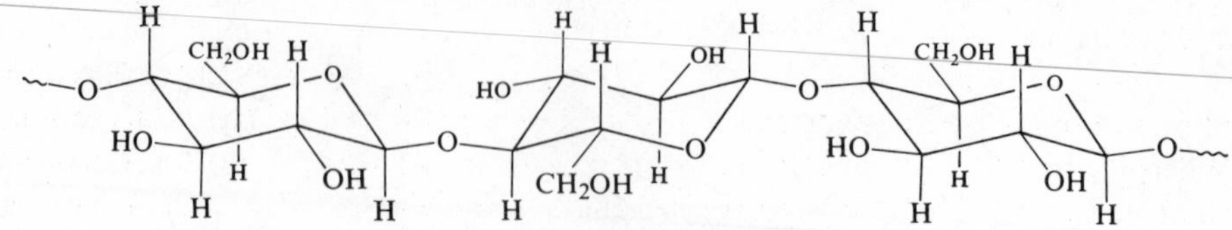

H CH₂OH H O HO H OH H H H O HO H CH₂OH H OH H H H CH₂OH H O HO H OH H

Cellulose

linkage. Upon treatment with acetic anhydride and sulfuric acid, cellulose yields octa-O-acetylcellobiose; there is evidence that all glycoside linkages in cellulose, like the one in (+)-cellobiose, are *beta* linkages.

Physical methods give molecular weights for cellulose ranging from 250,000 to 1,000,000 or more; it seems likely that there are at least 1500 glucose units per molecule. End group analysis by both methylation and periodic acid oxidation gives a chain length of 1000 glucose units or more. X-ray analysis and electron microscopy indicate that these long chains lie side by side in bundles, undoubtedly held together by hydrogen bonds between the numerous neighboring —OH groups. These bundles are twisted together to form rope-like structures, which themselves are grouped to form the fibers we can see. In wood these cellulose "ropes" are embedded in lignin to give a structure that has been likened to reinforced concrete.

35.11 Reactions of cellulose

We have seen that the glycoside linkages of cellulose are broken by the action of acid, each cellulose molecule yielding many molecules of D-(+)-glucose. Now let us look briefly at reactions of cellulose in which the chain remains essentially intact. Each glucose unit in cellulose contains three free —OH groups; these are the positions at which reaction occurs.

These reactions of cellulose, carried out to modify the properties of a cheap, available, ready-made polymer, are of tremendous industrial importance.

35.12 Cellulose nitrate

Like any alcohol, cellulose forms esters. Treatment with a mixture of nitric and sulfuric acids converts cellulose into *cellulose nitrate*. The properties and uses of the product depend upon the extent of nitration.

Guncotton, which is used in making smokeless powder, is very nearly completely nitrated cellulose, and is often called *cellulose trinitrate* (three nitrate groups per glucose unit).

Pyroxylin is less highly nitrated material containing between two and three nitrate groups per glucose unit. It is used in the manufacture of plastics like celluloid and collodion, in photographic film, and in lacquers. It has the disadvantage of being flammable, and forms highly toxic nitrogen oxides upon burning.

35.13 Cellulose acetate

In the presence of acetic anhydride, acetic acid, and a little sulfuric acid, cellulose is converted into the triacetate. Partial hydrolysis removes some of the acetate groups, degrades the chains to smaller fragments (of 200–300 units each), and yields the vastly important commercial *cellulose acetate* (roughly a *di*acetate).

Cellulose acetate is less flammable than cellulose nitrate and has replaced the nitrate in many of its applications, in safety-type photographic film, for example. When a solution of cellulose acetate in acetone is forced through the fine holes of a spinnerette, the solvent evaporates and leaves solid filaments. Threads from these filaments make up the material known as *acetate rayon*.

35.14 Rayon. Cellophane

When an alcohol is treated with carbon disulfide and aqueous sodium hydroxide, there is obtained a compound called a *xanthate.*

$$\mathrm{RONa} + \mathrm{S{=}C{=}S} \longrightarrow \underset{\text{A xanthate}}{\mathrm{RO{-}\underset{\underset{\displaystyle S}{\|}}{C}{-}SNa}} \xrightarrow{\mathrm{H^+}} \mathrm{ROH} + \mathrm{CS_2}$$

Cellulose undergoes an analogous reaction to form *cellulose xanthate*, which dissolves in the alkali to form a viscous colloidal dispersion called *viscose.*

When viscose is forced through a spinnerette into an acid bath, cellulose is regenerated in the form of fine filaments which yield threads of the material known as *rayon.* There are other processes for making rayon, but the viscose process is still the principal one used in the United States.

If viscose is forced through a narrow slit, cellulose is regenerated as thin sheets which, when softened by glycerol, are used for protective films (Cellophane).

Although rayon and Cellophane are often spoken of as "regenerated cellulose," they are made up of much shorter chains than the original cellulose because of degradation by the alkali treatment.

35.15 Cellulose ethers

Industrially, cellulose is alkylated by the action of alkyl chlorides (cheaper than sulfates) in the presence of alkali. Considerable degradation of the long chains is unavoidable in these reactions.

Methyl, ethyl, and benzyl ethers of cellulose are important in the production of textiles, films, and various plastic objects.

PROBLEMS

1. (+)-*Gentiobiose*, $C_{12}H_{22}O_{11}$, is found in the roots of gentians. It is a reducing sugar, forms an osazone, undergoes mutarotation, and is hydrolyzed by aqueous acid or by emulsin to D-glucose. Methylation of (+)-gentiobiose, followed by hydrolysis, gives 2,3,4,6-tetra-O-methyl-D-glucose and 2,3,4-tri-O-methyl-D-glucose. What is the structure and systematic name of (+)-gentiobiose?

2. (a) (+)-*Trehalose*, $C_{12}H_{22}O_{11}$, a non-reducing sugar found in young mushrooms, gives only D-glucose when hydrolyzed by aqueous acid or by maltase. Methylation gives an octa-O-methyl derivative that, upon hydrolysis, yields only 2,3,4,6-tetra-O-methyl-D-glucose. What is the structure and systematic name for (+)-trehalose?

(b) (−)-*Isotrehalose* and (+)-*neotrehalose* resemble trehalose in most respects. However, isotrehalose is hydrolyzed by either emulsin or maltase, and neotrehalose is hydrolyzed only by emulsin. What are the structures and systematic names for these two carbohydrates?

3. *Ruberythric acid*, $C_{25}H_{26}O_{13}$, a non-reducing glycoside, is obtained from madder root. Complete hydrolysis gives *alizarin* ($C_{14}H_8O_4$), D-glucose, and D-xylose; graded hydrolysis gives alizarin and *primeverose*, $C_{11}H_{20}O_{10}$. Oxidation of primeverose with bromine water, followed by hydrolysis, gives D-gluconic acid and D-xylose. Methylation of primeverose, followed by hydrolysis, gives 2,3,4-tri-O-methyl-D-xylose and 2,3,4-tri-O-methyl-D-glucose.

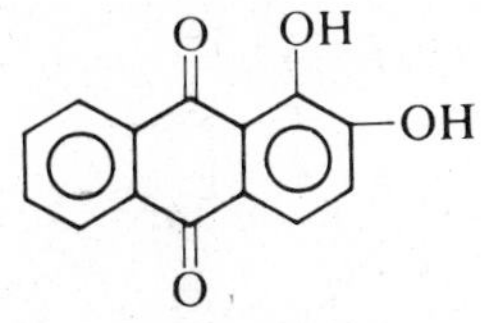

Alizarin

What structure or structures are possible for ruberythric acid? How can any uncertainties be cleared up?

4. (+)-*Raffinose*, a non-reducing sugar found in beet molasses, has the formula $C_{18}H_{32}O_{16}$. Hydrolysis by acid gives D-fructose, D-galactose, and D-glucose; hydrolysis by the enzyme α-galactosidase gives D-galactose and sucrose; hydrolysis by invertase (a sucrose-splitting enzyme) gives D-fructose and the disaccharide *melibiose*.

Methylation of raffinose, followed by hydrolysis, gives 1,3,4,6-tetra-O-methyl-D-fructose, 2,3,4,6-tetra-O-methyl-D-galactose, and 2,3,4-tri-O-methyl-D-glucose.

What is the structure of raffinose? Of melibiose?

5. (+)-*Melezitose*, a non-reducing sugar found in honey, has the formula $C_{18}H_{32}O_{16}$. Hydrolysis by acid gives D-fructose and two moles of D-glucose; partial hydrolysis gives D-glucose and the disaccharide *turanose*. Hydrolysis by maltase gives D-glucose and D-fructose; hydrolysis by another enzyme gives sucrose.

Methylation of melezitose, followed by hydrolysis, gives 1,4,6-tri-O-methyl-D-fructose and two moles of 2,3,4,6-tetra-O-methyl-D-glucose.

(a) What structure of melezitose is consistent with these facts? What is the structure of turanose?

Melezitose reacts with four moles of HIO_4 to give two moles of formic acid but no formaldehyde.

(b) Show that the absence of formaldehyde means either a furanose or pyranose structure for the fructose unit, and either a pyranose or septanose (7-membered ring) structure for the glucose units.

(c) How many moles of HIO_4 would be consumed and how many moles of formic acid would be produced if the two glucose units had septanose rings? (d) Answer (c) for one septanose ring and one pyranose ring. (e) Answer (c) for two pyranose rings. (f) What can you say about the size of the rings in the glucose units?

(g) Answer (c) for a pyranose ring in the fructose unit; for a furanose ring.

(h) What can you say about the size of the ring in the fructose unit?

(i) Are the oxidation data consistent with the structure of melezitose you gave in (a)?

6. The sugar, (+)-*panose*, was first isolated by S. C. Pan and co-workers (at Joseph E. Seagram and Sons, Inc.) from a culture of *Aspergillus niger* on maltose. Panose has a mol. wt. of approximately 475–500. Hydrolysis gives glucose, maltose, and an isomer of maltose called isomaltose. Methylation and hydrolysis of panose gives 2,3,4-tri-, 2,3,6-tri-, and 2,3,4,6-tetra-O-methyl-D-glucose in essentially equimolar amounts. The high positive rotation of panose is considered to exclude the possibility of any β-linkages.

(a) How many monosaccharide units make up a molecule of panose? In how many ways might these be arranged?

(b) Oxidation of panose to the aldonic acid, followed by hydrolysis, gives *no* maltose; reduction of panose to panitol, followed by hydrolysis, gives glucitol and maltitol (the reduction product of maltose). Can you now draw a single structure for panose? What must be the structure of isomaltose?

(c) Panose and isomaltose can be isolated from the partial hydrolysis products of amylopectin. What bearing does this have on the structure of amylopectin?

7. Cellulose can be oxidized by N_2O_4 to $[(C_5H_7O_4)COOH]_n$. (a) What is the structure of this product? (b) What will it give on hydrolysis of the chain? What is the name of this hydrolysis product?

(c) The oxidation product in (a) is readily decarboxylated to $(C_5H_8O_4)_n$. What will this give on hydrolysis of the chain? What is the name of this hydrolysis product? Is it a D or L compound?

8. Suggest structural formulas for the following polysaccharides, neglecting the stereochemistry of the glycoside linkages:

(a) An *araban* from peanut hulls yields only L-arabinose on hydrolysis. Methylation, followed by hydrolysis, yields equimolar amounts of 2,3,5-tri-O-methyl-L-arabinose, 2,3-di-O-methyl-L-arabinose, and 3-O-methyl-L-arabinose.

(b) A *mannan* from yeast yields only D-mannose on hydrolysis. Methylation, followed by hydrolysis, yields 2,3,4,6-tetra-O-methyl-D-mannose, 2,4,6-tri-O-methyl-D-mannose, 3,4,6-tri-O-methyl-D-mannose, and 3,4-di-O-methyl-D-mannose in a molecular ratio of 2:1:1:2, together with small amounts of 2,3,4-tri-O-methyl-D-mannose.

9. When a *xylan* (see Problem 35.17, p. 1127) is boiled with dilute hydrochloric acid, a pleasant-smelling liquid, *furfural*, $C_5H_4O_2$, steam-distills. Furfural gives positive tests with Tollens' and Schiff's reagents; it forms an oxime and a phenylhydrazone but not an osazone. Furfural can be oxidized by $KMnO_4$ to A, $C_5H_4O_3$, which is soluble in aqueous $NaHCO_3$.

Compound A can be readily decarboxylated to B, C_4H_4O, which can be hydrogenated to C, C_4H_8O. C gives no tests for functional groups except solubility in cold concentrated H_2SO_4; it gives negative tests for unsaturation with dilute $KMnO_4$ or Br_2/CCl_4.

Prolonged treatment of C with HCl gives D, $C_4H_8Cl_2$, which on treatment with KCN gives E, $C_6H_8N_2$. E can be hydrolyzed to F, $C_6H_{10}O_4$, identifiable as adipic acid.

What is the structure of furfural? Of compounds A through E?

10. Give a likely structure for each of the following polysaccharides:

(a) *Alginic acid*, from sea weed, is used as a thickening agent in ice cream and other foods. Hydrolysis yields only D-mannuronic acid. Methylation, followed by hydrolysis, yields 2,3-di-O-methyl-D-mannuronic acid. (Mannuronic acid is $HOOC(CHOH)_4CHO$.) The glycoside linkages in alginic acid are thought to be *beta*.

(b) *Pectic acid* is the main constituent of the *pectin* responsible for the formation of jellies from fruits and berries. Methylation of pectic acid, followed by hydrolysis, gives only 2,3-di-O-methyl-D-galacturonic acid. The glycoside linkages in pectic acid are thought to be *alpha*.

(c) *Agar*, from sea weed, is used in the growing of microorganisms. Hydrolysis yields a 9:1:1 molar ratio of D-galactose, L-galactose, and sulfuric acid. Methylation, followed by hydrolysis, yields 2,4,6-tri-O-methyl-D-galactose, 2,3-di-O-methyl-L-galactose, and sulfuric acid in the same 9:1:1 ratio. What uncertainties are there in your proposed structure?

11. The main constituent of the capsule surrounding the Type III pneumonococcus, and the substance responsible for the specificity of its antigen–antibody reactions, is a polysaccharide (mol. wt. about 150,000). Hydrolysis yields equimolar amounts of D-glucose and D-glucuronic acid, $HOOC(CHOH)_4CHO$; careful hydrolysis gives cellobiuronic acid (the uronic acid related to cellobiose). Methylation, followed by hydrolysis, gives equimolar amounts of 2,3,6-tri-O-methyl-D-glucose and 2,4-di-O-methyl-D-glucuronic acid.

What is a likely structure for the polysaccharide?

12. Draw structures of compounds G through J:

amylose + HIO_4 $\longrightarrow$ G + a little HCOOH and HCHO
G + bromine water $\longrightarrow$ H
H + H_2O, H^+ $\longrightarrow$ I ($C_4H_8O_5$) + J ($C_2H_2O_3$)

13. (a) Show what would happen to cellulose when treated with HIO_4. (b) How could this reaction be used to determine chain length? (c) If oxidation by HIO_4 of 203 mg of a sample of cellulose yields 0.0027 millimoles of HCOOH, what is the chain length of the cellulose?

Chapter 36 Amino Acids and Proteins

36.1 Introduction

The name **protein** is taken from the Greek *proteios*, which means *first*. This name is well chosen. Of all chemical compounds, proteins must almost certainly be ranked first, for they are the substance of life.

Proteins make up a large part of the animal body, they hold it together, and they run it. They are found in all living cells. They are the principal material of skin, muscle, tendons, nerves, and blood; of enzymes, antibodies, and many hormones.

(Only the nucleic acids, which control heredity, can challenge the position of proteins; and the nucleic acids are important because they direct the synthesis of proteins.)

Chemically, proteins are high polymers. They are polyamides, and the monomers from which they are derived are the α-amino carboxylic acids. A single protein molecule contains hundreds or even thousands of amino acid units; these units can be of twenty-odd different kinds. The number of different combinations, that is, the number of different protein molecules that are possible, is almost infinite. It is likely that tens of thousands of different proteins are required to make up and run an animal body; and this set of proteins is not identical with the set required by an animal of a different kind.

In this chapter we shall look first at the chemistry of the amino acids, and then briefly at the proteins that they make up. Our chief purpose will be to see the ways in which the structures of these enormously complicated molecules are being worked out, and how, in the last analysis, all this work rests on the basic principles of organic structural theory: on the concepts of bond angle and bond length, group size and shape, hydrogen bonding, resonance, acidity and basicity, optical activity, configuration and conformation.

36.2 Structure of amino acids

Table 36.1 gives the structures and names of 26 amino acids that have been found in proteins. Certain of these (marked *e*) are the *essential* amino acids, which must be fed to young animals if proper growth is to take place; these particular amino acids evidently cannot be synthesized by the animal from the other materials in its diet.

We see that all are *alpha*-amino carboxylic acids; in two cases (proline and hydroxyproline) the amino group forms part of a pyrrolidine ring. This common feature gives the amino acids a common set of chemical properties, one of which is the ability to form the long polyamide chains that make up proteins. It is on these common chemical properties that we shall concentrate.

In other respects, the structures of these compounds vary rather widely. In addition to the carboxyl group and the amino group *alpha* to it, some amino acids contain a second carboxyl group (e.g., aspartic acid or glutamic acid), or a potential carboxyl group in the form of a carboxamide (e.g., asparagine); these are called *acidic amino acids*. Some contain a second basic group, which may be an amino group (e.g., lysine), a guanidino group (arginine), or the imidazole ring (histidine); these are called *basic amino acids*. Some of the amino acids contain benzene or heterocyclic ring systems, phenolic or alcoholic hydroxyl groups, halogen or sulfur atoms. Each of these ring systems or functional groups undergoes its own typical set of reactions.

36.3 Amino acids as dipolar ions

Although the amino acids are commonly shown as containing an amino group and a carboxyl group, $H_2NCHRCOOH$, certain properties, both physical and chemical, are not consistent with this structure:

(a) In contrast to amines and carboxylic acids, the amino acids are non-volatile crystalline solids which melt with decomposition at fairly high temperatures.

(b) They are insoluble in non-polar solvents like petroleum ether, benzene, or ether, and are appreciably soluble in water.

(c) Their aqueous solutions behave like solutions of substances of high dipole moment.

(d) Acidity and basicity constants are ridiculously low for —COOH and $—NH_2$ groups. Glycine, for example, has $K_a = 1.6 \times 10^{-10}$ and $K_b = 2.5 \times 10^{-12}$, whereas most carboxylic acids have K_a's of about 10^{-5} and most aliphatic amines have K_b's of about 10^{-4}.

All these properties are quite consistent with a dipolar ion structure for the amino acids (I).

$$^{+}H_3N—CHR—COO^{-}$$

I

Amino acids: *dipolar ions*

The physical properties—melting point, solubility, high dipole moment—are just what would be expected of such a salt. The acid–base properties also become

Table 36.1 NATURAL AMINO ACIDS

Name	Abbreviation	Formula
(+)-Alanine	Ala	$CH_3CH(^+NH_3)COO^-$
(+)-Arginine[e]	Arg	$H_2NC(=^+NH_2)NHCH_2CH_2CH_2CH(NH_2)COO^-$
(−)-Asparagine	Asp(NH₂)	$H_2NCOCH_2CH(^+NH_3)COO^-$
(+)-Aspartic acid	Asp	$HOOCCH_2CH(^+NH_3)COO^-$
(−)-Cysteine	CySH	$HSCH_2CH(^+NH_3)COO^-$
(−)-Cystine	CyS—SCy	$^-OOCCH(^+NH_3)CH_2S—SCH_2CH(^+NH_3)COO^-$
(+)-3,5-Dibromotyrosine		HO-(3,5-$Br_2C_6H_2$)-$CH_2CH(^+NH_3)COO^-$
(+)-3,5-Diiodotyrosine		HO-(3,5-$I_2C_6H_2$)-$CH_2CH(^+NH_3)COO^-$
(+)-Glutamic acid	Glu	$HOOCCH_2CH_2CH(^+NH_3)COO^-$
(+)-Glutamine	Glu(NH₂)	$H_2NCOCH_2CH_2CH(^+NH_3)COO^-$
Glycine	Gly	$CH_2(^+NH_3)COO^-$
(−)-Histidine[e]	His	(imidazolyl, N—, NH)-$CH_2CH(^+NH_3)COO^-$
(−)-Hydroxylysine	Hylys	$^+H_3NCH_2CH(OH)CH_2CH_2CH(NH_2)COO^-$
(−)-Hydroxyproline	Hypro	4-HO-pyrrolidinium ($^+NH_2$, H H)-2-COO^-

Table 36.1 NATURAL AMINO ACIDS (*continued*)

Name	Abbreviation	Formula
(+)-Isoleucine[e]	Ileu	$CH_3CH_2CH(CH_3)CH(^+NH_3)COO^-$
(−)-Leucine[e]	Leu	$(CH_3)_2CHCH_2CH(^+NH_3)COO^-$
(+)-Lysine[e]	Lys	$^+H_3NCH_2CH_2CH_2CH_2CH(NH_2)COO^-$
(−)-Methionine[e]	Met	$CH_3SCH_2CH_2CH(^+NH_3)COO^-$
(−)-Phenylalanine[e]	Phe	$C_6H_5CH_2CH(^+NH_3)COO^-$
(−)-Proline	Pro	pyrrolidine ring, $\overset{+}{N}H_2$, COO^-
(−)-Serine	Ser	$HOCH_2CH(^+NH_3)COO^-$
(−)-Threonine[e]	Thr	$CH_3CHOHCH(^+NH_3)COO^-$
(+)-Thyroxine		$HO(C_6H_2I_2)O(C_6H_2I_2)CH_2CH(^+NH_3)COO^-$
(−)-Tryptophane[e]	Try	indole (N–H) ring–$CH_2CH(^+NH_3)COO^-$
(−)-Tyrosine	Tyr	$HOC_6H_4CH_2CH(^+NH_3)COO^-$
(+)-Valine[e]	Val	$(CH_3)_2CHCH(^+NH_3)COO^-$

[e] Essential amino acid

understandable when it is realized that the measured K_a actually refers to the acidity of an ammonium ion, RNH_3^+,

$$\underset{\text{Acid}}{^+H_3NCHRCOO^-} + H_2O \rightleftharpoons H_3O^+ + H_2NCHRCOO^-$$

$$K_a = \frac{[H_3O^+][H_2NCHRCOO^-]}{[^+H_3NCHRCOO^-]}$$

and K_b actually refers to the basicity of a carboxylate ion, $RCOO^-$.

$$\underset{\text{Base}}{^+H_3NCHRCOO^-} + H_2O \rightleftharpoons {}^+H_3NCHRCOOH + OH^-$$

$$K_b = \frac{[^+H_3NCHRCOOH][OH^-]}{[^+H_3NCH_2COO^-]}$$

In aqueous solution, the acidity and basicity of an acid and its conjugate base (CH_3COOH and CH_3COO^-, or $CH_3NH_3^+$ and CH_3NH_2, for example) are related by the expression $K_a \times K_b = 10^{-14}$. From this it can be calculated that a K_a of 1.6×10^{-10} for the —NH_3^+ of glycine means $K_b = 6.3 \times 10^{-5}$ for —NH_2: a quite reasonable value for an aliphatic amine. In the same way, a K_b of 2.5×10^{-12} for the —COO^- of glycine means $K_a = 4 \times 10^{-3}$ for —COOH: a quite reasonable value for a carboxylic acid containing the strongly electron-withdrawing (acid-strengthening) —NH_3^+ group.

When the solution of an amino acid is made alkaline, the dipolar ion I is converted into the anion II; the stronger base, hydroxide ion, removes a proton from the ammonium ion and displaces the weaker base, the amine.

$$\underset{\substack{\text{I}\\ \text{Stronger}\\ \text{acid}}}{^+H_3NCHRCOO^-} + \underset{\substack{\text{Stronger}\\ \text{base}}}{OH^-} \rightleftharpoons \underset{\substack{\text{II}\\ \text{Weaker}\\ \text{base}}}{H_2NCHRCOO^-} + \underset{\substack{\text{Weaker}\\ \text{acid}}}{H_2O}$$

When the solution of an amino acid is made acidic, the dipolar ion I is converted into the cation III; the stronger acid, H_3O^+, gives up a proton to the carboxylate ion, and displaces the weaker carboxylic acid.

$$\underset{\substack{\text{I}\\ \text{Stronger}\\ \text{base}}}{^+H_3NCHRCOO^-} + \underset{\substack{\text{Stronger}\\ \text{acid}}}{H_3O^+} \rightleftharpoons \underset{\substack{\text{III}\\ \text{Weaker}\\ \text{acid}}}{^+H_3NCHRCOOH} + \underset{\substack{\text{Weaker}\\ \text{base}}}{H_2O}$$

In summary, the acidic group of a simple amino acid like glycine is —NH_3^+ not —COOH, and the basic group is —COO^- not —NH_2.

Problem 36.1 In quite alkaline solution, an amino acid contains two basic groups, —NH_2 and —COO^-. Which is the more basic? To which group will a proton preferentially go as acid is added to the solution? What will the product be?

Problem 36.2 In quite acidic solution, an amino acid contains two acidic groups, —NH_3^+ and —COOH. Which is the more acidic? Which group will more readily give up a proton as base is added to the solution? What will the product be?

Problem 36.3 Account for the fact that *p*-aminobenzoic acid or *o*-aminobenzoic acid does not exist appreciably as the dipolar ion, but *p*-aminobenzenesulfonic acid (*sulfanilic acid*) does. (*Hint:* What is K_b for most aromatic amines?)

We must keep in mind that ions II and III, which contain a free $—NH_2$ or $—COOH$ group, are in equilibrium with dipolar ion I; consequently, amino acids undergo reactions characteristic of amines and carboxylic acids. As ion II is removed, by reaction with benzoyl chloride, for example, the equilibrium shifts to supply more of ion II so that eventually the amino acid is completely benzoylated.

$$\underset{\text{II}}{H_2NCHRCOO^-} \underset{OH^-}{\overset{H^+}{\rightleftarrows}} \underset{\text{I}}{^+H_3NCHRCOO^-} \underset{OH^-}{\overset{H^+}{\rightleftarrows}} \underset{\text{III}}{^+H_3NCHRCOOH}$$

Where feasible we can speed up a desired reaction by adjusting the acidity or basicity of the solution in such a way as to increase the concentration of the reactive species.

Problem 36.4 Suggest a way to speed up (a) esterification of an amino acid; (b) acylation of an amino acid.

36.4 Isoelectric point of amino acids

What happens when a solution of an amino acid is placed in an electric field depends upon the acidity or basicity of the solution. In quite alkaline solution,

$$\underset{\text{II}}{H_2NCHRCOO^-} \underset{OH^-}{\overset{H^+}{\rightleftarrows}} \underset{\text{I}}{^+H_3NCHRCOO^-} \underset{OH^-}{\overset{H^+}{\rightleftarrows}} \underset{\text{III}}{^+H_3NCHRCOOH}$$

anions II exceed cations III, and there is a net migration of amino acid toward the anode. In quite acidic solution, cations III are in excess, and there is a net migration of amino acid toward the cathode. If II and III are exactly balanced, there is no net migration; under such conditions any one molecule exists as a positive ion and as a negative ion for exactly the same amount of time, and any small movement in the direction of one electrode is subsequently canceled by an equal movement back toward the other electrode. The hydrogen ion concentration of the solution in which a particular amino acid does not migrate under the influence of an electric field is called the **isoelectric point** of that amino acid.

A monoamino monocarboxylic acid, $^+H_3NCHRCOO^-$, is somewhat more acidic than basic (for example, glycine: $K_a = 1.6 \times 10^{-10}$ and $K_b = 2.5 \times 10^{-12}$). If crystals of such an amino acid are added to water, the resulting solution contains more of the anion II, $H_2NCHRCOO^-$, than of the cation III, $^+H_3NCHRCOOH$. This "excess" ionization of ammonium ion to amine ($\text{I} \rightleftarrows \text{II} + H^+$) must be repressed, by addition of acid, to reach the isoelectric point, which therefore lies somewhat on the acid side of neutrality (pH 7). For glycine, for example, the isoelectric point is at pH 6.1.

Problem 36.5 (a) Will the isoelectric point be on the acid or alkaline side of pH 7 (neutrality) for a monoamino dicarboxylic acid? (b) For a diamino monocarboxylic acid? (c) Compare each of these isoelectric points with that for glycine.

An amino acid usually shows its lowest solubility in a solution at the isoelectric point, since here there is the highest concentration of the dipolar ion. As the solution is made more alkaline or more acidic, the concentration of one of the more soluble ions, II or III, increases.

Problem 36.6 Account for the fact that sulfanilic acid dissolves in alkalies but not in acids.

Problem 36.7 Suggest a way to separate a mixture of amino acids into three fractions: monoamino monocarboxylic acids, monoamino dicarboxylic acids (the acidic amino acids), and diamino monocarboxylic acids (the basic amino acids).

36.5 Configuration of natural amino acids

From the structures in Table 36.1, we can see that every amino acid except glycine contains at least one chiral center. As obtained by acidic or enzymatic hydrolysis of proteins, every amino acid except glycine has been found optically active. Stereochemical studies of these naturally occurring amino acids have shown that all have the same configuration about the carbon atom carrying the *alpha*-amino group, and that this configuration is the same as that in L-(−)-glyceraldehyde.

COO^- ; $H_3\overset{+}{N}$— C —H ; R

L-Amino acid

CHO ; HO— C —H ; CH_2OH

L-Glyceraldehyde

Problem 36.8 Draw all possible stereoisomeric formulas for the amino acid threonine. Naturally occurring threonine gets its name from its relationship to the tetrose *threose*; on this basis which is the correct configuration for natural threonine?

Problem 36.9 Besides threonine, there are four amino acids in Table 36.1 that can exist in more than two stereoisomeric forms. (a) What are they? (b) How many isomers are possible in each case? Indicate enantiomers, diastereomers, any *meso* compounds.

36.6 Preparation of amino acids

Of the many methods that have been developed for synthesizing amino acids, we shall take up only one: **amination of α-halo acids.** Considered in its various modifications, this method is probably the most generally useful, although, like any of the methods, it cannot be applied to the synthesis of all the amino acids.

Sometimes an α-chloro or α-bromo acid is subjected to **direct ammonolysis** with a large excess (Why?) of concentrated aqueous ammonia. For example:

$$\underset{\text{Propionic acid}}{CH_3CH_2COOH} \xrightarrow{Br_2,\ P} \underset{\alpha\text{-Bromopropionic acid}}{CH_3\underset{Br}{\underset{|}{C}}HCOOH} \xrightarrow{NH_3\ (excess)} \underset{\substack{\text{Alanine} \\ 70\%\ yield}}{CH_3\underset{NH_3^+}{\underset{|}{C}}HCOO^-}$$

The necessary α-halo acids or esters can be prepared by the Hell-Volhard-Zelinsky halogenation of the unsubstituted acids (Sec. 18.19), or by a modification of the **malonic ester synthesis**, the usual route to the unsubstituted acids. For example:

$$\underset{\text{Sodiomalonic ester}}{Na^+[CH(COOC_2H_5)_2]^-} \xrightarrow{C_6H_5CH_2Cl} \underset{\text{Ethyl benzylmalonate}}{HC(COOC_2H_5)_2{-}CH_2C_6H_5} \xrightarrow[heat]{KOH} \xrightarrow{HCl} \underset{\text{Benzylmalonic acid}}{H{-}C(COOH)_2{-}CH_2C_6H_5}$$

$$\xrightarrow{Br_2,\ ether,\ reflux} Br{-}C(COOH)_2{-}CH_2C_6H_5 \xrightarrow{heat} C_6H_5CH_2\underset{Br}{\underset{|}{C}}HCOOH \xrightarrow{NH_3\ (excess)} \underset{\substack{\text{Phenylalanine} \\ 35\%\ overall\ yield}}{C_6H_5CH_2\underset{NH_3^+}{\underset{|}{C}}HCOO^-}$$

Better yields are generally obtained by the **Gabriel phthalimide synthesis** (Problem 11, p. 744); the α-halo esters are used instead of α-halo acids (Why?). A further modification, the **phthalimidomalonic ester method**, is a combined malonic ester–Gabriel synthesis.

$$\underset{\text{Potassium phthalimide}}{C_6H_4(CO)_2N^-K^+} + \underset{\text{Ethyl chloroacetate}}{ClCH_2COOC_2H_5} \longrightarrow C_6H_4(CO)_2NCH_2COOC_2H_5 \xrightarrow{HCl,\ H_2O} \underset{\substack{\text{Glycine hydrochloride} \\ 89\%\ overall\ yield}}{Cl^-\,{}^+H_3NCH_2COOH} + \text{phthalic acid}$$

Potassium phthalimide + $Br{-}CH(COOC_2H_5)_2$ (Ethyl bromomalonate) $\longrightarrow$ $C_6H_4(CO)_2N{-}CH(COOC_2H_5)_2$ (Pthalimidomalonic ester) $\xrightarrow{Na}$ $C_6H_4(CO)_2N{-}C^-(COOC_2H_5)_2\,Na^+$

$\xrightarrow{\text{excess } ClCH_2COOC_2H_5\text{, heat}}$ $C_6H_4(CO)_2N{-}C(COOC_2H_5)_2{-}CH_2COOC_2H_5$ $\xrightarrow{\text{conc. HCl, heat}}$ $\xrightarrow{\text{base}}$ $HOOCCH_2CH(NH_3^+)COO^-$

Aspartic acid
43% overall yield

These synthetic amino acids are, of course, optically inactive, and must be resolved if the active materials are desired for comparison with the naturally occurring acids or for synthesis of peptides (Sec. 36.10).

Problem 36.10 Various amino acids have been made in the following ways:
Direct ammonolysis: glycine, alanine, valine, leucine, aspartic acid
Gabriel synthesis: glycine, leucine
Malonic ester synthesis: valine, isoleucine
Phthalimidomalonic ester method: serine, glutamic acid, aspartic acid

List the necessary starting materials in each case, and outline the entire sequence for one example from each group.

Problem 36.11 Acetaldehyde reacts with a mixture of KCN and NH_4Cl (**Strecker synthesis**) to give a product, $C_3H_6N_2$ (What is its structure?), which upon hydrolysis yields alanine. Show how the Strecker synthesis can be applied to the synthesis of glycine, leucine, isoleucine, valine, and serine (start with $C_2H_5OCH_2CH_2OH$). Make all required carbonyl compounds from readily available materials.

Problem 36.12 (a) Synthesis of amino acids by **reductive amination** (Sec. 22.11) is illustrated by the following synthesis of leucine:

$$\text{ethyl isovalerate} + \text{ethyl oxalate} \xrightarrow{NaOC_2H_5} A\ (C_{11}H_{18}O_5)$$

$$A + 10\%\ H_2SO_4 \xrightarrow{\text{boil}} B\ (C_6H_{10}O_3) + CO_2 + C_2H_5OH$$

$$B + NH_3 + H_2 \xrightarrow{\text{Pd, heat}} \text{leucine}$$

(b) Outline the synthesis by this method of alanine. Of glutamic acid.

36.7 Reactions of amino acids

The reactions of amino acids are in general the ones we would expect of compounds containing amino and carboxyl groups. In addition, any other groups that may be present undergo their own characteristic reactions.

Problem 36.13 Predict the products of the treatment of glycine with:

(a) aqueous NaOH
(b) aqueous HCl
(c) benzoyl chloride + aqueous NaOH
(d) acetic anhydride
(e) $NaNO_2$ + HCl
(f) $C_2H_5OH + H_2SO_4$
(g) benzyl chlorocarbonate (carbobenzoxy chloride), $C_6H_5CH_2OCOCl$

Problem 36.14 Predict the products of the following reactions:

(a) N-benzoylglycine (*hippuric acid*) + $SOCl_2$
(b) product of (a) + NH_3
(c) product of (a) + alanine
(d) product of (a) + C_2H_5OH
(e) tyrosine + $Br_2(aq)$
(f) asparagine + hot aqueous NaOH
(g) proline + methyl iodide
(h) tyrosine + methyl sulfate + NaOH
(i) glutamic acid + one mole $NaHCO_3$
(j) glutamic acid + excess ethyl alcohol + H_2SO_4 + heat

Problem 36.15 The reaction of primary aliphatic amines with nitrous acid gives a quantitative yield of nitrogen gas, and is the basis of the **Van Slyke determination of amino nitrogen.** What volume of nitrogen gas at S.T.P. would be liberated from 0.001 mole of: (a) leucine, (b) lysine, (c) proline?

Problem 36.16 When a solution of 9.36 mg of an unknown amino acid was treated with excess nitrous acid, there was obtained 2.01 cc of nitrogen at 748 mm and 20°. What is the minimum molecular weight for this compound? Can it be one of the amino acids found in proteins? If so, which one?

36.8 Peptides. Geometry of the peptide linkage

Peptides are amides formed by interaction between amino groups and carboxyl groups of amino acids. The amide group, —NHCO—, in such compounds is often referred to as the *peptide linkage*.

Depending upon the number of amino acid residues per molecule, they are known as *dipeptides*, *tripeptides*, and so on, and finally *polypeptides*. (By convention, peptides of molecular weight up to 10,000 are known as polypeptides and above that as proteins.) For example:

$$^{+}H_3NCH_2CONHCH_2COO^{-}$$

Gly.Gly
Glycylglycine
A dipeptide

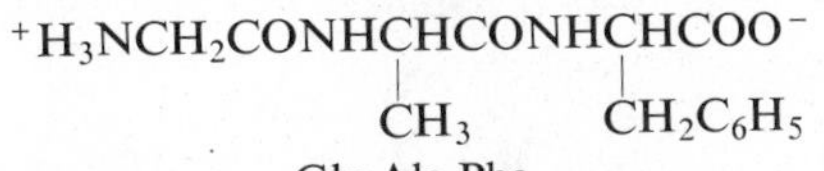

$$^{+}H_3NCH_2CONH\underset{\displaystyle CH_3}{\underset{|}{C}}HCONH\underset{\displaystyle CH_2C_6H_5}{\underset{|}{C}}HCOO^{-}$$

Gly.Ala.Phe
Glycylalanylphenylalanine
A tripeptide

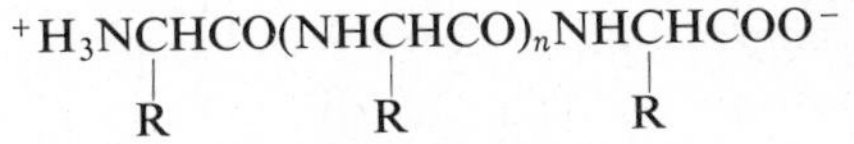

$$^{+}H_3N\underset{\displaystyle R}{\underset{|}{C}}HCO(NH\underset{\displaystyle R}{\underset{|}{C}}HCO)_nNH\underset{\displaystyle R}{\underset{|}{C}}HCOO^{-}$$

A polypeptide

A convenient way of representing peptide structures by use of standard abbreviations (see Table 36.1) is illustrated here. According to convention, the **N-terminal amino acid residue** (having the free amino group) is written at the left end, and the **C-terminal amino acid residue** (having the free carboxyl group) at the right end.

X-ray studies of amino acids and dipeptides indicate that the entire amide group is flat: carbonyl carbon, nitrogen, and the four atoms attached to them all lie in a plane. The short carbon–nitrogen distance (1.32 A as compared with 1.47 A for the usual carbon–nitrogen single bond) indicates that the carbon–nitrogen bond has considerable double-bond character (about 50%); as a result the angles of the bonds to nitrogen are similar to the angles about the trigonal carbon atom (Fig. 36.1).

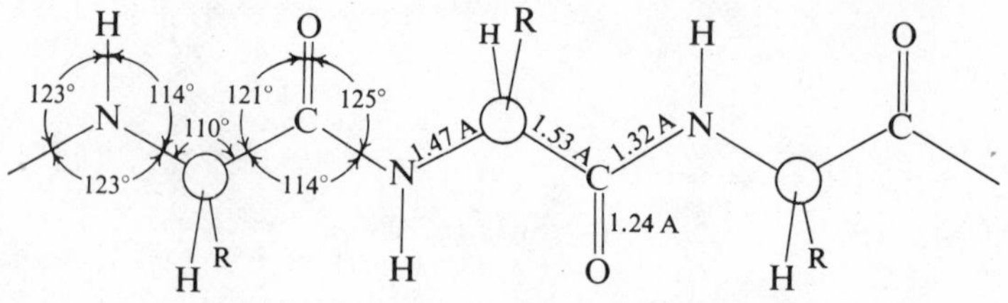

H O H R H O
123° 114° 121° 125°
N 110° C 1.47 A 1.53 A 1.32 A N C
123° 114° N C
1.24 A
H R H O H R

Figure 36.1. Geometry of the peptide link. Carbon–nitrogen bond has much double bond character. Carbonyl carbon, nitrogen, and atoms attached to them lie in a plane.

Problem 36.17 (a) What contributing structure(s) would account for the double-bond character of the carbon–nitrogen bond? (b) What does this resonance mean in terms of orbitals?

Problem 36.18 At room temperature, N,N-dimethylformamide gives the following nmr spectrum:

a singlet, δ 2.88, 3H
b singlet, δ 2.97, 3H
c singlet, δ 8.02, 1H

As the temperature is raised, signals *a* and *b* broaden and coalesce; finally, at 170°, they are merged into one sharp singlet. (a) How do you account for these observations? (b) What bearing do they have on the structure of the peptide linkage? (*Hint:* See Sec. 13.13.)

Peptides have been studied chiefly as a step toward the understanding of the much more complicated substances, the proteins. However, peptides are extremely important compounds in their own right: the tripeptide *glutathione*, for example, is found in most living cells; the nonapeptide *oxytocin* is a posterior pituitary hormone concerned with contraction of the uterus; *α-corticotropin*, made up of 39 amino acid residues, is one component of the adrenocorticotropic hormone ACTH.

$^{+}H_3N\underset{\displaystyle COO^-}{CH}CH_2CH_2CONH\underset{\displaystyle CH_2SH}{CH}CONHCH_2COOH$ *or* Glu.CySH.Gly

Glutathione
(Glutamylcysteinylglycine)

```
 CH3      OH
 |        |
 CH2     (C6H4)            NH2
 |        |                 |
CH3CH     CH2               |
 |        |                 |
 CHNHCOCHNHCOCH
/                \
O=C               CH2
|                  |
|                  S
|                  |
NH                 S
 \                 |
  \                CH2
   \              /
   CHCONHCHCONHCHCON(pyrrolidine ring)CONHCHCONHCH2CONH2
   |      |                                 |
   CH2    CH2                               CH2
   |      |                                 |
   CH2    CONH2                             CH
   |                                       /  \
   CONH2                                 H3C   CH3
```

or

```
Ileu.Tyr.CyS.
|          |
Glu.Asp.CyS.Pro.Leu.Gly(NH2)
|    |
H2N  NH2
```

Oxytocin

```
 Ser.Tyr.Ser.Met.Glu.His.Phe.Arg.Try.Gly.Lys.Pro.Val.┐
┌Gly.Lys.Lys.Arg.Arg.Pro.Val.Lys.Val.Tyr.Pro.Ala.Gly.┘
└Glu.Asp.Asp.Glu.Ala.Ser.Glu.Ala.Phe.Pro.Leu.Glu.Phe
```

α-Corticotropin (sheep)

We shall look at two aspects of the chemistry of peptides: how their structures are determined, and how they can be synthesized in the laboratory.

36.9 Determination of structure of peptides. Terminal residue analysis. Partial hydrolysis

To assign a structure to a particular peptide, one must know (a) what amino acid residues make up the molecule and how many of each there are, and (b) the sequence in which they follow one another along the chain.

To determine the composition of a peptide, one hydrolyzes the peptide (in acidic solution, since alkali causes racemization) and determines the amount of each amino acid thus formed. One of the best ways of analyzing a mixture of amino acids is to separate the mixture into its components by chromatography—sometimes, after conversion into the methyl esters (Why?), by gas chromatography.

From the weight of each amino acid obtained, one can calculate the number of moles of each amino acid, and in this way know the relative numbers of the various amino acid residues in the peptide. At this stage one knows what might be called the "empirical formula" of the peptide: the relative abundance of each amino acid residue in the peptide.

Problem 36.19 An analysis of the hydrolysis products of *salmine*, a polypeptide from salmon sperm, gave the following results:

	g/100 g salmine
Isoleucine	1.28
Alanine	0.89
Valine	3.68
Glycine	3.01
Serine	7.29
Proline	6.90
Arginine	86.40

What are the relative numbers of the various amino acid residues in salmine; that is, what is its empirical formula? (Why do the weights add up to more than 100 g?)

To calculate the "molecular formula" of the peptide—the actual number of each kind of residue in each peptide molecule—one needs to know the molecular weight. Molecular weights can be determined by chemical methods and by various physical methods: osmotic pressure or light-scattering measurements, behavior in an ultracentrifuge, x-ray diffraction.

Problem 36.20 The molecular weight of salmine (see the preceding problem) is about 10,000. What are the actual numbers of the various amino acid residues in salmine; that is, what is its molecular formula?

Problem 36.21 A protein was found to contain 0.29% tryptophane (mol. wt. 204). What is the minimum molecular weight of the protein?

Problem 36.22 (a) Horse hemoglobin contains 0.335% Fe. What is the minimum molecular weight of the protein? (b) Osmotic pressure measurements give a molecular weight of about 67,000. How many iron atoms are there per molecule?

There remains the most difficult job of all: to determine the sequence in which these amino acid residues are arranged along the peptide chain, that is, the structural formula of the peptide. This is accomplished by a combination of terminal residue analysis and partial hydrolysis.

Terminal residue analysis is the identifying of the amino acid residues at the ends of the peptide chain. The procedures used depend upon the fact that the residues at the two ends are different from all the other residues and from each other: one, the *N-terminal residue*, contains a free *alpha* amino group and the other, the *C-terminal residue*, contains a free carboxyl group *alpha* to a peptide linkage.

A very successful method of identifying the N-terminal residue (introduced in 1945 by Frederick Sanger of Cambridge University) makes use of 2,4-dinitrofluorobenzene (DNFB), which undergoes nucleophilic substitution by the free amino group to give an N-dinitrophenyl (DNP) derivative. The substituted peptide

$$O_2N\text{-}C_6H_3(NO_2)\text{-}F + H_2N\underset{\displaystyle R}{\underset{|}{C}}HCONH\underset{\displaystyle R'}{\underset{|}{C}}HCO\sim \xrightarrow[\text{medium}]{\text{alkaline}} O_2N\text{-}C_6H_3(NO_2)\text{-}NH\underset{\displaystyle R}{\underset{|}{C}}HCONH\underset{\displaystyle R'}{\underset{|}{C}}HCO\sim$$

2,4-Dinitrofluorobenzene (DNFB) — Peptide — Labeled peptide

$$\downarrow \text{aq HCl, heat}$$

$$O_2N\text{-}C_6H_3(NO_2)\text{-}NH\underset{\displaystyle R}{\underset{|}{C}}HCOOH + {}^{+}H_3N\underset{\displaystyle R'}{\underset{|}{C}}HCOOH, \text{ etc.}$$

N-(2,4-Dinitrophenyl)amino acid (DNP.AA) — Unlabeled amino acids

is hydrolyzed to the component amino acids, and the N-terminal residue, labeled by the 2,4-dinitrophenyl group, is separated and identified.

In its various modifications, however, the most widely used method of N-terminal residue analysis is one introduced in 1950 by Pehr Edman (of the University of Lund, Sweden). This is based upon the reaction between an amino group and phenyl isothiocyanate to form a substituted thiourea (compare Sec. 32.7). Mild hydrolysis with hydrochloric acid selectively removes the N-terminal residue as the phenylthiohydantoin, which is then identified. The great advantage of this

$$C_6H_5NCS + H_2N\underset{\displaystyle R}{\underset{|}{C}}HCONH\underset{\displaystyle R'}{\underset{|}{C}}HCO\sim \xrightarrow[\text{medium}]{\text{alkaline}} C_6H_5\overset{\displaystyle H}{\overset{|}{N}}-\underset{\displaystyle S}{\underset{\|}{C}}-NH\underset{\displaystyle R}{\underset{|}{C}}HCONH\underset{\displaystyle R'}{\underset{|}{C}}HCO\sim$$

Phenyl isothiocyanate — Peptide — Labeled peptide

$$\downarrow H_2O, HCl$$

$$\underbrace{C_6H_5N\!\!<\!\!\begin{matrix} C(=S) - NH \\ C(=O) - CHR \end{matrix}}_{\text{ring}} + H_2N\underset{\displaystyle R'}{\underset{|}{C}}HCO\sim$$

A phenylthiohydantoin — Degraded peptide *One less residue*

method is that it leaves the rest of the peptide chain intact, so that the analysis can be repeated and the *new* terminal group of the shortened peptide identified. In 1967, Edman reported that this analysis could be carried out *automatically* in his "protein sequenator," which is now available in commercial form. Ideally, residue after residue could be identified until the entire sequence had been determined. In actual practice, this is not feasible; after about the first 40 residues, there is interference from the accumulation of amino acids formed by (slow) hydrolysis during the acid treatment.

Problem 36.23 Edman has also devised the highly sensitive "dansyl" method in which a peptide is treated with 5-dimethylaminonaphthalenesulfonyl chloride, followed by acidic hydrolysis. A derivative of the N-terminal residue is obtained which can be followed during its analysis by virtue of its characteristic fluorescence. What is the derivative? Why does it survive the acid treatment that cleaves the peptide bonds?

The most successful method of determining the C-terminal residue has been enzymatic rather than chemical. The C-terminal residue is removed selectively by the enzyme *carboxypeptidase* (obtained from the pancreas), which cleaves only peptide linkages adjacent to *free alpha*-carboxyl groups in polypeptide chains. The analysis can be repeated on the shortened peptide and the *new* C-terminal residue identified, and so on.

In practice it is not feasible to determine the sequence of all the residues in a long peptide chain by the stepwise removal of terminal residues. Instead, the chain is subjected to partial hydrolysis (acidic or enzymatic), and the fragments formed—dipeptides, tripeptides, and so on—are identified, with the aid of terminal residue analysis. When enough of these smaller fragments have been identified, it is possible to work out the sequence of residues in the entire chain.

To take an extremely simple example, there are six possible ways in which the three amino acids making up glutathione could be arranged; partial hydrolysis to the dipeptides glutamylcysteine (Glu.CySH) and cysteinylglycine (CySH.Gly) makes it clear that the cysteine is in the middle and that the sequence Glu.CySH.Gly is the correct one.

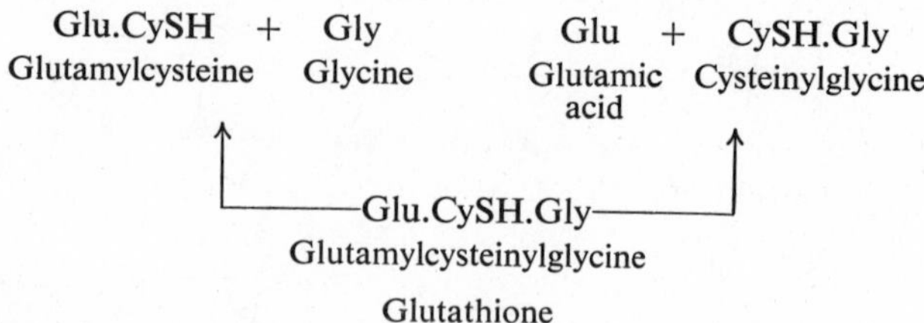

It was by the use of the approach just outlined that structures of such peptides as oxytocin and α-corticotropin (see p. 1143) were worked out. A milestone in protein chemistry was the determination of the entire amino acid sequence in the insulin molecule by a Cambridge University group headed by Frederick Sanger, who received the Nobel Prize in 1958 for this work. (See Problem 12, p. 1162.) Since then the number—and complexity—of completely mapped proteins has grown rapidly: the four chains of hemoglobin, for example, each containing 140-odd amino acid residues; chymotrypsinogen, with a single chain 246 units long; an immunoglobulin (*gamma*-globulin) with two chains of 446 units each and two chains of 214 units each—a total of 1320 amino acid residues.

As usual, final confirmation of the structure assigned to a peptide lies in its synthesis by a method that must unambiguously give a compound of the assigned structure. This problem is discussed in the following section.

Problem 36.24 Work out the sequence of amino acid residues in the following peptides:

(a) Asp,Glu,His,Phe,Val (commas indicate unknown sequence) *gives* Val.Asp + Glu.His + Phe.Val + Asp.Glu.
(b) CySH,Gly,His_2,Leu_2,Ser *gives* CySH.Gly.Ser + His.Leu.CySH + Ser.His.Leu.
(c) Arg,CySH,Glu,Gly_2,Leu,Phe_2,Tyr,Val *gives* Val.CySH.Gly + Gly.Phe.Phe + Glu.Arg.Gly + Tyr.Leu.Val + Gly.Glu.Arg.

36.10 Synthesis of peptides

Methods have been developed by which a single amino acid (or sometimes a di- or tripeptide) can be polymerized to yield polypeptides of high molecular weight. These products have been extremely useful as model compounds: to show, for example, what kind of x-ray pattern or infrared spectrum is given by a peptide of known, comparatively simple structure.

Most work on peptide synthesis, however, has had as its aim the preparation of compounds identical with naturally occurring ones. For this purpose a method must permit the joining together of optically active amino acids to form chains of predetermined length and with a predetermined sequence of residues. Syntheses of this sort not only have confirmed some of the particular structures assigned to natural peptides, but also—and this is more fundamental—have proved that peptides and proteins are indeed polyamides.

It was Emil Fischer who first prepared peptides (ultimately one containing 18 amino acid residues) and thus offered support for his proposal that proteins contain the amide link. It is evidence of his extraordinary genius that Fischer played the same role in laying the foundations of peptide and protein chemistry as he did in carbohydrate chemistry.

The basic problem of peptide synthesis is one of *protecting the amino group*. In bringing about interaction between the carboxyl group of one amino acid and the amino group of a different amino acid, one must prevent interaction between the carboxyl group and the amino group of the same amino acid. In preparing glycylalanine, for example, one must prevent the simultaneous formation of glycylglycine. Reaction can be forced to take place in the desired way by attaching to one amino acid a group that renders the $—NH_2$ unreactive. There are many such protecting groups; the problem is to find one that can be removed later without destruction of any peptide linkages that may have been built up.

$$^{+}H_3N\underset{R}{\underset{|}{C}}HCOO^- \longrightarrow Q—NH\underset{R}{\underset{|}{C}}HCOOH \longrightarrow Q—NH\underset{R}{\underset{|}{C}}HCOCl$$

Protection of amino group

$$Q—NH\underset{R}{\underset{|}{C}}HCOCl + {}^{+}H_3N\underset{R'}{\underset{|}{C}}HCOO^- \longrightarrow Q—NH\underset{R}{\underset{|}{C}}H\underset{O}{\underset{\|}{C}}—NH\underset{R'}{\underset{|}{C}}HCOOH$$

Formation of peptide linkage

$$Q—NH\underset{R}{\underset{|}{C}}H\underset{O}{\underset{\|}{C}}—NH\underset{R'}{\underset{|}{C}}HCOOH \longrightarrow \underset{\text{Peptide}}{{}^{+}H_3N\underset{R}{\underset{|}{C}}H\underset{O}{\underset{\|}{C}}—NH\underset{R'}{\underset{|}{C}}HCOO^-}$$

Removal of the protecting group

We could, for example, benzoylate glycine ($Q = C_6H_5CO$), convert this into the acid chloride, allow the acid chloride to react with alanine, and thus obtain benzoylglycylalanine. But if we attempted to remove the benzoyl group by hydrolysis, we would simultaneously hydrolyze the other amide linkage (the peptide linkage) and thus destroy the peptide we were trying to make.

Of the numerous methods developed to protect an amino group, we shall look at just one: **acylation by benzyl chlorocarbonate**, also called **carbobenzoxy chloride.** (This method was introduced in 1932 by Max Bergmann and Leonidas Zervas of the University of Berlin, later of the Rockefeller Institute.) The reagent, $C_6H_5CH_2OCOCl$, is both an ester and an acid chloride of carbonic acid, HOCOOH; it is readily made by reaction between benzyl alcohol and phosgene (carbonyl chloride), $COCl_2$. (In what order should the alcohol and phosgene be mixed?)

$$CO + Cl_2 \xrightarrow{\text{active carbon, }200^\circ} \underset{\substack{\text{Phosgene}\\\text{(Carbonyl chloride)}}}{Cl{-}\underset{\substack{\|\\O}}{C}{-}Cl} \xrightarrow{C_6H_5CH_2OH} \underset{\substack{\text{Carbobenzoxy chloride}\\\text{(Benzyl chlorocarbonate)}}}{C_6H_5CH_2O{-}\underset{\substack{\|\\O}}{C}{-}Cl}$$

Like any acid chloride, the reagent can convert an amine into an amide:

$$C_6H_5CH_2O{-}\underset{\substack{\|\\O}}{C}{-}Cl + \underset{\text{Amine}}{H_2NR} \longrightarrow \underset{\text{An amide}}{C_6H_5CH_2O{-}\underset{\substack{\|\\O}}{C}{-}NHR}$$

Such amides, $C_6H_5CH_2OCONHR$, differ from most amides, however, in one feature that is significant for peptide synthesis. The carbobenzoxy group can be cleaved by reagents that do not disturb peptide linkages: catalytic hydrogenation or hydrolysis with hydrogen bromide in cold acetic acid.

$$C_6H_5CH_2O{-}\underset{\substack{\|\\O}}{C}{-}NHR \xrightarrow{H_2,\ Pd} C_6H_5CH_3 + \underset{\substack{\text{A carbamic acid}\\\textit{Unstable}}}{\left[HO{-}\underset{\substack{\|\\O}}{C}{-}NHR\right]} \longrightarrow CO_2 + RNH_2$$

$$C_6H_5CH_2O{-}\underset{\substack{\|\\O}}{C}{-}NHR \xrightarrow{\text{HBr, cold HOAc}} C_6H_5CH_2Br + \left[HO{-}\underset{\substack{\|\\O}}{C}{-}NHR\right] \longrightarrow CO_2 + RNH_2$$

The carbobenzoxy method is illustrated by the synthesis of glycylalanine (Gly.Ala):

$$\underset{\substack{\text{Carbobenzoxy}\\\text{chloride}}}{C_6H_5CH_2OCOCl} + \underset{\text{Glycine}}{{}^+H_3NCH_2COO^-} \longrightarrow \underset{\text{Carbobenzoxyglycine}}{C_6H_5CH_2OCONHCH_2COOH}$$

$$\downarrow SOCl_2$$

$$\underset{\text{Acid chloride of carbobenzoxyglycine}}{C_6H_5CH_2OCONHCH_2COCl}$$

$$C_6H_5CH_2OCONHCH_2COCl + {}^+H_3N\underset{\underset{\text{Alanine}}{CH_3}}{\underset{|}{C}}HCOO^- \longrightarrow C_6H_5CH_2OCONHCH_2CONH\underset{CH_3}{\underset{|}{C}}HCOOH$$

Carbobenzoxyglycylalanine

$$C_6H_5CH_2OCONHCH_2CONH\underset{CH_3}{\underset{|}{C}}HCOOH \xrightarrow{H_2,\ Pd} {}^+H_3NCH_2CONH\underset{CH_3}{\underset{|}{C}}HCOO^-$$

Glycylalanine

Gly.Ala

$$+ C_6H_5CH_3 + CO_2$$

Problem 36.25 (a) How could the preceding synthesis be extended to the tripeptide glycylalanylphenylalanine (Gly.Ala.Phe)?

(b) How could the carbobenzoxy method be used to prepare alanylglycine (Ala.Gly)?

Methods like this can be repeated over and over with the addition of a new unit each time. In this way the hormone oxytocin (p. 1143) was synthesized by Vincent du Vigneaud of Cornell Medical College, who received the Nobel Prize in 1955 for this and other work. In 1963, the total synthesis of the insulin molecule—with the 51 amino acid residues in the sequence mapped out by Sanger—was reported.

But the bottleneck in such syntheses is the need to isolate and purify the new peptide made in each cycle; the time required is enormous, and the yield of product steadily dwindles. A major break-through came with the development of *solid-phase* peptide synthesis by R. Bruce Merrifield at Rockefeller University. Synthesis is carried out with the growing peptide *attached* chemically to polystyrene beads; as each new unit is added, the reagents and by-products are simply washed away, leaving the growing peptide behind, ready for another cycle. The method was automated, and in 1969 Merrifield announced that, using his "protein-making machine," he had synthesized—in *six weeks*—the enzyme ribonuclease, made up of 124 amino acid residues.

Problem 36.26 Give formulas for compounds A–G, and tell what is happening in each reaction.

$$\text{polystyrene} + CH_3OCH_2Cl \xrightarrow{SnCl_4} A + CH_3OH$$

$$A + C_6H_5CH_2OCONHCH_2COO^-\,{}^+NHEt_3 \longrightarrow B + Et_3NHCl$$

$$B + \text{dil. HBr} \longrightarrow C + C_6H_5CH_2Br + CO_2$$

$$C + \text{carbobenzoxyalanyl chloride} \longrightarrow D$$

$$D + \text{dil. HBr} \longrightarrow E + C_6H_5CH_2Br + CO_2$$

$$E + HBr \xrightarrow{CF_3COOH} F\ (C_5H_{10}O_3N_2) + G$$

36.11 Proteins. Classification and function. Denaturation

Proteins are divided into two broad classes: **fibrous proteins**, which are insoluble in water, and **globular proteins**, which are soluble in water or aqueous solutions of acids, bases, or salts. (Because of the large size of protein molecules, these solutions are colloidal.) The difference in solubility between the two classes is related to a difference in molecular shape, which is indicated in a rough way by their names.

Molecules of fibrous proteins are long and thread-like, and tend to lie side by side to form fibers; in some cases they are held together at many points by hydrogen bonds. As a result, the intermolecular forces that must be overcome by a solvent are very strong.

Molecules of globular proteins are folded into compact units that often approach spheroidal shapes. The folding takes place in such a way that the hydrophobic parts are turned inward, toward each other, and away from water; hydrophilic parts—charged groups, for example—tend to stud the surface where they are near water. Hydrogen bonding is chiefly intramolecular. Areas of contact between molecules are small, and intermolecular forces are comparatively weak.

Molecular and intermolecular structure determines not only the solubility of a protein but also the general kind of function it performs.

Fibrous proteins serve as the chief structural materials of animal tissues, a function to which their insolubility and fiber-forming tendency suit them. They make up: *keratin*, in skin, hair, nails, wool, horn, and feathers; *collagen*, in tendons; *myosin*, in muscle; *fibroin*, in silk.

Globular proteins serve a variety of functions related to the maintenance and regulation of the life process, functions that require mobility and hence solubility. They make up: all enzymes; many hormones, as, for example, *insulin* (from the pancreas), *thyroglobulin* (from the thyroid gland), *ACTH* (from the pituitary gland); antibodies, responsible for allergies and for defense against foreign organisms; *albumin* in eggs; *hemoglobin*, which transports oxygen from the lungs to the tissues; *fibrinogen*, which is converted into the insoluble, fibrous protein *fibrin*, and thus causes the clotting of blood.

Within the two broad classes, proteins are subdivided on the basis of physical properties, especially solubility: for example, albumins (soluble in water, coagulated by heat), globulins (insoluble in water, soluble in dilute salt solutions), etc.

Irreversible precipitation of proteins, called **denaturation**, is caused by heat, strong acids or bases, or various other agents. Coagulation of egg white by heat, for example, is denaturation of the protein egg albumin. The extreme ease with which many proteins are denatured makes their study difficult. Denaturation causes a fundamental change in a protein, in particular destroying any physiological activity. (Denaturation appears to involve changes in the secondary structure of proteins, Sec. 36.16.)

Only one other class of compounds, the *nucleic acids* (Sec. 37.7), shows the phenomenon of denaturation. Although closely related to the proteins, polypeptides do not undergo denaturation, presumably because their molecules are smaller and less complex.

36.12 Structure of proteins

We can look at the structure of proteins on a number of levels. At the lowest level, there is the *primary* structure: the way in which the atoms of protein molecules are joined to one another by covalent bonds to form chains. Next, there is the *secondary* structure: the way in which these chains are arranged in space to form coils, sheets, or compact spheroids, with hydrogen bonds holding together different chains or different parts of the same chain. Even higher levels of structure are gradually becoming understood: the weaving together of coiled chains to form ropes, for example, or the clumping together of individual molecules to form larger aggregates. Let us look first at the primary structure of proteins.

36.13 Peptide chain

Proteins are made up of peptide chains, that is, of amino acid residues joined by amide linkages. They differ from polypeptides in having higher molecular

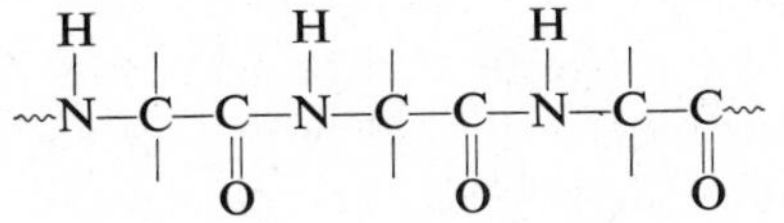

$$\sim N(H)-\overset{|}{\underset{|}{C}}-C(=O)-N(H)-\overset{|}{\underset{|}{C}}-C(=O)-N(H)-\overset{|}{\underset{|}{C}}-C(=O)\sim$$

weights (by convention over 10,000) and more complex structures.

The peptide structure of proteins is indicated by many lines of evidence: hydrolysis of proteins by acids, bases, or enzymes yields peptides and finally amino acids; there are bands in their infrared spectra characteristic of the amide group; secondary structures based on the peptide linkage can be devised that exactly fit x-ray data.

36.14 Side chains. Isoelectric point. Electrophoresis

To every third atom of the peptide chain is attached a side chain. Its structure depends upon the particular amino acid residue involved: —H for glycine, $—CH_3$ for alanine, $—CH(CH_3)_2$ for valine, $—CH_2C_6H_5$ for phenylalanine, etc.

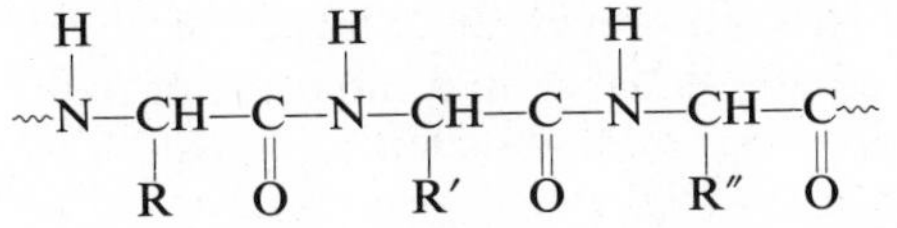

$$\sim N(H)-CH(R)-C(=O)-N(H)-CH(R')-C(=O)-N(H)-CH(R'')-C(=O)\sim$$

Some of these side chains contain basic groups: $—NH_2$ in lysine, or the imidazole ring in histidine. Some side chains contain acidic groups: —COOH in aspartic acid or glutamic acid. Because of these acidic and basic side chains, there are positively and negatively charged groups along the peptide chain. The behavior

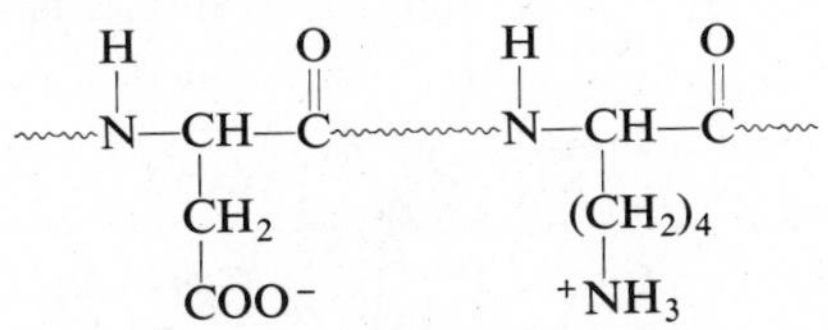

$$\sim N(H)-CH(CH_2COO^-)-C(=O)\sim\sim N(H)-CH((CH_2)_4\,{}^+NH_3)-C(=O)\sim$$

of a protein in an electric field is determined by the relative numbers of these positive and negative charges, which in turn are affected by the acidity of the solution. At the isoelectric point, the positive and negative charges are exactly balanced and the protein shows no net migration; as with amino acids, solubility is usually at a minimum here. On the acid side of the isoelectric point, positive charges exceed negative charges and the protein moves to the cathode; on the basic side of the isoelectric point, negative charges exceed positive charges and the protein moves to the anode.

While all proteins contain the peptide backbone, each protein has its own characteristic sequence of side chains, which gives it its characteristic properties. Different proteins have different proportions of acidic and basic side chains, and hence have different isoelectric points. In a solution of a particular hydrogen ion concentration, some proteins move toward a cathode and others toward an anode; depending upon the size of the charge as well as upon molecular size and shape, different proteins move at different speeds. This difference in behavior in an electric field is the basis of one method of separation and analysis of protein mixtures: **electrophoresis.**

Side chains affect the properties of proteins not only by their acidity or basicity, but also by their other chemical properties and even by their sizes and shapes. Hydroxyl and sulfhydryl (—SH) groups can form esters; amino nitrogen is not only basic but nucleophilic. It seems likely that the "permanent" waving of hair depends upon changes in disulfide (—S—S—) cross-linkages provided by cysteine side chains; that much of the difference between silk and wool is related to the small side chains, —H and —CH_3, that predominate in silk fibroin; that the toughness of tendon is due to the flatness of the pyrrolidine ring and the ability of the —OH group of hydroxyproline to form hydrogen bonds. Replacement of *one* glutamic acid side chain in the hemoglobin molecule (300 side chains in all) by a valine unit is the cause of the fatal sickle-cell anemia.

The sequence of amino acids in hemoglobin has been used to study evolution, in the new science called *chemical paleogenetics*. In the *beta*-chain of hemoglobin, for example, the horse differs from man at 26 of the 146 sites; a pig, at 10 sites; and the gorilla at just *one* site. It has been estimated that, on the average, it takes roughly ten million years for one successful amino acid substitution to occur—that is, a substitution that improves the chances of survival. (Such a change is due to a change in the base sequence in a molecule of nucleic acid, Sec. 37.8.)

36.15 Conjugated proteins. Prosthetic groups. Coenzymes

Some protein molecules contain a non-peptide portion called a **prosthetic group**; such proteins are called *conjugated proteins*. The prosthetic group is intimately concerned with the specific biological action of the protein.

The prosthetic group of hemoglobin, for example, is *heme*. As we see, heme

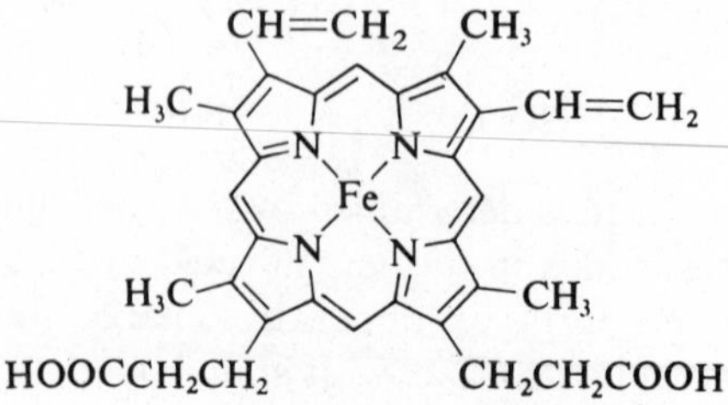

CH=CH_2 CH_3
H_3C CH=CH_2
N N
Fe
N N
H_3C CH_3
$HOOCCH_2CH_2$ CH_2CH_2COOH

Heme

contains iron bound to the pyrrole system known as *porphin* (compare with the structure of chlorophyll, p. 1004). It is the formation of a reversible oxygen–heme complex that enables hemoglobin to carry oxygen from the lungs to the tissues. Carbon monoxide forms a similar, but more stable, complex; it thus ties up hemoglobin, prevents oxygen transport, and causes death. Heme is held to the peptide portion (*globin*) of the protein by a combination of forces: coordination of iron by histidine nitrogen of the protein, hydrogen bonding, and van der Waals forces between hydrophobic parts of the two molecules.

Many enzymes require *cofactors* if they are to exert their catalytic effects: metal ions, for example. Organic cofactors are called **coenzymes** and, if they are covalently bonded to the enzyme, these too are prosthetic groups.

The coenzyme *nicotinamide adenine dinucleotide* (NAD), for example, is associated with a number of dehydrogenation enzymes. This coenzyme we see, is

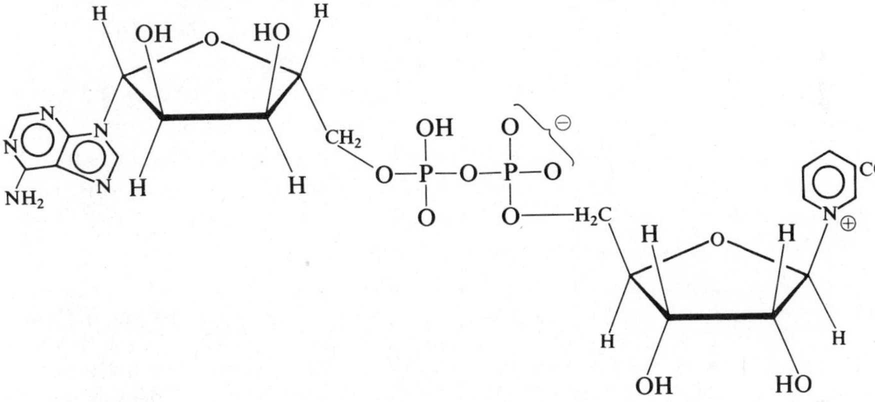

Nicotinamide adenine dinucleotide (NAD)
(Diphosphopyridinenucleotide)

made up of two molecules of D-ribose linked as phosphate esters, the fused heterocyclic system known as *adenine*, and nicotinamide in the form of a quaternary ammonium salt. In some systems one encounters *nicotinamide adenine dinucleotide phosphate* (NADP), in which the —OH on C-2 of the left-hand ribose unit of NAD has been phosphorylated. The characteristic biological function of these dehydrogenation enzymes (see, for example, Sec. 37.5) involves conversion of the nicotinamide portion of NAD or NADP into the dihydro structure.

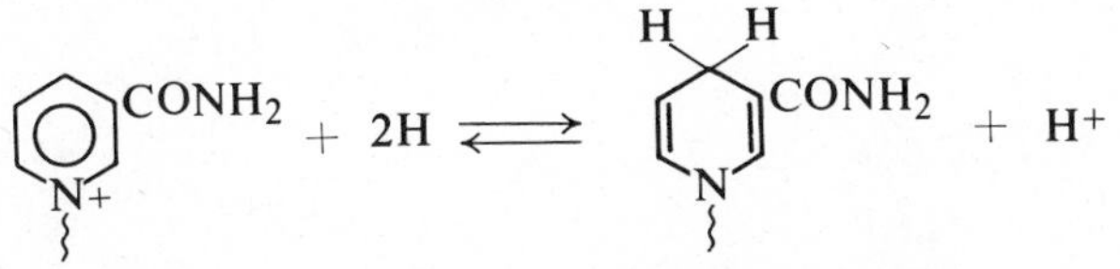

$$\text{NAD} + 2\text{H} \rightleftharpoons \text{Reduced NAD} + \text{H}^+$$

Like nicotinamide, many molecules making up coenzymes are **vitamins**, that is, substances that must be supplied in the diet to permit proper growth or maintenance of structure. Undoubtedly it is for their coenzyme activity that these substances are needed.

36.16 Secondary structure of proteins

It seems clear that proteins are made up of polypeptide chains. How are these chains arranged in space and in relationship to each other? Are they stretched out side by side, looped and coiled about one another, or folded into independent spheroids?

Much of our understanding of the secondary structure of proteins is the result of x-ray analysis. For many proteins the x-ray diffraction pattern indicates a regular repetition of certain structural units. For example, there are *repeat distances* of 7.0 A in silk fibroin, and of 1.5 A and 5.1 A in α-keratin of unstretched wool.

The problem is to devise structures that account for the characteristic x-ray diffraction patterns, and are at the same time consistent with what is known about the primary structure: bond lengths and bond angles, planarity of the amide group, similarity of configuration about chiral centers (all L-family), size and sequence of side chains. Of key importance in this problem has been recognition of the stabilizing effect of hydrogen bonds (5–10 kcal per mole per hydrogen bond), and the principle that the most stable structure is one that permits formation of the maximum number of hydrogen bonds. On the basis of the study of simpler compounds, it has been further assumed that the N—H---O bond is very nearly linear, hydrogen lying on, or within 20° of, the line between nitrogen and oxygen. In all this work the simultaneous study of simpler, synthetic polypeptides containing only a single kind of amino acid residue has been of great help.

The progress made on a problem of this size and difficulty has necessarily been the work of many people. Among them is Linus Pauling, of the California Institute of Technology, who received the Nobel Prize in 1954. In 1951 Pauling wrote: "Fourteen years ago Professor Robert B. Corey and I, after we had made a vigorous but unsuccessful attack on the problem of formulating satisfactory configurations of polypeptide chains in proteins, decided to attempt to solve the problem by an indirect method—the method of investigating with great thoroughness crystals of amino acids, simple peptides, and related substances, in order to obtain completely reliable and detailed information about the structural characteristics of substances of this sort, and ultimately to permit the confident prediction of precisely described configurations of polypeptide chains in proteins." (Record Chem. Prog., *12*, 156–7 (1951).). This work on simple substances, carried on for more than 14 years, gave information about the geometry of the amide group that eventually led Pauling and his co-workers to propose what may well be the most important secondary structure in protein chemistry: the α-helix.

Let us look at some of the secondary structures that have been proposed.

As a point of departure, it is convenient to consider a structure (perhaps hypothetical) in which peptide chains are fully extended to form flat zig-zags:

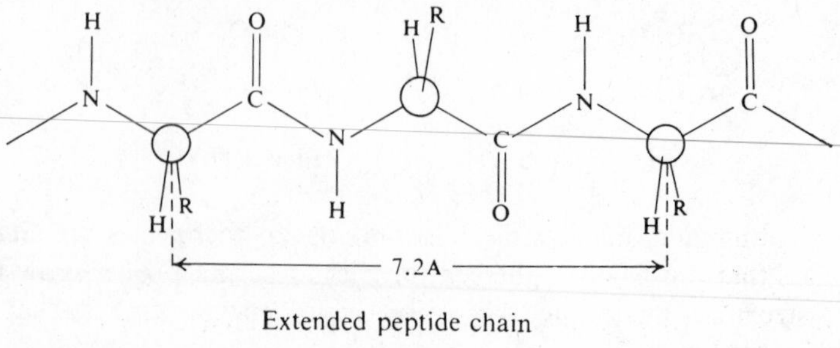

Extended peptide chain

These chains lie side by side to form a *flat sheet*. Each chain is held by hydrogen bonds to the two neighboring chains (Fig. 36.2). This structure has a repeat distance of 7.2 A, the distance between *alternate* amino acid residues. (Notice that alternate side chains lie on the same side of the sheet.) However, crowding between side chains makes this idealized flat structure impossible, except perhaps for synthetic polyglycine.

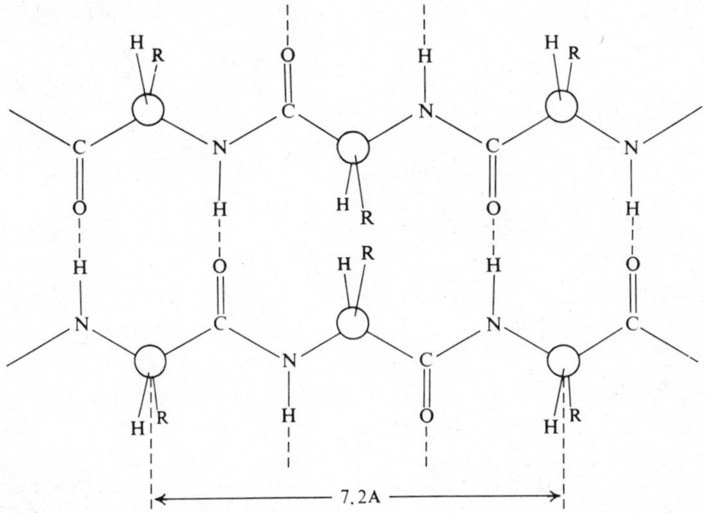

7.2A

Figure 36.2. Hypothetical flat sheet structure for a protein. Chains fully extended; adjacent chains head in opposite directions; hydrogen bonding between adjacent chains. Side chains (R) are crowded.

Room can be made for small or medium-sized side chains by a slight contraction of the peptide chains:

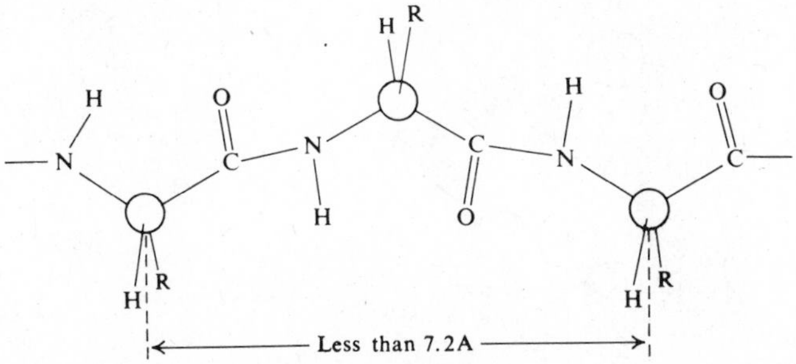

Less than 7.2A

Contracted peptide chain

The chains still lie side by side, held to each other by hydrogen bonds. The contraction results in a *pleated sheet*, with a somewhat shorter distance between alternate amino acid residues (see Fig. 36.3). Such a structure, called the **beta**

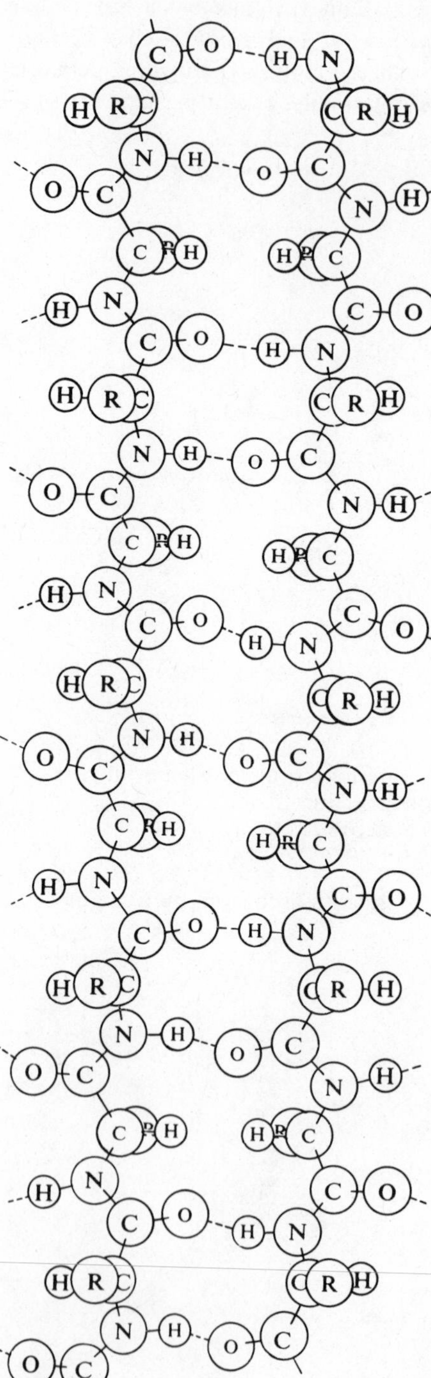

Figure 36.3. Pleated sheet structure (*beta arrangement*) proposed by Pauling for silk fibroin. Chains contracted to make room for small side chains. Adjacent chains head in opposite directions; hydrogen bonding between adjacent chains.

arrangement, has been proposed for silk fibroin, which has a repeat distance of 7.0 A and most closely approaches the fully extended, flat-sheet structure. It is significant that, although 15 kinds of amino acid residue are found in silk fibroin, 46% of the residues are glycine, which has no side chain, and another 38% are alanine and serine with the small side chains $—CH_3$ and $—CH_2OH$.

When the side chains are quite large, they are best accommodated by a quite different kind of structure. Each chain is coiled to form a *helix* (like a spiral

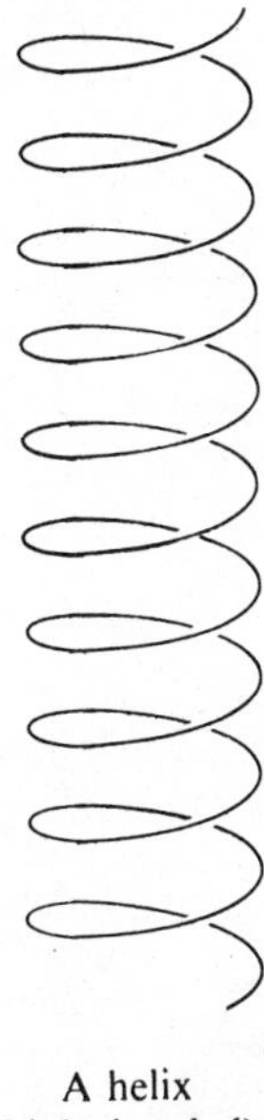

A helix
(right-handed)

staircase). Hydrogen bonding occurs between different parts of the *same* chain, and holds the helix together. For α-keratin (unstretched wool, hair, horn, nails) Pauling has proposed a helix in which there are 3.6 amino acid residues per turn (Fig. 36.4). Models show that this 3.6-helix provides room for the side chains and allows all possible hydrogen bonds to form. It accounts for the repeat distance of 1.5 A, which is the distance between amino acid residues measured along the axis of the helix. To fit into this helix, all the amino acid residues must be of the same configuration, as, of course, they are; furthermore, their L-configuration requires the helix to be *right-handed*, as shown. It is becoming increasingly clear that the **alpha helix**, as it is called, is of fundamental importance in the chemistry of proteins.

(To account for the second repeat distance of 5.1 A for α-keratin, we must go to what is properly the *tertiary structure*. Pauling has suggested that each helix can itself be coiled into a superhelix which has one turn for every 35 turns of the *alpha* helix. Six of these superhelixes are woven about a seventh, straight helix to form a seven-strand cable.)

When wool is stretched, α-keratin is converted into β-keratin, with a change in the x-ray diffraction pattern. It is believed that the helixes are uncoiled and the chains stretched side by side to give a sheet structure of the *beta* type. The hydrogen

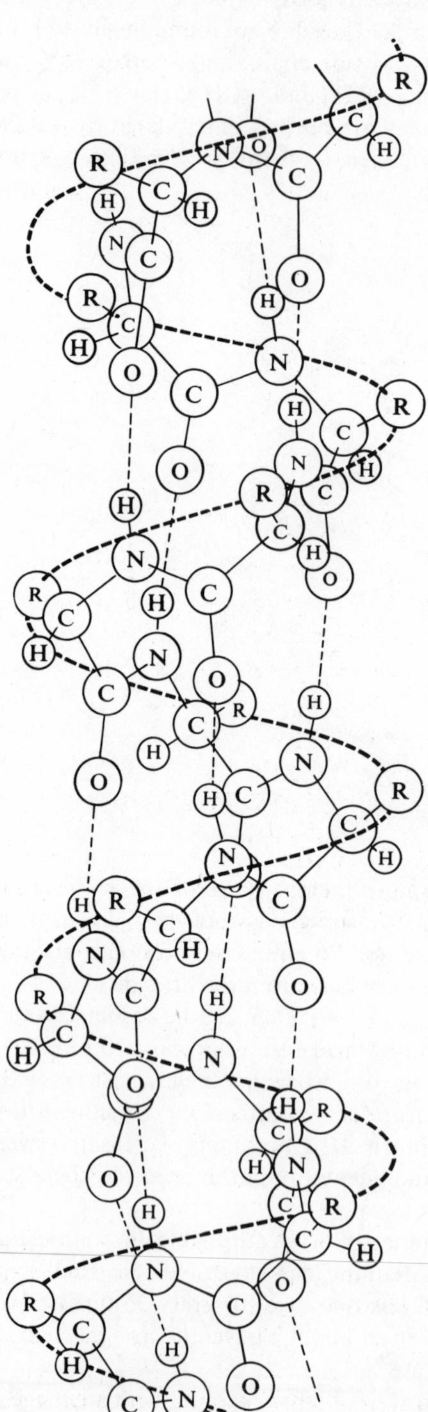

Figure 36.4. Alpha helix structure proposed by Pauling for α-keratin. Makes room for large side chains. Right-handed helix with 3.6 residues per turn; hydrogen bonding within a chain.

bonds within the helical chain are broken, and are replaced by hydrogen bonds between adjacent chains. Because of the larger side chains, the peptide chains are less extended (repeat distance 6.4 A) than in silk fibroin (repeat distance 7.0 A).

Besides the x-ray diffraction patterns characteristic of the *alpha*- and *beta*-type proteins, there is a third kind: that of *collagen*, the protein of tendon and skin. On the primary level, collagen is characterized by a high proportion of proline and hydroxyproline residues, and by frequent repetitions of the sequence Gly.Pro.Hypro. The pyrrolidine ring of proline and hydroxyproline can affect

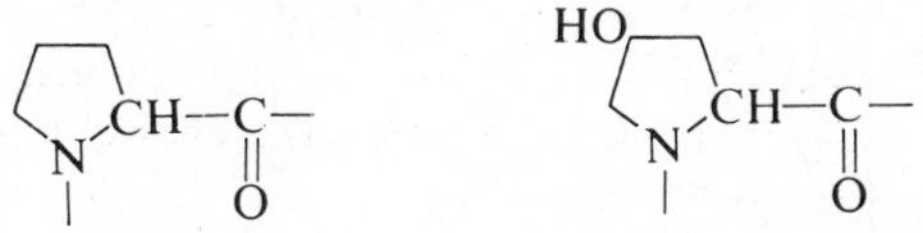

Proline residue

Hydroxyproline residue

the secondary structure in several ways. The amido nitrogen carries no hydrogen for hydrogen bonding. The flatness of the five-membered ring, in conjunction with the flatness of the amide group, prevents extension of the peptide chain as in the *beta* arrangement, and interferes with the compact coiling of the *alpha* helix.

The structure of collagen combines the helical nature of the *alpha*-type proteins with the inter-chain hydrogen bonding of the *beta*-type proteins. Three peptide chains—each in the form of a left-handed helix—are twisted about one another to form a three-strand right-handed superhelix. A small glycine residue at every third position of each chain makes room for the bulky pyrrolidine rings on the other two chains. The three chains are held strongly to each other by hydrogen bonding between glycine residues and between the —OH groups of hydroxyproline.

When collagen is boiled with water, it is converted into the familiar water-soluble protein *gelatin*; when cooled, the solution does not revert to collagen but sets to a gel. Gelatin has a molecular weight one-third that of collagen. Evidently the treatment separates the strands of the helix, breaking inter-chain hydrogen bonds and replacing them with hydrogen bonds to water molecules.

Turning from the insoluble, fibrous proteins to the soluble, globular proteins (e.g., hemoglobin, insulin, *gamma*-globulin, egg albumin), we find that the matter of secondary structure can be even more complex. Evidence is accumulating that here, too, the *alpha* helix often plays a key role. These long peptide chains are not uniform: certain segments may be coiled into helixes or folded into sheets; other segments are looped and coiled into complicated, irregular arrangements. Look, for example, at α-chymotrypsin in Fig. 37.1 (p. 1166).

This looping and coiling appears to be random, but it definitely is *not*. The sequence of amino acids is determined genetically (Sec. 37.7) but, once formed, the chain *naturally* falls into the arrangement that is *most stable* for that particular sequence.

We find all our kinds of "intermolecular" forces at work here—but acting between different parts of the same molecule: van der Waals forces, hydrogen bonds, interionic attraction (or repulsion) between charged groups. There is chemical cross-linking by disulfide bonds. The characteristic feature of these globular proteins is that hydrophobic parts are turned inward, toward each other and away from water—like the hydrophobic tails in a soap micelle.

In their physiological functions, proteins are highly specific. We have encountered, for example, an enzyme that will cleave α-glucosides but not β-glucosides, and an enzyme that will cleave only C-terminal amino acid residues in polypeptides. It seems clear that the biological activity of a protein depends not only upon its prosthetic group (if any) and its particular amino acid sequence, but also upon its molecular shape. As Emil Fischer said in 1894: "... enzyme and glucoside must fit together like a lock and key. . . ." In Sec. 37.2 we shall see how one enzyme is believed to exert its effect, and how that effect depends, in a very definite and specific way, on the shape of the enzyme molecule.

Denaturation uncoils the protein, destroys the characteristic shape, and with it the characteristic biological activity.

In 1962, M. F. Perutz and J. C. Kendrew of Cambridge University were awarded the Nobel Prize in chemistry for the elucidation of the structure of hemoglobin and the closely related oxygen-storing molecule, myoglobin. Using x-ray analysis, and knowing the amino acid sequence (p. 1146), they determined the shape—in three dimensions—of these enormously complicated molecules: precisely for myoglobin, and very nearly so for hemoglobin. They can say, for example, that the molecule is coiled in an *alpha* helix for sixteen residues from the N-terminal unit, and then turns through a right angle. They can even say *why*: at the corner there is an aspartic acid residue; its carboxyl group interferes with the hydrogen bonding required to continue the helix, and the chain changes its course. The four folded chains of hemoglobin fit together to make a spheroidal molecule, 64 A × 55 A × 50 A. Four flat heme groups, each of which contains an iron atom that can bind an oxygen molecule, fit into separate pockets in this sphere. When oxygen is being carried, the chains move to make the pockets slightly smaller; Perutz has described hemoglobin as "a breathing molecule." These pockets are lined with the hydrocarbon portions of the amino acids; such a non-polar environment prevents electron transfer between oxygen and ferrous iron, and permits the complexing necessary for oxygen transport.

PROBLEMS

1. Outline all steps in the synthesis of phenylalanine from toluene and any needed aliphatic and inorganic reagents by each of the following methods:

(a) direct ammonolysis
(b) Gabriel synthesis
(c) malonic ester synthesis
(d) phthalimidomalonic ester method
(e) Strecker synthesis
(f) reductive amination

2. (a) Give structures of all intermediates in the following synthesis of proline:

potassium phthalimide + bromomalonic ester $\longrightarrow$ A

A + $Br(CH_2)_3Br$ $\xrightarrow{NaOC_2H_5}$ B ($C_{18}H_{20}O_6NBr$)

B + potassium acetate $\longrightarrow$ C ($C_{20}H_{23}O_8N$)

C + NaOH, heat; then H^+, heat $\longrightarrow$ D ($C_5H_{11}O_3N$)

D + HCl $\longrightarrow$ [E ($C_5H_{10}O_2NCl$)] $\longrightarrow$ proline

(b) Outline a possible synthesis of lysine by the phthalimidomalonic ester method.

3. Give structures of all intermediates in the following syntheses of amino acids:

(a) ethyl acetamidomalonate [$CH_3CONHCH(COOC_2H_5)_2$] + acrolein $\xrightarrow{\text{Michael}}$ F ($C_{12}H_{19}O_6N$)

F + KCN + acetic acid $\longrightarrow$ G ($C_{13}H_{20}O_6N_2$)
G + acid + heat $\longrightarrow$ H ($C_{13}H_{18}O_5N_2$)
H + H_2, catalyst, in acetic anhydride $\longrightarrow$ [I ($C_{13}H_{24}O_5N_2$)]
I $\xrightarrow{\text{acetic anhydride}}$ J ($C_{15}H_{26}O_6N_2$)
J + OH^-, heat; then H^+; then heat $\longrightarrow$ (±)-lysine

(b) acrylonitrile + ethyl malonate $\xrightarrow{\text{Michael}}$ K ($C_{10}H_{15}O_4N$)
K + H_2, catalyst $\longrightarrow$ [L ($C_{10}H_{19}O_4N$)] $\longrightarrow$ M ($C_8H_{13}O_3N$)
M + SO_2Cl_2 in $CHCl_3$ $\longrightarrow$ N ($C_8H_{12}O_3NCl$)
N + HCl, heat $\longrightarrow$ O ($C_5H_{10}O_2NCl$)
O $\xrightarrow{\text{base}}$ (±)-proline

(c) Glutamic acid has been made from acrolein via a Strecker synthesis. Show how this might have been done. (*Hint:* See Sec. 27.5.)

4. Using the behavior of hydroxy acids (Sec. 20.15) as a pattern, predict structures for the products obtained when the following amino acids are heated:

(a) the α-amino acid, glycine $\longrightarrow$ $C_4H_6O_2N_2$ (*diketopiperazine*)
(b) the β-amino acid, $CH_3CH(NH_2)CH_2COOH$ $\longrightarrow$ $C_4H_6O_2$
(c) the γ-amino acid, $CH_3CH(NH_2)CH_2CH_2COOH$ $\longrightarrow$ C_5H_9ON (*a lactam*)
(d) the δ-amino acid, $H_2NCH_2CH_2CH_2CH_2COOH$ $\longrightarrow$ C_5H_9ON (*a lactam*)

5. (a) Show how the particular dipolar structure given for histidine in Table 36.1 is related to the answer to Problem 20 (b), p. 1026.

(b) Draw the two possible dipolar structures for lysine. Justify the choice of structure given in Table 36.1. (c) Answer (b) for aspartic acid. (d) Answer (b) for arginine. (*Hint:* See Problem 20.24, p. 686.) (e) Answer (b) for tyrosine.

6. (a) *Betaine*, $C_5H_{11}O_2N$, occurs in beet sugar molasses. It is a water-soluble solid that melts with decomposition at 300°. It is unaffected by base but reacts with hydrochloric acid to form a crystalline product, $C_5H_{12}O_2NCl$. It can be made in either of two ways: treatment of glycine with methyl iodide, or treatment of chloroacetic acid with trimethylamine.

Draw a structure for betaine that accounts for its properties.

(b) *Trigonelline*, $C_7H_7O_2N$, is an alkaloid found in coffee beans; it is also excreted from the body as a metabolic product of nicotinic acid. It is insoluble in benzene or ether, and dissolves in water to give a neutral solution. It is unaffected by boiling aqueous acid or base. It has been synthesized as follows:

nicotinic acid + CH_3I + KOH $\longrightarrow$ P ($C_8H_{10}O_2NI$)
P + Ag_2O + H_2O, warm $\longrightarrow$ trigonelline + AgI + CH_3OH

What structure for trigonelline is consistent with these properties?

7. Addition of ethanol or other organic solvents to an aqueous "solution" of a globular protein brings about denaturation. Such treatment also tends to break up micelles of, say, soap (Sec. 33.3). What basic process is at work in both cases?

8. An amino group can be protected by acylation with phthalic anhydride to form an N-substituted phthalimide. The protecting group can be removed by treatment with hydrazine, $H_2N—NH_2$ without disturbing any peptide linkages. Write equations to show how this procedure (exploited by John C. Sheehan of the Massachusetts Institute of Technology) could be applied to the synthesis of glycylalanine (Gly.Ala) and alanylglycine (Ala.Gly).

9. An elemental analysis of *Cytochrome c*, an enzyme involved in oxidation–reduction processes, gave 0.43% Fe and 1.48% S. What is the minimum molecular weight of the enzyme? What is the minimum number of iron atoms per molecule? Of sulfur atoms?

10. A protein, *β-lactoglobulin*, from cheese whey, has a molecular weight of 42020 ± 105. When a 100-mg sample was hydrolyzed by acid and the mixture was made alkaline, 1.31 mg of ammonia was evolved. (a) Where did the ammonia come from, and approximately how many such groups are there in the protein?

Complete hydrolysis of a 100-mg sample of the protein used up approximately 17 mg of water. (b) How many amide linkages per molecule were cleaved?

(c) Combining the results of (a) and (b), and adding the fact that there are four N-terminal groups (four peptide chains in the molecule), how many amino acid residues are there in the protein?

11. The complete structure of *Gramicidin S*, a polypeptide with antibiotic properties, has been worked out as follows:

(a) Analysis of the hydrolysis products gave an empirical formula of Leu,Orn,Phe,Pro,Val. (*Ornithine*, Orn, is a rare amino acid of formula $^{+}H_3NCH_2CH_2CH_2CH(NH_2)COO^{-}$.) It is interesting that the phenylalanine has the unusual D-configuration.

Measurement of the molecular weight gave an approximate value of 1300. On this basis, what is the molecular formula of Gramicidin S?

(b) Analysis for the C-terminal residue was negative; analysis for the N-terminal residue using DNFB yielded only DNP—$NHCH_2CH_2CH_2CH(\overset{+}{N}H_3)COO^{-}$. What structural feature must the peptide chain possess?

(c) Partial hydrolysis of Gramicidin S gave the following di- and tripeptides:

Leu.Phe	Phe.Pro	Phe.Pro.Val	Val.Orn.Leu
Orn.Leu	Val.Orn	Pro.Val.Orn.	

What is the structure of Gramicidin S?

12. The structure of beef insulin was determined by Sanger (see Sec. 36.9) on the basis of the following information. Work out for yourself the sequence of amino acid residues in the protein.

Beef insulin appears to have a molecular weight of about 6000 and to consist of two polypeptide chains linked by disulfide bridges of cystine residues. The chains can be separated by oxidation, which changes any CyS—SCy or CySH residues to sulfonic acids ($CySO_3H$).

One chain, A, of 21 amino acid residues, is acidic and has the empirical formula

$$\text{GlyAlaVal}_2\text{Leu}_2\text{Ileu(CySH)}_4\text{Asp}_2\text{Glu}_4\text{Ser}_2\text{Tyr}_2$$

The other chain, B, of 30 amino acid residues, is basic and has the empirical formula

$$\text{Gly}_3\text{Ala}_2\text{Val}_3\text{Leu}_4\text{ProPhe}_3\text{(CySH)}_2\text{ArgHis}_2\text{LysAspGlu}_3\text{SerThrTyr}_2$$

(Chain A has four simple side-chain amide groups, and chain B has two, but these will be ignored for the time being.)

Treatment of chain B with 2,4-dinitrofluorobenzene (DNFB) followed by hydrolysis gave DNP.Phe and DNP.Phe.Val; chain B lost alanine (Ala) when treated with carboxypeptidase.

Acidic hydrolysis of chain B gave the following tripeptides:

Glu.His.Leu	Leu.Val.CySH	Tyr.Leu.Val
Gly.Glu.Arg	Leu.Val.Glu	Val.Asp.Glu
His.Leu.CySH	Phe.Val.Asp	Val.CySH.Gly
Leu.CySH.Gly	Pro.Lys.Ala	Val.Glu.Ala
	Ser.His.Leu	

Many dipeptides were isolated and identified; two important ones were Arg.Gly and Thr.Pro.

(a) At this point construct as much of the B chain as the data will allow.

Among the numerous tetrapeptides and pentapeptides from chain B were found:

His.Leu.Val.Glu	Tyr.Leu.Val.CySH
Ser.His.Leu.Val	Phe.Val.Asp.Glu.His

(b) How much more of the chain can you reconstruct now? What amino acid residues are still missing?

Enzymatic hydrolysis of chain B gave the necessary final pieces:

Val.Glu.Ala.Leu His.Leu.CySH.Gly.Ser.His.Leu
Tyr.Thr.Pro.Lys.Ala Tyr.Leu.Val.CySH.Gly.Glu.Arg.Gly.Phe.Phe

(c) What is the complete sequence in the B chain of beef insulin?

Treatment of chain A with DNFB followed by hydrolysis gave DNP.Gly; the C-terminal group was shown to be aspartic acid (Asp).

Acidic hydrolysis of chain A gave the following tripeptides:

CySH.CySH.Ala Glu.Leu.Glu
Glu.Asp.Tyr Leu.Tyr.Glu
Glu.CySH.CySH Ser.Leu.Tyr
Glu.Glu.CySH Ser.Val.CySH

Among other peptides isolated from acidic hydrolysis of chain A were:

CySH.Asp Tyr.CySH Gly.Ileu.Val.Glu.Glu

(d) Construct as much of chain A as the data will allow. Are there any amino acid residues missing?

Up to this point it is possible to arrive at the sequences of four parts of chain A, but it is still uncertain which of the two center fragments, Ser.Val.CySH or Ser.Leu.Tyr, etc., comes first. This was settled by digestion of chain A with pepsin, which gave a peptide that contained no aspartic acid (Asp) or tyrosine (Tyr). Hydrolysis of this peptide gave Ser.Val.CySH and Ser.Leu.

(e) Now what is the complete structure of chain A of beef insulin?

In insulin the cysteine units (CySH) are involved in cystine disulfide links (CyS—SCy). Residue 7 of chain A (numbering from the N-terminal residue) is linked to residue 7 of chain B, residue 20 of chain A to residue 19 of chain B, and there is a link between residues 6 and 11 of chain A.

There are amide groups on residues 5, 15, 18, and 21 of chain A, and on residues 3 and 4 of chain B.

(f) Draw a structure of the complete insulin molecule. (*Note:* The disulfide loop in chain A is a 20-atom, pentapeptide ring, of the same size as the one in oxytocin.)

In the analysis for the N-terminal group in chain B of insulin, equal amounts of *two* different DNP derivatives of single amino acids actually were found. One was DNP.Phe; what could the other have been?

(g) What would have been obtained if that second amino acid had been N-terminal?

Chapter 37 Biochemical Processes

Molecular Biology

37.1 Biochemistry, molecular biology, and organic chemistry

In the past four chapters, we have learned something about fats, carbohydrates, and proteins: their structures and how these are determined, and the kind of reactions they undergo in the test tube. These, we said, are biomolecules: they are participants in the chemical process we call life. But just what do they *do*? What reactions do they undergo, not in the test tube, but in a living organism?

Even a vastly simplified answer to that question would fill—and does—a book as big as this one. Having come this far, though, we cannot help being curious. And so, in this chapter, we shall take a brief glance at the answer—or, rather, at the kind of thing the answer entails.

We shall look at just a few examples of biochemical processes: how one enzyme—of the thousands in our bodies—may work; what happens in one of the dozens of reactions by which carbohydrates are oxidized to furnish energy; how one kind of chemical compound—fatty acids—is synthesized. Finally, we shall learn a little about another class of biomolecules, the nucleic acids, and how they are involved in the most fascinating biochemical process of all—heredity.

The study of nucleic acids has become known as "molecular biology." Actually, of course, all of these processes are a part of molecular biology—biology on the molecular level—and they are, in the final analysis, organic chemistry. And it is as organic chemistry that we shall treat them. We shall see how all these vital processes—even the mysterious powers of enzymes—come down to a matter of molecular structure as we know it: to molecular size and shape; to intermolecular and intramolecular forces; to the chemistry of functional groups; to acidity and basicity, oxidation and reduction; to energy changes and rate of reaction.

Since catalysis by enzymes is fundamental to everything else, let us begin there.

37.2 Mechanism of enzyme action. Chymotrypsin

Enzymes, we have said, are proteins that act as enormously effective catalysts for biological reactions. To get some idea of how they work, let us examine the action of just one: *chymotrypsin*, a digestive enzyme whose job is to promote hydrolysis of certain peptide links in proteins. The sequence of the 245 amino acid residues in chymotrypsin has been determined and, through x-ray analysis, the conformation of the molecule is known (Fig. 37.1). It is, like all enzymes, a soluble globular protein coiled in the way that turns its hydrophobic parts inward, away from water, and that permits maximum intramolecular hydrogen bonding.

The action of chymotrypsin has been more widely explored than that of any other enzyme. In crystalline form, it is available for studies in the test tube under a variety of conditions. It catalyzes hydrolysis not only of proteins but of ordinary amides and esters, and much has been learned by use of these simpler substrates. Compounds modeled after portions of the chymotrypsin molecule have been made, and their catalytic effects measured.

To begin with, it seems very likely that chymotrypsin acts in two stages. In the first stage, acting as an alcohol, it breaks the peptide chain. We recognize this as alcoholysis of a substituted amide: nucleophilic acyl substitution. The products are an amine—the liberated portion of the substrate molecule—and, as we shall

(Stage 1)

$$\sim\overset{\overset{\displaystyle O}{\|}}{C}\!-\!NH\sim \;+\; E\!-\!OH \;\longrightarrow\; \sim\overset{\overset{\displaystyle O}{\|}}{C}\!-\!O\!-\!E \;+\; NH_2\sim$$

Protein	Enzyme	Acyl enzyme	Part of protein chain
Amide	*Alcohol*	*Ester*	*Amine*

(Stage 2)

$$\sim\overset{\overset{\displaystyle O}{\|}}{C}\!-\!O\!-\!E \;+\; H_2O \;\longrightarrow\; \sim COOH \;+\; E\!-\!OH$$

Acyl enzyme		Rest of protein chain	Regenerated enzyme
Ester		*Carboxylic acid*	*Alcohol*

see shortly, an ester of the enzyme. In the second stage, the enzyme ester is hydrolyzed. This yields a carboxylic acid—the other portion of the substrate molecule—and the regenerated enzyme, ready to go to work again.

What is the structure of this intermediate ester formed from the enzyme? The answer has been found by use of simple esters as substrates, *p*-nitrophenyl acetate, for example. An appreciable steady-state concentration of the intermediate ester builds up and, by quenching of the reaction mixture in acid, it can be isolated. Sequence analysis of the enzyme ester showed that the acetyl group from the substrate was linked to *serine-195*. It is, then, at the —OH group of this particular amino acid residue that the enzyme reacts.

$$HOCH_2\underset{\underset{\displaystyle {}^{+}NH_3}{|}}{C}HCOO^-$$

Serine

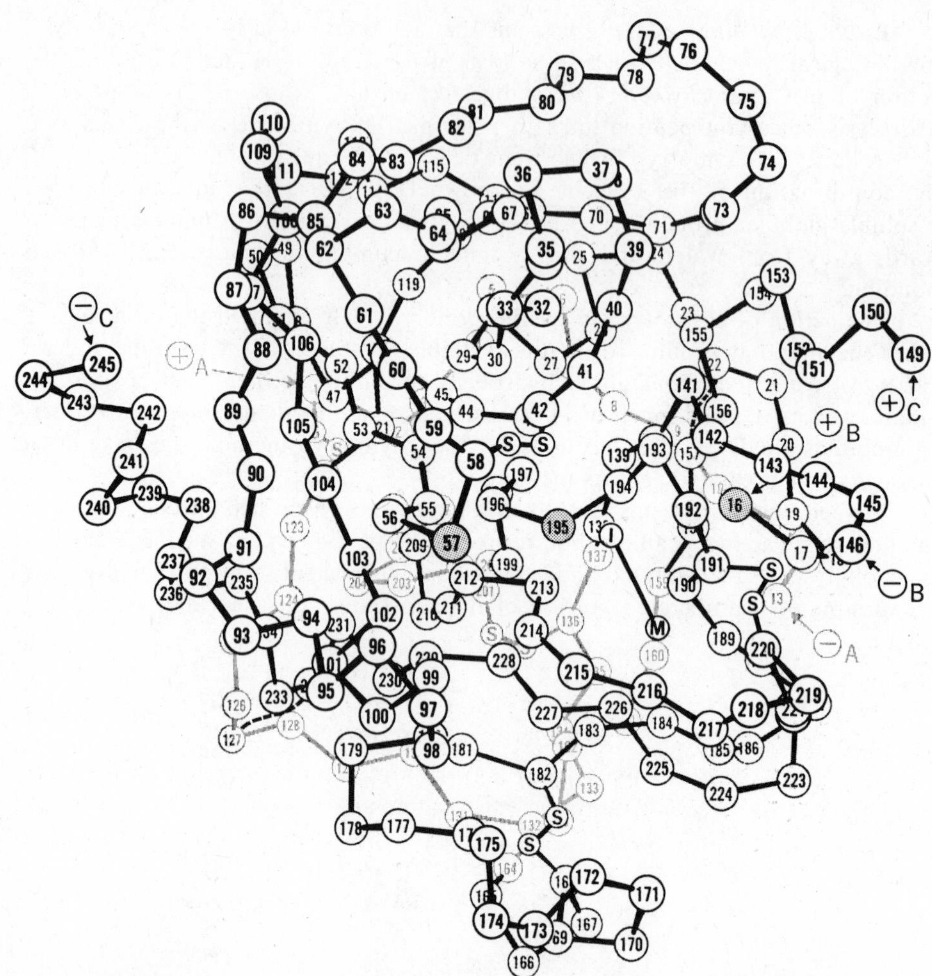

Figure 37.1. Three-dimensional structure of α-chymotrypsin. Histidine-57, serine-195, and isoleucine-16 are shaded. The hydrophobic pocket lies to the right of histidine-57 and serine-196, where M is marked; it is bounded by residues 184–191 and 214–227.

The ⊕ and ⊖ signs show the N-terminal and C-terminal ends of chains A, B, and C. The M and I stand for the methyl and sulfonyl parts of the inhibitor, a tosyl group held as an ester of serine-195.

We can see one short segment of α-helix at residues 234–245; another (mostly hidden) lies at 164–170. There is a hint of a twisted sheet beginning with residues 91–86 and 103–108, and extending to their right.

But evidence shows that certain other amino acid residues are also vital to enzyme activity. The rate of enzyme-catalyzed hydrolysis changes as the acidity of the reaction medium is changed. If one plots the rate of hydrolysis against the pH of the solution, one gets a bell-shaped curve: as the pH is increased, the rate rises to a maximum and then falls off. The rate is fastest at about pH 7.4 (fittingly, the physiological pH) and slower in either more acidic or more basic solution. Analysis of the data shows the following. Hydrolysis requires the presence of a free base, of K_b about 10^{-7}, and a protonated base, of K_b about 3×10^{-5}. At low pH (acid solution), both bases are protonated; at high pH (alkaline solution), both bases are free. Hydrolysis is fastest at the intermediate pH where the weaker base is mostly free and the stronger base is mostly protonated.

The K_b of the weaker base fits that of the imidazole ring of histidine, and there is additional evidence indicating that this is indeed the base: studies involving

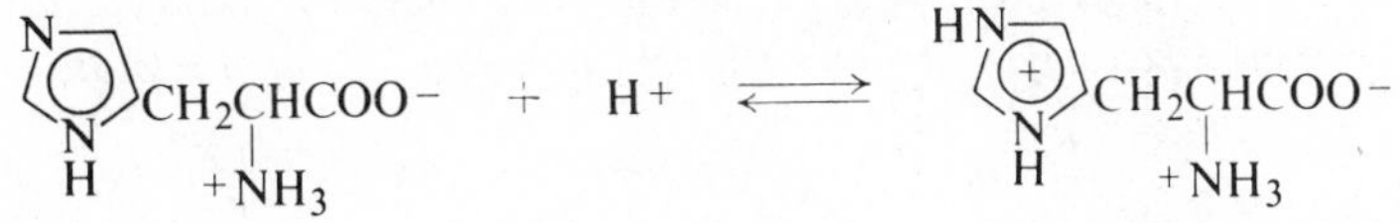

$$\text{(imidazolyl)}CH_2\underset{+NH_3}{CHCOO^-} + H^+ \rightleftharpoons \text{(protonated imidazolyl)}CH_2\underset{+NH_3}{CHCOO^-}$$

Histidine
Base

Protonated histidine
Acid

catalysis by imidazole itself, for example. Now, examination of the conformation of chymotrypsin (Fig 37.1) shows that very close to serine-195 there *is* a histidine residue. This is *histidine-57*, and it is believed to be the one involved in enzyme activity.

What about the stronger base which, according to the kinetics, is involved in its protonated form? Its K_b fits the α-amino group of most amino acids—an α-amino group, that is, which is not tied up in a peptide link. But all the (free) amino groups in chymotrypsin—*except one*—may be acetylated without complete loss of activity. The exception is *isoleucine-16*, the N-terminal unit of chain B.

$$CH_3CH_2CH(CH_3)\underset{^+NH_3}{CHCOO^-}$$

Isoleucine

Presumably, then, this amino group cannot be acetylated, but must be free to be protonated and do its part of the job.

Now, what is the job of each of these key units in the enzyme molecule? It is clear what serine-195 does: it provides the —OH for ester formation. What does isoleucine-16 do? The descending leg of the bell-shaped rate curve was attributed to protonation of this unit. But something else happens as the pH is raised above 7.4: the optical activity of the solution decreases—evidently due to a change in conformation of the enzyme molecule—and in a way that parallels the decrease in rate of hydrolysis. It is believed, then, that, through hydrogen bonding or electrostatic attraction, the $—NH_3^+$ on isoleucine-16 helps hold the enzyme chain in the proper shape for it to act as a catalyst: to keep histidine-57 near serine-195, among other things. At higher pH the $—NH_3^+$ is converted into $—NH_2$, and the chain changes its shape; with the change in shape goes loss of catalytic power and a change in optical rotation.

Next, we come to the question: what is the role of histidine-57? We are observing an example of *general acid–base catalysis:* catalysis not just by hydroxide ions and oxonium ions, but by all the bases and conjugate acids that are present, each contributing according to its concentration and its acid or base strength.

Let us look at this concept first with a simple example: hydrolysis of an ester catalyzed by the simple heterocyclic base, imidazole. Catalysis by hydroxide ions

$$\mathrm{RCOOR'} + \mathrm{H_2O} \xrightarrow{\text{imidazole}} \mathrm{RCOOH} + \mathrm{R'OH}$$

we understand: these highly nucleophilic ions are more effective than water at attacking acyl carbon. Imidazole generates some hydroxide ions by reaction with water, but these are already taken into account. We are talking now about hydrolysis that is directly proportional to the concentration of the base itself: imidazole. What seems to be involved in such reactions is something like the following. In step (1), water adds to acyl carbon with *simultaneous loss of a proton to the base*;

$$(1)\qquad \mathrm{R(R'O)C{=}O} + \mathrm{H{-}O{-}H} + \ddot{\mathrm{B}} \rightleftharpoons \mathrm{R{-}C(OH)(OR'){-}O^-} + \mathrm{H{:}B}$$

$$(2)\qquad \mathrm{R{-}C(OH)(OR'){-}O^-} + \mathrm{H{-}B} \rightleftharpoons \mathrm{(HO)(R)C{=}O} + \mathrm{R'OH} + \mathrm{B{:}}$$

reaction is fast because, in effect, the attacking nucleophile is not just water, but an incipient hydroxide ion. In step (2), transfer of the proton from the protonated base is simultaneous with loss of the ethoxy group; again reaction is fast, this time because the leaving group is not the strongly basic ethoxide ion, but an incipient alcohol molecule.

Reactions like (1) and (2) need not involve unlikely three-body collisions among the reactive molecules. Instead, there is prior hydrogen bonding between the base and water or between the protonated base and ester; it is these double molecules that collide with the third reagent and undergo reaction, with the dipole–dipole attraction of the hydrogen bonding being replaced by a covalent bond.

Figure 37.2 depicts the action of chymotrypsin, with the imidazole group of histidine-57 playing the same role of general base as that just described—and with protonated imidazole necessarily acting as general acid. There is general acid–base catalysis of both reactions involved: first, in the formation of the acyl enzyme, and then in its hydrolysis.

Chymotrypsin is not, as enzymes go, very specific in its action; it hydrolyzes proteins, peptides, simple amides, and esters alike. There is one structural requirement, nevertheless; a relatively non-polar group in the acyl moiety of the substrate, typically an aromatic ring. Now, turning once more to Fig. 37.1, we find that at the reactive site in the enzyme there is a pocket; this pocket is lined with hydrophobic substituents to receive the non-polar group of the substrate and thus hold the molecule in position for hydrolysis. It is the size of this pocket and the nature of its lining that gives the enzyme its specificity; here we find, in a very real sense, Emil Fischer's lock into which the substrate key must fit.

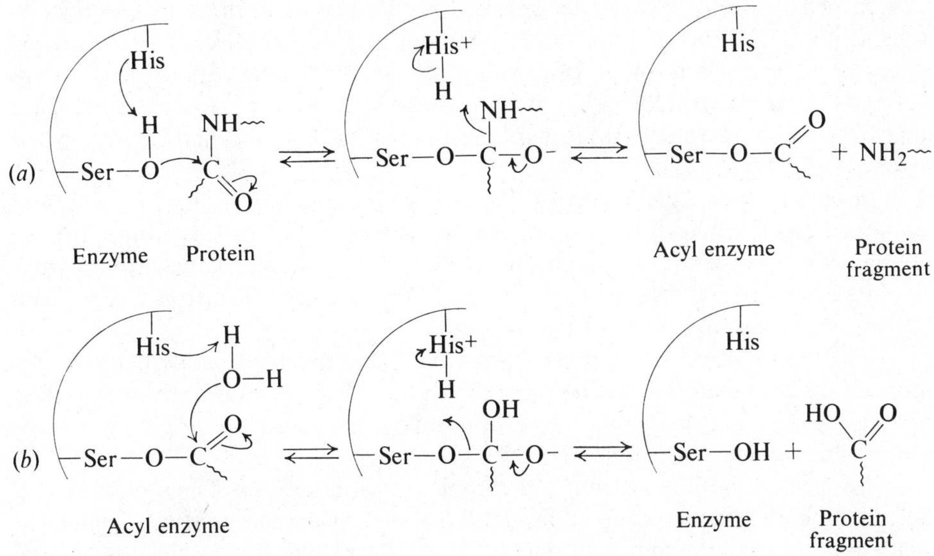

(*a*) His, H, NH~, —Ser—O, C, O ⇌ His+, H, NH~, —Ser—O—C—O⁻ ⇌ His, —Ser—O—C=O + NH_2~

Enzyme Protein Acyl enzyme Protein fragment

(*b*) His, H, O—H, —Ser—O—C=O ⇌ His+, H, OH, —Ser—O—C—O⁻ ⇌ His, —Ser—OH + HO—C=O

Acyl enzyme Enzyme Protein fragment

Figure 37.2. Catalysis by the enzyme chymotrypsin of the cleavage of one peptide bond in a protein: a proposed mechanism. Histidine and protonated histidine act as general base and acid in two successive nucleophilic substitution reactions: (*a*) cleavage of protein with formation of acyl enzyme and liberation of one protein fragment; (*b*) hydrolysis of acyl enzyme with regeneration of the enzyme and liberation of the other protein fragment.

We see, then, some of the factors that give enzymes their catalytic powers. The substrate is bound to a particular site in the enzyme, where the necessary functional groups are gathered: here, hydroxyl of serine and imidazole of histidine. In most cases, there are other functional groups as well, in molecules of cofactors—reagents, really—bound by the enzyme near the reactive site. In the enzyme–substrate complex, these functional groups are *neighboring groups*, and in their reactions enjoy all the advantages we listed (Sec. 28.1) for such groups. *They are there*, poised in just the right position for attack on the substrate. They need not wait for the lucky accident of a molecular collision; in effect, concentration of reagents is very high. Orientation of reacting groups is exactly right. There are no clinging solvent molecules to be stripped away as reaction occurs.

All these factors are important, and can be shown independently to speed up the rate of reactions, and very powerfully, too—but they do not seem to be nearly enough to account for the enormous activity of enzymes. Perhaps other factors are involved. It has been suggested, for example, that the pocket in which reaction occurs fits the transition state better than it fits the reactants, so that relief of strain or an increase in van der Waals attractions provides a driving force. Perhaps the correct factors are being considered but, in extrapolation to enzyme systems, their power has been underestimated.

37.3 The source of biological energy. The role of ATP

In petroleum we have a fuel reserve on which we can draw for energy—as long as it lasts. We burn it, and either use the heat produced directly to warm ourselves or convert it into other kinds of energy: mechanical energy to move things about; electrical energy, which is itself transformed—at a more convenient place than where the original burning happened—into light, or mechanical energy, or back into heat.

In the same way, the energy our bodies need to keep warm, move about, and build new tissue comes from a food reserve: carbohydrates, chiefly in the form of starch. (We eat other animals, too, but ultimately the chain goes back to a carbohydrate-eater.) In the final analysis, we get energy from food just as we do from petroleum: we oxidize it to carbon dioxide and water.

This food reserve is not, however, a limited one that we steadily deplete. Our store of carbohydrates—and the oxygen to go with it—is constantly replenished by the recombining, in plants, of carbon dioxide and water. The energy for recombination comes, of course, from the sun.

We speak of both petroleum and carbohydrates as sources of energy; we could speak of them as "energy-rich molecules." But the oxygen that is also consumed in oxidation is equally a source of energy. What we really mean is that the energy content of carbohydrates (or petroleum) plus oxygen is greater than that of carbon dioxide plus water. (In total, the bonds that are to be broken are weaker—contain more energy—than the bonds that are to be formed.) These reactants are, of course, energy-rich only in relation to the particular products we want to convert them into. But this is quite sufficient; in our particular kind of world, these *are* our sources of energy.

The body takes in carbohydrates and oxygen, then, and eventually gives off carbon dioxide and water. In the process considerable energy is generated. But in what form? And how is it used to move muscles, transport solutes, and build new molecules? Certainly each of our cells does not contain a tiny fire in which carbohydrates burn merrily, running a tiny steam engine and over which a tiny organic chemist stews up his reaction mixtures. Nor do we contain a central power plant where, again, carbohydrates are burned, and the energy sent about in little steam pipes or electric cables to run muscle-machines and protein-and-fat factories.

In a living organism, virtually the whole energy system is a chemical one. Energy is generated, transported, and consumed by way of chemical reactions and chemical compounds. Instead of a single reaction with a long plunge from the energy level of carbohydrates and oxygen to that of carbon dioxide and water—as in the burning of a log, say—there are long series of chemical reactions in which the energy level descends in gentle cascades. Energy resides, ultimately, in the molecules involved; as they move through the organism, they carry energy with them.

Constantly appearing in these reactions is one compound, *adenosine triphosphate* (ATP). It is called by biochemists an "energy-rich" molecule, but there

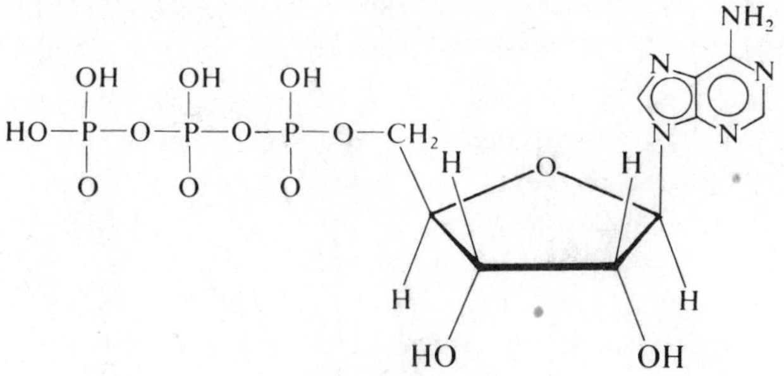

Adenosine triphosphate
ATP

is nothing magical about this. ATP does not carry about a little bag of energy which it sprinkles on molecules to make them react. Nor does it undergo hydrolysis alongside other molecules and in some mystical way make this energy available to them. ATP simply undergoes reactions—only one reaction, really. It *phosphorylates*, that is, transfers a phosphoryl group, $-PO_3H_2$, to some other molecule. For example:

$$\underset{\text{Adenosine triphosphate}}{\text{ATP}} + \underset{\text{An alcohol}}{\text{R—OH}} \longrightarrow \underset{\text{Adenosine diphosphate}}{\text{ADP}} + \underset{\text{A phosphate ester}}{\text{R—OPO}_3\text{H}_2}$$

ATP is called a "high-energy phosphate" compound, but this simply means that it is a fairly reactive phosphorylating agent. It is exactly as though we were to call acetic anhydride "high-energy acetate" because it is a better acetylating agent than acetic acid. And, indeed, there is a true parallel here: ATP is an anhydride, too, an anhydride of a substituted phosphoric acid, and it is a good phosphorylating agent for much the same reasons that acetic anhydride is a good acetylating agent.

When ATP loses a phosphoryl group to another molecule, it is converted into ADP, *adenosine diphosphate*. If ATP is to be regenerated, ADP must itself be phosphorylated, and it is: by certain other compounds that are good enough phosphorylating agents to do this. The important thing in all this is not really the energy level of these various phosphorylating agents—so long as they are reactive enough to do the job they must—but the fact *that the energy level of the carbohydrates and their oxidation products is gradually sinking to the level of carbon dioxide and water*. These compounds—and oxygen—are where the energy is, and ATP is simply a chemical reagent that helps to make it available.

We have seen that very often factors that stabilize products also stabilize the transition state leading to those products, that is, that often there is a parallel between ΔH and E_{act}. To that extent, the energy level of the various phosphorylating agents may enter in, too: less stable phosphorylating agents—less stable, let us say, relative to phosphate anion—may in general tend to transfer phosphate to more stable phosphorylating agents. In addition, of course, if any of the phosphate transfers should be too highly endothermic, this would require a prohibitively high E_{act} for reaction (see Sec. 2.17).

In following sections, we shall see some of the specific reactions in which ATP is involved.

37.4 Biological oxidation of carbohydrates

Next, let us take a look at the overall picture of the biological oxidation of carbohydrates. We start with glycogen ("sugar-former"), the form in which carbohydrates are stored in the animal body. This, we have seen (Sec. 35.9), is a starch-like polymer of D-glucose.

The trip from glycogen to carbon dioxide and water is a long one. It is made up of dozens of reactions, each of which is catalyzed by its own enzyme system. Each of these reactions must, in turn, take place in several steps, most of them unknown. (Consider what is involved in the "reaction" catalyzed by chymotrypsin.) We can divide the trip into three stages. (a) First, glycogen is broken down into its component D-glucose molecules. (b) Then, in *glycolysis* ("sugar-splitting"), D-glucose is itself broken down, into three-carbon compounds. (c) These, in *respiration*, are converted into carbon dioxide and water. Oxygen appears in only the third stage; the first two are anaerobic ("without-air") processes.

The first stage, **cleavage of glycogen**, is simply the hydrolytic cleavage of acetal linkages (Sec. 34.16), this time enzyme-catalyzed.

$$\underset{\text{Glycogen}}{(C_6H_{10}O_5)_n} + nH_2O \xrightarrow{\text{enzyme}} \underset{\text{D-Glucose}}{nC_6H_{12}O_6}$$

The second stage, **glycolysis**, takes eleven reactions and eleven enzymes. The sum of these reactions is:

$$\text{D-glucose} + \underset{\text{Phosphate}}{2HPO_4^{--}} + 2ADP^{3-} \longrightarrow \underset{\text{Lactate}}{2CH_3CHOHCOO^-} + 2H_2O + 2ATP^{4-}$$

No oxygen is consumed, and we move only a little way down the energy hill toward carbon dioxide and water. What is important is that a start has been made in breaking the five carbon–carbon bonds of glucose, and that two molecules of ADP are converted into ATP. (ATP is required for some of the steps of glycolysis, but there is a *net* production of two molecules of ATP for each molecule of glucose consumed.)

The third stage, **respiration**, is a complex system of reactions in which molecules provided by glycolysis are oxidized. Oxygen is consumed, carbon dioxide and water are formed, and energy is produced.

Let us look at the linking-up between glycolysis and respiration. Ordinarily, the energy needs of working muscles are met by respiration. But, during short periods of vigorous exercise, the blood cannot supply oxygen enough for respiration to carry the entire load; when this happens, glycolysis is called upon to supply the energy difference. The end-product of glycolysis, lactic acid, collects in the muscle, and the muscle feels tired. The lactic acid is removed by the blood and rebuilt into glycogen, which is ready for glycolysis again.

The last step of glycolysis is reduction of pyruvic acid to lactic acid. (The reducing agent is, incidentally, an old acquaintance, reduced nicotinamide adenine

$$CH_3COCOO^- + NADH + H^+ \longrightarrow CH_3CHOHCOO^- + NAD^+$$

Pyruvate	Reduced nicotinamide adenine dinucleotide		Lactate	Nicotinamide adenine dinucleotide

dinucleotide, Sec. 36.15.) Most of the time, however, glycolysis does not proceed to the very end. Instead, pyruvic acid is diverted, and oxidized to acetic acid in the form of a thiol ester, $CH_3CO–S–CoA$, derived from *coenzyme A* and called "acetyl CoA."

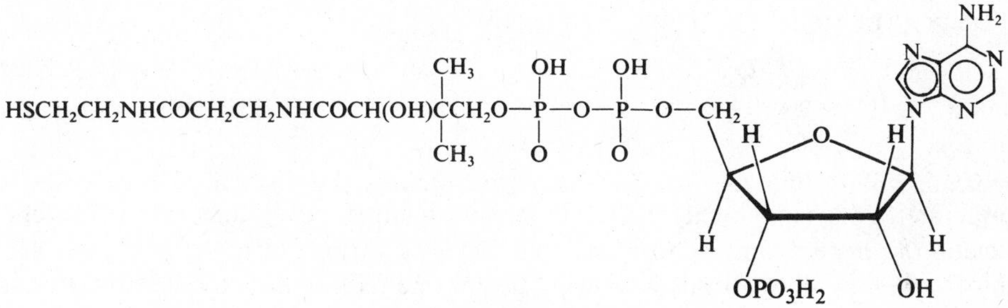

$HSCH_2CH_2NHCOCH_2CH_2NHCOCH(OH)C(CH_3)_2CH_2O—P(OH)(O)—O—P(OH)(O)—O—CH_2$ (ribose bearing adenine, with H, OPO_3H_2, OH)

Coenzyme A
CoA

It is as acetyl CoA that the products of glycolysis are fed into the respiration cycle.

The acetyl CoA that is fuel for respiration comes not only from carbohydrates but also from the breakdown of amino acids and fats. It is thus the common link between all three kinds of food and the energy-producing process. (Acetyl CoA is even more than that: as we shall see, it is the building block from which the long chains of fatty acids are synthesized.)

Thiols are sulfur analogs of alcohols. They contain the sulfhydryl group, —SH, which plays many parts in the chemistry of biomolecules. Easily oxidized, two —SH groups are converted into disulfide links, —S—S—, which hold together different peptide chains or different parts of the same chain. (See, for example, oxytocin on p. 1143.) Thiols form the same kinds of derivatives as alcohols: *thio*ethers, *thio*acetals, *thiol* esters. Thiol ester groups show the chemical behavior we would expect—they undergo nucleophilic acyl substitution and they make α-hydrogens acidic—this last more effectively than their oxygen counterparts.

37.5 Mechanism of a biological oxidation

Now let us take just one of the many steps in carbohydrate oxidation and look at it in some detail.

Although there is no *net* oxidation in glycolysis, certain individual reactions do involve oxidation and reduction. About mid-way in the eleven steps we arrive

$H_2O_3P—O—CH_2CHOHCHO$	$H_2O_3P—O—CH_2CHOHCOOH$
D-Glyceraldehyde-3-phosphate	3-Phosphoglyceric acid

at D-glyceraldehyde-3-phosphate and its oxidation to 3-phosphoglyceric acid. In the course of this conversion, a phosphate ion becomes attached to ADP to generate a molecule of ATP.

Two reactions are actually involved. First, D-glyceraldehyde-3-phosphate is oxidized, but not directly to the corresponding acid, 3-phosphoglyceric acid.

$$\underset{\text{D-Glyceraldehyde-3-phosphate}}{^{--}O_3POCH_2CHOHCHO} + NAD^+ + \underset{\text{Phosphate}}{HPO_4^{--}} \rightleftharpoons \underset{\text{1,3-Diphosphoglycerate}}{^{--}O_3POCH_2CHOH{-}\overset{\overset{\displaystyle O}{\|}}{C}{-}O{-}PO_3^{--}} + NADH + H^+$$

$$\underset{\text{1,3-Diphosphoglycerate}}{^{--}O_3POCH_2CHOH{-}\overset{\overset{\displaystyle O}{\|}}{C}{-}O{-}PO_3^{--}} + ADP^{3-} \rightleftharpoons \underset{\text{3-Phosphoglycerate}}{^{--}O_3POCH_2CHOHCOO^-} + ATP^{-4}$$

Instead, a phosphate ion is picked up to give the mixed anhydride, 1,3-diphosphoglycerate. This is a highly reactive phosphorylating agent and, in the second reaction, transfers a phosphoryl group to ADP to form ATP.

Now, how does all this happen? The enzyme required for the first reaction is *glyceraldehyde-3-phosphate dehydrogenase* ("enzyme-that-dehydrogenates-glyceraldehyde-3-phosphate"). Its action is by no means as well understood as that of chymotrypsin, but let us look at the kind of thing that is believed to happen. A sulfhydryl group (—SH) of the enzyme adds to the carbonyl group of glyceraldehyde-3-phosphate. Thiols are sulfur analogs of alcohols, and the product is a

$$\underset{\text{Enzyme}}{E{-}SH} + \underset{\text{Aldehyde}}{RCHO} \longrightarrow \underset{\text{Hemithioacetal}}{E{-}S{-}\underset{\displaystyle OH}{\overset{\displaystyle H}{C}}{-}R}$$

hemiacetal: more precisely, a hemi*thio*acetal. Like other acetals, this is both an ether (a *thio* ether) and an alcohol. Such an alcohol group is especially easily oxidized to a carbonyl group (Problem 11, p. 649).

The oxidizing agent is a compound that, like ATP, constantly appears in these reactions: *nicotinamide adenine dinucleotide* (NAD). The functional group here, we remember (Sec. 36.15), is the pyridine ring, which can accept a hydride ion to form NADH. Like the hemiacetal moiety, NAD is bound to the enzyme, and in a position for easy reaction (Fig. 37.3).

Oxidation converts the hemithioacetal into a thiol ester—an acyl enzyme. Like other esters, this one is prone to nucleophilic acyl substitution. It is cleaved, with phosphate ion as nucleophile, to regenerate the sulfhydryl group in the enzyme. The other product is 1,3-diphosphoglycerate. The molecule is (still) a phosphate ester at the 3-position, and has become a mixed anhydride at the 1-position.

The anhydride phosphoryl group is easily transferred; in another enzyme-catalyzed reaction, 1,3-diphosphoglycerate reacts with ADP to yield 3-phosphoglycerate and ATP. The 3-phosphoglycerate goes on in the glycolysis process.

The ATP is available to act as a phosphorylating agent: to convert a molecule of D-glucose into D-glucose-6-phosphate, for example, and help start another molecule through glycolysis; to assist in the synthesis of fatty acids; to change the cross-linking between molecules of *actin* and *myosin*, and thus cause muscular contraction.

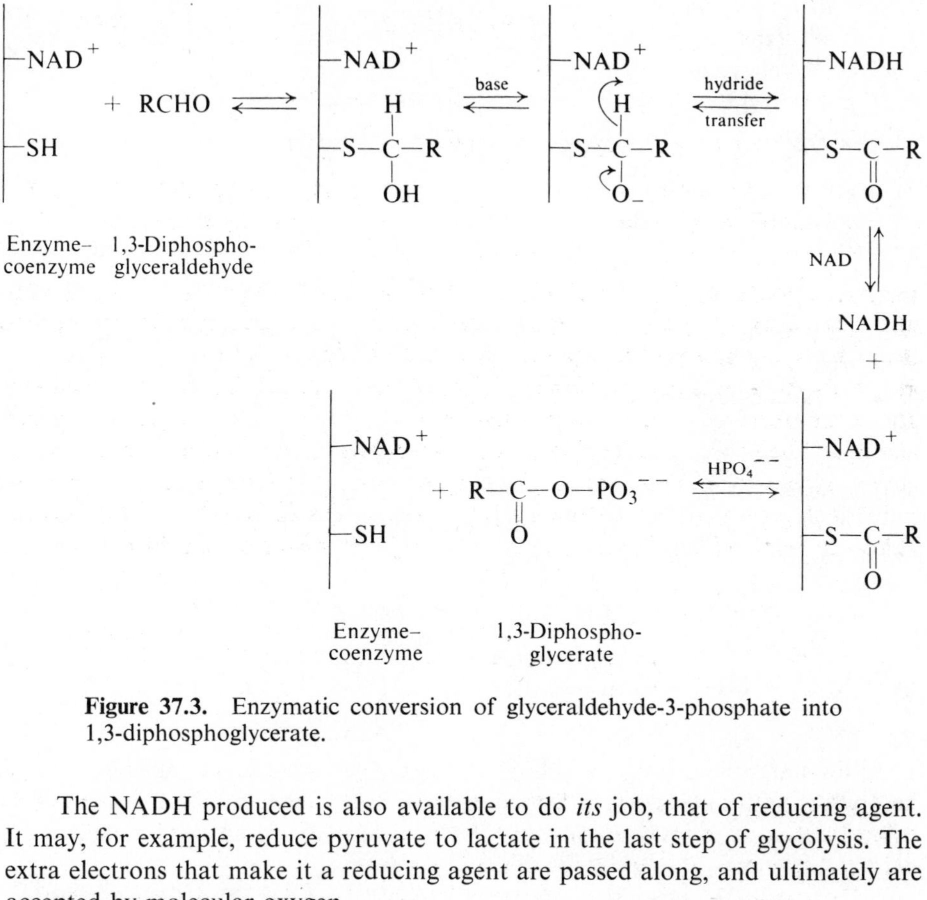

$$\text{Enzyme–coenzyme}\begin{cases}-\text{NAD}^+\\-\text{SH}\end{cases} + \underset{\text{1,3-Diphosphoglyceraldehyde}}{\text{RCHO}} \rightleftharpoons \begin{cases}-\text{NAD}^+\\-\text{S}-\underset{\text{OH}}{\overset{\text{H}}{\text{C}}}-\text{R}\end{cases} \xrightleftharpoons{\text{base}} \begin{cases}-\text{NAD}^+\\-\text{S}-\underset{\text{O}^-}{\overset{\text{H}}{\text{C}}}-\text{R}\end{cases} \xrightleftharpoons[\text{transfer}]{\text{hydride}} \begin{cases}-\text{NADH}\\-\text{S}-\underset{\text{O}}{\overset{}{\text{C}}}(=\text{O})-\text{R}\end{cases}$$

$$\begin{cases}-\text{NADH}\\-\text{S}-\text{C}(=\text{O})-\text{R}\end{cases} \xrightleftharpoons{\text{NAD}} \text{NADH} + \begin{cases}-\text{NAD}^+\\-\text{S}-\text{C}(=\text{O})-\text{R}\end{cases} \xrightleftharpoons{HPO_4^{--}} \underset{\text{Enzyme–coenzyme}}{\begin{cases}-\text{NAD}^+\\-\text{SH}\end{cases}} + \underset{\text{1,3-Diphosphoglycerate}}{\text{R}-\text{C}(=\text{O})-\text{O}-\text{PO}_3^{--}}$$

Figure 37.3. Enzymatic conversion of glyceraldehyde-3-phosphate into 1,3-diphosphoglycerate.

The NADH produced is also available to do *its* job, that of reducing agent. It may, for example, reduce pyruvate to lactate in the last step of glycolysis. The extra electrons that make it a reducing agent are passed along, and ultimately are accepted by molecular oxygen.

We are in a strange, complex chemical environment here, but in it we recognize familiar kinds of compounds—hemiacetals, esters, anhydrides, carboxylic acids—and familiar kinds of reactions—nucleophilic carbonyl addition, hydride transfer, nucleophilic acyl substitution.

37.6 Biosynthesis of fatty acids

When an animal eats more carbohydrate than it uses up, it stores the excess: some as the polysaccharide glycogen (Sec. 35.9), but most of it as fats. Fats, we know (Sec. 33.2), are triacylglycerols, esters derived (in most cases) from long straight-chain carboxylic acids containing an *even number* of carbon atoms. These even numbers, we said, are a natural consequence of the way fats are synthesized in biological systems.

There are even numbers of carbons in fatty acids because the acids are built up, two carbons at a time, from acetic acid units. These units come from acetyl CoA: the thiol ester derived from acetic acid and coenzyme A (Sec. 37.4). The acetyl CoA itself is formed either in glycolysis, as we have seen, or by oxidation of fatty acids.

Let us see how fatty acids are formed from acetyl CoA units. As before, we must realize that every reaction is catalyzed by a specific enzyme and proceeds by several steps—steps that in some direct, honest-to-goodness chemical way, involved the enzyme.

First, acetyl CoA takes up carbon dioxide (1) to form malonyl CoA. (To illustrate the point made above: this does not happen directly; carbon dioxide combines

(1) $\underset{\text{Acetyl CoA}}{CH_3CO\text{–}S\text{–}CoA} + CO_2 + ATP \rightleftharpoons \underset{\text{Malonyl CoA}}{HOOCCH_2CO\text{–}S\text{–}CoA} + ADP + \text{phosphate}$

with the prosthetic group of the enzyme—*acetyl CoA carboxylase*—and is then transferred to acetyl CoA.) Just as in the carbonation of a Grignard reagent, the *carbanionoid* character of the α-carbon of acetyl CoA must in some way be involved.

In the remaining steps, acetic and malonic acids react, not as CoA esters, but as thiol esters of *acyl carrier protein* (ACP), a small protein with a prosthetic group quite similar to CoA. These esters are formed by (2) and (3), which we recognize as examples of transesterification.

(2) $CH_3CO\text{–}S\text{–}CoA + ACP\text{–}SH \rightleftharpoons \underset{\text{Acetyl–S–ACP}}{CH_3CO\text{–}S\text{–}ACP} + CoA\text{–}SH$

(3) $HOOCCH_2CO\text{–}S\text{–}CoA + ACP\text{–}SH \rightleftharpoons \underset{\text{Malonyl–S–ACP}}{HOOCCH_2CO\text{–}S\text{–}ACP} + CoA\text{–}SH$

Now starts the first of many similar cycles. Acetyl–S–ACP condenses (4) with malonyl–S–ACP to give a four-carbon chain.

(4) $CH_3CO\text{–}S\text{–}ACP + HOOCCH_2CO\text{–}S\text{–}ACP \rightleftharpoons \underset{\text{Acetoacetyl–S–ACP}}{CH_3COCH_2CO\text{–}S\text{–}ACP} + CO_2 + ACP\text{–}SH$

At this point we see a strong parallel to the malonic ester synthesis (Sec. 26.2). The carbon dioxide taken up in reaction (1) is lost here; its function was to generate malonate, with its highly acidic α-hydrogens, its carbanionoid α-carbon. Here, as in test tube syntheses, the formation of carbon–carbon bonds is all-important; here, as in test tube syntheses (Sec. 26.1), carbanionoid carbon plays a key role. In the malonic ester synthesis, decarboxylation follows the condensation step; here, it seems, the steps are concerted, with loss of carbon dioxide providing driving force for the reaction.

The next steps are exact counterparts of what we would do in the laboratory: reduction to an alcohol (5), dehydration (6), and hydrogenation (7). The reducing agent for both (5) and (7) is reduced nicotinamide adenine dinucleotide phosphate, NADPH (Sec. 36.15).

(5) $CH_3COCH_2CO\text{–}S\text{–}ACP + NADPH + H^+ \rightleftharpoons$

$$\text{D-}CH_3CHOHCH_2CO\text{–}S\text{–}ACP + NADP^+$$
D-β-Hydroxybutyryl–S–ACP

(6) $\text{D-}CH_3CHOHCH_2CO\text{–}S\text{–}ACP \rightleftharpoons$ *trans*-$CH_3CH{=}CHCO\text{–}S\text{–}ACP + H_2O$
Crotonyl–S–ACP

(7) *trans*-$CH_3CH{=}CHCO\text{–}S\text{–}ACP + NADPH + H^+ \rightleftharpoons$

$$CH_3CH_2CH_2CO\text{–}S\text{–}ACP + NADP^+$$
n-Butyryl–S–ACP

We now have a straight-chain saturated fatty acid, and with this the cycle begins again: reaction of it with malonyl–S–ACP, decarboxylation, reduction, dehydration, hydrogenation. After seven such cycles we arrive at the 16-carbon acid, palmitic acid—and here, for some reason, the process stops. Additional carbons can be added, but by a different process. Double bonds can be introduced, to produce unsaturated acids. Finally, glycerol esters are formed: triacylglycerols, to be stored and, when needed, oxidized to provide energy; and phosphoglycerides (Sec. 33.8) to help make up cell walls.

Enzymes are marvelous catalysts. Yet, even with their powerful help, these biological reactions seek the easiest path. In doing this, they take advantage of the same structural effects that the organic chemist does: the acidity of α-hydrogens, the leaving ability of a particular group, the ease of decarboxylation of β-keto acids.

37.7 Nucleoproteins and nucleic acids

In every living cell there are found **nucleoproteins**: substances made up of proteins combined with natural polymers of another kind, the **nucleic acids**. Of all fields of chemistry, the study of the nucleic acids is perhaps the most exciting, for these compounds are the substance of heredity. Let us look very briefly at the structure of nucleic acids and, then, in the next section, see how this structure may be related to their literally vital role in heredity.

Although chemically quite different, nucleic acids resemble proteins in a fundamental way: there is a long chain—a backbone—that is the same (except for length) in all nucleic acid molecules; and attached to this backbone are various groups, which by their nature and sequence characterize each individual nucleic acid.

Where the backbone of the protein molecule is a polyamide chain (a polypeptide chain), the backbone of the nucleic acid molecule is a polyester chain (called a *polynucleotide* chain). The ester is derived from phosphoric acid (the acid portion) and a sugar (the alcohol portion).

```
  base     O     base      O
   |       |      |        |
~sugar—O—P—O—sugar—O—P—O~
           |               |
           O               O
```
Polynucleotide chain

The sugar is D-ribose (p. 1086) in the group of nucleic acids known as ribonucleic acids (RNA), and D-2-deoxyribose in the group known as deoxyribonucleic acids (DNA). (The prefix *2-deoxy* simply indicates the lack of an —OH group at the 2-position.) The sugar units are in the furanose form, and are joined to phosphate through the C–3 and C–5 hydroxyl groups (Fig. 37.4).

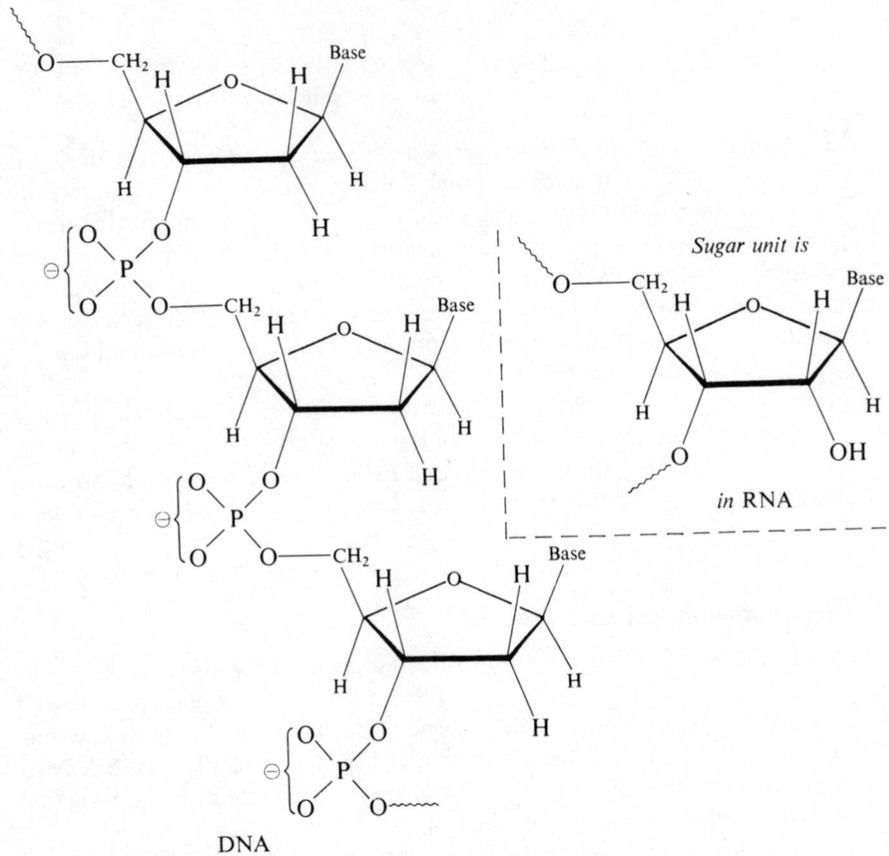

Sugar unit is *in* RNA

DNA

Figure 37.4. Deoxyribonucleic acid (DNA) and ribonucleic acid (RNA).

Attached to C–1 of each sugar, through a β-linkage, is one of a number of heterocyclic bases. A base–sugar unit is called a *nucleoside*; a base–sugar–phosphoric acid unit is called a *nucleotide*. An example of a nucleotide is shown in Fig. 37.5.

The bases found in DNA are *adenine* and *guanine*, which contain the purine ring system, and *cytosine*, *thymine*, and *5-methylcytosine*, which contain the pyrimidine ring system. RNA contains adenine, guanine, cytosine, and *uracil*. (See Fig. 37.6.)

The proportions of these bases and the sequence in which they follow each other along the polynucleotide chain differ from one kind of nucleic acid to another. This primary structure is studied in essentially the same way as the structure of proteins: by hydrolytic degradation and identification of the fragments. In this way, and after *seven years* of work, Robert W. Holley and his collaborators at

Cornell University determined the exact sequence of the 77 nucleotides in the molecule of one kind of transport RNA (p. 1181).

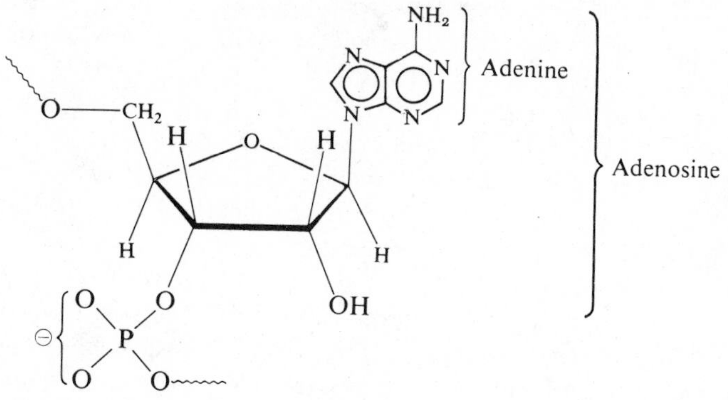

Figure 37.5. A nucleotide: an adenylic acid unit of RNA. Here, the nucleoside is adenosine, and the heterocyclic base is adenine.

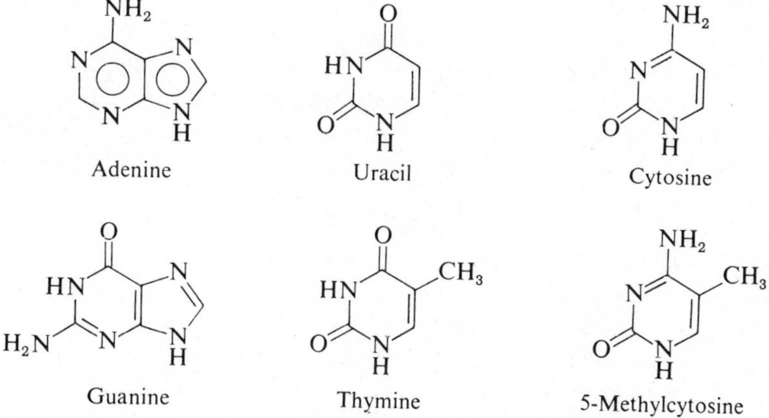

Figure 37.6. The heterocyclic bases of DNA and RNA.

What can we say about the secondary structure of nucleic acids? The following picture of DNA fits both chemical and x-ray evidence. Two polynucleotide chains, identical but heading in opposite directions, are wound about each other to form a double helix 18 A in diameter (shown schematically in Fig. 37.7). Both helixes are right-handed and have ten nucleotide residues per turn.

Figure 37.7. Schematic representation of the double helix structure proposed for DNA. Both helixes are right-handed and head in opposite directions; ten residues per turn. Hydrogen bonding between the helixes.

The two helixes in DNA are held to each other at intervals by hydrogen bonding between bases. From study of molecular models, it is believed that these hydrogen bonds can form only between adenine and thymine and between guanine and cytosine; hydrogen bonding between other pairs of bases would not allow them to fit into the double helical structure. In agreement with this idea, the adenine:thymine and guanine:cytosine ratios are found to be 1:1.

Less is known about structures of the various kinds of RNA, although here, too, helixes are involved. In 1973, the precise shape of one RNA molecule—the transport RNA that delivers phenylalanine—was reported: two short segments of double helix at right angles to each other and held together by two loops, the whole making a sort of four-leaf-clover pattern.

So far we have discussed only the nucleic acid portion of nucleoproteins. There is evidence that in one nucleoprotein (found in fish sperm), a polyarginine chain lies in one of the grooves of the double helix, held by electrostatic forces between the negative phosphate groups of the polynucleotide (which face the outside of the helix) and the positive guanidium groups of the arginine residues.

37.8 Chemistry and heredity. The genetic code

Just how is the structure of nucleic acids related to their function in heredity? Nucleic acids control heredity *on the molecular level*. The double helix of DNA is the repository of the hereditary information of the organism. The information is stored as the sequence of bases along the polynucleotide chain; it is a message "written" in a language that has only four letters, A, G, T, C (adenine, guanine, thymine, cytosine).

DNA must both *preserve* this information and *use* it. It does these things through two properties: (a) DNA molecules can duplicate themselves, that is, can bring about the synthesis of other DNA molecules identical with the originals; and (b) DNA molecules can control the synthesis, in an exact and specific way, of the proteins that are characteristic of each kind of organism.

First, there is the matter of self-duplication. The sequence of bases in one chain of the double helix controls the sequence in the other chain. The two chains fit together (as F. H. C. Crick of Cambridge University puts it) like a hand and a glove. They separate, and about the hand is formed a new glove, and inside the glove is formed a new hand. Thus, the pattern is preserved, to be handed down to the next generation.

Next, there is the matter of guiding the synthesis of proteins. A particular sequence of bases along a polynucleotide chain leads to a particular sequence of amino acid residues along a polypeptide chain. A protein has been likened to a long sentence written in a language of 20 letters: the 20 different amino acid residues. But the hereditary message is written in a language of only four letters; it is written in a *code*, with each word standing for a particular amino acid.

The genetic code has been broken, but research continues, aimed at tracking down the lines of communication. DNA serves as a template on which molecules of RNA are formed. It has been suggested that the double helix of DNA partially uncoils, and about the individual strands are formed chains of RNA; the process thus resembles self-duplication of DNA, except that these new chains contain ribose instead of deoxyribose. The base sequence along the RNA chain is different from that along the DNA template, but is determined *by it*: opposite each adenine of DNA, there appears on RNA a uracil; opposite guanine, cytosine; opposite thymine, adenine; opposite cytosine, guanine. Thus, AATCAGTT on DNA becomes UUAGUCAA on RNA.

One kind of RNA—called, fittingly, *messenger RNA*—carries a message to the ribosome, where protein synthesis actually takes place. At the ribosome, messenger RNA calls up a series of *transport RNA* molecules, each of which is loaded with a particular amino acid. The order in which the transport RNA molecules are called up—the sequence in which the amino acids are built into the protein chain—depends upon the sequence of bases along the messenger RNA chain. Thus, GAU is the code for aspartic acid; UUU, phenylalanine; GUG, valine. There are 64 three-letter code words (*codons*) and only 20-odd amino acids, so that more than one codon can call up the same amino acids: CUU and CUC, leucine; GAA and GAG, glutamic acid.

A difference of a single base in the DNA molecule, or a single error in the "reading" of the code can cause a change in the amino acid sequence. The tiny defect in the hemoglobin molecule that results in sickle-cell anemia (p. 1152) has been traced to a single gene—a segment of the DNA chain—where, perhaps, the codon GUG appears instead of GAG. There is evidence that antibiotics, by altering the ribosome, cause misreading of the code and death to the organism.

Thus, the structure of nucleic acid molecules determines the structure of protein molecules. The structure of protein molecules, we have seen, determines the way in which they control living processes. Biology is becoming more and more a matter of shapes and sizes of molecules.

At the beginning of this book, we said that the structural theory is the basis of the science of organic chemistry. It is much more than that: the structural theory is the basis of our understanding of life.

PROBLEMS

1. Carbon dioxide is required for the conversion of acetyl CoA into fatty acids. Yet when carbon dioxide labeled with ^{14}C is used, none of the labeled carbon appears in the fatty acids that are formed. How do you account for these facts?

2. Taken together, what do these two facts show about chymotrypsin action? (a) The two esters, *p*-nitrophenyl acetate and *p*-nitrophenyl thiolacetate, *p*-$NO_2C_6H_4SCOCH_3$, undergo chymotrypsin-catalyzed hydrolysis at the same rate and with the same pH-dependence of rate, despite the fact that —SR is a much better leaving group than —OR. (b) There is no oxygen exchange (Sec. 20.17) in chymotrypsin-catalyzed hydrolysis of an ester RCOOR′.

3. In DNA, the bases are bonded to deoxyribose at the following positions (that is, a hydrogen in Fig. 37.6, p. 1179, is replaced by C–1 of the sugar): adenine and guanine, NH in the five-membered ring; cytosine and thymine, the lower NH.

(a) Draw structures to show likely hydrogen bonding between adenine and thymine; between guanine and cytosine. (b) Can you account for the fact that guanine and cytosine pairs hold the chains together more strongly than do adenine and thymine pairs?

4. For each enzyme-catalyzed reaction shown in the following equations, tell what fundamental organic chemistry is involved.

(a) So that acetyl CoA can get through the membrane from the mitochondria where it is formed to the cytoplasm where fatty acids are made, it is converted into citric acid.

$$\underset{}{CH_3CO{-}S{-}CoA} + \underset{\text{Oxaloacetic acid}}{HOOCCOCH_2COOH} \rightleftharpoons \underset{\text{Citric acid}}{HOOCCH_2\overset{\overset{OH}{|}}{\underset{\underset{COOH}{|}}{C}}CH_2COOH} + CoA{-}SH$$

(b) Cholesterol is made up of isoprene units derived from isopentenyl pyrophosphate (Sec. 8.26), which is, in turn, formed from mevalonic acid.

$$CH_3CO{-}S{-}CoA + CH_3COCH_2CO{-}S{-}CoA \rightleftharpoons HOOCCH_2\overset{\overset{CH_3}{|}}{\underset{\underset{OH}{|}}{C}}CH_2CO{-}S{-}CoA + CoA{-}SH$$

$$HOOCCH_2\overset{\overset{CH_3}{|}}{\underset{\underset{OH}{|}}{C}}CH_2CO{-}S{-}CoA + 2NADPH + 2H^+ \rightleftharpoons \underset{\text{Mevalonic acid}}{HOOCCH_2\overset{\overset{CH_3}{|}}{\underset{\underset{OH}{|}}{C}}CH_2CH_2OH} + 2NADP^+ + CoA{-}SH$$

5. Three of the bases found in nucleic acids are *uracil*, *thymine*, and *cytosine*. (See p. 1179 for their structures.) They have been synthesized as follows:

(a) urea + ethyl acrylate $\xrightarrow{\text{Michael}}$ [A ($C_6H_{12}O_3N_2$)] $\longrightarrow$ B ($C_4H_6O_2N_2$) + C_2H_5OH

B + Br_2 in acetic acid $\longrightarrow$ C ($C_4H_5O_2N_2Br$)
C + boiling pyridine $\longrightarrow$ uracil ($C_4H_4O_2N_2$)

Give structures of A, B, and C.

(b) Thymine ($C_5H_6O_2N_2$) has been made in the same way, except that ethyl methacrylate, $CH_2{=}C(CH_3)COOC_2H_5$, is used instead of ethyl acrylate. Write equations for all the steps.

(c) uracil + POCl, heat $\longrightarrow$ D ($C_4H_2N_2Cl_2$), chlorine atoms on different carbon atoms

D + NH_3(alc), 100° $\longrightarrow$ E ($C_4H_4N_3Cl$) and F ($C_4H_4N_3Cl$)
E + $NaOCH_3$ $\longrightarrow$ G ($C_5H_7ON_3$)
G + HCl(aq) $\longrightarrow$ cytosine ($C_4H_5ON_3$)

Give structures of D through G.

(d) Six tautomeric structures for uracil have been considered. What are they?

6. In 1904, Franz Knoop outlined a scheme for the biological oxidation of fatty acids that was shown—50 years later—to be correct. In his key experiments, he fed rabbits fatty acids of formula $C_6H_5(CH_2)_nCOOH$. When the side chain ($n + 1$) contained an even number of carbons, a derivative of phenylacetic acid, $C_6H_5CH_2COOH$, was excreted in the urine; an odd number, and a derivative of benzoic acid was excreted. What general hypothesis can you formulate from these results?

7. In the actual cleavage reaction of glycolysis, D-fructose-1,6-diphosphate is converted into D-glyceraldehyde-3-phosphate and dihydroxyacetone, $CH_2OHCOCH_2OH$. What kind of reaction is this, basically? Sketch out a possible mechanism, neglecting, of course, the all-important role of the enzyme. (*Hints:* The enzyme required is called *aldolase*. See Problem 21.14, p. 711.)

8. Interpret each of the following facts.

(a) In the presence of the proper enzyme, nicotinamide adenine dinucleotide (Sec. 36.15) can oxidize ethanol reversibly to acetaldehyde. When D_2O is the solvent, the re-

$$CH_3CH_2OH + NAD^+ \rightleftharpoons CH_3CHO + NADH + H^+$$

duced NAD formed (NADH) contains no deuterium. When CH_3CD_2OH is oxidized, the reduced NAD formed (NADD) contains one atom of deuterium per molecule.

(b) Enzymatic reoxidation by acetaldehyde of the NADD of part (a) gives NAD^+ that contains no deuterium.

(c) If the NADD of part (a) is oxidized enzymatically by D-glucose, all of the original deuterium remains in the NAD^+.

(d) NADD can also be prepared non-enzymatically by chemical reduction ($Na_2S_2O_4$ in D_2O) of NAD^+. This, too, contains one atom of deuterium per molecule. When it is oxidized enzymatically by acetaldehyde, the NAD^+ formed still contains 0.44 atom of deuterium per molecule.

(e) Acetaldehyde is reduced enzymatically by the NADD of part (a) to give ethanol X. Labeled acetaldehyde, CH_3CDO, is reduced enzymatically by NADH to give ethanol Y. Both X and Y contain one atom of deuterium per molecule. On enzymatic oxidation by (unlabeled) NAD^+, ethanol X gives NADD and unlabeled CH_3CHO, whereas ethanol Y gives unlabeled NADH and CH_3CDO.

Suggested Readings

General

G. W. Wheland, *Advanced Organic Chemistry*, 3rd ed., Wiley, New York, 1960.

J. Hine, *Physical Organic Chemistry*, 2nd ed., McGraw-Hill, New York, 1962.

C. K. Ingold, *Structure and Mechanism in Organic Chemistry*, 2nd ed., Cornell University Press, Ithaca, 1969.

J. March, *Advanced Organic Chemistry*, McGraw-Hill, New York, 1968.

L. P. Hammett, *Physical Organic Chemistry*, 2nd ed., McGraw-Hill, New York, 1970.

E. S. Gould, *Mechanism and Structure in Organic Chemistry*, Holt, New York, 1959.

P. Sykes, *A Guidebook to Mechanism in Organic Chemistry*, 3rd ed., Wiley, New York, 1970.

R. Breslow, *Organic Reaction Mechanisms*, W. A. Benjamin, New York, 1965.

H. O. House, *Modern Synthetic Reactions*, W. A. Benjamin, New York, 1965.

V. Gold., ed., *Advances in Physical Organic Chemistry*, Academic Press, New York; a series starting in 1963.

A. Streitwieser and R. W. Taft, ed., *Progress in Physical Organic Chemistry*, Wiley, New York; a series starting in 1963.

S. Patai, ed., *The Chemistry of Functional Groups*, Wiley, New York; a series starting in 1964.

E. H. Rodd, ed., *Chemistry of Carbon Compounds*, Elsevier, New York; a series starting in 1951, 2nd ed. starting in 1964.

Organic Reactions, Wiley, New York; a series starting in 1942. Each chapter discusses one reaction ("The Clemmensen Reduction," "Periodic Acid Oxidation," etc.) with particular emphasis on its application to synthesis.

Note: Some of the above books will be referred to later by abbreviated names, e.g., O. R. III-2 for *Organic Reactions*, Vol. III, Ch. 2.

Molecular structure and intermolecular forces

G. W. Wheland, *Adv. Org. Chem.*, Ch. 1, 3.

L. N. Ferguson, *The Modern Structural Theory of Organic Chemistry*, Prentice-Hall, Englewood Cliffs, N. J., 1963.

C. K. Ingold, *Struct. and Mech.*, Ch. I, II, IV.

L. Pauling, *The Nature of the Chemical Bond*, 3rd ed., Cornell University Press, Ithaca, 1960

G. W. Wheland, *Resonance in Organic Chemistry*, Wiley, New York, 1955.

C. A. Coulson, "The Meaning of Resonance in Quantum Chemistry," Endeavour, **6**, 42 (1947).

M. J. S. Dewar, *Hyperconjugation*, Ronald Press, New York, 1962.

P. E. Verkade, "August Kekulé," Proc. Chem. Soc., 205 (1958).

W. Baker, "The Widening Outlook in Aromatic Chemistry," Chemistry in Britain, **1**, 191, 250 (1965).

R. Breslow, "Aromatic Character," Chem. Eng. News, June 28, 1965, p. 90.

G. W. A. Fowles, "Lone Pair Electrons," J. Chem. Educ., **34**, 187 (1957).

E. Cartmell and G. W. A. Fowles, *Valency and Molecular Structure*, 3rd ed., Butterworths, London, 1966.

M. Orchin and H. H. Jaffé, *The Importance of Antibonding Orbitals*, Houghton-Mifflin, Boston, 1967.

R. B. Woodward and R. Hoffmann, *The Conservation of Orbital Symmetry*, Academic Press, New York, 1970.

J. J. Vollmer and K. L. Service, "Woodward-Hoffmann Rules: Electrocyclic Reactions," J. Chem. Educ., **45**, 214 (1968): "Woodward-Hoffmann Rules: Cycloaddition Reactions," J. Chem. Educ., **47**, 491 (1970).

R. G. Pearson, "Molecular Orbital Symmetry Rules," Chem. Eng. News, Sept. 28, 1970, p. 66.

Isomerism and stereochemistry

E. L. Eliel, *Stereochemistry of Carbon Compounds*, McGraw-Hill, New York, 1962.

E. L. Eliel, *Elements of Stereochemistry*, Wiley, New York, 1969.

G. W. Wheland, *Adv. Org. Chem.*, Ch. 2, 6–9.

K. Mislow, *Introduction to Stereochemistry*, W. A. Benjamin, New York, 1965.

E. L. Eliel, N. L. Allinger, S. J. Angyal, and G. A. Morrison, *Conformational Analysis*, Interscience-Wiley, New York, 1965.

R. S. Cahn, "An Introduction to the Sequence Rule," J. Chem. Educ., **41**, 116 (1964).

E. L. Eliel, "Recent Advances in Stereochemical Nomenclature," J. Chem. Educ., **48**, 163 (1971).

D. F. Mowery, Jr., "The Cause of Optical Inactivity," J. Chem. Educ., **29**, 138 (1952).

J. M. Bijvoet, "Determination of the Absolute Configuration of Optical Antipodes," Endeavour, **14**, 71 (1955).

M. L. Wolfrom, "Optical Activity and Configurational Relations in Carbon Compounds," Rec. Chem. Progr. (Kresge-Hooker Sci. Lib.), **16**, 121 (1955).

Acids and bases

G. N. Lewis, "Acids and Bases," J. Franklin Inst., **226**, 293 (1938).

G. W. Wheland, *Adv. Org. Chem.*, Ch. 5.

C. A. VanderWerf, *Acids, Bases, and the Chemistry of the Covalent Bond*, Reinhold, New York, 1961.

R. P. Bell, *The Proton in Chemistry*, Cornell University Press, Ithaca, 1959.

J. Hine, *Phys. Org. Chem.*, Ch. 2, "Acids and Bases."

C. K. Ingold, *Struct. and Mech.*, Ch. XIV.

Nomenclature and pronunciation

A. M. Patterson, L. T. Capell, and M. A. Magill, "Nomenclature of Organic Compounds," Chem. Abs., **39**, 5875–5950 (1945).

E. J. Crane, "The Pronunciation of Chemical Words," Ind. Eng. Chem., News Ed., **12**, 202 (1934).

Free radicals

M. Gomberg, "An Instance of Trivalent Carbon: Triphenylmethyl," J. Am. Chem. Soc., **22**, 757 (1900).
G. W. Wheland, *Adv. Org. Chem.*, Ch. 15.
J. Hine, *Phys. Org. Chem.*, Ch. 18–23.
W. A. Pryor, *Free Radicals*, McGraw-Hill, New York, 1965.
W. A. Pryor, *Introduction to Free Radical Chemistry*, Prentice-Hall, Englewood Cliffs, N. J., 1965.
E. S. Huyser, *Free-Radical Chain Reactions*, Wiley, New York, 1970.
C. Walling, *Free Radicals in Solution*, Wiley, New York, 1957.
M. C. R. Symons, "The Identification of Organic Free Radicals by Electron Spin Resonance," Vol. 3, p. 284, of *Adv. in Phys. Org. Chem.*

Carbonium ions

F. C. Whitmore, "Alkylation and Related Processes of Modern Petroleum Practice," Chem. Eng. News, **26**, 668 (1948).
C. K. Ingold, *Struct. and Mech.*, Ch. VII.
D. Bethell and V. Gold, "The Structure of Carbonium Ions," Quart. Revs. (London), **12**, 173 (1958).
P. D. Bartlett, *Nonclassical Ions*, W. A. Benjamin, New York, 1965.
H. C. Brown, "The Norbornyl Cation—Classical or Non-classical?", Chemistry in Britain, **2**, 199 (1966).
G. A. Olah and P. v. R. Schleyer, ed., *Carbonium Ions*, Wiley, New York; in four volumes; Vol. I, 1968.

Carbanions and tautomerism

J. Hine, *Phys. Org. Chem.*, Ch. 10.
C. K. Ingold, *Struct. and Mech.*, Ch. X.
G. W. Wheland, *Adv. Org. Chem.*, Ch. 14.
D. J. Cram, *Fundamentals of Carbanion Chemistry*, Academic Press, New York, 1965.
D. C. Ayres, *Carbanions in Synthesis*, Elsevier, New York, 1966.

Nucleophilic aliphatic substitution

J. Hine, *Phys. Org. Chem.*, Ch. 6–7.
C. K. Ingold, *Struct. and Mech.*, Ch. 7.
J. March, *Adv. Org. Chem.*, Ch. 10.
C. A. Bunton, *Nucleophilic Substitution at a Saturated Carbon Atom*, Elsevier, New York, 1963.
A. Streitwieser, Jr., *Solvolytic Displacement Reactions*, McGraw-Hill, New York, 1962.
E. R. Thornton, *Solvolysis Mechanisms*, Ronald Press, New York, 1964.
H. C. Brown, C. J. Kim, C. J. Lancelot, and P. v. R. Schleyer, "Product-Rate Correlation in Acetolysis of *threo*-3-Aryl-2-butyl Brosylates. Supporting Evidence for the Existence of Two Discrete Pathways," J. Am. Chem. Soc., **92**, 5244 (1970).

Electrophilic aromatic substitution

J. Hine, *Phys. Org. Chem.*, Ch. 16.
C. K. Ingold, *Struct. and Mech.*, Ch. VI.
J. March, *Adv. Org. Chem.*, Ch. 11.
R. O. C. Norman and R. Taylor, *Electrophilic Substitution in Benzenoid Compounds*, Elsevier, New York, 1965.

R. M. Roberts, "Friedel-Crafts Chemistry," Chem. Eng. News, Jan. 25, 1965, p. 96.
G. A. Olah, ed., *Friedel-Crafts and Related Reactions*, Wiley, New York, four volumes; Vol. I, 1963.

Nucleophilic aromatic substitution

J. Hine, *Phys. Org. Chem.*, Ch. 17.
J. March, *Adv. Org. Chem.*, Ch. 13.
J. F. Bunnett, "Mechanism and Reactivity in Aromatic Nucleophilic Substitution Reactions," Quart. Revs. (London), **12**, 1 (1958).
J. F. Bunnett, "The Chemistry of Benzyne," J. Chem. Educ., **38**, 278 (1961).
J. Miller, *Aromatic Nucleophilic Substitution*, Elsevier, New York, 1968.
R. W. Hoffmann, *Dehydrobenzene and Cycloalkynes*, Academic Press, New York, 1967.
F. Pietra, "Mechanisms for Nucleophilic and Photonucleophilic Aromatic Substitution," Quart. Revs. (London), **23**, 504 (1969).
J. F. Bunnett, "The Base Catalyzed Halogen Dance, and Other Reactions of Aryl Halides," Accounts Chem. Res., **5**, 139 (1972).

Addition to carbon–carbon multiple bonds

J. Hine, *Phys. Org. Chem.*, Ch. 9, "Polar Addition to Carbon–Carbon Multiple Bonds."
C. K. Ingold, *Struct. and Mech.*, Ch. XIII, "Additions and Their Retrogression."
J. March, *Adv. Org. Chem.*, Ch. 15.
P. B. D. de la Mare and R. Bolton, *Electrophilic Additions to Unsaturated Systems*, Elsevier, New York, 1966.
O. R. XIII-3, C. Walling, and E. S. Huyser, "Free Radical Additions to Olefins and Acetylenes to Form Carbon–Carbon Bonds"; XIII-1, G. Zweifel and H. C. Brown, "Hydration of Olefins, Dienes, and Acetylenes via Hydroboration"; XIII-2, W. E. Parham and E. E. Schweizer, "Halocyclopropanes from Halocarbenes."
H. C. Brown, *Hydroboration*, W. A. Benjamin, New York, 1962.

Elimination

J. Hine, *Phys. Org. Chem.*, Ch. 8, "Polar Elimination Reactions."
C. K. Ingold, *Struct. and Mech.*, Ch. IX, "Olefin-forming Eliminations."
J. March, *Adv. Org. Chem.*, Ch. 17.
D. V. Banthorpe, *Elimination Reactions*, Elsevier, New York, 1963.

Oxidation

H. O. House, *Mod. Syn. Reactions*, Ch. 4.
R. Stewart, *Oxidation Mechanisms*, W. A. Benjamin, New York, 1964.
K. B. Wiberg, ed., *Oxidation in Organic Chemistry*, Academic Press, New York; Part A, 1965.
J. March, *Adv. Org. Chem.*, Ch. 19.
O. R. VII-7, D. Swern, "Epoxidation and Hydroxylation of Ethylenic Compounds with Organic Peracids"; II-8, E. L. Jackson, "Periodic Acid Oxidation."

Reduction and hydrogenation

H. O. House, *Mod. Syn. Reactions*, Ch. 1–3.
J. March, *Adv. Org. Chem.*, Ch. 19.
O. R. I-7, E. L. Martin, "The Clemmensen Reduction"; IV-8, D. Todd, "The Wolff-Kishner Reduction"; II-5, A. L. Wilds, "Reduction with Aluminum Alkoxides

(the Meerwein-Ponndorf-Verley Reduction)"; VI-10, W. G. Brown, "Reductions by Lithium Aluminum Hydride."
A. J. Birch, "Reduction of Organic Compounds," Quart. Revs. (London), **4**, 69 (1950).
A. J. Birch and H. Smith, "Reduction by Metal-Amine Solutions," Quart. Revs. (London), **12**, 17 (1958).
G. C. Bond, "Mechanism of Catalytic Hydrogenation and Related Reactions," Quart. Revs. (London), **8**, 279 (1954).

Rearrangements

F. C. Whitmore, "The Common Basis of Intramolecular Rearrangements," J. Am. Chem. Soc., **54**, 3274 (1932).
G. W. Wheland, *Adv. Org. Chem.*, Ch. 12–13.
J. Hine, *Phys. Org. Chem.*, Ch. 14, 15, 23.
C. K. Ingold, *Struct. and Mech.*, Ch. X–XII.
J. March, *Adv. Org. Chem.*, Ch. 18.
P. de Mayo, ed., *Molecular Rearrangements*, Interscience, New York, 1963.
O. R. III-7, E. S. Wallis and J. F. Lane, "The Hofmann Reaction"; III-9, P. A. S. Smith, "The Curtius Reaction"; III-8, H. Wolff, "The Schmidt Reaction."
C. J. Collins, "The Pinacol Rearrangement," Quart. Revs. (London), **14**, 357 (1960).
C. J. Collins, "Reactions of Primary Aliphatic Amines with Nitrous Acid," Accounts Chem. Res., **4**, 315 (1971).

Acyl compounds

C. K. Ingold, *Struct. and Mech.*, Ch. XXV.
J. Hine, *Phys. Org. Chem.*, Ch. 12–13.
H. O. House, *Mod. Syn. Reactions*, Ch. 7, "The Alkylation of Active Methylene Compounds."
O. R. I-9, C. R. Hauser and B. E. Hudson, Jr., "The Acetoacetic Ester Condensation and Related Reactions"; IV-4, S. M. McElvain, "The Acyloins"; II-4, W. S. Johnson, "The Formation of Cyclic Ketones by Intramolecular Acylation"; VIII-2, D. A. Shirley, "The Synthesis of Ketones from Acid Chlorides and Organometallic Compounds of Magnesium, Zinc, and Cadmium."
D. P. N. Satchell, "An Outline of Acylation," Quart. Revs. (London), **17**, 160 (1963).
A. G. Davies and J. Kenyon, "Alkyl-Oxygen Heterolysis in Carboxylic Esters and Related Compounds," Quart. Revs. (London), **9**, 203, (1955).
M. L. Bender, "Mechanisms of Catalysis of Nucleophilic Reactions of Carboxylic Acid Derivatives," Chem. Revs., **60**, 53 (1960).

Carbonyl compounds

C. K. Ingold, *Struct. and Mech.*, pp. 994–1015.
J. Hine, *Phys. Org. Chem.*, Ch. 11.
H. O. House, *Mod. Syn. Reactions*, Ch. 8, "The Aldol Condensation and Related Reactions."
O. R. XVIII-1, M. J. Jorgenson, "Preparation of Ketones from the Reaction of Organolithium Reagents with Carboxylic Acids"; XVI, A. J. Nielsen and W. J. Houlihan, "The Aldol Condensation"; IV-5, W. S. Ide and J. S. Buck, "The Synthesis of Benzoins"; II-3, T. A. Geissman, "The Cannizzaro Reaction"; V-6, N. N. Crounse, "The Gattermann-Koch Reaction"; I-10, F. F. Blicke, "The Mannich Reaction."

α,β-Unsaturated carbonyl compounds

C. K. Ingold, *Struct. and Mech.*, pp. 1015–1037.
R. C. Fuson, *Reactions of Organic Compounds*, Wiley, 1962. Ch. 17, "Nucleophilic Addition Reactions of Unsaturated Compounds."

O. R. IV-1, M. C. Kloetzel, "The Diels-Alder Reaction with Maleic Anhydride"; V-2, H. A. Bruson, "Cyanoethylation"; X-3, E. D. Bergmann, D. Ginsburg and E. Pappo, "The Michael Reaction."

Nitrogen compounds

H. Zollinger, *Azo and Diazo Chemistry*, Interscience, New York, 1961.
J. H. Ridd, "Nitrosation, Diazotisation, and Deamination," Quart Revs. (London), **15**, 418 (1961).
L. A. Paquette, *Principles of Modern Heterocyclic Chemistry*, W. A. Benjamin, New York, 1968.
A. G. Cook, *Enamines: Synthesis, Structure, and Reactions*, Dekker, New York, 1969.
E. Adams, "Barbiturates," Sci. American, Jan. 1958, p. 60.

Polymers and polymerization

L. R. G. Treloar, *Introduction to Polymer Science*, Springer-Verlag, New York, 1970.
R. B. Seymour, *Introduction to Polymer Chemistry*, McGraw-Hill, New York, 1971.
C. E. H. Bawn, "New Kinds of Macromolecules," Endeavour, **15**, 137 (1956).
G. Natta, "How Giant Molecules Are Made," Sci. American, Sept. 1957, p. 98.
G. Natta, "Precisely Constructed Polymers," Sci. American, Aug. 1961, p. 33.
C. E. H. Bawn and A. Ledwith, "Stereoregular Addition Polymerization," Quart. Revs. (London), **16**, 361 (1962).
M. Szwarc, *Carbanions, Living Polymers and Electron Transfer Processes*, Wiley, New York, 1968.

Natural products

C. S. Hudson, "Emil Fischer's Discovery of the Configuration of Glucose," J. Chem. Educ., **18**, 353 (1941).
R. D. Guthrie and J. Honeyman, *An Introduction to the Chemistry of Carbohydrates*, 3rd ed., Clarendon Press, Oxford, 1968.
I. L. Finar, *Organic Chemistry*, Longmans, Green, New York, Vol. II, 2nd ed. 1959.
L. F. Fieser and M. Fieser, *Steroids*, Reinhold, New York, 1959.
W. Klyne, *The Chemistry of the Steroids*, Wiley, New York, 1957.
J. Simonsen, *The Terpenes*, Cambridge University Press. Vols. I–III, 2nd ed., 1947. Vols. IV–V, with W. C. J. Ross, 1957.
J. B. Hendrickson, *The Molecules of Nature*, W. A. Benjamin, New York, 1965.

Amino acids and proteins

R. E. Dickerson and I. Geis, *The Structure and Action of Proteins*, Harper and Row, New York, 1969.
K. D. Kopple, *Peptides and Amino Acids*, W. A. Benjamin, New York, 1966.
Bio-organic Chemistry: Readings from Scientific American, Freeman, San Francisco, 1968, Sect. I.
L. Pauling, "The Configuration of Polypeptide Chains in Proteins," Rec. Chem. Progr. (Kresge-Hooker Sci. Lib.), **12**, 155 (1951).
P. Doty, "Proteins," Sci. American, Sept. 1957, p. 173.
L. Pauling, R. B. Corey, and R. Hayward, "Structure of Protein Molecules," Sci. American, Oct. 1954, p. 54.
F. Sanger and L. F. Smith, "The Structure of Insulin," Endeavour, **16**, 48 (1957).
F. Sanger, "The Chemistry of Insulin (Nobel lecture)," Chemistry and Industry, 104 (1959).
C. H. Li, "The ACTH Molecule," Sci. American, July 1963, p. 46.
A. L. Lehninger, *Biochemistry*, Worth, New York, 1970.

Chemistry of biological processes

F. Wold, *Macromolecules: Structure and Function*, Prentice-Hall, Englewood Cliffs, New Jersey, 1971.

Bio-organic Chemistry: Readings from Scientific American, Freeman, San Francisco, Calif., 1968, Sects. II and III.

The Molecular Basis of Life: Readings from Scientific American, Freeman, San Francisco, Calif., 1968.

J. D. Watson, *Molecular Biology of the Gene*, W. A. Benjamin, New York, 1965.

R. B. Clayton, "Biosynthesis of Sterols, Steroids, and Terpenoids," Quart. Revs. (London), **19**, 168, 201 (1965).

A. L. Lehninger, *Bioenergetics*, W. A. Benjamin, New York, 1965.

W. P. Jencks, *Catalysis in Chemistry and Enzymology*, McGraw-Hill, New York, 1969.

T. C. Bruice and S. Bentkovic, *Bioorganic Mechanisms*, 2 vols., W. A. Benjamin, New York, 1966.

Use of isotopes

D. A. Semenow and J. D. Roberts, "Uses of Isotopes in Organic Chemistry," J. Chem. Educ., **33**, 2 (1956).

J. G. Burr, *Tracer Applications for the Study of Organic Reactions*, Interscience, New York, 1957.

C. J. Collins, "Isotopes and Organic Reaction Mechanisms," Vol. 2, p. 3, in *Adv. in Phys. Org. Chem.*

H. Zollinger, "Hydrogen Isotope Effects in Aromatic Substitution Reactions," Vol. 2, p. 163, in *Adv. in Phys. Org. Chem.*

L. Melander, *Isotope Effects on Reaction Rates*, Ronald Press, New York, 1960.

J. A. Bassham, A. A. Benson, and M. Calvin, "Isotope Studies in Photosynthesis," J. Chem. Educ., **30**, 274 (1953).

J. A. Bassham, "The Path of Carbon in Photosynthesis," Sci. American, June 1962, p. 88.

W. F. Libby, *Radiocarbon Dating*, 2nd ed., University of Chicago Press, Chicago, 1955.

Analysis

R. L. Shriner, R. C. Fuson, and D. Y. Curtin, *Systematic Identification of Organic Compounds*, 5th ed., Wiley, New York, 1964.

N. D. Cheronis and J. B. Entrikin, *Semimicro Qualitative Organic Analysis*, 3rd ed., Interscience-Wiley, New York, 1963.

R. M. Silverstein and G. C. Bassler, *Spectrometric Identification of Organic Compounds*, 2nd ed., Wiley, New York, 1967.

J. R. Dyer, *Applications of Absorption Spectroscopy of Organic Compounds*, Prentice-Hall, Englewood Cliffs, N. J., 1965.

R. T. Conley, *Infrared Spectroscopy*, 2nd ed., Allyn and Bacon, Boston, 1972.

K. Nakanishi, *Infrared Absorption Spectroscopy*, Holden-Day, San Francisco, 1964.

F. A. Bovey, "Nuclear Magnetic Resonance," Chem. Eng. News, Aug. 30, 1965, p. 98.

J. D. Roberts, *Nuclear Magnetic Resonance*, McGraw-Hill, New York, 1959.

L. M. Jackman, *Applications of NMR Spectroscopy in Organic Chemistry*, 2nd ed., Pergamon, New York, 1969.

R. H. Bible, Jr., *Interpretation of NMR Spectra, An Empirical Approach*, Plenum Press, New York, 1965.

H. Budzikiewicz, C. Djerassi, and D. H. Williams, *Interpretation of Mass Spectra of Organic Compounds*, Holden-Day, San Francisco, 1964.

K. Biemann, *Mass Spectra and Organic Chemical Applications*, McGraw-Hill, New York, 1962.

Answers to Problems

Chapter 1

1.1 Ionic: a, e, f. **1.6** Linear. **1.7** (a) Expect zero; (b) expect $NF_3 > NH_3$. **1.8** a, e, f. **1.9** (a) $CH_3OH > CH_3NH_2$; (b) $CH_3SH > CH_3OH$; (c) $H_3O^+ > NH_4^+$. **1.10** (a) H_3O^+; (b) NH_4^+; (c) H_2S; (d) H_2O. **1.11** (a) $CH_3^- > NH_2^- > OH^- > F^-$; (b) $NH_3 > H_2O > HF$; (c) $SH^- > Cl^-$; (d) $F^- > Cl^- > Br^- > I^-$; (e) $OH^- > SH^- > SeH^-$. **1.12** $CH_3NH_2 > CH_3OH > CH_3F$. **1.13** (a) $OH^- > H_2O > H_3O^+$; (b) $NH_2^- > NH_3$; (c) $S^= > HS^- > H_2S$. **1.14** $NH_3 > NF_3$.

1. Ionic: a, d, e, g. **4.** Octahedral. **10.** (a) H_3O^+; (b) HCl; (c) HCl in benzene.

Chapter 2

2.1 (a) -8 kcal; (b) $+13$ kcal; (c) -102 kcal. **2.2** (a) $+46$, $+16$, -24 kcal; (b) $+36$, $+33$, -20 kcal; (c) $+38$, -32, -70 kcal. **2.5** (a) (%C + %H) < 100%; (b) 34.8.% **2.6** (a) 69.6% Cl; (b) 70.4% Cl; (c) 24.85 mg; (d) 26.49 mg; (e) 27.44 mg. **2.7** (a) CH_3; (b) $C_3H_6Cl_2$. **2.8** C_6H_6. **2.9** $C_4H_8O_2$.

1. A, 93.9% C, 6.3% H; B, 64.0% C, 4.5% H, 31.4% Cl; C, 62.0% C, 10.3% H, 27.7% O. **2.** (a) 45.9% C, 8.9% H, 45.2% Cl; (b) 52.1% C, 13.1% H, 34.8% O; (c) 54.5% C, 9.1% H, 36.3% O; (d) 41.8% C, 4.7% H, 18.6% O, 16.3% N, 18.6% S; (e) 20.0% C, 6.7% H, 26.6% O, 46.7% N; (f) 55.6% C, 6.2% H, 10.8% O, 27.4% Cl. **3.** (a) CH_2; (b) CH; (c) CH_2O; (d) C_2H_5OCl; (e) $C_3H_{10}N_2$; (f) $C_3H_4O_2Cl_2$. **4.** $C_{20}H_{21}O_4N$. **5.** $C_{14}H_{14}O_3N_3SNa$. **6.** (a) 85.8% C, 14.3% H; (b) CH_2; (c) C_6H_{12}. **7.** $C_2H_4O_2$. **8.** CH_2O. **9.** $C_{16}H_{10}O_2N_2$. **10.** (a) 942; (b) 6. **11.** (a) -130; (b) -44; (c) -26; (d) -2; (e) -13; (f) -8; (g) -1; (h) 1st step $+46$; 2nd steps $+10$, -3, 0; 3rd steps -23, -5, -1. **12.** $+58$, $+20$, -45; (b) E_{act} of a chain-carrying step ≥ 20 kcal. **13.** (b) Highly improbable, since E_{act} for reaction with Cl_2 is much smaller.

Chapter 3

3.2 No. **3.3** Van der Waals repulsion between "large" methyls. **3.9** (a) and (b) C_3H_8; (c) $CH_3CH_2CH_2D$ and CH_3CHDCH_3. **3.10** (a) 3; (b) 4; (c) 2; (d) 1. **3.14** (a)

44% 1-Cl, 56% 2-Cl; (b) 64% 1°, 36% 3°; (c) 55% 1°, 45% 3°; (d) 21% 1-Cl, 53% 2-Cl, 26% 3-Cl; (e) 28% 1-Cl-2-Me, 23% 2-Cl-2-Me, 35% 3-Cl-2-Me, 14% 1-Cl-3-Me; (f) 45% 1-Cl-2,2,3-triMe, 25% 3-Cl-2,2,3-triMe, 30% 1-Cl-2,3,3-tri-Me; (g) 33% 1-Cl-2,2,4-triMe, 28% 3-Cl-2,2,4-triMe, 18% 4-Cl-2,2,4-triMe, 22% 1-Cl-2,4,4-triMe. **3.15** (a) 4% 1-Br, 96% 2-Br; (b) 0.6% 1°, 99.4% 3°; (c) 0.3% 1°, 99.7% 3°; (d) 1% 1-Br, 66% 2-Br, 33% 3-Br; (e) 0.3% 1-Br-2-Me, 90% 2-Br-2-Me, 9% 3-Br-2-Me, 0.2% 1-Br-3-Me; (f) 0.6% 1-Br-2,2,3-triMe, 99% 3-Br-2,2,3-triMe, 0.4% 1-Br-2,3,3-triMe; (g) 0.5% 1-Br-2,2,4-triMe, 9% 3-Br-2,2,4-triMe, 90% 4-Br-2,2,4-triMe, 0.3% 1-Br-2,4,4-triMe. **3.16** 40:1. **3.17** 1.15:1. **3.22** 2,2-Dimethylhexane.

5. (e) 6. **6.** One monochloro, three dichloro, four trichloro. **7.** c, b, e, a, d. **10.** (a) 1-, 2-, and 3-chlorohexane; (b) 1-, 2-, 3-, and 4-chloro-2-methylpentane, and 1-chloro-4-methylpentane; (c) 1-, 3- and 4-chloro-2,2,4-trimethylpentane, and 1-chloro-2,4,4-trimethylpentane; (d) 1- and 3-chloro-2,2-dimethylbutane, and 1-chloro-3,3-dimethylbutane. **11.** Order of isomers as in Problem 10: (a) 16, 42, 42%; (b) 21, 17, 26, 26, 10%; (c) 33, 28, 18, 22%; (d) 46, 39, 15%. **16.** (a) 2650 g; (b) 8710 kcal; (c) 170 g. **17.** Carius: mono, 45.3% Cl; di, 62.8% Cl. Mol.wt.: mono, 78.5; di, 113. **19.** (a) Methane gas; 1.49 mg CH_3OH; (b) 59, *n*-propyl or isopropyl alcohol; (c) 3; $CH_2OHCHOHCH_2OH$.

Chapter 4

4.1 2 (mirror images). **4.2** (a) 3; (b) 2; (c) 3 (2 are mirror images); (d) 1. **4.3** (a) −39.0°; (b) −2.4°; (c) −0.6°. **4.4** Use a shorter or longer tube, measure rotation. **4.5** Chiral; b, d, f, g, h. **4.6** (b) 3 of 5 are chiral. **4.7** (d) Mirror images: a, b. **4.9** 3°, 2°, 1°, Me. **4.15** (b) Neither active: one is achiral, other is a racemic modification.

3. Equal but opposite specific rotations; opposite R/S specifications: all other properties the same. **4.** (a) Screw, scissors, spool of thread; (b) glove, shoe, coat sweater, tied scarf; (c) helix, double helix; (d) football (laced), golf club, rifle barrel; (e) hand, foot, ear, nose, yourself. **5.** (a) Sawing; (b) opening milk bottle; (c) throwing a ball. **7.** (a) and (b) 3-Methylhexane and 2,3-dimethylpentane. **8.** a, b, e, k, 2 pairs enantiomers; c, d, h, 1 pair enantiomers + 1 *meso*; f, 4 pairs enantiomers; g, 1 pair enantiomers + 2 *meso*; i, 2 diastereomers; j, 1 pair enantiomers. **11.** Attractive dipole–dipole interaction. **12.** 12% *gauche* (as non-resolvable racemic modification), 88% *anti.*

Chapter 5

5.5 (g) None. **5.7** (g) None. **5.9** (a) $(CH_3)_2C{=}CHCH_3$; (b) $(CH_3)_2C{=}CHCH_3$; (c) $(CH_3)_2C{=}C(CH_3)_2$.

3. b, d, g, h, i, k (3 isomers). **4.** (b) 4 show geometric isomerism. **5.** Differ in all except (h); (l) dipole moment would tell. **11.** Both solutions contain isopropyl cation. **12.** (a) $(CH_3)_2C{=}CHCH_3$ (major product) and $CH_2{=}C(CH_3)C_2H_5$.

Chapter 6

6.1 (c) 1-Butene 649.8, *cis*-2-butene 648.1, *trans*-2-butene 647.1. (d) 1-Pentene 806.9, *cis*-2-pentene 805.3, *trans*-2-pentene 804.3. **6.2** (a) H_3O^+; HBr; (b) HBr; (c) HBr. **6.10** Orlon, $CH_2{=}CH{-}CN$; Saran, $CH_2{=}CCl_2$; Teflon, $CF_2{=}CF_2$. **6.12** React with relatively scarce HCl, with a minimum E_{act} of 15 kcal. **6.20** A, alkane; B, 2° alcohol; C, alkyl halide; D, alkene; E, 3° alcohol.

6. 3° radical more stable than 2° radical, forms faster. **10.** (d) Steps (2) and (4) are too difficult with HCl. **16.** 3-Hexene.

Chapter 7

7.2 (a) 4; (c) none. **7.4** c, d, e, g. **7.6** −0.89°. **7.7** (f) R,R:*meso* = 29:71. **7.8** (a) 5 fractions, two inactive, others active; (b) 5, all inactive; (c) 6, all inactive; (d) 2, both active. **7.9** Rapidly inverting pyramid. **7.11** (a) Racemic; *meso*; (b) *syn*; (c) *anti*. **7.12** Racemic modification, a, c, d; *meso*, b.

1. (a) 3; (b) 5; (c) 7 (5 active); (d) 7 (6 active); (e) 1; (f) 3; (g) 2 (1 active); (h) 2. **3.** A, (S,S); B, (R,S); C, (S,S); D, (S,S); E, (2R,3S)-4-bromo-1,2,3-butanetriol; F, (R,R); G, (R,S). **5.** *Anti*; intermediate chloronium ion.

Chapter 8

8.4 (a) Propane. **8.5** Calcium acetylide. **8.6** H goes to terminal C. **8.7** 1,3-Hexadiene. **8.8** (a) 56–60 kcal. **8.11** (c) Position of equilibrium. **8.15** Head-to-tail polymer of isoprene.

7. No reaction: g through n. **8.** No reaction: g through n. **14.** (a) −42.2 kcal. **15.** (a) Two CH_2 planes perpendicular to each other. **19.** Geometric isomers. **23.** Cyclohexene. **24.** 1,3,5-Hexatriene. **25.** (b) Myrcene, $(CH_3)_2C{=}CHCH_2CH_2C({=}CH_2)CH{=}CH_2$. **26.** (a) Dihydromyrcene, $(CH_3)_2C{=}CHCH_2CH_2C(CH_3){=}CHCH_3$; (b) 1,4-addition. **27.** (c) 2 farnesyl units, head-to-head, form squalene skeleton.

Chapter 9

9.4 *Trans* is resolvable. **9.7** (a) 0 kcal; (b) 2.7 kcal; (c) 1.8 kcal + undetnd. methyl–methyl interaction; (d) 0 kcal; (e) 0 kcal; (f) 3.6 kcal. **9.8** (b) 3.6 kcal. **9.9** (a) *cis* > *trans*; (b) *trans* > *cis*; (c) 1.8 kcal/mole in each case. **9.10** More than: (a) 3.2 kcal; (b) 6.8 kcal; (c) 2.3 kcal. **9.11** Resolvable: b, d. *Meso*: c (e and f do not contain chiral centers). **9.12** (a) e; (b) a; (c) c, f; (d) d; (e) b; (f) none. **9.13** Pairs of enantiomers: a, b, c, d. No *meso* compounds. None are non-resolvable racemic modifications. **9.17** (e) For the same degree of unsaturation, there are two fewer hydrogens for each ring. **9.18** All are C_6H_{12}; no information about ring size. **9.19** 2, 2, 1, none.

4. (a) 4; (b) 6; (c) 7; (d) 9; (e) 5; (f) 2; (g) all-equatorial, **5.** A, *cis*-dimethyl; B, *trans*-dimethyl. **8.** (d) In the *trans*-isomer, both large substituents (the other ring) are equatorial; (e) high energy barrier (E_{act}) between decalins since bond must be broken. **11.** (a), (b), (c) 2 (1 active); (d) 2. **14.** (a) 2; (b) 4; (c) 1; (d) 3; (e) 4. **15.** (a) 1; (b) 2; (c) 1; (d) 3; (e) 3; (f) 5; (g) 4.

16. (b) 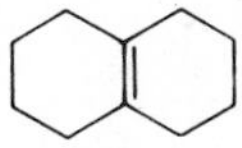 **17.** (c) 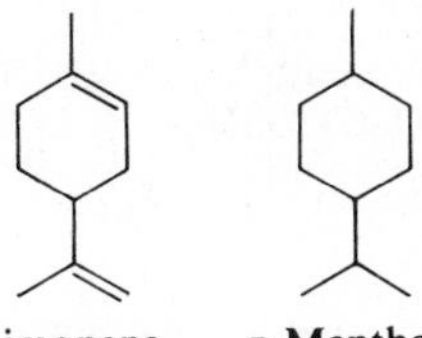 **18.** (b)

Limonene *p*-Menthane α-Terpinene

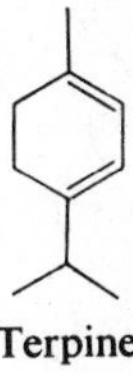

Chapter 10

10.1 (a) +5.6 kcal; (b) −26.8 kcal. **10.2** (a) 824.1 kcal; (b) 35.0 kcal greater. **10.8** *Ortho*, +6°; *meta*, −7°; *para*, +87°. **10.10** 26.0%. **10.11** 22.8%. **10.12** 18.5%. **10.13** 25.9%, 22.9%, 18.6%.

2. (a) 3; (b) 3; (c) 3; (d) 6; (e) 10; (f) 6. **3.** (a) 2, 3, 3, 1, 2; (b) 5, 5, 5, 2, 4 (neglecting stereoisomers; (c) none. **4.** (a) 2; (b) 3; (c) 1; (d) 4; (e) 4; (f) 2; (g) 4; (h) 4; (i) 2; (j) 1; (k) 3; (l) 2. **5.** (a) 1; (b) 1; (c) 2; (d) 1; (e) 2; (f) 3; (g) 2. **6.** Yes. **7.** (c) No, the *ortho* isomer would be chiral, and enantiomers would be possible. **8.** *Ortho*, 104°; *meta*, 63°; *para*, 142°. **9.** (a) For $n = 3, 5, 7, 9$; $n = 5$ has poor geometry; (b) $C_9H_9^-$. **11.** (a) $C_6H_6Cl_6$.

Chapter 11

11.3 (d) Carbonium ion mechanism. **11.6** Large size of complex. **11.8** (a) $RC{\equiv}O^+$; (b) ArN_2^+; (c) NO^+. **11.9** (b) D^+. **11.10** (a) 2.05; (b) 1.02 moles HCl:1 mole DCl.

11.11 [structure: ring cation with CH_3 at top, CH_3, CH_3 and H, H at the sp^3 carbon] BF_4^- **11.12** (a) 6.77; (b) yes; (c) no; (d) yes. **11.13** (a) CH_3-$CHCl^+$; (b) $CH_3CH_2^+$; (c) $CH_3CH_2^+$; (e) $^+CH_2CH_2Cl$; (f) CH_3CHCl^+; (h) inductive; (i) resonance.

Chapter 12

12.9 (a) Similar to Fig. 2.3, with E_{act} = 19 kcal, and ΔH = +11 kcal; (b) 8 kcal; (c) steric hindrance to combination.

6. ~$CH_2C_6H_4CH_2C_6H_4CH_2C_6H_4$~. **17.** 2-, 3-, 4-, 5-, and 6-phenyldodecane.

26. [structures]

Indene Indane

27. X and Y, racemic and *meso*-$C_6H_5CH(CH_3)$-$CH(CH_3)C_6H_5$; Z, $[C_6H_5C(CH_3)_2\text{-}]_2$.

Chapter 13

13.1 (a) $(CH_3)_3C^+$; $CH_2{=}CH{-}CH_2^+$; $CH_3CH_2^+$; $CH_2{=}CH^+$.

13.2 [structure: $(\text{H}_3\text{C}, \text{CH}_3$ ring with CH_3, chain with two CH_3 groups, $\text{CH=})_2$]

β-Carotene

13.3 (a) A, 1,4-pentadiene; B and C, *cis*- and *trans*-1,3-pentadiene. **13.4** (a) 2, 1; (b) 1, 2, 3, 4(1,2-dibromopropane); (c) 3, 2; (d) 2, 4, 3; (e) 3, 1; (f) 2, 4, 3, 5; (g) 2, 4; (h) 3, 1, 5. **13.6** 1 signal. **13.7** Electron release by methyl groups. **13.9** (a) Neopentylbenzene; (b) isobutylene bromide, $(CH_3)_2CBr$-CH_2Br; (c) benzyl alcohol, $C_6H_5CH_2OH$. **13.12** (a) Ethylbenzene; (b) 1,3-dibromopropane; (c) *n*-propyl bromide. **13.16** (a) $CH_3\cdot$; (b) $CH_3\dot{C}HCH_3$, $CH_3CH_2\dot{C}HCH_3$; (c) $Ph_3C\cdot$. **13.17** Cyclohexane.

1. (a) $CHCl_2CHClCHCl_2$; (b) $CH_2ClCCl_2CH_3$; (c) $(CH_3)_2CHCH_2Cl$; (d) $C_6H_5C(CH_3)_3$; (e) $C_6H_5CH_2CH(CH_3)_2$; (f) indane (see answer to Prob. 26, Ch. 12;) (g) $C_6H_5CH_2CCl(CH_3)_2$ (actually) or $C_6H_5C(CH_3)_2CH_2Cl$; (h) 1-phenyl-1-methylcyclopropane; (i) $C_6H_5CH_2CH_2CH_2Br$; (j) $CH_2ClCF_2CH_3$. **2.** X, *trans*-1,3-dibromo-*trans*-1,3-dimethylcyclobutane; Y, the *cis,cis*-isomer. **3.** See answer to Prob. 11.11. **4.** 1,2-Dimethylcyclopropene. **5.** See Sec. 29.6. **6.** B, [structure] C, [structure]

7. (a) eeeeee, eeeaaa; (b) eeeeea; (c) eeeeaa, eeaeea; (d) eeeeee, no change; eeeaaa, split into two peaks of equal area. **8.** (a) H on C–1; (b) equatorial H downfield from axial H. **9.** 82% equatorial —Br (axial H on C–1). **10.** (a) Isopropylbenzene; (b) isobutylene; (c) phenylacetylene. **11.** (a) Isobutylbenzene; (b) *tert*-butylbenzene; (c) *p*-cymene (*p*-isopropyltoluene). **12.** (a) α-Phenylethyl bromide, $C_6H_5CHBrCH_3$; (b) *tert*-pentylbenzene; (c) *sec*-butyl bromide. **13.** D, α-Methylstyrene, $C_6H_5C(CH_3){=}CH_2$.

Chapter 14

14.4 (a) 1.9%; (b) 16.4%; (c) 66.2%; (d) 95.1%; (e) 99.0%. **14.5** (a) Optical purity: bromide, 60%; alcohol, 40%. (b) 33% racemization, 67% inversion; (c) 17% front-side attack, 83% back-side attack. **14.7** Me, 300; Et 24; *i*-Pr, 1; *t*-Bu 1410. **14.13** *Anti.* **14.14** All —Cl atoms equatorial.

16. A, (1R,2S;1S,2R)-1,2-dichloro-1-phenylpropane; B, (1R,2R;1S,2S)-1,2-dichloro-1-phenylpropane. **20.** 1,1-Dimethylcyclopropane; 1,1-dimethylcyclopropane-2-d. **21.** (b) $(CH_3)_3C^+$, $(CH_3)_2CH^+$. **22.** (a) 1-Methylcyclopropene; (b) cyclopropene.

23. C, 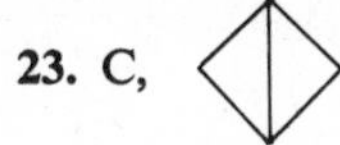

Chapter 15

15.1 Intramolecular H-bond in *cis*-isomer (see Sec. 24.2). **15.5** (a) Leucine → isopentyl alcohol; isoleucine → active amyl alcohol. **15.8** $C_6H_5CH(OCH_3)CH_3$. **15.10** *syn*-Addition, retention; or *anti*-addition, inversion. **15.12** *syn*-Addition, retention.

7. Intramolecular H-bond between —OH and —G. **8.** Coprostane-3β,6β-diol, by *cis*-hydration at more hindered "top" face of molecule. (b) *cis*-Hydration from beneath gives *alpha* —OH at C–11. **9.** (b) e,e; (c) a,a. **10.** Twist-boat. **12.** *anti*-Elimination.

Chapter 16

16.2 Free radical chlorination of neopentane. **16.9** With 100% inversion. **16.11** $1HIO_4$, a, b, c, e; $4HIO_4$, f, g; no reaction, d. **16.12** A, $(CH_3)_2C(OH)CH_2OH$; B, 1,2-cyclohexanediol; C, 2-hydroxycyclohexanone; D, HOOCCHOHCHOHCOOH; E, $HOCH_2CHOHCHOHCH_2OH$; F, $HOCH_2CHOHCOCHO$; G, $HOCH_2(CHOH)_4$-CHO. **16.13** Change concentration. **16.14** (a) 1°, triplet; 2°, doublet; 3°, singlet.

1. (a) Two give iodoform; (c) one gives negative test. **11.** (a) *anti*-Elimination. **13.** B, $HOCH_2CH_2OH$; D, $HOCH_2COOH$; E, OHC(CHOH)CHO; F, $CH_3(CH_2)_7CH{=}CH(CH_2)_7COOH$; J, $CH_2{=}CHCOOH$; M, $HOCH_2C{\equiv}CH$; O, CH_3COCH_3; S, CH_3-COONa; U, diacetate of *cis*-1,2-cyclohexanediol; W, triacetate of glycerol; AA, 3-methylbiphenyl, *m*-$CH_3C_6H_4C_6H_5$; GG, active 2,4,6,8-tetramethylnonane; HH, *meso*-2,4,6,8-tetramethylnonane. **15.** (a) Protonated alcohols; (b) and (c) *tert*-butyl cation. **16.** (a) R_3C^+, stabilized by overlap of empty *p* orbital with π clouds of rings. (b) Methyls located unsymmetrically; plane of methyls and trigonal carbon perpendicular to and bisecting ring. **20.** NN, $C_6H_5CH_2CHOHCH_3$; OO, $C_6H_5CH(CH_3)CH_2OH$. **21.** PP, 1,2,2-triphenylethanol; QQ, 1,1,2-triphenylethanol. **22.** (a) *sec*-Butyl alcohol; (b) isobutyl alcohol; (c) ethyl ether. **23.** (a) α-Phenylethyl alcohol; (b) β-phenylethyl alcohol; (c) benzyl methyl ether. **24.** RR, 2-methyl-2-propen-1-ol; SS, isobutyl alcohol. **25.** TT, 3,3-dimethyl-2-butanol. **26.** Geraniol, $(CH_3)_2C{=}CHCH_2CH_2C(CH_3){=}CHCH_2OH$. **27.** (a) Same as Prob. 26; (b) geometric isomers; (c) in geraniol, —H and —CH_3 are *trans*.

Chapter 17

17.7 (a) Configuration of (−)-ether same as (−)-alcohol; (b) maximum rotation is −17.6°. **17.8** (a) Practically complete inversion. **17.10** Trifluoroacetate is weaker base, weaker nucleophile, does not compete with alcohol. **17.18** (f) None. **17.26** 4.

6. Polyisobutylene. **9.** A, 3-bromo-4-methoxytoluene; B, *o*-methoxybenzyl bromide; C, *o*-bromophenetole. **15.** M, $(CH_2{=}CH)_2O$; N, $ClCH_2CHOHCH_2OCH_3$, retention;

O, CH_3OCH_2COOH; **P,** $CH_3OCH_2CH—CH_2$ (epoxide, bridged by O); Q, $CH_2—CH_2$ / $CH_2—O$ (four-membered ring); R, $C_6H_5CH_2CH_2OH$;

S, 2-methyltetrahydrofuran (ring O with CH_3); T, CH_3CHO; U, racemic *trans*-2-chlorocyclohexanol, inversion;

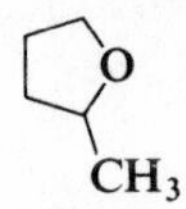

V, racemic *trans*-methyl-1,2-cyclohexanediol, inversion; W, racemic and *meso*-$HOCH_2CHOHCHOHCH_2OH$; X, racemic 2,3-butanediol; Y, *meso*-2,3-butanediol. **16.** *m*-Methylanisole. **17.** K, anisyl alcohol. **18.** (a) *tert*-Butyl ethyl ether; (b) *n*-propyl ether; (c) isopropyl ether. **19.** L, *p*-methylphenetole; M, benzyl ethyl ether; N, 3-phenyl-1-propanol.

Chapter 18

18.1 91 at 110°, 71 at 156°; association occurs even in vapor phase, decreasing as temperature increases. **18.2** (b) 2-Methyldecanoic acid; (c) 2,2-dimethyldodecanoic acid; (d) ethyl *n*-octylmalonate, *n*-$C_8H_{17}CH(COOEt)_2$. **18.3** (b) 2-Methylbutanoic acid. **18.4** (a) *p*-Bromobenzoic acid; (b) *p*-bromophenylacetic acid. **18.7** (a) F > Cl > Br > I; (b) electron-withdrawing. **18.19** *o*-Chlorobenzoic acid. **18.20** (a) 103; (b) ethoxyacetic acid. **18.21** (a) Two, 83; (b) N.E. = mol.wt./number acidic H per molecule; (c) 70, 57. **18.22** Sodium carbonate.

19. A and B, racemic and *meso*-2,3-dibromobutanoic acid; C, *meso*-HOOCCHOH-CHOHCOOH; F, *cis*-$HOOCCH_2CH—CHCH_2COOH$ (ring bridged by CH_2). **20.** G, $HC{\equiv}CMgBr$; J, $OHCCH_2COOH$. **26.** N.E. 165; *o*-nitrobenzoic acid. **27.** Q, *m*-ethylbenzoic acid; U, 3,5-dimethylbenzoic acid. **28.** Tropic acid, $C_6H_5CH(CH_2OH)COOH$; atropic acid, $C_6H_5C(=CH_2)COOH$; hydratropic acid, $C_6H_5CH(CH_3)COOH$. **29.** (a) CH_3-$CHClCOOH$; (b) $ClCH_2COOCH_3$; (c) $BrCH_2COOCH_2CH_3$; (d) $CH_3CH_2CHBrCOOH$; (e) $CH_3CH_2OCH_2COOH$. **30.** (a) Crotonic acid; (b) mandelic acid; (c) *p*-nitrobenzoic acid.

Chapter 19

19.1 $RCH(OH)_2$. **19.4** (a) Acetic, propionic, and *n*-butyric acids; (b) adipic acid. **19.5** (a) 1; (b) 1; (c) 1; (d) 2 (both active); (e) 2; (f) no change. **19.9** (a) Williamson synthesis of ethers; (b) acetals (cyclic). **19.17** Internal "crossed" Cannizzaro reaction.

10. B, 3,3-dimethyl-1-indanone (structure: benzene ring fused to five-membered ring bearing CH_3, CH_3 and C=O) C, 1,1-dimethylindane (structure: benzene ring fused to five-membered ring bearing CH_3, CH_3) D, $PhCH_2CH_2C(OH)(CH_3)_2$.

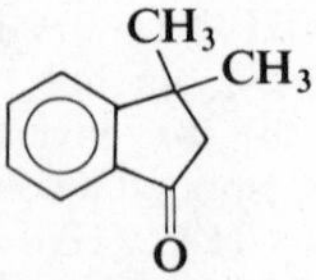

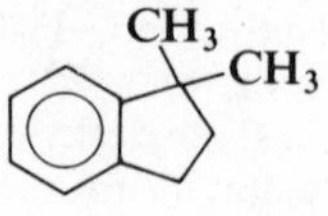

11. Hemiacetal is oxidized by mechanism of Sec. 19.9. **12.** See Fig. 34.6, Sec. 34.14. **13.** (a) Cyclic ketal. **18.** Hydride transfer from Ph_2CHO^- to excess PhCHO. **20.** Protonated aldehyde is electrophile, double bond is nucleophile. **21.** Chair: in E, all $—CCl_3$ equatorial; in F, two equatorial, one axial. **23.** (b) *trans*-Isomer: intramolecular H-bonding between —OH and ring oxygen. **26.** $(CH_3)_2C{=}CHCH_2CH_2C(CH_3){=}CHCHO$, citral *a* (H and CH_3 *trans*), citral *b* (H and CH_3 *cis*); dehydrocitral, $(CH_3)_2C{=}CHCH{=}CHC(CH_3){=}CHCHO$. **27.** Carvotanacetone, 5-isopropyl-2-methyl-2-cyclohexene-1-one. **28.** (a) 2-Butanone; (b) isobutyraldehyde; (c) 2-buten-1-ol. **29.** (a) 2-Pentanone; (b) methyl isopropyl ketone; (c) methyl ethyl ketone. **30.** P, *p*-anisaldehyde; Q, *p*-methoxyacetophenone; R, isobutyrophenone.

Chapter 20

20.3 Maleic acid is *cis* and fumaric acid is *trans*-butenedioic acid, HOOCCH=CH-COOH. **20.4** G, naphthalene. See Fig. 30.2, p. 987. **20.5** Final product is 1-phenylnaphthalene. **20.6** 9,10-Anthraquinone. See Sec. 30.18. **20.7** *o*-(*p*-Toluyl)benzoic acid (p. 993). **20.8** (a) *cis*-Acid: the only one that can form a cyclic anhydride. **20.16** Basicity of leaving group: $Cl^- < RCOO^- < OR^- < NH_2^-$. **20.17** Structure II in Sec. 20.17. **20.21** (a) Formic acid. **20.22** 1-Octadecanol and 1-butanol. **20.27** (b) Nucleophilic addition. **20.28** (a) RCOCl; (b) $RCOO^-NH_4^+$, $RCONH_2$, RCN, amides of low mol.wt. amines; (c) $RCOO^-NH_4^+$; (d) $(RCO)_2O$; (e) RCOOR′. **20.29** (a) 102; (c) 4; (d) no. **20.30** (a) Two, 97; (b) S.E. = mol.wt./number ester groups per molecule; (c) 297.

10. (a) $HOCH_2CH_2CH_2CONH_2$; (b) $HOCH_2CH_2CH_2CH_2OH$; (c) $HOCH_2CH_2CH_2COOEt$. **11.** Second step is S_N2 attack by benzoate anion. **17.** A *meso*; B, racemic. **18.** C, $CO_3^=$; D, $C_2H_5OCONH_2$; M, indene (see Chapter 12, Problem 26); O, *trans*-2-methylcyclohexanol. **19.** Progesterone. **20.** AA, 1,3-propanediol; BB, 1,2-propanediol; CC, 2-methoxyethanol; DD, dimethoxymethanol (dimethylacetal of formaldehyde); EE, α-hydroxypropionaldehyde; FF, hydroxyacetone; GG, β-hydroxypropionaldehyde; HH, propionic acid; II, ethyl formate; JJ, methyl acetate; KK, *cis*-1,2-cyclopropanediol; LL, MM, $CH_2—CH—CH_2OH$ (epoxide, O bridging). **21.** (a) Methyls are *trans* in NN, PP; *cis* in OO, QQ, RR; (b) NN is resolvable. **22.** See p. 1087. **23.** (a) Ethyl acetate; (b) methacrylic acid; (c) phenylacetamide. **24.** (a) *n*-Propyl formate; (b) methyl propionate; (c) ethyl acetate. **25.** SS, benzyl acetate; TT, methyl phenylacetate; UU, hydrocinnamic acid, $PhCH_2CH_2COOH$. **26.** Ethyl anisate. **27.** VV, vinyl acetate. **28.** (a) Ethyl adipate; (b) ethyl ethylphenylmalonate; ethyl acetamidomalonate.

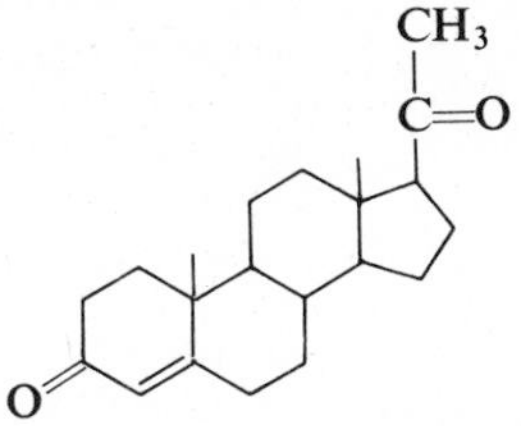

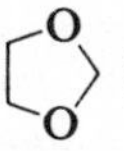

Chapter 21

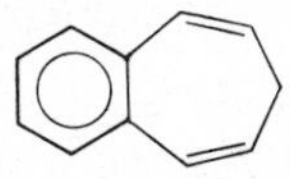

21.1 III, in which the negative charge resides on oxygen, the atom that can best accommodate it. **21.3** Order of decreasing delocalization of the negative charge of the anion. **21.6** (b) Hard to generate *second* negative charge. **21.7** Expect rate of racemization to be twice as fast as exchange. **21.8** (a) Both reactions go through the same slow step (2), formation of the enol. **21.9** (a) HSO_4^-; (b) D_2O. **21.11** Gives a mixture of aldol products. **21.12** Electrophile is protonated aldehyde; nucleophile is enol. **21.14** *Retro* (reverse) aldol condensation. **21.20** (a) γ-Hydrogen will be acidic. **21.23** Elimination → 1- and 2-butene. **21.26** A, $Ph_3P=CHOPh$; B, $C_2H_5(CH_3)=CHOPh$; C, $C_2H_5CH(CH_3)CHO$; a general route to aldehydes. **21.27** D, 1-phenylcyclopentene; E, $Ph_3P=CHCH_2CH=PPh_3$; F, **21.30** (a) Intramolecular Claisen condensation leading to cyclization; (b) 2-carbethoxycyclohexanone; (d) ethyl 2,5-dioxocyclohexane-1,4-dicarboxylate. **21.32** (b) 2,4-Hexanedione; (c) 1,3-diphenyl-1,3-propanedione (dibenzoylmethane); (d) 2-(EtOOCCO)cyclohexanone. **21.33** (a) PhCOOEt and $PhCH_2COOEt$; (b) EtOOCCOOEt and ethyl glutarate; (c) ethyl phthalate and CH_3COOEt. **21.36** C, citric acid, $(HOOCCH_2)_2C(OH)COOH$.

1. (e) Allylbenzene. **2.** (e) Methylenecyclohexane. **3.** (a) No reaction; (m) $PhCH{=}CHCH{=}CH_2$; (n) $PhCH{=}CHOPh$; (o) $PhCH_2CHO$. **13.** (b) Iodoform test. **15.** Triple aldol cond., followed by crossed Cannizzaro reaction. **18.** Dehydrocitral, $(CH_3)_2C{=}CHCH{=}CHC(CH_3){=}CHCHO$, formed by aldol cond. on γ-carbon of α,β-unsaturated aldehyde.
20. $CH_3COCH_2COOEt + CH_3MgI \rightarrow CH_4\uparrow + (CH_3COCHCOOEt)^-Mg^{++}I^-$.
21. (b) C=C conjugated with second C=O; (c) intramolecular H-bonding.

22. A, (structure: CH_3, O, O, H) B, (structure: O, O, O, CH_3), a triketone. **23.** (a) *a*, enol $—CH_3$; *b*, keto $—CH_3$; *c*, keto $—CH_2—$; *d*, enol $—CH{=}$; *e*, enol —OH. Ratios *a*:*b* and 2*d*:*c* are equal (5.5 and 5.6) and show 85% enol. (b) All enol; conjugation with ring.

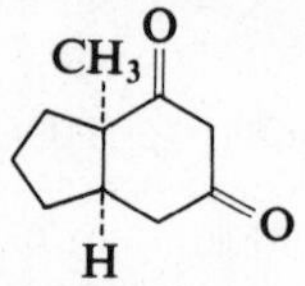

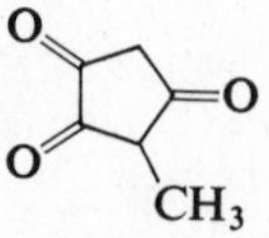

Chapter 22

22.4 $R:^-$ undergoes rapid inversion.

6. (a) Putrescine, 1,4-diaminobutane; (b) cadaverine, 1-5-diaminopentane. **9.** Pair of enantiomers: a, c, e, f; one inactive compound, b; inactive *cis-trans* pair, d. **11.** C, $CH_3CH_2CH_2NH_2$. Gabriel synthesis gives 1° amines free from 2° and 3°.

Chapter 23

23.2 $(CH_3)_3\overset{+}{N}:\overset{-}{B}F_3$. **23.6** 1,3-Pentadiene (from thermal isomerization of 1,4-pentadiene); 2-methyl-1,3-butadiene (isoprene). **23.8** Attack at acyl carbon less hindered than at sulfur; sulfonate better leaving group than carboxylate. **23.9** Free amine is much more reactive. **23.11** (a) *n*-Butyl cation. **23.12** (b) 2-Methyl-2-butene, 2-methyl-1-butene, *tert*-pentyl alcohol. **23.13** Leaving groups $Cl^- > H_2O > OH^-$. **23.18** (a) Electron withdrawal makes diazonium ion more electrophilic. **23.21** (a) 2′-Bromo-4-hydroxy-3,4′-dimethylazobenzene. **23.22** Reduction of azo compound formed by coupling N,N-dimethylaniline with some diazonium salt (usually $^-O_3SC_6H_4N_2^+$ from sulfanilic acid). **23.24** (a) That unknown is 3°; (b) separate, test solubility in acid.

13. See Sec. 32.7. **14.** Poor leaving group (OH^-) converted into a good leaving group (OTs^-). **15.** Reaction of PhN_2^+ is S_N1-like; reaction of p-$O_2NC_6H_4N_2^+$ is S_N2-like. **20.** Choline, $HOCH_2CH_2N(CH_3)_3{}^+OH^-$; acetylcholine, $CH_3COOCH_2CH_2N(CH_3)_3{}^+OH^-$. **21.** Novocaine, p-$H_2NC_6H_4COOCH_2CH_2N(C_2H_5)_2$. **22.** I, N-methyl-N-phenyl-*p*-toluamide. **23.** P, 1,3,5,7-cyclooctatetraene. **24.** Pantothenic acid, $HOCH_2C(CH_3)_2CHOHNHCH_2CH_2COOH$. **25.** W, $PhNH_3{}^+Cl^-$. **26.** (a) *n*-Butylamine; (b) N-methylformamide; (c) *m*-anisidine. **27.** (a) α-Phenylethylamine; (b) β-phenylethylamine; (c) *p*-toluidine. **28.** X, *p*-phenetidine (*p*-ethoxyaniline); Y, N-ethylbenzylamine; Z, Michler's ketone, *p,p′*-bis(dimethylamino)benzophenone.

Chapter 24

24.1 Intramolecular H-bond in *o*-isomer unaffected by dilution. **24.4** Benzene, propylene, HF. **24.9** *p*-Bromophenyl benzoate, p-$BrC_6H_4OOCC_6H_5$. **24.12** (a) The $—SO_3H$ group is displaced by electrophilic reagents, in this case by nitronium ion. **24.17** N.E.

5. No reaction: b, c, f, n. **6.** Reaction only with: c, p, r, s, t, u. **7.** Reaction only with: c, h, i, j, k, l, n. **13.** (a) Nucleophilic aliphatic substitution; (b) electrophilic aromatic substitution. **16.** Phenacetin, *p*-$CH_3CONHC_6H_4OC_2H_5$; coumarane, 3-cumaranone, carvacrol, 5-isopropyl-2-methylphenol; thymol, 2-isopropyl-5-methylphenol; hexestrol, 3,4-bis(*p*-hydroxyphenyl)hexane. **17.** Adrenaline, 1-(3,4-dihydroxyphenyl)-2-(N-methylamino)ethanol. **18.** Phellandral, 4-isopropyl-3,4,5,6-tetrahydrobenzaldehyde. **19.** Y, *m*-cresol. **20.** Z, *p*-allylanisole; AA, *p*-propenylanisole. **21.** BB, isopropylsalicylate. **22.** Chavibetol, 2-methoxy-5-allylphenol. **23.** GG, C_6H_5NHOH; HH, *p*-$HOC_6H_4NH_2$; (d) 2-methyl-4-aminophenol. **23.** Piperine,

CH=CHCH=CHC(=O)—N **24.** Hordinene, *p*-$HOC_6H_4CH_2CH_2N(CH_3)_2$ or *p*-$HOC_6H_4CH(CH_3)N(CH_3)_2$ (actually the former). **25.** α-Terpineol, 2-(4-methyl-3-cyclohexenyl)-2-propanol. **26.** Coniferyl alcohol, 3-(4-hydroxy-3-methoxyphenyl)-2-propen-1-ol. **27.** (a) UU, a ketal and lactone. **28.** AAA, piperonal; BBB, vanillin; CCC, eugenol; DDD, thymol; EEE, isoeugenol; FFF, safrole.

Chapter 25

25.4 (b) Nucleophilic aromatic substitution; (c) electron withdrawal.

1. No reaction: b, c, d, e, f, g, k, l, n, o. **2.** No reaction: h, i, j, k, m, n, o. **5.** (o) C_6H_6 + $HC{\equiv}CMgBr$. Racemic modifications: f, h, k. Optically active: n. **13.** Inductive effect, $o \gg m > p$. **14.** $—N_2^+$ activates molecule toward nucleophilic substitution.

15. $ArF + R_2NH \leftrightarrows \overset{-}{Ar}(F)(\overset{+}{N}HR_2) \xrightarrow{:B} \overset{-}{Ar}(F)(NR_2) \longrightarrow ArNR_2 + F^-$. **18.** (a) 28, N_2; 44, CO_2; 76, benzyne, C_6H_4; 152, biphenylene.

Biphenylene

(b) Anthranilic acid.

19. Tetraphenylmethane. **21.**

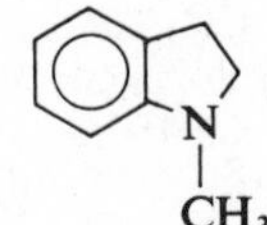

23. $Ar^\ominus + Ar'—Br \rightleftarrows Ar—Br + Ar'^\ominus$. Only carbanions with negative charge *ortho* to halogen are involved.

Chapter 26

26.3 Ethyl benzalmalonate, $PhCH{=}C(COOEt)_2$. **26.6** Nucleophilic substitution (S_N2); 1° > 2° ≫ 3° (or none); aryl halides not used. **26.7** (a) $CH_3COCH_2CH_2COOH$, a γ-keto acid; (b) $PhCOCH_2COCH_3$, $CH_3COCH_2CH_2COCH_3$, both diketones. **26.9** A, $EtOOCCOCH(CH_3)COOEt$. **26.11** (a) Charged end loses CO_2. **26.12** Gives relatively stable anion, 2,4,6-$(NO_2)_3C_6H_2$:$^-$. **26.15** Gives relatively stable anion, $PhC{\equiv}C$:$^-$.

26.17, B, ethyl 3-hydroxynonanoate. **26.18** E, Ph—(cyclohexene)—COOEt. **26.22** B, 2-benzalcyclopentanone; F, 3-phenyl-2,2-dimethylpropanal.

3. Cyclopentanone. **4.** C, 1,3-cyclohexanedicarboxylic acid; F, 1,4-cyclohexanedicarboxylic acid; H, succinic acid; J, 1,2-cyclobutanedicarboxylic acid. **5.** K, 1,5-hexadiene; O, 2,5-dimethylcyclopentanecarboxylic acid. **7.** (b) Intramol. aldol cond.; (d) gives 3-methyl-2-cyclohexen-1-one. **11.** (a) *Retro* (reverse) Claisen condensation.

13. S, 1-phenyl-3-nonanone. **16.** V, 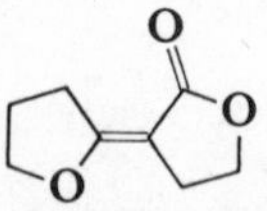 ; W, $ClCH_2CH_2CH_2COCH_2CH_2$-CH_2Cl. **17.** Nerolidol, $RCH_2C(CH_3)(OH)CH{=}CH_2$. **18.** Menthone, 2-isopropyl-5-methylcyclohexanone. **19.** Camphoric acid, $HOOCCH_2C(CH_3)(COOH)C(CH_3)_2COOH$.

20. Terebic acid, 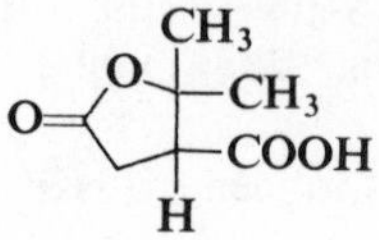

Terpenylic acid, 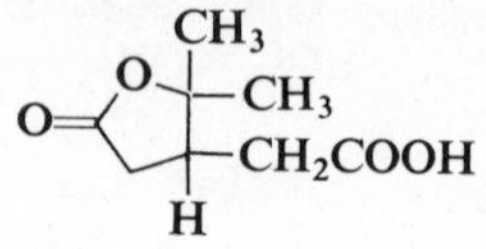

21. Phosphate ion, $H_2PO_4^-$, a better leaving group than OH^-.

Chapter 27

27.2 A, $PhCH_2CH_2CHO$; B, $PhCH_2CH_2CH_2OH$; C, $PhCH{=}CHCH_2OH$. **27.4** (d) ~~$CH_2CH(CN)$~~ (Orlon), ~~$CH_2CH(COOMe)$~~ (Acryloid), ~~$CH_2C(CH_3)(COOMe)$~~ (Lucite, Plexiglas) **27.6** All less stable than I. **27.7** An amide.

27.10 B, $CH_3CH(CH_2COOH)_2$; D, δ-ketocaproic acid; E, $CH_3COCH_2CH_2CH(COOEt)_2$; F, $PhCH(CH_2COPh)_2$; H, $H_2C{=}CHCH(COOH)CH_2CH_2COOH$; I, $EtOOCCH{=}C(COOEt)CH(COOEt)COCH_3$; J, $HOOCCH{=}C(COOH)CH_2COOH$. **27.11** (a) K, $H_2C{=}C(COOEt)_2$; (c) glutaric acid. **27.15** 1,4-Diphenyl-1,3-butadiene + maleic anhydride; 1,3-butadiene + 2-cyclopentenone; 1,3-butadiene (2 moles). **27.16** (a) 3-Ethoxy-1,3-pentadiene + *p*-benzoquinone; (b) 5-methoxy-2-methyl-1,4-benzoquinone + 1,3-butadiene. **27.18** (a) Ease of oxidation; (b) ease of reduction. **27.19** *p*-Nitrosophenol undergoes keto–enol tautomerization to give the mono-oxime.

3. (a) $C_6H_5COCH_2CH(C_6H_5)CH(CN)COOC_2H_5$; (f) $CH_3COCH_2C(CH_3)_2CH(COOEt)COCH_3$; (h) $(EtOOC)_2CHCH_2CH(COOEt)_2$; (j) $O_2NCH_2CH_2CH_2COOMe$; (l) $O_2NC(CH_2CH_2CN)_3$; (m) $Cl_3CCH_2CH_2CN$. **5.** A, $(EtOOC)_2CHCHPhCH_2COCH_2CHPhCH(COOEt)_2$; B, $(EtOOC)_2CHCHPhCH_2COCH{=}CHPh$; C, 4,4-dicarbethoxy-3,5-diphenylcyclohexanone. **6.** (d) 4-Acetylcyclohexene; (g) 5-nitro-4-phenylcyclohexene; (h) 1,4-dihydro-9,10-anthraquinone. **7.** (a) 1,3,5-Hexatriene + maleic anhydride; (b) 1,4-dimethyl-1,3-cyclohexadiene + maleic anhydride; (c) 1,3-butadiene + benzalacetone; (d) 1,3-butadiene + acetylenedicarboxylic acid; (e) 1,3-cyclopentadiene + *p*-benzoquinone; (f) 1,1′-bicyclohexenyl (see Problem 6 (b)) + 1,4-naphthoquinone (see Problem 6 (h)); (g) 1,3-cyclopentadiene + crotonaldehyde; (h) 1,3-cyclohexadiene + methyl vinyl ketone; (i) 1,3-cyclopentadiene (2 moles). **8.** *syn*-Addition. **9.** (a) Racemic modification; (b) *meso*; (c) 2 *meso*; (d) *meso*. **11.** Conjugate addition of OH^-, then *retro*-aldol condensation. **12.** $C_6H_5CH(C_2H_5)CH_2COCH_3$, 4-phenyl-2-hexanone. **14.** N, glyceraldehyde; P, aconitic acid, $HOOCCH{=}C(COOH)CH_2COOH$; R, tricarballylic acid, $HOOCCH(CH_2COOH)_2$; S, "tetracyclone", tetraphenylcyclopentadienone; U, tetraphenylphthalic anhydride; W, pentaphenylbenzene; BB, $(CH_3)_2C(CH_2COOH)_2$; DD, $CH_3CHOHC{\equiv}CCH_3$; EE, $CH_3COC{\equiv}CCH_3$; FF, acetylacetone; GG, $(CH_3)_2C{=}CHCOOH$; JJ, $HOOCCH{=}C(CH_3)CH_2COOH$; MM, 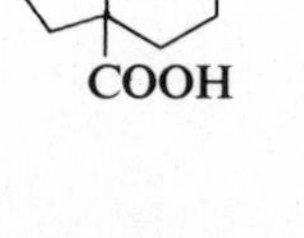

QQ, 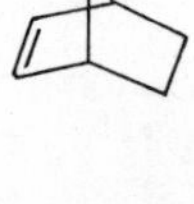

17. IV is correct. **18.** UU, [structure: cyclohexadiene with exocyclic CH_2] **19.** 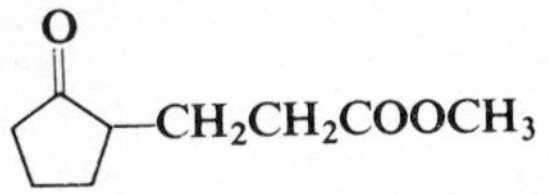 [structure: cyclopentanone with $-CH_2CH_2COOCH_3$]

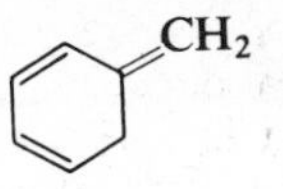

20. (b) 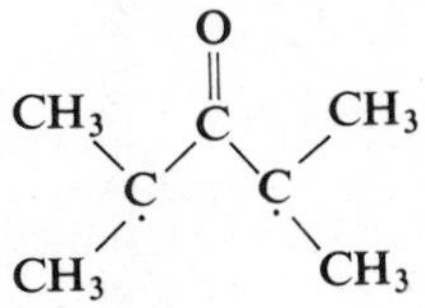 $(CH_3)_2\dot{C}-CO-\dot{C}(CH_3)_2$ is intermediate. **22.** Intermediate aryne: dehydrocyclopentadienyl anion.

Chapter 28

28.1 (a) Ammonium ion; (b) sulfonium ion; (c) protonated epoxide; (d) epoxide; (e) bromonium ion; (f) benzenonium ion; (g) oxonium ion; (h) ketone (dienone); (i) cyclopropylcarbinyl cation. **28.2** N_2 is leaving group. **28.3** Goes with retention, since only *cis* amino acid can form lactam. **28.4** If reaction (2), Sec. 28.6, occurs, it is not reversible; in view of substituent effect, then, (2) and (3) are concerted. **28.5** (a) *p*-Methylbenzaldehyde formed by migration of H; *p-cresol* (and formaldehyde), by migration of *p*-tolyl; (b) H migrates somewhat faster than *p*-tolyl. **28.6** H migrates much faster than alkyl. **28.7** Carbonium ion undergoes pinacol-like rearrangement. **28.9** Competition between solvent attack and rearrangement independent of leaving group; hence reaction is S_N1-like, with intermediate carbonium ion. **28.10** Intermediate is carbonium ion, which recombines with water faster than it rearranges. **28.14** Intermediate is an α-lactone. **28.15** Oxygens carry charge by *sharing* electrons. **28.16** Neighboring *trans*-Br and *trans*-I give anchimeric assistance. **28.18** α-Phenylethyl cation, by H-shift.

1. Successive H-shifts occur. **2.** $CH_3COCH_2CH_2CH_2CH_2CH_2OH$, formed by migration of ring carbon. **4.** (a) Analogous to Hofmann rearrangement, with $R'COO^-$ leaving group instead of X^-. **5.** Vinyl migrates predominantly, to give adipaldehyde, most of which undergoes intramolecular aldol to cyclopentene-1-carboxaldehyde. **6.** A, PhCONHPh; B, $PhNH_2$; C, PhCOOH. **8.** (b) *p*-Methoxyphenol and benzophenone; phenol and *p*-chlorobenzophenone. **10.** Two successive H-shifts. **11.** R undergoes 1,2-shift, with retention of configuration, from B to O in intermediate $R_3B—OOH$, with displacement of OH^-. **13.** Tosylate poorer leaving group than N_2, requires assistance from phenyl. **14.** (a) Neighboring —OH; (b) hydrolyzed. **15.** With p-CH_3OPh, nearly all reaction via (symmetrical) bridged ion; with p-NO_2Ph, most reaction via open cation; with Ph, about 50:50. **17.** Assistance by π electrons to give following intermediates (in (b), may be nonclassical ion):

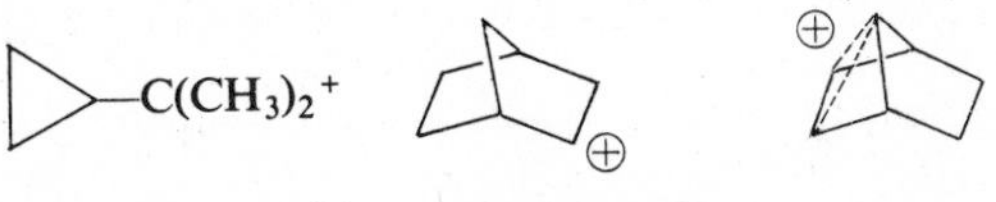

$-C(CH_3)_2^+$ ⊕ ⊕

(a) (b) (c)

19. Nucleophilic attack on acyl carbon of XXII by Z to give *tetrahedral intermediate*:

XXV ← [cyclic tetrahedral intermediate: benzofuranone ring carbon bearing O, Z, O^-; Ph, CH_2CH_2Br] → [phenoxide O^- with $C(Ph)(CH_2CH_2Br)COZ$] → XXIII or XXIV

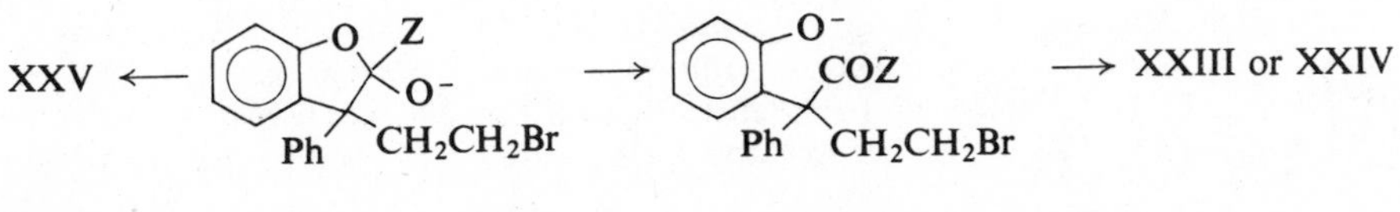

Chapter 29

29.1 First, monocation; then aromatic dication with 2 π electrons. **29.2** (a) Aromatic, with 2 π electrons:

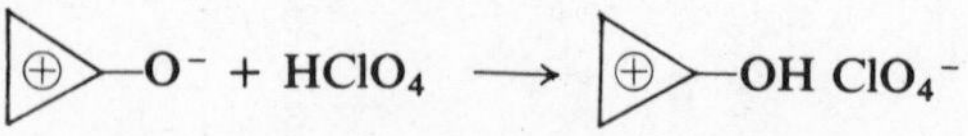

$$\text{(}\oplus\text{)}\!\!>\!\!-O^- + HClO_4 \longrightarrow \text{(}\oplus\text{)}\!\!>\!\!-OH\ ClO_4^-$$

29.3 (a) *Con* closure; I or III → *trans*; II → *cis*; (b) *dis* closure; I or III → *cis*; II → *trans*. **29.4** (a) ψ_1; 2 π electrons; (b) $4n + 2$; *dis* (thermal); (c) $4n$, *con* (thermal); (d) cation, $4n$, *con*; anion, $4n + 2$, *dis*. **29.5** (a) *Dis* opening; (b) *dis* closure; (c) *dis* closure; *con* opening; *dis* closure; (d) *con* opening (4 e); *dis* closure (6 e); (e) *dis* opening of cation (2 e), then combination with water; (f) protonated ketone like a pentadienyl cation, with 4 π electrons; *con* closure. **29.6** Via the cyclobutene, with *con* closures and openings. **29.7** (a) *cis*-3,6-Dimethylcyclohexene; [4 + 2]; (c) Ph's are *cis* to each other (*syn* addition) and *cis* to anhydride bridge (*endo* reaction); (d), (e), (f) all are tetramethylcyclobutanes; in D, one methyl is *trans* to other three. **29.8** (a) Diels-Alder; *retro*-Diels-Alder; (b) *endo* not *exo*. **29.9** (a) [4 + 2], not [6 + 2]; (b) photochemical (intramolecular) *supra,-supra* [2 + 2]; (c) *supra,supra* [6 + 4]; (d) *supra,supra* [8 + 2]; (e) *supra,antara* [14 + 2]. **29.10** (a) *supra* [1,5]-H to either face of trigonal carbon; (b) [1,5]-D, not [1,3]-D or [1,7]-D; (c) [1,3]-C (*supra*) with inversion at migrating C.

1. (a) Phenols; no; (b) dipolar structure is aromatic with 6 π electrons (compare answer to Problem 29.2); (d) intramolecular H-bond. **2.** (a) *Con* opening (4 e); [1,5]-H *supra*; (b) *con* opening (4 e); *dis* closure (6 e); (c) [1,7]-C *supra* and *dis* closure (4 e); [1,7]-H *supra*; (d) [4 + 4] *supra,supra*; *retro* [4 + 2] *supra,supra* (presumably thermal); (e) allylic cation (2 π electrons) undergoes [4 + 2] cycloaddition, followed by loss of proton; (f) bridge walks around the ring in a series of *supra* [1,5]-C shifts. **3.** (a) A, *trans*-7,8-dialkyl-*cis,cis,cis*-cycloocta-1,3,5-triene; (b) C, $(CH_3)_2C{=}C(CH_3)C({=}CH_2)C(CH_3){=}CH_2$; (c) D, 9-methyl-9-ethyl-*trans,cis,cis,cis*-cyclonona-1,3,5,7-tetraene; the *dis* closure takes place with both possible rotations; (d) E, *cis*-bicyclo[5.2.0]nona-8-ene; F, *cis,trans*-cyclonona-1,3-diene; G, *trans*-bicyclo[5.2.0]nona-8-ene. **4.** Symmetry-allowed *con* opening impossible on geometric grounds for bicyclo compound; reaction is probably not concerted. **5.** K, *cis*-bicyclo[4.2.0]octa-2,4-diene; L, Diels-Alder adduct, which undergoes *retro*-Diels-Alder. **6.** (a) [1,2] *supra* sigmatropic shift; π framework is a vinyl radical cation; HOMO is π; predict retention in migrating group; (b) π framework is diene radical cation; HOMO is ψ_2; predict inversion in migrating group.

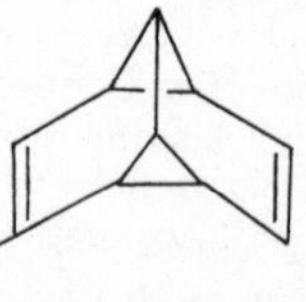

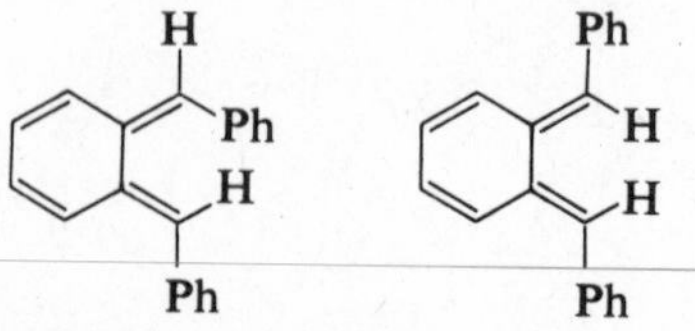

7. Symmetry-forbidden. **8.** [structure with D] **9.** (a) [4 + 2] cycloaddition of benzyne and diene; (b) [2 + 2] thermal cycloaddition symmetry-forbidden; reaction non-concerted, probably via diradicals. **10.** [structures: H, Ph, H, Ph; Ph, H, H, Ph] **11.** (a) *Meso* dibromide gives *cis*-VII (Fig. 29.26); racemic dibromide gives *trans*-VII; *cis*-VII contains four non-equivalent olefinic hydrogens; *trans*-VII, two equivalent pairs. **12.** (a) M and N, position isomers, both from *syn exo* addition; O and P, position isomers; (b) *retro*-Diels-Alder. **13.** (a) (Numbering from left to right in Fig. 29.19). Overlap between lobe of C–3 of diene and C–3 of ene, carbons to which bonds are not being formed; (b) lobes corresponding to those in (a) are of opposite phase.

14. (a)

Y (structure with CH_3, CH_2OH, H)

Z (structure with CH_3, O, H)

(b) intramolecular solvomercuration possible only for *cis* isomer. **15.** (a) Allowed thermal *con* opening (4 e) would give impossibly strained *cis,cis,trans*-cyclohexa-1,3,5-triene; (b) allowed *antara* [1,3]-H impossible on geometric grounds. **16.** (a) *Con* opening (6 e); [1,7]-H *antara*; (c) *dis* closure. **17.** (a) Via *cis,cis,cis,cis,cis*-cyclodeca-1,3,5,7,9-pentaene; (b) 10 π electrons fits Hückel rule, but evidently not very stable for steric reasons.

Chapter 30

30.1 2; 10; 14. **30.3** (b) *trans*-Decalin more stable; both large groups (the other ring) on each ring are equatorial; (c) *syn*-addition, rate control; *anti*-addition, equilibrium control. **30.4** Benzylic substitution; elimination of HBr to give conjugated alkenylbenzene; benzylic-allylic substitution; elimination to give aromatic ring. **30.5** (a) Cadalene, 4-isopropyl-1,6-dimethylnaphthalene; (b) cadinene has same carbon skeleton as cadalene, follows isoprene rule. **30.8** (a) Via aryne; (b) direct displacement of —F by amine; (c) both direct displacement and elimination-addition occur. **30.9** 1,2,4-Benzenetricarboxylic acid; 1,2,3-benzenetricarboxylic acid. **30.17** Deactivating acyl group transformed into activating alkyl group. **30.19** Phenanthrene (see Sec. 30.19, and Fig. 30.3, p. 995). **30.20** 23 kcal/mole; 31 kcal/mole. **30.22** (a) Most stable tetrahydro product; (b) reversible sulfonation yields more stable product. **30.24** (a) 1-Nitro-9,10-anthraquinone; (b) 5-nitro-2-methyl-9,10-anthraquinone (with some 8-nitro isomer).

30.29 Pyrene,

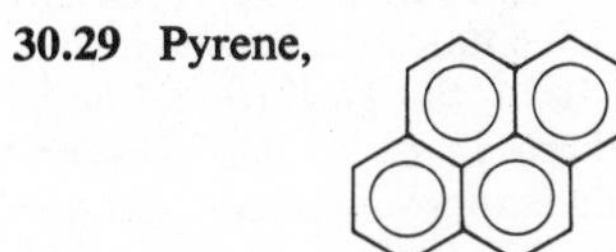

3. 1-, 5-, and 8-nitro-2-methylnaphthalene. **5.** F, phenanthrene. **7.** G, 1,2-benzanthracene; H, chrysene. **8.** α-Naphthol. **9.** (a) Diels-Alder; (c) J, *meso*; K, racemic modification. **10.** (d) β-Tetralone (2-oxo-1,2,3,4-tetrahydronaphthalene). **11.** (a) 1,6-Cyclodecanedione; (b) bicyclic unsaturated ketone, one 7-ring and one 5-ring. **12.** (a)

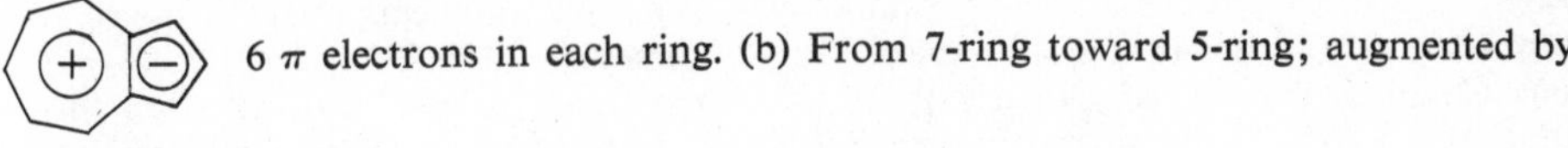

6 π electrons in each ring. (b) From 7-ring toward 5-ring; augmented by

Azulene

C—Cl dipole. **13.** (a) Aromaticity of 7-ring preserved.

(b) Protonation at C–1; azulene upon neutralization. (c) Deuteration via electrophilic substitution at C–1 and C–3, and deuteration again at C–1 comparable to the protonation in (b); expect 1,3-dideuterioazulene upon neutralization; (d) at C–1. **14.** Nucleophilic substitution in the 7-ring, at C–4; H Z aromaticity of 5-ring preserved, conjugation in 7-ring. **15.** Eudalene, 7-isopropyl-1-methylnaphthalene. **16.** Y, 2,2′,3,3′,5,5′-hexachloro-6,6′-dihydroxydiphenylmethane; CC, 3,4′-dimethylbiphenyl; FF, compound I, p. 393; HH, tetraphenylmethane; II, 1,3,5-triphenylbenzene. **17.** $—N_2^+$ activates molecule toward nucleophilic aromatic substitution. **18.** (a) JJ, methylene bridge between 9- and 10-positions of phenanthrene; (b) random insertion of methylene into *n*-pentane; (c) three insertion products and one addition product. **19.** KK,

Each ring contains 6 π electrons. **20.** (a) Via an aryne; (b) direct displacement accompanies elimination-addition. Fluoride least reactive toward benzyne formation (p. 838), most reactive toward direct displacement (Sec. 25.12). Piperidine shifts equilibrium (1) toward left, tends to inhibit benzyne formation. **21.** UU is aromatic, with 14 π electrons. Methyl protons are *inside* aromatic ring; see Fig. 13.4, p. 419.

CH_3 CH_3

UU

Chapter 31

31.1 B, $[—CH(COOEt)COCH_3]_2$. **31.3** —COOH deactivates ring. **31.4** Two units of starting material linked at the 5-positions through a $—CH_2—$ group. **31.5** Sodium furoate and furfuryl alcohol (Cannizzaro reaction). **31.10** Hygrine, 2-acetonyl-N-methylpyrrolidine; hygrinic acid, N-methyl-2-pyrrolidinecarboxylic acid. **31.11** Orientation ("*para*") controlled by activativg $—NH_2$ group. **31.13** Amine > imine > nitrile. **31.18** Piperidine, a 2° amine, would itself be acylated. **31.23** (a) 8-Nitroquinoline; (b) 8-hydroxyquinoline (8-quinolinol); (c) 4,5-diazaphenanthrene; (d) 1,5-diazaphenanthrene; (e) 6-methylquinoline. **31.28** Electrophilic aromatic substitution or acid-catalyzed nucleophilic carbonyl addition, depending upon viewpoint.

1. No reaction: c, h, i, j. **3.** Pyrroline has double bond between C–3 and C–4. **4.** C, acetonylacetone. **5.** Porphin, with same ring skeleton as in hemin, page 1152. **6.** D, 2–COOH; E, 3–COOH; F, 4–COOH. **7.** (a) 5- or 7-methylquinoline; (b) G, 7-methylquinoline. **9.** (e) Perkin reaction; (g) Reimer-Tiemann reaction. **10.** (See below for parent ring systems.) I, 2,4,6-trihydroxy-1,3-diazine; K, 3,6-dimethyl-1,2-diazine; L, 3,5-dimethyl-1,2-diazole; M, 2,3-dimethyl-1,4-diazanaphthalene; N, 1,3-dioxolan-2-one (ethylene carbonate); P, 3-indolol; R, 2,5-dimethyl-1,4-diazine; S, 1,3-diazolid-2-one (2-imidazolidone, ethyleneurea); T, 4,5-benzo-2-methyl-1,3-diazole (2-methylbenzimidazole); W, 2,4-dihydroxyquinoline; BB, 1,2-diazolid-3-one (3-pyrazolidone); CC, 4,5-diazaphenanthrene; GG, two indole units fused 2,3 to 3′,2′; HH, N-methyl-1,2,3,4-tetrahydroquinoline; II, 2-phenylbenzoxazole; JJ, the benzene ring of II completely hydrogenated.

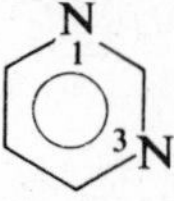

1,3-Diazine
(Pyrimidine)

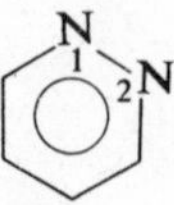

1,2-Diazine
(Pyridazine)

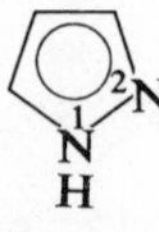

1,2-Diazole
(Pyrazole)

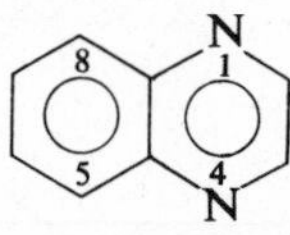

1,4-Diazanaphthalene
(Quinoxaline)

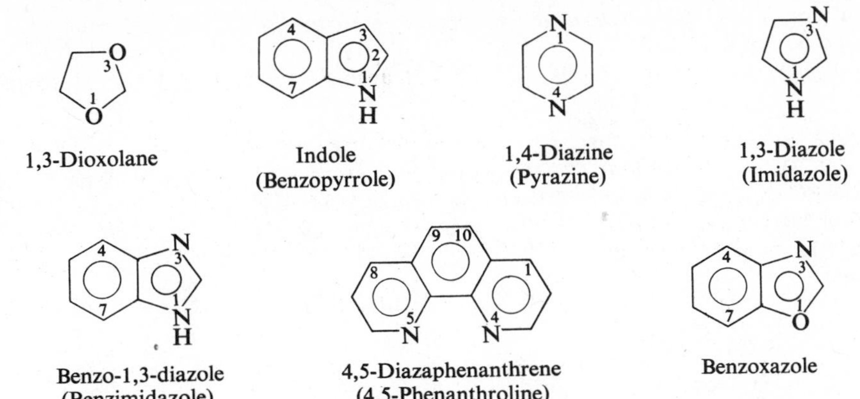

11. LL, 3,4-$(CH_3O)_2C_6H_3CH_2CH_2NH_2$; NN, 3,4-$(CH_3O)_2C_6H_3CH_2COCl$; OO, amide; PP, a 1-substituted-7,8-dimethoxy-3,4-dihydroisoquinoline; papaverine, the corresponding substituted isoquinoline. **12.** VV, $(C_2H_5)_2NCH_2CH_2CH_2CHBrCH_3$; XX, 8-amino-6-methoxyquinoline; Plasmochin, 8-amino group of XX alkylated by VV. **13.** Nicotine, 2-(3-pyridyl)-N-methylpyrrolidine. **14.** DDD, *o*-hydroxybenzalacetophenone; (c) oxygen contributes a pair of electrons to complete an aromatic sextet. **15.** Tropinic acid, 2–COOH–5–CH_2COOH–N–methylpyrrolidine. **17.** Pseudotropine has equatorial —OH, is more stable. **18.** (a) Guvacine, 1,2,5,6-tetrahydro-3-pyridinecarboxylic acid; arecaidine, N-methylguvacine; (b) nicotinic acid. **19.** UUU, one enantiomer of ethyl-*n*-propyl-*n*-butyl-*n*-hexylmethane; chirality does not necessarily lead to measurable optical activity (see Sec. 4.13). **20.** Aliphatic NH_2 > "pyridine" N > "pyrrole" NH. **21.** Dipolar ion loses CO_2.

Chapter 32

32.1 (a) Amide; see Sec. 32.7; (b) amide; 6-aminohexanoic acid; (c) ether; ethylene oxide; (d) chloroalkene; 2-chloro-1,3-butadiene; (e) chloroalkane; 1,1-dichloroethene. **32.2** (a) Amide; (b) ester; (c) acetal; (d) acetal. **32.3** 1,2- and 1,4-addition. **32.4** Combination. **32.6** Polymer is transfer agent. **32.9** (a) Chain-transfer.

2. Dehydration, polymerization. **4.** Nucleophilic carbonyl addition. **5.** Hydrolysis gives amine, alcohol, and carbon dioxide. **9.** ~$OCH_2CH_2COOCH_2CH_2COO$~; chain-reaction. **10.** Growing anion abstracts proton from solvent. **11.** *Some* head-to-head polymerization. **12.** (a) ~$NHCH_2(CH_2)_4CO$~; (b) chain-reaction. **13.** Cyclohexanone. **15.** Compounds are ionic, due to stability of benzylic anions. **16.** A, *meso*, resembles isotactic; B, racemic, resembles syndiotactic. **18.** Monomer acts as chain-transfer agent. **19.** Cross-linking by oxygen between allylic positions. **21.** F, syndiotactic; G, isotactic.

Chapter 33

33.1 Decarboxylation. Fatty acids could be precursors of petroleum hydrocarbons. **33.2** (a) Isoprene unit. (b) Likely that petroleum comes from green plants. **33.4** Alkoxide is poor leaving group. **33.5** Preserves semiliquidity of membranes in colder part of body.

1. Nervonic acid, *cis*- or *trans*-$CH_3(CH_2)_7CH{=}CH(CH_2)_{13}COOH$ (actually, *trans*). **2.** Transesterification to more random distribution of acyl groups among glyceride molecules. **3.** Hybrid (allylic) free radical is intermediate. **5.** Spermaceti, *n*-hexadecyl *n*-hexadecanoate. **6.** Cleavage of monoanion as dipolar ion (or with simultaneous transfer of proton) easiest because of (a) protonation of alkoxy group and (b) double negative charge on other oxygens:

$$R{-}\overset{+}{\underset{|}{\underset{H}{O}}}{-}PO_3^{--} \xrightarrow{H_2O} ROH + H_2PO_4^{-}$$

7. Vaccenic acid, *cis*-$CH_3(CH_2)_5CH{=}CH(CH_2)_9COOH$. **8.** Corynomycolenic acid, *cis*-*n*-$C_{13}H_{27}CH_2CH(COOH)CHOH(CH_2)_7CH{=}CHC_6H_{13}$-*n*. **9.** Tuberculostearic acid, 10-methyloctadecanoic acid. **10.** C_{27}-phthienoic acid, $CH_3(CH_2)_{17}CH(CH_3)CH_2CH$-$(CH_3)CH{=}C(CH_3)COOH$. **11.** CC, octadecanoic acid; DD, 2-methylheptadecanoic acid. **12.** Juvenile hormone,

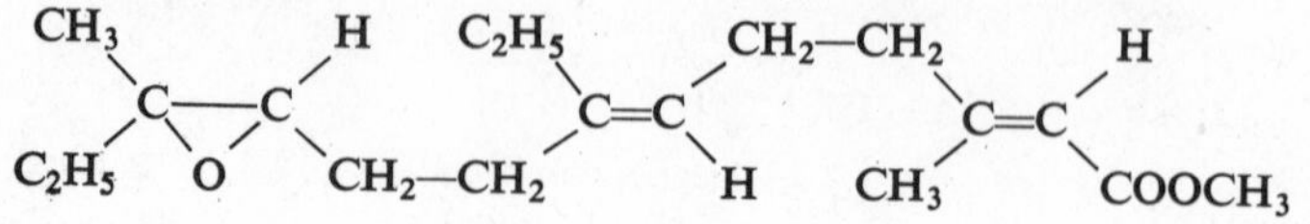

$$\underset{C_2H_5}{\overset{CH_3}{>}}C\overset{O}{\text{—}}\underset{}{C}\overset{H}{<}CH_2{-}CH_2{-}\underset{}{\overset{C_2H_5}{C}}{=}\underset{H}{C}{-}CH_2{-}CH_2{-}\underset{CH_3}{C}{=}\overset{H}{C}{-}COOCH_3$$

Chapter 34

34.3 (a) 3; (b) 8. **34.4** Glucose + $5HIO_4 \rightarrow 5HCOOH + HCHO$. **34.5** A, gluconic acid; B, glucitol; C, glucaric acid; D, glucuronic acid. **34.6** Fructose. Aldose → osazone → osone → 2-ketose. **34.7** Identical in configuration at C–3, C–4, and C–5. **34.8** Alditol. **34.9** (a) 2 tetroses; (b) 4 pentoses, 8 hexoses (see Problem 34.2); (c) D; (d) L. **34.11** I, (+)-allose; II, (+)-altrose; VI, (−)-idose; VII, (+)-galactose; VIII, (+)-talose. **34.15** (a) R; (b) R; (c) S; (d) R. **34.16** (S)-(+)-2-butanol. **34.17** (a) S,S-; (b) R,R-; (c) R,S-. **34.18** (b) 1:3; (c) the isomer favored in the L-series will be the mirror image of the isomer favored in the D-series. **34.19** L-(+)-Gulose. **34.20** (a) 36.2% α, 63.8% β. **34.23** (a) CH_3OH, HOOCCHO, and D-glyceric acid. **34.24** HCHO instead of HCOOH. **34.25** (a) Six-membered ring; (b) HCOOH, OHC—CHO, and $HOCH_2CHO$. **34.26** (a) Six-membered ring; (b) enantiomer. **34.27** (a) Five-membered ring; (b) optically active, L-family; (c) enantiomer.

4. E and E′, allitol and galactitol; F, glucitol (or gulitol); H, glucitol (or gulitol); I and I′, allitol and galactitol; N, ribitol; O, arabitol (or lyxitol). **5.** (a) P, glycoside of glucuronic acid; (d) $HOCH_2(CHOH)_3$-COCOOH. **6.** Rate-determining step involves OH^- before reaction with Cu^{++}: probably abstraction of proton leading to formation of enediol. **7.** (a) 5 carbons, five-ring; (b) C–1 and C–4; (c) Q, methyl α-D-arabinofuranoside. **8.** Salicin, *o*-(hydroxymethyl)phenyl β-D-glucopyranoside. **9.** Bio-inonose, the pentahydroxycyclohexanone in which successive —OH groups are *trans* to each other. **11.** (a) T, D-ribose; U, D-arabinose; (b) 3-phosphate. **12.** Z and AA are ketals: Z, furanose with acetone bridging C–1 to C–2 and C–5 to C–6; AA, pyranose, with acetone bridging C–1 to C–2. **14.** S_N1-type, with separation of relatively stable oxonium ion (see Sec. 19.15). **15.** (a) Proton on C–1 most deshielded by two oxygens. (b) JJ, β-anomer; KK, α-anomer; (c) LL, β-anomer; MM, α-anomer; (d) NN, α-mannose; OO, β-mannose; PP, β-glucose; QQ, α-glucose. **16.** L-(−)-Mycarose,

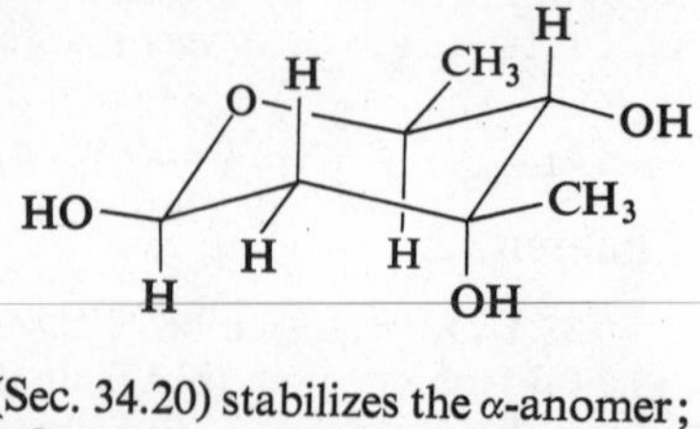

[Chair structure: ring O; HO, H at C–1; H at C–2; H at C–3; CH₃, H at C–4 (sic as drawn: H, CH₃, OH, CH₃, OH, H substituents)]

(e) α-glycoside; (f) β-anomer. **17.** (a) Anomeric effect (Sec. 34.20) stabilizes the α-anomer; (b) anomeric effect stabilizes diaxial chlorines. **18.** (a) On steric grounds, neither; anomeric effect would favor axial OAc on C–1. (b) Tells nothing: in either conformation two OAc are equatorial, two are axial. (c) The e:a peak area ratio would be 2:1 if C–1 OAc were all axial, 1:1 if half axial, 0.5:1 if none axial. Ratio of 1.46:1.00 shows C–1 OAc is axial in 79% of molecules.

Chapter 35

35.4 D-Glucose and D-erythrose; indicates attachment to other ring is at C–4. **35.8** D-Galactose and D-erythrose. **35.10** $C_{12}H_{20}O_{10}$, non-reducing. **35.11** Sucrose is an α-glucoside. **35.13** 1 (0.0025%); 3 (0.0075%); 9 (0.022%). **35.14** (a) A large group in an axial position. **35.15** (a) 3 molecules of HCOOH per molecule of amylose; (b) moles HCOOH/3 = moles amylose; wt. amylose/moles amylose = mol.wt. amylose; mol.wt. amylose/wt. (of 162) per glucose unit = glucose units per molecule of amylose; (c) 980. **35.16** A poly-α-D-glucopyranoside; chain-forming unit, attachment at C–1 and C–6; chain-linking unit, attachment at C–1, C–3, and C–6; chain-terminating unit, attachment at C–1. **35.17** A poly-β-D-xylopyranoside; chain-forming unit, attachment at C–1 and C–4; chain-linking unit, attachment at C–1, C–3, and C–4; chain-terminating unit, attachment at C–1.

1. Gentiobiose, 6-O-(β-D-glucopyranosyl)-D-glucopyranose. **2.** (a) Trehalose, α-D-glucopyranosyl α-D-glucopyranoside; (b) isotrehalose, α-D-glucopyranosyl β-D-glucopyranoside; neotrehalose, β-D-glucopyranosyl β-D-glucopyranoside. **4.** Raffinose, α-D-galactosyl unit attached at C–6 of glucose unit of sucrose; melibiose, 6-O-(α-D-galactopyranosyl)-D-glucopyranose. **5.** (a) Melezitose, α-D-glucopyranosyl unit attached at C–3 of fructose unit of sucrose; turanose, 3-O-(α-D-glucopyranosyl)-D-fructofuranose. **6.** Panose, α-D-glucopyranosyl unit attached at C–6 of non-reducing moiety of maltose; isomaltose, 6-O-(α-D-glucopyranosyl)-D-glucopyranose. **7.** D-Glucuronic acid; (c) D-xylose. **12.** I, D-$CH_2OHCHOHCHOHCOOH$; J, HOOCCHO. **13.** (a) 3 molecules of HCOOH per molecule of cellulose; (c) 1390 glucose units.

Chapter 36

36.1 $—NH_2 > —COO^-$; proton goes to $—NH_2$ to form $^+H_3NCHRCOO^-$. **36.2** $—COOH > —NH_3^+$; —COOH gives up proton to form $^+H_3NCHRCOO^-$. **36.5** (a) On acid side; (b) on basic side; (c) more acidic and more basic than for glycine. **36.8** 4 isomers. **36.9** CyS–SCy, Hylys, Hypro, Ileu. **36.11** Intermediate for Ala is $CH_3CH(NH_2)CN$. **36.12** A, $(CH_3)_2CHCH(COOEt)COCOOEt$; B, $(CH_3)_2CHCH_2COCOOEt$. **36.15** (a) 22.4 cc; (b) 44.8 cc; (c) no N_2. **36.16** Minimum mol.wt. = 114; could be valine. **36.19** Salmine, $AlaArg_{50}Gly_4IleuPro_6Ser_7Val_3$. **36.20** Same as empirical formula (preceding problem). **36.21** 70300. **36.22** (a) 16700; (b) 4. **36.23** A sulfonamide, which is more resistant to hydrolysis than carboxamides (see Sec. 23.6). **36.24** (a) Phe.Val.Asp.Glu.His; (b) His.Leu.CySH.Gly.Ser.His.Leu; (c) Tyr.Leu.Val.CySH.Gly.-Glu.Arg.Gly.Phe.Phe. **36.25** (a) Cbz.Gly.Ala, $SOCl_2$; Phe; H_2, Pd. (b) $PhCH_2OCOCl$, Ala; $SOCl_2$; Gly; H_2, Pd. **36.26** In A, polystyrene has $—CH_2Cl$ groups attached to rings; in G, $—CH_2Br$ groups.

2. D, $HOCH_2CH_2CH_2CH(NH_3^+)COO^-$. **3.** (a) F, $CH_3CONHC(COOC_2H_5)_2CH_2CH_2CHO$; J, $CH_3CONHC(COOC_2H_5)_2CH_2(CH_2)_2CH_2NHCOCH_3$. (b) K, $NCCH_2CH_2CH(COOC_2H_5)_2$; O, $^+H_3NCH_2(CH_2)_2CHClCOO^-$. **4.** (a) Diketopiperazine, cyclic diamide; (b) unsaturated acid; (c) γ-lactam, 5-ring amide; (d) δ-lactam, 6-ring amide. **6.** (a) Betaine, $^+(CH_3)_3NCH_2COO^-$; (b) trigonelline, N-methylpyridinium-3-carboxylate (dipolar ion). **7.** Polarity of solvent lowered; hydrophobic parts of organic molecules come out of their huddle. **9.** Minimum mol.wt. = 13000; minimum of one Fe atom and six S atoms. **10.** (a) Approx. 32 $—CONH_2$ groups; (b) 395–398 peptide links plus $—CONH_2$ groups; (c) 367–370 amino acid residues.

11.

```
Val.Orn.Leu.Phe
·Pro·          ·Pro·
Val.Orn.Leu.Phe   (printed inverted)
```

Gramicidin S

Cyclic decapeptide

12. Beef insulin:

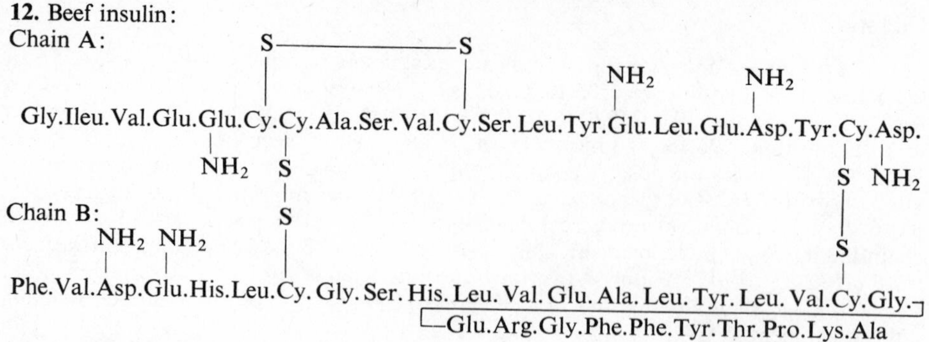

(g) DNP.NH$(CH_2)_4$CH$(NH_3{}^+)COO^-$ from ϵ-amino group of Lys. If Lys had been terminal, would have gotten a double DNP derivative of it, and no DNP.Phe.

Chapter 37

1. CO_2 becomes the —COOH of malonyl–CoA in reaction (1), Sec. 37.6; this is the carbon lost in reaction 4. **2.** Slow (rate-determining) formation of a tetrahedral intermediate (see Sec. 20.17) followed by fast loss of OR or SR. **3.** (b) Guanine and cytosine, 3 H-bonds per pair; adenine and thymine, only 2. **4.** (a) Aldol-like condensation between ester and keto group of oxaloacetate; (b) aldol-like condensation between ester and keto group of acetoacetyl–CoA; reduction of ester to 1° alcohol by hydride transfer. **5.** A, $C_2H_5OOCCH_2CH_2NHCONH_2$; B, a dihydroxydihydro-1,3-diazine (see p. 1206 for parent diazine ring system); C, a dihydroxydihydro-5-bromo-1,3-diazine; E, 2-chloro-4-amino-1,3-diazine; F, 4-chloro-2-amino-1,3-diazine. **6.** Biological oxidation of fatty acids removes 2 carbons at a time, starting at carboxyl end: "*beta*-oxidation." **7.** *Retro* (reverse) aldol condensation. **8.** (a) Direct transfer of a hydride ion from C–1 of ethanol to C–4 of pyridine ring of NAD^+. There are now two hydrogens on C–4 (see Sec. 36.15) and, if one of them is D, C–4 is chiral center. (Because of chirality of rest of NADD molecule, these are *diastereotopic* hydrogens; see Sec. 13.7.) Of the two hydrogens on C–4, *only* the one originally received from ethanol in part (a) is transferred back to aldehyde, indicating transfer in both directions is stereospecific. (c) Transfer to D-glucose of only the *other* hydrogen on C–4, indicating stereospecificity opposite to that in (b). (d) Chemical reduction is not stereospecific, and gives mixture of diastereomeric NADD molecules. (e) X and Y are the two enantiomers of CH_3CHDOH. Transfer is stereospecific not only with regard to which hydrogen on C–4 is transferred, but with regard to which face of acetaldehyde it becomes attached to. If D becomes attached to that face, X is formed; if H becomes attached, Y is formed.

Index

A

C

D

E

F

G

H

J

K

L

M

O

P

Q

R

T

Periodic Table of the Elements

Group	I	II											III	IV	V	VI	VII	O
Period																		
1	H 1																	He 2
2	Li 3	Be 4											B 5	C 6	N 7	O 8	F 9	Ne 10
3	Na 11	Mg 12					**Transition elements**						Al 13	Si 14	P 15	S 16	Cl 17	Ar 18
4	K 19	Ca 20	Sc 21	Ti 22	V 23	Cr 24	Mn 25	Fe 26	Co 27	Ni 28	Cu 29	Zn 30	Ga 31	Ge 32	As 33	Se 34	Br 35	Kr 36
5	Rb 37	Sr 38	Y 39	Zr 40	Nb 41	Mo 42	Tc 43	Ru 44	Rh 45	Pd 46	Ag 47	Cd 48	In 49	Sn 50	Sb 51	Te 52	I 53	Xe 54
6	Cs 55	Ba 56	* 57–71	Hf 72	Ta 73	W 74	Re 75	Os 76	Ir 77	Pt 78	Au 79	Hg 80	Tl 81	Pb 82	Bi 83	Po 84	At 85	Rn 86
7	Fr 87	Ra 88	† 89–103															
*** Lanthanide series**				La 57	Ce 58	Pr 59	Nd 60	Pm 61	Sm 62	Eu 63	Gd 64	Tb 65	Dy 66	Ho 67	Er 68	Tm 69	Yb 70	Lu 71
† Actinide series				Ac 89	Th 90	Pa 91	U 92	Np 93	Pu 94	Am 95	Cm 96	Bk 97	Cf 98	Es 99	Fm 100	Md 101	No 102	Lr 103